D0204132

Auto Body Repair Technology

Auto Body Repair Technology

Sixth Edition

James E. Duffy

Australia • Brazil • Mexico • Singapore • United Kingdom • United States

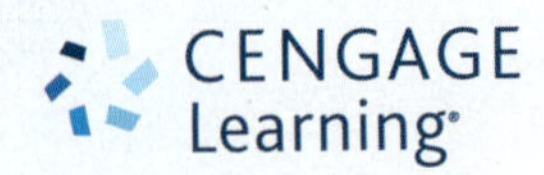

Auto Body Repair Technology Sixth Edition
James E. Duffy

SVP, GM Skills & Global Product Management: Dawn Gerrain

Product Team Manager: Erin Brennan

Senior Director, Development: Marah Bellegarde

Senior Product Development Manager: Larry Main

Senior Content Developer: Sharon Chambliss

Product Assistant: Maria Garguilo

Vice President, Marketing Services: Jennifer Ann Baker

Senior Marketing Manager: Jennifer Barbic

Senior Production Director: Wendy Troeger

Production Director: Andrew Crouth

Senior Content Project Manager: Cheri Plasse

Senior Art Director: Bethany Casey

Cover image(s): ©Dmitriy Bryndin/Shutterstock

Library of Congress Control Number: 2014936013

ISBN: 978-1-133-70285-6

Cengage Learning
20 Channel Center Street
Boston, MA 02210
USA

Cengage Learning is a leading provider of customized learning solutions with office locations around the globe, including Singapore, the United Kingdom, Australia, Mexico, Brazil, and Japan. Locate your local office at:
www.cengage.com/global

Cengage Learning products are represented in Canada by Nelson Education, Ltd.

To learn more about Delmar, visit **www.cengage.com**

Purchase any of our products at your local college store or at our preferred online store **www.cengagebrain.com**

Printed in the United States of America.
Print Number: 01 Print Year: 2014

Contents

Chapter 15
DOOR, ROOF, AND GLASS SERVICE 449

Chapter 16
PASSENGER COMPARTMENT SERVICE 481

SECTION 4
MAJOR BODY/FRAME REPAIRS 502

Chapter 17
BODY/FRAME DAMAGE MEASUREMENT 504

Chapter 18
UNIBODY/FRAME REALIGNMENT 556

Chapter 19
WELDED PANEL REPLACEMENT 592

Chapter 20
RESTORING CORROSION PROTECTION 635

SECTION 5
MECHANICAL AND ELECTRICAL REPAIRS 658

Chapter 21
CHASSIS SERVICE AND WHEEL ALIGNMENT 660

SECTION 7 COLLISION REPAIR PROFESSIONALISM 984

Preface

Welcome to the rewarding world of auto body repair! Whether you simply want to make minor repairs on your own car or to become an ASE-certified collision repair technician, you will find this book a valuable resource.

Auto Body Repair Technology details how to properly restore a damaged vehicle to a like-new condition from start to finish. It is designed to help you work on any make and model passenger car, pickup truck, van, or sport utility vehicle (SUV).

Auto body repair, also called collision repair, is an exciting area to study. With millions of vehicles on the road today, there is a strong demand for well-trained collision repair technicians. Thousands of technicians are needed in the industry every year. Just look in the newspaper and you will find numerous openings for skilled people to work in body shops. With today's high-tech vehicles and varied construction methods and repair techniques, competent collision repair takes well-trained, knowledgeable professionals.

Section 1 provides an introduction to the industry. It explains all of the basic information that pertains to the industry in general. Section 2 covers estimating and has new material on electronic estimating and shop management software. Section 3 covers minor repairs—the kinds of repair tasks that can be done by a novice or trainee. Section 4 explains major body/frame repairs, including topics such as vehicle frame damage measurement and repair. Section 5 summarizes mechanical and electrical repairs to the steering, suspension, and brake system parts, which are often damaged in a major accident. Section 6 details refinishing and how to prep and paint the vehicle. The last section explains how professionalism will help you prosper as a collision repair expert.

The text flows more like work actually does in the industry. For example, estimating is covered right after the fundamentals section because estimating is the first task when a vehicle needs repair. The estimate outlines what must be done to repair the vehicle. Proper estimating helps technicians plan and execute quality repairs.

Writing is all about communication between the writer and the reader. This edition of *Auto Body Repair Technology* is the most readable edition ever published. We have worked hard to make the book easier to understand without diminishing its technical completeness. Textbooks are just one part of a vocational technical education; it is our sincere desire that this book adds something positive to that education.

Unless you are experienced in collision repair, you will have to learn hundreds of new technical terms, which make up the language of collision repair personnel. These vocabulary terms are highlighted so that you know that the word must be learned and understood before reading on. Every effort has been made to help you identify new key terms and explain them on first use.

Auto Body Repair Technology has been used by thousands of students and continues to this day to be a market leader. All of the people involved in the publishing of this book hope that you will appreciate its improvements. Most of all, we hope that the students hoping to become collision repair professionals will take from this book the information they need to get started the right way.

New to this Edition

- A new chapter has been added. Chapter 30, Collision Repair for Hybrid and Electric Vehicles.
- Updated throughout to ensure all necessary NATEF competencies have been addressed.
- Video sequences demonstrating shop practices are included within the MindTap offering. Look for select videos within the instructor resource material.
- Every effort has been made to make this book accurate. Collision repair teachers, experts in the field, and others were brought in to review this book and ensure its accuracy. We worked with all of their comments and criticisms.

Safety is emphasized throughout the text. Safety cautions and warnings appear frequently, and we worked to make sure that all the illustrations represent safe practices.

Throughout this book, there are three special notes labeled "Shop Talk," "Warning," and "Danger." Each one has a specific purpose:

1. "Shop Talk" notes give examples of typical conversations between two technicians. Sometimes they show how someone did something wrong, which

damaged a part or caused injury. Shop Talk notes also provide "tricks of the trade" for doing better, more efficient repair work.

2. "Warning" notes are given to help prevent technicians from making errors that can damage a vehicle or a tool. These notes provide information about unsafe practices that can cause repair problems, waste time, and cost shops and technicians money for repairs.
3. "Danger" notes remind technicians to be especially careful of those tasks where carelessness can cause personal injury.

Remember to read these special notes carefully!

We are anxious to know what you think of the effort to update and upgrade *Auto Body Repair Technology*. Send letters or call the publisher so I can hear what you think.

We hope you will find this book a useful resource for many years to come.

Sincerely,

James E. Duffy
A Fellow Educator

Acknowledgments

REVIEWERS

The author and publisher would like to thank the following instructors for their help with reviewing this and previous editions of the book.

Cliff Ashton
Diman Regional Vocational Technical High School
Fall River, MA

Tom Bergstrom
Highland Community College
Freeport, IL

Rodney Bolton
Center of Applied Technology North
Severn, MD

Russell Butler
Idaho State University
Pocatello, IA

W. Jack Charles
Lakeshore Technical College
Cleveland, WI

Douglas Correll
Central Piedmont Community College
Huntsville, NC

Michael Crandell
Carl Sandburg College
Galesburg, IL

Craig Dickerson
Kentucky Community and Technical College
Paducah, KY

Eddie Ellis
Holmes Community College
Goodman, MS

James Friedel
Sussex Technical High School
Georgetown, DE

H. James Gaugler Jr.
Lehigh Career and Technical Institute
Schnecksville, PA

Don Headley
Wheeling Park High School
Adena, OH

Daniel Hodges, Jr.,
Florida State College- Jacksonville
Jacksonville, FL

Jim Ingles
Hawkeye Community College
Waterloo, IA

Tim Ingram
College of Central Florida
Ocala, FL

Terry Lindley
Union County Career Center
Monroe, NC

Robert Magee
Bergen County Technical High School
Teterboro, NJ

Eric Mason
Allan Hancock College
Santa Maria, CA

Josh Nelms
Technology Center of Dupage
Addison, IL

Gary Sanger
Des Moines Area Community College
Ankeny, IA

Rob Schultz
Iowa Lakes Community College
Emmetsburg, IA

Joe Youngwirth
Des Moines Area Community College
Des Moines, IA

The publisher would like to especially acknowledge W. Jack Charles, from Lakeshore Technical College, Cleveland, WI, for his contributions.

CONTRIBUTING COMPANIES

American Isuzu Motors Inc.
Atlantic Pneumatic, Inc.
Automotive Service Association
Automotive Service Excellence
BASF
Babcox Publications
Badger Air-Brush Co.
Bee Line Co.
Bondo Corporation
CRC Chemicals
Car-O-Liner Company
CEBORA/Cebotech, Inc.
Champion, A Gardner Denver Company
Chicago Pneumatic/Automotive Business Unit
Chief Automotive Technologies
Chrysler LLC
Daimler AG
Dana Corporation
Danaher Tool Group
Delphi Corporation
DeVair Inc.
Dorman Products
DuPont Automotive Finishes
Dynabrade
Eastwood Company
Equalizer Industries, Inc.
Ford Motor Company
GretagMacbeth
Henning Hansen Incorporated
Hunter Engineering Company
Hyundai Motor America
ITW Automotive Refinishing-DeVilbiss
KD Tools
Laser Mate, USA
Lisle Corp.
LORS Machinery, Inc.
Marson Corporation and Alcoa Fastening Systems
Mattson Spray Equipment
Mazda Motors of America, Inc.
Miller Electric Manufacturing Co.
Mitchell 1
Mitsubishi Motor Sales of America, Inc.
Morgan Manufacturing Inc.
N.A.D.A. Official Used Car Guide ® Company
National Detroit, Inc.
Nissan North America, Inc. and Charles Hopkins Photography
Noram, Inc.
Norco Industries, Inc.
Norton/brand of Saint-Gobain Abrasives
PBR Industries
PPG Industries, Inc.
Porsche Cars North America, Inc.
Porsche AG
Pull-it Corporation
S&G Tool Aid Corporation
S&H Industries
S&R Photo Acquisitions, LLC
SPX/OTC Service Solutions
Saab Cars USA, Inc.
Sartorius North America Inc.
Saturn Corporation
Snap-on Collision
Snap-on Tools Company
Stanley Works
Style-Line Corporation International
Subaru of America Inc.
Swan River Software
TRW Fasteners Division
TECNA/Cebotech, Inc.
Team Blowtherm
Tech-Cor, Inc.
3M Company
Toyota Motor Sales USA, Inc.
Urethane Supply Company Inc.
Volkswagen of America, Inc.
Volvo Cars of North America
Wedge Clamp

Features of the Text

OBJECTIVES

Each chapter begins with a list of cognitive and performance-based *objectives*. The objectives state the knowledge and skills that should be learned after reading the entire chapter. This feature is an excellent way for the reader to gain a quick overview of the essential material within the chapter that will be learned before the reading begins.

KEY TERMS

The *Key Terms* are the most important technical words you will learn in the chapter. These are listed at the end of each chapter and appear in bold print where they are first defined. For added study, you can write the key terms on a sheet of paper with their definitions to make sure you can explain the terms. These terms are also given in the Glossary at the back of the book.

CHAPTER 1

Collision Repair: Introduction and Careers

OBJECTIVES

After studying this chapter, you should be able to:

- Describe what happens to a motor vehicle during a collision.
- Summarize the basic steps needed to repair a vehicle damaged in an accident.
- Explain the major work areas of a typical collision repair facility.
- Summarize the work flow through a typical body shop.
- Describe the types of positions or jobs available in the collision repair industry.
- Identify the setup and inner workings of a typical collision repair shop.
- Answer ASE-style review questions relating to collision repair.
- Understand fundamental terms used in the collision repair industry.

INTRODUCTION

This chapter will explore the challenging world of auto body repair, also called collision repair. It will give you the basic knowledge needed to fully grasp the more detailed information in later chapters. You will follow damaged vehicles through a typical repair process from beginning to end: estimating damage; metal straightening, filling, sanding, masking, painting, and detailing; and final delivery to the customer.

Considering that the United States is a "nation on wheels" and auto accidents happen, you have selected an excellent area for further study. The collision repair skills you learn will remain in high demand as long as there are people driving automobiles. The collision repair industry is vast. Body shops do an astounding $10 billion of repair work per year. This enormous amount of work requires a vast number of well-trained technicians and related personnel.

If you are just beginning your study of collision repair, you have much to learn. Only highly skilled, knowledgeable professionals can properly repair today's vehicles. Study the material in this textbook carefully and you will be on your way to a successful career in collision repair.

HISTORY NOTE

Believe it or not, the very first "motor car" ever built was in an accident the first time it was driven on public roads. The huge, crude car looked something like a steam locomotive. It was being test-driven down a country road. Chugging over a hill, the steel-wheeled, "smoke-belching monster" almost ran over a horse-drawn carriage. The car's startled driver had to swerve off the narrow dirt road to avoid running over the horses, and he plowed into a brick wall. After the first car crash, plow horses had to be used to tow the "horseless carriage" back to the barn for body shop repairs.

1.1 WHAT IS COLLISION REPAIR?

A **collision**, nicknamed a *crash* or a *wreck* in laypersons' terms, is damage caused by an **impact** (hit) on the vehicle. This impact might be from another vehicle or from another object. The collision might be minor enough to only scratch the paint, or it might be severe enough to cause thousands of dollars' worth of damage to numerous metal and plastic parts.

A **head-on collision** results when two vehicles driving toward each other accidentally collide. It is often

4

Figure 1–1 During a major collision, like this head-on crash test, tremendous energy is dissipated into the body/frame structures of both cars. Modern vehicles are designed to absorb this energy by controlled collapse of structural body/frame members. The front and rear ends of the vehicles are designed to crush or collapse while keeping the passenger compartment intact. With today's front and side air bags and seat belts, the drivers and passengers would suffer little or no injury in this violent collision.

Figure 1–2 If the accident results in major damage and the vehicle cannot be driven, a wrecker is called to tow the car. Proper tow methods are needed to avoid further damage when moving the wrecked car.

the worst type of accident because the speed of both vehicles must be combined to determine the force of the impact. For example, if both vehicles are travelling 70 miles per hour (mph), the impact is similar to one vehicle smashing into a stationary brick wall at an astounding 140 mph. Refer to Figure 1–1.

Because cars can weigh well over a ton (2,000 to 3,000 pounds or more) and some trucks and sport utility vehicles (SUVs) can weigh over 2 tons (4,000 pounds or more), body parts are often crushed, bent, and torn in a collision. Major body/frame parts can even be forced out of alignment due to the tremendous energy dissipation produced when such heavy objects strike each other.

Auto body repair, also called **collision repair**, involves fixing a vehicle that has been damaged in an accident. With minor damage, this might only involve sanding and repainting a fender. With more serious damage, a large section of the frame or body might have to be straightened and numerous parts replaced.

Accident Scene

An **accident report** summarizes what happened during a collision and lists information about the drivers, their vehicles, and their insurance companies. Even after a minor accident, the police are usually called so that an accident report, or crash report, can be filled out. The police officer will ask the drivers questions about the accident and write down what happened so that the insurance companies can determine what caused the accident and who might be at fault. The car owners will also call their insurance companies to report the accident.

A **wrecker** is a truck equipped with special lifting equipment for raising and transporting a damaged vehicle away from an accident scene. Most wreckers have a power-operated cable or a large hydraulic arm for lifting one end of the damaged vehicle for towing. A flat-bed wrecker has a winch for pulling the whole vehicle up and onto the bed of the truck for transport.

If the car is no longer driveable, a wrecker will tow the damaged vehicle to a body shop or **salvage yard**, a business that resells parts from collision-damaged vehicles. Look at Figure 1–2.

Body Shop

A body shop, or **collision repair facility**, has well-trained technicians, specialized tools, and heavy equipment for restoring damaged vehicles to their pre-accident condition. There are several ways to classify a body shop. A few of the most common are discussed in the following section (Figure 1–3).

Figure 1–3 The modern body shop is equipped with specialized equipment and well-trained personnel. It can be a safe and enjoyable place to work if safety rules are followed. If not, it can be a very dangerous place!

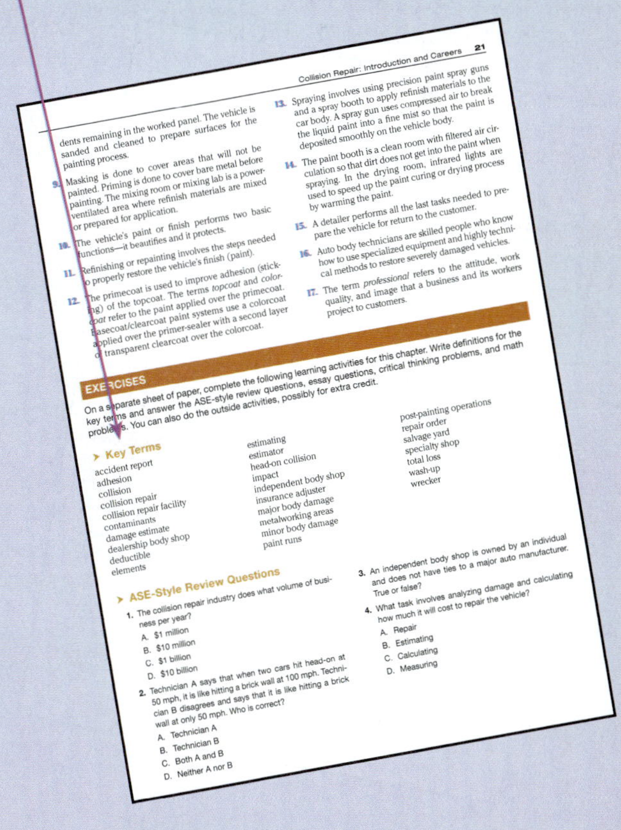

Collision Repair: Introduction and Careers 21

dents remaining in the worked panel. The vehicle is sanded and cleaned to prepare surfaces for the painting process.

9. Masking is done to cover areas that will not be painted. Priming is done to cover bare metal before painting. The mixing room or mixing lab is a power-ventilated area where refinish materials are mixed or prepared for application.
10. The vehicle's paint or finish performs two basic functions—it beautifies and it protects.
11. Refinishing or repainting involves the steps needed to properly restore the vehicle's finish (paint).
12. The primecoat is used to improve adhesion (sticking) of the topcoat. The terms *topcoat* and *colorcoat* refer to the paint applied over the primecoat. Basecoat/clearcoat paint systems use a colorcoat applied over the primer-sealer with a second layer of transparent clearcoat over the colorcoat.
13. Spraying involves using precision paint spray guns and a spray booth to apply refinish materials to the car body. A spray gun uses compressed air to break the liquid paint into a fine mist so that the paint is deposited smoothly on the vehicle body.
14. The paint booth is a clean room with filtered air circulation so that dirt does not get into the paint when spraying. In the drying room, infrared lights are used to speed up the paint curing or drying process by warming the paint.
15. A detailer performs all the last tasks needed to prepare the vehicle for return to the customer.
16. Auto body technicians are skilled people who know how to use specialized equipment and highly technical methods to restore severely damaged vehicles.
17. The term *professional* refers to the attitude, work quality, and image that a business and its workers project to customers.

EXERCISES

On a separate sheet of paper, complete the following learning activities for this chapter. Write definitions for the key terms and answer the ASE-style review questions, essay questions, critical thinking problems, and math problems. You can also do the outside activities, possibly for extra credit.

Key Terms

accident report
adhesion
collision
collision repair
collision repair facility
contaminants
damage estimate
dealership body shop
deductible
elements
estimating
estimator
head-on collision
impact
independent body shop
insurance adjuster
major body damage
metalworking areas
minor body damage
paint runs
post-painting operations
repair order
salvage yard
specialty shop
total loss
wash-up
wrecker

ASE-Style Review Questions

1. The collision repair industry does what volume of business per year?
   A. $1 million
   B. $10 million
   C. $1 billion
   D. $10 billion
2. Technician A says that when two cars hit head-on at 50 mph, it is like hitting a brick wall at 100 mph. Technician B disagrees and says that it is like hitting a brick wall at only 50 mph. Who is correct?
   A. Technician A
   B. Technician B
   C. Both A and B
   D. Neither A nor B
3. An independent body shop is owned by an individual and does not have ties to a major auto manufacturer. True or false?
4. What task involves analyzing damage and calculating how much it will cost to repair the vehicle?
   A. Repair
   B. Estimating
   C. Calculating
   D. Measuring

DANGER

Danger notes summarize critical safety rules. They alert you to operations that could hurt you or someone else. They do not only appear in the safety chapter; you will find them throughout the text where they apply. Read and remember all dangers. Your health is invaluable.

WARNINGS

Warnings provide important information to help prevent the kinds of accidents that can damage parts or tools. They are common mistakes that should be avoided. They appear throughout the text where they apply to the instructions being given.

SHOP TALK

Shop Talk notes give added information to help you complete a particular procedure successfully or to make a task easier. They are hints to help you work more efficiently and profitably.

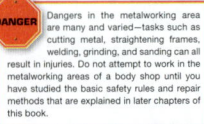

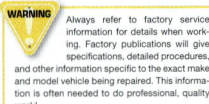

8 Chapter 1

Figure 1–7 Before starting repairs, body shops usually wash the vehicle thoroughly to remove road dirt. Keeping the shop clean is important to refinishing quality because it keeps paint contaminants out of the shop work areas.

Figure 1–8 To start work, you must normally remove damaged outer body parts to gain access to hidden parts that require straightening or replacement. Here the technician is removing a front fender.

wiping the body down with wax and grease remover. These steps will remove mud, dirt, wax, and water-soluble **contaminants** (unwanted substances). These substances must be cleaned off before starting bodywork because they could contaminate the paint and cause paint problems later. The car or truck should be completely dry before being moved to the repair area. Look at Figure 1–7.

Metalworking Areas

The **metalworking areas** are shop locations where damaged parts are removed, repaired, and installed. Such damage can result from either a collision or part

DANGER Dangers in the metalworking area are many and varied—tasks such as cutting metal, straightening frames, welding, grinding, and sanding can all result in injuries. Do not attempt to work in the metalworking areas of a body shop until you have studied the basic safety rules and repair methods that are explained in later chapters of this book.

WARNING Always refer to factory service information for details when working. Factory publications will give specifications, detailed procedures, and other information specific to the exact make and model vehicle being repaired. This information is often needed to do professional, quality work!

deterioration. The metalworking area is where the shop performs most of the collision repair tasks. Because of all of the grinding, sanding, and welding, this area tends to become dusty and dirty quickly.

Before starting work in this area, the body technician must first study and diagnose the damage that has occurred. Body technicians use the information on the RO to determine what repairs are needed. The technician may need to consult with the estimator before proceeding. It is then up to the technician to decide how to accomplish the repairs outlined on the estimate and RO.

Once the damage and repair methods are analyzed, the repairs must be completed in a systematic manner. For example, if a panel is creased, torn, or caved in, it can be straightened using hammers, hydraulic jacks, and other body shop tools. If the panel is badly crushed and folded, it must be replaced. If the unibody or frame is damaged, it must be straightened or parts replaced according to factory recommendations.

A *stall* is a work area for one vehicle, often marked off with painted lines on the floor. Also termed a *bay*, each stall is large enough so that the technician has room to work all the way around the vehicle.

Often, one of the first steps completed in the metalworking area is part removal. Badly damaged bolt-on parts must be removed to gain access to hidden damage. This might involve unfastening the bumper, grille, or fenders, for example Figure 1–8.

Vehicle Measurement

Vehicle measurement helps determine the extent and direction of major damage. If the vehicle has been in a serious accident, vehicle measurement is often done to find out whether the frame/unibody has been forced out of alignment. Specialized measuring tools are used to measure across specific reference points on the vehicle to find out whether body damage exists.

Measurement systems are specialized tools and equipment that allow the technician to check for frame

20 Chapter 1

shop manager is usually the same person. In large operations or dealerships, the owner might hire a shop manager. In all cases, the person in charge should understand all of the work done in the shop as well as its business operations.

The *shop supervisor* is in charge of the everyday operation of the shop. This job involves communication with all personnel who contribute to the body shop's success.

The *parts manager* is in charge of ordering all parts (both new and salvaged), receiving all parts, and seeing that they are delivered to the right technician. Because not every collision repair shop has a parts manager, the task of ordering parts can fall on all employees at one time or another.

The *bookkeeper* keeps the shop's books, prepares invoices, writes checks, pays bills, makes bank deposits, checks bank statements, and takes care of tax payments. Many shops hire an outside accountant to help perform these tasks.

The *office manager's* duties include various aspects of the business—handling letters, estimates, and receipts, for example. In many small shops, the office manager also acts as the parts manager and bookkeeper.

A *receptionist* is sometimes employed to greet customers, answer the phone, route messages, and do other tasks.

Larger shops generally have trained *tow truck operators* to operate their wrecker(s). Rather than own these expensive pieces of equipment, many smaller shops depend on independent towing services or farm out such work to other repair garages.

SHOP TALK It is important for body shop personnel to have good communication and cooperation skills. Everyone in the shop should work as a team member to make the facility profitable, enjoyable, and safe. "A chain is only as strong as its weakest link"—this saying applies to the smooth operation of a body shop. If anyone—from the estimator to the painter—does not do the job right, everyone suffers. Customers will not return to the shop, and everyone's paycheck will be affected in the long run.

There are other career openings in the collision repair field. Some of these include:

- Insurance adjuster or appraiser
- Vocational/technical instructor
- Salvage yard technician
- Dealership parts counterperson
- Paint company representative
- Auto manufacturer representative
- Equipment salesperson

To research these and other career opportunities, talk to your guidance counselor, visit your local library, or search the Internet. There you can obtain more detailed information on the qualifications and training requirements for each position.

SUMMARY

1. A collision, nicknamed a *crash* or a *wreck*, is damage caused by an impact (hit) on the vehicle.
2. Auto body repair, also called collision repair, involves fixing a vehicle that has been damaged in an accident. The vehicle is first brought into the shop, where a damage estimate is prepared to calculate the cost of repairs.
3. Minor body damage only requires that a few parts be replaced or repaired before the car is repainted. Major body damage usually must be corrected by replacing, repairing, and straightening large body parts before repainting. A collision can also be severe enough to cause a total loss, where repairs would be more expensive than buying another car.
4. Repair instructions are summarized on a printed repair order (RO), and the repairs are carried out according to those instructions. The metalworking areas are shop locations where damaged parts are removed, repaired, and installed. A stall is a work area for one vehicle, often marked off with painted lines on the floor.
5. Vehicle measurement is done to determine the extent and direction of major damage. Measurement systems are specialized tools and equipment that help the technician to check and measure for frame or body misalignment resulting from a high-impact collision.
6. Frame straightening equipment (also called a frame rack) uses a large steel framework, pulling chains, and hydraulic power to pull or force the frame or body back into its original position.
7. A panel is a large metal or plastic body part—a fender, hood, truck, or roof panel, for example. Structural panel replacement involves welding a new panel in place of an old, damaged one. Panel adjustment involves moving or shifting a part to properly align with other parts or panels. When parts are replaced, corrosion protection coatings must be applied to prevent the new parts from rusting.
8. Panel straightening involves using various hand tools and equipment to force the panel back into its original shape. Body filler is used to cover any small

20 Chapter 1

shop manager is usually the same person. In large operations or dealerships, the owner might hire a shop manager. In all cases, the person in charge should understand all of the work done in the shop as well as its business operations.

The *shop supervisor* is in charge of the everyday operation of the shop. This job involves communication with all personnel who contribute to the body shop's success.

The *parts manager* is in charge of ordering all parts (both new and salvaged), receiving all parts, and seeing that they are delivered to the right technician. Because not every collision repair shop has a parts manager, the task of ordering parts can fall on all employees at one time or another.

The *bookkeeper* keeps the shop's books, prepares invoices, writes checks, pays bills, makes bank deposits, checks bank statements, and takes care of tax payments. Many shops hire an outside accountant to help perform these tasks.

The *office manager's* duties include various aspects of the business—handling letters, estimates, and receipts, for example. In many small shops, the office manager also acts as the parts manager and bookkeeper.

A *receptionist* is sometimes employed to greet customers, answer the phone, route messages, and do other tasks.

Larger shops generally have trained *tow truck operators* to operate their wrecker(s). Rather than own these expensive pieces of equipment, many smaller shops depend on independent towing services or farm out such work to other repair garages.

SHOP TALK It is important for body shop personnel to have good communication and cooperation skills. Everyone in the shop should work as a team member to make the facility profitable, enjoyable, and safe. "A chain is only as strong as its weakest link"—this saying applies to the smooth operation of a body shop. If anyone—from the estimator to the painter—does not do the job right, everyone suffers. Customers will not return to the shop, and everyone's paycheck will be affected in the long run.

There are other career openings in the collision repair field. Some of these include:

- Insurance adjuster or appraiser
- Vocational/technical instructor
- Salvage yard technician
- Dealership parts counterperson
- Paint company representative
- Auto manufacturer representative
- Equipment salesperson

To research these and other career opportunities, talk to your guidance counselor, visit your local library, or search the Internet. There you can obtain more detailed information on the qualifications and training requirements for each position.

SUMMARY

1. A collision, nicknamed a *crash* or a *wreck*, is damage caused by an impact (hit) on the vehicle.
2. Auto body repair, also called collision repair, involves fixing a vehicle that has been damaged in an accident. The vehicle is first brought into the shop, where a damage estimate is prepared to calculate the cost of repairs.
3. Minor body damage only requires that a few parts be replaced or repaired before the car is repainted. Major body damage usually must be corrected by replacing, repairing, and straightening large body parts before repainting. A collision can also be severe enough to cause a total loss, where repairs would be more expensive than buying another car.
4. Repair instructions are summarized on a printed repair order (RO), and the repairs are carried out according to those instructions. The metalworking areas are shop locations where damaged parts are removed, repaired, and installed. A stall is a work area for one vehicle, often marked off with painted lines on the floor.
5. Vehicle measurement is done to determine the extent and direction of major damage. Measurement systems are specialized tools and equipment that help the technician to check and measure for frame or body misalignment resulting from a high-impact collision.
6. Frame straightening equipment (also called a frame rack) uses a large steel framework, pulling chains, and hydraulic power to pull or force the frame or body back into its original position.
7. A panel is a large metal or plastic body part—a fender, hood, truck, or roof panel, for example. Structural panel replacement involves welding a new panel in place of an old, damaged one. Panel adjustment involves moving or shifting a part to properly align with other parts or panels. When parts are replaced, corrosion protection coatings must be applied to prevent the new parts from rusting.
8. Panel straightening involves using various hand tools and equipment to force the panel back into its original shape. Body filler is used to cover any small

CHAPTER SUMMARY

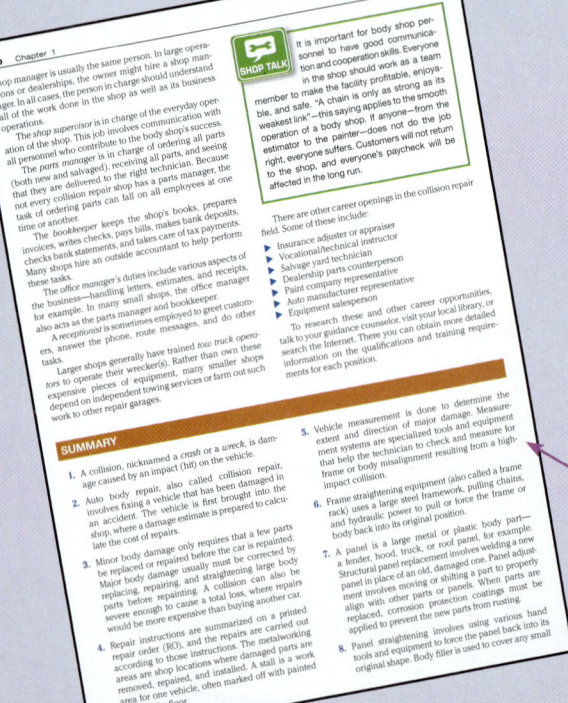

Each *Chapter Summary* gives a brief list of the most important information in the chapter. It will help you review and understand which points were the most important.

REVIEW QUESTIONS

Review Questions will help you to measure the skills and knowledge you learned in the chapter. To check your "brain power," different types of questions are given: ASE, Essay, Critical Thinking Problems, and Math Problems. The ASE questions will help you prepare to pass auto body repair certification tests.

ACTIVITIES

These are practical, hands-on activities that challenge the student to apply the skills learned in the chapter to real-life experiences. *Activities* often include going to a body shop or other workplaces to gather some type of research for analysis. This feature is a wonderful way to teach students how and where they will eventually use their skills.

PHOTO SUMMARIES

Several step-by-step photo summaries illustrate common auto body repair procedures.

Collision Repair: Introduction and Careers **21**

dents remaining in the worked panel. The vehicle is sanded and cleaned to prepare surfaces for the painting process.

9. Masking is done to cover areas that will not be painted. Priming is done to cover bare metal before painting. The mixing room or mixing lab is a power-ventilated area where refinish materials are mixed or prepared for application.
10. The vehicle's paint or finish performs two basic functions—it beautifies and it protects.
11. Refinishing or repainting involves the steps needed to properly restore the vehicle's finish (paint).
12. The primecoat is used to improve adhesion (sticking) of the topcoat. The terms *topcoat* and *colorcoat* refer to the paint applied over the primecoat. Basecoat/clearcoat paint systems use a colorcoat applied over the primer-sealer with a second layer of transparent clearcoat over the colorcoat.
13. Spraying involves using precision paint spray guns and a spray booth to apply refinish materials to the car body. A spray gun uses compressed air to break the liquid paint into a fine mist so that the paint is deposited smoothly on the vehicle body.
14. The paint booth is a clean room with filtered air circulation so that dirt does not get into the paint when spraying. In the drying room, infrared lights are used to speed up the paint curing or drying process by warming the paint.
15. A detailer performs all the last tasks needed to prepare the vehicle for return to the customer.
16. Auto body technicians are skilled people who know how to use specialized equipment and highly technical methods to restore severely damaged vehicles.
17. The term *professional* refers to the attitude, work quality, and image that a business and its workers project to customers.

EXERCISES

On a separate sheet of paper, complete the following learning activities for this chapter. Write definitions for the key terms and answer the ASE-style review questions, essay questions, critical thinking problems, and math problems. You can also do the outside activities, possibly for extra credit.

Key Terms

accident report
adhesion
collision
collision repair
collision repair facility
contaminants
damage estimate
dealership body shop
deductible
elements
estimating
estimator
head-on collision
impact
independent body shop
insurance adjuster
major body damage
metalworking areas
minor body damage
paint runs
post-painting operations
repair order
salvage yard
specialty shop
total loss
wash-up
wrecker

ASE-Style Review Questions

1. The collision repair industry does what volume of business per year?
   A. $1 million
   B. $10 million
   C. $1 billion
   D. $10 billion
2. Technician A says that when two cars hit head-on at 50 mph, it is like hitting a brick wall at 100 mph. Technician B disagrees and says that it is like hitting a brick wall at only 50 mph. Who is correct?
   A. Technician A
   B. Technician B
   C. Both A and B
   D. Neither A nor B
3. An independent body shop is owned by an individual and does not have ties to a major auto manufacturer. True or false?
4. What task involves analyzing damage and calculating how much it will cost to repair the vehicle?
   A. Repair
   B. Estimating
   C. Calculating
   D. Measuring

Collision Repair: Introduction and Careers **23**

Math Problems

1. A technician is paid $45 per hour. The repair will take 11.5 hours. Parts for the repair will cost $176.25 plus 7 percent tax. What is the total of the estimate?
2. An older car is valued at $2,500. The parts for the repair will cost $1,500. How many hours of labor at $35 per hour can be completed before the car should be declared totaled?
3. An independent repair shop made $33,507 total in a one-month period. The cost of labor was $45 per hour with a total of 200 hours of labor. Parts cost the repair shop $14,000. Power and property bills cost a total of $3,500. After these costs are taken into account, what did the repair shop make in profits for the month?

Activities

1. Take a field trip to a local collision repair facility. Have the shop owner or supervisor give you a guided tour of the repair operation. Discuss what you learned during the field trip in class the next day.
2. Ask a technician or shop owner to visit your classroom to describe their duties and answer questions.

FULL-COLOR CHAPTERS

Every chapter in *Auto Body Repair Technology* is now in full color, bringing the text to life with more detailed and realistic illustrations of collision repair technologies.

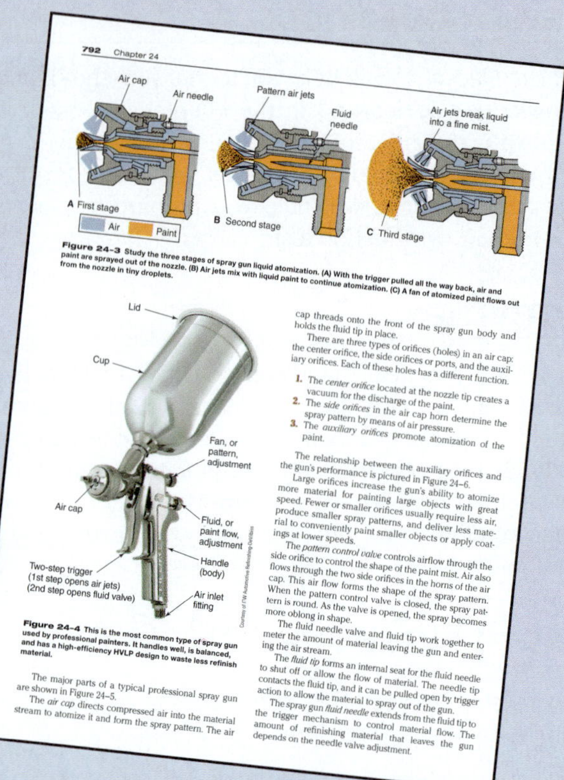

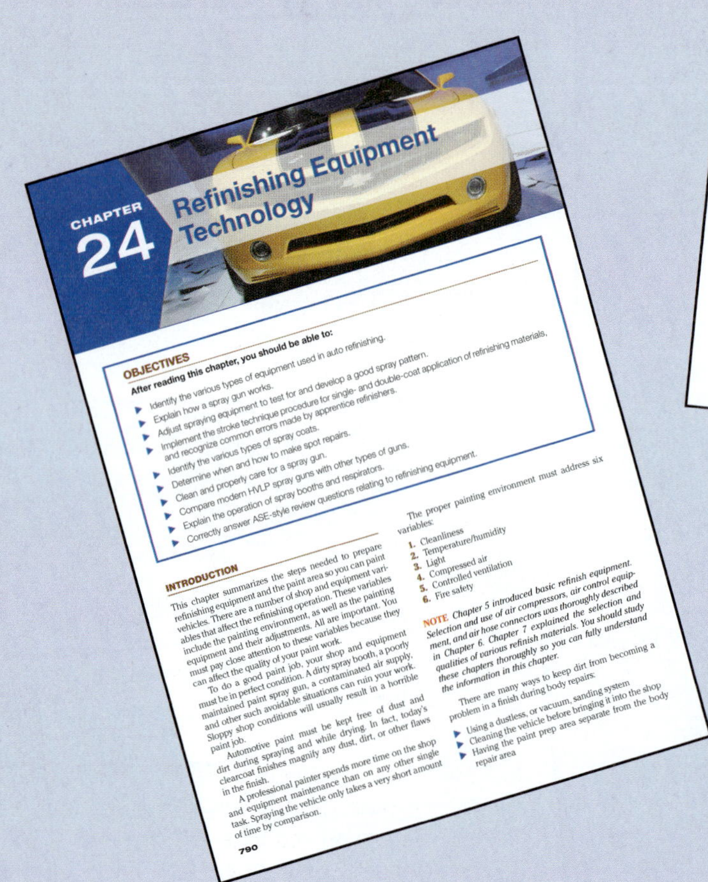

Supplements

Student Technician's Manual

Written by an experienced automotive collision repair teacher the Student's Technician's Lab Manual has been developed as a supplemental learning tool. The job sheets provide detailed directions for doing hands-on learning activities. They outline and question students as they complete competency-based learning tasks.

Instructor Resources CD

Carefully prepared, the Instructor Resources CD brings together several time-saving tools that allow for effective, efficient instruction. The Instructor Resources CD contains the following components:

- **PowerPoint®** lecture slides, which present the highlights of each chapter.
- An **Image Gallery**, which offers a database of hundreds of images in the text. These can easily be imported into the PowerPoint® presentations.
- An **Answer Key** file, which provides the answers to all end-of-chapter questions.
- **NATEF Correlations**
- **End-of-Chapter Review Questions**, which are provided in MS Word format.

Instructor Companion Website

The Instructor Companion Website, found on cengagebrain.com, includes the following components to help minimize instructor preparation time and engage students:

- **PowerPoint®** lecture slides, which present the highlights of each chapter.
- An **Image Gallery**, which offers a database of hundreds of images in the text. These can easily be imported into the PowerPoint® presentations.
- An **Answer Key** file, which provides the answers to all end-of-chapter questions.
- **NATEF Correlations**
- **End-of-Chapter Review Questions**, which are provided in MS Word format.

Cengage Learning Testing Powered by Cognero is a flexible, online system that allows you to:

- Author, edit, and manage test bank content from multiple Cengage Learning solutions.
- Create multiple test versions in an instant.
- Deliver tests from your LMS, your classroom, or wherever you want.

MindTap for Auto Body Repair Technology

MindTap is a personalized teaching experience with relevant assignments that guide students to analyze, apply, and improve thinking, allowing you to measure skills and outcomes with ease.

- *Personalized Teaching*: Becomes YOURS with a Learning Path that is built with key student objectives. Control what students see and when they see it—match your syllabus exactly by hiding, rearranging, or adding your own content.
- *Guide Students*: Goes beyond the traditional "lift and shift" model by creating a unique learning path of relevant readings, multimedia and activities that move students up the learning taxonomy from basic knowledge and comprehension to analysis and application.
- *Measure Skills and Outcomes*: Analytics and reports provide a snapshot of class progress, time on task, engagement and completion rates.

Introduction

CHAPTER

1 Collision Repair: Introduction and Careers

OBJECTIVES

After studying this chapter, you should be able to:

- Describe what happens to a motor vehicle during a collision.
- Summarize the basic steps needed to repair a vehicle damaged in an accident.
- Explain the major work areas of a typical collision repair facility.
- Summarize the work flow through a typical body shop.
- Describe the types of positions or jobs available in the collision repair industry.
- Identify the setup and inner workings of a typical collision repair shop.
- Answer ASE-style review questions relating to collision repair.
- Understand fundamental terms used in the collision repair industry.

INTRODUCTION

This chapter will explore the challenging world of auto body repair, also called collision repair. It will give you the basic knowledge needed to fully grasp the more detailed information in later chapters. You will follow damaged vehicles through a typical repair process from beginning to end: estimating damage; metal straightening, filling, sanding, masking, painting, and detailing; and final delivery to the customer.

Considering that the United States is a "nation on wheels" and auto accidents happen, you have selected an excellent area for further study. The collision repair skills you learn will remain in high demand as long as there are people driving automobiles. The collision repair industry is vast. Body shops do an astounding $10 billion of repair work per year. This enormous amount of work requires a vast number of well-trained technicians and related personnel.

If you are just beginning your study of collision repair, you have much to learn. Only highly skilled, knowledgeable professionals can properly repair today's vehicles. Study the material in this textbook carefully and you will be on your way to a successful career in collision repair.

Believe it or not, the very first "motor car" ever built was in an accident the first time it was driven on public roads. The huge, crude car looked something like a steam locomotive. It was being test-driven down a country road. Chugging over a hill, the steel-wheeled, "smoke-belching monster" almost ran over a horse-drawn carriage. The car's startled driver had to swerve off the narrow dirt road to avoid running over the horses, and he plowed into a brick wall. After the first car crash, plow horses had to be used to tow the "horseless carriage" back to the barn for body shop repairs.

1.1 WHAT IS COLLISION REPAIR?

A **collision**, nicknamed a *crash* or a *wreck* in laypersons' terms, is damage caused by an **impact** (hit) on the vehicle. This impact might be from another vehicle or from another object. The collision might be minor enough to only scratch the paint, or it might be severe enough to cause thousands of dollars' worth of damage to numerous metal and plastic parts.

A **head-on collision** results when two vehicles driving toward each other accidentally collide. It is often

Figure 1–1 **During a major collision, like this head-on crash test, tremendous energy is dissipated into the body/frame structures of both cars. Modern vehicles are designed to absorb this energy by controlled collapse of structural body/frame members. The front and rear ends of the vehicles are designed to crush or collapse while keeping the passenger compartment intact. With today's front and side air bags and seat belts, the drivers and passengers would suffer little or no injury in this violent collision.**

Figure 1–2 **If the accident results in major damage and the vehicle cannot be driven, a wrecker is called to tow the car. Proper tow methods are needed to avoid further damage when moving the wrecked car.**

the worst type of accident because the speed of both vehicles must be combined to determine the force of the impact. For example, if both vehicles are travelling 70 miles per hour (mph), the impact is similar to one vehicle smashing into a stationary brick wall at an astounding 140 mph. Refer to Figure 1–1.

Because cars can weigh well over a ton (2,000 to 3,000 pounds or more) and some trucks and sport utility vehicles (SUVs) can weigh over 2 tons (4,000 pounds or more), body parts are often crushed, bent, and torn in a collision. Major body/frame parts can even be forced out of alignment due to the tremendous energy dissipation produced when such heavy objects strike each other.

Auto body repair, also called **collision repair**, involves fixing a vehicle that has been damaged in an accident. With minor damage, this might only involve sanding and repainting a fender. With more serious damage, a large section of the frame or body might have to be straightened and numerous parts replaced.

Accident Scene

An **accident report** summarizes what happened during a collision and lists information about the drivers, their vehicles, and their insurance companies. Even after a minor accident, the police are usually called so that an accident report, or crash report, can be filled out. The police officer will ask the drivers questions about the accident and write down what happened so that the insurance companies can determine what caused the accident and who might be at fault. The car owners will also call their insurance companies to report the accident.

A **wrecker** is a truck equipped with special lifting equipment for raising and transporting a damaged vehicle away from an accident scene. Most wreckers have a power-operated cable or a large hydraulic arm for lifting one end of the damaged vehicle for towing. A flatbed wrecker has a winch for pulling the whole vehicle up and onto the bed of the truck for transport.

If the car is no longer driveable, a wrecker will tow the damaged vehicle to a body shop or **salvage yard**, a business that resells parts from collision-damaged vehicles. Look at Figure 1–2.

Body Shop

A body shop, or **collision repair facility**, has well-trained technicians, specialized tools, and heavy equipment for restoring damaged vehicles to their pre-accident condition. There are several ways to classify a body shop. A few of the most common are discussed in the following section (Figure 1–3).

Figure 1–3 **The modern body shop is equipped with specialized equipment and well-trained personnel. It can be a safe and enjoyable place to work if safety rules are followed. If not, it can be a very dangerous place!**

An **independent body shop** is owned and operated by a private individual. The shop is not associated with other shops or companies.

A *franchise body shop* is tied to a main headquarters that regulates and aids the operation of the shop. The shop logo, materials used, fees, and so on are all set by the corporate headquarters, and the franchisee must follow these guidelines.

A **dealership body shop** is owned and managed under the guidance of a new car dealership, such as General Motors, Chrysler, Lexus, Toyota, Jaguar, or Ford. This type of shop often concentrates on repairs of the specific make of cars sold by the dealership.

A *progression* or *production shop* often has an assembly line organization with specialists in each area of repair. One person might do nothing but heavy frame repair work. Another technician might be good at "building the body," or installing parts and panels. The shop might have a wheel alignment technician, prep people, painter, and cleanup specialists. The vehicle will move from one area and specialist to the next until fully repaired.

A **specialty shop** performs only specific types of repairs. For example, the body shop might send a radiator with a small hole in it to a specialty radiator shop for repair with specialized equipment.

A body shop that provides *complete collision services* might do wheel alignments, cooling system repairs, electrical system diagnosis and repair, suspension system work, and other repairs. Today, more and more collision repair shops are offering complete collision services. They have both a body shop area and a mechanical repair area.

Damage Estimating

A vehicle involved in a collision is first brought into the shop, where a **damage estimate** is prepared to calculate the cost of repairs. The labor, parts, and materials must be added to find the total cost of vehicle repair (Figure 1–4).

Estimating involves analyzing damage and calculating how much it will cost to repair a vehicle. It is critical that the quote on the repair be neither too high nor too low. If the estimate is too high, another shop with a lower bid will usually get the job. If too low, the profits may not be enough to cover the cost of repairs, and the shop could lose money.

In most shops, a well-trained **estimator** makes an appraisal of vehicle damage and determines what must be done to repair the vehicle. This person must be well versed in how cars and trucks are made and be good with numbers, computers, and communicating with people. An estimator at work is shown in Figure 1–5.

Minor body damage requires that only a few parts be replaced or repaired before being refinished or painted. Minor damage is often due to a low-speed "fender bender" in which two cars hit at low speed or one car runs into something. Such damage might be as minor as a tiny "door ding" that occurs when someone

A The estimator must study the damage carefully to determine the cost of repairs. Today's estimators often use an electronic storage unit and electronic camera so that data about damage can be input into the shop's computer.

B The vehicle interior is often damaged in a major collision. Replacing deployed air bags is expensive and must be included in the estimate.

C Damage to the engine, engine mounts, pulleys, and underbody parts must be accounted for in an estimate. Sometimes the vehicle is raised on a lift for undercarriage inspection.

Figure 1–4 Estimating is done to evaluate the extent of the damage and to arrive at a cost for repairing the vehicle.

A Data is downloaded into a personal computer in the business office.

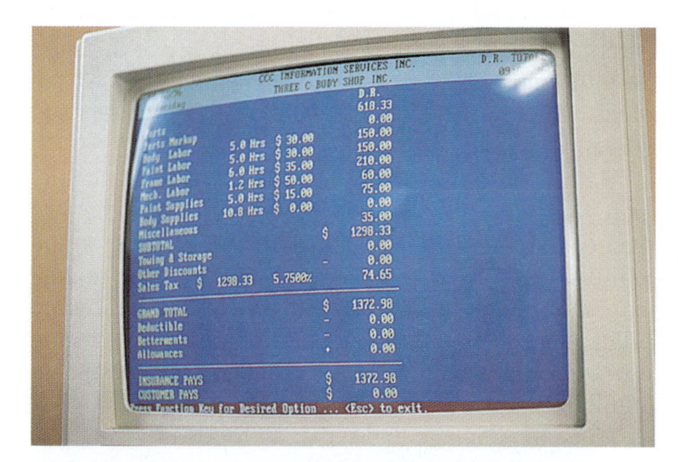

B A special estimating computer program will speed tabulation of labor rates and parts costs.

Figure 1–5 Once damage has been inspected, a computer is used to streamline estimating and ordering parts.

accidentally opens a car door into the side of another car, making a small paint chip.

Major body damage must usually be corrected by replacing, repairing, and straightening large body parts before refinishing. Parts might have to be cut off and new ones welded into place. Though severe, repairing the damage is less than the cost of vehicle replacement or less than the value of the vehicle.

A collision can also be severe enough to cause a **total loss**, in which repairs would be more expensive than buying another car. In this case, the insurance company does not pay for repairs but instead gives the driver enough money to purchase another similar year, make, and model vehicle.

Remember that nearly any damaged automobile can be restored to a safe, driveable condition if the vehicle owner or insurance company is willing to pay for the repair. It is this cost that is the major consideration (Figure 1–6).

The estimate is usually given to the customer, who submits it to the insurance company. **The insurance adjuster** reviews the estimates and determines which one best reflects how the vehicle should be repaired. The

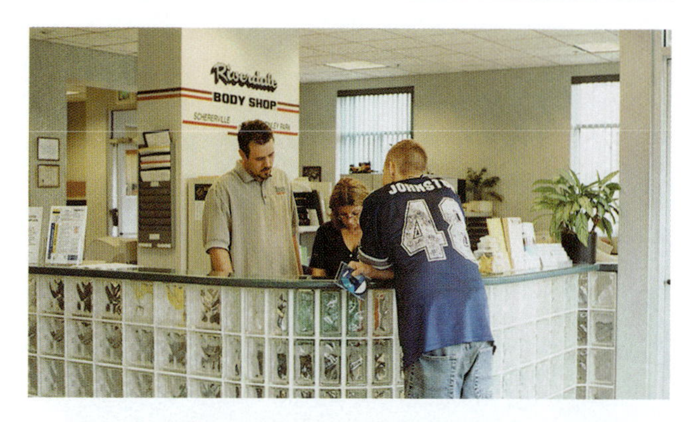

Figure 1–6 Once parts, labor, and materials have been tabulated, a printout or hard copy of the estimate summarizing what must be done to repair the vehicle and the repair costs is given to the customer and/or insurance adjuster.

adjuster may inspect the wrecked car to determine that the repairs will be done in a cost-effective fashion. The insurance company usually writes a check to the owner of the damaged vehicle to cover the cost of the repair, minus any **deductible** (amount the owner agrees to co-pay on the insurance policy).

1.2 BODY SHOP REPAIRS

Once the owner and the insurance company approve the repairs, the vehicle is turned over to the shop supervisor. The shop supervisor, sometimes with the help of a technician, will then review the estimate to determine how to do the repair.

Repair instructions are summarized on a printed **repair order** (RO), and the repairs are carried out according to these instructions. Once the RO is received in the shop, body shop repair procedures follow a general sequence. The basic repair sequence for a vehicle that has major frame/body damage is as follows:

1. Clean vehicle before moving it into repair area.
2. Study the RO and vehicle damage to determine repair procedure.
3. Remove badly damaged bolt-on parts.
4. Measure damage.
5. Straighten frame/unibody damage on frame rack.
6. Replace badly damaged welded-on parts.
7. Straighten minor body damage.
8. Apply body filler and coarse-sand repair area.
9. Apply a primer-filler around body filled area.
10. Fine-sand repair area and all parts to be refinished.
11. Mask areas not to be painted.
12. Clean surfaces to be painted.
13. Refinish (prime, seal, paint) damaged body parts.
14. Detail vehicle (unmask, clean, and polish) as needed.

Wash-Up Area

When a car is brought into the shop, the first step is usually wash-up. **Wash-up** involves a thorough cleaning of the vehicle with soap and water. This is followed by

Figure 1-7 **Before starting repairs, body shops usually wash the vehicle thoroughly to remove road dirt. Keeping the shop clean is important to refinishing quality because it keeps paint contaminants out of the shop work areas.**

Figure 1-8 **To start work, you must normally remove damaged outer body parts to gain access to hidden parts that require straightening or replacement. Here the technician is removing a front fender.**

wiping the body down with wax and grease remover. These steps will remove mud, dirt, wax, and water-soluble **contaminants** (unwanted substances). These substances must be cleaned off before starting bodywork because they could contaminate the paint and cause paint problems later. The car or truck should be completely dry before being moved to the repair area. Look at Figure 1–7.

Metalworking Areas

The **metalworking areas** are shop locations where damaged parts are removed, repaired, and installed. Such damage can result from either a collision or part deterioration. The metalworking area is where the shop performs most of the collision repair tasks. Because of all of the grinding, sanding, and welding, this area tends to become dusty and dirty quickly.

Before starting work in this area, the body technician must first study and diagnose the damage that has occurred. Body technicians use the information on the RO to determine what repairs are needed. The technician may need to consult with the estimator before proceeding. It is then up to the technician to decide how to accomplish the repairs outlined on the estimate and RO.

Once the damage and repair methods are analyzed, the repairs must be completed in a systematic manner. For example, if a panel is creased, torn, or caved in, it can be straightened using hammers, hydraulic jacks, and other body shop tools. If the panel is badly crushed and folded, it must be replaced. If the unibody or frame is damaged, it must be straightened or parts replaced according to factory recommendations.

A *stall* is a work area for one vehicle, often marked off with painted lines on the floor. Also termed a *bay*, each stall is large enough so that the technician has room to work all the way around the vehicle.

Often, one of the first steps completed in the metalworking area is part removal. Badly damaged bolt-on parts must be removed to gain access to hidden damage. This might involve unfastening the bumper, grille, or fenders, for example Figure 1–8.

DANGER Dangers in the metalworking area are many and varied—tasks such as cutting metal, straightening frames, welding, grinding, and sanding can all result in injuries. Do not attempt to work in the metalworking areas of a body shop until you have studied the basic safety rules and repair methods that are explained in later chapters of this book.

WARNING Always refer to factory service information for details when working. Factory publications will give specifications, detailed procedures, and other information specific to the exact make and model vehicle being repaired. This information is often needed to do professional, quality work!

Vehicle Measurement

Vehicle measurement helps determine the extent and direction of major damage. If the vehicle has been in a serious accident, vehicle measurement is often done to find out whether the frame/unibody has been forced out of alignment. Specialized measuring tools are used to measure across specific reference points on the vehicle to find out whether body damage exists.

Measurement systems are specialized tools and equipment that allow the technician to check for frame

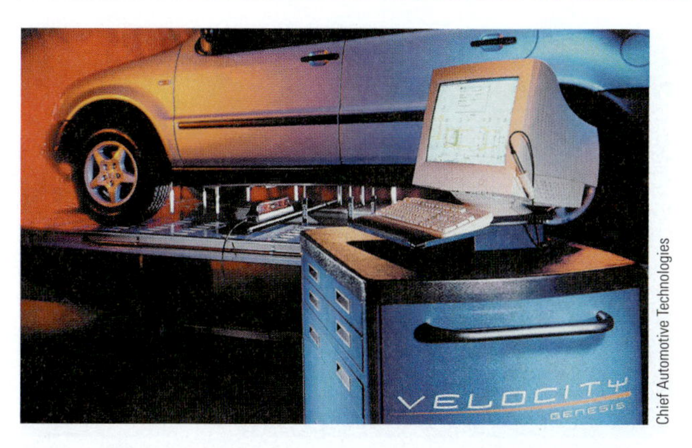

Chief Automotive Technologies

A To use this modern computerized measuring system, targets are hung from specific points on a vehicle. Laser light is reflected off the targets to make quick, accurate readings of the distance between targets.

Chief Automotive Technologies

B Computers can quickly compare known good specifications with laser measurements to calculate whether any section of a vehicle has been pushed out of alignment by a collision.

Figure 1–9 A measuring system is used to determine whether major body parts have been forced out of alignment by a major collision.

or body misalignment resulting from a collision. Various types of gauges and measuring devices can be used to compare *body specifications* to actual measurements taken from the damaged vehicle. The measurements will help determine what must be done to straighten any frame or body misalignment. If any measurement is not within factory specifications, the frame or unibody must be forced back into alignment using powerful hydraulic equipment (Figure 1–9).

Frame/Unibody Straightening

Once the extent and direction of frame misalignment are known, frame (unibody) straightening equipment can be used to pull the frame or body structure back into alignment.

Frame *straightening equipment* (also called a *frame rack*) uses a large steel framework, large steel towers, pulling chains, and hydraulic rams to pull the frame or body back into its original position (Figure 1–10).

Chief Automotive Technologies

Figure 1–10 A frame rack is a heavy steel framework with hydraulic equipment for forcing major structural parts back into alignment. The vehicle is anchored to the rack so it cannot move. Then pulling towers and chains can be attached to the damaged section of the vehicle to pull out the structural damage.

The vehicle frame or unibody is clamped down onto the frame equipment so it cannot move. Clamps and chains are then fastened to the damaged portion of the vehicle. Tremendous hydraulic pulling force is then applied to the chains to force the frame or body in the opposite direction of the collision impact. Refer to Figure 1–11.

After pulling, more measurements are taken to determine whether everything has returned to specification (Figure 1–12).

Panel Replacement

A *panel* is a large metal or plastic body part—a fender, hood, deck lid, or roof panel, for example. A vehicle body is made up of numerous panels welded, chemically bonded, or bolted together.

Panel replacement involves removing a panel or body part that is too badly damaged to be fixed. The new part has to be properly fit and fastened in place on the vehicle. This takes considerable skill (Figure 1–13).

Structural panel replacement involves welding a new panel in place of an old, damaged one. First, the badly damaged part is removed by drilling or grinding off its spot welds. Then the new parts are fitted into place while part locations are measured. Clamping pliers or self-taping screws are used to hold the new panel in place while welding. First, a welder is used to tack weld and fuse the part in place. Before final welding, measurements are again taken to make sure the new part is aligned properly. Look at Figure 1–14.

Panel adjustment involves moving or shifting a part to properly align it with other parts or panels. Accurate adjustment of body assemblies, such as hoods, deck lids, and doors, is often made by the technician. For example, if a door is not adjusted correctly, it may be difficult to close or may rattle when the car is driven over rough roads. The poorly adjusted door might also leak air and water. Such failures by the technician are bound to cause customer complaints.

When parts are replaced, *corrosion protection* coatings must be applied to prevent the new parts from rusting.

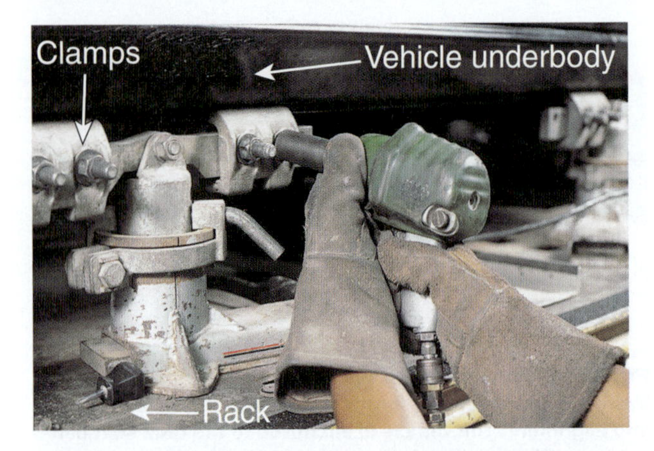

A First, the vehicle is clamped down to the frame rack using large clamps on the bottom of the unibody or frame. The vehicle must be anchored securely before pulling. Here pinch weld clamps are being installed to secure the vehicle while applying straightening force.

B After the vehicle is anchored, pulling clamps and chains are attached to the area to be straightened.

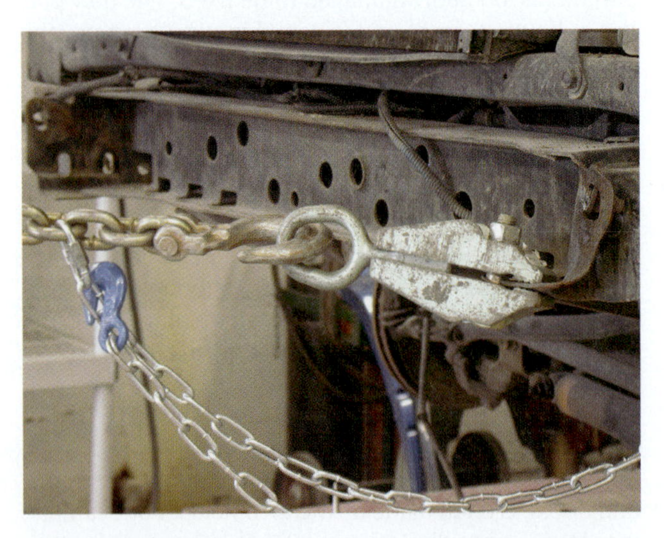

C Pulling power is applied to remove damage opposite to how it occurred. When pulling damage, you must be careful not to overpull. Metal will only flex back a small amount after releasing pulling force. If you overpull, it may be impossible to repair without part replacement.

Figure 1-11 Note how the technician is using a frame rack during a major repair.

Figure 1-12 Constantly measure as you pull out damage. This will let you pull in the right direction with the right amount of power. Here the tram gauge has been set to vehicle specs and compared to reference points on the damaged vehicle. If the tram gauge does not align with reference points on the vehicle, further pulling is needed until the points are lined up to specifications.

A Here a technician uses a plasma arc cutter to slice off a part for replacement.

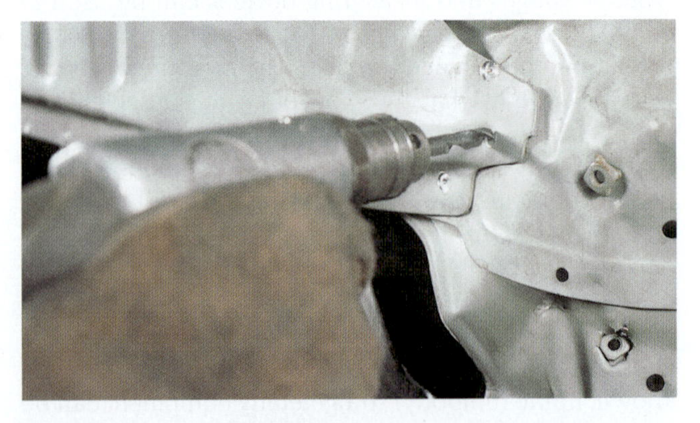

B Many parts or panels are attached with spot welds. A spot weld remover, a specialized drill, is often used to cut out each spot weld for part removal.

Figure 1-13 After the frame or unibody is straightened, you can then cut off damaged parts that require replacement.

A New structural parts must be temporarily held in place with clamping pliers or screws.

B First, tack weld parts in place, using only a few welds to hold the parts in alignment.

C After tack welding a part, measure its location before final welding. You may need to adjust a part's position while it is still tacked in place.

D If aligned properly, weld the structural parts into their final position following manufacturer recommendations. Proper welding skills are needed to do major collision damage repair. You may want to sign up for a welding class to further your career opportunities.

Figure 1–14 Replacing welded-on structural parts takes skill and training.

Weld-through primer and other materials are applied to the new parts to protect them from the **elements**—moisture, road salt, mud, and so on (Figure 1–15).

A technician must be able to hang the door in proper alignment and also transfer all internal parts (window mechanism, door handle, latch, and trim panel) to the new door. Look at Figure 1–16.

An air bag can cause injury if it fires accidentally. A technician replacing a side air bag should stand to one side of the bag in case it accidentally deploys. Refer to Figure 1–17.

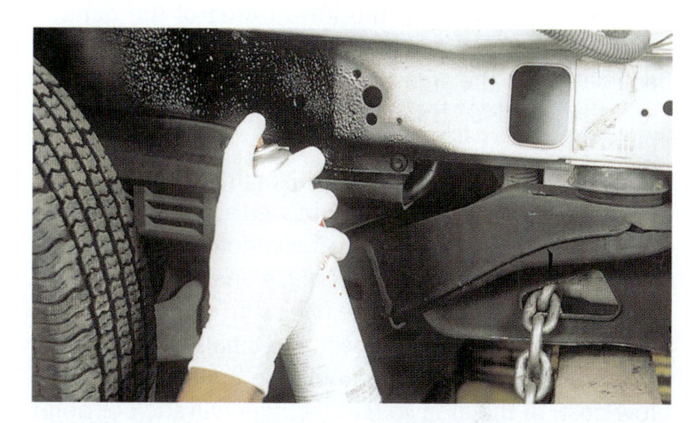

Figure 1–15 You must restore corrosion protection to all repair panels. Various products are available to prevent corrosion or rusting of new panels.

Figure 1–16 Here the entire door frame and skin had to be replaced because it was too damaged to be repaired.

Figure 1–17 A technician replaces a small air bag located on the inside of the door frame. Note how the technician stays to one side in case the air bag accidentally deploys. Air bags deploy with a "bang" and can cause injury to your arms, hands, and face.

Panel Straightening

Panel straightening involves using various hand tools and equipment to force a damaged panel back into its original shape. Body hammers, body filler, and sanders are a few of the tools and materials used to repair panel damage. A somewhat different approach is needed with plastic panels. Small parts (trim pieces, emblems, wiper blades, and so on) are removed before panel straightening to protect them.

With damage to a metal panel, a grinder is first used to remove the paint from the damaged area. Then small pull pins are welded to the lowest area of the dent. A puller can then be attached to the pins to force the dent back out. Once the area is almost back to its original contour, the welded pull pins can be ground off. See Figure 1–18.

When you can access the back of the panel, leave the paint on until the panel is straightened as close to its original contour as possible. The reflection of light off the shiny paint surface will reveal highs and lows when you use straightening tools on the back side of the damage.

Body filler is used to cover any small imperfections remaining in the worked panel. Use compressed air to blow dust off the area so that the filler will stick or bond to the area securely. Mix *hardener* into the body filler to make the material *cure*. Use a metal or plastic *spreader* to apply the body filler over the repair area on the panel. The filler cures or hardens quickly, so skill is needed to properly apply the body filler.

When partially cured, a coarse file is often used to knock off the high spots in body filler. This will save sanding time. After the filler fully cures, the area is coarse sanded to the original panel contour.

The mixing, application, filling, and sanding of body filler is shown in Figure 1–19.

Coarse sanding is done by the metalwork technician using an abrasive coated paper to level and smooth the body surface that is being repaired. Coarse, rough sandpaper may be used to level body filler.

Paint Preparation

In *paint preparation*, the vehicle is fine sanded and cleaned to prepare surfaces for the refinishing process. Fine, smooth sandpaper is used to lightly scuff old paint so the new paint will stick. The vehicle is moved from the metalworking area to a prep area in the shop. Dust is blown off the vehicle with compressed air (Figure 1–20).

Masking is used to cover areas that are not going to be painted to protect them from paint spray and physical damage. Parts of the car that are not to be

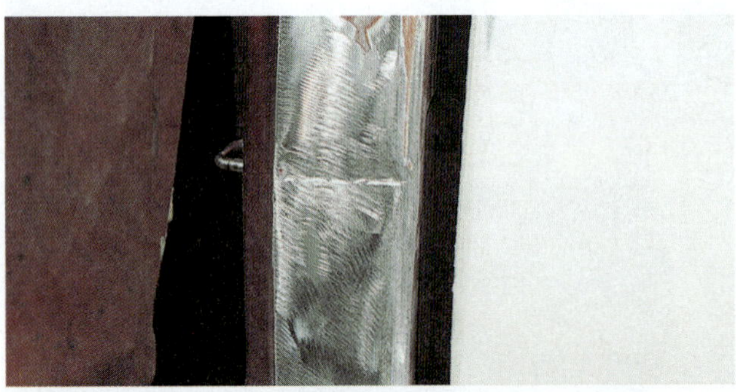

A

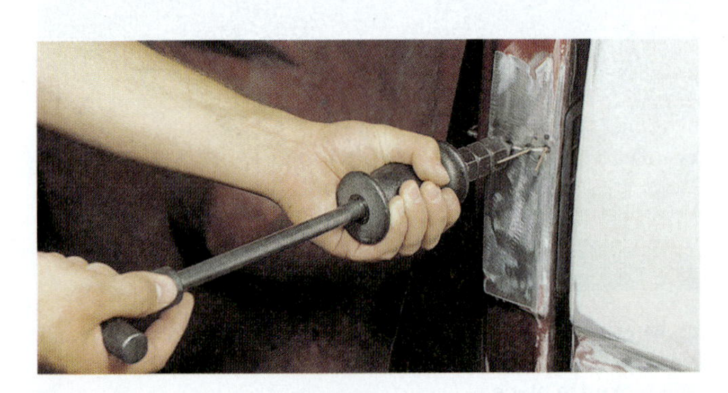

C

B

D

Figure 1–18 Note the basic steps for removing a dent from a metal part. (A) Often panels have minor damage that can be repaired. The cost of labor and materials must be weighed against the cost of new or salvaged parts. (B) When a dent cannot be hammered out from inside the panel, a nail weld gun can be used. The gun welds metal pull pins onto the damaged panel. (C) With pull pins welded into the dent, attach a puller to each pin for forcing out the low spot. Metal must be straightened to within 1/8 inch of the original contour so that the body filler is not too thick. (D) After pulling damage, cut and grind off the pins until they are almost flush with the panel surface. A slight depression is needed for body filler.

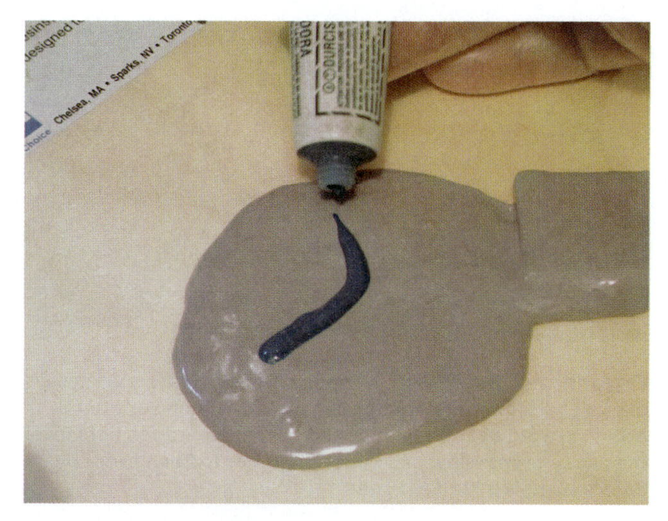

A

C

B

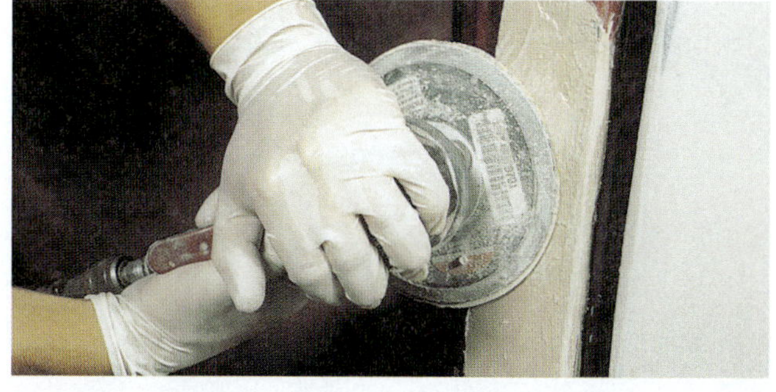

D

Figure 1-19 Body filler is commonly needed to smooth and level metal that has been straightened. (A) Mix body filler and hardener in correct proportions. As a general rule, for each golf ball-size lump of filler, use a 1-inch (25 mm) bead of hardener. If the amount of filler is as "big as a baseball," use a 6-inch (152 mm) bead of hardener. Do not stir hardener into body filler or you can cause air bubbles and pin holes. Wipe back and forth to mix ingredients completely. (B) Apply the body filler to the clean body surface right away. It will start to harden in a few minutes. Try to spread it out to match the body shape to reduce filing and sanding times. (C) As soon as filler starts to solidify, use a coarse "cheese grater" file to rough out the shape. File in multiple directions until the filler is contoured properly. Do not file too much. You want the filler to stick up above the panel surface a little so it can be sanded smooth. (D) After grating the filler, use an air sander and coarse grit sandpaper to cut the filler down more. Finish with a medium grit sandpaper. Featheredge the filler and old paint for a smooth transition into the panel.

Figure 1-20 Before masking and priming, use a blowgun to clean the vehicle of dust and debris. Aim the blowgun down into all gaps between parts that could hold dust and dirt. Do this before final masking and moving the vehicle into the paint booth.

painted—windows, chrome, lights, and so on—are covered with masking material, such as paper, plastic, tape, or a water-based spray-on coating (Figure 1–21).

In large shops, the sanding jobs are handled by the prep person, whereas the masking operation is performed by the masker or painter. In small shops, the final prep jobs are usually done by the painter or a helper.

Primer is applied to bare metal and body filler before painting. *Priming* is necessary because paint alone will not properly adhere to or stick to bare metal. A spray can or spray gun is used to apply primer to all exposed metal surfaces (Figure 1–22).

Fine sanding is done to remove scratches left from coarse sanding and also to scuff unrepaired areas so that the new paint will not peel off. All surfaces to be painted must be fine sanded or scuffed to provide a good bond between the old paint and the new, fresh paint. Refer to Figure 1–23.

Wax and grease remover is then wiped over the surfaces to be refinished to remove any remaining contaminants. A rag soaked in the cleaning agent is wiped over the body surfaces. A second clean, dry rag is used

A Here a technician uses masking tape and masking paper to cover the area not to be painted.

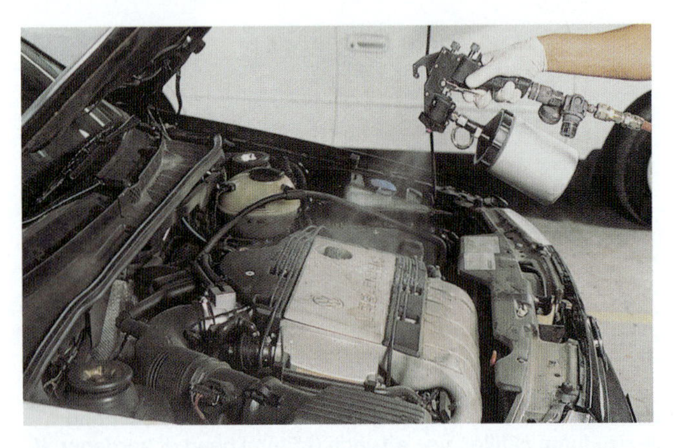

B Spray-on masking material is handy when masking a large area, such as underhood components. It is water-based masking coating and can be washed off with soap and water after painting.

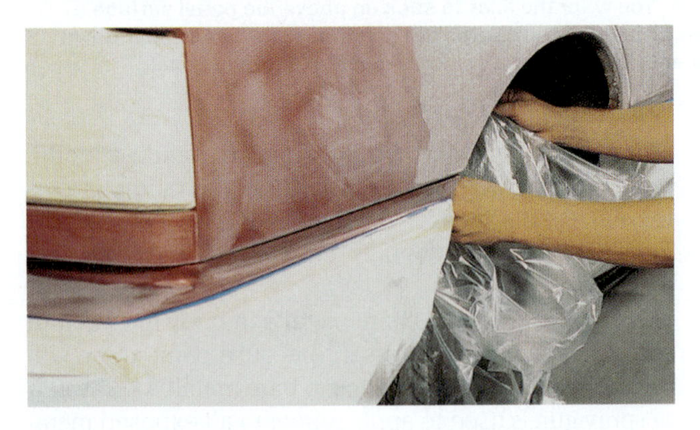

C Plastic wheel masks are installed over tires to protect them from paint spray.

Figure 1–21 Masking is used to protect parts of the car that are not going to be painted.

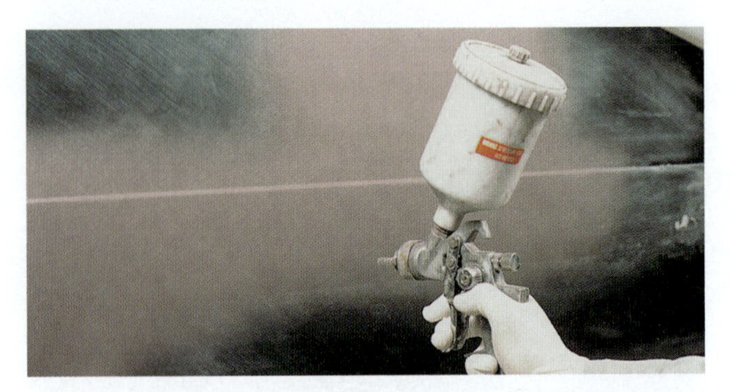

Figure 1–22 Self-etch or epoxy primer must be sprayed over all bare metal to bond to the metal. Primer-surfacer is often sprayed over body filler and the repair area to help level the repair area. Allow each coat of primer to flash properly before applying the next coat.

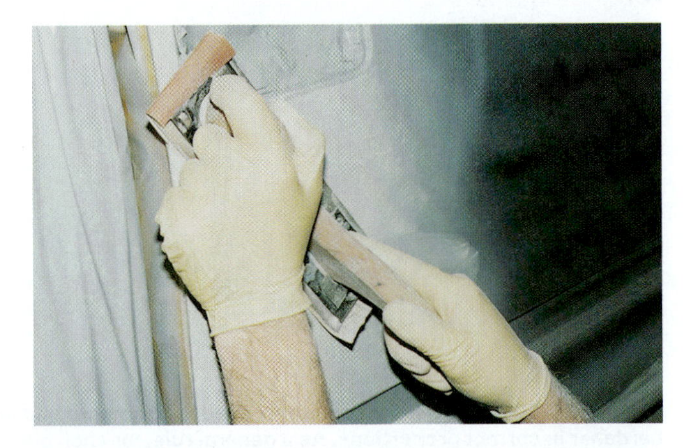

A A large sanding board is commonly used to smooth and level the repair area with the surrounding undamaged areas.

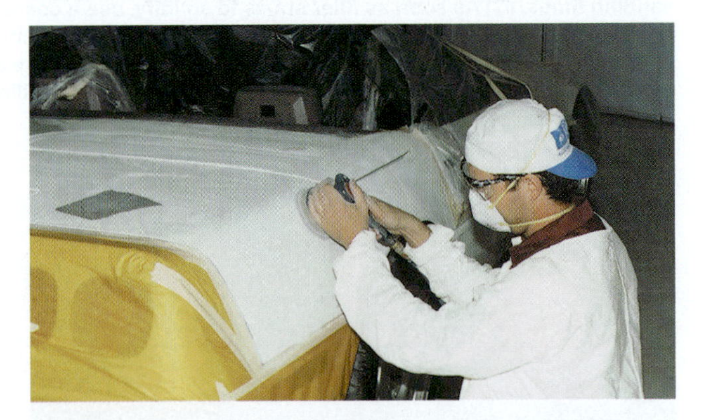

B An orbital sander or hand sanding with fine sandpaper is needed to scuff the old paint to prepare it to accept and bond with the new paint.

Figure 1–23 Finish sanding is needed between priming and painting to final smooth out the repair area.

to wipe off the solvent while the surface is still wet. Look at Figure 1–24.

Mixing Room

The mixing room, or *mixing lab*, is a power-ventilated area where refinish or paint materials are mixed and prepared for application. Most large body shops have an in-house mixing room. Smaller shops have their materials mixed by an outside paint supplier (Figure 1–25).

Paint mixing involves carefully measuring out the correct amount of refinish ingredients following manufacturer instructions. For example, when mixing a paint

Figure 1–24 Before painting, use two rags to wipe the repair area clean. First, use a rag soaked with wax and grease remover to remove any surface contaminants. Then use a clean rag to wipe the surface while the solvent is still wet.

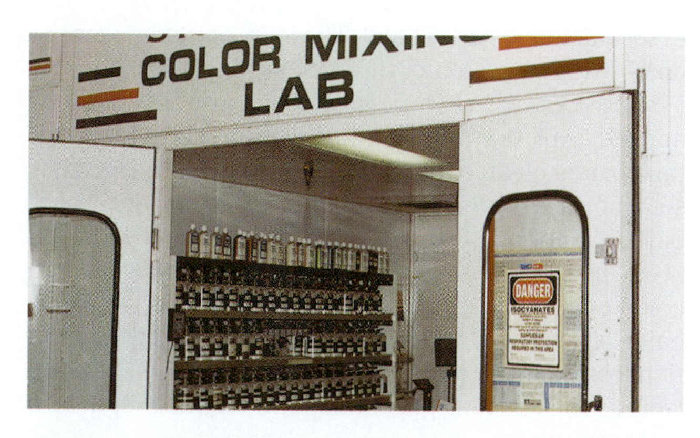

Figure 1–25 Most large body shops have their own paint mixing room. It has all color pigments and other ingredients for mixing paint materials. Mixing rooms also have forced ventilation to prevent buildup of paint fumes.

color, it is critical that the correct color pigments be added so that the new paint matches the existing color on the vehicle. Mixing usually involves adding reducer to make the paint the right viscosity (thickness or fluidity) for spraying. With water-based paints, distilled water is used to reduce the paint to a sprayable consistency. Mixing might also involve pouring in a recommended amount of *catalyst* (hardener) to make the top clearcoat cure or dry (Figure 1–26).

Refinishing

The vehicle's paint, or *finish*, performs two basic functions—it beautifies and it protects. Can you imagine what a car body would look like without paint? For a day or so, it would be the drab, steel gray of bare sheet metal. Then, as rusting eats into the metal, the body would turn an ugly, reddish brown. This degeneration, or oxidation, would continue until the body was solidly coated with rust.

The term *paint* generally refers to the visible topcoat. The most elementary painting system consists of a primer and final topcoat over the *substrate* (body surface

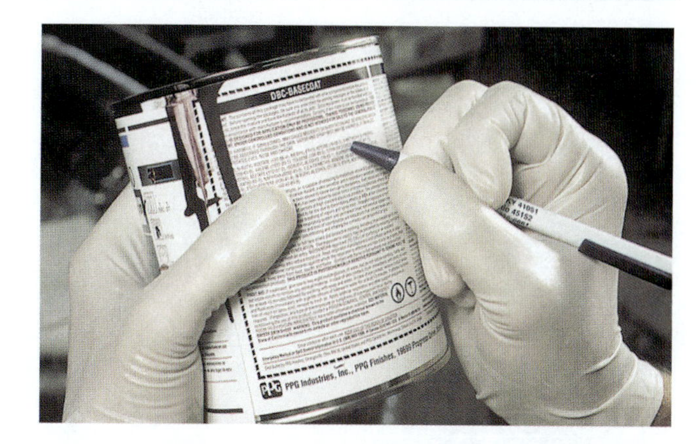

A Mix paint according to label directions or instructions given on the manufacturer procedure sheet. The label or procedure sheet gives ratios for reducer, catalyst, and paint.

B A computerized electronic scale aids in mixing paint ingredients. Some will give an electronic readout that tells you how much of each ingredient to add while mixing.

C Filter everything that goes into the spray gun cup. One piece of dirt or cured paint can ruin a paint job.

Figure 1–26 Paint mixing is like working in a chemistry lab. All ingredients must be added accurately to produce the correct color and chemical content.

material). This process can vary considerably and generally is more complex, as you will learn later.

Refinishing, or *repainting*, involves the steps needed to properly restore the vehicle's finish. Refinishing is a very important part of the auto repair process.

Damage caused by both minor and major accidents usually requires some painting. However, many older automobiles are repainted to simply enhance their beauty.

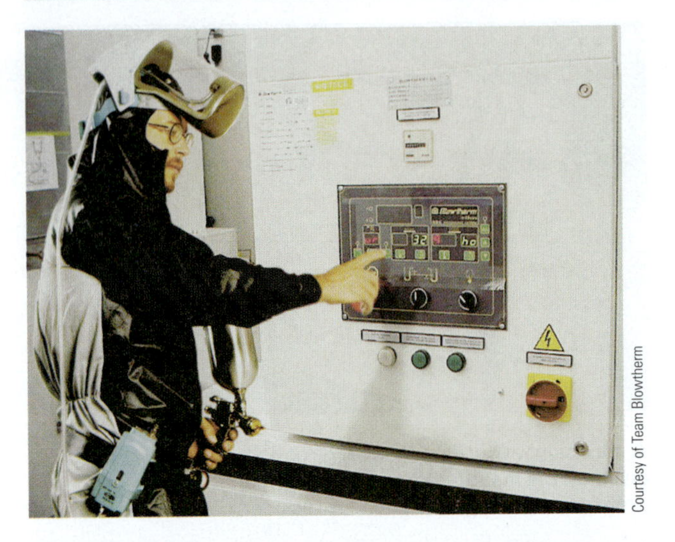

Courtesy of Team Blowtherm

Figure 1-27 **Refinish technicians must wear safety gear and be highly skilled at using painting tools and equipment. Here the technician is turning on the fans in a paint spray booth.**

Figure 1-28 **The spray gun is the main tool of a body shop paint technician. It must be adjusted properly to produce a smooth, even paint film on the vehicle.**

New and used-car dealers repaint automobiles to attract buyers. Sometimes an owner simply gets tired of looking at the same old color.

With today's high-tech paint materials, it is critical that painters wear proper safety gear (Figure 1–27).

A car's finish consists of several coats of two or more different materials. The most basic finish consists of:

1. Primecoat
2. Topcoat

The *primer* is used to improve **adhesion** (sticking) of the topcoat. It is the first coat applied. Paint alone will not stick or adhere as well as a primer. If you apply a topcoat to bare substrate, the paint will peel, flake off, or look rough. This is why you must "sandwich" a primecoat between the substrate (car body) and the topcoat. Primer-sealers also prevent any chemicals from bleeding through and showing in the topcoats of paint.

The terms *topcoat* and *colorcoat* refer to the paint applied over the primer-sealer coat. This usually consists of several thin coats of paint. The topcoat is the "glamour coat" because it features the eye-catching gloss.

Basecoat/clearcoat paint systems use a colorcoat applied over the primer-sealer with a second layer of transparent clearcoat over the colorcoat. The basecoat is sometimes a water-based paint. To protect the water-based paint from the elements, a conventional oil-based urethane clear topcoat is sprayed over the waterborne color coats. Basecoat/clearcoat is the most common paint system used today. The clear paint brings out the richness or shine of the underlying color and also protects it. The resulting *gloss* is superior to that of standard paint systems.

A *spray gun* uses compressed air to break the liquid paint into a fine mist so that the paint is deposited smoothly on the vehicle body. The *paint booth* is a clean room with filtered air circulation so that dirt does not get in the paint when spraying (Figure 1–28).

Spraying is the physical application of paint using an air pressure-powered paint spray gun. The technician methodically moves the spray gun next to the body while

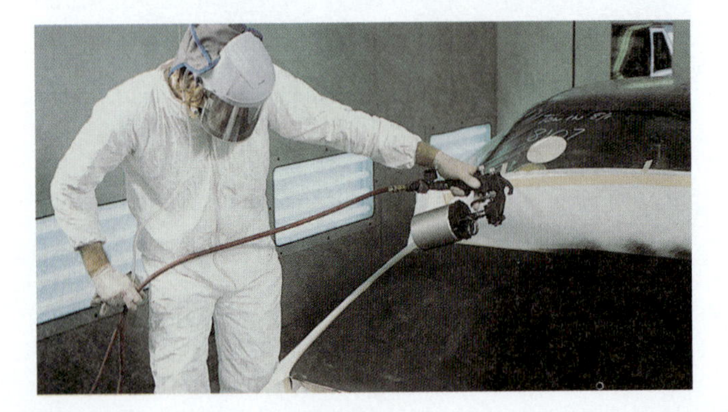

A

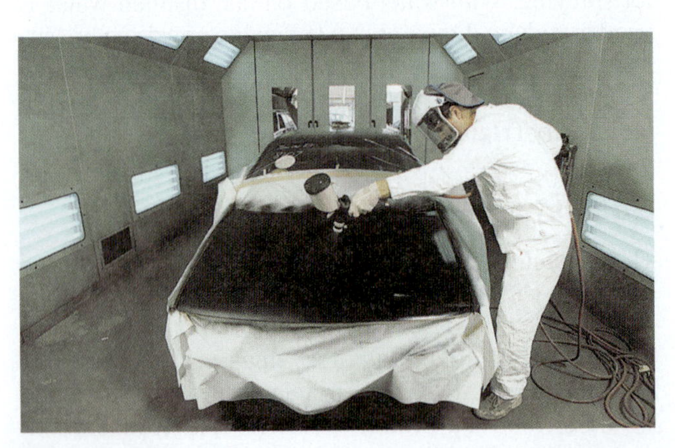

B

Figure 1-29 **Excellent hand-eye coordination and specialized knowledge is needed to be a professional painter. (A) When painting a vehicle, keep the gun the correct distance from the surface. Also keep the gun parallel with the surface. (B) When painting, try to move the spray gun smoothly and evenly. Use straight line paint strokes that overlap evenly. If you move the gun at different speeds or move it further or closer to the surface, paint problems will result.**

Figure 1-30 Here a refinish technician is spraying the colorcoat along the bottom and side of the car body. This will be followed by spraying the clearcoat to bring out the gloss or shine.

spraying accurate layers of paint onto it. This requires a high degree of skill to prevent **paint runs** (when excess paint flows down or runs) and other painting problems. Refer to Figure 1–29 and Figure 1–30.

In addition to being able to apply the new finish properly, the painter or refinishing technician must have knowledge of paint products and how to mix and match them. If the refinishing job looks good and the color matches well, the customer will usually be satisfied with all of the other repair work.

REMEMBER *The car owner notices the quality of the metal straightening and paint job more than any other aspect of collision repair.*

ECO-TECH Eco-Tech stands for ecology technology. Eco-tech deals with how a body shop can conserve materials to reduce the amount of air, water, and ground pollutants resulting from vehicle repairs. As you learn the information in this textbook, you will be given many tips for avoiding material waste and for recycling all useful materials. The information in this book will stress the recycling of unused paint solvents, metal body panels, plastic parts, air conditioning refrigerant, paper, antifreeze, motor oil, and other materials.

Drying

Drying involves using different methods to cure the fresh paint. If only partially dry when returned to the customer, the new paint can be easily damaged.

Air drying is done by simply letting the paint dry in the atmosphere. This can take a long time with enamel paint if a hardening agent is not added before applying the paint.

Forced drying uses special heat lamps or other equipment to speed the paint curing process. Most shops use drying equipment to speed up drying.

Post-Painting Operations

Post-painting operations include the tasks that must be done after painting but before returning the vehicle to the customer. These tasks include removing masking tape, reinstalling parts, and cleaning the vehicle.

Once the paint is hard enough so that it will not be easily damaged by being touched, any parts removed for repairs or painting must be reinstalled. This is done carefully, often by the original technician (Figure 1–31).

If the painted surface has a rough, textured surface like that of an orange peel, you may have to extra-fine sand and polish the hardened paint. This will remove the surface roughness of the paint.

Wet sanding involves using a water resistant, ultrafine sandpaper and water to level the paint. It can sometimes be done to fix small imperfections (orange peel, dirt, runs, and so on) in the paint. Dry ultrafine sanding uses 1000 grit or finer sandpaper on an air sander to repair minor paint imperfections.

Compounding, or buffing, may be required to smooth newly painted surfaces, after wet sanding, for example, or to remove a thin layer of old, dull paint. Compounding makes the paint smooth and shiny. An air or electric buffing machine equipped with a soft pad is used to apply buffing compound to the paint. The abrasive action cuts off a thin layer of topcoat to brighten the color and shine the paint (Figure 1–32).

Detailing is a final cleanup and touch-up on the vehicle. It involves washing any unpainted body sections, cleaning and vacuuming the interior, and touching up any chips in sections not painted. Detailing can also involve tasks such as polishing chrome, cleaning glass, installing trim, and cleaning the vinyl top and tires. This will ready the vehicle for return to the customer.

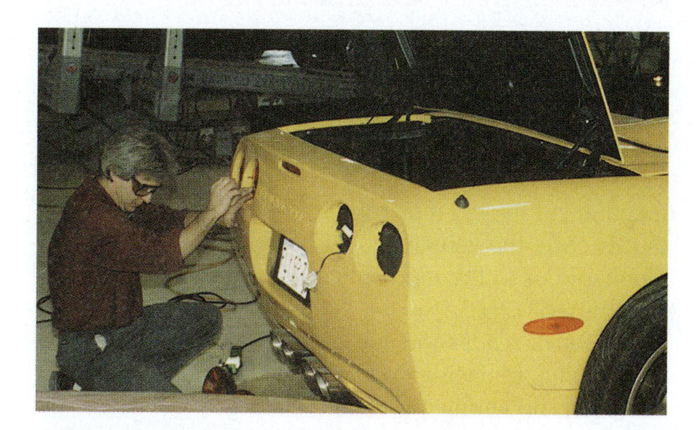

Figure 1-31 After the car body is painted, parts must often be reinstalled. Here a technician reinstalls taillights on this late model sports car.

A After paint cures properly overnight, it may be possible to wet sand or ultrafine dry sand any imperfections, such as a small protrusion or orange peel. Use a small sanding block and ultrafine (1000 or finer grit) sandpaper, and be careful not to cut through the clearcoat, or repainting will be required. A little dishwashing liquid in water will help keep the sandpaper from sticking to the surface.

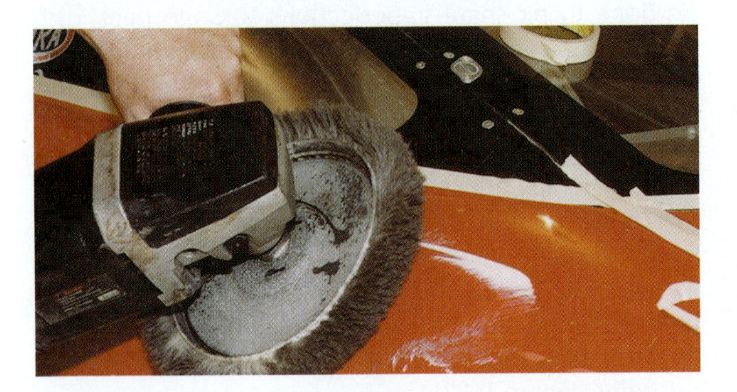

B After ultrafine wet or dry sanding, compound the area to revive the paint's gloss. When machine buffing, be careful not to cut through paint on body contours or panel edges. Most technicians like to apply masking tape to sharp edges to protect the finish from burn-through.

Figure 1–32 Minor problems in the paint, such as small dust or dirt particles, can often be removed without repainting.

Mechanical/Electrical Repairs

Mechanical repairs include tasks such as replacing a damaged water pump, radiator, or engine bracket. Mechanical components like these are often damaged in a major collision. Many mechanical parts are easy to replace and can be done by the auto body technician. However, other mechanical repairs may require special skills and tools. In this case, the vehicle would be sent to a professional mechanic or to the mechanical repair area of the shop (Figure 1–33).

Electrical repairs include tasks such as repairing severed wiring, replacing engine sensors, and scanning for computer or wiring problems. During a collision, the impact on the vehicle body and the resulting metal deformation can easily crush wires and electrical components.

Courtesy of Hunter Engineering Company.

A A technician uses special equipment to adjust wheel alignment. This must be done if suspension parts are damaged in an accident.

Courtesy of Hunter Engineering Company.

B Mechanical skills are helpful to the auto body technician.

Figure 1–33 Mechanical–electrical repairs are sometimes done in larger body shops.

For this reason, today's auto body technician must have the basic skills needed to work with and repair electrical/electronic components.

Complete Collision Services

More and more body shops are offering complete collision services, such as:

- Wheel alignment
- Cooling system repairs
- Electrical repairs
- Suspension repairs
- Air conditioning repairs
- Glass replacement

Many of these repairs are still done by auto specialty shops, which only do one type of repair (glass, air conditioning, and so forth). However, the expanding scope of

Figure 1-34 **A final cleanup of the interior and exterior of a repaired vehicle is important. You want the first impression customers have of your work to be pleasing so they will return to your shop again someday.**

the body shop has made it necessary for the body technician to have some knowledge of these repairs.

As an auto body technician, it is critical that you learn to repair a damaged vehicle properly. Your reputation as a professional technician and the safety of the occupants of the repaired vehicle depend on how well you do your work (Figure 1–34).

NOTE *Several chapters in this book are devoted to helping you learn to work on mechanical and electrical systems.*

1.3 AUTO BODY CAREERS

Auto body technicians are skilled, knowledgeable people who know how to use specialized equipment and highly technical methods to restore severely damaged vehicles. An auto body technician must be a "jack-of-all-trades." They must have basic skills in the following eight areas:

1. *Metalworker*—An auto body technician must be able to do all types of metalworking to properly form and shape sheet metal car bodies after an accident.
2. *Welder*—An auto body technician must weld and cut steel, aluminum, and plastic efficiently during major body repairs.
3. *Auto mechanic*—An auto body technician must remove and install mechanical systems; this requires some skill in auto mechanics.
4. *Sculptor*—An auto body technician must form body filler to match body shape, much like a sculptor shapes a piece of art.
5. *Plumber*—An auto body technician must work with numerous lines, hoses, and fittings during power steering, brake system, and fuel system service.
6. *Electrician*—An auto body technician must be good at testing and repairing wiring and electrical components after damage. Being able to find shorts, opens, and other wiring problems is essential with today's computer-controlled vehicles.
7. *Air-conditioning technician*—An auto body technician may be required to work on air-conditioning systems.
8. *Computer technician*—An auto body technician should be able to scan and repair computer problems stemming from collision damage.

Obviously, today's auto body technician must be a highly trained professional. The days of the "mud builder" or "parts replacer" are over. Individuals who lack specialized skills can no longer survive and earn a living with modern, complex automotive technology.

The term *professional* refers to the attitude, work quality, and image that a business and its workers project to customers. It is interesting to note that in Europe a shop cannot open without the presence of a "meister," or master craftsperson. A professional technician:

- Is customer oriented
- Is up-to-date on vehicle developments
- Keeps up with advancements in the repair industry
- Pays attention to detail
- Ensures that their work is up to specification
- Belongs to trade associations

Some shops have specialized technicians who concentrate their knowledge and skills in one area of collision repair.

Metalworking technicians are skilled at part removal/replacement, part repair, welding, and the use of body filler. They can also use frame straightening and measuring equipment to repair vehicles with frame damage.

Refinish technicians, also called painters or refinishers, specialize in vehicle prep and spraying. They are versed in paint mixing, color matching, and the use of spray equipment. These people must have good color perception and excellent hand–eye coordination.

Helpers work under the supervision of one or more professional technicians. They might help a technician mask a car before painting, install a hood, run for parts, or help clean up the work area at the end of the day. The helper is often an *apprentice* who is learning the trade by working with an experienced technician.

Other Collision Repair Personnel

Foremost in any collision repair business are the collision repair and paint technicians. However, there are other jobs that must be done as well. This section describes other personnel who work in and with the staff of a collision repair shop.

The *shop owner* must be concerned with all phases of work performed. In smaller shops, the owner and shop manager is usually the same person. In large operations or dealerships, the owner might hire a shop manager. In all cases, the person in charge should understand all of the work done in the shop as well as its business operations.

The *shop supervisor* is in charge of the everyday operation of the shop. This job involves communication with all personnel who contribute to the body shop's success.

The *parts manager* is in charge of ordering all parts (both new and salvaged), receiving all parts, and seeing that they are delivered to the right technician. Because not every collision repair shop has a parts manager, the task of ordering parts can fall on all employees at one time or another.

The *bookkeeper* keeps the shop's books, prepares invoices, writes checks, pays bills, makes bank deposits, checks bank statements, and takes care of tax payments. Many shops hire an outside accountant to help perform these tasks.

The *office manager's* duties include various aspects of the business—handling letters, estimates, and receipts, for example. In many small shops, the office manager also acts as the parts manager and bookkeeper.

A *receptionist* is sometimes employed to greet customers, answer the phone, route messages, and do other tasks.

Larger shops generally have trained *tow truck operators* to operate their wrecker(s). Rather than own these expensive pieces of equipment, many smaller shops depend on independent towing services or farm out such work to other repair garages.

It is important for body shop personnel to have good communication and cooperation skills. Everyone in the shop should work as a team member to make the facility profitable, enjoyable, and safe. "A chain is only as strong as its weakest link"—this saying applies to the smooth operation of a body shop. If anyone—from the estimator to the painter—does not do the job right, everyone suffers. Customers will not return to the shop, and everyone's paycheck will be affected in the long run.

There are other career openings in the collision repair field. Some of these include:

- Insurance adjuster or appraiser
- Vocational/technical instructor
- Salvage yard technician
- Dealership parts counterperson
- Paint company representative
- Auto manufacturer representative
- Equipment salesperson

To research these and other career opportunities, talk to your guidance counselor, visit your local library, or search the Internet. There you can obtain more detailed information on the qualifications and training requirements for each position.

SUMMARY

1. A collision, nicknamed a *crash* or a *wreck*, is damage caused by an impact (hit) on the vehicle.
2. Auto body repair, also called collision repair, involves fixing a vehicle that has been damaged in an accident. The vehicle is first brought into the shop, where a damage estimate is prepared to calculate the cost of repairs.
3. Minor body damage only requires that a few parts be replaced or repaired before the car is repainted. Major body damage usually must be corrected by replacing, repairing, and straightening large body parts before repainting. A collision can also be severe enough to cause a total loss, where repairs would be more expensive than buying another car.
4. Repair instructions are summarized on a printed repair order (RO), and the repairs are carried out according to those instructions. The metalworking areas are shop locations where damaged parts are removed, repaired, and installed. A stall is a work area for one vehicle, often marked off with painted lines on the floor.
5. Vehicle measurement is done to determine the extent and direction of major damage. Measurement systems are specialized tools and equipment that help the technician to check and measure for frame or body misalignment resulting from a high-impact collision.
6. Frame straightening equipment (also called a frame rack) uses a large steel framework, pulling chains, and hydraulic power to pull or force the frame or body back into its original position.
7. A panel is a large metal or plastic body part—a fender, hood, truck, or roof panel, for example. Structural panel replacement involves welding a new panel in place of an old, damaged one. Panel adjustment involves moving or shifting a part to properly align with other parts or panels. When parts are replaced, corrosion protection coatings must be applied to prevent the new parts from rusting.
8. Panel straightening involves using various hand tools and equipment to force the panel back into its original shape. Body filler is used to cover any small

dents remaining in the worked panel. The vehicle is sanded and cleaned to prepare surfaces for the painting process.

9. Masking is done to cover areas that will not be painted. Priming is done to cover bare metal before painting. The mixing room or mixing lab is a power-ventilated area where refinish materials are mixed or prepared for application.

10. The vehicle's paint or finish performs two basic functions—it beautifies and it protects.

11. Refinishing or repainting involves the steps needed to properly restore the vehicle's finish (paint).

12. The primecoat is used to improve adhesion (sticking) of the topcoat. The terms *topcoat* and *colorcoat* refer to the paint applied over the primecoat. Basecoat/clearcoat paint systems use a colorcoat applied over the primer-sealer with a second layer of transparent clearcoat over the colorcoat.

13. Spraying involves using precision paint spray guns and a spray booth to apply refinish materials to the car body. A spray gun uses compressed air to break the liquid paint into a fine mist so that the paint is deposited smoothly on the vehicle body.

14. The paint booth is a clean room with filtered air circulation so that dirt does not get into the paint when spraying. In the drying room, infrared lights are used to speed up the paint curing or drying process by warming the paint.

15. A detailer performs all the last tasks needed to prepare the vehicle for return to the customer.

16. Auto body technicians are skilled people who know how to use specialized equipment and highly technical methods to restore severely damaged vehicles.

17. The term *professional* refers to the attitude, work quality, and image that a business and its workers project to customers.

EXERCISES

On a separate sheet of paper, complete the following learning activities for this chapter. Write definitions for the key terms and answer the ASE-style review questions, essay questions, critical thinking problems, and math problems. You can also do the outside activities, possibly for extra credit.

Key Terms

accident report
adhesion
collision
collision repair
collision repair facility
contaminants
damage estimate
dealership body shop
deductible
elements
estimating
estimator
head-on collision
impact
independent body shop
insurance adjuster
major body damage
metalworking areas
minor body damage
paint runs
post-painting operations
repair order
salvage yard
specialty shop
total loss
wash-up
wrecker

ASE-Style Review Questions

1. The collision repair industry does what volume of business per year?
   A. $1 million
   B. $10 million
   C. $1 billion
   D. $10 billion

2. Technician A says that when two cars hit head-on at 50 mph, it is like hitting a brick wall at 100 mph. Technician B disagrees and says that it is like hitting a brick wall at only 50 mph. Who is correct?
   A. Technician A
   B. Technician B
   C. Both A and B
   D. Neither A nor B

3. An independent body shop is owned by an individual and does not have ties to a major auto manufacturer. True or false?

4. What task involves analyzing damage and calculating how much it will cost to repair the vehicle?
   A. Repair
   B. Estimating
   C. Calculating
   D. Measuring

5. Technician A says that a collision can be severe enough to cause a total loss. Technician B says that the repairs would be more expensive than buying another car. Who is correct?
   A. Technician A
   B. Technician B
   C. Both A and B
   D. Neither A nor B
6. Body shop repairs are carried out according to the instructions on which form?
   A. Repair order
   B. Estimate
   C. Quote
   D. Data sheet
7. Technician A says that a damaged vehicle does not have to be cleaned before starting repairs. Technician B says that wash-up will remove mud, dirt, wax, and water-soluble contaminants that could affect the paint job. Who is correct?
   A. Technician A
   B. Technician B
   C. Both A and B
   D. Neither A nor B
8. Technician A says that vehicle measurement is done to determine the extent of major damage. Technician B says that it is also done to determine the direction of damage. Who is correct?
   A. Technician A
   B. Technician B
   C. Both A and B
   D. Neither A nor B
9. Technician A says that various types of gauges and measuring devices can be used to measure vehicle damage. Technician B says that you can compare known good body specifications (normal measurements from an undamaged vehicle) with the actual body measurements on the damaged vehicle. Who is correct?
   A. Technician A
   B. Technician B
   C. Both A and B
   D. Neither A nor B
10. Technician A says that a frame rack can be used to do a wheel alignment. Technician B says that a frame rack uses mechanical power to pull or force the frame or body back into its original position. Who is correct?
    A. Technician A
    B. Technician B
    C. Both A and B
    D. Neither A nor B
11. Technician A says that panel replacement involves removing a panel or body part that is too badly damaged to be fixed. The new part would have to be bolted or welded in place on the vehicle. This takes considerable skill. Technician B says that structural panel replacement involves welding a new panel in place of the old, damaged one. Who is correct?
    A. Technician A
    B. Technician B
    C. Both A and B
    D. Neither A nor B
12. Technician A says that panel straightening involves using various hand tools and equipment to force the panel back into its original shape. Technician B says that body hammers, body filler, and sanders are a few of the tools and materials used to fix panel damage. Who is correct?
    A. Technician A
    B. Technician B
    C. Both A and B
    D. Neither A nor B

Essay Questions

1. What happens to a vehicle during a collision?
2. Explain the difference between a door ding and a total loss.
3. What is estimating?
4. Describe the differences between panel straightening and panel replacement.
5. Define the term *collision*.
6. Name two contaminants that must be cleaned off before starting bodywork.
7. Why is estimating the cost of a repair such an important step to consider for a body shop?

Critical Thinking Problems

1. How would you determine whether a car is a total loss?
2. Write the 14 steps of a basic repair sequence so you can memorize what is done when a vehicle enters a body shop.

Math Problems

1. A technician is paid $45 per hour. The repair will take 11.5 hours. Parts for the repair will cost $176.25 plus 7 percent tax. What is the total of the estimate?
2. An older car is valued at $2,500. The parts for the repair will cost $1,500. How many hours of labor at $35 per hour can be completed before the car should be declared totaled?
3. An independent repair shop made $33,507 total in a one-month period. The cost of labor was $45 per hour with a total of 200 hours of labor. Parts cost the repair shop $14,000. Power and property bills cost a total of $3,500. After these costs are taken into account, what did the repair shop make in profits for the month?

Activities

1. Take a field trip to a local collision repair facility. Have the shop owner or supervisor give you a guided tour of the repair operation. Discuss what you learned during the field trip in class the next day.
2. Ask a technician or shop owner to visit your classroom to describe their duties and answer questions.

CHAPTER 2

Vehicle Construction Technology

OBJECTIVES

After studying this chapter, you should be able to:

- Explain the past and present designs of motor vehicles.
- Summarize the various types of frames commonly used on modern cars, trucks, vans, and SUVs.
- Compare and contrast modern body-over-frame and unibody construction technology.
- Locate the major parts of a perimeter frame.
- Locate the major parts of a unibody frame.
- Compare a conventional full frame with modern hydroformed frames.
- Identify the major structural components, sections, and assemblies of a motor vehicle.
- Explain how simulated and actual crash tests are used to evaluate the structural integrity of a motor vehicle.
- Describe the layperson's names for body shapes used on passenger vehicles.
- Answer ASE-style review questions relating to vehicle construction.

INTRODUCTION

The term *vehicle construction* refers to how a passenger car, truck, van, or sport utility vehicle (SUV) is assembled at the factory. A typical car has over 15,000 parts that all work together to provide a safe, dependable means of transportation. As you will discover, the modern automobile is one of the most amazing engineering feats ever devised by humans (Figure 2–1).

This chapter familiarizes you with the auto parts vocabulary needed to become a successful auto body technician. You will learn to locate and describe the major body panels of an automobile.

Knowledge of vehicle construction will help you answer questions such as: What is the name of that part? How are the parts fastened together? What is that part made of? Does the vehicle use a full perimeter frame or does it have unibody construction?

To accurately repair a damaged car, you must fully understand the construction methods used during the manufacture of the particular make or model vehicle. You must accurately identify all damaged components and select the proper repair options from those available. To do this, you must know what kinds of materials are used and understand how these materials affect the repair process (Figure 2–2).

The goal of collision repair is to restore the vehicle to its preaccident condition. During repairs, you should use repair methods that closely duplicate how the car was manufactured on its assembly line.

2.1 BODY AND CHASSIS

To avoid confusion, the major parts of a vehicle can be categorized as parts of the body, the chassis, or the frame. You must understand each major division.

Vehicle Body

The *vehicle body* provides a protective outer hull, or "skin," around the outside of an automobile. The body is an attractive, colorful covering over the other parts. Body parts may also contribute to the structural integrity (safety and strength) of the vehicle.

A **Modern manufacturing facilities and dedicated people are needed to assemble modern cars, trucks, vans, and SUVs.**

B **Unibody vehicles move down an assembly line as automated robots weld the various body panels together.**

Figure 2–1 **Today's auto manufacturers design and build safe and efficient motor vehicles.**

A **Factory workers are inspecting the build quality of the Model T Ford.**

B **Modern vehicles are far superior to the cars and trucks of just a few years ago.**

Figure 2–2 **Great changes have occurred in automotive design since the early years of the automobile.**

The vehicle body can be made from steel, aluminum, fiberglass, plastic, or composite (a combination of materials like carbon fiber). The body is normally painted to give the vehicle its appealing, shiny color and appearance. Refer to Figure 2–3.

Vehicle Chassis

The **vehicle chassis** includes the frame, engine, suspension system, steering system, and other mechanical parts with the body removed. The body and chassis are two major categories used to classify the repair areas of a vehicle (Figure 2–4).

2.2 VEHICLE FRAME

The *vehicle frame* is a high-strength structure used to support all other parts of the vehicle. Besides bolt-on body panels, the frame holds the engine, transmission, suspension, and other parts in position. Frames are usually made of steel or aluminum and sometimes composite materials. The frame can be separate from the body or integrated into the body shell, as in the case of unibody design (Figure 2–5).

Body-Over-Frame

Body-over-frame construction has a separate body structure bolted to a thick steel framework. The engine and other major assemblies forming the chassis are mounted on the frame. Rubber body mounts fit between the frame and body structure to reduce road noise (unwanted sounds entering the passenger compartment from outside the vehicle). See Figure 2–6.

A *full frame* has a thick metal box or U-shaped stampings or rails welded and/or riveted together. The main structural members are two side rails connected by a series of cross members. For high load-carrying capabilities, the separate frame is made of much heavier gauge steel than the body panels.

A **This minivan with rounded body and large fenders resembles the vehicles built in the 1940s.**

B **Modern pickup trucks perform like performance cars and are built to last hundreds of thousands of miles.**

C **This "super car" has the looks and performance of a jet fighter.**

Figure 2-3 **Vehicle designs vary more today than ever.**

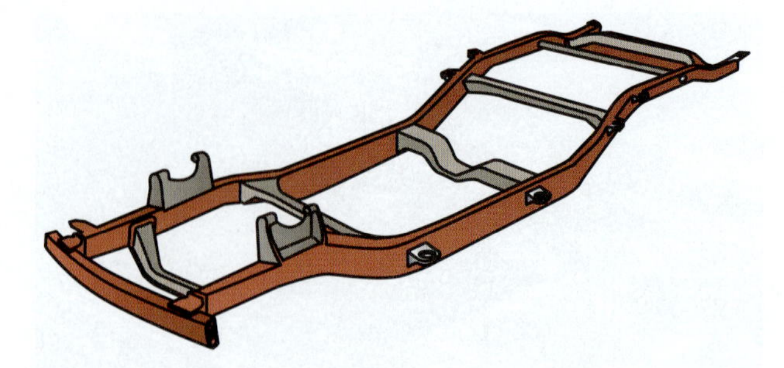

Figure 2-4 **The oldest and strongest frame design is the perimeter frame. A thick steel framework supports the chassis and body panels.**

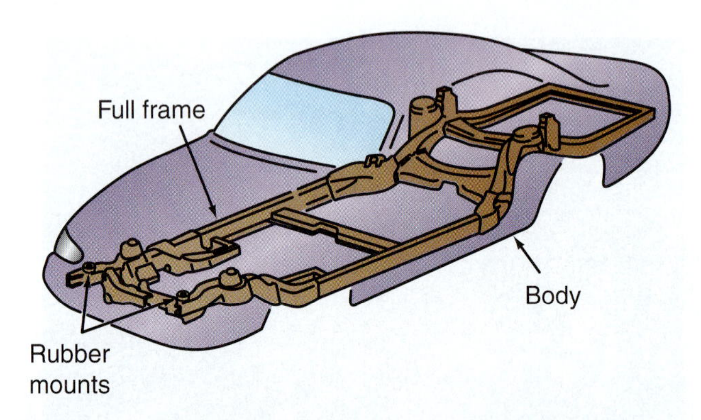

Figure 2-5 **Illustration of a vehicle with body panels removed shows chassis construction. Study the part names.**

The full frame rails extend the entire length of the vehicle. The frame cross members run sideways to secure the frame rails together. Body-over-frame or full frame construction is commonly used on pickup trucks, SUVs, and most full-size vans. Some larger luxury cars still use traditional body-over-frame construction.

A **hydroformed frame** is manufactured by using water under high pressure to force straight box-extruded frame rails into the desired shape or contour. A hydroformed frame is made of a thinner gauge metal than a conventional perimeter frame. Hydroformed frames are lighter, almost as strong, and equally as stiff as conventional heavy-gauge steel or aluminum frames.

Unibody

Unibody construction uses body parts welded or adhesive-bonded (glued) together to form an integral (built-in) frame. The body structure is designed to secure other chassis parts. No separate heavy-gauge steel frame under the body is needed.

Today's vehicles are manufactured using both unibody and body-over-frame construction. Refer to Figure 2–7.

Unibody construction is a totally different concept in vehicle design that requires more complex assembly techniques, new materials, and a completely different

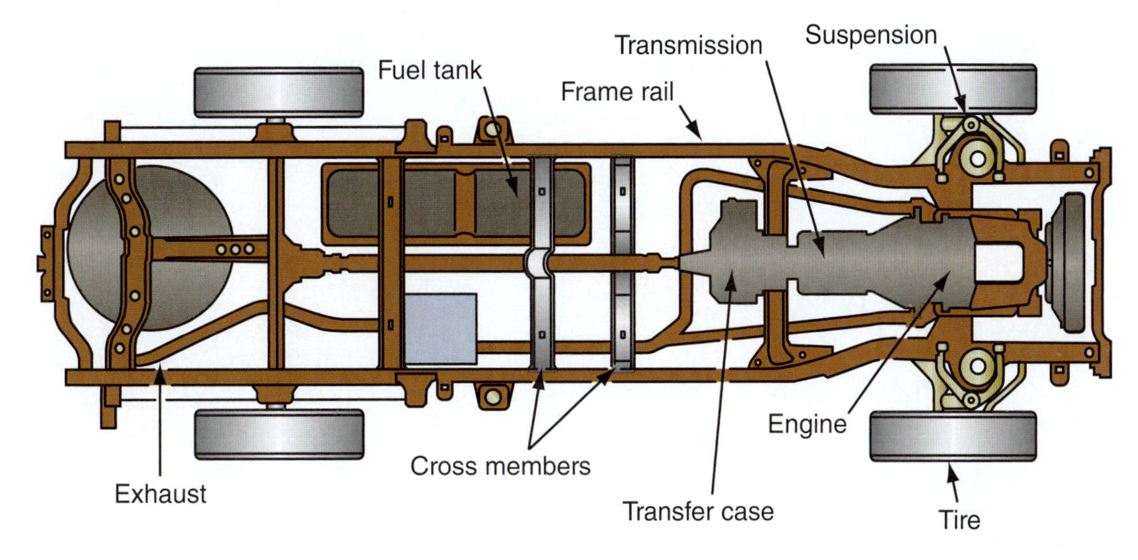

Figure 2-6 Rubber "biscuits," or mounts, fit between frame and body to reduce noise and vibration.

Figure 2-7 Unibody vehicles consist of small body panels welded together to serve as the vehicle frame. Unibody vehicles are light yet very strong. They are built for controlled collapse during a collision to help keep the passenger compartment intact.

approach to repairs. In unibody designs, heavy-gauge, cold-rolled steels have been replaced with lighter, thinner, high-strength steel or aluminum alloys. This requires new handling, straightening, and welding techniques.

2.3 MAJOR BODY SECTIONS

For simplicity and to help communication in auto body repair, a vehicle is commonly divided into three body sections—front, center, and rear. You should understand how these sections are constructed and which parts are included in each. See Figure 2–8.

Front Section

The *front section*, also called the nose section, includes everything between the front bumper and the fire wall. The bumper, grille, frame rails, front suspension parts, and the engine are a few of the items included in the front section of a vehicle.

The nickname **front clip**, or "doghouse," is used to refer to the front body section. It is often purchased from an automotive recycler or salvage yard and cut off from a wreck in one piece. The empty engine compartment forms the doghouse.

Center Section

The vehicle's *center section*, or midsection, typically includes the body parts that form the passenger compartment. A few parts in this section are the floor pan, roof panel, cowl, doors, door pillars, glass, and related parts. The center section is nicknamed the "greenhouse" because it is surrounded by glass.

Rear Section

The *rear section* (the tail section, or rear clip) commonly consists of the rear quarter panels, trunk or rear floor pan, rear frame rails, trunk or deck lid, rear bumper, and related parts. Also called the "cathouse," it is often sectioned or cut off of a salvaged vehicle to repair severe rear impact damage.

When discussing collision repair, body shop personnel often refer to these sections of the vehicle. It simplifies communication because everyone knows which parts are included in each section.

Vehicle Left and Right Sides

The left and right sides of a vehicle are determined by standing behind the vehicle or sitting in the driver's seat behind the steering wheel. In either position, the vehicle's left side is to your left; the right side is to your right.

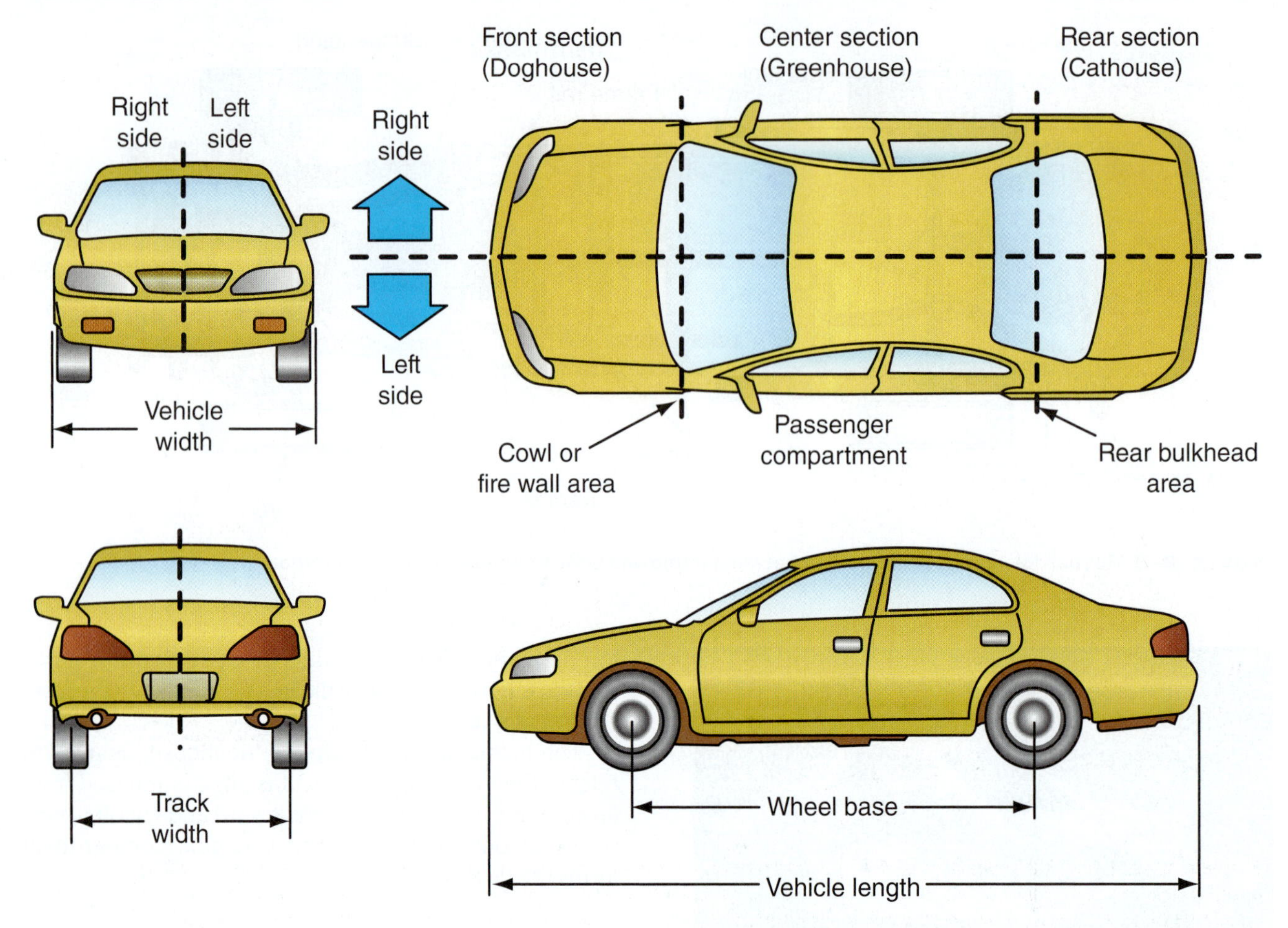

Figure 2-8 Study the locations of the basic sections of an automobile carefully.

Panels and parts are often named for the left or right side of the vehicle.

Note that vehicles built for American roads have the steering wheel on the left. Vehicles built for use in other countries may have the steering wheel on the right side of the passenger compartment.

Driveline Configuration

Driveline configuration refers to how power is transmitted from the engine to the drive wheels. There are six basic drivetrain designs: front-wheel drive; rear-wheel drive; rear-engine, rear-wheel drive; mid-engine, rear-wheel drive; four-wheel drive; and all-wheel drive.

The vast majority of unibody vehicles on the road today are front-wheel drive with the engine in the front. These variations affect vehicle construction and repair methods.

A **transverse engine** mounts sideways in the engine compartment. Its crankshaft centerline extends toward the right and left of the body. Both front-engine and rear-engine vehicles use this configuration (Figure 2–9A).

A **longitudinal engine** mounts the crankshaft centerline front to rear when viewed from the top. Front-engine, rear-wheel drive vehicles use this type of engine mounting (Figure 2–9B).

A *front-engine, front-wheel drive* (FWD) vehicle has both the engine and transaxle in the front. Drive axles extend out from the transaxle to power the front drive wheels. This is one of the most common configurations. The heavy drivetrain adds weight to the front drive wheels for good traction on slippery pavement.

A *front-engine, rear-wheel drive* (RWD) vehicle has the engine in the front and the drive axle in the rear. The transmission is usually right behind the engine, and a drive shaft transfers power back to the rear axle (Figure 2–9C).

A *rear-engine, rear-wheel drive* (RRD) vehicle has the engine in the back, and a transaxle transfers power to the rear drive wheels. Traction upon acceleration and cornering is good because more of the weight of the drivetrain is over the rear drive wheels.

A *mid-engine, rear-wheel drive* (MRD) vehicle has the engine centrally located, right behind the front seat. This helps to place the center of gravity in the middle so that the front and rear wheels hold the same amount of weight, which improves cornering ability (Figure 2–9D).

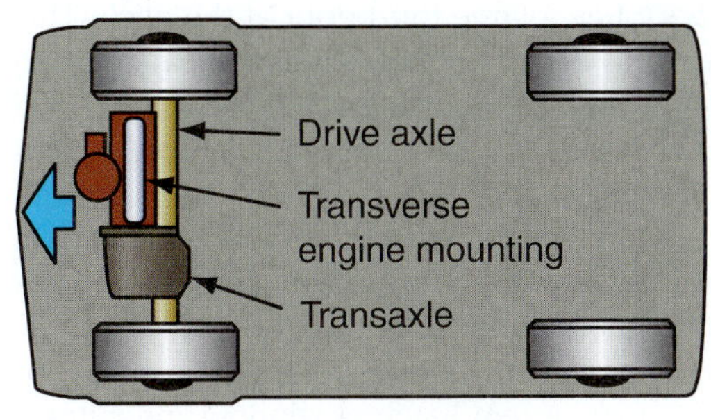

Front-engine, front-wheel drive (transverse engine)

A

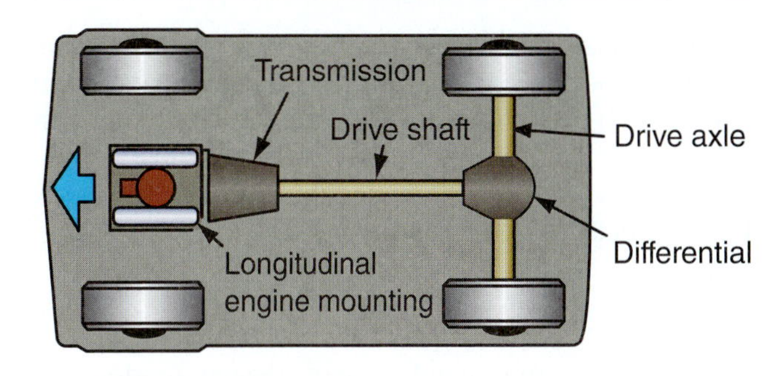

Front-engine, rear-wheel drive

C

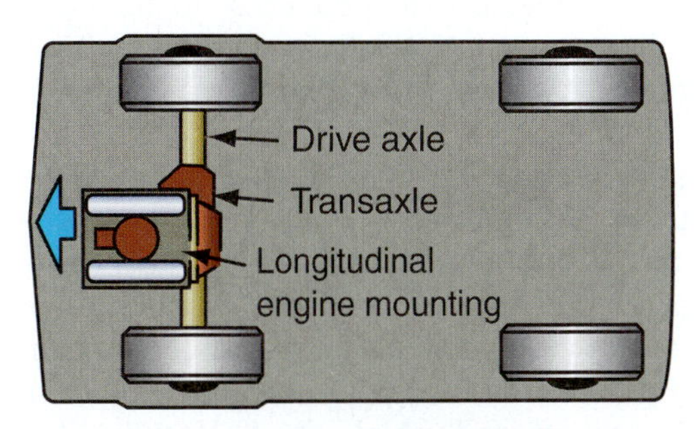

Front-engine, front-wheel drive (longitudinal engine)

B

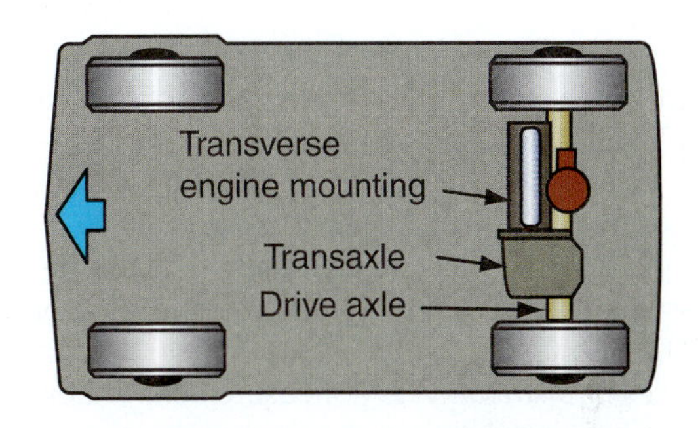

Mid-engine, rear-wheel drive

D

Figure 2-9 Compare the layout of common driveline configurations. The type of driveline used affects vehicle construction and repair methods. (A) Front-engine, front-wheel drive with a transverse engine mounting is the most common driveline configuration. The heavy engine and transaxle are in the front to place more weight over the front drive wheels. (B) This front-wheel drive arrangement places the engine longitudinally in the engine compartment. It is not as common as the transverse engine arrangement. (C) The front-engine, rear-wheel drive setup is still common on many sports sedans. This design steers and handles well on dry pavement. It does not provide as much traction on snow as front-wheel drive, however. There is less engine and transmission weight over the drive wheels. (D) A mid-engine, rear-wheel drive layout is the least common. It can be found on a few sports cars.

All-wheel drive (AWD) uses two differentials to power all four drive wheels. This is a relatively new design used on several makes of passenger vehicles.

Four-wheel drive (4WD) systems use a transfer case to send power to two differentials and all wheels. The transfer case can be engaged and disengaged to select two- or four-wheel drive as desired. It is common on off-road vehicles.

NOTE *For more information on vehicle construction (air bags, drivelines, and so on) and related subjects, refer to the Index. For example, Chapter 21 explains the operation and repair of mechanical systems in more detail. Air bag operation and service are explained in Chapter 23.*

2.4 BODY CLASSIFICATIONS

Various methods of classifying vehicles exist: engine type (gas or diesel), fuel system type (carburetor or injection), driveline type (automatic or manual transmission, front-wheel versus rear-wheel drive), and so forth.

The body classifications most recognized by consumers are car size, body shape, seat arrangement, number of doors, and so on. You should be familiar with these to communicate well on the job.

Car Size

A *compact car*, also called an economy car, is the smallest body classification. It normally uses a small, four-cylinder engine, is very light in weight, and gets the highest gas mileage.

A *micro car* is a tiny passenger vehicle for one or two adults. This type of car is often powered by a motorcycle engine. Micro cars are very popular in Europe. They are very light and have a very small cross-sectional area for low drag and wind resistance.

An *intermediate car* is medium in size. It often uses a four-, six-, or eight-cylinder engine and has average weight and physical dimensions. It usually has unibody construction, but a few older vehicles have body-over-frame construction.

A *full-size car*, or luxury car, is the largest classification of passenger car. It is larger and heavier and often uses a high-performance V8 engine. Full-size cars can have either unibody or body-over-frame construction. Full-size cars get lower fuel economy ratings, primarily because of their increased mass.

General Vehicle Data

Auto manufacturers or auto makers design, model, test, and build passenger vehicles in automated factories. The major auto companies include:

Vehicle curb weight is the total mass of the vehicle with a full tank of gas and no driver. Vehicle curb weights vary.

Generally, pickup trucks and SUVs are the heaviest vehicles and weigh 2.5 tons or approximately 5,000 lb. (2,267.96 kg). Full-size cars, station wagons, and full-size vans weigh about 2 tons or 4,000 lb. (1,814.369 kg). Economy cars have a low curb weight of about 3,000 lb. (1,360.777 kg). Small sports cars and micro cars are the lightest passenger vehicles, weighing in at about a ton or 2,000 lb. (907.185 kg).

Vehicle weight distribution is a measurement of how much force is pushing down on the front and rear tires of the vehicle. An ideal weight distribution of economy and corning ability is 50/50 or 50 percent front and 50 percent rear, as is found with many high-performance sports cars. Front-wheel drive vehicles have about a 70 percent front/30 percent rear weight distribution for good two-wheel drive traction at both front-axle hubs.

Vehicle wheelbase is the distance from the centerline of the front wheels to the centerline of the rear wheels.

Vehicle track is the distance between the right and left wheel or tire centerlines. The trend is to make vehicles wider so that they can corner more quickly without a rollover accident. A typical track rating might be 62 inches (0.0015748 km) front and 64 inches (0.0016256 km) rear.

Vehicle length is a measurement from the front of the front bumper to the back of the rear bumper. *Vehicle width* is a measurement between the two widest points on the right and left sides of the body. *Vehicle height* is measured from the ground to the top of the highest point on the roofline.

Roof Designs

There are several basic body shapes or roof designs in use today:

- A *sedan* refers to a body design with a center pillar that supports the roof. Sedans come in both two-door and four-door versions.
- A *hardtop* does not have a center pillar to support the roof, so the roof must be reinforced to provide enough strength. A hardtop is also available in both two- and four-door versions.
- A *hatchback* has a large third door at the back. This design is commonly found on small compact cars so that more rear storage space is available.
- A *convertible* uses a retractable canvas roof with a steel tube framework. The top folds down into an area behind the seat. Some convertibles use a removable hardtop. Some newer vehicles have gone back to the use of retractable hardtops.
- A *station wagon* extends the roof straight back to the rear of the body. A rear hatch or window and tailgate open to allow access to the large storage area.
- A crossover vehicle is a mixed design using traits from both a station wagon and an SUV.
- A *van* has a large, box-shaped body to increase interior volume or space. A full-size van normally is front-engine, rear-wheel drive with a full perimeter frame. A minivan is smaller and often uses front-engine, front-wheel drive with unibody construction.
- An *SUV*, or sport utility vehicle, has four-wheel drive and room for multiple passengers. This all-weather vehicle generally sits higher than passenger cars for increased ground clearance on rough terrain. Often classified as an off-road vehicle, the SUV is ideal for driving through snow and mud.
- A *pickup truck* normally has a separate cab and bed. Most pickup trucks use a front-engine, rear-wheel drive setup. Some are four-wheel drive.

The basic vehicle body shapes are shown in Figure 2–10.

Part/Panel Nomenclature

A *part*, also called a *component*, generally refers to an individual unit used to build a vehicle. Several parts fastened together form an assembly. For example, a steering column **assembly** is made up of the steering wheel, trim cover, air bag, turn signal mechanism, and other parts.

Stationary parts, such as the floor, roof, and quarter panels, are permanently welded or adhesive bonded into place. *Hinged parts* (the hood or trunk lid) and *hinged assemblies* (doors) can open and close.

Fastened parts are held together with various fasteners (bolts, nuts, clips, adhesive, and so forth). Many parts, such as the fenders, hood, and grille, are fastened or bolted into place.

Welded parts are permanently joined by heat fusing the material so that it flows together and bonds when cooled. Both metal and plastic parts can be welded.

Press-fit, or snap-fit, parts use clips or an interference (friction) fit to hold parts together. This assembly method is becoming more common to reduce manufacturing costs.

Adhesive-bonded parts use a high-strength epoxy or special glue to hold the parts together. Both metal and plastic parts can be joined with adhesive. *Structural adhesive* can also be used to bond parts together.

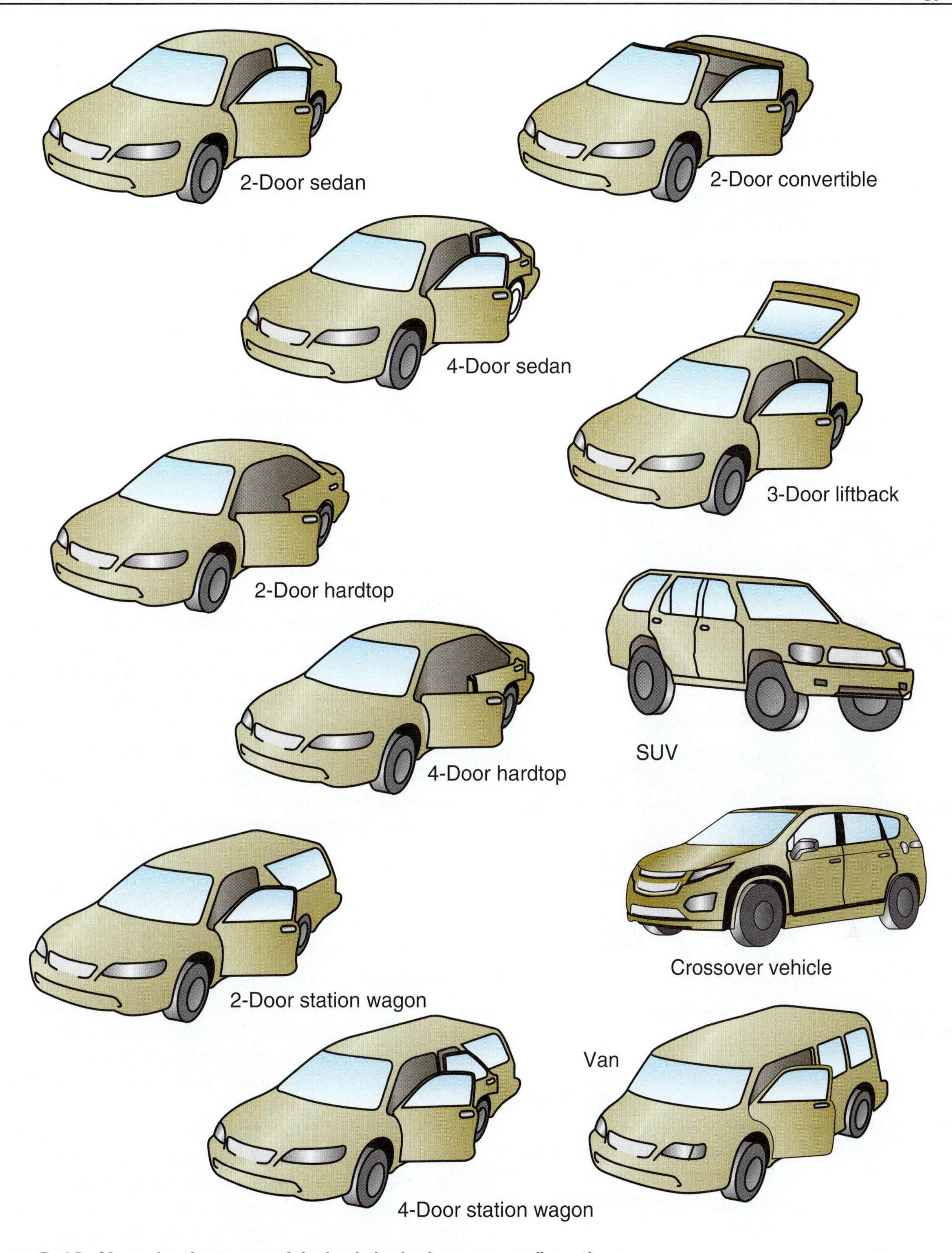

Figure 2–10 **Memorize the names of the basic body shapes or configurations.**

Body Panels

A **panel** is a steel, aluminum, or plastic sheet stamped or molded into a body part. Various panels are used in a vehicle. Usually, the name of the panel is self-explanatory: hood panel, fender panel, trunk lid panel, or roof panel.

Study the names and locations of each part or panel carefully. The major outer panels of a vehicle are shown in Figure 2–11.

During manufacturing, these complexly contoured panels are often stamped out of sheet metal, using a huge multi-ton drop forger. The giant machines crush the thin,

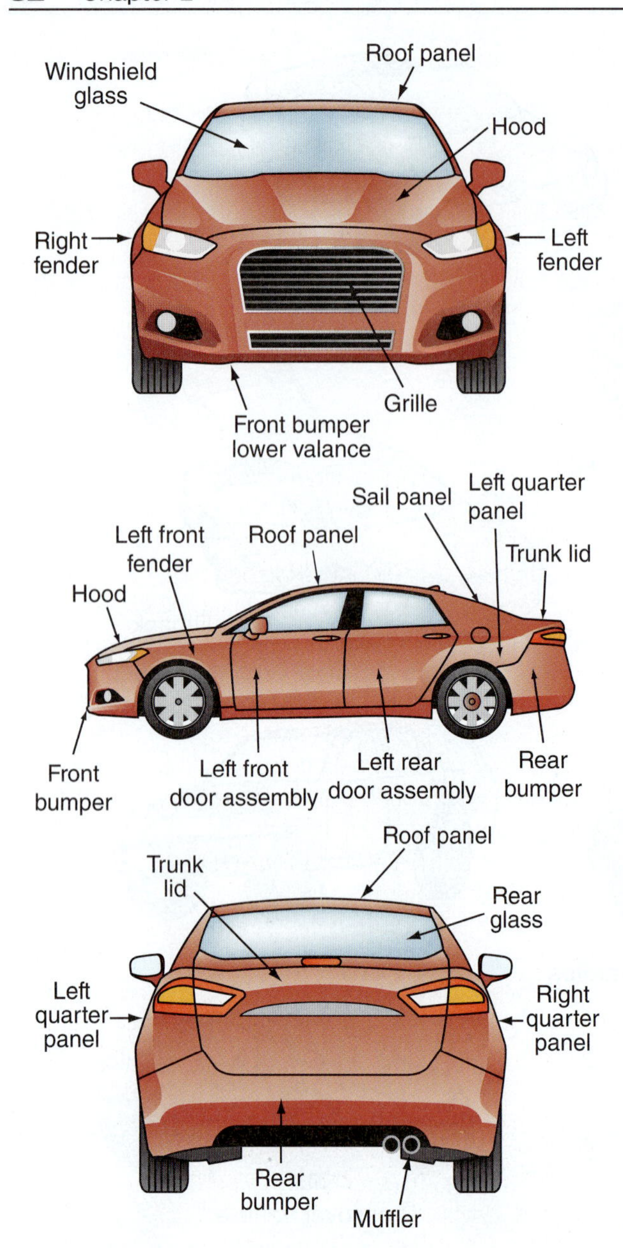

Figure 2-11 Study the major outer body panels, viewed from outside.

flat sheet metal in a die machined to match the shape of the desired body panel.

Body panels usually have compound curves formed to increase stiffness. With unibody construction, these panels are welded together to form the unibody frame at the vehicle assembly plant.

DANGER Make sure you fully understand correct repair procedures and construction technology before working on a vehicle. It can be costly and dangerous, even deadly, if you do not understand how a car or truck is made and should be repaired.

2.5 AUTO BODY REPAIR HISTORY

To provide background about how the passenger vehicle has changed over the years, this section of the chapter summarizes the history of automobile construction.

The First Motor Cars

The early automobile manufacturing companies originated in various ways. They evolved from bicycle makers, carriage and wagon makers, and other types of industry.

The very first motor cars were nothing more than a horse buggy with an engine. They had large, spoked wheels and a long bar for steering the front wheels. The small gas engine used a chain and sprockets to drive the rear axle. Other early motor cars used steam engines or electric motors and batteries for power.

Henry Ford developed the Model T, the first car mass-produced on an assembly line. Affectionately called the "Tin Lizzie," this car was introduced and sold in 1908. The Model T was deemed "the car for the multitude" and was an instant success, upon which Ford succeeded in building one of the largest auto manufacturing companies in the world.

During manufacturing, the Model T's rolling chassis was built first. The axles, wheels, and tires were attached to the frame, then the chassis was pulled down the assembly line by a floor conveyor. As the chassis reached each workstation, another part or assembly was installed. The body was one of the last units lowered onto the chassis. It had a 4-cylinder, 20-horsepower engine mounted in the front. To operate the planetary transmission, you had to use foot pedals on the floor.

The Model T was painted only one color—black. This and other mass-production steps helped Henry Ford keep the cost of his car down, low enough for the average American worker to afford. The first Model Ts cost about $850. However, by the early 1920s, the least expensive model sold for only $290.

Other car makers came along and imitated many of Henry Ford's assembly line techniques. American automobile companies used parts made by independent suppliers. These outsourced parts were shipped to the factory for assembly.

In the mid- to late 1920s, the two largest manufacturers were Ford and General Motors. The other major manufacturers at the time were Packard, Hudson, Maxwell, and Chrysler. Most of these companies were located in Detroit, Michigan, which became the center for world automotive production.

General Motors overtook Ford in sales volume in 1925 by producing vehicles more refined than the Model T. Because of this, the last Model T was manufactured in 1927.

The Sixties

Up through the 1960s, American automobiles were manufactured in pretty much the same way, with similar characteristics:

- Body-over-frame construction
- Rear drive, with solid axle housing
- Independent front suspension

By the 1960s, people wanted speed, power, and styling in their cars. Americans wanted cars that had low elapsed times in the quarter mile. They craved big, loud, "muscle cars" with huge V8 engines, such as the legendary 427 Chevy Big Block, 428 Ford Cobra Jet, 426 Dodge Maxi Wedge, Pontiac 421, and Oldsmobile 455. The "King Kong" 426-inch Chrysler Hemi Head engine with 2 four-barrel carburetors and 425 horsepower was stuffed into the small Plymouth Challenger. Chevrolet shoehorned a 454-cubic inch big block into a Chevelle. Ford installed a 429 Hemi in a Mustang. The Corvette had a 427 with 3 two-barrel carburetors, or for even more power, you could order "2-fours." These classic American icons are still increasing in value on the collector market.

The Seventies

In 1974, a variety of events took place that rocked the foundations of the automobile industry. First, the government placed strict fuel economy and emission control laws and standards on automobile manufacturers. This was because many large cities were suffering from smog caused by air pollution. This increased the public demand for safe as well as cleaner running vehicles. Second, the Arab oil embargo also occurred at this time. The price of gasoline went up quickly, and people often had to wait in long lines to buy fuel. Consumers started purchasing cars with increased fuel efficiency (smaller bodies and engines) to save money on gasoline. This meant that American automotive manufacturers had to start building cars with better fuel economy.

To combat one source of pollution and reduce our consumption of petroleum, the fuel economy standard known as **Corporate Average Fuel Economy (CAFE)** was set by the Environmental Protection Agency. The **Environmental Protection Agency (EPA)** was established in 1970 as an independent agency in the executive branch of the U.S. government. The purpose of the EPA is to coordinate government and industry action to protect the Earth's environment from chemical damage.

Foreign car makers, who had always manufactured smaller, lighter, more fuel-efficient vehicles, captured an increasing share of the domestic new car market. American auto makers reacted to the competition by producing smaller, more efficient cars. This sped the development of the millions of unibody cars on the roads today.

Third, because of the construction of interstate highways and higher speed limits, along with more high-performance cars on the road, accidents and deaths from auto accidents had increased dramatically. As a result of this growing problem, federal laws were passed to regulate safety standards for motor vehicles. Numerous regulations mandated the installation of seat belts, safety glass windshields, head restraints to prevent whiplash, tire standards, body strength, and other design factors to improve passenger safety.

In 1979, the first driver's side air bag was introduced in the passenger car. It worked in conjunction with the seat belts to protect the driver from injuries during a frontal collision. If the car was in a major head-on collision, the air bag deployed like a big cushion to keep the driver's body and head from hitting the steering wheel and other interior parts. Air bags were made mandatory in motor cars produced after 1990.

Modern Motor Vehicles

With the advancement of technology, modern motor vehicles are engineering marvels. Today's passenger cars, light trucks, vans, and SUVs are more powerful, cleaner running, and safer than ever. On-board computers add to driver control of everything from engine management to antiskid braking systems. On-board navigation and automation provide today's drivers with luxury and convenience.

It is amazing how much more refined present-day automobiles are compared to just a few years ago. Sports car icons like the Corvette and Dodge Viper produce over 400 horsepower while emitting far fewer pollutants than their muscle car counterparts of the past. High-performance, "race-bred" parts are found in every major system of today's "super cars."

To avoid a collision, computer-controlled disc brake systems let drivers almost "stop on a dime," while maintaining steering control without skidding. If drivers have to swerve to avoid a collision, computer-assisted handling packages sense vehicle speed, cornering forces, tire loss of traction, and other factors to help cars "corner as if on rails."

Today's small cars have set remarkable records for fuel efficiency. Fifty miles per gallon, once said to be impossible, is now a reality. Small but powerful engines squeeze more energy out of every drop of fuel. Lighter body/frame structures and sleeker body shapes have all helped reduce fuel consumption, conserving our natural resources.

The technology in a typical late-model vehicle has also helped to prevent thousands of highway deaths per year. High-speed auto accidents that used to be fatal in yesterday's low-tech cars now result in less severe injuries. People now walk away from car crashes that earlier would have killed them.

This improved safety record is primarily due to the superior structural body/frame designs as well as more advanced computer control of various safety systems. Air bags, front and rear structural crush zones, stronger pillars going up to the roof, and reinforced passenger compartment areas have contributed the most to these improved highway safety statistics.

At present, the largest U.S. auto manufacturers are:

- General Motors Corporation (Chevrolet, Pontiac, Cadillac, GMC)
- Ford Motor Company (Ford, Mercury Jaguar)
- Chrysler Corporation (Dodge, Plymouth, Jeep)

Ford, General Motors, and Chrysler are often called the "Big Three" because they are the largest U.S. auto manufacturers. Other auto manufacturers include Porsche, Audi, Volkswagen, BMW, Honda, Acura, Infinity, Lexus, Nissan, Saab, Subaru, Toyota, and Volvo.

The automotive industry is becoming more global in nature, with the major manufacturers owning many of the smaller auto makers. This aids technology transfer from country to country and from corporation to corporation, which in turn improves vehicle quality while reducing fuel consumption and emissions (pollution).

Other than the United States, countries with large automotive industries include Japan, Canada, France, Italy, Sweden, Korea, and Russia. The places in the world with the most motor vehicles are North America, Western Europe, Japan, Australia, and New Zealand. These countries have ratios of approximately one automobile for every three people.

Hybrid Motor Vehicles

The hybrid car is a glimpse into the future. The hybrid has tremendous potential to reduce fuel consumption and exhaust emissions. Hybrid vehicles are now being built and sold by several auto manufacturers.

The *hybrid vehicle* uses two power sources: an engine and an electric motor. When driving, the hybrid car uses batteries and its large electric motor to accelerate up to cruising speed. Then as the batteries become discharged, the gas or diesel engine starts up to generate electricity to keep the car moving and also to recharge the batteries. By using regenerative braking, the hybrid car saves even more fuel. Hitting the brakes actually generates and stores energy for the next acceleration. The hybrid car now holds the record for fuel economy at around 50 miles per gallon.

The Future

What will future motor vehicles be like? Will our automobiles be made of all plastics or carbon fiber or some yet to be invented material? Will car unibodies have fewer parts? Could the unibody be made of one large injection-molded piece of plastic? Will cars drive themselves? Engineers are studying self-steering cars that follow a strip of metal embedded in the highway. Will automatic braking systems sense an impending crash and apply the brakes by computer control?

Construction Methods Change

Passenger cars, trucks, vans, and SUVs use one of two types of construction:

1. Conventional body-over-frame
2. Unitized or unibody

The five construction areas in which domestic automobiles have changed since the mid-1970s are:

ECO-TECH Hybrid vehicles are now the "Gas Mileage Champions" of the auto industry. Plug-in hybrids can be charged overnight and then driven up to about 50 miles on full electric propulsion. The gasoline or diesel engine does not have to start and run at all. Only when the high-voltage battery becomes almost fully discharged does the gas engine start and run to propel the car. Many manufacturers claim that their hybrids produce zero emissions while driving in full electric mode. With millions of hybrid vehicles on the road today, this accounts for a substantial reduction in air pollution.

1. Body/frame construction (more unibody than full frame)
2. Weight (average vehicle weight has decreased)
3. Part composition (more use of thin, high-strength steels, aluminum, plastics, and composites)
4. Suspension/steering (more use of independent suspension, rack-and-pinion steering, and four- or all-wheel drive)
5. Engine location/drive (more front-engine, front-wheel drive vehicles)

In 1977, most new cars still used a full frame. They averaged around 4,500 lb. (2,038 kg) and used comparatively heavy, thick, 18-gauge, mild-strength steel in body panels. They were still predominantly front-engine, rear-wheel drive vehicles.

After that, body weight began to decrease, and thinner-gauge metal was used. Also, the first American-made transverse (sideways-mounted) engine, front-wheel drive, strut suspension cars were introduced.

By 1981, unibodies were used in almost half of American-made cars. Average weight decreased 600 lb. (272 kg), and 22-gauge high-strength steel was used in construction. At the same time, there was a shift toward rack-and-pinion steering, to the MacPherson strut–type suspension, and from rear- to front-wheel drive.

At the present time, most unibodies are constructed of thin, 24-gauge, high-strength steel; have an average weight of 900 lb. (407 kg) less than in 1980; and feature MacPherson struts, rack-and-pinion steering, and front-wheel drive. Today, most passenger cars on U.S. roads are unibodies.

Repair Methods Change

As design innovations and the construction of vehicles have changed over the years, so too has the collision repair profession. The job of repairing vehicles has also become more complex.

In the early days of Model T automobiles, there were no specialized shops for automobile collision repair.

When a car was brought in for repair, the damaged part was usually removed and replaced with a new one that was either forged from steel or cut from wood. This method was expensive and time-consuming. Many times there was a long wait for parts and to complete repairs. Most of the early body/frame technicians were carpenters or blacksmiths.

As automobile body/frame designs became more complex, it became more practical to repair than replace, even though early repair procedures often involved days of hammering on parts to straighten them. Early body repair was delicate and time-consuming because parts had to be straightened without using body filler. (None existed at this time.)

When the Nash Company introduced its unitized body in 1940, a whole new set of collision repair problems arose. Because there was no frame to apply pressure against, the technique of internal body and frame pushing was of little value. There was not enough material in any one place to push against without bending the body and causing further damage.

The basic repair technique of pushing out damaged sections evolved to one of pulling out damaged sections. Out of necessity, the portable body and frame puller was developed and was soon accepted on a worldwide basis.

The manufacturers of stationary frame equipment also had to modify their equipment. The change to a pull technique was made by adding adjustable pull towers. These units remained functional but grew more massive, complicated, and expensive.

It is important to know which type of auto body construction is used. Detailed in later chapters, repair work is different for each type of vehicle construction. Modern body technicians need a great deal more knowledge than their counterparts in the era prior to the advent of the unibody. Now technicians must know how to repair both full frame and unibody vehicles with great precision. The safety of the driver and passengers relies on the skill of the auto body technician who repairs their vehicle.

2.6 UNIBODY PANELS

To become a competent collision repair technician, it is important that you be able to quickly locate and identify the major panels of a motor vehicle. As you will learn, some are the same for both unibody and full frame vehicles, whereas others differ significantly depending on vehicle construction. Please refer to Figure 2–12.

Front Section Parts

The **frame rails** are the box members extending out near the bottom of the front section. They are usually the strongest part of a unibody. Frame members, or rails, are normally welded to the fire wall and to the bottom of the fender aprons. They usually have crush zones in them to absorb collision energy from a frontal impact.

The **cowl** is the assembly of panels at the rear of the front section, right in front of the windshield. This assembly includes the top cowl panel and side cowl panels.

The *front fender aprons* are inner panels that surround the wheels and tires to keep out road debris. They are often bolted or welded to the frame rails and cowl. They also add to the structural integrity of the front end (Figure 2–13).

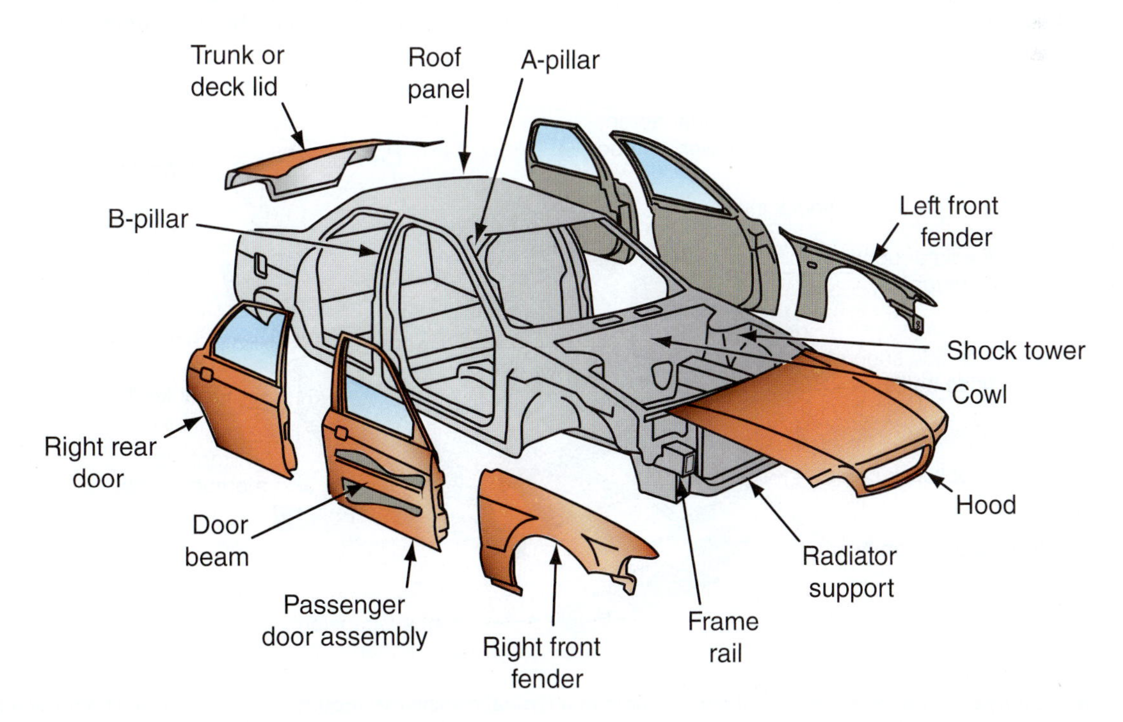

Figure 2–12 Here is a front-drive unibody vehicle with the bolt-on parts removed. Study the names and locations of the parts and panels.

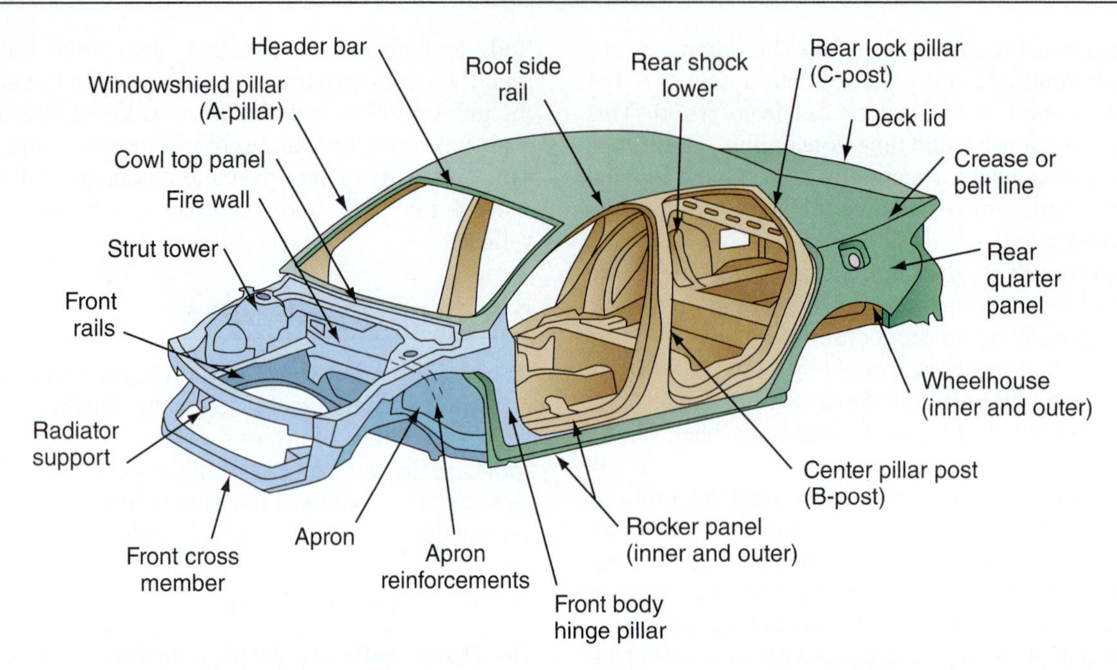

Figure 2–13 **Memorize unibody terminology. You will have to know the names of these parts when working in a body shop.**

The **shock towers**, or strut towers, are reinforced body areas for holding the upper parts of the suspension system. The coil springs and strut, or shock, absorbers fit up into the shock towers. They are normally formed as part of the inner fender apron.

The *radiator core support* is the framework around the front of the body structure for holding the cooling system radiator and related parts. It often fastens to the frame rails and inner fender aprons (Figure 2–14).

The *hood* is a hinged panel for accessing the engine compartment (front-engine vehicle) or trunk area (rear-engine vehicle). Hood hinges, bolted to the hood and cowl panel, allow the hood to swing open. The hood is normally made of two or more panels welded or bonded together to prevent flexing and vibration. Some hoods also hinge at the radiator support.

The *dash panel*, sometimes called the fire wall or front bulkhead, is the panel dividing the front section from the center passenger compartment section. It normally welds in place. The fire wall helps protect people in the car in the event of a fuel leak and resulting engine fire.

The *front fenders* extend from the front doors to the front bumper. They cover the front suspension and inner

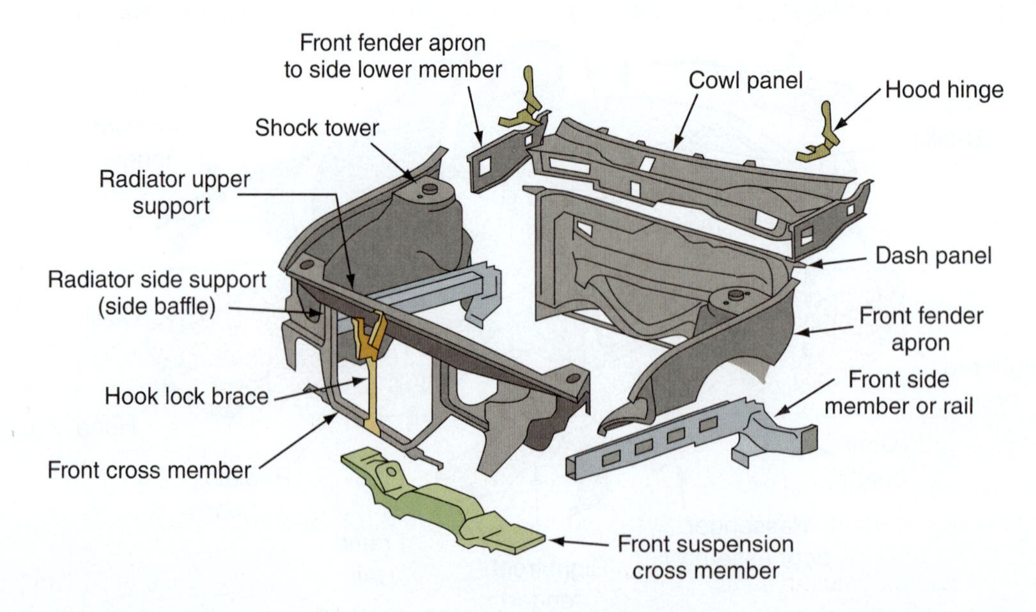

Figure 2–14 **Study the front body structural components of a typical transverse-mounted engine compartment of a front-engine, front-wheel drive vehicle. Because these components must not only hold the weight of the engine but also the transaxle, drive axles, steering, and suspension, they must be heavy duty in construction.**

A

C

B

D

Figure 2–15 Cutaway view shows what is inside modern unitized body structure. (A) Cutaway of a hood shows reinforcements to strengthen a panel. (B) Cutaway of a dash panel shows beams and brackets that keep the structure from collapsing into the passenger compartment. (C) Cutaway of rocker panels shows how the floor and pillars are strengthened. (D) View of the upper rear sail panel with rear glass removed shows how the roof is strengthened to withstand rollover damage.

aprons. They normally bolt into place around their perimeter.

The *bumper* assembly bolts to the front frame horns or rails to absorb minor impacts. The *grille* is the center cover over the radiator support. It generally has an opening for airflow through the radiator. Refer to Figure 2–15.

Center Section Parts

The *floor pan* is the main structural section in the bottom of the passenger compartment. It is often stamped as one large piece of steel.

Pillars are vertical body members that hold the roof panel in place and protect the passenger compartment in case of a rollover accident. The three types of pillars are shown in Figure 2–16.

The front pillars extend up next to the edges of the windshield. They must be strong to protect the passengers. Also termed **A-pillars**, they are steel box members that extend down from the roof panel to the main body section.

Center pillars or **B-pillars** are the roof supports between the front and rear doors on four-door vehicles. They help strengthen the roof and provide a mounting point for the rear door hinges.

Rear pillars extend up from the quarter panels to hold the rear of the roof and rear window glass. Also called **C-pillars**, their shape can vary with body style.

Rocker panels, or *door sills*, are strong beams that fit at the bottom of the door openings. They normally are welded to the floor pan and to the pillars, kick panels, or quarter panels. The kick panels are small panels between the front pillars and rocker panels.

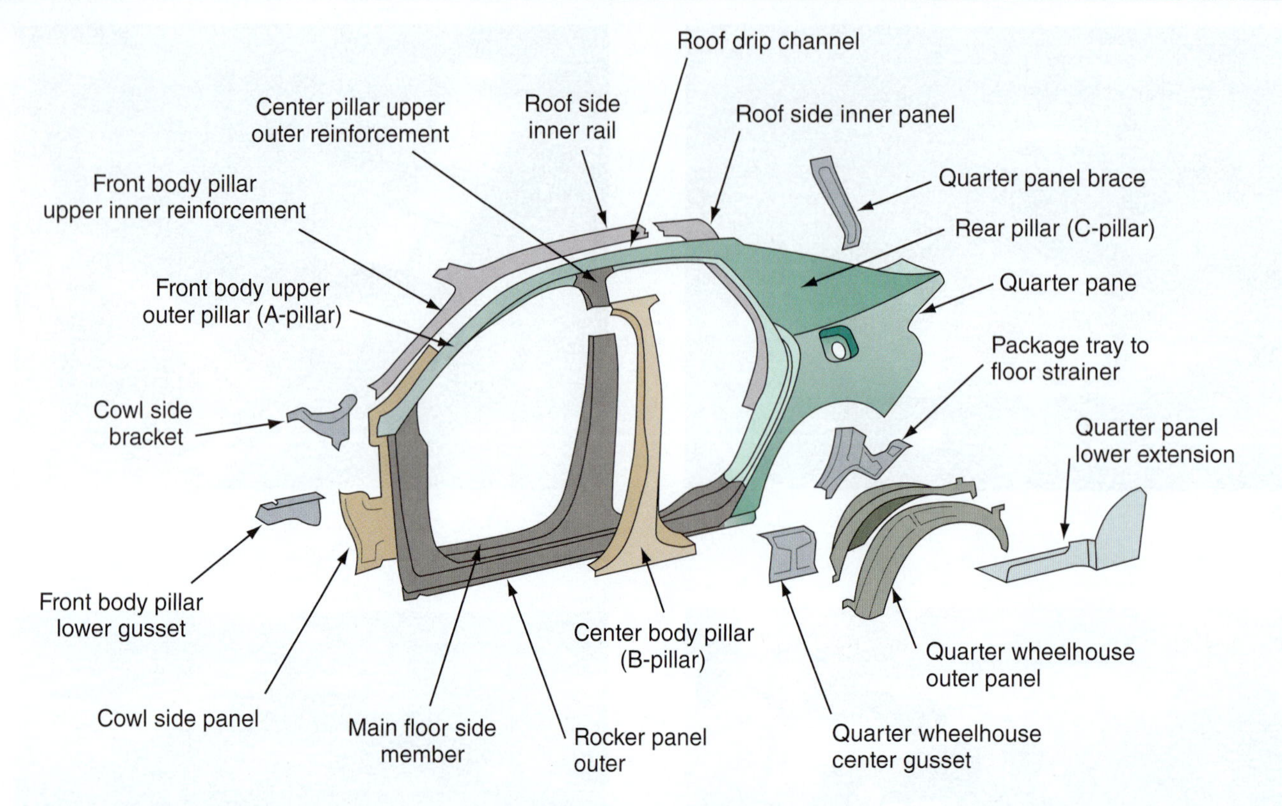

Figure 2–16 **Note the side body structural components of the typical vehicle. These parts must protect passengers in a side hit or collision. They also support the roof during a rollover accident. The pillars serve as a type of roll cage.**

The *rear shelf,* or *package tray,* is a thin panel behind the rear seat and in front of the back glass. It often has openings for the rear stereo speakers. The rear bulkhead panel separates the passenger compartment from the rear trunk area.

The doors are complex assemblies made up of an outer skin, inner door frame, door panel, window regulator, glass, and related parts. Door hinges are bolted or welded between the pillars and door frame.

The roof panel is a large panel that fits over the passenger compartment. It is normally welded to the pillars. Sometimes it includes a sunroof or removable top pieces, called *T-tops.* A headliner fastens to the roof panel for appearance and to deaden sound in the passenger compartment.

The *dash assembly,* sometimes called the *instrument panel,* is the assembly including the soft dash pad, instrument cluster, radio, heater and air-conditioning controls, vents, and similar parts. The dash assembly can be damaged in a major collision, when people fail to wear their seat belts, and by air bag deployment.

Rear Section Parts

The *rear frame rails* are strong boxed structures that give strength to the rear of the vehicle.

The *trunk floor panel* is a stamped steel part that forms the bottom of the rear storage compartment. Quite often the spare tire fits down into this stamped panel. It is typically welded to the rear rails, inner wheelhouses, and lower rear panel (Figure 2–17).

The *deck lid,* or trunk lid, is a hinged panel over the rear storage compartment. A *rear hatch* is a larger panel and glass assembly hinged for more access to the rear of the vehicle.

The **quarter panels** are the large, side body sections that extend from the side doors back to the rear bumper. They are welded in place and form a vital part of the rear body structure.

The *rear body panel* fits behind the rear bumper and between the quarter panels.

Rear shock towers hold the top of the rear suspension. The inner and outer *wheel housings* surround the rear wheels and weld to the quarter panels.

The *upper rear panel* is the area between the back glass and trunk lid.

Cross Members and Bracing

Cross members are thick-gauge supports that extend across the frame rails of both unitized and full frame vehicles. They provide added strength for holding various chassis parts, including the engine, transmission, and rear end (Figure 2–18).

Body bracing fastens between two body panels to increase stiffness and strength. It may be welded or bolted in place.

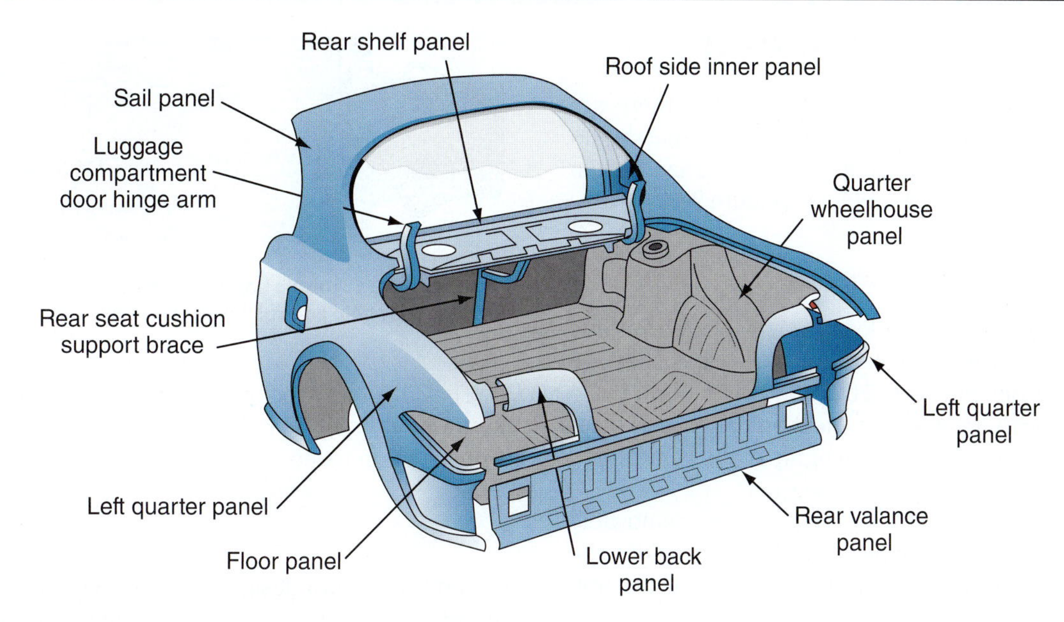

Figure 2-17 Study the rear body structural components of a typical unibody sedan.

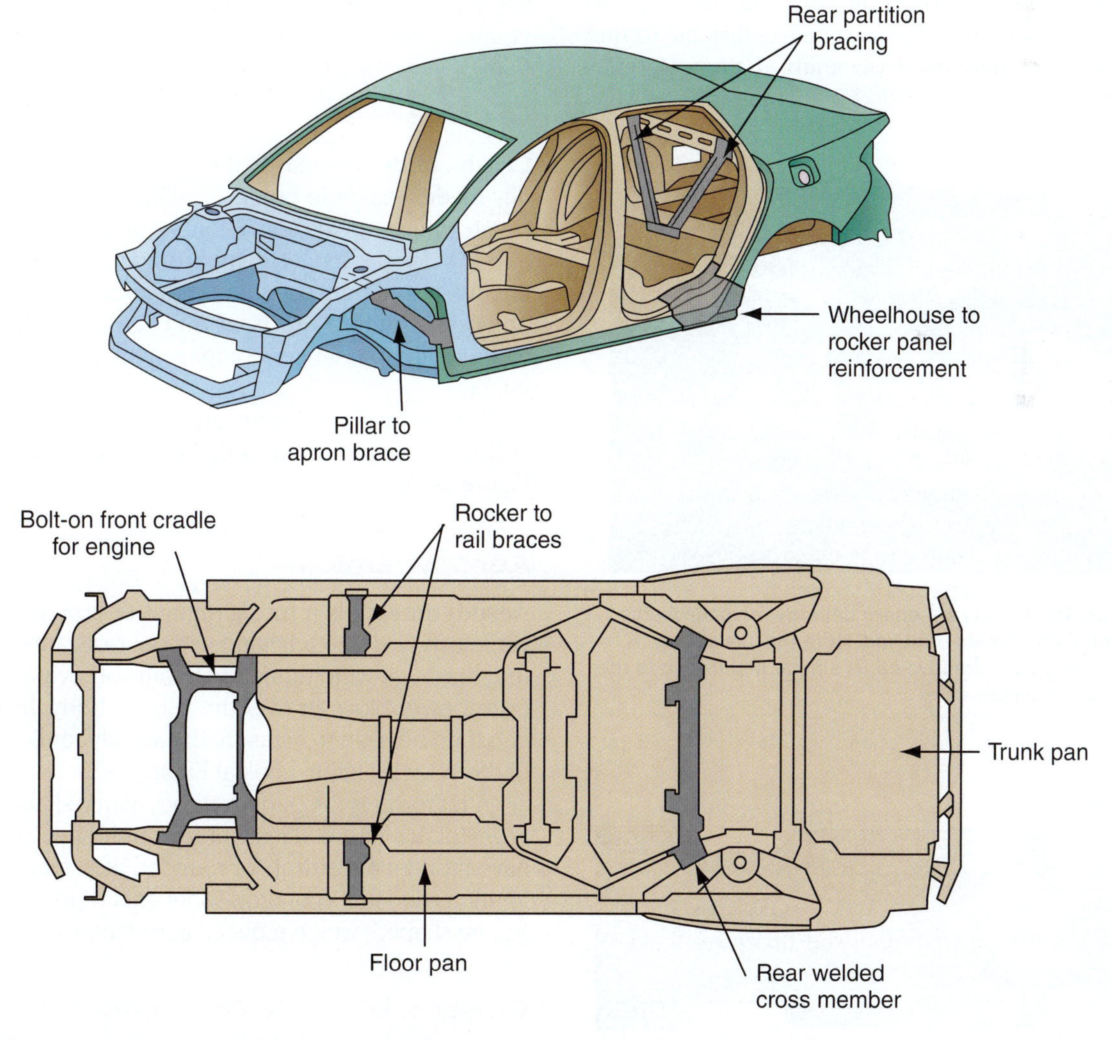

Figure 2-18 Various cross members and braces are welded or bolted to the unibody structure. Some add to the strength of the body. Others allow the mounting of mechanical parts.

2.7 UNIBODY DESIGN FACTORS

When learning to repair damaged vehicles, there are many other design and construction variables that should be understood. The following will summarize design considerations for unibody vehicles (Figure 2–19).

Semi-Unitized or Platform Frame Body

A *semi-unitized frame*, also known as a platform frame design, uses heavier gauge steel "stub" rails that are bolted to the front and rear of the body or platform structure. These heavy rails secure mechanical chassis parts and also add to the structural integrity of the vehicle during a collision (Figure 2–20).

Many small pickup trucks use a bolt-on removable stub frame under the engine. This design provides the advantages of both unitized and full frame construction on the same vehicle.

Unibody Torque Boxes

Unibody torque boxes allow some controlled twisting and crushing of the structure during severe collisions. Torque boxes are sometimes used on unitized frames/bodies (Figure 2–21).

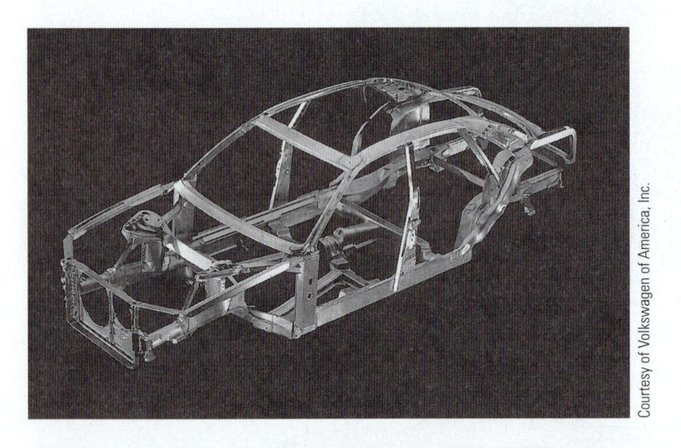

Figure 2–19 Here is a modern unibody or space frame with all of the flat panels removed. Note the use of box channels to form a protective cage, similar in purpose to the roll cages used in race cars.

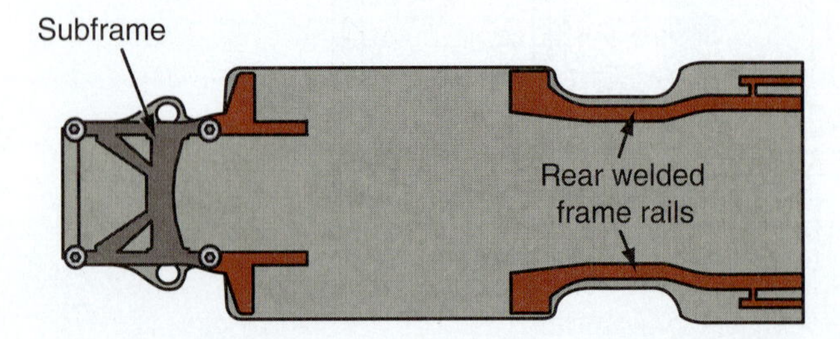

Figure 2–20 A bolt-on stub frame or carriage can be found on some vehicles. It simplifies repair because parts can be removed easily for replacement.

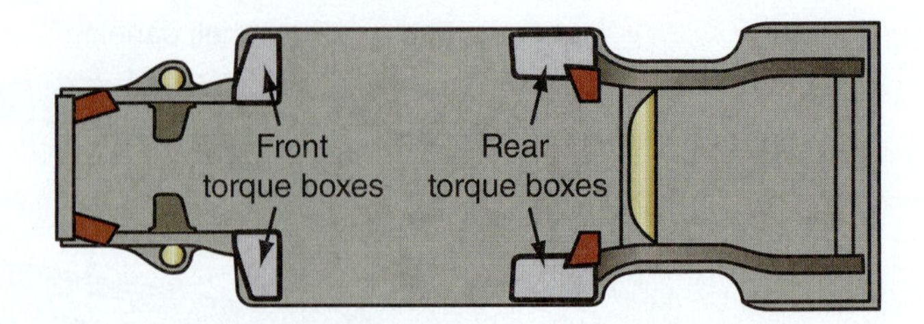

Figure 2–21 Torque boxes are used in some unibody designs. They help control flexing, twisting, and impact damage.

They also help secure the passenger compartment to the frame rails by increasing the surface area where the center section meets the front and rear sections.

Stress Hull Design

The unibody was a design concept used for the bodies of aircraft. This type of structure is often compared to an eggshell. Even when pressing hard on an eggshell lengthwise, it is difficult to crush. This is because the full force is not concentrated in one place but is dispersed effectively through the entire shell. In mechanics, this action creates a "stressed hull structure."

Crush Zones

Crush zones are areas in the unibody that are intentionally made weaker in order to collapse during a collision. Crush zones provide some control of secondary damage and a safer passenger compartment, because they are engineered to collapse in a predetermined fashion. In a unibody structure the front and rear areas will crush, whereas the passenger compartment area tends to stay intact.

The major crush zones and the flow of energy from collisions to the front, side, and rear are shown in Figure 2–22.

Aerodynamics

Aerodynamics is a measurement of how well a motor vehicle moves through wind without resistance. To measure this, huge wind tunnels with fans of several thousand horsepower blow air over the vehicle body. Instruments can then be used to measure the aerodynamic efficiency of the vehicle shape. Look at Figure 2–23.

A teardrop is the perfect aerodynamic shape. The box is one of the least efficient aerodynamic shapes. New cars have more of a teardrop, or rounded aerodynamic shape, for increased fuel economy. Older cars were more boxy or square shaped, which reduced aerodynamic efficiency.

General Unibody Characteristics

As mentioned, a vehicle body that integrates the frame and body into one assembly is called a unibody. It has the following characteristics:

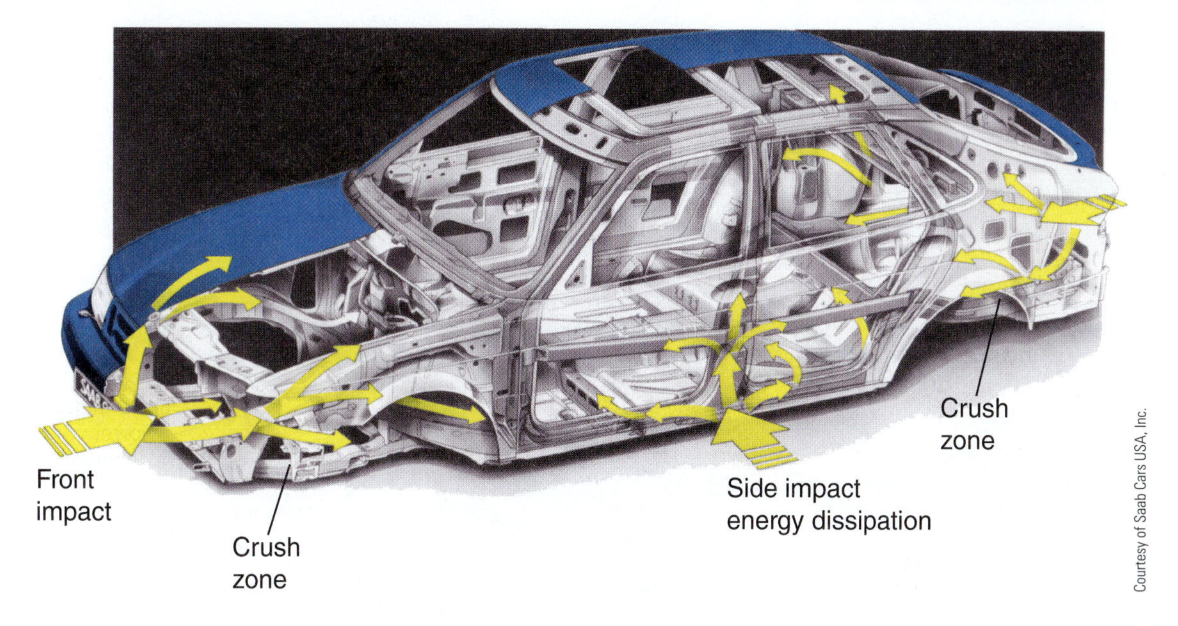

Figure 2-22 Arrows show how impact energy is designed to flow into and through a unibody structure. Crush zones are designed into the front and rear sections to cushion deceleration of the passenger compartment during a high-speed crash. Thick box pillars and door beams help avoid intrusion into the passenger compartment in the event of a side hit or collision.

Figure 2-23 Modern vehicles are designed to be as aerodynamic as possible. This helps them cut through the wind while driving to reduce fuel consumption. Even the bottom panels of new vehicles are made flat to avoid air turbulence, which resists vehicle movement.

- A unibody structure is made by combining pieces of thin sheet metal pressed to form panels of various shapes and joined into an integrated structure by spot welding. This lightweight structure is highly rigid and resistant to bending or twisting.
- The bulk or space taken up by a separate frame can be used to make the car more compact (smaller) and lighter.
- Vibration and noise from the drivetrain and suspension enter the floor pan and are amplified by the body, which acts as an acoustic chamber. This makes it necessary to add extra sound-deadening materials to the inside of the body to quiet the passenger compartment.

With the thin sheet metal body close to the road surface, adequate measures must be taken to prevent rusting during repairs. The thin sheet metal is structural, and severe rust can affect vehicle safety. **Anticorrosion materials** are used to prevent rusting of metal parts. Various types of anticorrosion materials are available (weld-through primer, sealers, rubberized undercoating, and so on). When performing repairs, you must restore all corrosion protection to keep the car safe to drive for extended periods of use.

Sound-deadening materials are used to help quiet the passenger compartment. They are insulation materials that prevent engine and road noise from entering the passenger area. Some metal panels sandwich plastic or wood between two metal sheets to quiet road noise in the passenger compartment.

The stiffer sections used with unibody design tend to transmit and distribute impact energy throughout more of the vehicle, causing misalignment in areas remote from the impact point. Even sections that are buckled or torn loose might have passed along heavy force before deforming. Worse still, much of this remote damage can easily be overlooked in a casual inspection but still be sufficient to cause handling or powertrain problems later.

The side body is joined to the front body and roof panel to form the passenger compartment. These panels distribute the loads from the underbody to the upper part of the vehicle and prevent bending of the left and right sides during side impacts or collisions.

The side body members also serve as door supports and maintain the integrity of the passenger compartment if the vehicle should overturn. Because the sides are weakened by the large door openings, they are reinforced by joining the inner and outer panels, which forms a very strong box-shaped structure.

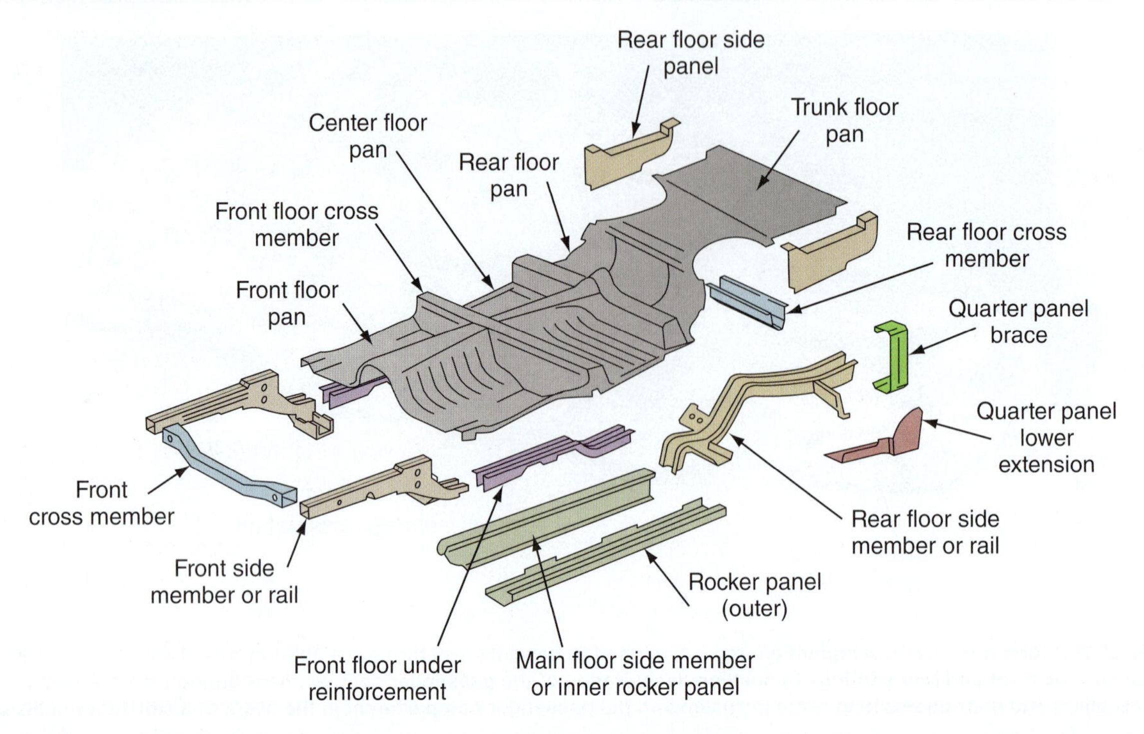

Figure 2-24 Note the underbody structural components of a typical unibody vehicle. These parts weld together to increase body stiffness, and they must also keep out the elements to avoid corrosion.

The unibody underbody includes the floor pan, trunk pan, rocker panels, cross members, front and rear members, and related parts. This large assembly must resist rusting and add strength to the unibody structure. The underbody is critically important to the structural integrity of a unibody vehicle (Figure 2–24).

Because the front side members and front cross members of the front underbody front section directly affect front-wheel alignment, they are often formed into a boxed section (Figure 2–25).

To prevent the collapse of the passenger compartment in a head-on collision, the front and rear frame rails are made with a kick-up area to bend, crush, and absorb shock loads before they reach the passenger compartment (Figure 2–26).

There are four basic unibody structures: front-engine, rear drive (FR); front-engine, front drive (FF); mid-engine, rear drive (MR); and (RR) rear-engine, rear-wheel drive.

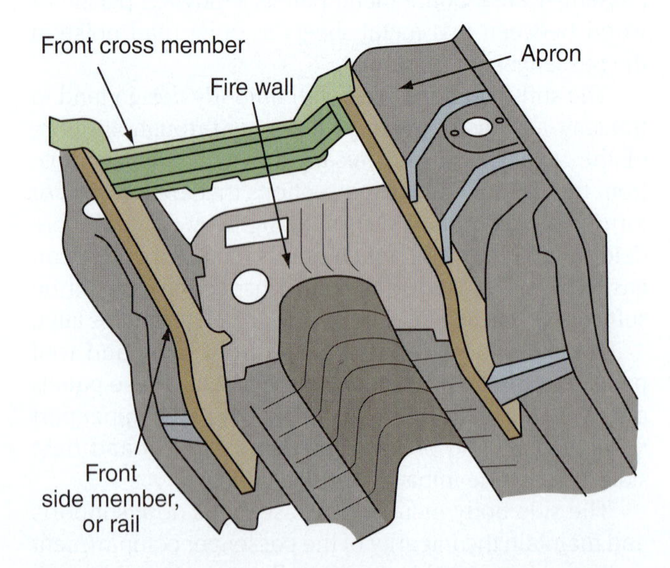

Figure 2-25 The underbody front section has strong side rails and cross members to support the engine and suspension system. Welded-on box rails are the main unibody structural members that replace full frame rails.

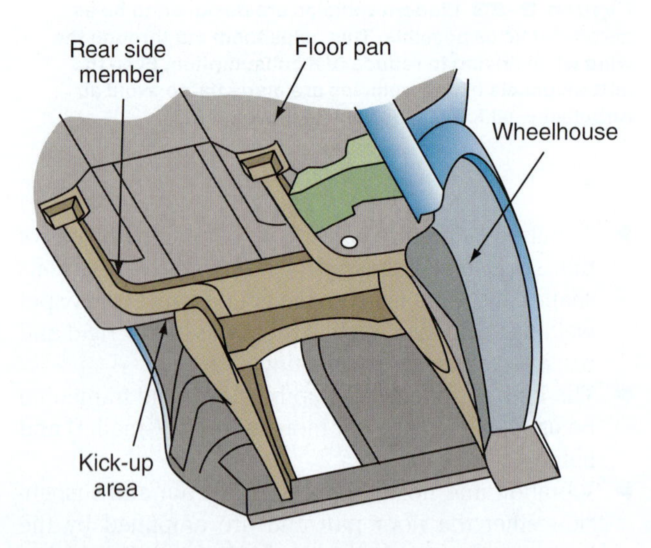

Figure 2-26 Note the underbody rear section. The kick-up area will bend during a major collision to help dissipate impact energy and cushion the shock to the passenger compartment.

FF Vehicle Unibody Structure

In a front-engine, front-wheel drive (FF) vehicle, the engine is mounted between the front rails and the transaxle drives the front wheels. It is also called a front-wheel drive (FWD) vehicle. The engine of an FF vehicle can be mounted either longitudinally or transversely.

In the space ordinarily taken up by the rear axle, the FF passenger compartment can be enlarged and the rear suspension simplified. This results in substantial weight reduction. Because the engine, transaxle, front suspension, and steering equipment are all located in the front body section, the methods of reinforcement are much different from those used in FR vehicles.

FF vehicles are characterized by the following:

- The transmission and differential are combined, and the drive shaft is eliminated, providing a substantial weight reduction.
- Overall noise and vibration are reduced because they are confined to the front of the vehicle.
- Because the engine and transmission are located in the front, the load on the front suspension and tires is increased.

Front-wheel drive vehicles suffer somewhat from torque steer. **Torque steer** is a problem in which engine torque is transmitted into the steering wheel under hard acceleration.

The interior of the FF vehicle is larger because there is no need for a propeller shaft or rear-drive axle. Because the fuel tank can be placed under the center of the vehicle, the luggage compartment can be large and flat.

Because of the location of the engine and transaxle, there is greater frontal mass and protection in a head-on collision. Therefore, engine-mounting components are reinforced accordingly.

The FF front body components—the engine hood, front fenders, radiator upper support, radiator side support, front cross member, front side members, front fender apron, and dash panel—are stamped from thin sheet metal.

FR Vehicle Unibody Structure

With front-engine rear-drive (FR) vehicles, the differential, rear axle housing, and rear suspension are mounted in the rear body/frame structure.

The engine, transmission, and differential are separate assemblies, so more weight can be distributed uniformly between the front and rear wheels, lightening the force needed to turn the steering wheel. Torque steer is also avoided because no power is flowing to the front wheels. Because it is possible to remove and install the engine, drive shaft, differential, and suspension independently, body restoration and repair workability are good.

FR vehicles require a tunnel in the floor pan. The **floor tunnel** provides a space under the floor pan for the rear-wheel drive shaft. (FF floor pans are flat and do not need this tunnel.) The floor tunnel forms the hump in the center of the floor of FR drive passenger compartments (Figure 2–27).

Because the engine's output is transmitted to the rear wheels by the drive shaft and differential, vibration and noise from the drivetrain are widely distributed over the length of the vehicle.

With the exception of outer shell parts, such as the engine hood, front fenders, and front valance panel (installed with nuts and bolts), all other exterior parts are welded together, reducing body weight and increasing body strength.

The side body is joined to the front, rear, and roof panels to form a strong passenger compartment. These panels distribute the loads from the underbody to the upper part of the vehicle to prevent body flex. The side body members also serve as door supports and maintain the integrity of the passenger compartment if the vehicle should overturn or be hit from the side.

Because the sides are weakened by large door openings, the side pillars are designed into a very strong box-type structure. The doors and pillars work together to prevent intrusion into the passenger compartment during a side collision.

The upper back panel and rear seat cushion support brace in sedans are joined at the side body and floor pan. The back panel prevents the body from twisting.

In station wagons and liftbacks, body rigidity is enhanced by adding enlarged roof side inner rear panels and a back window upper frame and by extending the roof side inner panels to the quarter panels (Figure 2–28).

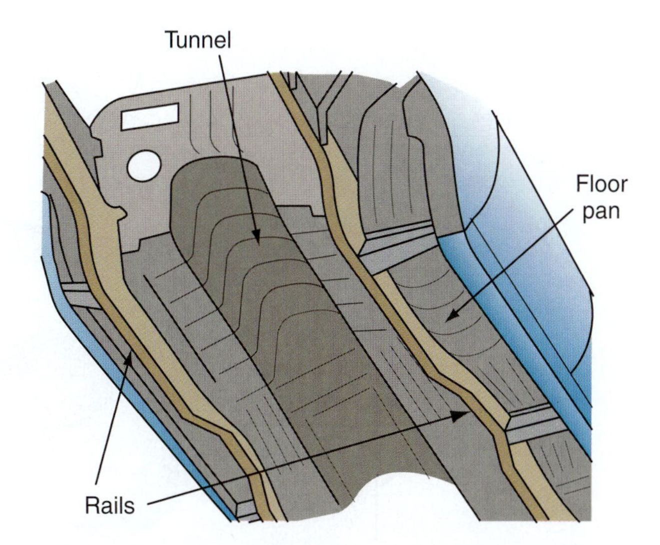

Figure 2-27 The bottom view of a front-engine, rear-wheel drive vehicle shows the floor pan and tunnel for the transmission and drive shaft. You can see this tunnel as a hump in the floor of the passenger compartment. Besides providing clearance for the drive shaft, the tunnel adds to the structural stiffness of the center section.

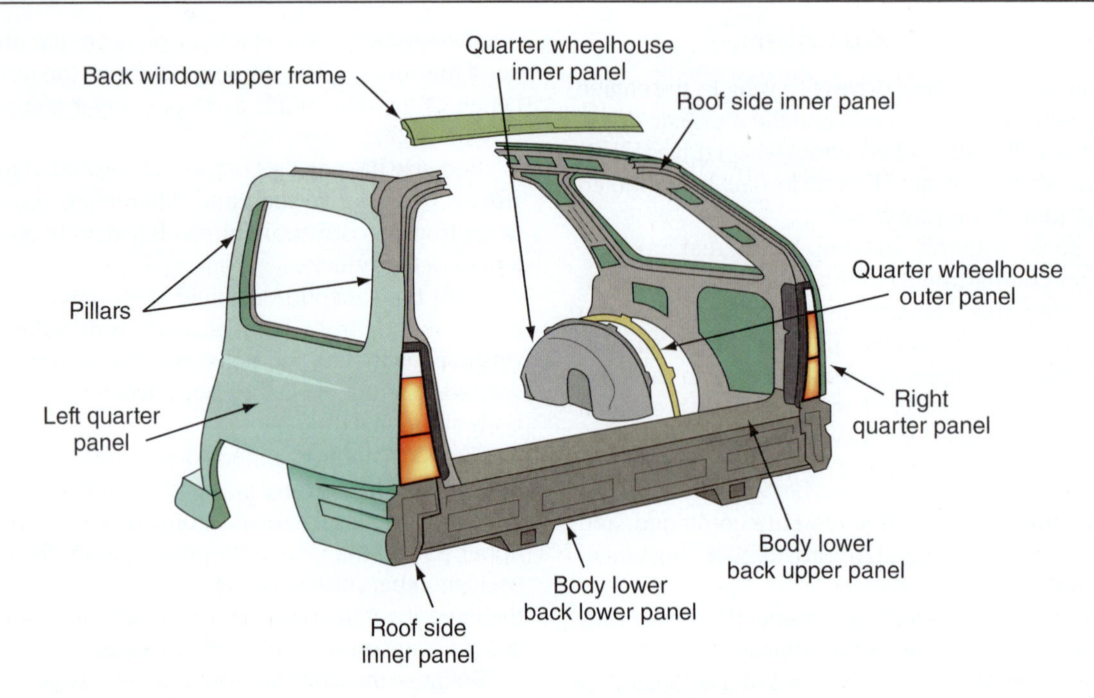

Figure 2-28 **Study the rear body structural components for a typical FR station wagon.**

MR Vehicle Unibody Structure

MR is an abbreviation for a mid-engine, rear-drive vehicle. The term *mid-engine* refers to the central positioning of the engine and powertrain between the passenger compartment and the rear axle. Mid-engine vehicles are usually high-performance sports cars, designed to handle and corner well. The Honda NSX, some Ferraris, the Toyota MR2, and a few other sports cars use this engine configuration. Look at Figure 2–29.

Due to its unique engine placement, a mid-engine vehicle has a lower front profile and often has a lower center of gravity than FF and most FR vehicles. Because this type of vehicle has the majority of its heavy components near the rear-center of the vehicle, the strength of the rear structure must be higher than normal.

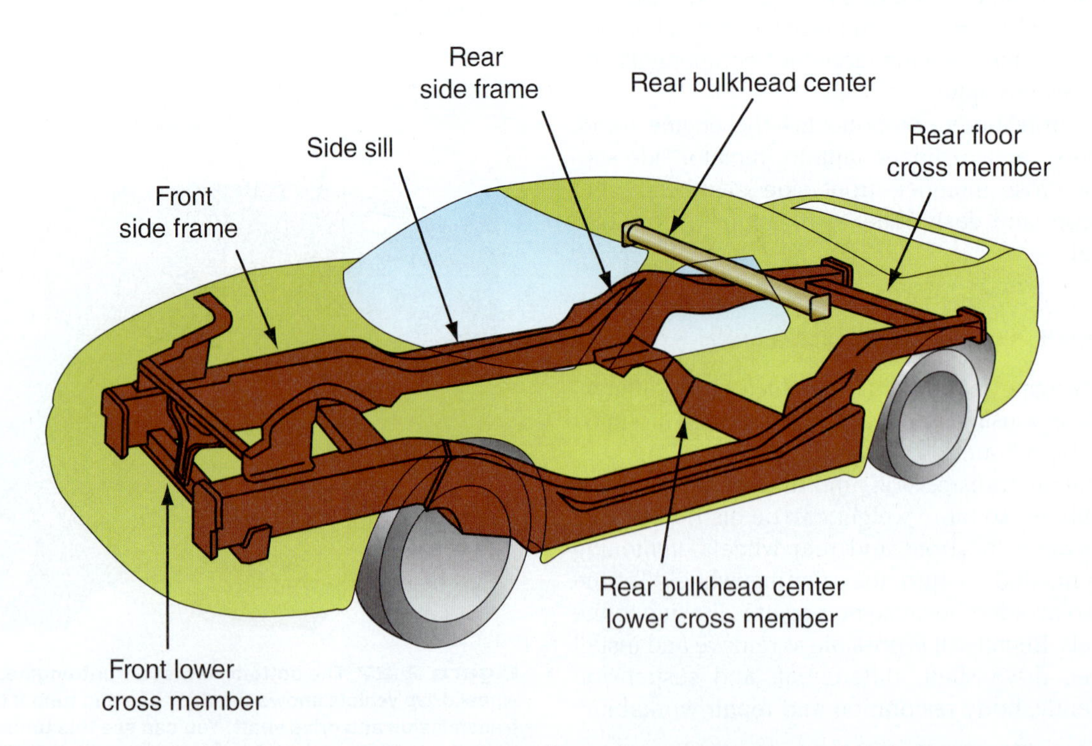

Figure 2-29 **Note the design of this rear-engine sports car. It uses huge, square box sections to form a unibody-type perimeter structure around the passenger compartment. Because the engine mounts in the rear, extra panels are welded in the front for head-on collision survivability.**

MR vehicles are not quite as safe in a head-on collision as a vehicle with the engine in the front. The mass of the engine and transaxle are in the rear, which tends to deform more of the body during a frontal impact. To counteract this safety deficiency, mid- and rear-engine vehicles must have a well-designed front structure. The front rails, aprons, and radiator support must be strong enough to absorb front impact energy without intrusion into the passenger compartment. The front of an MR or RR vehicle must be specially designed to protect occupants in head-on collisions (Figure 2–30).

Because the engine is in the rear of the vehicle, the front hood can be sloped downward, improving aerodynamics, lowering the center of gravity, and improving the driver's field of vision. Engine access and cooling efficiency are reduced because the engine is located between the passenger compartment and the rear axle assembly. This also results in less airflow over the engine and transaxle.

The fire wall is located behind the passenger compartment. A barrier must be constructed between the engine and the rear of the passenger compartment. The fire wall between the passenger compartment and the engine compartment is sometimes a three-layered structure to keep out noise, vibration, and heat. Study the rear section construction of a mid-engine vehicle in Figure 2–31.

The radiator and air-conditioning condenser are normally mounted in the front body section. Long coolant lines and air-conditioning lines must run from the rear to the front of the vehicle. The independent front suspension is supported by the front fender apron and front side members or frame rails. Because of the engine's unique location, there is room for a front luggage compartment.

Various removable parts, such as the front fenders, hood, and front valance panel, are bolted on. The other body panels in the front end of an MR vehicle are spot welded together for strength. The front hull must be made strong enough to survive front engine collisions without the mass of the powertrain.

The mid-engine vehicle rear body consists of the quarter panels, luggage compartment door, engine hood, body lower back panel, rear floor pan, room partition panel, rear floor partition panel, and rear side member.

The engine and rear luggage compartment are divided by the rear floor partition panel. The rear floor pan, room partition panel, and rear floor partition panel are reinforced with a deep bead structure and, together with the rear side members, form a rigid body.

The engine may be positioned transversely or longitudinally. Motor mounts are often located at four points: on the left and right rear side members and at cross members. Also, because an independent strut suspension is used for the rear suspension, the body structure is made to maintain body accuracy for components that have an influence on rear-wheel alignment, such as the rear floor side members and quarter wheelhousings.

RR Vehicle Unibody Structure

A rear-engine, rear-wheel drive vehicle, or RR, is similar to a mid-engine car, but the engine is located behind the

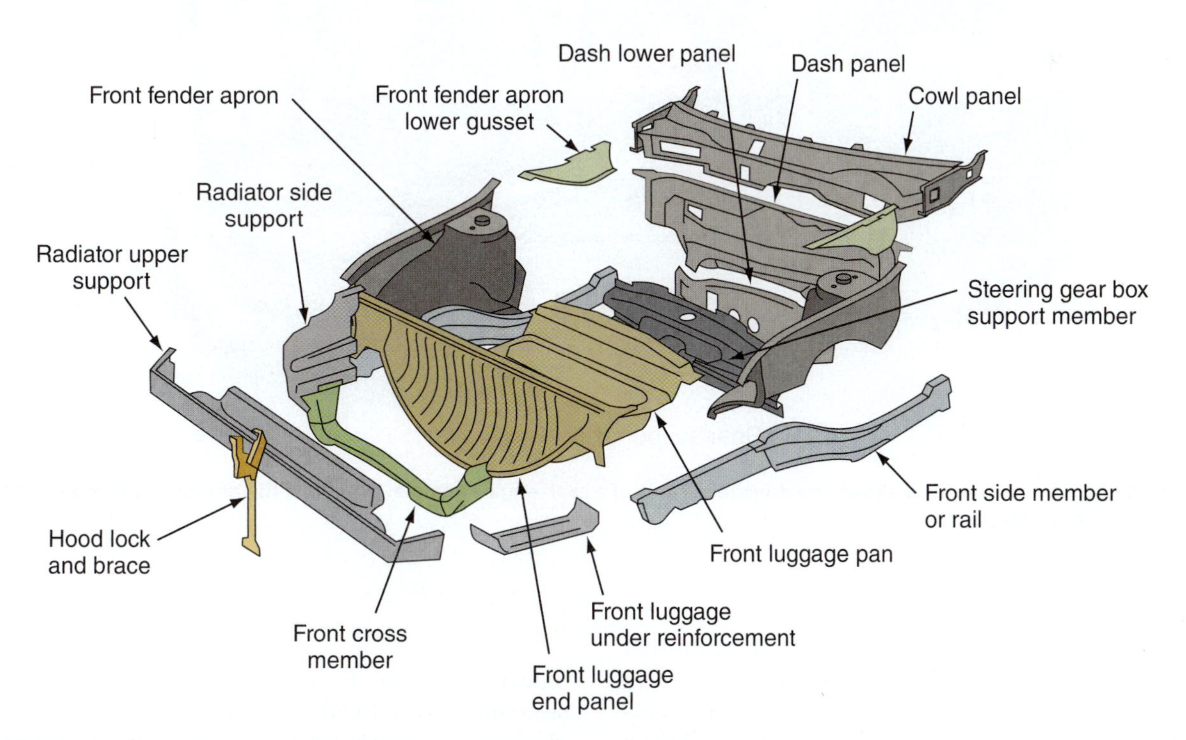

Figure 2–30 **Study the front body structural components of a typical mid-engine vehicle. Fender aprons and the luggage compartment front panel are strong to resist head-on collisions.**

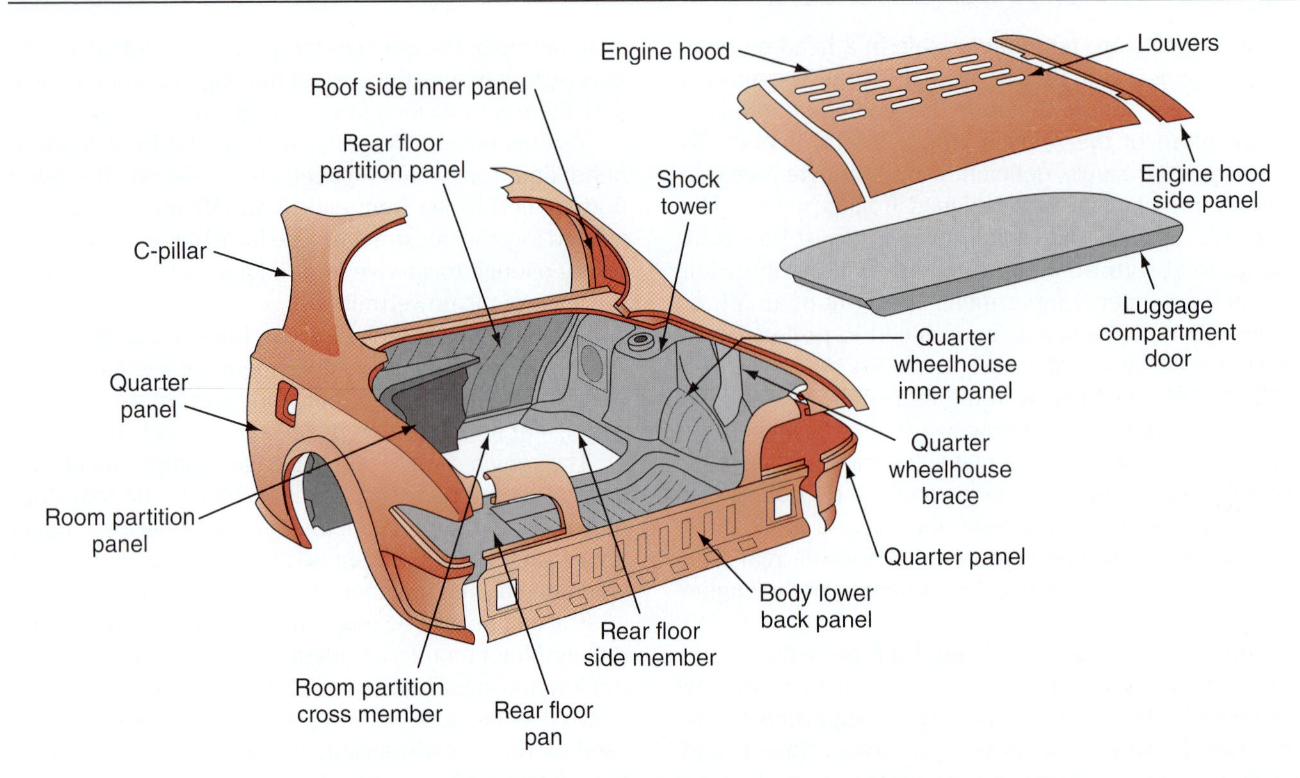

Figure 2–31 Note the rear body structural components of a typical mid-engine vehicle. Because the engine mounts in back, the body structure must allow for support of the engine and transaxle.

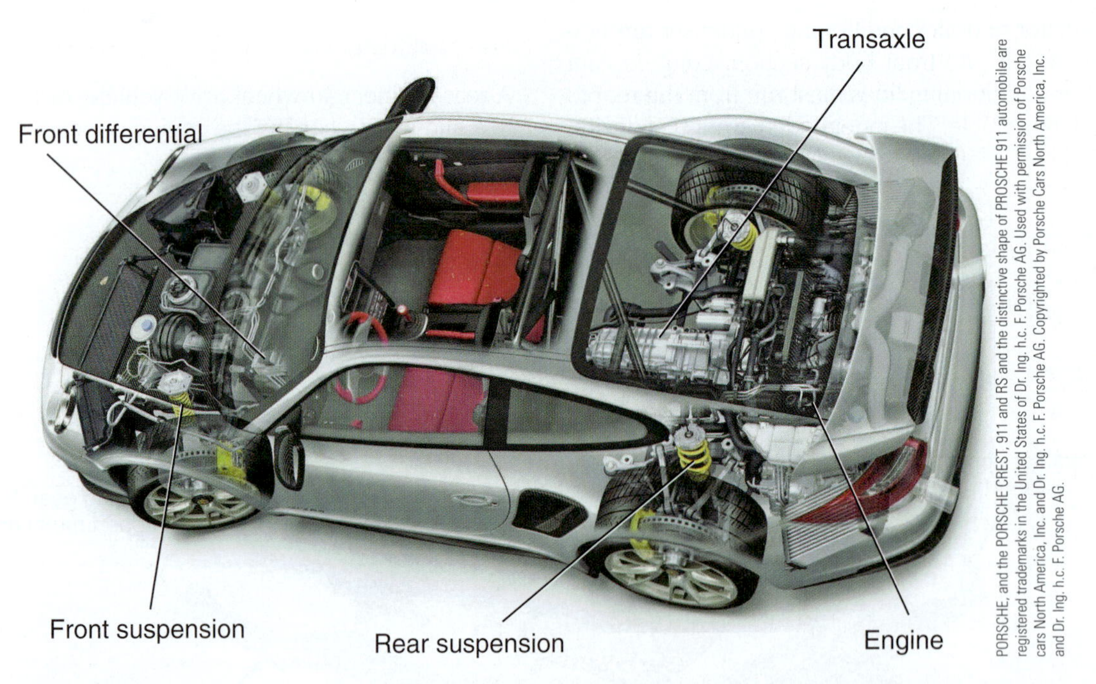

Figure 2–32 A phantom view shows the driveline layout of a rear-engine vehicle. Note how the engine mounts behind the transaxle and drive wheels.

rear axle. This moves the center of gravity even further back for improved traction. It also reduces frontal weight and steering effort. Older Volkswagen and some late model Porsche vehicles use this engine-transaxle layout (Figure 2–32).

Space Frame

Similar to a unibody, a **space frame** has a metal body structure covered with an outer skin of plastic or fiberglass panels. It is a relatively new type of vehicle construction. A space frame design is currently used on

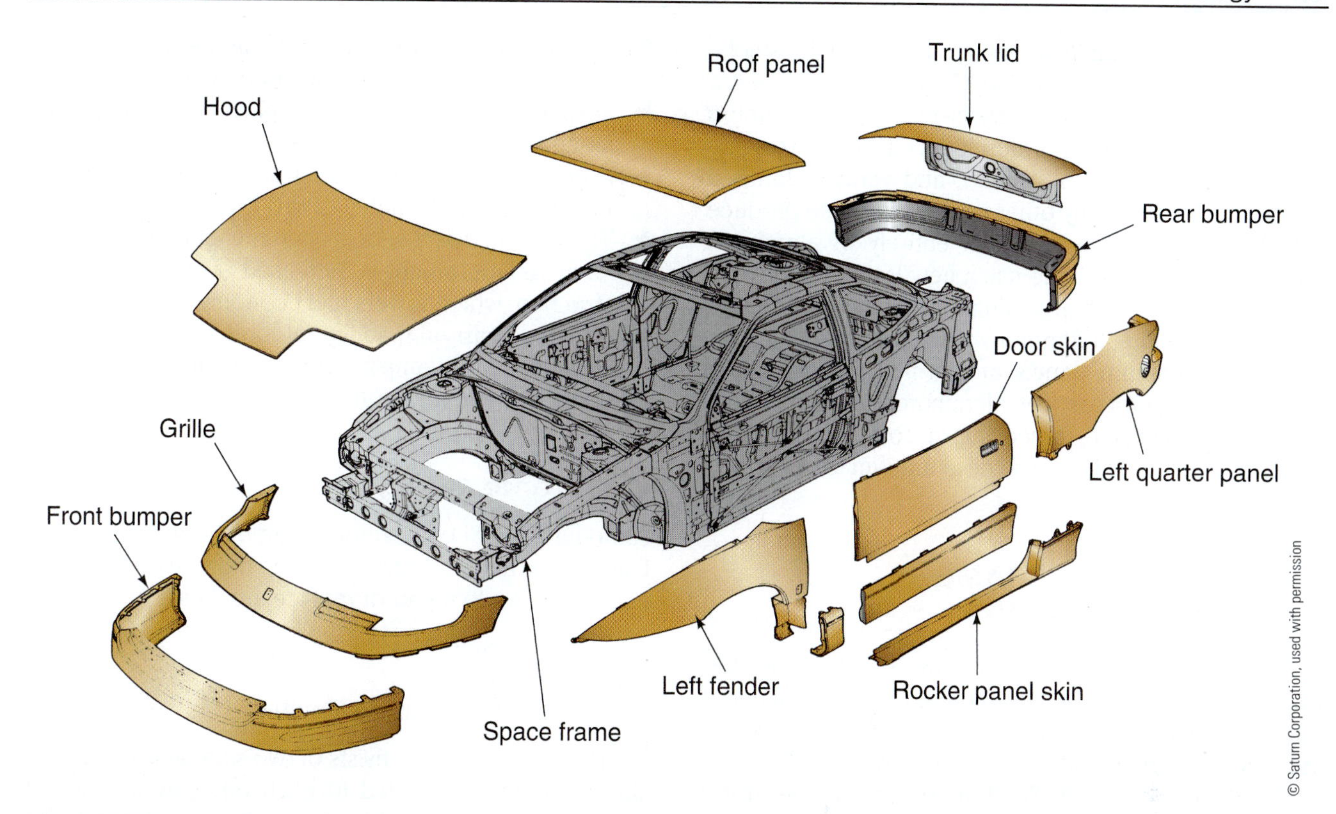

Figure 2–33 **A space frame has a metal frame structure covered with plastic or fiberglass body panels. This car's door panels are flexible and will not dent easily.**

some vans and economy vehicles. Quite often, the roof and quarter panels are not welded to the structure as they are with traditional unibodies. Exterior body panels are attached with mechanical fasteners or adhesives. One is shown in Figure 2–33.

After a collision, a space frame is more likely to have hidden damage because plastic panels can hide more severe damage. Corrosion protection is also important, since the plastic body panels may look good, but the hidden metal frame structure may become damaged and begin to deteriorate.

Plastic Parts and Panels

Plastic is a polymer material made from the processing of crude oil. It is a very light, strong, flexible, corrosion-resistant material. Plastics can be used in the manufacture of body parts and panels. Plastic parts are being used in the construction of most new vehicles for various smaller body and trim parts. Some large body panels can also be made of plastic.

Fiberglass, also called glass fiber, is used to strengthen plastic parts. It is a glass filament or fine strands of glass that can be added to plastic to increase its stiffness and strength.

Sheet molded compound refers to reinforced plastic materials that are formed into body panels. It is a form of fiberglass with materials added to strengthen the panel.

Carbon fiber is a composite of fiberglass cloth reinforced with strands of graphite. Carbon fiber body panels are now used on many exotic cars. Carbon fiber panels are very light and strong, almost as strong as steel, yet lighter than aluminum.

Composite Unibody

A **composite unibody**, or space frame, is made of specially formulated plastics and other materials, such as carbon fiber, to form the vehicle. These parts are adhesive-bonded (structurally glued) to each other. The frame is made almost totally of plastics, keeping metal parts to a minimum. The use of lightweight plastics lowers vehicle weight to improve vehicle performance and fuel economy. Although not mass-produced, several manufacturers are experimenting with composite unibody construction.

Aluminum Vehicle Construction

Aluminum is much lighter than steel and can reduce the vehicle's curb weight to reduce fuel consumption while increasing performance. Aluminum will also resist rust and corrosion much longer than steel.

Aluminum body panels and structural members are now utilized by almost all auto manufacturers. Aluminum hoods, fenders, roofs, and other panels are commonly fastened to conventional steel unibody vehicles. Many high-performance sports cars are now made almost entirely of aluminum.

The new ZO6 Corvette uses an all-aluminum perimeter frame strengthened and lightened with aluminum castings, magnesium castings around the windshield, and carbon fiber parts. The Ford GT mid-engine supercar also used an aluminum space frame made of 35 aluminum extrusions, seven complex aluminum castings, two semi-solid formed aluminum castings, and various stamped aluminum panels. Many other auto makers also produce vehicle unibodies made almost entirely of aluminum panels and castings. Some use a new laser-hybrid welding process that joins large aluminum panels with more efficiency, strength, and rigidity than riveting.

Some aluminum panels are formed using a super plastic forming process. The aluminum is heated to near its melting point and forced into a complex die using high-pressure air. This allows the aluminum to be formed into very complex shapes.

2.8 BODY-OVER-FRAME CONSIDERATIONS

In the conventional body-over-frame construction, the frame is the vehicle's foundation. The body and all major parts of a vehicle are attached to the frame. It must provide the support and strength needed by the assemblies and parts attached to it. The frame must also be strong enough to keep the other parts of the car in alignment should a collision occur. To the body technician, the frame is the most important part of the vehicle.

The conventional frame is an independent, separate component, because it is not welded to any of the major units of the body shell. The body is generally bolted to the frame. Large, specially designed rubber "biscuits," or mounts, are placed between the frame and body structure to reduce noise and vibration from entering the passenger compartment. Quite often, two layers of rubber are used in the mounting pads to provide a smoother ride.

Today, the strong steel frame side members of the modern conventional design are normally made of U-shaped channel sections or box-shaped sections. Cross members of the same material reinforce the frame and provide support for the wheels, engine, and suspension systems. Various brackets, braces, and openings are provided to permit installation of the many parts that make up the automotive chassis. The various cross members, brackets, and braces are welded, riveted, or bolted to the frame side rails.

Most conventional frames are wide at the center and narrow at the front and rear. The narrow front construction enables the front wheels to turn. A wide frame at the rear provides better support of the body.

Other characteristics of body-over-frame vehicles are:

- Fewer road-induced vibrations are transferred to the body via the frame, thus resulting in a smooth, soft ride.
- Rubber mountings between the body and frame insulate the body from vibrations and noise, providing a quiet interior.
- High amounts of energy can be absorbed by the frame during some collisions.
- Undersurfaces of the thin body panels are protected over rough roads by the thick framework.
- Suspension and powertrain parts can be quickly assembled on the basic frame.
- A heavy frame made of thick sheet metal is approximately 3/64-to 1/8-inch (1.2 to 3.1 mm) thick.
- The vehicle ride height is often higher off the ground.
- Total vehicle weight is increased over unibody construction, which lowers fuel economy and handling.
- The load-carrying ability of body-over-frame construction, for example, with a pickup truck, is normally higher than for unibody construction.

Full Frame Designs

Although several conventional frame designs have been used by the auto industry, the three that the body technician may have to repair or replace are these:

1. Ladder frame
2. Perimeter frame
3. X-frame (or backbone frame)

The *ladder frame* consists of two side rails, not necessarily parallel, connected to each other by a series of cross members like a ladder. In fact, some of the early car frames were perfect ladder shapes. Although the ladder frame design is no longer used for passenger vehicles because of its harsh, "log wagon" ride, it still can be found on some trucks because of its stiffness and strength (Figure 2–34).

As vehicle designs improved, the early ladder frames were modified and improved to absorb impact energy more efficiently (Figure 2–35).

Figure 2-34 An old-fashioned ladder frame actually looked something like a simple step ladder. Also note how the square shape of the old Model T was far from aerodynamic.

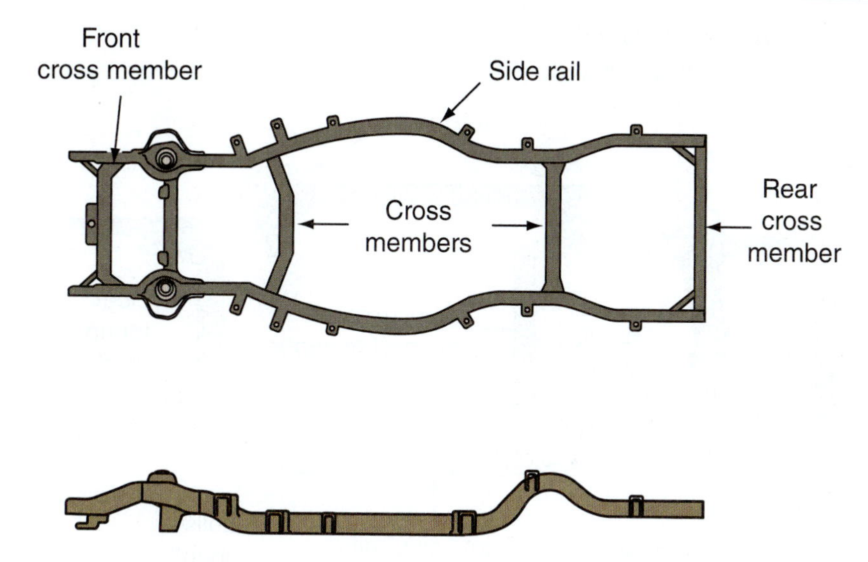

Figure 2–35 The ladder frame evolved into the perimeter frame, which can be found on pickup trucks. Note the part names.

The *perimeter frame* is similar in construction to the ladder frame. The full-length side rails support the body at its greatest width, which provides more protection to the passengers in the event of a side impact to the body. The areas behind the front wheels and in front of the rear wheels are stepped to form a torque box structure. Refer to Figure 2–36.

In a head-on collision, the front of the full frame absorbs most of the energy. In a side-impact collision, the passenger compartment is protected from collapse because the center frame rail is near the front floor side member. In rear-end collisions, the rear of the frame rails often kicks up or bends upward to help absorb the shock. As for twisting and bending, strategic areas are reinforced with cross members to try to avoid major frame damage (Figure 2–37).

The *X-frame* narrows in the center, giving the vehicle a rigid structure that is designed to withstand a high degree of twist. A heavy front cross member is used to support the upper and lower suspension control arms and coil springs. The X-frame has not been used since the late 1960s and can now be considered obsolete (Figure 2–38).

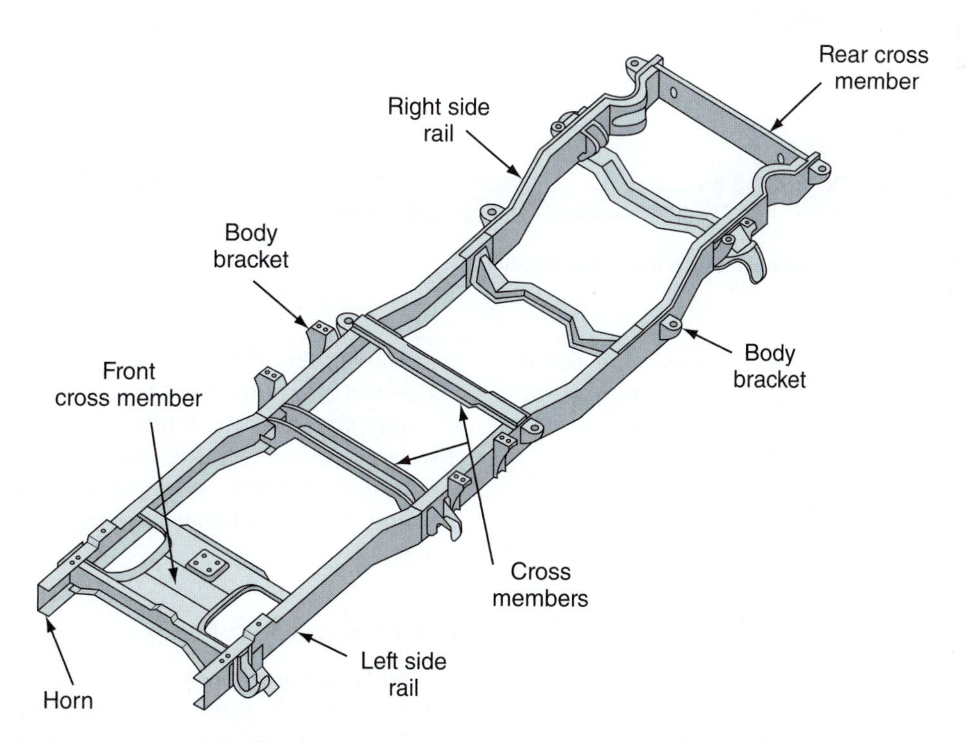

Figure 2–36 Study the basic parts of a full frame. It is made of thick steel for carrying the weight of a luxury car, pickup truck, or SUV.

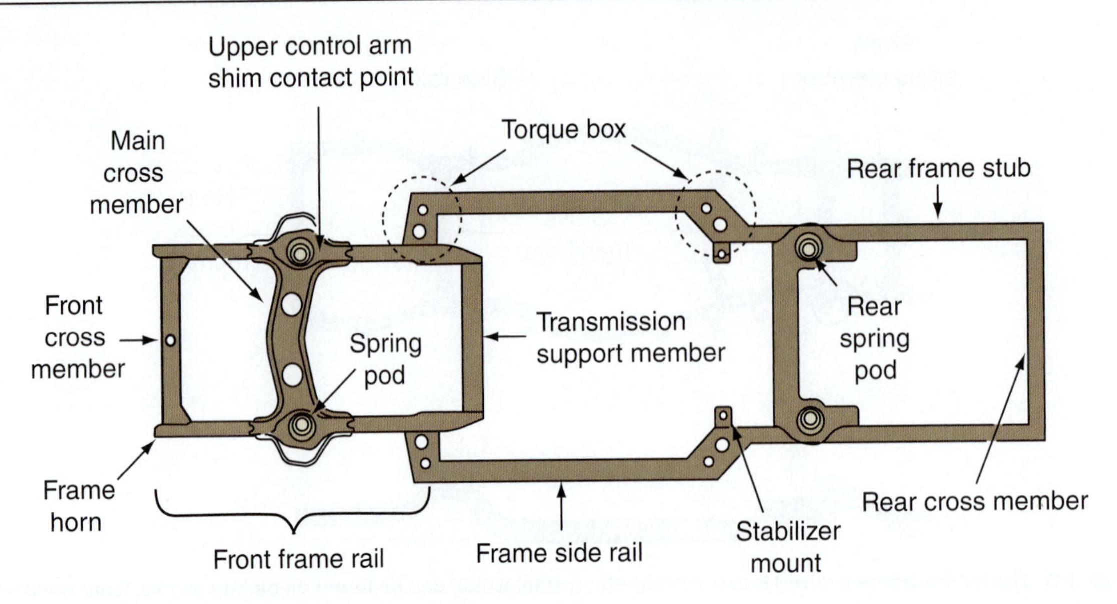

Figure 2-37 **Torque boxes in a perimeter frame help control the flow of impact damage during a high-speed collision. They help control twisting and collapse of the frame during a major impact.**

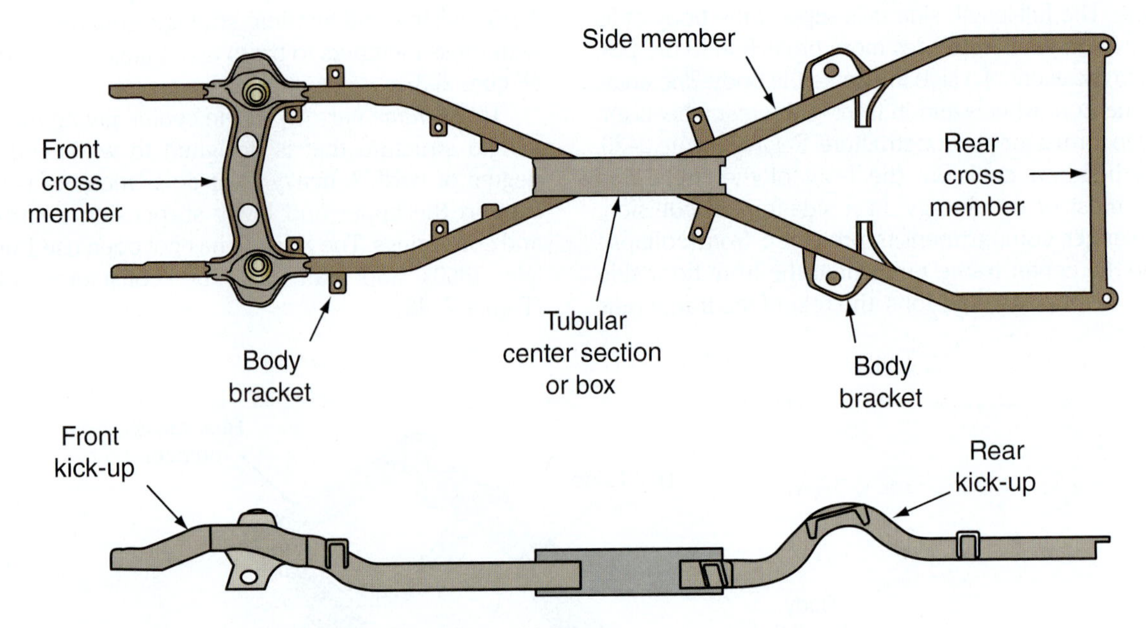

Figure 2-38 **The X-frame is still used on a few older vehicles.**

A variation of the X-frame is the *backbone frame*, which has a single thick beam or box section in the center section. It can be found on some sports cars.

Full Frame Front Section

The *full frame front section* is made up of the radiator support, front fenders, and front fender aprons. These components are installed with bolts and form an easily disassembled structure.

The radiator support is made up of the upper support, lower support, and left and right side supports welded together to form a single structure. The front fender of the separate frame-type vehicle differs from the front fender of the unibody. The panels in the upper inside and rear ends of the fender are spot welded. This not only increases the fender's strength and rigidity, it also works along with the front fender apron to reduce vibration and noise. In addition, it helps to prevent damage to the suspension and engine from side impacts.

Full Frame Main Body

The *full frame main body section* is made up of the dash panel, underbody, roof, and so on, to form the passenger and luggage compartments. It is similar in structure to that of a unibody (Figure 2–39).

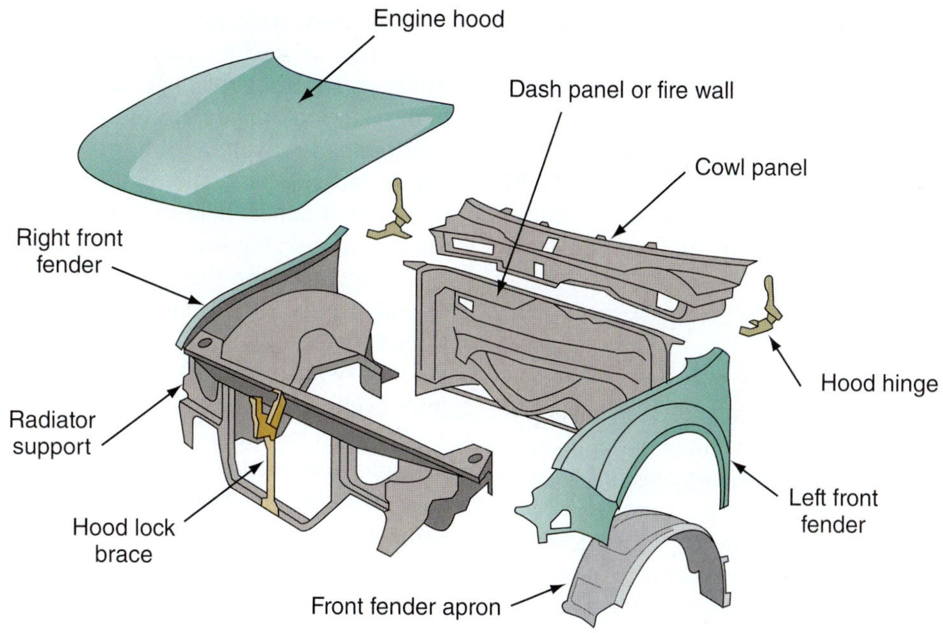

A **Many front body parts bolt in place to the frame.**

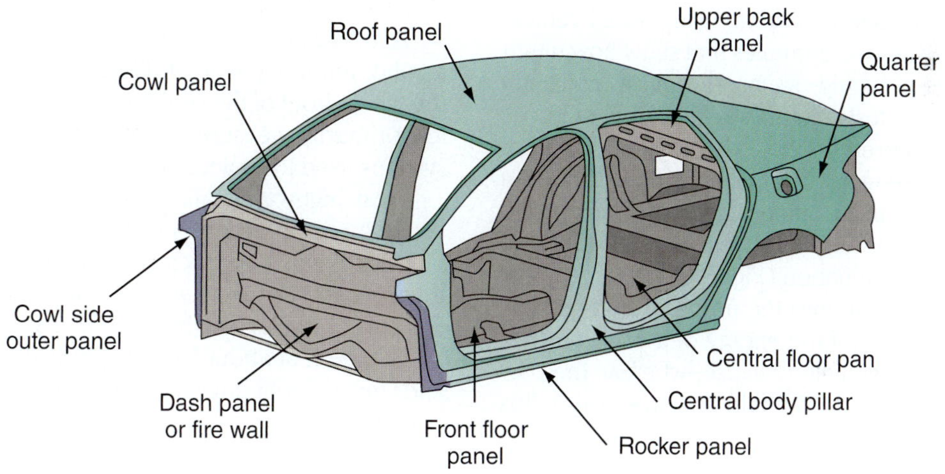

B **The main body structure bolts to the frame.**

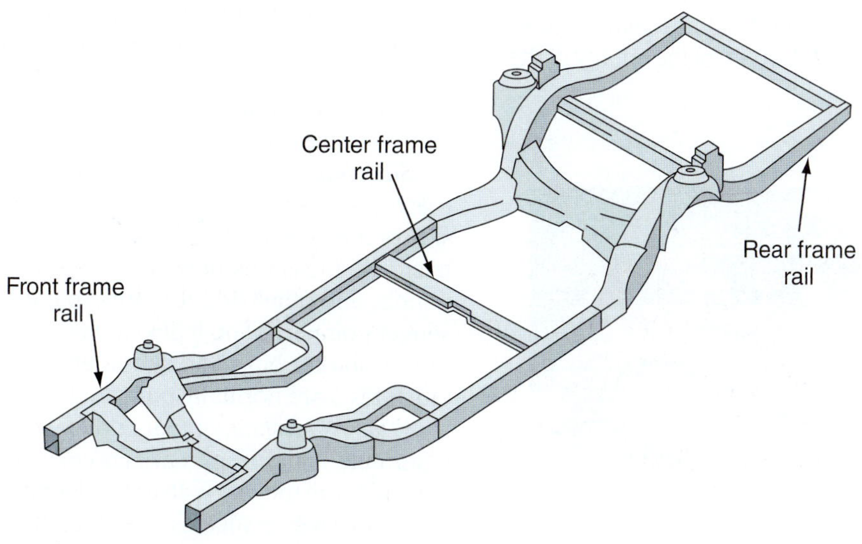

C **The shaded areas show torque box structures.**

Figure 2-39 **Note the body and frame parts with a body-over-frame vehicle.**

The dash panel, sometimes called the fire wall or front bulkhead, is the panel dividing the front section and the center, passenger compartment section. It normally welds in place.

The front of the underbody has a drive shaft floor tunnel built into it. In addition, the front, back, and left and right sides of the floor pan are made uneven in the stamping process, increasing the rigidity of the floor pan itself, which reduces vibration.

2.9 CRASH TESTING

Automobile manufacturers are challenged by having to design vehicles that are lightweight, but at the same time strong and safe. Modern passenger vehicles are, pound for pound, the safest ever made.

Computer-simulated crash testing helps to determine how well a new vehicle might survive a crash. Computer-simulated testing is used before building a prototype or the first real vehicle to find weak or faulty structural areas prior to investing in mass production.

Certified crash tests are done using a real vehicle carrying sensor-equipped dummies that show how much impact passengers would suffer during a collision. Sensors in the crash test dummies record impact forces (rapid changes in inertia) to the chest, legs, and head of the driver and passengers. Computer readings from the sensors in the crash test dummies give feedback in the form of raw data about each crash test for body structure and potential injury evaluation (Figure 2–40).

Crush zones are built into the frame or body to collapse and absorb some of the energy of a collision. The front and rear of the vehicle collapse, whereas the passenger compartment tends to retain its shape. This helps reduce the amount of force transmitted to the occupants.

It is critical that the passenger compartment be strong enough to help prevent injury to the driver and passengers. It is also important that the front and rear sections collapse upon impact to absorb some of the energy of the collision.

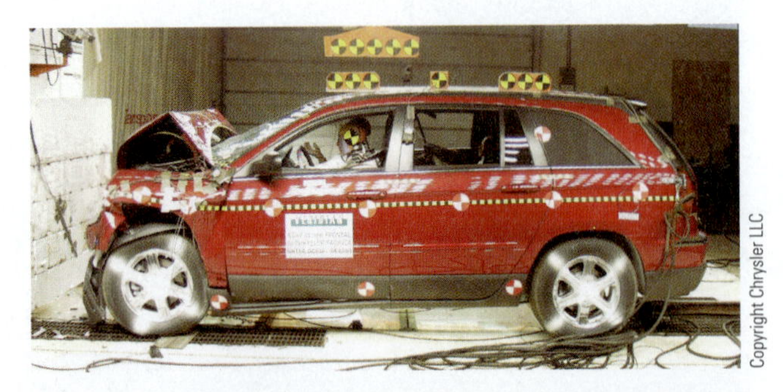

Figure 2–40 During a crash test, the vehicle is actually rammed into a brick or steel-reinforced concrete wall. This is done to measure how well the vehicle survives the collision. Different makes and models of vehicles are crash tested and compared.

A modern passenger vehicle is designed something like a race car, which has a five-point seat belt for the driver and a strong roll cage welded to body/frame members. The body pillars around the doors, the reinforced roof panel, fire wall, and rear body members all act as strong, stiff roll bars. The reinforced passenger compartment metal panels help to protect those inside the vehicle during even high-speed collisions.

During a wreck, the outer body is designed to crush or collapse around the stronger passenger compartment to protect the driver and passengers from injury. In conjunction with front and side air bags, the typical new motor vehicle is safer than ever.

For vehicles with large luggage areas, such as vans or liftbacks, the structural members and reinforcements designed into the body measurably affect overall torsional rigidity.

Types of Crash Tests

There are several types of crash tests to evaluate the structural integrity and safety of a vehicle.

A *full frontal crash test* pulls the vehicle squarely into a solid, immobile wall at 35 mph (56.327 km/h) so that the whole front of the vehicle hits the barrier. This type of test primarily measures the effectiveness of the restraint systems used to protect the driver and passengers.

Most highway fatalities occur in head-on collisions. For this reason, frontal crash tests are the most important. Hitting a solid brick wall at 35 mph (56.327 km/h) theoretically reenacts a head-on collision. When two vehicles of similar weights travelling at similar speeds collide head-on at about 50 mph (80.4672 km/h), their front ends hit as if colliding with a stationary object at 100 mph (160.934 km/h).

During an *offset frontal crash test*, the vehicles are pulled at almost 40 mph (64.3738 km/h) into a solid barrier that only impacts the driver's side of the front end. This test is better at evaluating the structural integrity of the vehicle's bumper, left front frame rail, left fender, left apron, and related parts. It measures how much damage the front end section can withstand before impact force enters the passenger compartment.

Side impact crash testing, nicknamed the "T-bone crash test," is done by moving a 3,000-lb. sled into the side of the vehicle at 38.5 mph (61.95974 km/h). This test basically measures how well the doors, pillars, rocker panels, and other side structural panels protect the passenger compartment. It also measures the effectiveness of the side air bags and shoulder harness. Side impact crash tests are normally only done on the driver's side.

Whiplash testing, or rear-end crash testing, is done to determine injury from rear-end collisions. The test vehicle is hit from the rear with a sled. Using crash test dummy data, this test evaluates the seat headrests as good, acceptable, marginal, or poor in terms of protection from neck injuries.

Bumper crash testing involves a 5-mph (8.04672 km/h) impact to determine the amount of damage to the front and rear ends of vehicles. This test does not evaluate injuries but rather the cost of repairs from very minor collisions. Today's vehicles must pass the 5-mph (8.04672 km/h) bumper crash test without significant damage.

Crash Test Agencies

There are two major U.S. agencies that conduct standardized crash testing of new cars:

1. National Highway Traffic Safety Administration (NHTSA)
2. Insurance Institute for Highway Safety (IIHS)

These organizations use crash test data to establish their own standards for vehicle safety.

NCAP Ratings

The NHTSA's **New Car Assessment Program (NCAP)** uses stars to indicate how well a vehicle does on the various types of crash tests. Their frontal crash test pulls the vehicle at about 35 mph into a flat, solid barrier, which impacts the whole front of the vehicle.

Insurance Rating System

An **insurance rating system** uses a number scale to rate various vehicles' accident survivability during a collision. It indicates how well the vehicle will survive a partial barrier frontal crash and how much it will cost to repair the damage at a body shop.

The rating system was developed by evaluating different vehicles during crash testing under controlled conditions. Negative five is the worst collision rating (most damage) and positive five is the best collision rating (least damage). A zero rating would be average.

For more information on crash test results, visit the following websites:

- New Car Assessment Program at http://www.nhtsa.gov
- Insurance Institute for Highway Safety at http://www.iihs.org

Try to learn something new about the paint and auto body repair industry every day! Besides studying this book and doing all of the hands-on workbook activities, read automotive magazines! Surf the Net or watch motorsports television programs. Try to become a wiser, more knowledgeable mechanic or body technician with very little effort!

SUMMARY

1. The term *vehicle construction* refers to how a passenger car, truck, van, or SUV is assembled at the factory. The vehicle body provides a protective outer hull or "skin" around the outside of a motor vehicle.
2. The vehicle chassis includes the frame, engine, suspension system, steering system, and other mechanical parts with the body removed.
3. The vehicle frame is a high-strength structure used to support all other parts of the vehicle. The perimeter frame is similar in construction to the ladder frame.
4. Body-over-frame construction has a separate body structure bolted to a thick steel framework.
5. Unibody construction uses body parts welded or adhesive-bonded (glued) together to form an integral (built-in) frame.
6. The front section, also called the nose section, includes everything between the front bumper and the fire wall. The center section, or mid-section, typically includes the body parts that form the passenger compartment. The rear section (tail section or rear clip) commonly consists of the rear quarter panels, trunk or rear floor pan, rear frame rails, trunk or deck lid, rear bumper, and related parts.
7. A panel is a steel or plastic sheet stamped or molded into a body part.
8. The frame rails are the box members extending out near the bottom of the front and rear sections of a unibody.
9. The cowl is the panel at the rear of the front section, right in front of the windshield.
10. The shock towers or strut towers are reinforced body areas for holding the upper parts of the suspension system.
11. The radiator core support is the framework around the front of the body structure for holding the cooling system radiator and related parts.
12. The floor pan is the main structural section in the bottom of the passenger compartment.
13. Pillars are vertical body members that hold the roof panel in place and protect the vehicle in case of a rollover accident.

14. Rocker panels, or door sills, are strong beams that fit at the bottom of the door openings.
15. The doors are complex assemblies made up of an outer skin, inner door frame, door panel, window regulator, glass, and related parts. Door hinges bolt or weld between the pillars and door frame.
16. The quarter panels are the large, side body sections that extend from the side doors back to the rear bumper.
17. Cross members are thick-gauge supports that extend across the frame rails of both unitized and full frame vehicles.
18. Crush zones are areas in the unibody that are made weaker to intentionally collapse during a collision.
19. Aerodynamics is a measurement of how well a motor vehicle moves through wind without resistance.
20. Anticorrosion materials are used to prevent rusting of metal parts. Sound-deadening materials are used to help quiet the passenger compartment.
21. Similar to a unibody, a space frame has a metal body structure covered with an outer skin of plastic or fiberglass panels. A composite unibody is made of specially formulated plastics and other materials, such as carbon fiber, to form the vehicle.
22. Certified crash tests are done using a real vehicle and sensor-equipped dummies that show how much impact passengers would suffer during a collision. An insurance rating system uses a number scale to rate the vehicle's accident survivability during a collision.

EXERCISES

On a separate sheet of paper, complete the following learning activities for this chapter. Write definitions for the key terms and answer the ASE-style review questions, essay questions, critical thinking problems, and math problems. You can also do the outside activities, possibly for extra credit.

Key Terms

A-pillar
aerodynamics
anticorrosion materials
assembly
B-pillar
body-over-frame construction
C-pillar
certified crash test
composite unibody
Corporate Average Fuel Economy (CAFE)
cowl
crush zone
Environmental Protection Agency (EPA)
fiberglass
floor tunnel
frame rail
front clip
hydroformed frame
insurance rating system
longitudinal engine
New Car Assessment Program (NCAP)
panel
plastic
quarter panel
rocker panel
shock tower
sound-deadening material
space frame
torque steer
transverse engine
unibody
vehicle chassis

ASE-Style Review Questions

1. Which type of vehicle construction uses a frame only in areas requiring extra support and a strong attachment point?
   A. Combination frame construction
   B. Semi-unitized stub rail construction
   C. First-generation unitized perimeter frame construction
   D. Fully unitized construction
2. Which parts of a unibody vehicle structure help to keep passengers safe in the event of a collision?
   A. Torque boxes.
   B. Cross members.
   C. Crush zones.
   D. all of the above
3. Which of the following is not an advantage of unitized vehicle design?
   A. Increased passenger compartment safety
   B. Reduced vehicle weight
   C. Higher fuel efficiency
   D. Localized collision damage to components
4. Which of the following mechanical components are commonly found on newer unitized constructed vehicles?
   A. MacPherson strut suspensions
   B. Rack-and-pinion steering
   C. Front-wheel drive
   D. All of the above

5. In front-engine, rear-wheel drive unitized vehicles, the engine is mounted
   A. longitudinally.
   B. transversely.
   C. between the passenger compartment and rear axle.
   D. either A or B.
6. The ___ are the large, side body sections that extend from the side doors back to the rear bumper. They are welded in place and form a vital part of the rear body structure.
   A. fenders
   B. quarter panels
   C. cross members
   D. sail panels
7. ___ construction uses body parts welded together to form an integral frame.
   A. Body-over-frame
   B. Unibody
   C. Welded
   D. Bonded
8. Technician A gives a more thorough damage analysis to a unibody vehicle than to a conventional frame vehicle. Technician B says the conventional frame vehicle requires the more thorough inspection. Who is correct?
   A. Technician A
   B. Technician B
   C. Both A and B
   D. Neither A nor B
9. Which of the following are designed to stiffen a unibody structure?
   A. Torque boxes
   B. Frame horns
   C. Cross members and bracing
   D. Stone deflectors
10. Which of the following frame designs are no longer used in automobile manufacturing?
    A. Perimeter
    B. Stub
    C. Hourglass
    D. Ladder
11. In a front-engine, front-wheel drive unibody structure, what panel supports the top of the MacPherson struts?
    A. Front cross member
    B. shock towers
    C. Side rails
    D. Radiator support
12. Technician A says that a unibody structure directly affects vehicle wheel alignment. Technician B says that a vehicle's unibody structure does not affect vehicle wheel alignment. Who is correct?
    A. Technician A
    B. Technician B
    C. Both A and B
    D. Neither A nor B
13. What type of automobile structure is welded and/or bonded into one unit?
    A. Frame body
    B. Unibody
    C. Stub frame
    D. Nose frame

Essay Questions

1. What is the difference between body-over-frame construction and unibody construction?
2. Explain the terms *part* and *assembly*.
3. Describe the three body sections of a vehicle.
4. Aluminum has what advantages over sheet steel when used in automobile construction?

Critical Thinking Problems

1. Why would you have to use different repair methods to repair a collision-damaged full frame car versus a unibody car?
2. From a repair standpoint, what are the pros and cons of space frame construction?
3. A passenger vehicle will typically have more than ___ parts.
4. The majority of the unibody cars on the road today feature what drivetrain?
5. Body panels are commonly made of what three materials?
6. Bumper crash testing involves a ___ mph test to determine the amount of damage to the front and rear ends of vehicles.

Math Problems

1. If a car weighs 4,000 lb. (1,816 kg) and each tire has the same amount of weight on it (equal weight distribution), how much weight is on each tire in pounds and kilograms?
2. In the previous question, if a tire has 4 square inches touching the road, how many pounds of weight would be pushing down per square inch?

Activities

1. Examine a few of the shop-owned vehicles to find information on how they are constructed. See how many parts you can identify. What kind of construction is used with each: unibody, body-over-frame, or space frame?
2. Have a class discussion comparing the advantages and disadvantages of conventional frame versus unibody construction.
3. Discuss front- and rear-wheel drive vehicles. What are the advantages and disadvantages of each?
4. Name the makes and models of vehicles that use steel, aluminum, and plastic or fiberglass construction. Discuss each.

CHAPTER 3

Service Information, Specifications, and Measurements

OBJECTIVES

After studying this chapter, you should be able to:

- Use printed and computer-based shop service information efficiently.
- Quickly find vehicle-specific service instructions in printed and computer-based shop manuals.
- List the different kinds of printed service data.
- Compare auto maker, aftermarket, tool manufacturer, and material manufacturer service data.
- List the different kinds of shop software used in a collision repair facility.
- Explain the symbols and abbreviations used in service information.
- Summarize the use of computer-based service information.
- Compare the different kinds of illustrations used to describe vehicle repairs.
- Utilize charts and graphs during collision repair.
- Explain the many types of measurements needed in collision repair.
- Make accurate linear, angle, pressure, volume, and other measurements.
- Compare U.S. and metric measuring systems.
- Identify and use basic measuring tools common to auto body repair.
- Use conversion charts.
- Answer ASE-style review questions relating to service information, specifications, and measurements.

INTRODUCTION

Today's motor vehicles are built to exacting standards using varied manufacturing methods, which results in different methods of repair. Because it is impossible to memorize all of the service details needed to properly rebuild damaged vehicles, it is imperative that you know how to use vehicle-specific (year, make, and model) service information efficiently.

This chapter describes how to utilize both printed and computer-based service data. It summarizes the various kinds of service information available and explains the purpose of each type. You will learn how to use service charts, symbols, abbreviations, and technical illustrations. The chapter also reviews most of the basic types of measurements you will have to make when employed in collision repair. This chapter will prepare you for more advanced repair topics.

A good technician must have a complete understanding of commonly used terms that identify parts and assemblies of a vehicle. The technician who is not familiar with this language will have difficulty ordering parts and reading repair orders.

Measurements are number values that help control repair processes in collision repair. For example, measurements are needed to measure structural damage, straighten frame damage, mix paint, adjust spray gun pressure, do a wheel alignment, torque a bolt, and numerous other repair tasks.

Vehicle manufacturers give **specifications** (measurements) for numerous repair procedures: body straightening dimensions, bolt or nut torque values, material thicknesses, electrical values, and other critical information. When working, you will have to refer to factory specifications to ensure competent repairs.

This textbook is designed to prepare you to use vehicle-specific service information. Instructions written for the exact year, make, and model vehicle are needed to do competent repairs. The information given throughout this textbook is general and does not apply to every repair situation. Always refer to vehicle manufacturer or aftermarket publisher service information when in doubt about any repair task.

Courtesy of ASE

Figure 3–1 Today's collision repair technician must be capable of using both printed and computerized service information. With modern vehicle repair methods, accessing accurate service information is more important than ever.

3.1 SERVICE INFORMATION

Service information includes written instructions and technical illustrations to help you properly repair a damaged vehicle. Service information is published by vehicle manufacturers (Jaguar, Chrysler, General Motors, Toyota, and so on) and aftermarket publishers (Mitchell Manuals, Motor Manuals, and Chilton Manuals, for example).

Printed Service Information

Printed service information places service instructions, vehicle dimensions, estimating data, and technical illustrations on paper in bound books. Books and professional magazines are another excellent way to retrieve service instructions and to improve repair knowledge and skills.

Computer-Based Service Information

Computer-based service information places service manuals, dimension manuals, estimating manuals, refinish material guides, mechanical repair procedures, and other data on compact discs (CDs). This allows one to use a personal computer (PC) to quickly look up and print the desired service repair information. Most PCs used in the collision repair industry are Windows-based (Figure 3–1).

There are various kinds of software or computer programs used in the collision repair industry. As a technician, or perhaps someday a shop owner, you should understand the purpose of these useful shop tools.

Most high-volume shops now access their service manuals with a PC. A huge volume of service data can be kept on CDs or retrieved online over the Internet. A PC allows more efficient handling of all shop operations. Estimating, parts ordering, bookkeeping, finding vehicle-specific service instructions, and the whole shop business operation can be more closely monitored and controlled by computer.

A PC can quickly access thousands of illustrations for fast and easy identification. By ordering new CDs every year, the computer-based information you have can be kept current, which means you will never be at a loss for the most up-to-date parts and service information.

Electronic media or computer-based service information provides potential advantages over print media by enabling the technician to:

- More quickly look up parts and labor information
- More efficiently cross-reference and validate parts and prices
- More easily create, store, and e-mail parts and labor worksheets and part orders
- Print customized parts, assemblies, and repair illustrations for shop use

Shop Publications

There are several publications, both printed and computer based, that all body shop personnel should become familiar with.

All automobile companies publish yearly **service manuals** that describe the construction and repair of their vehicles. These manuals give important details on repair procedures and part construction and assembly. Also called *shop manuals*, they give instructions, specifications, and illustrations for their specific cars, trucks, vans, and SUVs. Service manuals have both mechanical and body repair information. Refer to Figure 3–2.

The contents page of a service manual lists the broad categories in the manual and gives page numbers. Each service manual section concentrates on describing one area of repair: body parts, interior parts, suspension system, brakes, and so on (Figure 3–3).

The index in the back of the service manual contains the page numbers for hundreds of repair topics. It is used instead of the contents when you want information on a specific part of a vehicle.

Because it would be too expensive to purchase factory service manuals from every auto maker, most collision

Figure 3-2 **Part or panel removal and installation instructions can be found in factory and aftermarket reference materials (for example, service manuals or computer-based information).**

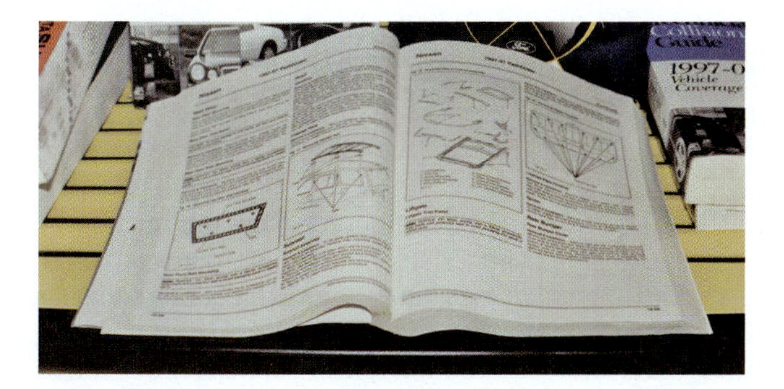

Figure 3-3 **A typical page from a shop manual gives written instructions and illustrations to clarify how repairs should be done.**

repair shops purchase aftermarket service information in both book and computer formats (Figure 3–4).

Aftermarket repair manuals, such as Mitchell Manuals, Motor Manuals, and Chilton Manuals, are sold by publishing companies. They are not as detailed as factory service materials, but they can give enough information for most body shop repairs.

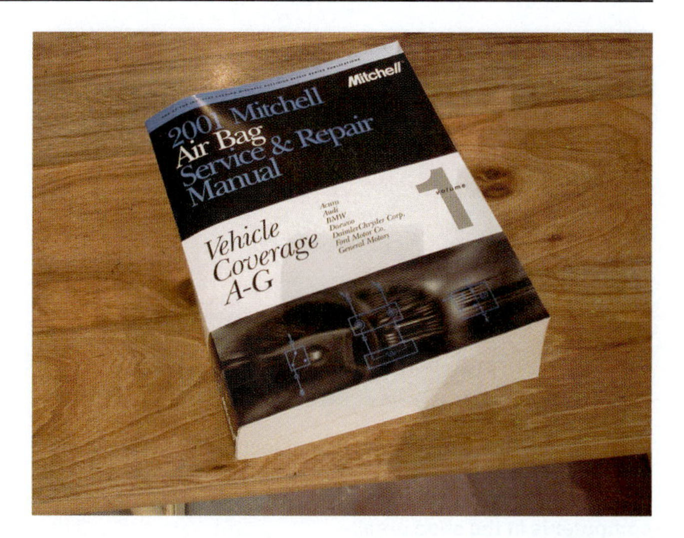

Figure 3-4 **This is an air bag service manual. It gives specific instructions for safely servicing the air bags on specific makes and models of vehicles.**

Service Information Categories

There are several types or categories of service manuals or computer-based service data used in collision repair. You should be familiar with each type.

Shop management software is designed to help enhance the collision repair facility's productivity, profitability, and customer service by coordinating all aspects of collision repair. It converts electronic estimates into repair orders, making it easy to track and order parts, generate work orders and productivity reports, and deliver the final bill. With management software, the shop owner or manager can see what is really going on in the facility while sitting at a computer.

With a *word processing program* in the management software, you can also create professional-looking follow-up letters and thank-you letters.

Parts ordering software enables you to generate a new parts order and fax it to a vendor. The returned vendor information can then be entered and used to maintain the parts status on vehicles. After entering the part information received from the vendor, the software will instantly generate a status report showing which parts have been received, back-ordered, or returned.

Collision repair guides give instructions, safety warnings, and technical illustrations for specific makes and models of vehicles. They are available in both printed and computerized formats. Collision repair guides will summarize procedures for removing parts, cutting and welding structural body panels, and similar types of repair operations.

Some shops have a personal computer in the repair area. This makes the task of retrieving service information quicker and easier. Today's collision repair guide CDs are packed with new, critical collision repair information from vehicle manufacturers. Computerized repair guides will help you master the latest collision repair procedures (Figure 3–5).

Chief Automotive Technologies

Figure 3-5 **A shop manager and technicians discuss the use of new computerized measuring systems for checking the amount and direction of vehicle damage. Note that a computer is in the shop area.**

Collision estimating guides provide information for calculating the cost of collision repairs. They give part numbers, part prices, labor charges per repair task, and other data to help the estimator calculate the cost of repairs.

NOTE *Electronic or computer-based estimating is fully discussed in Chapter 10.*

Refinishing materials information will help you to allocate the right amounts for refinishing materials and calculate their costs for each job. Modern refinishing information is based on the use of HVLP (high volume, low pressure), high-efficiency spray equipment that consumes less material than older, low-efficiency spray equipment.

Refinishing guides explain paint codes, types of paints, and how to apply and buff paints, and also describe other paint-related information. Refinishing guides can be printed or in electronic form.

Refinishing software typically allows you to calculate painting costs for approximately 5,000 paint codes from several paint manufacturers, including four types of paint. Refinishing software will help you develop summary reports and detailed breakdowns of materials costs.

A computer CD containing refinishing materials information is shown in Figure 3–6.

Refinishing guides and software will typically include information on:

- Paint code locations
- Paint code explanations
- Bodywork materials
- Plastic and fiberglass materials
- Sanding and buffing materials
- Blending paints
- Low volatile organic compound (VOC) information

There are several basic steps to using refinishing materials information effectively:

1. Look up the paint code tag location. This will show you where the paint code tag is located on the vehicle and how to use the information on the vehicle's paint code tag.

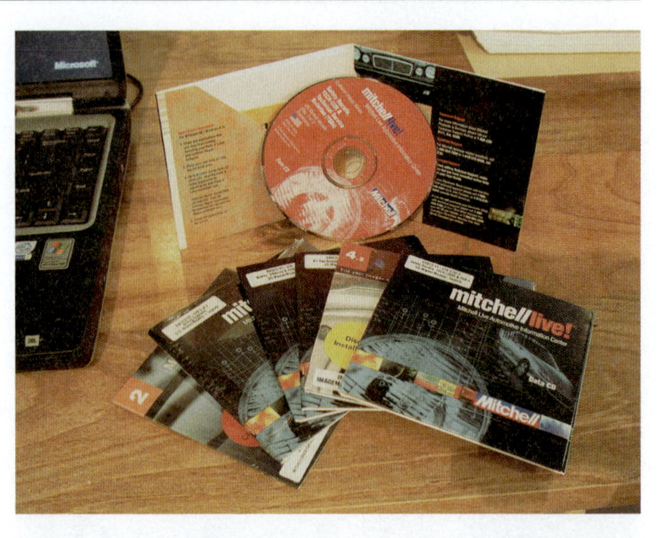

A Various types of CDs are available for collision repair.

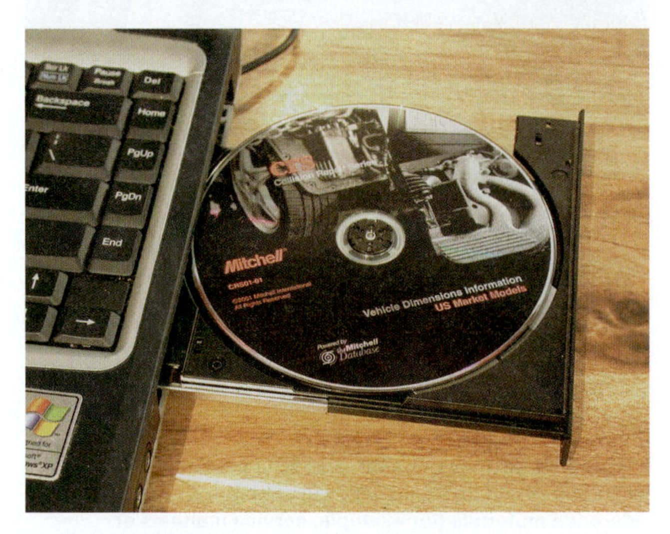

B By inserting the CD into your computer, you can access thousands of pages of technical information.

Figure 3-6 **Most body shops now purchase compact discs that contain essential service information so that service data can be quickly retrieved and used on a personal computer.**

2. Look up the paint code/color page information in the reference material for the vehicle being repaired. The manual or program will give directions on using the repair information.
3. Read the cost information for the specific vehicle being repaired. The information gives the typical costs for the type of paint being used, including single-state color, colorcoat/clearcoat, and three-stage finishes.
4. Refer to any additional information for refinish time, two-tone paint, blending, and so on.
5. Follow auto maker-specific instructions. For example, with GM vehicles you must select the WA number on the ID tag for the correct color selection.

A **vehicle dimension manual** gives the body and frame measurements of undamaged vehicles. Dimensions

A

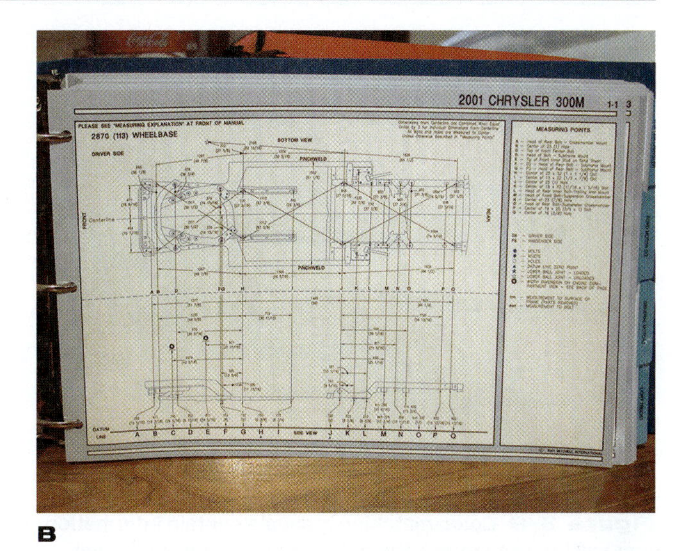

B

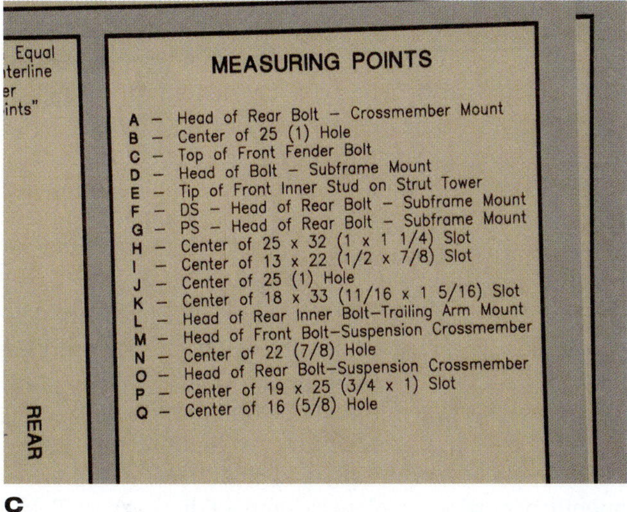

C

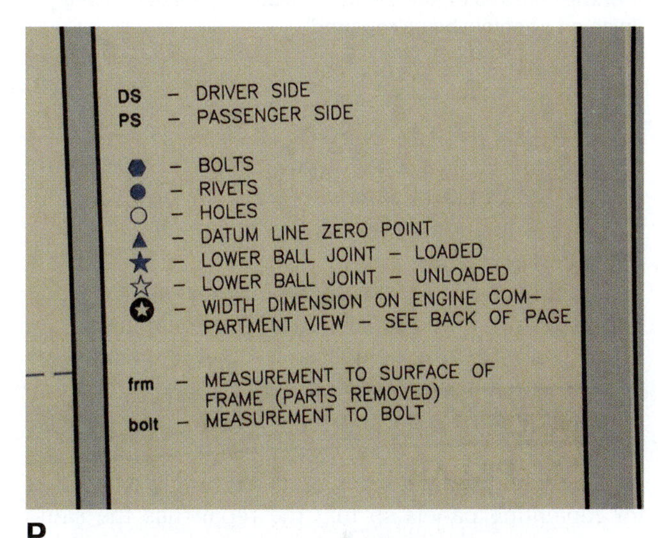

D

Figure 3-7 Vehicle dimension manuals give the measurement values across specific points on a damaged car, truck, van, or SUV. You can measure damage and compare to published values to find the direction and extent of major damage. (A) Both domestic and import dimension manuals are available. (B) A page from a dimension manual shows measurement points and specifications for one make and model of vehicle. (C) Measurement points are given as letters on the dimension page. (D) Symbols on the dimension page tell you exactly what must be measured to see whether the vehicle has major structural damage.

are given for every make and model of car, van, truck, and SUV. These known good dimensions can be compared with actual measurements taken of a wrecked car or truck to find out how badly the vehicle is damaged and what must be done to straighten it (Figure 3–7).

A vehicle dimension manual or software will usually contain the following repair information:

- Complete upper body measurements
- Wheel alignment specifications
- Engine compartment measurements
- Inside passenger compartment measurements
- Front windshield measurements
- Door opening measurements
- Deck lid opening measurements
- Torque values for suspension and steering
- Front suspension measurements
- Rear suspension measurements

Vehicle dimension manuals are usually issued once each year, with new model updates distributed at midyear. Vehicle dimension information is available for U.S. and European market models, each on a separate CD.

Center-of-hole dimensions in inches and millimeters are normally provided so that the measurement specifications are applicable for use with any model of frame equipment on the market.

A technician who needs the dimensions for a specific vehicle being straightened on a frame rack can quickly pull up this information on the computer, because all of the data about the vehicle (year, make, model, and so on) has already been entered into the computer system by the estimator.

Figure 3-8 **Color-matching manuals contain information that will help you match the color of an original or existing finish on a vehicle being repaired.**

Vector graphics-based software allows you to zoom in on any area of a dimension graphic without losing image quality or line definition. Vector graphics also print at high resolution so any printed illustrations are clear and easy to see.

A printout of the vehicle dimensions can be made and taken out to the repair area. Measurements on the printout can then be compared to actual points on the car. If the actual measurements do not match the numbers on the dimension drawing, then the car has major structural damage resulting from the collision impact.

Color-matching guides contain information needed for repainting panels so that the repair has the same appearance as the old finish. Color-matching guides have paint code information, color chips, blending and tinting data, tinting procedures, and other information. Refer to Figure 3–8.

Some color-matching manuals have charts that compare the paints manufactured by different companies.

Paint manufacturer guides give detailed instructions for using specific refinish or paint materials (DuPont, PPG, BASF, Sikkens, and so on). These guides describe specific products, including tips for use, mixing ratios, pot life, additives, application instructions, spray gun setup recommendations, and spray gun pressures. An example of a DuPont application guide is shown in Figure 3–9.

Air bag service guides give procedures, safety warnings, component locations, and other facts for working on active restraint systems. Because air bags often deploy during serious collisions, you must know how to properly service and repair these important vehicle safety devices.

For example, an air bag service guide typically contains the following information:

- Identification and graphics
- Description and operation
- System operational checks
- Removal and installation
- Disposal procedures

Parts interchange guides list which model vehicles from the same manufacturer use the same parts. For example, a part for a Chevrolet might fit a part for a Pontiac because both are manufactured by General Motors Corporation. These guides will help you locate hard-to-find salvage parts. See Figure 3–10.

Salvage parts software provides easy access to a user-customized database of recycled parts and suppliers in North America and Canada. This type of computer program often supports an online accessible parts database from a growing network of thousands of recyclers. Salvage parts software allows you to quickly search for and find quality recycled parts. You can then include them on the electronic estimate.

Mechanical repair manuals give instructions, torque specifications, diagrams, and part illustrations for servicing chassis and powertrain parts. This type of service information will help when you have to remove an engine, service an air-conditioning system, or work on the steering and suspension systems of a wrecked vehicle.

An example of a mechanical repair guide computer screen showing how to service an air-conditioning system compressor appears in Figure 3–11.

Technical service bulletins explain mechanical and structural problems found on specific makes and models of vehicles. They are available in print or computerized form. Bulletins can be published by the auto maker or an aftermarket service information company.

Service bulletins give factory vehicle recalls, symptoms, troubleshooting tips, and repair methods for frequently occurring problems. Compact discs are available that give you access to over 100,000 technical service bulletins and a complete library of recall information. The information is normally organized by system or symptom, making it very easy to find what you need.

Vehicle restoration guides list hard-to-find parts for older collector cars and trucks. Original or authentic aftermarket body parts and mechanical components are listed in this guide. Complete original equipment manufacturer (OEM) part numbers, part availability (shows current or last available price and indicates whether discontinued), part diagrams, and labor times are included in modern vehicle restoration guides.

Tool owner manuals often give service part numbers for power tools and equipment. An example is shown in Figure 3–12.

This information will help you order replacement parts if an air tool breaks or wears out. An exploded or phantom view illustration of the tool with the names of all parts is normally included.

Special service tools (SSTs) are tools designed to be used for specific repair applications. They include special pullers, drivers, and similar tools. Auto makers normally list special tools designed to be used on their make and model vehicles (Figure 3–13).

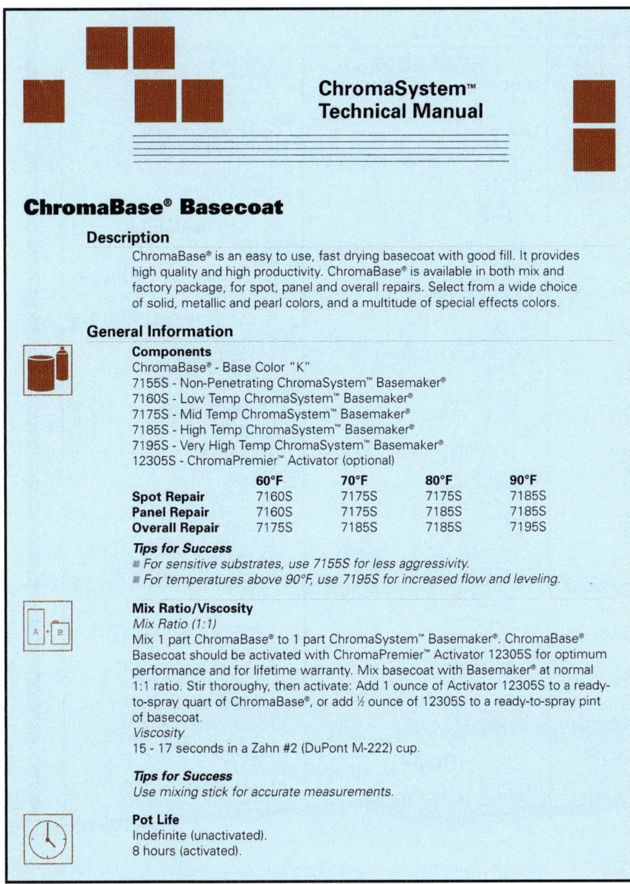

ChromaSystem™
Technical Manual

ChromaBase® Basecoat

Description

ChromaBase® is an easy to use, fast drying basecoat with good fill. It provides high quality and high productivity. ChromaBase® is available in both mix and factory package, for spot, panel and overall repairs. Select from a wide choice of solid, metallic and pearl colors, and a multitude of special effects colors.

General Information

Components
ChromaBase® - Base Color "K"
7155S - Non-Penetrating ChromaSystem™ Basemaker®
7160S - Low Temp ChromaSystem™ Basemaker®
7175S - Mid Temp ChromaSystem™ Basemaker®
7185S - High Temp ChromaSystem™ Basemaker®
7195S - Very High Temp ChromaSystem™ Basemaker®
12305S - ChromaPremier™ Activator (optional)

	60°F	70°F	80°F	90°F
Spot Repair	7160S	7175S	7175S	7185S
Panel Repair	7160S	7175S	7185S	7185S
Overall Repair	7175S	7185S	7185S	7195S

Tips for Success
- *For sensitive substrates, use 7155S for less aggressivity.*
- *For temperatures above 90°F, use 7195S for increased flow and leveling.*

Mix Ratio/Viscosity
Mix Ratio (1:1)
Mix 1 part ChromaBase® to 1 part ChromaSystem™ Basemaker®. ChromaBase® Basecoat should be activated with ChromaPremier™ Activator 12305S for optimum performance and for lifetime warranty. Mix basecoat with Basemaker® at normal 1:1 ratio. Stir thoroughly, then activate: Add 1 ounce of Activator 12305S to a ready-to-spray quart of ChromaBase®, or add ½ ounce of 12305S to a ready-to-spray pint of basecoat.
Viscosity
15 - 17 seconds in a Zahn #2 (DuPont M-222) cup.

Tips for Success
Use mixing stick for accurate measurements.

Pot Life
Indefinite (unactivated).
8 hours (activated).

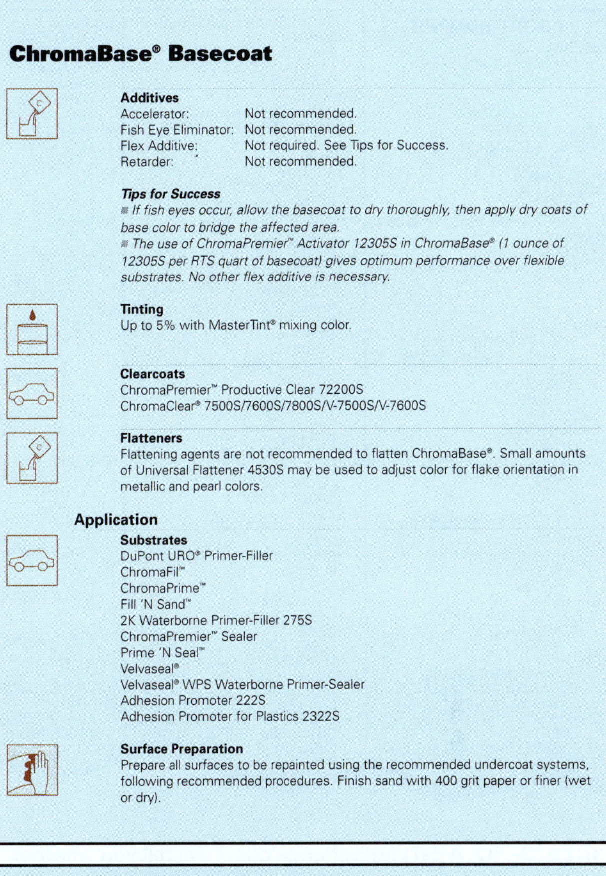

ChromaBase® Basecoat

Additives

Accelerator:	Not recommended.
Fish Eye Eliminator:	Not recommended.
Flex Additive:	Not required. See Tips for Success.
Retarder:	Not recommended.

Tips for Success
- *If fish eyes occur, allow the basecoat to dry thoroughly, then apply dry coats of base color to bridge the affected area.*
- *The use of ChromaPremier™ Activator 12305S in ChromaBase® (1 ounce of 12305S per RTS quart of basecoat) gives optimum performance over flexible substrates. No other flex additive is necessary.*

Tinting
Up to 5% with MasterTint® mixing color.

Clearcoats
ChromaPremier™ Productive Clear 72200S
ChromaClear® 7500S/7600S/7800S/V-7500S/V-7600S

Flatteners
Flattening agents are not recommended to flatten ChromaBase®. Small amounts of Universal Flattener 4530S may be used to adjust color for flake orientation in metallic and pearl colors.

Application

Substrates
DuPont URO® Primer-Filler
ChromaFil™
ChromaPrime™
Fill 'N Sand™
2K Waterborne Primer-Filler 275S
ChromaPremier™ Sealer
Prime 'N Seal™
Velvaseal®
Velvaseal® WPS Waterborne Primer-Sealer
Adhesion Promoter 222S
Adhesion Promoter for Plastics 2322S

Surface Preparation
Prepare all surfaces to be repainted using the recommended undercoat systems, following recommended procedures. Finish sand with 400 grit paper or finer (wet or dry).

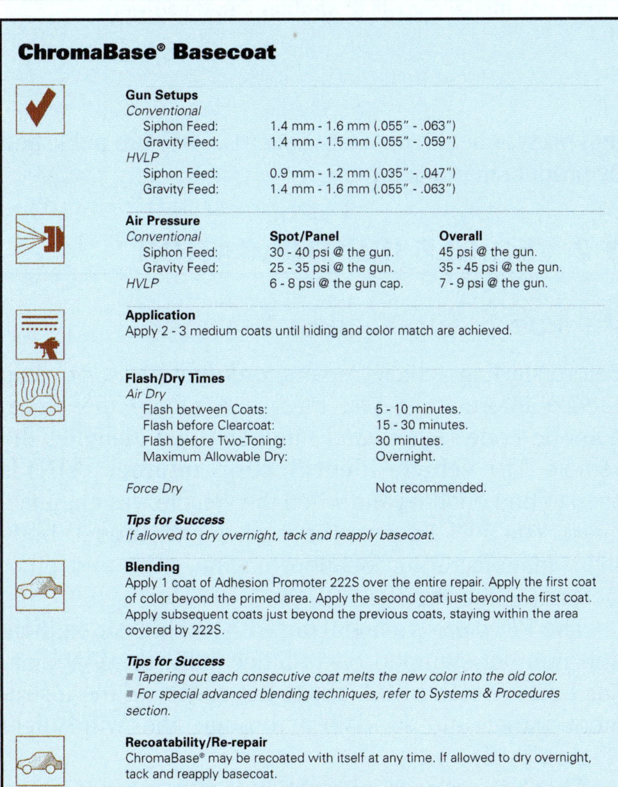

ChromaBase® Basecoat

Gun Setups

Conventional	
Siphon Feed:	1.4 mm - 1.6 mm (.055" - .063")
Gravity Feed:	1.4 mm - 1.5 mm (.055" - .059")
HVLP	
Siphon Feed:	0.9 mm - 1.2 mm (.035" - .047")
Gravity Feed:	1.4 mm - 1.6 mm (.055" - .063")

Air Pressure

	Spot/Panel	**Overall**
Conventional		
Siphon Feed:	30 - 40 psi @ the gun.	45 psi @ the gun.
Gravity Feed:	25 - 35 psi @ the gun.	35 - 45 psi @ the gun.
HVLP	6 - 8 psi @ the gun cap.	7 - 9 psi @ the gun.

Application
Apply 2 - 3 medium coats until hiding and color match are achieved.

Flash/Dry Times

Air Dry	
Flash between Coats:	5 - 10 minutes.
Flash before Clearcoat:	15 - 30 minutes.
Flash before Two-Toning:	30 minutes.
Maximum Allowable Dry:	Overnight.
Force Dry	Not recommended.

Tips for Success
If allowed to dry overnight, tack and reapply basecoat.

Blending
Apply 1 coat of Adhesion Promoter 222S over the entire repair. Apply the first coat of color beyond the primed area. Apply the second coat just beyond the first coat. Apply subsequent coats just beyond the previous coats, staying within the area covered by 222S.

Tips for Success
- *Tapering out each consecutive coat melts the new color into the old color.*
- *For special advanced blending techniques, refer to Systems & Procedures section.*

Recoatability/Re-repair
ChromaBase® may be recoated with itself at any time. If allowed to dry overnight, tack and reapply basecoat.

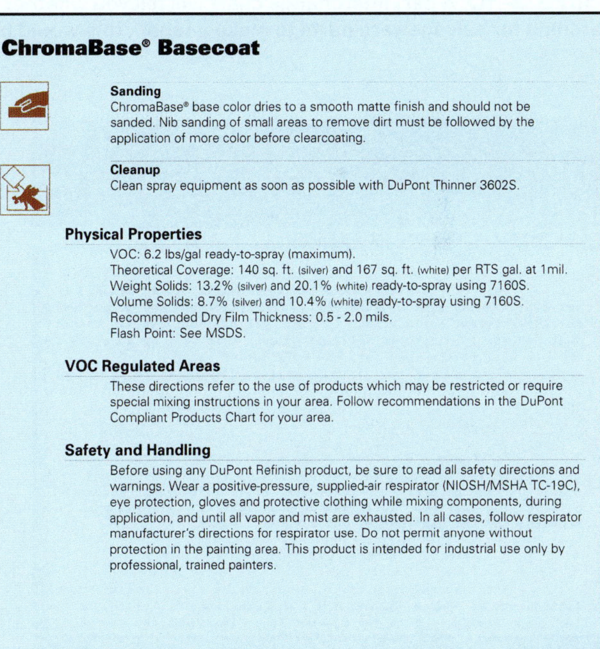

ChromaBase® Basecoat

Sanding
ChromaBase® base color dries to a smooth matte finish and should not be sanded. Nib sanding of small areas to remove dirt must be followed by the application of more color before clearcoating.

Cleanup
Clean spray equipment as soon as possible with DuPont Thinner 3602S.

Physical Properties

VOC: 6.2 lbs/gal ready-to-spray (maximum).
Theoretical Coverage: 140 sq. ft. (silver) and 167 sq. ft. (white) per RTS gal. at 1mil.
Weight Solids: 13.2% (silver) and 20.1% (white) ready-to-spray using 7160S.
Volume Solids: 8.7% (silver) and 10.4% (white) ready-to-spray using 7160S.
Recommended Dry Film Thickness: 0.5 - 2.0 mils.
Flash Point: See MSDS.

VOC Regulated Areas

These directions refer to the use of products which may be restricted or require special mixing instructions in your area. Follow recommendations in the DuPont Compliant Products Chart for your area.

Safety and Handling

Before using any DuPont Refinish product, be sure to read all safety directions and warnings. Wear a positive-pressure, supplied-air respirator (NIOSH/MSHA TC-19C), eye protection, gloves and protective clothing while mixing components, during application, and until all vapor and mist are exhausted. In all cases, follow respirator manufacturer's directions for respirator use. Do not permit anyone without protection in the painting area. This product is intended for industrial use only by professional, trained painters.

Courtesy of DuPont Automotive Finishes

Figure 3-9 This is a technical document published by one paint manufacturer. Note all of the useful information given for this specific paint product. Updated application guides can be obtained from the DuPont Web site: http://www.dupont.com/finishes/.

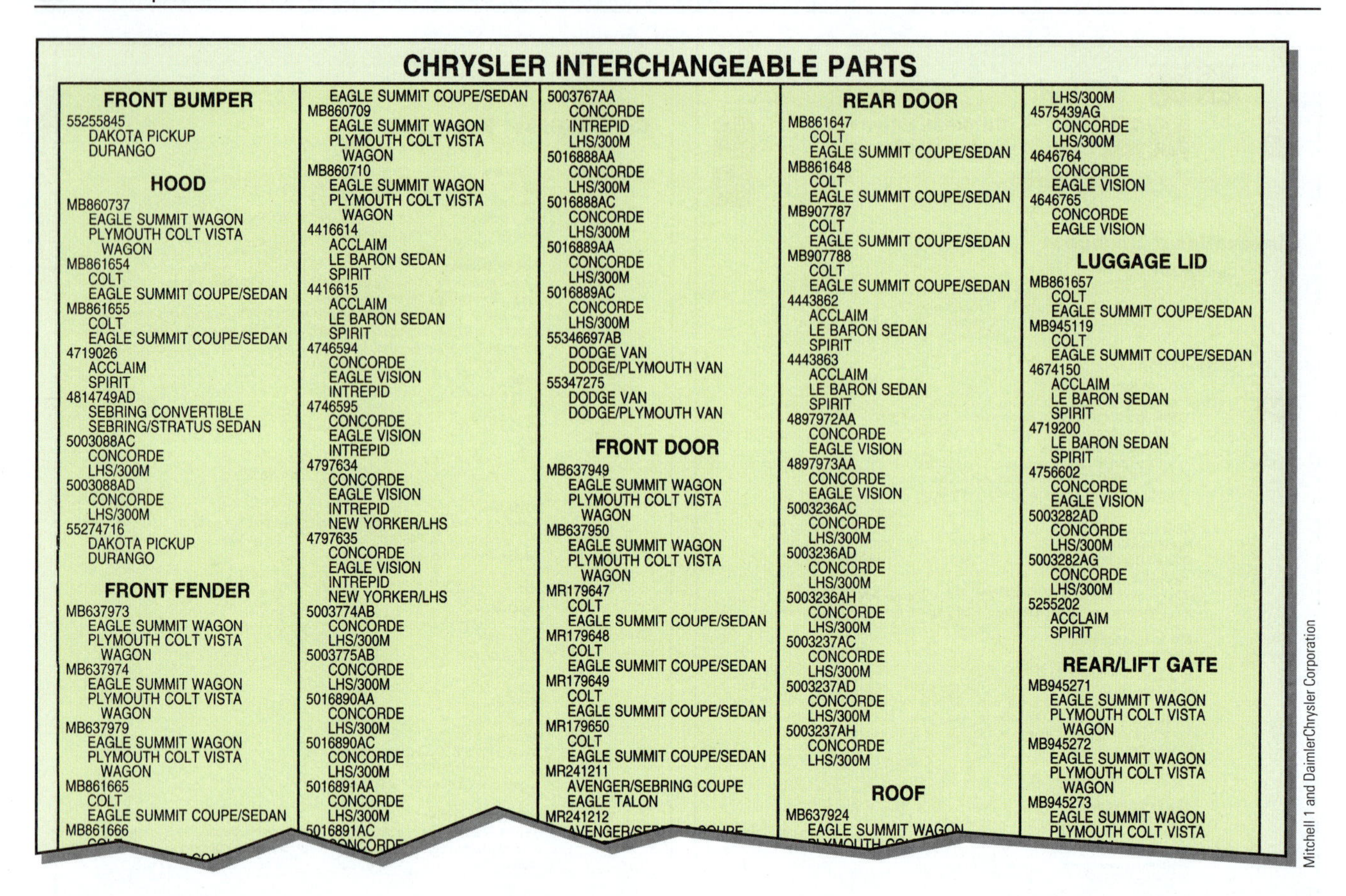

CHRYSLER INTERCHANGEABLE PARTS

FRONT BUMPER
55255845
DAKOTA PICKUP
DURANGO

HOOD
MB860737
EAGLE SUMMIT WAGON
PLYMOUTH COLT VISTA
WAGON
MB861654
COLT
EAGLE SUMMIT COUPE/SEDAN
MB861655
COLT
EAGLE SUMMIT COUPE/SEDAN
4719026
ACCLAIM
SPIRIT
4814749AD
SEBRING CONVERTIBLE
SEBRING/STRATUS SEDAN
5003088AC
CONCORDE
LHS/300M
5003088AD
CONCORDE
LHS/300M
55274716
DAKOTA PICKUP
DURANGO

FRONT FENDER
MB637973
EAGLE SUMMIT WAGON
PLYMOUTH COLT VISTA
WAGON
MB637974
EAGLE SUMMIT WAGON
PLYMOUTH COLT VISTA
WAGON
MB637979
EAGLE SUMMIT WAGON
PLYMOUTH COLT VISTA
WAGON
MB861665
COLT
EAGLE SUMMIT COUPE/SEDAN
MB861666

EAGLE SUMMIT COUPE/SEDAN
MB860709
EAGLE SUMMIT WAGON
PLYMOUTH COLT VISTA
WAGON
MB860710
EAGLE SUMMIT WAGON
PLYMOUTH COLT VISTA
WAGON
4416614
ACCLAIM
LE BARON SEDAN
SPIRIT
4416615
ACCLAIM
LE BARON SEDAN
SPIRIT
4746594
CONCORDE
EAGLE VISION
INTREPID
4746595
CONCORDE
EAGLE VISION
INTREPID
4797634
CONCORDE
EAGLE VISION
INTREPID
NEW YORKER/LHS
4797635
CONCORDE
EAGLE VISION
INTREPID
NEW YORKER/LHS
5003774AB
CONCORDE
LHS/300M
5003775AB
CONCORDE
LHS/300M
5016890AA
CONCORDE
LHS/300M
5016890AC
CONCORDE
LHS/300M
5016891AA
CONCORDE
LHS/300M
5016891AC

5003767AA
CONCORDE
INTREPID
LHS/300M
5016888AA
CONCORDE
LHS/300M
5016888AC
CONCORDE
LHS/300M
5016889AA
CONCORDE
LHS/300M
5016889AC
CONCORDE
LHS/300M
55346697AB
DODGE VAN
DODGE/PLYMOUTH VAN
55347275
DODGE VAN
DODGE/PLYMOUTH VAN

FRONT DOOR
MB637949
EAGLE SUMMIT WAGON
PLYMOUTH COLT VISTA
WAGON
MB637950
EAGLE SUMMIT WAGON
PLYMOUTH COLT VISTA
WAGON
MR179647
COLT
EAGLE SUMMIT COUPE/SEDAN
MR179648
COLT
EAGLE SUMMIT COUPE/SEDAN
MR179649
COLT
EAGLE SUMMIT COUPE/SEDAN
MR179650
COLT
EAGLE SUMMIT COUPE/SEDAN
MR241211
AVENGER/SEBRING COUPE
EAGLE TALON
MR241212

REAR DOOR
MB861647
COLT
EAGLE SUMMIT COUPE/SEDAN
MB861648
COLT
EAGLE SUMMIT COUPE/SEDAN
MB907787
COLT
EAGLE SUMMIT COUPE/SEDAN
MB907788
COLT
EAGLE SUMMIT COUPE/SEDAN
4443862
ACCLAIM
LE BARON SEDAN
SPIRIT
4443863
ACCLAIM
LE BARON SEDAN
SPIRIT
4897972AA
CONCORDE
EAGLE VISION
4897973AA
CONCORDE
EAGLE VISION
5003236AC
CONCORDE
LHS/300M
5003236AD
CONCORDE
LHS/300M
5003236AH
CONCORDE
LHS/300M
5003237AC
CONCORDE
LHS/300M
5003237AD
CONCORDE
LHS/300M
5003237AH
CONCORDE
LHS/300M

ROOF
MB637924

LHS/300M
4575439AG
CONCORDE
LHS/300M
4646764
CONCORDE
EAGLE VISION
4646765
CONCORDE
EAGLE VISION

LUGGAGE LID
MB861657
COLT
EAGLE SUMMIT COUPE/SEDAN
MB945119
COLT
EAGLE SUMMIT COUPE/SEDAN
4674150
ACCLAIM
LE BARON SEDAN
SPIRIT
4719200
LE BARON SEDAN
SPIRIT
4756602
CONCORDE
EAGLE VISION
5003282AD
CONCORDE
LHS/300M
5003282AG
CONCORDE
LHS/300M
5255202
ACCLAIM
SPIRIT

REAR/LIFT GATE
MB945271
EAGLE SUMMIT WAGON
PLYMOUTH COLT VISTA
WAGON
MB945272
EAGLE SUMMIT WAGON
PLYMOUTH COLT VISTA
WAGON
MB945273
EAGLE SUMMIT WAGON
PLYMOUTH COLT VISTA

Mitchell 1 and DaimlerChrysler Corporation

Figure 3-10 A part interchange chart will tell you whether parts from one model car will fit another car. If you were searching for salvage yard parts to make a repair, this would be useful.

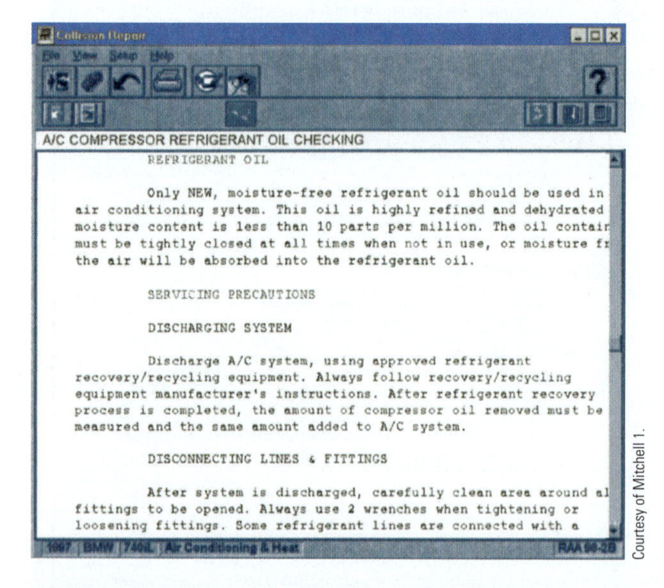

Courtesy of Mitchell 1.

Figure 3-11 A computer screen shows mechanical repair information on an air-conditioning system. You can quickly scroll down through the instructions to find the needed data.

Material Safety Data Sheets

Discussed fully in Chapter 9, a *material safety data sheet* (MSDS) is a printed summary of the chemical composition, handling, and disposal precautions for products that present health or safety hazards. They are published by product manufacturers.

3.2 VEHICLE IDENTIFICATION

Vehicle Identification Number

Before starting vehicle repairs, ordering parts, or using service information, you must first record the alphanumeric code (letters and numbers) that identifies the vehicle. The **vehicle identification number (VIN)** is used to find out how and when the vehicle was manufactured. The VIN denotes the year, make, model, body style, manufacturing location, engine, trim, and other facts about the vehicle.

The *VIN plate* is a metal tag with a vehicle identification number stamped on it. Since 1981, the VIN plate has been riveted to the upper left corner of the instrument panel and is visible through the windshield (Figure 3–14).

Check the service information for the location of the VIN, vehicle certification label, or body plate for vehicles made prior to 1981 and for foreign vehicles.

Service manuals and crash-estimating guides contain all of the necessary VIN decoding information. *VIN decode software* is a computer program that will

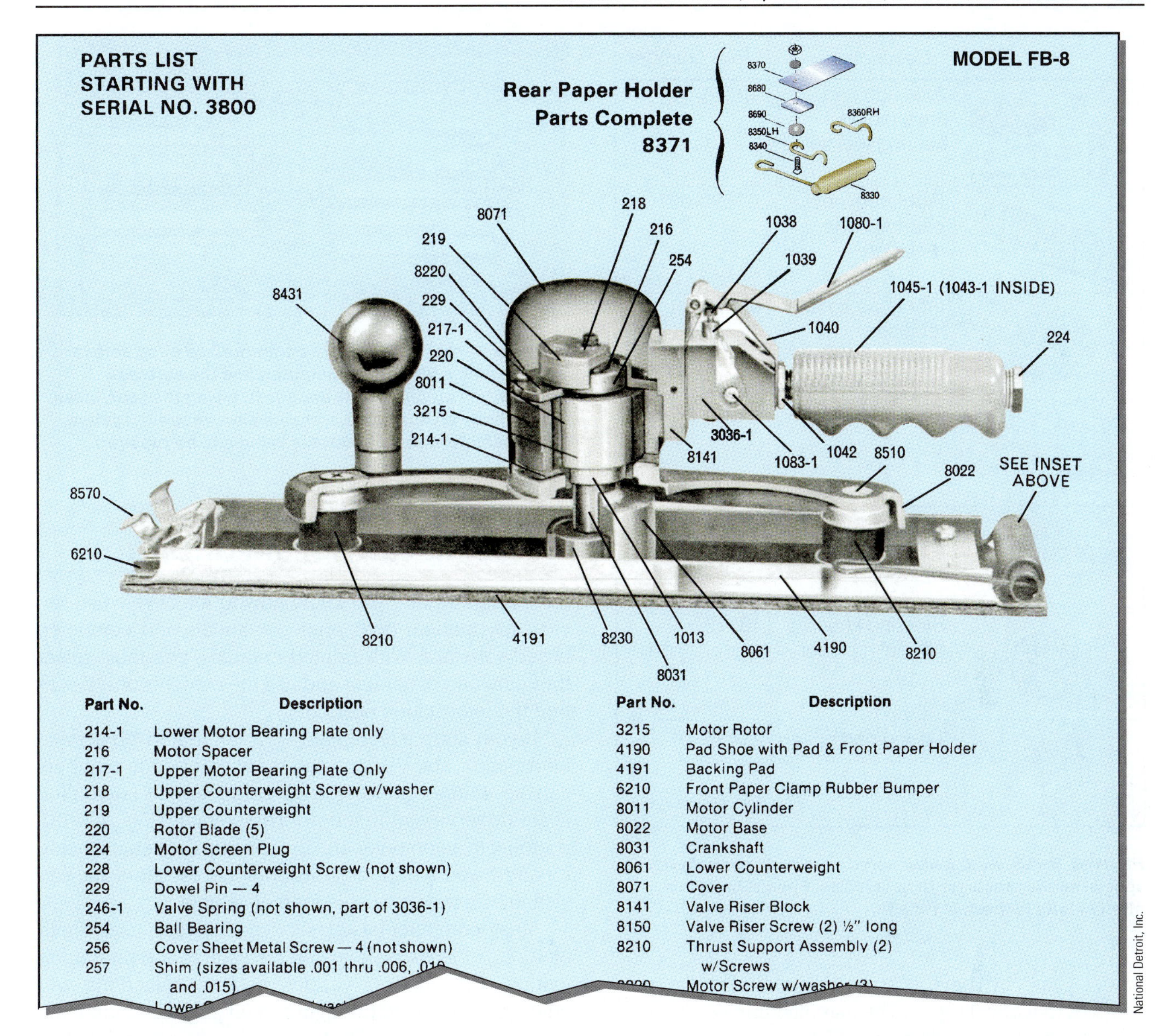

Part No.	Description
214-1	Lower Motor Bearing Plate only
216	Motor Spacer
217-1	Upper Motor Bearing Plate Only
218	Upper Counterweight Screw w/washer
219	Upper Counterweight
220	Rotor Blade (5)
224	Motor Screen Plug
228	Lower Counterweight Screw (not shown)
229	Dowel Pin — 4
246-1	Valve Spring (not shown, part of 3036-1)
254	Ball Bearing
256	Cover Sheet Metal Screw — 4 (not shown)
257	Shim (sizes available .001 thru .006, .010 and .015)
	Lower [illegible]

Part No.	Description
3215	Motor Rotor
4190	Pad Shoe with Pad & Front Paper Holder
4191	Backing Pad
6210	Front Paper Clamp Rubber Bumper
8011	Motor Cylinder
8022	Motor Base
8031	Crankshaft
8061	Lower Counterweight
8071	Cover
8141	Valve Riser Block
8150	Valve Riser Screw (2) ½" long
8210	Thrust Support Assembly (2) w/Screws
[illegible]	Motor Screw w/washer (3)

Figure 3–12 Tool owner's manuals often provide illustrations with part numbers. This allows you to order repair parts for your air tool or equipment.

automatically look up and convert the VIN into year, make, model, body style, engine, transmission, restraint system, and other vehicle-specific information. By typing in the alphanumeric VIN code, the program will automatically decipher and fill in the needed information about the vehicle. The computer screen for VIN decode software is shown in Figure 3–15.

Access to VIN information also allows shop personnel, locksmiths, fire departments, police departments, and car rental agencies to save valuable time and money by having the right tools for lockouts or lock replacements.

Vehicle Paint Code

The **paint code** is also an alphanumeric code that states what type and color of paint was used during vehicle manufacturing. The **body ID number**, or *service part label*, gives information about the finish (paint) and the trim or moldings used on a specific vehicle. It gives numbers for ordering the right type and color of paint, including the lower and upper body colors for a two-tone paint job. The body ID number will be on the body ID plate in various locations on the vehicle. Paint code locations are shown in Figure 3–16.

Become familiar with each vehicle manufacturer's method of vehicle identification and the specific information it contains. Remember that you should obtain all of the information possible about the vehicle being worked on to ensure quality repairs. See Figure 3–17.

Note that there are frequent "running changes" made by the vehicle manufacturers. Vehicles built in the same

	Description	Part Number
	Axle hub and drive pinion bearing tool set	12345-01
	Front hub inner bearing cone replacer	67890-02
	Front hub outer replacer	09876-03
	Oil seal puller	54321-04
	Rear axle bearing replacer	24680_01
	Steering knuckle oil seal replacer	13579_02
	Tie rod end puller	21435_01

Figure 3-13 **Auto maker service manuals normally list special service tools for their vehicles. Special tools are often helpful for certain repairs.**

model year can have one or several individual part changes that affect repair. The manufacturer or aftermarket publisher will usually note midyear changes in its service materials.

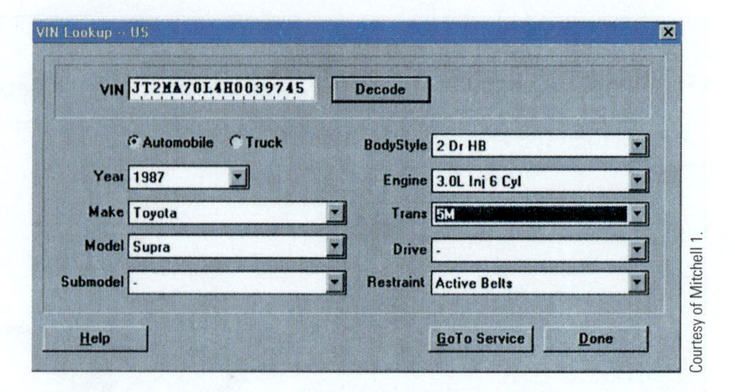

Courtesy of Mitchell 1.

Figure 3-15 **With modern computerized shop software, you can type a VIN into a computer, and the software program will automatically decode it, giving the year, make, model, body style, engine, transmission, restraint system, and other information about the vehicle to be repaired.**

3.3 USING SERVICE INFORMATION

It is important that you know how to effectively use service information, both printed manuals and computer-based software. With printed manuals, you must select the right kind of manual and use the contents or index to find the information needed.

If your shop is equipped with computerized service information, the VIN and the body paint code numbers can be entered into the shop computers to streamline access to service information for that vehicle (Figure 3–18). By going to a computer and bringing up the vehicle being repaired, you can almost instantly access all of the service information available for that vehicle.

Most computer-based service software uses small pictures, or *icons*, to help you navigate to the right information. Clicking on various content selections will quickly move you to the right screen page for finding the needed information. Modern shop software is intuitive and easy to use (Figure 3–19).

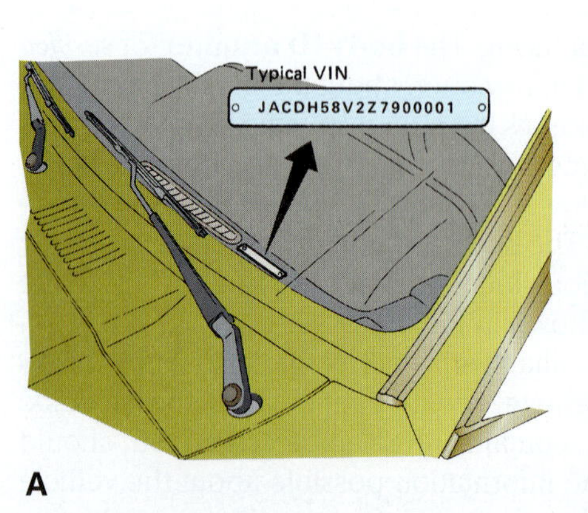

A

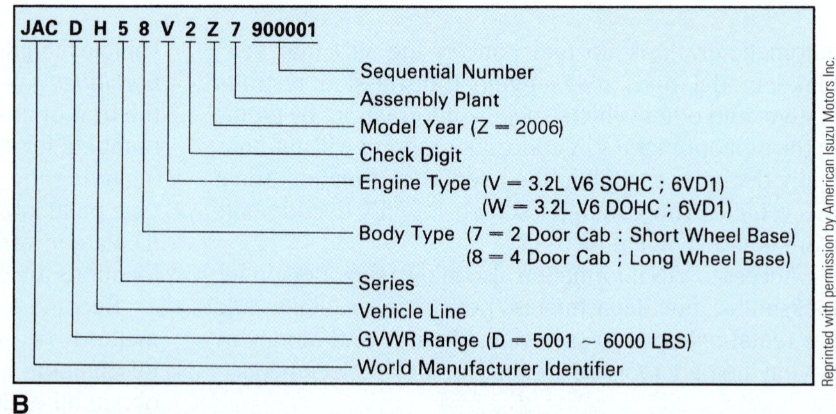

B

Reprinted with permission by American Isuzu Motors Inc.

Figure 3-14 **Study how to decipher a typical VIN. (A) The VIN is normally located on the dash and can be read through the windshield. (B) Note the meaning of the alphanumeric VIN for this car.**

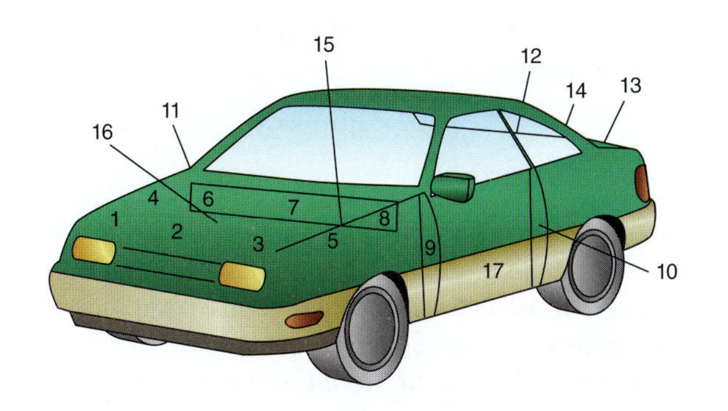

MODEL	POSITION
Acura	9
Alfa Romeo	4, 13
AMC	9, 10
Audi	12, 13
Austin Rover	17
BMW	4, 5
Chrysler	3, 5, 16
Chrysler Corp	3, 5
Caravan/Voyager/RamVan	6
Chrysler imports	1, 2, 4
Colt Vista	16
Conquest	7
Daihatsu	1, 6, 7
Datsun	2
Dodge D50	3
Ford	10
Ford Motor Co	10
General Motors	
A, J and L Bodies	14
E and K Bodies	12
B, C, H and N Bodies	13
GM imports	2, 12,13,14
Honda	8, 10
Hyundai	6, 7
Isuzu	2, 10
Lexus	7, 8
Mazda	1, 2, 3, 4, 6, 8
Mercedes	2, 7, 9
Mitsubishi	7
Montero/pickup	3
Cordia/Tredia	4
Others	1, 2, 3
Nissan	1, 3, 4, 6, 8, 15, *
Peugeot	2, 3, 4, 5, 8
Porsche	9
Renault	1, 3, 4, 5, 8
Rover	1, 3, 4, 5
Saab	5, 6, 8
Subaru	2
Suzuki	7, 11
Toyota passenger	7, 8, 14
Truck	4
Volkswagen	2, 11
Volvo	6, 7, 8
Yugo	

*Under right front passenger seat

Figure 3–16 **The paint code is needed before refinishing a vehicle. Note the paint code locations for the major auto manufacturers.**

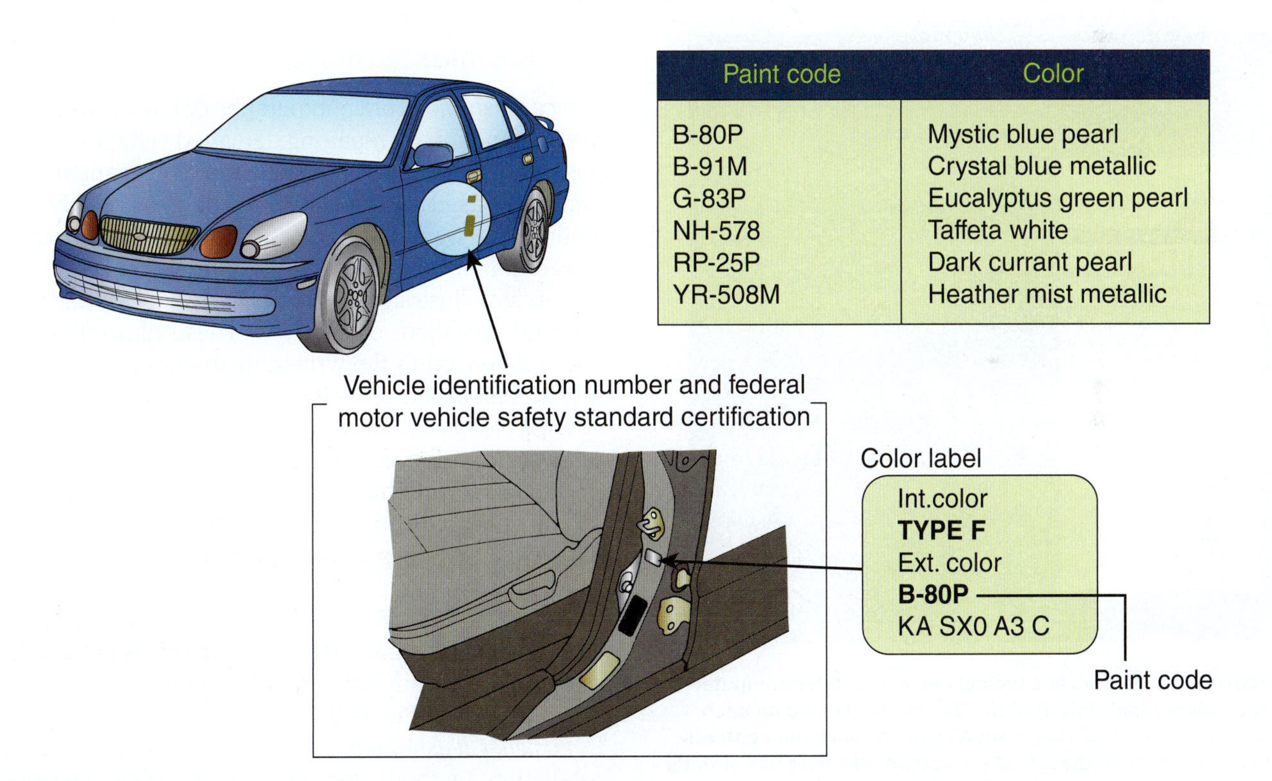

Paint code	Color
B-80P	Mystic blue pearl
B-91M	Crystal blue metallic
G-83P	Eucalyptus green pearl
NH-578	Taffeta white
RP-25P	Dark currant pearl
YR-508M	Heather mist metallic

Figure 3–17 **Here a factory service manual explains the meaning of the paint code. The color label is just above other labels on the C-pillar of this vehicle.**

Service Abbreviations

Service abbreviations are short series of letters that represent technical terms or words. They save space in the service literature and reduce reading time. Each manufacturer uses slightly different abbreviations. A sample of some of the most common body shop abbreviations appears in Figure 3–20.

NOTE *Refer to the tables in the back of this book for a more extensive list of service abbreviations.*

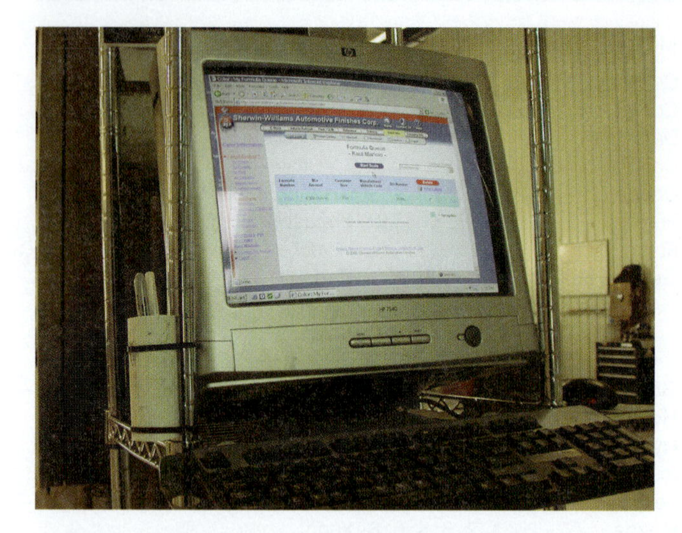

Figure 3-18 **The industry-wide trend is toward computer-based service information. Sometimes the technicians look up their own information; in other shops, the shop manager may look up repair data for the technicians and print it out.**

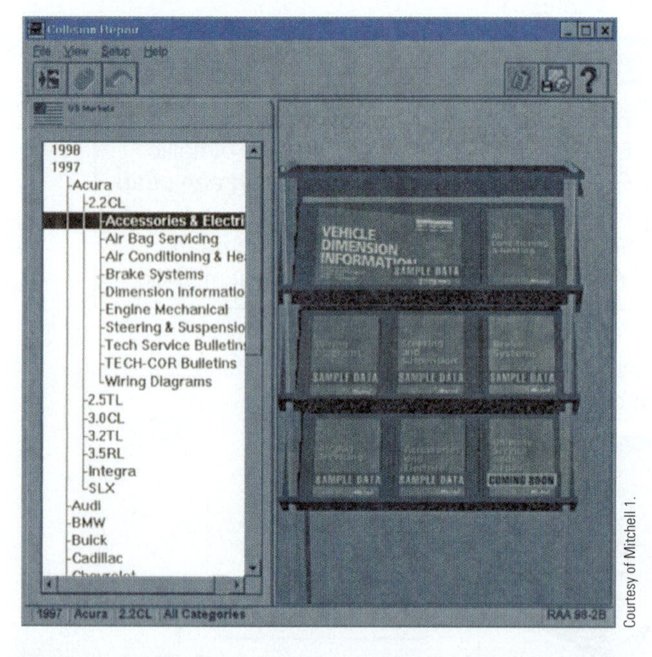

Courtesy of Mitchell 1.

Figure 3-19 **This is a typical main screen for computer-based shop repair information. By simply clicking on each icon for a type of service manual, you can view the contents of each manual on the left of the screen. VIN data has already been entered, so you will go directly to the service data needed for that make and model of vehicle.**

Service Symbols

Service symbols are small pictures that represent a location, part, procedure, safety warning, measurement point, or other aspect of repair (Figure 3–21).

Symbol use varies from publisher to publisher. Normally the contents or front matter of a service manual or computer software will explain its particular use of symbols. Figure 3–22 shows the symbols used in an air bag service manual.

Service Charts

Repair charts give diagrams that guide you through logical steps for making repairs. Though they vary in content and design, most use arrows and icons (graphic symbols) to represent repair steps.

Diagnosis charts, or *troubleshooting charts*, give logical steps for finding the source of problems. Mechanical, body, electrical, and other types of troubleshooting charts are provided in service manuals and computer-based service information. These charts give the most common sources of problems for the symptoms.

Paint reference charts list comparable paints manufactured by different companies. This will help you match the color of the paint to be used during refinishing with the paint already on the vehicle. These charts are manufactured by different companies and have different application qualities.

Service Illustrations

Service illustrations are photographs or line drawings that give information on repairing damaged vehicles. Service illustrations are used in both service manuals and computer-based service software. You can go to the publication or program index to find pictorial views of the parts or panels being repaired (Figure 3–23).

Service illustrations show how specific parts are removed, installed, or repaired. These illustrations are often referenced in the written instructions in a manual or computer-based service information (Figure 3–24).

Exploded view illustrations show each part or panel removed from the vehicle or assembly. Exploded views are helpful in finding hidden fasteners, part names for ordering new parts, the types of fasteners used, and how complex assemblies come apart or fit back together (Figure 3–25).

An exploded view of a fuel tank is shown in Figure 3–26. Sometimes, *detailed illustrations* giving more specific service information will accompany exploded and other views (Figure 3–27).

Fastener illustrations show the shapes and removal/installation methods for metal and plastic fasteners. These illustrations might show how to use a trim tool to pop out a press fit fastener (Figure 3–28).

Body illustrations show how the major welded body panels, fastened panels, plastic panels, unibody or perimeter frame, trim, lights, and other related parts are assembled on the vehicle. They are commonly used if the technician is not sure of some aspect of the disassembly or reassembly of a major section of the vehicle.

Mechanical illustrations show how chassis and drivetrain parts are serviced and repaired. As a body technician, you will often have to remove and install mechanical

AAM	All Active Module
AAS	Automatic Adjusting Suspension
ABS	Antilock Braking System
ACE	Active Cornering Enhancement
A/C	Air Conditioning
AIR	Air Injection Reactor
Add'l	Additional
Adj	Adjust/Adjustable
ADS	Adaptive Damping System
Alum	Aluminum
Alt	Alternator
AOT	Automatic Overdrive Trans
ASC	Auto Stability Control
ASC&T	Auto Stability Control plus Traction Control
ASD	Auto Locking Differential
ASR	Automatic Slip Control
Assy	Assembly
ATC	Automatic Temp Control
Attach	Attachments
Aux	Auxiliary
AWD	All Wheel Drive
BAS	Brake Assist System
Batt	Battery
BPMV	Brake Pressure Modulator Valve
BPT	Back Pressure Transducer
Brg	Bearing
Brkt	Bracket
BSC	Battery Safety Clamp
Bush	Bushing
"C"	Cleveland Eng
CB	Citizens Band Radio
CC	Cubic Centimeters
CCC	Computer Command Control
CCT	Computer Controlled Timing
Ch	Chassis
CNG	Compressed Natural Gas
Comb	Combination
Comm	Communication
Comp	Compression,Compressor
Compt	Compartment
Cond	Condenser or Conditioning
Cont	Control
Conv	Converter or Convertible
CPI	Central Port Injection
Cr/Cont	Cruise Control
Ctr	Center
CV	Constant Velocity
d	Discontinued
D & A	Disassemble & Assemble
D & C	Disconnect & Connect
DEFI	Digital Electronic Fuel Injection
Defl	Deflector
DERM	Diagnostic Energy Reserve Module
Diaph	Diaphragm
Diff'l	Differential
DME	Digital Motor Electronics
DOHC	Dual Over Head Cam
DPMS	Driving Position Memory System
DRL	Daytime Running Lights
DSTC	Dynamic Stability Traction Control
EATX	Electronic Automatic Transmission Control
EBCM	Electronic Brake Control Module
EBTC	Electronic Brake and Traction Control
ECM	Electronic Control Module
ECS	Emission Control System and Electronic Crash Sensor
EEC	Electronic Engine Control
EFC	Electronic Fuel Control
EFE	Early Fuel Evaporation
EFI	Electronic Fuel Injection
EGR	Exhaust Gas Recirculation
EGS	Elect Trans Gear System
Elec	Electric
Elect	Electronic
EPR	Evaporator Pressure Regulator
EPS	Electrical Power Steering
Equip	Equipment
ESP	Electronic Stability Program
EST	Electronic Spark Timing
ETR	Emergency/Energy Tension Reactor
ETS	Electronic Traction System
Evap	Evaporator
Ev-Em	Evaporative Emission Control
Exh	Exhaust
Ext	Extended or Extension
FI	Fuel Injection
Frt	Front
Ft	Foot
Fwd	Forward
FWD	Front Wheel Drive
Gal	Gallon
GDO	Garage Door Opener
Gskt	Gasket
GVW	Gross Vehicle Weight
HCU	Hydraulic Control Unit
HD	Heavy Duty
HEI	High Energy Ignition
HFM	Hot Film Management
HICAS	High Capacity Suspension
HID	High Intensity Discharge
H/Lamp	Headlamp
HO	High Output
HP	Horsepower
Hsg	Housing
HSLA	High Strength Low Alloy Steel
HSS	High Strength Steel
HUD	Head Up Display
HV	High Voltage
HVAC	Heater Ventilation Air Conditioning
Hvy	Heavy
Hyd	Hydraulic
ID	Identification or Inside Diameter
Ign	Ignition
Illum	Illuminated
IMA	Integrated Motor Assisted
Inr	Inner
Inst	Instrument
IVD	Interactive Vehicle Dynamics
Inter	Intermediate
IOH	Included in Overhaul
ITS	Inflatable Tubular Structure
km	Kilometer
KPH	Kilometer Per Hour
L	Litre or Left
Lb	Pound
Lic	License
LSD	Limited Slip Differential
Lwr	Lower
"M"	Modified Eng
Man	Manual
MAP	Manifold Absolute Pressure
MCU	Microprocessor Control Unit
M.D.	Medium Duty
MDM	Motor Drive Module
Med	Medium
MFI	Multi-Point Fuel Injection
Mldg	Moulding
mm	Millimeter
MPFI	Multi-Port Fuel Injection
Mtd	Mounted
Mtg	Mounting
N.A.	Not Available
Nav	Navigation
NOx	Nitrogen Oxide Emission Control
NP	New Process
OBD	On-Board Diagnostic
OD	Outside Diameter
O/D	Overdrive
O/H	Overhaul
OHC	Over Head Cam
OHV	Over Head Valve
OPDS	Occupant Position Detector Sensor
ORVR	On-Board Refueling Vapor Recovery
Otr	Outer
Opng	Opening
PAIR	Pulse Air Injection Reactor
PCM	Powertrain Control Module
Pass	Passenger or Passive
Perf	Performance
Pnl	Panel
Pos	Positive
PPD	Passenger Presence Detection
Press	Pressure
PTO	Power Take Off
Pwr	Power
Qtr	Quarter
R	Right
Rad	Radiator
Rect	Rectangular
Refrig	Refrigeration
Reg	Regulator or Regular
Reinf	Reinforcement
Resv	Reservoir
R & I	Remove & Install
R & R	Remove & Replace
RPO	Regular Production Option
RWD	Rear Wheel Drive
SAM	Signal Acquisition Module
SBEC	Single Board Engine Control
SDM	Sensing & Diagnostic Module
Sect	Section
Sed	Sedan
SEFI	Sequential Electronic Fuel Injection
Ser	Series
SFI	Sequential Fuel Injection
Sgl	Single
SIPS	Side Impact Protect System
SIR	Supplemental Inflatable Restraint
SISM	Side Impact Sensing Module
SLS	Self-Leveling Suspension
SOHC	Single Over Head Cam
Speedo	Speedometer
SRS	Supplemental Restraint System
Stab	Stabilizer
Std	Standard
Stpd	Stamped
Strg	Steering
Supt	Support
Susp	Suspension System
Tach	Tachometer
TBI	Throttle Body Injection
TCS	Traction Control System
TDC	Top Dead Center
TDI	Turbo Direct Injection
Temp	Temperature
TEMS	Toyota Electronic Modulated Suspension
T/Gate	Tailgate
TH	Turbo Hydra-Matic
TLEV	Transitional Low Emission Vehicle
Trans	Transmission
Transf	Transfer
Upr	Upper
Vac	Vacuum
VIN	Vehicle Identification Number
VIR	Valve in Receiver
VSC	Vehicle Skid Control
VTEC	Valve Timing Electronic Control
"W"	Windsor Eng
Warn	Warning
WB	Wheel Base
Whlhse	Wheelhouse
W/S	Windshield
W/Strip	Weatherstrip
X-Cool	Extra Cooling
Xmbr	Crossmember

-Misc.-

4x2, 2 WD	2-Wheel Drive
4x4, 4 WD	4-Wheel Drive
2 WS	2-Wheel Steering
4 WS	4-Wheel Steering

Figure 3-20 **Study these common abbreviations found in service manuals and computer-based service materials.**

parts, such as the brakes, suspension, steering, drive axles, and other similar parts, during major structural repairs. Figure 3–29 is a mechanical illustration showing how to service the front axle hub assembly.

Wiring diagrams are drawings that show the location of electrical components. They show how wires are routed to electronic parts (motors, solenoids, or sensors), fuses or circuit breakers, electronic control units (ECUs), and other components. During a major collision, the vehicle wires and electronic parts can be cut and damaged. Air bags can deploy and require replacement (Figure 3–30). This makes it very important that you know how to use wiring diagrams.

Wiring diagram CDs give you access to the most accurate, state-of-the-art wiring diagrams available. This type of CD often includes schematics, charts, electrical

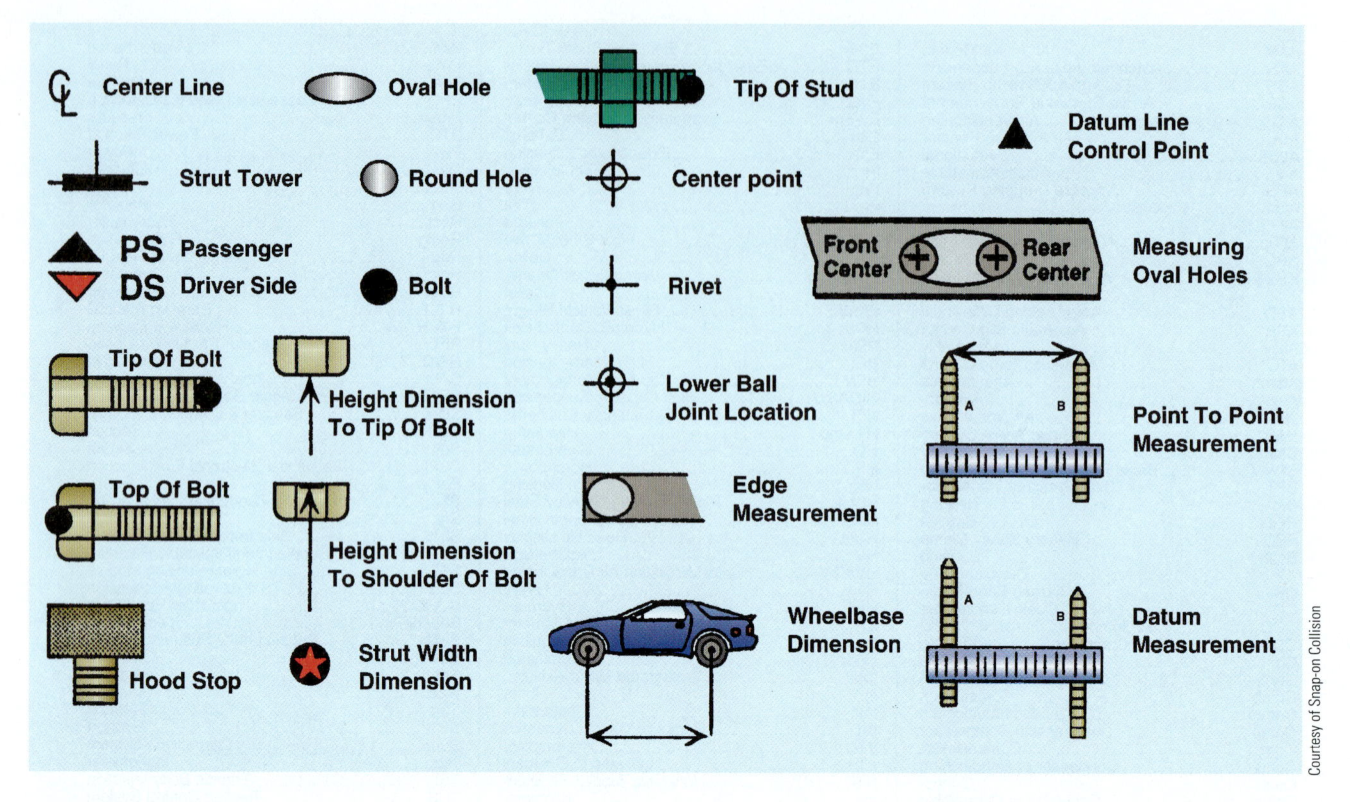

Figure 3-21 Various symbols are used in service information. A key at the front of the manual or computer contents page will explain each symbol's meaning.

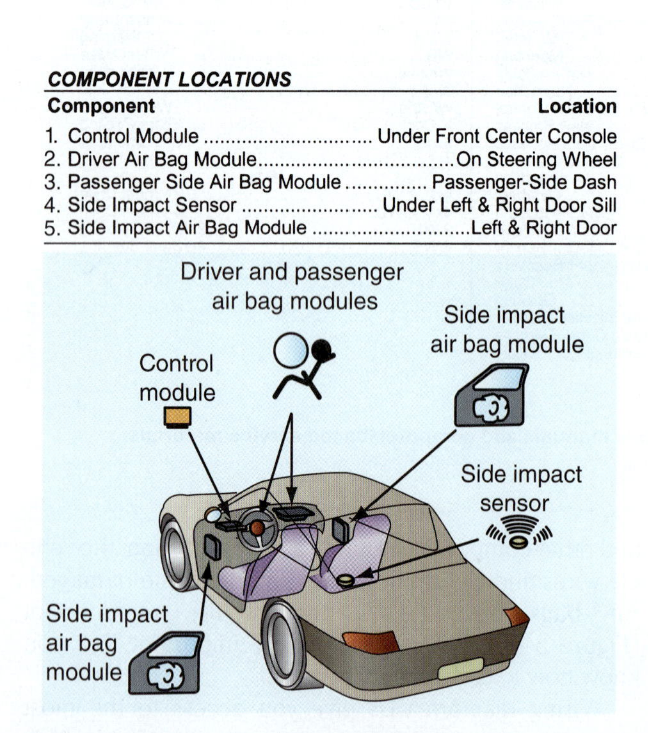

COMPONENT LOCATIONS

Component	Location
1. Control Module	Under Front Center Console
2. Driver Air Bag Module	On Steering Wheel
3. Passenger Side Air Bag Module	Passenger-Side Dash
4. Side Impact Sensor	Under Left & Right Door Sill
5. Side Impact Air Bag Module	Left & Right Door

Figure 3-22 Here are symbols used in one air bag service guide.

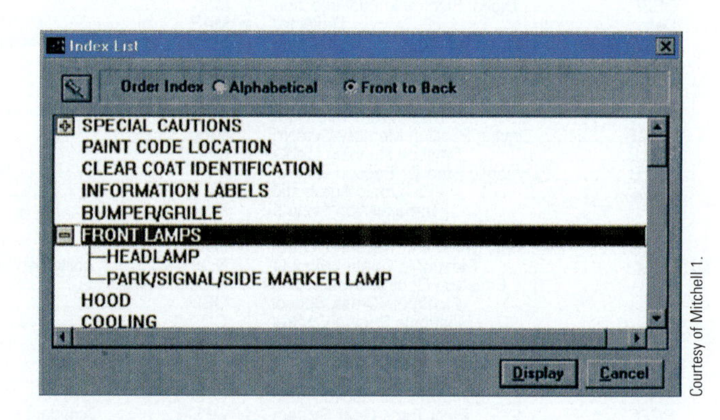

Figure 3-23 By clicking your mouse button on a contents list of parts, your computer will quickly go to that repair information.

connector drawings (Figure 3–31), and other service information for all major electrical systems, including:

- Starting/charging systems
- Air-conditioning and cooling fans
- Antilock brake system
- Cruise control
- Power window and door locks
- Interior and exterior lights
- Power and ground distribution
- Computer data lines
- Other circuits

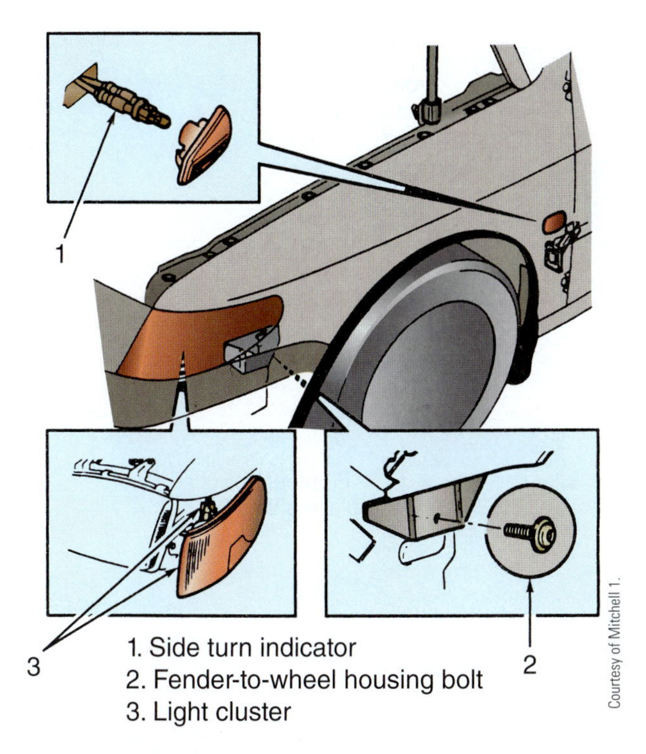

Figure 3–24 **This service illustration shows how the lights are held on a front fender. Repair-specific illustrations can save time.**

Flow diagrams are similar to wiring diagrams but show how various fluids or air move through a *hydraulic* (oil-filled) or *pneumatic* (air-filled) circuit. Various vehicle systems (power steering, air conditioning, automatic transmission, or transaxle) use hydraulic circuits for operation. Shop air tools use a pneumatic system or a high-pressure air supply system for power.

Part location illustrations help you to quickly locate hard-to-find components. They will often have large arrows that pinpoint part locations, and they sometimes show an enlarged second view of the part detail (Figure 3–32).

Figure 3–33 shows the computer screen with service instructions and a service illustration for an air bag assembly.

As mentioned earlier, **vehicle dimension illustrations** give measurements across specific points on a vehicle to help determine the extent of damage. Look at Figure 3–34. Note how the dimension data gives measurements for a door opening. You would measure across the door opening on the vehicle as shown. If any measurements are not correct, you could determine how the impact affected the door opening. Frame straightening and/or portable hydraulic equipment could then be used to push or pull the opening back into proper alignment.

A vehicle dimension manual gives body and frame measurements from undamaged vehicles. Dimensions are provided for every make and model of car and truck. These known good dimensions can be compared to actual measurements taken off a wrecked car or truck. This will let you know how badly the vehicle is damaged and what must be done to repair it (Figure 3–35).

Panel welding illustrations give the factory-recommended weld count, weld type, and weld locations for replacing structural panels. They show manufacturer-recommended details for removing and replacing a major body panel. They denote the types of welds recommended. These illustrations also show the number and location of all the welds required for the specific vehicle (Figure 3–36).

Trim illustrations show how molding and trim pieces attach to the vehicle. They are often shown for the passenger compartment of various vehicles (Figure 3–37).

3.4 COLLISION REPAIR MEASUREMENTS

Remember that modern vehicles are precision-manufactured using the latest technology. Most vehicle assembly at the factory uses automated conveyors and robotics, which results in motor vehicles that are built to closer tolerances than ever before. This requires auto body repair technicians who can measure accurately and repair vehicles precisely. In fact, accurate measurement is one of the most important aspects of today's collision repair industry.

English and Metric Systems

The *English measuring system* was first developed using human body parts as the basis for measurements. For example, the length of the human arm was used to standardize the yard.

The English system uses fractions and decimals for number values. Fractions are acceptable when precision is not critical. They divide the inch into 32nds, 16ths, and larger parts of an inch. Decimals are used when high precision is important. Decimals can be used to divide an inch into tenths, hundredths, thousandths, or even smaller divisions.

The *metric measuring system*, also called the *scientific international (SI) system*, uses a power of 10 as its base. It is a more precise and simple system than the conventional U.S. system. This is because multiples of metric units are related to each other by factors of 10.

Even after graduating from school, you should always continue your training in collision repair. Vehicle designs and repair methods constantly change. Only those who continue to learn more about the profession will excel. Smart technicians constantly try to learn more about their jobs (Figure 3–38).

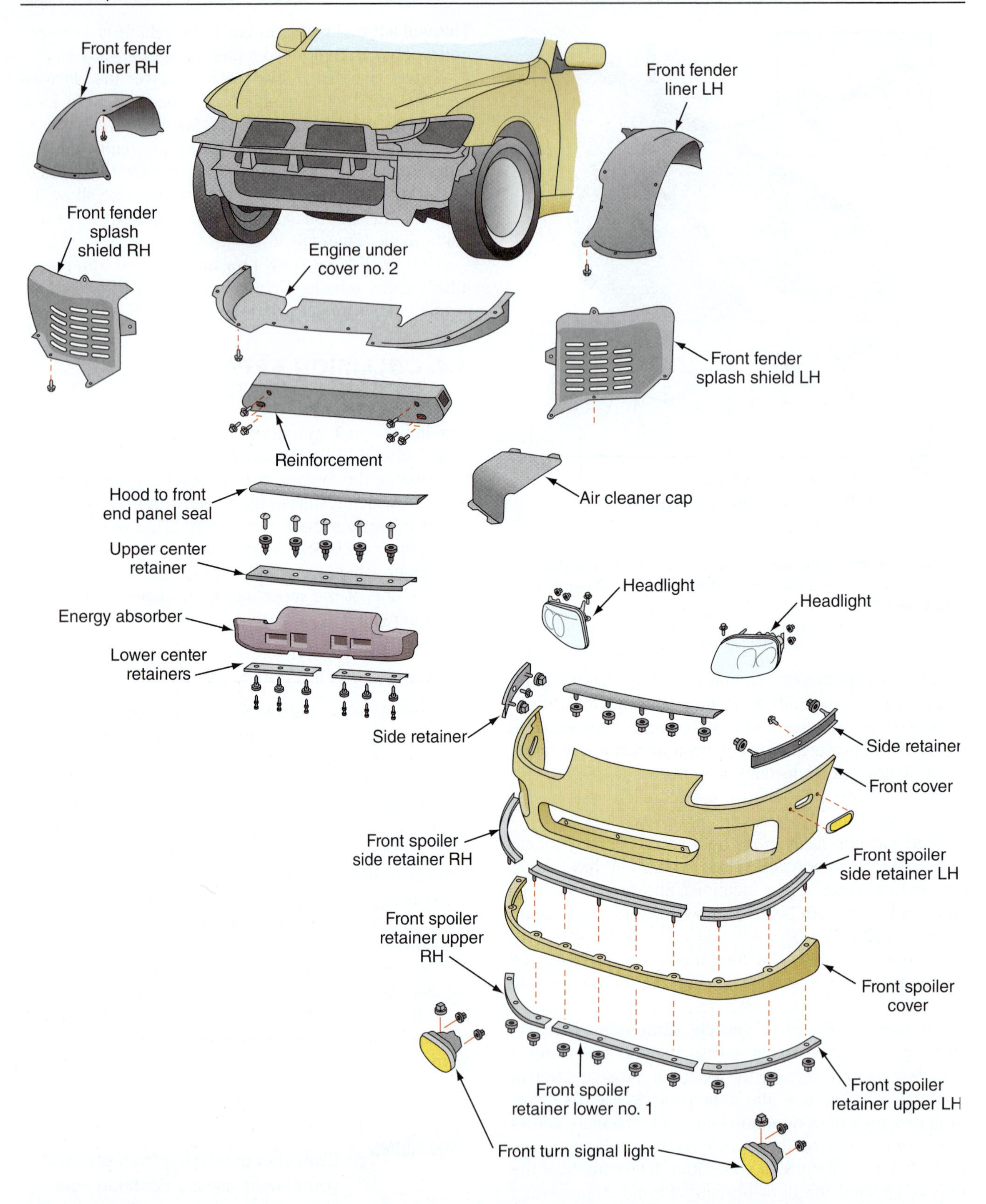

Figure 3-25 This service manual exploded view would be helpful when disassembling and reassembling a car that has been in a front-end or head-on collision. It shows all of the fasteners and which parts must be removed together.

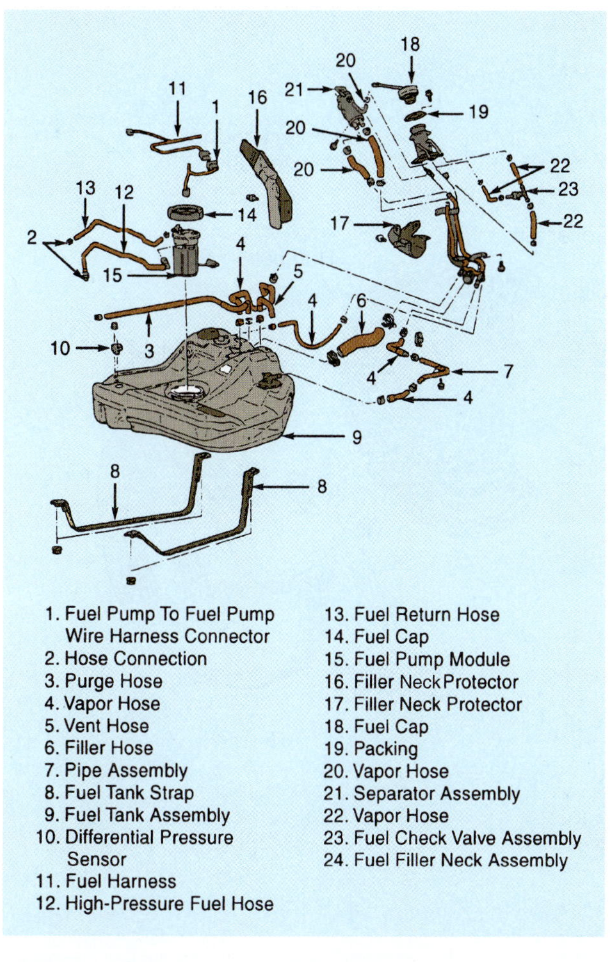

Figure 3-26 Vehicle fuel tanks must often be removed during auto body repairs. This exploded view shows how straps hold the tank to the unibody.

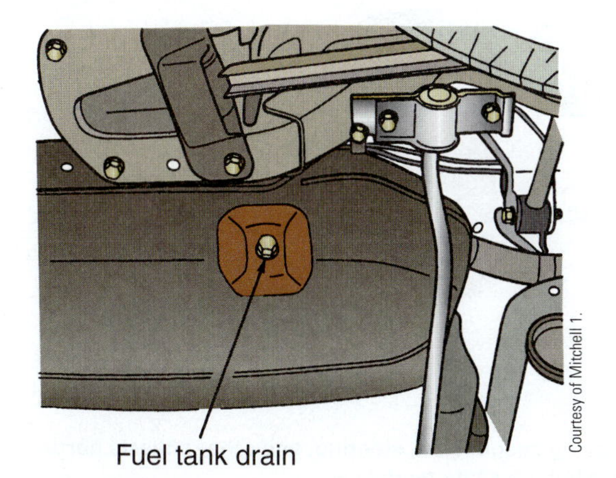

Figure 3-27 This illustration shows that the specific fuel tank has a drain. Fuel tanks are very heavy when full and should be drained before removal to prevent them from being crushed or starting a major fire.

Every metric unit can be multiplied or divided by 10 to get larger units (multiples) or smaller units (submultiples). There is less chance of math errors with the metric system.

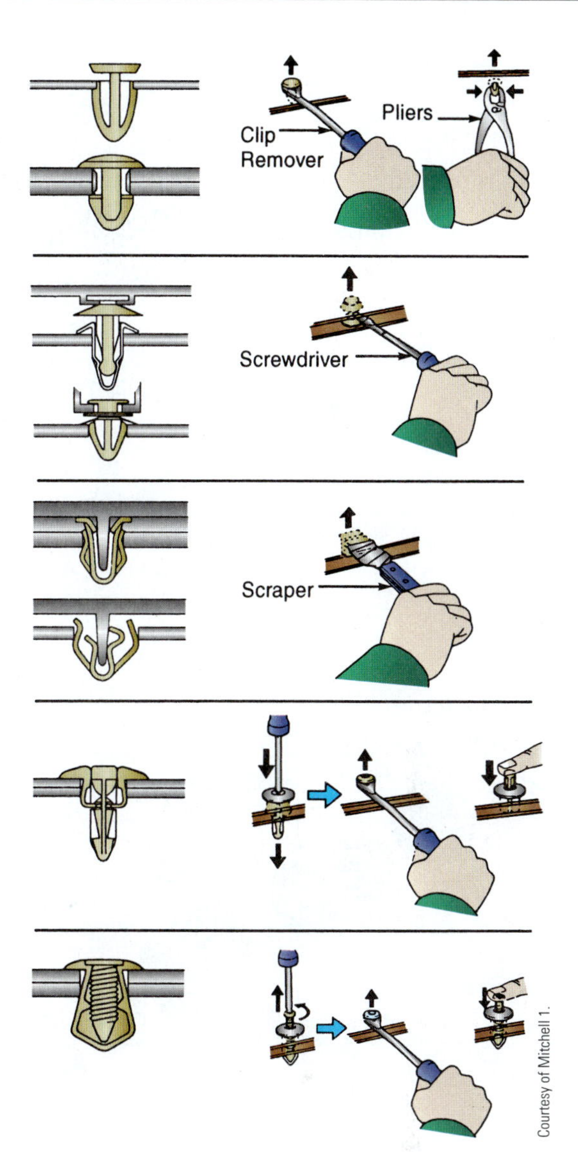

Figure 3-28 This service illustration shows all of the types of clips used on one make of vehicle. It also gives tips for removing clips without damaging them or the surrounding panel.

The English measuring system is primarily used in the United States, whereas other countries use the metric system. Service specifications are usually given in both English and metric values, which makes it important for you to understand both systems. The United States passed the Metric Conversion Act, which committed the country to a voluntary and gradual transition from its customary system to the metric system. To date, only a few industries—including the automotive industry—have moved toward adopting the metric system.

Conversion Charts

Conversion charts are handy for changing from one measuring system to another or from one value to another.

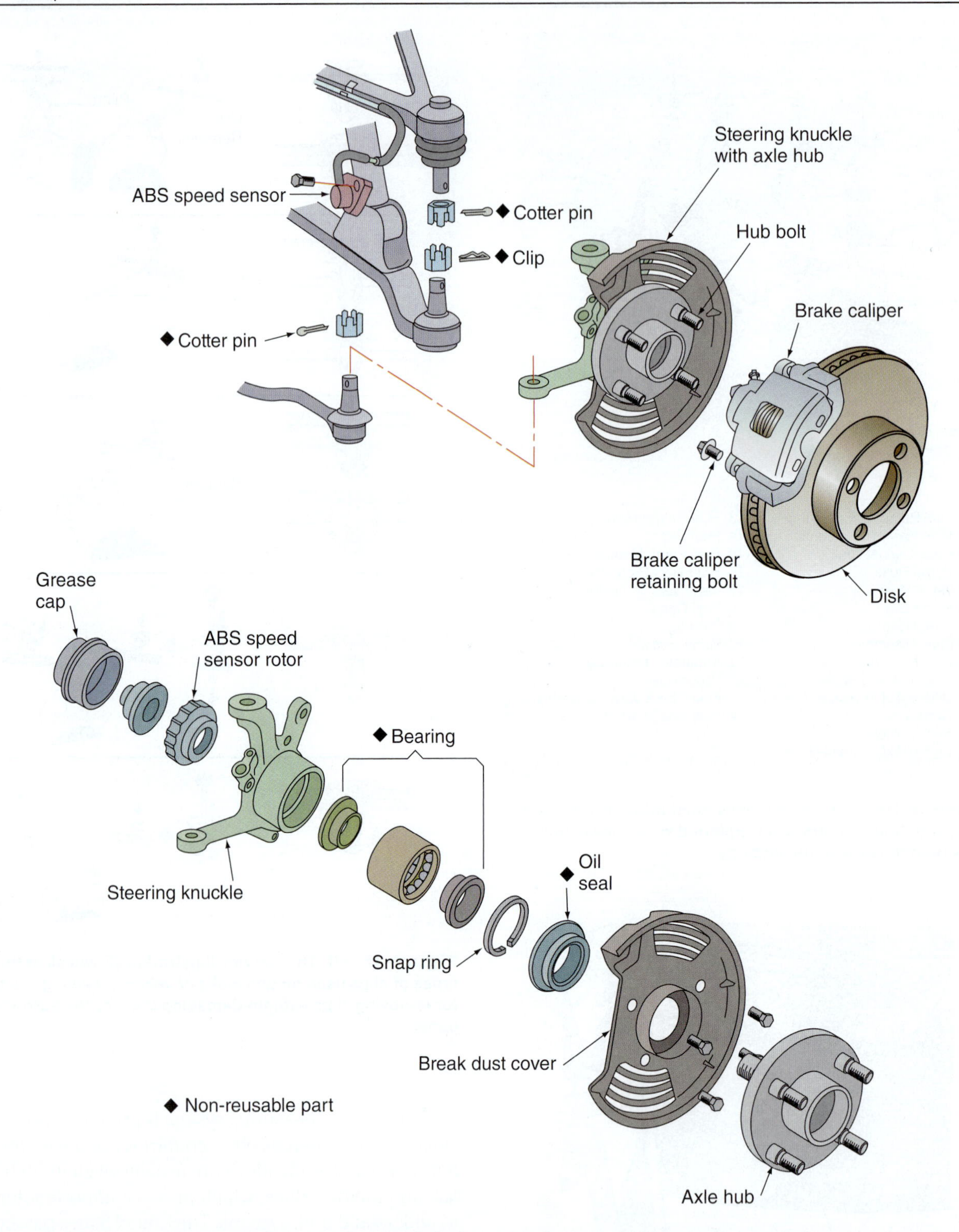

Figure 3-29 **Mechanical illustrations are often needed when removing suspension, steering, and other chassis parts. This illustration shows how to disassemble and reassemble a front axle hub on a late model car.**

An *English–metric conversion chart* allows you to change numbers from one system to the other. It gives multipliers that can be used to convert from English to metric or from metric to English values. For example, if a value is given in metric and you want to measure with a conventionally marked measuring tool, you could use this conversion chart (Figure 3–39).

A *decimal conversion chart* allows you to quickly change from fractions to decimals to millimeters. This chart is used for various tasks, such as selecting drill bits. One is shown in Figure 3–40.

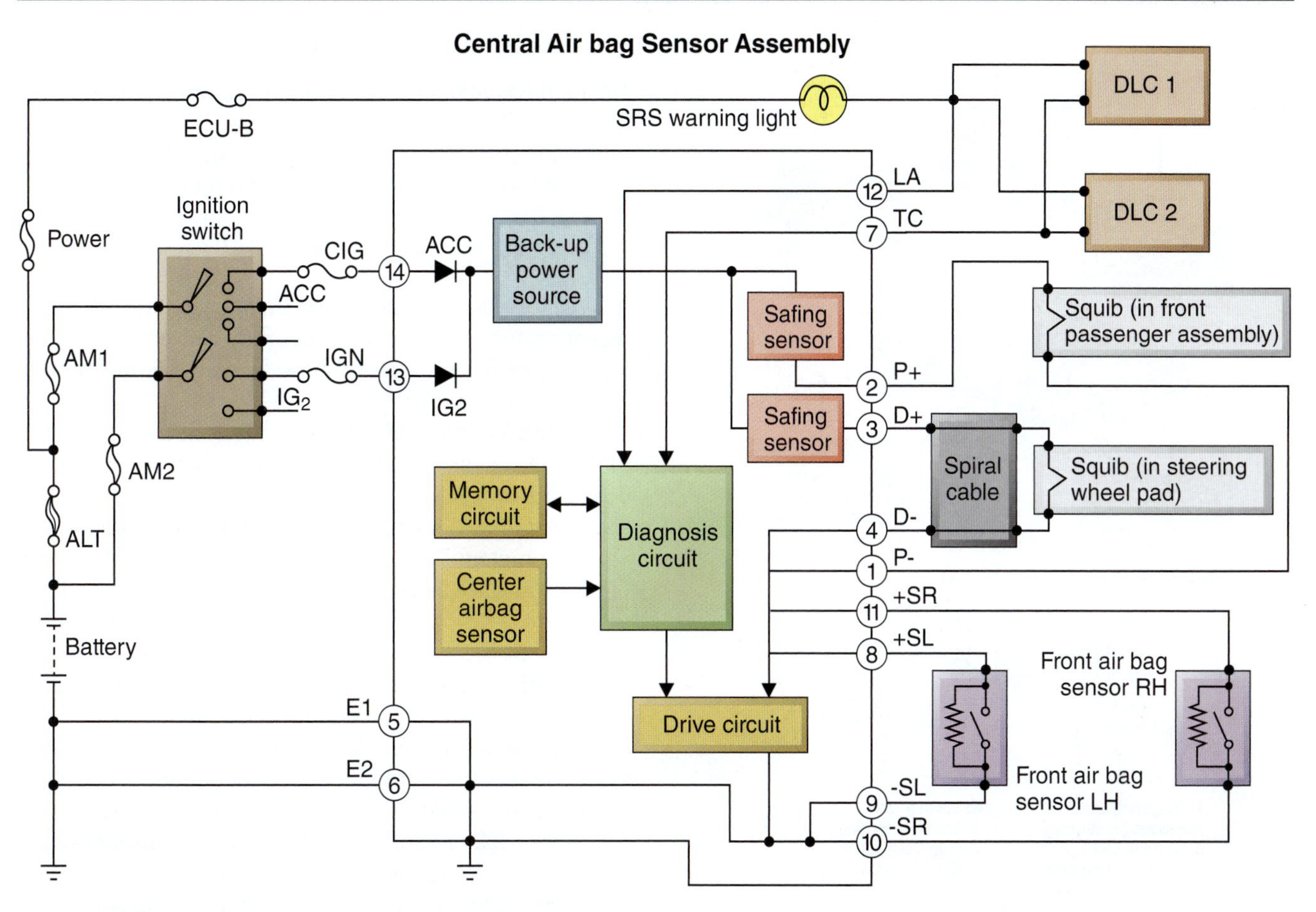

Figure 3-30 Wiring diagrams give wire color codes, wiring routing, and other information for electrical repairs.

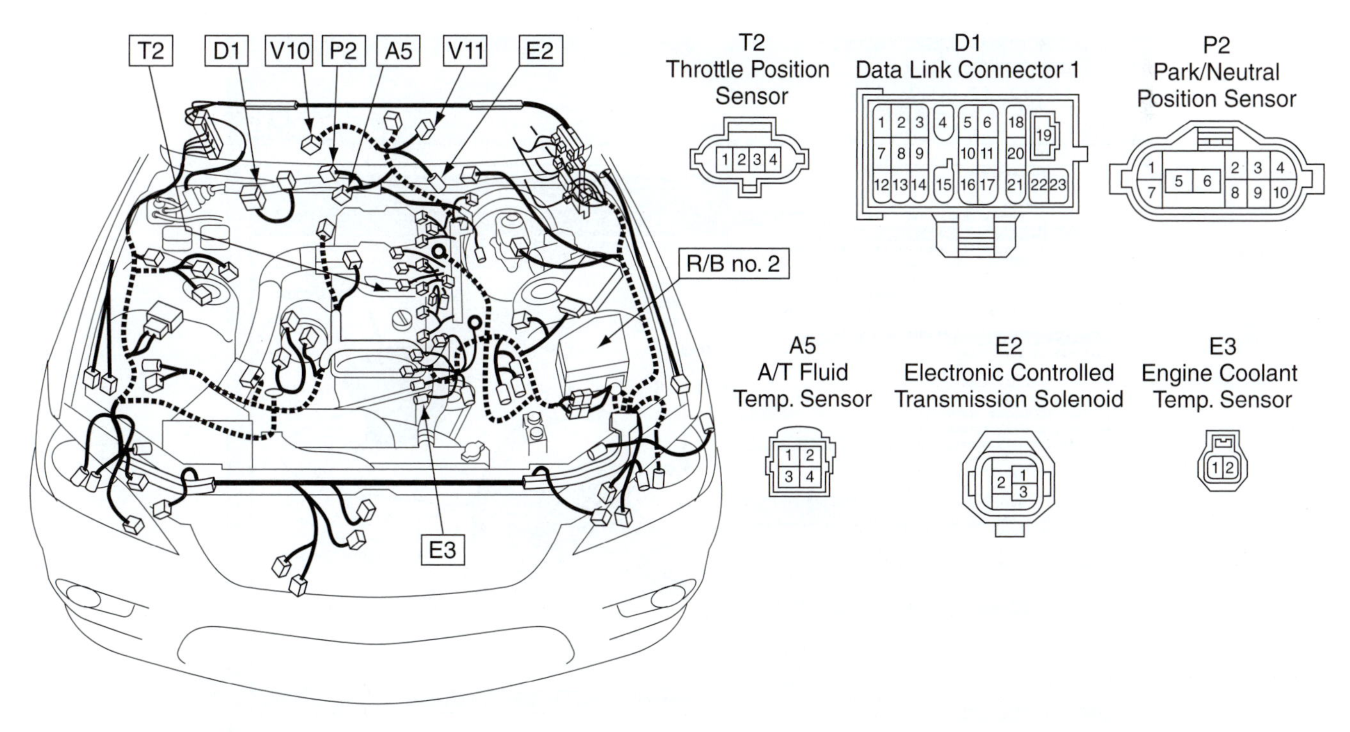

Figure 3-31 This is a connector location chart for one make and model of car. You must often disconnect and replace wiring harnesses damaged in collisions.

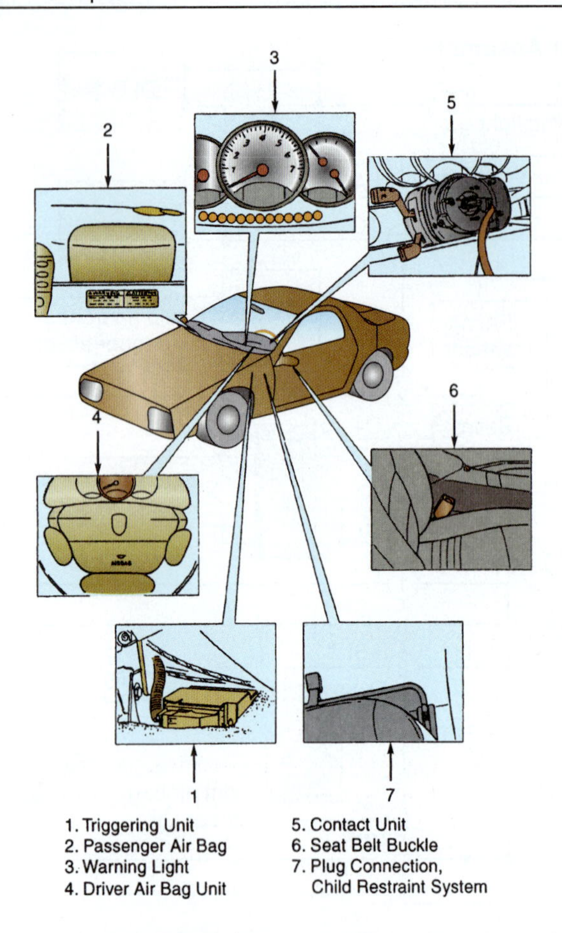

Figure 3-32 **This service manual illustration shows the locations for major air bag components.**

Linear Measurements

Linear measurements are straight-line measurements of distance. They are commonly used when evaluating major structural damage after a collision. There are many types of tools used for linear measurements.

Rules or Scales

A *scale*, or *ruler* (or rule) is the most basic tool for linear measurement. It has an accuracy of approximately 1⁄64 inch, or 0.5 mm. Study how to read a conventional ruler in Figure 3–41.

An *English ruler* often has markings in fractions of an inch (1⁄2, 1⁄4, 1⁄3, or 1⁄16) or in decimal parts of an inch (0.10, 0.20, 0.30, or 0.40).

A *metric ruler* or *meter stick* is marked in millimeters and centimeters. The numbered lines usually equal 10 mm (1 cm). In Figure 3–42, note how a metric ruler is read.

Figure 3–43 compares fractional, metric, and decimal inch scales.

Parallax error results when you read a rule or scale from an angle instead of looking at it straight down. Viewing at an angle makes you read the wrong line on the scale. Always look straight down when reading a rule.

A *pocket scale*, or *pocket rule*, is small (typically 6 inches or 152 mm long). It will clip into your shirt pocket and is handy for numerous small measurements. A yardstick or meter stick may be used for larger linear measurements.

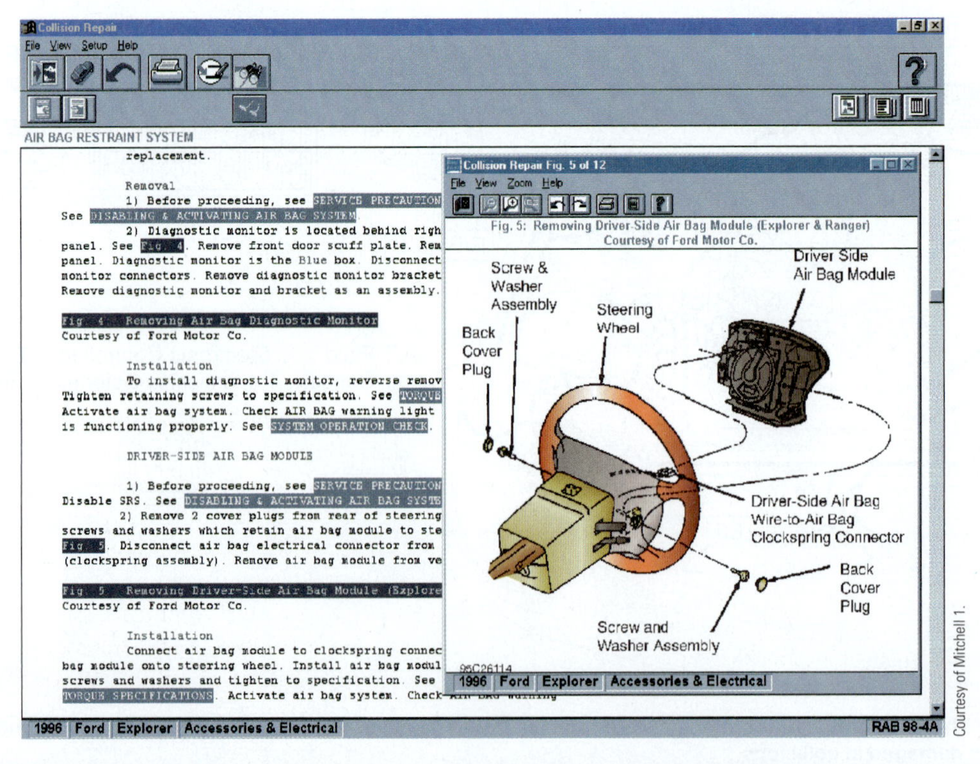

Figure 3-33 **This computer screen shows instructions and technical illustrations for the replacement of an air bag. Air bags often deploy during a frontal impact and must be replaced by the auto body technician.**

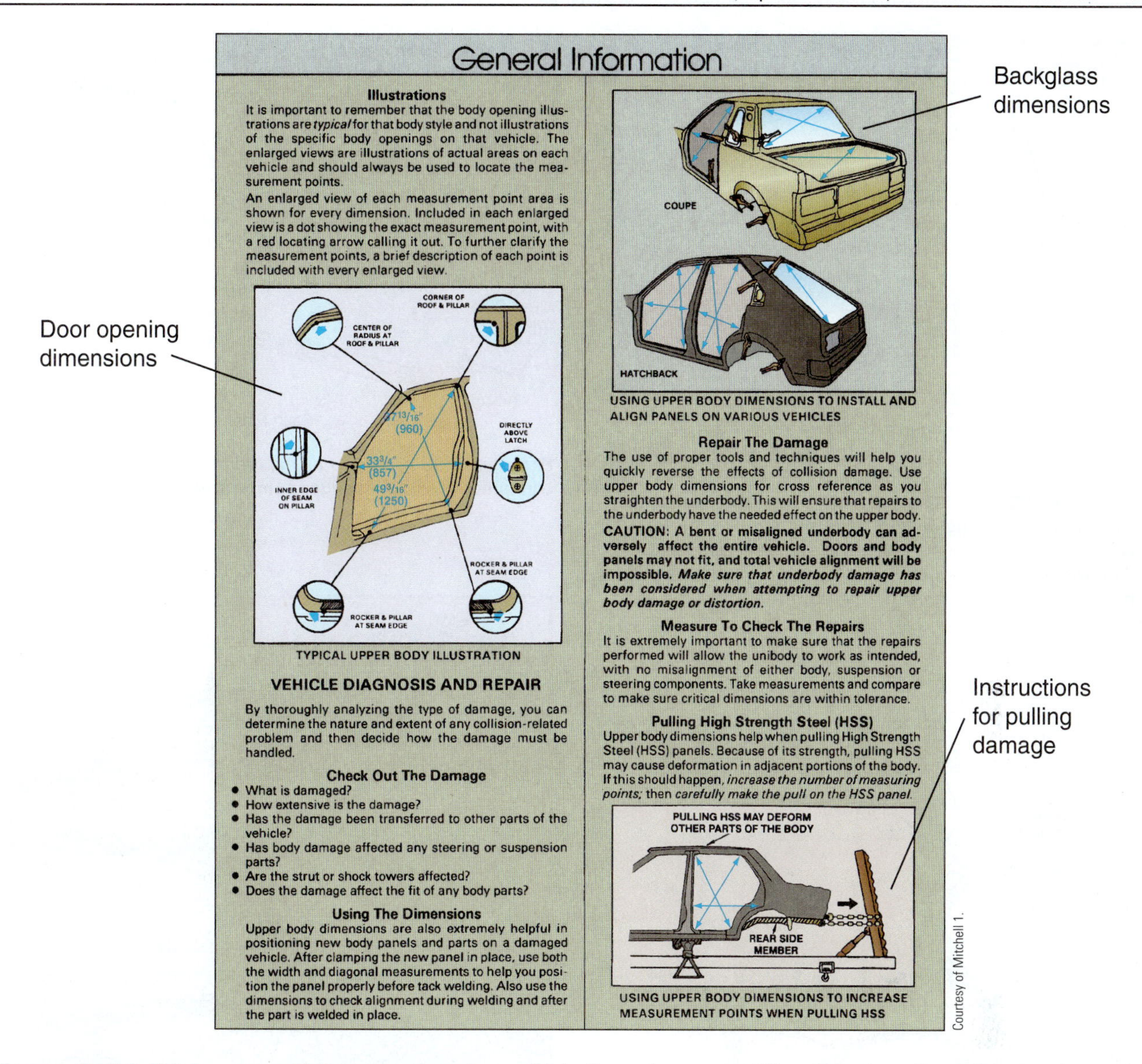

General Information

Illustrations

It is important to remember that the body opening illustrations are *typical* for that body style and not illustrations of the specific body openings on that vehicle. The enlarged views are illustrations of actual areas on each vehicle and should always be used to locate the measurement points.

An enlarged view of each measurement point area is shown for every dimension. Included in each enlarged view is a dot showing the exact measurement point, with a red locating arrow calling it out. To further clarify the measurement points, a brief description of each point is included with every enlarged view.

TYPICAL UPPER BODY ILLUSTRATION

VEHICLE DIAGNOSIS AND REPAIR

By thoroughly analyzing the type of damage, you can determine the nature and extent of any collision-related problem and then decide how the damage must be handled.

Check Out The Damage

- What is damaged?
- How extensive is the damage?
- Has the damage been transferred to other parts of the vehicle?
- Has body damage affected any steering or suspension parts?
- Are the strut or shock towers affected?
- Does the damage affect the fit of any body parts?

Using The Dimensions

Upper body dimensions are also extremely helpful in positioning new body panels and parts on a damaged vehicle. After clamping the new panel in place, use both the width and diagonal measurements to help you position the panel properly before tack welding. Also use the dimensions to check alignment during welding and after the part is welded in place.

USING UPPER BODY DIMENSIONS TO INSTALL AND ALIGN PANELS ON VARIOUS VEHICLES

Repair The Damage

The use of proper tools and techniques will help you quickly reverse the effects of collision damage. Use upper body dimensions for cross reference as you straighten the underbody. This will ensure that repairs to the underbody have the needed effect on the upper body.

CAUTION: A bent or misaligned underbody can adversely affect the entire vehicle. Doors and body panels may not fit, and total vehicle alignment will be impossible. *Make sure that underbody damage has been considered when attempting to repair upper body damage or distortion.*

Measure To Check The Repairs

It is extremely important to make sure that the repairs performed will allow the unibody to work as intended, with no misalignment of either body, suspension or steering components. Take measurements and compare to make sure critical dimensions are within tolerance.

Pulling High Strength Steel (HSS)

Upper body dimensions help when pulling High Strength Steel (HSS) panels. Because of its strength, pulling HSS may cause deformation in adjacent portions of the body. If this should happen, *increase the number of measuring points;* then *carefully make the pull on the HSS panel.*

USING UPPER BODY DIMENSIONS TO INCREASE MEASUREMENT POINTS WHEN PULLING HSS

Figure 3-34 **This is a general information page for a vehicle dimensions manual. It explains how dimensions can be used to determine the extent and direction of damage to a vehicle. Hydraulic equipment can then be used to straighten damage to measurements that are within specifications.**

A *tape rule*, or *tape measure*, will extend out to make very long measurements. A tape measure is commonly used to make large distance measurements during initial body damage evaluation. Because many body measurements are made at holes in the body, the tip of the tape rule must be small or ground to a point. This will prevent measurement errors.

Tram Gauge

A **tram gauge** is a special body dimension measuring tool (Figure 3–44). It is usually a lightweight frame with pointers. The pointers can be aligned with body dimension reference points to determine the direction and amount of body misalignment damage.

NOTE *Tram gauges are detailed in Chapter 17 on measuring systems.*

Dividers and Calipers

Dividers have straight, sharp tips for taking measurements or marking parts for cutting. In auto body repair, dividers are sometimes used for layout or marking cut lines. They will scribe circles and lines on sheet metal and plastic. They will also transfer and make surface measurements. Dividers are sometimes used when fabricating repair pieces for rust repair (Figure 3–45).

Outside calipers will make rough external measurements where 1/64 inch (0.40 mm) is accurate enough. The calipers are placed over the outside of parts and adjusted to just touch the parts. When placed next to a rule, part size can be determined.

Inside calipers are designed for measuring the inside of parts and are used in the same way as outside calipers. They are also accurate to about 1/64 inch (0.40 mm).

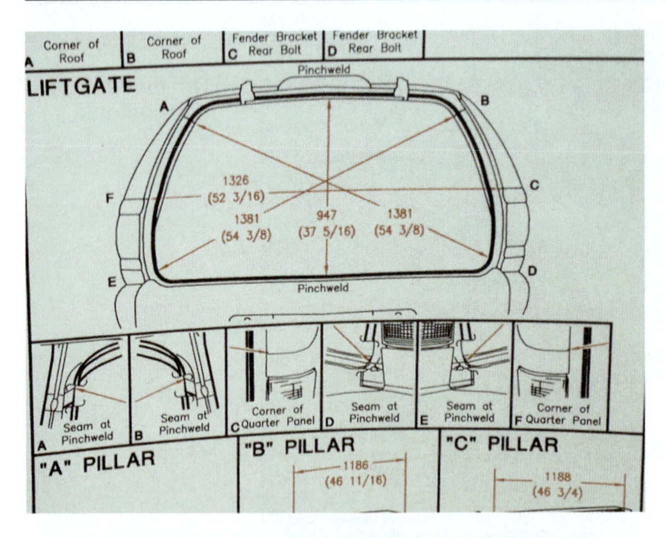

Figure 3-35 Here is an example of the dimensions given for the lift gate opening on one make and model van. Note how dimensions or measurements are given in both English and metric units. Instructions are to specific points (rivets, bolts, or holes) on the vehicle. Details are given at the bottom.

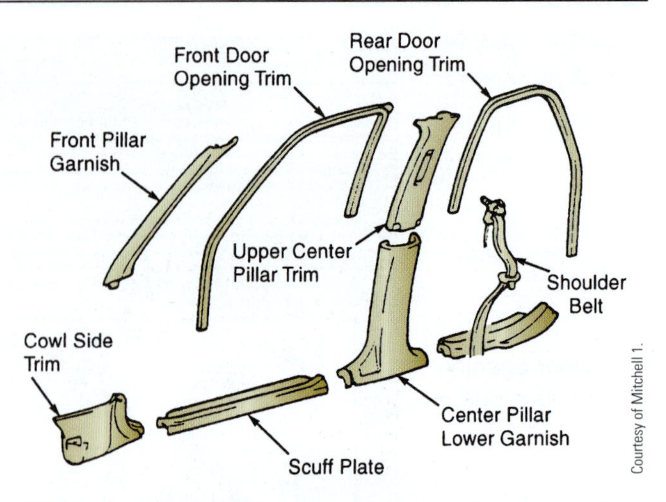

Figure 3-37 Here is an exploded view showing interior trim pieces for a passenger compartment.

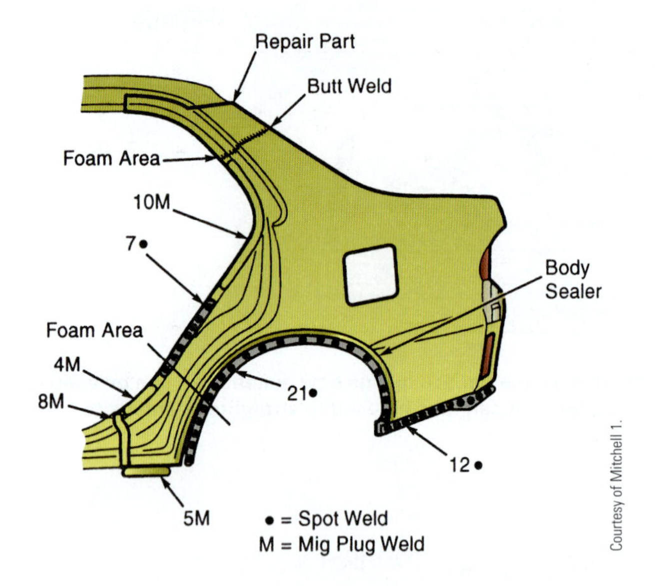

Figure 3-36 This service illustration shows that the factory recommends their quarter panel be welded onto the unibody structure. Note how spot welds and plug welds are both recommended. A letter M is used to denote MIG plug welds, and a dot is used to denote resistance spot welds. Also note how foam is recommended inside panels at specific locations.

Figure 3-38 Smart technicians know that they never know enough. Professionals are always trying to learn more about their professions. Here a technician is studying new wheel alignment methods.

Hermaphrodite calipers have one straight tip and one curved tip. They can measure cylindrical shapes easily.

Sliding calipers will measure inside, outside, and depth dimensions with high accuracy. Most sliding calipers measure to an accuracy of 0.001 inch (0.025 mm). This is a very convenient tool, especially digital readout electronic calipers, when high accuracy is necessary (Figure 3–46).

Micrometers

Micrometers are sometimes used to measure mechanical parts when high precision is important. For example, you might use one to measure the thickness of a brake system rotor. If worn too thin, this would tell you to replace the rotor. Often called a "mike," a micrometer can easily measure to a thousandth of an inch (0.001 in.) or a hundredth of a millimeter (0.01 mm).

CUSTOMARY	CONVERSION	METRIC
Multiply	by	to get equivalent number of:
LENGTH		
Inch	25.4	millimeters (mm)
Foot	0.3048	meters (m)
Yard	0.9144	meters
Mile	1.609	kilometers (km)
AREA		
Inch^2	645.2	millimeters^2 (mm^2)
	6.45	centimeters^2 (cm^2)
Foot^2	0.0929	meters^2 (m^2)
Yard^2	0.8361	meters^2
VOLUME		
Inch^3	16.387.	mm^3
	16.387	cm^3
	0.0164	liter (l)
Quart	0.9464	liters
Gallon	3.7854	liters
Yards^3	0.7646	meters (m^3)
MASS		
Pound	0.4536	kilograms (kg)
Ton	907.18	kilograms (kg)
Ton	0.907	tonne (t)
FORCE		
Kilogram	9.807	newtons (N)
Ounce	0.2780	newtons
Pound	4.448	newtons
TEMPERATURE		
Degree Fahrenheit	(T°F–32) ÷ 1.8	Degree Celsius

°F	–40F	0	32	60	98.6	140	180	212
°C	–40C	–20	0	20	37	60	80	100

CUSTOMARY	CONVERSION	METRIC
Multiply	by	to get equivalent number of:
ACCELERATION		
Foot/sec^2	0.3048	meters/sec^2 (ms/s^2)
Foot/sec^2	0.0254	meters/sec^2
TORQUE		
Pound-inch	0.11298	newton-meters (N-m)
Pound-foot	1.3558	newton-meters
POWER		
Horsepower	0.746	kilowatts (kW)
PRESSURE OR STRESS		
Inches of water	0.2491	kilopascals (kPa)
Pounds/sq. in.	6.895	kilopascals
ENERGY OR WORK		
Btu	1055.05	joules (J)
Foot-pounds	1.3558	joules
Kilowatt-hours	3,600,000 or 3.6 x 10^6	joules (J = one W's)
LIGHT		
Foot candle	1.0764	lumens/meter^2 (lm/m^2)
FUEL PERFORMANCE		
Miles/gal	0.4251	kilometers/liters (km/l)
Gas/mile	2.3527	liter/kilometer (l/km)
VELOCITY		
Miles/hour	1.6093	kilometer/hr. (km/h)

Figure 3–39 This conversion chart can be used to change back and forth from customary English values to metric values. What are the metric equivalents of the following customary values: 1 inch, 1 yard, 1 pound?

To use a micrometer, you hold the frame in the palm of your hand. Rotating the thimble with your thumb and finger, you adjust the tool down until it drags lightly on the part, but do *not* tighten it. You then read the micrometer.

Dial Indicators

A *dial indicator* is sometimes used to measure part movement and out-of-round in thousandths of an inch (or hundredths of a millimeter). For example, the indicator might be used to check for wheel damage by mounting it against the wheel or tire. When the tire is rotated, the dial indicator will show how much the wheel is bent. This would let you know that the wheel should be replaced (Figure 3–47).

To use a dial indicator, mount the tool base so it will not move. A magnetic base will stick to metal parts or a metal plate placed on the floor. Position the tool arm and indicator stem against the part. Turn the outside of the

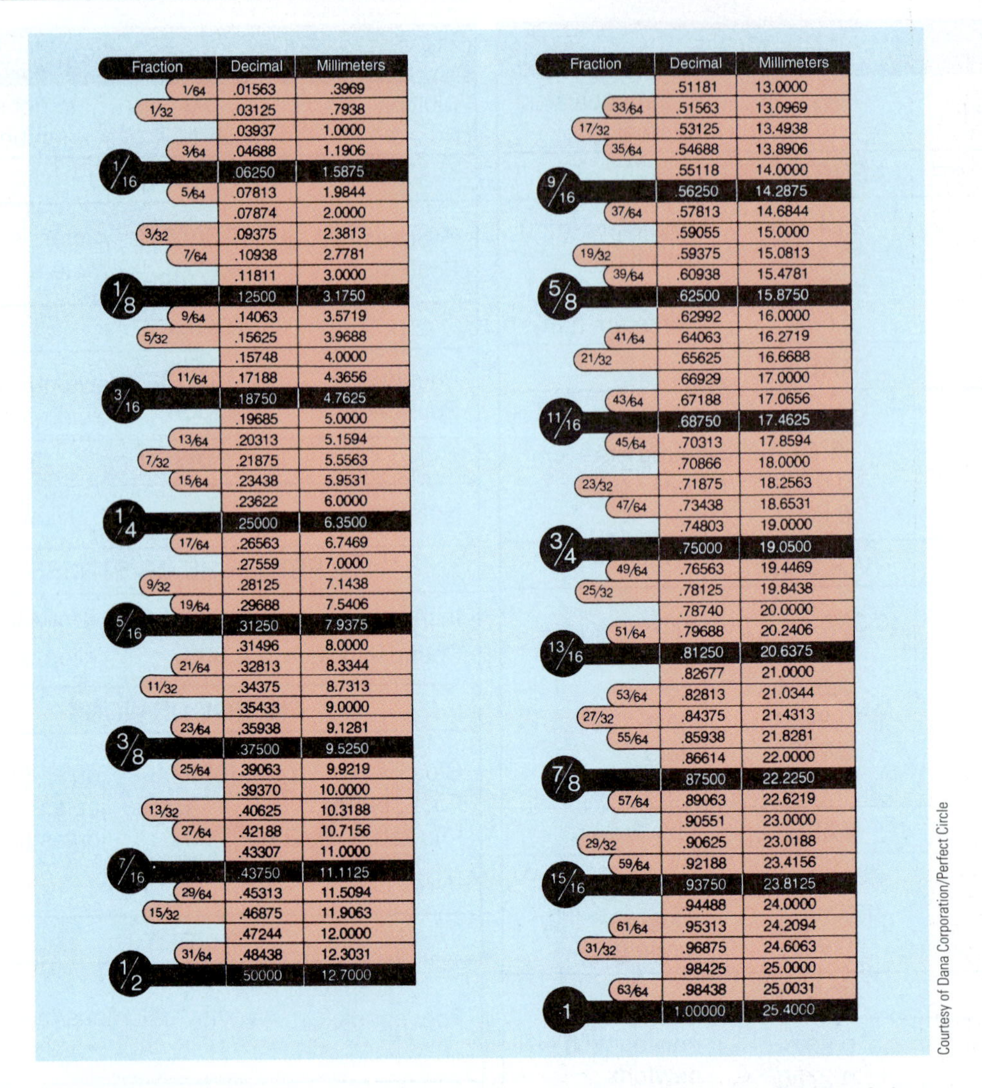

Fraction	Decimal	Millimeters
1/64	.01563	.3969
1/32	.03125	.7938
	.03937	1.0000
3/64	.04688	1.1906
1/16	.06250	1.5875
5/64	.07813	1.9844
	.07874	2.0000
3/32	.09375	2.3813
7/64	.10938	2.7781
	.11811	3.0000
1/8	.12500	3.1750
9/64	.14063	3.5719
5/32	.15625	3.9688
	.15748	4.0000
11/64	.17188	4.3656
3/16	.18750	4.7625
	.19685	5.0000
13/64	.20313	5.1594
7/32	.21875	5.5563
15/64	.23438	5.9531
	.23622	6.0000
1/4	.25000	6.3500
17/64	.26563	6.7469
	.27559	7.0000
9/32	.28125	7.1438
19/64	.29688	7.5406
5/16	.31250	7.9375
	.31496	8.0000
21/64	.32813	8.3344
11/32	.34375	8.7313
	.35433	9.0000
23/64	.35938	9.1281
3/8	.37500	9.5250
25/64	.39063	9.9219
	.39370	10.0000
13/32	.40625	10.3188
27/64	.42188	10.7156
	.43307	11.0000
7/16	.43750	11.1125
29/64	.45313	11.5094
15/32	.46875	11.9063
	.47244	12.0000
31/64	.48438	12.3031
1/2	.50000	12.7000

Fraction	Decimal	Millimeters
	.51181	13.0000
33/64	.51563	13.0969
17/32	.53125	13.4938
35/64	.54688	13.8906
	.55118	14.0000
9/16	.56250	14.2875
37/64	.57813	14.6844
	.59055	15.0000
19/32	.59375	15.0813
39/64	.60938	15.4781
5/8	.62500	15.8750
	.62992	16.0000
41/64	.64063	16.2719
21/32	.65625	16.6688
	.66929	17.0000
43/64	.67188	17.0656
11/16	.68750	17.4625
45/64	.70313	17.8594
	.70866	18.0000
23/32	.71875	18.2563
47/64	.73438	18.6531
	.74803	19.0000
3/4	.75000	19.0500
49/64	.76563	19.4469
25/32	.78125	19.8438
	.78740	20.0000
51/64	.79688	20.2406
13/16	.81250	20.6375
	.82677	21.0000
53/64	.82813	21.0344
27/32	.84375	21.4313
55/64	.85938	21.8281
	.86614	22.0000
7/8	.87500	22.2250
57/64	.89063	22.6219
	.90551	23.0000
29/32	.90625	23.0188
59/64	.92188	23.4156
15/16	.93750	23.8125
	.94488	24.0000
61/64	.95313	24.2094
31/32	.96875	24.6063
	.98425	25.0000
63/64	.98438	25.0031
1	1.00000	25.4000

Courtesy of Dana Corporation/Perfect Circle

Figure 3-40 A decimal conversion chart is handy if you need to change fractions to decimals and vice versa. What are the decimal and metric equivalent for 1/16; inch, 1/8 inch, 1/2 inch, and 1 inch?

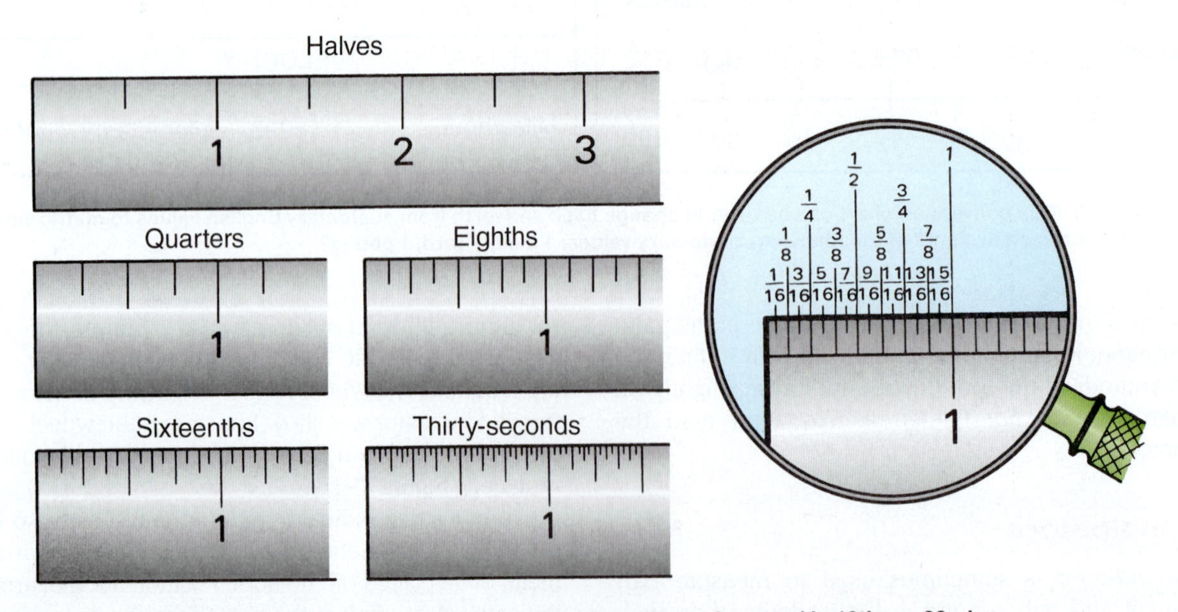

Figure 3-41 If needed, review the basic divisions of a ruler. Most shop rulers read in 16ths or 32nds.

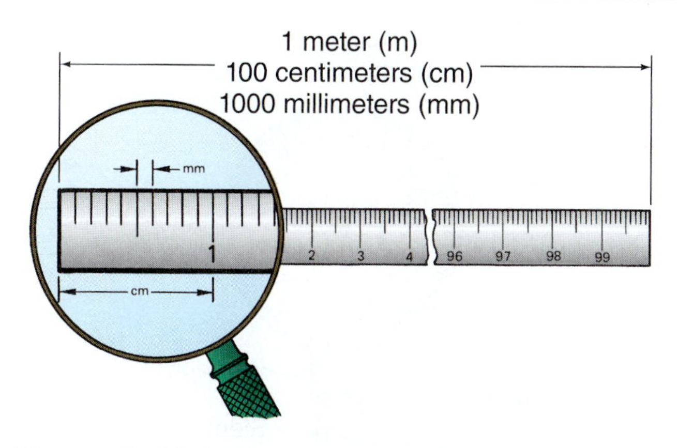

Figure 3-42 Small divisions of a metric ruler equal millimeters; numbered lines are centimeters.

Figure 3-43 Tape measure, yard or meter stick, and small pocket scale are commonly used to do measurements in collision repair.

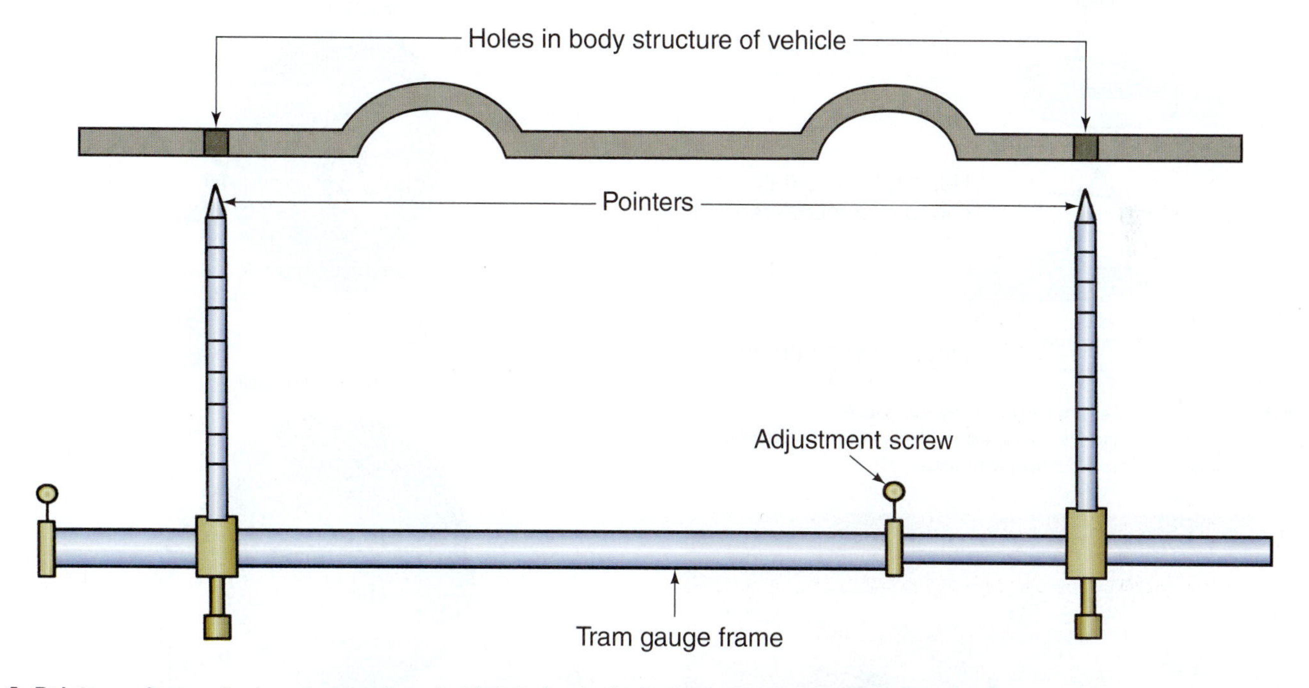

A Points can be touched on parts or inserted into holes in the body to make dimension measurements.

B The technician is using a tram gauge to check the amount of damage on the front of a pickup truck.

Figure 3-44 A tram gauge is handy when you have to measure across points on a vehicle to find major damage. Adjust the gauge to the correct dimension from the manual and then compare it to actual points on the damaged vehicle.

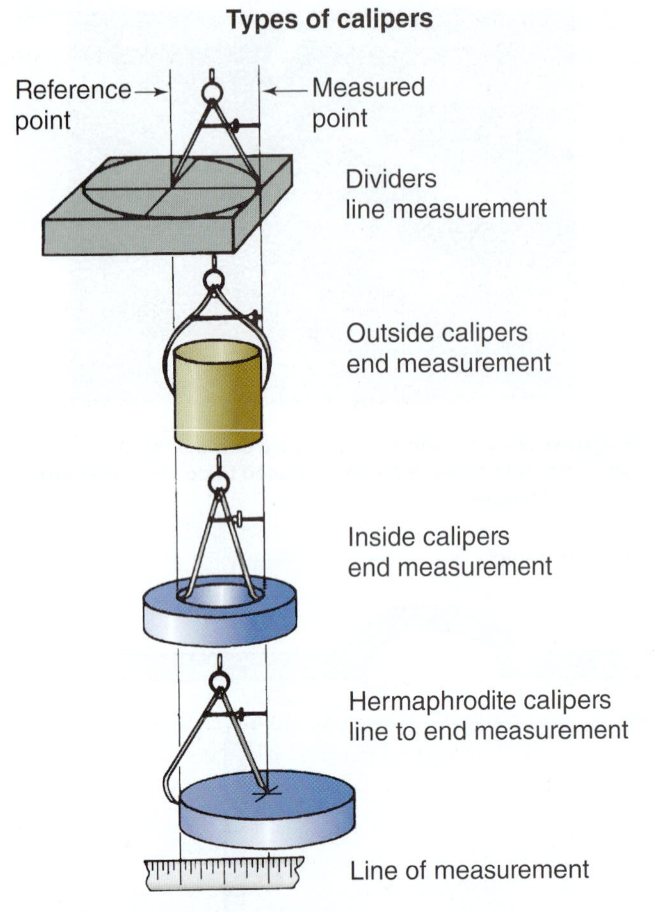

Figure 3-45 Calipers are handy for some measurements. Dividers can be used to scratch mark or lay out sheet metal when cutting it to size.

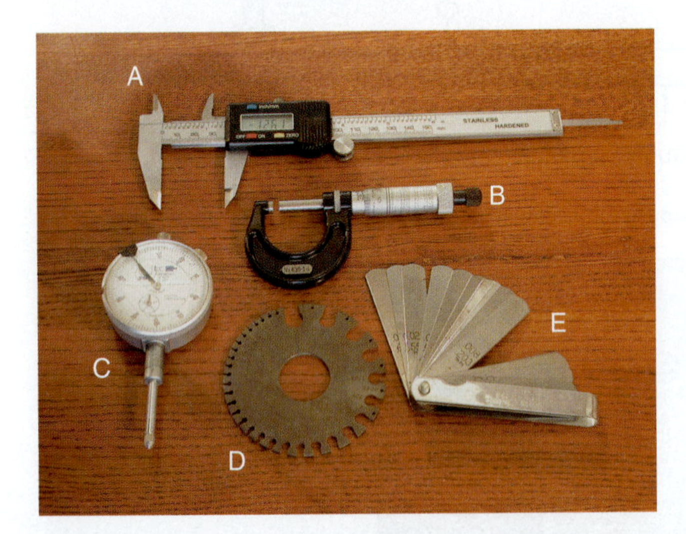

Figure 3-46 These measuring tools are used for very accurate measurements. (A) Sliding calipers measure inside, outside, and depth to within one-thousandth of an inch. (B) A micrometer produces a measurement even smaller than one thousandth of an inch. (C) A dial indicator will measure runout or motion as small as one thousandth of an inch. (D) A gauge will measure sheet metal thicknesses. (E) A feeler gauge will measure the opening between two parts. Sizes are printed on each blade.

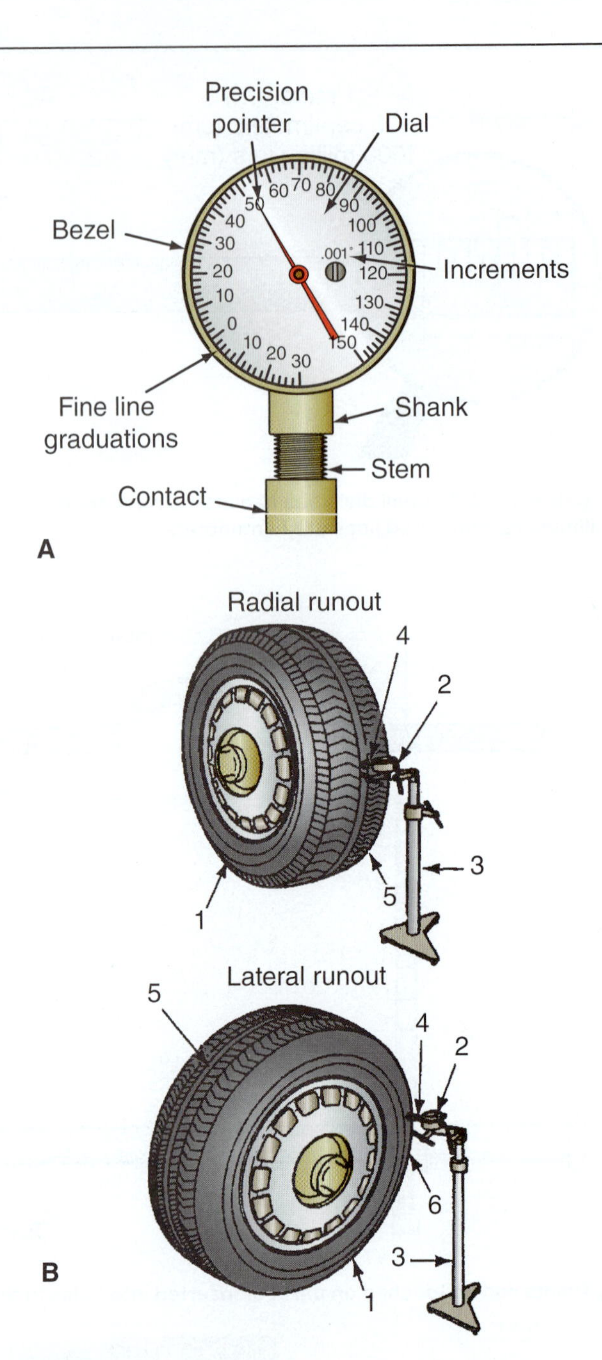

Figure 3-47 (A) A dial indicator will check part runout or wobble to within a thousandth of an inch or a hundredth of a millimeter. (B) Dial indicators are often used to check for wheel runout from collision damage.

dial to zero the needle. Move or rotate the part and take your reading.

Feeler Gauges

Feeler gauges measure small clearances inside parts. Blade thickness is given on each blade in thousandths of an inch (0.001, 0.010, or 0.020) or in hundredths of a millimeter (0.01, 0.07, or 0.10).

Flat feeler gauges are for measuring between parallel surfaces. Wire feeler gauges are round for measuring slightly larger distances between nonparallel or curved surfaces, such as spark plug gaps.

Spring Scale

A *spring scale* will measure pulling force. It is often used to check the force needed to turn a steering wheel. This will help determine whether there is damage to the steering mechanism from a collision.

To use a spring scale, hook the tool over the part (steering wheel, for example) and pull. Read the force or weight needed to move the part off the tool. Compare your measurement to known good specifications.

Pressure and Flow Measurements

You will have to make pressure and flow measurements when working. Pressures are important for air tools, spray guns, and other equipment. Adequate air flow is important when working in a paint spray booth, for example.

A *pressure gauge* reads in pounds per square inch (psi) or kilograms per square centimeter (kg/cm^2). A few pressure gauges also show or measure vacuum pressure. Note the two scales given on the pressure gauge shown in Figure 3–48.

Air pressure gauges are used on the shop's air compressor, on pressure regulators, or at the spray gun. Tire pressure gauges are needed to check for proper tire inflation. Hydraulic pressure gauges can be found on hydraulic presses and other similar equipment.

REMEMBER *Excessive pressure can be dangerous or can damage parts. Low air pressure may keep the tool from working properly.*

A *vacuum gauge* reads vacuum, negative pressure, or suction. It reads in inches of mercury (in. Hg) or kilograms per square centimeter (kg/cm^2).

A *flow meter* measures the movement of air, gas, or a liquid past a given point. In Figure 3–49, note the magnahelic gauge or air flow meter designed for use in a paint booth. Air flow meters can register inches of mercury, in feet per minute or meters per minute. Liquid flow meters often register in gallons per hour or liters per hour.

Angle and Temperature Measurements

Angle measurements basically divide a circle into 360 or more parts called degrees. You will need to read angles in degrees when doing wheel alignment, for example. This will be explained in later chapters.

Temperature Measurements

Temperature is usually measured or given in degrees Fahrenheit (F) or Celsius (C). In the body shop, temperatures are important to the drying of various materials, such as primers, paints, and adhesives.

Thermometers are often used to measure temperature in the paint booth and on the vehicle surface when refinishing. For example, manufacturers often recommend

Figure 3–48 Pressure gauges are needed to measure shop air and hydraulic pressure. Air tools and hydraulic equipment are designed to operate with specified pressure limits. Most pressure gauges have both customary and metric scales.

Figure 3–49 Adequate air flow is needed to pull toxic fumes out of the booth work area. Manometers and magnahelic gauges measure air flow by checking vacuum or negative pressure in an enclosed system. They are commonly used to check air flow in a paint spray booth.

different paint mixing values and different paint reducers for specific ambient (indoor or outdoor) temperatures.

A *mercury thermometer* encloses liquid mercury in a hollow glass rod to indicate temperature. Any change in temperature makes the mercury expand or contract. To read a thermometer, view the level of mercury next to the scale.

A *dial thermometer* is similar, but it has a needle that shows temperature. This type of thermometer is often used to check the output of an air-conditioning system. It is inserted into the center vent, with the system and blower on high. If the reading is too warm (above specifications), the system is not cooling properly.

Before applying any type of material during collision repairs, the vehicle and material to be used must be at room temperature. A vehicle's surface temperature must be brought up to room temperature prior to applying fillers and paint materials. Failure to do so may result in the formation of condensation on the surface.

Surface thermometers are often used to measure body temperature when painting. They are available in several different designs, including magnetic thermometers, paper thermometers, and digital thermometers.

Magnetic surface thermometers stick to metal body panels for measuring temperatures during paint drying or curing. When using infrared heaters after painting, keep in mind that magnetic surface thermometers have more mass than the panel they are attached to. This may result in a lower temperature reading than the panel alone would produce.

Disposable paper thermometers show the highest temperature a surface reached, but not necessarily the current temperature of the panel. They work well with infrared heaters.

A *digital thermometer* is an electronic instrument for accurately measuring temperature. It has a digital readout that will show temperature in Fahrenheit or Celsius. Special surface types are often used in collision repair.

Most digital thermometers have several different probe tips. A round, hollow tip is used for air temperature measurements. A flat, blunt tip is used for surface temperature measurements. A tip with a strand of wire is used for checking the temperatures of fluids.

Paint-Related Measurements

Mixing scales are used by paint suppliers and technicians to weigh the various ingredients when mixing paint materials. Scales will precisely weigh out each ingredient. Using paint manufacturer formulations, mixing scales are used to add different pigments, reducers, or other materials to make the paint the desired color and viscosity, or thickness. Look at Figure 3–50.

Standard mixing scales simply weigh each paint ingredient as it is poured into the mixing cup on the scale. You must look up how much of each material is needed from the paint formula and carefully pour out that amount while watching the scale. Most manual scales have been replaced by computerized scales.

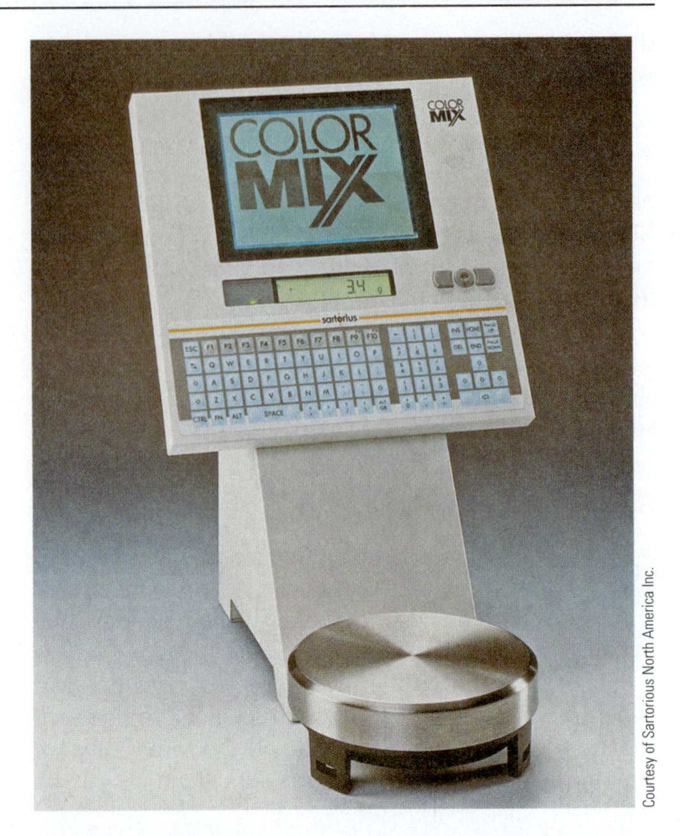

Courtesy of Sartorius North America Inc.

Figure 3–50 **Electronic or computerized scales are now used in most body shops. An intelligent computerized paint mixing scale prompts the user to add certain ingredients at the right time and also allows the system to maintain inventory and track volatile organic compounds (VOCs).**

Computerized intelligent scales allow automated paint mixing. After programming in the amount of paint needed for the area to be refinished, a computerized scale will state how much of each material to pour into the mixing cup. It will state how much pigment, reducer, and hardener to use by weight. As you pour each ingredient to the desired weight, the scale will prompt you to either add the next ingredient, or ask you if you would like to recalculate the mix in the event of an accidental overpour. Once it is zeroed, you have added the correct amount of that ingredient. The computerized scale will then prompt you to add a specific amount of the next ingredient. This will allow you to more quickly produce the correct mix of pigments and other ingredients for the specific color of paint.

When mixing and using paint and solvents or other additives, you must measure and mix their contents accurately. This is essential to doing good paint and bodywork. You must be able to properly mix reducers or thinners, hardeners, and other additives into the paint, or serious paint problems will result.

A *graduated pail* has lines denoting volume for pouring and measuring liquid materials. Its measurement lines are like those on a kitchen measuring cup. Liquid is poured into the pail until it is next to the scale for the amount needed.

Mixing instructions are normally given on the materials' labels. This might involve a percentage or parts of one ingredient compared to the other. See Figure 3–51.

A percentage reduction means that each material must be added in certain proportions, or parts. For instance, if a paint requires 50 percent reduction, this means that one part reducer or thinner (solvent) must be mixed with two parts of paint.

Mixing by parts means that for a specific volume of paint or other material, a specific amount of another material must be added. If you are mixing a gallon of paint, for example, and directions call for 25 percent reduction, you would add one quart of reducer. There are four quarts in a gallon and you want one part, or 25 percent, reducer for each four parts of paint.

Proportional mixing numbers denote the amount of each material needed by volume. The first number is usually the parts of paint needed. The second number is usually the solvent (thinner or reducer). A third number might be used to denote the amount of hardener or other additive required to mix the materials properly.

For example, a proportional mixing number of 2:1:1 means add one part solvent and one part hardener with two parts of paint. For a gallon of paint, you would add a quart of solvent and a quart of hardener for a 2:1:1 mix. Because mixing instructions vary, always follow the exact directions on the material's container.

A *mixing chart* converts a percentage into how many parts of each material must be mixed (Figure 3–52). Graduated *paint mixing sticks* have conversion scales that allow you to easily convert ingredient percentages into part proportions. They are used by painters to help mix paints, solvents, catalysts, and other additives right before spraying (Figure 3–53). Paint mixing sticks should not be confused with paint stirring sticks (wooden sticks) for mixing the contents after they are poured into the spray gun cup or a container.

Viscosity Cup Measurement

A *viscosity cup* is used to measure the thickness or fluidity of the mixed materials, usually paint. It is a small, stainless steel cup attached to a handle (Figure 3–54).

To use a viscosity cup, dip it into the mixed paint until submerged. Lift the cup out and hold it over the paint container. As soon as the cup is lifted out, start timing how long it takes for the cup to empty. The paint will leak out of a small hole in the bottom of the cup. Use the second hand on your watch or a stopwatch to time draining. When the paint stream breaks into drops, note how much time elapsed. This equals the paint viscosity in seconds.

The paint manufacturer will give a recommended viscosity value in viscosity cup seconds. It will vary between 17 and 30 seconds, depending upon the type of paint and type of cup used. Refer to the paint specifications for an exact value.

A Mixing container has several scales for mixing different proportions of each material.

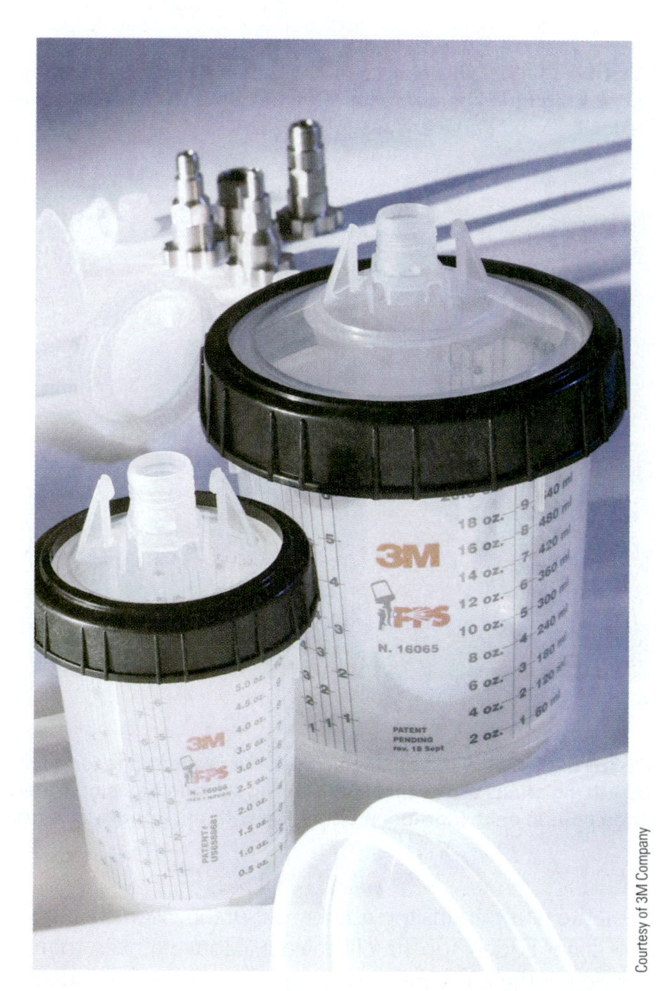

Courtesy of 3M Company

B This spray gun cup can be used for mixing paint, solvent, and hardener.

Figure 3-51 Graduated containers and spray gun cups will let you mix ingredients quickly and easily. Simply pour in paint, solvent, or hardener until it is even with the correct markings on the side of the container.

Reduction percentage	Reduction proportions	Paint (color)	Solvent
20%	= 5 parts paint / 1 part solvent		20%
25%	= 4 parts paint / 1 part solvent		25%
33%	= 3 parts paint / 1 part solvent		33%
50%	= 2 parts paint / 1 part solvent		50%
75%	= 4 parts paint / 3 parts solvent		75%
100%	= 1 part paint / 1 part solvent		100%
125%	= 4 parts paint / 5 parts solvent		125%
150%	= 2 parts paint / 3 parts solvent		150%
200%	= 1 part paint / 2 parts solvent		200%
250%	= 2 parts paint / 5 parts solvent		250%

Figure 3-52 This chart shows a range of mixing percentages and converts each into parts for each material. Compare various reduction percentages with the percent of solvent or reducer that must be added to the paint to obtain the correct viscosity, or paint thickness.

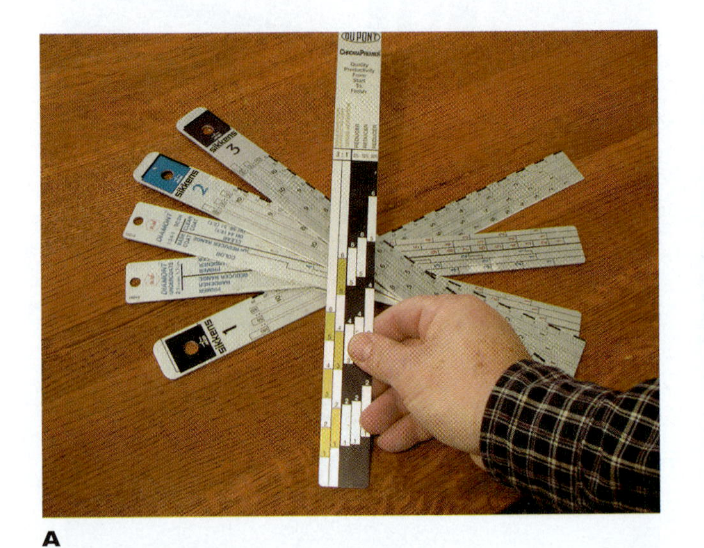

A

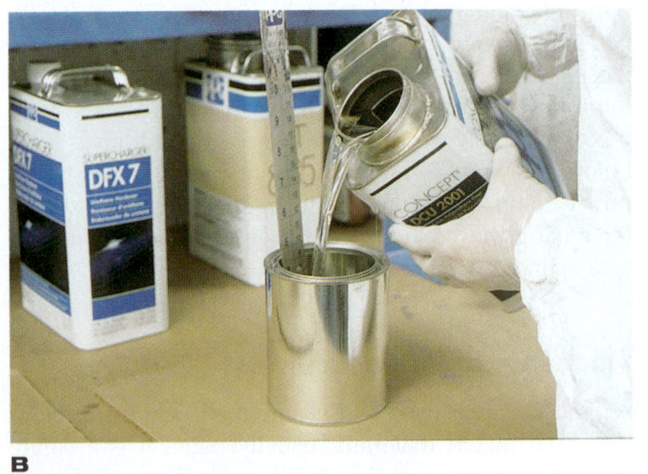

B

Figure 3-53 Paint mixing sticks have graduations for each material to be mixed together: color, reducer, and hardener. (A) Mixing sticks are available for mixing various types of materials. (B) Place the mixing stick in a straight-sided container. Then pour in each ingredient until even with the appropriate graduation on the stick.

If the paint drains too quickly out of the viscosity cup, you have added too much solvent. More paint is then needed. If the cup drains too slowly, you have not added enough solvent. Remix the paint until it passes the viscosity cup test.

If the paint is *too thick*, your paint will develop orange peel or a rough film. If the paint is *too thin*, excess solvent can cause the paint to provide poor coverage and other problems.

Paint Thickness Measurement

Paint thickness is measured in mils or thousandths of an inch (hundredths of a millimeter). Original equipment manufacturer (OEM) paints are typically about 4 to 8 mils thick. With basecoat/clearcoats, the basecoat is approximately 1 to 2 mils thick, which is approximately the thickness of a piece of typing paper. The clearcoat is about 2 to 4 mils thick. The primer is another 1 to 2 mils.

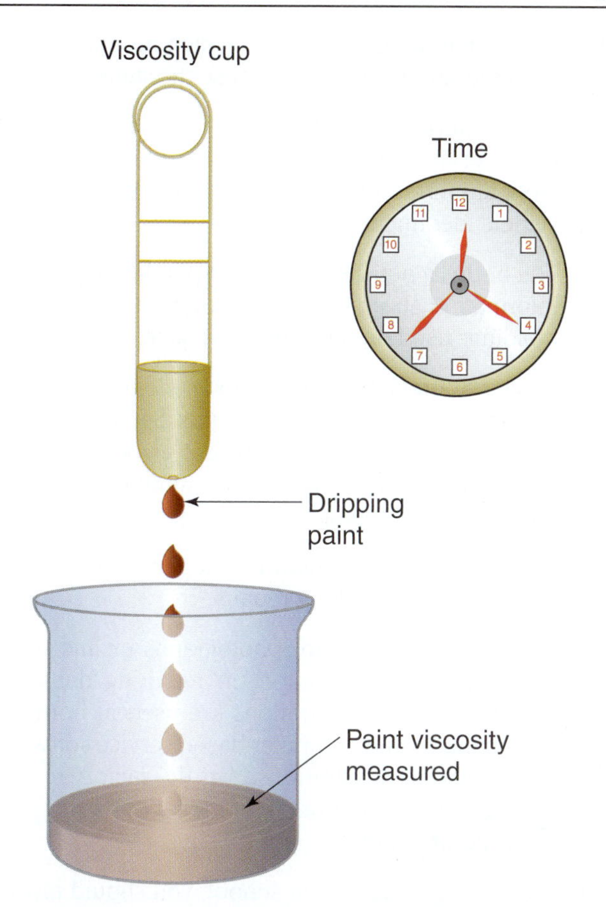

Figure 3-54 A viscosity cup measures the consistency, or fluidity, of the material to be sprayed. Paint manufacturers give a viscosity cup value in seconds. If you have mixed the paint product properly, it will take the recommended number of seconds for the paint to flow out of the small hole in the bottom of the cup.

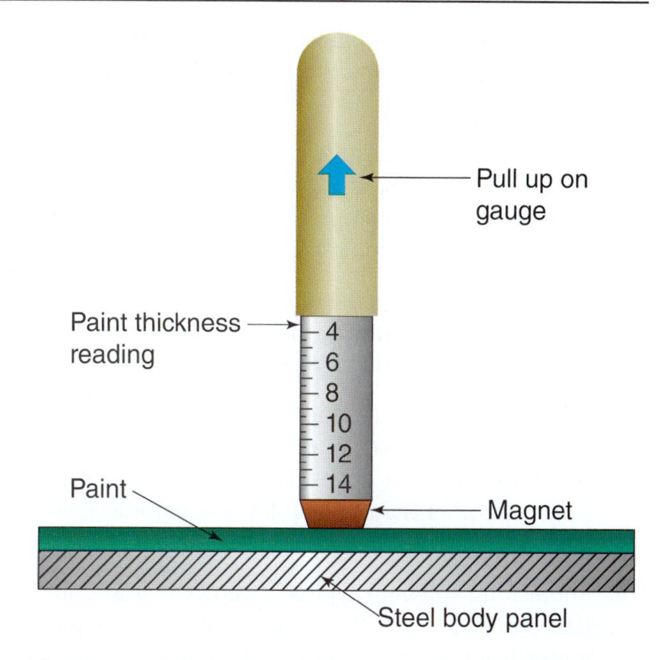

Figure 3-55 Paint must not be too thick or the paint surface will crack. A paint thickness gauge measures how much paint is on a vehicle body. A pencil-type paint thickness gauge uses a spring-loaded magnet to measure paint film thickness. Place the magnet on the paint and pull it straight out while reading the scale that indicates paint thickness in mils.

If a panel has been repainted, paint thickness will increase. If too much paint is already on the vehicle, it may have to be removed prior to refinishing. Paint buildup should be limited to no more than 12 mils. The OEM finish and one refinish usually equal just under 12 mils. Exceeding this paint thickness could cause cracking in the new finish. Chemical stripping or blasting would be needed to remove the old paint buildup.

A **mil gauge** can be used to measure the thickness of the paint on the vehicle. This can be done before refinishing, after refinishing, and during other finishing operations (Figure 3–55).

There are three types of mil gauges—pencil, gauge, and digital types. The pencil-type mil gauge uses a spring-loaded magnet and scale with markings for mil thickness. The magnetic tip is placed against the painted metal body. The pencil mil gauge is then pulled away from the body while a reading is taken. A mil gauge scale indicates how thick the paint is in mils.

The electronic mil gauge is similar but has a digital readout showing paint thickness. When placed and held on the surface, the tool will automatically register paint thickness.

SUMMARY

1. Today's motor vehicles are built to exacting standards using varied manufacturing methods, which result in different methods of repair.
2. Measurements are number values that help control repair processes in collision repair.
3. Vehicle manufacturers provide specifications (measurements) for numerous body dimensions and mechanical parts. You must refer to factory specifications to ensure a competent repair.
4. Service information includes written instructions and technical illustrations to help you properly repair a damaged vehicle.
5. Computer-based service information places service manuals, dimension manuals, estimating manuals, refinishing material guides, mechanical repair procedures, and other data on compact discs (CDs). This allows a personal computer (PC) to be used to more quickly look up and print the desired service–repair information.

6. All automobile companies publish yearly service manuals that describe the construction and repair of their vehicles.
7. Shop management software is designed to help enhance the collision repair facility's productivity, profitability, and customer service by coordinating all aspects of collision repair.
8. Collision repair guides give instructions, safety warnings, and technical illustrations for the make and model of vehicle being worked on.
9. Collision estimating guides provide information for calculating the cost of collision repairs. They give part numbers, part prices, labor charges per repair task, and other data to help the estimator to calculate the cost of repairs.
10. Refinishing software typically allows you to calculate painting costs for approximately 5,000 paint codes from several paint manufacturers, including four types of paint.
11. A vehicle dimension manual gives the body and frame measurements of undamaged vehicles. Dimensions are given for every make and model car, van, truck, and SUV. These known good dimensions can be compared with actual measurements taken off a wrecked car or truck to find out how badly the vehicle is damaged and what must be done to straighten it.
12. Color-matching guides contain information needed for repainting panels so that the repair has the same appearance as the old finish.
13. Air bag service guides give procedures, safety warnings, component locations, and other facts for working on active restraint systems.
14. Parts interchange guides list which model vehicles from the same manufacturer use the same parts.
15. Before starting vehicle repairs, ordering parts, or using service information, you must first record the alphanumeric code (letters and numbers) on the vehicle identification plate.
16. The body ID number, or service part label, provides information about the finish (paint) and the trim or moldings used on the specific vehicle.
17. Service abbreviations are a short series of letters that represent technical terms or words. Service symbols are small pictures that represent a location, part, procedure, safety warning, measurement point, or other aspect of repair.
18. Repair charts give diagrams that guide you through logical steps for making repairs.
19. Service illustrations are photographs or line drawings that give information on repairing damaged vehicles. Service illustrations are used in both service manuals and computer-based service software. Panel welding illustrations give the factory-recommended weld count and weld locations for replacing structural panels.
20. After you graduate from school, you should always continue your training in collision repair. Vehicle designs and repair methods are constantly changing.
21. Modern vehicles are precision-manufactured using the latest technology. Most vehicle assembly at the factory uses automated conveyors and robotics that result in motor vehicles that are built to closer tolerances than ever before. This requires auto body repair technicians who can measure accurately and repair vehicles precisely.

EXERCISES

On a separate sheet of paper, complete the following learning activities for this chapter. Write definitions for the key terms and answer the ASE-style review questions, essay questions, critical thinking problems, and math problems. You can also do the outside activities, possibly for extra credit.

Key Terms

body ID number
collision estimating guide
collision repair guide
color-matching guide
conversion chart
diagnosis chart
flow diagram
mil gauge
mixing by parts
mixing scale
paint code
paint thickness
panel welding illustration
parallax error
parts interchange guide
proportional mixing numbers
refinishing guide
salvage parts software
service manual
shop management software
specifications
surface thermometer
tram gauge
vehicle dimension illustration
vehicle dimension manual
vehicle identification number (VIN)
wiring diagram

➤ ASE-Style Review Questions

1. This type of manual provides information for calculating the cost of collision repairs.
   A. Service manual
   B. Dimensions manual
   C. Estimating manual
   D. Paint code manual
2. This tool is used to measure part movement and out-of-round in thousandths of an inch (hundredths of a millimeter).
   A. Outside micrometer
   B. Inside micrometer
   C. Feeler gauge
   D. Dial indicator
3. Paint buildup should be limited to no more than
   A. 12 mils.
   B. 2 mils.
   C. 50 mils.
   D. 100 mils.
4. Technician A says that a personal computer is the most efficient way to look up and print the desired service–repair information. Technician B says it is more cost-effective to purchase service manuals from all auto manufacturers. Who is correct?
   A. Technician A
   B. Technician B
   C. Both A and B
   D. Neither A nor B
5. This type of software is designed to enhance a collision repair facility's productivity, profitability, and customer service by coordinating all aspects of collision repair.
   A. Estimating software
   B. Dimensional software
   C. MSDS software
   D. Shop management software
6. Technician A says that collision repair guides give instructions, safety warnings, and technical illustrations for the make and model of vehicle being worked on. Technician B says that refinishing guides explain paint codes, types of paints, how to apply paints, buff paints, and other paint-related information. Who is correct?
   A. Technician A
   B. Technician B
   C. Both A and B
   D. Neither A nor B
7. Technician A says that panel welding illustrations give the factory-recommended weld count and weld type. Technician B says that they also give weld locations for replacing structural panels. Who is correct?
   A. Technician A
   B. Technician B
   C. Both A and B
   D. Neither A nor B
8. Technician A says that salvage parts software provides easy access to a database of recycled parts and suppliers. Technician B says that this type of computer program often supports an accessible online parts database from a growing network of thousands of recyclers. Who is correct?
   A. Technician A
   B. Technician B
   C. Both A and B
   D. Neither A nor B
9. These publications explain mechanical and structural problems found on specific makes and models of vehicles.
   A. Service manuals
   B. Refinish manuals
   C. Technical service bulletins
   D. MSDS
10. Technician A says that you have to go to the service manual to decipher VIN information. Technician B says that VIN decode software will automatically look up and convert the VIN into year, make, model, body style, engine, transmission, restraint system, and other vehicle-specific information. Who is correct?
    A. Technician A
    B. Technician B
    C. Both A and B
    D. Neither A nor B
11. Technician A says that you must look up the paint code to find out what type and color of paint was used during vehicle manufacturing. Technician B says that the VIN always gives this information. Who is correct?
    A. Technician A
    B. Technician B
    C. Both A and B
    D. Neither A nor B

➤ Essay Questions

1. How do you avoid parallax error when reading a scale?
2. What can happen if the paint on a vehicle is too thick?
3. Describe four steps for using refinish materials information.
4. Describe some of the information provided by vehicle dimension manuals or software.

Critical Thinking Problems

1. Explain how you might use conversion charts.
2. A damaged vehicle enters the shop, and the paint mil gauge shows a paint thickness of 18 mils. What should be done to properly repair the panel before painting?

Math Problems

1. If a paint requires 50 percent reduction, how would you mix the paint and its solvent?
2. When mixing paint materials, what might the numbers 2:1:1 mean?

Activities

1. Find the VIN on several vehicles. Write down the location of the VIN for several cars. Look up the VIN data in a service manual.
2. Use a paint thickness gauge to check for excess paint layers on several cars. Did different panels have more paint than others? How many vehicles had paint thicker than recommended? Could you see any paint cracking or other surface problems when the paint was too thick?

Hand Tool Technology

CHAPTER 4

OBJECTIVES

After studying this chapter, you should be able to:

- List the most common hand tools used in auto body repair.
- Identify common body shop hand tools.
- Explain the importance of having a wide range of hand tools.
- Choose the correct tool for the job at hand.
- Use each tool correctly and in a safe manner.
- Answer ASE-style review questions relating to hand tools.

INTRODUCTION

Hand tools are general tools used by both auto mechanics and auto body technicians. They include wrenches, screwdrivers, pliers, and other common tools. Basic hand tools are needed to remove parts, fenders, doors, and similar assemblies. The full range of bodyworking hand tools includes general purpose tools, metalworking tools, and body surfacing tools.

Hand tools are extensions of the human body. They allow you to do tasks otherwise impossible with your bare hands. By knowing how to select the right tool for the job, you will do higher quality work in less time. Hand tool knowledge is a sign of an experienced technician. Without the right tools, even the best body technician cannot do quality bodywork.

When purchasing tools, get high-quality tools from a reputable manufacturer. Most offer lifetime guarantees against failure. Never buy cheap, low-grade tools. Cheap tools slow down your work rate and efficiency because they are heavier, clumsier, and break more easily. You usually get what you pay for. Good tools will pay for themselves in a short period of time. See Figure 4–1.

This chapter explains which hand tools an auto body repair technician must have and how they are used. This is a vital chapter that gives you the basics for understanding procedures given throughout this book and in service manuals.

4.1 GENERAL PURPOSE TOOLS

Many of the tools a body technician uses every day are common, general purpose hand tools: wrenches, screwdrivers, pliers, and so forth. An apprentice auto body repair technician will probably already have many of these in a tool collection. The less familiar tools are designed for specific types of industrial fasteners often encountered in bodywork. Refer to Figure 4–2 and Figure 4–3.

Tool Storage

It is very important that you store tools properly. Tool care and storage will increase your tools' life span and speed your work. If you cannot find a tool quickly or if your tools are in bad condition, work quality will suffer.

A tool chest is a large roll-around cabinet for keeping larger tools. The bottom drawers are taller for holding

Never open more than one toolbox drawer at the same time because a box can flip over. Death or injury could result. Close each drawer before opening the next one.

Courtesy of Snap-on Tools Company (www.snapon.com)

Figure 4–1 **A professional set of tools and a toolbox are a big investment! Always purchase quality tools with a lifetime guarantee against failure. High-quality tools will save you time and effort and help you get more work done so you can earn more money.**

Courtesy of Snap-on Tools Company (www.snapon.com)

A A tool cart with casters will let you take needed tools to the vehicle. The lower tray can be used to hold small parts during repairs.

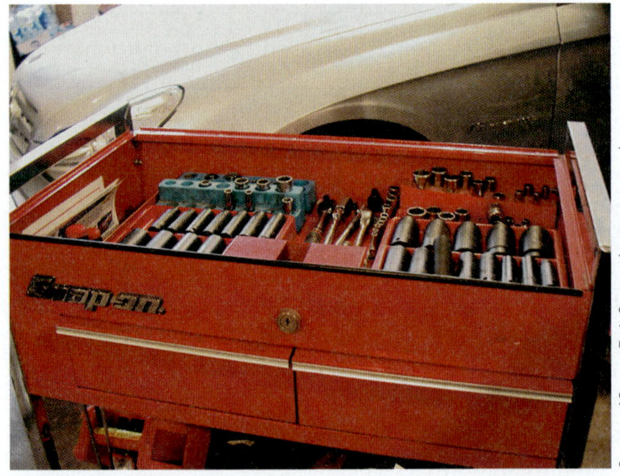

Courtesy of Snap-on Tools Company (www.snapon.com)

B The technician has organized common hand tools in a locking roll-around cart for working more efficiently.

Figure 4–2 **A tool cart or tray is very handy and will save time when working.**

power tools. A toolbox is a smaller cabinet that sits on top of the tool chest. Smaller tools, like hand wrenches, screwdrivers, and sockets, are often placed in the upper drawers of a toolbox.

A tool cart is handy for taking tools to the vehicle being repaired. A basic set of hand tools can be organized in the cart to help you work more efficiently. You will waste less time by not having to walk back and forth from your tool chest to get basic hand tools.

A creeper allows you to lie down and roll under a vehicle while doing underbody repairs. It has caster wheels for moving around and a padded area for lying on the floor in comfort. A stool creeper is a short chair with caster wheels for working at about waist level. You can place tools and fasteners in the tray of a stool creeper while working.

Wrenches

A complete collection of wrenches is indispensable for the auto body technician. A variety of auto body parts, accessories, and shop equipment utilizes common bolts and nuts. Fasteners can be standard SAE or metric size. A well-equipped auto body technician will have both metric and SAE wrenches in a variety of sizes and styles. See Figure 4–4.

The word *wrench* means "twist." A wrench is a tool for twisting and/or holding bolt heads and nuts. The width of the jaw opening determines **wrench size**. For example, a ½-inch wrench has a jaw opening (from face-to-face) of one inch (25.4 mm). The actual size is really slightly larger than its nominal size so that the wrench fits around the fastener head of equal size.

Larger wrench sizes are longer for more leverage. For example, a ¼-inch wrench is typically 4½ inches long.

A A creeper allows you to lie under the vehicle while working.

B A stool creeper allows you to sit while working, saving your legs and knees from strain.

Figure 4–3 Creepers allow you to lie down or sit while working.

A ¼-inch wrench probably is 10 to 12 inches long. The extra length provides the user with more leverage to turn the larger size nut or bolt. See Figure 4–5, Figure 4–6, and Figure 4–7.

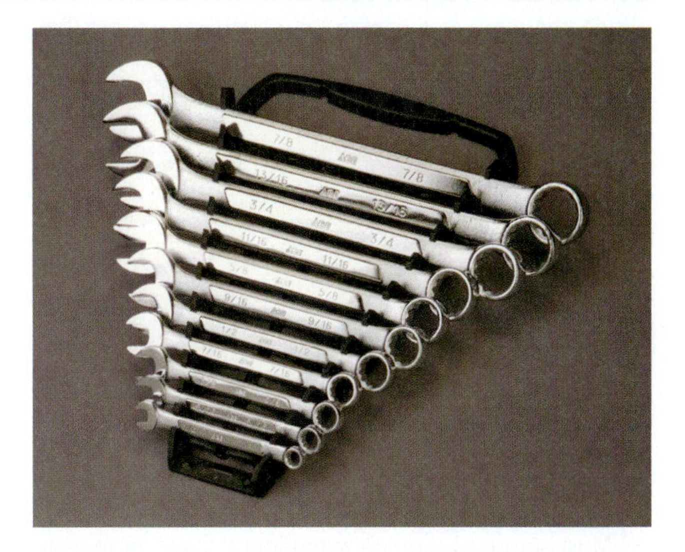

Figure 4–4 These are combination wrenches. Note the different head openings.

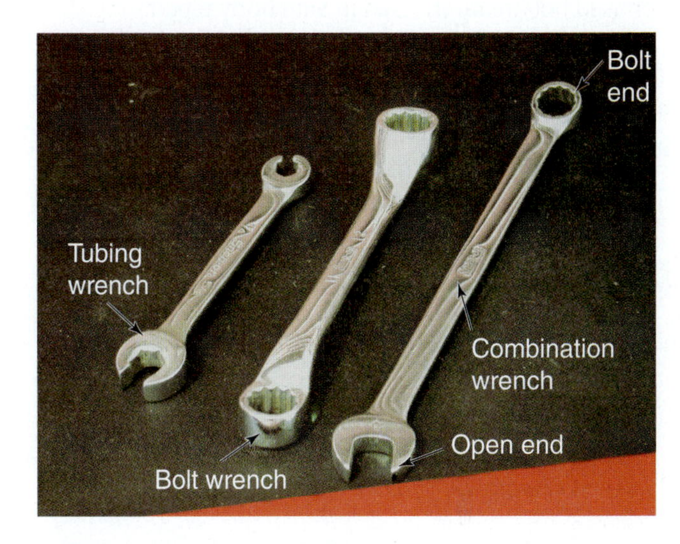

Figure 4–5 Note the three basic wrench configurations.

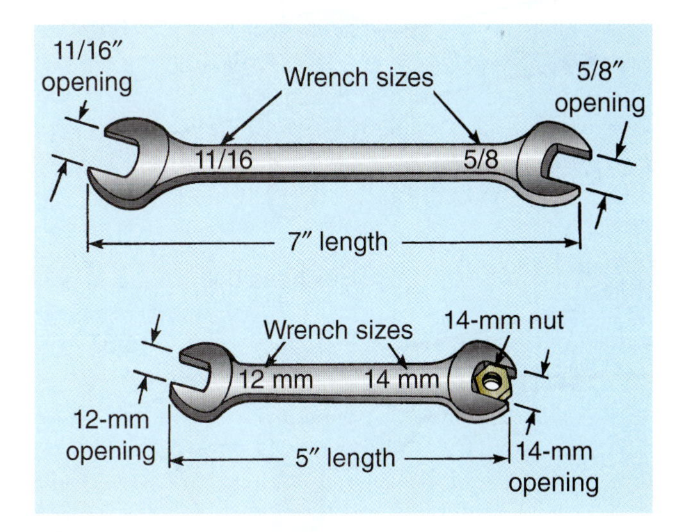

Figure 4–6 Wrench size is the size of the bolt or nut a wrench will fit properly, or the measurement across jaw openings.

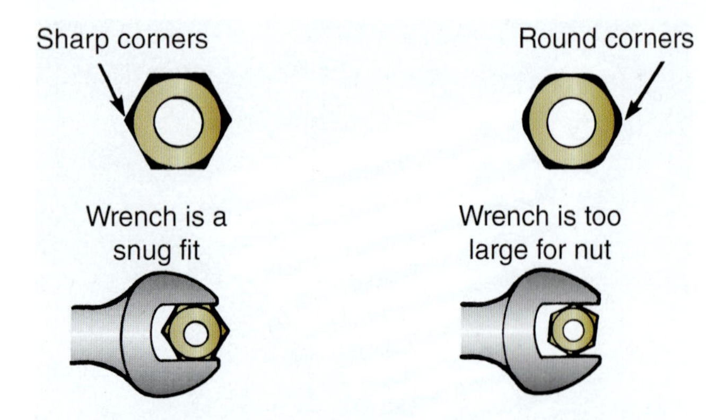

Figure 4-7 **Use the right size wrench to avoid rounding off flats on a fastener.**

Most standard wrench sets include sizes from ⅜ to ³⁄₁₆ inch. Metric sets usually include 6- to 19-millimeter wrenches. Smaller and larger wrenches can be purchased but are rarely used in auto body repair.

Metric and SAE-size wrenches are not interchangeable. For example, a ⁹⁄₁₆-inch wrench is ³⁄₁₀ millimeter larger than a 14-millimeter nut (see the conversion chart in Appendix C). If the ⁹⁄₁₆-inch wrench is used to turn or hold the 14-millimeter nut, the wrench will probably slip, rounding the points on the nut and possibly skinning knuckles as well.

Open-End Wrenches

Every tool chest should have a set of open-end or combination wrenches. The jaws of the **open-end wrench** will slide around bolts or nuts where there might be insufficient clearance above or on one side to accept a box wrench (Figure 4–8).

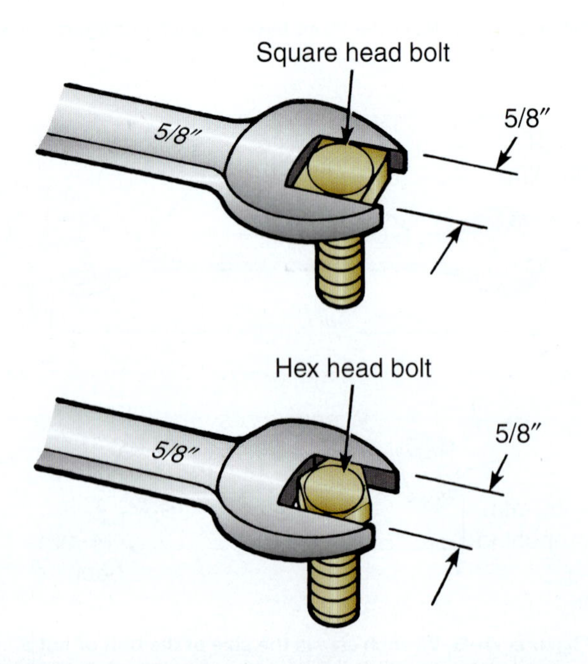

Figure 4-8 **An open-end wrench grasps only two faces on a fastener and can slip off easily.**

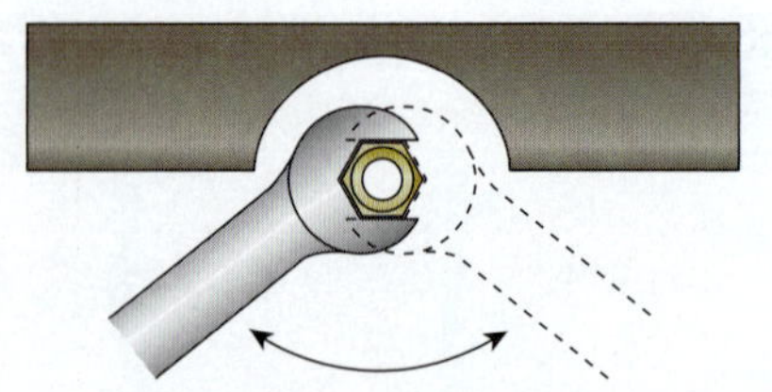

In tight places, flip wrench over after each turn of nut or bolt.

Figure 4-9 **An offset wrench can increase the turning radius and allows the wrench to fit over the fastener in an obstructed area.**

The open-end wrench fits both *square head* (four-cornered) and *hex head* (six-cornered) nuts. The disadvantage of using open-end wrenches is that **only two faces of the nut are gripped by the jaws, therefore creating a greater tendency for it to slip off the bolt or nut**. Rounded nuts and injured hands are too often the result. Because of this, the open-end wrench should be used for holding when there is not sufficient room for a box-end wrench to be used.

Open-end wrenches are often angled 15–80 degrees at both ends. The offset helps turn a bolt or nut that is recessed or in a confined area. Flipping the wrench over after each turn maximizes the turning arc (Figure 4–9).

Box-End Wrenches

The end of the **box-end wrench** is closed rather than open, for better holding power. The jaws of the wrench fit completely around a bolt or nut, gripping each point on the fastener (Figure 4–10).

The box-end wrench is thus the safest to use. More force can be applied without slippage and rounding bolt

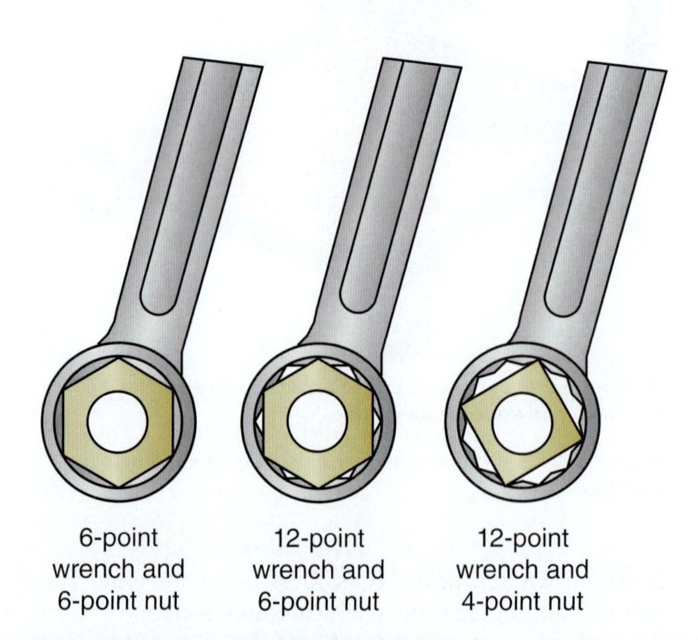

Figure 4-10 **Note how the contact areas of a 6-point wrench are greater than those of a 12 point.**

or nut heads. The handle of many box-end wrenches is offset to provide hand clearance. Each end is usually a different size. The box-end wrench does have limitations. There must be sufficient clearance for the jaws to fit over and around the head or nut. The box-end wrench must also be lifted off the head or nut and rotated to a new position for each pull.

Box-end wrenches are available in 6, 8, or 12 points. The 6-point wrench is the strongest because it completely surrounds the hex nut and brings force to bear on all six sides and points. The 12-point wrench also grips the six points but does not bear on the face surfaces of a hex nut, so there is a greater potential for slippage. The advantage of a 12-point wrench is that the additional engagement points increase the possible turning radius. The 8-point box-end wrench is seldom used because it fits only square head nuts.

The handle of a box-end wrench is often offset 10–60 degrees (Figure 4–11), which allows recessed fasteners to be reached more easily.

Combination Wrenches

The **combination wrench** has an open-end jaw on one end and a box-end on the other. Both ends are the same size. Every auto body repair technician should have two sets of wrenches: one for holding and one for turning. The combination wrench is probably the best choice for the second set. It complements either open-end or box-end sets. Combination wrenches are available with 6-, 8-, or 12-point box ends and with or without offset open ends and handles (Figure 4–5).

Adjustable Wrenches

An **adjustable wrench** has one fixed jaw and one movable jaw (Figure 4–12). The wrench opening can be adjusted by rotating a helical adjusting screw that is mated to teeth in the lower jaw. The jaw opening can be adjusted from fully closed to its maximum open width (1¾ inches). The auto body tool chest should have a set of adjustable wrenches in lengths from 4 to 24 inches.

Besides the obvious advantage of fitting various size bolt heads and nuts, the adjustable wrench has the same advantages and disadvantages of an open-end wrench.

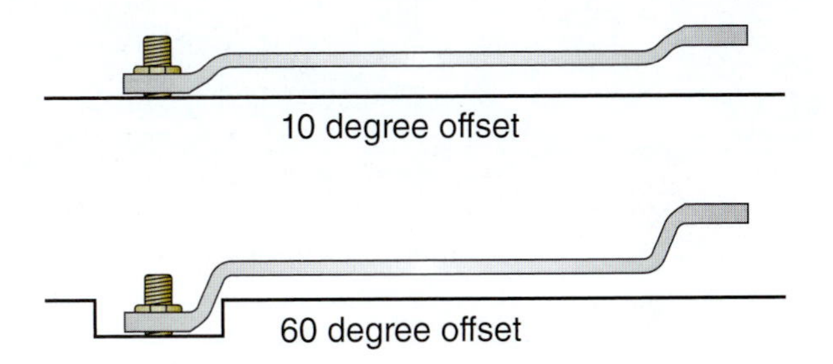

Figure 4–11 An offset box wrench will help you reach down into an area without the handle hitting on a part.

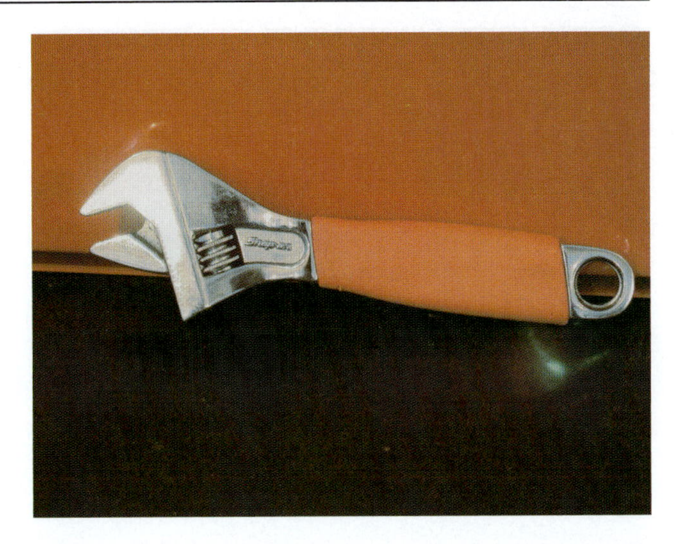

Figure 4–12 An adjustable wrench has a movable jaw for fitting different-size fasteners.

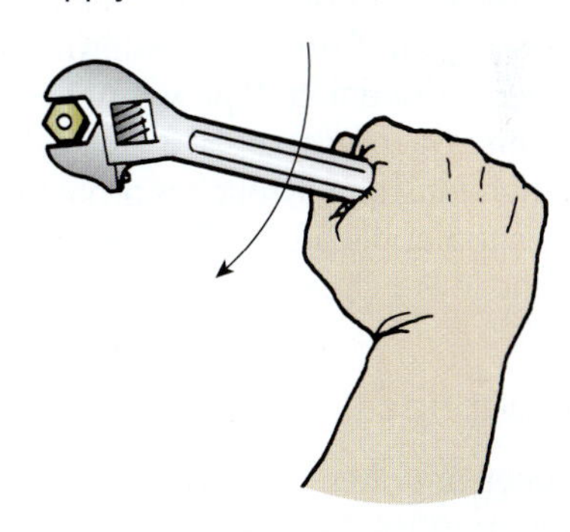

Figure 4–13 Pull on the wrench so that force bears against the fixed jaw.

It can be slipped around bolts or nuts in tight places, but it also bears against only two faces. It, too, slips more easily than does the box-end wrench.

As the adjustable wrench wears with use, the adjusting screw will lose some of its holding power. The jaws develop a tendency to loosen as force is applied and then slip off the nut or bolt. Therefore, use an adjustable wrench only when a suitable box-end or open-end wrench is not available. Use it to hold rather than turn when it must be used. Tighten it securely. Hold it flush, and pull the handle so that the force bears on the fixed jaw (Figure 4–13).

Pipe Wrenches

Another type of wrench that is occasionally used in the auto body shop is the pipe wrench. The **pipe wrench** gets its name from its most common use—turning pipes and pipe fittings. The advantage of the pipe wrench over other wrenches is that it will grab and turn round objects, such as pipes and studs (Figure 4–14).

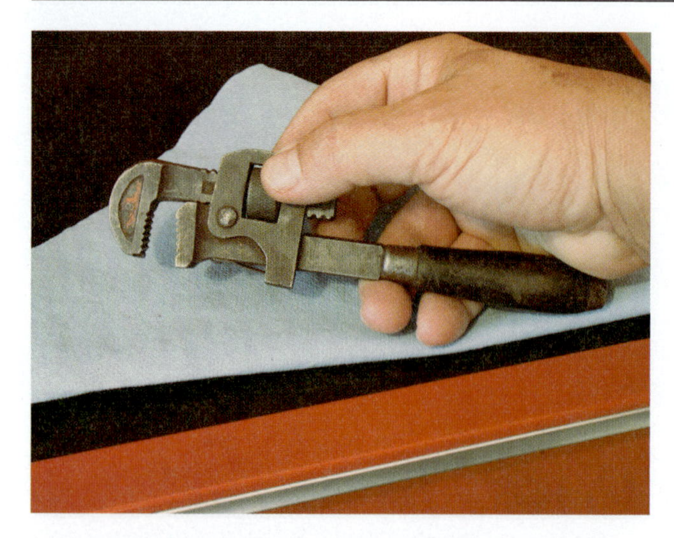

Figure 4-14 **A pipe wrench will grasp and turn round objects with great force. However, it will usually mar the part surface as the jaws dig into it.**

Like the adjustable wrench, the pipe wrench opening is adjustable. The top or hook jaw is threaded through a stationary adjusting nut. Turn the adjusting nut to increase or decrease the jaw opening. Pipe wrenches are available in maximum openings of 3/8–8 inches in various lengths. A 10-inch-long pipe wrench is probably adequate for most body shop applications. The best have replaceable teeth on the lower (heel) jaw.

Special Wrenches

Special wrenches are often needed to work on air tools. The small lugs or dowels formed on *a spanner wrench* engage slots or holes in threaded fasteners. Spanner wrenches are often used to service grinding discs or buffing pads, for example. Thin steel open-end and box-end wrenches are also provided with some air tools so you can loosen and tighten the chuck on the air tool. Special wrenches are shown in Figure 4–15.

Figure 4-15 **Here are some special slim or thin steel wrenches designed to be used on air tools.**

Hex and Torx Wrenches

An **Allen wrench** is a hex, hexagon, or six-sided wrench. It will install or remove setscrews. Setscrews are often used on pulleys, gears, rear view mirrors, and knobs (Figure 4–16).

A set of hex head wrenches, or Allen wrenches, should be in every toolbox. Hex head sockets are also available for removing large setscrews (Figure 4–17).

The **Torx fastener** is a 6-point fastener that has the strongest grip and drive torque without fastener damage.

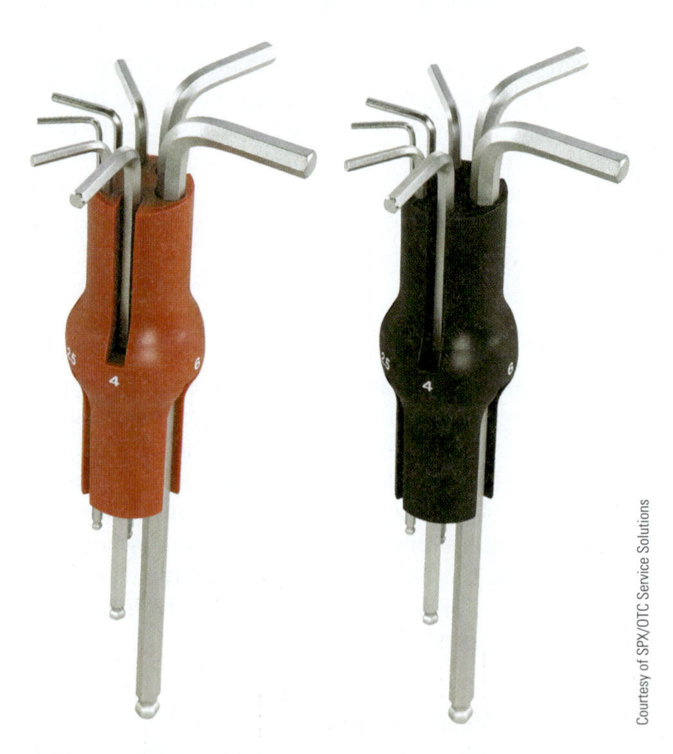

Courtesy of SPX/OTC Service Solutions

Figure 4-16 **Conventional and metric Allen wrench sets are needed on setscrews and other fasteners with a hex (six-sided) socket configuration.**

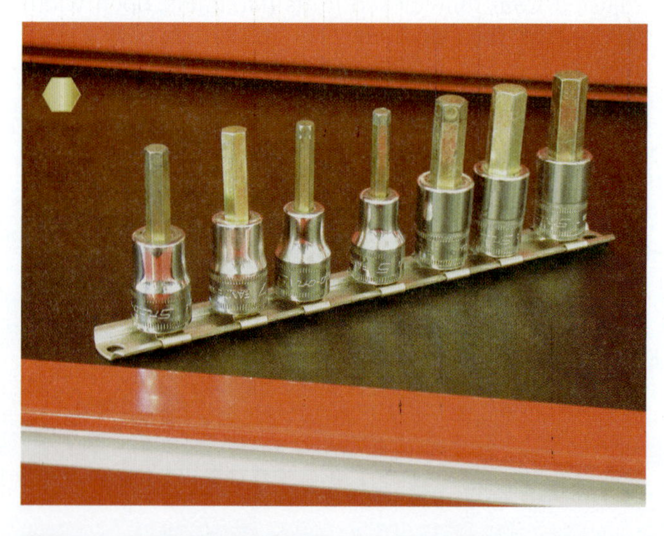

Figure 4-17 **A hex socket set is much easier to use because the ratchet will rapidly turn the tool.**

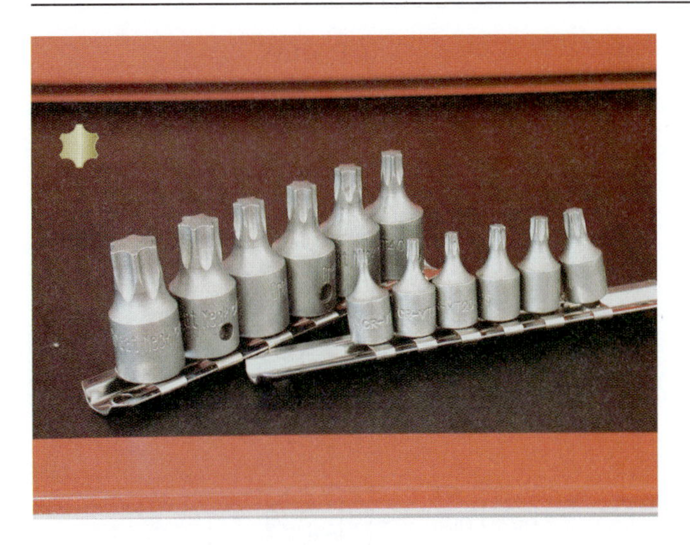

Figure 4–18 Note the shape of Torx head sockets.

Sometimes called a *star-fastener*, this type of wrench is used on most late model cars. On many vehicles, Torx fasteners are used in luggage racks, headlights, taillight assemblies, mirror mountings, door strikers, seat belts, and exterior trim. Torx wrenches or drivers are sold in sets of five or seven popular metric sizes (Figure 4–18).

Socket Wrenches

In many situations, a **socket wrench** is much faster and easier to use than an open-end or box-end wrench. Some applications absolutely require one. The auto body repair technician should have several sets of socket wrenches. See Figure 4–19.

The socket is closed on one end. The closed end has a square hole that accepts a square lug or drive on the socket handle (Figure 4–20). One handle fits all the sockets in a set. The size of the lug indicates the drive size of the socket wrench. On better-quality handles, a spring-loaded ball in the square lug fits into a depression in the socket. The ball holds the socket to the handle. See Figure 4–21.

The basic *socket wrench set* consists of a handle and several barrel-shaped sockets. The socket fits over and around a given size nut or wrench (Figure 4–22). Inside it is shaped like a box-end wrench. Sockets are available in 6, 8, or 12 points (Figure 4–23).

A 6-point socket gives the tightest hold on a hex nut. The face-to-face fit minimizes slippage and rounding of the fastener's points. The 8-point socket, like the 8-point box-end wrench, fits only square head fasteners and is thus limited in its usefulness. The 12-point socket does not

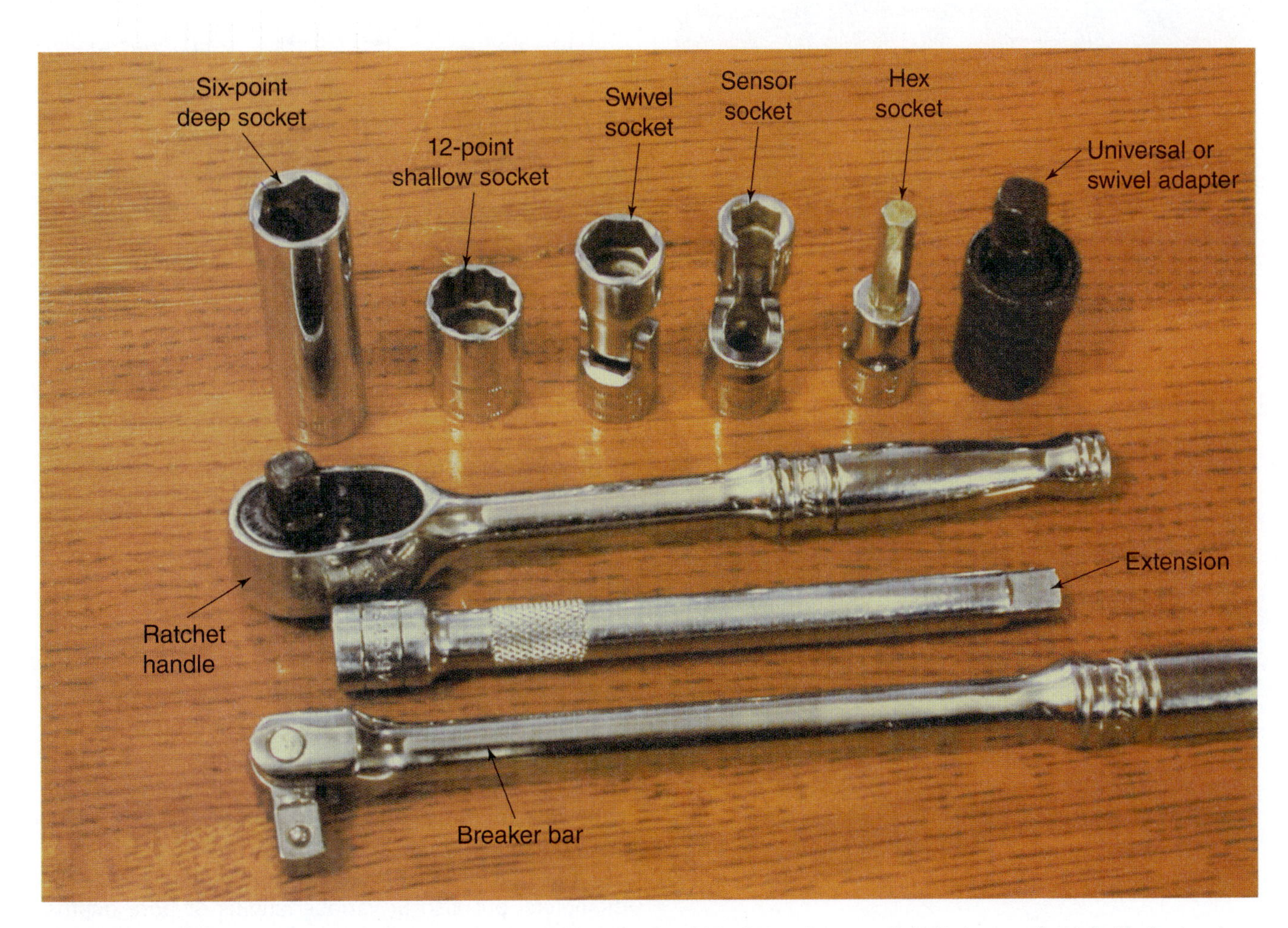

Figure 4–19 Study the basic tools in a socket set. A breaker bar is for loosening very tight fasteners. A ratchet is for turning bolts and nuts more quickly. An extension allows you to reach the fastener head.

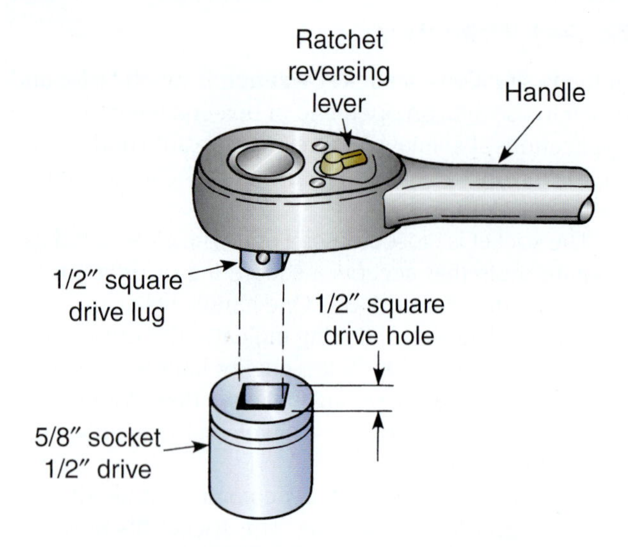

Figure 4–20 Socket drive size is the measurement across the square hole for the drive lug on the handle. Make sure it is large and strong enough not to break from excessive torque.

Figure 4–21 Compare three socket drive sizes. One-half inch is for larger bolts and nuts. One-quarter inch is for very small fasteners, as on the dash, trim, or moldings of vehicles. Three-eighths inch is for medium-size fasteners.

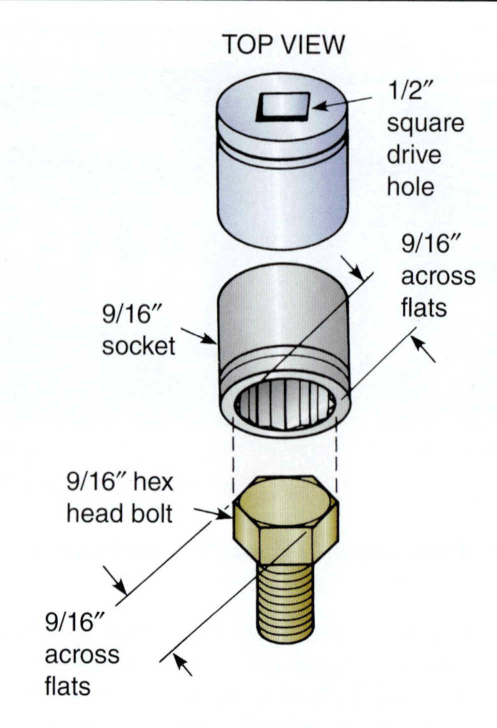

Figure 4–22 Socket size is almost the same as bolt or nut size. It is the measurement across flats.

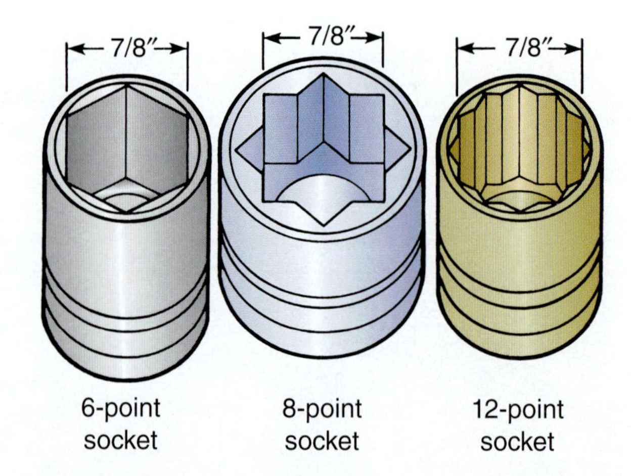

Figure 4–23 Compare 6-, 8-, and 12-point sockets. A 6-point socket should be used on very tight or rusted bolts and nuts to avoid stripping off flats.

have the holding power of the 6-point socket, but its numerous positions maximize the possible turning radius.

Deep sockets are longer for reaching over stud bolts. Swivel sockets have a universal joint between the drive end and socket body. *Impact sockets* are thicker and case-hardened for use with an air-powered impact wrench. Impact sockets are often black in color. *Conventional sockets*, or nonimpact sockets, are usually chrome-plated. See Figure 4–24 and Figure 4–25.

Socket Sizes

The size of the individual sockets depends on the drive size of the set as well as the size of the fastener head it fits. The socket size is slightly larger than the face-to-face dimension of the fastener. A ¼-inch set has sockets ranging from 3/16 to ½ inch or 4 to 13 mm. A ⅜-inch drive set has sockets ranging in size from ⅜ to ¾ inches or 9 to 25 mm. A good ½-inch socket wrench set has sockets ranging in size from 7/16 to 1¼ inches or 11 to 30 mm.

Both standard SAE and metric socket wrench sets are necessary for bodywork. A large percentage of the vehicles sold in the United States require metric tools.

Sockets are available not only in standard face-to-face diameters, but also in various lengths or bore depths. Normally, the larger the socket size, the deeper the well. Deep-well sockets are made extra-long to fit over bolt

Figure 4–24 These are hardened deep-impact sockets. Note the head sizes printed clearly on the sides of the sockets.

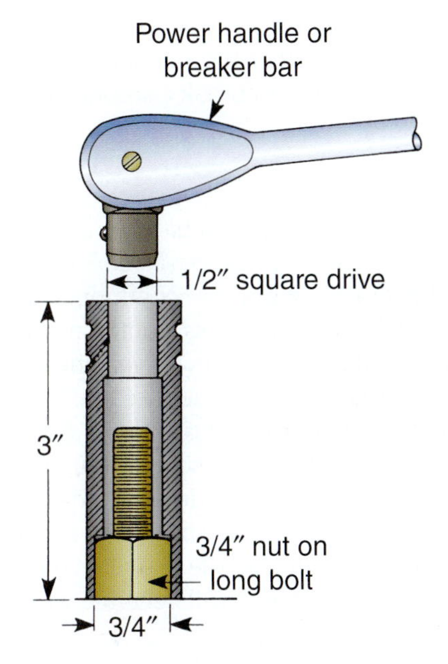

Figure 4–25 A deep-well socket will slide down over the studs to reach the nuts.

ends or studs to reach a nut. A spark plug socket is an example of a deep-well socket. Deep-well sockets are useful in a body shop for removing bumper bolts and also for reaching nuts or bolts in limited access areas. Deep-well sockets should not be used when a regular size socket will do the job. The longer socket develops more twisting torque and tends to slip off the fastener. Special sockets are shown in Figure 4–26.

Drive Sizes

Socket wrench sets can be purchased in ¼-, ⅜-, ½-, ¾-, and 1-inch **drive sizes**. The small drive sizes are used for

Figure 4–26 Here are two specialty sockets. The one on the left is for engine sending units, and the other is a sensor socket so wires or terminals on the sensor will clear the socket body.

turning small fasteners on emblems and trim where little torque is required. The larger drive sizes with longer handles are used where greater torque is needed. A ¾- or 1-inch socket wrench is useful for truck and heavy equipment repair.

An auto body repair technician will need a set of ¼-, ⅜-, ½-inch drive sockets. The ¼-inch drive set is a must for disassembly and removal of interior trim components. The ⅜-inch drive sockets will fit almost all sheet metal bolts and nuts found on a vehicle. The sockets in a ½-inch drive set are useful in removing exhaust systems, suspension parts, bumpers, and other related automotive parts commonly removed in body repair.

Always use the appropriate drive size for the job at hand. If the drive size is too small for the fastener, you can break the drive or socket. If the drive is too large, you will waste time trying to handle the clumsy tool.

Socket Wrench Accessories

Accessories multiply the usefulness of a socket wrench. A good socket wrench set has a variety of the following:

- Ratchet handles
- Spinner
- Ratchet adapter
- Breaker bars
- Sliding T-handle

- Speed handles
- Drive adapter
- Extension bars

Handles

Several different types of handles are available. One is the **breaker bar** or power handle. Held at a 90-degree angle, the extra-long handle provides the torque needed to loosen stubborn fasteners. After breaking the bolt loose, swing the handle straight out and you can turn the handle with your fingers to quickly remove the nut or bolt.

A *speed handle* is a bit-and-brace-type handle that can quickly turn off a nut or bolt. Its use requires sufficient clearance for turning the handle.

The **ratchet handle** is probably the most commonly used handle. The ratchet handle allows removing or tightening without removing the socket from the fastener. A reversing lever allows the ratchet mechanism to slip (ratchet) in one direction and turn the socket in the other. The turning direction can be changed by turning the ratchet lever position.

Some ratchet handles are equipped with a quick release push button for unlocking the socket from the handle drive. This allows the socket to be easily removed from the handle. Ratchet handles are available not only with flexheads, but also with offset handles to help reach obstructed areas.

A *T-handle* or *slide bar* is another handle that comes in handy when access is limited. The T-handle is similar to a long extension, but a slide bar fits in a hole in the upper end. The slide bar can be centered in the hole and gripped with both hands. The push–pull effort helps loosen stubborn fasteners with less likelihood of slippage. The slide bar can also be slid to one side and used as a breaker bar. See Figure 4–27.

Many socket wrench sets contain sockets in two drive sizes, such as ¼- or ⅜-inch and ½-inch drives. An adapter is often provided so that the larger drive handle can be used with the smaller drive sockets.

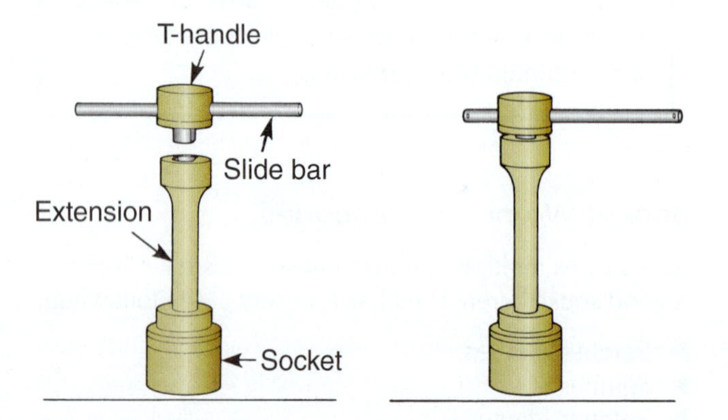

Figure 4–27 T-bars can be gripped with both hands or used as breaker bars.

Most socket wrench sets also contain extensions and universal joints. The **extensions** (3 and 6 inch are common lengths, with up to 36 inch available) reach into otherwise inaccessible places. A universal joint allows the work to be done at an angle to the fastener. With a universal joint adapter, one can reach around obstacles and use a socket wrench.

Flexockets

A *flexocket* is a combined socket and universal joint (see Figure 4–19). The **universal joint** allows the handle to be held at an angle other than 90 degrees to the fastener. Flexockets are normally ⅜ and ¼-inch drives.

Screwdriver Attachments

Screwdriver attachments are also available for use with a socket wrench.

These socket wrench attachments are very handy when a fastener cannot be loosened with a regular screwdriver. The leverage that the ratchet handle provides is often just what it takes to break a stubborn screw loose.

Screwdrivers

A variety of threaded fasteners used in the automotive industries are driven by a *screwdriver*. Some, like the self-tapping sheet metal screw, are common. Others, like the Torx or clutch head, are less common. Each fastener requires a specific kind of screwdriver. The well-equipped technician will have several sizes of each. See Figure 4–28.

All screwdrivers, regardless of the type of fastener they are designed for, have several things in common. The size of the screwdriver is determined by the length of the shank or the blade. The size of the handle is important, too. The larger the handle diameter, the better grip it has and the more torque it will generate when turned.

Do not use a screwdriver as a chisel, punch, or pry bar. Screwdrivers were not made to withstand bows or bending pressures. When misused in such a fashion, the tips will wear, become rounded, and tend to slip out of the fastener. Its usefulness will be impaired, and a defective tool is a dangerous tool.

Standard Tip Screwdriver

A **standard screwdriver** has a single blade that fits a screw with a slotted head. The blade and lengths should match the job. The blade tip width and thickness should fit the screw head perfectly (Figure 4–29).

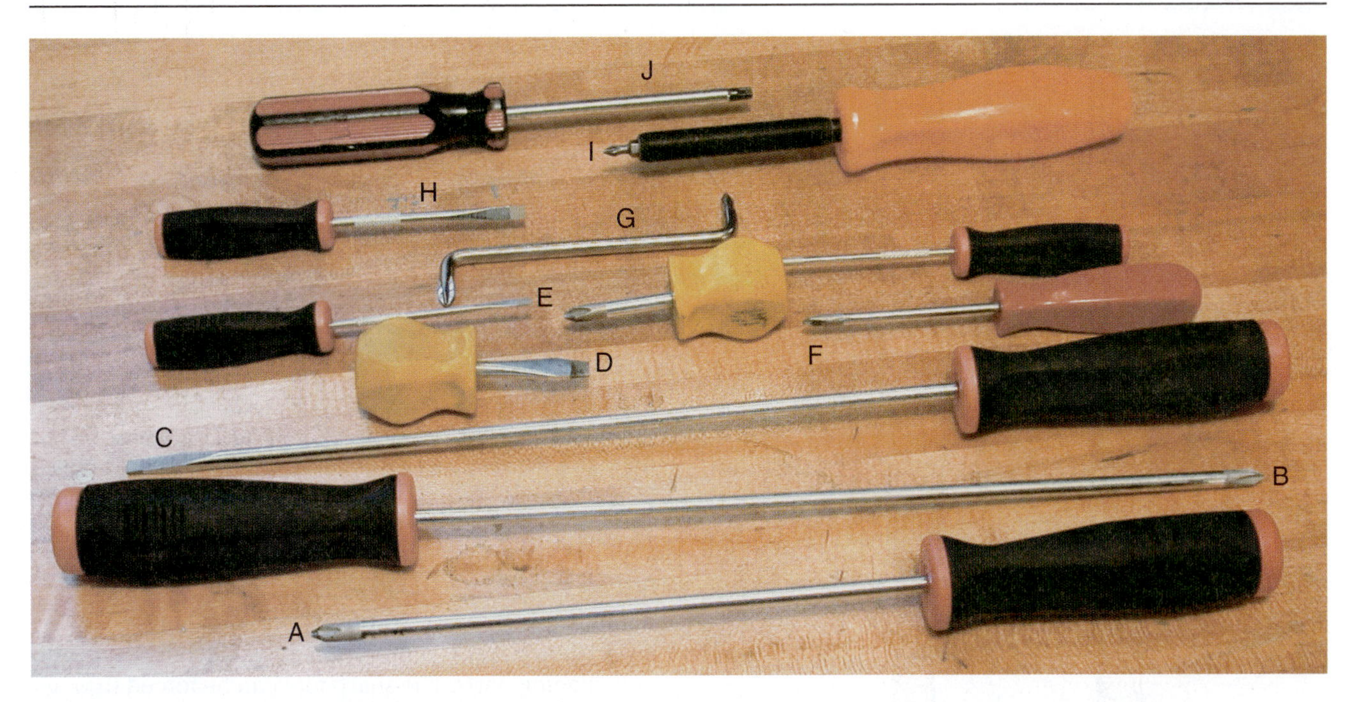

Figure 4–28 Study the different kinds of screwdrivers: (A) Phillips screwdriver, (B) long Phillips, (C) long standard, (D) stubby standard, (E) stubby Phillips, (F) small Phillips, (G) offset screwdriver, (H) small standard, (I) insulated electrical, (J) Torx screwdriver.

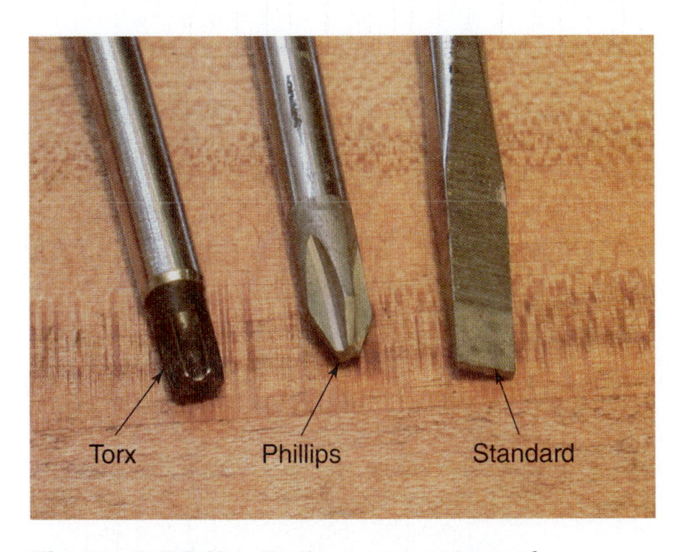

Figure 4–29 Note the three common types of screwdriver heads.

A good set of standard tip screws will have five to seven screwdrivers from the 1½-inch stubby to a 10-inch driver with a heavy-duty 5⁄16-inch blade.

Phillips Screwdrivers

The tip of a **Phillips screwdriver** has four prongs that fit the four slots in the Phillips screw head (Figure 4–29). This type of fastener is often used in the automotive field. Not only does it look nicer than the slot head screw, but it also is easier to install and remove. The four surfaces enclose the screwdriver tip so there is less likelihood that the screwdriver will slip off the fastener. Phillips screws, unlike the standard tips, can also be installed by automated machinery. This is the primary reason they are used on vehicles today.

A set of three Phillips screwdrivers with number 1, number 2, and number 3 tips will handle most body shop requirements. Purchase a set with large, insulated handles. They are easier and safer to use.

Phillips screwdrivers have one disadvantage. The prongs of the tip tend to wear and round off. Unlike a standard tip screwdriver, a worn Phillips cannot be sharpened. Therefore, it should be discarded and replaced.

Specialty Screwdrivers

A number of specialty fasteners have been replacing slot and Phillips head screws. These new breeds of fasteners are designed to improve transfer of torque from screwdriver to fastener, slip less, result in less work fatigue, and offer some tamper resistance. Most of these screwdriver bits will prove useful in auto body repair. Three of the most often used screwdrivers are the clutch head, Pozidriv, and Torx.

Clutch Head Screwdriver

There are two kinds of clutch head screws, the older G-style and the newer A-style. The G-style clutch type screwdriver tip (Figure 4–30) has an hourglass profile. It fits into a similarly shaped slot in a clutch head screw. Used most frequently by General Motors, this type of

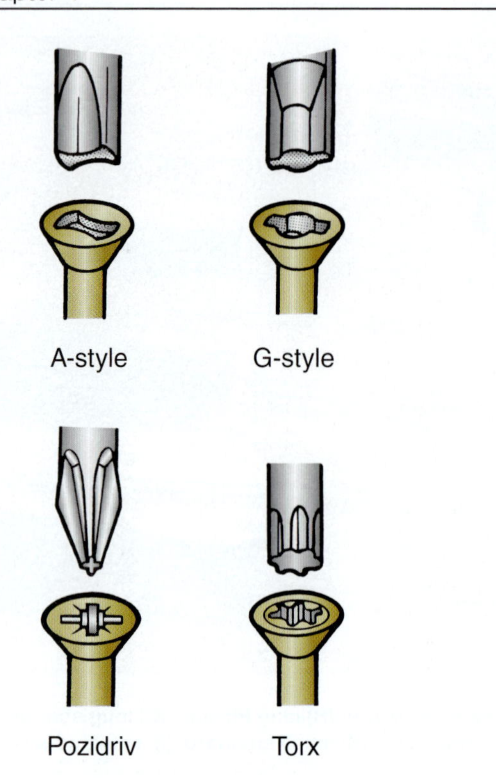

Figure 4–30 Note the less common types of screwdriver heads.

fastener system provides a more positive engagement with less slippage.

Pozidriv Screwdriver

This screwdriver is also like a Phillips but with a tip that is flatter and blunter. The square tip grips the screw head and slips less than a Phillips screwdriver. Less slippage results in less aggravation and fatigue, and lengthens the life of the screwdriver as well.

Torx Screwdriver

The Torx fastener is becoming more and more common. It is used in a variety of industries, including the automotive industries. Many American-made automobiles use the star-shaped Torx fastener to secure headlight assemblies, mirrors, and luggage racks. Not only does the six-prong tip provide greater turning power and less slippage, but the Torx fastener also provides a measure of tamper resistance. The popularity of Torx fasteners makes having a complete set of Torx screwdrivers a necessity for today's auto body repair technician.

A *prying screwdriver* has a very thick, strong shank and an angled tip. It can be used to move parts and panels into position. For example, a prying screwdriver might be used to force an engine belt tight for adjustment. See Figure 4–31.

Pick tools have points on their ends for moving small clips and other parts into alignment. Picks with curved

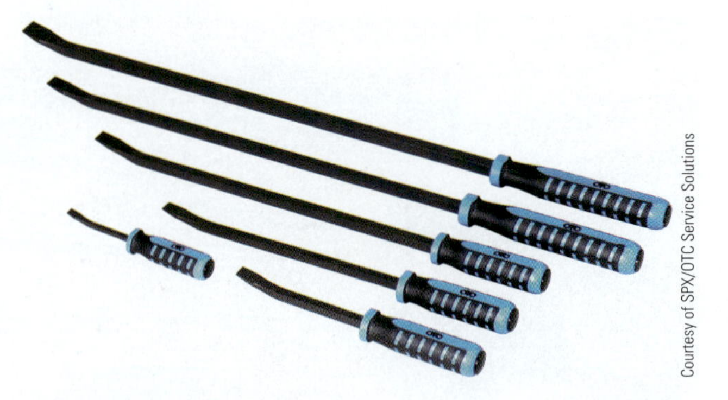

Figure 4–31 Thick shank screwdrivers are designed for prying. An angled tip lets you pry or move parts easily. Do not pry with a conventional screwdriver, because it could break or snap off.

tips are handy for reaching into parts, for example, when removing washers or bearings. Look at Figure 4–32A.

A *scratch awl* is a hand tool for scratching marks on and aligning parts. The sharp tool can be forced through the holes in several parts or panels to help align them during installation. Refer to Figure 4–32B.

Pliers

Pliers are an all-around grabbing and holding tool for working with wires, clips, and pins. The auto body repair technician must own several types: standard pliers for common parts and wires, needle nose for the really small parts, and large adjustable pliers for heavy-duty work, including bending sheet metal. See Figure 4–33.

Combination Pliers

Combination pliers are the most common type of pliers. The jaws have both flat and curved surfaces for holding flat or round objects. Also called *slip-joint pliers*, combination pliers have two jaw-opening sides. One jaw can be moved up or down on a pin attached to the other jaw to change the opening.

Adjustable Pliers

Adjustable pliers, commonly called *channel-locks*, have a multiposition slip joint that allows for many jaw opening sizes. Adjustable pliers are useful for grasping objects of varying sizes. The long handle provides plenty of turning leverage. Adjustable pliers are available with flat or curved jaws.

Needle Nose Pliers

Every auto body repair technician should have at least one 6- or 8-inch pair of needle nose pliers. **Needle nose pliers** have long, tapered jaws. They are indispensable for grasping small parts or for reaching into tight spots. Many needle nose pliers also have wire cutting edges

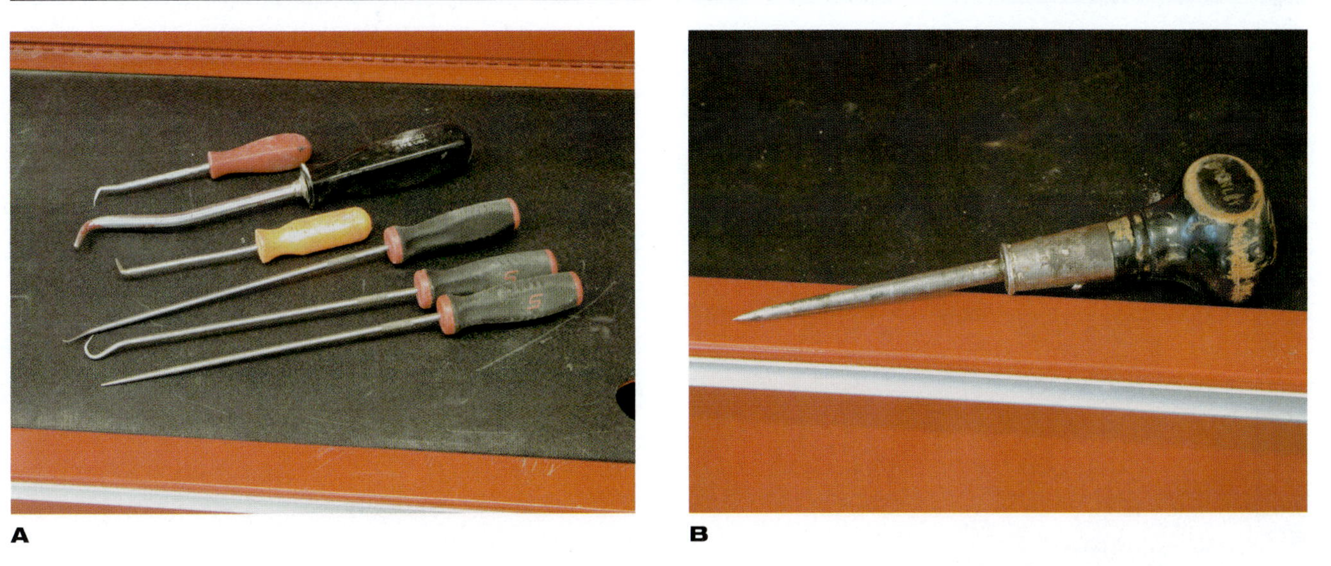

A B

Figure 4-32 Sharp or pointed tools can be used to mark sheet metal or move parts into alignment. (A) Pick tools are handy for aligning small parts, like body clips. (B) A scratch awl has a sharp, hardened point for marking panels or parts to be cut.

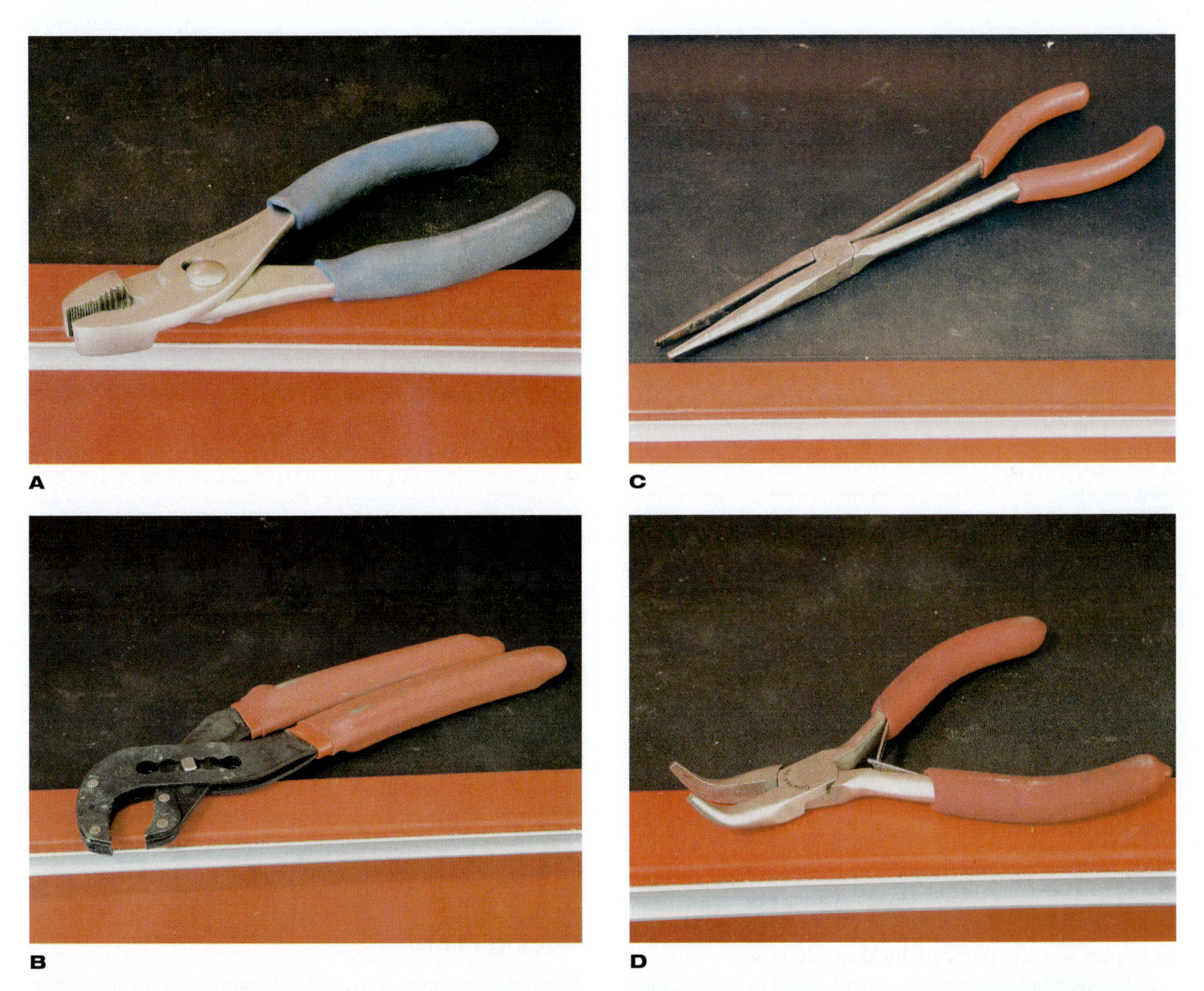

A C B D

Figure 4-33 Study plier types and their uses. (A) Slip-joint combination pliers are commonly used to grasp and hold parts while working. (B) Adjustable pliers have strong holding power but should not be used in place of a wrench. They will damage bolt and nut heads. (C) Needle nose pliers will reach into tight places and hold small parts. (D) Offset needle nose pliers will reach around corners to hold parts. (E) Snap ring pliers are used to reach into small holes to service snap rings in mechanical parts. (F) Fender pliers allow you to reach behind the lip on a fender or quarter panel to grasp or bend the inner lip on panels.

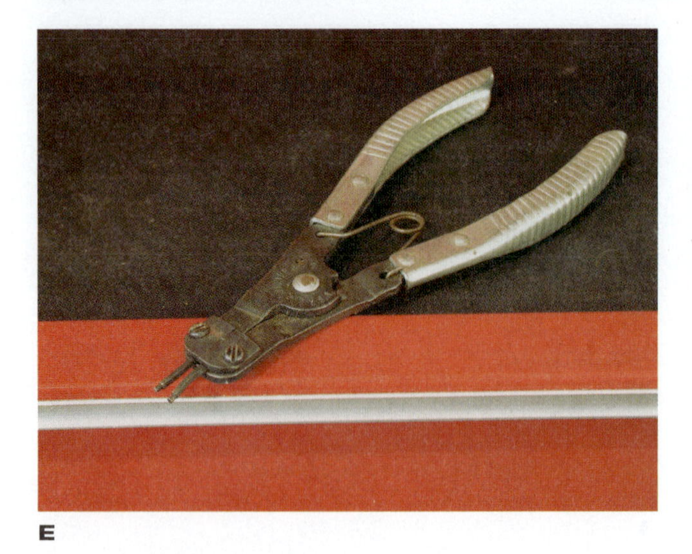

E

F

Figure 4–33 ***(continued)***

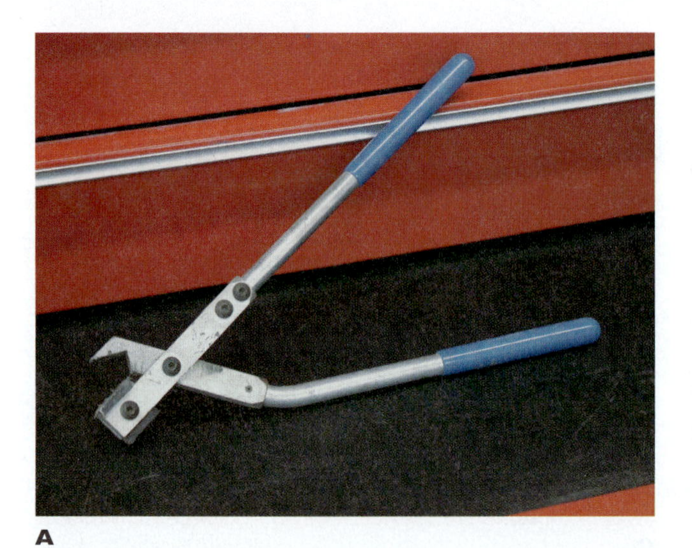

A

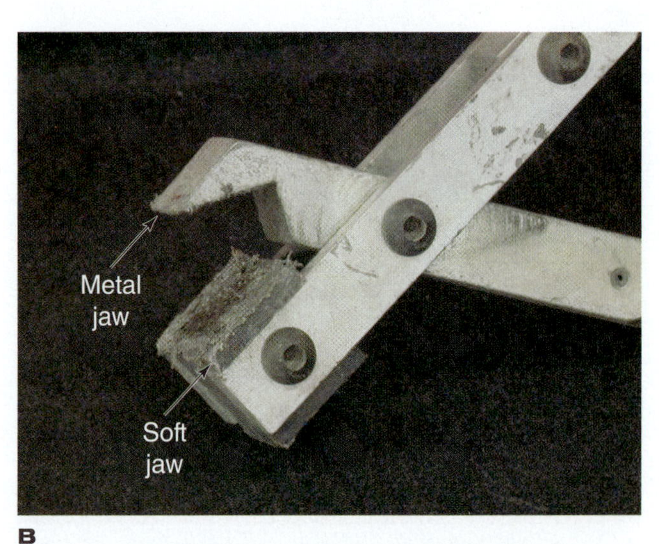

B

Figure 4–34 These folding pliers have one soft jaw and a metal jaw for crimping flanges down during panel repairs. They can be used when installing a new sheet metal door skin, for example. They will crimp the edge of the skin down around the door frame.

and a wire stripper. These are very handy for electrical work (headlights, for example). A needle nose with a 90-degree bend in the jaws is also handy for reaching behind or around obstacles. Figure 4–34 and Figure 4–35 show specialized plier types.

Locking Pliers

Locking pliers, or **Vise-Grips®**, are similar to standard pliers except that they lock closed with a very tight grip. They are extremely useful for holding parts together. For example, several pairs of locking pliers will come in handy when holding a replacement panel in position for spot welding (Figure 4–36). Locking pliers are also useful for getting a firm grip on a badly rounded, or stripped off fasteners on which wrenches and sockets are no longer effective. Locking pliers come in several sizes and jaw configurations for use in many auto body jobs (Figure 4–37).

C-clamp, welding, and duckbill types are among those frequently used. The C-clamp type is handy for reaching over and clamping pieces with flanges or beads. The welding Vise-Grip® has a special shaped jaw for gripping and aligning the weld joints in brazing or welding operations. The duckbill pliers have wide jaws for holding and bending sheet metal (Figure 4–38).

Cutting Pliers

Cutting pliers have sharp jaws for snipping through wire and other objects. See Figure 4–39.

Diagonal cutting pliers have sharp angled jaws for cutting though wire, cotter pins, and other metal or plastic parts.

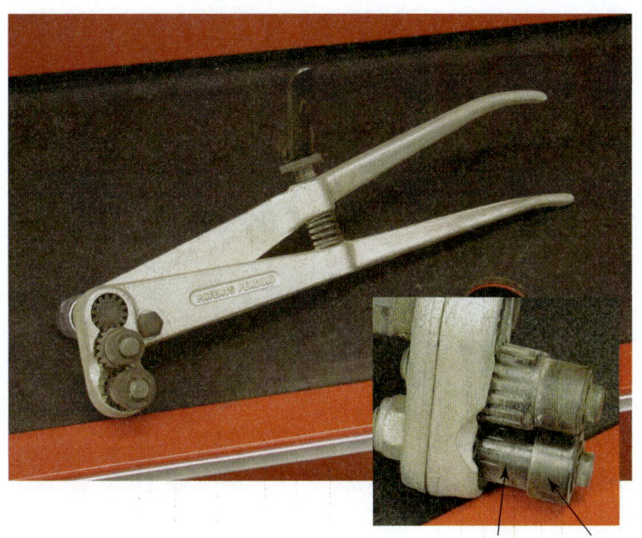

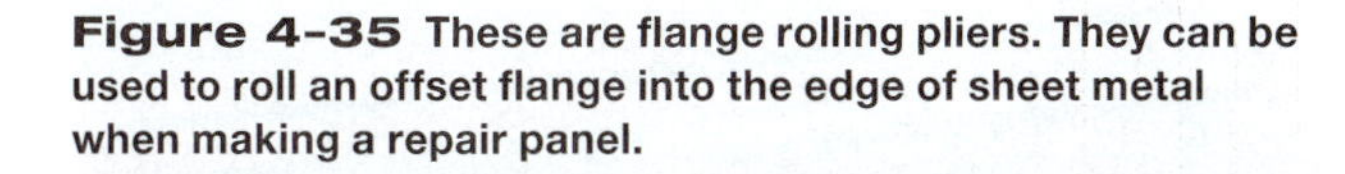

Figure 4-35 These are flange rolling pliers. They can be used to roll an offset flange into the edge of sheet metal when making a repair panel.

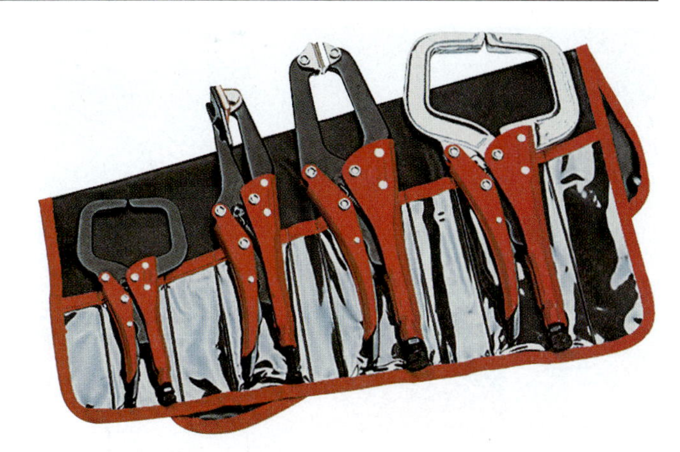

Figure 4-38 These locking pliers are commonly used to hold metal body panels in place while welding. They will fit around and clamp panel flanges together.

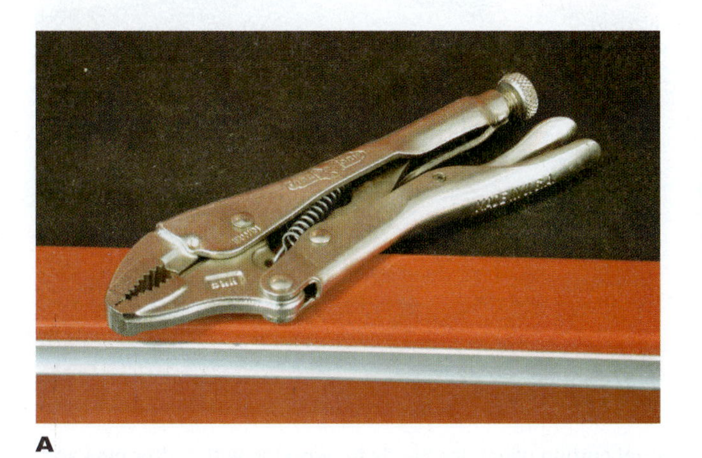

A

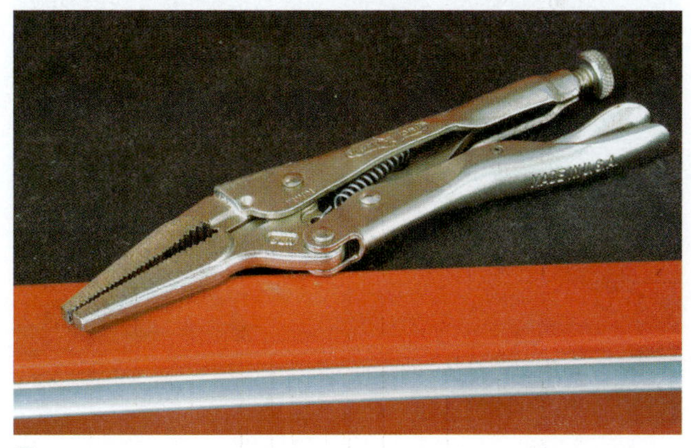

B

Figure 4-36 Locking pliers will lock onto a part, freeing your hands for other tasks. (A) Vise-Grips® pliers will lock securely onto parts. (B) These are long nose Vise-Grips® pliers.

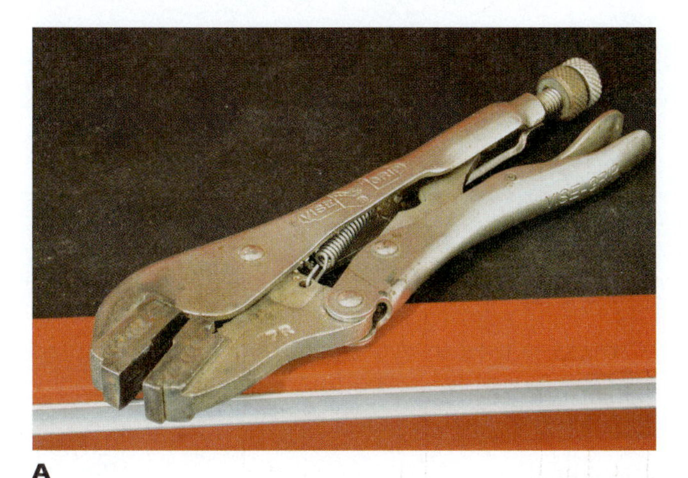

A

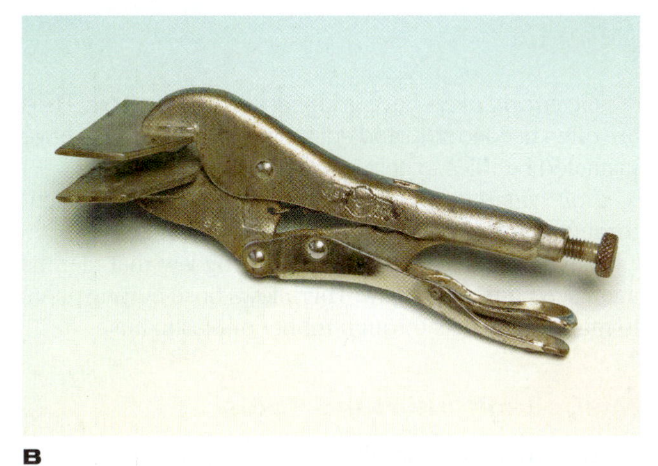

B

Figure 4-37 These Vise-Grips® pliers have special offset jaws for working sheet metal. (A) These locking pliers have offset jaws for forming a lip on sheet metal. (B) By clamping jaws down over sheet metal, an offset lip is formed so the repair piece can be welded flush with the existing panel.

A

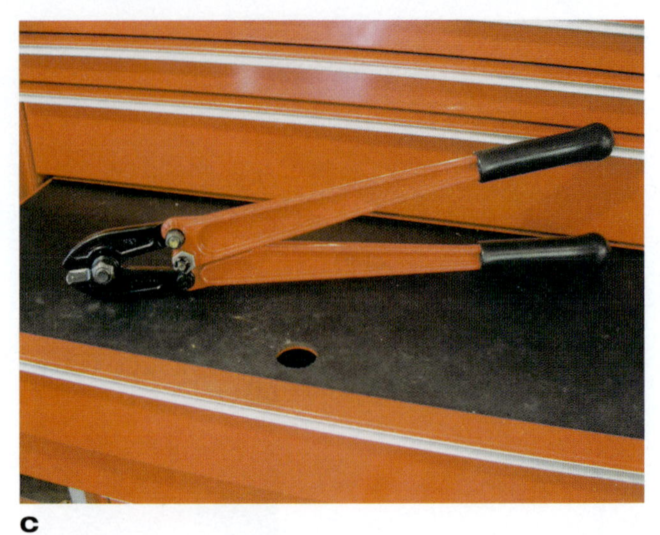

C

B

D

Figure 4–39 **Study the various types of cutting pliers. (A) Diagonal cutting pliers are handy for working with cotter pins and wires. They will bend and cut cotter pins and wires easily. (B) Electrician pliers have flat jaws for holding and cutter blades for snipping off wires. Handles are insulated to protect against electrocution. (C) Bolt-cutting pliers will produce tremendous cutting force. (D) Hose-cutting pliers will snip off rubber and plastic hose ends squarely.**

Electrician pliers have gripping and cutting jaws. They are often used to strip and cut wiring. They normally have insulated handles to help prevent electrocution.

Bolt cutters are very large and can be used to cut through thick steel fasteners and other objects.

Hose cutters have one sharp cutting jaw that clamps against a blunt or dull jaw. This allows hose-cutting pliers to make clean cuts through rubber or plastic hose.

Miscellaneous Hand Tools

A variety of other, miscellaneous hand tools will be useful from time to time. Many are inexpensive; most have a variety of uses. A few are very expensive and will probably be provided by the shop.

Utility Knife

A *utility knife* with a retractable razor blade is useful for general purpose cutting or trimming. Most come with extra blades stored in the handle. See Figure 4–40.

Scrapers

A *scraper* is used to remove old body filler, paint, and adhesive as well as to apply body filler and glazing putty when a plastic spreader or squeegee is not appropriate.

Wire Brush

Wire brushes are often used for cleaning, such as cleaning the weld joints of welding flux. A wire brush should

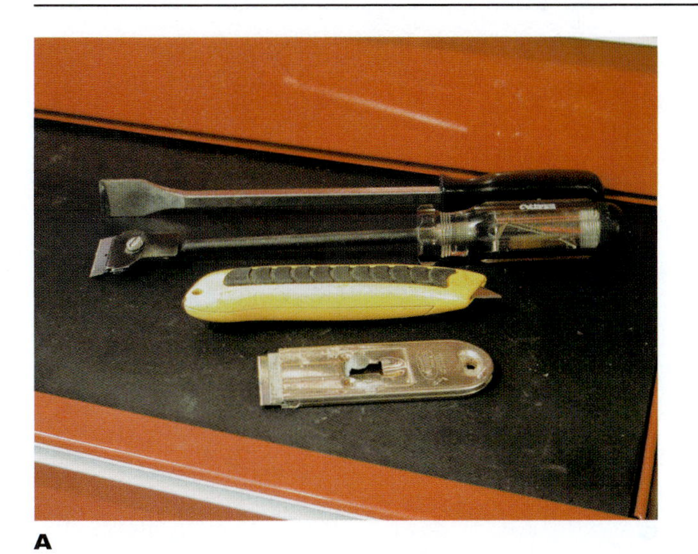

A

B

Figure 4–40 Note some of the scraping and sharp cutting tools used in collision repair. (A) Putty knives can be used for mixing and light scraping of soft materials. (B) Razor blade scrapers and knives are super sharp and will cut or scrape efficiently.

be used sparingly on bare metal because it can leave scratches in the metal.

C-Clamps

When a third hand is needed, a *C-clamp* is often the answer to hold parts in place while you are working. C-clamps come in various sizes, and a variety will be very useful in a body shop.

Clamping objects in a bench vise can damage them. Use soft lead or plastic jaw covers when there might be a chance of part damage.

Vise

A medium-duty *bench vise* is normally a shop-provided tool used for holding metal objects while grinding, bending, or welding. Most have a swivel base and serrated jaws. Clamping the workpiece in a vise frees both hands and keeps the workpiece stationary (Figure 4–41).

Rim Wrench

The *rim wrench*, or lug *wrench*, is used to remove wheel lugs (nuts or bolts holding the wheel to the axle or hub flange). A *four-way lug wrench* has four hex-shaped wrenches, or sockets. Each lug socket is sized to fit one of the four popular sizes of wheel lug nuts available. Metric sizes are also available.

Figure 4–42 shows various specialized tools used to collision repair.

Figure 4–41 A bench-mounted vise will secure parts during disassembly, drilling, cutting, and so on.

Tap and Die Set

A **tap** is a tool that cleans and rethreads holes. A **die** straightens damaged threads on bolts or studs. A tap and die set will perform most rethreading tasks in the body shop.

4.2 BODYWORKING TOOLS

Bodyworking tools include some familiar, general purpose metalworking tools as well as specialized tools used only in auto body repair. The following is a description of the most commonly used bodywork tools.

Hammers

A number of different hammers are useful in the body shop. Many are specially formed for a specific metal shaping operation.

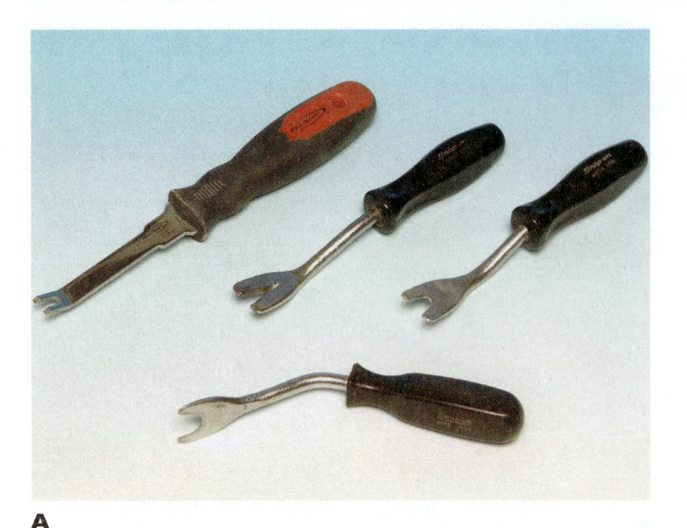

A

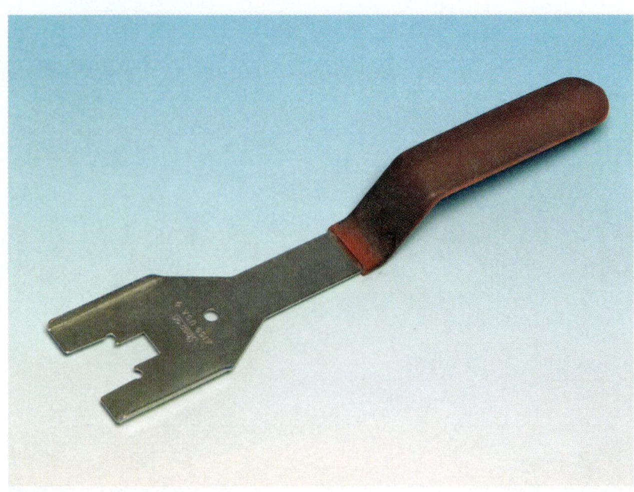

B

Figure 4-42 These tools are used to remove body clips by reaching behind a part to access the clip without damage to the part or the trim panel.

Ball Peen Hammers

The **ball peen hammer** is a useful, multipurpose tool for all kinds of work with sheet metal. Heavier than the body hammer, it is used for straightening bent underpinnings, smoothing heavy gauge parts, and roughly shaping body parts. It is sometimes used before work with a body hammer and dolly begins. Several ball peen hammers of different weights will see a lot of use in a body shop (Figure 4–43).

Sledgehammer

A light **sledgehammer** is an essential tool for the first stages of re-forming damaged thicker metal parts. Those with short handles can be used in tight places. The sledgehammer can be used to clear away damaged metal when replacing a panel (Figure 4–44).

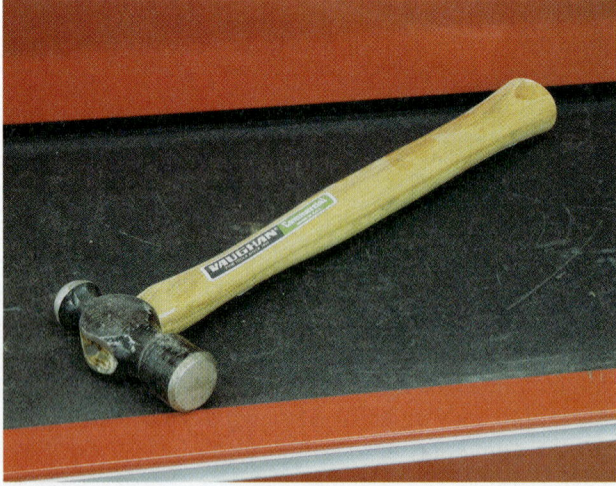

Figure 4-43 A ball peen hammer has a flat head and a rounded head. The flat head is for hammering on flat surfaces and the round head is for concave surfaces. The size of the hammer head should match the amount of driving force needed for the job.

Mallets

The *rubber mallet* (Figure 4–45A) gently bumps sheet metal without damaging the painted finish. It is often used with the suction cup on soft cave-in–type dents. While you pull upward on the cup, the mallet is used to tap lightly all around the surrounding high spots. A popping sound occurs as the high spots drop and the low spot springs back to its original contour.

A steel hammer with rubber tips is another mallet useful in bodywork. The *soft-faced hammer*, as it is sometimes called, is used to work chrome trim and other delicate parts without marring the finish.

Figure 4-44 The heavier mini-sledge hammer will produce a much more powerful blow than a small hammer.

A

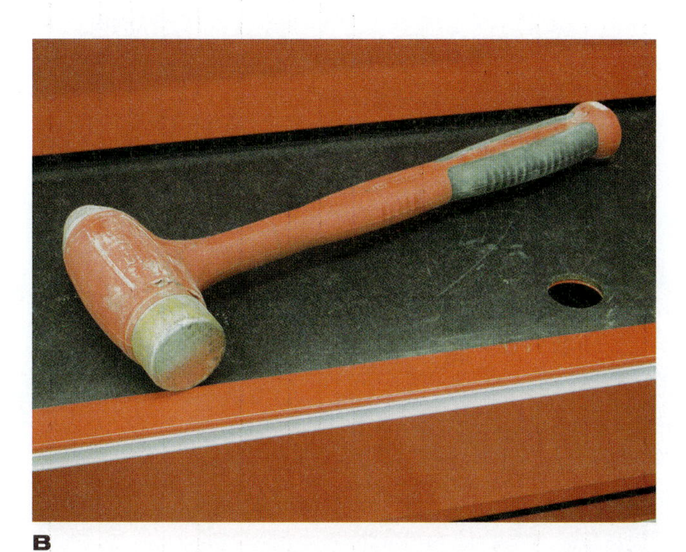

B

C

Figure 4–45 (A) Rubber mallets has a rubber face and is used on soft cave-in type dents. (B) This dead blow hammer has a metal face, but the head is filled with lead shot. This keeps the hammer from rebounding after striking an object. (C) This is a plastic-faced dead blow hammer. It will not mar or damage surfaces as easily as a metal-faced hammer head.

A **dead blow hammer** has a metal face filled with lead shot (balls) to prevent rebounding. It will not bounce back up after striking (Figure 4–45B and C).

Body Hammers

Body hammers are the basic tools for working sheet metal back into shape. They come in many different designs. As shown in Figure 4–46, they have flat, square, rounded, or pointed heads. Each style is designed for a special purpose.

Picking Hammers

The **picking hammer** has a pointed tip on one end and usually a flat head on the other. It will take care of many small dents. The pointed end is used to raise low spots from the inside. A gentle tap in the center usually does it (Figure 4–47).

The flat end is for hammer-and-dolly work to remove high spots and ripples. Picking hammers come in a variety of shapes and sizes. Some have long picks for reaching behind body panels. Some have sharp pencil points; others have blunted bullet points. Select the head best suited for the job. See Figure 4–48.

Be careful when using the pick hammer. If swung forcefully, the pointed end can pierce the lighter sheet metals used in late model cars. Use the pick only on small dents, and control impact force.

Bumping Hammers

Larger dents require the use of a **bumping hammer**. Bumping hammers can have round faces or square faces that are almost flat. The faces are large so that the force of the blows is spread over a large area. These hammers are used for initial straightening on dented panels or for working inner panels and reinforced sections that require more force but not a finish appearance. See Figure 4–49.

Sharp concave surfaces, such as the reverse curves on quarter panels, headlights, doors, and so on, require the use of a reverse curve light bumping hammer. The faces of these hammers are crowned—one in the opposite direction of the other. The tight curve of the faces allows concave contours to be bumped without the danger of stretching the metal. Remember that the contour of the hammer must be smaller than the contour of the panel to avoid stretching the metal.

Figure 4–46 A body hammer and dollies are often needed to take minor dents out of sheet metal. Body hammers have specially shaped heads for working sheet metal. Dollies are specially shaped blocks of steel for straightening sheet metal. Body hammers are the primary striking tools used in collision repair. (A) A body hammer has the head shape for working sheet metal. This one has large flat heads for flattening sheet metal. (B) This body hammer has rounded heads for forcing a curve into sheet metal. (C) This body hammer head is flat and smooth for working damage out of sheet metal. (D) The serrated body hammer head will shrink metal after it has been stretched from collision damage.

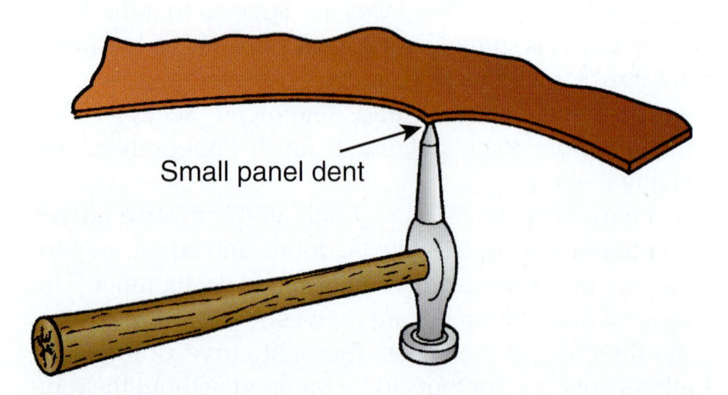

Figure 4–47 The pick hammer's pointed end can be used to raise a dent in sheet metal.

Finishing Hammers

After the bumping hammer is used to remove the dent, final contour is achieved with the **finishing hammer**. The faces on a finishing hammer are smaller than those of the heavier bumping hammer. The surface of the face is crowned to concentrate the force on top of the ridge or high spot (Figure 4–50).

A *shrinking hammer* is a finishing hammer with a serrated or cross-grooved face. This hammer is used to shrink spots that have been stretched by excessive hammering.

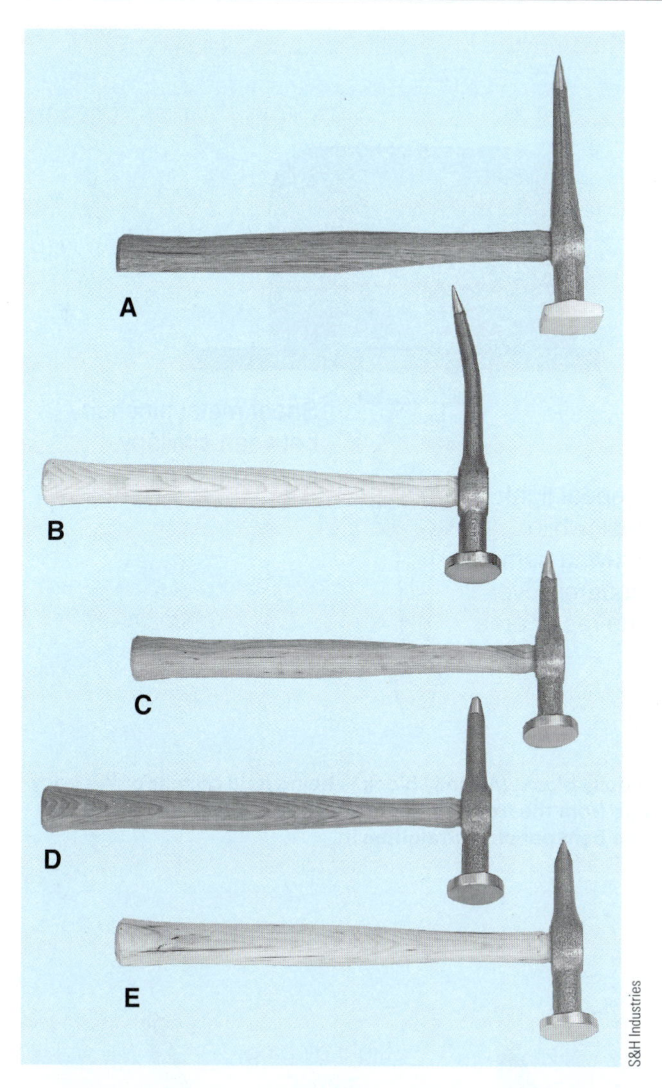

Figure 4–48 Note pick hammer names: (A) long pencil point, (B) long curved pencil point, (C) short pencil point, (D) short bullet point, and (E) short chisel point.

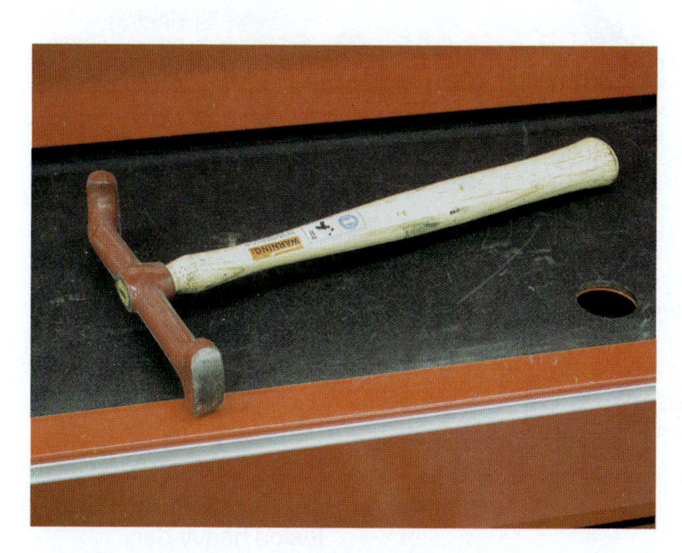

Figure 4–49 This body bumping hammer has a very large, long head on it for working obstructed areas on damaged panels.

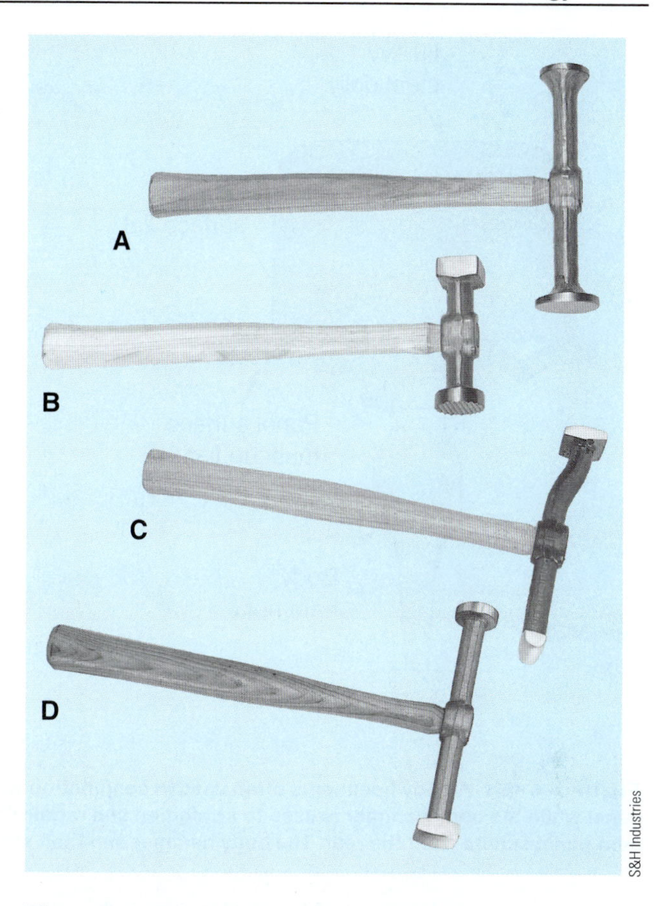

Figure 4–50 Here are several finishing hammers: (A) double round, (B) shrinking hammer, (C) offset bumping, and (D) dinging hammer.

Dollies

The **dolly** or *dolly block* is used like a small anvil while body damage is worked out. It is generally held on the back side of a panel being struck with a hammer. Together the hammer and dolly work high spots down and low spots up (Figure 4–51).

There are many different shapes of dollies (Figure 4–52). Each shape is intended for specific types of dents and body panel contours—high crowns, low crowns, flanges, and others. It is very important that the dolly fit the contour of the panel. If a flat dolly or one with a low crown is used on a high crown panel, additional dents will be the result.

A *general purpose* dolly has many contours. It can be used in most situations. A rail-type dolly is another commonly used dolly with many contours. Toe and heel dollies are used for bumping in tight places. The flat right angle edge is used for straightening flanges.

Spoons

Body spoons are another class of bodyworking tools used like a hammer or a dolly. They are available in a

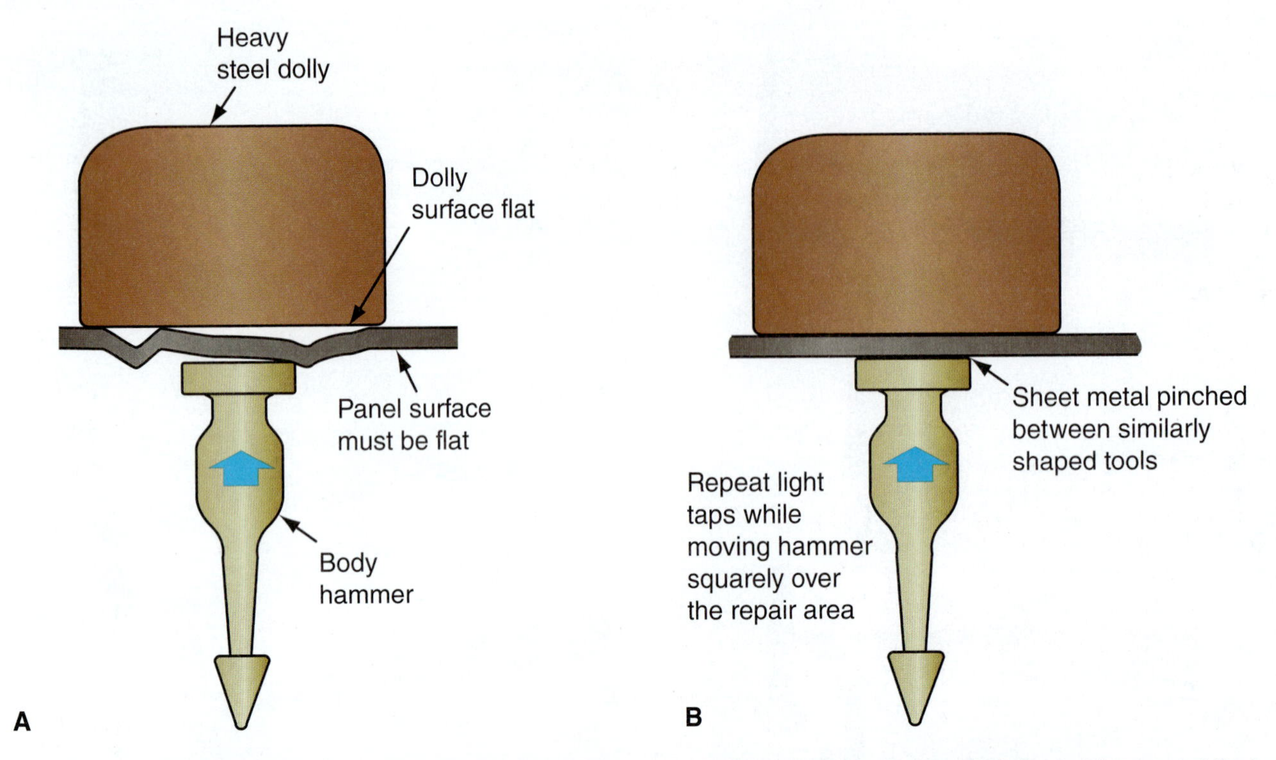

Figure 4-51 A body hammer is often used in conjunction with a dolly block. (A) Dolly block is being held on rear of the body panel while the body hammer is used to straighten and repair damage from the front. (B) Dolly acts as a small anvil to hold the bent panel secure from the rear. The body hammer can then strike the bent panel to straighten it.

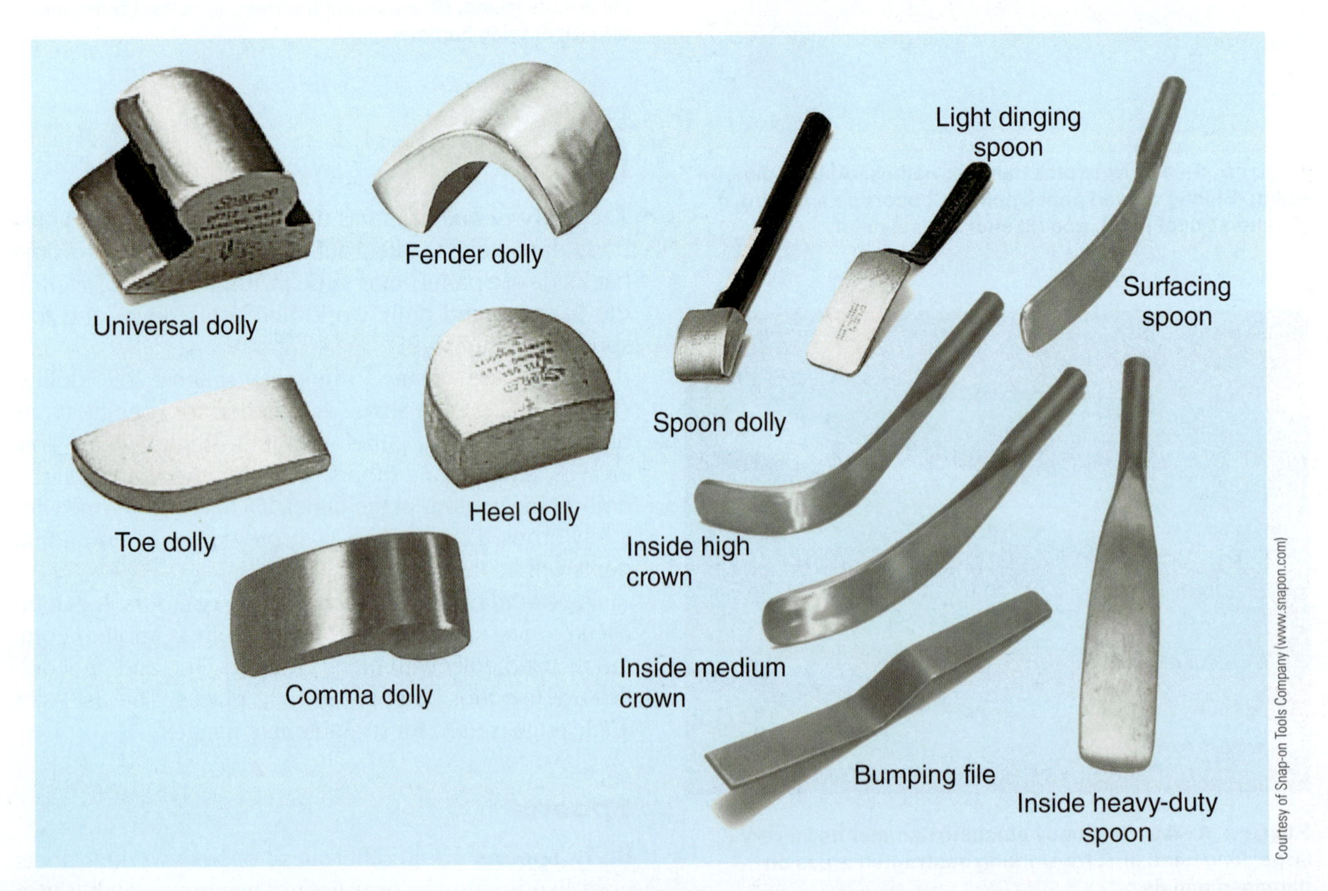

Figure 4-52 Study the various dolly block and spoon shapes. The shape should match the contour of the body panel being straightened.

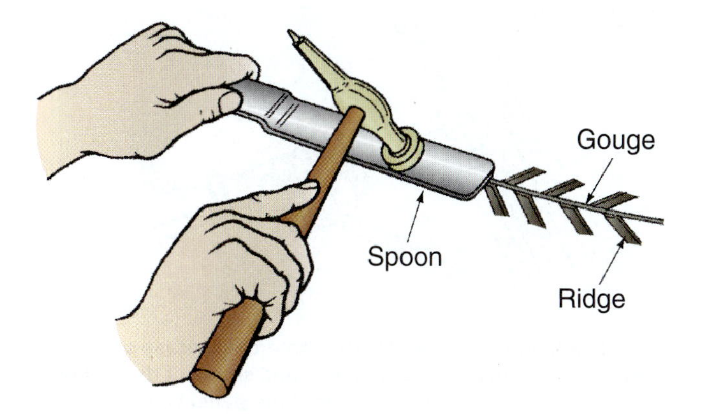

Figure 4-53 A minor ridge or bump in a sheet metal body panel can be lowered by placing the spoon over a high spot and hitting it with a hammer. The spoon increases the surface area so that hammering will not produce dents.

variety of shapes and sizes to match various panel shapes. The flat surfaces of a spoon distribute the striking force over a wide area (Figure 4–53).

Spoons are particularly useful when metalworking creases and ridges. A spoon dolly can be used as a dolly where the space behind a panel is limited. A dinging spoon is used with a hammer to work down ridges. Inside spoons can be used to pry up low places, or they can be struck with a hammer to drive up dents. Bumping files have serrated surfaces and are used to slap ridges or the underside of creases to bump the metal back to its original shape.

Picks

Picks, like spoons, are used to reach into confined spaces. The pick is used only to pry up low spots. Picks vary in length and shape, and most have a U-shaped end that serves as a handle. They are commonly used to raise low spots in doors, quarter panels, and other sealed body sections.

Picks are often preferred to slide hammers and pull rods because they do not require drilling holes into the metal and subsequently welding them shut after the repair (Figure 4–54).

Picks are sometimes used during *paintless dent removal* (removing small body dings or dents without painting the panel).

NOTE *Straightening tools and techniques are discussed further in other chapters. Refer to the index for additional information.*

Dent Pullers and Pull Rods

Creases in sealed body panels that cannot be reached from the back side can be pulled out with a *dent puller* (Figure 4–55). Never drill or punch holes in the crease for the puller or pull rod. Now, the common practice is to

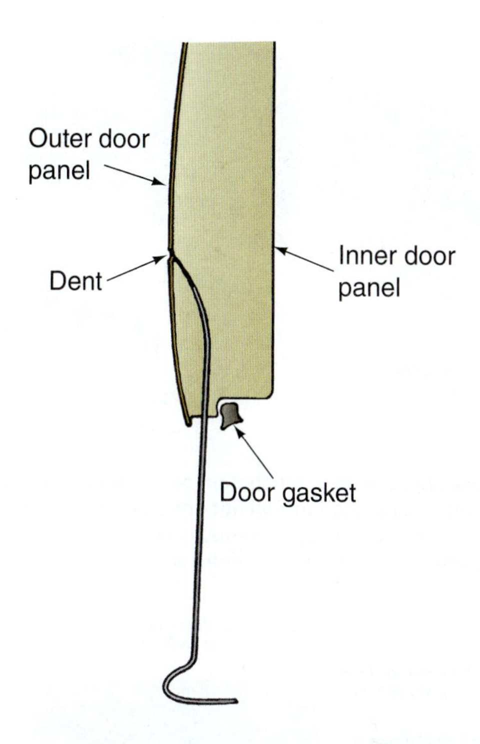

Figure 4-54 A pick can be inserted into obstructed areas, like inside a door, to help pry out small dents. This is the basis for paintless dent removal. If you carefully position and work the pick, you may not have to paint the panel.

weld a bracket or pull pin onto the surface instead of drilling. Either will give the rod something to grab and pull on.

A dent puller usually comes with a threaded tip and a hook tip. Either tip is inserted in the drilled hole or welded rod or bracket (Figure 4–56). Then the slide hammer is pulled back and struck against the handle. Tapping the slide hammer against the handle slowly pulls up the low spot (Figure 4–57).

A small dent or crease can be pulled up with a single pull rod (Figure 4–58).

Suction Cups

The *suction cup* (Figure 4–59) will pull out shallow dents quickly if they are not locked in by a crease in the metal. Simply attach the suction cup to the center of the dent and pull. The dent might come right out with no damage to the paint and no refinishing required. It is an easy tool to use and can make a simple repair. However, once a dent is locked in, some hammer-and-dolly work will be necessary to smooth the metal. Even so, the suction cup method is usually worth a try.

Punches and Chisels

A good set of punches and chisels (Figure 4–60) is absolutely necessary in every bodyworking tool chest.

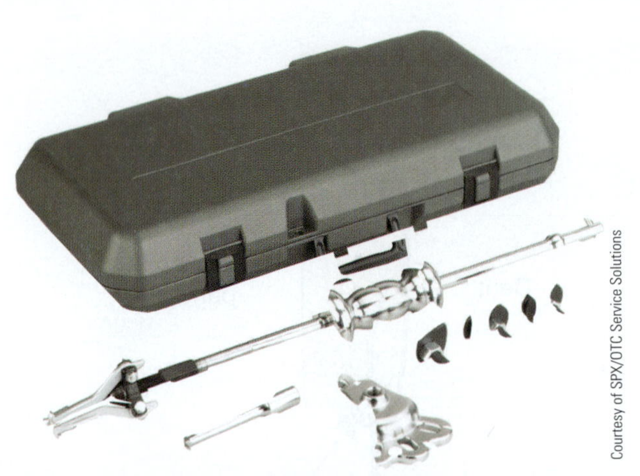

A Depending upon the attachment on the end of a slide hammer, it can be used to pull out drive axles or dents in the panels. A dent puller set has various attachments for hooking over a damaged panel to pull dents out.

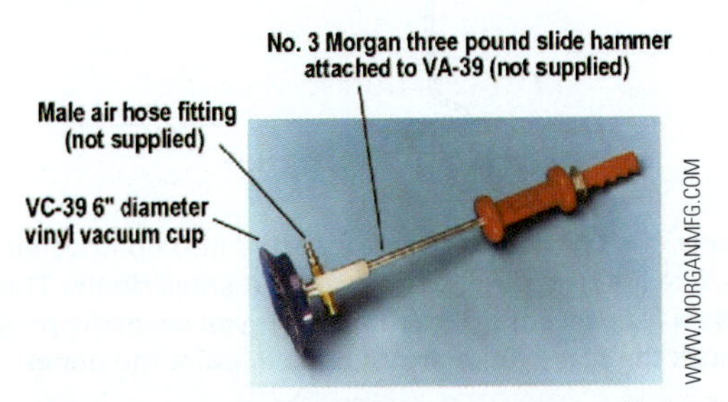

B A suction cup, when attached to an air line, will produce a powerful holding force for pulling large but shallow dents in sheet metal. Puller equipped with a suction cup head. It is attached to a compressed air source to form a powerful vacuum for popping out more stubborn dents.

Figure 4-55 Slide hammer set will produce a powerful pulling force on parts and panels.

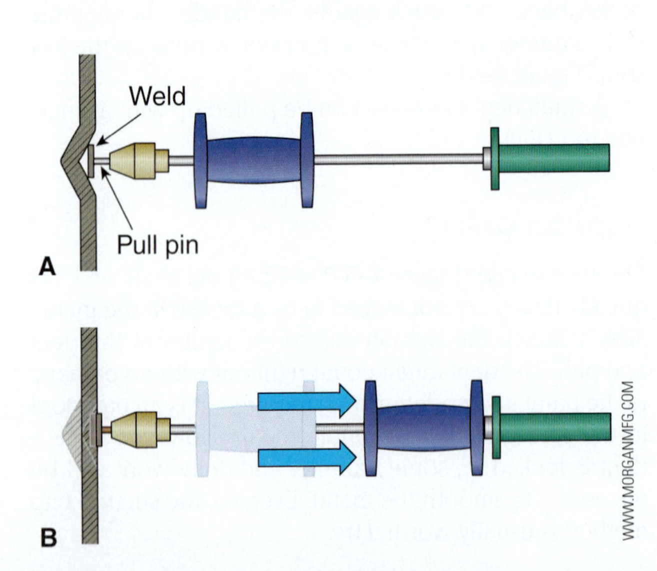

Figure 4-56 To pull a dent with a dent puller, (A) resistance weld a bracket, puller head, or pull pin to the surface of the repair area. (B) Slide the hammer back to pull out the dent.

A A slide hammer with hook attachment will grasp and pull the edges of panels easily. The size and shape of the pull tip should roughly match the contour of the panel being repaired.

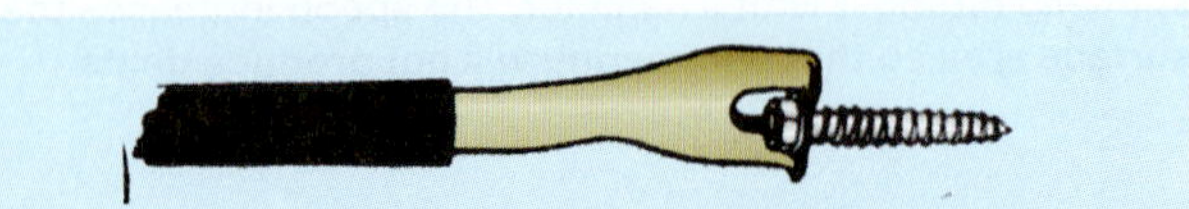

B A screw tip should only be used to pull panels if factory holes already exist in the panel or if the panel is going to be replaced, after frame straightening, for example. Never drill holes in panels for pulling with screws if the panel is being repaired. Drilled holes would have to be welded shut, and the weld can burn off corrosion protection on the back side of the panel, which can cause rusting of the repaired panel.

C A suction cup attachment on a slide hammer will sometimes pop out very minor, shallow stress dents.

D A cutter attachment will slice difficult-to-access panels' flanges to aid in removal.

Figure 4-57 Study the varied uses of a slide hammer puller. By pulling back briskly on the heavy weighted handle, a powerful outward pulling blow is exerted. Be careful not to pinch your hand between the two handles! The heavy striker handle can severely injure your hand.

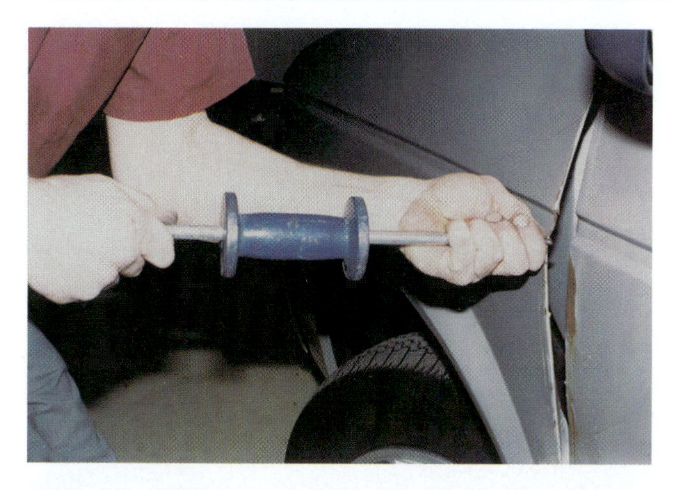

Figure 4–58 Here the technician is using a slide hammer to pull out and straighten the bent edge of a front fender.

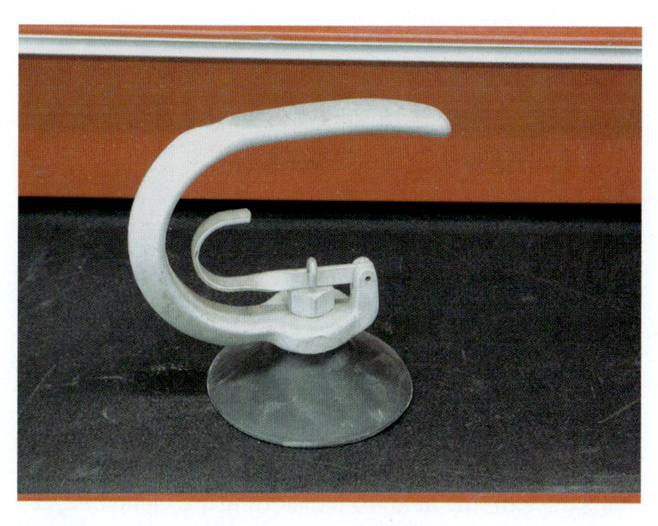

Figure 4–59 This hand-held suction cup tool will grasp and pull on glass or smooth body panels.

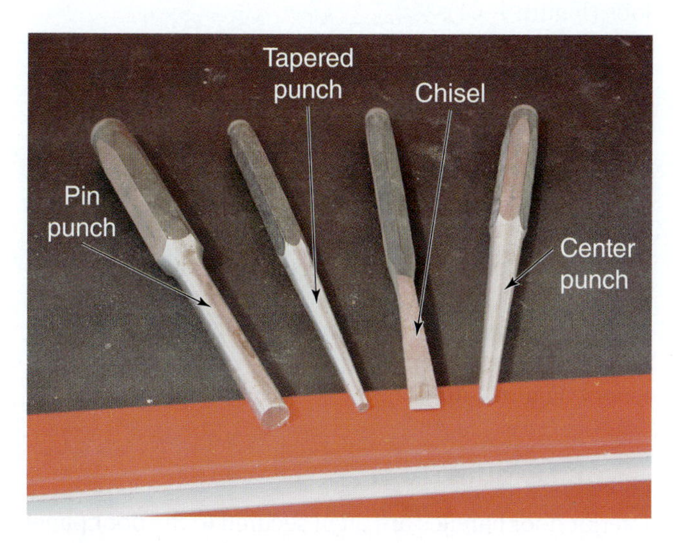

Figure 4–60 Study the names of punches and chisels.

DANGER Keep the end of a chisel or punch ground properly. If the end is mushroomed or enlarged from hammering, grind it down. A mushroomed end could cut your hand, and metal fragments could fly into your face.

Center punches are used to mark parts before they are removed and to mark a spot for drilling. The punch mark keeps the drill bit from wandering.

A *drift* or *starter punch* has a tapered point with a flat end used to drive out rivets, pins, and bolts. A *pin punch* is similar to the drift except its shaft is not tapered, so it can be used to drive out smaller rivets or bolts.

An **aligning punch** is a long, tapered punch used to align body panels for welding and for starting bolts. For example, one might be used to align fender bolt holes and a bumper.

A **chisel** is a steel bar with a hardened cutting edge for shearing steel. These chisels come in various sizes, and a set is necessary for both light- and heavy-duty work. The cold chisel is used to split frozen nuts, shear off rusted bolts, cut welds, and separate body and frame parts.

Scratch Awl

A *scratch awl* is very similar in appearance to an ice pick, but the pointed steel shank is heavier. A scratch awl is used to pierce holes in light-gauge metal when a specific size hole is not required. It is also used to mark metal for cutting, drilling, or fastening. Keep the awl ground to a sharp point so it can be used effectively and safely in every job.

Metal-Cutting Shears

Most body repair technicians have at least one pair of shears or tin snips. *Snips* are used to trim panels or metal pieces to size. Several types of metal cutters are useful.

Tin Snips

Tin snips (Figure 4–61) are perhaps the most common metal-cutting tool. They can be used to cut straight or curved shapes in sheet metal and aluminum.

Metal Cutters

Metal cutters, also called aviation snips, are used to cut through metal panels. The narrow profile of the jaws allows the snips to slip between the cut metal. The jaws are serrated to cut through the tough metal.

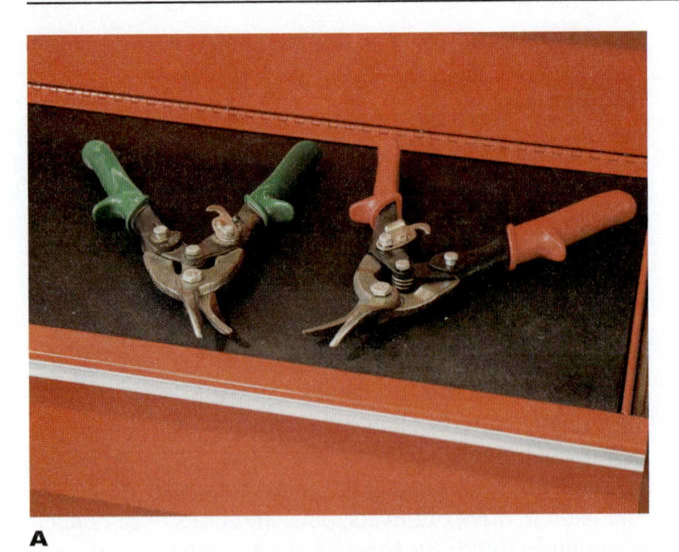

A

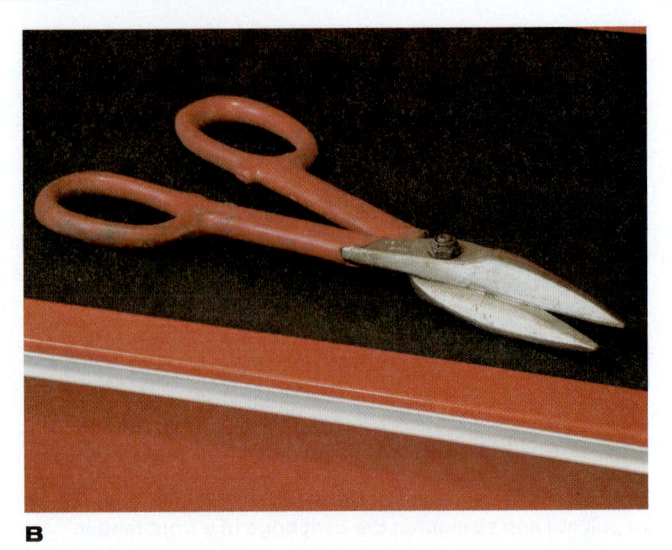

B

Figure 4-61 Study the types of snips often used in collision repair. (A) Sheet metal cutting pliers come in right- and left-hand configurations for cutting in different directions. (B) Straight-jaw sheet metal cutting pliers are handy when you have plenty of room for cutting.

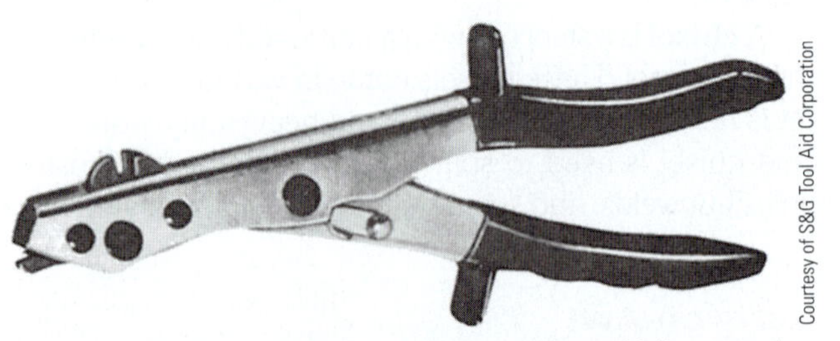

Courtesy of S&G Tool Aid Corporation

A A panel cutter, or two-way nibbler, will cut thicker panels easily.

Courtesy of The Eastwood Company, www.eastwood.com.

B The center jaw pinches between two stationary jaws to shear a thin strip of sheet metal away for cutting.

Figure 4-62 A panel nibbler will cut thick sheet metal easily even when making curved cuts.

Panel Cutters

Panel cutters are special snips used to cut through body sheet metal. These are used to make straight or curved cutouts in panels that require spot repair for rust or damage. They are designed to leave a clean, straight edge that can be welded easily (Figure 4–62).

Rivet Gun

Pop rivets are sometimes used to hold panels in place while repairs are made. They can be inserted into a blind hole through two pieces of metal and then drawn up with a riveting tool. This locks the pieces of metal together (Figure 4–63). There is no need to have access to the back of the rivets. They are used as temporary fasteners before the replacement sheet metal is welded. This prevents the extreme heat from distorting the metal or creating a safety hazard (such as around the gas tank).

A good rivet gun does not cost much. The most commonly used rivets in bodywork are ⅛ inch and 3⁄16 inch. A few others of assorted sizes might be needed for special jobs (Figure 4–64).

A heavy-duty riveter is used to rivet hard-to-reach places and heavier mechanical assemblies, such as a window glass regulator. It has long handles, a long nose, and sets 3⁄16- to ¼-inch blind rivets.

Trim and Upholstery Tools

Any repair work that requires removing interior trim and some body moldings will be facilitated with an *upholstery tool*. This prong-shaped prying tool is used to slip under and pry up upholstery tacks, springs, clips, and other fasteners (Figure 4–65).

Door Handle Tool

Interior door handles are often secured to the door panel by wire spring clips. These clips, shaped like horseshoes, fit over the handle shaft and hold the handle tightly against

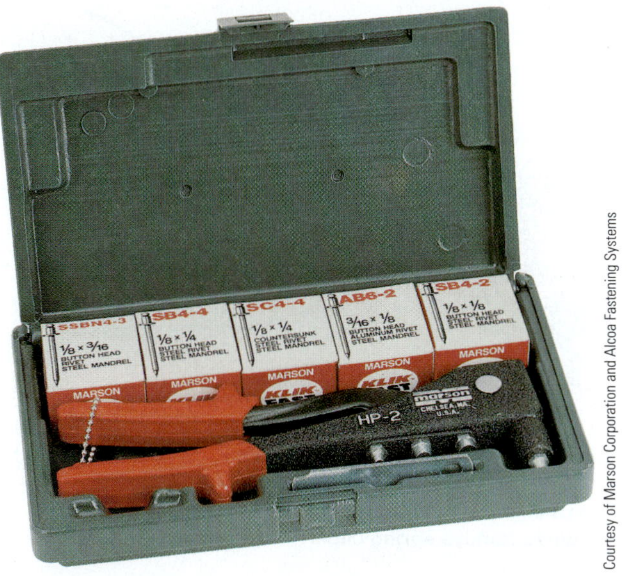

A A hand riveter in kit form for blind rivets.

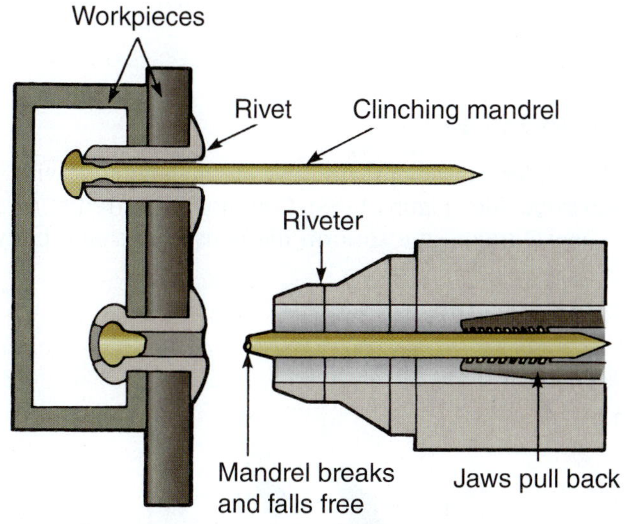

B When the rivet gun handle is squeezed, it will pull and expand the back side of the rivet head to secure parts together.

Figure 4–63 Note how rivets can hold two panels together. Factory-installed rivets are becoming common again.

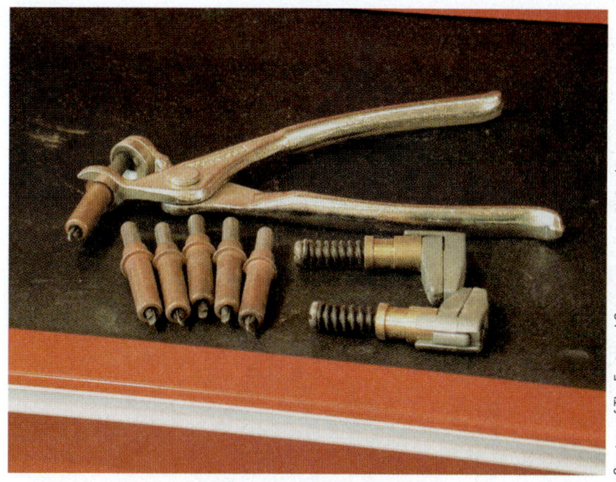

A These spring-loaded rivets and clamps can be installed and removed easily using special pliers.

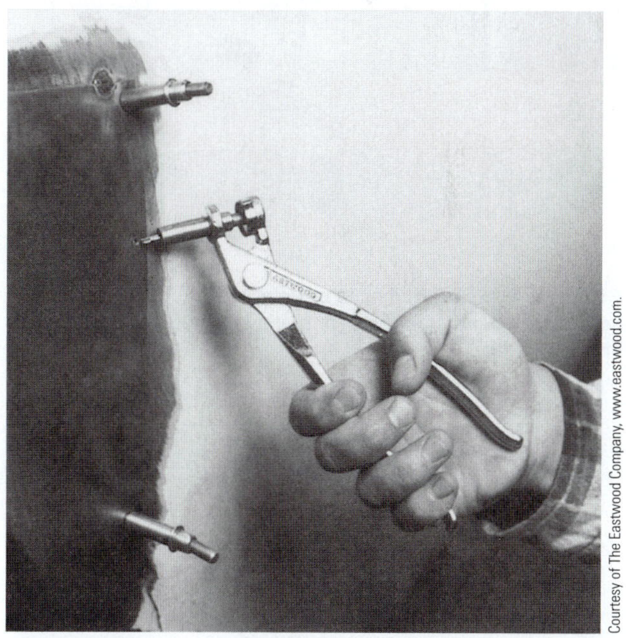

B These spring-loaded rivets can be installed and removed easily. They are ideal for holding panels together while welding.

Figure 4–64 Panel holding tools will temporarily align parts before welding.

the interior panel trim. **Clip pullers**, or *door handle tools* (Figure 4–66), are needed to reach inside the door and remove the clip. Some door handle tools pull the clip out; others push the clip off the shaft. Figure 4–67 shows an assortment of window and door tools.

4.3 BODY SURFACING TOOLS

A number of surfacing tools are used to give a repair its final shape and contour. Some are used to shape the repaired metal. Others are used to apply and shape plastic body filler and putty.

Metal Files

After working a damaged panel back to its approximate original contour, a *metal file* is used to mark (scratch) the metal to find high and low spots. Two special files are necessary for most bodywork.

Reveal File

The *reveal file* is a small file that is available in numerous shapes. Generally, it is curved to fit tightly crowned areas such as around windshields, wheel openings, and other panel edges. The reveal file is pulled, not pushed,

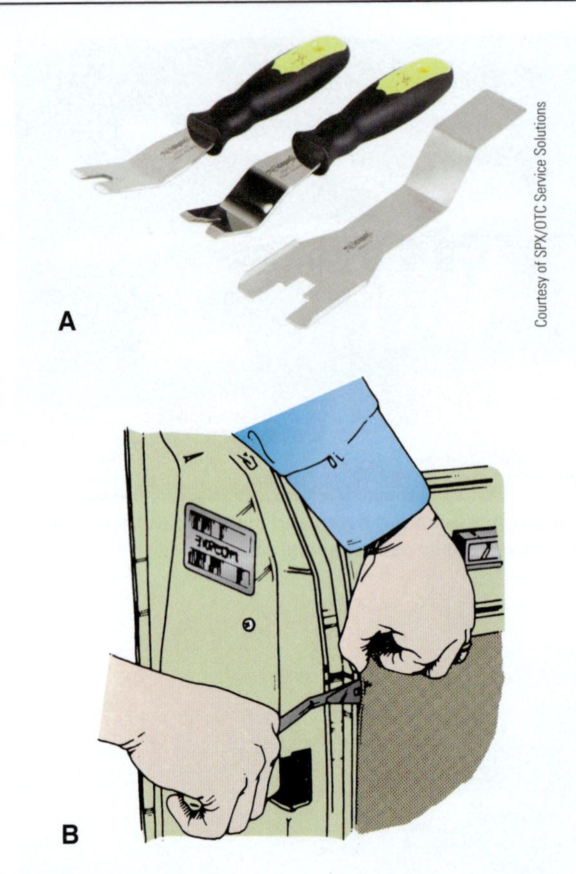

Figure 4–65 A body clip or upholstery tool will reach behind trim panels or parts for pulling out clips without breaking them.

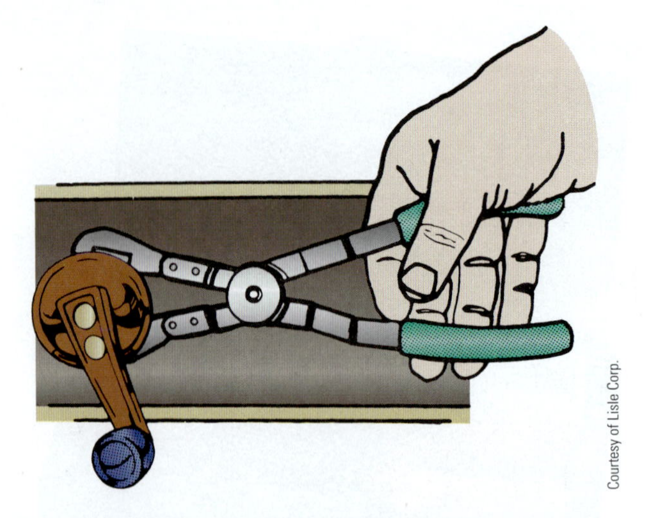

Figure 4–66 A door handle tool will reach behind the handle to remove spring clips.

when used. Pushing causes the file to chatter, resulting in nicks and an uneven surface.

Surform File

Body filler can be cut level and to rough contour with a **surform file** (Figure 4–68). Commonly referred to as a "cheese grater," the surform file is used to shape body

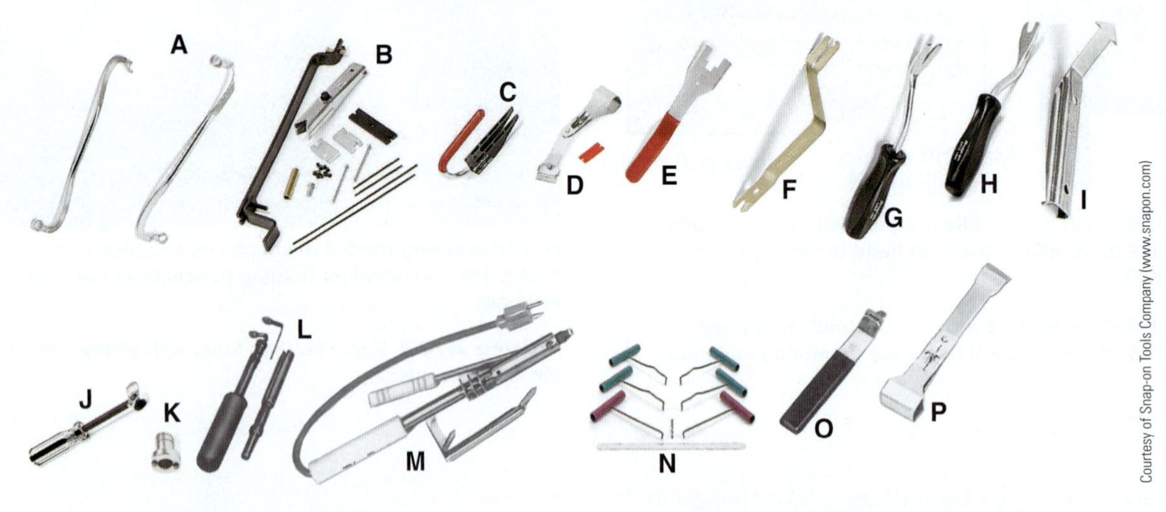

A. Door hinge bolt wrenches
B. Door removal kit
C. Door panel remover (GM and Ford)
D. Door panel remover
E. Door handle tool (GM, some Fords)
F. Door handle tool (Chrysler)
G. Trim pad remover (GM, Ford, Chrysler)
H. Trim pad remover (GM)
I. Window molding release tool
J. Windshield locking strip installation tool
K. Window sash nut spanner socket
L. Windshield remover
M. Hot-tip windshield removing kit
N. Windshield wiper removal tool
O. Windshield wiper tool
P. All-purpose window scraper

Figure 4–67 Study the names of window and door tools.

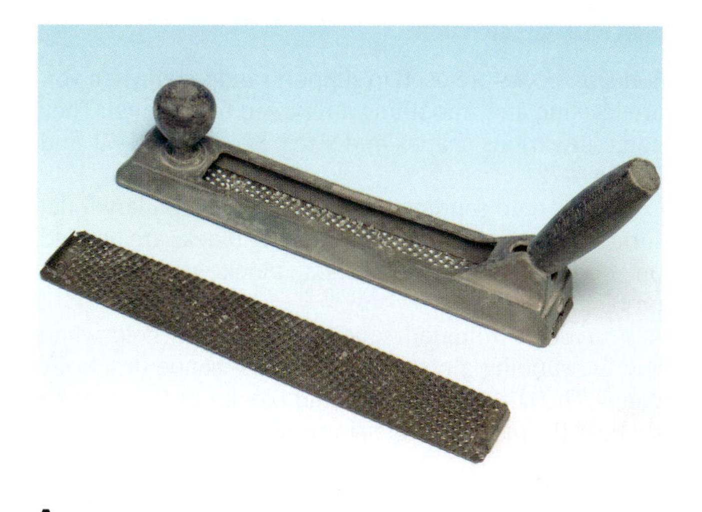

A

B

Figure 4-68 A body file is often used to rough-cut high spots in body filler after partial curing. This saves sanding time and does not produce sanding dust. Large teeth in the body file will scrape off semi-hard body filler quickly and easily.

filler while it is semihard. Shaping the filler before it hardens reduces sanding and shortens the waiting period while the filler cures. A dirt nib file is used to remove minor paint imperfections (Figure 4–69).

Figure 4-69 A dirt nib file is a small, finger-held tool used to carefully cut off dirt and dust particles in new paint.

Sanding Board

Once the body filler has hardened, the repair can be shaped and leveled with a *sanding board.* The long sanding board is a rigid wooden holder about 17 inches long and 2¾ inches wide. Also sometimes called a *flatboy*, the sanding board allows a repair area to be sanded quickly with long, level strokes, using 36- or 40-grit sandpaper. This eliminates waves and uneven areas (Figure 4–70).

The extra-long length helps avoid creating a wavy surface. The sander also flexes to match the panel contour. Adhesive-backed sandpaper is applied from a roll.

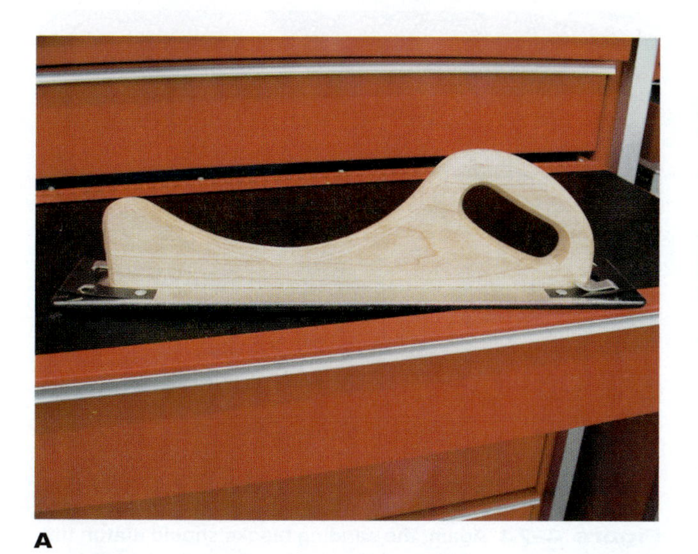

A

B

Figure 4-70 Hand-sanding boards are the most important sanding tools of a body technician. It is almost impossible to make a large body repair area perfectly flat without them. (A) A long sanding board will quickly level and smooth large areas of body filler on flat body parts being repaired. (B) A short sanding board is needed to level and smooth body filler on smaller repair areas.

Sanding Blocks

Sanding blocks are used to support sandpaper when you are leveling and smoothing a repaired body panel. They come in various shapes and sizes. See Figure 4–71 and Figure 4–72.

Stiff or hard sanding blocks are needed on larger flat surfaces (Figure 4–73). Softer sanding blocks are needed on curved surfaces or when block sanding paint problems.

Curved or rounded sanding blocks will help when you are sanding a contour or rounded shape in a body panel. The shape of the sanding block must match the shape of the part or panel being repaired. See Figure 4–74.

A Soft rubber sanding block with handle.

B Thin, soft sanding blocks are used for final sanding.

Figure 4–71 Soft sanding blocks are used on surfaces that have already been sanded level. They will support sandpaper but flex to follow the contour of the body panel.

Figure 4–73 Hard rubber sanding blocks, like these, are used when the surface is still rough or high and must be leveled and shaped flat.

Figure 4–72 The size of the sanding block must match the job. On the left is a tiny sanding block for repairing minor dirt damage in freshly applied paint. On the right is a large round sanding block that will accept Velcro-type sanding discs.

Figure 4–74 Again, the sanding blocks should match the shape or contour of the panel being repaired. On the left is a rounded sanding board for shaping long, curved body lines. On the right is a round rubber sanding block for shaping smaller curved surfaces.

A

B

Figure 4–75 This is a shapeable sanding block. (A) The sanding block has movable parts that can be moved to match shape of body panel. (B) The sanding block can be pressed against the undamaged surface on a body panel and then locked into that shape so the damaged area can be sanded to the right contour.

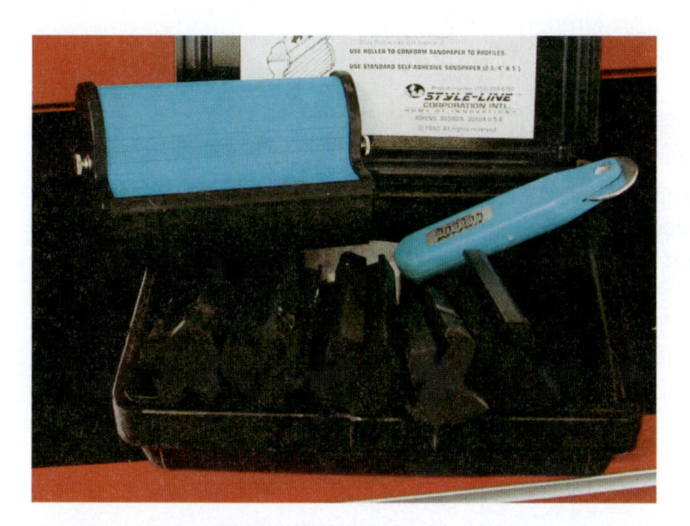

Figure 4–76 This sanding block has interchangeable shapes. You can select a shape that matches the body panel to repair complex body lines.

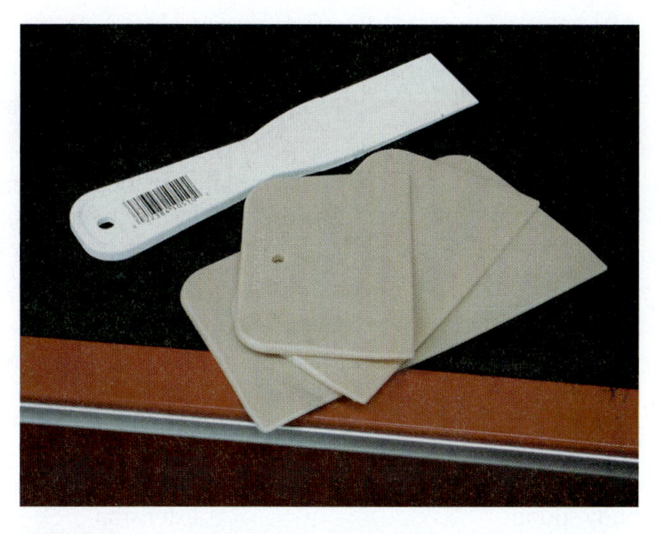

Figure 4–77 Upper left, a plastic putty knife should be used to mix body filler and hardener. Lower right, spreaders are used to apply body filler after mixing.

A contour sanding block has a shapeable or movable surface or a preformed irregular surface for sanding body lines. Look at Figure 4–75 and Figure 4–76. By matching the profile of the contour sander to the body shape on the vehicle, you can quickly repair irregular profiles in body filler.

Spreaders and Squeegees

Spreaders and squeegees are two important tools used in auto body resurfacing. **Spreaders** are used to apply body filler; they are made of rigid plastic and are available in various sizes (Figure 4–77). Be sure to use one that is large enough to apply plastic filler over the complete repair area smoothly before the filler begins to set.

A **squeegee** is a flexible rubber pad or block used to apply glazing putty and light coats of body filler. It is also used to skim water and sanding grit from the repair area when wet sanding.

4.4 HAND TOOL SAFETY

Hand tool safety begins with purchasing quality tools. Quality tools may require a greater investment in money, but the dividends—safety and durability—are worth the expense. In the long run, they cost less than cheap tools. Many quality tools are warranted to ensure their quality and protect the investment.

The second step in safe tool usage is knowledge. Read the manufacturer's instructions and use the tool only for the tasks it was designed to do. Misuse of a tool causes accidents, and it wears or weakens the tool, increasing the likelihood of slipping, chipping, or shattering.

The final step in tool safety is maintenance. Body shop tools must be kept clean, rust-free, sharp, and safely organized in a tool cabinet or chest. A tool should be maintained as close to the original condition as possible. Damaged and broken tools should never be used.

SUMMARY

1. Hand tools are extensions to the human body. They allow you to do tasks otherwise impossible with your bare hands.
2. When purchasing tools, get high-quality tools from a reputable manufacturer.
3. A complete collection of wrenches is indispensable for the auto body technician.
4. In many situations, a socket wrench is much faster and easier to use than an open-end or box-end wrench.
5. The ratchet handle is probably the most commonly used handle. The ratchet handle allows removing or tightening without removing the socket from the fastener.
6. A variety of threaded fasteners used in the automotive industries are driven by a screwdriver.
7. Pliers are an all-around grabbing and holding tool for working with wires, clips, and pins.
8. Body hammers are the basic tools for pounding sheet metal back into shape. They come in many different designs.

EXERCISES

On a separate sheet of paper, complete the following learning activities for this chapter. Write definitions for the key terms and answer the ASE-style review questions, essay questions, critical thinking problems, and math problems. You can also do the extra activities, possibly for extra credit.

Key Terms

adjustable wrench
aligning punch
Allen wrench
ball peen hammer
body spoon
box-end wrench
breaker bar
bumping hammer
center punch
chisel
clip puller
combination pliers
combination wrench
dead blow hammer
deep socket
die
dolly
drive sizes
extension
finishing hammer
needle nose pliers
open-end wrench
Phillips screwdriver
picking hammer
pick
pipe wrench
pliers
ratchet handle
sledgehammer
socket wrench
spreader
squeegee
standard screwdriver
surform file
tap
Torx fastener
universal joint
Vise-Grips®
wrench size

ASE-Style Review Questions

1. Which socket drive size should be used on large bolts or lug nuts?
   A. ¼ inch
   B. 5/16 inch
   C. ⅜ inch
   D. ½ inch
2. A striking tool should be discarded or serviced when it is
   A. enlarged from hammering.
   B. mushroomed.
   C. chipped.
   D. all of the above.

3. Technician A says that a ½-inch drive socket set is required to disassemble and remove interior trim components. Technician B says that a 1-inch drive socket set is useful for truck and heavy equipment repair. Who is correct?
   A. Technician A
   B. Technician B
   C. Both A and B
   D. Neither A nor B
4. Technician A says that a wire brush scratches bare metal and thus should be used sparingly on it. Technician B uses a wire brush on weld joints to clean off welding flux. Who is correct?
   A. Technician A
   B. Technician B
   C. Both A and B
   D. Neither A nor B
5. Which of the following wrenches provides the safest grip on a fastener?
   A. Open-end wrench
   B. Box-end wrench
   C. Adjustable wrench
   D. None of the above
6. Technician A sometimes uses a screwdriver as a chisel. Technician B sometimes uses a screwdriver as a pry bar. Who is correct?
   A. Technician A
   B. Technician B
   C. Both A and B
   D. Neither A nor B
7. Technician A says that a tap cuts external threads. Technician B says that a die cuts internal threads. Who is correct?
   A. Technician A
   B. Technician B
   C. Both A and B
   D. Neither A nor B
8. Which of the following wrenches is used with a ratchet handle?
   A. Combination
   B. Allen
   C. Socket
   D. Box
9. Which of the following screwdriver tips best prevents stripping out a fastener head?
   A. Standard
   B. Torx
   C. Phillips
   D. Slotted
10. Which of the following pliers is often used in electrical work?
   A. Adjustable pliers
   B. Needle nose pliers
   C. Retaining ring pliers
   D. Snap ring pliers
11. Which of the following hammers would be best used for initial straightening or working inner panels?
   A. Bumping
   B. Ball peen
   C. Plastic
   D. Pick

Essay Questions

1. Explain when you might use each drive size: ¼, ⅜, and ½ inch.
2. Describe the construction and organization of a typical tool chest.
3. Summarize the utilization of three types of punches.
4. What is the difference between a spreader and a squeegee?
5. Color matching guides contain information needed for what?

Critical Thinking Problems

1. What could happen if you used a body file excessively on an aluminum body panel?
2. A bolt is badly rusted and rounded off. How could you get it out?
3. What type of tool will remove small dust particles from new paint?

Math Problems

1. If a breaker bar is 2 feet long and you exert 125 lb. of force on the end of the handle, approximately how many pounds of force are exerted at the drive head?

Activities

1. Order catalogs from tool manufacturers. Compare the quality, price, and warrantees among manufacturers. A good place to search the Web for tool information is http://www.snapon.com.
2. Visit an outside body shop. Talk with other technicians about the tools they use. Discuss with the class anything you learned about new tools or techniques.
3. Tour your shop. Note the location of all hand tools. Prepare a report on your findings. Are any tools missing or damaged? Would you buy any new tools?

Power Tool and Equipment Technology

CHAPTER 5

OBJECTIVES

After studying this chapter, you should be able to:

- Identify and explain the operation of electric and air-powered tools.
- Name and select air and electric power tools most common to body shops.
- Describe the hydraulic equipment used in auto body repair.
- Maintain and use shop power equipment and tools.
- Answer ASE-style review questions relating to power tools.

INTRODUCTION

This chapter summarizes how to properly select and use power tools and equipment. Most of the tools explained here are general purpose collision repair power tools. More specialized tools and their uses are described in the appropriate chapters.

Power tools use air pressure (*pneumatic*), oil pressure (*hydraulic*), or electrical energy to effect repairs. This classification includes air wrenches, air and electric drills, sanders, and similar tools. Body shop technicians and painters must have a wide variety of power tools to make their tasks easier.

DANGER

The use of any air-driven or electric power tool requires that you wear safety glasses or a face shield. For hazardous operations, wear both. Also, never wear loose clothing that might get caught in a tool.

Power tools and equipment can be very dangerous if not used correctly. Always follow the instructions given in the owner's manual for the specific tool or piece of equipment. The information in this chapter is general and cannot cover all tool variations. If in doubt about any tool or piece of equipment, ask your instructor or shop foreman for a demonstration.

5.1 AIR-POWERED TOOLS

The automotive industry was one of the first industries to see the advantages of air-powered tools. Today they are known as "the tools of the professional technician."

Although electric drills, wrenches, grinders, polishers, drill presses, and heat guns are found in body and refinishing shops, the use of pneumatic (air) tools is a great deal more common. Pneumatic tools (Figure 5–1) have four major advantages over electrically powered equipment:

1. **Flexibility**. Air tools run cooler and have the advantage of variable speed and torque; damage from overload or stalling is eliminated. They can fit in tight spaces.
2. **Lightweight**. Air tools are lighter in weight and lend themselves to a higher rate of production with less fatigue.
3. **Safety**. Air tools reduce the danger of fire in some environments where the sparking of electric power tools can be a problem. Air tools also do not use electricity, so the danger of electrocution is reduced.
4. **Low-cost operation and maintenance**. Due to fewer parts, air tools require fewer repairs and less preventive maintenance. The original cost of air-powered tools is usually less than the equivalent electric type.

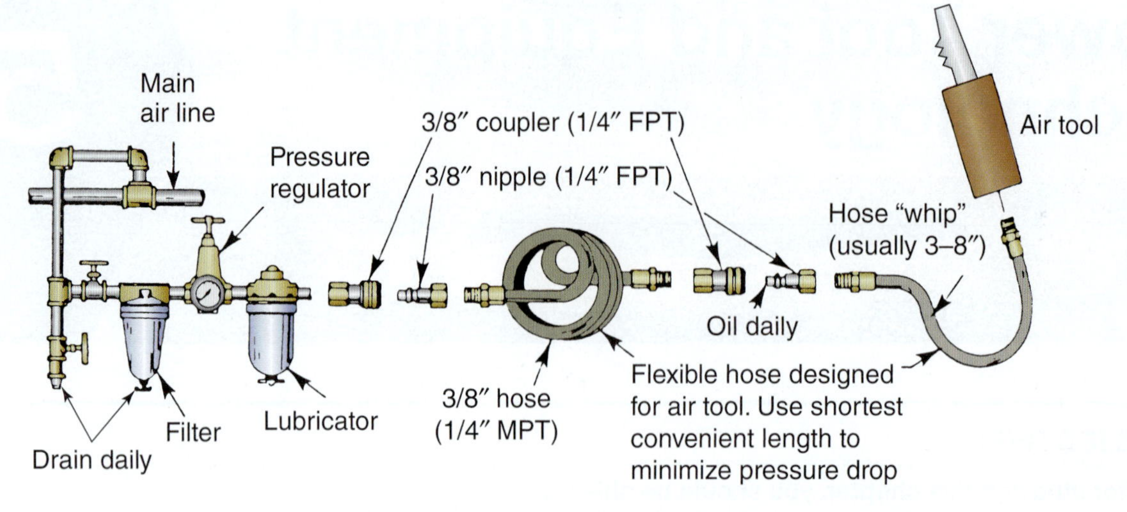

Figure 5-1 Study the parts of a typical body shop compressed air supply system for power air tools.

There are pneumatic equivalents for nearly every electrically powered tool, from sanders to drills, grinders, impact wrenches, and screwdrivers. Furthermore, there are some pneumatic tools with no electrical equivalent, such as the chisel, ratchet wrench, grease gun, and various auto tools.

Hoists, lifts, and frame and panel straighteners can be used in conjunction with a compressed air system. However, in most cases, these pieces of equipment are hydraulically operated. Described later in this chapter, hydraulic tools play a very important role in any body shop operation.

Spray Guns

Spray guns are used to apply sealer, primer, paint, and other liquid finishing materials to a vehicle (Figure 5–2). Spray guns must atomize the liquid, often paint, so that it flows onto the body surface smoothly and evenly.

A spray gun *atomizes* a liquid by breaking it into a fine mist of droplets. This requires sufficient pressure and volume at the gun, which can be powered by air or electric energy, although air is more common (Figure 5–3).

Spray Gun Parts

To use, service, and troubleshoot a spray gun properly, you must understand the operation of its major parts.

The *gun body* holds the parts that meter air and liquid. The body holds the spray pattern adjustment valve, fluid control valve, air cap, fluid tip, trigger, and related parts (Figure 5–4).

The *spray gun cup* attaches to the gun body to hold the material to be sprayed. The cup often fits against a rubber seal to prevent leakage. Another seal is mounted or formed in the lid to prevent leakage around the top of the cup.

The spray gun's *fluid control valve* can be turned to adjust the amount of paint or other material emitted. It consists of a thumbscrew or knob, needle valve, and spring. Turning the knob affects how far the trigger pulls the needle valve open. The *fluid needle valve* is seated in the fluid tip to prevent flow, or it can be pulled back to allow flow. Refer again to Figure 5–4.

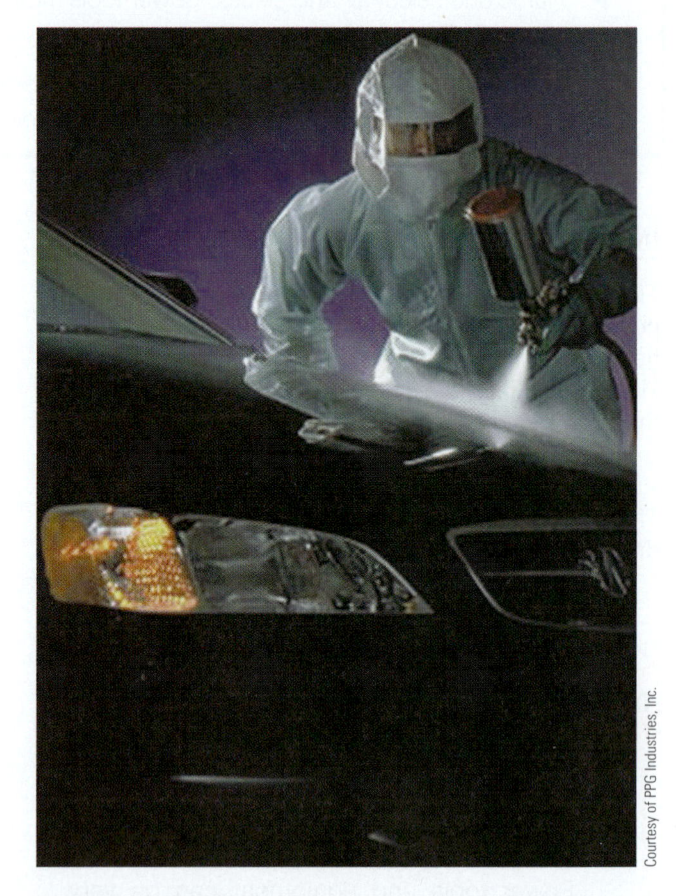

Courtesy of PPG Industries, Inc.

Figure 5-2 Here the painter is using a state-of-the-art spray gun to refinish a vehicle. The spray gun relies on a steady source of clean air for proper operation.

The spray gun's *air control valve*, or *spreader valve*, controls how much air flows out of the air cap side jets. It has an *air needle* that can be slid back and forth to open or close the air valve.

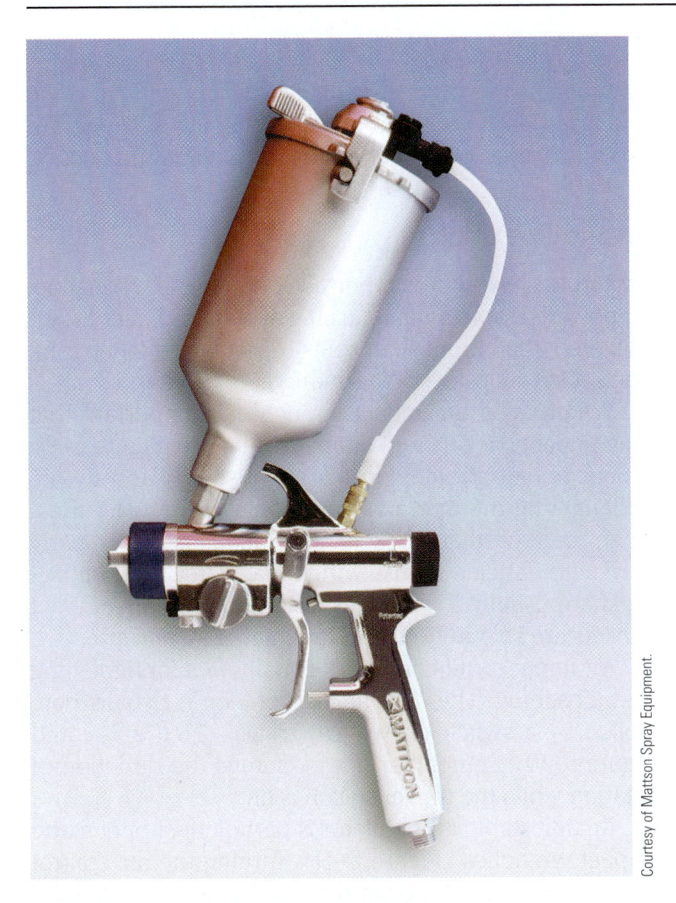

Figure 5–3 Spray guns are used to apply a paper-thin film of primer and paint to the vehicle body. They are precision tools that must be handled and cared for properly. This is a modern, high-efficiency gravity-fed spray gun.

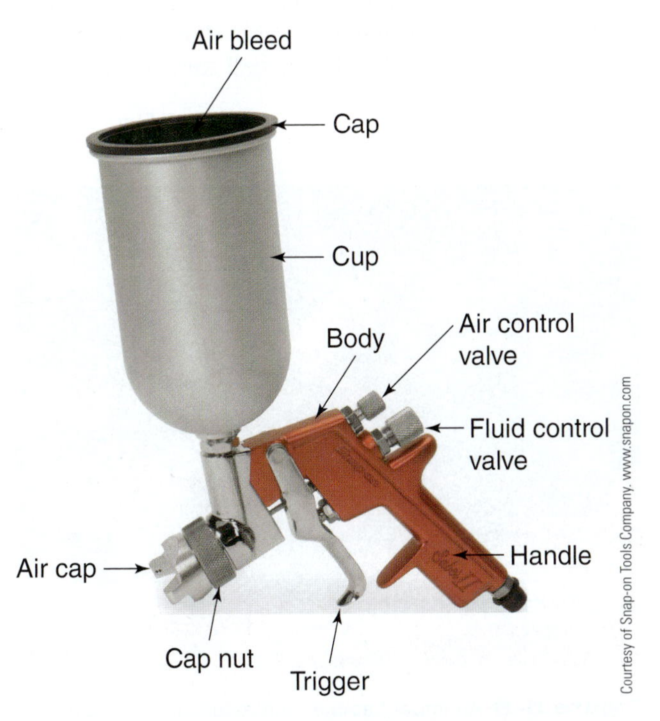

Figure 5–4 A high-volume, low-pressure (HVLP) spray gun uses lower air pressure to output a higher volume of primer or paint. This reduces unwanted overspray mist that does not stay on the vehicle, so less paint is wasted.

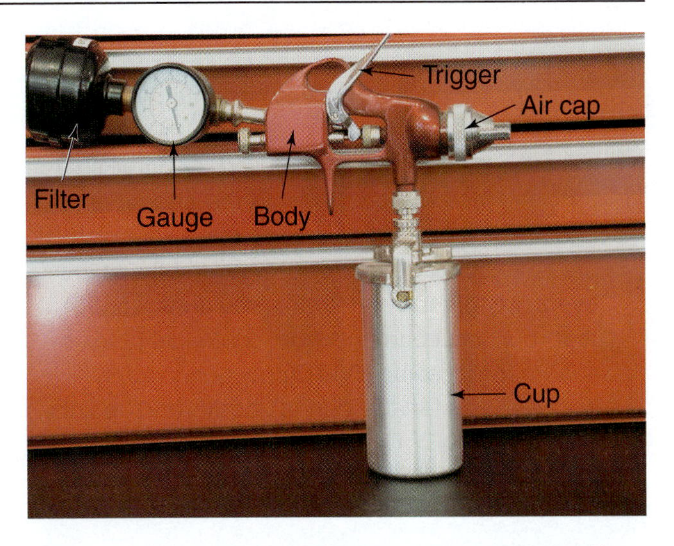

Figure 5–5 This is a suction-fed touch-up spray gun for painting small areas. Suction-fed spray guns can be high efficiency like gravity-fed guns.

The *spray gun trigger* can be pulled to open both the fluid and air valves. It uses lever action to pull back on the needle valves.

The *spray gun air cap* works with the air valve to control the spray pattern of the paint. It screws over the front of the gun head.

The *spray pattern* is the shape of the atomized spray when it hits the body. Little airflow out of the side jets on the cap produces a very round, concentrated spray pattern. When the airflow is adjusted up, the cap jets narrow and better atomize the paint flowing out of the gun.

The spray gun is probably the most used air-powered tool in the body/paint shop. It is used to do most of the refinishing work. Spray guns are also among the most efficient of all pneumatic tools. See Figure 5–5.

A conventional atomizing air spray gun is a precision tool that uses compressed air to atomize sprayable material. Replacing the conventional system in many body shops is the high-volume, low-pressure (HVLP) gun. Air and paint enter both types of guns through separate passages and are mixed and ejected at the air nozzle to provide a controlled spray pattern (Figure 5–6).

Pulling back slightly on the trigger opens the air valve to allow use of the gun as a blowgun. In this position the trigger does not actuate the fluid needle and no fluid flows. As the trigger is further retracted, it unseats the needle in the fluid nozzle and the gun begins to spray. The amount of paint leaving the gun is controlled by the pressure on the container, the viscosity of the paint, the size of the fluid tip, and the fluid needle adjustment (Figure 5–7). In industrial finishing where pressure tanks or pumps are used, the fluid needle adjustment should normally be fully opened. In suction cup operation, the needle valve controls the flow of paint.

Complete details on the operation of the various types of spray guns and their uses can be found in Chapters 24 through 28.

ECO-TECH Today, the HVLP is required by law to reduce the amount of overspray (spray mist not applied to body surface but emitted into the atmosphere and the air surrounding the vehicle). Older high-pressure spray guns produce several times the overspray of modern HVLP spray guns. To protect our environment from potentially toxic vapors, only use an HVLP spray gun to apply all spray-on materials during body and paint repairs.

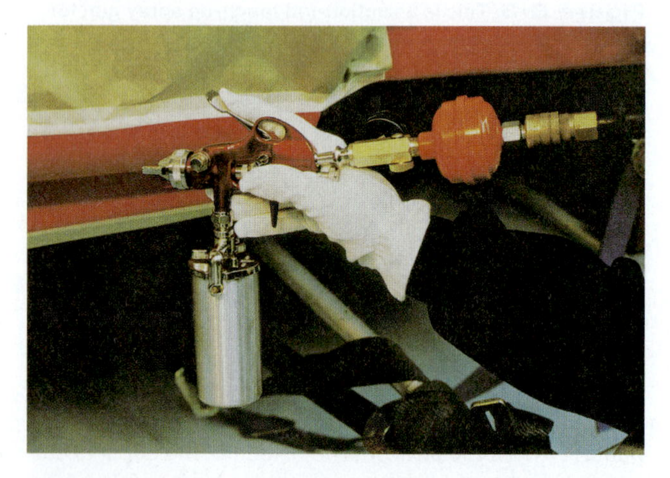

Figure 5-6 This small touch-up spray gun is good for painting small areas or for doing custom paint work.

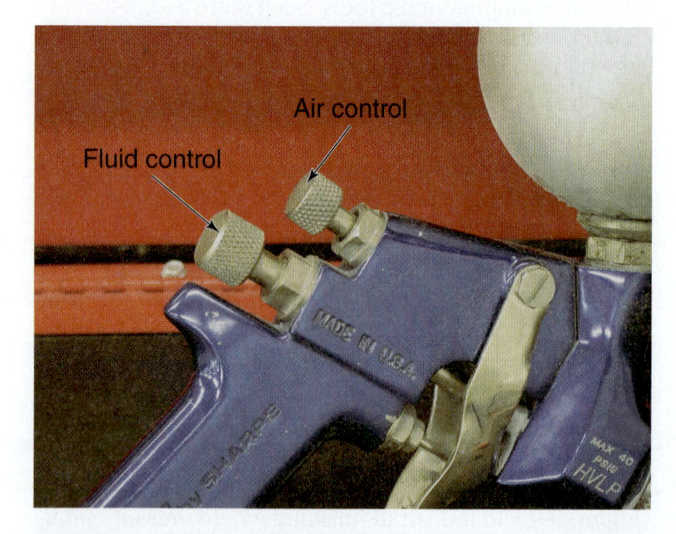

Figure 5-7 Note the two main controls on a typical spray gun. The fluid control changes the quantity of primer or paint coming out of the gun. Unscrewing the fluid knob increases the paint flow. The air control knob changes how the airflow fans out a pattern of spray coming out of the gun. Turning the knob in tightens the fan, and unscrewing the knob increases the fan width.

Air-Powered Wrenches

Any job that involves threaded fasteners can be done faster and easier with air-powered wrenches. There are two basic types of wrenches: the impact and the ratchet.

Impact Wrenches

An **impact wrench** is a portable, hand-held, reversible, air-powered tool for rapid turning of bolts and nuts. When triggered, the output shaft spins freely at over 2,000 rpm. The socket snaps over the square drive head and shaft.

When using an impact wrench, it is important that only impact sockets and adapters be used (Figure 5–8). Conventional chrome sockets might shatter and fly off, endangering the operator and others in the work area. Make certain that sockets and adapters are clearly marked or labeled for use with impact wrenches. Impact tools are usually a flat, black color, not chrome. Impact sockets are sold in both SAE and metric system sizes.

Air impact wrenches work equally well for tightening and loosening. The direction of rotation is usually controlled by a switch or two-way trigger (Figure 5–9 and Figure 5–10). Remember not to change the direction of rotation while the trigger switch is on.

An adjustable air regulator is part of most pneumatic impact wrenches (Figure 5–11). Turning the air control knob allows you to adjust the tool's speed and torque. Input air pressure above the usual 90 to 125 psi (620–861 kPa) range can be fed to the inlet without excessive tool wear.

To attach socket chucks and adapters to air impact wrenches, merely push them onto the output shaft as far as they will go. Power wrench output shafts come in two common variations: the detent ball and the retaining ring. See Figure 5–12–Figure 5–15.

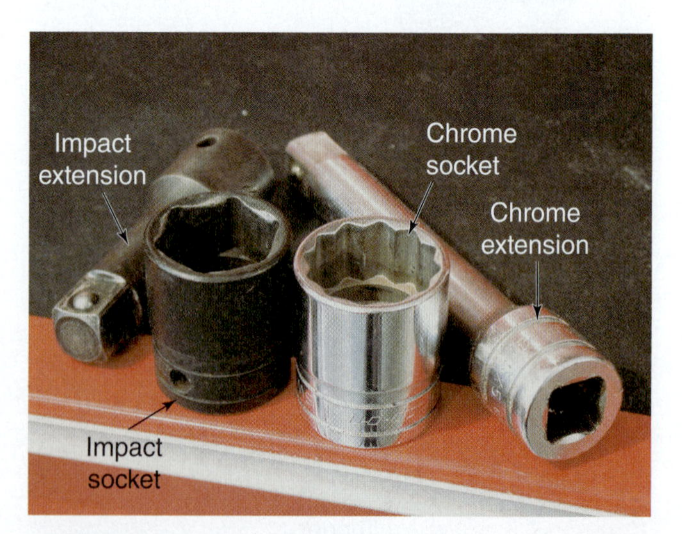

Figure 5-8 An impact socket and extension are tougher than the normal chrome sockets and extensions. The impact socket has thicker walls for a stronger, 6-point design. Impact tools are usually a flat, black color, indicating more hardening of the steel.

Courtesy of Snap-on Tools Company. www.snapon.com

Figure 5–9 An air impact wrench can produce tremendous torque to quickly loosen or tighten bolts and nuts.

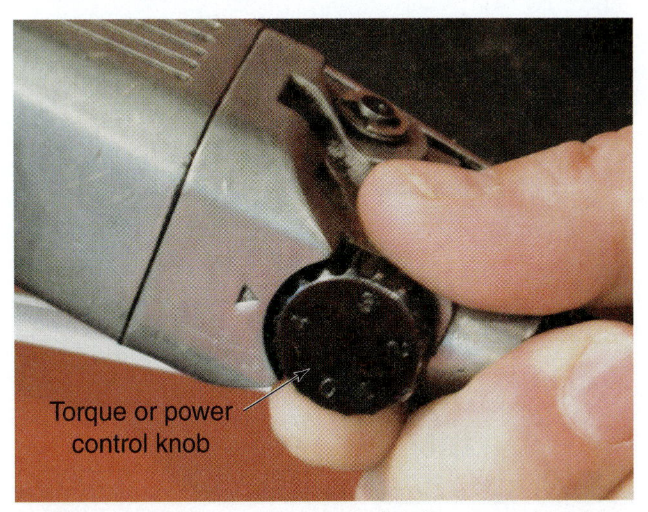

A The small numbers on this knob give the user an idea how much power will be produced by the impact wrench.

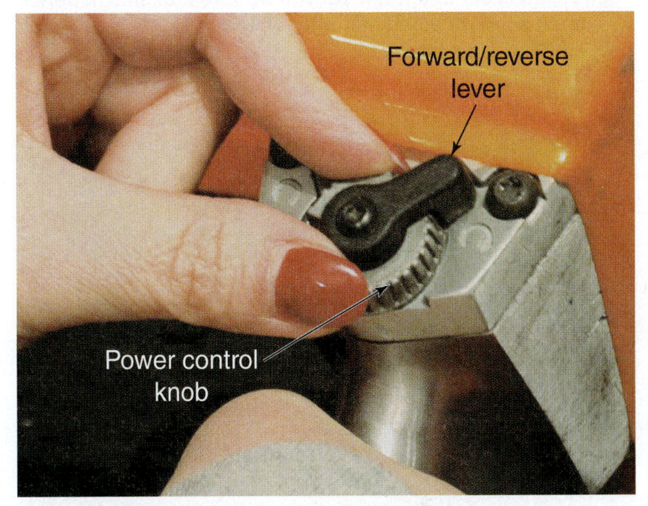

B This power control knob is built inside the forward/reverse lever.

Figure 5–11 The air regulator knob on a typical impact wrench can be adjusted to control speed and torque output.

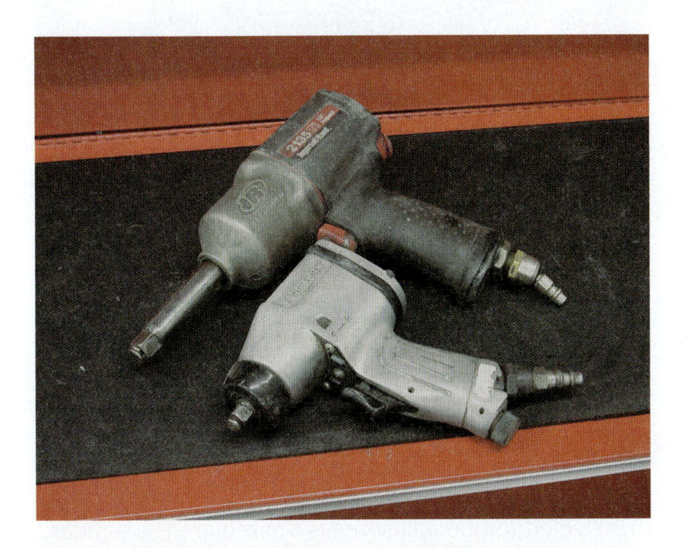

Figure 5–10 Note the different output shafts on these impact wrenches. The longer shaft is designed for use on lug nuts to remove and install wheels on vehicles.

To remove conventional threaded fasteners, set the switch for a left-hand counterclockwise rotation. Place the socket over the nut or fastener head. Exert forward pressure on the wrench while depressing the trigger switch. As soon as the nut or fastener becomes loosened, relax the forward pressure on the wrench to let it spin the nut or fastener free (Figure 5–16).

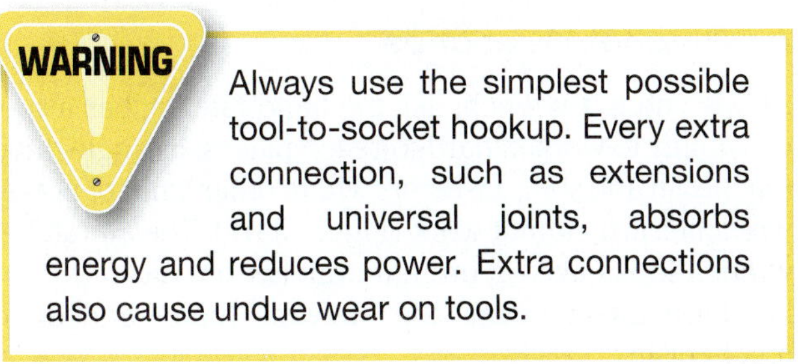

WARNING

Always use the simplest possible tool-to-socket hookup. Every extra connection, such as extensions and universal joints, absorbs energy and reduces power. Extra connections also cause undue wear on tools.

WARNING

Do not use an air impact wrench to final-torque fasteners. When important, final torque should be done by hand with a torque wrench. This will prevent broken fasteners, warped parts, and similar troubles.

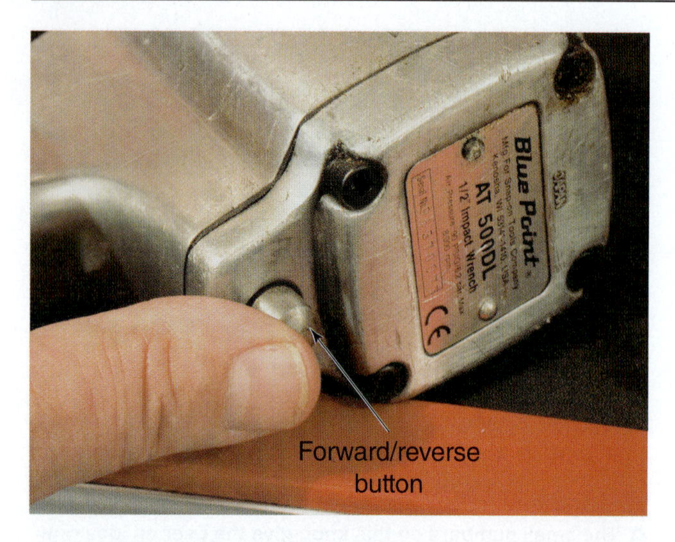

Figure 5–12 Always make sure the forward/reverse lever is set in the correct direction before triggering the gun. If the impact gun is set to tighten but you want to loosen a fastener, you could snap off a bolt instantly, which could be a costly or time-consuming mistake.

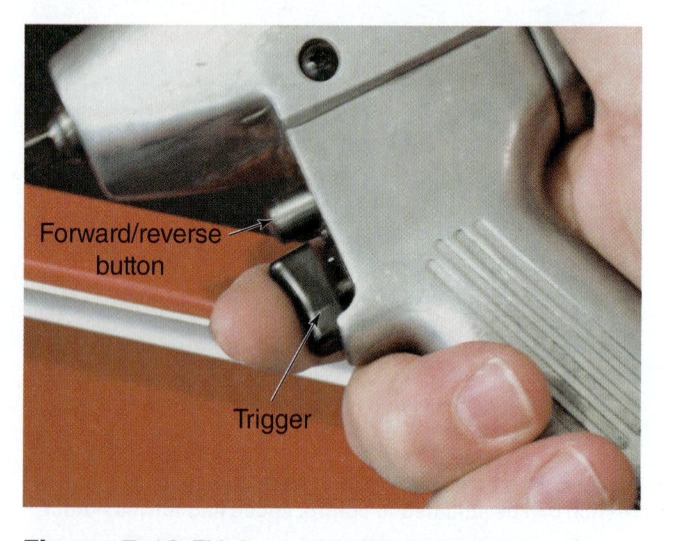

Figure 5–13 This is a push button-type trigger on an impact gun. Make sure the socket is completely over the nut or bolt and that the direction is correct before triggering.

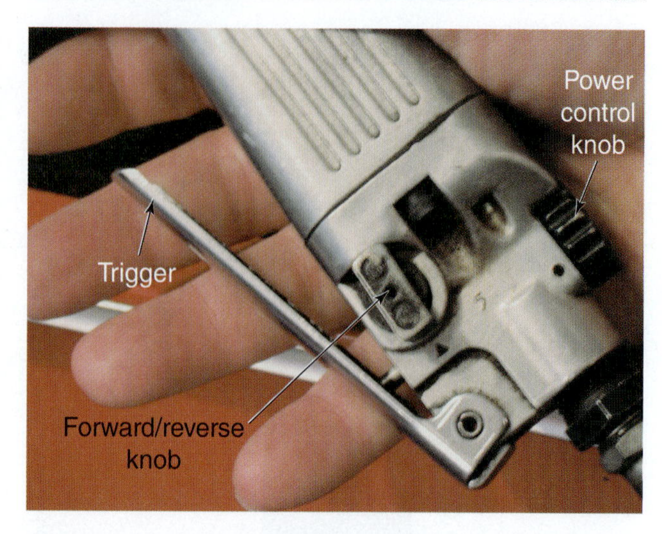

Figure 5–14 This is a lever-type trigger on a smaller impact gun. The direction control knob is right next to the trigger on the side of this gun.

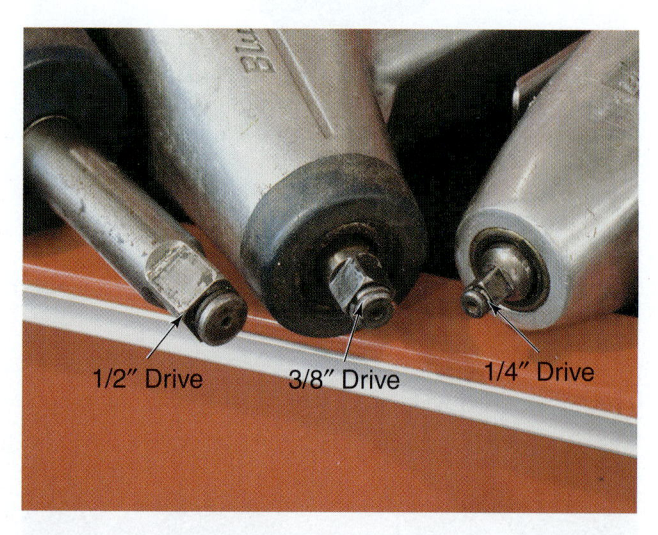

Figure 5–15 Compare the three common sizes of impact guns, starting from the left: ½-inch impact, ⅜-inch impact, and small ¼-inch impact. Larger ¾-inch impact guns are available, but they not commonly used in auto body shops.

To install fasteners, you normally set the switch for a right-hand, clockwise rotation. Start the nut on the stud or the bolt on the threads by hand to avoid cross threading. Place the socket over the nut or fastener head. Depress the trigger switch to spin the fastener in. As soon as the fastener bottoms, release the trigger. Then press the trigger quickly one or two more times to snug the fastener down.

For all their torquing power, air impact wrenches have practically no recoil. Holding one is rather easy, but it can be misleading. The flow of power through the socket is very strong. Therefore, the wrench should be held tightly.

Remember that there is no consistently reliable adjustment with any impact wrench. Where accurate preselected torque adjustments are required, a standard torque wrench should be used. The air regulator on air-powered wrenches can be employed to adjust torque to be slightly below the needed tightness of a known fastener.

Air Ratchet Wrenches

An **air ratchet** wrench, like the hand ratchet, has a special ability to work in hard-to-reach places. Its angle drive reaches in and spins fasteners where other hand or power wrenches just cannot work (Figure 5–17). The air ratchet wrench looks like an ordinary ratchet but has a thicker handgrip that contains the air vane motor and drive mechanism.

Figure 5–16 Here the technician is using a ½-inch impact gun to install a wheel on the vehicle. A torque wrench should be used to adjust lug nut tightness after running the nuts down lightly with an impact gun.

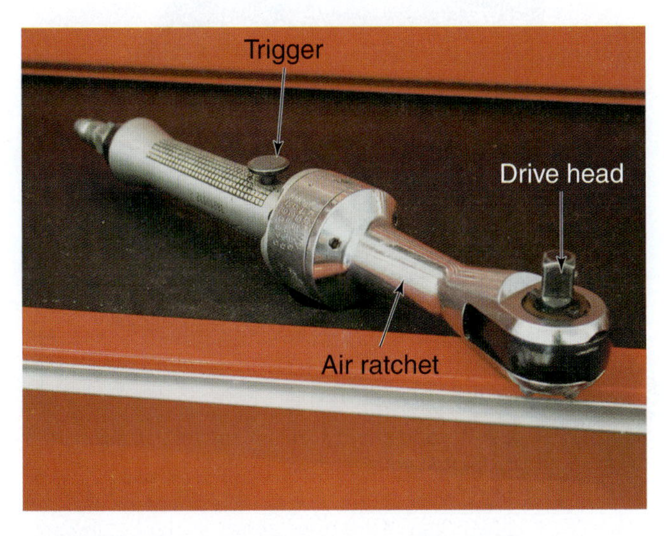

Figure 5–17 An air ratchet wrench is good for small fasteners requiring low torque.

When using an impact wrench, keep the following pointers in mind:

- Always make sure the impact wrench socket is properly retained. To prevent injury, use an impact-type socket. If the impact wrench has a pin retainer, do not substitute a bent nail or piece of wire.
- If an air wrench fails to loosen a bolt in 3 to 5 seconds, use a breaker bar. Soak large rusted nuts with penetrating oil before using the wrench.

Actual torque on a fastener is directly related to joint hardness, tool speed, the condition of the socket, and the time the tool is allowed to impact.

When tightening a fastener, run in the bolt or nut under power and finish tightening it by hand-pulling. Remember that an air ratchet has little torque, and hand-tightening and loosening are important. Look at Figure 5–18.

If your hand ever gets wedged between an air ratchet and a part, kink the air hose to relieve pressure and free your hand.

Air Drills

Air drills use shop air pressure to spin a drill bit. They can be adjusted to any speed and are more commonly used than electric drills. They are usually available in ¼-, ⅜-, and ½-inch sizes. Air drills are smaller and lighter than electric drills. Their compactness makes them a great deal easier to use for most tasks, especially for drilling operations in auto work (Figure 5–19).

Drill bits of different sizes fit into the chuck on drills for making holes in parts. Drill bits come in various sizes. The size is usually stamped on the upper part or shank of each bit (Figure 5–20).

A *key* is used to tighten the chuck. The *chuck* has movable jaws that close down and hold the bit. The key has a small gear that turns a gear on the chuck.

Figure 5–18 An air ratchet is good when working in tight quarters. Note how it will reach between this tire and fender apron when other air tools would not fit.

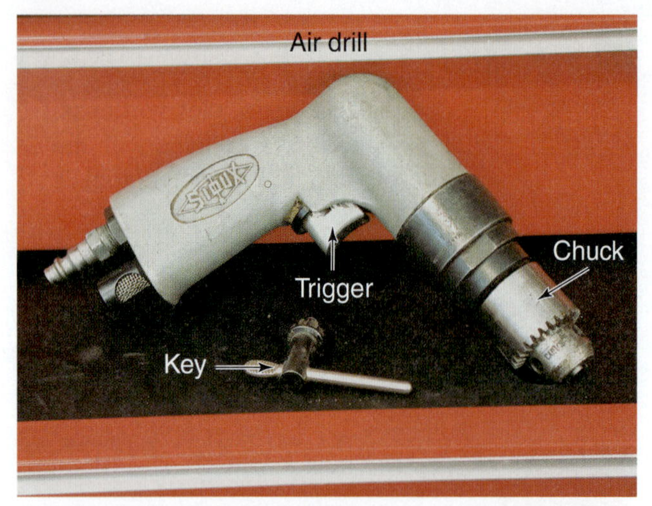

A A pistol grip air drill is handy for various drilling, brushing, and grinding operations. A key is needed to loosen and tighten the chuck on this air drill.

B This straight grip air drill has a hand-operated chuck so you can more quickly change drill bits or other attachments.

C A cordless drill is handy when prolonged high-power output is not required.

Figure 5-19 Drills are commonly used power tools in collision repair.

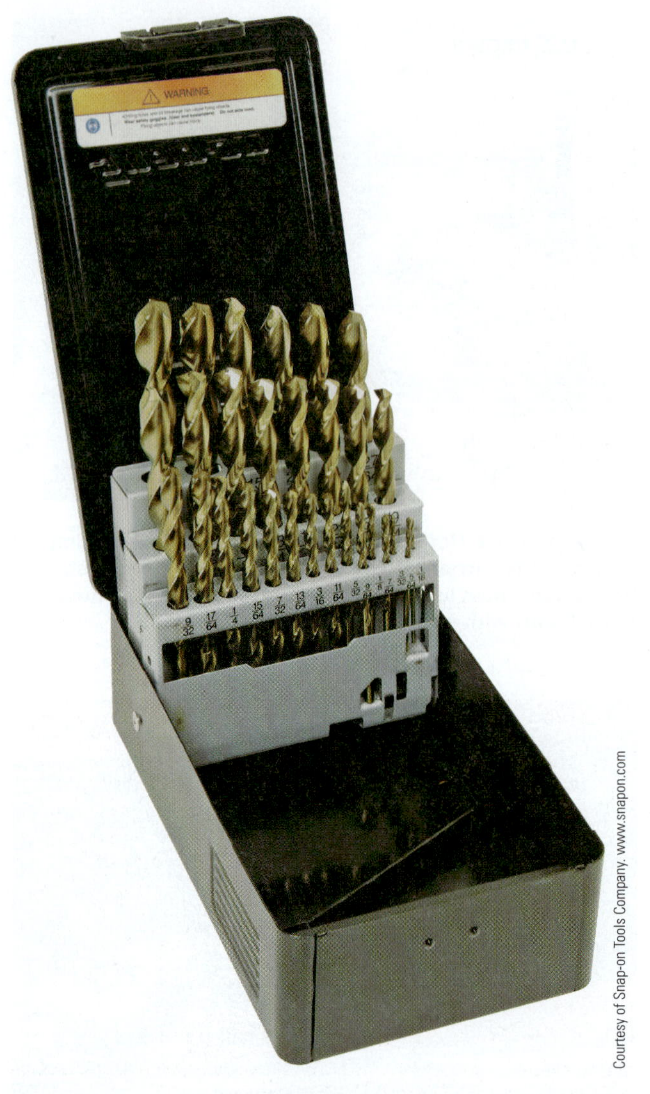

Courtesy of Snap-on Tools Company. www.snapon.com

Figure 5-20 A good quality drill bit set with all sizes of drills should be kept in your toolbox.

Never leave a key in a drill when not tightening. Always unhook the air hose when installing a bit to prevent injury. This also applies to operating a large drill press.

When using an air tool to drill into any material, the following general procedures should be followed:

1. Accurately locate the position of the hole to be drilled. Mark the position distinctly with a center punch or an awl to provide a seat for the drill point and to keep it from "walking" away from the mark when pressure is applied.
2. Always know what is on the other side. Do not drill a wiring harness or through a trim panel.
3. Unless the workpiece is stationary or large, fasten it in a vise or clamp. Holding a small item in the hand may lead to injury if it is suddenly seized by the bit and whirled out of grip. This is most likely to happen just before the bit breaks through the hole at the underside of the work.

4. Carefully center the drill bit in the jaws while securely tightening the chuck. Place the drill bit tip on the exact point to drill the hole, then start the motor by pulling the trigger switch. Never apply a spinning drill bit to the work.
5. Except when it is desirable to drill a hole at an angle, hold the drill perpendicular to the face of the work.
6. Align the drill bit and the axis of the drill in the direction of the hole. Changing the direction of this pressure will distort the dimensions of the hole. It could snap a small bit. To avoid stressing the drill bit, place only light pressure right above the housing on the drill body. If you push down with the handle, side loading could break the bit or affect the hole roundness.
7. Use just enough steady and even pressure to keep the drill cutting. Guide the drill; do not force it. Too much pressure can cause the bit to break or the tool to overheat. Too little pressure will keep the bit from cutting and dull its edges due to the excessive friction created by sliding over the surface.
8. When drilling deep holes, especially with a twist drill, withdraw the drill several times to clear the cuttings. Keep the tool running when pulling the bit back out of a drilled hole. This will help prevent jamming.
9. Reduce the pressure on the drill just before the bit cuts through the work.

There are special *spot remover air drills* and/or attachments that are used for cutting out spot welds (Figure 5–21). When cutting out spot welds, the drill can be fastened to the weld area with a clamp attachment that makes operation easier (Figure 5–22). The cutter will not deviate from the weld center during cutting.

There are two types of spot cutters available that can be mounted in an air drill for cutting out spot welds:

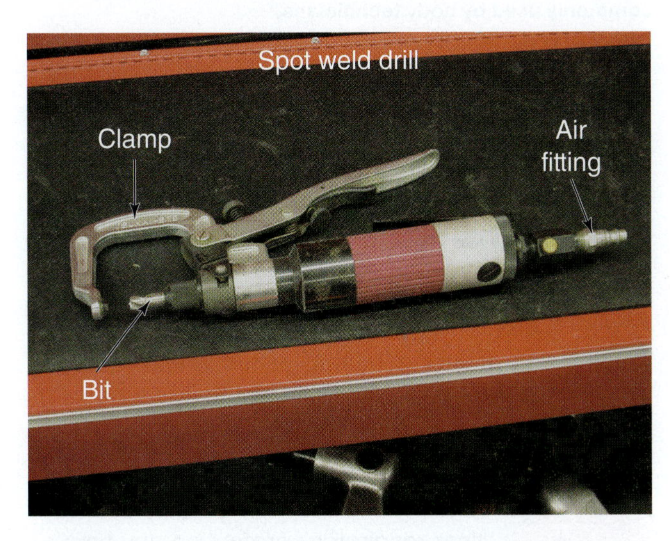

Figure 5–21 Spot weld drill is needed when removing welded-on panels. It will clamp over small spot weld nuggets to drill them out for panel removal.

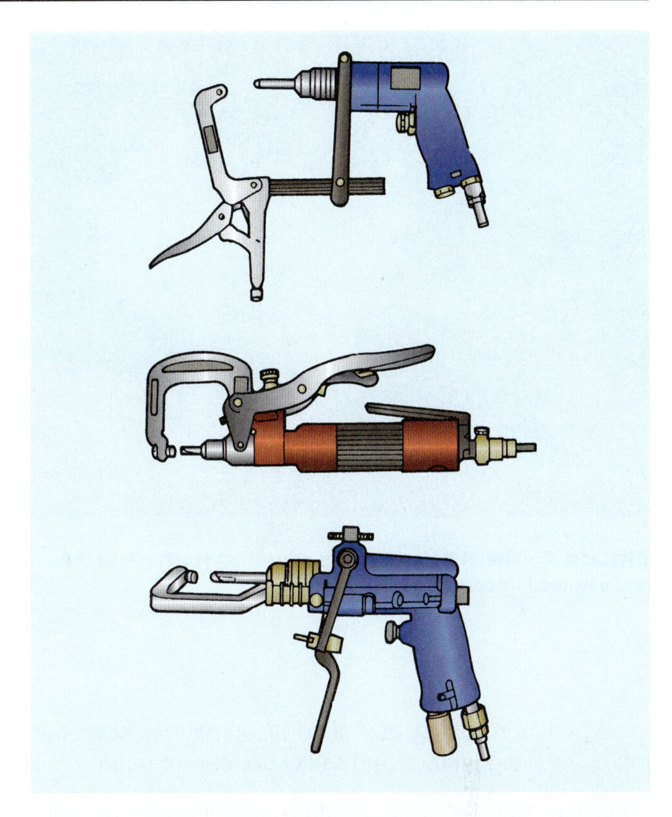

Figure 5–22 Several types of spot weld remover air drills are shown. They have plier attachments for helping to force the bit into the spot weld for removal.

- **Drill type** (Figure 5–23). This type does not damage the bottom panel, nor does it leave a nib in the bottom panel, so finishing is easy.
- **Hole saw type**. The cutting depth of this type can be adjusted so that the bottom panel is not damaged. It is necessary to sand off the remaining portion of the weld.

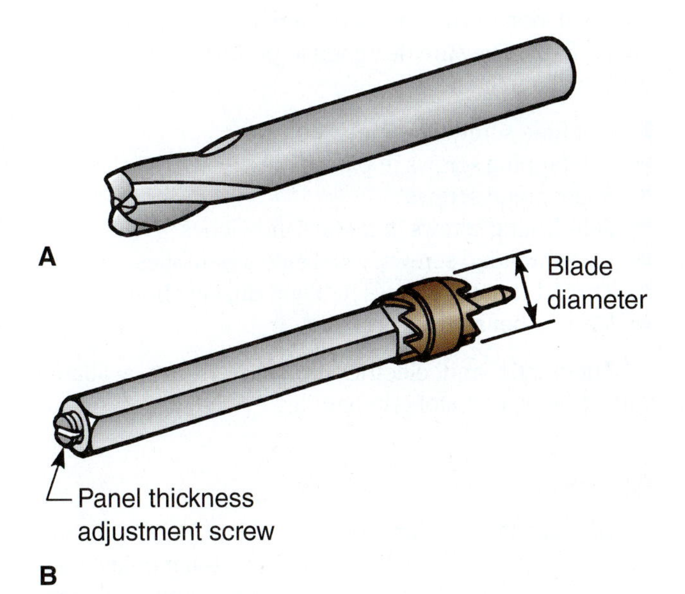

Figure 5–23 Note the two types of spot cutters: (A) drill type and (B) hole saw type.

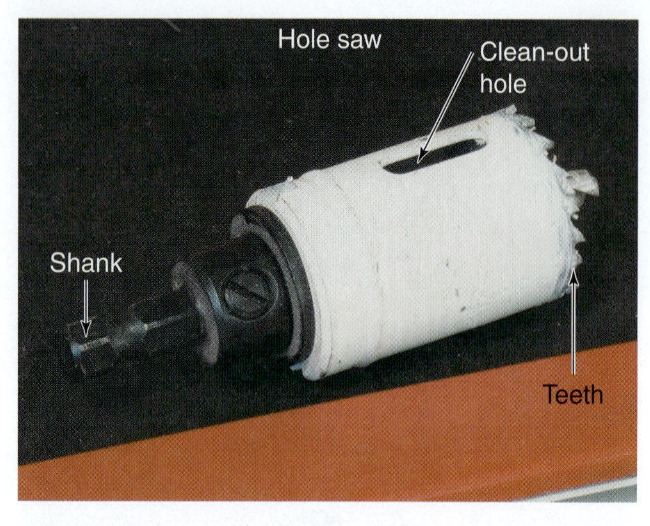

Figure 5–24 A hole saw is handy when you have to drill large holes in panels.

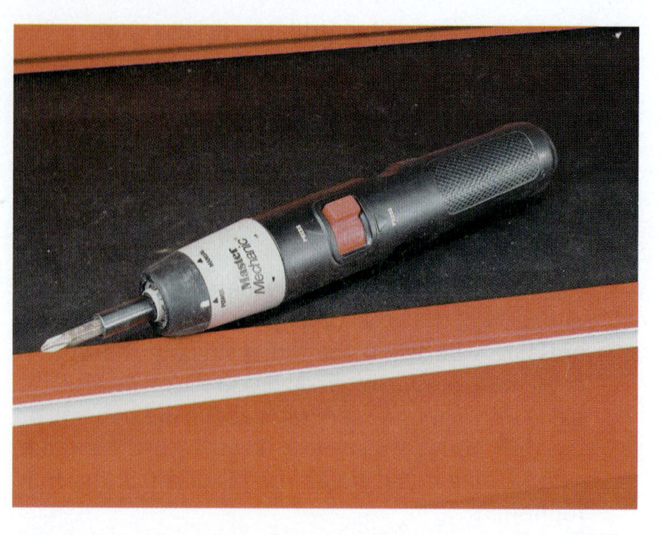

Figure 5–25 An electric screwdriver will work on small screws, but its torque is very limited.

When performing any air drill operation, keep the following maintenance and safety pointers in mind:

- Clean chuck jaws occasionally to prolong service life. Use sharp drills; they require far less effort and put less strain on the tool.
- To avoid the danger of breakage from high breakthrough torque, use properly sharpened twist drills and reamers, and select the proper drill speed.
- Start drilling at a low speed and gradually increase it. Avoid kickback by easing up at breakthrough. Figure 5–24 shows a hole saw.

Power Screwdrivers

Unlike electric screwdrivers, air screwdrivers run cool and will not burn out, even under constant use. See Figure 5–25. They are designed to perform well in a wide variety of applications:

- Machine screws in tapped holes
- Self-tapping screws in plastic
- Sheet metal screws
- Self-drilling screws in metal sandwiches
- Fine-threaded screws in delicate assemblies
- Thread forming screws in blend die-cast holes
- Many other similar shop jobs

Pneumatic and electric screwdrivers are available with straight- or pistol-grip handles.

Air Sanders

An **air sander** uses an abrasive action to smooth and shape body surfaces. Sandpapers of differing coarseness can be attached to the pad on the sander. *Coarse sandpaper* removes material more quickly. *Fine sandpaper* produces a smoother surface finish. Air sanders are among the most commonly used air tools in auto body repair. See Figure 5–26.

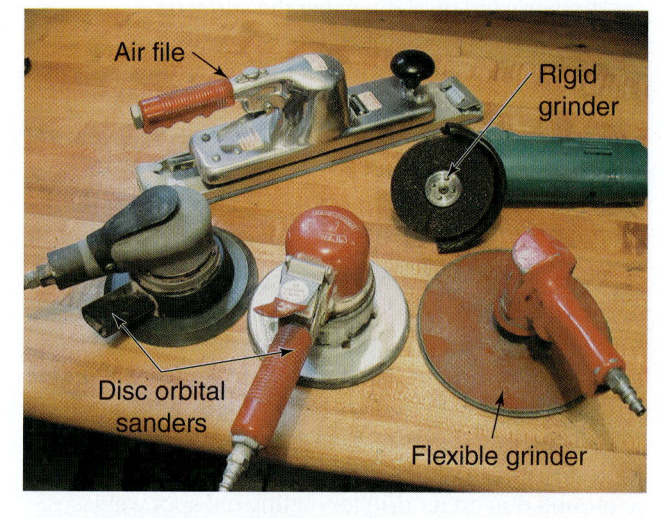

Figure 5–26 Various sander and grinder designs are commonly used by body technicians.

There are two basic types of air sanders: disc and orbital (finishing). Most rough sanding done in automotive work is done with a *disc sander* (Figure 5–27) or its counterpart, the *dual-action* (DA) orbital sander (Figure 5–28). The orbital sander oscillates while it is rotating and creates a buffing pattern rather than the swirls and scratches often caused by the disc sander.

Wear respiratory protection any time a sander is being operated.

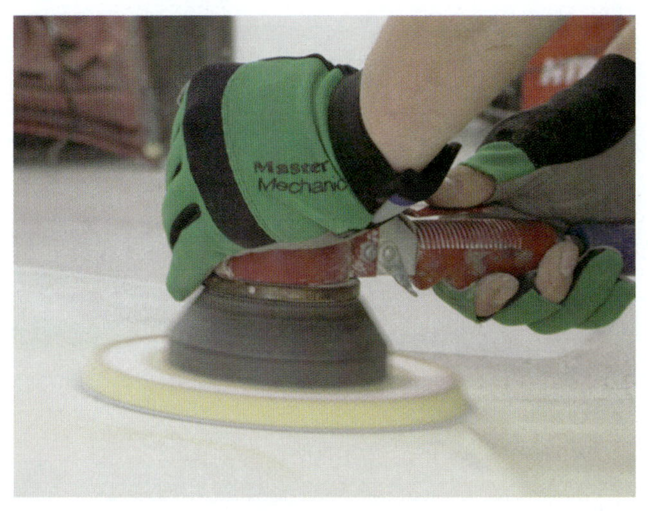

Disc sander

Motion of abrasive

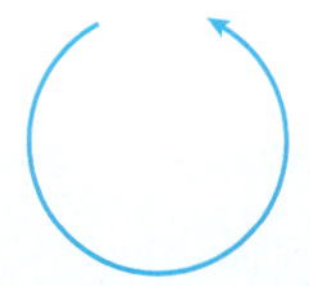

Single-action (circular)

Figure 5–27 A single-action sander can be used for rapid material removal of old paint.

Dual-action sander

Motion of abrasive

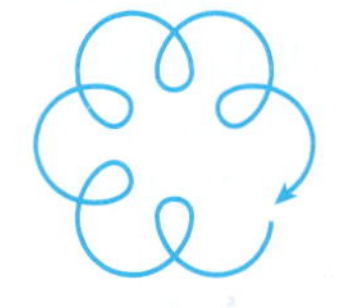

Dual-action (orbital)

Figure 5–28 The dual-, or double-action, sander and its motion are demonstrated for smoother material removal.

Finishing orbital sanders, also called "pad" or "jitterbug" sanders (Figure 5–29), are designed for fine finish sanding. It is possible to use a wider variety of abrasives with finish sanders. For the most part, the best work is done with comparatively fine grit abrasive paper. Finish sanders are also specially designed for hard-to-reach places and tight corners.

The **sander pad** is a soft mounting surface for the sandpaper. *Disc adhesive*, self-stick paper, or hook and latch paper holds the sandpaper onto the pad. *Disc adhesive* is special nonhardening glue that comes in a tube. It can be placed on the pad to adhere the paper to the sander. *Self-stick paper* already has a nonhardening glue on it.

A This 5-inch dual-action sander is good for most featheredging tasks.

B This small diameter sander will work well in confined spaces or on small parts.

Figure 5–29 Air sanders have different pad or sandpaper sizes.

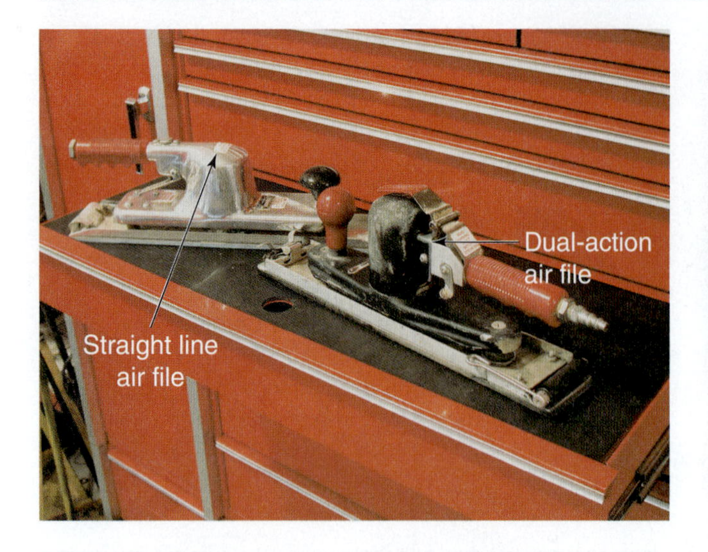

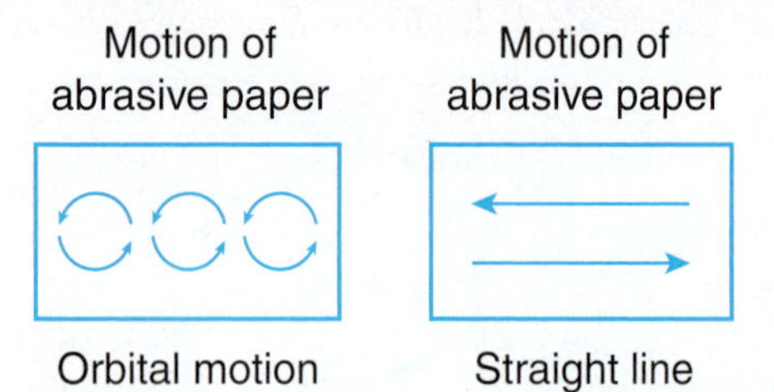

Figure 5–30 Here are two types of air files, or board sanders. They are available in straight line or orbital configurations.

Figure 5–31 Note the various types of grinders used in auto body repair work.

Figure 5–32 A grinder will quickly cut down to bare metal for paint removal. Take care not to catch projections when using a grinder, or the tool could fly into your body or face and cause serious injury. Always wear face protection.

Hook and latch paper is similar to Velcro®. One advantage is that you can take hook and latch paper off the tool without damage and reuse it.

An **air file** is a long, thin board sander for working large flat surfaces on panels. It is handy when you must "true" or flatten a large repair area. It will plane down filler so that a large area is flat. An air file is often used for rough shaping operations (Figure 5–30). It operates in either an orbital or straight-line motion.

Details on operating sanders can be found in Chapters 7 and 12.

Air Grinders

Grinders are used for fast removal of material. They are often used to smooth metal joints after welding and to remove paint and primer. They come in various sizes and shapes. Refer to Figure 5–31 and Figure 5–32.

An air grinder can cut metal, and it can cut bone and flesh as well. Wear leather gloves and a full face shield when using any air grinder.

The most commonly used portable air grinder in collision repair and refinishing shops is the *disc-type grinder.* It is operated like the single-action disc sander. An air grinder should be used carefully. It can quickly thin down and cut through body panels, causing major problems. See Figure 5–33 and Figure 5–34.

There are, of course, several other grinders used in the body shop. The more common ones include the following:

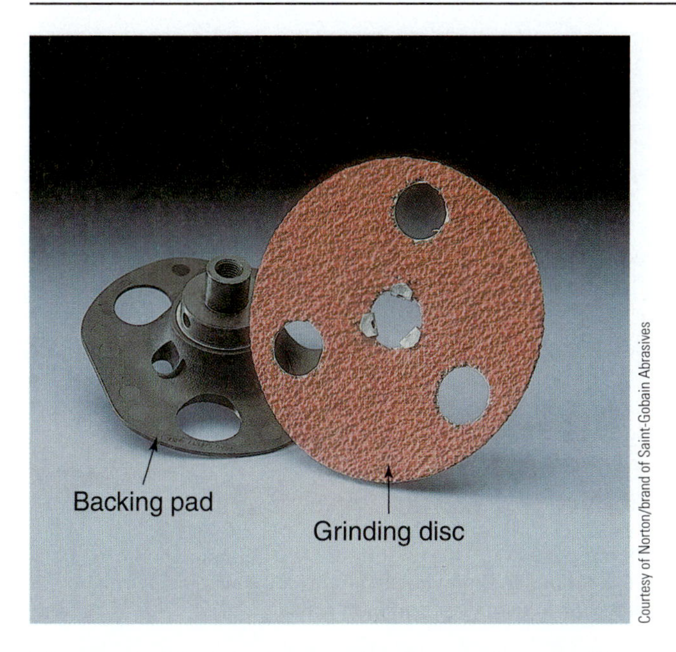

Figure 5-33 A flexible grinding disc requires a flexible backing pad. The flexible grinding pad and disc will help you avoid grinding too deeply into the body part being repaired. The patented AVOS disc from Norton allows you to see the work surface through the disc.

- A *horizontal grinder* is used for heavy-duty grinding.
- The *vertical grinder* is a larger version of the disc grinder. With a sanding pad, this grinder can be converted into a disc sander. Most vertical grinders can be used with both straight wheels and cup wheels.
- An *angle grinder* is used primarily for smoothing, deburring, and blending welds (Figure 5–35).
- A *small wheel grinder* is used with abrasive wheels and wire brushes (Figure 5–36 and Figure 5–37).
- The *die grinder* is used with mounted points and carbide burrs for a variety of applications, such as weld cleaning, deburring, blending, and smoothing. It is available in both straight and angle head designs. Look at Figure 5–38.
- A *cutoff grinder* is used to cut through muffler clamps and hangers with ease. It also slices through sheet metal and radiator hose clamps (Figure 5–39).

Figure 5–40 shows how to change a grinding wheel. Figure 5–41 shows a technician using a cutoff wheel.

Other Air Tools

An *air saw* uses a reciprocating action and a hacksaw-type blade to quickly cut through metal. An air saw is handy when a cutoff wheel will not reach into a body panel or the area to be cut.

When using an air saw, do not apply too much pressure to the blade or you will overheat the saw teeth and dull the blade. See Figure 5–42.

An *air scraper* produces a reciprocating action on a hardened steel scraper blade. The sharp air scraper blade

A Screw and tighten the pad mounting adapter onto the air tool shaft.

B Screw the flexible pad onto the adapter and tighten.

C Twist and lock the flexible grinding disc onto the pad.

Figure 5-34 Note the basic steps for installing a flexible grinding pad and disc.

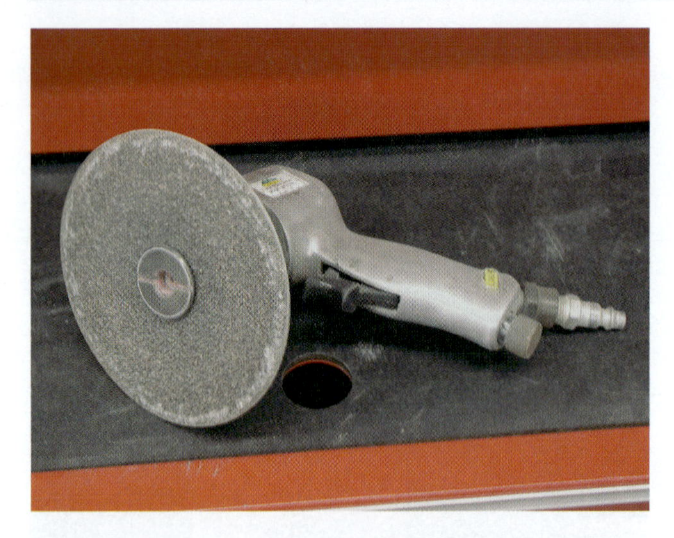

Figure 5-35 This pistol grip angle grinder has a large-diameter flexible grinding disc attached. It is often used for paint removal.

Figure 5-36 This air grinder has a thick, stiff grinding wheel attached. It is often used for metal grinding after welding.

Figure 5-37 This angle grinder is handy for removing paint over spot welding when replacing welding on panels. Note the scuff pad on the left and the sanding disc on the right.

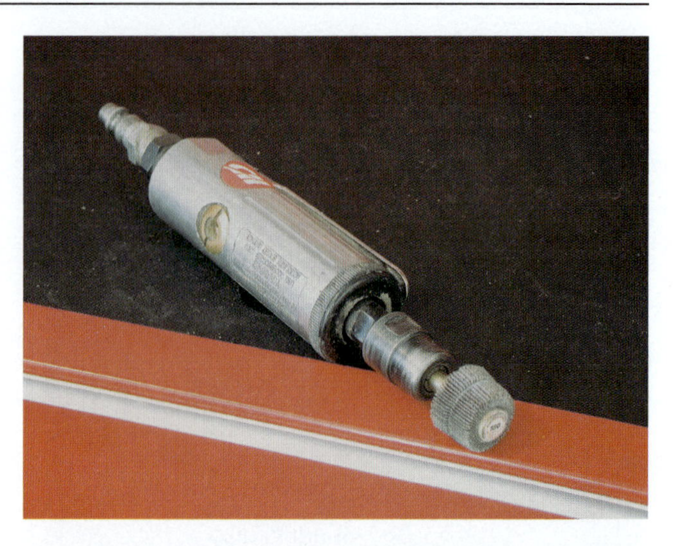

Figure 5-38 This die grinder is a very small air tool that will accept sanding wheels and grinding wheels. It can be used on very small or confined areas.

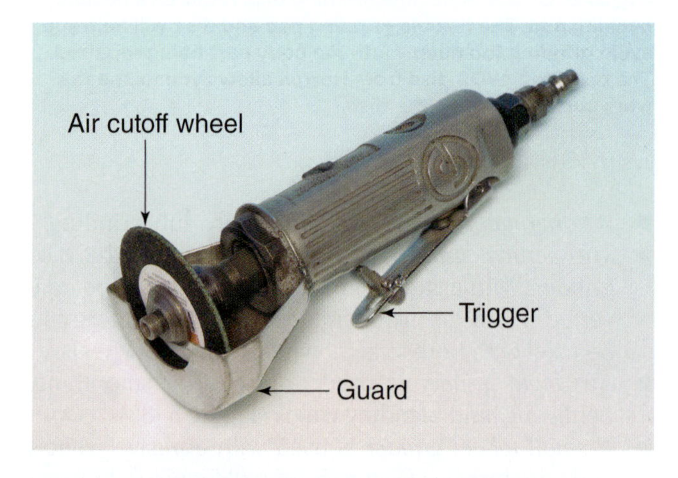

Figure 5-39 This small air grinder is equipped with a cutoff wheel that will quickly grind and cut metal.

Figure 5-40 Here the technician is replacing a damaged cutoff wheel. A wrench is used to hold the tool shaft stationary while an Allen wrench is used to loosen the bolt securing the cutoff wheel.

Figure 5-41 A speed cutoff wheel with its thin blade will quickly and accurately abrade through steel body panels during major structural repairs.

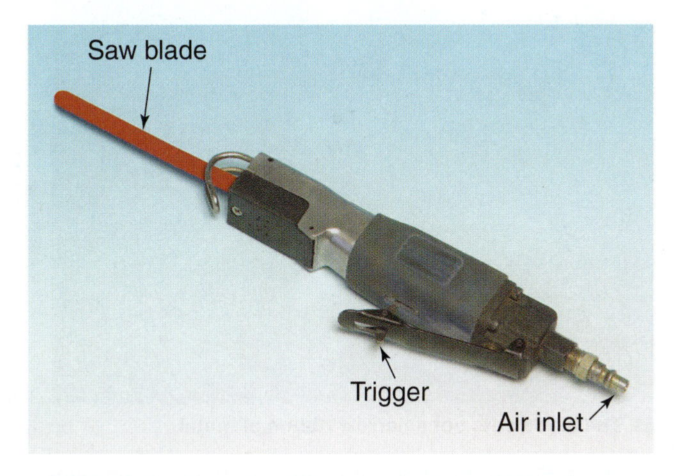

Figure 5-42 An air hacksaw is handy for cutting in obstructed locations where a cutoff wheel will not reach.

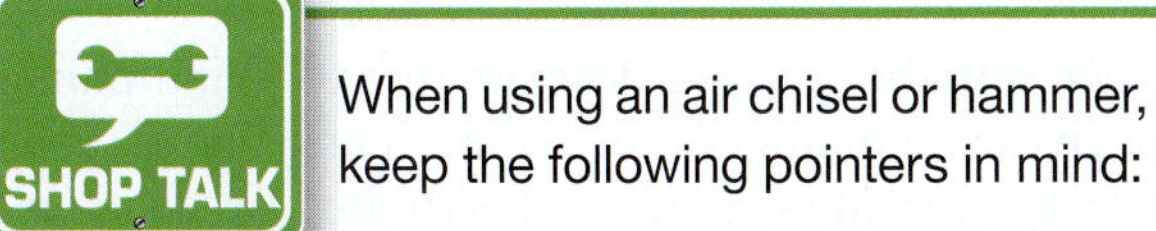

When using an air chisel or hammer, keep the following pointers in mind:

- Always use a chisel retainer when operating an air hammer.
- Position the tool by starting slowly, then increasing power. Avoid running into hardware, frames, and so forth, with sheet metal cutting tools.
- Check chisel shanks periodically for peening, and grind a new chamfer when required.
- Do not let the chisels ride out of the air hammer.
- Keep cutters sharp.
- Wear thick gloves, a face shield, and hearing protection when power-cutting metal.

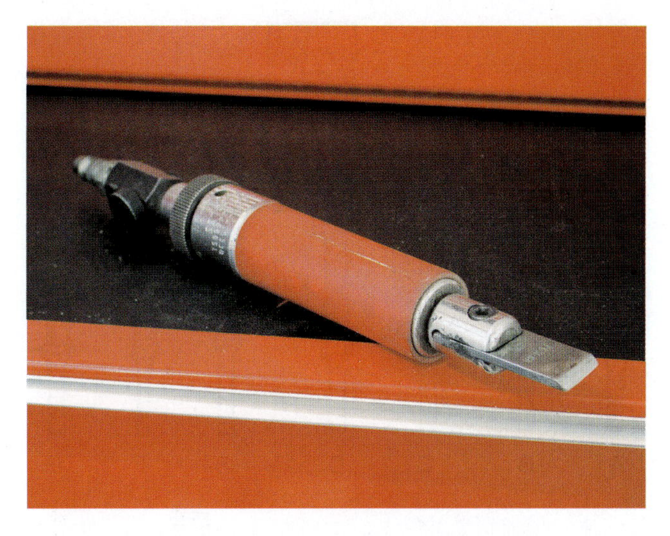

Figure 5-43 An air scraper can sometimes be convenient when removing undercoating and body sealer during major repairs.

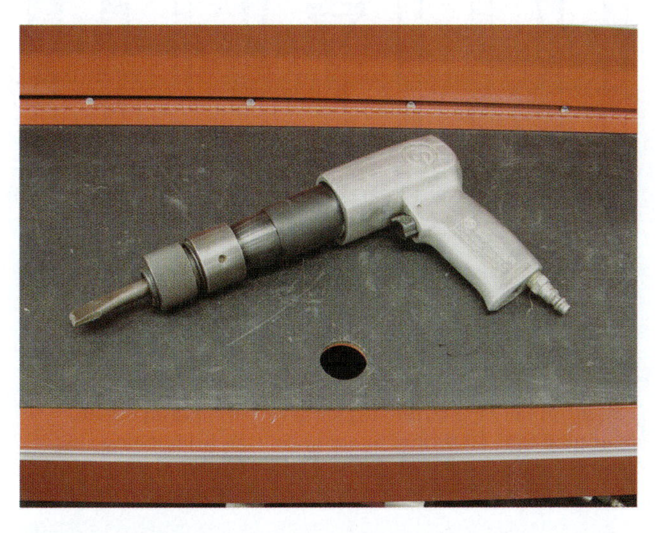

Figure 5-44 An air chisel will make rough cuts in metal quickly. If a panel is badly damaged and is crushed around fasteners, you might have to use an air chisel to cut and remove the panel.

will quickly peel off old sound-deadening material and sealer when you are removing damaged body panels (Figure 5–43).

An *air chisel* is used for various driving and cutting operations (Figure 5–44). Different types of chisels or blades can be installed in the air chisel body to complete various tasks (Figure 5–45).

Air shears use pneumatic power to move a sharp sheet metal-cutting blade back and forth. When the air shears are pushed into sheet metal, they will quickly slice through the metal. The shears can be moved to cut irregular shapes in sheet metal (Figure 5–46).

Air nibblers, as implied, use a punch and die head to produce a rapid cutting motion to "nibble" or cut small bits of metal. The air tool drives the die up and down so

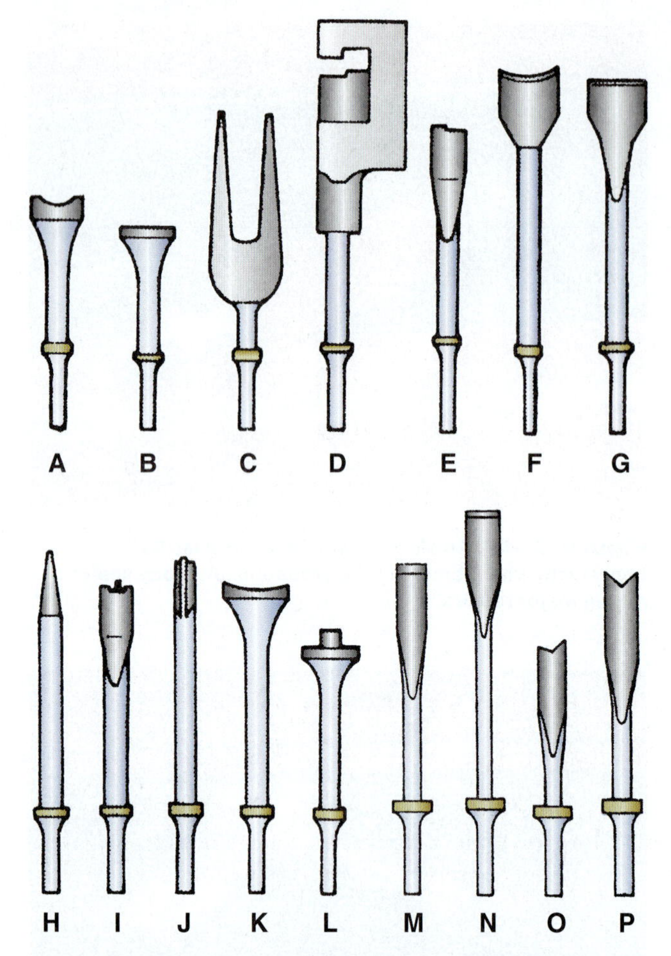

Figure 5-45 Typical air chisel accessories include: (A) universal joint and tie rod tool, (B) smoothing hammer, (C) ball joint separator, (D) panel crimper, (E) shock absorber chisel, (F) tail pipe cutter, (G) scraper, (H) tapered punch, (I) edging tool, (J) rubber bushing splitter, (K) bushing remover, (L) bushing installer, (M) rivet cutter, (N) flat chisel, (O) sheet metal cutter, and (P) spot weld breaker.

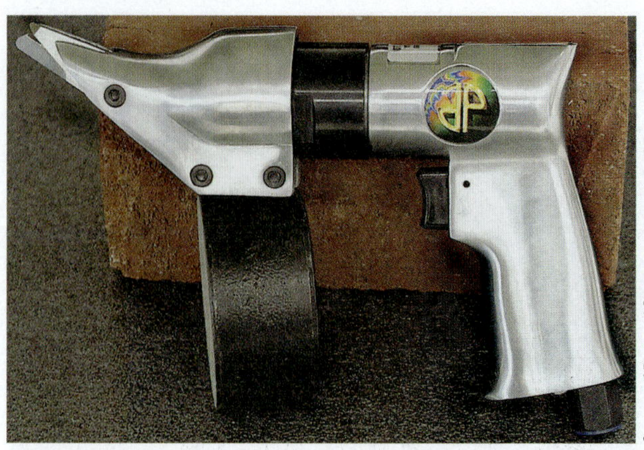

A Sharp blade is forced into sheet metal for rapid cutting.

Courtesy of Eastwood Company, www.eastwood.com

B The blade cuts out a narrow ribbon of metal.

Courtesy of Eastwood Company, www.eastwood.com

Figure 5-46 Air shears operate like automatic tin snips.

that small bits of sheet metal are nibbled away. An air nibbler will make a very clean cut and is ideal for cutting relatively flat sheets of metal (Figure 5–47).

An *air nozzle* directs a blast of compressed air out a small orifice to clean or dust off objects. When the trigger is depressed, line air pressure flows out the small orifice. The blast of air is commonly used to blow sanding dust off panels, cool parts after welding, or help dry parts off (Figure 5–48).

An *air chuck* is used to fill tires with compressed air. The chuck has a small valve that is opened when forced over the valve stem on a tire. This allows compressed air to flow into and fill the tire (Figure 5–49A).

A *tire gauge* is used to measure the amount of pressure in a tire. When pressed over the valve stem on the tire, the tire gauge will read inflation pressure in psi or kPa (Figure 5–49B). Figure 5–49C shows a combination tire chuck and tire gauge in one assembly. It is handy because it can read tire pressure as you inflate the tire.

Polisher/Buffers

An **air polisher** is used to smooth and shine painted surfaces by spinning a soft buffing pad (Figure 5–50). Polishing or buffing compound is rubbed over the paint with the polisher pad. This removes minor paint imperfections to increase *paint gloss* (shine).

A *polishing pad* is a thick cotton, synthetic cloth, lamb's wool, or foam cover that fits over the polisher's rubber backing plate. Sometimes the pad and backing plate are *integral* (one-piece). An integral pad has plastic backing with a foam pad. The pad may be smooth or wavy.

A *pad cleaning* tool is a metal star wheel and handle that will clean dried polishing compound out of the pad. The wheel is held onto the spinning pad to clean out and soften the cloth material. It will help prevent **swirl marks** (round or curved lines) in the paint caused by buffing. Note that a cleaning tool should not be used on foam pads because it damages the pad.

Figure 5–47 An air nibbler will cut flat sheet metal very accurately. It has a hardened steel punch that "nibbles" or presses out tiny round holes to make a cut in sheet steel and aluminum.

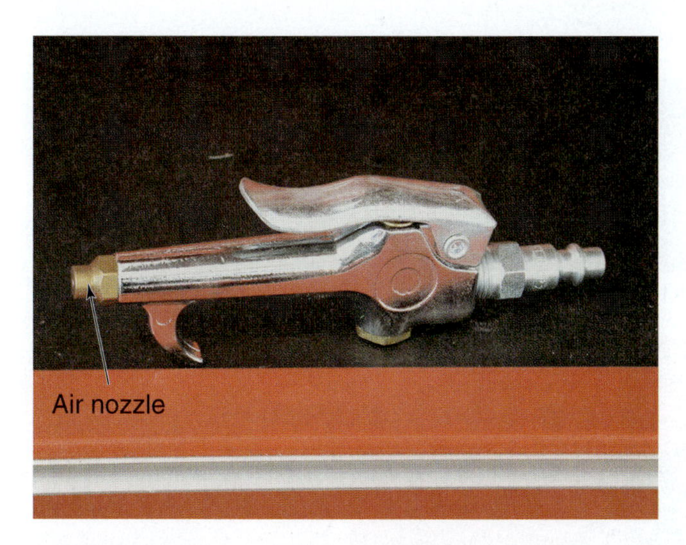

Figure 5–48 An air nozzle is commonly used to blow dust and debris off parts and panels being repaired.

One of the most important considerations when operating a polisher/buffer is the selection of the proper buffing pad. Here are some points to consider when making a selection:

- Match the pad to the needs of the job. Low pile heights work best for the early stages of cutting and compounding. High pile heights are better for light compounding and critical jobs. Thicker pads are good for touch-ups and blending where raised body lines demand cushioning. For final finishing and waxing, use a clean lamb's wool bonnet or foam pad. These will run the coolest and offer the most polishing action. For further protection, consider using pads with rounded-up edges.
- Let the pad do the work. Using the design of a pad to its best advantage means changing pads at the various stages of the job.
- Be sure that the pad does not load up too fast, does not burn (a rounded-up edge helps prevent edge burns), and is constructed tightly enough to prevent wool particles from flying out.
- Remember that 100 percent wool pads are best for automotive finishes. Wool runs cooler, cushions more, and lasts longer than synthetics because wool breathes, and its fibers retain their natural spring longer.
- It is poor practice to intermix the use of a buffer with a grinder. A buffer should operate at about 1,200 to 2,200 rpm. Employ a buffer for buffing only or surface scratches can result.
- Grinders should operate at about 5,000 rpm (Figure 5–51).
- Soft foam pads are used after wool or cotton pads to remove swirl marks.

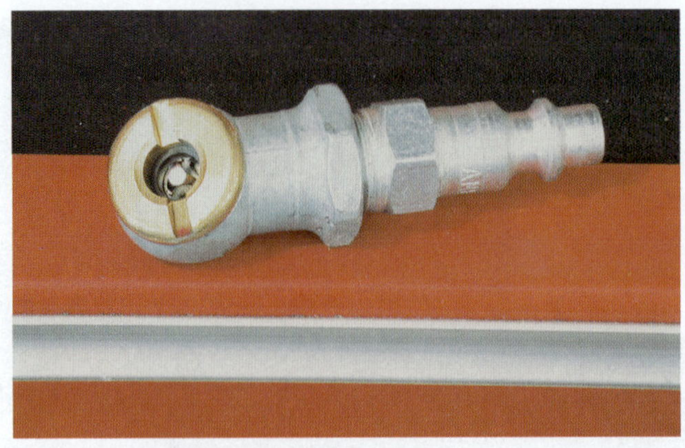

A The tire air chuck fits over the valve stem to increase inflation pressure.

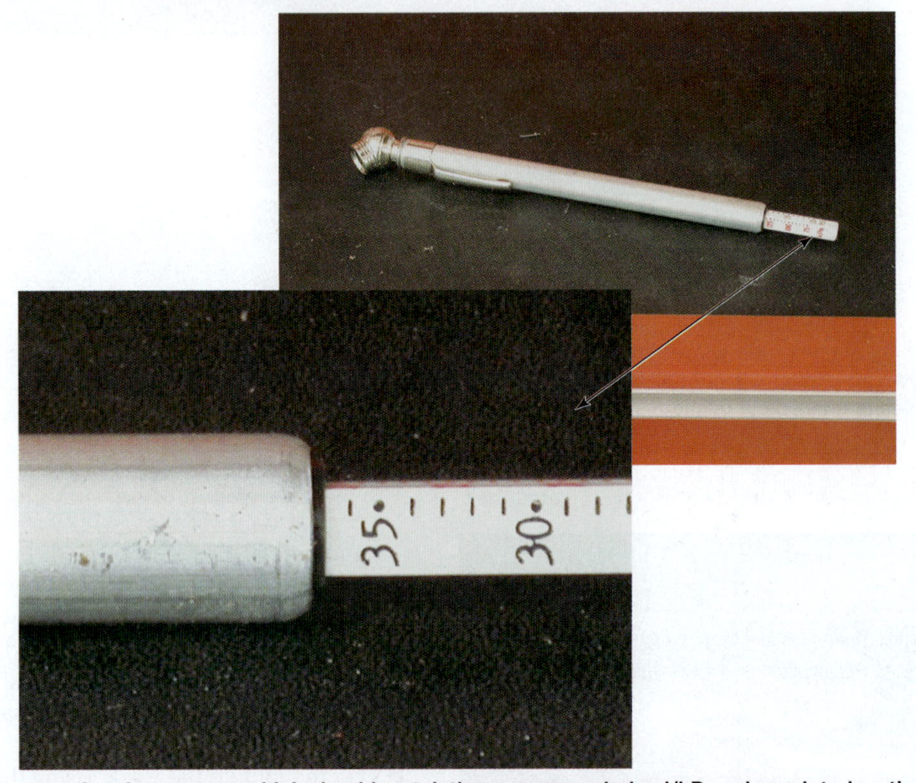

B A tire gauge will measure tire air pressure, which should match the recommended psi/kPa value printed on the side of the tire.

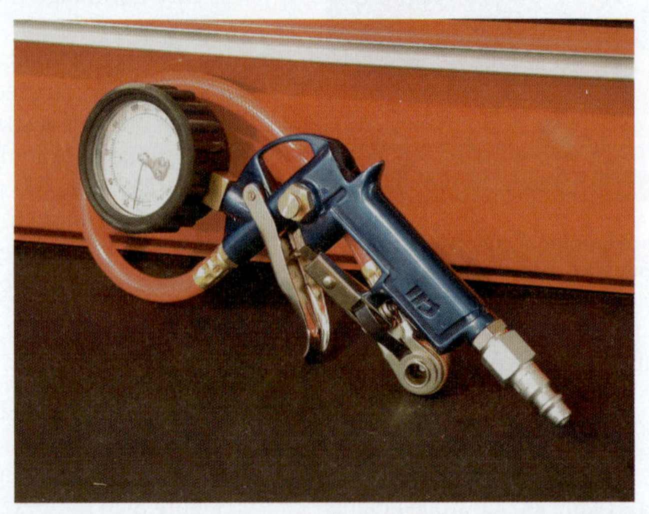

C This tire inflator/gauge assembly is handy because you can check the inflation pressure as you add air to the tire.

Figure 5-49 Study tire-related air tools.

A

C

B

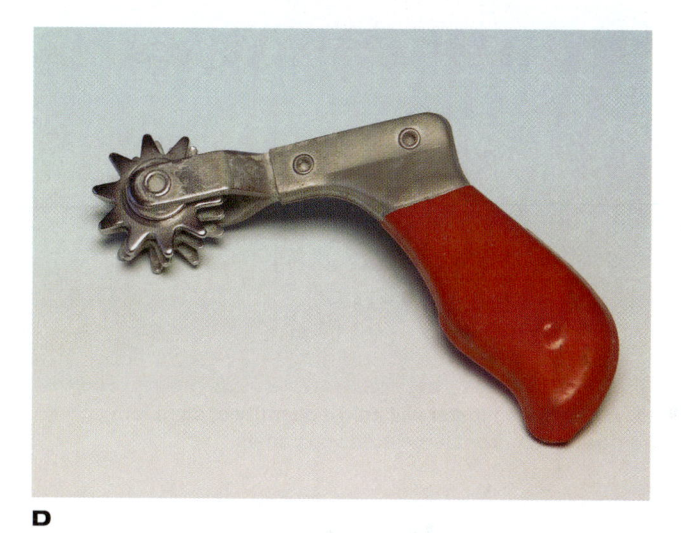

D

Figure 5-50 Buffers are used to polish old dulled paint or new paint that requires wet sanding, to remove minor imperfections. (A) A high-speed electric buffer is often used first because it will polish down imperfections in the clearcoat very quickly. (B) A dual-action buffer is used after a high-speed buffer to help remove swirl marks or microscopic scratches left from high-speed buffing. A compound is a fine buffing abrasive. (C) A cotton buffing pad is often used before softer foam rubber pads. (D) A star wheel is spun against a clogged cotton buffing pad to clean and soften it.

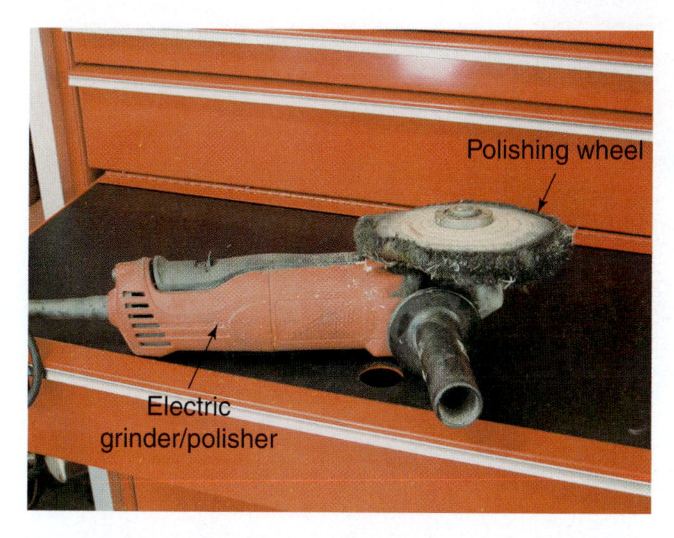

Figure 5-51 This electric power tool has been equipped with a buffing wheel for removing surface imperfections from chrome and aluminum part surfaces.

Media Blasters

Media blasting, often called sandblasting, is the most effective way to remove all finishes from any vehicle. Care must be taken, however, not to damage the underlying substrates. Media blasting operations are usually done at 60–100 psi (414–690 kPa). See Figure 5–52.

The use of plastic media blasting has replaced sand in many body shops. The process is very similar in principle to sandblasting. Instead of using hard silica sand, much softer reusable plastic particles are used at low blasting pressures of 20–40 psi (138–276 kPa). At the lower pressures, the plastic medium removes paint coatings without damaging the underlying substrates, including thin aluminum, steel, fiberglass, and even plastics.

Soda blasters are another new method of stripping paint. Baking soda is used, either wet or dry, under air pressure to strip away the old finish. As with plastic

A A tank-type blaster will hold a quantity of sand for removing rust from large areas on panels.

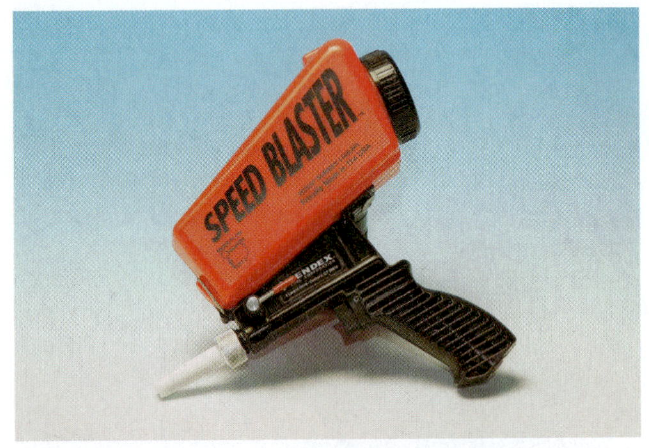

B A hand-held blaster is handy when only a small area is coated with rust or corrosion. It will blast off scale down to bare metal.

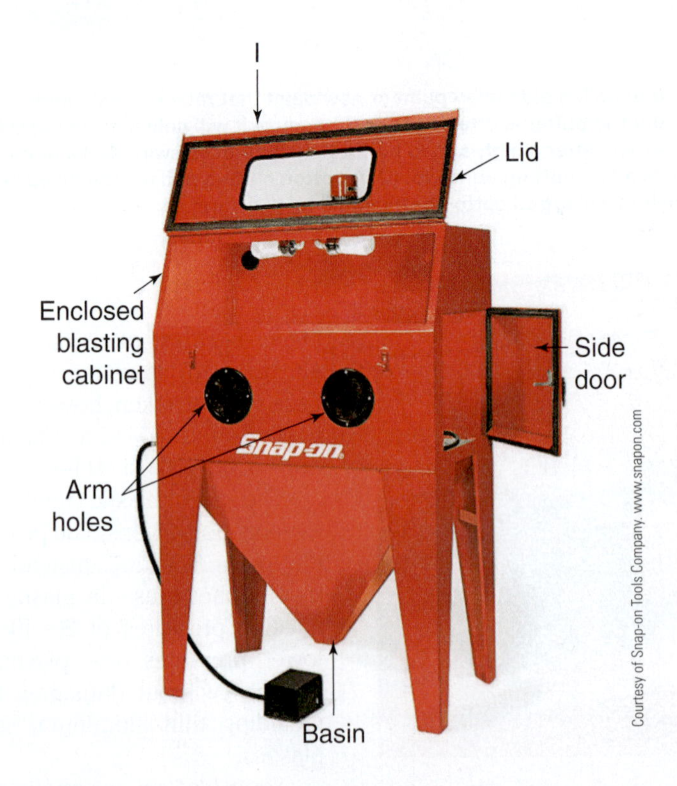

C A captive media sandblaster can be used inside the shop, because it prevents particles from flying all over.

Figure 5-52 Compare the types of sand or media blasting equipment.

media blasting, soda can be safely used on almost any substrate.

There are two basic types of sandblasters:

1. Standard sandblaster
2. Captive sandblaster

The standard sandblaster is usually operated outdoors, while a captive unit can be used indoors. The indoor sandblaster has a nozzle assembly that confines the blasting action. A vacuum in the machine sucks the abrasive and debris back into the machine. Plastic can be used in both types of blasters.

Pneumatic Tool Maintenance

Air tools need little upkeep. However, you will have problems if basic maintenance is not performed. For example, moisture gathers in the air lines and is blown into tools during use. If a tool is stored with water in it, rust will form and the tool will wear out quickly.

Lubricate your air tools periodically. A few drops of special, noncontaminating oil (spray gun-type oil) should be used in your air sanders, chisels, grinders, and other air tools. Do not use motor oil or transmission fluid to lubricate body shop air tools. Conventional oil could contaminate the body surface and cause painting problems.

Squirt a couple of drops of oil into the air inlet or into special oil holes on the tool before and after use (Figure 5–53). This will prevent rapid wear and rusting of the vane motor and other parts in the tool. Run the tool after adding the oil. Wipe excess oil off the tool to keep it from the body parts.

An **inline oiler** is an attachment that will automatically meter oil into air lines for air tools. It can be used on lines used exclusively for air tools but not for spray guns.

Remember not to use inline oilers in the paint area. Never over-oil sanders, grinders, and other air tools. You could contaminate the vehicle's surface with oil.

The most common causes for any pneumatic or air tool to malfunction are:

- Lack of proper lubrication
- Excessive air pressure or lack of it
- Excessive moisture or dirt in the air lines

All air tools have recommended air pressures (Table 5–1). If a tool is overworked, it will wear out sooner. If something goes wrong with a tool, it should be fixed promptly. A tool with worn parts will use more air pressure. The air compressor, in turn, will then become overworked and put out air that is not as clean or dry and is shot right back into the tools. Air tool troubleshooting procedures are given in Table 5–2.

Full information on pneumatic air system operation is provided in Chapter 6.

5.2 ELECTRIC-POWERED TOOLS

As mentioned earlier in this chapter, shop tools such as sanders, polishers, impact tools, and drills can also be powered by electric motors. The most important electric-only tools are drill presses, bench grinders, vacuum cleaners, heat guns, and plastic welders.

Complete data on metal welders is given in Chapter 8, while plastic welders are described in Chapter 13.

Other than these specialized electrically driven power tools, electric drills, polishers, sanders, and so on perform the same shop tasks as their pneumatic counterparts.

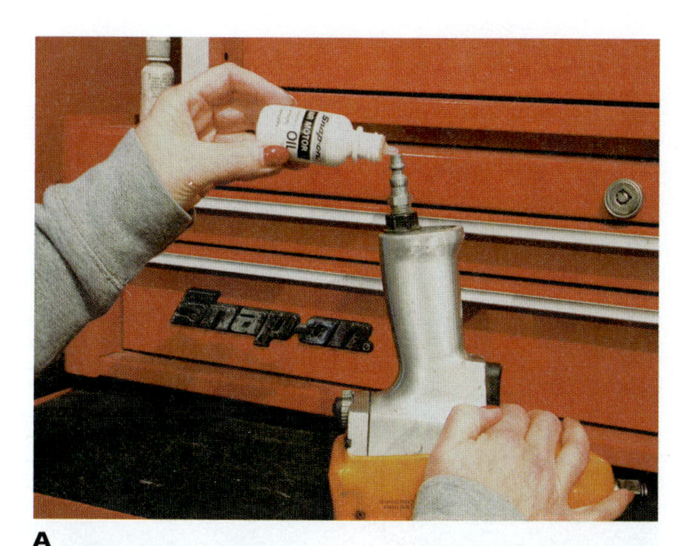

A

B

Figure 5–53 You must hand-oil air-powered tools frequently. It is best to oil them before putting them away at night. (A) Place a few drops of oil in the air inlet and then run the tool to circulate the oil through the air motor. (B) To help keep contaminants off the vehicle being painted, use spray gun oil instead of motor oil or transmission fluid in your air tools.

TABLE 5–1 AIR CONSUMPTION CHART†

Tool	Scfm*	Psi*	Tool	Scfm*	Psi*
Air brush	1	10–50	Material tank	1.8	10–50
Air chisel	4	70–90	Media blaster	2.2–8	30–90
Air filter cleaner	3	70–90	Media blaster/ hopper	2–6	40–90
Air hammer	6–10	70–90	Needle scaler	3–4	70–90
Blowgun	1–2.5	40–90	Nibbler	8	70–90
Brake tester	3.5	70–90	Nutsplitter	6–7.5	70–90
Burring tool	5	70–90	Paint sprayer	0.7–5	10–70
Car washer	8.5	40–90	Panel saw	4–8	70–90
Cut-off grinder	4–8	80–90	Polisher	12	70–90
Drill 3/8 inch	4–6	70–90	Ratchet wrench	4	70–90
Engine cleaner	4–6	70–90	Riveter	4.5–5.5	70–90
Grease gun	3–4	90–150	Sander, disc	4–6	60–80
Grinders, die	4–6	70–90	Sander, double action	6–8	60–80
Grinders, vertical	6–12	70–90	Sander, finish	6–8	60–80
Hacksaw	6–8	70–90	Sander, straight line	6–8	70–90
Hoist (1 ton)	1	70–90	Screwdriver	2–6	70–90
Hydraulic lift	5–7	90–150	Shears	5–8	70–90
Impact wrench 1/4 inch	1.4	70–90	Spark plug cleaner	5	70–90
Impact wrench 3/8 inch	3	70–90	Tire changer	1	90–150
Impact wrench 1/2 inch	4	70–90	Tire chuck	1.5	10–50

*Scfm: Standard cubic feet per minute

Psi: Pounds per square inch

†Always check with the tool manufacturers for the actual air consumption of the tools being used. These figures are based on averages and should not be considered accurate for any particular make of tool.

Drill Press

Some large auto repair shops use a permanently mounted drill press (Figure 5–54). It can be floor mounted or bench mounted. All drill work is performed on a table attached to the stand. The worktable can be adjusted up or down. The drill speed is variable for different materials and thicknesses.

Figure 5–54 The bench-type drill press is handy for drilling smaller, thicker parts.

Bench Grinder

This electric power tool (Figure 5–55) is generally bolted to one of the shop's workbenches. A bench grinder is classified by wheel size, with 6- to 10-inch wheels the most common in auto repair shops. Three types of wheels are available with this bench tool:

1. *Grinding wheels* are used for a wide variety of grinding jobs, from sharpening cutting tools to deburring.

TABLE 5–2 TROUBLESHOOTING AIR TOOLS

Problem	Probable Cause	Recommended Action
		Air Drills
Tool will not run or runs slowly; air flows slightly from exhaust; spindle turns freely.	Motor or throttle plugged with dirt	1. Check for dirt in air inlet. 2. Pour liberal amount of air tool oil in air inlet. 3. Operate trigger in short bursts. 4. Disconnect air supply, then turn empty and closed drill chuck by hand. Reconnect air supply. 5. If still not functional, tool should be checked by an authorized service center.
Tool will not run; air flows freely from exhaust; spindle turns freely.	Rotor vanes stuck with dirt or varnish	1. Pour liberal amount of air tool oil in air inlet. 2. Operate trigger in short bursts. 3. Disconnect air supply, then turn empty and closed drill chuck by hand. Reconnect air supply. 4. If still not functional, tool should be checked by an authorized service center.
Tool locked up; spindle will not turn.	Broken motor vane Gears broken or jammed by foreign object	Tool should be checked by an authorized service center.
Tool will not shut off.	Throttle valve O-ring blown off seat	See parts list for part number and replace O-ring or send tool to an authorized service center.
		Air Hammers
Tool will not run.	Cycling valve or throttle valve clogged with dirt or sludge Piston stuck in cylinder bore by rust or dirt	1. Pour liberal amount of air tool oil in air inlet (check for dirt). 2. Operate trigger in short bursts (chisel in place against solid surface). 3. If not free, first disconnect air supply, then tap nose or barrel lightly with plastic mallet, reconnect air supply, and repeat above steps. 4. If still not free, disconnect air supply, insert a 6-inch piece of ⅜-inch diameter rod in nozzle and lightly tap to loosen piston in rearward direction. Reconnect air supply, and repeat Steps 1 and 2.
Chisel stuck in nozzle.	End of shank peened over	Tool should be sent to an authorized service center.
		Air Ratchets
Motor runs; spindle doesn't turn or turns erratically.	Worn teeth on ratchet or pawl Weak or broken pawl pressure spring Weak drag springs fail to hold spindle while pawl advances for "another bite."	Replacement parts should be installed by an authorized service center.
Motor will not run; ratchet head indexes crisply by hand.	Dirt or sludge in motor parts	1. Pour liberal amount of air tool oil into air inlet. 2. Operate throttle in short bursts. 3. With socket engaged on bolt, alternately tighten and loosen bolt by hand. 4. If motor remains jammed, tool should be checked by an authorized service center.

(*continued*)

TABLE 5–2 TROUBLESHOOTING AIR TOOLS (*continued*)

Problem	Probable Cause	Recommended Action
		Air Wrenches
Tool runs slowly or not at all; air flows only slightly from exhaust.	Airflow blocked by accumulation of dirt Motor parts jammed with dirt particles Power regulator might have simply vibrated to closed position	1. Check air inlet strainer for blockage. 2. Pour liberal amount of air tool oil into air inlet. 3. Operate tool in short bursts, quickly reversing rotation back and forth. 4. Repeat as needed. 5. If this fails to improve performance, tool should be serviced at an authorized service center.
Tool will not run; exhaust air flows freely.	One or more motor vanes stuck due to sludge or varnish buildup Motor jammed due to rust	1. Pour a liberal amount of air tool oil into air inlet. 2. Operate tool in short bursts of forward and reverse rotation. 3. Tap motor housing lightly with plastic mallet. 4. Disconnect air supply, then attempt to free motor by rotating drive shank manually. (Some clutches will not engage sufficiently for this operation.) 5. If tool remains jammed, it should be serviced by an authorized service center.
Sockets will not stay on.	Worn socket retainer ring or soft backup ring	1. Wear safety glasses. 2. Disconnect air supply. 3. Using external retaining ring pliers, expand old retaining ring and remove; or if retaining ring pliers are not available, clamp tool "lightly" in soft jaw vise. 4. Holding square drive with appropriate open-end wrench, pry old retainer ring out of groove with small screwdriver. 5. Always pry off ring away from body; it can be propelled outward at high velocity. 6. Replace backup O-ring and retainer ring with correct new parts. (See parts list that accompanied tool.) 7. Place retaining ring on table; press tool shank into ring in a rocking motion. Snap into groove by hand.
Tool shows premature shank wear.	Use of chrome sockets or excessively worn sockets	Discontinue use of chrome sockets. Remember that chrome sockets have a hard surface and relatively soft core. Drive hole will become rounded, but still be very hard. Besides the danger of splitting, they will wear out wrench shanks prematurely.
Tool gradually loses power but still runs at full free speed.	Clutch parts worn, perhaps due to lack of lubricant Engaging cam of clutch worn or sticking due to lack of lubricant	*Oil lubed* 1. Check for presence of clutch oil (where oil is specified for clutch) and, removing oil fill plug, tilt to drain all oil from clutch case. Refill with 30-weight SAE oil or that recommended by manufacturer, but only the amount specified. 2. Check for excess clutch oil. Clutch cases need only be 50 percent full. Overfilling can cause drag on high-speed clutch parts. A typical ½-inch, oil-lubed wrench only requires ½ ounce of clutch oil. *Grease lubed* Vibration and heat usually indicate insufficient grease in the clutch chamber. The average greasing interval is specified in parts list. Severe operating conditions might require more frequent lubrication. 1. Check for excess grease by rotating drive shank by hand; it should turn freely. Excess grease is usually expelled automatically. 2. If disassembly is required for greasing, it should be done carefully to maintain orientation of mating parts.
Tool will not shut off.	Throttle valve O-ring broken or out of position Throttle valve stem bent or jammed with dirt particles	1. Remove assembly and install new O-ring. 2. Lubricate with air tool oil and operate trigger briskly. If operation cannot be restored, tool should be checked by an authorized service center.

2. *Wire wheel brushes* are used for general cleaning and buffing, removing rust and scale, paint removal, deburring, and so forth.
3. *Buffing wheels* are used for general purpose buffing, polishing, light cutting, and finish coloring operations.

When using a bench grinder, remember to:

- Always use a wheel with a rated speed equal to or greater than the grinder's.
- Always use a wheel guard.
- Inspect wheels for wear or cracks before using them. Use the correct wheel for the job and mount it properly.
- Always wear safety goggles or a face shield and make sure the eye shields of the grinder operator are in position.
- Adjust the rest as needed whenever the gap between it and the grinding wheel exceeds ⅛ inch.

These safety rules also apply to portable air grinders.

Vacuum Cleaner

A must in every body and refinishing shop is a *vacuum cleaner* for removing dust and debris from the vehicle interior. It should be one of the first tools used when a vehicle comes in for refinishing. An incoming vehicle should be completely washed and vacuumed before it is prepared for refinishing. This will greatly reduce the chance of dirt getting into the complete job. The vehicle should also be vacuumed after refinishing to ensure customer satisfaction.

There are two basic types of shop vacuum cleaners: the dry pickup type and the wet/dry unit. For interior vehicle cleaning, the portable vacuum cleaner is popular.

Vacuum tool attachments allow a vacuum cleaner to be attached to a power tool. They are available for sanders and other shop tools. For example, when on a sander, the vacuum cleaner will keep most of the sanding dust out of the shop, which is a plus for your health and the paint job.

Power Washers

Power washers can be used in exterior car preparation, engine cleaning, undercarriage cleaning, shop degreasing and cleaning, and snow and salt removal from vehicles. Figure 5–56 shows a typical unit. Before using a power washer, check the Occupational Safety and Health Administration (OSHA) regulations.

To keep dust out of the refinishing shop, some power orbital sanders—both electric and air-powered—have a sanding dust pickup arrangement. This helps keep the

A Place the object securely in the tool rest and wear thick leather gloves when using a powerful bench grinder.

B Close-up shows how the angle V-block will hold the tool at the correct angle while you are sharpening it to a point.

Figure 5-55 A bench grinder is often used to sharpen tools and dress smaller parts.

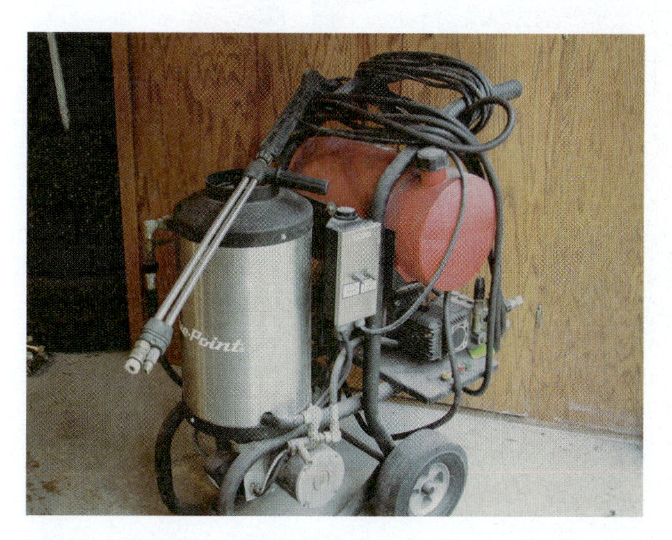

Figure 5-56 A pressure washer/steam cleaner will wash contaminants off the vehicle or the engine compartment before the vehicle enters the shop for body repairs.

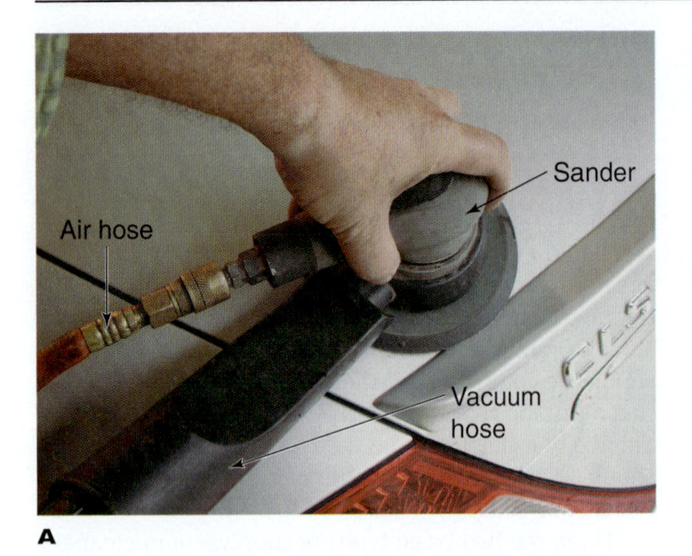

A

B

Figure 5-57 **Dust collection systems will help keep paint dust out of the shop. (A) A sander equipped with a dust collecting hose. (B) A shop vacuum with a good filter will capture sanding dust to make your work environment safer.**

shop cleaner. It also increases air cleanliness for safer working conditions (Figure 5–57).

Heat Gun

Heat guns have many uses in the auto body shop, where controlled heat is needed (Figure 5–58). They are used in almost all vinyl roof repairs as well as other plastic repairs. They can be used in some panel shrinking jobs as well as to speed up drying times. The overall use of heat guns can be found in several chapters of this book.

Courtesy of SPX/OTC Service Solutions

Figure 5-58 **A heat gun can be used to apply controlled heat to objects, such as when you are removing stick-on decals and pinstripes.**

Always allow a heat gun to run in the cool air position for a minute or two after use. This will protect the heating element from damage.

Battery Charger

A *battery charger* converts 120 volts AC into 13–15 volts DC for recharging drained batteries. One is shown in Figure 5–59.

To use a battery charger, connect red to positive and black to negative. The red lead on the charger goes to the positive terminal of the battery. The black lead goes to ground or the negative battery terminal.

After connecting the charger, adjust its settings as needed (12-volt battery, fast or slow charge, and so on). If a battery is low, it is best to slowly charge the battery for several hours. A fast charge for a few minutes will not restore battery charge properly.

A battery can explode if you connect a charger to a battery with the charger running.

Figure 5-59 A battery charger is often needed to recharge dead batteries in a body shop. Connect the red lead to the battery positive and the black lead to the battery negative or to the frame ground away from the battery.

Here are some safety tips to keep in mind when using extension cords:

- Always plug the cord of the tool into the extension cord before the extension cord is inserted into a convenience outlet. Always unplug the extension cord from the receptacle before the cord of the tool is unplugged from the extension cord.
- Extension cords should be long enough to make connections without being pulled taut, creating unnecessary strain and wear.
- Be sure that the extension cord does not come in contact with sharp objects.
- Before using an extension cord, inspect it for loose or exposed wires and damaged insulation. If a cord is damaged, it must be replaced. This advice also applies to the tool's power cord.
- Extension cords should be checked frequently while in use to detect unusual heating. Any cable that feels more than comfortably warm to the bare hand, which is placed outside the insulation, should be checked immediately for overloading.
- To prevent the accidental separation of a tool cord from an extension cord during operation, make a knot as shown in Figure 5–60A, or use a cord connector as shown in Figure 5–60B.

Electric Power Tool Safety

To protect the operator from electric shock, most power tools are built with an *external ground plug*. This is a wire that runs from the motor housing, through the power cord, to a third prong on the power plug. When this third prong is connected to a three-hole electrical outlet, the grounding wire will carry any current that leaks past the electrical insulation of the tool away from the operator and into the shop's wiring. In most modern electrical systems, the three-prong plug fits into a three-prong, grounded receptacle.

Some of the new electric power tools are self-insulated and do not require grounding. These tools have only two prongs since they have a nonconducting plastic housing. In shop operations, never connect a three-prong to a two-prong adapter plug.

Extension Cords

If an extension cord is used, it should be kept as short as possible. Very long or undersized cords will reduce

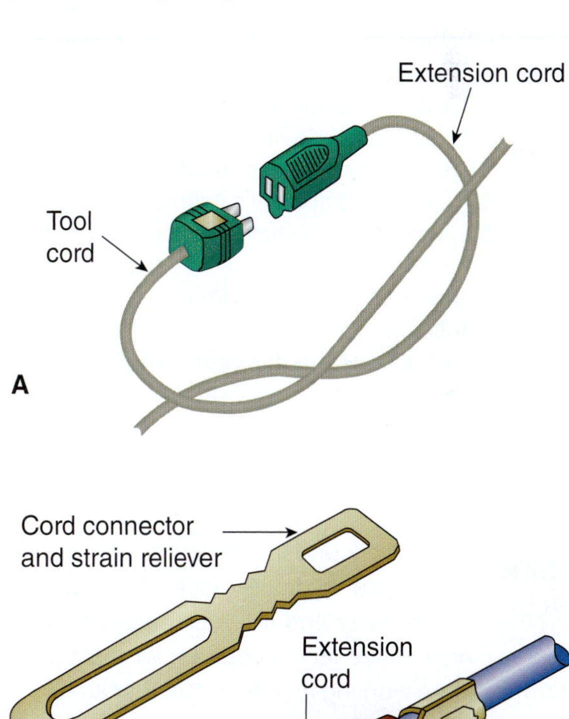

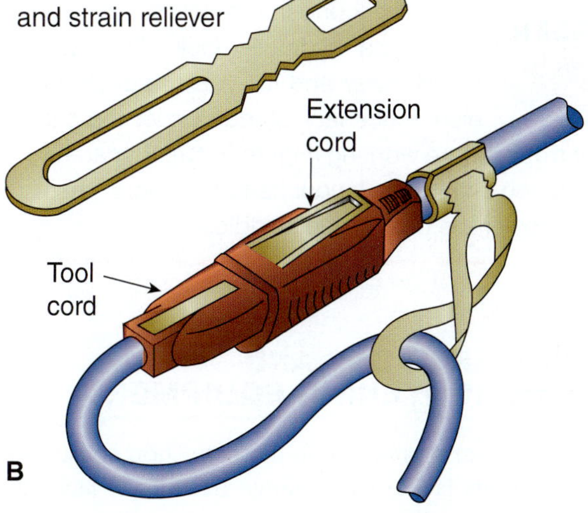

Figure 5-60 (A) A knot will prevent the extension cord from accidentally pulling apart from the tool cord during operation; (B) a cord connector will serve the same purpose effectively.

operating voltage and thus reduce operating efficiency, possibly causing motor damage. An extension cord should be used only as a last resort.

Tools with three-prong, grounded plugs must be used only with three-wire grounded extension cords connected to properly grounded, three-wire receptacles.

Many insurance companies now require automatic shut-off switches on electrical equipment to prevent them from inadvertently being left on.

Cordless Tools

As an alternative, cordless tools—cordless drills and sanders—have made their way into the auto body shop. These tools require no air hose or electric cord, but they do require recharging. Cordless tools are handy for smaller jobs where battery life is not a concern.

5.3 HYDRAULICALLY POWERED SHOP EQUIPMENT

Hydraulic body shop equipment uses an oil-like liquid, called hydraulic fluid, to develop the pressure necessary to operate it. This pressure is achieved manually by pumping on a handle or lever to build up the fluid pressure or by a small motor—either air- or electrically driven—that provides the pressure needed to force the hydraulic fluid into the equipment's cylinder. The cylinder then causes the tool to operate when a button or a lever is turned.

Hydraulic power equipment is usually classified as manual, air-over-hydraulic, or electric-over-hydraulic. *Air- or electric-over-hydraulic* means that either an air-powered or an electric-powered motor is used to force the hydraulic fluid into the tool's cylinder.

DANGER Never work under a vehicle supported only by a hydraulic jack. The jack could lower and kill you or someone else. Place jack stands under the car or truck before working (Figure 5–61). Be careful that you do not pinch your hand or fingers when you lower the heavy steel saddles.

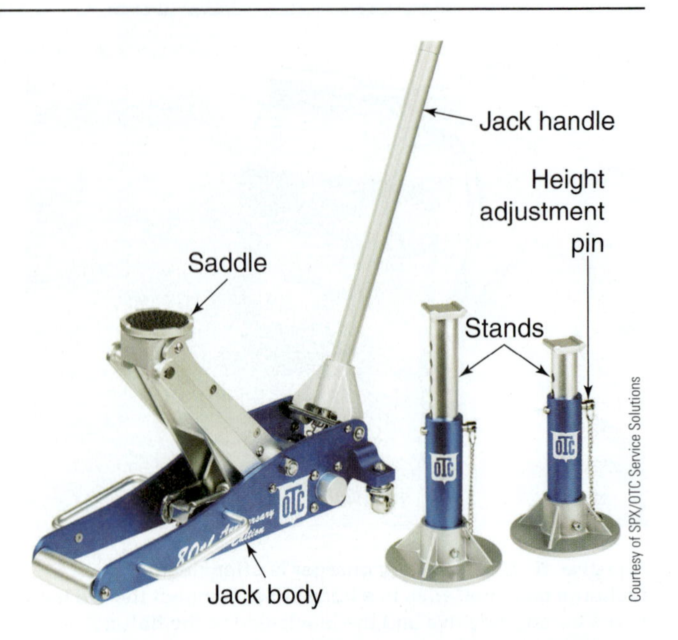

A **Study the parts names of the floor jack and jack stands.**

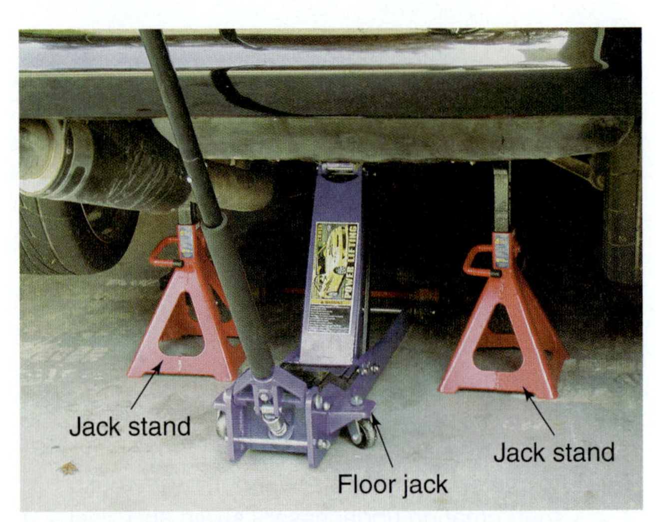

B **A floor jack can be used to raise a vehicle, but then jack stands must be placed under the car before working. Jacks can fail and drop a vehicle if jack stands are not used.**

Figure 5–61 **Hydraulic jacks are often needed during collision repair work to lift a heavy vehicle off the floor.**

5.4 POWER JACKS AND STRAIGHTENING EQUIPMENT

Hydraulic power equipment in the auto shop is used to operate various jacks. These range from body jacks to large frame jacks.

The average shop has approximately a dozen jacks, either air, hydraulic, or a combination of both, depending on the preference of the technicians. (Manual jacks are practically obsolete except for use in confined spaces.) The most popular jacks are:

- *Hand or bottle jacks* (Figure 5–62A). These tubular-shaped jacks are not specialized. Rather, they perform a variety of functions and range from 1½ tons to 100 tons of lifting capacity. They are useful when a service jack is too much.
- *Service jacks*. These four-wheeled jacks with a pump handle are by far the most commonly used jack in the body shop. Ranging in lifting capacity from 1½ tons to 5 tons, these jacks are easily "dollied" around the shop and rolled under a car to lift a section of it, as opposed to the entire structure. Compact and

portable, these jacks were developed for a variety of in-shop uses on full-size, intermediate, compact, or subcompact cars. They are also used for road service calls. Service jacks are recommended for all automotive, agricultural, and light truck repair facilities.

- *End lifts*. These jacks are either air-over-hydraulic or manual. As the name implies, they lift only a section of the vehicle by adhering to the bumper. They do not lift the sides of the vehicle. Lifting capabilities range from 1½ to 7 tons.
- *Transmission jacks* (Figure 5–62B). Often it is necessary to remove the transmission, engine, or drivetrain from a unibody before servicing a repair. This jack was developed specifically for this purpose. The lifting capacity ranges from ¼ to 1 ton and the jacks are mechanical, air-over-hydraulic, or manual.

A **body jack** (also known as a hydraulic power set portable power unit) is a hydraulic hand pump and ram for minor body and frame straightening operations. It can be used with frame/panel straighteners or it can be used by itself. To perform the many different straightening operations involving in pushing, pulling, or holding a panel to straighten or align metal, a large assortment of attachments (Figure 5–63) can be obtained. The hydraulic body jack is usually sold in sets or kits for general work. There is also a bodywork set, a mechanical set, and even a rescue set, which is used to help free people trapped in a vehicle after a bad accident.

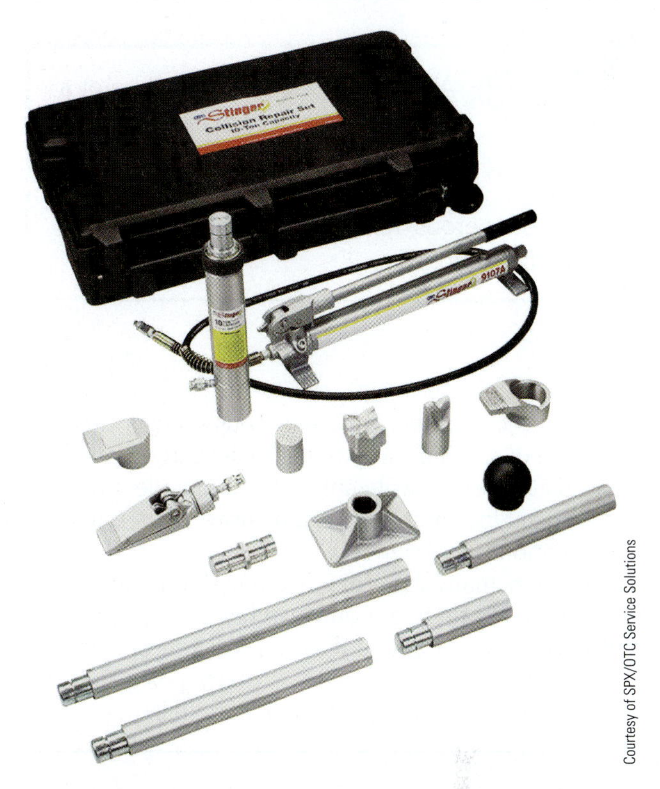

Figure 5-63 A portable power unit or hydraulic straightening unit is equipped with various attachments for forcing damaged body panels back into shape.

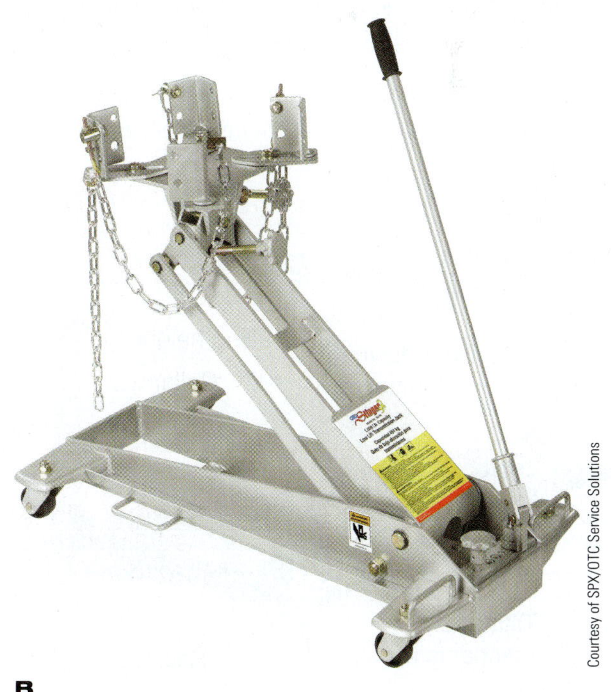

Figure 5-62 Here are two other kinds of jacks found in a body shop. (A) This small hand or bottle jack can exert tons of force to lift or bend panels. (B) The transmission jack has a clamping saddle for keeping a large transmission or other assembly from falling.

Always use the rated tonnage of a jack for the tons specified. If a jack is rated for 2 tons, do not attempt to use it for a job requiring 20 tons. It is dangerous for both the body technician and the vehicle.

The basic *hydraulic jack* unit consists of a manual, electrical, or air pump; a hydraulic hose; and a ram. Rams are available in various lengths. Included under rams are two wedge-type or spreader rams used for getting into tight locations.

There are many styles of frame/panel straighteners on the market, but there are only two basic types: portable and stationary. Portable units are less expensive to purchase, but they cannot make as many push and pull actions at one time as can the stationary units.

Frame Rack

A *frame rack* is a large electro/hydraulic machine for holding a vehicle while pulling out major structural damage. It consists of the following major parts:

1. The rack is a thick steel framework that supports and holds the vehicle secure while pulling out damage (Figure 5–64).
2. Pulling towers contain hydraulic rams and pulling chains that attach to the damaged area of the vehicle (Figure 5–65 and Figure 5–66).
3. Hydraulic rams mounted on or in the pulling towers attach and pull on the chains attached to the vehicle.

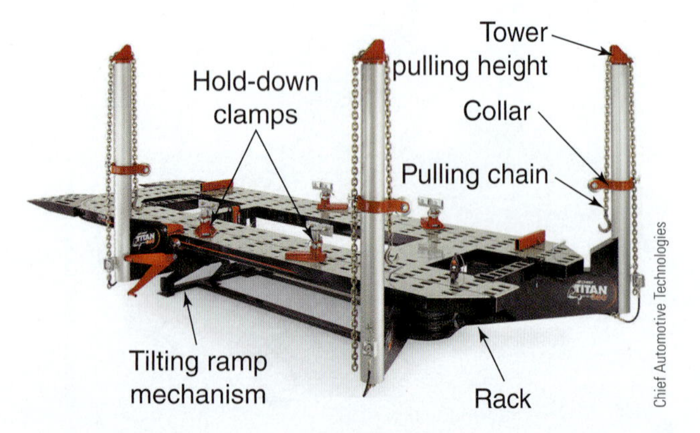

Figure 5–64 A frame/unibody machine is also called a "frame rack." It can exert tons of force to pull out structural damage after a major collision.

Figure 5–65 Always read the specific operating instructions before using a frame rack. This unit is equipped with several pulling towers for straightening major frame and unibody damage quickly and efficiently.

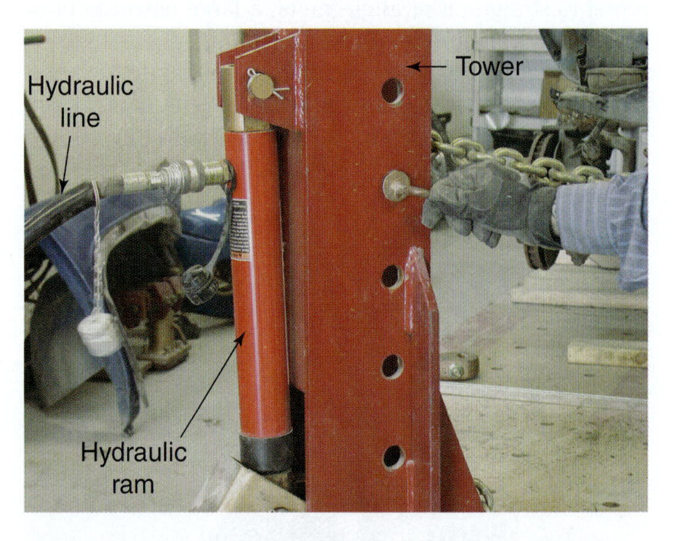

Figure 5–66 The hydraulic ram provides the tremendous power of a frame rack. When turned on, the ram can extend and place tons of pulling force on its pulling chain.

Measuring Systems

A measuring system is equipment used to compare known specifications to actual measurements taken off a damaged vehicle. If your measurements are different than specs, then you know the vehicle needs to be straightened on a frame rack or with a hydraulic ram. Look at Figure 5–67. It shows a state-of-the-art computerized measuring system.

Body shop technicians must understand how to make hookups for correcting body or frame damage with power jacks. Complete information on using body jacks and frame/panel straighteners is given in Chapters 9 and 10.

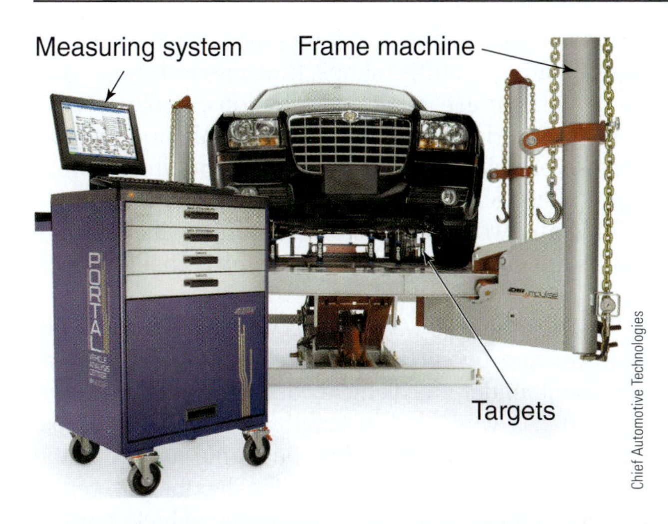

Figure 5-67 **A measurement system will check the dimensions of a damaged vehicle so you can determine how it should be repaired. Lasers or acoustics are normally used on computerized measuring systems.**

5.5 HYDRAULIC TOOL CARE

Hydraulic tools also require preventive maintenance to avoid failure at a critical moment. Just because a hydraulic tool is filled with hydraulic fluid does not mean it has been lubricated properly. Moving parts should be lubricated regularly with 30- or 40-weight oil. These parts include the moving mechanism, pump roller, universal joint, handle socket, pivot pins, wheels, and bearings. Also be sure to grease the fittings and sliding points used in pumping.

As with air tools, dirt can be a problem. Dirt is an abrasive that can scratch the bore of the ram. Where do dirt and debris come from? The auto repair shop is full of body filler and metal shavings. If a hydraulic tool has an air motor, the dirt may have come from the air hose sitting in dust. After the hose is connected, the dirt shoots right into the tool. The solution is to clean it just before connection.

Avoid overfilling a hydraulic tool. A certain amount of air is supposed to be left in the reservoir. If it is completely filled with fluid, too much vacuum is created and the fluid will not move out of the reservoir. When disconnecting a hydraulic line in a system, make sure the system and rams are fully retracted so oil pressure has been released.

Table 5–3 outlines the major problems associated with hydraulic tools and equipment operation, which include the following:

- *Spongy effect.* Air trapped in the hydraulic system easily compresses under pressure and causes sponginess. To bleed the system, place the pump at a higher elevation than the hose and ram. The objective is to "float" the air bubbles uphill and back to the reservoir where they belong. Close the valve and extend the unit as far as possible. Open the valve fully, allowing the oil and air to return to the reservoir. Repeat this

TABLE 5–3 TROUBLESHOOTING CHART

	Probable Cause								
Problem	**Reservoir low on hydraulic fluid**	**Reservoir too full**	**Air in system**	**Bent plunger**	**Release valve not fully closed**	**Dirt in release valve**	**Dirt in check valve**	**Loose dirt or air bubble in valve system**	**Damaged quick-coupler**
Spongy effect			X						
Tool will not extend all the way	X	X							
Tool will not retract		X		X					X
Tool leaks under pressure					X	X			
Handle kickback							X		
Works properly one time but not the next								X	

procedure until the tool starts to extend on the first stroke of the handle. Usually two or three times will do the trick.

- *Tool will not extend all the way.* This is usually a sign of low hydraulic fluid. Fill to the mark on the dipstick. Do not overfill. The tool should be fully retracted to check the oil level. If the tool does not have a dipstick, refer to the manufacturer's manual for filling procedure and fluid level. The hydraulic unit needs the prescribed amount of air chambered in the reservoir because it works on a partial vacuum to avoid venting to the outside.
- *Tool will not retract.* Usually this means there is too much oil and/or air in the system. Bleed the system and fill to the proper level. If that does not correct the problem, inspect for a bent plunger. If neither is the cause, the quick-coupler is probably damaged and needs to be replaced.
- *Tool leaks under pressure.* Make sure that the release valve is fully closed. If it still leaks, there is probably dirt in the return. Check the ball valve. Flush the system with mineral spirits or kerosene. If the problem still exists, the valve is damaged and the unit should go to a repair facility.
- *Handle kickback.* Dirt is in the check valve; flush the system. A damaged check valve should be taken to a repair facility.
- *Works properly one time, but not the next.* There is loose dirt or an air bubble in the valve system. Flush and refill the system.

5.6 HYDRAULIC LIFTS

Another important piece of equipment is the **hydraulic lift** used to raise the whole vehicle in the air for easier working conditions (Figure 5–68). The traditional, stationary, in-the-ground unit was usually found only in service stations, muffler shops, transmission shops, and tire dealers for use in oil and lube jobs, brake service, and other underbody repairs. Today, all body shops, because of unibody construction, are looking for ways to get the vehicle off the ground to estimate and repair. Technicians work better with cars on a lift. Damage reports are easier to prepare on a vehicle that has been up on a lift.

Figure 5-68 **A lift is commonly used to raise vehicles high in the air for suspension and underbody repairs.**

There are several ways to get a vehicle off the ground. Four-post and two-post hoists are neat and allow total movement under the vehicle, but they take up more space than a work stall in length and width. Side post hoists are great for estimating, but some access to the sides of the vehicle is impaired. Finally, the old center post hoists make some areas under the vehicle hard to reach but take up less space.

The use of lifts makes the job much less physically and mentally demanding on the body technicians, because they can work at a level that is comfortable for them. But a lift is a specialized piece of equipment, and a technician using one needs to know exactly what is to be done, especially if there are any vehicle weight problems. It is important to note that the quality of today's lifts, as well as the number of safety devices on each model, makes them very safe to operate.

The maintenance on above-ground lifts is minimal, but important. Depending on the lift, pulleys, pivoting lift links, and wheels should be greased. Bearings, pins, and other moving parts should be oiled, and cables and chains should be checked for worn or frayed areas, following the manufacturers' directions.

General maintenance on lifts includes inspecting the lift pads and bumper cushions regularly and replacing them if necessary. It is advisable to appoint a knowledgeable person to inspect the jacks and lifts daily, especially if the units are subjected to abnormal loads or shocks.

If any jack or lift appears to be damaged, worn, or operates improperly, remove it from service until the necessary repairs are made. A falling vehicle can cripple or kill!

Lift Safety

Raising a vehicle on a lift or a hoist requires special care. Drive-on lifts are fairly safe. However, it is important to make sure that vehicles equipped with a catalytic converter have enough clearance between the hoist and the

A

C

B

D

Figure 5-69 Note the basic steps for using a typical lift, but, again, refer to brand-specific lift operating instructions. Most important, make sure you have the weight of the vehicle centered on the lift. In a front-engine vehicle, lift arms should be more to the front of the vehicle. In a mid- or rear-engine vehicle, lift arms should be more to the rear to center the weight on the lift. Vehicles can easily tip and slide off the lift with deadly consequences if the weight is not centered on the lift arms. (A) Position the lift arms under the proper lift points on the frame, rails, cross members, control arms, or rear axle. (B) Raise the vehicle an inch or so off the floor and double-check your lift arms. (C) With the vehicle in the air a little, make sure that its weight is centered and that it will not fall off the lift. (D) Raise the vehicle fully and engage the mechanical locks so the lift cannot lower if the hydraulic system fails.

exhaust system components before driving the vehicle onto the ramps (Figure 5–69).

Adapters and hoist plates must be positioned correctly on twin-post and rail-type lifts to prevent damage to the underbody of the vehicle. The catalytic converter, tie rod, rod bracket, and shock absorbers are some of the components that could be damaged if the adapters and hoist plates are incorrectly placed.

There are specific contact points to use where the weight of the vehicle is evenly supported by the adapters or hoist plates. The correct lifting points can be found in a vehicle's service manual (Figure 5–70).

Engine Crane

An engine crane uses hydraulic power to raise heavy powertrains (engine and transmission) from a vehicle during major repairs. The engine and transmission may have to come out to replace frame rails or other structural panels. The engine crane has a large lift arm that can be

Figure 5-70 **The lift is a safe piece of equipment if used properly. However, thousands of cars and trucks have fallen off the lift because of improper use.**

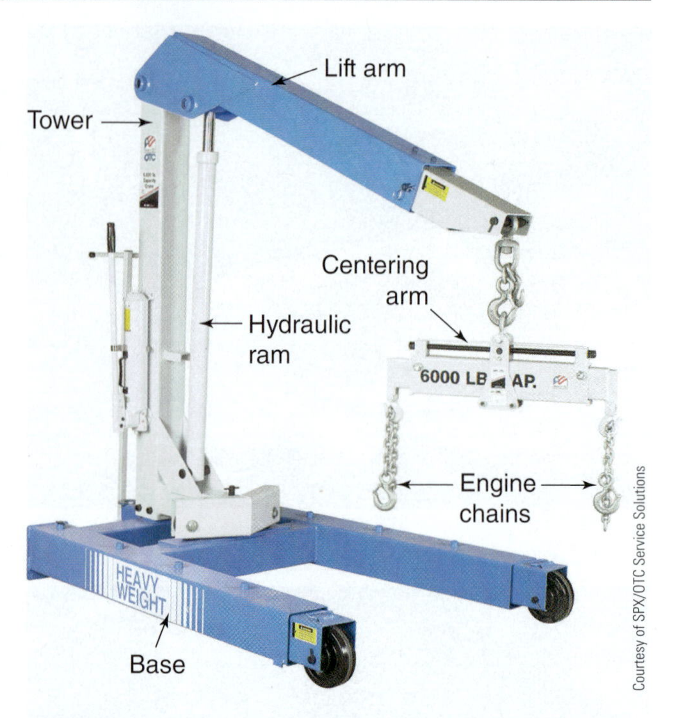

Figure 5-71 **An engine crane is sometimes needed to remove the powertrain from the vehicle during major frame or unibody repairs. Note the parts of the engine crane.**

raised by pumping the handle on the hydraulic cylinder. The ram will then extend to lift the arm and any assembly attached to it (Figure 5–71).

Here are some other lift safety tips that should always be kept in mind and followed when raising a vehicle up in the air:

- Never overload the lift. The manufacturer's rated capacity is shown on the nameplate affixed to the lift.
- Employees should stand to one side of a vehicle when directing it into position over a lift. Do not allow customers or bystanders to operate the lift or be in the lift area during its operation.
- Positioning of the vehicle and operation of the lift should be done only by trained and authorized personnel.
- Always keep the lift area free from obstructions, grease, oil, trash, and other debris.
- Operating controls are designed to close when released. Do not block them open or override them.
- Before driving a vehicle over a lift, position the arms and supports to provide unobstructed clearance. Do not hit or run over lift arms, adapters, or axle supports. This could damage the lift or the vehicle.
- Load the vehicle on the lift carefully. Check to make sure adapters or axle supports are in secure contact with the vehicle, per the manual instructions, before raising the lift to the desired working height. Remember that unsecured loads can be dangerous.
- Make sure the vehicle's doors, hood, and trunk are closed prior to raising the vehicle. Never raise a car with passengers inside.
- Position the lift supports to contact at the vehicle manufacturer's recommended lifting points. Raise the lift until the supports contact the vehicle. Check supports for secure contact with the vehicle and raise the lift to the desired working height.
- After lifting a vehicle to the desired height, always lower the unit onto mechanical safeties.
- Note that with some vehicles, the removal (or installation) of components can cause a critical shift in the center of gravity and result in raised vehicle instability. Refer to the vehicle manufacturer's service manual

(continued)

for recommended procedures when vehicle components are removed.

- Make sure tool trays, stands, and so forth are removed from under the vehicle. Release locking devices as per instructions before attempting to lower the lift.
- Before removing the vehicle from the lift area, position the arms, adapters, or axle supports to ensure that the vehicle or lift will not be damaged.
- Inspect the lift daily. Never operate it if it malfunctions or if it has broken or damaged parts. It should be removed from service and repaired immediately.

A lift requires immediate attention if it:

Jerks or jumps when raised
Slowly settles down after being raised
Slowly rises when not in use
Slowly rises when in use
Comes down very slowly
Blows oil out of the exhaust line
Leaks oil at the packing gland

- Repairs should be made with OEM parts only.

SUMMARY

1. Power tools use air pressure (pneumatic), oil pressure (hydraulic), or electrical energy to effect repairs.
2. An impact wrench is a portable, hand-held, reversible, air-powered tool for rapid turning of bolts and nuts.
3. Do not use an air impact wrench to final-torque fasteners. Fastener or part failure could result. This could endanger the driver and passengers.
4. Drill bits of different sizes fit into the chuck on drills for making holes in parts.
5. There are special spot remover air drills and/or attachments that are used for cutting out spot welds.
6. An air sander uses an abrasive action to smooth and shape body surfaces. Sandpapers of different coarsenesses can be attached to the pad on the sander.
7. Grinders are used for fast removal of material. They are often used to smooth metal joints after welding and to remove paint and primer.
8. An air polisher is used to smooth and shine painted surfaces by spinning a soft buffing pad.
9. Media blasting is the most effective way to remove all finishes from any vehicle. The use of plastic media or soda blasting has replaced sand in many body shops.
10. Air tools need little upkeep. However, you will have problems if basic maintenance is not performed.
11. The most common causes for any pneumatic tool to malfunction are lack of proper lubrication, excessive air pressure or lack of it, and excessive moisture or dirt in the air lines.

EXERCISES

On a separate sheet of paper, complete the following learning activities for this chapter. Write definitions for the key terms and answer the ASE-style review questions, essay questions, critical thinking problems, and math problems. You can also do the outside activities, possibly for extra credit.

Key Terms

air drill
air file
air polisher
air ratchet
air sander
body jack
grinder
heat gun
hydraulic lift
impact wrench
in-line oiler
power tool
sander pad
spray gun
swirl marks

ASE-Style Review Questions

1. To locate the position of a hole to be drilled, Technician A uses an awl. Technician B says that the distinct mark left by the awl provides a seat for the drill point. Who is correct?
   A. Technician A
   B. Technician B
   C. Both A and B
   D. Neither A nor B
2. Technician A uses a low-pile-height buffing pad for light compounding, touch-ups, and blending. Technician B uses a screwdriver to clean dried compound off the pad. Who is correct?
   A. Technician A
   B. Technician B
   C. Both A and B
   D. Neither A nor B
3. Which of the following operations can the air hammer perform?
   A. Bushing installer
   B. Bushing remover
   C. Shock absorber chisel
   D. All of the above
4. Which type of wheel is used for sharpening cutting tools?
   A. Grinding wheel
   B. Wire wheel brush
   C. Buffing wheel
   D. Cutting wheel
5. Which of the following is used to lift the entire automobile?
   A. Floor jack
   B. Hydraulic lift/hoist
   C. Hydraulic press
   D. All of the above
6. When working with a piece of hydraulic equipment that will not extend all the way, Technician A says a likely problem is low hydraulic fluid. Technician B says a likely problem is dirt in the check valve. Who is correct?
   A. Technician A
   B. Technician B
   C. Both A and B
   D. Neither A nor B
7. What causes a pneumatic tool to fail?
   A. Lack of lubrication
   B. Excessive air pressure
   C. Lack of air pressure
   D. All of the above
8. Which air sander creates a buffing pattern?
   A. Disc sander
   B. Dual-action orbital sander
   C. Long board sander
   D. All of the above
9. Technician A says that a cotton buffing pad is often used before a softer foam rubber pad. Technician B says that a star wheel is spun against a clogged cotton buffing pad to clean and soften it. Who is correct?
   A. Technician A
   B. Technician B
   C. Both A and B
   D. Neither A nor B
10. Which type of sandblaster can be used indoors?
    A. Standard
    B. Captive
    C. Both A and B
    D. None of the above
11. What is the recommended air pressure for a disc sander?
    A. 60 to 80 psi
    B. 70 to 90 psi
    C. 40 to 90 psi
    D. 10 to 50 psi

Essay Questions

1. Describe the eight major parts of a typical paint spray gun.
2. How does an air impact tool operate?
3. Why should you use a torque wrench to final tighten fasteners and not an impact wrench?
4. Explain the differences between sandblasting and plastic media blasting.
5. List some of the advantages of using air-powered tools over electric tools.

Critical Thinking Problems

1. If you have rough-filed a large area of body filler and the material is hard, how would you continue rough straightening?
2. What can happen to air tools if they are not oiled periodically?

3. It is recommended to wear loose-fitting clothing when operating a power tool. True or False?
4. An impact wrench can be used to make accurate adjustments. True or False?

Math Problems

1. If both air wrenches are made the same, how much stronger than a ¼-inch drive head is a ½-inch drive head?
2. A small hydraulic jack has a pumping plunger that is 1 square inch in diameter. The ram piston has a diameter of 5 square inches. If you exert 500 lb. of pressure on the pumping plunger through the handle's lever arm, how much weight can you lift with the jack?

Activities

1. Visit an auto repair shop. Watch technicians as they use power tools and equipment. Have a class discussion on how technicians were using these tools.
2. Walk through your shop and note the location of all power tools and equipment. Discuss with the class any questions you might have on power tool or equipment use.
3. Disassemble and reassemble a shop spray gun. Note the condition of all parts. Make a report on the condition of the spray gun.
4. Inspect your shop air compressor. Find the tank drain valve, shutoff valve, and other parts. Did water come out of the drain valve when it was opened?

CHAPTER 6 Compressed Air System Technology

OBJECTIVES

After studying this chapter, you should be able to:

- Identify the various types of air compressors used in a body shop.
- Explain the operation of the various air compressors.
- Describe the function of air and fluid control equipment.
- Use the different accessory equipment of the air compressor.
- Maintain an air supply system properly.
- List the air system safety rules.
- Answer ASE-style review questions relating to compressed-air equipment.

INTRODUCTION

The **compressed air supply system** is designed to provide an adequate supply of clean, dry air at a predetermined pressure to ensure efficient operation of all pneumatic equipment in the body shop. The system can vary in size from small portable units to large in-shop installations (Figure 6–1).

The basic requirements for compressed air systems are the same for most body shops (Figure 6–2):

- An air compressor, sometimes referred to as an air pump, can be one compressor or a series of compressors. The compressor is the heart of the system. The power source is generally an electric motor. (Portable gasoline-driven compressors are available for work outside the shop.)
- A control or set of controls is necessary to regulate the operation of the compressor and motor.
- Air intake filters/silencers are designed to muffle intake noises as well as filter out dust and dirt.
- The air tank or receiver must be properly sized. It cannot be too small or it will cause the compressor to cycle too often, thus causing excessive load on the motor. It should not be too large because of space constraints and should not have unnecessary capacity.
- The *distribution system* includes the pipes and hoses, or "arteries," that link the compressed air system.

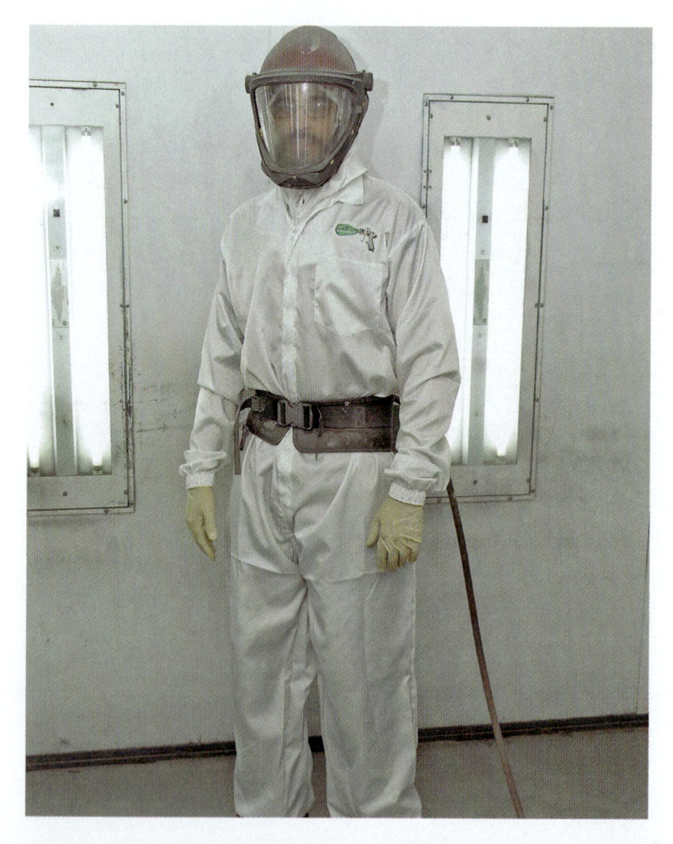

Figure 6–1 The compressed air system in a collision repair shop must provide a clean supply of air pressure for both the air tools and breathing masks in the paint booth.

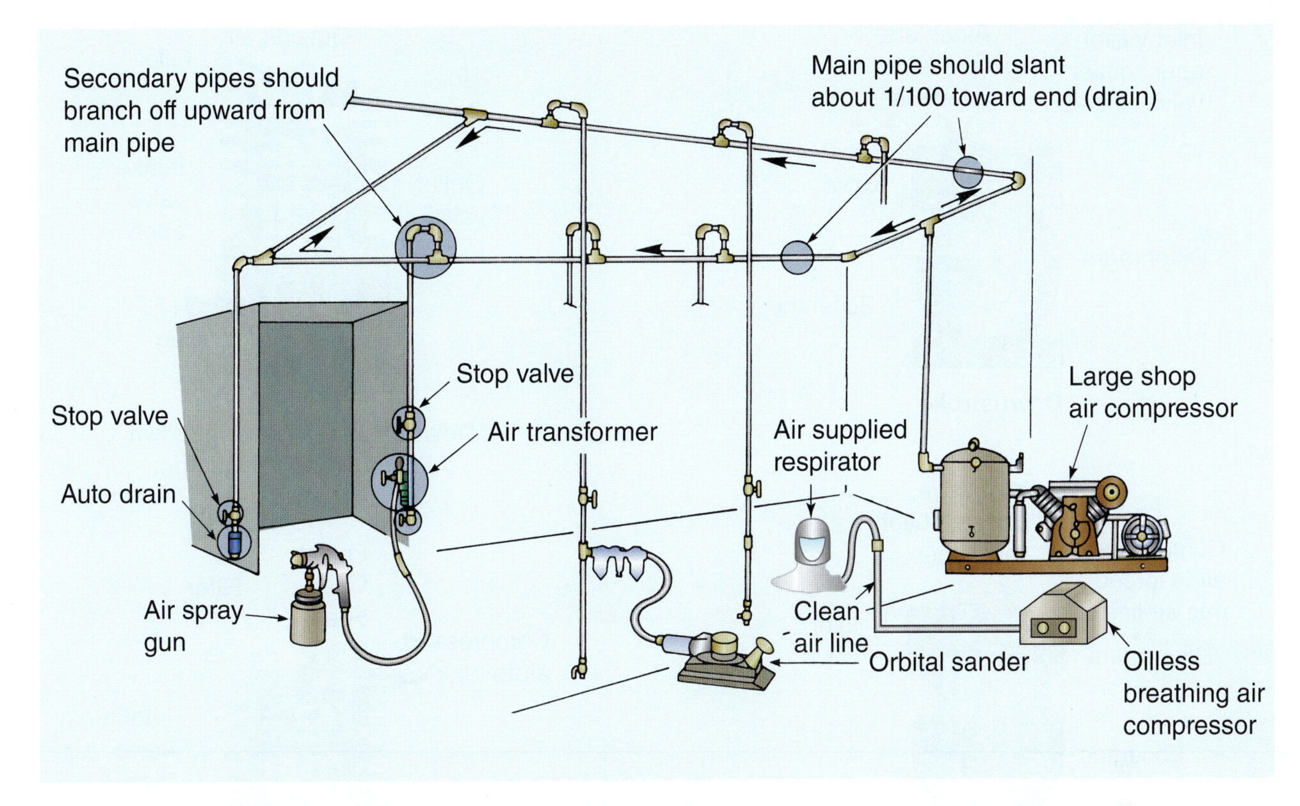

Figure 6–2 **Study the typical piping arrangement found in a body/paint shop. The large compressor is for all of the air tools in the shop. The small oilless air compressor is for breathing air in the paint booth.**

The hose and piping from the air receiver to distribution points requiring compressed air are included in the system. The distribution system must consist of properly sized hoses or pipes, fittings, valves, air filters, oil and water extractors, regulators, gauges, lubricators, and so on for the effective operation of specific air tools and spraying equipment.

6.1 THE AIR COMPRESSOR

The **air compressor** is designed to raise the pressure of air from normal atmospheric to some higher pressure. This pressure is measured in **pounds per square inch (psi)** or metric kilopascals (kPa). While normal atmospheric pressure is about 14.7 psi (101 kPa) at sea level, a compressor is capable of delivering air at pressures up to 200 psi, or 1,378 kPa.

An air compressor is usually made up of an electric motor, air pump, and large air storage tank. The motor spins the air pump, which works like a small reciprocating piston engine. The air pump piston action pushes the air into a large, thick, steel storage tank. The air compressor is an important part of any paint or body shop.

Compressor Types

There are three basic types of air compressors: the diaphragm type, the piston type, and the rotary screw type.

Diaphragm Compressor

The *diaphragm compressor* uses a flexible synthetic rubber membrane to produce a pumping action (Figure 6–3). This type of compressor is often used on very small compressors to power small air brushes for doing custom painting and to provide air to respirators.

A durable diaphragm is stretched across the bore of a very shallow compression chamber. An **eccentric** (egg-shaped part), mounted on the motor shaft, acts on the diaphragm plate to pull the diaphragm up and down.

Figure 6–3 **This diaphragm compressor is designed to run a small air brush for doing custom paint work.**

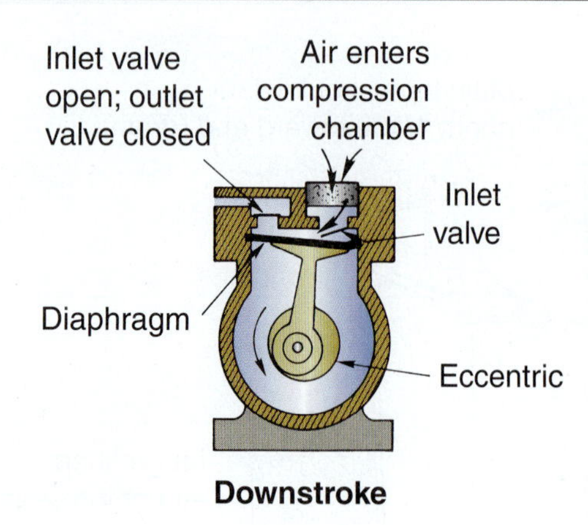

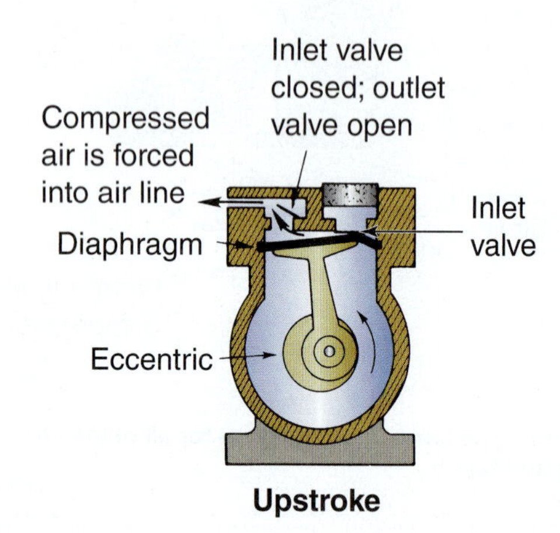

Figure 6–4 **The operation of a diaphragm compressor is illustrated. As the eccentric rotates, it pulls up and down on the flexible diaphragm. Valves then direct the flow in and out of the pump to produce pressure.**

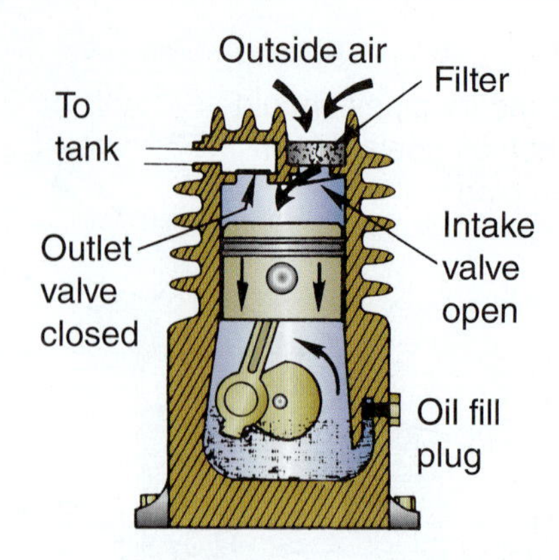

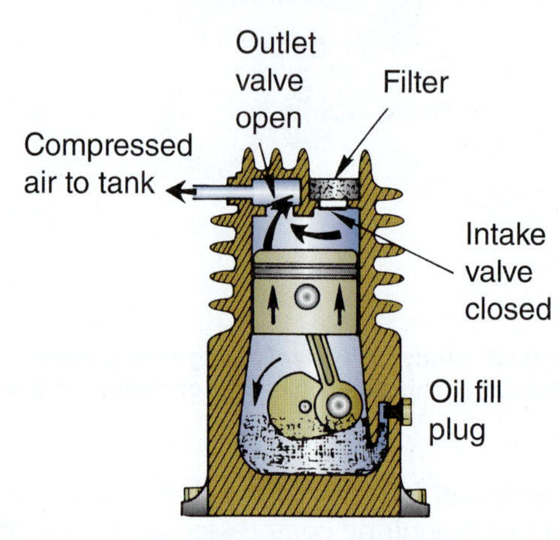

Figure 6–5 **Study the operation of a piston-type compressor. As the piston slides up, the inlet valve closes and the outlet valve opens; this forces air out of the cylinder under pressure.**

As the diaphragm is pulled down, air is drawn into the small space above the diaphragm. When the diaphragm is thrust upward, the air trapped in the compression chamber is squeezed and forced out into the delivery chamber and supply lines (Figure 6–4). Only a very small amount of air, in the 30 to 35 psi range, is compressed during each cycle. However, the pumping action is very rapid—in excess of 1,500 strokes per minute.

Piston Compressor

The *piston air compressor* pump develops compressed air pressure through the action of a reciprocating piston. The piston, which is actuated by a crankshaft, moves up and down inside a cylinder. This is very much like a piston in an automobile engine.

On the downstroke, air is drawn into the compression chamber through a one-way valve. On the upstroke, as the air is compressed by the rising piston, a second one-way valve opens (Figure 6–5). The air is forced into a pressure tank or receiver. As more and more air is forced into the tank, the pressure inside the tank rises.

Piston compressor pumps are available in single or multiple cylinder and single- or two-stage models. Selection depends on the volume and pressure required.

SHOP TALK

The displacement of a two-stage compressor is always given for that of the first-stage cylinder or cylinders only. This is because the second stage merely rehandles the same air the first stage draws in and cannot increase the amount of air discharged.

When air is drawn from the atmosphere and compressed in a single stroke, the compressor is referred to as a **single-stage compressor**. Single-stage units normally are used in pressure ranges up to 125 psi (861 kPa) for intermittent service. Most single-stage compressors are rated at *50 percent duty cycle* (half the time on, half the time off). They are available in single- or multicylinder compressors.

In a **two-stage compressor**, air is first compressed to an intermediate pressure and then further compressed to a higher pressure (Figure 6–6). Such a compressor has cylinders of unequal bore. The first stage of compression takes place in the large-bore cylinder. In the second stage, the pressurized air passes through an intercooler. Air is then compressed for a second time to a higher pressure in the smaller-bore cylinder (Figure 6–7).

Two-stage compressors are usually more efficient, run cooler, and deliver more air for the power consumed, particularly in the 100 to 200 psi (689–1,378 kPa) pressure range. This range of pressure is enough for most body or finishing applications.

Compressor oil plugs are provided for filling and changing air pump oil. The oil level in the compressor should be checked and changed periodically. Single-weight, nondetergent oil or the oil recommended by the manufacturer is normally used.

In recent years, an oilless or oil-free piston compression system has been introduced that employs self-lubricating materials that do not require an oil lubricant. Until recently, most oilless compressors, like the diaphragm type, were considered compact and were limited in both output and pressure. However, there are oilless compressors now on the market of up to approximately 5 horsepower that will nearly equal, in output and pressure, oil-lubricated compressors of the same horsepower. When kept in good condition, oilless compressors produce clean air output.

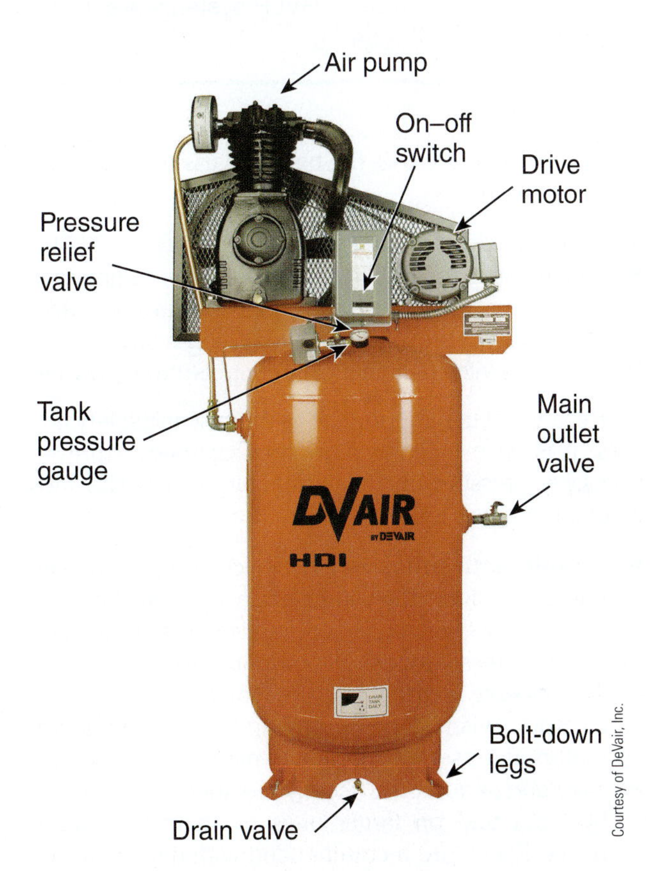

Figure 6–6 This is a typical two-stage compressor.

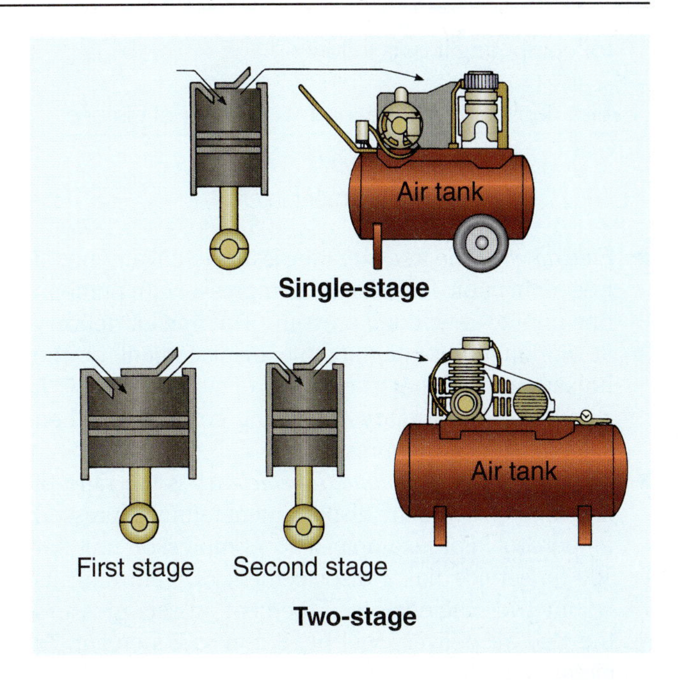

Figure 6–7 Compare single- and two-stage compressors.

Rotary Screw Air Compressor

Rotary screw air compressors have been a standard in other industries. The rotary screw air compressor is a highly efficient and dependable machine.

How Compressors are Rated

The following terms are used to measure the performance of a compressor.

- *Horsepower (HP)*. Horsepower is a measure of work. In a compressor, it is the capacity of the motor or engine that drives the compressor. Compressors found in body and paint shops usually range from 3 to 25 HP. As a general rule, the greater the horsepower, the more powerful the compressor. Also, in most cases, as the horsepower increases, so will the other compressor ratings that follow.
- **Cubic feet per minute (cfm)**. Cubic feet per minute is the volume of the air being delivered by the compressor to the air tool. This spec is used as a measure of the compressor's capabilities. Compressors with higher cfm ratings provide more air through the hose to the tool, thus making higher cfm outfits more practical for larger jobs. Actually, compressors have two cfm ratings.
- *Displacement cfm*. **Displacement** is the theoretical amount of air in cubic feet that the compressor can pump in one minute. It is a relatively simple matter to calculate the air displacement of a compressor if the piston diameter, length of stroke, and rpm are known. For example, the area of the piston multiplied by the length of the stroke and the shaft revolutions per minute equals the displacement volume. The formula

for computing it is as follows:

$$\frac{\text{Area of piston} \times \text{stroke} \times \text{rpm} \times \text{number of pistons}}{1{,}728} = \text{Displacement in cfm}$$

- *Free air cfm.* The free air rating is the actual amount of free air in cubic feet that the compressor can pump in one minute at working pressure. The free air delivery at working pressure, not the displacement or the horsepower, is the true rating of a compressor. It should be the primary cfm rating considered when selecting an air compressor.
- The *compressor's volumetric efficiency* is the ratio of free air delivery to the displacement rating, expressed in percent. For example, if a compressor unit for 100 lb. service has a displacement of 8 cfm and its volumetric efficiency is 75 percent, at this pressure the free air delivery will be: 8 cfm × 75 percent, or 6 cfm.
- *Pressure (psi).* Pounds per square inch (psi) is the measure of air pressure or force delivered by the compressor to the air tool. This is usually expressed as:

1. Normal or continuous working pressure
2. Maximum pressure

SHOP TALK

If the cfm of a compressor is too low for demand, the pump will not be able to keep up. The speed of the air tool will slow down as air pressure is consumed from the tank, and you will have to wait while the compressor builds pressure again. Always purchase a compressor that will handle maximum air consumption.

Tank Size

The **compressor air tank** is a heavy-gauge steel tank for holding an extra supply of compressed air. Working pressure is not available until the tank pressure is above the required psi or kPa rating of the air tool. The compressor puts more air pressure into the tank than is required for application. The larger the tank, the longer a job can be done at the required pressure before a pause is needed to rebuild pressure in the tank.

Air tanks or receivers are typically of a cylindrical shape. The compressor motor and pump are usually mounted on top. Tanks can be purchased with either horizontal or vertical stationary mountings, or they can be mounted horizontally on wheels for portability.

An **air tank shutoff valve** is a hand valve that isolates the tank pressure from shop line pressure. It should be closed at night or when the compressor is not going to be used. If it is not closed, the compressor would run all night if a hose leaked or ruptured.

A **compressor drain valve** on the bottom of the tank allows for water to be drained off. The compression of the air tends to make moisture condense. This moisture must be drained periodically to prevent it from entering the air lines.

Compressor Outfits

There are two types of compressor outfits used in body shops: portable and stationary. A portable outfit is designed for easy movement and is equipped with handles, wheels, or casters and usually a small air receiver or pulsation chamber.

A stationary outfit is one that is permanently installed. It is usually equipped with a larger air receiver than the portable type. It might have a pressure switch or an automatic unloader as found on larger industrial units. Larger stationary models are generally equipped with a centrifugal pressure release.

Another source of compressed air found in many body repair/paint shops is the high-volume, low-pressure (HVLP) system's turbine generator (Figure 6–8). HVLP systems are fully described in Chapter 28.

Shown in Figure 6–9, the typical parts of a stationary compressor are:

- Air compressor pump (pressurizes air)
- Electric motor or gasoline engine (powers compressor)
- Air receiver or storage tank (holds compressed air)
- Check valve (prevents leakage of stored air)
- Pressure switch (automatically controls the air pressure)

Because of the importance of the system's safety controls, it is wise to know how they operate to prevent excessive pressure and electrical problems. The most common of these are:

- An **automatic unloader** is a device designed to maintain a supply of air within given pressure limits on compressors when it is not practical to start and stop the compressor's electric motor during operations.
- A **pressure switch** is a pneumatically controlled electric switch for starting and stopping electric motors at preset minimum and maximum pressures.
- *Overload protection* is usually provided on small units by fuses and on larger ones by thermal overload relays. Relays are recommended with time delay features so that circuits will not be opened by short

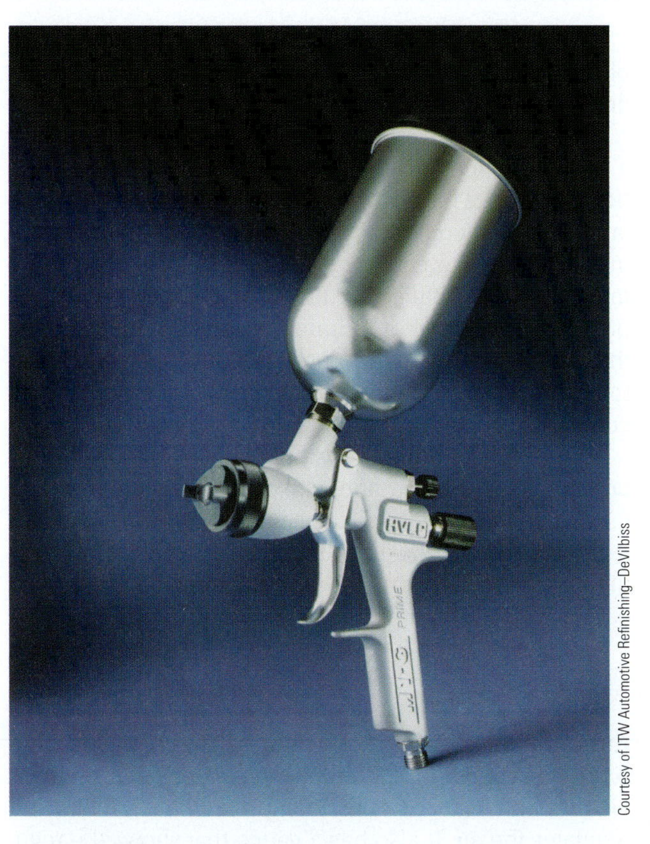

Courtesy of ITW Automotive Refinishing–DeVilbiss

Figure 6-8 A spray gun requires the cleanest supply of air of any air tool. Any oil or debris in the air supply system will cause dirt, fish eyes, and other problems with paint work.

duration surges that are not harmful to the motor. Overload protection should be employed on all compressor installations.

- A *centrifugal pressure release* is a device that allows a motor to start up and gain momentum before engaging the load of pumping air against pressure. When the compressor slows down to stop, rotating the crankshaft more slowly, steel balls move toward the center where they wedge against a cam surface, forcing the cam outward. This opens a valve, bleeding air from the line connecting to the check valve. With air pressure bled from the pump and aftercooler, the compressor can start up free of backpressure until it gets up speed. When normal speed is reached, the balls move out by centrifugal force, releasing the cam, closing the valve, and allowing air to again be pumped into the air receiver.
- A *fused disconnect switch* is a knife-type off–on switch containing the proper size fuse. This should be used at or near the compressor unit with the line going from the fused disconnect to the starter. Fuses should be the size recommended by the compressor's manufacturer.

A qualified electrician should always hook up an air compressor to meet all electrical codes. If something is wired incorrectly or connected improperly, severe part damage can result.

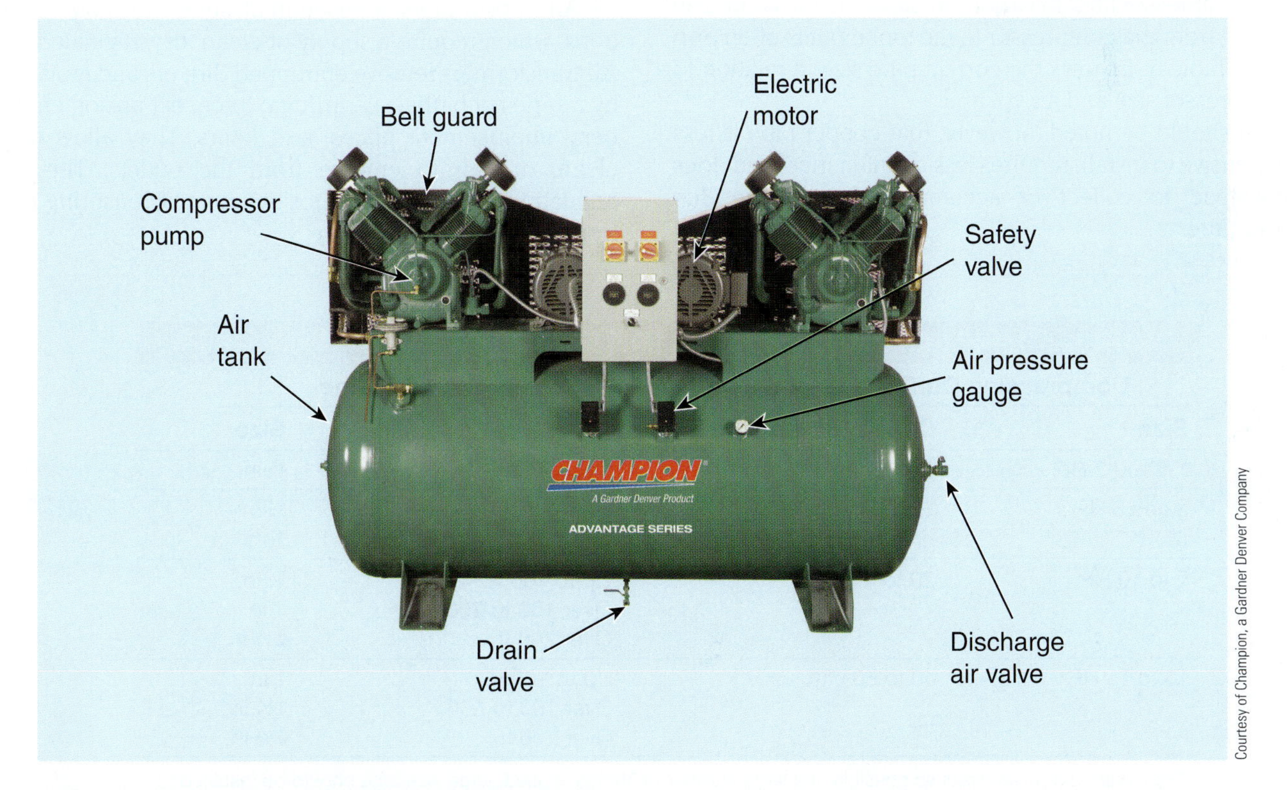

Courtesy of Champion, a Gardner Denver Company

Figure 6-9 Study the parts of a stationary compressor.

6.2 AIR AND FLUID CONTROL EQUIPMENT

The control of the volume (amount), pressure, and cleanness of the air going to pneumatic tools, especially spray guns, is of critical importance.

The **intake air filter** located on the compressor inlet is very important because all the air going into the compressor passes through this filter. The filter element must be made of fine mesh or felt material to ensure that small particles of grit and abrasive dust do not pass into the cylinders. The intake filter must prevent excessive wear on cylinder walls, piston rings, and valves.

If possible, the compressor should be placed where it can receive an ample supply of clean, cool, dry air. If necessary, connect the air intake to the outside of the building. Distance between the intake and the compressor should be as short as possible for highest efficiency. The outside intake should be protected from the elements with a hood or suitable weatherproof shield. The compressor air intake should not be located near steam outlets or other moisture-producing areas.

It must be remembered that unfiltered air piped directly from a compressor is of little use to the body or refinishing shop. The air contains small but harmful quantities of water, oil, dirt, and other contaminants that will lessen the quality of the sprayed finish. And the air will likely vary in pressure during the job.

Distribution System

The *air distribution system* carries compressed air from the compressor tank to various locations in the shop. The piping from the compressor to the tool input is often iron pipe. Table 6–1 shows the correct pipe size in relation to compressor size and air volume.

It should be noted, however, that copper pipe is less expensive to install, requires less maintenance, and does not tend to collect or accumulate harmful residue over time.

The compressor should be located as near as possible to operations requiring compressed air. Lengthy air lines can cause needless pressure drops.

A double loop or circle is accomplished by installing a tee in the line and then running the loop or circle in both directions back to the air tank. For this type of installation, it is recommended that an extra air tank be installed at the far end to balance out peak loads. All piping should be installed so that it slopes toward the compressor air receiver, or a drain leg should be installed at the end of each branch to provide for drainage of moisture from the main air line. This line should not run adjacent to steam or hot-water piping.

In the air distribution system, there should be a *shut-off valve* on the main line, close to the storage receiver tank. This valve is used to shut off the air at the air receiver. Keeping the air shut off at the storage tank overnight ensures a full tank of air when the shop is opened each day.

Air Transformer

An **air transformer**, sometimes called a moisture *separator/regulator*, is a multipurpose device. It removes oil, dirt, and water from the compressed air. It filters and regulates the air. It also has a gauge that shows the regulated air pressure. The transformer may also provide multiple air outlets for spray guns, blow guns, air-operated tools, and so on.

Figure 6–10 illustrates a typical air transformer. Some air transformers are equipped with a second gauge that indicates main line pressure.

Air transformers are used in all spray finishing operations, which require a supply of clean, dry, regulated air. Air transformers remove entrapped dirt, oil, and moisture by a series of baffles, centrifugal force, expansion chambers, impingement plates, and filters. They allow only clean, dry air to emerge from the outlets. The air-regulating valve provides positive control, ensuring uniformly constant air pressure.

TABLE 6–1 MINIMUM PIPE SIZE RECOMMENDATIONS*

Compressing Outfit		Main Air Line	
Size	**Capacity**	**Length**	**Size**
1½ and 2 HP	6 to 9 cfm	Over 50 ft.	¾ in.
3 and 5 HP	12 to 20 cfm	Up to 200 ft. Over 200 ft.	¾ in. 1 in.
5 to 10 HP	20 to 40 cfm	Up to 100 ft. Over 100 to 200 ft. Over 200 ft.	¾ in. 1 in. 1¼ in.
10 to 15 HP	40 to 60 cfm	Up to 100 ft. Over 100 to 200 ft. Over 200 ft.	1 in. 1¼ in. 1½ in.

*Piping should be as direct as possible. If a large number of fittings is used, large size pipe should be installed to help overcome excessive pressure drop.

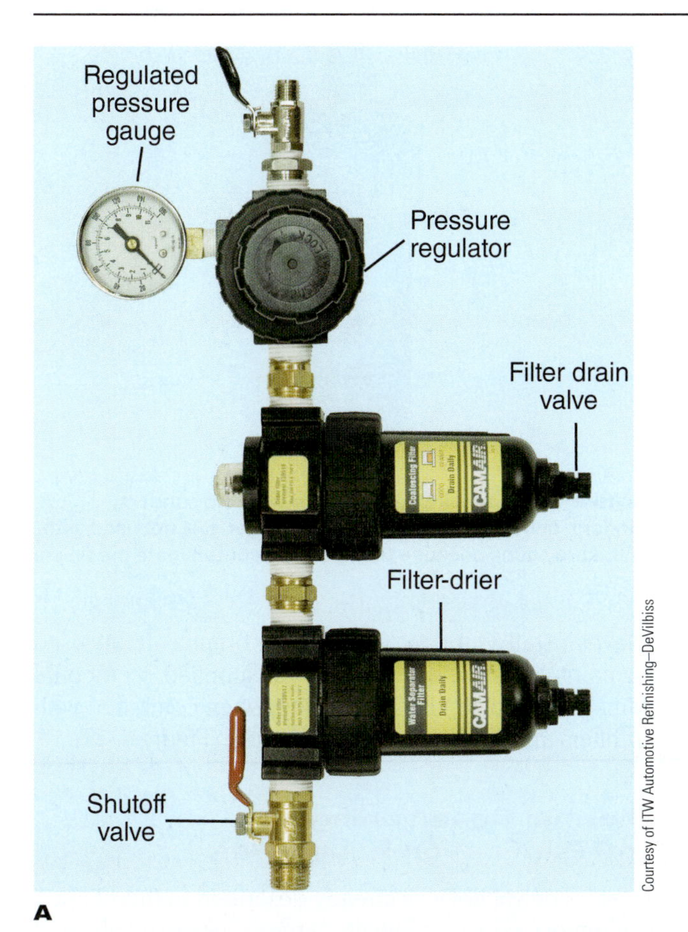

B

Figure 6–10 **(A) Air transformers, or regulators, are needed to reduce pressure to some air tools, such as a spray gun. A filter-drier is needed to purify air so that no contaminated air passes through the air hoses to the tool. (B) Open the drain on the filter periodically to remove water or moisture.**

Gauges indicate regulated air pressure and, in some cases, main line pressure as well. Outlets with valves allow compressed air to be distributed where it is needed. The *drain valve* provides for elimination of sludge consisting of oil, dirt, and moisture. The air transformer should be installed at least 25 ft. (7.6 m) from the compressing unit for efficient operation.

Air Condenser or Filter

An *air condenser* is basically a filter that is installed in the air line between the compressor and the point of use. It separates solid particles, such as oil, water, and dirt, out of the compressed air. No pressure regulation capability is supplied by this device. A typical air condenser is illustrated in Figure 6–11.

Air Pressure Regulator

An **air pressure regulator** is a device for reducing the main line air pressure as it comes from the compressor. It automatically maintains the required air pressure with minimum fluctuations. Regulators (Figure 6–12) are used in lines already equipped with an air condenser or other type of air filtration device. Air regulators are available in a wide range of cfm and psi capacities, with and without pressure gauges, and in different degrees of sensitivity and accuracy. They have main line air inlets and regulated air outlets (Figure 6–13).

Lubricator

Certain types of air-operated tools and equipment described in Chapter 5 require a very small amount of oil mixed in the air supply that powers them. An automatic **air line lubricator** (Figure 6–14) should be installed on a leg or branch line furnishing air to pneumatic tools.

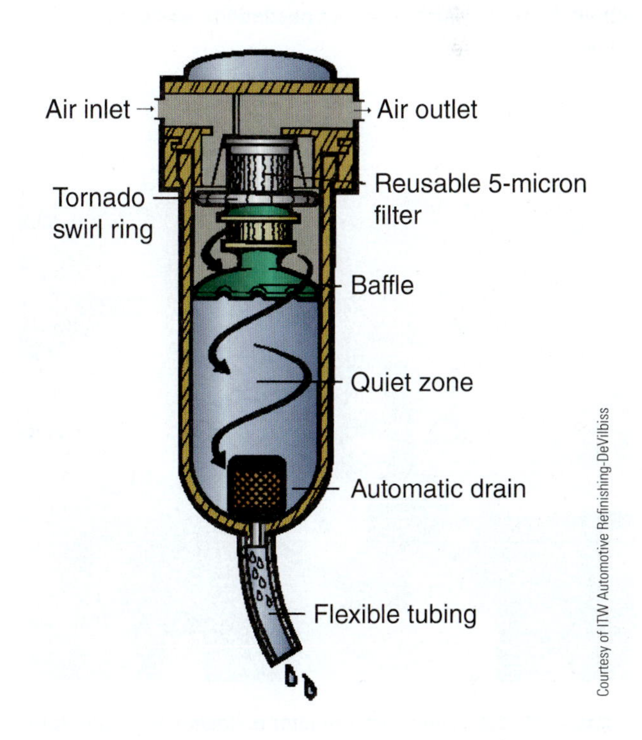

Figure 6–11 **This air filter removes debris and moisture from air that is entering the system.**

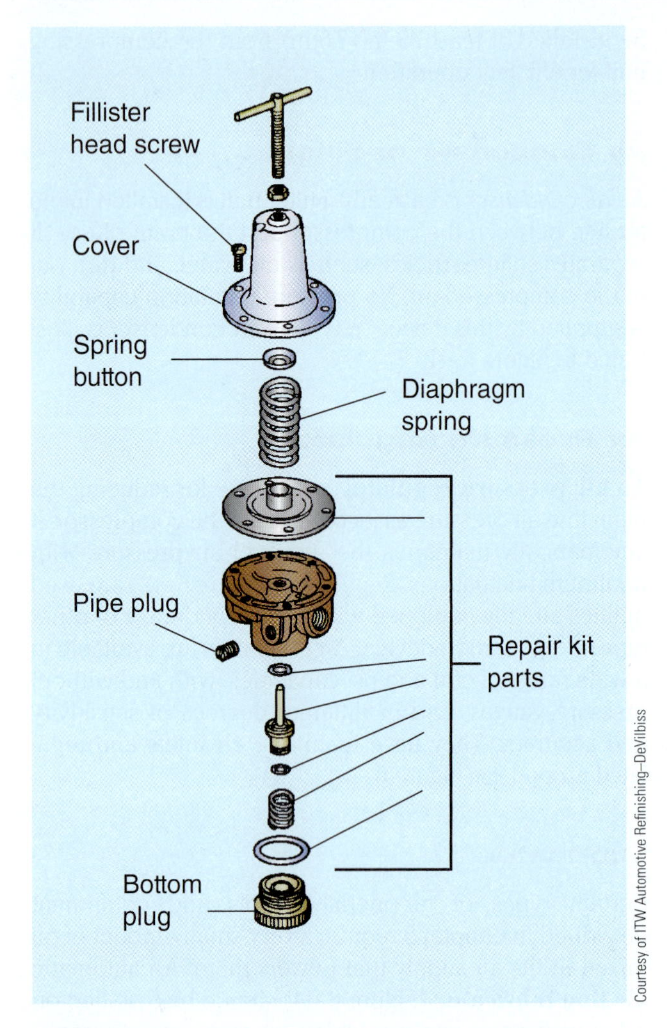

Figure 6–12 This exploded view shows the inner parts of an air pressure regulator. Quality units can be rebuilt. Note the parts repair kit needed for servicing the unit.

Figure 6–13 A filter/air regulator is designed for multiple air tools. Air pressure should be adjusted to match the requirements of the air tool.

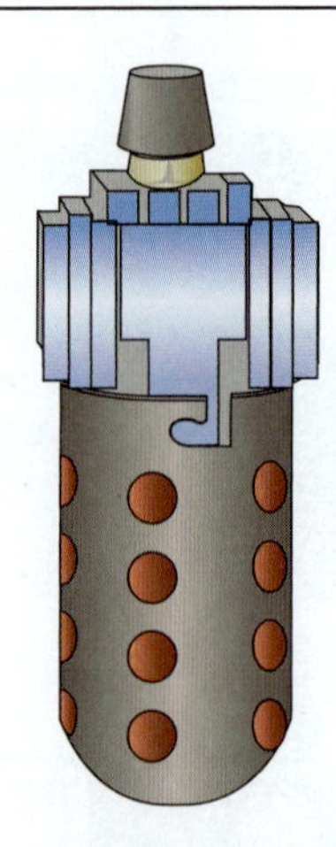

Figure 6–14 An air line lubricator will help sanders, grinders, and similar air tools to last longer. It is not used with paint spray guns because the oil would contaminate the finish.

(Never install a lubricator on a leg or branch air line used for paint spraying, because the oil supplied by it could damage the finish.) Lubricators are often combined with air filters and regulators in a single unit (Figure 6–15).

Thermal Conditioning and Purification Equipment

The air control devices already described in this chapter will remove contaminants from the compressed air most satisfactorily. However, there are some special problems related to heat, dampness, and dirt that require special thermal conditioning and purification equipment. This equipment is usually installed between the compressor and the air storage receiver tank (Figure 6–16). It includes the following:

- *Aftercooler.* An **aftercooler** is used to reduce the temperature of compressed air. Heat, as well as some impurities, can be removed by installing an aftercooler in the system. Aftercoolers are very efficient in lowering air temperature and removing most of the oil and water. The residue of oil and water is removed before it enters the air receiver. There are several different designs or types of aftercoolers available. The most common is the water-cooled "air tube" design in which air passes through small tubes. Recirculating water is directed back and forth across the tubes by means of baffles to reverse the direction of airflow. This cross-flow principle is accepted as the most efficient means of heat transfer.
- *Automatic dump trap.* An automatic dump trap, installed at the lowest point below the air receiver, collects condensed moisture. It opens automatically to discharge a predetermined volume. Due to the air pressure behind the water, the trap opens and closes with a snap action that ensures proper seating. A small line strainer should be installed ahead of any automatic device to keep foreign particles from clogging the working parts.

Courtesy of ITW Automotive Refinishing–DeVilbiss

A

Wet air in

Dry air out

Built-in screen prevents desiccant from going down tube on refill

Quick remove flange

Blue silica gel desiccant beads turn pink when saturated

Sintered bronze element prevents any desiccant dust from travelling downstream

Built in sightglass allows easy monitoring of desiccant color change

B

Figure 6–15 **(A) Air line filtration equipment ranges from the very basic gun-mounted disposable filter to a three-stage desiccant air drying system that removes water, dirt, oil, and water vapor from an air line. (B) The cutaway view shows silica gel used to remove moisture from air.**

- *Air dryers*. Good aftercoolers remove the greatest percentage of water vapor. However, to remove any remaining residue, an air dryer (Figure 6–17) is often used. There are many designs of air dryers available. Among these are chemical, desiccant (drying agent), and refrigeration types. All dryers are designed to remove moisture from the compressed air supply to prevent condensation in the distribution system under normal working conditions.

Breathable Air Systems

A breathable air system is designed to provide purified air to painters so they do not inhale toxic fumes and paint mists. These systems can use the main shop air compressor or a separate, small oilless air compressor.

Figure 6–18 shows a breathable air filtration system that uses the main shop air compressor to feed purified air to a breathing mask. It consists of a three-stage filtration system with a desiccant dryer to prepare the air for human consumption through a breathing hood or mask.

Figure 6–19 shows a breathable air compressor and filter system that works independently of the main shop air compressor. The box contains filters, dryers, and a small oilless air compressor. The unit is mounted outside the paint booth, and the breathing hose is routed through the wall of the paint booth and connects to the breathing hood or mask.

Figure 6–20 shows a painter wearing a fresh-air hood. Note the breathing air hose coming into the bottom of the hood. This setup provides the best protection from inhaling harmful paint fumes, especially from catalyzed paint products, which can be very harmful if inhaled. The fresh breathing air supply is forced into the hood to surround the painter's mouth and nose with clean fresh air. This keeps the painter cool as well as healthy.

Figure 6–21 shows a belt-mounted fresh air supply welding respirator. The small belt-mounted compressor forces clean, filtered air into the welding helmet to keep welding gases and fumes from being inhaled.

ECO-TECH

To avoid ill health when you get older, always wear an appropriate, properly fitting respirator when applying water-based and oil-based primers and paints. It would be wise to invest in a fresh-air supply respirator. It will filter out more airborne toxins than static respirators. An air-supplied respirator will also prevent air and vapor leakage between the rubber mask and your face when you inhale. Remember that even nonpolluting water-based paint can cause lung disease and breathing problems if inhaled.

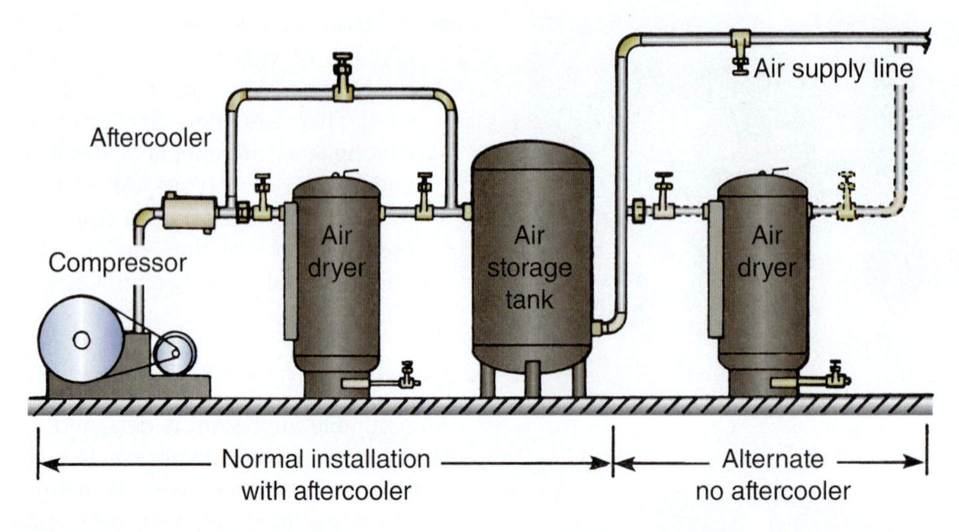

Figure 6-16 Note the arrangement of a large system where dry air supply is required.

Figure 6-17 Shown here are typical air dryers used in paint shops.

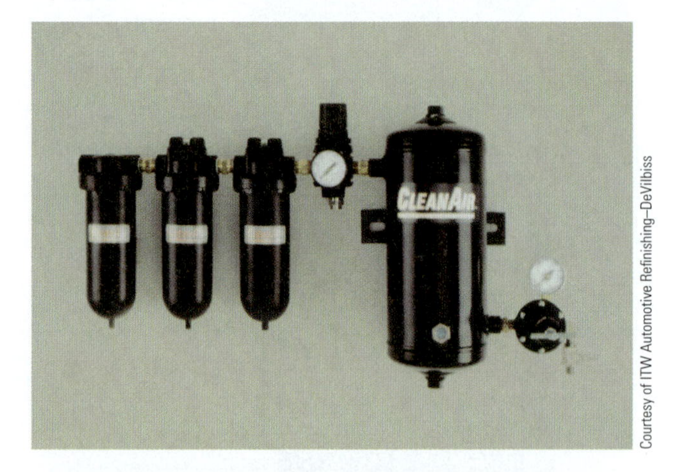

Figure 6-18 This is a filter-dryer system designed for breathable air going to a painter's hood or mask. It has three-stage filtering with a drying unit to provide very clean air for human consumption.

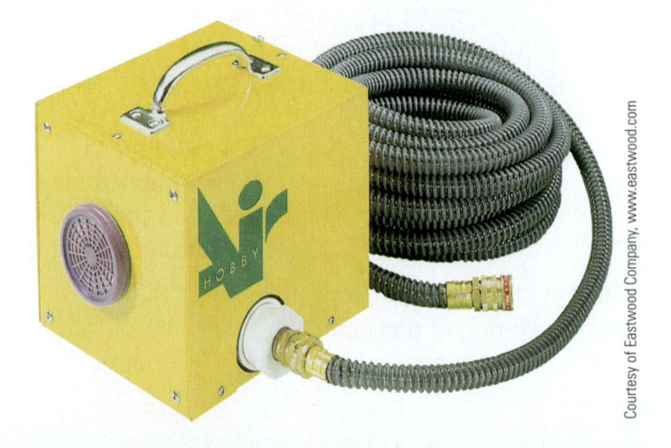

Air
HOBBY
Courtesy of Eastwood Company, www.eastwood.com

Figure 6-19 This is an independent breathing air supply system. The box, mounted outside of the paint booth, contains a small oilless air compressor and filters. A large breathing hose feeds into the paint booth and to the breathing hood or mask.

6.3 COMPRESSOR ACCESSORIES

The various types of hoses used to carry compressed air and fluid to the spray gun and other power tools are important parts of the system. Improperly selected or maintained hoses can create a number of problems.

Figure 6-20 The technician is wearing an air-supplied hood. Fresh air is pumped into the hood from outside of the paint booth. Cool, clean air is forced into the mask so no toxic fumes are inhaled.

Figure 6-21 This welder is wearing a fresh air-supplied welding helmet. Small pump and filter on belt feed clean, breathable air into helmet so that toxic gases from welding are not inhaled.

An air hose should never be used for solvent-based paints. Solvents can corrode the hose material and cause hole failure and rupture as well as paint problems.

Hose Types

There are three types of hoses found in a body shop compressed air system: air hose, breathing hose, and fluid hose. Each is designed differently, and they must never be interchanged.

High-pressure air hose is usually red or yellow rubber. Low-pressure air hose might be orange or black. Fluid hose may be black or brown rubber and can be found on pot- or tank-type spray guns. Breathing hose is a large-diameter plastic hose with a metal coil of wire spiraled inside the hose material to keep it from collapsing.

The air hose is usually a simple braid-covered hose that consists of rubber tubing (1) reinforced and covered by a woven braid (2), as shown in Figure 6–22A. The single-braid, rubber-covered hose (Figure 6–22B) consists of an inner tube (1), a braid (2), and an outside cover (3), all vulcanized into a single unit. The double-braid hose, illustrated in Figure 6–22C, consists of an inner tube (1), a braid (2), a separator or friction layer (3), a second layer of braid (4), and an outer rubber cover (5), all vulcanized into one unit. Double-braid hose has a higher working pressure than single-braid hose.

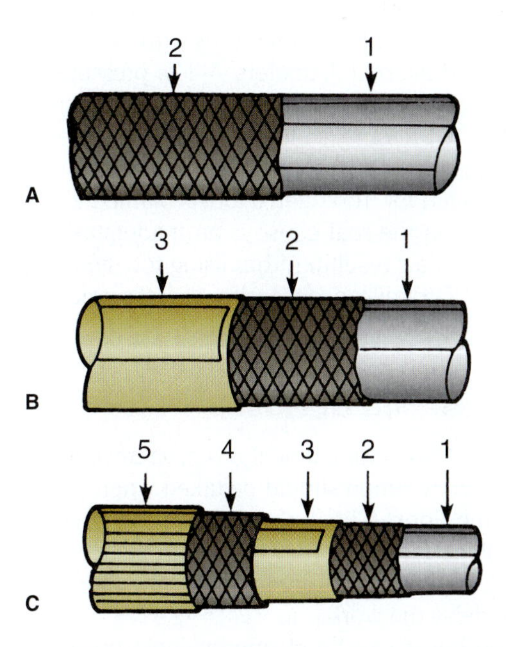

Figure 6-22 Study the construction of the hoses used in a compressed air system: (A) braid-covered hose, (B) single-braid hose, and (C) double-braid hose.

TABLE 6-2 AIR PRESSURE DROP

	Air Pressure Drop			
Size of Air Hose (ID)*	**5-Ft. Length**	**15-Ft. Length**	**25-Ft. Length**	**50-Ft. Length**
¼ Inch	**PSIG**	**PSIG**	**PSIG**	**PSIG**
@ 40 PSIG†	0.4	7.5	10.5	16.0
@ 60 PSIG	4.5	9.5	13.0	20.5
@ 80 PSIG	5.5	11.5	16.0	25.0
5⁄16 Inch				
@ 40 PSIG	0.5	1.5	2.5	4.0
@ 60 PSIG	1.0	3.0	4.0	6.0
@ 80 PSIG	1.5	3.0	4.0	8.0
⅜ Inch				
@ 40 PSIG	1.0	1.0	2.0	3.5
@ 60 PSIG	1.5	2.0	3.0	5.0
@ 80 PSIG	2.5	3.0	4.0	6.0

*ID: Inside diameter

†PSIG: Pounds per square inch gauge

Hose Size

It is important to use the proper size and type of hose to deliver the air from the compressor and the material from its source to the air tools and guns. When air is compressed and must travel a long distance, its pressure begins to drop. However, for a distance of up to 100 ft., pressure drop can be minimized when the proper diameter hose and fittings are used. Use only a hose constructed for compressed air use and with a rating of at least four times that of the maximum psi being used.

Table 6–2 indicates just how much pressure drop can be expected at different pressures with hoses of varying lengths and internal diameters. At low pressure and with short lengths of hose, this drop is not particularly significant. As the pressure is increased and the hose is lengthened, the pressure drop rapidly increases and must be compensated for. Too often a tool is blamed for malfunctioning when the real cause is an inadequate supply of compressed air resulting from using too small an *inside diameter* (ID, or distance measured across inside surfaces) hose.

Maintenance of Hoses

A hose will last a long time if it is properly cared for and maintained. Caution should be taken when it is dragged across the floor. A hose should never be pulled around sharp objects, run over by vehicles, kinked, or otherwise abused. A hose that ruptures in the middle of a job can ruin or delay the work.

Fluid hoses can be cleaned using a hose cleaner, a device that forces a mixture of solvent and air through the fluid hose and spray gun, ridding them of paint residue. A valve stops the flow of solvent and allows air to dry the equipment. Clean the fluid hose internally with the proper solvent when the gun is cleaned.

The outside of both air and fluid hoses should be wiped down with solvent at the end of every job. Wrap them into a large coil and hang them to store.

Connectors

Connections are needed between the compressor, the ends of hoses, and the air tools. Of the many different types used, the most common are the threaded and quick-connect types (Figure 6–23). The screw-type fitting is usually connected and disconnected with a wrench. The quick-connect is readily attached and detached by hand by pulling back on its sleeve.

Both types of connections may use the compression ring system to mount the fittings to the air or fluid hose. They may also use pipe fittings.

To install a compression ring connection, slip the sleeve and the compression ring over the end of the hose. Hold the body of the connection in a vise, and push the hose into the body as far as it will go. Slide the compression ring up to the body, bring the sleeve over the ring, and thread it on by hand. Tighten it with a wrench.

Most paint spray guns require either ¼- or 5⁄16-inch (6.3 or 7.9 mm) hoses; air hoses for pneumatic tools usually have 5⁄16- to ⅜-inch (7.9–9.5 mm) inside diameters. A few of the air tools described in Chapter 5 require hoses of specified inside diameters, which the tool manufacturers usually supply. A minimum of ⅜-inch (9.5 mm) diameter hose is recommended for HVLP paint guns; anything less will "starve" the equipment.

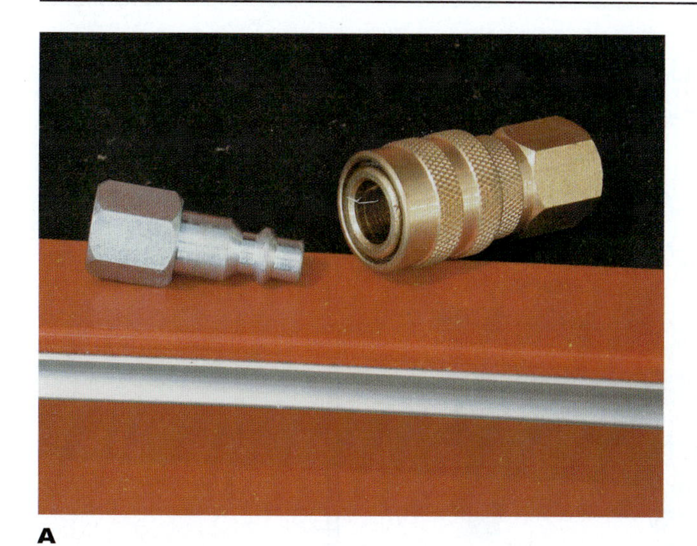

A

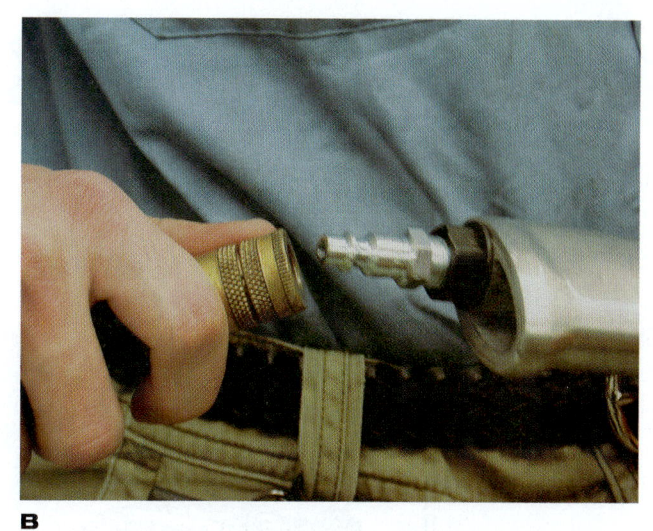

B

Figure 6–23 (A) The male hose end inserts into the female quick-disconnect coupler. (B) By pulling back on the sleeve of the couple, you can quickly connect the air tool and air hose.

Adapters and Couplings

An **adapter** is a type of connection that is male on one end and female on the other. It is used to convert the connections on the hose and other equipment from one thread size to another. Adapters are available in a very wide variety of sizes and threads.

A **coupling** is a type of connection that is male on both ends. It is used to couple two pieces of hose or pipe together or to convert a female connection of one size thread to a male connection of another size thread.

6.4 AIR SYSTEM MAINTENANCE

The manufacturer's maintenance schedule for the air supply system, given in the owner's manual, should be followed exactly. For example, if you fail to service a compressor properly, it will cut the unit's service life and affect the quality of your paint work. If you fail to change compressor oil at specific intervals, compressor parts will wear and leak. Excess oil can then enter the air lines and air tools. Some of this oil could get into the spray equipment or on the vehicle surface, ruining your paint work.

In general, all air systems require the periodic maintenance outlined in the following sections.

Daily Maintenance

- Drain the air receiver and the moisture separator/regulator or air transformer. If the weather is humid, drain them several times a day. See Figure 6–24.
- Check the level of the oil in the crankcase. Although it should be kept at full level, do not overfill. Overfilling causes excessive oil usage.
- SAE 10W-30, a multigrade oil, can be used as a substitute when SAE 10- or 20-weight oil is not readily available. Multigrade oils do contain additives that can cause harmful carbon residue and varnish. Detergent-type oils are satisfactory if used before hard carbon deposits have developed. Before changing to a detergent-type oil, pistons, rings, valves, and cylinder heads should be cleaned because the detergent oil may loosen hard carbon deposits that can plug passages and damage cylinders and bearings.
- Check for and immediately repair air leaks and couplings. One small pinhole leak will result in a loss of 30 cfm daily.

Weekly Maintenance

- Pull the ring on the safety valve and unseat it. If the valve is working properly, it will release air. Reseat the safety valve by pushing the stem down with your finger. If the valve sticks or fails to seat, repair or replace it immediately.
- Clean air filters. Felt and foam air filters should be washed in nonexplosive solvent, allowed to dry, and reinstalled. A dirty air filter decreases compressor efficiency and will increase oil usage (Figure 6–25).
- Clean or blow off fins on cylinders, heads, intercoolers, aftercoolers, and any other parts of the compressor or outfit that collect dust or dirt. A clean compressor runs cooler and provides longer service.
- Check the oil filter in the air line and change the filter element if necessary.

Monthly Maintenance

Add or change the compressor crankcase oil. Under clean operating conditions, the oil should be changed at the end of 500 running hours or every six months,

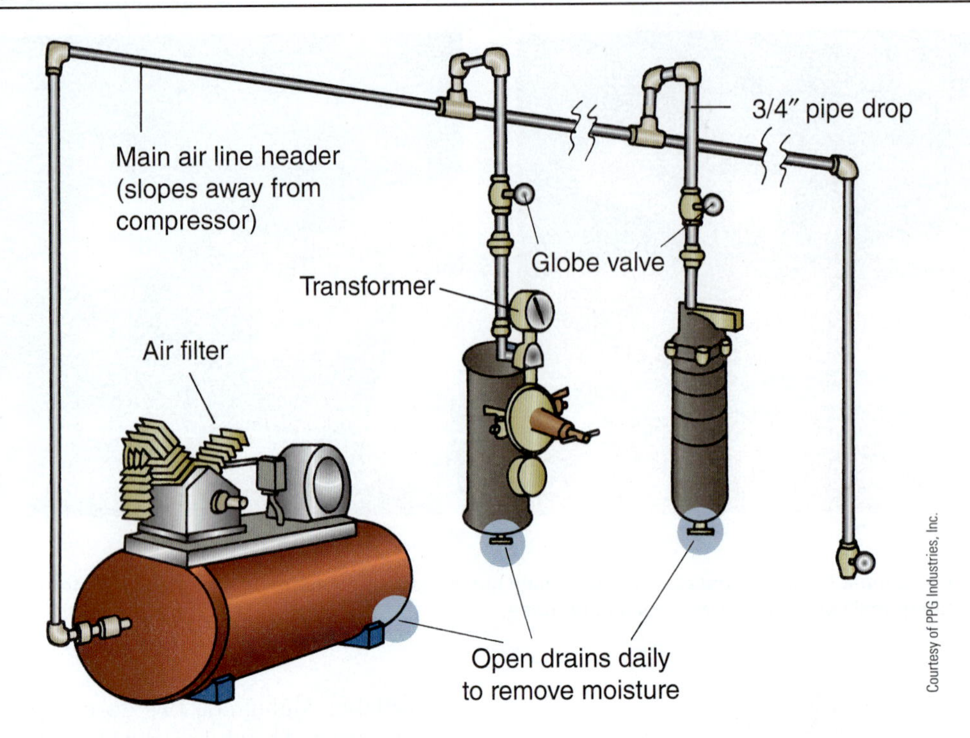

Figure 6-24 **Note the drain valve locations on a typical compressed air system. These should be drained on a daily basis.**

whichever occurs first. If operating conditions are not clean, change oil more frequently.

- Adjust the pressure switch cut-in and cut-out settings if needed.
- Check the relief valve for exhausting of head pressure each time the motor stops.
- Tighten belts to prevent slippage. A heated motor pulley is a sign of loose belts. Be careful not to overtighten belts, which can cause motor overload or premature failure of motor and compressor bearings.
- Check and align a loose motor pulley or compressor flywheel. It will be necessary to remove the front section of the enclosed belt guard to do so.
- Tighten all valve plugs and covers on the compressor head to ensure that each valve does not become loose and damage the valve or piston.
- Check for air leaks on the compressor outfit and air piping system.
- Check compressor pump-up time when the air receiver outlet valve is closed.
- Listen for unusual noises.
- Check and correct oil leaks.

6.5 AIR SYSTEM SAFETY

An air compressor system is a very safe arrangement to operate. Accidents seldom happen, but the few that do occur can usually be traced to human error.

- *Read the instructions.* Learn what each part of the compressor does by carefully reading the owner's manual that comes with the unit.
- *Inspect the unit before each use.* Carefully check the hoses, fittings, air control equipment, and overall appearance of the compressor before each use. Never operate a damaged unit.
- *Use proper electrical outlets.* Electrical damage often results from using improperly grounded outlets. Use only a properly grounded outlet that will accept a three-prong plug.
- *Always run the compressor on a dry surface.* The compressor should be located where there is a circulation of clean, dry air. Avoid getting dust, dirt, and paint spray on the unit.
- *Be aware of starts and stops.* Most compressors start and stop automatically. Never attempt to service a unit that is connected to a power supply.
- *Keep hands away.* Fast-moving parts can cause injury. Keep fingers away from the compressor while it is running. Do not wear loose clothing that could get caught in the moving parts. Unplug the compressor before working on it.
- *Keep the belt guard on.* Use all the safety devices available and keep them in operating condition. Also, remember that compressors become hot during operation. Exercise caution before touching the unit.
- *Release air slowly.* Fast-moving air will stir dust and debris. Be safe! Release air slowly by using a pressure regulator to reduce pressure to that recommended for the tool.
- *Keep air hose untangled.* Keep air hoses and power and extension cords away from sharp objects, chemical spills, oil spills, and wet floors. All of these can cause injury.

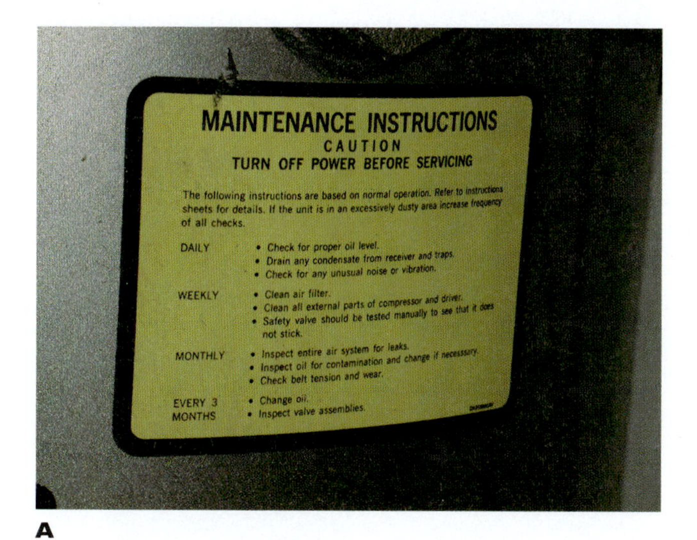

A

B

C

D

Figure 6-25 **Air compressor maintenance is important to compressor life and the quality of air supplied to the shop. (A) The label on an air compressor gives maintenance instructions. (B) The shutoff valve can be closed to isolate compressor tank pressure for air lines and hoses. (C) Small air filters and the inlet to the compressor pump should be cleaned or replaced at recommended intervals. (D) The drain on the bottom of a compressor tank must be opened daily to remove water condensation. Many compressor tanks have automatic drains.**

- *Depressurize the tank.* Be sure the pressure regulator gauge reads zero before removing the hose or changing the air tools. The quick release of high-pressure air can cause injury.
- *Disinfect a fresh air supply system periodically to prevent the buildup of germs.* Remove the breathing hose and wash it out in dishwashing liquid and water. Submerse the hose in disinfectant solution so that the inside diameter of the hose is washed clean.

SUMMARY

1. The compressed air supply system is designed to provide an adequate supply of clean, dry air at a predetermined pressure to ensure efficient operation of all pneumatic equipment in the body shop.
2. The air compressor, the heart of the system, is designed to raise the pressure of air from normal atmospheric to some higher pressure. There are three basic types of air compressors: the diaphragm type, the piston type, and the rotary screw type.
3. The piston air compressor pump develops compressed air pressure through the action of a reciprocating piston.

4. Single-stage units normally are used in pressure ranges up to 125 psi (861 kPa) for intermittent service.
5. Two-stage compressors are usually more efficient, run cooler, and deliver more air for the power consumed, particularly in the 100 to 200 psi (689–1,378 kPa) pressure range.
6. Compressor oil plugs are provided for filling and changing air pump oil. Pounds per square inch (psi) is the measure of air pressure or force delivered by the compressor to the air tool.
7. The compressor air tank is a heavy-gauge steel tank designed to hold an extra supply of compressed air. The typical parts of a stationary compressor are:

   Air compressor pump (pressurizes air)

   Electric motor or gasoline engine (power compressor)

   Air receiver or storage tank (holds compressed air)

   Check valve (prevents leakage of stored air)

   Pressure switch (automatically controls the air pressure)

   Air transformer (removes oil, dirt, and water from the compressed air)
8. A breathing air system uses large-diameter hose and highly filtered air to feed clean, fresh air to a painter's hood or mask. A three-stage filter and drying unit are needed to prepare shop air for breathing, or a separate oilless air compressor and filter outside the shop can feed clean breathing air to the painter.

EXERCISES

On a separate sheet of paper, complete the following learning activities for this chapter. Write definitions for the key terms and answer the ASE-style review questions, essay questions, critical thinking problems, and math problems. You can also do the outside activities, possibly for extra credit.

Key Terms

adapter
aftercooler
air compressor
air line lubricator
air pressure regulator
air tank shutoff valve
air transformer
automatic unloader
compressed air supply system
compressor air tank
compressor drain valve
compressor oil plugs
coupling
cubic feet per minute (cfm)
displacement
eccentric
gauges
intake air filter
pounds per square inch (psi)
pressure switch
single-stage compressor
two-stage compressor

ASE-Style Review Questions

1. Which type of air compressor would be inadequate as the main type of compressor in a body shop?
   A. Reciprocating piston compressor
   B. Diaphragm compressor
   C. One-stage compressor
   D. Two-stage compressor
2. The two-stage piston compressor provides
   A. less air pressure than the diaphragm compressor.
   B. more air pressure than the diaphragm compressor.
   C. enough pressure for most body or finishing applications.
   D. both B and C.
3. This is the actual amount of free air in cubic feet that the compressor can pump in one minute at working pressure.
   A. cfm
   B. Displacement cfm
   C. Free air cfm
   D. psi
4. Technician A says an airline lubricator helps paint guns function smoothly. Technician B says it is unnecessary to lubricate air tools. Who is correct?
   A. Technician A
   B. Technician B
   C. Both A and B
   D. Neither A nor B
5. The three basic types of air compressors are:
   A. Diaphragm, Rotary and Screw
   B. Diaphragm, Piston and Dual Stage
   C. Diaphragm, Single and Dual Stage
   D. Diaphragm, Rotary Screw and Piston

6. Which safety control is designed to maintain a supply of air within given pressure limits on gasoline and electrically driven compressors when it is not practical to start and stop the motor?
   A. Pressure switch
   B. Automatic unloader
   C. Centrifugal pressure release
   D. Overload protection
7. What is the recommended fuse rating for a fused disconnect switch?
   A. Equal to the current rating stamped on the motor
   B. Twice the current rating stamped on the motor
   C. Size recommended by manufacturer
   D. None of the above
8. The intake air filter must be made of
   A. fiberglass.
   B. fine mesh.
   C. felt material.
   D. both B and C.
9. How far away from the compressor unit should an air transformer be installed?
   A. At least 5 ft. (1.52 m)
   B. At least 15 ft. (4.56 m)
   C. At least 25 ft. (7.6 m)
   D. At least 35 ft. (10.6 m)
10. Which of the following is often combined with an air filtration system in a single unit?
   A. Thermal conditioning equipment
   B. Air dryer
   C. Air pressure regulator
   D. Both B and C

Essay Questions

1. List and explain the basic requirements for a compressed air supply system.
2. How is air pressure measured in an air supply system?
3. How does a piston-type compressor operate?
4. Explain cfm.
5. What is an air transformer?
6. What are some daily maintenance tasks on a compressed air supply system?

Critical Thinking Problems

1. What will happen if the oil level is too low or dry in a piston compressor?
2. What are some of the possible reasons that an air tool might slow down and run slowly?
3. Name and explain the major types of air compressors.
4. In general, a small hose causes a greater pressure drop than larger-diameter hoses. True or False?

Math Problems

1. If a compressor piston has an area of 5 square inches and a 3-inch stroke, runs at 2,000 rpm, and has one piston, what is its theoretical cfm?
2. How much pressure is dropped at 60 psi (408 kPa) when using a 50-foot (15.2 mm) length, ¼-inch (6.3 mm) inside diameter air hose?
   A. 20.5
   B. 6.0 psig
   C. 4.5 psig
   D. 1.5 psig

Activities

1. Inspect your shop air supply system. Identify as much of the system as you can (compressor types, number of air outlets, system pressure, size of fittings, and so on). Write down your findings in chart form.
2. Talk to shop technicians and people from other trades to find out if they know of anyone hurt with compressed air. Have they ever heard of air tools injuring anyone? Report your findings to the class.

CHAPTER 7

Body Shop Materials and Fastener Technology

OBJECTIVES

After studying this chapter, you should be able to:

- Select the right repair materials for a particular job.
- Explain the basic purpose of primers, sealers, surfacers, and other refinish materials.
- Compare the use of similar shop materials.
- Summarize when to use different kinds of filler.
- Know how to select the right type of primer and paint.
- Understand the importance of using a complete paint system.
- Identify the various fasteners used in body construction.
- Remove and install bolts and nuts properly.
- Explain when specific fasteners are used in body construction.
- Explain bolt and nut torque values.
- Summarize the use of chemical fasteners.
- Answer ASE-style review questions relating to shop materials and fasteners.

INTRODUCTION

When customers look at a vehicle's paint job, they often see only a shiny, bright color. They seldom understand all of the technology involved in producing that long-lasting, tough, durable, high-gloss finish. There is much hidden technology under the surface of the paint. A professional technician comprehends all of the "chemistry" and skill needed to do a good repair. This chapter will introduce you to the materials needed to do competent paint and bodywork (Figure 7–1).

Body shop materials are more than just refinishing or paint materials. They include the various fillers, primers, sealers, adhesives, sandpapers, and other compounds common to a body shop. It is critical that you understand their selection and use.

This first part of the chapter will explain the purpose of these basic types of materials. The second section of the chapter summarizes fasteners. This information will prepare you for later chapters that explain how to use materials and fasteners in more detail.

7.1 REFINISHING MATERIALS

A vehicle body is protected by a complete finishing system. All parts of the system work together to protect the vehicle from ultraviolet radiation, weathering, pollutants, and corrosion.

Refinishing materials is a general term referring to the products used to repaint a vehicle. Refinishing material chemistry has changed drastically in the past few years. New paints last longer but require more skill and safety measures for proper application (Figure 7–2).

The *substrate* is the metal, fiberglass, or plastic material used in the vehicle's construction. It will affect the selection of refinishing materials.

The term **paint** generally refers to the visible topcoat. The most elementary painting system consists of primer

Figure 7–1 Over the years, product manufacturers have developed better products for refinishing a vehicle. The results are paints and related products that last longer and look better.

Figure 7–2 Auto body technicians must work with many types of chemicals and learn how to properly and safely handle these materials.

coats and final topcoats of paint over the substrate. This process can vary considerably and is usually more complex, as you will learn later.

Primecoats and Topcoats

A basic finish consists of several coats of two or more different materials. The most basic finish consists of:

- Primecoats (primer, primer-surfacer, sealer, etc.)
- Topcoats (colorcoat or basecoat/clearcoat)

The **primer** improves adhesion of the topcoat to bare metal or plastic. It is the first coat applied to the substrate. Paint alone will not stick or adhere as well as a primer. If you apply a topcoat to bare substrate (metal for example), the paint will peel, flake off, or look rough in a matter of weeks. This is why you must "sandwich" a layer of primer between the substrate and the topcoats if clear. Primer-sealers also prevent any chemicals from bleeding through and showing in the topcoats of paint (Figure 7–3).

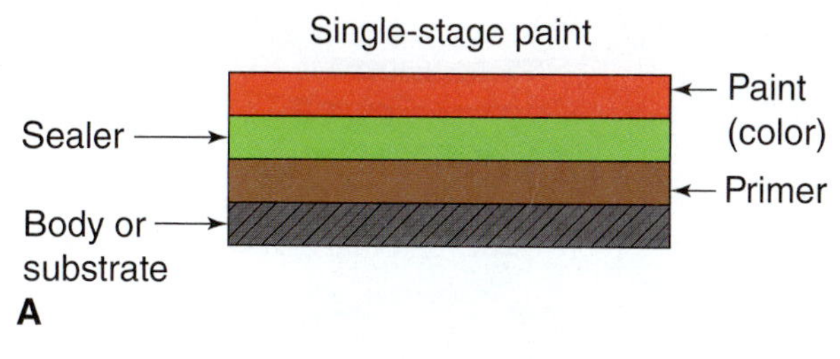

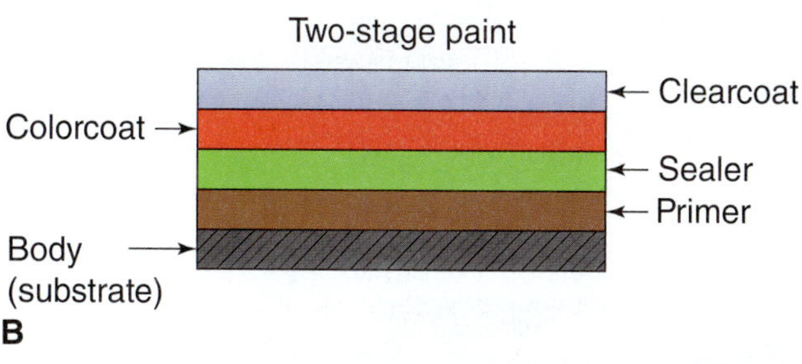

Figure 7–3 Compare single- and two-stage paints. Both require a primer and sealer over the substrate. (A) Single-stage paint has the color and clear formulated together so the color goes on with a medium gloss. (B) Two-stage paint uses clearcoats applied over colorcoats to make the finish extremely high gloss. Most shops use a two-stage paint to better match the original OEM finish.

The term **colorcoat** refers to the color paint sprayed over the prime and seal undercoats. It is usually several light coats of dull color of pigmented paint. The colorcoat is the "glamour coat" because it features the eye-catching color, color effects like "candy apple," or "metal flakes," that make a new paint job eye catching and appealing.

Clearcoat refers to the transparent layers of paint sprayed over the colorcoats of paint. The clear paint brings out the gloss and color of the basecoats. The coats of clear paint also protect the colorcoats from weather and sun damage.

Basecoat/clearcoat paint systems spray several thin layers of colorcoat over the primer, and then a second layer of clearcoat over the colorcoat. This is the most common paint system used today. The clear paint brings out the richness of the underlying color and also protects it. It makes the paint shine more than a single layer of color without a clearcoat (Figure 7–4).

Paint Types

There are three general types of paint:

1. Lacquer
2. Enamel (urethane)
3. Waterbase

As you will learn, there are variations within these categories. It is important that you know what types of finishes manufacturers use, because there are slightly different methods required for refinishing them.

Lacquer is an older paint that dries quickly because of solvent evaporation. Lacquers have been phased out and replaced by the more durable urethane enamel paints

A

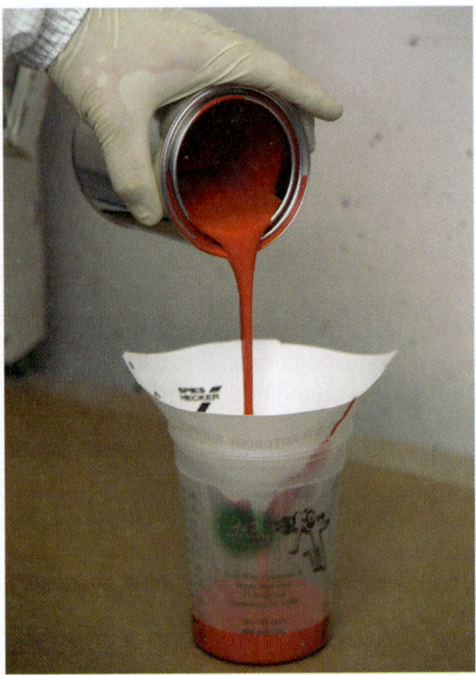

B

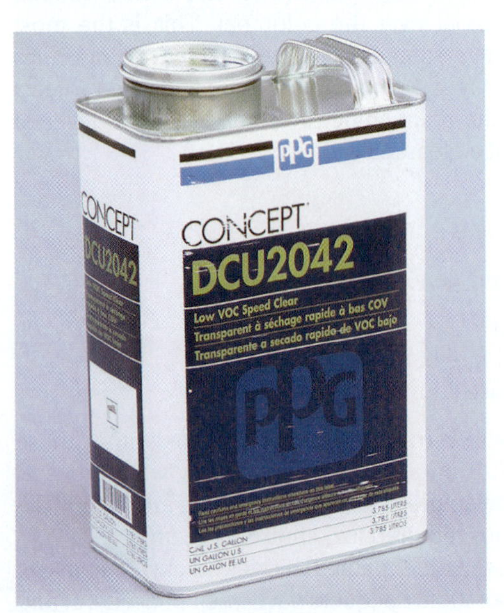

C

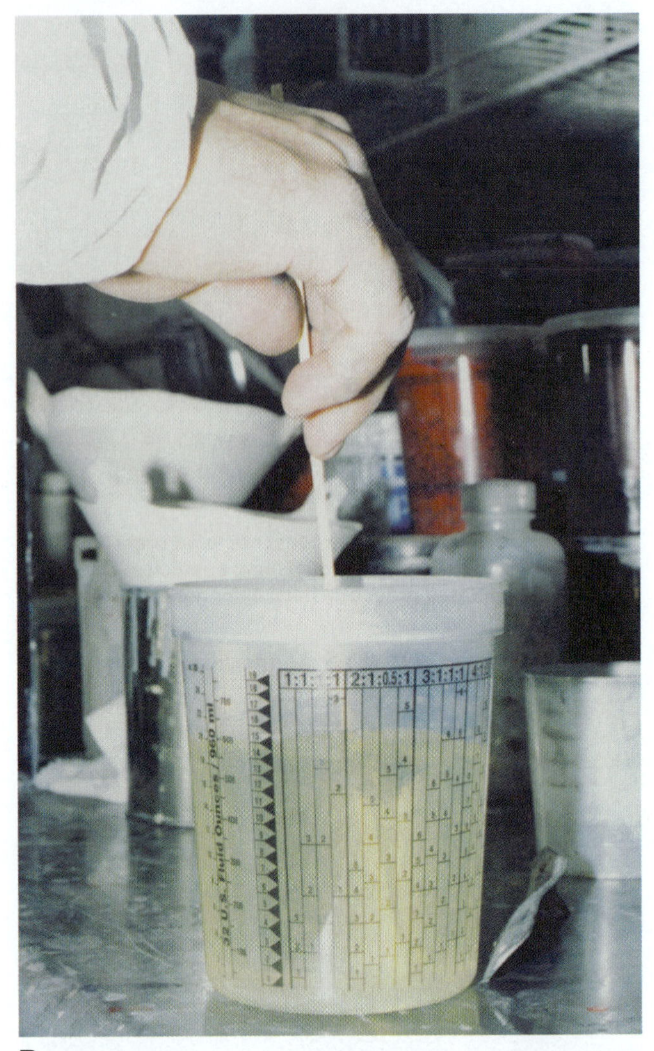

D

Figure 7-4 **When doing paint or refinish work, always use a complete matching system manufactured by one company. You are then sure that all ingredients are compatible and designed to be used together without problems.**

by both original equipment manufacturers (OEMs) and body shops. Only during the restoration of an older vehicle is lacquer paint still used.

Lacquer topcoats usually must be compounded or rubbed with a compound or polish to bring out their gloss.

Enamel finishes dry with a gloss and do not require rubbing or polishing, because the enamel dries through a chemical change rather than by solvent evaporation.

Never spray lacquer over the top of enamel. The enamel can lift and cause problems. Lacquer paint has been outlawed in many states and counties because of its high content of volatile organic compounds (VOCs) and air pollution.

Because enamels generally dry more slowly, there is more of a chance for dirt and dust to stick in the finish. Although there is usually a slight amount of surface roughness (orange peel) in an enamel film, too much will cause a lower gloss.

Two-stage paints consist of two distinct layers of paint: basecoat and clearcoat. Basecoat/clearcoat enamel is now the most common system used to repaint cars and trucks. First, a layer of color is applied over the primer-sealer. Next, while the basecoat is still not fully cured hard, a second coat of clear is sprayed over the color basecoat.

Acrylic urethane and urethane enamel are two specific types of catalyzed or hardened primers or paints that are commonly used in the industry. Urethane enamels are more durable than plain acrylics or enamels. Each type is available in a variety of colors and clears.

It can be difficult to tell the difference between lacquer and enamel, especially if the vehicle has been refinished. As you become more familiar with automotive refinishing, it will be easier to distinguish between them. This will be discussed later.

Waterbase paint, as implied, uses water to carry the pigment. It dries through evaporation of the water. Some manufacturers and repair shops are starting to use waterbase paints to reduce VOC emissions. The use of distilled water as a reducer, instead of mineral spirits, helps satisfy stricter emission regulations in some areas. The basecoat of color is waterbase. Then an enamel topcoat is applied over the waterbase paint to protect it from the environment.

Waterbase primers have been used for years as a fix for lifting problems. Waterbase primers serve as an excellent barrier coat when there are paint incompatibility problems.

OEM Finishes

Today's passenger car and light truck *OEM finishes* (factory paint jobs) are either "thermosetting" acrylic enamel (paint is oven-baked and hardened at the factory), new high-solids basecoat/clearcoat enamels, or sometimes waterbase, low-emission paints. Common enamel finishes are baked in huge ovens to shorten the drying times and cure the paint. This is done before installing the interior and other nonmetal parts.

Vehicle manufacturers use several different types of finish materials, coating processes, and application processes. Each type of finish requires different planning and repair steps.

The four most common types of OEM coating processes are:

1. Single-stage (color and clear premixed)
2. Two-stage (basecoat/clearcoat)
3. Three-stage paint (tri-coat for pearl effect)
4. Multistage (color change with viewing angle)

These topics will be fully explained in Chapter 24 through Chapter 28 of this textbook.

Contents of Paint

A paint's *chemical content* includes the following:

- Pigments
- Binders
- Solvents
- Additives

Each of these ingredients has a specific function within the paint formula.

Paint Pigments

The **pigments** are fine powders that impart color, opacity, durability, and other characteristics to the primer or paint. The main purpose of the pigment is to hide everything under the paint.

The size and shape of the pigment particles are important. Pigment particle size affects hiding ability and appearance. In addition, pigment shape affects strength. Pigment particles may be nearly spherical, rod-like, or plate-like. Rod-shaped particles, for example, reinforce paint film like iron bars in concrete.

Medium-size reflective pigment particles, such as mica, are added to **pearl paints** to give the paint a luster or shine that tends to change color with viewing angle. As you will learn later, pearlescent paints are now common and are the most difficult to match when repainting.

Large, reflective pigment flakes are added to **metallic paints**. The size, shape, color, and material in the flakes can vary. Often called *metal flakes*, they can be made of tiny but visible bits of metal or polyester. When light strikes the flakes, they reflect the light at different angles, making them look like tiny glittering stars inside the paint.

Paint solids are the nonliquid contents of the paint or primer. *High-solids paints* are needed to reduce air pollution or emissions when painting. *Low-solids paints* are less desirable because they contain more solvents that pollute the air.

Paint Binders

The **paint binder** is the ingredient in a paint that holds the pigment particles together. It is the backbone, or film-former, of the paint. The binder helps the paint to stick to the surface being painted. Various materials are used in the binder. The binder type determines the kind of paint—lacquer or enamel—because it contains the drying mechanism.

Paint binders are generally made of a natural resin (such as rosin), drying oils (such as linseed or cottonseed), or a manufactured plastic and are usually modified with plasticizers and catalysts. They improve such properties as durability, adhesion, corrosion resistance, mar resistance, and flexibility.

Paint Solvents

The **paint solvent** is the liquid solution that carries the pigment and binder so the mixture can be sprayed.

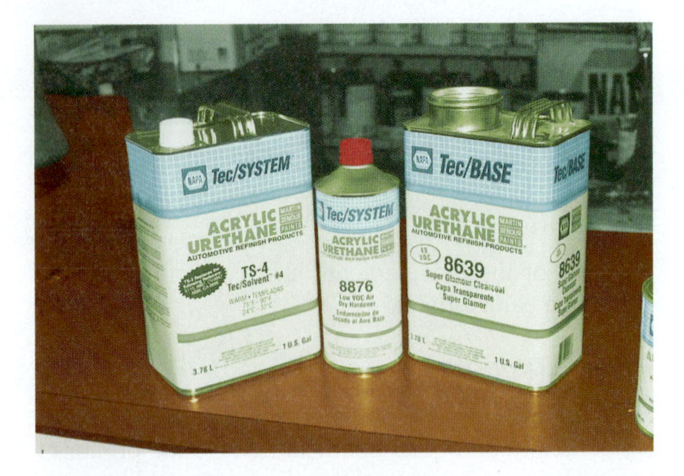

Figure 7-5 **When mixing modern paints, you must usually add reducer (solvent) and a hardener (catalyst) to the paint. The reducer will make the paint material the right thickness, or viscosity, for spraying. The hardener will make the paint material cure (dry) more quickly.**

Thinners and reducers are composed of one or several solvents and provide a transfer medium. Solvents are the volatile part of a paint in non–water-based systems. They are used to reduce a paint for spraying and give the paint its flow characteristics, evaporating as the paint dries (Figure 7–5).

Most solvents are made from crude oil or petroleum. However, waterbase paints are increasing in use to meet strict pollution regulations. The solvent reduces or thins the binder and transfers the pigment and binder through the spray gun to the surface being painted.

When used with lacquer, the solvent is called a *thinner.* When used with enamel, the solvent is called a **reducer**. This is an important distinction. The respective products are so labeled.

It is important to remember that you "thin" lacquer and "reduce" enamel. Thinning and reducing are needed to make the mixture the right thickness, or *viscosity*, to flow smoothly onto the surface.

Waterbase paints come *premixed* (ready to spray), and they are not normally reduced. In an emergency, distilled water can be added to make a thinner, more liquid solution of water-based colorcoats.

When using waterbase materials, it is important to remember that the water used for equipment cleaning:

- Contains hazardous materials and should be disposed of as hazardous waste
- Cannot be poured down the drain for disposal
- Must not be combined with other waste solvents, such as reducers or thinners; keep a separate container for storing water wastes

Due to clean air regulations, some solvents are no longer being used. To meet these regulations, traditional solvents are being replaced by water or other solvents. Be sure to check local ordinances before choosing products.

Paint Additives

Paint additives are ingredients added to modify the performance and characteristics of the paint (Figure 7–6). Additives are used to:

- Speed up or slow down the drying process
- Lower the freezing point of a paint
- Prevent the paint from foaming when shaken
- Control settling of metallics and pigments
- Make the paint more flexible when dry
- Increase gloss or shine
- Decrease gloss or shine

To assure compatability, the additive should be manufactured by the same company.

Drying and Curing

Drying is the process of changing a coat of paint or other material from a liquid to a solid state. Drying is due to the evaporation of solvent (Figure 7–7A), a chemical reaction of the binding medium (Figure 7–7B), or a combination of these causes.

The term *drying* refers to a paint material that evaporates its solvent to harden. The term **curing** refers to a chemical reaction in paint or other material that causes hardening.

Flash is the first stage of drying, where some of the solvents evaporate from the surface, which dulls the surface from a high gloss to a normal gloss.

A **retarder** is a slow-evaporating thinner or reducer used to retard, or slow, drying. It is a slow-drying solvent often used in very warm weather. If a paint dries too quickly, problems such as blushing can result. If used in cooler temperatures, a retarder can keep a paint from truly drying for several days.

An **accelerator** is a fast-evaporating thinner or reducer for speeding the drying time. It is needed in very cold weather to make the paint dry in a reasonable amount of time. The term *accelerator* also refers to a hardener or catalyst in most paint systems. Refer to Figure 7–6B.

A general rule to follow in selecting the proper thinner or reducer is: The faster the shop drying conditions, the slower drying the thinner or reducer should be. In hot, dry weather, use a slow-drying thinner or reducer. In cold, wet weather, use a fast-drying thinner or reducer.

A **catalyst** is a substance that causes or speeds up a chemical reaction. When mixed with another substance, it speeds the reaction but does not change itself. Catalysts are used with many types of materials—primers, paints, putties, fillers, fiberglass, and plastics.

A *paint catalyst, drier,* or *hardener* is an additive used to make enamel paints into urethane that cures quickly. Seldom is an enamel paint used alone. A hardener is almost always used by professional painters. The hardener speeds curing and makes the paint more durable (Figure 7–5).

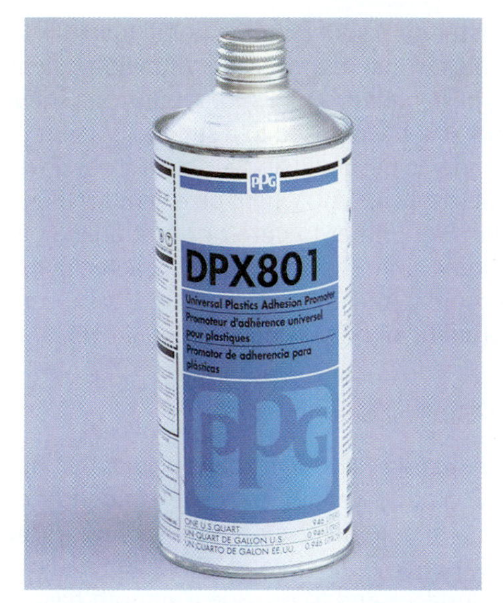

A

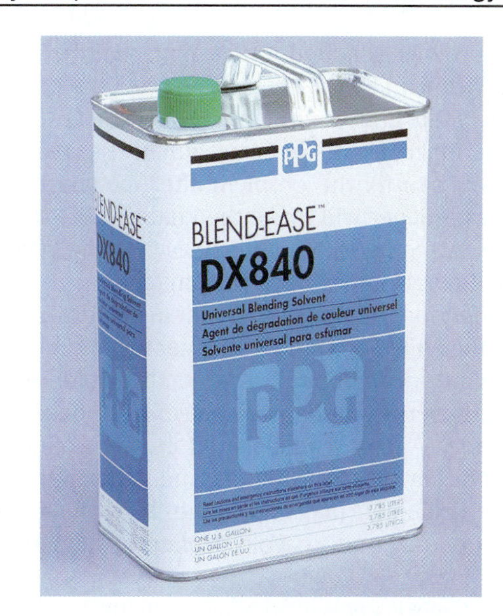

C

B

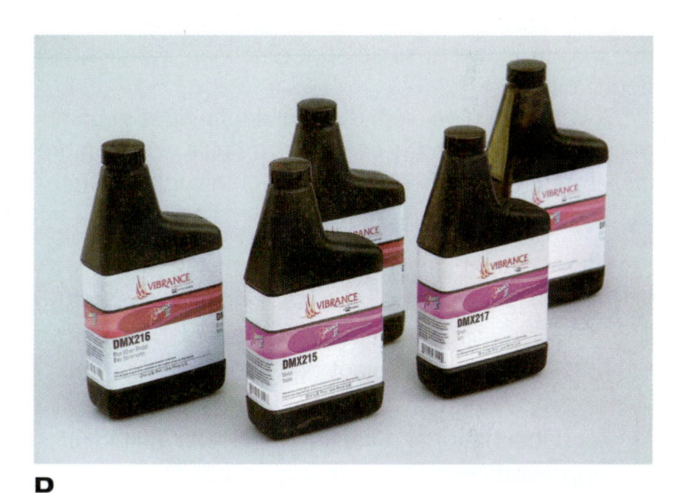

D

Figure 7-6 Many types of paint additives are available. (A) An adhesion promoter will make the paint adhere or bond to the vehicle more securely. (B) An accelerator will make the paint material cure more quickly when desired. (C) Blending solvent will help dissolve new paint into existing paint when blending color. (D) Toner can be added to paint to help color match more closely.

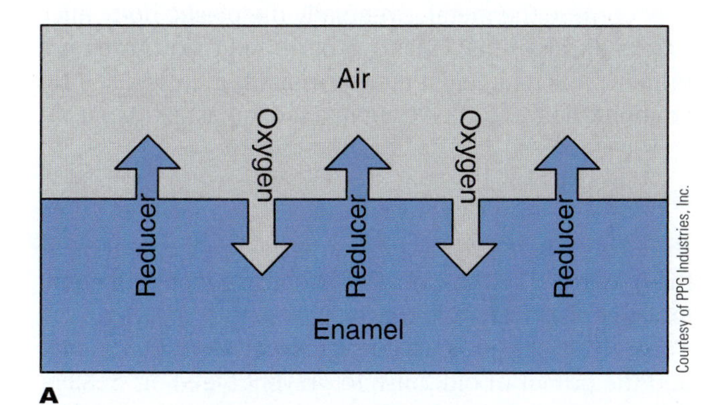

Courtesy of PPG Industries, Inc.

A

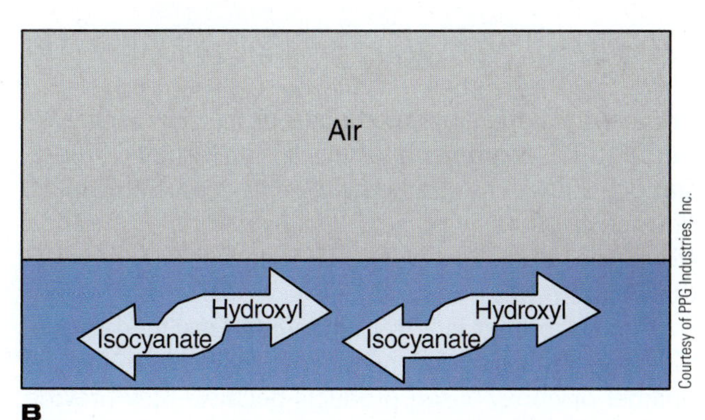

Courtesy of PPG Industries, Inc.

B

Figure 7-7 Automotive materials (primers, sealers, surfacers, and paints) harden in one of two ways. (A) One-part, noncatalyzed products, such as enamel paints, dry by the evaporation of solvent or reducer first, then by oxidation. Controlled heat speeds this chemical process. Most refinish materials are no longer one-part or noncatalyzed. (B) Urethane and polyurethane products cure by molecular cross-linking. However, oxygen does not have to be present for these catalyzed materials to solidify. For example, a paint with an added hardener or catalyst will solidify inside a spray gun cup if you forget to clean it after use.

The hardener is added to the paint right before it is sprayed. When an enamel catalyst is used, the paint can be wet sanded and compounded (polished) the next day. If you make a mistake (paint run, dirt in paint, and so on), you can fix the problem after the short curing time. The hardener will make the enamel cure in just a few hours. Also, the car can be released to the customer sooner with less chance of paint damage.

An *adhesion promoter* is a chemical designed to increase the ability of a paint or primer to bond to its substrate (Figure 7–6A). A paint adhesion promoter is sometimes used over OEM paints that have been baked hard at the factory. A plastic adhesion promoter is often used when repairing plastic bumpers and other parts. The adhesion promoter melts into the base material to increase bonding and to avoid peeling or separation of the repair materials.

A *blending solvent* is used when spraying a spot repair area to make the new paint melt into the existing paint more thoroughly (Figure 7–6C). It has powerful solvents that melt the existing paint so the new and old paints flow into each other and mix more thoroughly. A blending solvent makes a small spot repair on a panel less noticeable.

Toners are colors added in small quantities to paints to make them match the existing color on a vehicle more closely. They are high-solids colors for matching paint.

DANGER

Isocyanate resin is a principal ingredient in some urethane hardeners. Because this ingredient has toxic effects, you must always wear a National Institute for Occupational Safety and Health (NIOSH) approved respirator. Usually, a positive pressure or fresh air–supplied respirator is required when spraying an isocyanate product.

Primers and Sealers

Primers come in many variations—primer, primer-sealer, primer-surfacer, primer-filler, etc. It is important to understand the functions of subcoating or undercoat materials. You must follow manufacturer's instructions. Deviation from these directions will result in unsatisfactory work. See Figure 7–8.

A plain primer is a thin undercoat designed to provide good adhesion for the topcoat. Primers are usually enamel or epoxy-based products because they provide better adhesion and corrosion resistance than older lacquers.

A **self-etching primer** has acid in it to treat bare metal so that the primer will adhere properly. Quality primer-sealers and primer-surfacers have an etching material in them.

UV primer is a spot repair material that fully cures in 2–5 minutes when exposed to ultraviolet light. This primer is designed for small spot repairs of a square foot or less that can be exposed to a strong UV lamp.

UV primer is normally applied with several wet coats from an aerosol can. A UV light on a stand is then placed next to the wet primer to cure or dry it quickly. The UV primer should glow, showing it is being acted upon by the UV light source. After a few minutes of UV exposure, this type of primer can be sanded or painted over.

Primer-Surfacers

A **primer-surfacer** is a high-solids primer that fills small imperfections and usually must be sanded. It is often used after a filler to help smooth the surface. Primer-surfacers are used to build up and level featheredged areas or rough surfaces and to provide a smooth base for topcoats (Figure 7–9).

A *primer-filler* is a very thick form of primer-surfacer. It is sometimes used when a very pitted or rough surface must be filled and smoothed quickly. It might be used on a solid but badly rusted and pitted body panel, for example.

The industry trend is to use combination materials (primer-sealers or primer-surfacers) over single-purpose materials (sealer, primer, and so on). This saves time and money and helps improve the quality of the work.

Epoxy Primers

An **epoxy primer** is a two-part primer that cures fast and hard. Some material manufacturers recommend epoxy primer prior to the application of body fillers. Using an epoxy primer greatly increases body filler adhesion and corrosion resistance over bare metal. Epoxy primers most closely duplicate the OEM primers used for corrosion protection.

Body fillers, once mixed with the appropriate catalyst, start a chemical reaction, which in turn causes heat. The heat on bare metal tends to create condensation that may corrode the metal. Eventually the plastic body filler cracks and loosens, leaving a corroded area. An epoxy primer protects against moisture entrapment caused by condensation and can result in a longer-lasting repair.

Sealers

Bleeding, or *bleedthrough*, is a problem when colors in the undercoat or old paint chemically seep into the new topcoats. This can discolor the new paint.

A **paint sealer** is an innercoat between the topcoat and the primer or old finish to prevent bleeding. Sealers differ from primer-sealers in that they cannot be used as primers. Sealers are sprayed over a primer or primer-surfacer or a sanded old finish. Sealers do not normally need sanding, but some are sandable (Figure 7–10).

Sealers are often used over repair areas sprayed with primer or when a sharp color difference is visible after

A

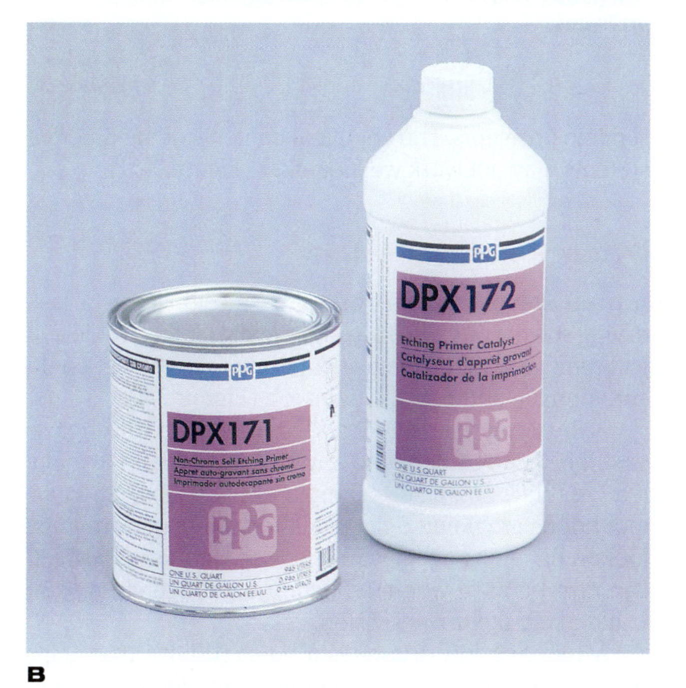

B

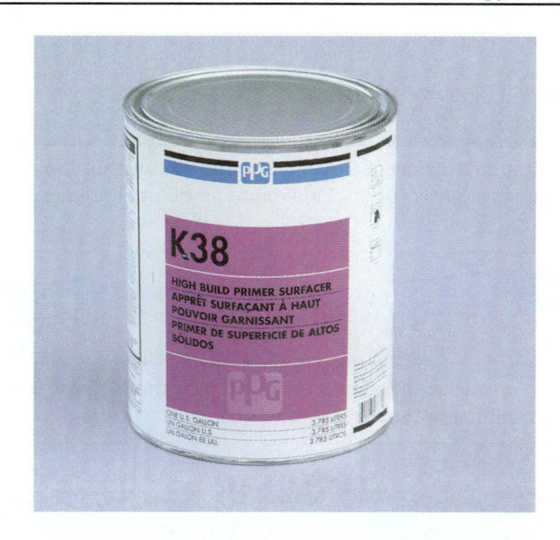

C

D

E

Figure 7-8 There are many types of primer in use today. Here are a few of the most common. (A) Plain primer is handy for rapidly priming small repair areas when the area is very smooth and ready for paint. (B) Etching primer is applied to bare metal. It contains acid that will eat into and bond to sheet metal. (C) High-build primer-surfacer is a very thick primer and is ideal for spraying over body filler repair areas. It will build up quickly and is like a spray-on body filler. (D) Corrosion resistor or epoxy primer can be used in place of self-etch or acid-type primer. It will bond to bare metal while resisting corrosion or rust. Epoxy primer-surfacer is similar and is a commonly used product. (E) Water-based primer is designed to replace OEM water-based products. All other paint products must be designed to be used with water-based primer.

A A magnified view of a cross-section shows that the surface is slightly rough. This might be due to initial rough sanding.

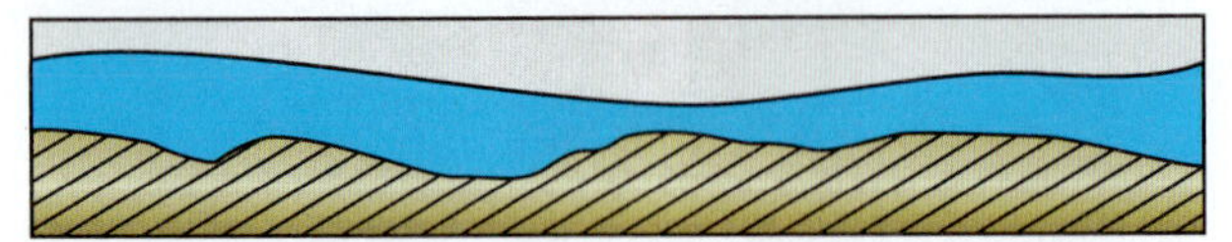

B Primer-surfacer has been sprayed over the surface. It has a high solids content that flows and fills indentations.

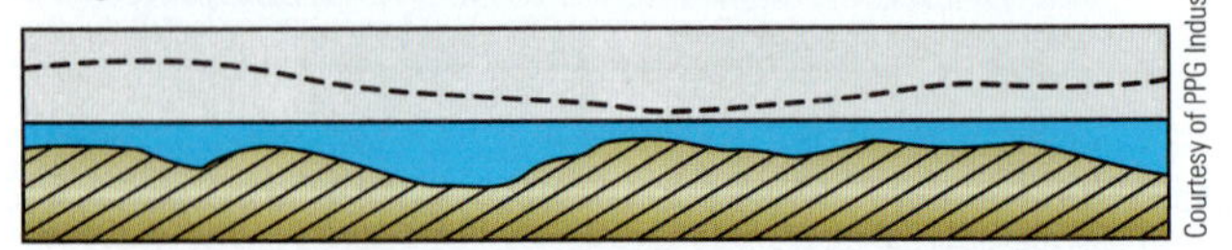

C Sanding the primer-surfacer will quickly level and smooth the surface. This readies the surface for topcoats.

Courtesy of PPG Industries, Inc.

Figure 7–9 Primer-surfacer is now the workhorse of the auto body repair industry.

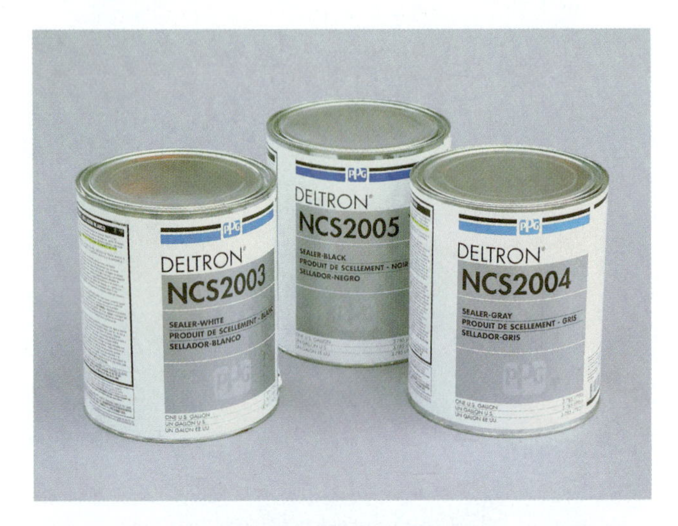

Figure 7–10 Sealer is normally applied over the primer, primer-surfacer, and repair areas. It will help final smooth the repair area and prevent bleedthrough of other materials into new paint.

Courtesy of PPG Industries, Inc.

Figure 7–11 When doing paint or refinish work, always use a complete matching system manufactured by one company. You are then sure that all ingredients are compatible and designed to be used together without problems.

sanding a repair area. Sealer can be applied over small areas of bare metal (even the size of a 50-cent piece) when final block sanding has cut through the primer.

A primer-sealer improves adhesion and also provides a barrier coat between the repair materials and the topcoat. It will solve two potential problems—adhesion and bleedthrough—in one application.

Complete Paint Systems

Remember to always use a complete paint system. A *paint system* means all materials (primers, catalysts, paints) are compatible and manufactured by the same company. They are designed to work properly with each other. If you mix materials from different manufacturers, you can run into problems. The chemical contents of the different systems may not work well together (Figure 7–11).

Other Paint Materials

A **prep solvent**, or wax and grease remover, is a fast-drying solvent often used to chemically clean a vehicle. It will remove wax, oil, grease, and other debris that could contaminate and ruin the paint job (Figure 7–12).

A **flattener** is an agent added to paint to lower gloss or shine. It can be added to any color gloss paint to make it a *flat* (dull) color. For example, some factory and custom hoods are painted flat black for a high-performance look. A flattening agent would be used in this instance. It can also be used where reflection off a high-gloss paint could affect the driver's vision.

A **fish-eye eliminator** is a paint additive that helps smooth the paint when small craters or holes in the paint film are a problem. Contaminants (usually silicone) make the paint flow away from small debris in the paint film. Fish-eye eliminator is an oil-based material that makes the paint flow over the top of the contaminant. It should be used only when absolutely necessary. However, always have fish-eye eliminator on hand for emergencies.

Figure 7–12 Cleaning agents are needed to remove contaminants from the body surface before painting.

A **flex agent** is an additive that allows primers and paints to flex or bend without cracking. It is commonly added to paints being applied to plastic bumper covers. Also called an *elastomer*, it is a manufactured compound with flexible and elastic properties that can be added to primers and paints.

Antichip coating, also called *gravel guard, chip guard*, or *vinyl coating*, is a rubberized material used along a vehicle's lower panels and on the front edges of hoods and fenders. It is designed to be flexible or rubbery to resist chips from rocks and other debris flying up off the tires. Antichip coatings are usually applied with a special spray gun.

Many manufacturers are using special chip-resistant coatings in areas that are exposed to stones and gravel. These coatings are generally between the E-coat (factory or manufacturer electrostatic applied primer) and the topcoats. Some chip-resistant coatings are clear and can be applied over the topcoat. If a vehicle has chip-resistant coatings, they must be replaced during the refinishing process.

Rubberized undercoat is a synthetic-based rubber material applied as a corrosion or rust-preventive layer. It can be applied using a production gun or a spray can (Figure 7–13).

A **metal conditioner** is phosphoric acid used to etch bare sheet metal before priming. It is a chemical cleaner that removes rust and corrosion from bare metal and helps prevent further rusting.

Remember the following important points about metal conditioners:

1. Acid cleans the metal.
2. It dissolves light surface rust.
3. It etches metal, improving adhesion.
4. It needs to be completely neutralized with water after applying.
5. It may have to be diluted; follow product directions.

Figure 7–13 Rust converters change rust into an inactive solid coating over the metal.

6. It is always followed by conversion coating.
7. It is necessary to wear rubber gloves, a respirator, and eye protection when using a metal conditioner.

A *conversion coating* is a special metal conditioner or primer used on galvanized steel, uncoated steel, and aluminum to prevent rust. It is applied after acid etching or metal conditioning.

Corrosion is a chemical reaction of air, moisture, or corrosive materials on a metal surface. Corrosion of steel is usually referred to as *rusting*, or oxidation.

Paint stripper is a powerful chemical that dissolves paint for fast removal of an old finish. If the old paint is cracking or peeling, you may have to use a chemical stripper. It is applied over the old paint. After it soaks into and lifts the paint, a plastic scraper is used to remove the softened paint.

Chemical paint strippers are not environmentally friendly. Check local and national regulations regarding the use of chemical strippers. Plastic media and soda blasting are alternate, less polluting means of paint removal.

A **tack cloth** can be used to remove dust and lint from the surface right before painting. It is a cheesecloth treated with nondrying varnish to make it tacky. A tack cloth must be wiped gently over the surface to keep the varnish from contaminating the paint.

Body Fillers

A **filler** is any material used to fill or level a damaged area. There are several types of filler. You should understand their differences.

Courtesy of Bondo Corporation

Figure 7–14 Body filler comes in cans, buckets, and plastic bags. A dispenser will save time and keep the filler uncontaminated.

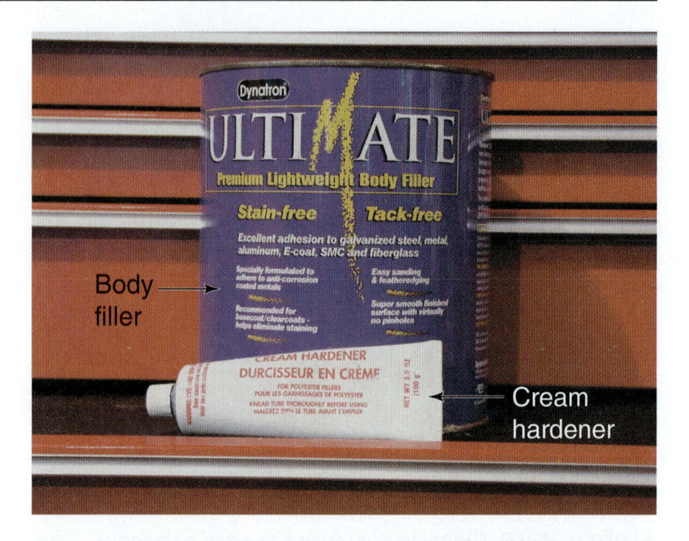

Figure 7–15 Cream hardeners are added to filler to make it cure. Note that hardeners can differ and can deteriorate with age. Always use the type recommended by the filler manufacturer.

Body filler or *plastic filler* is a heavy-bodied plastic material that cures very hard for filling small dents in metal. It is a compound of resin and plastic used to fill dents on car bodies (Figure 7–14).

Body fillers come in cans and in plastic bags. When in a plastic bag, a dispenser is used to force the filler onto your mixing board. This keeps the filler clean. A *mixing board* is the surface (metal, glass, or plastic) used for mixing the filler and its hardener.

Light body filler is formulated for easy sanding and fast repairs. It is used as a very thin topcoat of filler for final leveling and can be spread thinly over large surfaces for block or air tool sanding.

Fiberglass-reinforced body filler has fiberglass material added to the body filler and is used for rust repair or where strength is important. It can be applied on both metal and fiberglass substrates. Because fiberglass-reinforced filler is very difficult to sand, it is usually used under a conventional, lightweight body filler.

Short-strand fiberglass filler has tiny particles of fiberglass in it. It works and sands like a conventional filler, but is much stronger. *Long-strand fiberglass filler* has long strands of fiberglass for even more strength. It is commonly used for repairing holes in metal or fiberglass bodies.

Cream hardeners are used to cure body fillers and usually come in tubes. Once the hardening cream is mixed in, the body filler will heat and harden (Figure 7–15). Too much cream hardener can cause problems with adhesion and pinholing.

Fiberglass

Fiberglass resin is another form of plastic body repair material. It is a thick resin liquid that comes in a can. The fiberglass resin must be mixed with its own special type of hardener to cure. If you accidentally use cream hardener in fiberglass resin, it will not harden and you will have a mess to clean up.

Fiberglass mat is a series of long fiberglass strands irregularly distributed to form a patch. It is used to strengthen and form a shape for the resin liquid (Figure 7–16).

Fiberglass cloth is made by weaving the fiberglass strands into a stitched pattern.

Fiberglass tape will stick to the repair surface to help form the body shape and is a fast method of applying fiberglass for repairs. Resin can be applied over the tape.

Glazing Putty

Glazing putty is a material made for filling small holes or sand scratches. It is similar to primer-surfacer, but it has a higher solids content. Putty is applied over the undercoat of primer-sealer or primer-surfacer to correct small surface imperfections. The purpose of glazing putties is to fill imperfections that cannot be filled with a primer-surfacer.

Spot putty is the same as glazing putty except it has even more solids. Spot putty can be used for scratches or nicks up to 1⁄16 inch (1.5 mm) deep. It should not be used to fill large surface depressions. For larger depressions, use body filler or catalyzed putty.

A **Clear resin cures to a hard, brittle solid. A special catalyst is needed; do not use cream hardener. Fiberglass cloth, mat, or tape is used with resin. Cut pieces to size, coat them with resin, and apply them over the repair area.**

B **Fiberglass-reinforced body filler is much stronger than regular body filler and is also waterproof to resist rusting.**

Figure 7–16 **Fiberglass resin and mat or cloth is often used to repair fiberglass body panels and parts. It is sometimes used for rust repair.**

One-part putty often comes in a tube. A rubber spreader is used to work the putty into small holes in the primer. After fully curing for 24 hours or more, the putty is sanded smooth. Only the small pinholes or scratches remain filled with the putty. One-part putty is seldom used in production shops because it can shrink and cause low spots under the paint.

Two-part putty comes with its own hardener for rapid curing, which is the main advantage of two-part putty. If you paint over partially cured putty, it can shrink and cause problems (Figure 7–17). Two-part, or catalyzed, putty has replaced slow-drying one-part putty in most shops.

Figure 7–17 **Two-part spot putty can be used to fill tiny chips in existing paint and small pinholes in body filler. It is more runny than body filler and will flow down into small surface imperfections.**

A common mistake is to try to use spot putty like a filler. Spot putty is very expensive and may not adhere to metal properly. Only use two-part putty over the top of primer or paint to fill small pinholes or sand scratches. Don't be a "putty builder," or your work will be very weak.

Masking Materials

Masking materials are used to cover and protect body parts from paint overspray. *Overspray* is unwanted paint spray mist floating around from a spray gun. It can stick to glass and body parts and take considerable time to clean off (Figure 7–18).

Figure 7–18 **Masking is needed to keep overspray off panels and parts that are not being painted.**

Figure 7-19 Masking paper and plastic are used to cover and protect areas not to be painted on the vehicle.

Masking paper is a special paper designed to be used to cover body parts that are not to be painted. It comes in a roll and, when mounted on a masking machine, masking tape is automatically applied to one edge of the paper, which speeds the work (Figure 7–19).

Many collision repair shops use two kinds of masking paper: primer masking paper and paint masking paper.

Primer masking paper is a less expensive, more porous masking paper. It is suitable for masking off primer spray, but not paint spray. If you use it to mask off for painting, the paint can bleed through the green paper and get onto the unpainted surfaces.

Paint masking paper is a nonporous paper designed for masking off paint spray. It is a more expensive paper that will keep paint from bleeding through to the masked surface (Figure 7–20).

Masking plastic is used just like paper to cover and protect parts from overspray. It also comes in rolls and can cover large body areas more easily than paper. Masking plastic is used to cover areas of the vehicle away from the area being painted. Plastic should not be used right next to the area being sprayed because the plastic will not absorb the paint, and it can cause paint to drip down onto the body surface (Figure 7–21).

Heat-resistant masking paper is designed to protect a vehicle body from damage caused by hot weld spatter and grinding sparks. This specially thick masking paper will protect glass and paint like a welding blanket. It can be used to protect a large surface area or the sides of a vehicle when a welding blanket would be difficult to use (Figure 7–22).

Courtesy of 3M Company

A A paint technician is masking the front of the car before painting.

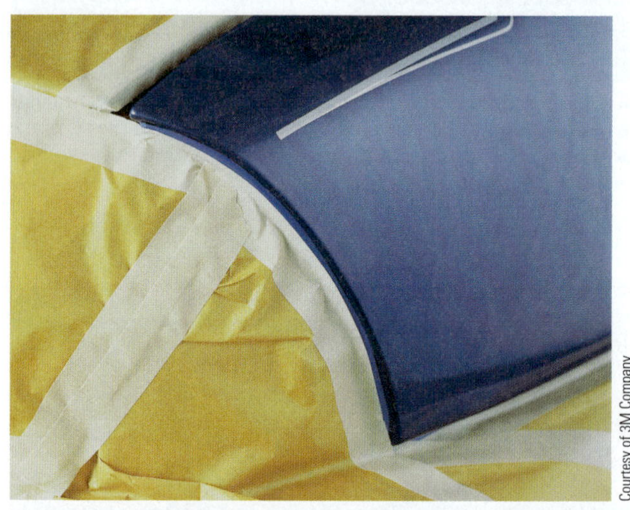

Courtesy of 3M Company

B A close-up shows how masking tape and masking paper have been applied around the area to be painted.

Figure 7-20 Masking materials must be installed carefully to clean surfaces.

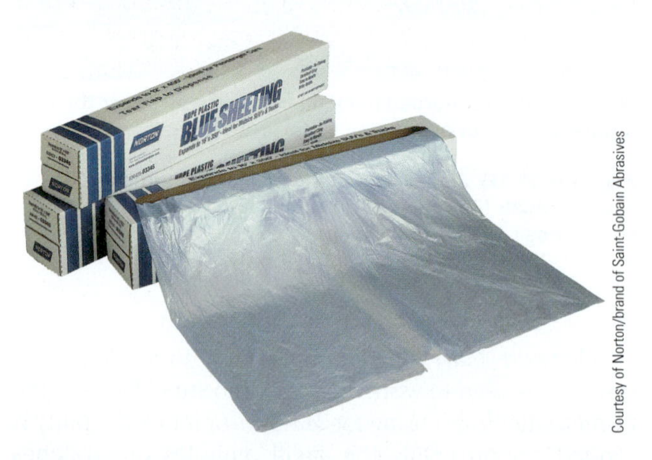

Courtesy of Norton/brand of Saint-Gobain Abrasives

Figure 7-21 Corona Treated Masking plastic is used to mask a vehicle and can be used right next to the critical edge of the painted area.

Figure 7–22 Welding masking paper is thick and fire-resistant and should be applied to the vehicle if grinding and welding could damage the paint or parts.

Masking tape is used to hold masking paper or plastic in position. It is a high-tack, easy-to-tear tape. It comes in rolls of varying widths, ¾-inch (19 mm) wide being the most common. Refer to Figure 7–23. To save time and improve the quality of the work, always purchase quality masking tape.

Fine line masking tape is a very thin, smooth-surface plastic masking tape (Figure 7–24). Also termed *flush masking tape*, it can be used to produce a better *paint part edge* (edge where old paint and new paint meet). When the fine line tape is removed, the edge of the new paint will be straighter and smoother than if conventional masking tape were used.

Fine line tape can help a painter mask flush-mounted parts. It may be better to mask these parts than to remove them. Door handles, side trim, mirror mounts, and moldings are a few examples. Fine line tape is also used where two different paint colors come together, as when painting stripes or two-tone finishes.

Wheel masks are preshaped plastic or cloth covers that fit over the vehicle's wheels and tires (Figure 7–25). Plastic wheel covers are disposable and should only be used once. Cloth covers are reused but should be cleaned off periodically. Preshaped plastic antenna, headlamp, and mirror covers are also available.

Masking rope is round, self-stick masking foam often used to mask inside body openings. The masking rope can be adhered inside of body gaps to prevent overspray from entering the openings in the body. The round shape of the masking foam tends to feather overspray to avoid a visible sharp paint edge from forming on the inside edges of panels. See Figure 7–26.

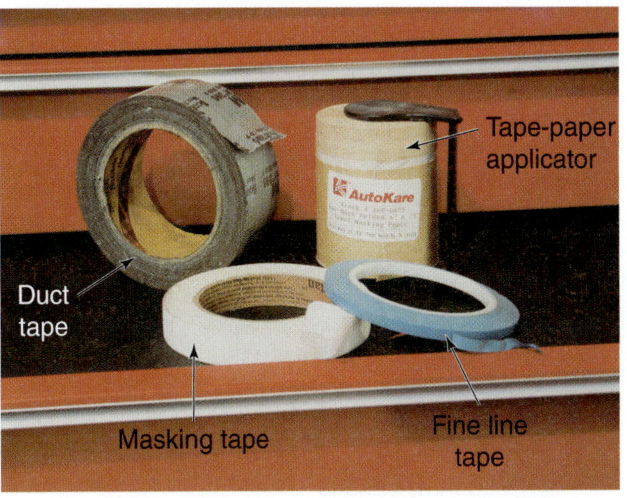

A

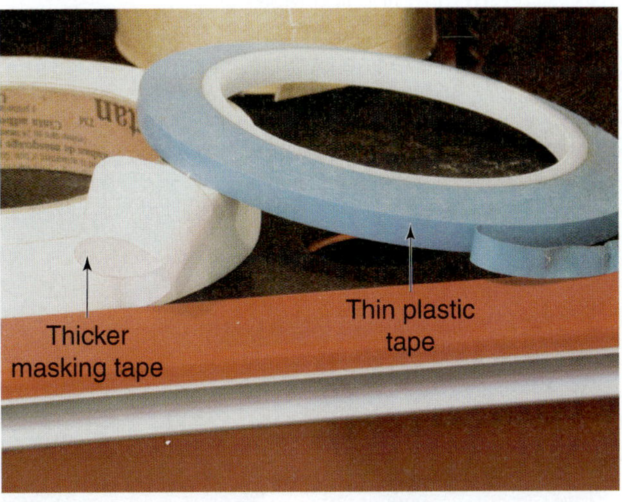

B

Figure 7–23 (A) Masking tapes come in various widths, ¾-inch wide being the most common. Masking tape has a paper backing. Fine line tape is made of thin plastic for making a smoother paint edge. A tape-paper dispenser is handy for masking narrow or small areas. (B) A close-up shows the difference between regular masking tape and fine line masking tape.

Masking liquid, also called *masking coating*, is usually a water-based sprayable material for keeping overspray off body parts. Some are solvent based. Masking liquids come in large, ready-to-spray containers or drums. These materials are sprayed on and form a paint-proof coating over the vehicle (Figure 7–27).

Some masking coatings are tacky and are used only during priming and painting. They form a film that can be applied when the vehicle enters the shop. Others dry to a hard, dull finish.

Masking coatings can be removed when the vehicle is ready to be returned to the owner. They wash off with soap and water. Local regulations may require that liquid masking residue be captured in a floor drain trap and not put into the sewer system.

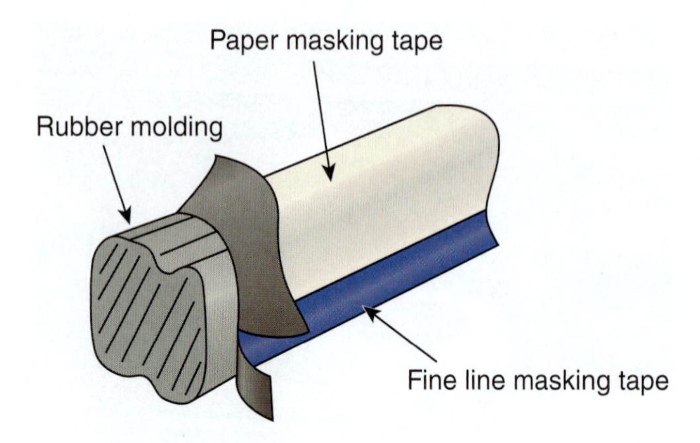

1. Apply masking tape to within 1/8″ (3 mm) of edge
2. Apply masking paper
3. Mask to edge of part with 1/4″ (6 mm) fine line tape
4. Leave a tail or handle
5. Remove fine line tape while paint is still wet

Figure 7–24 Note the example of how fine line tape is used along rubber molding to produce a good paint edge. Paper masking tape is then used to hold the paper in place next to the fine line tape.

Figure 7–26 Masking rope is self-stick foam that will quickly back mask along openings in body panels to block overspray and produce a feathered spray edge.

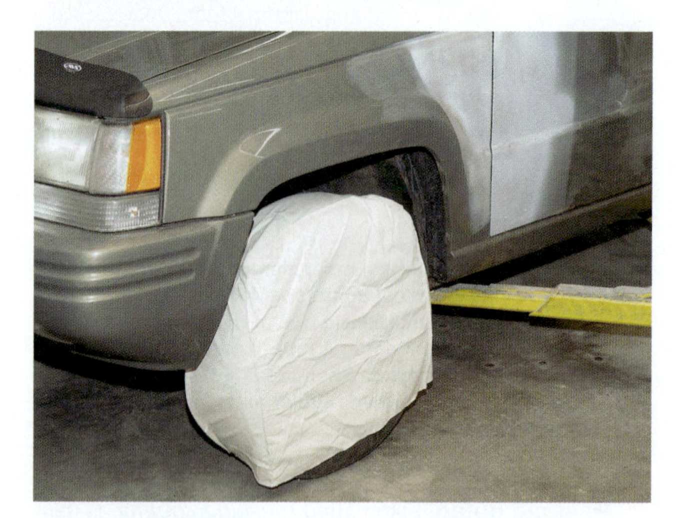

Figure 7–25 Disposable wheel masks are handy for protecting the wheel and tire from overspray.

Figure 7–27 Spray masking materials are applied with a spray gun to protect surfaces from paint overspray. They can then be washed off with water.

Duct tape is a thick tape with a plastic body that is sometimes used to protect parts from damage when grinding, sanding, or blasting. Duct tape is thicker than masking tape and provides more protection for the surface under the tape (Figure 7–23).

Abrasives

An **abrasive** is any material, such as sand, crushed steel grit, aluminum oxide, silicon carbide, or crushed slag,

Figure 7-28 **Sandpaper comes in many shapes, sizes, and grits.**

used for cleaning, sanding, smoothing, or material removal. Many types of abrasives are used by the auto body technician (Figure 7–28).

Grit refers to a measure of the size of particles on sandpaper or discs. *A coarse sandpaper* has large grit; *fine sandpaper* has smaller grit.

Grit Ratings

A **grit numbering system** denotes how coarse or fine the abrasive is. For example, 16 grit would be one of the coarsest and 1500 grit would be one of the finest. The grit number is printed on the back of the paper or disc. Figure 7–29 shows grit classifications and their applications in a body shop.

Very coarse grit of 16–24 is generally used for fast material removal. It will quickly remove paint down to bare metal. This grit is commonly used on grinding discs and air files for rapid cutting.

A *coarse grit* of 36–60 is basically used for rough sanding and smoothing operations. This coarseness might be used to get the general shape of a large body filler area.

Medium grit of 80–120 is often used for sanding body filler high spots and for sanding of old paint.

Fine grit of 150–180 is normally used to sand bare metal and for smoothing existing painted surfaces. This is also used for final sanding of body filler and to featheredge paint.

Very fine and *ultrafine grits* range from 220 to about 1500 and are used for numerous final smoothing operations. Larger grits of 220–320 are for sanding primer-surfacers and old paint prior to being finished. Finer grits of 2000–3000 are for colorcoat sanding and sanding before polishing or buffing. Very fine grits are usually wet sandpaper, which is required to keep the paper from becoming clogged or filled with paint.

When starting your work, use the coarsest grit practical. This will remove and smooth the area quickly. Then gradually go to finer paper to achieve the desired surface smoothness. This will be detailed in later chapters.

Open- and Closed-Coat Grits

Sandpaper and discs come in either open-coat or closed-coat types of grit.

With an *open coat*, the resin that bonds the grit to the paper touches only the bottom of the grit. About 50–70 percent of its surface is covered by grit materials. Open-coat grit will not clog as quickly.

With a *closed coat*, the resin completely covers the grit. This bonds the grit to the paper or disc more securely. About 90 percent of the surface is covered by grit materials. Closed coat will clog faster than open coat.

Grinding Discs

Grinding discs are round, very coarse abrasives used for initial removal of paint, plastic, and metal (weld joints). Some are very thick and do not require a backing plate. Others are thinner and require a *disc backing plate* mounted on the grinder spindle. They are used for material removal operations, with 24 grit being the most common (Figure 7–30).

Grinding disc size is measured across the outside diameter. The most common grinding disc sizes are 7 and 9 inch. The hole in the center of the disc must match the shaft on the grinder or sander.

Sandpapers

Sandpaper is a heavy paper coated with an abrasive grit. It is the most commonly used abrasive in auto body repair. There are several kinds, shapes, and grits of sandpaper.

Sanding discs are round and are normally used on air-powered orbital sanders. They might have self-stick coating or require the use of disc adhesive to hold the sandpaper onto the tool pad.

Sanding sheets are rectangular and can be cut to fit sanding blocks. Long sheets are also available for use on air files.

Dry sandpaper is designed to be used without water. Its resin is usually an animal glue, which is not water resistant and will dissolve when wet, ruining the sandpaper.

Dry sandpaper is often used for coarse- to medium-grit sanding tasks, like shaping and smoothing body filler. One example is 80-grit dry sandpaper, which is often used on body filler. It quickly cuts the filler down but does not leave deep sand scratches in the paint surrounding the filler (Figure 7–31 and Figure 7–32).

Wet sandpaper, as implied, can be used with water for flushing away sanding debris that would otherwise clog fine grits. Wet sandpaper comes in finer grits, from

Abrasive grading scales for sandpaper						
	CAMI (U.S. std) (see note 1)	FEPA (P-scale) (see note 2)	Finishing scale	Average grit particle size		Auto body use
				Microns	Inches	
	1200					
	1000	P2000		9.6	0.00042	Finish sanding before buffing or polishing; wet or dry paper.
	800	P1500		12.3	0.00051	
Finishing (fine)		P1200	A16	15.8	0.00060	
	600			16.0	0.00062	
		P1000		18.3	0.00071	Finish sanding of paint runs or other imperfections in new part; wet or dry paper.
	500			19.7	0.00077	
		P800	A25	21.8	0.00085	
	400		A30	23.6	0.00092	
		P600	A35	25.8	0.00100	Finish sanding before priming and painting; wet or dry paper.
	360			28.8	0.00112	
		P500		30.2	0.00018	
		P400	A45	35.0	0.00137	
	320			36.0	0.00140	
		P360		40.5	0.00158	
	280			44.0	0.00172	
		P320	A60	46.2	0.00180	Final sanding of good paint; wet or dry paper.
		P280		52.5	0.00204	
	240		A65	53.5	0.00209	
		P240	A75	58.6	0.00228	
Smoothing (medium)		P220	A90	65.0	0.00254	
	220			66.0	0.00257	Final sanding of body filler and old paint; dry sanding.
	180	P180	A110	78.0	0.00304	
	150		A130	93.0	0.00363	
		P150		97.0	0.00378	Initial smoothing and final leveling of body filler; dry paper.
	120			116.0	0.00452	
		P120	A160	127.0	0.00495	
	100			141.0	0.00550	
		P100	A200	156.0	0.00608	
	80			192.0	0.00749	
		P80		197.0	0.00768	Rough sanding of body filler and plastics; dry paper.
		P60		260.0	0.01014	
	60			268.0	0.01045	
		P50		326.0	0.01271	
	50			351.0	0.01369	
		P40		412.0	0.01601	Coarse grinding and sanding to remove paint; dry paper.
	40			428.0	0.01669	
Roughing (coarse)		P36		524.0	0.02044	
	36			535.0	0.02087	
		P30		622.0	0.02426	
	30			638.0	0.02488	
	24			715.0	0.02789	
		P24		740.0	0.02886	

Notes:

1. CAMI = Coated Abrasives Manufacturers Institute (North America)
   (Allows a wide tolerance range of particle sizes within the definition of a particular grit)

2. FEPA = Federation of European Producers Association
   (More consistent sized grit particles than CAMI)

Figure 7-29 Note how sandpaper is made and classified by coarseness.

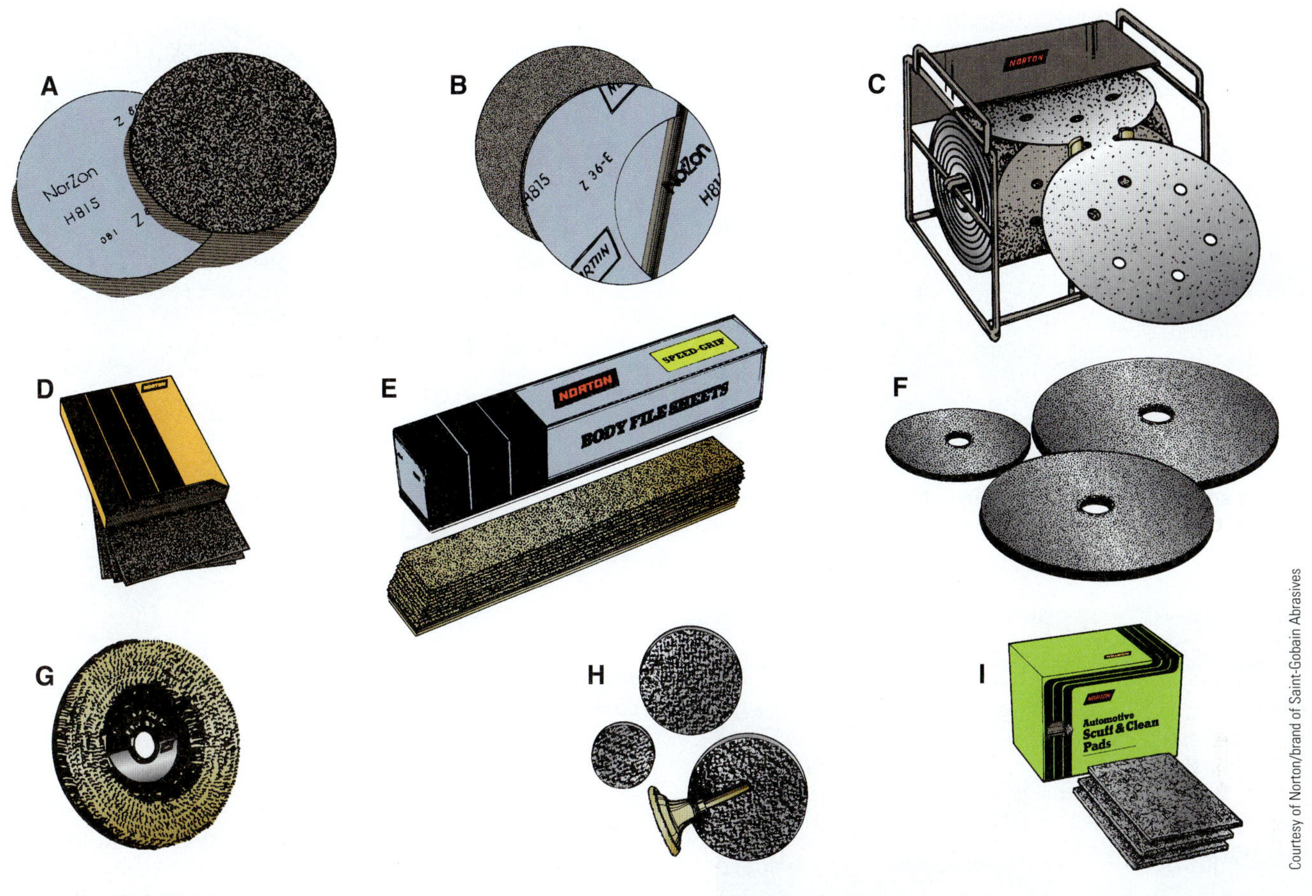

Figure 7–30 These are the most commonly used abrasive materials utilized by a collision repair facility: (A) Coarse sanding discs are used for rapid leveling of filler and paint. (B) Fine sanding discs are used for texturing surfaces for good adhesion and to ensure that sanding scratches will not show. (C) These sanding discs have holes for use in a vacuum system. (D) Some types of air sanders take rectangular sandpaper. (E) Air files for leveling large areas use these long, narrow sandpaper sheets. (F) Grinding discs are for rapid material removal, often down to bare metal. (G) Wire and synthetic brushes are handy for material removal in pockets where sandpaper or a grinding disc will not reach. (H) Small scuff wheels are often used to remove paint over spot welds during structural panel replacement. (I) Scuff pads are used for texturing good paint so the new finish will adhere properly. They are often used in door jambs and other hard-to-sand areas.

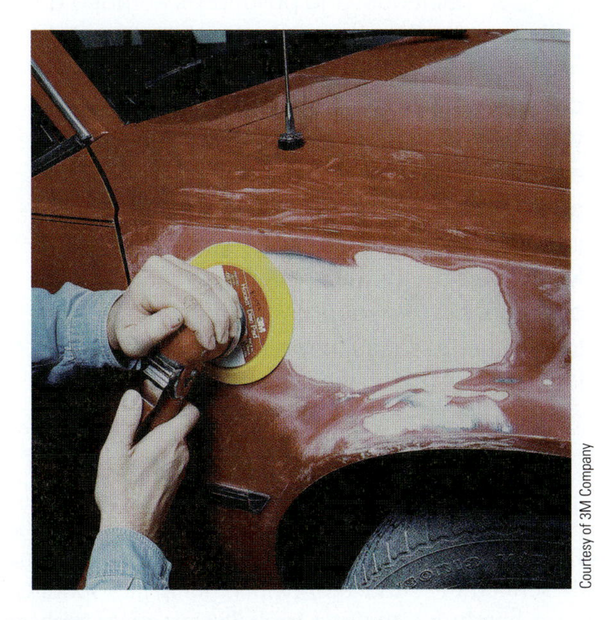

Figure 7–31 Here the body technician is using a smoothing, or medium grit, sanding disc to level and smooth body filler.

about 220 to 2000, for final smoothing operations before and after painting (Figure 7–33).

Wet sandpaper is commonly used to block-sand paint before compounding or buffing. Wet sanding will knock down any imperfections (color sanding) in the paint film. Buffing or compounding is then needed to make the paint shiny again.

When using sandpapers, remember these important points:

1. Sand in one direction only; this is usually along the line of sight. If several grades of sandpaper are used on one area, cross sand to eliminate scratches.
2. Use the finest abrasive possible to do the job.
3. Start with as fine a grade as possible. If too fine, go to a coarser grade, then work back to the finer grade.
4. Adjust one grade finer for hand versus machine sanding.
5. Support the abrasive with a block or pad to avoid finger marks and crowning.

Courtesy of 3M Company

A Disc adhesive is an older method of holding sandpaper on the pad, but the adhesive can contaminate the body surface.

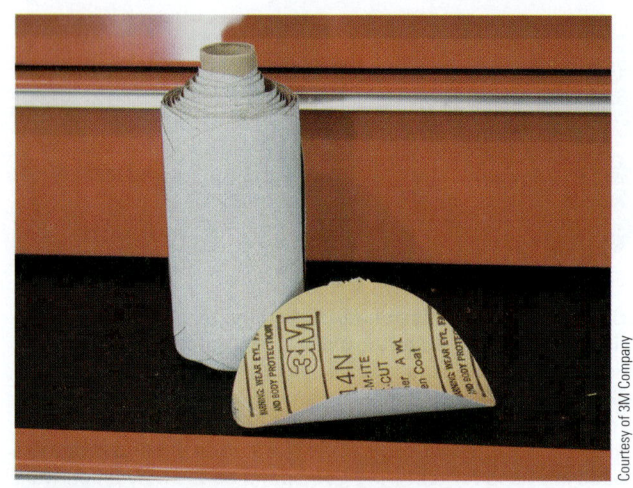

Courtesy of 3M Company

B Self-stick sandpaper has a controlled thickness of adhesive on the back side of the sanding discs. This type is still used.

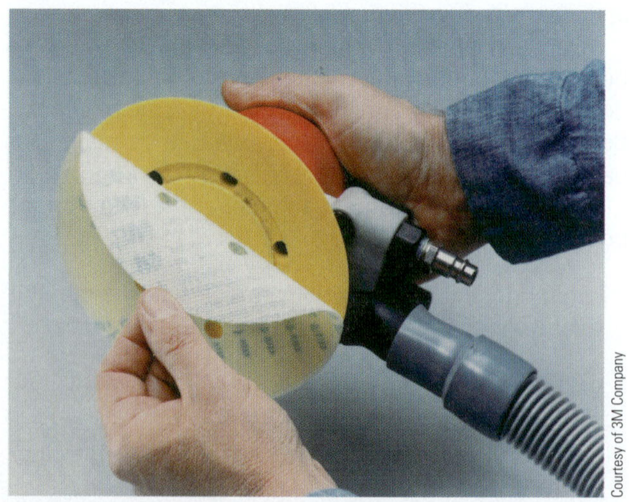

Courtesy of 3M Company

C Velcro, or hook and loop, sandpaper can be easily installed and removed from the pad on an air sander.

Figure 7-32 Note three methods that can be used to hold the sandpaper onto the air sanding pad.

Figure 7-33 Wet sandpaper will not dissolve when soaked in water. Wet sanding is very fast and does not produce sanding dust.

6. Adjust two grades finer when wet sanding, due to faster-cutting abrasives.
7. Choose one manufacturer's line so you learn its cutting characteristics.
8. Follow the manufacturer's recommendations for use.

ECO-TECH

When power sanding or grinding paint off a body panel, a large amount of paint dust and pollution is abraded off the panel and ejected into the shop area. This dust is laden with the harmful chemicals used to formulate the paint. Never sand without wearing a dust respirator or even better, an air-supplied respirator. Just imagine the quantity of toxic debris that could be inhaled into your lungs after 10 or 20 years on the job! Many retired nonprofessional auto body technicians have suffered a painful, early death from lung failure because they refused to wear a "silly looking dust mask" while sanding and grinding.

Scuff Pads

Scuff pads are tough synthetic pads used to clean and lightly scratch the surface of paints. Being like sponges, they are handy for scuffing irregular surfaces, such as door jambs, around the inside of the hood and deck lids, and other obstructed areas. They clean and lightly scuff these areas so the paint will stick. Scuff pads are also used to lightly scuff exterior surfaces prior to blending. This light roughening allows the paint and clearcoat to adhere without showing any scratches.

Compounds

Compounding involves using an abrasive paste material to smooth and bring out the gloss of the applied topcoat. Compounding can be done by hand or with a polishing wheel on an air tool. A compound has a fine volcanic pumice or dust-like grit in a water-soluble paste. When rubbed on a painted surface, a thin layer of paint is removed. It will remove the very top layer of old, weathered paint, leaving a new, fresh surface of paint (Figure 7–34).

A hand compound is designed to be applied by hand with a rag or cloth. *Machine compound* is formulated to be applied with an electric or air polisher. It will not cut as fast and will not break down with the extra friction and heat of machine application.

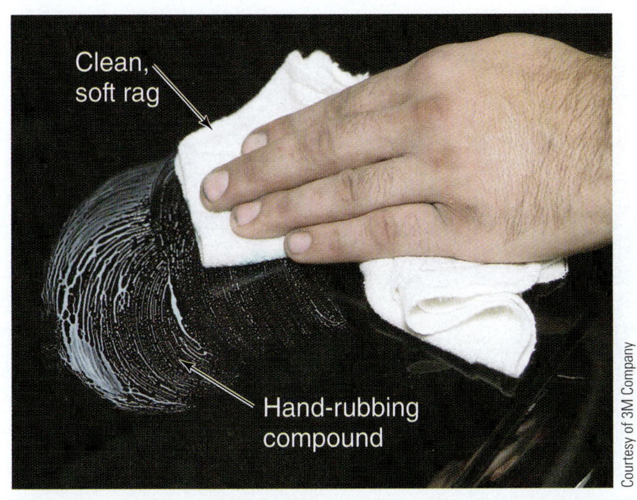

A Hand-rubbing compound is a coarser abrasive and should usually only be applied by hand.

B Machine compound is a finer polishing paste and is designed for use with a buffing machine.

Figure 7–34 Compounds are abrasive pastes used to final smooth paint.

Rubbing compound is the coarsest type of hand or machine compound. It will rapidly remove paint but will leave visible scratch marks. Rubbing compound is usually designed for hand application, but machine application is possible with some materials. It is often used on small areas to treat larger imperfections in the paint surface.

Polishing compound, or *machine glaze*, is a fine grit compound designed for machine application. A polisher is used to carefully run the compound over the cured or dried paint. Polishing compound is often used after a rubbing compound or after wet (color) sanding. It will make the paint shiny and smooth.

Hand glazes are used for final smoothing and shining of the paint. They are the last process used to produce a professional finish and are applied by hand, using a circular motion, like a wax.

Besides these, other compounds come in various formulations. Read the label on the compound to learn about its use. A painter will generally use the following compounds:

- Rubbing compounds to remove surface imperfections in paint
- Machine glazes to restore paint gloss after wet or color sanding
- Hand glazes to remove swirl marks after machine buffing

Adhesives

Adhesives are special glues designed to bond parts to one another. Various types are available.

Weatherstrip adhesive is designed to hold rubber seals and similar parts in place. Weatherstrip adhesive dries to a hard, rubber-type consistency. This makes it ideal for holding door seals, trunk seals, and other seals onto the vehicle body (Figure 7–35).

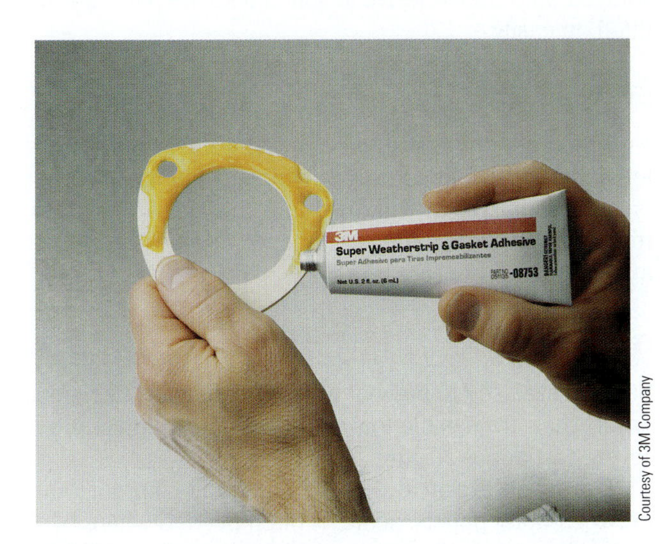

Figure 7–35 Weatherstrip adhesive is commonly used to install rubber seals around doors and trunks. Black color adhesive is best on black rubber seals.

Courtesy of 3M Company

Figure 7–36 **Plastic adhesive dries harder than weatherstrip adhesive. It is often used on trim pieces that bond directly to paint.**

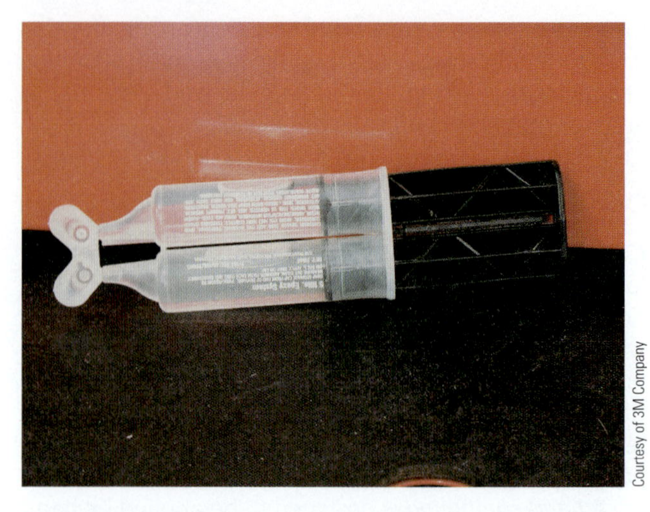
Courtesy of 3M Company

Figure 7–37 **Epoxy is a two-part adhesive that produces a powerful bond between parts. Numerous types of two-part epoxies are used to help bond parts. Make sure you use the type recommended by the manufacturer.**

Plastic adhesive, or emblem adhesive, is designed to hold hard plastic and metal parts. It is used to install various types of emblems and trim pieces onto painted surfaces (Figure 7–36). Before using emblem adhesive, make sure the paint is fully cured or dry, and use tape to hold the emblem in place until the adhesive dries.

Vinyl adhesive is designed to bond a vinyl top to the vehicle body. This type of adhesive is often used when installing or repairing vinyl tops, interior roof liners (headliners), and similar parts.

An *adhesive release* agent is a chemical that dissolves most types of adhesives. It is used when you want to remove a part without damage. The release agent is sprayed onto the adhesive to soften it so the part can be lifted off easily.

Epoxies

An **epoxy** is a two-part glue used to hold various parts together. The two ingredients are mixed together in equal parts. This makes the mixture cure through a chemical reaction. Always use the type of epoxy suggested by the auto manufacturer (Figure 7–37).

Sealers

Sealers are used to prevent water and air leaks between parts. They are flexible to prevent cracking and come in several variations (Figure 7–38).

Seam sealers are designed to make a leakproof joint between body panels. Sealer is often needed where two panels butt or overlap each other. Seam sealers come in different forms and each is applied differently. Read the directions.

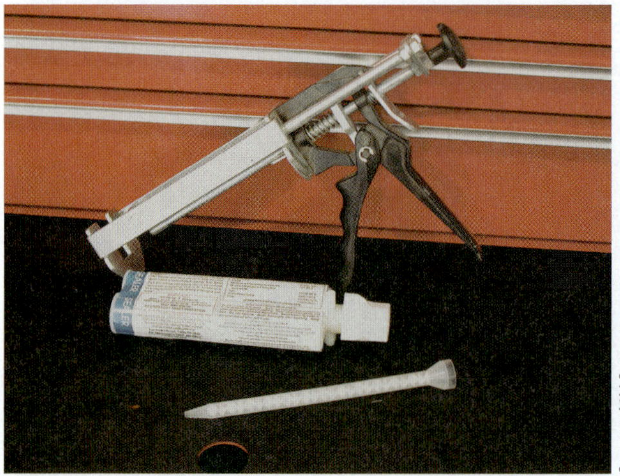
Courtesy of 3M Company

A

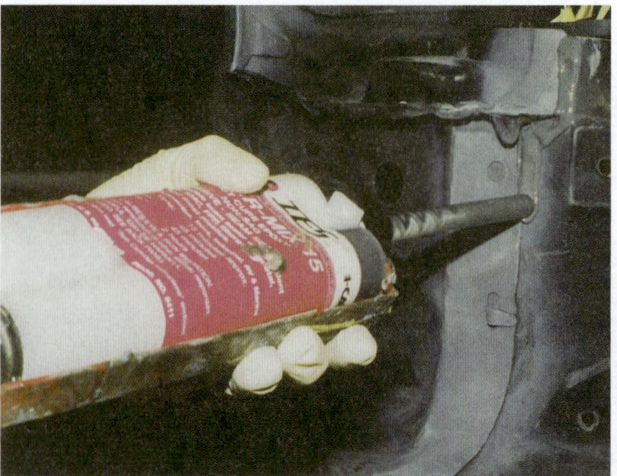
Courtesy of 3M Company

B

Figure 7–38 **(A) Various types of sealers are used to prevent leakage between body parts. (B) The technician is using a caulking gun to seal a newly installed body panel.**

Tube sealers are applied directly from the tube or by using a caulking gun. They squirt out like toothpaste and cure in a few hours.

Apply primer before applying seam sealer. Seam sealers are paintable but may need to be reprimed if the product directions specify. Silicone sealers are not paintable and should not be used in auto body repair. Follow instructions on the product for finishing sealers.

Ribbon sealers come in strip form and are applied by hand. They are thick sealers that must be worked onto the parts with your fingers.

NOTE *The proper use or application of these materials will be explained later in this book. Refer to the index if you need more information now.*

7.2 FASTENERS

Fasteners are the thousands of bolts, nuts, screws, clips, and adhesives that literally hold a vehicle together.

As an auto body technician, you will constantly use fasteners when removing and installing body parts. This is why it is important for you to be able to identify and use fasteners properly.

Remember that each fastener is engineered for a specific application. Always replace fasteners with exactly the same type that was removed from the OEM assembly. Never try to reengineer the vehicle. Keep in mind that using an incorrect fastener or a fastener of inferior quality can result in failure and possible injury to the vehicle occupants.

Bolts

A *bolt* is a metal shaft with a head on one end and threads on the other. A *cap screw* is a term that describes a high-strength bolt. Bolts and cap screws are usually named after the body part they hold, such as fender bolt or hood hinge bolt. Their shape and head drive configurations also help name them.

Bolt Terminology

To work with bolts properly, you must understand basic bolt terminology (Figure 7–39).

The *bolt head* is the top and is used to torque or tighten the bolt. A socket or wrench fits over the head, which enables the bolt to be tightened. Some English and metric sockets are very close in size. It is important not to use metric sizes for English bolts or English sizes for metric bolts, as the heads can be rounded and damaged.

Bolt length is measured from the end of the threads to the bottom of the bolt head. It is *not* the total length including the bolt head.

Bolt diameter, sometimes termed bolt size, is measured around the outside of the threads. For example, a 1/2-inch bolt has a thread diameter of 1/2 inch, whereas its head or wrench size would be 3/4 inch.

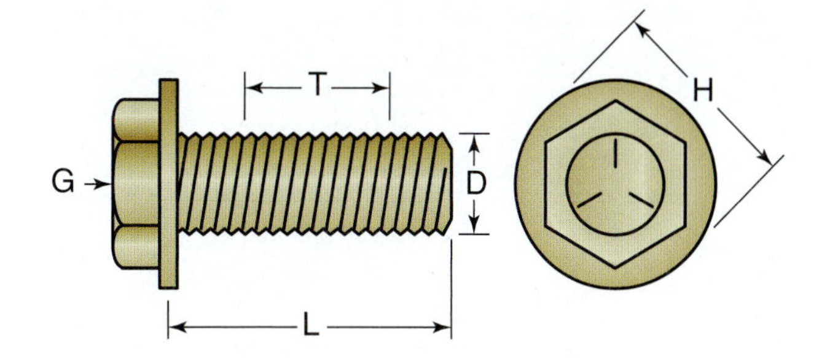

H Head
G Grade marking (bolt strength)
L Length (inches)
T Thread pitch (thread/inch)
D Nominal diameter (inches)

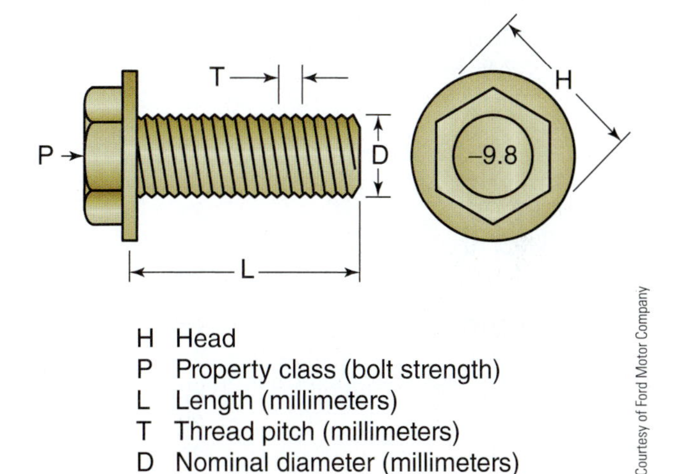

Figure 7-39 **You need to know bolt measurements when you are working. Study each dimension of both English and metric bolts.**

Bolt head size is the distance measured across the flats of the bolt head. In the United States Customary System (USCS), head size is given as fractions, just like wrench size. A few common sizes are 7/16, 1/2, and 9/16 inch. In the metric system, 3-, 4-, and 5-mm head sizes are typical (Figure 7–40).

Bolt thread pitch is a measurement of thread coarseness. Bolts and nuts can have coarse, fine, and metric threads. Bolt threads can be measured with a *thread pitch gauge*.

A number of classifications have been used over the years to identify the various types of threads. The terms used in the automotive trade—the United States Standard Institute (USSI), the American National Standard (ANS), and the Society of Automotive Engineers Standard (SAE)—have all been replaced by the Unified National Series.

The two common metric threads are coarse and fine and can be identified by the letters SI (Système International or International System of Units) and ISO (International Standards Organization).

Common English USC/SAE head sizes	Common metric head sizes
Wrench size (inches)*	Wrench size (millimeters)*
3/8	9
7/16	10
1/2	11
9/16	12
5/8	13
11/16	14
3/4	15
13/16	16
7/8	17
15/16	18
1	19
1-1/16	20
1-1/8	21
1-3/16	22
1-1/4	23
1-5/16	24
1-3/8	26
7/16	27
1-1/2	29
	30
	32

*The wrench sizes given in this chart are not equivalents, but are standard head sizes found in both inches and millimeters.

Figure 7-40 **These are common bolt head and wrench sizes. Never use a standard wrench on metric bolts and vice versa. This will round off the bolt head.**

Do *not* accidentally interchange thread types or damage will result. It is easy to mistake metric threads for English threads. If the two are forced together, either the bolt or the part threads will be ruined.

Bolts and nuts are also available in right- and left-hand threads. *Right-hand threads* must be turned clockwise to tighten. Less common *left-hand threads* must be rotated in a counterclockwise direction to tighten the fastener. Left-hand threads may be denoted by notches or the letter *L* stamped on them.

When ordering bolts, it is necessary to designate the bolt diameter, thread pitch, and length. An example would be ¼ inch–20 × 1 inch.

Bolt Strengths or Grades

Bolt strength indicates the amount of torque or tightening force that should be applied. Bolts are made from different materials having various degrees of hardness. Softer or harder metal can be used to achieve different hardnesses and strengths for use in different situations.

Bolt grade markings are lines or numbers on the top of the head to identify bolt hardness and strength. The hardness or strength of metric bolts is indicated by using a property class indicator on the head of the bolt.

Bolt strength markings are given as lines (Figure 7–41). The number of lines on the head of the bolt is related to the strength. As the number of lines increases, so does the strength.

Metric bolt strength markings are given as numbers—the higher the number, the stronger the bolt. These markings apply to both bolts and nuts (Figure 7–42).

SAE grade markings					
Definition	No lines: unmarked indeterminate quality SAE grades 0-1-2	3 lines: common commercial quality Automotive & AN bolts SAE grade 5	4 lines: medium commercial quality Automotive & AN bolts SAE grade 6	5 lines: rarely used SAE grade 7	6 lines: best commercial quality NAS & aircraft screws SAE grade 8
Material	Low carbon steel	Med. carbon steel tempered	Med. carbon steel quenched & tempered	Med. carbon alloy steel	Med. carbon alloy steel quenched & tempered
Tensile strength	65,000 psi	120,000 psi	140,000 psi	140,000 psi	150,000 psi

Figure 7-41 **Bolt tensile strengths are denoted on the heads of bolts. Slash marks [Society of Automotive Engineers (SAE)] or numbers (metric) are used. Always replace bolts with equal or higher rated bolts to prevent failure.**

Grade	Identification	Class	Identification
Hex nut grade 5	3 dots	Hex nut property class 9	Arabic 9
Hex nut grade 8	6 dots	Hex nut property class 10	Arabic 10
Increasing dots represent increasing strength		Can also have blue finish or paint dab on hex flat. Increasing numbers represent increasing strength	

Figure 7-42 **Quality nut strengths are also denoted. More dots or a high number means more strength.**

Tensile strength is the amount of tension per square inch the bolt can withstand just before breaking when it is pulled apart. The harder or stronger the bolt, the greater the tensile strength.

Never replace a high-grade bolt with a bolt having a lower grade marking, because the weaker bolt could break. Using a lower grade bolt in the steering or suspension could seriously endanger the passengers of the vehicle.

Bolt Torque

Bolt torque is a measurement of the turning force applied when installing a fastener. It is critical that bolts and nuts are torqued or tightened properly. Overtightening will stretch and possibly break the bolt. Undertightening may allow the bolt or nut to loosen and fall out.

Torque specifications are tightening values for a specific bolt or nut. They are given by the manufacturer. A torque wrench must be used to measure values when installing critical bolts and nuts.

If you cannot find the factory torque specification for a bolt, you can use a *general bolt torque chart.* It will give a general torque value for the size and grade of bolt. Such a chart is shown in Figure 7–43. Normally the bolt threads should be lubricated to get accurate results. Refer to the chart to see whether the threads should be lubricated or dry.

A *tightening sequence*, or **torque pattern**, ensures that parts are clamped down evenly by several bolts or nuts. Tightening fasteners in a crisscross pattern pulls the part down evenly, preventing warpage. This is commonly recommended for wheels, as shown in Figure 7–44.

Tighten the fastener in steps to about half torque, three-fourths torque, and then full torque, at least twice.

Be careful when tightening bolts and nuts with air wrenches. It is easy to stretch or break a bolt instantly. An air wrench can spin a bolt or nut so fast that it can hammer the fastener past its yield point. This can strip threads or snap off the bolt.

Metric Standard							SAE Standard / Foot Pounds						
Grade of Bolt		5D	.8G	10K	12K		Grade of Bolt	SAE 1 and 2	SAE 5	SAE 6	SAE 8		
Min. Tensile Strength		71,160 PSI	113,800 PSI	142,200 PSI	170,679 PSI		Min. Ten. Strength	64,000 PSI	105,000 PSI	133,000 PSI	150,000 PSI		
Grade Markings on Head		5D	8G	10K	12K	Size of Socket on Wrench Opening	Markings on Head					Size of Socket or Wrench Opening	
Metric		Foot Pounds				Metric	U.S. Standard	Foot Pounds				U.S. Regular	
Bolt Dia.	U.S. Dec. Equiv.					Bolt Head	Bolt Dia.					Bolt Head	Nut
6 mm	.2362	5	G	8	10	10 mm	1/4	5	7	10	10.5	3/8	7/16
8 mm	.3150	10	16	22	27	14 mm	5/16	9	14	19	22	1/2	9/16
10 mm	.3937	19	31	40	49	17 mm	3/8	15	25	34	37	9/16	5/8
12 mm	.4720	34	54	70	86	19 mm	7/16	24	40	55	60	5/8	3/4
14 mm	.5512	55	89	117	137	22 mm	1/2	37	60	85	92	3/4	13/16
16 mm	.6299	83	132	175	208	24 mm	9/16	53	88	120	132	7/8	7/8
18 mm	.709	111	182	236	283	27 mm	5/8	74	120	167	180	15/16	1
22 mm	.8661	182	284	394	464	32 mm	3/4	120	200	280	296	1-1/8	1-1/8

Figure 7-43 **If factory specs are not available, use this general bolt torque chart. It gives different values for each bolt tensile strength rating.**

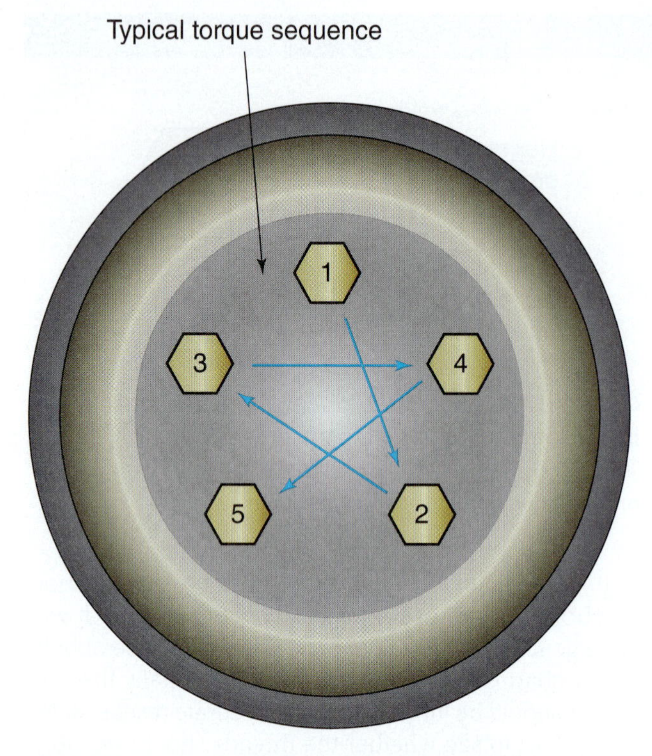

Figure 7–44 When tightening several bolts that hold one part—a wheel, for example—always use a crisscross pattern. This will prevent part warpage and runout problems.

When torque is not critical, do not use the air impact to run the nut full speed onto the bolt. Instead, run it up slowly until it contacts the work. Then mark the socket and watch how far it turns. Smaller air-powered speed wrenches do not produce the severe force of impact wrenches and are much safer to use.

Nuts

A *nut* uses internal (inside) threads and an odd-shaped head that often fits a wrench. When tightened onto a bolt, a strong clamping force holds the parts together. Many different nuts are used by the automotive industry. Several are shown in Figure 7–45.

Castellated, or *slotted, nuts* are grooved on top so that a safety wire or cotter pin can be installed into a hole in the bolt. This helps prevent the nut from working loose. For example, castellated nuts are used with the studs that hold wheel bearings in position. Slotted nuts are also used on steering and suspension parts for safety.

Self-locking nuts produce a friction or force fit when threaded onto a bolt or stud. The top of the nut can be crimped inward. Some have a plastic insert that produces a friction fit to keep the nut from loosening. Locking nuts may need to be replaced after removal. Front-wheel-drive spindles sometimes use self-locking nuts.

Jam nuts are thin nuts used to help hold larger, conventional nuts in place. The jam nut is tightened down against the other nut to prevent loosening.

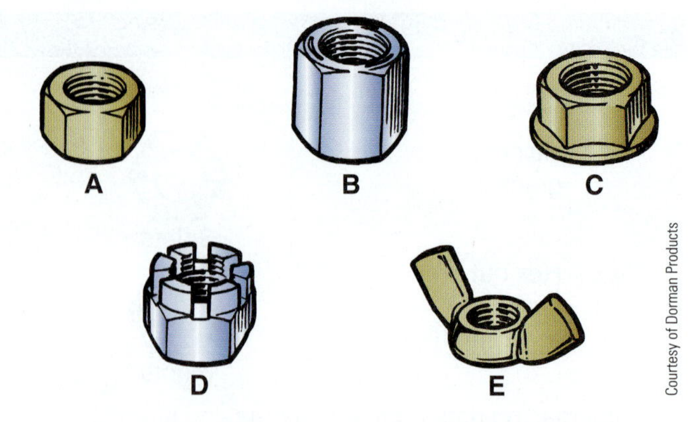

Figure 7–45 Memorize nut types: (A) hex nut, (B) high, or deep, nut, (C) flange nut, (D) castle, or slotted, nut, and (E) wing nut.

Wing nuts have two extended arms for turning the nut by hand. They are used when a part must be removed frequently for service or maintenance. Air cleaners sometimes use wing nuts.

Acorn nuts are closed on one end for appearance and to keep water and debris off the threads. They can be used when they are visible and looks are important.

Special types of nuts are used to hold parts onto the vehicle. Sometimes a washer is formed onto the nut. Termed *body nuts*, the flange on the nut helps distribute the clamping force of the thin body panel or trim piece to prevent warpage (Figure 7–46).

Washers

Washers are used under bolts, nuts, and other parts. They prevent damage to the surfaces of parts and provide better holding power (Figure 7–47).

Flat washers are used to increase the clamping surface area. They prevent the smaller bolt head from pulling through the sheet metal or plastic.

A *wave washer* adds a spring action to keep parts from rattling and loosening.

Body or *fender washers* have a very large outside diameter for the size hole in them. They provide better holding power on thin metal and plastic parts.

Copper or *brass washers* are used to prevent fluid leakage, as on brake line fittings. *Spacer washers* come in specific thicknesses to allow for adjustment of parts. *Fiber washers* will prevent vibration or leakage but cannot be tightened very much.

Finishing washers have a curved shape for a pleasing appearance. They are used on interior pieces.

Split lock washers are used under nuts to prevent loosening by vibration. The ends of these spring-hardened washers dig into both the nut and the work to prevent rotation. The lock washers should be placed next to the bolt head or nut.

Shakeproof or *teeth lock washers* have teeth or bent lugs that grip both the work and the nut. Several designs, shapes, and sizes are available. An external type has

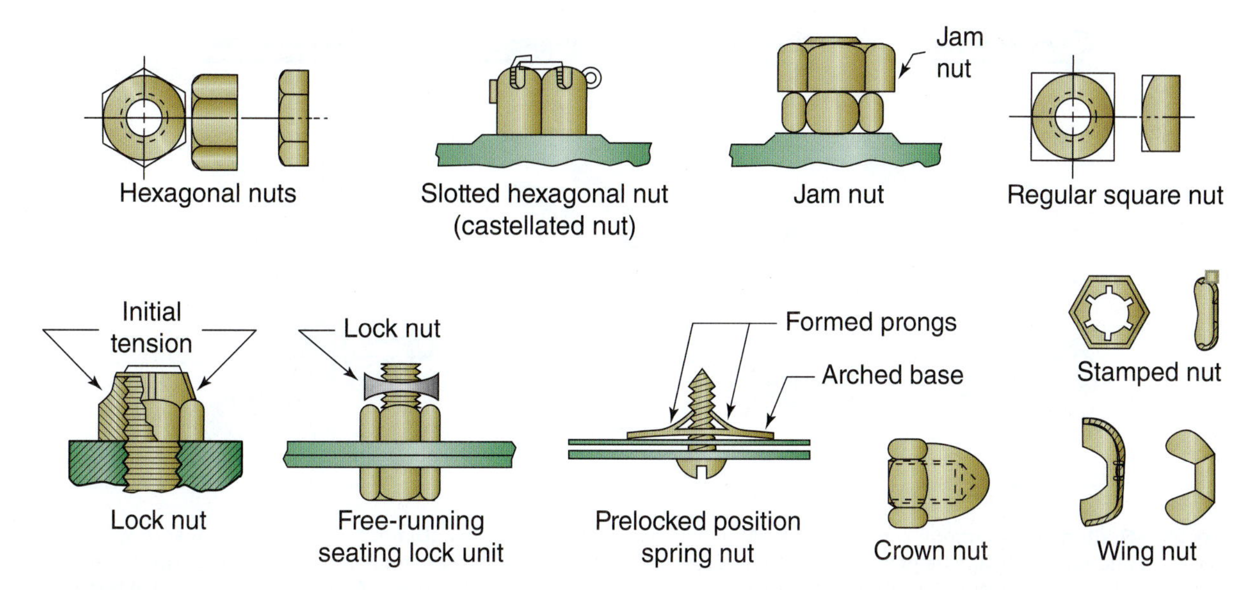

Figure 7-46 Body nuts are specially designed for specific holding applications.

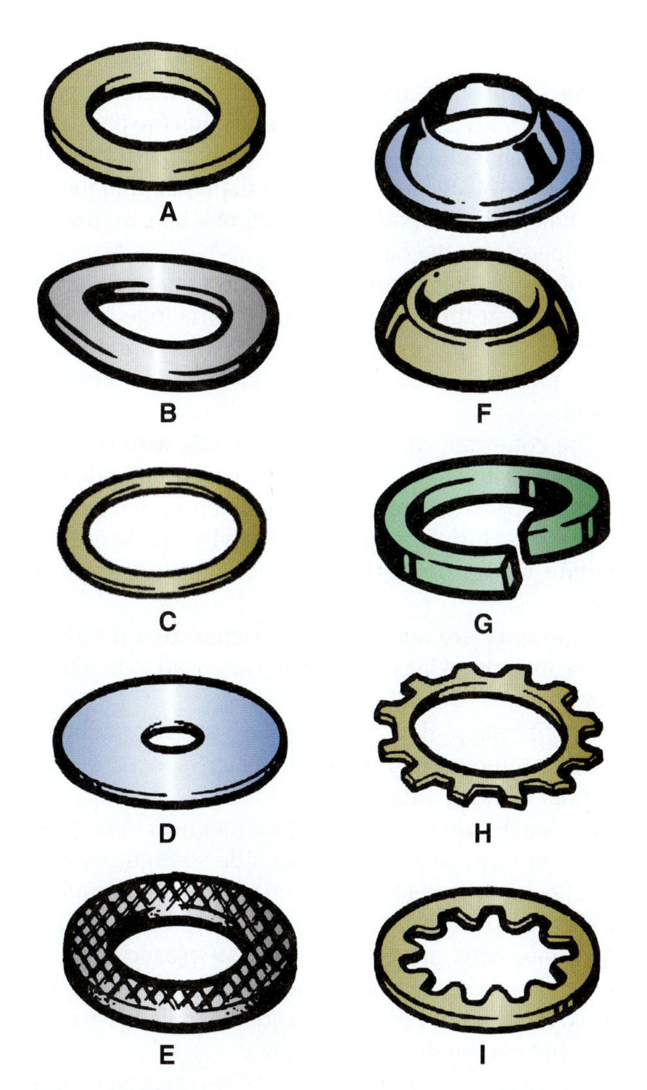

Figure 7-47 Study the washer types: (A) plain flat washer, (B) wave, or spring, washer, (C) spacer washer, (D) fender washer, (E) fiber washer, (F) finishing washers, (G) split lock washer, (H) external lock washer, and (I) internal lock washer.

teeth on the outside, and an internal type has teeth around the inside. Lock washers are extremely hard and tend to break under severe pressure.

A rule of thumb is if the part did not come with a lock washer, do not add one. If the bolt or nut has a lock washer, use one. The manufacturer would not use one if it did not have a purpose.

Screws

Screws are often used to hold nonstructural parts on a vehicle (Figure 7–48). Trim pieces, interior panels, and so on, are often secured by screws.

Machine screws are threaded their full length and are relatively weak. They come in various configurations and will accept a nut.

Set screws frequently have an internal drive head for an Allen wrench and are used to hold parts onto shafts.

Sheet metal screws have pointed or tapered tips; they thread into sheet metal for light holding tasks.

Self-tapping screws have a pointed tip that helps cut new threads in parts.

Trim screws have a washer attached to them, which improves appearance and helps keep the trim from shifting.

Headlight aiming screws have a special plastic adapter mounted on them. The adapter fits into the headlight assembly. Different design variations are needed for different makes and models of vehicles.

Nonthreaded Fasteners

As implied, *nonthreaded fasteners* do not use threads. They include keys, snap rings, pins, clips, adhesives, and so on. Various keys and pins are used by equipment manufacturers to retain parts in alignment. It is important

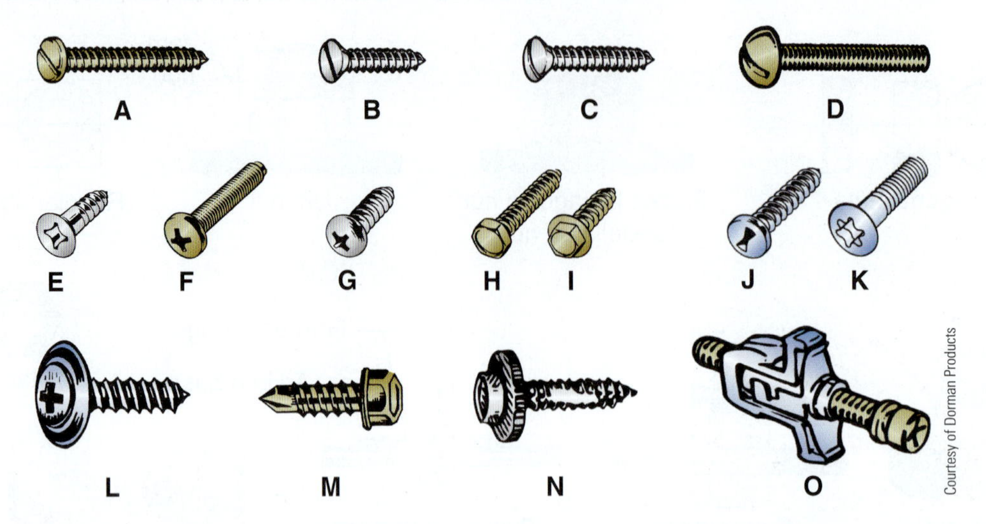

Figure 7-48 Study the common screw names: (A) pan head sheet metal screw, (B) flat head sheet metal screw, (C) oval head, (D) round head, (E) Phillips head screw, (F) threaded screw, (G) oval head sheet metal screw, (H) hex, or nutdriver, screw, (I) hex screw with flange or integral washer, (J) clutch head, (K) Torx head, (L) trim screw, (M) self-taping screw, (N) body screw, and (O) headlight aiming screw.

to be able to identify these keys and pins to order replacements.

Square keys and *Woodruff keys* are used to prevent hand wheels, gears, cams, and pulleys from turning on their shafts. These keys are strong enough to carry heavy loads if they are fitted and seated properly (Figure 7–49).

Figure 7–50 shows the various types of pins. Round *taper pins* have a larger diameter on one end than the other. They are used to locate and position matching parts. They can also be used to secure small pulleys and gears to shafts.

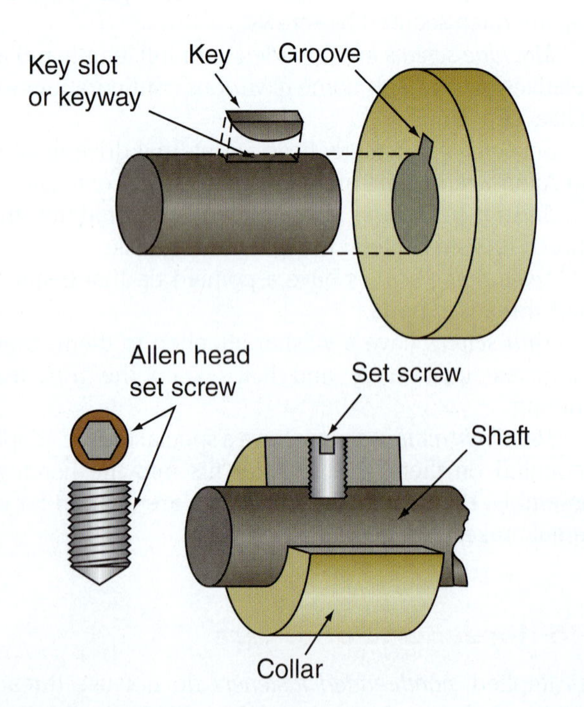

Figure 7-49 Keys and set screws are both used to align parts on shafts: (A) key and keyway and (B) set screw application.

Dowel pins have the same general diameter their full length. They are used to position and align the parts of an assembly. One end of a dowel pin is chamfered, and it is usually 0.001–0.002 inch greater in diameter than the size of its hole. When replacing a dowel pin, be sure that it is the same size as the old one.

Cotter pins help prevent bolts and nuts from loosening, or they fit through pins to hold parts together. They are also used as stops and holders on shafts and rods. All cotter pins are used for safety and should never be reused.

The cotter pin should fit into the hole with very little side play. If it is too long, cut off the extra length. After insertion, bend it over in a smooth curve with needle nose pliers. Sharply angled bends invite breakage. Final bending of the prongs can be done with a soft-faced mallet.

Snap rings are nonthreaded fasteners that install in a groove machined into a part. They are used to hold parts on shafts.

Special snap ring pliers are designed to flex and install or remove snap rings. They have special tips that will hold the snap ring.

Body clips are specially shaped retainers to hold trim and other body pieces requiring little strength. The clip often fits into the back of the trim piece and through the body panel (Figure 7–51).

Push-in clips are usually made of plastic and they force fit into holes in body panels. Push-in clips are used to hold interior door trim panels, for example. They install easily but can be difficult to remove.

Pop rivets can be used to hold two pieces of sheet metal together. They can be inserted into a blind hole through two pieces of metal and then drawn up with a riveting tool or gun. This will lock the pieces together. There is no need to have access to the back of the rivets.

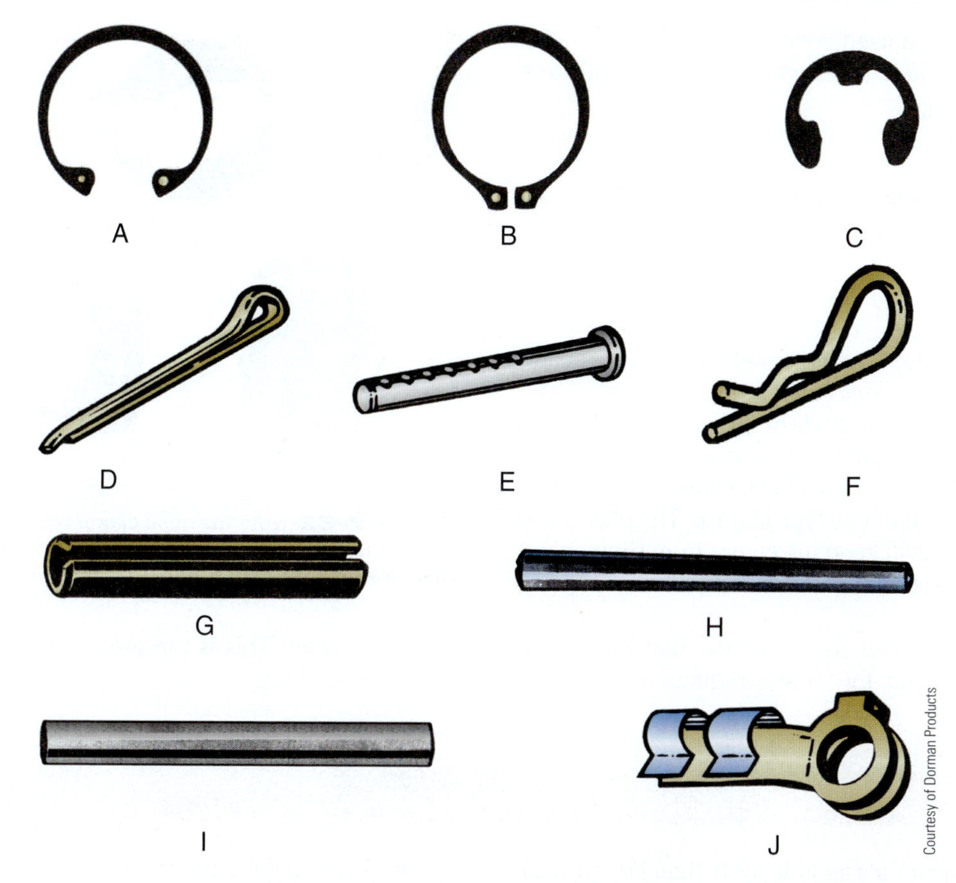

Figure 7-50 Learn these names of nonthreaded fasteners: (A) internal snap or retaining ring, (B) external snap ring, (C) E-clip or snap ring, (D) cotter pin, (E) clevis pin, (F) hitch pin, (G) split rollpin, (H) tapered pin, (I) straight dowel pin, and (J) linkage clip.

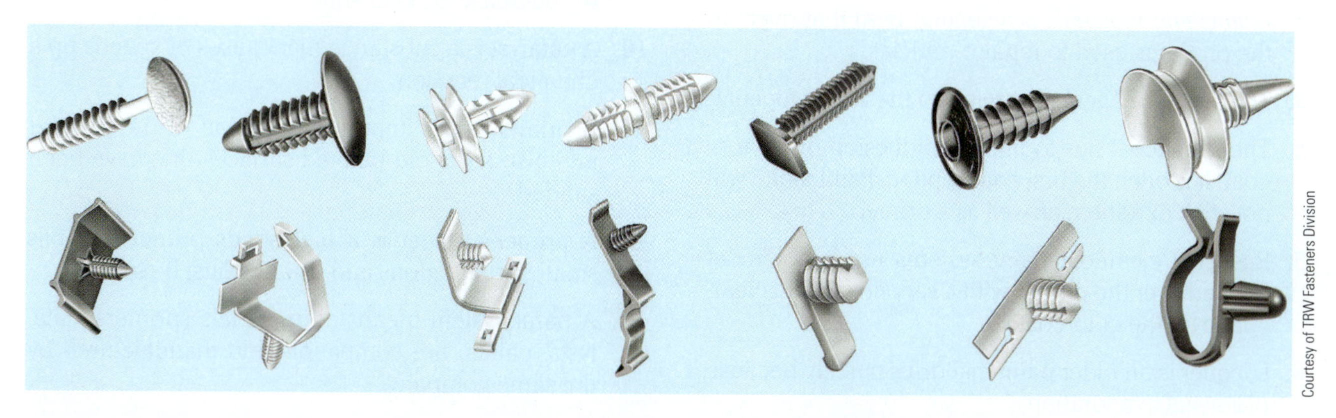

Figure 7-51 These are a few of the special plastic retainers available. These types are often used in interiors. They quickly press into a hole. To remove them, you must carefully pry next to the retainer with a flat, forked trim tool.

Pop rivets should not be used in areas that are subject to excessive vibration or for structural panels. The rivets can work loose and weaken the repair.

Wear safety glasses when removing or installing snap rings. Because they are spring steel, they can shoot out with great force.

Replacing Fasteners

When replacing fasteners, observe the following precautions:

1. Always use the same number of fasteners.
2. Use the same diameter, length, pitch, and type of fasteners.
3. Observe the OEM recommendations given in the service manual for tightening sequence, tightening steps (increments), and torque values.

4. Always replace a used cotter pin.
5. Replace stretched fasteners or fasteners with any signs of damage.
6. Use the correct washers and pins as specified by the OEM.
7. Always replace "one-time" fasteners. They can be found in suspension and steering assemblies.

Hose Clamps

Hose clamps are used to hold radiator hoses, heater hoses, and other hoses onto their fittings (Figure 7–52).

A *spring hose clamp* is made of spring steel with barbs on each end. Squeezing the ends opens and expands the clamp. Special hose clamp pliers should be used to remove or install round wire-type clamps. The pliers have a deep groove that will keep the clamp from slipping out of the jaws. Conventional pliers will work fine on spring strap clamps.

A *worm hose clamp* uses a screw that engages a slotted band. Turning the screw reduces or enlarges clamp diameter. This is the most common replacement type of hose clamp.

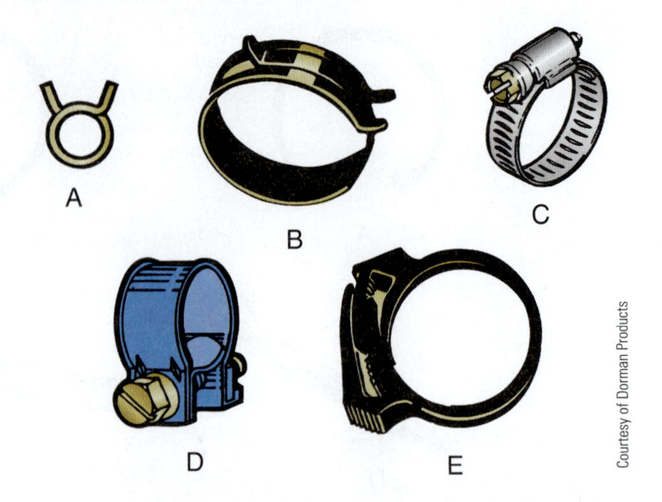

Figure 7–52 Note the hose clamp types: (A) wire spring hose clamp, (B) wire strap hose clamp, (C) worm hose clamp, (D) screwnut hose clamp, and (E) plastic hose clamp.

SUMMARY

1. Body shop materials include more than just refinishing or paint materials. They include the various fillers, primers, sealers, adhesives, sandpapers, and other compounds common to a body shop.
2. *Refinishing materials* is a general term that refers to the products used to repaint vehicles.
3. The term *paint* generally refers to the visible topcoat.
4. The primecoat has to improve adhesion of the topcoat. It is often the first coat applied. Paint alone will not stick or adhere as well as a primer.
5. Basecoat/clearcoat paint systems use a colorcoat applied over the primer with a second layer of clearcoat over the colorcoat.
6. Lacquer is an older paint that dries quickly because of solvent evaporation.
7. Enamel finishes dry with a gloss and do not require rubbing or polishing.
8. A paint's chemical content includes the following:
   - Pigments
   - Binders
   - Solvents
   - Additives
9. Paint additives are ingredients added to modify the performance and characteristics of the paint. Additives are used to:
   - Speed up or slow down the drying process
   - Lower the freezing point of a paint
   - Prevent the paint from foaming when shaken
   - Control settling of metallics and pigments
   - Make the paint more flexible when dry
   - Increase gloss or shine
   - Decrease gloss or shine
10. A catalyst is a substance that causes or speeds up a chemical reaction.
11. A primer-sealer improves adhesion of the topcoat and also seals old painted surfaces that have been sanded.
12. A primer-surfacer is a high-solids primer that fills small imperfections and usually must be sanded.
13. A paint system means all materials (primers, catalysts, paints) are compatible and manufactured by the same company.
14. A prep solvent or wax and grease remover is a special fast-drying agent often used to chemically clean a vehicle's body surfaces before painting.
15. A metal conditioner is used to etch bare sheet metal before priming.
16. Body filler or plastic filler is a heavy-bodied plastic material that cures very hard for filling small dents in metal.
17. Glazing putty is a material made for filling small holes or sand scratches.
18. Masking materials are used to cover and protect body parts from paint overspray.

19. Grit refers to a measure of the size of particles on sandpaper or discs. A coarse sandpaper would have large grit. A fine sandpaper would have smaller grit.

20. Compounding involves using an abrasive paste material to smooth and bring out the gloss of the applied topcoat.

21. Adhesives are special glues designed to bond parts to one another.

22. Sealers are used to prevent water and air leaks between parts.

23. Fasteners include the thousands of bolts, nuts, screws, clips, and adhesives that literally hold a vehicle together.

24. Bolt torque is a measurement of the turning force applied when installing a fastener. It is critical that bolts and nuts are torqued or tightened properly.

EXERCISES

On a separate sheet of paper, complete the following learning activities for this chapter. Write definitions for the key terms and answer the ASE-style review questions, essay questions, critical thinking problems, and math problems. You can also do the outside activities, possibly for extra credit.

Key Terms

abrasive
accelerator
adhesives
antichip coating
basecoat/clearcoat
bleeding
body filler
bolt grade markings
catalyst
colorcoat
compounding
corrosion
cream hardener
curing
drying
dry sandpaper
enamel
epoxy
epoxy primer
fasteners
fiberglass cloth
fiberglass mat
fiberglass-reinforced body filler
fiberglass resin
filler
fine line masking tape
fish-eye eliminator
flattener
flash
flex agent
glazing putty
grinding disc
grit
grit numbering system
light body filler
masking liquid
masking material
masking paper
masking plastic
masking tape
metal conditioner
metallic paint
paint
paint binder
paint masking paper
paint sealer
paint solvent
paint stripper
pearl paint
pigment
polishing compound
prep solvent
primer
primer masking paper
primer-surfacer
reducer
refinishing material
retarder
rubbing compound
sandpaper
scuff pad
sealer
seam sealer
self-etching primer
tack cloth
torque pattern
torque specifications
two-part putty
two-stage paint
UV primer
waterbase paint
wet sandpaper
wheel mask

ASE-Style Review Questions

1. Which of the following terms refers to a factory-applied finish?
   A. Lacquer
   B. Enamel
   C. OEM paint
   D. ASE paint

2. Which primer coat improves adhesion of the topcoat and prevents bleed through of old painted surfaces that have been sanded?
   A. Primer
   B. Primer-sealer
   C. Primer-filler
   D. Primer-surfacer

3. Which bolt would be the strongest?
   A. Three-head markings
   B. Two-head markings
   C. No-head markings
   D. One-head marking
4. When installing a wheel on an older car, no service manual can be found for getting a factory torque specification. Technician A says to use an impact wrench on the medium setting. Technician B says that a crisscross pattern should be followed when tightening the wheel nuts. Who is correct?
   A. Technician A
   B. Technician B
   C. Both A and B
   D. Neither A nor B

Essay Questions

1. What are two functions of a vehicle's paint or finish?
2. List and explain the three basic types of automotive paints.
3. What are the differences between thinner, reducer, and premix?
4. What is a catalyst?
5. What is corrosion or rust?
6. What is masking liquid or masking coating?
7. In detail, explain the sandpaper grit numbering system and how it is used.
8. What general sequence should be used when tightening a series of bolts or nuts?
9. Define the terms *primecoat* and *topcoat*.
10. Name the four ingredients in a paint.
11. Define the term *cap screw*.
12. What is bolt thread pitch and how is it measured?
13. If you turn right-hand threads clockwise, what will happen?

Critical Thinking Problems

1. If lug bolts on a wheel are improperly tightened, what might happen?
2. When repairing a wrecked car, what are some of the fasteners that you might have to torque while working?

Math Problems

1. If a paint is supposed to be reduced one part paint to one part thinner, what is the percentage of reduction?
2. If a bolt torque spec is 100 ft-lb. of torque, what is the metric torque value in newton-meters?

Activities

1. Inspect a car or truck. Write down the various types of fasteners you can locate. Make a chart giving their names and applications or locations.
2. Read the directions on a few types of body shop materials (primers, paints, and so on). Write a short report on their uses.
3. Visit a body shop supply house or hardware store. Inspect the various types of fasteners available. How much does each type of fastener cost?

Welding Equipment Technology

CHAPTER 8

OBJECTIVES

After studying this chapter, you should be able to:

- Identify the three classes of welding.
- Explain how to use a MIG welding machine.
- Name the six basic welding techniques employed with MIG equipment.
- Describe differences between MIG electrode wires.
- Determine where and how to use resistance spot welding.
- Identify oxyacetylene welding equipment and techniques.
- Explain general brazing and soldering techniques used in a body shop.
- Describe plasma arc cutting of body panels.
- Explain plasma cutting techniques.
- List safety procedures important in each welding operation.
- Answer ASE-style review questions relating to welding equipment.

INTRODUCTION

With major collision repair work, many of the panels on a vehicle must be replaced and welded into place. As you will learn, this requires considerable skill and care. The structural integrity of the vehicle is dependent on how well you weld and install repair panels.

If not already trained, you may want to consider taking a metal inert gas (MIG) welding course in school or through another agency. Ask your instructor or guidance counselor for more information on welding courses in your school or area. Welding is an essential skill if you plan on becoming a master auto body technician.

8.1 JOINING METALS

There are three basic methods of joining metal together in the automobile assembly:

1. Mechanical (metal fastener) methods (Figure 8–1)
2. Chemical (adhesive fastening) methods
3. Welding (molten metal fusion) methods (Figure 8–2)

Welding uses extreme heat to join or fuse pieces of metal together. Welding can be divided into three main categories:

1. **Pressure welding**. The metal is heated to a softened state by electrodes. Pressure is applied, and the metal is joined. Of the various types of pressure welding, electric resistance welding (spot welding) is an indispensable method used in automobile manufacturing and, to a lesser degree, in repair operations.
2. **Fusion welding**. Pieces of metal are heated to the melting point, joined together (usually with a filler rod), and allowed to cool.
3. *Braze welding*. Metal with a melting point lower than the base metal to be joined is melted over the joint of the pieces being welded (without fusing pieces of base metal). Braze welding is classified as either soft or hard brazing, depending on the temperature at which the brazing material melts. *Soft brazing* is done with brazing material that melts at temperatures below 850°F (455°C). *Hard brazing* is done with brazing materials that melt at temperatures above 850°F (455°C).

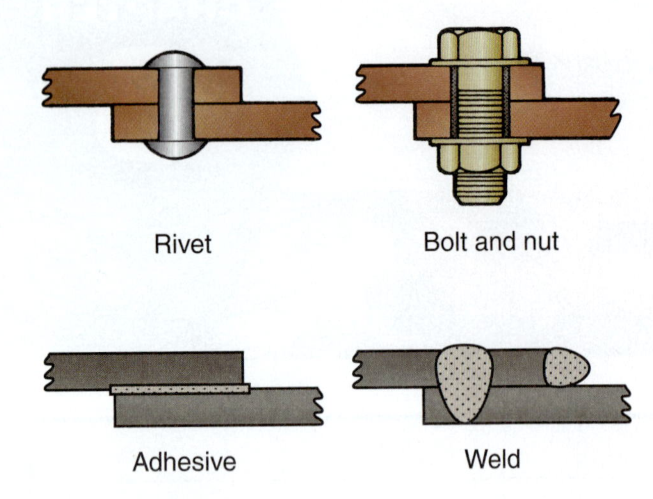

Figure 8–1 Mechanical joining methods use threaded or nonthreaded fasteners, which are heavier and less dependable than an adhesive-bonded or welded joint.

Figure 8–2 Welding is a fast, strong method of permanently joining metal parts.

Shown in Table 8–1, there are distinct welding methods within each respective category. Many of these methods can be used in the auto body shop. Gas metal arc welding is the preferred method.

Weld Terminology

The **weld root** is the part of the joint where the wire electrode is directed. The **weld face** is the exposed surface of the weld on the side that has been welded.

Visible **weld penetration** is indicated by the height of the exposed surface of the weld on the back side. Full weld penetration is needed to assure maximum weld strength.

A **burn mark** on the back of a weld is an indication of good weld penetration. **Burn-through** results from penetrating too much into the lower base metal, which burns a hole through the back side of the metal.

Fillet weld parts include the following: The **weld legs** are the width and height of the weld bead; the **weld throat** refers to the depth of the triangular cross section of the weld.

Joint fit-up refers to holding work pieces tightly together, in alignment, to prepare for welding. It is critical to the replacement of body parts.

Welding Characteristics

Joint welding is indispensable in the restoration of collision-damaged vehicles. The characteristics of welding can be summarized as follows:

- Because the shape of welding joints is limitless, it is the perfect method for joining a vehicle structure while still maintaining body integrity.
- Weight can be reduced (no fasteners are necessary).
- Air and water tightness are excellent.
- Production efficiency is very high.
- The strength of a welded joint is greatly influenced by the level of skill of the operator.
- Surrounding panels will warp if too much heat is used.

Welding in the Auto Body Shop

Auto body shop welding is quite different and often more demanding than that of many other welding professions. Welding the ultrathin, complexly shaped unibody panels of a modern motor vehicle requires great skill and knowledge. Professional collision repair technicians must be master welders. The quality, safety, and integrity of vehicles depends on their skills.

New welding techniques and equipment have entered the auto body repair picture, replacing the once-popular arc and oxyacetylene processes (Figure 8–3). New steel alloys used in today's cars cannot be welded properly by these two processes. At present, gas metal arc welding (GMAW)—better known as **metal inert gas (MIG) welding**—offers more advantages than other methods for welding high-strength steels (HSS) and high-strength, low-alloy (HSLA) steel component parts used in modern cars. Most of the applications of HSS and HSLA steels are confined to body structures, reinforcement gussets, brackets, and supports, rather than large panels or outer skin panels.

Common Auto Body Welding Techniques

Several types of welding, from all positions (horizontal, vertical, flat, and overhead), are recommended by auto makers during collision repair of their high-tech vehicles. Auto body welders often have to do the following three welding operations:

1. MIG, also called *wire feed arc*, is the most common type of welding for steel unibody panels, medium

TABLE 8–1 WELDING METHODS

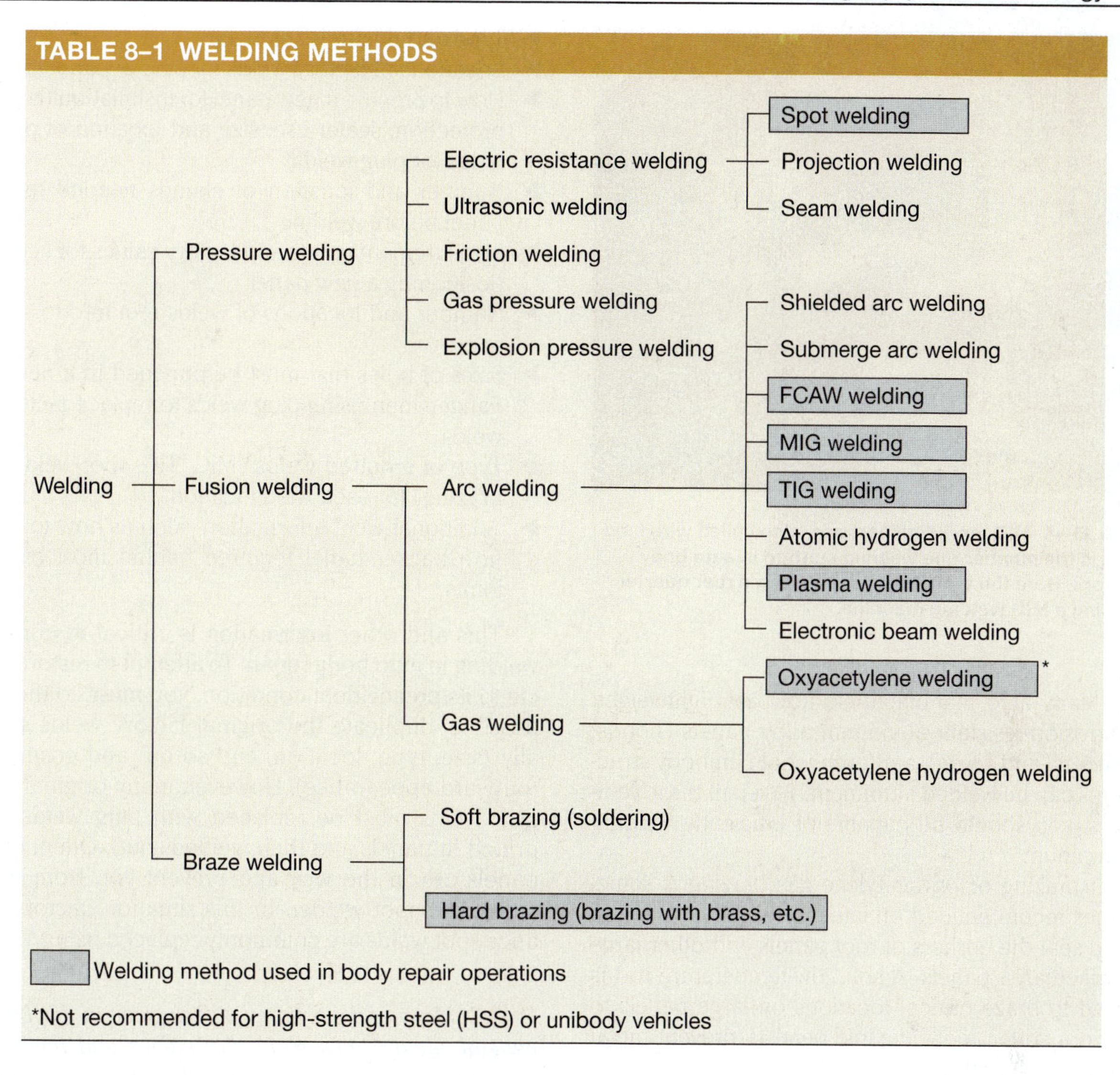

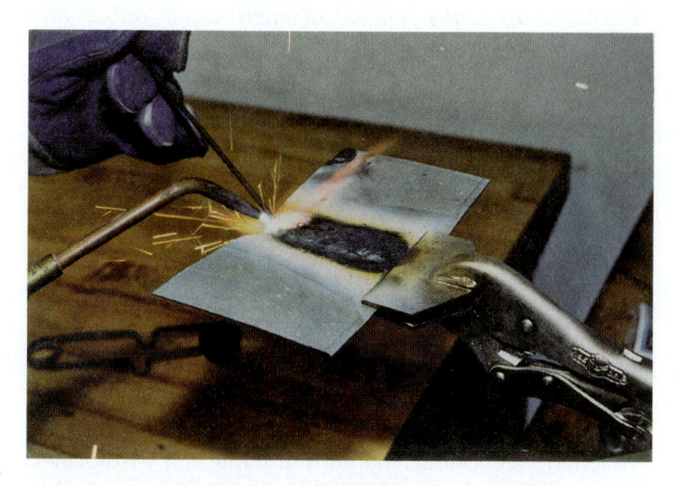

Figure 8–3 Oxyacetylene welding is not used with today's thin, easily warpable, HSS unibodies; note the warpage of two sheet metal pieces from high heat transfer. Brazed or soft torch weld joints are still sometimes recommended to seal and bond certain low structural joints between panels—for example, between the roof and the rear quarter panel—to prevent water leakage.

thickness hydroformed frames, and heavy-gauge full frames. Because a vast majority of new vehicles use HSS construction, MIG welding is by far the most common method used in collision repair facilities (Figure 8–4).

2. **Tungsten inert gas (TIG)**, which uses a hand-held rod and gas-shielded arc, is often recommended when welding aluminum alloy body panels. However, MIG welding with a larger 0.047-inch (1.2 mm) aluminum wire can also be used.

Many experts agree that TIG welding aluminum is superior to MIG for several practical reasons. Large diameter, soft aluminum wire does not feed through a MIG welder as well as steel wire. The soft wire tends to jam in the MIG feed mechanism more often. It is easier to control the aluminum weld puddle and prevent burn-through with a hand-held rod. You can move the TIG rod in and out of the weld puddle to control weld heat better than an automatic feed MIG wire.

Figure 8–4 **MIG, or metal inert gas, also called wire feed welding, is the number one welding method in auto body repair work. Here the technician is replacing a rear quarter panel using a MIG welding machine.**

Many auto manufacturers now use lightweight, corrosion-resistant aluminum alloy panels (hoods, fenders, and so on), and even whole unibody structures can be welded aluminum. To repair these vehicles, you should be capable of efficiently welding aluminum.

3. Soft brazing or oxyacetylene gas brazing is sometimes recommended on late model vehicles to join and seal the corners of roof panels and other large surface area panels. A soft, low-temperature rod is used to braze critical locations on large panels to form a softer, more flexible bead to prevent metal cracks and water leakage.

   Remember that oxyacetylene welding is not recommended on today's thin, high-strength steels; only soft brazing in conjunction with MIG welding is allowed.

Factory Weld Specifications

Always follow auto manufacturer instructions and specifications when welding panels during collision repair. The company that designed and built the vehicle knows the correct weld procedures for properly repairing its own vehicles after an accident. These factory specifications and detailed instructions are the "repair standard."

The auto maker establishes its repair information by considering the exact metal gauge (thickness), metal type (cold-rolled steel, HSS, aluminum, and so on), as well as the original "weld specs" that were used during vehicle manufacturing. As a result, welding methods often vary with the year, make, and model of car, truck, van, or SUV being repaired.

When welding a new panel on a vehicle, you should know the following information before starting to work:

- Type and thickness of metal
- Locations of factory welds for proper panel removal
- How to prepare a new panel for installation (corrosion protection, sealer use, size and location of punched holes for plug welds)
- Number and location of clamps needed to secure panel before welding
- Measurement points and factory values for accurately positioning a new panel
- Number and locations of welds required to secure a panel
- Sizes of holes that must be punched in a new panel flange when using plug welds to replace factory spot welds
- Type of required welds (MIG, TIG, spot weld, or soft brazing) for each weld location
- Additional weld information, such as how to cut and fit a backer panel required behind most butt weld joints

This and other information is critical to competent welding in auto body repair. To attempt to restore a vehicle to its preaccident condition, you must, to the extent possible, duplicate the original factory welds structurally (size, type, location, and so on) and cosmetically (outward appearance). However, many original factory spot welds must be replaced with plug welds (holes drilled in panels and then welded shut). Often, existing panels are in the way and prevent you from using a resistance spot welder. In this situation, factory resistance spot welds are commonly replaced using MIG plug welds.

NOTE *Many of the welding values given in this book are general. Always refer to factory values and methods for repairing a specific year, make, and model vehicle. The auto manufacturer or aftermarket reference materials (Mitchell, Motor, and so forth) provide detailed specifications and instructions for the exact vehicle being repaired and should be consulted.*

Chapter 19 explains how to replace a welded body panel in more detail. For information on welding plastics materials, refer to Chapter 13.

Typical Auto Body MIG Wire Sizes

Today's steel unibody panels are commonly welded using 0.023-inch (0.58 mm) MIG wire. On superthin, ultralight steel unibodies, the manufacturer may recommend an even smaller MIG wire size. Be sure to check published specifications if in doubt.

When sectioning medium thickness hydroformed frames or frame horns on full perimeter frames, higher weld settings and a slightly larger 0.03-inch (0.76 mm) MIG wire is sometimes better.

Very thick body-over-frame steel members may require larger 0.035-inch (0.89 mm) MIG welding wire. When welding the frames on many full-size cars and trucks, you will have to set the MIG welder up for thicker gauge steel and install the larger wire and gun tip in the weld machine.

When MIG welding aluminum alloy body parts, most service manuals recommend 0.030- to 0.035-inch (0.76–0.89 mm) wire sizes. The actual thickness of an aluminum panel or part will determine which wire size would result in the best weld, so check factory specifications if needed.

Typical Auto Body Shielding Gases

When MIG welding steel, manufacturers often recommend using C-25 inert shielding gas, which is 25 percent carbon dioxide and 75 percent argon. But, again, refer to published details from the auto maker for the exact aluminum thickness and metallurgical content of the alloy.

When welding an aluminum body structure, either a MIG welder with aluminum wire or a TIG welder with an aluminum rod may be recommended. When welding aluminum alloy, the most common MIG and TIG shielding gas is 100 percent argon.

The advantages of MIG welding (Figure 8–4) over conventional stick electrode arc welding (Figure 8–5) are so numerous that manufacturers now recommend it almost exclusively. MIG welding is recommended by all OEMs, not only for HSS and unibody repair, but for *all* structural collision repair. This recommendation extends also to independent collision repair shops.

Ultralight steel auto bodies (ULSAB) are now being used by many manufacturers. ULSAB panels are even thinner than HSS. You must adjust your welder for thinner gauge metal when welding this ultrathin steel. Refer to the auto manufacturer's specifications for wire size and type, shielding gas, number and size of welds, and so on for this ultralightweight metal.

Heat Effect Zone

The **heat effect zone** is the area around the weld that becomes adversely hot. The heat effect zone should always be kept to a minimum to prevent panel warpage and part damage.

Heat sink compound (clay-like material) can be placed on the body panel around the weld area to reduce the heat effect zone. Heat will easily flow into heat sink compound to keep the surrounding area cool. Water-soaked rags can also be used to cover parts or assemblies to keep them cool and safe from excessive welding heat.

Welding blankets are thick covers made of fire-resistant cloth for protecting vehicle surfaces from heat, sparks, and weld spatter. Weld blankets should be placed over painted surfaces, glass, upholstery, exposed plastic parts, and any surface that could be harmed by the welding. Glass will pit and damage badly if weld spatter containing molten metal particles lands on the glass.

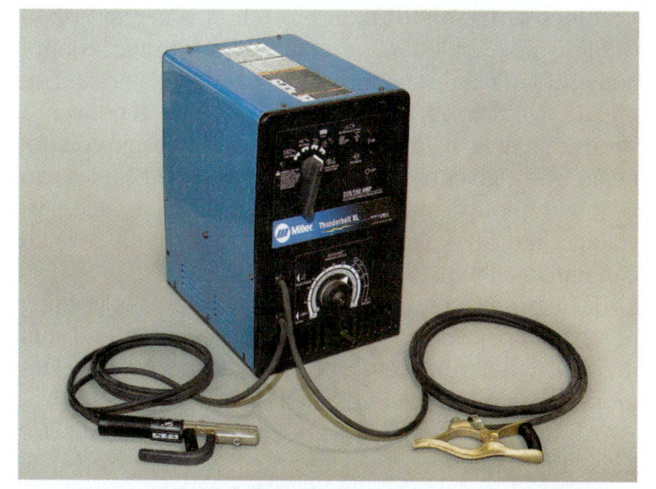

A A stick welding machine uses a large-diameter welding rod and high current flow to produce an electric arc that melts the rod down into the joint.

Courtesy of Miller Electric Manufacturing Co.

B Stick welding is needed on very thick steel parts. It will badly warp or burn through the thin sheet metal used in body panels.

Figure 8–5 Conventional stick electrode, or shielded arc, welding spreads too much heat into the part for thin metal body panels.

DANGER

An excessive heat effect zone can ignite undercoating or sound-deadening materials, causing a toxic, smoky fire. Make sure you take steps to reduce heating of the area around the weld joint. This is extremely important when making continuous welds that transfer a great deal of heat into the weld area.

Electronic shielding or protection is needed when welding near on-board computers (electronic control units [ECUs]) and sensor wiring. The high electrical current flowing through the welding cables generates a strong magnetic field that can damage electronic components. The high current flow through the weld cables can generate enough magnetic flux to introduce excessive voltage into low-voltage electronic components, which can damage them.

If electronic units are too close to the arc, the welding heat can also overheat electronic components and ruin them. Do not run weld cables or place your welder ground clamp near electronic circuits.

Always follow manufacturer recommendations for protecting electronic devices when welding. You may have to completely remove the ECU or other computer system, or you may have to simply unbolt it and wrap it in a welding blanket.

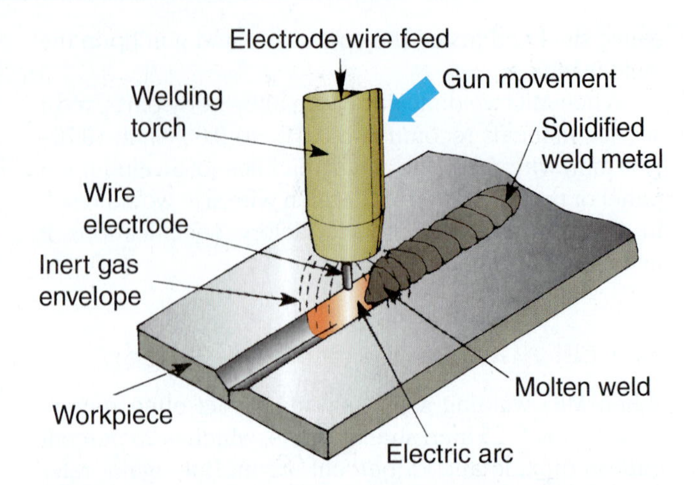

Figure 8–6 Study the basic MIG welding process. Wire is automatically fed down through the nozzle and into the welding arc. The arc melts the wire and fills the gap between parts with molten metal. Inert gas also flows out of the nozzle to protect the weld from contamination.

Auto Body MIG Welding

Here are some potential advantages of MIG welding from an auto body technician's viewpoint:

- MIG welding is easier to learn than arc or gas welding. The typical welder can learn to use MIG welding equipment with proper training. Moreover, experience shows that even an average MIG welder can produce higher quality welds faster and more consistently than a highly skilled welder using older stick electrode welds.
- MIG welding produces 100 percent fusion in the parent metals. This means MIG welds can be dressed or ground down flush with the surface (for cosmetic reasons) without loss of strength.
- Low current can be used for thin metals. This prevents heat damage to adjacent areas, which can cause strength loss and warping.
- The arc is smooth and the weld puddle small, so it is easily controlled (Figure 8–6). This ensures maximum metal deposit with minimum splatter.
- MIG welding is more tolerant of gaps and misfits. Several gaps can be spot welded immediately (no slag to remove) by making several spots on top of each other. Therefore, the area can be easily refinished.
- Almost all steels can be welded with one common type of weld wire. What is in the machine is generally right for any job.
- Metals of different thicknesses can be welded with the same diameter wire. Again, what is in the machine is right for almost any job.
- The MIG welder can control the temperature of the weld and the time the weld takes place.
- With MIG welding, the small area to be welded is heated for a short period of time, thereby reducing metal fatigue, warpage, and distortion of the panel. Vertical and/or overhead welding is possible because the metal is molten for a very short time.
- Portable resistance spot welding is also recommended today for some repairs (Figure 8–7). This type of equipment is used to form spot weld attachments like the production welds. To use this kind of spot welding equipment, you must install the proper extensions and electrodes on the welder to provide access to the area being welded. The clamping force on a squeeze-type resistance spot welder must be properly adjusted. On some equipment, the amperage, current flow, and timing are all made with one adjustment. After the adjustments are made, position the spot welder over the panels being joined, making sure the electrodes are directly opposite to each other. Squeeze the trigger and the spot weld takes place.

The resistance spot welder, which probably requires the least skill to operate, provides very fast, high-quality welds while maintaining the best control of temperature buildup in adjacent panels and structure.

Be sure to consult the car manufacturer's recommendations in the vehicle's service manual before welding.

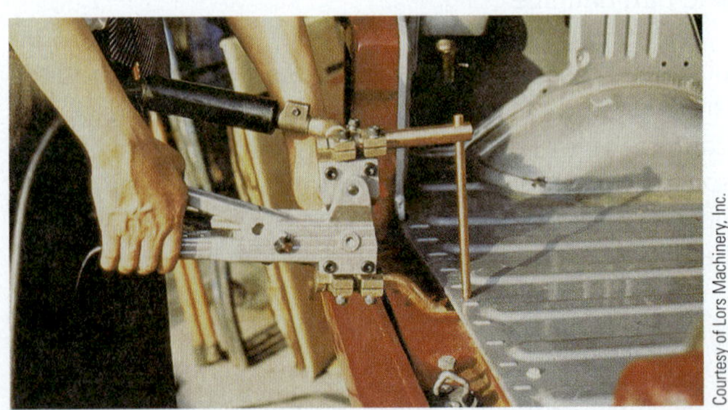

Figure 8–7 The technician is using a portable squeeze-type resistance spot welding gun to install a new floor panel on this vehicle.

> **WARNING**
>
> Always follow service manual recommendations when welding. This will ensure structural integrity.
>
> It is essential to use appropriate welding methods that do not reduce the original strength and durability of the body when making repairs. This is accomplished if the following basic points are observed:
>
> - Try to use either spot welding or MIG/MAG (metal inert gas/metal active gas) welding.
> - Do not braze any body components other than those brazed at the factory.
> - Do not use an oxyacetylene torch for welding late model auto bodies.

When replacing body panels, all the new welds should be similar in size to the original factory welds. The number of replacement welds should be the same as the original number of welds in production.

The manufacturer decides what is the most appropriate welding method (Figure 8–8) by first determining the intended use, the physical characteristics, and the location of the part.

8.2 MIG WELDING

MIG welding became popular in body shops when auto manufacturers began using thin-gauge HSLA steels. Car makers insisted that the only correct way to weld HSLA and other thin-gauge steel was with MIG (or the similar GMAW system). Once the MIG welder was in place, it was easy to see that it provided clean, fast welds for all applications.

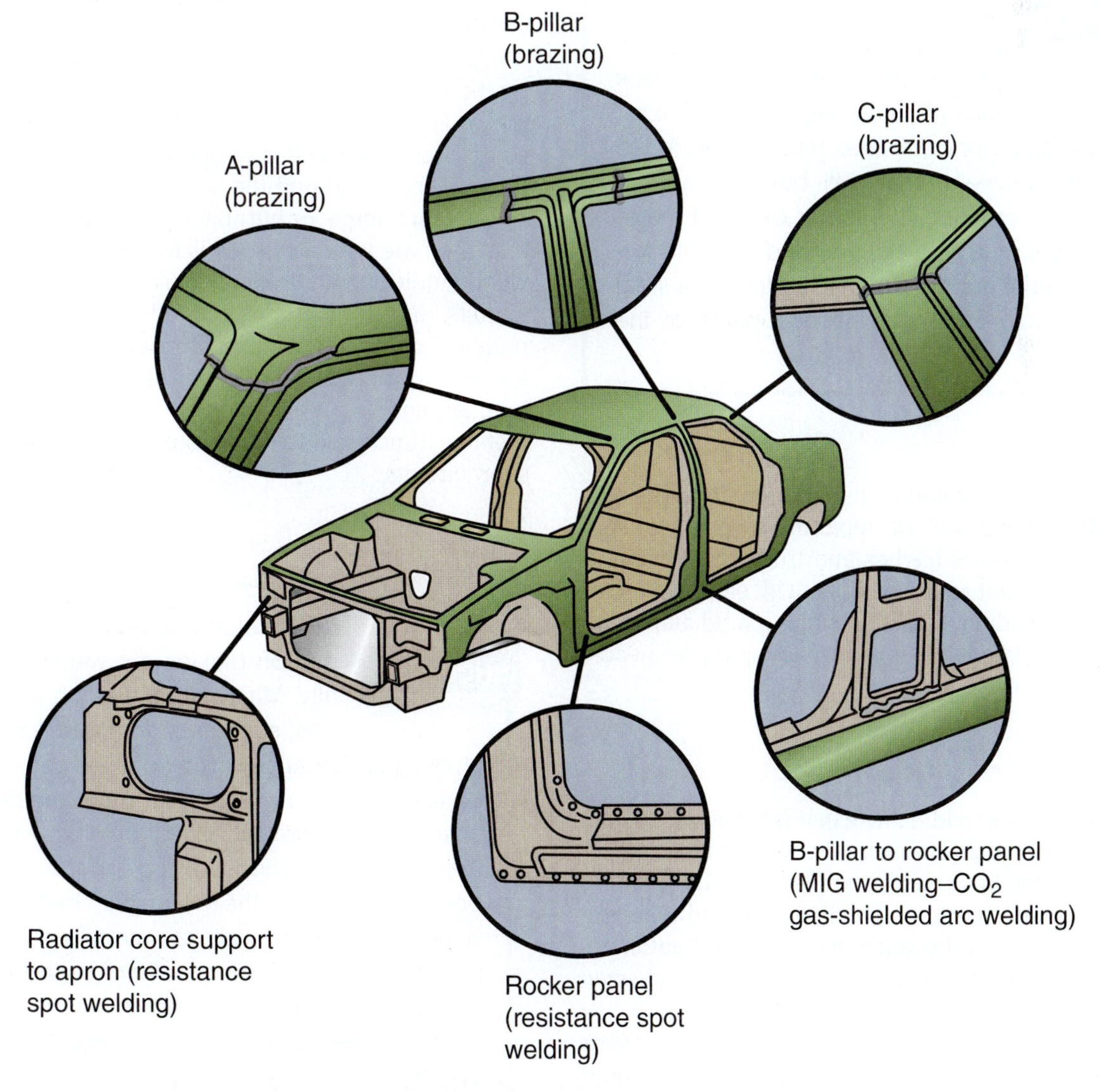

Figure 8–8 Compare the welding methods used in vehicle production.

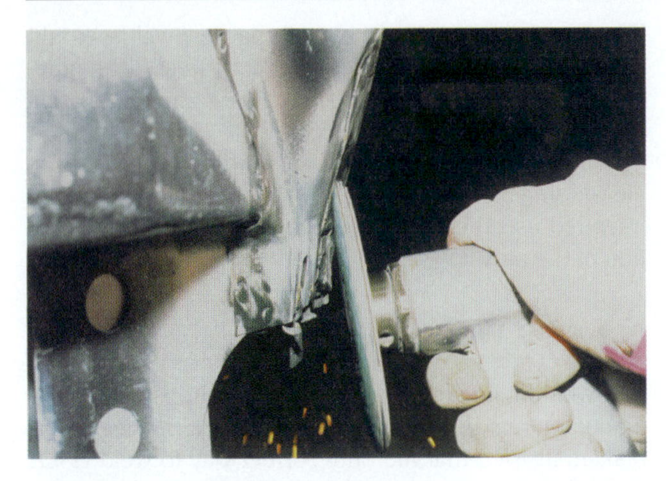

Figure 8-9 Make sure the surfaces to be welded are completely free of rust and scale contaminants loosened by grinding, sanding, or media blasting.

Regardless of the type of welding, you must properly clean the surface before starting the weld. Remove all surface materials back to the bare metal (Figure 8-9). When dirt, rust, sealers, paint, or other materials are left in the area of the weld, they will burn during the application of heat. The ash or oxidized material can become a part of the weld. Dirt and foreign material can weaken the weld. If the weld fails, it could endanger the occupants of the vehicle.

MIG welding is not limited to body repairs alone. It is also ideal for exhaust work, repairing mechanical supports, installing trailer hitches and truck bumpers, and any other welds that would be done with either an arc or a gas welder. In addition, it is possible to weld aluminum castings, like cracked transmission cases, cylinder heads, and intake manifolds.

MIG Principles and Characteristics

MIG welding uses a welding wire that is fed automatically at a constant speed as an electrode. A short arc is generated between the base metal and the wire. The resulting heat from the arc melts the welding wire and joins the base metals together. Because the wire is fed automatically at a constant rate, this method is also called *semiautomatic arc welding*.

During the welding process (Figure 8–10), either *inert gas* or active gas shields the weld from the atmosphere and prevents oxidation of the base metal. The type of inert or active gas used depends on the base material to be welded. For most steel welds, carbon dioxide (CO_2) is used as the shield gas (Figure 8–11).

With aluminum, either pure argon gas or a mixture of argon and helium is used, depending on the alloy and the thickness of the material. It is even possible to weld stainless steel by using argon gas with a little oxygen (between 4 and 5 percent) added.

MIG *flux core wire* has its own flux contained in a tubular electrode and does not require a shielding gas. As with stick welding, the flux forms slag that must be chipped off. Flux core electrode wire is not convenient for most collision repair work. It takes more time to clean the weld.

MIG welding uses the *short circuit arc method*—a unique method of depositing molten drops of metal onto the base metal. Welding of thin sheet metal for automobiles can cause welding strain, blow holes, and warped panels. To prevent these problems, it is necessary to limit the amount of heat near the weld. The short circuit arc method uses very thin welding rods, a low current, and low voltage. By using this technique, the amount of heat introduced into the panels is kept to a minimum and penetration of the base metal is quite shallow.

As shown in Figure 8–12, the end of the wire is melted by the heat of the arc and forms into a drop. The drop then comes in contact with the base metal and creates a short circuit. When this happens, a large current flows through the metal and the shorted portion is torn away by the pinch force, or burnback, which reestablishes the arc. The bare wire electrode is fed continuously into the weld puddle at a controlled, constant rate where it short-circuits, and the arc goes out. While the arc is out, the puddle flattens and cools, but the wire continues to feed, shorting to the workpiece again. This heating and cooling happens on an average of 100 times a second. The metal is transferred to the workpiece with each of these short circuits.

MIG welding is sometimes called carbon dioxide arc welding. Actually, MIG welding uses a fully inert gas, such as argon or helium, as a shield gas. Because carbon dioxide gas is not a completely inert gas, it is more accurately called MAG (metal active gas) welding. Although most auto body shop welding is done with carbon dioxide gas as the shield gas, the term MIG is used to describe all gas metal arc welding processes. In fact, many welders on the market can use carbon dioxide (a semiactive gas) or argon (an inert gas) by simply changing the gas cylinder.

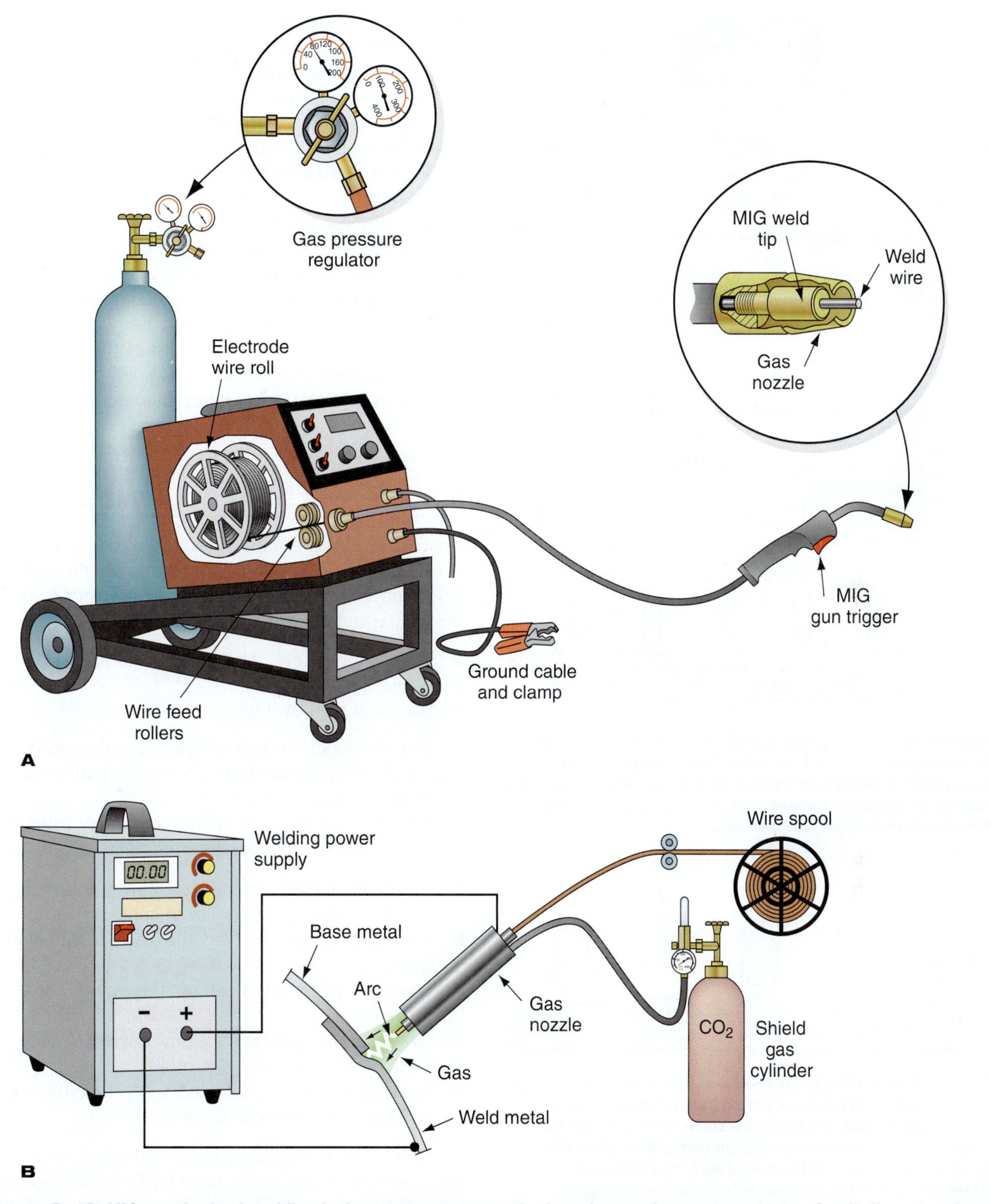

Figure 8–10 **MIG, or wire feed, welding is the most common method used on modern motor vehicles. Small-diameter wire will make very precise welds with localized heat to avoid panel warpage. (A) Study the parts of a basic MIG welder. Note that the welding wire and gas move through the welding gun whenever the trigger is depressed. (B) Study the principles of MIG welding. If shielded by inert gas flow, the welding arc is maintained by wire feeding into the arc at the right speed.**

If current is flowing through a cylindrical-shaped fluid (in this case molten metal) or current is flowing through an arc, the current is pulled toward the weld. This works as a constricting force in the direction of the center of the cylinder. This action is known as the pinch effect, and the size of the force is called the *pinch force*.

In summary, the MIG welding process works as follows:

- At the weld point, the wire undergoes a split-second sequence of short circuiting, burnback, and arcing (Figure 8–13).

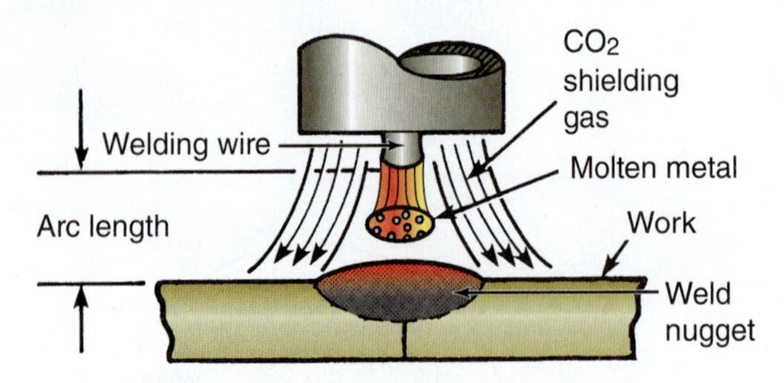

Figure 8-11 Carbon dioxide (CO_2) from the welding machine tank protects molten metal from contamination by the atmosphere.

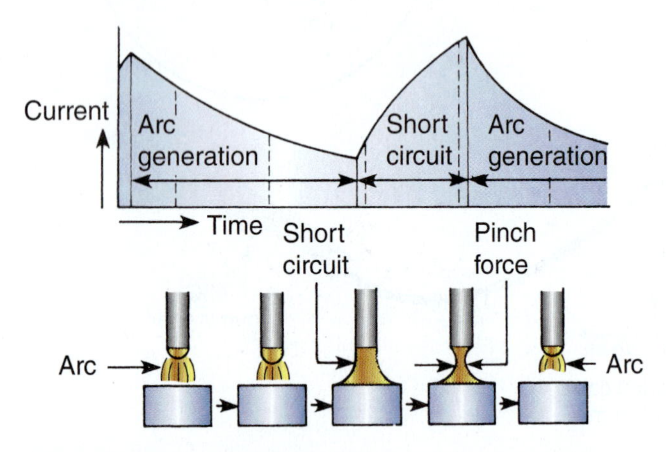

Figure 8-12 The graph and illustrations show how the short circuit arc method operates. Note how the current draw varies with the formation of molten metal flow. Because each cycle occurs in milliseconds, you can hear this as a "hissing sound" when MIG welding.

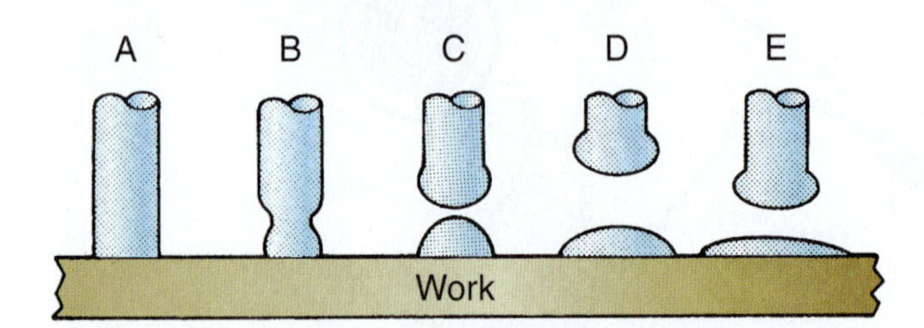

Figure 8-13 Here is the typical action of welding wire as it burns back from the work during MIG welding. (A) Wire touches workpiece. (B) Weld wire starts to liquefy. (C) Tiny nugget of molten metal deposits on workpiece. (D) Molten metal in workpiece and from weld wire flow together. (E) Weld wire and metal in workpiece form weld nugget.

- Each sequence produces a short arc transfer of a minute drop of electrode metal from the tip of the wire to the weld puddle.
- A gas curtain or shield surrounds the wire electrode. This gas shield prevents contamination from the atmosphere and helps stabilize the arc.
- The continuously fed electrode wire contacts the work and sets up a short circuit, and resistance heats the wire and the weld site.
- As the heating continues, the wire begins to melt and thin out or neck down.
- Increasing resistance in the neck accelerates the heating in this area.
- The molten neck burns through, depositing a puddle on the workpiece and starting the arc.
- The arc tends to flatten the puddle and burn back the electrode.
- With the arc gap at its widest, it cools, allowing the wire feed to move the electrode closer to the work.
- The short end starts to heat up again, enough to further flatten the puddle, but not enough to keep the electrode from recontacting the workpiece. This extinguishes the arc, reestablishes the short circuit, and restarts the process.
- This complete cycle occurs automatically at a frequency ranging from 50 to 200 cycles a second.

8.3 MIG WELDING EQUIPMENT

Most MIG welding equipment for collision repair work is considered semiautomatic. This means that the machine's operation is automatic, but the gun is hand-controlled. Before starting to weld, the operator sets

- Voltage for the arc
- Wire speed
- Shielding gas flow rate

Then the operator has complete freedom to concentrate entirely on the weld site, the molten puddle, and the welding technique that is used.

Regardless of the type of MIG equipment used, it will comprise the following basic components (Figure 8–14):

- Supply of shielding gas with a flow regulator to protect the molten weld pool from contamination
- Wire/feed control to feed the wire at the required speed
- Spool of electrode wire of a specified type and diameter
- Power supply to control welding current
- Work cable and clamp assembly
- Welding gun and cable assembly that the welder holds to direct the wire to the weld area

Fine diameter welding wires are used in collision repair and typically range from 0.023 inch (0.584 mm) through 0.030 inch (0.792 mm). This is roughly equivalent in diameter to ultrafine leads in today's mechanical pencils. A wire that is becoming more commonly used today is 0.023 inch (0.584 mm). Once a specialty wire, it is now stocked by most wire manufacturers. These small diameter wires can be used at low currents (10–20 amps) and voltages (120 volts), thus greatly reducing heat input to the base material. The welding wire must carry a minimum specification of AWS-ER70-6.

Because of the power demands in this process, it is necessary to use a constant potential, constant voltage

Figure 8-14 Study the basic parts of a MIG welding machine.

power source (Figure 8–15). The controls are a voltage adjustment and wire feed speed adjustment (Figure 8–16). Some optional controls available on this type of equipment are a spot control and pulse control.

MIG spot welding is termed consumable spot welding because the welding wire is consumed in the weld puddle. Consumable spot welds can be made in a variety of methods and in all positions using various nozzles equipped with this option.

When you are spot welding different thicknesses of materials, the lighter gauge material should always be spotted to the heavy material.

Spot welding usually requires greater heat to the weld than continuous or pulse welding. It is best to use sample materials when setting the controls for spot welding. To check a spot weld, pull the two pieces apart. A good weld will tear a small hole out of the bottom piece. If the weld pulls apart easily, increase the weld time or heat. After each spot is complete, the trigger must be released and then pulled for the next spot.

MIG spot welding has the advantage of producing an easily grindable crown. The procedure does not leave any depression requiring a fill.

The pulse control allows continuous seam welding on the material with less chance of burn-through or distortion. This is accomplished by starting and stopping the wire for preset times without releasing the trigger.

A

B

Figure 8–15 **(A) MIG welder has a roll of fine welding wire that is automatically fed through the welding machine and out the nozzle. (B) The purge button can be pushed one way to feed a new roll of wire through the machine. If pushed the other way, you can purge air out the welding nozzle after changing tanks.**

The weld "on" time and weld "off" time can be set for the operator's preference and the metal thickness.

The burnback control on most MIG welders gives an adjustable burnback of the electrode to prevent it from sticking in the puddle at the end of a weld.

In MIG welding, the polarity of the power source is important in determining the penetration to the workpiece. DC power sources used for MIG welding typically use DC reverse polarity. **DC reverse polarity** means the wire (electrode) is positive and the workpiece is negative. Weld penetration is greatest using this connection.

Weld penetration is also greatest using CO_2 gas. However, CO_2 gives a harsher, more unstable arc, which leads to increased spatter. So when welding on thin materials, it is preferable to use argon/CO_2 (Figure 8–17).

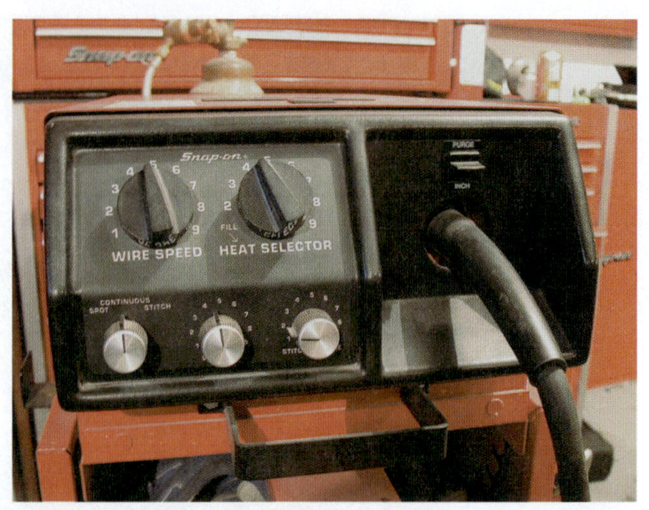

A **The most important controls on a MIG welder are the heat and wire feed controls. They must be adjusted so a hissing sound is produced for good weld penetration.**

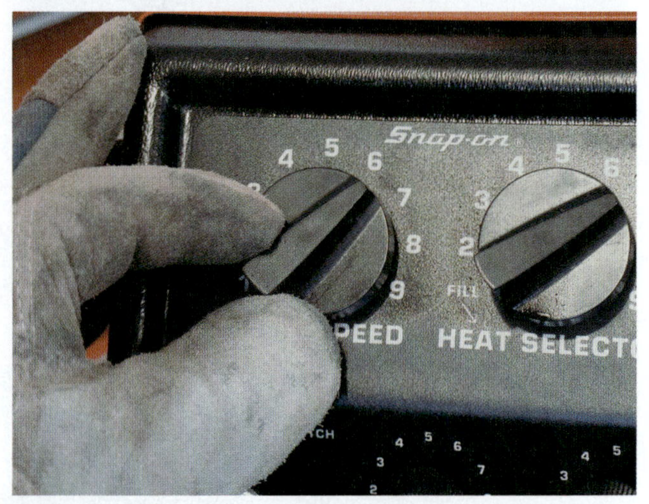

B **Wire speed must be matched with the heat or current control. More current and heat are needed when you increase wire feed speed for thicker metals.**

Courtesy of TECNA/Cebotech, Inc.

C **This inverter spot welder is programmable. You can set the machine for different types of metal (high-strength steel, galvanized, boron steel, etc.), set metal thickness, adjust for one, two, or more layers of metal, and preset other values to help you make good welds.**

Figure 8–16 **Control panels on inverter resistance spot welders can vary.**

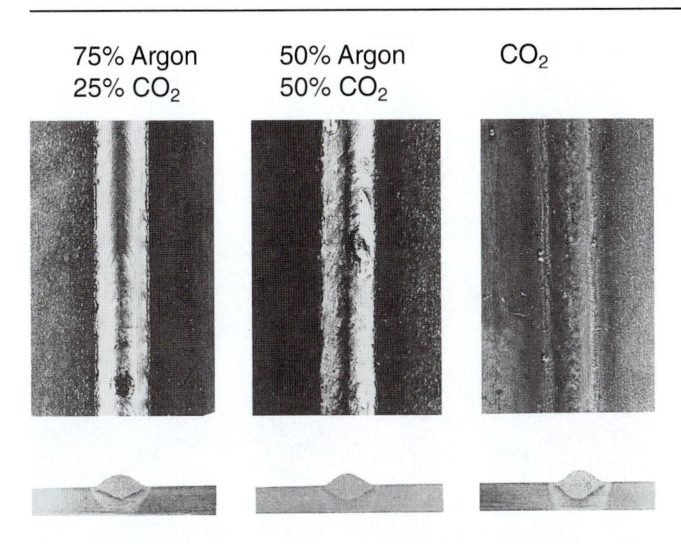

Figure 8–17 Note how the type of inert gas affects weld penetration.

Voltage adjustment and wire feed speed must be set according to the diameter of the wire being used and metal thickness. It should be noted that when setting these parameters, the manufacturer's recommendations should be followed to reach approximate settings. When rough parameters are selected, change only one variable at a time until the machine is fine-tuned for an optimum welding condition. MIG welders can be tuned in using both visual and audio signals.

Welding Lens

A **welding filter lens**, sometimes called a *filter plate*, is a shaded glass welding helmet insert for protecting your eyes from ultraviolet burns. The lenses are graded with numbers, from 4 to 12; the higher the number, the darker the filter. The American Welding Society (AWS) recommends Grade 9 or 10 for MIG welding steel (Figure 8–18).

Note that there are *self-darkening filter lenses* available that instantly turn dark when the arc is struck. There is no need to move the face shield up and down.

A A modern welding helmet has a self-darkening lens. You can see work easily through an almost clear lens at first. Then, when you strike an arc, the lens almost instantly darkens so you can see the weld arc.

B Welding goggles with a see-through shaded lens can be worn when gas heating and cutting, because the heat and light intensity is lower than for an electric arc.

Figure 8–18 Proper eye protection must be worn when welding. If you look at a welding arc without a proper shaded lens, severe eye burns and permanent eye damage can result.

8.4 MIG OPERATION METHODS

To match MIG welding power to the available input voltage, follow the procedure prescribed on the machine or in the manufacturer's manual (Figure 8–19).

Handle the cylinder of shielding gas with care. It might be pressurized to more than 2,000 psi (13,800 kPa). Chain or strap the cylinder to a support sturdy enough to hold it securely to the MIG machine (Figure 8–20). Install the regulator, making sure to observe the recommended safety precautions (Figure 8–21).

When the clamp is attached to clean metal on the vehicle (Figure 8–22) near the weld site, it completes the welding circuit from the machine to the work and back to the machine. This clamp is not referred to as a ground cable or ground clamp. The ground connection is for safety purposes and is usually made from the machine's case to the building ground through a third wire in the electric input cable.

Ground the welding machine as close to the weld area as possible. This will make for a better electrical connection and helps avoid unwanted current flow through other body parts. Keep the welding machine as far away from vehicle computers as possible to avoid electronic circuit damage. When welding near a vehicle computer, remove the computer from the vehicle to avoid

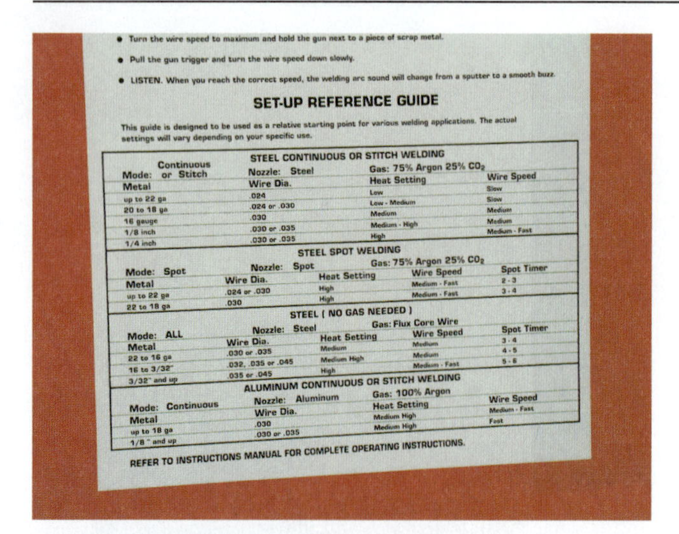

- Turn the wire speed to maximum and hold the gun next to a piece of scrap metal.
- Pull the gun trigger and turn the wire speed down slowly.
- LISTEN. When you reach the correct speed, the welding arc sound will change from a sputter to a smooth buzz.

SET-UP REFERENCE GUIDE

This guide is designed to be used as a relative starting point for various welding applications. The actual settings will vary depending on your specific use.

STEEL CONTINUOUS OR STITCH WELDING

Mode: Continuous or Stitch	Nozzle: Steel	Gas: 75% Argon 25% CO_2	
Metal	Wire Dia.	Heat Setting	Wire Speed
up to 22 ga	.024	Low	Slow
20 to 18 ga	.024 or .030	Low - Medium	Slow
16 gauge	.030	Medium	Medium
1/8 inch	.030 or .035	Medium - High	Medium
1/4 inch	.030 or .035	High	Medium - Fast

STEEL SPOT WELDING

Mode: Spot	Nozzle: Spot		Gas: 75% Argon 25% CO_2	
Metal	Wire Dia.	Heat Setting	Wire Speed	Spot Timer
up to 22 ga	.024 or .030	High	Medium - Fast	2 - 3
22 to 18 ga	.030	High	Medium - Fast	3 - 4

STEEL (NO GAS NEEDED)

Mode: ALL	Nozzle: Steel		Gas: Flux Core Wire	
Metal	Wire Dia.	Heat Setting	Wire Speed	Spot Timer
22 to 16 ga	.030 or .035	Medium	Medium	3 - 4
16 to 3/32"	.032, .035 or .045	Medium High	Medium	4 - 5
3/32" and up	.035 or .045	High	Medium - Fast	5 - 6

ALUMINUM CONTINUOUS OR STITCH WELDING

Mode: Continuous	Nozzle: Aluminum	Gas: 100% Argon	
Metal	Wire Dia.	Heat Setting	Wire Speed
up to 18 ga	.030	Medium High	Medium - Fast
1/8" and up	.030 or .035	Medium High	Fast

REFER TO INSTRUCTIONS MANUAL FOR COMPLETE OPERATING INSTRUCTIONS.

Figure 8-19 **Check the manufacturer's manual before hooking up equipment. This label on the inside lid of a welder summarizes wire sizes, heat settings, and wire speeds for different metal thicknesses. Note the use of 75 percent argon/25 percent CO_2 gas, found in most body shops.**

Figure 8-20 **Make sure the chain or strap is in place so the heavy welding tank cannot fall over.**

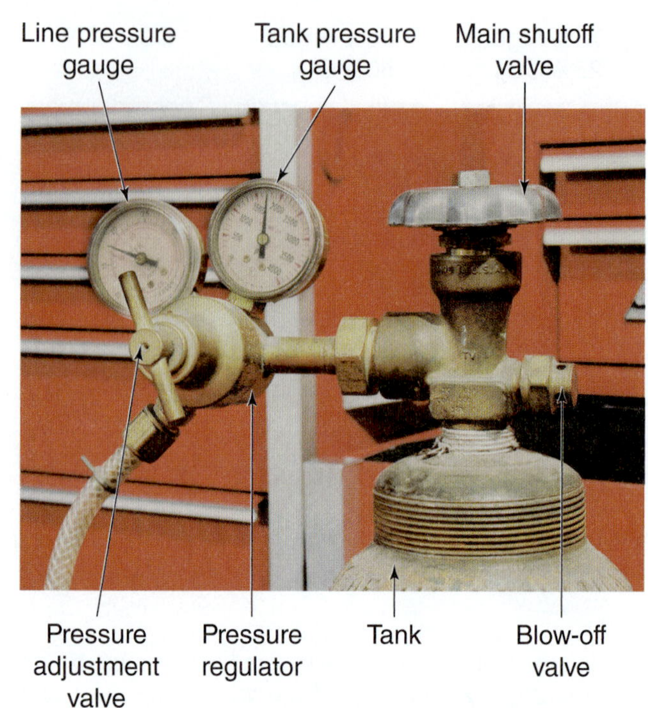

Figure 8-21 **Gauges on an MIG welding tank read pressure in the tank and pressure being fed to the nozzle. Make sure fittings are snug and that you close the main tank valve when not in use.**

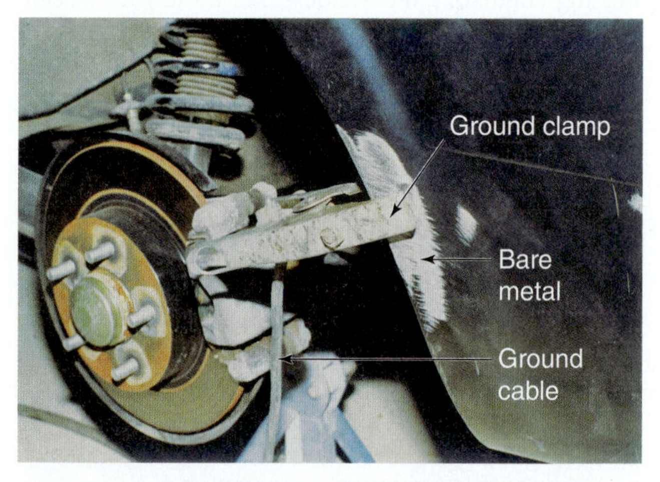

Figure 8-22 **Attach the welder's ground clamp to a clean metal surface to prevent electrical resistance from affecting weld current.**

electronic circuit damage. Disconnect the vehicle's positive battery cable when welding to avoid unwanted current flow through the vehicle's circuitry.

Consult the manufacturer's manual for the specific procedure to assemble, install, and adjust the wire feeder components. In general, the adjustment of the wire feeder can be done as follows:

- Mount the wire. Feed the wire manually for about 12 inches (305 mm), making sure that it travels freely through the gun assembly.
- A correct setting on the drive rollers will ensure just enough pressure on the wire to pull it off the wire spool and through the gun/cable assembly. The tension must be set so that the wire will slip at the rollers when the wire is stopped at the nozzle, but tight enough to withstand a 30- to 40-degree deflection. If too much pressure is applied, the wire will be deformed, creating a spiral effect through the liner and erratic feed.
- Stopping the wire at the tip with this much pressure will also cause the wire to bird-nest between the rollers and cable entrance. The tension on the wire spool spindle should also be set so that the wire can be pulled off easily, but just tight enough to stop the spool from freewheeling when the trigger is released.

The proper handling of any welding equipment is an essential ingredient in successful welding. When tuning

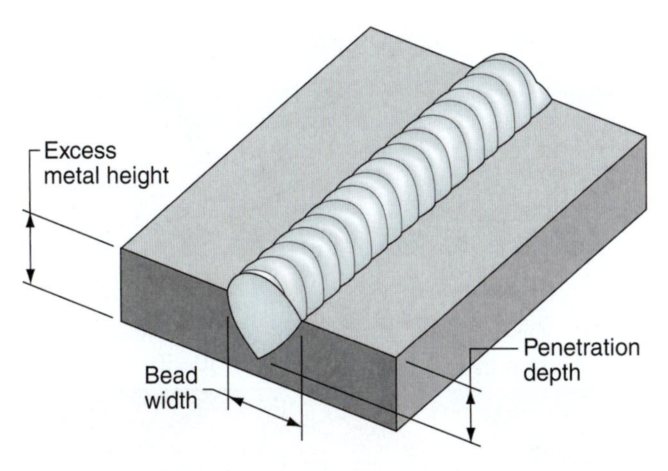

Figure 8–23 These are the three most important variables for a good weld bead: penetration depth, excess metal height, and bead width.

the MIG welder for any given welding job, you have to deal with a number of parameters, meaning values that are variable: input voltage to the welding equipment, welding current, arc voltage, tip-to-base metal distance, torch angle, welding direction, shield gas flow volume, welding speed, and wire speed. Most manufacturers of MIG welders provide tables that show the variable control parameters that apply to their machines.

MIG Welding Current

Welding current affects the base metal penetration depth (Figure 8–23), the speed at which the wire is melted, arc stability, and the amount of weld spatter. As the electrical current is increased, the penetration depth, excess metal height, and bead width also increase (Table 8–2).

MIG Arc Voltage

Good welding results depend on a proper arc length. The length of the arc is determined by the arc voltage. When the arc voltage is set properly, a continuous light hissing or cracking sound is emitted from the welding area.

When the arc voltage is high, the arc length increases, the penetration is shallow, and the bead is wide and flat.

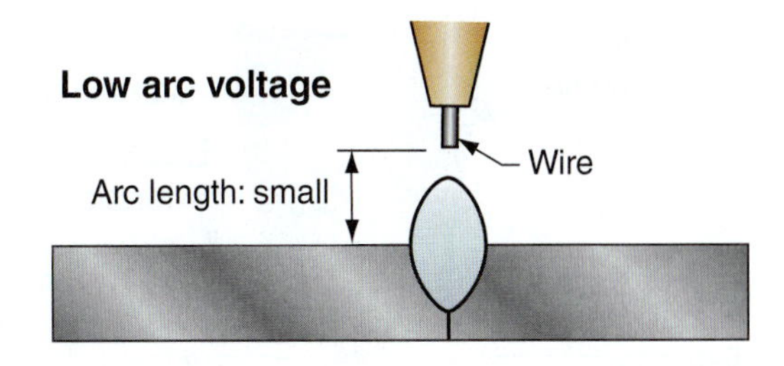

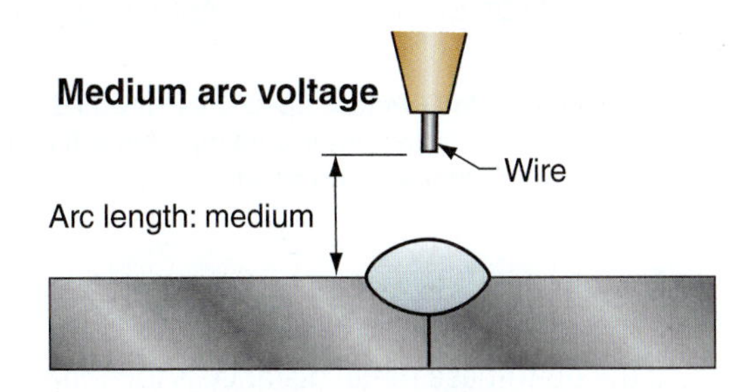

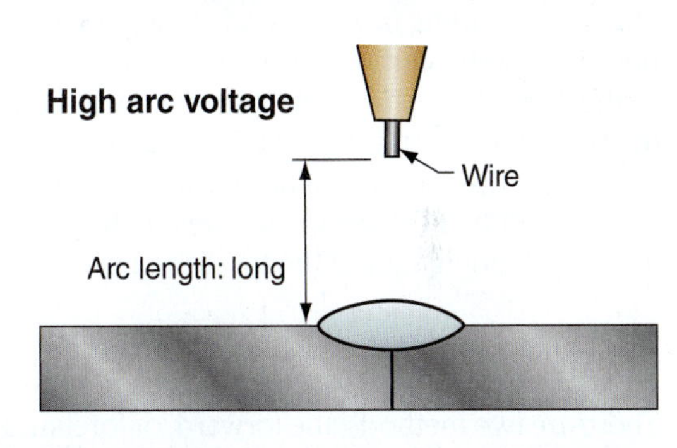

Figure 8–24 Note how arc voltage or heat adjustment affects the bead shape.

When the arc voltage is low, the arc length decreases, penetration is deep, and the bead is narrow and dome shaped (Figure 8–24).

Because the length of the arc depends on the amount of voltage, voltage that is too high will result in an overly long arc and an increase in the amount of weld spatter. A sputtering sound and no arc means that the voltage is too low.

MIG Tip-To-Base Metal Distance

The tip-to-base distance (Figure 8–25) is also an important factor in obtaining good welding results. The standard

TABLE 8–2 RELATIONSHIP BETWEEN WIRE DIAMETER, PANEL THICKNESS, AND WELDING CURRENT

	Panel Thickness						
Wire Diameter	**1/64 in.**	**1/32 in.**	**Less Than 3/64 in.**	**3/64 in.**	**1/16 in.**	**3/32 in.**	**1/8 in.**
1/64 in.	20–30 A	30–40 A	40–50 A	50–60 A	–	–	–
1/32 in.	–	–	40–50 A	50–60 A	60–90 A	100–120 A	–
More than 1/32 in.	–	–	–	–	60–90 A	100–120 A	120–150 A

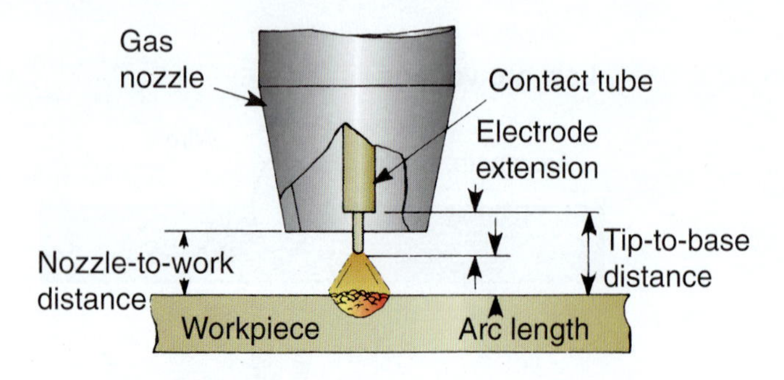

Figure 8–25 MIG welder tip-to-base metal distance should be kept constant when welding. This will help maintain a consistent arc and bead.

MIG welding tip-to-base distance is approximately ¼ to ⅝ inches (6.3–40 mm).

If the tip-to-base metal distance is too long, the length of wire protruding from the end of the gun increases and becomes preheated, which increases the melting speed of the wire. Also, the shield gas effect will be reduced if the tip-to-base metal distance is too long.

If the tip-to-base metal distance is too short, it becomes difficult to see the progress of the weld because it will be hidden behind the tip of the gun.

MIG Gun Angle and Welding

There are two methods: the forward, or forehand, method and the reverse, or backhand, method (Figure 8–26). With the forward method, the penetration depth is shallow and the bead is flat. With the reverse method, the penetration is deep and a large amount of metal is deposited (Figure 8–27). The gun angle for both methods should be between 10 and 30 degrees.

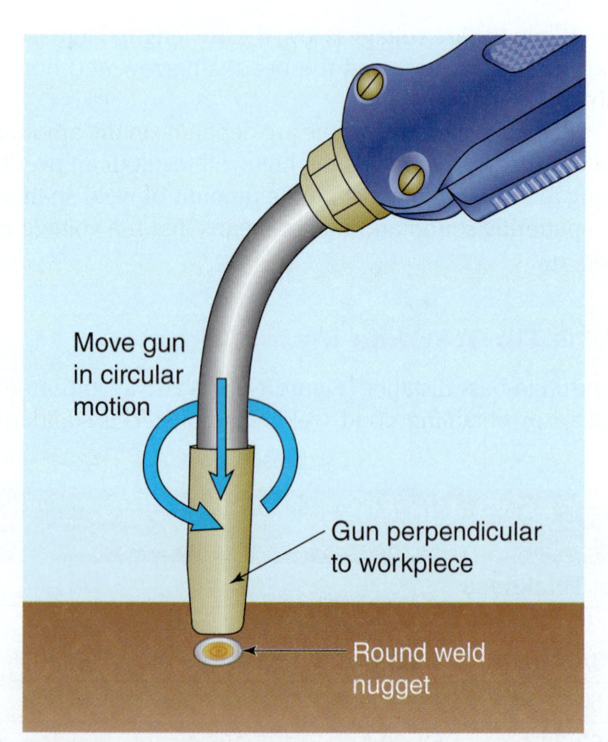

A Here the gun angle is perpendicular to, or straight down into, the workpiece. The gun should be held perpendicular when making a plug weld hole during body panel replacement.

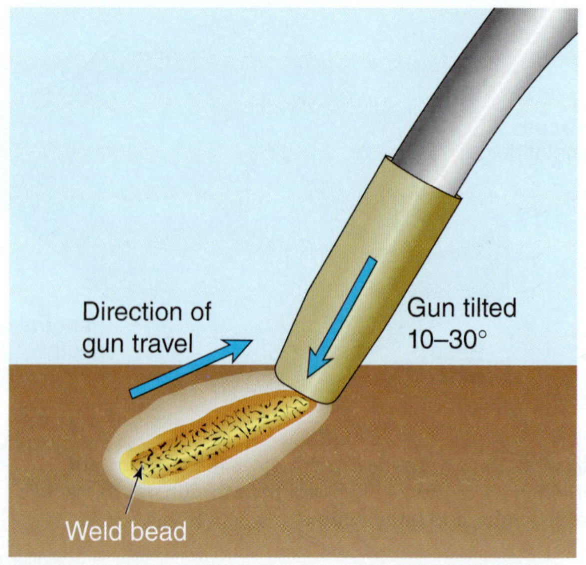

B Trailing or dragging the torch is the most common direction used to make continuous, flat welds. The torch is tilted 10–30 degrees and slid along in front of the weld bead.

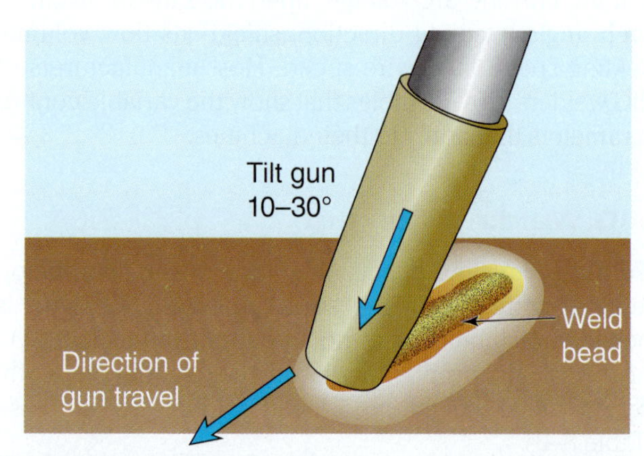

C Leading or pushing the torch is sometimes preferred by some welders. The gun is again tilted but is pushed over workpieces.

Figure 8–26 Note the various ways to angle and move the MIG welding gun.

MIG Shield Gas Flow Volume

Precise gas flow is essential to a good weld. If the volume of gas is too high, it will flow in eddies and reduce the shield effect. If there is not enough gas, the shield effect will also be reduced. Adjustment is made in accordance with the distance between the nozzle and the base metal, the welding current, welding speed, and welding environment (nearby air currents). The standard flow volume is approximately 1-3/8–1-1/2 cubic inches (0.022–0.024 liters)

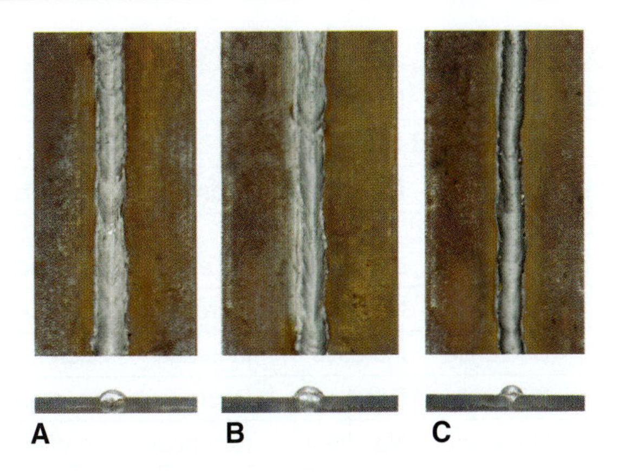

Figure 8-27 Study the penetration and weld pattern for three travel methods: (A) forehand, (B) transverse angle of 90 degrees, and (C) backhand (reverse).

per minute or 15–25 cubic feet (420–700 cubic liters) per hour.

MIG Welding Speed

If you weld at a rapid pace, the penetration depth and bead width decrease, and the bead is dome shaped. If the speed is increased even faster, **undercutting**—producing a weld surface level lower than base metal—can occur. Welding at too low a speed can cause burn-through holes. Ordinarily, welding speed is determined by base metal panel thickness and/or voltage of the welding machine (Table 8–3).

MIG Wire Speed

An even, high-pitched buzzing sound indicates the correct wire-to-heat ratio, producing a temperature in the 9,000°F (4,986°C) range. Visual signs of the correct setting occur when a steady reflected light starts to fade in intensity as the arc is shortened and wire speed is increased.

If the wire speed is too slow, a hiss and a plop sound will be heard as the wire melts away from the puddle. The visual signal will be a much brighter reflected light.

Too much wire speed will choke the arc. More wire is being deposited than the heat and puddle can absorb. The result is spitting and sputtering as the wire melts into tiny balls of molten metal that fly away from the weld. The visual signal is a strobe light arc effect.

TABLE 8–3 WELDING SPEED

Panel Thickness	Welding Speed (in./min.)
1/32 in.	41-11/32–45-9/32
More than 1/32 in.	39-3/8
3/64 in.	35-7/16–39-3/8
1/16 in.	31-1/2–33-15/32

Before this critical ratio can be obtained, a thorough understanding of what is happening to produce these signals is essential.

When the trigger is first activated, a solid steel wire makes its initial contact with a solid steel plate. Prior to contact, the wire has been charged with current and the gas flow has been started. The first contact produces tiny sparks of oxide being burned off the wire and base metal.

Immediately after the oxide sparks, tiny molten balls are produced as the wire melts prior to forming a molten puddle that will absorb them. Once the heat creates the puddle, the balls stop. A consistent transfer and sound with only oxide sparks are present as they burn off the wire and base metal during the weld process.

In slow motion, after the arc transfer has been started, an on–off action occurs. Every time the metal is deposited, a plop is heard. When it pulls away, a hiss is heard. Speeded up to 200 plops and hisses per second, it creates a smooth buzz, like the sound of bacon frying in a pan.

When welding overhead, the danger of having too large a puddle and ball is obvious. The ball is pulled by gravity down onto the contact tip or into the gas nozzle where it can create serious problems. Therefore, overhead welding should always be done using a higher wire speed, with the arc and ball kept tiny and close together. Pressing the gas nozzle against the work ensures that the wire is not moved out of the puddle. If it is moved out, the balls are produced by melting wire until a new puddle is formed to absorb them.

Normal buildup of oxide sparks in the gas nozzle area must be carefully removed before they fall inside and short out the nozzle. Balls caused by too slow a wire speed must also be removed before a short is formed.

As a summary, Table 8–4 outlines the various effects of several welding parameters and the changes necessary to alter a variety of weld characteristics.

MIG Gun Nozzle Adjustment

The guns used on automotive MIG welders serve two main functions:

1. To provide proper gas protection
2. To feed the wire into the arc at the right speed

If the insulation is bypassed by accidentally grounding the body of the gun, the power intended for the wire is transferred to the gas nozzle, causing the nozzle to burn up. Welding on dirty or rusty material can cause heavy bombardment into the nozzle and will require immediate cleaning if proper welding performance is to be achieved. To successfully weld on a poor, rusty surface, slow the wire speed. Set the burnback control to its maximum and tap the trigger, floating the ball on and off the material.

Of the four main components in a MIG welder, the nozzle area is the most crucial. The wire feed delivery is second. A clogged or damaged liner will cause erratic wire speed and produce molten balls that, in turn, will short out the gas nozzle.

TABLE 8–4 ADJUSTMENTS IN WELDING PARAMETERS AND TECHNIQUES

Welding Variables to Change	Desired Changes							
	Penetration		Deposition Rate		Bead Size		Bead Width	
	Increase	Decrease	Increase	Decrease	Increase	Decrease	Increase	Decrease
Current and wire feed speed	Increase	Decrease	Increase	Decrease	Increase	Decrease	No effect	No effect
Voltage	Little effect	Little effect	No effect	No effect	No effect	No effect	Increase	Decrease
Travel speed	Little effect	Little effect	No effect	No effect	Decrease	Increase	Increase	Decrease
Stickout	Decrease	Increase	Increase	Decrease	Increase	Decrease	Decrease	Increase
Wire diameter	Decrease	Increase	Decrease	Increase	No effect	No effect	No effect	No effect
Shield gas percent CO_2	Increase	Decrease	No effect	No effect	No effect	No effect	Increase	Decrease
Torch angle	Backhand to 25 degrees	Forehand	No effect	No effect	No effect	No effect	Backhand	Forehand

To summarize the basic adjustment procedure of the gas nozzle, proceed as follows:

1. *Arc generation.* Position the tip of the gun near the base metal. When the gun switch is activated (Figure 8–28), the wire is fed at the same time as the shield gas. Bring the end of the wire in contact with the base metal and create an arc. If the distance between the tip and the base metal is shortened a little, it will be easy to generate an arc. If the end of the wire forms a large ball, it will be difficult to generate an arc, so quickly cut off the end of the wire with a pair of wire cutters (Figure 8–29).
2. *Spatter treatment.* Remove weld spatter promptly. If it adheres to the end of the nozzle, the shield gas will not flow properly and a poor weld will result. Antispatter compounds are available that reduce the amount of spatter that adheres to the nozzle (Figure 8–30). Weld spatter on the tip will prevent the wire from moving freely. If the wire feed switch is turned on and the wire is not able to move freely through the tip, the wire will become twisted inside the welder. Use a suitable tool, such as a file, to remove spatter from the tip and then check to see that the wire comes out smoothly.
3. *Contact tip conditions.* To ensure a stable arc, the tip should be replaced if it has become worn. For a good current flow and stable arc, keep the tip properly tightened (Figure 8–31).

Heat Buildup Prevention

Too much heat during welding distorts and weakens the metal. Always make sure you do not allow excess heat to transfer into any area of a panel.

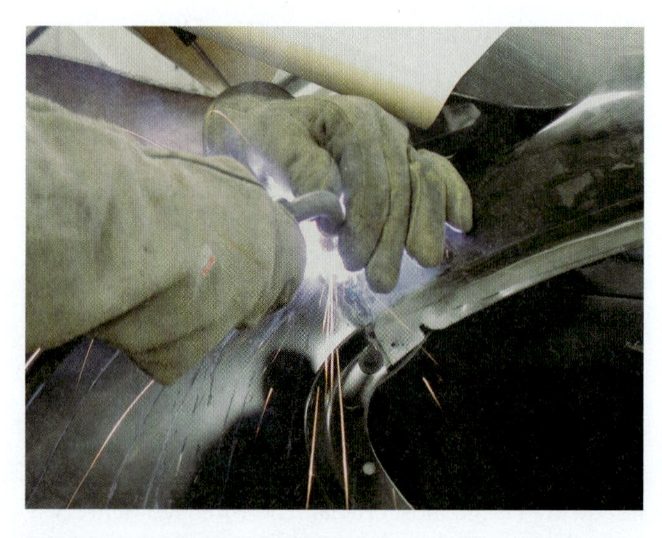

Figure 8–28 Activating the gun switch starts the wire feed. Welding gloves protect against high heat while allowing you to guide the gun accurately.

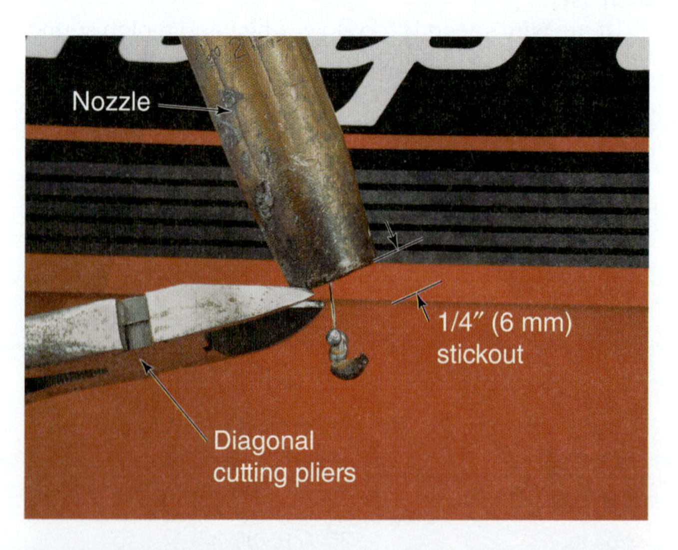

Figure 8–29 Cut off new or damaged welding wire, leaving about a ¼-inch stickout.

Figure 8–30 Spraying antispatter compound into a MIG nozzle will help protect the tip and prevent the wire from sticking in the gun.

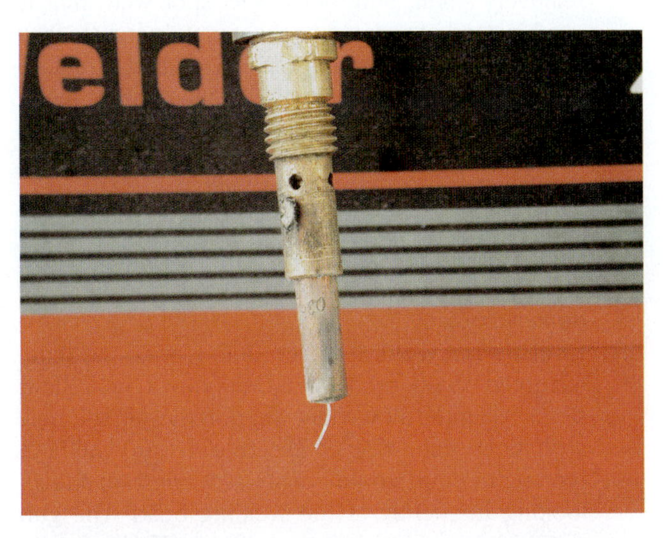

Figure 8–31 Remove the gas nozzle and check the gun tip condition periodically. Replace the tip if it is burned or worn down. When feeding new wire into the machine, you might have to remove the tip so the wire will feed through the machine without catching on the inside of the tip and bending.

Stitch and skip welding will prevent costly and time-consuming panel warpage. Another method of preventing heat buildup is heat sink compound.

Heat sink compound is a paste that can be applied to parts to absorb heat and prevent warpage. It comes in a can and can be applied and reused. Heat sink compound is sticky and can be placed on the panel next to the weld. Heat will flow into the compound and out of the metal to prevent heat damage.

Heat crayons or thermal paint can also be used to warn you when a panel is becoming too hot. They are commonly used on aluminum, which does not change color with heat and will be discussed later.

Clamping Tools for Welding

Locking jaw (vise) pliers, C-clamps, sheet metal screws, tack welds, or special clamps, described in Chapter 4, are necessary tools for good welding practices. Anybody can clamp panels together (Figure 8–32), but clamping panels together correctly to guarantee a sound weld will require close attention to every detail. As shown in (Figure 8–33), a hammer can often be used to fit panels closely together in places that cannot be clamped.

Many times clamping both sides of a panel is not possible. In these cases, sheet metal screws or pop rivets can be employed to gain proper clamping during welding operations. To clamp panels together with sheet metal screws, punch or drill holes through the panel. In the case of plug welding, every other hole is filled with a sheet metal screw. The empty holes are then plug welded using proper plug welding techniques. After the original holes are plug welded, the sheet metal screws are

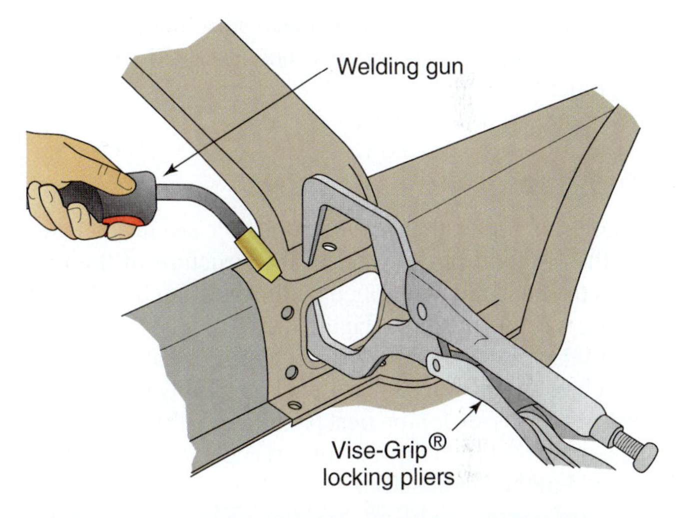

Figure 8–32 Clamping pliers will help hold the body panel in place during welding.

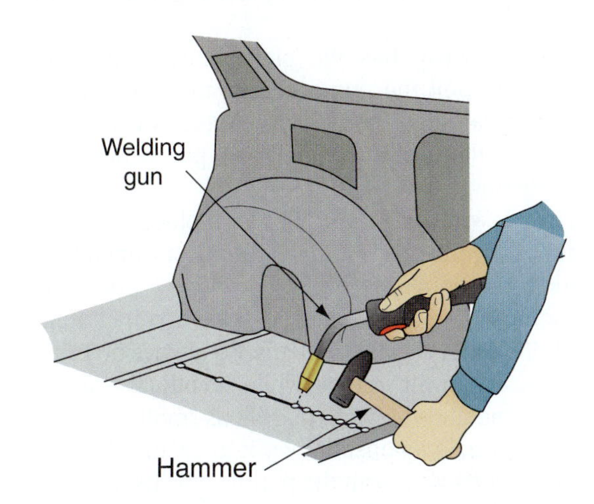

Figure 8–33 If needed, use hammer blows to move the panels together for a tighter joint before welding at next location on panel.

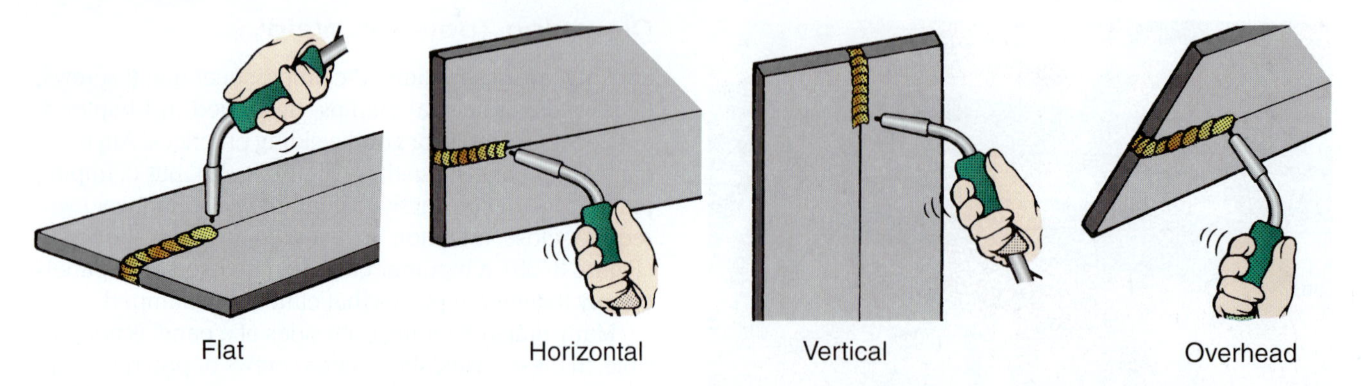

Figure 8–34 Note the four basic welding positions. Flat welding is the easiest and overhead is the most difficult.

removed and the holes left from the sheet metal screws are then plug welded.

Fixtures can also be used in some cases to hold panels to be welded in proper alignment. Fixtures alone, however, should not be depended on to maintain tight clamping force at the welded joint. Some additional clamping will be required to make sure that panels are tightly clamped together and not just held in proper alignment.

Welding Position

In collision repair, the *welding position* is usually dictated by the location of the weld in the structure of the car. Both the heat and wire speed parameters can be affected by the welding position (Figure 8–34).

Flat welding means the pieces are parallel with the bench or shop floor. Flat welding is generally easiest and fastest and allows for the best penetration. When welding a member that is off the car, try to place it so that it can be welded in the flat position.

Horizontal welding has the pieces turned sideways. Gravity tends to pull the puddle into the bottom piece. When welding a horizontal joint, angle the gun upward to hold the weld puddle in place against the pull of gravity.

Vertical welding has the pieces turned upright. Gravity tends to pull the puddle down the joint. When welding a vertical joint, the best procedure is usually to start the arc at the top of the joint and pull downward with a steady drag.

Overhead welding has the workpieces turned upside down. Overhead welding is the most difficult. In this position, the danger of having too large a puddle is obvious; some of the molten metal can fall down into the nozzle, where it can create problems. So always do overhead welding at a lower voltage, while keeping the arc as short as possible and the weld puddle as small as possible. Press the nozzle against the work to ensure that the wire is not moved away from the puddle. It is best to pull the gun along the joint with a steady drag.

To avoid painful burns to your ear canals, wear ear plugs when overhead welding.

8.5 BASIC WELDING TECHNIQUES

As shown in Figure 8–35, there are six basic welding techniques employed with MIG equipment.

Tack weld. The tack weld is exactly that: a tack—a relatively small, temporary MIG spot weld that is used instead of a clamp or sheet metal screw to tack and hold the fit in place while proceeding to make a permanent weld. And like the clamp or sheet metal screw, a tack weld is always a temporary device. The distance between each tack weld is determined by the thickness of the panel. Ordinarily, a length of 15–30 times the thickness of the panel is appropriate Figure 8–36. Temporary welds are very important in maintaining proper panel alignment and must be done accurately.

4. **Continuous weld**. In a continuous weld, an uninterrupted seam or bead is laid down in a slow, steady, ongoing movement. Support the gun securely so it does not wobble. Use the forward method, moving the gun continuously at a constant speed, looking frequently at the welding bead. The gun should be inclined between 10 and 15 degrees to obtain the best bead shape, welding line, and shield effect (Figure 8–37). Maintain proper tip-to-base metal distance and correct gun angle. If the weld is not progressing well, the problem might be that the wire length is too long. If this is the case, penetration of the metal will not be adequate. For proper penetration and a better weld, bring the gun closer to the base metal. If the gun handling is smooth and even, the bead will be of consistent height and width with a uniform, closely spaced ripple.
5. **Plug weld**. A plug weld is made in a drilled or punched hole through the outside piece (or pieces). The arc is directed through the hole to penetrate the inside piece. The hole is then filled with molten metal.
6. **Spot weld**. In a MIG spot weld, the arc is directed to penetrate both pieces of metal, while triggering a timed impulse of wire feed.
7. **Lap weld**. In the MIG lap spot technique, the arc is directed to penetrate the bottom piece and the puddle is allowed to flow into the edge of the top piece.

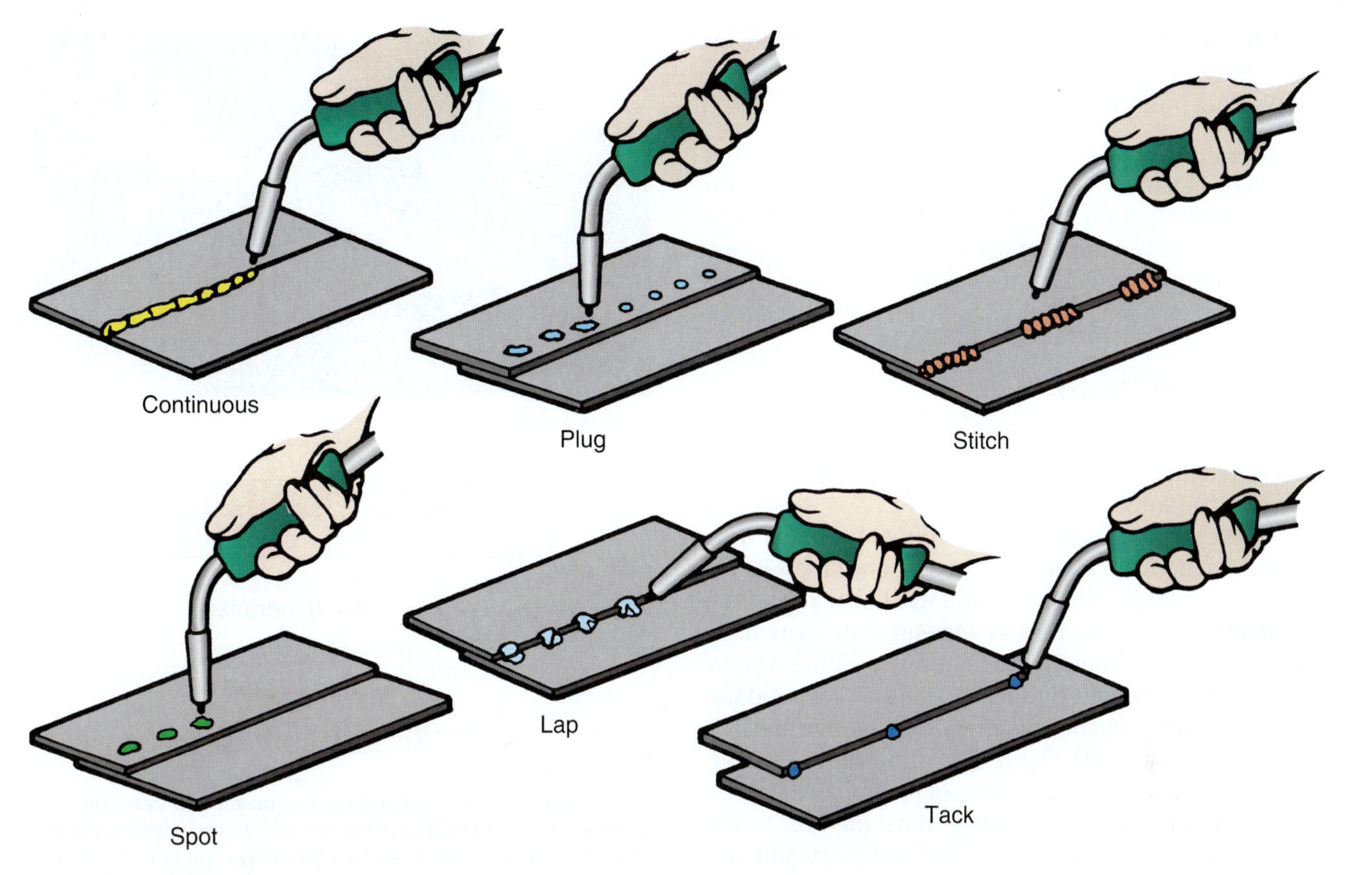

Figure 8-35 **Study and memorize the basic welding techniques. Compare the orientation of two weld pieces and the placement of the weld.**

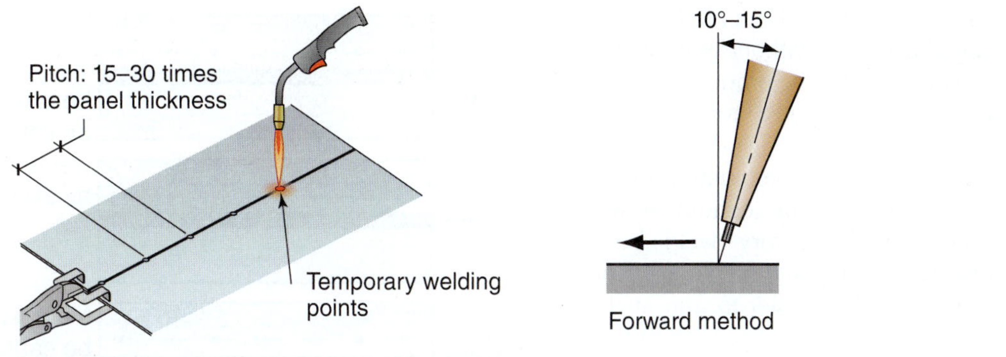

Figure 8-36 **Temporary, or tack, welding is commonly used to hold parts in place before the final continuous weld.**

8. **Stitch weld**. A stitch weld is a series of connecting or overlapping MIG spot welds, creating a continuous seam.

Basic Welding Methods

Each type of joint can be welded by several different techniques. The technique used depends mainly on the given welding situation:

- Thickness or thinness of the metal
- Condition of the metal

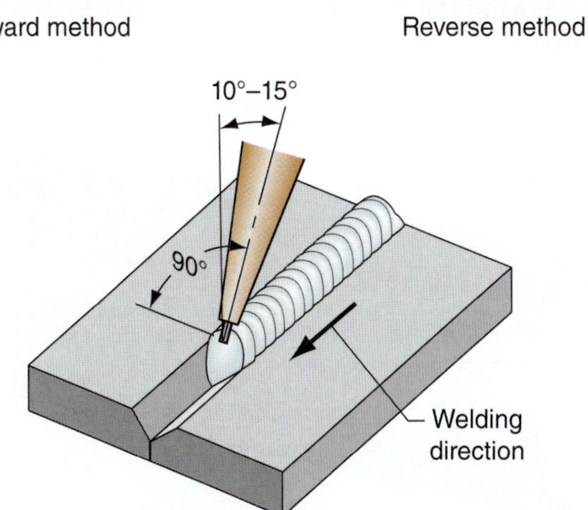

Figure 8-37 **Continuous welding, if done properly, produces a joint stronger than the base metal. You must be careful of excess heat and panel warpage, however.**

- Amount of gap, if any, between the pieces to be welded
- Welding position

For example, the butt joint can be welded with the continuous technique or the stitch technique. And it can be tack welded at various points along the joint to hold the parts in place while completing the joint with a permanent continuous weld or a stitch weld. Lap and flange joints can be made using all six welding techniques.

Making Butt Welds

Butt welds are formed by fitting two edges of adjacent panels together and welding along the mating or butting edges of the panels.

In butt welding, especially on thin panels, it is wise not to weld more than ¾ inch (19 mm) at one time. Closely watch the melting of the panel and welding wire and the continuity of the bead. Be sure the end of the wire does not wander away from the butted portion of the panels. If the weld is to be long, it is a good idea to tack weld the panels in several locations (stitch weld) to prevent panel warpage (Figure 8–38). Figure 8–39 shows how to generate an arc a short distance ahead of the point where the weld ends and then immediately move the gun to the point where the bead should begin. The bead width and height should be uniform at this time.

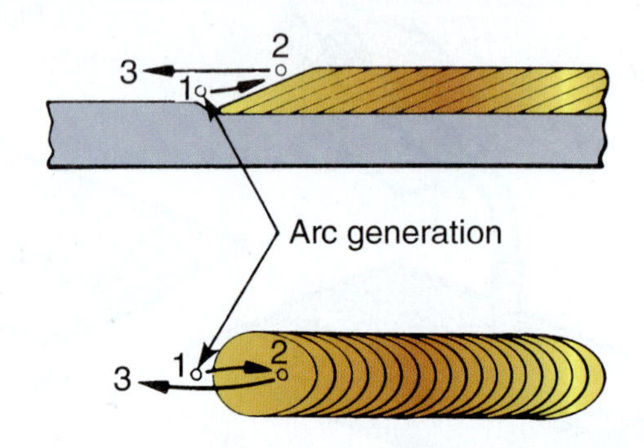

Figure 8–39 If gun handling is smooth and even, the bead will be of a consistent height and width, with a uniform, closely spaced ripple. Note how you move the gun in and out of the molten metal to produce overlapping nuggets that form a continuous bead.

Weld in a sequence that allows an area to cool before the next area is welded (Figure 8–40). To fill the spaces between intermittently placed beads, first grind the beads along the surface of the panel. Then fill the space with metal. If weld metal is placed without grinding the surface of the beads, blowholes can be produced.

When welding thin panels that are 1/32 inch (0.79 mm) or less, an intermittent or a stitch welding technique is a *must* to prevent burn-through. The combination of the proper gun angle and correct cycling techniques will enable you to achieve a satisfactory weld bead (Figure 8–41). The reverse welding method can be used for moving the gun because it is easier to aim at the bead.

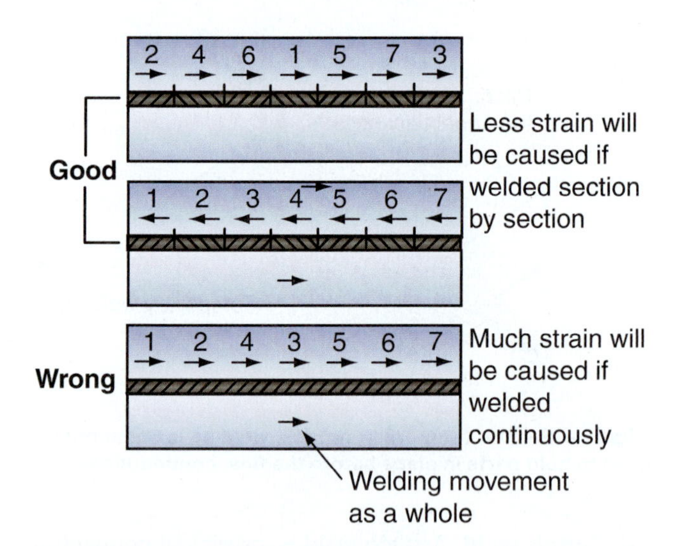

Figure 8–40 Compare the right and wrong welding sequences.

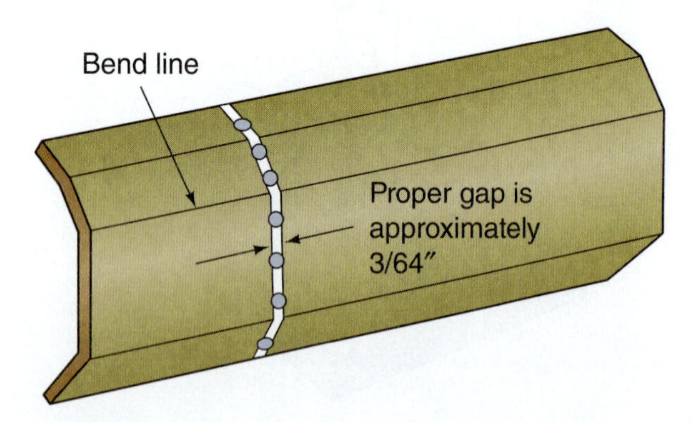

Figure 8–38 Tack welding of panels will also help prevent warpage because the area around the weld has time to cool.

Figure 8–42 shows a typical butt welding procedure for installing a replacement panel. If the desired results are not obtained, the cause can be that the distance between the tip of the gun and the base metal is too great. Weld penetration decreases as the distance between the tip and the base metal increases. Try holding the tip of the gun at several distances away from the

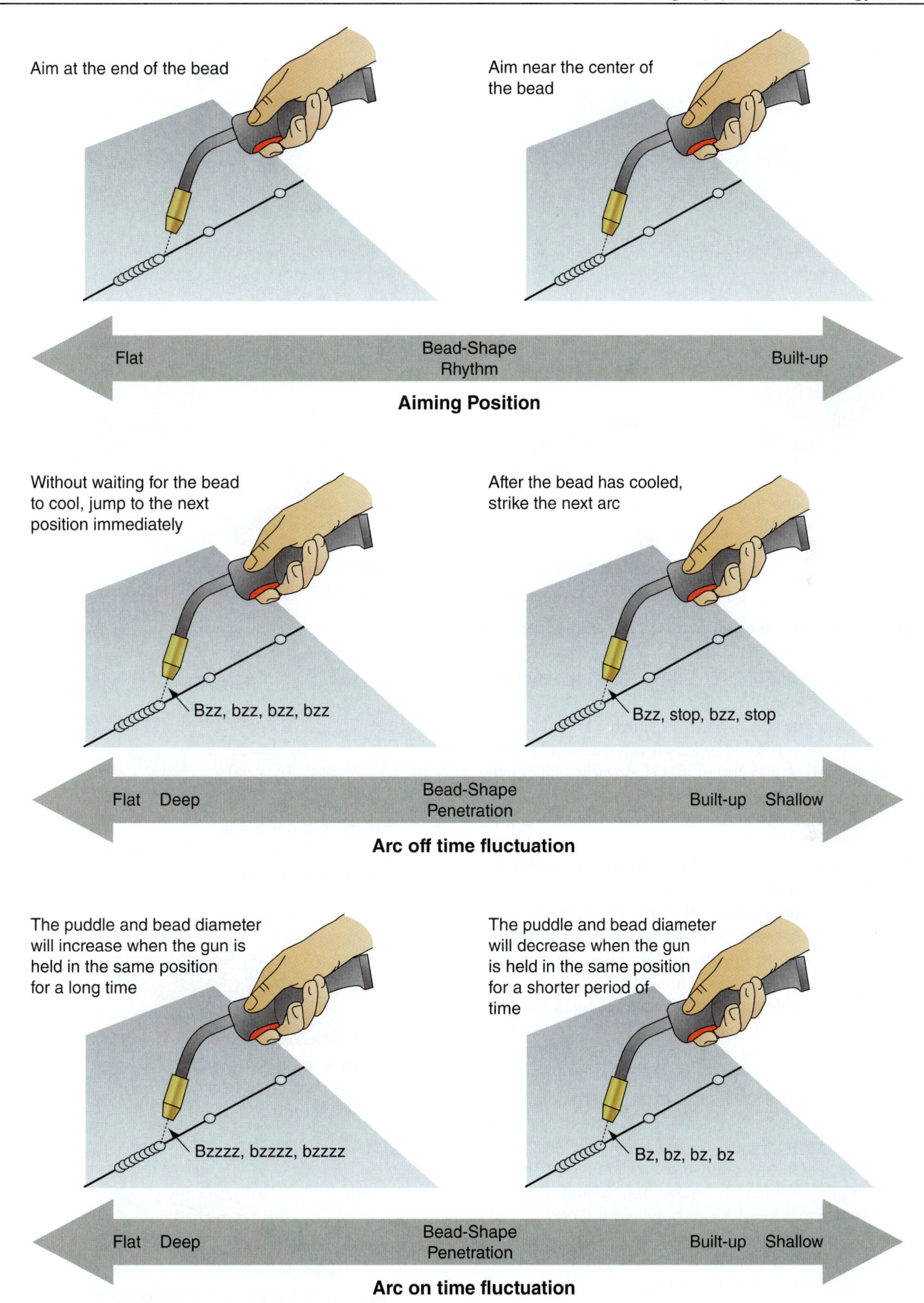

Figure 8-41 Study the steps in achieving a proper bead.

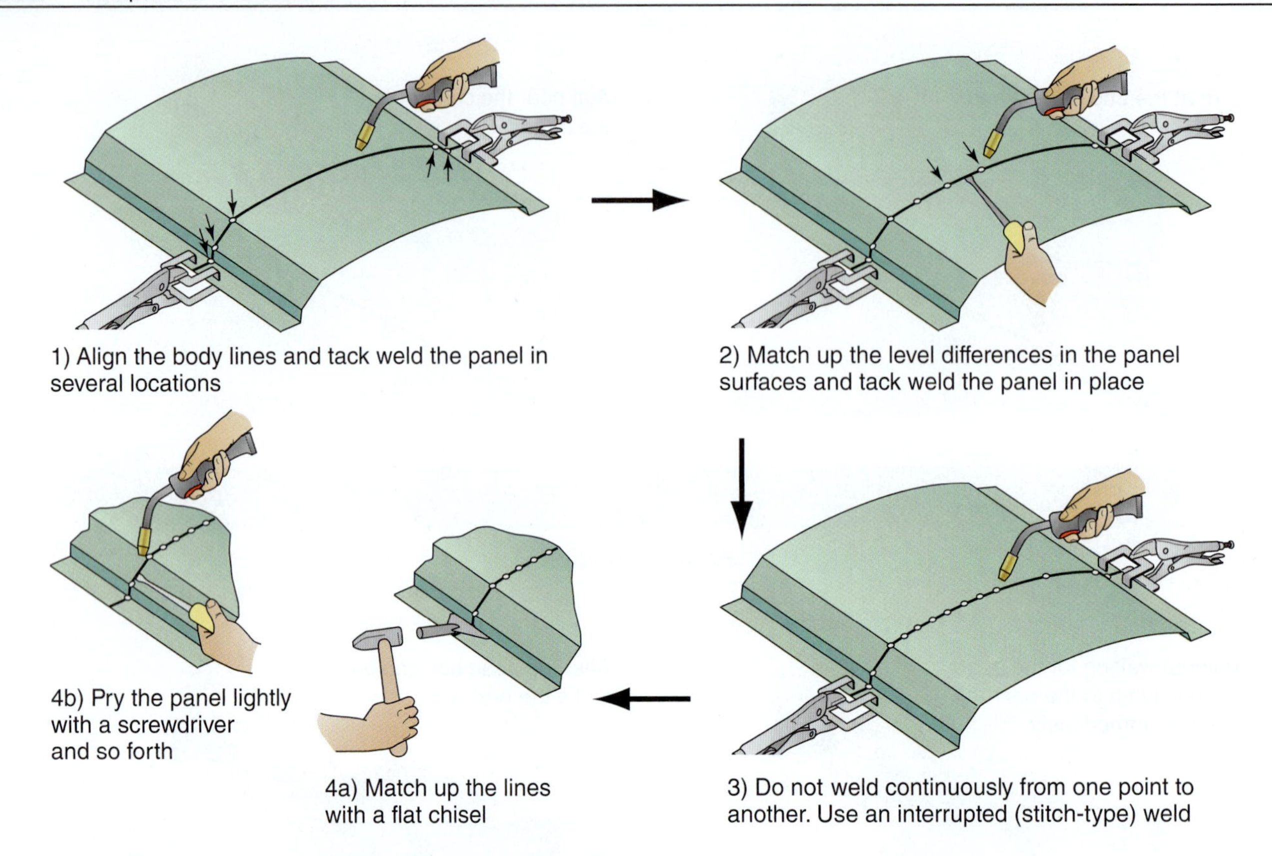
1) Align the body lines and tack weld the panel in several locations
2) Match up the level differences in the panel surfaces and tack weld the panel in place
3) Do not weld continuously from one point to another. Use an interrupted (stitch-type) weld
4a) Match up the lines with a flat chisel
4b) Pry the panel lightly with a screwdriver and so forth

Figure 8–42 Note the procedure for butt welding sectional areas.

base metal until the proper distance gives the desired results (Figure 8–43).

Moving the gun too fast or too slowly (Figure 8–44) will produce poor welding results (even if speed of wire feed is constant). A gun speed that is too slow will cause melt-through. Conversely, a gun speed that is too fast will cause shallow penetration and poor weld strength (Figure 8–45).

Even if a proper bead is formed during butt welding, panel warpage can result if the weld is started at or near the edge of the metal (Figure 8–46A). Therefore, to prevent warpage, disperse the heat into the base metal by starting the weld in the center of the panel. Frequently change the location of the weld area (Figure 8–46B). The thinner the panel thickness, the shorter the bead length.

When welding a butt joint, be sure the weld penetrates all the way through to the back side of the joint. Where the metal thickness at a butt joint is 1⁄16 inch (1.59 mm) or more, a gap should be left to ensure full penetration. If it is not practical to leave a gap, grind a V-groove in the joint (Figure 8–47) so the weld can penetrate to the back side.

MIG *butt welds* are often used to make two joints when sectioning frame rails, rocker panels, and door pillars. For butt welds keep a gap between the two pieces the thickness of one piece. This helps weld penetration and prevents expansion and contraction problems. Also, hold the gun at 90 degrees to the joint.

An **insert**, or *backing strip*, made of the same metal as the base metal can be placed behind the weld. The backing contributes to proper fit, helps align the joint,

Incorrect

Insufficient penetration. Weld strength is poor and the panel could separate when the panel is finished with a grinder

Incorrect

There is good penetration, but finish grinding will be both difficult and time-consuming

Correct

Good penetration and easy to grind

Figure 8–43 Compare the bead shapes in the cutaway diagrams.

Figure 8-44 Compare the welds from left to right. There is too much penetration on the left and not enough on the right beads.

and gives the joint the same strength and rigidity as the original structure.

Making Lap and Flange Welds

Lap and flange welds are made with identical techniques. They are formed by welding or fusing two surfaces to be joined at the top edge of two overlapping surfaces. This is similar to butt welds except only the top surface has an edge. Lap and flange welds should be made only in repairs where they replace original factory lap or flange welds or where outer panels, but not structural panels, are involved. These welds should not be used to join more than two thicknesses of material together.

The same technique used for temperature control in butt welding should be followed for lap and flange welding. Welds should never be made continuously but should be sequenced to allow for natural cooling and to prevent temperature buildup in the welding area.

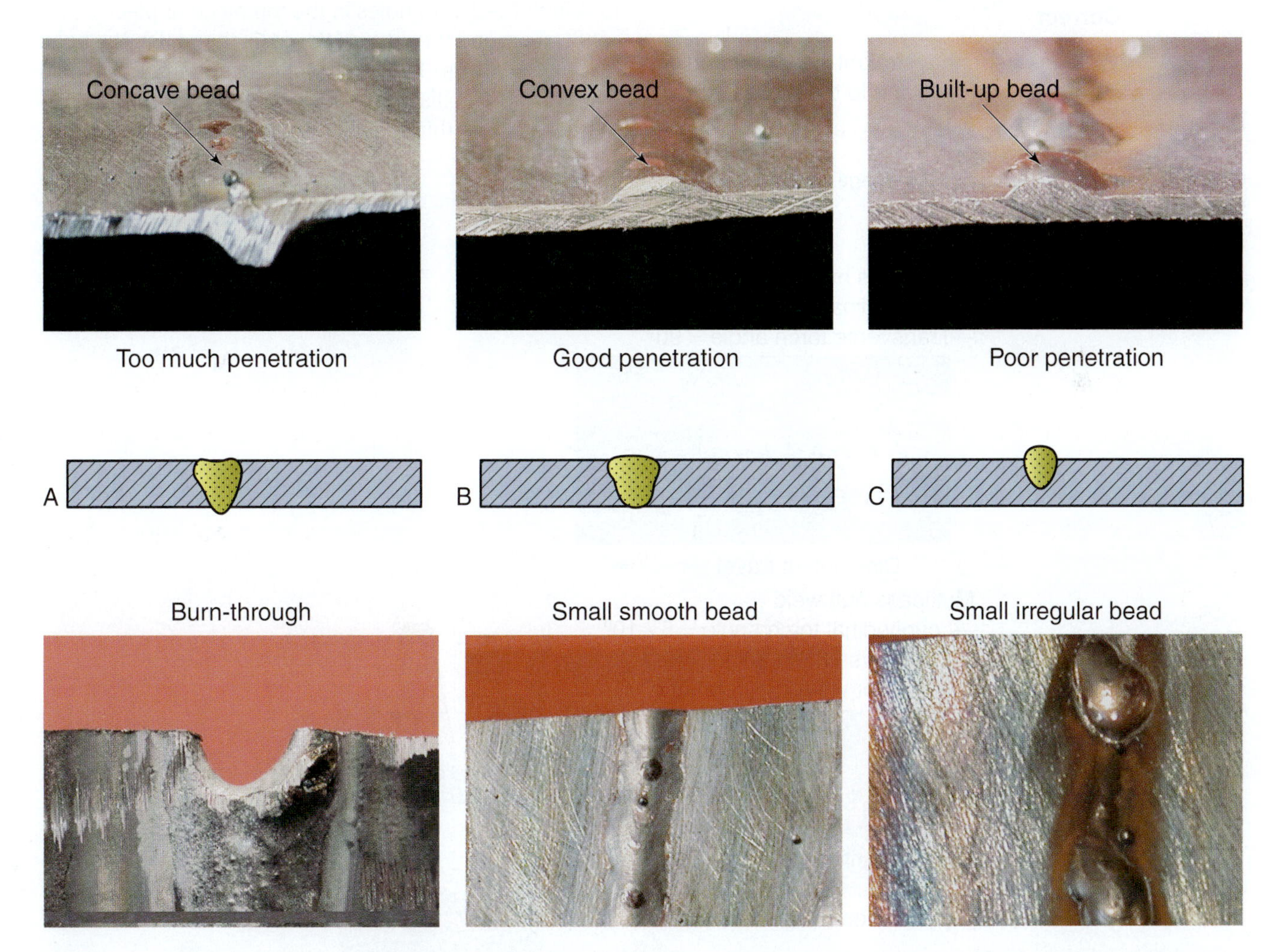

Figure 8-45 Compare good and bad weld beads. Gun movement, welding speed, and welder adjustments affect the bead shape and penetration. (A) Too much weld penetration is often caused by heat or by setting the current too high or the gun movement too slow. (B) A good weld on thin sheet metal resulted when the MIG welder setting was correct and the gun was moved properly. (C) This weld did not penetrate the sheet metal. It resulted from the welder current or heat adjustment being too low or from moving the gun too quickly.

Incorrect

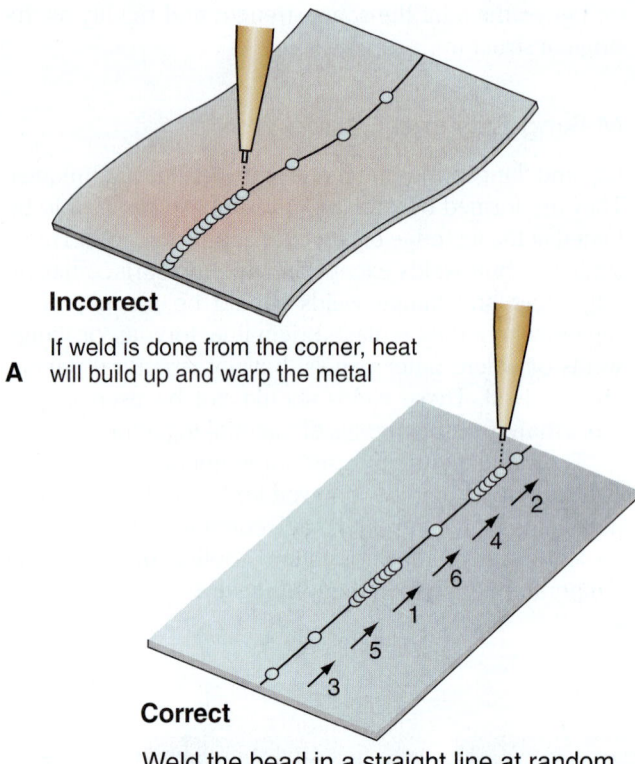

If weld is done from the corner, heat will build up and warp the metal

A

Correct

Weld the bead in a straight line at random lengths, always moving to the coolest area for the next weld

B

Figure 8-46 **Preventing panel warpage can be done by careful placement of each weld.**

Making Plug Welds

The plug weld is the body shop alternative to the OEM resistance spot weld made at the factory, because it can be used anywhere in the body structure that the factory used a resistance spot weld. Its use is not restricted. It has ample strength for welding load-bearing structural members. It can also be used on cosmetic body skins and other thin-gauge sheet metal (Figure 8–48).

A plug weld is formed by drilling or punching (Figure 8–49) a hole in the outer panel being joined (Figure 8–50). The materials should be tightly clamped together. Holding the torch at right angles to the surface, aim the electrode wire in the hole, and trigger the arc while moving the gun in a circular motion around the hole (Figure 8–51). The puddle fills the hole and solidifies.

When plug welding, try to duplicate the number and the nugget size of the original factory spot welds. The hole that is punched or drilled should not be larger in diameter than the factory weld nugget. Following the vehicle manufacturer's specifications, drill or punch 3/16- to 3/8-inch (5–9 mm) holes in the top piece or pieces.

Start around the edge of the hole, then fill in the hole. A 5/16-inch (8 mm) hole is a typical standard and works well for most collision repair. A 3/16-inch (5 mm) hole is better with very thin metals (24 gauge and lighter), and a

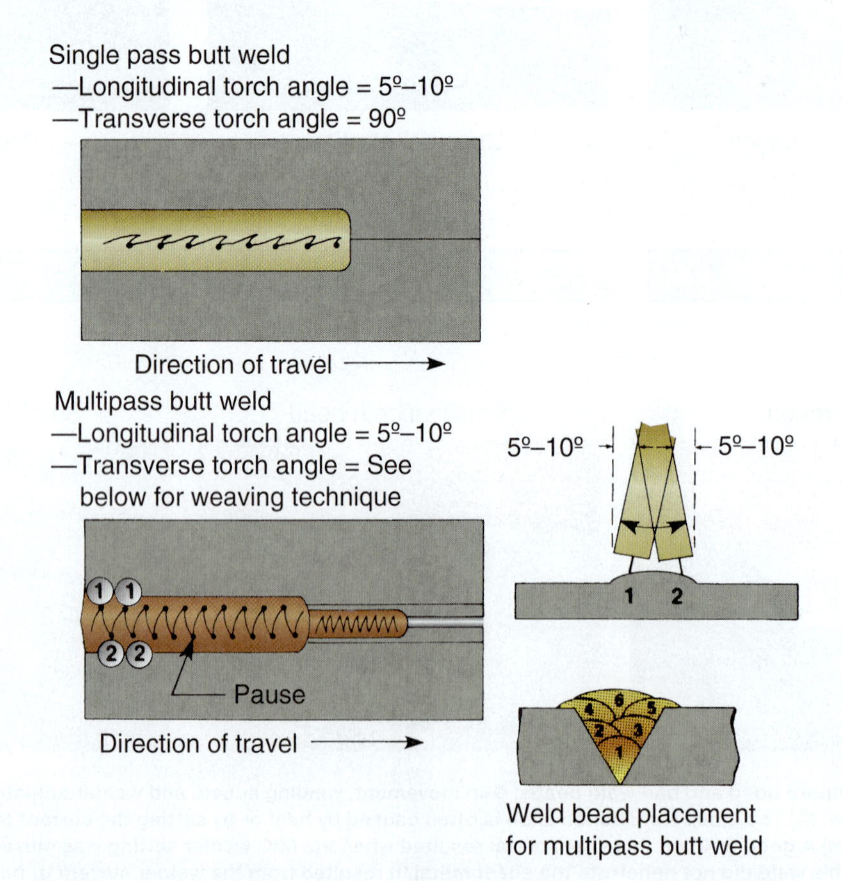

Figure 8-47 **Study torch movement for flat position butt welding.**

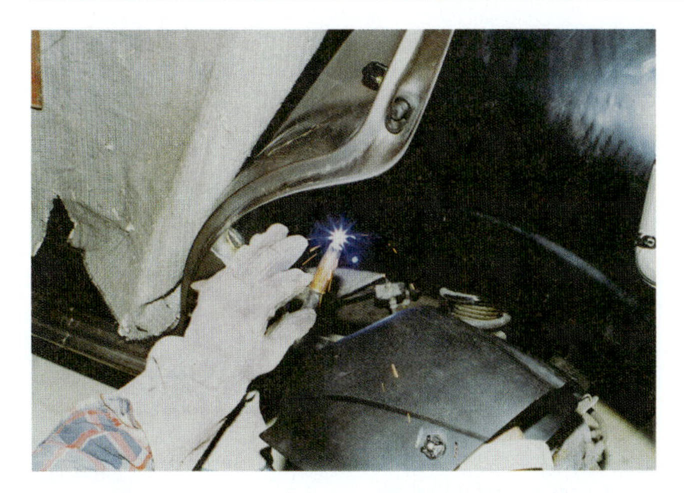

Figure 8–48 Welding takes great skill, knowledge, and patience.

Figure 8–49 Plug welds are formed by drilling or punching a hole in the outer panel to be welded. Here, the technician is using an air punch to make a series of holes along the flange of a new replacement panel.

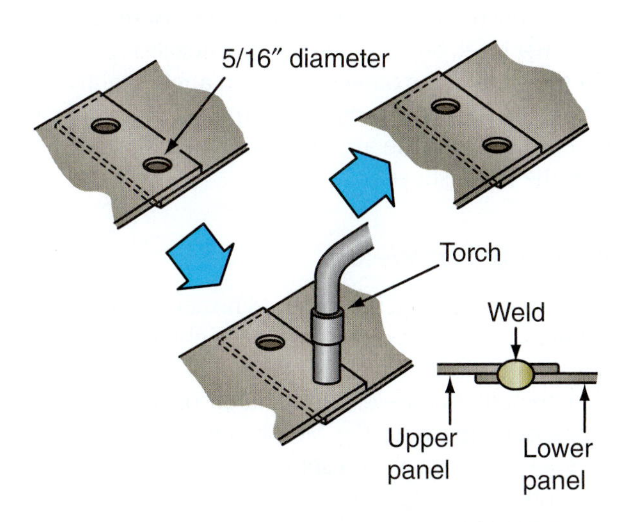

Figure 8–50 Study the steps in making a plug weld.

Figure 8–51 Position welding wire next to the edge of the hole in the top panel and press the welding gun trigger. Work your way around the hole until it is full of molten metal.

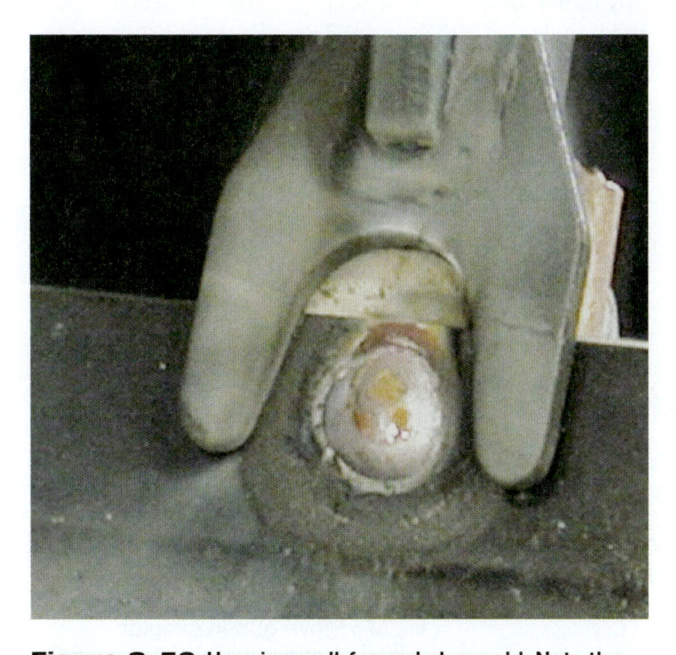

Figure 8–52 Here is a well-formed plug weld. Note the use of special clamping pliers that press the panel together next to the plug weld.

⅜-inch (9 mm) hole is better with heavier metal (14 gauge and heavier). See Figure 8–52.

When plug welds are used to join three or more panels together, holes are punched or drilled in every piece except the bottom piece. The holes are made progressively smaller from the top down. This is done to get better fusion of each layer to the adjacent one.

Typical hole size combinations are as follows: With three layers of metal, use 5⁄16-inch (8 mm) and ⅜-inch (9 mm) holes. With four layers, use ¼-inch (6 mm), 5⁄16-inch (8 mm), and ⅜-inch (9 mm) diameter holes.

When MIG plug welds are used to replace factory spot welds, remember these two points:

1. Follow the manufacturer's recommendations for number, size, and location of plug welds.
2. If this information is not available, duplicate the number, size, and location of the original factory welds.

Proper welding wire length is an important factor in obtaining a good plug weld. If the length of the wire protruding out of the end of the gun is too long, the wire will not melt properly, causing inferior weld penetration. The weld will improve if the gun is held closer to the base metal. Be sure the weld penetrates into the lower panel. Round, dome-shaped protrusions on the underside of the metal are good indicators of proper weld penetration.

The area welded should be allowed to cool naturally before any adjacent welds are made. Areas around the weld should not be force cooled using water or air. It is important that they be allowed to cool naturally. Slow, natural cooling without using water or air will minimize any panel distortion and keep the strength designed into the panels.

Plug welds can also be used to join more than two panels together. A hole is punched in every panel except the lower panel (Figure 8–53). The diameter of the plug weld hole in each panel being joined should be smaller than the diameter of the plug weld hole on top. Likewise, if panels of different thicknesses are being joined, a larger hole is punched in the thinner panel to ensure that the thicker panel is melted into the weld first. When welding panels of different thicknesses using the plug weld method, the thinner panel should be on top.

A plug weld using a MIG welder can be accomplished in a minimum amount of time, creating less temperature buildup in adjacent panels. While adjacent welds should not be made immediately, the area being welded will cool in a very short period of time.

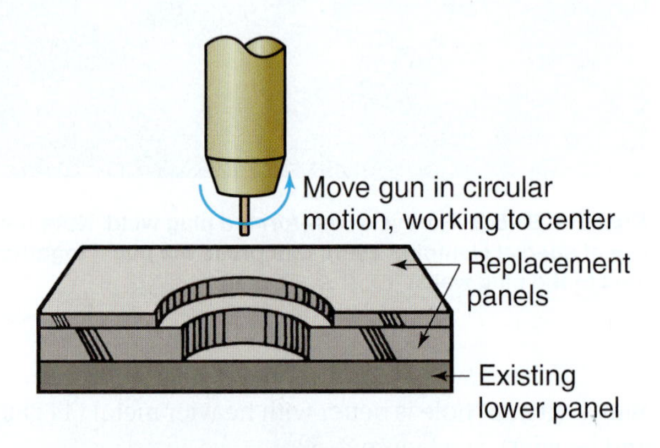

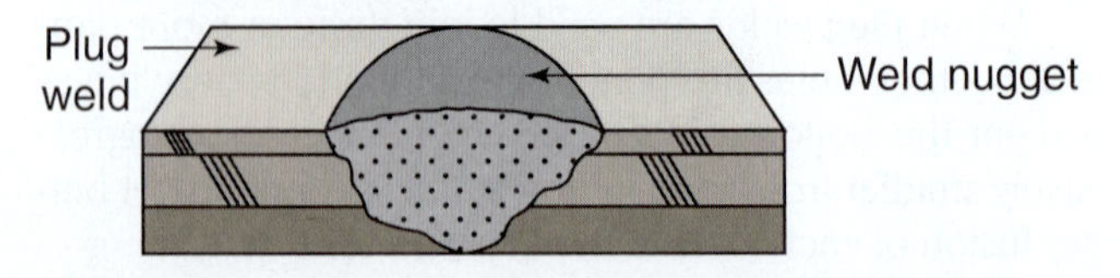

Figure 8-53 Note the method for welding two new panels over one existing panel. (A) Use a smaller hole on a lower panel when plug welding. (B) Weld all three panel flanges together with good penetration.

Considerations important to high-quality plug welds are proper current flow, wire feed adjustment, proper penetration, panel fitup, shielding gas flow rate, and filler rod type.

Making Spot Welds

Most MIG machines that are designed for collision repair work have built-in timers that shut off the wire feed and welding arc after the time required to weld one spot. Some MIG equipment also has a burnback time setting. It can be adjusted to prevent the wire from sticking in the puddle. The setting of these timers depends on the thickness of the workpiece. This information can usually be found in the machine's owner's manual.

For MIG spot welding, a special welding nozzle must replace the standard nozzle. Once in place and with the spot timing, welding heat, and backburn time set for the given situation, the spot nozzle is held against the weld site and the gun is triggered. For a very brief period of time, the timed pulses of wire feed and welding current are activated, during which the arc melts through the outer layer and penetrates the inner layer. After this, the automatic shutoff goes into action and no matter how long the trigger is squeezed, nothing will happen. The trigger must be released and then squeezed again to obtain the next spot pulse.

Because of varying conditions, the quality of a MIG spot weld is difficult to determine. On load-bearing members, therefore, MIG plug welding is the preferred method.

The MIG lap spot technique is a popular one for the quick, effective welding of lap joints and flanges on thin-gauge, nonstructural sheets and skins. Here again the spot timer is set, but this time the spot nozzle is positioned over the edge of the outer sheet at an angle slightly off 90 degrees. This will allow contact with both pieces of metal at the same time. The arc melts into the edge and penetrates the lower sheet.

Making Stitch Welds

In MIG stitch welding, the standard nozzle is used, not the spot nozzle. To make a stitch weld, combine spot welding with the continuous welding technique. To do this, set the automatic timer—either a shutoff or pulsed interval timer—depending on the MIG machine (Figure 8–54). The spot weld pulses and shutoffs occur with automatic regularity: weld-stop-weld-stop-weld-stop, as long as the trigger is held in.

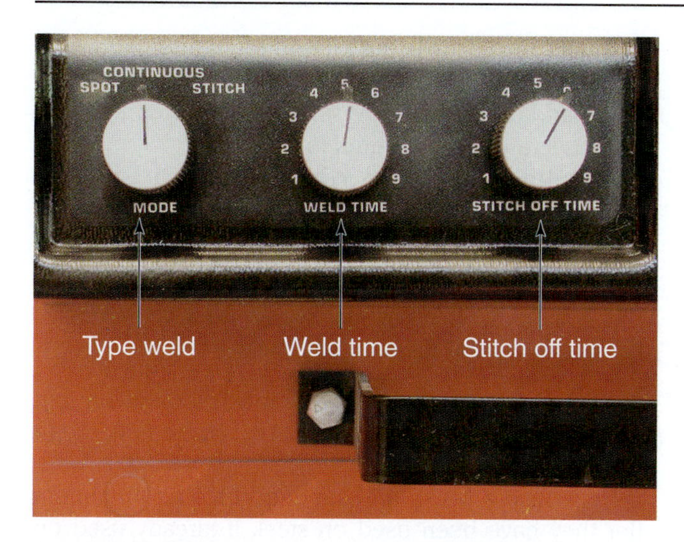

Figure 8-54 **Spot and stitch welding can be accomplished with most MIG machines by adjusting these controls.**

The arc off period allows the last spot to cool slightly and start to solidify before the next spot is deposited. This intermittent technique means less distortion and less melt-through or burn-through. These characteristics make the stitch weld preferable to the continuous weld for working thinner gauge cosmetic panels.

The intermittent cooling and solidifying of the stitch weld also makes it preferable to continuous welding on vertical joints where distortion is a problem. The welder does not have to contend with a continuous weld puddle that gravity is trying to pull down the joint ahead of the arc. Stitch welding is also preferable in the overhead position. There is virtually no weld puddle for gravity to pull.

If the MIG machine does not have automatic stitch modes, the spot and stitch welds can be made manually. The operator merely has to be capable of triggering the gun on–off, on–off, on–off—the same way the automatic system does.

8.6 MIG WELDING GALVANIZED METALS AND ALUMINUM

When MIG welding galvanized or zinc-metallized steels, also called zinc-coated steels, do *not* remove the zinc. If zinc is ground away, the thickness of the metal is reduced and so is its strength. And when a zinc-free area around the weld site is created, it is an inviting target for corrosion.

With galvanized or zinc-coated steels, use a slower gun travel speed than when welding uncoated steels. This is because the zinc vapors tend to rise into the arc zone and interfere with arc stability. A slower travel speed allows the zinc to burn off at the front of the weld pool. How much to reduce the gun travel speed will depend on the thickness of the zinc coating, the joint type, and the welding position. Experience is the best teacher with these variable conditions.

Because there is slightly less weld penetration with galvanized or zinc-coated steels than with uncoated steels, a slightly larger gap in square edge butt welds is needed. To prevent burn-through or excessive penetration of the wider gap, the welding gun should be handled with a side-to-side weaving motion.

It must be remembered that there is more spatter when welding galvanized or zinc-coated steels than with uncoated steels. Therefore, it is a good idea to apply anti-spatter compound inside the gun nozzle and to clean the nozzle frequently.

Always wear a welding respirator when fusing galvanized metals. The fumes can cause serious lung or respiratory illness.

Welding Aluminum

Several vehicles now have body, frame, and chassis parts made of aluminum. In fact, entire bodies made of aluminum are now available. As a result, the need for information on welding aluminum is growing.

Aluminum is light and relatively strong. It is naturally corrosion resistant. Aluminum forms its own corrosion barrier of aluminum oxide when exposed to air. A disadvantage of aluminum is its high cost.

There are some major differences to keep in mind when working with aluminum as opposed to steel, particularly when it comes to welding. Pure aluminum is lightweight and useful more for its ability to be formed than for its strength. When used on vehicles, it is alloyed with other elements and heat treated for additional strength.

Concerning welding, pound for pound aluminum is the best conductor of electricity. It conducts heat three times faster than steel. Aluminum becomes stronger when welded and is not brittle in extreme cold. It is also easily recyclable. Aluminum conducts heat faster than steel and also spreads the heat faster. Therefore, it requires special attention when welding.

Aluminum looks similar to magnesium, which, if welded, could start a flash fire. To make sure the part is aluminum, brush the part with a stainless steel brush. Aluminum turns shiny; magnesium turns dull gray.

When welding aluminum, be sure to protect wire harnesses and electronics from potential damage caused by spreading heat. Aluminum takes more voltage and amperage for the same thickness of material than steel (Figure 8–55).

If the part is found to be magnesium, do not weld it. It can start to burn with tremendous heat.

Courtesy of Miller Electric Manufacturing Co.

Figure 8-55 **TIG, or tungsten inert gas, is being used more in body shops now because of the aluminum construction and aluminum body panels used by many auto manufacturers. A hand-held aluminum rod is fed into the electric arc and inert gas.**

Use the following ten guidelines when MIG welding aluminum:

1. Match the wire to the aluminum alloy.
2. Set the wire speed faster than with steel.
3. Hold the gun closer to vertical. Tilt it only about 5–10 degrees from the vertical in the direction of the weld.
4. Use only the forward welding method. Always push; never pull. When making a vertical weld, start at the bottom and work up.
5. Set the tension of the wire drive roller lower to prevent twisting. But do not lower the tension too much or the wire speed will not be constant.
6. Because there tends to be more spatter, use an antispatter compound to control buildup at the end of the nozzle and contact tip. Only apply the compound to the nozzle, and clean off any excess to keep it out of the weld puddle. Antispatter compound must be kept off all welding surfaces because it will contaminate the weld.
7. Shop squeeze-type resistance spot welders do not have enough amperage for aluminum. Do not use resistance spot welders.
8. Always use skip and stitch welding techniques to prevent heat warping. Set wire speed slightly faster. Hold the gun closer to vertical than you do when welding steel.
9. Use about 50 percent more shielding gas.
10. Use 100 percent argon for the shield gas.

A MIG gun for welding aluminum can be either the standard equipment or self-contained with a motor in the handle and aluminum wire spool mounted on top. The nozzle is straight, not tapered in at the end.

Aluminum electrode wire is classified by series, according to the metal or metals the aluminum is alloyed with and whether the aluminum is heat treated. The series are set up by the Aluminum Association, not the AWS. The number does not indicate the strength of the electrode.

Special cleaning instructions are needed for aluminum. Remove all aluminum oxide with a stainless steel brush before welding. The metal might look clean, but aluminum oxide always needs to be brushed off. Clean the metal until shiny.

Never use a brush and sanding discs on aluminum after they have been used on steel. If already used on steel, iron powder will remain on the surface of the aluminum and contaminate the weld.

The typical procedure for MIG welding aluminum is as follows:

1. Clean the weld area completely, both front and back. Use wax and grease remover and a clean rag.
2. If the pieces to be welded are coated with a paint film, sand a strip about ¾-inch (19 mm) wide to the bare metal, using a disc sander and a No. 80 disc. Do not press too hard or the sander will heat up and peel off aluminum particles, clogging the paper.
3. Clean the metal until shiny with a stainless steel wire brush.
4. Load 0.030 aluminum wire into the welder. Trigger it to extend about an inch (25 mm) beyond the nozzle.
5. Set the voltage and wire speed according to the instructions supplied with the welding machine. Remember that the wire speed must be faster than for steel. Make a practice weld on scrap pieces.

   Position the two pieces together and lay a bead along the entire joint. The distance between the contact tip and the weld should be 5⁄16–9⁄16 inch (8–14 mm).
6. If the arc is too large, turn down the voltage and increase the wire speed. The bead should be uniform on top, with even penetration on the back side.

The high heat conduction property of aluminum means that the technician must protect against warpage. There are two methods for doing this: stitch welding and center out welding, which were explained earlier.

Parts made of aluminum are usually 1½–2 times as thick as steel parts. When damaged, aluminum feels harder or stiffer to the touch because of work hardening.

8.7 TESTING THE MIG WELD

Repair welds should be tested from time to time on every job. This can be done simply with test panels. Before welding on a vehicle, make some welds on pieces of scrap sheet metal like the panels that are going to be installed on the vehicle. If the proper settings on the MIG welder are obtained on the test pieces, the quality of the weld on the car can be ensured.

8.8 MIG WELD DEFECTS

Defects in MIG welds and their causes are summarized in Table 8–5. Proper welding techniques ensure good welding results. If welding defects should occur, think of ways to change your procedures to correct them.

When making any MIG repairs, the materials and panels must be similar enough to allow mixing when they are welded together. The melting and flowing of metals can be accomplished by many methods, depending on the materials being joined. The combinations of cleanliness of the welded area, the mixing of proper metals, and the right heat application will result in a good MIG weld.

A *welding problem* causes a weak or cosmetically poor joint that reduces quality. Eight common weld problems include:

1. *Weld porosity* (holes in the weld)
2. *Weld cracks* (cracks on the top or inside the weld bead)
3. *Weld distortion* (uneven weld bead)

TABLE 8–5 WELDING PRECAUTIONS

Defect	Defect Condition	Remarks	Main Causes
Pores/pits		There is a hole made when gas is trapped in the weld metal.	1. There is rust or dirt on the base metal. 2. There is rust or moisture adhering to the wire. 3. Improper shielding action; the nozzle is blocked or wind or the gas flow volume is low. 4. Weld is cooling off too fast. 5. Arc length is too long. 6. Wrong wire is selected. 7. Gas is sealed improperly. 8. Weld joint surface is not clean.
Undercut		Undercut is a condition where the overmelted base metal has made grooves or an indentation. The base metal's section is made smaller and, therefore, the weld zone's strength is severely lessened.	1. Arc length is too long. 2. Gun angle is improper. 3. Welding speed is too fast. 4. Current is too large. 5. Torch feed is too fast. 6. Torch angle is tilted.
Improper fusion		This is an unfused condition between weld metal and base metal or between deposited metals.	1. Check torch feed operation. 2. Is voltage lowered? 3. Weld area is not clean.
Overlap		Overlap is apt to occur in a fillet weld rather than in a butt weld. Overlap causes stress concentration and results in premature corrosion.	1. Welding speed is too slow. 2. Arc length is too short. 3. Torch feed is too slow. 4. Current is too low.
Insufficient penetration		This is a condition in which there is insufficient deposition made under the panel.	1. Welding current is too low. 2. Arch length is too long. 3. The end of the wire is not aligned with the butted portion of the panels. 4. Groove face is too small.
Excess weld spatter		Excess weld spatter occurs as speckles and bumps along either side of the weld bead.	1. Arc length is too long. 2. There is rust on the base metal. 3. Gun angle is too severe.

(continued)

TABLE 8–5 WELDING PRECAUTIONS (*continued*)

Defect	Defect Condition	Remarks	Main Causes
Spatter (short throat)		Spatter is prone to occur in fillet welds.	1. Current is too great. 2. Wrong wire is selected.
Vertical crack		Cracks usually occur on the top surface only.	1. There are stains on the welded surface (paint, oil, rust).
Nonuniform bead		This is a condition in which the weld bead is misshapen and uneven rather than streamlined and even.	1. The contact tip hole is worn or deformed and the wire is oscillating as it comes out of the tip. 2. The gun is not steady during welding.
Burn-through		Burn-through is the condition of holes in the weld bead.	1. The welding current is too high. 2. The gap between the metal is too wide. 3. The speed of the gun is too slow. 4. The tip-to-base metal distance is too short.

4. *Weld spatter* (drops of electrode on and around weld bead)
5. *Weld undercut* (groove melted along either side of the weld and left unfilled)
6. *Weld overlap* (excess weld metal mounted on top and either side of the weld bead)
7. *Too little penetration* (weld bead sitting on top of the base metal)
8. *Too much penetration* (burn-through beneath the lower base metal)

8.9 FLUX-CORED ARC WELDING

Flux-cored arc welding (FCAW) is an electric arc welding process that uses a tubular wire with flux inside. With the development of 0.030 self-shielded, flux-cored wire, the flux-cored welding process has proven to be valuable for work on HSS (coated or uncoated). The FCAW process uses the same type of constant potential power source as MIG. It also uses the electrode feed system, contact tube, electrode conduit, welding gun, and many other pieces of equipment that are used in MIG. Nevertheless, the process itself differs somewhat from MIG.

There is no external shielding gas in FCAW. As the flux within the wire melts in the heat of the arc, the created gases shield the weld puddle, stabilize the arc, help control penetration, and reduce porosity. The melted flux also mixes with the impurities on the metal surface and brings them to the top of the weld where they solidify as slag. The slag can then be chipped or brushed away.

Two very important advantages of the FCAW process over MIG are its ability to tolerate surface impurities (thus requiring less precleaning) and to stabilize the arc. Other beneficial characteristics of the process include the following:

- High deposition rate
- Efficient electrode metal use
- Requires little edge preparation
- Welds in any position
- Welds a wide range of metal thicknesses with one size of electrode
- Produces high-quality welds
- Weld puddle is easily controlled and its surface appearance is smooth and uniform, even with minimal operator skill
- Produces a weld with less porosity than MIG when welding galvanized steels

While the FCAW process has a number of advantages over MIG, it has the following drawbacks:

- FCAW wires are more expensive than MIG hard wires. However, the cost is quickly recovered through higher productivity.
- The flux from the wire changes to slag as it cools. Until it does cool, the slag is sharp and hot and should be considered an eye and skin hazard. Once it cools, this slag must be removed prior to the application of fillers, seam sealers, primers, or paint.
- Spatter is worse when using flux-cored wires. Use nozzle gel and keep the nozzle scraped clean. Spatter

buildup in the gun nozzle can jam the wire in the contact tip; it can also fall off during welding and mix with the molten puddle, diminishing the quality of the weld.

- Excessive tension on the drive rollers or using the incorrect style of drive rollers can collapse the tubular wire. Check the owner's manual for flux-cored wire requirements.
- Only ferrous metals can be welded.

If a machine is used for both MIG and FCAW, the welder must have polarity switching capabilities. FCAW with 0.030- or 0.035-inch (0.76 or 0.89 mm) wire uses straight polarity, while 0.023-, 0.030-, and 0.035-inch (0.58, 0.76, or 0.89 mm) hard wires for MIG use DC reverse polarity. Many of the gas metal arc welding machines sold over the past few years were originally designed to run DC reverse. Without going inside the machine to change polarities, which is difficult and time-consuming, this type of machine will not run DC straight polarity. Check the owner's manual for polarity reversing capabilities.

- FCAW wires are more expensive per pound than hard wires for GMAW.
- The 0.030-inch (0.58 mm) self-shielded cored wire contains fluoride compounds. *Use adequate ventilation.*
- The flux in the core of the wire changes to slag upon cooling. This slag must be removed prior to the application of fillers, seam sealers, primers, or paint.
- In addition, the slag is sharp and hot until it cools, so it must be considered an eye and skin hazard.
- Spatter is worse when using cored wires. Use nozzle gel and keep the nozzle scraped clean. Spatter buildup in the gun nozzle can jam the wire in the contact tip or fall off during welding, mixing with the molten puddle and contributing to a poor quality weld.
- Wire feed problems for FCAW are similar to those encountered with GMAW, but with one important difference. Because cored wires are not solid, excessive tension on the drive rolls or the incorrect style of drive rolls may collapse the tubular wire, which leads to feeding problems. Again, check with the owner's manual for correct drive rolls and tension requirements for flux-cored wire.

8.10 TIG WELDING

Tungsten inert gas (TIG) welding, another form of GMAW, uses a nozzle-fed shielding gas and a hand-held filler rod. It has somewhat limited use in body shop repair applications. In a general auto repair or engine rebuilding, however, it does things that make it a valuable tool.

MIG welders lay down weld beads at the average of 25 inches (635 mm) per minute. TIG welding is much slower, with weld speeds ranging between 5 and 10 inches (127–254 mm) per minute. However, this slower speed allows for much more control, and the end result is the best-looking weld obtainable. A TIG unit can be used to repair cracks in aluminum cylinder heads and reconstruct combustion chambers and other automotive components that need to be welded.

Like MIG (Figure 8–56), TIG welders use an inert gas, such as argon or helium, to surround the weld area and prevent oxygen and nitrogen in the atmosphere from contaminating the weld. But instead of having a wire feed welding electrode like MIG units, TIG machines use a tungsten electrode with a very high melting point (about 6,900°F) to strike an arc between the welding gun and the work.

Because the tungsten electrode has such a high melting point, it is not consumed during the welding process. A filler rod must be used for welding thicker materials (Figure 8–57).

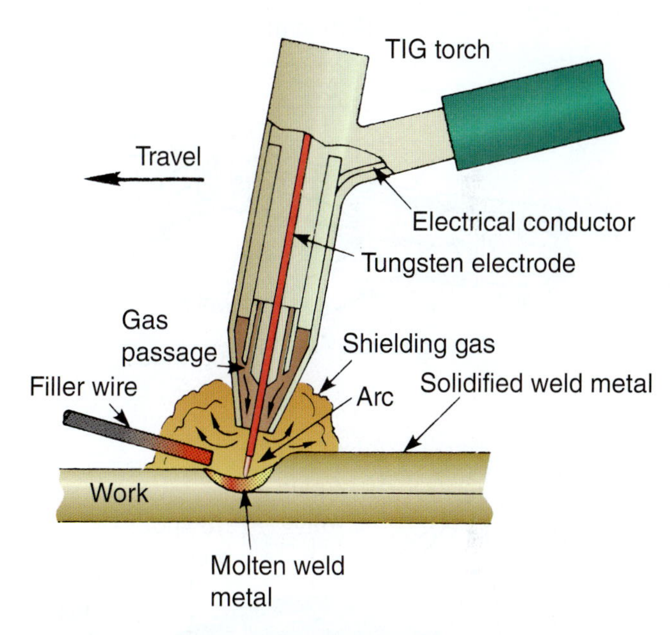

Figure 8–56 Study the principles of the TIG process. Filler metal is fed into the pool of molten aluminum from a separate filler rod. You can control the bead of aluminum much more easily with TIG compared to MIG.

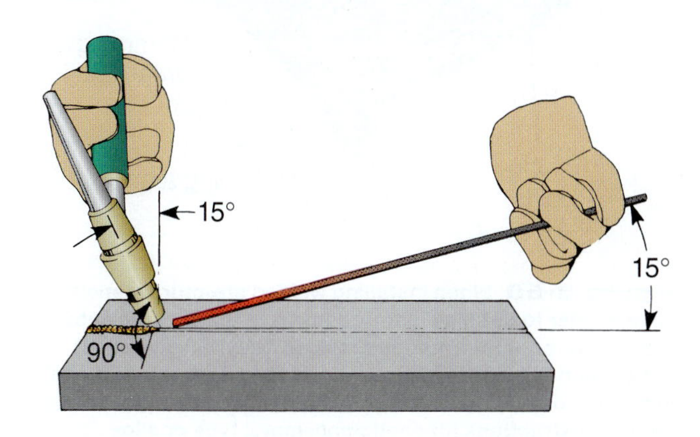

Figure 8–57 Proper position of the torch and filler rod for manual TIG welding.

8.11 RESISTANCE SPOT WELDING

Resistance spot welding is the most important welding process used by automobile manufacturers. It is used on their assembly lines to make many of the OEM welds on unibody cars (Figure 8–58). It is estimated that between 90 and 95 percent of all factory welds in a unibody structure are spot welds. In this country, it is also widely used in the automotive aftermarket for sunroof installations and vehicle conversions, including recreational vehicles (RVs) and stretch limousines.

Because resistance spot welding is now specified by a growing number of automobile manufacturers for repair welding their vehicles, the repair specialist must know how to use a resistance spot welding gun.

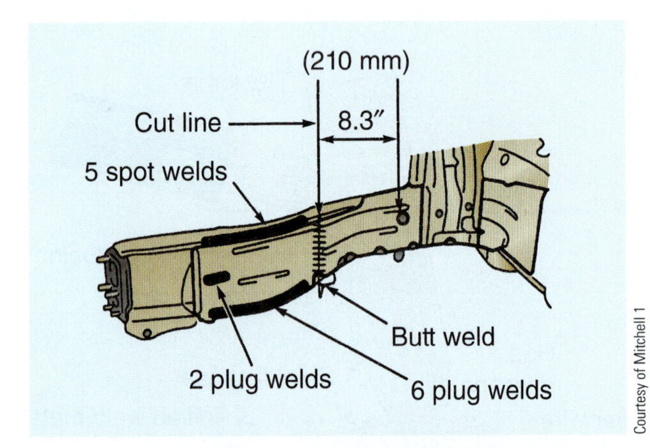

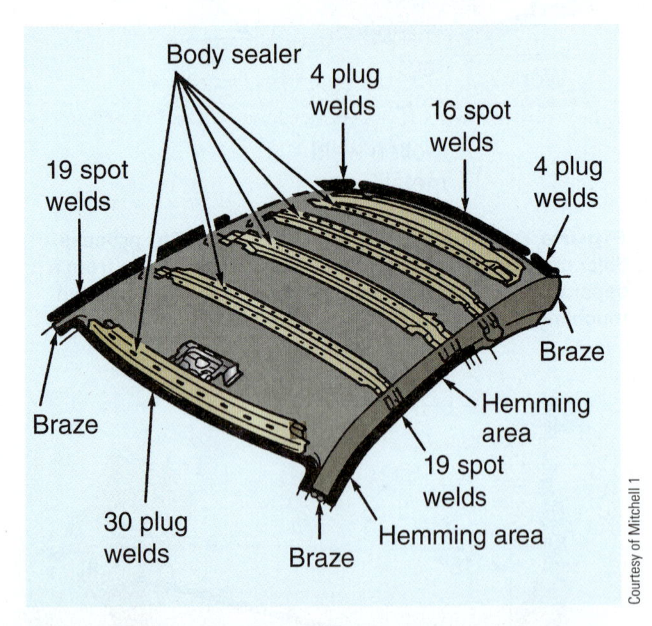

Figure 8-58 When installing welded structural panels, always refer to detailed instructions from the specific auto manufacturer or information provider. They will give exact cut or section locations, types of welds (plug, spot, or braze), locations for sealing joints, and other information. Never use general instructions for sectioning; metal type or alloy composition, thickness, number of pieces welded together, and load-carrying requirements all affect procedures. Note the three types of welds recommended for correct installation of these panels.

The squeeze-type resistance spot welder (Figure 8–59) is ideal for repair welding many of the unibody's thin-gauge sections that require good weld strength and no distortion. Typical applications include roofs, window and door openings, rocker panels, and many exterior panels (Figure 8–60). Due to the strength requirements of unibody repairs, it is often important that a squeeze-type resistance spot welder be used and that the repair specialist know how to set it up, make test welds, and use it.

Resistance spot welding has several advantages:

1. It reduces welding costs.
2. No consumable filler wire, rod, or gas is required.
3. It is clean with no smoke or fumes.
4. It allows the use of weld through conductive zinc primers to restore corrosion protection to repair joints.
5. It duplicates OEM factory weld appearance.
6. It eliminates the need for grinding of welds.
7. It is fast; weld times of a second or less make strong welds on HSS and HSLA steels as well as mild steels, with a very small heated zone, eliminating distortion of metal.

How Resistance Spot Welding Works

Resistance spot welding relies on the resistance heat generated by low-voltage electric current flowing through two pieces of metal held together, under pressure, by the squeeze force of the welding electrodes. Thus, the three important factors in the operation of resistance spot welding are

1. *Pressurization.* The mechanical welding bond between two pieces of sheet metal is directly related to the amount of force exerted on the sheet metal by the welding tips. As the tips squeeze the sheet metal together, an electrical current flows from the tips through the base metal, causing the metal to melt and fuse together. Weld spatter (internal or external) is the result of low pressure on the tip or excessive electrical current flow. A high tip pressure causes a small spot weld (Figure 8–61) and a reduced mechanical bond of the weld. In other words, the high tip pressure forces the tip into the softened area, thinning and weakening the weld.
2. *Current flow.* When pressure is applied to the metal, a high electric current flows through the electrodes and through the two pieces of metal. The temperature rises rapidly at the joined portion of the metal where the resistance is greatest (Figure 8–62A). If the current continues to flow, the metal melts and fuses together (Figure 8–62B). If the electrical current becomes too great or the pressure too low, internal spatter will result. However, if the current is decreased or the pressure is increased, weld spatter will be held to a minimum. As can be seen, there is a mutual relationship between the electrical current and the pressure applied to the spot weld.

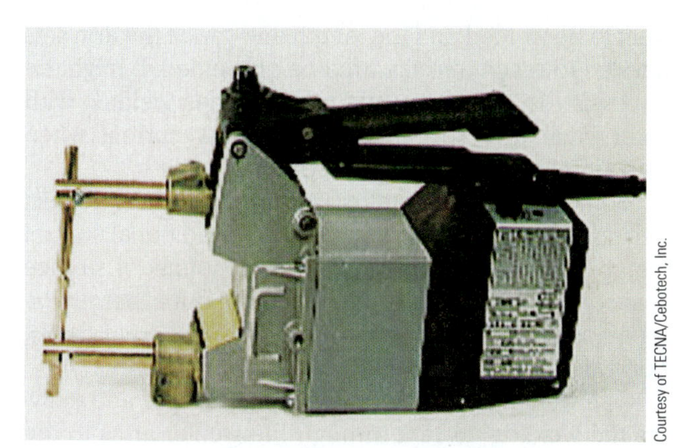

A This is a hand-held squeeze-type resistance spot welder. It can be used on easy-to-get-at flanges on panels.

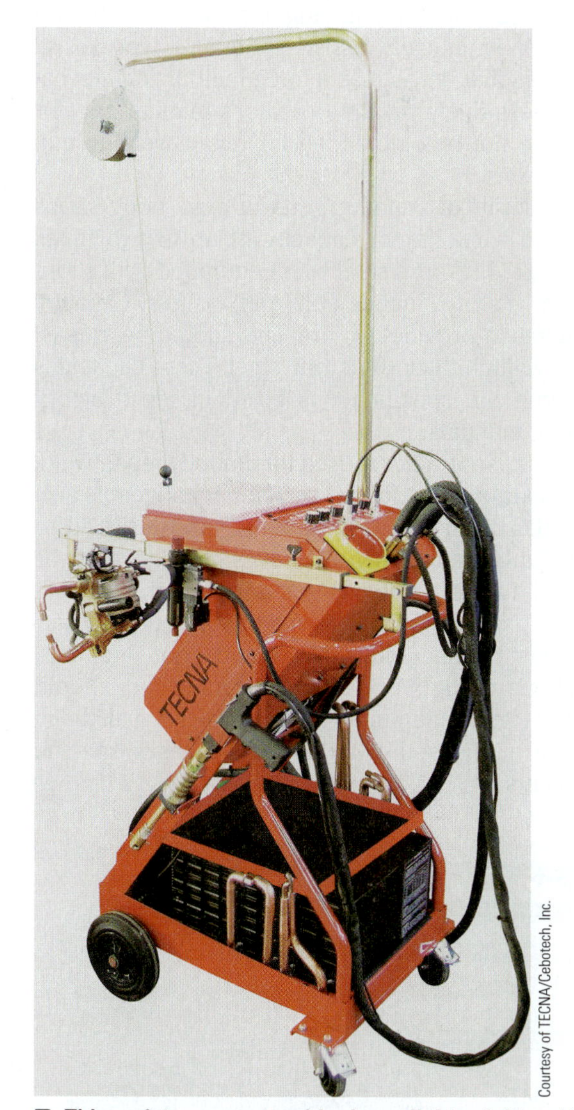

B This resistance spot welder has a lightweight welding head with various tip lengths. This type of welder will reach most flanges during panel replacement.

Figure 8-59 When replacing badly damaged body panels, most manufacturers recommend resistance spot welding over plug welding. It is faster and will result in a more factory-type weld. When a car is built in a factory, resistance spot welds are used, not plug welds.

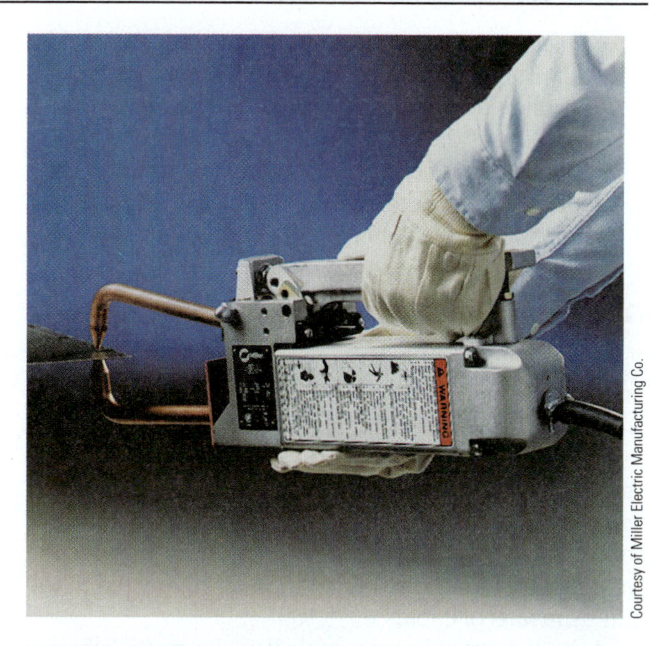

Figure 8-60 When placed around two pieces of sheet metal, a resistance spot weld sends a current through the workpieces to join them together without a filler rod. This results in a clean, factory looking spot weld.

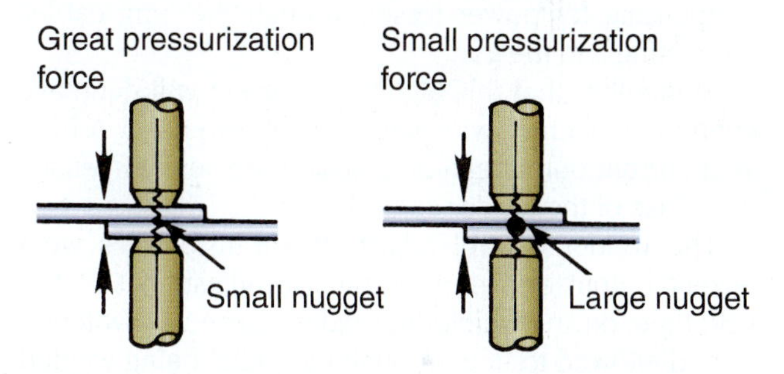

Figure 8-61 Electrode (tip) pressure affects the spot weld.

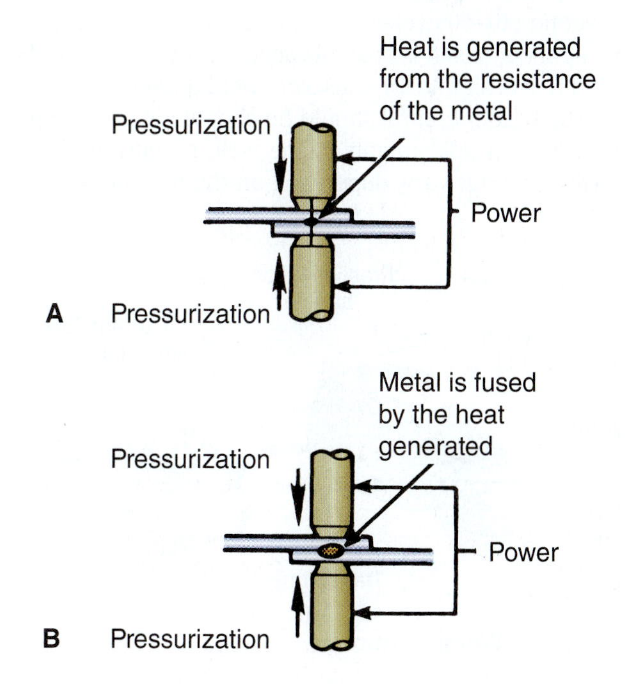

Figure 8-62 Note how electrical current (amperage) forms a weld.

3. *Holding.* If the current flow is stopped, the melted portion begins to cool and forms a round, flat bead of solidified metal (nugget). This structure becomes very dense due to the pressurization force, and its subsequent mechanical bonding is excellent. Pressurization time is very important. Do not use less time than specified in the operator's manual.

Resistance Spot Welding Components

The components of a resistance spot welder (Figure 8–63) are the welding transformer, the welder control, and the welding gun with interchangeable arm sets.

The transformer converts low-amperage, 240-volt shop line current to high secondary amperage, low-voltage (2–5 volts) welding current, safe from electrical shock. The welder transformer can either be built into the welding gun or mounted remotely and connected to the gun by means of cables.

A built-in transformer is electrically more efficient because there is little or no loss of welding current between the transformer and the gun. A remote transformer must be larger and draw more shop line current to compensate for power losses through the long cables connecting it to the gun.

Remember that this high weld current will decrease when long-reach or wide-gap arm sets are used. A high weld current output can be adjusted to a lower intensity by the use of the welder control.

The welder control adjusts the transformer's weld current output and permits precise adjustment of the weld time, during which the welding current is switched on and allowed to flow through the metal being welded and then switched off. It is desirable to have a range of timing adjustment from approximately 1⁄6 of a second to 1 second (10–60 cycles) for typical collision repair welding applications. A repeatable accuracy of at least 1⁄10 of a second is desirable for consistent weld quality.

The welder control should be capable of providing a full range of adjustments to the welding current. Weld current settings vary, depending on the thickness of the steel to be welded and the length and gap of the arm sets needed to reach into the area being welded. It might be necessary to decrease weld current when welding with short-reach arm sets, or increase weld current when using long-reach or wide-gap arm sets.

Some manufacturers of resistance spot welders designed for unibody repair work offer additional control features that compensate for small amounts of surface scale or slight rust on the metal. Such features permit the repair specialist to determine when a poor weld condition exists.

The welding gun applies the squeeze force and delivers the welding current through the welder arms to the metal being welded. Most resistance spot welders are designed with a force multiplying mechanism to produce the high electrode force required for consistent weld quality. These force multiplying mechanisms can be spring or pneumatically assisted. Squeeze-type resistance welders that do not use a force multiplying mechanism and rely solely on the operator's manual grip for pressure are not recommended for repair welding unibody structures.

The majority of welding guns in auto body shops should have a maximum capacity of up to two times 5⁄64-inch-thick (1.98 mm) steel when equipped with short-reach arm sets of 5 inches (127 mm) or less. Capacity with long-reach or wide-gap arm sets should be at least two times 1⁄32-inch-thick (0.79 mm) steel. These capacities comply with the specifications listed in most factory body repair manuals.

Resistance spot welders used for unibody repair welding are available with a full range of interchangeable arm sets. Standard arm sets (Figure 8–64) are designed to reach difficult areas on most makes of cars, such as wheel well flanges, drip rails, taillight openings, and other tight pinch weld areas, as well as floor pan sections, rocker panels, and window and door openings. Repair shops doing work for new car dealers should check the factory repair manuals and look for availability of special arm sets for the hard-to-reach areas on specific makes of cars.

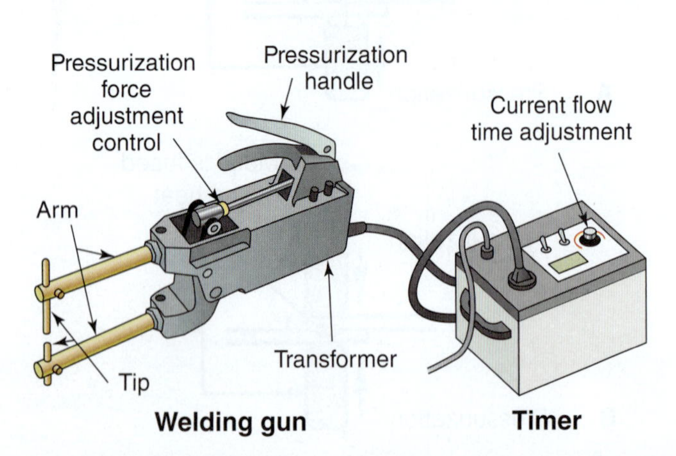

Figure 8–63 Study the components of a resistance spot welding system.

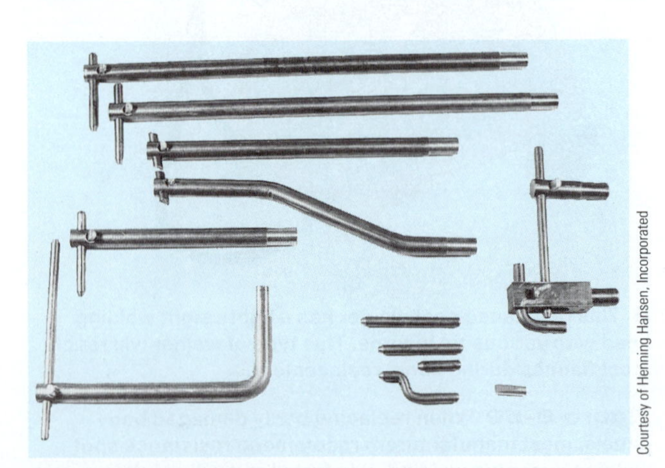

Figure 8–64 Various accessory arms are needed to reach around the panels to be welded.

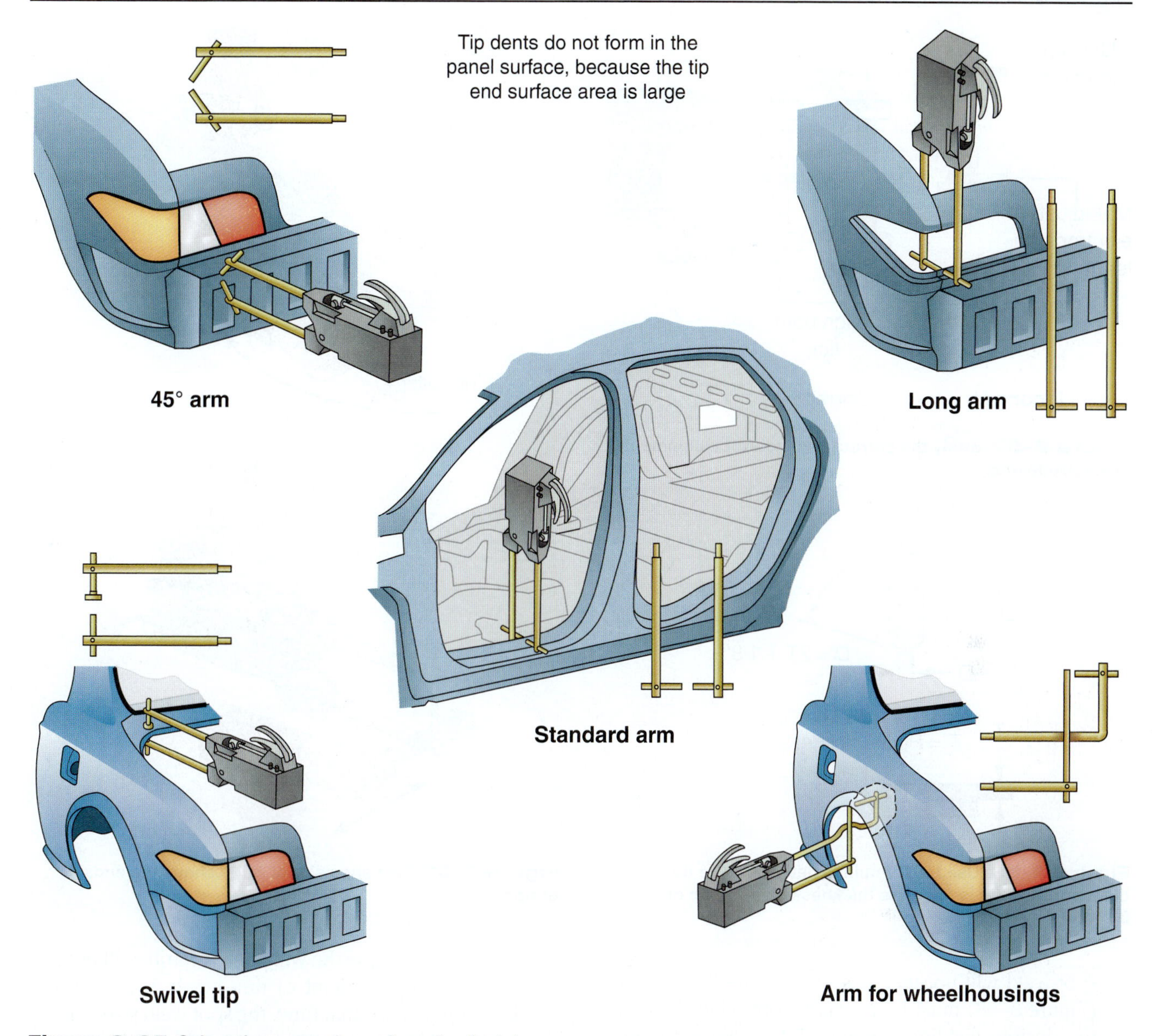

Figure 8–65 Select the proper type of arm for the job.

Spot Welder Adjustments

To obtain sufficient strength at the spot-welded portions, perform the following checks and adjustments on the squeeze-type resistance spot welding gun before starting:

- *Arm selection.* It is important to select the arm according to the area to be welded (Figure 8–65).
- *Adjustment of arm.* Keep the gun arm as short as possible to obtain the maximum pressure for welding (Figure 8–66). Securely tighten the gun arm and tip so that they will not become loose during the operation.
- *Alignment of electrode tips.* Align the upper and lower electrode tips on the same axis (Figure 8–67). Poor alignment of the tips causes insufficient pressurizing, and this results in insufficient current density and insufficient strength at the welded portions.

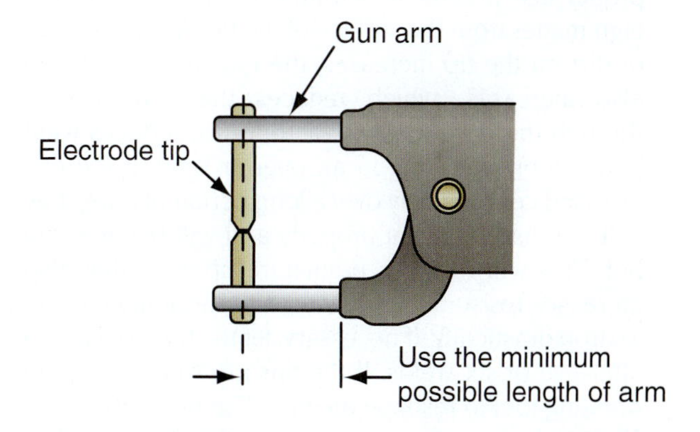

Figure 8–66 Adjust the gun arms for proper alignment.

- *Diameter of electrode tip.* The diameter of the spot weld decreases as the diameter of the electrode tip increases. Also, if the electrode tip is too small, the

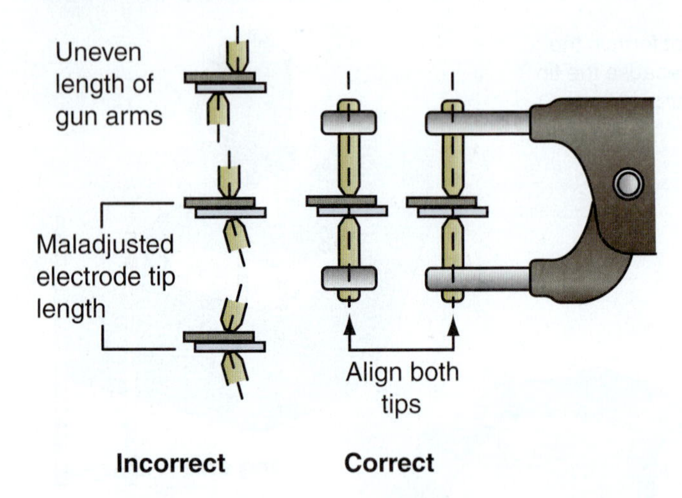

Figure 8-67 Study the correct and incorrect alignment of electrode tips.

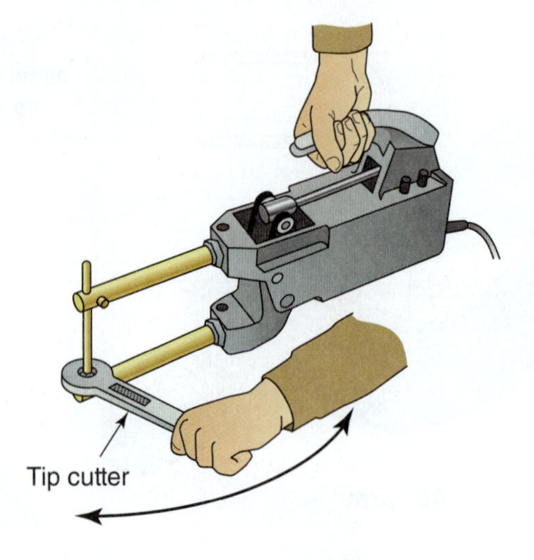

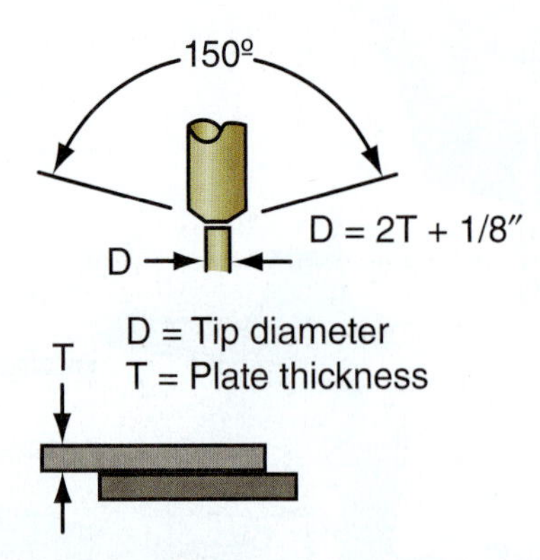

Figure 8-68 Use the formula to determine the tip diameter: two times the plate thickness plus ⅛ inch or 3.2 mm.

Figure 8-69 This special tool will reshape the ends of tips.

spot weld will not increase in size. The tip diameter (Figure 8–68) must be properly controlled to obtain the desired welding strength. Before starting operation, make sure that the tip diameter (D) is kept the proper size, and file it cleanly to remove burnt or foreign matter from the surface of the tip. As the amount of dirt on the tip increases, the resistance at the tip also increases, which reduces the current flow through the base metal, which in turn reduces weld penetration, resulting in an inferior weld. If the tips are used continuously over a long period of time, they will not dissipate heat properly and will become red hot. This will result in premature tip wear that also increases resistance and causes the welding current to drop drastically. If necessary, let the tips cool down after five or six welds. If the tips are worn, use a tip dressing tool to reshape the tips (Figure 8–69).

- *Electrical current flow time.* Current flow time also has a relationship to the formation of a spot weld. When the electrical current flow time increases, the heat that is generated increases the spot weld diameter and penetration. The amount of heat that is dissipated at the weld increases as the current flow time increases. Because the weld temperature will not rise after a certain amount of time, even if the current flows longer than that time, the spot weld size will not increase. However, tip pressure marks and heat warping might occur.

The pressurization force and welding current of many spot welders cannot be adjusted, so the current value might be low. However, welding strength can be ensured by lengthening the current flow time (letting low current flow for a long time).

While spot welding galvanized or zinc-coated steel panels used in auto bodies, offset the drop in current density by raising the current value 10–20 percent above that for ordinary steel panels. Because the current value cannot be adjusted in spot welders ordinarily used for body repairs, lengthen the current flow time a little.

The best welding results can be obtained by adjusting the arm length or welding time according to the thickness of the panels. Although the welder instruction manual has these values listed inside, it is best to test the quality of the weld using the methods described later.

Operating a Squeeze-Type Resistance Spot Welder

Hold the welding gun and position it so that the welder arm electrodes contact the body parts to be welded. Then use the squeeze mechanism to apply weld force to both sides of the metal being welded. As force is applied and maintained on the metal, the force mechanism initiates an electrical signal to the welder control that switches on the flow of weld current for a preset time and then switches it off. Because the weld time is usually less than a second, the entire process is very fast.

Other important operational considerations when using a squeeze-type resistance spot welder are:

- *Clearance between welding surfaces.* Any clearance between the surfaces to be welded causes poor current flow (Figure 8–70). Even if welding can be done without removing such a gap, the welded area would become smaller, resulting in insufficient strength. Flatten the two surfaces to remove the gap and clamp them tightly before welding.
- *Metal surface to be welded.* Paint film, rust, dust, and any other contamination on the metal surfaces to be welded cause insufficient current flow and produce poor results. Remove such foreign matter from the surfaces to be welded (Figure 8–71).
- *Corrosion.* Coat the surfaces to be welded with an anticorrosion agent (see Chapter 20) that has higher conductivity. It is important to apply the agent uniformly to the end face of the panel (Figure 8–72 and Figure 8–73).
- *Performance of spot welding operations.* When performing spot welding operations, be sure to:

  Use the direct welding method. For the portions to which direct welding cannot be applied, use plug welding by MIG welding.

  Apply electrodes at a right angle to the panel. If the electrodes are not applied at right angles, the current density will be low, resulting in insufficient welding strength.

  For the portion where three or more metal sheets are overlapping, spot welding should be done twice.
- *Number of points of spot welding.* The capacity of spot welding machines available in a repair shop generally

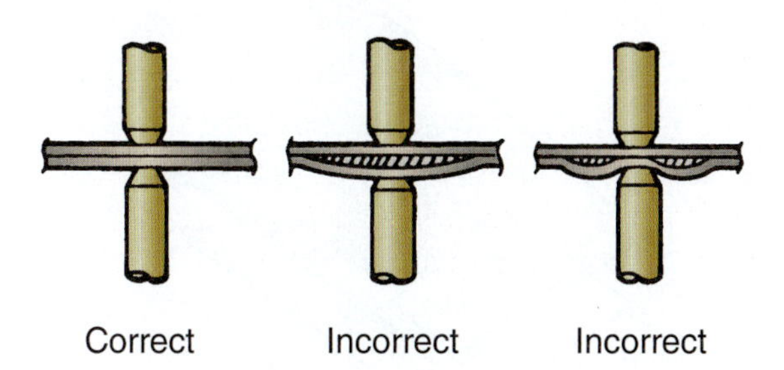

Figure 8–70 Compare the correct and incorrect clearance between welding surfaces.

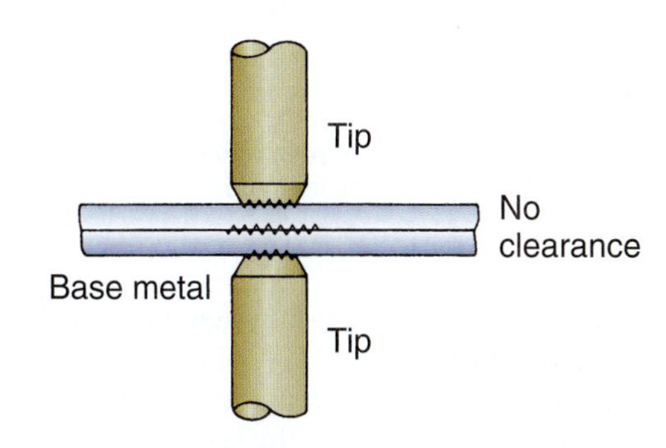

Figure 8–71 The condition of base metal surfaces is critical to weld quality.

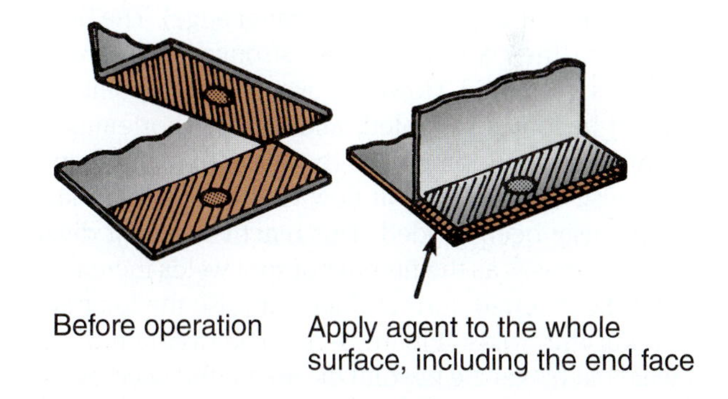

Figure 8–72 Apply an anticorrosion agent to metal surfaces requiring protection from rust. The service manual will give locations of the areas needing an agent.

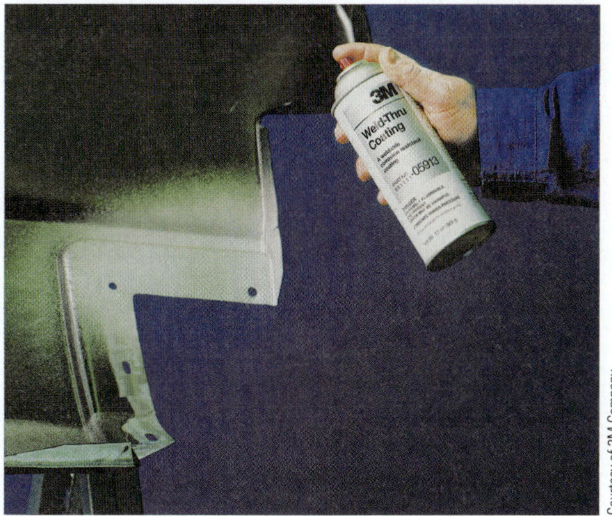

Figure 8–73 During major repairs and panel replacement, weldthrough primer must be applied to flanges of panels to be welded. It will help protect from corrosion and allow for a good weld.

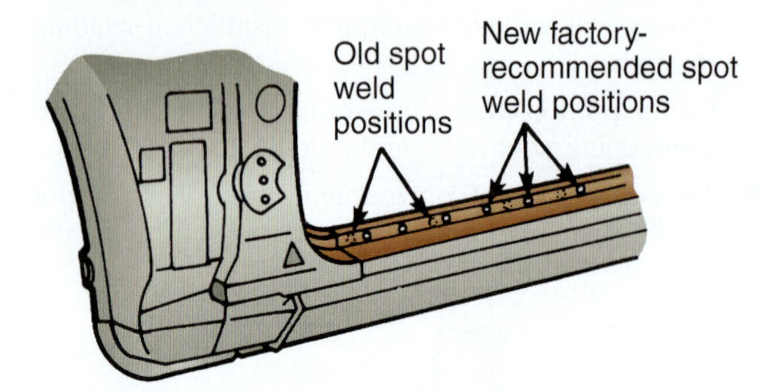

Figure 8-74 **The number of points to spot weld will also be given in the vehicle's service manual.**

is smaller than that of welding machines at the factory. The number of points of spot welding should be increased accordingly by 30 percent in a service shop compared to spot welding in the factory (Figure 8–74).

- *Minimum welding pitch.* The strength of individual spot welds is determined by the *spot weld pitch* (the distance between spot welds) and *edge distance* (the distance of the spots from the panel edge). The bond between the panels becomes stronger as the weld pitch is shortened. However, over a certain point, the metal becomes saturated, and further shortening of the pitch will not increase the strength of the bond, because the current will flow to the spots that have previously been welded. This reactive current diversion increases as the number of spot welds increases, and the diverted current does not raise the temperature at the welds (Figure 8–75). The distance of the weld pitch must be beyond the area influenced by the reactive current diversion. In general, the values given in Table 8–6 should be observed.
- *Position of welding spot from edge and end of panel.* The edge distance is also determined by the position of the welding tip. Even if the spot welds are normal, the welds will not have sufficient strength if the edge distance is insufficient. When welding near the end of a panel, observe the values for the distance from the end of a panel given in Table 8–7. A distance that is too small results in insufficient strength and also in a strained panel.

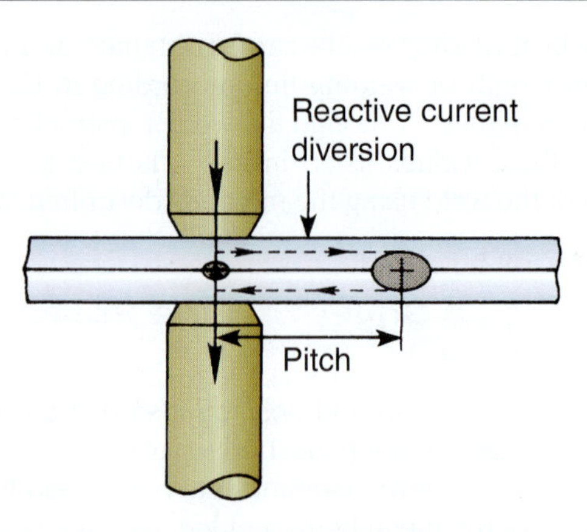

Figure 8-75 **Minimum welding pitch is spec for how far apart the welds should be located. This information is given by the auto manufacturer in its body repair manuals.**

- *Spotting sequence.* Do not spot continuously in one direction only. This method provides weak welding due to the shunt effect of the current (Figure 8–76). If the welding tips become hot and change their color, stop welding and allow them to cool.
- *Welding corners.* Do not weld the corner radius portion (Figure 8–77). Welding this portion results in concentration of stress that leads to cracks. The following three locations require special consideration:

1. Upper corner of front and center pillars
2. Front upper portion of the quarter panel
3. Corner portion of front and rear windows

Inspection of Spot Welds

Spot welds are inspected either by outward appearance (visual inspection) or destructive testing. Destructive testing is used to measure the strength of a weld, and a visual inspection is used to judge the quality of the outward appearance.

TABLE 8–6 SPOT WELDING POSITIONS

Panel Thickness	Pitch S	Edge Distance P
1/64 in.	7/16 in. or more	13/64 in. or more
1/32 in.	9/16 in. or more	13/64 in. or more
Less than 3/64 in.	11/16 in. or more	1/4 in. or more
3/64 in.	7/8 in. or more	9/32 in. or more
1/16 in.	1-9/64 in. or more	5/16 in. or more

TABLE 8–7 POSITION OF WELDING SPOT FROM END OF PANEL

Thickness (t)	Minimum pitch (ℓ)
1/64 in.	7/16 in. or over
1/32 in.	7/16 in. or over
Less than 3/64 in.	15/32 in. or over
3/64 in.	9/16 in. or over
1/16 in.	5/8 in. or over
1/64 in.	11/16 in. or over

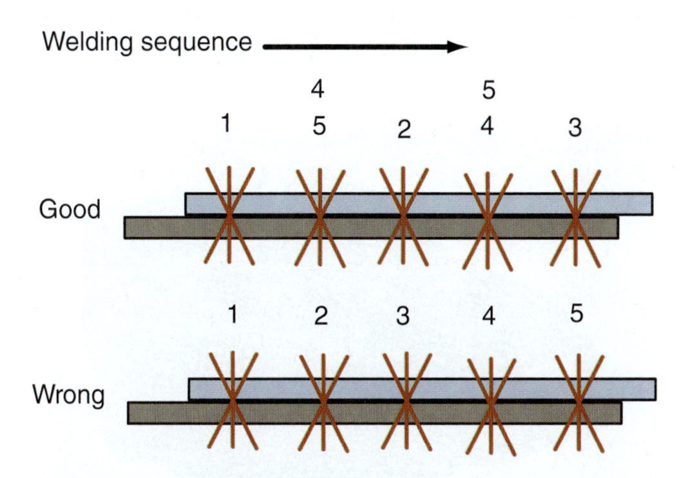

Figure 8–76 **Memorize the proper welding sequence to avoid heat buildup and panel warpage.**

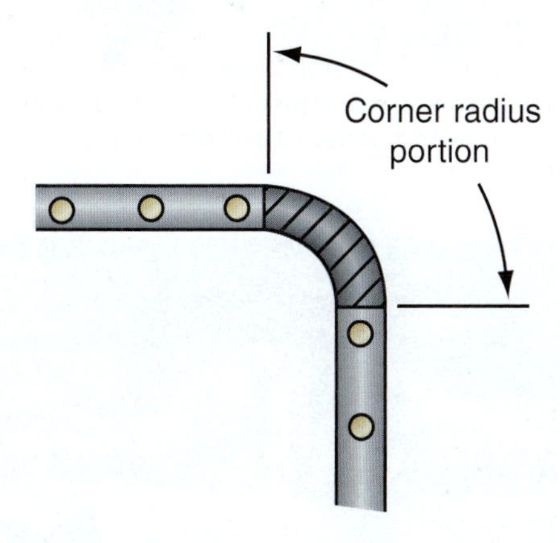

Figure 8–77 **Often, you do not weld corners; just weld right up to them or as directed in the manual.**

Appearance Inspection

Check the finish of the weld visually and by touching. The items to check are:

- *Spot position*. The spot weld position should be in the center of the flange, with no tip holes, and have no spot welds overriding the edge. As a rule, an old spot position should be avoided.
- *Number of spots*. There should be 1.3 or more times the number made by the manufacturers. (For example, 1.3 times 4 original factory spot welds equals roughly 5 new repair spot welds.)
- *Pitch*. It should be a little shorter than that of the manufacturer and spots should be uniformly spaced. The minimum pitch should be at a distance where reactive current diversion will not occur.
- *Dents (tip bruises)*. There should be no dents on the surfaces that exceed half the thickness of the panel.
- *Pinholes*. There should be no pinholes that are large enough to see.
- *Spatter*. A glove should not catch on the surface when rubbed across it.

Destructive Testing

Most destructive tests require the use of sophisticated equipment, a requirement that most body shops are unable to meet. For this reason, the two simpler methods described here have been developed for general use in body shops.

1. **Destructive test**. A test piece of the same metal as the welded piece and with the same panel thickness is made and welded in the positions shown in Figure 8–78. Force is then applied in the direction of the arrow and the spots are separated. How cleanly the weld breaks determines whether it is satisfactory. If the weld pulls out cleanly, like a cork from a bottle, the weld is judged to be good. It should be noted that since the weld performance cannot be exactly duplicated by this test, the results should only serve as a reference.
2. **Nondestructive test**. To confirm a spot weld after it has been made, use a chisel and hammer and proceed as follows:

Insert the tip of a chisel between the welded plates (Figure 8–79) and tap the end of the chisel until a clearance of ⅛–5⁄32 inches (3.2–3.97 mm, when the plate thickness is approximately 1⁄32 inch [0.79 mm]) is formed between the plates. If the welded portions remain normal, it indicates that the welding has been done properly. This clearance varies with the

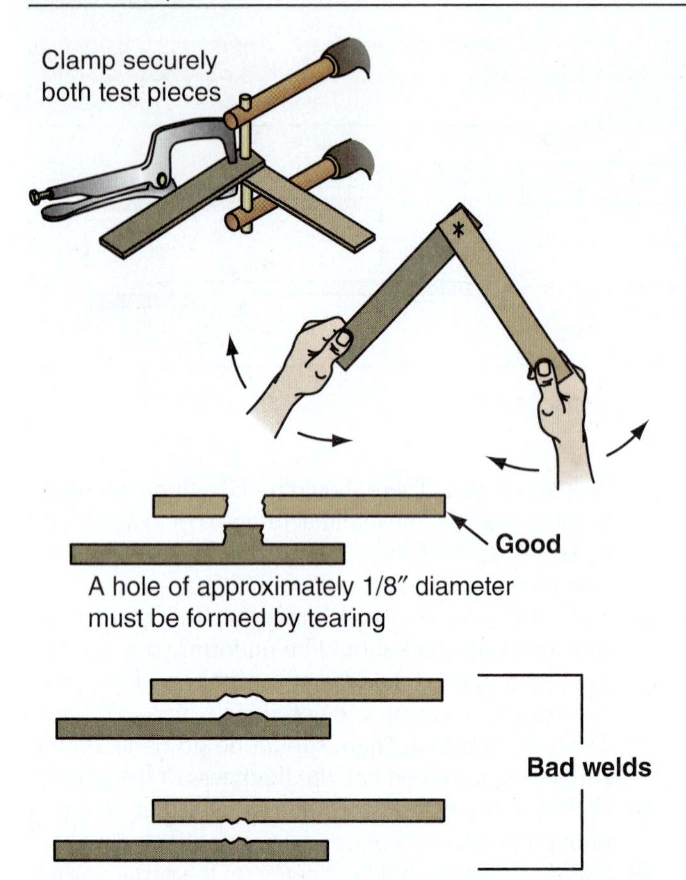
Clamp securely
both test pieces
Good
A hole of approximately 1/8″ diameter
must be formed by tearing
Bad welds

Figure 8-78 **When performing a destructive check, force the test weld apart and inspect it**

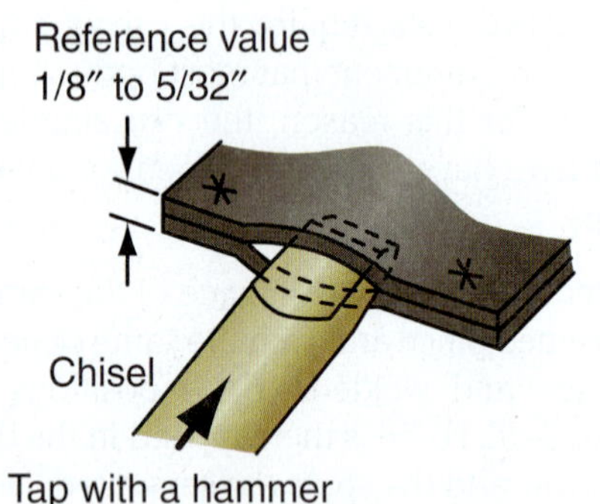
Reference value
1/8″ to 5/32″
Chisel
Tap with a hammer

Figure 8-79 **Perform a nondestructive test on metal identical to that being welded on the vehicle. This will enable you to double-check welder adjustments and welding technique.**

location of the welded spots, length of the flange, plate thickness, welding pitch, and other factors. Note that the values given here are only reference values.

If the thickness of the plates is not equal, the clearance between the plates must be limited to 1⁄16–5⁄64 inches (1.58–1.98 mm). Note that further opening of the plates can become a destructive test.

Be sure to repair the deformed portion of the panel after inspection.

8.12 OTHER SPOT WELDING FUNCTIONS

Although the squeeze-type welding gun is the most used in the repair shop, there are other types of guns used with spot welding equipment. With the proper gun attachment, a spot welder can be used as a panel spotter, stud welder, spot shrinker, and mold rivet welder.

8.13 STUD SPOT WELDING FOR DENT REMOVAL

Studs used in dent removal can be resistance welded with a special stud welder (Figure 8–80) or a panel spotter equipped with stud welding attachments. With either

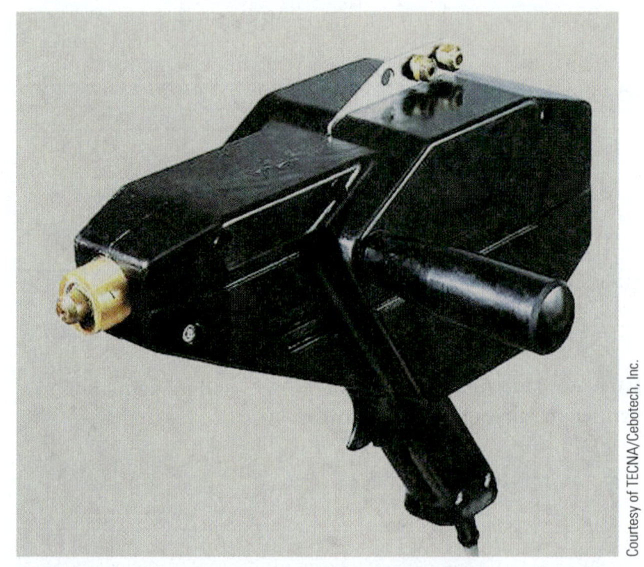

Courtesy of TECNA/Cebotech, Inc.

A By inserting a small metal pull pin into the tip of the stud welder, you can weld a series of pull pins to the low area in a panel dent.

CEBORA

Courtesy of CEBORA/Cebotech, Inc.

B Here a technician is using special aluminum studs and a stud welder to remove a small dent from an aluminum body panel. Note framework for pulling on studs to remove dent in aluminum.

Figure 8-80 **Stud welders can be used to weld pull pins on dented panels so you can pull out damage from the front.**

method, a stud pulling kit containing all the necessary items (including a slide hammer) is a must for dent removal.

To remove a dent properly with either a stud or stud spot welder, a good quality stud is necessary. The stud should offer the necessary combination of pull strength and tensile strength, while remaining extremely flexible. The flexibility allows the stud to be bent out of the way when working on adjacent studs, then bent back when required. The importance of this stud is to minimize the heat required and, therefore, maintain the flexibility of the steel when being applied and removed. Complete details on using stud or panel welding for dent removal can be found in Chapter 11.

8.14 OXYACETYLENE WELDING

Oxyacetylene welding is a type of fusion welding. Acetylene and oxygen are mixed in a chamber, ignited at the tip, and used as a high-temperature heat source (approximately 5,400°F, or 2,984°C) to melt and join the welding rod and base metal together (Figure 8–81).

Because it is difficult to concentrate the heat in one area, the heat affects the surrounding areas and reduces the strength of steel panels. Because of this problem, auto makers do not recommend the use of oxyacetylene to repair damaged vehicles. Although oxyacetylene is in disfavor with most automobile manufacturers—with good reason—it has some use in the body shop. The oxyacetylene flame is still used to repair other damaged auto bodies and for some heat-shrinking operations, brazing, soldering, surface cleaning, and cutting of nonstructural parts. Oxyacetylene should not be used to cut structural parts of any vehicle unless special care is taken.

Figure 8-81 An oxyacetylene welder is seldom used to weld body panels because of the heat warpage that results.

Welding and Cutting Equipment

In general, an oxyacetylene welding and cutting outfit (Figure 8–82) consists of the following:

- *Steel tanks (cylinders)* filled with:

  Oxygen

  Acetylene
- *Regulators* (Figure 8–83) that reduce the pressure coming from the tanks to the desired level and maintain a constant flow rate of:

  Oxygen pressure: 15–100 psi (103–689 kPa)

  Acetylene: 3–12 psi (21–83 kPa)
- *Hoses* from the regulators and cylinders connect the oxygen and acetylene to the torch.
- *Torch.* The torch body mixes the oxygen and acetylene from the tanks in the proper proportions and produces a heating flame capable of melting steel. There are two main types of torches:

  Welding torch

  Cutting torch

The low-pressure torch is generally used for acetylene welding. This torch can be used at an extremely low acetylene pressure and has an injector nozzle. The gases are mixed by the discharge of oxygen from the center nozzle (Figure 8–84).

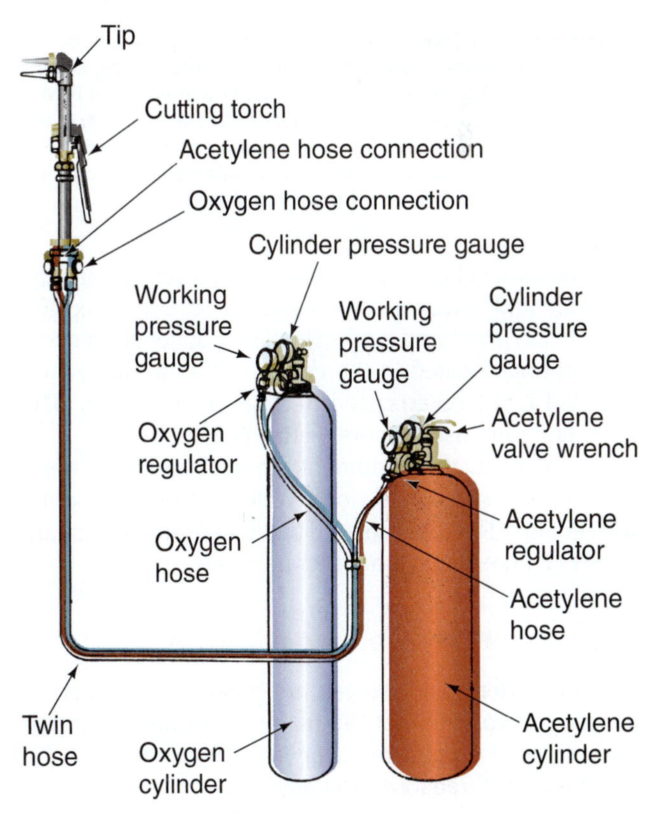

Figure 8-82 Study the parts of an oxyacetylene welding and cutting outfit.

Figure 8-83 Oxyacetylene equipment is still used for numerous body shop tasks, from heating metal during stress relieving to soft brazing of panel joints. You must know how to use this equipment. Study the major controls and parts of oxyacetylene equipment. For safety reasons, always shut off the main tank valves right after use. Gas can still leak out of the hoses if you only shut off the torch valves.

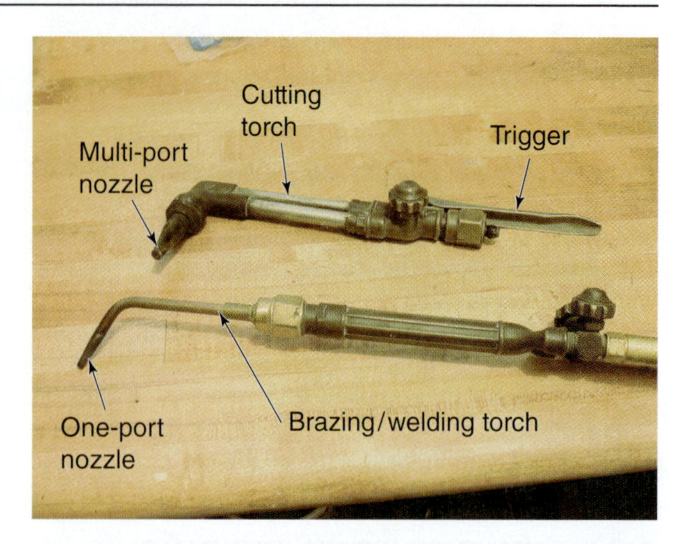

Figure 8-84 Compare gas cutting and welding torches. Study this close-up of an oxyacetylene cutting torch. When you press down on the trigger, a high-volume blast of pure oxygen comes out of the torch to blow away the molten metal preheated by the gas flame.

As shown in Figure 8–85, the cutting torch has an oxygen tube and valve for conducting high-pressure oxygen attached to a welding torch. The flame outlet has a small oxygen hole located in the center of the tip that is surrounded by holes arranged in a spherical pattern. The outer holes are used for preheating (Figure 8–86).

To round out the equipment, the safety gear described in Chapter 9 should be worn. Welding should be done with either a Number 4, 5, or 6 tinted filter shade. A spark lighter (Figure 8–87) is another necessity.

Types of Flame and Adjustment

When acetylene and oxygen are mixed and burned in the air, the condition of the flame varies depending on the volume of oxygen and acetylene (Figure 8–88).

There are three forms of flame:

1. **Neutral flame**. The standard flame is said to be a neutral flame. Acetylene and oxygen mixed in a 1 to 1 ratio by volume produces a neutral flame. As shown in Figure 8–89A, this flame has a brilliant white cone surrounded by a clear blue outer flame.
2. **Carburizing flame**. The carburizing flame, also called a surplus or reduction flame, is obtained by mixing slightly more acetylene than oxygen. Figure 8–89B shows that this flame differs from the neutral flame in that it has three parts. The cone and the outer flames are the same as the neutral flame, but between them there is an intermediate, light-colored acetylene cone enveloping the cone. The length of the acetylene cone varies according to the amount of surplus acetylene in the gas mixture. For a double surplus flame, the oxygen–acetylene mixing ratio is about 1 to 1.4 (by volume). A carburizing flame is used for welding aluminum, nickel, and other alloys.
3. **Oxidizing flame**. The oxidizing flame is obtained by mixing slightly more oxygen than acetylene. The oxidizing flame (Figure 8–89C) resembles the neutral flame in appearance, but the acetylene cone is shorter and its color is a little more violet compared to the neutral flame. The outer flame is shorter and fuzzy at the end. Ordinarily, this flame oxidizes melted metal, so it is not used in the welding of mild steel, but it is used in the welding of brass and bronze.

The acetylene line pressure must never exceed 15 psi (103 kPa). Free acetylene has a tendency to dissociate at pressure above 15 psi (103 kPa) and could cause an explosion.

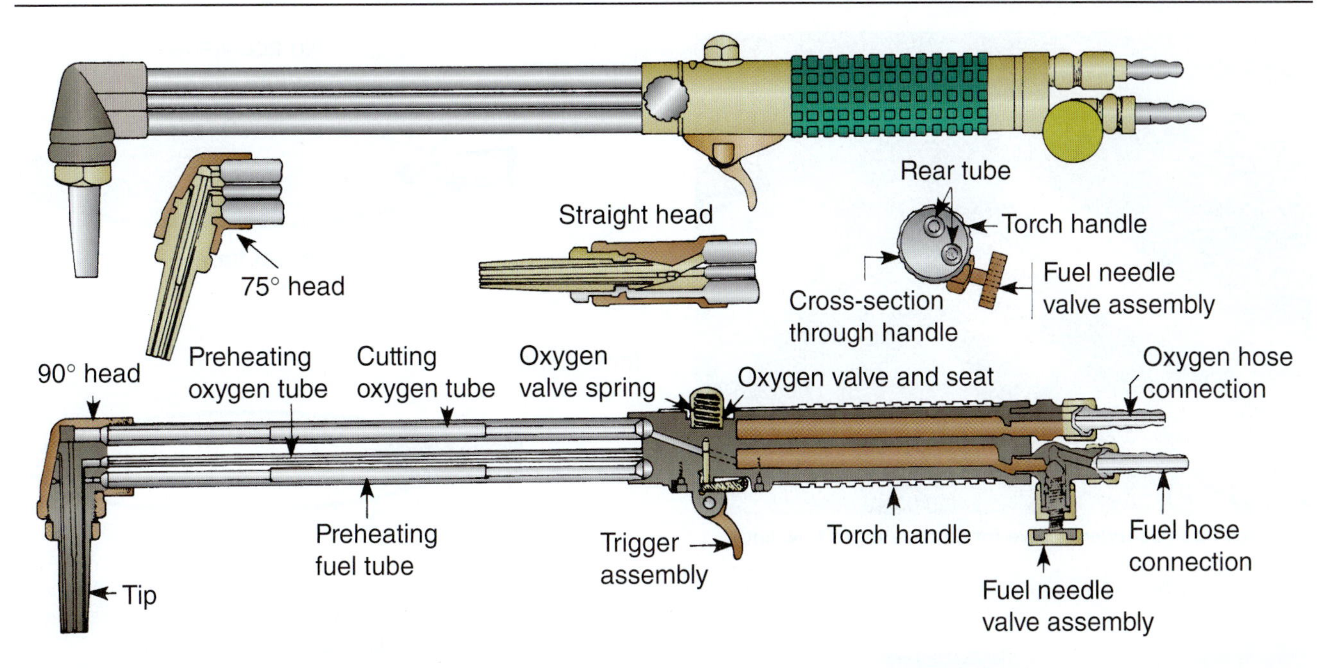

Figure 8-85 Study the parts of a typical cutting torch.

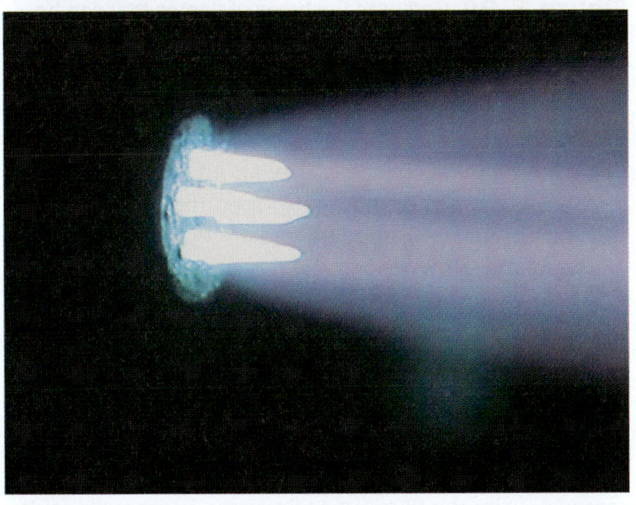

Figure 8-86 To use a cutting torch, adjust oxygen and acetalyene valves to produce a good flame as shown. Note the bright cones of flame coming out of the cutting torch nozzles.

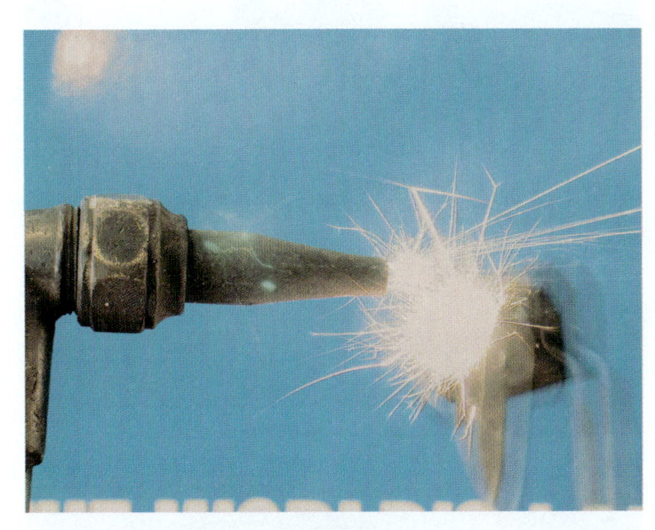

Figure 8-87 By squeezing the handle on a spark lighter, you can light the flame on the torch.

Welding Torch Flame Adjustment

As stated in the overview of welding, oxyacetylene welding is not used for welding modern auto bodies, but it is used for brazing certain nonstructural panels at factory-brazed seams. When using a welding torch, proceed as follows:

1. Attach the appropriate tip to the end of the torch. Use the standard tip for sheet metal (each torch manufacturer has a different system for measuring the size of the tip orifice).
2. Set the oxygen and acetylene regulators at the proper pressure:

   Oxygen = 8–25 psi (55–172 kPa)

   Acetylene = 3–8 psi (21–55 kPa)

A Crack open acetylene valve for moderate gas flow and light torch.

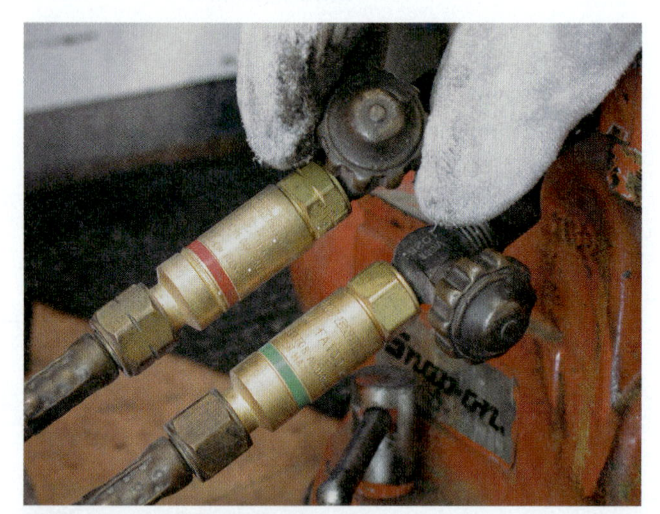

B Then adjust oxygen for a good flame.

C A good flame will show a bright cone at the outlet of the nozzle.

Figure 8–88 Study the technique for lighting a gas welding torch. Note spark arrestors in gas and oxygen feed lines. They prevent accidental gas burnback into lines.

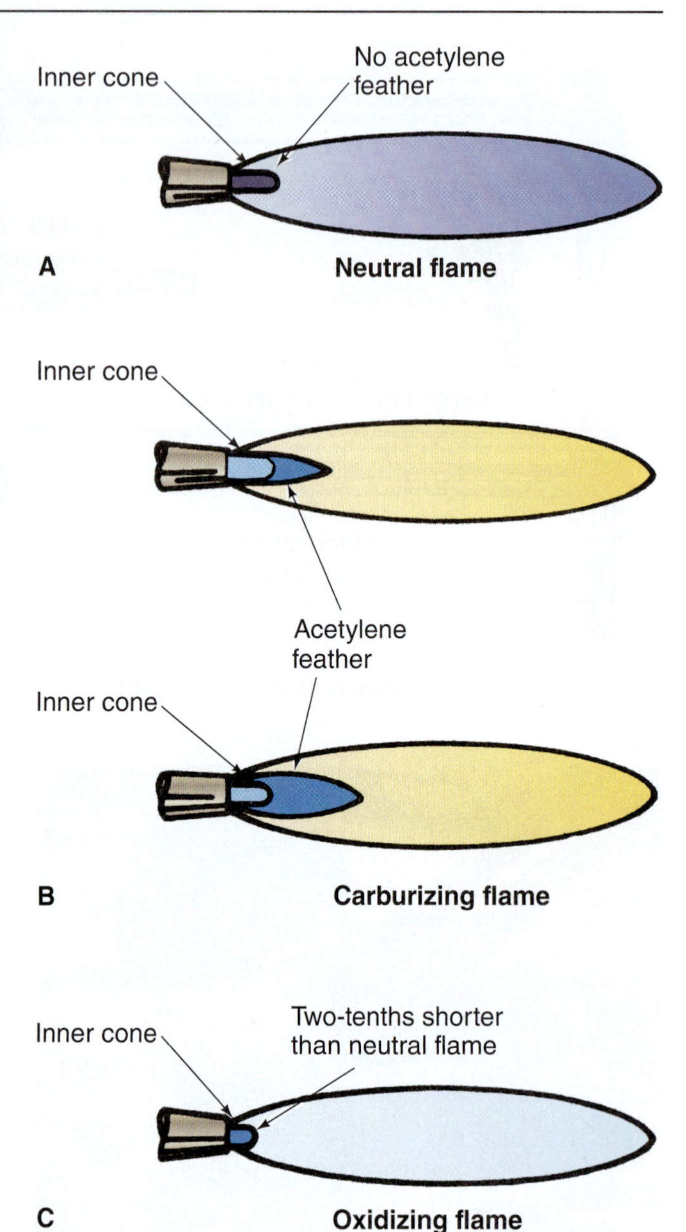

Figure 8–89 Compare the types of flames that can be produced with a gas torch.

3. Open the acetylene valve about half a turn and ignite the gas. Continue to open the valve until the black smoke disappears and a reddish yellow flame appears. Slowly open the oxygen valve until a blue flame with a yellowish white cone appears. Further open the oxygen valve until the center cone becomes sharp and well defined. This type of flame is called a neutral flame and is used for welding mild steel (other than automobile bodies).

If acetylene is added to the flame or oxygen is removed from the flame, a carburizing flame will result.

If oxygen is added to the flame or acetylene is removed from the flame, an oxidizing flame will result.

Figure 8–90 shows a torch tip being cleaned.

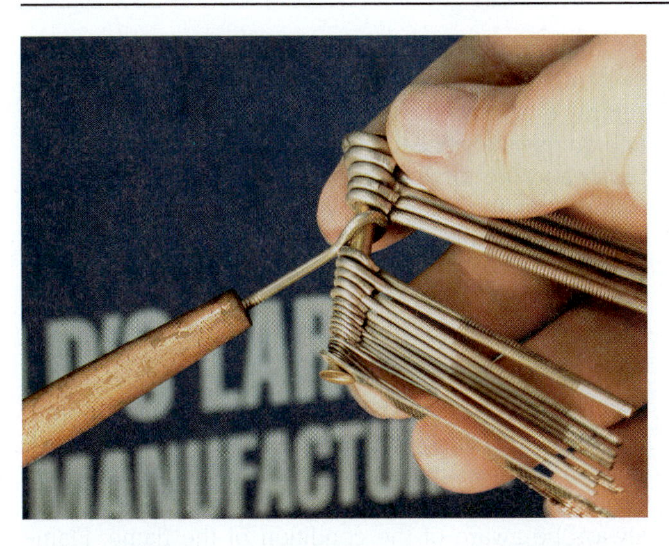

Figure 8–90 If a gas torch nozzle becomes restricted, use a nozzle cleaning tool or small round files to remove debris.

Gas Cutting Torch Flame Adjustment

The cutting torch is sometimes used in collision repair shops to rough cut damaged panels. Gas cutting torch flame adjustment and cutting procedures are as follows:

1. Adjust the oxygen and acetylene valves for a preheating neutral flame.
2. Open the preheating oxygen valve slowly until an oxidizing flame appears. This makes it difficult for melted metal to remain on the surface of the cut panel, allowing for clean edges.
3. *Thick panel cutting method.* Heat a portion of the base metal until it is red hot. Just before it melts, open the high-pressure oxygen valve and cut the panel. Advance the torch forward while making sure the panel is melting and being cut apart. This method is widely used for thick panels when there are several pieces overlapped together or for a side member, even when there is an internal reinforcement.
4. *Thin panel cutting method.* Heat a small spot on the base metal until it is red hot. Just before it melts, open the high-pressure oxygen valve and incline the torch to cut the panel. When cutting thin material, incline the tip of the torch so that the cut will be clean and fast (this prevents unwanted panel warpage).

As soon as the cutting operation is completed, quickly turn off the high-pressure oxygen flow used for cutting and pull the torch away from the base metal. This action prevents sparks from entering the tip and igniting the oxygen–acetylene mixture in the torch handle. In extreme cases the ensuing fire could melt the torch handle.

Cutting HSS for Salvage Purposes

Salvage components must be cut with a grinding wheel disc, an air chisel and/or metal cutting saw, or with a plasma cutter. If the use of a gas torch is necessary when cutting HSS sheet metal components for salvage purposes or cutting a body structure for a front/rear clip, factory engineers advise the following approach:

1. Cut the metal structure at least 2 inches (51 mm) away from the desired cut line. Sheet metal within the heat-affected area will lose strength when subjected to the high heat levels of a torch.
2. After torch cutting, use a grinding wheel disc, an air chisel, or a metal saw to make the final cut at the originally intended dimension line. HSS damage will then be cut out of the salvaged part.

As stated previously, oxyacetylene equipment should not be used on HSS components for welding or cutting. Vehicle design engineers stress this point. There is just too much heat buildup that can reduce structural strength. However, in some instances an oxyacetylene torch can be used to heat HSS components or parts ("hot working"), provided the critical 1,200°F temperature is not exceeded. (Check the manufacturer's shop manual on this point, because some state that 1,000°F is the critical temperature.)

High-strength steels should be exposed to high temperatures from an oxyacetylene torch for only a very short period of time. Two minutes is the recommended maximum time span for exposing HSS to a 1,200°F (538°C) temperature to reduce the amount of scaling that normally takes place on the metal surface. High-temperature exposure causes discoloration.

To determine and control temperatures of HSS parts and components being "heat worked" with oxyacetylene equipment, it is necessary to use a temperature-indicating crayon (Figure 8–91).

Heat Crayons

With steel, the use of heat is avoided whenever possible to avoid reducing the strength of the metal. With aluminum, heat must be used to restore flexibility caused by work hardening. If not, it will crack when straightening force is applied.

Before straightening, heat is often applied to the damaged area of the aluminum. It is easy to apply too much heat, because aluminum does not change color with high temperatures. It also melts at a relatively low 1,220°F (660°C). Careful heat control is very important.

Heat crayons or *thermal paint* can be used to determine the temperature of the aluminum or other metal being heated. They will melt at a specific temperature and warn you to prevent overheating (Figure 8–91).

The crayon or paint is applied next to the aluminum area to be heated. The mark will begin to melt when the crayon's or paint's melting point is almost reached.

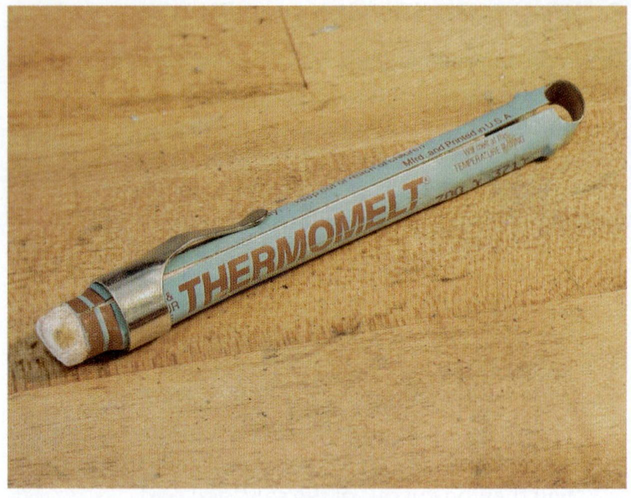

A

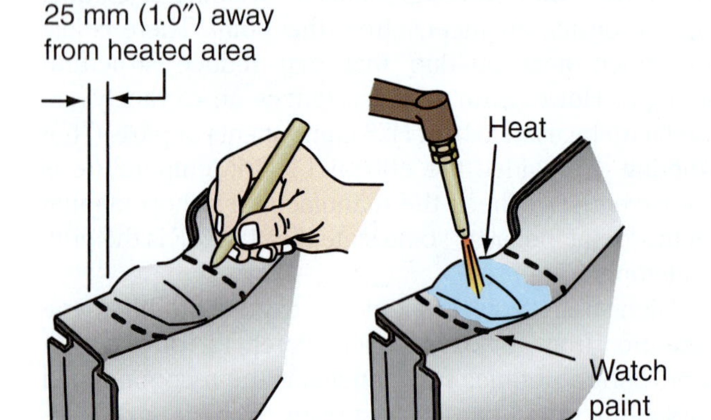

B

Figure 8-91 **(A) A typical heat-indicating crayon will melt at a specific temperature and help prevent overheating of parts while welding, cutting, or heating. (B) A heat crayon or thermal paint will melt at a predetermined temperature; this is especially helpful when heating or welding aluminum, which does not change color with heat. This aluminum frame is being straightened.**

The melting will let you know that you are about to reach the melting point of the aluminum.

The metal should be marked closely adjacent to the area being worked with a crayon rated no higher than 1,200°F (538°C). Using such a crayon will indicate to the welder whether an excessive amount of heat is being applied. Thus metal temperatures can be controlled within safe levels and HSS damage easily prevented.

Cleaning With a Torch

Before starting any weld, the surfaces to be joined must be thoroughly clean. The weld site must be completely free of any foreign material that might contaminate the weld. Otherwise, the finished weld is quite likely to be brittle, porous, and of poor integrity.

To remove heavy undercoating, rustproofing, tars, caulking, sealants, road dirt, and primers, first use a scraper to remove the loose material. Then use a scraper and an oxyacetylene torch. If needed, follow with a wire brush and the torch, using a carburizing flame. In any event, keep the torch at a very low, controlled heat to prevent part damage. Use just enough heat to get the job done.

Flame Abnormalities

When changes occur during gas welding, such as overheating of the flame outlet, adhesion of spatter, or fluctuations in the gas adjustment pressure, the result will be variations in the flame and weld. Therefore, you must always be aware of the condition of the flame. Flame abnormalities, their causes, and remedies are described in Table 8–8.

8.15 BRAZING

Brazing is applied only to places for sealing. This is a method of welding in which a nonferrous metal with a lower melting point (temperature) than that of the base metal is melted without melting the base metal (Figure 8–92). Brass brazing is frequently applied to automotive bodies.

Brazing is similar to joining two objects with adhesives; melted brass sufficiently spreads between the base metals to form a strong bond. Braze joint strength is less than that of the base metal, but the same as the melted brass. Therefore, never use brazing as a structural joint unless recommended by the vehicle manufacturer.

There are two types of brazing:

1. Soft brazing (soldering)
2. Hard brazing (brass or nickel)

Ordinarily, the term *brazing* refers to hard brazing. The basic characteristics of brazing are:

- The pieces of base metal are joined together at a relatively low temperature so that the base metal does not melt. Therefore, there is a lower risk of distortion and stress in the base metal.
- Because the base metal does not melt, it is possible to join otherwise incompatible metals.
- Brazing metal has excellent flow characteristics; it penetrates well into narrow gaps and it is convenient for filling gaps in body seams.
- Because there is no penetration and the base metal is joined only at the surface, it has very low strength to resist repeated loads or impacts.
- Brazing is a relatively easy skill to master.

Automobile assembly plants sometimes use arc brazing to join the roof and quarter panels together (Figure 8–93). Arc brazing uses the same principles as MIG welding. However, argon is used with brazing metal

TABLE 8–8 FLAME ABNORMALITIES AND REMEDIES

Symptom	Cause	Remedy
Flame fluctuations	1. Moisture in the gas; condensation in hose. 2. Insufficient acetylene supply.	1. Remove the moisture from the hose. 2. Adjust the acetylene pressure and have the tank refilled.
Explosive sound while lighting the torch	1. Oxygen or acetylene pressure is incorrect. 2. Removal of mixed-in gases are incomplete. 3. The tip orifice is too enlarged. 4. The tip orifice is dirty.	1. Adjust the pressure. 2. Remove the air from inside the torch. 3. Replace the tip. 4. Clean the orifice in the tip.
Flame cutoff	1. Oxygen pressure is too high. 2. The flame outlet is clogged.	1. Adjust the oxygen pressure. 2. Clean the tip.
Popping noises during operation	1. The tip is overheated. 2. The tip is clogged. 3. The gas pressure adjustment is incorrect. 4. Metal is deposited on the tip.	1. Cool the flame outlet (while letting a little oxygen flow). 2. Clean the tip. 3. Adjust the gas pressure. 4. Clean the tip.
Reversed oxygen flow (Oxygen is flowing into the path of the acetylene.)	1. The tip is clogged. 2. Oxygen pressure is too high. 3. Torch is defective. (The tip or valve is loose.) 4. There is contact with the tip and the deposit metal.	1. Clean the tip. 2. Adjust the oxygen pressure. 3. Repair or replace the torch. 4. Clean the orifice.
Backfire (There is a whistling noise and the torch handle grip gets hot. Flame is sucked into the torch.)	1. The tip is clogged or dirty. 2. Oxygen pressure is too low. 3. The tip is overheated. 4. The tip orifice is enlarged or deformed. 5. A spark from the base metal enters the torch, causing an ignition of gas inside the torch. 6. Amount of acetylene flowing through the torch is too low.	1. Clean the tip. 2. Adjust the oxygen pressure. 3. Cool the tip with water (while letting a little oxygen flow). 4. Replace the tip. 5. Immediately shut off both torch valves. Let torch cool down. Then relight the torch. 6. Readjust the flow rate.

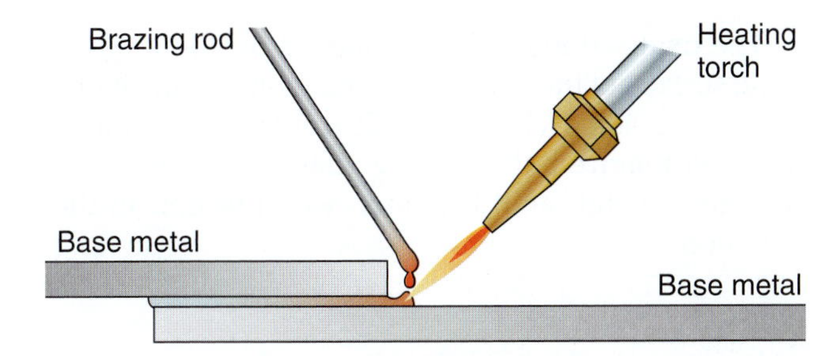

Figure 8–92 The brazing principle involves heating the base metal until the molten metal sticks to its surface. The base metals do not become molten as in welding.

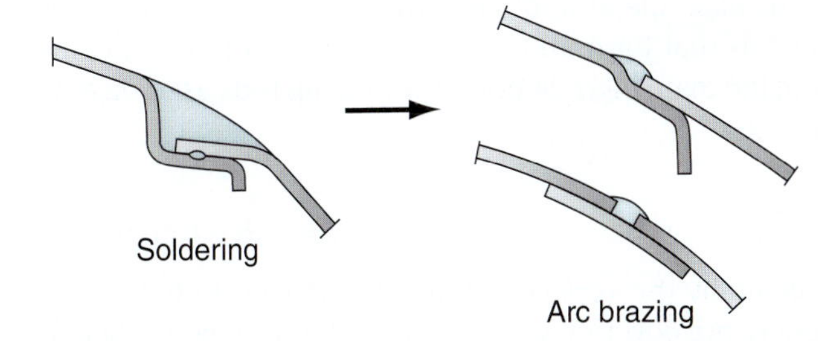

Figure 8–93 Note the typical body construction using solder or arc brazing.

instead of CO_2 or an argon/CO_2 mixture (Figure 8–94). Special brazing wire is also required. Because the amount of heat applied to the base metal is low, overheating is minimized. There is little distortion or warpage of the base metal. Compared to flame brazing, arc brazing shortens both the time for making the weld and finishing. Also, there is no danger of lead poisoning.

In the body shop, the brazing equipment is usually about the same as oxyacetylene welding. For brazing, an

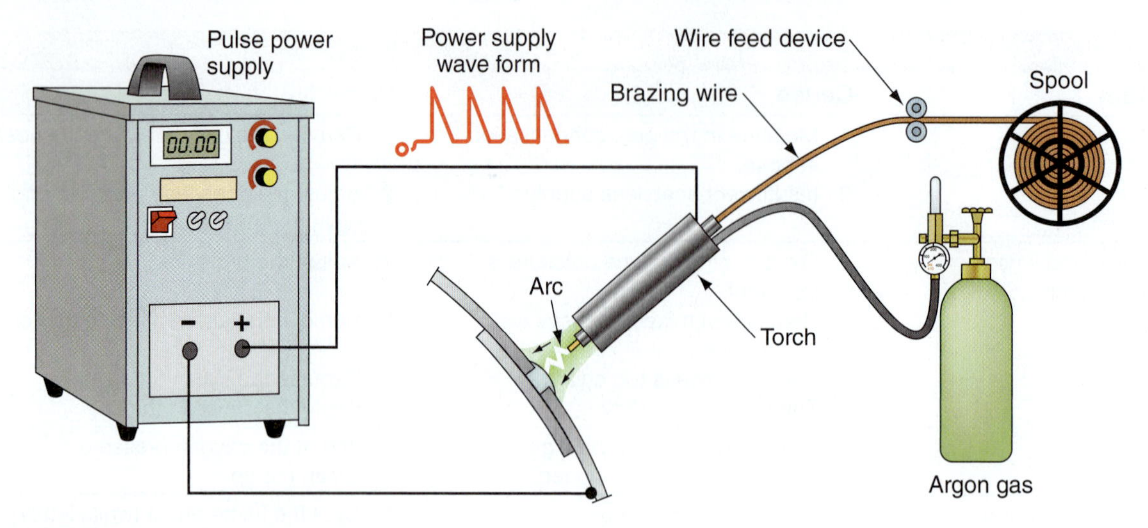

Figure 8–94 Study the principles of arc brazing.

TABLE 8–9 ALLOY-BASED BRAZING MATERIALS

Types of Brazing Materials	Main Ingredients
Brass brazing metal	Copper, Zinc
Silver brazing metal	Silver, Copper
Phosphor copper brazing metal	Copper, Phosphorus
Aluminum brazing metal	Aluminum, Silicon
Nickel brazing metal	Nickel, Chrome

oxyacetylene torch, brass filler rods, flux welding goggles, gloves, and a torch lighter are needed. Although the oxyacetylene torch can be used in soft brazing (soldering), it is best to use one designed specifically for soldering.

A brazing material with good qualities, such as flow characteristics, melting temperature, and compatibility with base metal and strength, is made of two or more metals that form an *alloy* (Table 8–9). Copper and zinc are the main ingredients of the brazing rods used on auto bodies.

Interaction of Flux and Brazing Rods

Generally the surfaces of metals exposed to the atmosphere are covered with an oxidized film, which, if heat is applied, thickens. *Flux* removes this oxidized film and prevents the metal surface from reoxidizing. It also increases the bond between the base metal and the brazing material.

If a brazing material is melted over a surface that has an oxidized film and foreign matter adhering to it, the brazing material will not adequately bond to the base metal. Surface tension will cause the brazing material to ball up and not stick to the base metal (Figure 8–95A).

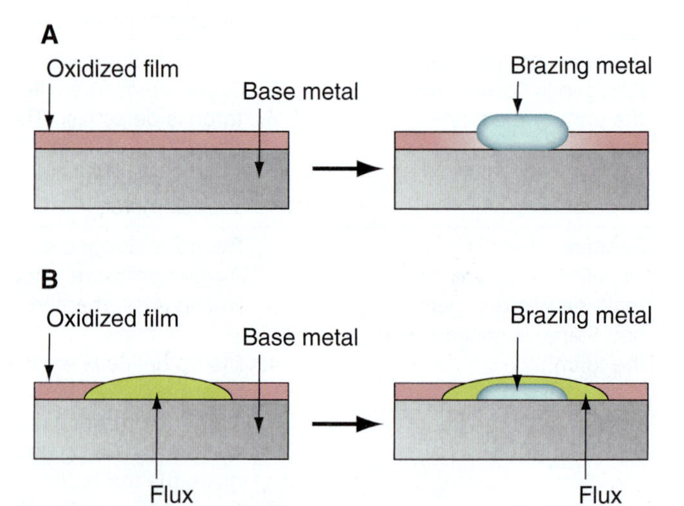

Figure 8–95 (A) Compare when flux is not used and (B) when flux is used.

The oxidized film can be removed by applying flux to the surface of the base metal and then heating it until it becomes liquid (Figure 8–95B). After the oxidation has been removed, the brazing material will adhere to the base metal and the flux will prevent further oxidation.

Brazing Joint Strength

Because the strength of the brazing material is lower than that of the base metal, the shape of the joint and the clearance of the joint are extremely important. Joint strength is dependent on the surface area of the pieces to be joined. Therefore, make the joint overlap as wide as possible.

Even when the items being joined are of the same material, the brazed surface area must be larger than that of a welded joint. As a general rule, the overlapping portion must be *three* or more times wider than the panel thickness.

Brazing Operations

General brazing procedure is as follows:

1. *Cleaning the base metal.* If there is oxidation, oil, paint, or dirt on the surface of the base metal, clean the surfaces before brazing. These contaminants, if allowed to remain on the surface, can cause eventual joint failures. Even though flux acts to remove oxidized film and most contaminants, it is not strong enough to completely remove everything. Therefore, first clean the surface mechanically with a wire brush.
2. *Flux application.* After the base metal is thoroughly cleaned, apply flux uniformly to the brazing surface. If a brazing rod with flux in it is used, this operation is not necessary.
3. *Base metal heating.* Heat the joining area of the base metal to a uniform temperature capable of accepting the brazing material (Figure 8–96). Adjust the torch flame so that it is a slight carburizing flame. By watching the melting flux, you can estimate the proper temperature for the brazing material.
4. *Base metal brazing operation.* When the base metal has reached the proper temperature, melt the brazing material onto the base metal (Figure 8–97). Let the braze metal flow out naturally. Stop heating the area when the brazing material has flowed into the gaps of the base metal.

Other points to consider are:

- Because brazing material flows easily over a heated surface, it is important to remember to heat the entire joining area to a uniform temperature.
- Do not melt the brazing material prior to heating the base metal, because the brazing material will not adhere to the base metal.

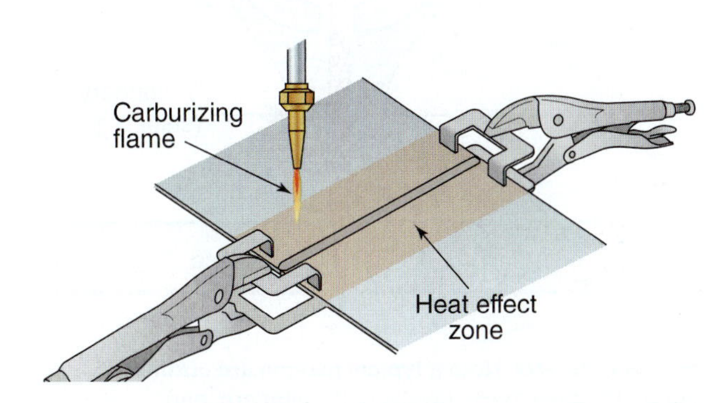

Figure 8-96 **Use a carburizing flame to heat the base metal.**

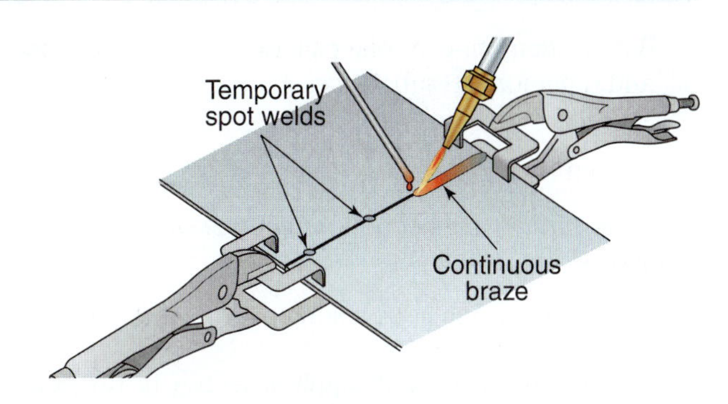

Figure 8-97 **Once the base metal is heated, brazing metal can be applied to surfaces.**

- If the surface temperature of the base metal becomes too high, the flux will not clean the base metal, resulting in a poor brazing bond and inferior joint strength.

The following additional precautions should be taken when brazing:

- Brazing temperature must be higher than the melting point of brass by 50°F–190°F (10°C–89°C).
- The size of the torch tip must be slightly larger than the thickness of the panel.
- Preheat the panel to deposit brazing filler metal more efficiently.
- Secure the panel to prevent the base metal from moving and the brazing zone from breaking.
- Evenly heat the portion to be welded without melting the base metal.
- Control the heat by tilting the torch more horizontally (flatter to surface) or by removing the flame and allowing the area to cool briefly.
- The brazing time must be as short as possible to avoid lowering weld strength.
- Avoid brazing the same place again.

Treatment After Brazing

Once the brazed portion has cooled down sufficiently, rinse off the remaining flux sediment with water. Scrub the surface with a stiff wire brush. Baked and blackened flux can be removed with a sander or a sharp-pointed tool. If the remaining flux sediment is not adequately removed, the paint will not adhere properly. Corrosion and cracks might form in the joint.

8.16 SOLDERING (SOFT BRAZING)

Soldering is not used to reinforce the panel joints. It is used only for final finishing, such as in leveling the panel surface and correcting the surface of the welded joints. Because soldering functions by "capillary phenomenon," it has outstanding sealing ability.

Before attempting to solder a joint, remove paint, rust, oil, and other foreign substances.

Soldering Procedure

After the surface has been thoroughly cleaned, proceed as follows:

1. Heat the portion to be soldered. Wipe it with a cloth after heating.
2. Stir solder paste well. Apply it with a brush to an area 1–1½ inches (25.4–38 mm) larger than the built-up area.
3. Heat it from a distance.
4. Wipe the solder paste from the center to the outside.
5. Make sure the soldered portion is silver gray. If it is bluish, it is due to overheating. If any spot is not soldered, reapply the paste for soldering.

When soldering, keep the following points in mind:

- It is desirable to use a special torch for soldering. If a gas welding torch is used, the oxygen and acetylene gas pressures must be 4.3–5 psi (29.7–34 kPa).
- The solder must contain at least 13 percent zinc.
- Maintain the appropriate temperature.
- Move the torch so that the flame evenly heats the entire portion to be soldered (without heating a single spot only).
- When the solder begins to melt, remove the flame and start finishing with a spatula.
- When additional solder is required, the previously built-up solder must be reheated.

8.17 PLASMA ARC CUTTING

Plasma arc cutting creates an intensely hot air stream, which melts and removes metal over a very small area. Extremely clean cuts are possible with plasma arc cutting. Because of the tight focus of the heat, there is no warpage, even when cutting thin sheet metal.

Plasma arc cutting is replacing oxyacetylene as the best way to cut metals. It cuts damaged metal effectively and quickly but does not destroy the properties of the base metal.

In plasma arc cutting, compressed air is often used for both shielding and cutting. As a shielding gas, air covers the outside area of the torch nozzle, cooling the area so the torch does not overheat.

Figure 8–98A shows that there are two areas for gas flow. In air plasma arc cutting, compressed air is used for both shielding and cutting. As a shield gas, air shields the outside area of the torch nozzle, cooling the area so the torch does not overheat. Air also becomes the cutting gas. The air swirls around the electrode as it heads toward the nozzle opening. The swirling action helps constrict

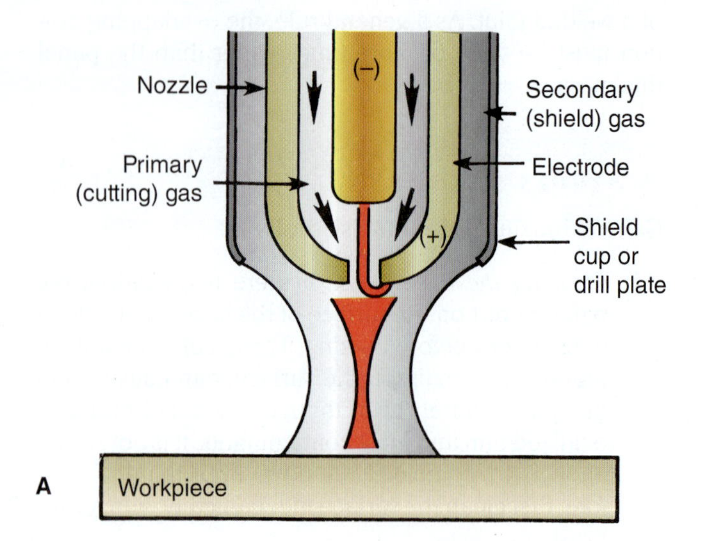

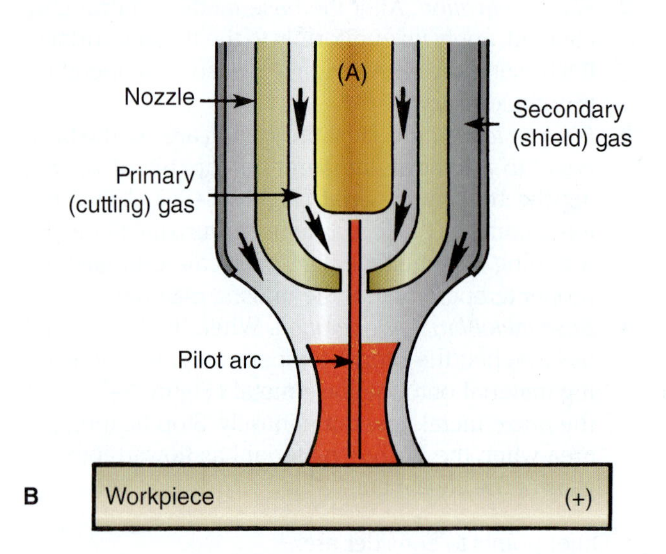

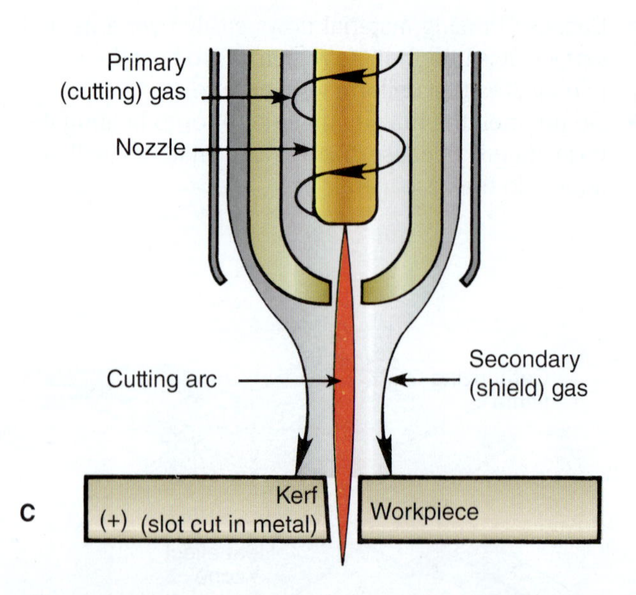

Figure 8-98 Note a typical plasma arc cutting setup: (A) basic parts involved, (B) pilot arc, and (C) cutting arc.

and narrow the gas. When the machine is turned on, a pilot arc is formed between the nozzle and the inner electrode (Figure 8–98B). When the cutting gas reaches this pilot arc, it is superheated—up to 60,000°F.

The gas is now so hot it ionizes and becomes capable of carrying an electrical current (ionized gas is actually the plasma). The small, narrow opening of the nozzle accelerates the expanding plasma toward the workpiece. When the workpiece is close enough, the arc crosses the gap, with the electrical current being carried by the plasma (Figure 8–98C). This is the cutting arc.

The extreme heat and force of the cutting arc melt a narrow path through the metal. This serves to dissipate the metal into gas and tiny particles. The force of the plasma literally blows away the metal particles, leaving a clean cut.

A 10–15 amp plasma arc cutter is generally adequate for mild steel up to 3/16-inch (5 mm) thick, a 30 amp unit can cut metal up to 1/4-inch (6 mm) thick, and a 60 amp unit will slice through metal up to 1/2 inch (13 mm) thick.

Controls are usually quite simple. Plasma arc cutters made specifically for thinner metals might only have an on–off switch and a ready light. More elaborate equipment can include a built-in air compressor, variable output control, on-board coolant, and other features.

On some units, a switch is provided that allows you to alter the current mode depending on the surface being cut. When cutting painted or rusty metal, a continuous, high-frequency arc is best.

Two critical parts of the torch are the cutting nozzle and the electrode. These are the only consumables (besides air) in plasma arc cutting. If either the nozzle or the electrode is worn or damaged, the quality of the cut will be affected. They wear somewhat with each cut. Moisture in the air supply, cutting thick materials, or poor technique will make them fail more quickly. Keep a supply of electrodes and nozzles on hand and replace them when needed.

Today's plasma arc cutters do an excellent job using clean, dry compressed air. The air can be supplied through an external or built-in air compressor or by using a cylinder of compressed air. Cylinders of air can be expensive, whereas shop air is almost free. To reduce contaminants, use a regulator with a filter.

Also, check the air pressure regularly. Using the wrong pressure can reduce the quality of the cuts, damage parts, and decrease the cutting capacity of the machine.

Operating a Plasma Arc Cutter

To operate a typical plasma arc cutter (Figure 8–99), proceed as follows:

1. Connect the unit to a clean, dry source of compressed air with a minimum line pressure of 60 psi (413 kPa) at the air connection.

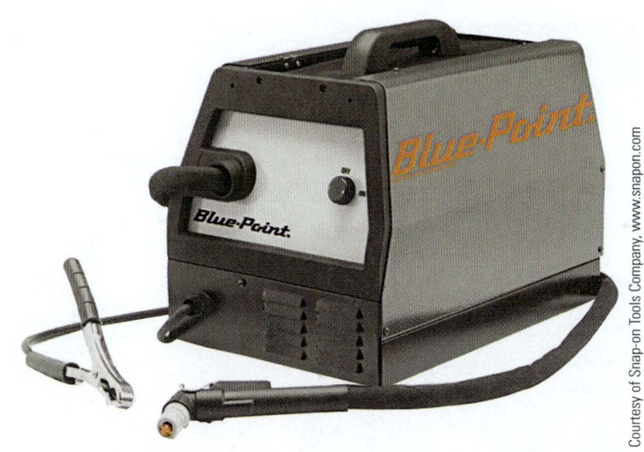

Courtesy of Snap-on Tools Company, www.snapon.com

A A plasma arc cutter uses an electric arc and a blast of air to cut metal quickly.

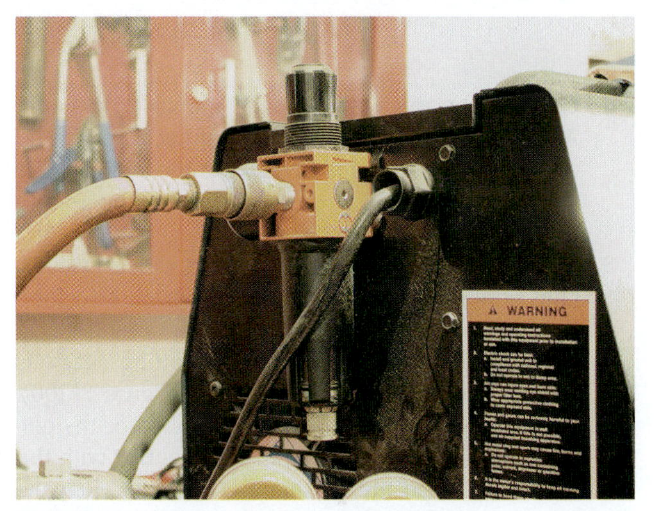

B A shop air hose connects to a filter/drier on back of the plasma arc cutting machine.

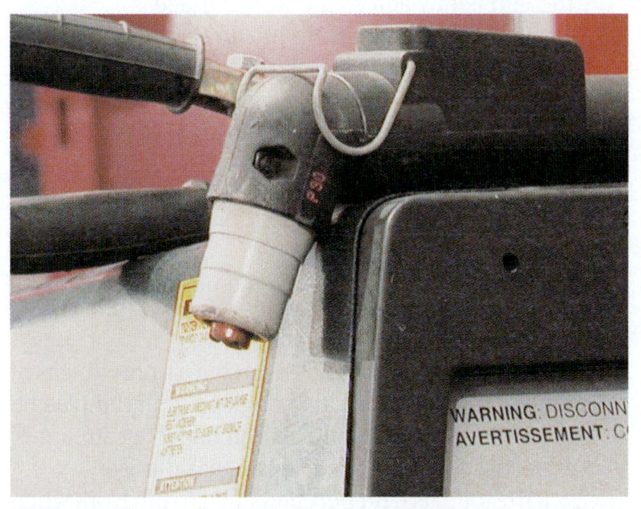

C A metal tip produces an electric arc and tremendous heat. The air blast blows molten metal away from the arc to make a cut in the metal.

Figure 8–99 A typical modern plasma arc cutter will make rapid, smooth cuts in metal.

Figure 8–100 **When using a plasma arc cutter, be careful of molten metal spray on the back side of the cut. It could ignite and burn the interior parts of the vehicle.**

2. Connect the torch and ground clamp to the unit. After plugging the machine in, connect the ground clamp to a clean metal surface on the vehicle. The clamp should be as close as possible to the area to be cut.
3. Move the cutting nozzle into contact with an electrically conductive part of the work. This must be done to satisfy the work safety circuit.
4. Hold the plasma torch so that the cutting nozzle is perpendicular to the work surface (Figure 8–100). Push the plasma torch down. This will force the cutting nozzle down until it comes in contact with the electrode. Then the plasma arc will start. Release downward force on the torch to let the cutting nozzle return to its normal position. While keeping the cutting nozzle in light contact with the work, drag the gun lightly across the work surface.
5. Move the plasma torch in the direction the metal is to be cut. The speed of the cut will depend on the thickness of the metal. If the torch is moved too fast, it will not cut all the way through. If moved too slowly, it will put too much heat into the workpiece and might also extinguish the plasma arc (Figure 8–100).

Other points that should be remembered when using a plasma arc cutter are:

- When piercing materials ⅛ inch (3 mm) thick or more, angle the torch at 45 degrees until the plasma arc pierces the material. This will allow the stream of sparks to shoot off away from the gas diffuser.

  If the torch is held perpendicular to the work when piercing heavy-gauge material, the sparks will shoot back up at the gas diffuser. The molten metal will collect on the diffuser. This might plug the air holes and shorten the life of the diffuser.
- Torch cooling is important to extend the life of the electrode and nozzle. At the end of a cut, the air continues to flow for several seconds. This prevents the nozzle and electrode from overheating. Some equipment suppliers also recommend idling the unit for a couple of minutes after the cut is made.
- When making long, straight cuts, use a metal straightedge as a guide. Simply clamp it to the work to be cut. For elaborate cuts, make a template out of thin sheet metal or wood and guide the tip along that edge.

When angling the torch, be aware that the sparks can shoot as far as 20 feet (6 m) away. Be sure that there are no combustibles or other workers in the area.

- When cutting ¼-inch (6 mm) materials, start the cut at the edge of the material.
- When making rust repairs on cosmetic panels, it is possible to piece the new metal over the rusted area and then cut the patch panel at the same time that the rust is cut out. This process also works when splicing in a quarter panel.
- Be aware of the fact that the sparks from the arc can damage painted surfaces and can also pit glass. Use a welding blanket to protect these surfaces.
- Make sure there is nothing behind the panel that can be damaged. Check for wiring, fuel lines, sound-deadening materials, and other objects that could cause a fire.

Air ordinarily will not conduct electricity, but with very high voltage the air molecules ionize and become electrically conductive. The air becomes superheated and forms a path along which voltage can easily flow.

Remember that these three variables will have a bearing on cut quality:

1. *Travel speed.* The thicker the material, the slower the speed. Travel is faster for thin material.
2. *Parts wear.* The tip and electrode will erode with use. The more wear, the poorer the quality of the cut.
3. *Air quality.* Moist or oil-contaminated air will contribute to a poor quality weld.

Courtesy of CEBORA/Cebotech, Inc.

Figure 8-101 The technician is using a plasma arc cutter to remove an old aluminum door frame during major repairs.

Plasma Air Cutter

Some equipment has a built-in safety protection system to protect the operator. This type of system cuts output power automatically if the safety cup is removed from the torch, if the tip and electrode are accidentally short-circuited because of insufficient air pressure, or if the duty cycle is exceeded. The open circuit voltage of plasma cutting equipment can be very high (in the range of 250–300 volts), so insulated torches and internally connected terminals are also essential.

On some units, a switch is provided that allows you to alter the current mode when cutting bare or painted metal. When cutting painted or rusty metal, a continuous, high-frequency arc is best for punching through the nonconductive surface layer and for keeping the arc going while cutting. When cutting bare metal, a high-frequency arc is needed only to start the arc. Once the torch starts to cut, a direct current pilot arc is all that is needed to keep things going. The bare metal position gives the longest electrode and nozzle life.

Figure 8–101 shows a plasma arc cutter designed for aluminum.

SUMMARY

1. There are three basic methods of joining metal together in the automobile assembly:
    - Mechanical (metal fastener) methods
    - Chemical (adhesive fastening) methods
    - Welding (molten metal) methods

    Welding is one method of repair in which heat is applied to the pieces of metal to fuse them together into the shape desired.
2. Visible weld penetration is indicated by the height of the exposed surface of the weld on the back side. Full weld penetration is needed to ensure maximum weld strength.
3. MIG welding is recommended by all OEMs, not only for HSS and unibody repair, but for all structural collision repair.
4. The resistance spot welder provides very fast, high-quality welds while maintaining the best control of temperature buildup in adjacent panels and structure.
5. Always follow service manual recommendations when welding to ensure structural integrity.
6. During the welding process, either inert gas or active gas shields the weld from the atmosphere and prevents oxidation of the base metal.
7. Flat welding means the pieces are parallel with the bench or shop floor.
8. Horizontal welding has the pieces turned sideways. Gravity tends to pull the puddle into the bottom piece. When welding a horizontal joint, angle the gun upward to hold the weld puddle in place against the pull of gravity.
9. Vertical welding has the pieces turned upright. Gravity tends to pull the puddle down the joint. When welding a vertical joint, the best procedure is usually to start the arc at the top of the joint and pull downward with a steady drag.
10. Overhead welding has the workpieces turned upside down.
11. The tack weld is a relatively small, temporary MIG spot weld that is used instead of a clamp or sheet metal screw to tack and hold the fit-up in place while a permanent weld is made.

12. In a continuous weld, an uninterrupted seam or bead is laid down in a slow, steady, ongoing movement.

13. A plug weld is made in a drilled or punched hole through the outside piece (or pieces).

14. If welding defects should occur, think of ways to change your procedures to correct them.

EXERCISES

On a separate sheet of paper, complete the following learning activities for this chapter. Write definitions for the key terms and answer the ASE-style review questions, essay questions, critical thinking problems, and math problems. You can also do the outside activities, possibly for extra credit.

Key Terms

aluminum electrode wire
brazing
burn mark
burn-through
carburizing flame
continuous weld
DC reverse polarity
destructive testing
electronic shielding
flat welding
fusion welding
heat crayon
heat effect zone
heat sink compound
horizontal welding
insert
joint fit-up
lap weld
metal inert gas (MIG) welding
neutral flame
nondestructive test
overhead welding
oxidizing flame
plasma arc cutting
plug weld
pressure welding
spot weld
stitch weld
tack weld
tungsten inert gas (TIG)
undercutting
vertical welding
weld face
weld legs
weld penetration
weld root
weld throat
welding
welding blanket
welding current
welding filter lens

ASE-Style Review Questions

1. Technician A uses a forward gun angle to achieve a deep penetration in the metal. Technician B uses the reverse gun angle to achieve a flat bead. Who is correct?
   A. Technician A
   B. Technician B
   C. Both A and B
   D. Neither A nor B
2. Technician A says that the main function of the gun nozzle is to provide gas protection. Technician B says that if the insulation in the gun nozzle area is bypassed, the current will ignite the inert shielding gas. Who is correct?
   A. Technician A
   B. Technician B
   C. Both A and B
   D. Neither A nor B
3. Welding current affects which of the following?
   A. Base metal penetration depth
   B. Arc stability
   C. Amount of weld spatter
   D. All of the above
4. When MIG welding, what happens if the tip-to-base metal distance is too long?
   A. The shield gas effect is reduced
   B. The wire protruding from the end of the gun increases and becomes preheated
   C. The melting speed of the wire increases
   D. All of the above
5. Technician A starts a butt weld in the center of the metal. Technician B says that it is wise not to weld more than ¾ inch (19 mm) at one time. Who is correct?
   A. Technician A
   B. Technician B
   C. Both A and B
   D. Neither A nor B
6. Which of the following welds is the body shop alternative to the OEM resistance spot welds made at the factory?
   A. Spot
   B. Plug
   C. Stitch
   D. All of the above

7. What determines the length of a tack weld?
   A. Operator preference
   B. Thickness of the panel
   C. Type of base metal being welded
   D. Type of shielding gas being used
8. When using a resistance welder, Technician A installs a larger diameter electrode tip to increase the diameter of the spot weld. Technician B says that when the tips are worn, a tip dressing tool can be used to reshape the tips. Who is correct?
   A. Technician A
   B. Technician B
   C. Both A and B
   D. Neither A nor B
9. Which of the following statements concerning plasma arc cutting is incorrect?
   A. The plasma arc process cuts mangled metal effectively
   B. Plasma cutting is an extension of the TIG process
   C. The nozzle must come in contact with an electrically conductive part of the work before the arc can start
   D. When piercing material that is more than ⅛-inch (3.1 mm) thick, hold the torch perpendicular to the work
10. The typical acetylene pressure for oxyacetylene welding is
   A. 15–100 psi (103–689 kPa).
   B. 3–12 psi (21–83 kPa).
   C. 3–25 psi (21–173 kPa).
   D. 30–120 psi (207–827 kPa).
11. Mixing slightly more acetylene than oxygen will obtain what type of flame?
   A. Neutral
   B. Standard
   C. Carburizing
   D. Oxidizing
12. Which of the following is not characteristic of brazing?
   A. Relatively high strength
   B. Can join parts of varying thickness
   C. Greater risk of distortion in the base metal
   D. Can join otherwise incompatible metals
13. When operating a MIG welder, which of the following indicates the correct wire-to-heat ratio?
   A. An even, high-pitched buzzing sound
   B. A steady, reflected light of low intensity
   C. Both A and B
   D. Neither A nor B
14. Technician A says that all steels can be MIG welded with one common type of weld wire. Technician B says that metals of different thicknesses can be MIG welded with the same diameter wire. Who is correct?
   A. Technician A
   B. Technician B
   C. Both A and B
   D. Neither A nor B

Essay Questions

1. Summarize the MIG welding process.
2. Describe the basic guidelines to follow when MIG welding aluminum.
3. Describe plasma arc cutting.
4. When welding a repair panel, what information should be known before starting?

Critical Thinking Problems

1. If undercutting occurs while MIG welding, what should you do?
2. What can be done to prevent heat buildup during welding?
3. __________ is the most common type of welding used in collision repair facilities.

Math Problems

1. When setting up a typical plasma arc cutter, the air pressure gauge shows only 20 psi (138 kPa). How much should this pressure be changed?
2. During a butt weld on a thin panel, the technician has welded ¼ inch (6 mm). How much farther can the technician go before stopping to allow cooling?

Activities

1. Visit your guidance counselor. Ask about welding courses offered in your area. Give a report to the class about your findings.
2. Practice destructive tests of welds. Make MIG and spot welds on scrap metal. Tear them apart with a hammer and chisel. Visually inspect the welds and write a report on your findings.
3. Visit a body shop. While wearing all necessary protective gear and a helmet with an approved lens, observe experienced auto body welders working. Write a report on what you learned.

CHAPTER 9

Shop Safety and Efficiency

OBJECTIVES

After studying this chapter, you should be able to:

- List the general rules regarding personal safety while working.
- Summarize the major shop areas and safety rules that apply to each.
- Summarize the importance of wearing a respirator when airborne contaminants are present.
- Explain how to fit and adjust a respirator.
- List the types of safety gear needed when working in a body shop.
- Review precautions for using hand tools and power equipment.
- List the types of accidents that can occur in an auto shop.
- Describe how to prevent auto shop accidents.
- Explain what to do in case of a shop fire.
- Summarize important methods of handling hazardous waste materials found in a body shop.
- Explain Right-to-Know Laws.
- Answer ASE-style review questions relating to shop safety and efficiency.

INTRODUCTION

In previous chapters, you learned how a damaged vehicle moves through a typical body shop during repairs. You learned the names of the major parts of a vehicle. You also learned the proper selection and use of hand tools, power tools, compressed air systems, and shop repair/repaint materials. Now that you understand the basic language of an auto body repair professional, you are ready to learn specific safety rules critical to your work as a collision repair technician.

This chapter summarizes the types of hazards present in a body shop. You will study how to best avoid these potential injuries and how to work with hazardous materials and equipment.

True professionalism and shop safety begins with how technicians perform their given work tasks. A professional understands the ever-present dangers in a body shop and strives to avoid making dangerous mistakes.

NOTE *A body shop can be a very safe and enjoyable place to work if everyone follows standard safe work procedures (Figure 9–1). Conversely, a body shop can be a very dangerous, even deadly place to work, if you or another worker fails to follow basic safety rules.*

The book stresses the importance of abiding by **Occupational Safety and Health Administration (OSHA)** standards. OSHA was established by Congress in the Occupational Safety and Health Act of 1970. This law was enacted to ensure that every American worker had safe and healthy working conditions. OSHA has a staff of over 2,000 people and an annual budget of several hundred million dollars.

OSHA works with the **National Institute for Occupational Safety and Health (NIOSH)**. NIOSH does scientific and technical research to establish basic safety standards. OSHA then enforces these standards by doing surprise inspections of workplaces to make sure established safety standards are being followed.

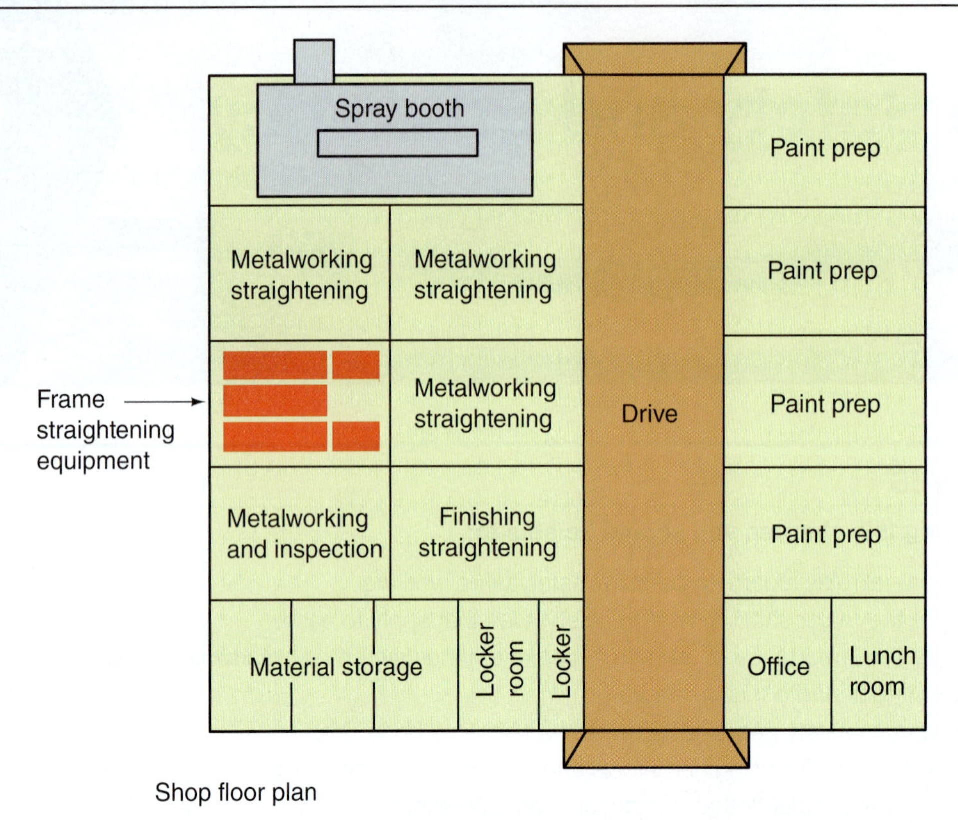

Figure 9–1 **As soon as you get a job in a collision repair facility, become familiar with the shop's layout. Know the location of fire exits, fire extinguishers, a first-aid kit, telephones, and the eye flushing station.**

9.1 SHOP ACCIDENTS

Shop accidents are unplanned mishaps that hurt people, damage parts or tools, and result in other adverse effects on the shop and its employees. Because a body shop has so many potential sources of danger, safety must be everyone's primary concern (Figure 9–2).

The most important action taken in any body shop is accident prevention. Carelessness and the lack of safety habits frequently cause accidents that can result in serious injury and even death. Thousands of auto body technicians are injured or killed on the job each year. Ignoring and/or violating safety rules caused most of these accidents.

ECO-TECH In the collision repair industry, safety and efficiency go "hand in hand"! By working efficiently and avoiding waste, you make your work environment safer for yourself and your co-workers. By not mixing too much and wasting paint, by not wasting sandpaper or oversanding, by adhering to all safety and procedural rules, you reduce shop overhead costs while also improving your working conditions and work environment.

Safety Program

Accidents have far-reaching effects, not only on the victim, but also on the victim's family and society in general. Therefore, all shop employees must foster, develop, and promote a safety program to protect the health and welfare of those involved.

A *safety program* is a written shop policy designed to protect the health and welfare of all employees. It provides rules on everything from equipment use to disposal of hazardous chemicals.

Manufacturer's Instructions

Manufacturer's instructions are the detailed procedures for products provided by the manufacturer. They are written to guide you in the use of a specific item, whether it is a piece of equipment or a certain type of paint.

Generally, instructions take the form of the owner's or instruction manual for tools and equipment. With paint and chemicals, they are normally printed on the container label. Always refer to manufacturer's instructions when in doubt about any task.

NOTE *Other chapters in this text give more specific safety rules and cautions. However, these safety rules are generic and cannot cover every challenge you will encounter in the body shop. For this reason, you should always refer to manufacturer's instructions for specific tools and make/model vehicles before you undertake repairs.*

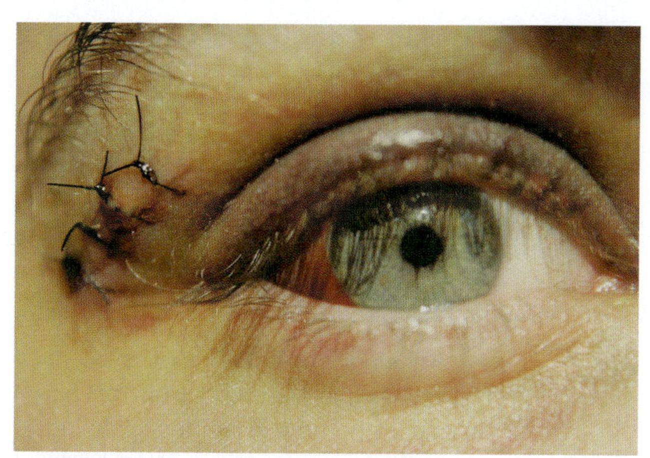

A This technician almost lost an eye when a small cutting disc shattered and shot debris into his eye. Safety glasses would have prevented this injury.

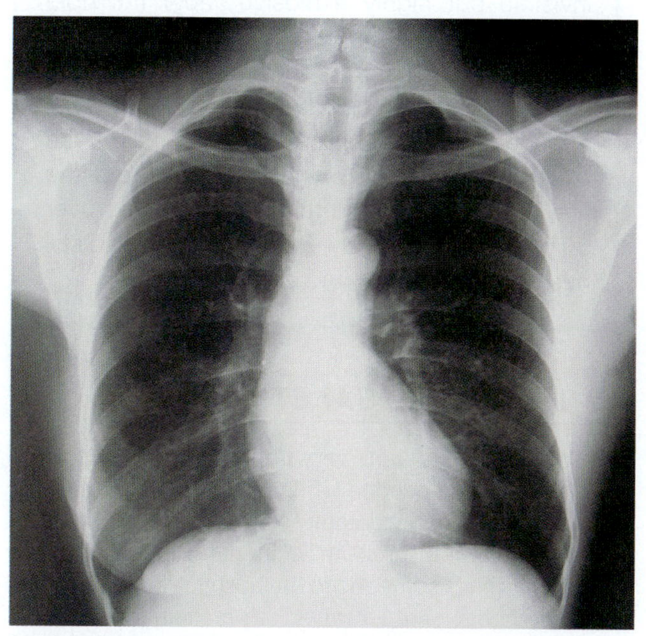

B Breathing problems and even lung, throat, and nose cancer can result if you fail to wear an approved respirator to keep toxic materials out of your respiratory system.

Figure 9–2 A smart, professional technician understands the importance of wearing appropriate protective gear when working in a body shop. Every shop area, except the office, requires the use of safety clothing, eye protection, respirators, and so on.

In Case of an Emergency

You should know what to do in the event of an emergency, whether it be an injury, a shop fire, or another accident.

All shops should have a list of *emergency telephone numbers* clearly posted next to all telephones. These important phone numbers include the police, the fire department, poison control, an ambulance service, a doctor, the nearest hospital, and so on.

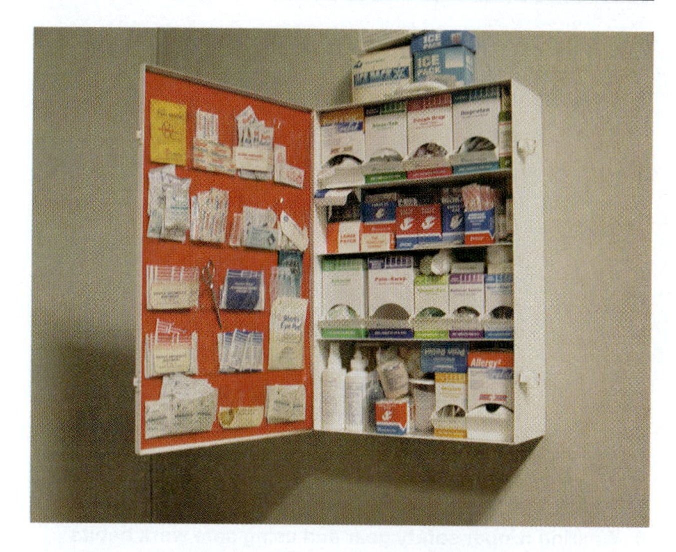

Figure 9–3 The shop's first-aid kit will have antiseptics, bandages, and other medical items for treating minor injuries. Know its location and help keep it stocked with supplies. Inform the boss if any item needs to be reordered.

Know the location of your shop's first-aid kit. A **first-aid kit** includes many of the medical items needed to treat minor shop injuries. It will have sterile gauze, bandages, scissors, antiseptics, and other items to help treat minor cuts and burns. A fully stocked first-aid kit should be kept in a handy location, usually near the office or rest room (Figure 9–3).

An **eye flushing station** is equipped with fresh water and nozzles for washing out your eyes if you get anything in them (Figure 9–4). You should know where the eye flushing station is located and how to use it.

Safety signs provide information that helps to improve shop safety. Signs often mark the locations of fire exits and fire extinguishers, dangerous or flammable chemicals, and other information. Safety signs should be located throughout the shop. Make sure you read and understand them.

Types of Accidents

Technicians can help to prevent several kinds of accidents—asphyxiation, chemical burns, electrocution, fires, and explosions.

- **Asphyxiation** refers to anything that prevents normal breathing. There are many mists, gases, dusts, and fumes in a body shop that can damage the lungs and affect the ability to breathe. As you will learn, there are several things you can do to protect yourself from these airborne dangers.
- *Chemical burns* result when a corrosive chemical comes into contact with the skin or eyes. These can result from various chemicals in a body shop: paint removers, part cleaners, refinish materials, and other common shop chemicals.
- **Electrocution** results when electricity passes through a person's body. Severe injury or death can

A Wearing proper safety gear and using safe work habits helps prevent having to use this safety equipment.

B With your eye held open, allow the pure flushing water to flow over your eyeball to wash away chemicals.

Figure 9-4 This eye flushing station is used if chemicals splash into a technician's eyes.

result. Electric power tools, drop lights, and other equipment operating on either 120 or 240 volts can all cause electrocution. You will learn several ways to avoid electrocution.

- A *fire* is rapid oxidation of a flammable material, producing high temperature. A burn from a fire can cause horrible, permanent scar tissue and death. There are numerous combustibles (paints, thinners, reducers, gasoline, and dirty rags) in a body shop, any of which can quickly cause a fire.
- *Explosions* are fast-moving air pressure waves that result from extremely rapid burning. For example, if you were to weld on or near a car's gas tank, the fumes in the tank could ignite and explode. The metal tank would act like a powerful bomb that could send metal shrapnel into the work area. There are many other objects in the body shop that can explode if mishandled: welding tanks, propane tanks, etc.
- *Physical injury* is a general category that includes cuts, broken bones, strained backs, and similar injuries. To prevent these painful injuries, you should constantly evaluate repair techniques—always think about what you are doing and try to do it better.

A young auto body repair student was helping a friend work on his car at home. He started cleaning greasy suspension system parts with gasoline. He noticed that the gasoline seemed very cool to the touch. The gasoline felt cool because it was evaporating into an invisible but highly explosive vapor (invisible cloud of gasoline fumes).

The pilot light from a hot water heater ignited the gasoline fumes and the student was engulfed in flames. His friend grabbed a fire extinguisher, but before he could put out the fire, the pupil suffered third-degree burns over 50 percent of his body. He almost died and still has terrible physical and emotional scars.

9.2 PERSONAL SAFETY

This section details some very important personal safety rules that must be heeded while working.

Horseplay is unacceptable! Proper, professional conduct can also help prevent accidents. Horseplay is not fun when it sends someone to the hospital. Such things as air nozzle fights, creeper races, or practical jokes do not have any place in the shop. Stay away from anyone who does not take shop work seriously. Remember, a joker is "an accident waiting to happen."

Dress like a technician. Remove rings, bracelets, necklaces, watches, and other jewelry. These items can get caught in engine fans, belts, and drive shafts, causing severe and permanent injury. Also, roll up long sleeves and secure long hair, which also can get caught in spinning parts.

Work like a professional. It is easy to get excited about your work, but never let this excitement prompt you to work too fast. Then you might overlook an important repair procedure or safety rule. In the end, you won't save time by trying to work too fast.

There is usually a "best" tool for each repair task. Always use the right tool for the job; using the wrong tool will slow your work and can result in inferior repairs. It can also be dangerous. It pays to consider whether one tool will work better than another, especially when you run into difficulty.

An apprentice mechanic was using an air impact wrench to tighten large clamps on a frame rack. This technician mistakenly used a non-impact-type universal joint on the extension and socket setup. When high torque from the wrench was applied, the universal joint broke and almost cut off one of the mechanic's fingers. The extension then flew off and broke the windshield of the pickup truck parked in the next stall.

By using the wrong tool, the apprentice was badly injured and he had to pay for the broken windshield.

Remember, customers and all nonemployees should never be allowed in any of the shop's work areas.

Air Passage and Lung Safety

There are many airborne materials in the body shop that can cause health problems, particularly with repeated exposure. These airborne contaminants include:

- Abrasive dusts from sanding and grinding
- Welding flux and soldering fumes
- Vapors from caustic solutions, primers, paints, and solvents
- Spray mists from undercoats and painting materials

All these materials present dangers to your air passages and lungs, especially when you are surrounded by them day in and day out.

Respirators

Respirators are often needed in the collision repair facility to keep airborne materials from being inhaled into your nose, throat, and lungs. Respirators are recommended even when adequate ventilation (outside airflow) is provided in the shop area.

Dust Respirator

A paper filter that fits over your nose and mouth to block small airborne particles is called a **dust respirator** (Figure 9–5). This type of respirator will help keep dangerous dust particles from entering your nose, throat, and lungs. A dust respirator should be worn when sanding, grinding, or blowing off dirty panels with a blow nozzle (Figure 9–6).

A paper or cloth dust respirator is not designed to stop paint mists and fumes.

Sanding operations create dust that can cause bronchial irritation and possible long-term lung damage. Protection from this health hazard is often overlooked. Just because sanding dust does not cause immediate symptoms does not mean that there will not be problems later in life. An approved dust respirator should be worn anytime as particles could be in your breathing air.

Follow the instructions provided with the dust respirator to ensure a proper fit. Bend or shape it so that air

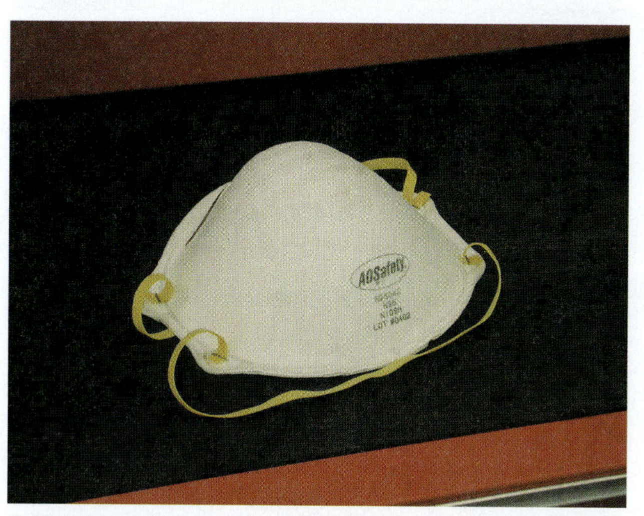

A This type of inexpensive, disposable respirator should be worn when power sanding and grinding. The filter will trap harmful paint dust and dirt that would otherwise enter your lungs.

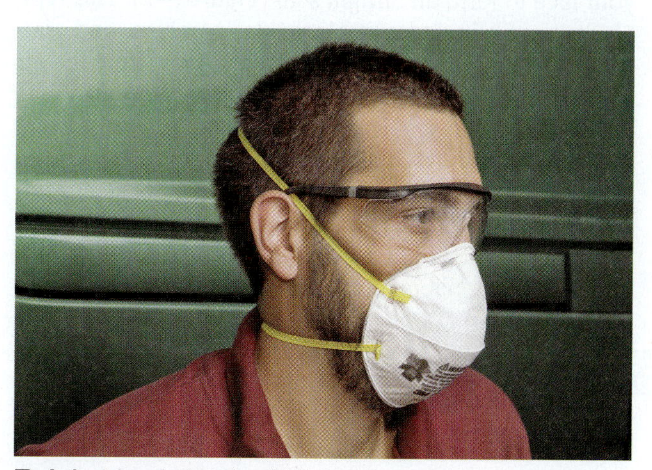

B A dust respirator should fit snugly on the face to keep debris from going around outside of the mask.

Figure 9–5 A dust respirator is like an "air filter" for your nose, mouth, throat, and lungs.

Figure 9–6 Always put on a dust particle mask before grinding or sanding. The dust can be very toxic and harmful with prolonged exposure.

Figure 9–7 A cartridge-type respirator will filter fine particles, mists, and some fumes that a dust respirator will not stop.

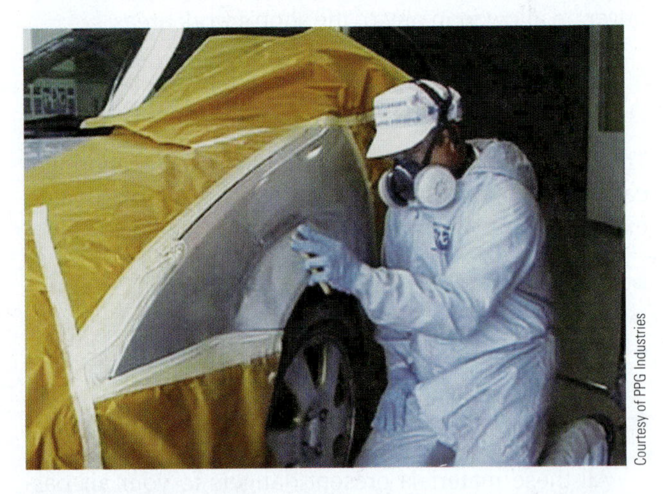

Figure 9–8 Cartridge-type respiratory masks will filter finer particles than fiber dust masks. They provide adequate protection from most primers, sealers, and non-catalyzed paints.

cannot leak around your face, nose, and mouth. Note that a dust respirator will not work properly if you have a large beard or facial hair.

Cartridge Respirator

A **cartridge filter respirator** protects against vapors and spray mists of one-part enamel (no hardener added), lacquers, and other nonisocyanate materials. Cartridge filter respirators should be used only in well-ventilated areas. They must not be used in environments containing less than 19.5 percent oxygen.

The cartridge filter respirator typically consists of a rubber face piece designed to conform to the contour of your face to form an airtight seal (Figure 9–7). This type of respirator includes replaceable prefilters and cartridges that remove solvent and other vapors from the air (Figure 9–8). The cartridge filter respirator also has intake and exhaust valves, which ensure that all incoming air flows through the filters.

Cartridge filter respirators are available in several sizes. Purchase and use one that fits your face. To maintain the cartridge filter respirator, keep it clean and change the prefilters and cartridges as directed by the manufacturer.

Welding Respirator

A **welding respirator** has special cartridge inserts designed to trap welding fumes. Welding fumes are most dangerous when the metals being fused are zinc-coated steel for improved rust resistance.

Air-Supplied Respirator

An approved **air-supplied respirator** typically consists of a hood with a clear visor and an external air supply hose. Clean, breathable air is pumped through the hose from a separate air source and into the hood or helmet (Figure 9–9).

An air-supplied respirator provides protection from inhaling extremely dangerous airborne materials: paint with added hardener, isocyanate paint vapors and mists, and hazardous solvent vapors. An air-supplied respirator is the safest type of protection and is highly recommended when spraying all types of primers, sealers, and paints (Figure 9–10).

Figure 9–9 **This refinish technician is wearing a state-of-the-art, air-supplied respirator while painting a car. It uses a lightweight hood with a clear lens. The hood keeps the head clean and also allows circulation of fresh breathing air.**

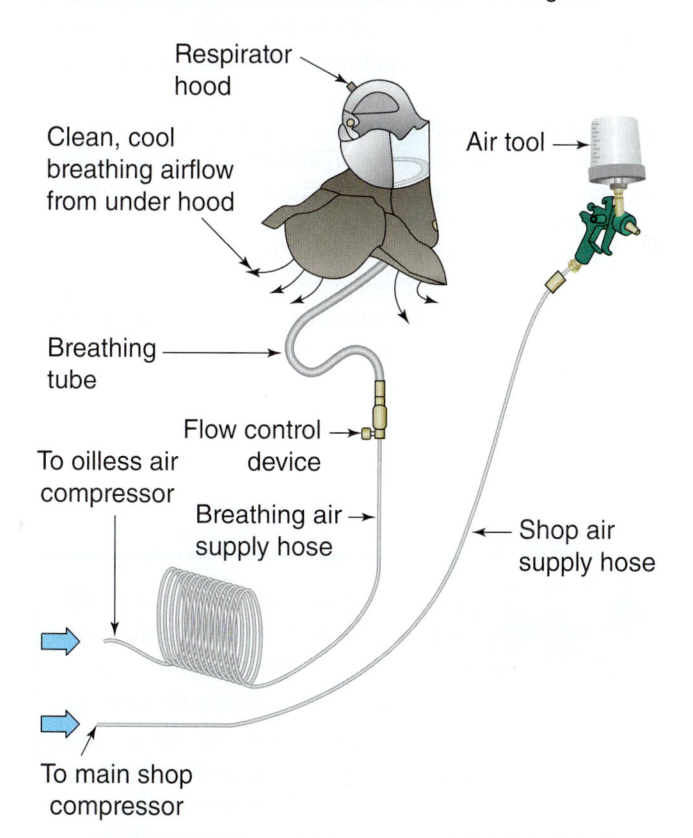

Figure 9–10 **Note parts of a fresh-air respirator that is fed from its own oilless air compressor located outside of the paint booth. This one uses an air visor that affords protection but still allows for good vision. Cool breathing air circulates around your face and head. A shop air line can only be used for air tools because it is not pure enough for breathing.**

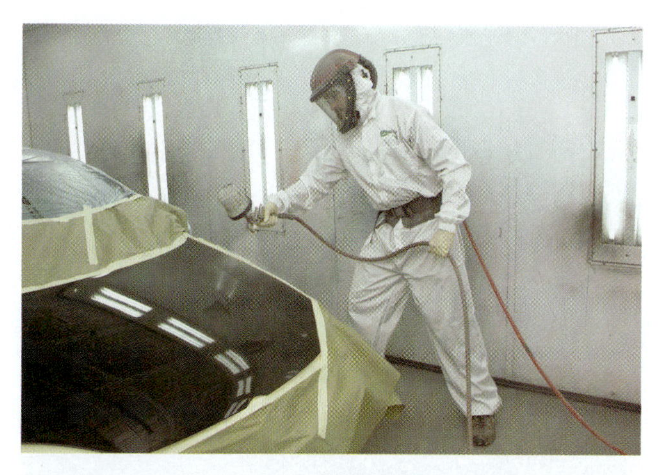

Figure 9–11 **Today's multistage, catalyzed refinish materials are much more harmful to your lungs than old lacquer and single-stage enamels. You should wear an air-supplied visor, hood, or respirator when spraying these materials.**

An air-supplied respirator is comfortable to wear and does not require fit testing (Figure 9–11).

The air-supplied respirator may include a self-contained oilless pump to supply air to the hood or half-mask respirator. The shop may also have additional filters to prepare, dry, and clean main shop air pressure for use in air-supplied respirators (Figure 9–12).

The pump's air inlet must be located in a clean air area. Some shops mount the pumps on an outside wall, away from the dust and dirt generated by shop operations. If shop compressed air is used, it must be filtered

Figure 9–12 **When mixing paints, you should work in a well-ventilated area or a power-ventilated mixing room and wear a cartridge respirator.**

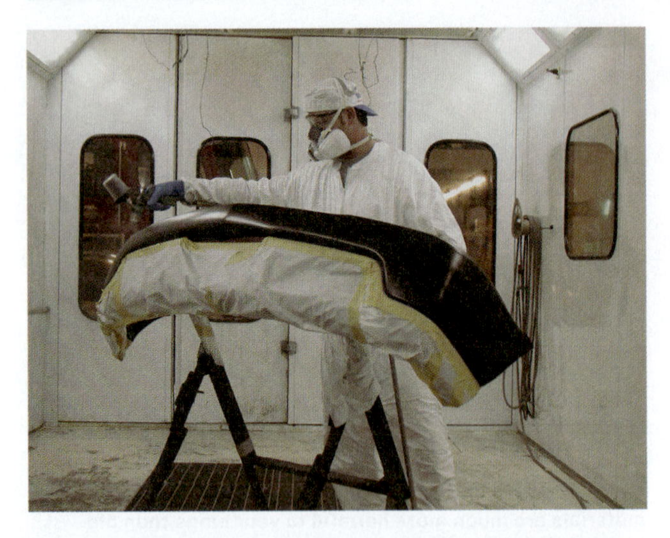

Figure 9–13 You can sometimes breathe a large quantity of toxic materials without feeling immediate illness. The resulting sickness will show up as major health problems after a few years of not wearing an approved respirator. Look and feel professional by wearing all necessary safety gear when working.

with a trap and carbon filter to remove oil, water, scale, and odor.

The air supply must have a valve to match air pressure to respirator equipment specs and an automatic control to sound an alarm or shut down the compressor in the event of overheating and contamination. Overheating causes carbon monoxide contamination of the air supply (Figure 9–13).

DANGER Isocyanate resin is a principal ingredient in some urethane hardeners. Because this ingredient has toxic effects, you must always wear a NIOSH-approved respirator. Hardened primers and paints can actually cure when inside your lungs!

Respirator Testing and Maintenance

It is very important that an air-purifying cartridge filter respirator or half-mask fit securely around your face. This will prevent contaminated air from bypassing the filters and entering your lungs.

To check for air leaks, a **respirator fit test** should be done prior to using the respirator. Perform both negative and positive pressure checks.

To make a *negative pressure test*, place the palms of your hands over the cartridges and inhale. A good fit will be evident if the face piece collapses onto your face when breathing normally.

To perform a *positive pressure test*, cover up the exhalation valve and exhale or breathe out. A proper fit is evident if the face piece billows out without air escaping from the mask when breathing normally.

Another form of fit testing consists of exposing amyl acetate (banana oil) near the seal around the face. If no odor is detected, the mask fits properly. Follow the manufacturer's instructions to ensure proper maintenance and fit of respirators.

Replace the prefilters when it becomes difficult to breathe through the respirator or after manufacturer-recommended intervals. Replace the cartridge(s) at the first sign of solvent odor. Regularly check masks to make sure they do not have any cracks or deformities. Store respirators in airtight containers or resealable plastic bags to keep them clean.

Remember that facial hair may prevent an airtight seal, which presents a health hazard. Therefore, refinishing technicians with facial hair should use an air-supplied respirator.

Head Protection

Be sure to tie long hair securely behind your head before beginning to work on a vehicle. Hair that becomes tangled in moving parts or air tools can be torn out and mutilation can occur.

Your hair should also be protected from dust and spray mists. To keep hair clean and healthy, wear a cap at all times in the work area. Wear a protective painter's stretch hood in the spray booth.

There are numerous objects in the shop that can cause painful head injuries: car body panels, frame rack towers and chains, and other heavy equipment. Be careful not to accidentally hit your head on these objects while working in the body shop.

Several years ago, a newly hired apprentice was helping a technician find an abnormal driveline noise. The car was on a lift running in drive while he listened for the odd noise. The apprentice accidentally got his hair caught in the rotating drive shaft. The shaft pulled his head and face up against the floor pan before ripping out a huge chunk of hair and skin, resulting in the young man being "scalped." It took several hundred stitches and hours of surgery to reattach his scalp and hair, and the scars are still obvious. Remember to keep long hair tied up or under a cap.

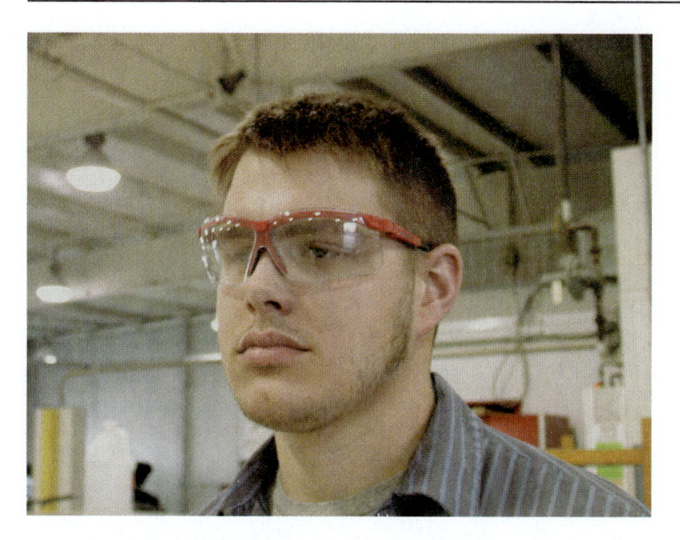

Figure 9–14 Dress like a professional when at work. This technician is wearing lightweight and attractive safety glasses to protect his eyesight, something that cannot be replaced!

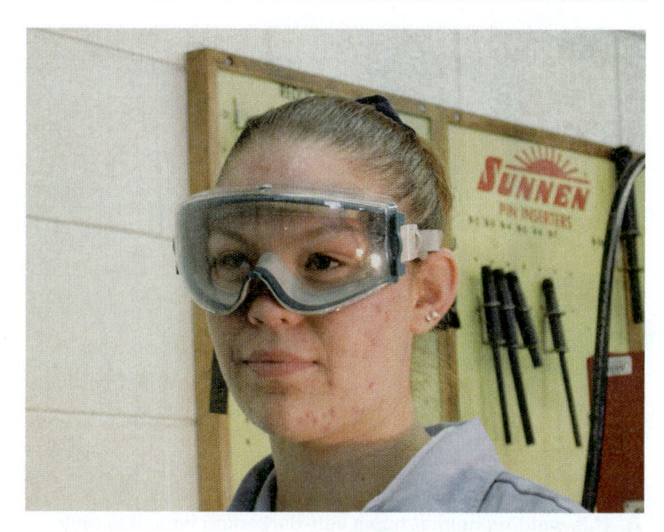

Figure 9–15 Goggles are better than safety glasses when a spill of a chemical might splash into your eyes. These are vital tools that can protect you from eye injuries and even blindness.

A body technician should consider wearing a padded bump cap or welder's cap when working beneath hoods, under cars, or on a frame rack.

Eye and Face Protection

Eye protection—safety glasses, goggles, face shields, hoods, or helmets—is required at all times in most shops to comply with OSHA or insurance company requirements. Some type of eye protection should be worn at all times when working in the shop (Figure 9–14).

Safety glasses are sufficient eye protection when doing tasks such as hammering, drilling, and cutting. Safety glasses are not sufficient when doing tasks that could lead to major injuries to your face. Also remember to wear safety glasses when others in the shop are grinding, sanding, and so forth.

Goggles should be worn when handling fluids that could burn your eyes. For example, wear goggles when working with solvents that could splash into your eyes. See Figure 9–15 and Figure 9–16.

Welding goggles with a shaded lens are often used when heating and cutting with a gas torch outfit (Figure 9–16). The flame from the oxyacetylene gas is not as bright as that produced when electric arc welding.

When doing more dangerous operations, wear a **full face shield** to protect your eyes and your face from injury. For example, wear a full face shield when using an air grinder or bench grinder. It is possible that the grinding disc or wheel could break and shoot large pieces of abrasive into your face. A face shield will prevent deep lacerations and scars.

REMEMBER *When you are in the shop, there is always the possibility that flying objects, dust particles, or splashing liquids could injure your eyes. Not only can this be painful, but it can also cause loss of sight. Your eyes are irreplaceable. Get in the habit of wearing safety goggles, glasses, or face shields in all work areas.*

Figure 9–16 These tinted goggles should be worn when flame heating and torch cutting. Their shaded lens is dark enough for the brightness of a welding flame, but not dark enough to protect your eyes when electric welding.

A **welding helmet** that covers your whole head has a dark, shaded lens that must be worn when electric welding (Figure 9–17). The helmet will protect your face from extreme heat and flying pieces of molten metal. The shaded lens will protect your eyes from the extremely bright, harmful rays generated by the electric arc (Figure 9–18 and Figure 9–19).

NOTE *Sunglasses are not adequate protection when gas welding, cutting, or heating! As shown in Figure 9–20, ultraviolet light can cause severe burns and permanent eye damage.*

A This welding helmet has a self-darkening lens. It is only lightly shaded until you start welding and then it darkens enough to protect your eyes.

B Controls on the inside of the self-darkening helmet allow you to adjust how fast and dark the lens gets when you are welding.

Figure 9-17 A welding helmet is needed to protect your eyes and face from the extreme light of an electric arc when MIG or TIG welding.

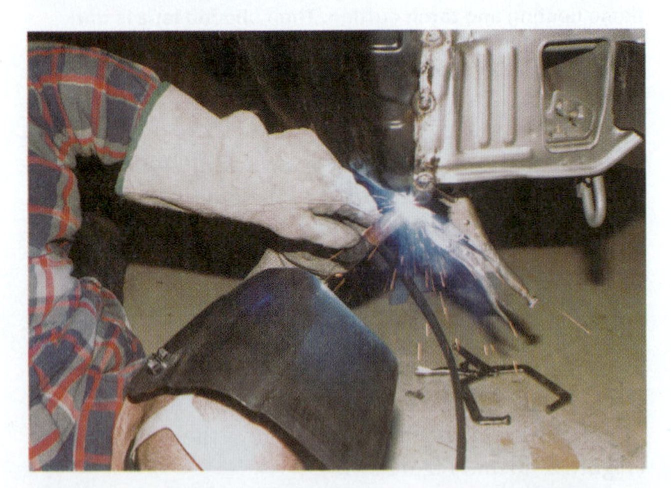

Figure 9-18 This technician is wearing a welding-type cartridge respirator and an automatic dimming welding helmet while making plug welds during panel replacement. Both are needed for protection when arc welding today's steel and aluminum panels during structural body repair.

Figure 9-19 Here a body technician is welding during major repairs of this sports car. Note the brightness of the MIG welding arc.

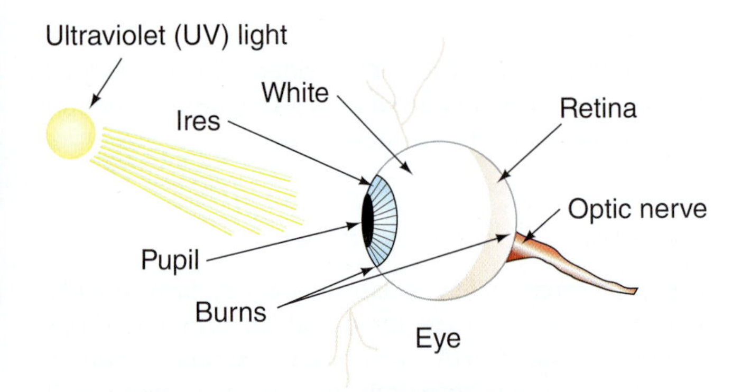

Figure 9-20 Electric welding generates powerful ultraviolet rays that can badly burn your eyes. Even though its light does not hurt immediately, never watch a weld arc without proper eye protection.

DANGER Do *not* weld any parts made of lightweight magnesium. Magnesium burns with tremendous heat. In fact, large, 10-foot-long "burning rods" made of magnesium are used to cut up and scrap the heavy plate steel used in U.S. Navy ships. Once magnesium starts to burn, it is very difficult to put out the fire.

Ear Protection

Ear protection—ear plugs or earmuffs—is needed when *decibel* (sound) levels are high, for example, when using air-powered cutting tools, such as an air chisel (Figure 9–21). The level of noise in a body shop—the noise of

Figure 9-21 Many shop tools produce very high decibel levels. This air chisel being used to remove a badly rusted exhaust system is a good example. Its sound level is loud enough to cause permanent hearing loss.

panel beating, the piercing sound of sanding, even a radio playing full blast—may be enough to deafen a person if proper precautions are not taken. When in metalworking areas, wear earmuffs to protect the eardrums from damaging noise levels. Earmuffs can also help prevent molten metal from getting into the inner ear when performing overhead welding.

Body Protection

Loose clothing, unbuttoned shirt sleeves, dangling ties, and shirts hanging out are very dangerous in a body shop. Instead, wear approved shop coveralls or jumpsuits (Figure 9–22).

Keep clothing away from moving parts when an engine or a machine is running. Any loose or hanging clothing (shirttails, ties, cuffs, scarves, and so on) presents a risk for catching in moving parts. This can cause serious bodily injury. Also be sure to remove all jewelry before working.

Your pants should always be long enough to cover the tops of your shoes. When welding, for example, this will prevent hot sparks or beads of molten metal from going down into your shoes, causing painful burns to your feet. For added safety, leather welder's pants, leggings, or spats are often worn to prevent molten metal from burning through your clothing. Upper body protection should include either a welder's jacket or leather apron.

A clean jumpsuit or lint-free coveralls should be worn when you are in the spray area or paint booth. Clean coveralls will protect the vehicle paint from dirt contamination and will also protect your body from chemical contamination.

If you spill chemicals (cleaning solvents, reducers, thinners, paint removers, and so forth) on your clothing, remove the clothing right away. Dirty, solvent-soaked clothing will hold these chemicals against your skin, causing irritation, a rash, or even a severe chemical burn.

Hand Protection

The harmful effects of liquids, undercoats, and finishes on the hands can be prevented by wearing work gloves.

Impervious gloves, such as the latex type or synthetic rubber gloves, should be used when working with any chemical that can be harmful if exposed to your skin, such as solvents or two-part primers and topcoats. Chemical impervious gloves offer special protection from the materials found in two-component refinish or paint systems (Figure 9–23).

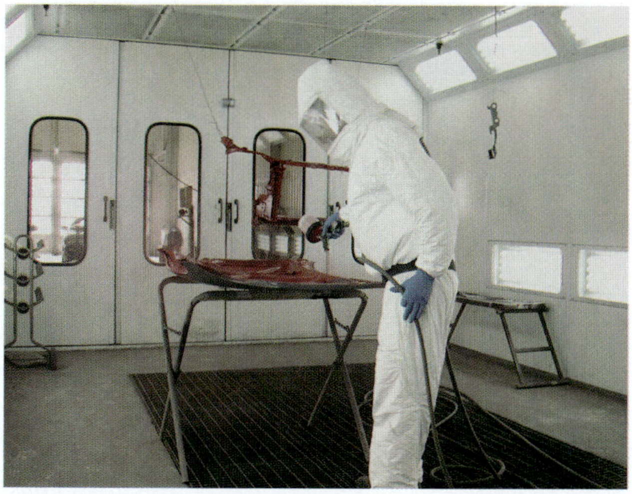

Figure 9-22 A smart technician knows that taking the time to put on safety gear is time well spent. Health is something often taken for granted until an illness hits with debilitating or deadly force.

Figure 9-23 When working with chemicals, take the time to slip on plastic or rubber gloves. Make sure the plastic gloves are a type that will not be dissolved by the chemicals you are using.

NOTE *Some thin, inexpensive plastic gloves dissolve when exposed to certain chemicals. Make sure the gloves you are wearing are resistant to the solution you plan to handle. Material safety data sheets (MSDS) and product labels often give glove recommendations.*

Wear rubber or safety gloves to prevent chemical burns while handling paints and thinners. If any of these materials get on the skin, promptly wash the affected area with soap and water. Cleaners, some paints, refrigerants, and solvents can cause skin and eye burns. Make sure you wear long sleeves for complete protection (Figure 9–24).

Chemical paint remover is very powerful. When using remover, exercise extreme caution and wear all recommended protective gear for your hands, body, face, eyes, and lungs.

Thick, strong work gloves should be worn in the prep area to avoid cuts or abrasions. Sheet metal can cut your skin as easily as a knife, and if an air grinder easily cuts metal, it can do the same to flesh and bone. Always wear leather gloves and a full face shield when grinding.

Keep hands away from the moving parts of a running tool. Never clear chips or debris when a tool such as a drill or power saw is turned on. Stop the tool or use a brush to remove debris from the workpiece.

Stay away from engine fans, which have been compared to a "spinning knife." An engine fan can inflict serious cuts and major injuries. Parts or tools dropped into a running fan can fly out and hit a person with incredible force. Remember that electric engine fans can turn on even with the ignition key off!

Always remember to wash your hands thoroughly before leaving the shop area. This provides protection from ingesting any harmful elements that may have been touched. Wash your hands with a proper hand cleaner. At the end of a day's work, it is wise to oil or moisten your skin a little by applying a good skin cream.

Do not use paint thinner or reducer as a hand cleaner. Many chemicals can be absorbed into the skin, causing illnesses later on.

Foot Protection

It is best to wear *safety shoes* that have metal toe inserts and nonslip soles. The steel inserts protect the toes from falling objects; the soles help prevent falls. In addition, good work shoes provide arch support and comfort while standing and walking all day at work (Figure 9–25).

Never wear gym shoes or dress shoes, which do not provide adequate protection in a body shop. It is easy to drop heavy tools and parts and break bones in your feet.

When working in the paint booth, many technicians wear disposable shoe covers. They help keep work shoes clean. In general, disposable garments (coveralls, shoe covers, and slip-on hoods) are becoming more common in the body shop.

Knee pads are good when you have to kneel on the floor while working (Figure 9–26).

Back Protection

It is very easy to injure your back when working in a body shop. Once your back is injured, recovery can take a long time (Figure 9–27).

Courtesy of PPG Industries, Inc.

Figure 9–24 This technician wisely wears rubber gloves and long sleeves when mixing chemicals in the paint mixing room.

Figure 9–25 Quality leather work shoes will protect your feet much better than tennis shoes. Many shops will not let their technicians wear tennis shoes, to avoid foot injuries from falling objects.

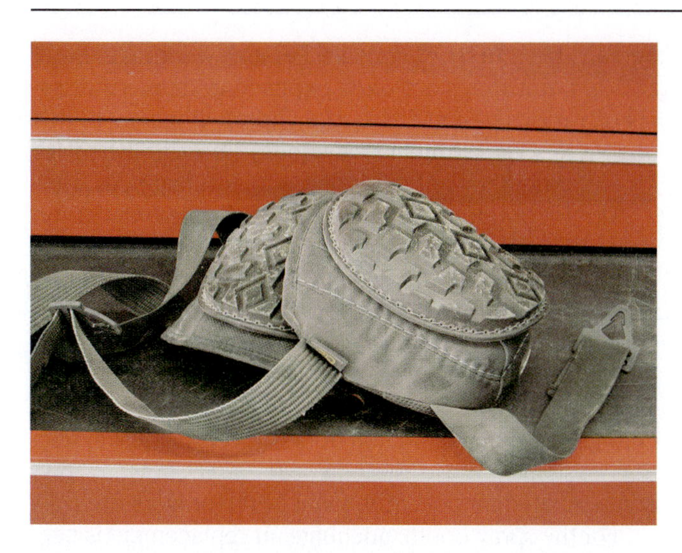

Figure 9–26 Knee pads are often worn when you have to kneel next to a vehicle being repaired. They are much softer than the concrete shop floor and will prevent knee injury over time.

Never overreach when working on a vehicle. Overreaching can easily injure your back. Maintain a balanced stance to help avoid slips and falls.

When you have to lean over a fender or panel, rest your arm or arms on the panel to avoid back strain. If you have to lean over while doing a repair operation, stand up and stretch now and then to avoid back spasms.

Figure 9–27 When lifting and carrying heavy objects in the shop, stay balanced and keep your back straight to avoid strain.

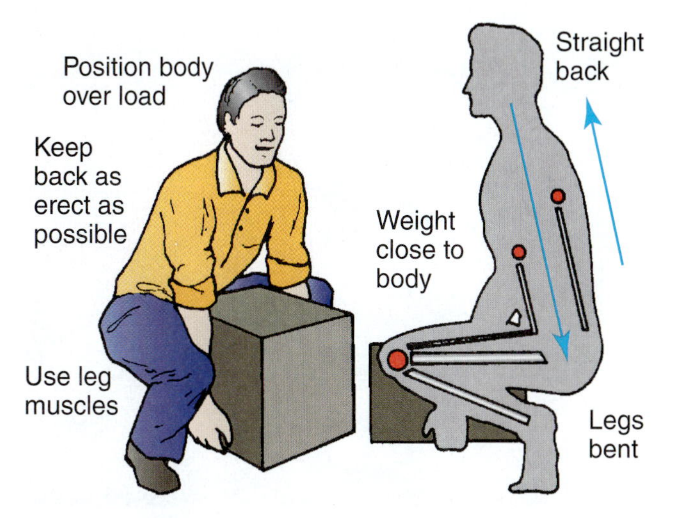

Figure 9–28 Study the proper way to lift objects to prevent back injury.

A *back support* is an elastic brace that wraps around your back and stomach to prevent injury to the lower spine. The back brace should be stretched tightly around your torso and secured with Velcro straps. The brace then lifts up on your chest cavity to remove body weight from your back. It also helps hold your spine in proper alignment while working. If you ever feel back pain, a back brace is a good investment.

Lifting improperly can cause long-term back problems! When lifting and carrying objects, bend with the knees, not the back. Keep your back straight when picking up objects. Bending from the waist when lifting increases your risk for back injury (Figure 9–28).

Heavy objects should be lifted and moved with the proper equipment for the job (floor jack, roll-around cart, engine crane, forklift, and so on). There are many assemblies that are extremely heavy (doors, hoods, deck lids, front clips, transmissions, rear axles, transaxles). These heavy, and sometimes clumsy, parts or assemblies require equipment or the help of another technician during removal and installation.

9.3 GENERAL SHOP SAFETY PROCEDURES

In addition to personal safety, body/paint technicians must be aware of general shop safety procedures. This section outlines some of the rules and precautions that should be observed.

Electrical Safety

Electrocution results when a small amount of electric current passes through the human body. This can affect heart and brain functions, possibly causing death.

Disconnect electrical power supplies before performing any service on a machine or tool (Figure 9–29).

Figure 9–29 Know the location of all power shutoff switches in the shop in case of an emergency.

Keep all water off the floor. Water conducts electricity and presents a serious shock hazard if a live wire falls into a puddle in which a person is standing. Floors must be dry where electric power tools are used.

Make sure electric tools and equipment are properly grounded. If the third, round, ground prong on the tool cord is broken off, replace it before using the tool. Check for cracks in wiring insulation, as well as for bare wires.

To prevent serious injury, make sure the switch is off before plugging in any electric tool. When you are not using an electric tool, turn it off.

Environmental Safety

Environmental safety includes those procedures that protect people and the Earth's resources (land, water, and air) from **toxic** chemicals. Persons working in body and/or paint shops are often exposed to dangerous levels of various gases, dusts, and vapors. Because of this exposure, control measures should be established for air contaminants and other hazardous substances.

Do not breathe contaminated air! Proper ventilation is very important in areas where caustics, degreasers, undercoats, sanding dust, and finishes are used. The vapors from thinners used in most paints have a narcotic effect, and long-term exposure can cause serious illness.

Ventilation can be provided by means of an air-changing system, extraction floors, or central dust extraction systems. These systems use large fans to pull contaminated air out of the paint booth or work area.

For the spray booth, adequate air replacement is necessary not only to promote evaporation and drying of paint materials, but also to remove harmful mists and vapors (Figure 9–30). Always remember to turn on the air exchange system when working in the paint booth or paint mixing room.

Dustless Sanding Systems

Use a *dustless sanding system* to minimize exposure to toxic airborne dust particles created by sanding automotive paints and primers.

A dustless sanding system uses a blower or air pump to draw airborne dust into a storage container, much like a vacuum cleaner. This action pulls airborne sanding dust through holes in a special sanding pad and sandpaper or through a shroud that surrounds the sanding pad.

Some dustless system manufacturers claim that their machines can trap over 99 percent of the toxic dust created by sanding operations.

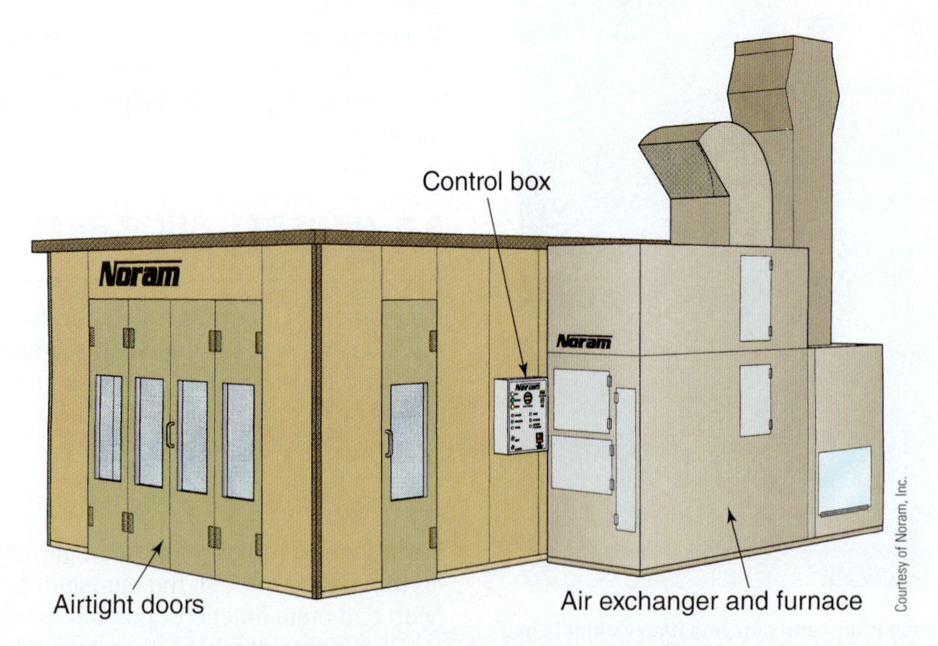

Figure 9–30 This paint booth has a large air exchange unit. It pulls fresh air through the booth for breathing and pulls airborne contaminants away from fresh paint. Make sure the air exchanger is turned on anytime you work in a paint booth.

Figure 9–31 Whenever a vehicle is in a stall with the engine running, connect a shop exhaust hose(s) to the tailpipe(s). This will keep invisible, odorless, but deadly, carbon monoxide gas from entering the work area.

Carbon monoxide (CO) is an invisible, odorless, but deadly gas! Car and truck engine exhaust produces harmful CO gas. CO poisoning from engine exhaust fumes can cause drowsiness and vomiting; it can even be fatal.

Operate engines only in a well-ventilated area to avoid the danger of CO poisoning. Connect a shop vent hose to the tailpipe of any vehicle being operated in the shop. Look at Figure 9–31.

Space heaters used in some shops can also be a major source of CO. These heaters should be inspected periodically to ensure that they are adequately vented and have not become blocked.

Never use automotive-type paints on household items such as toys or furniture. Automotive paints pose a dangerous health hazard to children who might ingest some of the toxic paint.

Asbestos was used in the manufacture of older brake and clutch assemblies. **Asbestos dust** contains cancer-causing agents. Never blow this dust into the shop. Use a vacuum system while wearing a filter mask to safely clean off asbestos dust.

Vehicle Safety in the Shop

When driving a vehicle during repairs, keep the following safety rules in mind.

Drive carefully! Drive slowly and keep one window rolled down when in the shop. It is easy for someone to walk in front of a vehicle you are driving. If you have the window down, you can more easily hear instructions or warnings from co-workers (Figure 9–32).

Look carefully in front of the vehicle. Make sure no one is working under a car with their legs sticking out into your stall. When moving a vehicle around the shop, be sure you look in all directions to make certain that nothing is in your way (Figure 9–33).

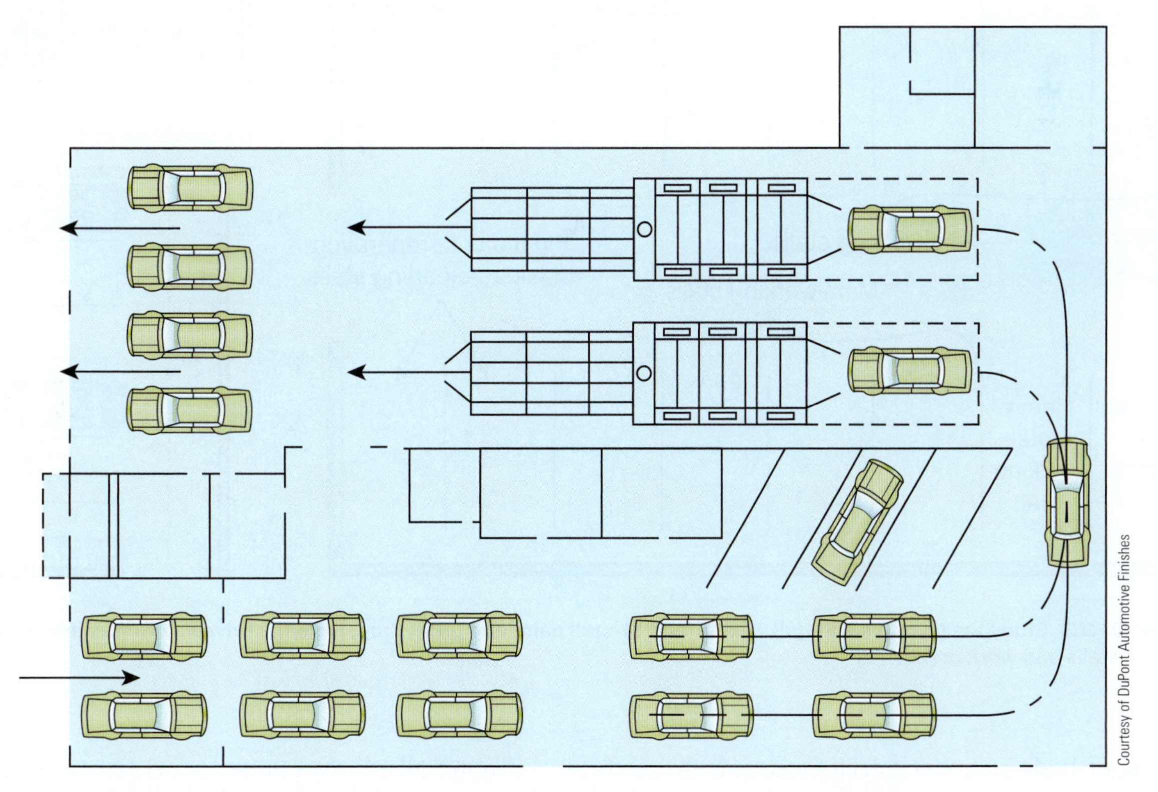

Figure 9–32 Vehicles are constantly being moved around in a body shop. Normally, it is best to use established routes through the shop, such as the one shown in this paint prep and refinishing area of a typical shop.

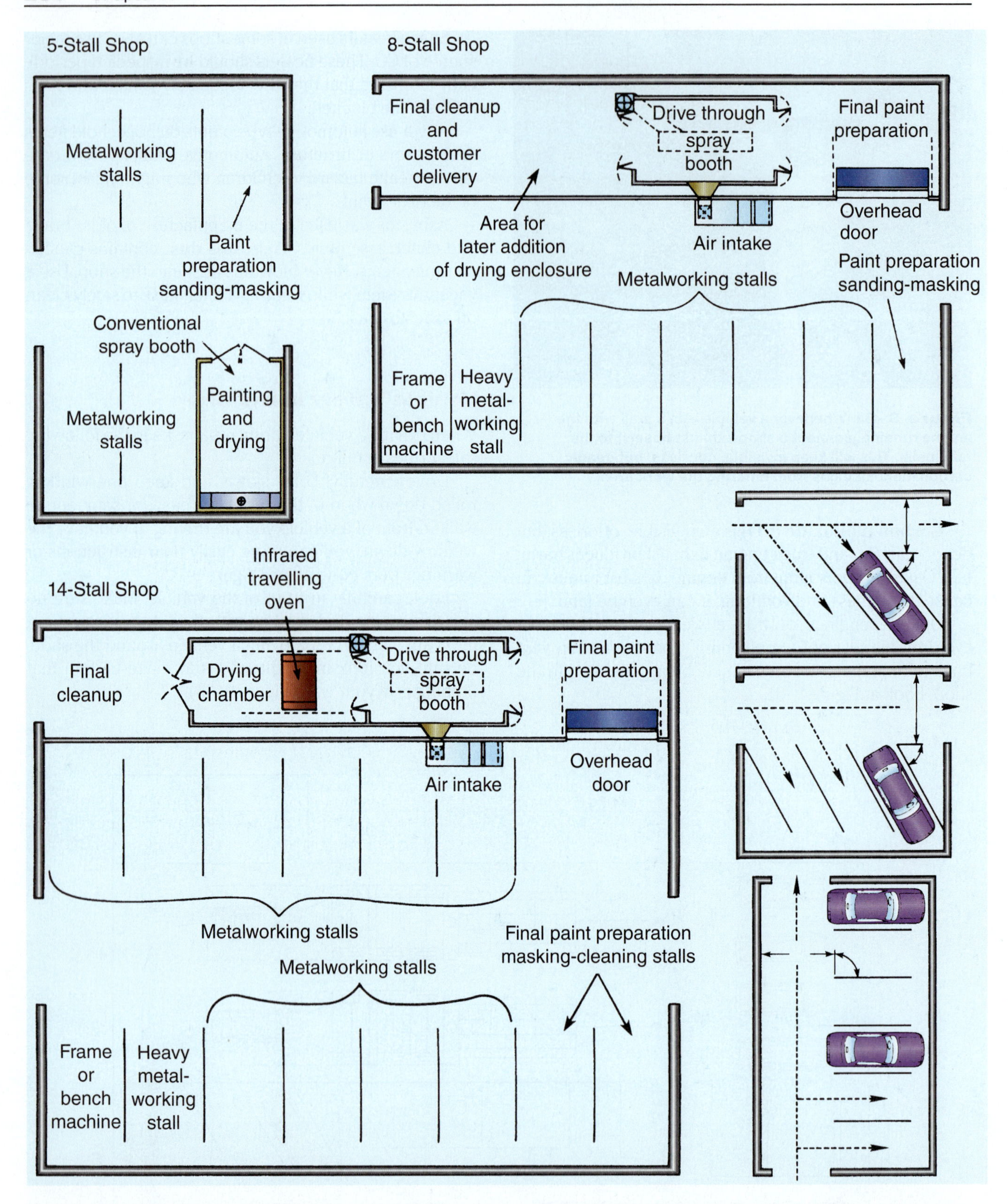

Figure 9-33 **Study the layout of a 5-stall, 8-stall, and 14-stall paint and body shop. Note the driveway widths and locations of various stalls and work areas.**

In larger body shops, a forklift is commonly used to move large assemblies around the shop. Be aware of potential hazards at all times when operating a forklift.

Secure all vehicles properly. Set the parking brake when you are working on a vehicle. If the car has an automatic transmission, set it in park. If the vehicle has a

Figure 9–34 Be careful when driving through the shop or pulling up onto equipment. Have someone guide you onto lifts and frame racks to keep the vehicle centered on the equipment.

manual transmission, it should be in reverse (engine off) or neutral (engine on) (Figure 9–34).

Avoid hot or moving parts. Each year, thousands of technicians are injured by touching hot parts or spinning parts on vehicles. To prevent serious burns, avoid contact with hot parts, such as the radiator, exhaust manifold, tailpipe, catalytic converter, muffler, or any parts being welded.

Keep your clothing and body away from moving parts when an engine is running, especially the radiator fan blades and belts. Belts can easily pull your hand or fingers into the pulley, breaking fingers or inflicting severe lacerations.

Check to see that the ignition key switch is always in the off position when working, unless otherwise required for the procedure. If the key is on, the engine could start if you accidentally rotate the engine crankshaft by bumping the vehicle or rolling the vehicle when in gear.

Reinstall all fasteners properly. All bolts, nuts, lock rings, and other fastening components are crucial to the safe operation of a vehicle. Failure to use specified items could cause extensive damage and injuries. Manufacturer's torque specifications must be also followed on critical assemblies. Use a torque wrench to final tighten suspension, steering, and wheel nuts and bolts.

Vehicle batteries often explode! Hydrogen gas floating around the battery can cause it to explode like a small bomb. Flying chunks of plastic and sulfuric acid can shoot out of the battery, blinding, maiming, or even killing technicians. Always charge batteries in a well-ventilated area.

Modern injection systems can retain fuel pressure. You must release fuel pressure on many fuel-injected engines before disconnecting a fuel hose or line. A fitting on the engine can be used to relieve fuel pressure. You can also pull the fuse for the fuel pump and crank the engine to relieve fuel pressure in the system. Disconnect the battery before continuing work on a fuel system. Hold a rag around an engine's fuel line fitting to prevent any fuel spray or loss when disconnecting it.

Hood and door hinge springs are exceptionally strong and can cause painful injury! Wear eye protection. Also, keep your fingers away from springs when they are stretched. Your fingers could be severely pinched and cut by powerful springs.

9.4 TOOL AND EQUIPMENT SAFETY

Do not risk injury through lack of knowledge; use shop tools or perform repair operations only after receiving proper instruction. Note that safety tips and important safety procedures are given throughout this book. It is imperative that you observe the following hand and power tool safety guidelines (Figure 9–35).

Hand Tool Safety

Do not use hand tools for any job other than that for which they were specifically designed. For example, never hammer on a file or screwdriver. The tool could break and cause injury.

Hand tools should be kept clean and in proper working condition. Greasy, oily hand tools can easily slip out of one's grasp, causing skinned knuckles or broken fingers. Wipe tools with a shop rag before putting them away. Dirty tools are unprofessional and dangerous (Figure 9–35)!

Pull, don't push, on a wrench. If the wrench accidentally slips off the fastener, you are likely to smash your hand. If you have to push, use the palm of your hand with your fingers open.

Never open several toolbox drawers at once. Your toolbox can flip over and cause serious injury, or even death. A full tool chest can weigh as much as a ton

Courtesy of Chief Automotive Technologies

Figure 9–35 Always take the time to keep your work area and tools clean. In the long run, this will save time.

(2,000 pounds)! Remember to slide each drawer shut before opening the next.

Check all hand tools for cracks, chips, burrs, broken teeth, or other dangerous conditions before using them. If any tools are defective, repair or replace them.

Be careful when using sharp or pointed tools that can slip and cause cuts and contusions (bruises).

Keep chisels and punches properly ground and shaped. Chisel cutting edges should be sharp and square. The head of a chisel or punch can become "mushroomed" (deformed and enlarged) after prolonged hammering. If this happens, a piece of metal could fly off the head when struck with a hammer. Use a bench grinder to remove any mushroom to form a chamfer (tapered edge) on the tool head. This will keep you from cutting your hand.

Do not carry screwdrivers, punches, or other sharp hand tools in your pockets. It is easy to stab yourself or damage a vehicle with sharp tools in your pockets.

Do not wear contact lenses when grinding, sanding, or handling solvents.

Store all parts and tools properly by putting them away neatly where other workers will not trip over them. This will increase safety and reduce time wasted looking for a misplaced part or tool.

Never leave a creeper lying on the floor. Stand it up on edge when not in use. If someone steps on a creeper, he or she could fall and be seriously injured.

Power Tool and Equipment Safety

Do not operate a power tool without its guard(s) in place. Disconnect the tool's air hose or electrical cord before performing any service or maintenance on the tool. Refer to Figure 9–36.

Do not attempt to use a power tool beyond its stated capacity. For example, grinding discs and stones usually have a maximum revolutions per minute (rpm) stamped on them. Make sure the power tool does not exceed the rpm for the grinding disc, brush, or other tool. The disc or brush could explode and throw off chunks of abrasive or steel wire with great force.

Figure 9-36 **This technician is using an air sander to repair a plastic bumper. How many safety devices is he wearing?**

When grinding to maintain a tool, grind slowly to avoid overheating the case-hardened metal. If you grind fast enough to turn the metal a blue color, you have probably ruined the tool. Too much heat removes the case hardening from the metal and softens the metal in the tool.

When using power equipment on a small part, never hold the part with your hand. The part could slip, causing serious hand injury. Always use clamping pliers or a bench vise to hold small parts when grinding, drilling, or sanding.

When working with a hydraulic press, make sure that hydraulic pressure is applied safely. It is generally wise to stand to the side when operating the press. Always wear a full face shield when using a hydraulic press. Parts can shatter and fly out.

Never drop a welding tank or gas cylinder. The head could break off and fly out. Shut off the main gas valve on top of welding/cutting tanks after use, which can prevent a major explosion or loss of expensive gases in the tank if a hose leak develops.

Using Compressed Air

Be careful when using shop air pressure! Shop air pressure is usually around 100–120 psi (689–1,034 kPa). This is enough pressure to severely injure or kill.

Use the utmost caution with compressed air. Pneumatic tools must be operated at the pressure recommended by their manufacturers.

The downstream pressure of compressed air used for cleaning purposes must remain at a pressure level below 30 psi whenever the nozzle is a non-OSHA-type blowgun (Figure 9–37). OSHA-approved blow nozzles use air pressure release holes for safety. An OSHA-approved nozzle can be operated off higher pressures. Refer to the equipment manufacturer's maximum allowable pressure limit rating for each tool.

Be careful when cleaning with compressed air! Eye protection and particle masks should always be worn when using an air blowgun to clean door jambs and other hard-to-reach places.

Do not use compressed air to clean clothes. Even at low cleaning pressure, compressed air can cause dirt particles to become embedded in the skin, which can result in infection.

Never direct a blowgun at your skin. If you force air into your bloodstream through a cut, an air bubble could cause instant death.

Using a Vehicle Lift

If the shop has a hydraulic lift, be sure to read the instruction manual before using it. Refer to service information for the specific vehicle to find recommended vehicle lift points.

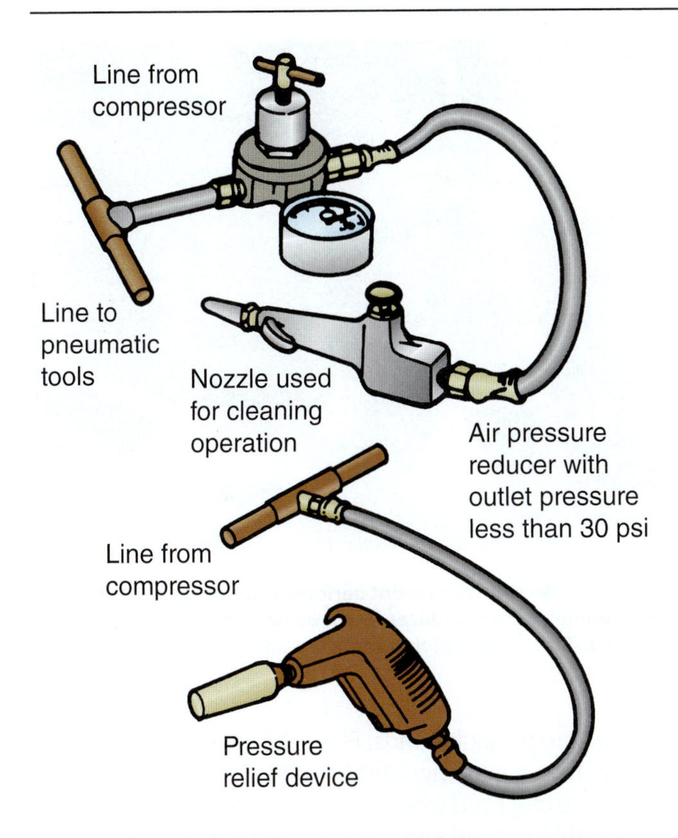

Figure 9-37 Air pressure-reducing regulators are needed to keep the air pressure of some tools at recommended levels for safe operation. The blow nozzle at top does not have OSHA-recommended holes in its tip and requires pressure reduction. The lower nozzle has an approved pressure relief device so full shop pressure can be used with the tool.

Vehicle lift points are the exact locations for positioning the lift pads of a lift or a floor jack to safely raise the vehicle off the ground. Using vehicle lift points locates the vehicle's center of gravity over the center of the lift to keep the vehicle from falling. Check the pads to see that they are making proper contact with the frame or unibody pinch welds or other recommended lift points (Figure 9–38).

Center a vehicle on the lift following service information specifications for that make and model. Raise the lift slowly.

Raise the vehicle about 6 inches and shake it to make sure it is well balanced on the lift. If there are any rattling or scraping sounds, it means that the vehicle is not locked properly in place. If this happens, lower the lift and realign the pads to the vehicle. Test it again as previously described.

After lifting a vehicle to full height, double-check that the safety catch is engaged before working underneath the vehicle. Never work under the lift without the catch locked. Even if the lift's hydraulics fail, the mechanical catch will keep the lift and vehicle from crushing and killing you.

Unless absolutely necessary for problem diagnosis, avoid permitting anyone, either technician or customer, to remain in the car while it is being lifted.

SAFETY TECH! *Before using a lift to raise any vehicle for the first time in the shop, make sure that your instructor has given you a demonstration. Your instructor may also want you to ask for permission before initially raising a car or truck on a lift.*

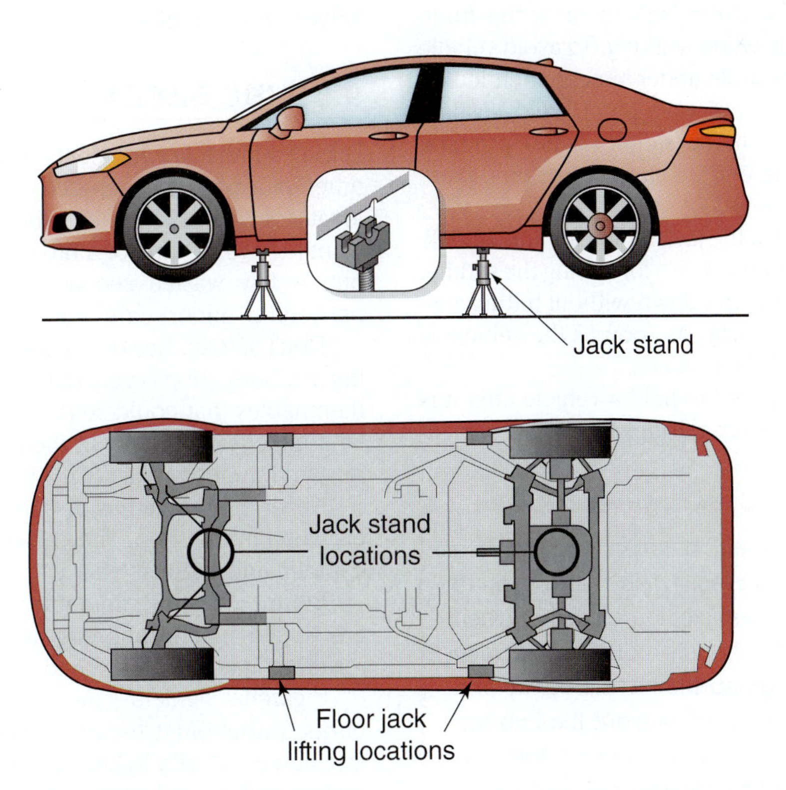

Figure 9-38 Here is an example of the recommended lift points on a modern passenger car. Note the different positions for a floor jack, a vehicle lift, and jack stands.

Center
Center the jacking
bracket in the middle
of the jacking lift platform

Figure 9–39 **Besides the pinch weld area below the rocker panels on the sides of a vehicle, a service manual or CD will also give other jacking points like at these cross members or beams under the unibody. A common mistake is to jack a car by its floorpan, crushing it upward into the passenger compartment, a time-consuming mistake.**

Make sure you have enough ceiling clearance before raising trucks and campers. The vehicle roof must not hit overhead pipes, lights, or the ceiling.

Using Floor Jack and Stands

Technicians often use a **floor jack** to raise the front, sides, or rear of a vehicle while working. To avoid vehicle damage, place the jack saddle under a recommended lift point under the vehicle (frame rail, pinch weld, suspension arm, or rear axle). If the saddle is not properly located, it is easy to dent and damage sheet metal parts in the underbody (Figure 9–39).

Normally rotate or turn the jack handle or knob clockwise to close off the hydraulic valve for raising the saddle. Then pump the handle up and down without hitting anything. Raise the vehicle slowly. As soon as the vehicle is high enough, secure it on jack stands.

Jack stands are required to hold a vehicle after it is raised by a floor jack. After raising, lower the vehicle down onto the jack stands. Place the vehicle in park, then apply the emergency brake and block the wheels. Be aware that even when the car is resting on jack stands, it can still be jolted enough to bump it off the stands. Use jack stands to support a vehicle whenever you must work under it (Figure 9–40).

DANGER Many floor jack handles tend to catch or stick. As a result, it is easy for you to lower the vehicle too quickly. When turning the handle counterclockwise to lower a vehicle, grasp the handle tightly and turn it very slowly. This will prevent the two-ton vehicle from slamming down rapidly and violently, possibly causing damage or serious injury!

Figure 9–40 **To prevent serious injury or death, follow recommended procedures when using a floor jack and jack stands. This weight of this truck would easily kill or cripple.**

Never trust hydraulic jacks alone; they are for raising vehicles, not holding them. Many people have been crushed and killed when a vehicle fell on them.

To lower a floor jack, slowly turn the pressure relief valve counterclockwise.

9.5 FIRE SAFETY

During a fire, a few moments of time can seem like a lifetime. To be properly prepared to deal with a fire, know the locations of all fire extinguishers in the shop. You don't want to have to hunt for a fire extinguisher if a fire breaks out. A few wasted seconds can mean the difference between a "minor scare" and a "major catastrophe."

Don't smoke! Besides being a health concern, smoking in a body shop is very dangerous. There are too many flammables that could start a fire in a body shop. Never light matches or smoke in the paint spraying area! Paint mist and fumes can act like a bomb ready to go off.

Never carry matches or butane lighters in your pockets when in the shop. They have been known to burn or explode and seriously injure technicians.

Butane is highly flammable and can explode. Matches are tipped in phosphorous and can burn all at once, resulting in a large, hot flame.

Cigarette lighters can cause serious third-degree burns. Burns often happen when technicians arc weld and flame cut with lighters in their pants pockets. If hot, molten metal touches the lighter, it can melt the plastic housing and cause the rapid release of the butane.

When depressurized, liquid butane changes into a highly explosive gas vapor. The small lighter can then produce a hot, blowtorch-like flame.

Keep heat sources away from materials that can ignite. Do not use torches or welding equipment near the paint mixing bench/room or in the paint booth area. Sound-deadening material is combustible and must be removed from any body panel to be welded or cut with a torch or plasma arc cutter.

Never attempt to cut or weld magnesium parts of a vehicle. If overheated, magnesium will burn with tremendous heat and flame. Magnesium will burn with a brilliant white light if heated above 800°F. Many lightweight structural parts on new vehicles are made of magnesium: radiator supports, windshield frames, cross members, and so forth. Also, avoid grinding magnesium because the small chips and fine dust generated are also highly flammable and pose a serious fire risk if not properly handled. Refer to the vehicle's service manual to find out which parts are made of magnesium, if needed.

Before spraying refinish materials, be sure to remove any portable lamps and turn on the ventilation system. Spray areas must be free from hot surfaces, such as heat lamps. The spray area must be kept clean of combustible residue. Always leave the ventilation system on while the paint is drying.

When welding and cutting, remember that very high heat and sparks can travel a long distance. Never weld or cut near paints, thinners, or other flammable liquids or materials. Cover open containers or move them to a safe area. Never cut or weld a container before checking what material was originally in that container.

Never weld or grind near a battery. The battery charging operation produces hydrogen gas, and a battery explosion can result.

Fuel tanks should be drained and removed, if necessary, to repair panels next to them. It is easy to cut and rupture a tank with body tools. When welding or grinding near fuel filler lines, close them tightly and cover them with wet rags.

When welding or cutting near car interiors, remove seats and floor mats or cover them with a water-dampened cloth or a welding blanket. Always have a pail of water handy and a fire extinguisher nearby. Other welding safety tips are given in Chapter 8.

Don't short out vehicle wiring while working! **Electrical fires** result when excess current causes wiring to overheat, melt, and burn. This often results when a wire or wires are shorted to ground while doing electrical work or because wire insulation has been cut as a result of a collision. To prevent electrical fires, always disconnect the battery when doing electrical work or when body damage may have cut through wires.

Disconnecting the battery also can prevent a costly vehicle fire from excess current flow. This is necessary before disconnecting any wires or if you suspect wire damage (Figure 9–41).

Figure 9–41 Be careful when working with a car battery. Never connect jumper cables or a charge backward or you can damage electronic circuits in the vehicle. Always disconnect the battery if you will be disconnecting wiring or welding on the vehicle.

Remember what to do in case of a fire! A good rule is to make sure someone calls the fire department before attempting to extinguish the fire. Stay low to avoid inhaling the smoke. If it gets too hot or too smoky, get out. Remember, never go back into a burning building for anything.

NOTE *More detailed paint spraying safety precautions are given in Chapters 26, 27, and 28.*

Using Fire Extinguishers

A fire can be extinguished by depriving it of its essential ingredients, which are heat, fuel, and oxygen. A fire extinguisher is a metal tank filled with a chemical agent designed to quickly smother and stop a fire.

During a fire, a few seconds' time can be a "lifetime" for someone. Know where all shop fire extinguishers are located and make sure you know how to use them (Figure 9–42).

Every body shop must have fire extinguishers. Because fires are classified as A, B, C, and D, there are different types of extinguishers specially designed for each particular class of fire.

Table 9–1 gives the common classes of fire that are found in body shops and methods of containing them. Some extinguishers are capable of being used on more than one type of fire.

A multipurpose *dry chemical fire extinguisher* will put out ordinary combustible, flammable liquid, and electrical fires. Most shops are equipped with multipurpose fire extinguishers.

Operating instructions are printed on each extinguisher. However, during an emergency there might be no time to read the label. You must know how to use the fire extinguisher before any emergency takes place.

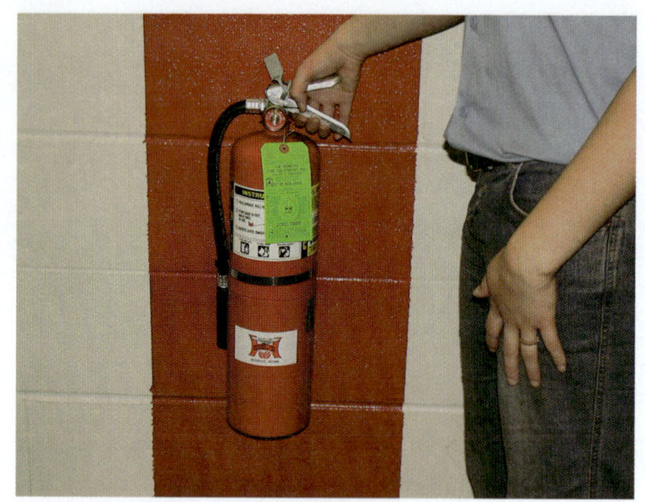

A If a fire breaks out in your shop, quickly walk to the closest fire extinguisher and lift it off of the wall.

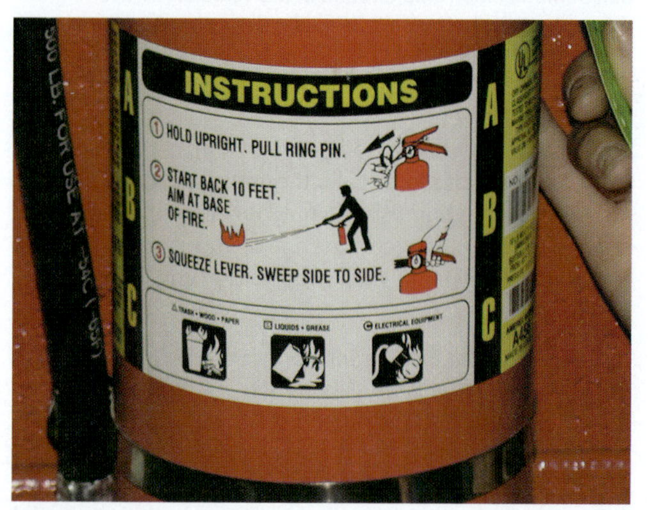

B Pull the pin out of the fire extinguisher handle and it should be ready for use.

Figure 9-42 **Know where all fire extinguishers are located and how to use them. A few minutes' time during a fire can be a lifetime.**

To use a fire extinguisher, first pull out the safety pin on the handle. Aim the nozzle at the base of the flames. Then squeeze the handle and direct the agent into the flames to extinguish them.

Stand 6–10 feet (2–3 meters) from the fire, holding the extinguisher firmly in its recommended position. Use a side-to-side motion, sweeping the entire width of the fire (Figure 9–43).

Most extinguishers work by cooling the fire and removing oxygen. If the fire extinguisher is going to be used effectively, it must be aimed at the base of the flame where the fuel is located.

During a fire, never open doors or windows unless it is absolutely necessary to escape. The extra air draft will only make the fire worse.

Fire extinguishers should be recharged regularly and placed at strategic, well-marked shop locations (Figure 9–44).

Handling Flammable Material

Body/paint technicians work with many kinds of flammable materials (paints, sealers, primers, reducers, solvents, cleaning agents, and so on). Some of these materials are also explosive and toxic. Gasoline and refinish material fumes, in particular, can ignite explosively.

Gasoline is a highly flammable petroleum or crude oil-based liquid that vaporizes and burns rapidly. Always keep gasoline and diesel fuel in approved safety cans and in a fireproof cabinet. Remember never to use automotive fuels to wash your hands, parts, or tools.

Store Flammable Products Properly

Store paints, thinners, solvents, pressurized containers, and other combustible materials in approved and designated fireproof metal storage cans, cabinets, or rooms (Figure 9–45).

All ignition sources should be carefully controlled and monitored to avoid any possible fire hazard where a high concentration of vapor from flammable liquids might be present.

Oily, greasy, or paint-soaked rags should be stored in an approved metal container with a lid. Keep chemically soiled paper towels and other paper products in a separate covered container, which should be emptied every day (Figure 9–46).

When chemically soiled rags or paper towels are left lying in a pile in the shop, they are prime candidates for **spontaneous combustion** (fire that starts by itself). A pile of oily or solvent-soaked rags or towels can start burning without an outside source of heat, spark, or flame. This is because when certain chemicals mix together, they can chemically generate heat and ignite on their own.

Storage or mixing rooms should have adequate ventilation. Never leave flammable materials in the work area.

Keep partially used containers tightly closed. Solvent fumes in the bottom of these containers are prime ignition sources.

Do not light matches or smoke in the painting or mixing areas or anywhere in the shop. Make sure that your hands and clothing are free from solvent when lighting matches or a gas torch and when leaving the shop.

Before lighting a gas torch or striking an arc with a welder, double-check that the work area is free of flammables. Sweep up any oil absorbent sprinkled over spills. The oil or fuel-soaked absorbent can flash into a large fire with the slightest spark.

Transfer flammable liquids properly! Use only approved explosion-proof equipment in hazardous locations. An Underwriters Laboratories (UL)-approved drum

TABLE 9–1 GUIDE TO EXTINGUISHER SELECTION

	Class of Fire	Typical Fuel Involved	Type of Extinguisher
Class A Fires (green)	**For Ordinary Combustibles** Put out a class A fire by lowering its temperature or by coating the burning combustibles.	Wood Paper Cloth Rubber Plastics Rubbish Upholstery	Water*[1] Foam* Multipurpose dry chemical[4]
Class B Fires (red)	**For Flammable Liquids** Put out a class B fire by smothering it. Use an extinguisher that gives a blanketing, flame-interrupting effect; cover the whole flaming liquid surface.	Gasoline Oil Grease Paint Lighter fluid	Foam* Carbon dioxide[5] Halogenated agent[6] Standard dry chemical[2] Purple K dry chemical[3] Multipurpose dry chemical[4]
Class C Fires (blue)	**For Electrical Equipment** Put out a class C fire by shutting off power as quickly as possible and by always using a nonconducting extinguishing agent to prevent electric shock.	Motors Appliances Wiring Fuse boxes Switchboards	Carbon dioxide[5] Halogenated agent[6] Standard dry chemical[2] Purple K dry chemical[3] Multipurpose dry chemical[4]
Class D Fires (yellow)	**For Combustible Metals** Put out a class D fire of metal chips, turnings, or shavings by smothering or coating with a specially designed extinguishing agent.	Aluminum Magnesium Potassium Sodium Titanium Zirconium	Dry powder extinguishers and agents only

*Cartridge-operated water, foam, and soda-acid types of extinguishers are no longer manufactured. These extinguishers should be removed from service when they become due for their next hydrostatic pressure test.

Notes:

(1) Freezes in low temperatures unless treated with antifreeze solution, usually weighs over 20 pounds, and is heavier than any other extinguisher mentioned.

(2) Also called ordinary, or regular, dry chemical (sodium bicarbonate).

(3) Has the greatest initial fire-stopping power of the extinguishers mentioned for class B fires. Be sure to clean residue immediately after using the extinguisher so sprayed surfaces will not be damaged (potassium bicarbonate).

(4) The only extinguisher that fights A, B, and C classes of fires. However, they should not be used on fires of liquefied fat or oil of appreciable depth. Be sure to clean residue immediately after using the extinguisher so sprayed surfaces will not be damaged (ammonium phosphates).

(5) Use with caution in unventilated, confined spaces.

(6) May cause injury to the operator if the extinguisher agent (a gas) or the gases produced when the agent is applied to a fire are inhaled.

transfer pump along with a drum vent should be used when transferring flammable chemicals. Keep all solvent containers clearly labeled and closed, except when pouring.

Handle all solvents or liquids with care to avoid spills. Extra caution should also be used when transferring flammable materials from bulk storage. Remember to make sure the drum is grounded and that a bond wire connects the drum to a safety can. Otherwise, static electricity can build up and create a spark that could cause an explosion (Figure 9–47).

9.6 HAZARDOUS MATERIAL SAFETY

Work intelligently around hazardous materials! Be informed; read the warnings on the product labels and in manufacturers' literature. If more information is desired, get copies of the material safety data sheets (MSDS) for specific products from the shop's office or from the material suppliers.

Material safety data sheets (MSDS) contain information on hazardous ingredients and protective measures that the technician should use. MSDS, available from

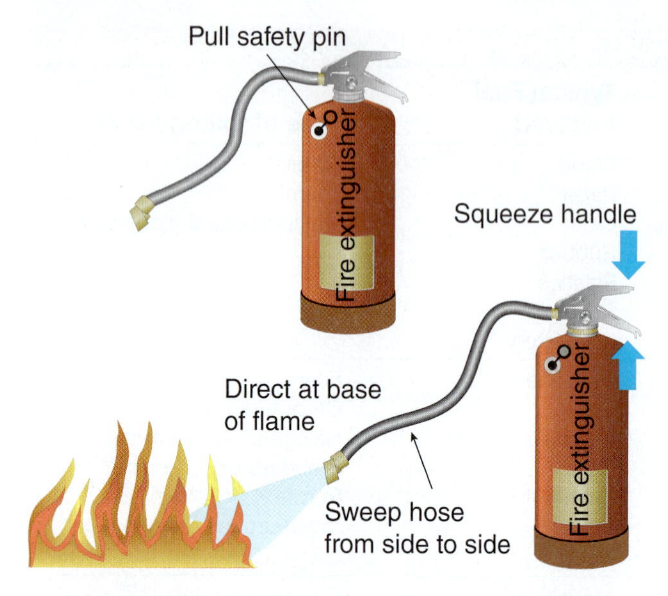

Figure 9-43 Note the basic steps for using a fire extinguisher.

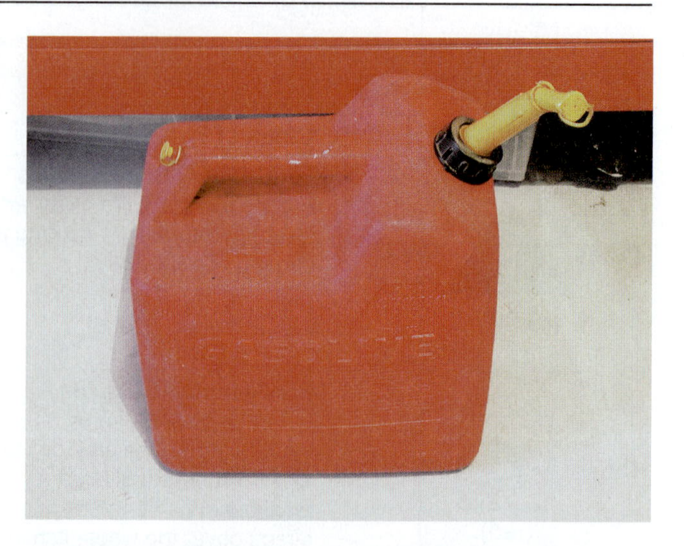

Figure 9-45 All flammable materials must be stored in a metal safety cabinet. Gasoline is by far one of the most flammable materials in a shop. When storing gasoline, make sure vent cap and spout cap are in place so it cannot evaporate.

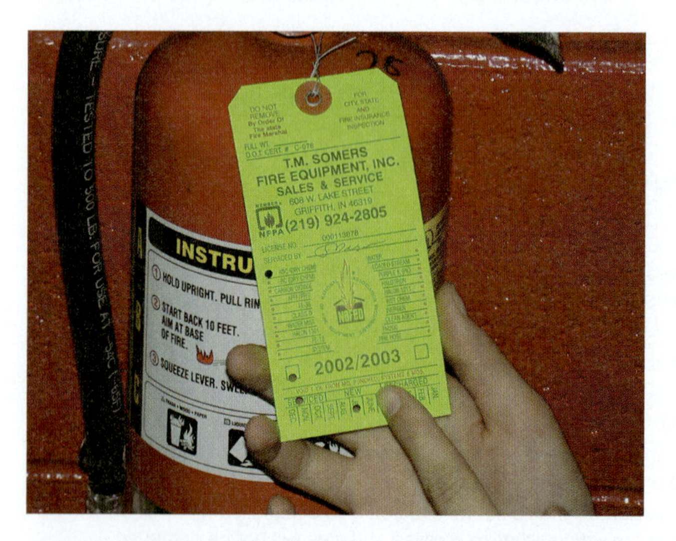

Figure 9-44 Someone in the shop must inspect the dates on all fire extinguishers. They must be recharged at periodic intervals.

Figure 9-46 Soiled or dirty rags and towels must be placed in a metal can with a lid. If spontaneous combustion occurs, the metal lid will smother the fire.

all product manufacturers, detail chemical composition and precautionary information for all products that can present a health or safety hazard. An example of an MSDS is shown in Figure 9–48.

Hazardous waste, as determined by the Environmental Protection Agency (EPA), is a solid or liquid that can harm people and the environment. If the waste is on the EPA list of known harmful materials or has one or more of the following characteristics, it is considered hazardous.

1. *Ignitability* means the material or waste fails the ignitability test if it is a liquid with a flash point below 140°F or a solid that can spontaneously ignite.
2. *Corrosiveness* means a material or waste is considered corrosive if it dissolves metals and other materials or burns the skin. It is an aqueous solution with a pH of 2 and below, or 12.5 and above. Acids have the lower value and alkalis have the higher value.
3. *Reactivity* means a material reacts violently with water or other materials or releases cyanide gas, hydrogen sulfide gas, or similar gases when exposed to low pH solutions (acid). This also includes material that generates toxic mists, fumes, vapors, and flammable gases.
4. *Toxicity* means a material leaches one or more heavy metals in concentrations greater than 100 times

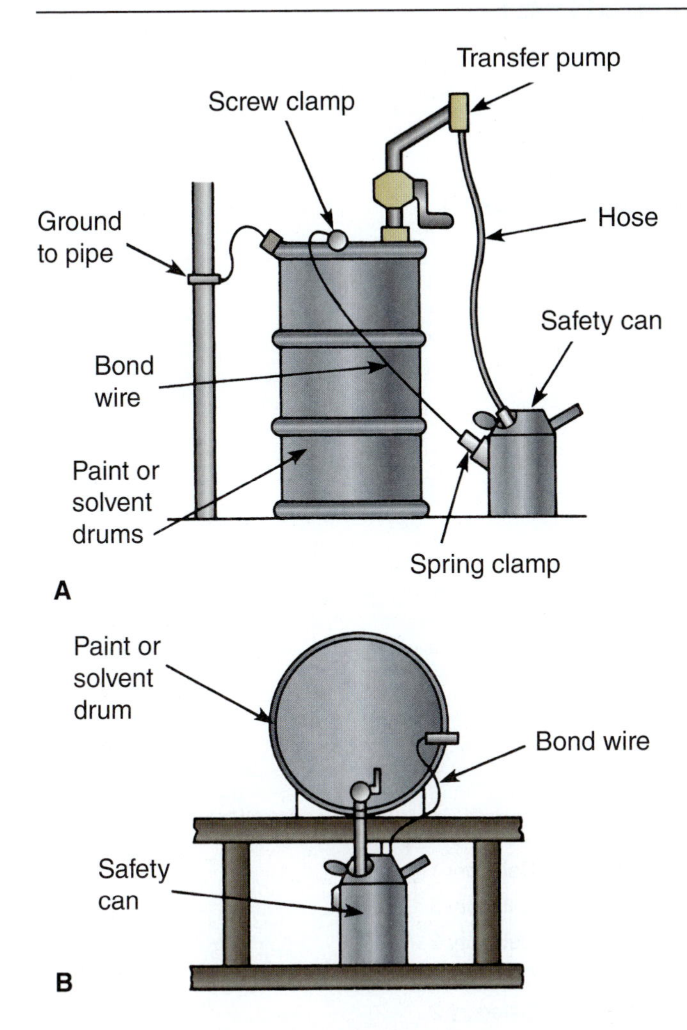

Figure 9-47 Two safe methods of moving flammable liquids from a drum to a portable safety can are shown here.

primary drinking water standard concentrations. These heavy metals include lead, cadmium, chromium, and arsenic.

Complete EPA lists of hazardous wastes can be found in the *Code of Federal Regulations*. Materials and wastes of most concern to the body/paint technician are organic solvents that contain heavy metals, especially lead. During disposal, all hazardous waste must be handled according to the appropriate regulations (Figure 9–49 and Figure 9–50).

All collision repair shop employees must know the general uses, protective equipment, accident or spill procedures, and other information for safe handling of hazardous materials. This training must be given to employees as part of their job orientation.

The best way to protect yourself when using paint and body repair products is to be familiar with the correct application procedures. Follow all safety and health precautions found on MSDS and on product labels.

Metal conditioners can be very toxic and corrosive! Metal conditioners contain phosphoric acid. Breathing these chemicals or allowing them to come in direct contact with the skin, eyes, or clothing may cause irritation or injury. Wear safety glasses or, preferably, goggles to prevent these kinds of materials from splashing into your eyes. In addition, wear coveralls, rubber gloves, and a NIOSH-approved organic vapor respirator when using these products. If coveralls become soaked for any reason, be sure to change right away or soak the spill with water to dilute the chemicals.

Warnings and Regulations

Right-to-Know Laws specify essential information and stipulations for safely working with hazardous materials. They started with OSHA's Hazard Communication Standard. This document was originally intended for chemical companies and manufacturers that require employees to handle potentially hazardous materials in the workshop.

Since then, the majority of states have enacted their own Right-to-Know Laws. The federal courts have decided that these regulations should apply to all companies, including the auto collision repair and refinishing professions.

The general intent of such laws is that employers provide their employees with a safe workplace. Specifically, there are three areas of employer responsibility:

1. Obtain proper hazardous materials training. All employees must be trained about their rights under the legislation. They must learn the nature of the hazardous chemicals, the labeling of chemicals, and the information about each chemical. This information should be posted on MSDS. MSDS summarize all chemicals used in the shop products. These information sheets should be made readily available to employees. Employees must be familiarized with the general uses, characteristics, protective equipment, and accident or spill procedures associated with major groups of chemicals. This training must be given to employees annually and provided to new employees as part of their job orientation.
2. Study information about potentially hazardous chemicals. All hazardous materials must be properly labeled, indicating what health, fire, or reactivity hazard they pose. Information about what protective equipment is needed when handling each chemical must be given. This safety data must be read and understood by the user before application.
3. Shops must keep hazardous material records. Shops must maintain documentation on the hazardous chemicals in the workplace. They also must provide proof of training programs, records of accidents and/or spill incidents, satisfaction of employee requests for specific chemical information via the MSDS, and a general Right-to-Know compliance procedure manual used within the shop.

SAFETY DATA SHEET

1. Identification

Product identifier	**Brakleen® Brake Parts Cleaner - Non-Chlorinated**
Other means of identification	
Product code	05088
Recommended use	Brake parts cleaner
Recommended restrictions	None known.

Manufacturer/Importer/Supplier/Distributor information

Manufactured or sold by:

Company name	CRC Industries, Inc.
Address	885 Louis Dr. Warminster, PA 18974 US
Telephone	
General Information	215-674-4300
Technical Assistance	800-521-3168
Customer Service	800-272-4620
24-Hour Emergency (CHEMTREC)	800-424-9300 (US) 703-527-3887 (International)
Website	www.crcindustries.com

2. Hazard(s) identification

Physical hazards	Flammable aerosols	Category 1
Health hazards	Acute toxicity, oral	Category 1
	Acute toxicity, dermal	Category 3
	Skin corrosion/irritation	Category 2
	Serious eye damage/eye irritation	Category 2A
	Reproductive toxicity (the unborn child)	Category 2
	Specific target organ toxicity, single exposure	Category 3 narcotic effects
	Specific target organ toxicity, repeated exposure	Category 1
	Aspiration hazard	Category 1
OSHA defined hazards	Not classified.	

Label elements

Signal word	Danger
Hazard statement	Extremely flammable aerosol. Fatal if swallowed. May be fatal if swallowed and enters airways. Toxic in contact with skin. Causes skin irritation. Causes serious eye irritation. May cause drowsiness or dizziness. Suspected of damaging the unborn child. Causes damage to organs through prolonged or repeated exposure.
Precautionary statement	
Prevention	Obtain special instructions before use. Do not handle until all safety precautions have been read and understood. Keep away from heat/sparks/open flames/hot surfaces. - No smoking. Do not spray on an open flame or other ignition source. Pressurized container: Do not pierce or burn, even after use. Do not breathe gas. Do not breathe mist or vapor. Wash thoroughly after handling. Do not eat, drink or smoke when using this product. Use only outdoors or in a well-ventilated area. Wear protective gloves/protective clothing/eye protection/face protection. Extinguish all flames, pilot lights and heaters. Do not apply while equipment is energized. Vapors will accumulate readily and may ignite. Use only with adequate ventilation; maintain ventilation during use and until all vapors are gone. Open doors and windows or use other means to ensure a fresh air supply during use and while product is drying. If you experience any symptoms listed on this label, increase ventilation or leave the area.

Figure 9-48 A typical MSDS will give information on hazardous materials. Study the information on this sample MSDS carefully.

Figure 9–49 Many automotive service operations, including body shops, are hiring outside contractors to handle their waste materials.

Figure 9–50 When washing a shop floor, make sure all electrical cords are off the ground. Also, never wash chemical spills down the shop drain.

Blood and Tissue Hazard

Pathogens are disease-producing microorganisms or agents. Pathogens can enter your body by being inhaled or eaten or through open cuts in the skin. In collision repair, a common source of dangerous pathogenic organisms is contaminated blood remaining in wrecked vehicles. You must take several safety precautions to protect yourself from pathogens.

Contaminated blood and small pieces of human tissue often remain in the passenger compartment of vehicles entering the body shop. Even with air bags and seat belts, the driver and their passengers are sometimes badly hurt when vehicles crash at high speed. Blood and tiny pieces of flesh can be found on seats, carpets, trim panels, the dash, and other interior parts.

Always treat this human body material as hazardous. If anyone injured in the vehicle was suffering from a communicable disease, such as HIV (human immunodeficiency virus), handling this material could make you sick. Generally, if the blood and tissue are still moist or wet, they are more dangerous than when the material has had enough time to dry completely.

If blood or tissue is found in an unrepaired vehicle, put on coveralls, plastic or rubber gloves, and a respirator before working. An air-supplied painter's suit and hood would provide the best protection, especially if the bloodstains are still wet. This safety gear offers protection from potential infection when cleaning vehicle interiors or removing contaminated trim parts.

In the washup area of the shop, the passenger compartment should be thoroughly washed with an antiseptic cleaning agent to help kill any pathogens on blood stains. Washup personnel should also wear recommended safety gear.

9.7 GOOD SHOP HOUSEKEEPING

It is very important that the work area be kept clean and safe. This should be a team effort of all shop employees. Here are some simple good housekeeping precautions that should be followed in every shop.

DANGER When working on the interior or trunk area of a vehicle, be aware of the danger of contaminated needles or drug apparatus. Contaminated medical syringes used to inject legal and illegal drugs could prick your skin and enable pathogens (AIDS or similar disease) to enter your bloodstream. Use a drop light to look under seats or down along hidden areas in the trunk before reaching into these areas while working. Never run your hand down inside the gap between interior seat cushions!

REMEMBER *A clean, well-organized shop is much safer to work in than a dirty, unorganized shop.*

All work areas and surfaces should be kept clean, dry, and orderly. Any oil, coolant, or grease on the floor can cause slips or falls that could result in serious injuries. Items stacked on benches and tables can easily fall, and they generally cause a cluttered work area, which invites accidents (Figure 9–51).

Hang tools up or put them away in your toolbox when not in use. Roll up air hoses when not in use so they do

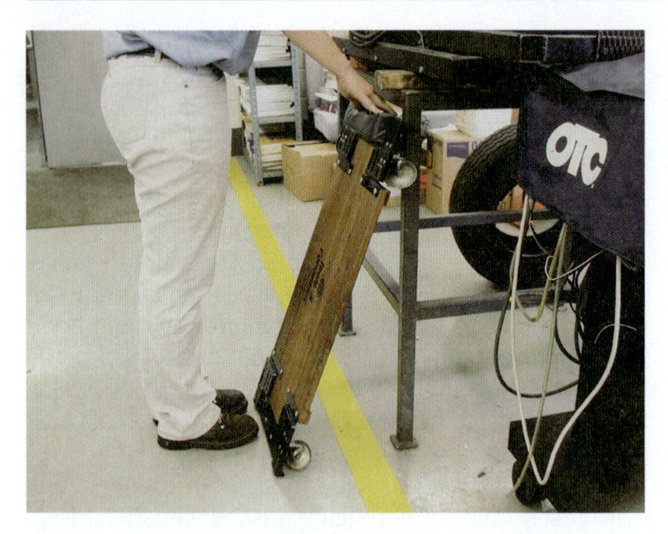

Figure 9-51 **Keep all walkways clear, and never leave a creeper lying on the floor. If someone were to step on a creeper, a serious fall and injury could result.**

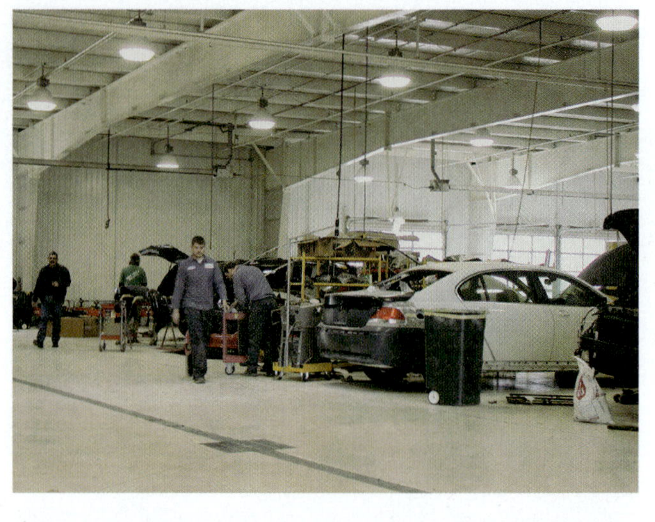

Figure 9-52 **If everyone takes part in shop maintenance and safety, the body shop should be a fun, rewarding place to work.**

not create traffic hazards. Floor jacks, bumper jacks, jack stands, and creepers should be kept in their designated area, out of aisles and walkways.

To clean up an oil spill on the shop floor, use a commercial *oil absorbent.* Spread the absorbent on the spill. Rub it with your foot in a circular motion. Then use a dust pan and broom to pick up the absorbent right away, especially if the spill was a flammable liquid.

Trash and rubbish should be removed from the shop area regularly. If not, serious fire dangers can result.

Clean up broken glass and remove parts with jagged metal edges from the shop area right away. Recycle bins should be provided for recycling metal and glass. If you fail to clean up broken glass properly, someone could be cut and injured. Also, if you fail to clean out dash ducts, glass can blow into people's eyes when the heater or air conditioning is turned on. Wear gloves and a full face shield.

Make sure that aisles and walkways are kept clean and are wide enough for safe clearance. Organize your work space so there is room to walk around all machines and equipment. Cluttered walking areas are a common source of injuries.

Keep all shop floor drain covers snugly in place. Open drains have caused many toe, ankle, and leg injuries.

Make sure that hazardous materials are not discharged through floor drains or other outlets leading to public waterways.

Keep gas cylinders away from sources of heat, such as a furnace or room heater. Check and service furnaces and water heaters in the shop at least once every six months.

Before leaving the shop, store toxic materials properly! Solvents, chemicals, and other materials can contaminate clothing and wind up on the hands when you remove personal protective equipment or put away the refinishing tools. Wash your hands thoroughly before handling food. See Figure 9–52.

9.8 AIR BAG SAFETY

Air bags can be very dangerous because they inflate with tremendous force when deployed. Air bags can reach speeds of over 100 miles per hour during deployment. Air bags can break arms, hands, fingers, or even cause death if someone's head is near the bag during accidental deployment.

When working around air bags, use service manual directions and caution to prevent accidental deployment. Never install or connect a new air bag module until all wiring is checked with a scan tool and repaired, if needed. A short in the wiring to an impact sensor or to the air bag module could make the new air bag deploy as soon as it is connected to its control circuit.

Refer to the vehicle's service manual to learn the precautions that must be followed when servicing an air bag. It will give the details needed to work safely.

For details on air bag service, please refer to Chapter 24.

Always remember to follow all safety rules. When you cut corners and break safety rules, you put the health and safety of yourself and your coworkers in danger.

SUMMARY

1. True professionalism and shop safety begins with how technicians perform their work tasks. Professionals understand the ever-present dangers in a body shop and strive to avoid making dangerous mistakes.
2. Manufacturer's instructions are detailed procedures for a specific product, whether that be a piece of equipment or a certain type of paint.
3. You should instantly know what to do in case of an emergency, whether it involves a bad injury, a shop fire, or another accident.
4. Know the location of your shop's first-aid kit.
5. Horseplay in the shop is unacceptable. Proper conduct can help prevent accidents.
6. There is usually a "best tool" for each repair task. Using the correct tool can help prevent injuries.
7. Sanding operations create dust that can cause bronchial irritation and possible long-term lung damage.
8. A dust respirator should be worn when sanding, grinding, or blowing off dirty panels with a blow nozzle.
9. An air-supplied respirator provides protection from the dangers of inhaling extremely dangerous airborne materials: paint with a hardener, isocyanate paint vapors and mists, and hazardous solvent vapors.
10. Respirators are recommended even when adequate ventilation (outside airflow) is provided in the shop area.
11. Some type of eye protection should be worn at all times when working in the shop.
12. The harmful effects of liquids, undercoats, and finishes on the hands can be prevented by wearing work gloves.
13. Loose clothing, unbuttoned shirt sleeves, dangling ties, jewelry, and shirts hanging out are very dangerous in a body shop.
14. Car and truck engine exhaust produces harmful carbon monoxide (CO) gas. Operate engines only in a well-ventilated area to avoid the danger of CO poisoning.
15. Do not risk injury through lack of knowledge; use shop tools or perform repair operations only after receiving proper instruction.
16. Store all parts and tools properly by putting them away neatly where other workers will not trip over them. This will increase safety and reduce time wasted looking for a misplaced part or tool.
17. Shop air pressure is usually around 100–150 psi (689 to 1,034 kPa), which is sufficient to severely injure or kill.
18. If the shop has a hydraulic lift, read the instruction manual before using it. Refer to service information for the specific vehicle to find recommended vehicle lift points.
19. To be properly prepared for a shop fire, know the location of all fire extinguishers in the shop, and know how to use them.
20. To prevent electrical fires, always disconnect the battery when doing electrical work or when body damage may have cut through wires.
21. Body/paint technicians often work with flammable liquids. Fumes, in particular, can ignite explosively. Do not light matches or smoke in the spraying and painting area. Make sure your hands and clothing are free from solvent when lighting matches or a gas torch.
22. Read the warnings on the product labels and in manufacturers' literature. If more information is desired, get copies of the material safety data sheets (MSDS) for specific products.
23. Hazardous wastes, as determined by the Environmental Protection Agency (EPA), are solids or liquids that can harm people and the environment.

EXERCISES

On a separate sheet of paper, complete the following learning activities for this chapter. Write definitions for the key terms and answer the ASE-style review questions, essay questions, critical thinking problems, and math problems. You can also do the outside activities, possibly for extra credit.

Key Terms

air-supplied respirator
asbestos dust
asphyxiation
carbon monoxide (CO)
cartridge filter respirator
dust respirator
electrical fire
electrocution
environmental safety
eye flushing station
first-aid kit
floor jack
full face shield
goggles
hazardous waste
impervious gloves
jack stand
manufacturer's instructions
material safety data sheets (MSDS)
National Institute for Occupational Safety and Health (NIOSH)
Occupational Safety and Health Administration (OSHA)
pathogens
respirator fit test
Right-to-Know Laws
spontaneous combustion
toxic
vehicle lift points
welding helmet
welding respirator

ASE-Style Review Questions

1. Which of the following present dangers to the air passages and lungs of body shop technicians?
   A. Dust
   B. Vapors from caustic solutions and solvents
   C. Spray mists from undercoats and finishes
   D. All of the above
2. Which respirator covers the entire head and neck area?
   A. Cartridge filter respirator
   B. Dust respirator
   C. Air-supplied respirator
   D. None of the above
3. Technician A and Technician B spray prime then paint materials continuously for extended periods of time. Technician A changes the pre-filters when it becomes difficult to breathe through the respirators. Technician B performs a fit test prior to using the respirator. Who is correct?
   A. Technician A
   B. Technician B
   C. Both A and B
   D. Neither A nor B
4. Which respirator is commonly used to protect against dust from sanding and grinding?
   A. Hood respirator
   B. Organic vapor-type respirator
   C. Air-supplied respirator
   D. None of the above
5. Eye protection should be worn when using
   A. Drills.
   B. Cutting tools.
   C. Hammers.
   D. all of the above.
6. Which of the following should not be worn in a body/paint shop?
   A. Jumpsuit
   B. Loose clothing
   C. Cap
   D. Both B and C
7. By what means can ventilation be achieved in the body/paint shop?
   A. Extraction floors
   B. Central dust extraction
   C. Air-changing system
   D. All of the above
8. Which of the following products should be stored in designated fireproof areas or containers?
   A. Paints
   B. Thinners and solvents
   C. combustible materials
   D. All of the above
9. Technician A discards all empty solvent containers. Technician B keeps empty solvent containers in the paint mixing area. Who is correct?
   A. Technician A
   B. Technician B
   C. Both A and B
   D. Neither A nor B
10. Which type of extinguisher can be used on all classes of fire?
    A. Water
    B. Foam
    C. Class D
    D. None of the above
11. Which respirator can be used when applying nonisocyanate paint materials?
    A. Dust mask
    B. Cartridge respirator
    C. Welding respirator
    D. None of the above

12. Technician A will operate a tool beyond its stated capacity in order to get the job done faster. Technician B uses appropriate tools for each task even if it takes longer to complete the job. Who is correct?

A. Technician A
B. Technician B
C. Both A and B
D. Neither A nor B

13. Technician A wears a dust mask while spraying a clearcoat with hardener, whereas Technician B wears a cartridge filter respirator. Who is correct?

A. Technician A
B. Technician B
C. Both A and B
D. Neither A nor B

14. When transferring solvents into smaller containers, Technician A leaves flammable materials in the work area. Technician B has built a separate wooden storage facility for solvents. Who is correct?

A. Technician A
B. Technician B
C. Both A and B
D. Neither A nor B

15. Technician A says that shop employees should receive proper hazardous materials training each year. Technician B says that new shop employees should receive proper hazardous materials training as part of their job orientation. Who is correct?

A. Technician A
B. Technician B
C. Both A and B
D. Neither A nor B

Essay Questions

1. Why is a cluttered, unorganized shop dangerous?
2. How do you use a fire extinguisher?
3. Can a battery explode? Explain.
4. How can an air bag cause injuries?
5. Summarize how to safely use a vehicle lift.
6. Describe the types of accidents that could happen in a typical shop setting.

Critical Thinking Problems

1. Why should you pull rather than push on a hand wrench?
2. Fire breaks out in the shop and a co-worker's clothes catch fire. What should you do?
3. What can you do to make your shop a safer place in which to work?
4. A person with a bad safety track record might face what problem when looking for a new job?
5. The first response in an emergency should always be ________.
6. List ways to prevent these accidents from happening.

Math Problems

1. After working in an auto body shop for 30 years, an auto body technician came down with terminal lung cancer and had to retire. Through all of his years of work, he failed to wear a respirator mask while sanding and grinding. This resulted in 1 gram of airborne material per day entering his mouth, nose, and lungs. If the technician worked a total of 7,800 days, how many pounds of this material could have been inhaled?

Activities

1. Inspect your shop for safety violations. Write a report on how shop safety could be improved or better maintained.
2. Organize a field trip to a local body shop. Ask the owner if your class can observe shop personnel at work. Make a report on any safety violations observed.
3. Invite a professional auto body technician to visit your class. Ask them to lecture on the importance of shop safety and the kinds of accidents that can happen in a body shop.
4. Visit a salvage yard to evaluate vehicle damage. Ask permission from the owner to inspect wrecked vehicles for a report. Try to determine the direction of impact damage.

SECTION

2

Estimating

CHAPTER

10 Estimating Repair Costs

OBJECTIVES

After reading this chapter, you should be able to:

- Explain the general purpose of damage estimates.
- Manually and electronically prepare an estimate.
- Outline the sequence for evaluating vehicle damage.
- Describe the method of determining the repairability of a damaged vehicle.
- Use gathered information, determine whether damaged parts should be repaired or replaced with new ones.
- Explain the difference between flat-rate labor time and overlap labor time when estimating repair costs.
- Calculate material costs based on a refinishing materials list.
- Use computer-based estimating programs.
- Describe the benefits of using a digital camera, hand-held computer, personal computers, and estimating software when preparing estimates.
- Answer ASE-style review questions relating to the estimating process.

INTRODUCTION

An *estimate*, also called a *damage report* or *damage appraisal*, calculates the cost of parts, materials, and labor for repairing a collision-damaged vehicle. Developed by the estimator and equipment available, it is a handwritten or printed summary of the collision repairs needed. The estimate lets the customer, insurance company, shop management, and technician know what must be done to repair a vehicle.

Estimates must be accurately written. Repair costs are a major consideration for both the owner and the collision repair shop. The profit margins for a collision repair shop depend heavily on the accuracy of estimates. Insurance companies can also write their own estimates.

This chapter will help you understand how both handwritten and electronic, or computerized, estimates are prepared (Figure 10–1).

Courtesy of Liss CarStar Collision Center

Figure 10–1 Estimators must have a thorough knowledge of vehicle construction, collision repair methods, and computer use to develop accurate estimates of repair costs.

10.1 THE ESTIMATE

To understand computer-based estimating, you must first understand how a handwritten estimate is done. Basically, a *handwritten estimate* involves using a pen and paper or a printed form to list all of the information needed to calculate and bill the customer or insurance company for doing the repairs.

Before any decision involving the repair of a damaged vehicle can be made, a detailed repair estimate must be calculated and printed. All of the parts and procedures that must be completed are listed on the estimate. Then part prices and labor charges can be tabulated.

Although printed estimating guides are still used in the industry, the trend is to use computer-based estimating tools (Figure 10–2).

Before writing the formal damage report, preliminary information that identifies the owner

REPAIR ESTIMATE INFORMATION

Customer Info:

Name: | Date:
Address:
City: | State: | Zip:
Home #: | Work #: | Cell #:
Fax #: | E-mail Address:
Year: | Make: | Model: | Style:

Insurance/Payment Info:

Who will be responsible for your repairs? | ☐ My Ins. | ☐ Their Ins. | ☐ Other:
Has the Insurance Co. written an estimate? | ☐ Yes | ☐ No | Claim #:
Name of your Insurance Co.: | Deductible:
Name of your Insurance Agent: | Date of loss:
Name of other party's Insurance Co.:
Is this your ☐ 1st ☐ 2nd ☐ 3rd estimate? | Would you like us to repair your vehicle? | ☐ Yes | ☐ No
Will you require a rental car during the repairs? | ☐ Yes | ☐ No
Do you have any special concerns?
Estimator: | VIN:
Year: | Make: | Model: | Production Date:
Lic. Number: | Mileage: | Color: | Trim Code:
Options: | Picture Code: | Paint Code:

Repair	Replace	R & I	Repair Description	Labor: Body	Labor: Refinish	Labor: Frame	Labor: Mech.

Prior Damage

Figure 10–2 A form like this one is often used when making a handwritten or computer-based estimate. Much of it can be filled out by the customer. Other information about the damaged vehicle is filled in by the body shop estimator.

of the vehicle and who will pay the bill must be obtained. This includes the VIN, make, year, body style or type, license plate number, mileage, and date.

All parts that need to be replaced or repaired must be identified. A detailed description of all the labor operations that must be performed, with a listing of parts and materials needed to make the necessary repairs, can then be prepared.

A copy of the estimate is then given to the customer. Estimate information—pricing of labor, parts and materials, and their totals—is then compared to estimates from competing body shops. A written estimate is needed to prevent any misunderstanding between the shop, the insurance company, and the vehicle owner.

At least three copies of the written estimate should be made. One is kept by the shop, one is given to the insurance company, and the other is given to the customer. An estimate is an *approximate bid* for a given period of time—usually for 30 days. The reason for a specified time period is obvious: part prices change and damaged parts can deteriorate.

Most body shops provide damage reports or estimates free of charge. Because estimate preparation can involve a considerable amount of time and paperwork, some body shops charge a flat fee for the estimate.

Direct Repair Programs

Direct repair programs (DRPs) are made up of cooperating insurance companies and body shops. An insurance company will approve a body shop for repairing vehicles insured by that insurance company. The insurance company will have its customers take their damaged vehicles to one of the approved body shops for repairs. This eliminates the time needed to obtain and approve one of several estimates.

The estimate is considered authorization to complete the repair work as listed, but only when it is agreed upon and signed by the vehicle owner or by the insurance company appraiser. The estimate explains the legal conditions under which the repair work is accepted by the collision repair shop. It also protects the shop against the possibility of undetected damage that might be revealed later as repairs progress. With direct repair, work is often done for a reduced hourly rate.

Insurance Deductibles

Many insurance policies contain a **deductible clause**, which means that the owner is responsible for a given amount of the bill, usually the first $100–$500. The remaining cost is paid by the insurance company. In such cases, both the customer and insurance company should authorize the estimate.

Estimate Time Factors

There are additional factors to be considered when writing an estimate. Such added time is usually negotiated between the estimator and the insurance company or customer, and includes the following:

- Setting the vehicle on a frame machine and making a damage diagnosis
- Pushing, pulling, cutting, and so on to remove collision-damaged parts; called **access time**
- Straightening or aligning related parts
- Removing undercoating, tar, grease, and similar materials
- Repairing rust or corrosion damage to adjacent parts
- Freeing rusted or frozen parts
- Drilling for ornamentation or mounting holes
- Filling or plugging unneeded holes in new parts
- Repairing damaged replacement parts prior to installation
- Checking suspension and steering alignment/toe-in
- Removing shattered glass
- Rebuilding, reconditioning, and installing aftermarket parts, not including refinishing time
- Applying sound-deadening material; undercoating, caulking, and painting of the inner areas
- Restoring corrosion protection
- Removing and installing (R&I) main computer module when excessive temperatures (above 176°F or 80°C) are necessary in repair or paint-drying operations
- Removing and installing wheel or hub cap locks
- Replacing accessories such as trailer hitches, sunroofs, and fender flares

Work Orders

Another important function of the estimate is that it serves as a basis for writing the work order or operational plan. It is usually prepared from the damage appraisal of the estimator (using the written estimate) and a visual inspection by the shop supervisor and/or a technician.

The **work order** is a printed form that outlines the procedures that should be taken to return the vehicle to its preaccident condition. It is a valuable tool for the technician, estimator, and shop supervisor. The work order summarizes the actual methods necessary to do the repairs on a specific vehicle. Technicians are given the vehicle work order before starting repair work. The work order is often clipped on the vehicle or on the technician's toolbox for easy reference.

Using Estimating Guides

Discussed briefly in Chapter 3, *collision estimating guides* help with filling out the estimate by listing part and labor costs. Whether in printed or electronic form, estimating guides contain:

- Lists of vehicle makes and models in the contents or index
- Section indexes for major part assemblies for each vehicle
- Illustrated parts breakdowns

- Part names and numbers
- Flat rate times
- Part prices
- Other information

Collision estimating guides, also known as collision damage manuals, are essential tools of the estimator. They contain such items as vehicle identification information, the price of new or recycled parts, the amount of time needed to install the parts, identification of almost any part of the car from the front bumper to the rear bumper, and refinishing data, such as paint code references (Figure 10–3).

Estimating guides are often published and updated at different times of the year when manufacturers change prices or make model revisions. The information in these guides is of value to body technicians and refinishers. You should be able to read and understand them (Figure 10–4).

Collision estimating guides can be used as a reference for pricing parts. However, they should never be used to determine the final estimated total. The prices in estimating guides are factory-suggested list prices. Parts that have been discontinued are usually listed and noted with the letter D. The price that appears in the guide is the latest one available at the time of publication.

Each collision estimating guide will have **procedure pages**. Sometimes called *pages*, these procedure pages provide important information such as:

- Arrangement of material in the guide
- Explanation of symbols used in the guide
- Definitions of terms used in the guide
- How to read and use the parts illustrations
- Procedure explanations, including which operations are included and which are *not* included together (Figure 10–5)
- How discontinued parts information is displayed
- How interchangeable part information is displayed
- Additions to labor times
- Labor times for overlap items
- How to identify structural operations
- How to identify mechanical operations

Service manuals and estimating guides also contain exploded views of parts. Such illustrations are useful in determining which parts must be replaced during repairs.

When using crash estimating guides, you must be familiar with a few basic technical terms found in estimating guides:

- **Remove and install (R&I)** means the item is removed as an assembly, set aside, later reinstalled and finally aligned for a proper fit. This is generally done to gain access to another part or metal panel to be fixed. For example, "R&I bumper assembly" means that the complete bumper assembly (front cover, backing pad, steel frame, and energy absorbing shock absorbers) must be removed from the chassis and frame.
- **Remove and replace (R&R)** means to take out the old parts, transfer necessary items to new parts, replace, and align.
- **Overhaul (O/H)** means to remove an assembly from the vehicle; disassemble, clean, inspect, and replace parts as needed; then reassemble, install, and adjust (except wheel and suspension alignment). Overhaul time should be used only if the time for repairing the individual parts (less overlap) is more than the overhaul time.

Looking at Figure 10–6, how many hours of labor are listed to R&R and refinish the front fender?

Appendix B in this textbook lists other terms that are accepted by most estimating guide publishers and estimators when filling out written forms. Chapter 3 also introduced basic shop abbreviations.

10.2 PART PRICES

Once the vehicle is judged repairable, the next decision is whether the collision damage requires new parts or the repair and straightening of existing parts. The estimator must have the ability to compare the cost of repairs against the cost of new parts or units.

As a rule of thumb, repair costs should never exceed replacement costs. If there is some doubt that repairs and straightening will not produce a quality job, then new parts should be used. Remember, sheet metal parts usually offer the most opportunities for repair and straightening. As a result, sheet metal repairs, replacement, and refinishing of panels generally account for the largest number of estimate dollars.

New parts are manufactured by the auto maker or an aftermarket company. They are usually more expensive than used parts and sometimes require more time to prepare them for installation and painting. New parts often do not have the factory undercoating and corrosion protection of a salvage part.

To reduce parts costs, many insurance appraisers and some customers might want the body shop to use salvage parts. **Salvage parts** are used parts in good condition that were removed from totally wrecked vehicles by salvage yards.

Some customers might object to used parts in their repaired cars. Inform them that used parts might be preferable to new parts. Explain that new parts do not always have factory corrosion protection and that factory rust protection is difficult and time-consuming to match in the shop.

Many salvage yard dealers offer a free computer parts location service. They use shortwave radio or computer messaging to request your needed parts from salvage yards all over the country.

Remember that salvage parts could have been previously damaged and repaired. For this reason, carefully inspect all salvage parts before they are installed to be sure that they are structurally sound.

SUNFIRE 1995-02

SECTION INDEX

SPECIAL CAUTIONS

AIR BAG

Refer to Procedure Explanation 29 for Supplemental Restraint/ Air Bag Special Cautions.

ANTILOCK BRAKE SYSTEM

Certain components in the Antilock Brake System are not intended to be serviced individually. Attempting to remove or disconnect certain system components may result in personal injury and/or improper system operation. Only those components with approved removal and installation procedures should be serviced.

SEAT BELTS

Replace belts, retractors and hardware in use during all but a minor collision. Restraint systems should be replaced and anchorages properly repaired if they were in areas damaged by a collision, whether the belt was in use or not. If there is any question, replace the belt system. Damage, whether visible or not, could result in serious personal injury in the event of an accident.

PAINT CODE LOCATION

Paint code located on spare tire cover in trunk.

CLEAR COAT IDENTIFICATION

All colors are clear coat.

INFORMATION LABELS

001-12029

Part	Year/Side	Part No.	Price
1 **Label, A/C Charging¶**		22640332	9.55
¶Order by Application			
2 **Label, Belt Routing**			
2.2L Eng	95-97	10171018	9.55
	98-01	24575714	9.55
	02	N.A.	0.00
3 **Label, Child Lock**	R/L	22578972	9.55
4 **Label, Fan Blade Caution¶**		10120010	9.55
¶Order by Application			
5 **Label, SIR Systems Caution**		15001737	9.25
6 **Label, SIR Systems Info**	95-96	15001739	9.25
7 **Label, SIR Sensor Service**		10247726	9.55
8 **Label, Jack Caution¶**		21011879	9.55
¶Order by Application			
9 **Label, Information**			
Spare Storage	95-99	22614521	9.25
Jack Usage	98-01	22614521	9.25
	02	N.A.	0.00
10 **Label, Emission¶**			
2.2L Eng			
Man Trans			
w/Calif Emissions		24575930	0.00
w/o Calif Emissions		24575620	0.00
Auto Trans			
w/Calif Emissions		24576253	9.25
w/o Calif Emissions		24576000	9.55
2.3L Eng, 2.4L Eng			
Man Trans			
w/Calif Emissions		24576237	9.25
w/o Calif Emissions		24575319	8.33
Auto Trans			
w/Calif Emissions		24575339	8.33
w/o Calif Emissions		24575318	8.33
¶Order by Application			
11 **Label, Lifting Instruction**	95-01	22588328	9.55
	02	22679118	9.25
12 **Label, Odometer Service**		10443718	9.25

FRONT BUMPER

W/GT MODEL 1995-99

Refinish Front Cover 2.5
R&I Bumper Cover Assy .... #1.5
#Includes R&I/R&R Lower Radiator Air Baffle & Splash Shields
O/H Bumper Cover Assy (Includes R&I) .... #2,5
#Includes R&I/R&R Lower Radiator Air Baffle, Splash Shields, Park/Signal Lamps & Side Marker Lamps

NOTE: All Parts in this section are included in overhaul unless noted otherwise.

001-11524

Part	Side	Part No.	Labor	Price
1 **Cover, Front (P)**		22597555	#1.9	400.00
(P) Paint to Match				
#Includes R&I/R&R Lower Radiator Air Baffle, Splash Shields, Park/Signal Lamps & Side Marker Lamps				
2 **Brkt, License**		22594636	.2	3.45
3 **Grille, Front Cover**	R/L	22599196-7	.2	5.35
4 **Retainer, Push-In (4)**		20699808		.39
5 **Retainer, Outer Cover**	R	22574624		7.20
	L	22629293		7.10
6 **Absorber, Impact**		16516749	IOH	105.00
7 **Bar, Impact**		22608945	IOH	284.00
8 **Brkt, Impact Bar**	R/L	22595724-5	#.8	13.10
#w/Bumper Assy Removed, Not Included in O/H, Included in R&R Side Lower Rail Assy				
9 **Plate, Stud (2/Side)**	R/L	22598028		3.45

W/GT MODEL 2000-02

Refinish Front Cover 2.6
Refinish License Plate Brkt. 5
R&I Bumper Cover Assy .... #1.5
#Includes R&I/R&R Lower Radiator Air Baffle & Splash Shields
O/H Bumper Cover Assy (Includes R&I) .... #2.8
#Includes R&I/R&R Lower Radiator Air Baffle, Splash Shields, Park/Signal Lamps & Side Marker Lamps & Fog Lamps

NOTE: All Parts in this section are included in overhaul unless noted otherwise.

001-17023

Part	Side	Part No.	Labor	Price
1 **Cover, Front (P)**		12335340	#2.2	234.00
#Includes R&I/R&R Lower Radiator Air Baffle, Splash Shields, Park/Signal Lamps, Side Marker Lamps & Fog Lamps				
2 **Brkt, License Plate (P)**		12365279	.2	24.20
(P) Paint to Match				
3 **Emblem (Adhesive)**		22658036	.2	6.80
4 **Retainer, Push-In (4)**		10121502		.25
5 **Retainer, Outer Cover**	R	22574624		7.20
	L	22629293		7.10
6 **Retainer, Push-In (4)**		20699808		.39
7 **Absorber, Impact**		22600628	IOH	119.00
8 **Bar, Impact**		22608945	IOH	284.00
9 **Plate, Stud (2/Side)**	R/L	22598028		3.45
10 **Brkt, Impact Bar**	R/L	22595724-5	#.8	13.10
#w/Bumper Assy Removed Not Included in O/H, Included in R&R Side Lower Rail Assy				

W/O GT MODEL 1995-99

Refinish Front Cover 2.5
R&I Bumper Cover Assy .... #1.5
#Includes R&I/R&R Lower Radiator Air Baffle & Splash Shields
O/H Bumper Cover Assy (Includes R&I) .... #2.3
#Includes R&I/R&R Lower Radiator Air Baffle, Splash Shields & Park/Signal Lamps

NOTE: All Parts in this section are included in overhaul unless noted otherwise.

001-11525

Part	Side	Part No.	Labor	Price
1 **Cover, Front (P)**		22597554	#1.8	214.00
(P) Paint to Match				
#Includes R&I/R&R Lower Radiator Air Baffle, Splash Shields & Park/Signal Lamps				
2 **Brkt, License**		22594635	.2	10.50
3 **Grille, Front Cover**				
Coupe	R/L	22598286-7	.2	5.65
Convertible	R/L	22599196-7	.2	5.35
4 **Retainer, Push-In (3)**	R/L	15958694		.37
5 **Retainer, Outer Cover**	R	22574624		7.20
	L	22629293		7.10

Figure 10-3 Study this actual page from an estimating guide book. This information will help you do both manual and computer estimates. The section index lists major parts, assemblies, or systems that may have to be repaired on a specific year, make, and model vehicle. The numbers on the part illustrations correspond with the numbers in the column list of parts.

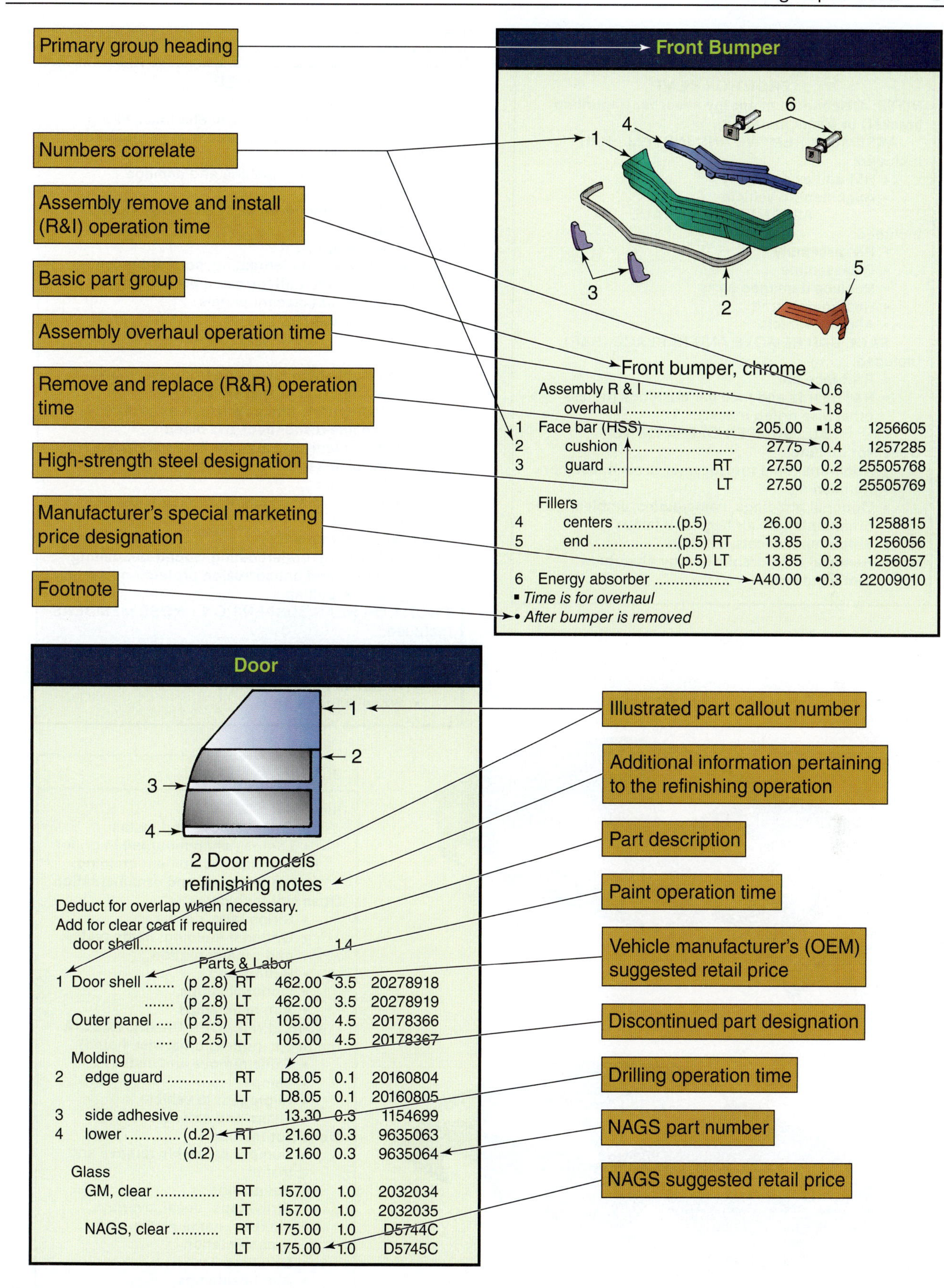
Primary group heading
Front Bumper
Numbers correlate
Assembly remove and install (R&I) operation time
Basic part group
Assembly overhaul operation time
Remove and replace (R&R) operation time
High-strength steel designation
Manufacturer's special marketing price designation
Footnote
1
2
3
4
5
6
Front bumper, chrome
Assembly R & I 0.6
overhaul 1.8
1 Face bar (HSS) .... 205.00 ▪1.8 1256605
2 cushion 27.75 0.4 1257285
3 guard .... RT 27.50 0.2 25505768
LT 27.50 0.2 25505769
Fillers
4 centers ....(p.5) 26.00 0.3 1258815
5 end ....(p.5) RT 13.85 0.3 1256056
(p.5) LT 13.85 0.3 1256057
6 Energy absorber .... A40.00 •0.3 22009010
▪ Time is for overhaul
• After bumper is removed
Door
1
2
3
4
2 Door models refinishing notes
Deduct for overlap when necessary.
Add for clear coat if required
door shell...................... 1.4
Parts & Labor
1 Door shell .... (p 2.8) RT 462.00 3.5 20278918
.... (p 2.8) LT 462.00 3.5 20278919
Outer panel .... (p 2.5) RT 105.00 4.5 20178366
.... (p 2.5) LT 105.00 4.5 20178367
Molding
2 edge guard .... RT D8.05 0.1 20160804
LT D8.05 0.1 20160805
3 side adhesive 13.30 0.3 1154699
4 lower ....(d.2) RT 21.60 0.3 9635063
(d.2) LT 21.60 0.3 9635064
Glass
GM, clear .... RT 157.00 1.0 2032034
LT 157.00 1.0 2032035
NAGS, clear .... RT 175.00 1.0 D5744C
LT 175.00 1.0 D5745C
Illustrated part callout number
Additional information pertaining to the refinishing operation
Part description
Paint operation time
Vehicle manufacturer's (OEM) suggested retail price
Discontinued part designation
Drilling operation time
NAGS part number
NAGS suggested retail price

Figure 10-4 Study how an estimating guide gives part pricing and labor time information.

1. BUMPER

FRONT OR REAR

NOTE: Disconnect at energy absorber (mounting bracket) or frame mounting.

ASSEMBLY REMOVE AND INSTALL (R&I)

Included:
- R&I unit as assembly
- Alignment to vehicle

ASSEMBLY OVERHAUL

Included:
- R&I assembly
- Disassemble
- Replace damaged parts
- Reassemble unit
- Align to vehicle

FACE BAR REMOVE AND REPLACE (R&R)

Included:
- R&R face bar
- R&I guards and face bar cushions (unless otherwise noted in text)

ALL BUMPER OPERATIONS

Does Not Include:
- Additional time for frozen or broken fasteners
- Refinishing
- Optional moldings, name plates, emblems, and ornamentation; air bags and lamps
- Stripe tape, decals, overlays
- Removal of hydraulic energy absorbers
- Aim headlamps

2. FRAME

UNITIZED

Included:
- Welding as necessary and electrical wiring
- Floor mats, insulation, and trim (if required)

Does Not Include:
- Setup on frame machine and damage diagnosis
- R&I of all bolted-on parts and body sheet metal
- Wheel alignment
- Refinishing, undercoating, sound-deadening material, and anticorrosion protection
- Removal of adjacent panels
- Time for pulling

CONVENTIONAL

Included:
- R&I front sheet metal and body assembly
- Front and rear suspension parts
- Steering parts and powertrain as assembly
- Brake line disconnect and bleed
- R&I fuel tank and bumper assembly
- Control linkage and electrical wiring

Does Not Include:
- Setup on frame machine and damage diagnosis
- Wheel alignment
- Refinishing, undercoating, sound-deadening material, and anticorrosion protection
- Time for pulling

FRONT OR REAR SUSPENSION CROSS MEMBERS

Included:
- Welding as necessary

Does Not Include:
- R&I of all bolted-on parts
- Wheel alignment

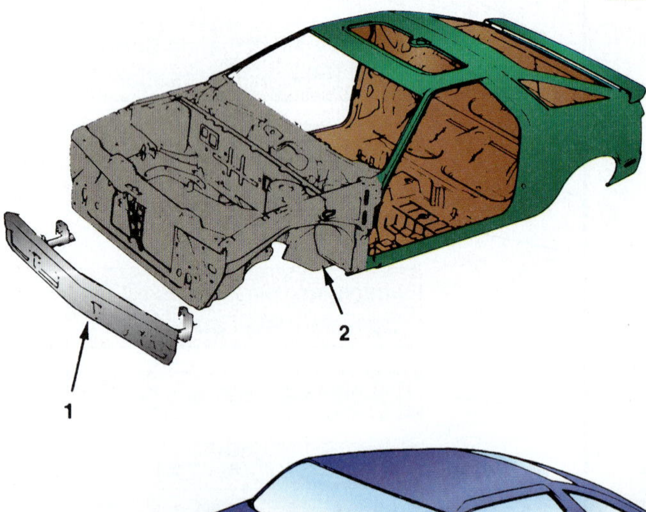

3. GRILLE

Included:
- Grille remove and install
- Lamps (when mounted in grille)
- Standard equipment molding, name plates, and ornamentation

Does Not Include:
- Stripe tape, decals
- Optional molding, name plates, and ornamentation
- Refinishing
- Optional lamps
- Aim headlamps

GRILLE/HEADER PANEL/FASCIA

Included: (unless otherwise noted)
- Grille remove and install
- Lamps
- Alignment to vehicle
- Fillers and extensions

Does Not Include:
- Bumper assembly remove and install
- Refinishing
- Stripe tape, decal, overlays
- Molding, name plates, and ornamentation
- Drill time
- Aim headlamps

Figure 10-5 Note how an estimating guide summarizes operations for repairs that are included and those that are not included. Labor time would have to be added for repair operations not included for rust, frozen parts, moldings, aiming headlights, and other labor tasks.

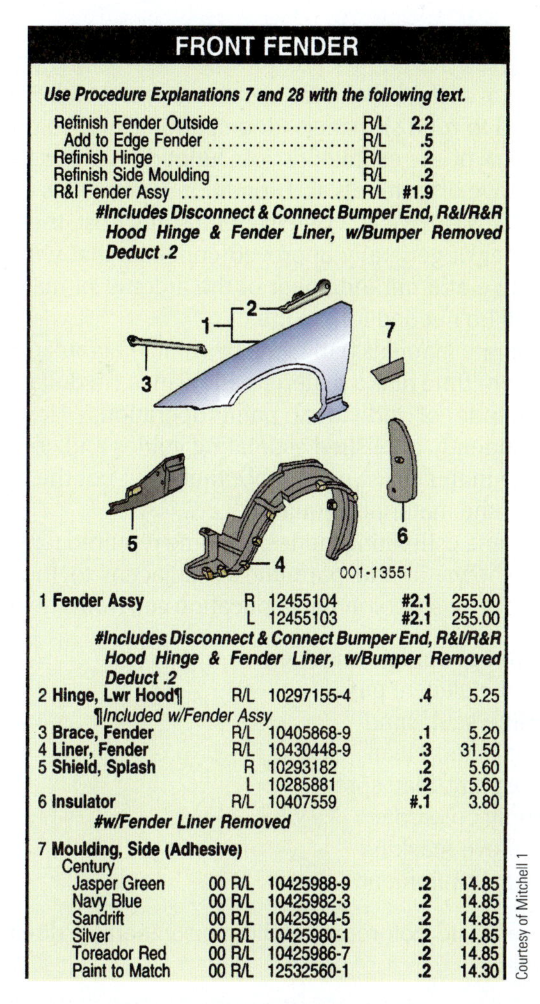

FRONT FENDER

Use Procedure Explanations 7 and 28 with the following text.

Operation	Side	Time
Refinish Fender Outside	R/L	2.2
Add to Edge Fender	R/L	.5
Refinish Hinge	R/L	.2
Refinish Side Moulding	R/L	.2
R&I Fender Assy	R/L	#1.9

#Includes Disconnect & Connect Bumper End, R&I/R&R Hood Hinge & Fender Liner, w/Bumper Removed Deduct .2

Part	Side	Part No.	Time	Price
1 **Fender Assy**	R	12455104	#2.1	255.00
	L	12455103	#2.1	255.00
#Includes Disconnect & Connect Bumper End, R&I/R&R Hood Hinge & Fender Liner, w/Bumper Removed Deduct .2				
2 **Hinge, Lwr Hood¶**	R/L	10297155-4	.4	5.25
¶Included w/Fender Assy				
3 **Brace, Fender**	R/L	10405868-9	.1	5.20
4 **Liner, Fender**	R/L	10430448-9	.3	31.50
5 **Shield, Splash**	R	10293182	.2	5.60
	L	10285881	.2	5.60
6 **Insulator**	R/L	10407559	#.1	3.80
#w/Fender Liner Removed				
7 **Moulding, Side (Adhesive)**				
Century				
Jasper Green	00 R/L	10425988-9	.2	14.85
Navy Blue	00 R/L	10425982-3	.2	14.85
Sandrift	00 R/L	10425984-5	.2	14.85
Silver	00 R/L	10425980-1	.2	14.85
Toreador Red	00 R/L	10425986-7	.2	14.85
Paint to Match	00 R/L	12532560-1	.2	14.30

Courtesy of Mitchell 1

Figure 10-6 **Study how columns in an estimating guide give typical part prices and labor times in hours. For example, how much does the front fender cost and how many hours of labor would it take to replace it?**

Each damaged car poses different problems that must be answered to arrive at a repair versus replacement decision. The most numerous and difficult questions arise from major collision wreckage.

For example, let us say that you are estimating a car with a "hard hit" in the right front. You might have to decide whether to install a partial or complete right front frame rail. Do you want to section the damaged rail and splice on a partial front-half section? Or should you remove and replace the complete rail? Which would save time and money while still producing a solid structural repair? These kinds of questions take time and thought to answer. Therefore, you must know correct repair procedures to write an estimate properly.

Generally, clues to body distortion (twist, sag, or side sway) are apparent if any cracking of stationary glass (windshields/back window) is noted. **Stress cracks** indicate minor panel damage and movement. Always use a shop light to look around affected areas to find stress cracks between panel joints. Sealer and undercoating cracks offer hints of damage.

In some situations, such as a severely damaged front end, an estimator might want to consider a complete front clip. The **front clip assembly** generally includes all body parts from the front bumper to the rear of the fenders.

Part prices given in estimating guides usually do not include the cost of state and local taxes; shipping from the supplier; items such as bolts, rivets, screws, nuts, washers, clips, and fasteners; body repair materials; and refinishing costs, unless otherwise noted. These costs must be added to the estimate.

Once the part prices, material costs, and repair time have been determined, they can be entered into the estimate.

10.3 LABOR COSTS

The **flat rate** is a preset amount of time and money charged for a specific repair operation. Estimating guides provide an explanation of what the flat rate labor time includes and does not include. For example, replacing a panel or fender includes transfer of the part attached to the panel. It does not include the installation of moldings, antennas, refinishing, pin striping, decals, or other accessory parts. Also, it does not consider rusted bolts, undercoating, and alignment or straightening of damaged adjacent parts or bolts. You should add a nominal amount of time to cover these types of unwritten repair operations.

The flat rate labor time reported in collision damage manuals should be used *only* as a guide. It is based primarily on data reported by vehicle manufacturers who have arrived at these estimates by repeated performance of each operation under normal shop conditions. Estimating guides usually provide an explanation of the established requirements for the average mechanic, working under average conditions, and following procedures outlined in their service manuals.

The labor times do not apply to cars with equipment other than that supplied by the car manufacturer as standard or regular production options. If other equipment is used (body spoilers, ground effects, and so forth), the time must be increased to compensate for these added variables.

Job Overlap

Job overlap means that replacement of one part duplicates some labor operations required to replace an adjacent or attached part. With job overlap, reductions in an estimating guide's flat-rate labor times must be considered. For example, when replacing a quarter panel and a rear body panel on the same vehicle, the area where these two components join is considered overlap. Where a labor overlap condition exists, less time is required to replace adjoining components collectively than is required when they are replaced individually.

Overlap labor estimating information is generally included at the beginning of each group. In those instances where overlap information is not given, appropriate allowances should be negotiated after an on-the-spot evaluation.

Included Operations

Another labor cost reduction is known as included operations. **Included operations** are jobs that can be performed individually but are also part of another operation. For example, the suggested time for replacing a door includes the replacement of all parts attached to the door. Unless a salvage door is used, it would be impossible to replace the new door without transferring these parts. Consequently, the time involved to transfer these parts is considered "included operations" and is disregarded. This is because the times for the individual items are already included in the door replacement time.

An experienced estimator will not overlook the removal of exterior trim and body sheet metal hardware. These parts must often be removed prior to repair or painting operations. They must then be replaced after these operations.

Once all the repairs and labor times have been entered on the estimate, the computer will total your figures for you. If you do not have computerized estimating, it is necessary to refer to a conversion table. A conversion table can be used to convert flat-rate labor time into dollars to fit local hourly labor or operating rates. With a computer estimating system, this form is built into the software and is calculated automatically.

When establishing flat labor rates, shop overhead (including such items as rent, management and supervision, supplies, and depreciation on equipment) must be determined. Then the actual labor cost of all employees (including office help) and the profit required to keep the business operating must be added to the shop overhead to obtain a dollar flat rate for repairs. This flat-rate cost is usually figured on an hourly basis.

Labor times shown in all collision estimating guides are listed in hours and tenths of an hour. If a vehicle requires a new right front fender, a new wheel opening molding, and a new name plate ornament, the total labor replacement time, according to a leading estimating guide, is:

Front right fender	3.0 hours
Wheel opening molding	0.2 hours
Installation of nameplate	0.2 hours
Total labor	3.4 hours

If the shop's dollar per hour operating rate is $25.00, the total estimated labor cost would be $87.50. When the labor costs are determined, they are written on the estimate form.

On the whole, flat-rate operations comprise all work in which components must be removed and replaced. Estimated work is work done to straighten and/or repair body members. However, it is not always that simple. Sometimes there are combinations. A good example of this is when body members must first be straightened before cutting them off, to make sure the new member will line up prior to welding. This operation, even though it is a flat rate R&R, must have some additional estimated time added to the total time for the procedure.

10.4 REFINISHING TIME

Making a correct estimate of the amount of labor time required to refinish panels, doors, hoods, and so on is a vital part of an estimator's job function. Although the wide range of materials and conditions sometimes makes it difficult to arrive at a precise refinishing cost, there are a number of generally approved concepts that will help you arrive at a fair judgment of the amount of materials that will be needed for the job.

Flat-rate manuals published by vehicle manufacturers list a labor time plus a materials allowance (in dollars) for a multitude of individual paint operations. However, independently published estimating guides and collision damage manuals list paint labor times but not the dollar value of the materials required.

In some estimating guides, the time required for painting is shown in the parentheses adjacent to the part name. The basic colorcoat application generally includes:

- Clean panel/light sanding
- Mask adjacent panels
- Prime/scuff sand
- Final sand/clean
- Mix paint/load sprayer
- Apply colorcoat
- Remove masking
- Clean equipment

The basic colorcoat application generally does not include:

- Cost of paints or materials
- Matching and/or tinting color
- Grinding, filling, and smoothing welded seams
- Blending into adjacent panels
- Removal of protective coatings
- Spatter paint
- Custom painting
- Undercoating
- Anticorrosion materials
- Sound-deadening materials
- Edging panel
- Underside of hood or trunk lids
- Covering entire vehicle prior to refinishing if necessary
- Protective coatings
- Additional time to produce custom, non-OEM finishes

Painting times given in most estimates are for one color on new replacement parts' outer surfaces only. Additions to paint times are usually made for the following operations:

- Underside of hood — Add 0.6
- Underside of trunk lid — Add 0.6
- Edging new part
  1. First panel — Add 0.5
  2. Each additional — Add 0.3
- Anticorrosion coating — Add 0.3
- Two-tone operations (unless otherwise specified in text)

1. First panel — Add 0.6
2. Each adjacent — Add 0.4

- Stone chip (protective material)
  1. First panel — Add 0.5
  2. Each additional — Add 0.3
- Clearcoat (basecoat/clearcoat) after deduction for overlap
  1. First major panel — Add 50%
  2. Each additional — Add 25%

Reductions to paint times can be considered for the following:

- Overlap (adjacent parts)
  1. First major panel — Full time
  2. Each additional (except extensions) — Deduct 0.4
  3. Extensions — Deduct 0.2
- Overlap (nonadjacent parts)
  1. First major panel — Full time
  2. Each additional — Deduct 0.2

The labor allowance given in a collision estimating guide does not include any material costs. These must be estimated using a locally compiled refinishing materials list or one that is accepted on a national or regional basis.

The use of a guide permits shop owners and estimators to place a fair evaluation on the refinish materials actually used. Figure 10–7 shows a table for tabulating repair times for single-stage, two-stage, and three-stage paints.

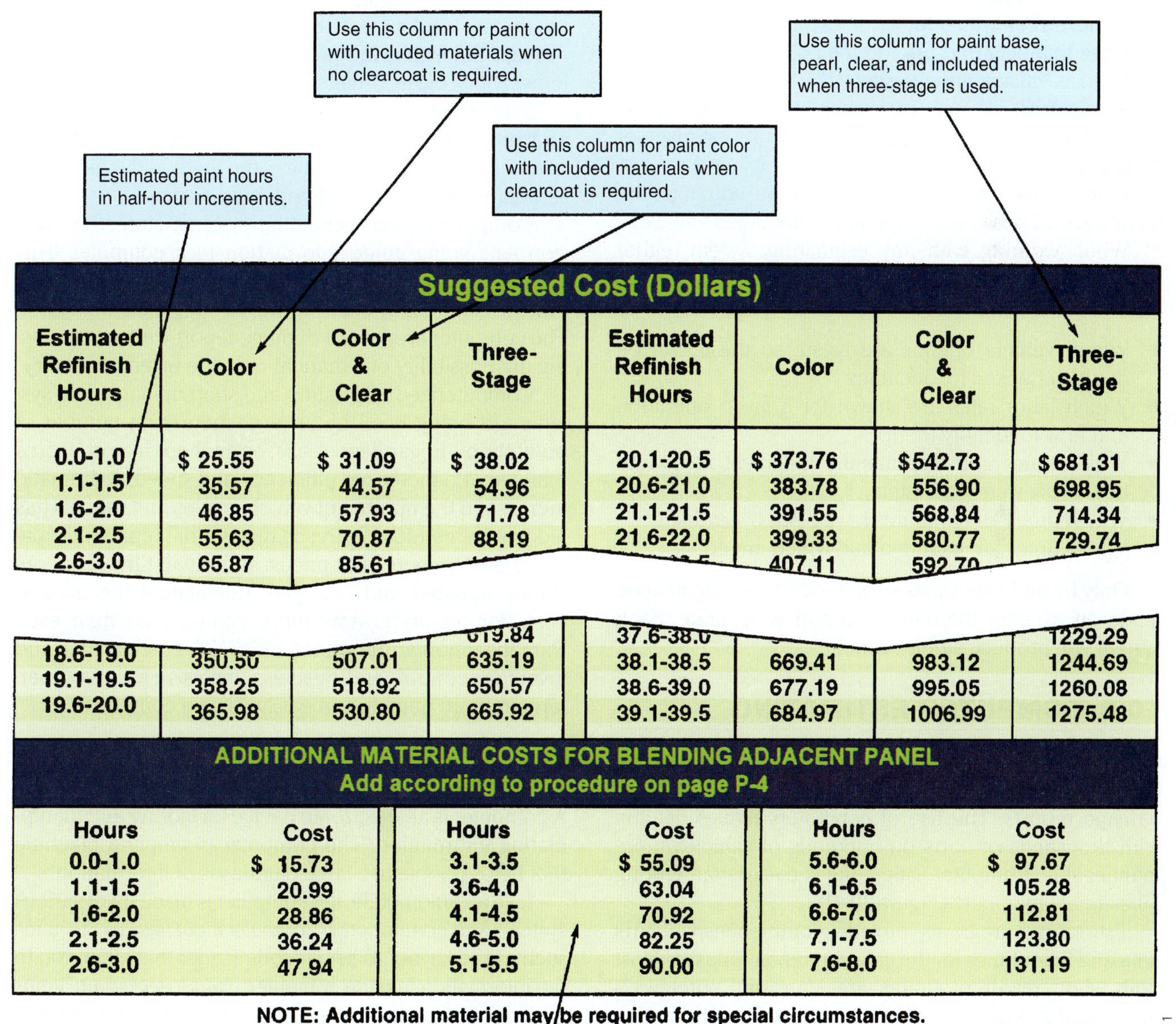

Suggested Cost (Dollars)

Estimated Refinish Hours	Color	Color & Clear	Three-Stage	Estimated Refinish Hours	Color	Color & Clear	Three-Stage
0.0-1.0	$ 25.55	$ 31.09	$ 38.02	20.1-20.5	$ 373.76	$542.73	$681.31
1.1-1.5	35.57	44.57	54.96	20.6-21.0	383.78	556.90	698.95
1.6-2.0	46.85	57.93	71.78	21.1-21.5	391.55	568.84	714.34
2.1-2.5	55.63	70.87	88.19	21.6-22.0	399.33	580.77	729.74
2.6-3.0	65.87	85.61			407.11	592.70	
			619.84	37.6-38.0			1229.29
18.6-19.0	350.50	507.01	635.19	38.1-38.5	669.41	983.12	1244.69
19.1-19.5	358.25	518.92	650.57	38.6-39.0	677.19	995.05	1260.08
19.6-20.0	365.98	530.80	665.92	39.1-39.5	684.97	1006.99	1275.48

ADDITIONAL MATERIAL COSTS FOR BLENDING ADJACENT PANEL
Add according to procedure on page P-4

Hours	Cost	Hours	Cost	Hours	Cost
0.0-1.0	$ 15.73	3.1-3.5	$ 55.09	5.6-6.0	$ 97.67
1.1-1.5	20.99	3.6-4.0	63.04	6.1-6.5	105.28
1.6-2.0	28.86	4.1-4.5	70.92	6.6-7.0	112.81
2.1-2.5	36.24	4.6-5.0	82.25	7.1-7.5	123.80
2.6-3.0	47.94	5.1-5.5	90.00	7.6-8.0	131.19

NOTE: Additional material may be required for special circumstances.

Figure 10-7 This chart lets you add correct costs for different types of refinish materials. Three-stage refinish materials cost more than single-stage paint.

10.5 ESTIMATE TOTAL

Once all cost information is entered (parts, labor, and materials), the amounts can be added together for a subtotal. To this figure, the estimator adds any extra charges, such as wrecker and towing charges, storage fees, and state/local taxes. These figures, added to the subtotal, compose the grand total estimate of the repair (Figure 10–8).

Although more and more larger body shops are doing repair jobs such as wheel alignment, rustproofing, and tire replacement, smaller shops still "farm out," or "sublet," these jobs to others. When this is done, the specialty shop bills the body shop for the work. Shops that farm out work usually have a column—Sublet—where the retail labor cost is marked. This sublet figure is added to others to obtain the grand total of the estimate.

If the owner wants to have "extra" work performed (damage that occurred prior to the collision), this should be noted as "customer-requested (C/R)" repairs. Insurance companies will not pay for customer-requested repairs that are not a result of the wreck. A separate estimate for these extra repairs must be made for the customer.

From the estimate, a work order can be written for the technician who is going to repair the damaged vehicle.

When working with any estimating system, either manual or computer, those working with the damage report must understand:

- Where vehicle options are listed, so these can be double-checked for accuracy
- Which labor rates are shown for a given operation and how to identify them
- Which times and amounts are judgment items and how these were generated
- Which times and amounts have been overridden
- Which times are included and which are not

Only by understanding all of these can a judgment be made on whether the damage report is accurate. Each system will produce a unique look.

10.6 COMPUTER ESTIMATING

Computer-based estimating systems using a personal computer (PC) provide more accurate and consistent damage reports. The use of computers makes dealing with thousands of parts on hundreds of vehicles more manageable. Computer estimating is easier than writing an estimate manually (Figure 10–9).

Estimating programs provide comprehensive parts data and labor times for the collision repair industry. Most body shop repair programs are Microsoft Windows®-based and are used on IBM-compatible PCs. Estimating programs are organized just like collision estimating guides and are based on a comprehensive database of parts and labor information.

Estimating software can also help save time on searching for parts. Estimating software automatically narrows the qualified parts selection range to the specific vehicle displayed. Comprehensive VIN decoding (right down to the drivetrain information) helps the estimator select the right parts for the right car.

Estimating software makes it easy to prepare estimates while interacting with on-screen parts graphics. You simply "point and click" on the description or graphic of the part that needs repair or replacement, the database is automatically accessed for the part, and then it is calculated into your estimate.

The part lister displays the information contained within the collision estimating software, including important part and labor information notes, in the same front-to-rear sequence as the guide. Part price and new vehicle updates are provided regularly by software publishers.

Estimating software takes care of important details, such as refinishing parts and accounting for additional labor on optional equipment. Estimating software guides you through the estimating process to help present your customer with a clean, professional estimate in just minutes.

Estimating software handles complex calculations and procedures, which means fewer supplements and additions. It virtually eliminates the math and logic errors sometimes found in handwritten estimates.

Computer-based estimating systems store the collision estimating guide information in a computer. This eliminates a lot of time in looking up parts and labor times and manually entering and totaling them on a form. The computer prepares a damage report while still allowing the possibility of a manual override when necessary.

Computerized estimating and shop management systems are being used by an ever increasing number of body shops. Regardless of size, body shops face concerns that demand increased efficiency and speed. A dramatic increase in the number of vehicle makes and models has led to a tremendous increase in essential repair information. Part numbers, part prices, and repair times are constantly updated and changed. Calculating the needed costs for repairs is now more complicated than ever. Damage report writing, accounting, job costing, time and inventory control, and business analysis can all be performed more quickly and easily using a computer.

Computer Components and Jargon

A computer is an electronic device for storing and manipulating information. The computer is simply a machine that helps people use large databases of information.

Like an automobile, the computer is made up of various systems and components. *Computer input devices* are items such as the keyboard or bar code reader that allow you to put information into the computer. *Output devices* are items such as cathode ray tubes (CRTs) or monitors and printers that allow you to read information inside the computer.

An awareness of computer jargon is essential to understanding the computer. The following is a list of definitions to acquaint the damage report writer with computer terminology, to increase an understanding of the computer, and to serve as a reference guide.

B&J

Collision Estimating Services

ESTIMATE OF REPAIRS

Nº 002128

SHEET NO. 1 OF 1 SHEETS

NAME	ADDRESS	PHONE	DATE
KAREN Miller	1143 RAILROAD ST. CRESSONA, PA. 17929	HOME 395-2719 BUS. 623 7347	12-17-08

YEAR	MAKE	MODEL	LICENSE NO.	MILEAGE	SERIAL/V.I. NO.
2006	ford	P.V.	MAE 917	14864	1FTDF15YSGNA69994

INSURANCE COMPANY	TYPE OF INSURANCE	ADJUSTER	PHONE	CAR LOCATED AT
Amerisure	COL			mf6. 4/07

PARTS NECESSARY AND ESTIMATE OF LABOR REQUIRED	PAINT COST ESTIMATE	PARTS COST ESTIMATE	LABOR COST ESTIMATE
(1) FRONT FACE BAR Chrome NO gds OR PAdS		205 82	.5
(1) " STONE deflector	1 0	40 50	.5
(1) Left HeadLAMP door (with ARgENT GRILL)		52 32	2
(1) " " Shield		3 20	
(1) " fRONT fender	3 1	133 00	1 6
(1) " " " APRON		60 47	1 0
(2) wheeLS 15" 55.60 ea.		111 20	6
(2) stems and Balance		2 50	5
(2) HUB CAPS		43 46	—
RePAIR Radiator Support		—	2 0
ALign fRONT End		—	1 5
(1) Left door trim Panel		73 40	
STRIPE Left fRONT fender		15 00	.5
LABOR 25.0 HRS @ 23.00		575 00	
(note MAY Be fRONT suspension damage)			
PAINT MAT.		89 10	
UNdeRCOAT		15 00	
TOTALS		1,419 .97	

INSURED PAYS $________ INS. CO. PAYS $________ R. O. NO.________

INS. CHECK PAYABLE TO________

The above is an estimate, based on our inspection, and does not cover additional parts or labor which may be required after the work has been opened up. Occasionally, after work has started, worn, broken or damaged parts are discovered which are not evident on first inspection. Quotations on parts and labor are current and subject to change. Not responsible for any delays caused by unavailability of parts or delays in parts shipment by supplier or transporter.

ESTIMATOR________

AUTHORIZATION FOR REPAIRS. You are hereby authorized to make the above specified repairs to the car described herein.

SIGNED X ________ DATE ________ 19

GRAND TOTAL	1,419.97
TOWING & STORAGE	
TAX	85.20
TOTAL OF ESTIMATE	$ 1,505.17

Figure 10-8 This is an example of a handwritten damage estimate or damage appraisal for parts and labor. Note the columns for tabulating paint cost, parts cost, and labor cost.

Figure 10-9 Most collision repair shops are now computer based to streamline and speed the estimating process. The estimator has taken a laptop computer and estimating software out to a vehicle with minor damage. Parts and labor needed can be input with a few mouse clicks.

Figure 10-10 A laptop computer is a very efficient way of inputting estimate information about vehicle damage.

Central processing unit (CPU) The portion of the computer that directs the sequence of operations by electrical signals and governs the actions of the units that make up the computer. The CPU is the "computer brain" where calculations take place.

Command An instruction to the computer to perform a predefined operation. It can be given by typing on the keyboard or installing data from another device.

Compact disc (CD) A disc that stores optical data instead of magnetic data. It can hold several times the information of a floppy. A CD is a "read-only" disc. You cannot erase or alter data on a CD.

Cursor A movable, blinking marker or pointer on the monitor screen that shows where the next point of entry or change will be made.

Drive A device that holds a disc, cassette tape, or CD so that the computer can read data from and write data onto it.

Erase To remove information or data from memory or from disc storage.

Floppy disc A thin magnetic disc (in protective paper or plastic jacket) that stores data or programs. It is erasable and reusable and comes in several sizes. They are often used to back up data in case the information is lost on the main hard drive.

Hard disc A device, usually inside the computer, to store large amounts of data magnetically. Unlike a floppy disc, it is rigid and not readily interchangeable but can store and retrieve data faster than a floppy.

Hardware The physical parts of a computer, such as the keyboard, mouse, screen, disc drives, printer, and so on.

I-Pad A medium-size-screen personal computer with touch screen interface for the keyboard, and other functions.

Laptop computer A small portable computer with a flat display screen for working in remote locations (Figure 10–10); also called a notebook computer.

Mainframe computer A powerful computer capable of storing and processing extremely large amounts of data; also called the "host" computer in some systems.

Memory Computer chips that hold information electronically. This information is lost when the computer is shut off, unless the information is "saved" to make it permanent.

Menu A displayed list of options from which the user selects an action to be performed.

Modem A device for sending computer data over phone lines for use by another computer at a remote location.

Monitor The computer screen for viewing information.

Peripherals Hardware that can be added to a basic computer, such as a printer, disc drive, CD drive, or modem.

Program Disc information that loads into memory to allow the computer to do a specific task.

Save To write data from memory to disc for permanent storage.

Software Instructions written in a computer language that tell the computer what the user wants to do. Smaller programs come on floppy discs. Larger, more complex programs come on a CD; also known as programs.

System software Disc information that allows the computer to start up and operate.

NOTE *For more information on computer systems used in a body shop, refer back to Chapter 3 on Service Information.*

Types of Computer Estimating Systems

Computers are revolutionizing the art of estimating by making enormous amounts of data readily accessible and user-friendly. There are basically two types of computer estimating systems: dial-up systems and resident, or in-shop, systems.

The *dial-up estimating system* relies on data stored in a mainframe computer at a remote location. The data in the mainframe is usually accessed over the Internet via a modem. The shop may be charged a fee whenever data from the mainframe is accessed. This is a seldom used system.

A Late model cars will have a bar code that represents the VIN number.

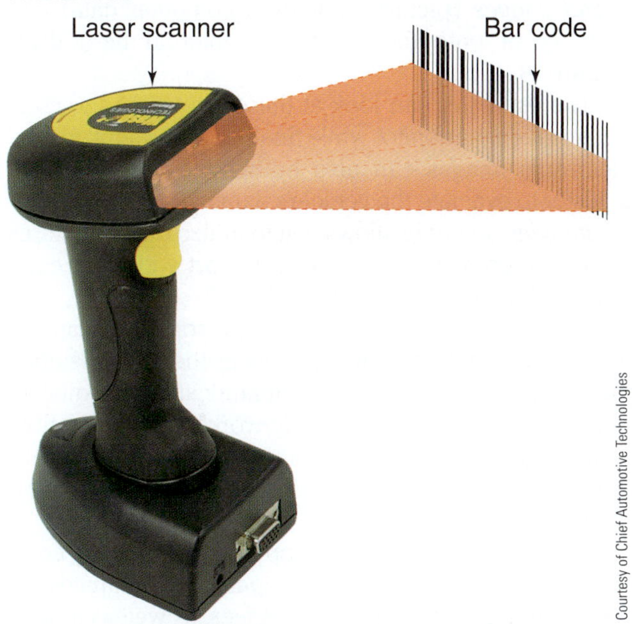

B By moving the bar code scanner over the label on the vehicle body, it will automatically read VIN data so you do not have to type in a long number.

Figure 10-11 **A bar code reader can be used to input the VIN, or the vehicle identification number.**

Even with the best computerization, you must still use a logical, sound inspection sequence to locate all damaged parts. If you miss a part or two, you or the shop's profits will suffer.

An *in-shop estimating system* has all of the data needed at the shop office, usually on CDs. The system does not require a fee for each estimate. However, the shop must purchase current CDs or subscribe and pay an annual fee to get updated CDs. Modern computer estimating systems store an electronic database containing up-to-date part prices and labor times that can be retrieved from the shop's computer system.

With the use of these sophisticated programs, estimators can now use the stored data to make virtually error-free estimates rapidly at a very low cost.

Downloading means to hardwire or link two information holding devices (scanner, computer, laptop, and so on) so they can exchange digital data or information. A *VIN reader* will automatically read and download an alphanumeric code for the vehicle to be repaired (Figure 10–11). The bar code scanner is used to gather and store data for later downloading into the shop's PCs.

10.7 COMPUTER DATABASE

A body shop **computer database** contains all service information (estimating guides, labor rates, and part and material vendor information) in electronic form.

Once damage information has been gathered at the vehicle, the estimator will take this data into the office. The damage data is then electronically downloaded (entered) into the PC. This allows the estimator to use the speed and convenience of a computer and estimating program to make up the estimate. There is no need to type in the names and numbers of damaged parts.

Estimating Programs

An **estimating program** is software (computer instructions) that will automatically help to find the parts needed and the labor rates and calculate the total cost for the repairs. The estimating program is often on computer floppy discs or on a CD.

To use the program, you must load the data into the PC's memory. This will then let you "pull up" screen data for everything relating to making the estimate (Figure 10–12).

By entering the VIN into the computer, the estimating program will decode this number into year, make, and model information. You can then click on the part assemblies that must be repaired or replaced (Figure 10–13).

The estimating program can access a huge database of information. The computer database includes part numbers, part illustrations, labor times, labor rates, and other data for filling out the estimate (Figure 10–14).

The most common and modern way of storing an estimating database is the CD. One CD can hold all of the crash estimating guides, dimension manuals, and other information for every make and model vehicle. The CD database contains the same basic information found in a written collision estimating guide.

When a special drive is wired to the shop's computers, PCs can then access this huge amount of data. In other words, CD data is pulled into the PC's memory for manipulation.

Estimating programs can display exploded views of the parts requested. By clicking on more specific parts, the program will pull up and show part prices and labor repair times for each part.

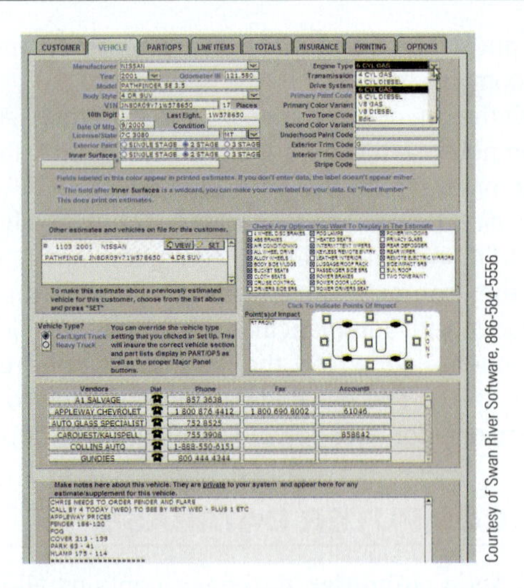

Courtesy of Swan River Software, 866-584-5556

Figure 10–12 This is a typical vehicle information window from computerized estimating software. You must enter make, model, paint, and other information. Note how it gives commonly needed phone numbers of salvage yards, glass installers, insurance agents, and other sources.

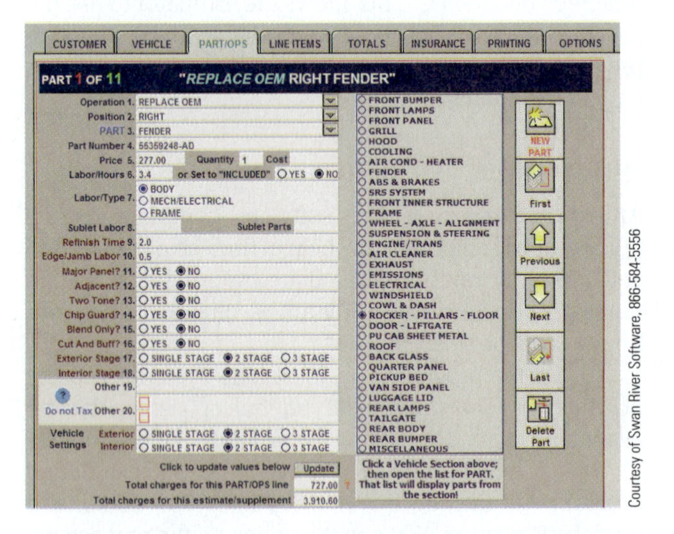

Courtesy of Swan River Software, 866-584-5556

Figure 10–13 The parts and operations window from this estimating software lets you input what repairs and parts are needed for the damaged vehicle. By clicking on buttons along the left side of the page, you can view parts and add them to the estimate.

Wireless Estimating Systems

Wireless estimating systems streamline the process even more. A wireless laptop computer can be taken out to the damaged vehicle. The laptop can then use wireless communication to access the main shop computer for pulling up part illustrations right at the vehicle.

Imaging Software

Advanced computer estimating systems now enable photos of wrecked vehicles to be downloaded to a shop's computers.

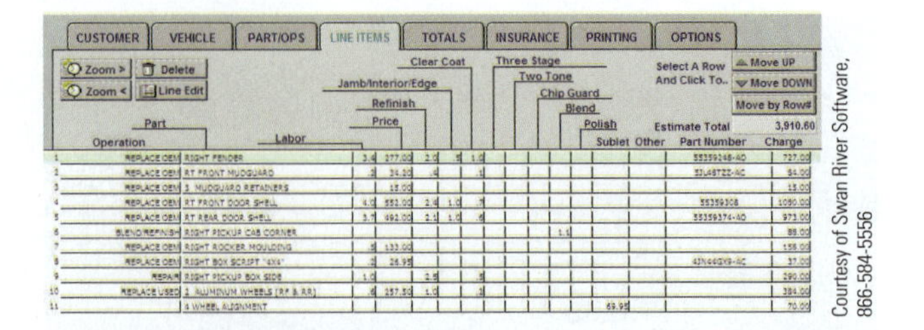

Courtesy of Swan River Software, 866-584-5556

Figure 10–14 This line items window shows what parts will be replaced, repaired, painted, or removed and installed, with labor and costs tabulated automatically.

A *digital camera* can be used to take photos of a vehicle's damage and store them as digital data. The camera can be connected to the computer to download these *digital images* (pictures stored as computer data) into memory or onto the computer's internal hard drive (Figure 10–15).

The estimator can look at these photos on the computer screen or monitor while finalizing the estimate. The electronic images can also be sent to the insurance adjuster to evaluate the vehicle's damage.

Imaging software allows you to utilize digital pictures of vehicle damage or of a specific part with your computer estimating system (Figure 10–16).

Imaging software allows shop personnel to capture and retrieve data and images using the most flexible, time-saving, and reliable communications available. Plus, imaging software normally works with compatible estimating software to give your shop a complete, easy-to-use solution for handling repair and estimating information (Figure 10–17).

Digital imaging frees the shop from the ongoing costs of traditional photography development. It eliminates the cost of film, processing, and print fees, as well as associated costs in time. This software provides instant access to any photo ever entered into the computer system. Most software organizes photos into "job folders" right along with the estimate and other service information. Look at Figure 10–18.

A printer can be used to make a *hard copy* (printed images on paper) for the customer, insurance adjusters, and shop personnel. You can quickly print vehicle dimension drawings, part views, or the actual estimate.

With computer-based estimating, printouts or Internet files with pictures of the damaged vehicle can be sent out to an insurance adjuster for approval (Figure 10–19).

Computer Part Lists

Part and labor prices are also stored in the computer database. This allows new or salvage parts to be ordered quickly and easily. Vendors and their part prices can be pulled up on the computer screen to get actual prices. Vendor lists for new and used parts can be stored in the database for repeated access by the estimator.

A While holding the camera very still, take several photos of the damage from different angles.

B When you are back in the office, connect the camera to a personal computer and download files to the hard drive.

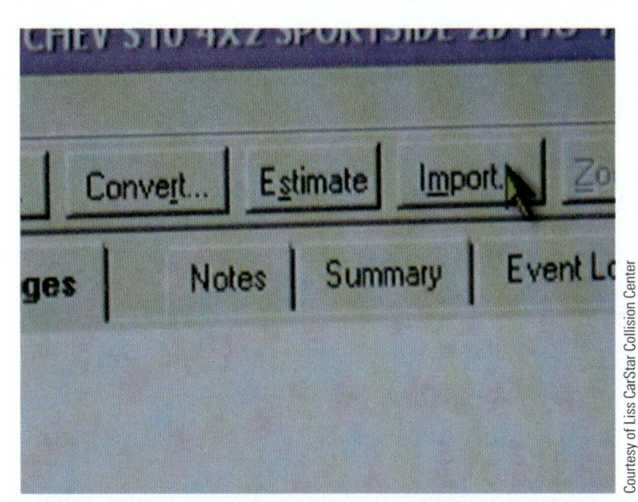

C With the correct file or estimate open, click download to bring the electronic photos into the program.

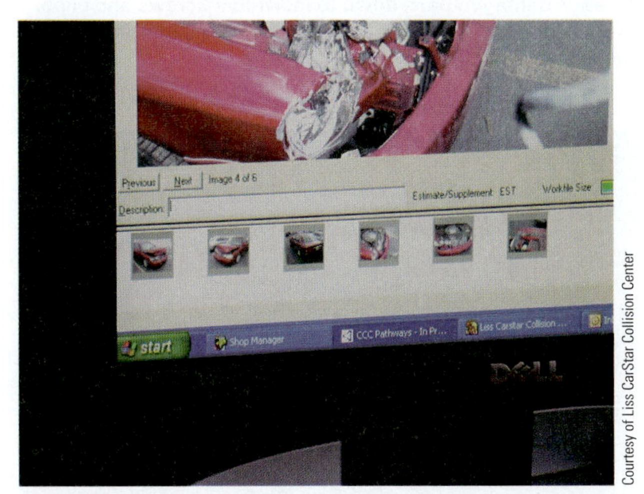

D By clicking on the small thumbnails of the damage photos, you can open them for analysis.

Figure 10-15 Study the major steps for loading digital photos of damage into a computer.

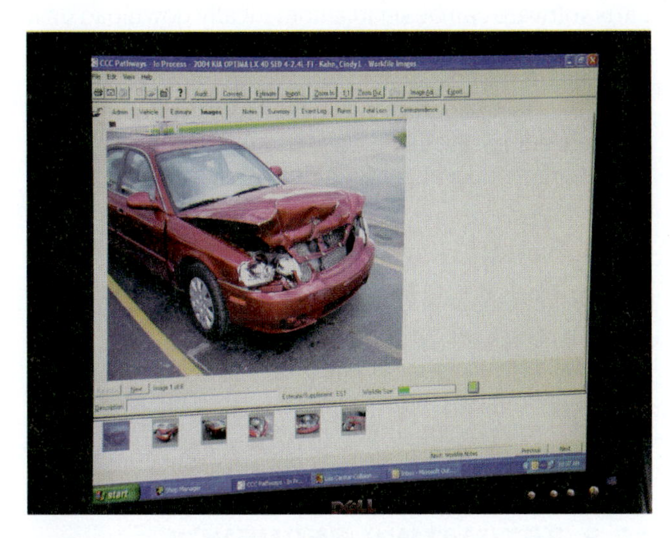

Figure 10-16 Viewing the high-resolution photos full screen will let you double-check that all damaged parts have been found and the estimate is written accurately. You can also e-mail the photos to the insurance adjuster, if needed.

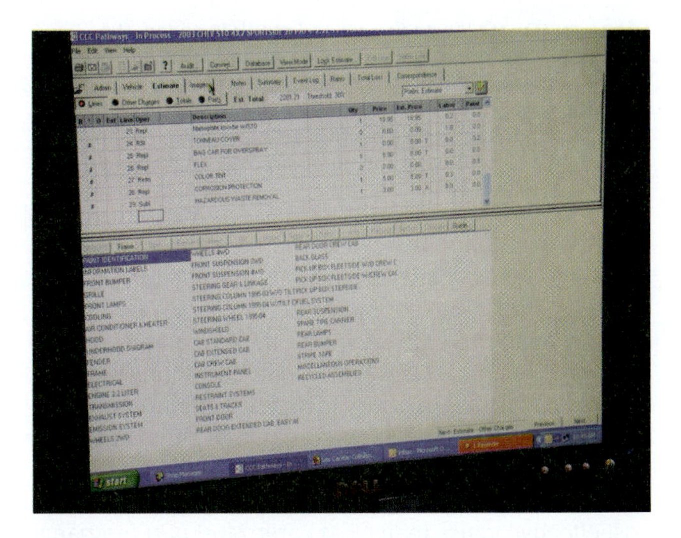

Figure 10-17 In this estimate window, you can add repair operations, a description of the repair, part prices, and labor costs for the damaged sections of the vehicle. By clicking on one of the categories along the bottom, you will be shown exploded assembly views.

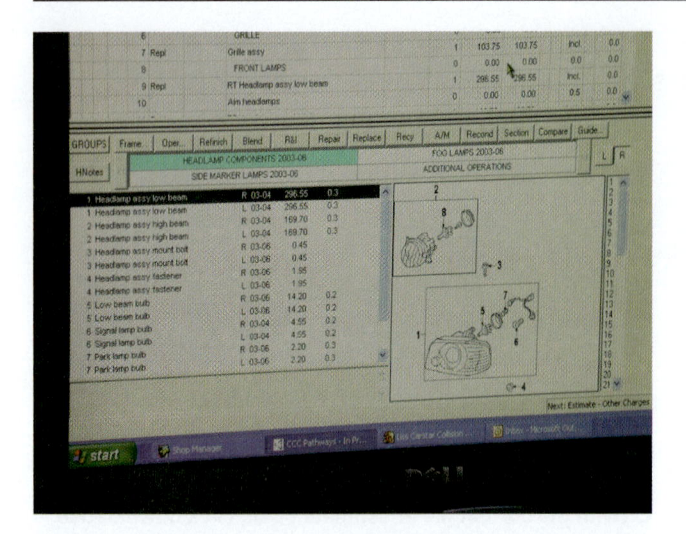

Figure 10–18 In the exploded view window, you can click on each damaged part, down to individual screws and clips. This will load each part name, its price, and labor time into the estimate for that vehicle.

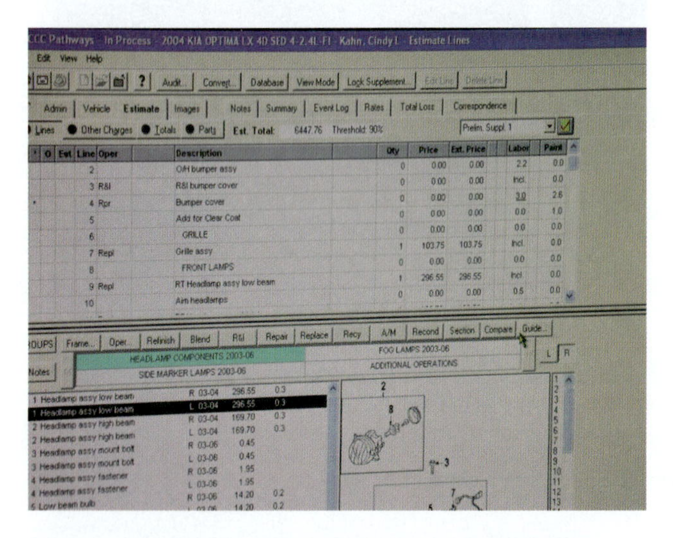

Figure 10–19 As each part and repair is added, the estimate will automatically add the repair operation (replace, repair, remove and install), a description of the operation, quantity, part price, repair labor time in hours, and paint labor time in hours. Note how this estimate gives a labor time of 2.6 hours to paint the bumper cover on this car.

A *part category* in service data contains information on a single type of part. For example, the radiator category contains only complete radiators. Other categories can contain many parts, such as the front suspension category, which contains A-arms, shocks, springs, knuckles, and so forth. A category includes either new parts or used parts (rechromed, recored, or OEM-remanufactured parts).

Aftermarket and remanufactured parts can be more cost-effective to use than OEM parts. Most part programs are aimed at reducing the cost of automotive collision repairs. Many times, not using new parts makes the difference between doing repairs or having the insurance company classify the damaged vehicle as a total loss.

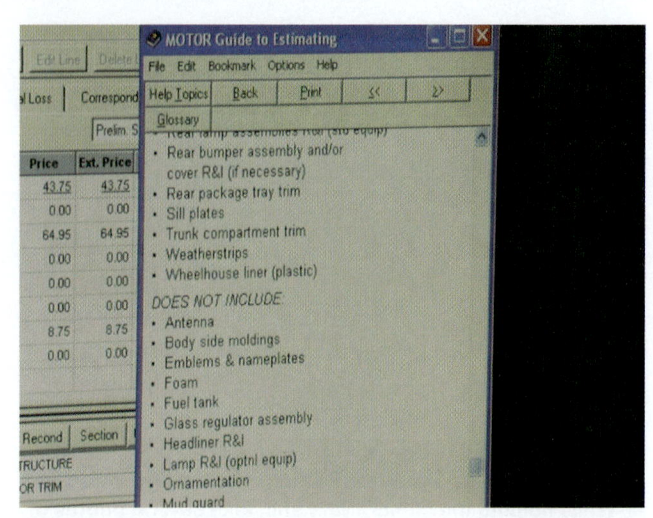

Figure 10–20 Another window is usually provided for figuring which operations are included and not included. Included operations automatically add parts that must be removed to access the main part to be replaced.

Modern service software will help you select the lowest cost parts available from the selected suppliers. The suppliers and parts information prints at or near the bottom of the estimate, detailing the supplier's address, phone number, and part number and price.

Salvage parts software provides easy access to a user-customized database of recycled parts and suppliers in North America. This type of computer program often supports an online accessible parts database from a growing network of thousands of recyclers. Salvage parts software allows you to quickly search and find quality recycled parts. You can then include them on the electronic estimate.

The salvage parts database is often supplied by the Automotive Recyclers Services Corporation. With some software, these recyclers update their inventory of replacement parts into the shop's Internet server. Salvage parts software can be set to automatically download on a predetermined schedule, every day if desired. The shop's computer hard drive will then have an information database on recycled parts. Having the most current recycled parts inventory available ensures that you will quickly find the part you want at the price listed.

Automatic Overlap Deduction

Most estimating programs automatically deduct for overlap. The program notes operations as included if they are part of a larger operation. The program may also identify judgment times (Figure 10–20).

Labor time information for a given operation is taken from the same sources as in the collision estimating guides.

10.8 ESTIMATING SEQUENCE

When estimating any type of damage to a vehicle, whether minor or major, a logical sequence must be followed. For an estimate to clearly establish a true cost

Figure 10–21 When analyzing damage, methodically view all body surfaces and individual parts and panels. Sight down the side of the vehicle to look for damage.

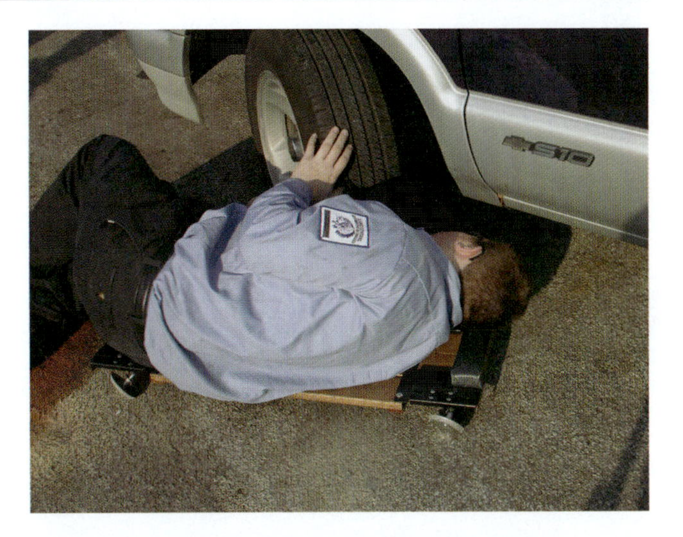

Figure 10–22 If the vehicle sustained a hard hit or had tire and wheel damage, always look under it to find undercarriage damage.

of repairs, a thorough damage analysis must be performed on the wrecked vehicle (Figure 10–21).

Before making an estimate, the estimator should make a visual inspection of the entire vehicle, paying special attention to damaged subassemblies and parts that are mounted to (or part of) a damaged component. The estimator must consider basically the same points as the technician does before making any decision on the repair work. That is, most estimators start from the outside of the car and work inward, listing everything—by car section—that is found bent, broken, crushed, or missing (Figure 10–22).

The estimator should develop a basic pattern or system for analyzing damage. This pattern should be used on every vehicle to prevent missed parts or overlooked problems.

Figure 10–23 The front of the vehicle scraped lightly on a guard wall on an interstate highway. Because the bumper and fender were only scratched, all of these parts can be repaired.

Types of Damage

When writing an estimate, the following are some of the conditions to look for:

- Direct outer body damage starting at point(s) of impact (Figure 10–23 and Figure 10–24)
- Improper fit of doors and movable panels (Figure 10–25), which shows major movement of body panels
- Hidden structural damage under outer body panels
- Cracked sealer and undercoating, which indicate structural damage (Figure 10–26).
- Indirect damage from shock waves travelling through body parts
- Mechanical damage to engine parts, steering and suspension parts, and the drivetrain
- Damage to interior of vehicle, air bag deployment, or dash damage, for example
- Bent wheels and damaged tires
- Broken glass and mirrors

As an example, if the front grille and some of the related parts are damaged, the repairs or replacements needed would be listed as follows:

Front grille .... Replace
Front grille .... Refinish
Opening panel .... Replace
Deflector (or valance panel) .... Replace
Headlamp door................................ Replace
Grille opening panel .... Refinish

Notice that the parts to be repaired, straightened, replaced, or refinished are listed in a definite sequence

Figure 10-24 The rear of the van contacted the wall with greater force, knocking out rear glass and badly crushing a quarter panel. The quarter panel and rear bumper cover will require replacement. The rear hatch and side door are repairable.

Figure 10-25 When analyzing damage, look for uneven gaps around doors and panels. They indicate major movement of body parts from collision impact and will add to repair costs.

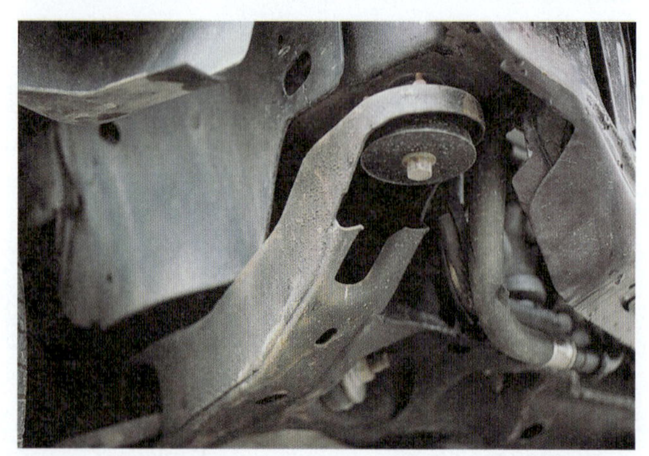

Figure 10-26 When looking under the vehicle or inside wheel openings, check for cracked sealer, missing undercoat, and movement of parts. This engine cradle shows movement, indicating major front-end damage.

according to factory disassembly operations or exploded views as provided in shop manuals or crash estimating guides.

You must find all damage, no matter how minor (Figure 10–27). After noting obvious exterior damage, look for hidden damage. Doors out of alignment and cracked stationary glass are often a solid clue to hidden damage to the frame or underbody structure. If applicable, open the hood and check in the engine compartment for damage (Figure 10–28).

When entering mechanical damage for an estimate, you will often have to designate mechanical entering after the operation. Then the computer will know to charge a different rate.

Look for cracked sealer or undercoating, which indicates the metal panels have been pushed out of alignment.

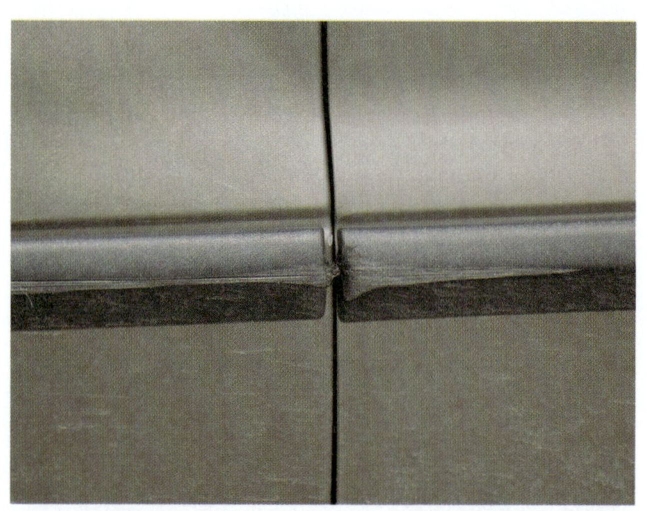

A These body moldings were deeply scratched and would require replacement.

B This door handle was badly abraded and would also require replacement.

Figure 10-27 Make sure you check all major parts and surfaces for damage.

A **The air-conditioning condenser and engine radiator are commonly crushed, requiring repair or replacement.**

B **These aluminum air-conditioning lines were fractured during collision. Part replacement and recharging the air conditioner with refrigerant would have to be added to the estimate.**

C **This power steering pump was bent and had to be replaced. Also check the water pump and all engine pulleys for damage.**

Figure 10–28 **After a front-end collision, open the hood and check for these kinds of problems.**

Find Indirect Damage

The estimator must be on constant guard against accidentally missing related damage. Damage that happens to a vehicle during the moment of impact or immediately after impact is referred to as related, or *indirect, damage.*

Often, in the case of unibody cars, related damage is not near the area of impact, but some distance away. Buckles in large panels like the roof and quarter panels are often indirect damage. Always look for indirect damage when writing an estimate (Figure 10–29).

Hidden or secondary damage, as stressed many times in this book, can occur almost anywhere or to any part or component on a car that has been involved in a collision. For example, the snout of an engine crankshaft can snap from frontal impact forces. Badly damaged fan blades and pulleys can clue you in to a damaged engine water pump. Check for coolant, oil, and other fluid leakage. This would indicate ruptured hoses, lines, or physical damage to fluid reservoirs.

Transmission cooler lines running from the engine radiator to the automatic transmission case might have become crushed or pinched. Steering linkage might have become bent and yet, because of its unusual configuration, misalignment can be difficult to detect at first glance.

Radiators are commonly damaged in front collisions. Always check for smashed radiator fins or a leaking radiator when estimating damage to the front of a vehicle.

Front motor mounts can be sheared even though the engine resettles back into position, hiding this type of damage. Castings of the engine, transmission, or bell housing may crack, leaving only a hairline fracture that is difficult to detect even on close examination.

Leaf springs can crack and be hard to spot, or the parking pawl, located inside the automatic transmission, may have snapped in two if the car was hit with the shift lever in the park position.

Finding Frame Damage

The estimator must look for buckles in frame members. Search for bolts, flat washers, or other types of fasteners that have moved or shifted out of position, leaving unpainted, bare, or shiny metal showing. Also look for displaced, cracked, or fractured undercoating underneath the body structure. Check crush zones on the frame rails for buckles and signs of major frame rail damage, which would add considerable cost to the repairs.

Wheel and Suspension Damage

Always inspect the inside and outside lips of wheels for cracks and damage. New wheels should be added to the estimate if needed. Tires should also be closely inspected on both the outside and inside of the sidewall. If the estimator finds extensive or deep cuts, the tire is unsafe. A new tire must be listed in the estimate. Rotate any suspect wheels to make sure that they are not bent (Figure 10–30).

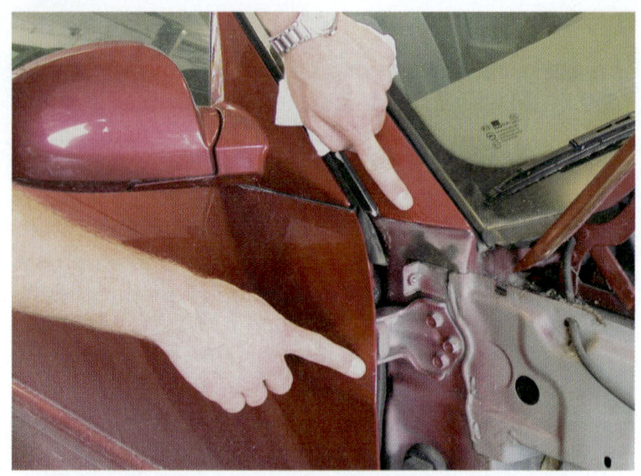

A A fender can often be pushed back into the door and A-pillar. Check edges of these panels for paint nicks that would require repair and repainting.

B This truck was "rear-ended." This often pushes the bed of the truck into the cab, a common location of secondary damage.

C Sight down large panels and rub them with your hand. You must find any indirect panel warpage away from the point of impact.

Figure 10-29 With front and rear hits, check for secondary damage where large panels meet and along long, smooth surfaces of panels. Indirect damage is common. Impact force can travel through the body structure and buckle panels at the other end of a vehicle.

A Note how the rear axle of this truck is way out of alignment. Minor wheel alignment problems are more difficult to catch.

B Here a camber angle gauge is being used to quickly check for wheel misalignment. If not within specs, major repairs and a wheel alignment would have to be added to the estimate.

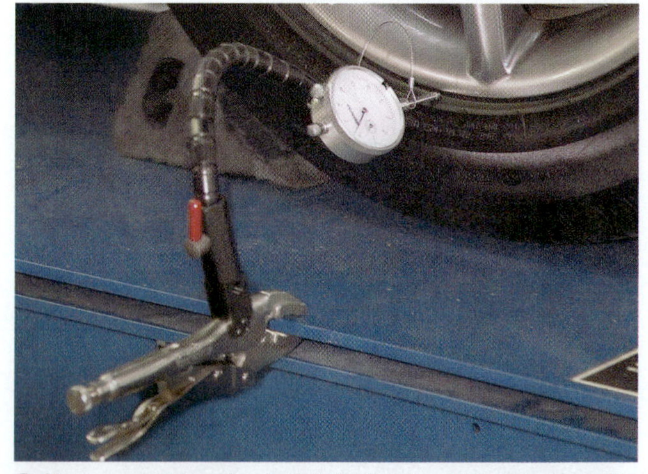

C If you suspect a wheel might be bent, a dial indicator will quickly measure runout and damage. Aluminum wheels are very susceptible to damage. If a wheel is badly damaged, the tire probably has cut and damaged internal cords, and should be replaced.

Figure 10-30 Sight down wheels and tires to find major wheel alignment problems.

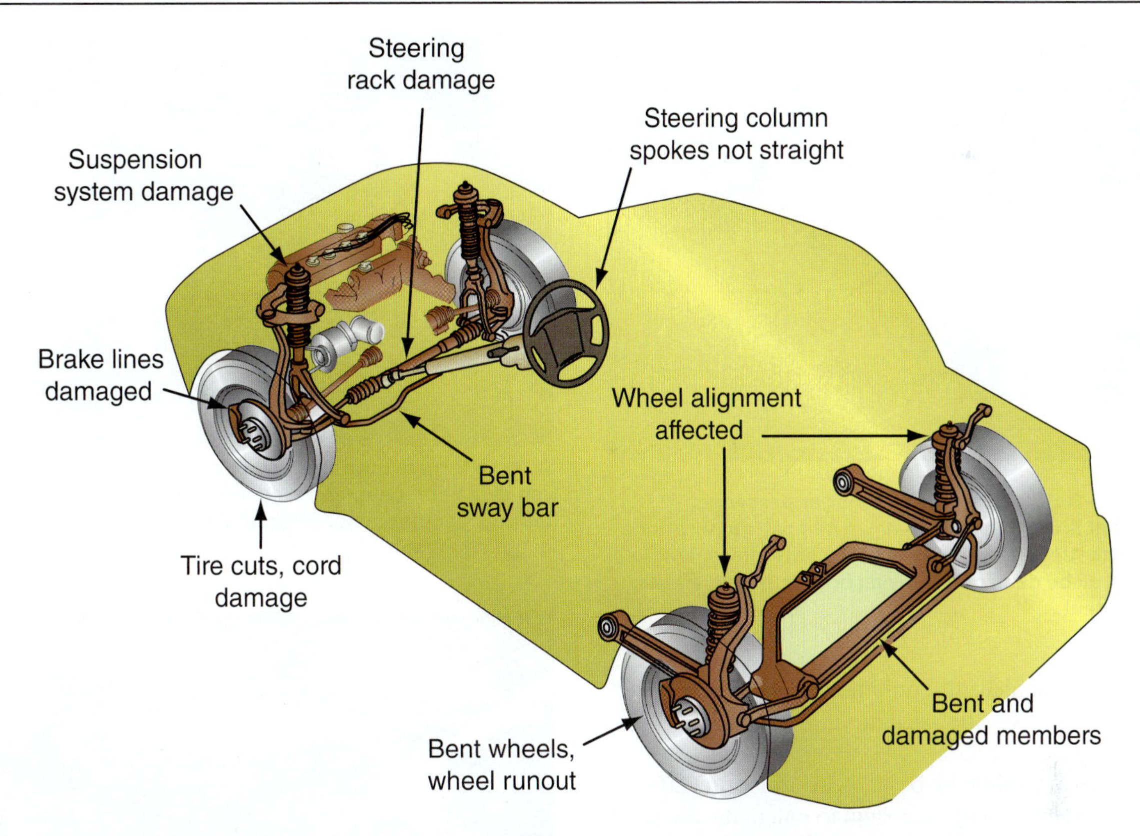

Figure 10–31 Note the types of damage that often happen to suspension, brake, and steering systems.

Sometimes, wheel damage occurs and yet the tire looks undamaged. Keep in mind that an impact that damaged a wheel will often cut cords inside the tire. This will cause tire runout and vibration when returned to service. In such instances, have someone dismount the tire and make an inspection of the inner surfaces.

Figure 10–31 shows some of the chassis parts that are often damaged in major collisions.

During your inspection of vehicle damage, also look at and include damaged wiring harnesses in your estimate. Wires are often cut and smashed during a major collision (Figure 10–32).

With major damage, it is best to raise the vehicle on a lift so you can inspect under the vehicle. This will let you check frame rails, cross members, and other underbody components more closely.

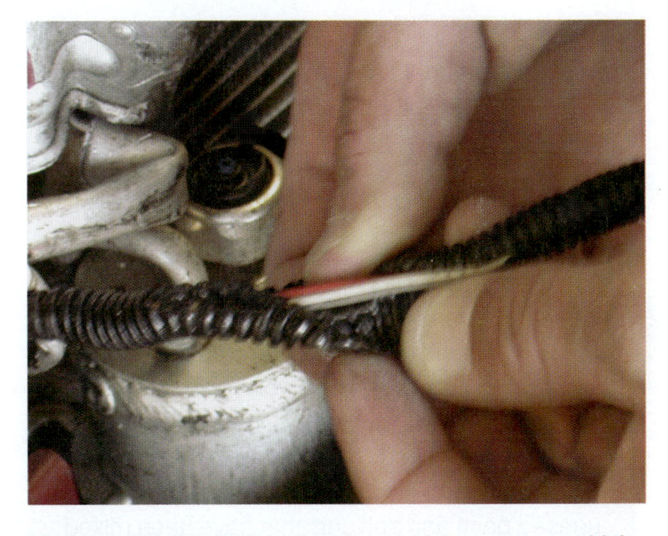

Figure 10–32 With major structural damage to a vehicle, wiring harnesses are often damaged. The cost of purchasing and installing new wiring harnesses would have to be added to the estimate.

Passenger Compartment Damage

The interior of the vehicle must also be checked closely for damage to be repaired. In modern vehicles, air bags commonly deploy from the steering wheel, dash, seats, or door panels. Air bag deployment can damage parts such as the windshield or trim pieces (Figure 10–33).

Measuring Damage During Estimating

As mentioned, severe collisions from any direction often cause the frame or unitized body to distort. With today's auto construction, the "eyeballing" technique is not enough to detect frame misalignment. It is far better to use measuring equipment (Figure 10–34) to check for misalignment. However, unless the estimator is thoroughly skilled in this area of estimating, it is wise to consult with a frame/body technician for an appraisal to determine the labor time necessary to get the frame back into proper alignment.

To evaluate major damage, you might need to print out vehicle dimension specifications. This would let you

Figure 10–33 The vehicle's interior or passenger compartment is another area that can be costly to fix after a bad accident. Air bags, the dash panel, seats, glass, and carpets can all be damaged by flying bodies, blood, and air bag deployment.

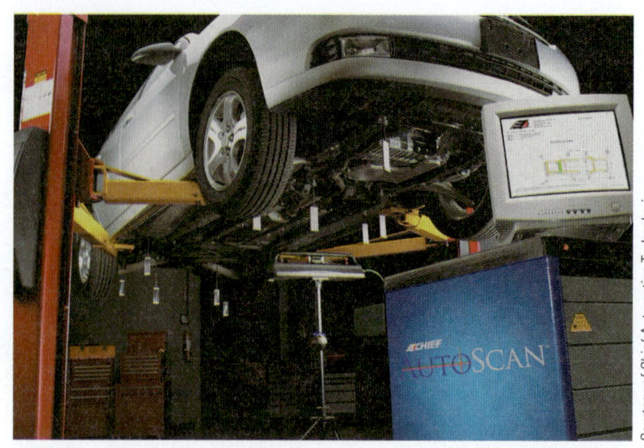

Courtesy of Chief Automotive Technologies

A Here targets have been placed on specific points of the vehicle. A computerized measuring system can instantly tell you whether major structural parts are damaged and out of alignment.

Courtesy of Chief Automotive Technologies

B Here the estimator is taking measurements with an electronic tram gauge to check for major damage that would affect the estimate.

Figure 10–34 Inspections of major damage are often done on a lift or bench, which helps you more easily inspect under the vehicle to measure damage.

use the shop's measuring equipment to precisely determine the extent of the structural damage. Look at (Figure 10–35).

After gathering all of the damage information you can, use the computer estimating system to calculate and print out the damage report (Figure 10–36 and Figure 10–37).

Hidden Damage

Some estimates do include a so-called "hidden damage clause" that permits added charges to the original estimate. When hidden damages are discovered after exterior panels or parts are removed for repairs, this is called an *estimate supplement.*

ECO-TECH When estimating paint repair, you must consider and add any costs of recycling materials. If your shop does not have a solvent or paint recycling machine, your shop may have to pay an outside company to pick up, process, and recycle unused paint and solvents that have been mixed but not consumed during the repair. This is another way that you, as a body/paint technician, can help protect our environment while making a little more money to cover recycling or reprocessing costs.

10.9 VEHICLE TOTAL LOSS

A *total loss* occurs when the cost of the repairs exceeds the value of the vehicle. The insurance company or customer will normally *not* want the vehicle repaired in that case. Instead, the insurance company will write a check to cover the cost of a replacement vehicle. An equivalent year, make, and model vehicle can then be purchased by the customer.

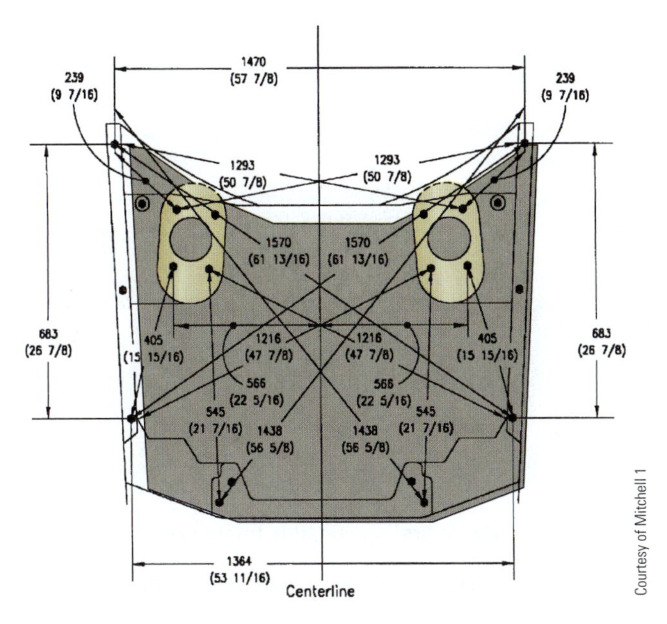

Courtesy of Mitchell 1

Figure 10-35 With major structural damage, a technician will often measure vehicle damage to help the estimator calculate repair costs. With a computerized estimating system, you can quickly pull up vehicle dimensions so they can be measured and compared to specs. This lets you know whether frame-straightening costs must be added to the estimate.

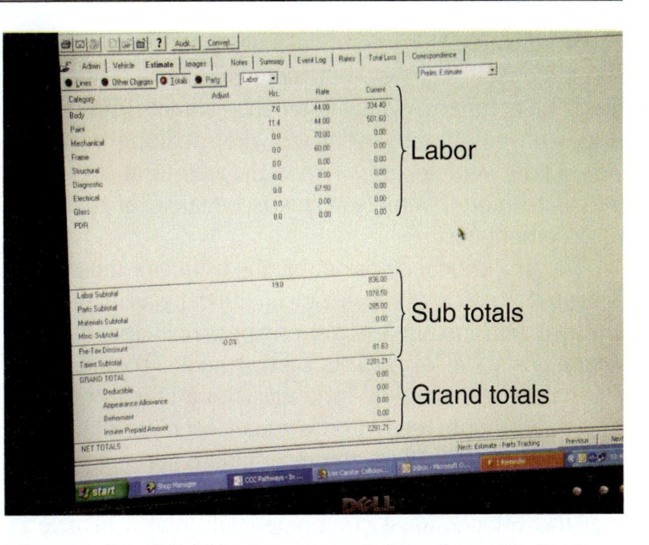

Figure 10-36 This computer window shows the final computerized estimate. The labor subtotal is $836.00. The parts subtotal is $1,078.58. The repair materials subtotal is $285.00. Tax subtotal on parts is $81.63. The grand total is $2,281.21.

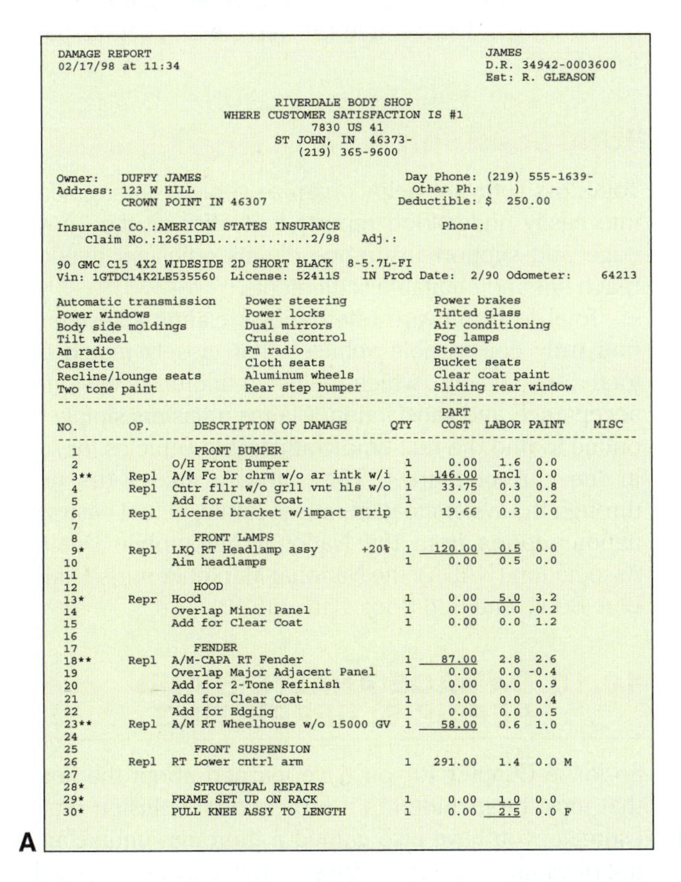

DAMAGE REPORT
02/17/98 at 11:34

JAMES
D.R. 34942-0003600
Est: R. GLEASON

RIVERDALE BODY SHOP
WHERE CUSTOMER SATISFACTION IS #1
7830 US 41
ST JOHN, IN 46373-
(219) 365-9600

Owner: DUFFY JAMES
Address: 123 W HILL
CROWN POINT IN 46307

Day Phone: (219) 555-1639-
Other Ph: ( ) - -
Deductible: $ 250.00

Insurance Co.:AMERICAN STATES INSURANCE Phone:
Claim No.:12651PD1.............2/98 Adj.:

90 GMC C15 4X2 WIDESIDE 2D SHORT BLACK 8-5.7L-FI
Vin: 1GTDC14K2LE535560 License: 52411S IN Prod Date: 2/90 Odometer: 64213

Automatic transmission	Power steering	Power brakes
Power windows	Power locks	Tinted glass
Body side moldings	Dual mirrors	Air conditioning
Tilt wheel	Cruise control	Fog lamps
Am radio	Fm radio	Stereo
Cassette	Cloth seats	Bucket seats
Recline/lounge seats	Aluminum wheels	Clear coat paint
Two tone paint	Rear step bumper	Sliding rear window

NO.	OP.	DESCRIPTION OF DAMAGE	QTY	PART COST	LABOR	PAINT	MISC
1		FRONT BUMPER					
2		O/H Front Bumper	1	0.00	1.6	0.0	
3**	Repl	A/M Fc br chrm w/o ar intk w/i	1	146.00	Incl	0.0	
4	Repl	Cntr fllr w/o grll vnt hls w/o	1	33.75	0.3	0.8	
5		Add for Clear Coat	1	0.00	0.0	0.2	
6	Repl	License bracket w/impact strip	1	19.66	0.3	0.0	
7							
8		FRONT LAMPS					
9*	Repl	LKQ RT Headlamp assy +20%	1	120.00	0.5	0.0	
10		Aim headlamps	1	0.00	0.5	0.0	
11							
12		HOOD					
13*	Repr	Hood	1	0.00	5.0	3.2	
14		Overlap Minor Panel	1	0.00	0.0	-0.2	
15		Add for Clear Coat	1	0.00	0.0	1.2	
16							
17		FENDER					
18**	Repl	A/M-CAPA RT Fender	1	87.00	2.8	2.6	
19		Overlap Major Adjacent Panel	1	0.00	0.0	-0.4	
20		Add for 2-Tone Refinish	1	0.00	0.0	0.9	
21		Add for Clear Coat	1	0.00	0.0	0.4	
22		Add for Edging	1	0.00	0.0	0.5	
23**	Repl	A/M RT Wheelhouse w/o 15000 GV	1	58.00	0.6	1.0	
24							
25		FRONT SUSPENSION					
26	Repl	RT Lower cntrl arm	1	291.00	1.4	0.0 M	
27							
28*		STRUCTURAL REPAIRS					
29*		FRAME SET UP ON RACK	1	0.00	1.0	0.0	
30*		PULL KNEE ASSY TO LENGTH	1	0.00	2.5	0.0 F	

A

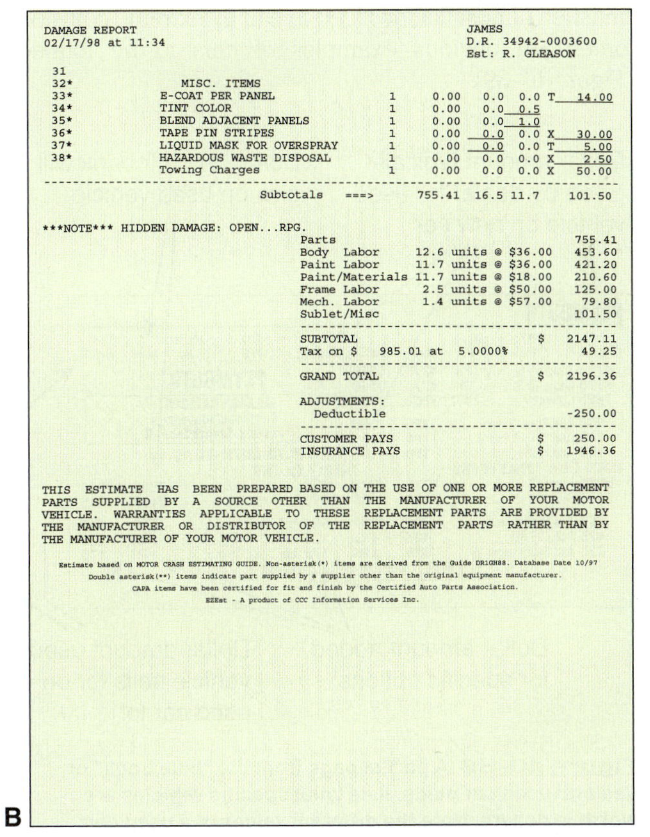

DAMAGE REPORT
02/17/98 at 11:34

JAMES
D.R. 34942-0003600
Est: R. GLEASON

NO.	DESCRIPTION	QTY	PART COST	LABOR	PAINT	MISC
31						
32*	MISC. ITEMS					
33*	E-COAT PER PANEL	1	0.00	0.0	0.0	T 14.00
34*	TINT COLOR	1	0.00	0.0	0.5	
35*	BLEND ADJACENT PANELS	1	0.00	0.0	1.0	
36*	TAPE PIN STRIPES	1	0.00	0.0	0.0	X 30.00
37*	LIQUID MASK FOR OVERSPRAY	1	0.00	0.0	0.0	T 5.00
38*	HAZARDOUS WASTE DISPOSAL	1	0.00	0.0	0.0	X 2.50
	Towing Charges	1	0.00	0.0	0.0	X 50.00
	Subtotals ===>		755.41	16.5	11.7	101.50

NOTE HIDDEN DAMAGE: OPEN...RPG.

Parts			755.41
Body Labor	12.6 units @ $36.00		453.60
Paint Labor	11.7 units @ $36.00		421.20
Paint/Materials	11.7 units @ $18.00		210.60
Frame Labor	2.5 units @ $50.00		125.00
Mech. Labor	1.4 units @ $57.00		79.80
Sublet/Misc			101.50
SUBTOTAL		$	2147.11
Tax on $ 985.01 at 5.0000%			49.25
GRAND TOTAL		$	2196.36
ADJUSTMENTS:			
Deductible			-250.00
CUSTOMER PAYS		$	250.00
INSURANCE PAYS		$	1946.36

THIS ESTIMATE HAS BEEN PREPARED BASED ON THE USE OF ONE OR MORE REPLACEMENT PARTS SUPPLIED BY A SOURCE OTHER THAN THE MANUFACTURER OF YOUR MOTOR VEHICLE. WARRANTIES APPLICABLE TO THESE REPLACEMENT PARTS ARE PROVIDED BY THE MANUFACTURER OR DISTRIBUTOR OF THE REPLACEMENT PARTS RATHER THAN BY THE MANUFACTURER OF YOUR MOTOR VEHICLE.

Estimate based on MOTOR CRASH ESTIMATING GUIDE. Non-asterisk(*) items are derived from the Guide DR1GH88. Database Date 10/97
Double asterisk(**) items indicate part supplied by a supplier other than the original equipment manufacturer.
CAPA items have been certified for fit and finish by the Certified Auto Parts Association.
EZEst - A product of CCC Information Services Inc.

B

Figure 10-37 This is a printed computer-based estimate. Study it closely to see how all part prices, material prices, labor charges, hidden charges, and other data have been included on a damage report, or estimate. A The first page of a computerized estimate lists information on the vehicle. The lower part of the estimate gives a description of the damage, parts cost, labor cost, and paint cost for each task. B The second page of the estimate lists costs for tinting color, blending paint into another panel, hidden damage to body, paint, frame, and mechanical parts, and it gives the subtotal and grand total. It also gives the amount the customer and insurance company must pay the shop, which depends upon the insurance deductible amount.

The insurance company usually determines whether a vehicle is a total loss, but the customer can negotiate this. The company will evaluate the estimate and market prices for comparable vehicles when making this decision. Older vehicles are written up as a total loss more than late model vehicles. This is because of their low replacement cost.

The first critical decision that the estimator must make is the repairability of the vehicle. If the car was involved in a severe collision, there is a strong possibility that the total repair costs will exceed its market value. For example, repairs are not practical for a car with a market value of $3,150 (if undamaged) and a repair estimate that totals $3,710. More than likely the insurance company adjuster will agree with such a decision and would authorize "totaling" the vehicle. Most companies will total a vehicle at 70–80 percent of the vehicle cost. If the vehicle's value is $20,000, the insurance company may total it at $14,000.

Figure 10–39 At first glance, this totaled vehicle may seem repairable at reasonable cost, but hidden and indirect damage make repairs too costly for the value of the vehicle.

Blue Book

The auto industry "blue book" (Figure 10–38) lists the market values for used vehicles in various conditions. Refer to this type of publication when decisions to "repair or scrap" badly damaged vehicles must be made.

Of course, cars that have been wrapped around trees, smashed almost flat, or sliced in half by extreme collision forces are obvious examples of cars to be totaled (Figure 10–39).

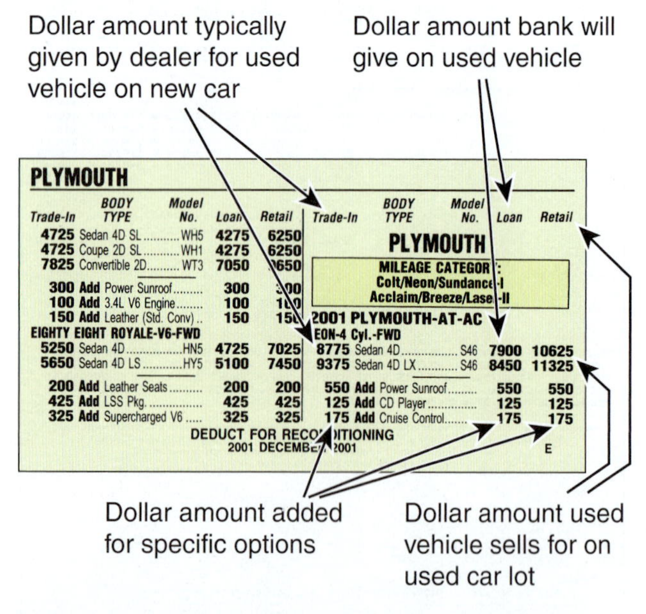

Figure 10–38 A partial page from the "blue book" or dealers' used car guide, lists what specific vehicles are worth in dollars. Note the different values of a used car: trade-in, wholesale, and retail. Retail is the highest dollar value and is what the vehicle might sell for on a used car lot. Trade-in and loan values are lower. A blue book is used to determine whether a vehicle is a total loss. If the cost of repairs is higher than the value of a vehicle, the insurance company would not approve repairs but would give the customer a check to purchase another vehicle.

Some totaled vehicles have a *salvage value*, which is the price a junkyard will pay for the totaled vehicle. Parts of a wrecked car or truck, such as the engine, transmission, drive shaft, and other body parts, might be usable and saleable. The totaled vehicle will usually be auctioned or sold to a salvage yard or recycler. The recycler will then disassemble the vehicle and sell its parts for a profit.

Total Loss Software

Total loss software helps organize vehicle damage data into easily understood reports that offer all the knowledge and support insurance claims adjusters need to reach fair and equitable settlements.

Total loss software often allows claims adjusters to find truly comparable vehicles that closely match the total loss vehicle, which helps to increase customer acceptance and satisfaction. Claims adjusters simply go online to find the real-time comparable vehicles for sale in the current market. The software guides the user through the system to generate accurate, actual cash valuation reports from the National Automobile Dealers Association (NADA), the National Market Reports (NMR) Blue Book, and Red Book Canada.

10.10 DAMAGE ANALYSIS AND REPAIR BLUEPRINTING

So far in Chapter 10 you have learned about the items that must be included in a professional collision repair estimate; you have also gained a thorough understanding of estimating terminology. In this part of Chapter 10 we will focus on customer relations, how to properly assess vehicle damage, and how to find hidden damage that could easily be missed by new estimators. In order to do a thorough damage analysis there are several vehicle systems that we must carefully inspect. We will also look at labor judgment items and different choices for replacement parts.

Customer Relations

The vehicle owner is usually the first person you will encounter in the repair process. It is important to treat customers with a high level of professionalism throughout the entire collision repair process. The customer can be an excellent resource when it comes to vehicle damage analysis. In the estimating process it is important to ask the customer what occurred in the accident. Ask questions like: Does the vehicle drive the same as before the accident? Or have you noticed anything different since the accident occurred? Examples like the vehicle hitting a curb and bending suspension parts or leaving the road and damaging exhaust system parts could be types of damage that could be discovered easily by talking to the customer. Finding hidden damage early in the process leads to faster repair times and increased overall customer satisfaction.

Damage Analysis

After gathering details of the accident/incident from the customer, the first step in a thorough damage analysis is to completely wash the vehicle and dry it with a soft, clean chamois. Walk around the vehicle with the customer and take pictures of all four sides. Estimators typically use a tablet or other hand-held electronic device to input data while discussing repairs with the customer. At this time it is important to document previous repairs or previous damage that may not be accident related. Most estimating systems have headnote and footnote options as convenient locations to write specific information about the vehicle. Give the customer the option to repair any preexisting damage and explain that it will make the entire repair area look better if preexisting door dings or scratches are repaired along with the collision damage. If the customer decides not to repair previous damage, this should be marked on the vehicle with water-based marker so the repair technician will be aware of it. Ask the customer if the vehicle has any aftermarket or add-on equipment. Examples of add-on equipment could be step bars for pickup trucks or ground effects for cars. If any add-on equipment is damaged and requires replacement, it may take longer to get than normal factory replacement parts, so aftermarket add-on parts should be ordered as early as possible in the repair process.

The VIN and paint code should always be included on the estimate. The VIN can identify almost everything about the vehicle as it was equipped from the factory and is necessary in order to get the correct factory parts for repairs. It is important to record the paint code on the estimate because sometimes repair facilities may paint bumper covers or other parts for a vehicle that is not on site. In this scenario all the refinish work could be completed while the customer drives his or her own vehicle, only to come to the repair facility briefly for the installation of the new painted parts.

The vehicles for which you will write estimates will have varying levels of damage. Light damage may be easy to identify, but if vehicles have moderate or severe damage or you suspect hidden damage, there is another important level of damage analysis that requires more than just looking at the outside of the damaged areas.

Repair Blueprinting

Repair blueprinting (sometimes referred to simply as blueprinting) is a process used in the collision repair industry to perform a thorough damage analysis and make a repair plan. In order to make a good repair blueprint, a complete teardown (disassembly) of damaged or affected areas of the vehicle is necessary. Blueprinting helps repair technicians to be efficient as they work through the complicated processes necessary for a complete repair.

Blueprinting requires a disassembly of all the affected areas of the vehicle. Remove bumper covers, sheet metal, lights, and any other components that might conceal hidden damage. It is a common practice during this stage of blueprinting to lift the damaged vehicle using a two-post hoist and inspect the undercarriage. While the vehicle is suspended on the hoist and the safety locks are engaged, it is important to identify any structural damage. Structural damage can be easily documented by using a three-dimensional measuring system such as the Car-O-Tronic Vision or PointX. These systems will provide a printout of damaged areas of the vehicle in order to document what structural repairs will need to be completed in the repair process. If major structural repairs are needed, such as the replacement of body or frame components, it is important to gather the specific auto manufacturer recommendations during the blueprinting process. Although challenging, finding vehicle-specific information is vital for a complete repair. I-Car maintains a list of vehicle maker websites at www.i-car.com, and there is also a repairability tech site located at rts.i-car.com.

Specific Systems to Inspect During the Blueprinting Process

In the blueprinting process it is important to pay close attention to various vehicle systems. Chapter 30 has some specific things to look for while performing damage analysis on hybrid vehicles, but on any vehicle we need to carefully inspect the following systems:

- Safety systems
- Electrical systems
- ABS, traction, and stability control systems
- A/C systems
- Steering and suspension systems

Safety systems include air bags or supplemental inflatable restraints (SIRs). SIR technology has evolved a great deal in recent years. Where vehicles were once equipped only with standard driver's side and passenger air bags, now they have many air bags in locations throughout the vehicle. The 2015 Chevy Spark is one example of a vehicle which has 10 air bags. These air bag modules are in different places around the vehicle passenger compartment. If damage exists in the passenger compartment area, we must make sure that none of the undeployed air bags have become damaged. Any damage to undeployed air bag modules would require replacement. Check all the seat belts in a vehicle after an accident. Some vehicles have pyrotechnic charges in the seat belt tensioners that tighten during an accident; after these tensioners are deployed they must be replaced. Some vehicle makers recommend replacing all the air bag sensors after a deployment. It is important to refer to specific vehicle maker repair information while working on these types of advanced restraint systems.

Although not a restraint system, another vehicle safety system that requires inspection is the blind-spot detection system (BSDS). The BSDS will illuminate a light or make a sound if there is another vehicle in a blind spot during an attempted lane change. The BSDS generally has sensors located behind the rear tires under the rear bumper cover. Damage to the bumper cover in the area of a BSDS sensor requires bumper cover replacement rather than a repair because creating a repair area in close proximity to the BSDS sensor (which acts like a radar system) could cause a malfunction.

Electrical systems can easily become damaged during collisions. Check the 12-volt battery, and if it is damaged it must be replaced. If the battery is leaking acid, the spill must be cleaned up according to local, state, and federal requirements, and the battery should be disposed of properly. Damaged wiring harness insulators are a sign of possible wiring damage. Orange-colored cables indicate high voltage and are not repairable, so please refer to Chapter 30 for more information about hybrid and electric vehicle repairs. If low-voltage wiring is damaged, check for available voltage and diagnose electrical problems with a digital multimeter (DMM). Use vehicle wiring diagrams and check for voltage drop and resistance in electrical circuits and components to help identify electrical system damage.

ABS, traction, and stability control systems share many of the same components and must be inspected after a collision. If the computer that controls the ABS, traction, and stability control systems senses a fault or malfunction, a DTC (diagnostic trouble code) will be displayed. If a DTC is displayed, use a diagnostic tool to determine what repairs are needed. A physical inspection of the ABS parts will require the wheels to be removed from the vehicle; the best time for this inspection is while the car is suspended on a two-post lift. Look for bent dust shields, broken or cracked housings, or missing teeth from the ABS sensor ring.

A/C system inspection includes checking the drive belts and pulley alignment. In a collision it is possible for pulleys to get pushed back while in spinning motion. This phenomenon makes the pulley run "true" but with an improper amount of set-back; for this reason it is important to look carefully at belt alignment. A/C compressors are commonly damaged in collisions, so check for damaged housings or signs of leaking oil or refrigerant. A/C condensers are often damaged in frontal collisions because of their vulnerable location behind the front impact reinforcement. While you are inspecting the A/C condenser it is also a good time to check auxiliary oil or fluid coolers for damage; generally these coolers will be close to the A/C condenser. Removal and replacement of the A/C condenser will require the removal of the radiator in most cases (they are usually connected and installed as an assembly in the manufacturing process), so checking the vehicle's cooling system for damage as part of the A/C system inspection will help save time. Check cooling system electrical and mechanical fans and their shrouds for damage; this can be done by spinning the fans by hand while looking for bent or broken blades or misalignment.

Steering and suspension systems can also be inspected easily while the vehicle is suspended from a two-post lift. As in the case of measuring body structural damage, a three-dimensional measuring system such as the Car-O-Tronic Vision or PointX can identify bent or damaged (out of spec) steering and suspension components. This is a fast and easy way to document which parts require replacement early in the repair process. Subframes and suspension parts may have one-time use fasteners that require replacement after removal. These bolts may have a special coating that helps keep them tight. Refer to specific vehicle maker information to know which fasteners are one-time use.

Judgment Items

When writing an estimate, you will be required to make "judgment" decisions for repair operations. Estimating software has predetermined labor times for removing and replacing a vehicle fender, but there is no predetermined amount of labor for a softball-size dent. Part of your job as an estimator is to assign a value to judgment items. Generally the amount of dent damage that can be hidden by a dollar bill would be between 2 and 3 hours of body labor. Scratches are a little different; a scratch that would be covered with a sheet of notebook paper would generally be a 1-hour repair. A dent the size of a dime, such as a parking lot ding, would be ½ to 1 hour of body labor. A dent the size of a sheet of notebook paper would be 6 to 8 hours of body repair labor. Judgment items must take type of material, damage location, and severity of damage into consideration. The more complicated or extensive the damage, the more time or body labor is required for a complete repair. Labor operations include body, structural, nonstructural, mechanical, and refinish labor. You must select which labor operation will be performed for each repair line that is entered when writing an estimate.

Replacement Parts

There are different part sources that include factory new parts, recycled parts, and aftermarket parts. Factory new parts are generally preferred by customers and repair technicians because they are clean, free from previous damage or corrosion, and recommended by vehicle manufacturers for quality repairs. In some cases new replacement parts are unavailable or their cost may make the needed repairs too expensive. Every vehicle has a total loss threshold, so there is a limit to what a repair can cost before the vehicle is deemed a total loss. Alternatives to factory new parts include recycled and aftermarket parts. Recycled parts are undamaged parts that are from damaged or salvaged vehicles. Aftermarket parts are new because they have never been on a vehicle before, but they are manufactured by sources outside of the vehicle maker's supply chain, so quality of fit and finish can vary with aftermarket parts. When selecting parts for your repair estimate, it is important to verify availability, compatibility, and condition. A good way to do this is when the part arrives, inspect it before accepting it. Repair times can be greatly reduced if the right parts are ordered early in the repair process and any mistakes resolved as soon as they are discovered. Repair blueprinting can help determine every part that is needed for a complete repair, from sheet metal parts to hidden clips and fasteners that may break during teardown.

Use a Checklist

Pilots walk around their aircraft using a preflight checklist before starting the engine and taxiing down the runway. Checklists help people remember things that are important but may be overlooked because of redundancy. A checklist can be helpful through the many processes of collision repair. Checklists are commonly used for collision repair estimating, repairs, and even vehicle cleanup. Make a checklist to help you with repair blueprinting, and organize it with the damage analysis operations and the inspection of the vehicle systems we have learned about in this chapter.

SUMMARY

1. An estimate, also called a damage report or an appraisal, calculates the cost of parts, materials, and labor for repairing a collision-damaged vehicle.
2. An estimate is a firm bid for a given period of time—usually 30 days.
3. Computerized estimating systems using a personal computer (PC) provide more accurate and consistent damage reports.
4. An in-shop estimating system has all of the data needed at the shop office, usually on CDs.
5. A total loss occurs when the cost of the repairs exceeds the value of the vehicle.
6. The flat rate is a preset amount of time and money charged for a specific repair operation.

EXERCISES

On a separate sheet of paper, complete the following learning activities for this chapter. Write definitions for the key terms and answer the ASE-style review questions, essay questions, critical thinking problems, and math problems. You can also do the outside activities, possibly for extra credit.

Key Terms

access time
computer-based estimating
computer database
deductible clause
estimating program
flat rate
front clip assembly
included operations
job overlap
overhaul (O/H)
procedure pages
remove and install (R&I)
remove and replace (R&R)
salvage parts
stress crack
work order

ASE-Style Review Questions

1. Technician A uses the flat-rate labor times reported in collision damage manuals only as a guide for estimating. Technician B says that flat-rate labor times are based on data reported by vehicle manufacturers. Who is correct?
   A. Technician A
   B. Technician B
   C. Both A and B
   D. Neither A nor B
2. How long does an estimate usually remain a firm bid?
   A. 1 year
   B. 6 months
   C. 90 days
   D. 30 days
3. Which of the following abbreviations found in collision estimating guides means that the item in question should be removed as an assembly, set aside, and later reinstalled?
   A. R&R
   B. R&I
   C. O&H
   D. O/H
4. Many insurance policies contain a ______ clause, which means that the owner is responsible for a given amount of the estimate (usually the first $500–$1000).
   A. premium
   B. deductible
   C. review
   D. final
5. Which of the following conditions is a good sign of minor body distortion?
   A. Stress cracking
   B. Front clip
   C. Side clip
   D. All of the above
6. This is used to take photos of a vehicle's damage and store them in a computer estimating system.
   A. Scanner
   B. Digital camera
   C. Film camera
   D. Bar code reader
7. What term describes jobs that can be performed individually but are also part of larger procedures?
   A. R&R
   B. R&I procedures
   C. Included operations
   D. Flat rate
8. What is the smallest typical increment in which labor times are listed in crash estimating guides?
   A. Hours
   B. Half-hours
   C. Quarter hours
   D. Tenths of an hour
9. Technician A says estimates are used in place of work orders in most shops. Technician B says estimates are used to help write work orders. Who is correct?
   A. Technician A
   B. Technician B
   C. Both A and B
   D. Neither A nor B
10. Technician A says a totaled vehicle has no salvage value. Technician B says that a totaled vehicle often has salvage value. Who is correct?
    A. Technician A
    B. Technician B
    C. Both A and B
    D. Neither A nor B
11. Which of the following is not included in the refinishing times for painting a new panel?
    A. Removing moldings
    B. Featheredging body putty
    C. Masking handles
    D. All of the above
12. Which of the following refinishing operations is generally considered "included" or part of the job in estimating software and flat-rate manuals?
    A. Clean panel/light sanding of the repair area
    B. Masking the complete vehicle to prevent overspray damage
    C. Tinting the colorcoat for a quality color match
    D. All of the above
13. Printed collision estimating guides can be used as a reference for ____________.
    A. final estimated total.
    B. checking total-loss threshold.
    C. pricing parts.
    D. all of the above.
14. What process used in the collision repair industry requires a vehicle tear-down, thorough damage analysis and repair planning?
    A. Estimating
    B. Blueprinting
    C. Damage analysis
    D. All of the above

Math Problems

1. A repair will require 3 hours at $40 per hour. How much will the total labor cost be for this job?
2. A gallon of paint is $95 and you only need a quart for the repair. How much should you charge the customer for the paint?

Activities

1. Manually write an estimate for a vehicle with minor damage. Then use a computer estimating system to write another estimate for the same vehicle. Compare your two estimates. Note any differences in calculations.
2. Arrange a visit to watch a shop estimator at work. Write a report on what you learned from the professional estimator.

Essay Questions

1. List the advantages of a resident data system.
2. When writing an estimate, what are some of the problems to look for?
3. Doors that are out of alignment and cracked stationary glass are often solid clues to what?
4. What does *salvage value* mean?

Critical Thinking Problems

1. If you make your estimate too low or too high, how could it affect the shop operation?
2. A car was driven over a cement median curb at high speed by a drunken driver. You find cracked sealer around the front frame rails. What does this tell you about the vehicle's damage?
3. What does "deductible clause" mean?

Photo Summary

Estimating Repair Costs

Figure P10-1 The estimator must understand all aspects of collision repair to make accurate calculations of repair costs.

Figure P10-2 The estimator must inspect for obvious signs of damage to the outer body structure and must also be knowledgeable enough to find hidden damage.

Figure P10-3 Today's estimators use PCs and specialized software to automate the estimating process.

SECTION

3

Minor Repairs

CHAPTER

11 Working Sheet Metal

OBJECTIVES

After reading this chapter, you should be able to:

- Describe different types of metals used in vehicle construction.
- Explain the strength ratings of metals.
- Summarize the deformation effects of impacts on steel.
- Use a hammer and dolly to straighten metal.
- Explain how to bump dents with spoons.
- List the steps for shrinking metal.
- Summarize the procedures for paintless dent removal.
- Answer ASE-style review questions relating to sheet metal work.

INTRODUCTION

This chapter will introduce you to basic metalworking methods. It will explain how to analyze minor damage to sheet metal before showing you how to work out or straighten that damage. Good metalworking skills and the use of proper sheet metal tools are critical to your success as an auto body technician (Figure 11–1).

Figure 11–1 Many special tools are needed to efficiently straighten damaged sheet metal.

To do quality sheet metal repairs, you must first know how to return the sheet metal to its original shape. You can then use a thin layer of plastic body filler to raise any low spots in the metal. By carefully feather sanding the body filler area level and smooth, the body surface repair area can stand level with the undamaged portion of the body panel. The straightened body panel can next be sprayed with primer-surfacer or sealer and primer to ready the metal panel for several coats of color and clear paint (Figure 11–2).

Metalworking skills are probably the most important craft a collision technician can bring to a shop. They are also probably some of the most neglected skills. An untrained worker can spend more time shaping and sculpting an excessive layer of body filler than would be spent properly working the damaged metal. Not only does this waste valuable shop time and materials, but repair quality also suffers.

The estimate will normally state whether a part should be repaired or replaced. With minor damage, a sheet metal part can usually be straightened. However, if the labor to straighten a part is more than the cost of a new part, then the damaged part is normally replaced (Figure 11–3).

Table 11–1 summarizes the basic steps for correcting minor sheet metal damage. Study it carefully.

Figure 11-2 Damage to sheet metal can be difficult to repair if you are not properly trained. Some panels on the wrecked car would require replacement, but some panels would have to be straightened and repainted.

Figure 11-3 The estimate or work order will normally state whether a part or panel should be straightened or replaced. The doors on this truck would require repair, whereas the fenders and front-end parts would require replacement.

TABLE 11-1 MINOR DAMAGE REPAIR PROCEDURE

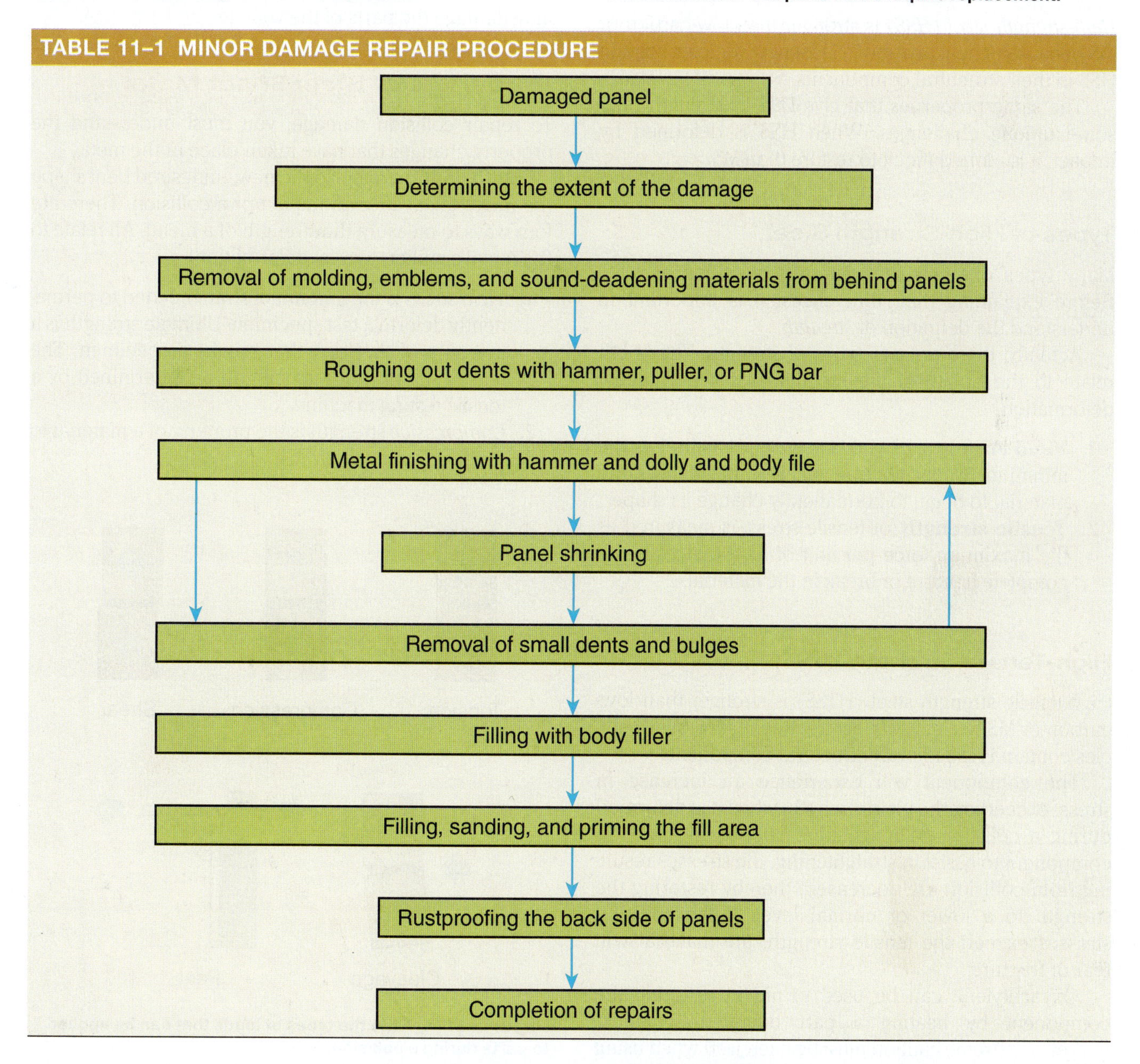

11.1 AUTOMOTIVE SHEET METAL

To do body work, you must have a basic understanding of metals and their properties.

Low-Carbon Steels

Low-carbon steel, or *mild steel (MS)*, has a low level of carbon and is relatively soft and easy to work. The sheet metal panels on older or antique vehicles are sometimes low-carbon. Mild steel can be safely welded, heat shrunk, and cold worked without seriously affecting its strength.

Because MS is easily deformed and relatively heavy, vehicle manufacturers have begun using high-strength steels (HSS) in load-carrying parts of the vehicle.

High-Strength Steels

High-strength steel (HSS) is stronger than low-carbon or MS because of heat treatment. Most new vehicles contain HSS in their structural components.

The same properties that give HSS its strength offer some unique challenges. When HSS is deformed by impact, it is more difficult to restore than MS.

Types of High-Strength Steel

Many types of steels are generally classified as HSS. Before explaining their differences, it is important to understand the definition of *strength*.

Actually, there are two types of strength. They both relate to the ability of the metal to resist permanent deformation.

1. **Yield strength**, or yield stress, is measured as the minimum force per unit of area that causes the material to begin to permanently change its shape.
2. **Tensile strength**, or tensile stress, is measured as the maximum force per unit of area that causes a complete fracture or break in the material.

High-Tensile Strength Steel

High-tensile strength steel (HTSS) is stronger than low-carbon or MS because of heat treatment. Most new vehicles contain HTSS in body structural components.

The component will experience an increase in stress, exceeding the yield strength, when it is deformed during a collision. When heat is applied to an HTSS component to assist in straightening, the stresses resulting from collision are decreased, thereby restoring the strength to a lower or normal level. If the collision stresses exceed the tensile strength, the material will tear or fracture.

Oxyacetylene can be used to aid in restoring the component by heating a part being straightened. However, extreme caution must be exercised when using oxyacetylene. Thermal crayons or paint should be applied around the area to be heated with an oxyacetylene torch to restrict temperatures to 1,200°F (649°C).

Part Loading

Loading refers to the type of force applied to a part to damage it. As shown in Figure 11–4, the five basic types of loading are:

1. *Tension* A load that tries to pull parts straight apart.
2. *Compression* A load that forces the parts straight into each other.
3. *Shear* A load that pulls parts sideways.
4. *Cleavage* A load that tries to force parts apart from an angle.
5. *Peel* A load that pulls parts straight away from each other.

During a collision, one or more of these types of loads may damage the parts of the vehicle.

Properties of Steel Sheet Metal

To repair collision damage, you must understand the property changes that have taken place in the metal.

Deformation refers to the new, undesired bent shape the metal takes after an impact or a collision. There are four ways to measure the strength of a metal. All relate to the metal's ability to resist deformation.

1. *Yield stress* is the amount of strain needed to permanently deform a test specimen. Ultimate strength is a measure of the load that breaks a specimen. The tensile strength of a metal can be determined by a tensile testing machine.
2. *Compressive* strength is the property of a material to resist being crushed.

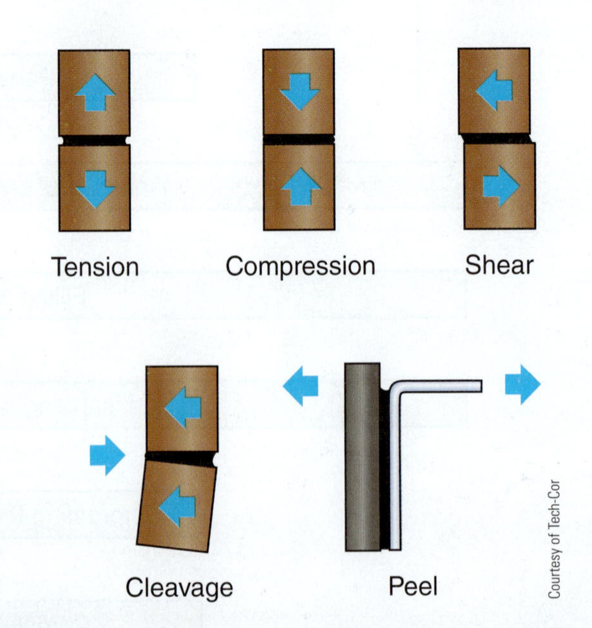

Figure 11–4 Note the types of loads that can be applied to parts during a collision.

3. *Shear strength* is a measure of how well a material can withstand forces acting to cut or slice it apart.
4. *Torsional strength* is the property of a material that withstands a twisting force.

Strength is expressed in pounds per square inch (psi) or kilograms per square millimeter (kPa).

When flat sheet steel is formed into a shape for a part, it takes on certain properties that harden it. For example, a roof panel is relatively flat. If hit lightly in the center, the panel will usually bend and then pop back to its original shape. However, if you hit a panel with a curved shape, the panel will hardly move. Although both are the same steel, the one that has been changed the most will be stronger and more resistant to bending.

The same is true for panels whose shape has been changed during a collision. The structure of the metal in the affected areas has changed, causing the metal to become harder and more resistant to corrective forces.

Effect of Impact Forces

The grain pattern (how atoms line up and link together) of a metal will determine how it reacts to force. Sheet metal's resistance to change has three properties: elastic deformation, plastic deformation, and work hardening. All of these properties are related to the yield point.

Yield point is the amount of force that a piece of metal can resist without tearing or breaking.

Spring-back is the tendency for metal to return to its original shape after deformation. It will occur in any area that is still relatively smooth. Many such areas will spring back to shape if they are released by relieving the distortion in the buckled areas.

Elastic Deformation

Elastic deformation (Figure 11–5) is the ability of metal to stretch and return to its original shape. For example, take a piece of sheet metal and gently bend it to form a slight arc. When released, it will spring back to its original shape. This spring-back tendency makes it necessary for you to recognize elastic deformation in damaged panels. This will enable you to plan your work to take advantage of any tendency of the damaged metal to spring back.

Plastic Deformation

Plasticity is the ability of metal to be bent or formed into different shapes. Plasticity is important to the collision repair technician because both stretching and permanent deformation take place in various areas of most damaged panels. When metal is bent beyond its elastic limit, it will have a tendency to spring back. However, it will *not* spring all the way back to its original shape. This is because the grain structure has been changed (Figure 11–6). *Plastic deformation* occurs when the metal has permanently changed shape.

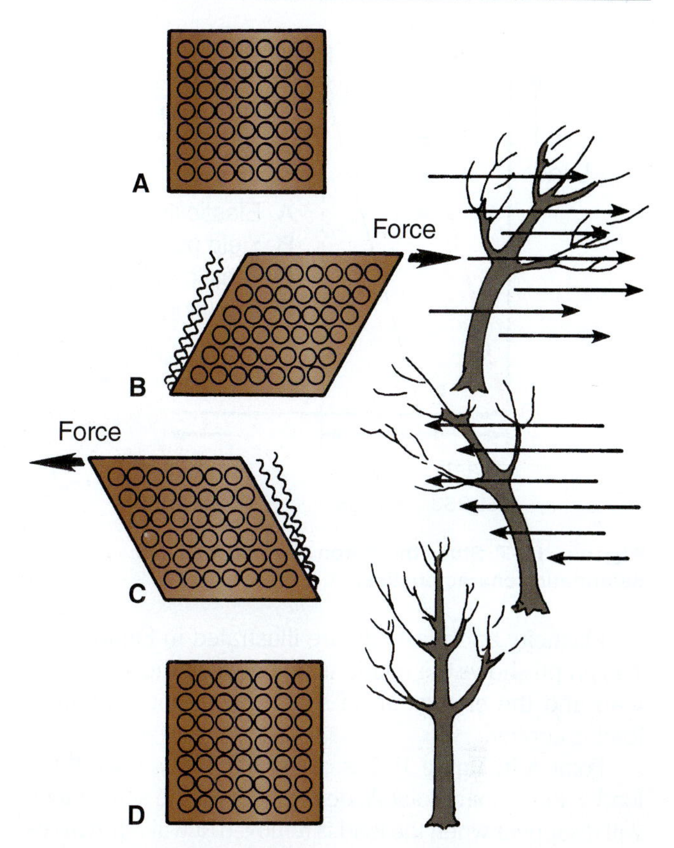

Figure 11–5 Elastic deformation. (A) Metal at rest. (B) Metal under pressure bends like a tree in the wind. (C) When the force is released, metal rebounds. (D) Finally, metal returns to its original shape if not overly bent, just like the tree.

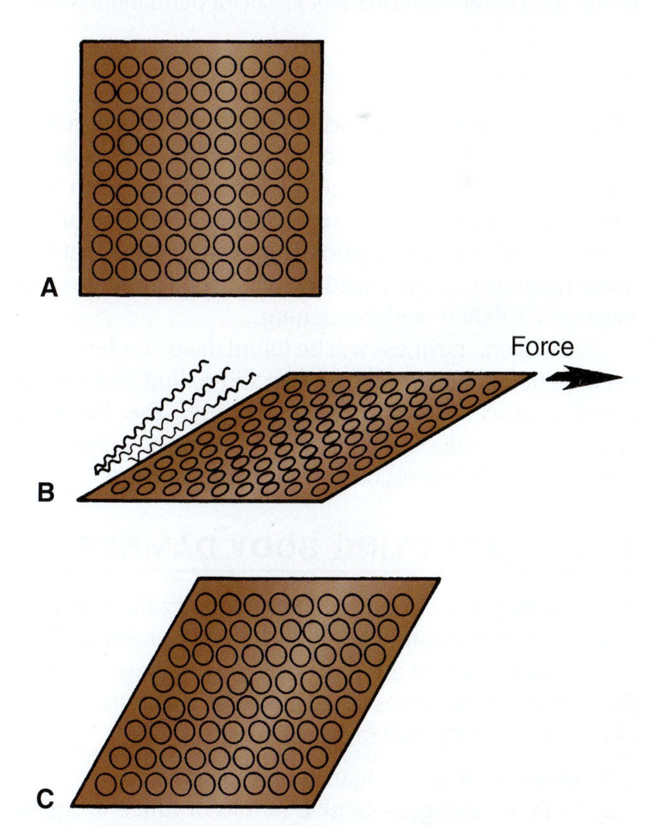

Figure 11–6 Plastic deformation occurs when the grain structure of metal is forced beyond its elastic limit and takes on a new set. (A) Metal at rest. (B) Metal bent beyond its elastic limit. (C) Metal takes on a new shape.

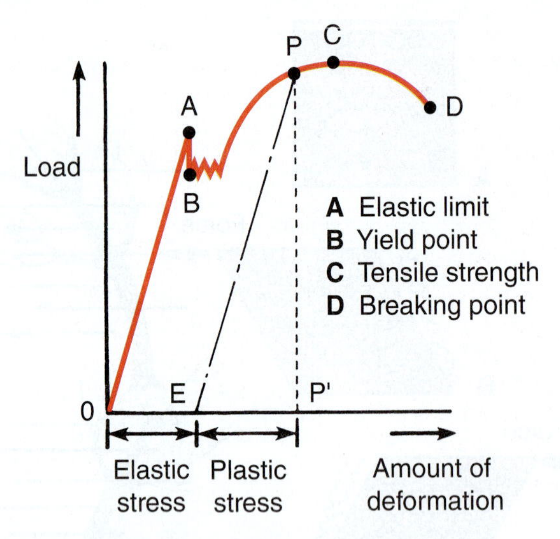

Figure 11-7 Study this graph showing load and deformation characteristics.

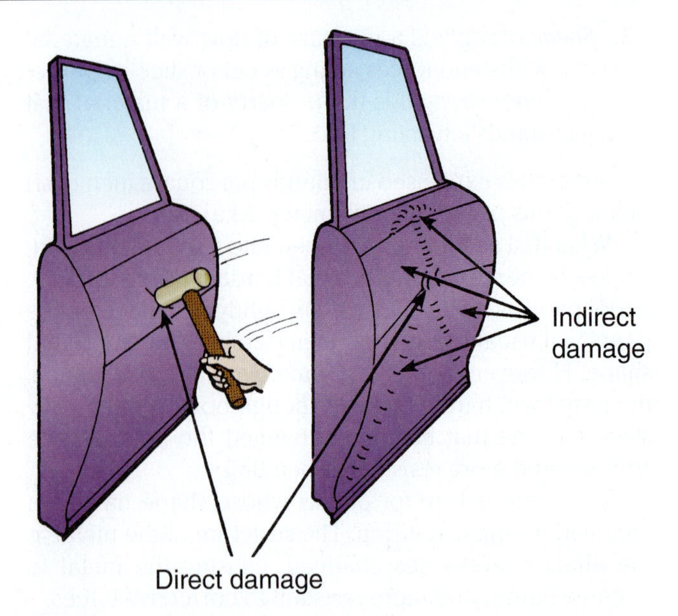

Figure 11-8 Direct damage is a result of impact by another object. Indirect damage results from movement of metal, which affects other areas.

Elasticity and plasticity are illustrated in Figure 11–7. The graph shows the relationship between the size of the load and the elongation of sheet metal when a tensile load is applied.

Point A in Figure 11–7 is called the elastic limit. If the load is lower than Point A, deformation of the sheet metal will disappear when the load is removed. It will return to its original shape. This is called elastic stress. If the load exceeds Point A, even if the load is removed, the deformation will remain. The panel will not return to its original shape. This is called permanent plastic or permanent stress.

Work Hardening

Work hardening occurs when plastic deformation has caused the metal to become very hard in the bent area. For example, if a welding rod is bent back and forth several times, a fold or buckle will appear at the point of the bend. The plastic deformation has been so great that the metal became hard and brittle where bent. This increased hardness is called work hardening.

Some work hardness will be found in any undamaged body panel. It is the result of the cold-rolling and panel-forming process during manufacturing. The bending caused by a collision adds still more work hardening where the body panels have been badly damaged.

11.2 CLASSIFYING BODY DAMAGE

The first step in auto body repair is analyzing the damaged area. A number of conditions that the body technician must recognize are present in any damaged panel. Each of the three items listed below is a condition that occurs when metal is damaged by impact.

1. Direct damage—a tear, gouge, or scratch
2. Indirect damage—buckle (a fold or hinge in metal due to damage or tension) or pressures (unwanted force due to impact damage)
3. Work hardening—normal and impact created

Direct Damage

Direct damage is simple, visible damage that is easy to find, such as a gouge, a tear, or a scratch. It is the damaged portion of the panel that came in direct contact with the object that caused the impact (Figure 11–8).

Direct damage is usually about 20 percent of the total damage. Direct damage repair at the point of impact is limited. Metal used in today's cars is often too thin to be reworked. Straightening is time-consuming and usually not practical on areas containing direct damage. Direct damage usually requires some body filler or, on rare occasions, lead, after all indirect damage has been handled. Direct damage varies from job to job.

Indirect Damage

Indirect damage is caused by the shock of collision forces travelling through the body and inertial forces acting on the rest of the unibody. Indirect damage can be more difficult to completely identify and analyze. It may be found anywhere on the vehicle. Indirect damage represents, on average, 10 to 20 percent of the overall damage.

11.3 ANALYZING SHEET METAL DAMAGE

In the fender illustrated in Figure 11–9, there are "soft" areas (unshaded) and "hard" areas (shaded). The shaded areas (crowns and ridges) are harder to damage. However, when damaged, they are more difficult to straighten.

Work hardening is in all sheet metal panels of a car to varying degrees. It is important to know where the metal was hardest and softest before it was damaged.

To demonstrate how work hardening affects the repair process, imagine a piece of steel about 12 inches (25 mm)

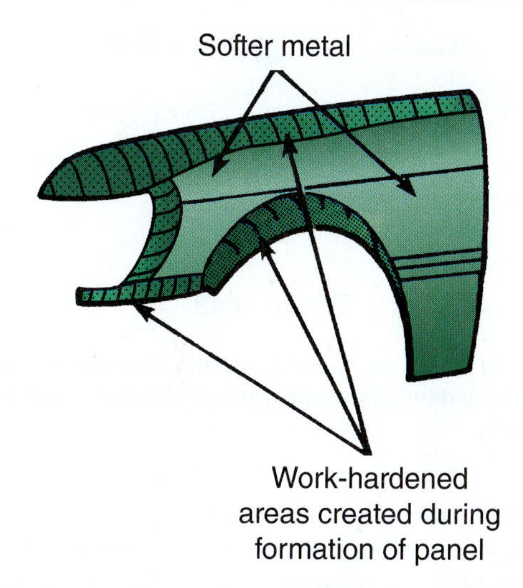

Figure 11-9 **Factory-formed, work-hardened areas are often built into parts.**

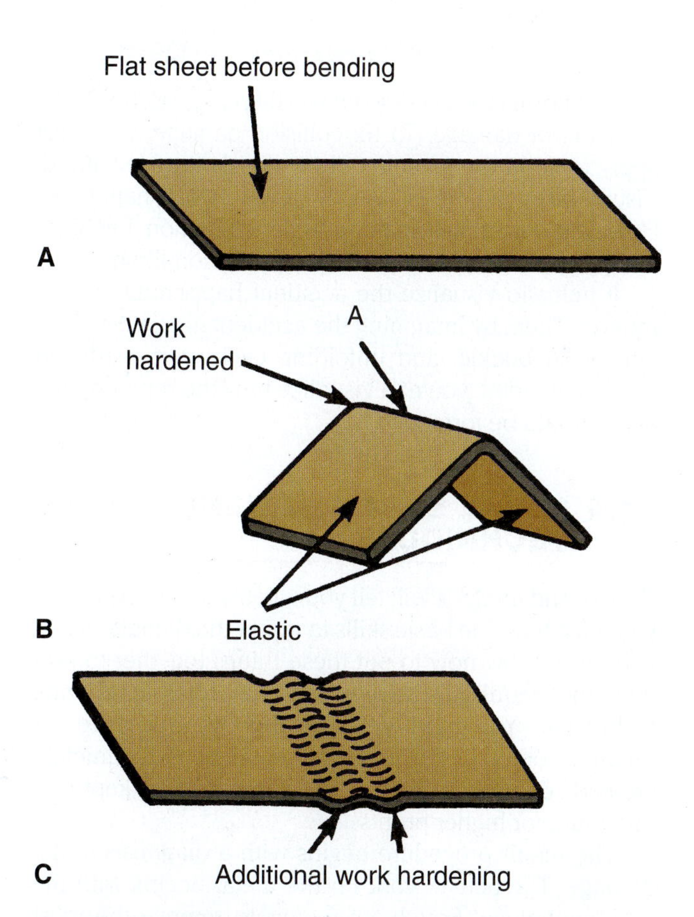

Figure 11-10 **Additional work hardening is created by trying to bend work-hardened metal back to shape. (A) An undamaged piece of sheet metal. (B) The sheet metal has been bent severely, exceeding its elastic limit. (C) If you try to bend metal straight, the work-hardened areas in the bend will not straighten, whereas unhardened areas on each side of the bend will bend. The result is more damage to the panel.**

long and 6 inches (12.5 mm) wide (Figure 11–10). One can bend this metal strip slightly and it will return to its original shape. If bent past a certain limit (elastic limit), the metal takes a set called a buckle. The metal surrounding the new bend returns to a straight condition, but at the point of the bend work hardening has set in. If an attempt is made to bend the metal back to its original shape, two additional buckles (work hardened) are created adjacent to the original bend because the bend will not open up. It is too hard.

Buckles caused by impact create additional work hardening in an automotive sheet metal panel, but bent metal is not necessarily buckled metal. When metal is bent but returns to its original shape later, it is not buckled metal. It is important to recognize the difference between bent and buckled areas as they play an important role in determining a sound repair procedure.

Buckles

As mentioned, **buckles** are a result of bending metal past its elastic limit. Bent beyond this limit, metal will not return to its original shape. New work hardening has occurred and a new shape is formed.

Simple Hinge Buckles

The simple *hinge buckle* bends like a hinge equally along its entire length. The buckle usually causes little stretching or shrinking. If straightened incorrectly, however, it will cause considerable trouble. When severe, it should be "pulled" out rather than pushed out.

Any metal that is bent to form an angle is considered a box section. In today's unibody cars there are quite a number of complete boxed sections. Boxed structural rails, rocker panels, windshield pillars, center pillars, and roof rails are just a few. Some boxed sections, such as the door assembly, are quite large.

Late model cars have a great number of ridges and flanges in them. These are all areas where work hardening is built in, and they are considered "partial boxed sections." Entire fenders can be thought of as partial boxed sections. Just as the complete boxed section in Figure 11–11 collapsed, partial boxed sections can also collapse. Improper straightening produces the same results as with the complete box—overall shortage in dimensions.

Pressure Forces

The terms *pressure* and *tension* are commonly used to describe the conditions in metal after damage. These conditions are also referred to as high spots and low spots.

A high spot or *sheet metal bump* is an area that sticks up higher than the surrounding surface. A low spot or *sheet metal dent* is just the opposite: It is recessed below the surrounding surface. Minor low and high spots in sheet metal can often be fixed with metalworking tools.

The term *sheet metal raising* means to work a dent outward or away from the body. The term *sheet metal lowering* means to work a high spot or bump down or into the body.

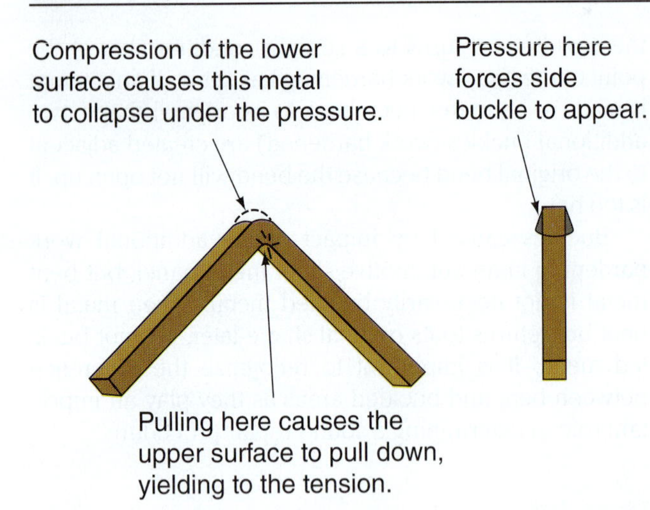
Compression of the lower surface causes this metal to collapse under the pressure.
Pressure here forces side buckle to appear.
Pulling here causes the upper surface to pull down, yielding to the tension.

Figure 11–11 Study this collapsed box section.

There are pressures and tensions (stresses and strains) within the metal before any damage occurs. Metal that is forced up has a new pressure applied to it. The work-hardened buckles are holding the pressure there. If they were to suddenly disappear, the metal would return to its original shape. Consider the metal as being without pressure or tension before it was damaged when evaluating the changes that have taken place and the corrections that must be made.

The metal that is pushed up is called a *pressure area*. Areas that are pushed down are called *tension areas*. It is important to remember that the repair procedure and the application of tools are determined by whether the area is under tension or pressure. A hammer should never strike in a tension area; the dolly should never strike the underside of a pressure area. Power hook-ups are determined by the direction of pressure forces. No body filling can be done when there are pressure areas present. This is because the filler could pop off from movement of the pressure area.

Single Crown Panels

A single crown panel is flat in one direction (left to right) and crowned in another (90 degrees or crosswise). The damage has tension in one direction and pressure in another.

Door Dings

"Parking lot" dings are a good example of pressure and tension areas (Figure 11–12). The impact creates a shallow tension area surrounded by a ridge or pressure area. Sometimes you can correct small dings by using a pick from inside the door. Carefully position the pick and push while watching where pressure is being applied. When you are on the center of the ding, use enough pushing force to move the ding out level with the panel. You may not even have to repaint the door.

For deeper dings with more serious tension areas, you may have to use a hammer and dolly to work the ding.

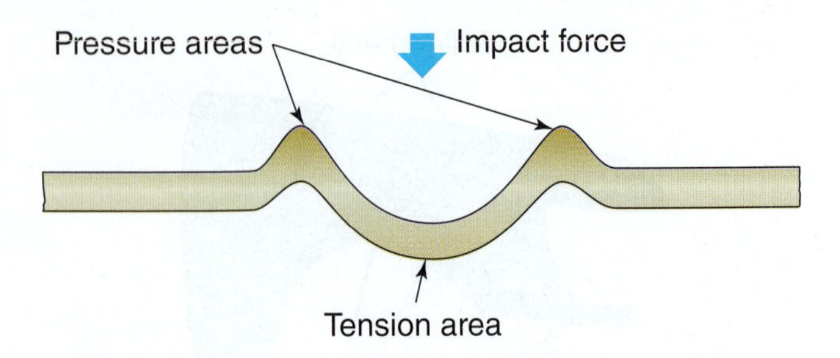
Pressure areas
Impact force
Tension area

Figure 11–12 Exaggerated drawing of a simple door ding, or small dent, shows pressure and tension areas in the damage.

The ridge around the ding must be tapped down level with the panel and the low area must be pushed out with a dolly that has the correct contour. Then the area might need to be leveled with body filler and refinished.

Determining the Direction of Damage

All the information given so far will help you determine the direction of damage. To fix collision damage, you must apply force in the reverse of how the damage occurred. Visual inspection can usually reveal what happened. However, sometimes making a determination becomes complicated when there are overlapping conditions.

It helps to visualize the accident happening in slow motion. Then, by imagining the accident in reverse, studying each buckle, and unfolding each work-hardened section of metal, you can visualize how the repair operations should be formed.

11.4 METAL STRAIGHTENING TECHNIQUES

Theory and analysis will tell you what is wrong. After that you must have the basic skills to repair the damage. Then you must know how to put these things together to produce the overall results required of a professional body technician. You must develop a good procedure for repair. Good procedure saves a great deal of "technician-created" damage so that overall repair time is kept to a minimum for higher profits.

The repair procedure begins with a diagnosis of the damage. The actual work on the metal begins with the rough-out stage. Rough-out means to remove the most obvious damage so that it has almost its original part or panel shape. It must be done properly if finishing operations are to succeed. When finishing operations are started too soon, it becomes difficult to do a good job.

Roughing out the damage can be as simple as using a rubber or plastic hammer on the edge of a door. Carefully placed hammer blows on the back of the panel may be all that is needed to straighten minor damage. When using hammer blows, always grasp the end of the handle and make sure the hammer head strikes the metal squarely (Figure 11–13).

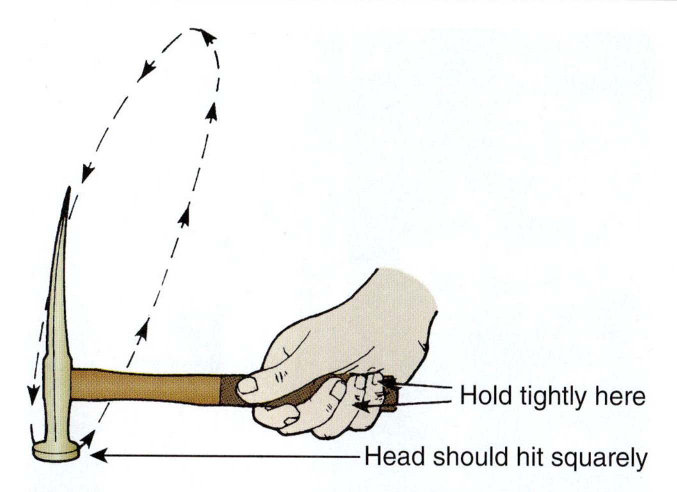

Figure 11–13 Hold a hammer with your third and fourth fingers and swing it with a circular motion to hit the damaged area squarely.

Figure 11–14 Here a body technician is using a body hammer to lower the sheet metal area where the quarter panel meets the roof panel.

The rough-out operations change with each damage, with each vehicle, and with each location of the damage on the car. In other words, the rough-out is very important to the particular vehicle being worked on.

The rest of the chapter is devoted to explaining some of the common skills used by body technicians from the rough-out stage of repair up to the body filling stage.

The buckles and creases in a dented panel can be unlocked in a variety of ways. On panels where the back side of the panel is accessible, hammers and dollies or spoons are used for the initial roughing out. On areas where the back side of the panel is difficult to reach, slide hammers, picks, and welded studs can be used to reverse the damage.

Always remember the rule: "First damage in, last damage out," or "Work indirect damage first, work direct damage last." This means that you must repair the damage away from the point of impact before finally removing the worst damage at the point of impact. If you are using a hammer and dolly to work a small dent, start working around the perimeter of the dent and gradually work your way in to the deepest part to remove it last. If you work backward, trying to hammer out the deepest point of the dent first, you will not be able to work the dent out as smoothly.

Using Body Hammers

The **body hammer** is designed to strike sheet metal and rebound off its surface as a means of straightening minor bumps and dents. It is not designed to be driven down to dent the metal. A too forceful, driving action would create additional damage (small hammer head dents) in the sheet metal (Figure 11–14).

The secret of metal straightening is to hit the right spot, at the right time in the repair sequence, with the right hammer, and using the correct amount of hammering force. When using a body hammer, swing the hammer in a circular motion at your wrist. Do not swing the hammer with your whole arm and shoulder. Hit the part squarely and let the hammer rebound off the metal. Space each blow ⅜–½ inch (9.5–13 mm) apart until the damaged metal is level.

The face of the hammer must fit the contour of the panel. Use a flat face on flat or low-crown panels. Use a convex-shaped or high-crown face when bumping inside curves. Use repeated light hammer taps or blows to progressively work the metal body panel back to its original, undamaged contour (Figure 11–15).

Heavy body hammers should be used for roughing out the damage quickly. *Finishing hammers*, or dinging hammers, should be used for final shaping. The secret to finish hammering is light, rapid taps. It is also important to hit squarely. Hitting with the edge of the hammerhead will put additional "half moon" dents in the metal.

When you do not want to chip paint on a panel, a soft-faced plastic or rubber-faced hammer head could be used as shown in Figure 11–15.

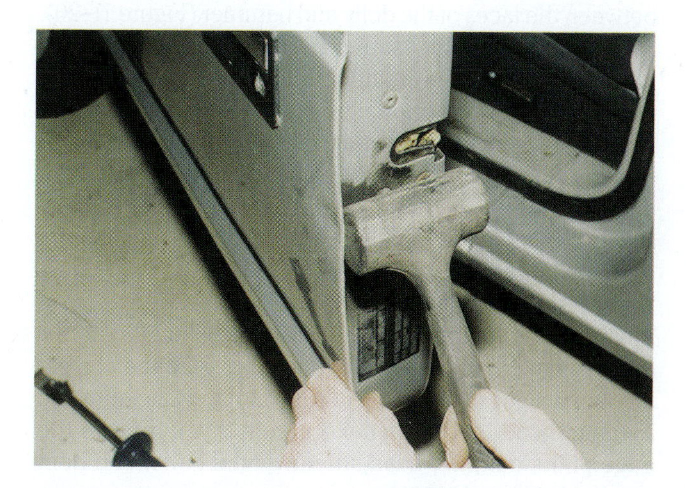

Figure 11–15 Select a body tool that will work best to remove damage. To remove a small dent from the rear edge of this door, the technician is using a plastic mallet.

Bumping Dents with Dollies

A *dolly* is a heavy steel block with various shapes on each side for straightening sheet metal. In the rough-out phase, a heavy steel dolly block is sometimes used as an impact tool. A dolly is often used as a striking tool on the back of panels. Sometimes you can reach into obstructed areas with a steel dolly more easily than you can with a hammer. You can strike the back side of a dented panel with the dolly to raise low areas and to unroll buckles (Figure 11–16).

The contour of the dolly must fit the contour of the back side of the damaged area (Figure 11–17). This will make the blows from the dolly force the metal back into the original contour. If the wrong surface hits the panel (the sharp edge of the dolly, for example), you will further damage the panel (Figure 11–18).

Use accurate hammer blows. Start out with light blows from the dolly while watching the front of the panel. Make sure you are hitting exactly where needed. Gradually increase the force of your blows to raise the damage. It is normally better to use several moderate blows than to use a few hard blows. Numerous well-placed blows with the dolly will let you better control how you work the metal back into shape.

As you hit the panel, the dolly tends to rebound slightly. This creates a secondary lifting action on the metal. You can increase rebound blows by releasing pressure as soon as the dolly hits the panel. Using a large dolly will also increase impact and rebound forces on the panel (Figure 11–19).

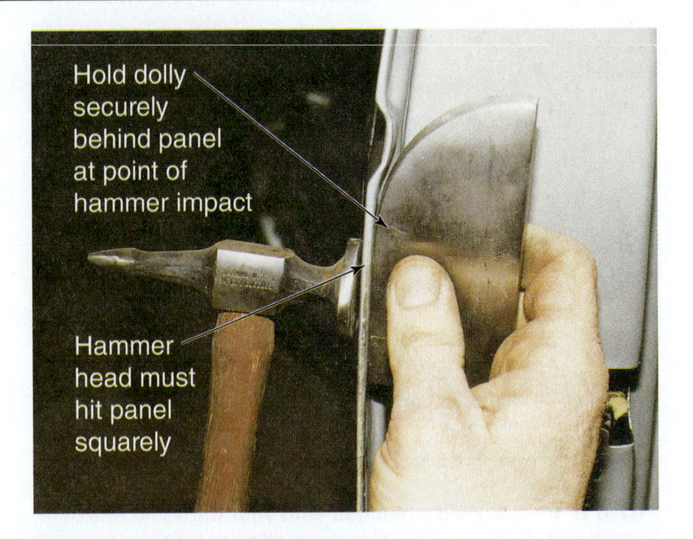

Figure 11–16 A hammer and dolly will pinch or crush the damaged sheet metal between tools and force it into the shape of the dolly. Here the flat side of the dolly is being used to flatten and straighten damage on a door.

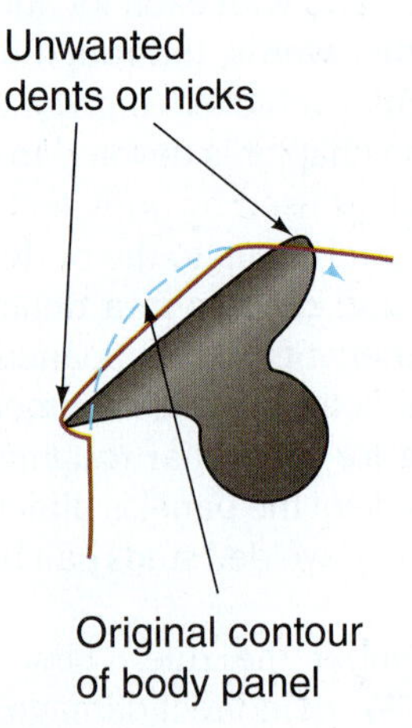

Figure 11–17 Using a dolly whose contour does not fit the contour of the panel will result in additional damage to the panel.

Hammer-On-Dolly Method

Hammer-on-dolly is a method used to exert a smoothing force to a small area on a damaged panel. The dolly is held against the back of the damage and the hammer hits the metal right over the top of the dolly. This exerts a pinching force on the metal between the dolly and hammer head. A small area of damaged metal is crushed and flattened between the faces of the dolly and hammer (Figure 11–20).

Hammer-on-dolly straightening requires you to repeatedly move the point of hammer impact and the dolly slightly. Each blow should overlap the next. By repeatedly moving hammer-on-dolly blows, you can steadily smooth and level the panel damage. Try to work out the damage methodically. Generally start at the outside of a dent and gradually work toward the center of the damage (Figure 11–21).

When learning hammer-on-dolly straightening, you might want to practice on an old scrap of metal or a discarded panel. Practice making light blows to the correct locations right over the top of the dolly. Make sure the hammer head hits the panel squarely. If you hit with the edge of the hammer head, an unwanted half-moon dent or "ding" will be formed.

A proper hammer-on-dolly blow will make a high-pitched "ping" sound. The force of the blow goes into the

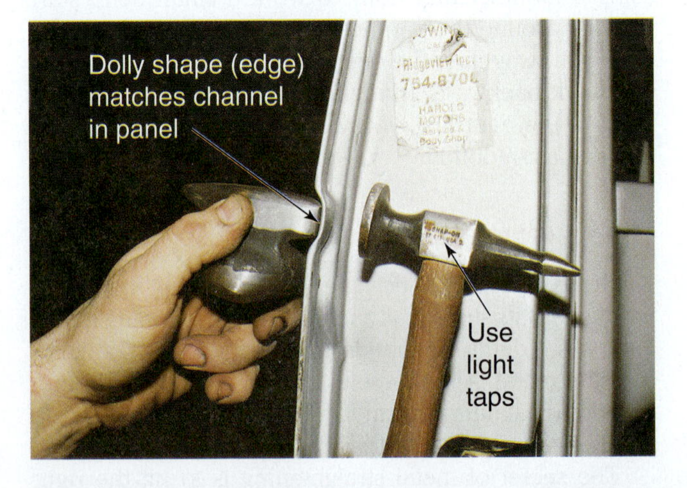

Figure 11–18 Notice how the technician has selected a surface on the steel dolly block that matches the shape of the door panel. The groove or body line formed in the panel is almost the same shape as the dolly surface. Hammer blows can be used to restore the panel's original contour.

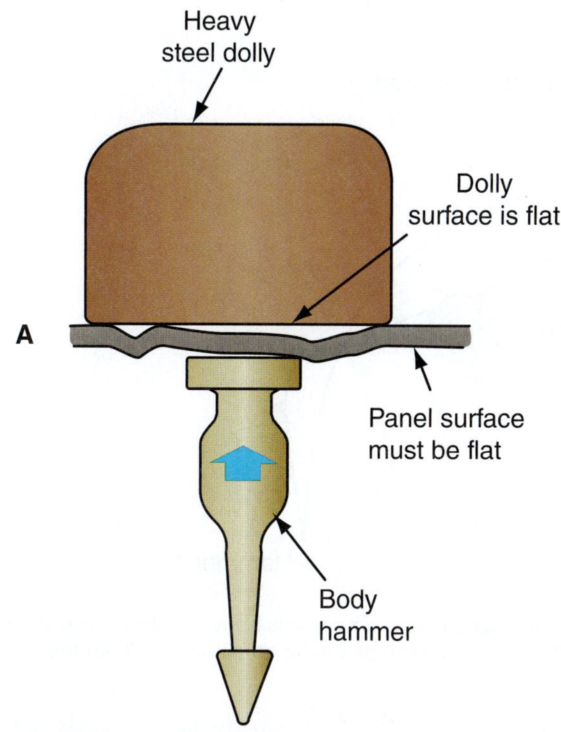

A **Press the dolly against the back of the damage. Use light hammer blows from the front to remove damage from sheet metal.**

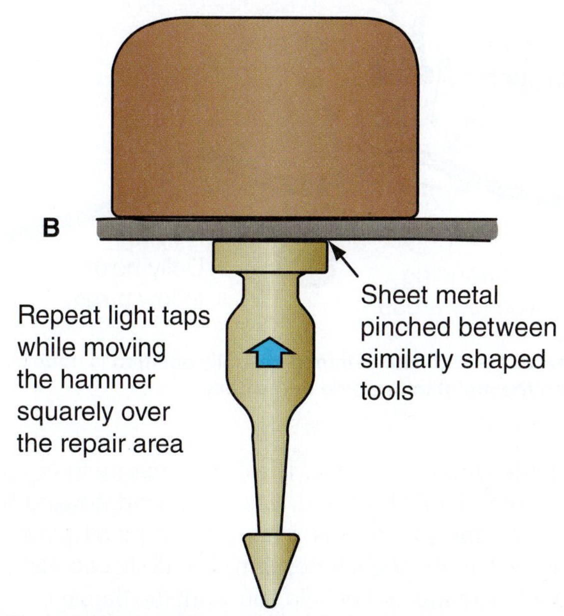

B **Keep striking areas of damage until the sheet metal is almost perfectly straight.**

Figure 11-19 **Study the basic methods for using a hammer and dolly.**

panel and then into the heavy steel block. Hitting the heavy dolly block makes the pinging sound. If you accidentally miss the spot backed by the dolly, a more dull or dead sound is produced as only the metal is hit. If you miss the dolly with a hammer tap, a small unwanted dent is often produced in the panel.

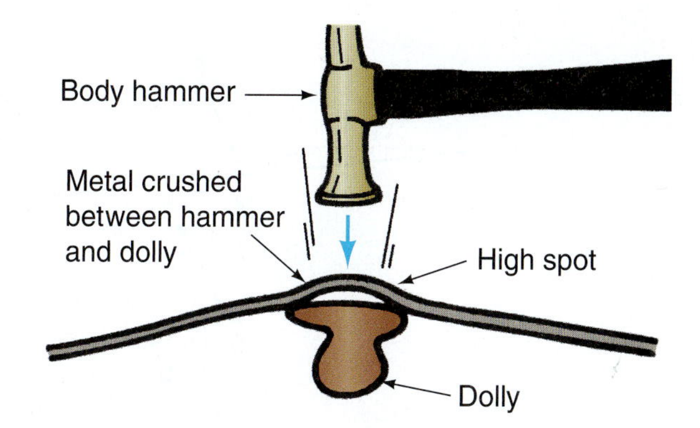

Figure 11-20 **Hammer-on-dolly repairing is done by hitting the panel right over the dolly.**

With the hammer-on-dolly method, the shapes of the dolly and hammer head must match the desired shape of the panel. If the area to be straightened is flat, the dolly surface and hammer head must be flat. If the panel is curved, the dolly and hammer head must also be curved to match the panel's shape. When you bump or hit the damage with the hammer, the metal is flattened against the dolly and a tiny area is formed into the shape of the hammer face and dolly face.

Always start out with light hammer blows. A common mistake is to use excessively hard or poorly aimed hammer blows that dent, stretch, and damage the panel. By starting light and working up to stronger blows, you can better control the movement of the metal to avoid unwanted dents. Carefully observe the results of each blow to make sure you slowly reshape the metal as desired.

Hold the dolly securely against the back of the panel. Hit the area lightly so that the hammer bounces back. Light hammer-on-dolly blows are used to smooth small, shallow dents and bulges. Hard hammer-on-dolly blows can be used to stretch the metal.

To lower a bulge, place the dolly against the back side of the panel directly behind the bulge and use a hammer from the front side. There will be a slight rebound as your hammer hits the dolly. The dolly will then hit the back side of the panel. As the force of the dolly pressing against the panel is increased, the flattening action will also increase.

With hard blows using hammer-on-dolly, the metal is smashed between the hammer and dolly. This tends to crush the metal thinner and make it stretch out to fill a slightly larger surface area. All blows that are designed to stretch should be hard and accurate. Remember that an inaccurately placed hard blow can damage the panel.

Keep in mind that light hammer blows are for straightening, not stretching. In other words, when using the hammer-on-dolly technique for stretching, hit hard and do not miss!

Hammer-on-dolly is used only if there is access to the back side of the panel. If there is not, dent pullers and filling are used. The hammer-on-dolly method can also be used to stretch metal.

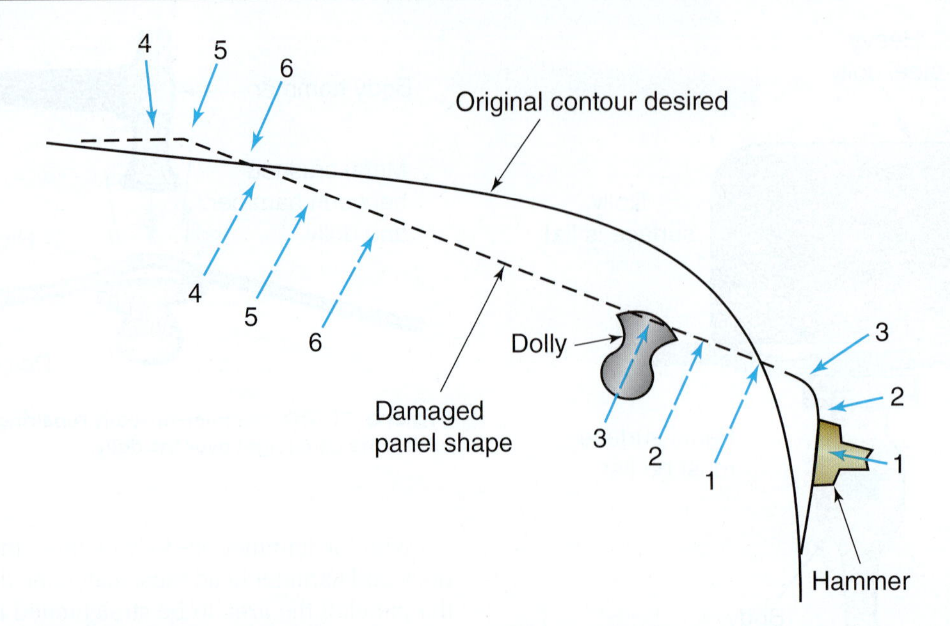

Figure 11–21 Note how the hammer and dolly can be used in a specific sequence to remove this body damage. First, start on the end of the damage at the right, with hammer-off-dolly blows. Then work the damage on the other side. Roll out the damage toward the middle.

REMEMBER *You can control the effects of hammer-on-dolly straightening by:*

- *Changing the shape of the hammer face*
- *Using a different-shaped dolly face*
- *Altering how hard you hit the metal*
- *Increasing the force of the dolly against the back of the panel*

Hammer-Off-Dolly Method

The **hammer-off-dolly** method is used to raise low spots and lower high spots simultaneously. The hammer hits the panel slightly to one side of where the dolly is being held. It is often used to rough out or shape large areas of damage during initial straightening. In this procedure, you hold the dolly under the lowest area on the back of the panel, then hit any high area right next to the dolly with your hammer. Hammer off to one side of the dolly, not directly on top of the dolly (Figure 11–22).

Generally, use the hammer and dolly to roll out the damage in the reverse order from which it was formed. Normally, the damage must be rolled out working toward the center. Start at the outer perimeter of the damage and work to the middle of the damage.

If a panel has a large buckle, you can use the hammer-off-dolly method. Place the dolly on the low spot at the back of the panel, then hit a high spot with your hammer. This will lower the high spot and raise the low spot without stretching the metal. The hammer blow will push the high spot down and the rebound of the dolly will force the low spot up.

If the panel has a raised ridge of damage, you can also use the dolly-off method. Use a flat-faced dinging hammer to direct light to medium blows at the outer ends of the ridge. The blows from the hammer gradually force down the ends of the ridge. The dolly pressure forces the end of the channel upward. Gradually work toward the center. As the pressure is released, the metal tends to move back to its original position. The dolly can also be used as a driving tool to help you work the damage.

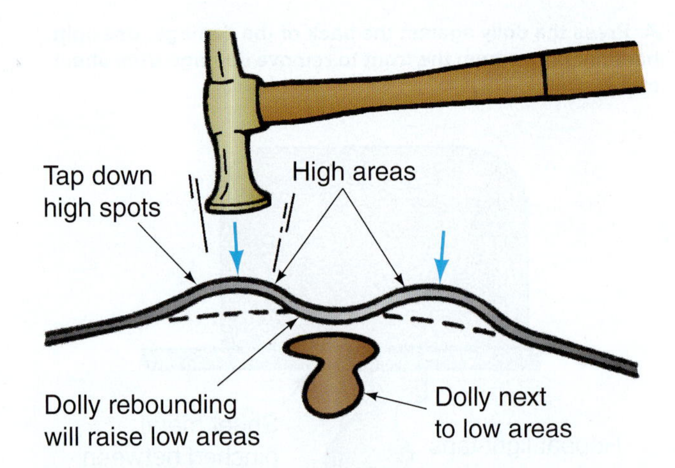

Figure 11–22 The hammer-off-dolly position is done by hitting the metal to one side of the dolly.

Once the area has been brought back to its basic shape, use the hammer-on-dolly method to smooth and level smaller damaged areas. You are then ready for metal finishing or plastic filling procedures.

REMEMBER *You can control hammer-off-dolly straightening by:*

- *Altering how hard you hit the panel. Start out with light blows and then increase their force as needed to lower high spots in the panel.*
- *Changing how hard to push the dolly against the back of the panel. Pushing harder tends to increase lifting action to raise low spots.*

- *Adjusting how far away the dolly is from the hammer blows. Moving the dolly farther away tends to spread out the lowering/raising force to a larger area on the panel. Moving the dolly next to the hammer blow tends to concentrate the effect of the lowering/raising force.*

Picking Dents

There are several methods of picking up metal with the use of a pointed (not necessarily sharp) tool. Picking dents often involves final straightening of very small areas of damage with the pointed end of a body hammer or with a long rod that has a curved, pointed tip (Figure 11–23).

The pick on a body hammer is often used to lower any small, high spots in the repair area. Very light, carefully placed blows with the point of the hammer will lower any dimples still sticking up in the repair area.

Long picking tools can also be used to pry up metal in areas that cannot be reached with a dolly or spoon. A car door is a good example. A pick can sometimes be inserted through a drainage hole or a hole drilled behind the door gasket. This eliminates the need to remove the inside door trim or to drill holes in the outer panel for pulling the dent. Picks are used during paintless dent removal (removing small body dings or dents without painting the panel).

When prying with a pick, be careful not to stretch the metal by exerting too much pressure. Deep creases should be straightened by starting from the shallow area and working toward the deep area. Start with the original point of contact or the lowest point. Slowly pry the crease up. On larger dents, use a flat blade pick rather than a pointed one. Tap down pressure areas while prying up low-tension areas.

Unlocking Dents with a Hammer and Dolly

A minor dent (Figure 11–24A) is often straightened by using hammer and dolly to "roll out" the metal in reverse of the order in which the damage happened.

Figure 11–23 A pointed head or pick on a hammer is often used to lower small high spots during final straightening of a panel. All high spots must be lowered to be even or slightly below the original contour to allow application of body filler.

To remove the dent, roll out the damage from the outside, working toward the center in the reverse order of damage. Hold the dolly tightly under the channel at the outer end where there is the least amount of damage (Figure 11–24B). A flat-faced dinging hammer might be used to direct light to medium blows at the outer ends of the ridge closest to the dolly (off-dolly blows). The blows from the hammer will gradually force down the ends of the ridges. Your hand or arm pressure on the heavy dolly will force the end of the channel upward. The same procedure is then repeated on the other end of the channel and the adjacent ridges (Figure 11–24B).

The off-dolly method is gradually worked toward the center or where the greatest degree of bend exists in the ridges and channels. As the pressure is released in the ridges and channels, the surrounding elastic metal tends to move back to its original position. The dolly can also be used as a driving tool to work the channel upward (Figure 11–24C). However, if the dolly does not move when the channel is hit upward, there is still too much pressure on either or both the ridges and/or channel. More dollying must be done to relieve the tension (Figure 11–24D).

Once the area has been brought back to its basic shape, use the light on-dolly method to smooth and level the area (Figure 11–24E). It is then ready for either the metal finishing or filling procedure.

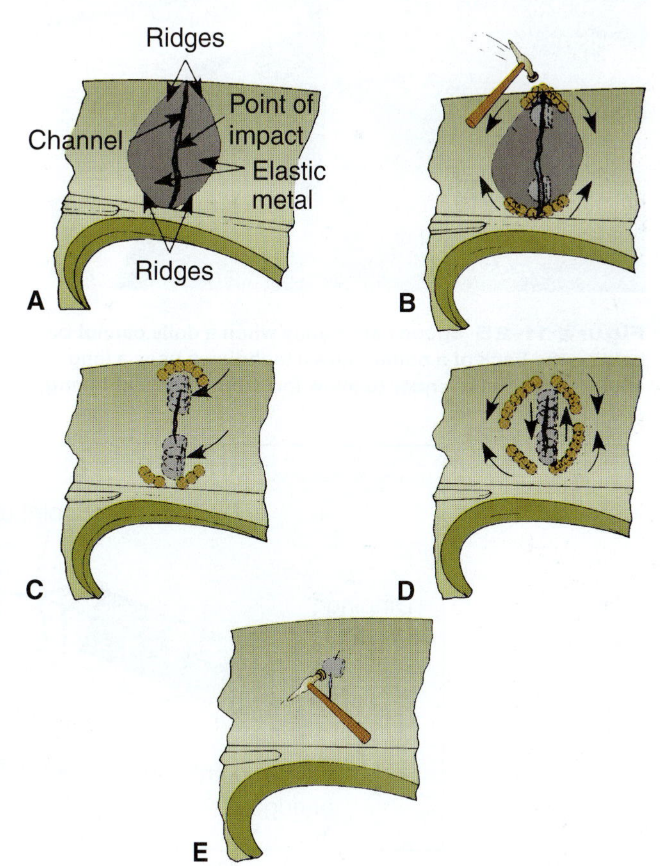

Figure 11–24 Study the steps in repairing a dented panel with a hammer and dolly.

Straightening Dents with Spoons

Spoons can be used in a number of ways to straighten sheet metal. They can be used to pry out dents, and certain kinds can be struck with a hammer to drive out dents. In hard-to-reach areas, a spoon can be used as a dolly (Figure 11–25). Some are even designed to be used in place of a hammer.

Spring hammering is commonly done with a hammer and a dinging spoon. The dinging spoon is lightweight and has a low crown. When used, it is held firmly against the high ridge or crease. The spoon is then struck with a ball peen or bumping hammer. The force of the blow is distributed by the spoon over a large area of the crease or ridge. This reduces the likelihood of stretching the metal.

Always keep firm pressure on the spoon when spring hammering. It must never be allowed to bounce. Part of the corrective force is the pressure of the spoon. Begin at the ends of a ridge (hinge buckle) and work toward the high point on the ridge, alternating from side to side (Figure 11–26).

Slapping spoons are sometimes used instead of body hammers. They can be driven down harder and more often without damaging the panel. They can be used with a dolly (Figure 11–27).

Remember that hammer blows on top of a panel can be a corrective force only when they are placed on pressure (high) areas of the damage. A bumping file can also be used to "slap" down ridges. It has a serrated surface that shrinks the stretched metal.

Spoons can be used to back up the hammer or in combination with a slapping spoon. With a long body spoon, you can often reach into places inaccessible to a hammer or dolly. Pressure can be applied to tension areas with the spoon, while high areas are bumped down (Figure 11–28A).

Spoons can also be used to pry metal up in the rough-out stage or to drive deep dents out. Figure 11–28B shows a spoon being used to pry out a dent in a door panel. The door is supported on blocks of wood to provide clearance for the door panel to move.

Figure 11-25 **Spoons are handy when a dolly cannot be held on the back of a panel. Here a technician uses a long spoon to pry out a fender to allow for hammer straightening.**

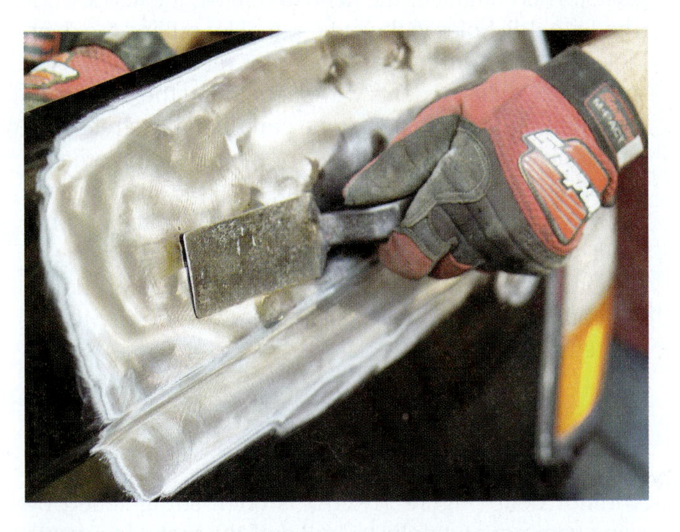

Figure 11-27 **A slapping spoon can be hit on a sheet metal area to produce a driving, straightening force over larger areas. A dolly would normally be held on the back of the area being hit by the slapping spoon.**

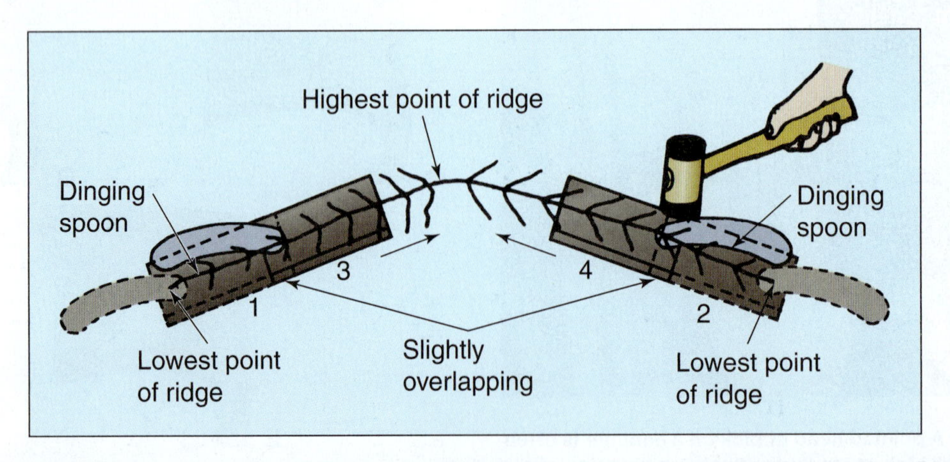

Figure 11-26 **A large dinging spoon can be used to lower a ridge formed in damaged sheet metal. Start at the outer ends of the damage and work your way inward or to the middle of the damage.**

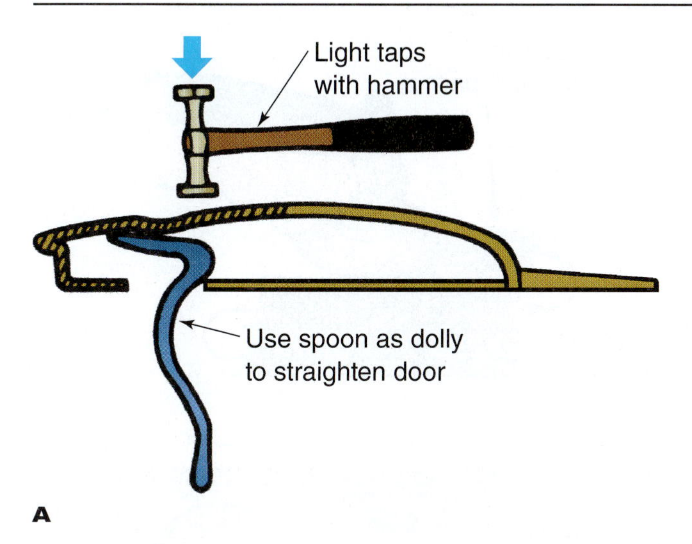

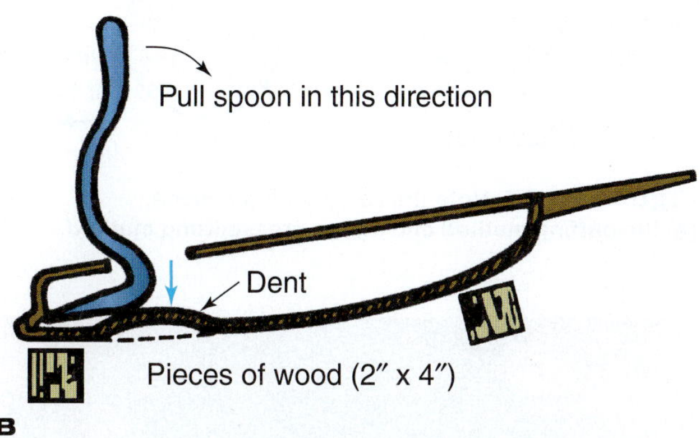

Figure 11-28 (A) Using a spoon as a dolly: (1) Use a hammer to work the metal from the front while (2) holding a spoon at the rear of the panel. (B) Using a spoon to pry out a dent in a door panel.

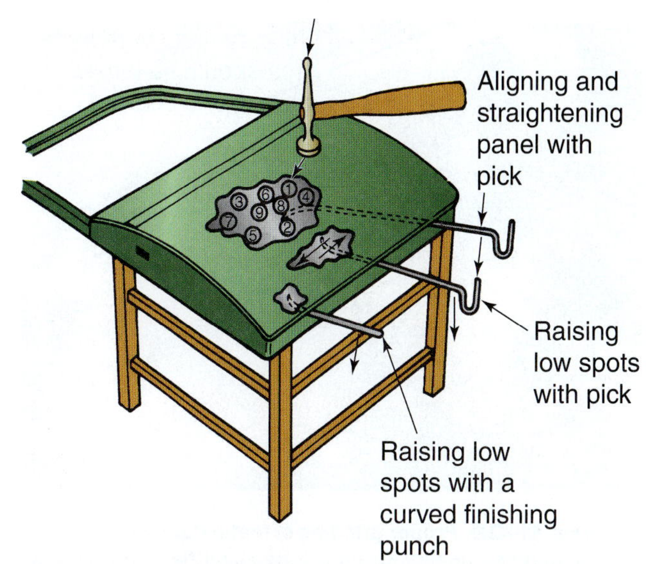

Figure 11-29 Pry picks are sometimes used to remove small dents in hard-to-reach areas. Pry on the back of low spots with the pick while hitting next to the area with a body hammer. Picks will reach through small drain holes in the door or through restricted areas in other panels.

Other Metal Straightening Methods

When a hammer and dolly will not reach or fit behind the damaged body panel, long steel picks can often be used to reach and push out dents. As shown in Figure 11–29, insert the correct length pick through openings on the back of the dented panel. In doors, for example, go through drain holes or the inner door frame to access the back of the damage. Then pry directly on the back of all low spots while hammering on them from the front. Reshape the sheet metal until smooth enough for a thin skim coat of body filler.

Rubber air bladders are designed to push out larger dents from the back of panels (Figure 11–30). The air bladder can be installed inside doors, quarter panels, and other unibody areas to push out large dents (Figure 11–31).

Figure 11-30 Air bladders can be used to push out large dents from the back of some panels.

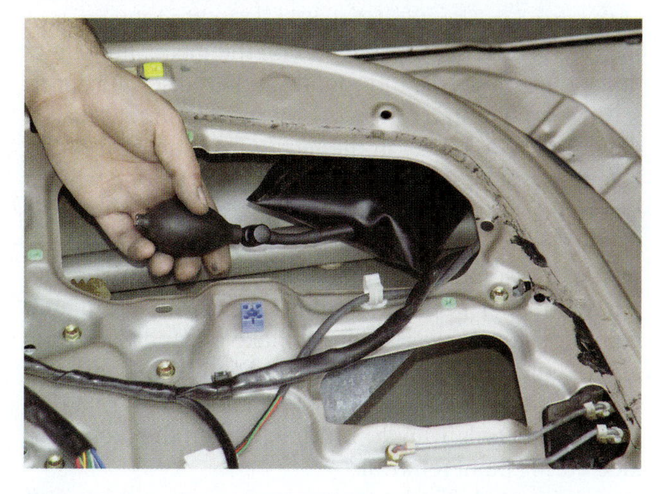

Figure 11-31 Here an air bladder has been placed inside a panel. When shop air is forced into the bladder, the bladder will expand and push out the dent from the rear of the panel.

Paint Removal

Depending on the amount and type of damage, you will often have to sand or grind off the paint when straightening sheet metal damage. Usually this is done with a disc grinder (Figure 11–32).

Figure 11-32 **Proper grinding of metal during repairs is critical to repair quality. Grind the area as little as possible to reach bare metal. An area slightly larger than the repair area should be ground to remove the finish. Body filler should not be applied over paint.**

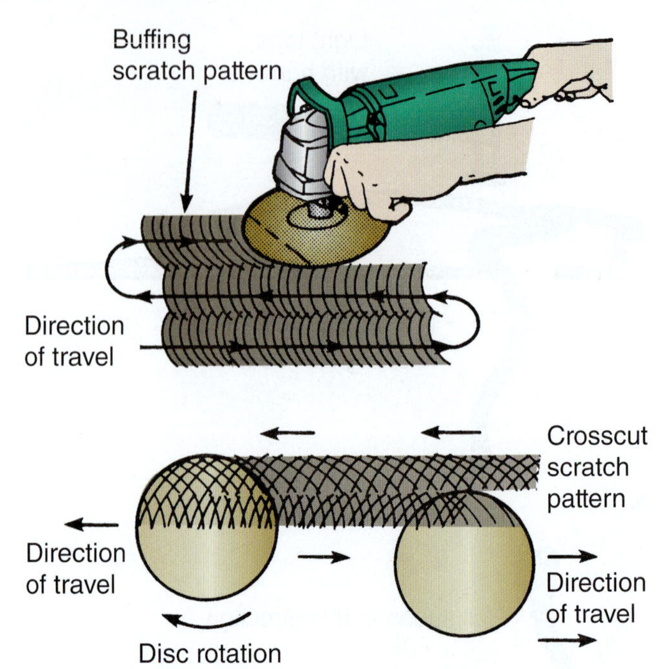

Figure 11-33 **Note the two grinding actions: (A) the buffing method and (B) the crosscutting method.**

On today's vehicles, the thin metal almost always requires the use of a 36 grit disc or a hard synthetic scuff pad. Two types of backing pads are used. A nonflexible pad is used for removing metal, such as weld beads. A softer backup pad should be used when removing paint or polishing metal. The softer pad allows the disc to "roll" and give with the metal.

There are two different ways to use a grinder. Called buffing, the first is used to remove paint and smooth body filler with little overlap of grinding marks. The second is called crosscutting (grinding marks overlap) and is used to remove metal (Figure 11–33).

When using a grinder, only the top 1.5–2 inches (38–51 mm) should contact the surface. Do not use excessive pressure. The weight of the grinder should be just about enough.

On vertical surfaces, use pressure equal to the weight of the grinder. The grinder should be held so the back of the disc is raised 10–20 degrees off the metal. It is sometimes difficult to use the round sanding disc in a sharp reverse crown area. The edge of the disc can cause a deep groove to be cut in the metal. This can be avoided by cutting the edge of the disc into points to form what is commonly called a star disc.

In Figure 11–34, a body technician is using a large-diameter disc orbital sander equipped with very coarse grit sandpaper to strip this hood down to bare metal. Paint was failing, was starting to flake off, and would not featheredge smoothly. It is easier to control the depth of a cut into sheet metal with a soft foam sanding pad or backer. You only want to brush lightly on the metal surface during paint removal.

In Figure 11–35, the technician is using a thick, stiff scuff pad to remove paint with a high-speed single-action grinder. A scuff pad helps protect the body panel because it will not cut into steel as easily as a flexible grinding disc or very coarse sandpaper.

Figure 11-34 **Here the technician is using a large air sander with coarse grit sandpaper to remove paint from a hood. A foam backing pad on the sander will help prevent gouging and damage to sheet metal.**

When grinding in deep body contours, try to hold the abrasive wheel so that it fits down into curved surfaces without gouging the panel. Be extremely careful when grinding on high crowns or ridges in panels; a grinding wheel will concentrate its abrasive force in a tiny area and can easily cut through the metal body panel. When working on crowns or ridges, lift up on the grinder and guide it lightly and quickly over the pointed or raised body surface (Figure 11–36).

Figure 11-35 Here a stiff scuff pad on a high-speed grinder is being used to remove paint from the repair area.

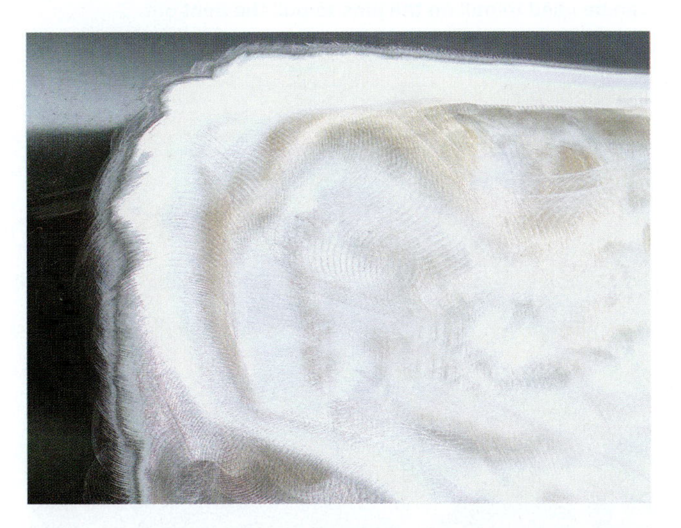

Figure 11-36 When grinding down to bare metal, be extremely careful not to grind too deep into the metal. Late model HSS panels are very thin gauge and can be burned through and damaged if you grind too much in one location.

Figure 11–37 shows a close-up of a body panel with its paint properly ground off for repairs. The sand scratches barely cut through the final coating of OEM primer and into the bare metal.

Pulling Dents

Dents can be pulled out with a number of tools: suction cups, dent pullers, and spot weld dent pullers. The purpose of a dent puller is to lift out simple dents that cannot be reached easily or lifted out by other means.

The dent puller is probably one of the body technician's most frequently used tools. One reason is the growing complexity of automobile body construction and corrosion protection. Access to the inside of many panels is blocked by welded-in inner panels and window mechanisms. Using a dent puller and welded-on nail or suction cup, you can often repair a simple dent in less time than would be required for the disassembly necessary to start the repair from the inside.

Figure 11-37 Here is a close-up of a properly ground surface ready for body filler or primer. A small amount of factory corrosion protection or zinc has been left on the sheet metal.

As discussed, a hook tip on a dent puller or slide hammer is handy for straightening panel edges. Both will reach around the edge of the panel easily.

Flat and round tips are available for dent pullers. Always select a tip that matches the shape of the part being straightened. For example, in Figure 11–38, the technician is straightening the edge of a fender with a curved contour. A round tip has been installed in the dent puller to match the shape of the panel.

A suction cup can be used to pull out large, shallow dents. Wet the area and install the cup. If handheld, pull straight out on the cup's handle. If mounted on a slide hammer, use a quick blow to pop out the dent.

To use a slide hammer or dent puller, thread the appropriate tip on the end of the puller. Hold the handle of the dent puller in one hand and slide the weight straight back against the handle. This will exert a powerful pulling force on the damaged panel.

A vacuum suction cup uses a remote power source (separate vacuum pump or air compressor airflow) to produce negative pressure (a vacuum) in the cup. This increases the pulling power, because the cup is forced tightly against the panel. Larger, deeper dents can be pulled with a vacuum suction cup.

Avoid drilling or punching holes in panels that are going to be reused. Only make holes when the panel is going to be removed and replaced or under unusual circumstances. As you rough straighten or pull badly damaged panels, you may want to weaken a highly stressed area, in a deep crease, for example. The holes will weaken the damaged area so that you can pull the panels back into alignment before removal. Drilling or punching weakens the panel and should be avoided.

When pulling damage and stress relieving on a frame rack, technicians will punch or drill holes in the damaged

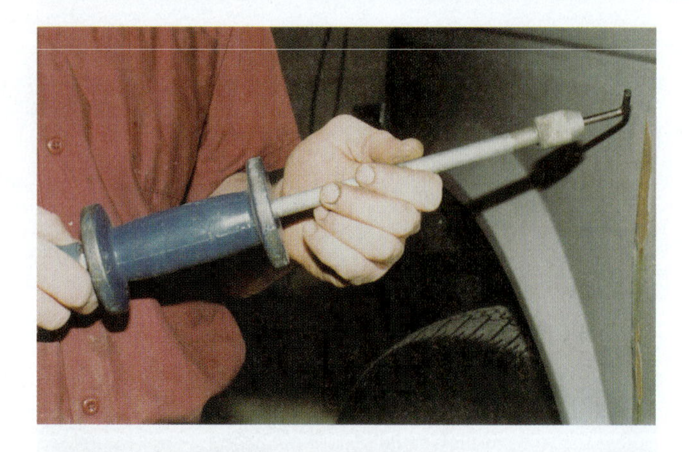

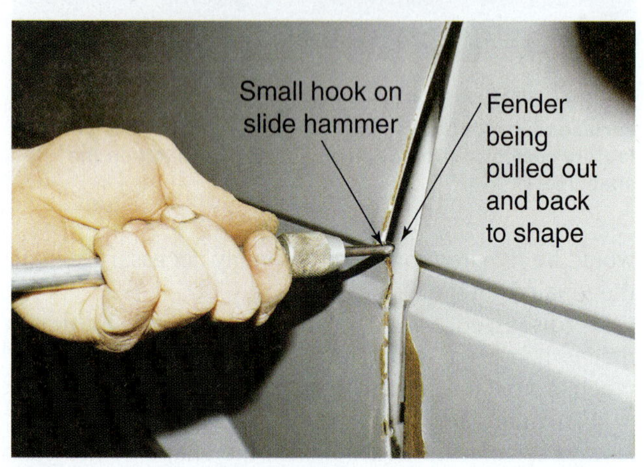

Figure 11–38 Note how the dent puller is being used to pull out minor damage along a lip in the fender. A round hook tip has been installed in the slide hammer. The hook fits into and matches the curve in the panel groove to efficiently pull out damage.

panels to be replaced to weaken them. The old panels are pulled and straightened just to help align adjoining parts. In this case, punching or drilling holes is acceptable.

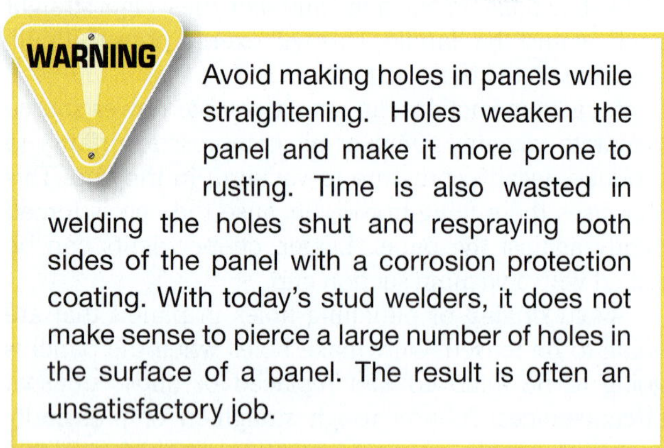

WARNING Avoid making holes in panels while straightening. Holes weaken the panel and make it more prone to rusting. Time is also wasted in welding the holes shut and respraying both sides of the panel with a corrosion protection coating. With today's stud welders, it does not make sense to pierce a large number of holes in the surface of a panel. The result is often an unsatisfactory job.

Spot Weld Dent Pullers

A spot weld dent puller can be used to remove dents in steel panels from the front without drilling holes (Figure 11–39). The heat from the resistance spot welder

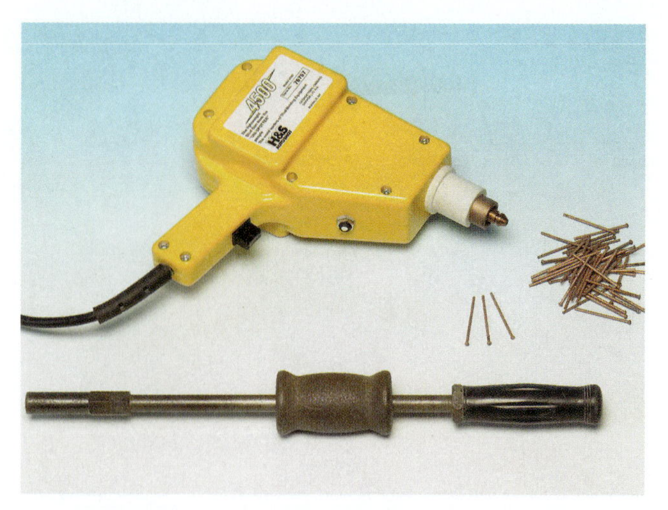

Figure 11–39 A weld-on stud or pin dent puller is commonly used in the industry. A spot welder will weld small pull pins to the dented area. Then a slide hammer can be used to pull on the pins to pull the dent out.

Figure 11–40 Grind the area to be pulled down to bare metal.

fuses a metal pull tip or pin to the steel body. This allows the damage to be pulled out from the front without accessing the back of the panel. Refer to Figure 11–40.

Before the invention of spot weld dent pullers, holes sometimes had to be drilled in the panel. Then screws or hook pull rods were inserted into these holes to pull out the dent. This damaged the panel and its corrosion protection. The holes had to be welded shut before using body filler.

Resistance spot weld dent pullers are a far superior method of straightening damaged sheet metal (steel, but not aluminum panels). There are two basic types of spot weld dent pullers: reusable tip and disposable tip types.

To use a spot weld dent puller, attach the ground cable near the area to be repaired. Grind a bare metal spot next to the dent, if needed (Figure 11–41). If a magnetic ground cable is used, just place it over the panel in contact with bare metal. If a clip-type ground cable is

Figure 11-41 **Attach the welder ground cable to the ground area of the bare metal. Install the pin in the welder head. Press the pin against the dented area and pull the welder trigger. This will weld a small pull pin to the damaged area.**

Figure 11-42 **If the dent is very small, you may be able to pop it out by pulling on the welded pin with locking pliers.**

Figure 11-43 **If the dent is larger, attach a slide hammer to the pulling pins. By sliding the hammer on the tool outward, a strong driving or pulling force is applied to each pulling pin.**

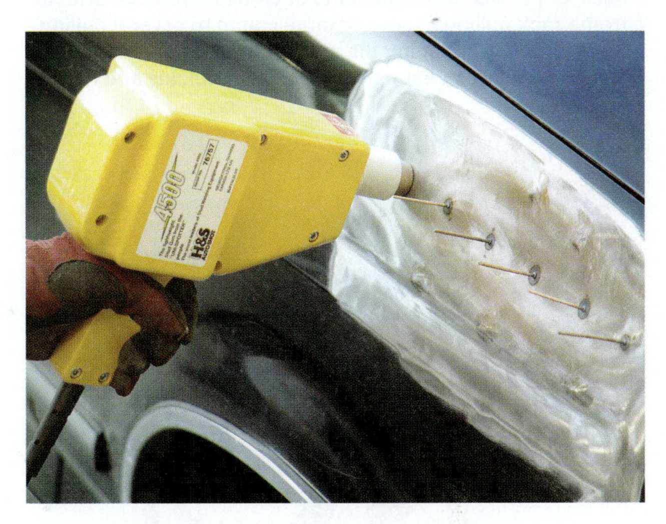

Figure 11-44 **On a long crease, several pulling pins can be welded to the low spot in the crease. Start pulling on the ends of the crease and work your way inward to the center, the deepest area of the crease.**

used, make sure it is making a good electrical connection or the spot welding will not work properly.

Push the spot welding gun and rivet up tight against the damaged body panel. While wearing welding gloves, press the trigger and the metal pin will be spot welded to the vehicle. With the steel pins welded to the dented area, various means can be used to pull out the damage. In Figure 11–42, a technician has attached locking pliers to a welded-on pull pin. Pulling by hand with pliers will often pop out shallow dents. For deeper dents, a slide hammer can be attached to a pulling pin. Slide hammer blows will easily pull the sheet metal back out and almost level with the surrounding surfaces (Figure 11–43).

Several pins can be welded to each low spot or a larger dent in the body panel for added pulling power (Figure 11–44). Carefully locate the pins along the valley of low spots for pulling.

Figure 11–45 illustrates how several pulling pins lined up in a row can be used to pull out a crease or valley in a damaged panel. Weld the pins in a row along the valley of the crease. Then clamp a frame rack pulling chain to all of the pulling pins. When hydraulic power is energized, the ram, chain, and clamp will pull on all of the pins at once. This will quickly and easily pull out a long, low spot in a sheet metal repair area.

After the dent has been pulled out straight, cut off the pull pins and grind the area flush. Finish straightening the panel with a body hammer and dolly.

The *reusable tip spot weld dent puller* has a permanent weld tip and slide hammer attachment mounted on the spot welder (Figure 11–46). The star or triangular tip can be quickly welded to the bare steel panel by pushing the trigger button on the slide hammer attachment (Figure 11–47). Then, by sliding the heavy handle outward, low spots in the dent can be pulled out (Figure 11–48).

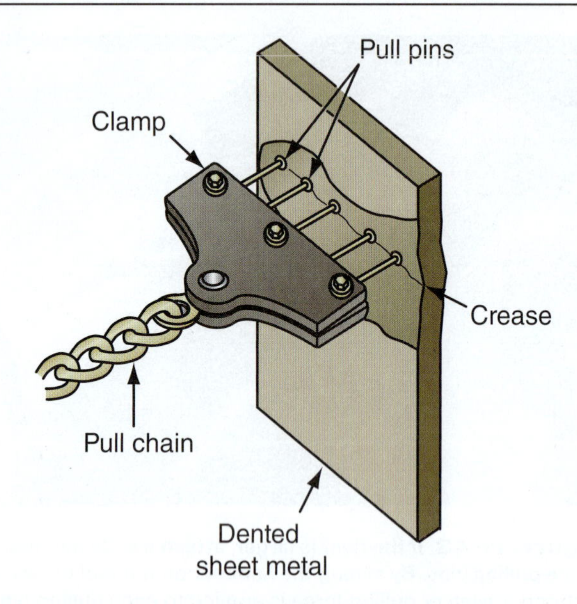

Figure 11-45 On large dents or creases, you can attach a frame rack pulling chain and large clamp to several pulling pins at once. By engaging the pulling tower chain on the frame rack, a large dent can be pulled out quickly and easily.

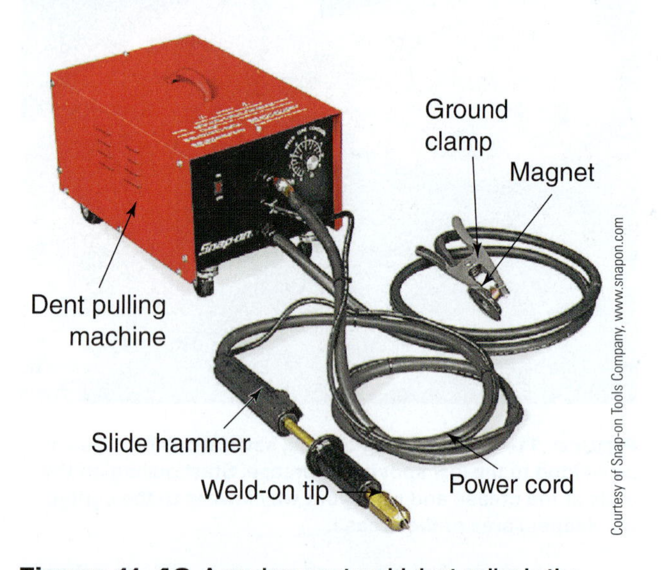

Figure 11-46 A modern spot-weld dent puller is the quickest way of removing small dents from sheet metal body panels. This machine has a magnetic ground clamp and a replaceable welding tip with a slide hammer built onto the welding machine.

The advantage of the reusable tip dent puller is that it removes small dents more quickly. Also, no weld pins remain on the panel. You do not have to cut off pins or grind off their large weld nuggets after pulling out the dent.

Set the weld time on the dent puller to the correct time in seconds for the thickness of the metal on the vehicle body.

Press the slide hammer puller tip against the sheet metal. Press the trigger on the spot welder handle and the pull tip will be fused to the panel. Blows from the slide hammer will then force the dent out. To remove the puller tip, twist or rotate the tool, and the tip will break off the

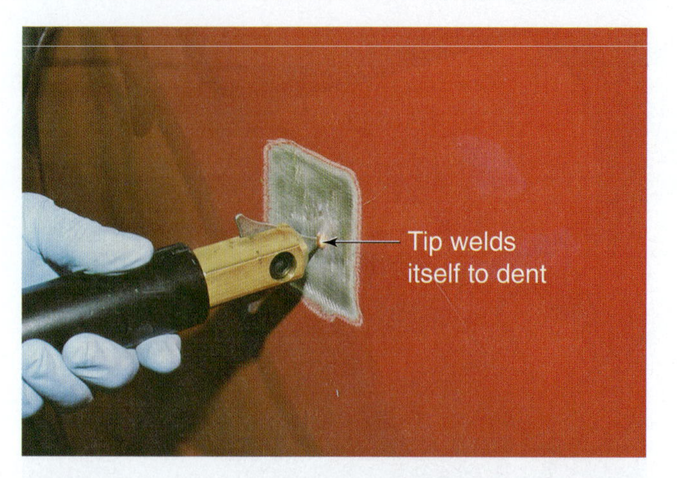

Figure 11-47 Press the spot-weld dent puller tip against the dented body surface. Press the trigger and the machine will automatically weld its tip to the body panel. Set the weld timer knob to low and then turn it up until fusing is strong enough for pulling out dents.

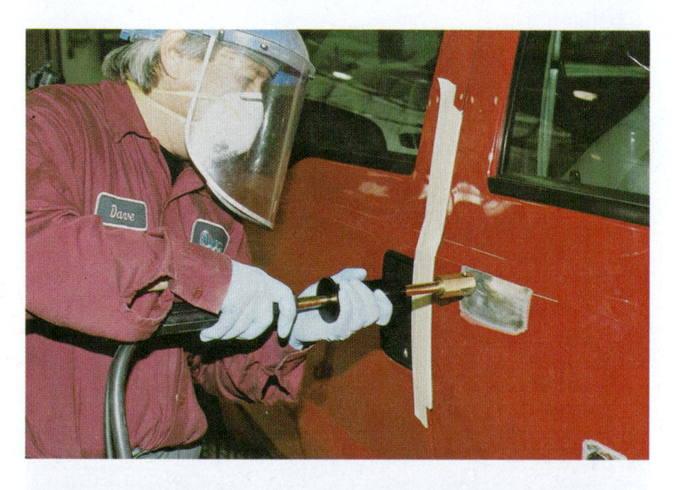

Figure 11-48 Once the tip is welded to the damaged panel, you can force the slide hammer back to pull out the small dent. Twist on the handle to free the weld tip from the body for another pull.

panel. Very little of the weld nugget will remain on the panel. Another pull can be made with the same tip.

To pull out a crease-shaped dent, start on each end and work your way to the middle. On a round dent, start at the perimeter of the dent and work your way to the middle in a circular pull pattern. Do not try to pull a dent out all at once or you will stretch the metal. Make several small pulls to work the metal back out evenly without stretching the panel.

A *disposable tip spot weld dent puller* uses small pull pins, studs, or "nails" that weld to the damaged panel. Pull pins are inserted into the tip of the spot welder. They are then resistance spot welded to the dent. After that, a slide hammer or pull chain can be attached to the pull pins to force the dent back out. After pulling, the pins are cut off with side cut pliers and the weld nuggets are ground flush with the panel surface.

The advantage of the disposable pin spot welder is that several pull pins can be welded to the dent. By attaching one or more pulling clamps and chains to the pins, a larger dent can then be pulled at one time.

When pulling dents out of thicker metal or double- or triple-thick layers of metal, you may want to use an MIG welder to fuse large metal objects to aid the pull. The stud can be a large metal flat washer, a pull tab, a piece of angle iron, or a bolt.

DANGER Make sure you are wearing shaded goggles or safety glasses when resistance spot welding. Normally, there will not be an electric arc. However, if the spot welder loses electrical contact, the resulting electric arc flash could burn your eyes! Also, do not operate electric welders if you have a heart pacemaker.

11.5 METAL SHRINKING, STRESS RELIEVING

Shrinking metal is needed to remove strain or tension on a damaged, stretched sheet metal area. During impact, the metal can be stretched. When pulled or hammered straight, the area can still have tension or strain on it. This is because the stretched metal no longer fits in the same area. The metal will tend to pop in and out when you try to final straighten it.

If a strained area is filled with body filler, road vibrations can cause the panel to make a popping or flapping noise. After prolonged movement of the strained area, the filler can crack or fall off. Eventually, you will be required to spend extra time correcting work that should have been done properly in the first place.

Stretched Metal

Stretched metal has been forced thinner in thickness and larger in surface area by impact. When metal is severely damaged in a collision, it is often stretched in badly buckled areas. Sometimes these same areas are also stretched slightly during the straightening process. Most of the stretched metal will be found along ridges, channels, and buckles in the direct damage area. When there are stretched areas of metal, it is impossible to correctly straighten the area back to its original contour. The stretched areas can be compared to a bulge on a tire. There is no place for the area to fit within the correct panel contour.

Before shrinking, dolly the damaged area back as close as possible to its original shape. Then you can accurately determine whether there is stretched metal in the damaged area. It will usually pop in and out if stretched. If stretched, you must shrink the metal (Figure 11–49).

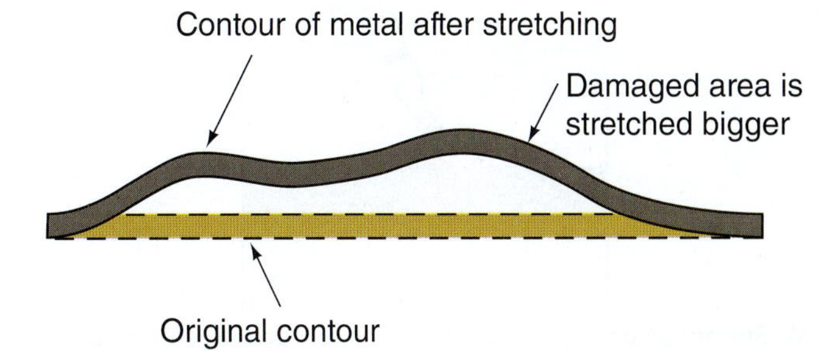

Figure 11–49 Stretched metal must be shrunk to relieve stress so the damaged area will lie flat again.

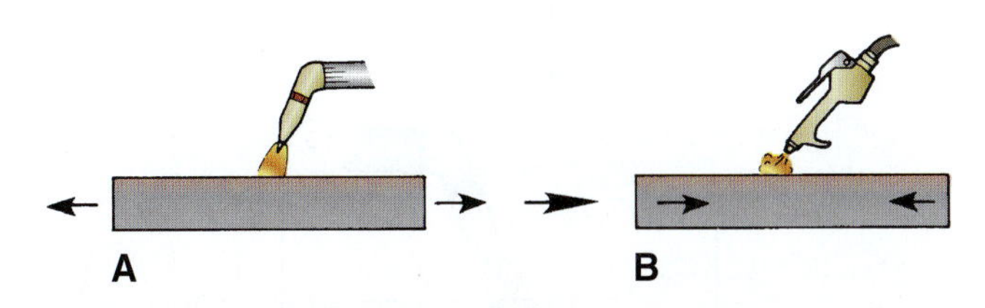

Figure 11–50 (A) Heat causes metal to expand. (B) Cooling causes metal to contract.

Principles of Shrinking

Figure 11–50 shows that a steel bar, with both ends free to expand or contract, will expand when heated and contract to its original length when cooled.

When heated, the steel bar tries to expand (Figure 11–51A), but because it is prevented from expanding at both ends, a strong compression load is generated inside the bar.

When the temperature is increased even more, and the steel becomes red-hot and soft, the compression load concentrates in the red-hot area and is relieved as the diameter of the red-hot area increases (Figure 11–51B). If the steel bar is suddenly cooled down, the steel contracts and the length of the bar is shortened (Figure 11–51C).

This principle of shrinking steel also applies to the shrinking of a warped area in a piece of sheet metal. A small spot in the center of the warped area of steel is heated to a dull red. When the temperature rises, the heated area of the steel panel swells and attempts to expand outward toward the edges of the heated circle (circumference). Because the surrounding area is cool and hard, the panel cannot expand, so a strong compression load is generated.

If heating continues, the expansion of the metal is centered in the soft, red hot portion, pressing it out. This causes it to thicken, thus relieving the compression load. If the red hot area is suddenly cooled while in this state, the steel will contract and the surface area will shrink to less than its area before heating.

A variety of pieces of welding equipment can be used to heat metal for shrinking. Attachments are available for spot and MIG welding equipment to transform them into shrinking equipment. The oxyacetylene torch with a #1 or #2 tip can also be used to heat-shrink sheet metal panels made of steel, but it is more difficult to control the heat to avoid damage to HSS.

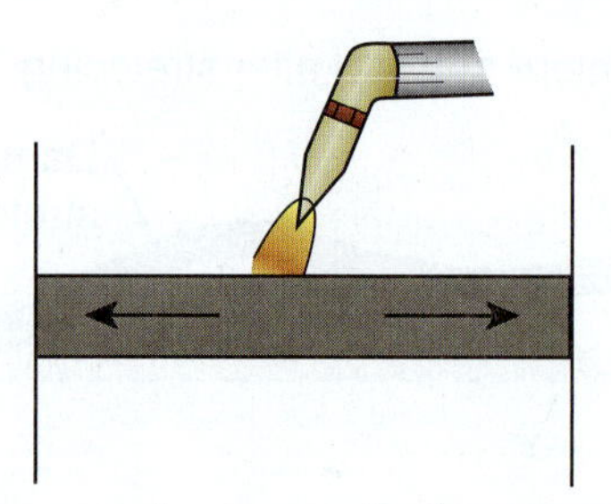

A Shrinkage occurs when expansion forces are restricted by panel rigidity.

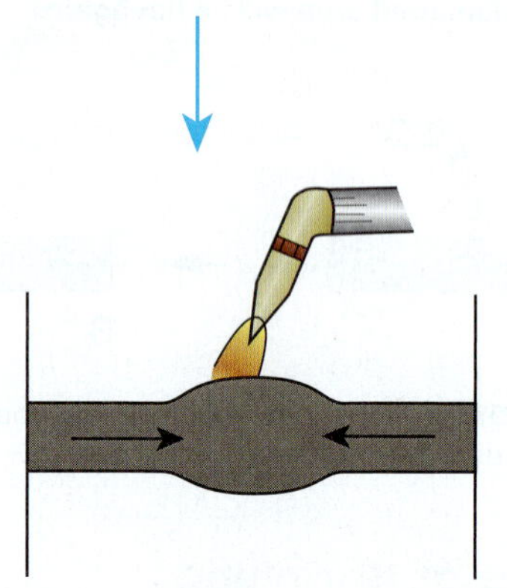

B This causes the heat-softened metal to expand and thicken.

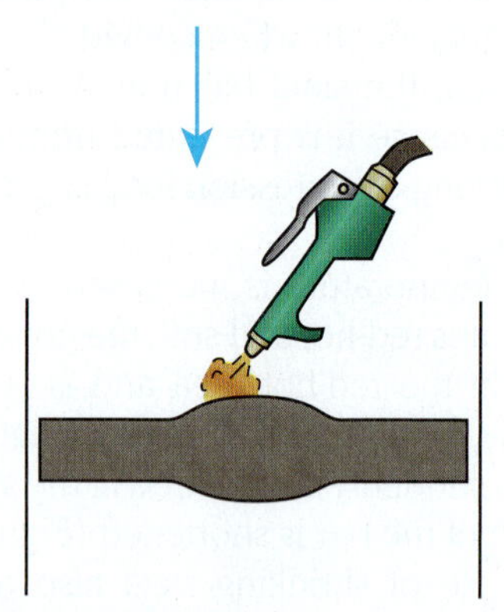

C When the metal cools, the panel contracts and, due to the increased area of the hot spot, shrinks to an area smaller than its original size.

Figure 11–51 Study the basic principles of heat shrinking damaged, stretched sheet metal.

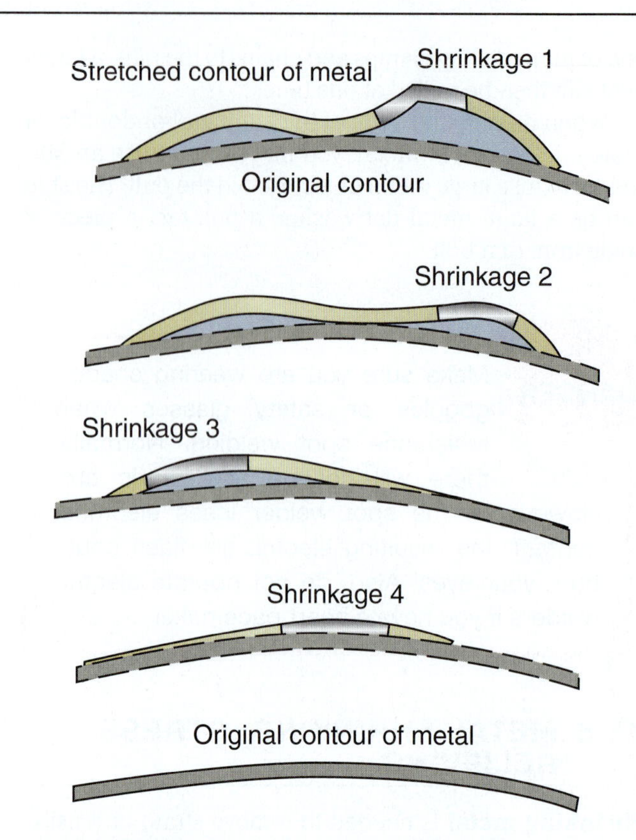

Figure 11–52 Shrinking stretched metal usually requires heating more than one spot. Always heat the highest or lowest spot first.

Shrinking Steel Panels with Heat

To shrink a damaged area with heat from an electric welder or gas torch, a small spot in the middle of the stretched area of sheet steel is heated to a "cherry-red" color. The shrink is placed in the highest spot of the stretched area, then in the next highest spot, and so on. This is repeated until the area has been shrunk back to its proper position (Figure 11–52).

The size of the shrink or hot spot is determined by the amount of excess metal in the area to be shrunk. The shrinks can be anywhere in size from a silver dollar down to the head of a thumbtack. The larger the hot spot, the harder the heat is to control. An average-sized shrink is usually about the size of a dime. Small shrinks should always be used on flat panels, because panels tend to warp easily.

A very small hot spot would be used to take an oil can-size bulge out of a flat panel. The term *oil can* is used to describe an area of a panel that is stretched very slightly. It can be pushed in. However, as soon as the pressure is released, the area will pop back out again, just as the bottom of an oil can does.

A neutral flame and a #1 or #2 tip are often used to heat the hot spots. The point of the cone is brought straight down to within ⅛ inch (3.2 mm) of the metal and held steady until the metal starts to turn red. The torch is then slowly moved outward in a circular motion until the complete hot spot is cherry-red (Figure 11–53).

Many spot welders and weld-on dent pullers can also be used to heat and shrink metal. They are better than a torch on HSS, because you can control the amount of heat that enters the sheet metal better.

As the heat enters the small spot in the panel, the heated metal expands. The cooler metal surrounding the hot spot resists the expansion forces. As the temperature increases, the heated metal becomes softer. This soft metal piles up and forms a bulge in the hot spot.

The metal usually bulges up instead of down due to the crown in the panel and because the top of the metal is heated first. When it starts to bulge, the rest of the metal in the hot spot follows.

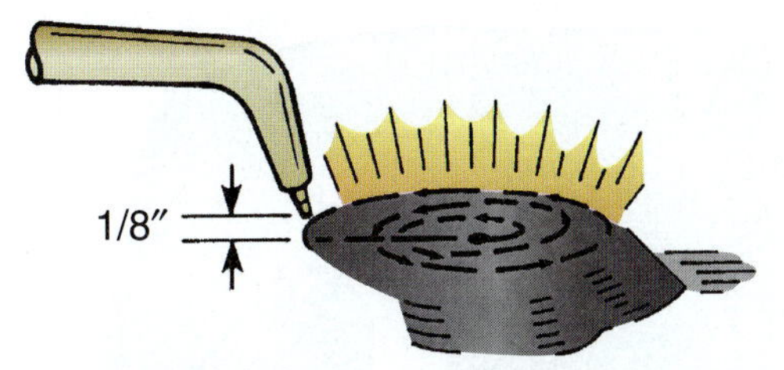

A Keep the flame cone ⅛ inch (3.2 mm) from the metal and move the torch in a circular motion from the center out.

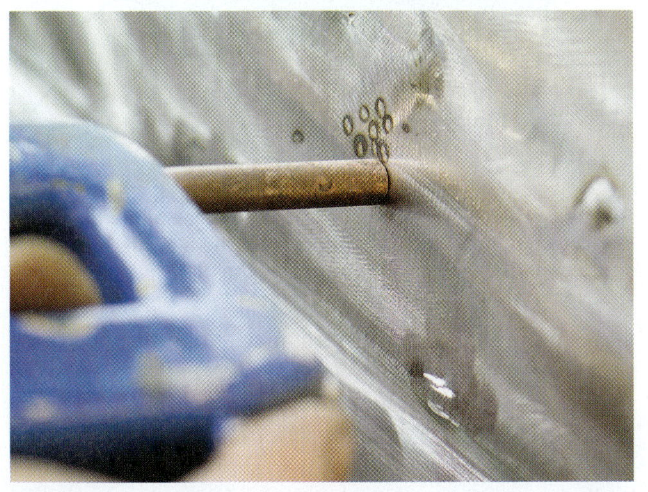

B Hold the electric welder shrinking tip next to the stretched area of metal. Move in a circular motion around metal until the steel panel glows cherry-red.

Figure 11–53 Note how either a torch or electric welder shrinking tip can be used to shrink metal.

After the deformed area on the body panel has been carefully heated, use controlled hammer/dolly blows on the hot area first. Because the metal is more pliable when heated, hitting this area will force the molecules in the stretched metal back together to lower the hump or high spot on the panel (Figure 11–54).

After lowering the heated area, methodically work your way around the damaged area. Progressively hammer out smaller and smaller high spots in the metal. Tap lightly and hit accurately to reshape the metal without making hammer indentations in the metal. Work the repair area until it is smooth enough for body filler (within ⅛ inch of original panel shape).

During this procedure, it is not necessary to support the metal with the dolly unless the metal collapses. If it is necessary to support the panel, the dolly should only be held lightly under the metal. As soon as the redness disappears, off-dolly blows and light on-dolly blows are used to smooth and level the area around the hot spot.

It is very hard to determine accurately the amount that each hot spot will shrink. One shrink can remove far more excess metal in one area than the same size would in another. It is not uncommon to find that an area has been overshrunk once the shrink has completely cooled. When an area has been overshrunk, the metal in the area of the last shrink is usually collapsed or pulled flat. Sometimes the metal surrounding the shrink area can even be pulled out of the proper contour.

Overshrinking is corrected by using hard, hammer-on-dolly blows to stretch the last shrink. The last shrunken area is usually the direct cause of overshrinking.

Metalworking must be done precisely so that only a thin coat of body filler is required. Taking the time to properly straighten sheet metal is the sign of a professional—a technician who cares about the quality of his/her work!

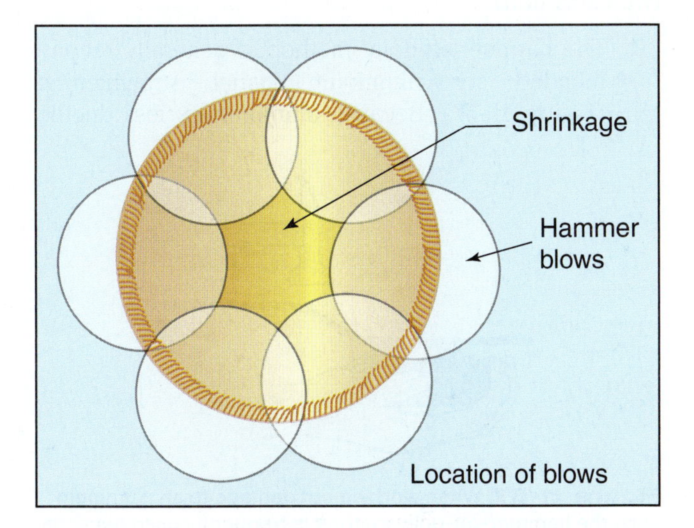

Figure 11–54 Hammer around the hot bulge to shrink it.

KINKING

Kinking involves using a hammer and dolly to create pleats, or kinks, in the stretched area to shrink its surface area. Instead of using heat to shrink the metal, kinking is another way to deal with stretched metal (Figure 11–55).

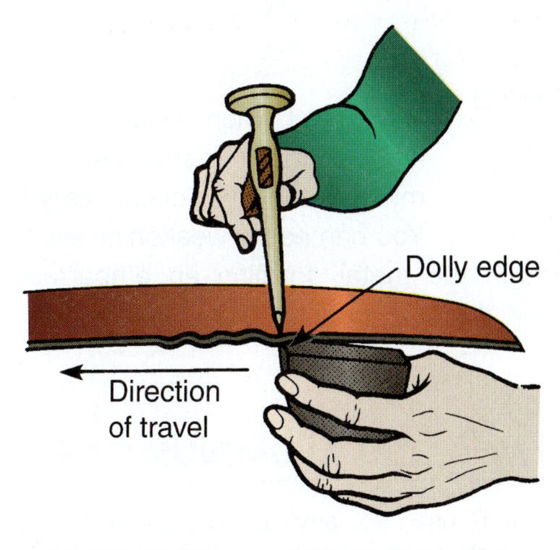

Figure 11–55 Shrinking metal by kinking a high spot can only be done when heat shrinking is not practical.

Kinking the metal will lower the area slightly below the rest of the panel. The low spot should be filled with body filler, then filed and sanded level with the panel.

Shrinking a Gouge

A **gouge** is caused by a focused impact that forces a sharp dent or crease into a panel. A gouge causes the metal to be stretched. Gouges must be shrunk to their original size to properly repair the damage. Simply picking up the low area only distorts the panel. Filling the gouge with filler without restoring the panel's original contour leaves tension in the panel that can cause the filler to crack or pop off.

Follow this procedure for shrinking gouges:

1. Heat the lowest point of the gouge with a gas torch until the metal is a dull red.
2. Use a dolly to hammer up the hot spot. This will increase the tension on the soft spot, forcing it to swell and return to its original position.
3. While the metal is still hot, hold the dolly directly under the groove and tap down the ridges that will have developed on either side of the groove. This will not only drive down the ridges but will also bump up the gouged metal.
4. If the gouge is a long one, this process will have to be repeated several times to raise the whole length of the gouge. Only heat as much of the gouge as can be worked before the metal cools.

Filing the Repair Area

When the damaged area has been bumped and pulled as level and smooth as possible, a body file can be used to locate any remaining high and low spots. High spots will be scratched more than flush surfaces (Figure 11–56).

The scratch pattern created by the file identifies any low spots. You then "pick" up the low spots and bump down the high spots. This process ensures that the repair area is ready for body filler.

Be careful when filing or grinding high spots in metal panels. The metal on new vehicles is very thin. You can easily weaken or even cut through the metal, forming an embarrassing hole in the body panel.

11.6 WORKING ALUMINUM PANELS

Aluminum is used for a variety of automotive panels, such as hoods, fenders, and roof panels. A few passenger cars have their entire unibody made of an aluminum alloy. Aluminum's natural resistance to rusting and corrosion is its primary advantage.

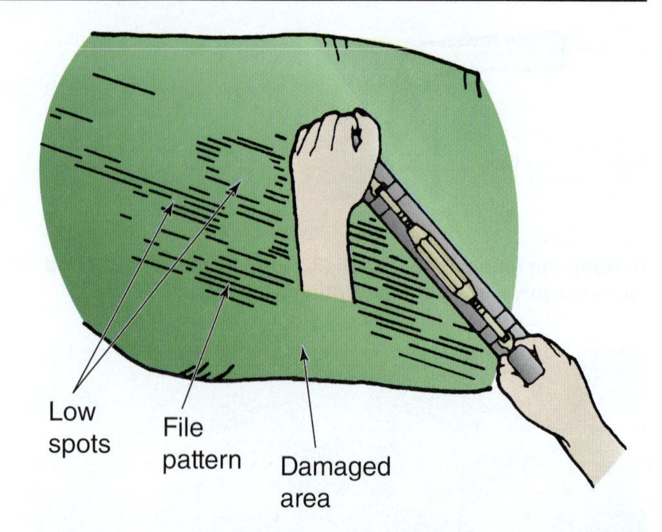

Figure 11-56 Push the file at a 30-degree angle on flat or low-crown panels. Scratch marks from the file will reveal low spots because they will not be cut by the file.

The repair of aluminum panels requires much more care than working steel panels. Aluminum is much softer than steel, yet it is more difficult to shape once it becomes badly bent and work hardened. It also melts at a lower temperature and readily distorts when heated.

Aluminum panels are usually thicker than today's HSS panels. It is important to keep in mind that aluminum body and frame parts are usually 1.5–2 times as thick as steel parts. When damaged, aluminum feels harder or stiffer to the touch because of work hardening. These characteristics must be considered when working damaged aluminum panels.

Working Aluminum with Hammer and Dolly

Straightening aluminum with a hammer and dolly is basically the same process as that for steel, with the following two exceptions:

1. The hammer-off-dolly method is generally recommended for aluminum panel straightening (Figure 11–57). Because aluminum is less ductile

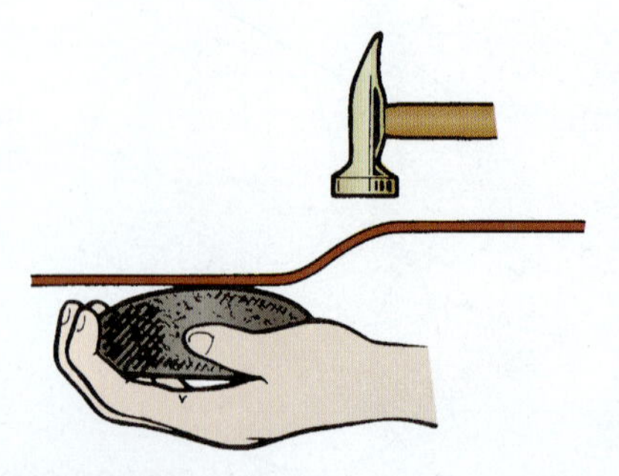

Figure 11-57 When working out damage to an aluminum part, the hammer-off-dolly method is frequently used because aluminum is softer than steel.

than steel, it does not readily bend back to its original shape after being buckled by an impact. Therefore, aluminum does not respond well to off-dolly hammering. Care must also be exercised to avoid additional damage when attempting to lower ridges with hammer and dolly blows.

2. Aluminum alloys bend much more quickly when the panel is sandwiched between the hammer and dolly, as with the hammer-on-dolly method. When it is necessary to hammer on dolly, hammering too hard or too much can stretch the soft aluminum panel. It is better to use many light strokes than a few heavy blows.

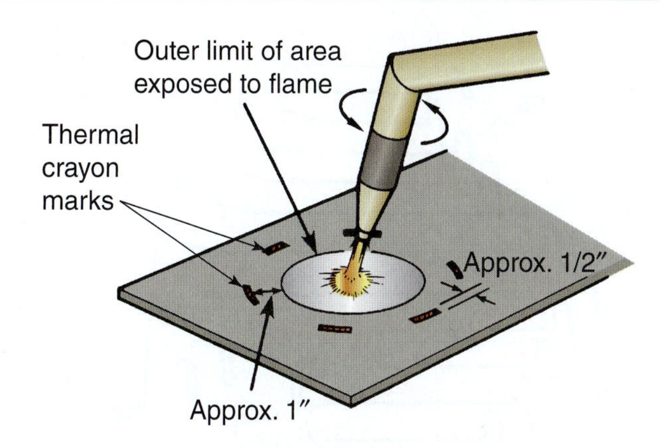

Figure 11–58 Use a temperature-sensitive crayon to prevent excessive heat buildup in soft aluminum. It is very easy to burn through aluminum because there is little reddening of hot metal to warn against burn-through.

Straightening Aluminum with a Hammer

Shrinking hammers used for working steel should not be used with aluminum, because they can cause cracking. That is, separate sets of tools should be used on steel and aluminum.

When hammer picking aluminum, work slowly and methodically. Raising small dents with a pick hammer or pry bar is an excellent way to repair aluminum panels. However, be careful not to raise the panel too far, stretching the soft aluminum.

Spring hammering with hammer and spoon is an excellent way to unlock stresses in high-pressure areas in aluminum. The spoon distributes the force of the blow over a wider area of the soft aluminum, minimizing the possibility of creating additional dents in any unyielding buckles.

Filing and Grinding Aluminum

Because aluminum is so soft, you should reduce hand pressure on the body file when marking high and low spots. Use a file with rounded edges to avoid scratching and gouging the metal. Soft aluminum cuts much more easily than steel.

Grinding must be done very carefully on aluminum panels. A coarse grit disc on a high-speed grinder can quickly burn through the soft metal. The heat from the grinding operation can also quickly warp the panel.

You can use a #36 grit open coat disc, but grind carefully in order to remove only paint and primer, not the metal. Make two or three passes. Then, quench the area with a wet rag to cool the metal and minimize heat gain. Grinding small areas and featheredging should be done with a dual-action sander or an electric polish machine that rotates at less than 2,500 rpm. Use #80 or #100 grit paper and a soft, flexible backing pad.

Heat-Shrinking Aluminum

There is one major difference with straightening aluminum by heat shrinkage. With steel, use heat *only* when the metal is stretched and cannot be straightened by other means. With aluminum, heat *must* be used to restore the flexibility that was reduced by work hardening. Without the use of heat, the aluminum may crack when straightening force is applied.

Before attempting to straighten aluminum, heat the damaged metal with a torch. It is easy to apply too much heat because aluminum does not change color with high temperatures. Aluminum melts easily, so careful heat control is very important. Use a temperature-sensitive paint or a heat-sensitive crayon made to change color at about 750°F (417°C). Follow these steps:

1. Apply a temperature-sensitive paint or crayon in a circular pattern around the area that will be exposed to the flame (Figure 11–58). Aluminum does not glow cherry-red like steel does right before it melts.
2. Heat the area, moving the flame constantly.
3. Stop heating when the paint or crayon color changes. The surface temperature at the center of the heated area will be between 750° and 800°F, a safe margin from aluminum's melting point. A lack of caution will result in a melted panel. Also, the shrink spot must be very slowly quenched to avoid distorting the panel by excessive contraction.

11.7 PAINTLESS DENT REMOVAL

Paintless dent removal involves removing small dents using specialized tools: long picks, soft-faced hammers, plastic blocks, and other equipment (Figure 11–59).

You can save repair time and retain the factory baked-on finish when you use picks to remove small dents, because you will not have to repaint the panel. Paintless dent removal only works with very small dents that do not damage the finish. It is commonly used on small door dings and hail damage (Figure 11–60).

First, all small dents should be marked with a crayon. Place a bright light to one side of the panel so that light reflects off from the side to help make the dents more visible (Figure 11–61).

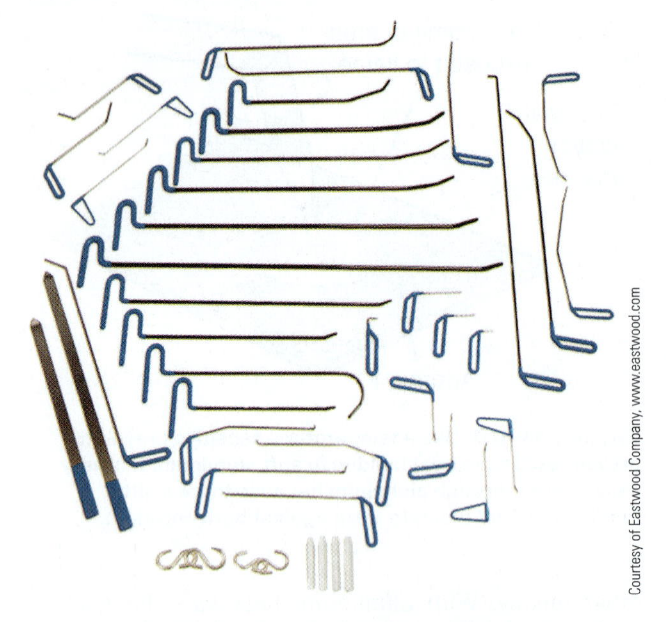

Figure 11-59 This set of long picks is used for paintless dent removal. They can be used to push out small dents from the back of panels.

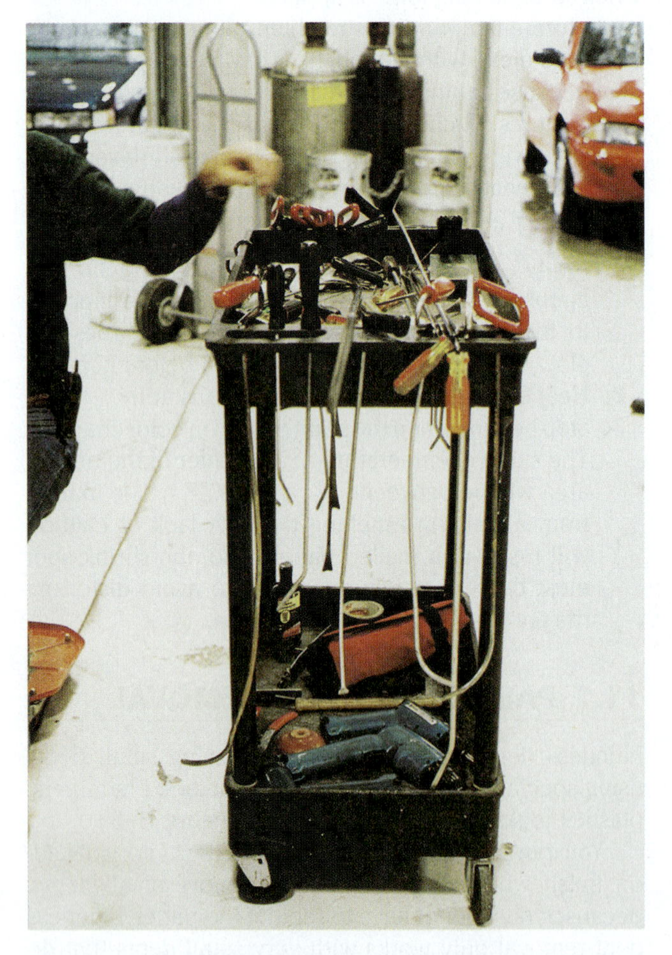

Figure 11-60 Note the various specialty tools commonly used during paintless dent removal.

With your eyes almost level with the panel, sight across the panel. Small dents can be hard to see. In Figure 11–62, the technician is marking damage on a car hood.

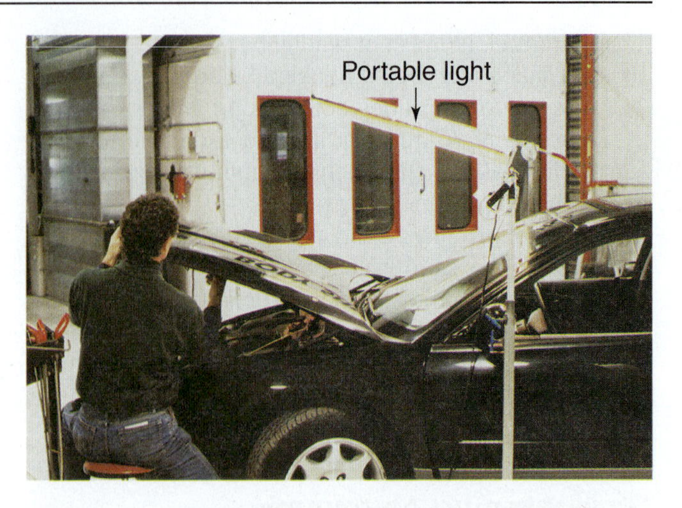

Figure 11-61 A large, portable light is needed to see clearly during paintless dent removal. Metalworking must be extremely precise so as not to damage the paint film when straightening metal damage.

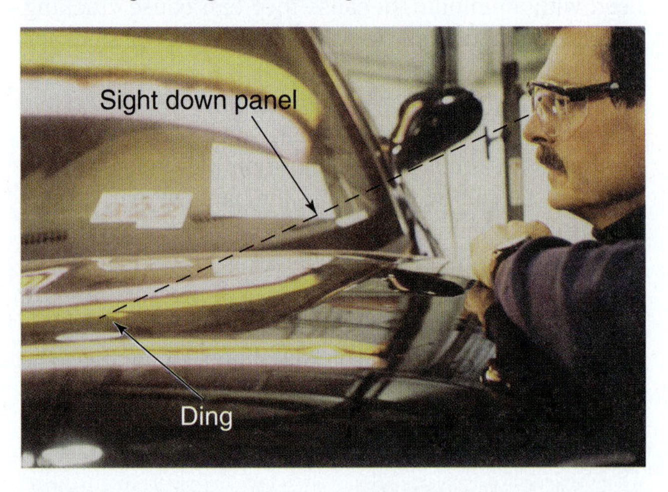

Figure 11-62 Sight down a panel to look for the shadows of any dings. This will make small dings more visible.

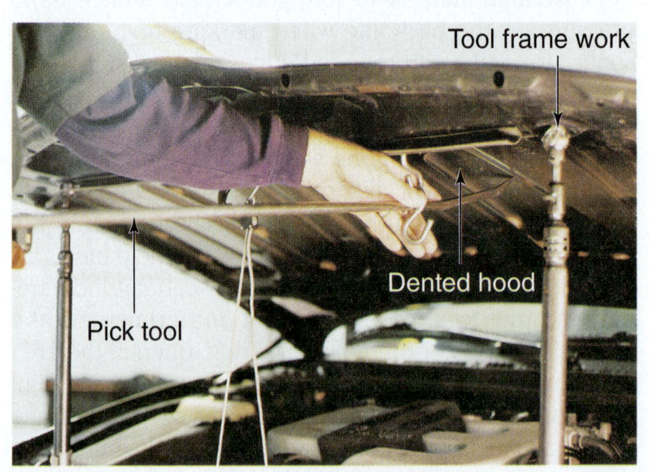

Figure 11-63 A large metal hook allows this technician to use a long pick to pry up small dents from the bottom side of the hood.

Figure 11–63 shows how you would use a large hook attached to the bottom of a hood in a paintless repair. The hook allows you to pry upward on the bottom of the hood right under small dents.

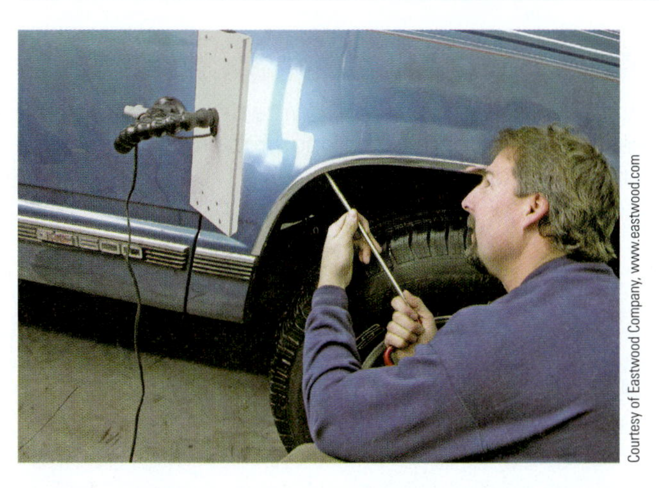

Figure 11–64 Here a paintless dent removal technician is prying against a car tire to push out a small door ding in a quarter panel.

Push up on the panel while watching for the pushing action of the pick. The metal will flex upward at the point of the pick. Move your pick until it is next to the ding or dent. Then, slowly and carefully pry up on the dent. Circle around larger dents to work them out a little at a time.

When prying with a pick, be careful not to stretch the metal by exerting too much pressure. Start with the point of impact or the lowest point. Slowly pry up the damaged area. On the larger areas, use a flat blade pick rather than a pointed one. Tap down pressure areas while prying up low-tension areas.

Figure 11–64 shows how a technician has pushed prying picks up inside a wheel well for removing a door ding in a quarter panel. The picks are pushed up against the tire to pry outward on the back side of the panel. When feasible, paintless dent removal provides a superior repair, because the original finish does not have to be repainted.

To lower high spots next to small dings, place a plastic dolly or dowel over the high spot. Very light taps from a plastic hammer or mallet will force the high spot down without damaging the paint film. Light, careful blows from a tiny metal dolly will also lower the perimeter of dings, without requiring repainting (Figure 11–65).

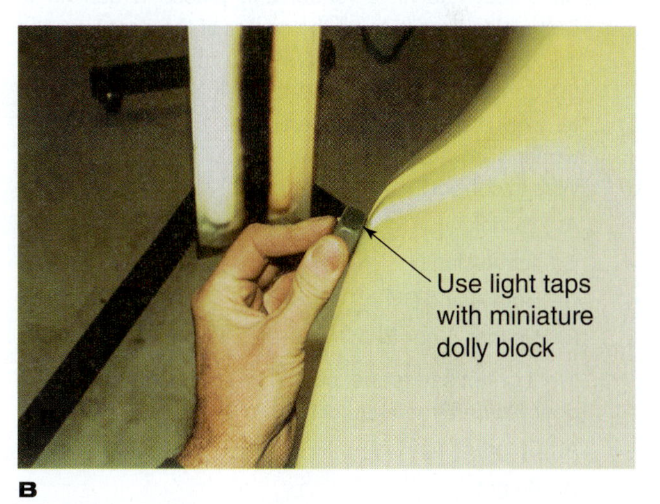

Figure 11–65 Lowering high spots during paintless dent removal takes skill and patience. (A) Here a plastic rod is being used with light blows from a plastic mallet to lower high spots without creating paint damage. (B) Light taps with a miniature dolly with rounded edges will final lower high spots without damage to the paint film.

SUMMARY

1. Metalworking skills are probably the most important craft a body technician can bring to a shop.
2. To do quality sheet metal repairs, you must first know how to return the sheet metal to its original shape.
3. To repair collision damage, you must also understand what property changes have taken place in the metal.
4. Spring-back is the tendency for metal to return to its original shape after deformation.
5. Work hardening is the upper limit of plastic deformation that causes the metal to become very hard in the bent area.
6. The term *raising* means to work a dent outward or away from the body. The term *lowering* means to work a high spot or bump down or into the body.
7. Hammer-on-dolly repairs are used to smooth small, shallow dents and bulges and to stretch metal so that it can return to its original shape.
8. The hammer-off-dolly method is used to straighten metal just before the final stage of straightening.
9. Aluminum is used for a variety of automotive panels, such as hoods and roof panels.

EXERCISES

On a separate sheet of paper, complete the following learning activities for this chapter. Write definitions for the key terms and answer the ASE-style review questions, essay questions, critical thinking problems, and math problems. You can also do the outside activities, possibly for extra credit.

Key Terms

body hammer
buckles
deformation
gouge
hammer-off-dolly
hammer-on-dolly
shrinking metal
spring-back
stretched metal
tensile strength
work hardening
yield strength

ASE-Style Review Questions

1. Which type of steel is most often used in new vehicle structural components?
   A. Mild steel
   B. High-strength steel
   C. Both A and B
   D. Neither A nor B
2. An object has hit a fender and placed a dent in it. Technician A says to hit the center of the dent from the back with a dolly first to drive out the deepest part of the dent. Technician B says to use a dolly and hammer to work the dent out, starting at the perimeter and working inward to the deepest part of the dent last. Who is correct?
   A. Technician A
   B. Technician B
   C. Both A and B
   D. Neither A nor B
3. What grit of sandpaper should be used to remove paint down to bare metal quickly?
   A. 36 grit
   B. 80 grit
   C. 120 grit
   D. 180 grit
4. Which kind of spoon can be used instead of a hammer?
   A. Dinging spoon
   B. Slapping spoon
   C. Both A and B
   D. Neither A nor B
5. A dented panel cannot be accessed from the rear for straightening with a hammer and dolly. Technician A says to drill holes in the panel and use screws to pull out the damaged sheet metal. Technician B says to avoid drilling holes in panels, because this will damage the panel and its corrosion resistance. Who is correct?
   A. Technician A
   B. Technician B
   C. Both A and B
   D. Neither A nor B
6. Technician A says that aluminum has a lower melting point than steel. Technician B says that you should use a temperature-sensitive crayon when heating aluminum since it does not glow cherry-red right before it melts. Who is correct?
   A. Technician A
   B. Technician B
   C. Both A and B
   D. Neither A nor B
7. Technician A says that 20 percent of sheet metal damage is direct damage. Technician B says that bent metal is not necessarily buckled metal. Who is correct?
   A. Technician A
   B. Technician B
   C. Both A and B
   D. Neither A nor B

Essay Questions

1. Explain, in your own words, how to use a body hammer to straighten metal.
2. Describe the hammer-on-dolly method.
3. Summarize the method for removing a gouge in metal.

Critical Thinking Problems

1. An aluminum panel has a bulge after initial straightening. The bulge pops in and out with hand pressure. What should you do?
2. After shrinking a steel panel, the technician finds a flat area under tension and lower than the rest of the panel. What is wrong?
3. Please name four ways to measure the strength of a metal.
4. *Indirect damage* is caused by what?

Math Problems

1. If a fender is 1/16 inch (1.6 mm) thick and you grind away 0.031 inch (0.79 mm) of metal, how thick is the remaining panel?
2. If a 12-inch (25 cm) metal rod is heated and expands 2 percent in size, how long has it become?

Activities

1. Obtain a damaged panel (fender, hood, door, or lid) from a salvage yard. Use a hammer to place a small dent in a flat area of the panel. Using the information in this chapter, remove the dent.
2. Write a report on the steps taken to work out the dent in your panel. Describe the tools used, how much time it took, any difficulties, and so forth.
3. Use the Periodic Table of the Elements to list all of the metals that can be used in the manufacture of a modern automobile. Describe why a part should be made of each type of metal discussed. Also summarize each metal's melting point, hardness, ductility, and other characteristics.

Working Sheet Metal

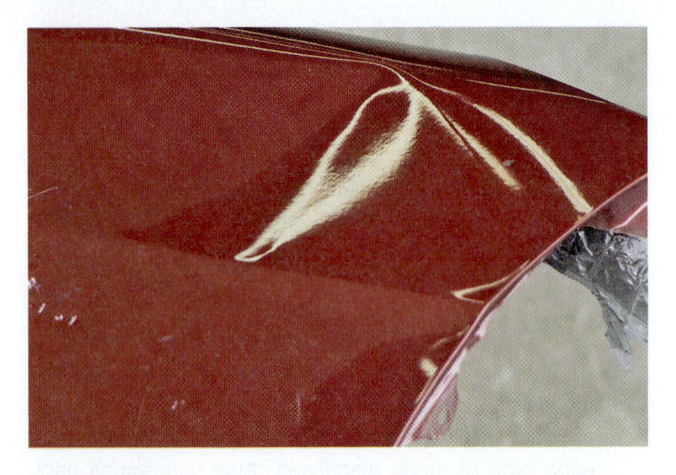

Figure P11-1 A dent in a contoured part forms a crease. Proper metalworking methods are needed to remove the dent efficiently and without further stretching the metal.

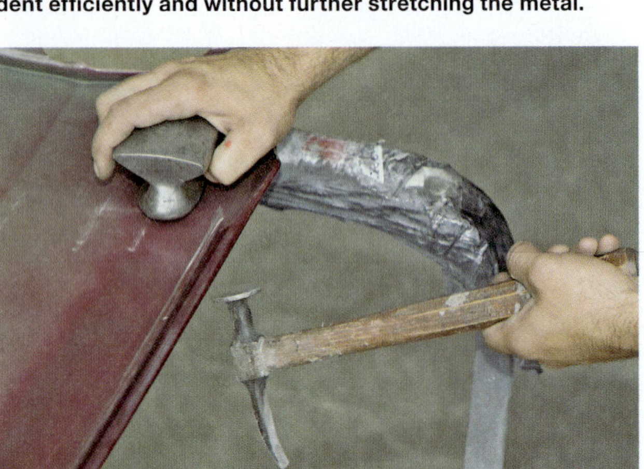

Figure P11-2 Using a properly shaped dolly, start working the dent from the ends of the crease, not from the middle. Flatten the curve at the ends of the dent so that the metal will not be stretched as the center is moved back out.

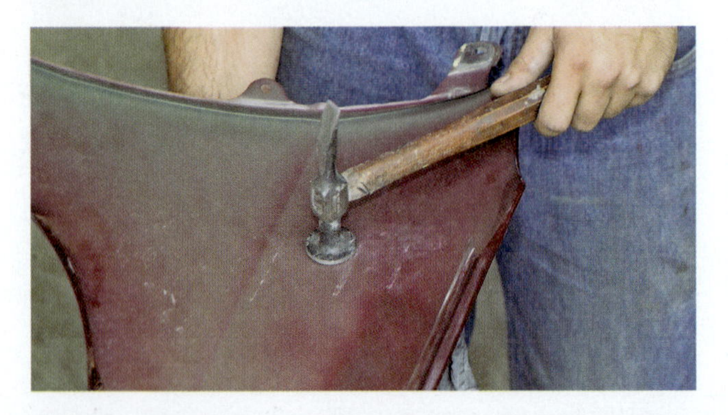

Figure P11-3 Next, move the center of the crease out part way. You will need the edge of a dolly with a larger contour to match the larger contour of the center area of the crease.

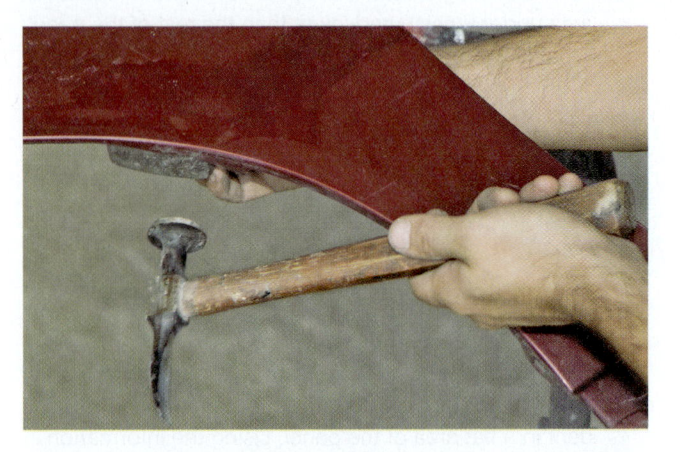

Figure P11-4 Then, go back and work the ends. Try to remove the damage either as it occurred or all at once.

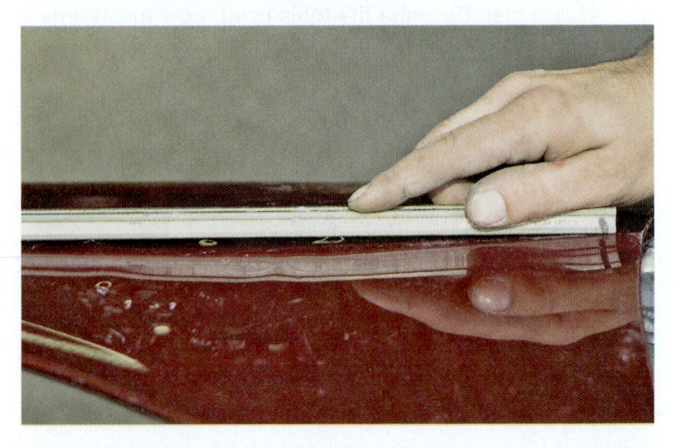

Figure P11-5 Use a straightedge to check your progress.

Figure P11-6 Light hammer blows will help to lower any surfaces that have been raised too high. The repair area should be within ⅛ inch (3.1 mm) of level so it can be filled with body filler.

Using Body Fillers

CHAPTER 12

OBJECTIVES

After reading this chapter, you should be able to:

- List the different types of body fillers and glazes.
- Choose the correct body filler for a particular repair job.
- Identify the correct way to mix filler and hardener.
- Correctly apply filler.
- Use recommended methods for shaping filler.
- List the common mistakes made when using filler and spot putty.
- Describe how to use special sanding aids.
- Explain how to repair scratches, nicks, dings, and surface rust with body filler and glazing putty.
- Properly repair rustout damage.
- Answer ASE-style review questions relating to body filler.

INTRODUCTION

The metalworking techniques discussed in Chapter 11 are fundamental to any repair job. Damaged metal must be restored to its original contour using industry-accepted methods. Metal straightening must be done to unfold buckles, relieve stresses in work-hardened ridges, shrink stretched metal, and stretch shrunken areas. Repair quality depends on sound metalworking techniques. Proper metalworking will require a very thin coat of plastic body filler to level and straighten the repair area in the sheet metal.

Two-part plastic body filler is the finishing touch for most sheet metal repairs. Minor surface irregularities can be quickly filled and smoothed with a thin coat of plastic body filler. Restoring bent and stretched HSS to its exact original shape is almost impossible in many instances. It is critical that you properly prepare the surface and apply the correct type fillers. If not, the filler can crack or pop off, or the topcoat of paint may be adversely affected.

This chapter details the procedures for the final repair of minor damage to sheet metal. You will learn how to properly select, mix, apply, file, sand, and shape body filler to match the contour of a body panel. Repairing small surface scratches and dings using two-part spot putty will be described. This chapter also summarizes how to professionally repair rust damage without complete panel replacement, using a welded metal patch and waterproof filler (Figure 12–1).

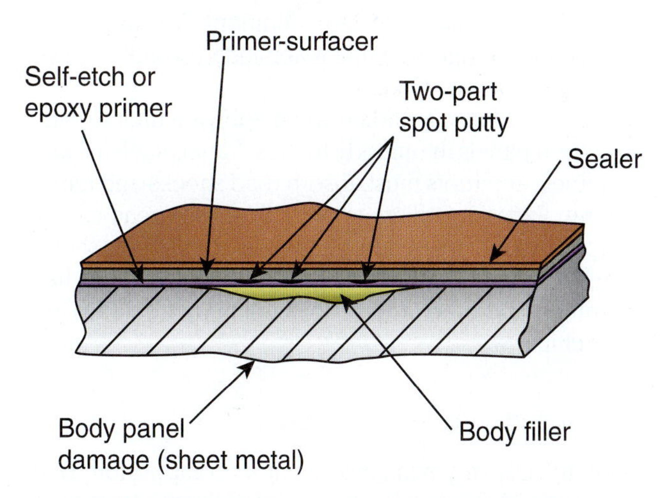

Figure 12–1 Study the basic sequence of a minor repair after metalworking. First, body filler goes over metal. Follow with self-etch primer, epoxy primer, or primer-surfacer. Then apply spot putty before spraying on sealer and paint or topcoat.

12.1 BODY FILLERS

Body filler, or *plastic filler*, is a heavy-bodied plastic material that cures very hard for filling small dents in metal. This compound of resin and plastic is regularly used to fill minor dents or low spots on car and truck bodies.

Most auto body repairs require some application of body filler. Modern body filler is a fast, inexpensive way to restore the final contour of a damaged panel.

Only after the damaged panel has been bumped, pulled, pried, and dinged to within at least ⅛ inch (3.1 mm) of the original contour can filler be applied. Then you can fill, shape, and smooth the repair area with a thin layer of body filler.

Unfortunately, some body shops fail to properly do sheet metal repairs and simply hide the damage under a thick layer of filler. Body fillers were never meant to replace proper metalworking techniques. Before any fillers are applied, the damaged panel should be returned to its correct shape and dimension by bumping, stretching, and pulling. After the panel has been filed to locate low spots, low areas should be bumped or picked up so that no area is more than ⅛ inch (3.1 mm) below the original contour of the panel.

Before any filler is applied, all holes, cracks, and joint gaps must be welded. Some body fillers are hygroscopic, which means they absorb moisture when exposed to humid conditions. Unless filled with a waterproof pigment, these fillers will absorb moisture through holes or cracks in the metal. The moisture will penetrate to the metal, where rust will begin to form. Eventually, the rust will destroy the bond between the filler and the metal.

Body fillers and putties (discussed later in this chapter) can also be used to repair minor surface defects, such as dings, stone chips, and surface rust. Be aware, though, that body fillers have limitations. Large panels such as hoods, deck lids, and door panels tend to vibrate violently under normal road conditions. Vibrations can crack and dislodge filler that is applied too thickly or over too large an area.

Care must also be taken when applying filler to semistructural panels in unibody frames. Panels such as quarter panels and roofs must absorb road shocks and torque flexing. Excessive fillers applied in these areas can be popped off by stresses in the panels. Body filler should also be used sparingly on rocker panels, lower rear wheel openings, and other areas subjected to flying stones and rock chips.

Body Filler Ingredients

Body filler is very similar to paint in composition. Both are made of resins, pigments, and solvents. Most body fillers have a polyester resin that acts as a binder. When the filler is applied and the solvents evaporate, the binders hold the pigments together in a tough, durable film. The basic pigment or material in conventional fillers is talc.

Figure 12–2 Body filler is a two-part plastic repair material for sheet metal. It is in a semisolid state until hardener is added. If the filler has been in storage, it should be mixed with a back-and-forth motion of a mixing stick or putty knife.

Talc, also used in baby powder, absorbs moisture. That might be good for a baby, but it is bad for a car, so proper steps must be taken to shield the filler from moisture. If holes in the metal or cracks in the paint expose the filler to the atmosphere, the talc in the filler absorbs moisture, which attacks the metal substrate and forms rust. The rust destroys the filler-to-metal bond, causing the filler to fall off. Waterproof fillers are available. Fiberglass strands or metal particles are used instead of talc as pigments.

Like some paints, body fillers harden by chemical action. Hardening, or curing, produces a molecular structure that will not shrink or soften. The chemical reaction is set off by oxygen. Consequently, if a container of plastic filler is left open and exposed to the air, it will slowly harden. Body fillers come in various types of containers (Figure 12–2).

Two-part body fillers may produce a *waxy coating*, or *paraffin*, on their surface. The purpose of paraffins in the filler is to form a film that prevents oxygen absorption. The paraffins are suspended in the filler solvent and are carried to the surface when the solvents evaporate. The paraffins either must be removed with a wax and grease remover before being sanded or else filed off with a surform "cheese grater."

Body Filler Hardener

To speed up the drying process, a chemical catalyst is provided by the manufacturer. The catalyst, in liquid or cream form, is called **hardener**. Hardener is basically a chemical compound called peroxide. The oxygen in the peroxide drastically speeds up the curing process of the body filler (Figure 12–3).

Table 12–1 shows how quickly filler will become too stiff or thick to spread after adding hardener at various temperatures. For example, if you add the correct amount of hardener to the body filler at 85°F, it will only take

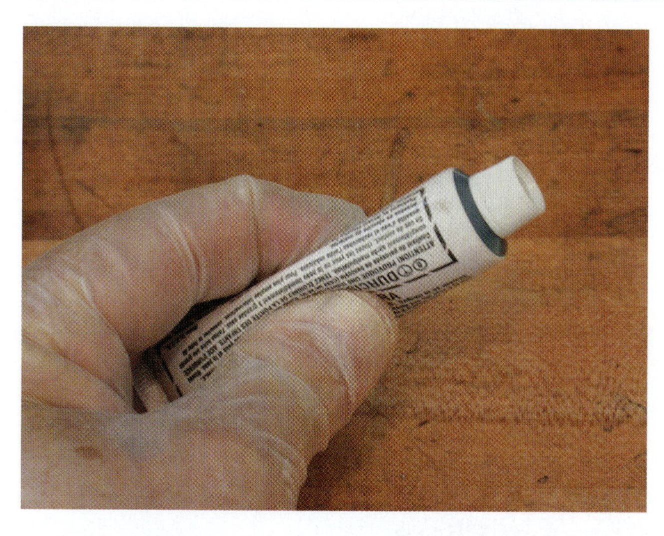

Figure 12-3 **Hardener is a catalyst for body filler. It comes in a tube. To knead or mix hardener after storage, move your fingers back and forth over the tube.**

TABLE 12-1 EFFECT OF TEMPERATURE ON WORKING TIME

Temperature	Working Time
100°F	3 to 4 minutes
85°F	4 to 5 minutes
77°F	6 to 7 minutes
70°F	8 to 9 minutes

As filler cures and hardens, the chemical reaction produces a tremendous amount of heat. For this reason, unused filler should *not* be discarded in trash cans containing solvent-soaked paper or cloths. A serious shop fire could result!

4–5 minutes for the filler to set up too much to be worked over a damaged area.

Types of Body Fillers

There are many types of body fillers on the market today. Each has slightly different qualities or uses. You should understand each type so that you know which kind of body filler should be used for specific repairs.

Body Filler History

During the first 50 years of auto body repair, blemishes in sheet metal panels were corrected by applying lead filler. *Lead filler*, or solder, is an alloy of lead and tin. A welding torch was used to soften the solder and bond it to the body sheet metal.

Before World War II, automobiles were made with heavy-gauge steel panels that were unaffected by the heat used in the "tinning" operation, but changes began to take place in automotive construction in the late 1940s and early 1950s. During the economic boom following World War II, Americans began demanding larger and fancier cars. Manufacturers responded by producing vehicles made with thinner, larger, and more complex body panels. The thinner metals, however, made the old lead repair methods almost obsolete. The heat required for the lead filler warped the thin panels, and hammer-and-dolly work stretched metals too thinly for filing. There was a real need for an inexpensive, time-saving substitute.

In the early 1950s, epoxy-based fillers were developed. Usually mixed with aluminum powder, epoxy fillers cured very slowly and did not harden at all if applied too thickly.

In the mid-1950s, the first polyester resin-based body fillers were developed. These fillers were made from the same resin used to make fiberglass boats and required mixing with a liquid hardener and accelerator. Because fiberglass resin is very brittle when cured and depends on cloth or matte for flexibility, the early polyester body fillers were also very brittle and hard.

The first successful filler was named "Bondo," and many body technicians still refer to body fillers as Bondo. The early fillers were composed of approximately 40 percent polyester resin and 60 percent talc (by weight). The Bondo Corporation is still in existence and produces advanced, quality body repair products.

When early fillers and hardener were mixed together, the fillers dried or cured very hard. Early fillers were difficult to file and had to be levelled with a grinder, resulting in choking clouds of dust that blanketed the shop. Low-dust, straight-line air fillers had not been developed yet.

As the technology developed, body fillers became softer, easier to apply, and easier to shape. Fillers soon appeared in black, red, gray, white, and yellow. Cream hardeners in contrasting colors—red, white, green, and blue—were also developed to provide a mixing reference for the various colored polyester fillers. The softer fillers could also be grated while still semicured, thus reducing the amount of sanding required. Note that the addition of color does not affect the working characteristics of the filler.

Today's body fillers have over 30 years of development backing them. The premium fillers use very fine-grained talc to provide superior workability, sandability, and featheredging. High-quality resins ensure excellent adhesion, flexibility, and quick curing properties. Most plastic fillers can be grated and rough shaped within 5–10 minutes. They can be sanded within 10–15 minutes.

Fiberglass Fillers

Fiberglass body filler has fiberglass material added to the plastic filler material. Because it does not absorb moisture, it is used for rust repair or where strength is important. It can be used on both metal and fiberglass substrates. Because fiberglass-reinforced filler is very difficult to sand, it is usually used under conventional, lightweight plastic filler. After welding in a repair patch, waterproof fiberglass body filler is used over the weld joint first and then covered with easily sandable lightweight filler.

As thinner-gauge sheet metal replaced heavy-gauge steel, rust became a problem, especially in areas of the country where road salts are used in winter. A product was needed to repair surface rust and rustout repairs. Because talc-filled body fillers absorb moisture readily, the available fillers did not provide long-lasting protection when used over surface rust or a rustout.

To meet this demand for waterproof filler, fiberglass-reinforced fillers were developed. Fiberglass fillers use fiberglass strands rather than talc as a bulking agent. These fillers are more flexible and stronger than conventional fillers.

Fiberglass fillers are available in two basic forms. One is formulated with short strands of fiberglass. The other is made with long strands.

1. **Short-hair fiberglass filler** has tiny particles of fiberglass in it. Short-hair filler works and sands a little harder than conventional filler but is much stronger.
2. **Long-hair fiberglass filler** has long strands of fiberglass for even more strength. It is waterproof but very hard to sand. Again, it can be used where filler strength is vital to the repair. Long-hair filler is often covered with a thin coat of easily sandable lightweight body filler.

Aluminum Fillers

Aluminum fillers actually contain particles of aluminum. The first premixed, 100 percent aluminum auto body fillers were introduced in 1965. These products were waterproof, used a red-tinted liquid hardener, and had a fairly good shelf life. Due to their very high relative costs, the 100 percent aluminum body fillers are used sparingly for special applications, such as restoring antique cars. Today there are several similar aluminum products available. Metal fillers are nonshrinking, waterproof, and very smooth. Metal fillers have the look of lead but are easier to work. When cured, they are harder than talc or fiberglass-filled body fillers.

Note that some paint and vehicle manufacturers recommend applying epoxy primer to bare metal (usually aluminum) before applying a filler. Aluminum will not bond to filler as well as to primer. Refer to the paint or vehicle manufacturer's instructions if in doubt.

Lightweight Fillers

Light body filler is formulated for easy sanding and fast repairs. It is used as a very thin topcoat of filler for final leveling of damaged metal body surfaces. It can be spread thinly over large surfaces for block sanding or air tool sanding the panel level. Lightweight fillers were formulated by replacing about 50 percent of the talc in the filler with tiny glass spheres. The resulting higher resin content dramatically improved the filing and sanding characteristics of the filler as well as the filler's adhesion and water resistance.

Most lightweight fillers are homogenous. The glass bubbles remain suspended in the resin and do not settle to the bottom of the can. This homogenous composition allows lightweight fillers to be packaged in plastic bags or cans and dispensed with rollers or compressed air or squeezed out with a plastic spreader. The plastic bags keep filler fresh and eliminate much of the waste sometimes associated with canned fillers.

Nationally, lightweight fillers represent more than 80 percent of the total filler used in body shops.

Premium Fillers

In the mid-1980s, filler manufacturers starting taking advantage of technology to produce premium quality fillers. Premium fillers have superior performance qualities that go beyond the capabilities of conventional lightweight fillers. Premium fillers are moist and creamy. They spread easily yet will not sag on vertical surfaces. They dry tack-free without pinholing.

Spot Putties

Spot putty, also called *glazing putty*, is used to fill minor surface imperfections on tiny sections of the repair area. Because body fillers usually have tiny pinholes and sand scratches, spot putty provides a quick way of leveling these imperfections with minimum effort.

One-part spot putties are applied directly out of the tube and cure slowly. They were developed to fill minor imperfections, producing a smoother surface. They are being phased out for quicker curing, more chemically stable two-part putties.

Although one-part glazing putties featheredge very nicely, they do not develop the hardness of a body filler. When coated with primer or paint, putties absorb paint solvents and swell. Sufficient time must be allowed for the putty to fully cure again before finish sanding of the finish coats. If sanded too soon, scratches will appear in the finish as the putty dries completely and shrinks below the sanded surface.

New basecoat/clearcoat paint systems caused problems with conventional one-part putties. The rich solvents and multicoats required for these "trick" paint jobs caused the pigment to "bleed" and stain the finish on light colors, usually after several days of exposure to sunlight. The widespread use of basecoat/clearcoat products and other multicoat systems has made this staining problem a more frequent occurrence.

Spot putties should be used only to fill very shallow sand scratches and pinholes. Maximum filling depth of most putties is only 1⁄32 inch (0.794 mm).

Conventional one-part glazing putty does not dry below its surface very quickly. It takes several hours, sometimes days, for one-part putty to dry all the way through. Even though the putty may sand normally, it still may not be completely dry at the bottom of the scratch or pit. If paint is applied over the partially cured putty, the surface may sink or shrink, as well as bleed, and cause a flaw in the paint or topcoat.

Polyester Glazing Putty

To solve these problems, body filler manufacturers have developed fine-grained, catalyzed **polyester glazing putty**, or *two-part spot putty.* Like body filler, a hardener or catalyst must be mixed with the putty to initiate and speed curing.

Polyester glazing putty does not shrink, has excellent dimensional stability, and resists solvent penetration (the cause of bleedthrough). When applied over traditional body fillers, polyester glazing putties effectively solve the bleedthrough problem.

Many two-part spot putties can be applied over small dents or "dings" in the paint. All you have to do is scuff the surface of the paint, blow the surface clean, and wipe on a thin coat of putty over the indention. You can then featheredge sand the putty level with the surrounding paint. This saves considerable time. You also do not have to sand down to bare metal, which cuts through factory corrosion protection under the paint.

12.2 APPLYING BODY FILLER

Body fillers are designed to cover up minor depressions that cannot be removed by metal straightening alone. As discussed, there are several types of filler for specific tasks.

Table 12–2 summarizes the ingredients, characteristics, and applications of the currently available body fillers and putties. Study it carefully so you can use the product best suited for a particular job.

WARNING Always refer to manufacturer instructions and recommendations before using products. Taking shortcuts when mixing and applying fillers might save time at first, but it affects the quality of the repair. Fillers that are improperly used will eventually crack, lose adhesion, permit rust to form, or even fall off the panel. Sooner or later, the repair will have to be redone. The loss in time, money, and reputation can be considerable.

Preparing Surface For Filler

One of the most important steps in applying body fillers is surface preparation. As emphasized in earlier chapters, always begin repairs by washing the vehicle to remove dirt and grime. Then clean the area with wax and grease remover to eliminate wax, silicones, road tar, and grease that can contaminate the area.

Wash brazed joints between panels with soda water to neutralize the acids in the flux. Do not grind these areas before neutralizing the acids. Grinding simply drives the acids deep into the metal. New cars still have brazed joints between highly flexible panels.

Grind the area to remove the old paint see (Chapter 11). Remove the paint for 3 or 4 inches (76–101 mm) around the area to be filled. If filler overlaps any of the existing finish, the paint film will absorb solvents from the new primer and paint, destroying the adhesion of the filler. The filler will lift, cracking the paint and allowing moisture to seep under the filler. Rust will then form on the metal.

Use a #24 or #36 grit grinding disc to remove the paint. This coarse grit removes paint and surface rust quickly and also etches the metal to provide better adhesion. You can also blast off the paint to prevent any removal of metal (Figure 12–4).

If applying filler over a metal patch, avoid hammering down the excess weld bead. Instead, grind it level with the surface. Hammering the weld distorts the metal, creates stress in the panel, and increases the thickness to be filled.

Use metal straightening methods to straighten the metal as much as possible to match its original contour. If the panel is flat, use a long straightedge to gauge the panel's straightness. Hold it over the repair area and look for high and low spots. You must shape the panel to within ⅛ inch (3.1 mm) of its original contour.

After grinding away the finish from the repair area, blow away the sanding dust with compressed air, and wipe the surface with a tack rag to remove any remaining dust particles.

A common mistake for the beginner is to apply filler over paint. This is an easy to do but serious mistake that will affect repair quality. Remember that body filler is designed to be applied to the body substrate, usually metal, not to paint.

Wear eye protection and a respirator when blowing off dust. It is very easy for particles to fly into your eyes. The dust can also contain materials that are harmful if inhaled.

TABLE 12–2 COMPARING FILLERS AND PUTTIES

Filler	Composition	Characteristics	Application
		Conventional Fillers	
Heavyweight fillers	Polyester resins and talc particles	Smooth sanding; fine featheredging; nonsagging; less pinholing than lightweight fillers	Dents, dings, and gouges in metal panels
Lightweight fillers	Microsphere glass bubbles; fine grain talc; polyester resins	Spreads easily; nonshrinking; homogenous; no settling	Dings, dents, and gouges in metal panels
Premium fillers	Microspheres; talc; polyester resins; special chemical additives	Sands quickly and easily; creamy and moist; spreads smooth without pinholes; dries tack-free; will not sag	Dings, dents, and gouges in metal panels
		Fiberglass Reinforced Fillers	
Short-strand	Small fiberglass strands; polyester resins	Waterproof; stronger than regular fillers	Fills small rustouts and holes; used with fiberglass cloth to bridge larger rustouts
Long-strand	Long fiberglass strands; polyester resins	Waterproof; stronger than short-strand fiberglass fillers; bridges small holes without matte or cloth	Cracked or shattered fiberglass; repairing rustouts, holes, and tears
		Specialty Fillers	
Aluminum filler	Aluminum flakes and powders; polyester resins	Waterproof; spreads smoothly; high level of quality and durability	Restoring classic and exotic vehicles
Finishing filler/polyester putty	High-resin content: fine talc particles; microsphere glass bubbles	Ultrasmooth and creamy; tack-free; nonshrinking; eliminates need for air dry type glazing putty	Fills pinholes and sand scratches in metal, filler, fiberglass, and old finishes
Sprayable filler/polyester primer-surfacer	High-viscosity polyester resins; talc particles; liquid hardener	Virtually nonshrinking; prevents bleedthrough; eliminates primer/glazing/primer procedure	Fills file marks, sand scratches, mildly cracked or crazed paint films, and pinholes; seals fillers and old finishes against bleedthrough

Mixing Filler

Body fillers come in cans and in plastic bags. When packaged in a plastic bag, hand pressure or a dispenser can be used to force the filler onto your mixing board. This keeps the filler perfectly clean.

A **mixing board** is a flat surface (metal, glass, or plastic) used for mixing the filler and its hardener. A handy type of mixing board has a pad of coated paper on it. After use, you can simply tear off the used sheet and use the new, clean one underneath (Figure 12–5).

Mix the can of filler or knead the bag of filler to a uniform and smooth consistency free of lumps. If the container has been shelved for a while, using a paint shaker will save time.

If it is not stirred thoroughly to a smooth and uniform consistency before use, the body filler in the upper portion of the can will be too thin. Concurrently, the filler in the lower portion of the can will be too thick, making it very coarse and grainy. Close the container of filler right away (Figure 12–6).

Improper and inadequate body filler mixing can result in:

- Runs and sags during application
- Slow and poor curing
- Gummy condition when sanded
- Poor featheredging
- Tacky, sticky surface
- Blistering and lifting
- Poor adhesion
- Rampant pinholing
- Poor color holdout

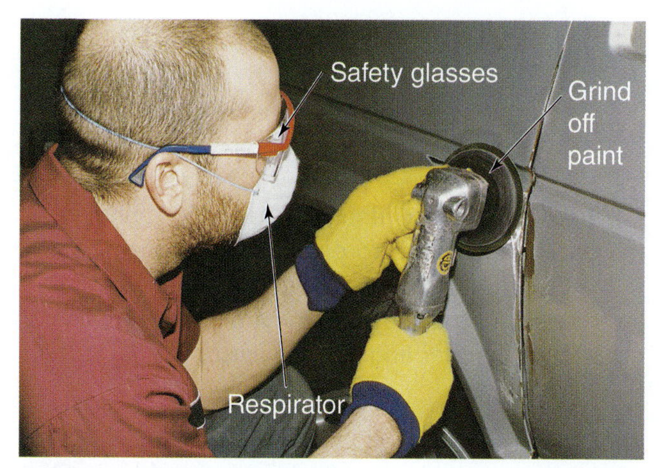

A Remove all paint several inches beyond the damaged area.

B Methodically grind from the top of the repair area to the bottom, removing all paint, but be careful not to grind too long in one location or you can thin and damage sheet metal.

C If needed, grind quickly along edges of panels to remove paint. Be careful that the grinding disc does not catch and dig into panel edges. A thin layer of body filler is commonly applied over an area where the metal has been worked. It is used to level and smooth a metal area that has been straightened.

Figure 12-4 To ready the damaged area on the vehicle for body filler, grind the area around the damage with a grinding disc.

A This mixing board has removable, or disposable, sheets. Tear off an old sheet to expose a new, clean sheet of mixing paper.

B Place the needed amount of body filler onto the mixing board.

Figure 12-5 A mixing board should be used when applying body filler. Never use cardboard, because it can be coated with waxes and other substances that can contaminate the repair area.

Figure 12-6 Always close the container of body filler right after it has been dispensed. Filler can dry out and become contaminated with dust and dirt if not sealed right away. Light taps with a plastic mallet will seal up the can of body filler.

Kneading the Hardener

Loosen the cap of the cream hardener tube to prevent the hardener from becoming air bound. **Hardener kneading** is done by thoroughly squeezing its contents back and forth inside the tube. This will ensure a smooth, paste-like consistency (like toothpaste) when it is squeezed out.

If you do *not* knead the hardener, the result can be the same problems as mentioned earlier for poor filler mixing. If the hardener feels gritty or is thin and watery, it has spoiled. Spoiled hardener should not be used, because it has chemically broken down. Hardener can spoil if frozen or if stored too long.

Mixing Filler and Hardener

Numerous problems can occur from the improper *catalyzing* (mixing) of *cream hardener* (filler catalyst) and filler. Before catalyzing, make sure the filler and hardener are compatible. They should be manufactured by the same company and recommended for use together.

The following tips will help eliminate problems relating to mixing filler and its catalyst. Open the can of filler without bending its lid. Using a clean putty knife or spreader, remove the desired amount of filler (Figure 12–7).

Place the filler on a smooth, clean mixing board. You can use a sheet metal, glass, hard plastic, or coated-paper mixing board. Mixing boards with a handle are also available commercially.

Add hardener according to label directions on the body filler container. Typically, use 2 percent hardener when mixing body filler. For example, for 100 parts of body filler, add 2 parts hardener. This would provide normal curing time for applying and shaping the filler.

Cardboard should *not* be used as a mixing board filler. It is porous and contains waxes for waterproofing. These waxes will be dissolved in the mixed filler and cause poor bonding. Cardboard also absorbs some of the chemicals in the filler and hardener, slightly changing the filler's curing quality. Cardboard fibers can also stick in the filler and ruin the finish.

Another example: For 4 tablespoons of body filler, add 1 teaspoon of hardener. This will result in approximately 2 percent hardener.

As a general rule, for each golf ball-size "glob" of filler, use a 1-inch (25 mm) bead of hardener. If the filler is as big as a baseball, use about a 6-inch (152 mm) bead of hardener. Use less in hot weather and a little more in cold temperatures. However, always refer to the manufacturer's instructions for exact mixing directions.

Filler overcatalyzation results when you use too much hardener for the amount of filler. Too much hardener produces excessive gases, resulting in pinholing. In addition to pinholing, adding too much hardener can cause paint color bleedthrough, poor adhesion, and poor sanding properties (Figure 12–8).

Filler undercatalyzation is caused by not using enough hardener in the filler. Too little hardener produces soft, gummy filler that will not cure or adhere properly to the metal. It will also not sand or featheredge cleanly. Figure 12–9 shows the results of adding too much and too little hardener.

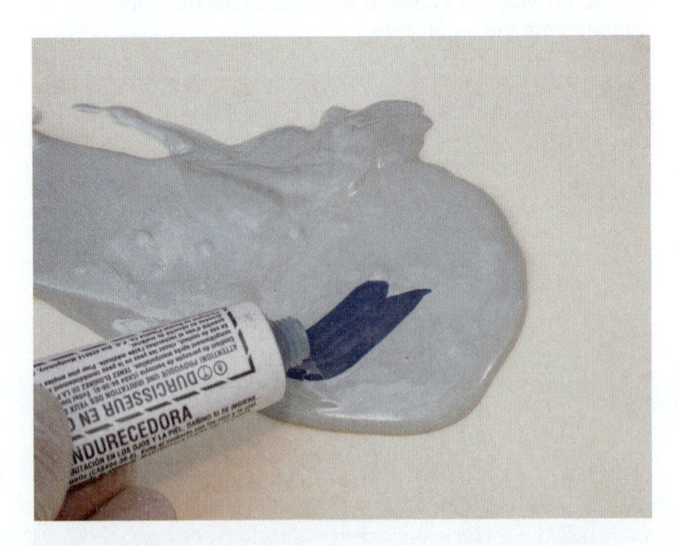

Figure 12–7 Body filler is a two-part material. Squeeze the tube to dispense a little hardener onto the body filler. Only add a small amount of hardener to avoid problems. For a golf ball-size glob of body filler, use about an inch-long bead of hardener.

Figure 12–8 Here is a common mistake made in auto body repair: Too much hardener has been mixed into the body filler. The body filler has cured solid in less than 5 minutes while still on the mixing board. There was not enough curing time to spread body filler on the damaged area of the vehicle.

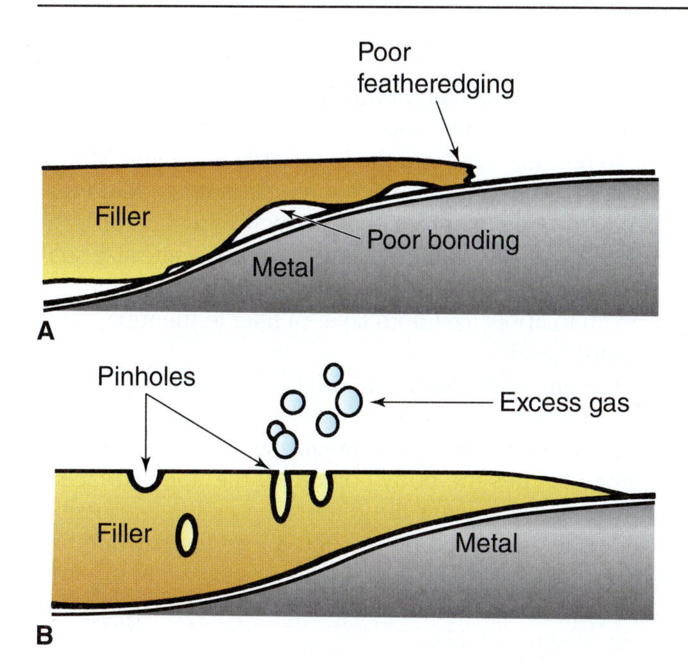

Figure 12–9 (A) Too little hardener often causes poor featheredging and weak bonding. (B) Too much hardener often results in pinholes. The body filler may also harden on your mixing board before you have time to apply it smoothly on the repair.

With a clean plastic putty knife, use a scraping motion (back and forth) to mix the filler and hardener together thoroughly and achieve a uniform color. Scrape filler off both sides of the knife and mix it in. After every few strokes back and forth, scrape the filler into the center of your mixing board by circling inward. Refer to Figure 12–10.

If the filler and hardener are not thoroughly mixed to a uniform color, soft spots will form in the cured filler. The result is an uneven cure, poor adhesion, lifting, and blistering (Figure 12–11).

Figure 12–10 Mix the filler and hardener with a scraping motion.

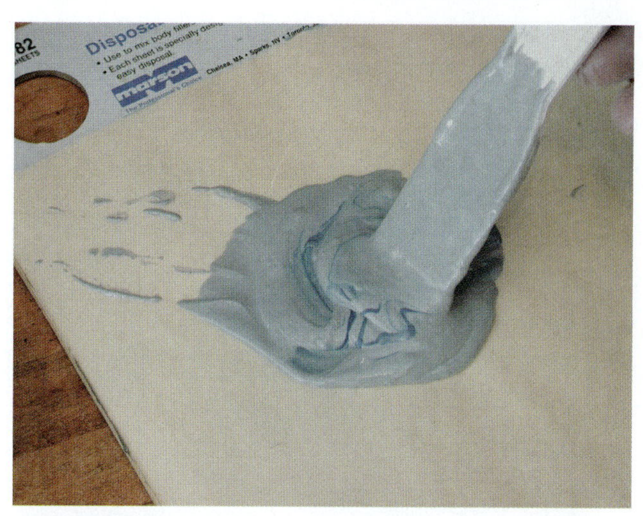

A Use a back and forth spreading motion to mix the filler and hardener.

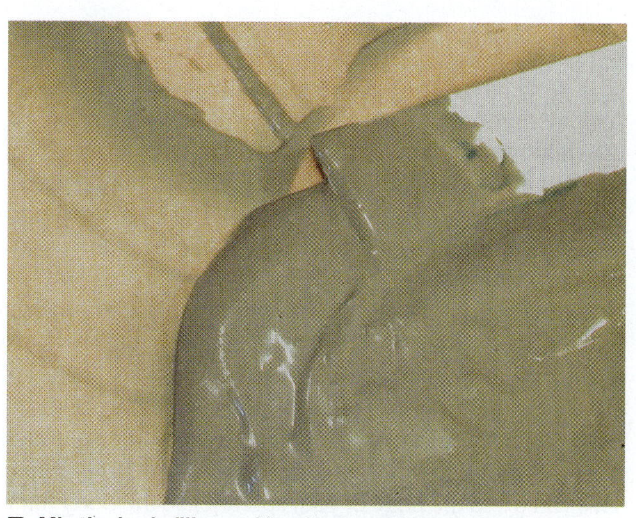

B Mix the body filler and hardener until it is a consistent, uniform color. Streaks or globs of filler and hardener must be completely mixed together to avoid problems.

Figure 12–11 Mix, do not stir, the body filler and hardener together or air bubbles and pinholes can form in the filler.

Do *not* stir the filler. Stirring whips air into the filler and causes air pockets and pinholes when it is applied.

Replace the cover on the can of filler promptly. This will keep out dust and dirt. It will also help prevent liquids in the filler from evaporating.

Always use clean tools when removing the filler from the can and mixing the filler and hardener together. *Do not* re-dip the knife, spreader, or mixed filler into the can. Over time, this causes the whole can of filler to harden. Hard lumps of filler might form in the can and/or applied filler, which will cause problems the next time you try to use the filler (Figure 12–12).

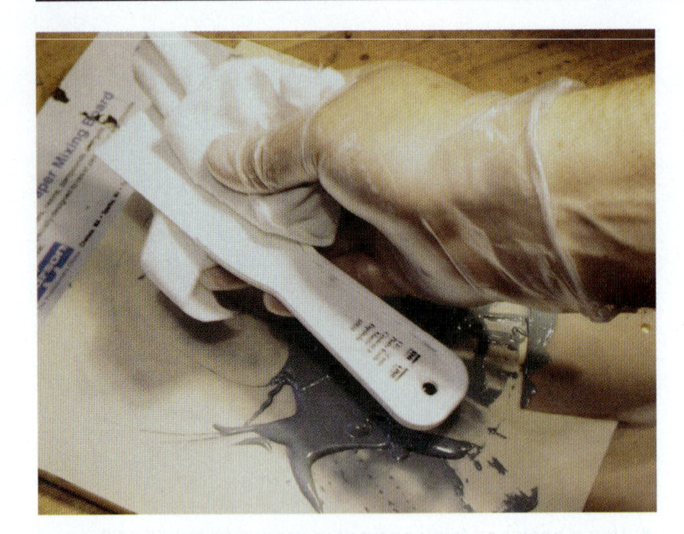

Figure 12–12 Wipe off the mixing stick or putty knife right after use to save clean-up time.

Use different spreaders to mix and to apply the filler. A small amount of unmixed filler will always remain on the mixing spreader. If any is applied, you will have soft spots in the cured filler, and the paint finish may peel.

Note that some spreaders have two sides. Only one side is for spreading or applying body filler, the one opposite the writing.

When working with a used plastic spreader, always make sure its edge is perfectly straight and smooth. If it is rough, gouged, grooved, or nicked, rub the spreader edge on a piece of #320 grit dry sandpaper. Lay the sandpaper on a smooth work surface and move the spreader over the abrasive while holding it at a 45-degree angle. This will dress the edge of your spreader so the body filler goes on smoothly.

If you mix or apply body filler in the direct sunlight, work time before the filler hardens will be shortened drastically. Avoid sunlight when using body filler.

Before applying body filler, mask off any panel openings, trim pieces, and other parts to help keep filler from sticking to or collecting on them. This will save time, because hardened body filler can be difficult to remove from parts.

The most common mistake of the apprentice is using *too much cream hardener*! The body filler will set up or harden in a couple of minutes or before you have time to spread it on the body. Literally, tons of body filler have been wasted because of hardener overuse. Using too much cream hardener can also cause problems with adhesion and result in pinholing.

Spreading Body Filler

Apply the mixed filler promptly to a thoroughly clean and well-sanded surface. Make sure you use an air blow nozzle to remove dust from the body surface (Figure 12–13).

A first tight, thin application is recommended. Press firmly to force filler into sand scratches to maximize the bond. Make sure you are using the appropriate size spatula. If the spatula is too big or too small, it will be difficult for you to apply a smooth layer of filler to the repair area (Figure 12–14).

Rough, improperly smoothed filler will take extra time to sand off. To properly apply body filler, you have to be a "sculptor." You must spread the soft filler so that its shape matches the body contour. If the body surface is almost flat, you want the filler to be as flat as possible while sticking up about 1/16 inch above the surrounding surface. If the surface is curved, you must wipe the filler so that it curves and matches the rounded shape on the body (Figure 12–15).

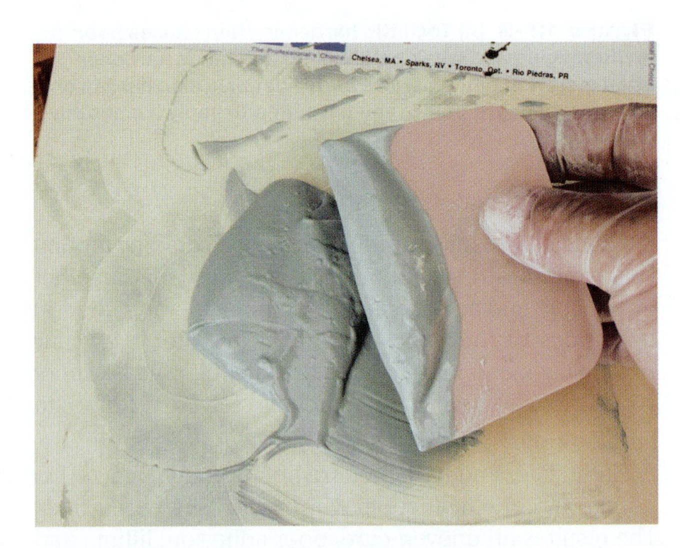

Figure 12–13 After mixing, use a clean spreader to center the mixed filler onto the mixing board.

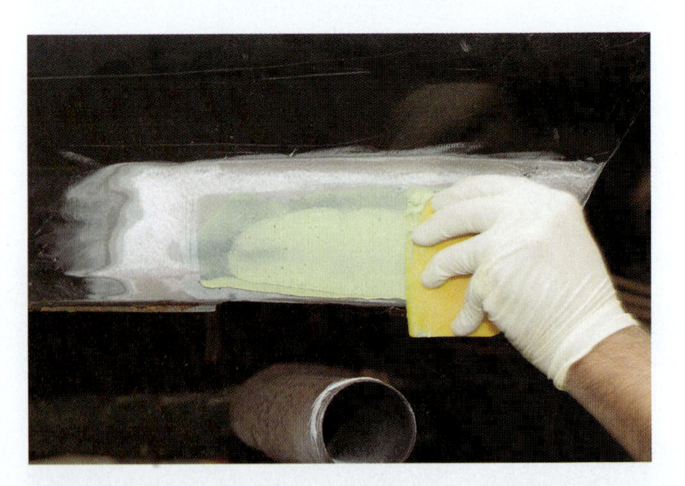

Figure 12–14 Go right to the vehicle and start applying body filler. Make sure you have blown off the repair area with compressed air so the body filler will stick and adhere properly.

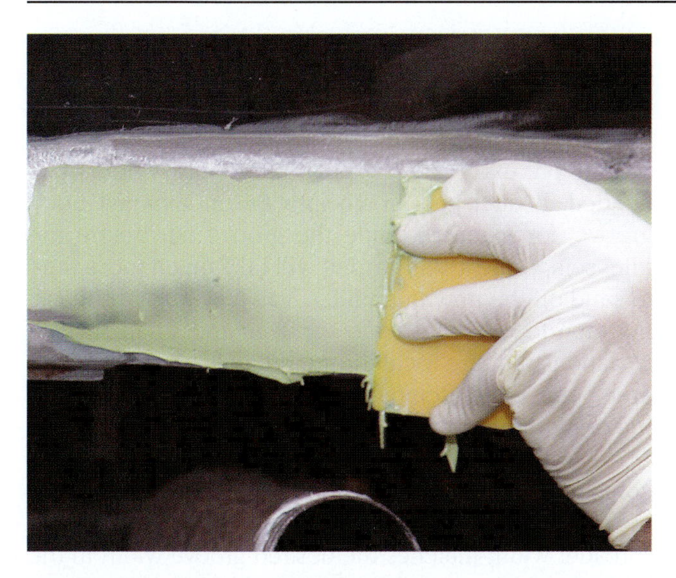

Figure 12–15 Apply filler to the repair area and spread it out several inches beyond the damaged area.

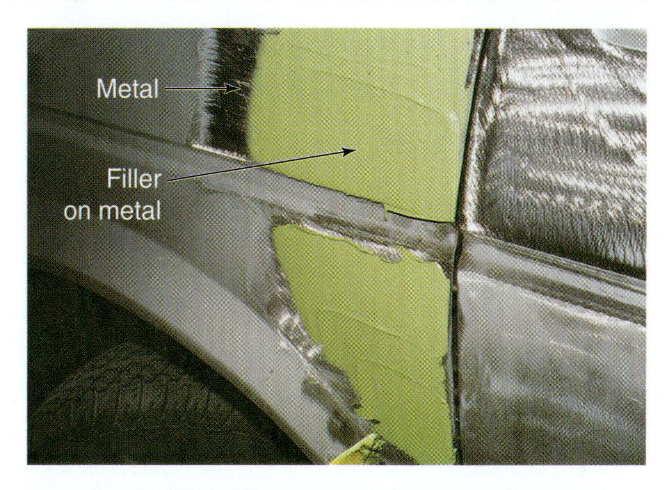

Figure 12–17 Apply body filler only where it is needed. Note how the body line is not damaged, so care was taken to keep filler out of the groove in the panel.

When applying filler, wipe it on one way and then the other, usually upward and then sideways, over the repair area. This will help work the filler down into the grind marks in the metal for good adhesion and a minimum of air pockets (Figure 12–16).

Never apply body filler in a thick, single coat. This is a common mistake that can make the filler crack, pinhole, and not bond properly to the body. When applying body filler, use one or more thin coats to fill remaining low spots in the body surface (Figure 12–17).

Make sure the metal surface is completely dry and free of dust before applying filler. If filler is applied over a layer of sanding dust or a wet surface (from cleaning agents or wet sanding), the body filler may not bond securely to the body.

Body filler will not bond to smooth, unground, or unsanded metal. The metal *must* be sanded so the body filler can "bite into" and adhere to the metal. Also remember to sand several inches beyond the area to be coated with body filler.

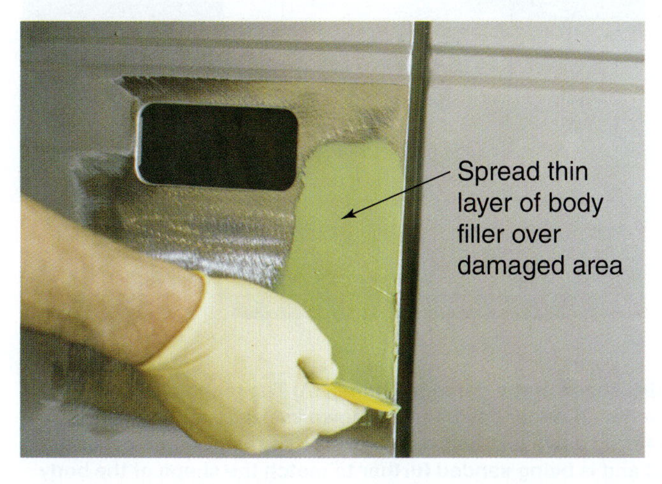

Figure 12–16 After wiping body filler over the repair area, use a steady wiping motion to spread it out smooth.

SHOP TALK

Always use a clean plastic spreader to apply the filler. *Do not* use the same spreader you used to mix the filler. A small amount of unmixed (uncatalyzed) filler will always remain on the mixing spreader. If any of this is transferred to the metal, there will be soft spots in the cured filler and the paint finish will often peel.

Use a straightedge to make sure there are no metal high spots in the repair area. The whole area to be filled with body filler must be slightly lower than the surrounding panel surface. If you apply filler to an area with a high spot of metal, you will have to use a picking hammer to lower the metal, possibly breaking the bond between the cured filler and the body panel.

Avoid using filler in cold temperatures. When the filler, shop, and/or body panel temperatures are cold, the body filler will not cure properly. This produces a filler that is too soft and results in a tacky surface and poor sanding properties. Tremendous pinholes can also be created. Filler should be stored at room temperature. If needed, a heat lamp can be used to warm cold surfaces on the vehicle body.

Avoid moisture on the repair area in conditions of high humidity in the shop. Again, use a heat lamp to warm and dry damp surfaces before applying filler. Failing to remove moisture accumulation on the surface can result in inadequate adhesion, poor featheredging, pinholing, and lifting when recoating with refinishing materials.

If you have to apply filler over two adjacent panels, make sure you have a parting line between the panels. Use the edge of your spreader to remove any filler

Figure 12–18 **Note how this technician has applied filler to the fender and door. Before the filler cures completely, the technician removes excess filler with the edge of the spreader. This will save sanding time.**

between body gaps. It is much easier to remove this excess while the filler is still soft (Figure 12–18).

While the filler is still soft, also remove any excess filler around the repair area. Use your spreader to scrape off extra filler. Apply filler only where it is needed. If filler gets down into an undamaged body groove, for example, wipe the groove out before the filler sets. Once it sets hard, it will take much more time to sand off the unwanted filler. You may also use masking tape to cover intricate surfaces and small surface variations that are located immediately adjacent to the area being filled. Remove the tape prior to the filler curing or hardening.

Figure 12–19 shows a technician filling a very complex, difficult shape on the rear of a truck bed. Note how he is using spreaders cut down in width. Each spreader width matches the desired groove width in the body panel.

A

C

B

D

Figure 12–19 **Study how body filler is being applied to a complex shape of the damaged panel. (A) The technician is filling a repair area in a very complex corner of a truck bed panel. Note how the spreader has been cut down in size to match the width of the body line. (B) A tiny high spot in the metal is being pick hammered down just below the surface so the filler will cover the whole repair area. (C) A second coat of body filler has been applied and is being sanded further to match the shape of the body panel. (D) Normally, never finger sand. Only sand with your fingers when your fingers match the shape you want to form in the body filler.**

12.3 GRATING AND SANDING BODY FILLER

After application, you must closely monitor how quickly the filler sets up or hardens. On larger repair areas, you should file or grate the filler before sanding. Grating removes excess filler sticking up beyond the desired level. Sanding is done to smooth and shape the filler.

Grating the Filler

Filler grating is done with a coarse body file to remove high spots or edges that stick up on the freshly applied filler. Allow the filler to cure to a semi-hard consistency. This usually takes 5–10 minutes.

If the filler leaves a firm, white track when scratched with a fingernail, it is ready to be filed. Filing is important because it greatly reduces sanding time and reduces the amount of dust generated during the repair.

The *surform*, or *cheese grater, file* is used to cut the excess filler to contour quickly. Its long length produces an even, level surface. The teeth in the file are open enough to prevent the tool from clogging. Grinders, sanders, and air files do not level well. They become loaded quickly, create too much dust, and waste a lot of sandpaper (Figure 12–20).

To use the grater, hold it at a 30-degree angle and pull it lightly across the semi-hard filler. Work the filler in several directions. Stop filing when the filler is slightly *above* the desired level. This will leave sufficient filler for sanding out the file marks and for feathering the edges. If the filler is undercut, additional filler must be applied (Figure 12–21).

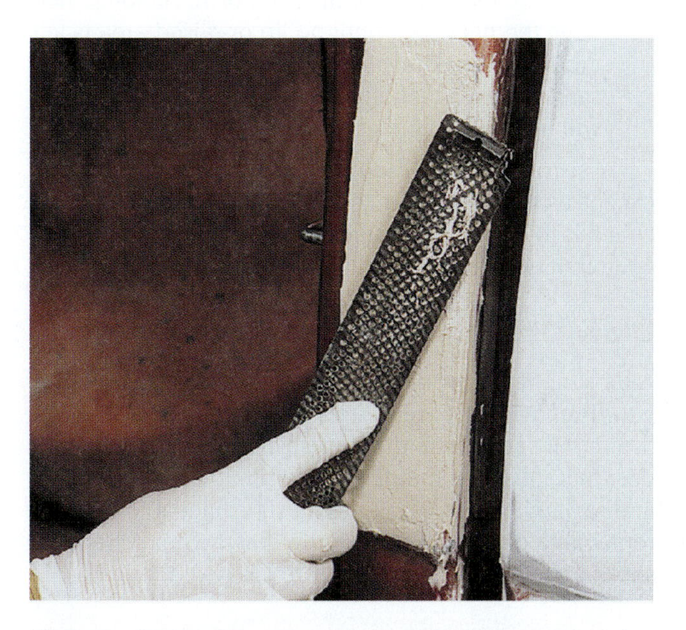

Figure 12–20 When the filler is only partially hardened, file off the high spot with a super-coarse cheese grater file. Do not cut too deeply into the semi-hard body filler or you will not be able to sand out the file marks.

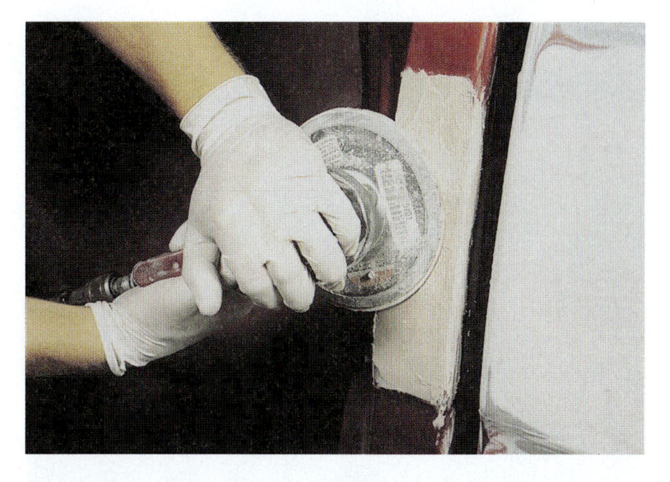

Figure 12–21 When initially sanding filler, use coarse sandpaper, like #80 grit, and as large a sander as will fit the repair area.

Coarse Sanding Filler

After grating, sand out all file marks with very coarse sandpaper. Use a block or air file on large, flat surfaces. Use a disc orbital sander, or **DA**, on smaller areas. An air file is often used on large, flat areas. Do not to try to sand out all imperfections in the first coat of filler (Figure 12–22).

Start with #36 or #40 grit sandpaper to quickly shape the filler. Only sand the first coat to get the general shape of the repair. Then, follow with a finer #80 grit sandpaper until all grating and coarse sandpaper scratches are removed. Final level and shape the body filler with #150 grit sandpaper. Sandpaper with #150 grit will cut fairly quickly but is fine enough that thick, high-solids primer-surfacer will fill and level the remaining sand scratches.

A common mistake is oversanding the first layer of filler below the desired level. Two or more coats of body

Figure 12–22 An air file is commonly used first on relatively flat body surfaces to quickly level body filler. Note that the technician is wearing a dust mask to avoid inhaling plastic dust particles.

Figure 12-23 After knocking off high spots in the body filler, block sand the repair area. Hand-block sanding is the best way to smooth and level a repair area properly.

filler are normally needed to get a good, smooth surface (Figure 12–23).

After sanding the first layer of filler to shape, blow off the area and apply a second layer of body filler. Work the filler in two directions to fill any imperfections or holes in the repair.

You must make the file or sander match the angle or shape of the panel. Tilt the air file upward to match the lip shape on the panel.

Some shops are equipped with a vacuum system that attaches to the air sander. Holes in the sandpaper allow the vacuum to pull the sanding dust off the surface so it does not float into the shop area.

WARNING The most common mistake is sanding too much in one place. If you hold the air sander in one location for too long, you will cut a flat area in the filler. Keep the sander moving while monitoring the shape of the filler. With coarse sandpaper, lightweight body fillers sand off very, very quickly.

DANGER Sanding generates a tremendous amount of plastic dust. It is critical that you wear a respirator or dust mask when sanding to protect your lungs. Even though the dust might not make you feel ill right away, the long-term effects of the plastic particles in your lungs could cause major health problems later.

Final smoothing should be done using #150 to #180 grit paper until all #80 grit scratches are removed. The DA or air file can again be used or a long speed file can be used.

Be careful not to oversand. This results in the filled area being lower than the desired level, which in turn makes it necessary to apply more filler. Oversanding is a common mistake for the novice body technician. Always sand a little and then check your work. Slowly cut the filler down until it is almost flush with the undamaged surface. On flat surfaces, you can use a straightedge to check filler straightness.

Generally, when sanding filler, use the following sanding procedure:

1. Start out sanding body filler with coarse sandpaper on a larger-size sander. Coarse grit is needed to rapidly knock off high spots and any deep grate (file) marks in the filler. Use a long air file or large-diameter DA sander to level the filler when possible.

   If less sanding is needed, hand or power sand using a medium sandpaper. Hand sanding with a long sanding board or sanding block is better than power sanding on smaller repair areas. The coarse to medium sandpaper will let you quickly remove and shape the filler to slightly above the desired level.
2. Stop sanding the body filler when the repair area is still a little high. There should still be a slight hump in the filler compared to the surrounding body surface. You must change to finer sandpaper as you continue to shape the repair area to avoid deep scratches in the filler.
3. To final level and feather the filler, hand block or power sand the surface down level with an even finer grit of sandpaper (#150 to #180 grit). Hand-sanding with a rubber block or a sanding board will allow better control of filler removal. Hand-featheredge the filler area until no lip or raised area is felt around the perimeter of the repair area. You must sand the area smooth enough so that primer coats will fill the remaining sand scratch marks.

A second coat of body filler is normally required, except on minor repairs to very small areas. A second coat fills low spots and builds up the filler to the existing contour if needed.

Blow Off Sanding Dust

After sanding, blow off the area with a high-pressure air gun. Then wipe the area with a tack cloth. This will make the repair more visible and also removes fine sanding dust that might be hiding surface pinholes. These holes and remaining sand scratches must be filled. Blowing the repair area clean will also ensure that the next filler coat sticks or adheres properly (Figure 12–24).

Checking Filler Repair

Inspect the filler closely. Run your hand over the surface while sanding to check for evenness. Do not trust only

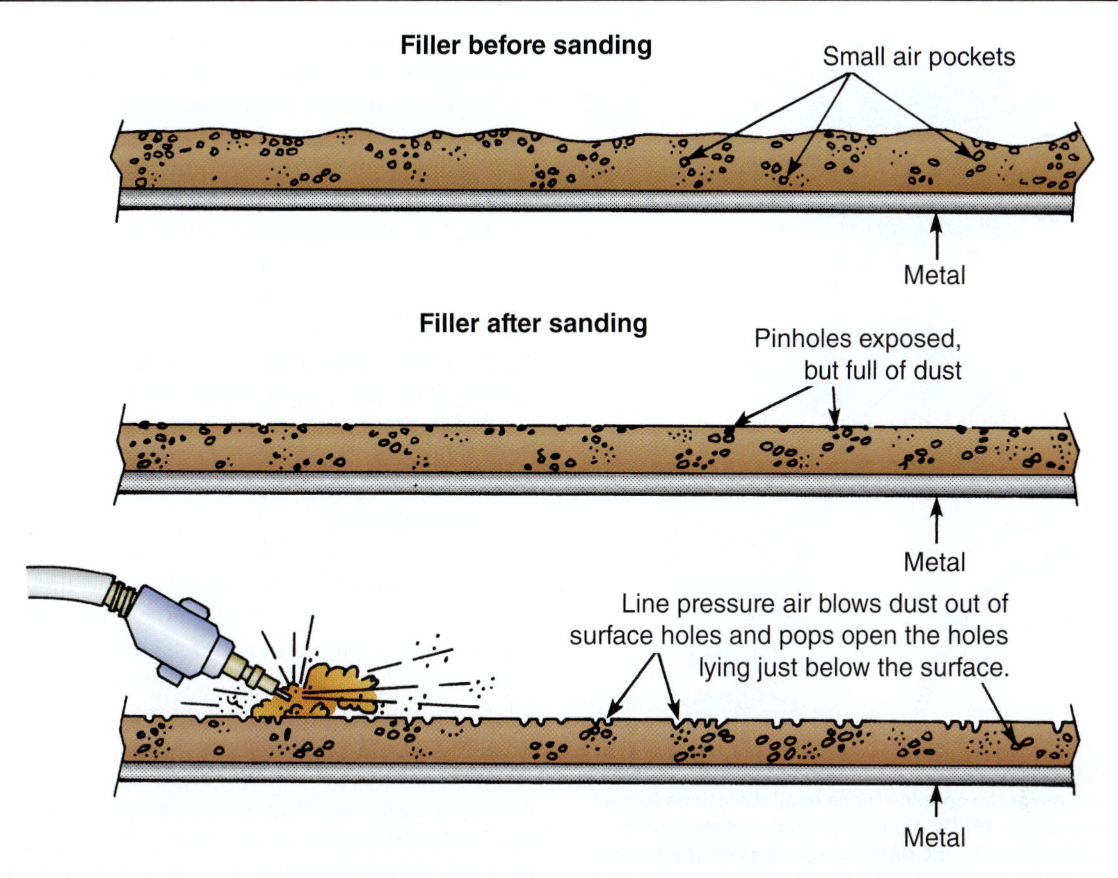

Figure 12–24 Blow filler dust out of pinholes before inspecting the repair area and before applying more filler.

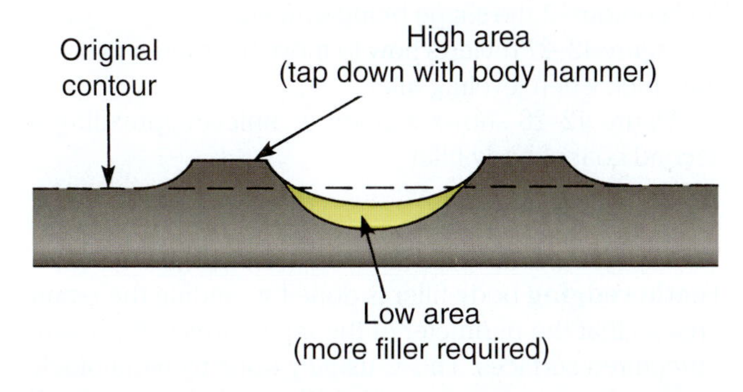

Figure 12–25 Feel carefully for high and low spots. Move your hand or a clean rag over the surface quickly in different directions. This will help you detect bumps and dips in the surface.

your eyes for accuracy. Remember, paint does not hide imperfections; it highlights them. Do not be satisfied until the repaired surface both looks *and* feels perfectly even. Inspect for craters, pits, scratches, and other surface imperfections that must be refilled.

Inspect for high and low spots in the filler. If you find small areas of metal exposed in the sander filler, the sheet metal is too high. Use soft blows from a pick hammer to lower these high spots in the metal, as illustrated in Figure 12–25.

Applying Second Filler Coat

A second filler coat is often needed to build up the level of the first coat or to fill remaining surface imperfections. Using a clean mixing board and spreader, apply the second coat over the first (Figure 12–26 and Figure 12–27).

Figure 12–28 shows a technician applying a second coat of body filler to the bottom of a door panel. A thin layer of filler has been applied to the whole bottom of the door. This will ensure that the panel is not wavy when finished.

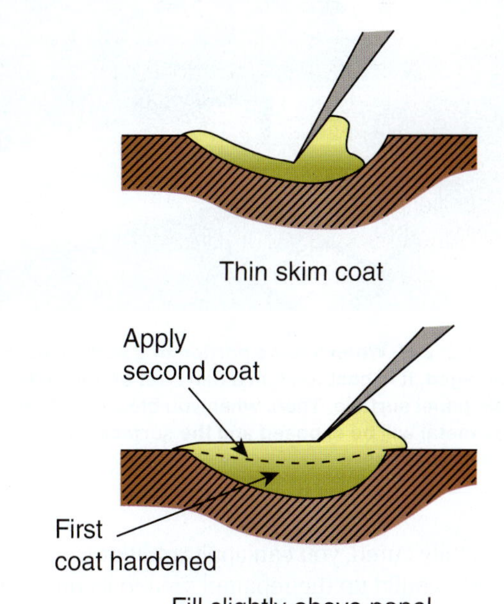

Figure 12–26 Build body filler thickness in several coats. Never apply a thick coat, or cracks can develop from heat expansion during curing.

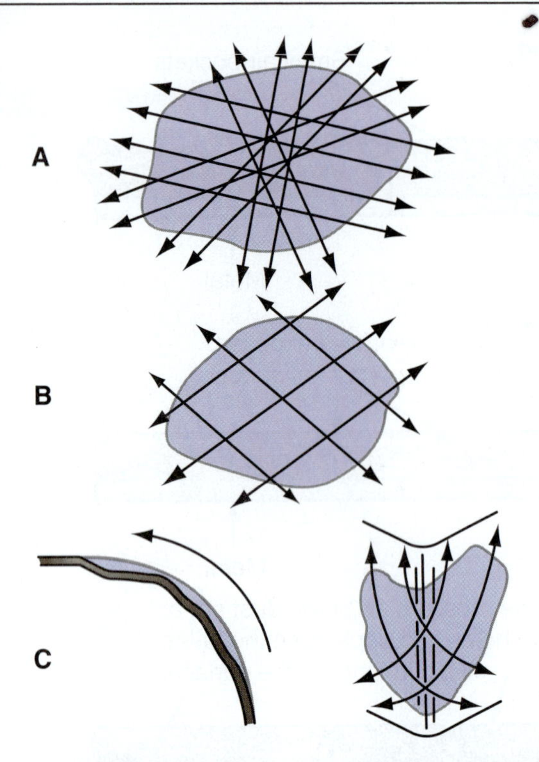

Figure 12-27 **Study the basic sanding methods. (A) On a flat surface, move the sanding block in all directions to cut an area down evenly. (B) On a gently curved surface, move the sanding block in specific directions across or at an angle to the curve. (C) On a sharply curved part, move the sanding block with a rolling action to move over the top of the curve at an angle.**

Figure 12-28 **When a large portion of a body panel has been damaged, it is best to skim a thin coat of body filler over the whole panel surface. Then, when you block sand the panel, no metal will be exposed and the surface can be sanded flat more easily.**

When fully cured, you can apply additional filler coats as needed to build up the repaired area to its proper contour. Allow each application to set up before applying the next coat of filler. Conventional body fillers should be built up slightly so that the waxy film that curing produces on the surface of the filler can be removed by sanding. It is seldom necessary to file or grate the second coat, because you do not want to cut too deeply into the surface.

To sand and level the second coat of body filler, start with #80 to #120 grit sandpaper. Then smooth the second coat of filler with #150 to #180 grit sandpaper to ready the area for a spray coat of epoxy primer-surfacer.

Always select sandpaper grit based on the amount of sanding that must be done. Use a coarser grit for fast material removal from a larger area. Use finer grits on smaller areas to avoid forming deep scratches in the body surface. Remember that coarser sandpapers leave deeper scratches. These scratches will have to be fine sanded to prepare for priming or primer-surfacers. However, if you use too fine a dry sandpaper on body filler (#400 and finer), the sandpaper abrasive will clog quickly with filler material.

To feather the body-filled area with the surrounding unrepaired area, hand-block or DA-sand the plastic surface down level with #100 to #150 grit sandpaper. Hand-sanding with a rubber block or sanding board will allow better control of filler removal. Fine sand and feather the filler area until no lip or raised area can be felt around the perimeter of the repair.

It is unlikely that you will often sand with your hands or fingers alone holding the sandpaper. You would only do so when the curve of the body panel exactly matches the curve of your hand or fingers. In Figure 12–19D, note how the width of the technician's finger equals the width and contour of the shape being sanded.

Figure 12–27 reviews how to move the sander or sanding block when leveling filler.

Figure 12–28 shows a body technician spreading a second coat of body filler.

Featheredging

Featheredging body filler is done by sanding the repair area so that the perimeter of the repair area is flush with unrepaired surfaces. This is usually done by hand-blocking over the repair area and into the undamaged area. If viewed from the side in a cutaway view, the filler should plane straight into the undamaged area of the body panel (Figure 12–29).

A trick to make sure you have a good featheredge is to be sure there are no sharp or distinct lines where different materials (paint, primer, filler, or metal) overlap. The lines showing different materials should look like a "bird's feather," with a soft, fluffy-looking edge, as shown in Figure 12–30.

When satisfied with the smoothness of the filler surface, clean the area with a tack cloth. A tack cloth picks up bits of filler dust that normal cleaning leaves behind. Remember that the tiniest particle will mar or ruin the paint job.

When the panel is perfectly smooth and level or contoured properly, the panel is ready to be primed, sealed, or sprayed with primer-surfacer to further prepare the surface for paint (Figure 12–31).

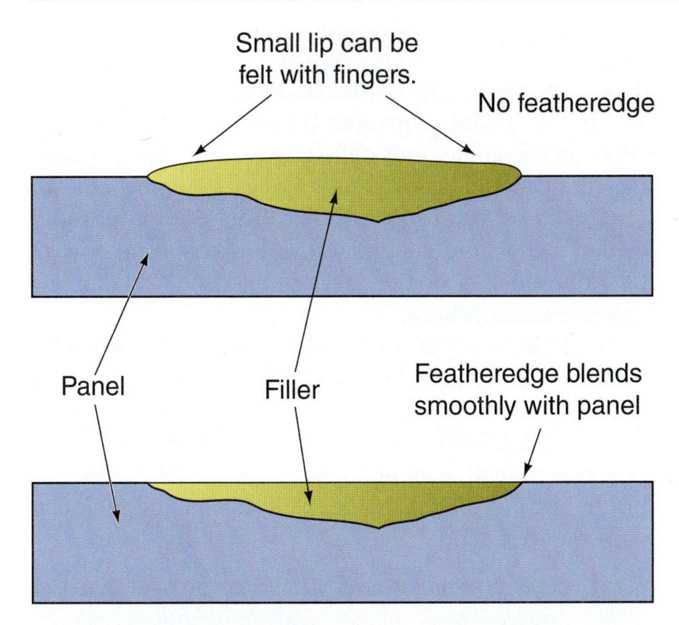

Figure 12–29 Note correct and incorrect featheredging. (A) The filler has not been featheredged properly; a lip still exists between filler and panel. (B) With proper featheredging, the filler and panel meet smoothly, or feather into each other.

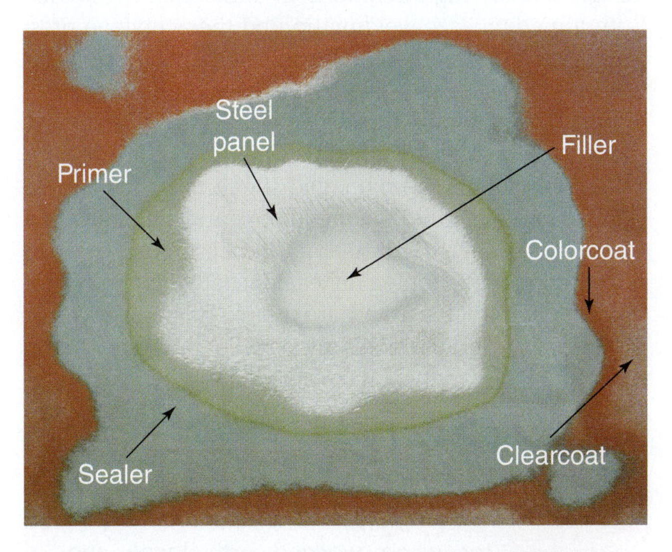

Figure 12–30 Study this properly featheredged repair area. Note how no sharp or distinct lines are visible where the different repair materials meet. One or two coats of thick primer-surfacer will fill the feathered repair area so it can be block sanded perfectly smooth.

Applying Filler to Body Lines

Many cars and trucks today have sharp body lines (grooves along the panel) in doors, quarter panels, hoods, and other body parts. Maintaining the sharpness of these lines when doing filler work is difficult, especially in recessed areas. There are several sanding aids to help you restore body lines using body filler.

A **contour sanding stick** has a specific shape to match the body line being sanded. An assortment of sanding stick shapes are provided in each set. This will save time when sanding these difficult areas (Figure 12–32).

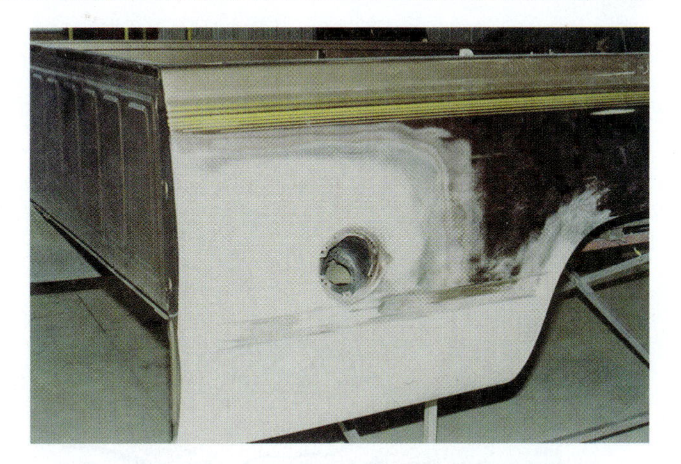

Figure 12–31 Note how this large area on a truck bed has been filled and feathered into undamaged areas. Spraying on primer-filler will allow further block sanding so no repair areas can be seen when it is painted.

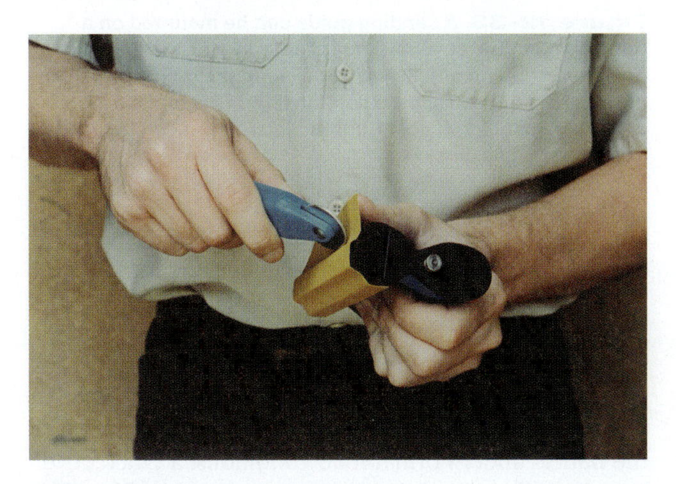

Figure 12–32 When sanding body lines, a specially shaped sanding block will save time.

A roller is often used to force sandpaper to match the shape of a contoured sanding stick. You can select from a variety of sanding stick shapes to match precisely the contour of the body line. Each stick indicates vehicle manufacturer and model with directional arrows for proper usage.

A **body line sanding guide** is a special tool for sanding straight, special contour lines on panels. The tool framework mounts on the panel. It has long slides that will hold the sanding block for making perfectly straight cuts in the filler. The sanding block is designed in segments so that it can be shaped to match the irregular shape of the edge. After fitting the block over a good section of the panel and moving its segments/rubber blades to match the contour, the rubber blades are locked in place. The block is then mounted in the guide and moved over the filler. This enables you to cut a straight contoured edge in the panel quickly and easily (Figure 12–33).

Applying Filler to Panel Joints

Many panels on unibody vehicles have joints that are factory-finished with a flexible mastic (seam sealer) or a

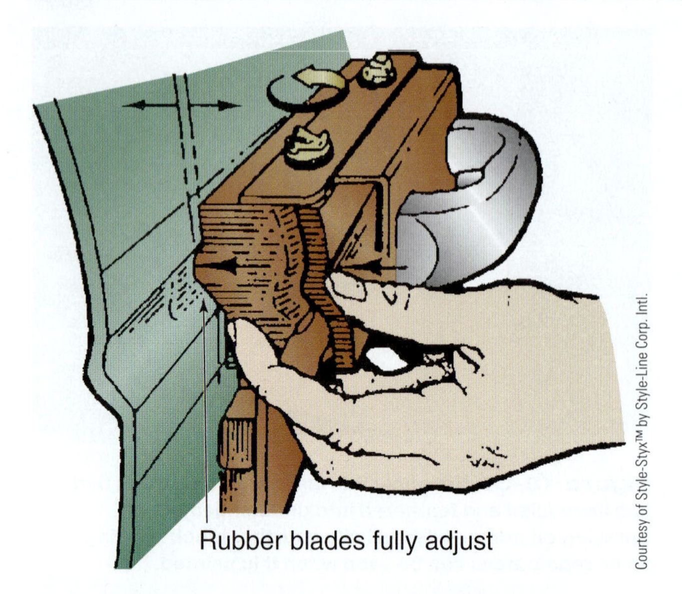

Figure 12–33 A sanding guide can be mounted on a panel to sand the straight-line contour in the panel. Specially designed sanding blocks will slide along the guide rails to make a perfectly straight cut in the filler. This tool has a shapeable sanding block. Rubber blades can slide to match irregular shapes.

brazed joint that allows the panel to flex and move. Many times both halves of the body joint suffer damage and require filling. Often, inexperienced technicians make the mistake of covering the damaged joint with body filler.

When the filler is subjected to the twisting action of the panels under normal road conditions, a crack develops that allows moisture to seep under the filler. This causes rust to form on the metal surface, which eventually results in the failure of the filler-to-metal bond and a weakened sheet metal joint (Figure 12–34).

The original flexibility of the joint can be preserved by taping its alternate sides. Tape is applied to one panel. Filler is then applied to the other panel and the tape is pulled up, removing the excess filler. The other panel is filled in the same way. After the filler in both panels is cured and shaped, a sealer is forced into the joint, as shown in Figure 12–35.

Applying Lead Filler

Most body shops use plastic body fillers exclusively in dent repair. Those shops that do use lead filler, or *body solder* as it is sometimes called, use it only when restoring antique and classic automobiles or doing custom work. Some shops use lead filler to fill door edges and welded seams, but generally lead work is done only at the request of a customer. It is a specialized skill that few body technicians have.

Lead filler is an alloy of lead and tin. Most lead solder used in body repair is 30 percent tin and 70 percent lead. Thus, it is often called 30/70 solder. At approximately 360°F (182°C), 30/70 body solder becomes soft or plastic. It becomes liquid at approximately 490°F (255°C). Within this heat range, body solder is plastic enough to be worked and shaped.

NOTE *Lead body filler was designed for use on older, thicker, mild steel body panels. Lead should not be used on modern thin gauge, HSS body panels because the heat can cause panel warpage.*

Use an oxyacetylene welding torch, or a specially designed soldering torch, to heat and soften the solder. With a medium-size welding tip and the acetylene pressure set at 4–5 psi (28–34 kPa), adjust the torch for a carburizing flame. The low heat and wide flame are adequate to heat the solder and the repair area without overstressing most mild steel sheet metal panels.

After bumping the damaged sheet metal back to its original contour, grind the metal bare and clean it with a metal conditioner. Then heat the repair area and brush on a tinning flux. Tinning fluxes are used to clean microscopic rust particles from bare metal. This prevents rust from forming and promotes adhesion of the solder to the sheet metal. A variety of fluxes is available.

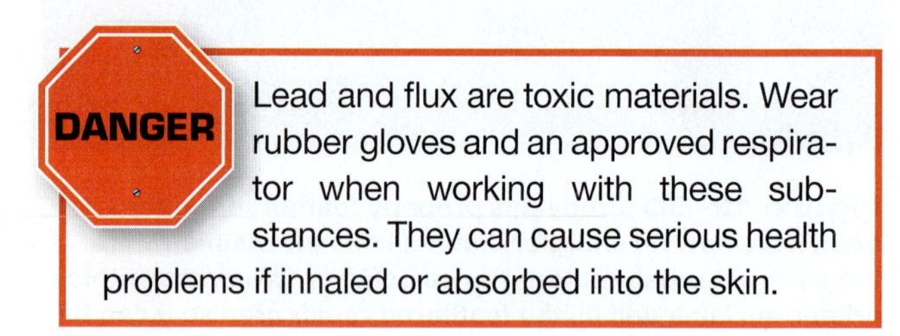

Lead and flux are toxic materials. Wear rubber gloves and an approved respirator when working with these substances. They can cause serious health problems if inhaled or absorbed into the skin.

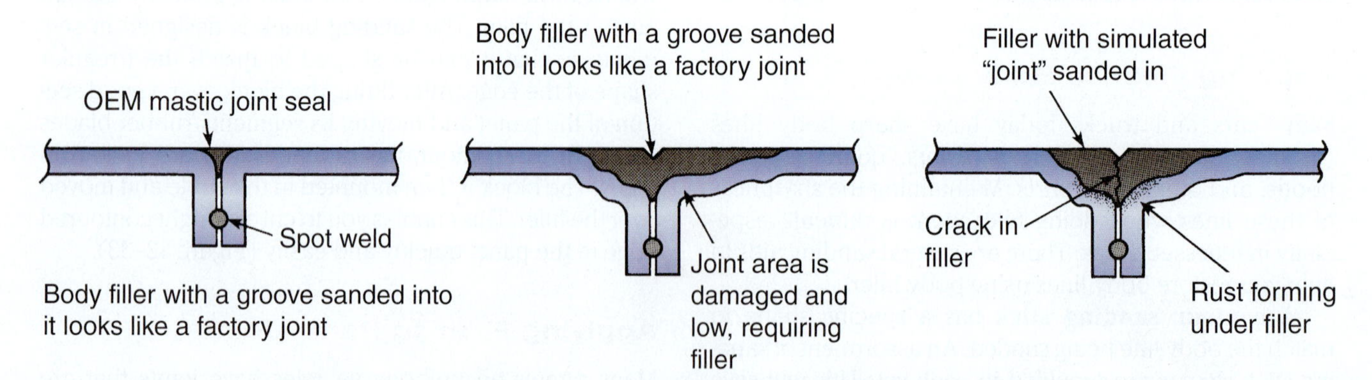

Figure 12–34 This body joint was accidentally covered with filler, which cracked.

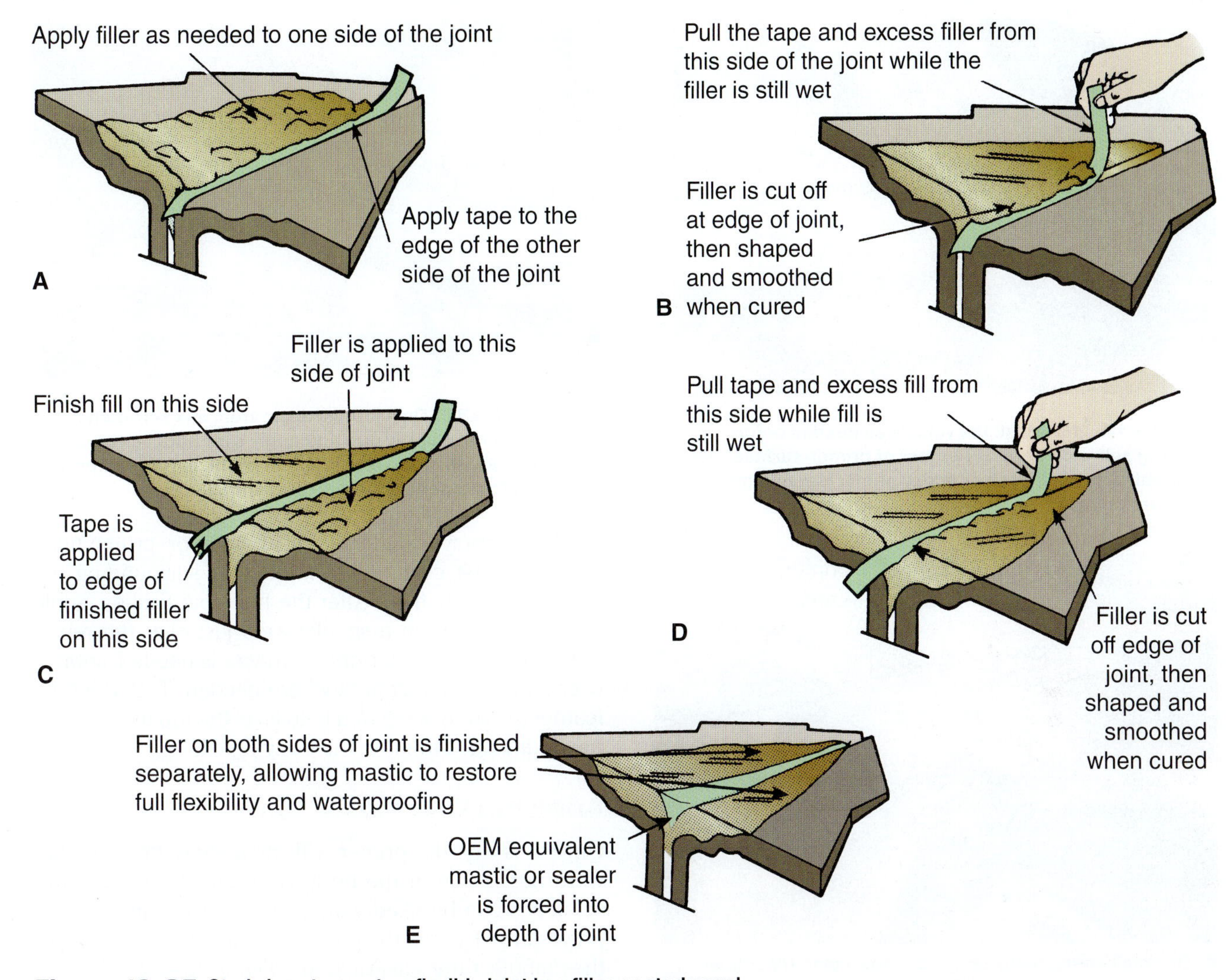
Apply filler as needed to one side of the joint
Apply tape to the edge of the other side of the joint
A
Pull the tape and excess filler from this side of the joint while the filler is still wet
Filler is cut off at edge of joint, then shaped and smoothed when cured
B
Filler is applied to this side of joint
Finish fill on this side
Tape is applied to edge of finished filler on this side
C
Pull tape and excess fill from this side while fill is still wet
Filler is cut off edge of joint, then shaped and smoothed when cured
D
Filler on both sides of joint is finished separately, allowing mastic to restore full flexibility and waterproofing
OEM equivalent mastic or sealer is forced into depth of joint
E

Figure 12-35 **Study how to create a flexible joint in a filler-coated panel.**

After applying the flux, heat the metal with the torch and rub the solder bar over the hot metal. This will deposit a thin layer of solder over the repair area. Although the solder is still plastic, wipe the area with a clean shop cloth. This will spread the solder over the bare metal and will remove any impurities. The tinned area will be silvery white.

Next, fill the area with solder. Do this by heating 1 inch (25 mm) of the solder rod or bar and the adjacent metal. Press the heat-softened bar against the hot metal. This will deposit the solder on the sheet metal. Do this as much as is necessary to fill the repair area.

Now, shape and smooth the solder. Heat the solder by moving the welding torch with a carburizing flame back and forth over the solder until it begins to sag. Then, with a soldering paddle, spread the solder and smooth it over the repair area. A flat paddle is used on flat and convex surfaces. A curved paddle is used on concave surfaces. The paddle must be clean and properly waxed so that solder will not stick to it. The solder level should be slightly above the panel.

While the metal is still hot, quench the area with cool water. Quenching cools the metal so that it can be filed and also relieves the heat stress in the panel.

Use a body file to level the solder with the panel. Do this with a dull file if one is available. A sharp file will cut the solder very quickly. File with long strokes, working from the edge across the middle in every direction. Sand the solder to final contour with #80 or #100 grit sandpaper and a speed file. Prep the metal before priming.

Priming Filler Area

When you are sure the filler has been smoothed and leveled properly, you should next prime the area. Detailed in later chapters, the vehicle must be masked to prevent primer overspray (Figure 12–36). All surfaces to be primed must be sanded or scuffed and cleaned. Usually, a self-etching primer, epoxy primer, or UV-primer is sprayed over the filler and bare metal areas (Figure 12–37).

Sometimes a self-etching primer and a thick primer-filler are sprayed over the area if it is badly pitted from surface rust. If the surface pits are shallow, a primer-surfacer or primer-filler will quickly help you sand the body surface smooth and level (Figure 12–38).

The trend is to use an epoxy primer-surfacer or primer-filler over body filler and repair areas. Epoxy

Figure 12–36 **Mask the vehicle as needed before spraying self-etch or epoxy primer or primer-surfacer over the repair area. Self-etch or epoxy primer is recommended over bare metal.**

A On small areas, many shops are now using UV-primer. It can be sprayed on from an aerosol can.

B By shining UV light on the repair area, the primer will fully dry in just a couple of minutes. This will speed repair time.

Figure 12–37 **Note the use of ultraviolet or UV-primer for small spot repairs.**

primer will bond to any surface and builds up easily to help level and smooth the repair areas quickly. Self-etch primer contains acid, has little film thickness, and does not dry fully for up to 24 hours.

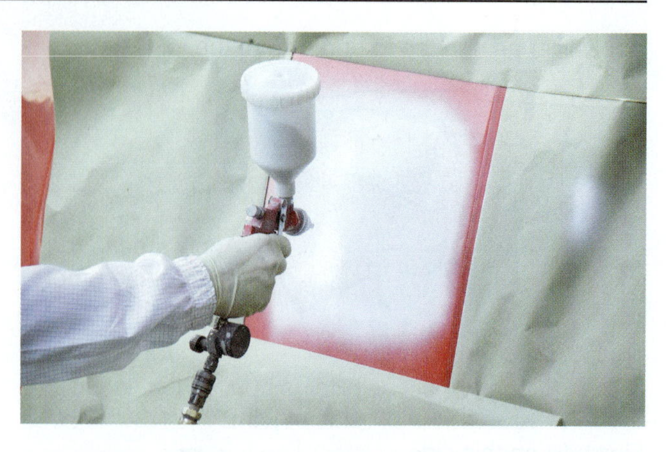

Figure 12–38 **On a larger repair area, mix self-etch or epoxy primer in a spray gun and apply over body filler and bare metal areas of the repair. Either of these types of primer is needed to properly adhere to bare metal.**

When spraying on primer-surfacer or primer-filler, apply the first coat over a large area surrounding the repair and body filler. After the first coat flashes, apply any second coat in a smaller area just over the repair area. If a third coat of primer-surfacer is needed, spray it over an even smaller area where needed. This will help feather the surface out and help level the repair area with the undamaged areas of the body panel.

Applying Glazing Putty

Once it is dry, the primer will often show tiny pits that were not visible in the body filler. Small pinholes and scratches can be filled with glazing putty. With modern polyester putty, mix the putty and hardener according to the manufacturer's instructions (Figure 12–39).

Place a small amount of putty onto a clean rubber squeegee. Apply a thin covering of putty over any pits or other imperfections in the filler. Use single strokes and a fast scraping motion. Use a minimum number of strokes when applying putties. They skim over or surface dry very fast. Repeated passes of the spreader might pull the putty away from the filler (Figure 12–40).

Allow the putty to dry completely before sanding smooth with #240 or finer grit sandpaper. Sanding the putty before the solvents in it have completely evaporated results in subsequent sand scratches in the finish. Only sand the putty until it is flush with the surface of the primer. Avoid cutting through the primer.

When properly mixed, applied, sanded, and primed, a quality body filler will have no imperfections. Excessive use of glazing putties is usually an indication of a lack of expertise on the part of the body technician. Remember that glazing putty is only designed for tiny pits and pinholes, *not* as filler.

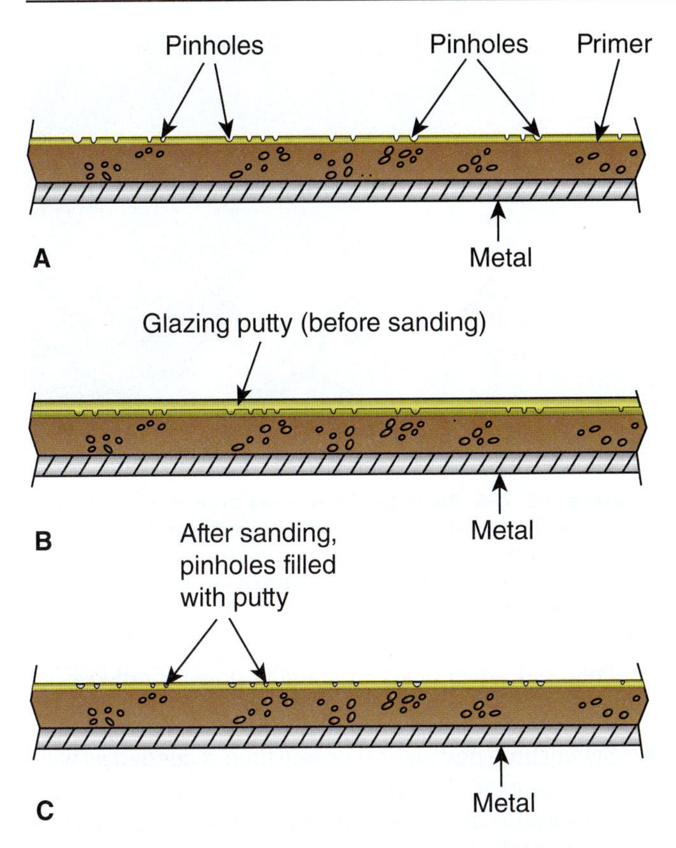

Figure 12–39 Glazing putty fills pinholes in primer and filler. (A) Pinholes can remain after sanding and priming. (B) Apply a thin layer of putty over pinholes. (C) Sand off putty until it is flush with the primer. Putty will remain and fill small holes or pits.

Using a Guide Coat

A **guide coat** is often used to check for high and low spots on your repair area. A guide coat is a thin layer of a different color primer or a special powder applied to the repair area. By watching what happens to the guide coat with light sanding, you can easily find low and high spots (Figure 12–41).

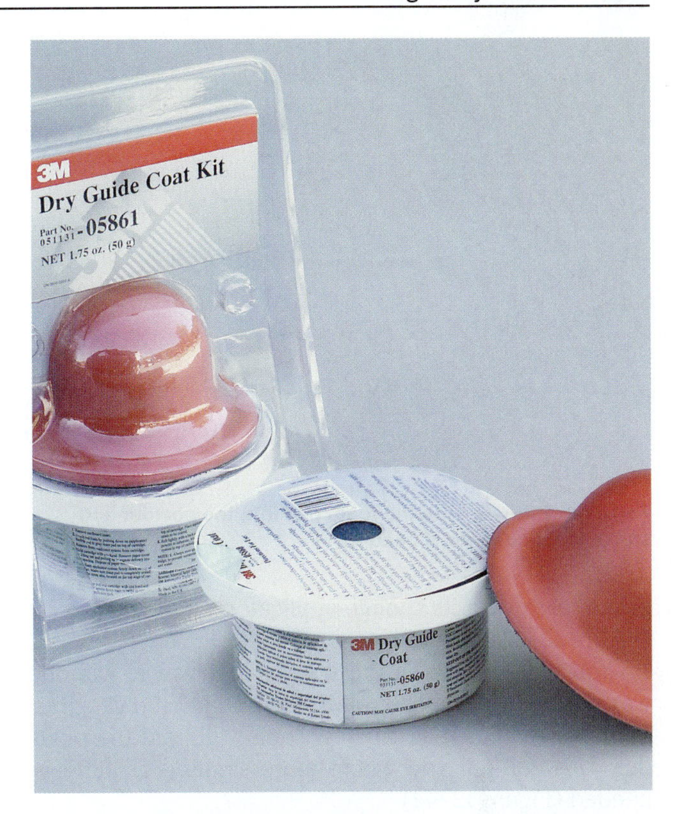

Figure 12–41 A powder guide coat is a fast, efficient way to make sure the repair area is flat and smooth enough for painting.

In the past, *primer guide coat* of a different color was used. A thin mist of primer was sprayed over the repair area. By sanding off the primer guide coat, you could easily detect filler high and low spots, small pits, and other surface problems. High spots sand off more quickly, while low spots leave the primer mist intact.

Today, technicians use a powder guide coat to final check their body work (Figure 12–42). The guide coat powder usually comes in a plastic container with a sponge applicator. The fine powder is spread over the

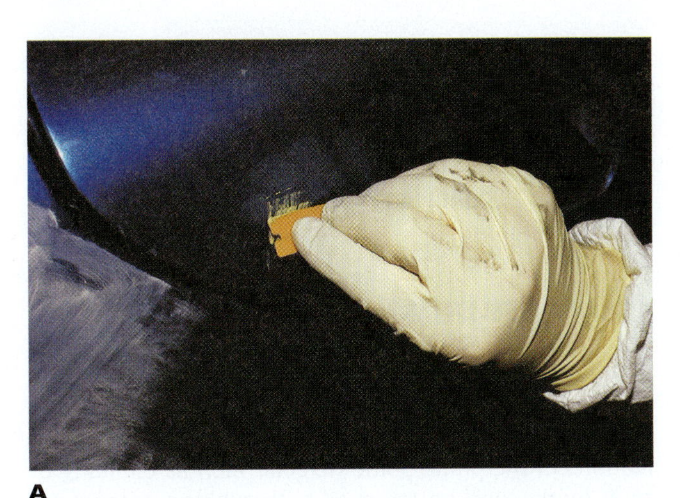
A

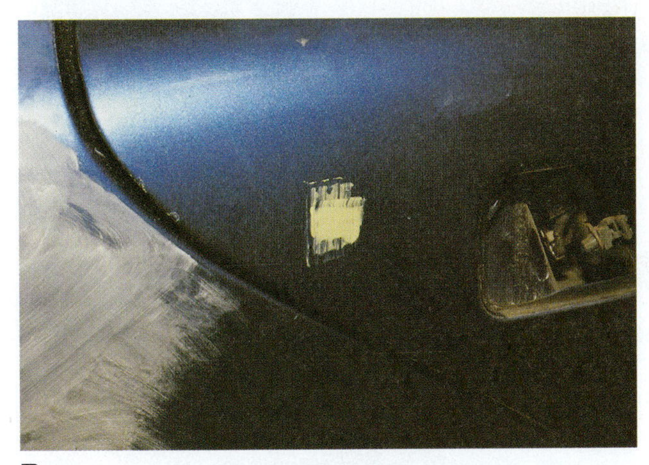
B

Figure 12–40 Most two-part spot putties can be applied to metal, primer, or even paint to save time. (A) A tiny spreader is being used to apply two-part spot putty to a tiny chip in the paint. (B) When spot putty cures, block sand putty down flush with existing paint. This saves you from having to deep sand and featheredge a large area around a small paint chip.

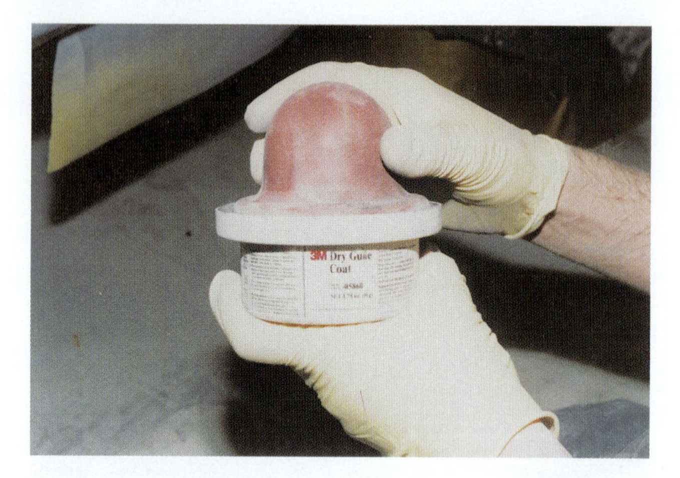

Figure 12-42 Rub the foam pad into the powder and shake the dispenser, if needed.

primer area with its sponge applicator. This is shown in (Figure 12–43).

You can then sand the powder right away without waiting for it to dry. The powder will sand off high spots immediately, whereas any low spots will hold the powder. This will tell you where more sanding or putty is needed (Figure 12–44).

Ideally, the second color primer or the dry guide coat powder should all sand off at the same time. This shows that the surface is flat and ready for sealer, a colorcoat, and other operations (Figure 12–45 and Figure 12–46).

Figure 12-43 Rub the entire repair area with guide coat power.

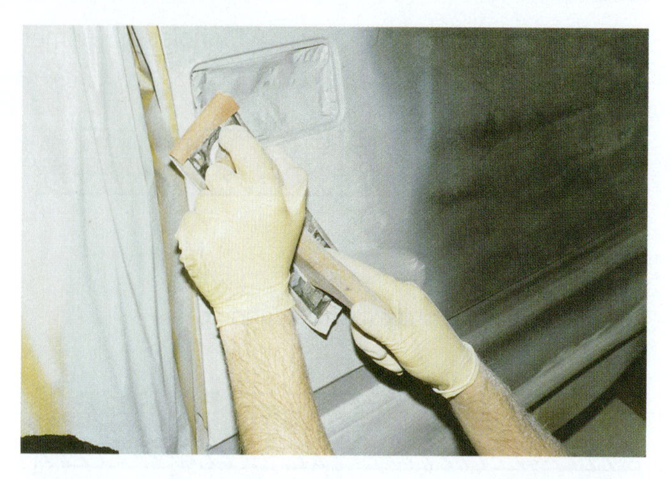

Figure 12-44 Block sand the repair area and watch the guide coat to see whether it comes off evenly. If the guide coat sands off too quickly, you have a high spot. If the guide coat does not sand off evenly, you have a low spot that must be filled or primed more.

Potential advantages of using a powder guide coat include:

- Saving time because no drying time is needed as with a spray-on guide coat.
- No solvents are sprayed into the air, so it is more environmentally friendly.
- It works well with dry and water sanding.
- Small amounts of the product are used for cost and environmental savings.

Final check all areas to be refinished. Look carefully around edges to find any remaining surface imperfections. Identify any surface problems now, before painting.

Make sure you hand-sand the edges on the repaired panels to prepare them for painting. Many technicians like to wet sand repair areas as a final check of the repair. When the surface is wet, it is much easier to notice minor imperfections in the surface. Wet sanding also produces a very smooth surface, free of scratches.

When working with body filler, it is advisable to wear gloves (neoprene or surgical) to keep the material from contacting your skin.

12.4 REPAIRING PAINT SURFACE IMPERFECTIONS

Most vehicles brought to an auto body shop for damage repair and/or refinishing have a variety of minor imperfections in the paint. Some defects, such as chalking or scuff marks, can be removed with rubbing compound. The abrasive compound removes the damaged surface paint and brings out the luster in the paint beneath without repainting.

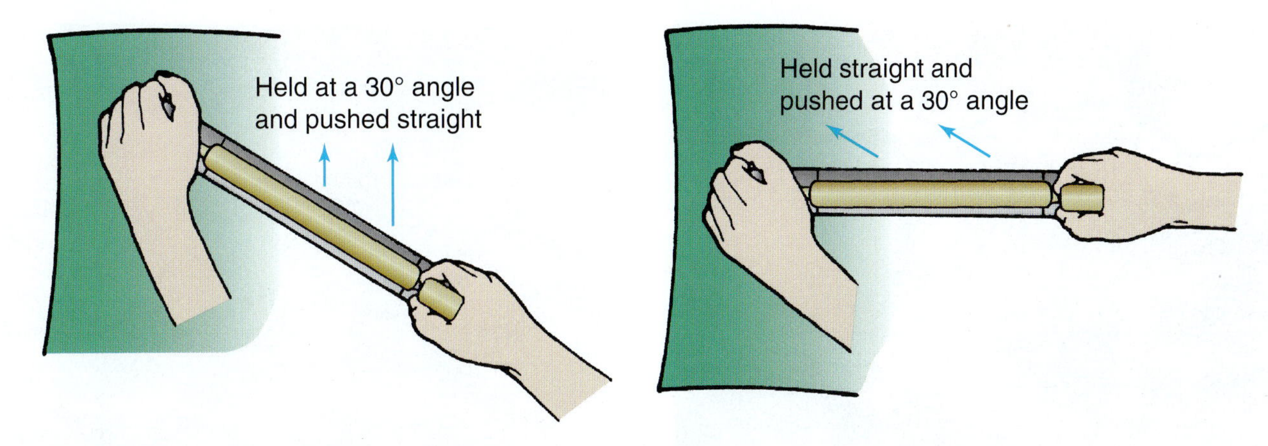

Figure 12–45 Block sanding is the most important skill of a sheet metal technician. Move a large sanding block at different angles to plane and sand the area level and smooth.

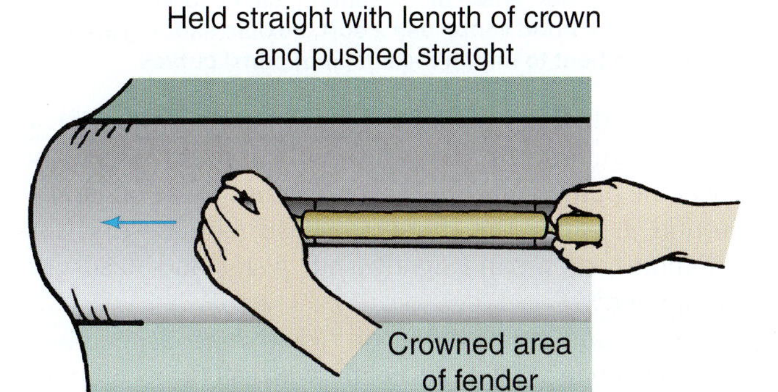

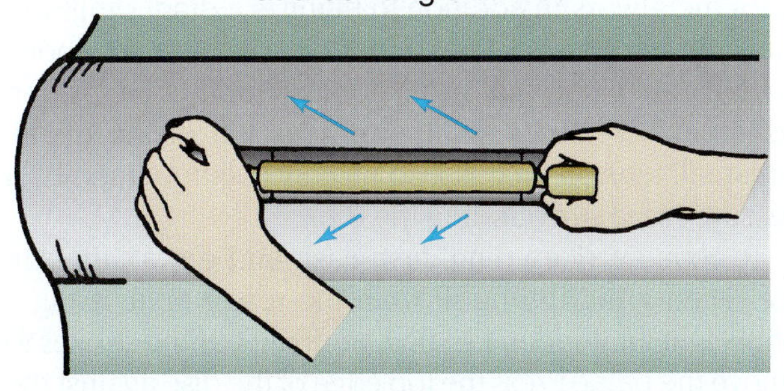

Figure 12–46 Note board sanding of a crowned shape on a body panel. Push the sanding board back and forth along the crown to sand it flat. Then move the sanding board at an angle down the curve, as shown. Twist the sanding board with your wrists to match the shape of the curve.

Compounding is discussed at length in later chapters. Other defects, such as a scratch, are too deep to buff out with rubbing compound. If the scratch penetrates through the primer and exposes the metal underneath, the scratch must be sanded, primed, and painted. On the other hand, shallow scratches that are too deep to be buffed out but do not reach the metal underneath the paint can often be filled with two-part putty designed to adhere to paint.

Repairing Paint Scratches

Wash and clean the repair area with a wax and grease removing solvent. Then lightly sand or scuff pad the whole panel to be refinished. Use a sanding block when sanding large areas. A normal size sheet of #150 or finer grit paper folded up is fine for small areas. Many technicians like to wet sand the area.

A light sanding will rough up the finish coat so the repair materials will adhere to the old finish. Do not press too hard on the sandpaper. Excessive pressure can result in low spots or a wavy surface that will require additional filling and sanding. After the rough sanding is complete, clean the sanded area with compressed air or a soft cotton cloth and wipe it with a tack cloth.

If the scratch is too deep to featheredge, apply two-part spot putty over it. Using moderate pressure, spread spot putty over the scratch. Apply spot putty with the rubber squeegee. Use a fast scraping motion.

Allow the putty to dry completely, following label directions. Drying time varies, but it usually takes between 20 and 60 minutes. Sanding the putty before it completely cures will result in sand scratches and shrinkage problems in the finish.

After the putty dries, sand the repair area with #150 or finer grit sandpaper. Use a sanding block to avoid making low spots with finger pressure (Figure 12–47).

When sanding, rub the palm of your hand over the puttied area to feel for high spots on the surface. When finished, rinse the sludge away and wipe the surface dry. Clean the repair area with a tack cloth.

Inspect the scratch for low spots and voids in the putty. If the scratch requires additional putty, repeat this procedure. When the surface of the previously scratched area is free of imperfections, it is ready for primer-surfacer, sealer, and refinishing.

Repairing Nicks

Minor bumps and scrapes often leave nicks and scratches in a car's finish. A stone thrown up by a passing vehicle can chip the paint, exposing the sheet metal beneath.

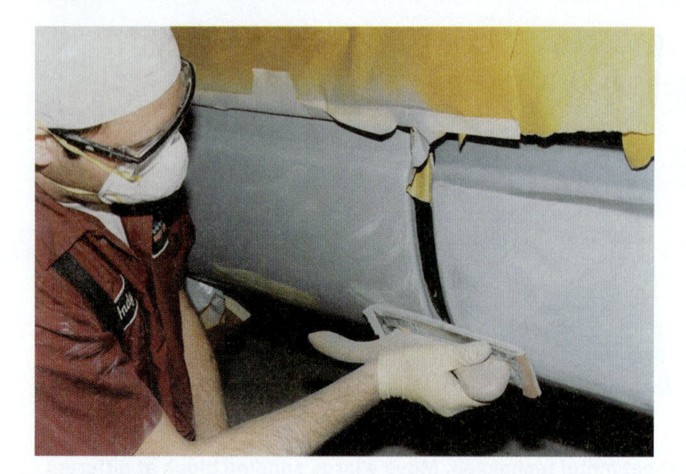

Figure 12–47 During final block sanding, all primer should sand evenly and deep sand scratches must be removed. The technician is final board sanding spot putty applied to primer-surfacer.

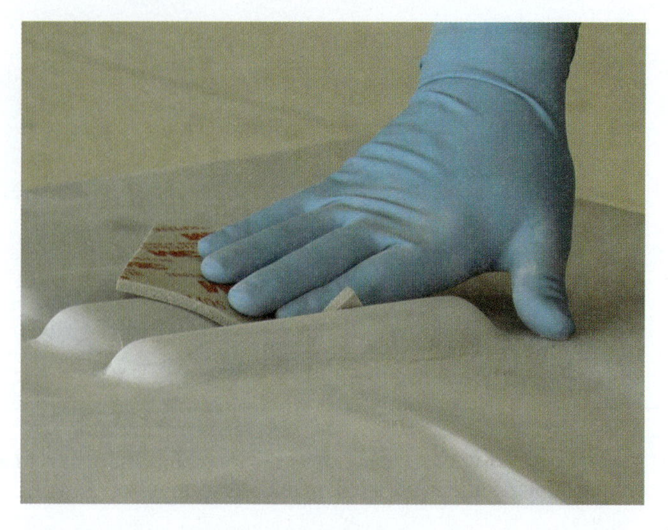

Figure 12–48 When block sanding small curved areas, as around the hood vents, use a soft, flexible sanding sponge that can be bent to match the shape of panel curves.

Side swipe collisions result in scrapes and gouges. Anytime bare metal is exposed to the air, the metal must be coated with self-etching or epoxy primer to prevent rerusting before the new paint finish is applied.

First, clean and dewax the repair area with an approved solvent. Wipe the area down with a clean towel or rag. Sand the ragged edges of the chipped paint to a smooth, feathered surface. This is commonly referred to as featheredging. As discussed, featheredging tapers the edges of the paint so that it gradually blends in with the metal surface.

Featheredging chips and nicks is done quickly with a #150 grit disc and a disc orbital (DA) sander. In tight spots, use a sanding block. Sand the edges of the old finish to a fine taper. When the sanded area is smooth to the touch, switch to #220 grit sandpaper. Sand any remaining scratches away.

Never leave bare metal surfaces exposed to air. Moisture in the air quickly encourages rust to form on the metal. The slightest film of rust will prevent primer and paint from properly adhering to the metal. Subsequent lifting and blistering will eventually ruin the paint. The area will have to be sanded down and refinished again. Priming the bare metal areas with a zinc chromate base primer inhibits rust formation and ensures good bonding of the finish paint.

Blow away any sanding dust and wipe the area with a tack cloth. Then, apply a coat of primer-surfacer to build up the area and fill in any uneven featheredging. After the primer-surfacer has dried, apply a mist coat of another color or contrasting color guide coat. Block sand the area to identify low spots. Apply two-part glazing putty to fill the low spots or pits if needed.

Final sanding and priming are necessary to achieve a super-smooth surface. Wet sand with #220 grit sandpaper and a sanding block. Sand in long, straight strokes to avoid creating low spots. When sanding curved surfaces, sand very lightly, holding the paper with the palm of the hand, or use a flexible sander (Figure 12–48).

Clean, dry, and wipe the sanded surface with a tack cloth. Then, spray the repair area with primer. Completely cover the puttied area and several inches of the old finish around it. Allow the primer to flash (surface dry) for 5 minutes, then sand lightly with water and #220 grit sandpaper.

Clean and prime once or twice more as needed. Between coats, wet sand lightly with #400 or #600 grit sandpaper to achieve an extremely smooth surface. The surface is now ready to be painted.

Repairing Dings

One kind of surface imperfection that sometimes requires minor metalwork in addition to filling is a ding. Dings are small dents often caused by carelessly opened doors. When a panel is struck by the edge of another car's door, the impact creates a shallow depression in the metal. This small tension area is also usually accompanied by a pressure ridge surrounding the ding.

To repair a deep ding, first wash and dewax the surface. Then grind the finish from the repair area, using a lightweight air grinder. Use up and down buffing strokes to remove the paint. Press the top edge of the disc against the metal and move the grinder up and down. Avoid removing too much metal, which thins and weakens the body panel. On smaller dings, a fiber abrasive stripping wheel in a small air tool or drill works well. It removes the paint over the ding without cutting too deeply into the metal.

After cleaning the panel, apply a skim coat of body filler. Allow the filler to harden. Then, block sand the filler smooth with #80 grit paper. Run your hand over the sanded plastic to feel for high and low spots. High metal areas might need to be tapped down. Low spots will need another application of body filler or two-part spot putty.

Once the ding has been properly filled and leveled, the surrounding paint edges must be featheredged. A dual action sander or a block sander can be used to featheredge the area, depending on the size of the repair.

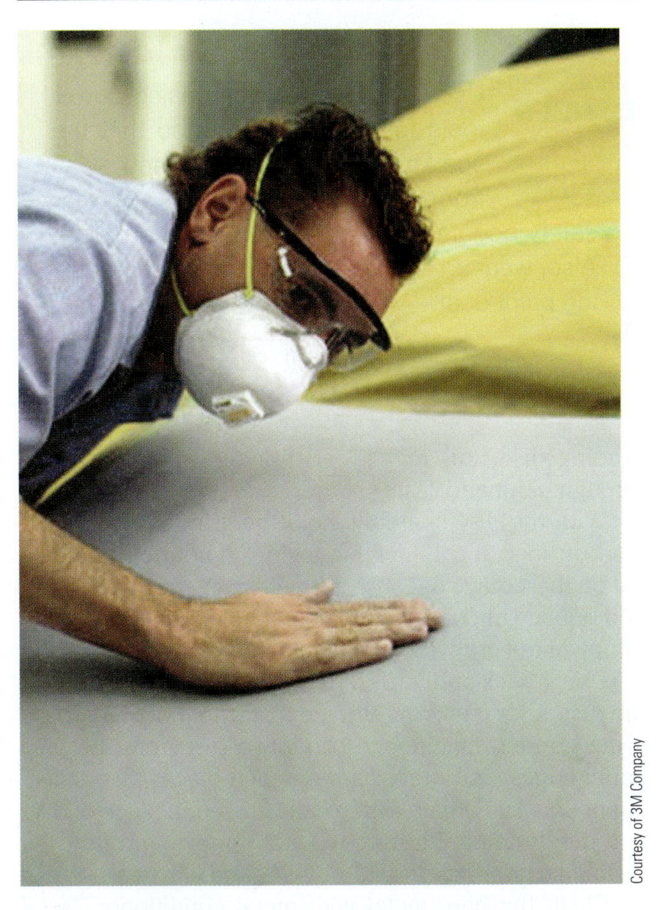

Figure 12–49 **A close inspection of all surfaces is critical before painting. High-gloss paint will magnify and show surface problems much more than dull primer.**

Note that some spot putties are formulated so that they can be applied over cured paint. This cuts repair time and avoids having to sand or grind down to bare metal, cutting through the factory corrosion protection.

To use spot putty to repair a small dent or ding, sand the surface of the paint inside and around the ding or small dent. Then, blow off the surface. Apply the manufacturer-approved, two-part spot putty over the small indentation in the panel. Wipe in two directions to get good adhesion.

To avoid shrinkage, allow the putty to cure the recommended time. Then, block sand the repair area flush with the panel. You can then spray primer over the putty. Afterward, wet or dry sand the primer with very fine sandpaper to prepare for painting (Figure 12–49).

12.5 REPAIRING RUST DAMAGE

In addition to minor collision damage, the body repair technician must be able to recognize and repair corrosion damage created by rusting sheet metal.

Rust is produced by a chemical reaction known as *oxidation*. Oxidation occurs when metal is exposed to moisture and air. The oxygen in the air, water, or other chemicals combines with the steel molecules to form iron oxide (Figure 12–50).

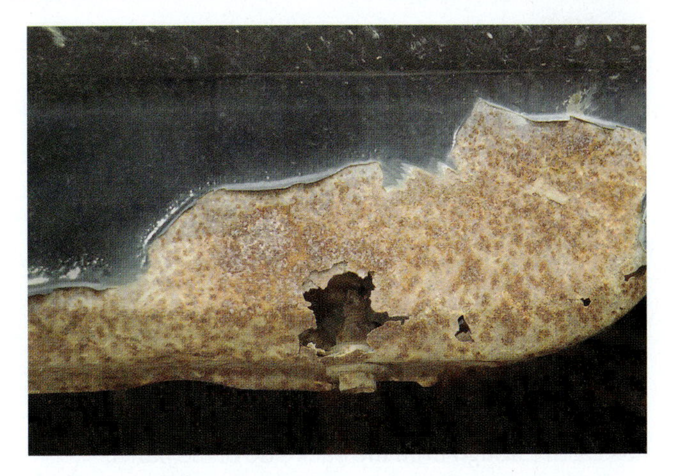

Figure 12–50 **Rust is a major problem along the lower edges of panels in older vehicles. The best way to repair rust is to replace the whole panel if one is available. If one is not available, as may be the case for an older collector car, you may have to make a repair panel out of sheet metal.**

Iron oxide is the reddish-brown compound commonly referred to as rust. By turning the metal into flaky or powdery iron oxide, rust can eat completely through a sheet metal panel if not treated.

If a crack in the car's finish allows moisture and air to seep under the paint film, the sheet metal beneath will begin to rust.

The same is true if a chip or nick exposes sheet metal to the air and no immediate action is taken to repair it. Left alone, rust will soon form on the metal surface and begin to eat pits into the sheet metal. These pits harbor rust and prevent sanding the surface rust away. If simply painted over, rust in the pits will eventually bubble up and break through the new paint (Figure 12–51).

Corrosion damage on a vehicle panel takes two forms: surface rust and rustouts.

Surface rust is an early stage of corrosion that has not penetrated the steel panel. It may simply leave surface pits and a reddish coating on the metal surface.

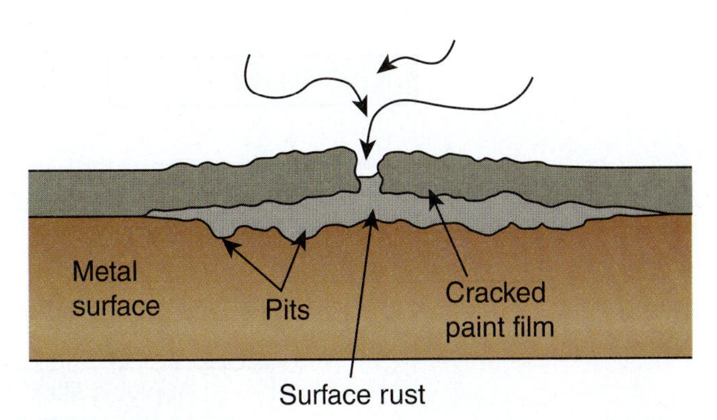

Figure 12–51 **Once the paint film is broken, moisture can seep under the paint to accelerate rust or oxidation of the steel body panel. This will initially lift and bubble paint above surface rust.**

Given time, surface rust will eventually develop into rust holes or larger rustouts.

A **rustout** is corrosion that has had time to penetrate completely through the steel panel. Rustouts often occur along the lower areas of a vehicle that are exposed to more road salt and debris. The lower part of the quarter panels, bottoms of doors, and fenders are the most common areas to suffer rustout damage.

Rust can form on either side of a metal panel. Rust that is present on the back side of a panel might go unnoticed until the paint begins to bubble and lift. By this time, the rust has eaten completely through the panel. Spots of surface rust or humps in the paint might be a sign that rust has eaten through from the back side of the panel.

Both types of rust damage require different repair procedures. These are outlined in Table 12–3 and are explained in the rest of this chapter.

Preparing Surface Rust

Repairing an area affected by light surface rust can be as simple as grinding or sandblasting the rust film away and chemically neutralizing the area with metal conditioner. However, if the metal is pitted, additional steps are required. The surface rust shown in Figure 12–51 requires minor metalworking and waterproof body filler.

Prepare the deteriorated area for sanding by first washing it with a mild detergent. Then clean it with a wax and grease remover solvent. Apply masking tape to nearby trim before grinding away paint and rust. The tape will protect the trim from sparks, chips, and accidental contact with the sanding disc.

A lightweight air grinder is ideal for doing minor rust repair. Used with a rigid backing plate and a #24 grit sanding disc, this high-speed grinder will quickly cut through paint and rust.

DANGER Use extreme care when using power tools for grinding. Grind so that the sparks and dust fly down and away from your face and eyes. Always wear safety goggles or a face shield when grinding. Also, wear an air-filtering mask to avoid breathing paint dust.

With the disc spinning, hold the grinder against the work surface at a 10-degree angle. Do not hold the disc flat on the surface because this will make the grinder skip and bounce uncontrollably. Do not hold the disc on edge or unwanted grooves will be cut into the metal.

Use a back-and-forth, crosscutting action to remove the rust. After removing surface rust with the grinder, use a die grinder attachment to remove rust from the pits, panel edges, and other hard-to-reach places. Sandblasting or plastic media blasting can be used in place of disc and die grinding. Blasting will not remove and thin the metal-like grinding.

Clean the bare metal with metal conditioner. *Metal conditioner* is an acid compound that neutralizes microscopic rust particles. The acid also etches the metal surface to improve the bond between the metal and repair materials. Always wear rubber gloves and safety glasses when handling conditioners. Follow the manufacturer's instructions carefully.

TABLE 12–3 RUST REPAIR PROCEDURES

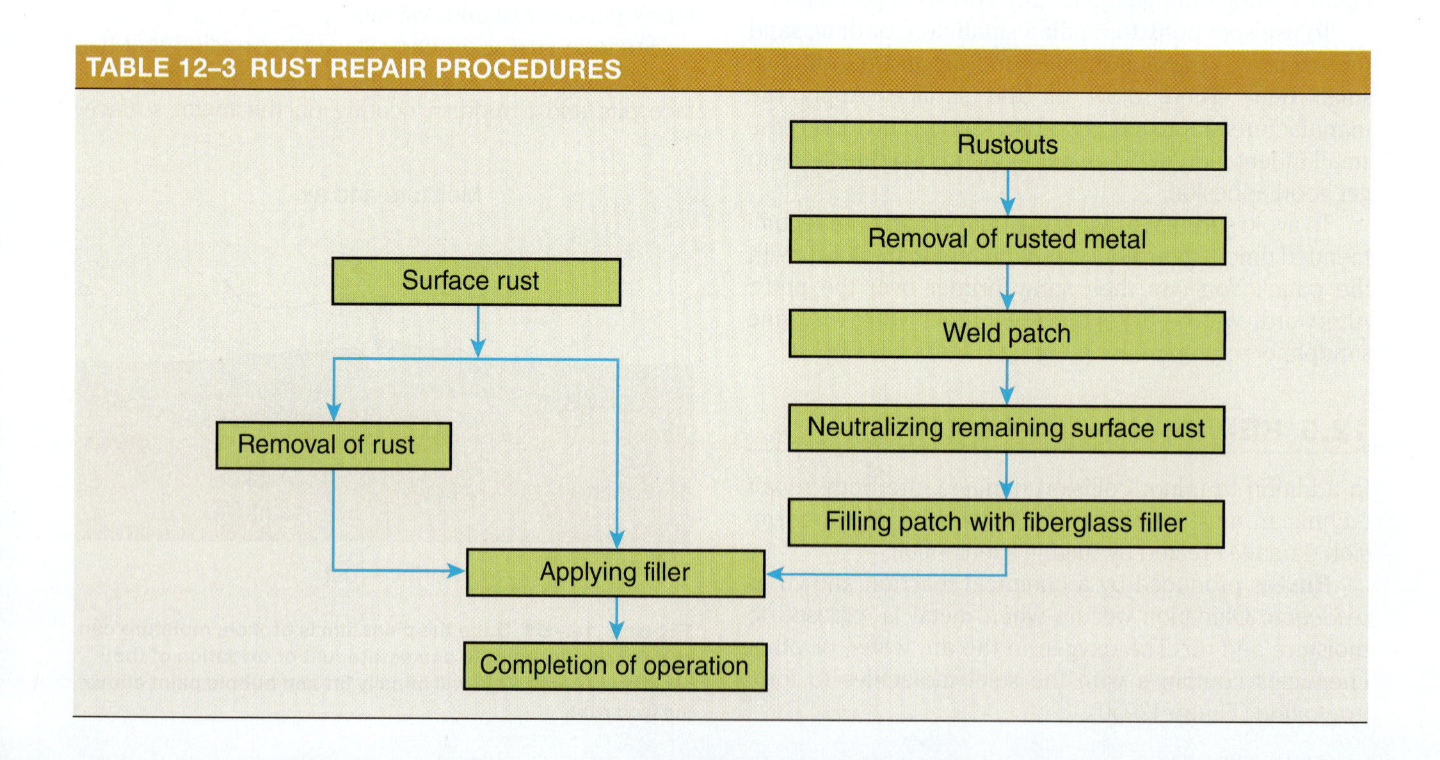

Never apply body filler directly to a conditioned metal surface. The plastic will not adhere to the metal and could crack when in service.

Make sure you wash off and neutralize any acid-type metal conditioner right after it has been applied. The industry trend is away from using acid metal conditioner for this reason. Metal conditioner is only used when surface rust is a major problem on the panel to be repaired.

Now, mix waterproof body filler and hardener together. Scoop some filler onto the edge of the spreader and spread a thin skim coat over the pitted area. Apply moderate pressure to force the body filler into the pits. Allow the filler to harden. Then, block sand with #80 grit sandpaper. Sand until the filler is level with the panel surface.

Always wear a dust mask while working with body fillers. Inhaling filler dust can be harmful.

Use compressed air to blow filler dust from the area. Spray the areas with a self-etching or epoxy primer. Primer-filler is often used next to help level the repair area. If necessary, apply two-part spot putty to any remaining pits. Follow the sanding and priming procedures already described in this chapter.

Repairing Rustouts

There are two ways to properly repair rustouts or rust holes in a steel body panel: Install a whole new panel (for example, entire whole quarter panel) or a partial repair panel (only the lower section of a quarter panel). In professional collision repair, it is not acceptable to cover rust holes with waterproof fiberglass or reinforced body filler. This is an inappropriate way of temporarily fixing the serious rust damage.

Treating Rusted Areas

When the surface rust and paint are ground or blasted away, small holes called rustouts will be uncovered (Figure 12–52). After grinding the area with an air grinder and a #24 grit rigid disc or media blasting the area, use a pick hammer to find thin spots in the metal. They will dent easily because they are very thin.

If the back side of the panel is accessible, remove accumulated dirt and undercoating. A wire brush,

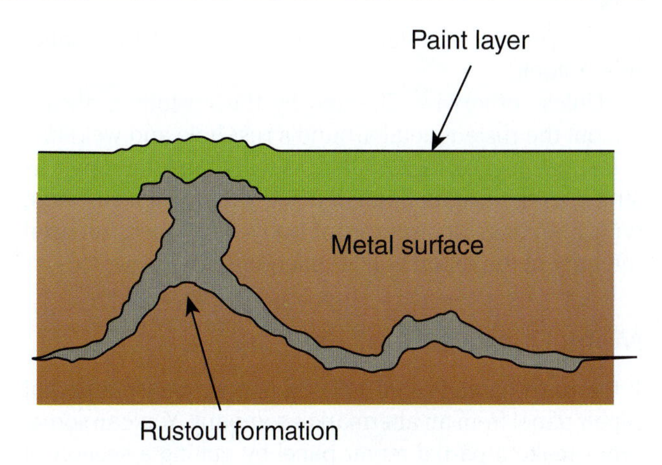

Figure 12–52 **Small spots or bumps on the paint surface often indicate rustout from the inside or back side of a steel body panel.**

scrapers, or blaster can be used to remove the rust on the back side of the repair area.

If difficult to remove, you might want to apply a commercially available rust deactivator. The chemical reacts with the rust to form a hard, black polymer coating over the rust. This coating seals out air and moisture, preventing any further oxidation.

After cleaning and sanding the repair area, apply metal conditioner or a rust deactivator to neutralize any rust remaining in pitted areas and around the edges of the rustout. Always wear rubber gloves when handling rust deactivator and metal conditioners. Shake the container well before using and follow instructions carefully.

Apply the rust deactivator to the back side of the panel too. With a wire brush or scraper, first clean dirt and undercoating from the back side of the repair area. Then apply the deactivator. Be sure to carefully follow the manufacturer's instructions. A black coating will soon develop over the rust. For maximum protection, this coating should have a solid black color. Wait 1–2 hours between coats. If the color is splotchy and uneven, apply additional coats. Two or three thin coats neutralize the rust better than one thick coat. If excess rust deactivator runs over the finish paint, wipe it off immediately using a cloth dampened with mineral spirits.

You may also want to blast the area to remove rust from pits. This may eliminate the need to use a rust converter.

Cutting Out Rust Holes

Cut out the hole and any metal around the hole that has been thinned by surface rusting. Use a straightedge to mark the lines for the area to be removed. Cut the area out with a cutoff wheel or a plasma arc cutter.

If only a small area of a panel is badly rusted, you can cut out only the damage area and weld in a new section. You can purchase a partial repair panel from an aftermarket panel manufacturer or cut out the needed section from a salvage yard panel. If a very small, flat surface is rusted, you can also fabricate a metal patch from sheet metal stock.

Unless otherwise directed by the customer, always cut out the rusted metal around a rust hole and weld in a metal patch (Figure 12–53). This will make the repair as strong as the original panel. If you simply fill a rust hole with fiberglass or waterproof body filler, the structural integrity of the repair will be inferior.

Metal Rustout Patches

If the rusted area is a complex shape, purchase a partial repair panel from an aftermarket company. You can sometimes make a partial repair panel by cutting a section of the needed panel off a wrecked vehicle at a salvage yard.

When possible, the metal patch panel should be slightly larger than the hole cut in the damaged vehicle. This will let you fold or crimp and lap the patch panel over or under the vehicle's panel so the two pieces overlap. Overlapping the repair panel and the existing panel will simplify welding (Figure 12–54).

To overlap a repair panel with an existing panel, use a panel recessing tool. A *panel recessing tool* has offset rollers that will bend and fold the edge of the panel down lower. When the repair panel is fit in place behind or in front of the hole in the panel, its lip will form an offset fit between the two panel edges. The repair panel surface will then be flush with the existing panel surface. See Figure 12–55, Figure 12–56, and Figure 12–57.

First, fit and clamp the patch panel in place on the vehicle. Use clamping pliers or sheet metal screws to hold the patch panel in alignment. Tack weld the center edges of the patch panel. Then continuous weld the joint between the patch panel and existing panel. Work your way to the corners of the patch when welding.

Fabricating a Metal Patch

As mentioned briefly, the best way to repair a large rustout is to make and weld in a metal patch. This makes the repair as strong as the original panel, so the repair will last longer. The patch can be made by cutting out the needed area from a salvage yard panel or it can be cut from sheet metal stock if the area is relatively flat (Figure 12–58, Figure 12–59, and Figure 12–60).

To install a small, simple metal patch in an antique car door, preferably begin by blasting the area to be repaired. Make sure the area around the repair is sound. It is easy for a large area to be thinned and damaged by rust from the inside out. To check for badly thinned metal, tap all over the area with the tip of a pick hammer. If the area dents easily, that area is rusted too thin.

After determining the amount of area damaged, grind or blast off the paint and primer. Mark around the area to be replaced. Then cut it out with a cutoff tool, power cutter, or plasma arc cutter.

Use the cutout piece of metal as a template to make a new repair piece. Use a scribe or scratch awl to mark the

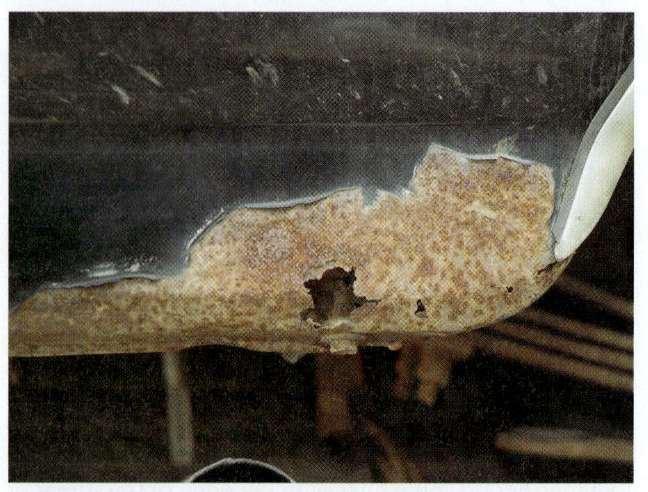

A This is a large rustout in the bottom of a rear quarter panel. Moisture, salt, and road debris tend to collect on the body behind rear tires, which promotes rusting.

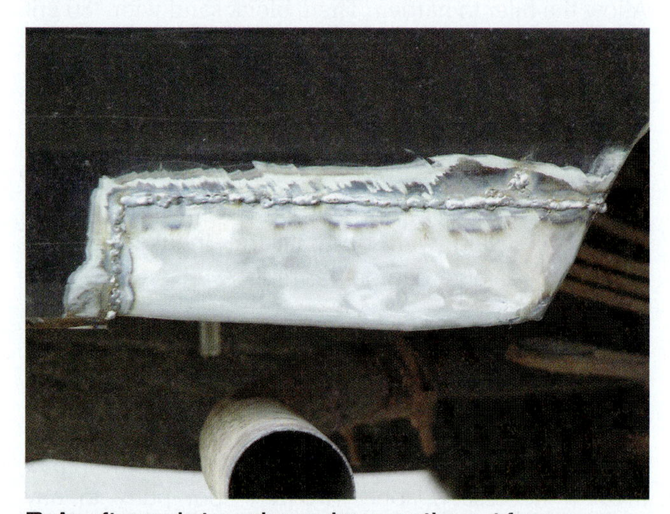

B An aftermarket repair panel or a section cut from a salvage yard vehicle can be cut and fitted into the rustout area. Ideally, the repair panel should be slightly larger than the hole so that it extends behind the panel on the vehicle. This allows you to use a lap weld.

C Waterproof filler should be used over a welded patch because the weld may not be moisture-tight.

Figure 12–53 **Note the basic steps to repair rustout without whole panel replacement.**

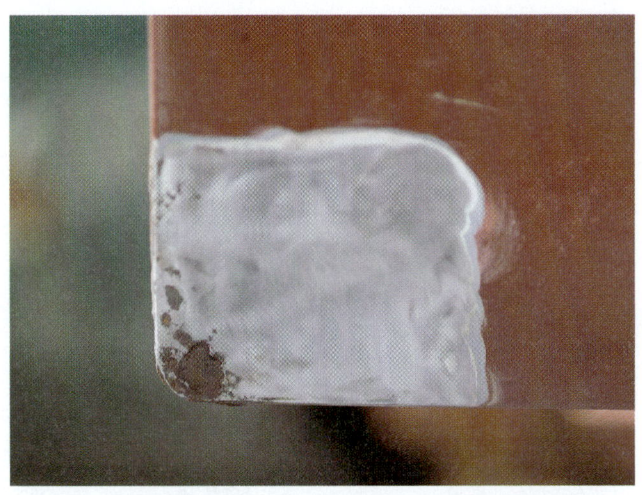

A The rusted area has been sandblasted to remove paint and rust.

B A large enough area is cut to make sure all rusted, thinned metal is removed. A small cutoff wheel or grinder quickly cuts a sheet metal area out of the door.

C Treat any rust inside of the door with conditioner. If you do not remove or treat all rust, corrosion will start up again quickly.

Figure 12-54 Note how a small area of rusted metal is removed from this collector car door.

Figure 12-55 Body shops that restore collector cars often have sheet metal equipment for making body repair panels. Here a craftsperson is using large shears to cut out a repair panel.

new metal piece to be cut. Make sure the repair piece is the same type metal and thickness as the original. Cut along your scribed line. If accessible from the rear, you might want to make the patch larger than the hole. You can fit the patch in from the back on over the front and make a lap weld.

Make sure that the labor for making and installing a metal patch is worth it. You may be able to purchase a whole replacement panel and install it for a lower total cost.

Fit and clamp the repair piece into place in the hole. Use clamps or Vise-Grip pliers to secure the repair piece in position.

MIG weld the metal patch. Start by spot welding the patch in the center of each seam. Then, stitch weld from the center out to each corner to prevent warpage. Finish by continuous welding along each edge to form a leakproof weld joint. You might want to use *heat sink compound* to help prevent panel warpage (Figure 12–61).

Grind the weld down flush with the panel. If applicable, apply undercoating or anticorrosion compound to the inside of the panel to protect from further rusting. Apply filler over the metal patch as summarized earlier.

Filling Rustout Patches

After welding a metal patch around the rust hole, grind the weld bead down nearly flush with the surface of the panel. Then cover the metal patch area with a waterproof, fiberglass-reinforced filler. Use a plastic spreader to force the filler into the weld and patch area.

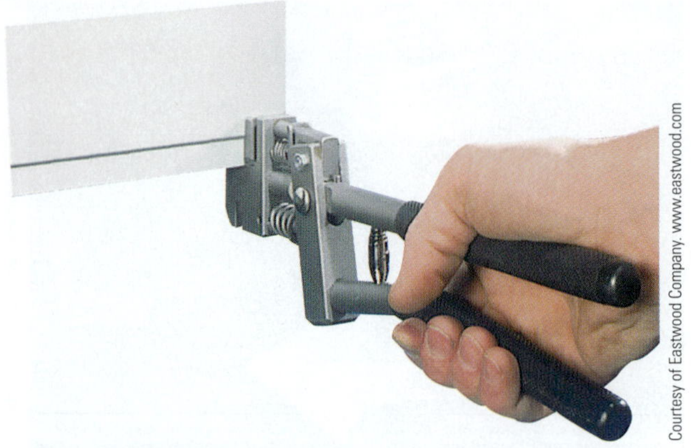

Courtesy of Eastwood Company. www.eastwood.com

A Folding pliers are being used to form a lip on a repair panel.

Courtesy of Eastwood Company. www.eastwood.com

B Pliers with folding rollers work better on a rounded body panel lip.

Courtesy of Eastwood Company. www.eastwood.com

C This air tool will quickly form an offset lip on a repair or patch panel.

Figure 12-56 Folding tools will bend the edge of a patch panel so it can be recessed behind the existing panel and welded.

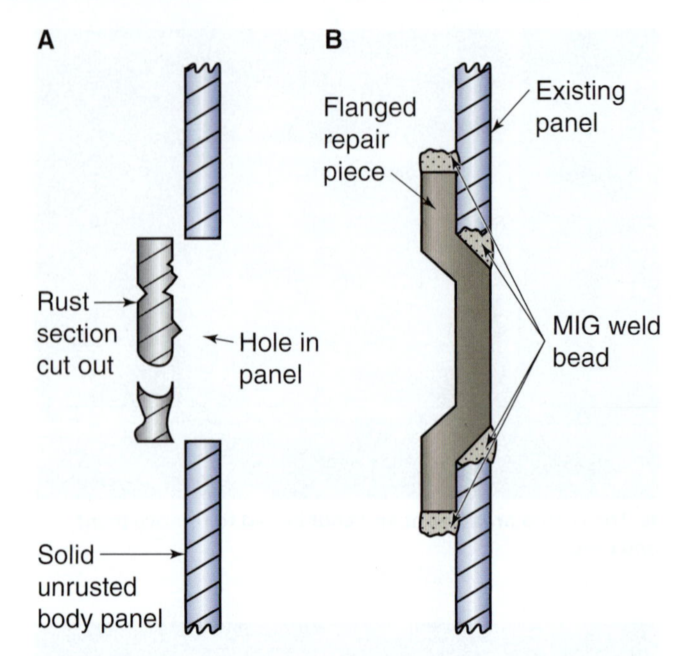

Figure 12-57 Cutaway view shows how to install a patch panel over a rusted-out area. (A) Mark the area with rust to be removed. Use air cutoff wheel to remove rusted-out metal. (B) Fit the repair panel behind the existing panel. Continuous MIG weld the repair panel into place. Grind the weld bead down flush. Then use waterproof fiberglass-reinforced body filler to cover the welds.

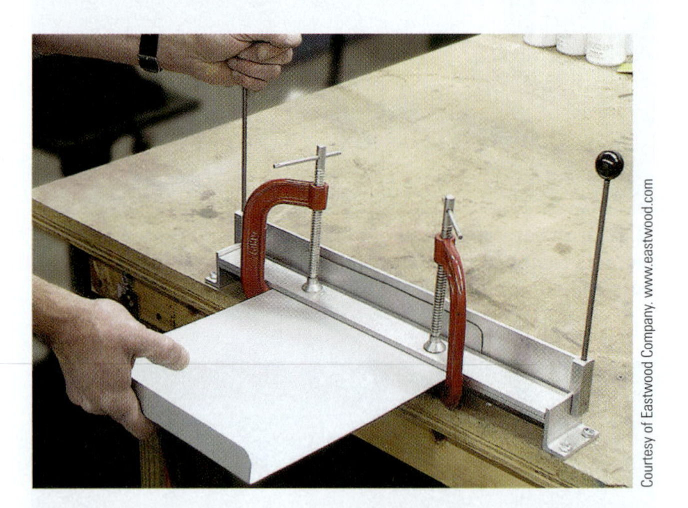

Courtesy of Eastwood Company. www.eastwood.com

Figure 12-58 Here a body technician is using a sheet metal brake to fold a flange into a repair panel.

REMEMBER *Regular body fillers with a talc bulking agent absorb moisture and are not suitable for filling over rust repair areas.*

Apply a waterproof filler over the welded repair patch. Do not use conventional filler, because it is not waterproof. Even the best welds can have tiny pinholes that allow moisture entry through the weld. Always use waterproof filler as the first coat over a rust repair area.

With one coat of waterproof filler over the metal patch, mix enough conventional body filler to fill the depressed area. Using a larger plastic spreader, apply a

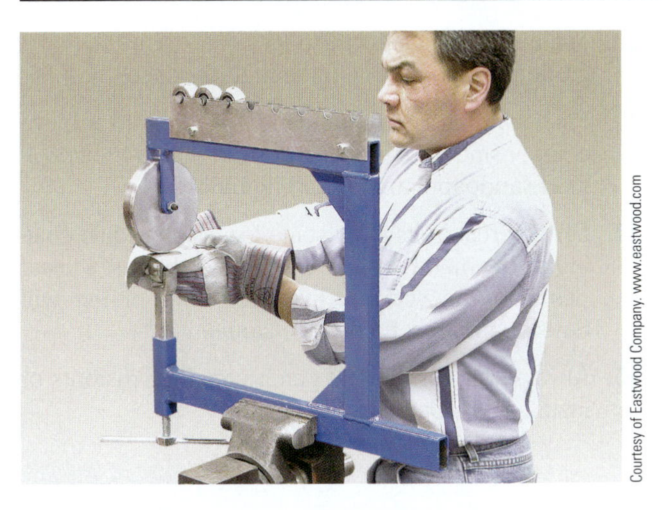

Courtesy of Eastwood Company. www.eastwood.com

Figure 12-59 Here the technician is using a forming tool to roll a curve into a small repair panel.

Figure 12-61 MIG welding can be used to install a repair or patch panel. This repair panel is being installed in a rusted-out wheel well.

Courtesy of Eastwood Company. www.eastwood.com

Figure 12-60 A rounded plastic mallet and leather shot-filled bag are being used to form a complex repair panel in sheet metal.

coat of filler over the repair area. Press hard on the spreader to force out air bubbles. Spread from the edges to the center.

Wet sand the filler smooth and blow the dust away. Cover the waterproof filler with easier-to-sand lightweight body filler. Then featheredge the repair area.

When the filler has been sanded to the desired level and smoothness, clean the area and dry it with an unsoiled cloth. Wipe the repair area with a tack cloth. Blow with compressed air to remove any remaining dust particles. Spray with self-etching primer or self-etching primer-surfacer. After the primer dries, fill any pinholes with two-part glazing putty. Apply it with a clear rubber squeegee using smooth strokes, and allow it to dry.

Using #150 to #240 grit sandpaper and long, even strokes with the sanding block, sand the glazed repair to a smooth surface. Spray primer-surfacer over the area and block sand the area.

Clean with a tack cloth and prime the repair area once more. The repaired rustout is ready to be painted.

More information on rust and corrosion protection can be found in Chapter 20.

A rustout in a structural panel should be repaired by replacing the part or by welding in a metal patch. The repair of the structural part is critical to the integrity of the vehicle. If you try to repair a large rustout hole with only fiberglass, the part will be weaker than when new.

SUMMARY

1. Body filler is the finishing touch for most sheet metal repairs.
2. Body filler is a heavy-bodied plastic material that cures very hard for filling small dents in metal.
3. As the filler cures and hardens, the chemical reaction produces a tremendous amount of heat. For this reason, unused filler should *not* be discarded in trash cans containing solvent-soaked paper or cloths.
4. Fiberglass body filler has fiberglass material added to the filler. It is used for rust repair or where strength is important.
5. Light body filler is formulated for easy sanding and fast repairs.
6. Body filler manufacturers have developed a fine-grained, catalyzed polyester glazing putty, or two-part spot putty. Like body filler, a hardener or

catalyst must be mixed with the putty to initiate and speed curing.

7. Hardener kneading is done by thoroughly squeezing the hardener back and forth inside the tube.
8. Add hardener according to the proportion indicated on the can, usually 2 percent hardener (2 parts hardener for each 100 parts of filler).
9. The most common mistake of the apprentice is using too much cream hardener!
10. The surform, or cheese grater, file is used to quickly cut the excess filler to size.
11. A sanding guide is a special tool for sanding straight, special contour lines on panels.
12. Featheredging tapers the edges of the paint so that it gradually blends in with the metal surface.
13. Final sanding and priming are necessary to achieve a super-smooth surface. Wet sand with #400 grit or finer sandpaper and a sanding block.
14. Surface rust in its early stage leaves a reddish coating on the metal surface. Given time, the rust will eat pits into the surface. Eventually the pitting will develop into rust holes, or rustouts.
15. Always weld in a metal patch panel for rustouts or install a new panel.

EXERCISES

On a separate sheet of paper, complete the following learning activities for this chapter. Write definitions for the key terms and answer the ASE-style review questions, essay questions, critical thinking problems, and math problems. You can also do the outside activities, possibly for extra credit.

Key Terms

body line sanding guide
contour sanding stick
DA
featheredging
filler grating
guide coat
hardener
hardener kneading
long-hair fiberglass filler
mixing board
one-part spot putties
polyester glazing putty
rust
rustout
short-hair fiberglass filler
spot putty
surface rust

ASE-Style Review Questions

1. Technician A uses body filler over rust holes in body panels. Technician B welds in a repair panel after cutting out the rust damage. Who is correct?
   A. Technician A
   B. Technician B
   C. Both A and B
   D. Neither A nor B
2. What type of spot putty should be used to cure quickly and prevent shrinkage when drying time is limited?
   A. One-part
   B. Two-part
   C. Lacquer-based
   D. Enamel-based
3. Fiberglass fillers are
   A. waterproof.
   B. available in 10 basic forms.
   C. used to bridge large rustouts.
   D. all of the above.
4. Lightweight fillers have improved
   A. filing characteristics.
   B. sanding characteristics.
   C. water resistance.
   D. all of the above.
5. Technician A says that plastic body filler is commonly used to final straighten sheet metal damage. Technician B says that lead is commonly used in place of plastic body filler. Who is correct?
   A. Technician A
   B. Technician B
   C. Both A and B
   D. Neither A nor B
6. If too little hardener is used, the filler
   A. will not adhere to the metal.
   B. will be subject to rampant pinholing.
   C. will be easier to handle.
   D. none of the above.

7. Technician A uses a body file to initially level high spots in body filler. Technician B says to use a high-speed grinder first. Who is correct?
   A. Technician A
   B. Technician B
   C. Both A and B
   D. Neither A nor B
8. The first step in repairing nicks is
   A. sanding the ragged edges of the paint to a smooth surface.
   B. featheredging.
   C. dewaxing.
   D. both A and B.
   E. none of the above.
9. Once a ding has been properly filled and leveled, Technician A uses a dual action sander to featheredge the surrounding paint edges. Technician B uses a block sander. Who is correct?
   A. Technician A
   B. Technician B
   C. Both A and B
   D. Neither A nor B
10. Metal conditioner is
    A. an acid compound.
    B. used to etch the metal surface.
    C. used to neutralize rust.
    D. both A and B.
    E. all of the above.
11. Technician A grinds a brazed joint before neutralizing the acids in the flux. Technician B neutralizes the acids, then grinds. Who is correct?
    A. Technician A
    B. Technician B
    C. Both A and B
    D. Neither A nor B
12. Technician A uses the same putty knife to mix and apply body filler. Technician B uses a separate knife for each step. Who is correct?
    A. Technician A
    B. Technician B
    C. Both A and B
    D. Neither A nor B

Essay Questions

1. How and why do you remove paint before using body filler?
2. How do you make and install a metal patch for a rustout?
3. What is a metal conditioner?
4. Describe how to mix and apply body filler.

Critical Thinking Problems

1. If you use spot putty to fill a large dent on bare metal, will the repair be sound? Why or why not?
2. Describe some of the problems that can result if body filler is not stirred up thoroughly to a smooth and uniform consistency.
3. To speed up the drying process of body filler, a chemical catalyst is added. What is this chemical catalyst called?
4. Heat or high temperature on or around the work area will have what affect on body filler?
5. During the first 50 years of auto body repair, blemishes in sheet metal panels were corrected by what?
6. Most plastic fillers can be grated and rough shaped within how many minutes? They can be sanded within how many minutes?

Math Problems

1. If a panel is 0.031 inch (0.79 mm) thick and you grind off 0.024 inch (0.61 mm) of metal, how thick is the remaining panel?
2. If maximum filler thickness is $\frac{1}{8}$ inch (31 mm) and you have already applied a thickness of $\frac{1}{32}$ inch (0.78 mm), how much more thickness is allowable?

Activities

1. Obtain a damaged panel (fender, hood, door, or lid) from a salvage yard. Use a hammer to place a small dent in a flat area of the panel. Use the information in this chapter to grind, fill, and smooth the dent using filler.
2. Featheredge, prime, and spot putty the area filled in Activity 1.
3. Read the mixing instructions on several brands and types of body filler. Report on your findings. What differences in filler characteristics and instructions did you find?

Photo Summary

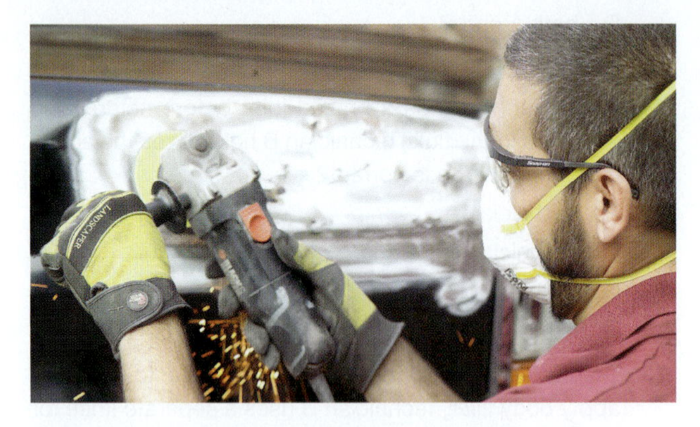

Figure P12-1 Body filler should not be placed over existing paint. To prepare the surface for adhesion, grind or sand off the paint in the area to be filled. Grind the paint carefully without cutting too deeply into the metal. Too much grinding will weaken today's thin metal panels.

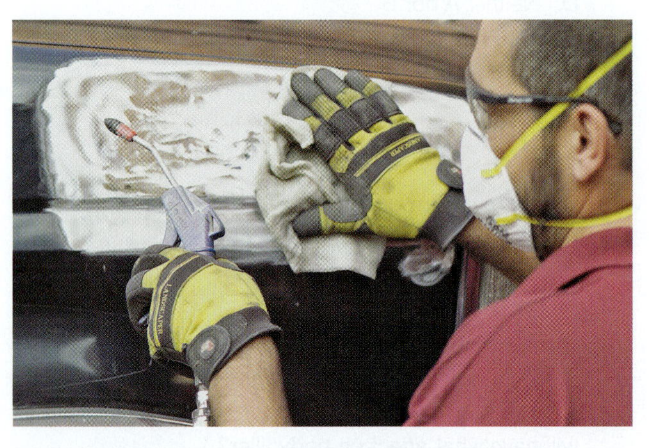

Figure P12-2 Blow and wipe off the area to remove debris. If dust is left on the repair area, the filler may not stick or adhere to the body. Vibration and flexing of the body could cause the filler to crack or pop off.

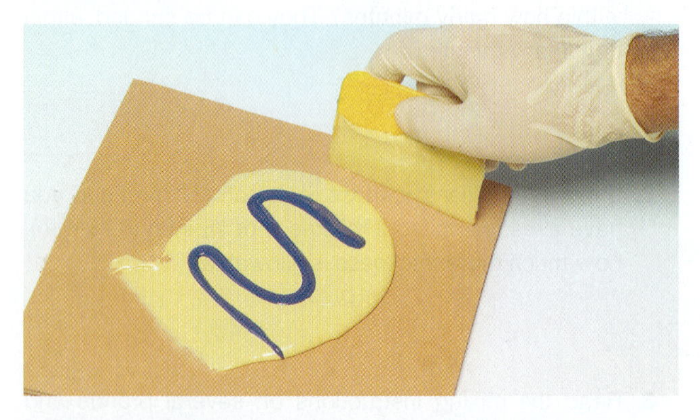

Figure P12-3 Mix quality body filler with catalyst or hardener. Use a 1-inch bead of hardener with a golf ball-size scoop of filler. Wipe the ingredients back and forth to mix them properly and to avoid stirring in air bubbles.

Figure P12-4 Place the filler on a spreader. Wipe the filler across the repair area. Use hand motions to make the filler match the general shape of the part contour. Try to make the filler slightly higher than the surrounding body surface.

Figure P12-5 When the filler is partially hardened, file it to rough shape with a super-coarse cheese grater file. Hold the grater at a 30-degree angle and pull it across the filler.

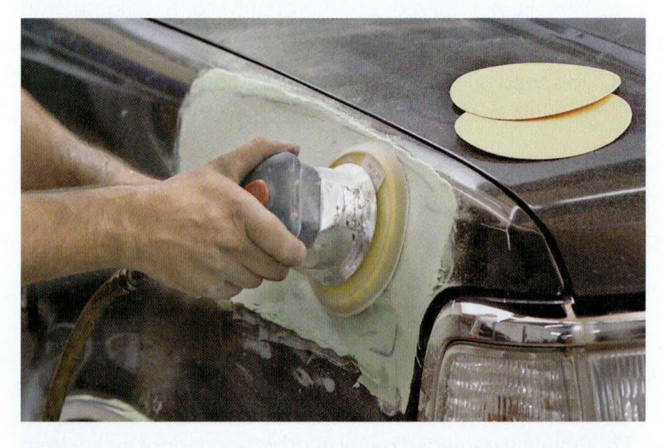

Figure P12-6 Now, use either a dual action sander or a hand block with medium grit sandpaper to sand the area. Knock down any remaining high spots as you featheredge, or taper the outer edge, of the filler into the undamaged body surface.

CHAPTER 13

Repairing Plastics

OBJECTIVES

After reading this chapter, you should be able to:

- Identify and explain the difference between the two major types of plastics used in automobiles. Identify unknown plastics.
- Repair minor cuts and cracks in plastics using adhesives.
- Repair gouges, tears, and punctures in plastics by means of a chemical bonding process.
- Explain the keys to good plastics welding.
- Describe the proper plastics welding repair sequence.
- Explain the safety precautions used when working with fiberglass.
- Explain how fiberglass is used in adhesives to reinforce the damaged surface.
- Make SMC and RRIM repairs.
- Answer ASE-style review questions relating to repairing plastics.

INTRODUCTION

The term *plastics* refers to a wide range of synthetic materials chemically compounded from synthetic or semi-synthetic organic solids that are moldable. Most plastics contain organic polymers that serve as "atomic chains" to make plastic parts so tough and durable. Unlike metals, plastics do not occur in nature and must be manufactured. Because plastic is much lighter in weight than sheet metal, it has become an important component of today's vehicles. Today, more and more plastic is being used in automobile manufacturing.

Plastic parts include bumpers, fender extensions, fascias, fender aprons, grille openings, stone shields, instrument panels, trim panels, fuel lines, door panels, quarter panels, and engine parts. Fuel-saving, corrosion prevention, and weight reduction programs by auto makers have made plastic parts more common.

Many of the new reinforced plastics are almost as strong and rigid as steel. Some are even more stable dimensionally. Plastic parts are also extremely corrosion resistant. Tests are being made on plastic engine blocks and plastic frame parts. Plastic suppliers are projecting increased use of plastic in floor pans, windows, steering shafts, springs, wheels, bearings, and other mechanical components.

This increasing use of plastic has resulted in new approaches to collision repair. Many plastic parts can be repaired more economically than they can be replaced, especially if the part does not have to be removed. Cuts, cracks, gouges, tears, and punctures are all repairable. When necessary, some plastics can also be re-formed back to their original shape after distortion. Because parts are not always available, this means less downtime for the vehicle and more profits for you and your shop.

Symbols are sometimes stamped on the back of plastic parts to denote the type of plastic used during manufacturing. However, if you cannot find a symbol, as is often the case, the chart in Figure 13–1 provides logical methods for identifying the type of plastic to be repaired.

13.1 TYPES OF PLASTICS

Two general types of plastics are used in automotive construction: thermoplastics and thermosetting plastics.

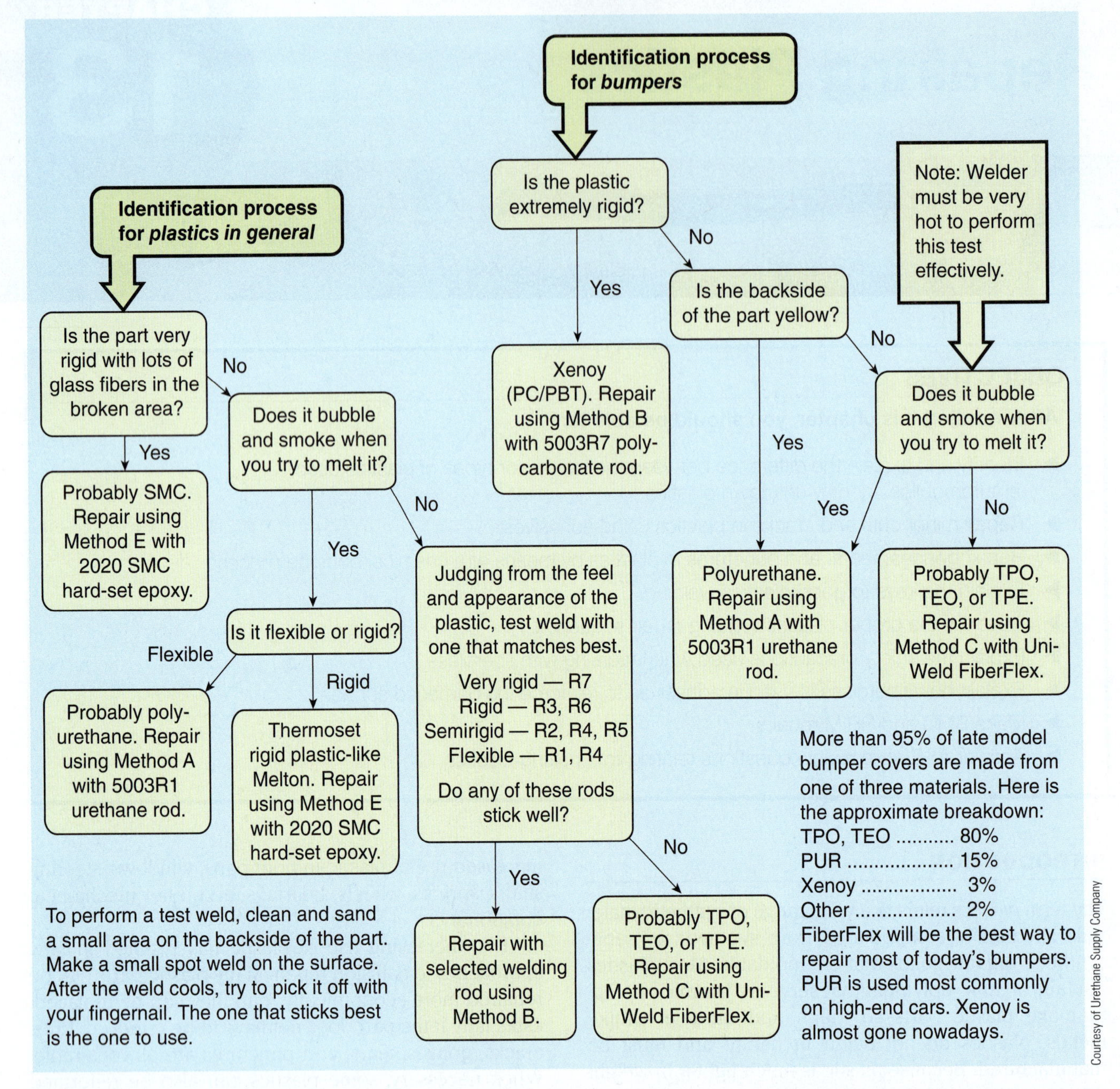

Figure 13–1 **If you cannot find a plastic ID symbol on the back of a part, you can use this basic sequence to help identify the type of plastic to be repaired.**

Thermoplastics can be repeatedly softened and reshaped by heating, with no change in their chemical makeup. They soften or melt when heated and harden when cooled. Thermoplastics are wieldable with a plastic welder or they can be adhesively repaired.

Thermosetting plastics, or *thermosets*, undergo a chemical change by the action of heating, a catalyst, or ultraviolet light. They are hardened into a permanent shape that cannot be altered by reapplying heat or catalysts. Thermosets are usually repaired with flexible parts repair materials. In general, chemical adhesive bonding is used to repair thermosetting plastics, and welding is used for thermoplastics.

Figure 13–2 explains more fully the effects of heat on the two types of plastics. Table 13–1 shows some of the more common plastics with their full chemical names, common names, and where on the vehicle they might be found. Their designations as thermosetting or thermoplastic are also noted.

Composite plastics, or *hybrids*, are blends of different plastics and other ingredients designed to achieve specific performance characteristics.

A good example of this change is the use of fiber reinforced composite plastic panels, commonly known as **sheet-molded compounds** (SMC). The reason for using SMC is simple. It is light, corrosion proof, dent resistant,

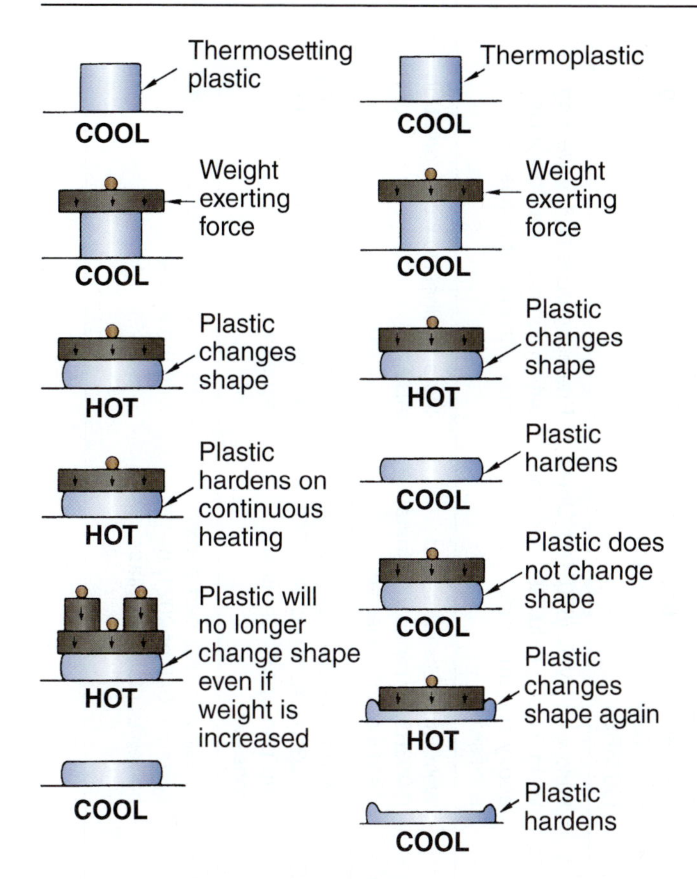

Figure 13-2 Here are the effects of heat on thermosetting plastic and thermoplastic.

and relatively easy to repair compared to more traditional materials.

The use of SMC and other *fiber-reinforced plastics* (FRP) is not new. They have been used in various applications on automobiles for years. The use of large external body panels of reinforced plastic is not unusual, either. What is new is that now, unlike the external panels on earlier vehicles, these panels are bonded to a metal space frame using structural adhesives, adding overall structural rigidity to the vehicle. SMC repair will be detailed later.

Table 13–2 contains additional information on types of automotive plastics and their repair methods.

Plastics Safety

Working with plastics and fiberglass requires you to think about safety at all times. Resins and related ingredients can irritate your skin and stomach lining. The curing agent or hardener can produce harmful vapors.

Read and understand the following safety points before using any of these types of products:

- Read all label instructions and warnings carefully.
- When cutting, sanding, or grinding plastics, dust control is important.
- Wear rubber gloves when working with fiberglass resin or hardener. A long-sleeved shirt with buttoned collar and cuffs is helpful to prevent sanding dust from getting on your skin. Disposable paint suits will keep dust away from clothes.
- A protective skin cream should be used on any exposed areas of the body.
- If the resin or hardener comes in contact with your skin, wash with borax soap and hot water or alcohol.
- Safety glasses are a necessity at all times.
- Always work in a well-ventilated area.
- Wear an approved respirator to avoid inhaling sanding dust and resin vapors.

Plastics Identification

There are several ways to identify unknown plastics. One way to identify plastics is by the *international symbols*, or *ISO codes*, that are molded into the parts. Many manufacturers use these symbols. The symbol or abbreviation is formed in an oval on the back of the part. One problem is that you usually have to remove the part to read the symbol.

If the part is not identified by a symbol, the body repair manual will give information about the plastics used on the vehicle. Body manuals often name the types of plastic used in a particular application.

The *burn test*, which is no longer recommended, involved using a flame and the resulting smoke to determine the type of plastic. However, the test was not always reliable. Many parts are now being manufactured from composite plastics that use more than one ingredient, and the burn test would be of no help in such cases. It is also environmentally unsound to burn plastics, which can produce **carcinogens** (cancer-causing agents).

A reliable means of identifying an unknown plastic is to make a weld rod adhesion test or a trial-and-error weld on a hidden or damaged area of the part. Try several different filler rods until one sticks. Most suppliers offer only a few types of plastic filler rods; the range of possibilities is somewhat limited. The rods are color coded. Once you find a rod that works, the base material is identified.

Another way to help identify a plastic part is the **plastic flexibility test**. To do a plastic flexibility test, use your hands to flex and bend the part and compare it to the flexibility of samples of plastic. Use the repair material that most closely matches the characteristics of the part's base material.

13.2 PLASTIC REPAIR

Plastic repair, like any other kind of body repair work, begins with the estimation process. At that time, it is determined whether a part should be repaired or replaced. A minor crack, tear, gouge, or hole in a nose fascia or large panel that is difficult to replace, costly, or not readily available probably indicates a repair should be considered. Extensive damage to the same component or damage to a fender extension or plastic trim item that is cheap and easy to replace would dictate replacement. In short, it is up to the repair person or estimator to decide whether it makes more sense to repair a plastic part than to replace it.

TABLE 13–1 COMMON AUTOMOTIVE PLASTICS IDENTIFICATION TABLE

Identifying Symbol	Chemical Composition	Typical Usage	Common or Trade Names	Suggested Repair Method	Repair Tips
PUR (RIM, RRIM)	Thermoset Polyurethane	Bumper covers, front and rear body panels, filler panels	Elastoflex, Bayflex, Specflex (Reaction Injection Molding)	Weld with urethane rod (5003R1) or Uni-Weld (5003R8).	Do not try to melt the base material; just melt rod into the V-groove.
TPU (TPUR)	Thermoplastic Polyurethane	Bumper covers, soft filler panels, gravel deflectors, rocker panel covers	Pellethane, Estane, Roylar, Texin, Desmopan	Weld with urethane rod (5003R1) or Uni-Weld (5003R8).	
TPO, EPM, TEO	Polypropylene + Ethylene Propylene rubber (at least 20 percent) (polyolefin)	Bumper covers, valence panels, fascias, air dams, dashboards, grilles	TPO (Thermoplastic Olefin), TPR (Thermoplastic Rubber), EPI, EPII, PTO, HiFax	Weld with Uni-Weld (5003R8) or TPO Blended Gray rod (5003R5).	Use adhesion promoter before applying filler or coating.
PP	Polypropylene (polyolefin)	Bumper covers, deflector panels, interior moldings, radiator shrouds, inner fenders	Profax, Oreflo, Marlex, Novolen, Carlona	Weld with Uni-Weld (5003R8) or Polypropylene Black rod (5003R2).	Use adhesion promoter before applying filler or coating.
PC + PBT	Polycarbonate + Polybutylene Terephthalate	Bumper covers (Ford)	Xenoy (GE)	Weld with Polycarbonate Clear rod (5003R7) or Uni-Weld (5003R8).	Preheat groove before welding with polycarbonate rod.
PPE + PA (PPO + PA)	Polyphenylene Ether + Polyamide	Fenders (Saturn, GM), exterior trim	Noryl GTX (GE)	Weld with Nylon (5003R6), Uni-Weld Ribbon (5003R8), two-part epoxy system, or instant adhesive.	Preheat groove before welding with nylon rod. Use fiberglass mat with instant adhesive.
ABS	Acrylonitrile Butadiene Styrene	Instrument clusters, trim moldings, consoles, armrest supports, grilles	Cycolac (GE), Magnum (Dow), Lustran (Monsanto)	Weld with ABS White rod (5003R3) or repair with Insta-Weld instant adhesive.	Instant adhesive works great on ABS.
PC + ABS	Polycarbonate + Acrylonitrile Butadiene Styrene	Door skins (Saturn), instrument panels	Pulse (Dow), Bayblend (Bayer), Cycoloy (GE)	Weld with Polycarbonate rod (5003R7), ABS rod (5003R3), two-part epoxy, or instant adhesive.	Preaheat groove before welding. Use fiberglass mat with instant adhesive.
UP, EP	Unsaturated Polyster, Epoxy (Thermoset)	Fender extensions, hoods, roofs, decklids, instrument housings	SMC, Fiberglass, FRP	Repair with two-part epoxy system (2020 or 2021) or polyester resin and glass cloth.	This material cannot be repaired with the welder.

PE	Polyethylene (polyolefin)	Inner fender panels, valences, spoilers, interior trim panels	Lacqtene, Lupolen, Dowlex, Hostalen	Weld with Polyethylene Opaque White rod (5003R4).	Use adhesion promoter before applying filler or coating.
PC	Polycarbonate	Interior rigid trim panels, valence panels	Lexan, Merlon, Calibre	Weld with Polycarbonate Clear rod (5003R7).	Preheat groove before welding.
PA	Polyamide	Radiator tanks, headlamp bezels, quarter panel extensions, exterior trim finish parts	Nylon, Capron, Celanese, Zytel, Rilsan, Orgamide, Vydyne, Minlon	Weld with Nylon Opaque White rod (5003R6) or Uni-Weld (5003R8).	Preheat groove before welding, especially on radiator tanks.
TEEE	Thermoplastic Ether Ester Elastomer	Bumper fascias (Bonneville SSE, Park Ave., 91–'96 Vette front), rocker panel covers (Camaro & Firebird)	Bexloy V (DuPont)	Weld with Uni-Weld (5003R8) or two-part epoxy system (2000 or 2020).	
PET	Polyethylene Terephthalate + Polyester	Fenders (Chrysler LH)	Bexloy K (DuPont), Vandar (Hoechst)	Weld with Uni-Weld (5003R8) or two-part epoxy system (2020 or 2021).	
EEBC	Ether Ester Block Copolymer	Rocker cover moldings, fender extensions ('91–'96 DeVille)	Lomod (GE)	Weld with Uni-Weld (5003R8) or two-part epoxy system (2010).	
EMA	Ethylene/ Methacrylic Acid	Bumper covers (Dodge Neon)	Bexloy W (DuPont)	Weld with Uni-Weld (5003R8).	

TABLE 13–2 PLASTIC REPAIR QUICK REFERENCE

Repair Method		A	B	C	D	E	F
Step 1	Identify Plastic	Thermoset Polyurethane	ABS, Polyurethane, Nylon, Polycarbonate	PP, TPO, TEO, TPE, PE, or Other	ABS, SMC, Fiberglass, PC Blend	SMC, UP, FRP, Fiberglass	ABS, SMC, Fiberglass, PC Blend
Step 2	Clean	Clean part with soap and water and plastic cleaner.					
Step 3	Repair	Thermoset Urethane Weld	Thermoplastic Fusion Weld	Uni-Weld FiberFlex	Insta-Weld Adhesive	Two-Part Epoxy Adhesive	Rigid Plastic Repair Kit
Step 4	Fill	Grind and apply filler to match the hardness of the substrate.					
Step 5	Prime	Prime with flexible primer, then seal with sealer.					
Step 6	Paint	Apply topcoat with flex additive.					

Symbol & Type	How to Identify	Typical Usage	Suggested Repair Method	Repair Tips
PUR, RIM, RRIM Thermoset polyurethane	Usually flexible, may be yellow or gray, bubbles and smokes when attempting to melt	Flexible bumper covers (esp. on domestics), filler panels, rocker panel covers, snowmobile cowls	Method A with urethane rod or Method C with Uni-Weld FiberFlex	Do not try to melt the base material! Just melt the rod into the V-groove like a hot melt glue.
SMC, UP, FRP Fiberglass	Rigid, polyester matrix reinforced with glass fibers, sands finely	Rigid body panels, fenders, hoods, deck lids, header panels, spoilers	Method E—two-part epoxy repair with fiberglass reinforcement.	Use backing plate over holes, layer in fiberglass cloth for extra strength and to approximate thermal expansion.
ABS (Acrylonitrile Butadiene Styrene)	Rigid, often white, but may be molded in any color, sands finely; very pungent odor when heated	Instrument panels, grilles, trim moldings, consoles, armrest supports, street bike fairings	Method B with ABS rod (5003R3), or Method D Insta-Weld adhesive repair, or Method E Two-Part epoxy repair	Weld repairs may be backed with epoxy for extra strength.
EEBC (Ether Ester Block Copolymer)	Flexible, off-white in color, similar in appearance to PUR (Lomod by GE)	Rocker cover moldings, bumper extensions ('91–'96 DeVille)	Method C with Uni-Weld FiberFlex (5003R10)	

EMA (Ethylene Meth-acrylic Acid)	Semirigid, molded in a variety of colors, unpainted (Bexloy W by DuPont)	Bumper covers (Dodge Neon first generation base model)	Method C with Uni-Weld FiberFlex (5003R10) or Method B with slivers cut from scrap	Sand entire bumper for refinishing, restore texture with Flex Tex (3800).
PA Polyamide (Nylon)	Semirigid or rigid, sands finely	Radiator tanks, headlamp bezels, exterior trim finish parts	Method B with nylon rod (5003R6)	Preheat plastic with heat gun before welding, mix rod completely with base material.
PC + ABS Pulse (Polycarbonate & ABS)	Rigid, sands finely, usually dark in color	Door skins (Saturn), instrument panels, street bike fairings	Method B with Polycarbonate rod (5003R7), or Methods D or E adhesive repairs	Preheat plastic with heat gun before welding with Method B.
PC + PBT Xenoy (Polycarbonate blend)	Rigid, sands finely, usually dark in color	Bumper covers (primarily Ford products, 84–95 Taurus, Aerostar, some Mercedes and Hyundai's)	Method B with polycarbonate rod (5003R7), Method C, or Method E adhesive repairs	Preheat plastic with heat gun before welding with Method B.
PE Polyethylene	Semiflexible, melts & smears when grinding, usually semi-translucent	Overflow tanks, inner fender panels, valences, interior trim panels, RV water storage tanks, gas tanks	Method B with polyethylene rod (5003R4) or Method C with FiberFlex (5003R10)	Applying filler or painting is nearly impossible.
PP Polypropylene	Semiflexible, usually black in color, melts & smears when grinding	Bumper covers (usually blended with EPDM), inner fenders, radiator shrouds, interior panels, gas tanks	Method C with Uni-Weld FiberFlex (5003R10) or Method B with polypropylene rod (5003R2)	Use 1060FP Filler Prep adhesion promoter when applying two-part epoxy filler.
PPO + PA Noryl GTX (Nylon blend)	Semirigid, sands finely, usually off-white in color	Fenders (Saturn & GM), exterior trim	Method B with nylon rod (5003R6), or Methods D or E adhesive repairs	Preheat plastic with heat gun before welding with Method B.
TEEE (Thermoplastic Ether Ester Elastomer)	Flexible or Semiflexible (Bexloy V by duPont)	Bumper covers (especially on domestics, filler panels, rocker panel covers)	Method C with Uni-Weld Fiberflex (5003R10) or Method B with slivers cut from scrap	
TPE Thermoplastic Elastomer	Semiflexible, usually black or gray, melts & smears when grinding	Bumper covers, filler panels, underhood parts	Method C with Uni-Weld FiberFlex (5003R10)	Use 1060FP Filler Prep adhesion promoter before applying two-part epoxy filler.
TPO, EPM, TEO Thermoplastic Olefin	Semiflexible, usually black or gray in color, melts & smears when grinding	Bumper covers, air dams, grilles, interior parts, instrument panels, snowmobile cowls	Method C with Uni-Weld FiberFlex (5003R10) or Method B with PP or TPO rod	Use 1060FP Filler Prep adhesion promoter when applying two-part epoxy filler.
TPU, TPU—Thermoplastic Polyurethane	Flexible, sands finely	Bumper covers, soft filler panels, gravel deflectors, rocker panel covers	Method B with urethane rod (5003R1) or Method C with Uni-Weld FiberFlex (5003R10)	

If repair is the answer, it must be determined whether the part needs to be removed from the vehicle. The entire damaged area must be accessible to do a quality repair. If it is not accessible, the part must be removed. Keep in mind that the part will also have to be refinished.

As mentioned earlier, there are two methods of repairing plastics:

1. By use of chemical adhesives
2. By plastic welding

13.3 CHEMICAL-ADHESIVE BONDING TECHNIQUES

Adhesive repair systems are of two types: cyanoacrylate (CA) and two-part. Two-part is the most commonly used.

Cyanoacrylates, or *CAs*, are one-part, fast-curing adhesives used to help repair rigid and flexible plastics. They are often used as a filler or to tack parts together before applying the final repair material. CAs are sometimes referred to as "super glues." They can be a valuable tool for the repair of plastic parts. CAs set up very quickly.

Two-part adhesive systems consist of a base resin and a hardener (catalyst). The resin comes in one container and the hardener in another. When mixed, the adhesive cures into a plastic material similar to the base material in the part. Two-part adhesive systems are an acceptable alternative to welding for many plastic repairs. They are also stronger than CAs.

Not all plastics can be welded, while adhesives can be used in all but a few instances. If adhesive repair is chosen, you must first identify the type of plastic. A good way to do this is the plastic flexibility test described earlier.

Repairs of Minor Cuts and Cracks

Adhesives are usually used for repairing minor cuts and cracks in plastic parts. First, wash the area thoroughly with soap and hot water. Second, wipe or wash the repair area clean with water and a plastic cleaner. The surfaces must be free of wax, dust, and grease. Allow the part(s) to warm to 70°F (21°C) before applying adhesives.

After cleaning, prepare the crack with an adhesive kit. The kit should have two elements: an accelerator and an adhesive. Spray one side of the crack with the accelerator, as shown in Figure 13–3. Then apply the adhesive to the same side of the crack.

Carefully position the two sides of the cut or crack in their original position, and quickly press them together with firm pressure. Hold for a full minute to achieve good bond strength. Then allow the repair to cure for 3–12 hours for maximum strength, or according to the instructions on the label. Note the precautions and instructions on the container for the adhesive used.

If the original paint was not damaged and the repair was properly positioned, painting may not be required. Where painting is required, special procedures are needed.

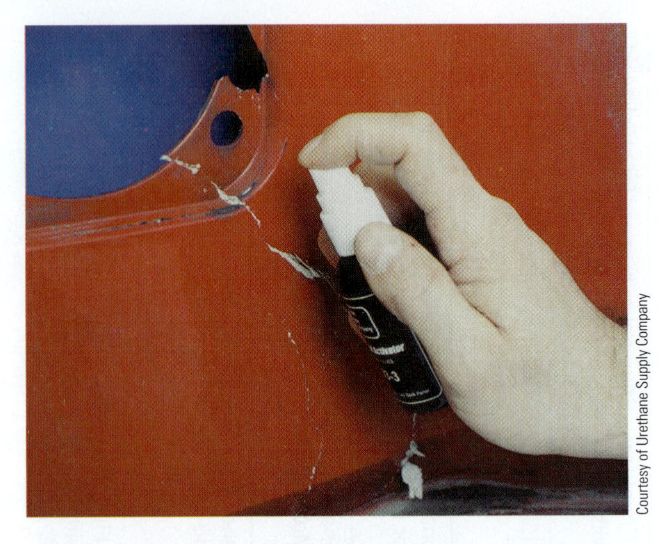

Courtesy of Urethane Supply Company

Figure 13–3 Spray the crack with the correct type of activator or accelerator to prepare it for the adhesive.

Repairs of Gouges, Tears, and Punctures

The procedure for repairs of gouges, tears, and punctures is somewhat more involved than for minor cuts but requires no special skills or tools.

First, wash the area thoroughly with soap and hot water. Then clean around the damaged area thoroughly with a wax-, grease-, and silicone-removing solvent applied with a water-dampened cloth. Wipe the area dry.

To prepare for the structural adhesive, bevel the edges of the hole back about ¼–⅜ inch (6.4–9.5 mm). The technician in Figure 13–4 is using a small grinding disc, with medium grit. Use a slow speed when grinding (2,000 rpm or less). In this repair, the beveling has left a coarse surface for good adhesion. In any repair, the mating surfaces should be scuffed to improve adhesion.

Featheredge the paint around the repair area. Use a finer grit disc. Remove the paint, but very little of the urethane plastic. Blend the paint edges into the plastic. Continue removing paint until there is a paint-free band around the hole—about 1–1½ inches (25–38 mm) wide. The repair material must not overlap the painted surface.

Carefully wipe off all paint and urethane dust. The repair area must be absolutely clean for proper bond strength. If recommended by the product manufacturer, flame treat the beveled area of the hole. This flame treatment improves the adhesion of some types of structural adhesive. Use any torch with a controlled flame, and develop a 1-inch (25 mm) cone tip.

The next step is to apply auto backing tape to the repair area. An aluminum foil with a strong adhesive on one side and a moisture-proof backing is recommended. Clean the inner surface of the repair area with silicone and wax remover. Then install the tape. Cover the hole completely, with about a 1-inch (25 mm) adhesion surface around the edges.

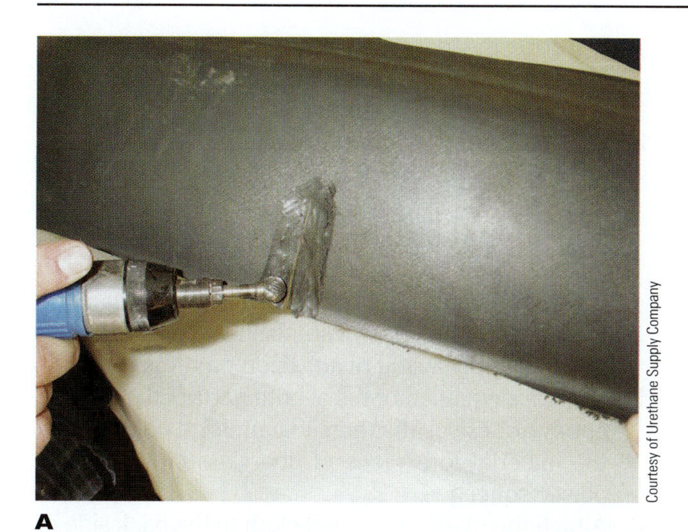

A

B

Figure 13-4 **(A) Bevel the damaged area with a 3-inch grinding disc to strengthen the repair. (B) Sometimes it will strengthen the repair if you drill holes next to the crack in the plastic part.**

Some technicians like to back the repair with cloth rather than tape. The cloth will remain in place and reinforce the repair. Saturate the entire cloth on both sides with the adhesive. This will make the cloth bond to the back of the plastic properly and also seal the cloth.

Prepare the repair adhesive material on a clean, flat, nonporous surface, such as metal or glass, as directed by the manufacturer. Most adhesive compounds come in two tubes. Squeeze out equal amounts of the repair mix. Then, with an even paddling motion to reduce air bubbles, completely mix the two components until a uniform color and consistency is achieved.

Paddle the structural adhesive into the hole, using a squeegee or plastic spreader (Figure 13–5). This must be done carefully and swiftly, as the structural adhesive material will begin to set in about 2–3 minutes. Two applications of the adhesive are usually required. The first application is used to fill the bottom of the hole. It is not necessary to worry about contour at this time. In the first application of patch material, try to fill the greater part of the hole's volume. Then cure for about an hour at room

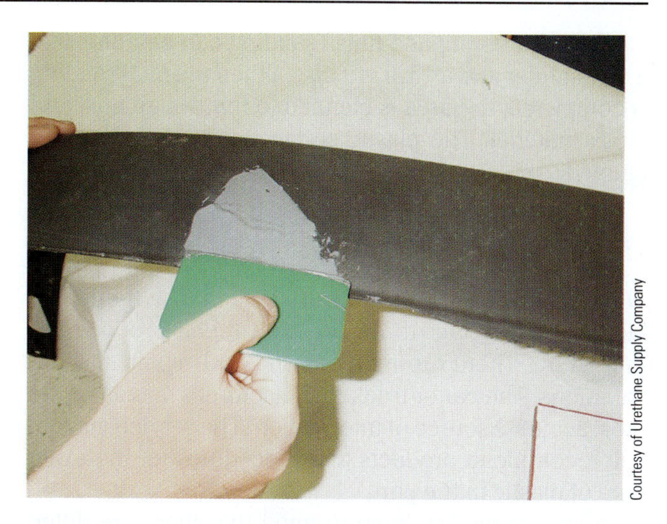

Figure 13-5 **Apply structural adhesive with a squeegee to match the shape of the part.**

temperature, or 20 minutes with a heat lamp or gun at 190°–200°F (88°–93°C) if the manufacturer's directions allow heat curing.

Before the final application of the adhesive, use a fine grit disc to grind down the high spots of the first application (Figure 13–6). Wipe the dust from the repair area.

After the first application is ground and wiped clean, mix the second application of the adhesive, squeezing the components together as before for about 2 minutes. Then apply the second adhesive mixture, paddling it into an overfill contour of the area. A flexible squeegee or spatula is useful in approximating the panel contours.

When the adhesive repair material has dried, establish a rough contour to the surrounding area with a #80 grit abrasive on a sanding block. You can also feather sand the area using a disc sander with #180 sandpaper followed by a #240 sandpaper to achieve an accurate level with the surface of the part.

Figure 13-6 **Grind down high spots on the adhesive as you would with conventional body filler. Never use conventional body filler on plastic parts, because it will not bond to plastic properly.**

Final feathering and finish sanding can be done with a disc sander and a #320 grit disc. When the final sanding is completed, the area is cleaned to remove all dust and loose material. The plastic surface is then ready to be painted.

Using the Right Adhesive

When working with an adhesive system, use the manufacturer's categories to decide on a repair product and procedure. There are many plastic and repair material variations. The car or truck manufacturer's manual is the most accurate source of information. The service manual will recommend products and procedures for the exact type of plastic in the part.

It is important to keep in mind that there are differences among manufacturers' repair materials. When using plastic repair adhesives, remember the following:

- Mixing product lines is not acceptable. Choose a product line and use it for the entire repair.
- Most product lines have two or more adhesives designed for different types of plastic.
- The product line usually includes an adhesion promoter, a filler product, and a flexible coating agent. Use each as directed.
- Some product lines are formulated for a specific base material. For example, one manufacturer offers individual products for use on each type of plastic (TPO, urethanes, or Xenoy, for example), regardless of plastic flexibility.
- Clean plastic parts with a plastic prep. Never use solvent-based cleaners.

A product line might use single flexible filler for all plastics, or there may be two or more flexible fillers designed for different types of plastic.

An **adhesion promoter** is a chemical that treats the surface of the plastic so the repair material will bond properly. Some plastics (TPO, PP, and E/P) require an adhesion promoter. There is a simple test that indicates whether the plastic will require an adhesion promoter. Lightly sand a hidden spot on the piece using a high-speed grinder and #36 grit sandpaper.

If the material gives off dust, it can be repaired with a standard structural adhesive system. If the material melts and smears or has a greasy or waxy look, then you must use an adhesion promoter. Many plastic fillers and adhesives contain an adhesion promoter. Check their labels.

Flexible Part Repair

Here is a typical procedure for using a two-part epoxy adhesive to repair a flexible part, a bumper cover in this example:

1. Clean the entire cover with soap and hot water. Wipe or blow-dry. Then clean the surface with a good plastic cleaner (Figure 13–7).
2. V-groove the damaged area. Then grind about a 1½-inch (38 mm) taper around the damage for good adhesion and repair strength.
3. Use a sander with #180 grit sandpaper to featheredge the paint around the damaged area. Then blow off the dust. Depending on the extent of the damage, the back may need reinforcement. To do this, follow steps 4 through 6.
4. To reinforce the repair area, sand and clean the back side of the cover with plastic cleaner. Then, if needed, apply a coat of adhesion promoter.
5. Dispense equal amounts of both parts of the flexible epoxy adhesive. Mix them to a uniform color. Apply the material to a piece of fiberglass cloth using a plastic squeegee.
6. Attach the plastic-saturated cloth to the back side of the bumper cover. Fill in the weave with additional adhesive material.
7. With the back side reinforcement in place, apply a coat of adhesion promoter to the sanded repair area on the front side. Let the adhesion promoter dry completely (Figure 13–8).
8. Fill in the area with adhesive material. Shape the adhesive with your spreader to match the shape of the part. Allow it to cure properly.
9. Rough grind the repair area with #80 grit sandpaper, then sand with #180 grit, followed by smoother #240 grit.
10. If additional adhesive material is needed to fill in a low spot or pinholes, apply a coat of adhesion promoter again.

13.4 PLASTIC WELDING

Plastic welding uses heat and sometimes a plastic filler rod to join or repair plastic parts. The welding of plastics is not unlike the welding of metals. Both methods use a heat source, welding rod, and similar techniques (butt joints, lap joints, and so on). Joints are prepared in much the same manner, and evaluated for strength. There are differences between welding metal and welding plastics, however.

When welding plastics, the materials are fused together by the proper combination of heat and pressure (Figure 13–9). Successful welds require that both pressure and heat be kept constant and in proper balance. Too much pressure on the rod tends to stretch the bead; too much heat will char, melt, or distort the plastic. With practice, plastic welding can be mastered as completely as metal welding.

13.5 HOT-AIR PLASTIC WELDING

Hot-air plastic welding uses a tool with an electric heating element to produce hot air (450–650°F or 232–345°C) that blows through a nozzle and onto the plastic. The air supply comes from either the shop's air compressor or a self-contained portable compressor that comes with the welding unit.

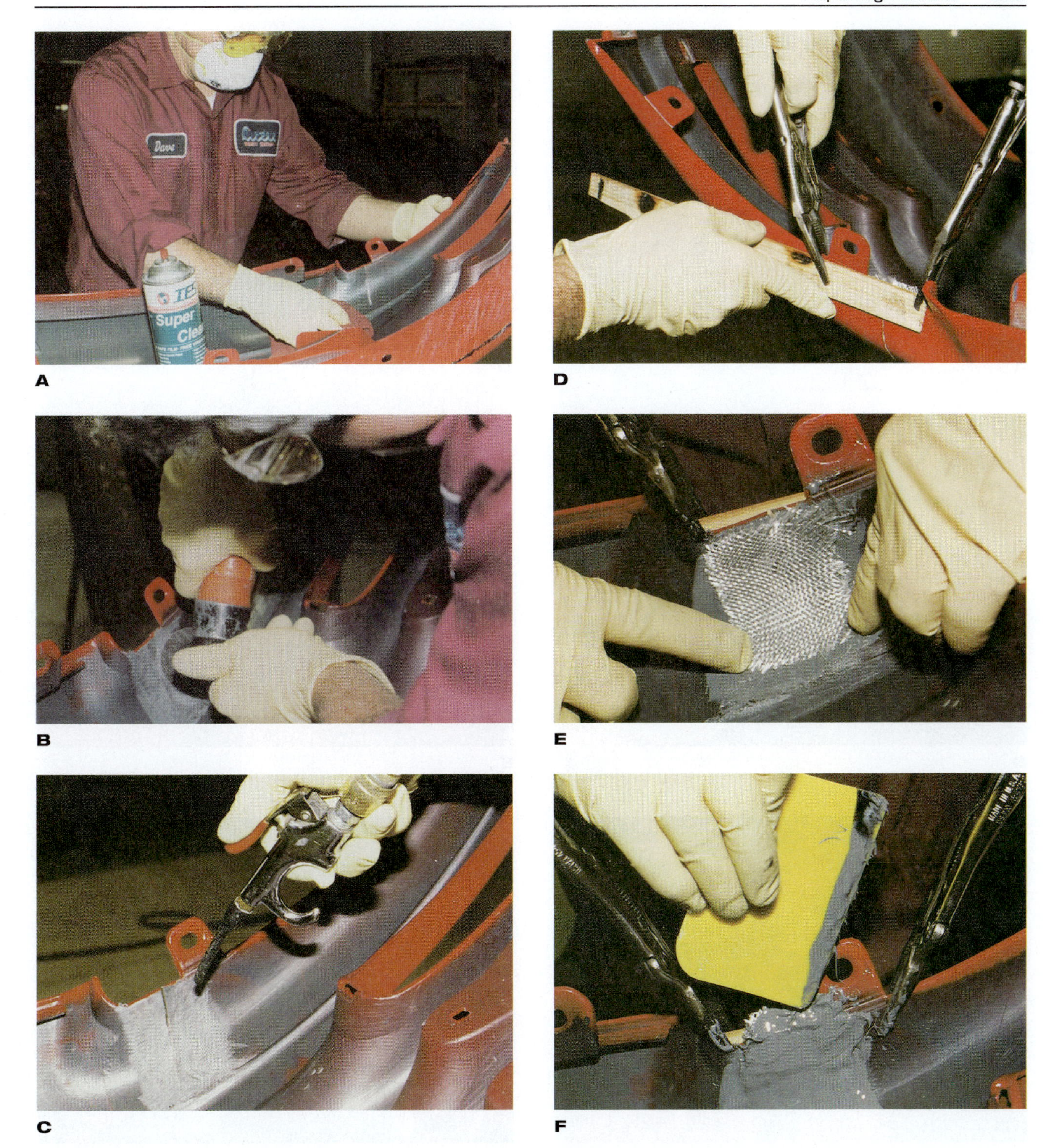

Figure 13-7 **When damage is all the way through a flexible part, it is best to repair and reinforce the back of the part first. (A) Clean the back of the part with an approved plastic cleaner. (B) Grind a groove in the crack or grind a bevel around the damaged area. (C) Blow off the plastic dust and apply an adhesion promoter. (D) Here the technician is clamping a wooden mixing stick to the bumper to keep it aligned while applying repair material. (E) Apply a two-part repair material and then a layer of fiberglass cloth to strengthen the repair. (F) Coat the fiberglass cloth with a layer of two-part repair material. Allow the material to cure before repairing the front of the part.**

One of the problems with hot-air welding is that the plastic welding rod is often thicker than the panel to be welded. This can cause the panel to overheat before the rod has melted. Using a smaller diameter rod can often correct such warpage problems.

Three types of welding tips are available for use with most hot-air plastic welding torches:

Tacking tips are shaped to tack-weld broken sections of plastic together before welding. If necessary, tack welds can be easily pulled apart for realigning.

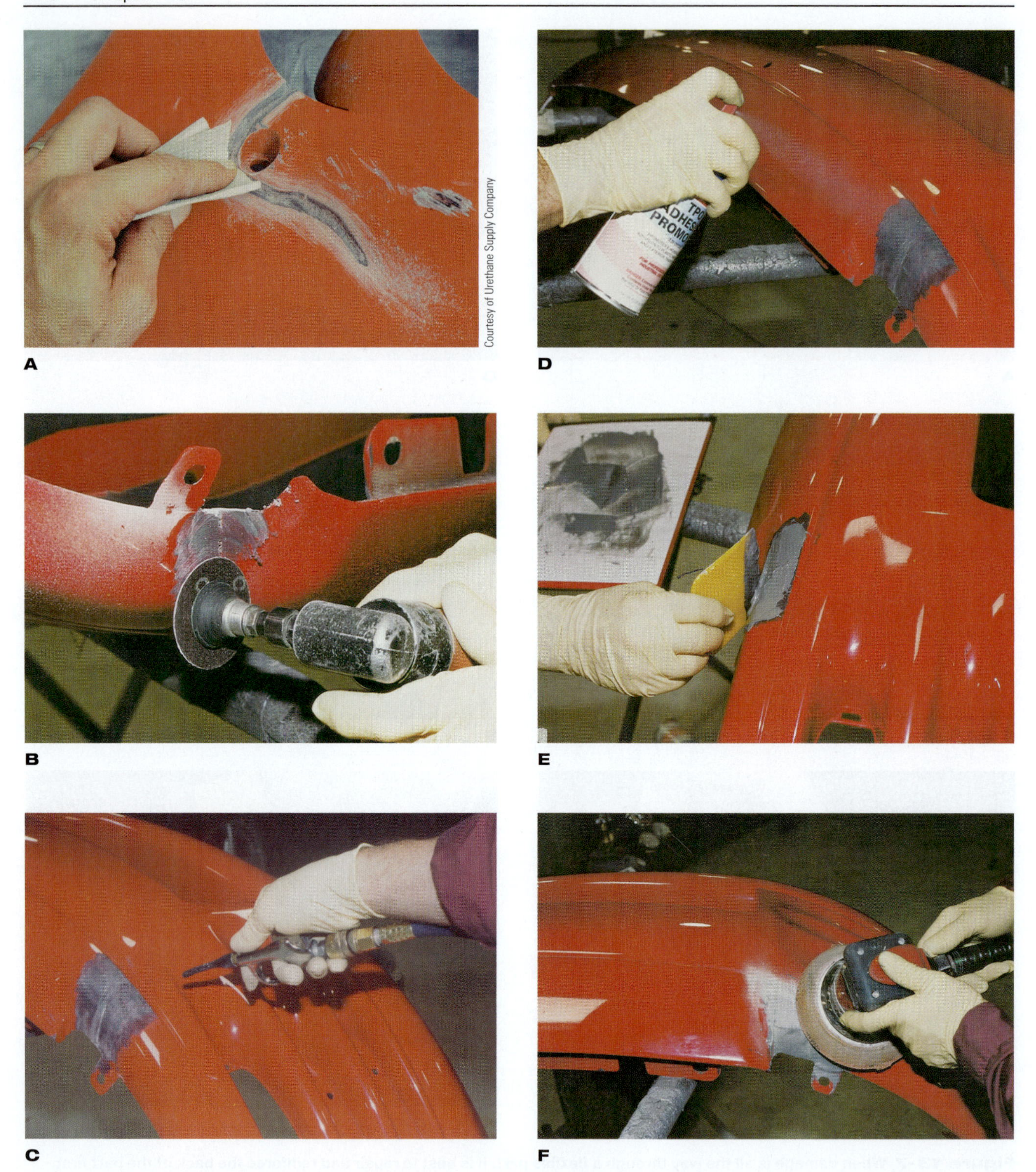

Figure 13-8 After repairing the rear of the part, the front can be repaired using a similar procedure. (A) Grind a groove in the crack and then feather an area to be coated with flexible part repair material. (B) Feather the ground area out with an air sander. (C) Blow off plastic dust so the repair material will bond securely. (D) Apply adhesion promoter to the repair area, if needed. (E) Apply a two-part repair material to the front of the damage. (F) Sand and feather the repair area as you would body filler.

Round tips are used to make short welds to weld small holes and to weld in hard-to-reach places and sharp corners.

Speed tips hold, feed, and automatically preheat the plastic welding rod. This design feeds the rod into the base material, thus allowing for faster welding speeds. They are used for long, fairly straight welds.

Some hot-air welder manufacturers have developed specialized welding tips and rods to meet specific needs. Check the product catalog for more information.

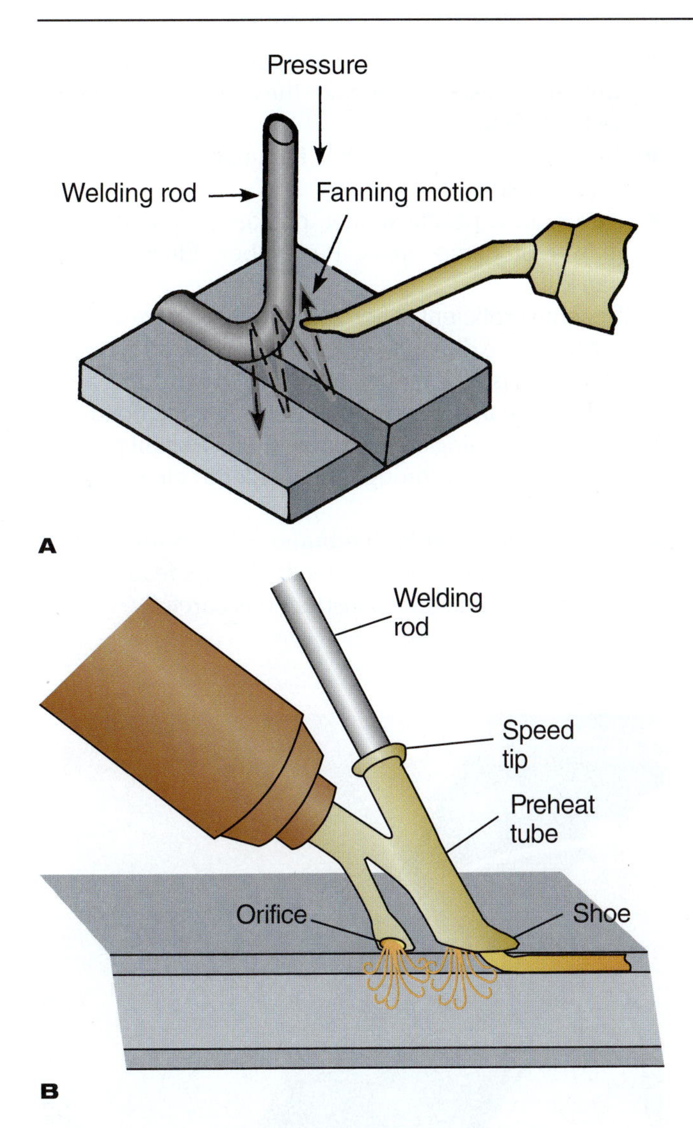

Figure 13-9 **(A) Successful plastic welding requires the proper combination of heat and pressure. (B) This plastic welder is equipped with a high-speed tip.**

High-Speed Welding

High-speed welding incorporates the basic methods utilized in hand welding except a specially designed and patented high-speed tip is used, which enables the welder to produce more uniform welds and work at a much higher rate of speed. As with hand welding, constant heat and pressure must be maintained.

Plastic Welder Setup, Shutdown, and Servicing

No two hot-air plastic welders are exactly alike; for specific instructions, always refer to the owner's manual and other material provided by the welder manufacturer.

Some manufacturers advise against using their welder on plastic thinner than ⅛ inch (3 mm) because of distortion. It is sometimes acceptable to weld thin plastics if they are supported from underneath while welding.

13.6 AIRLESS PLASTIC WELDING

Airless plastic welding uses an electric heating element to melt a smaller ⅛-inch (3 mm) diameter rod with no external air supply. Airless welding with a smaller rod helps eliminate two troublesome problems: panel warpage and excess rod buildup.

When setting up an airless welder, set the temperature dial at the appropriate setting. This will depend on the specific plastic being worked. It will normally take about 3 minutes for the welder to fully warm up.

Make sure the rod is the same material as the damaged plastic or the weld will be unsuccessful. Many airless welder manufacturers provide rod application charts. When the correct rod has been chosen, it is good practice to run a small piece through the welder to clean out the tip before beginning.

13.7 ULTRASONIC PLASTIC WELDING

Ultrasonic plastic welding relies on high-frequency vibratory energy to produce plastic bonding without melting the base material. Handheld systems are available in 20 and 40 kHz frequencies. They are equally adept at welding large parts and tight, hard-to-reach areas. Welding time is controlled by the power supply.

13.8 PLASTIC WELDING PROCEDURES

The basic methods for hot-air and airless welding are very similar. To make a good plastic weld with either procedure, keep the following factors in mind:

- Plastic welding rods are frequently *color-coded* to indicate their material. Unfortunately, the coding is not uniform among manufacturers. It is important to use the reference information provided. If the rod is not compatible with the base material, the weld will not hold.
- Too much heat will char, melt, or distort the plastic. Too little heat will not provide weld penetration between the base material and the rod.
- Too much pressure stretches and distorts the weld.
- The angle between rod and base material must be correct. If too shallow, a proper weld will not be achieved.
- Use the correct welding speed. If the torch movement is too fast, it will not produce a good weld. If the tool is moved too slowly, it can char the plastic.

The basic repair sequence is generally the same for both plastic welding processes:

1. Prepare the damaged area.
2. Align the damaged area.
3. Make the weld.
4. Allow it to cool.

5. Sand. If the repair area has pinholes or voids, bevel the edges of the defective area. Add another weld bead and resand.
6. Apply a topcoat.

General Welding Techniques

Welding plastic is not difficult when done in a careful and thorough manner. The following guidelines cannot be stressed enough:

- The welding rod must be compatible with the base material in order for the strength, hardness, and flexibility of the repair to be the same as the part.
- Always test a welding rod for compatibility with the base material. To do this, melt the rod onto a hidden side of the damaged part, let the rod cool, and then try to pull it from the part. If the rod is compatible, it will adhere.
- Pay close attention to the temperature setting of the welder; it must be correct for the type of plastic being welded (Figure 13–10).
- Never use oxygen or other flammable gases with a plastic welder.
- Never use a plastic welder, heat gun, or similar tool in wet or damp areas. Remember: Electric shock can kill.
- Become proficient at horizontal welds before attempting the more difficult vertical and overhead types.
- Make welds as large as they have to be. The greater the surface area of a weld, the stronger the bond.
- Before beginning an airless weld, run a small piece of the welding rod through the welder to clean out the torch tip.
- Consult a supplier for the brands of tools and materials that best fit the shop's needs. Always read and follow the manufacturer's instructions carefully.

Courtesy of Urethane Supply Company

A

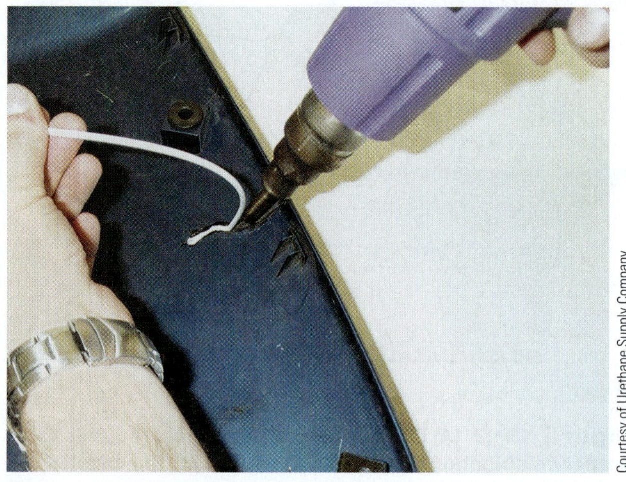

Courtesy of Urethane Supply Company

C

Courtesy of Urethane Supply Company

B

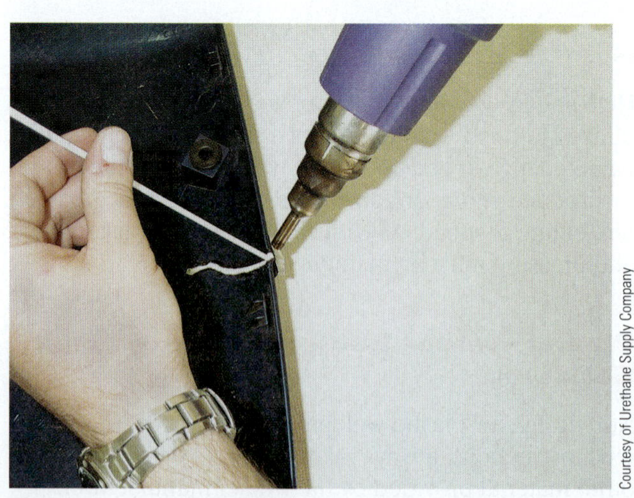

Courtesy of Urethane Supply Company

D

Figure 13-10 Note the basic steps for plastic welding. (A) Use a Dremel tool or small grinder to bevel crack to be plastic welded. (B) Begin heating joint and apply plastic welding rod. (C) Only heat part and plastic rod enough to make it stick to joint or crack. Overheating will prevent proper welding. (D) Finish the weld all the way to the end of the crack.

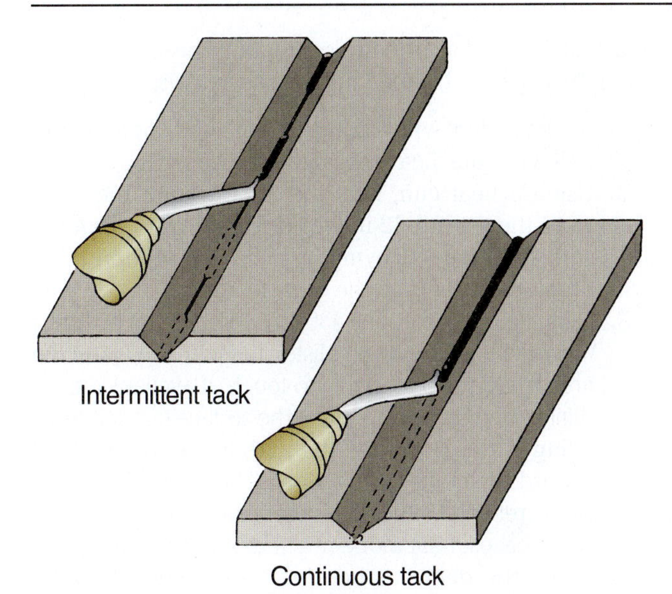

Figure 13-11 Two methods of making a tack weld are intermittent and shallow continuous.

Tack Welding

On long tears where backup is difficult, small *tack welds* can be made to hold the two sides in place before doing the permanent weld (Figure 13–11). For larger areas, a patch can be made from a piece of plastic and tacked in place.

To tack weld, proceed as follows:

1. Hold the damaged area in alignment with clamps or aluminum body tape.
2. Using a tacking welding tip, fuse the two sides to form a thin hinge weld along the root of the crack. This is especially useful for long cracks because it allows for easy adjustment and alignment of the edges.
3. Start tacking by drawing the point of the welding tip along the joint. Press the tip in firmly, making sure to contact both sides of the crack. Draw the tip smoothly and evenly along the line of the crack. No welding rod is used when tacking.
4. The point of the tip will fuse both sides in a thin line at the root of the crack. The fused parts will hold the sides in alignment. Then you can fuse the entire length of the crack.

Speed Welding Procedure

As mentioned, **plastic speed welding** uses a specially designed tip to produce a more uniform weld at a higher rate of speed. You must preheat both the rod and base material. The rod is preheated as it passes through a tube in the speed tip. The base material is preheated by a stream of hot air passing through a vent in the tip.

Following are some techniques essential for quality speed welding:

1. Hold the speed torch like a dagger. Bring the tip over the starting point a full 3 inches from the base material. You do not want the hot air to affect the part.
2. Cut the welding rod at a 60-degree angle. Insert it into the preheat tube. Immediately place the pointed shoe end of the tip on the base material at the starting point.
3. Hold the torch perpendicular to the base material. Push the rod through until it stops against the base material at the starting point. If necessary, lift the torch slightly to allow the rod to pass under the shoe.
4. Keep a slight pressure on the rod and only the weight of the torch on the shoe. Then pull the torch slowly toward you to start the speed weld.
5. In the first 1–2 inches (25–50 mm) of travel, push it into the preheat tube with slight pressure.
6. Once started, swing the torch to a 45-degree angle. The rod will now feed without the need for pressure. As the torch moves along, inspect the quality of the weld.

Airless Melt-Flow Plastic Welding

Melt-flow plastic welding is the most commonly used airless welding method. It can be utilized for both single-sided and two-sided repairs.

A typical melt-flow procedure is as follows:

1. With the welding rod in the preheat tube, place the flat shoe part of the tip in the V-groove.
2. Hold it in place until the rod begins to melt and flow out around the shoe.
3. A small amount of force is needed to feed the rod through the preheat tube. The rod will not feed itself, and care should be used not to feed it too quickly.
4. Move the shoe slowly. Crisscross the groove until it is filled with melted plastic.
5. Work the melted plastic well into the base material, especially toward the top of the V-groove.
6. Complete a weld length of about 1 inch (25 mm) at a time. This will allow smoothing of the weld before the plastic cools.

Plastic Stitch-Tamp Welding

Plastic stitch-tamp welding is used primarily on hard plastics, like ABS and nylon, to ensure a good base and rod mix.

After completing the weld using the melt-flow procedure, remove the rod. Turn the shoe over and slowly move the pointed end of the tip into the weld area to bond the rod and base material together. Stitch-tamp the entire length of the weld. After stitch-tamping, use the flat shoe part of the tip to smooth out the weld area.

Single-Sided Plastic Welds

Single-sided plastic welds are used when the part cannot be removed from the vehicle. To make a single-sided weld, proceed as follows:

1. Set the temperature dial on the welder for the plastic being welded. Allow it to warm up to the proper temperature.

2. Clean the part by washing with soap and hot water, followed by a good plastic cleaner.
3. Align the break using aluminum body tape.
4. V-groove the damaged area 75 percent of the way through the base material. Angle or bevel back the torn edges of the damage at least ¼ inch (6 mm) on each side of the damaged area. Use a die grinder or similar tool.
5. Clean the preheat tube and insert the rod. Begin the weld by placing the shoe over the V-groove and feeding the rod through. Move the tip slowly for good melt-in and heat penetration.
6. When the entire V-groove has been filled, turn the shoe over and use the tip to stitch-tamp the rod and base material together into a good mix along the length of the weld.
7. Resmooth the weld area using the flat shoe part of the tip, again working slowly. Then cool with a damp sponge or cloth.
8. Shape the excess weld buildup to a smooth contour, using a razor blade and/or abrasive sandpaper.

Two-Sided Plastic Welds

A *two-sided plastic weld* is the strongest type of weld because you weld both sides of the part. When making a two-sided weld, be sure to follow these steps:

1. Allow the welder to heat up. Then clean the preheat tube.
2. Clean the part with soap and hot water and plastic cleaner.
3. Align the front of the break with aluminum body tape, smoothing it out with a stiff squeegee or spreader.
4. V-groove 50 percent of the way through the back side of the panel.
5. Weld the back side of the panel using the melt-flow method. Move slowly enough to achieve good melt-in.
6. When finished, smooth the weld with the shoe.
7. Quick-cool the weld with a damp sponge or cloth.
8. Remove the tape from the front of the piece. V-groove deep enough so that the first weld is penetrated by the second V-groove.
9. Weld the seam, filling the groove completely.
10. Use a razor blade or slow-speed grinder to reshape the contour.

13.9 REPAIRING VINYL

Vinyl is a soft, flexible, thin plastic material often applied over a foam filler. Vinyl over foam construction is commonly used on interior parts for safety. Common vinyl parts are the dash pads, armrests, inner door trim, seat covers, and exterior roof covering. Dash pads or padded instrument panels are expensive and time-consuming to replace. Therefore, they are perfect candidates for repair.

Most dash pads are made of vinyl-clad urethane foam to protect people during a collision. Surface dents in foam dash pads, armrests, and other padded interior parts are common in collision repair. These dents can often be repaired by applying heat as follows:

1. Soak the dent with a damp sponge or cloth for about half a minute. Leave the dented area moist.
2. Using a heat gun, heat the area around the dent. Hold the gun 10–12 inches (254–305 mm) from the surface. Keep it moving in a circular motion at all times, working from the outside in.
3. Heat the area to around 130°F (54°C). Do not overheat the vinyl because it will blister. Keep heating until the area is too uncomfortable to touch. If available, use a digital thermometer to meter the surface temperature.
4. Using gloves, massage the pad. Force the material toward the center of the dent. The area might have to be reheated and massaged more than once. In some cases, heat alone might repair the damage.
5. When the dent has been removed, cool the area quickly with a damp sponge or cloth.
6. Apply vinyl treatment or preservative to the part.

Spraying Vinyl Paints

Vinyl repair paints are usually ready for spraying as packaged. Because application properties cannot be controlled with thinners or other additives, air pressure is an important factor.

Using Heat To Reshape Plastics

Many bent, stretched, or deformed plastic parts, such as flexible bumper covers and vinyl-clad foam interior parts, can often be straightened with heat. This is because of **plastic memory**, the tendency of a piece to keep or return to its original molded shape. If it is bent or deformed slightly, it will return to its original shape if heat is applied.

To reshape a distorted bumper cover, use the following procedure:

1. Thoroughly wash the cover with soap and hot water.
2. Clean with plastic cleaner. Carefully remove all road tar, oil, grease, and undercoating.
3. Dampen the repair area with a water-soaked rag or sponge.
4. Apply heat directly to the distorted area. Use a concentrated heat source, such as a heat lamp or high-temperature heat gun. When the opposite side of the cover becomes uncomfortable to the touch, it has been heated enough.
5. Use a paint paddle, squeegee, or wood block to help reshape the piece if necessary.
6. Quick-cool the area by applying cold water with a sponge or rag.

Do not overheat textured vinyl or you will damage the vinyl surface.

13.10 ULTRASONIC STUD WELDING

Ultrasonic stud welding uses high-frequency movement and friction to generate heat that bonds plastic parts together. It can be used to join plastic parts at a single point or at numerous locations. In many applications, a continuous weld is not required. The welding cycle is short, almost always less than half a second.

Ultrasonic stud welding is made along the circumference of the stud. Its strength is a function of the stud diameter and the depth of the weld. Maximum tensile strength is achieved when the depth of the weld equals half the diameter of the stud.

13.11 REINFORCED PLASTIC REPAIR

Reinforced plastic—including sheet-molded compound (SMC), fiber-reinforced plastic (FRP), and reinforced reaction injection-molded polyurethane (RRIM)—parts are being used in many unibody vehicles (Figure 13–12). They often provide a durable plastic skin over a steel unibody.

Table 13–3 provides an overview of reinforced plastic repair materials.

The damage that generally occurs in reinforced plastic panels includes:

- One-sided damage, such as a scratch or gouge
- Punctures and fractures

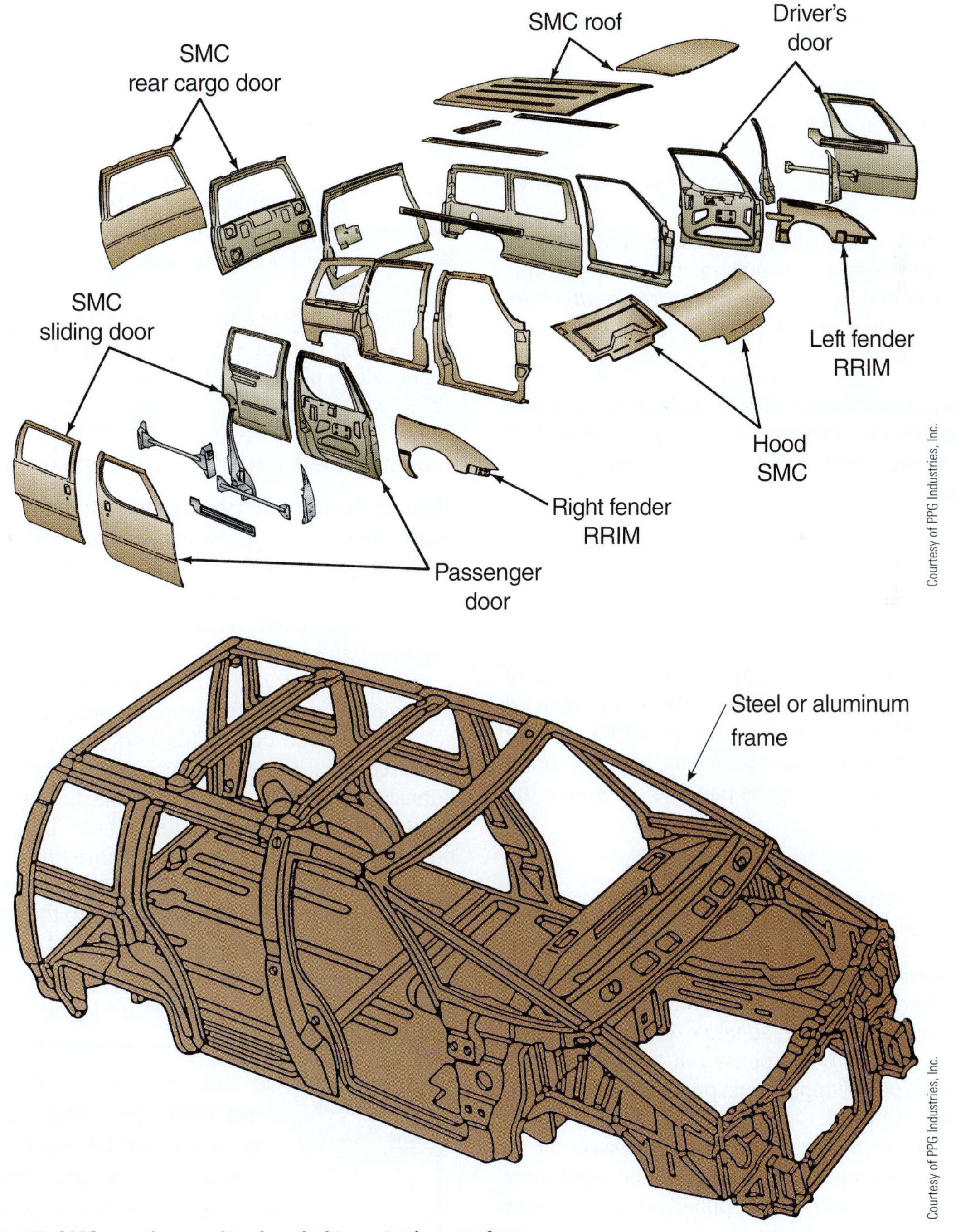

Figure 13–12 SMC panels are often bonded to a steel space frame.

TABLE 13–3 REINFORCED PLASTIC REPAIR MATERIAL SELECTION CART

Type of Repair	Applicable Repair Product Panel Adhesive	Patching Adhesive	Structural Filler	Cosmetic Filler	Glass Fiber Reinforcement
Panel replacement	X				
Panel sectioning	X		X[1]	X[1]	X
One-sided repairs				X[1]	
Two-sided repairs	X[2]	X[2]	X	X	X

Notes: (1) Some panel adhesives can also be used as structural and cosmetic fillers, depending on sanding characteristics.
(2) Panel adhesives can also be used as patching adhesives, but not vice versa.

- Panel separation, where the panel pulls away from the metal space frame
- Severe damage, which requires full or partial panel replacement
- Minor bends and distortions of the space frame, which can be repaired by pulling and straightening
- Severe kinks and bends to the space frame, which require replacement of that piece along factory seams or by sectioning

Combinations of these types of damage often occur on a single vehicle. Depending on the location and amount of damage, there are four different types of reinforced plastic repairs:

1. Single-sided repair
2. Two-sided repair
3. Panel sectioning
4. Full panel replacement

To select a repair method, thorough examination of the vehicle is required. Examine all affected reinforced plastic panels. First, check the entire panel for signs of damage. Also check all panel seams for adhesive bond failure. Examine the back of the panel to determine the extent of the damage.

Reinforced Plastics Repair Applicators

Most of the tools used for repairs of reinforced plastics should already be available in a well-equipped body shop. The *reinforced plastic adhesive applicator* allows two-part adhesives to be dispensed at a constant rate. There are two types of applicators: pneumatic and hand-operated (Figure 13–13).

The pneumatic applicator uses compressed air to force the materials out. The hand-operated applicator works like a caulking gun. Hand-applied pressure is used to force material out of the tubes.

To use either type of applicator, follow these simple rules:

- Follow the manufacturer's instructions.
- Check for proper product flow.
- Check for consistent mix of the two-component product.
- When changing cartridges, run a new test bead.
- If saving part of a cartridge, leave the static mixing nozzle in place.

Reinforced Plastic Adhesives

Many of the materials that are used for reinforced plastic repair are two-part adhesive products. Two-part adhesive means a base material and a hardener must be mixed to cure the adhesive. Each must be mixed together in the proper ratio. Both parts must be thoroughly mixed together before use.

Work life or *open time* is the time when it is still possible to work the adhesive and still have the adhesive set up for a good bond. This work life/open time will be provided by the manufacturer. The cure time of some adhesives used in reinforced plastic repair can be shortened with the application of heat. Temperature and humidity can affect work life and cure time.

After mixing, remember that each product has a work life or open time. If you move or disturb the adhesive as it starts to harden, you will adversely affect its durability.

Do not use fillers designed for sheet metal on reinforced plastic. The repair will be weak and will crack and fail quickly. This would be an embarrassing mistake!

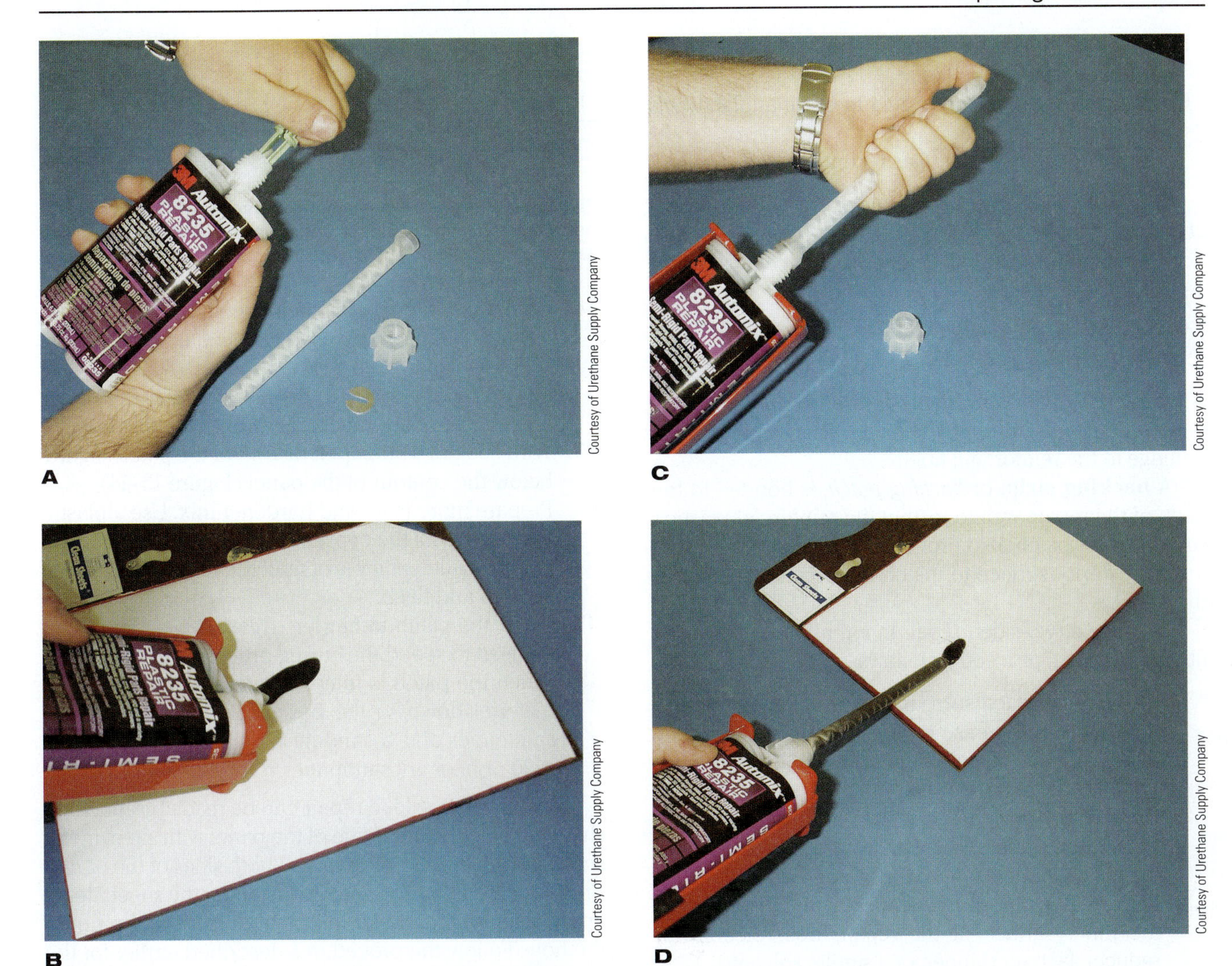

Figure 13-13 A dual cartridge applicator will dispense equal parts of two-part adhesive quickly. It can be either hand-operated (like this one) or pneumatic. (A) Remove cap and plug from plastic adhesive tubes. (B) Squeeze out a small amount of material to clean out tube ends. (C) Install the mixing tip and tighten the cap securely. (D) Before applying adhesive to the part, dispense out a small amount. It will be mixed while inside the tip.

Reinforced Plastic Fillers and Glass Cloth

Two filler products are specifically formulated for use on reinforced plastic. They are cosmetic filler and structural filler.

Cosmetic filler is typically a two-part epoxy or polyester filler used to cover up minor imperfections.

Structural filler is used to fill the larger gaps in the panel structure while maintaining strength. Structural fillers add to the structural rigidity of the part.

All two-part products will shrink to some degree. The use of heat will help to speed the drying time and will eliminate some of the shrinkage.

There are several different types of glass cloth available. Rovings and matting are not appropriate for reinforced plastic repair. Choose unidirectional cloth, woven glass cloth, or nylon screening. The cloth weave must be loose enough to allow the adhesive to fully saturate the cloth, leaving no air space around the weave.

Single-Sided Repairs of Reinforced Plastic and Fiberglass

Single-sided damage is surface damage that does not penetrate or fracture the back of the panel. Damage might pass all the way through a panel, but no pieces of the panel have broken away. If the break is clean and all of the reinforcing fibers have stayed in place, then a single-sided repair is adequate.

For a single-sided repair, you must bevel deep to penetrate the fibers in the panel. The broken fibers must come into contact with the adhesive.

The following is a typical single-sided repair procedure for reinforced plastic:

1. Clean the repair area with soap and hot water.
2. Clean again, using mild wax and grease remover.
3. Remove any paint from the surrounding area by sanding with #80 grit sandpaper.
4. Scuff sand the area surrounding the damage.

5. Bevel the damage to provide an adequate area for bonding.
6. Mix two-part filler according to the manufacturer's instructions.
7. Apply the filler and cure as recommended.

Once the filler has been sanded, apply additional coats as required and resand between coats. The product manufacturer will provide grit recommendations.

Two-Sided Repairs of Reinforced Plastic and Fiberglass

A two-sided repair is normally needed on damage that passes all the way through the panel. This would include damage to the reinforcing fibers.

A **backing strip**, or *backing patch*, is bonded to the back of the repair area to restore the reinforced plastic's strength. The patch also provides a foundation for forming the exterior surface to match the original contour of the panel.

To make a two-sided repair in a reinforced plastic or fiberglass panel, proceed as follows:

1. Clean the surface surrounding the damage with a good wax and grease remover. Use a #36 grinding disc to remove all paint and primer at least 3 inches (76 mm) beyond the repair area.
2. Grind, file, or use a hacksaw to remove all cracked or splintered material away from the hole on both the inside and outside of the repair area.
3. Remove any dirt, sound deadener, and the like from the inner surface of the repair area. Clean with reducer, lacquer thinner, or a similar solvent.
4. Scuff around the hole with #80 grit sandpaper to provide a good bonding surface.
5. Bevel the inside and outside edge of the repair area about 30 degrees to permit better patch adhesion.
6. Clean the repair area thoroughly.
7. Cut several pieces of fiberglass cloth large enough to cover the hole and the scuffed area. The exact number of pieces will depend on the thickness of the original panel.
8. Prepare a mixture of resin and hardener. Follow the label recommendations.
9. Using a small paintbrush, saturate at least two layers of the fiberglass cloth with the activated resin mix.
10. Apply the material to the inside or back surface of the repair area. Make sure the cloth fully contacts the scuffed area surrounding the hole.
11. Saturate three more layers of cloth with the mix. Apply it to the outside surface. These layers must also contact the inner layers and the scuffed outside repair area.
12. With all of the layers of cloth in place, form a saucer-like depression in them. This is needed to increase the depth of the repair material. Use a squeegee to work out any air bubbles.
13. Clean all tools with a lacquer thinner immediately after use.

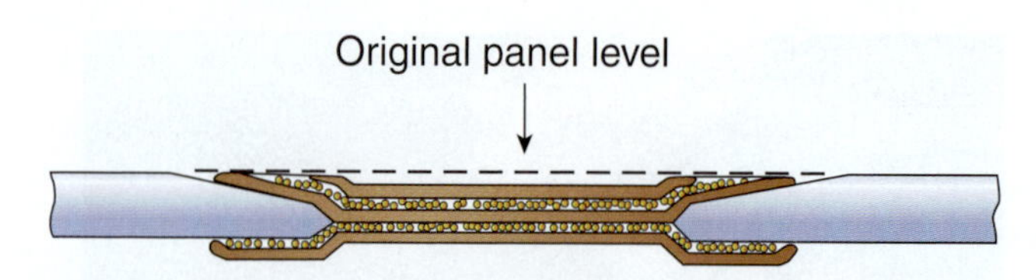

Figure 13–14 To repair a hole or crack in SMC, you can install a patch made of plastic adhesive and layers of fiberglass. Sand the patch slightly below the contour of the panel.

14. Let the saturated cloth patch become tacky. An infrared heat lamp can be used to speed up the process. If one is used, keep it 12–15 inches (305–381 mm) away from the surface. Do not overheat the repair area, because too much heat will cause distortion.
15. With #50 grit sandpaper, disc sand the patch slightly below the contour of the panel (Figure 13–14).
16. Prepare more resin and hardener mix. Use a plastic spreader to fill the depression in the repair area. You need a sufficient layer of material for grinding down smooth and flush.
17. Allow the patch to harden. Again, a heat lamp can be used to speed the curing process.
18. When the patch is fully hardened, sand the excess material down to the basic contour. Use #80 grit sandpaper and a sanding block. Finish sand with #120 or finer grit sandpaper.

This same two-sided repair can be made by attaching sheet metal to the back side of the panel with sheet metal screws. Sand the sheet metal and both sides of the part to provide good adhesion. Before fastening the sheet metal, apply resin and hardener mix to both sides of the rim of the hole. Follow the procedures described earlier for the remainder of the repair.

When the inner side of the hole is not accessible, apply a fiberglass patch to the outer side only. After the usual cleaning and sanding operations, apply several additional layers of fiberglass cloth to the outer side of the hole. Before it dries, make a saucer-like depression in the cloth to provide greater depth for the repair material.

Sectioning Reinforced Plastic and Fiberglass Panels

Proper sectioning requires that you understand what areas are most appropriate for sectioning. You must also know how to avoid problems with horizontal bracing, rivets, and concealed parts. The replacement panel used will depend on the amount and location of the damage (Figure 13–15).

Using the left rear quarter panel as an example, there are three possibilities. The entire panel can be ordered, or just a front or rear half (Figure 13–16).

Remember that reinforced plastic is a very forgiving and workable material. Just because quarter panels come split at the wheel well, the sectioning point does not have to be located there. With proper backing strips to reinforce the joints, sectioning can be done almost anywhere.

Mill and drill pads are used to help the factory hold panels in place while the adhesive cures. These mill and

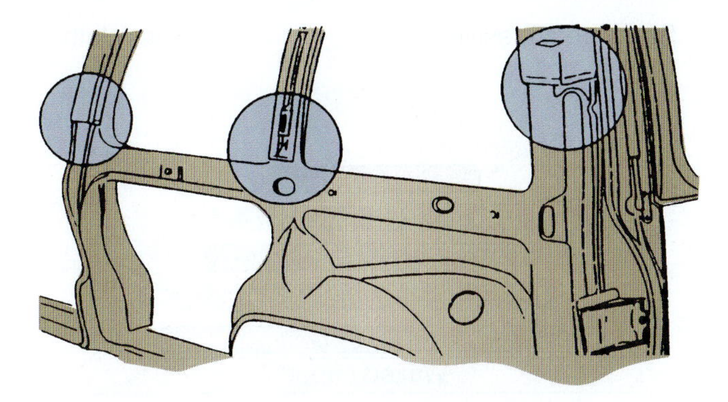

Figure 13–15 Sectioning locations are often given in the service manual and are typical of nonplastic parts.

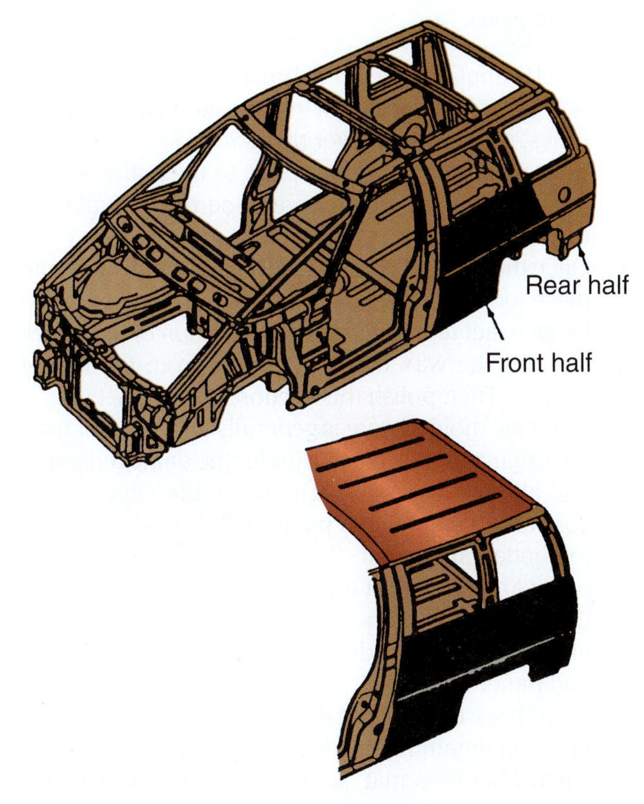

Figure 13–16 Depending on the extent of the damage, you may want to install a complete panel or section it down to replace only the damaged area.

drill pads also help you to hold, align, and level replacement panels. If a panel is to be sectioned, it should be done between the mill and drill pad locations (Figure 13–17).

Removing Reinforced Plastic Panels

First, remove the interior trim to expose the horizontal bracing and mill and drill pads. Examine the back of the panel to gauge the extent of panel damage. Also determine the location of the horizontal bracing, mill and drill pads, and electrical and mechanical components.

Once the interior trim is removed, a "window" can be cut. Controlling the depth of the cut is very important. Space frame components, as well as electrical lines and heating/cooling elements, may be located behind the

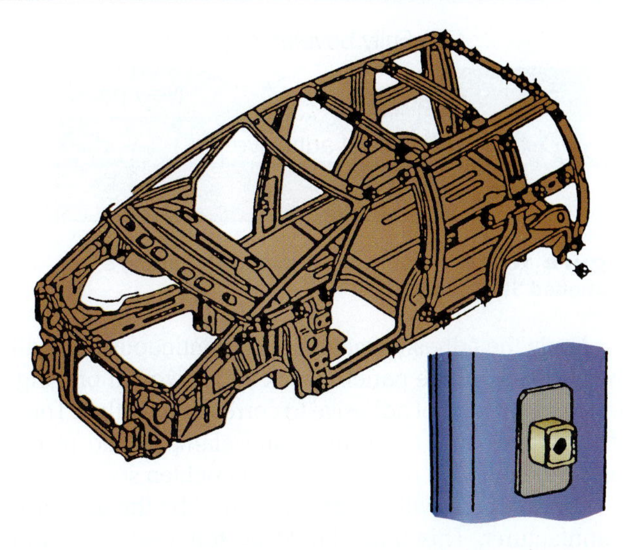

Figure 13–17 Mill and drill pad locations will be given in the service literature.

panels. Before cutting the window, be sure you know what is behind the panel you plan to cut, or limit the depth of the cut to ¼ inch (6.35 mm) to avoid doing damage.

Now that the window has been cut, the rest of the panel can be removed from the space frame. This can be done using heat and a putty knife, or by carefully using an air chisel. Choose a flat chisel, beveled on one side only. Be careful not to damage the space frame.

If the door surround panels are to be left attached to the vehicle, the air chisel method may not be the best choice for separating the seam between two panels. Use heat and a putty knife to separate the seams to avoid doing damage to the door surround pieces.

Fitting Reinforced Plastic Panels

After removing the scrap panel, prepare the space frame for the new panel. First, remove the old adhesive from the space frame. In some cases, the adhesive can be peeled away from the frame. You may also have to use heat and a putty knife or a sander. Remove all of the adhesive, because the heat will break down the adhesive.

Bevel the outside edges of the existing panel to a 20-degree taper. Sand and clean the back side of the panel where the backing strips will be attached. Backing strips are made using scrap material that duplicates the original panel contour as closely as possible. They should extend about 2 inches (50 mm) beyond either side of the sectioning location. Clean the backing strips. Remove the paint from those places where adhesive will be applied.

Measure the replacement panel for fit. Trim the panel to size. Check the fit again. Leave a ½-inch (13 mm) gap between the existing panel and the replacement panel. When proper fit has been established, the new panel can be prepared for the adhesive. Sand or grind bevels into the panel where they mate to the existing panel (Figure 13–18).

Bevel the mating edges of the new panel to a shallow taper, just like on the existing panel. Make sure to bevel all the way through the panel. Do not leave a shoulder.

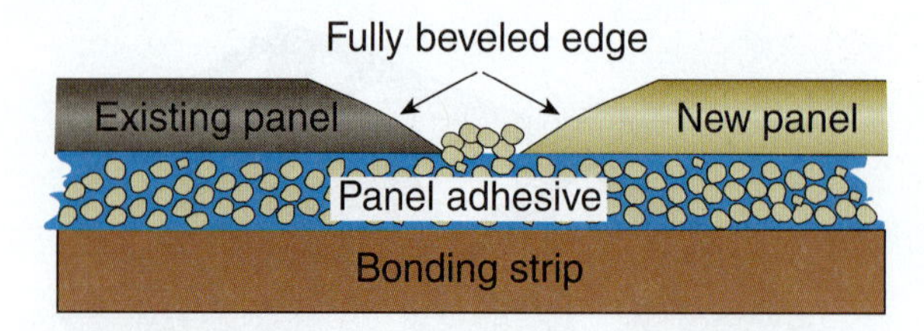

Figure 13–18 Note the basic method for fitting a sectioned SMC panel.

Apply the adhesive material in a continuous bead all the way around the panel. Check for horizontal bracing, and apply a bead of adhesive to correspond with it. Then fit the panel onto the vehicle and clamp it into place. Install the mill and drill pad nuts and tighten securely.

The work life will be recommended by the adhesive manufacturer. This time should be followed to ensure that the proper fit is achieved.

Using Molded Cores

A **molded core** is a curved body repair part made by applying plastic repair material over a part and then removing the cured material. Naturally, holes are much more difficult to repair in a curved portion of a reinforced plastic panel than those on a flat surface. Basically, the only solution (short of purchasing a new panel section) is to use the molded core method of replacement. This is often the quickest and cheapest way to repair a curved surface.

Although the following example relates to a rear fender section, its procedures can be applied to any type of curved reinforced plastic panel.

1. Locate an undamaged panel on another vehicle that matches the damaged one. It will be used as a model or pattern. A new or used vehicle can be used because it will not be harmed. You can also use a new or undamaged used panel removed from a vehicle.
2. On the model vehicle, mask off an area slightly larger than the damaged area. Apply additional masking paper and tape to the surrounding area, especially on the low side of the panel. This will prevent any resin from getting on the finish.
3. Coat the area with paste floor wax. Leave a wet coat of wax all over the surface. A piece of waxed paper can be substituted for the coat of wax. Make sure the waxed paper is taped firmly in place.
4. Cut several pieces of thin fiberglass mat in sizes larger than the area to be repaired.
5. Mix the fiberglass resin and hardener following the label instructions.
6. Starting from one corner of the mold area, place pieces of fiberglass mat on the waxed area so each edge overlaps the next one; use just one layer of matting (Figure 13–19).
7. Apply the resin/hardener to the matting with a paintbrush. Force the mixture into the curved surfaces and around corners with the tips of the bristles.
8. Use the smaller pieces of matting along the edges and on difficult curves. Additional resin/hardener can be applied if needed, brushing in one direction only to force the material into the indentations. In all cases, use only one layer of matting.

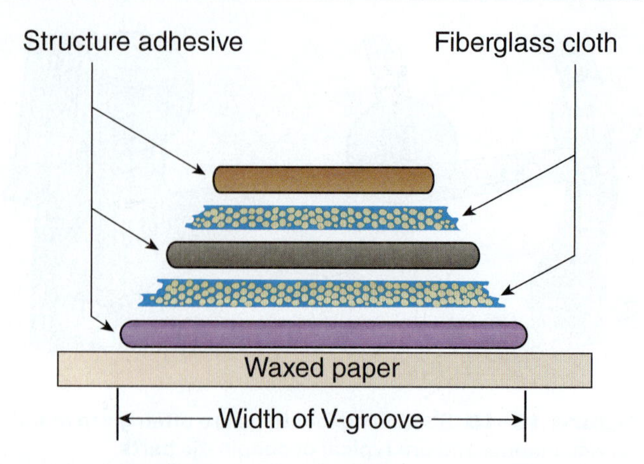

Figure 13–19 Note the method for making a filler patch for an SMC panel.

9. After matting has been applied to the entire waxed area, allow the molded core to cure a minimum of one hour.
10. Once the molded core has hardened, gently work the piece loose from the model vehicle. The core should be an exact reproduction of this section of the panel.
11. Remove the wax or waxed paper from the model vehicle. Then polish this section of the panel.
12. Since the molded core is generally a little larger than the original panel, place it under the damaged panel and align. If necessary, trim down the edges of the core and the damaged panel slightly where needed for better alignment. The edges of the damaged panel and core must also be cleaned.
13. Using fiberglass adhesive, cement the molded core in place on the inside of the panel. Allow the core and panel to cure.
14. Grind back the original damaged edges to a taper or bevel, maintaining the desired contour.
15. Lay a fiberglass mat, soaked in resin/hardener, on the taper or bevel and over the entire core. Once the mat has hardened, level it with a coat of fiberglass filler. Then prepare it for painting.

In some instances it might not be possible to place the core on the inside of the damaged panel. In this case, the damaged portion must be cut out to the exact size of the core. After the panel has been trimmed and its edges beveled, tabs must be installed to support the core from the inside. These tabs can be made from pieces of the panel or from fiberglass strips saturated in resin/hardener.

After cleaning and sanding the inside sections, attach the tabs to the inside edge of the panel and bond with fiberglass adhesive. Clamping pliers can be used to hold the tabs in place. Taper the edge of the opening and place the core on the tabs. Fasten the core to the tabs with fiberglass adhesive. Grind down any high spots so that layers of fiberglass mat can be added.

Place the saturated mats over the core, extending about 1½–2 inches (38–50 mm) beyond the damaged area in all directions. Work each layer with a spatula or squeegee to remove all air pockets. Additional resin/hardener can be added with a paintbrush to secure the layers. Allow sufficient curing time. Then sand the surface level. For a smooth surface, use fiberglass filler to finish the job before painting.

Figure 13–20 summarizes making a repair with a molded core.

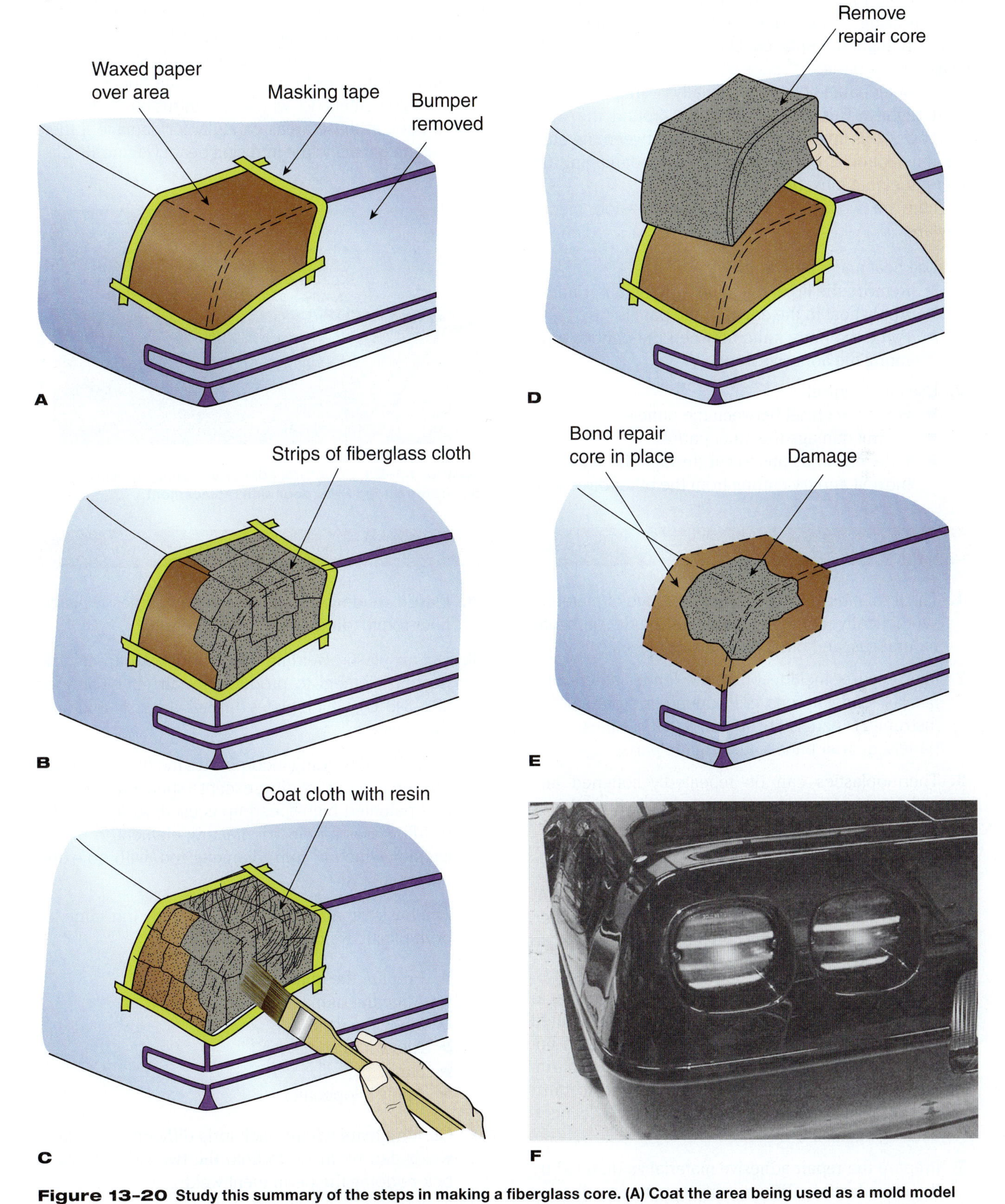

Figure 13-20 Study this summary of the steps in making a fiberglass core. (A) Coat the area being used as a mold model with paste floor wax or a piece of waxed paper. (B) Place pieces of fiberglass veil over the waxed or waxed paper surface. (C) Apply resin/hardener to the veil material. (D) Remove the mold core from the model. (E) Cement the core piece in place. (F) This view shows the completed job.

SMC Door Skin Replacement

Door skin replacement is a straightforward repair because most are made of an inner and an outer piece of SMC bonded together. The exterior of the door might be repaired using a single-sided or two-sided repair, or the door skin might be replaced. Outer door skins are available as service parts.

The outer panel of the door usually overlaps the inner panel slightly. This forms a little lip around the door. Grind away this lip to expose the joint between the outer and inner panels. Use caution to avoid damaging the inner panel that must be saved.

There are two methods of separating the door pieces. They are:

1. Use heat and a putty knife.
   - ▶ Remove the lip of the panel with an air grinder.
   - ▶ Apply heat to the edges of the panels.
   - ▶ Force the putty knife between the panels, separating the adhesive bond.
2. Use an air chisel.
   - ▶ Force the chisel between the panels.
   - ▶ Do not damage the inner panel.
   - ▶ If the chisel begins to cut through the panel, remove it and try cutting from the other direction.

Clean the mating edges of the inner panel of loose adhesive or SMC or fiberglass parts.

Usually there is an inner UHSS reinforcement that runs the length of both front doors and the sliding cargo door. There will also be some inner SMC reinforcements attached directly to the SMC outer door skin itself. Look for any inner reinforcements by removing the inner door skin. The leading and trailing edges inside the doors are bonded with a metal reinforcement to which the intrusion beams are bolted. Look at these areas carefully for damage if there has been an impact to the intrusion beam (Figure 13–21).

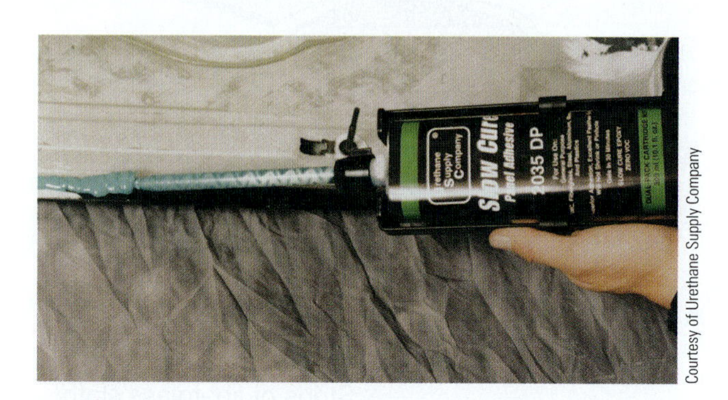

Courtesy of Urethane Supply Company

Figure 13–21 **Structural adhesive is being applied to a door frame during SMC door skin replacement.**

SUMMARY

1. The term *plastics* refers to a wide range of materials synthetically compounded from crude oil, coal, natural gas, and other natural substances.
2. Plastic parts include bumpers, fender extensions, fascias, fender aprons, grille openings, stone shields, instrument panels, trim panels, fuel lines, door panels, quarter panels, and engine parts.
3. Thermoplastics can be repeatedly softened and reshaped by heating, with no change in their chemical makeup.
4. Thermosetting plastics undergo a chemical change by the action of heating, a catalyst, or ultraviolet light.
5. Composite plastics, or hybrids, are blends of different plastics and other ingredients designed to achieve specific performance characteristics.
6. Working with plastics and fiberglass requires you to think about safety at all times. Resins and related ingredients can irritate your skin and stomach lining. Curing agents, or hardeners, can produce harmful vapors. It is environmentally unsound to burn plastics, which can produce carcinogens (cancer-causing agents).
7. Prepare the repair adhesive material as directed by the manufacturer. When working with an adhesive system, use the manufacturer's categories to decide on a repair product and procedure.
8. Plastic welding uses heat and sometimes a plastic filler rod to join or repair plastic parts.
9. Hot-air plastic welding uses a tool with an electric heating element to produce hot air (450–650°F or 232–345°C), which blows through a nozzle and onto the plastic.
10. High-speed welding incorporates the basic methods utilized in hand welding, except a specially designed and patented high-speed tip is used. Airless plastic welding uses an electric heating element to melt a smaller ⅛-inch (3 mm) diameter rod, with no external air supply.
11. The basic repair sequence is generally the same for both plastic welding processes:
    - ▶ Prepare the damaged area.
    - ▶ Align the damaged area.
    - ▶ Make the weld.
    - ▶ Allow it to cool.
    - ▶ Sand.
    - ▶ Apply a topcoat.
12. On long tears where backup is difficult, small tack welds can be made to hold the two sides in place before doing the permanent weld.
13. Vinyl is a soft, flexible, thin plastic material often applied over a foam filler.

14. Reinforced plastic—including sheet-molded compound (SMC), fiber-reinforced plastic (FRP), and reinforced reaction injection-molded polyurethane (RRIM)—parts are being used in many unibody vehicles.
15. Work life or open time is the time it is possible to work the adhesive and still have the adhesive set up for a good bond.
16. Cosmetic filler is typically a two-part epoxy or polyester filler used to cover up minor imperfections.
17. Structural filler is used to fill larger gaps in the panel structure while maintaining strength. Structural fillers add to the structural rigidity of the part.
18. A backing strip, or backing patch, is bonded to the rear of the repair area to restore the reinforced plastic's strength.

EXERCISES

On a separate sheet of paper, complete the following learning activities for this chapter. Write definitions for the key terms and answer the ASE-style review questions, essay questions, critical thinking problems, and math problems. You can also do the outside activities, possibly for extra credit.

Key Terms

adhesion promoter
backing strip
carcinogen
melt-flow plastic welding
mill and drill pad
molded core
plastic flexibility test
plastic memory
plastic speed welding
plastic stitch-tamp welding
plastic welding
reinforced plastic
sheet-molded compound
thermoplastics
thermosetting plastics
two-part adhesive systems
ultrasonic plastic welding
vinyl

ASE-Style Review Questions

1. When tack welding plastic parts, Technician A does not use welding filler rod. Technician B says that the tack welds will hold the sides of the part being welded in alignment. Who is correct?
   A. Technician A
   B. Technician B
   C. Both A and B
   D. Neither A nor B
2. Which type of welding tip is ideal for working in hard-to-reach places?
   A. Round
   B. Speed
   C. Both A and B
   D. Neither A nor B
3. The recommended welding rod size for airless welding is
   A. 3/16-inch diameter.
   B. 1/32-inch diameter.
   C. 1/2-inch diameter.
   D. 1/8-inch diameter.
4. Which of the following statements is incorrect?
   A. CAs work equally well on all automotive plastics
   B. PP and TPO are examples of plastics that require an adhesion promoter as part of the repair process
   C. The best way to identify a plastic for adhesion bonding is by using the flexibility test
   D. Both A and B
5. When Technician A grinds the base material, it melts and smears, so he or she uses an adhesion promoter to make the repair. Technician B says that if the plastic material gives off dust when sanded, it can be repaired with a standard structural adhesive system. Who is correct?
   A. Technician A
   B. Technician B
   C. Both A and B
   D. Neither A nor B
6. Technician A uses a pneumatic adhesive applicator when repairing a reinforced plastic. Technician B says that the pneumatic applicator uses hydraulic fluid to force the materials out. Who is correct?
   A. Technician A
   B. Technician B
   C. Both A and B
   D. Neither A nor B
7. What allows plastic bumper covers and other plastic parts to be straightened with heat?
   A. Plastic shrinking.
   B. Mold release agent.
   C. Plastic memory.
   D. All of the above.

Essay Questions

1. Summarize the basic procedures for making a good plastic weld.
2. How do you make a typical hot-air plastic weld?
3. Summarize how to make a two-sided plastic weld.

Critical Thinking Problems

1. How would you determine whether it is better to repair an SMC panel or replace it?
2. You have a tear in a plastic part. How can you tell what type of plastic it is?
3. Other than using the ISO code on the back of a plastic part, what are some other methods to identify a plastic?
4. Name the two ways of repairing plastics.
5. In general, two types of plastic are used in automotive construction: thermoplastics and thermosetting plastics. Can you name some ways they are different?

Math Problem

1. A technician made a bevel ⅛-inch wide, whereas specs call for a 6.4 mm-wide bevel. How incorrect is this groove?

Activities

1. Inspect plastic body parts. Try to find an identification code on the back of each part. Make a report on the type of plastic used for different parts.
2. Make a one-sided repair on a plastic part. Summarize the repair in a report. Describe the type of plastic, the damage, and the steps you followed to repair the part.
3. Visit a body shop. Talk to technicians about new methods of repairing plastics. Report your findings to the class.

Hood, Bumper, Fender, Lid, and Trim Service

CHAPTER 14

OBJECTIVES

After reading this chapter, you should be able to:

- Remove, install, and adjust front fenders and hood.
- List the various methods for adjusting mechanically fastened panels.
- Perform hood-to-hinge, hood height, and hood latch adjustments.
- Remove, install, and adjust deck lids.
- Remove, install, and adjust bumpers.
- List the precautions to follow when servicing impact energy absorbers.
- Replace grilles and other bolt-on body parts.
- Service trim pieces on the outside of body panels.
- Answer ASE-style review questions relating to replacement of hoods, bumpers, fenders, deck lids, and trim pieces.

INTRODUCTION

A collision-damaged vehicle can require a variety of repair operations. Repair steps depend on the nature and location of the damage. Panels with minor damage often can be straightened and filled with plastic. Minor bulges, dents, and creases can be fixed using the techniques discussed in earlier chapters. However, quite often the damage is too extensive, and part replacement is the only logical solution.

This chapter covers replacement procedures for hoods, fenders, bumpers, deck lids, trim pieces, and similar bolt-on parts. Many major parts bolt onto the vehicle.

Keep in mind that on-the-job experience is the only way to become competent at body part removal and replacement (*part R&R*). Sometimes you must remove one part at a time. In other instances, it is better to remove several parts as an assembly. This chapter will give you the background information to make this learning process easier (Figure 14–1).

Courtesy of Liss CarStar Collison Repair

Figure 14–1 Many major body panels are held on with bolts, nuts, screws, clips, or adhesives. You must know how to efficiently remove and install these parts during collision repair.

14.1 HOW ARE FASTENED PARTS SERVICED?

Fastened parts are held on a vehicle by bolts, nuts, screws, clips, and adhesives. The methods of fastening parts to cars and trucks have changed over the past few years. Many parts that were held on with bolts and screws in the past now snap-fit into place. Plastic retainers now hold these parts onto the vehicle. This was done to save time during vehicle manufacturing. The part is simply pressed or "popped" into position on the assembly line.

The panels and parts that typically fasten in place without welding are shown in Figure 14–2.

Refer to the estimate or shop work order to get guidance on how to start work. The estimator will have determined which parts need to be repaired and which should be replaced. Use this information and shop manuals to remove and replace parts efficiently.

The estimate is a critical reference tool when doing repairs, and it must be followed. The insurance company and estimator have both determined which parts to repair. If you fail to follow the estimate, the insurance company may not pay for your work.

The estimate is also used to order new parts. You should make sure all ordered parts have arrived. Compare new parts on hand with the parts list. If any parts are missing, have the parts person order them. This will save time and prevent your stall from being tied up while waiting for parts to be delivered.

Part Removal Sequence

Generally, you start by removing large, external parts first. For example, if the front end has a damage, you must remove the hood first. This gives you more room to access rear fender bolts. It also allows more light into the front to aid in finding and removing hidden bolts. Use this same kind of logic to remove parts efficiently.

If in doubt about how to remove a part, refer to the vehicle's service manual or computerized service information (Mitchell, for example). Factory service information normally has a body repair section. The body repair section of the manual or computerized information explains and illustrates how parts are serviced. It gives step-by-step instructions for the specific make and model vehicle, as well as bolt locations, torque values, removal sequences, and other important information. For example, look at Figure 14–3. The rear quarter panels on this car bolt in place. Most cars have welded quarter panels. This car also has aluminum body panels that must be repaired differently than steel.

NOTE *Chapter 3 provides detailed information on using service information when doing collision repairs. Refer back to this chapter if needed.*

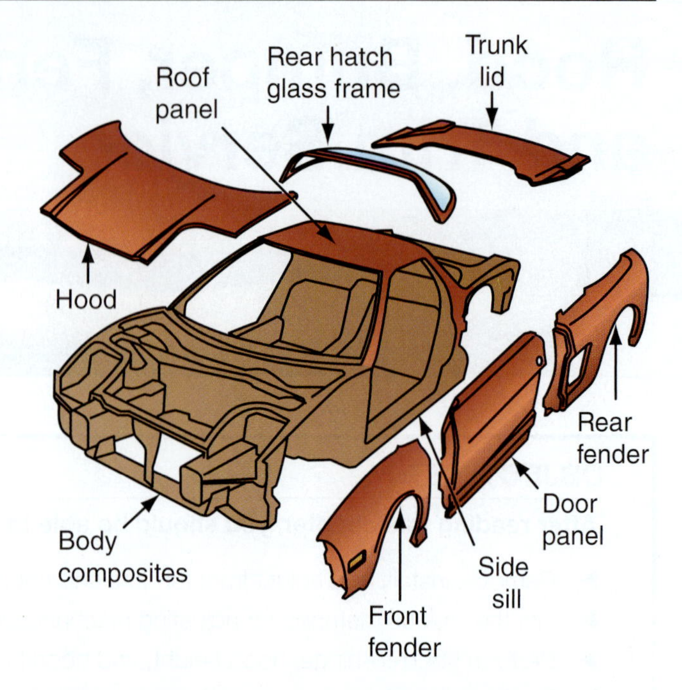

Figure 14–3 This high-performance sports car is a mid-engine design. Lightweight, rustproof aluminum panels, including the rear quarter panels, bolt to this advanced space frame.

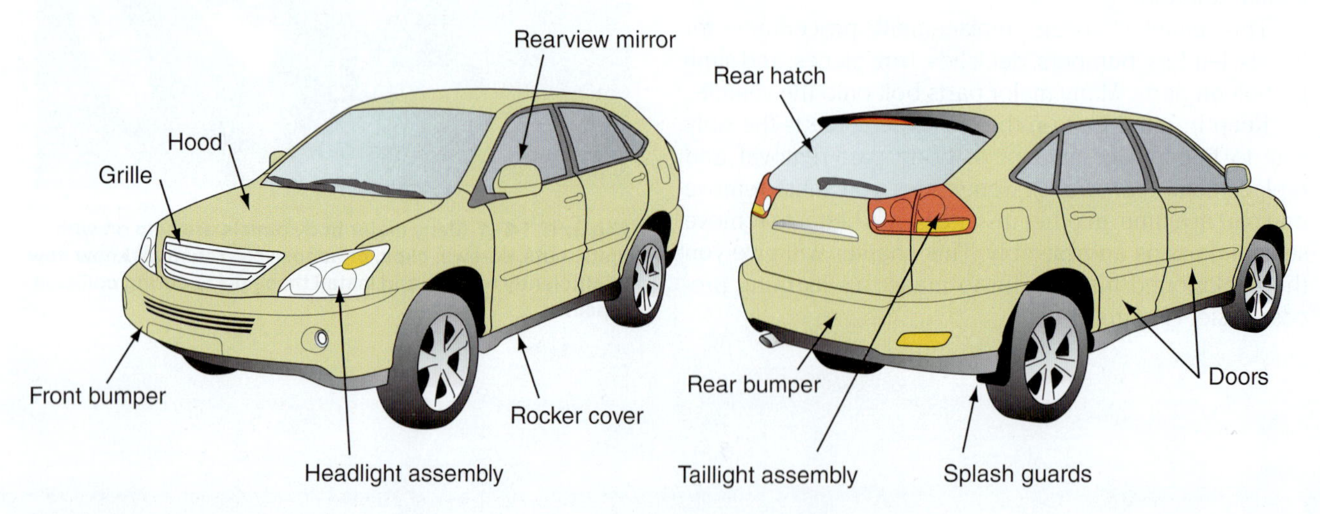

Figure 14–2 Note some of the panels and parts that are held onto a vehicle by fasteners.

14.2 HOOD SERVICE

The hood provides an external cover over the front of the vehicle. It is one of the largest, heaviest panels on a car or truck. In a front-engine vehicle, it provides access to the engine compartment. With a rear-engine car, it serves as a deck or trunk lid over a storage compartment (Figure 14–4).

Before removing the hood, analyze the condition of its parts. Open and close the hood. Check for binding and bent hinges. If applicable, inspect hood alignment with the fenders and cowl. This will help you determine what must be done during repairs.

Figure 14–4 This display vehicle has had all bolt-on parts removed, including the hood, fenders, bumper, doors, and grille, exposing the unibody structure and suspension.

Hood Removal

To remove a hood, first disconnect any wires and hoses. Wires often connect to an underhood light. Hoses might run to the hood for the windshield washer system.

A **hood prop tool** is a rubber-tipped extension rod for holding the hood open as you remove the hood shocks and other hood parts. It will help keep the hood from falling while you work.

A common mistake during hood removal is forgetting to disconnect windshield washer hoses and wires, which can break or snap small hose fittings or sever wires.

Hood struts, or shocks, are spring-loaded rods used to hold the hood open when working in the engine compartment. With the hood propped open, you next remove the struts. Various methods are used to secure the ends of the hood struts (clips, small nuts, and so on). Figure 14–5 gives a typical service manual example of how to remove the hood struts.

Next, remove the hood hinge bolts. If the hood is not badly damaged and will be reused, mark the hood hinge alignment. To mark the hood, scribe alignment marks around the sides of the hood hinge where it contacts the hood. You may also want to mark the hinge where it mounts on the body. You can then use these marks to rough adjust the hinges and hood during reinstallation.

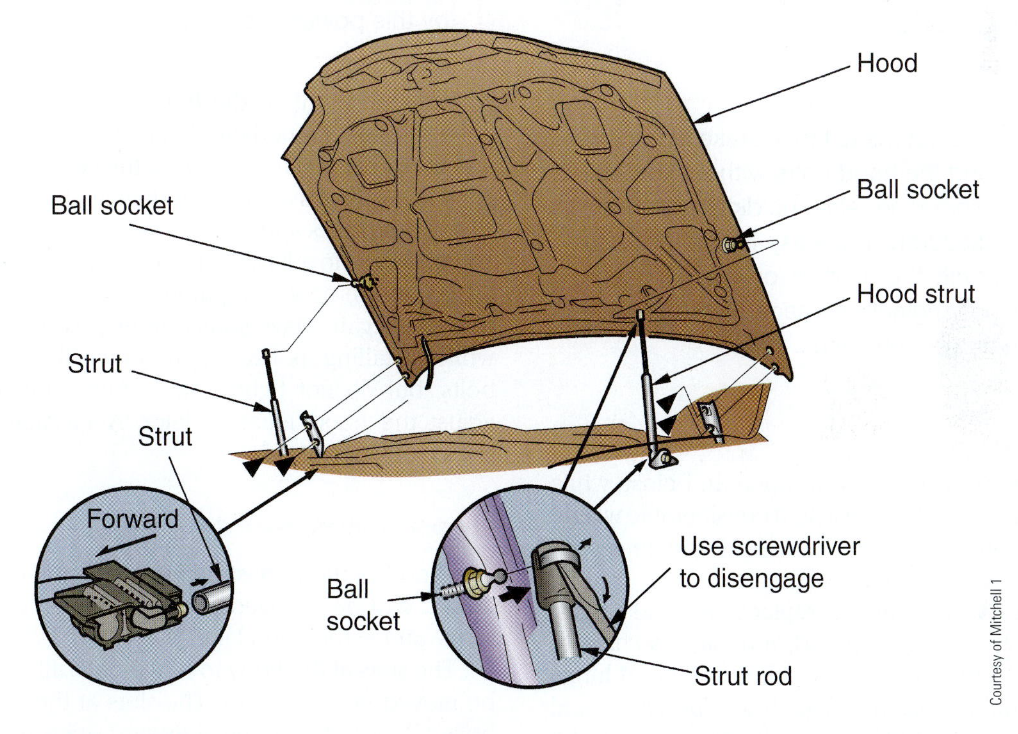

Figure 14–5 Most hood struts use a spring-clip mechanism. A small screwdriver is needed to pry out the clip to free the strut from its ball socket.

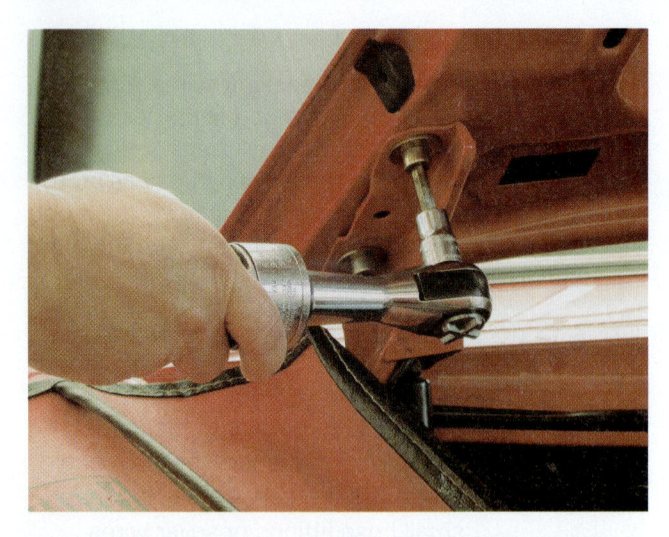

Figure 14–6 An air ratchet is handy for removing and installing hood hinge bolts. If they are not knocked out of alignment from collision, mark the hinge positions to simplify reassembly and realignment.

To prevent part damage, have someone help you hold the hood. Place your shoulder under the hood while holding the bottom edge of the hood with one hand. This will keep the hood from sliding down and hitting the windshield, cowl, or fenders. Use your shoulder to support the weight of the hood. With your free hand, remove the hood bolts. Your helper should do the same. Do not let the weight of the hood rest on the bolts as you loosen them (Figure 14–6).

Note the location of any body shims or spacers that help adjust the hood. If there is no major damage, you may need to reinstall the spacers in the same locations. Place the hood out of the way, where it cannot get hit, scratched, or knocked over.

WARNING Do not make the mistake of removing the hood bolts without you and someone else holding the hood securely. It is easy for the hood to slip and scratch the fenders or cowl. If these parts were not originally damaged, you will have to fix them on your own time.

Hood Hinge R&R

Hood hinges allow the hood to open and close while staying in alignment. They must hold considerable weight and keep the hood secure while driving. Hood hinges are often damaged in a frontal impact.

If bent badly, you will have to replace the hood hinges. If equipped with large coil springs, you may also have to install the old springs on the new hinges. A hood hinge spring tool should be used to stretch the springs off and on. It is a hooked tool that will easily pry the end of a spring off and onto its mount.

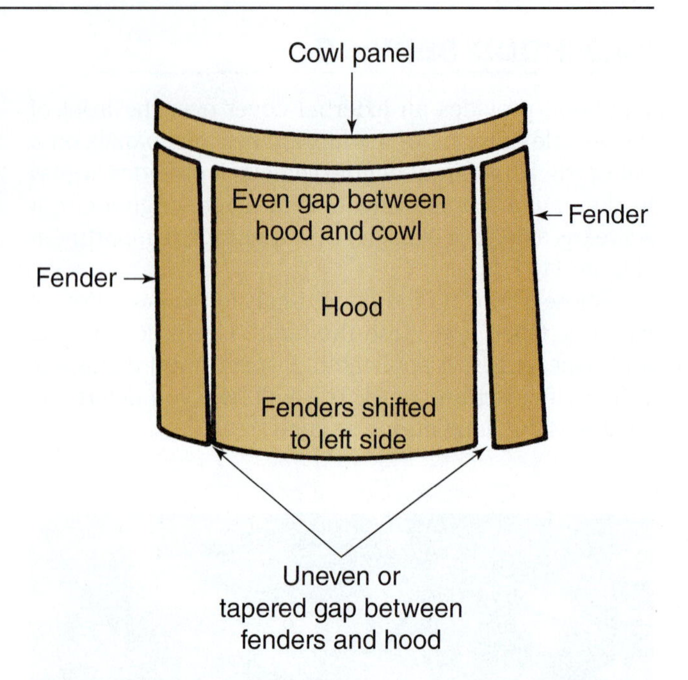

Figure 14–7 Gaps around a hood should be equal. This drawing exaggerates the misalignment of fenders. Generally, if the cowl is straight, first align the back of the hood with the cowl. If there are still gaps between the hood and fenders, the fender bolts must be loosened and the fenders moved into alignment with the hood.

DANGER Hood hinge springs are very strong. Wear eye protection when working on them. Also, keep your fingers away from the spring as it is stretched. Your fingers could be severely pinched and cut by this powerful spring.

If needed, unbolt the hinges from the inner fender panels. Again, mark their alignment if the unibody structure is not badly damaged from the collision.

If hood hinges are bent or damaged, install new ones. Align your marks and snug down the body-to-hinge bolts. You may have to adjust the hood height at these bolts later.

Install the new or repaired hood in reverse order of removal. Again, have someone help you hold the hood while installing its bolts. Snug down the hinge-to-hood bolts, but do not tighten them fully. You will need to adjust the hood location later. A misaligned hood is shown in Figure 14–7.

Hood Adjustment

The hood is the largest adjustable panel on most vehicles. It can be adjusted at the hinges, at the adjustable stops, and at the hood latch.

The slots at the body-to-hinge bolts allow the hood to be moved up and down. The slots at the hood-to-hinge bolts allow forward and rearward adjustments to align the hood with the fenders and cowl.

Make sure there is enough gap at the back edge of the hood to clear the cowl panel. Before opening the hood, check the clearance at the back of the hood. A common mistake is to have the hood hit the cowl, denting it or chipping the paint.

The hood should align with the cowl and fenders with an equal gap between them. The front edge of the hood should also be even with the front edge of the fender.

Many technicians remove the hood latch or leave it off while adjusting the position of the hood. If in place, the latch could kick the hood sideways and give a false impression of proper alignment. After positioning the hood to the cowl and fenders, you can install and adjust the latch.

Hood hinges bolted to either the cowl or the inner fender hold the rear of the hood to the vehicle. The holes in the hinges are slotted to allow the hinges to be adjusted. The hinges can be raised or lowered on the cowl or fender and moved forward or backward.

The front of the hood is held in place by a hood latch. The latch is used to secure the front of the hood so that it holds tightly and aligns with the fenders. Slotted holes are usually provided on the latch for alignment purposes.

A safety catch keeps the hood from flying open if the latch accidentally opens while the vehicle is being driven. The safety catch is simply a spring-loaded hook or lever that engages the hood.

Most vehicles have adjustable hood bumper stops mounted on the radiator support or along the inner edges of the fenders. These bumper stops provide up and down adjustment points to set the hood height in relation to the fenders. They also prevent hood flutter during driving. These stops help determine the position of the hood when it is closed.

Hood-to-Hinge Adjustment

To adjust a hood, slightly loosen the bolts attaching the hood to the hinges. Keep them tight enough to hold the hood during adjustment, but loose enough to allow you to shift the hood.

Close the hood and line it up properly. Shift it by hand until the gap around all sides of the hood is equal. Carefully raise the hood far enough for another technician to tighten the bolts. The front of the hood must align with the front of the fenders and any panel in front of the hood. Be sure there is sufficient clearance between the hood and cowl to allow the hood to be raised without rubbing the cowl.

If you cannot obtain the right clearance between the fenders and hood, the fenders may be out of alignment.

Hood Height Adjustments

To correct the alignment of the hood up and down at the rear, slightly loosen the bolts holding the hinges to the fenders or cowl. Then, slowly close the hood and raise or lower the rear edge of the hood as necessary. When the rear of the hood is level with the adjacent fenders and cowl, open the hood and tighten the bolts.

Once the rear of the hood is adjusted to the correct height, the adjustable stops must be checked. The rear stops must be adjusted to touch lightly against the hood. This eliminates hood movement and rattle. The front stops control the height of the front of the hood. Turn the stops in or out until the front of the hood is even with the top of the fenders. Be sure to retighten the locknut on the stop after adjustment.

When first closing a hood after installation, be sure to close it slowly. If it is not centered, it could hit and dent the fender. It might be a good idea to place tape or use fender covers over the fender edges and cowl to protect them (Figure 14–8).

Remember, hood adjustments are made at the hinges, at the adjustable stops, and at the hood latch. You can adjust the hood up or down, side to side, and forward or rearward. This allows you to align the hood vertically and horizontally with the fenders and cowl. Refer to Figure 14–9.

Hood hinge adjustments control the general position of the hood in relation to the fenders and the rear hood height. By loosening the hood-to-hinge bolts, you can move the front end of the hood right or left. You can also slide the hood to the front or back. Tighten them down when the hood is centered in the opening. There should

Figure 14–8 Be careful when first closing the hood. Never slam the hood down after realignment. Lower the hood slowly to make sure it is not going to hit the fenders or latch. It is easy to damage parts or the paint if the hood is still not perfectly aligned.

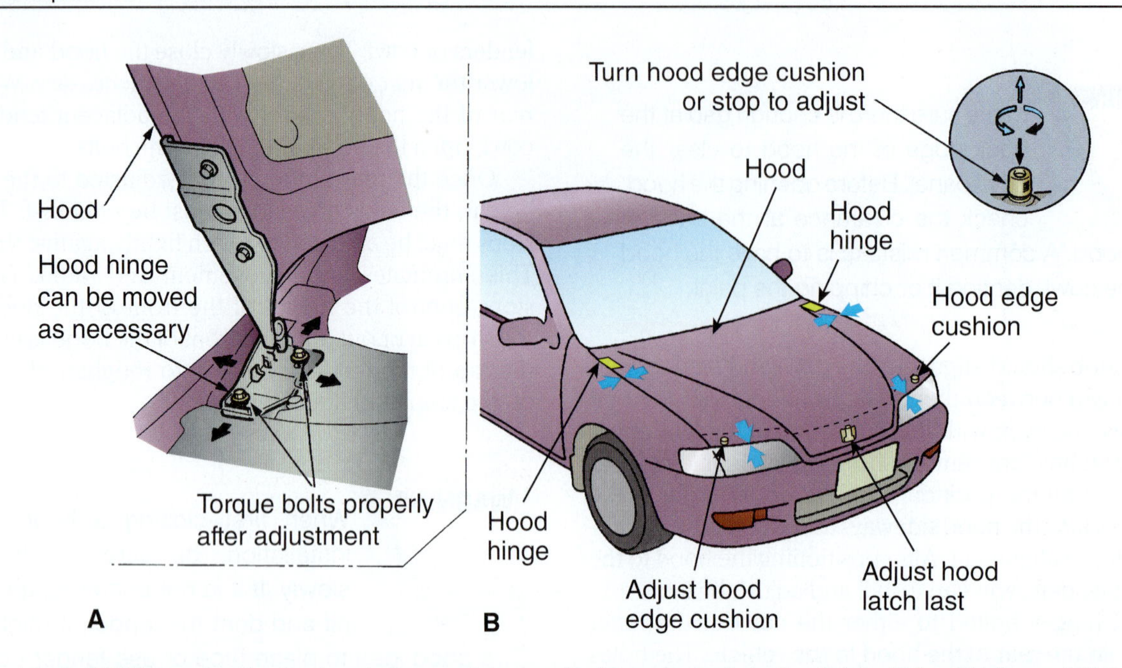

Figure 14–9 **(A) When adjusting the hood alignment, loosen the hinge-to-body bolts slightly. This will let you shift the hinge mounting in the elongated holes right to left and fore to aft. To avoid paint damage, slowly lower the hood while checking alignment. Tighten the hinge bolts when the hood is aligned with the cowl and fenders. (B) Hood edge cushions or stops must be adjusted so that the height of the hood is even with the fenders. You may want to remove the latch during hood adjustment.**

be an equal gap around the hood's perimeter. There should also be enough of a gap at the back edge to clear the cowl panel (Figure 14–10).

By partially loosening the hinge-to-body bolts, you can raise or lower the rear height of the hood. Do not loosen the bolts too much or the weight of the hood will push the hinge all the way down. Tap on the hinges with a mallet to shift them as needed. The back of the hood should be level with the fenders and cowl when fully closed.

Generally, adjust the hood hinges so that the hood is centered in the fenders. Adjust the hood to have the proper gap around its perimeter. A part gap or clearance is the distance measure between two adjacent parts, such as hood-to-fender gap. Then, adjust the hinges to raise or lower the back of the hood.

Hood stop adjustment controls the height of the front of the hood. Hood stops are usually rubber or plastic bumpers mounted in the top of the radiator. By rotating the hood stops, you can raise or lower the stops and the hood when it is in the closed position. Adjust the hood stops so that the hood is even with the front of the fenders and fascia.

Hood Latch Mechanisms

The **hood latch** mechanism keeps the hood closed and releases the hood when activated. Some hood latches

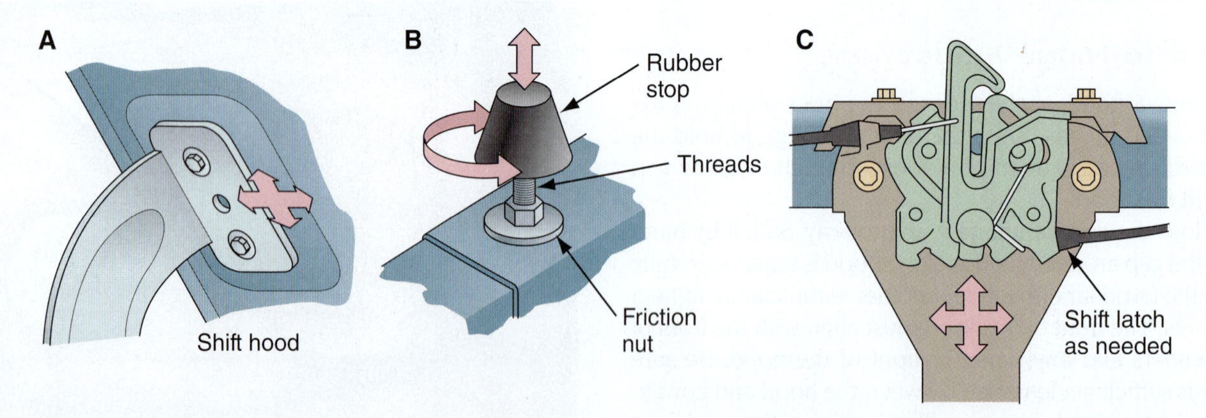

Figure 14–10 **Study the basic methods of adjusting a hood. (A) With the hood-to-hinge bolts loose, shift the hood as needed to position it in alignment with the cowl or windshield. Check that the proper gap is formed at the back of the hood. Also shift the hood sideways to center it with the cowl or windshield. To raise or lower the back of the hood, loosen the hinge-to-body bolts and raise or lower the hinges on the body. (B) Hood cushions or rubber stops control the height of the front of a hood. For example, if the hood edges are sticking up above the fenders, lower the stops by screwing them down. To keep the hood from rattling or bouncing up and down when the car is being driven, make sure the stops are lightly touching the bottom of the hood. (C) Make sure the hood striker moves straight down into the hood latch. If necessary, loosen the latch mounting bolts and slide the latch sideways or up and down as needed for proper engagement.**

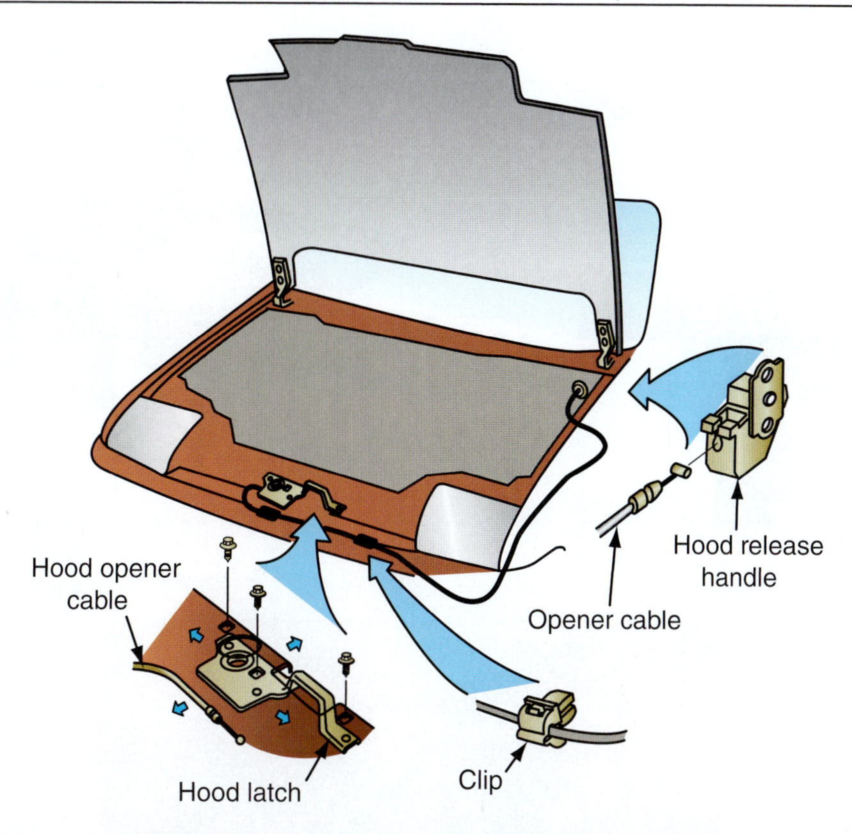

Figure 14–11 **Hood release mechanisms normally use a long steel cable running from the latch to a handle release in the passenger compartment. It can be damaged in a major collision.**

are opened by moving a lever behind the grille. Most have a cable release that runs into the passenger compartment. All have slots for adjustment (Figure 14–11).

A cable hood release consists of four major components:

1. The hood release handle in the passenger compartment, which is pulled to slide a cable running out to the hood latch. It is normally mounted on the lower left side of the passenger compartment under the dash.
2. The hood release cable, which is a steel cable that slides inside a plastic housing. One end fastens to the release handle and the other to the hood latch.
3. The hood latch, which has metal arms that grasp and hold the hood striker. The spring-loaded arms lock over the hood striker when the hood is closed. When the cable release is pulled, the arms release the striker so the hood can open.
4. The **hood striker**, which bolts to the hood and engages the hood latch when closed (Figure 14–12).

To remove a hood latch, scribe mark its location if needed. Remove its bolts and disconnect any cable. Slots in the latch mount provide up-and-down and side-to-side adjustment.

After an impact, you should always check the operation of the hood and trunk latches before closing a new lid. If the latch has been damaged in the collision, it may be very difficult to open the hood or deck lid using the release lever in the passenger compartment. Use a screwdriver to push down on the spring-loaded lever and engage the latch in the fully closed position. Then, pull on the release lever under the dash and make sure the latch lever pops back up freely. It should then be safe to fully close the hood.

Hood Latch Adjustments

After making height and position adjustments, test the hood for proper latching. Slowly lower the hood and make sure it engages the latch in the center. If the hood must be slammed excessively hard to engage the latch, the latch should be raised. If the hood does not contact the front stoppers when latched, the latch should be lowered. To adjust the hood latch, do the following:

1. Remove the hood latch assembly from the radiator support and lower the hood.
2. Check that all the gaps around the hood are properly aligned.
3. Reinstall the hood latch and lower the hood until it engages or contacts the first latch (auxiliary latch or safety catch).
4. Attempt to raise the hood. If it does open, adjust the safety catch so that it engages. Sometimes the hook can be shifted or bent until the auxiliary latch "catches."
5. Lower the hood slowly. Check to see whether the hood shifts to one side or the other when it is locked. The striker bar bolted to the hood should be centered in the "U" of the latch. When the hood is latched, it should be even with the surrounding sheet metal and fit tightly.

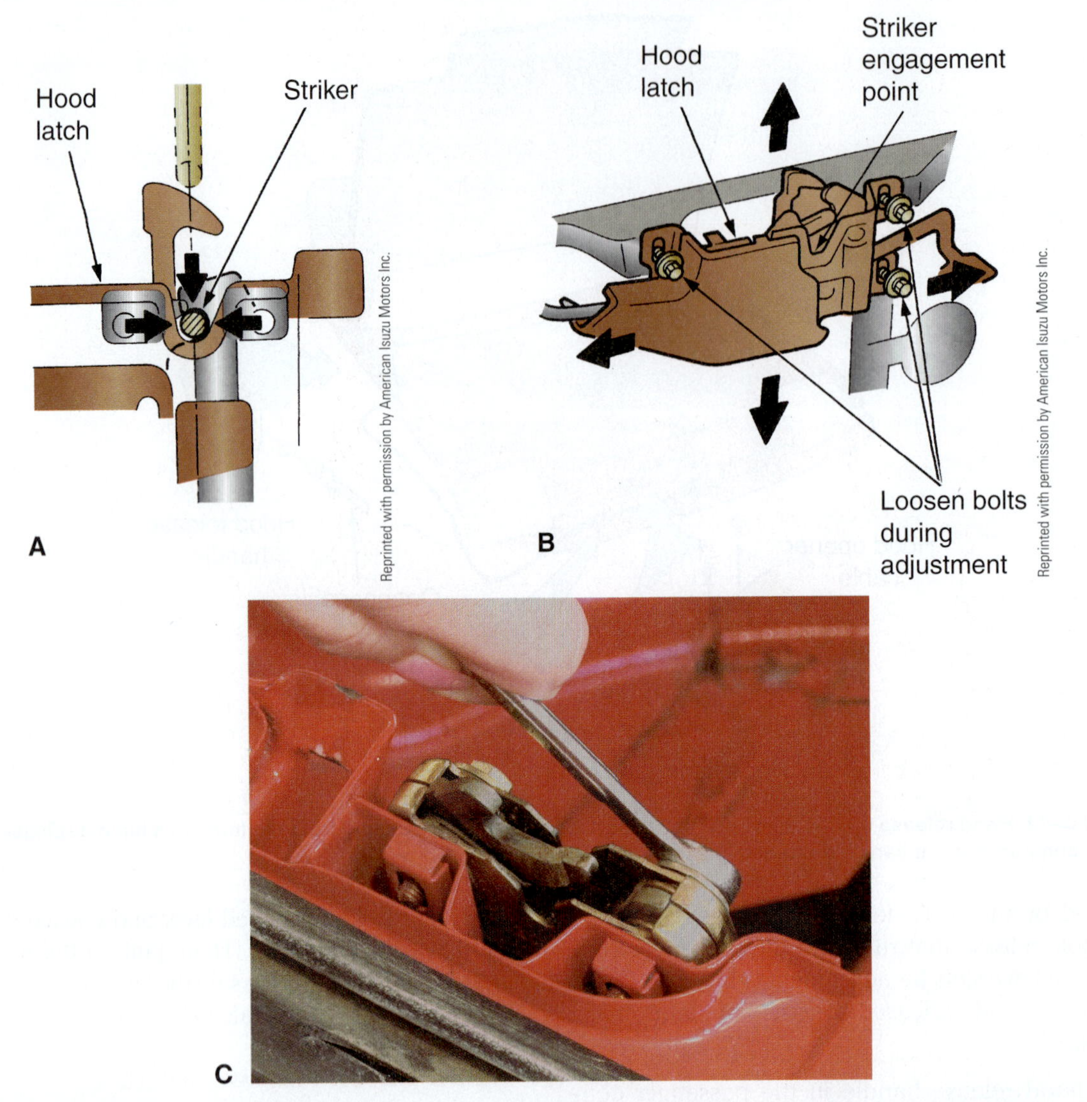

Figure 14-12 Note how to adjust a hood or deck lid latch. (A) With hood alignment and height adjusted, adjust the latch. Move the latch until the striker is centered in the latch as shown. (B) The latch should be positioned so that it places a slight downward pull on the hood when it is fully engaged with the striker. (C) After the hood is aligned with the fenders and cowl, reinstall the latch. The striker should engage the very center of the latch when closed. The latch will shift sideways and up and down when bolts are loose.

6. Loosen the hood latch just enough to maintain a tight fit, but with enough give to allow you to move the latch.
7. Move the latch from side to side to align it with the hood latch hook. Move the latch up or down as required to obtain a flush fit between the top of the hood and the fenders when an upward pressure is applied to the front of the hood.
8. Tighten the hood latch attaching hardware.
9. Open the hood and double-check its action.
10. Close the hood. Make sure it is still at the same height as the fenders. If necessary, again adjust the bumper stops to eliminate any looseness at the front of the hood and ensure a good, tight fit.
11. Tighten the attaching hardware on the bumper stops.
12. Check to see that the side bumper stops (if any) are in place and in good condition.
13. Make sure that the safety catch is working properly.

Hood latch adjustment controls how well the hood striker engages the latch mechanism. Basically, with the hood centered and set at the right height, adjust the latch for proper closing.

Slowly lower the hood while watching to see whether the striker centers itself in the latch. The hood should not pivot right or left when it engages the latch. If the hood moves to one side when closed, shift the latch right or left as needed.

The latch should also produce a slight downward compression of the rubber stops. This keeps the hood from bouncing up and down. Remember, if you must slam the hood down to engage the latch, you will need to raise the latch. If the hood moves up and down when latched, lower the latch.

After adjustment, tighten the latch bolts. Make sure the latch releases the hood properly. Always check the vehicle's service manual for specific hood adjustment procedures.

14.3 BUMPER SERVICE

New vehicle bumpers are designed to withstand minor impact without damage. Bumpers used to be made of heavy chrome-plated steel to protect other parts of the vehicle from light impact with sheer mass and bulk.

Bumpers are designed to protect the front and rear of a vehicle from damage during a low-speed collision. Most modern bumper covers are made of flexible plastic. Some bumpers are made of heavy-gauge spring steel plated with bright chromium metal. Many trucks are still equipped with chrome bumpers. Other vehicles, however, have aluminum bumpers, which are much lighter than the steel bumpers and maintain a bright finish without chrome plating. Painted steel bumpers are popular on many trucks.

Today, bumpers are designed to be light, yet strong. Many use an outer covering of flexible plastic with a heavy steel or aluminum inner bumper. Some have a plastic honeycomb structure behind their flexible cover. The trend is also to have a large one-piece cover over the lower front half of the vehicle nose. This is shown in Figure 14–13.

Bumpers on many late model cars are covered with urethane or other plastics. The use of urethane, polypropylene, or other plastic allows the bumper to be shaped to

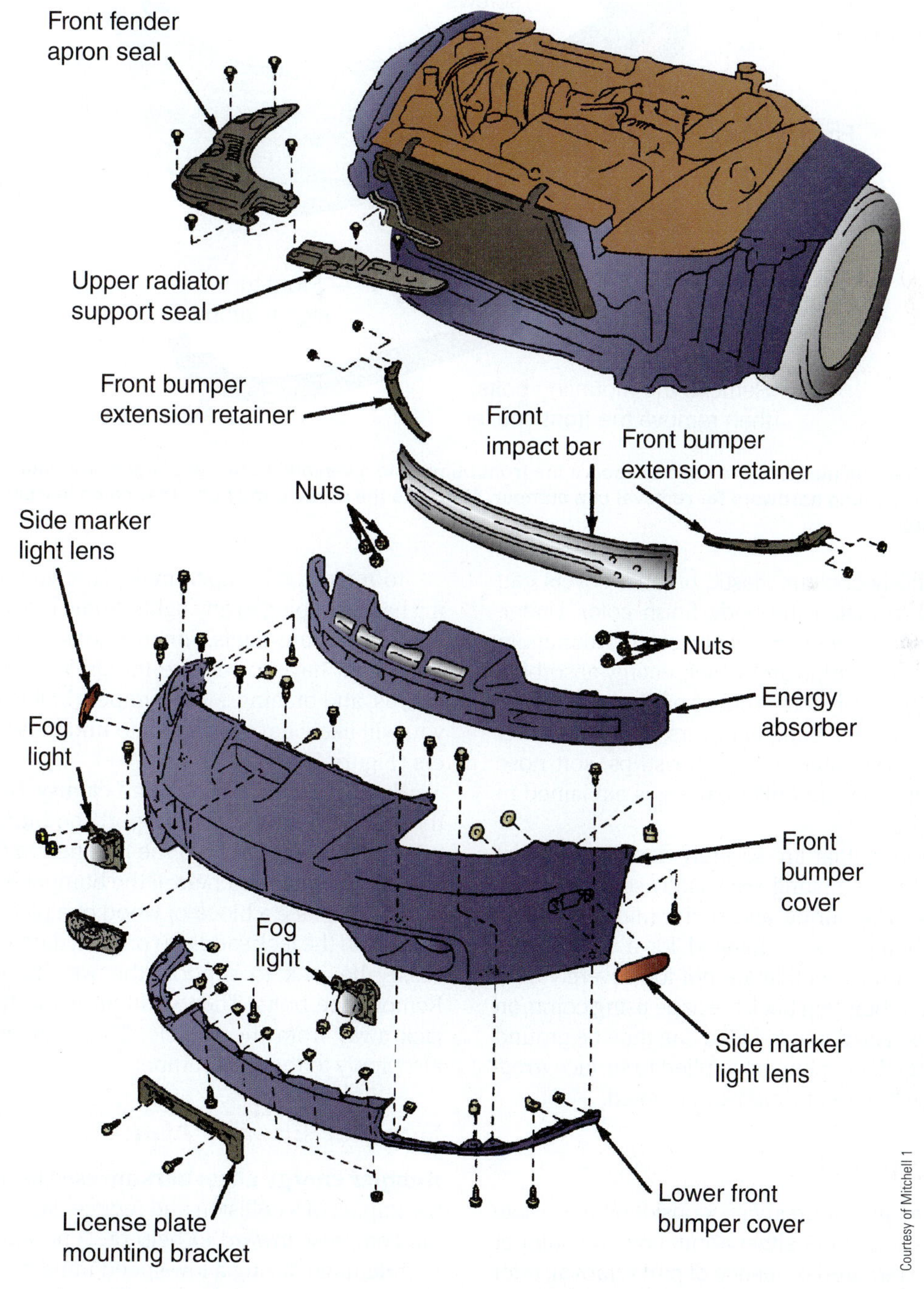

Figure 14–13 **Consult the service manuals or computer CDs for an exploded view of the bumper assembly of the vehicle you are working on. Note the impact bar, energy absorber, bumper cover, and location of all fasteners.**

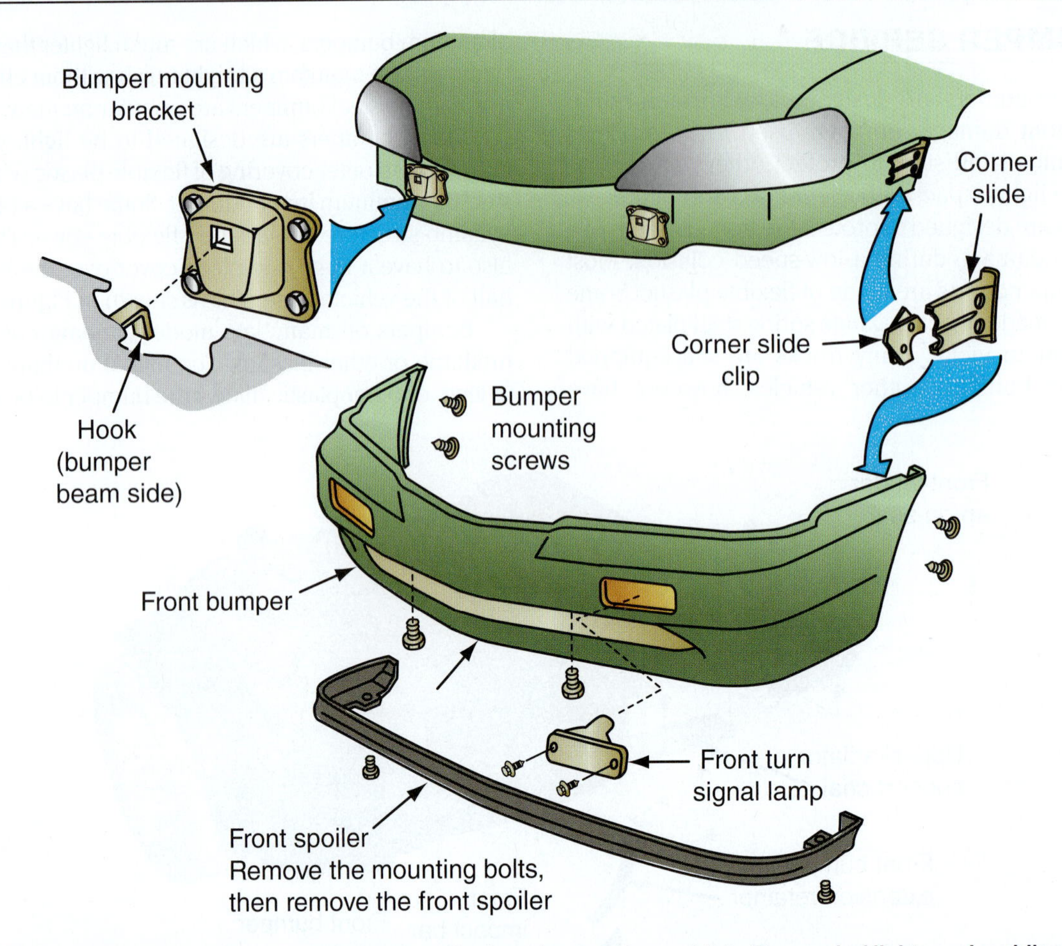

Figure 14–14 Various methods are used to secure the front bumper to a vehicle. Removal of lights and moldings often allows access to mounting hardware for removal of a bumper. Also note the screws and bolts that come in from the bottom and rear of a bumper.

blend with the body contour. Plastic bumper covers can also be painted to match the body finish color. Underneath the plastic covers there might be a steel or aluminum face bar or reinforcement bar and a thick, energy-absorbing pad made of high-density foam rubber or plastic.

The repair of flexible bumper covers, as well as other plastic bumper parts, such as bumper strips, soft nose parts, gravel deflectors, and filler panels, is explained in detail in Chapter 13.

Chrome bumpers that are severely damaged or that have chipped chrome plating are usually sent to a shop that specializes in repairing and rechroming bumpers. The damaged bumper is exchanged for a refurbished bumper. Painted bumpers that are not too severely bent can be pulled and bumped back to shape using common shop procedures. The damaged area can then be ground down to bare metal, filler can be applied to surface irregularities, and the bumper primed and painted.

Bumper Removal

Bumper removal procedures vary considerably on late model vehicles. If in doubt either about how the bumper is secured or the required sequence of part removal, refer to service information for the specific make and model vehicle.

To remove a bumper, make sure you unplug the wiring harness going to any lights. Some bolts can be hidden behind parking lights, inner fender panels, and so on. You might have to remove the grille, headlights, splash guards, and bracing. Most bumpers bolt to the fenders, so you will likely have to search to find these hidden fasteners (Figure 14–14).

Bumpers can be heavy and clumsy. Before removing the last mounting bolts, support the bumper on a floor jack. Get a helper to hold the bumper on the jack as you remove the final fasteners. If the bumper is to be repaired or reused, place a block of wood or a piece of thick foam rubber on the jack saddle to prevent damage to the finish. Raise the jack to support the weight of the bumper. Remove the bolts. Then you can work the bumper and jack away from the vehicle. These general procedures also apply to the rear bumper.

Bumper Energy Absorbers

Bumper energy absorbers are used to cushion some of the impact of a collision and reduce damage. The absorbers compress inward to help prevent bumper and other part damage during a low-speed impact.

Modern bumpers are designed to absorb the energy of a low-speed impact, minimizing the shock to both the vehicle frame and the occupants of the vehicle. A few years ago, car manufacturers were required by federal regulations to equip cars with bumpers that could withstand 5 mph (8 km/h) collisions without incurring damage to the vehicle. Later, the federal standards were relaxed to 2.5 mph (4 km/h).

In order to comply with the federal regulations, manufacturers fitted their bumpers with energy absorbers. Most energy absorbers are mounted between the bumper face bar or bumper reinforcement and the frame (Figure 14–15).

There are many types of energy absorbers. The most common is similar to a shock absorber. The typical bumper shock is a cylinder filled with hydraulic fluid. Upon impact, a piston filled with inert gas is forced into the cylinder. Under pressure, the hydraulic fluid flows into the piston through a small opening. The controlled flow of fluid absorbs the energy of the impact. Fluid also displaces a floating piston within the piston tube, which compresses the inert gas. When the force of the impact is relieved, the pressure of the compressed gas forces the hydraulic fluid out of the piston tube and back into the cylinder. This action forces the bumper back to its original position.

In another energy absorber design, upon impact, fluid flows from a reservoir through a metering valve into an outer cylinder. When impact forces are relieved, a spring in the absorber returns the bumper to its original position.

An isolater is another type of bumper energy absorber. In principle, it works like a motor mount. A rubber pad is sandwiched between the isolator and the frame. Upon impact, the isolator moves with the force, stretching the rubber pad. The "give" in the rubber absorbs the energy of the impact. When the force is relieved, the rubber retracts to its original shape (unless it is torn from its base by the impact) and returns the bumper to its normal position (Figure 14–16).

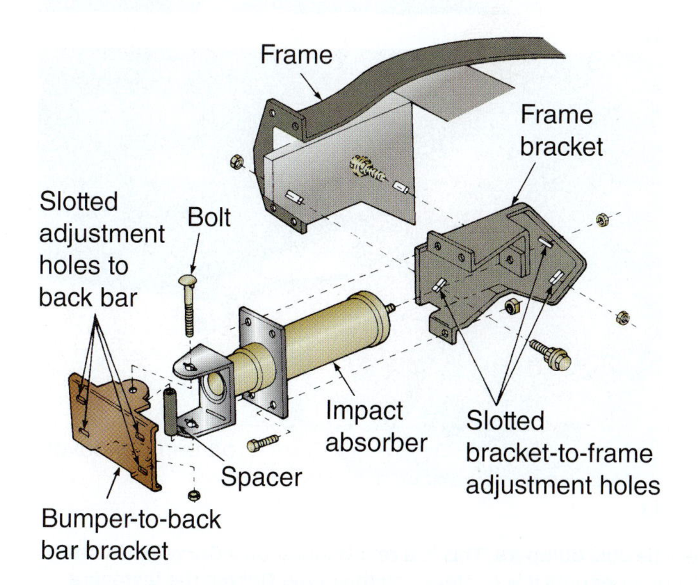

Figure 14–15 **A bumper shock absorber bolts between the body or frame and the bumper. Some are gas filled and/or spring-loaded.**

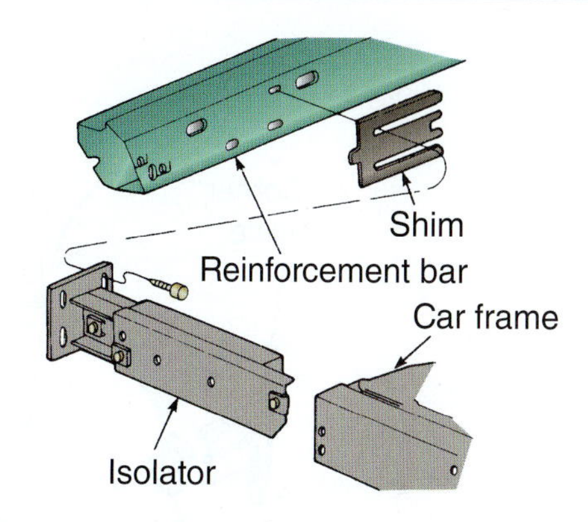

Figure 14–16 **This is a typical bonded isolator. It is often damaged in front-end collisions.**

Figure 14–17 **This truck is getting a new rear bumper. The bumper will be painted and then brackets will be transferred to it before installation.**

A different type of energy absorber is found on many late model vehicles. Instead of shock absorbers mounted between the frame and the face bar or reinforcement bar, a thick urethane foam pad or honeycomb-shaped absorber is sandwiched between an impact bar and a plastic face bar or cover. The pad is designed to give and rebound to its original shape in a 2.5-mph collision.

On some vehicles, the impact bar is attached to the frame with energy-absorbing bolts. The bolts and brackets are designed to deform during a collision in order to absorb some of the impact force. The brackets must be replaced in most collision repairs (Figure 14–17).

Bumper Replacement

Removing a bumper is basically a matter of removing the correct bolts. This job is made easier if the bumper is supported by a floor jack to hold its weight. On some vehicles, stone deflectors, parking lights, windshield washer hoses, and other items must be disconnected before the bumper can be removed from the vehicle (Figure 14–18).

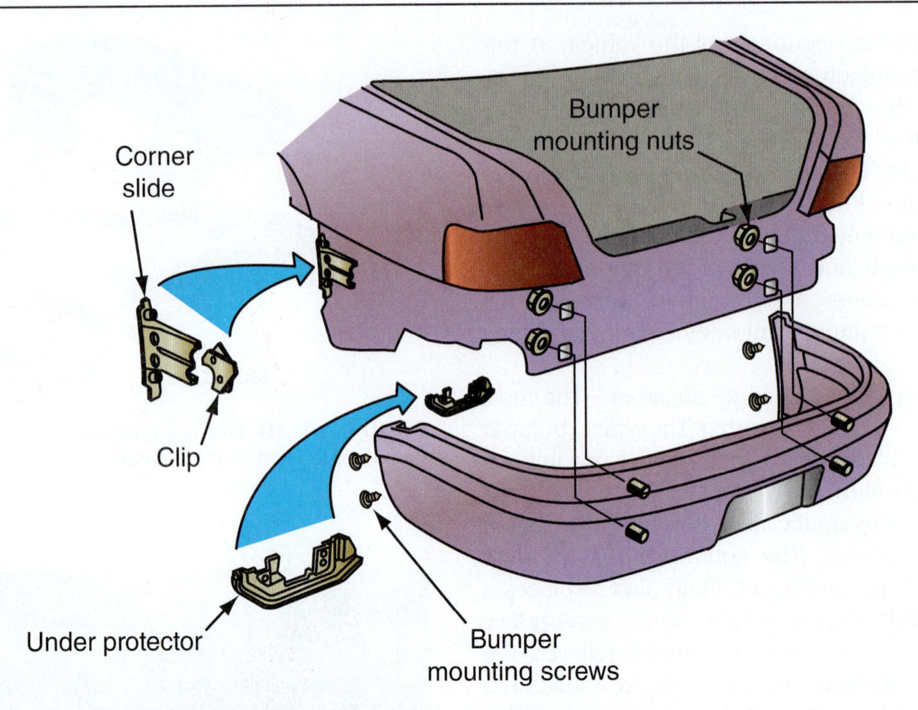

Figure 14-18 Note the typical fasteners used to secure the rear bumper assembly.

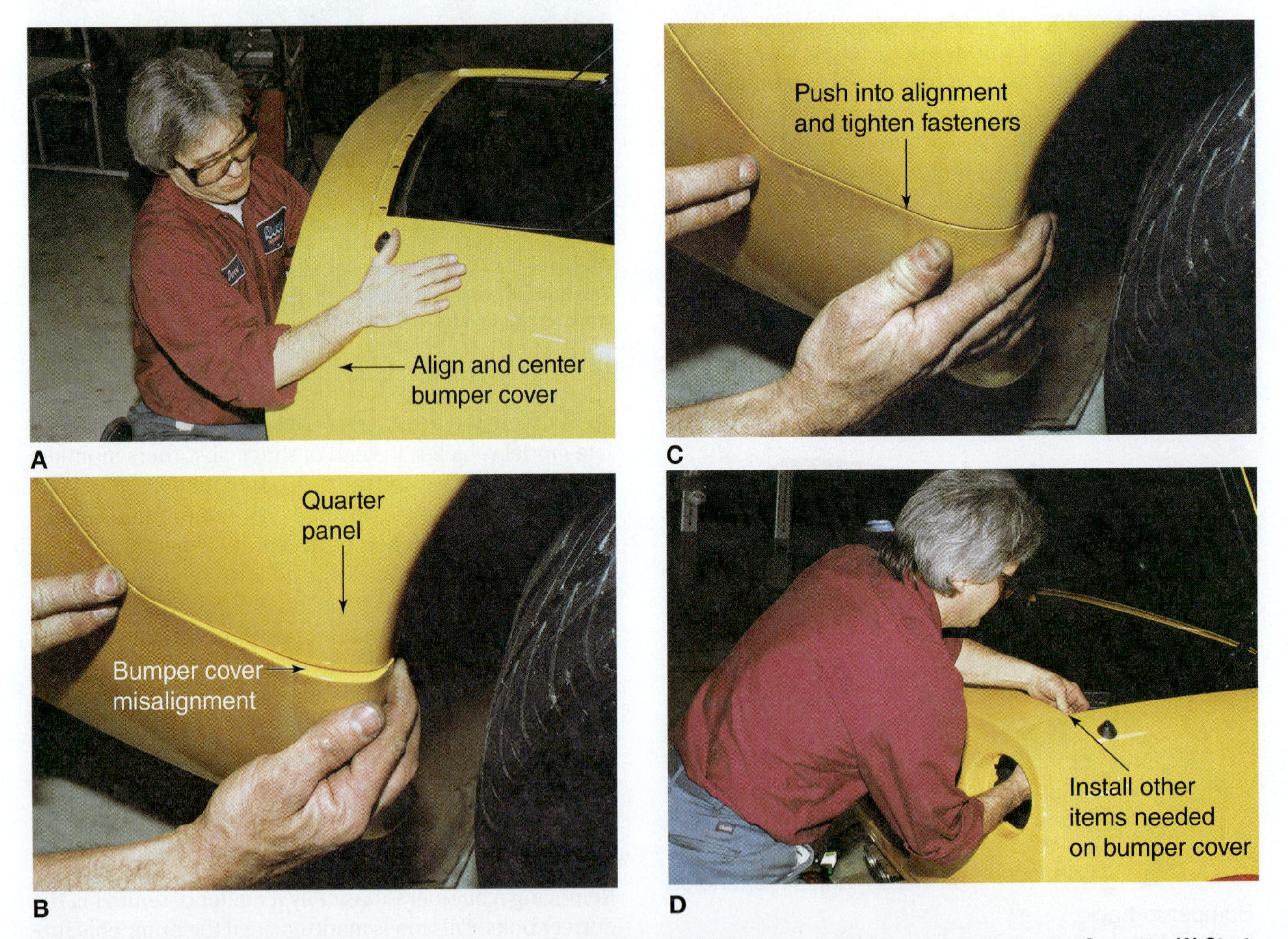

Figure 14-19 Follow a logical sequence when installing modern flexible bumpers. This is a rear bumper on a Corvette. (A) Start the fasteners in the bumper cover but do not tighten them. Shift the bumper until it is centered on the body. Tighten the fasteners along the top of the bumper. (B) Work your way down the sides of the flexible bumper, aligning and tightening fasteners as the edges are aligned. (C) Here the technician is pushing the rear bumper cover into alignment with the quarter panel before tightening the fasteners behind the cover. (D) The technician is reaching through the taillight lens to install clips that secure the wiring harness.

After bolting the bumper in place, it must be adjusted so that it is an equal distance from the fenders and front grille. The clearance across the top must be even. Adjustments are made at the mounting bolts. The mounting brackets allow the bumper to be moved up or down, side to side, and in and out. If necessary, steel shims can be added between the bumper and the mounting bracket to adjust the bumper alignment (Figure 14–19).

With major front end damage, it is sometimes best to remove the large assemblies, such as the front clip or bumper-spoiler assembly. This enables you to gain access to hard-to-reach parts on the assembly more easily. This may also allow you to service a damaged bumper, lights, brackets, and other front end parts in less time.

14.4 FENDER SERVICE

Fender service is commonly needed after frontal impacts. Fenders are among the most common parts damaged during collisions. You must be proficient at removing and installing fenders on various makes and models of vehicles.

Fender Removal

To remove a fender, find and remove all of the bolts securing it to the vehicle. Also remove any wires going to fender-mounted lights. Fenders are usually bolted to the radiator core support, inner fender panels, and cowl. Bolts are often hidden behind the doors, inner fender panels, and under the vehicle (Figure 14–20).

If the old fender has factory-installed fender shims or spacers, note their locations during disassembly. If there is no major unibody or frame damage, reinstalling the fender shims in their original locations will help you more quickly realign the fender after repairs.

With all of the bolts removed, carefully lift the fender off. Transfer any needed parts (trim, body clips, and so forth) from the old fender over to the new fender.

You will usually have to send the fender to the refinishing area for preinstallation paint edging. During *paint edging*, all ends, corners, edges, and sometimes the rear of the fender panel should be scuff sanded, primed, and painted. The surfaces of the fender are often painted before installation on the vehicle. These areas would be difficult or impossible to paint after the fender has been bolted to the vehicle.

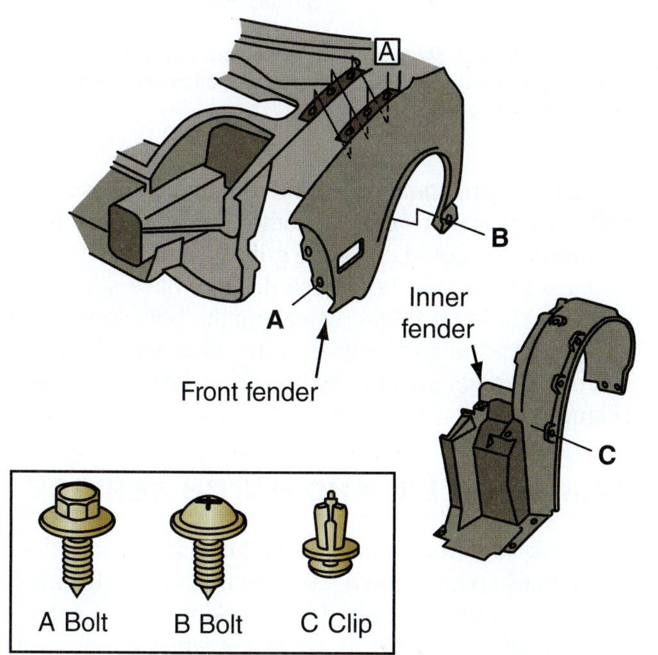

Figure 14–20 Fender bolts are typically located in these locations. Some are hidden behind the fender's inner splash guard.

Installing Fenders

Install the replacement fender in reverse order of removal. If the doors or cowl are undamaged, place masking or duct tape over their edges to protect from scratches when installing the fender (Figure 14–21).

When installing fenders, hand-start all of the fender bolts, but do not tighten them. Leave the bolts loose enough so that you can adjust the fender.

Fender Adjustments

Fenders are bolted to the radiator core support or to the grille, to the inner fender panel in the engine compartment, and to the cowl behind the door and under the car. When these bolts are loosened, the fender can be moved for adjustment.

Start adjusting and tightening the fender bolts at the rear, next to the top of the door. Obtain the correct fender-to-door gap and the correct fender-to-hood gap. Then

Figure 14–21 Fender R&R can be challenging or simple depending on the design of the vehicle. When installing a fender, fit it into place and hand-start the bolts. Then, shift the fender to produce the proper gaps between the door and hood. Tighten the bolts after the fender is properly aligned. After fender installation, open doors slowly to check for adequate clearance so you do not damage the fender or door. If there were shims under the fender, note their locations for reassembly. If the vehicle did not suffer major structural damage, shims can and should be used in original factory-assembled locations.

tighten the fender bolts, working your way to the front of the vehicle.

Shift the fender on its bolts so that it properly aligns with other body parts. Shift the fender forward or backward until the fender, door, and cowl have the correct spacing or gap. Also adjust the fender in and out so that it is flush with the door and parallel with the hood. Tighten the fender bolts only after you have the fender in alignment.

You must make sure the curvature of the fender matches the shape of the front door edge. Sometimes, a mounting bolt is provided at the center rear of the fender. It can be tightened when you have the correct curvature. If not, you will need to adjust the position of the upper and lower rear mounting holes (up and down) so the fender matches the door.

A **body shim** is another means of making an adjustment to a fender or other body panel. A body shim is a thin, U-shaped piece of metal for making part adjustments. By loosening a bolt, a shim can be slipped under the panel and around the bolt. When retightened, the position of the attached panel is raised or moved the distance equal to the thickness of the shim.

Shimming body panels was once a very common operation with full-frame vehicles. However, with the welded panels of today's unibody construction, there are few body panels that can be shimmed. Body panels can still be shimmed on full-frame trucks and full-size cars, however.

Fender shimming is an adjustment method that uses spacers under the bolts that attach the fender to the cowl or inner fender panel. By changing shim thicknesses, you can move the position of the fender for proper alignment.

The fender-to-door alignment can sometimes be made by shimming the two large bolts attaching the fender to the cowl. The top bolt is usually in the door pillar. The bottom bolt is either in the hinge post or under the car in the rocker panel. By shimming the top bolt, the upper fender can be moved out. Shimming the lower bolt will move the lower portion out. If the fender is in too far and not flush with the door, the protruding door edge will cause noisy wind turbulence when the vehicle is in motion.

Complete alignment can be achieved on many fenders without using any shims. Shims should be used only if alignment cannot be achieved without them.

These adjustments allow the fender, hood, and door to be properly aligned. Often the fender and hood adjustments must be made simultaneously to achieve a satisfactory result. The gap between fender and hood should be to factory specifications. The result will be even spacing all around the fender (Figure 14–22).

Figure 14–22 **The technician is using plastic feeler gauges to check quickly for perfect alignment or gaps between fender and door.**

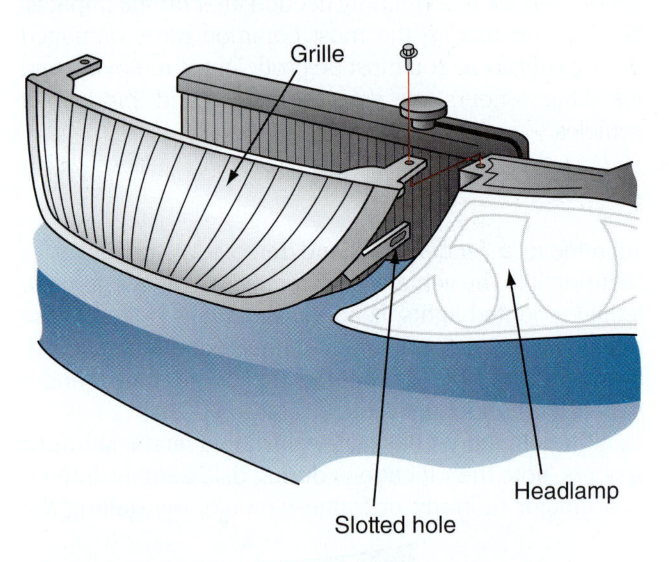

Figure 14–23 **Most modern grilles are made of lightweight plastic. Only a few screws or clips are used to secure today's grilles.**

14.5 GRILLE SERVICE

Grilles are often held in place with small screws and clips. You might have to remove a cover to access grille fasteners.

An air ratchet is handy for reaching down and unscrewing grille bolts. When installing a grille, make sure all clips are undamaged and installed. Because most grilles are plastic, be careful not to overtighten any bolts or screws. You could crack the grille.

Most grilles can be adjusted. They have slotted or oversized holes in them. By leaving the bolts loose, you can shift and align the grille with other vehicle parts. Once it is aligned, tighten the grille fasteners slowly (Figure 14–23).

14.6 DECK LID AND HATCH SERVICE

The deck lid is very similar to the hood in construction. Two hinges connect the deck lid to the rear body panel. The trailing edge is secured by a locking latch. Deck lid or hatch door removal and replacement are similar to hood R&R (Figure 14–24).

The deck lid must be evenly spaced between the adjacent panels. Slotted holes in the hinges and/or caged

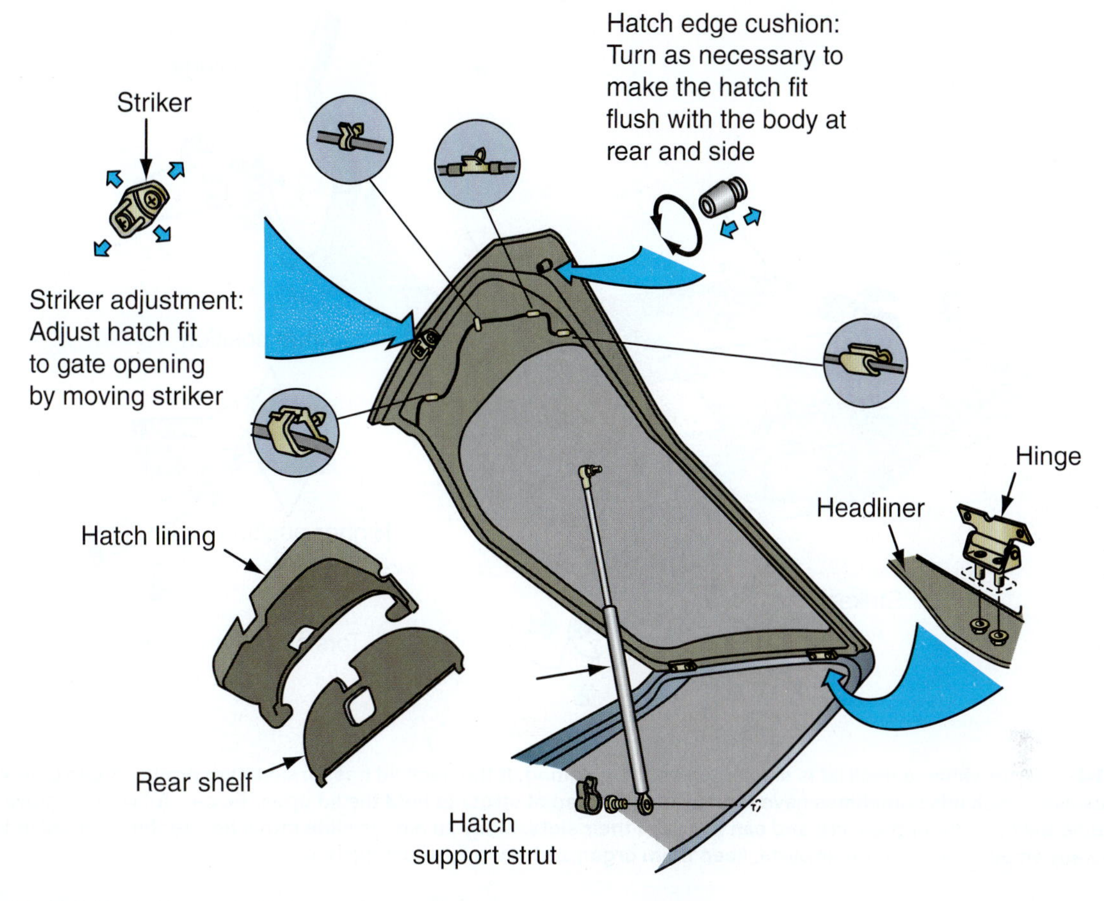

Figure 14–24 Study the parts that must be serviced to replace a rear hatch lid.

plates in the deck lid allow it to be moved. To adjust the deck lid forward or rearward, slightly loosen the bolts on both hinges. Close and adjust the deck lid as required. Then raise the lid and tighten the bolts.

Weatherstripping is a rubber seal that prevents leakage at the joint between the movable part (lid, hatch, or door) and the body. To prevent air and water leaks, the deck lid must contact the weatherstripping evenly when closed. The latch must be adjusted so that it holds the lid or hatch closed against the weatherstripping.

Deck lids and hatches usually do not have exterior or interior door handle mechanisms. They operate with a key (or instrument panel switch on powered units) and lock mechanism.

Lock cylinders contain a tumbler mechanism that engages the key so that you can turn the key and disengage the latch. When you insert your key into a door or lid, it engages the lock cylinder. The lock cylinder then transfers motion to the latch.

Lid torsion rods are spring steel rods used to help lift the weight of the lid. They extend horizontally across the body and engage a stationary bracket. Some torsion rod brackets have adjustment slots. You can change tension on the torsion rods by moving them in these slots (Figure 14–25).

Trunk Lid Adjustments

The lid must be evenly spaced between the adjacent panels. Slotted holes in the hinges and/or caged plates in the lid allow the trunk lid to be moved forward, rearward, and side to side. To adjust the lid forward or backward, slightly loosen the attaching hardware on both hinges. Close and adjust the lid as required. Then raise the lid softly and tighten the attaching hardware.

In some cases, it might be necessary to use shims between the bolts and the trunk lid to raise or lower the front edges. If the front edge must be raised, place the shim(s) between the hinge and the lid in the front bolt area. To lower the front edge of the lid, place the shim(s) at the back of the hinge.

Vehicles with hatchback-type trunk lids are usually difficult to align mainly because of their size. The hatchback types also use gas-filled door lift or shock assemblies, or springs, one at each upper corner of the lid.

Panel Alignment

After installing all new body parts, you must check overall panel alignment. Make sure that clearances between parts are equal. As shown in Figure 14–26, the gap around

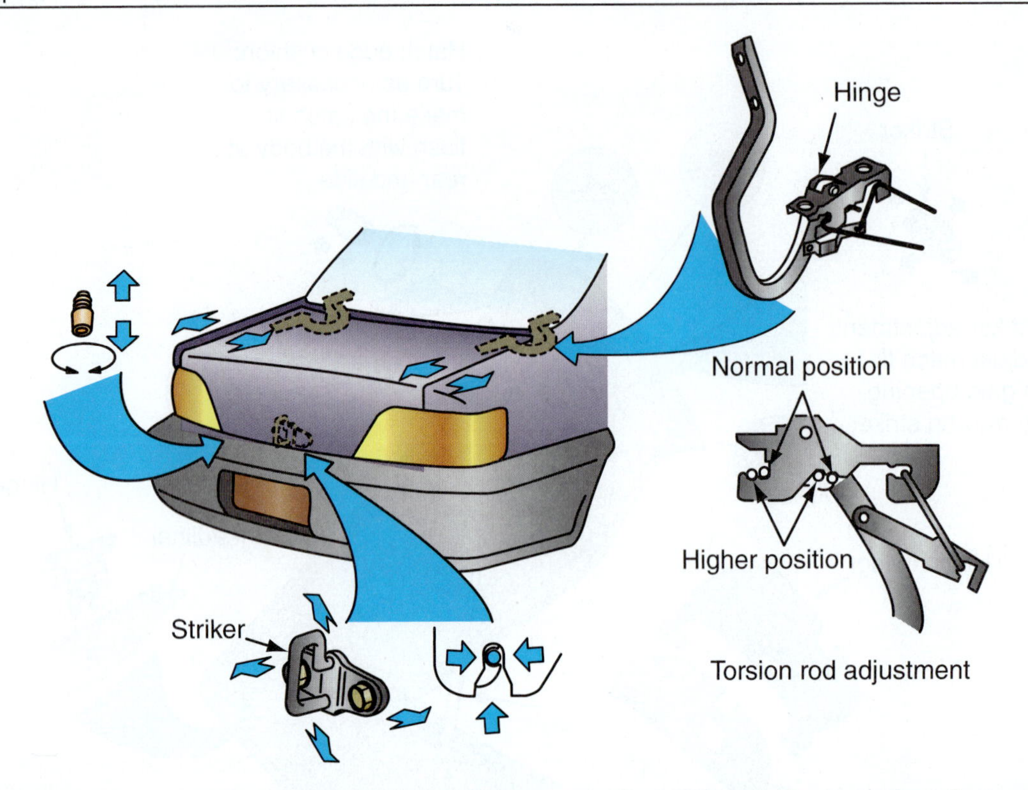

Figure 14-25 Servicing a deck lid is similar to servicing a hood. If the deck lid has torsion bar springs, it can complicate service, however. Deck lids sometimes have torsion rods instead of struts to hold the lid open. Be careful when removing these rods, because they are under pressure and can fly out of their slots. Many lid hinges slide into a box section formed in the panel. Do not lose any small spacers or other parts; keep them organized in a tray or plastic bag.

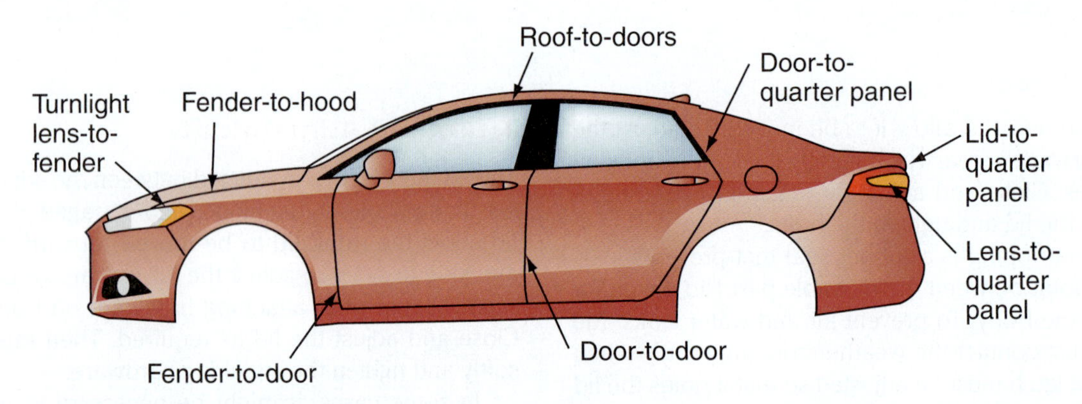

Figure 14-26 After installing major parts and panels, double-check all panel gaps. Some gaps could still be off due to indirect damage. Adjust panels until all gaps are within specifications.

all parts must be within specifications. Also check that the surfaces of all panels are even with each other. Take the time to double-check all panels to ensure good alignment. This is a sign of a professional technician.

14.7 TRUCK BED SERVICE

Truck beds are usually bolted to a full ladder frame. To remove the bed, simply remove the bolts that extend up through brackets on the frame. Keep track of bolt lengths and rubber mounting cushion locations. They must be reinstalled in their original positions (Figure 14–27).

Truck tailgates mount on two hinges. A steel cable often limits how far down the tailgate will open. Latches in the sides of the tailgate engage strikers on the body. The handle on the outside of the tailgate moves small linkage rods that run out to the latches.

The tailgate hinges must be adjusted as described earlier for other parts. Generally, leave the hinge bolts loose while shifting the tailgate as needed. Then, tighten the hinge bolts and recheck adjustment (Figure 14–28).

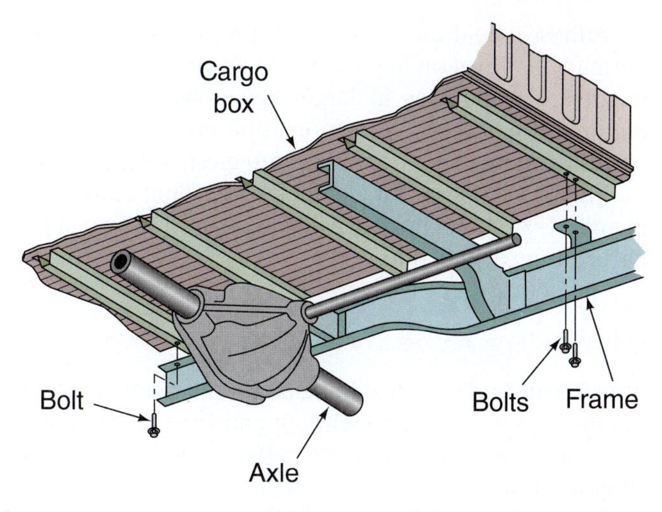

Figure 14-27 Truck beds normally bolt to the full ladder frame. Large bolts secure the truck bed to the vehicle.

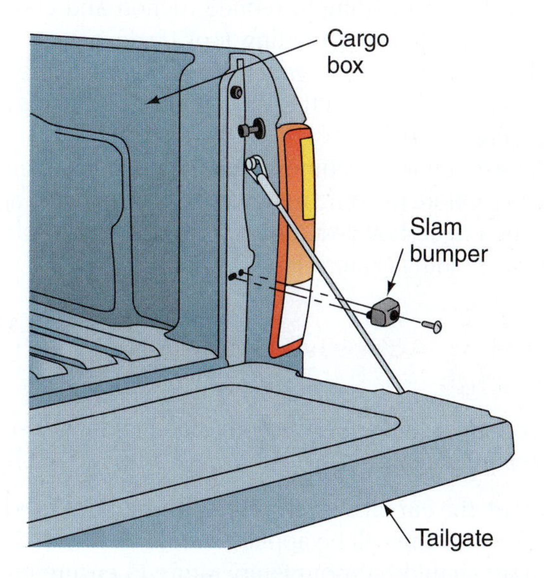

Figure 14-28 Truck tailgates are surprisingly heavy. Be careful when removing a tailgate. Ask someone to help you remove and install a tailgate. A tailgate must align with the bed and properly engage the striker.

Truck tailgates are surprisingly heavy. Ask someone to help you when removing or installing one. Tailgates can easily fall, injuring your legs and feet.

14.8 SOUND-DEADENING PADS

Sound-deadening pads are often bonded to the inside surface of trunk cavities and doors to reduce noise, vibration, and harshness. Sound-deadening material is made of a plastic or asphalt-based material. It helps to quiet the passenger compartment by preventing the thin sheet metal panels from acting and sounding like large metal drums. The original factory material is bonded and sometimes heat-formed to the surface.

Sound-deadening material is combustible and must be removed from any area to be welded or flame cut.

During collision repair, sound-deadening pads must be replaced to match preaccident performance. After repair, the area must be properly refinished to provide corrosion protection before the sound-deadening material is applied.

14.9 CUSTOM BODY PANELS

Custom body panels are aftermarket parts that alter the appearance of the vehicle. Customizing is a growing trend. When faced with the need to replace damaged panels, many vehicle owners spend the extra money to give their vehicles a performance face-lift. They have a body shop install custom parts.

Sometimes, aftermarket custom parts cost no more than original parts. A number of companies sell spoiler and air dam kits that enhance vehicle appearance. These are also known as ground effects.

Inform customers that they may be affecting their factory warranty by installing aftermarket parts.

Air dams improve aerodynamics by restricting air passing under the vehicle body, thus reducing turbulence and resistance to airflow. If designed properly, air dams can increase fuel economy.

A *rear spoiler* mounts in the trunk lid to alter the airstream at the rear of the body to increase body down force. This helps rear wheel traction at highway speeds.

Spoilers, side skirts, air dams, and other aftermarket parts are commonly made from molded polyurethane plastic or fiberglass. Before installing them, clean the parts with a wax and grease remover. As an added precaution, also wipe with isopropyl alcohol, particularly if they are fiberglass. Then, scuff sand the parts with #240 grit sandpaper followed by #400 grit paper to improve adhesion.

Most custom body panels utilize original fasteners. They simply bolt in place in the same way as the original panel. Double-sided adhesive tape is sometimes used to fit custom panels tightly against the original panels. To use double-sided tape, clean the area with wax and grease remover. Scuff sand the area to ensure a positive bond between the vehicle and the new panel.

When installing add-on panels, read and follow the kit instructions carefully. Then, if you have problems, the kit manufacturer is responsible.

The key to successful add-ons is proper alignment, careful positioning of the new fasteners, and preparation of the mating panel surfaces. Keep these points in mind when installing aftermarket body panels and parts:

1. Remove the lower body panel from under the front bumper.
2. Place the new air dam under the front bumper and align the air dam with the wheel wells. Make sure that the front top lip of the air dam is located inside the front panel.
3. Clamp the corners of the air dam to the wheel well with Vise-Grips.
4. Transfer the front body panel mounting holes with a scribe onto the air dam.
5. Use a scribe to transfer the mounting holes in the ends of the air dam to the wheel well.
6. Using the recommended size drill bit, drill the right amount of holes through the sheet metal and air dam.
7. Loosely bolt the air dam in position and check it for correct alignment.
8. Without overtightening them and causing part distortion, securely snug down all fasteners.

14.10 INSTALLING BODY TRIM AND MOLDINGS

Every vehicle has a variety of trim pieces or moldings. Moldings enhance the appearance of a vehicle by hiding panel joints, preventing door dings, and accenting body lines. They also help to weatherproof by channeling wind and water away from windows and doors. Moldings often must be replaced due to collision damage, or moldings can be added as a custom accessory (Figure 14–29).

Exterior moldings can be attached to the body using adhesives, self-stick backing, two-sided tape, and mechanical fasteners (clips, rivets, or screws). A variety of fasteners is used to secure moldings to a vehicle. The clips and bolts shown in Figure 14–30 are examples.

An older molding attachment method is a stud welded to the body. A clip fits over the stud and the molding slides over the clip. If a weld stud is bent or broken off, replace it with an oval head blind rivet, or weld a new stud in place with a stud welder equipped with a special rivet electrode.

Adhesive-held moldings, when properly applied, are permanent. Rivet-on moldings, while also permanent, might create buckles in large, low-crown panels and require drilling through the corrosion protection on the body. Instructions for both types are given here.

Figure 14–31 shows a technician installing body trim. Figure 14–32 shows how an awl can be used to help align parts during installation.

Removing Adhesive-Held Moldings

A **molding tool** or power knife has a thin blade that will cut through the adhesive without causing part damage. It is the best way to remove adhesive-held moldings (Figure 14–33).

To cut off an old molding, apply masking tape around the molding to protect the paint. Then adjust the tool blade depth to match the molding size. Apply a soaping solution to the molding to reduce friction and cool the cutter blade. Slip the molding tool blade between the molding and the body panel. While holding the tool firmly square, begin cutting slowly from one end to the other (Figure 14–34).

A heat gun is sometimes recommended to soften the adhesive before molding removal. Be careful not to apply too much heat, however, or you can blister the paint or ruin the molding (Figure 14–35).

Installing Adhesive Body Side Moldings

To properly install adhesive body side moldings, use the following procedure:

1. Park the car on a level surface. The surface to which the molding will be applied should be cleaned. The part should be room temperature to ensure proper adhesion.
2. Select the area to which the molding will be applied. For greatest protection, the molding should be applied to the outermost surface of the vehicle. If the car has an outermost ridge, install the molding ⅛ inch (3.1 mm) above or below the ridge, but not on the ridge itself. If the panel does not have a prominent ridge, select the outermost surface of the body contour.
3. After determining the best location for the molding, thoroughly clean the area with water and detergent. Then, use a clean rag wetted with a wax and grease remover to remove waxes and silicones. Use a clean cloth for each side of the vehicle.
4. If the body molding will not be aligned above or below a body ridge that can be used as a guide, mark the correct height of the molding with a steel rule and a soft lead pencil or a china marker. Mark the height at the rear and front of the car and at each door gap. Then, stretch a piece of masking tape from

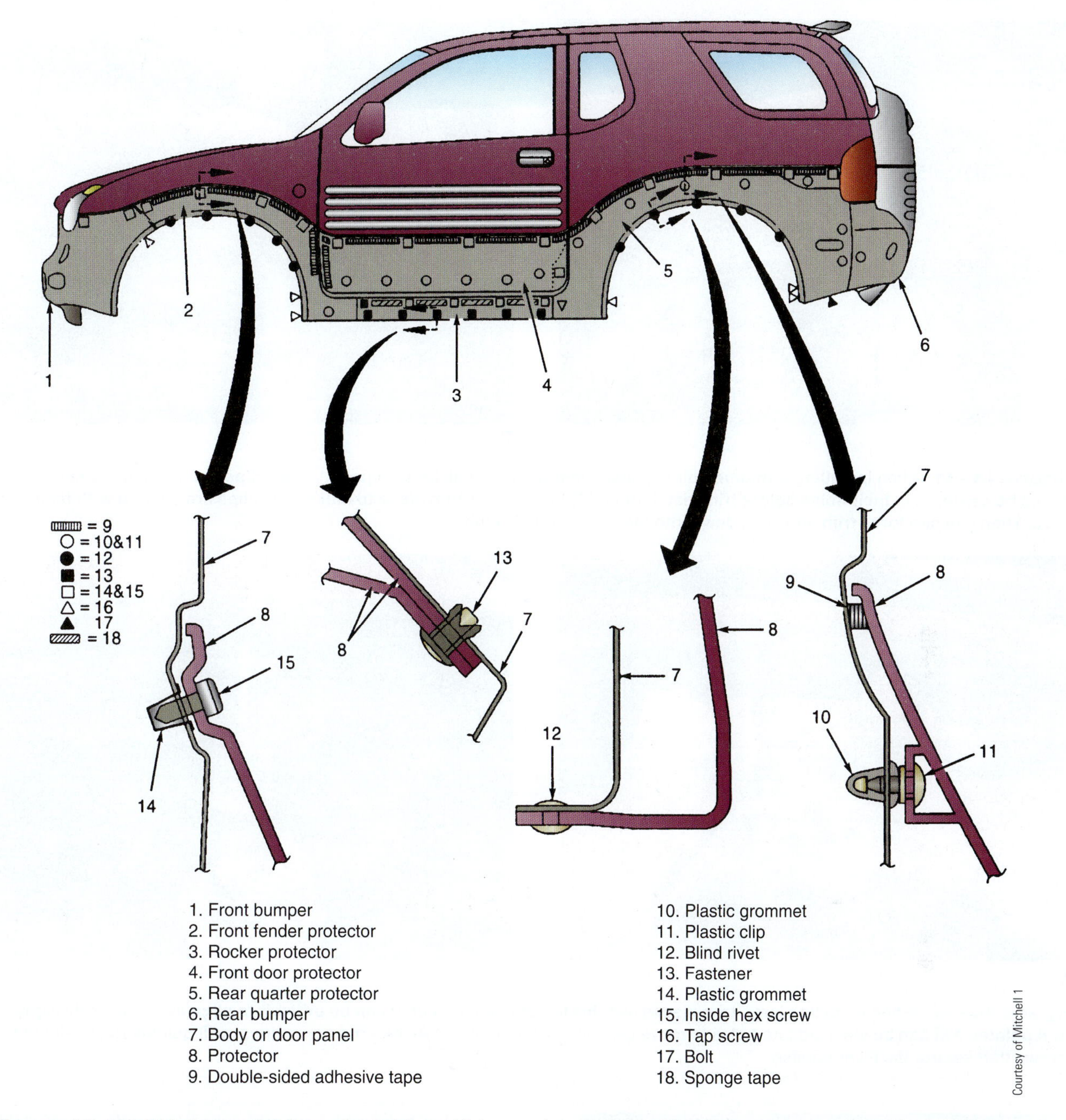

Figure 14-29 This late model, four-wheel-drive vehicle uses large, plastic trim pieces to protect metal body panels from road debris. Note how this section view details how the trim is fastened to the vehicle.

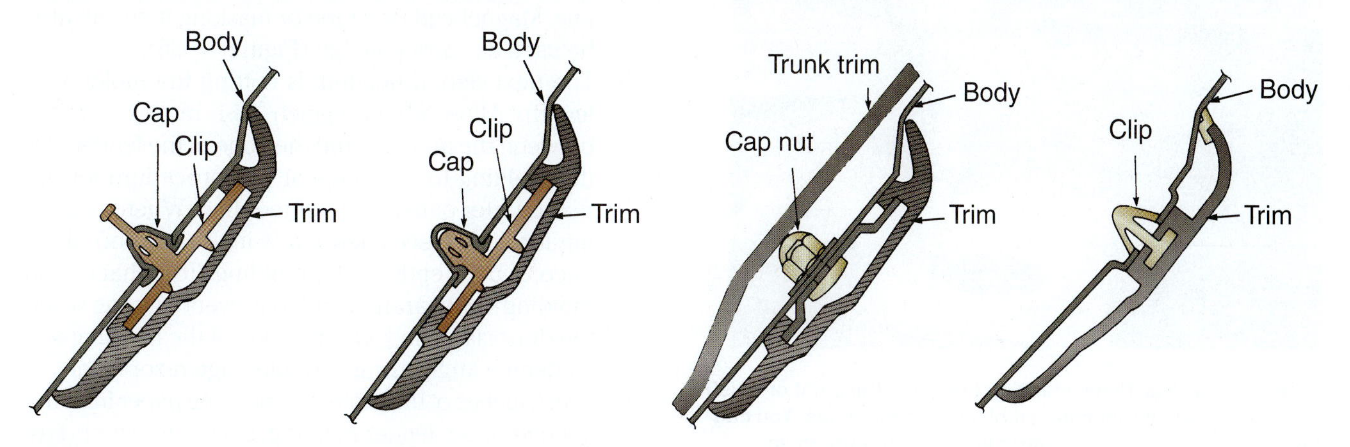

Figure 14-30 Here are a few molding attachment methods.

A

B

Figure 14–31 When installing trim or moldings, check installation of all body clips. (A) Body clips must not be broken and should be centered in their holes before trim installation. (B) Look under trim piece to make sure clips are aligned with their holes. Then you can force trim and clips down into place without difficulty.

A

B

Figure 14–32 When installing large trim pieces like this lower front spoiler, it can be difficult to align all of the body clips. (A) A pointed awl can be inserted into holes to move clips into alignment. (B) An electric screwdriver will quickly run in all of the screws that secure the front spoiler.

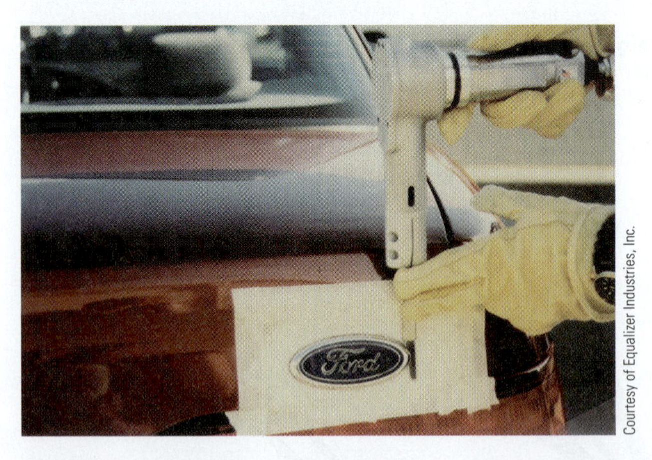
Courtesy of Equalizer Industries, Inc.

Figure 14–33 Place masking tape over the paint or finish to protect it when replacing emblems and moldings. You can then use a heat gun and electric knife to cut through an adhesive-held trim piece.

front to rear, connecting the marks. Keep the tape taut and sight along its length to ensure a straight line. Magnetic plastic tape or masking tape can also be used as a straightedge (Figure 14–36).

5. The next step, if needed, is cutting the molding to length. Allow about ⅛-inch (3.1 mm) clearance between the molding and the edge of the fender. Cut the molding to size. Repeat this procedure for the rear quarter panel piece of molding. When measuring for door pieces, leave a ⅛-inch (3.1 mm) clearance at both ends of the molding. To enhance the molding's appearance and to prevent binding when the door is opened, cut the ends of the molding at a 45-degree angle, using a single-edge razor blade.
6. Peel 6 inches of the protective backing paper from the cut end of the fender molding. Do not touch or dirty the exposed adhesive after removing the backing.

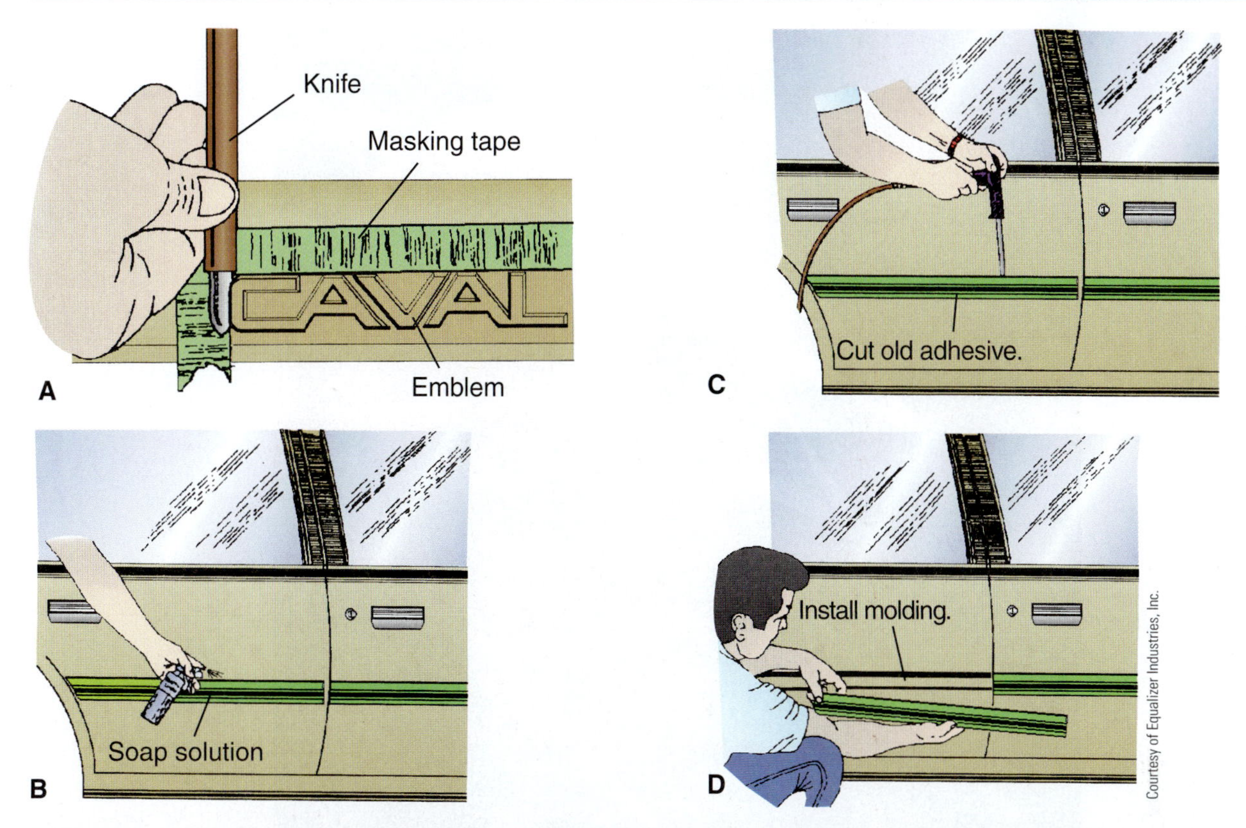

Figure 14–34 (A) Apply masking tape to protect paint before cutting. Adjust your blade length to match the width of the emblem or molding. (B) Apply a soapy solution to prevent the blade from overheating. (C) Hold the tool securely as you cut between the body and emblem. (D) Before installing new molding, repair the area as needed and clean it properly to provide good adhesion.

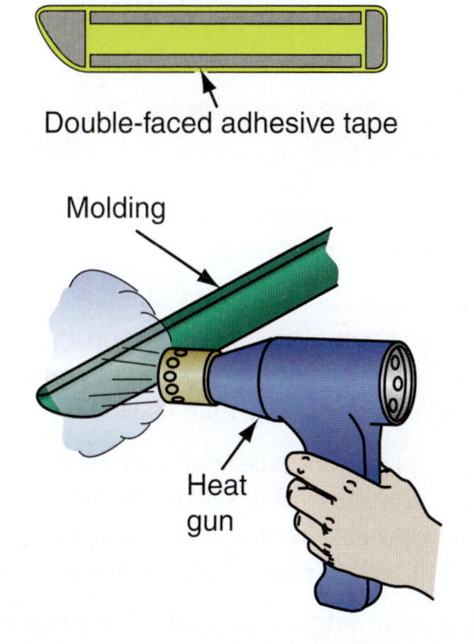

Figure 14–35 A heat gun is sometimes recommended to remove emblems and moldings. It will soften adhesive so the part can be pulled off easily. Do not apply too much heat, however.

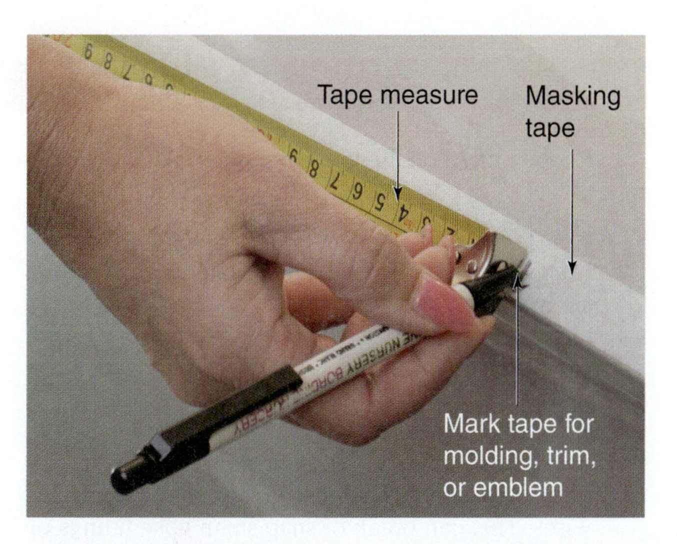

Figure 14–36 If needed, you can apply masking tape as a straightedge for aligning body moldings, trim, and emblems. Tape can be marked to serve as a guide for placing a part correctly on the body.

Begin installing the fender molding ⅛ inch (3.1 mm) from the fender rear edge. Align the molding with the top edge of the tape and lightly press against the panel. Progressively remove the backing and press the molding against the surface and along the edge of the tape. Do not attempt to reposition a piece after it is applied. After the whole length of molding is applied, press along the entire length with the heel of your hand or with a roller. See Figure 14–37.

Repeat this process with the door moldings and the rear quarter panel molding.

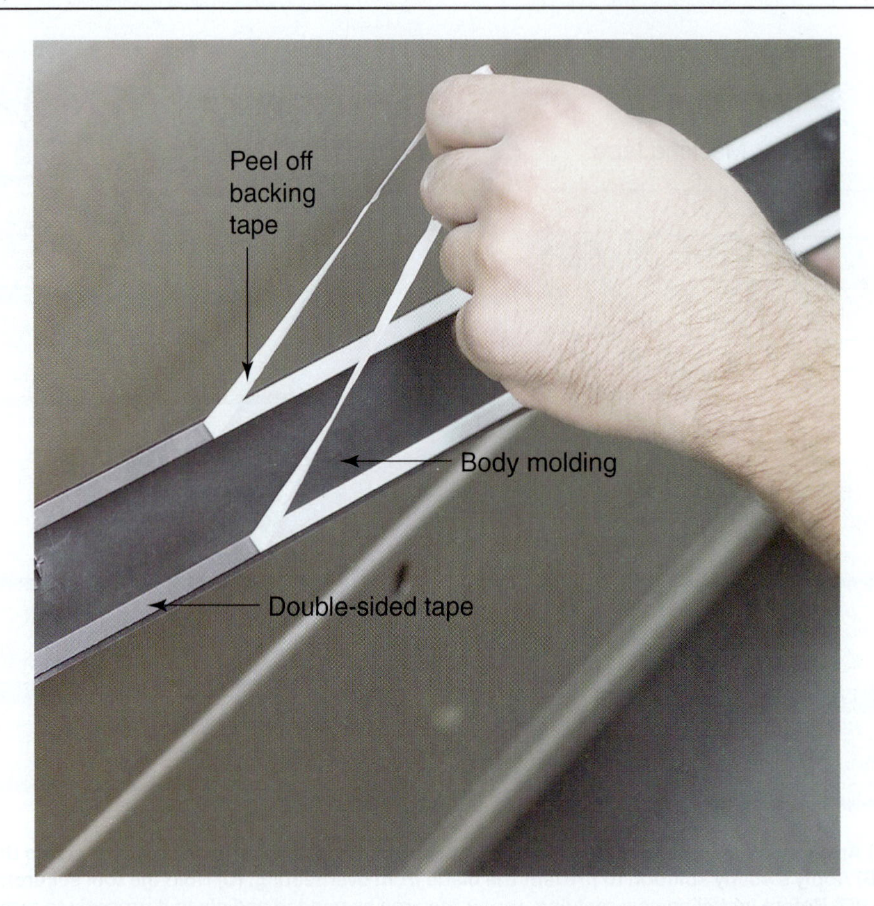

Figure 14-37 Some moldings, trim, and emblems are held on with double-sided tape. Peeling off the plastic tape will expose the adhesive so the part can be bonded to the body. The body surface must be perfectly clean for the tape to adhere properly.

SUMMARY

1. Keep in mind on-the-job experience is the only way to become competent and fast at body part R&R (part removal and replacement).
2. Fastened parts are held on the vehicle by bolts, nuts, screws, clips, and adhesives.
3. If in doubt about how to remove a part, refer to the vehicle's service manual or computerized service information.
4. A common mistake during hood removal is forgetting to disconnect windshield washer hoses and wires. This can break or snap small hose fittings or sever wires.
5. To prevent part damage, have someone help you hold the hood. The hood is the largest adjustable panel on most vehicles. It can be adjusted at the hinges, at the adjustable stops, and at the hood latch.
6. Most vehicles have adjustable hood bumper stops mounted on the radiator support or along the inner edges of the fenders. After first installing the hood, close the hood slowly. If the hood is not centered, it could hit and dent the fenders.
7. Bumpers on many late model cars are covered with urethane or other plastics.
8. Bumpers can be heavy and clumsy. Before removing the last mounting bolts, support the bumper on a floor jack. Get a helper to hold the bumper on the jack as you remove the final fasteners.
9. Major front end damage sometimes requires removal of the large assemblies like the front clip or the bumper-spoiler assembly. This will give you better access to hard-to-reach parts on the assembly.
10. During paint edging, all ends, corners, edges, and sometimes the rear of the fender panel should be scuff sanded, primed, and painted.
11. Weatherstripping is a rubber seal that prevents leakage at the joint between the movable part (lid, hatch, door) and the body. To prevent air and water leaks, the deck lid or door must contact the weatherstripping evenly when closed.
12. Truck beds are usually bolted to a full perimeter frame. To remove the bed, simply remove the bolts that extend up through brackets on the frame.

EXERCISES

On a separate sheet of paper, complete the following learning activities for this chapter. Write definitions for the key terms and answer the ASE-style review questions, essay questions, critical thinking problems, and math problems. You can also do the outside activities, possibly for extra credit.

Key Terms

body shim
bumper energy absorber
fastened part
hood hinges
hood latch
hood prop tool
hood stop
hood striker
hood strut
molding tool

ASE-Style Review Questions

1. Technician A says that many parts that were held with bolts and screws in the past now snap-fit into place. Technician B says that most parts now bond into place. Who is correct?
   A. Technician A
   B. Technician B
   C. Both A and B
   D. Neither A nor B
2. Technician A says that he can remove a hood by himself. Technician B says it is wiser to have a helper when removing or installing a hood. Who is correct?
   A. Technician A
   B. Technician B
   C. Both A and B
   D. Neither A nor B
3. Technician A says that all quarter panels are welded in place. Technician B says that a few makes of cars have bolt-on quarter panels. Who is correct?
   A. Technician A
   B. Technician B
   C. Both A and B
   D. Neither A nor B
4. When starting work on a damaged vehicle, Technician A says to refer to the estimate or work order. Technician B says to refer to factory service information for detailed instructions when in doubt. Who is correct?
   A. Technician A
   B. Technician B
   C. Both A and B
   D. Neither A nor B
5. Technician A says that factory service manuals normally have a body repair section that explains and illustrates how body parts are serviced. Technician B disagrees. Who is correct?
   A. Technician A
   B. Technician B
   C. Both A and B
   D. Neither A nor B
6. Technician A says if the hood will be reused, the hinge alignment should be marked before removal. Technician B says this is not necessary with major front end damage because the hood will be replaced. Who is correct?
   A. Technician A
   B. Technician B
   C. Both A and B
   D. Neither A nor B
7. Technician A says that hood adjustments are made at the hinges, at the adjustable stops, and at the hood latch. Technician B says that hood adjustments are made at the cowl. Who is correct?
   A. Technician A
   B. Technician B
   C. Both A and B
   D. Neither A nor B
8. A new hood is sticking up above the fenders. Technician A says to check the hood stops. Technician B says to check the radiator support. Who is correct?
   A. Technician A
   B. Technician B
   C. Both A and B
   D. Neither A nor B
9. A hood is being removed for replacement. Technician A says to check for bent hinges. Technician B says to use a hood prop tool as you remove the hood shocks. Who is correct?
   A. Technician A
   B. Technician B
   C. Both A and B
   D. Neither A nor B
10. Technician A says that new vehicle bumpers are designed to withstand moderate impacts without damage. Technician B says that bumpers are designed to protect the front and rear of a vehicle from damage during a low speed collision. Who is correct?
    A. Technician A
    B. Technician B
    C. Both A and B
    D. Neither A nor B

Essay Questions

1. In your own words, summarize how to remove a hood.
2. What cautions must be observed when removing bumpers with energy absorbers?

Critical Thinking Problems

1. If in doubt, how would you find out how to remove a front fender?
2. You have uneven gaps on each side of a new hood. How would you correct the problem?

Math Problem

1. When adjusting headlights, you find a front curb height measurement to be 12.6 inches (320 mm). Specs call for a minimum ride height of 13.2 inches (335 mm). How much is the vehicle out of spec?

Activities

1. Inspect several vehicles. Determine whether their bolt-on panels are adjusted properly. Is the gap between parts equal all the way around all panels? Make a report of your findings.
2. Using a service manual, summarize the procedures for removing a hood and fender from three different makes of cars and trucks. Write a report on how these procedures vary.
3. After getting instructor permission, loosen the bolts on a hood latch and hood. Move the hood and latch out of alignment. Adjust the hood and latch.

Door, Roof, and Glass Service

CHAPTER 15

OBJECTIVES

After reading this chapter, you should be able to:

- Describe windshield glass replacement procedures.
- Compare different methods used to secure windshield glass.
- Remove, replace, and adjust door assemblies.
- Describe how to service both manual and power window regulators.
- Summarize door glass replacement and adjustment.
- Properly replace rearview mirror glass and heating elements.
- Explain how to replace and repair vinyl roofs.
- Answer ASE-style review questions relating to glass, trim, and other service operations.

INTRODUCTION

This chapter will give you the general knowledge needed to service the glass in any make or model vehicle. By referring to factory service information and specifications, you will be well prepared to do competent glass replacement.

The chapter also explains how to service doors. You will learn how to remove, replace, and adjust door hinges and door latch mechanisms. The service of the inner door trim panel, window regulator, and related parts is explained in detail. Roof panel, convertible top, and vinyl roof service is also summarized.

In today's body shops, auto body repair technicians are frequently required to service vehicle glass. Technicians often service door glass and fastened stationary side glass. However, most body shops "farm out," or hire a glass technician to replace, windshields and rear passenger compartment glass. Passenger side air bag deployment frequently cracks the windshield, and the broken glass must be removed before doing repairs.

WARNING There are a wide variety of methods used to service the parts discussed in this chapter. Always refer to a manufacturer's service manual or their electronic service data when in doubt. Vehicle-specific service instructions with exact specifications are needed to do competent service of doors, roof panels, and glass.

15.1 VEHICLE GLASS TECHNOLOGY

Glass is a transparent substance manufactured by heating a mixture of sand, soda (sodium carbonate), limestone, and other materials to a temperature of about 2,400°F (1,300°C). Today's vehicles are built with a lot of glass for greater visibility. Frequently this glass is broken out or cracked as a result of a collision, air bag deployment, flying gravel, or vandalism.

DANGER Stationary glass strengthens the body structure and helps contain passengers during a serious collision. If you fail to install glass properly, you could be lowering the structural integrity of the vehicle and endangering the driver and passengers.

Glass is sometimes considered a structural component of the vehicle, especially the windshield and back glass. It is important for the body shop technician to be familiar with the various techniques to remove and install vehicle glass properly. Broken glass must also be removed before doing major structural repairs to the body.

Types of Glass

There are two types of glass used in today's vehicles: laminated and tempered. Both are considered safety glass. They may or may not be tinted (Figure 15–1).

Laminated plate glass consists of two thin sheets of glass with a thin layer of clear plastic (vinyl) between them. It is used to make all windshields and some side glass. The plastic or vinyl material is usually clear to provide an unimpeded view from all angles.

When laminated glass is broken, the plastic material helps to hold the shattered glass in place and prevent it from causing injury (Figure 15–2).

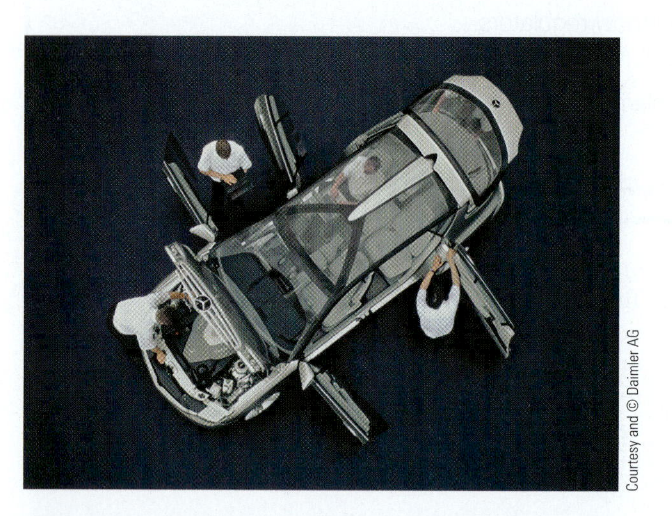

Courtesy and © Daimler AG

Figure 15–1 The doors and glass are important structural members in a modern vehicle. They must be serviced properly for the driver and passengers to be safe while driving.

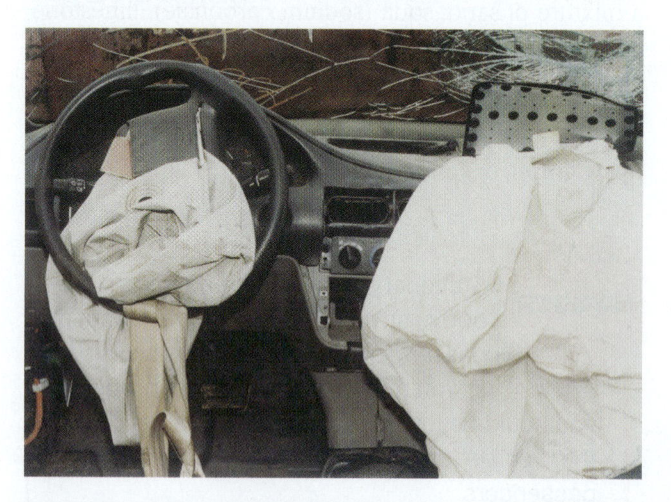

Figure 15–2 Lamination kept this windshield glass intact upon impact. Tempered glass would have shattered into small pieces. Air bag deployment often shatters windshield glass.

Tempered glass is used for side and rear window glass but rarely for windshields. It is a single piece of heat-treated glass and has more resistance to impact than regular glass of the same thickness. Unlike laminated glass, tempered glass shatters into tiny pieces when broken. The pieces of glass are small and have a granular texture.

Antilacerative glass is similar to conventional multilayered glass, but it has one or more additional layers of plastic affixed to the passenger side of the glass. This glass is used in the front windshield only and provides added protection against shattering and cuts during impact.

15.2 GLASS SERVICE

Outside *glass specialty shops* are often hired to install windshields and back glass. They have glass technicians who are specially trained to install bonded and rubber gasket-held glass on all makes and models of vehicles. However, some large shops do their own glass work, including windshield and back glass replacement.

Windshields and rear windows are usually secured in place by rubber weatherstripping or by an adhesive. Generally, moldings are used on the interior and exterior of the body around the glass opening.

Interior moldings around glass are called *garnish moldings*, and exterior moldings around glass are called *reveal moldings*. These moldings, the windshield wipers, cowl grille, and related parts often must be removed before the windshield will come out (Figure 15–3).

One of the first steps in glass service is to protect the vehicle's interior. Place seat covers over the cushions and fender covers over the dash as needed. To keep glass bits or slivers out of the defrost ducts, cover them with duct tape.

The replacement of a windshield and a rear window follow almost identical procedures, varying slightly for different makes of vehicles.

Replacement of windshield glass involves two different methods based on the materials used: rubber gasket installations or adhesive installations. The adhesive-type installation is further refined into two additional methods: the full cutout and the partial cutout method.

The *partial cutout method* takes advantage of the fact that if most of the adhesive is in good condition and of sufficient thickness, it can be utilized as a base for the application of new adhesive. When the original adhesive is defective or requires complete removal, the *full cutout method* must be used.

The gasket installation was more predominant in older vehicles but still finds use in present-day, high-end

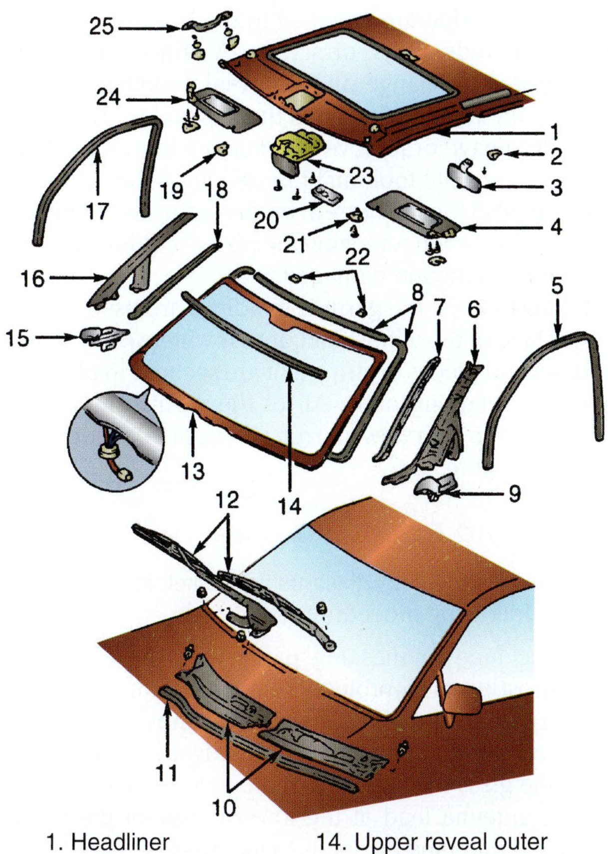

1. Headliner
2. Mirror cover
3. Rear view mirror
4. Sun visor
5. Front door opening trim
6. Front pillar garnish
7. Side reveal outer molding
8. Dam
9. Instrument side panel
10. Cowl louver
11. Weatherstrip
12. Wiper arm
13. Windshield glass
14. Upper reveal outer moulding
15. Instrument side panel
16. Front pillar garnish
17. Front door opening trim
18. Side reveal outer molding
19. Holder
20. Lens
21. Holder
22. Stopper
23. Map light assembly
24. Sun visor
25. Assist grip

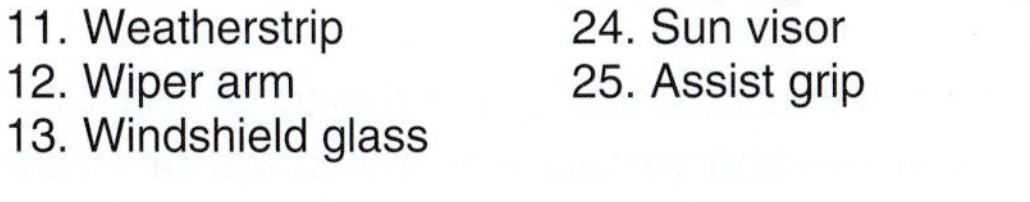

Courtesy of Mitchell 1

Figure 15–3 **Study the typical parts relating to windshield service.**

(expensive) vehicles. The gasket is grooved to accept the glass, the sheet metal pinch weld flange, and sometimes the exterior reveal molding (Figure 15–4).

DANGER

Litigation has resulted from improper installation of windshields. The windshield is very important to the protection of drivers and their passengers. You must install all windshields according to factory recommended procedures to avoid liability problems.

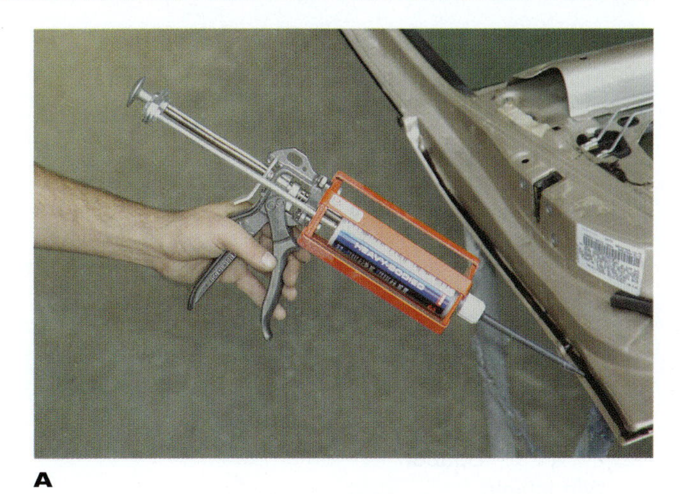

A

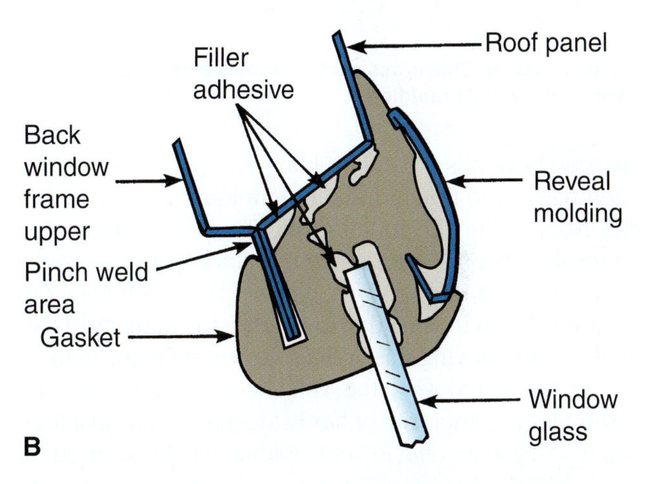

B

Figure 15–4 **This windshield glass is secured with a rubber gasket. (A) A soft rubber gasket locks around the windshield glass and body structure. No adhesive is required with some designs. (B) This cutaway view shows how rubber gaskets secure windshield glass.**

Removing Windshield Molding

The windshield and rear glass on many older model vehicles and some late model vehicles are held in place by a locking strip of rubber or metal. The locking strip fits into a groove between the glass and the pinch weld flange. The strip forces the rubber gasket tightly against the glass and the flange. This locking strip must be removed *before* the glass can be removed from the opening.

The adhesive-type installation, as the name implies, uses an adhesive material to secure the glass in place. The use of adhesive permits the windshield to be mounted flush with the roof panel, decreasing wind drag and noise. Adhesive-bonded windshields also increase the overall rigidity of the vehicle, minimizing body twist and helping to keep the glass in place during a collision. Rubber stops and spacers often separate the glass from the metal with adhesive installation (Figure 15–5).

Before removal of the windshield glass or rear window, the interior and exterior moldings must be removed. An older, but very common, metal reveal molding is made of polished aluminum. Millions of vehicles still on

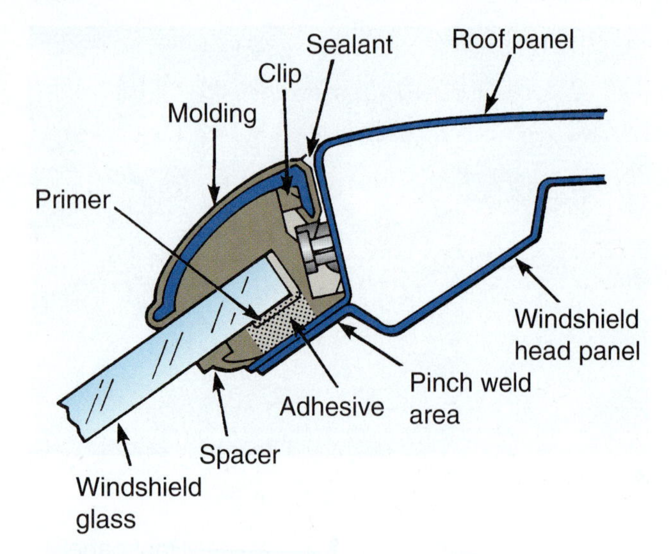

Figure 15-5 This cross section shows a bonded windshield with its molding.

the road have this type of molding. A *clip removal tool* is often needed to reach under the molding to release the clip from under the molding lip. If unavailable, a clip removal tool can be fabricated from banding strap steel.

With late model vehicles, a cowl grille often clips on to cover the windshield wipers and mechanism. One method of removing this grille is shown in Figure 15–6.

On the exterior of the vehicle, remove the reveal moldings and other trim or hardware (such as windshield wiper arms) if needed. Reveal molding can be secured in

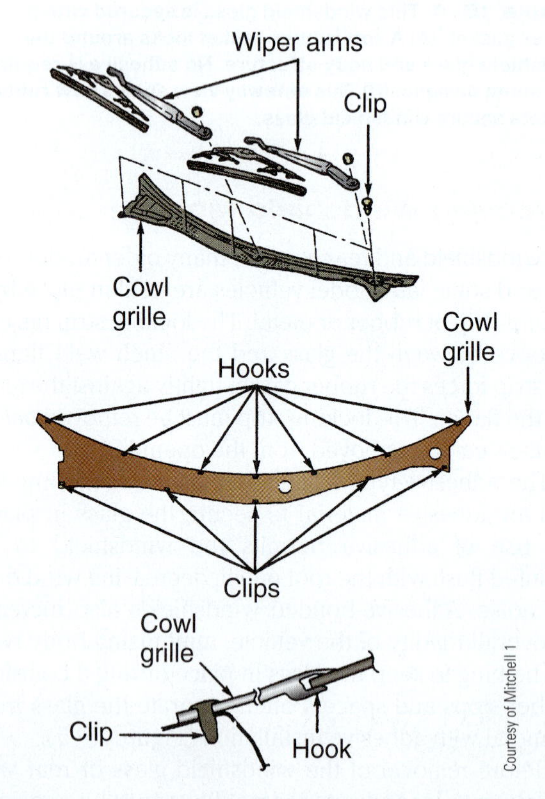

Figure 15-6 During windshield service, you must remove the wipers and sometimes a cowl grille. Clips or screws normally secure the grille.

place by clips that are attached to the body opening by welded-on studs, bolts, or screws. A projection on the clip engages the flange on the reveal molding, thereby retaining the molding between the clip and body metal.

To disengage or remove the molding from the retaining clips, a special tool might be required. Reveal moldings can also be anchored in the adhesive material. Exercise care when removing the reveal moldings so that they are not bent and damaged.

In most cases, the garnish moldings are used on the interior face of the windshield or rear window. They consist of several pieces or strips that are secured in place by screws or retaining clips. All of the garnish moldings, as well as the rearview mirror (if possible), should be removed (Figure 15–7).

Windshield Rubber Gasket Service

To replace windshield glass using gasket material, perform the following procedure:

1. Place tape or masking paper over the dash and defrost vents to protect them from damage and to keep debris from falling in the vent holes.
2. Be sure all moldings, trim, and hardware are removed.
3. If the glass has a built-in radio antenna, disconnect the antenna lead at the lower center of the windshield and tape the leads to the glass.
4. Locate the locking strip on the outside of the gasket. Pry up the tab and pull the tab to open the gasket all the way around the windshield glass.
5. Use a putty knife to pry the rubber channel away from the pinch weld inside and outside of the vehicle.
6. With an assistant, push out the windshield glass and gasket.
7. Clean the windshield body opening with an acceptable solvent to clear the area of dirt or residual sealant.
8. If the glass was not cracked and is to be reused, do not exert uneven pressure to the glass or strike it with tools. The technician should always wear safety goggles and gloves when replacing windshield glass or rear window glass—broken or not.
9. Place the removed glass on a suitable bench or table that is covered to protect the glass. If the glass was removed to accommodate body repairs, leave the gasket and moldings intact. If the glass was replaced because it was broken, remove the associated moldings and gaskets from the glass.
10. Cracks that develop in the outer edge of the glass are sometimes caused by low or high spots or poor spot welds in the pinch weld flange. Examine the pinch weld and correct the problem if applicable.
11. Apply a double layer of masking tape around the outside edge of the glass with ¼-inch (6.4 mm) overlap onto the inside of the glass. This will prevent chipping or breaking the glass.
12. Install stop blocks and spacers. If the original blocks are not available, cut pieces of used gasket for blocks.

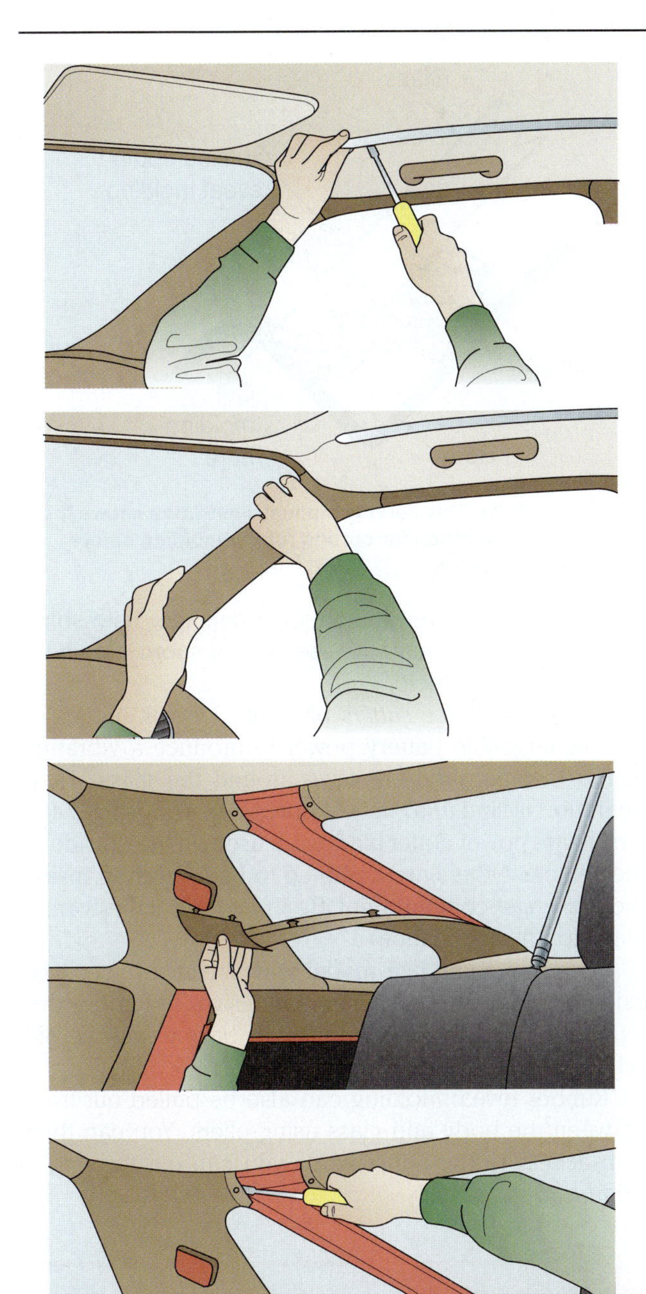

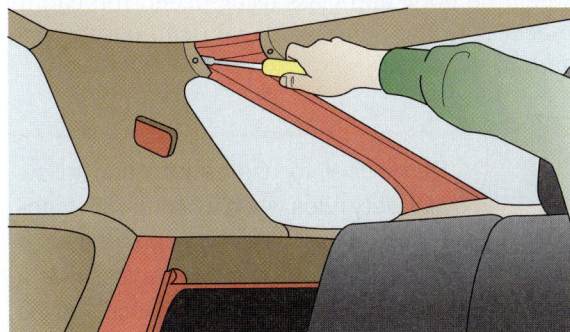

Figure 15-7 Interior garnish molding must sometimes be removed during glass service.

13. Carefully install the glass on the blocks. Center the glass and then check the gap between the glass and the pinch weld. The gap should be even around the entire pinch weld. Remove the masking tape around the edges of the glass.
14. Apply a bead of approved sealer in the glass channel and install the gasket on the glass.
15. Insert a cord (vinyl or nylon type) in the pinch weld groove of the gasket. Start at the top of the glass. The cord ends should meet in the lower center of the glass. Tape the ends of the cord to the inside surface of the glass. Squirt a soapy solution in the pinch weld groove to ease installation (Figure 15–8).

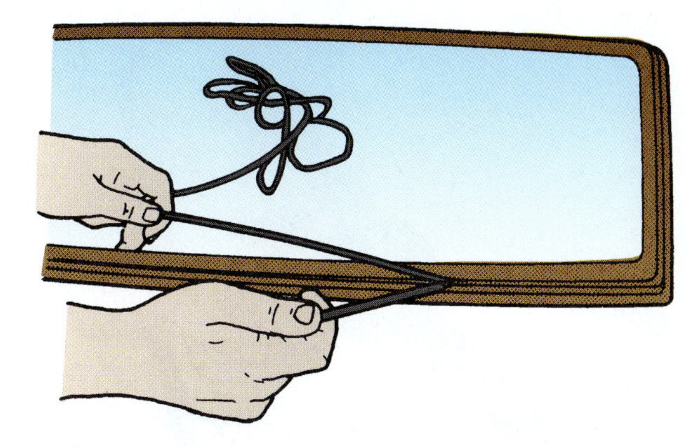

Figure 15-8 A cord in the pinch weld opening can be used to work the rubber gasket and windshield into position during installation.

16. Apply factory-recommended sealer to the base of the gasket.
17. With the aid of an assistant, install the glass and gasket assembly in the body opening and center it. Slip the bottom groove over the pinch weld.
18. Very slowly pull the cord ends so that the gasket slips over the pinch weld flange. Work the bottom section of the glass in first, then do the sides, and finally the top section. Be sure to work the sections evenly, because the glass might crack if the cord end is pulled from one side only (Figure 15–9).
19. Apply a small bead of sealer around the body side of the gasket.
20. Remove excess sealer with a suitable solvent that will not harm the paint.
21. Install the reveal and garnish moldings.
22. Check the windshield for water leaks using a low-pressure stream of water. Start at the bottom and

Figure 15-9 A hooked tool can be used to work any stuck part of the rubber gasket over the lip for complete installation.

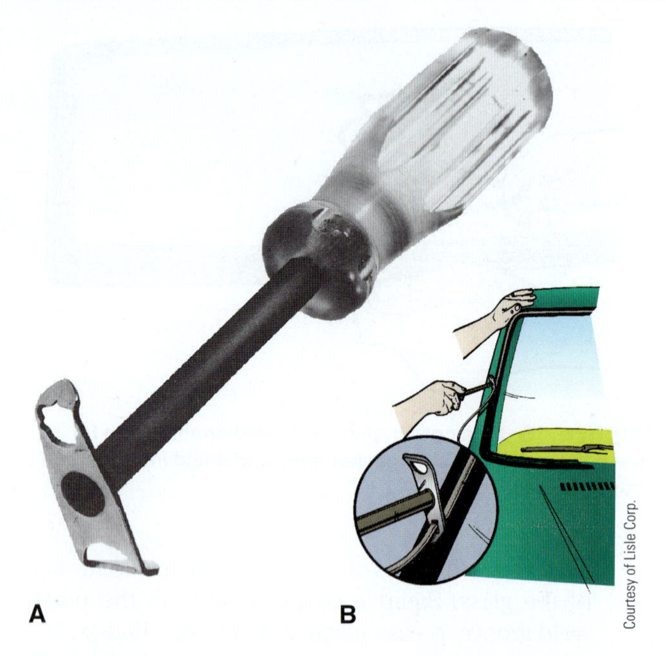

Figure 15-10 **(A) A windshield locking strip tool and (B) installing a locking strip.**

slowly work your way up each side. Do the top last to help isolate the location of the leak.

23. Place a soapy solution in the locking strip groove and, with a tool designed for the job, replace the locking strip. The wedge-shaped tool shown in Figure 15–10 spreads the groove and feeds the strip into the opening. The soapy solution lubricates the groove and makes it easier to slide the tool through the rubber groove.

NOTE *It is critical that the shear strength and tensile strength of the adhesive be within specifications. For this reason, most technicians use the windshield adhesive recommended by the manufacturer. Then there is no doubt that the adhesive is strong enough to properly secure the windshield.*

Glass Adhesive—Full Cutout Method

This method involves the complete removal of the old adhesive sealer. To remove the glass, all of the old adhesive must first be cut. Several devices are available to do this: a steel wire; a hot knife; a pneumatic knife; an electric knife; or a cold, fine sharp knife. Each device has advantages and disadvantages. The pneumatic knife with a thin steel blade is preferred by many technicians.

A 3-foot (914 mm) length of single-strand steel piano wire (smallest diameter available) is the safest to use to prevent glass breakage. The hot or cold knife can crack the glass in the areas where the reveal molding clips are very close to the glass.

Pneumatic windshield cutters use shop air pressure and a vibrating action to help cut the adhesive around the glass. There are a variety of blade designs available for use on different windshields. Many have blades with

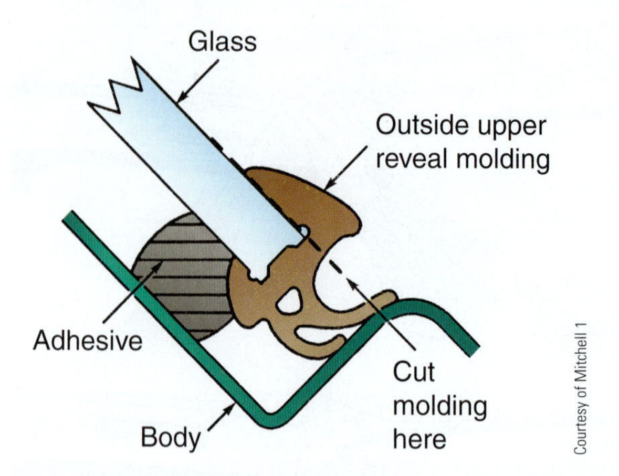

Figure 15-11 **This service manual illustration shows the recommended method for cutting rubber molding before cutting out adhesive.**

depth stops to prevent pinch weld damage. With some designs, you may have to use two or more blades to remove the windshield.

Electric adhesive cutters plug into 120-volt outlets or use rechargeable battery power to produce a vibrating action that cuts the adhesive around the glass. Their operation is similar to that of pneumatic windshield cutters. Some power cutter blades are used on the outside of the vehicle. Others are designed to be used from inside the passenger compartment. Be sure to select the correct blade for the application.

Note that some late model windshields are secured using both a rubber gasket and adhesive. To remove this type of windshield, you will usually have to cut the old rubber gasket, as shown in Figure 15–11.

Rubber reveal molding can also be pulled out from between the body and glass using pliers. You can then gain access to the adhesive so you can cut through it (Figure 15–12).

DANGER Remember to use caution and follow all safety rules when working with automotive glass. Wear eye protection to guard against flying bits of glass. Wear leather gloves to prevent cuts. Plastic gloves should be worn to keep adhesives off your skin.

Use the following procedure to replace an adhesive-bonded windshield:

1. If a steel wire is used to remove the glass, soften up the adhesive by using a heat gun. Cut excessive adhesive from the glass edge to the pinch weld with a sharp knife. Attach one end of the wire to a wooden handle. Force the other end of the wire through the adhesive and under the bottom of the glass. Also attach this end to a wooden handle. With one technician inside the vehicle, work the wire back and forth to cut through the sealant (Figure 15–13).

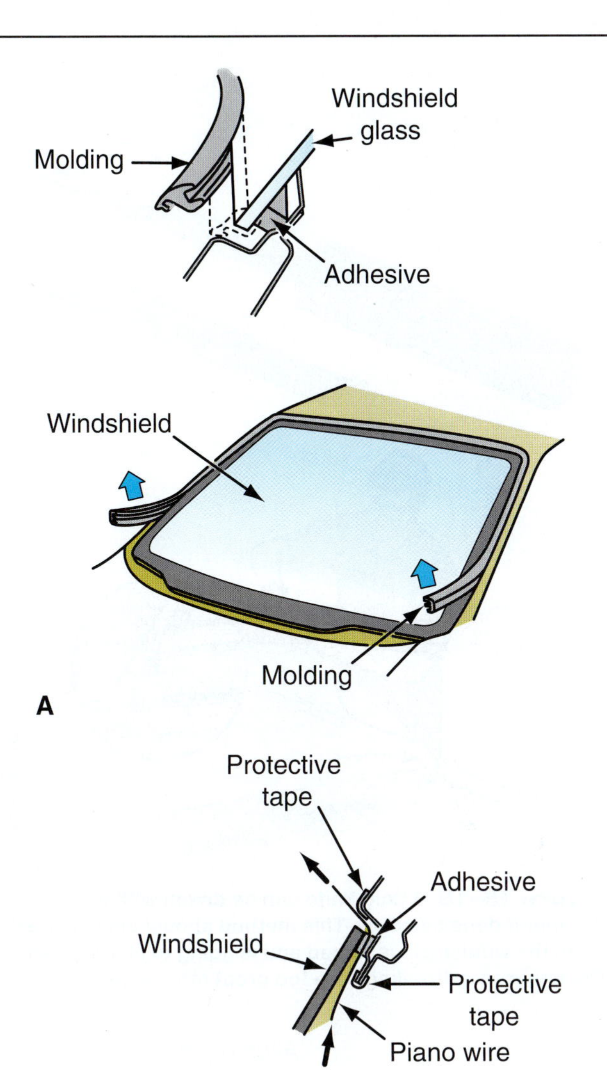

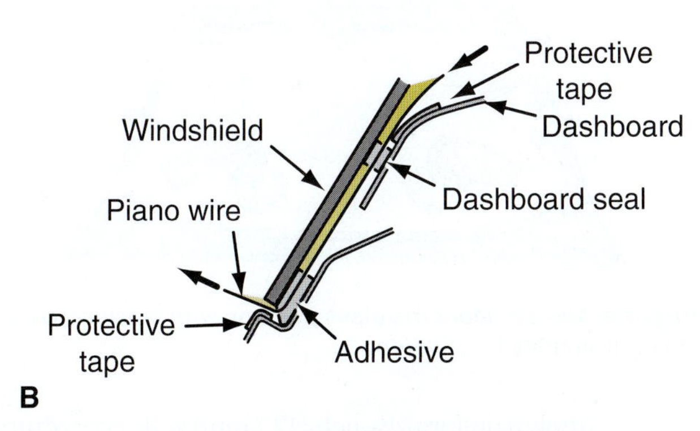

Figure 15-12 Note the technique for removing the windshield on this vehicle. (A) Molding can be pulled out of the slot between the glass and vehicle body. (B) This cutaway view shows how to cut through windshield adhesive. Note the use of protective tape.

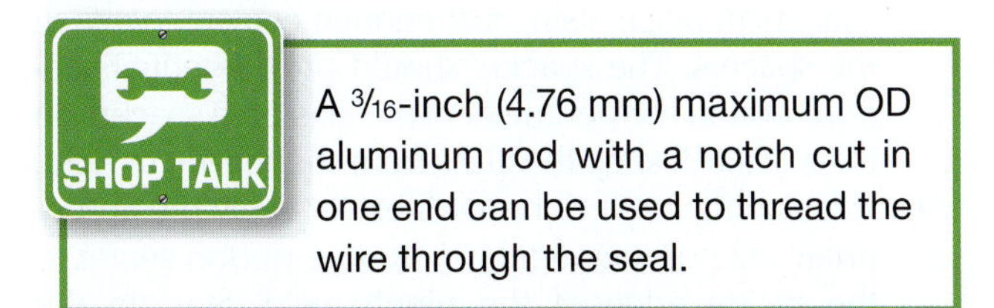

A 3/16-inch (4.76 mm) maximum OD aluminum rod with a notch cut in one end can be used to thread the wire through the seal.

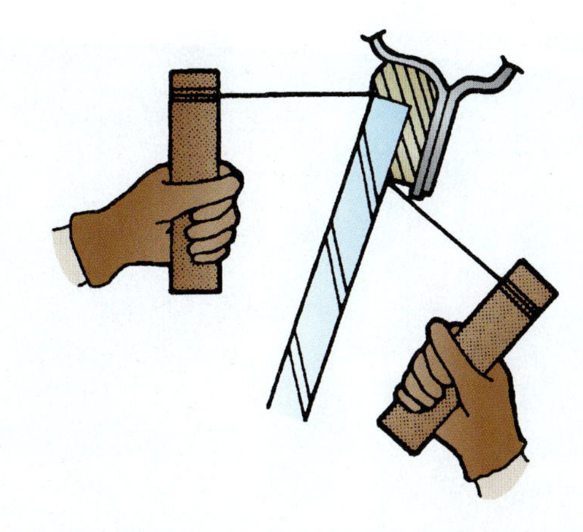

Figure 15-13 You can cut through windshield adhesive with piano wire. Be extremely careful not to cut your hand!

2. If a hot knife, such as the one shown in Figure 15–14, is used to remove the glass, cut excessive adhesive from the edge of the glass to the pinch weld. Insert a hot knife in the adhesive and keep it as close to the glass as possible. Cut around the entire perimeter of the glass. To cut the adhesive at the corners of the glass, move the handle of the tool as close to the corner as possible. Then rotate the tool to cut the adhesive seal. Be careful not to twist the blade of the knife, because it will break. Use wedges (wooden, plastic, and so on) if the adhesive tends to reseal itself after being cut. A power knife is handy along the bottom of the windshield where it is hard to reach the old adhesive (Figure 15–15).
3. If a cold knife is used to remove the glass, cut excess adhesive from the edge of the glass to the pinch weld. Soften up the adhesive by using a heat gun.

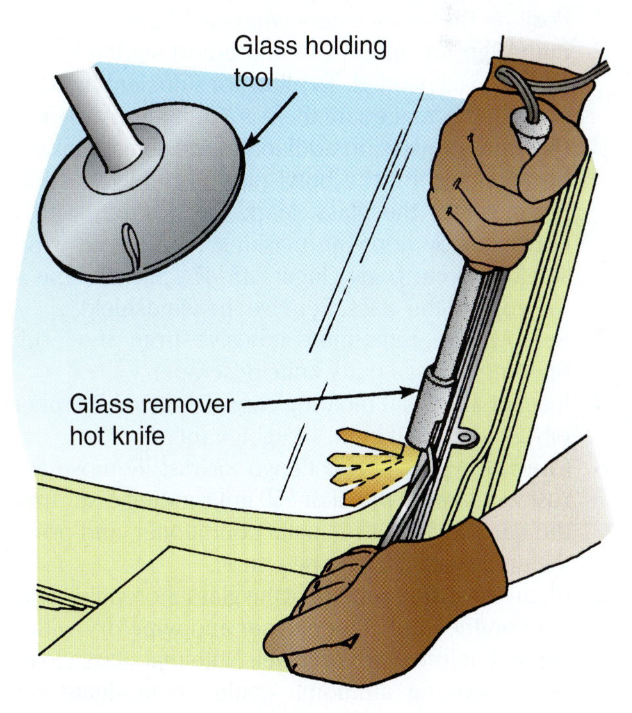

Figure 15-14 Windshield adhesive can also be cut with a hot knife.

Figure 15-15 This technician is using a power knife to cut out a windshield for replacement.

Insert the knife and pull it carefully through the sealant. Tip the knife slightly so that the forward edge of the blade scrapes along the glass surface. Cut around the entire perimeter of the glass. Sharpen the knife blade as required.

Figure 15–16 shows a cold knife designed to be driven with an air hammer. This knife cuts through even tough urethane sealants with ease.

4. When the adhesive has been cut, remove the glass and place it in a safe area if it is to be reused. If the glass is damaged, remove as required and discard. Be sure to wear safety goggles and gloves when handling glass.
5. Position the replacement windshield into the opening. Align for uniform fit and adjust setting blocks (spacers) as needed. To allow for sufficient bonding of urethane, make sure there is a minimum of ¼-inch (6.4 mm) of glass, in addition to the space that will be taken up by the butyl tape around the entire perimeter of the glass. Mark the position with a crayon or by applying masking tape to the windshield and car body (Figure 15–17). Slit the tape at the edge of the glass. Remove the windshield.
6. Remove the remaining adhesive from the body opening, using a putty knife or scraper.
7. Inspect all reveal molding clips. Replace all broken or rusted clips; if bent, straighten them.
8. Check the pinch weld flange for rust. Remove any rust with a wire wheel or #50 grit sanding disc. Treat the bare metal with a metal conditioner, and prime the areas with a urethane primer.
9. Clean the inside surface of the glass thoroughly with a recommended glass cleaner and wipe dry with a clean, lint-free cloth or towel. Note that glass cleaners containing ammonia could contaminate surfaces to be painted. Use a recommended product line so that all materials are compatible.

Figure 15-16 A cold knife can be driven with an air hammer if done carefully. This method should only be used when the windshield is ruined and is going to be replaced. The chances of breakage are too great otherwise.

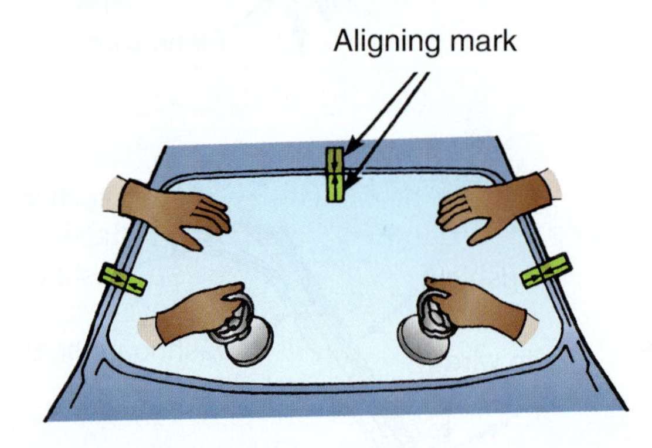

Figure 15-17 Mark the glass position with masking tape. This will simplify installation.

Apply a uniform, ½-inch (12.7 mm) wide coat of urethane primer to the inside edge of the glass (Figure 15–18). Allow primer to dry a few minutes. (See manufacturer's instructions for suggested drying time.)

10. Ensure that the glass supports or spacers are in place. Install new ones if necessary. Cement the flat rubber spacers in place, using just enough cement to attach the spacers. The spacers should provide equal support around the perimeter of the glass; the spacers on the sides will keep the glass from shifting left or right.
11. If replacing butyl ribbon adhesive, apply the appropriate size of rectangular adhesive ribbon sealer to the inside edge of the pinch weld. Start in the

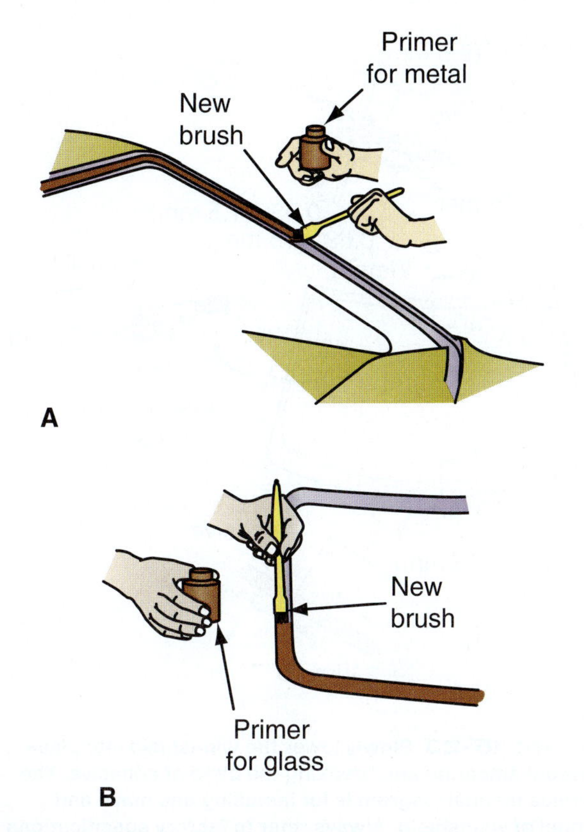

Figure 15-18 Primers are often recommended to increase adhesion. (A) Body primer is applied to the inner lip. (B) Glass primer is applied to the area that accepts adhesive.

bottom center of the window opening to help avoid leakage. Do not stretch the strip of sealer. Cut the ends at a 45-degree angle and butt together.

Figure 15–19 shows an air-powered caulk or adhesive gun. It will save time and help form a continuous bead of adhesive.

12. Apply a bead of urethane sealant around the glass or the perimeter of the pinch weld flange. Cut the cartridge nozzle at a 45-degree angle. The opening should produce a bead size slightly larger than the ribbon sealer. Apply the sealant directly behind the ribbon sealer dam on the pinch weld. Do not apply sealant on antenna lead wires, as shown in Figure 15–20.

Courtesy of Equalizer Industries Inc.

Figure 15-19 This is a powered caulk gun. It will speed and improve the application of adhesive.

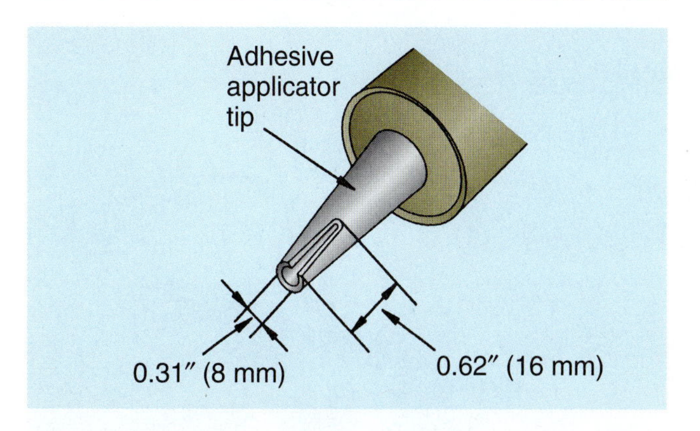

Figure 15-20 Always apply urethane adhesive following service manual directions. Cutting a notch in the tip of the adhesive tube will help form the proper bead of adhesive.

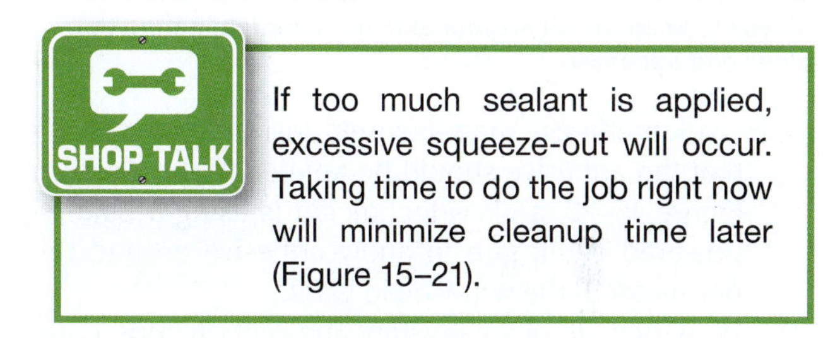

If too much sealant is applied, excessive squeeze-out will occur. Taking time to do the job right now will minimize cleanup time later (Figure 15–21).

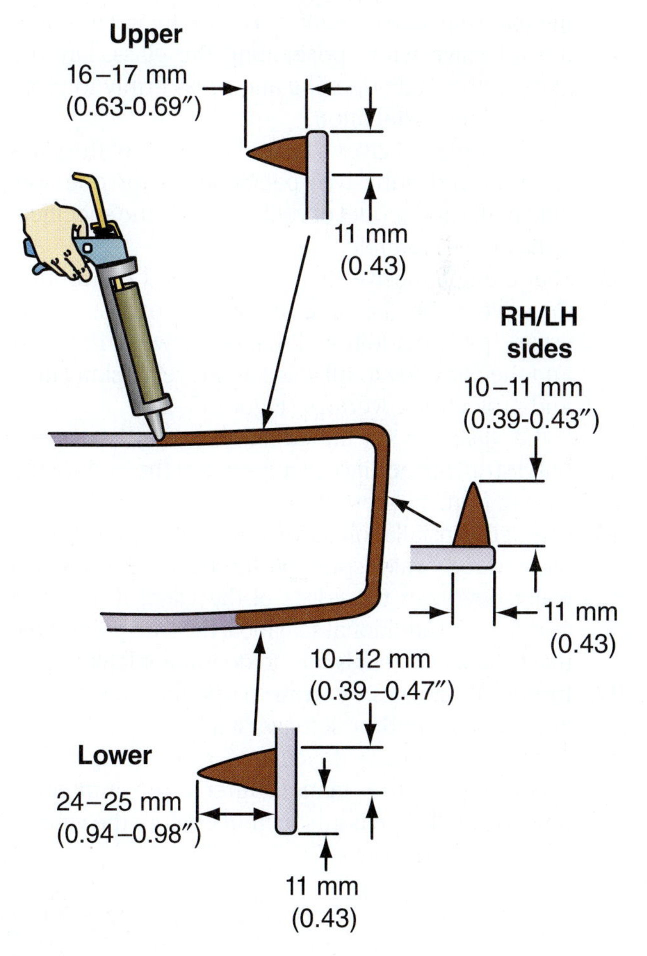

Figure 15-21 Here is a typical service manual illustration giving specifics for applying glass adhesive.

Figure 15-22 **A technician uses an electric caulk gun to apply adhesive to windshield glass. You must move the caulk gun consistently at the correct speed to produce a good bead of adhesive. Never touch the cleaned glass and wear gloves to keep the oil on your skin from contaminating the glass and adhesive.**

Sometimes, manufacturer instructions specify that the adhesive should be applied to the glass. In Figure 15–22, a glass technician is using a battery-powered caulk gun to apply adhesive around the perimeter of the windshield glass.

13. With the help of an assistant and suction cups, carefully position the glass in the body opening, using the masking tape as a guide. Be careful not to smear the adhesive when positioning the glass. Lay the glass in the body opening and press firmly to properly seal the installation.

    Figure 15–23 gives a typical example of the glass location and adhesive specifications for one specific make and model of vehicle. Study the specifications in this example.
14. Shape any adhesive that has squeezed out around the edge of the glass and remove any excess. If necessary, paddle additional sealant between the glass and the car body to fill voids. Remove masking tape and protective coverings (Figure 15–24).
15. If the glass has an imbedded antenna that uses a butyl strip, put additional adhesive at the ends of the strip to form a watertight seal.
16. Check the installation using a soft water spray. Do not use a direct water spray on the fresh adhesive. Let water flow over the edges of the glass. If a leak is found, apply additional sealant at the leak point. Note that Chapter 16 details how to do a water leak test.
17. Install all necessary trim parts and attach the antenna lead and/or defogger lead.
18. Allow the adhesive to cure at room temperature according to the manufacturer's recommended time, typically 6–8 hours or more, before the vehicle is returned to its owner.

Glass Adhesive—Partial Cutout Method

If the partial cutout method is to be used to install the glass, thoroughly inspect the remaining adhesive first, before attempting the procedure. There must be sufficient adhesive remaining in the pinch weld to give adequate clearance between the body and the glass. The remaining adhesive must also be tightly bonded to the pinch weld to provide a good base for the new adhesive.

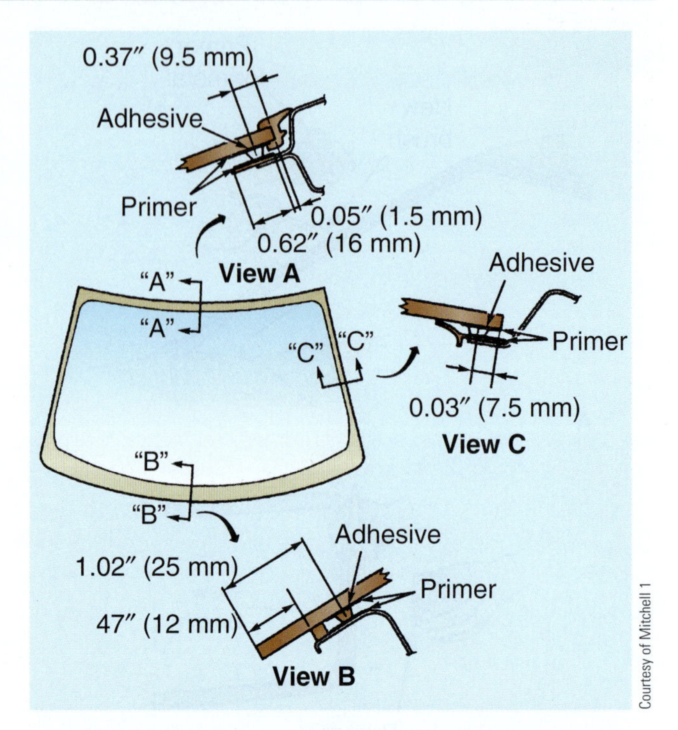

Figure 15-23 **Slowly lower the windshield into place without smearing and breaking the bead of adhesive. The service manual diagram is for installing one make and model of windshield. Always refer to factory specifications when in doubt.**

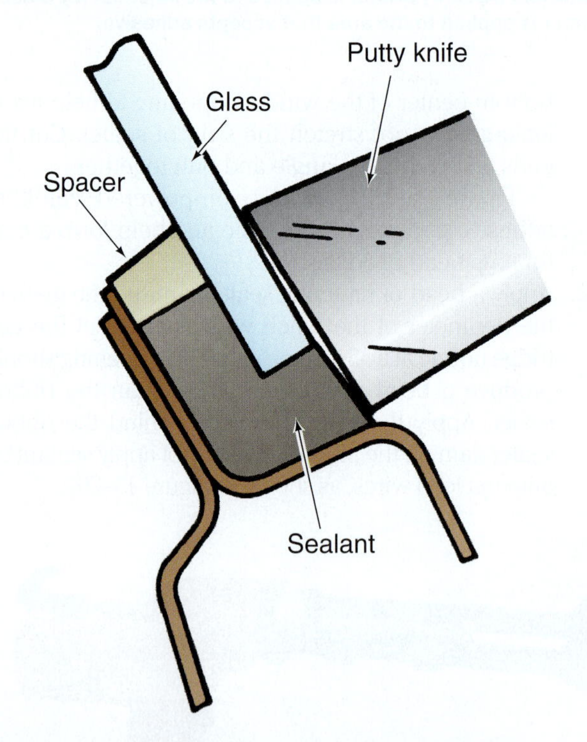

Figure 15-24 **Smooth the glass adhesive with a putty knife.**

If the original glass is to be reinstalled, you must remove all traces of adhesive from the glass. Use either denatured alcohol or an approved solvent to clean any residual adhesive from the edge of the glass.

To replace a windshield using the partial cutout method, do the following:

1. Place protective coverings on the vehicle to prevent damage to the paint or interior.
2. Remove windshield wiper arms, trim, antenna, and so on to expose the entire perimeter of glass.
3. Using a utility knife, make a cut into the existing urethane sealant around the entire perimeter of glass. Cut as close to the edge of the glass as possible.
4. Use a cutout knife or piano wire to cut out the glass, keeping the tool as close to the edge of the glass as possible. Remove the windshield. Trim any high spots on the urethane bed to ensure a flat surface. The remaining adhesive should be approximately 3⁄32-inch (2.38 mm) thick.
5. Inspect the reveal molding clips for damage. Replace any clips if necessary.
6. Select an adhesive that will be compatible with the adhesive used on the body pinch welds. Refer to manufacturer recommendations for the type and amount of adhesive to use.
7. Replace the lower glass supports or spacers where applicable.
8. With the help of an assistant, position the glass in the body opening. Ensure that the gap is equal on both sides and that there is ample clearance on the top. Lower or raise the lower supports or spacers as required to get the correct placement of the glass.
9. Apply two pieces of masking tape from the bottom portion of the glass to the body about 6–8 inches (152–203 mm) in from the corner. Repeat this procedure at the top of the glass.

   Use a razor blade or knife to cut the masking tape strips and remove the glass. The tape strips will help to align the glass when reinstalling it.
10. Using a clean, dry, lint-free cloth, clean the surface of the urethane sealer remaining on the pinch weld. Replace the butyl tape strip in the antenna area with a new piece of butyl tape.
11. Clean the inside surface of the windshield thoroughly with an ammonia-free, noncontaminating glass cleaner. Wipe dry with a clean, lint-free cloth or towel. Apply a uniform ½-inch (12.7 mm) wide coat of urethane primer to the inside edge of the glass. Allow the primer to dry 3–5 minutes. Also, apply adhesive primer to the existing or remaining adhesive.
12. If the windshield contains an antenna, place a piece of butyl tape about 8 inches (203 mm) from the antenna pigtail. Do not use urethane or primer near the pigtail, because it will interfere with radio reception.
13. Apply the new, manufacturer-recommended adhesive directly over the existing adhesive.
14. Apply masking tape about ¼ inch (6.4 mm) from the outer edge of the inside of the glass on the top and both sides. This will aid the cleanup process when the glass is installed. Apply a smooth bead of the adhesive around the outer end of the glass or to the pinch weld.
15. With the help of an assistant, install the glass into the body opening. Place the glass on the lower supports or spacers with the masking tape strips properly aligned.
16. Open the vehicle front doors. Place one hand inside the opening and gently lay the glass in position. Use suction cups to control glass movement. An alternate method is to rest the glass on the lower supports or spacers. Then, one technician can go inside the vehicle and help lay the glass in position.
17. Firmly press the glass in place to set the adhesive material.
18. If adhesive was placed in the pinch weld, a dark line in the glass will indicate a sealed area. The dark line should be completely around the glass. Any light spots that appear indicate improper sealing.
19. If cartridge-type adhesive was used, the adhesive can be smoothed out along the edge of the glass.
20. Check the installation using a fine water spray. Do not use a direct flow of water on the fresh adhesive. Correct leaks by adding additional adhesive in the applicable areas.
21. Install any necessary trim and moldings and connect the antenna and/or defogger pigtails as applicable.
22. Clean excess adhesive from the glass area or body.
23. Allow the adhesive to cure for 6–8 hours before moving the vehicle.

Windshield Wiper Service

The windshield wipers must often be serviced during collision repair. The wiper arms can be held in place by spring clips or nuts. With spring clips, pinch the spring and pull for removal. You might also have to lift up on the end of a spring under the arm. This will free the arm for removal.

A special tool is available for spring tension-mounted wiper arm removal. If a nut is used, a cover over the nut must be pivoted upward or popped off. Then you can remove the retaining nut and wiper arm.

When installing windshield wiper arms, make sure they are adjusted properly. Specs are usually given for the arms in the down position. By engaging the arm into different teeth on the drive shaft, you can change the adjustment. Operate the wipers to make sure they are adjusted correctly and do not sweep too far one way or the other (Figure 15–25).

If the wiper blades must be replaced, disengage them from the wiper arm. A small spring on the end of the blade often must be engaged into the arm. Before removing the wiper arms, you can mark their position with masking tape on the windshield to make reinstallation easier.

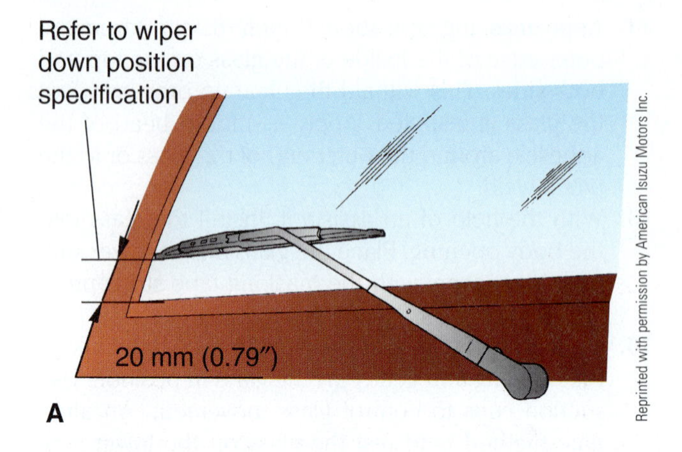

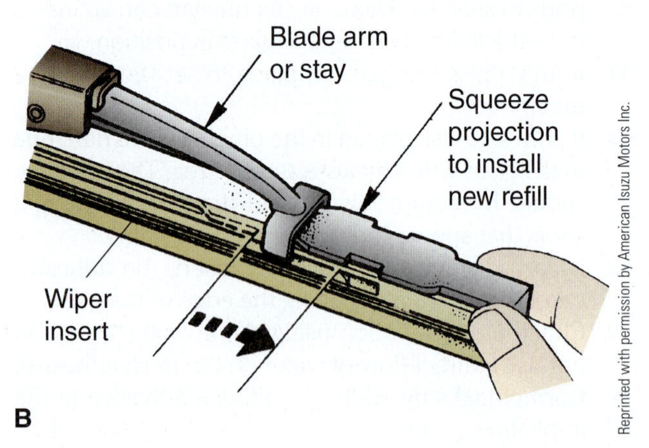

Figure 15–25 (A) When installing windshield wiper arms, make sure they are adjusted properly. (B) Most wiper refills are installed by compressing the small spring on the end of the arm.

Rear and Quarter Window Service

The removal and replacement procedures for rear windows and many stationary side windows are similar to windshield replacement. Methods can vary slightly for different vehicle makes. However, many of the operations that are applied to one make of vehicle can readily be applied to others.

Quarter windows can be secured with rubber molding, adhesive, fasteners, or a combination of methods. After removing interior and exterior moldings, the quarter window can be removed. Again, refer to vehicle service information for details about fastener locations and procedures (Figure 15–26).

Figure 15–27 shows a technician installing rear quarter glass.

15.3 SERVICING DOORS AND LIFTGATES

Doors are the most used and abused parts of a vehicle. They are opened and closed thousands upon thousands of times during the life of a car or truck. They must also remain strong enough to stay closed and protect the driver and passengers from injury during a collision.

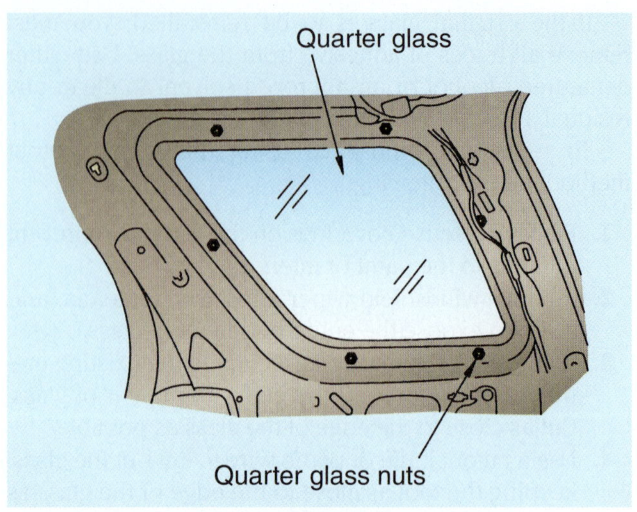

A Note how small nuts and adhesive secure glass.

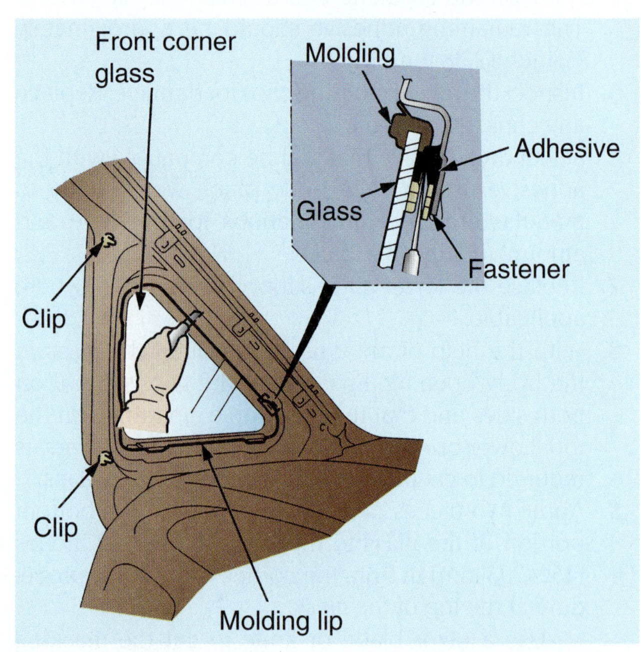

B A knife is used to cut the adhesive holding quarter glass.

Figure 15–26 Quarter glass is basically serviced like rear glass.

In addition, doors must seal out water and wind noise to keep the vehicle interior dry and quiet. Doors are frequently damaged in collisions.

Door Construction

With the variety of models available from different vehicle manufacturers, there are many variations on vehicle door designs. Although the doors on passenger cars and trucks have similar designs, it is common for minivans to have sliding doors that open and close themselves by pulling the handle or pushing a button on the key fob. These electric-powered doors can create unique challenges for collision repair technicians.

For passenger cars there are two basic door designs: framed doors and frameless doors. Framed doors

A Butyl tape is being applied to quarter glass. After tape is bonded to the glass, pull off the back tape to expose the adhesive.

B Align glass and press glass into place. Fasteners are also used to help hold quarter glass in place.

Figure 15-27 The technician is installing rear quarter glass.

surround the top and sides of the door glass with a metal frame. This frame helps keep the window glass in alignment. The door frame seals against the door opening when the door is closed. Sliding doors that have roll-down windows are considered framed doors.

Frameless doors have the glass extending up out of the door (when the glass is rolled up) without a frame around it. With frameless doors the glass itself must seal against the weatherstripping in the door opening.

Illustrated in Figure 15–28, the basic parts of a door include:

- The **door frame** is the main steel frame of the door. Other parts (hinges, glass, handle, and so on) mount on the door frame.
- The **door skin** is the outer panel over the door frame. It can be made of steel, aluminum, fiberglass, or plastic.
- The *door glass* must allow good visibility out the door.
- The *door glass channel* serves as a guide for the glass to move up and down. It is a U-shaped channel lined with a low friction material, felt, for example.
- The **window regulator** is a gear and arm mechanism for moving the glass. When you turn the window handle or press the window button, the regulator moves the glass up or down.
- The *door latch* engages the door striker on the vehicle body to hold the door closed.
- Inner and outer *door handles* use linkage rods to transfer motion to the door latch. This allows you to activate the latch to open the door.
- The door *trim panel* is an attractive cover over the inner door frame. Various parts (inner handle, window buttons, speakers) can mount inside the inner trim panel.
- A plastic or paper **door dust cover** fits between the inner trim panel and door frame to keep out wind noise.
- **Door weatherstripping** fits around the door or door opening to seal the door-to-body joint. When the door is closed, the weatherstripping is partially compressed to prevent air and water leaks.
- A rearview mirror is often mounted on the outside of the door frame. A remote mirror knob on the inner trim panel allows for mirror adjustment.

Manual and Power Regulators

Window regulators can be manually or electrically powered. Both types of regulators are very similar, the only difference being the handle crank mechanism on manual regulators and the electric motor-driven gear mechanism on powered regulators. The lift arms or mechanisms are the same for both types.

One or two lift arms can be used, depending on the make of the vehicle. If two lift arms are used in the window regulator, it is usually referred to as an *X-type regulator*. The X-design uses an auxiliary arm that is mounted into a cam or stabilizer channel that is adjustable. The cam adjustments allow the glass to be tilted or rocked so that it can be raised in a parallel position.

Checking Door Operation

Before door removal, check that the door and its related parts operate normally. Inspect the door assembly, its hinges, and the door opening in the body. Look for uneven or nonparallel gaps all the way around the door edge. **Nonparallel body gaps** indicate panel misalignment from structural damage, shifted panel fasteners, or worn mechanical parts (hinges or latches).

Look between the fender and door, rocker panel and door, quarter panel or rear door and front door, and between the roof rail and top of door. If you find gap misalignment, it may give you a clue about what needs to be done to repair the door, hinges, and door opening in the body.

You might find a front fender pushed back into the door by checking gaps. The fender might have to be adjusted back to allow the door to open. An even gap might be due to the A-pillars or rocker panel being deformed from the collision. This would tell you to measure the door opening to check for body damage. These kinds of problems must

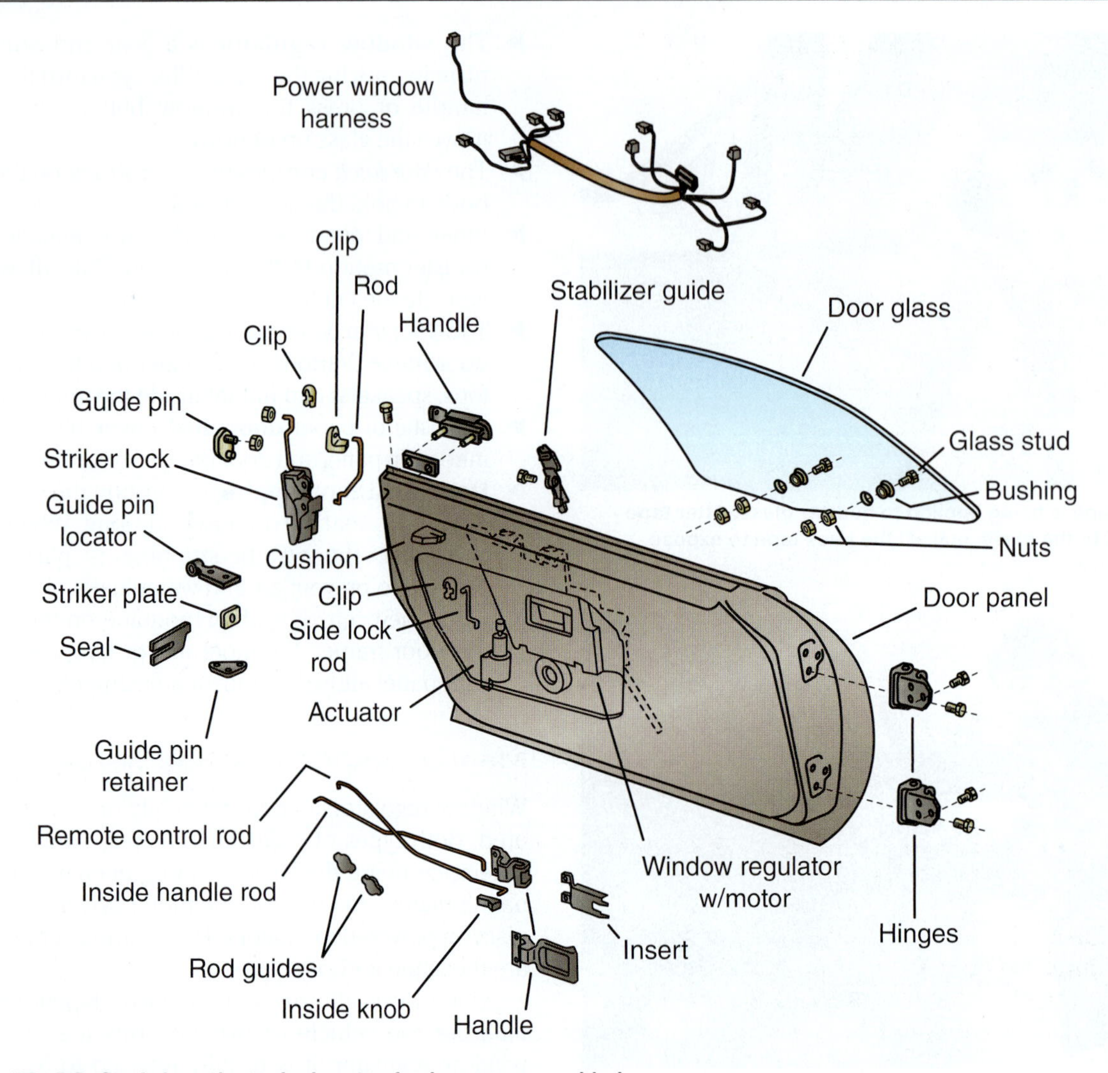

Figure 15-28 Study how the typical parts of a door are assembled.

be found before door removal so that they can be corrected during door service.

Door sagging results when the rear of the door is lower than the front. This is a common problem that is often due to badly worn hinge pins. Without periodic lubrication the hinge pins can wear, producing play in the hinges. This play allows the door to sag or drop on its unhinged end.

A **door hinge check** involves trying to move the door assembly up and down on its hinges. On lighter doors, you can try to raise and lower the door by hand. On larger doors, use a floor jack to move the door upward while watching the hinges.

Worn door hinges show up when the two halves of the hinge body shift, allowing up and down door movement. Good door hinges are evident when there is little or no play between the pin and hinge body. Worn door hinges should be replaced before reinstalling the door assembly.

A *door operational check* involves slowly opening and closing the door to check its latch, lock, hinge action, and other parameters (rattles from loose parts, squeaks from unlubricated parts, or binding from misaligned panels). Roll the windows up and down to check for binding or other troubles. With power windows, turn the ignition key on and activate all power window buttons. If a power window is inoperative, you want to find out now so it can be repaired while the door is apart.

As stated at the beginning of Section 15.3, electric-powered sliding doors can create unique challenges for collision repair technicians. These sliding doors operate on at least three tracks. If any of these tracks are damaged in a collision, the door may not function properly. For damage analysis the door should be switched to manual so that you can open and close it slowly and check for clearance. If the door requires replacement, the replacement door should be fit to the vehicle before being refinished to verify that it fits and functions properly. Make sure the repaired or replacement sliding door opens and closes easily by hand before using the automatic feature. It is also important to achieve the correct panel gaps for the closed position before operating the door in automatic mode.

Door Removal

Door removal is necessary for replacement and many repairs, such as door skin replacement or frame straightening. To remove a typical door, you must remove the two door hinge bolts or drive out the welded hinge pins (Figure 15–29).

Disconnect wiring going into the door frame and disconnect the door hinges. Some wiring is easy to disconnect on the outside of the door; other wiring requires complete door disassembly.

WARNING

If you fail to find and fix a structural problem during a repair, you might have to do the whole job all over again, without pay! For any job, always try to find and correct all problems relating to the parts being worked on. For example, if you find worn door hinges, they should be replaced for the safety of the people using the vehicle. However, be sure to get supervisor permission before doing extra work not listed on the work order or estimate.

Open the door about halfway. Place a floor jack under the door. Place a fender cover, rag, notched block of wood (a short piece of 2 × 4 works well), or door holding tool on the jack saddle to protect the painted edge of the door. A door holding tool is a rubber jack saddle insert that has a long groove to engage the bottom of the door flange (Figure 15–30).

With the saddle near the center of the door, raise the jack just enough to take most of the weight off the hinges. Be careful not to raise the jack too much, because it is easy to damage the door with the power of a hydraulic jack. You want the weight of the door balanced on the jack so the hinge bolts unscrew easily (Figure 15–31).

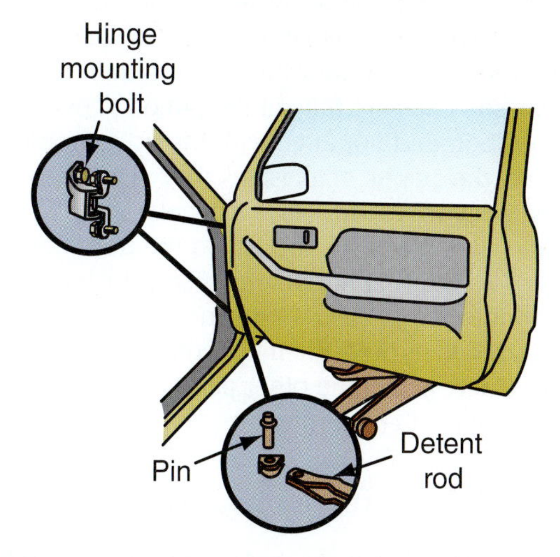

Figure 15-29 A floor jack covered with a shop towel or holding fixture is commonly used to support the weight of a door when loosening hinges. Have someone hold the door to keep it from falling when removing the last hinge bolt.

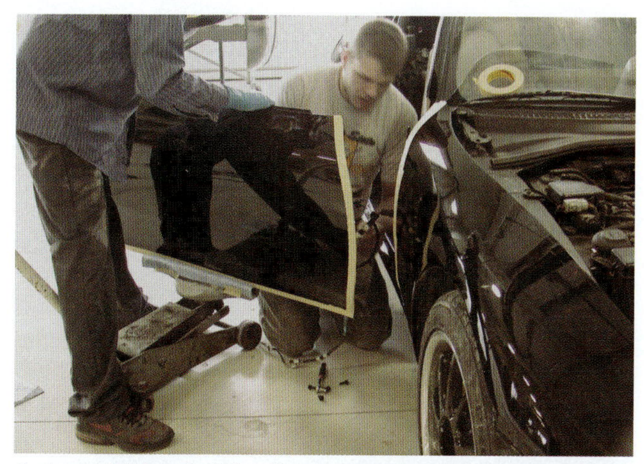

A A special door holding fixture has been placed on the floor jack. It has a rubber insert that will not damage paint. Masking tape has also been applied to door edges to protect against paint chips and damage.

B Raise or lower the jack so that the door hinges are at the right height. Slowly move the door into position while making sure it does not hit the fender. Hand-start and tighten the door hinge bolts. Then feed wiring harness into door.

Figure 15-30 Note how two technicians are working together to install a freshly painted door without damage.

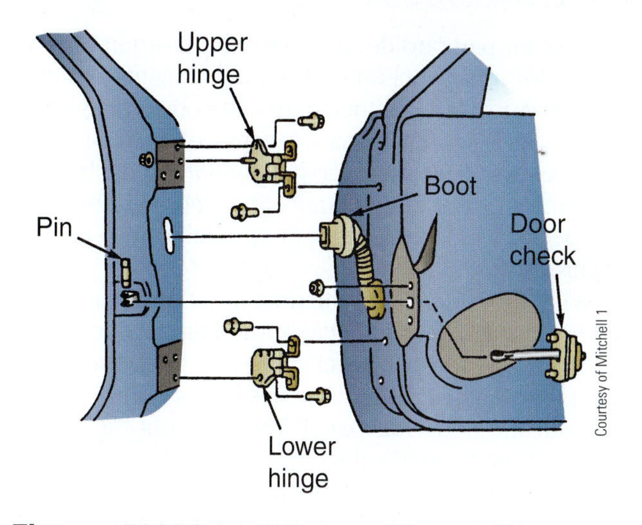

Figure 15-31 Note how hinges fasten to the door and vehicle body. Sometimes a wire harness can be unplugged on the outside of the door frame.

Figure 15-32 When loosening door hinge bolts, make sure the door is held so it does not fall off the jack.

Before removing the last bolt, ask a coworker to help hold the door and keep it from falling off the jack. The two of you can then move the door to a workbench or out of the way. Normally, place the door skin or outer panel down on the work surface. If the door does not have to be repainted, make sure you place a clean shop blanket on the work surface to prevent scratches in the finish (Figure 15–32).

Door Weatherstrip Service

With the door removed, inspect the rubber weatherstrip for deterioration or damage. The weatherstrip can be cut or worn. If you find any holes, splits, or cracks, remove and replace the door weatherstrip to prevent air and water leakage in the passenger compartment.

For more information on weatherstrip service refer to Chapter 16, which explains how to find and fix air and water leaks.

Door Inner Trim Panel R&R

To work on parts inside the door, you must remove the inner door trim panel and related components. Remove any screws that hold on the armrest and other trim pieces. You may have to pop out small decorative plugs over some of the screws. Refer to the service manual if in doubt.

Remove the window handle and door handle. They can be held on by a screw or by a clip from behind (Figure 15–33).

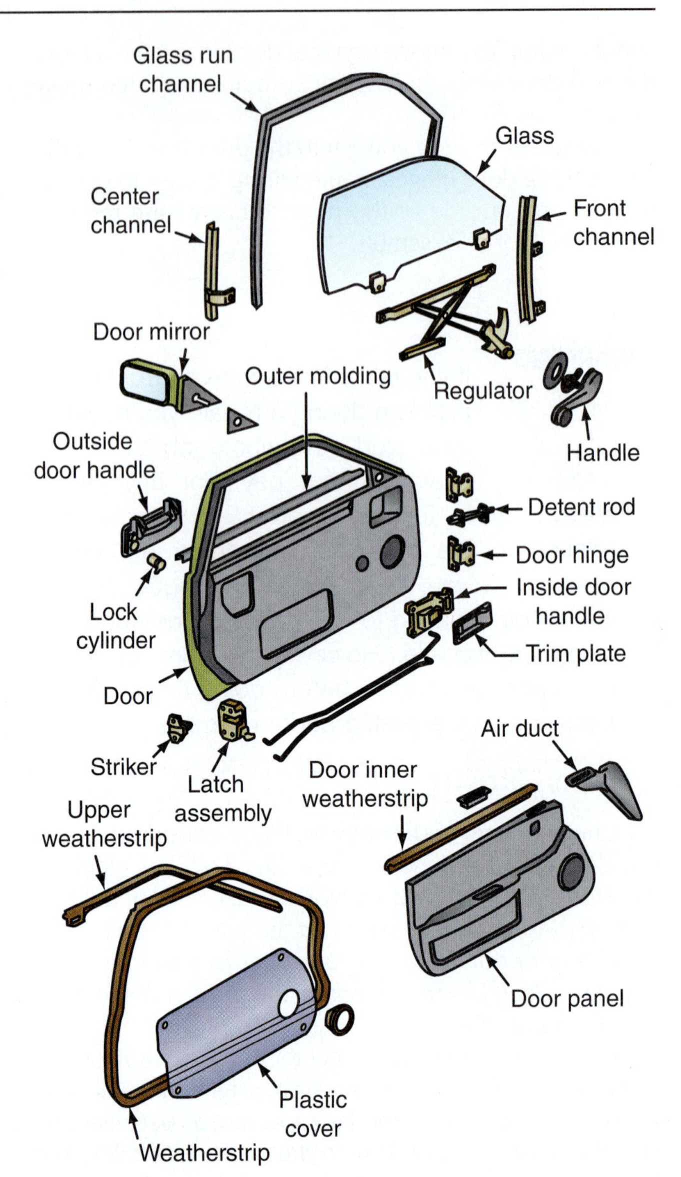

Figure 15-33 Study the basic parts of a door assembly.

With all screws out of the door inner trim panel, you will usually have to pop out a series of plastic clips. They install around the perimeter of the panel. Use a forked trim tool designed to remove clips in trim parts. Slide it between the door and panel. Then pry out the plastic clips without damaging the panel (Figure 15–34).

As you lift off the door inner trim panel, disconnect any wires going to the panel. Feed the wires through the panel.

With the door panel off, peel back any paper or plastic material over the door. Pull slowly so that you do not damage it. You now have access to the bolts and nuts securing the window regulator (Figure 15–35).

When replacing a door inner trim panel, check that all the plastic clips are fully in the panel. They can slide out of position and not align with the holes in the door. Re-adhere the paper or plastic cover back into place. Then feed any wires through the panel.

Fit the panel over the top lip on the door and down into position. Starting at one end, use your hand to pop each clip into the door. Lean down and check that each clip is started in its hole. Then use your hand to pop each clip into place. Install the other parts in reverse order of removal (Figure 15–36).

Door Window Regulator Service

A window regulator is a mechanism for raising and lowering the door glass. It consists of a set of gears, a window crank or electric motor, and sash channels. The sash channels guide the glass as it slides up and down (Figure 15–37).

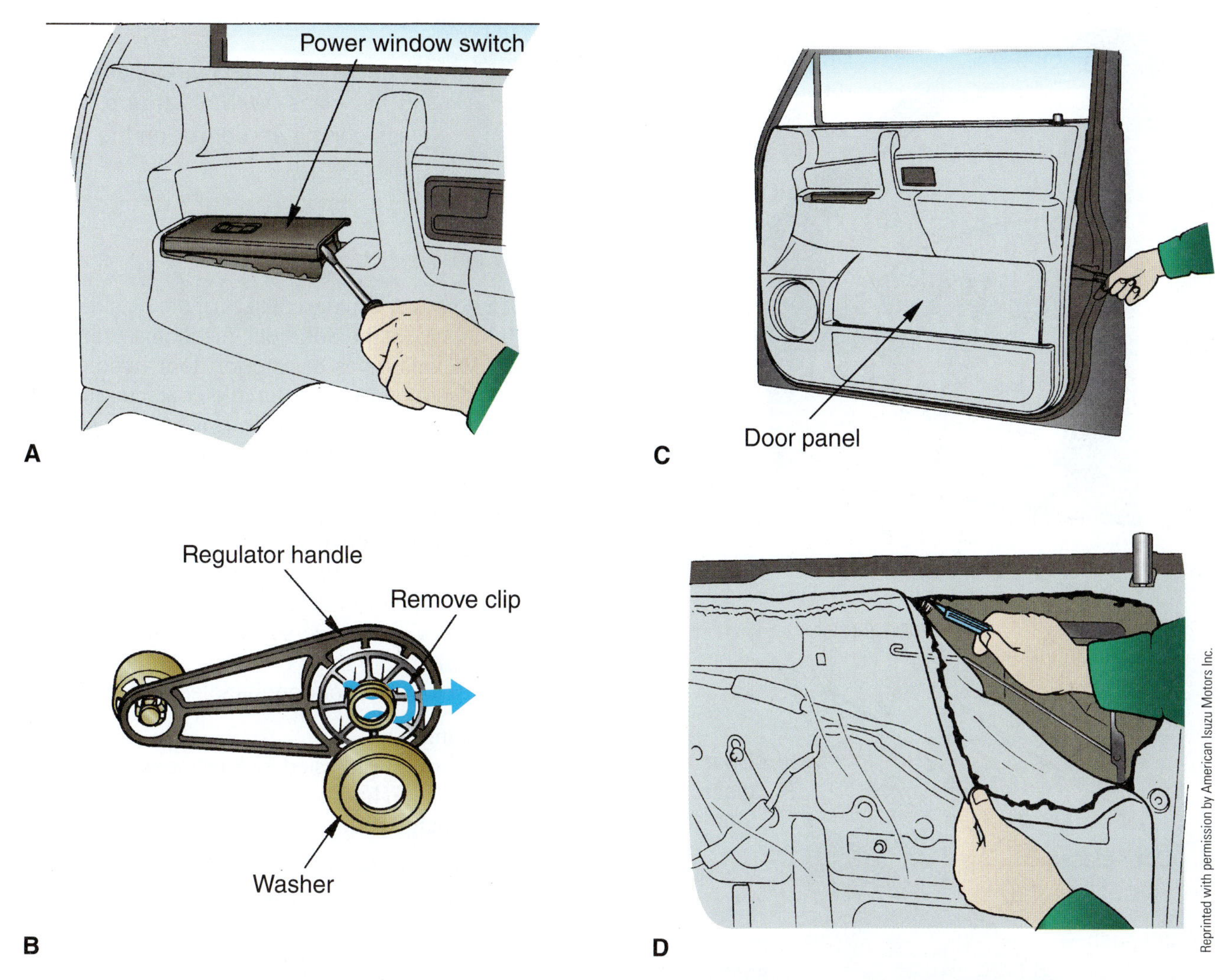

Figure 15-34 (A) To remove the inner door panel, first remove any snap-on parts, such as this window switch. (B) The door handle can be secured with a snap ring or screws. A hook can be used to pull out the snap ring. (C) After removing any screws around the perimeter of the panel, use a wide tool to pry out the clips. (D) Try not to tear the dust cover under the door panel. It keeps dirt, noise, and moisture out of the passenger compartment.

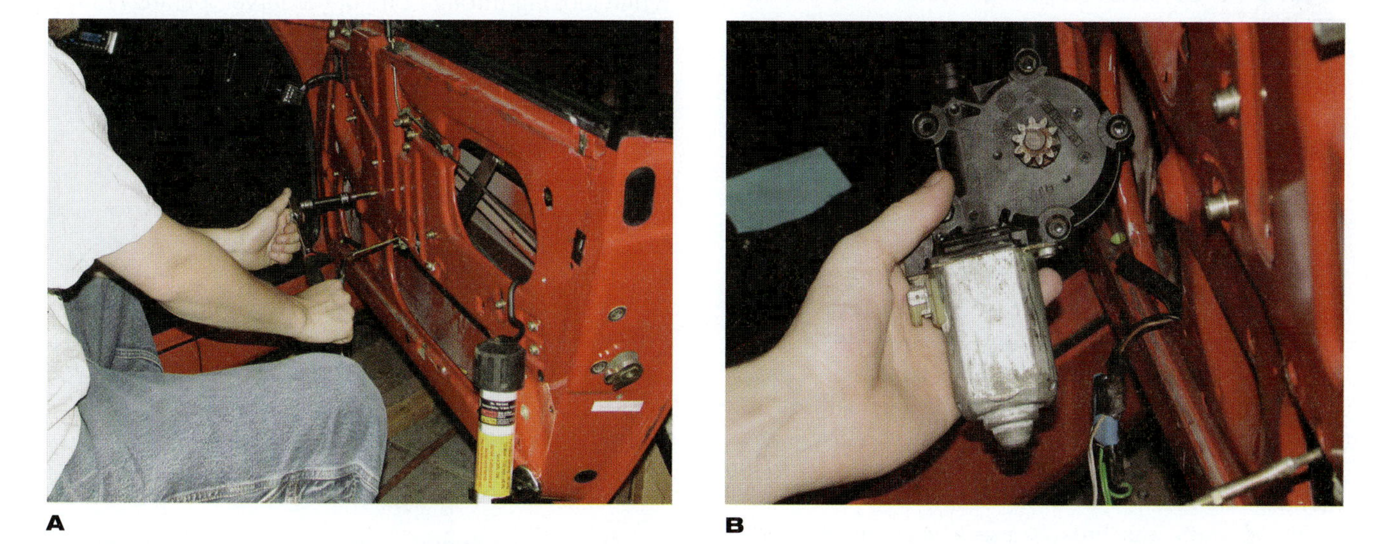

Figure 15-35 This technician is working on the inside of a door assembly. (A) A drop light is essential for working inside a door frame. (B) A power window motor is normally held engaged with the window regulator with three or four small machine screws.

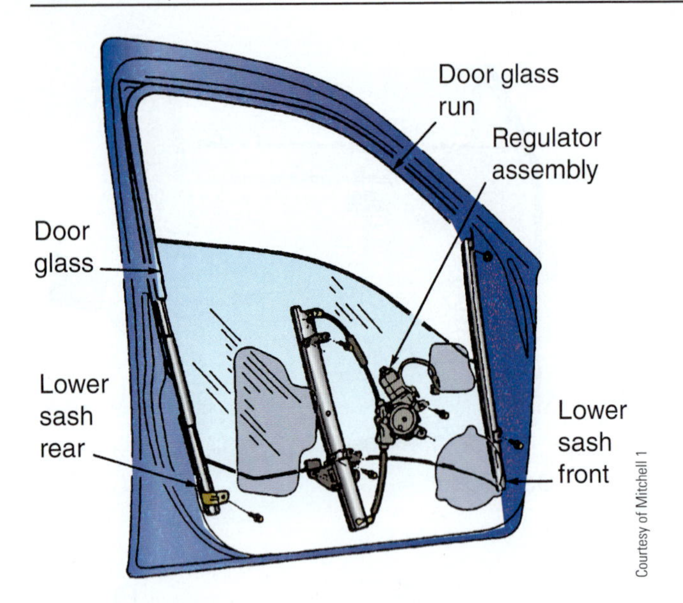

Figure 15-36 This is a modern power window regulator.

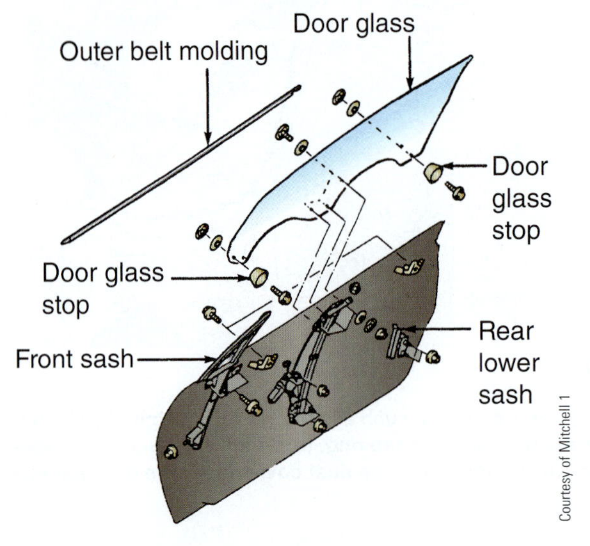

Figure 15-37 Note how glass is secured to the window regulator.

A *manual window regulator* uses a hand crank to turn the gears in the regulator. A *power window regulator* uses a small electric DC motor to spin the regulator gears. A station wagon tailgate window regulator is very similar to door window regulators.

Small nuts and bolts secure the window regulator and glass guides or tracks in position. Usually, the glass bolts to the upper arms of the regulator. Rivets can also be used to secure the glass to its regulator. On a few older vehicles, the glass may be held with a special adhesive or epoxy.

To remove the glass, unbolt it from the regulator. You must then remove any parts that prevent you from sliding the glass out of the door. This can vary, so refer to the manual if needed.

If the glass was broken, use a vacuum cleaner to remove all broken glass from inside the door. Install the new glass and bolt it to the regulator. Make sure you use all rubber, plastic, and metal washers.

Do not slam the door before adjusting the glass. With a hardtop, the glass can hit and break on the top door opening.

Door Lock and Latch Service

A *door lock assembly* usually consists of the outside door handle, linkage rods, the door lock mechanism, and the door latch. Various types of exterior door handles are available. The button or outer handle contacts the lock lever on the latch to open the door. However, most exterior door handles operate through one or more metal rods (Figure 15–38).

Door handles can be replaced by raising the window and removing the interior trim, panel, and water shield to gain access to the inside of the door. Exterior door handles are often held on with screws or bolts. A short screwdriver or small, ¼-inch drive socket is often needed to remove an outside door handle.

Some causes of exterior door handle problems are:

- Worn bushings
- Bent or incorrectly adjusted lock cylinder rods
- No lubrication on handle, linkage, or latch
- Worn or damaged latches

Inside door lock mechanisms are generally the pull handle type. The mechanisms are connected to the lock by one or more lock cylinder rods. Clips or bushings are used to secure the rods in place.

The door lock mechanism usually mounts through a hole in the door. A spring clip often fits over the inside of the lock to hold it in place. An arm on the lock transfers motion through a rod to the door latch.

To remove the lock, place a drop light inside the door. While looking through one of the large openings in the inside of the door, pop off the small clip holding the lock rod. Then pry the clip off the lock. You can use a

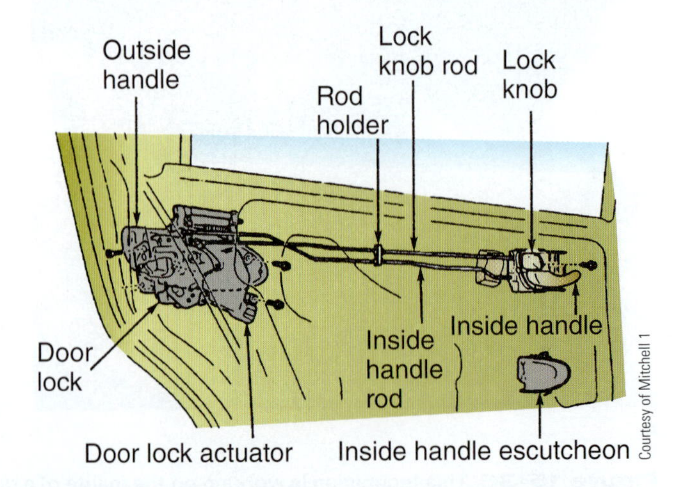

Figure 15-38 Note how small metal rods extend from the inner door handle to the door latch.

screwdriver or needle nose pliers depending on the amount of space around the lock. Slide the lock and washer off the outside of the door.

Electric door locks often use electric solenoids to move the linkage to the door locks. When you turn the key or press a door lock button, current flows to the solenoids in the doors. The solenoids convert the electrical energy into motion. The solenoid motion moves the lock linkages and latches. This locks or unlocks the doors.

NOTE *For more information on electrical systems, refer to Chapter 22.*

Door Reinforcements

All the doors of unibody vehicles have inner metal reinforcements at various locations. There are some other door frame reinforcements, such as at the hinge locations and the door lock plate.

Door intrusion beams are normally used inside side doors. A door intrusion beam is welded or bolted to the metal support brackets on the door frame to increase door strength.

15.4 LIFTGATES

Liftgates are common on SUVs, minivans, sport utility vehicles, and crossover vehicles. Liftgates swing up and out of the way for convenience in loading any type of cargo. Power liftgates are growing in popularity, especially with vehicle models also equipped with power sliding doors. Collision repair technicians commonly replace liftgates when they are damaged. Similar to the installation of a new door, it is important to fit the new liftgate to the vehicle before it is refinished. When a new, undamaged liftgate is installed on a vehicle that has been hit in the rear in a collision, it may not fit. Why is this the case? The reason is that the liftgate hinges or the vehicle itself is probably bent. More than likely the striker area where the liftgate latches is pushed in from impact damage. In some cases the roof where the liftgate hinges mount could be pushed up on one or both sides, causing a misalignment problem when the new gate is bolted in place. Damage to the vehicle structure must be repaired before the new liftgate will properly seal and latch to the vehicle. When installing a liftgate that closes automatically, make sure it closes properly and easily by hand before closing it using the automatic feature.

15.5 DOOR PANEL (SKIN) REPLACEMENT

Like other damaged panels, a door can be bumped or pulled into shape or replaced. The decision to repair or replace is determined by comparing the amount of time required to repair the door versus the cost of a replacement door.

The outer panel, or door skin, wraps around the door frame and is clinched to the pinch weld flange. The skin

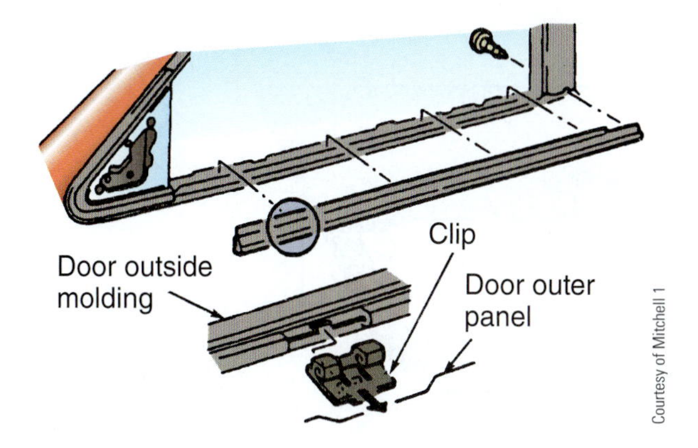

Figure 15-39 Outer door moldings are often held on by small screws or clips.

is secured to the frame either with plug or spot welds or with adhesives. Before servicing a door skin, you might have to remove any outer door moldings (Figure 15–39).

Replacing Welded Door Skins

Use the following procedure to replace welded door skins:

1. Before removing the door, check to see whether the hinges are sprung. Check the alignment of the door with respect to its opening.
2. Observe how the door skin is fastened to the door so you can tell how much interior hardware must be removed.
3. Remove the trim panel and disconnect the battery to isolate all door power accessories.
4. To prevent loss, place parts inside a container. Identify the parts and store the container in the trunk of the vehicle.
5. Use a hydraulic or body jack to remove some of the damage and possibly straighten or align the inner door frame.
6. Remove the door glass to prevent breakage. Now remove the door from the vehicle and move it to a suitable work area.
7. Remove all hardware from the door.
8. As a reference, apply tape to the door frame. Measure the distance between the lower line of the tape and the outer panel edge. Also measure the distance between the front or rear edge of the outer panel and the door frame (Figure 15–40). These measurements will be used to properly align and install the new door skin.
9. Use a scuff wheel, air chisel, or cutoff wheel to remove any corrosion protection material or brazed joints on the door skin flange.

   The quickest way to remove an exterior door panel is to grind off the edge of the hem flange. Only grind off enough metal so that the panel can be separated from the inner flange. Do not grind into the inner panel. Do not use a welding torch or power chisel to separate the panels. The inner panel can

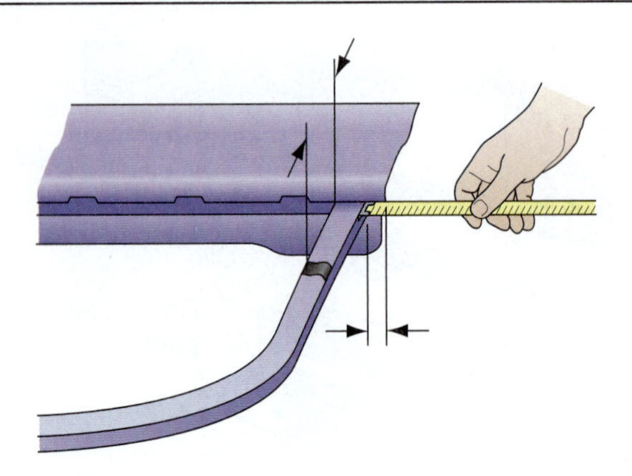

Figure 15-40 Measure the location of the old door panel on its frame. Write down the horizontal and vertical positions. Here a piece of tape has been placed on the frame for measurement reference.

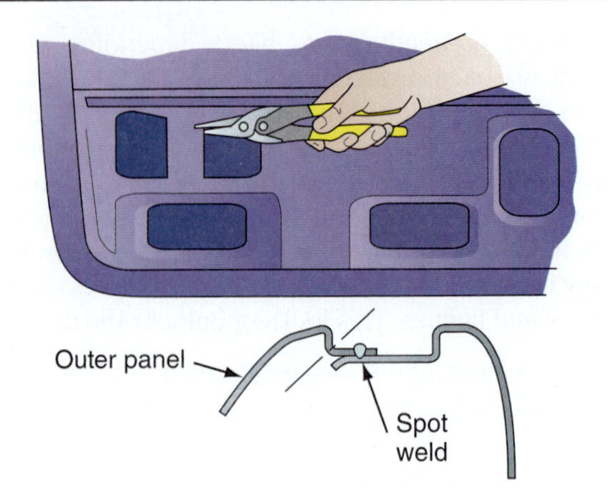

Figure 15-42 Cut around spot welds as needed.

become distorted or be accidentally cut. Separate the reinforcing strip on the top of the panel (if installed or used) (Figure 15–41).

10. Using a hammer and chisel, loosen the two panels. Use a pair of tin snips to cut around any spot welds that could not be drilled or ground off. When the exterior panel moves freely, remove the panel. Use Vise-Grips or pliers to remove what remains of the inner hem flange. Any remaining spot welds or brazing should be ground off using a disc grinder (Figure 15–42).
11. With the exterior panel removed, carefully examine the inner panel and door frame for damage. Straighten any minor door frame damage at this time. Remove dents on the inner flange with a hammer and dolly. Media blast away any rust.

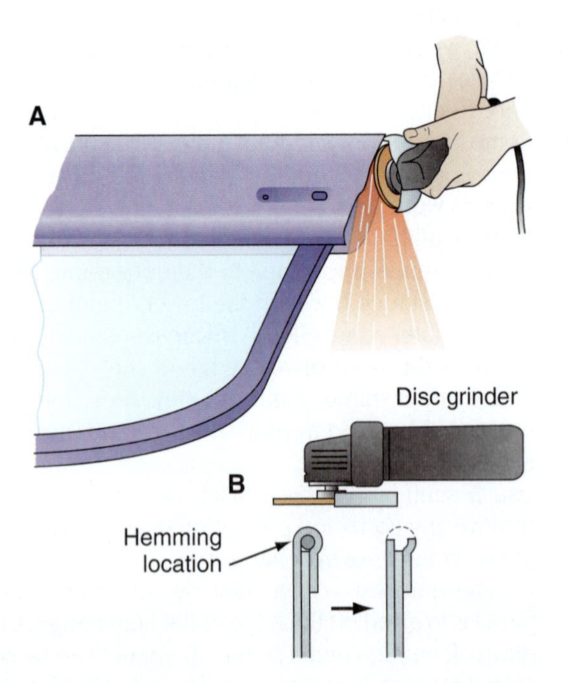

Figure 15-41 Cut off the old door panel or skin without damaging the frame. (A) Grind off the door hemming flange. (B) Note the hem flange cross section before and after grinding.

12. Apply weld-through primer to any areas to be spot or plug welded. Cover other bare metal areas with a rust-resistant primer or other manufacturer-recommended rust treatment.
13. Prepare the new panel for installation. Drill holes for any plug welds. Using a sander, remove the paint from the weld and braze locations. Apply weld-through primer to the bare metal seam areas and prime any other bare metal areas.
14. Some outer door panels are accompanied by a silencer pad that must be glued to the outer panel. To do this, clean the outer panel with an approved alcohol or solvent. Heat the outer panel and silencer pad with a heat lamp. Then glue the silencer pad to the outer panel.
15. Before installing the new panel, apply body sealer to the back side of the new panel. Apply the sealer evenly ⅜ inch (9.5 mm) from the flange in a ⅛-inch (3.2 mm) thick bead.
16. Using Vise-Grips, attach the new outer panel to the door. Align it properly, using the dimensions determined earlier. Refer to specific instructions from make and model service information so you use the correct number, placement, and types of welds. Model-specific service data will also specify where seam sealer and corrosion protection materials are needed (Figure 15–43).
17. Use a hammer and dolly to bend the outer panel flange. Cover the dolly face with cloth tape to avoid scarring the panel. Bend the hem gradually in three steps. Be careful not to tap the panel edge because that could throw the panel out of alignment. Do not create bulges or creases in the body lines of the outer panel (Figure 15–44).
18. After working the flange within 30 degrees of the inner panel, use a hemming tool to finish the hem. Again, finish the hem in three steps, being careful not to deform the outer skin. By placing a wooden paint stick between your hemming pliers and the door skin, you can help avoid denting the new door skin (Figure 15–45).

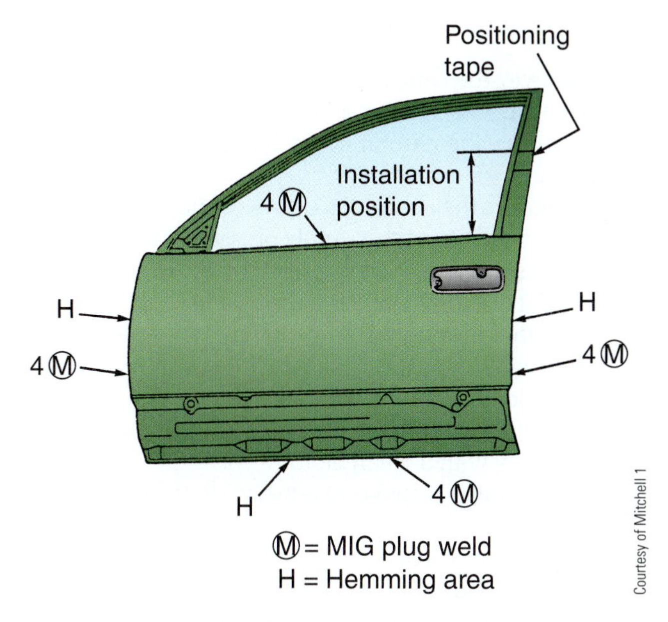

Figure 15–43 Factory or aftermarket service information will give details like these. Note the locations and number of recommended MIG plug welds and hemming areas.

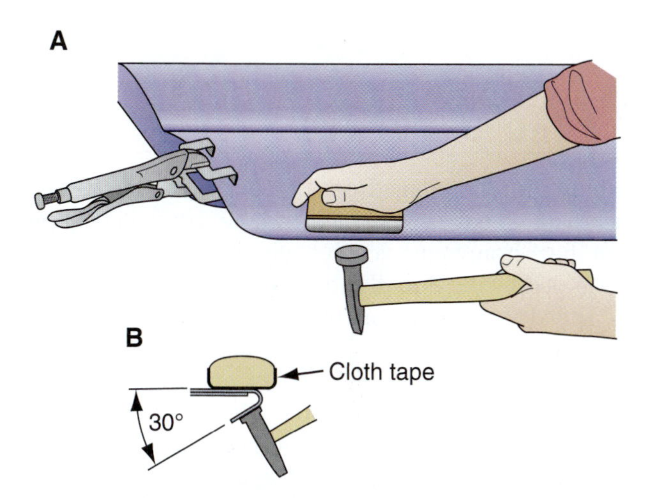

Figure 15–44 Carefully straighten the door frame flange after removing the old panel. (A) Dollying the panel flange. (B) A cross-sectional view.

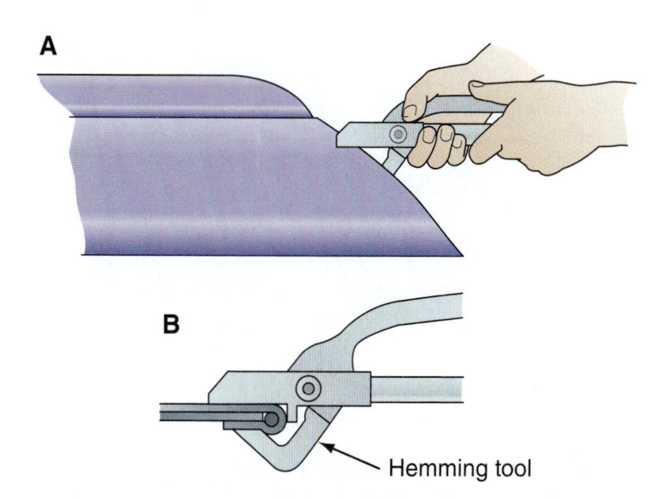

Figure 15–45 (A) Special tools are used when forming the panel hem. (B) A cross-sectional view.

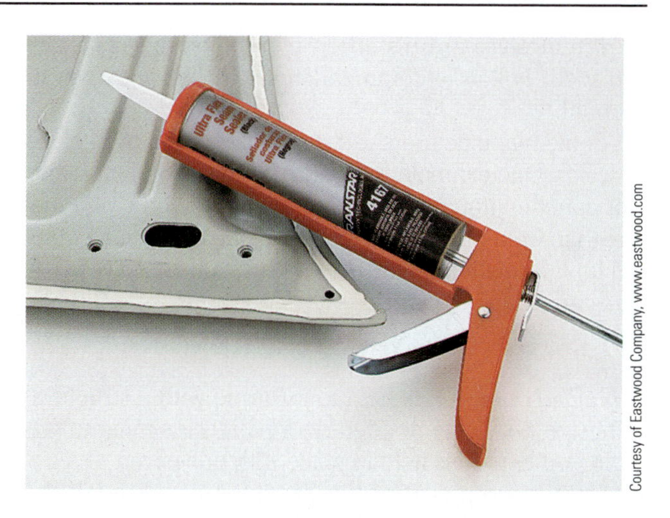

Figure 15–46 Apply sealer to the flange of the new door skin in a smooth, continuous bead.

19. Fit the door onto the vehicle and align it in the door opening. This will let you double-check door skin alignment before welding. With the door still on the vehicle, place a couple of welds to secure the door skin.
20. Remove the door and finish welding the skin in place, using the factory recommended number and types of welds.
21. Clean the welds and apply sealer to the hemming flange to make it waterproof. If needed, spray the inside of the door with anticorrosion or sound-deadening materials (Figure 15–46).
22. If needed, drill holes in the new door skin panel for moldings or trim pieces.
23. Prepare the door for painting and reinstall it. You may want to paint the door while it is still off the vehicle.
24. Before refinishing the door, be sure to install the door glass. This will prevent overspray from getting on the interior of the door and will avoid paint damage.
25. Be sure to align the door with all adjacent panels and check for panel gaps and proper closure or latching.

Wear leather gloves when working with door skins and other sharp sheet metal panels. Jagged metal edges can cause serious cuts.

15.6 PANEL ADHESIVE TECHNOLOGY

Some vehicle manufacturers use structural adhesives to install door skins and other panels. These two-part epoxy adhesives are sometimes called weld-bond adhesives because spot welds can be placed through the adhesive.

Weld-bond adhesives are used to add strength and rigidity to the vehicle body. They also improve corrosion protection in weld seams. Adhesives also help control noise and vibrations.

If adhesives are disturbed by repairs, they must be replaced. Follow recommendations in the body repair manual.

Some manufacturers use structural adhesives in place of welds. One example is around the wheel openings and sail panel reinforcement. This is a different type of adhesive than weld-bond adhesive. Check the body repair manual for information on the use of structural adhesives.

Replacing Bonded Door Skins

A *bonded door skin* is held to its frame with a structural adhesive. Servicing bonded door panels is similar to servicing other panels held in place with adhesives.

To begin removal of the damaged panel, mark the location of the old panel on the door frame. This will allow you to install the new panel in the proper location on the frame.

To gain better access to the inside of the door, cut out the center area of the door skin panel. Use power shears, a cutoff wheel, or an air chisel. With the center section of the door panel cut out, you can also inspect the door frame for damage. If the frame is damaged, you should consider purchasing a new or recycled door assembly.

Use a heat gun to soften the adhesive that holds the door panel to the door frame. As you heat the adhesive, it will soften so that you can wedge a putty knife or chisel into the joint. Be careful not to damage the door frame when separating the door panel. Keep working around the flange with heat and your tool until the entire panel is removed.

Clean off the remaining old adhesive with your putty knife. A very thin layer of adhesive can remain on the frame so long as it is not thick enough to affect new panel installation.

Remember that mating surfaces must be clean (no grease, oil, undercoating, and so forth). Some products require you to apply a primer to the surfaces before the adhesive. This will ensure a strong adhesive bond between the door skin, any remaining adhesive, and the door frame.

NOTE *It is a good idea to "dry test-fit" the door skin before applying adhesive.*

Apply a liberal amount of new adhesive in a continuous bead around the frame flange. Place the new panel down into place. Align your reference marks so that the new panel is positioned properly. Use clamping pliers and light pressure to secure the panel in place as the adhesive cures. Allow the adhesive to cure for the time suggested by the manufacturer before moving the door.

Replacing SMC Door Skins

Sheet molded compound (SMC) door skin panels are becoming popular. SMC is similar to fiberglass. Many doors are now made completely of SMC, except for steel door intrusion beams and steel lock and hinge reinforcements.

To replace an SMC door skin, proceed as follows:

1. Cut away the center of the skin. Air shears work well because you can easily control the cut depth. If you use a saw, be careful not to hit or cut internal door parts.
2. To remove the remaining door skin, heat the bonded areas with a heat gun. Apply pressure with a pry bar or chisel to remove the rest of the material. Be careful not to damage the door flange.
3. Sand the door frame flange to remove all remaining adhesive. Clean the bonded areas of the replacement door skin with soap and water. Allow it to dry. Sand the bonding areas to expose the SMC fibers. Wipe dry with a clean cloth.
4. Apply a bead of two-part adhesive to the door frame flange.
5. Set the door skin on the door frame and lightly clamp it. Do not squeeze too tightly. You want to leave adhesive between the skin and frame.

To complete the job, allow the adhesive to cure, paint the door, and then reassemble it and mount it on the vehicle.

15.7 DOOR AND DOOR GLASS ADJUSTMENTS

If there are undamaged panels next to the door, cover their painted edges with masking tape. This will help prevent them from being accidentally scratched and nicked if bumped by the door during installation.

The door installation procedure involves reversing the removal procedure. Have someone help hold the door on the floor jack. Raise the jack until the door hinges are the same height as their bolt holes in the body. Make sure you are holding the door level (Figure 15–47).

Slowly slide the door hinges against their bolt holes. Wiggle and shift the door until you can start the bolts

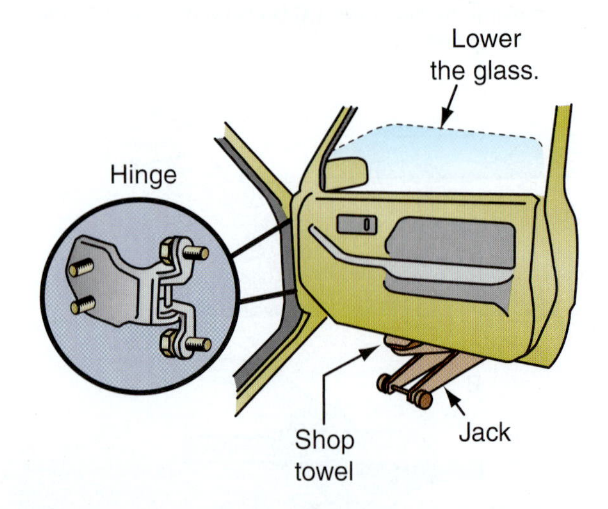

Figure 15–47 Use a floor jack to raise a door to the correct height for starting hinge bolts. The door should be level on the jack. Have someone help you hold the door on the jack.

with your fingers. Align the hinges to their original positions. Snug the bolts down but do not tighten them yet.

Door frame adjustment is needed to ensure that the door will close easily and not rattle or leak water and dust. This section will describe various door adjustments and door glass adjustments. Door glass adjustment is also needed to prevent air and water leakage into the passenger compartment.

Worn door hinges will have play that allows up and down movement of the rear of the door. If the hinge pins are worn out, you should replace the hinges. Some hinges use bushings around the hinge pins. When these bushings are worn out, replace them. This will retighten the pin in the hinges and also readjust the door to a certain extent. Make sure replacement hinge bushings are available.

If the hinges are to be removed, scribe a line around the hinge to mark its position on the body and door. This will simplify reinstallation and positioning of the new hinge. You might have to loosen the fender at the rear bottom edge to reach the hinge bolts.

Servicing Welded Door Hinges

Obviously, service methods are different for welded hinges. The large pin must be driven in and out of a welded hinge to service the door assembly. The bolted hinge can be easily adjusted forward, rearward, up, and down. The use of shims behind the hinge also allows the hinge to be moved as desired.

A specially designed pry bar can be used to adjust a door. The end of the bar hooks over the striker bar and a U-shaped bracket engages the latch.

As mentioned, some vehicles use a welded-on door hinge that has no adjustment provisions. A pin is provided to remove the door for servicing the hinges. The half of the hinge that is to be installed on the door is predrilled to permit a bolt-on installation with tapped caged plates and bolts. The half of the hinge on the hinge pillar must be rewelded on the pillar when it is replaced.

When removing the door hinge pins, use a special spring compressing tool. The spring must be seated properly in the tool before compressing it. Otherwise, the spring can slip and cause damage or personal injury. After the pin in each hinge is removed, the door can then be removed from the vehicle.

To replace the welded door side hinge, first scribe the outline of the hinge on the door. Center punch the spot welds and drill an ⅛ inch (3.2 mm) pilot hole completely through the welds. The weld is then drilled out with a larger bit (about ½ inch, or 12 mm), but only deep enough to penetrate the hinge base to release the hinge from the panel. Next, a chisel is driven between the hinge and the base to break it free from the panel (Figure 15–48).

The new part is installed on the door by drilling the recommended size holes into the attaching holes. The holes will allow for slight adjustment on the door assembly because the bolts are often smaller than the holes.

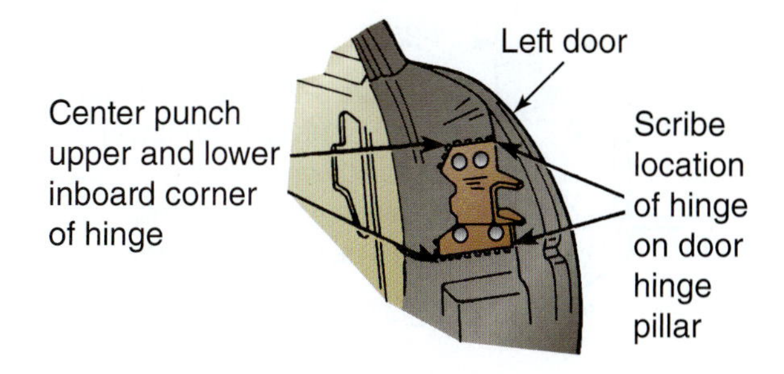

Figure 15–48 **Scribing the hinge location can help you to rough adjust a door when you are reinstalling it.**

To remove the body side hinge, scribe the hinge position. Then use a cutting torch to cut the tabs holding the hinge together. The door sill plate and carpet should be removed or covered with an asbestos sheet to protect them from the hot slag of the cutting operation.

Using a suitable tool, such as grip-type pliers, the welds holding the separated hinge tabs are twisted or rotated to break them. Once the tabs are removed, the pillar is ground smooth and prepared to receive the new part.

To install the new hinge strap, manufacturer measurements must be transferred to the new part. First, tack weld the hinges carefully in place and then hang the door to check its fit in the door opening and with surrounding panels, as shown in Figure 15–49.

If it fits properly, the door is removed and the hinge is welded completely around the upper and lower hinge tabs. The area is cleaned properly and a paintable sealer is applied around the perimeter of the hinge. The area is then refinished to the proper color before the door is reinstalled.

Bolted Door Hinge Adjustment

Doors must be accurately adjusted so that they close easily and do not rattle or leak. Basically, the door hinges must be adjusted to hold the door in the center of its opening when closed. The door striker must be adjusted to engage the latch smoothly. This section will describe various door adjustments.

Doors must fit their openings and align with the adjacent body panels. When the doors on a sedan need adjusting, start at the rear door. Because the quarter panel cannot be moved, the rear door must be adjusted to fit these body lines and the opening. Once the rear door is adjusted, the front door can be adjusted to fit the rear door (Figure 15–50).

Next, the front fender can be adjusted to fit the door. On hardtop models, the windows can then be adjusted to fit the weatherstripping. The windows are usually adjusted starting at the front and working toward the back. The front is adjusted to fit the front door pillar, and the front window is then adjusted to it. The rear door window is adjusted to the front window rear edge and the opening for the rear door assembly.

Some vehicles have rubber door stops that can be turned to adjust the closed door in or out. They are

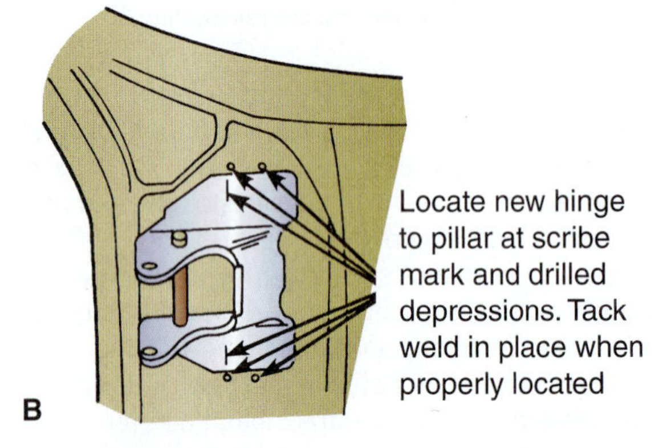
1-3/4″
Measure 1-3/4″ from forward flange of upper and lower hinge tab on service replacement hinge and scribe location on hinge
A
1-3/4″
Locate new hinge to pillar at scribe mark and drilled depressions. Tack weld in place when properly located
B

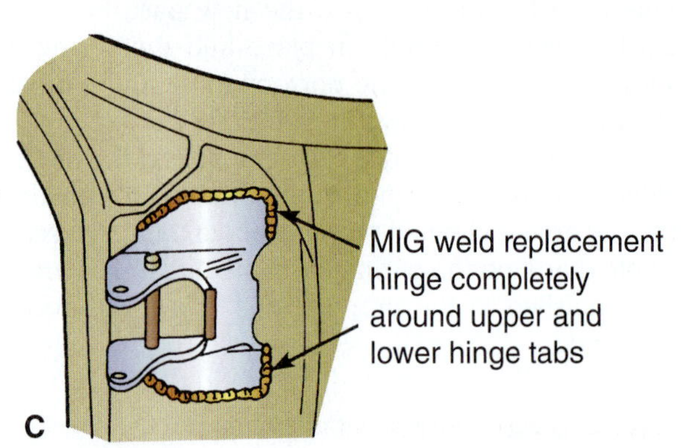
MIG weld replacement hinge completely around upper and lower hinge tabs
C

Figure 15–49 (A) Measure the hinge location, (B) transfer these measurements to the new hinge, and (C) weld the hinge tabs.

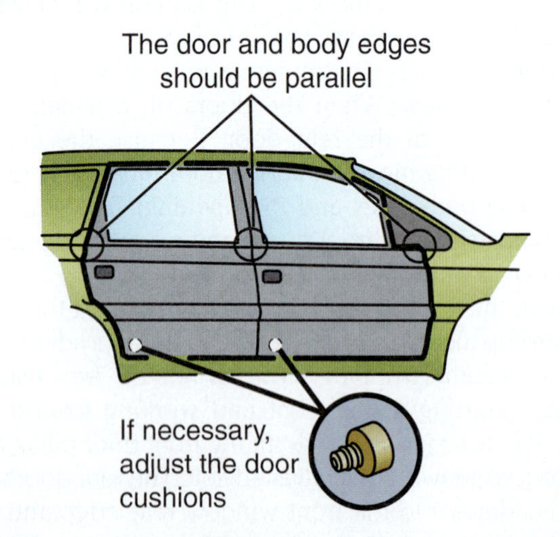
The door and body edges should be parallel
If necessary, adjust the door cushions

Figure 15–50 With a four-door vehicle, all body gaps or edges must be parallel and equal in width. Note the use of adjustable door stops on this vehicle.

similar to hood stops. You can rotate the door stops to screw them in or out so that the door panel is flush with the adjacent panels.

To adjust a door in its opening, follow these steps:

1. Remove the striker bolt so it will not interfere with the alignment process.
2. Determine which hinge bolts must be loosened to move the door in the desired direction. First, establish door height.
3. Loosen the hinge bolts just enough to permit movement of the door with a padded pry bar or jack and wooden block. On some vehicles, a special wrench must be used to loosen and tighten the bolts.
4. Move the door as needed. Tighten the hinge bolts. Then check the door fit to be sure there is no bind or interference with the adjacent panel.
5. Repeat the operation until the desired fit is obtained.
6. Install the striker bolt and adjust it so that the door closes smoothly and is flush with the rear door or quarter panel. Check that the door is in the full latched position, not the safety latch position (Figure 15–51).
7. On all hardtop models, the door and quarter glass must be checked to ensure proper alignment with the roof rail and weatherstrip.

In-and-out adjustments are also very important. The door must fit the opening and be aligned in and out to fit the body panels. The door must also provide a good seal between the weatherstripping and the body opening. The weatherstrip must be compressed sufficiently in the opening to prevent water, dust, drafts, and wind noises from entering the automobile.

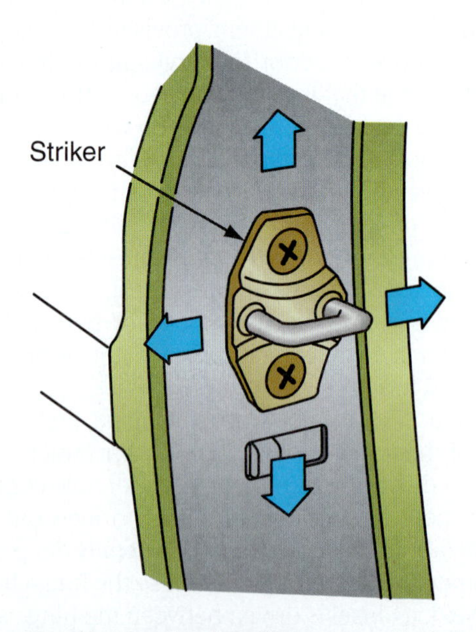
Striker

Figure 15–51 After the door hinges are adjusted, you may have to adjust the striker. When the door is closed, latch it to smoothly engage the striker. If the door is forced up or down, adjust the striker up or down as needed. If the door does not close enough or closes too much, adjust the striker in or out as needed.

Care must be taken when adjusting the in-and-out movement of the door. Moving a door out on the top hinge will not only affect the top of the door but also move the opposite bottom corner in. If the bottom of the door is moved in on the hinge, it will move the top opposite corner out. If the door is moved in or out equally on both hinges, however, it will only affect the front of the door because the amount of adjustment decreases toward the back of the door.

The center door post, striker bolt, and lock determine the position of the door. The front leading edge of the door should always be slightly in on the front edge from the rear of the other panel (usually the front fender). This will help to stop wind noises at the leading edge of the door panel. If the front edge is sticking out, wind noise will annoy the car owner and passengers.

SHOP TALK

The striker plate is not adjusted properly if the door rises or if it is forced down when the door is closed. The striker should slide and engage smoothly into the latch when the door is closed. The striker can be moved up and down, in and out, and back and forth.

Another common fastener that allows for door striker adjustments is the caged plate. A **caged plate** is a thick, heavy, steel plate with threaded holes to accept large bolts (Figure 15–52).

The threaded plate is often housed in a "cage," or box of thin sheet metal spot welded to the body panel. The cage is larger than the plate so that the plate can be moved around, but the cage prevents the plate from falling away from or down into the panel. Oversized holes in the panel allow the plate and bolts for the hinge or latch to be adjusted in any direction.

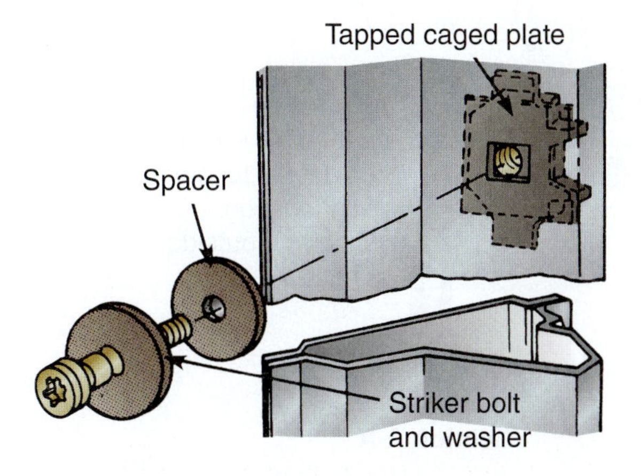

Figure 15–52 A caged plate allows for striker adjustments and provides a strong mounting point.

Figure 15–53 The first time you close the door after installation, close it slowly and softly while watching how the door fits. The striker should engage the latch smoothly without pushing up or down on the door.

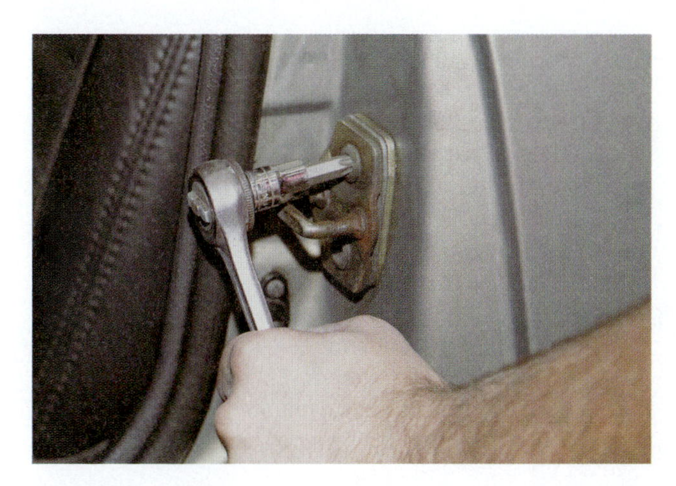

Figure 15–54 Here the technician is using a ratchet and Torx socket to install and tighten the door striker.

Caged plates are often used in doors and door pillars. In this application, the cluster of oversized bolt holes would weaken the panel without the reinforcement of the steel backing plate. See Figure 15–53 and Figure 15–54.

15.8 DOOR GLASS SERVICE

Various methods are used to attach the door glass to the window lifting or regulator mechanism:

- Bolt-through, or rivet, method
- Adhesive, or bonding, method
- Sash channel method

Two basic types of doors are used on today's vehicles: framed and hardtop. The framed door (sedan) uses an upper door frame structure that surrounds the glass. The hardtop door does not use a frame structure around the door glass. Both door types normally use one-piece door glass (Figure 15–55).

The bolt-through method utilizes bolts or nuts that have plastic or rubber gaskets to prevent direct contact

Figure 15-55 **This window glass has a frame around it. A hardtop door would not have a frame around the glass.**

with the glass. The fasteners pass through a hole in the window glass and secure it to the lift channel or bracket. Bolts are used to attach the lift channel or brackets to the glass. The bolts are inserted through the glass or clips bonded to the glass.

The use of adhesives is another method of securing the lower lift bracket to the glass. Usually a U-channel with insulator stays is used to prevent the glass from contacting the metal channel (Figure 15–56).

The oldest method of attaching glass to lift channels uses the sash channel. A rubber seal or tape is put on the lower edge of the glass. Then a channel is positioned and tapped onto the glass by using a rubber mallet. If the channel is too loose, tape can be used as a shim to tighten it. Usually the edges of the channel can be squeezed slightly for an even tighter fit. Be careful not to break the glass, however.

Methods for removing and servicing door glass vary. Door glass is secured in a channel with either bolts, rivets, or adhesive. Doors on sedans are basically the same as different makes of hardtop doors. Some hardtop doors require the removal of the upper window stops, lower lift brackets or bolts, front or rear glass run channel, upper glass stabilizers, and many other parts. If the glass is to be reinstalled, store it in a safe place.

Mark the position of the channel on the glass. The sash channel can then be unbolted or rivets drilled out. If the glass is glued into the channel, follow this procedure:

Remove the channel from the glass by applying heat from a welding torch with a #2 or #3 tip along the full bottom length of the channel. Slowly pass the tip back and forth for 60–90 seconds, then grip the channel with pliers and pull it loose. If the channel does not separate easily, repeat the heating operation.

8. Clean the replacement glass. If the original glass is to be used, scrape all traces of adhesive off with a sharp-bladed tool. If the original channel is to be reused, clamp it in a vise and burn out the remaining adhesive with a welding torch. While still hot, wire brush adhesive traces from the channel. After it has cooled, remove the remaining adhesive from the glass and channel with lacquer thinner. Complete the cleaning operation with water.

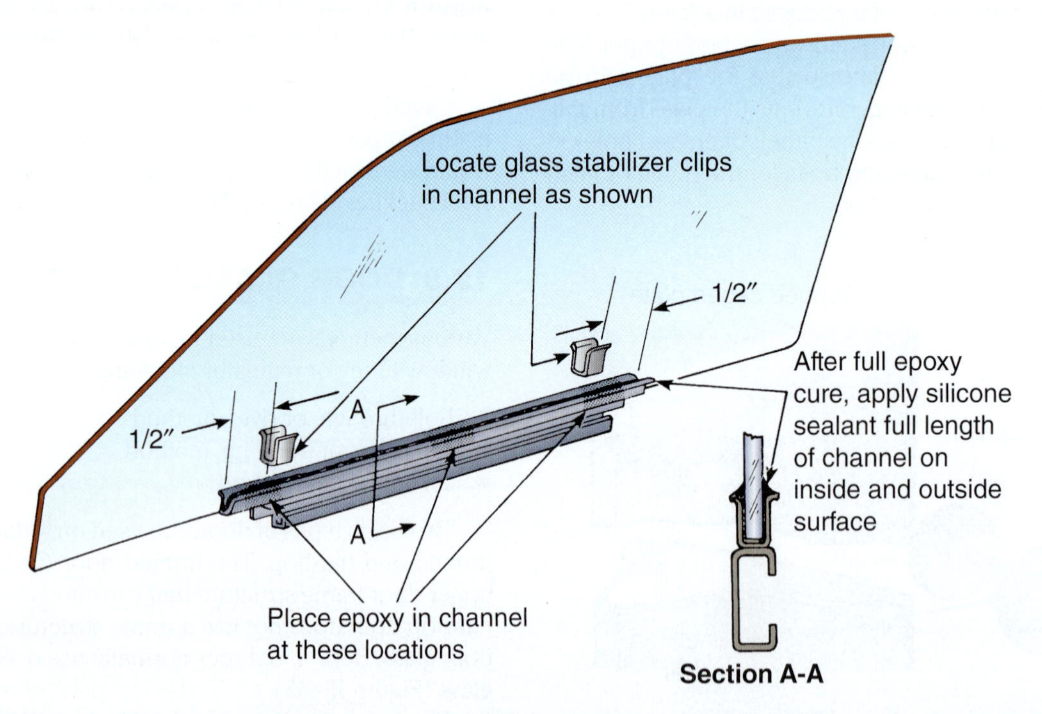

Figure 15-56 **This is adhesive-bonded glass in a bracket. You would need to refer to specific service information to get exact measurements for installing this glass.**

When the sash channel is clean and dry, the replacement glass can be glued in place. If using new glass, transfer the position markings from the original glass to the replacement glass. The vehicle manufacturer's service manual can be consulted for correct channel position.

To reinstall glass in a bolted or riveted channel, a strip of tape of the proper thickness should be applied to the bottom of the glass. Rest the top part of the glass on a piece of soft wood or carpeting. Then position the channel on the glass and use light blows to force the channel on the glass. If possible, use a rubber hammer. If the channel is loose on the glass, use a thicker piece of tape to close the gap in the channel and provide the proper width. Then, reattach the channel bolts or rivets and spacers.

After the channel has been installed on the glass, the glass and channel assembly can be positioned into the door and secured with the necessary attaching hardware. Install the lower glass stop and adjust it if necessary. Finally, reinstall all the hardware and trim panels.

Door Glass Adjustment

By far, framed doors require the least amount of adjustments to the door glass. The glass mechanism on framed doors is rather simple, because the upper frame serves as a guide and support for the door glass. Hardtop doors do not have this advantage, and glass generally rests against the roof edge or the top of the door opening. A soft, rubber, weatherstrip gasket is used in the door opening to protect the glass when it is fully raised or closed. The glass must have some means of support and height control when it is lowered or raised.

WARNING Hardtop door glass is prone to damage if it is slammed against the roof drip rail when the door is closed. Make sure the door glass is down or rough adjusted before closing the door. Never slam it until all adjustments are completed.

On some vehicle doors, the window glass is adjusted by loosening nuts or bolts that hold the channels and stops on the inside of the door frame. The window channels have slotted holes to control the forward and rearward window adjustments. The window stop brackets also have slotted holes to control how high the window raises.

Some vehicle windows utilize adjustable guide rollers in a movable channel. Still others have a center lift guide and can be adjusted forward and rearward, as well as in and out, by tilting the glass. These adjustments are controlled at the bracket that is attached to the lower sash channel or where the guide attaches to the inner door panel.

If the window binds or is stiff, check the channel or add approved lubricant to the glass runs or guide channels. A door window that tips forward (or rearward) and binds can be caused by improper adjustment of the lower sash brackets, a loose channel or cam roller, or a channel that is out of adjustment.

On some sedan doors, a full or partial length rubber glass run or channel is used. If the channels are too tight or lack proper lubricant, the glass or rubber will bind. To free up the glass, a dry silicone spray should be applied to the glass run.

Oil is not recommended for use on rubber or felt window channels. It can cause the rubber or felt to swell and deteriorate.

The regulator and its associated parts are sometimes riveted, rather than bolted, to the door structure. In this case, drill out the rivets in accordance with good shop practice and reinstall the necessary parts using the appropriate rivet gun and rivets.

If the regulator is to be removed without removing the glass, secure the glass in an up position to prevent it from dropping inside the door shell. Heavy cloth tape or a wedge can be used for this purpose. Always consult the manufacturer's manual to obtain the proper removal and installation procedures for the regulator.

Power regulators require additional installation care because of the use of counterbalance springs in their design. On some vehicle models, the counterbalance spring must be released before servicing the regulator motor or other associated parts.

Vehicle doors that use a full trim panel sometimes have a set of brackets at the top of the door. The trim panel is attached to these brackets. If they are set too far inward, the window glass will often bind on these brackets. Other items that can cause the window glass to bind when raised are the antirattle slides or other devices, if they are not set correctly.

The hardtop glass must also be properly tilted to make contact with the upper gasket. If it does not make the proper contact, the glass will leak water and dust. If it is tilted too far in, the door will be hard to close and the gasket can be damaged (Figure 15–57).

Door Trim Panel Installation

After the door and its glass have been adjusted, you must reinstall all of the door trim pieces and other parts. First, reposition and press the plastic dust and wind sheet into place on the door frame. Make sure all holes in the plastic align with their parts or holes.

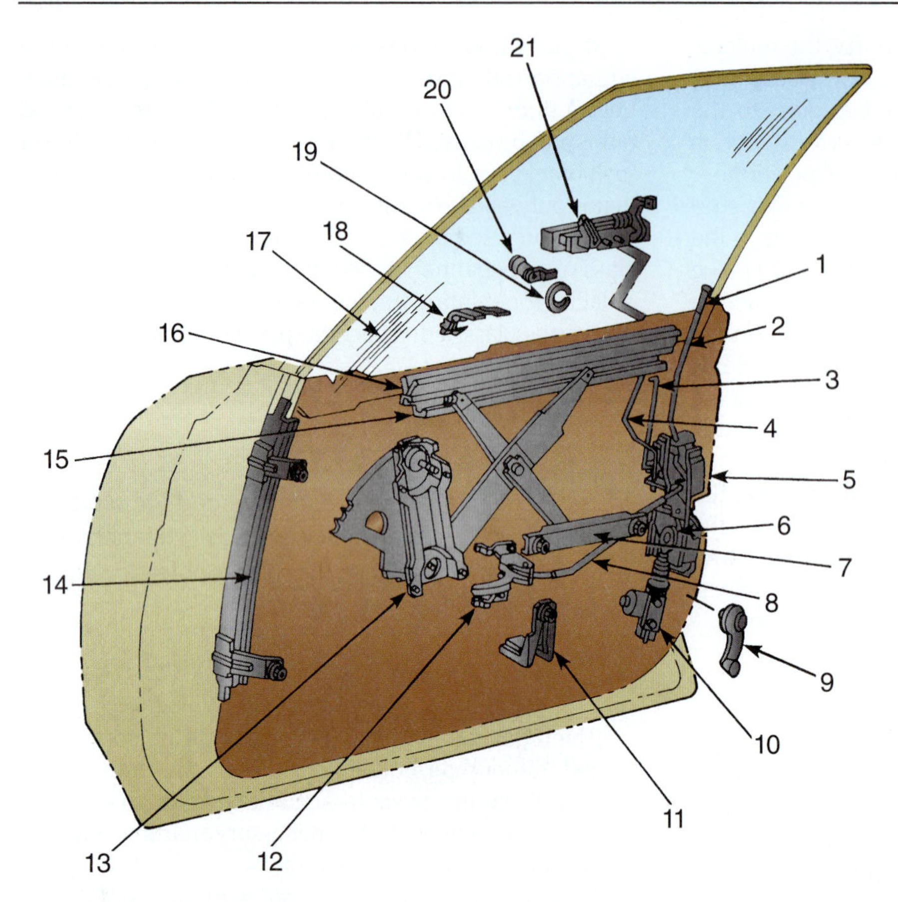

1. Inside locking rod knob
2. Inside locking rod
3. Lock cylinder to lock connecting rod
4. Outside handle to lock connecting rod
5. Door lock
6. Inside locking rod to electric actuator connecting rod
7. Inner panel cam
8. Inside remote handle to lock connecting rod
9. Manual window regulator handle
10. Power door lock actuator
11. Down-travel stop
12. Inside remote handle
13. Manual window regulator
14. Glass run channel retainer
15. Lower sash channel cam
16. Lower sash channel
17. Window glass
18. Lock cylinder retainer
19. Lock cylinder gasket
20. Lock cylinder assembly
21. Outside handle assembly

Figure 15-57 **Study the X-design window regulator.**

Connect wires to any lights or electric controls on the inner trim panel. Make sure all clips are properly positioned on the back of the door trim panel. Align these clips with the holes in the door frame. Then use your hand to force the clips into the door to secure the trim panel. Install any screws in the panel or on inner door handles or other parts.

When servicing a door for a keyless entry system, all wiring harness connectors must be disconnected and reconnected carefully. It is very easy to break the plastic harness connectors.

Tailgate Glass Service

Tailgate glass is generally secured to a regulator with screws or bolts. Removal or replacement of tailgate glass is similar to procedures presented for door glass. If in doubt, refer to the factory service manual for details.

Station Wagon Tailgate Adjustments

Many station wagons have a three-way tailgate that swings open like a door or drops down like a truck tailgate. Compact wagons are the exception. They often have a hatchback-style rear gate. The three-way tailgate has a unique hinge and locking arrangement. The three-way tailgate can be operated as a tailgate with the glass fully down or as a door with the glass up or down.

Glass Element Repair

Some heating elements built onto rear glass can be repaired when damaged. A special electrically conductive adhesive is used to bridge the gap in any breaks in the heating element (Figure 15–58).

You can reattach broken antenna and heating element wires by soldering. Refer to the service manual for details.

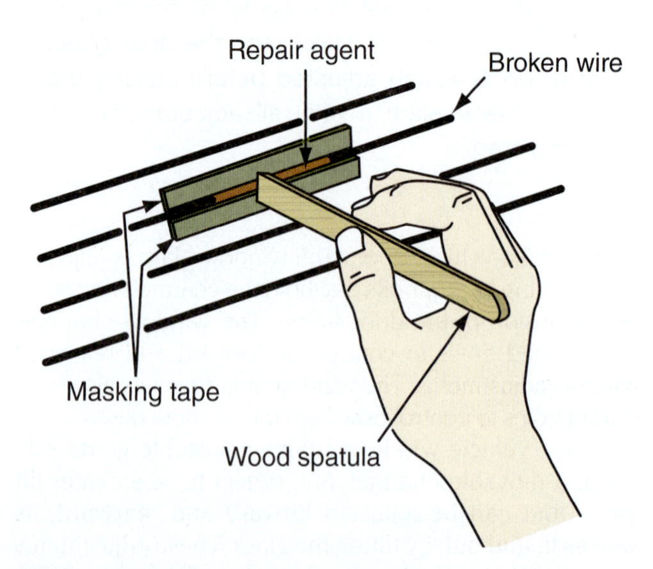

Figure 15-58 **Some glass defrost elements can be repaired using a special agent that conducts current. Masking tape is used as a guide for applying the agent.**

15.9 REARVIEW MIRROR SERVICE

Both outside and inside rearview mirrors can be damaged in collisions. Inside rearview mirrors normally attach to the windshield glass using a quick bond adhesive. Outside rearview mirror housings normally bolt to the door frame.

To service an inside rearview mirror, use a sharp putty knife to remove the old mirror mount. Apply heat to the mirror wedge. While it is warm, twist it back and forth with pliers. Use a single-edge razor blade to scrape the area clean on the inside of the windshield. Spray clear primer where the mirror is going to be mounted. Then place a few drops of clear adhesive on the windshield glass and the mounting surface for the mirror. Press and hold the mounting pad or mirror onto the glass without moving it. Hold the mirror or metal pad tight for about a minute. This will secure the mirror or its mounting pad. Sometimes a small Allen wrench is needed to tighten a small screw that holds the mirror to its mounting pad.

If the outside rear mirror housing is damaged, the whole assembly is normally replaced. After removing the door inner trim panel, you can remove the two or three nuts that secure the outer rearview mirror to the door. Make sure you position any rubber gasket properly between the mirror housing and the door skin (Figure 15–59).

Quite often, only the rearview mirror glass is broken. You can purchase and install just the mirror glass.

Sometimes the rearview mirror glass is held on with clips, but usually it is adhesive-bonded in place. Use a heat gun to soften the adhesive on the broken mirror. Twist the mounting pad 30 degrees each way until it releases from the glass (Figure 15–60).

Then use a putty knife or similar tool to pry off the old mirror. If the mirror has a heating element, it can be reused if still in working condition. Apply recommended adhesive to the back of the mirror and tape it in place. Allow the adhesive to cure for the recommended amount of time before tape removal.

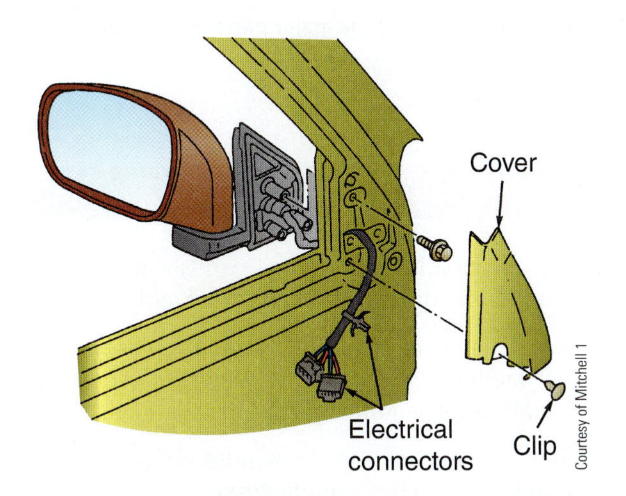

Figure 15-59 Outside rearview mirrors are usually held on by several small bolts or nuts. By removing the inner trim, you can gain access to these fasteners.

A Warm the glass with a heat gun to soften the adhesive.

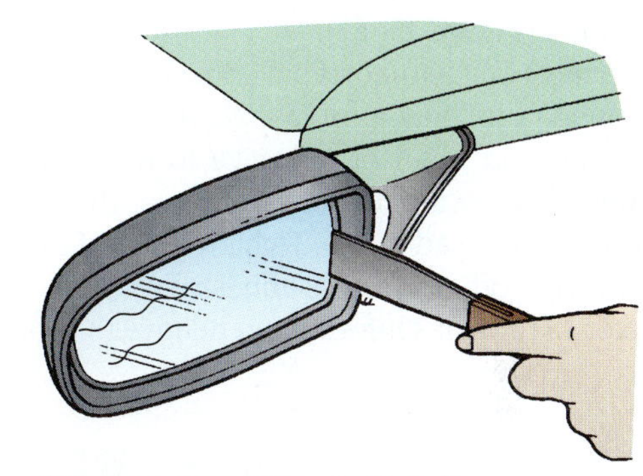

B Pry off the old glass with a putty knife.

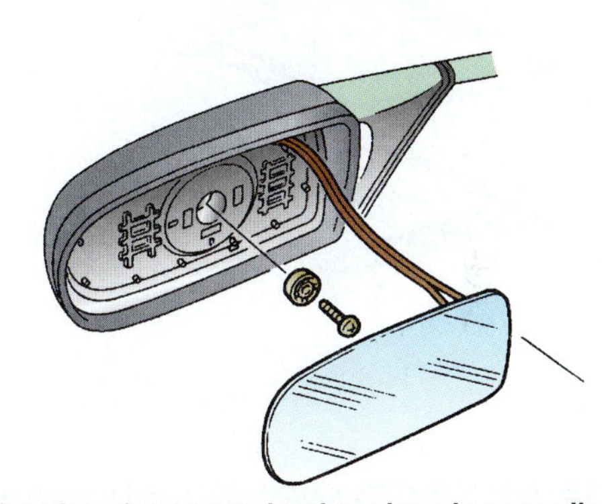

C The heating element under the mirror is normally held on by small screws.

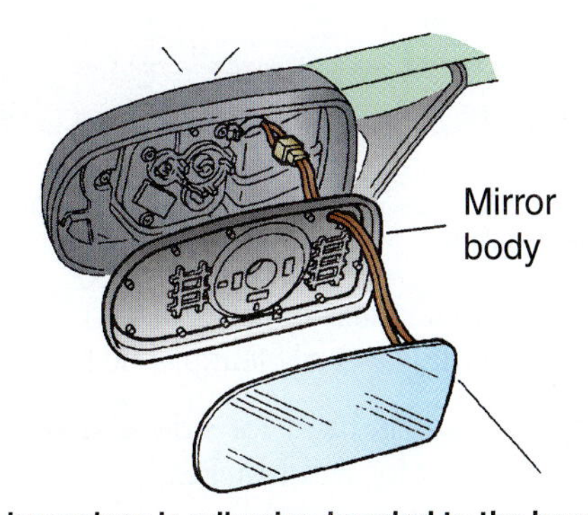

D New mirror glass is adhesive-bonded to the housing.

Figure 15-60 Note the basic steps for replacing rearview mirror glass.

15.10 ROOF PANEL SERVICE

Roof service includes repair and replacement of roof panels. Roof panels are often damaged in rollovers. You might also have to service a retractable hardtop roof. A retractable roof mechanism is very complex, and designs vary. You will need to refer to factory service information to properly repair a retractable hardtop.

Fastened Roof Panel Service

Some roof panels are not welded to the pillars. Instead, they are secured with fasteners and an adhesive. Their replacement involves removing the headliner and fasteners. Then the adhesive must be pried loose to remove the old roof panel (Figure 15–61).

The lip must be cleaned of old adhesive to prepare for installation of the new roof panel. Apply the recommended type and bead of adhesive to the mating surfaces. Then have someone help you lower the new roof panel down into the adhesive. Tighten the roof panel fasteners and wipe off any excess adhesive.

Chapter 19 provides information relative to roof panel service.

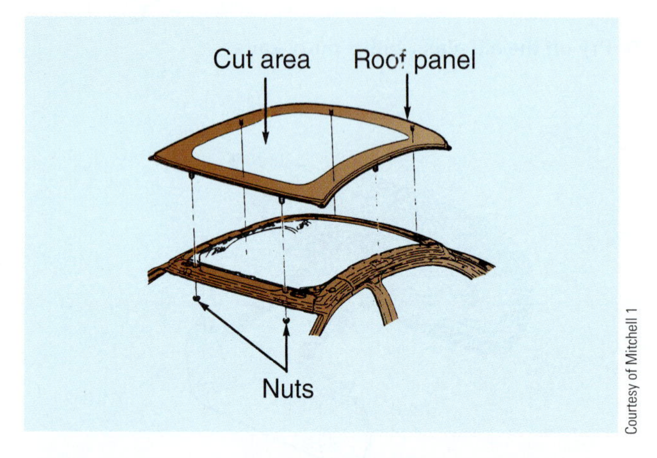

Figure 15–61 Many late model roof panels are not welded in place. They are held on with fasteners and structural adhesive.

Convertible Top Service

A convertible top consists of a retractable metal form covered with canvas material. It is also very complex. Convertible tops are often sent to an upholstery or specialty shop for service.

Sunroof Service

A *sunroof assembly* typically consists of a large glass sheet, frame assembly, and motor assembly for sliding the glass back into the rear of the roof panel. One is shown in Figure 15–62.

Because sunroof designs vary, you should refer to service information for the particular make and model vehicle. The instructions will be specific enough to guide you through the repair or replacement of the unit.

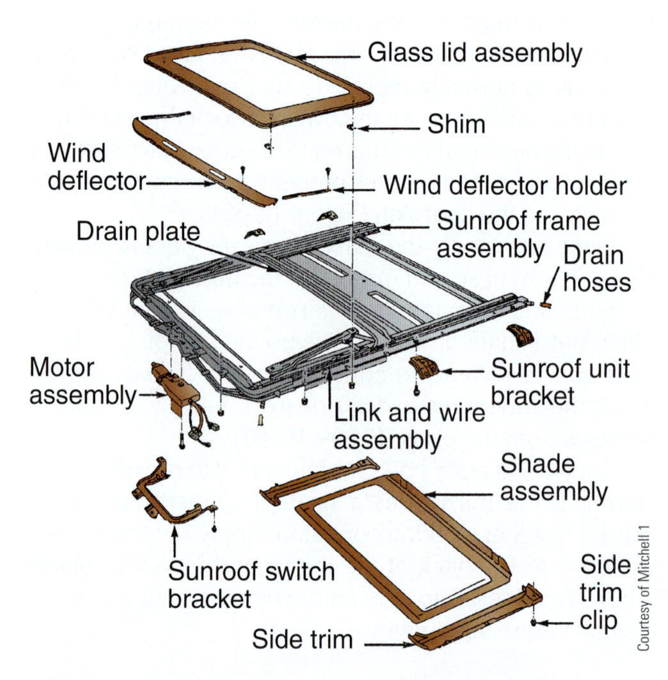

Figure 15–62 Study the major parts of a sunroof assembly.

SUMMARY

1. Today's vehicles are built with a lot of glass for greater visibility.
2. Laminated plate glass consists of two thin sheets of glass with a thin layer of clear plastic between them.
3. Tempered glass is used for side and rear window glass, but rarely for windshields.
4. Antilacerative glass is similar to conventional multilayered glass, but it has one or more additional layers of plastic affixed to the passenger side of the glass.
5. Interior moldings are called garnish moldings and exterior moldings are called reveal moldings.
6. Replacement of windshield glass involves two different methods based on the materials used: gasket installations or adhesive-type installations.
7. Most technicians use the windshield adhesive recommended by the manufacturer.
8. The outer panel, or door skin, wraps around the door frame and is clinched to the pinch weld flange.

9. Some manufacturers use structural adhesives in place of welds.
10. Door adjustment is needed so that doors close easily but do not rattle or leak water and dust.
11. Worn door hinges will have play that allows up and down movement of the rear of the door.
12. Oil is not recommended for use on the rubber window channels because it can cause the rubber to swell and deteriorate.
13. The window regulator is a gear mechanism that allows you to raise and lower the door glass.

EXERCISES

On a separate sheet of paper, complete the following learning activities for this chapter. Write definitions for the key terms and answer the ASE-style review questions, essay questions, critical thinking problems, and math problems. You can also do the outside activities, possibly for extra credit.

Key Terms

caged plate
door dust cover
door frame
door hinge check
door sagging
door skin
door weatherstripping
laminated plate glass
nonparallel body gaps
tempered glass
window regulator

ASE-Style Review Questions

1. Technician A says that a broken windshield will usually shatter into tiny slivers of glass that fall out of the body. Technician B says that some vehicles have laminated side glass to prevent shattering. Who is correct?
   A. Technician A
   B. Technician B
   C. Both A and B
   D. Neither A nor B
2. What is a common name for exterior moldings?
   A. Rubber moldings
   B. Plastic moldings
   C. Reveal moldings
   D. Garnish moldings
3. Tempered glass is rarely used for ______________.
   A. Windshields
   B. Rear windows
   C. Side windows
   D. Both A and B
4. Windshield wiper arms should be adjusted in this position.
   A. Up
   B. Down
   C. Left
   D. Right
5. What type of glass has more resistance to impact than regular glass of the same thickness?
   A. Modular
   B. Tempered
   C. Laminated
   D. Channel
6. Technician A says it is best to check the opening and closing of automatic doors in manual mode. Technician B says that automatic doors do not need to be adjusted. Who is correct?
   A. Technician A
   B. Technician B
   C. Both A and B
   D. Neither A nor B
7. Gasket glass installation is more common in ______________.
   A. Older vehicles
   B. Newer vehicles
   C. Vehicles with modular glass
   D. None of the above
8. Which of the following items is used in the full cutout windshield replacement method?
   A. Windshield adhesive
   B. Setting spacers
   C. Putty knife or scraper
   D. All of the above
9. In the partial cutout windshield replacement method, what serves as the base for the new adhesive?
   A. Butyl ribbon sealer
   B. Butyl tape
   C. Masking tape
   D. The old adhesive

Essay Questions

1. How do you adjust a door in its opening?
2. Summarize the replacement of an SMC door skin.

Critical Thinking Problems

1. A door is sagging and does not close during an operation check. What should you do next?
2. A door wiggles up and down on its hinges. What should you do to correct this problem?
3. What are the two types of glass used in today's vehicles?

Math Problems

1. If factory specs say that panel overlap should be ¾ inch and you only have a ¼-inch overlap, how much more is needed?
2. A foreign service manual gives a 12.7 mm spec for a drilled hole. What size U.S. or metric drill bit should be used?
3. Two holes have to be drilled 2 inches apart. How far apart will the holes be in millimeters?

Activities

1. Visit a salvage yard. Ask to inspect some wrecked vehicles. Write a report on the types of glass damage and injury that might have resulted from the collisions on some of the vehicles in the salvage yard.
2. Write a report on how glass is manufactured. Use the library or the Internet to do your research.

Passenger Compartment Service

CHAPTER 16

OBJECTIVES

After reading this chapter, you will be able to:

- Identify the major parts of a typical passenger compartment.
- List which interior parts and panels are often damaged in major collisions.
- Remove and replace passenger compartment trim pieces.
- Explain the parts of power and manual seat assemblies.
- Service front and rear seats.
- Remove and install a headliner.
- Remove and install a dash assembly and its instrument cluster.
- Describe methods of fastening interior parts.
- Find and correct air and water leakage into the passenger compartment.
- Answer ASE-style review questions relating to passenger compartment service.

INTRODUCTION

During a major auto accident, the interior or passenger compartment of the vehicle is often crushed in and damaged, requiring major structural repairs before repainting. Passenger compartment damage results from many factors. Damage to the windshield, dash, seats, and door trim can be caused by deployment of the air bags. Other damage such as cracked plastic, bent seat frames, and soiled seat cushions can result from the impact of human body parts with the interior of the vehicle. With side hits, the pillars and doors can be smashed into the passenger compartment, damaging interior parts.

During collision repairs, the body technician often has to remove and replace the parts of a passenger compartment. To be a competent technician, you must possess the skills and knowledge to service all interior components. This chapter will overview the most common repairs and problems that you will encounter on the job.

NOTE *Chapter 23 explains the replacement of air bags and seat belts. Refer to that chapter for more information on restraint systems, if needed.*

16.1 PASSENGER COMPARTMENT ASSEMBLIES

The interior of today's motor vehicles is much more luxurious than in the past. A multitude of new fastening methods, power seats, sound systems, and navigation systems has added to the cost and complexity of repairing interior damage (Figure 16–1).

The force of a car wreck is unbelievably powerful and destructive. Modern passenger compartments are designed for beauty and safety. Engineers and designers are aware that any protruding object on an interior surface can act as a knife, injuring people who may be thrown around violently inside the passenger compartment during a serious collision.

This has led auto manufacturers to adopt design and fastening methods that allow interior surfaces to act as soft, crushable surfaces if hit by parts of human bodies during a collision. Softer, rattle-free plastic clips are replacing many sharp metal fasteners, such as screws. The use of metal fasteners is limited to heavier parts mounted behind the plastic trim pieces.

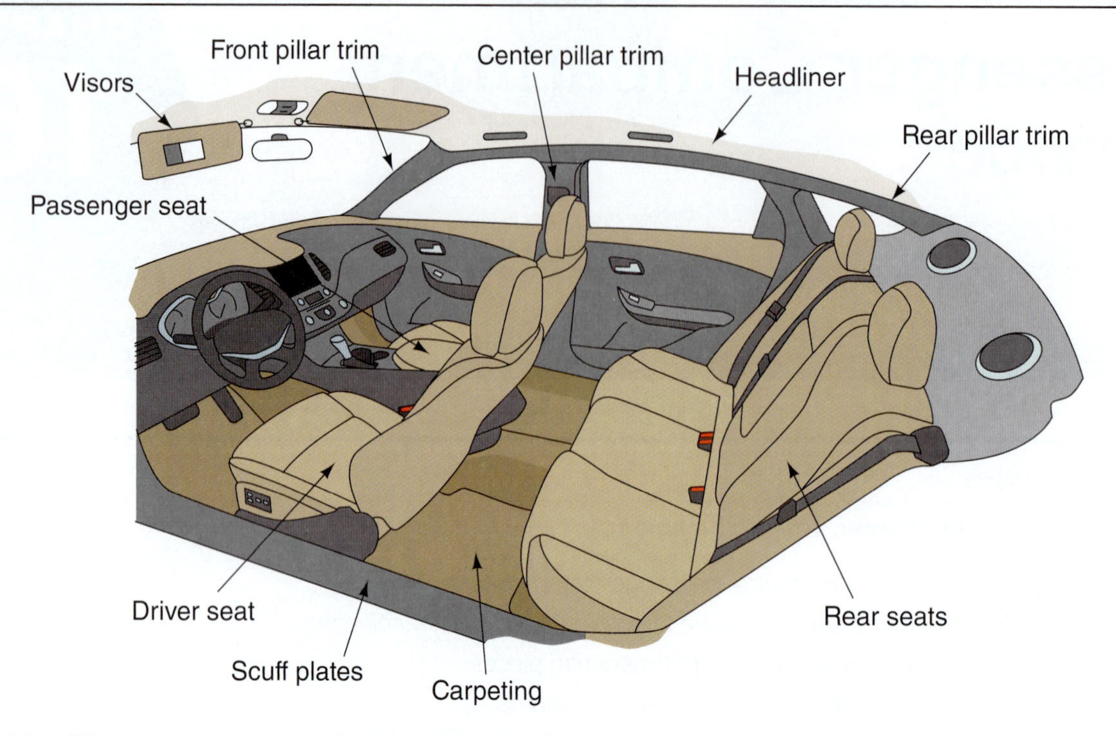

Figure 16–1 There are many parts and interior trim panels that can be damaged during a major auto accident. It is important to know their names, locations, and how to service them.

Besides the vehicle restraint system, doors, and glass, the major parts of a passenger compartment or interior include:

- The **dash assembly**, which includes the dash panel or cover, instrument cluster, heating and air-conditioning vents, stereo, and related parts.
- The **instrument cluster**, which fits in the dash assembly and normally contains the warning and indicator lights, gauges, and speedometer head.
- The **seat assemblies**, which include the seat tracks, seat cushions, headrests, trim pieces, and sometimes power seat accessories (seat motor-transmission assembly, heating elements, and so forth).
- **Interior trim**, which includes the upholstery as well as the plastic panels, covers, and moldings that fit over the pillars, headers, rockers, and other unattractive parts in the passenger compartment (Figure 16–2).
- The **steering column assembly**, which uses a long steel shaft to transfer steering wheel rotation to the steering gear assembly. The steering gear transfers this motion to the front wheels.

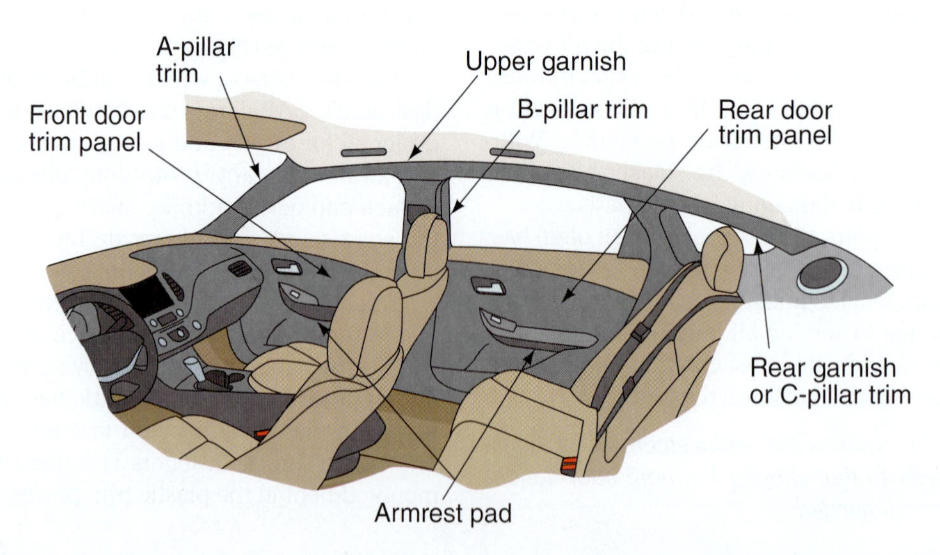

Figure 16–2 Pillar trim pieces, also called garnish trim, include plastic covers over pillars. Padded door trim panels help protect a passenger thrown sideways during a serious side or T-bone collision. Small trim pieces often cover bolt and nut heads, as on the upper shoulder belt anchor bolt and grab handles.

- The **headliner assembly**, which is a cloth or vinyl cover for the inside of the roof panel. It sometimes has grab handles, trim for interior lighting, and a sound deadening backing.
- The **carpeting**, which is a woven fabric cover, often with a sound-deadening backer, that fits over the floor panels.
- **Weatherstripping**, which surrounds the door openings to prevent air and water leakage around the doors.

NOTE *Door assemblies, restraint systems, and glass are also classified as structural members of the passenger compartment. These parts protect the driver and passengers during a major auto accident. Because of their complexity and importance, door assembly and glass service was detailed in Chapter 15. Restraint systems (air bags and seat belts) are covered in Chapter 23.*

16.2 INTERIOR TRIM

Mentioned briefly, various pieces of trim, also called **interior trim panels**, are used in the passenger compartment for appearance and safety. Most are held by plastic snap-in clips or small screws. Sometimes screw heads are covered by small plastic plugs or small plastic trim pieces. Screws can also be hidden under protruding parts. Vehicle-specific service reference materials will give locations for the fasteners holding interior trim parts (Figure 16–3).

The major interior trim panels include:

- **Pillar trim panels**, which fit over the upper section of the A-, B-, and C-pillars. The A-pillar trim fits next to the sides of the windshield. The B-pillar trim covers the center pillar on a four-door sedan. The C-pillar trim, also called quarter trim, covers the area on the sides of the back seat.
- The **dash panel**, which fits between the A-pillars, or windshield pillars, to hold the instrument cluster, air-conditioning system vents, passenger side air bag, glove box door, stereo, and other items.
- **Door trim panels**, which are padded trim pieces that fit over the door frames. They include provisions for armrests, inner door handles, stereo speakers, and other parts.
- **Glass trim panels**, which fit over the edges of the windshield and back glass.
- **Sill plates**, or *scuff plates*, which cover the rocker panels and hold the edges of the carpeting.
- Cloth-, vinyl-, or leather-covered *visors*, which swing down to block the sun from the upper area of the window glass.

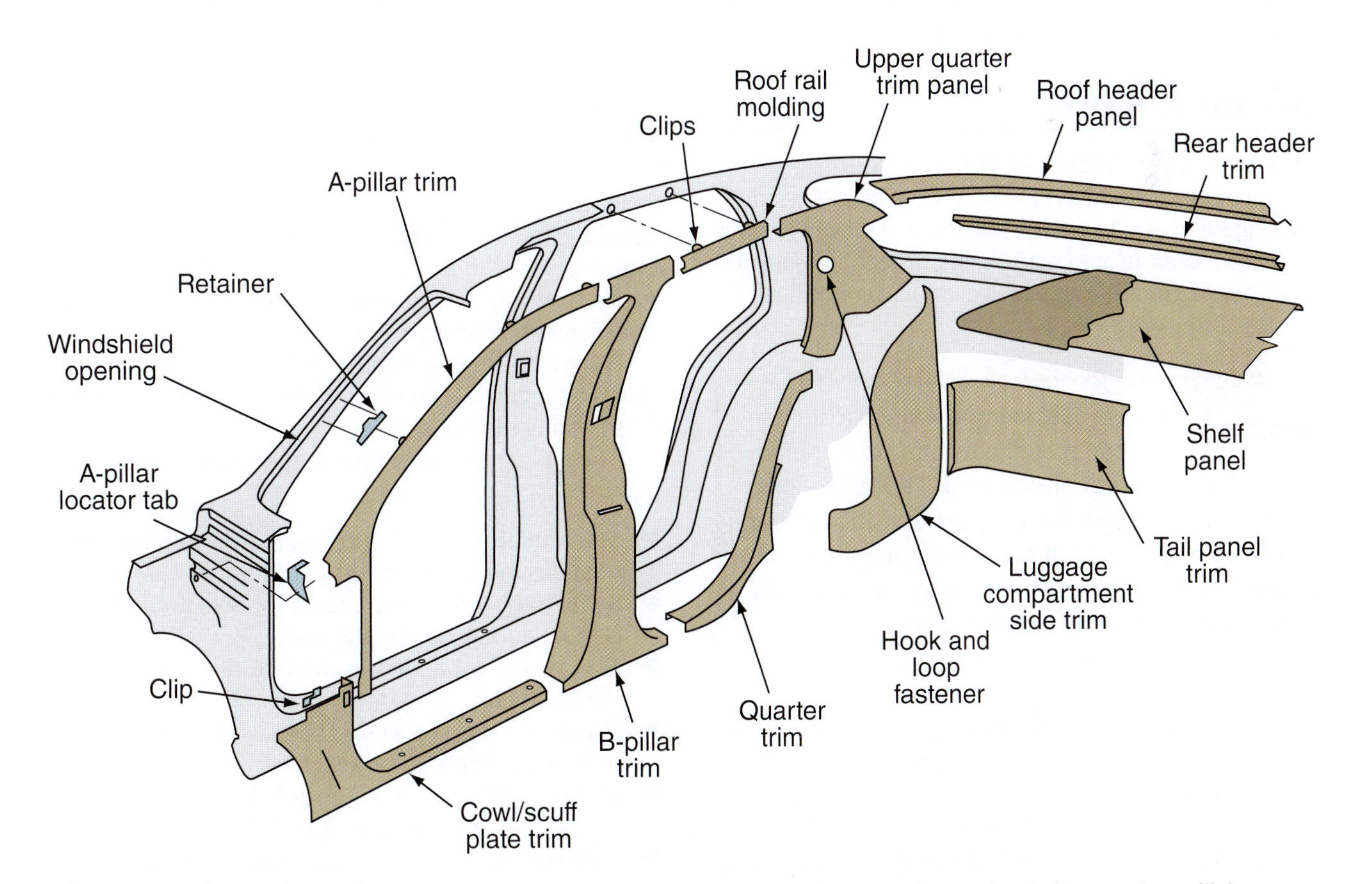

Figure 16-3 Study the location of fasteners for interior trim pieces that often must be serviced after a major collision.

Interior Trim Service

The key to servicing interior trim pieces is finding how they are fastened or held in the vehicle. For appearance's sake, most fasteners are hidden and hard to find. Screws or clips can be hidden under small pop-out plugs or cover plates.

When in doubt, always refer to service information for the specific vehicle interior to be disassembled. If needed, print out a copy of an illustration showing how all of the interior trim is fastened. Vehicle-specific illustrations will show fastener types and locations.

When servicing trim pieces, remember these basic rules:

- Make sure all metal screws are removed from hidden areas before popping out plastic clips.
- Be careful not to break brittle plastic trim pieces. Do not use excessive force, which could bend or even break the panel. Use a forked trim tool to reach behind trim pieces to pry out all plastic clips. Pulling force should be applied to or near the head of the clip. This will avoid trim panel breakage.
- Disconnect the battery before disconnecting any wires in the passenger compartment. This will prevent a possible short circuit that could quickly burn up wiring.
- Keep all fasteners organized in plastic bags or cans. This will help you keep track of their locations and speed reassembly.
- Use a logical sequence of trim panel removal. Usually one trim piece overlaps and helps hold an adjacent panel. Remove the top trim piece first.

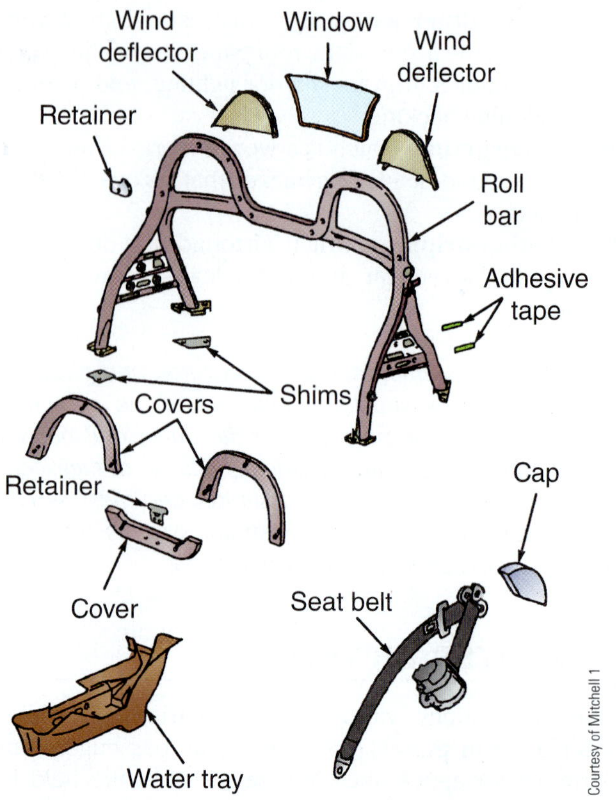

16.3 ROLL BARS

A **roll bar** is a steel framework designed to protect people in a vehicle with a convertible roof during a rollover accident (Figure 16–4). A *fixed roll bar* does not move. A *retractable roll bar* uses an automatic mechanism to raise the roll bar when electronic sensors detect a rollover accident.

DANGER

Always use a torque wrench to tighten roll bar mounting bolts to manufacturer-recommended specifications. This will help ensure that the roll bar performs properly. If the vehicle is in an accident that causes it to flip over, you want the bolts and roll bar to perform as designed.

The roll bar is often padded and covered with vinyl or plastic. Sometimes wind deflectors attach to the roll bar to quiet the passenger area in a convertible. Large case-hardened bolts are normally used to secure the roll bar assembly to a reinforced area of the frame or unibody structure.

Courtesy of Mitchell 1

Figure 16–4 **Several convertible sports cars and some trucks now come factory-equipped with a roll bar. If a vehicle flips over during an accident, a strong, structural roll bar will help protect the occupants from being crushed and killed. Always use a torque wrench to tighten the roll bar and seat belt anchor bolts.**

16.4 SEAT SERVICE

Seats are often damaged during a collision. A seat can be damaged by the inertia of its occupant, by side impact intrusion into the passenger compartment, or by bloodstains. You might also have to remove seats for carpet replacement or floor pan repairs.

A *bucket seat* is a single seat for one person. A *bench seat* is a longer seat for several people. Both require similar servicing methods.

The typical parts of a front seat, shown in Figure 16–5, include:

- **Seat cushion**—the bottom section of the seat, which includes the cover, padding, and frame
- **Seat back**—the rear assembly, which includes a cover, padding, and metal frame
- **Headrest**—a padded frame that fits into the top of the seat back
- *Headrest guide*—a sleeve that accepts the headrest post and mounts in the seat back
- *Recliner adjuster*—a hinge mechanism that allows adjustment of the seat back to different angles
- **Seat track**—a mechanical slide mechanism that allows the seat to be adjusted forward or rearward

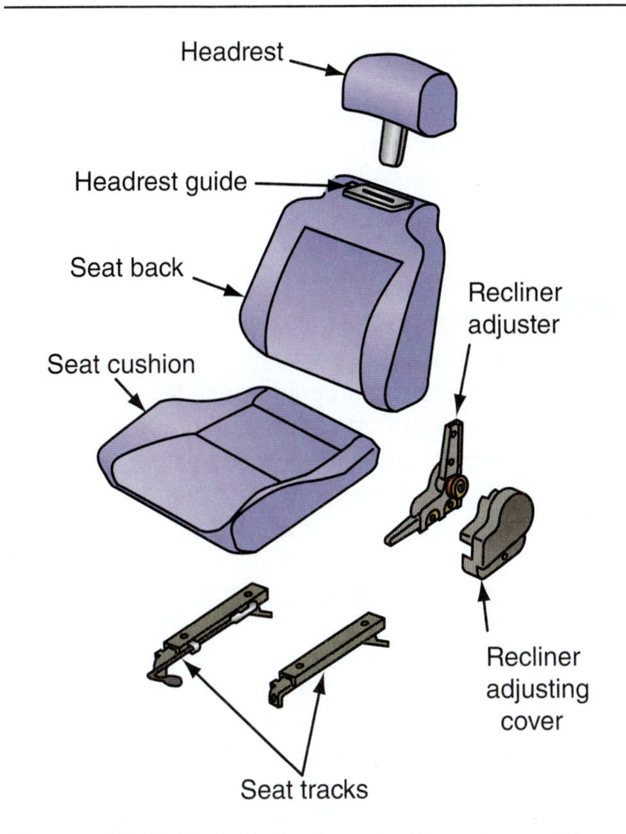

Figure 16-5 Study the basic parts of a seat assembly.

Front Seat Service

Seat anchor bolts are case-hardened cap screws that secure the seat track to the floor structure. Four bolts normally secure the seat to the floor. Sometimes the seat anchor bolts are covered with press-fit or screw-held plastic trim pieces. These pieces may have to be removed to access the seat anchor bolts (Figure 16–6).

To remove the front seat anchor bolts, slide the seat fully backward. This will allow easier access to the front bolts. Then, slide the seat forward to remove the two rear anchor bolts.

If you are repairing power or heated seats, unplug the wiring going to the seat. Tilt the seat to gain access to the harness connectors. Carefully lift the seat out of the vehicle and place it where it cannot be damaged. You might want to cover the seats with shop welding blankets.

When installing front seats, make sure all tools and items that could rattle (coins, fasteners, and so on) are cleaned out from the floor pan. Lift the seat into the interior. You might want a coworker to help you lift and position the seats, because they are heavy and clumsy to handle.

Reconnect any power seat wiring. Start the seat anchor bolts by hand. Run them down snug. Then, use a torque wrench to tighten the seat anchor bolts to factory specifications. If you must replace a seat anchor bolt, make sure you use bolts of equal or greater tensile strength (number of slash marks on bolt head). Weaker seat anchor bolts could break in an accident.

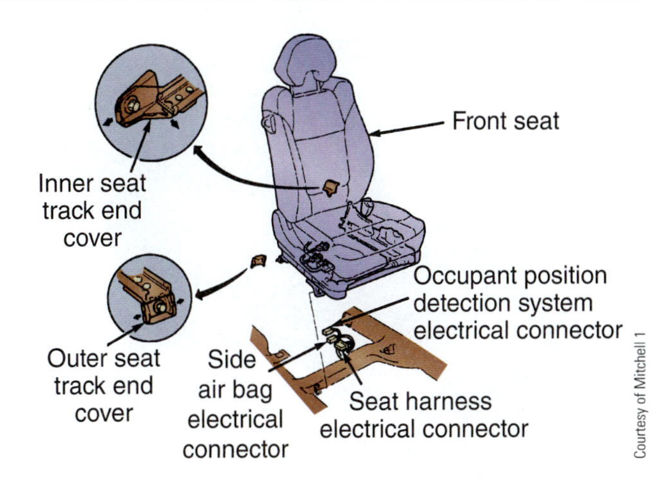

Figure 16-6 To remove a seat, first gain access to the front or rear anchor bolts by sliding the seat all the way forward to get at the rear bolts and all the way rearward to get at the front bolts. Remove the bolts and disconnect any wires to allow removal of the seat assembly.

DANGER If you fail to use a torque wrench to tighten seat anchor bolts, you are endangering the people who will use the vehicle. Loose or overtightened bolts can snap off during a severe collision. This would allow the seat assembly and passenger to fly around inside the passenger compartment. Severe injury or death could result if you fail to properly torque seat anchor bolts. This also applies to seat belt anchor bolts.

Rear Bench Seat Service

A rear bench seat is often held in position by screws or spring-loaded clips. The screws are normally at the front bottom of the seat cushion. You might have to lie down next to the rear seat to see the fastener (Figure 16–7).

When rear seat screws are removed, the seat can be pushed back and lifted up and out.

With spring clips, use your hands to force the seat down and back. You might have to punch the seat cushion back with the palm of your hand to free the spring clips. This will allow you to lift the bench seat out.

To install a seat with spring clips, place the seat in position. Use your knee and blows from the palm of your hand to push the seat down and back. This will engage most spring clip designs. If the seat is held by screws, start them before tightening them.

Seat Cover Service

The seat cover is a cloth, vinyl, or leather cover over the seat assembly. The cover may require replacement when

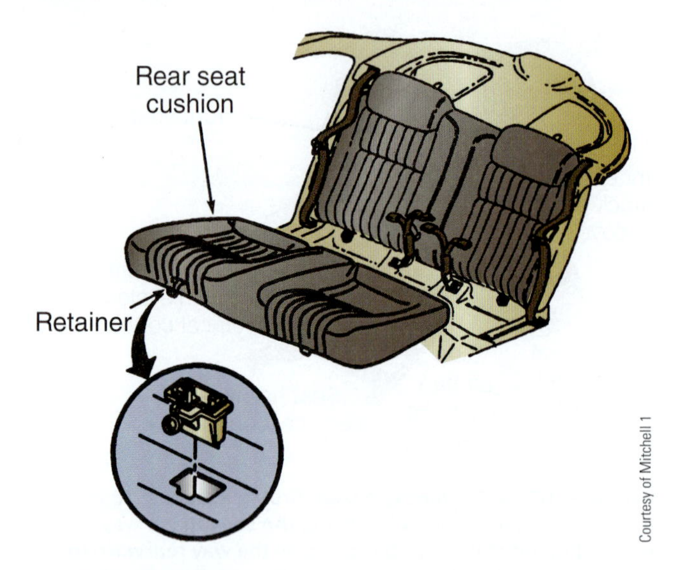

Figure 16–7 Rear seats are fastened by small screws, bolts, or spring clips. With spring clips, you must usually push down and back on the cushion to free the seat. Refer to vehicle-specific service literature if in doubt.

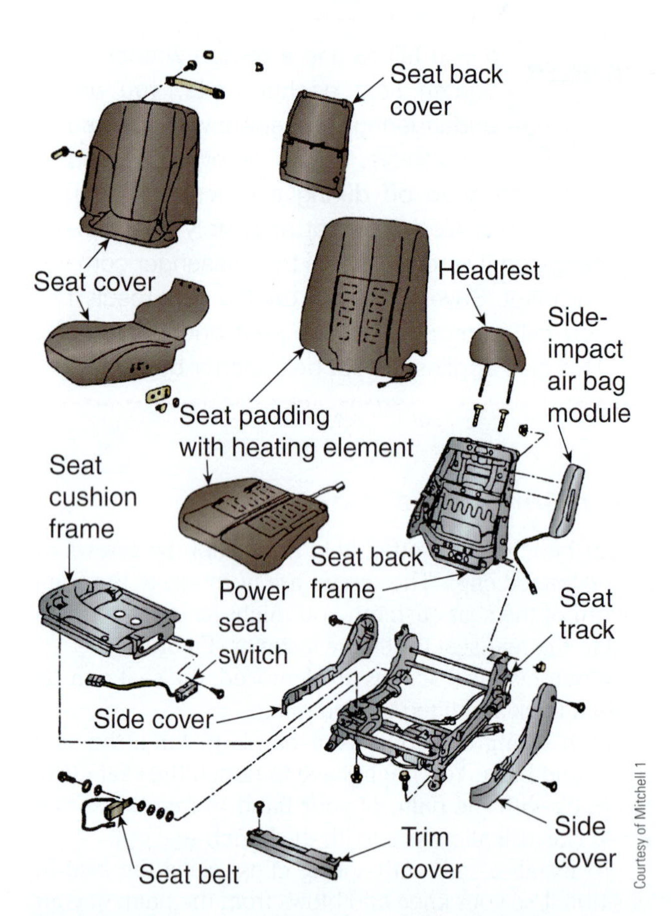

Figure 16–8 This exploded view shows the individual parts of a power seat assembly. Because it contains a side air bag, the battery would have to be disconnected and the air bag system disarmed before repairs.

damaged. With minor damage, an upholsterer can sometimes repair small holes and tears. You must disassemble the seat to replace the covers (Figure 16–8).

Hog rings and clips normally stretch and hold the seat cover over the seat frame and padding. They are located on the bottom of the seat cushion or rear of the seat back. Remove them and you can lift off the seat cover. The new cover can then be installed in reverse order of removal.

DANGER When servicing seats, refer to the manufacturer's manual for details. Procedures vary. Improper seat installation could endanger the vehicle's passengers. Always use a torque wrench and factory-specified torque values when tightening seat fasteners.

16.5 CARPETING SERVICE

Carpeting must often be removed during major structural repairs that involve welding in or next to the passenger compartment. Carpeting must also be replaced if damaged during the auto accident. Bloodstains and tears are a common reason for carpet replacement.

If a bloodstained carpet cannot be cleaned with a strong carpet cleaner, or if it is torn, the carpeting must be replaced. A skilled upholsterer can sew small tears and holes in carpeting. However, with most damage, installation of new carpeting is necessary.

Most carpeting comes in two or three sections. The carpet overlaps under the front seats. With a station wagon, van, or SUV, a separate rear carpet section fits over the back area of the passenger compartment. The major components of interior carpeting are shown in Figure 16–9.

To replace carpeting, you must remove the seats, seat belt anchors, trim pieces, rocker covers, and other parts mounted over the carpeting. These might include the console, any electronic control units, and any wiring harnesses bolted down to the carpet. Screws and clips hold these parts and the carpet down in place (Figure 16–10).

After removal, the new carpet is installed in the reverse order of its removal. Make sure the new carpet is stretched out smooth and is properly centered before installing any fasteners. An adhesive may be required between the carpet and floor in some locations. Refer to the service manual if in doubt (Figure 16–11).

DANGER To prevent possible exposure to communicable diseases, avoid direct contact with any human blood, whether wet or dry, in the passenger compartment. Wear plastic gloves and a respirator and use seat covers to keep the blood from contacting your skin.

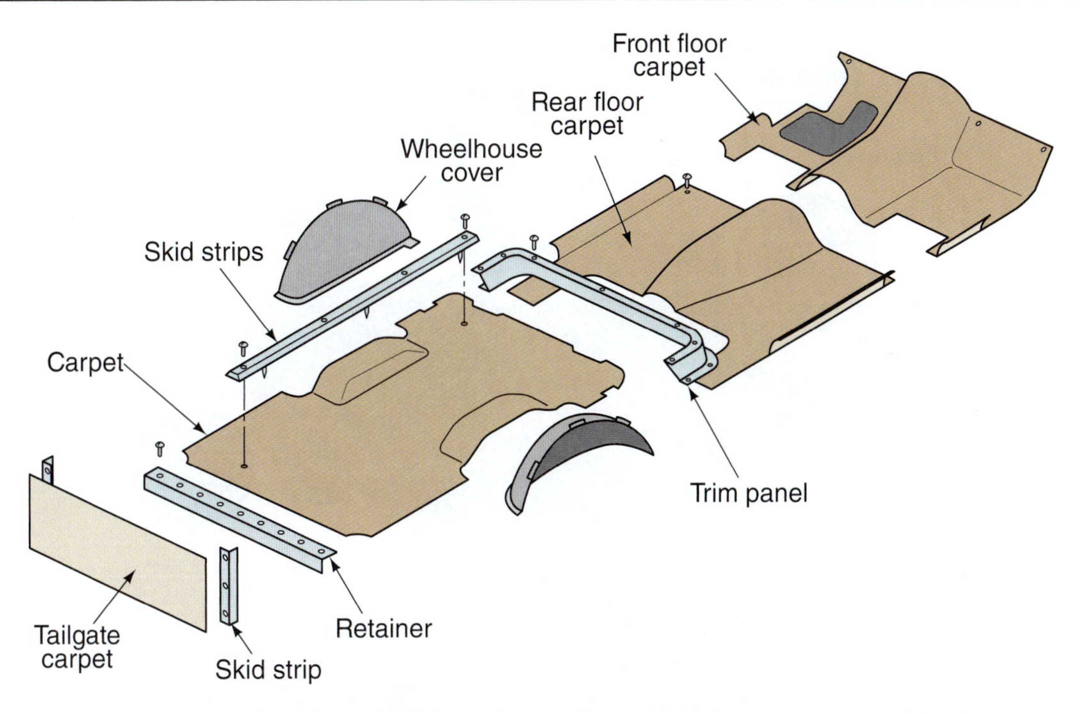

Figure 16-9 Carpeting often comes in sections. Only ruined carpet sections have to be replaced. Study carpet section names, skid strips, and wheel house cover panels. This carpeting is for a station wagon.

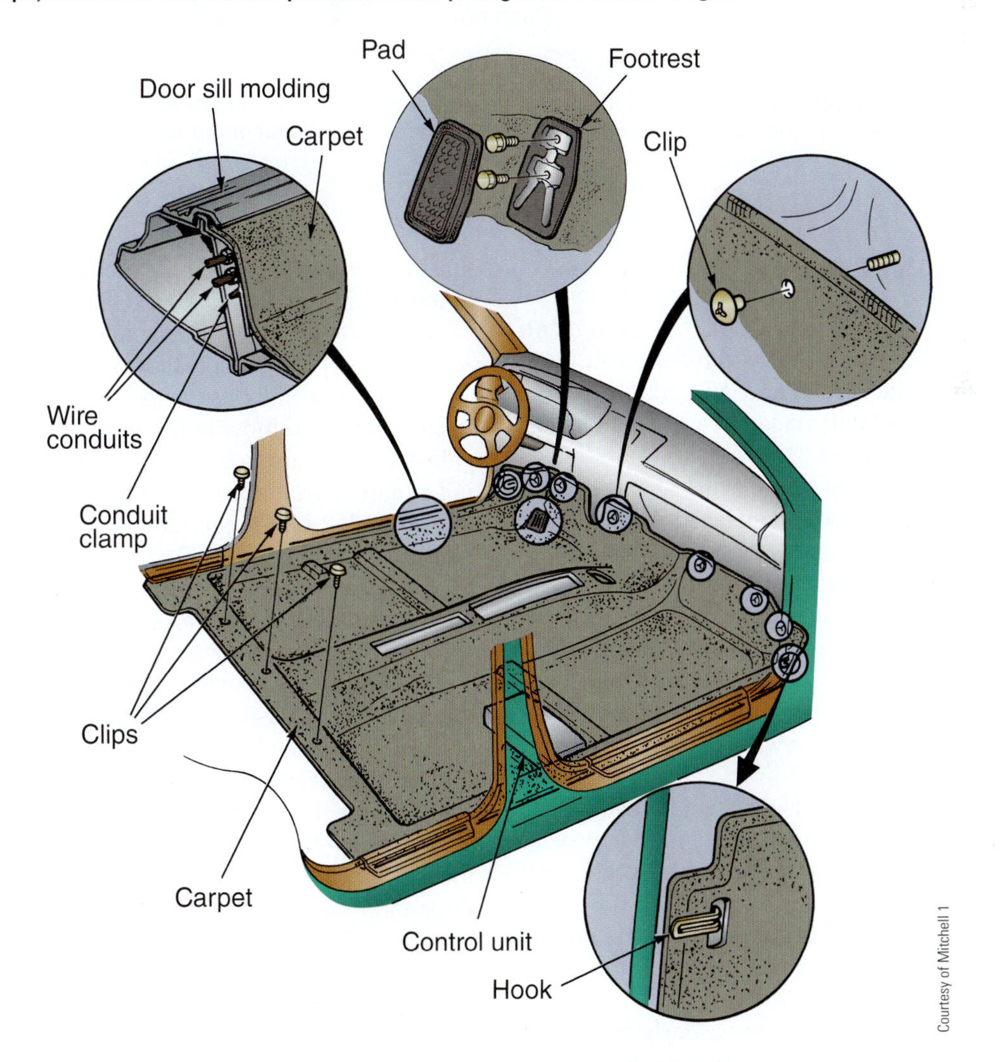

Figure 16-10 To remove carpeting, the seats and console must be removed first. The clips, screws, and parts securing the carpeting can then be removed.

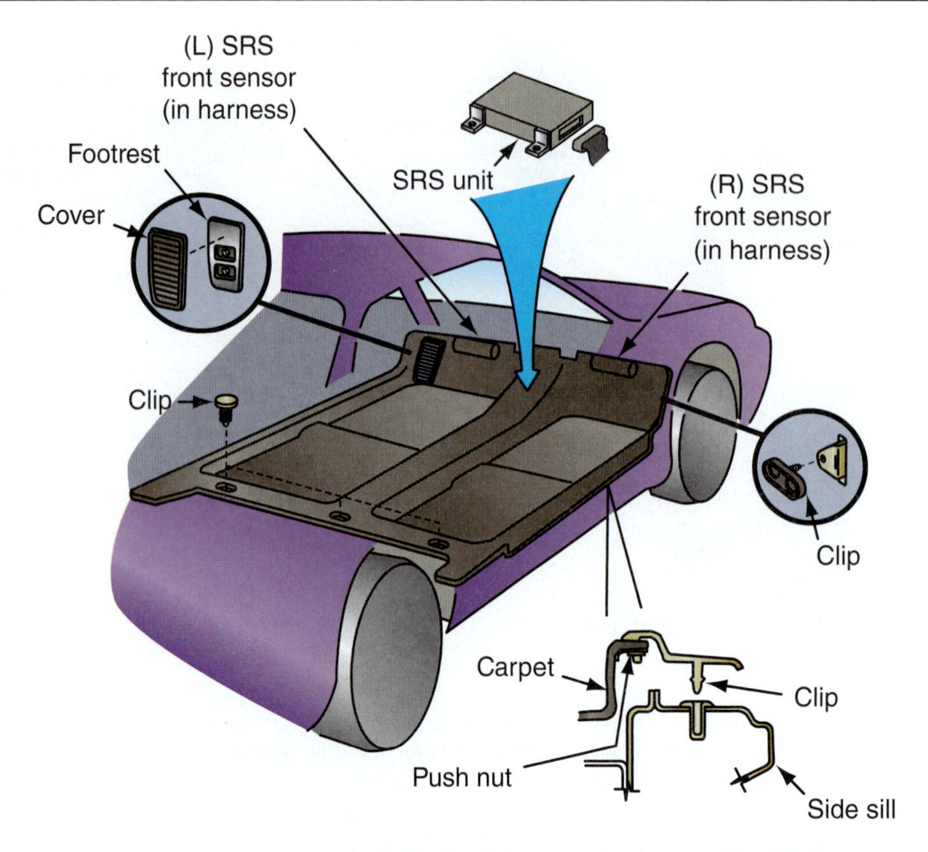

Figure 16-11 **Door sill plates, or skid plates, are the primary way to secure edges of carpeting. Carpeting should be spread out flat and attached starting in the middle. Work wrinkles out toward the sides and rear of the vehicle. Trim or cut the carpet as needed to fit.**

16.6 DASH PANEL SERVICE

The dash assembly, sometimes called the *instrument panel*, is the assembly that includes the soft dash pad, instrument cluster, radio, heater and AC controls, vents, and similar parts. It can be damaged in a collision by air bag deployment and the impact.

When parts of the instrument panel are damaged in a collision, they must be removed and replaced. An exploded view of a typical instrument panel is shown in Figure 16–12. Study the relationship of its various components.

Many instrument panel parts can be replaced without unbolting the dash pad. The instrument cluster, vents, and many trim pieces can be removed and replaced while the main part of the dash remains intact. Vents often snap into place. A thin screwdriver can be used to release and remove most vents.

Some of the screws and bolts that secure instrument panel parts can be difficult to find and remove, however. Some are placed along the bottom of the dash; others are on the sides (Figure 16–13).

A few fasteners can be inside openings in the instrument panel. You will have to remove parts to access these fasteners (Figure 16–14).

If in doubt about how to remove a dash assembly, always refer to vehicle-specific service data. The illustrations will show you the locations and types of fasteners used to secure the dash (Figure 16–15).

Figure 16–16 shows an example of the details given for removing part of a console that attaches to the lower dash.

You must usually use a specific sequence to remove dash parts. By removing the correct part first, you will gain access to hidden fasteners. For example, after removing a snap-on trim panel, you can sometimes gain access to screws that secure the radio and heater or AC control head. Refer to Figure 16–17.

Sometimes radios have small holes in them. You can insert a pointed, slender tool into these holes to release the clips for radio removal.

Heating and AC ducts route air from the blower fan to the vents in the dash panel, defrost vents, or floor vents. During dash assembly service, you must make sure all of the ducts are reconnected properly to provide airflow out of each vent (Figure 16–18).

Figure 16–19 shows how large bolts commonly secure the bottom of the dash assembly. You must lie on the floor and look up to find some of these bolts.

Figure 16–20 illustrates a reinforcement bar that is sometimes used behind the dash assembly. It can be bent and damaged in a severe collision. Note how this one bar bolts into place. It can only be serviced after complete removal of the dash.

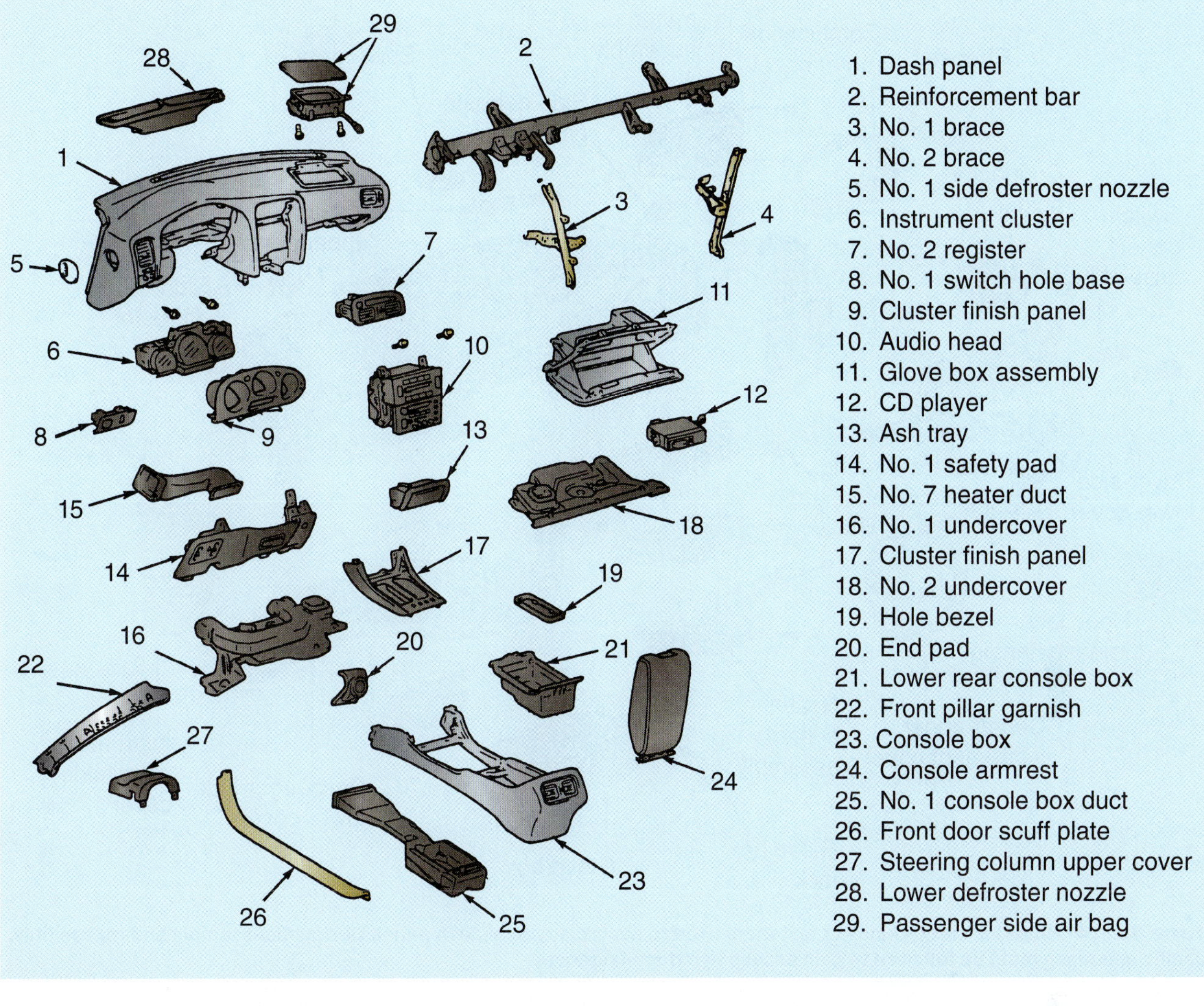

Figure 16-12 This exploded view shows the typical parts of a dash assembly. The dash panel is the core that holds other parts of the assembly. It is bolted to the body structure and reinforcing steel bar in this vehicle.

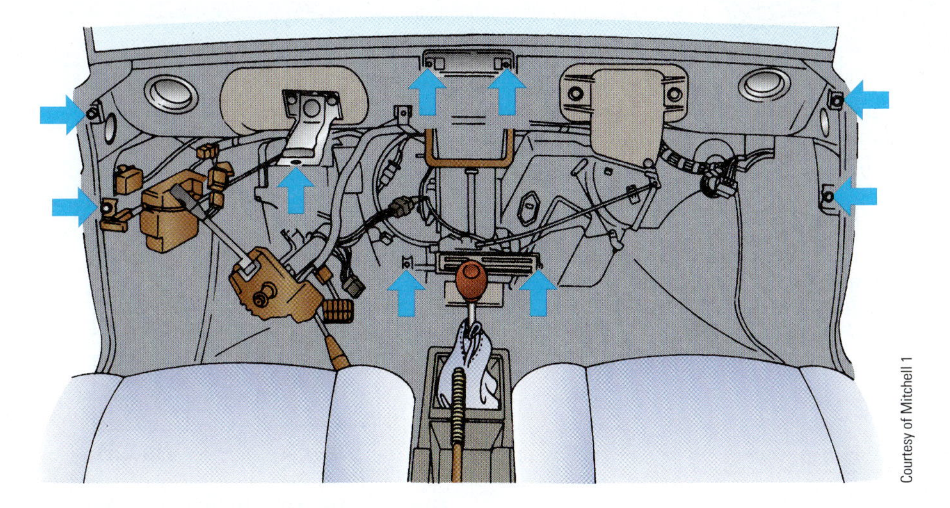

Figure 16-13 Some dash panels are difficult to remove because many fasteners are hidden from sight. Refer to vehicle-specific service information if needed.

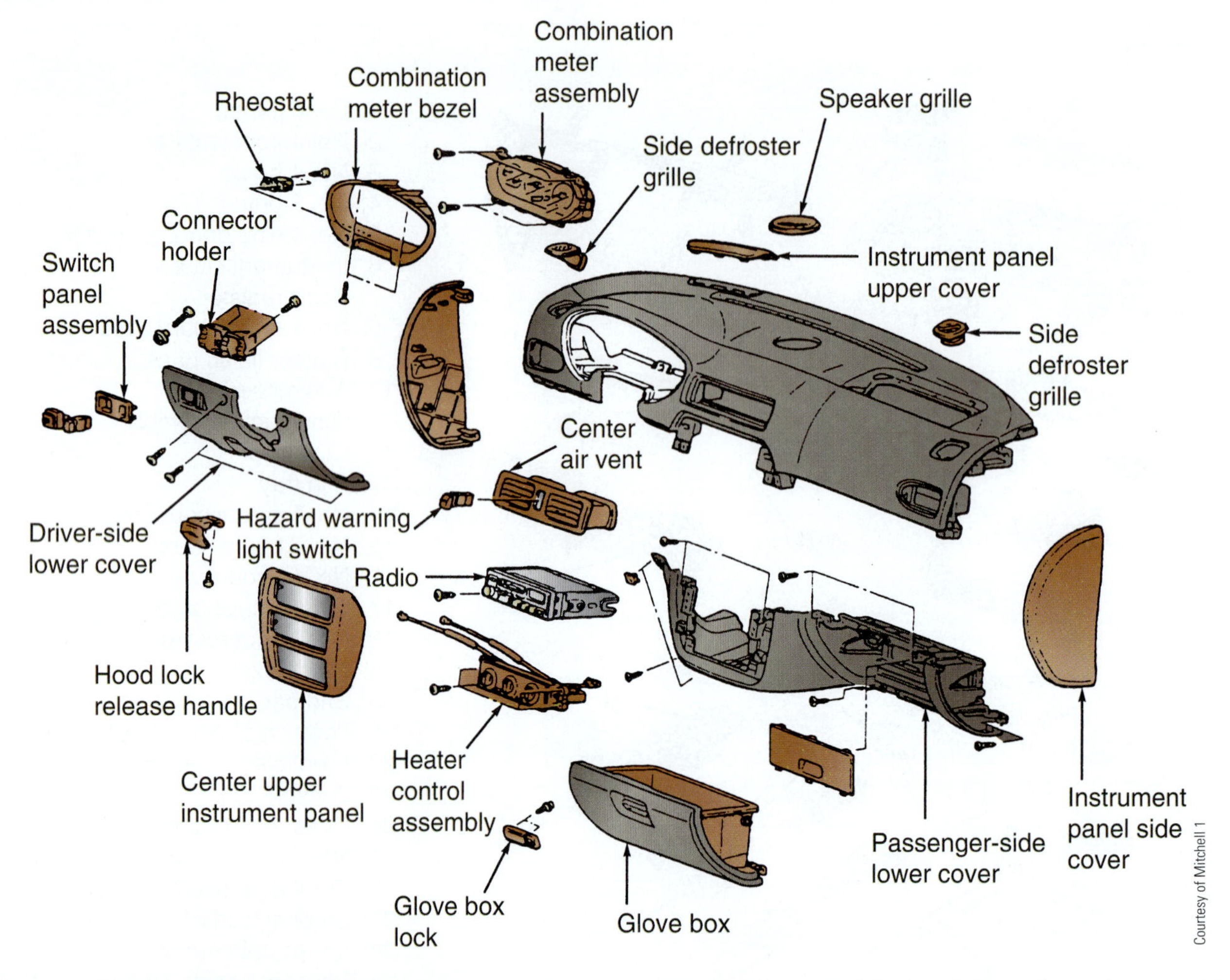

Figure 16-14 Note the various kinds of fasteners used to secure a typical dash panel. During disassembly and reassembly, a specific sequence must be followed to gain access to hidden fasteners.

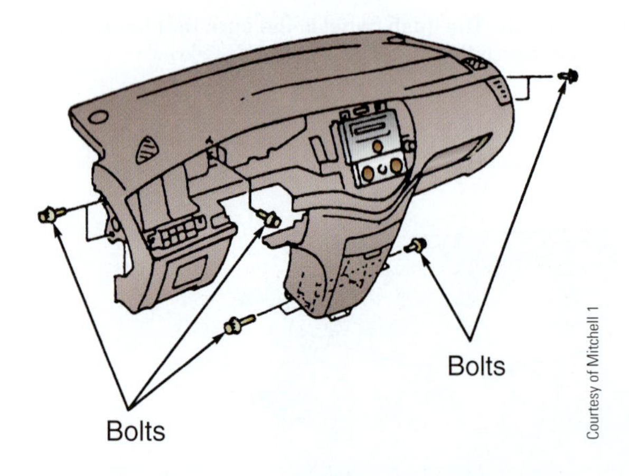

Figure 16-15 After removing the smaller dash assembly parts, you can remove large mounting or anchor bolts.

B C B C A Screws Screws Console Clip Clip Pin View A View B View C Courtesy of Mitchell 1

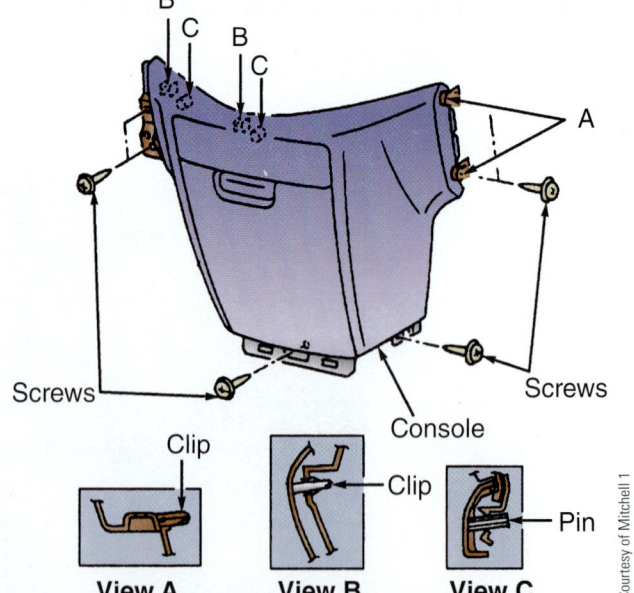

Figure 16-16 Study these cutaway views to see how dash parts are secured. Screws and small friction clips hold the lower console cover. Pry clips out carefully so that you do not crack the plastic.

16.7 CONSOLE SERVICE

Many sport vehicles use a **console** between the bucket seats to house the gearshift mechanism, electrical controls, a console lid and compartment, and other items.

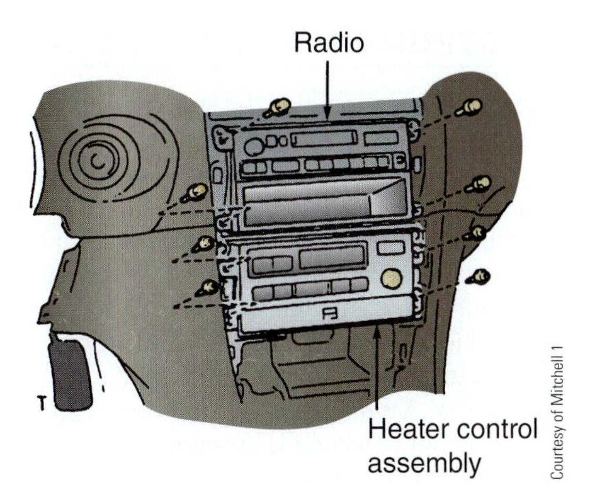

Figure 16-17 Once cover plates are removed, it can be seen that most sound and climate control heads are secured with small screws. However, because other methods of removing stereos exist, refer to vehicle-specific service data.

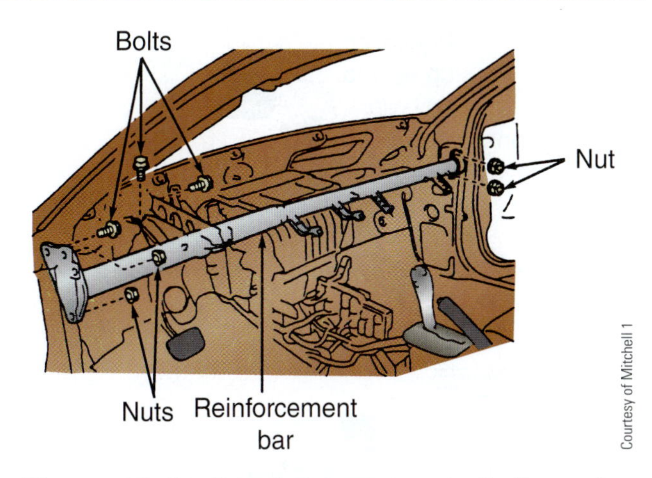

Figure 16-20 If the dash or cowl was badly damaged in an accident, measure the reinforcement bar to make sure it is not bent. Replace it if needed; torque fasteners to specifications.

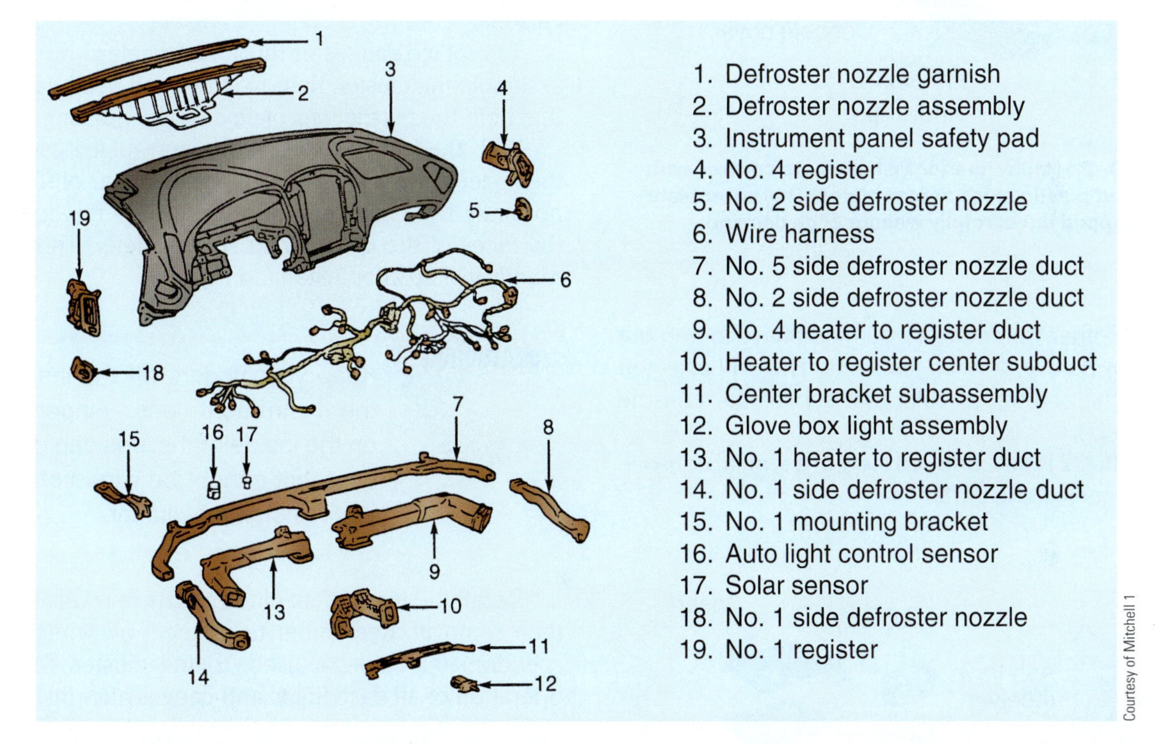

Figure 16-18 Plastic ducts carry airflow to vents in a dash panel. Make sure all ducts are pushed over each other properly as you install the dash panel. Wiring harnesses must also be located correctly before dash installation.

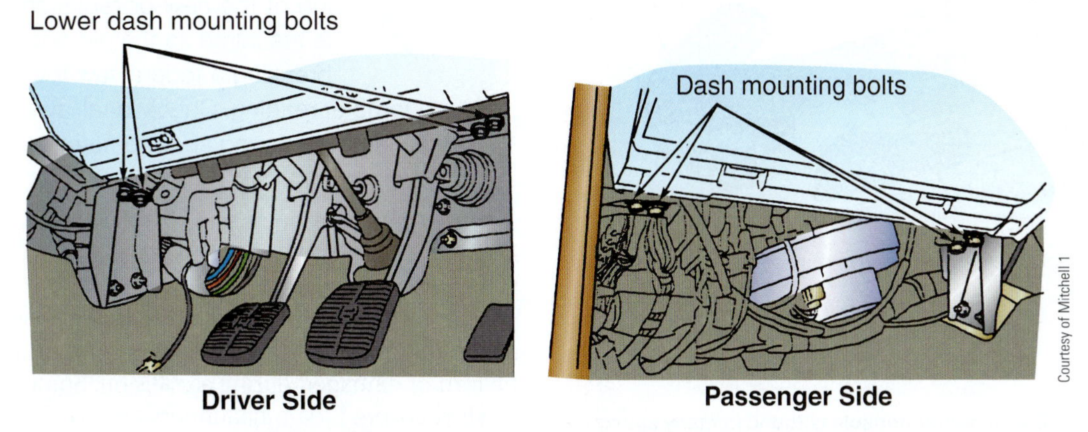

Figure 16-19 To see and service lower dash mounting bolts, you must lie down on the vehicle floor and look up.

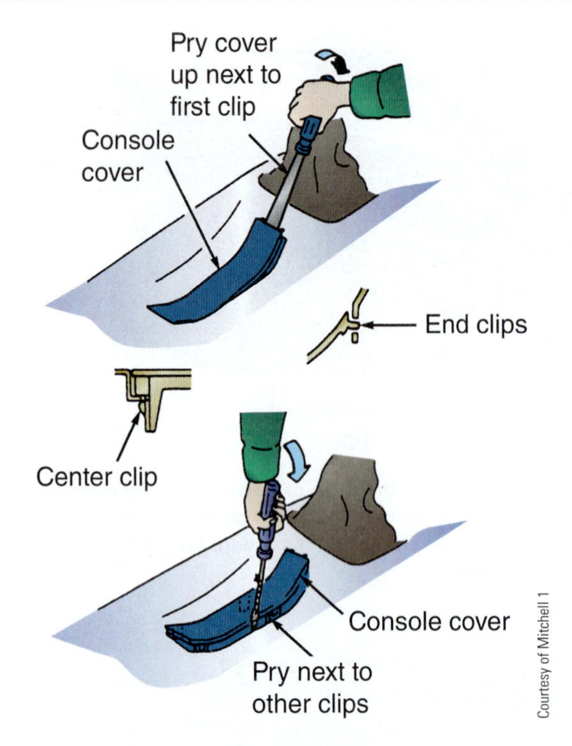

Figure 16–21 Many interior fasteners are covered with small pop-out plastic plugs or face plates. This cover plate has to be popped out carefully without edge damage.

As with other trim panels, you might have to pop out plugs or covers over hidden fasteners. This will give you access to the screws or bolts that secure the console (Figure 16–21).

Figure 16–22 is an exploded view of a typical console. Note the names of the parts.

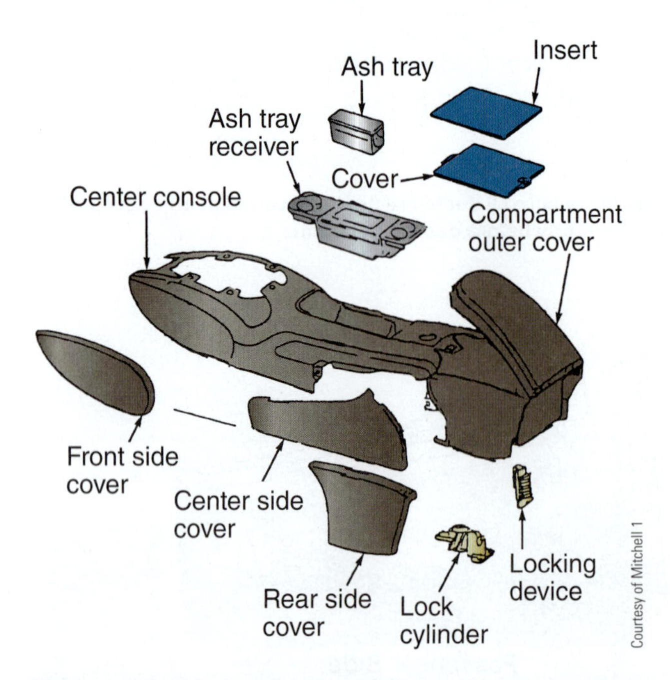

Figure 16–22 A center console is found in many sports vehicles. Note the names of parts.

16.8 INSTRUMENT CLUSTER SERVICE

An instrument cluster contains the speedometer, various gauges, indicating lights, and similar parts. It may require service when damaged in a collision or when parts are not working.

To service an instrument cluster, first disconnect the battery. This will prevent the possibility of an electrical short to ground. Also note that static electricity can damage the chips in a digital dash. You might want to use a static strap to ground your arm and prevent any buildup of static charges. A *static strap* is a soft metal strap that wraps around your wrist and is grounded to the vehicle.

Remove the instrument panel cover. Several screws secure it. Next, remove the screws that hold the cluster to the dash. Pull the cluster out far enough to disconnect the wires and speedometer cable. Then you can lift the cluster out. Bulbs can be replaced from the rear of the cluster.

To replace gauges or the speedometer, you must disassemble the cluster. Remove the small hex head screws that hold the plastic lens plate over the housing.

With the lens removed, you can replace gauges and the speedometer head. Screws on the rear of the cluster normally hold each unit in place. Keep fingerprints off the faces of the gauges and speedometer. They will be readily visible after installation.

Keep your fingers off the inside of the instrument lens. Fingerprints on the inside of the lens can collect dust that cannot be wiped off after you reinstall the cluster.

Install the instrument cluster parts in reverse order of their removal. Remember to connect all wires and the speedometer cable, if used, to the cluster. Check the operation of all dash lights and gauges after installation.

Make sure the speedometer reading of the new or replacement unit is the same as the old one. You are breaking a federal law if you alter a speedometer reading. Check local laws for rules pertaining to reporting speedometer service.

16.9 HEADLINER SERVICE

The headliner assembly is a cloth or vinyl cover over the inside of the roof in the passenger compartment. It can be torn or damaged during a collision. Some thick cloth- or vinyl-covered foam headliners are bonded directly to the roof panel. Others are thin vinyl suspended by metal

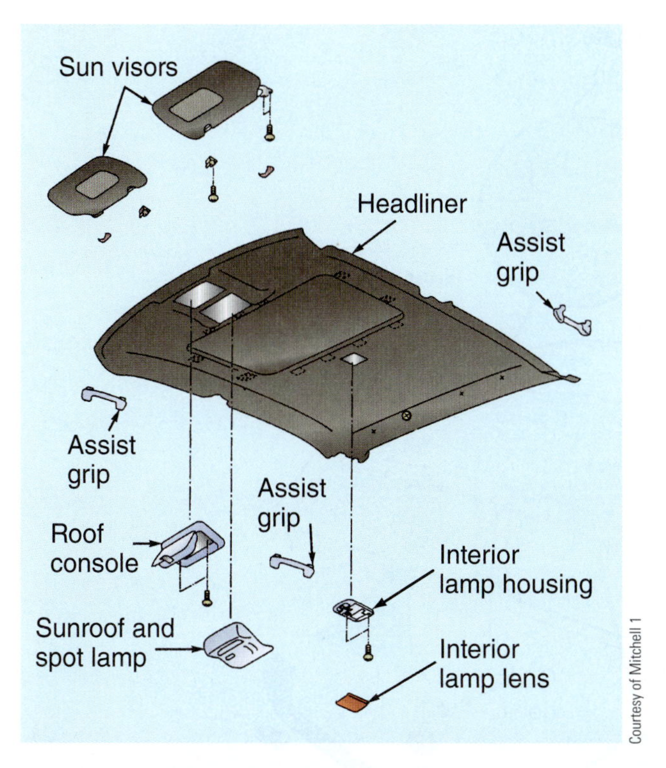

Figure 16-23 **Most of today's headliners are large, fabric-covered, sound-deadening panels. After removing the parts over a headliner (grips, roof console, lamp assemblies, sun roof switch, and visors), release the headliner mounting clips. Sometimes the seat and shoulder harness assemblies must also be removed. A large headliner must be slid through the front door if the windshield and rear glass are still intact.**

rods and bonded around the edges of the roof. One headliner assembly is shown in Figure 16–23.

To service a headliner, first remove all of the trim pieces around the edges of the roof. Various screws and clips secure the trim pieces. You may also have to remove the sun visors, grab handles, and other parts for headliner service (Figure 16–24).

When installing a foam-backed headliner, be careful not to overbend and kink it. Center it in position. Then install it in the reverse order of its removal. Again, refer to the vehicle-specific service manual if in doubt.

16.10 LOCATING AIR AND WATER LEAKS

Water leaks are evident when moisture or rain enters the passenger compartment and collects on the carpeting. **Air leaks** normally cause a whistling or hissing noise in the passenger compartment while the car is being driven. Both are frequent customer complaints when vehicles are brought to a body shop. Such problems are often difficult to locate.

Checking Drain Hoses

Figure 16–25 shows how some vehicles use a water drain system for the sunroof frame. The drain system routes any water that collects in the sunroof frame to the outside of the vehicle.

If the drain hoses become clogged with leaves and other debris, water can leak into the passenger compartment. You can sometimes clean out the drain hoses by directing a blast of air through the hose using a shop air nozzle.

Air-conditioning systems also have a drain hose to remove water condensation from the evaporator. The evaporator normally mounts behind the dash on the right or passenger side of the interior. If the evaporator drain hose becomes clogged, water will usually leak out onto the right floor carpeting. To clean this drain hose, you must normally raise the vehicle on a lift or on jack stands. The tip of the evaporator drain hose extends down the firewall. By pinching and opening the tip of the hose, it can usually be cleaned out without major part removal.

Water leaks frequently occur at panel joints and glass-to-metal joints due to cracked or insufficient sealer. Dust and water leaks also occur at doors, windows, trunk lids, and windshields whenever the weatherstripping becomes damaged or loose or when doors or window glass are improperly adjusted (Figure 16–26).

Wind noises are annoying, high-frequency swishing sounds heard when the vehicle is being driven. They are heard mainly around the door when the window is closed. This is generally due to loose, worn, or improperly applied weatherstripping or misaligned doors, which allows air to leak into the passenger compartment.

Wind noise is also produced when moving air hits a projection. This disturbance produces an eddy or swirl behind the object, thus creating a noise (the principle of flute and bugle sounds).

A loose body molding, a poorly aligned front fender, or an improperly adjusted hood are just a few causes of wind noise. A troubleshooting chart on how to identify and solve wind noise problems is given in Table 16–1.

If you suspect an air leak around a rear hatch or door seal, for example, perform a paper pull test on the rubber seal. Place a dollar bill or piece of paper between the rubber seal and body. Close the hatch or door completely. Then pull out the dollar bill or paper. As you pull it out, you should feel a slight drag, or resistance. This shows that the rubber seal is pressed normally against the body and should seal out air leaks. If the dollar or paper pulls out too easily, the hatch or door should be adjusted inward. If the dollar or paper is held too tightly under the seal, the door or hatch is moved in too much and is crushing the rubber seal too tightly.

Types of Leak Tests

The following principal methods are often used to locate air and water leaks (Figure 16–27):

- Spraying water on the vehicle
- Driving the vehicle over dusty terrain
- Directing a strong beam of light on the vehicle and checking for light leakage between the panels

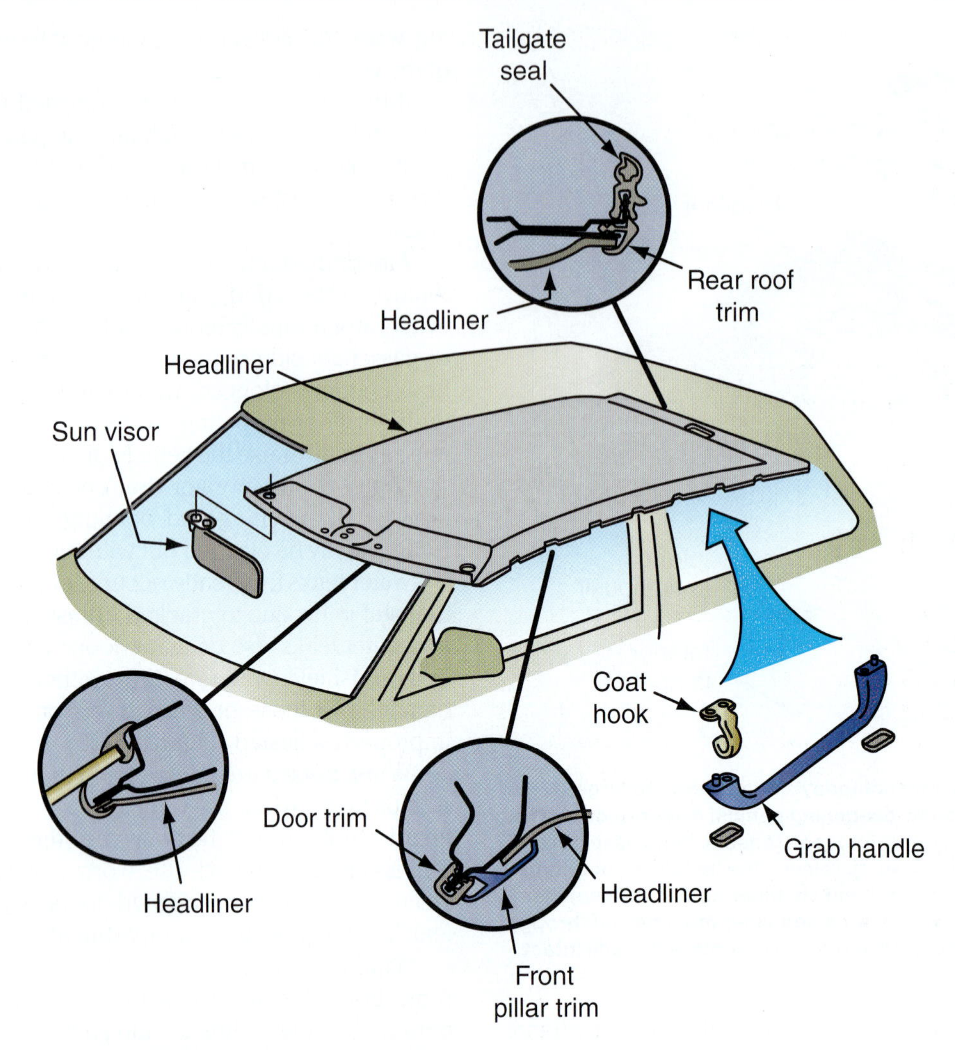

Figure 16-24 This cross section shows how a typical headliner is held on by clips and trim.

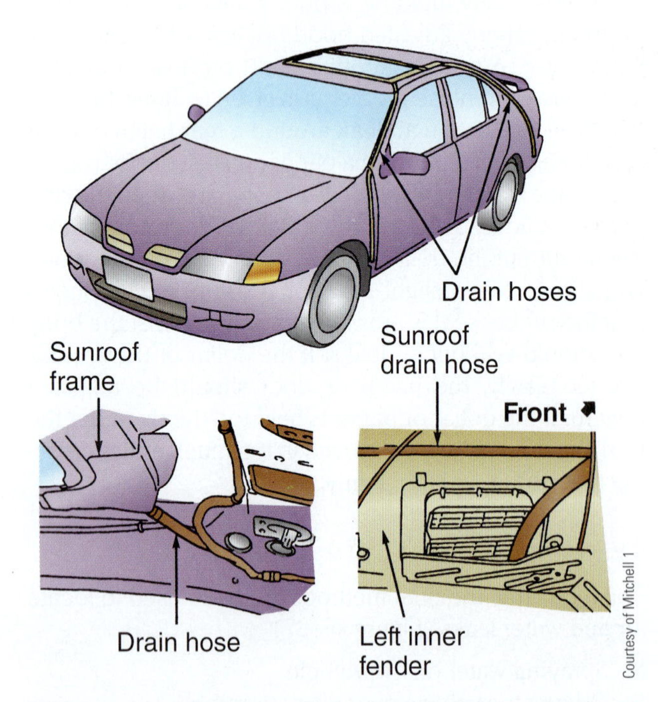

Figure 16-25 Some vehicles have drain hoses for the sunroof. Water leakage in areas around the hose could indicate a clogged drain or disconnected hose.

- Using soapy water and an air nozzle
- Using a listening device

Before making an actual leak test, remove all applicable interior trim from the general area of the reported leak. The spot where dust or water enters the vehicle might be some distance from the actual leak. Therefore, remove all trim, seats, and floor mats from areas that are suspected as possible sources of the leak. Entrance dust is usually noticed as a pointed shaft of dust or silt at the point of entrance. These points should be sealed with an appropriate sealing compound and then rechecked to verify that the leak is sealed.

Leak Checks Using Water

After all the applicable trim has been removed, have one person sit inside the vehicle with all the doors and windows closed. Then, spray the vehicle with a low-pressure stream of water in the suspected area of the leak (Figure 16–28). The person inside the vehicle should act as an observer to locate exactly where the water enters.

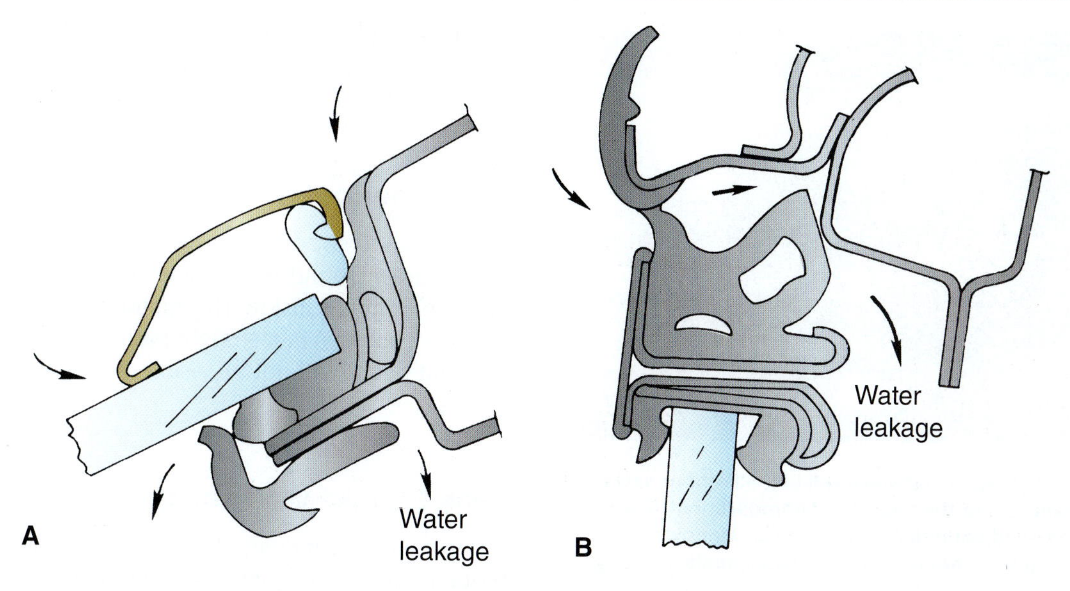

Figure 16-26 These are common locations for water leakage around (A) windshield glass and (B) door weatherstripping.

You can alter the flow from the hose by pressing the end lightly with the thumb, according to the condition of the panel joint. Water should be sprayed for several minutes because small leaks can take a long time before showing up.

Another way to discover water leaks around a windshield or back light is to apply a soapy solution around the outside edge of the window. Then, from inside the vehicle, apply compressed air to the window-to-panel joint. Any gap in the sealant will produce bubbling in the soap solution.

Air Nozzle Leak Test

An air nozzle can also be used to find water and air leaks. As in the windshield test just noted, coat the area where

TABLE 16–1 ELIMINATING NOISE LEAKS

Sources	Cause	Corrections
Weatherstrip	Imperfect adhesion to contact surface and improper contact of lip due to separation, breakage, crush, and hardening	Repair or replace weatherstrip.
Door sash and related parts	1. Improper weatherstrip contact due to a bent door sash 2. Gap caused by corner piece improperly installed 3. Gap caused by corner sash badly finished 4. Separation and breakage of the rubber on the door glass run	1. Repair 2. Install properly 3. Repair with body sealer and masking tape 4. Repair
Door assembly	Improper weatherstrip contact due to improper fitting door	Correct door fit
Door glass	Gap caused due to ill-fitting door glass	Align door glass
Body	Improper body finishing on contact surface for door weatherstrip (uneven panel joint, sealer installed improperly, and spot welding splash)	Repair contact surface
Drip molding	Rise and separation of molding	Repair or replace
Front pillar	Rise and separation of molding	Repair or replace
Waist molding	Door glass gap due to rise of molding and deformation of rubber seal	Repair

Figure 16-27 Rain and snow will test all rubber seals that keep water out of the passenger compartment. They must all be in good condition and all doors, windows, and rear hatches must be adjusted to fit against seals properly.

Figure 16-28 Low-pressure water spray should be directed over potential leakage points on the vehicle body. If water starts dripping inside the vehicle, you have found the source of trouble.

the leak is suspected with a soap and water solution. Then, blow air into the area with an air hose and nozzle. Any leaks will be evident when soap bubbles form (Figure 16–29).

Figure 16–30 shows cross-sectional views of how weatherstripping is often secured to the door openings.

Figure 16–31 shows how to replace weatherstripping.

Leak Checks Using Light

Simple leaks can often be located by moving a strong light source around the outside of the vehicle while an observer remains inside. This method is useful only if the leakage course is in a straight path. If the path deviates, the light beam will not be visible through the turns and curves.

Leak Checks Using a Listening Device

A stethoscope (a medical listening device) with the metal probe removed or a piece of vacuum hose can also help locate air leaks. While someone else drives the vehicle, move the hollow tube around potential leakage points. The sound of any air leakage will become very loud when you move the hose past the air leak point.

Special vacuum leak detectors that listen for the high pitch of an air or vacuum leak are also available. These devices emit a warning sound when a test probe moves near an air leak.

Electronic Leak Detector

An **electronic leak detector** uses a high-frequency sending unit and a receiver unit to find openings or leaks between parts. One technician sits in the interior with the receiver unit. Another technician moves the sending unit around possible leakage points. When a gap between parts or weatherstripping is found, the receiver unit will make an audible signal or illuminate an indicator light. This is shown in Figure 16–32.

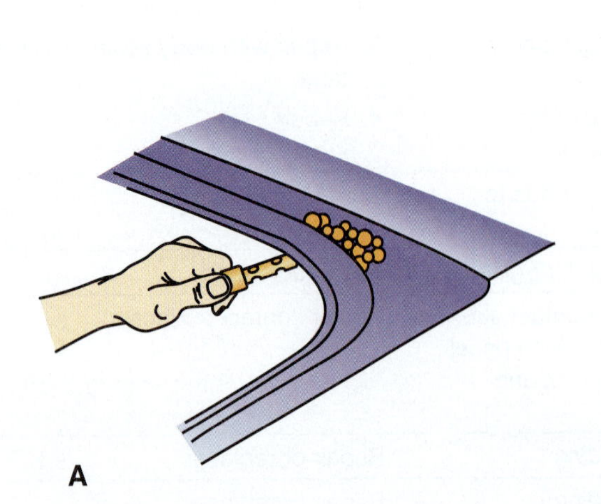

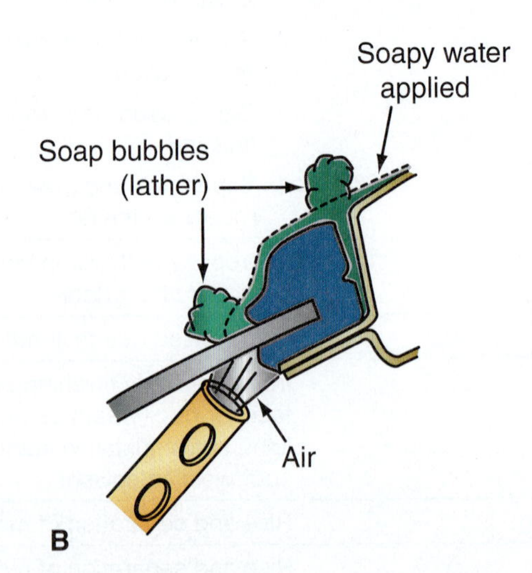

Figure 16-29 (A) You can use soapy water and compressed air to find leaks. Spread soapy water over possible leakage areas, then blow low-pressure air into part gaps. (B) This cutaway view shows how air pressure will form bubbles in the soap solution to pinpoint leaks.

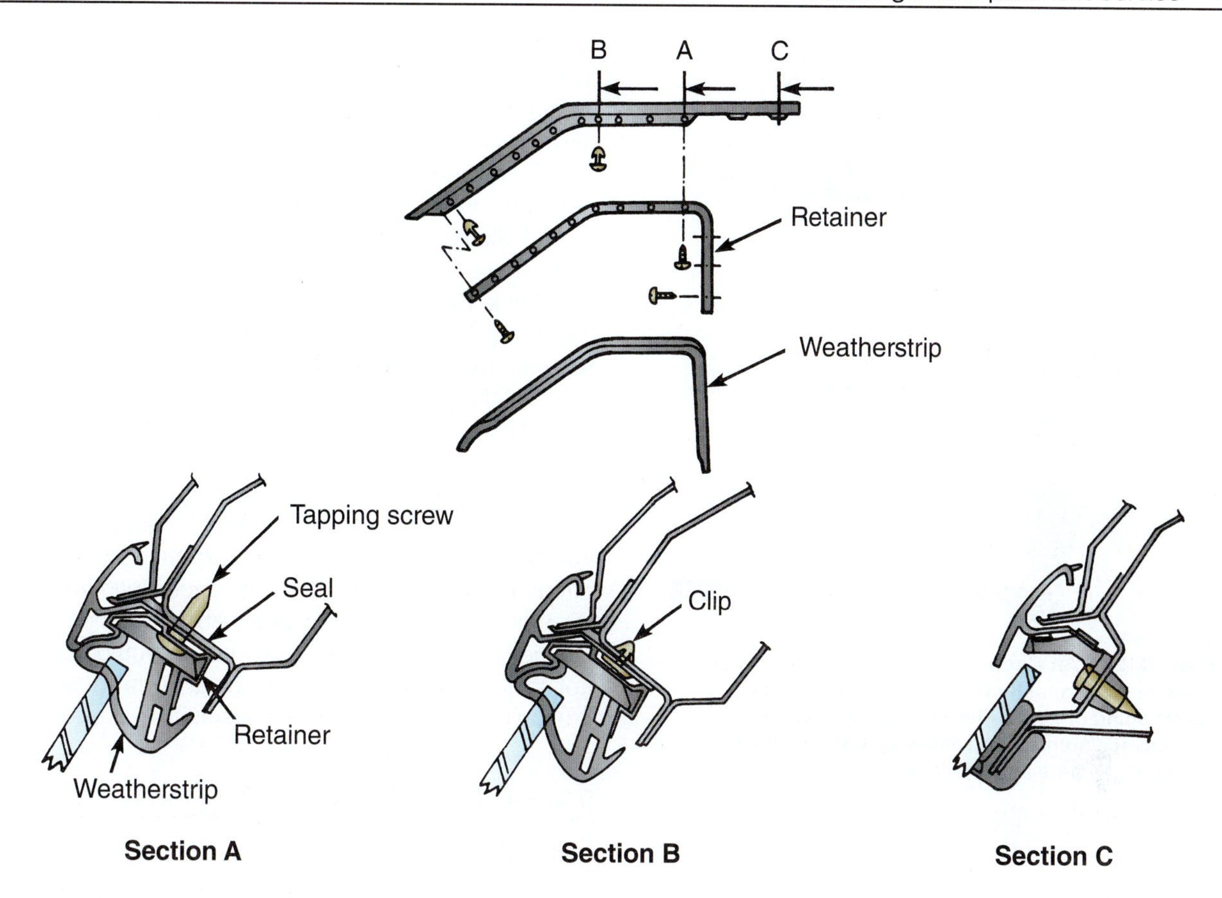

Figure 16-30 This weatherstripping is held on by a retainer.

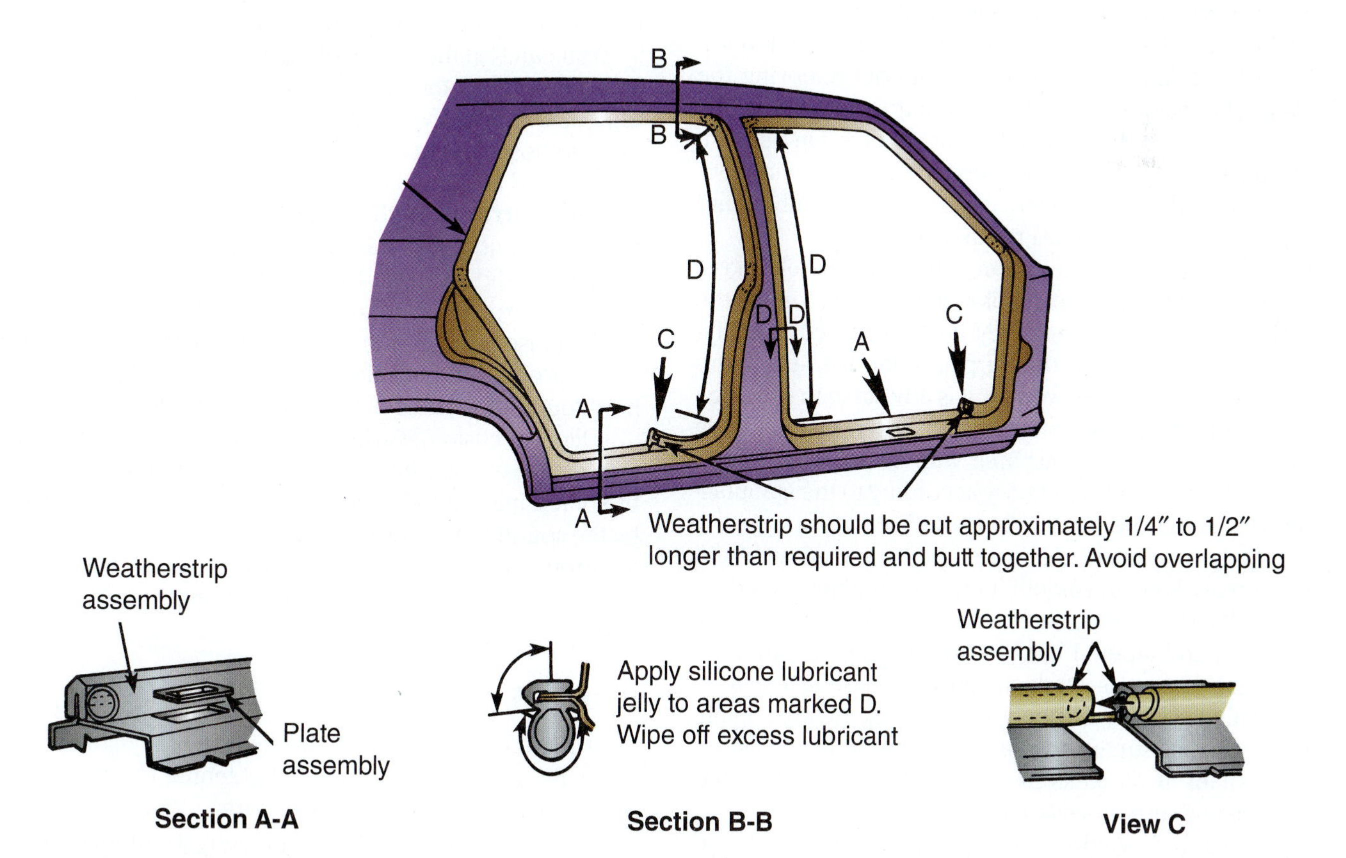

Figure 16-31 Note a typical door gasket, or weatherstrip, installation.

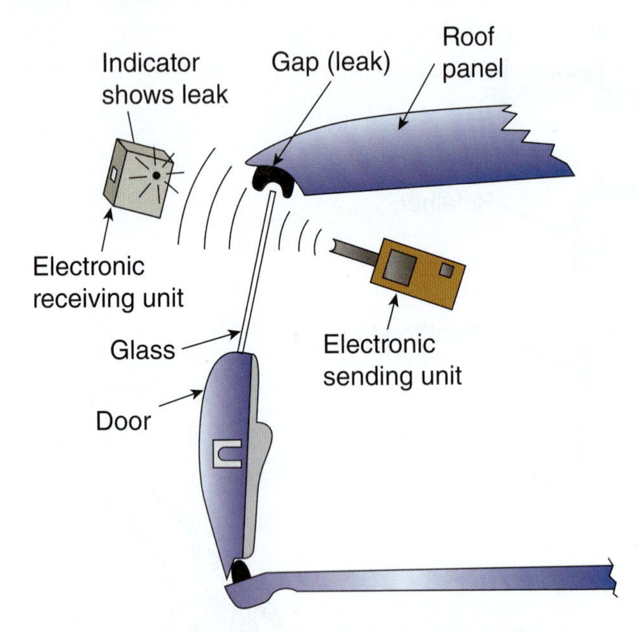

Figure 16–32 An electronic sending unit generates a high-frequency signal that will pass through gaps in weatherstripping or seals. Move the sending unit over possible leakage points. The receiving tool will pick up the signal and indicate leakage locations.

Repairing Leaks

Plugs and grommets are used in floor pans, dash panels, and trunk floors of a vehicle to keep out dust and water from the interior. These items should be carefully checked to ensure they are in good condition.

Vehicle windshields and rear windows often develop water leaks that can be repaired without removing the glass. The majority of leaks occur at the top of the windshield or top and bottom of the rear window. On a station wagon, the leaks usually occur on the rear side windows. If several leaks are detected, it is better to seal all around the area rather than at each leak point.

To repair a water leak through adhesive-held glass, first clean the leak area and blow it dry. Then, trim off the surplus adhesive that extends beyond the edge of the glass. After removing the surplus adhesive, again dry the area using compressed air. It is advisable to use a solvent to clean the area of any oil or grease that might be present. Prime the repair area with a urethane primer-sealer. Allow the primer to dry according to the manufacturer's instructions (generally about 5 minutes).

Apply windshield sealer along the cleaned area and use a putty knife to smooth it out. The sealant should be applied and spread so that it is even with the top edge of the glass and tapered back to the molding clip area. Be sure the sealant is worked into any existing crevices. While the sealant is still soft, water-check the area again. Use a very soft stream or spray of water so as not to disturb the sealant. If no leaks are detected, reinstall trim and remove any surplus sealant on the glass or vehicle.

Doors and windows are sealed against wind and water with rubber gaskets called weatherstripping. Weatherstripping usually fits over a pinch weld flange or inside a channel. The rubber gaskets can be glued on, held with screws or clips, or simply held securely by the design of the gasket.

When the weatherstripping on doors and trunk lids becomes loose, damaged, or deteriorated, dust and water leakage results. On some vehicles the weatherstripping used on doors or trunks is cemented in place.

Check the weatherstrip for correct positioning by placing a thin feeler gauge or a dollar bill between the weatherstrip and frame. If there is little or no resistance when withdrawing the gauge or card, the weatherstrip should be moved closer to the edges of the door or trunk or replaced completely.

When applying weatherstripping around a door or trunk, cut the strip ½ inch (12.7 mm) longer than required and butt the cut ends together. A sponge rubber plug is often used to hold the cut ends together. Some manufacturers require an application of silicone lubricant jelly to the base of the weatherstrip bulb. Be careful not to stretch the weatherstripping during installation. Pulling the strip too tightly will result in an improper seal.

16.11 RATTLE ELIMINATION

Rattles and squeaks are normally caused by loose or rubbing parts. These, in turn, are often caused by loose bolts and screws and improperly adjusted doors, hood, or body panels. Other rather simple things, such as a broken or loose exhaust mount or an improperly secured jack or tire, or even articles in the trunk, can also cause rattles.

You can find the source of rattles using a stethoscope. When you touch the metal probe near parts that are rattling, it will amplify the sound. Both conventional doctor's stethoscopes and electronic stethoscopes are now used to find abnormal noises.

An **electronic stethoscope** is handy when trying to find rattles and other mechanical noises. Some have alligator clips and long test leads that can be connected to auto parts. You can then drive the vehicle while listening to the noise through the tester. When you connect the alligator clip to a part and the noise is the loudest, you have found the source of the rattle (Figure 16–33).

Often, a customer will pinpoint a noise in a certain area of the vehicle when, in fact, the noise is being caused by something in another location. This effect is generated by the sound travelling through the vehicle body. Usually a thorough investigation and a test drive of the vehicle is recommended so that the rattle or noise can be accurately located.

Fixing Rattles

Most rattle or noise repairs involve the readjustment of parts, the replacement of parts, tightening loose attaching hardware, or welding broken parts.

If the noise is outside the passenger compartment near the front, check the hood for proper alignment at

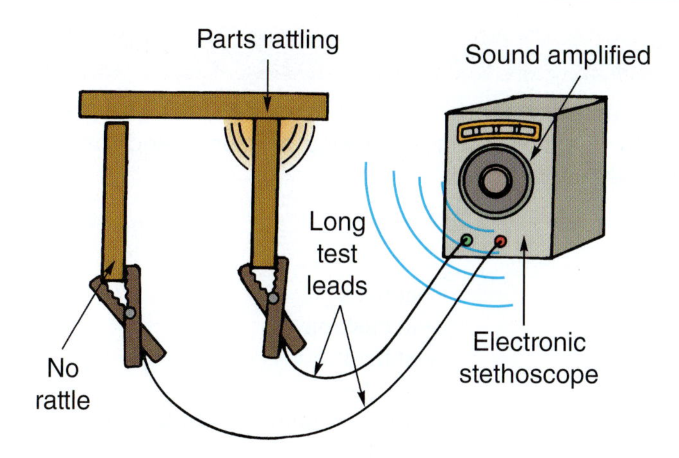

Figure 16-33 Electronic stethoscopes have long test leads with alligator clips. The clips are clamped to parts near the rattle or unwanted noise. A speaker in the tool amplifies noise so you can find which part is at fault.

the front and the back. If the paint is knocked off or scratched on one end, the hood is probably hitting another edge or rubbing against it. Check the hood latch pin, as well as the rubber hood bumpers, for looseness and proper fit. If the back of the hood flutters, readjust the hood so that it properly fits at the back seal. Also check the grille, wheelhousing, trim moldings, and bumper brackets for tightness.

Many areas on the body of the vehicle can also cause rattles, noises, and squeaks. The most susceptible areas are the dash, doors, steering column, and seat tracks. It is also possible for weatherstripping to squeak, especially when it becomes very dry. Lubricant should be applied to moving parts, such as door hinges.

All attaching hardware should be checked for tightness, especially in the area of the suspected noise source. Soft blows with a rubber mallet on structural parts can make parts rattle and help pinpoint the source of noises. Be careful not to damage anything being hit by the rubber hammer, however.

SUMMARY

1. The dash assembly includes the dash panel or cover, instrument cluster, heating and AC vents, stereo, and related parts.
2. The instrument cluster fits in the dash assembly and normally contains the warning and indicator lights, gauges, and speedometer head.
3. The seat assemblies include the seat tracks, seat cushions, headrests, trim pieces, and sometimes power seat accessories (seat motor-transmission assembly, heating elements, and so forth).
4. Interior trim includes the plastic panels, covers, and moldings that fit over the pillars, headers, rockers, and other unattractive parts in the passenger compartment.
5. The steering column assembly uses a long steel shaft to transfer steering wheel rotation to the steering gear assembly. The steering gear transfers this motion to the front wheels.
6. The pedal assembly is a metal frame that bolts under the dash to support and hinge the brake and clutch pedal (manual transmission or transaxle only) assemblies (pedal arms, bushings, pedal hinge shaft, brake light switch, and so on).
7. The headliner assembly is a cloth or vinyl cover for the inside of the roof panel. It sometimes has grab handles, trim for interior lighting, and a sound-deadening backing.
8. The carpeting is a woven fabric cover, often with a sound-deadening backer, that fits over the floor panels.
9. Weatherstripping surrounds the door and trunk openings to prevent air and water leakage.
10. Make sure all metal screws are removed from hidden areas before popping out plastic clips.
11. Be careful not to break brittle plastic trim pieces. Do not use excessive force that could bend or break the panel. Use a special trim tool to pry out all plastic press-in clips.
12. Disconnect the battery before disconnecting any wires in the passenger compartment. This will prevent a possible short circuit that could damage wiring.
13. Keep all fasteners organized in plastic bags or cans to keep track of their locations and speed reassembly.
14. Use a logical sequence of trim panel removal. Usually one trim piece will overlap and help hold an adjacent panel. Remove the top trim piece first.
15. Use a torque wrench to tighten the seat anchor bolts to factory specifications.
16. Water leaks are noticed when moisture or rain enters the passenger compartment and collects on the carpeting.
17. Air leaks normally cause a whistling or hissing noise in the passenger compartment while the vehicle is being driven.
18. The principal methods used to locate air and water leaks are spraying water on the vehicle, driving the vehicle over dusty terrain, checking for light leakage between the panels, using soapy water and air, and using a listening device.

EXERCISES

On a separate sheet of paper, complete the following learning activities for this chapter. Write definitions for the key terms and answer the ASE-style review questions, essay questions, critical thinking problems, and math problems. You can also do the outside activities, possibly for extra credit.

Key Terms

air leaks
carpeting
console
dash assembly
dash panel
door trim panels
electronic leak detector
electronic stethoscope
glass trim panels
headliner assembly
headrest
instrument cluster
interior trim
interior trim panels
pillar trim panels
roll bar
seat anchor bolts
seat assembly
seat back
seat cushion
seat track
sill plates
steering column assembly
water leaks
weatherstripping
wind noises

ASE-Style Review Questions

1. Technician A says that most damage to the interior of a vehicle is due to air bag deployment. Technician B says that passengers or unsecured items can also damage interior parts in a collision. Who is correct?
   A. Technician A
   B. Technician B
   C. Both A and B
   D. Neither A nor B
2. Which of these trim pieces would be next to the windshield?
   A. Garnish trim
   B. Rocker trim
   C. B-pillar trim
   D. A-pillar trim
3. What is the most common method of attaching passenger compartment trim pieces?
   A. Self-tapping screws
   B. Bolts
   C. Adhesive
   D. Plastic clips
4. Technician A says to torque roll bar mounting bolts to specifications. Technician B says it is acceptable to use lower-grade bolts to secure a roll bar. Who is correct?
   A. Technician A
   B. Technician B
   C. Both A and B
   D. Neither A nor B
5. If a rear seat is held by spring clips, what is the most common way to remove the bottom seat cushion?
   A. Pull forward and up before lifting
   B. Push down and back before lifting
   C. Pull forward and down before lifting
   D. Jerk up and down before lifting
6. Technician A says that you must normally remove the seat to replace carpeting. Technician B says that you must normally remove the vehicle dash panel to replace carpeting. Who is correct?
   A. Technician A
   B. Technician B
   C. Both A and B
   D. Neither A nor B
7. Technician A says you cannot contract disease from coming in contact with wet blood. Technician B says you can. Who is correct?
   A. Technician A
   B. Technician B
   C. Both A and B
   D. Neither A nor B
8. Where do the majority of passenger compartment water leaks occur?
   A. Roof area
   B. Glass to metal joints
   C. Floor seams
   D. Trunk area
9. A vehicle has an air leak somewhere around the door. Technician A says to remove the metal tip from a stethoscope to use it as an air leak listening device. Technician B says to use air pressure to blow through potential leak points while another technician sits in the vehicle to listen. Who is correct?
   A. Technician A
   B. Technician B
   C. Both A and B
   D. Neither A nor B

10. Technician A says to use soapy water and an air nozzle to find leaks. Technician B says you could use blows from a rubber mallet to find air leaks. Who is correct?

A. Technician A
B. Technician B
C. Both A and B
D. Neither A nor B

Essay Questions

1. How do you use an electronic stethoscope?

2. Describe the principal methods used to locate air and water leaks.

3. How do you remove front seats?

Math Problems

1. A new dash panel costs $295. If sales tax on the dash panel is 7 percent, what is the total cost for the part?

2. If a small trim piece costs $11, what is the total cost for this part with 5 percent sales tax?

Critical Thinking Problems

1. If you cannot find hidden fasteners and something is preventing dash removal, what should you do?

2. A customer complains of water leakage onto the passenger floor carpet. What should you check first?

3. List six parts of a front seat that will need to be inspected after a crash.

4. Why is it a must to use a torque wrench when tightening seat anchor bolts?

Activities

1. Inspect the interior of a few wrecked cars. Note which parts were damaged in the accident.

2. Look up the repair procedures for the dash of one make and model car. Write a report on what must be done to remove and replace the dash assembly.

SECTION

4

Major Body/Frame Repairs

CHAPTER

17 Body/Frame Damage Measurement

OBJECTIVES

After reading this chapter, you will be able to:

- Explain how impact forces are transmitted through both full frame and unibody construction.
- Describe how to visually determine the extent of impact damage.
- List the various types and variations of body measuring tools.
- Analyze damage by measuring body dimensions.
- Analyze impact damage to mechanical parts of the vehicle.
- Explain the importance of the datum plane and centerline concepts as related to unibody repair.
- Interpret body dimension information and locate key reference points on a vehicle, using body dimension manuals.
- Discuss the use of tram bars, self-centering gauges, and strut tower gauges.
- Diagnose various types of body damage, including twist, mash, sag, and side-sway.
- Locate and measure key points using a tape measure, tram bar, and self-centering gauges, when given a damaged vehicle and a body specification manual.
- Answer ASE-style review questions relating to vehicle measurement and damage analysis.

INTRODUCTION

Vehicle measurement involves using specialized tools and equipment to measure the location of reference points on a vehicle. These measurements are then compared to published dimensions from an undamaged vehicle. By comparing known good and actual measurements, you can determine the extent of damage. The difference in the two measurements indicates the direction and amount of frame or body misalignment.

When a car or truck is in a high-speed crash, powerful impact forces can bend the frame or unibody structure of the vehicle. The frame or body is designed to absorb some of the energy of the collision and protect its occupants. When a heavily damaged vehicle is brought to the shop, the extent of the damage must be carefully evaluated. Sometimes measurements are needed to help the estimator calculate the costs of the repairs.

After studying damage measurements, frame straightening equipment is used to pull the frame or body back into alignment. Straightening procedures are explained in the next chapter.

Table 17–1 illustrates the major collision repair processes right after estimating.

To repair any vehicle properly, you must accurately diagnose the collision damage. The severity and extent of damage to parts must be analyzed. Once the total damage has been determined, a plan can be made for repair or salvage of the vehicle.

The importance of a complete and accurate damage diagnosis cannot be overstressed. Any inaccurately diagnosed damage on a vehicle will be uncovered during repair. When this happens, the repair method or procedure must be changed. The finished product will be less than satisfactory, resulting in the need for further repairs. Therefore, the best person for the body technician to talk to is the person who prepares the estimate.

Physical damage is rarely missed during an inspection by a competent estimator and body technician. However, the effects of the damage on unrelated systems

TABLE 17–1 COLLISION REPAIR PROCESSES

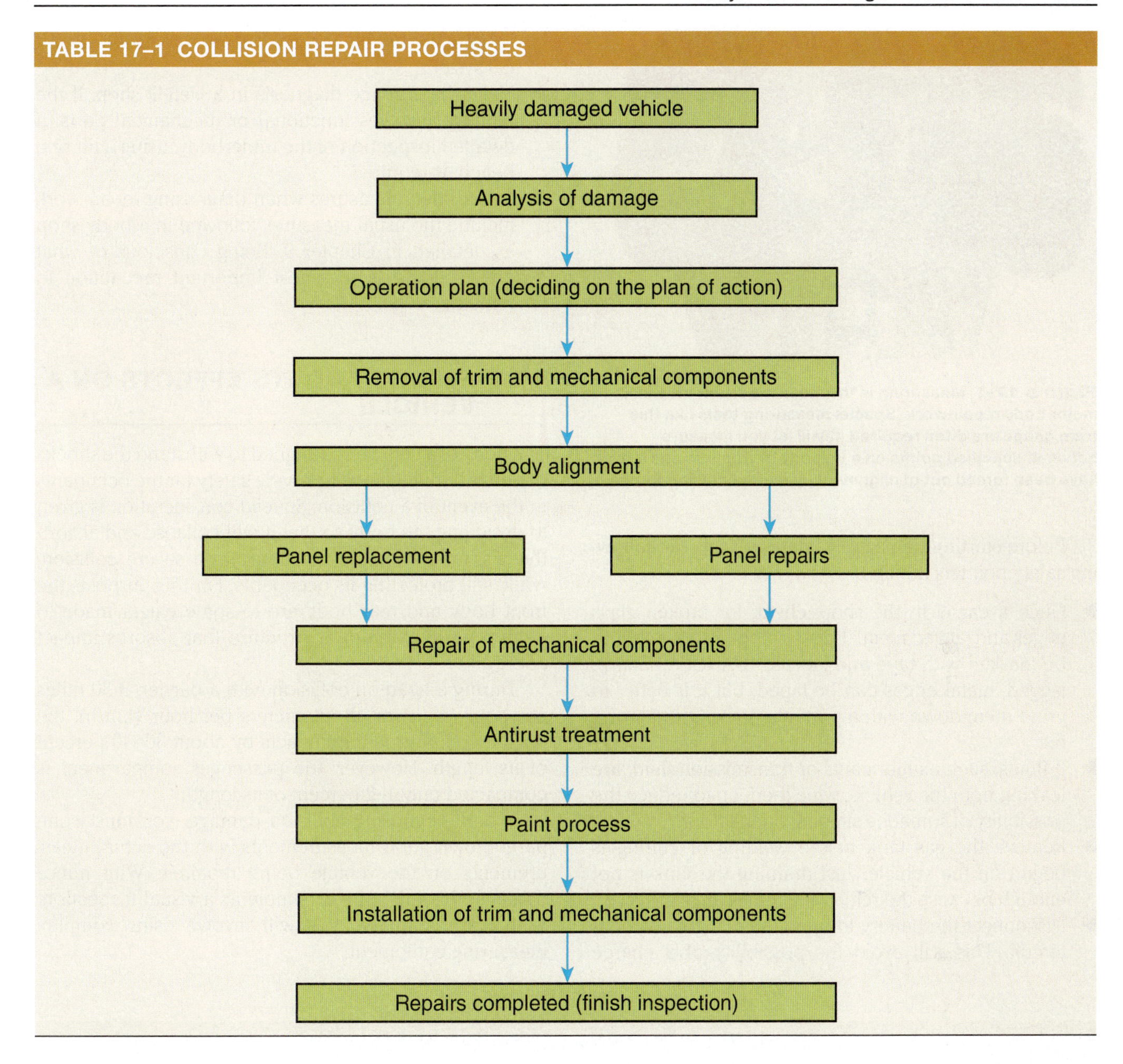

and damage occurring next to an impacted part are sometimes overlooked. A visual inspection alone is generally inadequate with modern vehicles. Accident damage should be assessed carefully by measurements with the proper tools and equipment.

The following is a basic, recommended diagnostic procedure:

1. Know the vehicle construction type.
2. Visually locate the point of impact.
3. Visually determine the direction and force of the impact; once determined, check for possible damage.
4. Determine whether the damage is confined to the body or whether it involves functional parts (wheels, suspension, engine, and so on).
5. Systematically inspect damage to the components along the path of the impact. Find the point where there is no longer any evidence of damage. For example, pillar damage can be determined by checking the door fitting conditions.
6. Measure the major components. Check **body dimensions** (known correct body measurements of an undamaged vehicle) by comparing the actual measurements with the values in the repair manual or body dimensions chart (Figure 17–1).
7. Check for suspension and overall body damage with the proper equipment.

Vehicle damage conditions are diagnosed from the procedures given in Table 17–2.

Figure 17-1 **Measuring is the most important step in major body repair work. Special measuring tools like this tram gauge are often required. It will let you measure between specified points on a unibody or frame to see if they have been forced out of alignment by major collision forces.**

Before starting damage evaluation, keep the following safety pointers in mind:

- Once a car is in the shop, check for broken glass edges and jagged metal. Edges of broken glass should be masked with tape and labeled "DANGER." Sharp, jagged, metal edges can be taped, but it is better to grind them down with a portable power grinder or a file.
- If fluids, such as lubricants or transmission fluid, are leaking from the vehicle, wipe them up to reduce the possibility of someone slipping on the floor.
- Remove the gas tank before welding or cutting is begun on the vehicle. Just draining the tank is not enough because the remaining fumes are explosive.
- Disconnect the battery to open the electrical system circuit. This will avoid the possibility of a charge igniting flammable vapors. It also protects the electrical system.
- Make the damage diagnosis in a well-lit shop. If the damage involves functional or mechanical parts, a detailed inspection of the underbody, using a lift or a bench, is required.
- Other safety measures when diagnosing repair work include the usual measures followed in a body shop as detailed in Chapter 9. Being conscious of what one is doing is the most important precaution to remember.

17.1 IMPACT AND ITS EFFECTS ON A VEHICLE

The body of a vehicle is designed to withstand the shocks of normal driving and to provide safety for the occupants in the event of a collision. Special consideration is given to designing the body so that it will collapse and absorb the maximum amount of energy in a severe collision, while still protecting its occupants. For this purpose, the front body and rear body are to some extent made to deform easily, forming a structure that absorbs impact energy.

During a head-on collision with a barrier at 30 miles per hour (mph) or 48 kilometers per hour (km/h), the engine compartment compacts by about 30–40 percent of its length. However, the passenger compartment is compacted only 1–2 percent of its length.

When diagnosing collision damage, you must compare known good measurements with the actual measurements on the vehicle being repaired. With minor damage, this might be as simple as a visual inspection. With major damage, this will involve using complex measuring equipment.

TABLE 17-2 FACTORS TO CONSIDER IN THE DIAGNOSIS OF COLLISION DAMAGE

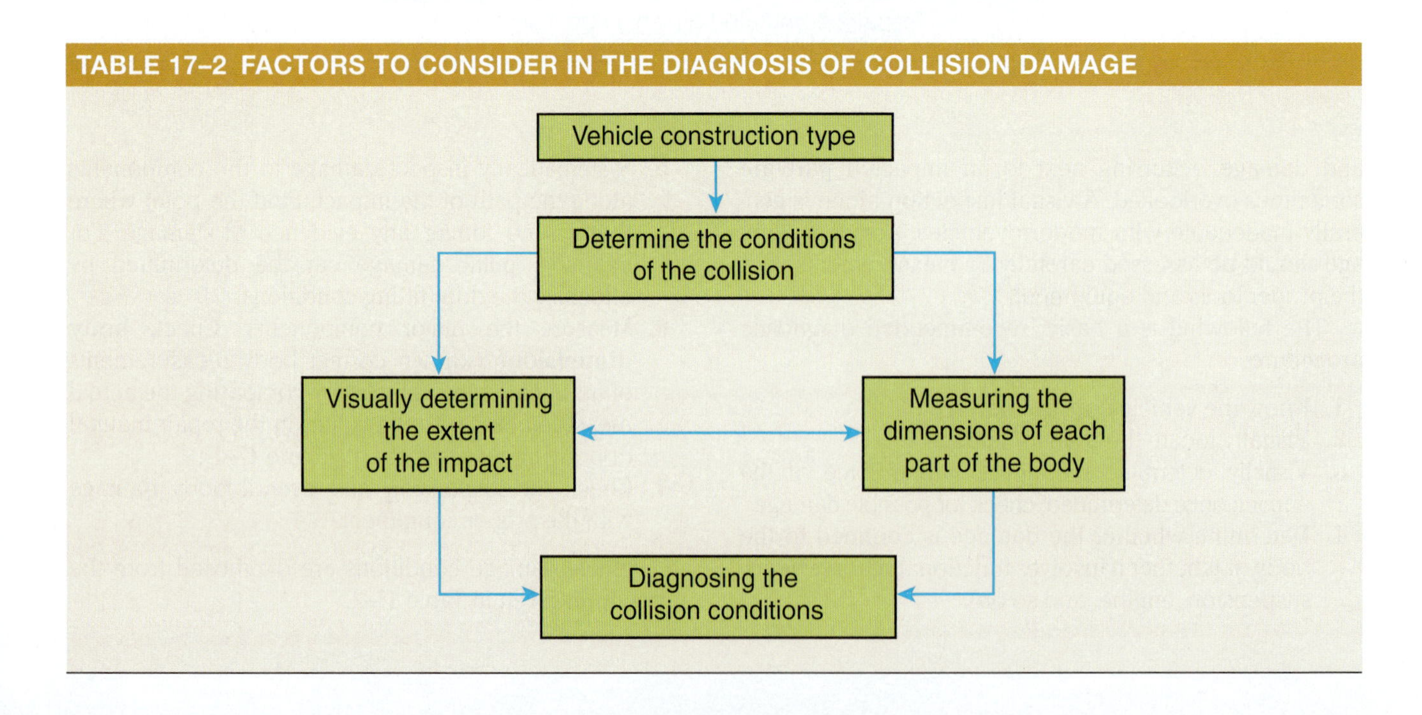

To correctly analyze damage on a unibody vehicle, the entire structure must be considered. To do this, it is necessary to be able to take proper measurements to locate damage. It will also help you plan where to pull.

Measurement gauges are special tools used to check specific frame and body points. They allow you to quickly measure the direction and extent of vehicle damage.

Specific points or locations on the frame or body are given by the manufacturer for making measurements. They might be holes, specific bolts, nuts, panel edges, or other locations on the vehicle. To repair a badly damaged vehicle, you must restore these reference points to their factory dimensions while reference points in the undamaged area remain in their correct locations.

Therefore, the collision repair technician must work with the whole vehicle. This is done by measuring and recording dimensional changes. The most widely accepted method of checking body dimensions is to use the charts supplied in the body dimension manuals.

When the collision damage has been identified using the proper identification and analysis procedures, anyone skilled in the mechanics of collision damage repair is capable of repairing the car or truck.

The terms *control point* and *reference point* have different meanings. The **control points** used in manufacturing are not necessarily the same as the reference points the collision repair technician uses to measure the vehicle. **Reference points** refer to the points, bolts, holes, and so on, used to give unibody and frame dimensions in body specification manuals. The distance between reference points can be measured with either a tram bar or a tape measure.

Determining the Conditions of the Collision

The extent of the damage differs depending on the conditions at the time of the accident. To put it another way, the damage can be partly determined by understanding how the collision occurred. To understand the circumstances of the collision, it would be necessary to contact persons directly involved or eyewitnesses. Such a task would undoubtedly be a waste of time. However, it is possible for the person responsible for making the estimate to get a direct response from the customer. This method of damage assessment is sometimes necessary in order to estimate the cost of the repair. Therefore, the body technician should talk to the estimator to help analyze the methods of repair needed.

The body technician should know the following items:

- The size, shape, position, and speed of the vehicles involved in the collision
- Speed of the vehicle at the time of the collision
- Angle and direction of the vehicle at the time of the impact
- The number of passengers and their positions at the time of the impact

A good body/frame or structural technician can usually determine what actually happened during a collision to cause the damage. Because of the predictable nature of a driver's reactions before a collision, certain types of damage almost invariably occur in a rather predictable pattern and sequence.

If a driver's first reaction (Figure 17–2) is to turn away from the danger, the vehicle will be forced to take the hit on the side. If the driver's reaction is to slam on the brakes the direction of impact will be frontal (Figure 17–3).

A frontal collision where the point of impact is high on the vehicle could cause the cowl and roof to move rearward and the rear of the vehicle to move downward. Or, if the point of impact is low at the front, the inertia of the body mass could cause the rear of the vehicle to distort upward, forcing the roof forward. This would leave an excessively large opening between the front upper part of the door and the roofline (Figure 17–4).

Even with vehicles of similar weights travelling at about the same speed, vehicle damage will vary significantly depending on what is struck, for example, a telephone pole or a wall. If the impact is spread over a larger area, such as a wall, the damage will be spread over a wide body surface area (Figure 17–5).

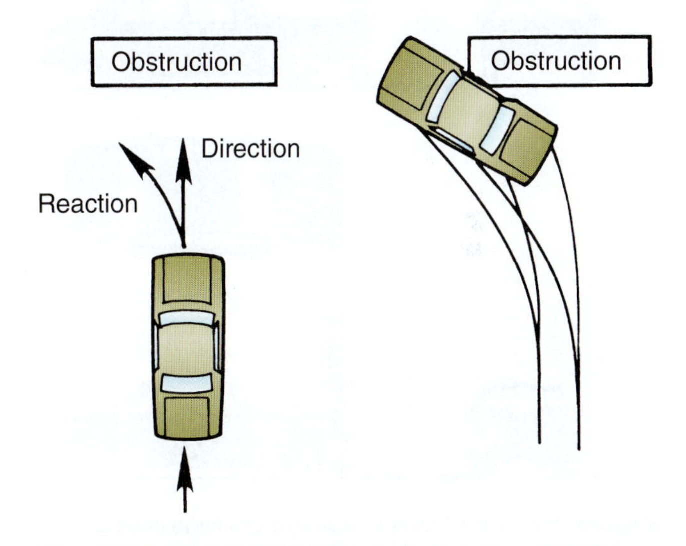

Figure 17–2 A driver's first reaction is to turn away from danger, forcing the hit to a side, causing side-sway damage.

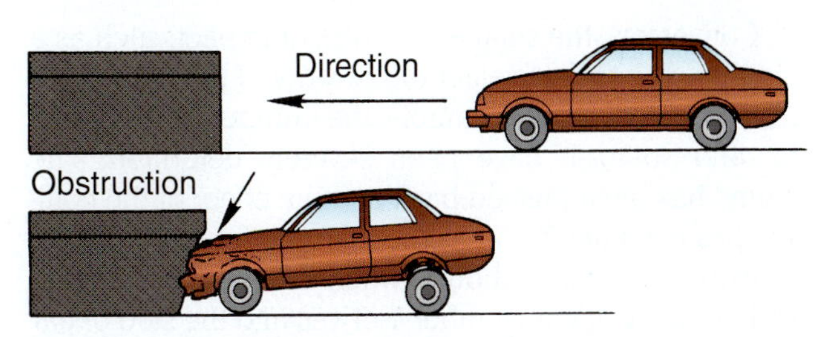

Figure 17–3 A driver's second reaction is to slam on the brakes, forcing the front end to drive down, causing sag.

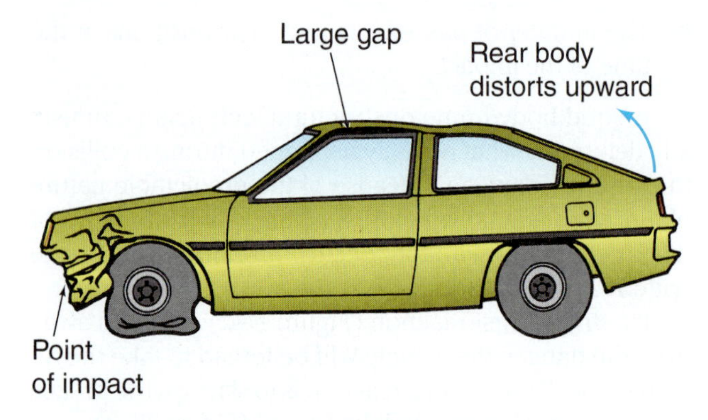

Figure 17–4 A hard frontal impact (primary damage) often causes secondary damage (for example, buckles in the roof).

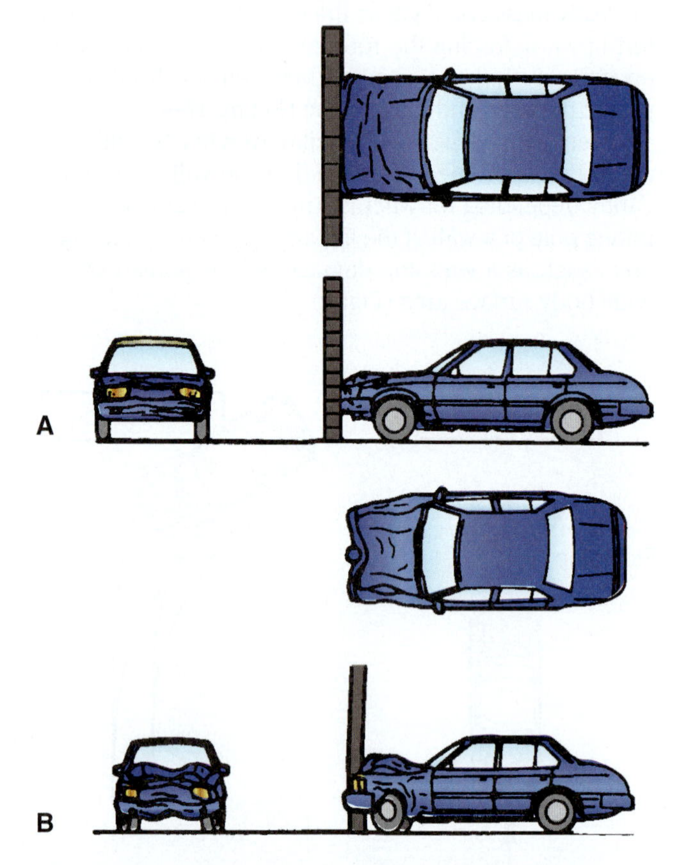

Figure 17–5 (A) This is an example of a large impact surface area (a brick wall) and (B) an example of a small impact surface area (a telephone pole).

Conversely, the smaller the area of impact, such as a telephone pole, the greater the severity of the damage in a smaller area. In this example, the bumper, hood, radiator, and so forth have been severely deformed. The engine has been pushed back and the effect of the collision has extended as far as the rear suspension.

Another consideration is when one car hits another while moving. If car Number 1 drives into the side of car Number 2 while Number 2 is moving, the motion of the first car will drive the front end of the car back.

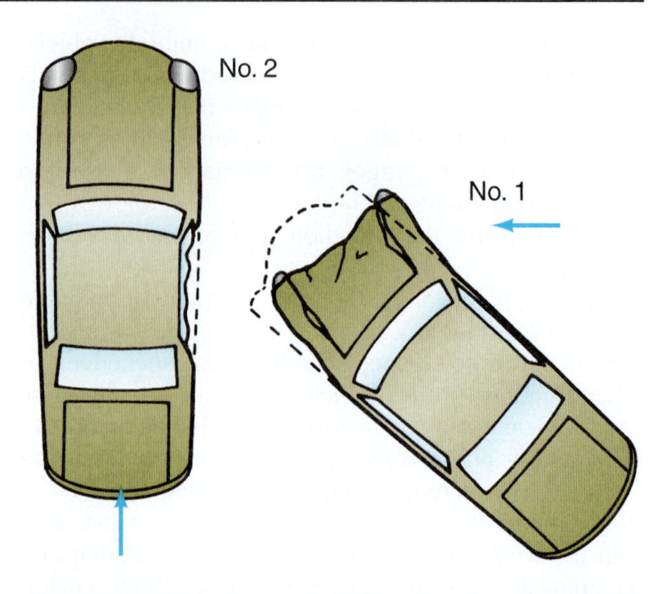

Figure 17–6 Note a typical broadside collision. If car No. 2 is moving, car No. 1 will have its front end mashed back and to one side.

Simultaneously, the motion of car Number 2 will also drag that same front end to the side. There is only one collision, but the damage is in two directions (Figure 17–6).

On the other hand, there might be two collisions in only one direction. This is a fairly common occurrence in freeway pile-ups. A car that collides with another car and then leaves the road to hit a pole or guard rail ends up with two completely separate types of damage.

There are many other variables and possible combinations of damage. It is important to determine what actually happened before an accurate diagnosis can be made. Get as many facts as possible. Combine them with physical measurements and centerline gauge readings to determine exactly what collision repair procedure should be taken.

A little extra time spent evaluating damage can save many hours in the overall repair time. "Think time" saves "work time" and increases profits.

Influence of Impact on a Body-Over-Frame Vehicle

Figure 17–7 illustrates a body with a perimeter frame with its built-in collapsible sections. The circled areas indicate the softer sections of the frame designed to absorb the major impact of a collision.

The body is attached to the frame by rubber mounts. The *rubber body mounts* reduce the effects of road shocks travelling from the frame to the body. This quiets the ride

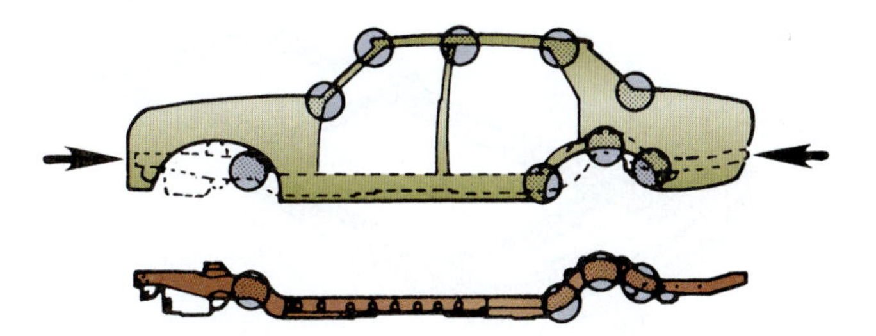

Figure 17-7 These are typical perimeter frame and body section collapsible sections.

in the passenger compartment. In the event of a large impact or collision, the bolts of the rubber mounts might bend, resulting in a gap between the frame and the body. Also, depending on the magnitude and direction of impact, the frame might experience damage while the body does not.

Frame deformation can be broken down into five categories:

1. **Side-sway damage**. Collision impacts that occur from the side often cause *side-sway damage* or a side-bending frame damage condition. Side-sway usually occurs in the front or rear of the vehicle. Generally, it is possible to spot side-sway by noting whether there are buckles on the inside of one rail and buckles on the outside of the opposite side rail (Figure 17–8 and Figure 17–9).

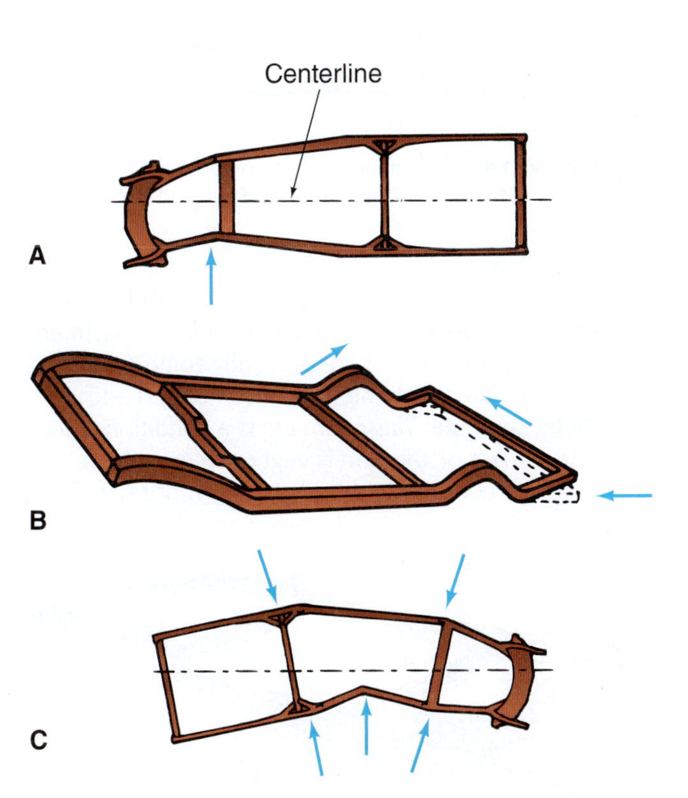

Figure 17-8 Study various kinds of side-sway damage: (A) sideways at the front of the frame caused by a front-end collision, (B) rear side-sway, and (C) double side-sway on the frame's outer section.

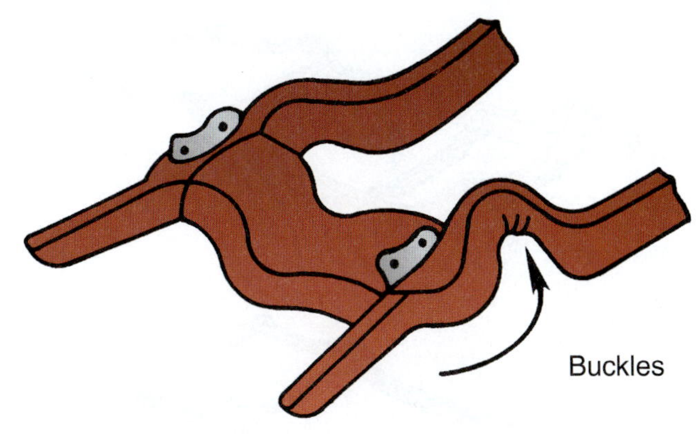

Figure 17-9 A good clue to frame misalignment is buckles in the crush zones of the frame rails.

Side-sway can be recognized by abnormalities, such as a gap at the door on the long side and wrinkles on the short side. Look for impact damage obvious from the side, such as the hood and deck lid not fitting into the proper opening (Figure 17–10).

2. **Sag damage**. *Sag damage* is a condition in which one area, often the cowl area, is lower than normal. The structure has a swayback appearance. Sag damage generally is caused by a direct impact from the front or from the rear. It can occur on one side of the vehicle or on both sides (Figure 17–11).

Sag can usually be detected visually by a gap between the fender and the door that is narrow at the top and wide at the bottom. Also look for the door appearing to hang too low at the striker. Sag is the most common type of damage and occurs in most vehicles that are involved in an accident. Enough sag can be present in a frame to prevent body panel alignment even though wrinkles or kinks are not visible in the frame itself (Figure 17–12).

Figure 17-10 Misalignment of doors gives clues to the extent and direction of the damage.

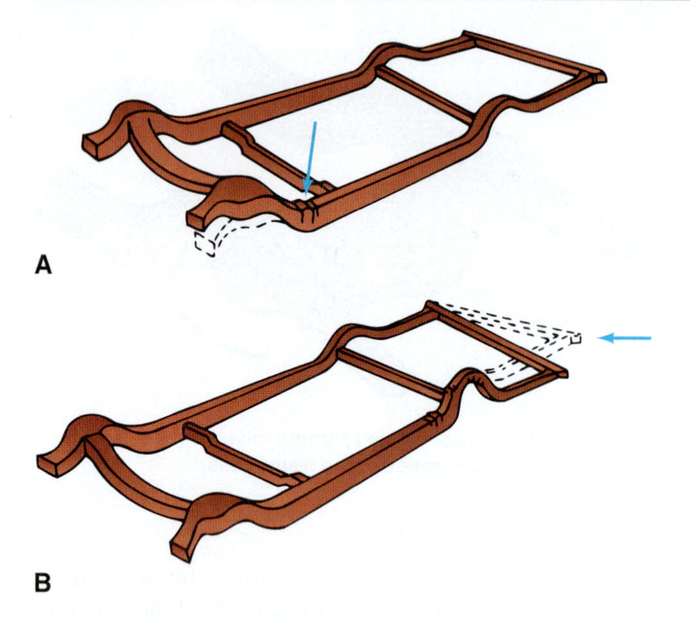

Figure 17–11 (A) Note the sag condition on the left front frame section and (B) rear-end sag.

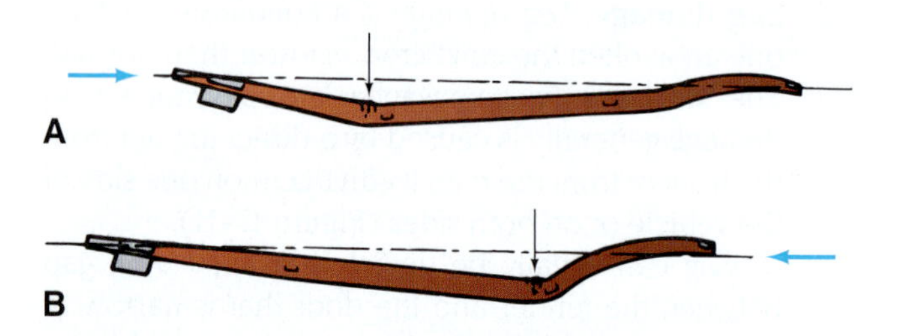

Figure 17–12 (A) This side rail sag resulted from a front-end collision; (B) this side rail sag resulted from a rear-end collision.

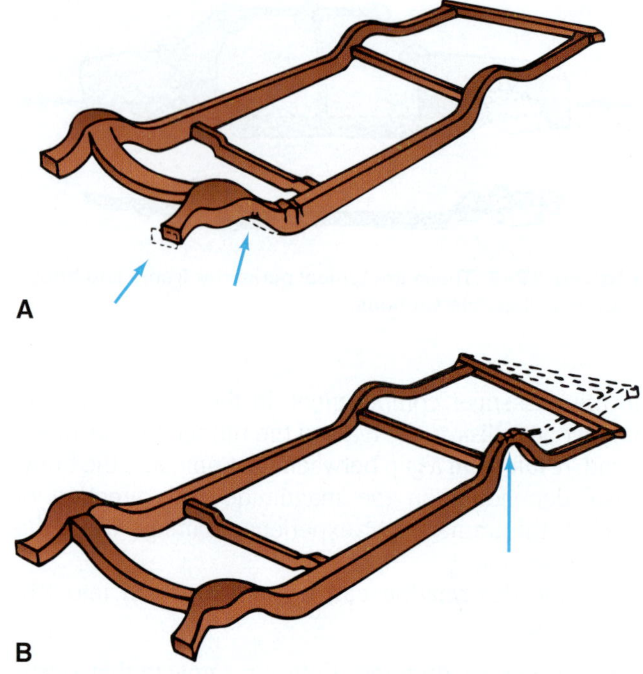

Figure 17–13 (A) Note the mash damage on the left front side rail; (B) mash damage on the left rear side rail.

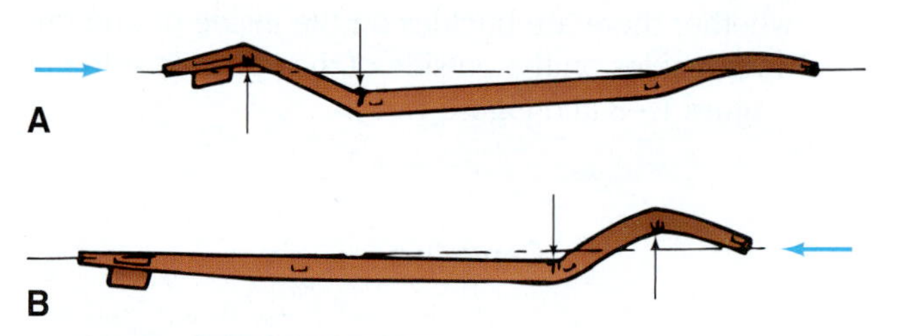

Figure 17–14 (A) A frame mashed and buckled from a front-end collision. (B) A frame mashed from a rear-end collision.

3. **Mash damage**. *Mash damage* is present when any section or frame member of the car is shorter than factory specifications. Mash is usually limited to the areas forward of the cowl and rearward of the rear window. Doors might fit well and appear to be undisturbed. Wrinkles and severe distortion will be found in fenders, the hood, and possibly frame rails or horns. The frame will rise upward at the top of the wheel arch, causing the spring housing to collapse. With mash damage, there is very little vertical displacement of the bumper. The damage results from direct front or rear collisions (Figure 17–13 and Figure 17–14).
4. **Diamond damage**. *Diamond damage* is a condition where one side of the car has been moved to the rear or front, causing the frame and/or body to be out of square. The resulting shape is a figure similar to a parallelogram and is caused by a hard impact on a corner or off-center from the front or rear. Diamond damage affects the entire frame, not just the side rails. Visual indications are hood and trunk lid misalignment. Buckles can appear in the quarter panel near the rear wheel housing or at the roof-to-quarter panel joint. Wrinkles and buckles often will appear in the passenger compartment and/or trunk floor. There is usually some mash and sag combined with the diamond (Figure 17–15).
5. **Twist damage**. *Twist damage* is a condition where one corner of the car is higher than normal; the opposite corner might be lower than normal. Twist

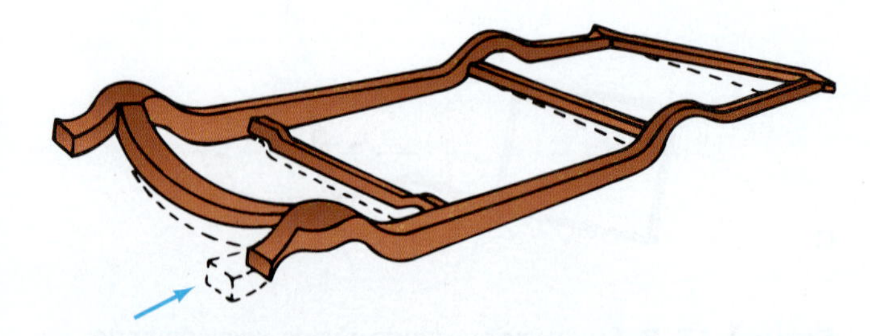

Figure 17–15 This diamond condition affecting the entire frame alignment resulted from a hard frontal impact, but only on one side.

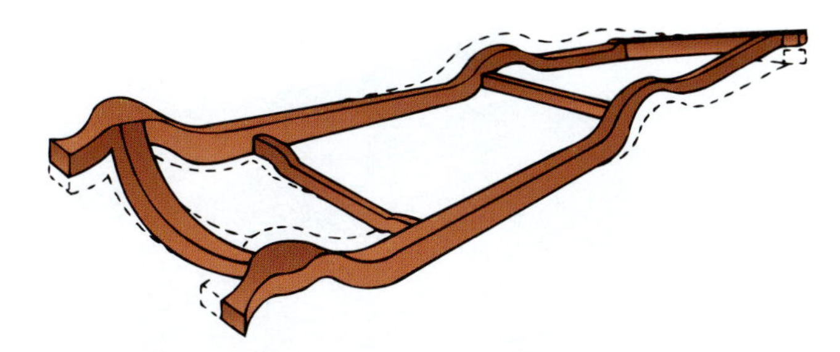

Figure 17–16 Twist conditions affect the entire frame alignment.

can happen when a car hits a curb or median strip at high speed. It is also common in rear corner impacts and rollovers (Figure 17–16).

A careful inspection reveals no apparent damage to the sheet metal. However, the real damage is hidden underneath. One corner of the car has been driven upward by the impact. Most likely, the adjacent corner is twisted downward. If one corner of the car is sagging close to the ground as though a spring is weak, the car should be checked for twist.

Diamond damage usually occurs when the vehicle is struck off-center. However, a frame will rarely experience deformation involving the whole frame (Figure 17–17).

The most frequent order of occurrence of damage is:

1. Side-sway
2. Sag
3. Mash
4. Diamond
5. Twist

As described in Chapter 18, the most important rule in body/frame alignment is *reverse direction and sequence*. This means, to correct collision damage on a conventional vehicle, the pulling or pushing of the damaged area must be done in the opposite direction of impact. The repair must also be made in the reverse sequence that it happened.

Unfortunately, most accidents result in a mix of one or more of these damage problems. Side-sway and sag frequently occur almost simultaneously. Some damage affects the frame's cross members, especially the front member. In a rollover accident, for example, the front cross member on which the motor mounts are attached is often forced out of shape because of the engine's weight. This will result in a sag of this cross member. While cross member damage is rather rare, it must be corrected because cross member alignment can affect the handling of the vehicle.

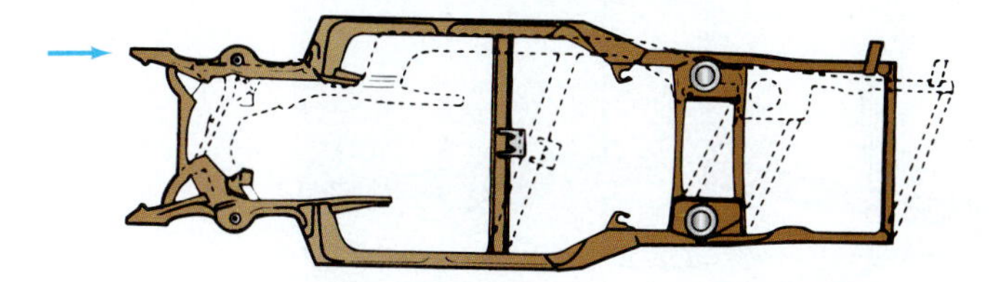

Figure 17–17 Study diamond conditions. Solid lines show the undamaged frame and dotted lines represent the damaged frame.

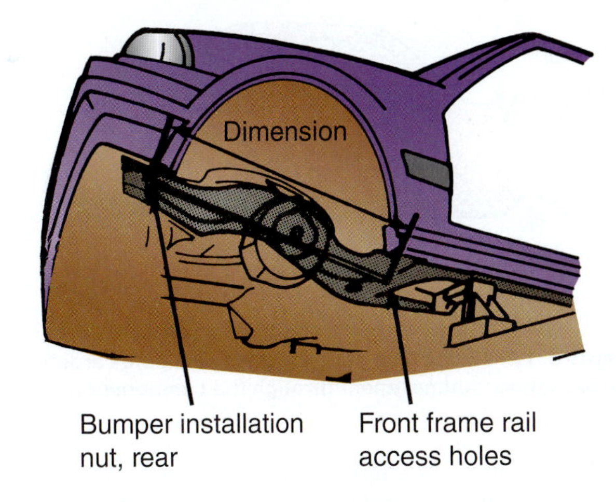

Figure 17–18 Measurements of undersurface dimensions are often needed.

Inspect a deformed frame by comparing the space between the body rocker panel and the front and back of the frame. Also compare the space between the front fender and the front and back of the wheel hub. To inspect front frame deformation, compare the left and right measurements from the rear hole for the front bumper to a point on the front frame rail (Figure 17–18).

Impact Effect on Unibody Vehicles

The damage that occurs to a unibody car as the result of an impact can best be described by using the *cone concept*. The unibody vehicle is designed to absorb a collision impact. When hit, the body folds and collapses as it absorbs energy. As the force penetrates the structure, it is absorbed by an ever increasing area of the unibody. This characteristic spreads the force until it is completely dissipated. Visualize the point of impact as the tip of the cone.

The centerline of the cone will point in the direction of impact. The depth and spread of the cone indicate the direction and area that the collision force travelled through the unibody. The tip of the cone is the **primary damage** area.

Because unibodies are structured entirely from the joining of pieces of thin sheet metal, the shock of a collision is absorbed by a large portion of the body shell. The effects of the impact shock wave as it travels through the body structure are called **secondary damage**. This damage is toward the inner structure of the unibody or toward the opposite end or side of the vehicle (Figure 17–19 and Figure 17–20).

To provide some control on secondary damage distortion and to provide a much safer compartment for passengers, a unibody vehicle is designed with crush zones or areas at the front and rear (Figure 17–21).

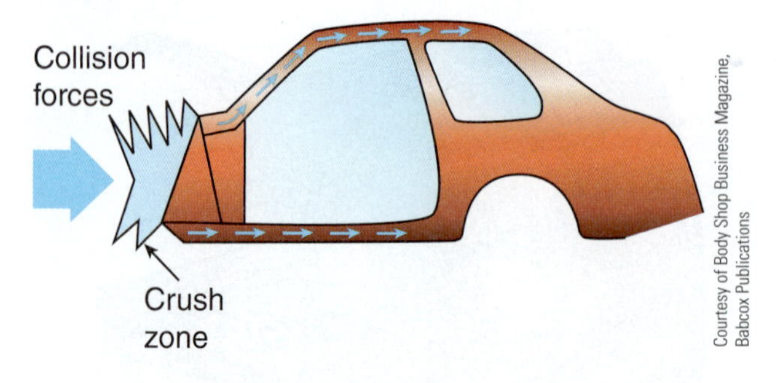

Figure 17-19 Collision energy often dissipates around the passenger compartment through the components.

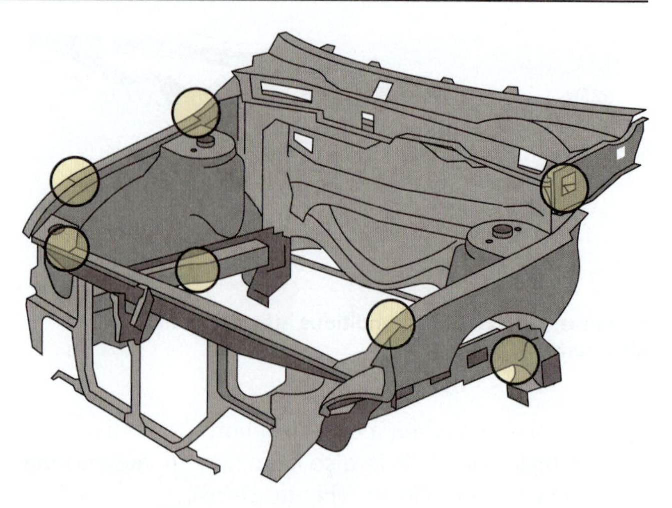

Figure 17-22 Here are common locations for unibody front crush zones. They should be inspected when analyzing damage.

Crush zones are engineered to collapse in a predetermined fashion to protect a vehicle's passengers and to localize damage. The effects of the impact shock wave to the body structure are reduced. In other words, front impact shocks are absorbed by the front body and crush zones (Figure 17–22).

Rear shocks are absorbed by the rear body and crush zones. Side shocks will be absorbed by the rocker panel, roof side frame, center pillar, and door (Figure 17–23).

Impact damages on unibody vehicles can be described in the following ways.

Frontal unibody damage results from a head-on collision with another object or vehicle. The impact of a collision depends on the vehicle's weight, speed, area of impact, and the source of impact. In the case of a minor impact, the bumper is pushed back, bending the front

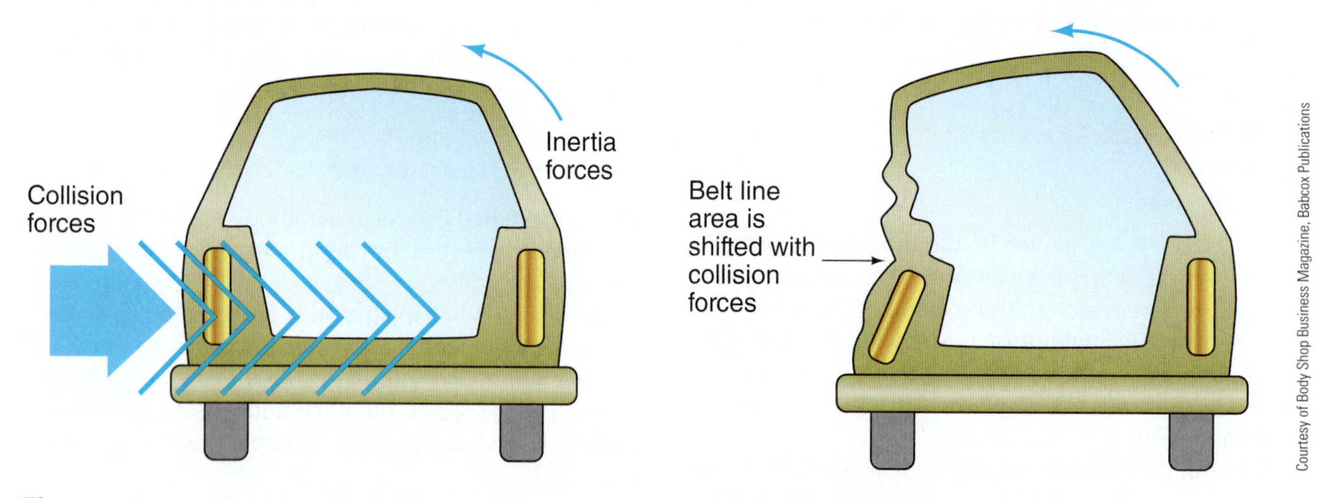

Figure 17-20 The roof shifted toward the side of impact because of weight/mass inertia.

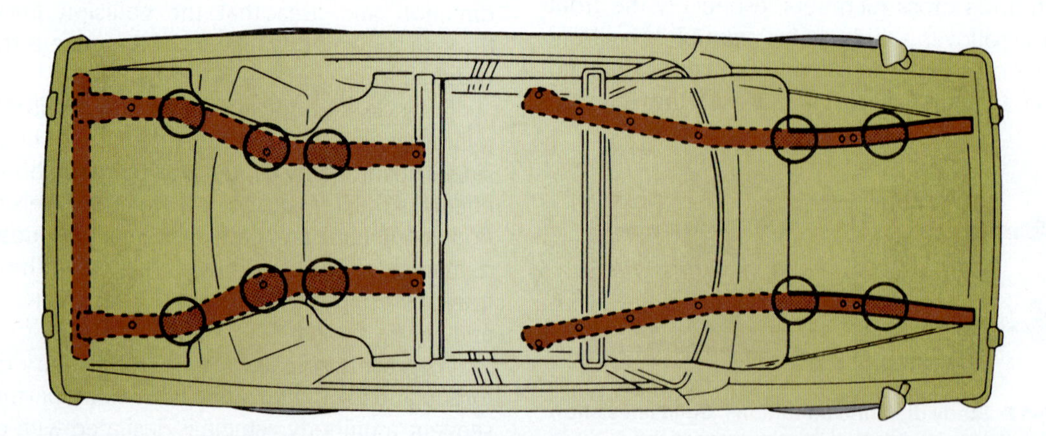

Figure 17-21 These are typical unibody impact-absorbing areas.

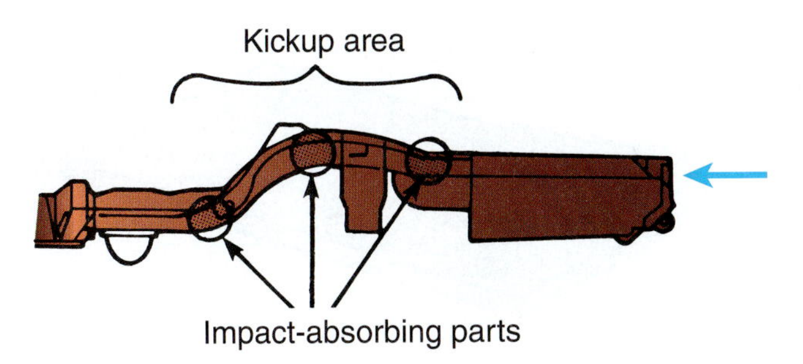

Figure 17-23 **The rear side member impact-absorbing areas should also be inspected for crumples.**

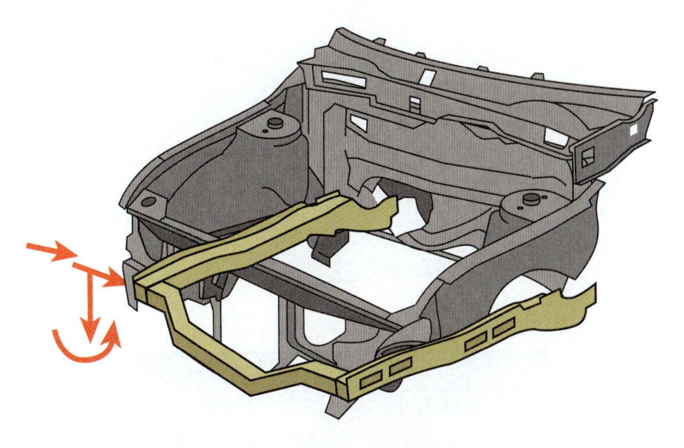

Figure 17-25 **Both lateral and vertical bending movement of a unibody vehicle often happen during a collision.**

side members, bumper stay or bracket, front fender, radiator support, radiator upper support, and hood lock brace. If the impact is further increased, the front fender will contact the front door. The hood hinge will bend up to the cowl top. The front side members may also buckle into the front suspension cross member, causing it to bend. If the shock is great enough, the front fender apron and front body pillar (particularly the front door hinge upper area) will be bent, which will cause the front door to drop down. In addition, the front side members will buckle and the front suspension member will bend. The dash panel and front floor pan may also bend to absorb the shock (Figure 17–24).

If a frontal impact is received at an angle, the attachment point of the front side member becomes a turning axis. Lateral as well as vertical bending occurs. Because the left and right front side members are connected together through the front cross member, the shock from the impact is sent from the point of impact to the front side member of the opposite side of the vehicle (Figure 17–25).

Rear unibody damage occurs when the vehicle is moving backward and hits something or is hit by another vehicle from behind. When the impact is comparatively small, the rear bumper, the back panel, trunk lid, and floor pan will be deformed. The quarter panels will also bulge out. If the impact is severe enough, the quarter panels

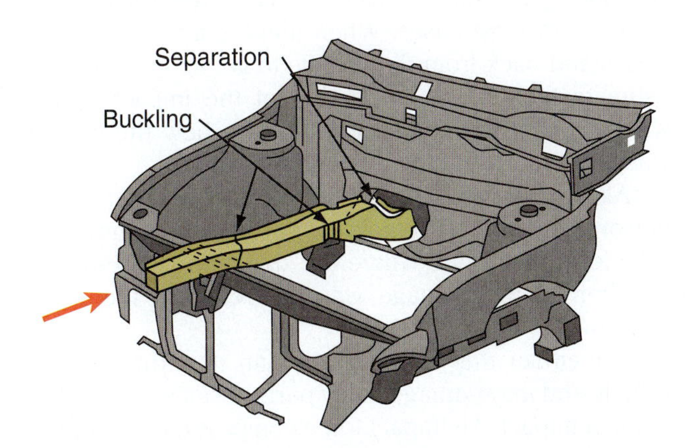

Figure 17-24 **Buckling and separation action in a unibody vehicle indicate major damage. Cracked undercoating is also an indicator of more serious damage.**

will collapse to the base of the roof panel. On four-door vehicles, the center body pillar might bend. Impact energy is absorbed by the deformation of these parts and by the deformation of the kickup of the rear side member.

Side unibody damage will cause the door, front section, center body pillar, and even the floor to deform. When the front fender or quarter panel receives a large perpendicular impact, the shock wave extends to the opposite side of the vehicle.

When the central area of the front fender receives an impact, the front wheel is pushed in. The shock wave extends from the front suspension cross member to the front side member. If the impact is severe, the suspension parts are damaged and the front wheel alignment and wheelbase may be changed. The steering gear or rack can also be damaged by side impacts.

Top impacts can result from falling objects or from a vehicle rollover. This type of damage involves not only the roof panel, but also the roof side rail, the quarter panels, and possibly the windows.

When a vehicle has rolled over and the body pillars and roof panels have been bent, the opposite ends of the pillars will be damaged as well. Depending on the manner in which the vehicle rolled over, the front or back sections of the body will be damaged, too. In such cases, the extent of the damage can be determined by the deformation around the windows and doors.

The typical collision damage sequence on a unibody structure is as follows, as illustrated in Figure 17–26:

1. *Bending.* In the first microseconds of impact, a shock wave attempts to shorten the structure, causing a lateral or vertical bending in the central structure. Most of the forces that broadcast impact shock to remote areas occur at this instant. Because the structure is stiff and springy, it tends to snap back to its original shape—at least momentarily. Bending is usually indicated by the height measurement being out of tolerance. This damage—similar to sag in a conventional structure—can occur on one side of the car and not the other (Figure 17–27).

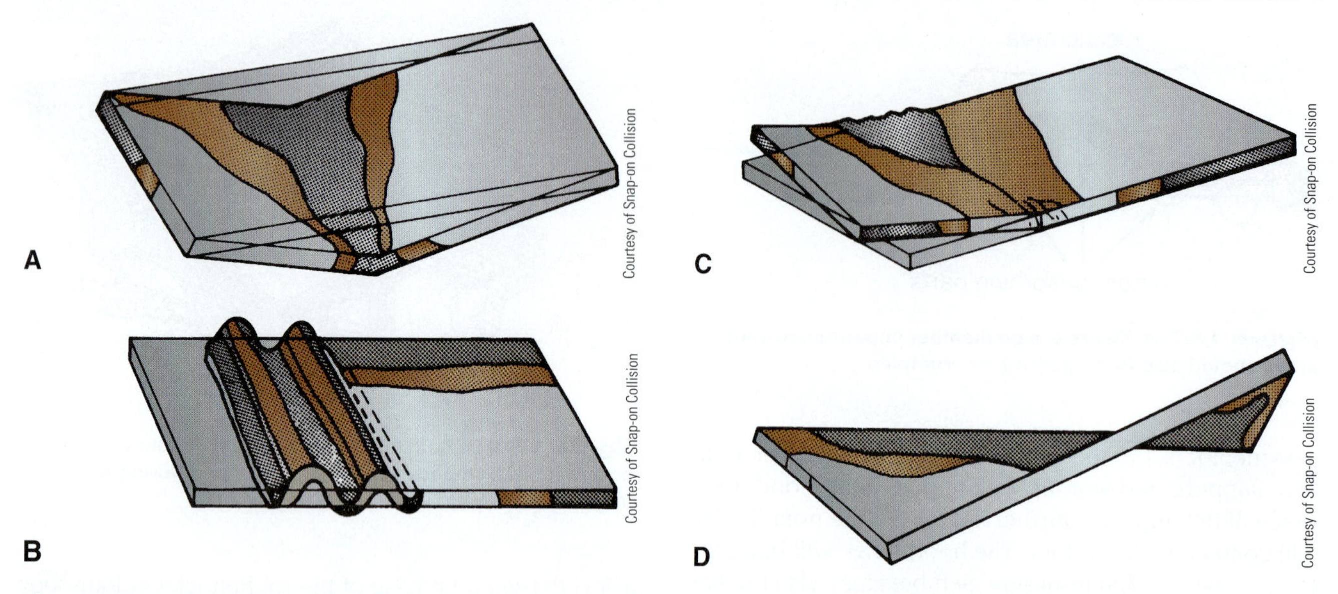

Figure 17-26 Study the types of unibody collision damage: (A) bending, (B) crushing or collapsing, (C) widening, and (D) twisting.

2. *Crushing or collapsing.* As the collision event continues, visible crushing occurs at the point of impact. Impact energy is absorbed in the deforming structure (helping protect the passenger compartment). Remote areas may buckle, tear, or pull loose. Crush damage, which is similar to mash on body-over-frame (BOF) vehicles, is indicated by the length measurement being out of tolerance.
3. *Widening.* In a well-designed unibody structure, impact forces reaching the passenger compartment cause the side structure to bow out away from the passengers (never in), distorting side rails and door openings. Widening is similar to side-sway damage in BOF vehicles and is indicated by the width measurement being out of tolerance.
4. *Twisting.* Even if the initial impact is dead center, the secondary impact can introduce torsional loads that cause a general twisting of the structure. Unibody structural twisting, like twisting of a conventional vehicle frame, is usually the last part of the collision event. It is indicated by combinations of height and width measurements being out of tolerance.

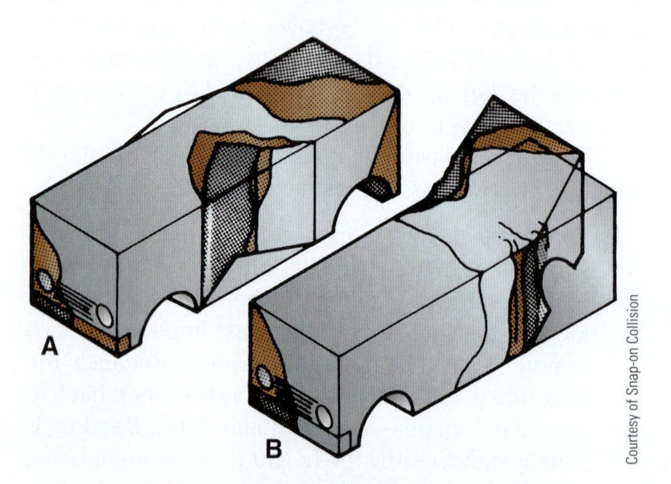

Figure 17-27 If a vehicle is viewed as a rectangular box, one can see how force moves through the structure: (A) center bending and (B) rear bending.

Although there is a great similarity between the types of damage that can occur on BOF and unibody vehicles, unibody damage is often more complex. Note that a severe collision will not normally cause diamond damage on unitized cars. Also like conventional aligning, pulling secondary damage (last-in) so that it is corrected first (first-out) is the best way to correct damage to a unibody car. Secondary damage is identified by accurate measurement.

17.2 VISUALLY DETERMINING THE EXTENT OF IMPACT DAMAGE

Damaged parts show signs of structure deformations or fractures in most cases. When making a visual inspection, stand back from the vehicle to get an overall view. Estimate the size and direction of the impact (place where impact was received). Estimate how the impact was propagated and the damage sustained.

Also investigate whether there is any twisting, bending, or slanting of the vehicle overall. Look over the entire vehicle to determine where the damage occurred and whether all the damage was the result of the same collision.

Remember that impact force can flow through the vehicle and may damage many parts besides those at the point of impact. An impact force can pass easily through the strong portions of the body, finally ending up in the weak portions, damaging them as well. Therefore, in searching for damage, inspection must be made along the path of the impact damage through the weak portions of the body.

You must look for the presence of strain, panel joint misalignment, cracks in and peeling of the paint film, cracked undercoat and sealer, and so on.

Damage can be detected easily by finding these types of symptoms:

- Areas where the cross sections of the components were suddenly deformed
- Parts that are broken or missing
- Gaps in strengthening materials, such as reinforcements or patches
- Part-to-part joints that are shifted
- Corners and edges of components that are misaligned

When surveying the extent of damage to the frame components, such as side members, it is easier to locate the damage on the concave side of the component. The concave side may appear as a sharp dent or kink rather than a minor bulge that appears on the opposite side of the member.

A body is designed so that the energy received during impact travels along a predetermined path. Energy flow starts at the point of impact and flows through the structure until all the energy has been dissipated. Therefore, the evidence of damage will usually be greater near the point of impact because the extent of damage is reduced as the energy is dissipated into the adjacent structure. However, in some cases, the energy is passed through the impact point (with little evidence of damage showing) and is propagated to a point that is deep within the body.

Inspecting Clearance and Fit of Each Part

Checking the door alignment makes it easy to determine if the body pillar has been damaged. Simply open and close the door, observing its alignment and action (Figure 17–28).

In the event of a front-end collision, it is important to check the clearances and level differences between the rear doors and quarter panels and rocker panels. Another good method is to compare the clearances on the left and right sides of the vehicle.

Hinges on a vehicle wear over a period of time, and doors tend to drop down. This is especially true of the door on the driver's side, which is opened and closed quite frequently. Inspection should be made with the vehicle on a level shop floor or drive-on rack. If you place the car on a lift or stands, the fit of the doors can be affected by the flexibility of the body.

Inspecting for Inertia Damage

Because of the presence of heavy objects, such as engines mounted on rubber mounts, inertia is a powerful force during a collision. Inspect for damage to the mounts or to surrounding parts and panels. During a collision, the powerful impact usually causes the body and frame to become misaligned, damaging the body isolator mountings.

Inspect all mounting hardware for signs of inertia damage. With a unibody vehicle, look for a dent in the roof. On pickup trucks, inspect for an out-of-parallel condition between the cab and the bed (Figure 17–29).

Inspecting for Damage from Passengers and Luggage

Passengers and luggage can cause secondary damage to the vehicle as a result of inertia during a collision. The damage will vary depending on the position of the passengers and the severity of the impact. Parts that are frequently damaged include the instrument panel, steering wheel, steering column, and seat backs. Luggage in the trunk has also been known to cause damage to the body quarter panels.

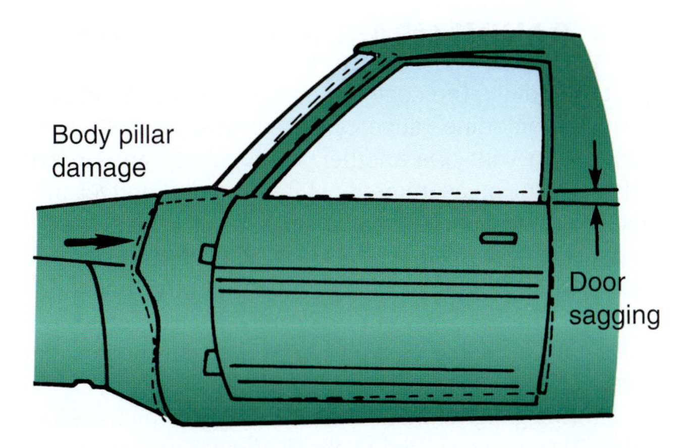

Figure 17–28 Do a careful visual inspection and make repair notes, because procedures for pulling damage can be complex. Inspecting of door alignment will give valuable information about damage.

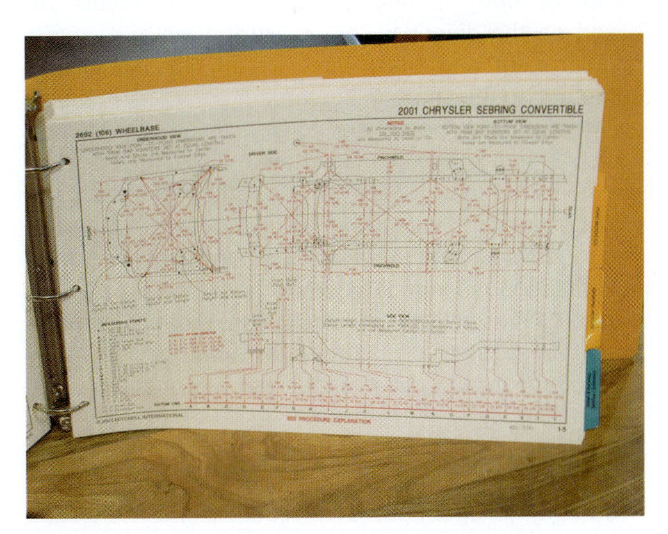

Figure 17–29 Look up in the dimension manual the measurements for the make and model vehicle being repaired.

17.3 MEASUREMENT OF BODY DIMENSIONS

Measuring is critical to the success of any major collision repair job regardless of the type of body structure. But with unibody vehicles, measurements are vital to successful repair because the steering and suspension are mounted to the body structure. In addition, some of the suspension geometry is built into the body structure. As a result, the angles of wheel alignment (caster and camber) often have a fixed (nonadjustable) value. Body damage often seriously affects suspension geometry. The rack-and-pinion control box for the steering assembly is also mounted to a panel, resulting in a fixed relationship to the steering arms. The mechanical components, engine, transmission, and differential are all mounted directly to body members or to cradles supported by body members (panels or integral rails).

A distortion of any of these measuring points will change steering or suspension geometry or misalign mechanical components. This can result in improper steering and handling, vibration and noise in the drivetrain, and excessive wear of tie rod ends, tires, rack-and-pinion assemblies, universal joints, or other drive or steering components. To maintain proper steering, handling, and drivability, repair tolerances must be held to within a maximum value of less than ⅛ inch, or 3 mm.

Body Dimensions Charts

Accurate damage assessment can be made at specific points on the body using a body dimensions chart. The **body dimensions chart** gives measurement points and measurement specifications for a specific type of vehicle. You need to use the chart for the specific make and model vehicle you are repairing. The chart information enables you to use measurement tools to compare the damaged vehicle to known good measurements.

In the body dimensions chart, measurements are based on the diagonal line measuring method engine compartment and body dimensional data should be compared to the chart and recorded.

Measurement points and tolerances are determined by inspection of the damaged area. Normally, in front-end collisions that cause slight amounts of door sag, the damage does not extend beyond the center of the vehicle, so measurement in the rear section is not necessary. In a situation where a large impact has occurred, many measurements must be taken to ensure proper alignment procedures. However, taking and recording too many measurements may cause unnecessary confusion.

Vehicle Measuring Basics

In unibody construction, each section should be checked for diagonal squareness by comparing diagonal lengths. Length and width should also be compared. The center section should be used as a base when reading structural alignment. All measurements and alignment readings should be taken relative to the center section.

Start measuring in the center, or middle, section. If it is not square, then move to the undamaged end of the vehicle to find three correctly positioned reference points.

Keep in mind that to accurately measure a vehicle, you must start with at least three reference points you know are right. The way to do this is to check the squareness of the vehicle. If the vehicle is not symmetrical, refer to the dimensions chart for correct measurements.

Measurement Importance

In the entire repair process of both conventional and unibody cars, it is not possible to overemphasize the importance of measuring. A vehicle cannot be satisfactorily repaired unless all of the major manufacturing control points in the damaged area are returned to the manufacturer's specifications. To achieve this, the body technician must:

- Measure accurately
- Measure often
- Recheck all measurements

Because of the importance of measuring, many kinds of equipment have been developed and marketed by automotive equipment manufacturers strictly for the purpose of providing the capability to measure quickly and accurately. While there are a number of styles of measuring equipment that can be found in body shops, most of it can be divided into five basic systems:

1. Gauge measuring system
2. Universal measuring system
3. Dedicated fixture system
4. Universal/laser
5. Computer/electronic

17.4 GAUGE MEASURING SYSTEMS

The tram gauge, the centering gauge, and the MacPherson strut centerline gauge can be used separately or in conjunction with one another. *Tram gauges* are scaled rods used for measurement, while *centering gauges* are metal rods used to check for misalignment. Supported by suspension system strut tower domes, centerline gauges allow visual alignment of the critical control points of unibody vehicles. The tram centering and strut centerline gauges are available as a unit or as separate diagnostic tools.

Another gauge similar to the tram type is the tracking gauge. This gauge is used to check alignment of the front and rear wheels. If the front and rear wheels are not in alignment, the vehicle will not handle properly.

17.5 TRAM GAUGES

The tram gauge (Figure 17–30) measures one dimension at a time. Each dimension must be recorded and must be cross-checked from two additional control points—at least one being a diagonal measurement. The best areas to select for tram gauge measurements are the attachment points for suspension and mechanical components, because these are critical to alignment. Throughout the repair operation, critical control points must be measured (and recorded) repeatedly with the tram gauge in order to monitor progress and to prevent overpulling.

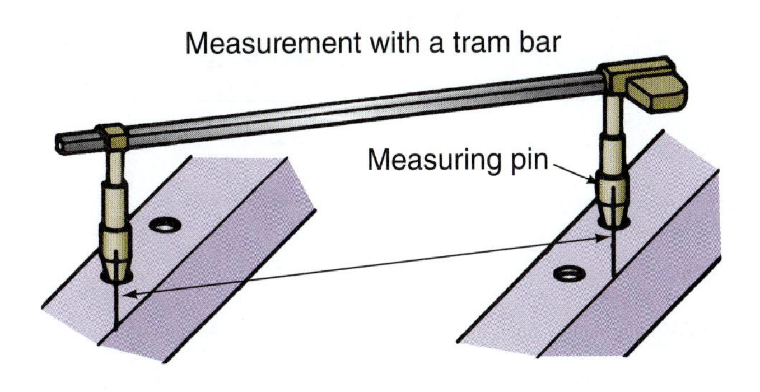

Figure 17–30 A tram gauge is simply a ridged tape measure with pointers. Place the pointer on the bolt or in the holes to measure between the reference points.

Because these control point tram measurements must be taken and written down several times in a repair operation, a method of tabulation must be devised. One of the ways to keep track of your frame or unibody measurements and analyze damage is to use or make a damage direction chart and/or diagram. Both are shown in Figure 17–31.

Note the different damage direction chart columns for manufacturer's specifications and your own measurements. Enter the known good dimension in the first column and your measurements while pulling in the next columns. As you pull out the damaged frame, improved dimension measurements will help you keep track of repair progress.

Study how you can make a copy or simple drawing of the vehicle's dimension diagram for keeping track of frame/unibody straightening. Write down your actual measurements next to the factory ones. Subtract the difference to find how much that area of the vehicle is out of alignment.

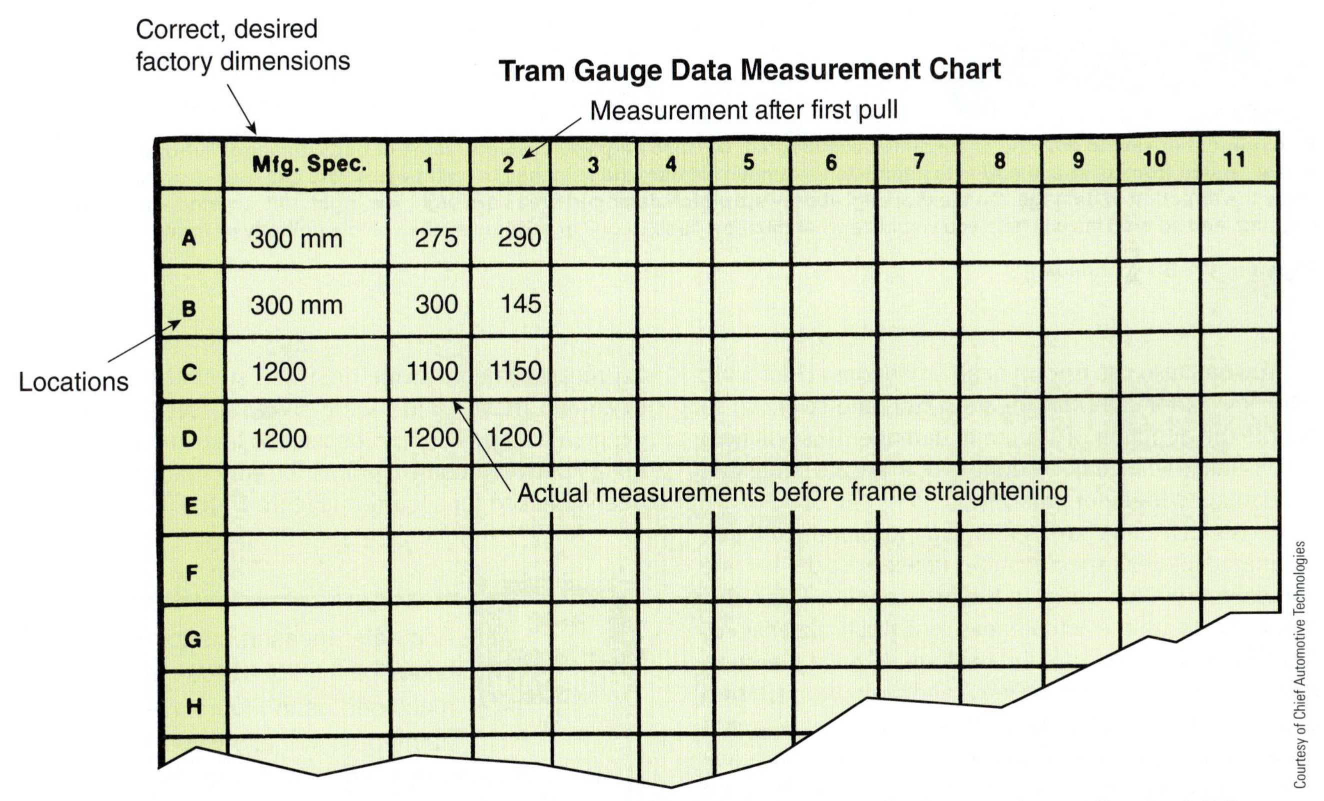

Tram Gauge Data Measurement Chart

	Mfg. Spec.	1	2	3	4	5	6	7	8	9	10	11
A	300 mm	275	290									
B	300 mm	300	145									
C	1200	1100	1150									
D	1200	1200	1200									
E												
F												
G												
H												

Courtesy of Chief Automotive Technologies

A Make measurements to crushed areas of the vehicle and write them down next to the factory values on the sheet. If the vehicle is not damaged, the factory specs and your measurements should be the same. The difference between them equals the amount of damage. Write down more measurements in the other columns as you use the frame rack to pull out the damage. Corresponding factory numbers should be equal when the vehicle is realigned.

Figure 17–31 Study how to easily chart and diagram vehicle structural damage to simplify frame-unibody straightening.

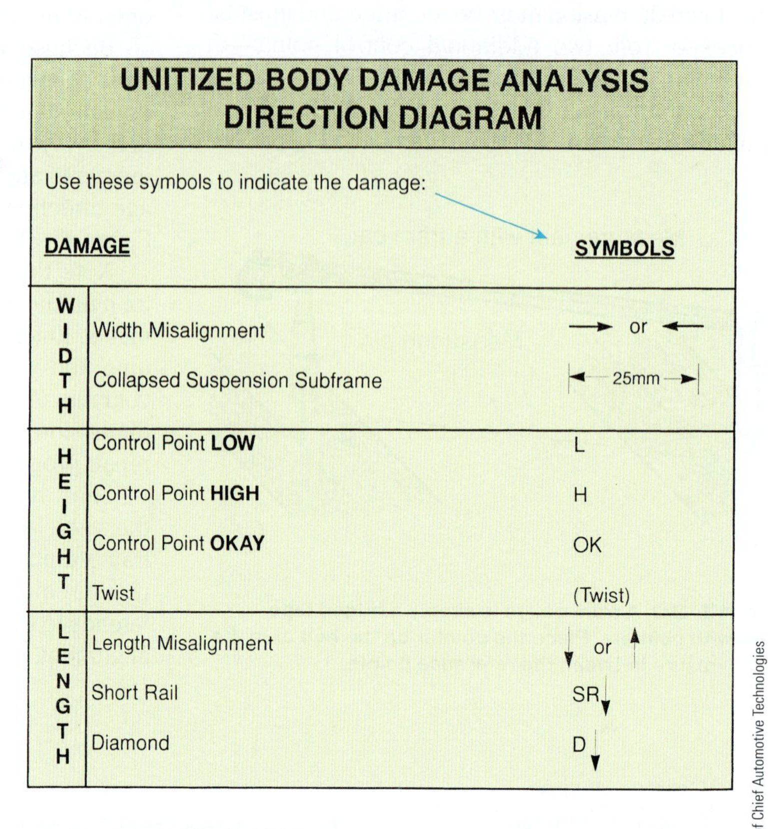

UNITIZED BODY DAMAGE ANALYSIS DIRECTION DIAGRAM

Use these symbols to indicate the damage:

	DAMAGE	SYMBOLS
WIDTH	Width Misalignment	→ or ←
	Collapsed Suspension Subframe	←25mm→
HEIGHT	Control Point **LOW**	L
	Control Point **HIGH**	H
	Control Point **OKAY**	OK
	Twist	(Twist)
LENGTH	Length Misalignment	↓ or ↑
	Short Rail	SR↓
	Diamond	D↓

Courtesy of Chief Automotive Technologies

B A printout or simple drawing of the dimension diagram will also help you evaluate frame or body damage. Subtract your measurement from its specification to calculate the amount of damage in inches or millimeters. You can then draw arrows to show the direction of damage. On the drawing, abbreviate which damaged areas are high, low, right, left, twisted, diamond, sagging, and so on. This will help you visualize what must be done to pull or straighten the vehicle unibody or frame.

Figure 17–31 ***(continued)***

You can also use arrows or abbreviations (H for high, L for low, R for right, SR for short rail, and so forth) to denote the direction of structural damage. This will help you visualize what must be done to pull out and straighten the damaged frame or unibody.

To use this data chart or a similar measurement sheet, the manufacturer's specifications taken from the service manual are written down in the first column. The A-B-C designations are the actual measuring point dimensions; the 1-2-3 designations are the readings taken at measurement Step 1, measurement Step 2, and so on. As each step of a restoration repair is made, the measurements should be recorded, including those dimensions that have just been corrected. This measurement data chart tells the body technician at a glance whether the job has succeeded in restoring the vehicle to its original state (Figure 17–32).

The tram gauge may have a scale superimposed on it. However, because almost all manufacturer specifications list measurements in metric, use a steel tape with both fractional inches and metric scales to set up the tram gauge. The tape can also be used to take quick measurements between control points. Be sure that the tape has been checked for accuracy (Figure 17–33).

Accurate measurements can be taken if the tip of a tape measure is machined as in Figure 17–34. The pointed tip will allow you to insert the measurement tape fully into the control measurement hole.

Most reference points are actually holes in the vehicle structure, and dimensions are center-to-center distances.

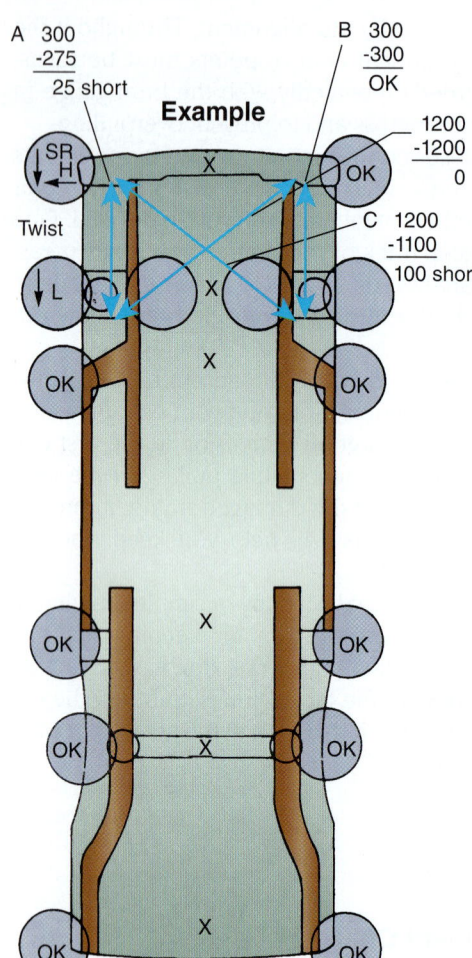

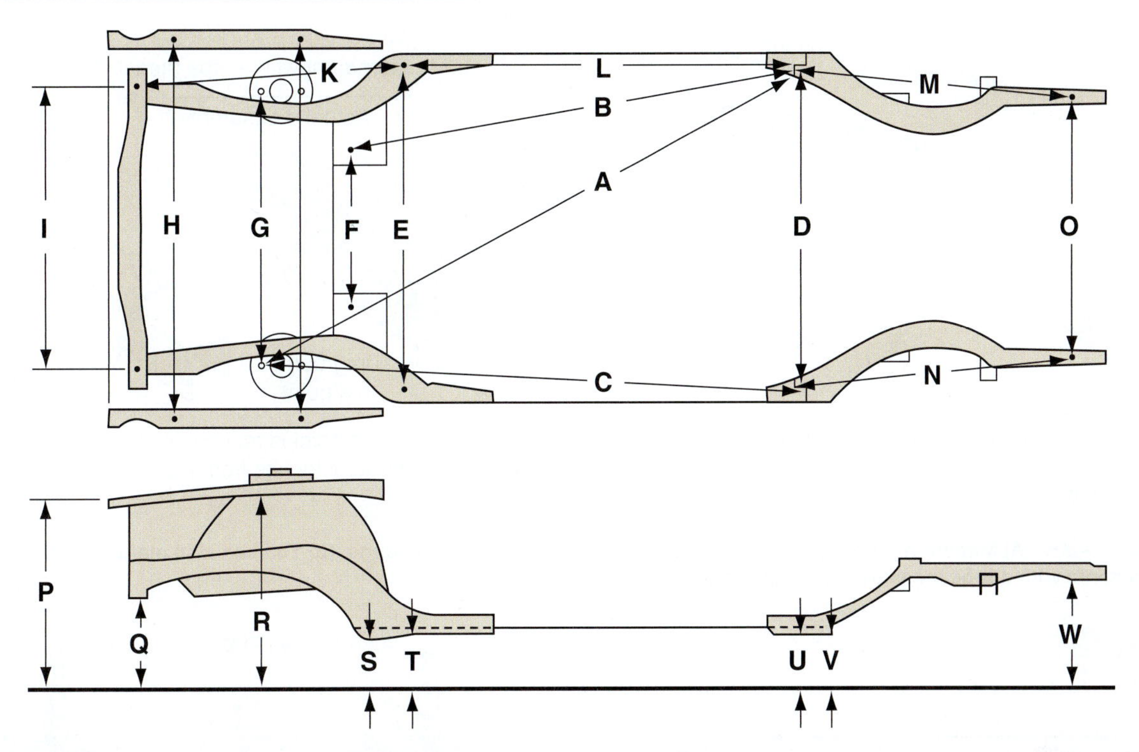

Figure 17-32 Many body technicians find it is easier to substitute letters for numbers when making up a tabulation chart. If the manual drawing has only letters, you need to refer to the accompanying chart that gives number values for each letter.

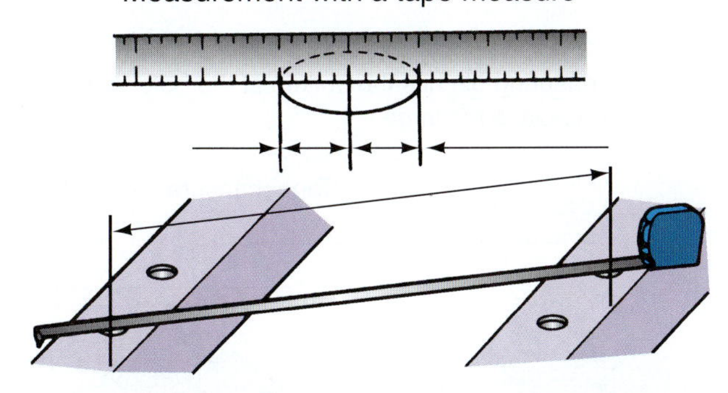

Figure 17-33 Quick measurement of reference points can be taken with a measuring tape. By grinding the tip sharp, you will get better readings in holes and on parts.

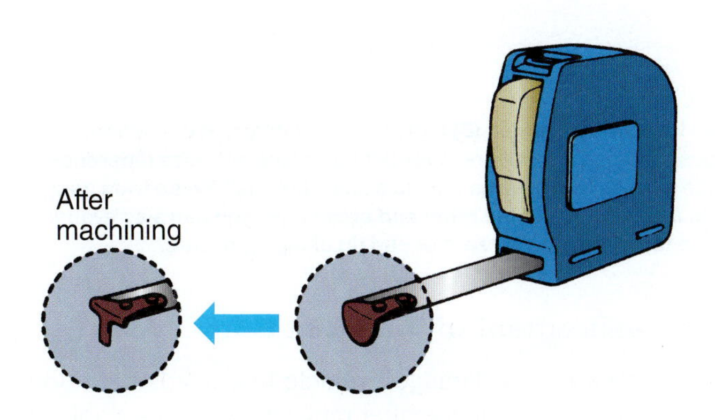

Figure 17-34 As shown above, machining the tape measure's tip gives more accurate measurements.

Control point holes are frequently larger in diameter than the tram gauge tip. To measure accurately with the tram gauge (when holes are same diameter), measure like edge to like edge. A few gauge manufacturers' spec books give the measurement based on the gauge's bar length. Always check the method used for specification measurements (Figure 17–35).

When the holes are not the same size, find the center-to-center measurement. They will usually be the same type of hole: round, square, oblong, and so on. In this case, measure inside edge to inside edge, then outside edge to outside edge. Add the results of the two measurements and divide by 2 (Figure 17–36).

For example, two round holes, one being ½ inch (13 mm) in diameter, the other 1½ inches (38 mm) in diameter, have an inside measurement of 30 inches (762 mm) and an outside measurement of 32 inches (813 mm). The center-to-center dimension is 30 inches + 32 inches ÷ 2 = 31 inches (787 mm). The 31 inches (787 mm) is the dimension for the tram gauge (Figure 17–37).

In using a tram gauge for measuring, the manufacturer's specifications for the vehicle are needed. They will enable you to accurately assess the damage and restore the vehicle structure to factory dimensions.

If the manufacturer's specifications are not available, use an undamaged vehicle of the same make, year, model, and body style as a source for correct dimensions. Frequently, if only one side of a vehicle is damaged, it is possible to take measurements on the undamaged side. You can then apply them to the damaged side for a comparison measurement (Figure 17–38).

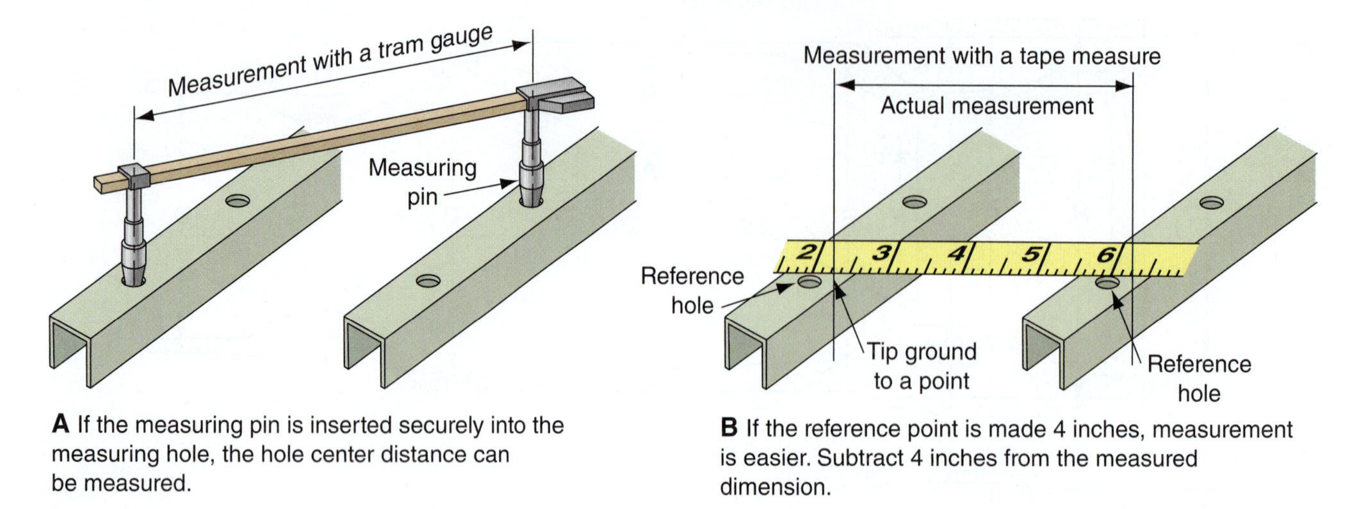

A If the measuring pin is inserted securely into the measuring hole, the hole center distance can be measured.

B If the reference point is made 4 inches, measurement is easier. Subtract 4 inches from the measured dimension.

Figure 17-35 **(A) Measuring the distance between hole centers with a tram gauge. (B) You can also make quick, accurate measurements with a tape measure if the holes are the same size.**

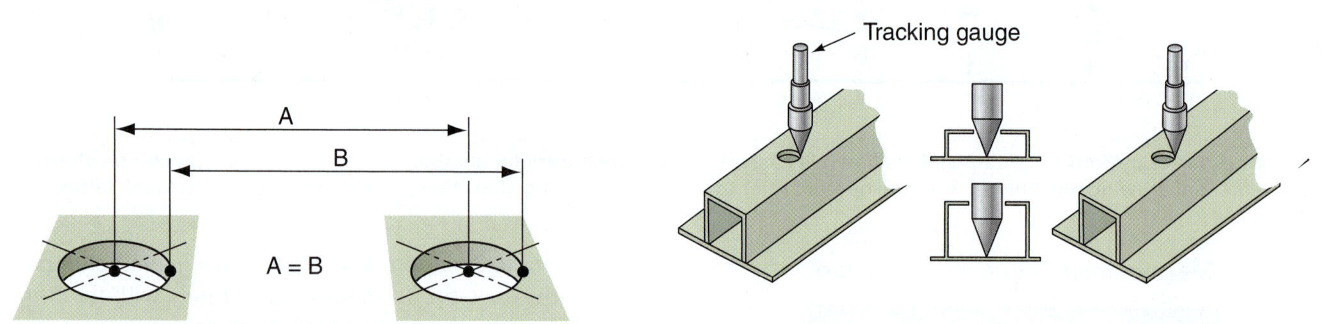

A In a situation where the hole diameters are the same

C The measuring pin hits the bottom of the hole, or the measuring hole is too large

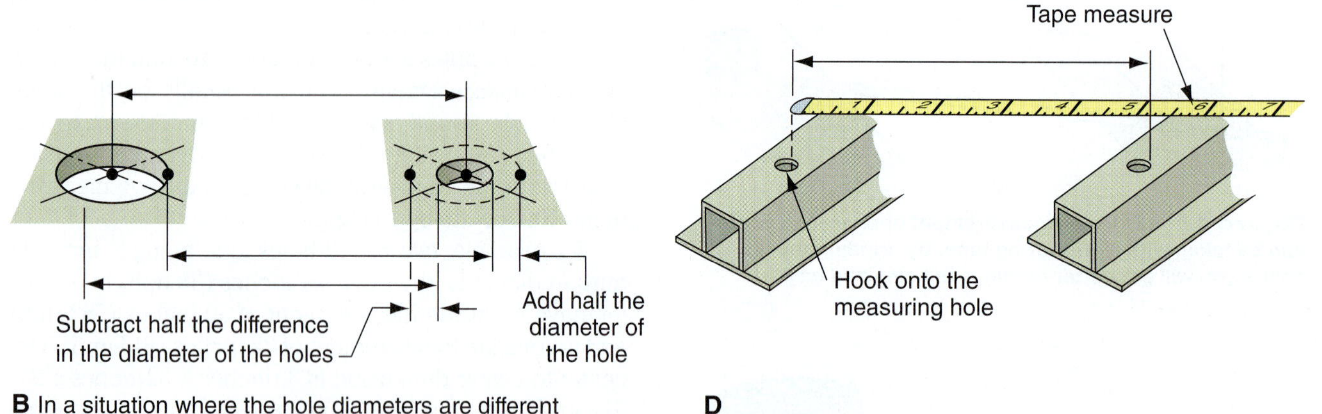

B In a situation where the hole diameters are different

D

Figure 17-36 **(A) Note two ways of measuring when the holes are the same diameter. (B) If the holes are different diameters, measure from the centers if possible. If measuring them from the edges with a tape measure, subtract half of the hole size difference if going from a small hole to a large hole. Add half the diameter difference if going from a large hole to a small hole. (C) Pin on tram gauge fully inserted to contact hole or inaccurate readings will result. (D) With a helper to hold the other end of the tape, you can also eyeball centers by reading the tape as shown. Subtract the starting point on the tape (since not at zero or end tip of tape measure.**

Upper Body Dimensioning

Upper body damage can also be determined with tracking trams and a steel measuring tape. Their use is basically the same as when they are used to do an underbody evaluation. Manufacturers furnish specifications on the most important upper body reference points.

Measurement of the Front Body

In the case of a damaged vehicle that needs the hood edge and front side member replaced, it is reasonable to take measurements before making the repair. Even if only the front right side of the body received the impact, the left side will usually be damaged as well. Therefore,

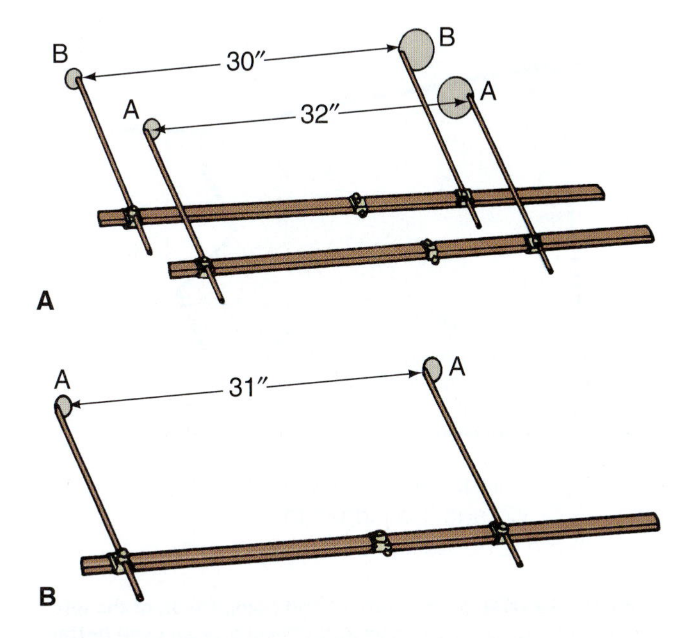

Figure 17–37 (A) Two round holes, one being ½ inch (13 mm) in diameter, the other 1 ½ inches (38 mm) in diameter, have an inside measurement of 30 inches (762 mm) and an outside measurement of 32 inches (813 mm). (B) Center-to-center dimension is 30 inches + 32 inches/2 = 31 inches (787 mm). The 31 inches (787 mm) is the dimension for the tram gauge.

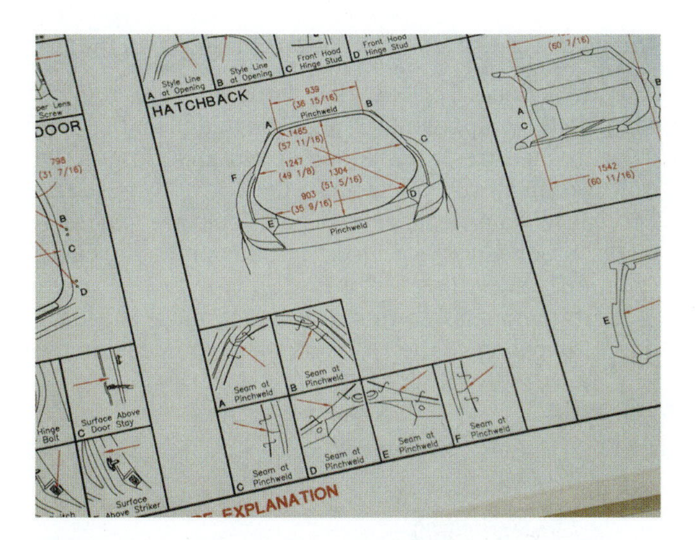

Figure 17–38 Study these examples of dimensions and specifications from a typical manual. Many body dimension charts are in metric measure or both U.S. and metric. Note how the main image shows how to measure a body opening, and details for the measurements are given on the bottom of the image.

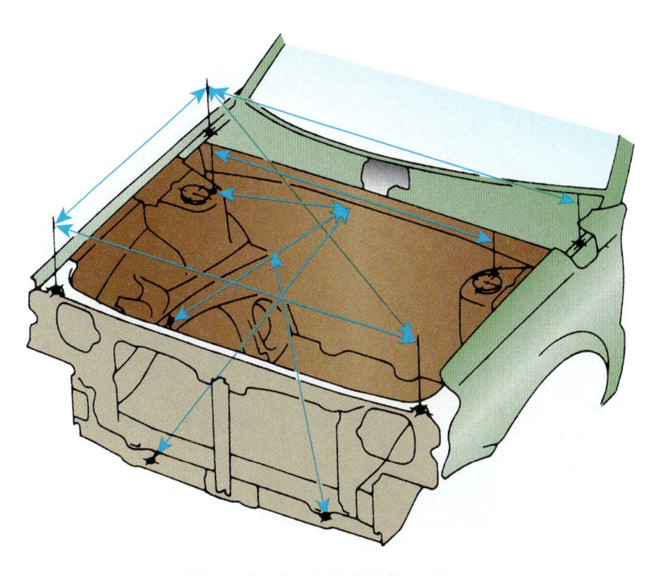

Figure 17–39 Front body measurement points are similar in all charts and manuals, but always use the correct one for the specific vehicle.

the extent of deformation must be checked before remeasuring.

Figure 17–39 shows the typical front body reference points, which can be checked against the manufacturer's body dimensions diagram. You will need to measure across these points to analyze damage. If any measurements are shorter or longer than specifications, you need to use frame straightening equipment to pull or push the body parts back into alignment.

When checking front-end dimensions, the best areas to select for the tram gauge measurements are the attachment points for suspension and mechanical components. These are critical to proper alignment. Each dimension should be checked from two additional reference points, with at least one reference point being a diagonal measurement. The longer the dimension, the more accurate the measurement. For example, a measurement from a lower cowl to the front engine mount cradle is better than from a lower cowl area to another lower cowl area. The longer dimension takes in a larger area of the vehicle. The use of two or more measurements from each reference point ensures greater accuracy. It helps you identify the extent and direction of any panel damage.

Measurement of the Body Side Panel

Any deformation of the body side structure can often be found by noting irregularities in the door when it is opened and closed. Depending on where the deformation is located, attention should be given to possible water and air leakage. It is important that accurate measurements be taken. The tracking tram gauge is primarily used to measure the body side panel (Figure 17–40).

Symmetrical means that the dimensions on the right side of the vehicle are equal to the dimensions on the left side of the vehicle. If the vehicle is asymmetrical, these dimensions are not the same.

A **left-to-right symmetry check** compares measurements on the undamaged side of the vehicle to the damaged side. Warping can generally be detected if the left-to-right symmetry is different in each side. Measure diagonal lines. Use this measuring method if the data on the engine compartment and underbody is missing, if

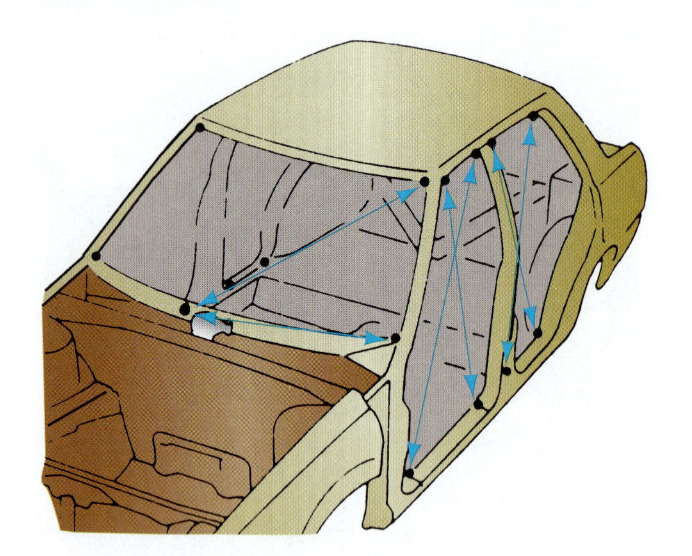

Figure 17–40 These are typical body side panel measurement points. Diagonal measurements across the door and windshield openings tell an important story about major center body damage.

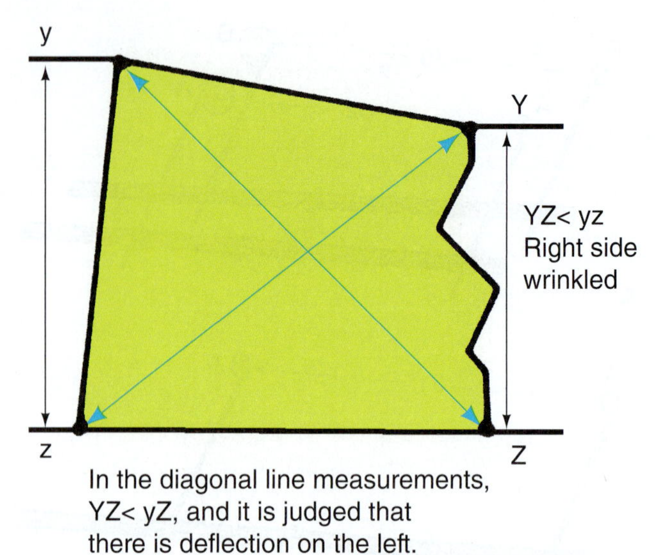

Figure 17–42 Measurement and comparison of the left and right lengths between yz and YZ will give an even better indication of damage conditions. This method should be used in conjunction with the diagonal line measurement. This method can be applied where there are parts that are symmetrical on the left and right sides.

there is no data available in the body dimensions chart, or if the vehicle has been severely damaged in a rollover (Figure 17–41).

The diagonal line measurement method is not adequate when inspecting damage to both sides of the vehicle or in the case of twisting. The left-to-right difference in the diagonal lines cannot be measured. If deformation is the same on the left and the right, a difference will not be apparent (Figure 17–41).

The measurement and comparison of the left and right lengths between yz and YZ will give an even better indication of damage conditions (this method should be used in conjunction with the diagonal line measurement method). It can be applied where there are parts that are symmetrical on the left and right sides (Figure 17–42).

Measurement of the Rear Body

Any deformation of the rear body can be roughly estimated by appearance and irregularities that are evident when the trunk lid is opened and closed. The trunk lid might rub or catch on the body. Because this can cause water leakage and paint damage, measurements are needed to check for the probable cause (Figure 17–43).

Furthermore, any wrinkle in the rear floor is usually due to buckling of the rear side member. Measure the

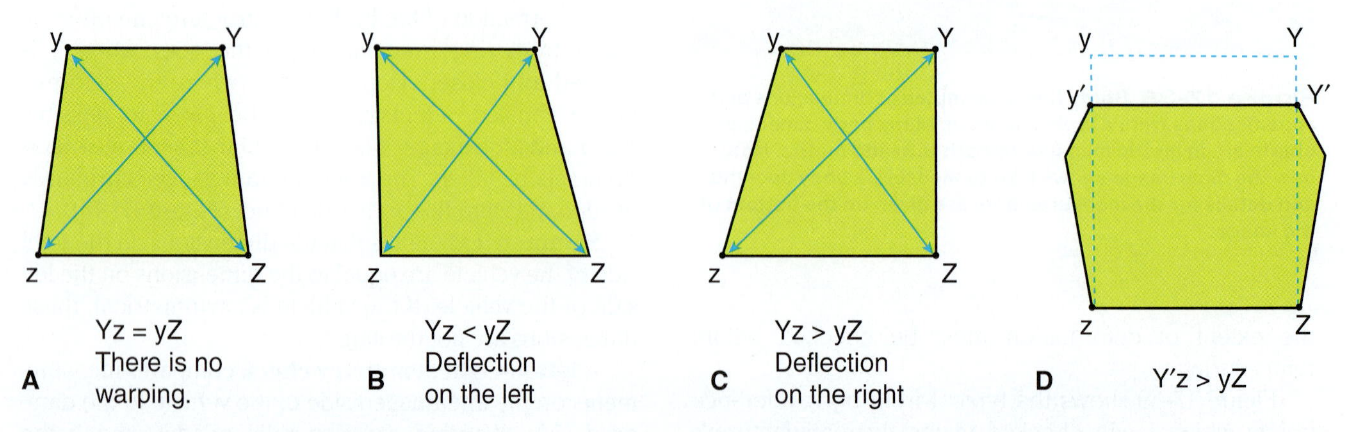

Figure 17–41 Carefully study the diagonal line measurement method. (A) No straightening needed or same as specs. (B) Damage to left, so pull to right for repair. (C) Door opening or other section has been damaged and pushed right. (D) With this damage, diagonal measures would be the same and might even be within specs. Other measurement points are needed to pull out damage.

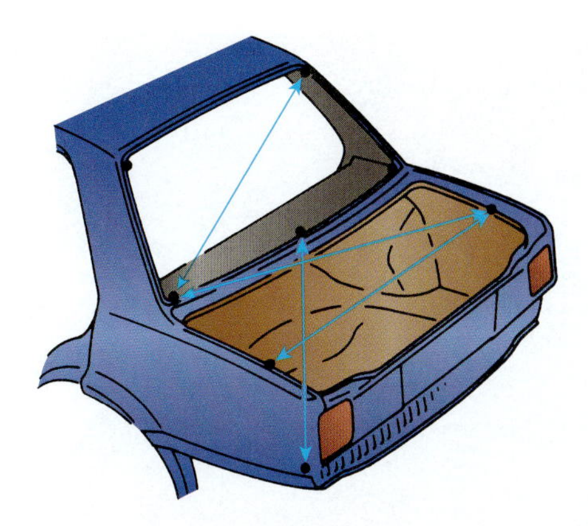

Figure 17-43 Typical rear body measurement points can be checked like other parts already discussed.

rear body together with the underbody. In this way, the straightening work can be performed effectively.

When using a tram gauge (Figure 17–44, Figure 17–45, Figure 17–46), be sure to keep the following methods in mind:

- Measurements are made to fixed points on the vehicle, such as bolts, plugs, or holes.
- A point-to-point measure is the direct actual measurement between two points (Figure 17–47).
- The tram bar should be parallel to the car body. This might require the pointers on the tram bar to be set at different lengths.
- Some body dimensions manuals show dimensions in bar length. Other dimension books show dimensions in point-to-point lengths. Some manuals use both. Keep this in mind when reading specs and measuring.
- Make all measurements on the damaged vehicle at the points specified in the body manual. The amount of damage can usually be determined by subtracting the actual measurement from the specified measurement.

Figure 17-44 The technician is looking up a dimension in the manual so he can set tram gauge to that dimension.

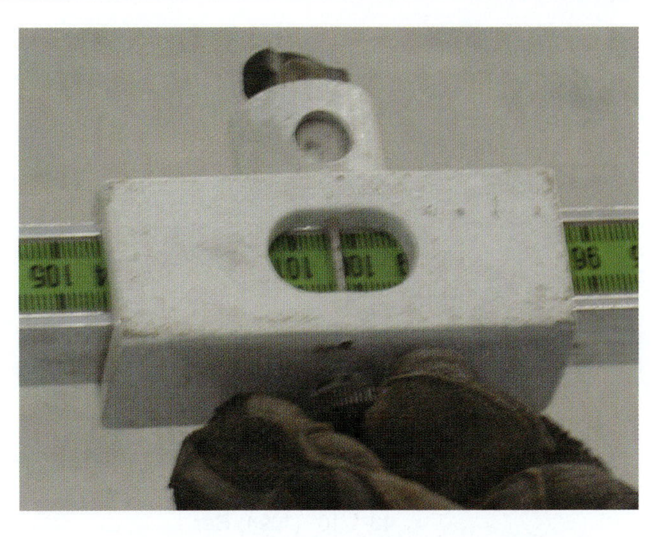

Figure 17-45 The scale and line on a tram gauge will show the length the tram gauge is able to measure.

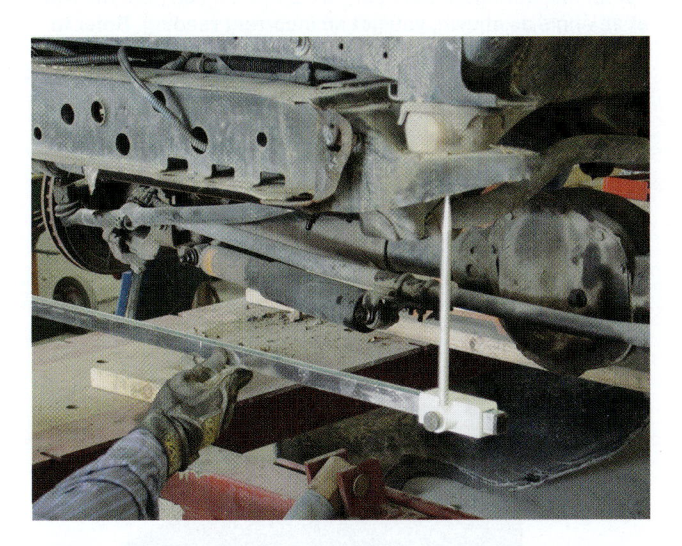

Figure 17-46 With the tram gauge length set correctly, pointers should touch the center of reference points. If the pointers do not align with the point on the vehicle, you have found major structural damage.

17.6 DIGITAL TRAM GAUGES

A **digital tram gauge** can slide to different lengths and will electronically measure its own length and show a numeric readout in inches or millimeters. Many can send a signal out to the measurement computer to help streamline the damage measuring process. See Figure 17–48.

Figure 17–49 shows a digital tram gauge used in conjunction with a small laptop computer. The laptop can store the vehicle dimensions for easy comparison with the digital tram gauge readings. This setup also has a small printer for making hard copies of dimension sheets to use as a reference when pulling the damage out on a frame rack.

Various rods and ends are provided with a digital tram gauge for measuring different kinds of references

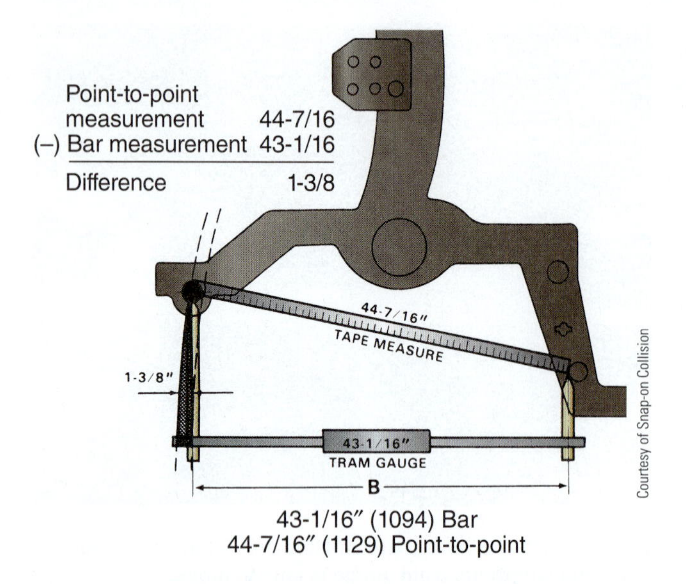

Figure 17–47 Point-to-point measurement is direct, actual measurement between two points. If you hold the tram at an angle as shown, you get an incorrect reading. Refer to spec chart footnotes for getting correct values and procedures.

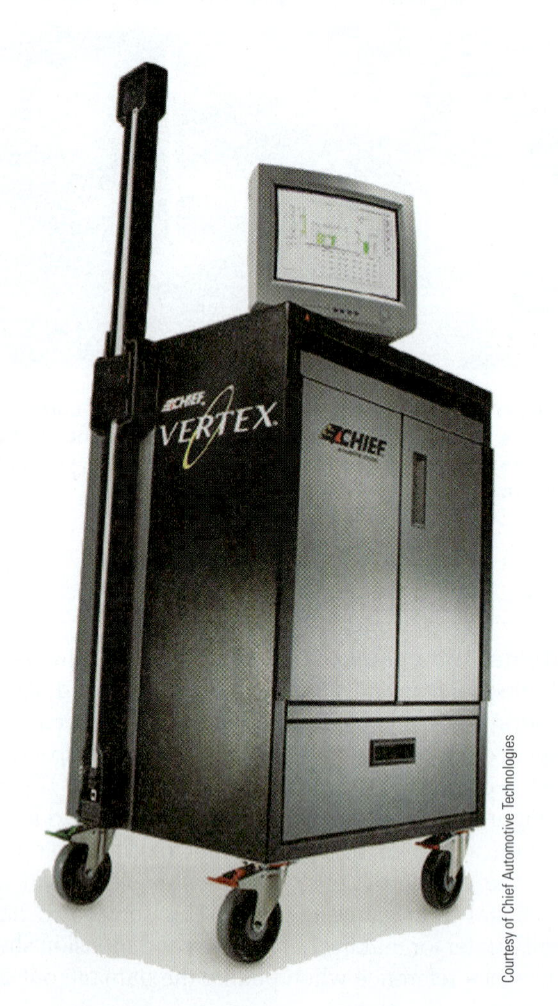

Figure 17–48 A digital or electronic tram gauge and computer is the modern method of measuring vehicle frame and unibody damage.

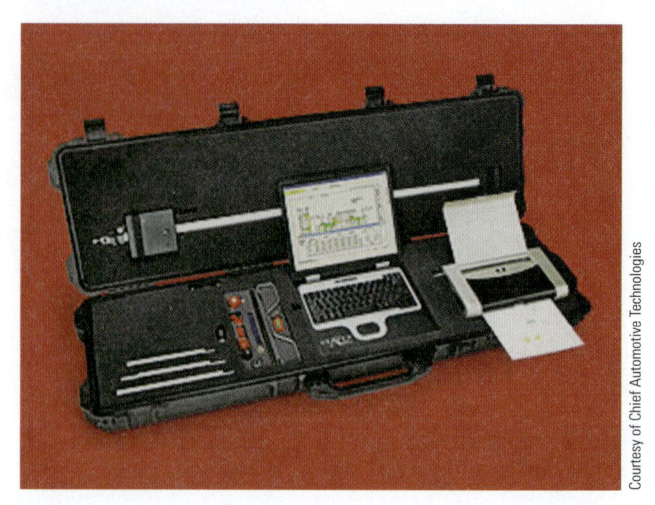

Figure 17–49 This digital tram gauge comes with its own laptop computer and printer. Some electronic tram gauges can communicate with a personal computer over a wireless network; some cannot.

points and locations (Figure 17–50). The user's manual for the unit helps you select the correct attachments needed for each measurement.

A close-up of one digital tram gauge (Figure 17–51) shows buttons for zeroing, holding a measurement, entering a measurement, and so on. Also note the digital readout.

By comparing your measurements to the correct factory specs for those reference points, you can determine whether the vehicle is badly damaged and in need of frame straightening (Figure 17–52).

Figure 17–53 shows the workstation for one type of digital tram gauge. Note the keyboard, computer display, and attachments for the gauge. You can quickly see the known good dimensions on the monitor and instantly compare them to your actual measurements to determine the extent and direction of damage.

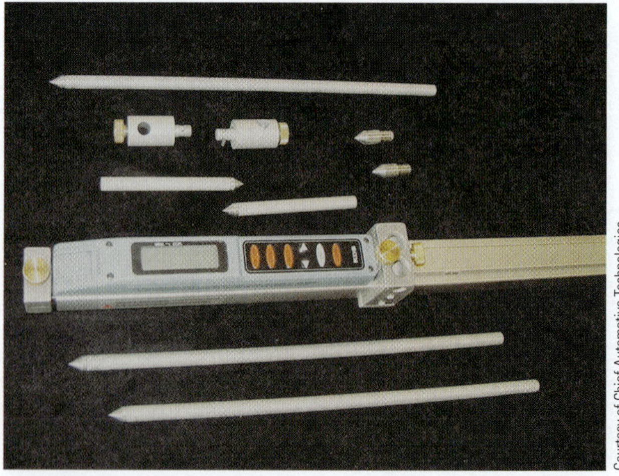

Figure 17–50 Various attachments come with a digital tram gauge.

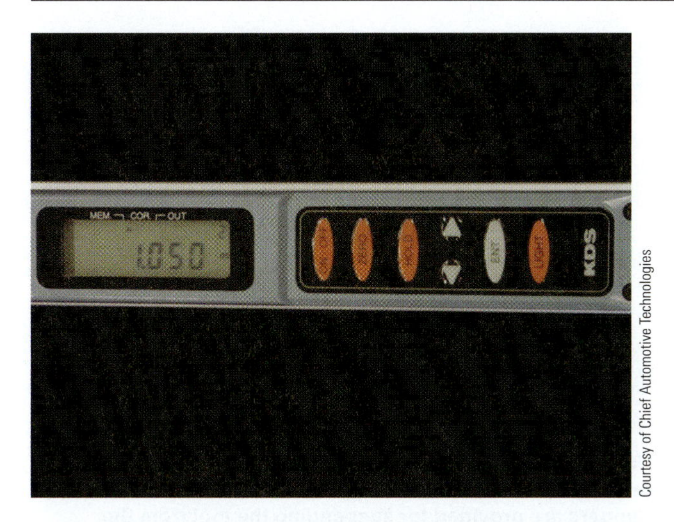

Courtesy of Chief Automotive Technologies

Figure 17-51 Note the controls and digital readout on this digital tram gauge. It has a built-in wireless sending unit that can communicate with a laptop or personal computer.

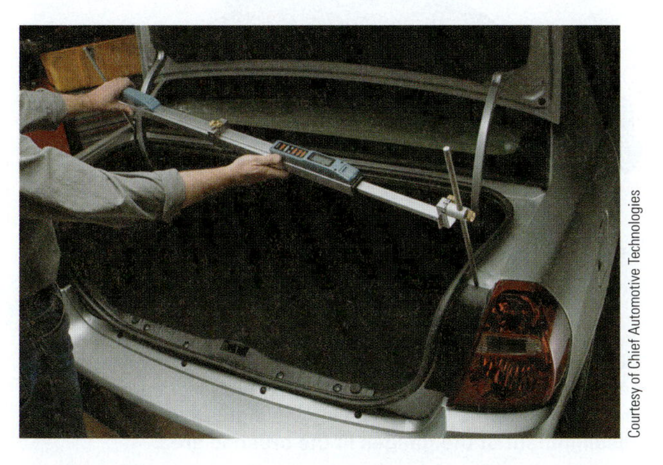

Courtesy of Chief Automotive Technologies

Figure 17-52 By sliding the digital tram gauge to fit the reference points on the vehicle and pressing Enter, the unit will store measurements and send them to a personal computer. The personal computer will then compare the measurements to specs to analyze damage.

In Figure 17–54, a technician is using a digital tram gauge to measure underbody dimensions. With the computer nearby, measurements can be downloaded quickly and viewed on the computer monitor.

Figure 17–55 shows a close-up view of a readout from a digital tram gauge. Note how it shows length and height at the same time.

By pressing *Enter* on some digital tram gauges, you can use a built-in wireless transmitter to send the measurement to a computer. The computer then automatically compares your measurements to specifications, for rapid damage analysis. See Figure 17–56.

17.7 CENTERING GAUGES

While self-centering gauges are closely related to the tram gauges; they do not measure, they show alignment or misalignment by projecting points on the vehicle's

Courtesy of Chief Automotive Technologies

Figure 17-53 The console for a digital tram gauge holds a monitor, mouse, attachments, and keyboard. By clicking on the monitor screen and selecting a measurement point, the tram can quickly tell whether a measurement is correct or indicates vehicle damage.

Courtesy of Chief Automotive Technologies

Figure 17-54 Here the technician is measuring under the vehicle and communicating with a personal computer to quickly find the amount and direction of damage.

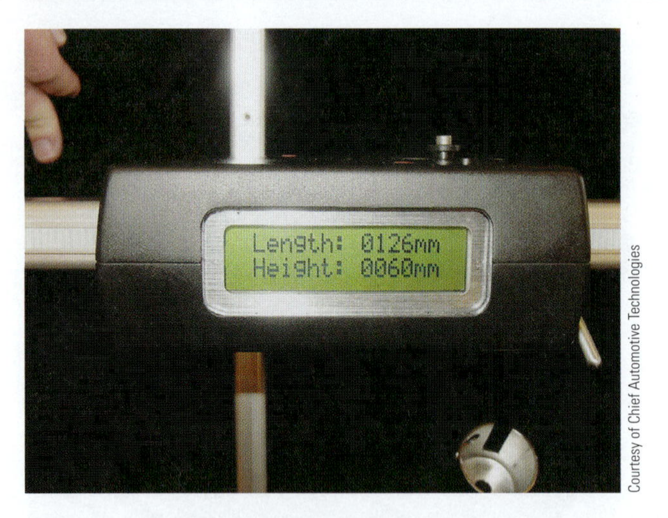

Figure 17-55 A close-up of this digital tram gauge shows how it reads both length and height at the same time.

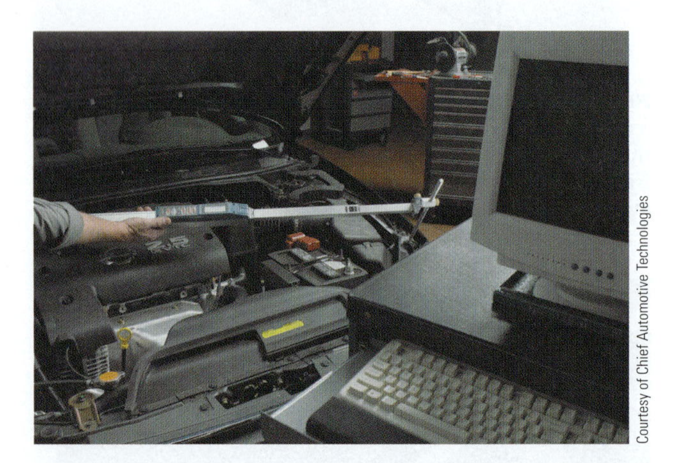

Figure 17-56 A digital tram gauge will save considerable time evaluating vehicle damage when estimating or before pulling out the damage.

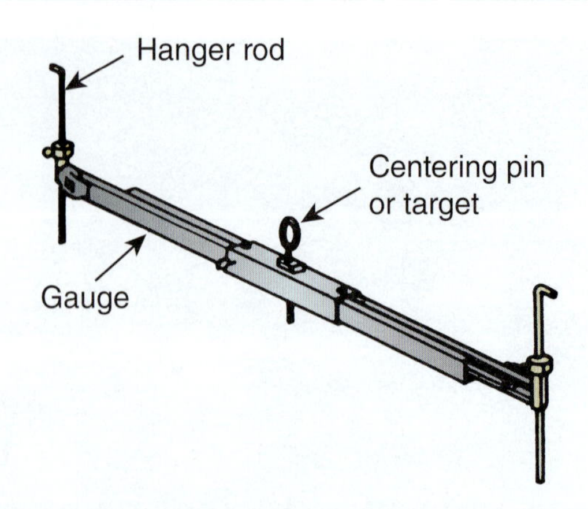

Figure 17-57 The typical self-centering gauge has a center pin or target for viewing the centerline of the vehicle. Hangers are provided for suspending the tool from the underbody of the vehicle.

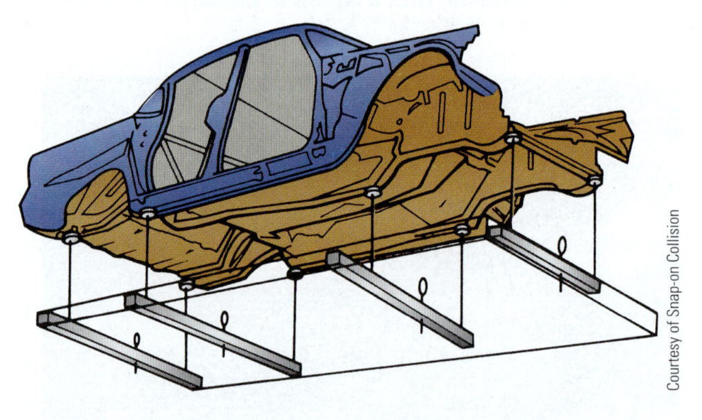

Figure 17-58 Typical starting locations for centering gauges. At least two gauges must be hung from undamaged areas on the vehicle. This will let you sight down and see any misalignment of the gauges in the crushed area.

structure into the technician's line of sight. They are installed at various control areas on the vehicle (Figure 17–57).

Self-centering gauges have two sliding horizontal bars that remain parallel as they move inward and outward. This action permits adjustment to any width for installation on various areas of the vehicle. After the gauges are hung on the car (usually three or four sets), the horizontal bar will be parallel to the portion of the structure to which it is attached.

Place one centering gauge at the front of the vehicle, one at the extreme rear, one just behind the front wheels (front torque box), and one forward of the rear wheels (rear torque box). The two gauges in the center are usually considered to be the baseline gauges (Figure 17–58).

When inspecting for collision damage, first hang centering gauges from two places where there is no visible damage, usually the center section of the body or frame. Then hang two more gauges where there is obvious damage. Then look or sight along the gauges and analyze how the self-centering gauges line up. Check for parallel misalignment of the gauges or misalignment of the centering pins. This will help determine the direction and extent of major body or frame damage (Figure 17–59).

Centering gauges are equipped with center pins or sights that always remain in the center of the gauge, regardless of the width of the horizontal bars. This allows the body technician to read the centerline throughout the length of the vehicle.

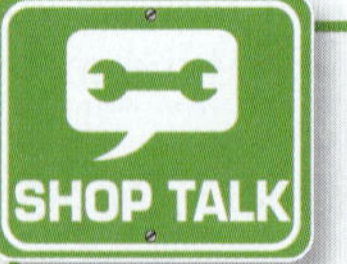

There should not be any deformation at the point where the gauge is installed. There are many instances where the alignment of the gauge holes has been deformed by previous collisions or other causes. Do not use deformed holes unless they can be repaired satisfactorily (Figure 17–60).

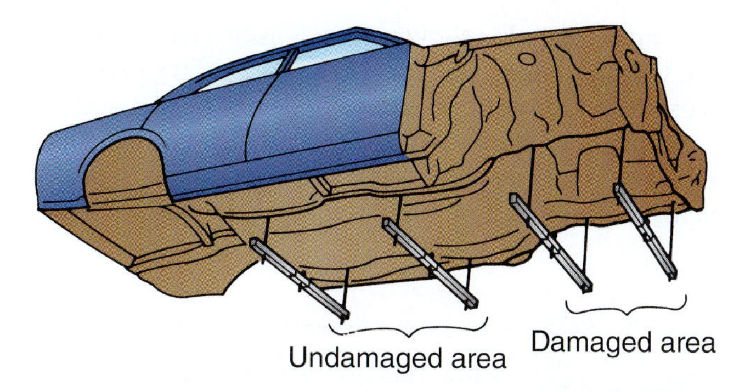

Figure 17-59 Note the placement of centering gauges to check a damaged area. For example, if the center section is undamaged, you can view and analyze damage to the front body/frame structure for straightening.

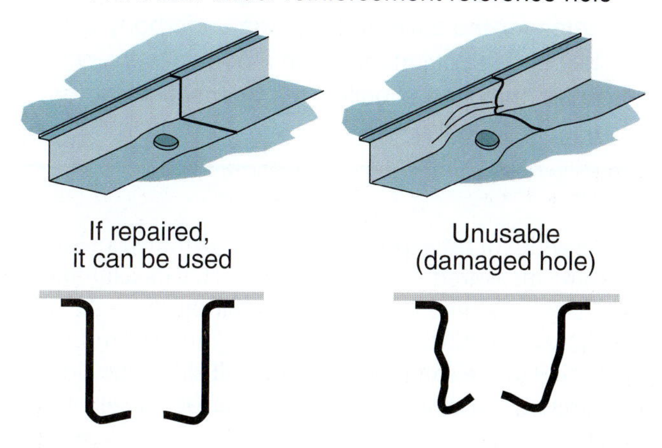

Figure 17-60 Never use body holes that are deformed or damaged. Fix or reshape them before measuring.

Each self-centering gauge accommodates two vertical scales—one on the left side, one on the right. These scales can be adjusted vertically, which ensures that the horizontal bars accurately reflect the true positions of the parts to which they are attached. Once hung in specific locations, these gauges generally remain on the car throughout the entire repair operation, unless one or all of them interfere with straightening or with tram gauge measurements.

Special centering gauges are available that can be used to check such items as body pillar damage. The same system of alignment is employed when using centering gauges to check underbody damage.

The centering gauge reading is the visual alignment of parallel bars and pins. The final objective is to achieve a level and centered structural alignment in the vehicle.

Here are some points to keep in mind when reading and using centering gauges:

- Assuming that the centering gauges have been installed properly, correct alignment will be achieved when all the gauges are parallel and the centering pins line up in a row. This indicates that the frame or unibody is level, not twisted or deformed.
- When sighting crossbars for parallel, always stand directly in the middle, scanning with both eyes. To ensure accuracy, readings should be made at the outer edge of the centering gauge, not in the middle.
- The farther one stands from the centering gauges while reading, the more accurate the reading will be. Standing close changes the line of sight to the front gauges so that an accurate reading is nearly impossible.
- Centering gauges should always be set at the same height or plane. Different heights will change the angle of sight and give a false reading.
- It is sometimes beneficial to sight over one gauge and under another. Going to the end of the vehicle, opposite the damage, to make readings will sometimes result in a more accurate reading. This is true because you are able to read the base gauges before sighting into the damaged area. With practice and a certain amount of experimentation, you can improve the damage analysis.
- The sighting of centerline pins must be done with one eye. Since the center section is always the base for gauging, the line of sight must always project through the pins of the base gauges. Observing pins in other sections of the frame will then reveal how much they are out of alignment.
- Never attach the centering gauges to any movable parts, such as control arms or springs.

Self-centering alignment gauges are used to read three major elements of collision damage: datum, center, and zero planes. Critical measurements—the fourth major element of analysis—are handled with a tape measure and tram gauge.

Dimensional References

Two major dimensional references are indicated in all body dimension manuals: the datum plane and centerline.

A *datum line*, or **datum plane**, is an imaginary flat surface parallel to the underbody of the vehicle at some fixed distance from the underbody. It is the plane from which all vertical or **height dimensions** are taken by the vehicle manufacturer. It is also the plane that is used to measure the vehicle during repair. The datum is normally shown on dimension charts from the vehicle's side view.

Using this line of reference, centering gauges can be strategically suspended under the vehicle from side to side at varying distances along the length of the chassis frame or unibody. First, place the base gauges at the main platform: one across the vehicle beneath the rear seat (rear torque box) and another under the cowl area (front torque box). Add two more gauges before and after the base gauges: one located at the front cross member and a second at the rear cross member. Additional gauging of the front cross member area and/or strut tower completes the picture.

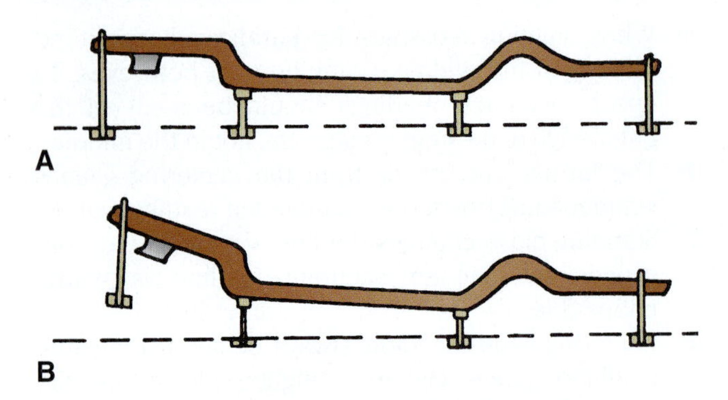

Figure 17–61 (A) Datum is correct and (B) datum is off at front. The front centering gauge would be too high when you sighted down the other three gauges.

To read datum, all gauges must be on the same plane, as indicated by the spec sheets. After hanging all four gauges, read across the top to determine whether datum is correct. If all four gauges are level at the top, the vehicle is on datum. If they are not level, the vehicle is off datum (Figure 17–61).

Because the datum line is an imaginary plane, datum heights can be raised or lowered to facilitate gauge readings. If the datum height is changed at one gauge location, all gauges must be adjusted an equal amount to maintain accuracy.

Although datum readings are usually obtained from centering gauges, there are individual gauges available for measuring datum heights. These datum gauges are usually held in position by magnetic holders. Remember that the dimensions that allow the vehicle to be level with the road are measured from the datum plane.

The Center Plane

The **center plane**, or *centerline*, divides the vehicle into two equal halves: the passenger side and the driver's side. Note how the center plane or line cuts up through the middle of the vehicle (Figure 17–62).

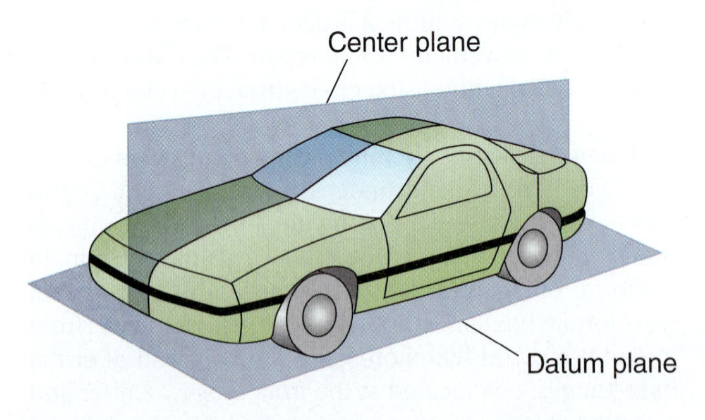

Figure 17–62 The center plane or centerline of a vehicle allows for horizontal measurements, whereas the datum plane allows for vertical measurements.

The centerline is shown on dimension charts in either the bottom or top views. It can be found on some vehicles in the form of body center marks.

Body center marks are often stamped into the sheet metal in both the upper and lower body areas of the vehicle. They can save time when taking measurements.

Most vehicles are built symmetrically. But if the vehicle is not symmetrical (if it is asymmetrical or has unequal measurements on each side), the self-centering gauges will not align and will not indicate a true center reference. In such a case, a centering gauge that compensates for the asymmetry of the underbody can be used in conjunction with a body dimensions chart that has a built-in compensation factor (Figure 17–63).

All width or **lateral dimensions** of symmetrical vehicles are measured from the center. The measurement from the centerline to a specific point on the right side will be exactly the same as the measurement from the centerline to the same point on the left side. One side of the structure should be a perfect mirror image of the other.

Zero Planes

It is usually necessary to think of the vehicle as a rectangle divided into three zero plane sections. The **three zero plane sections** break the vehicle into three areas—front, center, and rear. The torque box location is used as the dividing line. This three-section principle is a result of the vehicle's design and the way it reacts during a collision (Figure 17–64).

To check for centerline misalignment, all four centering gauges must be hung. To establish the true centerline, the center pin on the #2 gauge must be lined up with the center pin on the #3 gauge. Then the center pins of #1 and #4 can be read relative to the centerline of this base.

Of course, other damage conditions will affect a centerline reading. If a vehicle has a sway condition or an out-of-level condition, the centerline reading will be affected. Further inspection by gauging or measuring might be necessary to determine the presence or absence of these problems.

The controlling points of any car underbody are the front cross member, the cross member at the cowl, the cross member at the rear door, and the rear cross member (Figure 17–65). The center section or the area between the cowl and rear door cross member is the portion used when doing a major straightening operation.

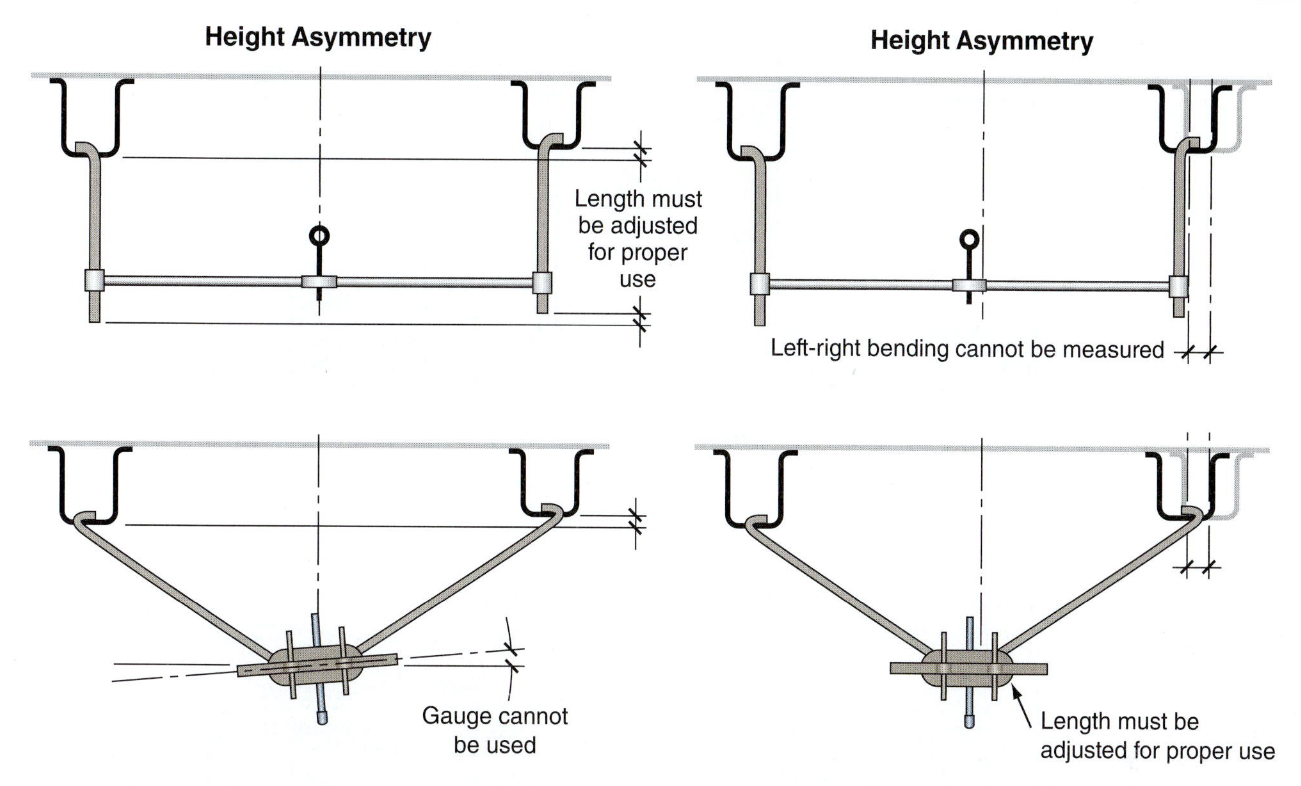

Figure 17-63 Note how these gauges are arranged when measuring asymmetric installation points.

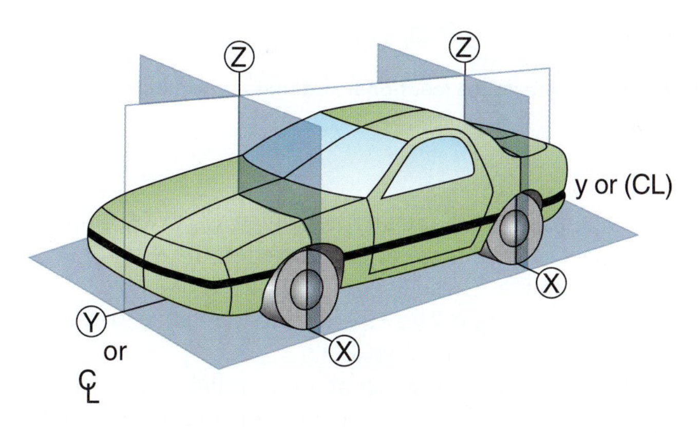

Figure 17-64 Three vehicle planes are combined in this illustration. This allows for three-dimensional, or 3D, measurements.

Level in a zero plane means the condition in which all areas of the vehicle are parallel to one another. Level refers to parallel conditions in the vehicle structure only and has nothing to do with any outside reference, such as the floor. Check for an out-of-level condition in the front or rear sections. When the #2 and #3 base gauges are hung, the center section is read for level. When these base gauges are parallel, no twist can exist. However, the front or rear sections could still be out of level. To check for this condition, hang #1 and #4 gauges and read relative to the nearest base gauges. If #1 hangs parallel to #2, the front section is level, relative to the base. If #4 hangs parallel to #3, the rear section is level, relative to the base. If an out-of-level condition exists in the front, #1 will not hang parallel to #2. This same type of reading should be done in the rear section.

17.8 DIAGNOSING DAMAGE USING GAUGE MEASURING SYSTEMS

As previously mentioned, the most common rule in body/frame alignment is: Reverse direction and sequence. To correct collision damage, pull or push the damaged area in the opposite direction of impact, and because the repair must be made in the reverse sequence, the damage must also be measured in the reverse sequence.

When measuring damage, keep in mind that a vehicle is similar to a building. If the foundation is not square and level, the rest of the structure also will be uneven. The vehicle's foundation, which is the center section of the vehicle, is measured for twist and diamond first. These two measurements will tell the collision repair technician whether the foundation is square and level. The remaining measurements use the foundation as a reference.

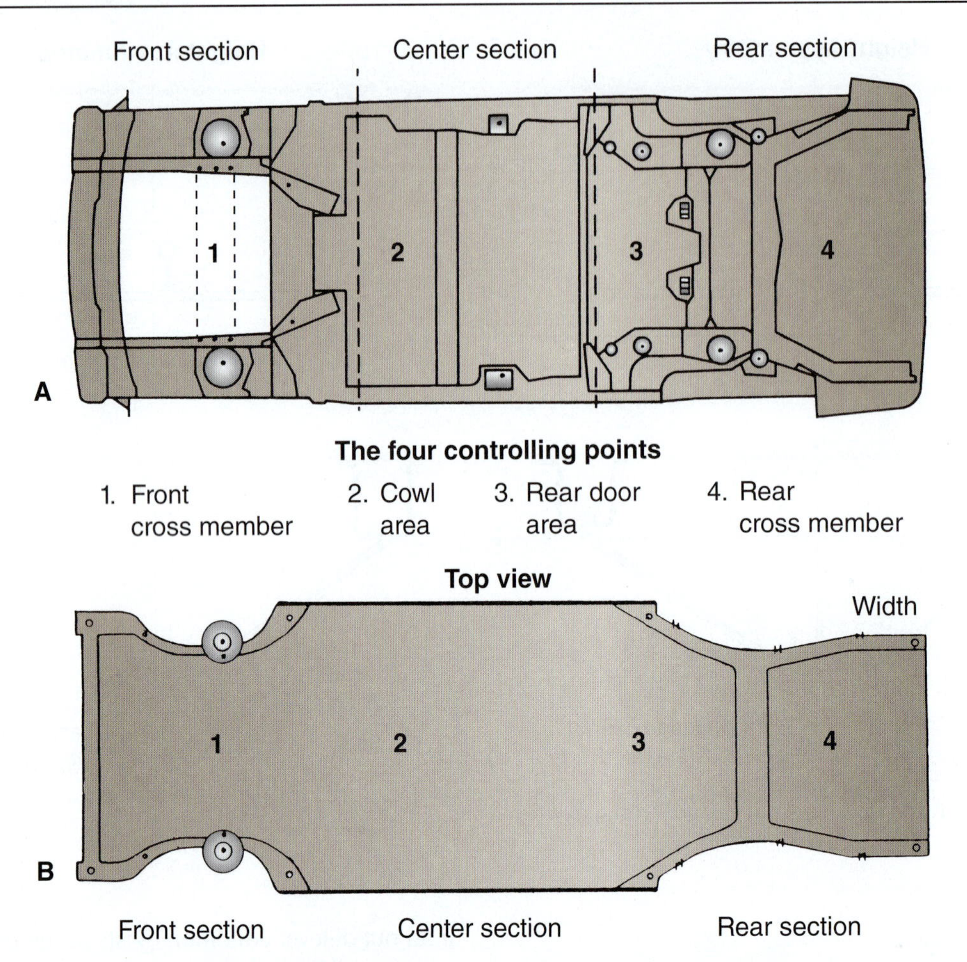

Figure 17-65 **During major repairs, the vehicle is normally divided into three zones (sections) and four controlling areas. Compare (A) a conventional frame vehicle and (B) a unibody car.**

Measuring Twist Damage

Twist damage is a condition in which one corner of the car is higher than normal; the opposite corner may be lower than normal.

The first damage condition to look for and measure is twist. Twist is the last damage condition to occur to the vehicle and exists throughout the entire vehicle.

To check for twist, sight down your properly hung self-centering gauges. Twist will show up when the gauges are not parallel.

Twist can be checked only in the center section; otherwise, additional misalignment in the front or rear may give an inaccurate reading of twist.

To check for twist, two base gauges must be hung. These base gauges are also referred to as the #2 (front center) and #3 (rear center) gauges. The #2 should be hung as far forward of the center section, up to the cowl, as possible. The #3 is hung as far rearward of the center section, toward the rear kickup, as possible. The #2 gauge is then read relative to #3.

If the gauges are parallel, no twist exists. If the gauges are not parallel, then a twist may exist. Remember that a true twist must exist throughout the entire structure. To check for a true twist versus an out-of-level condition in the center section, hang another gauge. Go to the undamaged section of the car and hang either the #1 (front) or #4 (rear) gauge. These gauges will be read relative to the nearest base gauge. The #1 will be read relative to #2 and the #4 will be read relative to #3.

If the front or rear gauge reads parallel to the nearest base gauge, true twist cannot exist, and there is an out-of-level condition in the center section. If a true twist exists, the gauges will read like those shown in Figure 17–66.

Measuring Diamond Damage

Diamonding is a condition in which one rail or rocker is pushed either forward or rearward of the opposite rail or

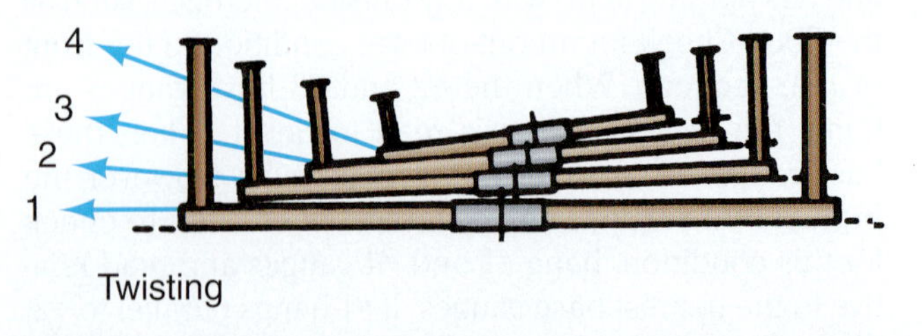

Figure 17-66 **If self-centering gauges read like this, the body or frame has twist damage.**

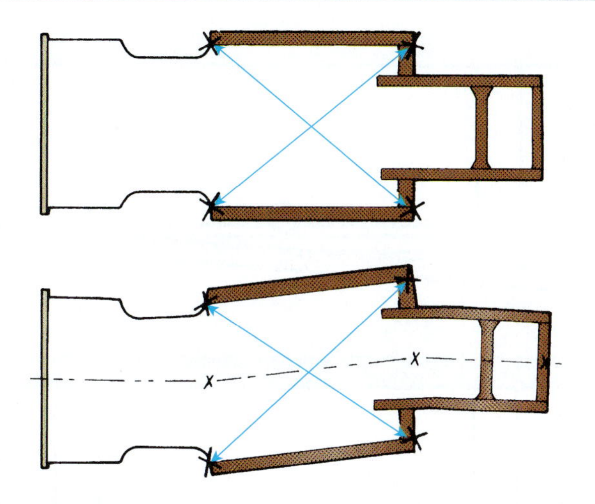

Figure 17-67 Unequal diagonal measurements show diamonding.

rocker. This condition is often found in conventional frames. Rarely do you find a diamond condition in unibody vehicles, but it is possible.

The check for diamonding is simple. Using a tram gauge, measure from the front corner of one rail or rocker to the rear corner of the opposite side. If one measurement is longer than the other, a diamond condition exists. If one measurement is 1 inch (25 mm) longer than the other, the condition is referred to as "1-inch diamond," or "25-millimeter diamond" (Figure 17–67).

Measuring Mashing (Crushing)

Mash can be measured with a tram gauge. Mash is present when any section or frame member of the vehicle is shorter than the factory specification. When using a tram gauge on a mash-damaged vehicle, be sure to make the measurement specified on the manufacturer's specification sheets or in the dimensions data book. The amount of mash is determined by subtracting the actual measurement from the specified measurement. The proper methods of measuring various impacts with a tram gauge are shown in Figure 17–68.

Measuring SAG Damage

Sag is a condition where the cowl or another area of the vehicle is lower than normal. Sag can also occur at the front cross member. The ends of the cross member will be closer than normal and the center will be too low.

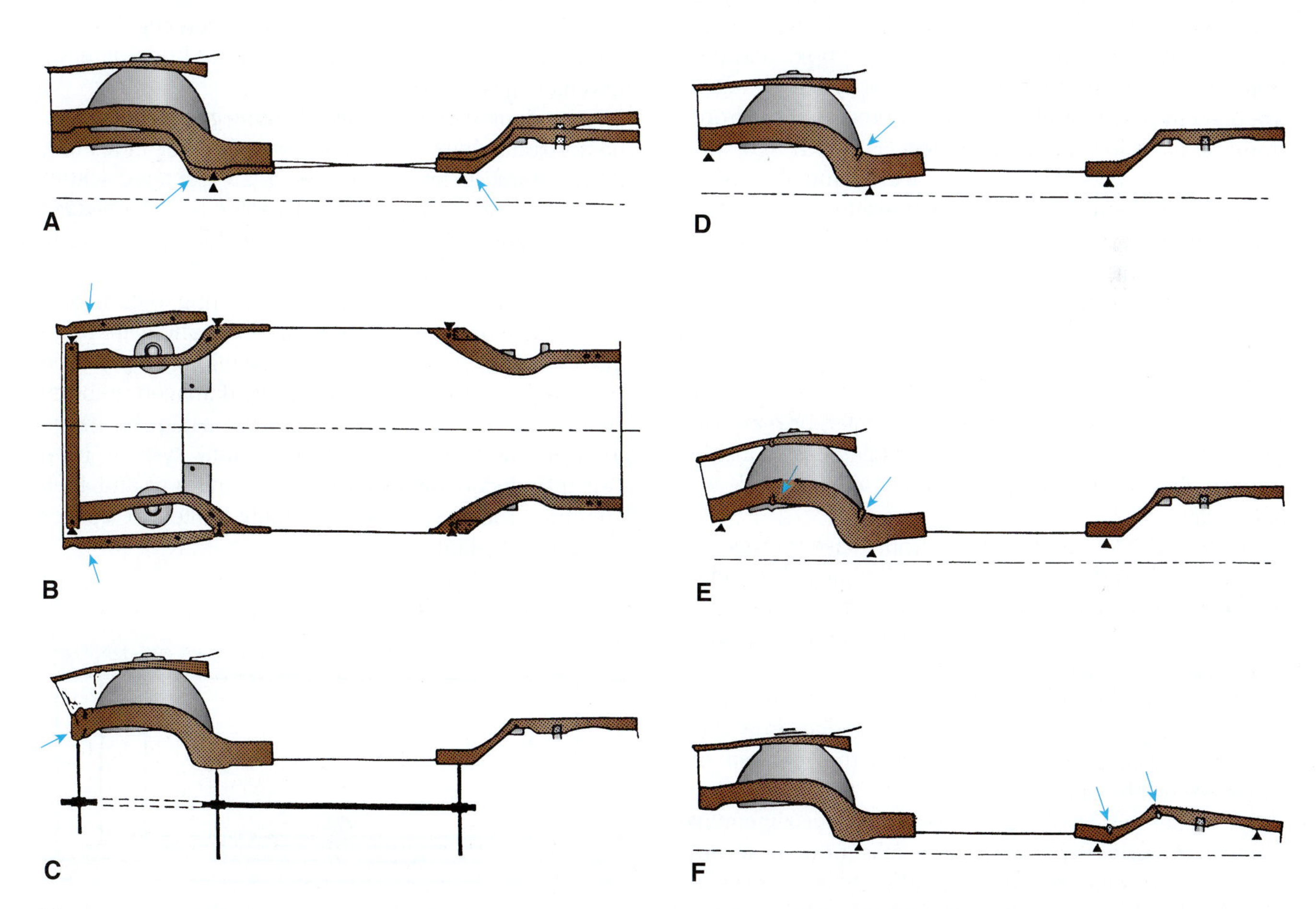

Figure 17-68 Study variations of unibody distortion under impact force. (A) High front impact with secondary damage to the rear of the assembly, (B) right front corner impact, (C) direct front impact, (D) low front impact, (E) high front impact, and (F) high rear impact.

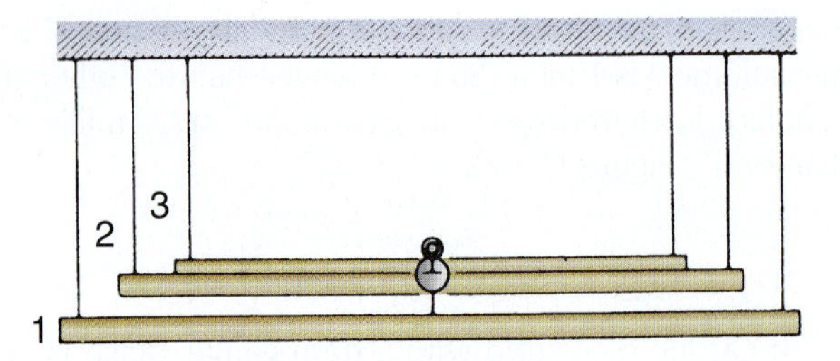

Figure 17–69 Cowl area sag is shown by the gauge pointer being lower in undamaged areas.

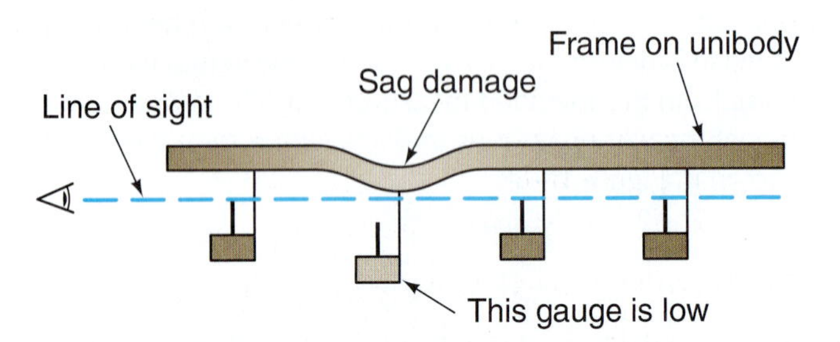

Figure 17–70 A side view gives a better idea of sag.

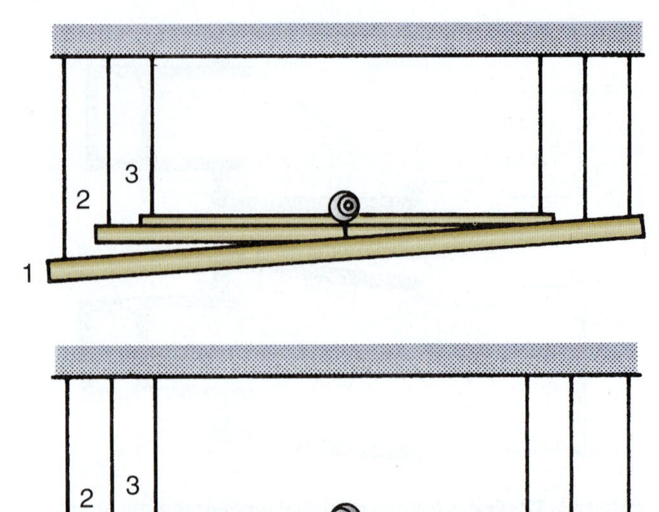

Figure 17–71 Self-centering gauges show side-sway damage.

Three centering gauges are used to check for a sag condition. One gauge is placed at the front cross member, the next one at the cowl area (torque box), and the third one at the rear door area (torque box). If the gauges are on-center and parallel with each other, but the front frame gauge is lower than the others (Figure 17–69), this indicates a sag condition at the front cross member area.

In Figure 17–70 observe the relationship of the centering gauges to each other. Note that the parallel bar in the cowl area is about 2 inches lower than the other bars. This means there is 2 inches of sag.

Measuring Side-Sway Damage

Side-sway is present when the front, center, or rear portion of the vehicle is pushed out of alignment by a side impact. Three centering gauges are used to check for side-sway.

If the vehicle was hit in the front, base gauges #2 (cowl area) and #3 (rear door area) are hung. The sighting gauge #1 is located in the front cross member area. If gauge #1 does not line up with the other two, front side-sway is present (Figure 17–71).

If the vehicle has been struck in the rear, the misalignment will appear on the self-centering gauges in a way similar to front side-sway, except that the rear pin or bull's-eye will be out of alignment.

A center hit on a vehicle causes a misalignment known as *double side-sway*. It results from a severe impact in the center section, but it affects the entire vehicle. The dimensions of both front and rear sections must be checked during the pulling of double side-sway damage.

While the self-centering and datum gauges give a total picture of frame and body damage, their functions can be adapted into a so-called frame gauge. By viewing body damage with a frame gauge arrangement, it is possible to measure the amount of frame or body damage the vehicle has incurred.

An undamaged frame gives a frame gauge indication. The horizontal bars are parallel to each other, indicating that the frame is level. The targets are centered within each other, indicating a perfect centerline. The gauges reveal horizontal and vertical alignment for certain body and frame damages, as shown in Figure 17–72.

Measuring gauges—tram, self-centering, datum, and frame—have been in use for many years and were originally designed for measuring conventional body-over-frame vehicles. However, structurally damaged unibody vehicles can be successfully repaired using the gauge measuring system. In recent years, new systems have been introduced for use with both unibody and BOF structures. Today, gauges are usually limited to light or medium body damage.

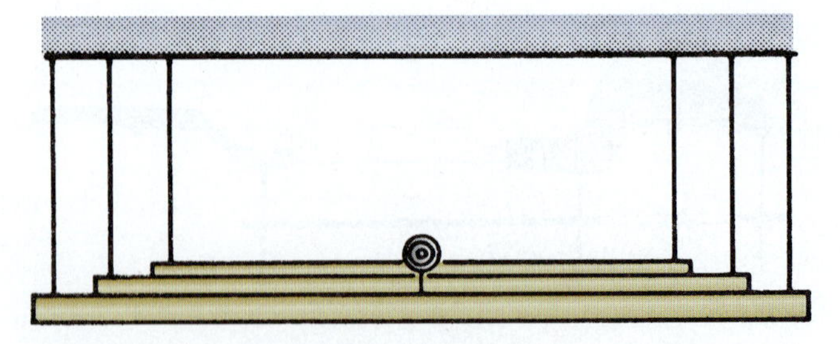

Figure 17–72 When the body/frame is fixed by power straightening, no deviation should be seen in the centering circles or pointers of gauges.

17.9 STRUT CENTERLINE GAUGE

A **strut tower gauge** shows misalignment of the strut tower/upper body parts in relation to the centerline plane and datum plane. The strut tower gauge allows visual alignment of the upper body area. It is usually mounted on the strut towers (Figure 17–73).

The strut tower gauge features an upper and a lower horizontal bar, each with a center pin. The upper bar is usually calibrated from the center out. Pointers, which are positioned in an adjustable housing on the upper horizontal bar, are used to mount the gauge to the strut tower/upper body locations (Figure 17–74).

Two types of pointers are provided: cone and reverse cone. The reverse cone is notched to provide additional means of mounting on the vehicle, for example, on ridged surfaces. The pointers are usually held in the housing by means of thumb screws (Figure 17–75).

Different length pointers are provided for situations when more length is needed to position the gauge. When using different length pointers to mount the gauge, remember that they change the scale reading.

The vertical scales that link the upper and lower horizontal bars are used to set the lower bar at the datum height of the mounting parts. The scales fasten in housings at the ends of the horizontal bars. Height adjustments are made at the housings of the upper horizontal bar (Figure 17–76).

By referring to a body dimensions chart, the technician can adjust the strut tower gauge to the correct dimensions, using the vertical scales that link the upper and lower horizontal bars to set the lower bar at the datum plane. With the lower bar set to align properly, the upper pointers should be located at reference points on the strut towers. If they are not, the strut towers are damaged and pushed out of alignment. This would tell the technician that straightening is needed so that the front suspension and the wheels can be aligned properly (Figure 17–77).

Figure 17–73 Here is typical strut centerline gauge in place. The tops of the strut towers is a common measurement point, because they are often moved out of alignment during frontal impact.

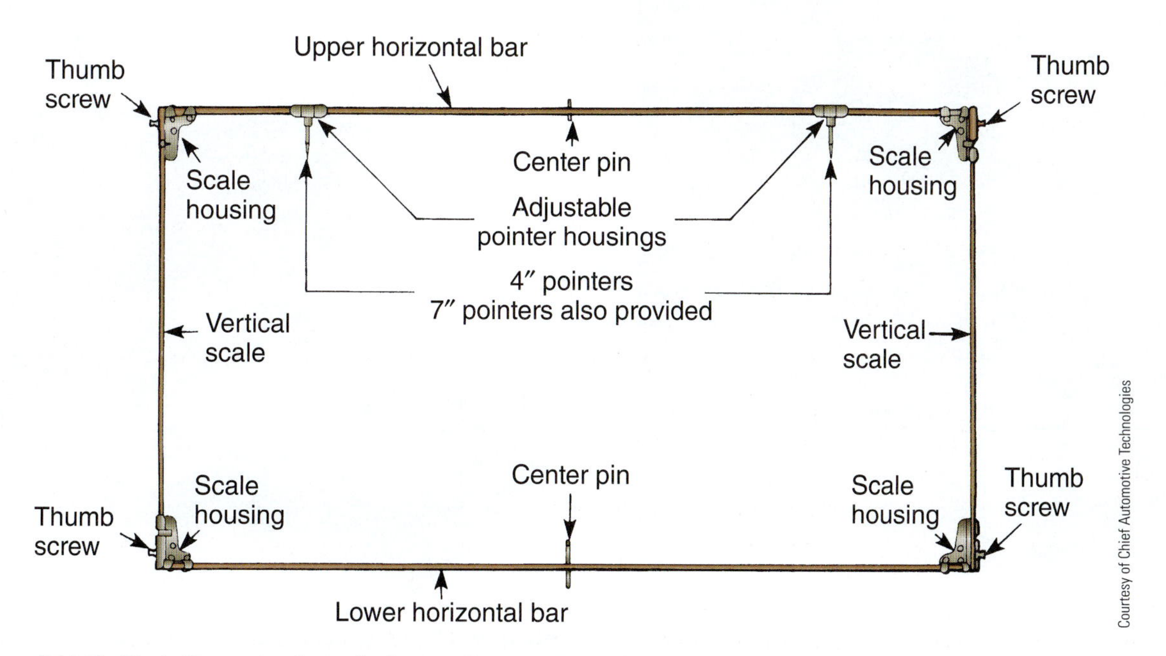

Figure 17–74 Study the parts of a typical centerline strut tower/upper body gauge.

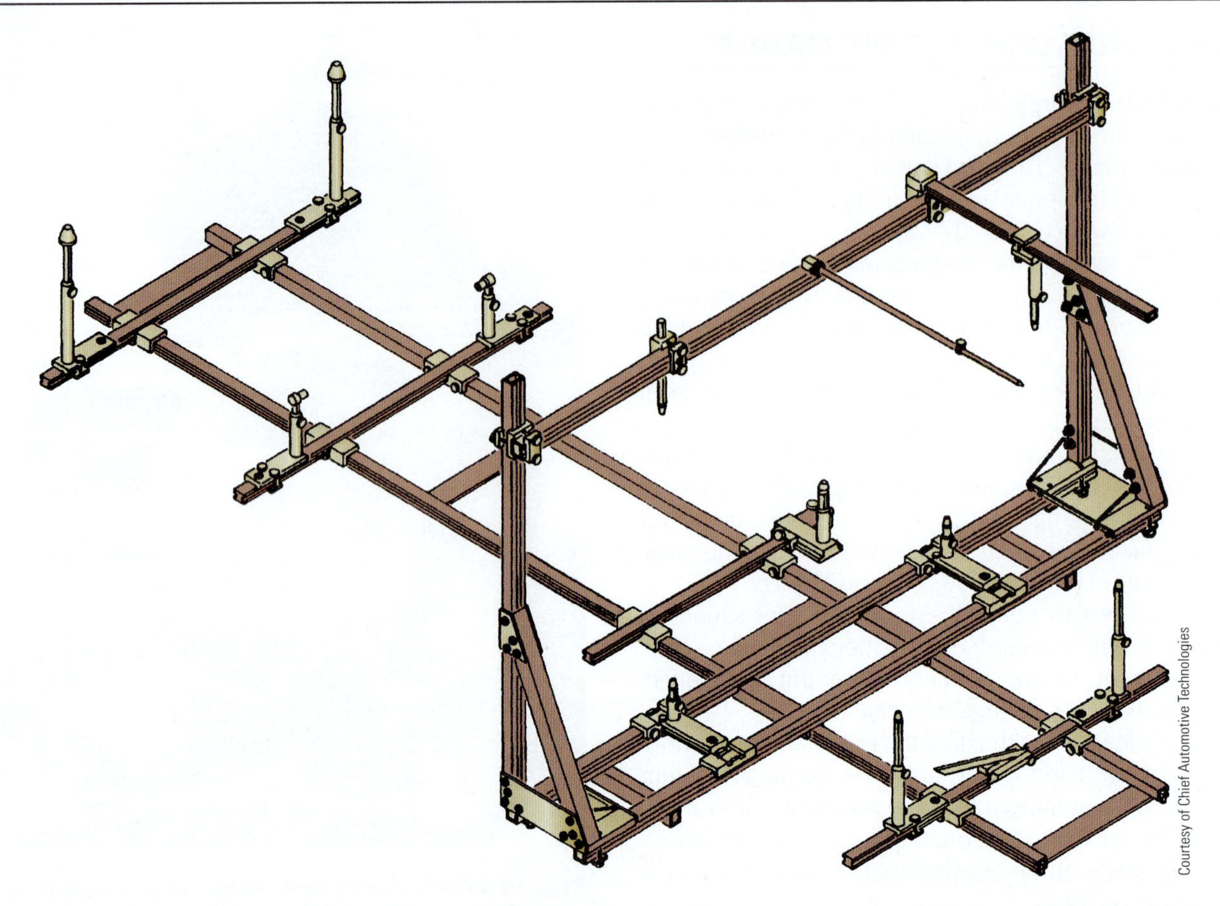

Courtesy of Chief Automotive Technologies

Figure 17-75 Universal measuring offers the advantages of measuring a third dimension for better collision damage analysis. This type of tool mounts to the vehicle, which allows measurements to be taken at any time during repair.

The strut tower upper body gauge is used most often to detect misalignment of the strut towers. However, it can also be used to detect misalignment of a radiator support, center pillar, cowl, quarter panel, and so on.

Photos courtesy of Car-O-Liner Company

Figure 17-76 A free-standing type universal bridge mounts next to or under the vehicle to set up a baseline for measurements.

Photos courtesy of Car-O-Liner Company

Figure 17-77 With modern universal systems, gauge adjustments are simple and readings are easy to make.

17.10 UNIVERSAL MEASURING SYSTEMS

Universal measuring systems are the most efficient application of tram centering gauge technology. They make parts of the measuring job much easier and more accurate, but still require a degree of skill and attention to detail. These systems can measure all the reference points at the same time. But to get the proper measurement reading, the equipment must be set to the manufacturer's specifications.

With a universal measuring system, all the reference points can be checked just by moving around the vehicle. You can quickly determine where each reference point on the vehicle is in comparison to the measuring system (Figure 17–78).

If a reference point on the vehicle is not in the same position as the dimension chart says it should be, the reference point on the vehicle is wrong. When the system is set up properly, you can monitor the key points by simply looking at the pointers. If the pointers are out of position, then the vehicle is not dimensionally correct. A reference point that is out of position must be brought back to preaccident specifications.

Before beginning any universal measuring operations, be sure to:

- Remove detachable damaged body parts, both mechanical and sheet metal body panels.
- If the damage is severe, perform rough straightening to the center section or foundation of the vehicle.
- If the mechanical parts are left in the vehicle and an overhang condition exists, this must be compensated for.

Photos courtesy of Car-O-Liner Company

Figure 17–78 Note the ruler scales on a universal measuring system. By setting the points at the correct locations, you can accurately tell whether a unibody reference point has been pushed out of alignment.

Universal measuring systems fall into three groups:

1. Mechanical systems
2. Laser systems
3. Sonic systems

Mechanical Measuring Systems

With most **mechanical measuring systems**, several mechanical pointers are attached to a precision measurement bridge. Freestanding-type bridges are available. The pointers on the measuring system are positioned on the bridge or framework according to the vehicle's correct factory specifications for horizontal and vertical dimensions. This allows simultaneous observation of a number of reference points on the damaged vehicle.

Care must be taken to ensure that the measurement bridge is not stressed or damaged during the repair process. The accuracy of this system depends on the location and precision of the pointers on the measurement.

The advantage of a universal system over a tram gauge is that readings are instantaneous; the pointers either align with the reference points or they do not (Figure 17–79).

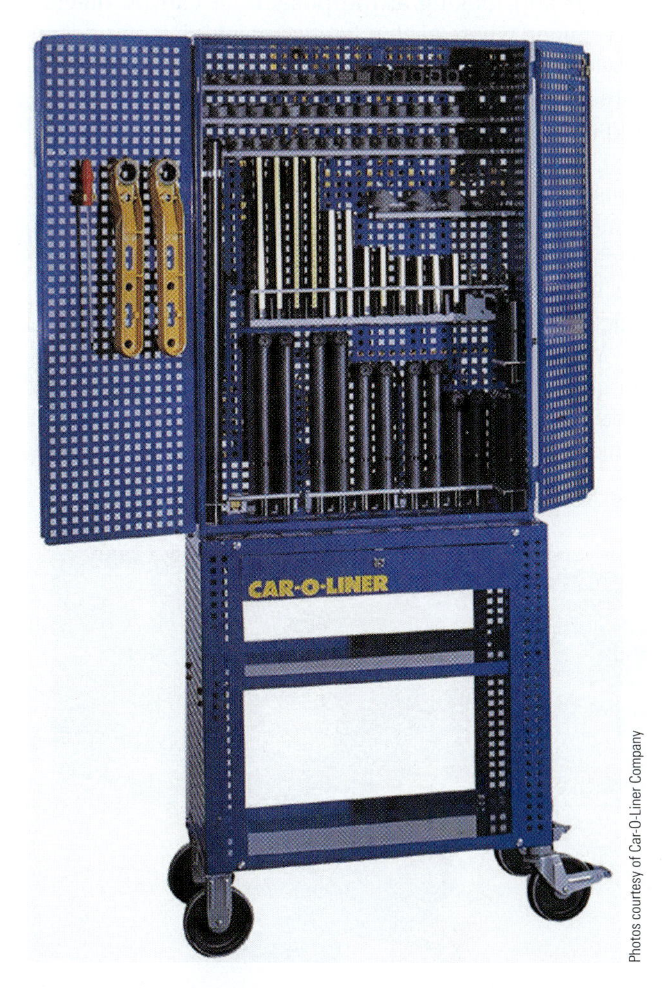

Photos courtesy of Car-O-Liner Company

Figure 17–79 Various attachments are available for universal measuring systems.

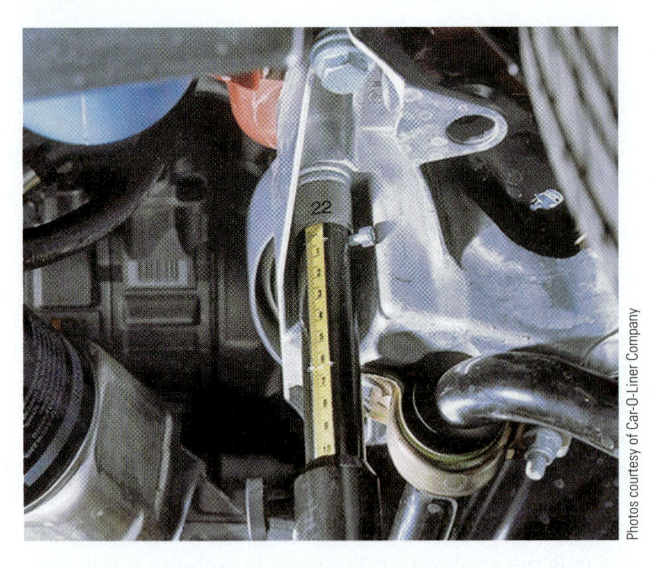

Photos courtesy of Car-O-Liner Company

Figure 17-80 **Here the pointer has been moved up into the reference hole on the underbody of the vehicle to check for damage.**

Photos courtesy of Car-O-Liner Company

Figure 17-82 **This U-attachment allows you to measure around obstructions like a sway bar, tie rod, or other part.**

In practice, a universal measurement system offers the technician the advantage of being able to visually inspect all the reference points just by walking around the car and looking at the pointers. It can be quickly determined where each reference point is in comparison to where it should be. If the reference point is not aligned with its pointer, the reference point on the car is wrong and must be straightened (Figure 17–80).

Universal measuring systems vary from complex units to simple ones. An example of the latter is a tram/centering gauge system that is fastened to the vehicle. Because equipment designs vary, you must read the owner's manual for each piece of equipment to learn the specific procedures for proper use (Figure 17–81).

Most mechanical universal measuring systems work on both unitized and conventional frame vehicles. They measure the lower and upper body reference points of a vehicle as identified in dimensions manuals and make comparison measurements of components from one side of a vehicle to the other. They measure all three dimensions of the vehicle: length, width, and height (Figure 17–82).

Photos courtesy of Car-O-Liner Company

Figure 17-81 **The technician is taking length measurements on this wrecked car.**

Figure 17–83 and Figure 17–84 show how measurements are made in a typical mechanical measuring system.

A universal mechanical measuring system assesses a damaged vehicle by showing how far components are out of alignment. It also remains on the vehicle to guide the technician and verify that components are back in their proper places when the repair is complete. As you use frame straightening equipment to pull or push the structure into alignment, you can watch the pointers to check your work.

A **pivot measure system** uses rotating rods and pointers to measure vehicle damage. Its use is similar to the ones just described. The main difference is how the pointers are mounted and positioned for measuring (Figure 17–85).

Basically, the poles and pointers are mounted and calibrated from vertical and horizontal reference posts. Then the bars can be used to measure reference points in the vehicle.

Figure 17–86 shows other sample measurements using a pivot measuring system.

When using any measuring system, record keeping is critical. A **damage analysis form** (Figure 17–87) is handy because it helps you organize specs, actual measurements, and differences in good and actual measurements. The form will help you more quickly and accurately determine what must be done to straighten the frame or unibody.

Laser Tram

A laser tram is a tool designed to save time checking symmetry when measuring damage or straightening a vehicle body. It will also help diagnose damage to mechanical components at the front and back of a vehicle. A laser

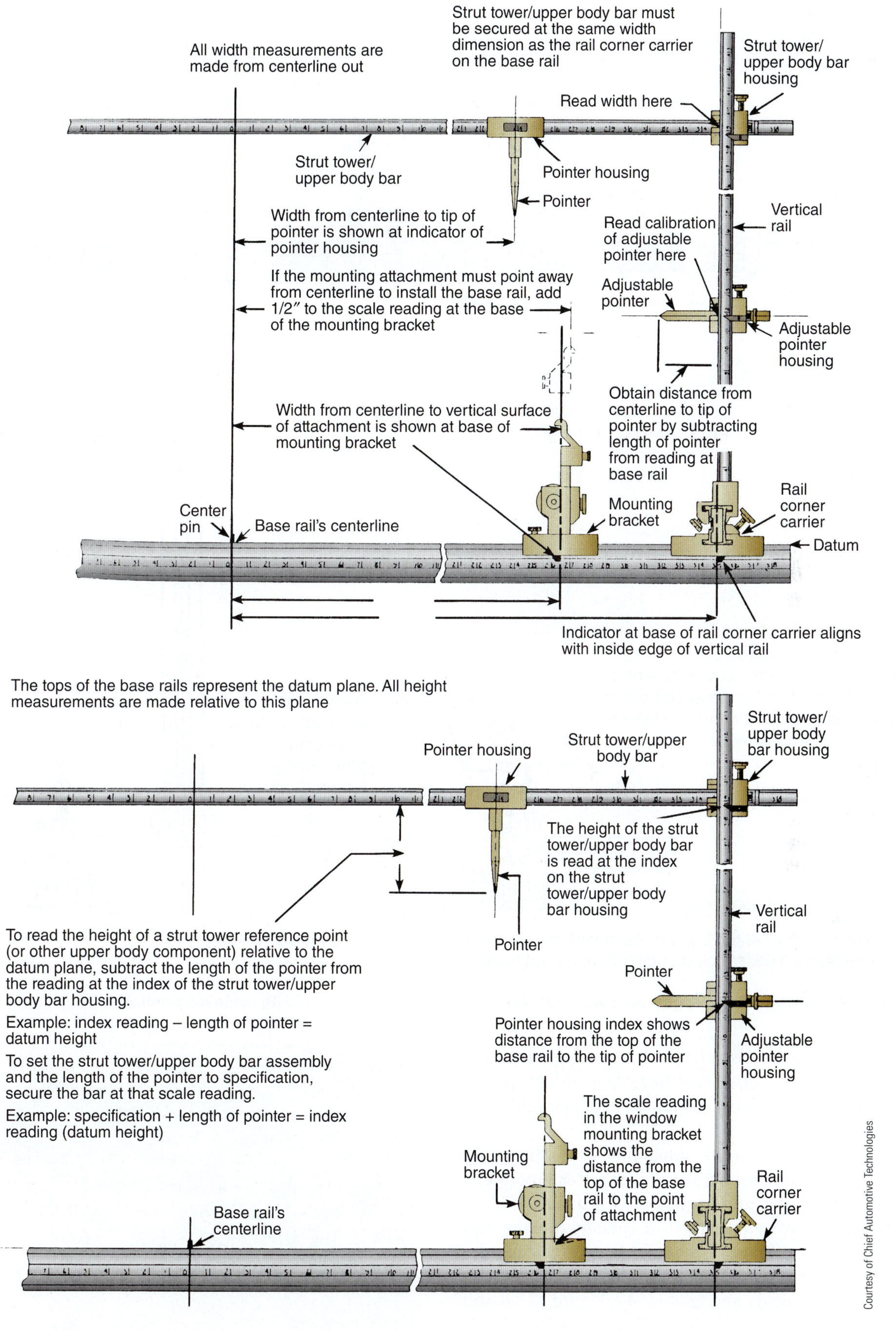

All width measurements are made from centerline out
Strut tower/upper body bar must be secured at the same width dimension as the rail corner carrier on the base rail
Strut tower/ upper body bar housing
Read width here
Strut tower/ upper body bar
Pointer housing
Pointer
Width from centerline to tip of pointer is shown at indicator of pointer housing
Read calibration of adjustable pointer here
Vertical rail
If the mounting attachment must point away from centerline to install the base rail, add 1/2″ to the scale reading at the base of the mounting bracket
Adjustable pointer
Adjustable pointer housing
Width from centerline to vertical surface of attachment is shown at base of mounting bracket
Obtain distance from centerline to tip of pointer by subtracting length of pointer from reading at base rail
Mounting bracket
Rail corner carrier
Center pin
Base rail's centerline
Datum
Indicator at base of rail corner carrier aligns with inside edge of vertical rail
The tops of the base rails represent the datum plane. All height measurements are made relative to this plane
Pointer housing
Strut tower/upper body bar
Strut tower/ upper body bar housing
The height of the strut tower/upper body bar is read at the index on the strut tower/upper body bar housing
Vertical rail
Pointer
To read the height of a strut tower reference point (or other upper body component) relative to the datum plane, subtract the length of the pointer from the reading at the index of the strut tower/upper body bar housing.
Example: index reading – length of pointer = datum height
To set the strut tower/upper body bar assembly and the length of the pointer to specification, secure the bar at that scale reading.
Example: specification + length of pointer = index reading (datum height)
Pointer
Pointer housing index shows distance from the top of the base rail to the tip of pointer
Adjustable pointer housing
The scale reading in the window mounting bracket shows the distance from the top of the base rail to the point of attachment
Mounting bracket
Rail corner carrier
Base rail's centerline
Courtesy of Chief Automotive Technologies

Figure 17-83 Study how widths and heights are measured on a typical mechanical universal system.

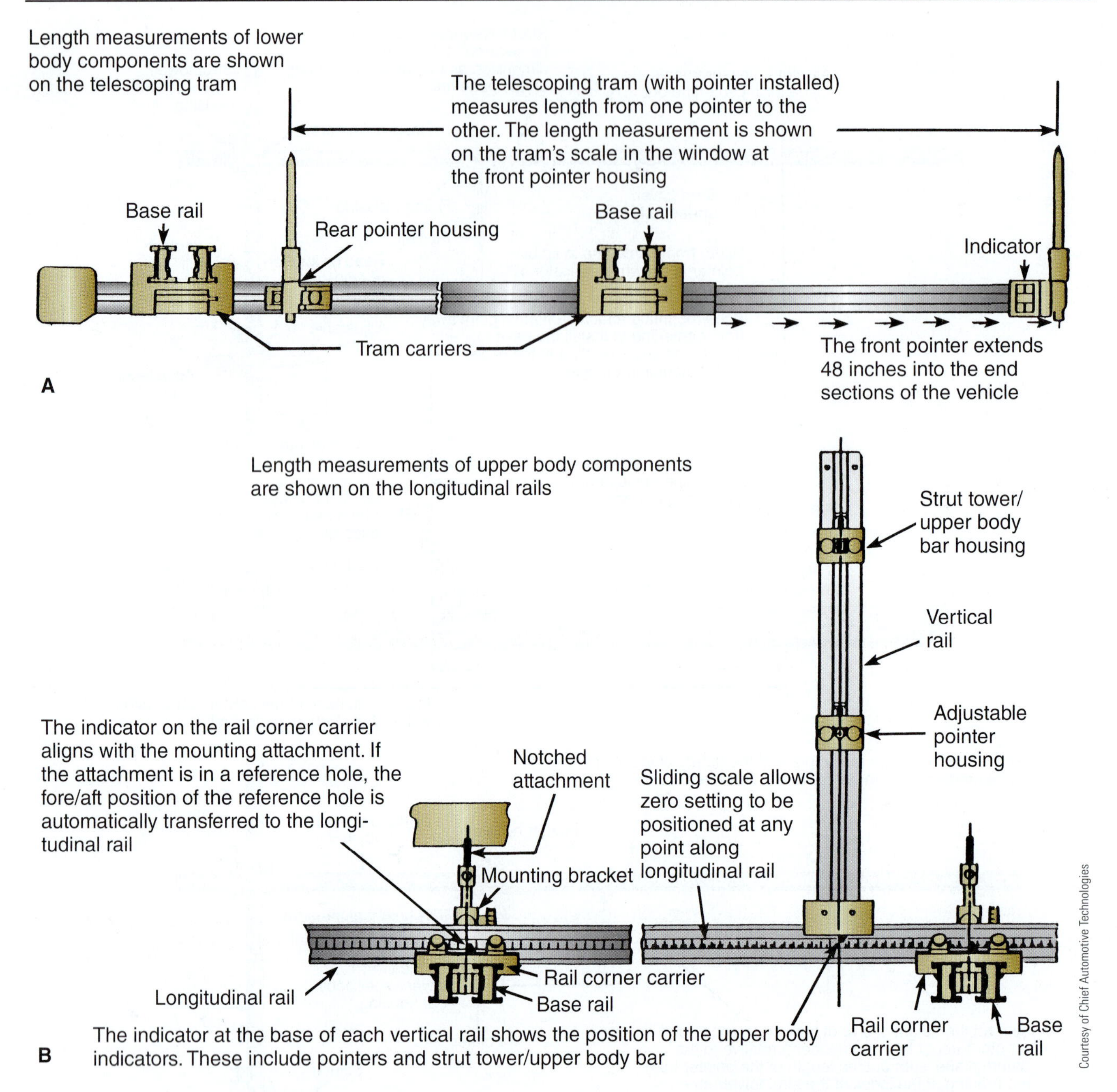
Length measurements of lower body components are shown on the telescoping tram
The telescoping tram (with pointer installed) measures length from one pointer to the other. The length measurement is shown on the tram's scale in the window at the front pointer housing
Base rail
Rear pointer housing
Base rail
Indicator
Tram carriers
The front pointer extends 48 inches into the end sections of the vehicle
A
Length measurements of upper body components are shown on the longitudinal rails
Strut tower/ upper body bar housing
Vertical rail
Adjustable pointer housing
The indicator on the rail corner carrier aligns with the mounting attachment. If the attachment is in a reference hole, the fore/aft position of the reference hole is automatically transferred to the longitudinal rail
Notched attachment
Sliding scale allows zero setting to be positioned at any point along longitudinal rail
Mounting bracket
Rail corner carrier
Longitudinal rail
Base rail
The indicator at the base of each vertical rail shows the position of the upper body indicators. These include pointers and strut tower/upper body bar
Rail corner carrier
Base rail
B
Courtesy of Chief Automotive Technologies

Figure 17-84 Study how lengths are measured (A) using a telescoping tram and (B) using longitudinal rails.

tram can mount on a mechanical measuring system, a dedicated bench system, or on the vehicle's body itself so you can quickly check the alignment of panels and parts (Figure 17–88).

A laser tram with its attachments is shown in Figure 17–89. The laser tram will send out perfectly straight laser beams from the body of the gauge.

The tram can be extended to specific lengths and centered on the vehicle. This allows you to measure the distance from the laser beam to various reference points on the vehicle quickly and accurately (Figure 17–90 and Figure 17–91).

Figure 17–92 shows the laser tram mounted and centered under the front of a vehicle. This would allow you to make measurements from the laser beam to various

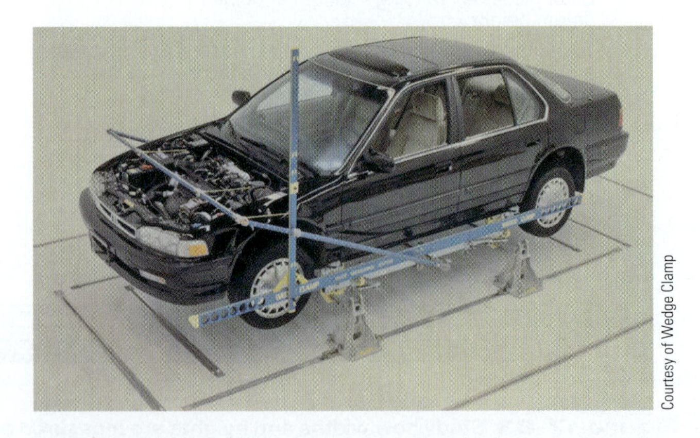
Courtesy of Wedge Clamp

Figure 17-85 This type of measuring system uses long rods that pivot to different angles to measure any location on the vehicle.

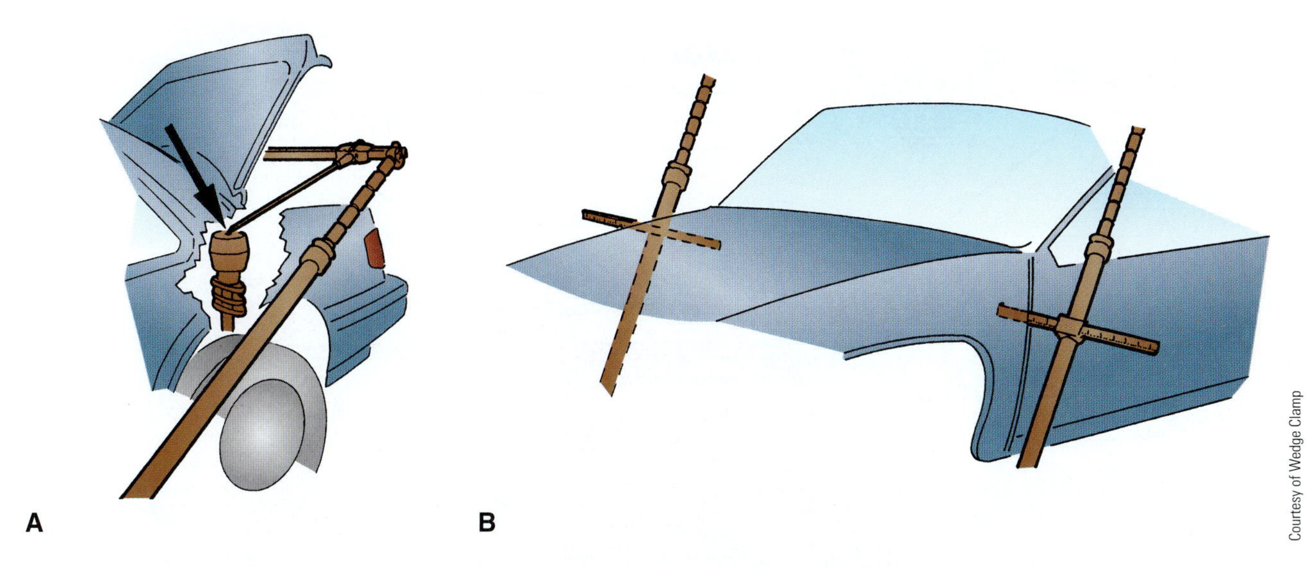

Figure 17-86 (A) A measurement is being taken at the rear strut towers, a hard-to-reach area, with a pivot-type system. (B) Here a measurement at one side of the cowl is being taken by extending the pointers through door gaps.

points along the side and top of the body. With a laser tram mounted inside the passenger compartment, you can quickly measure from the laser light beam to specific points on the pillars to determine the extent of upper body damage.

17.11 COMPUTERIZED MEASURING SYSTEMS

A **computerized measuring system** uses a personal computer and specialized electronic hardware to evaluate structural damage quickly and easily. Most computer measuring systems give "real-time" readings that change as the vehicle frame or unibody is pulled back into alignment. Larger, more advanced body shops tend to prefer the advantages that computerized measuring systems offer.

The dedicated PC or workstation stores the dimension manual information for the vehicle being repaired. A picture of the vehicle's dimensions can be pulled up and downloaded into the measuring system's software (program). This allows the system to automatically compare actual vehicle measurements with known good specifications. You do not have to thumb through a printed manual or write down measurements, because they are displayed on the computer monitor.

There are three basic types of computerized measuring systems:

1. Sonic measuring systems
2. Robot arm measuring systems
3. Laser measuring systems

Although the basic methods for measuring vehicle damage are similar, the exact operating instructions for different brands of systems vary. Most measuring equipment manufacturers have specific dimension charts for their equipment, often displayed on a computer monitor. These charts, one for each vehicle make and model, serve as guides to use before and during the repair.

Many equipment manufacturers' dimension charts are intended only for a specific piece of equipment. Because of this variation between systems, always refer to and read the equipment manual before measuring vehicle damage.

Most modern computer-based measuring systems denote measurements as plus or minus in millimeters for length, width, and height.

- *Negative length*—measurement point has been pushed in or is shorter than specs.
- *Positive length*—measurement points is too long or has been forced out from center of vehicle.
- *Negative width*—measurement point is pushed to center of vehicle or is too short.
- *Positive width*—measurement point is pushed out or is too wide.
- *Negative height*—measurement point has been pushed down or is too low.
- *Positive height*—measurement points has been pushed up or is too high.

For example, if the measuring system shows that a measurement point is –15 for length, +15 for width, and –13 for height, you would know that the collision has pushed the reference point inward 15 mm; sideways, or outward, 5 mm; and down 13 mm. This is the principle of reading a computerized measuring system. You would have to use a frame rack to pull this amount of damage back out in the opposite direction.

Sonic Measuring Systems

Sonic measuring systems use sound waves as a means of measuring vehicle dimensions. Because sound travels at a constant speed, the time it takes for sound waves to

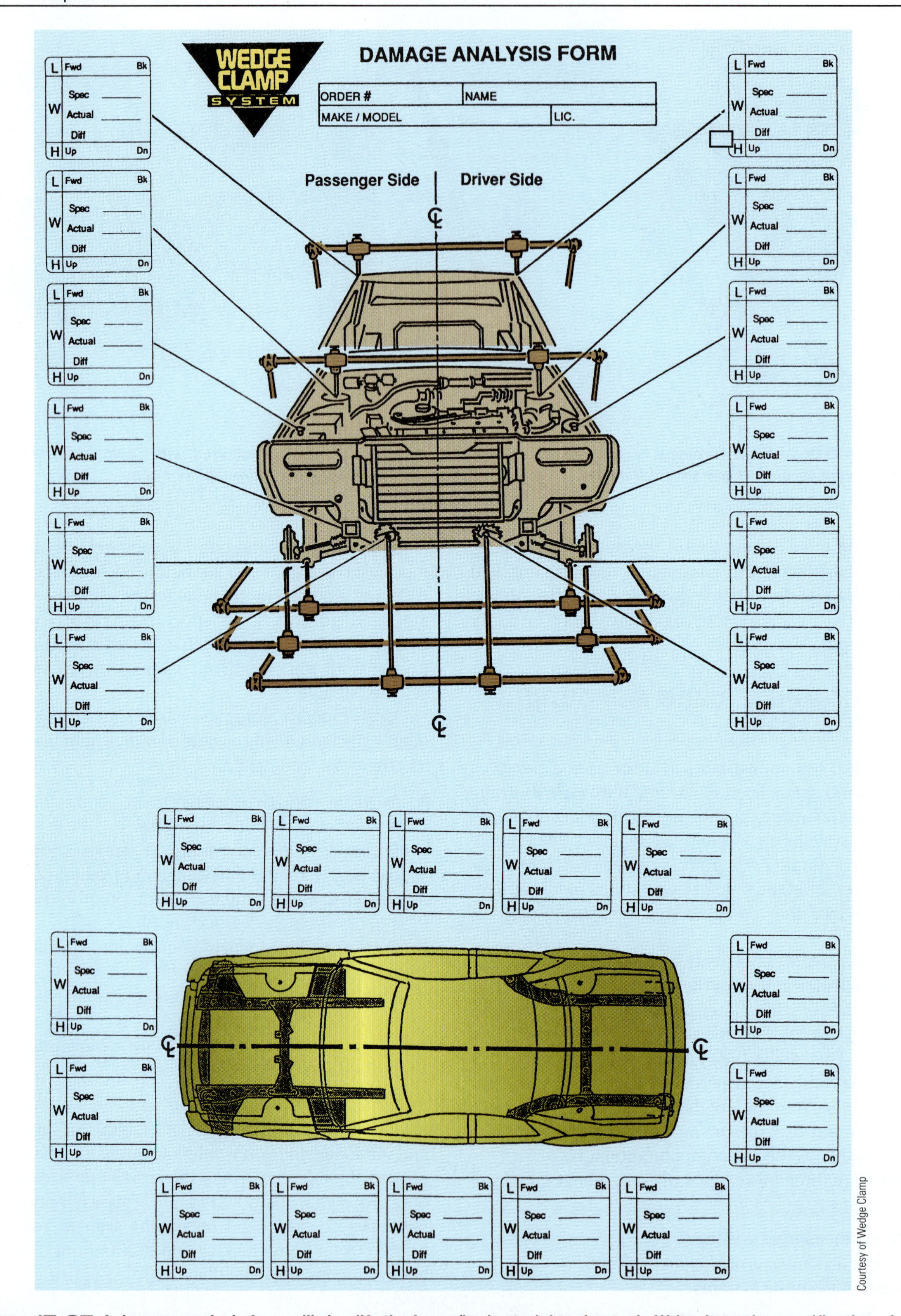

Courtesy of Wedge Clamp

Figure 17-87 A damage analysis form will simplify the frame/body straightening task. Write down the specifications from the manual. Then, measure and record values taken from the vehicle. By subtracting the two measurements, you can determine the direction and extent of damage.

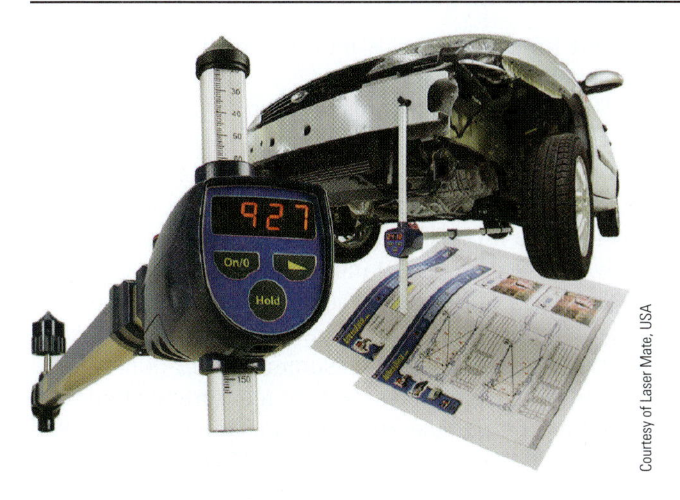

Figure 17–88 Using modern equipment and a basic knowledge of measuring methods, you should be able to learn how to analyze vehicle damage.

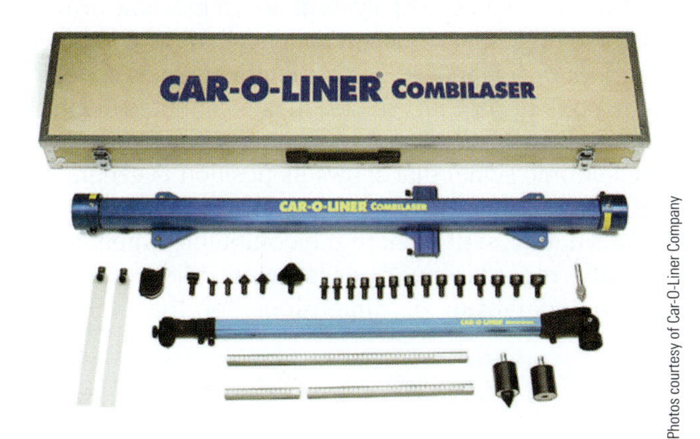

CAR-O-LINER COMBILASER
Photos courtesy of Car-O-Liner Company

Figure 17–89 This laser tram can be used in conjunction with a universal measuring system to help find damage more quickly.

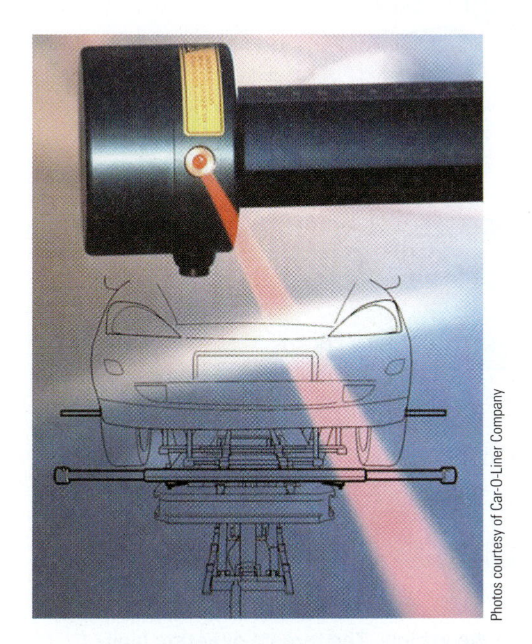

Figure 17–90 Laser light is emitted from the laser tram so that you can see it hit other points on the measuring system or vehicle. Note how the laser beam from the tram can be used to take width measurements.

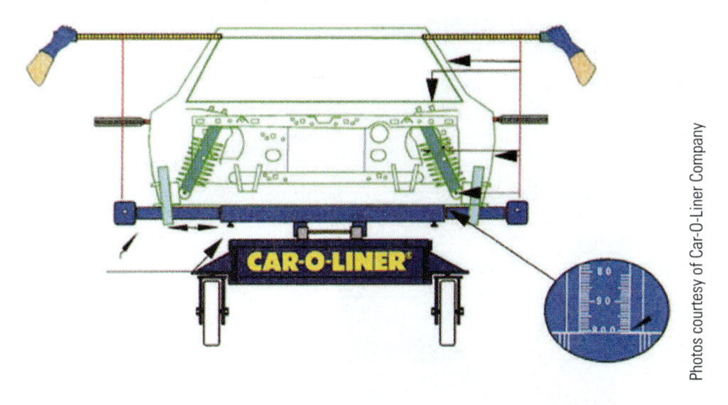

Figure 17–91 A laser tram is an additional measuring tool used to supplement a main measuring system.

Figure 17–92 Here a laser tram is being used to check for roof and pillar misalignment.

travel between different reference points on the vehicle can be measured quickly and accurately (Figure 17–93).

A sonic measuring system basically consists of:

- A personal computer to operate the system and store repair data
- Emitter targets that produce clicking sounds
- A receiving unit to detect incoming sound waves from targets (Figure 17–94).

To use a sonic measuring system, slide the lightweight aluminum receiving unit under the vehicle. Try to center it on the rack or bench. See Figure 17–95.

Open the computer measurement software. Pull up needed data about the vehicle to be straightened. This will let you know how and where to mount your emitter targets. Normally, four emitters are attached under the center section of the vehicle to establish a base. Other targets are placed around the damaged area of the vehicle.

Emitter targets can be attached to the control points using magnet or mechanical expander mounts. Plug the wires from the emitter targets into the receiving unit under the vehicle.

Figure 17-93 **The sonic measuring system uses a PC or workstation to analyze electronic data from probes and microphones. Sound instead of laser light is used to make accurate measurements of vehicle damage.**

When the unit is turned on, the emitters send out sound waves. The sound waves are heard as "clicking" sounds.

Microphones in the receiving unit then detect the sounds coming from each emitter. The microphones convert the sound waves into electrical signals for the computer system.

The computer compares how long it takes the sound from each emitter target to enter each microphone, which allows for computation of distance. Because the sonic system can measure how long it takes each sound wave to hit a receiver microphone, the system can accurately measure target distances.

Computer intelligence automatically calibrates the setup and zeroes the undamaged center section base reference points. Damage measurements are done by the computer. It also updates damage measurement number values as the vehicle structure and targets are pulled back into alignment. Refer to Figure 17–96.

Robot Arm Measuring System

A **robot arm measuring system** uses a movable arm to send electronic measurement data back to the computer The robot arm is moved so that its tip contacts specific control points on the vehicle. A rack-mounted control unit can then be activated to store each measurement. An electronic signal that shows its position is generated by the robot (Figure 17–97).

This data is sent back to the computer, which automatically records each measurement and compares it to the database of dimensions for that vehicle.

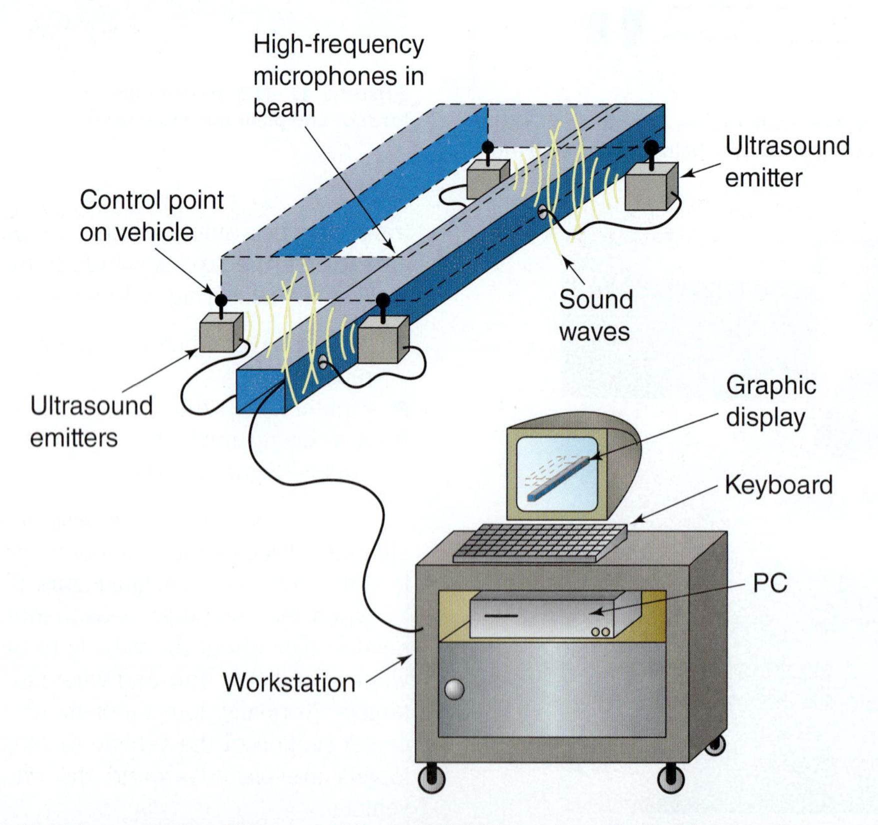

Figure 17-94 **The targets or sound emitters mount in control points on the vehicle. The targets generate high-frequency clicking sounds picked up by the microphones in the center beam. Because the speed of sound is constant, the computer system can then use the amount of time taken for the click sounds to reach the microphones to accurately calculate distance.**

Courtesy of Snap-on Tools Company, www.snapon.com

A

Courtesy of Snap-on Tools Company, www.snapon.com

C

B

Courtesy of Snap-on Tools Company, www.snapon.com

D

Courtesy of Snap-on Tools Company, www.snapon.com

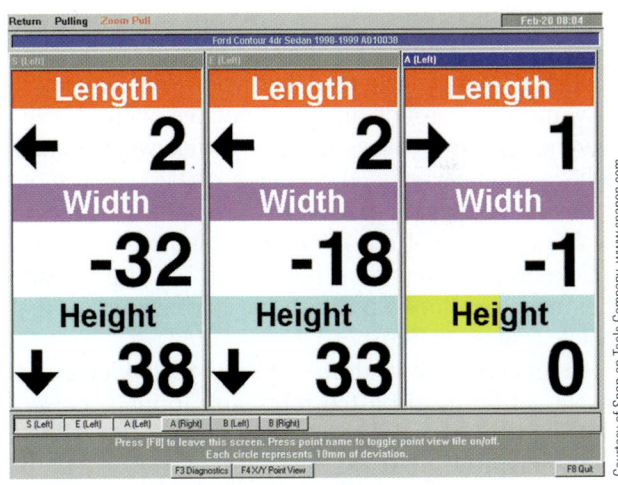

Figure 17-95 **Note the basic steps for using a sonic measuring system. (A) Slide the microphone beam under the vehicle and onto the rack. (B) The console will tell you where to hang probes on the vehicle. The probe is hung from the bolt for this vehicle reference point. (C) Using the denoted adapter, hang the probe from reference points on the vehicle. (D) Measurements will then be given for each measurement point. Note the length, width, and height measurements.**

SHOP TALK

Remember that the proper use of any measuring equipment is the secret of successful vehicle repair. With any measurement system, the key to correct pulling and straightening lies in accurately monitoring all measurements—before starting the pull, while pulling, and immediately after the pull has been made.

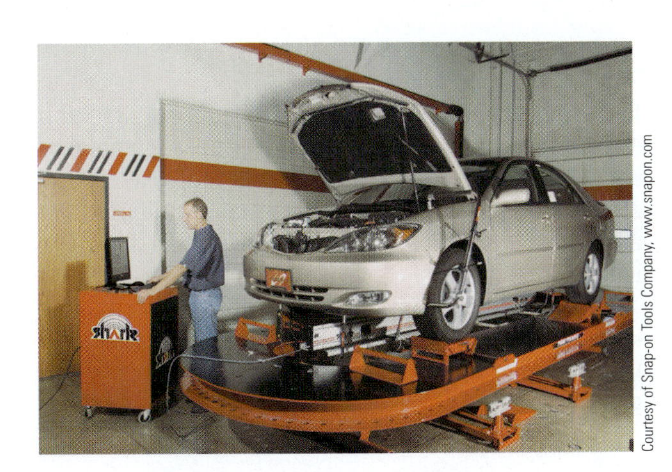

Courtesy of Snap-on Tools Company, www.snapon.com

Figure 17-96 **A sonic measuring system will provide accurate measurements with minimum setup time.**

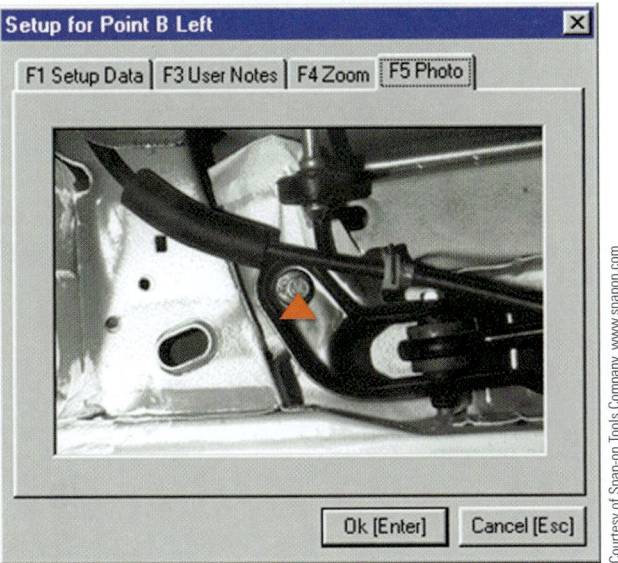

Figure 17-97 The robot arm measuring system is very accurate and will read height, width, and length all at one time for each reference point.

The robot arm is a very accurate but uncommon system (Figure 17–98).

Figure 17–99 shows a technician measuring the damage to the front of a late model vehicle. When you physically move the robot arm into contact with specific reference points on the vehicle, the system will automatically determine whether the points have been moved out of alignment or out of specs.

When you input the information about the specific vehicle on the computer screen, the robot arm measuring system can look up dimension information for that vehicle (Figure 17–100).

Other computer windows will show you which adapters should be fitted on the robot arm to make correct contact with the control or reference point on the vehicle. Look at Figure 17–101.

The dimension information and readings from the robot arm are read using the computer display window in Figure 17–102.

Other computer display windows may even show you where to measure next with the robot arm. You can then quickly input each measurement into the system for damage analysis (Figure 17–103).

Laser Measuring Systems

The **laser measuring system** uses beams of light and vehicle-mounted targets to measure misalignment of vehicle control points. Laser measuring systems are extremely accurate when properly installed and used. An early but still used laser measuring system is shown in Figure 17–104.

The word *laser* stands for "*l*ight *a*mplification by stimulated *e*mission of *r*adiation." A laser is good for measuring because it shoots out a perfectly straight beam of light.

Computerized Laser Measuring

A **computerized laser measuring system** uses light beams, reflective targets, a laser-receiver unit, and a PC to speed and simplify damage analysis. Modern computerized laser measuring systems are relatively easy to use and very precise. It is one of the most common types found in today's collision repair facility. A computerized laser measuring system is shown in Figure 17–105.

A A computer console is for comparing actual measurements with factory specifications.

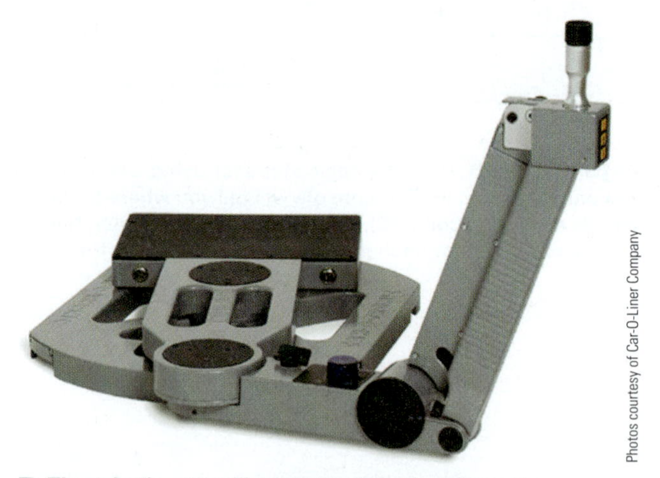

B The robotic arm will output a signal to inform the computer about the exact location of the pointer on the end of the arm.

Figure 17-98 Note the two major parts of the robot arm measuring system.

Photos courtesy of Car-O-Liner Company

Figure 17-99 By moving the robot arm pointer or adapter into a reference point on the vehicle, you can measure its exact location to determine damage.

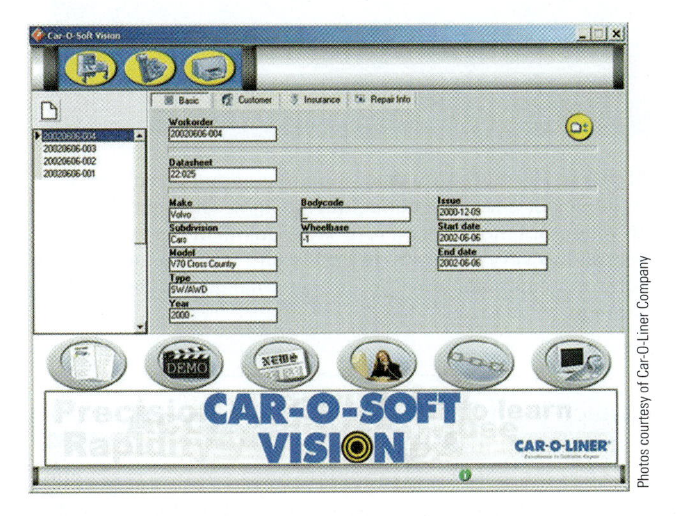

Photos courtesy of Car-O-Liner Company

Figure 17-100 This computer window allows you to input vehicle information so it can pull up correct dimensions.

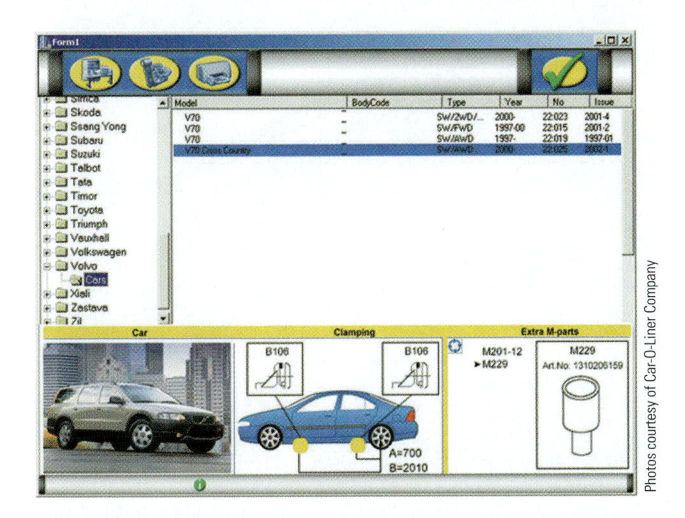

Photos courtesy of Car-O-Liner Company

Figure 17-101 This computer window shows information on clamping or securing a vehicle to the rack.

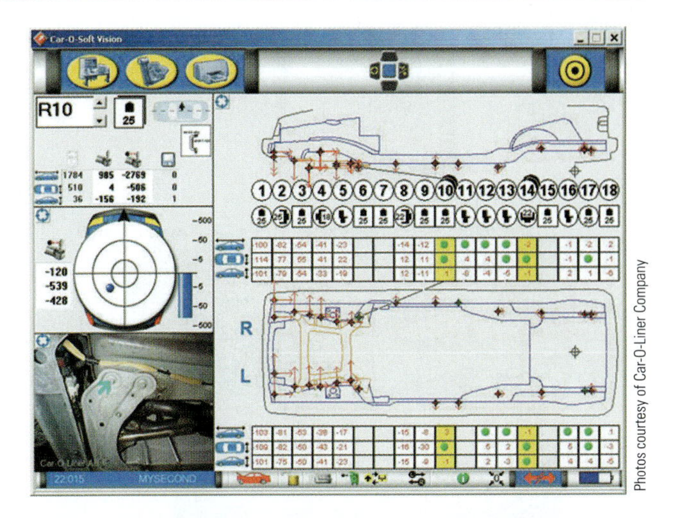

Photos courtesy of Car-O-Liner Company

Figure 17-102 This window shows measurement points, adapters needed, and gives a live reading on the right side of the window. In the lower left, it also gives pictures of measurement points.

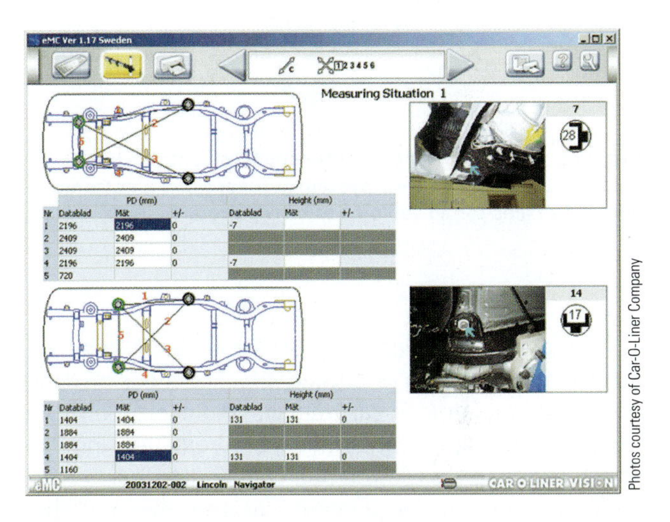

Photos courtesy of Car-O-Liner Company

Figure 17-103 This window gives measurement situations to help analyze damage.

Courtesy of Chief Automotive Technologies

Figure 17-104 Laser measuring systems are one of the most common types found in today's repair facilities. The lasers send accurate distance and alignment information back to the computer console for analysis.

Courtesy of Chief Automotive Technologies

Figure 17–105 **The computer console contains a PC loaded with the measuring system software and electronic dimension files for all vehicles to be repaired.**

Courtesy of Chief Automotive Technologies

Figure 17–106 **The laser mounts under a vehicle and emits power beams of concentrated light. The light bounces off reflectors mounted on reference points of the vehicle. This allows very accurate detection and measurement of any body or suspension point forced out of alignment by impact damage.**

Computer-based laser measuring systems combine the convenience of an electronic vehicle dimension database with real-time measurement feedback. You can watch the measurement numbers on the dimension diagram change to zero as you pull out the damage.

Today's systems use a central laser-receiver unit that generates light beams and also reads them (Figure 17–106). This is unlike older systems that mounted the laser on a perimeter rack system and required manual reading of ruler scales (Figure 17–107).

The multiple laser head sends out laser beams that strike and bounce off the reflective targets. Data on the speed and angle at which the laser beams return to the unit are evaluated by the computer. The computer is wired to the laser-receiver unit so that this data can be downloaded into the measurement software at the workstation (Figure 17–108).

When properly set up, most laser systems can remain in position during the entire repair operation unless the mounting hardware or the laser targets interfere with the operation of the pulling and straightening equipment. Laser equipment has fewer mechanical parts that can be bent and damaged while doing repairs.

Laser systems provide direct, instantaneous dimensional readings. Reference points in both the damaged and undamaged areas of the vehicle can be monitored continually during the pulling and straightening operation.

After the vehicle is mounted on the frame rack, slide the laser-receiver unit under the center section of the

Courtesy of Chief Automotive Technologies

Figure 17–107 **Never look directly into a laser light. The concentrated light beam can be powerful enough to cause eye damage.**

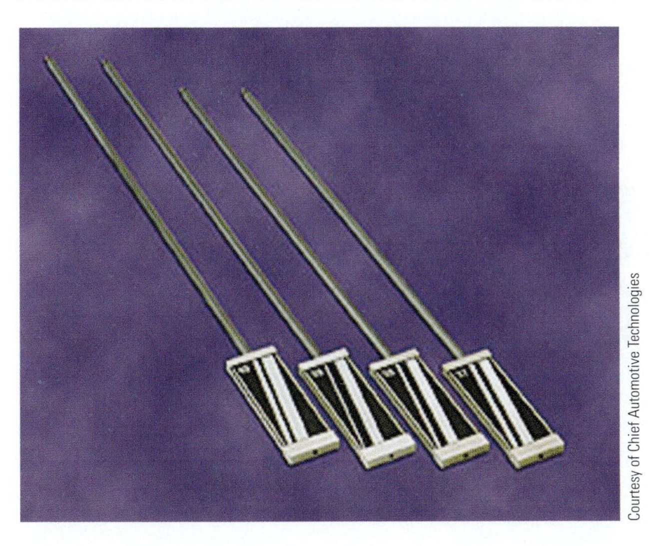

Courtesy of Chief Automotive Technologies

Figure 17-108 **Several targets are usually required to measure all areas of damage that must be straightened.**

vehicle. Plug the cable from this laser unit into the computer workstation.

Use the measurement software, mouse, and computer keyboard to pull up the electronic dimension diagram for the make and model vehicle to be repaired. This will download the correct unibody-frame illustration and dimensions for an undamaged vehicle. The dimension charts and diagrams usually provide one, two, or three views of the vehicle frame or unibody. Some charts also give underhood and upper body dimensions.

The diagrams will also help you find reference points for mounting the targets. The measurement software can be used to determine vehicle hole sizes for selecting target mounting adapters, target mounting rod lengths, and other information for the specific vehicle.

After reviewing this information, mount the targets on the vehicle. The targets and mounting hardware are normally stored in the bottom of the computer workstation. Sometimes large magnets are used to hold the targets on their designated locations or over reference points on the vehicle (Figure 17–109).

Spring-loaded or adjustable target mounts are also available. They will expand to fit into different-size holes in the vehicle body (Figure 17–110).

To measure upper body points, a special strut frame is mounted on the suspension towers formed in the inner fender aprons. With pointers touching specified points on the shock towers, reflective laser targets on the bottom of the frame can be read by the laser-receiver unit. Refer to Figure 17–111 and Figure 17–112.

With the laser unit and targets mounted on the vehicle, use the computer to calibrate the system and then read vehicle dimensions. By touching the screen, clicking the mouse on icons, or inputting keystrokes, you can make accurate measurements of structural damage (Figure 17–113).

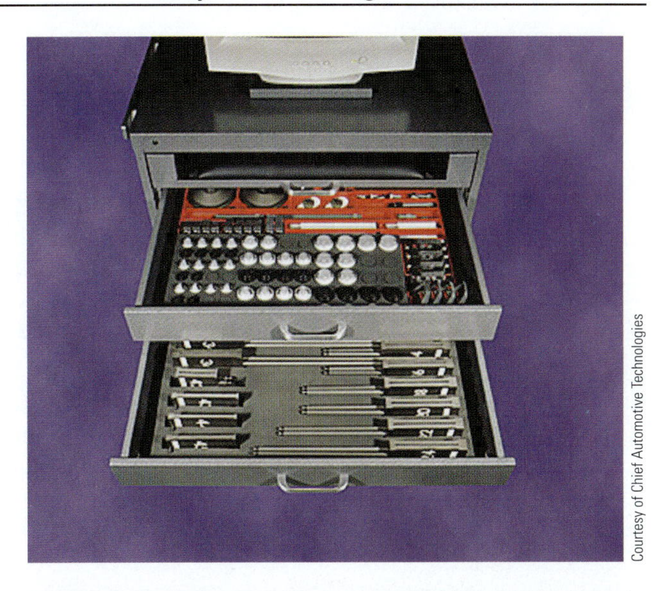

Courtesy of Chief Automotive Technologies

Figure 17-109 **A computer console will normally contain adapters and targets.**

Courtesy of Chief Automotive Technologies

Figure 17-110 **Hang reflective targets at points on the vehicle recommended in the computer diagram. Be sure to use the correct mounting method and rod length.**

Courtesy of Chief Automotive Technologies

Figure 17-111 **These laser targets will reflect laser beams back to a laser-receiver unit. The adapter on the end of the rod allows the target to be mounted on a vehicle. Some adapters are magnetic and others mechanically open to lock into reference holes on the vehicle.**

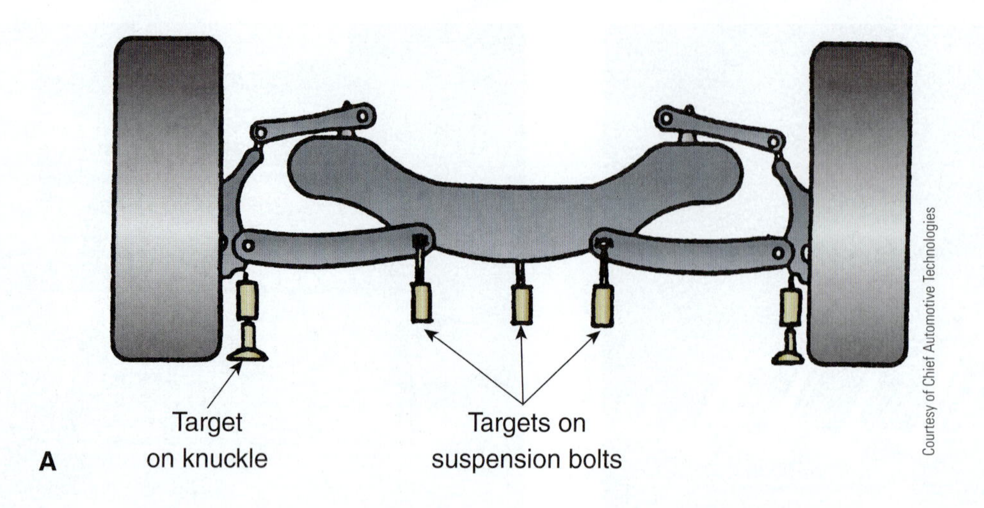

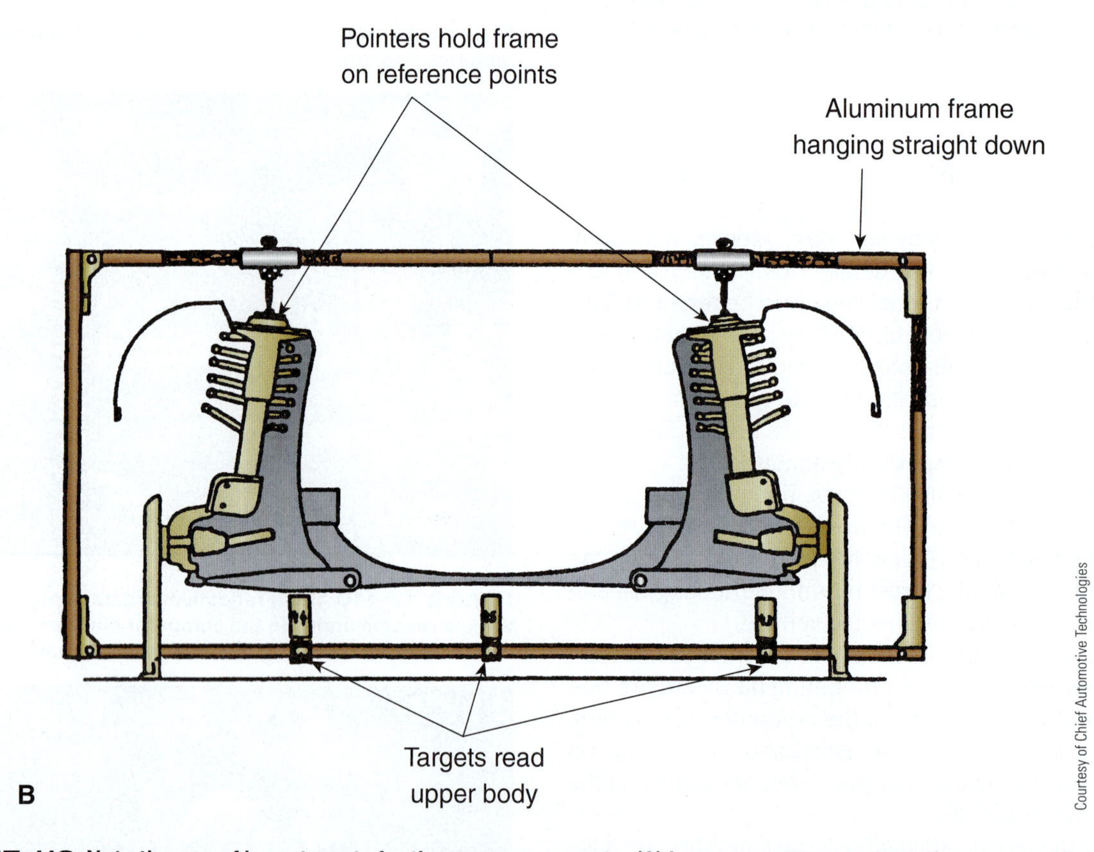

Figure 17–112 Note the use of laser targets for these measurements. (A) Laser targets are mounted on suspension bolts to measure cross member damage. (B) A strut tower gauge has laser targets mounted on it. The gauge hangs from pointers on specific locations of the strut towers. This allows the computerized laser system to measure upper body points.

Figure 17–114 shows the computer monitor screen for a typical laser measuring system. The window shows how you can navigate and use the measurement software. Note the menu bar, title bar, scanner view, tool bars, status bar, and vehicle dimension graphic.

First, calibrate the system by checking the center section base targets. They must be mounted on an undamaged section of the vehicle. They are normally mounted under the passenger compartment in holes in the underbody structure (Figure 17–115 and Figure 17–116).

The targets are represented by the small boxes with the zeros in them. The centerline for making width measurements is established by projecting a line between the right and left targets. A line perpendicular to the centerline for making length measurements runs through right and left base targets at the front or rear targets.

With most computerized laser measuring systems, numbers on the dimension diagrams are represented in millimeters. Some systems give both metric and U.S. (inch) units. Arrows next to these numbers show the

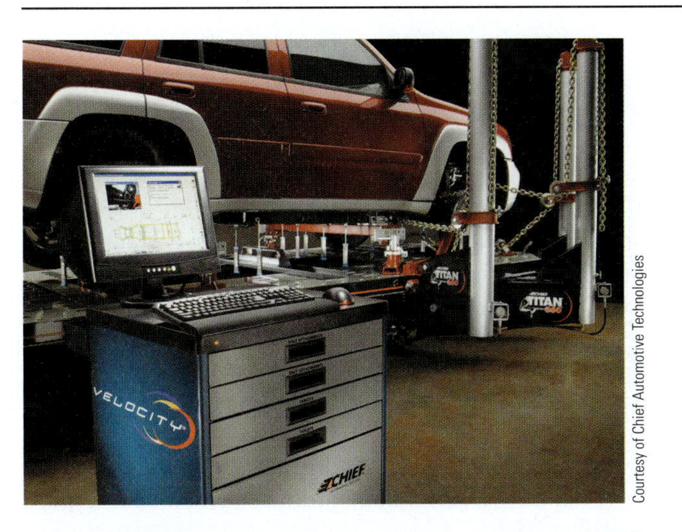
Courtesy of Chief Automotive Technologies

Figure 17-113 This is a state-of-the-art, computer-based laser measuring system. It uses a computer workstation that stores a dimension database and actual measurement data fed into the system by the laser system.

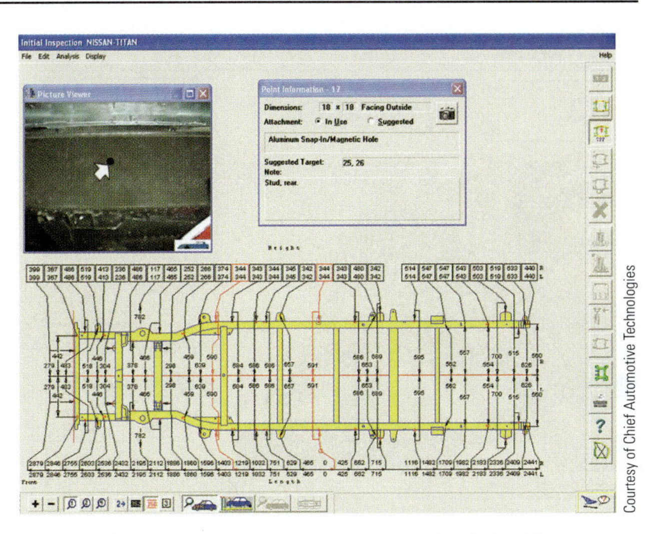
Courtesy of Chief Automotive Technologies

Figure 17-115 Here is another window that shows where to hang targets and which adapter is needed for each measurement point.

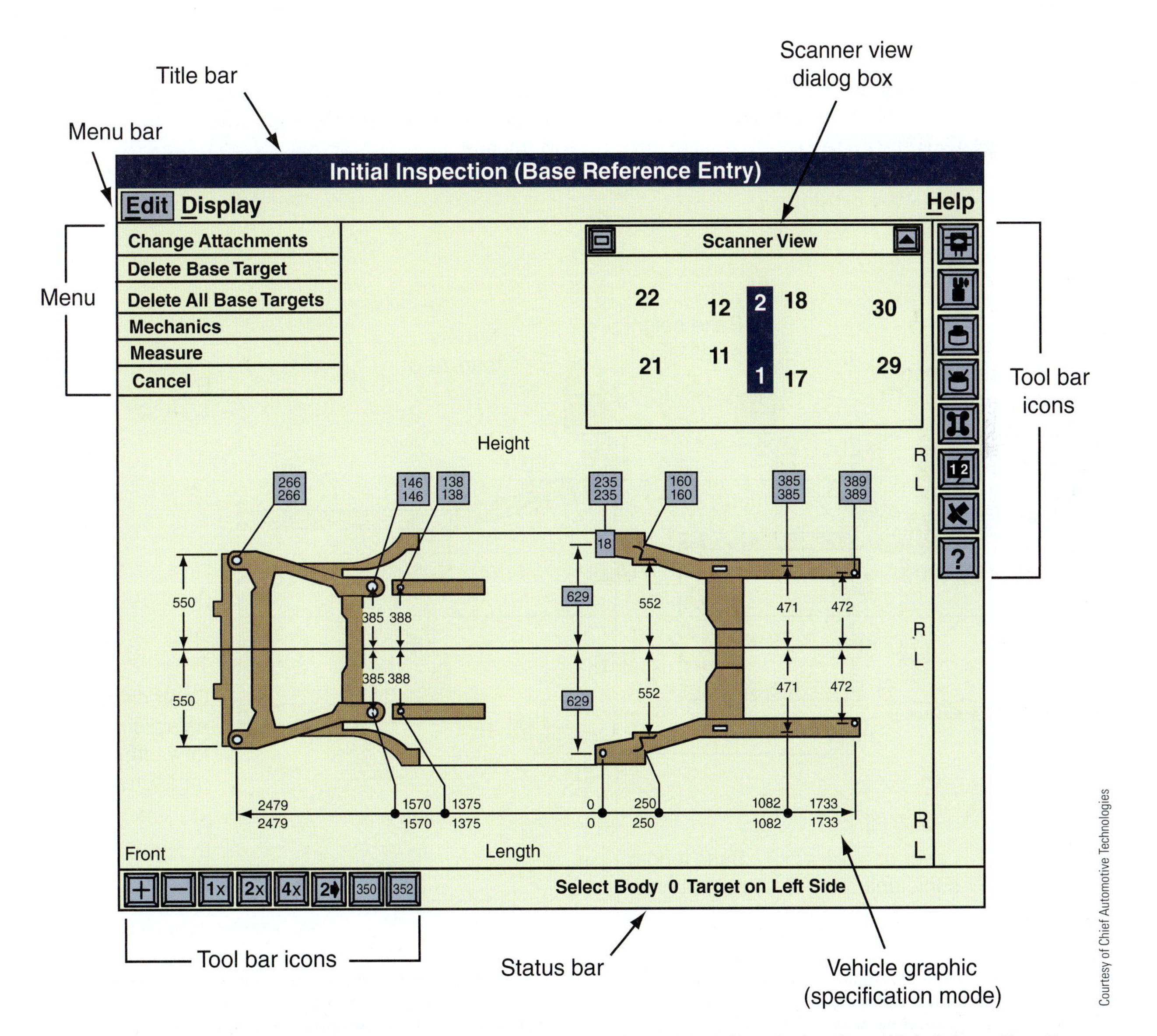

Courtesy of Chief Automotive Technologies

Figure 17-114 Study a typical window for a modern laser measuring system. By entering the vehicle information, the computer will show known good measurements for comparison with actual measurements on a damaged vehicle.

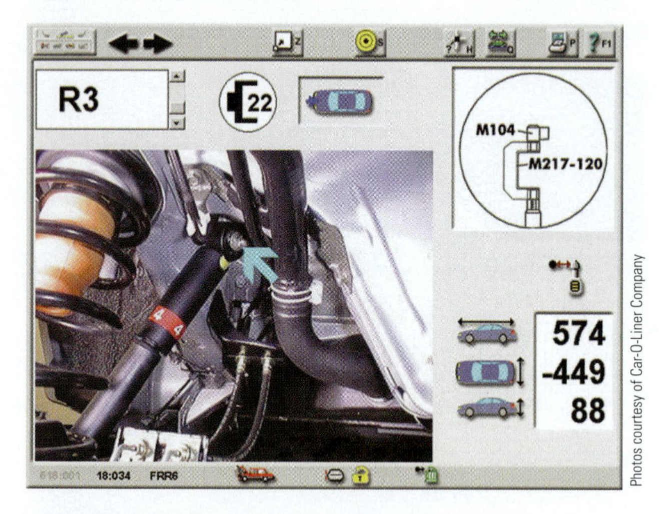

Figure 17–116 It is very important that you use the correct adapters when hanging targets. The arrow in the photo denotes measurement points. Upper left shows adapters needed. Bottom right shows measurements for length, width, and height.

direction of damage. Generally, zeros and ones show no damage because a 3-mm tolerance is given by most manufacturers (Figure 17–117).

Note how the height measurements are given along the top of the dimension diagram. Length dimensions are given along the bottom. Also note how this vehicle shows frontal damage indicated by the numbers not reading zero. The rear of the vehicle shows no damage, because all of the rear targets are right on spec (zero). The computer software automatically subtracts the correct dimension and the actual measurement to get a difference, which indicates structural damage (Figure 17–118).

Also note that upper body dimensions are given in the upper left of the diagram in Figure 17–118. A strut gauge with targets would be used to measure these points.

A close-up, partial view of a "live" computer readout for a vehicle with a damaged frame is provided by Figure 17–119.

Note how each box at the top of the drawing gives measurement information for height. The top number in each box is for the right side of the vehicle. The lower box number is the height measurement for the left side of the vehicle.

Widths are given in the center of the drawing. Length dimensions are at the bottom. Again, the upper length number is for the right and the lower is for the right side.

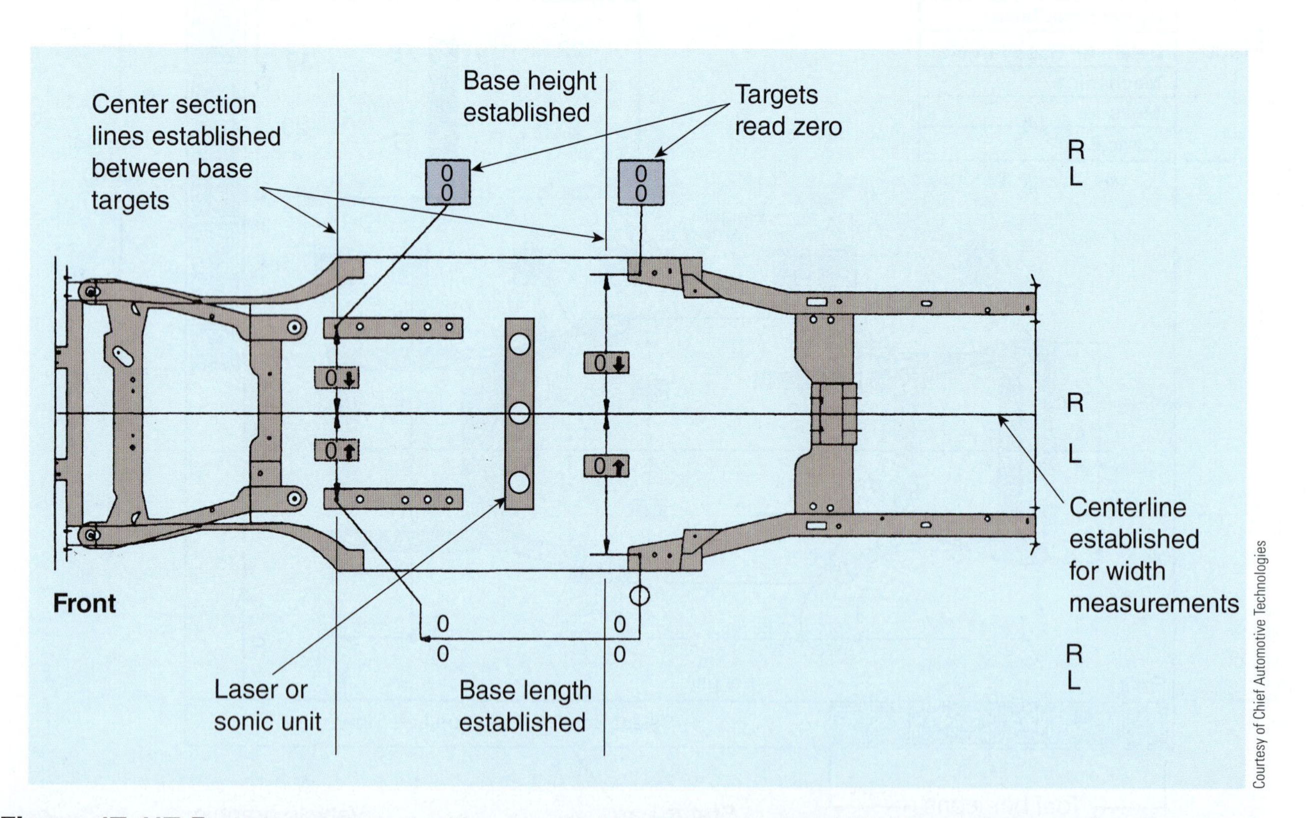

Figure 17–117 To start measurements with a computer-based laser system, establish base reference points. Four targets are usually hung under the passenger compartment in the center section. Any undamaged section can be used, however. Ideally, base targets should be at zero. This will let the computer establish a vehicle centerline (width measurements), a length reference line through the front and rear targets, and a height, or datum, reference.

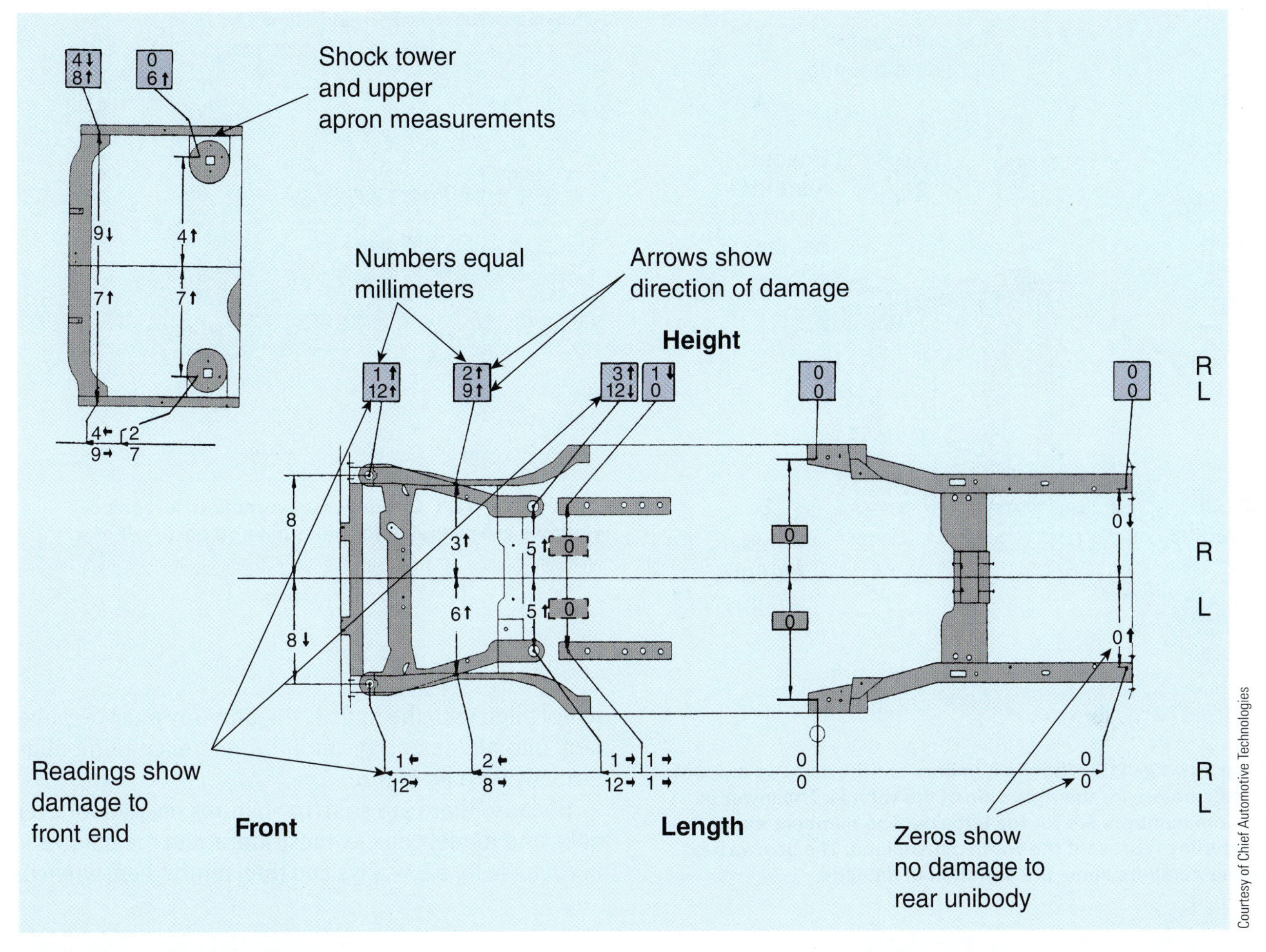

Figure 17-118 Note what the symbols and number values on this laser measuring system indicate. The numbers equal the damage amount in millimeters.

Study how the measurement values indicate the extent and direction of damage. The car has taken a hard hit to the left front frame rail in Figure 17–120.

The rail has been pushed back, as denoted by the 12-mm values with the arrows pointing to the rear. This shows that the frame rail on the left side has been pushed out of alignment up to 12 mm. The height values indicate that the left frame rail has been pushed up as well. The width dimensions also show collapsing inward.

When reading a damage measurement graphic such as this one, you must visualize how much and in what direction the frame has been forced out of alignment by the collision. This information will help you position the pulling equipment for straightening the frame or unibody damage. Basically, you will pull opposite the direction of the damage shown in the diagram.

Figure 17–121 shows a computer screen for measuring damage with major body parts still on the vehicle. Other tabs on the window will show measurements for major body parts removed, because these measurement points will become exposed.

A printout of the correct and incorrect dimensions on the damaged vehicle is often used as a quick reference when pulling out damage (Figure 17–122).

Dedicated Bench Measuring System

A **dedicated bench measuring system** uses large steel fixtures to determine whether the frame or unibody has been pushed out of alignment in a collision. Dedicated bench systems are gaining popularity again because some auto manufacturers require their use when repairing their makes of vehicles (Figures 17–123 and 17–124).

Several high-end manufacturers stipulate the use of a dedicated bench when repairing their vehicles.

A dedicated bench measuring system is basically a "go–no go system." This means that when the fixtures are bolted to the bench in their correct locations, the fixture aligns or engages that specific reference point on the vehicle. If a structural member and reference point

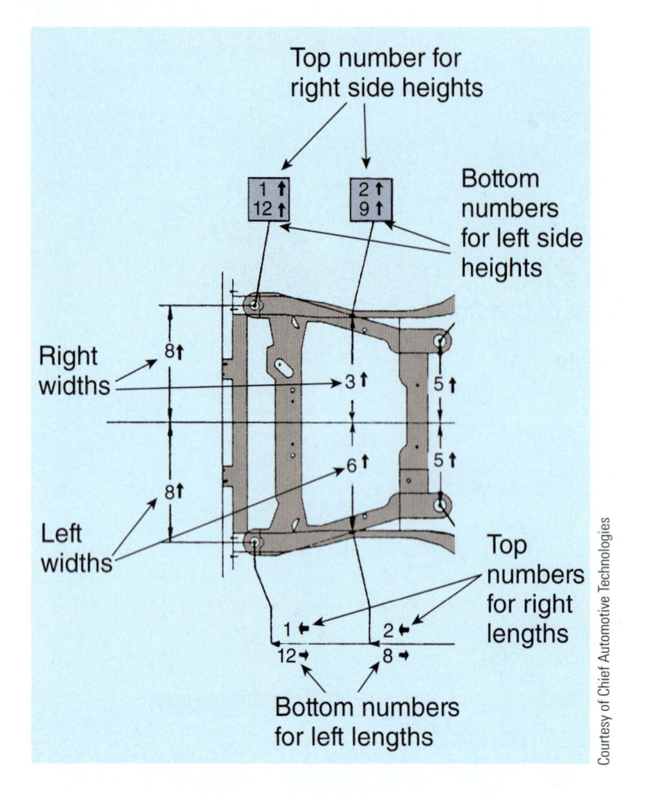

Figure 17-119 The top numbers are the damage amount in millimeters for the right side of the vehicle. The lower or bottom numbers are for the left side. The numbers along the centerline represent the width dimensions. The arrows next to the numbers show the direction of damage.

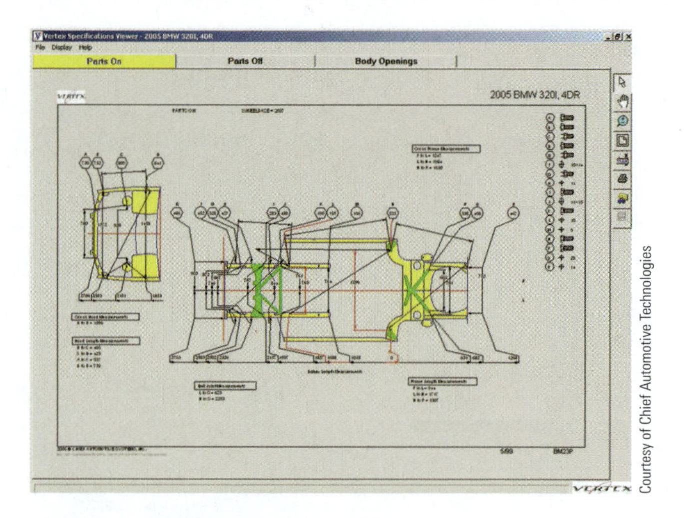

Figure 17-121 Optional measurement screens or windows are available for parts-on and parts-off of a vehicle.

do not align with the fixture, the unibody must be pulled back into alignment, or until the reference point aligns perfectly with its fixture.

Because there are so many fixtures needed for every make and model vehicle, most shops rent the fixtures for that vehicle for a few days and then return them when the

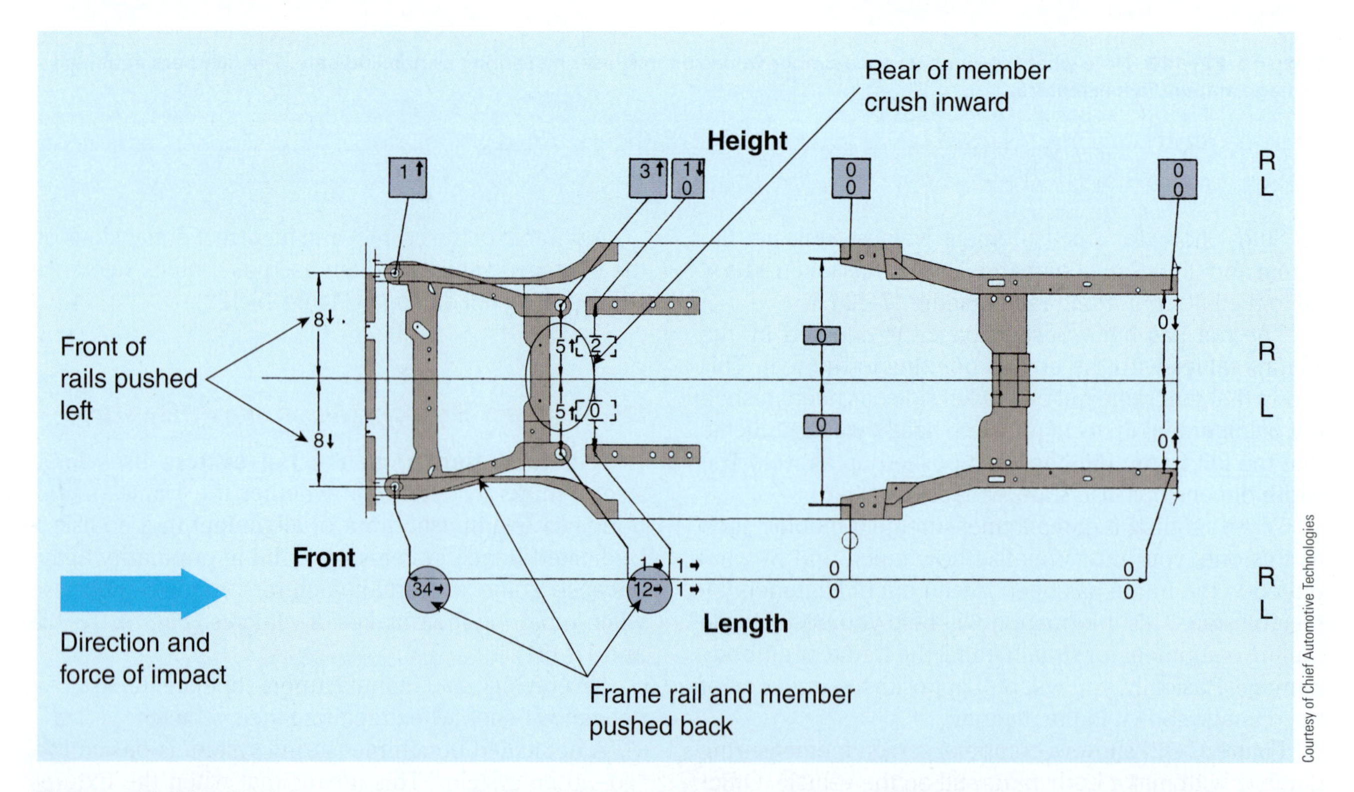

Figure 17-120 Study how a computer-displayed diagram shows the extent and direction of frame or unibody damage. This vehicle suffered a hard hit to the left front. Impact damage forced the left frame rail backward and up. The ends or tips of the frame rails are pushed sideways. The cross member is crushed inward and narrowed in the middle. You should be able to visualize the direction and extent of damage by looking at the dimensions' numbers and arrows. A frame rack would be used to pull out the damage until the reads were at or near zero.

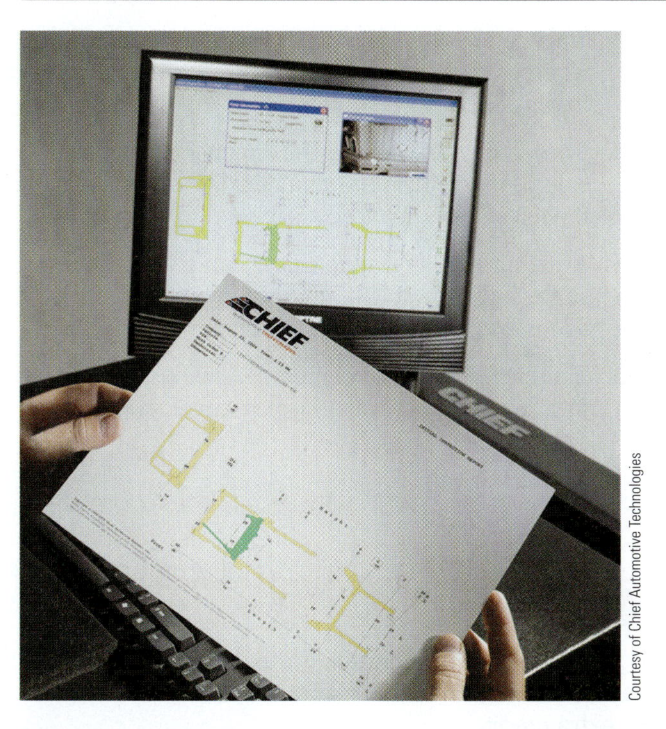

Courtesy of Chief Automotive Technologies

Figure 17-122 A printout of windows or measurements can be made at any time.

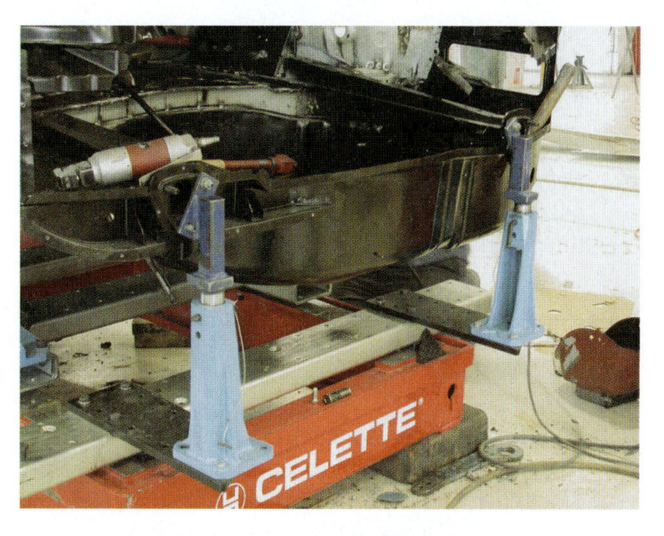

Figure 17-123 Some high-end automobiles require a dedicated bench measuring system. Reference points on the vehicle must align with fixtures. Fixtures can be rented for each make and model vehicle being repaired.

job is finished. Large shops or auto dealerships may have all of the fixtures for their makes of vehicles on hand at all times.

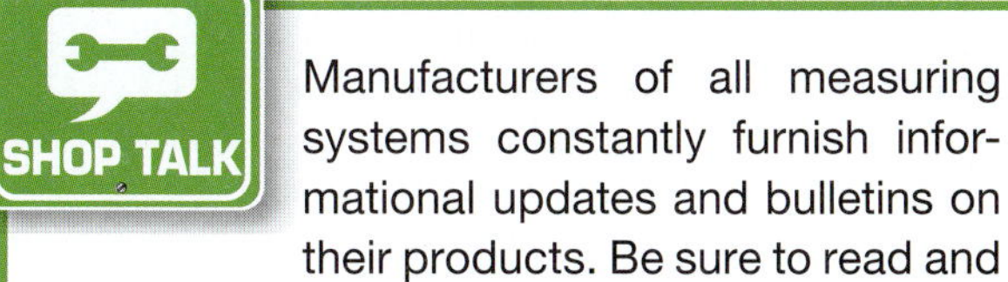

Manufacturers of all measuring systems constantly furnish informational updates and bulletins on their products. Be sure to read and study them because they will help to make the repair procedure easier.

Figure 17-124 The technician is using a dedicated bench measuring system to repair rear damage to a vehicle.

SUMMARY

1. Vehicle measurement involves using specialized tools and equipment to measure the location of reference points on the vehicle.
2. Check body dimensions (known good body measurements of the undamaged vehicle) by comparing the actual measurements with the values in the repair manual or body dimensions chart.
3. The control points used in manufacturing are not necessarily the same as the reference points the collision repair technician uses to measure the vehicle. Reference points refer to the points, bolts, holes, and so on used to give unibody and frame dimensions in body specification manuals.
4. Collision impacts that occur from the side often cause side-sway damage or a side-bending frame damage condition.
5. Sag damage is a condition in which one area, often the cowl area, is lower than normal.
6. Mash damage is present when any section or frame member of the car is shorter than factory specifications.
7. Diamond damage is a condition in which one side of the car has been moved to the rear or front, causing the frame and/or body to be out of square.

8. Twist damage is a condition in which one corner of the car is higher than normal; the opposite corner might be lower than normal.
9. The most important rule in body/frame alignment is: *Reverse direction and sequence.* This means that to correct collision damage on a conventional vehicle, the pulling or pushing of the damaged area must be done in the opposite direction of impact.
10. Crush zones are engineered to collapse in a predetermined fashion to protect the vehicle's passengers and to localize damage.
11. The body dimensions chart gives measurement points and measurement specifications for a specific type of vehicle.
12. Tram gauges are scaled rods used for measurement, whereas centering gauges use metal rods to check for misalignment.
13. Symmetrical means that the dimensions on the right side of the vehicle are equal to the dimensions on the left side of the vehicle.
14. A datum line, or datum plane, is an imaginary flat surface parallel to the underbody of the vehicle at some fixed distance from the underbody.
15. The center plane, or centerline, divides the vehicle into two equal halves: the passenger side and the driver's side.

EXERCISES

On a separate sheet of paper, complete the following learning activities for this chapter. Write definitions for the key terms and answer the ASE-style review questions, essay questions, critical thinking problems, and math problems. You can also do the outside activities, possibly for extra credit.

Key Terms

body center marks
body dimensions
body dimensions chart
center plane
computerized laser measuring system
computerized measuring system
control points
damage analysis form
datum plane
dedicated bench measuring system
diamond damage
digital tram gauge
height dimensions
laser measuring system
lateral dimensions
left-to-right symmetry check
mash damage
measurement gauge
mechanical measuring system
pivot measure system
primary damage
reference points
robot arm measuring system
sag damage
secondary damage
side-sway damage
sonic measuring system
strut tower gauge
symmetrical
three zero plane section
twist damage
vehicle measurement

ASE-Style Review Questions

1. Which type of full frame vehicle damage usually occurs first in a collision?
   A. Side-sway
   B. Sag
   C. Mash
   D. Diamond
2. In a unibody structure, which of the following occurs last in the typical collision damage sequence?
   A. Bending
   B. Widening
   C. Twisting
   D. Crushing
3. Technician A says that the tolerance of critical manufacturing dimensions must be held to within a maximum value of 5 mm. Technician B will check for door sag in the rear section of a vehicle after a front-end collision. Who is correct?
   A. Technician A
   B. Technician B
   C. Both A and B
   D. Neither A nor B
4. To accurately measure a vehicle, how many correct dimensions are required as a starting point?
   A. At least two
   B. At least three
   C. At least four
   D. One
5. Technician A says that any wrinkle in the rear floor is usually due to buckling of the rear side member. Technician B says that self-centering gauges are used to measure in a manner closely related to the tram gauges. Who is correct?
   A. Technician A
   B. Technician B

C. Both A and B
D. Neither A nor B

6. When proceeding through the measurement of the damage, Technician A measures the center section of the vehicle for twist and diamond first. Technician B measures for mash with a strut tower upper body gauge. Who is correct?
A. Technician A
B. Technician B
C. Both A and B
D. Neither A nor B

7. Technician A says that diamond damage can frequently occur in unibody vehicles. Technician B says that a visual inspection is sufficient to accurately assess impact damage. Who is correct?
A. Technician A
B. Technician B
C. Both A and B
D. Neither A nor B

8. Technician A leaves the laser system in position during the repair operation. Technician B keeps targets for a laser measuring system in a shared drawer with hammers and dollies. Who is correct?
A. Technician A
B. Technician B
C. Both A and B
D. Neither A nor B

9. How many gauges are usually hung to check for centerline misalignment?
A. 2
B. 3
C. 4
D. 5

10. Technician A frequently uses tram gauges and centering gauges in conjunction with one another. Technician B stands as close as possible to centering gauges in order to read them accurately. Who is correct?
A. Technician A
B. Technician B
C. Both A and B
D. Neither A nor B

Essay Questions

1. Explain the difference between primary and secondary damage.
2. Give seven steps in a basic collision damage diagnosis procedure.
3. What is diamond damage?
4. Explain the cone concept of damage.
5. What is a left-to-right symmetry check?
6. Before beginning any universal measuring operations, what three things should you do?

Critical Thinking Problems

1. If you see an accident, and car Number 1 is braking hard and hits a stationary car Number 2 center broadside, what damage would you expect to each vehicle?
2. During an inspection, you find the right front frame rail crush zone badly collapsed and pushed to the left. What was the direction of impact?

Math Problems

1. When measuring damage, you find that two round holes, one being ½ inch in diameter, the other 1½ inches in diameter, have an inside measurement of 32 inches and an outside measurement of 34 inches. What is this dimension?
2. A body dimension is 35.2 inches (890 mm). Your measurements show a reading of 37.1 inches (942 mm). If the tolerance is $\frac{1}{16}$ inch, what is the minimum amount you must straighten the body to be within tolerances?

Activities

1. Visit a local body shop. Ask the owner or manager whether you can watch a technician measuring damage on an actual vehicle. Write a report on what you observe.
2. Obtain and study operating manuals for several types of measurement systems. Write a report or summary of unique methods for using each type of equipment.
3. Using a body dimensions manual, locate the data for one specific make and model vehicle. Note the locations of the reference points. Make a few practice measurements on this vehicle.

CHAPTER

18 Unibody/Frame Realignment

OBJECTIVES

After reading this chapter, you should be able to:

- Summarize how different types of unibody/frame straightening equipment are set up and used.
- Describe the basic straightening and aligning techniques.
- Determine pull directions by analyzing damage.
- Summarize safety considerations to follow when using unibody/frame realignment equipment.
- Properly plan and execute collision repair procedures.
- Identify signs of stress/deformation on a unibody vehicle and make the necessary repairs.
- Compare the differences between the straightening of unibody vehicles and larger, full-frame vehicles.
- Explain why you might need to pull damaged parts before their removal.
- Correctly answer ASE-style review questions relating to unibody/frame realignment procedures.

INTRODUCTION

Vehicles with major damage must often have their frame or body structures straightened. Frequently called "frame straightening," body aligning, or pulling, is often thought to be a "rough and tough" physical operation. Actually, frame straightening involves using a logical, step-by-step procedure to methodically remove large dents and bends in sheet metal. This chapter will warn you of potential danger points and how to prevent common injuries that can occur during the frame straightening process. To pull out major structural and sheet metal damage, powerful hydraulic/electric frame straightening equipment applies tons of pulling, straightening force on the partially crushed vehicle to realign major panels.

An important requirement when straightening is accuracy. If the frame or unibody is not straightened exactly, panels will not fit and the wheels may not be alignable.

Unibody/frame straightening or realignment involves using high-powered hydraulic equipment, mechanical clamps, and chains to bring the full frame or unibody structure back into its original shape (Figure 18–1).

Improper straightening techniques lead to costly and time-consuming mistakes. Accurate vehicle alignment positively affects safety, repair time, repair quality, and the confidence of your customers. This chapter summarizes the most important methods for realigning a vehicle with major unibody or frame damage.

Photos courtesy of Car-O-Liner Company

Figure 18–1 Frame straightening equipment uses hydraulic power to pull or force a vehicle's body or frame back into its original shape.

18.1 REALIGNMENT BASICS

A **frame machine**, also called a *frame rack* or *frame bench*, is a large framework with hydraulic equipment for pulling out major structural damage. Even though equipment designs and setups vary, frame straightening equipment use is similar from machine to machine (Figure 18–2).

NOTE *Always follow realignment equipment operating instructions. The information in this book is general and does not apply to all equipment types and applications.*

Anchoring and Pulling

Vehicle anchoring involves clamping the vehicle down so that it will not move during the straightening process.

The term **pulling** refers to using hydraulic realignment equipment to stretch the damaged metal back out to its original shape. The vehicle is secured and held stationary by the equipment. Then, clamps and chains are attached to the damaged area. When the hydraulic system is activated, the chains slowly pull out the damage.

Measurements are made at body/frame reference points while pulling to return the vehicle to its original dimensions (Figure 18–3).

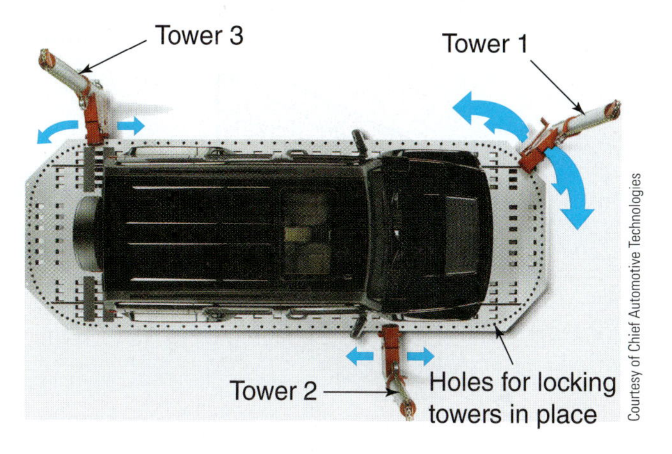

Figure 18–3 Note how the towers can be rotated around the rack and secured to different hole locations. This allows you to pull from a direction opposite to that of the force of impact damage.

Pulling Direction

When realigning a vehicle, a pull force, or **traction**, should be applied in the *opposite direction* from the force of the impact. When determining the direction of a pull, basically you must set the equipment to pull perpendicular to the damage.

Here is a simple example. If a vehicle with a simple box perimeter frame hits a brick wall directly head-on, how would you use hydraulic equipment to pull out the damage? You would simply have to pull straight out on the frame rails while holding or anchoring the frame. Vehicle damage and required pulling force are shown in Figure 18–4.

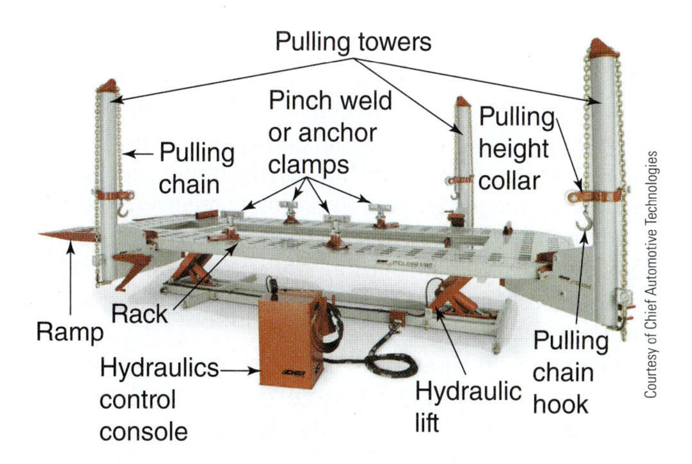

Figure 18–2 Study the major parts of a typical frame rack. Large towers can be swiveled into different pulling positions. Collars on towers allow the pulling height to be changed quickly. Vehicle anchor clamps or pinch weld clamps hold a vehicle on the rack so it does not move during the pulling operation.

When pulling, you must think of how the damage should be pulled to move all parts back into their original positions. Set your reference point on an imaginary line extending along the desired axis of the pull (Figure 18–5).

In reality, vehicles frequently hit objects that are not flat, today's vehicle frames are complex, and the direction of impact is often from an angle. This complicates the damage and pulling methods needed for repairs.

The pulling and straightening process must remove both direct and indirect damage. It must return all of the damaged metal back to preaccident dimensions. To do this, the equipment must reverse the direction and sequence in which the damage occurred. Generally, the damage that occurred last during the collision should be pulled out first (Figure 18–6).

Use the method that works the best for a given situation. Because applying force in only one place will usually not result in proper repairs, it is recommended to exert pulling force on many places at the same time. For convenience, the phrase *direction opposite to the input* will be used to describe the effective pulling direction.

$$X + Y = Z$$

$$X + Y' = Z'$$

$$X' + Y + Z''$$

Single and Multiple Pulls

The **single-pull method** only uses one pulling chain. This method works well with minor damage on one part. A small bend in a part can often be straightened with a single pull.

With major damage to several panels, a **multiple-pull method** with several pulling directions and steps works best.

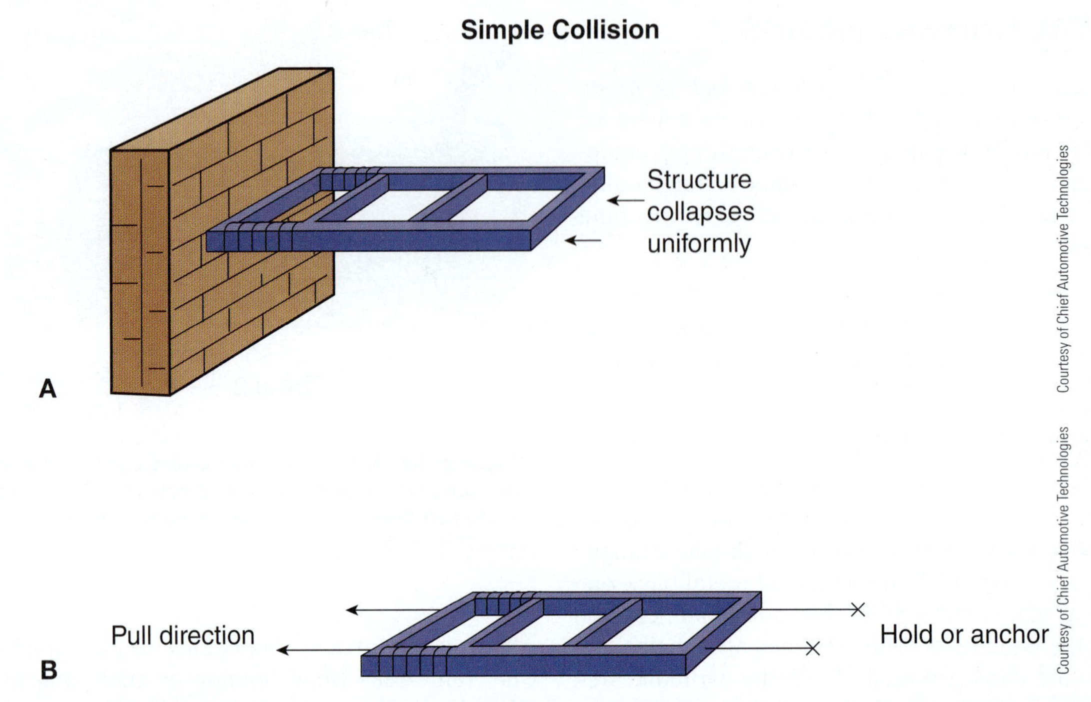

Figure 18-4 **(A) During a simple collision, the frame or unibody structure is crushed straight back with no deflection. (B) Pulling or frame straightening involves holding the frame while using chains to pull it straight back to reverse the direction of damage.**

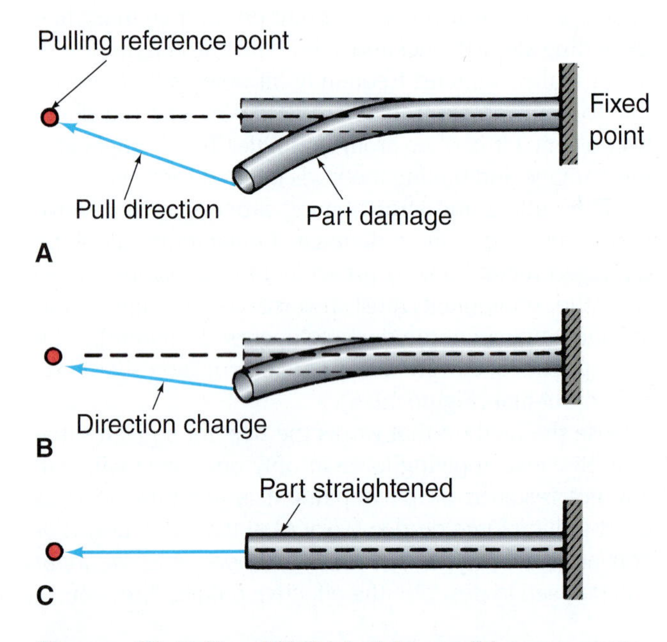

Figure 18-5 **(A and B) Think of the condition you want to achieve when repairs are completed. Set a reference point on an imaginary line extending along the desired axis from which to exert force and pull from that point. (C) When force is applied and the bend is repaired, the part will be straightened.**

With major damage, body panels are often deformed into complex shapes with altered strengths in the damaged areas.

To alter the direction while pulling, divide the pulling force into two or more directions. This will allow you to change the direction of the **composite force**—the force of all pulls combined. Refer to Figure 18–7.

With push–pull force, the straightening equipment is used to pull out damage while a portable ram or blocking device pushes out damage at the same time. This reduces the total force needed and restores the parts with less effort and reduced repair time.

The repair of a bent boxed part, such as a side member, is done by clamping the surface of the bent-in side and pulling. The pulling direction should be applied in an imaginary straight line extending through the original position of the part (Figure 18–8).

Energy-Absorbing Construction

Late model vehicles, both unibody and full frame, are designed to absorb much of the energy of a collision. The impact force travels into and through the frame/body structure to keep the passenger compartment relatively intact.

Figure 18–9 shows how impact force travels into a unibody during a frontal collision. Note how the front frame rails transfer some of the energy into the rocker panels and front pillars, which resist damage.

Figure 18–10 shows how impact force travels into a unibody during a side hit. Note that the rockers and pillars are designed to absorb the damage in this event.

Visualizing Front-End Collisions

A good way to determine how to remove or pull damage is to visualize what happened during an accident. Did the vehicle get hit straight on, get hit from an angle, or flip

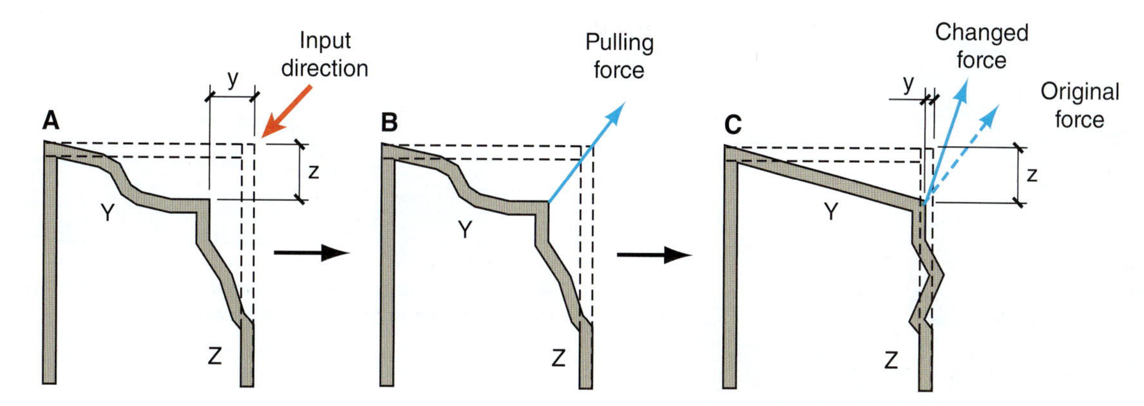

Figure 18-6 Study how to establish pulling direction to correct vehicle damage: (A) Input was in the direction of the back arrow, causing damage in directions Y and Z; (B) apply force in the direction opposite to the input force; and (C) if a difference in the degree of repair between Y and Z occurs, change the pulling direction accordingly.

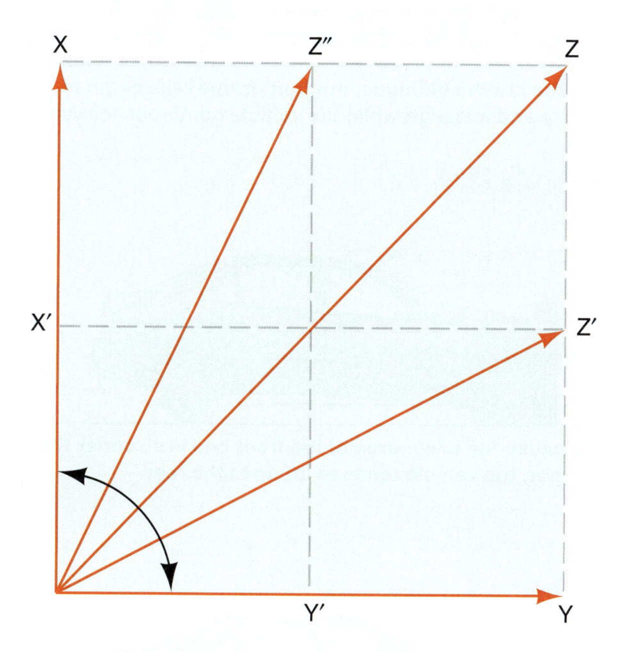

Figure 18-7 If the pulling force is divided between directions X and Y, composite force direction Z will change freely with adjustments to the force in the two directions.

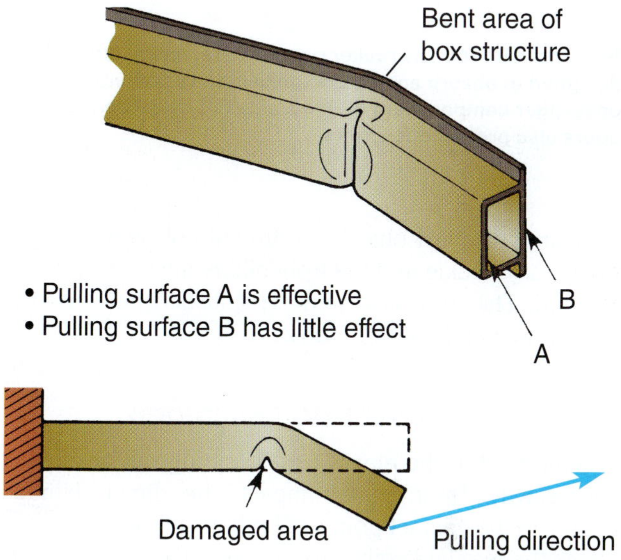

Bent area of box structure

- Pulling surface A is effective
- Pulling surface B has little effect

B

A

Damaged area

Pulling direction

Figure 18-8 Note the point of clamping and the direction of pull to straighten a box section like this one. The same would apply to frame rails, pillars, and similar unibody parts.

over? This kind of thinking will help you pull the damage in the opposite direction from which it occurred.

Front-end collisions result when a vehicle is moving forward and hits a stationary object or another vehicle. At the moment of impact, the parts at the front of the vehicle come to a complete stop and are crushed. The remainder of the vehicle continues to move forward from inertia.

As the front end continues to collapse, the frame rails begin to misalign upward due to the design of the frame. The rest of the vehicle continues forward (Figure 18–11).

As the frontal collision continues, additional collapsing and deflection of the rails results. Because the rails are stronger than the upper structure, the rear of the vehicle is forced upward. This may bend the rear of the frame or unibody upward. Door openings might be forced out of alignment.

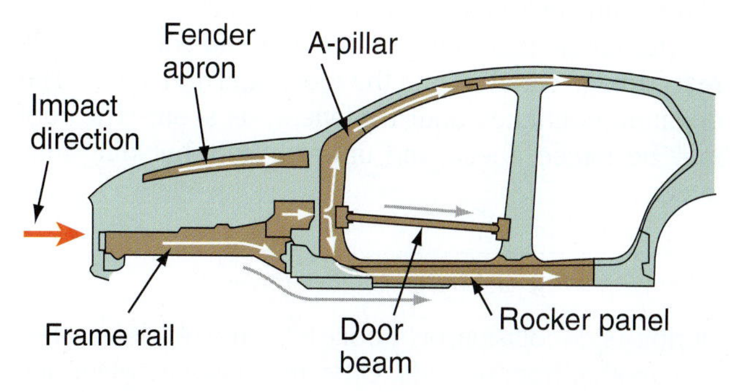

Figure 18-9 Note how impact force travels through a unibody structure. Most of the damage is absorbed by the front section. However, with a hard hit some damage can transfer into the front pillars, roof, and rockers.

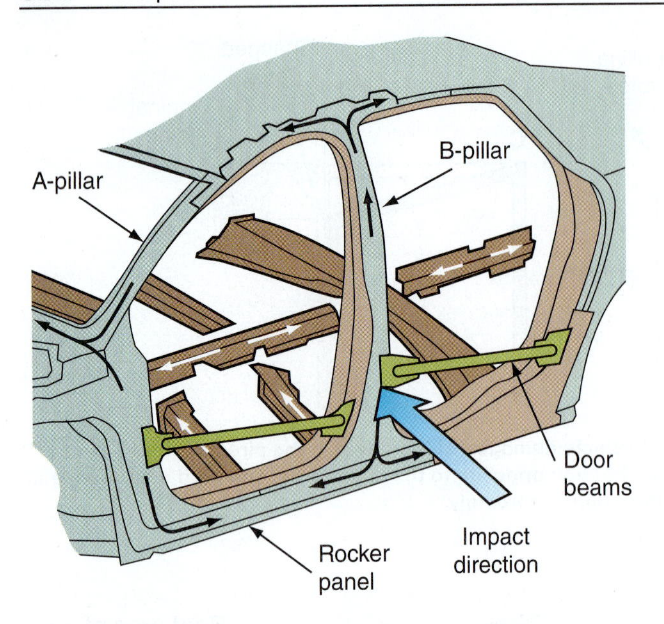

Figure 18–10 **The rocker panel, doors, and pillars are designed to absorb energy and prevent intrusion into the passenger compartment during a side hit. UHSS beams in doors also protect occupants.**

A At the moment of impact, the front section collapses while the rest of the vehicle moves forward.

B As the crash continues, the front frame rails begin to collapse and misalign while the vehicle continues forward.

C Because the lower area of the front end is stronger than the upper, the vehicle tends to lift up in the rear.

D If the speed is high enough, the front section completely crushes and the inertia of the rear of the vehicle causes secondary damage to the pillars and roof, causing door gap misalignment.

Figure 18–11 **Study the basic stages of frontal impact into a brick wall.**

During the last phase of a frontal collision, the roof panel may buckle and the front pillars may bend up and forward. The door also may be smashed between the front fender and quarter panel.

Visualizing Rear-End Collisions

A vehicle hit in the rear is usually sitting still or moving very slowly. Upon initial impact, the direct damage occurs to the contact points of both vehicles.

As the rear-end collision continues, the first vehicle is pushed forward, but inertia resists movement. The rear rails and floor pan start to collapse forward. Their energy-absorbing design also causes them to flex and deform. See Figure 18–12.

If vehicle speed is high enough, additional collapsing of the rear-end structure occurs. The strength of the center section of the vehicle resists this collapsing. The strength of the frame rails causes the rear to deflect upward.

During the final stages of the rear-end collision, indirect damage can flow into the roof panel and pillars. The crushing continues until the energy is spent. The roof may be forced ahead and upward. Roof buckling may occur.

Visualizing a Side Collision

During a side collision, or "T-bone hit," one vehicle smashes into another from the side. Often this happens when one vehicle runs a stop sign or signal and rams into the doors, fender, or quarter panel of another vehicle (Figure 18–13).

During the first stages of a side hit to the doors, the door skins are crushed inward. Steel reinforcement beams in the doors resist the initial impact. As the side hit continues, the front of the vehicle tries to move into the passenger compartment. The strength of the doors and pillars resists this intrusion.

Impact energy pushes the passenger compartment sideways and tries to shorten and pull in on the side of the vehicle being hit. Lateral deflection continues until the hit vehicle starts to slide sideways. The collapsing of

A The car in front is hit from behind by another vehicle. Initially, the bumpers are crushed.

B As the crash continues, the rear frame rails and trunk floor panel start to absorb impact energy. The area of the rail over the suspension starts to kick up, and the rear of the rail often kicks or bends down.

C Additional collapsing and deflection occur, and damage spreads to the center section and roof.

D With a sufficiently hard hit, the pillars are pushed forward and the rear section is forced upward.

Figure 18–12 Study the sequence of a typical rear-end collision.

the pillars and rocker panel normally shortens the side of the vehicle hit from the side.

Visualizing Rollover Damage

Rollover damage results when a vehicle flips over. The vehicle may have turned too sharply, been hit from the side, or slid into a curb before rolling onto its roof. If the vehicle rolls over several times, a series of collisions actually occurs. Refer to Figure 18–14.

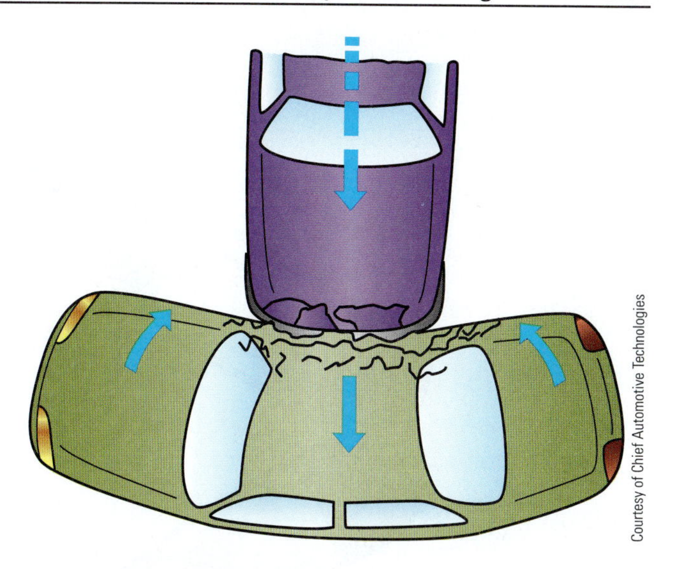

Figure 18–13 During a side collision or "T-bone hit," one vehicle rams another from the side. The rocker panel, doors, and pillars absorb impact energy. The vehicle that is hit on the side often takes on a "banana" shape because the front and rear sections are bent sideways.

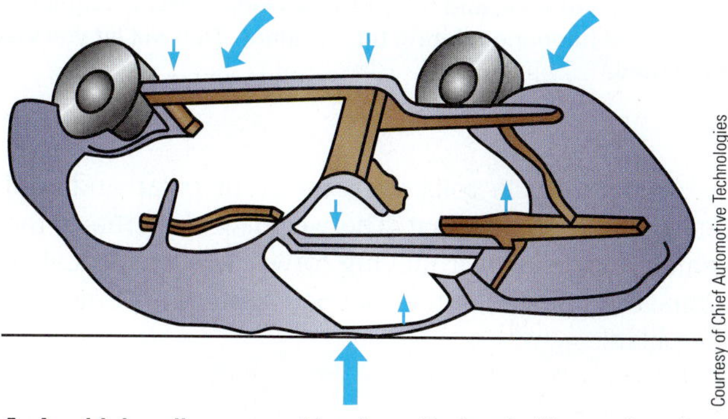

A A vehicle rolls over and lands on its front pillar and roof, pushing them down.

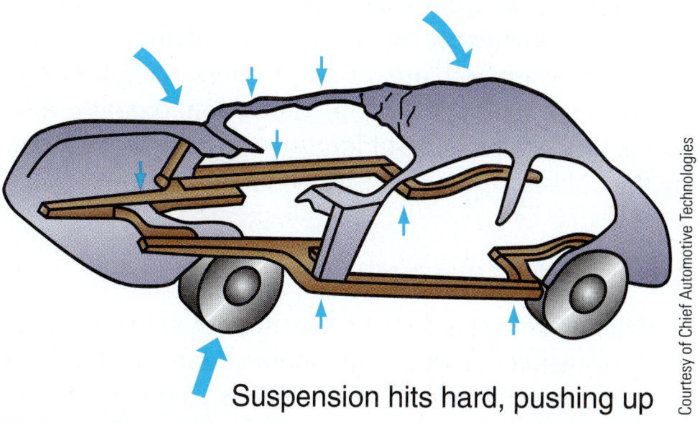

B If travelling fast enough, the vehicle may continue to flip and then land hard on its wheels, pushing the suspension and rails upward.

Figure 18–14 Study the sequence of events during a rollover accident. Rollover damage is more common with vehicles that have a high center of gravity, such as SUVs and vans.

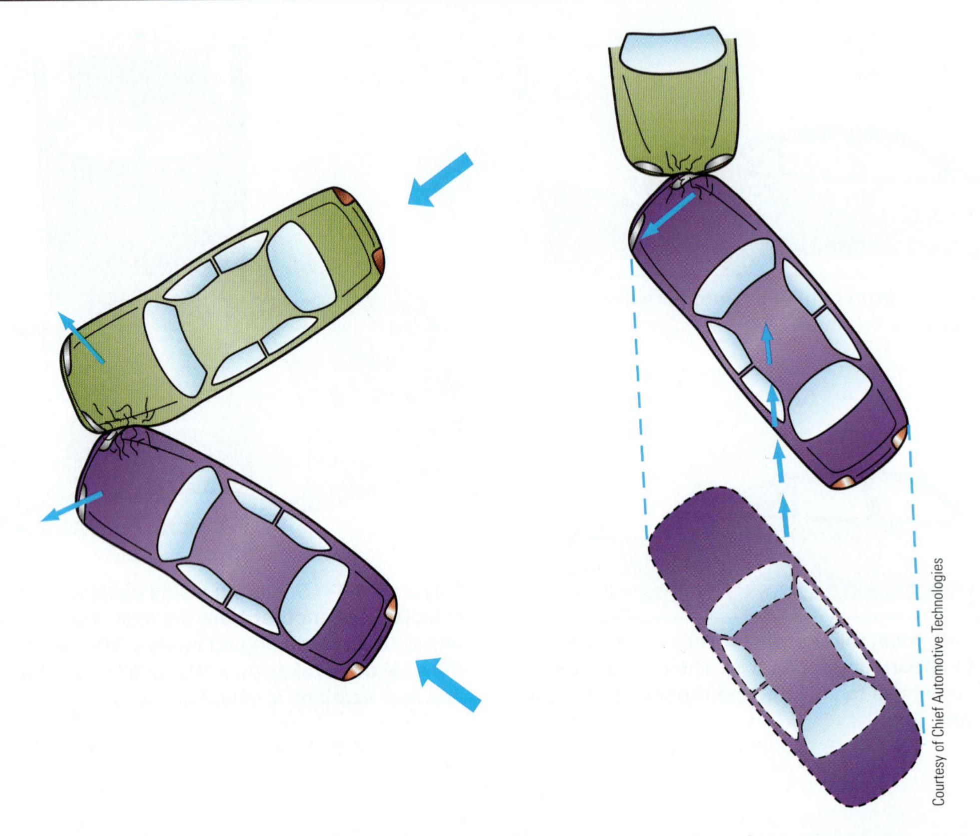

Figure 18–15 Deflection from the direction of travel causes damage at an angle from the centerline of vehicles. Drivers often hit the brakes, and their vehicle skids sideways right before collision. By studying the damage, you can visualize what might have happened during the accident. This will let you visualize how to pull out the damage in reverse sequence to how it happened.

As the vehicle rolls over, the front pillar and roof often hit the ground first. The weight and inertia of the vehicle try to keep it moving forward. A great deal of damage is inflicted to the roof panel and one or more of the pillars.

Visualizing Angled Impacts

Lateral deflection is sideways damage. This type of damage often results from a side hit or a hit from an angle. When two vehicles are moving in different directions and collide, as when both are turning corners, complex damage normally results. Lateral deflection damage also results if one vehicle is skidding or sliding sideways during the collision. This is illustrated in Figure 18–15.

Analyzing Damage

Analyzing damage is done before unibody/frame straightening by inspecting direct and indirect damage. Look for direct damage to frame rails and indirect damage to large surface area panels, such as the roof and quarter panels. Check for improper panel gaps and alignment. Look for cracked sealer or undercoating, bulges, buckles, and other signs of major damage (Figure 18–16).

Finding uneven gaps or spaces between panels can give you helpful information on how to pull out the

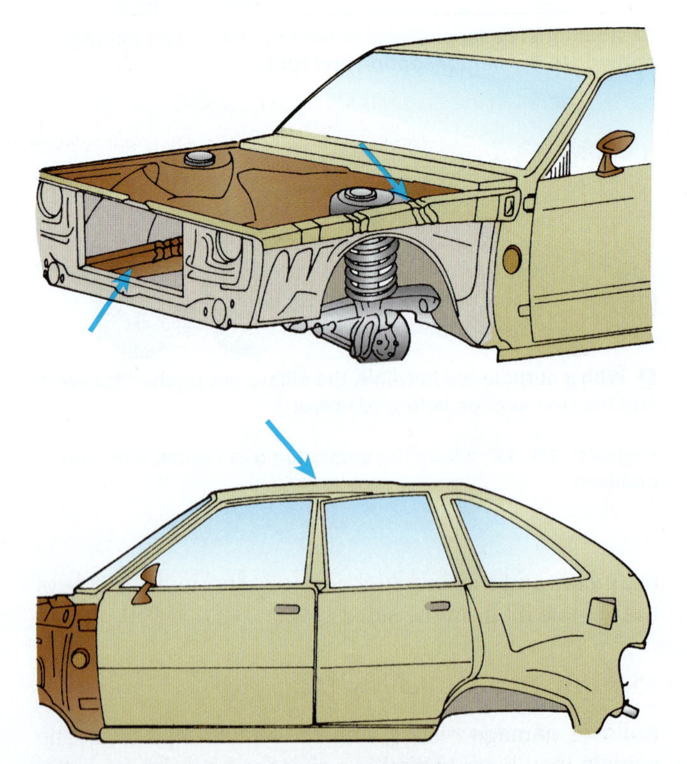

Figure 18–16 Signs of stress/deformation on unibody cars are cracked sealer or undercoating, bulges, buckles, and misaligned panels. Always look for these signs when planning your pulling sequence.

damage. For example, if the front of a quarter panel is pushed into the back door at the bottom, you would know that lower rear structure pulling is needed.

Regarding vehicle dimensions, there are three types or directions of unibody/frame damage:

1. **Length damage** normally results from a front or rear hit that pushes body panels toward the center of the vehicle. The length measurement will usually be shorter than specifications. However, it is sometimes possible for the length to be stretched longer than normal.
2. **Width damage** results when parts are pushed to the center by a side or angled impact. The width measurement will usually be shorter than factory dimensions.
3. **Height damage** results from impact damage that forces parts or panels up or down. A rollover accident will normally crush the roof down and cause the height dimension to be too short. If a frame rail or another panel is pushed up or down, the height dimensions will not be within factory specifications.

As you have read in previous chapters, vehicle measurements are compared to factory vehicle dimensions or specs to determine the type and direction of damage.

18.2 UNIBODY/FRAME STRAIGHTENING EQUIPMENT

Unibody/frame *straightening equipment* is used to apply tremendous force to move the frame or body structure back into alignment. Straightening equipment includes anchoring equipment, hydraulic pulling equipment, and other accessories (chains, clamps, and so forth) (Figure 18–17).

Anchoring equipment holds the vehicle stationary during pulling and measuring. Anchoring can be done by fastening the frame or unibody of the vehicle to anchors in the shop floor or to the straightening equipment rack, frame, or bench. The anchoring system is designed to hold a vehicle solidly in place while tremendous pulling forces are applied. To avoid damage, anchoring must also distribute pulling forces throughout the vehicle.

Traction direction refers to the direction of pulling force applied by the frame straightening equipment to remove the damage. Traction direction, or pulling direction, is normally opposite the direction that the damage was made during the impact (Figure 18–18).

Pulling equipment uses hydraulic power to force the body structure or frame back into position. Many different types of pulling equipment are available. Regardless of their design or operating features, each system uses the same basic pulling theory and is operated in a similar manner. You should be familiar with a variety of pulling systems and their general operation.

Hydraulic rams use oil pressure from a pump to produce a powerful linear motion. When you activate the system, oil is forced into the ram cylinder. The ram is then pushed outward with tremendous force. This pulls on the chain attached to the vehicle to remove the damage. The rams can be mounted in or on the pulling towers or posts, or between the vehicle and anchoring system.

Pulling posts or towers are strong steel members used to hold the pulling chains and hydraulic rams. Depending on equipment design, they can be positioned wherever needed to make the pull. They push against the rack or bench as the pull is made. This eliminates the need for separate anchoring to keep the pulling equipment from sliding under the rack or bench as the pull is made.

Pins or hardened bolts lock the tower to the rack. The tower can be rotated sideways for different pulling angles. Some will also tilt into an angle for even more pulling flexibility. Make sure all lock pins are securely in their holes before applying pulling power.

There are several different types of frame straightening equipment on the market that are suitable for both body-on-frame and unibody collision repair work.

The amount of pulling pressure required to remove damage should not be too much. If the pulling equipment is straining or spot welds begin to break during the pulling process, something is wrong. If this happens, stop pulling! Release tension, and reevaluate the setup to find the problem. If too much pressure is applied, parts or equipment can be damaged and serious injuries can result.

In-Floor Straightening Equipment

Two types of in-floor systems available are the anchor-pot system and the modular rail frame system.

In-floor straightening systems have anchor pots or rails cemented or mounted in the shop floor. Some use anchor pots or small steel cups in various locations in the shop floor. Others use a system of steel anchor rails in the floor so that a number of pulling–holding locations can be used. Both systems must be balanced both in direction and force of the pull.

Figure 18–19 shows how one type of in-floor system is used to anchor a vehicle before pulling.

An in-floor system is ideal for a small body shop. After the rams and other power accessories have been neatly stored away, the area can be used for other body shop purposes. In-floor systems also provide single or multiple pulls and positive anchoring without sacrificing shop space.

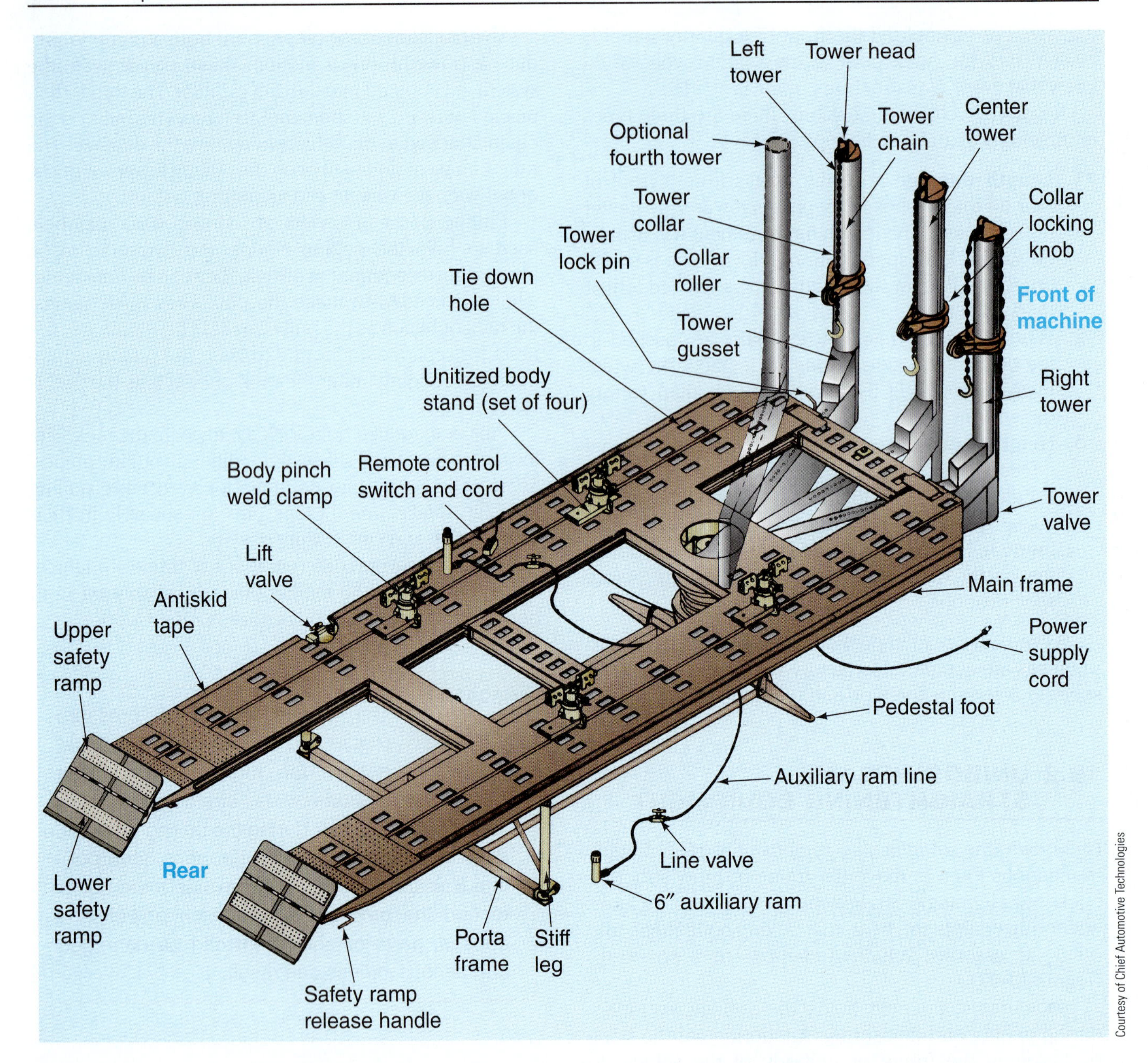

Figure 18-17 A frame rack or bench provides the most efficient method of correcting frame or unibody damage. Study the parts of a typical straightening system.

Portable Body and Frame Pullers

Portable body and frame pullers have a hydraulic pressure system installed between the removable main frame and the mast. This type of system is designed to extract damage using chains and clamps. It is often used for repair of minor damage.

Because these systems are easily movable, you can quickly set the traction direction or pulling direction opposite to the damage input. Many units of this type, however, are able to pull only in one direction. These units are more dangerous to use than rack or floor systems.

Rack Straightening Systems

Most *rack straightening systems* have a thick steel table to which pulling towers and anchoring clamps are attached. Angle ramps are provided for driving or winching a vehicle up onto the bench. This is one of the most common types of unibody/frame equipment found in modern body shops.

A stationary frame rack usually gives the technician infinite positioning possibilities to pull from any angle and height, 360 degrees around the vehicle. It is possible to make up-pulls or down-pulls. In fact, pulls can be made with the rack positioned at full rack height or flush

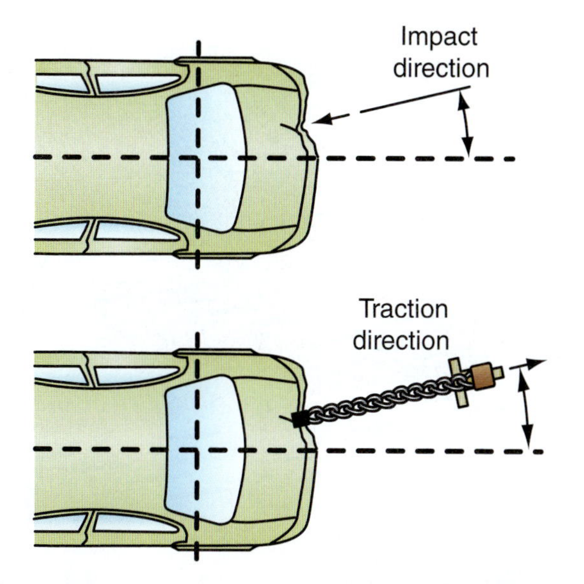

Figure 18-18 Basically, the traction direction, or direction of pull, should be opposite the damage direction. Body and frame pulling may seem complex and confusing at first. However, if you methodically analyze the directions of the damage, it is fairly easy.

to the floor. Most racks tilt hydraulically so that vehicles can either be driven on or pulled into position with the optional power winch, shown in Figure 18–20.

Bench Straightening Systems

A *bench straightening system*, similar to a rack, is generally a portable or stationary steel table for straightening severe vehicle damage. Some benches tilt. Others have drive-on ramps, like a rack, that can be raised up or down as needed. Alignment benches are available in fixed and movable types.

A bench-rack system is a hybrid machine that has features of both a bench and a rack. Its table will often tilt like a rack for quick loading of a vehicle. It also provides the accuracy and convenience of a bench once the vehicle is in place. This is the most modern type of system in use today.

Some bench-rack systems have a drive-on ramp with a rolling dolly system. This allows for fewer obstructions

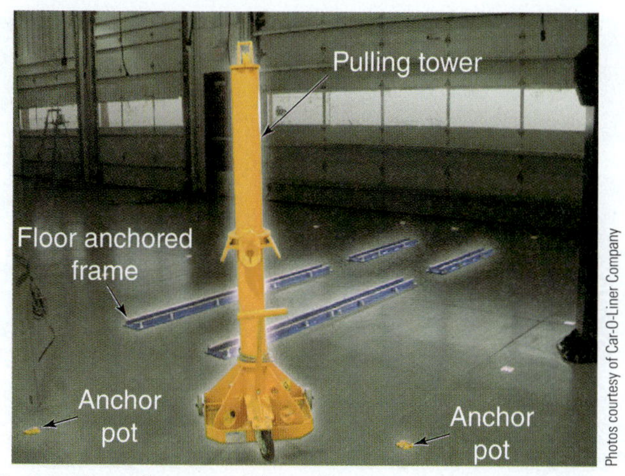

A Note the major parts of the pulling system using floor anchors.

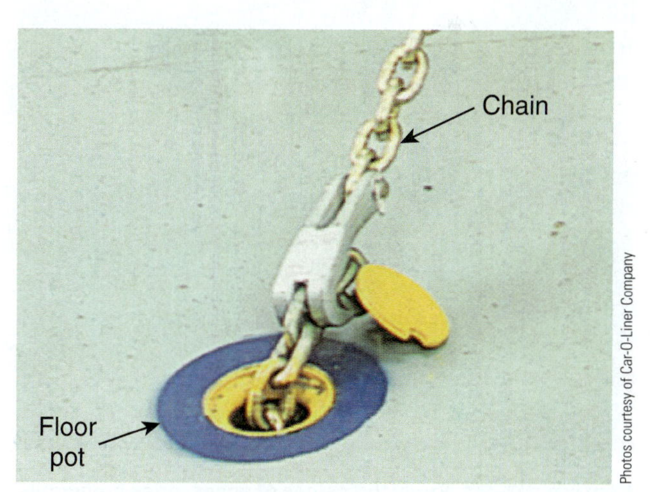

B A steel anchor pot is buried on the concrete floor of the shop. This allows it to hold when tremendous force is applied.

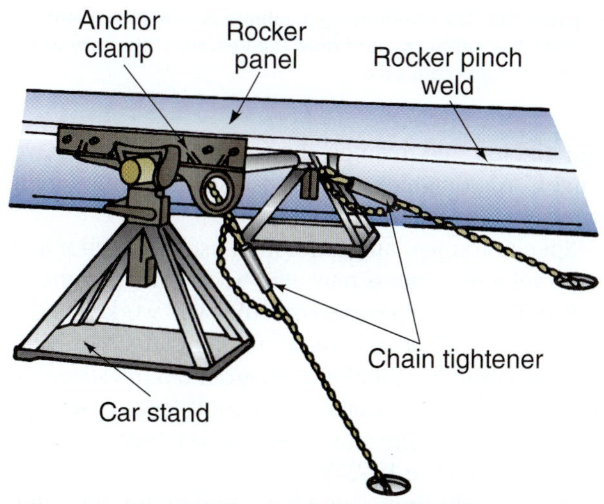

C With this unibody/frame straightening equipment, anchor chains extend from the body pinch weld clamps to floor pots to hold the vehicle while pulling damage.

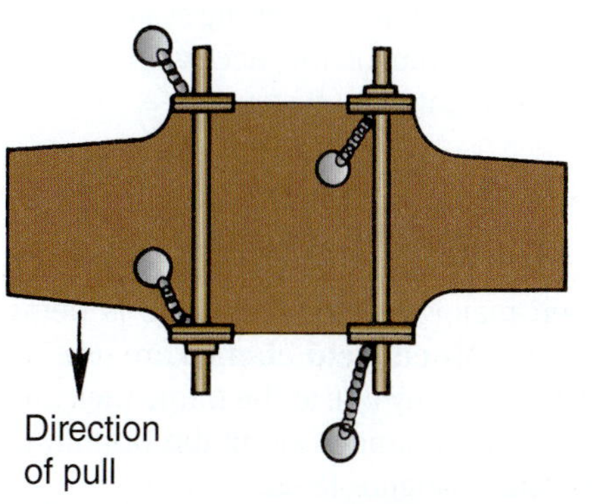

D If anchoring for a side-pull of a vehicle's front section, anchor chains must be positioned to resist the pulling action.

Figure 18-19 Note how to anchor using a floor-type straightening system.

Figure 18-20 Many racks tilt hydraulically so that vehicles can either be driven on or pulled into position with a power winch, as shown here.

Courtesy of Chief Automotive Technologies

A Pinch weld clamps are now the most commonly used method to anchor unibody vehicles. The bottom of the anchor clamp bolts to the bench or frame rack.

B Note how four large bolts and nuts secure the clamp to the flange under the rocker panel with this frame rack.

Figure 18-22 Anchoring a vehicle is very important because the anchors must hold the vehicle stationary when damage is being pulled out.

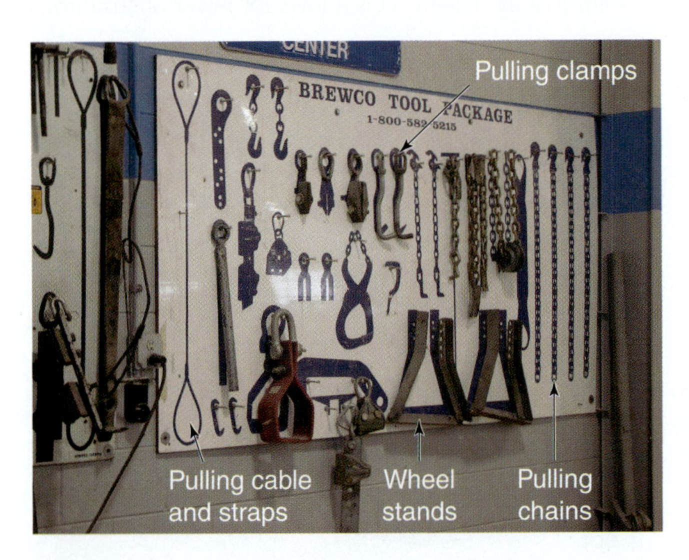

Figure 18-21 Study the names of accessories commonly used on unibody/frame straightening equipment.

under a vehicle and easier access to repair areas and reference points. The ramp may also be used as a stand while installing the vehicle on its anchoring system.

Figure 18–21 shows some of the accessories used with a rack or bench straightening system.

Anchoring the Vehicle

As noted earlier, a vehicle must be anchored so that it will not move when major structural damage is being pulled out. Today, large **pinch weld clamps** are used to anchor unibody vehicles. They bolt to the frame rack and to the vehicle's pinch weld flanges along the bottom of the rocker panels. Refer to Figure 18–22.

Raise the vehicle off the rack a few inches with an air jack or hydraulic jack. Then position four pinch weld clamps under the outer corners of the center section of the vehicle. Lower the vehicle down into the clamps, then tighten the clamps' bolts onto the pinch weld flanges. Also tighten the bolts, securing the anchor clamps to the rack or bench (Figure 18–23).

Special frame clamps are often used to anchor a full-frame vehicle. Spacers may also be available for the unibody pinch weld clamps so that they will work with a full frame. The anchor spacers allow the clamps to have a wider jaw opening for clamping around a thick steel box frame. With a light down- or up-pull, chains wrapped around the full frame may be all that is needed to anchor the vehicle (Figure 18–24).

When anchoring a vehicle in preparation for pulling, attempt "overanchoring" or "overclamping." An extra anchor point or two takes very little time, and it improves

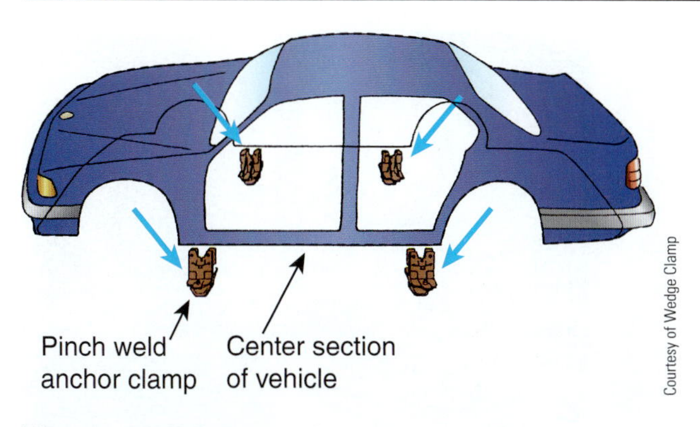

Figure 18-23 Four anchor or pinch weld clamps are normally mounted at the four corners of the center section of a vehicle.

Figure 18-24 To anchor this full-frame vehicle for a light up or down pull, anchor chains have been wrapped around the full frame and cinched down. Adapters are available so that you can use pinch weld clamps on the full perimeter frame for heavier pulling.

safety. However, there are many cases where a clamp cannot be fastened to the exact area of deformation. If the section is to be replaced, a piece of steel can be temporarily welded to the section to improve anchoring.

A full-frame vehicle can be anchored by placing a suitable plug hook in the fixture holes located on the bottom of the frame rail. Blocking should be used to keep the hook in line with the frame rail. If a hard pull is to be made, it is advisable to weld a washer around the hole as a reinforcement. Make identical hookups on both sides of the vehicle.

Pulling Clamps and Chains

Pulling clamps are attached to the damaged area of a vehicle to accommodate pulling chains. Various pulling clamps are used to allow the frame rack to force body parts back into alignment. The *pulling chains* are large, high-tensile-strength chains that attach between the clamps and pulling towers.

Pulling towers contain hydraulic rams for applying pulling force to the chains. They can be swiveled into different positions and locked to the rack (Figure 18–25).

Tower collars allow you to adjust the height of the traction directions. The collars can be slid up or down on the tower and clamped to establish the height of the chain and its pulling force.

Figure 18–26 shows a technician connecting a pulling chain from a tower to a clamp. When pulling, you must stand to one side of the pulling chain to avoid injury.

A Swivel the pulling tower into position so that it will pull in the correct direction to remove damage. Install the pin or engage the latch to secure the tower position.

B Once located properly, the pulling tower chain can be attached to the vehicle to be straightened.

Figure 18-25 With the vehicle anchored securely to the frame rack, this technician is ready to begin the pulling and straightening operation.

Figure 18–26 This technician is attaching the pulling chain to a large hook on the vehicle.

A If a clamp slips off or a chain breaks, serious injury can result.

B Here a safety chain has been attached to the pulling chain to stop it from flying out if its clamp slips off the vehicle.

Figure 18–27 Stand to one side and away from the chains when pulling damage. Tons of force can be applied.

DANGER Never stand in line with a pulling chain when applying traction. If the clamp were to slip off or a chain were to break, you could be killed by the flying chain. Place a safety chain or welding blanket over the pulling chain. Refer to Figure 18–27.

Figure 18–28 shows how various types of clamps can be secured to different parts of a unibody vehicle for pulling damage out. A single-bolt clamp is the common setup for pulling radiator supports, flanges, and other panels with minor damage.

An adapter is available to allow a single-pull chain to be used with two clamps for added power. A hinge plate that can be bolted to existing mounting holes on the vehicle is handy when you need to pull on an area that does not have a flange or lip for using conventional clamps. Also, hole adapters can be used to pull or anchor full-frame vehicles.

A scissor clamp is pictured in Figure 18–29. Pulling force causes this type of clamp to pinch down on the body part for pulling. It is handy and easy to install in some situations.

A pulling chain can be attached to a frame rail for a down-pull. The bolted-on angle and eye allow a slight lever action to be produced by the down-pull direction.

Pulling slings are nylon straps that can be wrapped around parts during pulling. Nylon straps help protect parts from damage. Because they wrap around parts and pull from the back of the part, they are desirable when pulling some parts.

A *chain pulley* is often used to pull down from under the vehicle. The pull chain is fed through a pulley, which is anchored to the rack. This allows the pulling angle from the tower to be changed into a down-pull (Figure 18–30).

Figure 18–31 shows how a chain pulley is being used to pull down on a front frame rail. Study the pulling setup.

Other Straightening Accessories

Various other adapters are often supplied with frame straightening equipment. Some are holding devices to clamp parts in place so that they do not move while pulling other parts.

Many times when pulling, you need to retain the movement of one area while straightening another. A **restraint bar** can be used to hold or maintain a dimension in an opening when pulling. It is an adjustable steel bar that can be slid out and locked into position.

A *door aligner* is designed to flex a door and its hinges for correcting alignment. It is a special bar that snaps onto the door lock and striker. By pulling up or down on the handle, you can quickly adjust minor door misalignment (Figure 18–32).

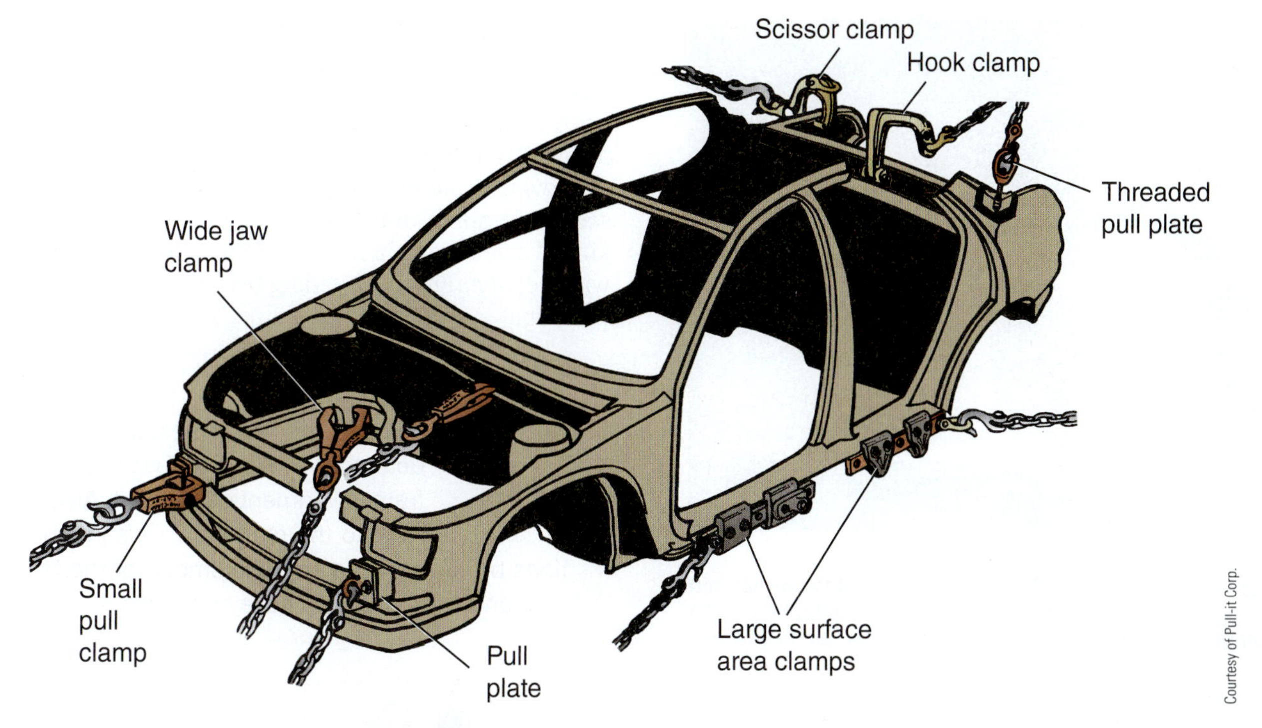

Figure 18–28 Note the various pulling clamps that are available. The size and shape of a pulling clamp should match its attachment point and the amount of pulling force required.

Figure 18–29 When you are installing a clamp like this one, the pulling chain end should be free to move around in the clamp. Then, when pulling force is applied, the clamp hinges down even tighter on the body part to avoid slippage.

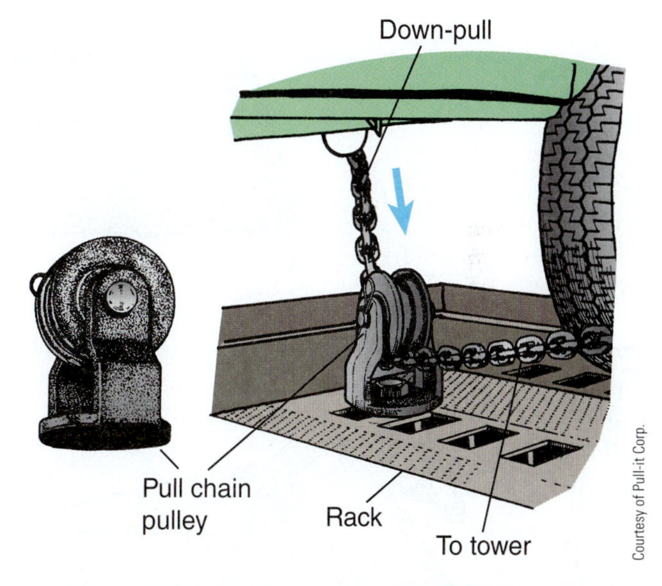

Figure 18–30 A pulley is often used for down-pulls. The pulley can be bolted to a bench or rack. This allows the pulling angle from the tower to be changed into a downward direction.

When it is necessary to remove the engine or transmission mounts, an *engine holder* can be used to support the engine. It rests on the inner fenders and is adjustable in width. An adjustable chain hook is used to hold the chain attached to the engine. This allows you to remove engine and transmission mounts or the cradle for repairs.

Portable hydraulic rams can be used to apply straightening force to hard-to-reach or enclosed areas on the vehicle. They are possibly the most versatile of all aligning tools. They can be used to push, spread, clamp, pull, and stretch (Figure 18–33). Figure 18–34 shows other pulling attachments.

Some pulling towers are designed to allow you to pull and push the damage at the same time. A portable power cylinder can be mounted between the damage to be pushed and the tower. It will push damage inward while pulling chains force the damage outward.

Figure 18-31 A single down-pull is used to correct minor damage to a front frame rail. Note how the pulley is used to change the direction of the pull to move the rail downward.

A *work platform* attaches to the rack so that you can stand and work on upper body areas. Many frame racks and benches come with a portable work platform.

A *holding device* can be clamped to the rack and adjusted out to keep one area of the vehicle from flexing or moving. Note how a holding device is being used to secure an apron and rail during a side pull. The rail is kinked sideways right in front of the holding device. This will keep the rail from bending behind the damage.

It is beyond the scope of this text to detail the use of every type of unibody/frame alignment equipment. Always refer to the manufacturer's instructions before you use any equipment for the first time and whenever you have questions.

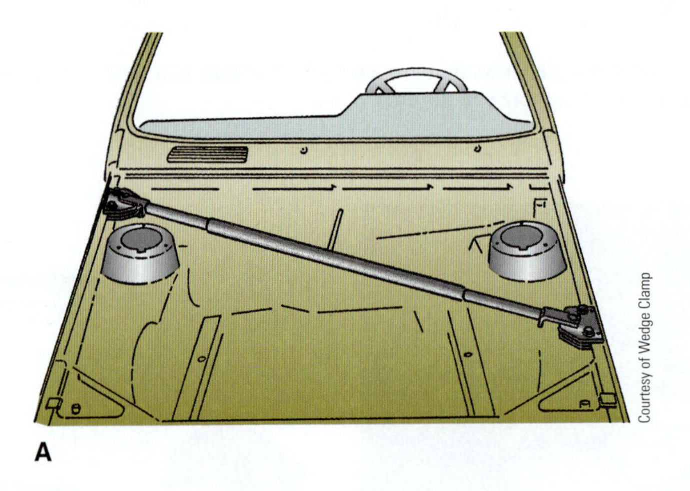

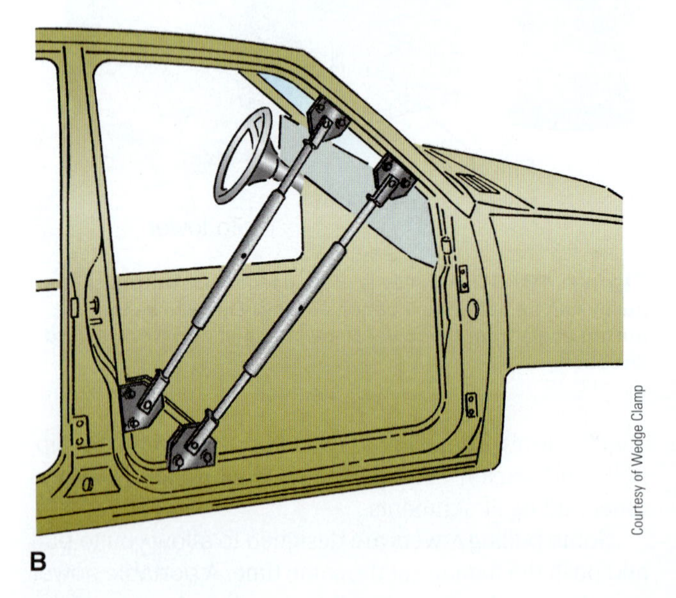

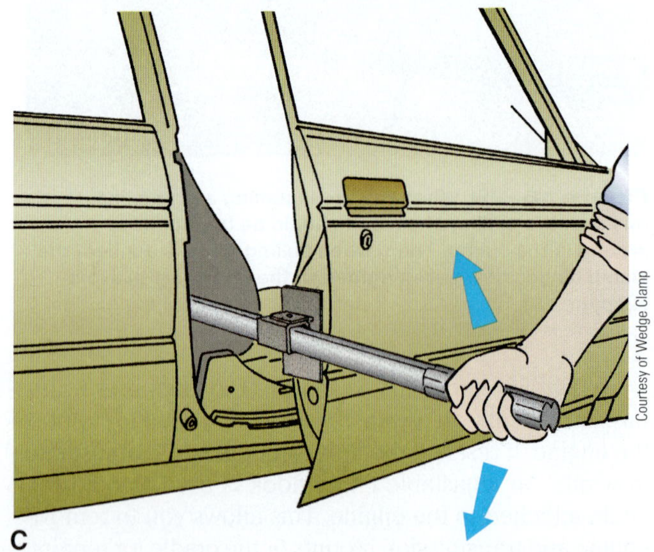

Figure 18-32 Study a few more frame rack accessories. (A) A single holding restraint is secured across the fender aprons to hold good dimension while pulling. (B) Two restraints are used to secure the door opening dimension. (C) A door aligner fits into the door lock and catch so you can flex the door up or down for minor adjustment.

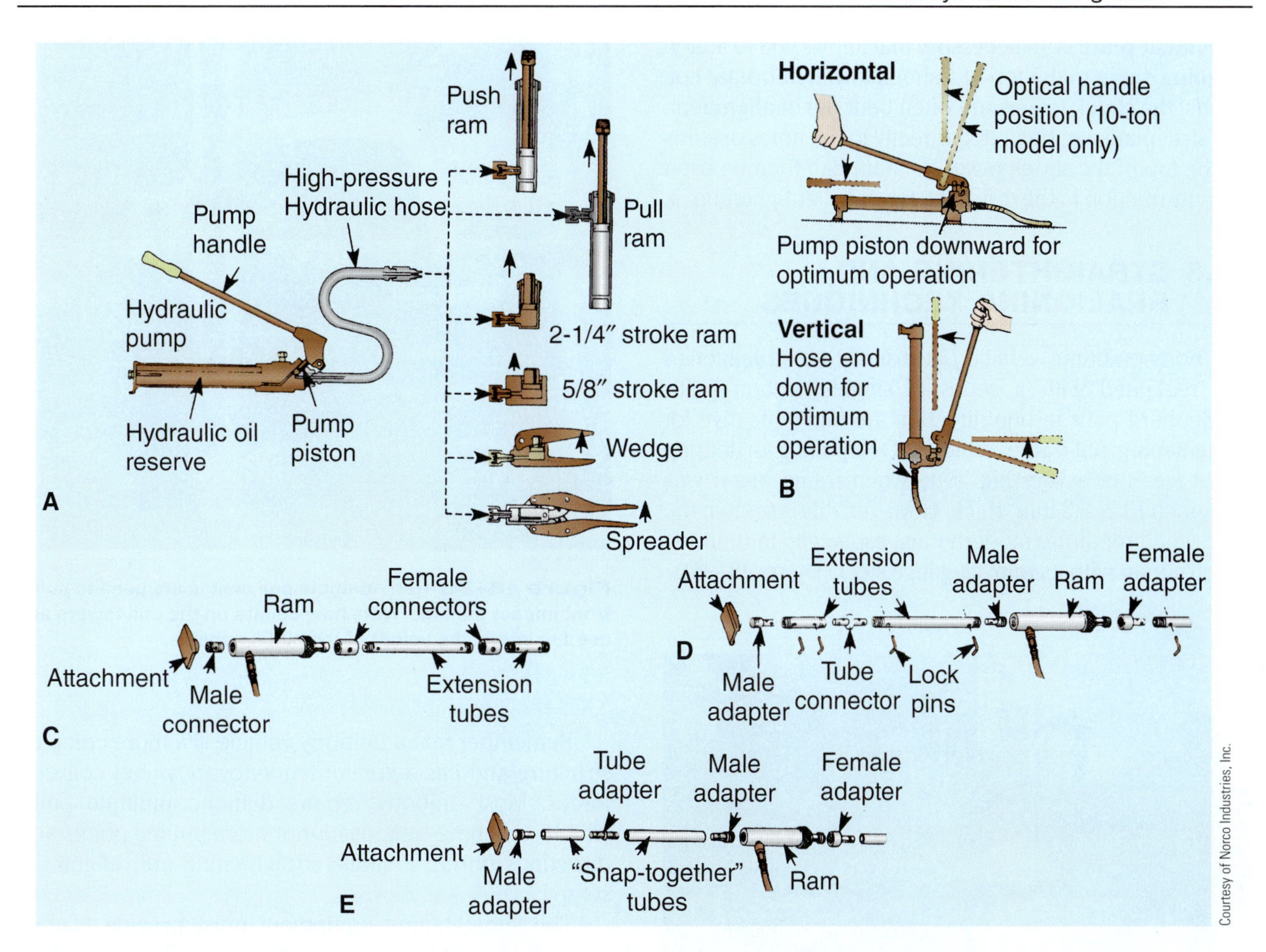

Figure 18-33 Study the major components of a portable hydraulic ram system: (A) the heart of the system—a portable pump, high-pressure hydraulic hose and hydraulic ram, and wedge, or spreader; (B) the positioning and operation of the pump; (C) the threaded connection; (D) the quick-fitting connection; and (E) the snap-together connection. The latter should not be used for pulling.

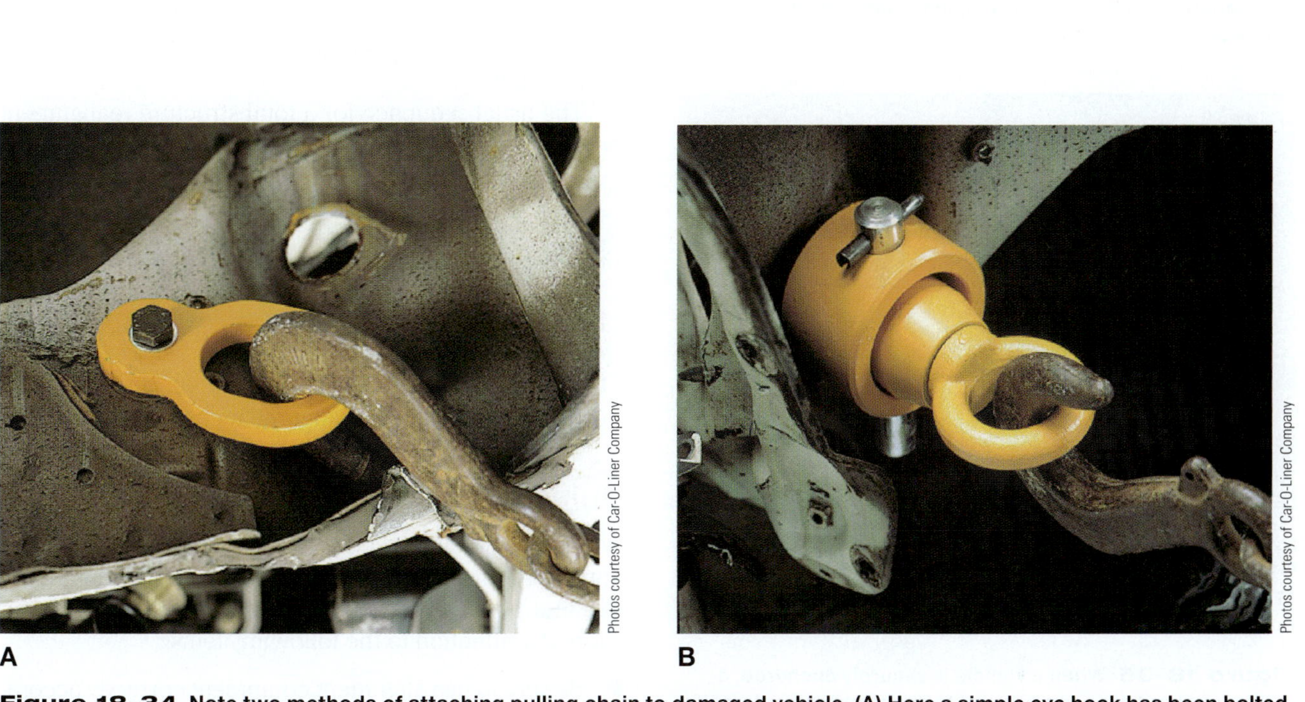

Figure 18-34 Note two methods of attaching pulling chain to damaged vehicle. (A) Here a simple eye hook has been bolted to an existing hole in the body structure to remove minor damage. (B) This pulling adapter fits over the bumper mount and can be quickly secured with a large steel pin.

A **strut plate** is an accessory that allows you to attach a pulling chain to the top of a shock tower. In frontal collisions, the shock towers are often bent out of alignment. The strut plate can be bolted directly to the holes or studs on the top of the shock tower. A chain can then be used to apply traction to the top of the tower for straightening it.

18.3 STRAIGHTENING AND REALIGNING TECHNIQUES

The body-on-frame vehicle can usually be straightened and realigned with a series of single-direction pulls. Single, hard pulls in one direction are fairly effective for straightening full-frame vehicles. Overpulling or tearing metal was not a very big issue when frame metal was $\frac{1}{8}$–$\frac{1}{4}$-inch (3.2–6.3 mm) thick. However, this is seldom the case when repairing modern cars, especially in unibody construction with its thin, 24-gauge steel (Figure 18–35).

A Carefully watch and listen as you pull damage. Watch for the tearing of sheet metal or popping sounds of breaking spot welds.

B As you press on the foot pedal, hydraulic force will be applied to the pulling chain.

Figure 18–35 When a vehicle is securely anchored, a pulling post or tower can be used to apply corrective force to damaged areas through pulling chains. You must constantly measure to determine the pulling direction and force needed.

Courtesy of Chief Automotive Technologies

Figure 18–36 Here multiple pull chains are used to pull front impact damage. Note how collars on the pull towers are used to adjust the height of the pulling chains.

Remember that a unibody vehicle is a more complex structure and has a greater tendency to spread collision forces. Most unibody repairs demand multiple pulls, which sometimes means four or more pulling points and directions during a single straightening and alignment setup.

The straightening equipment must provide clamps that will prevent further damage to the unibody structure during the pull. A single, hard pull in one direction on a unibody vehicle will usually tear the metal or crack spot welds before the area is pulled straight (Figure 18–36).

Figure 18–37 shows portable straightening equipment being used for multiple pulls to realign the rear of a unibody vehicle.

The usual sequence for a total structure realignment procedure is:

1. Understand the safety considerations of the alignment equipment.
2. Analyze damage to determine which areas require unibody/frame straightening.
3. Anchor the vehicle.
4. Attach and locate pulling chains.
5. Execute the planned pulling sequences with additional clamping and alignment checks.

Unibody/Frame Realignment Safety

When using aligning equipment, inadequate attention to any procedure can result in vehicle or equipment damage and possible physical injury to you or others in the shop. Pay attention to the following items:

- Be sure to use alignment equipment correctly according to the instruction manual prepared by the manufacturer.

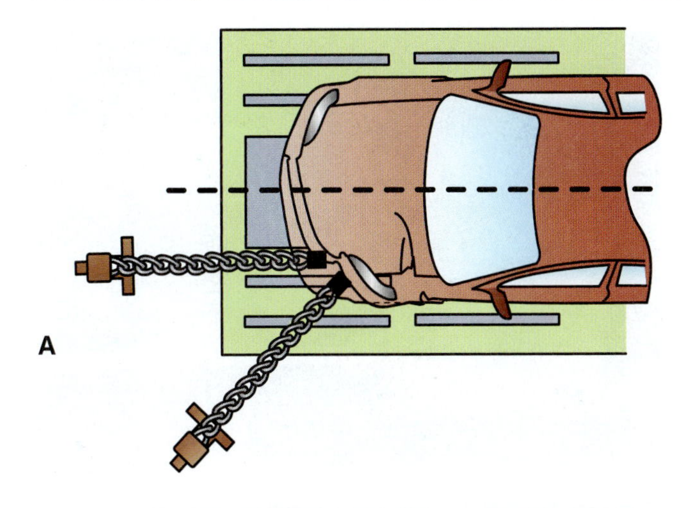

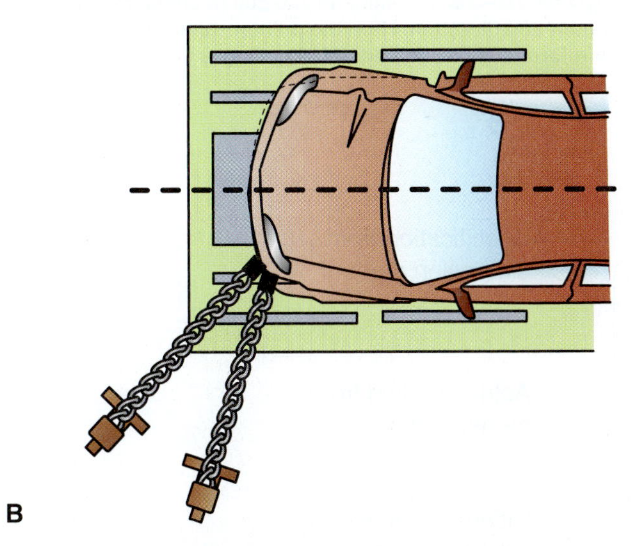

Figure 18–37 **Here multiple pulls are used to remove rear damage. (A) Note how chains are attached to pull opposite the force of impact. Damage is exaggerated. (B) Because the rear end has been pushed sideways during collision, two pulling chains are being used to force the rear of the vehicle back into alignment. Again, damage is exaggerated for clarity.**

- Never allow unskilled or improperly trained personnel to operate aligning equipment without supervision.
- Make sure the rocker panel pinch welds and chassis clamp teeth are tight. As you pull, check the clamps to make sure they are not slipping.
- Always anchor the vehicle securely before making a pull. Check that the chassis clamps and anchor bolts are tightened.
- Always use the size and grade (alloy) chain recommended for pulling and anchoring. Use only the chain and bolt grades supplied with the aligning equipment.
- Drawing chains must be securely attached to the vehicle and/or anchoring locations so that they will not come off during the pulling operation. Avoid placing chains around sharp corners.
- Before powerful side pulls are executed, apply counter supports to prevent the vehicle from being pulled off the straightening equipment.
- Never use a service jack for supporting the vehicle while working on or under it.
- Always use car stands for supporting the vehicle. Use only those stands recommended for the aligning equipment.
- A pull clamp can slip and cause a sheet metal tear. Prevent injuries and material damage by always using safety wires. Watch them closely while pulling.
- Never stand in line with a chain or clamp. Chain breakage, clamp slippage, or sheet metal tearing can cause injury or damage. Remember, it can be dangerous to work inside the vehicle at the same time pulls are being made outside.
- Cover pulling chains with a heavy blanket or attach a safety chain. If a chain breaks, this will keep the chain from being thrown across the shop.
- Wear leather gloves to prevent hand injuries.

Before doing any pulling work, protect the vehicle body and externally attached parts as follows:

- When welding or plasma arc cutting, cover glass, seats, instruments, and carpet with a heat-resistant material. Also, disconnect the battery and computer modules to protect them from welding or cutting current.
- After the initial pull, recheck the clamps. They may have rotated slightly with the initial pull and sometimes will bite into the metal, leaving some slack in the clamp tightness.
- When removing external parts (moldings and trim) attached to the body, apply cloth or protection tape to the body to prevent scratching.
- If the painted surface on an undamaged part is accidentally scratched, make a note to repair it. Even a small flaw in the painted surface may lead to corrosion and affect customer satisfaction.

18.4 MEASURING WHEN PULLING

Vehicle measurement is necessary before, during, and after pulling. Measurement of vehicle damage helps you determine how to pull. Measurement while pulling helps you check that you are properly pulling the damage out. Measurement after pulling enables you to check your work.

Measurements can be done with tape measures, tram bars, and electronic measuring systems. The industry trend is toward computerized laser or sonic measuring systems that give "real-time feedback" of the pulling operation. For the unibody/frame repair, the measuring equipment must show the amount and direction of misalignment.

You may have to make some general measurements with a tape measure or tram gauge. These include diagonal measurements to check for diamond damage and length measurements to check for mash damage. Try to get as good an idea as possible of where the damage begins and ends. Use all the dimension data available, including body/frame dimension books and vehicle manufacturers' manuals, or by checking against an undamaged car (Figure 18–38).

Figure 18–39 reviews how to use a tram bar to document vehicle damage. For more information on how to measure vehicle damage, refer back to Chapter 17.

If you are using a computerized measuring system, pull up the dimension data for the vehicle being repaired. Hang targets for three or four base reference points and for measurement points on the damaged section of the vehicle (Figure 18–40).

Most computerized measuring systems help you find reference points and automatically download known good dimensions for comparison to vehicle damage measurements. Figure 18–41 shows a fixture-type measuring system.

Figure 18–38 Measure as you pull to check the pulling amount and direction. Here the pointer does not quite fit into the reference hole, so additional pulling is needed.

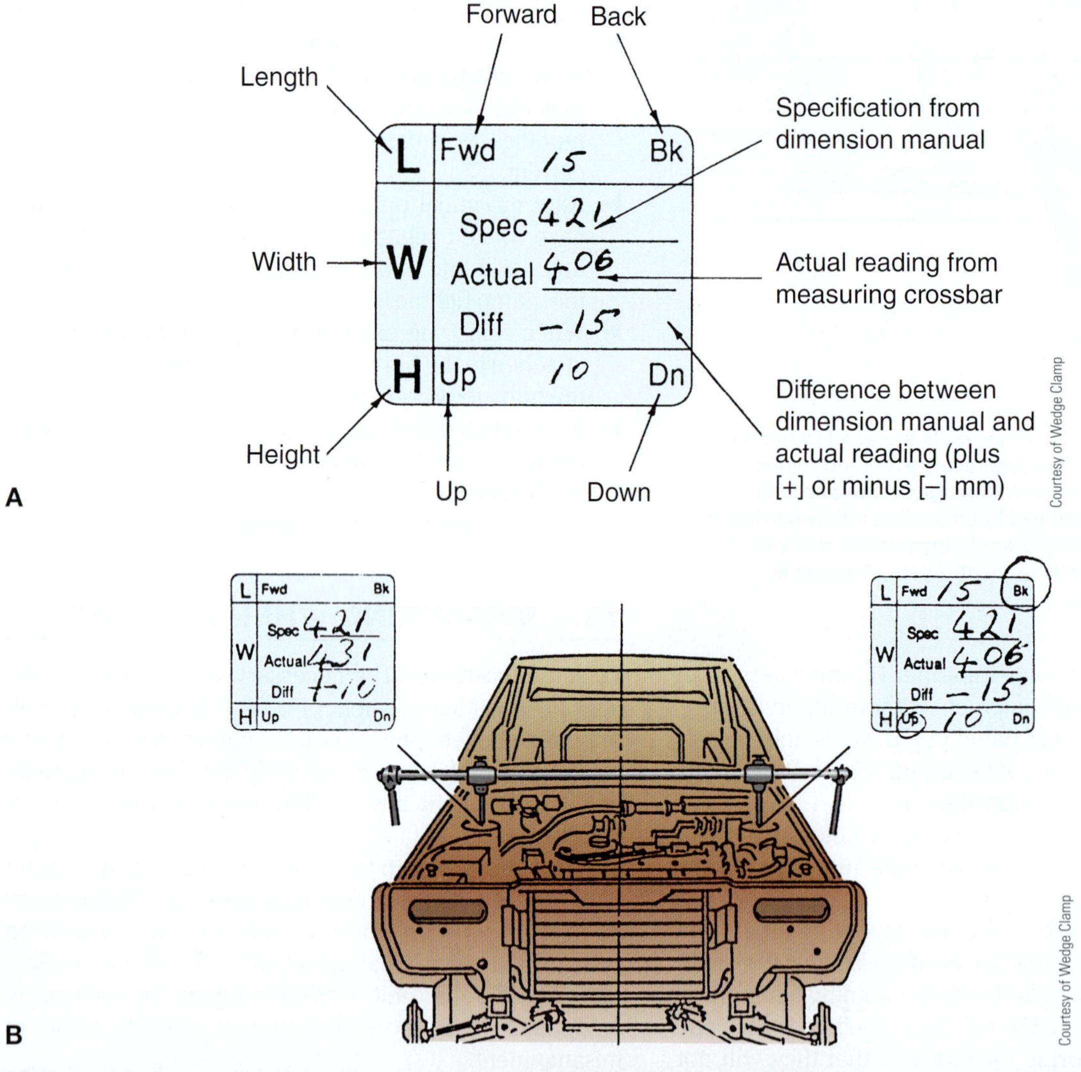

Figure 18–39 If measuring with a tram gauge, write down your measurements and compare them to specifications. (A) A form such as this one can be used to keep track of the pulling sequence. (B) Sample measurements of strut towers show that the driver's side varies in width by 15 mm toward the centerline. This point is also pushed back 15 mm and up 10 mm. The passenger side width also varies from the dimension manual by 10 mm. You would have to pull out, down, and sideways to make these measurements match dimension manual specs.

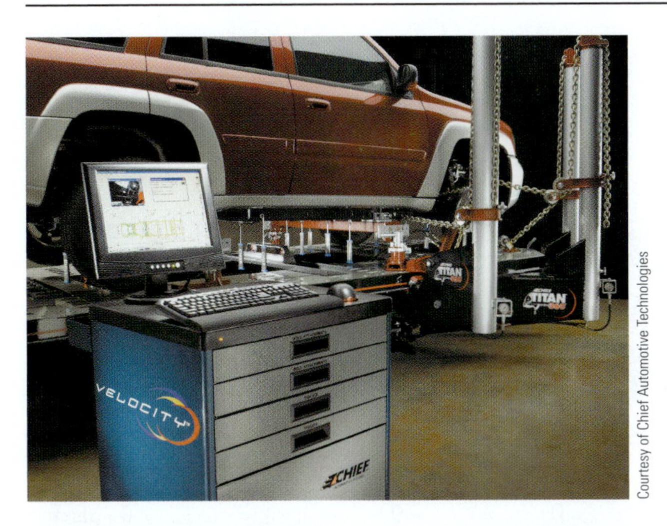

Figure 18–40 Computerized measuring systems greatly simplify the unibody/frame straightening operation. Monitor can be watched to observer results of pulling on frame rack in real time.

Figure 18–41 This dedicated bench is being used to check alignment during pull on the unibody. Fixtures should slide up and align with reference points when the frame is straightened.

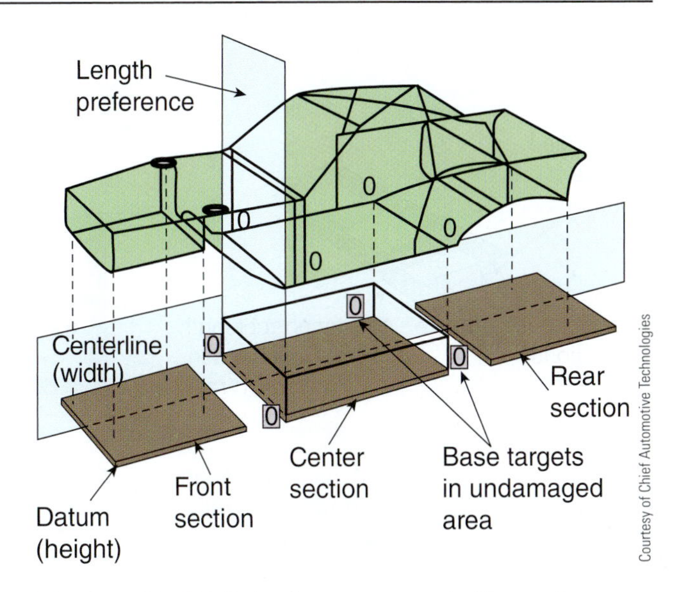

Figure 18–42 To begin measurement of frame damage with a computerized system, first establish your base. Targets in the undamaged center section of the vehicle are normally used to establish datum plane (height), centerline (width), and body zero (length) dimensions.

Ideally, if the center section of the vehicle is not damaged, you should have four reference points on which to base your measurements of unibody/frame damage. It is necessary to have at least *three reference points* on the undamaged part of the vehicle that can be used to set the vehicle up properly. These three locations form the datum plane on which all of the other measurements are based. If there are more than three locations that are undamaged, they can also be used for setup. Refer to Figure 18–42.

For situations in which there are *not* three undamaged reference points, such as a severe side impact, it might be necessary to do some rough straightening of the underbody until three points can be secured. You can also use undamaged rear or front section reference points, if needed, to set up your base.

If you are measuring frame or unibody damage with a ram gauge, you should write down your measurements on a copy of the dimension sheet for that vehicle. This will let you compare factory dimensions and your measurements of damage.

Shortened Frame Rail Damage

With computerized measuring systems, the workstation computer monitor will normally illustrate the extent and direction of damage. For example, in Figure 18–43 note how the computer diagram shows a collapsed frame rail.

All measurements in Figure 18–43 except the left front show little or no damage. The left front frame rail has been pushed straight back, indicated by the length dimensions not being zero. The arrows show the direction of damage. The numbers show the extent of damage in millimeters. Some computerized measuring systems also show English measurements as well.

Basically, to straighten a shorted frame rail, attach your clamp and chain to the rail. Adjust the pulling collar height to pull straight out on the crushed rail. This will stretch the rail back out. Because the metal will tend to flex back, use a slight overpull of a few millimeters. Then, release traction and check your measurements.

Measurement tolerance is a 2- to 3-mm difference between the spec dimensions and your actual measurement. A small tolerance is allowed by most auto manufacturers.

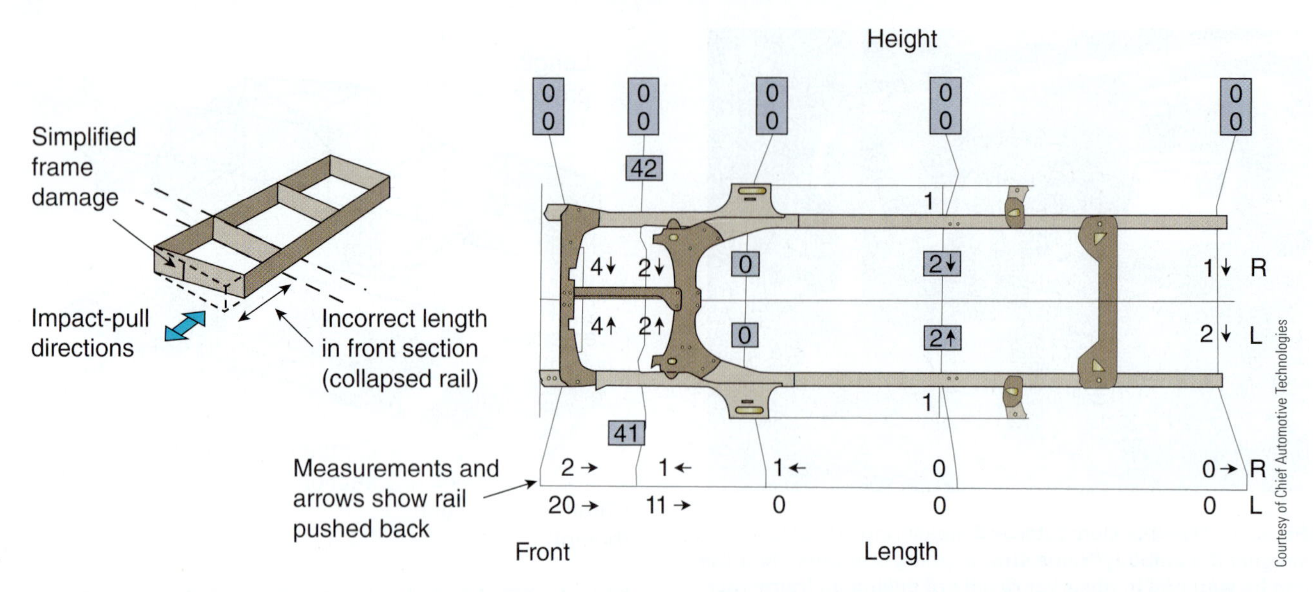

Figure 18-43 Compare this simple drawing of a collapsed rail with readings that would be given on a typical computerized laser measuring system. Arrows on the dimension drawing show the direction of the damage. Numbers equal the amount of damage in millimeters. The top numbers are for the right side of the vehicle and the lower numbers are for the left side.

Frame Rail Pushed Back and Up

Figure 18–44 shows a computerized measuring system screen for a front frame rail that is crushed backward and upward. Note that the length and height measurements are not correct. The left front rail's height dimensions show misalignment. The length dimensions for the left rail are also off by several millimeters.

To pull this damage, the pulling chains and collars would have to be located slightly below and in front of the rail. This would let you pull forward and down on the

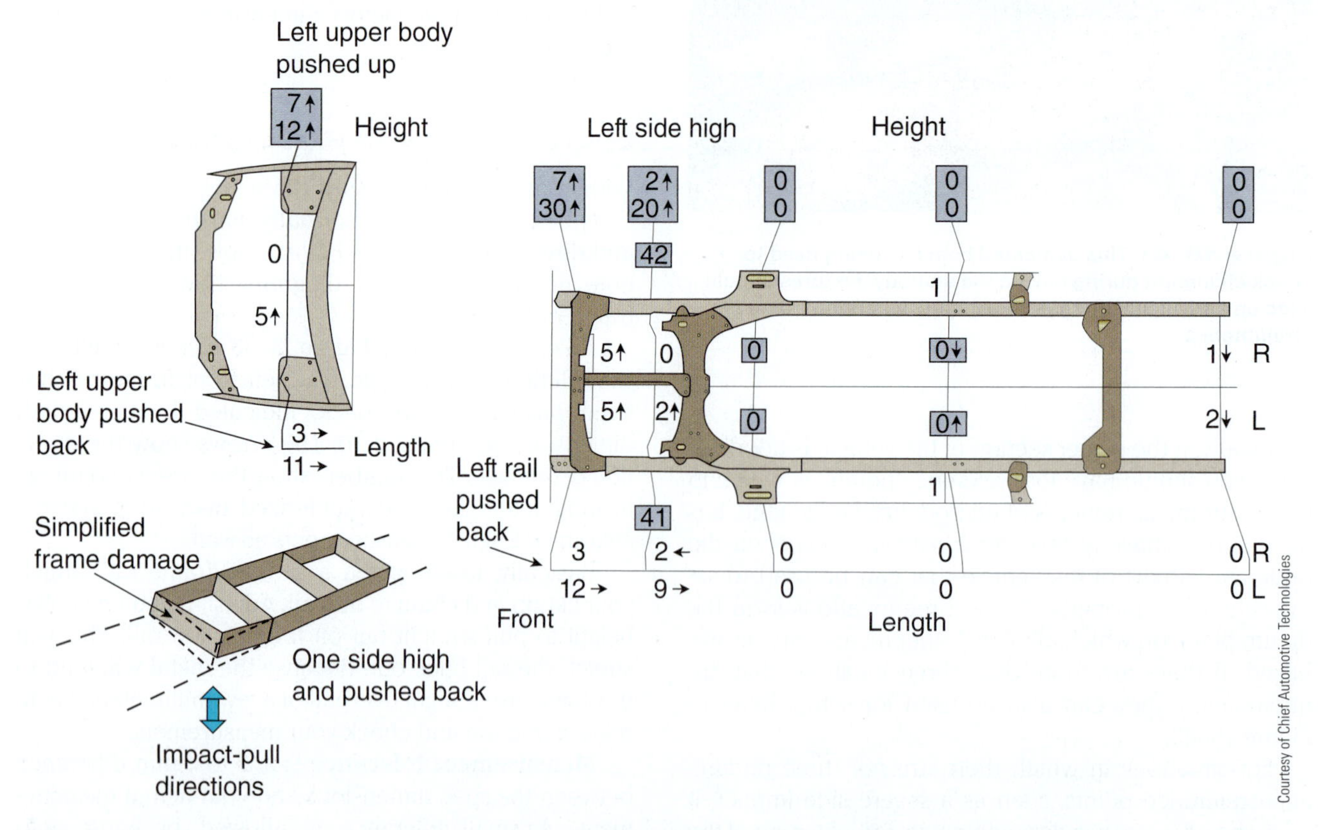

Figure 18-44 A computer monitor diagram shows how the left front frame rail has been pushed back and up. The pulling direction would be opposite to the directional arrows. Ideally, pull until all numbers read zero. With most systems, numbers out of spec appear in red and values within spec change to blue.

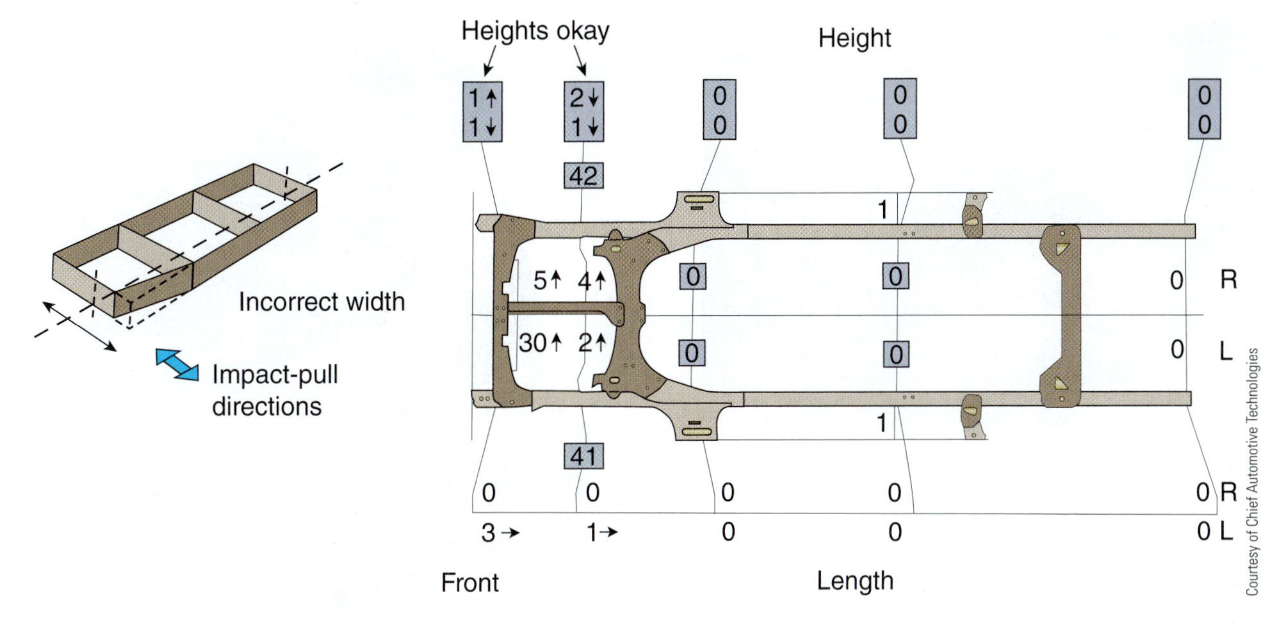

Figure 18–45 Here a side hit has pushed one side of a vehicle sideways. Compare the basic damage to the numbers and arrows on the computer monitor dimension diagram.

rail. As you pull, monitor the readings so that all of them move to zero almost equally.

One Rail Pushed Sideways

Figure 18–45 shows what your electronic measurements might look like if only one rail has been pushed sideways from a light side hit. Note how the width dimensions are off on both sides, but primarily on the left. The right rail has not been moved sideways or shortened. You would have to set up your frame rack for a side pull.

Front Section Pushed Sideways

Figure 18–46 shows sideways misalignment from a much harder side hit. After being hit hard on the left front, damage was transferred to the right frame rail. Note the width and length misalignments denoting the direction and extent of damage. Again, a multiple chain pull would be needed.

Part Removal When Pulling

As a general rule, only remove the parts that prevent you from getting to the area of the vehicle being repaired.

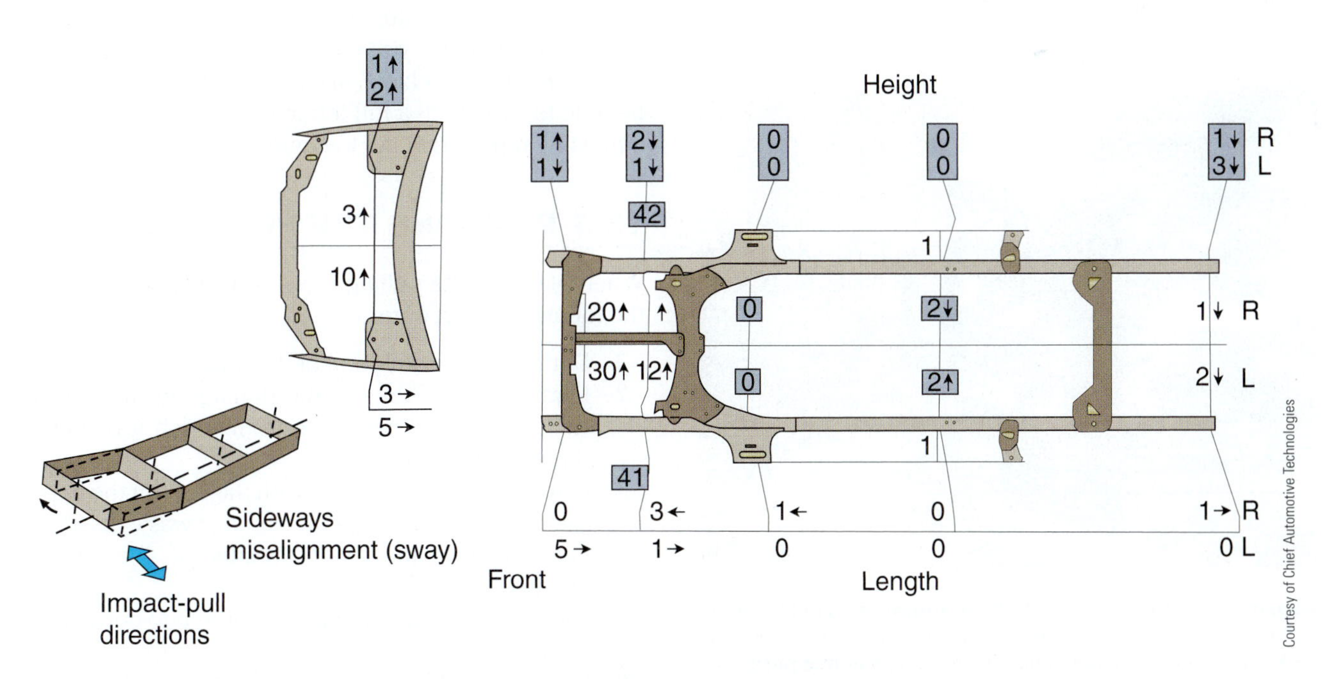

Figure 18–46 Here is an example of sway damage. The entire front section has been pushed to one side by impact forces. Note that width dimensions are way off, so most of the pull would be from the side.

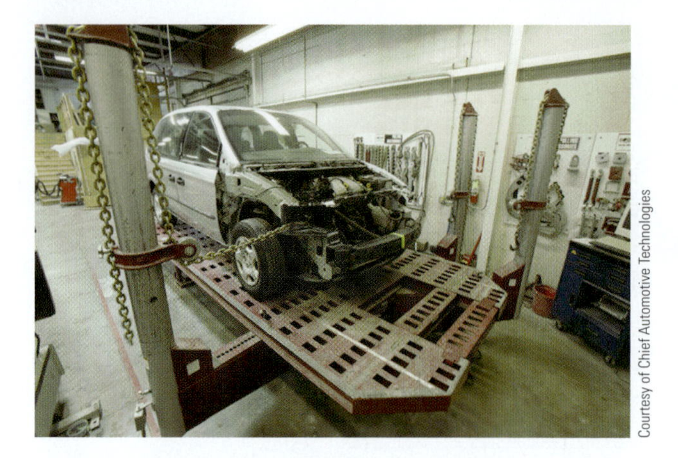
Courtesy of Chief Automotive Technologies

Figure 18–47 Bolt-on parts often have to be removed before you can properly pull out vehicle damage.

For example, you will often have to remove the fenders to access pulling points on the frame rails after a frontal collision (Figure 18–47).

Depending on the construction of the vehicle and the location and degree of damage, there will be cases in which it is more convenient to remove parts before proceeding with the repair. Carefully analyze the vehicle and the damage to determine what must be removed. It is sometimes best to remove parts before putting the vehicle on the rack. You might have better access to the fasteners.

Do not remove damaged structural parts that can be used for pulling. In many situations, you will want to pull on damaged welded-on parts even if they are going to be replaced. By pulling the badly damaged part, you can often help align other structural parts that are not going to be replaced. If you remove the damaged structural part first, it will be more difficult to pull and realign adjoining parts or panels. See Figure 18–48.

Figure 18–48 Do not cut off structural parts until you have finished pulling other parts into alignment. You can pull on badly crushed parts to help pull other repairable parts into alignment. Then you can cut off old parts or panels and weld in new ones.

Photos courtesy of Car-O-Liner Company

Figure 18–49 This pulling equipment is ideal for single pulls of minor frame or unibody damage. The vehicle, anchoring framework, and pulling tower can be raised off the ground so work is easier.

At one time, you often had to remove the suspension and driveline completely from a unibody vehicle before putting it on a straightening machine. With most of the current straightening systems and such accessories as the engine holder, this is no longer necessary. Major straightening operations can be done with *most* major mechanical parts intact (Figure 18–49).

Take the time to carefully study the locations of engine, transmission mounts, and suspension mounts and to determine whether these parts themselves are damaged.

By analyzing the direction and extent of damage from your measurements, you can determine what should be done to repair the vehicle. In extreme cases, you may not be able to straighten a full frame or a welded panel, and it must be replaced. Refer to Figure 18–50.

18.5 PLANNING THE PULL

When planning the pulling process, you should observe the following procedure:

1. Determine the direction of the pulls.
2. Find out how to repair the damage in the reverse (first-in, last-out) sequence from which it occurred during the collision.
3. Plan the pulling sequence with the pulls in the opposite direction from how the damage was caused.
4. Find correct attachment points of the pulling clamps.
5. Estimate the number of pulls required to correct the damage.
6. Determine which parts must be removed to make the pulls.

A The vehicle body has been unbolted and lifted off its chassis.

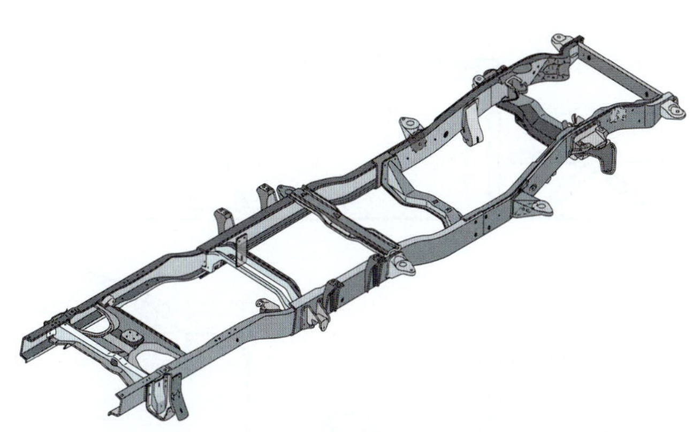

B A new perimeter frame uses fasteners to secure all other parts.

Figure 18-50 When a full frame is badly damaged, a whole new frame may have to be installed.

Many times it is best to draw out the repair plan prior to actually pulling the vehicle. This drawing should show OEM and actual dimensions, anchoring, and pulling locations.

The easiest way to determine where to pull from is to picture the damage being removed by pulling with "your bare hands." The pulling process will work in the exact same manner.

As a general rule, vehicle straightening is needed whenever the damage involves the suspension, steering, or powertrain mounting points or major damage to the center section of the vehicle. Determine whether a particular collision meets this rule either by eye, where there is obvious damage, or by making some general measurements with a tape measure or tram gauge. These would include diagonal measurements to check for diamond and length measurements to check for mash. Try to get as good an idea as possible of where the damage begins and ends. Use all the dimension data available, including body/frame dimension books, vehicle manufacturers' manuals, or measurements against an undamaged vehicle.

Once it is determined how far the damage travelled in the unibody structure and it is fully identified, the damaged area can be pulled and straightened. The corrected reference points provide a larger guide for subsequent pulls.

In planning the repair (pulling) sequence, remember the two basic guides to ensure that misalignment and damage will be corrected with minimum metalworking and without further damage:

1. Repair the damage in the reverse (first-in, last-out) sequence from which it occurred during the collision.
2. Plan the pulling sequence with the pulls in the direction opposite to the damage input.

18.6 MAKING PULLS

A single-pull setup is capable of making a simple directional pull on the damaged area of a vehicle. Single pulls are effective on primary damage to full-frame vehicles and minor damage to unibody vehicles.

Multiple-pull setups allow you to pull and hold when damage is in several directions (length, width, and height). Multiple pulls are often required to correct major damage to unibody vehicles. Multiple-pull setups provide the ability to exert a great deal of control over any pulling task. This improves the precision with which a pull can be made. Multiple-pull setups also eliminate the need for disconnecting and moving the power posts or towers.

The multiple-pull approach accomplishes these two objectives:

1. The exact desired direction of pull can easily be achieved from three or four points at one time. This gives the control needed in the repair of modern unibody construction.
2. The use of multiple-pull points reduces the amount of force required at any single point. This reduces the risk of tearing lightweight metals. Due to the design of today's vehicles, there simply is not enough strength available in any one place to transmit sufficient force to complete a repair. Again, as in the anchoring system, the pull load must be distributed through several attaching points.

Figure 18–51 shows a typical setup to straighten a hard side hit into the center section of a car.

A full-frame vehicle that has been hit hard from the side is shown in Figure 18–52. Note how multiple clamps and chains are being used to pull out the damage before part removal.

It is necessary to set the pulling clamp so that the line extending along the path of the pulling force passes through the middle of the teeth of the clamp. If this is not done, rotational force will act on the clamp to pull it off, further damaging the section (Figure 18–53).

Move the pulling towers into position. Attach the tower chain to the vehicle. Make sure the chains are not twisted. Repeat this on the other towers. If damage has pushed one area of the vehicle up and back, then you

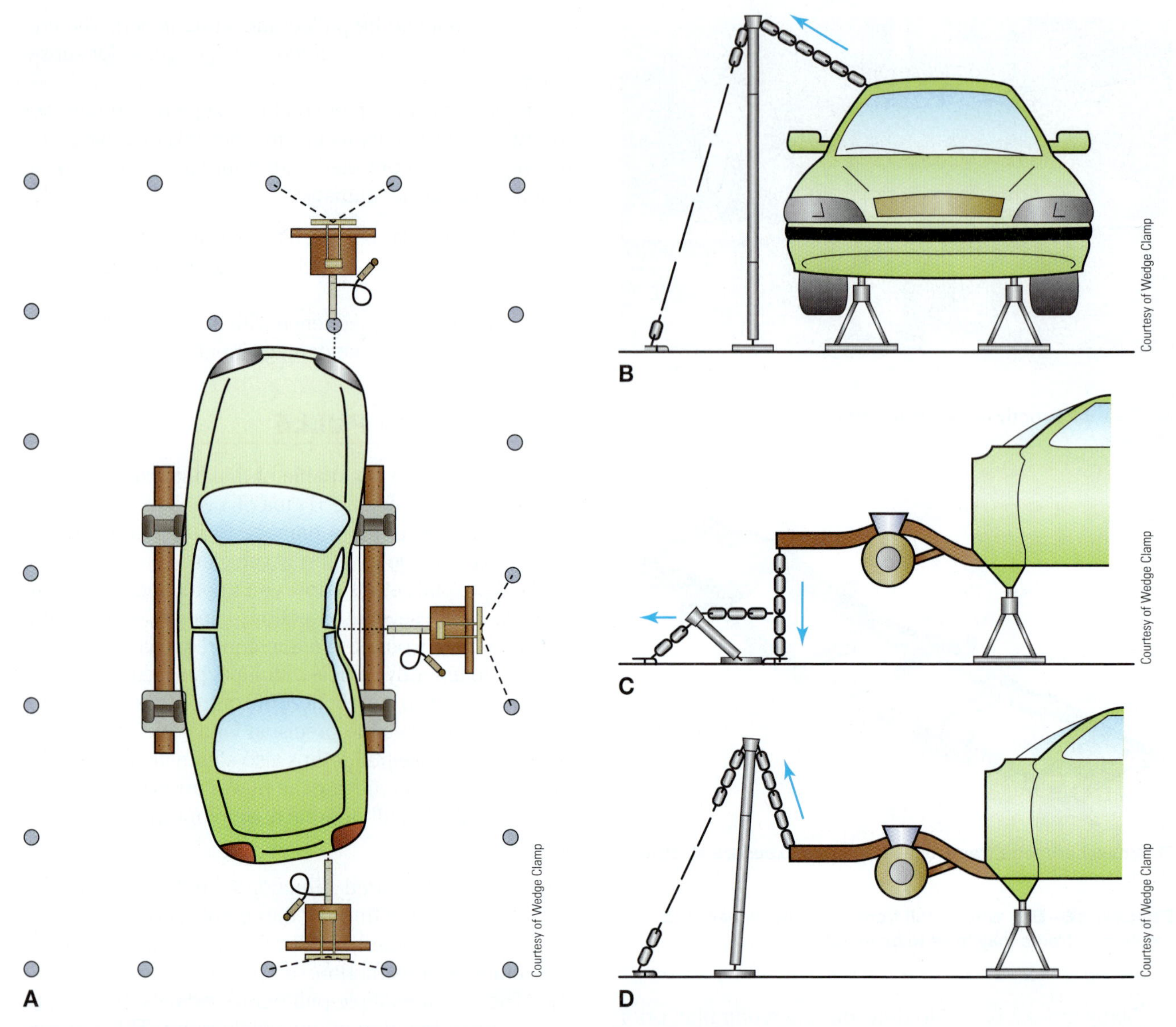

Figure 18-51 Note the examples of portable towers being used to fix damage. (A) Three towers are pulling out side impact damage. Two stretch the vehicle lengthwise while another pulls it sideways. (B) The tower is pulling up on the roof panel. (C) The tower is being used to make a downward pull on the front rail. (D) The tower is pulling up on the rail.

Figure 18-52 Straightening a full-frame vehicle is similar to straightening a unibody vehicle. However, more pulling force must be applied to the frame rails and less to the body structure.

need to position the tower and chains to pull down and out.

Collars on the towers allow you to raise or lower pulling points. Take the slack out of the chain and you are ready to pull.

When hooking up to make a pull on a unibody vehicle, consider these pointers:

- The unitized body has made multiple anchoring a must. At least four anchors are required, one on each of the body clamps. Depending on the vehicle's construction, additional anchoring might be required.
- Always look for the possibility of more than one hookup for both damage correction and restraints. Twin pulls and/or restraints allow twice the pull potential with less damage at the points of attachment.
- Use multiple hookups on structural members and on sheet metal sections to be worked. Today's metals

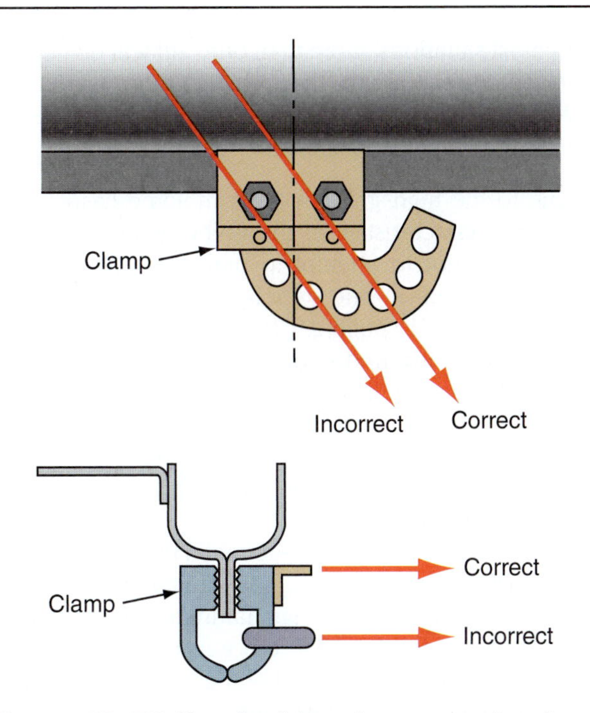

Figure 18–53 Note the right and wrong ways to set clamps.

shift, shrink, and stretch quite readily. This is why an incorrect (too localized) pull can cause more damage than it removes.

- Always install an additional security chain or chains to a substantial member on the vehicle chassis.
- Treat each damaged area as individually as possible because today's cars are manufactured for isolated collapse upon impact.
- Carefully observe the "last-in, first-out" rule in areas of primary as well as secondary damage. Occasionally this principle can be violated for initial pulls, but it nearly always holds true in the "fine-tuning" phases of unibody alignment.
- Use imagination in utilizing available clamps for multiple hookups, including the shaping of straps and other attaching devices.
- After you place a small amount of pressure on the pulling chains, the anchor clamps will be seated, and they should be retightened to prevent slippage.

18.7 EXECUTING A PULLING SEQUENCE

The progress toward alignment should be monitored during the pull. Because sheet metal has elasticity or flexibility, the body structure will, to a certain extent, flex back and return to its damaged condition even when the body has been pulled back to its prescribed dimensions. Therefore, it is important to estimate the amount of return or flex back in advance. This is why a controlled overpull is so important.

Because of the power of the rams, the metal will begin moving as soon as the chain slack is taken up. Always check your dimensions frequently to avoid overpulling too much.

Several attempts may be needed to get the damaged area to remain in the proper position. You may have to pull and release tension to see where the panel moves when tension is released. Then, repull and release to slowly move the part or panel to within specs. Each time this is done the panel will move a little closer to the desired position. Shocking the metal in an adjacent area will help relieve stress and keep the panel moving as needed.

Make the pulls a little at a time, relieve the stress, and then take a measurement. Typically, when straightening unibody/frame damage you should work from the center section outward, achieving the following sequence:

1. Length damage removal
2. Width damage removal
3. Correction of height damage

Approach the pulling operation as if you were going to do it with your bare hands. That is, determine how the metal should be moved to force it back into its original shape with your hands. How many areas could be moved at one time and in which directions? This is the key to effective pulling.

There are a number of setups when pulling or pushing. The pulling arrangement with the vector system is determined by a simple triangle. Figure 18–54 shows a triangular arrangement that will provide a fairly straight out-pull.

At no time should pulling continue if the chain between the ram and the anchor goes beyond perpendicular. Should this condition occur, the possibility of overloading a chain is increased because of the added stress placed on the anchored end of the chain. To avoid this condition, be sure that the chain lock head is not placed behind the chain anchor.

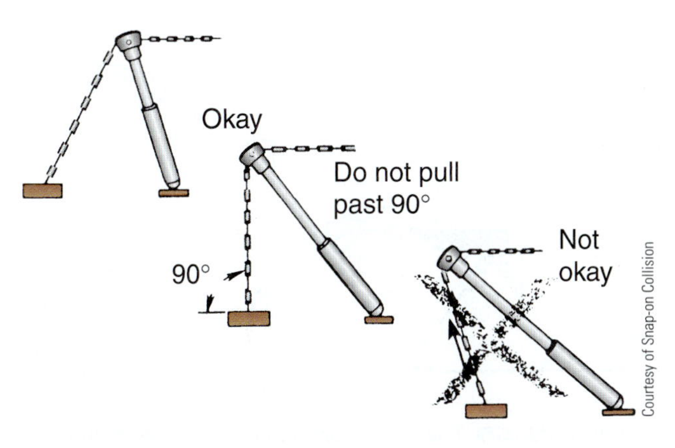

Figure 18–54 Study the right and wrong triangular arrangement for a straighter outward pull.

There are other basic single-pull setups when using a ram:

- For a high pull, more tubing is required. For an outward and down-pull, less tubing is needed. Another way to make a down-pull is to attach a chain between the vehicle and floor anchors. By pulling on the chain bridge, the structure is forced down.
- A horizontal pull on a rail can be accomplished by placing the ram at about a 45-degree angle.
- By adding tubing to the ram, a straight out-pull on the cowl can be accomplished.
- To pull straight out at the roofline, use the ram with extension tubes.
- Upward pulls are very easy to set up. In most cases, the ram is in a vertical position. This pull setup will produce an upward and slightly outward pull.
- The same type of setup can be used at roof height by adding extensions to the ram.
- Although pushing is not used to the extent it once was in damage repair, the capability to push is still important. The vector system provides push capability from any angle around the vehicle by means of a simple triangular setup.
- It is also possible to push from underneath the vehicle at whatever angle is needed. This push setup can be used to effectively remove sag at the cowl area (Figure 18–55).

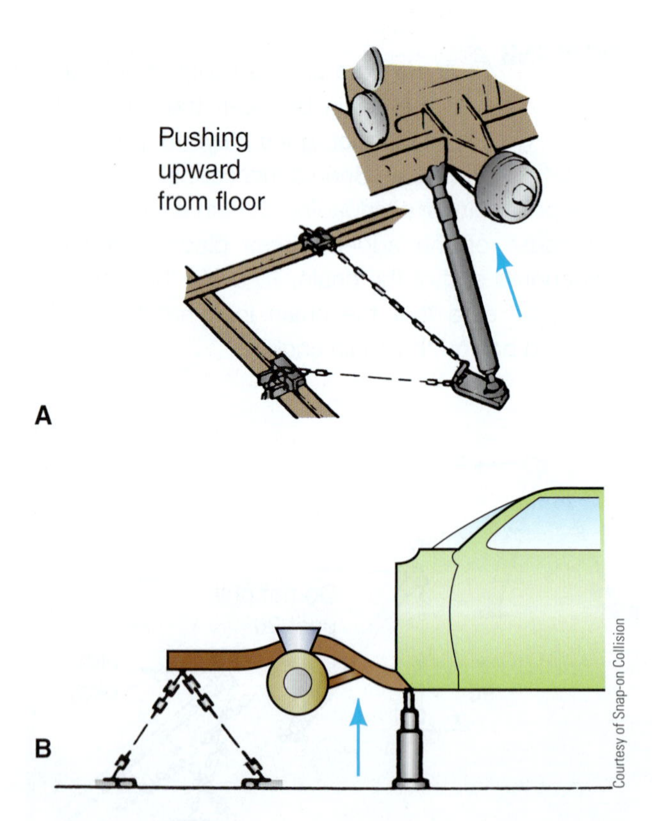

Figure 18–55 These are common pushing ram setups. (A) A ram anchored by chains will push up and sideways on the part. (B) A ram will push straight up.

- In most situations, more than one pull will be needed to effectively repair the vehicle, for a variety of reasons. Some important multiple-pull setups are discussed later in this chapter.

Due to the high-strength (and, in some cases, heat-sensitive) characteristics of the unibody structure, it is usually best not to attempt to make an alignment or straightening pull in one step. Instead, use a sequence that consists of (1) pulling, (2) hold the pull, (3) more pulling, (4) holding, and so on.

Start the hydraulics moving slowly and carefully. Watch the movement closely. Is it doing what it is supposed to do? If it is on the right track, keep going. If not, determine why and make the angle or direction adjustment, then try again.

Relieve stressed or locked-up metal by hammering as you pull out the damage. Pull the damaged metal, then loosen it up with hammering. Hammer bent or dented areas of the panel. Increase the tension and loosen it again. If you are not sure whether the metal is locked, try freeing it anyway with hammering

The repair of a bent, closed, cross-sectional structure such as a side member is done by clamping the surface of the bent-in side and pulling. The pulling direction should be such that force is applied in the direction of an imaginary straight line extending through the original position of the part. The minor dented portion of the part can often be repaired by welding on studs and pulling them with a sliding hammer.

If some of the buckles are folded so tightly that they threaten to tear, it might be necessary to use a little heat. However, use heat carefully and only on the corners and double panels. Applying heat on a low spot in the side of a frame rail or box section will only drive the damage deeper. Use heat carefully as a means to release locked-up metal, not as a means to soften up an area. Although a torch is not recommended on HSS metals, it can sometimes be used with care, as described in Chapter 8.

By bringing damaged metal back into shape slowly and carefully, a first-class, solid, and safe repair is easy. Although there will be exceptions, a general rule of thumb to follow is to achieve proper length, width, and height, in that order.

Overpulling

Overpulling is done by pulling the damage a few millimeters beyond its original dimension. If done in a controlled way, the metal will flex back slightly when tension is released. The unibody/frame reference points will then line up properly.

If overdone, however, overpulling may result in an irreversible error. Excessive overpulling is a mistake that can take days of work to correct. If the overpulled, stretched panels cannot be shrunk, new welded panels may have to be installed at your expense (Figure 18–56).

Figure 18-56 Slight overpulling is needed so that the metal can spring back to its original dimensions. If overpulling is excessive, expensive and time-consuming part replacement will often be needed.

Overpulling damage results from failing to measure accurately and often. To prevent overpull damage, measure the progress when pulling the damaged area.

Straightening Front-End Damage

The general repair method for front-end damage is best approached by examining a typical repair, in this case a moderate front hit to one side of the vehicle.

Even though the left frame rail and apron will be replaced, it is necessary to pull the badly damaged side member in the direction opposite to the damage input. This will help realign the parts that will not be replaced. Then repair the fender apron and side member on the repair side.

At the same time, repair the front fender apron and side member installation areas on the replacement side. There are many cases in which the entire fender apron or side member on the repair side is deflected left or right only. Because there is practically no warping in the lengthwise direction, repairs involve measuring the diagonal dimensions A and B, as shown in Figure 18–57.

This measurement would be compared to the frame/body manual to determine the extent of the damage.

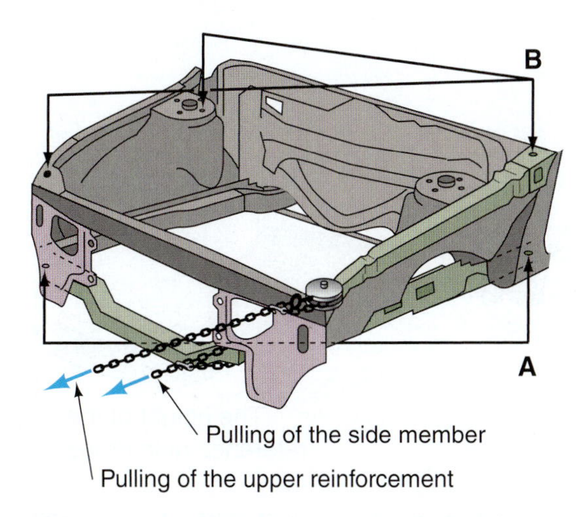

Figure 18-57 Constantly check the front dimensions as you pull out damage. Make sure the direction and amount of pull are correct.

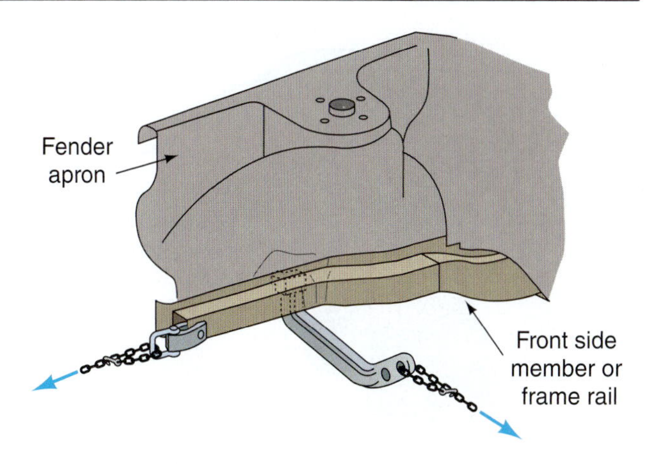

Figure 18-58 Minor front side member damage often requires pulling from the front and side at same time.

Correct this distance while keeping an eye on the repair condition. The operation can be done efficiently if the fender apron upper reinforcement is pulled at the same time as the side member.

If there is severe bending damage, it might be best to detach the front cross member and radiator upper support and repair or replace them separately. Grip the inside broken face of the side member, and while pulling it forward, pull the bent piece from the inside or push it from the outside. After repairing the bent portion, match up the dimensions to specs (Figure 18–58).

To repair the replacement side front fender apron and side member area, the main repairs are near the dash and cowl panels. If the impact was severe, the damage will extend into the front body pillar; the door will fit poorly in this case. Simply gripping the front edge of the side member of the fender apron and pulling will not repair the major damage to the front body pillar or the dash panel. In this case, cut the fender apron and side member near the installation area, clamp near the major panel damage, and pull while keeping an eye on the door fit conditions. Good results can be obtained using this method. Also, at the same time that the pillar is being pulled forward, pushing can be done from the interior side with a power ram (Figure 18–59).

During body aligning, confirm the degree of restoration by measuring critical dimensions. Reference holes in the underbody front floor and rear of the front fender installation hole are the standard reference points. Refer to Figure 18–60.

If the impact to the front side member structure is severe, there is a tendency for it to take on the shape shown in Figure 18–61.

The height of the standard measuring point might be distorted, so use caution during repairs of this nature. Further, the front side member used in vehicles has a reference point in the rear that has a tendency to be deflected upward when damaged (Figure 18–62).

To correct lateral unibody/frame damage of the front caused by side impact, you must sometimes block specific points opposite the pull direction to allow

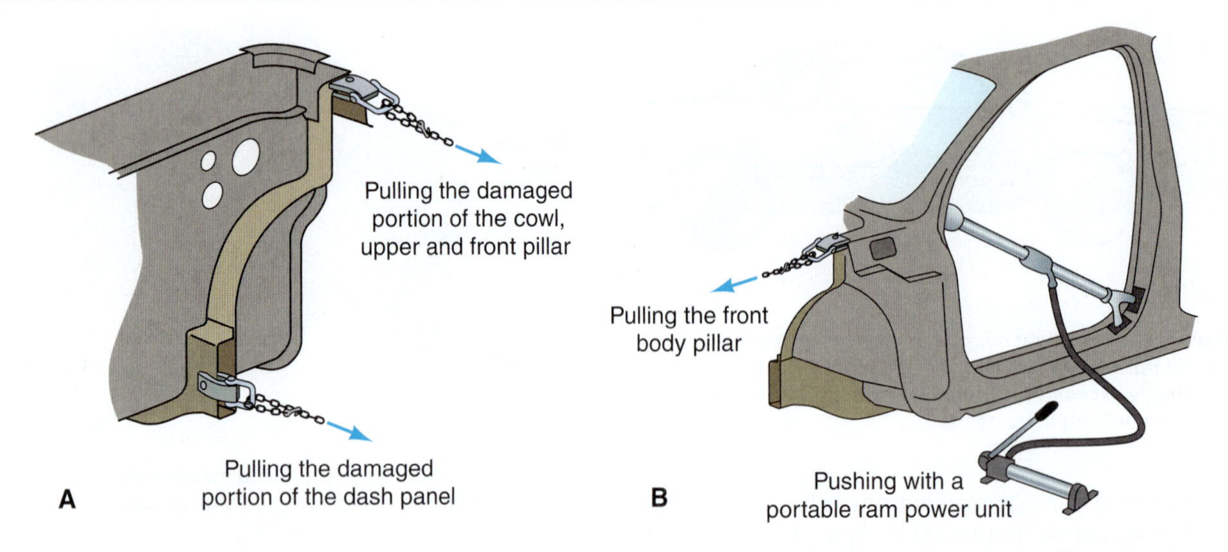

Figure 18-59 If major damage extends to the cage around the passenger compartment, this type of pulling might be needed: (A) Pulling on the cowl and rail at same time. (B) While pulling from the front, you may also need to use a portable power unit to spread the door opening.

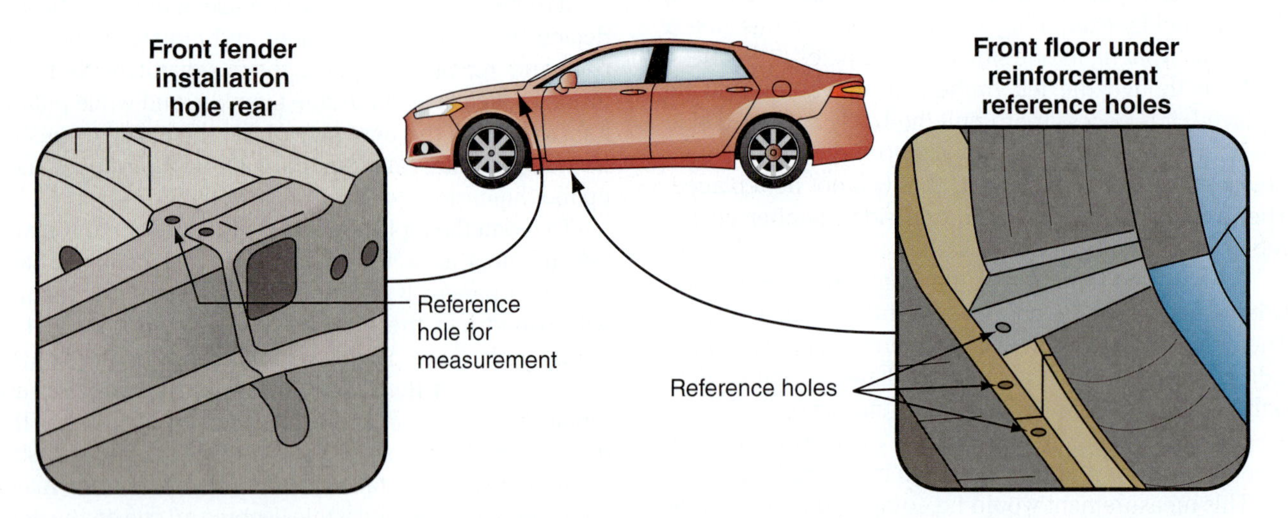

Figure 18-60 Measuring or reference points will vary from vehicle to vehicle. Refer to the specs for the exact make and model car, truck, van, or SUV being repaired to get the right dimensions or measurements.

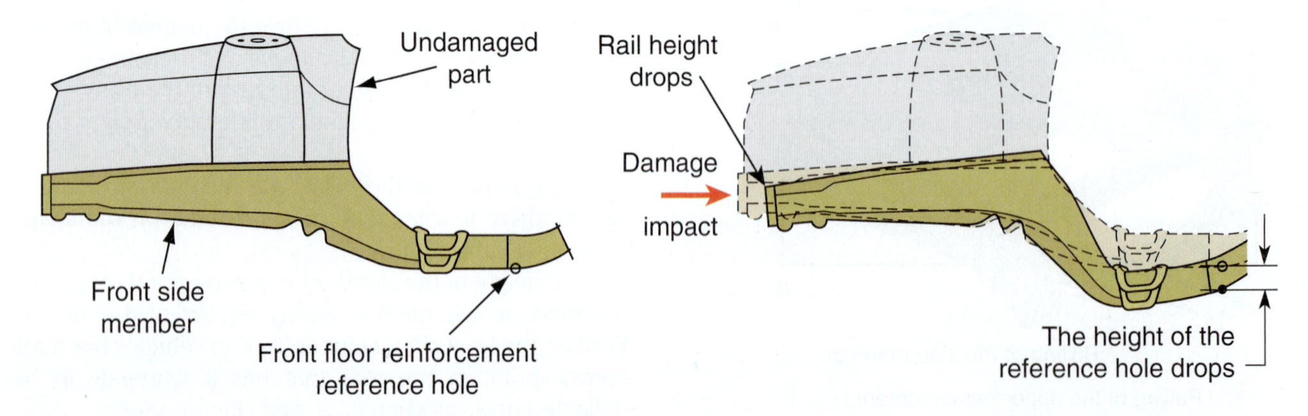

Figure 18-61 With front side member downward impact, damage may extend into the floor and affect its reference points. This is an example of a front-engine, rear-wheel-drive vehicle.

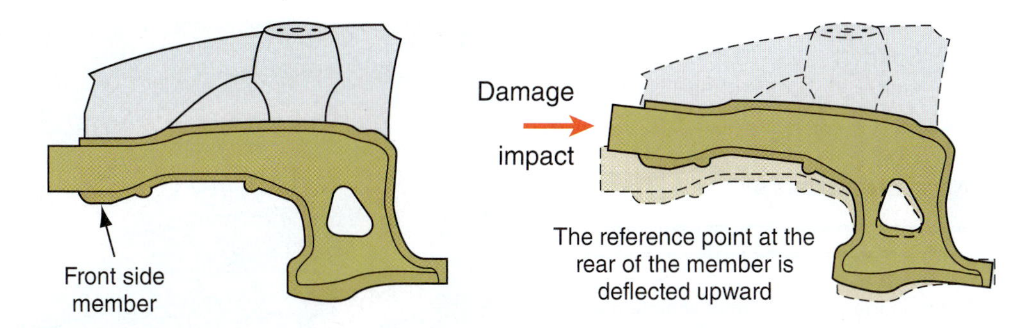

Figure 18–62 Note typical front-engine, front-wheel-drive vehicle front side member damage.

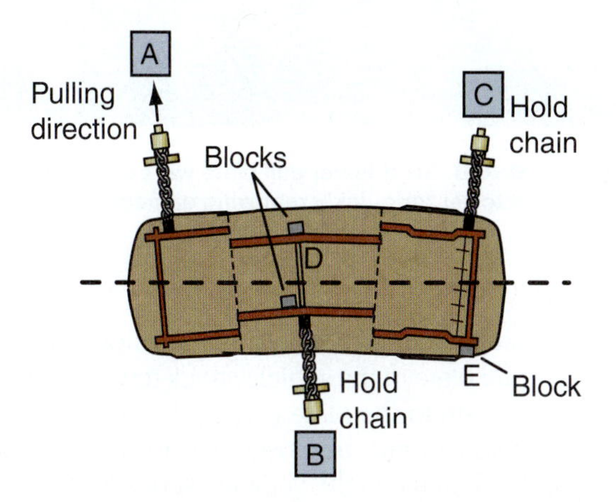

Figure 18–63 Here is a common setup for correcting lateral bending damage. B is the point of impact that pulls against A and C. Blocking devices at points D and E prevent part damage and help pull all damaged areas together.

straightening. The holding point receiving the greatest force is Point B, which must be clamped securely and also blocked to allow straightening (Figure 18–63).

Straightening Rear Damage

Because panel construction of the rear body is not as strong as it is for the front body, damage can be more complex and more extensive. The impact force will usually radiate through the ends of the rear side rails or nearby panels and cause damage to the kickup area. Next, the wheel housings will deform, causing the quarter panels to move forward, which in turn creates clearance problems between other components. If the impact is severe enough, the roof, door panels, and center body pillars will also be affected.

Attach clamps or hooks to the rear portion of the rear side member, rear floor pan, or quarter panel. Pull while measuring the dimensions of each part of the underbody. Determine the degree of repairs necessary by the conditions of panel fit and clearances.

NOTE *Do not clamp and pull a quarter panel that has little or no strain on it when there is major rear damage. When the rear side member is pushed into the wheel housing or there are clearance problems at the rear door, you should not pull on the quarter panel. Relieve the stress in the quarter panel by pulling on the side member only. If the wheel housing or the roof side inner panel is clamped and pulled along with the rear side member, the clearances with the door panel can be maintained properly.*

Straightening Side Damage

If there is a severe impact to the center of the rocker panel, the floor pan will deform and the entire body will take on a curved "banana" shape. To align this type of damage, use a method similar to straightening a piece of bent wire. The two ends of the body are pulled apart and the caved inside is pulled outward, employing three-way pulling as shown in Figure 18–64.

The portable beam and knee can be used as a side anchor with either inside or outside contact. It can be used as an added anchoring attachment for difficult pulls.

It is advisable to make an end-to-end stretch pull whenever pulling outward on the center section of a vehicle. If pulling high on the body, it will be necessary to tie the vehicle down on the opposite side. Pulling outward

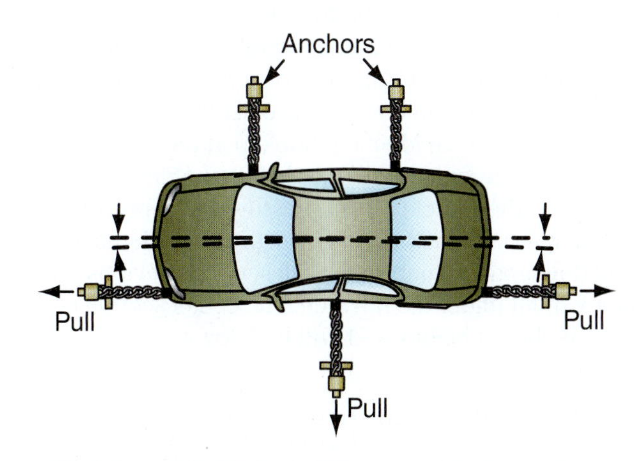

Figure 18–64 With hard side hit damage, it might be necessary to pull in three directions. You must pull and stretch the length of the unibody while pulling from the side. This will remove the "banana" shape.

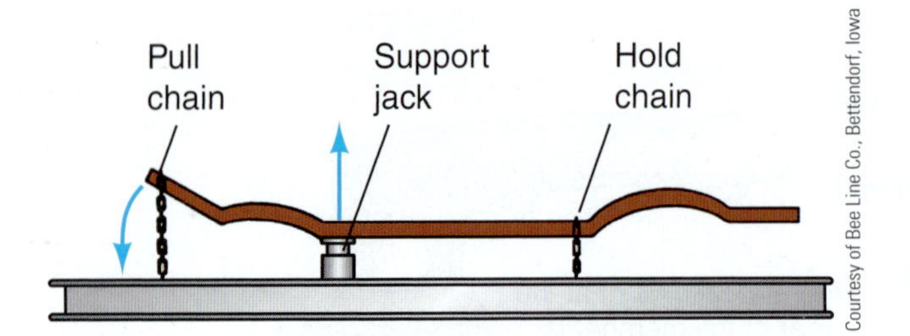

Figure 18-65 **Sag damage to a full-frame vehicle can be corrected by placing a jack under the frame while making a downward pull.**

on the center section of the vehicle can also be done with the portable beam and knee attached to the pulling tower or ram. The chain roller can be in the lowest position on the power tower.

Straightening Sag Damage

Blocking under the low area and pulling down on the high end will correct sag. The vehicle must also be tied down to the equipment at the opposite end. Anchoring the high portion of the vehicle to the rack with chains and pushing up at the low spot will also correct the datum line. When using the pulley and base for the downward pull, the tower pull chain must be in the lowest position (Figure 18–65).

Sag can also occur at the front frame cross member. The ends of the cross member will be closer than normal and the center will be too low. This condition can be corrected by using three hydraulic rams and two chains, plus an anchoring rail or pots. Check the repair with a tram gauge and compare the measurements to the specifications in the body manual or chart.

Straightening Twist Damage

Position and lock a ram or tower on the side of the platform or bench next to the low side of the vehicle. With the tower chain in the highest position, route the chain under the lower horn (on the isolator) and over the high horn. Attach a chain hook to the outside edge of the platform bed. Make an identical hookup at the opposite end of the vehicle or tie down and block under the center section of the vehicle. Apply pressure to the pull chain.

An alternate method of correcting twist damage is to pull down on the high side, as described previously, and block or lift under the low side. The center section of the vehicle should be blocked and tied down.

Straightening Diamond Damage

Place a pulling tower or ram on each end of the frame rack on opposite sides. Adjust the chain height and attach it to the vehicle as described for end pull corrections. Block or anchor one side of the vehicle to prevent side movement. Activate the pull ram.

Figure 18-66 **Strut tower pull plate will bolt directly to top of shock tower for quickly removing damage.**

Straightening Strut Tower Damage

To pull a MacPherson strut tower into position, attach the multihole pull plate to the vehicle shock tower. Then connect a pull chain to the plate (Figure 18–66). The frame rack pull chain should be positioned to traction (pull) opposite the damage. If the vehicle's shock tower is bent back and in equally, you must pull the damage forward and out at an angle from the front side of the vehicle.

If both towers are tipped left or right, they can be repositioned by mounting adapter plates to both towers and installing a strap to make the pull. After the pull is made, a dimension check should be performed using a strut measuring gauge.

Most MacPherson strut adapter plates or pull plates can also be used on frame horns, hinge mounts, or similar locations with threaded bolt holes or studs. Again, by checking or measuring critical reference points, you should be able to determine the number of pull chains needed. You can also determine the direction of pulls (Figure 18–67).

18.8 STRESS RELIEVING

Stress relieving uses hammer blows, and sometimes carefully controlled heat, to help return damaged metal to its original shape and state. It is important to note that *shape* and *state* do not necessarily mean the same thing. Something can be manipulated back into its original shape, while its original state is nowhere close. There are two separate problems in the pulling procedure:

1. Restoring the vehicle to its original shape.
2. Relieving all the stress accumulated in the metal when its shape was distorted in the accident. This is called *original state*.

To return something to its original state means to change it back to its original form. Metal has a "memory,"

Figure 18-67 By analyzing all critical reference points, you can determine multiple pull setup and methods.

or an elastic property, because it "knows" its original state. The metal will be "comfortable" once it is returned to that condition.

Unbent metal contains layers of grain or molecules, all in a relatively relaxed state. As a piece of metal is bent, these grains become slightly distorted, introducing stress. If a piece of metal is flexible enough once pressure is released, the grain will return to its original state. If the metal is bent too far, as in a collision, the grain on the outside of the bend is severely distorted by tension, while the grain of the inside is distorted by compression. Because a large amount of stress is present, the metal will remain in this shape.

Stress is defined in metallurgical terms as the internal resistance a material offers to being deformed when subjected to a specific load (force). In the collision repair industry, stress can be defined as the internal resistance a material offers to corrective techniques. This resistance or stress can be caused by:

- Original stamping of a curved part at the factory
- Deformation from impact damage
- Overheating from improper use of a torch
- Improper welding techniques when installing new panels
- Undesirable stress concentrations

Signs of stress/deformation on unibody cars are:

- Misaligned door, hood, trunk, and roof openings
- Dents and buckles in aprons and rails
- Misaligned suspension and motor mounts
- Damaged floor pans and rack and pinion mounts
- Cracked paint and undercoating
- Pulled or broken spot welds
- Split seams and seam sealer

Because the damaged area will frequently give greater resistance to alignment than adjacent areas, additional holds for clamping are sometimes required.

Hammer Stress Relieving

The use of controlled heat or a combination of heat and hammering on specific stress areas helps the metal grain relax back into its original state.

Figure 18–68 shows a technician marking, hammering, and heating a stress area while pulling out damage.

When stress relieving with a hammer, use light blows so that the hammer head bounces or springs up off the panel. This technique transfers pressure waves through the metal to relieve the stress.

Avoid hammering on the peak of a buckle when stress relieving because this can cause further stress. Instead, with pulling force applied, hammer around the point of stress so that the metal relaxes and straightens from the pulling traction. A block of wood is sometimes used when hammering out stress. It can help you access hard-to-reach areas when stress relieving during unibody/frame straightening and repair (Figure 18–69).

Once the damage has been analyzed and the angle and direction of the pull decided upon, tension is applied and spring hammering is used to relieve the stress. Spring hammering is usually done with a spoon or a block of wood to distribute the force of the blow over a large area. This releases the tension and allows the elasticity of the metal to help it return to its original size and shape. Spring hammering should also be applied to areas adjacent to the major damage.

A dolly or large wood block and hammer will work out a lot of stress. Most of the stress relieving will be "cold work." Not much heat will be used.

Figure 18-68 Stress relieving is important to repair quality. Here the technician is carefully using heat and hammer blows to remove stress while pulling.

Figure 18–69 If you overpull without stress relieving, you can have a major repair nightmare because you will damage parts that were not affected in the collision.

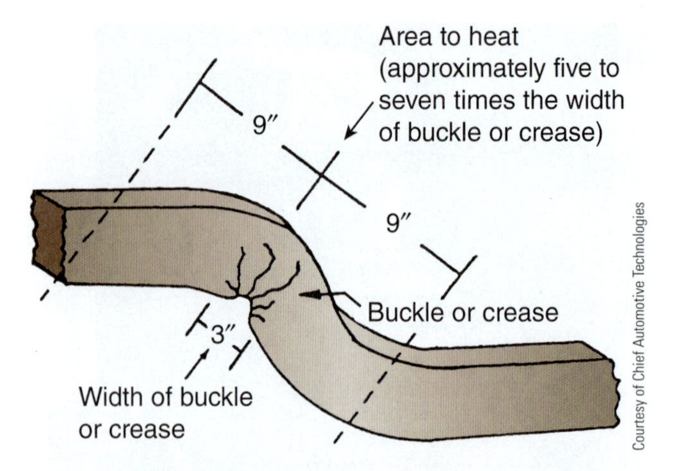

Figure 18–70 Generally, heat the area during stress relieving about five to seven times the width of the buckle. Use a heat crayon to avoid damaging the metal due to overheating.

Stress Relieving with Heat

Along with spring hammering, controlled heat can also be used to remove stress from a part or panel. Most late model vehicles have HSS panels that require special care when heating. Generally, all steel panels should be treated as HSS to avoid problems when heating.

Normally, when heat stress relieving, use the following procedure:

1. Follow manufacturer recommendations for heating metal. The type of steel or aluminum alloy will greatly affect the maximum temperatures that can be used without damaging the metal. Remember that aluminum melts at a much lower temperature than steel. Also, aluminum does not "glow red" as it starts to melt and liquify.
2. Allow the metal to cool naturally after stress relieving. Do not apply water to speed cooling because it could affect the metal's hardness. Rapid cooling can cause the metal to become hard and brittle.
3. Use heat crayons or a high-temperature noncontact thermometer to determine the exact temperature of the metal when heating. Different temperature heat crayons are available. The heat crayon mark will begin to melt when its temperature rating is reached.
4. Heat an area about 2–3 times greater than the width of the buckle or stress area. This is illustrated in Figure 18–70.
5. Do not heat UHSS parts. UHSS is often used in door guard beams and bumper reinforcements.

The best way to monitor heat application is with a *heat crayon*. Stroke or mark the cold piece with the crayon. When the stated temperature has been reached, the crayon mark will liquefy.

A quality repair can be achieved only if function, durability, and appearance have been restored. When stress is not removed, the following can occur:

- Fatigue is caused by loading and unloading of suspension and steering components.
- In the event of a second similar collision, less force is required to cause the same or greater damage, and it could endanger the occupants of the vehicle.
- The vehicle can dimensionally distort, causing handling problems.

Stress Concentrators

Stress concentrators are designed into unibody vehicles to control and absorb collision forces, minimize structural damage, and increase occupant protection. They also make damage more predictable and allow for easier detection when analyzing damage or estimating. Do not attempt to remove designed stress concentrators. Follow the car manufacturer's recommendations for straightening or replacement of parts that have been designated as stress concentrators. See Figure 18–71.

Figure 18–71 After the vehicle frame or unibody has been pulled back into alignment, you can cut off any badly damaged body panels that must be replaced with new ones.

SUMMARY

1. Vehicle straightening involves using high-powered hydraulic equipment, mechanical clamps, and chains to bring the frame or body structure back into its original shape.
2. With major damage to several panels, a multiple-pull method with several pulling directions and steps is needed.
3. Straightening equipment is used to apply tremendous force to move the frame or body structure back into alignment.
4. Anchoring equipment holds a vehicle stationary while you are pulling and measuring.
5. Hydraulic rams use oil pressure from a pump to produce a powerful linear motion.
6. It is necessary to have at least *three reference points* on the undamaged section of a vehicle that can be used for making measurements of damaged areas.
7. Straightening system accessories include the various chains, clamps, hooks, adapters, straps, and stands needed to mount various makes and models of vehicles.
8. Computerized equipment is helping to automate the measuring and straightening process.
9. When using aligning equipment, inadequate attention to any procedure can result in vehicle or equipment damage and possible physical injury.
10. Use heat carefully and as a way to release locked-up metal, not to soften an area.
11. Overpulling is done by pulling the damage a few millimeters beyond its original dimension. If done in a controlled way, the metal will flex back slightly when tension is released.
12. Stress relieving uses hammer blows, and sometimes carefully controlled heat, to help return damaged metal to its original shape and state.

EXERCISES

On a separate sheet of paper, complete the following learning activities for this chapter. Write definitions for the key terms and answer the ASE-style review questions, essay questions, critical thinking problems, and math problems. You can also do the outside activities, possibly for extra credit.

➤ Key Terms

composite force
frame machine
height damage
length damage
measurement tolerance
multiple-pull method
overpulling
pinch weld clamps
pulling
restraint bar
single-pull method
stress relieving
strut plate
tower collars
traction
vehicle anchoring
width damage

➤ ASE-Style Review Questions

1. Technician A says that in-floor systems provide fast hookup and positive anchoring without sacrificing space. Technician B says that in-floor systems cannot perform multiple pulls. Who is correct?
   A. Technician A
   B. Technician B
   C. Both A and B
   D. Neither A nor B
2. Technician A says a rack system should be used whenever the damage involves the suspension, steering, or powertrain mounting points. Technician B says that it is necessary to have at least three reference points on the undamaged part of the car that can be used to set up the vehicle properly on the frame straightening equipment. Who is correct?
   A. Technician A
   B. Technician B
   C. Both A and B
   D. Neither A nor B
3. Technician A completely removes the suspension and driveline from vehicles before putting them on frame straightening equipment. Technician B says that a single, hard pull in one direction will usually repair unibody damage. Who is correct?
   A. Technician A
   B. Technician B
   C. Both A and B
   D. Neither A nor B

4. What pulling technique may be necessary when repairing shortened frame rail damage?
   A. Side-sway removal
   B. A slight overpull of a few millimeters
   C. Height damage correction
   D. Repairs in the center section
5. When stress relieving, Technician A uses hammer blows and controlled heat. Technician B uses a plasma cutter or cutting torch to remove material from the vehicle structure. Who is correct?
   A. Technician A
   B. Technician B
   C. Both A and B
   D. Neither A nor B
6. When a vehicle has been hit from the side, ______ damage often results.
   A. tulip
   B. sway
   C. sag
   D. accordion
7. Which of the following is true?
   A. Cracked paint and undercoating is a sign of stress
   B. Most of the stress relieving will be "cold work"
   C. The best way to monitor heat applications is with a heat crayon
   D. All of the above
8. Technician A says to remove all badly damaged panels that are going to be replaced before pulling. Technician B says you should sometimes pull badly damaged components separately. Who is correct?
   A. Technician A
   B. Technician B
   C. Both A and B
   D. Neither A nor B
9. A computerized measuring system shows that a left front rail length dimension is pushed 10 mm to the rear. This vehicle was hit:
   A. From the side.
   B. At an angle.
   C. From the front.
   D. From the rear.
10. When using a computerized measuring system during frame straightening, how do you know that you are pulling out the damage properly?
   A. Directional arrows will disappear and the numbers will go to zero
   B. Directional arrows will appear and the numbers will go to zero
   C. Directional arrows will reverse and the numbers will go to zero
   D. Directional arrows will disappear and the numbers will go to infinity

Essay Questions

1. Explain the usual sequence for a total structure realignment.
2. Describe six things to remember when planning a pull.
3. List three steps that should be taken to protect the body and externally attached parts before doing any pulling work.

Critical Thinking Problems

1. A unibody car has been driven over a cement barrier, badly scraping and damaging its underbody. There are only two unbent flanges for installing clamps. What should you do?
2. During final inspection, you find a door badly out of alignment. How should you proceed?

Math Problems

1. If the manufacturer states that a 100 psi (689 kPa) reading on a pressure gauge equals 1 ton (0.9 metric ton) of pulling power, what would 63 psi (425 kPa) equal?
2. If a pulling chain is rated at 2,100 pounds (945 kg) and it should not be strained over 50 percent of its rating, what is the maximum force that should be applied to the chain?

Activities

1. Visit a body shop. Observe an experienced technician using unibody/frame straightening equipment. Report on your visit.
2. Read the operating manuals for different types of straightening equipment. Compare their differences and similarities.
3. Inspect several badly damaged vehicles. Write a report on what would have to be done to repair one of them.

Photo Summary

Unibody/Frame Realignment

Figure P18-1 Remove any parts that are in the way of clamping and pulling operations.

Figure P18-2 Anchor the vehicle so it will not move when pulling force is applied. Pinch weld clamps are often tightened down around the lower rocker panel flange to secure the vehicle.

Figure P18-3 Attach pulling chains at the locations of damage. Generally, you want to pull out damage in the reverse of the order in which it was formed by the collision impact forces.

Figure P18-4 Stand to one side of the chains as force is applied. Do not exceed the recommended equipment pressure.

Figure P18-5 Stop periodically to measure progress. Remove the pulling force and measure the area being pulled. Keep pulling until all reference points are within specifications.

CHAPTER

19 Welded Panel Replacement

OBJECTIVES

After reading this chapter, you should be able to:

- List the parts and panels of a vehicle considered to be structural.
- Use the information in a vehicle dimension manual to properly replace welded body panels.
- List the steps necessary for replacing a body panel along factory seams.
- Describe how factory spot welds are separated.
- Explain how new body panels should be positioned on a vehicle.
- List the steps for welding new body panels in place.
- Describe how to install foam panel fillers.
- Section rails, rocker panels, pillars, floor pans, and trunk floors.
- Answer ASE-style review questions relating to panel replacement.

INTRODUCTION

Structural panel replacement involves cutting, measuring, and welding a new body panel in place of a badly damaged one. With quarter panels, frame rails, and other welded body assemblies, you will have to find all factory welds. These welds must be drilled or ground out with power tools for removal of the damaged panel. Then, the new panel must be fitted in place, measured, and welded to the vehicle. This process takes considerable skill.

A collision-damaged vehicle can require a variety of repair operations. Repair steps will depend on the nature and location of the damage. Panels with minor damage can often be straightened and filled with plastic. Bent structural panels may have to be pulled and realigned using hydraulic equipment. Some panels, however, may be so badly damaged that replacement is the only practical and effective procedure for cost-effective repair.

Table 19–1 outlines the general procedure for replacing both bolted and welded panels.

Previous chapters explained how to repair minor panel damage and how to replace fastened or bolted panels. This chapter summarizes how to replace the many welded-on structural panels of modern vehicles. See Figure 19–1.

19.1 WELDED PANELS

Before you begin any repair process, refer to the estimate for guidance on how to proceed. The estimator will have determined which parts need to be repaired and which should be replaced. You should use this information and shop manuals to efficiently remove and replace parts.

The estimate or repair order is an important reference tool for doing repairs, and it must be followed. The insurance company and estimator have both agreed on which parts must be replaced or repaired. If you fail to follow the estimate or repair order, the insurance company may not pay for your work.

The estimate is also used to order new parts. You might want to make sure all ordered parts have arrived. Compare new parts on hand with the parts list. If anything is missing, have the shop's parts person order them. This will save time and prevent your work area or stall from being tied up while you wait for parts.

In general, to replace a welded-on structural panel you should use the following procedure:

1. Remove fastened parts that prohibit repairs to structural parts.
2. Measure the amount and direction of structural damage.

TABLE 19–1 TYPICAL PANEL REPLACEMENT PROCEDURE

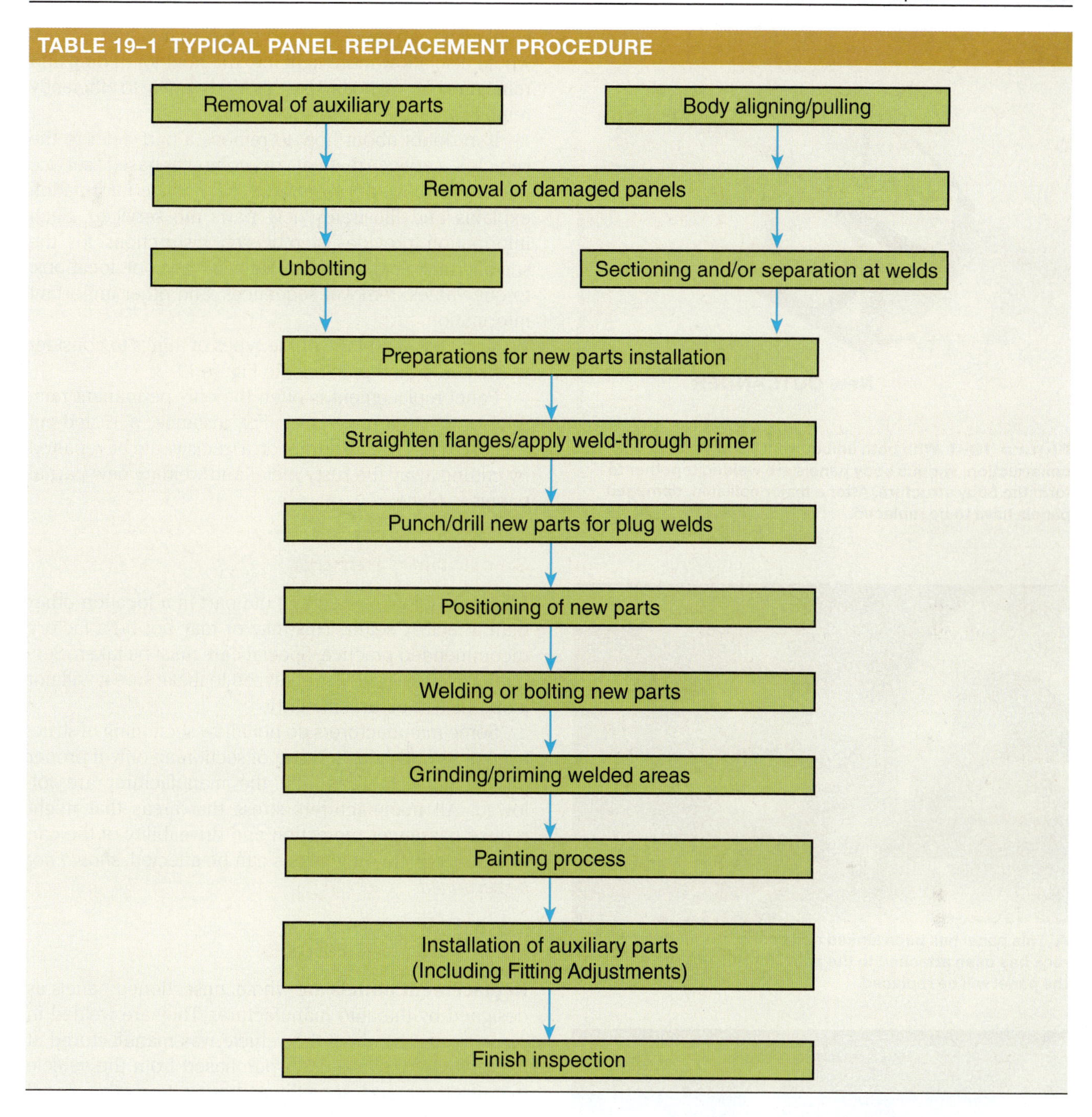

3. Use a frame rack to pull, realign, and straighten repairable panels (Figure 19–2).
4. Find and clean off flanges to find spot welds on panels to be replaced.
5. Grind or drill out spot welds holding panels on the vehicle.
6. Separate flanges and remove damaged panels.
7. Clean, grind, and straighten flanges on all panels to be welded.
8. Look up recommended weld types, counts, and locations in applicable service literature.
9. Drill or punch holes in new panels for plug welds.
10. Coat panel flanges with weld-through primer to prevent corrosion.
11. Fit and secure new panels to the vehicle with locking pliers or screws.
12. Measure the alignment of new panels.
13. Tack weld new panels and recheck alignment of adjacent panels.
14. Final weld new panels to the vehicle.
15. Apply sealer and anticorrosion materials to new panels as needed.

Starting Structural Repairs

Generally, you start by removing large, external, bolt-on parts that are badly damaged. For example, if the front end was hit hard, you might remove the hood first. This

New OUTLANDER
Courtesy of Mitsubishi Motors North America, Inc.

Figure 19-1 With both unibody and full-frame construction, various body panels are welded together to form the body structure. After a major collision, damaged panels have to be replaced.

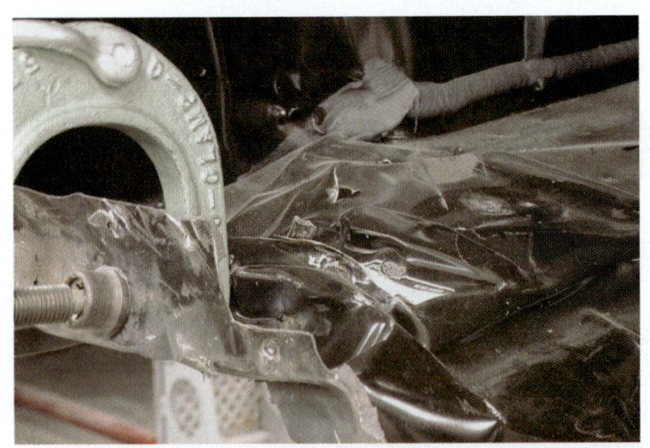

A This panel has been kinked and badly damaged. A frame rack has been attached to the panel for pulling, even though the panel will be replaced.

B Hammer blows with pulling traction will help relieve stress and aid in pulling other panels into alignment.

Figure 19-2 Before removing any badly damaged welded-on body panels, pull out as much damage as possible to realign adjacent panels that will be reused.

would give you more room to access rear fender bolts. It would also allow more light into the front for finding and removing hidden bolts. Use this kind of logic to efficiently remove parts.

If in doubt about how to remove a part, refer to the vehicle's service manual or computer-based service information. Factory or aftermarket service information explains and illustrates how parts are serviced. Such information provides step-by-step instructions for the specific make and model vehicle. It gives bolt locations, torque values, removal sequences, and other important information.

To give you an idea of the types of things to consider when starting a repair, refer to Figure 19–3.

Panel replacement is often the only permanent remedy for corrosion damage. For instance, a rusted-out rocker panel and cab corner on a truck would be repaired by cutting away the rusty metal and welding new partial panels in place.

Sectioning Panels

Sectioning involves cutting the part in a location other than a factory seam. This may or may not be a factory recommended practice. Special care must be taken. Sectioning a part should be analyzed to make sure it will not jeopardize structural integrity.

Some manufacturers do not allow sectioning of structural panels. Others approve of sectioning only if proper procedures established by the manufacturer are followed. All manufacturers stress that areas that might reduce passenger protection and driveability of the car, or where critical dimensions can be affected, *should not be sectioned.*

Replacement Panels

Replacement panels are whole, unsectioned panels as designed by the auto manufacturer. They are welded in place to simulate how the vehicle was manufactured at the factory. They are often purchased from the vehicle manufacturer and sometimes from aftermarket panel manufacturers (Figure 19–4).

OEM replacement panels are purchased from the original equipment manufacturer. If you are working on a Buick, for example, an OEM replacement panel would be obtained from General Motors Corporation.

Aftermarket replacement panels are manufactured by smaller companies, not the original equipment manufacturer (OEM). They cost less than OEM replacement panels but may not fit as well and may require more fit-up time.

Salvage replacement panels are undamaged panels cut off a wrecked vehicle at a salvage yard. Salvage replacement panels usually cost less than new parts, and they also have OEM corrosion protection. They can reduce the cost of parts and labor charges.

A An old, badly damaged panel must be cut off the vehicle.

B Reusable panels must be straightened and prepped for welding.

C A new panel is then fitted and welded into place.

Figure 19-3 Study the major steps for replacing a structural panel.

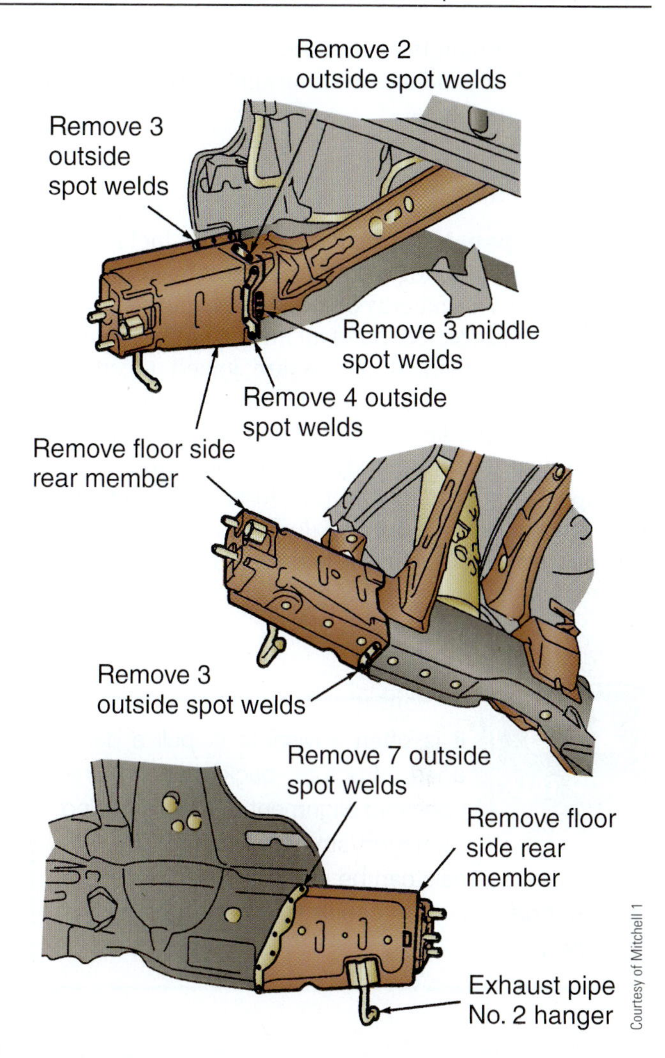

Figure 19-4 When replacing welded panels during collision repair, always refer to factory recommendations for cut locations, weld types, and weld locations. The auto manufacturer knows how to do proper repairs that meet crash test requirements. This is an example of recommendations on removing a section of a rear frame rail or structural member from one manufacturer.

Partial replacement panels are often designed to replace only a section or area of a large panel. They are ideal for body areas commonly subjected to rustout. Partial replacement panels are available from a number of aftermarket parts manufacturers, local salvage yards, or the OEM. For example, the lower area of a quarter panel can be cut off a wrecked vehicle at a salvage yard and welded in place to repair lower rustout. This saves considerable labor time over having to install a whole quarter panel.

Fabricated panels are handmade repair parts to fix small problems (for example, gouge or rusted holes) in panels when a new or partial panel is not available or practical. When making a fabricated panel, use the same metal type and thickness as that found on the vehicle. Cut out the damaged section of the part. Then, use it as a template to make the new part. Usually, a fabricated part is made larger than the cutout section so a lap joint can be formed. This produces a strong joint for welding in the repair section of metal.

Structural Panels

Structural panel is a general term that includes the body parts welded together to form the frame of a unibody vehicle. The integrity of the whole vehicle is dependent on the interconnection of all the individual structural panels. The individual panels are joined together at flanges or mating surfaces usually formed at the edges of the panels during factory production. Some examples of structural panels in unibody construction are the radiator support, inner fender aprons, front side members or rails, fender aprons, upper reinforcements, floor pan, pillars, quarter panels, rear side members, trunk floor panel, and rear panel.

Figure 19–5 You will often need to cut a window in a panel to be replaced. This will aid in getting at and removing spot welds along the flange of the old panel.

It is often advisable to pull a damaged structure back into factory-specified alignment before removing damaged structural panels. The old damaged panels can be used to pull and remove minor damage from panels that will remain on the vehicle.

The structural panels provide the foundation to which all the mechanical components are mounted, and all outer panels are attached to them. Therefore, all appearance fits and suspension alignments are determined by the accuracy of the positioning of the welded structural panels. Structural panels must be accurately positioned *before* final welding.

It is very important to always follow the manufacturer's recommendations when servicing structural panels. This is especially true when sectioning.

19.2 REMOVING STRUCTURAL PANELS

Structural body panels are often joined together in the factory by resistance spot welding. Therefore, removing panels mainly involves the separation of spot welds. Spot welds can be drilled out, blown out with a plasma torch, chiseled out, or ground out with a high-speed grinding wheel. The best method for removing a spot-welded panel is determined by the number and arrangement of mating panels and the accessibility of the weld.

A **panel replacement illustration** shows the type, number, and location of cuts and welds needed to properly install a structural panel. Factory service information is available for all major structural panels. Service illustrations show the locations of factory welds and where parts should be sectioned if necessary.

It is often necessary to cut a window into a panel that will be replaced (Figure 19–5). This will allow you to access the spot welds with a C-clamp–type drill (Figure 19–6).

Finding Spot Welds

It is usually necessary to remove the paint film, undercoat, sealer, or other coatings covering the joint area to find the locations of spot welds. To do this, remove the paint using a small grinding disc or use a coarse scuff wheel in a grinder. A coarse wire wheel or power brush can also be used to remove paint over spot welds. Scrape off thick portions of undercoating or wax sealer before trying to remove the paint. See Figure 19–7.

Avoid using an oxyacetylene or propane torch to remove paint, because they can overheat the metal. If you do use a torch, do not burn through the paint film so that the sheet metal panel begins to turn color. Heat the area only enough to soften the paint, and then brush or scrape it off.

NOTE *It is not necessary to remove paint from areas where the spot welds are visible through the paint film.*

In areas where the spot weld positions are not visible after the paint is removed, drive a chisel between the panels. Doing so will cause the outline of the spot welds to appear.

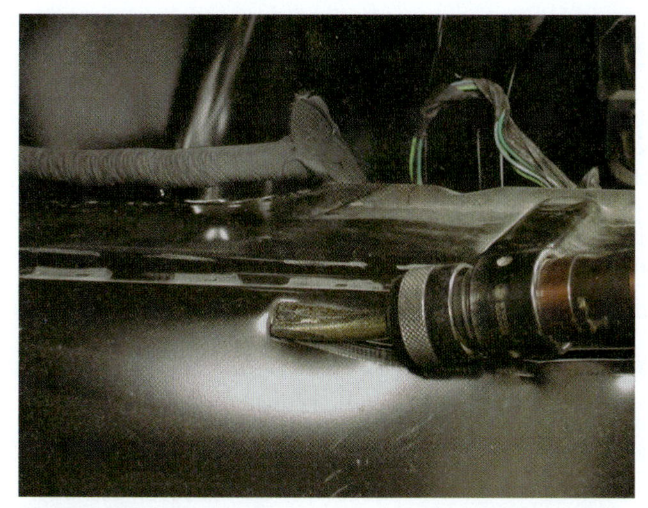

A Cut along outer perimeter of damaged panel close to the existing spot welds.

B Lift out the cut piece so you can gain access to the spot welds with a spot weld drill.

Figure 19–6 An air chisel is often used to cut away the center area of damaged panels.

A

B

Figure 19–7 (A) If needed, clean the part flanges to expose the spot welds so they can be drilled or ground out for panel removal. (B) To determine the spot weld locations, grind the flange on the panel to be replaced.

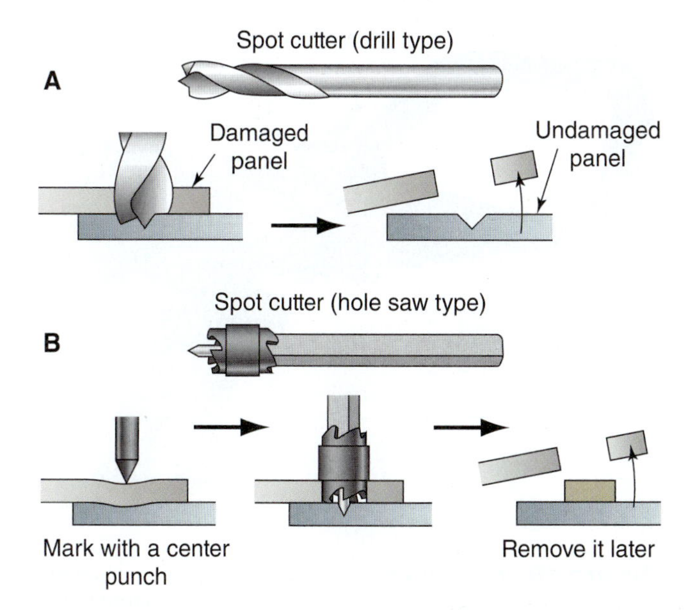

Figure 19–8 Note the types of spot weld cutters: (A) drill type and (B) hole saw type.

Separating Spot Welds

After the spot welds have been located, they can be drilled out and removed, using a spot weld cutter (Figure 19–8). Two types of cutting bits can be used: a drill type or a hole saw type.

Table 19–2 shows when either a cutter or drill should be used to remove factory spot welds. Be careful not to cut into the lower panel. Also, be sure to cut out the plugs precisely to avoid creating an excessively large hole.

A **spot weld drill** includes a lever and clamp for pushing the drill into the panel flange for rapid removal of panel welds. Drilling out the numerous spot welds in a panel can be tedious. To make the job easier, use a spot removing drill with an integral clamping mechanism. Hand pressure and lever action force the special rounded bit into the weld for faster cutting action (Figure 19–9, Figure 19–10, and Figure 19–11).

TABLE 19–2 SEPARATION OF SPOT WELDS

Type			Application Method	Characteristics
Spot cutter	Drill type	Small	Places where the replacement panel is between other panels and welding cannot be done from the backside Places where the replacement panel is on top and the weld is small	The separation can be accomplished without damaging the bottom panel Because the nugget is not left in the bottom panel, finishing is easy
		Large	When the replacement panel is on top When the panel is thick (places where nuggets are large) Places where the weld shape is destroyed	
	Hole saw Type		When the replacement panel is on top	Separation can be accomplished without damaging the bottom panel Because only the circumference of the nugget is cut, it is necessary to remove the nugget remaining in the bottom panel after the panels are separated
Drill			When the replacement panel is on bottom When the replacement panel is between and welding can be done from the back side (Select a drill diameter that is appropriate for the panel thickness and the weld diameter)	Lower cost A labor-saving spot weld removing tool has been developed that is easy to use and has a built-in attaching clamp

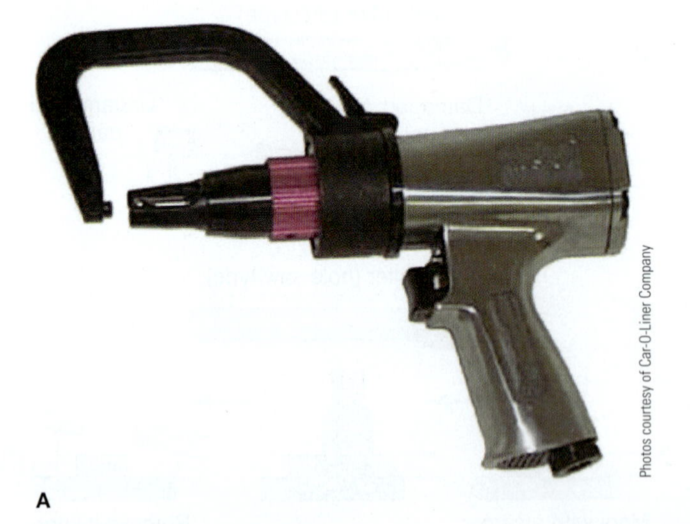

A

B

Figure 19-9 A spot weld drill has a clamp mechanism for holding and forcing the cutter bit into a resistance spot weld for panel removal. (A) This spot weld drill uses air pressure to force the bit down into the weld nugget when the trigger is pressed. (B) The collar around the drill bit can be adjusted to set the depth that the drill bit will enter the body panel. This avoids drilling into the lower panel that will be reused during repair.

Figure 19–10 Note how the spot weld drill clamps over the body flange to drill out the weld nugget.

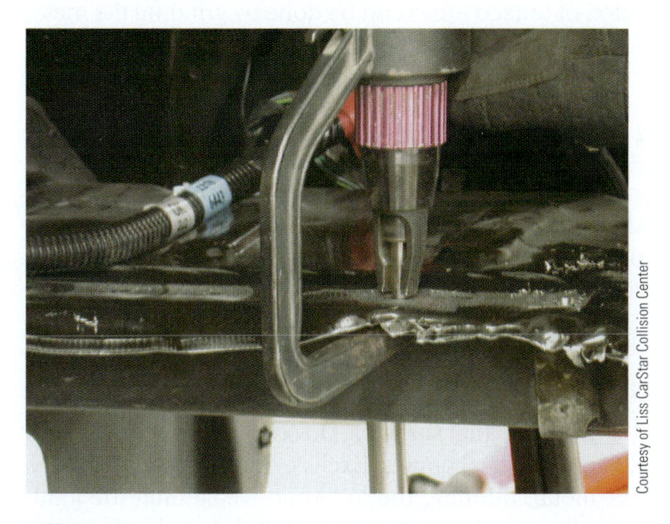

Figure 19–11 Here the panel has been cut so the frame of the spot weld drill can reach the weld nuggets.

A plasma arc torch is seldom recommended for removal of spot welds, although it can be done. The plasma torch will quickly blow a hole in all the thicknesses of metal at the same time. Obviously, the use of a plasma torch does not preserve the integrity of the underlying panels, and the less heat used, the better with today's metals. A plasma arc torch is used when neither of the panels will be reused (Figure 19–12).

WARNING When using a plasma torch, remember that tremendous heat is blown through the cut parts. This can burn wires, undercoating, and other parts. It can also start a serious fire that could lead to severe vehicle damage or even injury.

A high-speed grinding wheel can also be used to separate spot-welded panels. Use this technique only when

3200 kg
Courtesy of Liss CarStar Collision Center

Figure 19–12 Note how the major rear panels have been cut out of the vehicle for replacement.

the weld is not accessible with a drill, where the replacement panel is on top, or where a plug weld (from a previous repair) is too large to be drilled out (Figure 19–13).

DANGER Painful accidents and serious cuts can result from the sharp metal left from drilling out spot welds. Never run your hand over a drilled hole. It is also advisable to wear heavy leather gloves in case the drill or cutting tool slips.

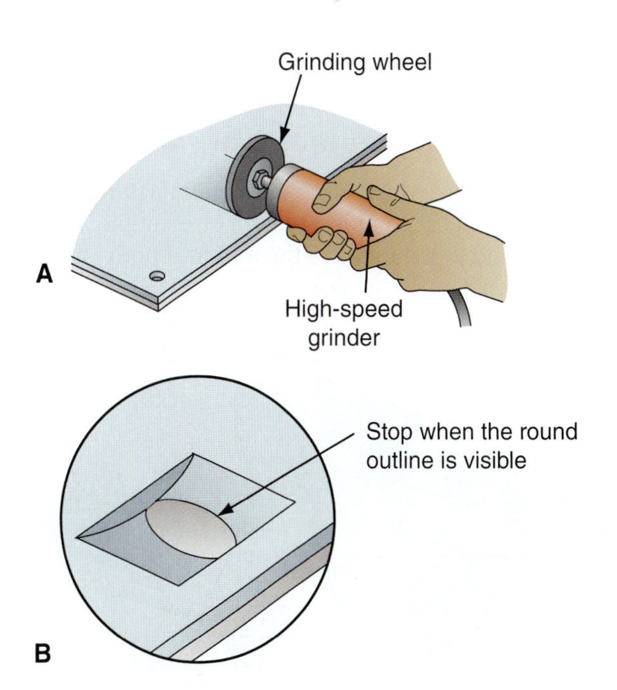

Figure 19–13 To use a grinder to remove spot welds, carefully remove the metal over the weld. (A) Hold the cutter securely to keep it from walking sideways. (B) Form a small pocket over the weld nugget.

Figure 19–14 An air chisel will quickly spread and separate panels after all of the factory spot welds have been cut out.

After the spot welds have been drilled out, blown out, or ground down, drive a chisel between the panels to separate them (Figure 19–14). Be careful not to cut or bend the undamaged panel.

Separating Continuous MIG Welds

Structural panels are sometimes joined by continuous MIG welding. Because the welding bead is long, use a grinding wheel or high-speed grinder to separate the panels. Cut through the weld without cutting into or through the panels. Hold the grinding wheel at a 45-degree angle to the lap joint. After grinding through the weld, use a hammer and chisel to separate the panels (Figure 19–15 and Figure 19–16).

Separating Brazed Joints

Brazing is still used by some auto manufacturers at the ends of large surface area panels. For example, brazing can be found at the joints of the roof and body pillars to

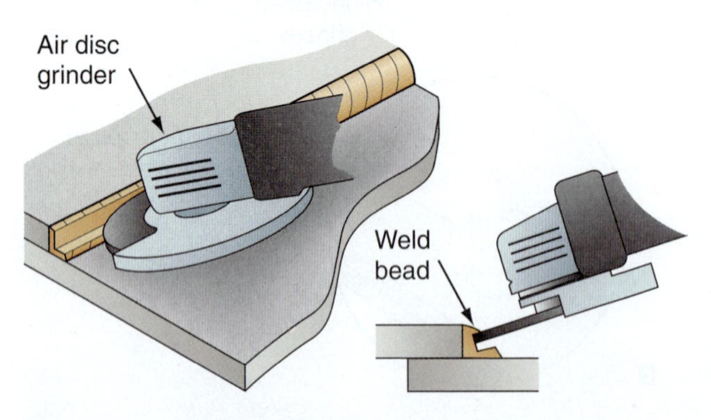

Figure 19–15 A continuous weld can be removed with a disc grinder. Again, guide the cutter carefully to prevent damage to the lower panel or the panel to remain on the vehicle.

Figure 19–16 A high-speed grinder will quickly remove a continuous weld. Be careful not to grind into panels that will be reused.

improve finish quality and the body seal. Generally, separation of brazed areas can be done by grinding the metal out from the joint.

19.3 PREPARING PANELS FOR WELDING

After removing the damaged panels, prepare the vehicle for installation of the new panels. To do this, follow these steps:

1. Grind off the welding marks from the spot welding areas. Use a wire brush or abrasive disc to remove dirt, rust, paint, sealers, zinc coatings, and so on from the joint surfaces. Do not overgrind the flanges of structural panels. Excessive grinding will remove metal, thinning the area and weakening the joint. Also, remove any paint and undercoating from the back sides of the panel joining surfaces on parts that will be spot welded during installation (Figure 19–17).

Figure 19–17 After the damaged panel has been cut off, grind off the remaining weld nuggets so the flange is smooth, and use a scuff wheel to clean the flange surface.

Figure 19–18 Use a hammer and dolly to straighten flanges on the panels to be reused. They must fit tightly against the new panel to be welded in place.

2. Smooth the dents and bumps in the mating flanges with a hammer and dolly (Figure 19–18).
3. Apply weld-through primer to areas where the base metal is exposed after the paint film and rust have been removed from the joining surfaces. It is very important to apply the antirust primer to joining surfaces or to areas where painting cannot be done in later processes (Figure 19–19).
4. To make sure that the MIG machine is correctly adjusted for the specific joint being welded, always do a test weld.

A **test weld** is done on scrap pieces of metal that are the same thickness and type of metal as the parts to be repaired. This is especially important on a closed section where the back side of the weld cannot be checked. A test weld is the only way to ensure that the welding techniques and machine adjustments will restore the original strength, integrity, and alignment of the panel.

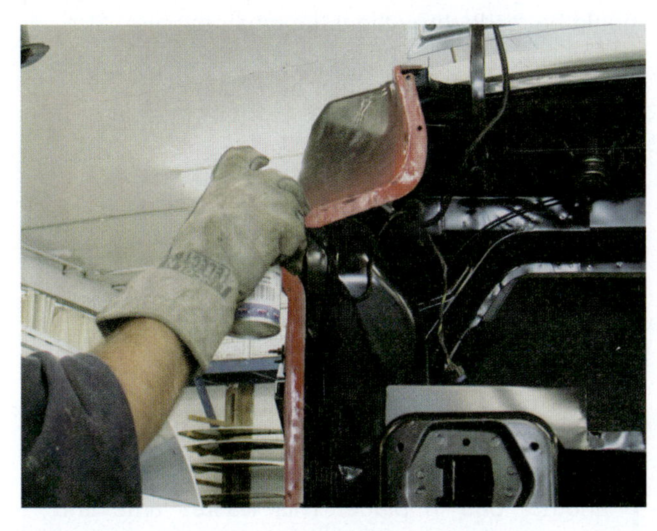

Figure 19–19 After flange grinding, cleaning, and straightening, spray weld-through primer onto the flanges.

Replacement Panel Preparation

New parts are coated with a primer. This coating must be removed from the mating flanges before welding to allow the welding current to flow properly during resistance spot welding. It is also necessary to drill holes for plug welds where spot welding is not possible. Use plug hole diameters that correspond to the thickness of the panels.

NOTE *For more information on panel welding, refer to Chapter 8. It details how to do spot welding, plug welding, and brazing.*

To prepare the new panel for welding, follow these steps:

1. Use a disc sander to remove the paint from both sides of the spot welding area. Do not grind into the steel panel. Do not heat the panel so that it turns blue or begins to warp.
2. Make holes for plug welding with a power punch or a drill. Always refer to the body repair manual or computer-based information for each type of vehicle to determine the number of holes needed and their sizes. Generally, you will have to make more holes than used on the factory assembly line.

   Be sure to make plug welding holes of the proper diameter. If the welding holes are too large or too small for the thickness of the panel, either the metal will melt through or the weld will be weak. Space the holes as directed in the factory service manual or computer-based service information. Refer to Figure 19–20.
3. Apply weld-through primer as an antirust treatment to the welding surfaces where the paint film was removed. Weld-through primer can be sprayed or brushed on.

   Apply the weld-through primer carefully so that it does not ooze out or run from the joining surfaces. If the primer does squeeze out, it will have a detrimental effect on painting, necessitating extra work. Remove any excess with a solvent-soaked rag.

If the new panel is sectioned to overlap any of the existing panels, rough cut the new panel to size using an air saw, cutoff grinding wheel, or similar tool. The edges should overlap the portion of the panel remaining in the sectioning area on the body by about ¾ –1 inch (–25 mm). If the overlap portion is too large, it will make matching the position of the panel more difficult during temporary installation. Again, refer to factory recommendations for panel overlap, because allowable overlap varies with the type and thickness of the panels.

Positioning New Panels

Aligning new parts with the existing body is a very important step in body repair. Improperly aligned panels will affect both the appearance and the driveability of the vehicle.

A This technician is using a drill to make holes in the new panel for plug welds.

B This technician is using an air-powered punch to make holes in the flange of a new body panel. This is much faster than drilling.

Figure 19–20 Normally, you will have to drill or power punch holes in replacement panels. Generally, use the same number of welds in the same locations as were on the old, damaged panel.

Panel clamping pliers have wide U-shaped jaws for reaching around and holding panels at their flanges. Clamping pliers are often used to hold the new panel in place while welding. Use as many panel clamping pliers as necessary to hold the panel securely in place (Figure 19–21).

Self-tapping screws can also be used to hold new panels in position while welding. A power drill with a small socket can be used to rotate and force the self-tapping screws through the panels. Screws should only be used when panel fit-up is difficult, as in restricted areas or when several panels must be held in alignment while being welded. One disadvantage of using screws to hold panels is that the holes left from the self-tapping screws will have to be welded shut.

Figure 19–21 Clamping pliers with wide jaws are often used to hold body panels while welding is being done. When clamps cannot be used to pull panel flanges together, small self-tapping screws can be driven through flanges to secure them while welding. After welding, remove all screws and weld the holes shut.

Measuring Panel Location

With major damage, use dimension measuring instruments to determine the correct part position. Either a tram gauge or computerized measuring system will ensure that the new panel is located within specifications before welding. With minor body damage, you can often visually find the correct panel position by examining the relationship of the new part to surrounding panels.

The dimensional accuracy of the engine compartment, fender aprons, front side members, rear side members, and similar rear structural parts have a direct effect on wheel alignment and driving characteristics. Therefore, when replacing structural panels in unibody vehicles, accurately measure part position. Whether structural or cosmetic panels, the emphasis should be on proper fit. You must often use both methods to ensure accuracy and the necessary fit for a high-quality repair.

When installing major structural parts, measurements should be made throughout the repair process. Also, all straightening must be done before replacing panels. Otherwise, proper alignment of the new panels will be impossible. As panels are fit into place and welded, they should be measured again (Figure 19–22 through Figure 19–31).

Failure to measure properly before welding new panels into place can be very embarrassing and expensive. If you make the mistake of welding a part in place out of alignment, you will probably have to cut off the new part and reweld another one in place. This wastes time and money and affects your reputation!

Here is a typical procedure for a vehicle being fitted with a front fender apron assembly, front cross member, and radiator support while measurements are being taken.

1. Match the assembly reference marks on the installation areas of the front fender apron and the side member. Fasten them in place with Vise-Grips. Parts that have no assembly reference marks should be installed in the same location as the old parts.
2. Match the length dimensions by setting the tracking gauge at the reference values. Adjust the length dimensions so they match those values. Temporarily install the front cross member. Use a hammer and block of wood to shift the parts in the desired direction (Figure 19–22).
3. If the length dimensions match the reference values, temporarily install the front floor reinforcement by tack welding one spot. Choose a spot weld location in an area where it will be easy to remove if necessary. Scribe a positioning line at the end of the part that is not welded and drill a small hole. Fasten the parts together with a sheet metal screw. Scribe a line on the apron installation area, but do not weld the panels together yet.
4. Use a centering gauge to match the height of the new components to the components on the opposite side of the vehicle. Support the new parts with a hydraulic jack so that the height does not change (Figure 19–23).
5. Match the diagonal and width dimensions. Then move the side member back and forth to match the dimension (Figure 19–24).
6. Confirm the height dimensions again.

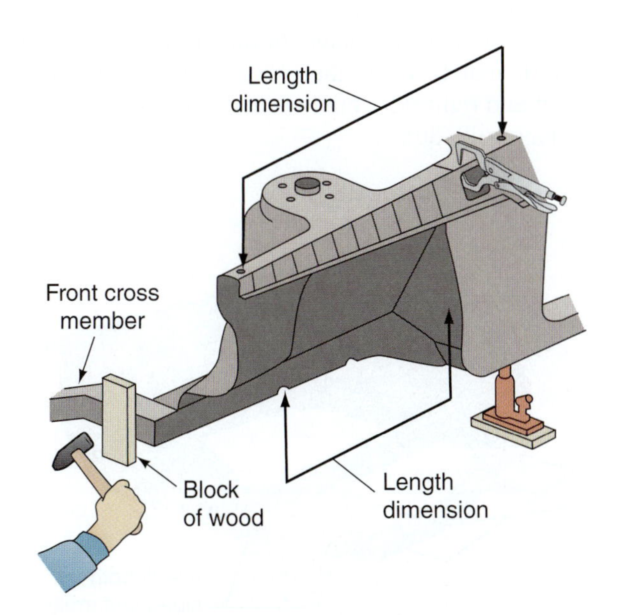

Figure 19-22 When making length adjustments, use a measuring system to compare the part location with the correct dimensions. Use a block of wood and light hammer blows to move the part as needed.

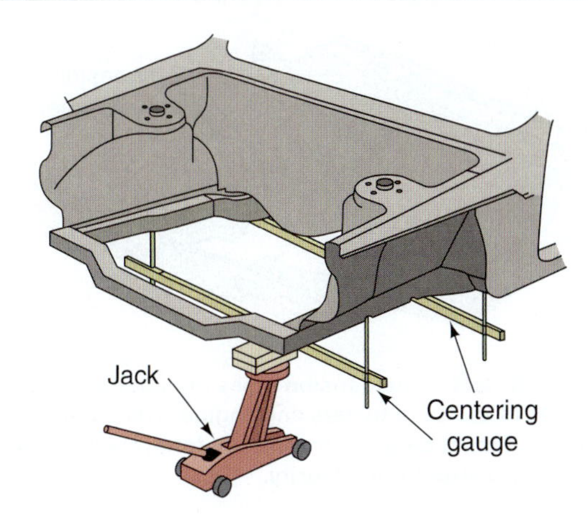

Figure 19-23 Here is an example of using centering gauges to check the height adjustments of new parts.

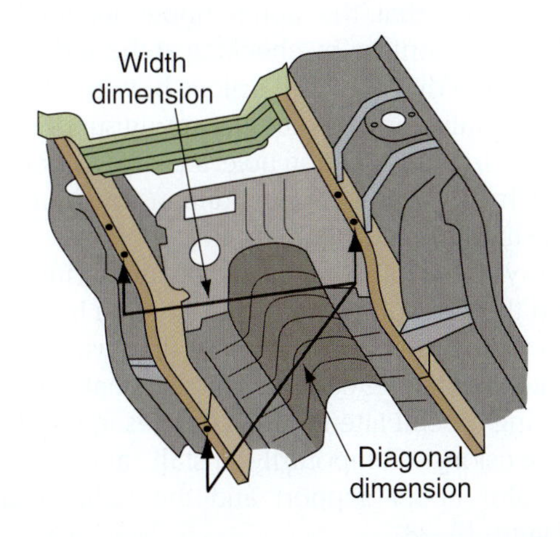

Figure 19-24 A diagonal measurement is made from an undamaged reference point to a point in the part being installed. Width dimension is checked from the undamaged to the damaged side.

7. Position the strut bar bracket and front cross member. Install the cross member so that both the left and right ends are uniform (Figure 19–25 and Figure 19–26).
8. Once the dimensions of the side member match the reference dimensions, secure the member in place. Use plug welds at several locations to fasten the side

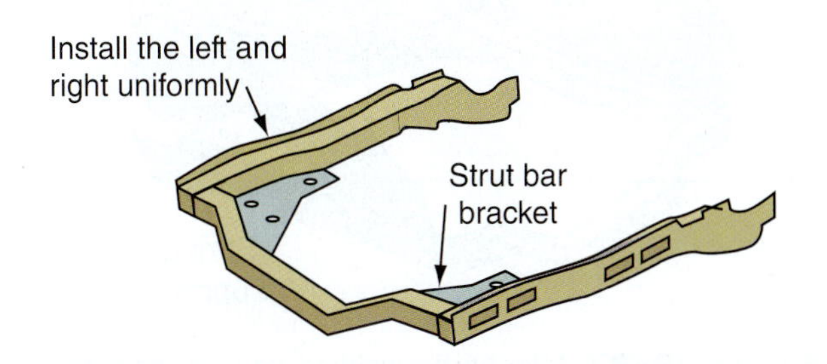

Figure 19-25 The correct position of a strut bar bracket is critical to the action of a suspension system. Measure it carefully before final welding.

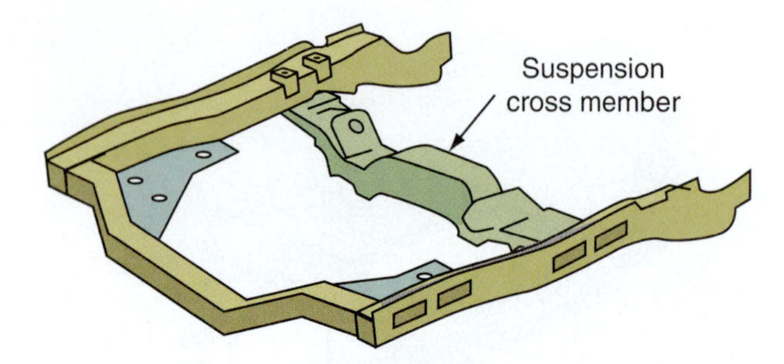

Figure 19-26 A suspension cross member supports the lower ends of the strut towers and engine. The panels supporting it must be welded correctly for proper wheel alignment and structural integrity.

member, the under reinforcement, and the side member to the front cross member.

9. Make sure that the apron upper length has not changed. Confirm by checking at the scribed line.
10. Match the diagonal dimensions between the fender rear installation hole and the spring support hole or fender front installation hole. It is also a good idea to match the spring support dimension from side to side at this time (Figure 19–27).
11. Verify the width dimension of the spring support and the front of the fender installation hole and fasten them together. If the width dimension does not match the reference value, make a small adjustment, paying careful attention to changes in the diagonal dimensions. Temporarily install and fasten the radiator upper support and the radiator support (Figure 19–28).
12. Match the radiator support width dimensions. Set the tracking gauge to the reference value and adjust the support so that the dimensions match that on the gauge. Lightly fasten it with Vise-Grips (Figure 19–29).

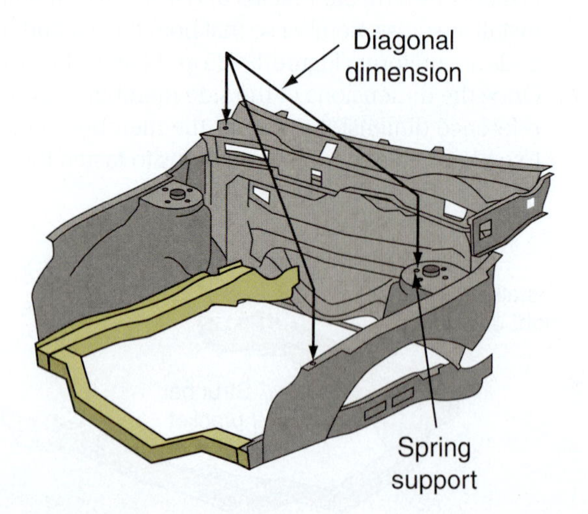

Figure 19-27 After partial welding, you may need to adjust the fender apron dimensions at the top. Check them continuously during the welding process.

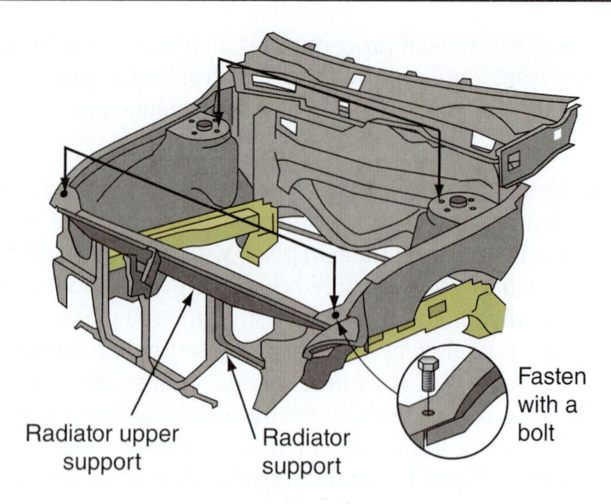

Figure 19-28 Check the fender apron width dimensions as you install the radiator support.

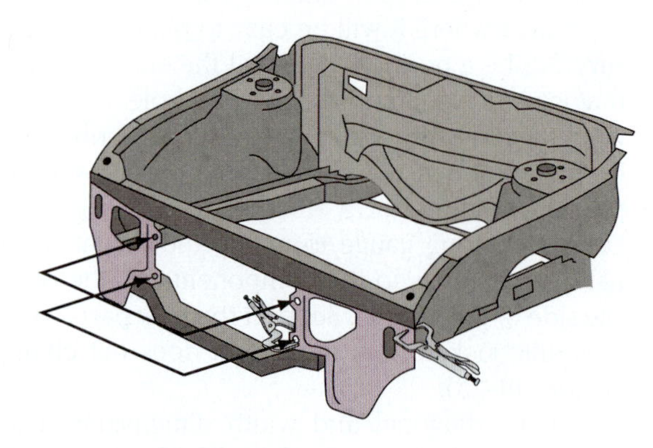

Figure 19-29 Match the width dimensions to make sure the front is aligned properly.

13. Match the diagonal dimensions for the radiator support. Be sure the diagonal dimensions of the supports match. Verify their height and make sure the left and right sides are installed in the same manner (Figure 19–30).

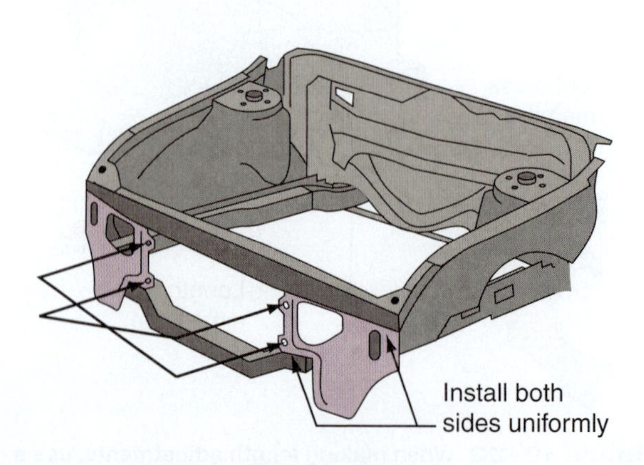

Figure 19-30 Also match the diagonal dimensions as you fit and weld parts back on the vehicle.

Figure 19–31 With the new panel perfectly aligned and jigged on a car bench, you can start final pulse MIG welding. This is an aluminum unibody structure that requires a different welder setup. The welder is an advanced programmable unit for making precise welds in aluminum.

14. Visually verify the left–right balance. Stand back and visually compare the new parts with existing ones.
15. Temporarily install the front fender and inspect it for proper fit with the door. If the clearance is not correct, it might be because the fender apron or the side member height is off on both the left and right sides. Also temporarily install the hood to check its alignment with the cowl and fenders.
16. Verify the overall dimensions once more before welding (Figure 19–31).

When using the tram and centering gauge method of component positioning, it is important to remember that measurement points for the new parts should be the same on the opposite side of the vehicle. If the dimensions do not match, the reference points must be verified and changed if necessary.

NOTE *Complete information on measurement and adjusting of panels is provided in Chapter 17.*

Positioning Panels Visually

Nonstructural outer panels can sometimes be visually aligned with adjacent panels without the precise measurements necessary in replacing structural panels. This is true of both mechanically fastened panels and welded panels. The emphasis here is on appearance. Body lines must be flush and aligned, and gaps between panels must be even, not tapered.

For example, when installing a hood, it is impractical to measure it during installation. You want it to fit in its opening properly. If the cowl, fenders, and radiator support are installed correctly, simply center the hood in its opening by aligning it with the cowl. Then install the fenders and align them with the hood. You must also make sure the hood latch and safety catch engage properly and the hood hinges operate smoothly.

Fitting a Quarter Panel

Quarter panels are often severely damaged in rear and side hits. They are large panels and must be replaced properly to maintain vehicle integrity. Make- and model-specific service information will detail cut locations, existing weld types and locations, and the recommended number and types of welds for replacement (Figure 19–32).

Following removal of the damaged quarter panels, check hidden parts and panels for damage. Any damage sustained under the quarter panel must be repaired before installing the new panel (Figure 19–33 and Figure 19–34).

For example, the wheel house or side brace may have been crushed from a side impact. These parts must be removed and rewelded as with other welded panels.

Prepare the new quarter panel as described earlier. Clean flanges and edges to be welded. Coat these surfaces with weld-through primer. The quarter panel top edge must normally be ground to accept a continuous weld. Refer to Figure 19–35 and Figure 19–36.

Next, fit and align the new quarter panel with adjacent body parts. Temporarily install the quarter panel

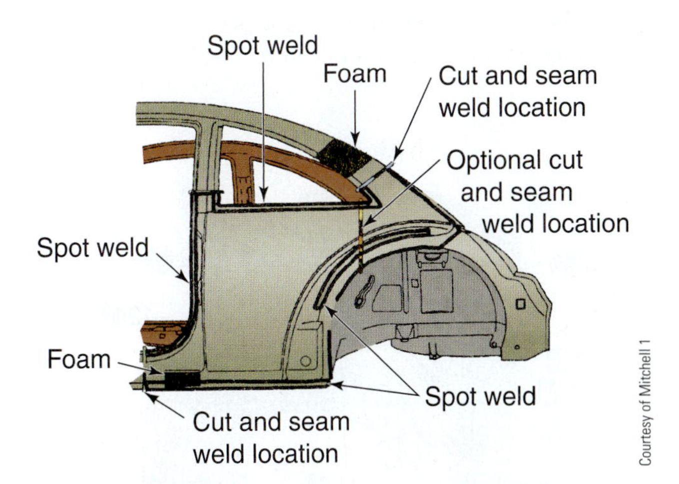

Figure 19–32 Service information provides panel cut or section locations for specific make and model vehicles. Always refer to such illustrations to make sure a new panel is correctly removed and installed.

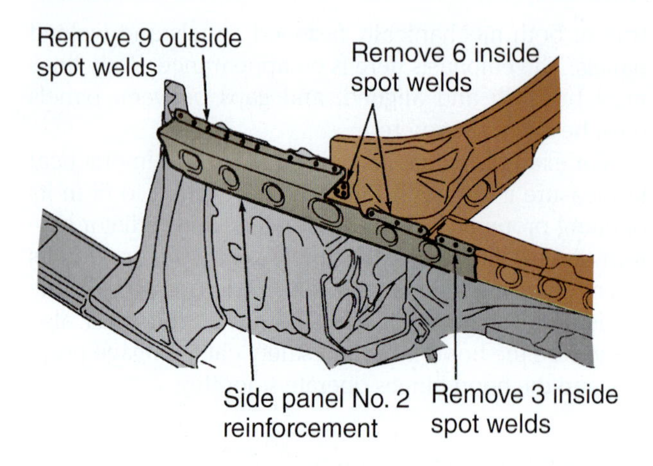

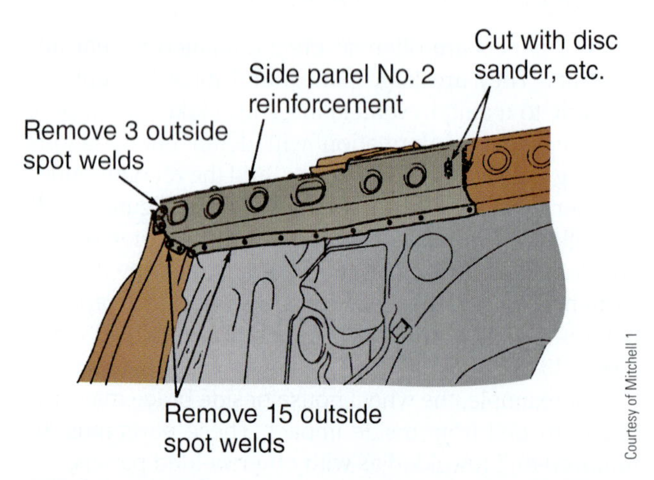

Figure 19-33 **A side panel acts as an upper structural member. Note the factory instructions for removing this part on this particular make of vehicle. Other makes of vehicle use similar methods.**

and fasten it at several points with Vise-Grips, screws, or another holding tool. Make sure that the panel ends and flange match up (Figure 19–37).

Carefully adjust the fit with the surrounding panels. Adjust the panel so that the clearances with the door and body lines match each other (Figure 19–38).

Then, install the trunk lid in its correct position and adjust the clearances and heights. Confirm that there is no left–right difference in the diagonal dimensions for the rear window opening. Match the rear glass to the opening to verify proper alignment (Figure 19–39).

After fitting the panel to the door and the trunk lid, fasten the panel with self-tapping screws. If it is fastened with Vise-Grips, the fit cannot properly be verified. With the trunk lid aligned with the upper rear body panel, you can align the quarter panel to the trunk lid.

Adjust the body line and panel overlap to match the lower back panel and the rear valance panel. Install the rear combination lamp and fit the panel to the lamp assembly (Figure 19–40).

When the clearance, body line, and height differences of each part have been adjusted, visually check for overall twisting or bending.

After the panel is properly positioned, cut the overlapping portion of the joining sail panel area with an air saw or cutoff grinding wheel. Be precise when making cuts in the sectioning area. If a gap opens up at the cut, welding will be difficult or impossible. Therefore, after matching the panel fit at every point, it is important that accurate cutting be done.

If there is sufficient overlapping, both panels can be cut simultaneously. If the overlapping is small, a line can be scribed at the end of the overlapping panel. When cut along the scribed line, the panels should fit snugly together with little or no gap.

After cutting the overlap to fit, remove the replacement panel. Clean off any metal chips and other foreign material from the inside of the panel before proceeding. Drill or punch holes for any plug welds and apply a weld-through primer.

Apply body sealer around the inside perimeter of the quarter panel. Install the panel and other parts with self-tapping screws in the same screw holes as before. Verify the fit once more.

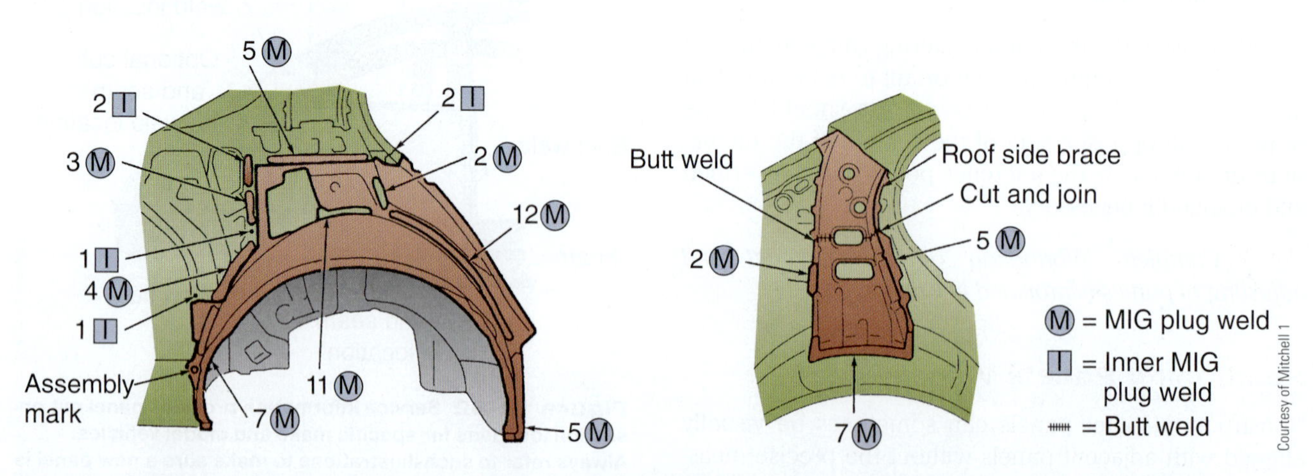

Figure 19-34 **Here are the factory recommendations for replacing an inner wheel house and roof brace, which can be damaged by a side hit to the quarter panel. Note the symbols for MIG plug welds, inner MIG plug welds, and butt welds.**

A Cuts or sections in existing panels should be made in locations recommended by the auto manufacturer.

B A new panel can be placed on the car and marked for cuts to match the section points on the vehicle.

C A new panel can then be cut to fit against the sectioned points made on the vehicle. The technician is using an air-powered cutoff wheel because it will make smooth, straight cuts in metal.

Figure 19-35 Precise cuts have been made in the B- and C-pillars to install a new quarter panel.

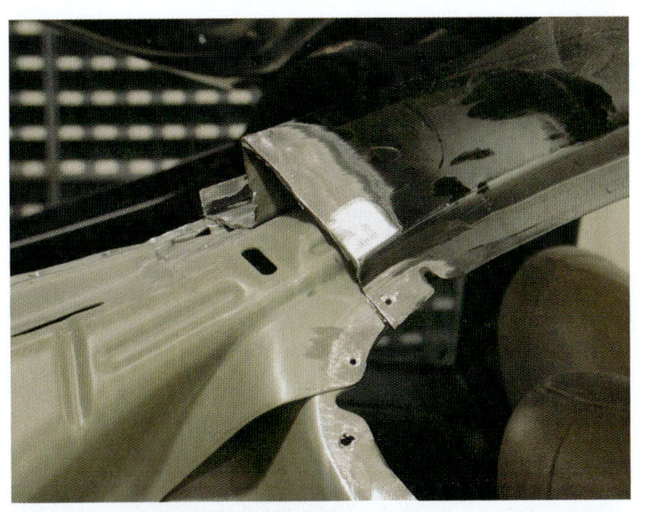

A A sail panel near the roof or upper quarter panel has been sectioned and cleaned.

B A small piece of the new quarter panel has been cut off and welded behind the butt joint and coated with weld-through primer. This will allow the butt weld to be made flush with the surface. Little grinding and body filler will be needed to smooth the area over the butt weld.

Figure 19-36 A small backer panel must often be made when making welded butt joints between panels.

Figure 19-37 Here a roof sail panel and the upper end of a new quarter panel have been fitted and checked before welding. Always cover glass and upholstery with a welding blanket or welding masking paper to avoid damage to these surfaces.

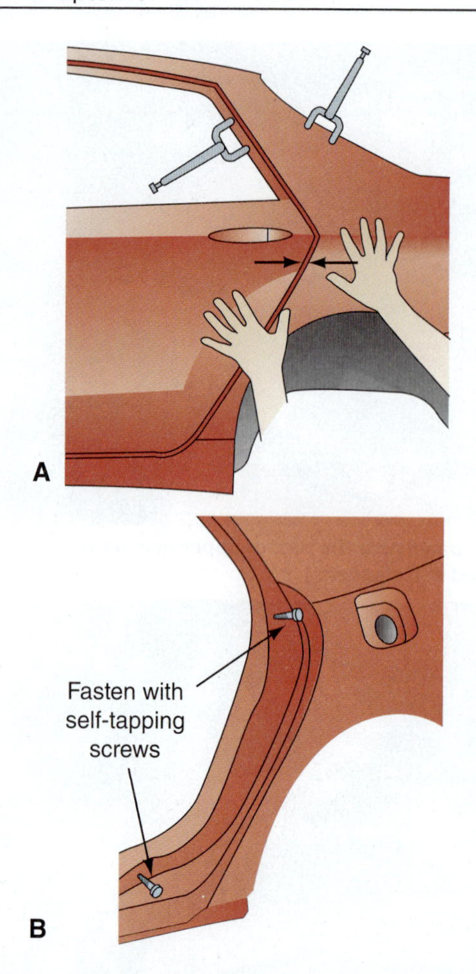

Figure 19–38 (A) Use clamps to position a quarter panel and check its alignment with the door panel. (B) Small screws can be used to secure the position of the panel.

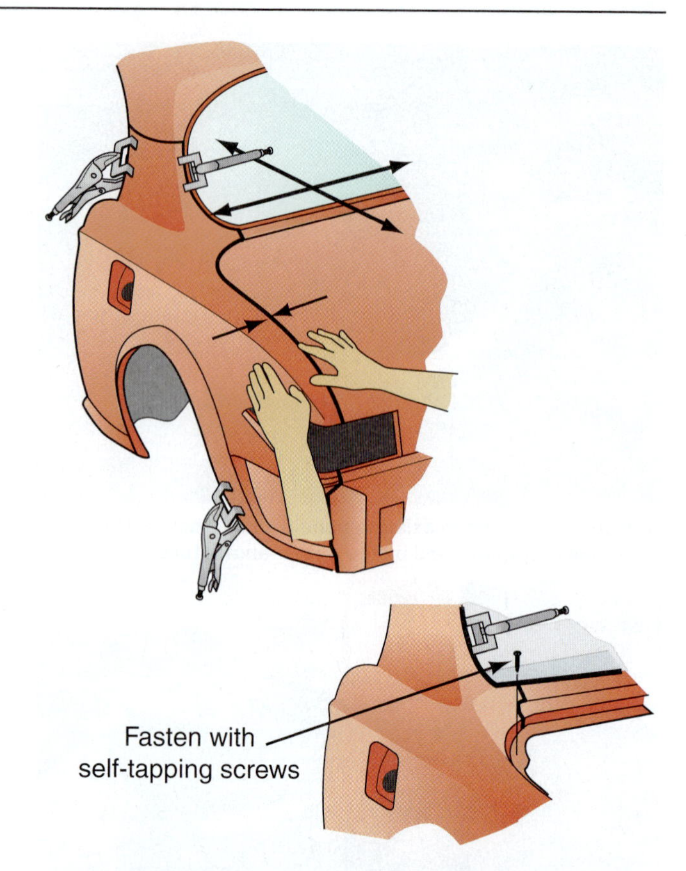

Figure 19–39 With the front of a panel secured, adjust the alignment to the rear window and deck lid. Install other sheet metal screws as needed to precisely hold the panel.

Welding Quarter Panels

Once the dimensions and position of the new part are correct, weld the quarter panel in place. Attach the welder ground to a clean, paint-free area on the panel (Figure 19–41). Adjust the MIG welder properly by doing a test weld as described earlier.

Refer to factory instructions for section or cut locations and number and types of welds. One example of a factory weld count and location illustration is given in Figure 19–42.

Tack weld the panel in place and then check quarter panel alignment. Remeasure the panel to double-check that it is fitted in place properly.

Typically, a continuous weld is used across the pillar (Figure 19–43). Spot welds are often used along its edge near the window (Figure 19–44). After welding, grind down the weld until it is flush with the panel surface. Be careful not to overgrind. Overgrinding creates a serious weakness and is a common mistake made by novice body technicians.

NOTE *Welding operations for new panels are fully explained in Chapter 8. The installation of fiberglass body panels is given in Chapter 13.*

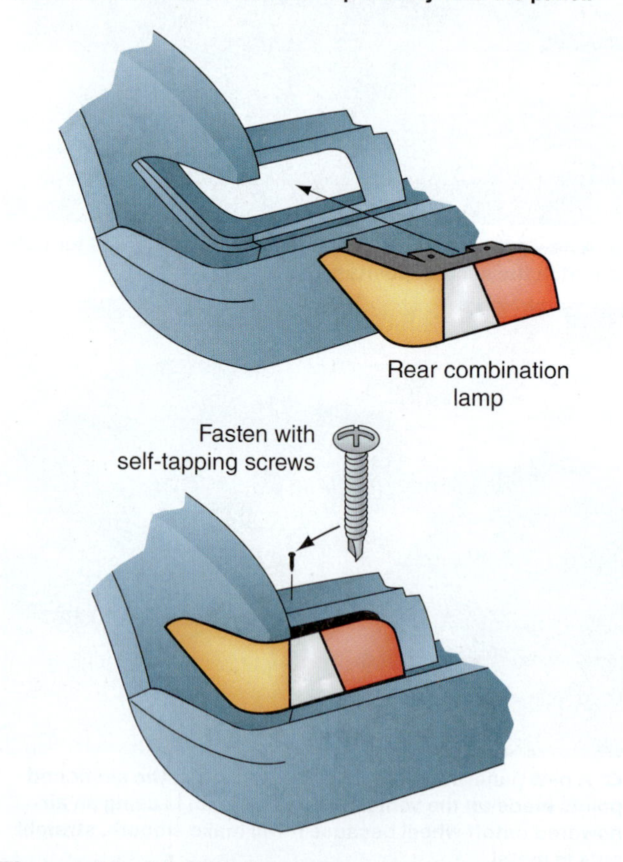

Figure 19–40 With the quarter panel dimensionally correct, clamp on the lower back panel and check the fit of the lamps.

Figure 19–41 After the new and existing panels have been cut, test fitted, and their flanges have been coated with weld-through primer, spot weld the panel in several locations. Then double-check panel alignment with the other panels or by using a measuring system before final welding.

19.4 STRUCTURAL SECTIONING

As mentioned earlier, sectioning involves cutting and replacing panels at locations other than factory seams. When body parts need to be replaced, replacing them at factory seams is the logical first choice. However, this is impractical when many seams have to be separated in undamaged areas. In some repairs, sectioning of parts such as rails, pillars, and rocker panels may be required to make their repair economically feasible.

Remember that sectioning requires precision, as well as strict adherence to manufacturer-recommended procedures. Always check the body repair manual or service data to obtain the manufacturer's procedure for the specific sectioning location.

As a collision repair technician, never forget that sectioning finally comes down to sound judgment. It is the technician who must ensure the quality of the repair through the proper application of tested and proven procedures.

Figure 19–45 shows some of the parts that are often sectioned: rocker panels, quarter panels, floor pan, front rails, rear rails, trunk floor, pillars, and similar parts.

Unibody parts that can be sectioned include these types of construction:

- Closed sections, such as rocker panels and A- and B-pillars
- Hat or open U-channels, such as rear rails
- Single layer or flat parts, such as floor pans and trunk floors

Closed sections are the most critical because they provide the principal strength of the unibody structure.

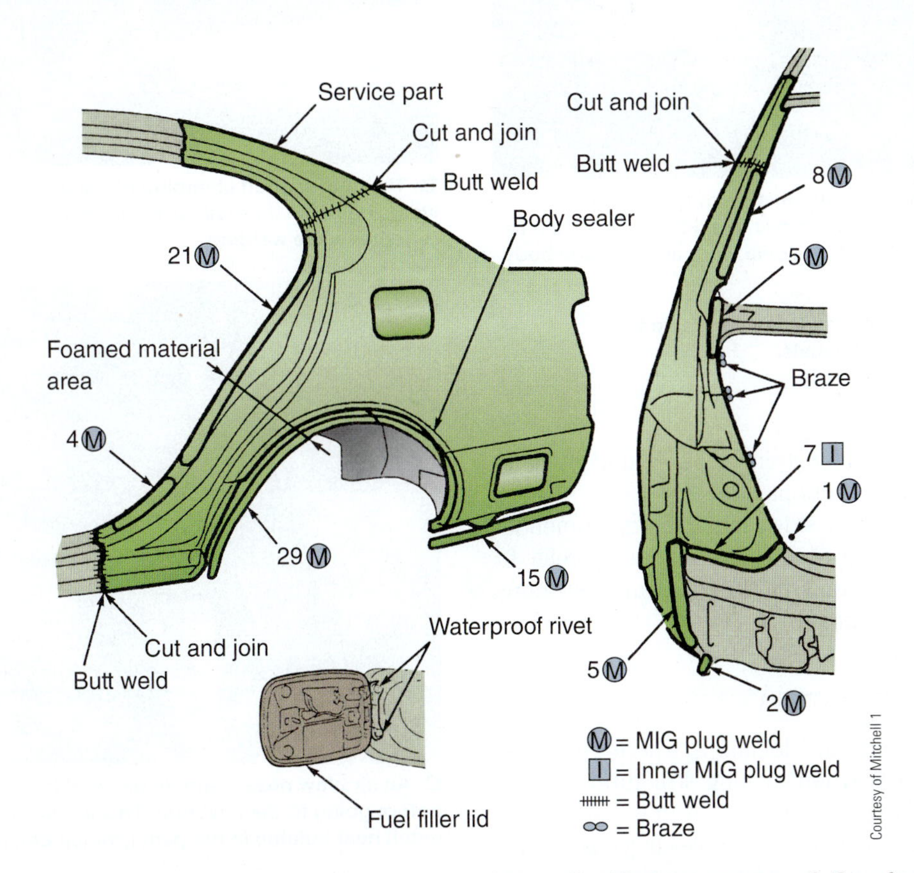

Figure 19–42 Although exact procedures vary, this is a typical example of factory recommendations for quarter panel replacement. Can you find the places on this late model vehicle where the manufacturer recommends braze joints?

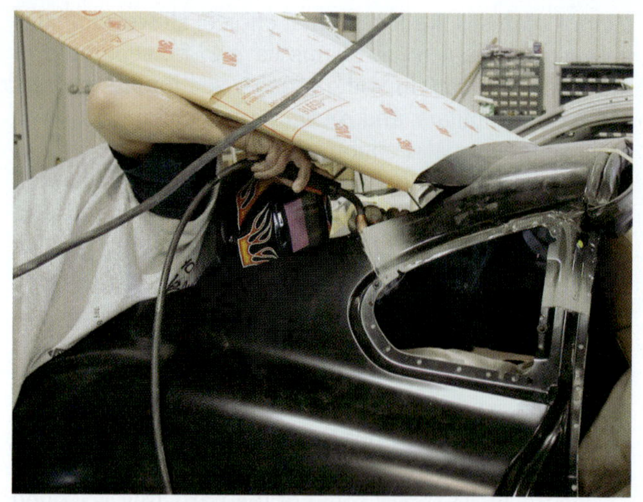

A After fitting and measuring the quarter panel location, weld one or two of the plug weld holes in the new panel. Then recheck alignment of the new panel.

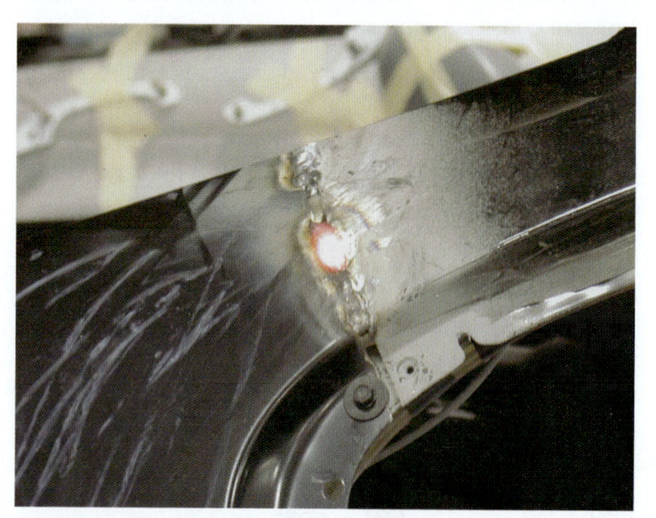

B When making a long continuous weld, stitch weld the joint. Weld an inch or so, then move to a different location to avoid excess heat buildup that could warp and damage body panels.

Figure 19-43 Proper welds are critical to the replacement of structural panels.

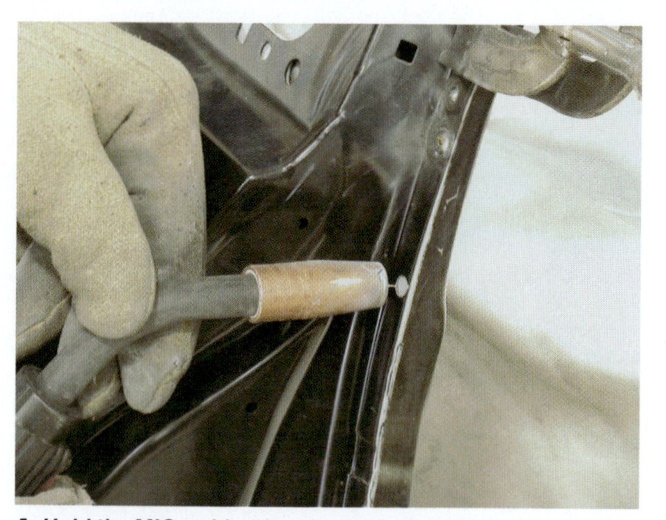

A Hold the MIG welder tip over the hole in the upper panel. Trigger the welder and fill up the hole with molten metal, using a circular motion of the welding gun.

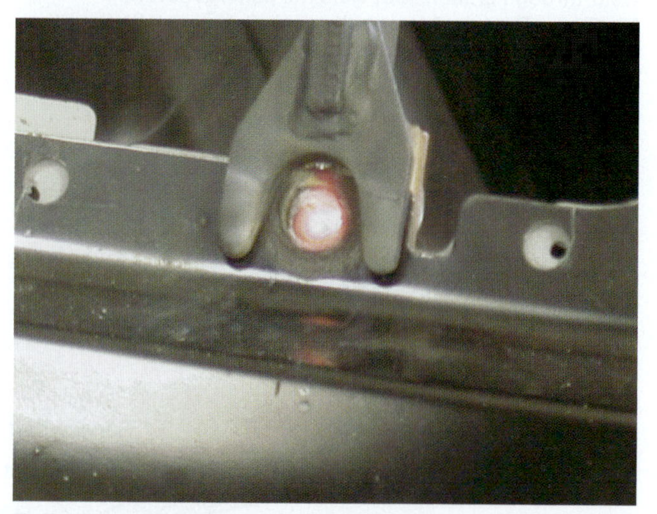

B These U-shaped clamping pliers are ideal for making plug welds because they will force the two panel flanges tightly together while welding.

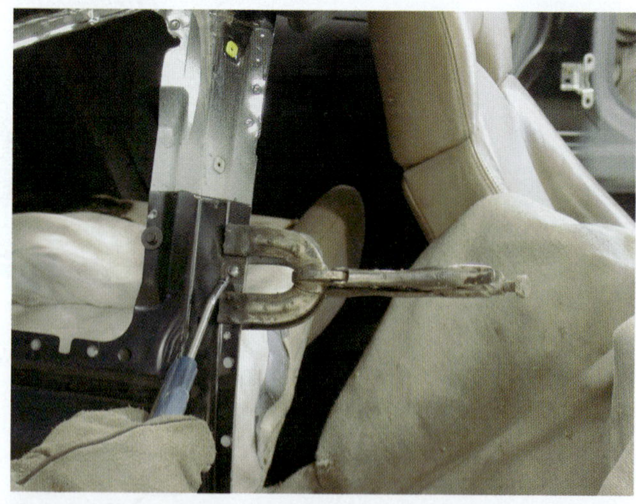

C An air blow nozzle should be used to cool each plug weld before going to the next one. This again will help avoid too much heat buildup in the panel, which can lead to warpage.

Figure 19-44 Plug welds are commonly used in body shops to replace factory resistance spot welds.

They possess much greater strength per pound of material than other types of sections.

Figure 19–46 shows the side sectioning recommendations for one make and model of car. Always refer to vehicle-specific instructions if you have any questions about how to section a panel.

Nonsectionable Areas

Nonsectionable areas are locations where you must not cut through parts when making structural repairs. There are several areas to stay away from when making sectioning cuts. Do not cut through holes in parts or panels, because they are often used for measuring, and welding will be difficult. Do not cut through any inner

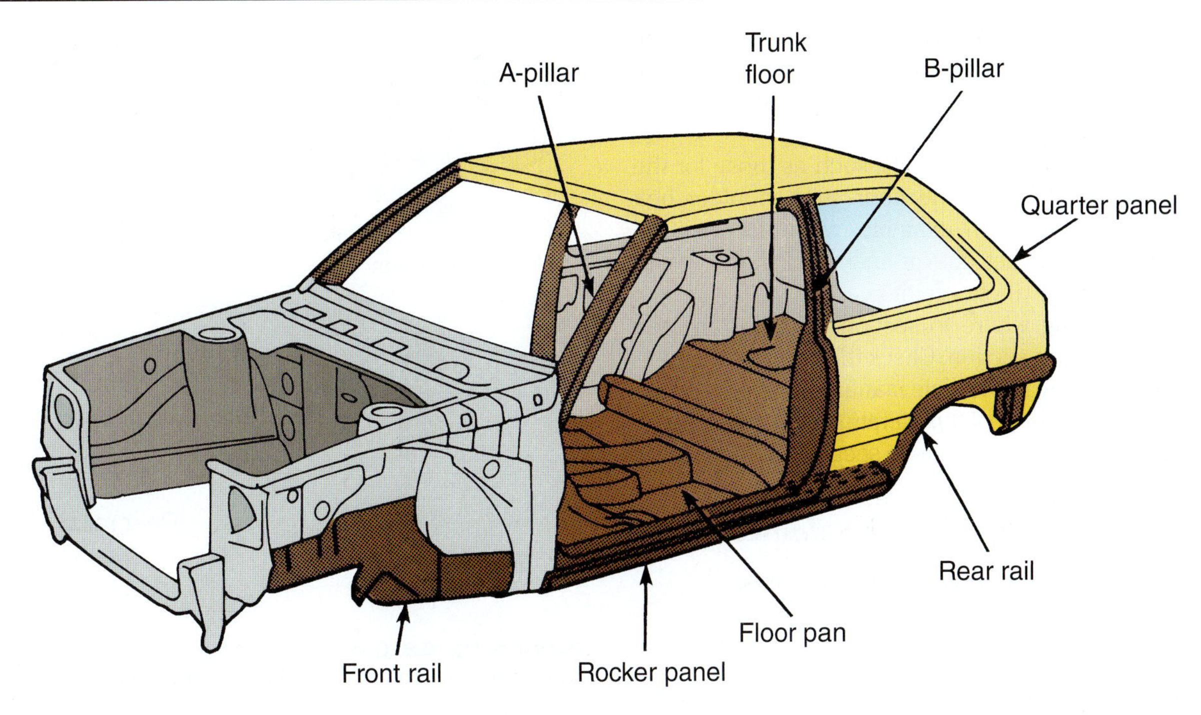

Figure 19-45 These are some other areas commonly sectioned on this body type because the whole replacement panel would be too difficult to install.

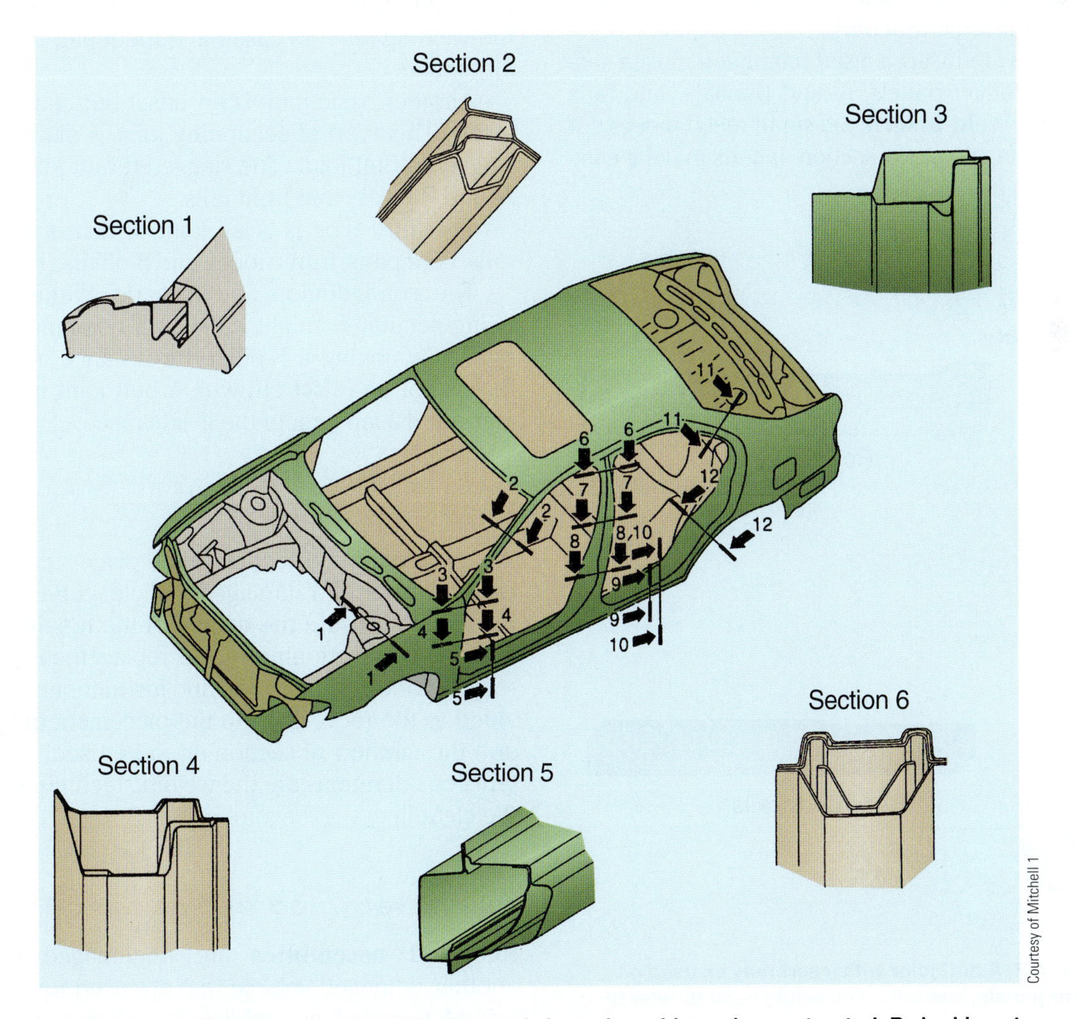

Figure 19-46 This illustration from a body service manual shows how this car is constructed. By looking at a cross section of panels, you can determine how best to cut or section each location.

reinforcements, meaning double layers in the metal. Careless cutting through a closed section with inner reinforcements may make it impossible to restore the area to preaccident strength.

Stay away from anchor points, such as those for the suspension, seat belts in the floor, and shoulder belt D-ring anchor points. For example, when sectioning a B-pillar, make an offset cut around the D-ring area to avoid disturbing the anchor reinforcement.

When deciding where to section, look for an area with a uniform cross section. Check the body repair information provided by the vehicle manufacturer or an aftermarket publisher. Much of this literature provides specific instructions on how and where to section.

Basic Types of Sectioning Joints

There are three basic types of sectioning joints:

1. Butt joint with insert
2. Offset butt joint
3. Lap joint

One of these joints, or a combination of these joints, will be used for all sectioning procedures. The type of joint used for a specific repair depends on the location and design of the structural part.

A butt joint with insert is used mainly on closed sections, such as rocker panels, A- and B-pillars, and rails (Figure 19–47). **Weld inserts** are small metal pieces cut to fit behind or inside a box section. Inserts make it easy to fit and correctly align the joints. They also help make the welding process easier and the repair more structurally sound.

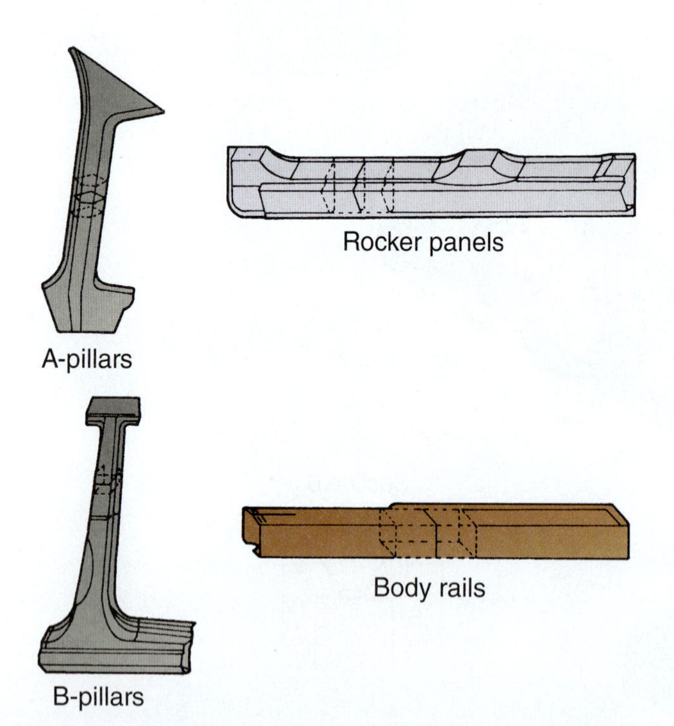

Figure 19–47 A butt joint with insert may be used on pillars, rocker panels, and rails. You would need to refer to the manual or use common sense to determine whether this type of joint would work best.

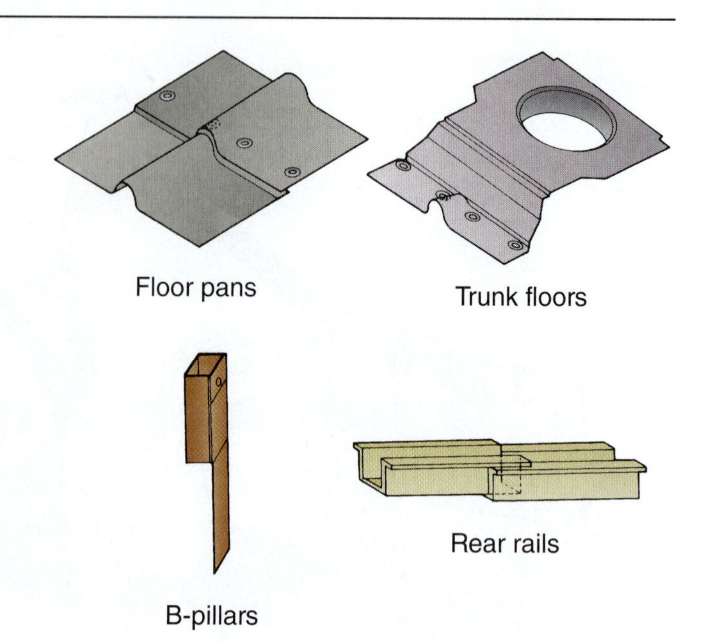

Figure 19–48 Overlap joints are common on floor and trunk pans and sometimes on pillars and rails.

Another basic joint is an offset butt joint without an insert. This type of sectioning joint is also known as a staggered butt joint. The staggered butt joint is used on A- and B-pillars and front rails.

The third type is a lap joint, which is used on rear rails, floor pans, trunk floors, and B-pillars (Figure 19–48).

The configuration and makeup of the component being sectioned might call for a combination of joint types. Sectioning a B-pillar, for instance, might require the use of an offset cut with a butt joint in the outside piece and a lap joint in the inside piece.

Preparing to Section

When preparing to section and replace a panel or structural member of a damaged vehicle, certain steps must be taken to ensure the quality of the repair. The first of these is the sectioning of the replacement part by the recycler or technician. Specific instructions must be provided to the recycler as to the placement of the section and the method of sectioning to be used. Other important considerations are the welding techniques used and the cleanliness of the joint metal.

Using Recycled or Salvaged Parts

Recycled assemblies are undamaged parts from another damaged vehicle that are used for repairs. The use of recycled assemblies in collision repair makes sense for four reasons:

1. Fewer welds need to be made when using recycled assemblies compared to new, separate parts.
2. Less factory corrosion protection is disturbed.
3. More measuring is required when welding separate new parts and attaching them to the vehicle.
4. There is an abundance of recyclable assemblies available in most areas.

When using recycled parts, tell the recycler exactly where to make the cuts. It is preferable to have the required part removed with a metal saw. But if the recycler uses a cutting torch, make sure that at least 2 inches (50 mm) of extra length is left on the part to ensure that the heat dispersion from the cut does not invade the joint area. Instruct the recycler to make the cut so that reinforcing pieces that are welded inside the component are not cut through.

When a recycled or salvaged part is received, examine it for corrosion. If it has a lot of rust, do not use it. Ask for another part. Before installing a recycled part, check it for possible damage and make sure it is dimensionally accurate.

Remember! Using quality replacement parts is a must to achieve a quality repair. If the recycled part is almost rusted through, the repair will be inferior and may even endanger the passengers of the repaired vehicle.

Careful joint preparation is another necessity for doing a proper job of structural sectioning used parts. Before starting to weld, be sure to thoroughly clean the surfaces to be joined. Use a scraper to remove thick materials (heavy undercoatings, rustproofing, tars, caulking and sealants, and road dirt). Be sure to remove rustproofing, lead, body filler, and other contaminants from the inside of structural closed sections when preparing them for welding. Then use a scuff wheel or a sander to remove thinner, less flammable primers and paint.

Make sure the surfaces to be welded are completely free of rust and scale. This is best removed by sanding or media blasting until there is a clean metal welding surface. In some cases, it is possible to do this with a power wire brush.

The weld site must be completely free of any foreign material that might contaminate the weld. Improper cleaning can result in a brittle, porous weld of poor integrity. In addition, you must attach the welder's cable clamp to a clean surface to have a trouble-free welding circuit. As mentioned earlier, application of a weld-through primer is also recommended.

To make sure complete penetration and full fusion are achieved, do test welds on sample pieces that duplicate the intended workpiece: the same types of welds on the same type and configuration of joint and the same gauge metal. The ideal way to do this is to use pieces of excess material from the components being joined on the car, as from the scrap cut off to make the fit-up. While making the test welds, adjust the MIG machine to suit the situation (Figure 19–49).

Figure 19–49 MIG welding is a must when sectioning and installing recycled parts to prevent warpage and for strong joints. Here the body technician is welding a rocker panel.

Sectioning Body Rails

Virtually all front and rear rails are closed sections, but the closures are of two distinct types. One is called a closed section. It comes from the factory or the recycler with all four sides intact. Sometimes it is referred to as a box section. The other type comes from the factory as an open "hat" channel and is closed on the fourth side by being joined to some other component in the body structure (Figure 19–50).

The butt joint with insert is commonly used for repairing a closed section rail. Most rear rails, plus various makes of front rails, are of the hat channel type. Some hat channel closures are vertical, such as a front rail joined to a side apron. Some of them are horizontal, such as a

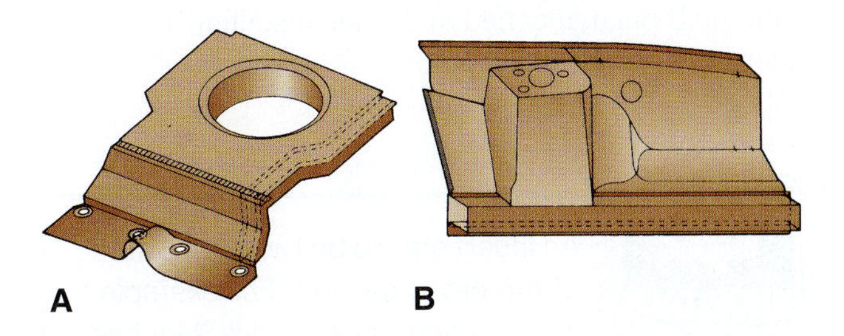

Figure 19–50 Hat channels have a complex shape. (A) Never section near holes or openings in panels. (B) The front inner fender panel is welded to the strut tower and the frame rail to form a strong but complexly shaped structure.

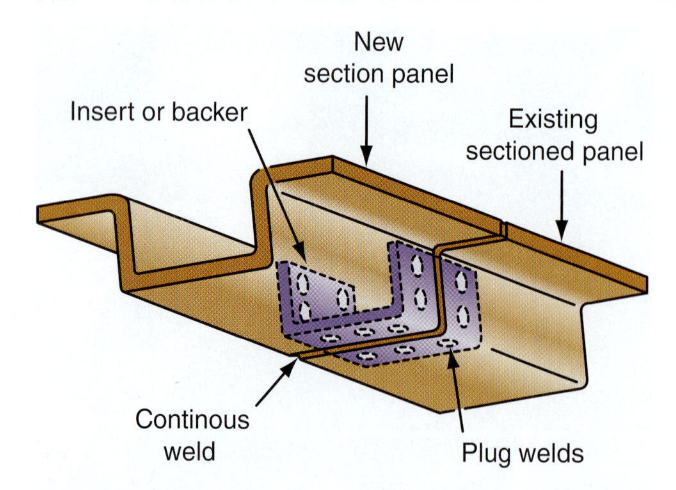

Figure 19-51 **Joining open rails is often done by using a backer or insert piece with plug welds and a continuous weld for a strong butt weld repair.**

rear rail joined to a trunk floor. In most cases, when sectioning an open hat channel rail, the procedure involves a lap joint with plug welds in the overlap areas and a continuous lap weld along the edge of the overlap (Figure 19–51).

Sectioning Rocker Panels

Rocker panels are constructed differently depending on the make and model of vehicle. The rocker panel might contain reinforcements. The reinforcements might be intermittent or continuous. Before starting work, you should know how the rocker panel is made.

Refer to factory recommendations for sectioning a specific make and model rocker panel. One example is shown in Figure 19–52. Note how the service illustration gives cut-and-join locations, weld removal locations, section dimensions, and new weld types and locations.

Depending on the nature of the damage, the rocker panel can be replaced with the B-pillar or without it (Figure 19–53).

To section and repair the rocker panel, a straight-cut butt joint with an insert can be used. The outside piece of the rocker panel can also be cut and the repair piece installed with overlap joints. Generally, a butt joint with insert is used when installing a recycled rocker panel with the B-pillar attached and when installing a recycled quarter panel (Figure 19–54).

An insert should be twice the width of the cross section. For example, if the widest dimension is 2 inches (51 mm), the insert should be 4 inches (102 mm) long.

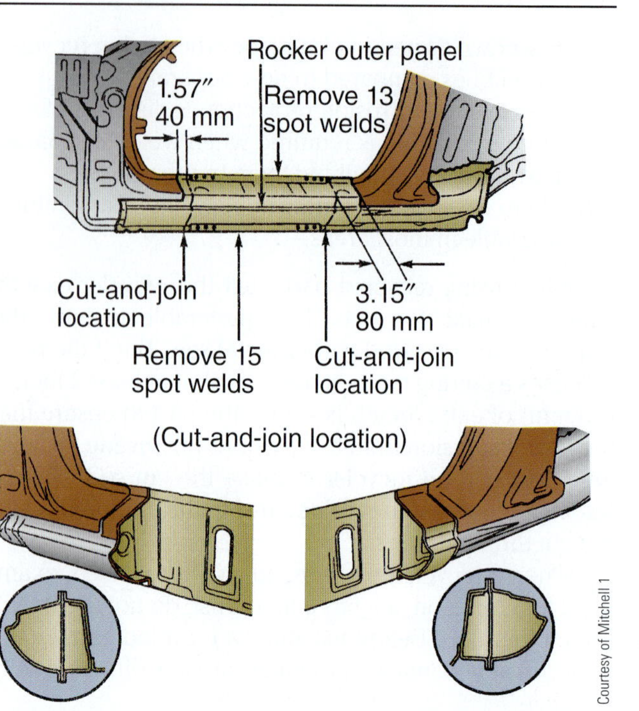

A The removal illustration shows the locations of cuts from reference points. It also gives the number and locations of factory spot welds.

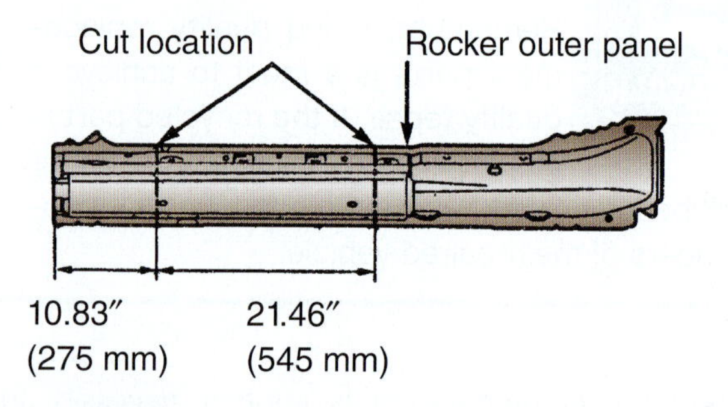

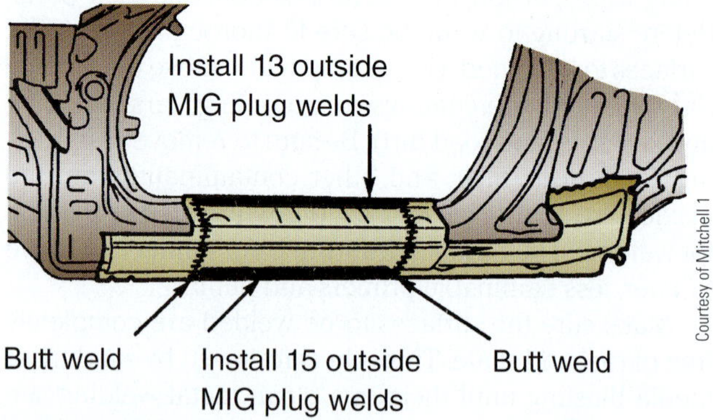

B This installation illustration shows how to weld one type of rocker panel.

Figure 19-52 **Service manual illustrations showing how to remove and weld a rocker panel for a specific make and model vehicle.**

To do a butt joint with an insert, cut straight across the panel. The insert is fashioned out of one or more pieces cut from the excess length on the repair panel or from the end of the damaged panel. It should typically be 4 to 6

Figure 19–53 After determining the internal structure of the rocker panel and how it will be repaired, cut it off. A power saw works well on a rocker panel. This vehicle has had the outer section of the rocker panel removed for replacement.

Figure 19–54 A properly welded rocker panel will be as strong as a new one. Its installation is critical to vehicle integrity. Note the use of a welding blanket to protect the interior.

inches long and should be cut lengthwise into two to four pieces, depending on the rocker panel's configuration (Figure 19–55).

Remove the pinch weld flange so that the insert will fit inside the rocker panel. With the insert in place, secure it with plug welds. For structural sectioning, 5⁄16 inch (8 mm) **plug weld holes** are typical to achieve an adequate nugget and acceptable weld strength. This 5⁄16 inch (8 mm) hole requires a circular motion of the gun to properly fuse the edge of the hole to the base metal.

When installing an insert in a closed section, whether it is a rocker panel, A- or B-pillar, or body rail, make sure the closing weld fully penetrates the insert. When closing the job with a butt weld, leave a gap wide enough to allow thorough penetration into the insert. See Figure 19–56.

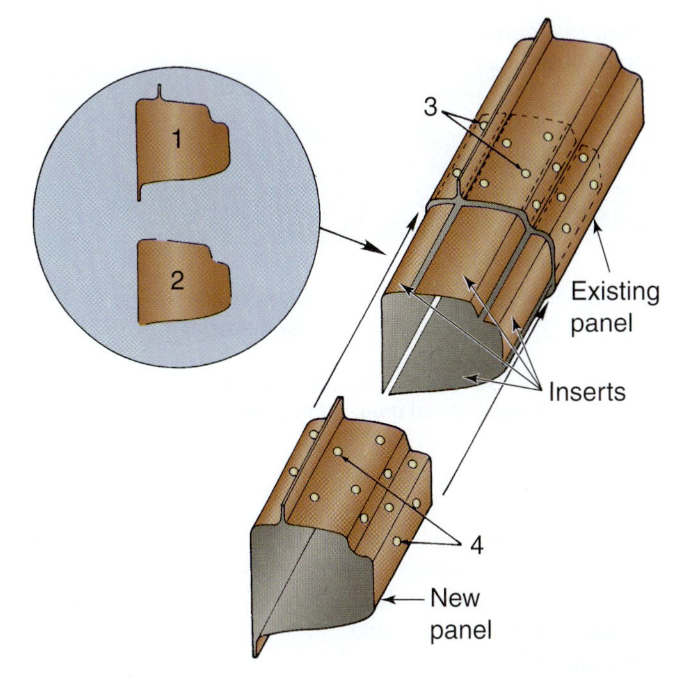

1. Cross section of rocker panel insert material before cutting lengthwise
2. Insert cut lengthwise into sections
3. Insert inside rocker panel, secured with plug welds or sheet metal screws
4. 5/16″ holes for plug welds

Figure 19–55 If practical, cut an insert to fit a rocker panel. You can make it from an old section of rocker panel or from an unused new section. By removing the flange from the insert, you can slide it inside the rocker to provide a strong repair.

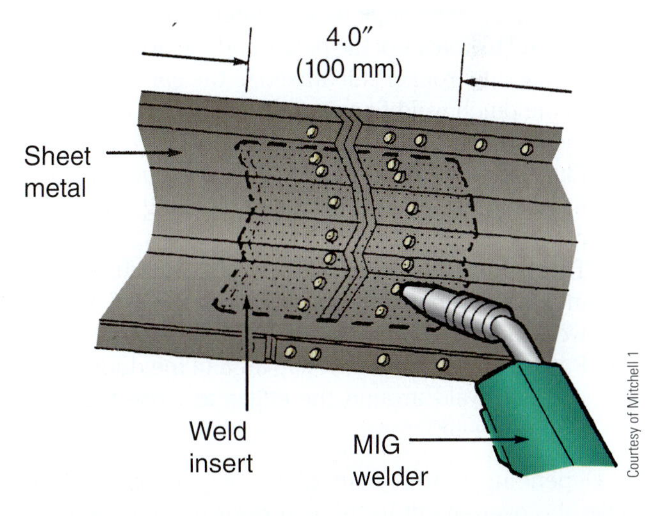

Figure 19–56 A weld insert is often required behind butted panel weld joints. Holes drilled in the outer panels allow you to plug weld the insert to two halves of the outer panel. Then, when you continuously weld a butt joint, there is less chance of burn-through and the joint will be much stronger and more level with surfaces.

The width of the gap depends on the thickness of the metal, but ideally it should not be less than 1⁄16 inch (1.59 mm), nor more than 1⁄8 inch (3.2 mm).

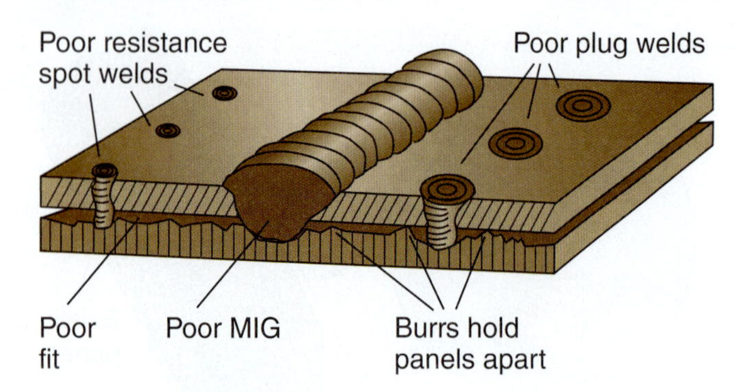

Figure 19–57 Burrs in a joint or along panel flanges will weaken the weld. Grind them off flush with the flange before fitting the parts.

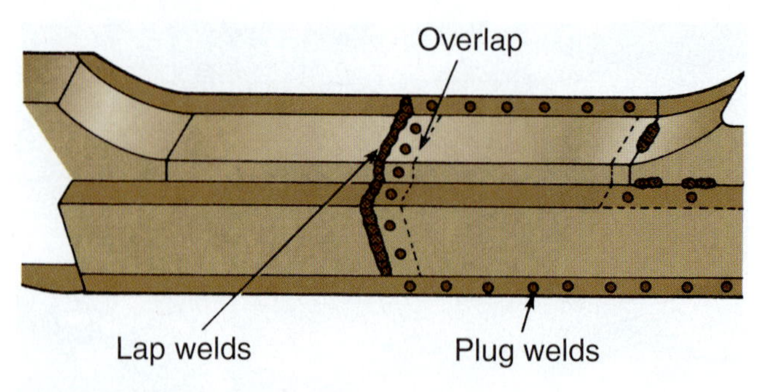

Figure 19–58 Plug welds and a lap weld will secure a rocker panel.

Be careful to remove the burrs from the cut edges before welding. Otherwise, the weld metal tends to travel around and up under the burr. This can create a flawed weld, resulting in cracks and weakening the joint (Figure 19–57).

In general, use the overlap procedure on a rocker panel when installing only the outer rocker or a portion of it. Leave the inner piece intact and cut only the outer piece. One way to make an overlap joint is to make the cut in the front door opening and allow for an overlap there when measuring.

When making this cut, stay several inches away from the base of the B-pillar to avoid cutting any reinforcement underneath it. Use the following procedure:

1. Cut around the bases of the B- and C-pillars, leaving overlap areas around each.
2. Cut out the new outer rocker panel so that it overlaps around the bases of the pillars and the original piece of the outer rocker still affixed to the car.
3. In the pinch weld flanges, use plug welds to replace the factory spot welds.
4. Plug weld the overlaps around the B- and C-pillars, using approximately the same spacing as in the pinch weld flanges.
5. Then lap weld the edges with about a 30 percent intermittent seam, that is, about ½ inch (12.7 mm) of weld in every 1½ inches (38 mm) of overlap edge.
6. Put plug welds in the overlap area in the door opening. Lap weld around the edges to close the joint (Figure 19–58).

Depending on the nature of the hit, you may need to make the overlap cut in the rear door opening and cut out and overlap around the bases of the A- and B-pillars. Use this same basic technique to replace the entire outer rocker. In this version, cut around the bases of all three pillars and overlap all three bases in the same way as before.

Figure 19–59 shows how service information for one particular vehicle recommends replacing a B-pillar with a section in the rocker panel. Note how the vehicle-specific instructions detail where to make cuts and how to place new welds.

Sectioning A-Pillars

The front pillars, or A-pillars, extend up next to the edges of the windshield. They must be strong to protect the passengers. They are steel box members that extend down from the roof panel to the main body section.

A-pillars can be either two-piece or three-piece components. They can be reinforced at the upper end or the lower end, or both. However, they are not usually reinforced in the middle. Therefore, A-pillars should be cut near the middle to avoid cutting through any reinforcing pieces. The middle is also the easiest place to work (Figure 19–60).

To section an A-pillar, use a straight-cut butt joint with an insert or an offset butt joint without an insert. The butt joint with insert repair is made in the same manner as already described for the rocker panel.

Figure 19–61 shows the details for replacing an A-pillar for one make and model vehicle. Note how each layer of metal must be cut at different locations. This will allow you to weld in each layer or section, one at a time.

The A-pillar insert should be typically 4–6 inches (102–152 mm) in length. After cutting the insert lengthwise and removing any flanges, tap the pieces into place. Secure the insert in place with plug welds. Then, close all around the pillar with a continuous butt weld (Figure 19–62).

To make the offset butt joint, cut the inner piece of the pillar at a different point than the other piece was cut, creating the offset. Whenever possible, try to make the cuts between the factory spot welds so that it will not be difficult to drill them out. Make the cuts no closer to each other than 2–4 inches (50–100 mm). But the sections together and close them all around with a continuous weld.

With rollover damage, it is often necessary to replace the front roof panel that fits between the two A-pillars. Plug welds often secure the bottom of the roof panel, and a braze joint is sometimes recommended along the top, as shown in Figure 19–63.

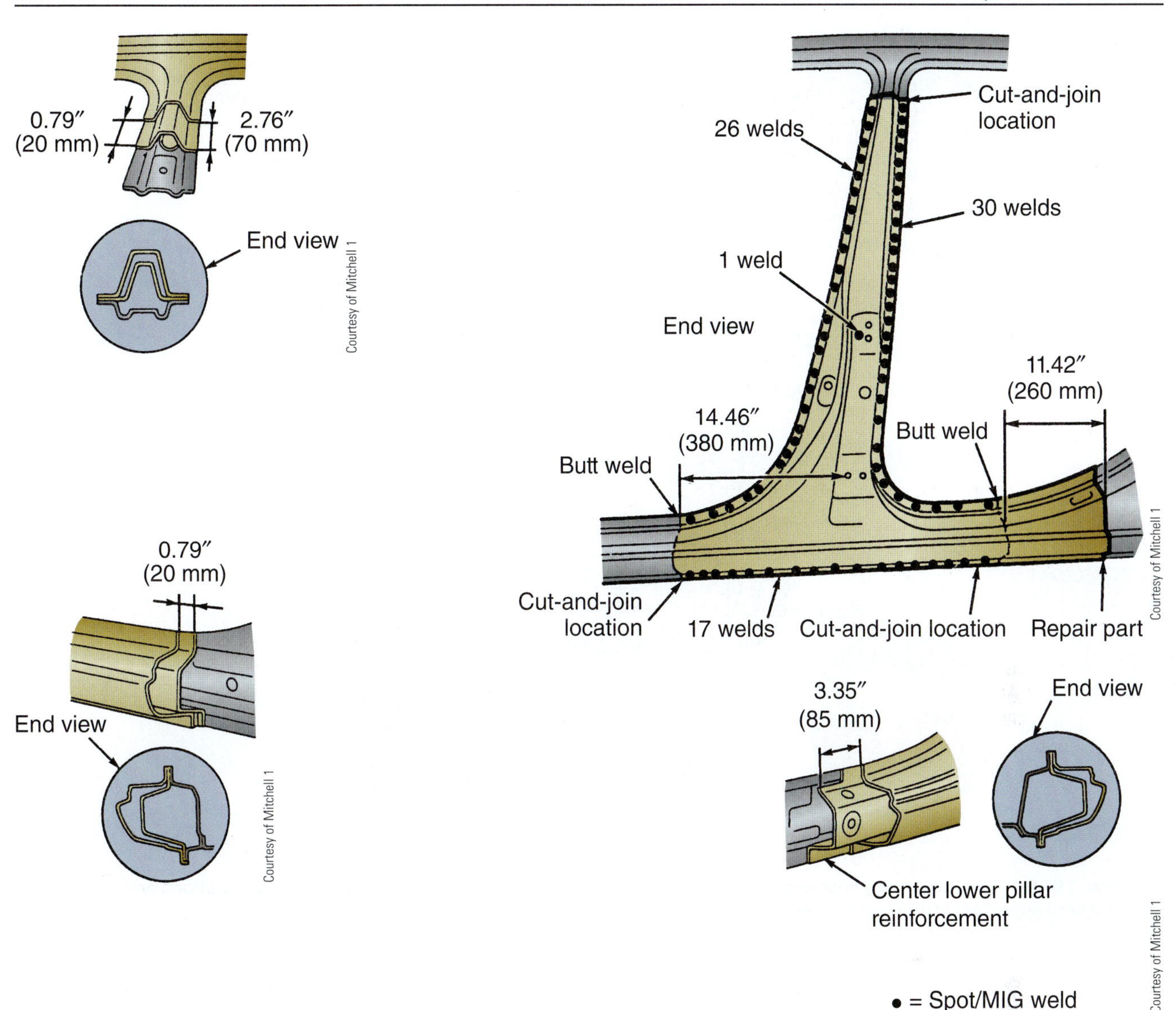

Figure 19-59 Here is a service manual illustration recommending weld procedures for a particular make and model vehicle.

Figure 19-60 When replacing important structural panels, always refer to the service manual for directions, because construction methods and procedures vary.

Sectioning B-Pillars

Center pillars, or B-pillars, are the roof supports between the front and rear doors on four-door vehicles. They help strengthen the roof and provide a mounting point for the rear door hinges.

For sectioning B-pillars, two types of joints can be used: the butt joint with insert and a combination of offset cut and overlap. The butt joint with insert is usually easier to align and fit up when the B-pillar is a relatively simple two-piece cross section without a lot of internal reinforcing members. The insert provides additional strength.

Be sure to cut below the seat belt D-ring mount low enough to avoid cutting through the D-ring anchor reinforcement. The majority of B-pillars have them. In the case of the B-pillar, use a channel insert only in the outside piece of the pillar. The D-ring anchor reinforcement welded to the inside piece prevents the installation of an insert there.

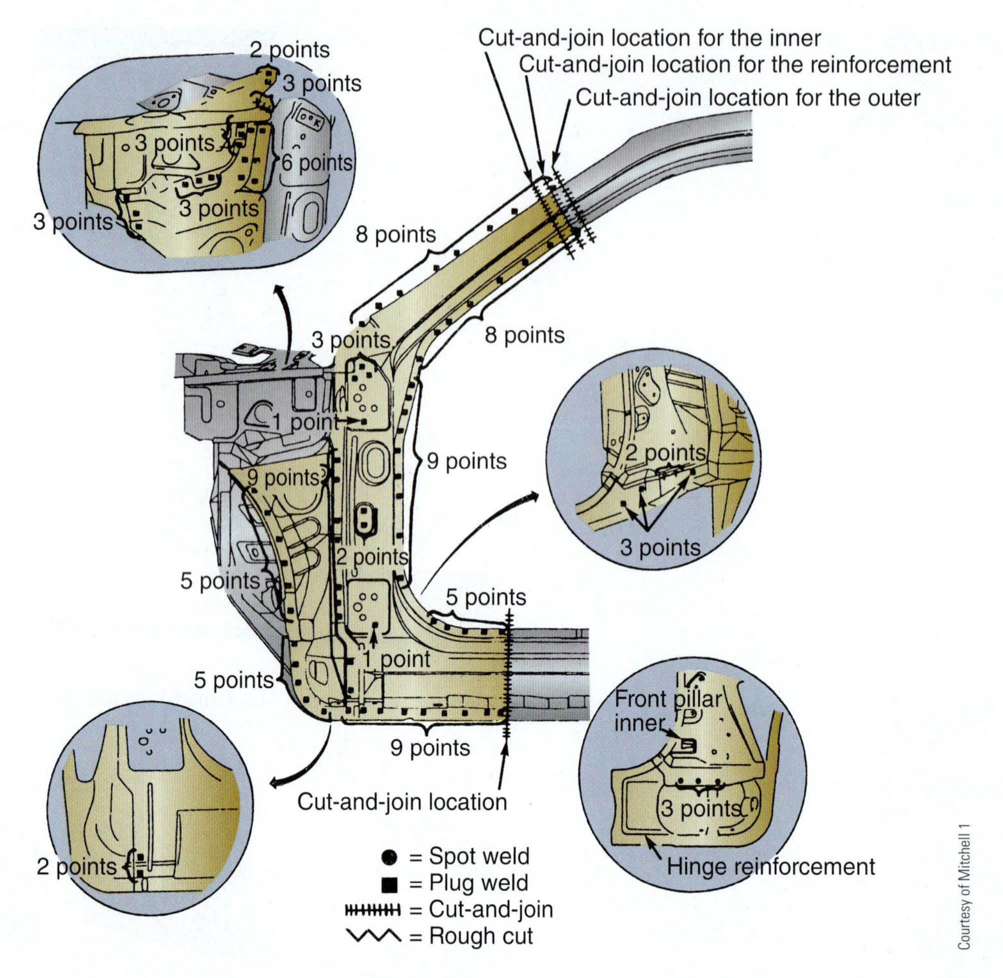

Figure 19–61 Note how many spot and plug welds are recommended on this front hinge pillar. Cut locations are also given. This is a very important structural area of the vehicle, and it must be installed to factory specifications.

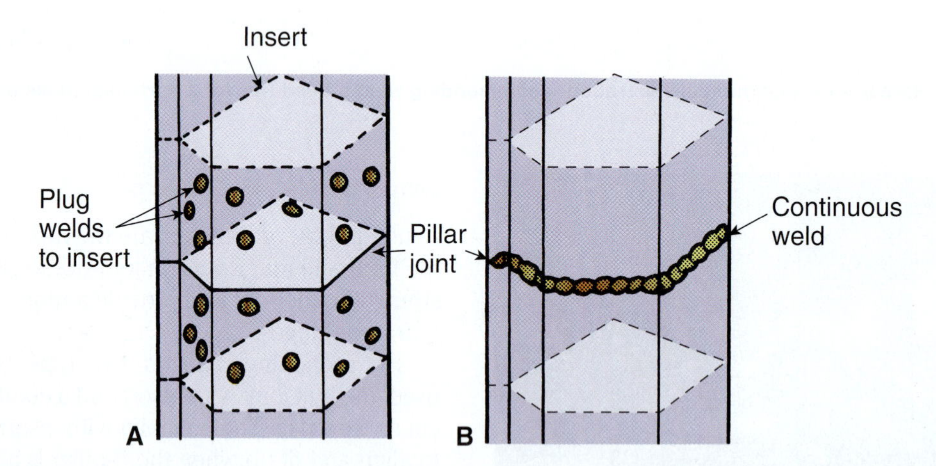

Figure 19–62 With many box or closed section welds, you must use an insert or backing to ensure structural integrity. (A) Plug welds are used to secure outer A-pillar panels to the insert. (B) A continuous butt weld is then used to join the two halves of the pillar together and to the insert.

Begin by overlapping the new inside piece on the existing one, rather than by butting them together, and lap weld the edge. Then, secure the insert in place with plug welds and close the joint with a continuous butt weld around the outer pillar (Figure 19–64).

You may want to obtain a recycled B-pillar and rocker panel assembly and replace them as a unit. This is because any time a B-pillar is hit so hard that it needs to be replaced, the rocker panel is almost invariably damaged as well. Install the upper end of the B-pillar with

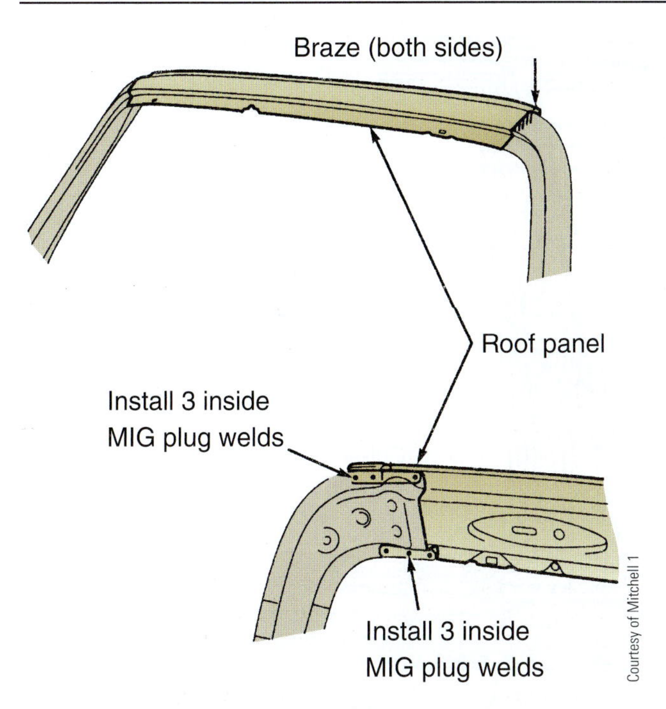

Figure 19-63 This manufacturer recommends plug welds and brazing to secure this front roof panel to the A-pillars.

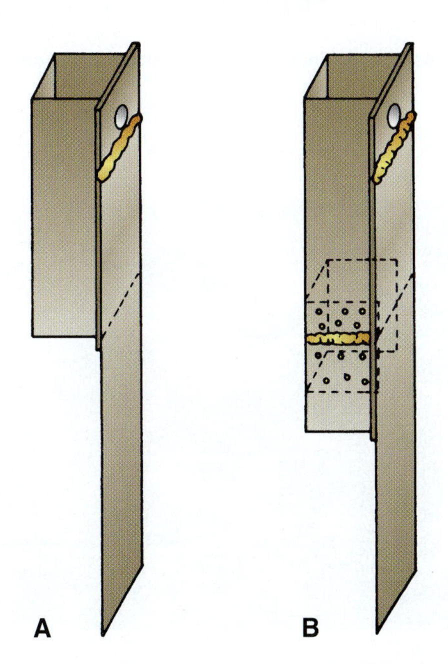

Figure 19-64 (A) Lap weld the inner panel; (B) plug and butt weld the outer panel.

either of the two approved types of joints and make a butt joint with insert in the rocker panel in the manner already shown.

If the main damage is in the rear door opening, make the butt joint with insert in the front door opening. Install the other end of the rocker in its entirety. If the main damage is in the front door opening, reverse the procedure.

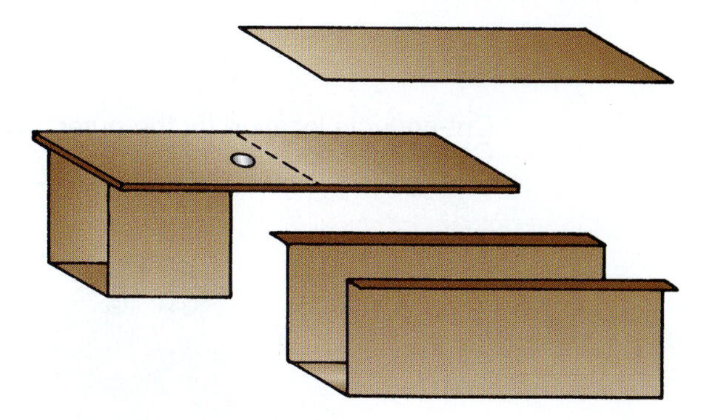

Figure 19-65 This is a combination offset and overlap joint.

Generally speaking, the combination offset and overlap joint is used more often when installing new parts and when working with separate inside and outside pieces (Figure 19–65).

Use the following procedure to make a combination offset and overlap joint:

1. Cut a butt joint in the outside piece above the level of the D-ring anchor reinforcement.
2. Make an overlap cut in the inside piece below the D-ring anchor reinforcement.
3. Install the inside piece first, with the new segment overlapping the existing segment.
4. Lap weld the edge.
5. Put the outside pieces in place, and make plug welds in the flanges. Close the section with a continuous weld at the butt joint (Figure 19–66).

Usually, it is best to use the offset and overlap joint on a B-pillar with three or more pieces in its cross section.

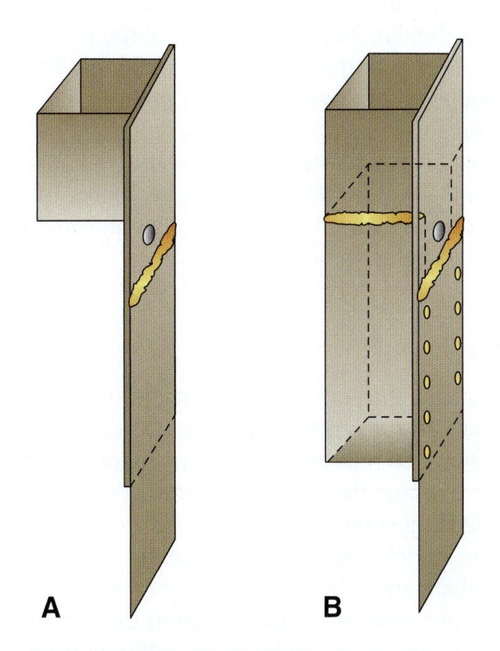

Figure 19-66 Here is an example of creating a combination offset and overlap joint: (A) lap welding inside and (B) plug and lap welding outside.

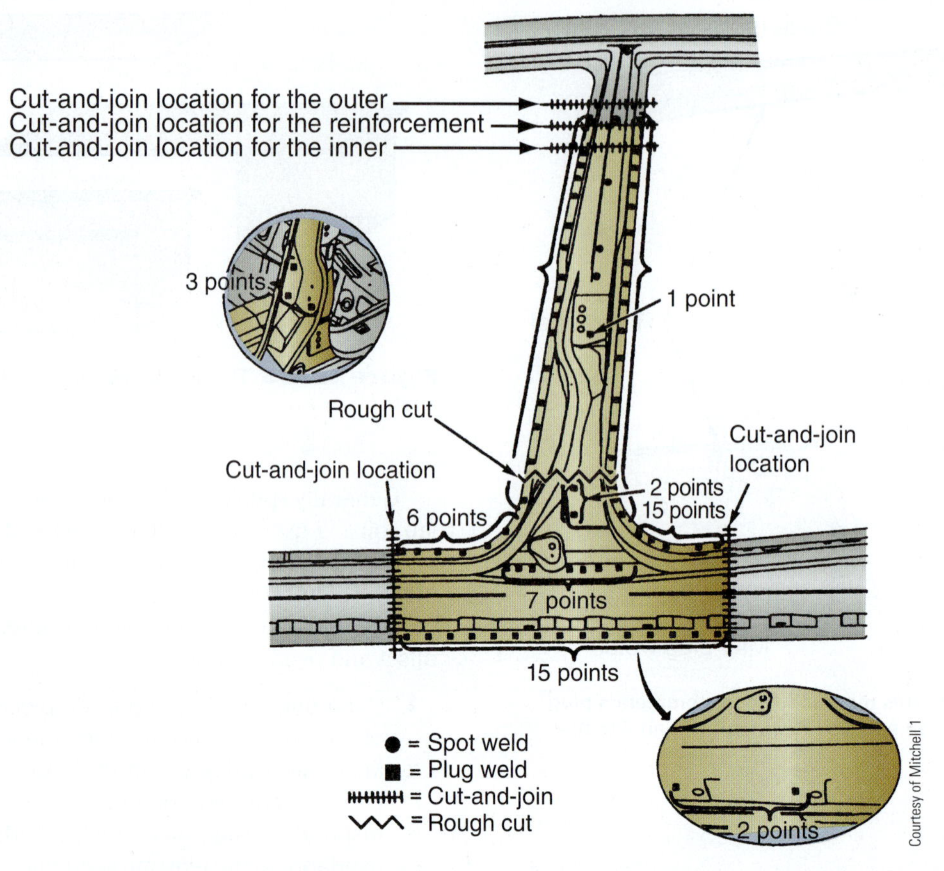

Figure 19-67 **Study details of the sectioning in a new center or B-pillar for one model car. Note the three cut locations at the top of the pillar section and the rough cut locations.**

This design would make it difficult to install an insert. In fact, sometimes the offset and overlap procedure is mandatory, because it is not possible to install an insert.

Figure 19–67 shows the factory instructions for sectioning a B-pillar. Note how the top of the pillar must be cut one layer at a time in different locations. This allows the installation and welding of the outer panel, the reinforcement, and the inner panel.

Again, always remember to measure and adjust structural panels before welding. In Figure 19–68, the technician is using a hydraulic ram to realign new and existing roof panels before final welding.

Figure 19-68 **This technician is using a portable hydraulic ram to adjust the location of the roof before final welding.**

Replacing Foam Fillers

Some manufacturers place foam inside of panels. **Foam fillers** are used to add rigidity and strength to structural parts. They also reduce noise and vibrations. Cutting and welding damage the foam. Consequently, replacing the foam fillers must be part of the repair procedure.

Some vehicle manufacturers use urethane foam in A- and B-pillars and other locations. The manufacturer may or may not consider the foam filler to be structural. The use and location of foam fillers are different from vehicle to vehicle. Follow manufacturer's recommendations for replacing or sectioning foam-filled panels.

Single-part urethane foams made for home use *cannot* be used for replacing automotive foam fillers.

Some OEM replacement parts come with the foam already in them. When parts come without foam filler or foam filler needs to be replaced, a product designed specifically for this application must be used to fill the panel.

When sectioning foam-filled A-pillars, the foam filler is removed in the repair area. It is then replaced after all welding is completed. Some manufacturers have specific recommendations for the type of replacement foam filler needed. The repair usually calls for a foam filler that, when cured, does not change in volume due to differences in temperature and humidity. There may also be specific requirements for foam density given in ounces per cubic inch (grams per cubic centimeter).

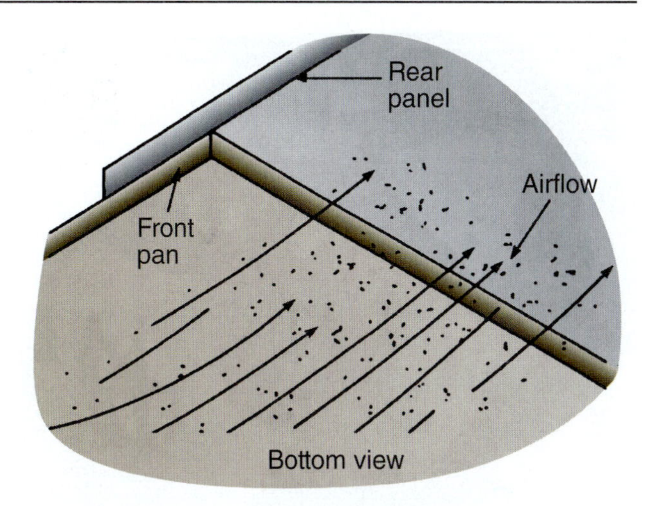

Figure 19–69 The rear section of a floor pan should overlap to shield the joint from airflow and water.

Sectioning Floor Pans

When sectioning a floor pan, do not cut through any reinforcements, such as seat belt anchors. Always make sure the rear section overlaps the front section. You want the edge of the bottom piece, under the car, to point rearward. This type of overlap prevents road debris and water splash from streaming into the joint between the parts (Figure 19–69).

To section a floor pan, use the following procedure:

1. Join all floor pan sections with an overlap.
2. Plug weld the overlap, putting the plugs in from the topside, downward (Figure 19–70).
3. Caulk the top, forward edge with a recommended, flexible body caulk.
4. On the bottom side, lap weld the floor pan edge with a continuous bead.
5. Cover the lap weld with a primer, a seam sealer, and a topcoat. The primer helps the sealer hold better, and the topcoat completes the protection. This ensures that there will be no deadly carbon monoxide intrusion from any leaking of the engine exhaust system through the joint and into the passenger compartment.

Sectioning Trunk Floors

When sectioning a trunk floor, in general, follow the basic procedures just described for the floor pan, with some variations (Figure 19–71):

In a collision that requires sectioning of the trunk floor, the rear rail usually requires sectioning.

1. There is generally some kind of a cross member under the trunk floor near the rear suspension. Whenever possible, section the trunk floor above the cross member's rear flange. Section the rail just rearward of the cross member.

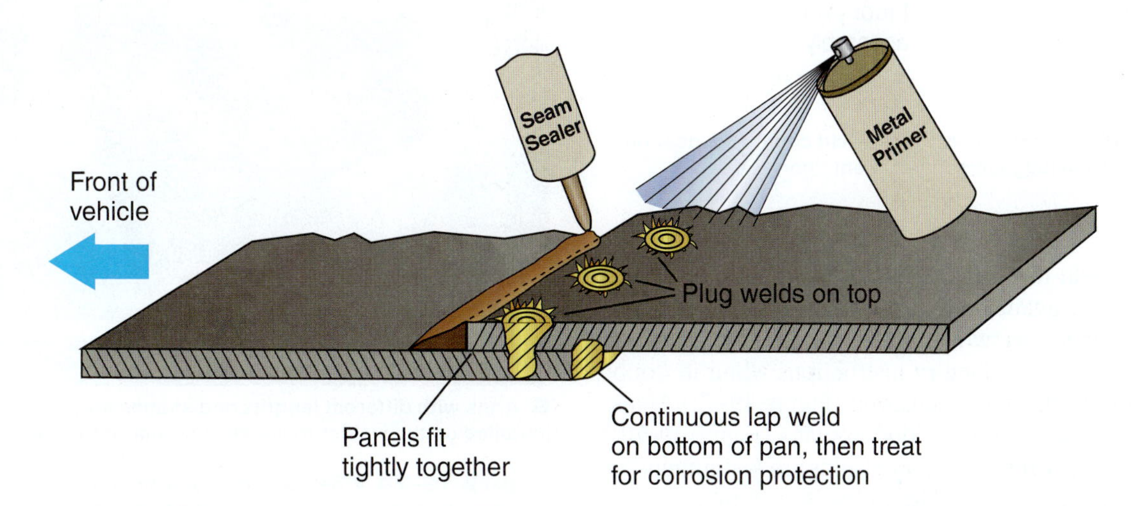

Figure 19–70 Note the basic method of installing floor pans. Plug weld from the top and lap weld the bottom edge; prime and seal the joint. Note that weld-through primer may require several hours to cure properly before you can apply seam sealer.

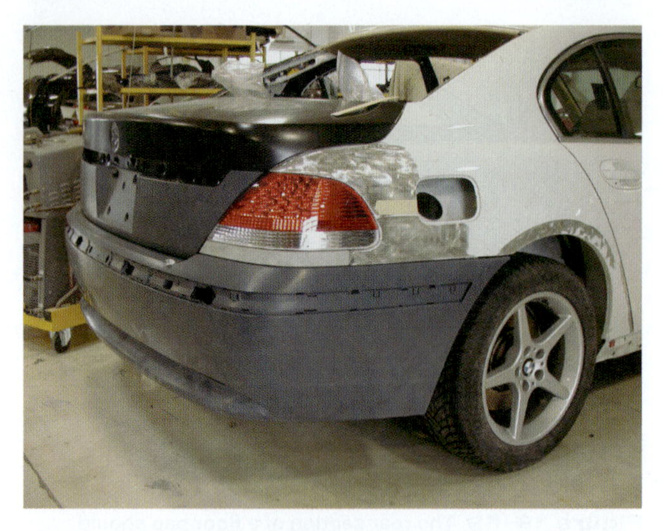

Figure 19–71 When replacing major structural panels, you should install all adjacent panels to make sure everything fits properly. Here the trunk lid, bumper, and even taillights have been test fitted before final welding of the rear rails and trunk pan.

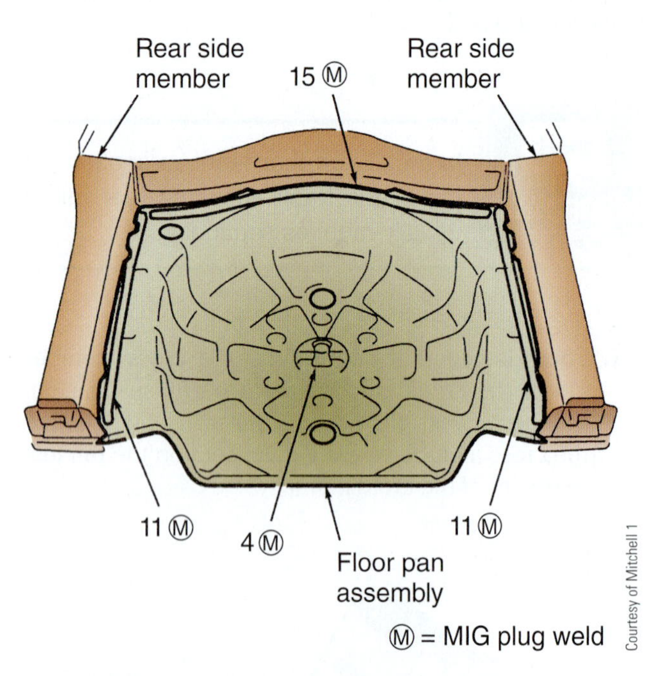

Courtesy of Mitchell 1

Figure 19–72 Here is a factory weld count and location illustration for a luggage compartment floor.

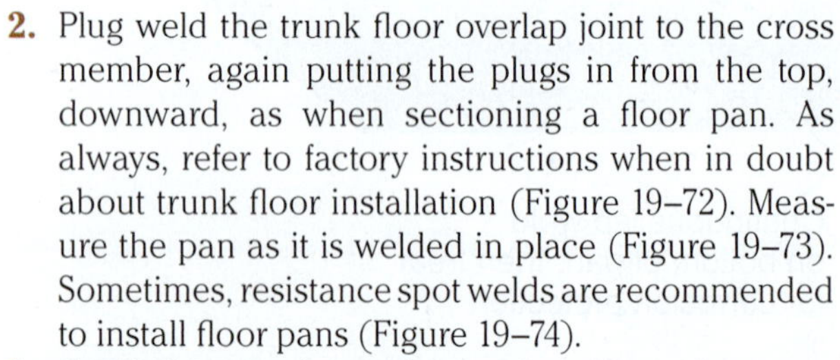

2. Plug weld the trunk floor overlap joint to the cross member, again putting the plugs in from the top, downward, as when sectioning a floor pan. As always, refer to factory instructions when in doubt about trunk floor installation (Figure 19–72). Measure the pan as it is welded in place (Figure 19–73). Sometimes, resistance spot welds are recommended to install floor pans (Figure 19–74).
3. Caulk the top, forward edge just like a floor pan seam (Figure 19–75).

Courtesy of Liss CarStar Collision Center

Figure 19–73 This technician is using a measuring system to check the location of new panels before and during the welding process.

Photos courtesy of Car-O-Liner Company

A A resistance spot welder clamps panel flanges together and then sends a high current through the electrodes to fuse the panels together without using a filler wire.

Photos courtesy of Car-O-Liner Company

B Arms with different lengths and shapes may have to be installed on the welder to access the rear of the panel.

Figure 19–74 When possible, you should use a resistance spot welder to install new body panels. When the vehicle was manufactured, resistance spot welds, not plug welds, were used to assembly the body structure.

A Use a caulking gun to apply a bead of seam sealer to the joint between panels. Make sure the primer has had enough time, up to 24 hours, to cure before applying seam sealer.

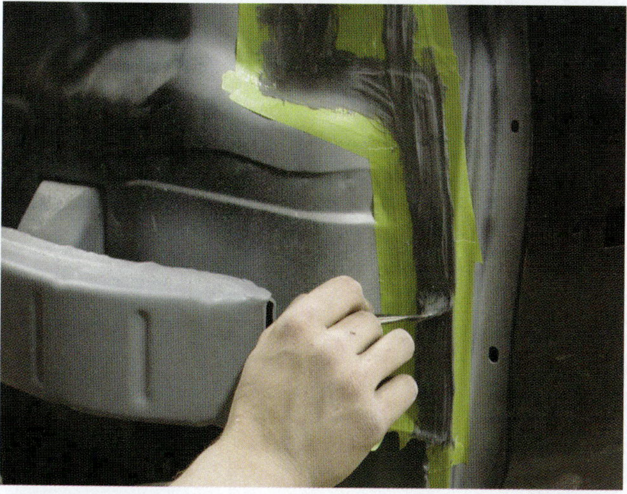

B Masking tape can be applied to each side of the seam. Then a brush can be used to smooth and level the seam sealer for a factory look.

C When the tape is removed, the seam sealer will keep moisture and air from entering the joint between the welded panels.

Figure 19–75 When instructed to do so in service literature, caulk the seams of newly welded panels to keep water and poisonous gas out of the exhaust system.

4. On the bottom side, a lap weld is not always necessary because of the strength provided by the cross member. However, on cars where the trunk floor section is not above a cross member, the lower edge must be lap welded.

In both cases, cover the bottom-side seam with a primer, a seam sealer, and a topcoat. Sealing against poisonous carbon monoxide gas is critical because of the proximity of the trunk floor to the tailpipe.

19.5 SECTIONING SIDE MEMBERS (FRAME RAILS)

Certain structural components have *crush zones*, or buckling points, designed into them for absorbing impact energy in a collision. This is particularly true of the front and rear members or rails, because they take the brunt of the impact in most collisions. Crush zones are now designed into all front and rear rails.

You can often identify crush zones by their appearance. Some appear convoluted or crinkled. Others have dents or dimples, and others have holes or slots so the rail will collapse at these points. Crush zones are ahead of the front suspension and behind the rear suspension (Figure 19–76).

Avoid cutting near crush zones. Sectioning procedures can change the designed collapsibility if improperly located. If a rail has suffered major damage it will be buckled in the crush zone, so the crush zone will usually be easy to locate. Where only moderate damage has occurred, be very careful. The hit might not have used up the entire crush zone. Be aware of other potential areas where designed-in collapse might occur.

Vehicle and part manufacturers often give recommendations about how to install frame rail replacements. Refer to these instructions (Figure 19–77).

They provide critical information about where and how to cut. They also specify the type of weld joint that

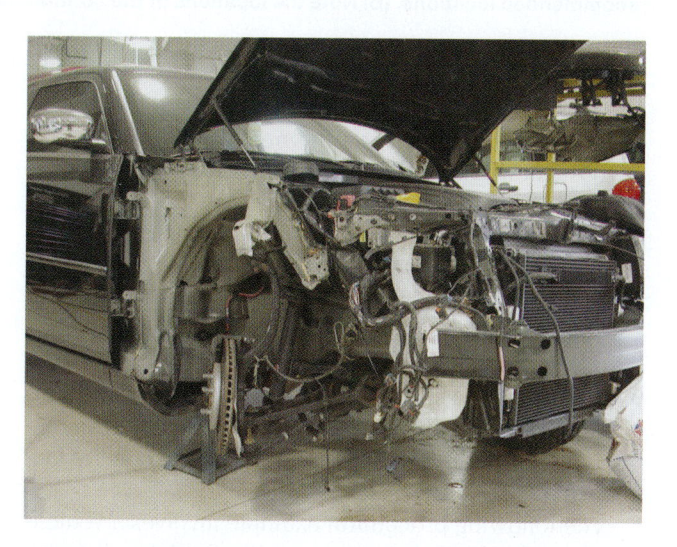

Figure 19–76 With a hard front hit, the front frame rails are often damaged and require partial or full replacement.

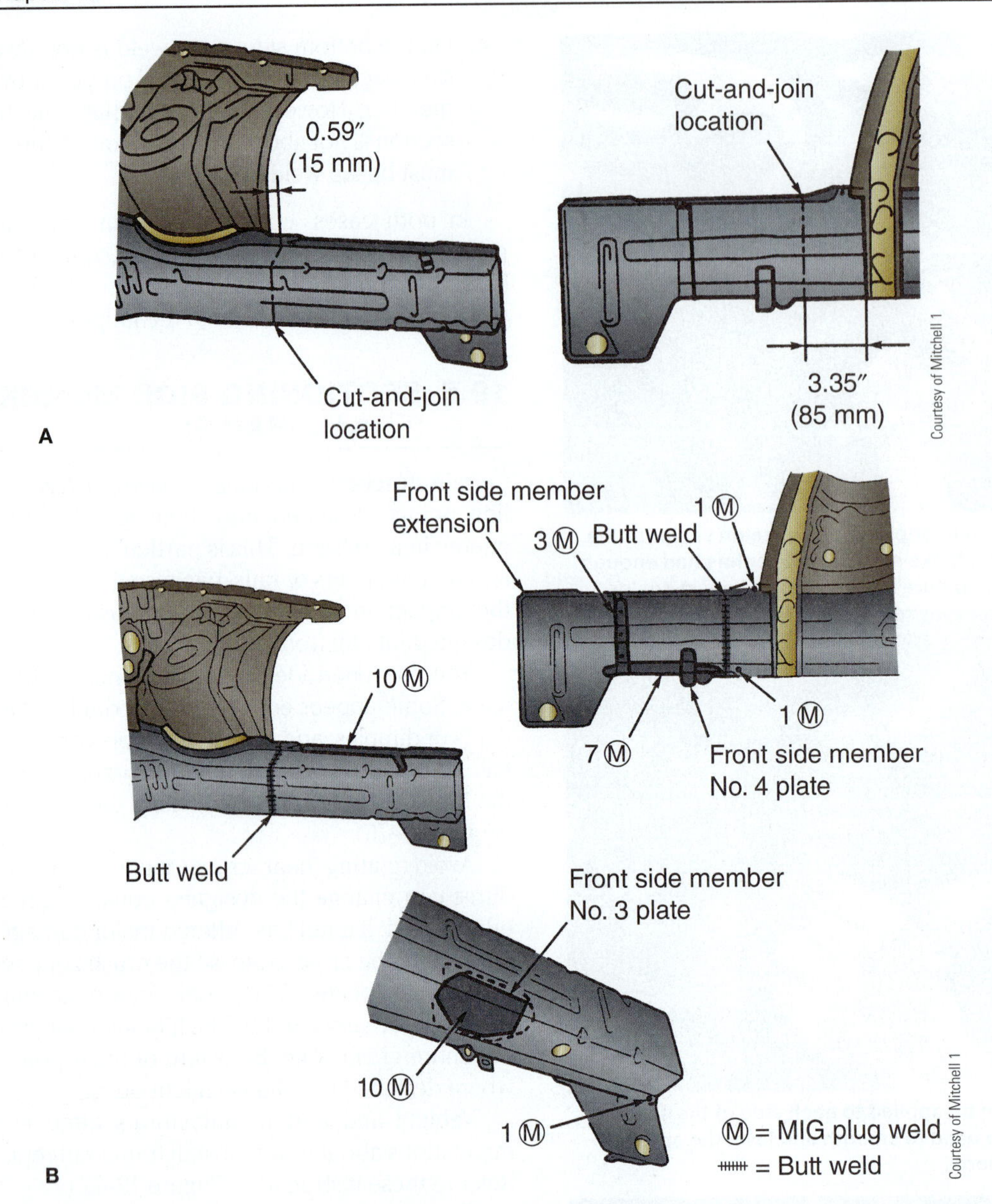

Figure 19-77 Rail and side member section locations vary with each make and model vehicle. Always refer to manufacturer recommendations in the service literature. Here is an example for one particular vehicle. (A) Cuts must be straight and at recommended locations. (B) Note the locations of the continuous butt welds.

will work best with the particular rail. In addition, these factory instructions tell you the best way to secure and align the new rail section.

Recent testing has determined that lap welding of front section frame rails and rocker panels can yield a tighter fitting section and superior corrosion protection when compared to inserts.

The new rail section must be cut to match and overlap the remaining section. Fit the new section in place with clamps. Measure the clamped section to make sure it is aligned properly. Use an MIG welder to make the recommended plug and seam welds as described in the factory service literature.

The following procedural example involves a vehicle that has sustained damage requiring the left rail to be sectioned:

1. Locate and drill out the factory spot welds that attach the upper rail to the cowl at the base of the windshield. A propane torch, scraper, and wire brush might be needed to remove sealant or caulk from the spot weld areas. Set the torch on low heat to avoid burning the seam sealer.
2. Remove the spot welds that secure the upper rail to the rear outer flange of the strut tower. They are normally visible through a hole at the rear portion of the upper rail (Figure 19–78).
3. Remove the spot welds that attach the strut tower to the rail extension panel at the base of the strut tower. Remove any seam sealer covering these welds inside the engine compartment. The undersurface is frequently coated with sealer and sound-deadening material.

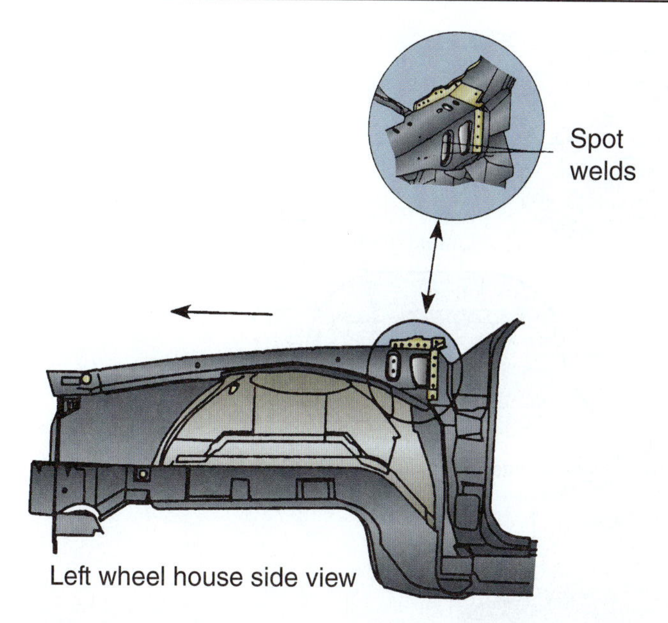

Figure 19-78 Spot welds that secure the upper rail to the strut tower can sometimes be seen through a hole at the rear of the upper rail.

4. The lower rail sectioning is often done forward of the center of the strut tower. The sectioning procedure uses a staggered cut of the inner and outer lower rail with a lap joint at both cut lines. There is usually an inner rail reinforcement located in the area of the section. Because of this reinforcement, this part of the rail is ideal for sectioning.
5. There are spot welds that attach the reinforcement to the inside of the lower rail. These must be removed before any cuts are made. They are visible on the wheel house side of the rail.
6. The location of the cut on the engine side is usually about 12 inches (305 mm) from the cowl (Figure 19–79). This will position the cut line near the end of the inner reinforcement. Refer to the service manual to get an exact dimension for making this cut.

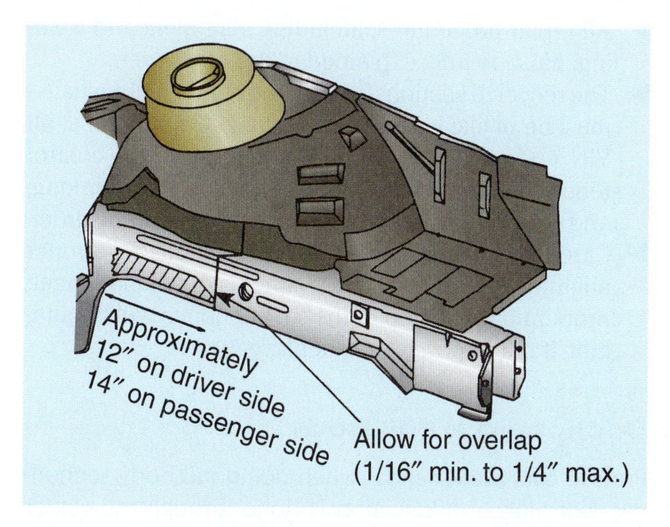

Figure 19-79 This is a typical sectioning cut on the engine side of the rail.

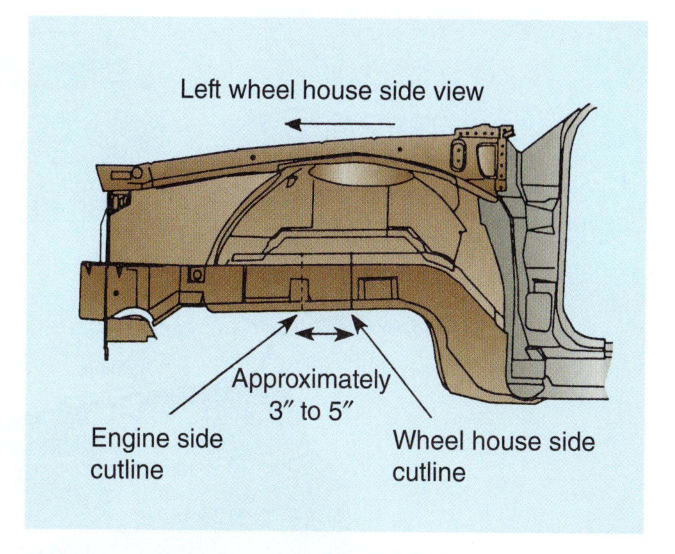

Figure 19-80 This sectioning cut on the outer rail is also typical. Refer to manufacturer directions for the exact location.

7. The outer rail cut line (wheel house side) is typically made 3–5 inches (76–127 mm) rearward of the engine side cut line (Figure 19–80).
8. To achieve the correct overlap, carefully "split" the corners on the original structure at the exposed end. These splits should not exceed ¼ inch (6.4 mm). Any part of the splits that is exposed after fit-up must be welded closed.
9. It is very important that the replacement structure is positioned over the original structure. This will allow the application of the corrosion protection to be more effective. The open portion of the joint will face the open end of the rail.
10. Separate the opposite side lower rail extension from the lower rail by first drilling the spot welds that secure the radiator support and apron extension panel. Next, carefully fold the apron extension panel upward to expose the other spot welds that attach the extension to the rail.
11. After all the spot welds have been removed and the offset cuts made, the damaged assembly can be removed from the vehicle.
12. Preparation of the used assembly (for example, spot weld removal and lower rail offset cut) is identical to that of the damaged assembly. It is very important to inspect, measure, and, if necessary, straighten the used assembly to the proper dimensions prior to installing it on the vehicle. Remember to add to the length measurement on the replacement rail to allow for the overlap.
13. All adjoining flanges and weld areas must be cleaned using a propane torch and wire brush. Do not grind or burn off any galvanized coatings. After cleaning and before welding, a weld-through primer must be applied to all bare metal mating surfaces.

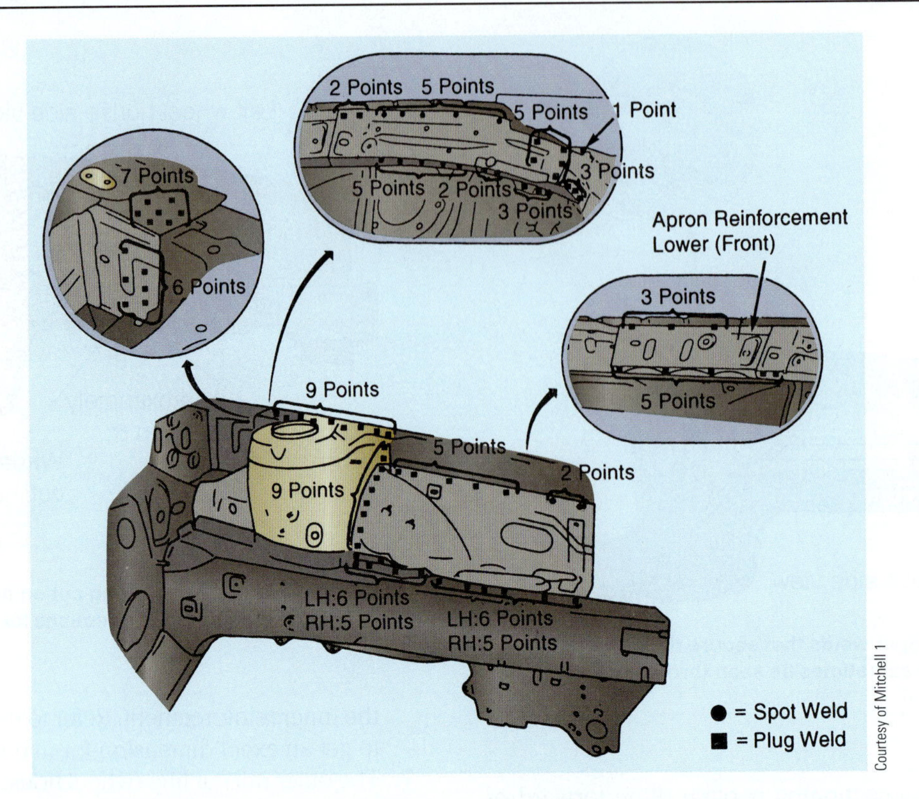

Figure 19–81 Here are weld locations for one type of front wheel apron. Note that both spot and plug welds are recommended.

14. After the used assembly is installed, measuring equipment can be used to check the correct position. Then the assembly can be clamped in place.
15. After checking that all dimensions are within tolerance, the assembly can be welded. Continuous welds should be made in alternating segments of ½–¾ inch (13–19 mm) to minimize distortion. Be sure to completely close all seams, and do not leave any gaps.
16. Corrosion protection, including refinishing of the replacement pieces, should be completed as detailed in Chapter 20.

Figure 19–81 shows how to weld in a new front apron for one make and model vehicle. The apron is normally welded in place after the frame rail. Again, clamp and tack weld the part and double-check panel alignment before final welding.

Figure 19–82 shows how to section the front cap of a full-frame vehicle. They are often damaged in frontal collisions.

Full Body Sectioning

One of the most drastic repairs that can be performed is full body sectioning. **Full body sectioning** involves replacing the entire rear section of a collision-damaged vehicle with the rear section of a salvage vehicle. It may be more economical than trying to rebuild the damaged vehicle using new parts. Full body sectioning requires the highest quality workmanship possible.

When the individual components are properly sectioned, aligned, and welded using the correct techniques and procedures, full body sectioning is a suitable and satisfactory procedure. Vehicles repaired by full body sectioning are completely crashworthy. This has been tested and proved time and again. Keep in mind, however, that full body sectioning is not a frequently required procedure, and full disclosure should be made to the car owner before repairs are started.

A discussion among the insurance representative, car owner, and repairer must be conducted and the following points covered:

- All repair procedures, including alignment and welding, must be fully explained to the car owner.
- The recycled sections—both body and mechanical—must be of like kind and quality. Always verify that all VIN code identifications and EPA emission control requirements are met and that all suspension, braking, and steering components are in proper working order.
- Carefully inspect front and rear sections for proper alignment before cutting. If either is out of alignment, proper fit and lineup of the section joints will be difficult, if not impossible, to achieve.

Sectioning a Full Body

Precise cuts are essential when doing full body sectioning. Normally, a full section is done through the center area or passenger compartment of the vehicle, with minimum reinforcements.

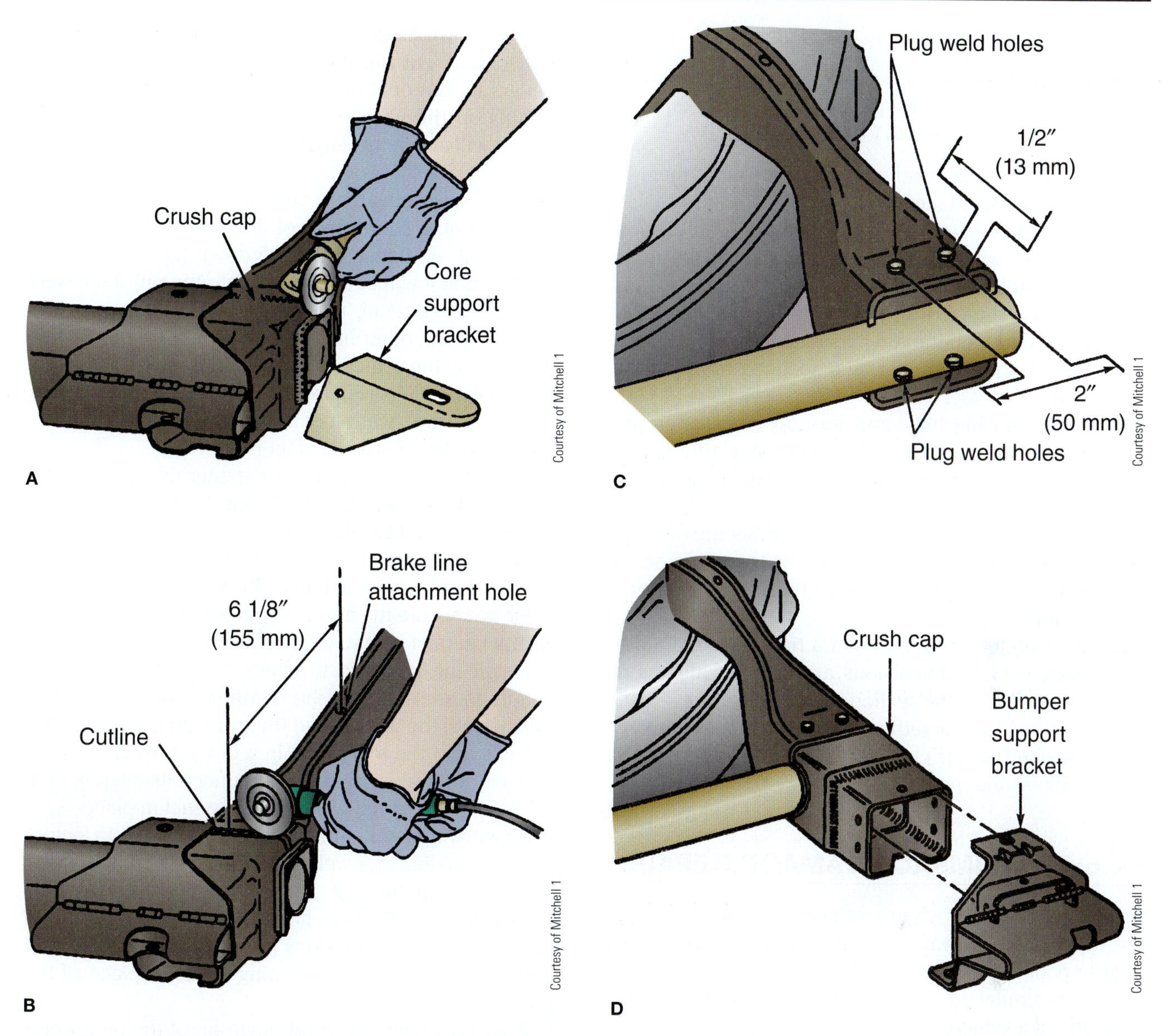

Figure 19-82 **Note the basic steps for sectioning a bracket on the front of a full frame. (A) Use a power cutoff wheel to remove the radiator core support bracket from the thick frame rail to access the crush cap. (B) Grind out continuous welds that secure the crush cap on the full-frame rail. (C) Drill out spot welds on the damaged crush cap. (D) The new part can then be welded to the full frame.**

Full body sectioning procedures often require sectioning the two A-pillars, two rocker panels, and the floor pan. When this procedure is properly performed, sectioned vehicles have been shown to be as strong and serviceable as an undamaged vehicle.

Full body sectioning is complicated by antilock brake systems (ABS). The replacement vehicle must be so equipped, or the ABS parts must be retrofitted. The location and function of vehicle computers may change, even within the same production year of a given vehicle. Check the locations of these computers on both the damaged vehicle and the salvage section.

The undamaged front half of one car must be cut and joined to the undamaged rear half of another. Butt joints with inserts are normally used in the middle of the A-pillars and in the two rocker panels. An overlap joint is normally used in the floor pan. The rocker panel and floor pan cuts are often made in the middle of the front door opening to avoid any brackets or reinforcements in the A- and B-pillars.

Remember that the floor pan might have reinforcements and brackets that need to be removed before sectioning. Reinforcements can be left on the replacement rear half to aid in alignment. Proper corrosion protection must be restored during and after sectioning.

The cut line on both vehicles must be carefully measured and marked. The cut lines must align perfectly for the sectioning to be done properly. A cutoff wheel and saber saw are normally used to cut the sections to size.

Joining the Full Body Sections

After the front and rear sections have been trimmed to fit, drilled for plug welds, and prepared with weld-through primer, follow these steps to join the sections:

1. Install the rocker and pillar inserts. Clamp them in place with sheet metal screws.
2. Place the A-pillar inserts in the upper or lower portion of the windshield pillar, depending on the angle and contour of the windshield.
3. Fit the two halves together by first joining the rocker panels and then the A-pillars. Clamp the rocker and pillar flanges to prevent the sections from pulling apart.
4. Check the windshield and door openings for proper dimension, using a tram gauge or a steel rule. If possible, install the doors and windshield to verify proper alignment.
5. When proper alignment is achieved, secure overlapping areas with sheet metal screws to pull the seam areas together and hold the sections together during welding.
6. Using centerline gauges and a tram gauge, double-check vehicle dimensions and section alignment before welding the sections together.
7. Weld sections together using techniques already described in this chapter for joining rocker panels, A-pillars, and the floor pan.

19.6 REAR IMPACT DAMAGE REPAIR

As discussed earlier, when a vehicle is hit from the rear, the bumper cover, rear panel, floor pan, rear rails, and quarter panels are often damaged. The repair of the rear section is similar to the procedures described for the front end of a vehicle.

Basically, you must remove the bumper cover and other fastened parts that are over the top of the structural damage (taillight lenses, bumper cover, trunk lid, and so on). Then you can measure and pull out damage to parts that will remain on the vehicle. For example, measure the dimensions of the rear frame rails or side members and compare your measurements to specifications. If pushed forward, you would have to use the frame rack to pull the rails and other panels back into alignment.

After pulling, badly damaged parts that must be replaced can then be removed. Minor damage to the trunk floor pan and quarter panels can often be straightened without replacement.

For example, the rear body panel is often badly damaged and must be replaced. Clean, find, and remove all of the spot welds holding the rear panel to the vehicle. Remove the rear panel. Then, straighten and smooth the flanges on the panels that remain on the vehicle.

Punch or drill the recommended number of holes for plug welds in the new rear panel. Clean the flanges on all parts to be welded and coat them with weld-through primer.

If you have to section a rear rail during trunk panel replacement, cut the damaged rail off or section it as needed so that only the undamaged portion of the rail remains on the vehicle. Measure and section the new rail so it will match the length of the existing sectioned rail (Figure 19–83).

To aid in welding the section together, make an insert by cutting off a short piece of the unused piece of new rail. Cut off the flange and slice down the center of the insert so that it will fit fully down into the rail. Tack weld the insert into the rail.

Then fit the other half of the new rail to the existing piece still on the vehicle. Clamp the rail sections together using locking pliers (Figure 19–84).

Measure the location of the new rail section to make sure it is located correctly (Figure 19–85).

Weld the rear rail sections together using a continuous weld along the butt ends (Figure 19–86). As you weld the sections together, double-check your measurements. Install and measure the new trunk floor panel. Then weld it into place as shown in Figure 19–87.

Fit and secure the new rear panel to the vehicle using locking pliers or screws (Figure 19–88).

Install the trunk lid to check the alignment of surrounding panels. Adjust the quarters and rear panel as needed until the gap around the trunk lid is acceptable.

Tack weld the rear panel in a few places around the perimeter of the panel and again check its alignment. If within tolerance, final weld the rear panel in place.

As with other structural panels, always refer to factory recommendations for weld type, count, and locations to correctly secure the rear panel.

In Figure 19–89, note how spot and plug welds are suggested. If your shop does not have a resistance spot welder, you will have to use plug welds in place of the spot welds.

After final welding, grind down all of the plug welds until they are almost flush with the panel flanges (Figure 19–90).

If screws were used to secure parts, weld them shut and grind them flush. Grind your plug welds down enough so that they look like factory spot welds. A small air grinder with a flexible backing and disc works well.

After grinding the welds, restore anticorrosion protections to the ground flanges and seams between parts (Figure 19–91).

19.7 ANTIRUST TREATMENTS

The application of antirust agents is necessary not only before welding but also before and after the painting process. Welded panel joints are treated with weld-through primer before they are fused together. Some weld joints must also be sealed with body sealer before finishing. Undercoating or an antirust treatment must be applied to the joints after finishing to seal out moisture and prevent rust formation.

Figure 19-83 Note the basic steps used by this technician to section a rear frame rail. (A) The piece cut off of the damaged rail is aligned, and the holes are placed over each other and used to mark the new rail for sectioning. (B) A plasma arc cutter is cutting a piece of rail to make an insert or backer. The flanges are being cut off. (C) The insert should be cut down the middle so it will fit fully down the inside rail sections to be welded. (D) The insert pieces have been tack welded inside the existing rail to serve as a backer for butt welding the two halves together.

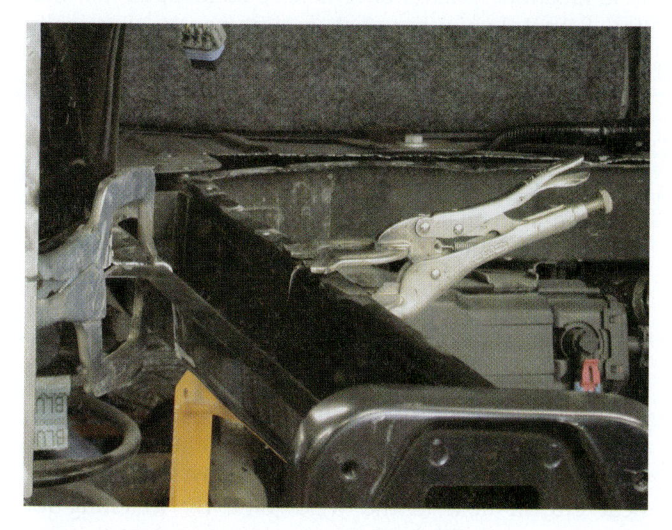

Figure 19-84 A new rear rail section has been clamped and butted against a piece of the existing rail.

Figure 19-85 Before starting to weld, use a measuring system to check rear rail length, width, and height locations.

Figure 19-86 Tack weld the rail into place. Then test-fit parts to make sure the rail is perfectly aligned before final welding.

Figure 19-87 Here the technician is using plug welds to install a new trunk pan or panel.

Figure 19-88 The rear panel is being fitted into place, clamped, and measured before welding.

Rustproofing and corrosion protection techniques are discussed fully in Chapter 20.

19.8 REPLACING PANELS WITH ADHESIVES

Some vehicle manufacturers use structural adhesives along certain weld seams. These two-part epoxy adhesives are sometimes called weld-bond adhesives, because spot welds are placed through the adhesive.

Weld-bond adhesives are used to add strength and rigidity to the vehicle body. They also improve corrosion protection in weld seams. In addition, adhesives help control noise and vibrations.

Parts most commonly weld-bonded are:

- A- and B-pillars
- Rocker panels
- Roof panels
- Rear quarter panels
- SMC door panels

If adhesives are disturbed by repairs, they must be replaced. Follow recommendations in the body repair manual or computer-based service information.

Some manufacturers use structural adhesives in place of welds. One example is around the wheel openings and sail panel reinforcement. This is a different type of adhesive than the weld-bond adhesive. Check factory recommendations for details on the use of structural adhesives.

To replace and repair fiberglass plastic door skins and other parts, see Chapter 13 and Chapter 15.

Replacing Aluminum Body Panels

Always refer to the manufacturer-published specifications for the specific make and model vehicle being repaired. It will give welding rod, shielding gas, cut and weld locations, and other vehicle-specific repair information. Some aluminum panels require you to chemically bond and electric weld body panels to the aluminum unibody structure.

Aluminum body panels are repaired or replaced using fundamental sheet metal and welding principles. However, if you are used to welding only steel, aluminum can be a little challenging at first. It is very easy to overheat aluminum, which causes it to melt and drip off of the surrounding body panel. If you accidentally burn a hole in a body panel, you will have to weld and refill the hole with aluminum.

Heat crayons are used to measure the temperature of a body panel during welding. Many auto body welders use heat crayons when welding aluminum, to measure heat buildup in the soft alloy panel. Draw a line with the heat crayon next to the weld joint. By watching for the heat crayon to melt, the welder can better judge

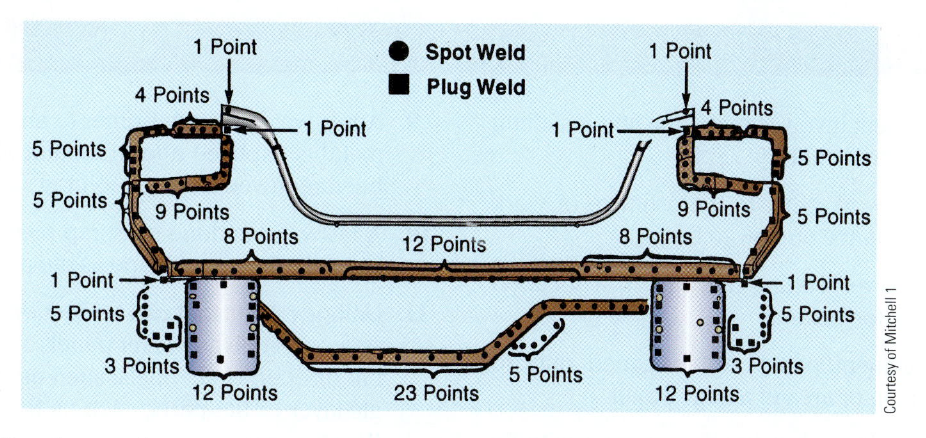

Figure 19-89 Here is a service manual illustration showing the weld count, types, and locations for a rear panel.

Figure 19-90 This vehicle has had new rear rails, a trunk pan, and a rear panel installed.

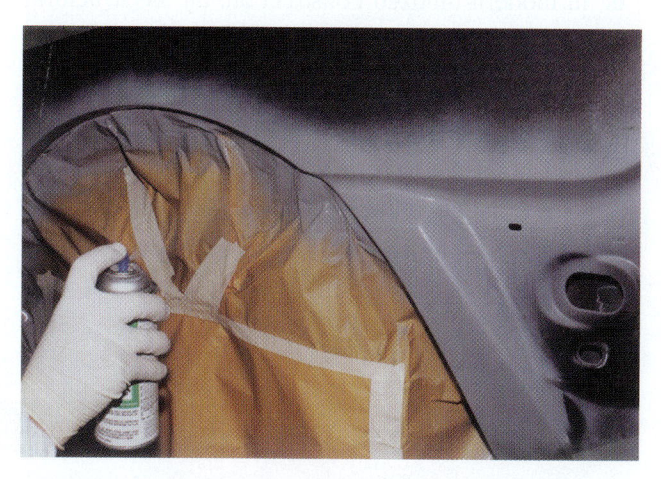

Figure 19-91 After grinding down plug weld nuggets, apply self-etch primer to bare metal. The vehicle is now ready to move into the paint prep area of the shop.

welding speed and heat buildup on the soft aluminum sheet metal.

Avoid using a stiff stone grinder on aluminum. A grinder with a coarse grit stone wheel can almost instantly gouge into and cut completely through modern, lightweight aluminum panels. Do not use grinders on aluminum fenders, door panels, hoods, lids, and quarter panels. A small grinder with flexible pad and coarse abrasive paper can be used gently on aluminum but with great care. It might be used to clean and bevel body panel flanges that must be welded.

Use 80-grit or finer sandpaper on aluminum. A coarse grit may accidentally abrade too much metal, making areas of the repair too thin and weak to be structurally sound. When sanding, use foam backing pads, not stiff backing pads.

Make sure points and picks are not too sharp for aluminum. You should have a sharp set of metalworking tools for steel body panels. You should also have a dull, more rounded set of metalworking tools for straightening soft, more flexible and bendable aluminum.

Make sure that hammer heads are free of steel shards, mushroomed edges, or other surface imperfections. It is best to use plastic tools to work aluminum when possible. Steel tools can impact and bend the soft aluminum too much.

Use **dye penetrant** to check for cracks in aluminum panels and components. First, spray a primer-cleaning agent over the suspect area on the part or panel. Then spray dye penetrant onto the possible crack. The dye will collect inside any crack or fracture in the part or panel for easy detection. Only weld to repair an aluminum part when a replacement or used assembly is not available.

Clean surfaces of aluminum with stainless steel bristle brushes only. Stainless steel will not contaminate the surface of aluminum, which could adversely affect welding.

SUMMARY

1. Panel replacement involves removing and installing a new panel or body part.
2. Before starting work, refer to the estimate or work order to get guidance on how to begin.
3. Sectioning involves cutting the part in a location other than a factory seam.
4. Partial replacement panels are designed only to replace a section or area of a large panel.
5. Fabricated panels are handmade repair parts to fix small problems in panels when a new or partial panel is not available or practical.
6. In modern unitized construction, all the structural panels, from the radiator support to the rear end panel, are welded together to make a one-piece frame structure.
7. It is usually necessary to remove the paint film, undercoat, sealer, or other coatings covering the joint area to find the locations of spot welds.
8. After the spot welds have been located, the welds can be drilled out and removed using a spot weld cutter.
9. Apply weld-through primer to areas where the base metal is exposed after the paint film and rust have been removed from the joining surfaces.
10. A test weld is done on scrap pieces of metal of the same thickness and type as the parts to be repaired.
11. Unibody parts that can be sectioned include closed sections, such as rocker panels and A- and B-pillars; hat or open U-channels, such as rear rails; and single-layer or flat parts, such as floor pans and trunk floors.
12. The three basic types of sectioning joints are the lap joint, the offset butt joint, and the butt joint with insert.
13. Recycled assemblies are undamaged parts from another damaged vehicle that are used for repairs.
14. Foam fillers are used to add rigidity and strength to structural parts, as well as to reduce noise and vibrations.
15. Full body sectioning involves replacing the entire rear section of a damaged vehicle with the rear section of a salvage vehicle.

EXERCISES

On a separate sheet of paper, complete the following learning activities for this chapter. Write definitions for the key terms and answer the ASE-style review questions, essay questions, critical thinking problems, and math problems. You can also do the outside activities, possibly for extra credit.

Key Terms

aftermarket replacement panels
dye penetrant
fabricated panels
foam fillers
full body sectioning
heat crayon
nonsectionable areas
OEM replacement panels
panel clamping pliers
panel replacement illustration
partial replacement panels
plug weld holes
recycled assemblies
replacement panels
salvage replacement panels
sectioning
spot weld drill
structural panel
structural panel replacement
test weld
weld inserts

ASE-Style Review Questions

1. Technician A sometimes positions nonstructural outer panels visually without making precise measurements. Technician B says that measurements must always be made. Who is correct?
   A. Technician A
   B. Technician B
   C. Both A and B
   D. Neither A nor B
2. Rocker panels and pillars are known as what type of section?
   A. Open surface
   B. Crush zone
   C. Closed
   D. Compound

3. When starting work, Technician A says to refer to the estimate or work order to know how to proceed. Technician B says this is not necessary. Who is correct?
   A. Technician A
   B. Technician B
   C. Both A and B
   D. Neither A nor B
4. Technician A says that sectioning involves cutting the part in a location other than a factory seam. Technician B says that pillars and rocker panels are often sectioned. Who is correct?
   A. Technician A
   B. Technician B
   C. Both A and B
   D. Neither A nor B
5. Technician A says that the number and location of plug welds should always be about an inch apart. Technician B says to refer to factory specifications for weld type, count, and locations. Who is correct?
   A. Technician A
   B. Technician B
   C. Both A and B
   D. Neither A nor B
6. What should be applied to the panel flanges before welding?
   A. Caulk
   B. Sealer
   C. Weld-through primer
   D. Paint
7. Where should A-pillars be cut when sectioning?
   A. At the upper end
   B. Near the middle
   C. At the lower end
   D. Both A and C
8. Technician A uses a spot weld drill to remove a structural panel. Technician B uses a smaller grinding wheel. Who is correct?
   A. Technician A
   B. Technician B
   C. Both A and B
   D. Neither A nor B
9. Which of the following statements concerning replacement at factory seams is incorrect?
   A. Replacing panels at factory seams is common in unibody repair
   B. Damaged rails and panels should not be returned to factory specifications until after they have been removed from the vehicle
   C. It is easy to destroy more of the factory welds than is necessary when doing the job
   D. All of the above
10. What is the best method for separating spot welds?
    A. Blowing them out with a plasma arc torch
    B. Drilling them out
    C. Blowing them out with an oxyacetylene torch
    D. Grinding them down with a high-speed grinding wheel
11. Technician A says to use a measuring system to check panel alignment before final welding. Technician B says to install adjacent parts or panels to visually check panel alignment before welding. Who is correct?
    A. Technician A
    B. Technician B
    C. Both A and B
    D. Neither A nor B
12. When preparing a vehicle for installation of replacement body panels, Technician A uses a disc sander to prepare the flanges of the structural panels, while Technician B does not. Who is correct?
    A. Technician A
    B. Technician B
    C. Both A and B
    D. Neither A nor B
13. Technician A says to use a scribe in order to mark the cut line on a quarter panel to roof joint. Technician B says that if there is sufficient overlapping both panels can be cut at the same time. Who is correct?
    A. Technician A
    B. Technician B
    C. Both A and B
    D. Neither A nor B
14. When visually positioning a nonstructural outer body panel, Technician A makes sure that the gaps between panels are tapered. Technician B makes sure that the gaps between panels are even. Who is correct?
    A. Technician A
    B. Technician B
    C. Both A and B
    D. Neither A nor B
15. Which of the following should be avoided when making sectioning cuts?
    A. Structural member mounts
    B. Compound member mounts
    C. Dimensional reference holes
    D. All of the above

Essay Questions

1. How would you use the estimate when starting work on a wrecked car?
2. Describe how to clean a typical part before removing its spot welds.
3. Why should you be careful when using a plasma arc torch to cut parts?
4. Why does the use of recycled assemblies in collision repair make sense?
5. Summarize how to fully section a car.

Critical Thinking Problems

1. You are working on a late model Porsche. You are not quite sure how to section a damaged front rail. What should you do?
2. After welding in a new quarter panel, you find that you cannot make the trunk lid fit properly in its opening. What are your options to fix this problem?
3. OEM stands for what?
4. In general, to replace a welded-on structural panel you should use what procedure?

Math Problems

1. If factory specs say that panel overlap should be ¾ inch and you only have 0.25-inch overlap, how much more is needed?
2. If a recycled panel has been cut 1.3 inches longer than what is needed for proper overlap and the total length is 2 feet, 1.5 inches, what should the part's final length be?
3. A foreign service manual specifies 12.7 mm for a drilled hole. What size English or U.S. drill bit should be used?
4. An automobile needs two panels costing $300 each to finish a repair (with an added sales tax of 7%). It will take eight hours to install, and paint the panels at $75 per hour. What is the total cost of the repair?

Activities

1. Inspect a badly damaged vehicle. List the structural parts that would require replacement.
2. For the same vehicle, summarize the procedures for structural part replacement.
3. Visit a body shop. Ask the owner to allow you to watch a technician replacing a structural part.

Restoring Corrosion Protection

CHAPTER 20

OBJECTIVES

After reading this chapter, you should be able to:

- Define *corrosion*, and describe the common factors involved in rust formation.
- Describe the anticorrosive materials used to prevent and retard rust formation.
- Explain the conditions and events that lead to corrosion on a vehicle body.
- Choose the correct anticorrosive application equipment for specific applications.
- Outline the correct corrosion treatment procedures for each of the four general corrosion treatment areas.
- List the four types of seam sealers, and explain where each should be used.
- Answer ASE-style review questions relating to corrosion-protection materials and methods.

INTRODUCTION

Corrosion protection involves using various materials to protect steel body parts from rusting. When doing repairs, you must always use recommended methods for protecting repair areas from rust damage. Rustproofing is often a joint task or effort of both the collision repair and paint technicians.

Anticorrosion materials are used to prevent rusting of metal parts. Various types of anticorrosion materials are available (weld-through primer, sealers, e-coat, wax, conversion coatings, rubberized undercoating, and so forth). When performing repairs, you must restore all corrosion protection to keep the car safe to drive for extended periods of use.

DANGER If you fail to restore proper corrosion protection, it can endanger the driver and passengers of the vehicle. After prolonged service, rust can weaken the body structure. This could cause failure of the body where it supports the suspension components. The vehicle could become unstable and dangerous to drive because of this rust.

Corrosion-protection materials are critically important to the structural integrity of the vehicle. This means that rust is not just an "eyesore." The unibody car has more welded joints in critical structural areas where corrosion can do serious damage. It is an ever-present danger to the unibody vehicle because rusting of structural panels and rails can affect the drivability of the car and the safety of its passengers.

20.1 WHAT IS CORROSION?

Corrosion, or rust, is the oxidation and chemical change of metal. When it occurs on steel, it is the product of a complex chemical reaction with serious and costly consequences. The formula for rust in a car body is:

Iron + Oxygen + Electrolyte = Rust (Iron Oxide)

Chemical corrosion requires three elements:

1. Exposed metal
2. Oxygen
3. Moisture (electrolyte)

There are three basic types of corrosion protection used on today's automobiles:

1. Galvanizing or zinc coating
2. Paint
3. Anticorrosion compounds

See Figure 20–1 through Figure 20–3.

Galvanizing is a process of coating steel with zinc. It is one of the principal methods of corrosion protection applied during the manufacturing process. On galvanized steel, the zinc forms a natural barrier between the steel and the atmosphere. As the zinc corrodes, a layer of zinc oxide will form on the surface. Unlike iron oxide, or rust, the zinc oxide adheres to the zinc coating tightly, forming a barrier between the zinc and the atmosphere. See Figure 20–4.

Figure 20–1 This close-up shows a car's number one enemy—rust, also called corrosion. Paint chips from rocks flying off tires exposes small areas of bare metal. Exposed to elements, metal quickly rusts.

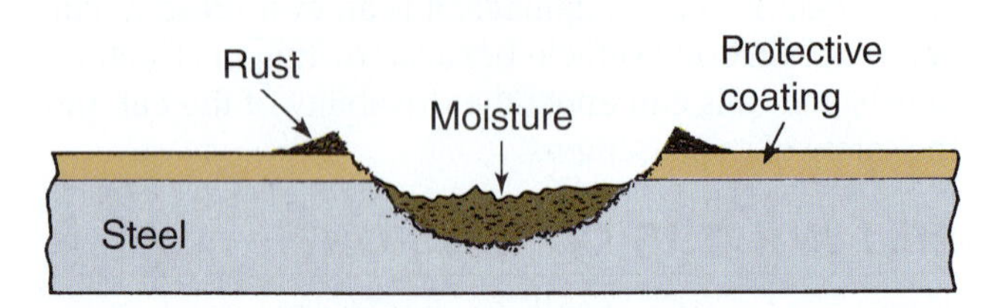

Figure 20–2 The breakdown of a vehicle's protective paint coating causes rapid rust corrosion formation.

When the surface of the vehicle's finish is damaged by a scratch or nick, the zinc coating undergoes corrosion, sacrificing itself to protect the iron under it. The resulting zinc oxide actually forms a protective coating and repairs the exposed area of the steel. Thus, zinc performs a twofold protective process. First, it provides chemical, galvanic protection, and second, it forms a repair over the exposed steel with a layer of zinc oxide.

A paint system, such as those described in later chapters of this book, also provides a barrier between the atmosphere and the steel surface. When this barrier is in place, the moisture and impurities in the air cannot interact with the steel surface, and the steel is protected from corrosion.

If the paint surface or barrier is broken by a stone chip or scratch, the steel in this area is no longer isolated from moisture and impurities in the air. Corrosion will then take place in this region. Corrosion will spread between the paint and steel surface. If the adhesion of the paint to the steel is poor, large sections of the paint can be separated from the steel. This will result in a large area of the steel being left unprotected. Severe rust in this region will quickly follow. If impurities are present between the paint and the steel, oxygen in the air can pass through the paint, reacting with the impurities and the steel to form rust. In this case, corrosion will take place on the steel surface, and the protective paint barrier will be destroyed. Paint, by itself, is effective only as long as the paint film remains intact.

Anticorrosion compounds are additional coatings applied over and under the paint film. Protective coatings can be applied either by the manufacturer or as an aftermarket process. The two most popular types of anticorrosion coatings are:

1. Petroleum-based compounds
2. Wax-based compounds

Anticorrosion compounds are primarily used in enclosed body sections and other rust-prone areas.

Auto manufacturers are increasing their corrosion-protection measures all the time. New processes and methods, including the use of coated steels, zinc-rich

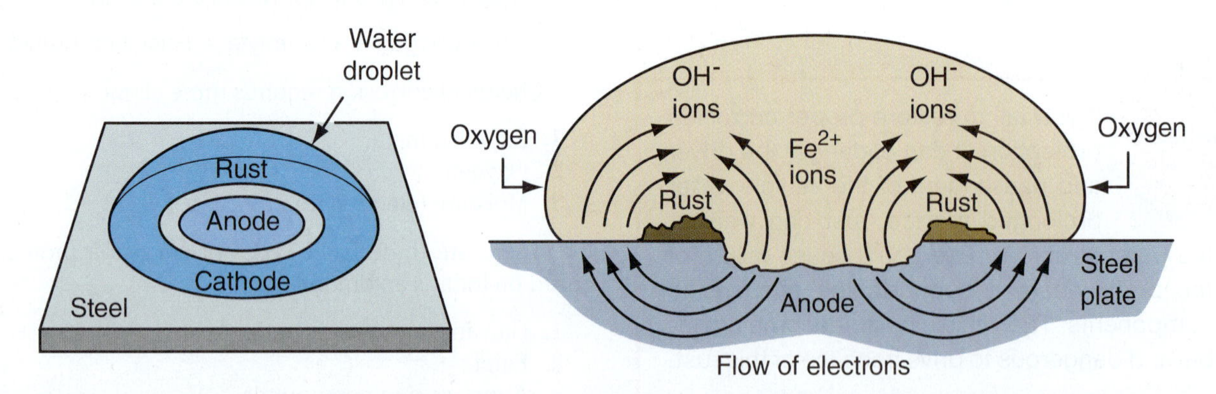

Figure 20–3 Note the chemical action during the rusting, or corrosion, process. An anode is set up near the middle of a water droplet and a cathode around the perimeter of water. The flow of electrons and oxygen then converts steel into rust.

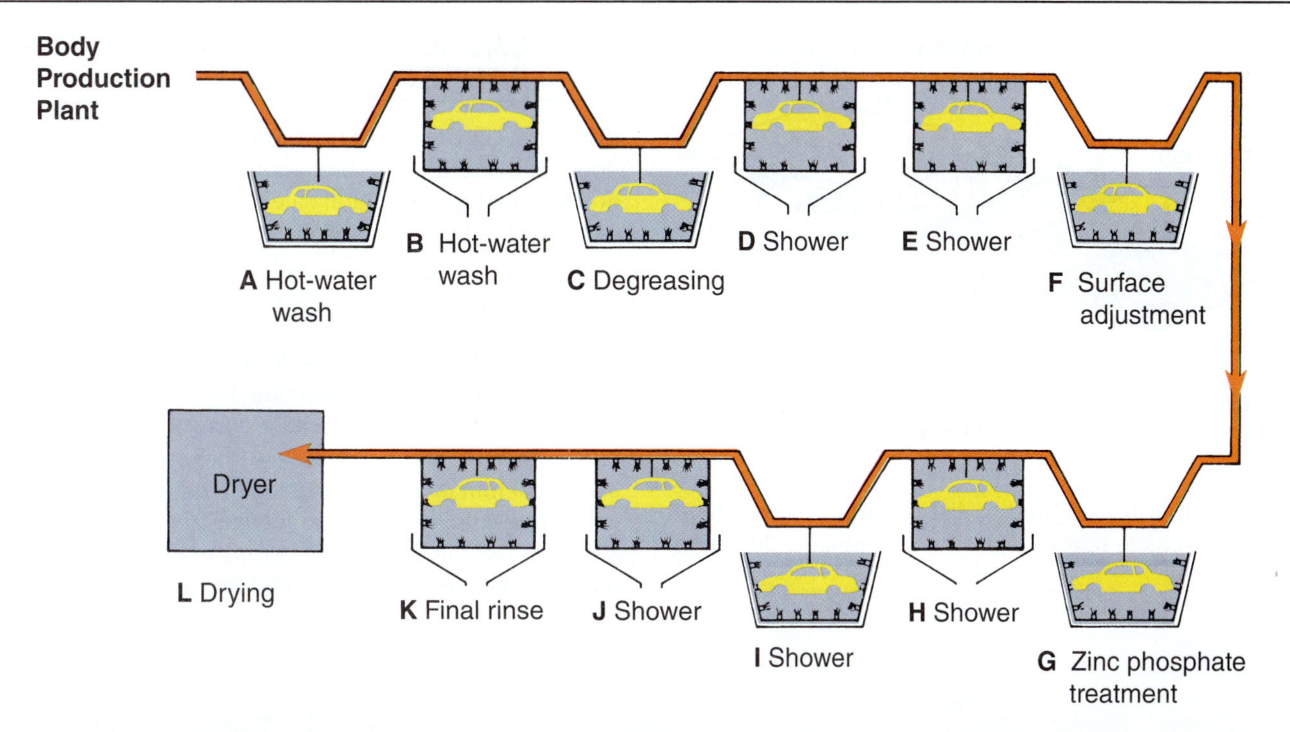

Figure 20-4 Study the typical zinc corrosion treatment, which uses the full unibody dip method, as done by a vehicle manufacturer: (A) and (B) Metal chips, dirt, and other foreign particles are washed off with hot water at 100–125°F (36–52°C). (C) Press oil and anticorrosion oil are removed with a weak alkali degreasing agent. (D) and (E) The degreasing agent is washed off with water in two stages. (F) The nucleus of the zinc phosphate film is adhered to the panel surfaces. (G) The body is dipped into a tank of zinc phosphate for crystallization. (H), (I), and (J) The zinc phosphate liquid is washed off by water in three stages. (K) The body is given a final rinse to prevent blistering. (L) The body is dried at 212–300°F (100–149°C).

primers, and more durable base coatings, have made it possible for modern cars to survive corrosive forces for longer periods than ever before (Figure 20–5).

The following is a typical new car finishing sequence used by major auto manufacturers:

1. Use coated or galvanized steel.
2. Chemically clean and rinse.
3. Apply conversion coating.
4. Apply epoxy primer.
5. Bake primer.
6. Seam seal process.
7. Apply primer-surfacer.
8. Apply colorcoats.
9. Bake colorcoats.
10. Apply anticorrosion materials.

Figure 20-5 Several anticorrosion materials are commonly used to protect rusted steel frame or unibody structure during repairs.

Because of these better procedures, corrosion-protection warranties of several years are now common. With the dramatic improvements in the performance of OEM products, the repair industry must rise to the challenge of producing corrosion resistance in repaired areas that matches or exceeds the durability of the original product. Repair work that does not stand up will draw attention to itself next to the outstanding durability of many original

Remember that you, the body shop technician, are responsible for the quality and durability of the completed repairs. Remember that the customer is entitled to a car restored to the way it was before the damage occurred.

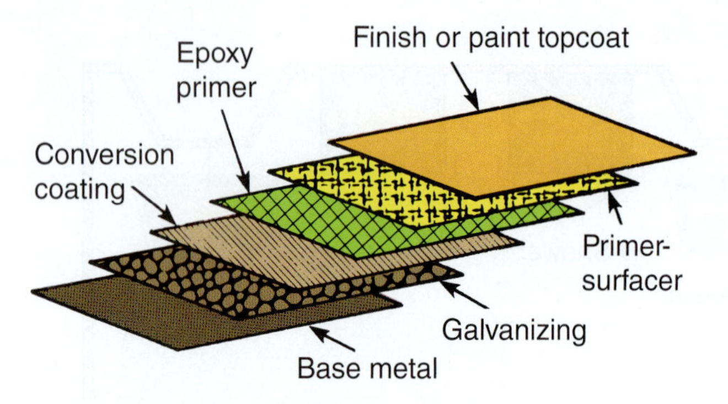

Figure 20-6 Note the typical makeup of corrosion prevention material used by car manufacturers.

finishes. It can also lead to liability challenges where issues of vehicle safety are involved. See Figure 20–6 and Figure 20–7.

20.2 CAUSES FOR LOSS OF FACTORY PROTECTION

Even with all the care taken to protect vehicles, breakdown still occurs. The breakdown of corrosion protection falls into three general categories:

1. Paint film failure
2. Collision damage
3. Repair process

The paint film is the result of the entire process of coatings, primers, and colorcoats that the manufacturer applies. When the paint film fails, corrosion begins. Stone chips, moisture, and improper surface preparation can all lead to film failure. See Figure 20–8.

During a collision, the protective coatings present on a car are usually damaged. Corrosion protection can be damaged not just in the areas of direct impact, but also at

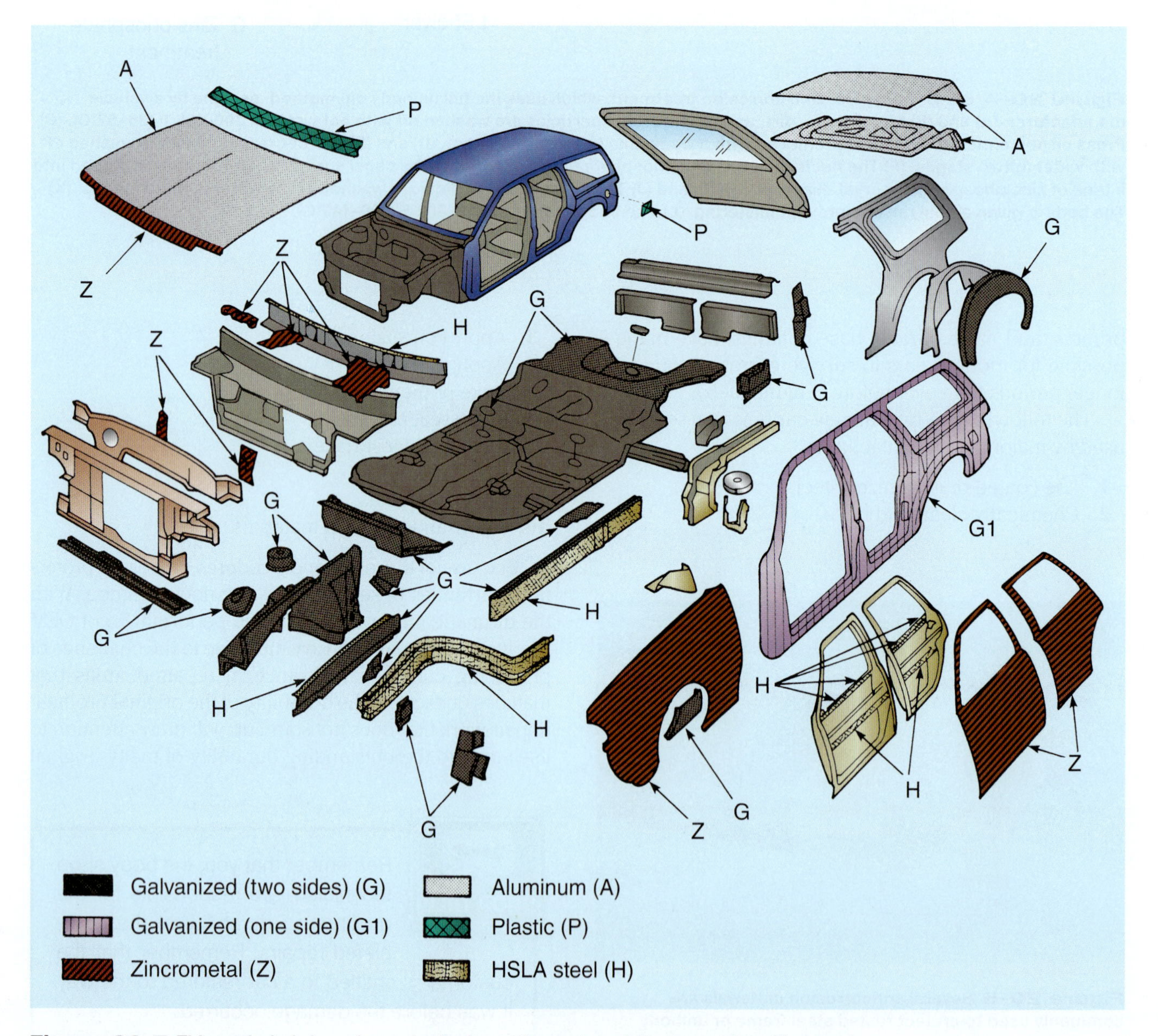

Figure 20-7 This exploded view of a car body shows the parts and the types of coating used.

Figure 20-8 Stone chips or door dings from other cars can break the paint film and lead to rust spots.

indirect damage zones. Seams pull apart, caulking breaks loose, and paint can crack and chip. Locating all affected areas and restoring corrosion protection to them remains a key challenge (Figure 20–9).

Vehicle repair is possibly one of the major causes of protective coating damage. For example, repair procedures often require cutting body panels and seams either mechanically or with a plasma torch. Even minor straightening and stress-relieving procedures can damage protective coatings. Normal welding temperatures cause zinc to vaporize and be lost from the weld area. Abrasive operations during repair and refinishing can also leave areas unprotected. After all welding and repair work has been completed, these damage points need careful attention to eliminate contaminants. Then steps must be taken to keep the atmosphere off the metal by sealing all surfaces thoroughly.

Figure 20-9 During a major collision, seam sealer is often cracked and broken, a sign of major structural damage. Note the small crack in the right side of the sealer. You must find and replace all cracked seam sealer during repairs.

Other precautions that should be taken to protect the factory corrosion protection are the following:

- Minimize the size or area where paint is removed, such as at cut and weld points.
- Be extremely careful not to scratch any part not being repaired.
- Place protective covers over adjacent painted surfaces and surrounding areas to protect them from the flame or sparks.
- Cover any opening in body sills and similar areas with masking tape to prevent metal chips from entering during the grinding, cutting, or welding operations.
- Remove any metal chips from inside the body. Use a vacuum cleaner, not compressed air, to remove metal chips. If compressed air is used, metal chips can be blown out and accumulate in other corner areas.

There are also some environmental and atmospheric conditions that influence the rate of corrosion:

- *Moisture.* As the water on the underside of the body increases, corrosion accelerates. Floor sections that have snow and ice trapped under the floor matting will not dry. Likewise, if holes at the bottom of the doors and side sills are clogged or sealed shut, water will accumulate. Remember, water is one of the requirements for rust (Figure 20–10).

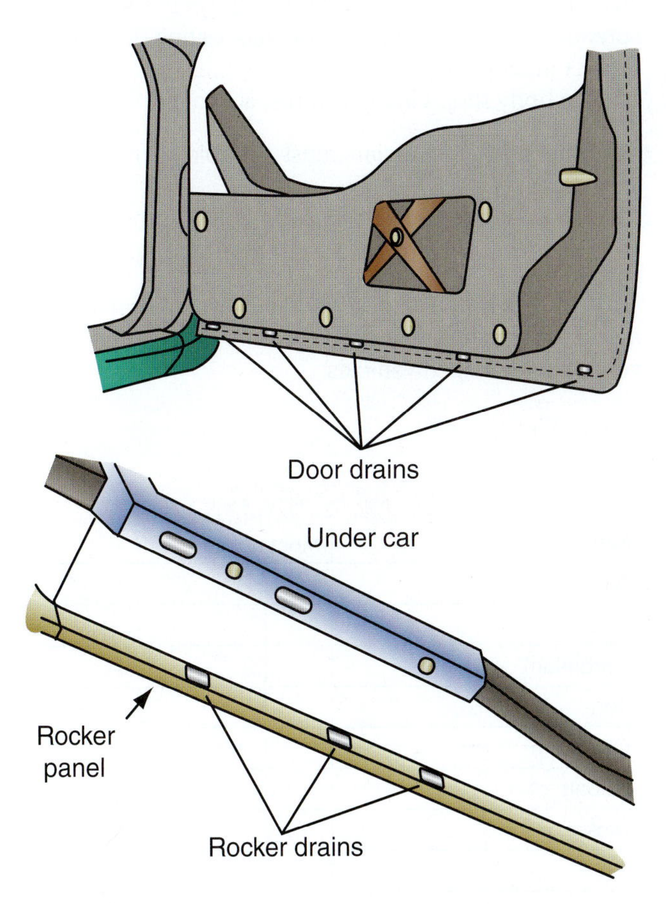

Figure 20-10 Keep drain holes open at the bottom of the doors, side sills, and so forth to avoid water accumulation inside panels.

- *Relative humidity.* Corrosion will be accelerated in areas of high relative humidity, especially those areas where the temperatures stay above freezing and where atmospheric pollution exists and road salt is used.
- *Temperature.* A temperature increase will accelerate the rate of corrosion to those parts that are not well ventilated.
- *Air pollution.* Industrial pollution, the salty air of coastal areas, and the use of heavy road salt accelerate the corrosion process. Road salt will also accelerate the disintegration of paint surfaces.

Another type of corrosion that must be considered when working on automobiles is known as galvanic corrosion. **Galvanic corrosion** occurs when two dissimilar metals are placed in contact with each other. The more chemically active of the two metals will corrode. As shown in Table 20–1, this is why zinc will sacrifice itself to protect steel. In the case of other metals, as mentioned later in the chapter, galvanic corrosion can cause problems.

Regardless of the cause, if corrosion prevention is not practiced, the cost to the body shop and insurer is "comebacks," or lost customers. Inadequate preparation that leaves dirt, grease, or acids on the metal causes the loss of adhesion. Rust will start, a little at first, creating corrosive hot spots. Surface failure will progress quickly, spreading under the surface coatings, eating deeper and deeper into the metal.

The body shop's interest in rust is twofold:

1. The body technician must be able to repair rust damage.
2. The body technician must be able to provide treatment that will prevent rust from recurring.

Chapters 11 and 12 described how to repair rust damage. This chapter is devoted to restoring corrosion prevention to damaged vehicles.

TABLE 20–1 RELATIVE ACTIVITY OF METAL

Metal	Activity
Magnesium	Most active ↑
Aluminum	
Zinc	
Chromium	
Iron	
Cadmium	
Cobalt	
Nickel	
Tin	
Lead	
Copper	↓ Least active

20.3 ANTICORROSION MATERIALS

The body and paint shop's efforts to protect metal from rusting should first focus on creating a clean, chemically neutral surface on the sheet metal. Then it should focus on sealing the material under layers of paint. Under certain conditions, a wax- or petroleum-based anticorrosion compound is used to exclude air and moisture from the metal surface.

More and more, new vehicles come off the assembly line with anticorrosive materials that are available to the body shop. Being able to replace or install these materials properly is a very important skill for today's auto body technician.

Corrosion prevention has not always been a popular body shop operation. The original rustproofing was called **undercoating**, and it was an asphalt-based product. Applying this "tar" was sheer agony because it smelled bad and was messy. But, worst of all, it did not work very well. In time, the solvents used would evaporate and the asphalt would harden and crack. The moisture that causes oxidation would actually become trapped under the undercoating.

Asphalt undercoats did have benefits in terms of sound deadening and preventing stone marks under fenders. And they are still useful today on fiberglass panels for the same reasons. Rubberized undercoating is now more common than asphalt-based undercoating (Figure 20–11).

Anticorrosion materials or agents can be divided into 11 broad categories:

1. *Undercoat.* Anticorrosion compound can undercoat, sound-deaden, and completely seal large surface areas from the destructive causes of rust and corrosion. It should be applied to the undercarriage and inside body panels so that it can penetrate into joints and body crevices to form a pliable, protective film.
2. *Sealant*, or body sealer. It prevents the penetration of water or mud into panel joints and serves the important role of preventing rust from forming between adjoining surfaces (Figure 20–12).
3. **Weld-through primer**. Weld-through primer is used to provide anticorrosion protection to weld zones. This primer must be applied to clean surfaces. Most weld-through primers have poor adhesion qualities. Do not overuse them, and always follow directions closely. Weld-through primer can be applied to galvanic mating surfaces where the coating was removed during repair. After welding, remove the excess primer (Figure 20–13).
4. *Self-etching primers.* Self-etching primers etch the bare metal to improve paint adhesion and corrosion resistance, while providing the priming and filling properties of primer-surfacers. Self-etching primers work best on lightly sanded surfaces where a slight-to-moderate amount of filling is required. They must be applied as directed by the manufacturer.

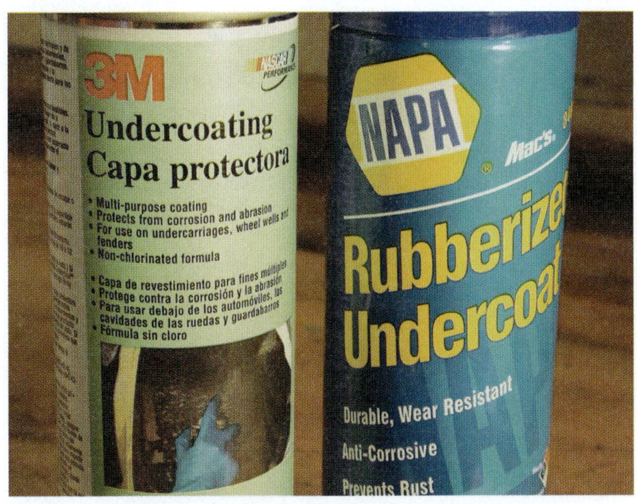

A Undercoating is available as a tar-based product or synthetic rubberized material. Use the type recommended by the auto manufacturer.

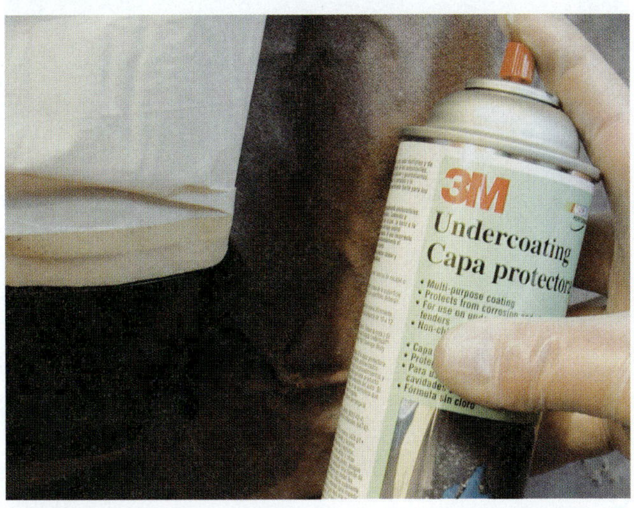

B Shake the can thoroughly before use. Undercoating is very thick and takes several minutes to mix completely. Mask areas not to be sprayed. Apply a smooth layer of undercoat to protect the area.

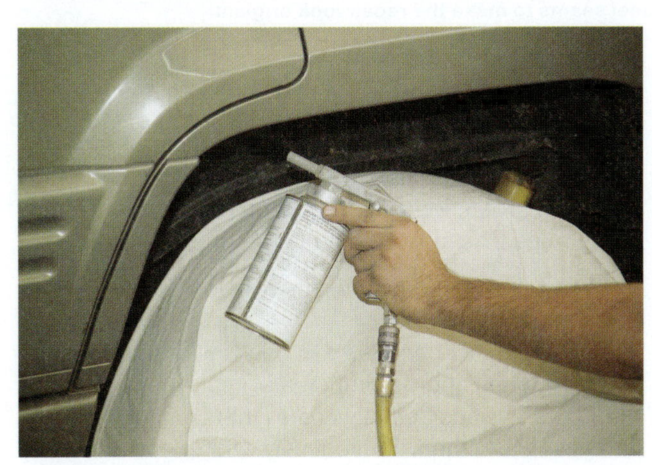

C Here the technician is using a large air-powered applicator to apply undercoat to the inner wheel housing.

Figure 20–11 Undercoating is often applied to lower areas of the body to protect them from road debris damage.

5. *Two-part epoxy primers.* Two-part epoxy primers provide very strong base coating with good adhesion to bare metal. They are mixed together and cure very quickly, which helps prevent corrosion by more tightly bonding the coating over the metal (Figure 20–14).
6. **Rust converters** change ferrous (red) iron oxide to ferric (black/blue) iron oxide. Rust converters also contain some type of latex emulsion that seals the surface after the conversion is complete. These products offer an interesting alternative for areas that cannot be completely cleaned (Figure 20–15).
7. *Chip guard.* Chip guard is a rubberized coating often used along the lower rocker panel to protect the paint and body from stone chips. It is a soft, flexible coating that can be applied over or under paint. Chip guard coatings can be clear or colored. By helping to avoid chips and metal exposure, they serve as an important corrosion-protection material. They should be reapplied if removed during a repair (Figure 20–16).
8. *Rust preventive paint.* This is a self-etch primer and paint combined in one product. It can be quickly applied to bare metal to treat minor surface rust and also to provide a protective paint film that seals out moisture. It is sometimes used on the underbody or frame of vehicles to help restore corrosion protection with minimum prep time (Figure 20–17).
9. *Hot-melt "wax" coatings.* These are thermoplastic corrosion-prevention compounds consisting of solvent or waterborne materials. Thermoplastic coatings have the ability to heal and reseal over small localized damage, such as that caused by small stone chipping. Some new vehicles come with hot-melt wax corrosion-protection coatings.
10. *Electrocoat (E-coat).* E-coating is a process wherein electrically charged paint particles are deposited out of a water suspension to coat a conductive metal body part. An E-coat provides a barrier to corrosive situations. Most new repair panels are E-coated in the factory.
11. *Powder coatings.* Powder coats are dry thermoplastic powders or finely ground particles applied to the surface of the object. The part is electrically charged to attract the particles to its surface. Heat (typically 180°C or 356°F for 10 minutes) is used to fuse the colored powder to the part to create a smooth,

Be sure to carefully read and follow the manufacturer's instructions on the container of any anticorrosion material. Failure to follow product directions can lead to failure of your repair!

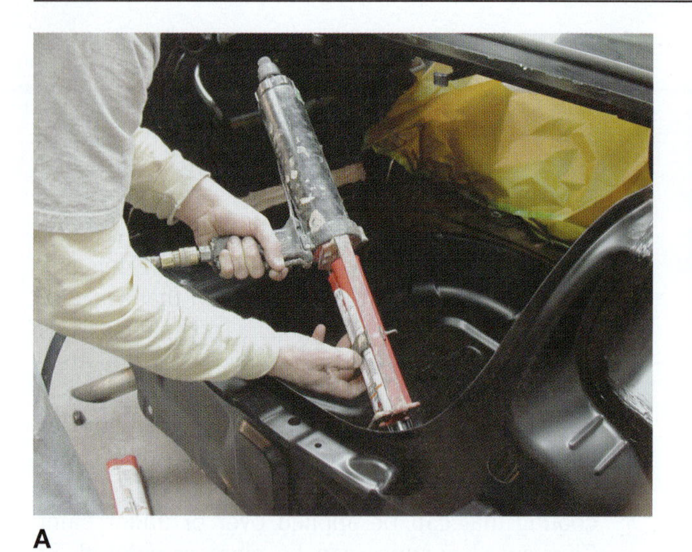

A

C

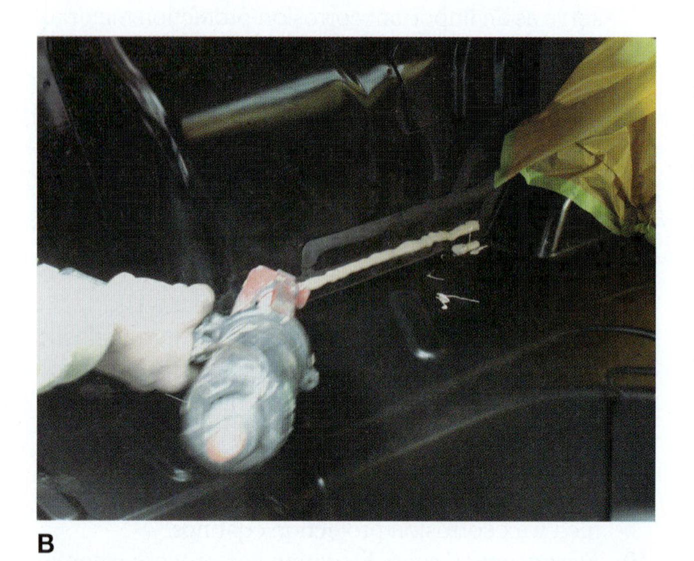

B

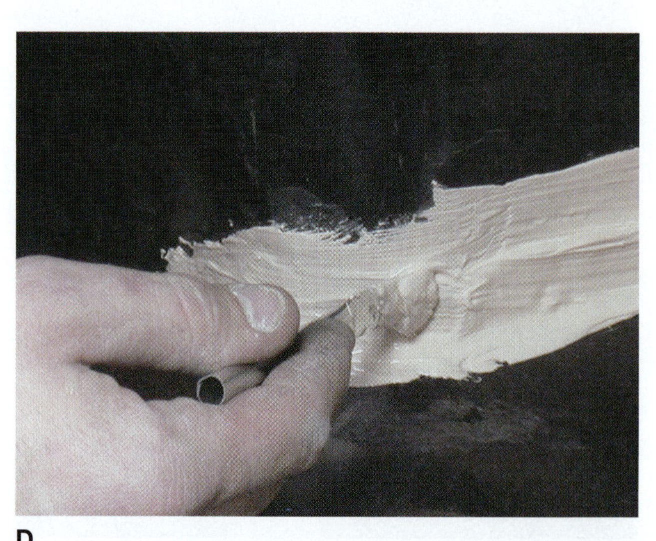

D

Figure 20–12 **The technician is sealing the trunk seams of a repaired vehicle. The area has already been sprayed with self-etch or an epoxy primer and allowed to fully dry. (A) New cartridges of two-part seam sealer have been installed in an air-powered application gun. (B) Press the trigger and move the gun tip to produce a continuous bead of sealer along panel edges. (C) An inner-mix nozzle mixes two ingredients of sealer and deposits the sealer on body panel joints. (D) After applying a bead of seam sealer, use a small brush to spread the sealer out over the panel seams to make the repair look original.**

continuous, plastic paint-like film bonded to the part. Powder coating can contain a wide range of compounds: acrylic, vinyl, epoxy, nylon, polyester, and urethane. See Figure 20–18.

NOTE! *The industry is moving from liquid paints to powdered colorcoats and clearcoats rapidly. Powdered topcoats resist acid rain, the sun's ultraviolet rays, and road and weather damage better than conventional paints. Several auto manufacturers are using powder-coated finishes on their new model vehicles. Powder coating can be used to prime as well as paint new vehicle bodies/frames on the assembly line. The repair industry should see more powder coating of parts in the future and should be ready to someday apply these durable coatings in-shop.*

Corrosion-Protection Safety

As with other materials used in collision repair, the use of corrosion-protection materials requires that you follow safety rules. The most basic rules are the following:

- Wear gloves and avoid skin contact. Epoxy systems can irritate skin.
- If skin contact occurs, wash hands with soap and hot water. Then apply a skin cream.
- If adhesive accidentally comes into contact with your eyes, wash them immediately with clean water for 15 minutes. Then consult a physician.
- Be sure to work in a well-ventilated area and wear a respirator. Spot welding in weld-bond joints can generate gases that can be harmful if inhaled.

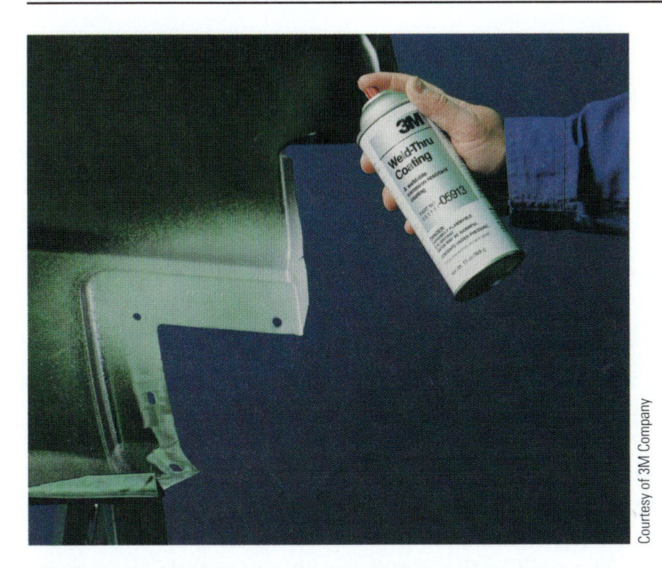

Courtesy of 3M Company

Figure 20-13 Always read the manufacturer's instructions and literature before using any chemical product. The technician is spraying a weld-through coating or primer to panel flanges to be butted together and welded.

Figure 20-14 Here a painter is applying primer to a large area of a quarter panel. A spray gun will work better than a spray can on large areas.

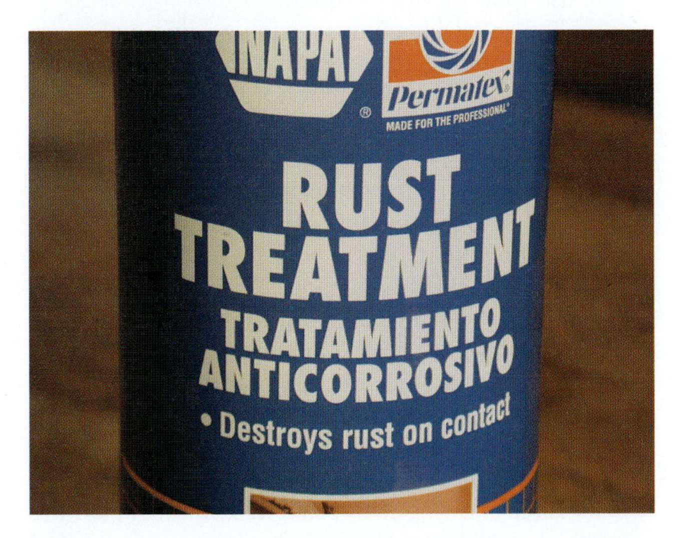

Figure 20-15 Rust treatment will neutralize and convert rust inside small pits in the metal into an inactive primer-like coating.

A A stone or chip guard comes in clear or flat black to match OEM coatings. You can apply stone guard under or over paint. Applying it over the paint works best.

B Here the technician is applying flat black stone guard to the rocker panel right in front of a rear tire. This matches factory process.

Figure 20-16 A stone or chip guard is a rubberized coating often needed along the bottom of a rocker panel to protect it from chips and metal exposure.

Figure 20-17 Chassis paint is a primer, rust converter, and paint product mixed together. It is often used when restoring vehicles that have severe rust pits in frame rails.

Courtesy of Eastwood Company, www.eastwood.com

A A special powder coat gun is needed to spray or dust parts with solid plastic particles.

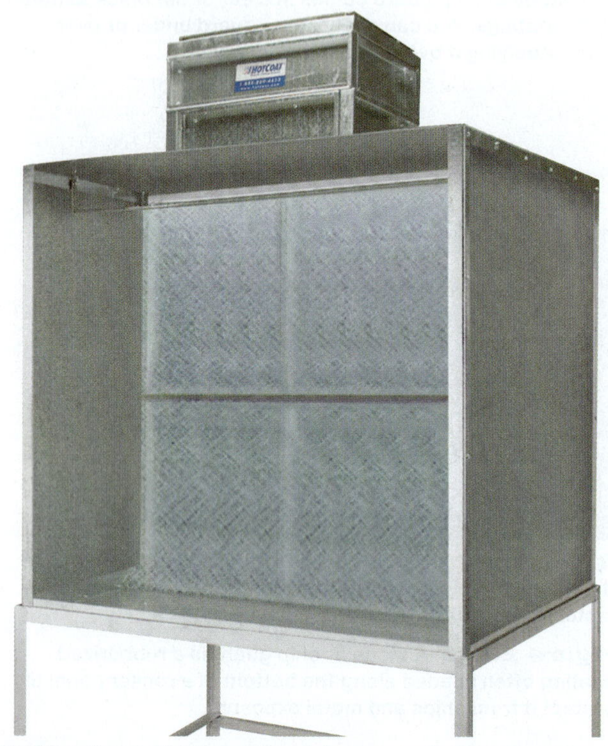

B When heated in an enclosed furnace or booth, powder melts and bonds to the metal part.

Figure 20-18 Powder coating of parts is becoming increasingly popular in the OEM and aftermarket. Powder coating is very durable.

20.4 BASIC SURFACE PREPARATION

Surface preparation is one of the most important steps in ensuring long-term corrosion resistance of body panels and other metal parts. Without the proper surface (especially bare metal), the rest of the repair procedure and refinishing efforts will be futile. A common system generally consists of the following four-step process called **metal treating**:

1. *Cleaning to remove contaminants.* Use a wax and grease remover to dissolve and float off oily, greasy film as well as other contaminants from the surface. Apply the remover with a clean, white cloth. Work small areas of no more than 2–3 square feet (0.6–0.9 square meters). Wet the surface liberally and, keeping it wet, use a second cloth to wipe the surface to remove any contaminants. Turn the cloth frequently while drying the surface.
2. *Cleaning with metal conditioner.* A metal conditioner is a phosphoric acid used to etch bare sheet metal before priming. It is a chemical cleaner that removes rust and corrosion from bare metal and helps prevent further rusting (Figure 20–19).

   Remember the following about metal conditioners:

- Acid cleans the metal.
- It dissolves light surface rust.
- It etches metal, improving adhesion.
- The acid in the metal conditioner needs to be completely neutralized after use. Follow product label directions to neutralize the acid; most recommend water to neutralize the acid.
- It may have to be diluted, following product directions.
- It is always followed by conversion coating.
- Wear rubber gloves and eye protection.

   Dilute the conditioner with water in a plastic bucket according to label instructions and apply it to the metal. A spray bottle is often recommended. Then rinse with clear water and wipe dry with a clean cloth.

3. *Applying conversion coatings.* The conversion coating forms a zinc phosphate coating that is chemically bonded to the metal. This layer makes an ideal surface for the primer and prevents rust from creeping under the paint. Use conversion coatings on galvanized and uncoated steels and aluminum. Be sure to use the correct product for each type of surface.

   There are two general types of rust conversion coatings: a wipe-and-rinse acid wash and an easier-to-use spray-on rust treatment.

   To use a wipe-on conversion coating, make sure you are wearing acid-tolerant rubber gloves, goggles, respirator, and a rubber apron, if available. The acid is very powerful and can cause severe chemical burns. Coarse sand or scuff the area to be treated to remove all loose debris from the surface. Remove as much rust as practical without thinning down the metal panel too much.

   Following label directions: Pour the rust remover into a bucket and mix it with the right amount of water (usually 1:1 or 1:2 acid:water). Either wipe or preferably spray the acid mix over the area with surface rust. Leave the rust remover on the surface for about 5 minutes. This allows the acid to dissolve any rust in small pits in the metal.

   After allowing it to soak, wash the acid off the vehicle with water. Use a garden hose or a bucket of clean water and a sponge. Wipe the area dry with a clean towel and allow to dry.

A Rust remover is powerful acid that should be used with care.

B Read the label directions because rust remover is a strong acid that can cause painful chemical burns. You must usually mix water and acid together.

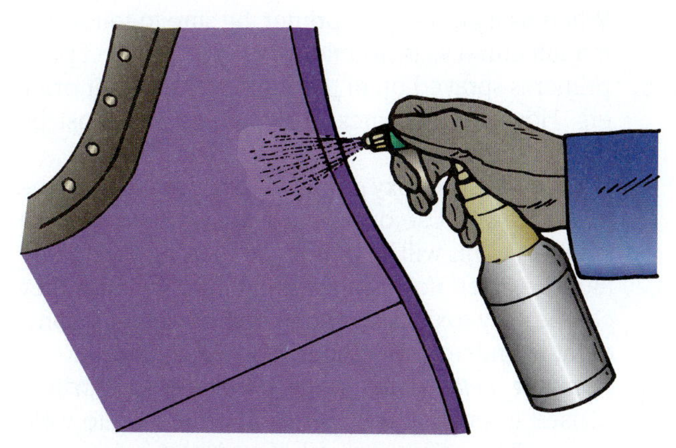

C Pour the acid/water solution into a spray bottle. While wearing rubber gloves and goggles, spray rust remover onto the surface rust. After letting it soak in for a few minutes, wash the acid off the panel with clean water.

Figure 20-19 **Rust remover or metal conditioner is a strong acid that literally eats surface rust off metal panels.**

To apply rust treatment with an aerosol can, grind or coarse sand the surface rust to remove loose material. Shake the can vigorously for several minutes to mix the ingredients. Spray a thin coat of rust treatment over the repair area. After a flash time of about 2 minutes between coats, apply two more coats of rust treatment. Allow the rust conversion coating to fully dry for at least 24 hours. You can then sand the area smooth before applying primer, primer-surfacer, and your topcoats of paint (Figure 20–20).

4. Use a self-etch primer on bare metal to increase adhesion. This type of primer will actually eat into the metal to produce a strong bond. It will help prevent paint peeling and other problems.

Apply the self-etch primer in two thin, wet coats. Wet coats help the material etch and bond with the metal. Avoid dry coats that will not adhere properly (Figure 20–21).

Note that most self-etching primers do not recommend the use of metal conditioners or conversion coatings before their application. Most self-etch primers require 24 hours for full drying. Do not apply seam sealer or other materials before full drying or they will not adhere to the self-etch primer properly.

20.5 CORROSION TREATMENT AREAS

The corrosion treatment areas that must be considered when body repair work is done can be grouped into four categories. They are the following:

1. *Enclosed interior surfaces*, which include body rails and rocker assemblies.
2. *Exposed interior surfaces*, including floor pan, apron, and hood sections.
3. *Exposed joints*, such as quarter-to-wheel housing and quarter-to-trunk floor joints.
4. *Exposed exterior surfaces*, such as fenders, quarter panels, and door skins.

The term *exposed* as used in this chapter refers to a panel surface that is accessible without having to remove a welded component.

20.6 CORROSION-PROTECTION PRIMERS

Of all the areas to be protected during a repair job, the enclosed interior surfaces are the most important. These include underbody structures such as front rails, rear rails, and rocker panels. The reason for the importance of these sections is that they represent the principal load-carrying members of the unibody car. Corrosion of these components can have a severe effect on the crashworthiness and durability of the vehicle.

Metal conditioners and conversion coatings are not recommended for use inside closed sections. The reason is that the chemicals and moisture might be difficult, if

A **Always mix ingredients thoroughly. If you do not want to shake the can by hand for up to five minutes, place it upside down on a paint shaker machine.**

B **Make sure the spray can nozzle is not partially obstructed by dried material.**

C **After use, hold the spray can upside down and spray until only propellent comes out. This will blow thick corrosion-prevention material out of the nozzle so it will be clean and ready for the next use.**

Figure 20–20 **Note the important steps when using an aerosol can of corrosion-prevention materials.**

Figure 20–21 **After welding a new body panel, exposed metal surfaces should be coated with self-etch or epoxy primer. Self-etch primer is commonly used because it comes in a spray can and does not have to be mixed and sprayed from a commercial spray gun.**

not impossible, to fully remove from the inside seams. Because stone chipping is not a problem here, the primer should develop adequate adhesion in these areas without conversion coating.

With this in mind, begin the process on closed sections with a thorough cleaning and degreasing. Because of the closed construction of these components, the cleaning must be done before the part is welded into the repair area.

After cleaning and degreasing, the enclosed metal surfaces must be protected with a primer. There are many different primers used by the auto trade. However, for corrosion protection, especially for enclosed interior surfaces, the three used most often are the following:

1. *Two-part epoxy primers* are recommended by most automobile makers in place of the standard primer. When using a two-part primer, be sure to follow the manufacturer's instructions to the letter. Epoxy primer is sprayed on in the same way as other primers. However, it comes in two parts and must be mixed properly.

   Reduce the epoxy primer as recommended by the manufacturer. Use one or two wet coats to saturate the metal. This will help produce a strong bond. Care must be taken not to inhale the fumes because they can be very toxic. Use epoxy primer when a strong, durable undercoating is desired (Figure 20–22).
2. *Applying weld-through primer.* Weld-through primer is used to provide anticorrosion protection to weld zones. This primer must be applied to clean surfaces. Most weld-through primers have poor adhesion qualities. Do not overuse them. Always follow directions closely. Weld-through primer can be applied to galvanic mating surfaces where the coating was removed during repair. After welding,

remove the excess primer. Be sure to select a weld-through primer and not just a galvanized spray coating, which may interfere with welding. Any overspray on the outside of panels should be removed before conventional priming and painting.

3. *Applying self-etch primer.* Self-etch primers contain acids that eat into the metal to produce better adhesion. This type of primer should be used when there is a concern about paint peeling or lifting. Mix and spray self-etch primer according to the primer manufacturer's directions. Wear an approved respirator to prevent fumes from entering your throat and lungs.

Do not use lacquer-based primers, because they do not provide enough adhesion under enclosed interior conditions.

The application of both the primer and anticorrosion materials to the inside of closed sections must be done only with the manufacturer's recommended equipment. This is normally the airless or pressure-feed type of spray gun, although some suction equipment is also recommended.

Aerosol or conventional spray gun equipment will not work in enclosed interior sections because you cannot spray the material directly on the surface. This work requires special wands to reach all the inside cavities and joints where the material must go.

The airless, or pressure-fed, spray gun uses compressed air behind the fluid to force the liquid through the wand and nozzle. The fluid is broken up into a very fine atomized state, sometimes called a fog. When the substance is sprayed inside the closed section in this atomized state, it spreads rapidly and evenly into all areas, including tiny crevices (Figure 20–23).

To use a spray wand, insert it into the cavity to the farthest point that the coat is needed. Begin the spray, and pull the wand out at a steady rate, coating the section of the cavity evenly as it moves along.

Failure to clean the gun and wands on a regular basis can result in clogging or total blockage of the wands, which will prevent proper function, cause work delays, and entail a time-consuming service of the tool.

Once a week or as needed:

- Fill the gun or tank with cleaning solvent or mineral spirits.
- Spray cleaning solvent through all wands.
- Hang wands to drain overnight.

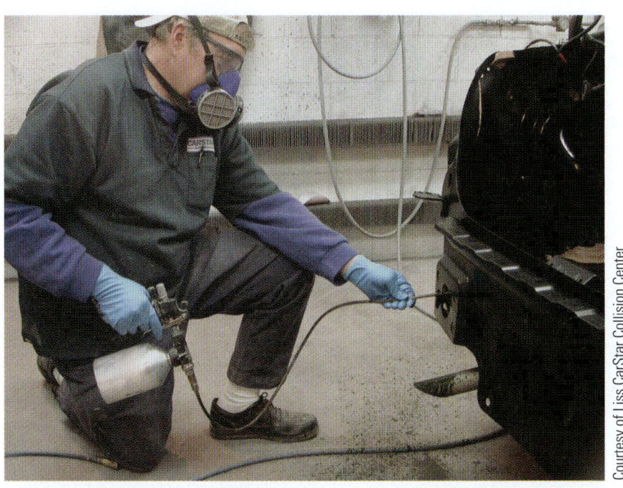

Courtesy of Liss CarStar Collision Center

A The technician is using an air-powered gun to spray epoxy primer inside a new rear frame rail. Do not press the gun trigger until the nozzle is fully inside the enclosed panel, or primer will spray out into the shop or paint booth.

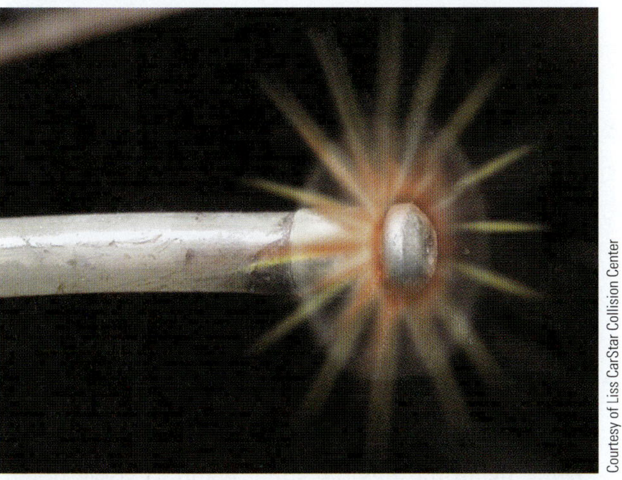

Courtesy of Liss CarStar Collision Center

B The nozzle on the end of a flexible plastic tube sprays out in a circular pattern to coat all surfaces inside the boxed or enclosed areas.

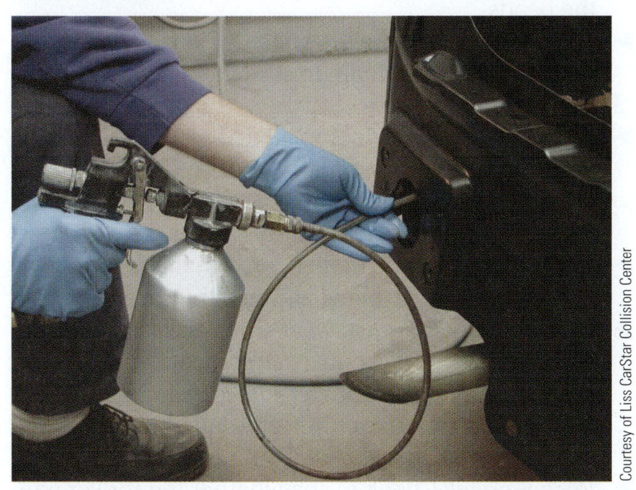

Courtesy of Liss CarStar Collision Center

C Use your hand to mark how far in the spray nozzle must reach to get at the inner weld inside the rail. Then slide the tube and nozzle that far into the rail. Press the gun trigger and slowly pull the nozzle out of the rail. Release the trigger before the nozzle is pulled out.

Figure 20-22 **Enclosed welds should be coated with epoxy primer to prevent rapid rusting inside closed panels.**

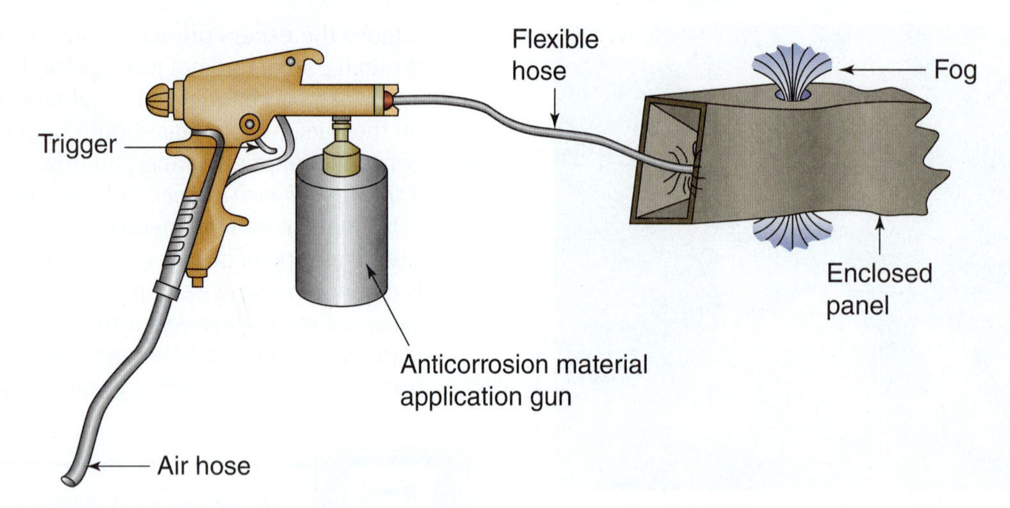

Figure 20–23 **A spray gun like this one is often used to apply material to enclosed areas. Note the "fog" coming out of the holes in the rail. The fog mist helps coat all interior surfaces, especially back sides of welds that are prone to rusting.**

There are several wand styles available. Before spraying with a wand, be sure that the spray pattern is checked and corrected and the pistol or barrel is filled with the desired primer or anticorrosion material.

The general corrosion restoration process for an enclosed interior surface is as follows:

1. Clean the enclosed interior surface with wax and grease remover.
2. Apply weld-through primer only to bare steel areas to be welded. Do not apply over paint, primer, or galvanized surfaces.
3. Apply only in the immediate weld area because this product has poor adhesion characteristics. After welding, thoroughly remove all welding residue and surplus primer from the joint area.
4. Research has shown that wire brushing is not the best way to clean a weld area. Wire brushing can leave scratches in the original primer that are not always filled by the new primer. The primer tends to "float" over the scratches, creating minute voids in which corrosion can start. A better way to clean the weld area is to use a plastic abrasive. Another way is to sandblast or plastic media blast with a captive blaster or with regular blasting equipment.
5. After the area is thoroughly cleaned, apply a primer. A two-part or self-etch epoxy primer is usually recommended for the inside area. Be sure to allow sufficient drying time according to the primer manufacturer's recommendations.
6. Apply anticorrosion compound according to the manufacturer's directions. The material is applied using service or access clip holes and drain holes. When the rustproofing material is dry, in approximately 1 hour, the water drain holes must be cleared. Refer to Figure 20–24.

To spray specific enclosed interior surfaces, special techniques may be necessary. These area considerations include the following:

Refer to the vehicle manufacturer's recommendations before drilling holes in panels for applying corrosion-protection materials. Holes in structural panels may weaken the vehicle and reduce its integrity.

- *Trunk.* Remove the spare tire, tools, floor mat, board, and padding on each side of the trunk. The rear quarter panel behind the wheels is coated from inside the trunk using the flexible spray wand to spray downward in the recess between the trunk and the quarter panel. Spray the back edge of the trunk, getting under the beads.
- Spray the trunk lid by inserting the flexible wand into the existing holes, making sure the material reaches the edges.
- When the spraying is completed, replace the padding, floor extensions, floor mats, tools, and tire. Wipe off overspray with a cloth dampened lightly with enamel reducer, solvent, or kerosene.
- *Doors.* Doors can be treated through their drain holes or after the interior panel has been removed.
- *Rear post and quarter panel.* The reverse side of the rear post and quarter panel can sometimes be sprayed from the trunk area. To ensure coverage, the front edge of the wheel well and the quarter panel area should be coated.
- *Rocker panel.* Check to see whether the rocker panels are boxed. If not, work from both ends. Before drilling any holes, check both ends of the rockers underneath for existing plugs. If satisfactorily located, these can be used to spray the entire length of the rocker with the flexible cone spray wand. If it is inconvenient or undesirable to do the rocker

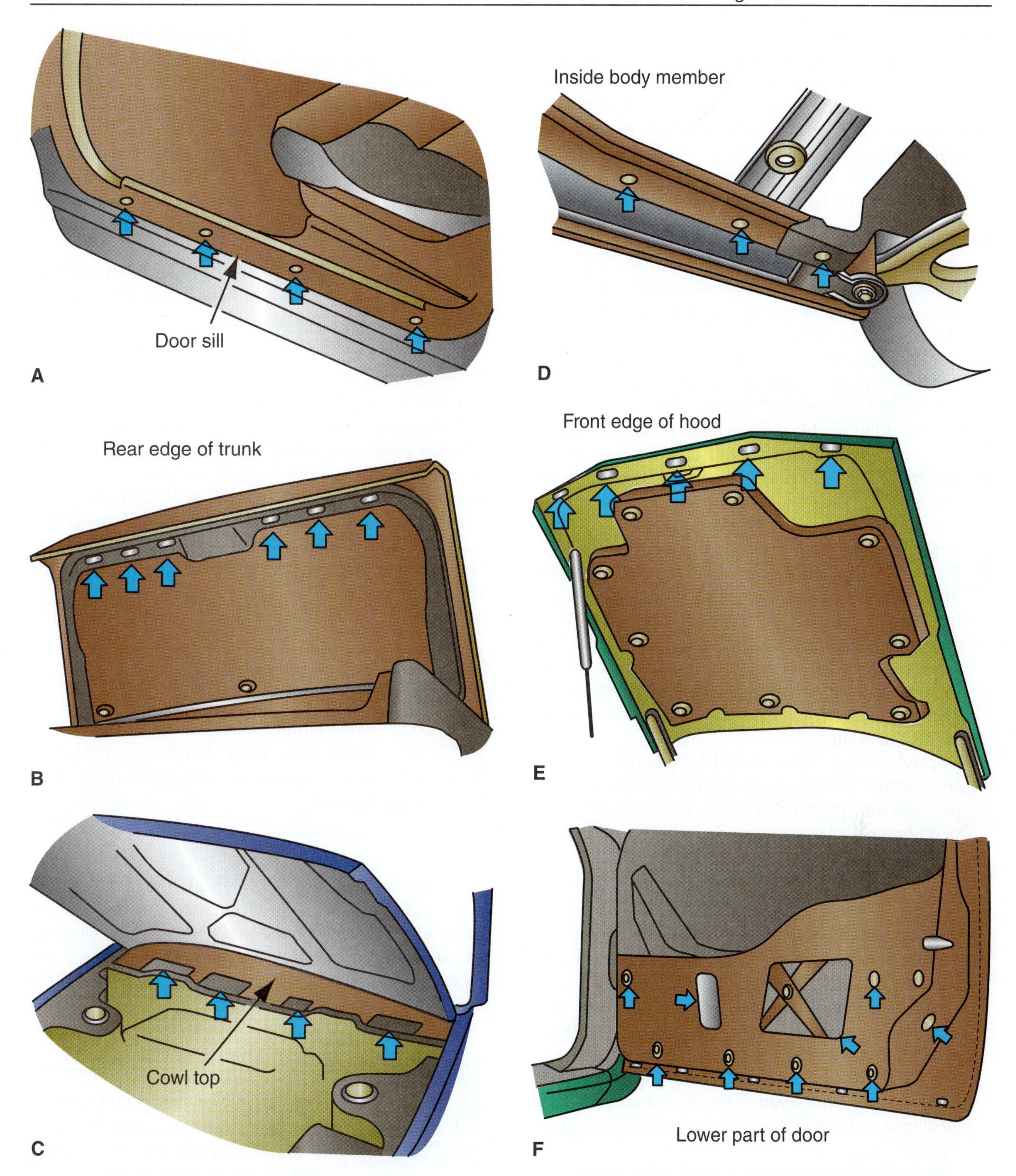

Figure 20-24 Inspect these typical service or access clip holes: (A) door sill, (B) rear edge of trunk lid, (C) inside cowl top, (D) inside body member, (E) front edge of hood, and (F) lower part of door. Keep the mechanical parts of the door free of compounds; windows should be in a closed position.

panel from above, drill a hole from below into the rocker panel at about the center and spray in both directions using the flexible cone spray wand. Be sure to spray on both sides of internal baffles, if present.

20.7 EXPOSED JOINTS

Body panel joints and seams require special attention. These areas are highly vulnerable to corrosion and must be protected correctly. This is because of the effect of welding on the metal as well as the tendency of water, snow, dirt, mud, and other contaminants to become trapped in the joint area. As a general rule, a body sealant must be applied over all the joints. The sealant must be applied so there are no gaps between the material and the panel surface.

The following are some factors to be considered when selecting and applying seam sealers:

- *Paintability.* All sealants must be paintable and have good adhesion to bare and primed metal. A seam sealer should be allowed to adequately dry before painting. The necessary drying time depends on the sealer itself, the thickness applied, and the temperature and humidity during the drying period. Normally, the lower the temperature is below 70°F (21°C), the longer the necessary drying time. The higher the humidity is above 50 percent relative humidity, the longer the necessary drying time.
- *Flexibility.* This is a critical issue with today's unibody automobiles. The sealer must be able to withstand the motion associated with the automobile. If it is not flexible, vibration could crack and damage the sealer.
- *Tooling sealants.* A finger wetted with solvent or water makes tooling easier and helps to keep the sealer from sticking to the finger of your plastic glove. This is a good application tip to help improve the finished seam's appearance. Brushable seam sealer should be tooled with a stiff bristle brush. It should be stroked in one direction only to help it match the original equipment appearance.
- *Silicone sealants.* These should not be used as a body seam sealer. They typically are not paintable, attract dust and dirt with time, and do not offer the adhesion of other types of sealants.

There are four types of seam sealers that are commonly used in auto corrosion protection work:

1. **Thin-bodied sealers** are designed to fill seams under ⅛ inch (3.2 mm) wide. This sealer will shrink slightly to provide definition to the joint, while remaining flexible to resist vibration. Adhesion is good to both primed metal and bare metal surfaces. Because many of the seams are on vertical surfaces, sag control is important so the sealer does not run out of the seam. Typically, thin-bodied sealants carry the generic names of drip-check or flow grade.
2. **Heavy-bodied sealers** are used to fill seams from ⅛ to ¼ (3.2–6.4 mm) inch wide. These sealers can be tooled to hide the seam or can be left in bead form. Shrinkage should be minimal, with good resistance to sagging and high flexibility to resist cracking in service. Heavy-bodied sealants are used on both coach joints and overlap seams. They are typically dispensed from cartridges. Some products are available in squeeze tubes.
3. **Brushable seam sealers** are used on interior body seams where appearance is not important. These seams are normally hidden and not seen by the customer. Brushable sealers are designed to hold brush marks and to resist salt and automotive fluids, such as gasoline, transmission fluid, and brake fluid. Any seams such as those under the hood and under the carriage that may be exposed to automotive fluids should have a brushable seam sealer. Applied with a brush, it normally has overlap seams.
4. **Solid seam sealers** containing 100 percent solids are used to fill larger voids at panel joints or holes. This product comes in strip caulking form, designed to be pressed into place with your thumb (Figure 20–25).

Be sure to carefully follow the manufacturer's instructions for the use of these versatile products.

The application sequence for exposed joints is basically the same as it is for interior exposed panels, with the addition of two steps. After the welded areas are thoroughly cleaned and primed, seal all body panel joints with a seam sealer. Finish the joints by applying another coat of primer over the seam sealers, then topcoat with the same material used over the rest of the repaired area.

DANGER

Some anticorrosion compounds can be harmful if they come into contact with human skin. Read the material directions. Wear plastic or rubber gloves when in doubt or when the material could be harmful.

It is important to use a nozzle with a small hole in the end to apply the sealer. Then spread out the bead of sealer with a fingertip. If the nozzle hole is small, the finish can be kept neat. If the nozzle hole is too large, the sealer will spread too wide and may cause a poor-looking finish.

Do not use latex-based seam sealers designed for home use! These may draw moisture into the joint and cause rapid rust formation and joint failure. Use only recommended sealers.

Automotive masking tape can also be used to make a parting line for the sealer application area. In some situations, this makes the repair look more OEM.

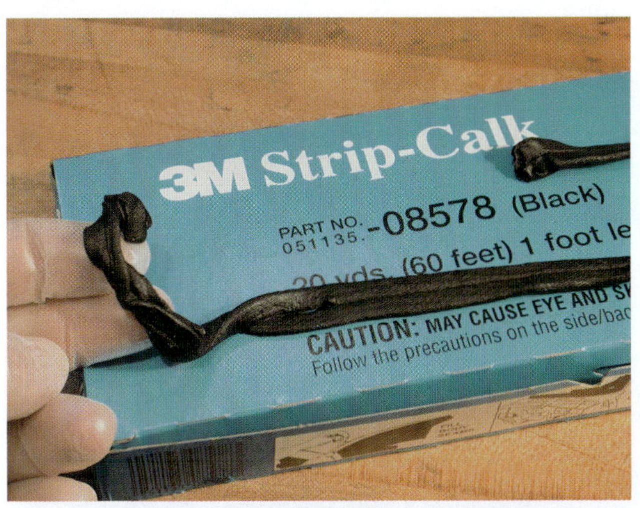

A Strip caulk comes in long beads. Auto manufacturers recommend this type of sealer in some body locations.

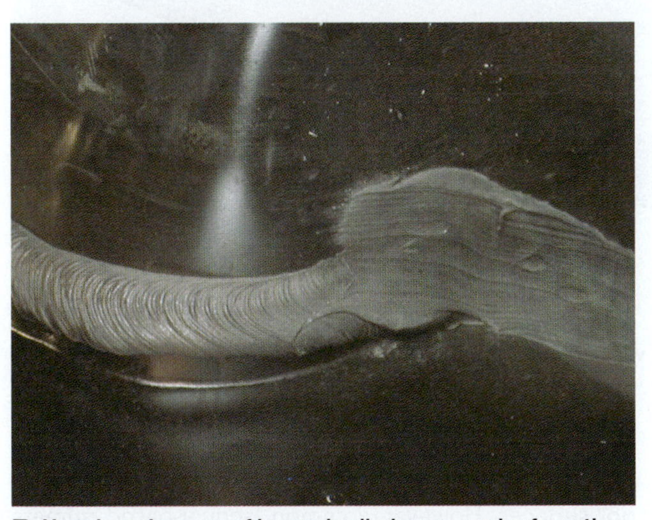

B Here is a close-up of heavy-bodied seam sealer from the factory.

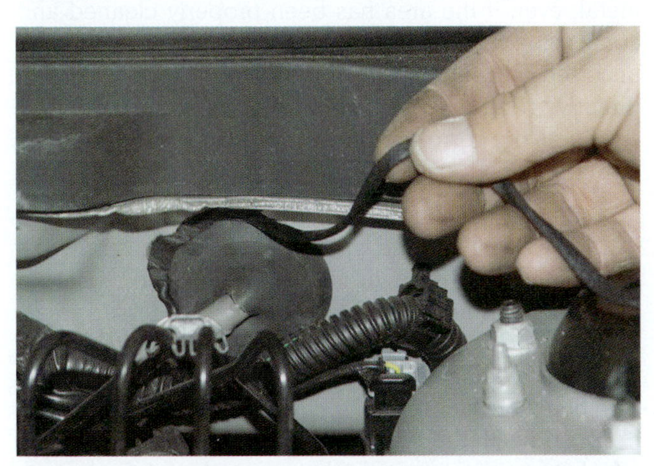

C Here heavy-bodied seam sealer is being applied where air-conditioning lines come through the firewall on the vehicle. Force the soft strip down into the joint. It will fill large gaps in panels and even holes.

Figure 20–25 Strip caulk is thick-bodied sealer that can be worked into panel and part joints to prevent water and air leakage. Solid seam sealer often comes in a box in individual strips.

When applying sealant, refer to the shop manual for the vehicle being repaired. Determine the sealer application area or look at the other side of the vehicle to see where the sealer is applied (Figure 20–26).

To summarize the corrosion protection process for exposed joints and seams, proceed as follows:

1. Thoroughly clean the joint or seam.
2. Apply primer or primer-sealer.
3. Seal the joints with seam sealer.
4. Apply a second coat of primer or primer-sealer.
5. Finish with a colorcoat in a spray booth.

More than one type of seam sealer can be used on any given joint. For example, a brushable seam sealer could be used over a thin-bodied sealer. The material instructions and service manual will give details for proper application.

20.8 EXPOSED INTERIOR SURFACES

The bottom surfaces of the underbody and inside of the wheel housing can be damaged by flying stones, causing rust to develop. These areas are given an undercoat treatment with a material such as shock-absorbing wax. Apply this treatment from below the underbody.

Do not spray anticorrosion material into the passenger compartment. Metal conditioners and conversion coatings also are not recommended for interior surface protection. There are three reasons for this:

1. These surfaces are not exposed to physical damage the way exterior surfaces are.
2. These areas contain joints and seams that should not be contaminated with etching and conversion-coating chemicals and are generally difficult to rinse clean.
3. They can cause harmful odors that remain in the passenger area.

The corrosion-protection process begins with a thorough cleaning with a wax and grease remover. Once the surface is completely air dry, spray the first coating of wax- or petroleum-based undercoat compound on all welded areas and panel joints. Then apply a second coat over the entire area. Cover places surrounding the application area with masking paper and/or tape to prevent the undercoating from sticking to areas where it is not wanted (Figure 20–27).

Although there are several different types available, the two-part epoxy primers most closely duplicate the baked-on electrode position coating, or e-coat, used by car manufacturers. Nearly any material can be applied over the epoxy primer.

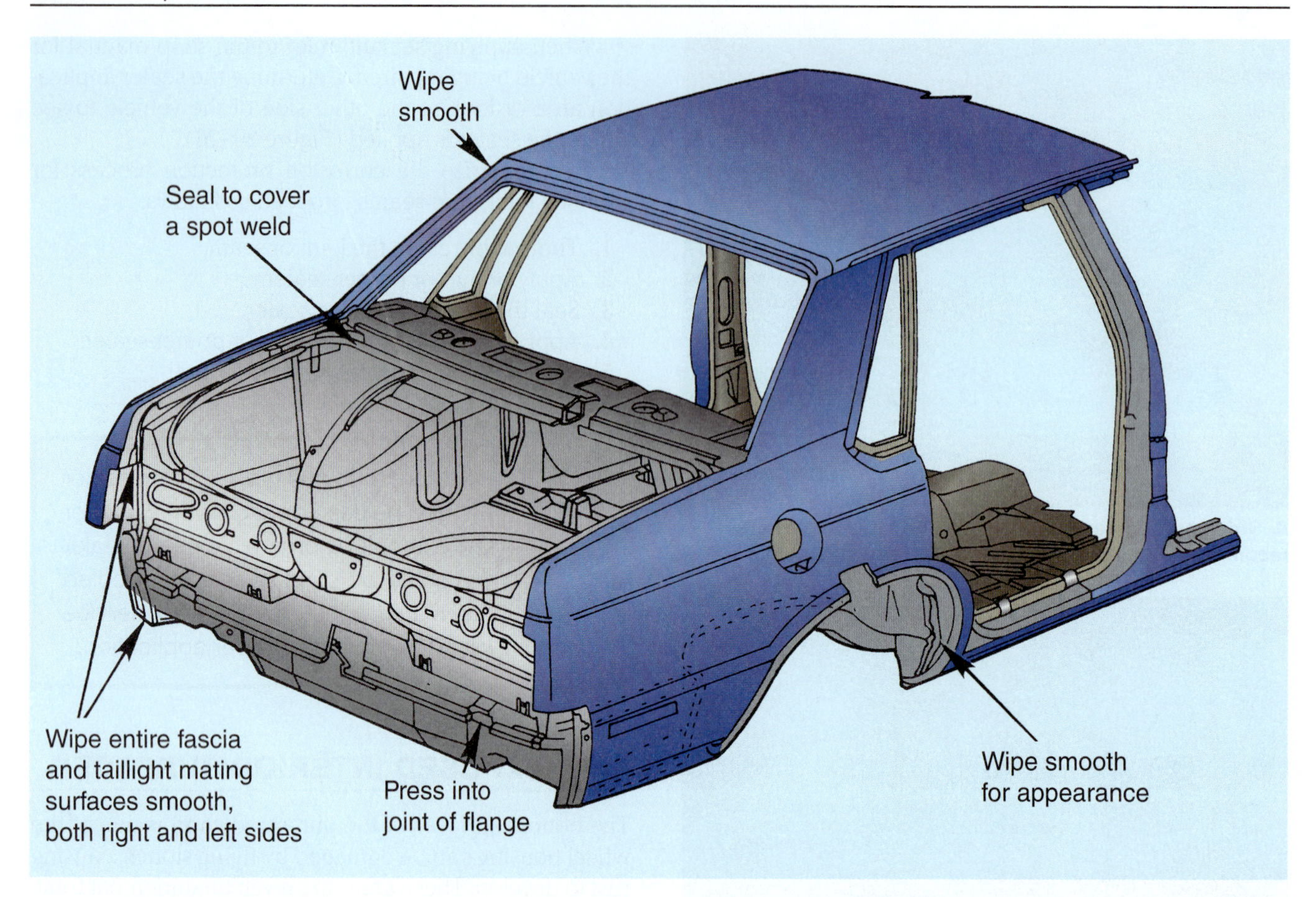

Figure 20-26 Service manual illustrations, such as this one, will give valuable information about seam sealers and other anticorrosion materials.

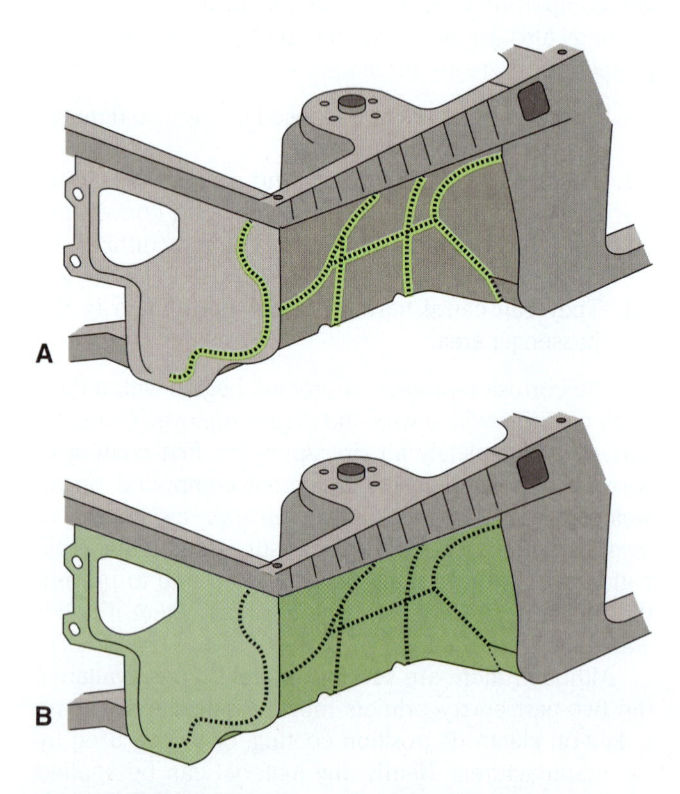

Figure 20-27 (A) Apply undercoat to all welded areas and panel joints, (B) then apply to the entire area.

Self-etch primers can also be used for exposed interior surfaces. However, common lacquer-based primers will not provide proper adhesion when used on bare metal, even if the area has been properly cleaned and conversion coating used. This point cannot be overemphasized. Lacquer-based primer should never be used directly on the bare metal of modern unibody cars.

To restore corrosion protection to specific areas such as under the hood, proceed as follows:

1. Lift the hood and spray the front fender or apron between it and the wheel well. Be sure to apply the material right down to the fender beads. Use the flexible cone spray wand to reach all recessed areas.
2. Cover the large open spaces with a 45-degree flat spray wand.
3. Spray the leading edge and the side channels of the hood with the flexible cone spray wand.
4. Loosen or remove the battery and coat the battery tray and surrounding areas.

The headlight areas on some cars might be reached from under the hood. On other makes, the headlight areas can be reached from existing holes under the hood by means of the flexible spray wand. If not accessible from under the hood, the headlight areas can be sprayed from under the car, working forward in the front wheel well when the car is put on a lift. This can be a baffled area with a rubber edge that can be depressed to insert a flexible cone spray or flat spray wand. Choose the best method to ensure complete coverage.

20.9 EXPOSED EXTERIOR SURFACES

Exterior surfaces are subjected to much greater exposure to chips and nicks than interior surfaces. Therefore, the use of etching and conversion coating agents is of critical importance on exterior surfaces. Conversion coating provides the kind of superior paint film adhesion that retards creeping rust from working its way under the paint when chips and nicks do occur.

Exposed exterior surfaces are of two types: cosmetic and underbody.

Anticorrosion procedures for exterior cosmetic surfaces are generally as follows:

1. Clean with a wax and grease remover.
2. Apply a metal conditioner. Rinse with water. Apply a conversion coating and allow to thoroughly air dry. Drying can be speeded with compressed air or a clean, white rag. Rinse with water.

   or

   Apply a self-etch primer that does not require the use of a metal conditioner. This can save time and effort.
3. Apply a primer-surfacer.
4. Apply a colorcoat or paint system.

If available, a lift makes underbody corrosion protection work easier. When corrosion-proofing the underbody, start by spraying the fenders and wheel wells, paying particular attention to the fender beads. On some cars, it will be necessary to remove the wheels to do an adequate spraying job (Figure 20–28).

Spray the remaining underbody and splash pans adjacent to the front and rear bumpers. Spray the underside of the floor pan, welded joints, frame, tank straps, and seams. Remove any loose debris or sound deadener, particularly around joints, before spraying. Loose sound-deadening materials or dirty surfaces prevent the rustproofing material from reaching the metal and will create pockets in which rust will form.

Anticorrosion procedures for exterior underbody surfaces are generally as follows:

1. Clean with a wax and grease remover.
2. Apply a metal conditioner.
3. Rinse with water.
4. Apply a conversion coating.
5. Rinse with water.
6. Apply a recommended self-etch or epoxy primer.
7. Apply anticorrosion compound and sound-deadening materials to restore to factory specifications.
8. Most undercoat overspray can be removed with enamel reducer, solvent, or kerosene and by washing.

Care is needed when applying anticorrosion compounds. Keep the material away from parts that conduct heat, electrical parts, labels, identification numbers, and moving parts. Avoid applying corrosion-protection materials to:

- Seat belt retractors and passive restraint guide rails
- Hidden headlamp assemblies
- Power window motors and cables
- Exhaust system
- Engine and accessories
- Air filter
- Air lift shock absorbers
- Transmission parts
- Shift linkages
- Speedometer cables
- Brake parts
- Locks, key cylinders, and door latches
- Power antennas
- Theft prevention labels
- Driveshaft

Make sure you mask or cover parts that you do not want to undercoat (exhaust system, brake rotors, fan belts, and so forth). For example, if undercoat gets on the exhaust system, it could cause heavy smoke and possibly a fire upon engine operation—an embarrassing mistake.

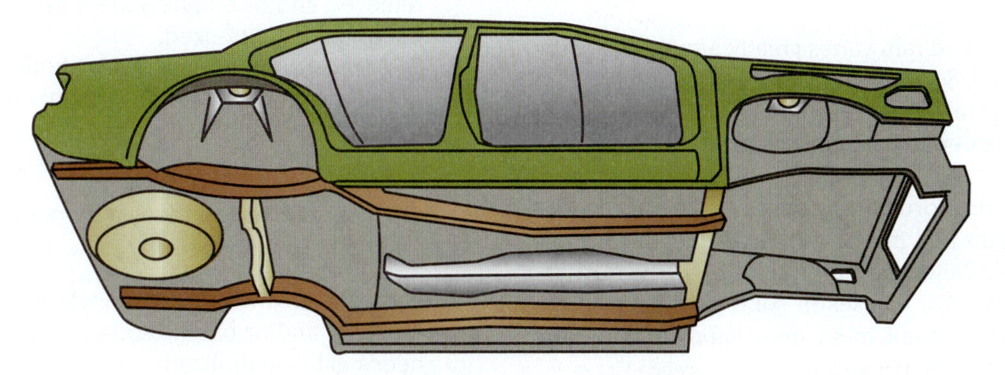

Figure 20-28 A shop manual will note those underbody panels that require protection.

20.10 EXTERIOR ACCESSORIES

To prevent corrosion, it is very important to install a barrier between dissimilar metal components, such as aluminum bumpers and stainless and aluminum body trim. Plastic or rubber isolating pads accomplish this effectively. Mounting stainless and aluminum body trim must be done correctly to avoid galvanic corrosion. For example, when mounting trim requires drilling holes in a new or repaired panel, drill all holes before applying the primer, coating the inside edges of all holes completely.

When using a kit for replacement trim, be sure to use all parts supplied with the kit. If parts are not purchased as a kit, duplicate the original assembly exactly. Clearly, there is a great variety of body trim and accessories requiring many different application techniques. In all cases, be sure to follow the manufacturer's recommendations to avoid problems in making these repairs.

20.11 ACID RAIN DAMAGE

As mentioned earlier, air pollutants can damage an automotive finish. Because most of such damage is done to exterior, finished surfaces, air pollutants are a major concern of the refinisher.

Acid rain and other pollutants have generated a lot of controversy in recent years. There has been some confusion as to their causes and effects. Sulfur dioxide or nitrogen oxides create acid rain when released into the atmosphere. They combine with water and the ozone to create either sulfuric or nitric acid. It is estimated that the United States alone pumps out 30 million tons (27 metric tons) of sulfur dioxide and 25 million tons (22.5 metric tons) of nitrogen oxides yearly. More than two-thirds of the sulfur is emitted from power plants burning coal, oil, or gas. Iron and copper smelters, automobile exhaust, and natural sources like volcanoes, wetlands, and forest fires account for most of the remaining pollutants.

The standard for measuring acid rain is the pH scale. It runs from zero to 14, with 7 being neutral or equal to distilled water. A pH reading of 4 is 10 times more acidic than a solution of acid and water with a pH of 5, and 100 times more acidic than a pH of 6. Once released into the ozone, these acids are readily dissolved into cloud droplets, which, if low enough in pH, can cause significant paint damage.

The level of acid rain varies greatly around the country. For example, South Carolina is reported to be one of the most acidic states in the nation. In Los Angeles, fog has been measured to have the acidic strength of lemon juice.

Rainfall in the northeastern states is extremely corrosive to car paints and finishes. For example, the average pH of rainfall in New Jersey is an acidic 4.3. Several manufacturers now have clauses in some of their new car warranties that exempt them from liabilities involving paint damage in high pH areas.

Acid rain damage most frequently affects the paint pigments, with lead-based pigments the most susceptible. Typically, the damage looks like water droplets that have dried on the paint and caused discoloration. Sometimes the damage appears as a white ring with a clear, dull center. Severe cases show pitting. Discoloration varies depending on the color. For example, acid rain damage to a yellow finish may appear as a white or dark brown spot. Medium blue may have a whitening look. A white finish may be a discolored pink, and a medium red may be purple.

Metallic finishes can be damaged because the acidic solution reacts with the aluminum particles and etches away the finish. A fresh finish is more easily damaged than an aged finish. Lacquers and uncatalyzed enamel finishes are most susceptible to damage, followed closely by catalyzed enamels.

Clearcoated finishes add a layer of protection against acid rain, so late model vehicles with two- and three-coat finishes are less susceptible to damage. A clearcoat protects the paint pigments from discoloration, but it is still possible for acid rain to create a peripheral etch, or ring, on the clearcoat.

Acid Rain Repairs

The procedure for restoring acid rain damage varies depending on the level and depth of the damage. When the problem has been corrected, stop at that stage. Remember that polishing or compounding removes part of the original finish and thereby reduces its overall life.

If the surface damage from acid rain is only into the clearcoat, proceed as follows:

1. Clean the area with soap and water, wax and grease remover, and/or baking soda and water if the acid needs to be neutralized.
2. Wet or dry sand the pitted clearcoat with #1500- to #2000-grit sandpaper.
3. Machine buff or compound the area to return its gloss.

If the acid rain damage is down into the colorcoat, you must:

1. Clean the area with soap and water, wax and grease remover, and/or baking soda and water if the acid needs to be neutralized.
2. Dry or wet sand the area with about #600- to #800-grit sandpaper to level and smooth the area.
3. If you did not cut through the colorcoat, respray the whole panel with clearcoat (Figure 20–29).

If the acid rain ate all the way through the clearcoat and colorcoat, you must:

1. Clean the area with soap and water, wax and grease remover, and/or baking soda and water if the acid needs to be neutralized.

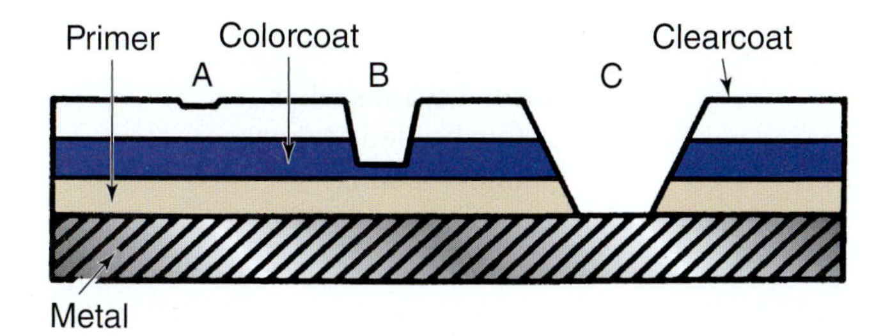

Figure 20-29 Different levels of acid rain damage will determine the repair methods required.

2. Dry or wet sand the area with #220- to #400-grit sandpaper to level and smooth the damaged paint. Then sand the whole panel with #600- to #800-grit sandpaper to prepare it for repainting.
3. Spray the damaged area with colorcoat as needed and blend the color into the surrounding undamaged areas on the panel.
4. Clearcoat the entire panel.

Industrial Fallout Surface Damage

Generally speaking, damage from **industrial fallout** is caused when small, airborne particles of iron fall on and stick to the vehicle's surface. The iron can eventually eat through the paint, causing the base metal to rust. Sometimes the damage is easier to feel than see. Sweeping a hand across the apparent damage will likely reveal a gritty or bumpy surface. Rust-colored spots might be visible, however, on light-colored vehicles.

The steps for repairing damage caused by industrial fallout are similar to those used when repairing acid rain damage, but with the following exception: After washing the car, treat the repair area with a "fallout remover," a chemical treatment product made especially for industrial fallout damage. Do not buff the damaged area before removing the fallout, because buffing will drive the particles into the paint surface. If the particles break loose and become lodged in the buffing pad, deep gouges can occur.

SUMMARY

1. Corrosion protection involves using various materials to protect steel body parts from rusting. When doing repairs, you must always use recommended methods of protecting repair areas from rust damage.
2. Failing to restore proper corrosion protection can endanger the driver and passengers of the vehicle.
3. Corrosion or rust is the oxidation and chemical change of metal.
4. Galvanizing is a process of coating steel with zinc. It is one of the principal methods of corrosion protection applied during the manufacturing process.
5. During a collision, the protective coatings present on a car are usually damaged.
6. Weld-through primer is used to provide anticorrosion protection to weld zones.
7. Self-etch primers etch bare metal to improve paint adhesion and corrosion resistance, while providing the priming and filling properties of primer-surfacer.
8. Two-part epoxy primers provide a very strong base coating with good adhesion to bare metal.
9. Rust converters change ferrous (red) iron oxide to ferric (black/blue) iron oxide.
10. Be sure to carefully read and follow the manufacturer's instructions for the use of anticorrosion compounds.
11. Refer to the vehicle manufacturer's recommendations for applying corrosion-protection materials before drilling holes in panels.
12. Some anticorrosion compounds can be harmful if they come into contact with human skin.
13. *Acid rain* is the term given to rain containing pollutants. It causes discoloration and even destruction of the paint surface and can lead to corrosion damage.
14. Generally speaking, damage from industrial fallout is caused when small airborne particles of iron fall on and stick to the vehicle's surface.

EXERCISES

On a separate sheet of paper, complete the following learning activities for this chapter. Write definitions for the key terms and answer the ASE-style review questions, essay questions, critical thinking problems, and math problems. You can also do the outside activities, possibly for extra credit.

Key Terms

acid rain
brushable seam sealers
corrosion prevention
corrosion protection
galvanic corrosion
galvanizing
heavy-bodied sealers
industrial fallout
metal treating
rust converters
solid seam sealers
thin-bodied sealers
undercoating
weld-through primer

ASE-Style Review Questions

1. Corrosion protection involves using various materials to protect steel bodies from ___________.
   A. Rust
   B. Corrosion
   C. Rustproofing
   D. All of the above
   E. Both A and B
2. Corrosion is accelerated in areas ___________.
   A. of high relative humidity
   B. where temperatures drop below freezing
   C. Both A and B
   D. None of the above
3. The higher the pH number rises above 6.0, the ___________.
   A. greater the chance of acid rain
   B. less likely the chance of acid rain
   C. Both A and B
   D. None of the above
4. When two dissimilar metals are placed in contact with each other, the more chemically active will corrode, protecting the other metal in the process. This is called ___________.
   A. zinc coating
   B. galvanic corrosion
   C. Both A and B
   D. None of the above
5. Technician A uses a conversion coating and then a metal conditioner. Technician B says that conversion coatings are usually mixed with water before using them. Who is correct?
   A. Technician A
   B. Technician B
   C. Both A and B
   D. Neither A nor B
6. Technician A uses a conversion coating on inside closed sections. Technician B states that inside closed sections must be cleaned and degreased before being welded into place. Who is correct?
   A. Technician A
   B. Technician B
   C. Both A and B
   D. Neither A nor B
7. Which sealant is not paintable and attracts dust and dirt with time?
   A. Silicone sealant
   B. Tooling sealant
   C. Thin-bodied sealant
   D. Heavy-bodied sealant
8. Heavy-bodied sealers are used to fill seams from ___________.
   A. $\frac{1}{16}$ to $\frac{1}{8}$ inch wide
   B. $\frac{1}{8}$ to $\frac{1}{4}$ inch wide
   C. $\frac{1}{4}$ to $\frac{1}{2}$ inch wide
   D. All of the above
9. Technician A uses conversion coating on aluminum. Technician B says that conversion coatings are applied using a spray bottle and rinsed off before drying. Who is correct?
   A. Technician A
   B. Technician B
   C. Both A and B
   D. Neither A nor B
10. When restoring corrosion protection to an enclosed interior surface, Technician A applies the weld-through primer over existing paint. Technician B applies the weld-through primer over galvanized surfaces. Who is correct?
    A. Technician A
    B. Technician B
    C. Both A and B
    D. Neither A nor B
11. Any seams that might be exposed to automotive fluid should have a ___________.
    A. thin-bodied sealer
    B. heavy-bodied sealer
    C. brushable seam sealer
    D. solid seam sealer
12. Corrosion protection materials should never be applied to ___________.
    A. The exhaust pipe or muffler
    B. Shock absorbers
    C. Drivetrain parts
    D. Brake drums
    E. Any of the above

Essay Questions

1. Summarize the corrosion process for exposed joints and seams.
2. Explain the four broad categories of anticorrosion materials.
3. Summarize four basic safety rules to follow when working with anticorrosion compounds.
4. What is weld-through primer?
5. How would you repair acid rain damage embedded in a vehicle's surface coat?
6. Define corrosion.

Critical Thinking Problems

1. An untrained worker fails to apply anticorrosion materials after major structural repairs to a car's frontal frame rail and shock tower areas. The car will be driven on salty roads in winter months. What can happen after a few years of service?
2. An inexperienced worker applies undercoating on a replaced floor pan without first cleaning the area properly. What could happen?
3. Name various types of anticorrosion materials.
4. What are the basic types of corrosion protection used on today's automobiles?
5. List the new car finishing sequence used by major auto manufacturers.

Math Problems

1. A can of sprayable undercoating will cover 10 square feet. You must spray panels that are 2×3 feet, 1×2.5 feet, and 1.1×1.1 feet. How many cans of undercoating will you need?
2. To neutralize an area damaged by acid rain, you need 3 gallons of solution. If you must mix 1 tablespoon of baking soda to 1 quart of water, how many tablespoons will you need?

Activities

1. Inspect a badly damaged vehicle. Find any corrosion-protection materials that would need replacement during the repair.
2. Read the instructions on various corrosion-protection materials.
3. Visit a body shop. Ask the owner to allow you watch a technician replacing corrosion-protection materials.
4. Search the Internet for information on vehicle manufacturing.

Mechanical and Electrical Repairs

Chapter 23 Restraint System Operation and Service

CHAPTER

21 Chassis Service and Wheel Alignment

OBJECTIVES

After reading this chapter, you should be able to:

- Explain the procedure for removing a powertrain from unibody and full-frame vehicles.
- Describe how suspension and steering systems work.
- List the elements of proper wheel alignment.
- Diagnose and service a steering system.
- List typical driveline variations.
- Name the service procedures for the major parts of a cooling system.
- Describe the service procedures for an air-conditioning system.
- Diagnose and service an emission control system.
- Answer ASE-style review questions relating to chassis service.

INTRODUCTION

Today's collision repair technicians often have to service mechanical chassis parts when doing major auto body repairs. To gain access to structural panels for replacement or during full-frame replacement, they must remove and install engines, transmission assemblies, drive axles, suspension systems, axle housings, fuel lines, emission control devices, brake assemblies, and other mechanical units.

Body technicians frequently complete minor mechanical repair tasks, such as replacing a damaged water pump, a smashed radiator, or a broken engine motor mount. The crushing force of a major collision often ruins the mechanical components.

In many cases, major driveline parts, such as engines, transmissions, and drive axles, are mounted directly to structural unibody panels or the frame. In other cases, these parts are mounted to supporting cross members, subframe assemblies, or cradle assemblies that are mounted to the unibody or frame (Figure 21–1).

Collision repair technicians must often remove and install these mechanical assemblies when doing major structural repair work. However, more complex mechanical repairs may require the special skills and tools of a certified auto mechanic. In such cases, vehicles are sent to another shop or to the mechanical repair area next to the body shop.

This chapter will give you the basic knowledge needed to service mechanical components in a body shop environment. You will learn which mechanical parts

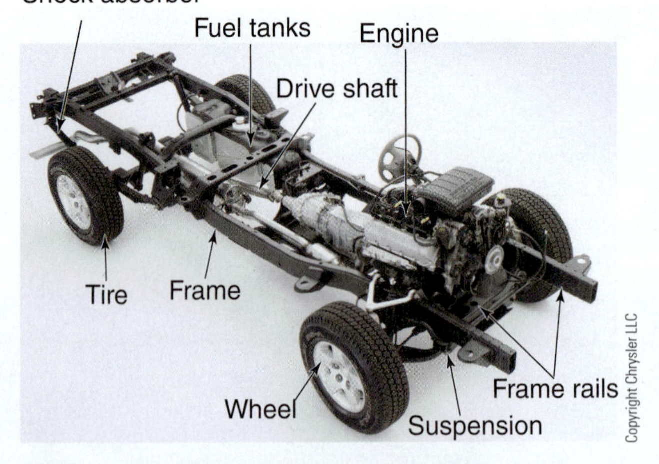

Figure 21–1 Chassis parts are often damaged in a major collision. The engine, transmission, axles, suspension, brakes, and other mechanical parts may need repair.

are typically damaged in a collision and how to replace them. You will learn when, why, and how mechanical assemblies are removed while doing collision repairs.

21.1 POWERTRAIN CONSTRUCTION

The **powertrain** includes all of the parts that produce and transfer power to the drive wheels to propel the vehicle. This includes the engine, transmission or transaxle, drive axle, differential, and other related parts. The *drivetrain* typically includes all of the assemblies that send power to the drive wheels, except the engine.

When diagnosing and repairing mechanical parts, always refer to factory or aftermarket service information for the specific vehicle. The computer-based repair information provides the detailed procedures and specifications essential for doing competent work.

Engine

The **engine** provides energy to move the vehicle and power all of its accessories. Most cars and light trucks use a gasoline engine; some use a diesel engine (Figure 21–2).

The basic parts of a typical internal combustion, reciprocating, piston engine include:

- The *engine block*, which is the foundation of the engine; all the other engine parts are either housed in or attached to the block. The cylinder is a round hole bored (machined) in the block that guides piston movement.
- The *engine piston*, which transfers the energy of combustion (burning of air–fuel mixture) to the connecting rod. Circular seals called rings are installed around the top, sides of the piston. They keep combustion pressure and oil from leaking between the piston and cylinder wall (cylinder surface). A connecting rod attaches the piston to the crankshaft.
- The *engine crankshaft*, which changes the reciprocating (up and down) motion of the piston and rod into a more useful rotary (spinning) motion. Power to turn the driving wheel comes from the rear of the crankshaft, and accessories are driven off the front.
- A *cylinder head*, which covers and seals the top of the cylinder. It contains valves, rocker arms, and, sometimes, the camshaft. The combustion chamber is a small enclosed area between the top of the piston and the bottom of the cylinder head. The burning of the air–fuel mixture occurs in the combustion chamber.
- *Engine valves*, which are flow control devices that open to allow the air–fuel mixture into and exhaust out of the combustion chamber. Valve springs hold

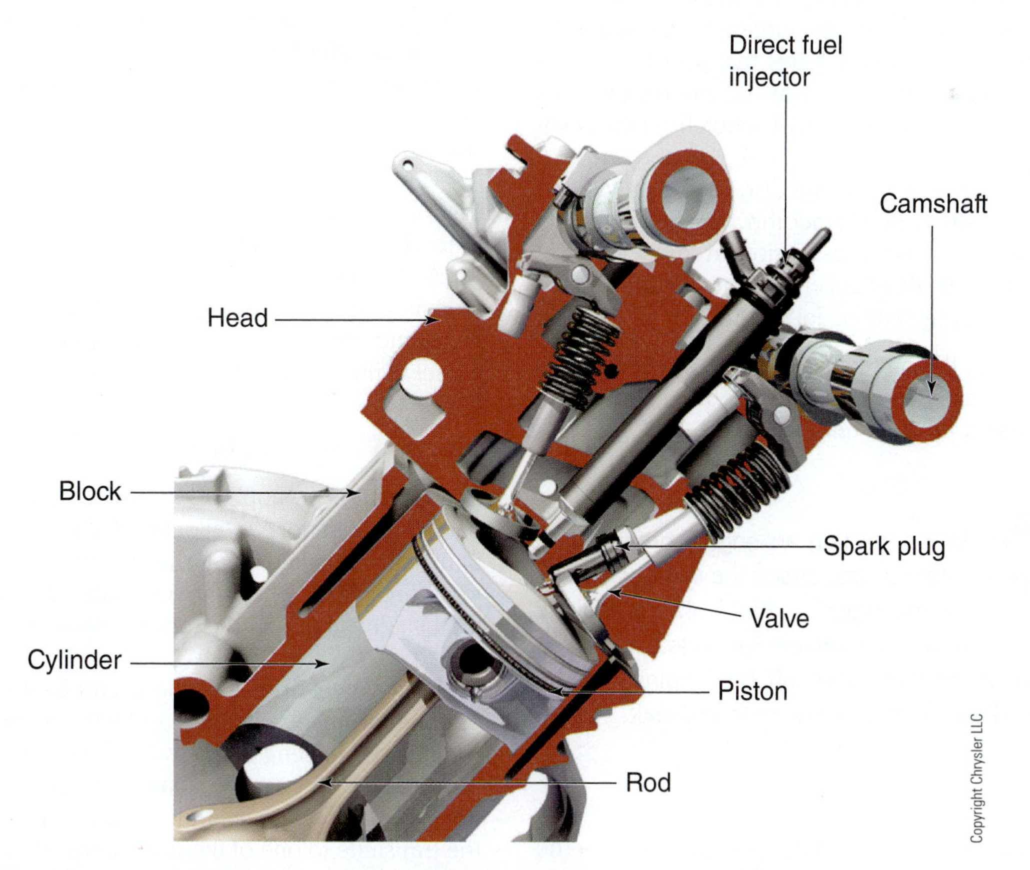

Figure 21-2 This cutaway view shows the internal parts of a modern engine. Study the names of the various parts.

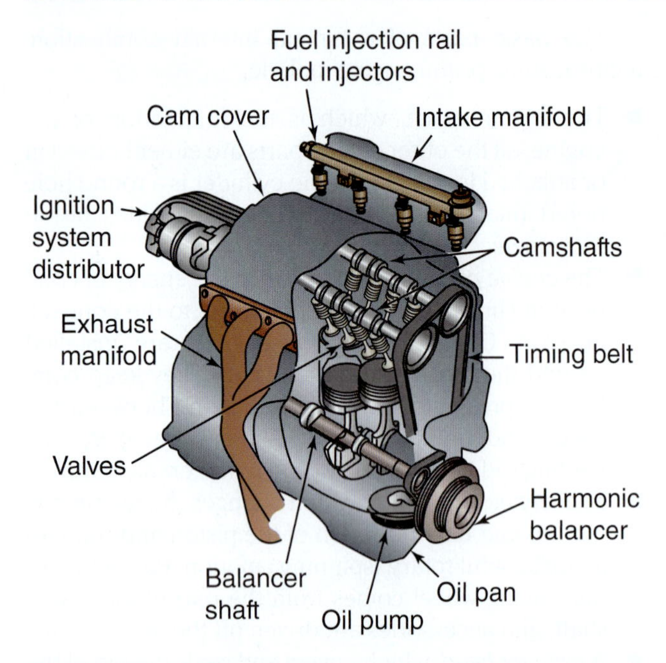

Figure 21-3 Even though as a body technician you may never work on the internal parts of an engine, it is interesting to know how an engine works.

the valves closed when they do not need to be open. They also return the valve train parts to the at-rest position.

- The *camshaft*, which controls the operation of valves. It can be located in the block or the cylinder head. The lifter is a cylindrical-shaped part that rides on the camshaft lobes and transfers motion to the pushrods. The pushrods are hollow tubes that transfer motion from the lifters to the rocker arms. The rocker arms are levers that transfer camshaft action from the pushrods to the valves. See Figure 21–3.
- The *flywheel* is a heavy metal disc used to help keep the crankshaft turning smoothly. It also connects engine power to the transmission. A larger gear on the outside of the flywheel engages the starting motor when cranking the engine for starting.

In a collision, the oil filter, oil pan, and related parts are sometimes damaged. They are made of thin metal and can be easily crushed and ruptured.

> **WARNING**
> When starting an engine before or after repairs, check the oil level with the oil dipstick. Also, always look under the vehicle for oil leakage. If you find an oil leak, shut off the engine right away. Find and fix the source of the oil leak.

Drivetrain Construction

The drivetrain includes everything after the engine—the clutch, transmission, drive shaft, and drive axles. Drivetrain designs vary. Some cars use a manual (hand-shifted) transmission. Others use an automatic transmission, which shifts gears automatically using internal fluid pressure.

The **transmission** is an assembly with a series of gears for increasing torque to the drive wheels so the car can accelerate properly. It provides high power for acceleration in lower gears and good gas mileage in higher gears. Look at Figure 21–4.

With an automatic transmission, a *torque converter* (fluid coupling) is used in place of a clutch.

A **transaxle** is a transmission and differential combined into a single housing or case. Both automatic and manual transaxles are available. After collision repairs, check the transmission or transaxle fluid level before test-driving the vehicle.

A **clutch** is a device used to couple and uncouple engine power to a manual transmission or transaxle. It uses a friction disc, pressure plate, flywheel face, and release bearing for activation.

Front-wheel-drive (FWD) vehicles use a transaxle to transfer engine torque to the front drive wheels. Constant-velocity axles, or **CV axles**, transfer torque from the transaxle to the wheel hubs. They can be found on vehicles with independent suspension at the drive wheels. Constant-velocity (CV) joints have overcome the design limitations of conventional universal joints (U-joints). They eliminate the vibration problem typical of older cross and roller-type U-joints.

Front-engine, rear-wheel-drive (RWD) vehicles use a conventional transmission, drive shaft, and rear axle assembly to transfer power to the rear drive wheels.

A **drive shaft** is a long tube that transfers power from the transmission to the rear axle assembly. It has U-joints at both ends that provide flexibility to the suspension while maintaining driving force.

The **rear axle assembly** is the housing that contains the ring gear, pinion gear, differential assembly, and axles. Rear suspension springs attach to the housing.

A **differential** assembly is a unit within the drive axle assembly that uses gears to allow different amounts of torque (turning force) to be applied to each drive wheel while the vehicle is making a turn.

Front-Wheel Drive

In a typical FWD application, two CV joints are used on each half shaft, for a total of four CV joints. Two outboard joints are installed near the wheels and two inboard joints are installed near the transaxle. The outboard joints are usually fixed and the inner ones are generally plunging types.

Front-wheel-drive half shafts can be solid or tubular, of equal or unequal length, and with or without damper weights. Equal-length shafts help reduce *torque steer*—the tendency to steer to one side as engine power is applied. In these applications, an intermediate shaft links the transaxle to one of the half shafts. At the outer end is a support bracket and bearing assembly (Figure 21–5).

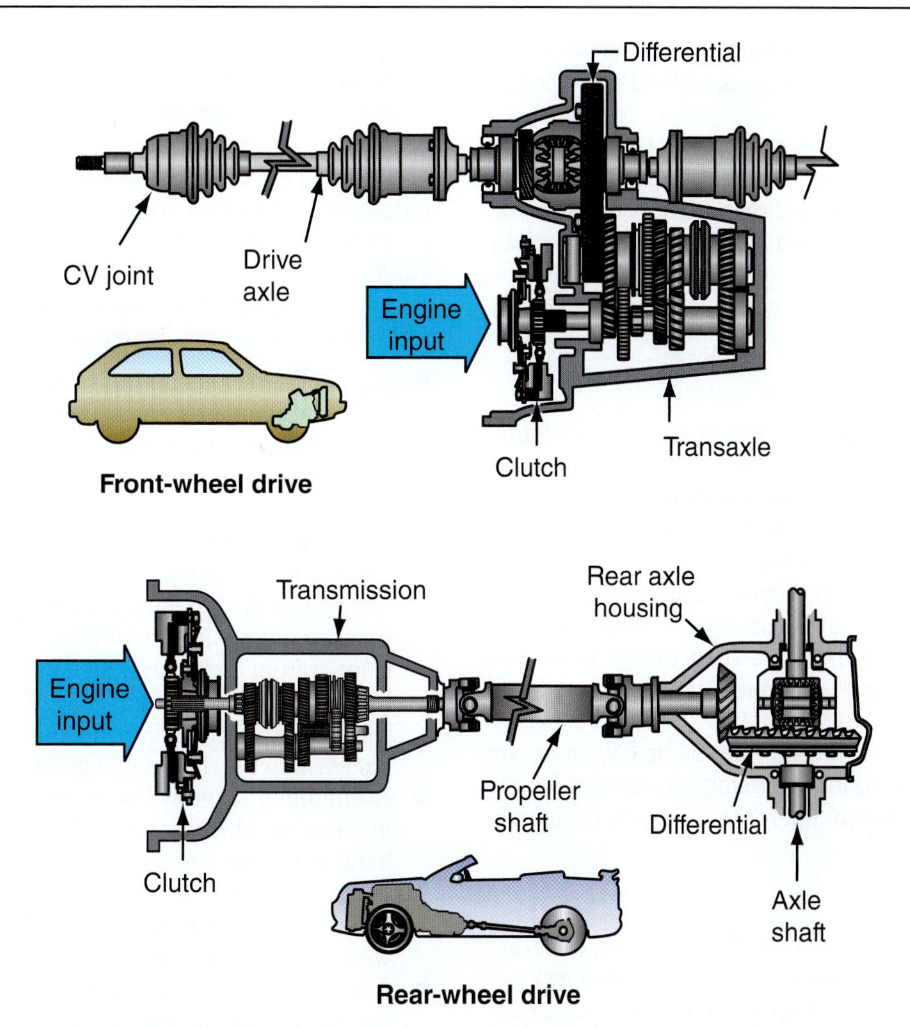

Figure 21-4 The transaxle or transmission bolts to the rear of an engine. The clutch or torque converter fits between the engine and the transmission. The differential is part of the transaxle. With a rear-drive transmission, the differential is in the rear axle assembly. With front-drive, CV axles transfer power to drive the wheels and tires.

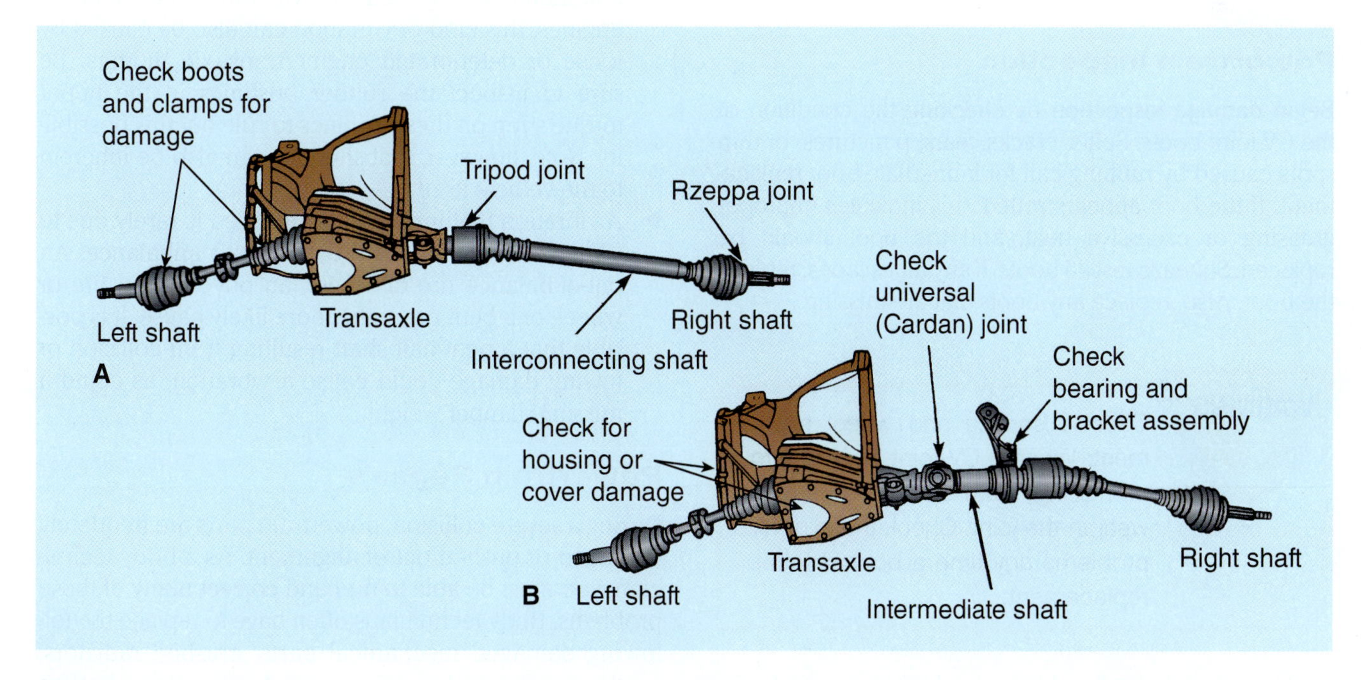

Figure 21-5 This car has (A) unequal- and (B) equal-length FWD half shafts. Always inspect these parts for collision damage and have them replaced if necessary.

Always inspect these drivetrain components because loose bearings and/or brackets can vibrate and fail in service.

Because the half shafts on a FWD vehicle turn at roughly one-third the speed of the drive shaft on a RWD vehicle, half shaft balance and runout are not very important. A bent drive shaft must be rebuilt by a specialty shop or replaced if bent or badly dented by the auto accident.

Rear-Wheel Drive

There are two basic types of CV-joint applications found on RWD unibody vehicles: independent rear suspension and solid axle housing. In RWD with independent rear suspension (IRS), CV joints can be found at both ends of the axle shafts (for a total of four).

The rear axle assembly contains a differential and two axles. The differential is a set of gears and shafts that transmits power from the drive shaft to the axles. The axles are steel shafts that connect the differential and drive wheels.

With a solid axle housing, no U- or CV joints are required. Solid rear axles are being phased out for the smoother riding independent rear suspension designs.

Four-Wheel Drive

On the typical four-wheel drive (4WD) vehicle, a *transfer case* is used to send power to the front and rear axles. It is mounted to the side, underneath, or the back of the transmission. A chain or gear drive within the case receives the power flow from the transmission and transfers it to two separate drive shafts leading to the front and rear axles.

Powertrain Inspection

Begin damage inspection by checking the condition of the CV-joint boots. Splits, cracks, tears, punctures, or thin spots caused by rubbing call for immediate boot replacement. If the boot appears rotted, this indicates improper greasing or excessive heat, and the boot should be replaced. Squeeze-test all boots. If any air escapes, replace the boot. Also, replace any boots that are missing.

Chassis service and wheel alignment: When a CV-joint boot is torn or missing, there is often damage or wear in the joint. Check the joint for problems any time a boot requires replacement.

The drive shafts should be checked for signs of contact against the chassis, or rubbing. Rubbing can be a symptom of a weak or broken spring or engine mount, or chassis misalignment.

On FWD transaxles with equal-length half shafts, inspect the intermediate shaft U-joint, bearing, and support bracket for looseness by rocking the wheel back and forth and watching for any movement.

Various drivetrain and suspension problems can be confused with symptoms produced by a bad CV joint. The following list of symptoms should help guide the technician to a proper diagnosis:

- A popping or clicking noise when turning signals a worn or damaged outer joint. Putting the car in reverse and backing in a circle aggravates the condition. If the noise gets louder, the outer joint(s) should be replaced.
- A "clunk" when accelerating, decelerating, or when putting the transaxle into drive can come from excessive play in the inner joint of FWD applications, either inner or outer joints in an RWD independent suspension, or from the drive shaft CV joints or U-joint in an RWD or 4WD powertrain. However, be aware that the same kind of noise can be produced by excessive backlash in the differential gears and transmission.
- A humming or growling noise is sometimes due to inadequate lubrication in either the inner or outer CV joint. It is more often due to worn or damaged wheel bearings, a bad intermediate shaft bearing on equal-length half shaft transaxles, or worn shaft bearings within the transmission.
- A shudder or vibration when accelerating indicates excessive play in either the inboard or outboard joints, but more likely the inboard plunge joint. These kinds of vibrations can also be caused by a bad intermediate shaft bearing on transaxles with equal-length half shafts. On FWD vehicles with transverse-mounted engines, this kind of vibration can also be caused by loose or deteriorated engine/transaxle mounts. Be sure to inspect the rubber bushings in the upper torque strap on these engines to rule out this possibility. Note, however, that shudder can also be inherent to the vehicle itself.
- A vibration that increases with speed is rarely due to CV-joint problems or FWD half shaft imbalance. An out-of-balance tire or wheel, an out-of-round tire or wheel, or a bent rim is the more likely cause. It is possible that a bent half shaft resulting from collision or towing damage could cause a vibration, as could a missing damper weight.

Powertrain Repairs

During a severe collision, powertrain parts are frequently damaged or pushed out of alignment. As a body technician, you must be able to find and correct many of these problems. Body technicians often have to replace the following damaged mechanical parts: crushed radiators, cut hoses, cracked water pumps, bent pulleys, broken drive belts, engine intake air tubes, engine covers, fan

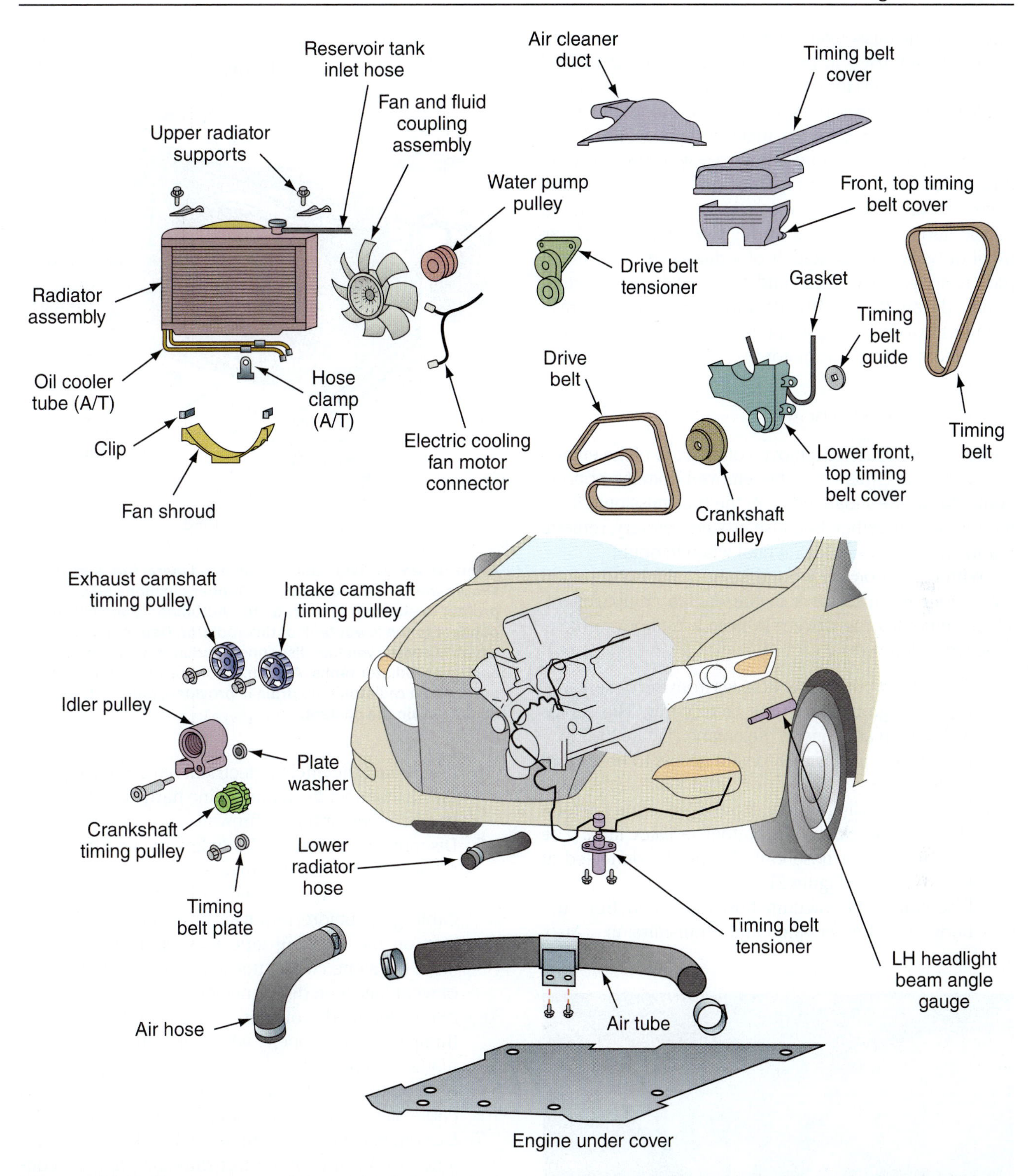

Figure 21-6 These are some of the mechanical parts that are crushed and must be replaced after a major frontal collision.

blades, and related parts. Some of these parts are shown in Figure 21–6.

Many mechanical parts such as engine mounts and transmission supports are through-bolted. The position of these mountings must be maintained parallel to each other to allow for the correct movement of the mechanical parts. When these mechanical mountings are not in proper alignment, free movement of parts can be restricted. For instance, misalignment in transmission linkages can easily cause erratic transmission performance. Proper drive shaft angles must be maintained to prevent vibration and chatter of the drive shaft and U-joints.

Motor mounts prevent minor engine vibrations and noise from being transferred into the body. Misalignment of motor mounts can cause vibrations to be transferred directly to the passenger compartment. Special fasteners are frequently used to provide the necessary structural support mountings for mechanical parts.

At times it is desirable to completely remove the drivetrain from the unibody to make repairs to the body. Removal of the drivetrain allows ready access to structural unibody panels for repair or replacement. In some cases, the time taken to remove the drivetrain pays off in considerable time savings during the repair or replacement of body panels. Repair of a damaged mechanical part is sometimes easier and faster after the piece is removed from the car. The decision to remove the drivetrain or to work around it in the car can be made by the repair technician or estimator.

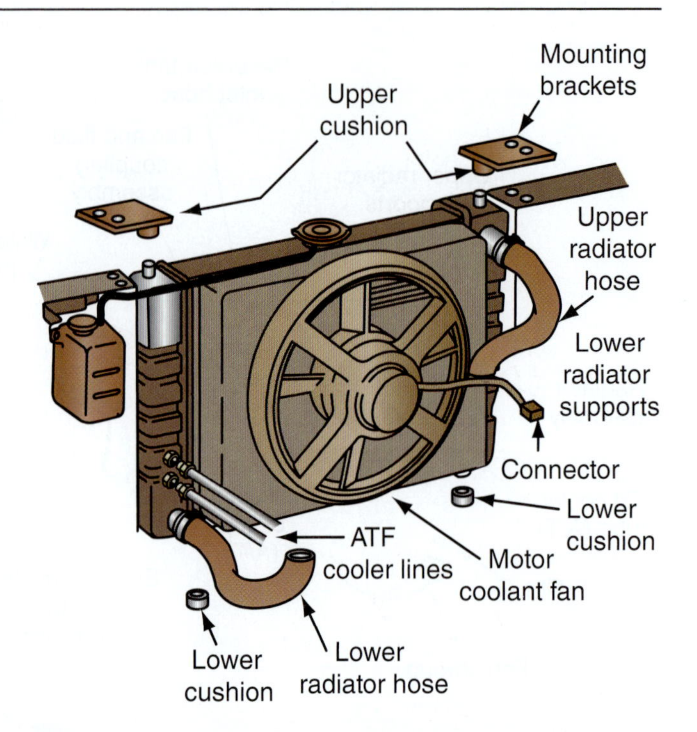

Figure 21-8 **Two rubber-cushioned metal brackets often secure the radiator to the body. Rubber cushions also protect the bottom of the radiator. Automatic transaxle lines connect to the lower tank of this radiator. Depending on which is easier, you can disconnect radiator hoses at the engine or radiator tanks. A drain is normally provided at the bottom of a radiator. If no drain is provided, pull the lower hose to drain the coolant.**

Powertrain Removal

Some engines must be removed out the top of the engine compartment. Others must be removed from the bottom. Some should be separated from the transmission; others are removed together. Because procedures vary, remember to refer to the service manual when in doubt.

With the car on the ground, remove the hood so you have more room to work in the engine compartment. When removing the drivetrain from a unibody vehicle, proceed as follows:

1. Disconnect both battery cables from the battery and the body ground from the battery tray. The cables will remain attached to the engine (Figure 21–7).
2. Remove the air cleaner to gain access to hoses and cables.
3. Drain the cooling system. If provided, open the radiator drain cock on the bottom radiator tank. The bottom radiator hose can also be disconnected at either end. See Figure 21–8.
4. Disconnect the vacuum hoses attached from the body to the engine and transmission. Also, disconnect the engine electrical harnesses. If possible, disconnect the main engine harness at the bulkhead connector on the firewall.
5. Disconnect the throttle body linkage and the transmission or transaxle linkage (Figure 21–9). Use masking tape to label where each hose, wire, or cable goes before removal. Print the name of the part on the masking tape to simplify reassembly. Mark the same code letter or number on both sides of what has been disconnected. Once all parts have been identified, you can feel assured that everything will work properly when all reconnections are made (Figure 21–10).
6. With a manual transaxle, disconnect the clutch cable, slave cylinder, or linkage. Refer to Figure 21–11.
7. Disconnect the speedometer cable. It is often fastened to the side of the transmission case. Disconnect the transaxle or transmission cooling lines at the radiator.
8. Disconnect the heater hoses. If connected to the heater core, be careful not to damage the heater core by twisting and pulling too hard.
9. If needed, disconnect the power steering pump lines where accessible. It may be easier to remove the pump from the engine.
10. Remove the air-conditioning (A/C) compressor. If the A/C system is not damaged and a recharge is not required, remove the A/C compressor from its

Figure 21-7 **Always disconnect the battery cables before starting any work on a vehicle. Impact damage can sever or short wires and possibly start an electrical fire.**

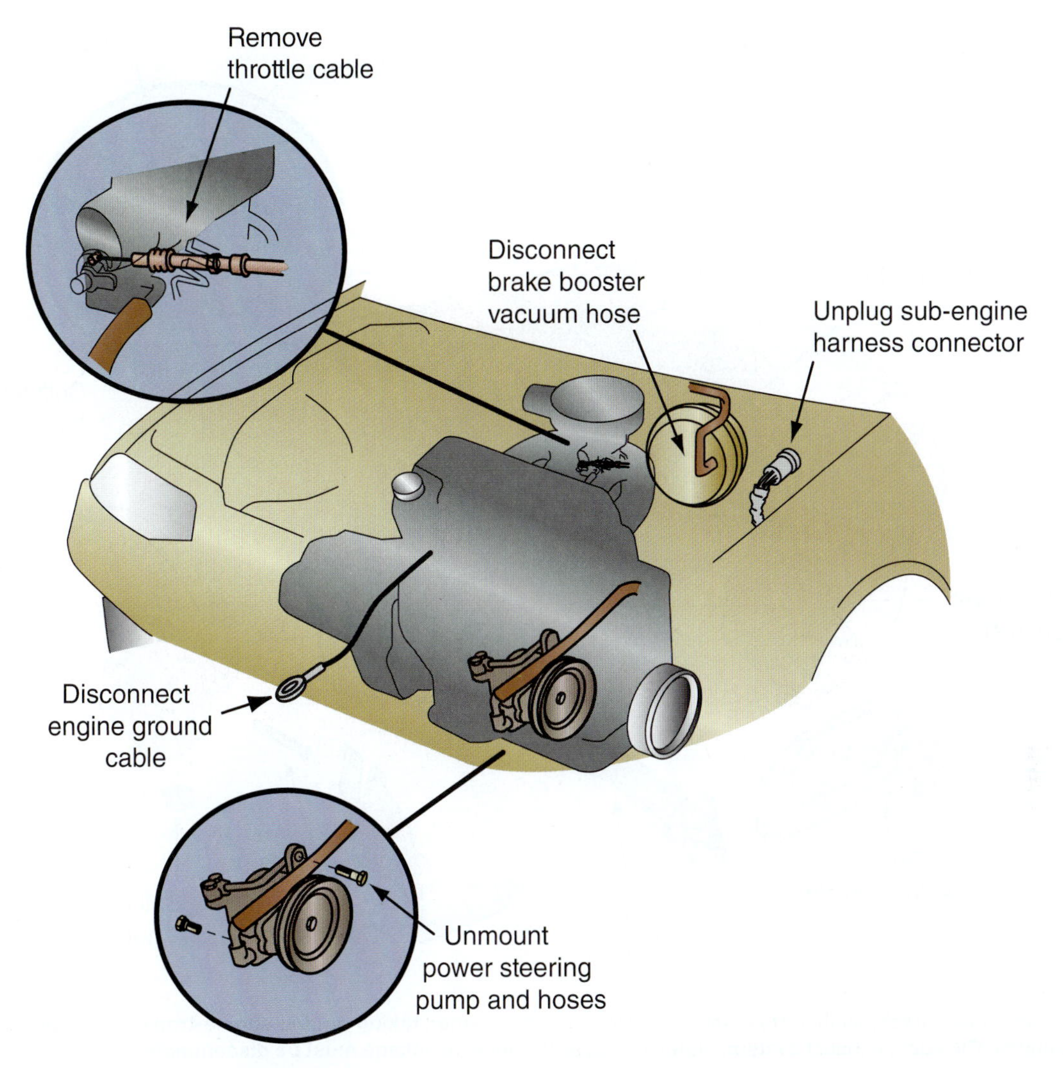

Figure 21-9 Service literature will detail how the throttle cable, wiring harness, lines, and hoses must be disconnected for powertrain removal.

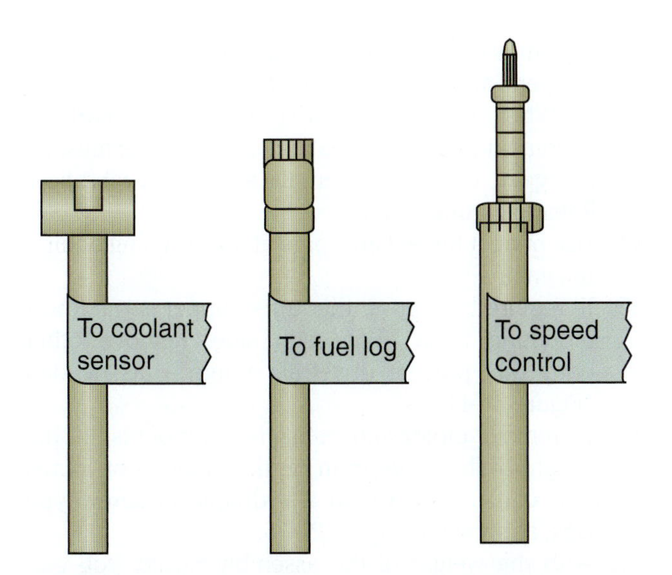

Figure 21-10 Label hoses, wires, and cables as they are disconnected to simplify and speed reassembly. This can save you time. If a wire goes to a sensor on the radiator or a specific location on the engine, write down where it goes on masking tape and wrap the tape around the wire, hose, or cable so you do not get confused during reassembly.

mounting bracket and leave it with the body. If there are open lines, they should be plugged. Remove the radiator fan and shroud for additional clearance if necessary. Double-check for any individual wires attached to the engine and traveling to the vehicle body.

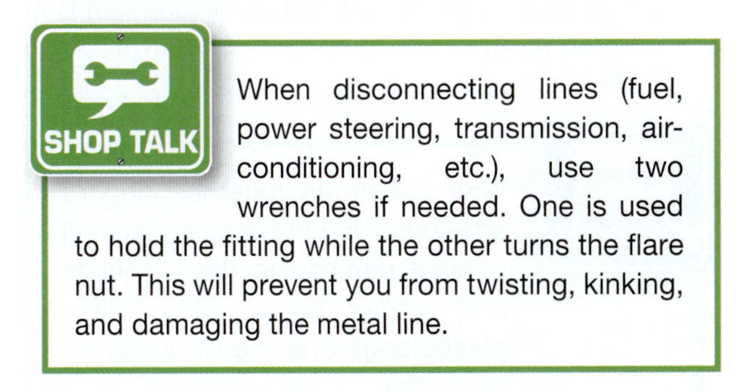

When disconnecting lines (fuel, power steering, transmission, air-conditioning, etc.), use two wrenches if needed. One is used to hold the fitting while the other turns the flare nut. This will prevent you from twisting, kinking, and damaging the metal line.

11. Disconnect the fuel lines between the engine and body. Plug the fuel lines to prevent leakage.
12. To get the vehicle up in the air, use frame-straightening equipment, if possible. Most racks and bench-racks

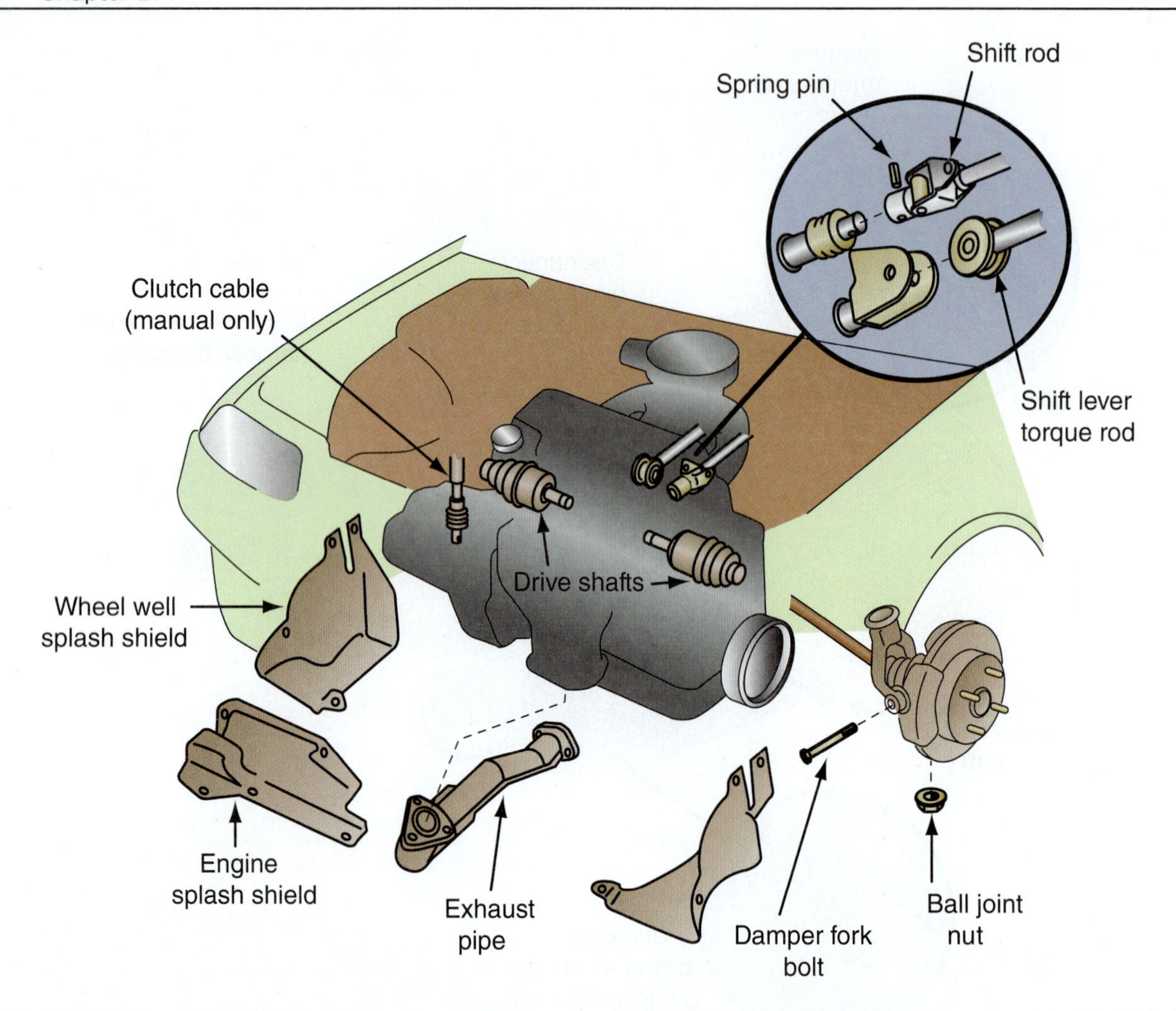

Figure 21–11 These mechanical parts must normally be removed when taking out the engine-transaxle assembly. Note how the splash shields, CV axles, exhaust system, clutch cable, and transaxle linkage must be disconnected.

allow you to quickly raise a vehicle using built-in jacks or a scissor lift inside the rack. The vehicle can be placed on stands, or a mechanical safety catch must be engaged on a scissors lift before working. Dolly wheels can also be used to move the vehicle to other areas in the shop for part removal.

If using older equipment, you might have to use a floor jack under the car and place jack stands in position on the cross tubes before removing some parts.

Wire the struts together inboard after removing the engine. This will prevent damage to the CV joints and rubber boots. Also, place plastic bags over the CV joints for protection.

13. Once the vehicle is in the air and on jack stands, remove any other parts that prevent powertrain removal. This might include the CV axles, exhaust system, steering linkage, suspension system, brake calipers, and so on (Figure 21–12).

    When removing a drive shaft joint, mark its alignment to maintain balance. Your marks must be realigned during reassembly to prevent vibration. Refer to Figure 21–13.
14. Disconnect the exhaust pipe at the coupling behind the engine.
15. If required, remove the three upper strut tower mounting bolts from each side. Remove other parts that prevent drivetrain removal, as needed (Figure 21–14).
16. To remove motor and transmission mounts, lift the weight of the powertrain off the unibody or frame. This can be done with a hydraulic or screw-type jack, as shown in Figure 21–15.
17. With the weight of the assembly raised, you can remove the cross members or mounts that secure the drivetrain. See Figure 21–16.
18. Double-check that everything is disconnected before trying to remove the engine/drivetrain assembly (Figure 21–17).

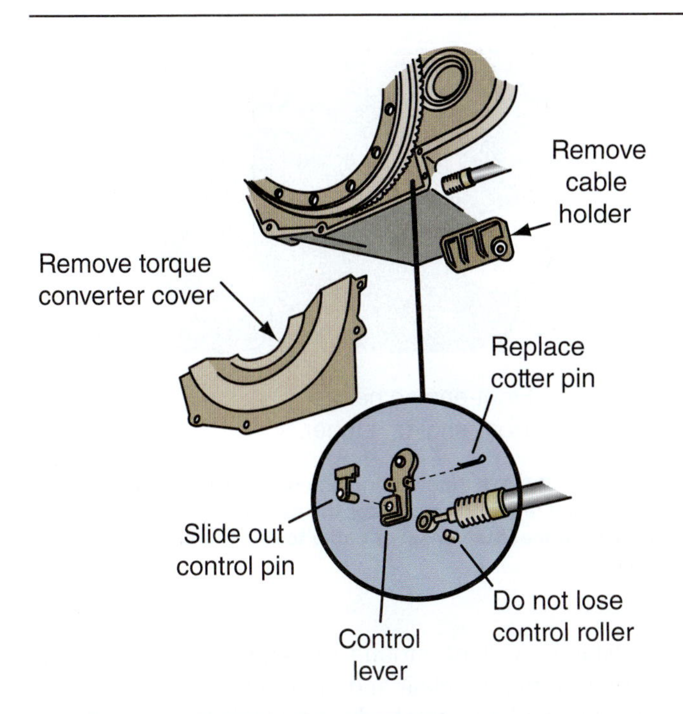

Figure 21-12 This vehicle must have the torque converter cover removed to disconnect the transmission shift control cable.

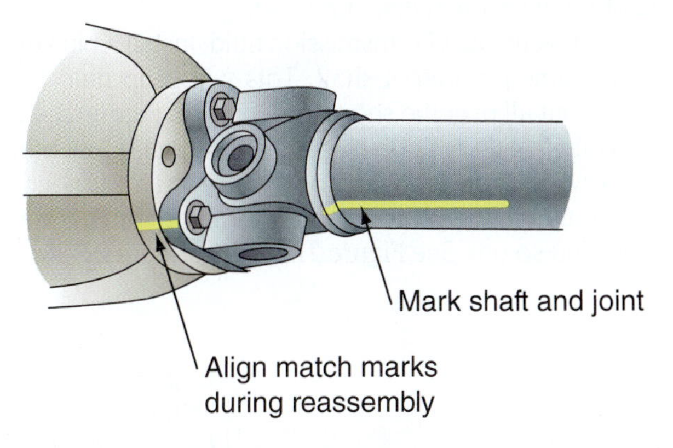

Figure 21-13 When disconnecting a drive shaft, scribe the alignment of parts. The drive shaft is often balanced and must be realigned to prevent possible vibration.

Figure 21-14 Refer to the factory service information when working on mechanical assemblies. It will give specific instructions, special tools needed, torque specs, and other critical information.

Figure 21-15 Here the technician is heating a tie-rod end to aid in part removal. Threads can become corroded. Heat will expand the part, making removal easier.

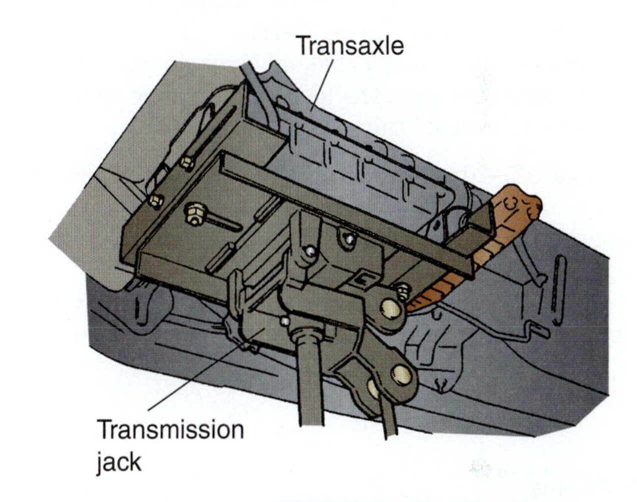

A A transmission jack is used to raise the transaxle to remove the transmission mount.

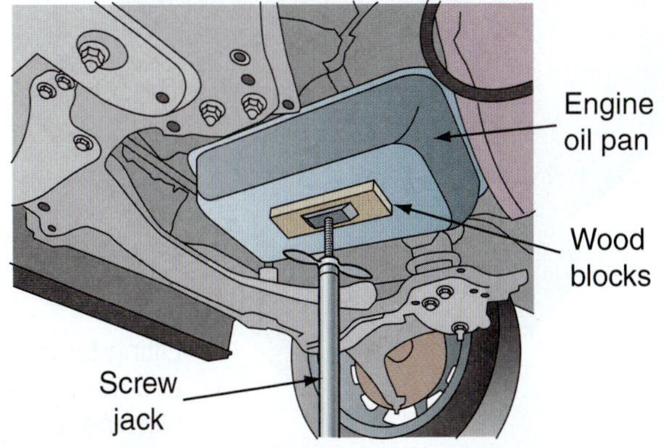

B Here a screw jack with a block of wood to protect the oil pan is used to raise the engine for removal of the motor mounts.

Figure 21-16 To service motor mounts, you must take the weight of the engine or transaxle off the mounts.

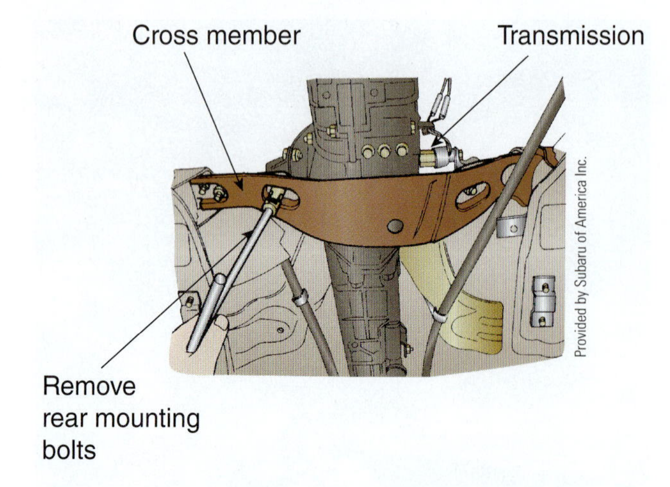

Figure 21-17 **With the engine and transmission supported, you can remove the rear cross member.**

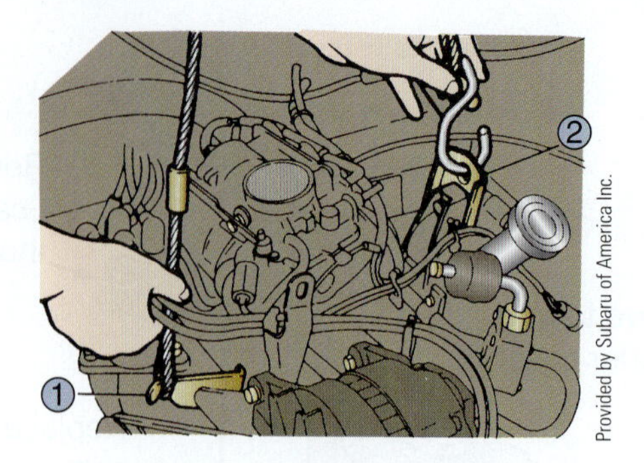

Figure 21-19 **When lifting an engine, keep your hands and feet out from under the heavy assembly at all times. Connect the chain or cable to recommended lift points.**

19. An engine support fixture is sometimes used during powertrain service. Sometimes the fixture is designed to mount under the engine, or it might be used on the top (Figure 21–18).
20. Attach the lifting cable or chain to the engine. Hangers are often provided on the engine for the lift chain. Look at Figure 21–19.

Figure 21–20 shows how one auto maker recommends connecting the lift chain and how to remove the motor mounts. Figure 21–21 shows how to lift engines with varying mounting configurations.

When starting to remove a drivetrain from any unibody, perform the actual separation very slowly, while constantly walking around the entire vehicle to make sure everything is clear and disconnected. Care must be taken when lifting the body away from the drivetrain or when lifting the powertrain out of the body. As the lifting continues, keep checking all sides for wires or hoses that might not have been disconnected.

To prevent loss of transmission fluid, install a plug over the transmission output shaft. This will keep fluid from leaking out all over the shop floor. Refer to Figure 21–22.

While the body is being repaired, an auto mechanic can service damaged parts of the drivetrain, if needed (bent engine pulleys, a broken alternator, damaged CV axles, and so on). See Figure 21–23 and Figure 21–24.

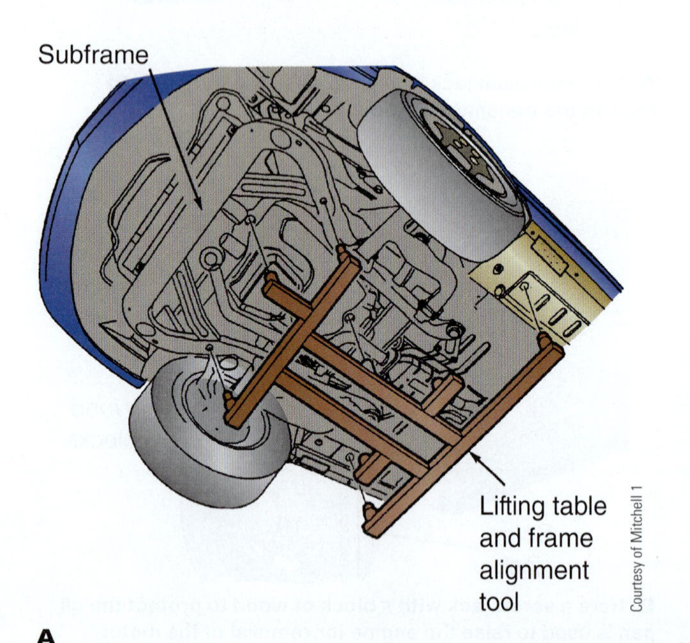

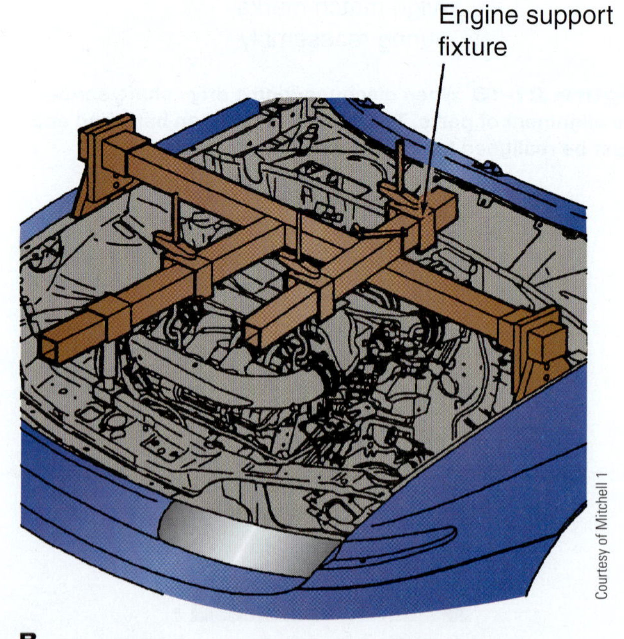

Figure 21-18 **A lifting table or an engine support fixture will support the weight of mechanical assemblies as they are serviced. (A) A lifting table is recommended during powertrain service on this make and model vehicle. (B) The engine support fixture mounts across the engine compartment to hold the engine-transaxle assembly while it is being worked on.**

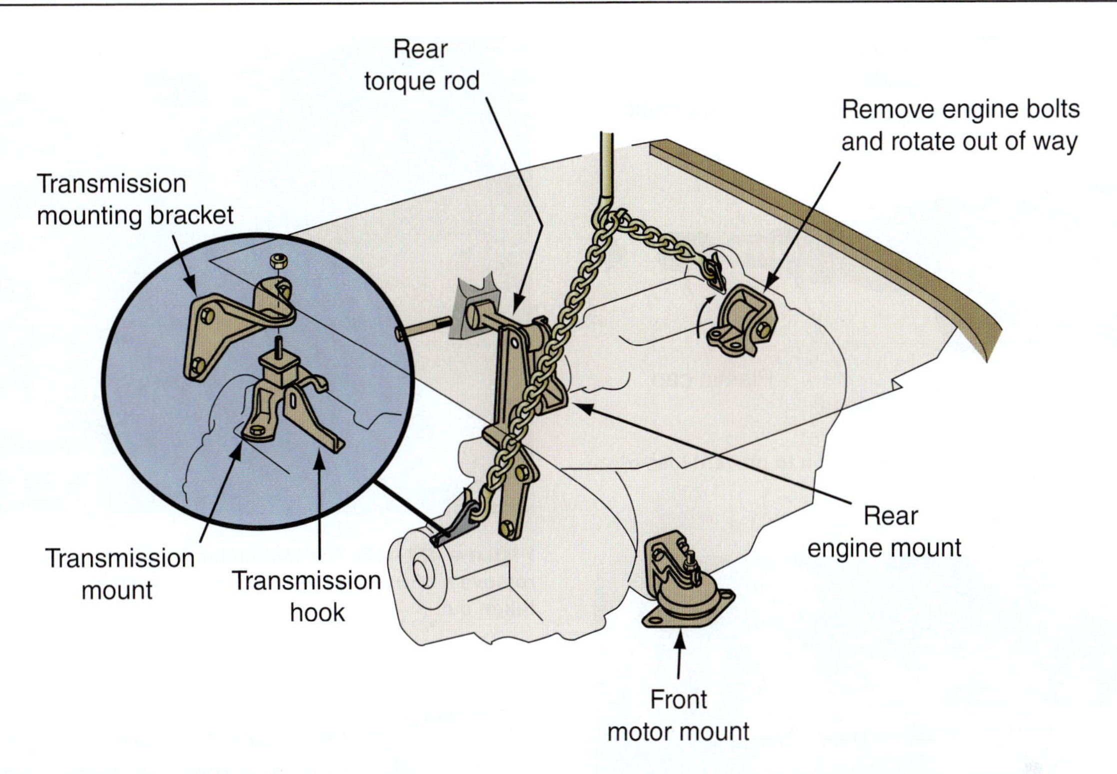

Figure 21-20 Note how the engine and transmission mounts secure this engine to the unibody structure.

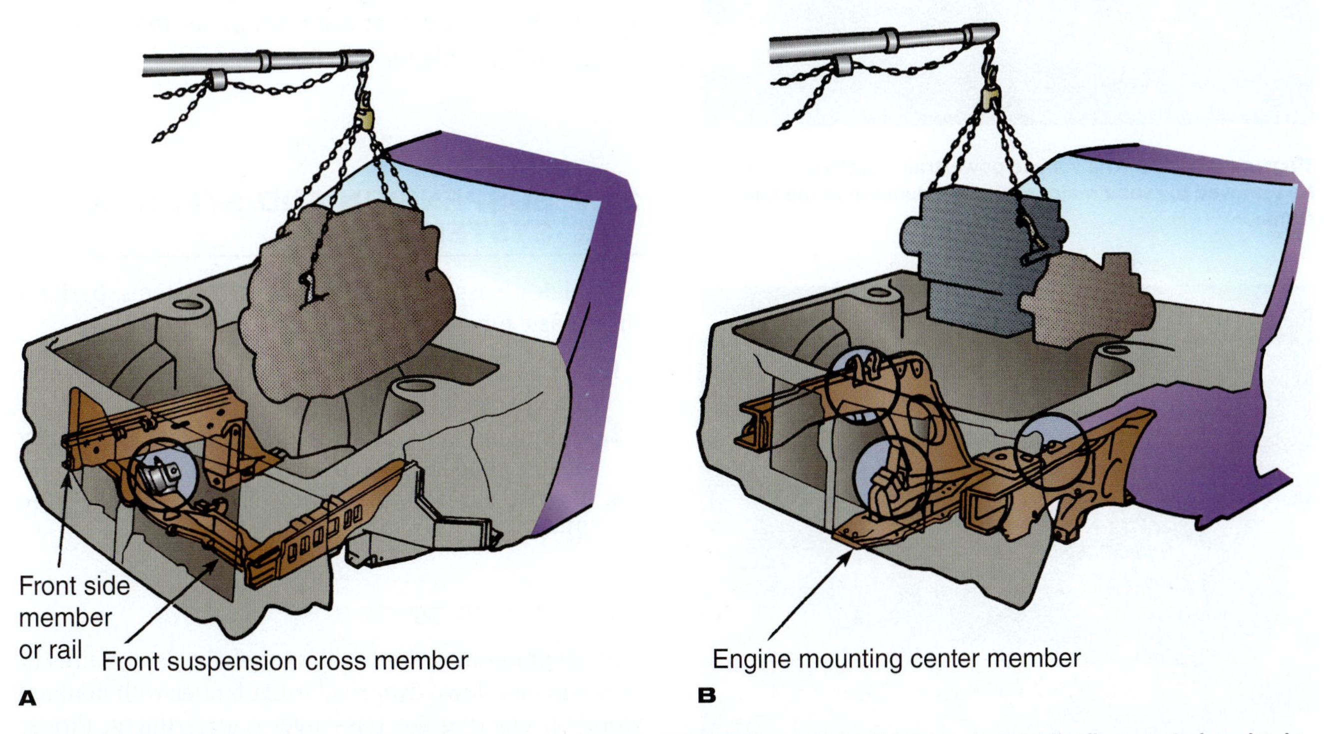

Figure 21-21 Note the mount locations for two engine mounting arrangements. (A) With a longitudinally mounted engine in a FWD vehicle, the mount locations can be seen with the engine raised out of the compartment. (B) A transversely mounted engine in a FWD vehicle has mounts in different locations than the longitudinal arrangement.

It is much easier to access and repair many structural body parts with the powertrain removed. Make sure you accurately measure body panel locations before final welding. Look at Figure 21–25.

Powertrain Installation

Reinstallation of the drivetrain can be accomplished by reversing the removal procedure. After the unibody structure has been accurately repaired, the cradle can be

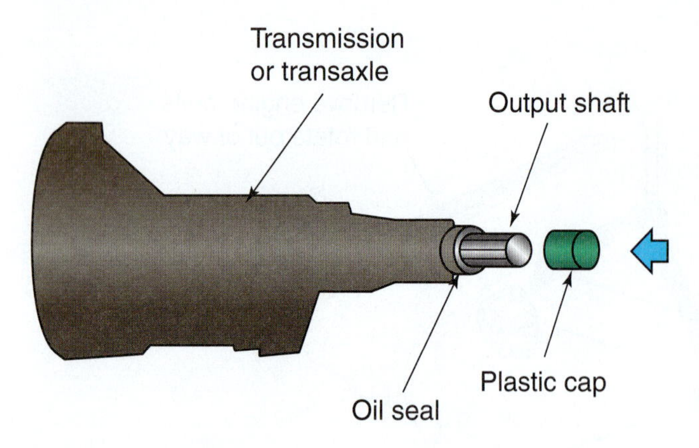

Figure 21-22 Plug the transmission to prevent loss of fluid.

Figure 21-23 The vehicle powertrain must sometimes be removed to repair major structural damage to the body or frame.

Figure 21-24 This axle assembly was removed to aid in structural repairs to the unibody.

quickly and correctly positioned by using the line-up holes located at the right front and right rear cradle mounting points. An incorrectly positioned cradle can give the customer a wheel alignment problem.

Figure 21-25 The technician is using a cutting torch to remove a structural member so the crushed radiator can be taken out.

An engine/transaxle or transmission assembly is very heavy. If dropped, it can easily sever toes and fingers or crush bones. Keep your hands and feet out from under the engine assembly while moving it.

21.2 SUSPENSION AND STEERING SYSTEMS

A vehicle's suspension and steering systems perform three basic functions:

1. Act as the overall connection between the wheels and the vehicle body
2. Damp and control the ride; that is, act to partially absorb road shock and sway
3. Provide directional control of the vehicle (Figure 21–26)

Suspension Systems

The **suspension system** allows the wheels and tires to move up and down over road irregularities with minimal vibration entering the passenger compartment. Proper collision repairs of suspension systems and supporting unibody structural panels must restore the ability of these panels to support the high dynamic loads experienced during operation of the suspension system. Most body repair technicians focus their attention during suspension system repairs on the ability to restore traditional wheel alignment angles to specification.

The major parts of a modern suspension system shown in Figure 21–27 include the following:

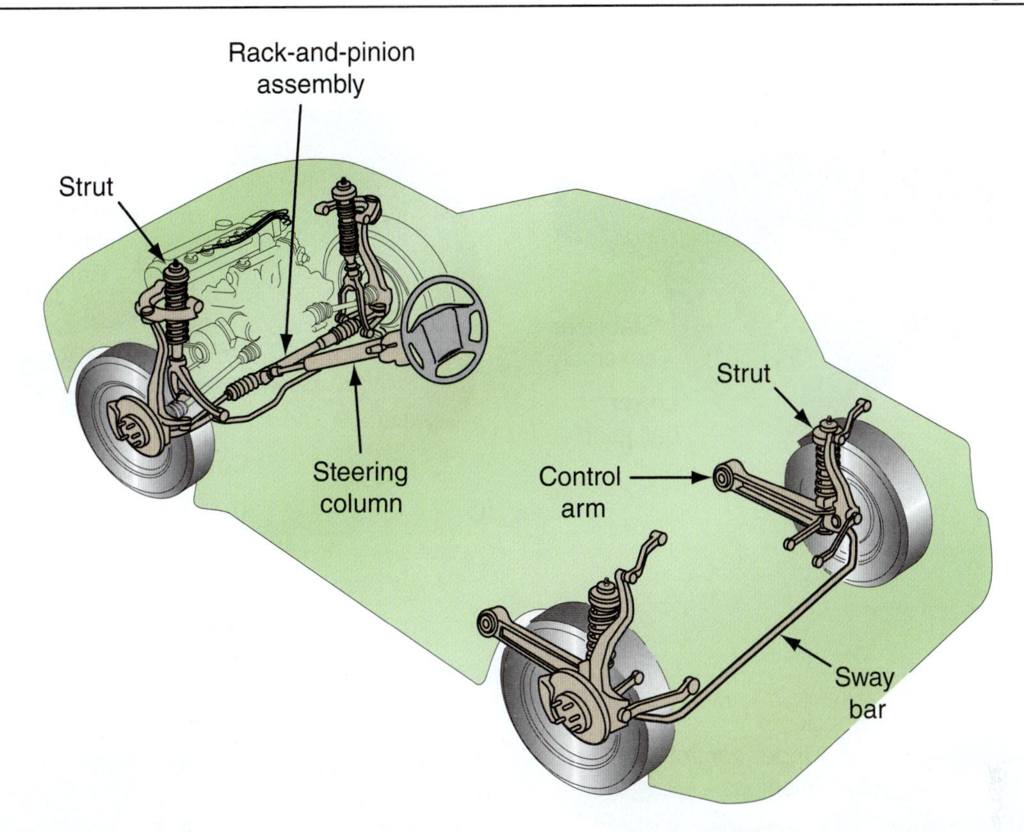

Figure 21-26 Here you can see the major parts of the steering and suspension systems, which often must be removed and replaced during repair of the surrounding body panels.

- *Control arms*, which mount on the frame and move up and down.
- *Ball joints*, which are located on the outer end of the control arms and allow the steering knuckles to swivel and turn.

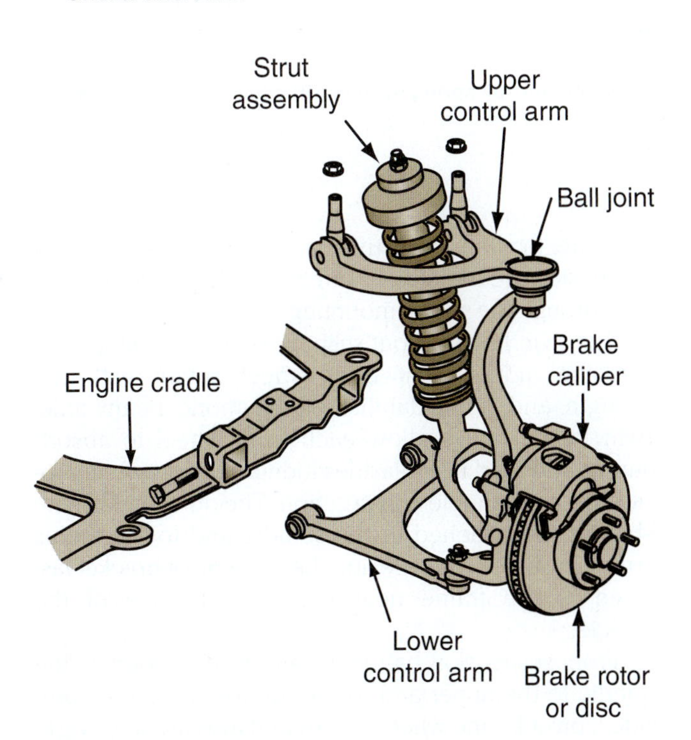

Figure 21-27 Study the major parts of a modern suspension system.

- *Steering knuckles*, which hold the wheel bearings and wheels.
- *Hubs*, which mount on the wheel bearings to hold the wheels or rims. The wheels hold the tires.
- *Suspension system springs*, which support the weight of the car and allow suspension flexing.
- **Shock absorbers**, which are dampening devices that absorb spring oscillations (bouncing) to smooth the vehicle's ride quality. They may be gas, oil, or air filled.

Front Suspension

There are several basic types of suspension systems used in passenger cars and light-duty trucks. Most frame bodies use either the coil spring, leaf spring, or torsion bar system. The strut suspension is widely used in unibody cars. Light-duty trucks sometimes use the twin I-beam system (Figure 21–28).

The coil spring suspension system often uses both an upper and a lower control arm. These arms are attached with pivots to a structural component, such as the frame or side rails. The outer ends of the control arms are attached to the spindle and steering knuckle assembly with ball joints. The spring is usually placed between the lower control arm and the frame. Some types place the spring above the upper control arm and others use torsion bars. A separate shock absorber is connected to one of the control arms and a structural member.

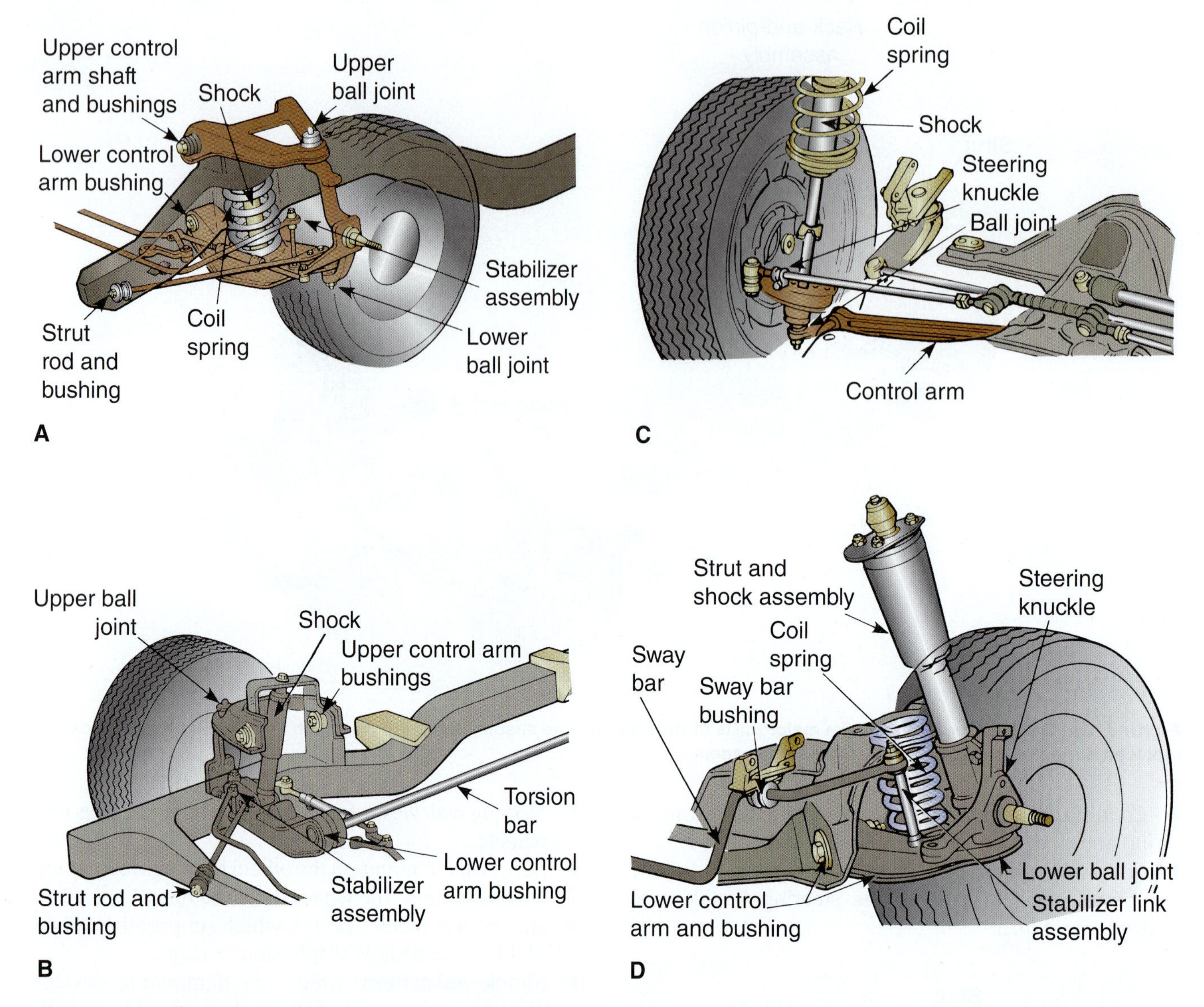

Figure 21-28 Study the various front suspension systems: (A) conventional coil spring system, (B) conventional torsion bar system, (C) conventional strut system, and (D) modified strut system.

The torsion bar suspension system uses torsion bars instead of coil springs. Vehicle weight is supported by a twisting action of the torsion bar. The front of the bar is attached to the lower control arm and the rear is attached to the frame. Torsion bars installed in this manner are commonly called "longitudinal" because they run lengthwise in the vehicle.

The most commonly used front suspension system for unibody vehicles is the MacPherson strut suspension and a modified version of it. The design and operation of a strut suspension system are simple compared with the parallel arm suspension system.

Like the parallel arm suspension, the MacPherson strut suspension has a lower control arm and spring. The strut replaces the upper control arm. The strut suspension system uses a coil spring that is part of the strut assembly. In some cases, the coil spring is placed between the lower control arm and the unibody structure. In either case, the loads generated by the strut suspension are transferred directly to the unibody structure through the spring mounting.

The twin I-beam front suspension was developed to combine independent front wheel action with the strength and dependability of the mono beam axle. Twin I-beam axles allow each front wheel to absorb bumps and road irregularities independently, while providing sturdy, simple construction. The outer ends of the I-beams are attached to the spindle and to the radius arms. The inner ends are attached to a pivot bracket fastened to the frame near the opposite side of the vehicle.

Front suspension ball joints are used to connect the spindle to the upper and lower control arms. They provide a pivot for the wheel to turn and also allow for vertical movement of the control arms as the vehicle moves over irregularities in the road (Figure 21–29).

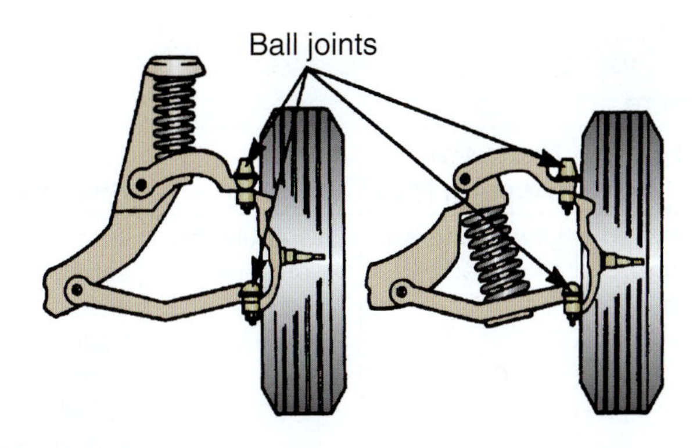

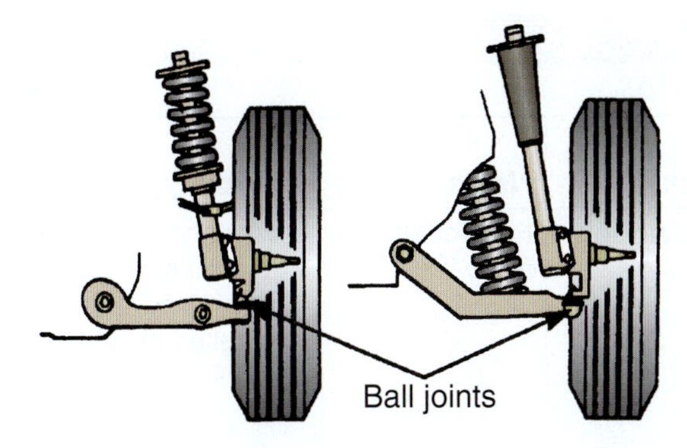

Coil spring or torsion bar mounted on upper control arm

Coil spring or torsion bar mounted on lower control arm

MacPherson strut

Coil spring mounted on lower control arm with modified strut

Figure 21–29 Study the ball joint location variations.

Rear Suspension

Generally, rear suspensions require no special service. Broken or worn parts should be replaced. Remember that rear wheels, just like front wheels, are affected by road shock, acceleration, and braking forces. Control arm or leaf spring bushings are constantly flexing. In addition, bushings keep the rear wheels in line with the front wheels and when worn or damaged can upset the settings of the entire suspension and drive shaft systems.

Loose, worn, or broken attaching parts allow the rear wheels to shift, causing premature tire wear as well as short U-joint service life. A metallic jingling sound when driving over small bumps or unusual tracking (sometimes called dog tracking) also indicate the need for inspection. Usually a visual inspection is enough to determine repair requirements.

The coil spring and leaf spring nonindependent rear suspensions are the most common today on rear-drive vehicles. The solid axle design will exhibit some of the same teeter-totter characteristics as noted with solid axle front systems. However, the effect is not nearly as dramatic because the rear wheels do not pivot (Figure 21–30).

Table 21–1 gives diagnoses of suspension problems. Various rear wheel suspension system variations are shown in Figure 21–31.

A

B

Figure 21–30 These are two of the more popular rear suspension systems: (A) leaf spring nonindependent and (B) coil spring nonindependent systems. A large hollow axle housing holds the differential, solid axles, wheel bearings, and rear brake assemblies.

Suspension System Service

The exact procedures for servicing a suspension system will vary with the make and model vehicle. Always refer to factory instructions when servicing a suspension system (Figure 21–32).

A coil spring compressor is often needed when removing and installing a coil spring. A *coil spring compressor* is a special tool for squeezing the spring coils to reduce their height. This will give you enough room to slide the spring out of the control arm. It will also keep the spring from shooting out when you disconnect the lower ball joint.

DANGER A compressed coil spring has a tremendous amount of stored energy. Use extreme caution when removing a coil spring. Never unbolt the ball joint without first compressing the coil spring. If the spring is not compressed, the lower control arm and spring could move downward, possibly causing serious injury or even death.

TABLE 21–1 SUSPENSION PROBLEM DIAGNOSIS

	Problem					
Check	**Noise**	**Instability**	**Pulls to One Side**	**Excessive Steering Play**	**Hard Steering**	**Shimmy**
Tires/wheels	Road or tire noise	Low or uneven air pressure; radials mixed with belted bias ply tires	Low or uneven air pressure; mismatched tire sizes	Low or uneven air pressure	Low or uneven air pressure	Wheel out of balance or uneven tire wear or overworn tires; radials mixed with belted bias ply tires
Shock dampers (struts/ absorbers)	Loose or worn mounts or bushings	Loose or worn mounts or bushings; worn or damaged struts or shock absorbers	Loose or worn mounts or bushings	—	Loose or worn mounts or bushings on strut assemblies	Worn or damaged struts or shock absorbers
Strut rods	Loose or worn mounts or bushings	Loose or worn mounts or bushings	Loose or worn mounts or bushings	—	—	Loose or worn mounts or bushings
Springs	Worn or damaged	Worn or damaged	Worn or damaged, especially rear	—	Worn or damaged	—
Control arms	Steering knuckle control arm stop; worn or damaged mounts or bushings	Worn or damaged mounts or bushings	Worn or damaged mounts or bushings	—	Worn or damaged mounts or bushings	Worn or damaged mounts or bushings
Steering system	Component wear or damage	Component wear or damage	Component wear or damage	Component wear or damage	Component wear or damage	Component wear or damage
Alignment	—	Front and rear, especially caster	Front, camber and caster	Front	Front, especially caster	Front, especially caster
Wheel bearings	On turns or speed changes front-wheel bearings	Loose or worn (front and rear)	Loose or worn (front and rear)	Loose or worn (front)	—	Loose or worn (front and rear)
Brake system	—	—	On braking	—	On braking	—
Other	Clunk on speed changes: transaxle; click on turns: CV joints; ball joint lubrication	—	—	—	Ball joint lubrication	Loose or worn friction ball joints

To remove a coil spring from most front suspensions, place the car on jack stands. Remove the shock absorber. Install the spring compressor, and then unbolt the lower ball joint.

Use a fork tool or ball joint separator and hammer blows to remove the lower ball joint from the steering knuckle. Special pullers and drivers are also available for ball joint separation. See Figure 21–33.

Make sure you also remove any other component (brake line, strut rod, or steering linkage) that could be damaged when the control arm is lowered. Pull the spring and compressor out as a unit (Figure 21–34).

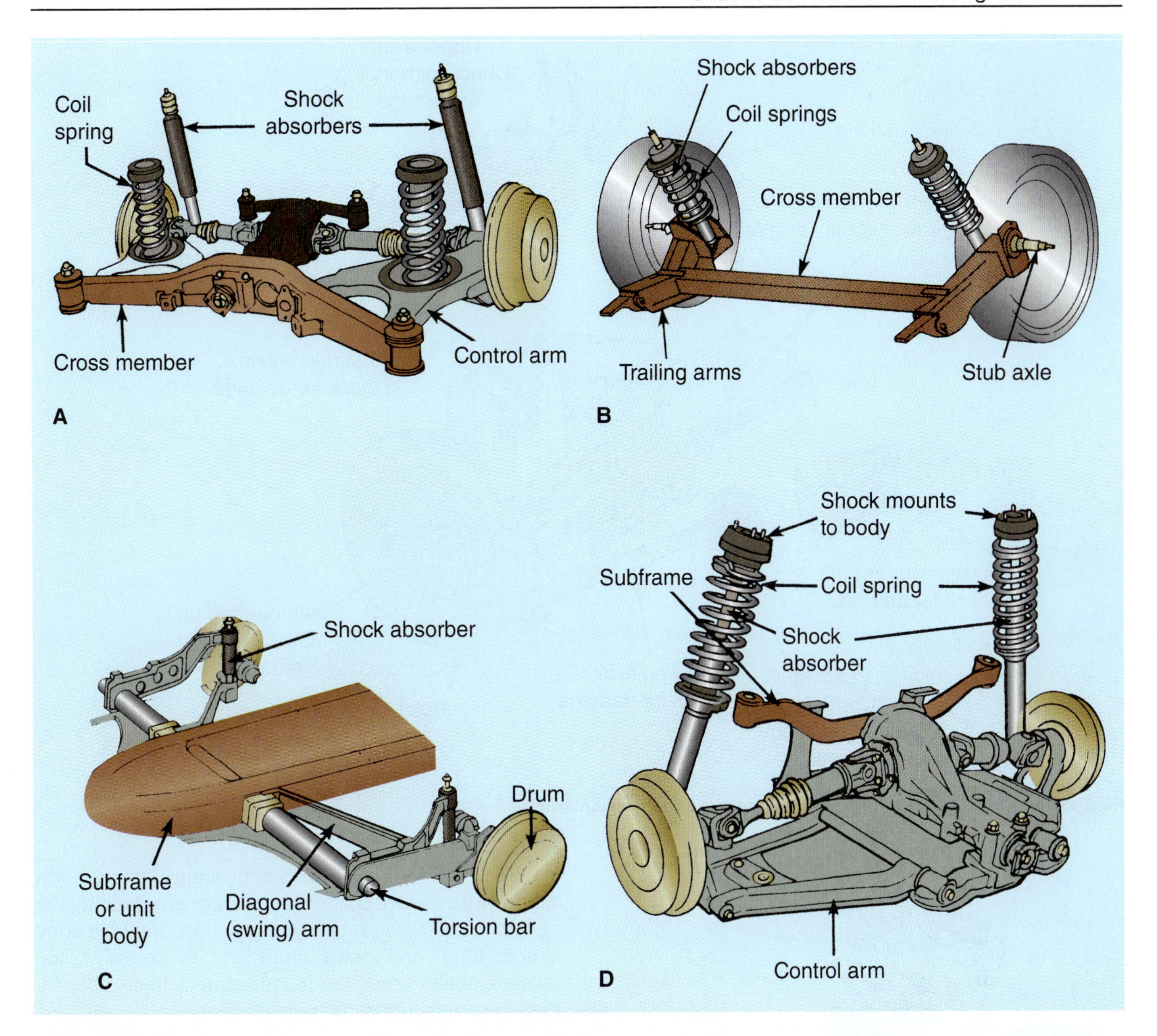

Figure 21-31 **Compare the rear suspension systems on FWD vehicles: (A) independent rear suspension, (B) independent rear axle suspension, (C) swing arm rear suspension, and (D) strut rear suspension.**

Install the compressor on the new spring. Slip the spring into place and position the coil ends in the same location as the old spring. Reassemble the ball joint and other components. Then, unscrew the spring compressor while guiding the coil into its seats. Keep your fingers out from under the spring!

An air ratchet can be used to run suspension system fasteners down. However, make sure you use a torque wrench to tighten them to recommended torque specifications.

Figure 21–35 shows how a floor jack can be used to lower and remove a rear axle assembly. Note how the vehicle is supported on jack stands for safety. The floor jack will support the weight of the heavy differential and axle housing.

Steering Systems

The steering system transfers steering wheel motion through gears and linkage rods to swivel the front wheels. When you turn the steering wheel, a steering shaft extends down through the steering column and rotates the steering gearbox.

The **steering gearbox**, either a worm or rack-and-pinion type, changes the wheel rotation into side movement for turning the wheels. A series of linkage rods connect the steering gearbox with the steering knuckles.

The **parallelogram steering system** is still used on many large, full-size pickup trucks (Figure 21–36).

A *pitman arm* attaches the steering box to the linkage rods. Steering action is relayed via the center link, again

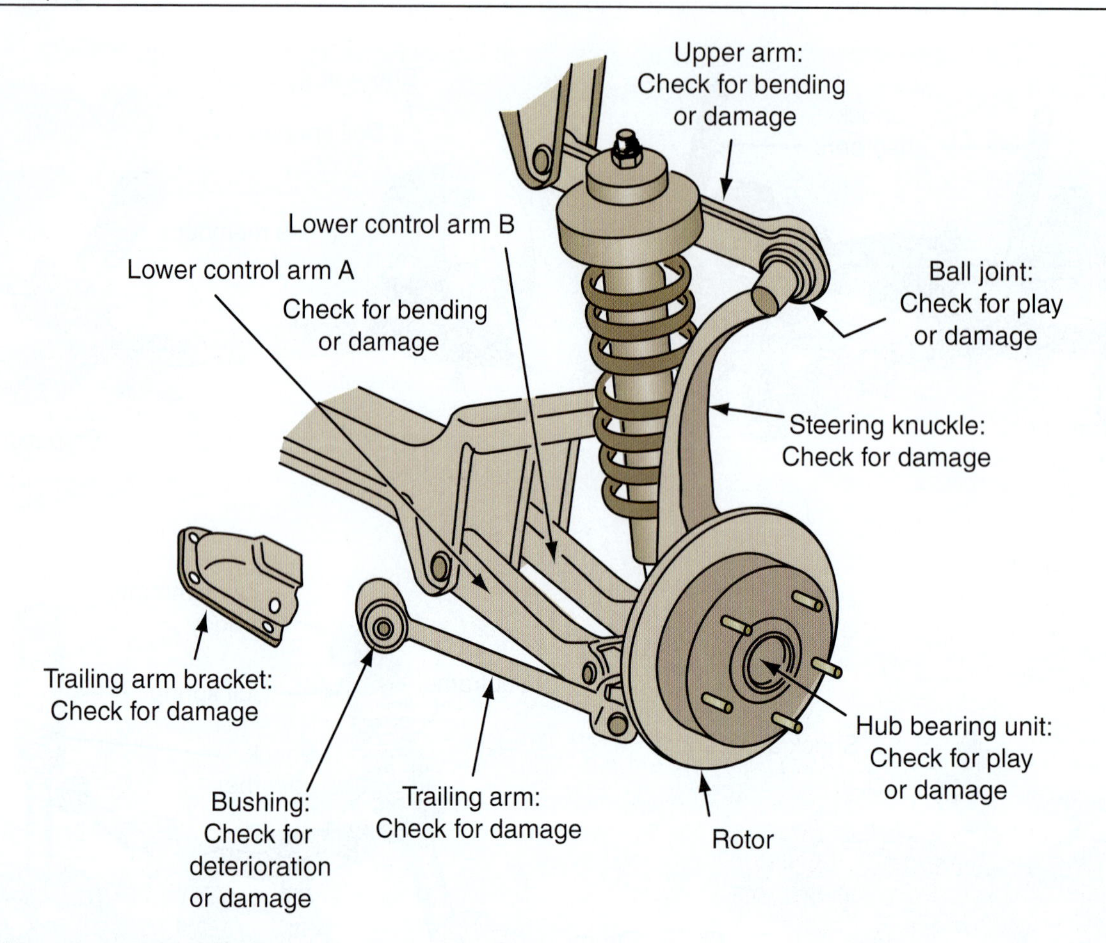

Figure 21-32 Study how this rear suspension system is constructed.

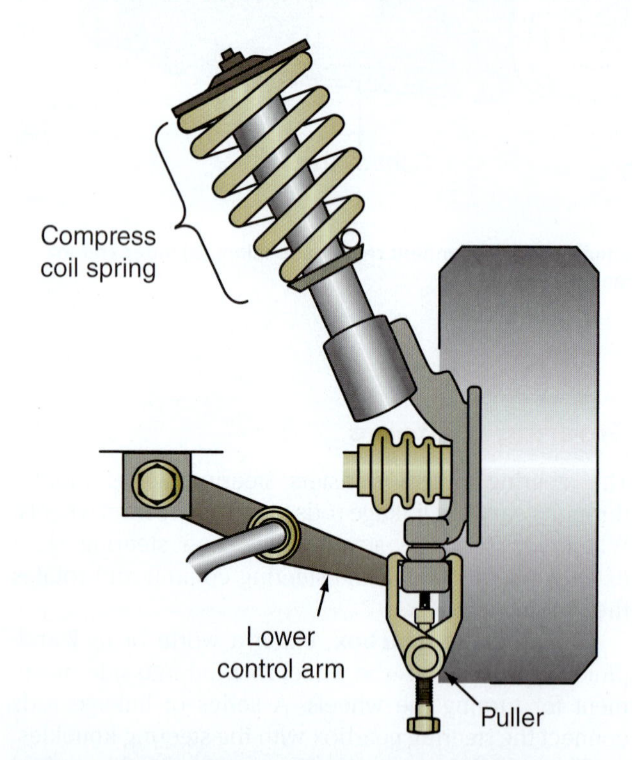

Figure 21-33 After compressing the coil spring or placing the jack under the lower control arm, a puller can be used to separate the ball joint from the control arm. Remember the danger if you forget to retain spring tension: The control arm could fly downward with lethal force.

attached by either ball sockets or bushings. The *idler arm* supports the center link at the opposite end, holding the system parallel and transmitting horizontal steering action. If up and down movement is excessive, toe change might exceed the manufacturer's limits, thereby creating premature and rapid tire wear.

Figure 21-34 Note the spring compressor mounted on a large coil spring. It is screwed down to compress the spring to allow its safe removal and installation. Be careful, because a compressed spring has tremendous stored energy and can fly out with deadly force.

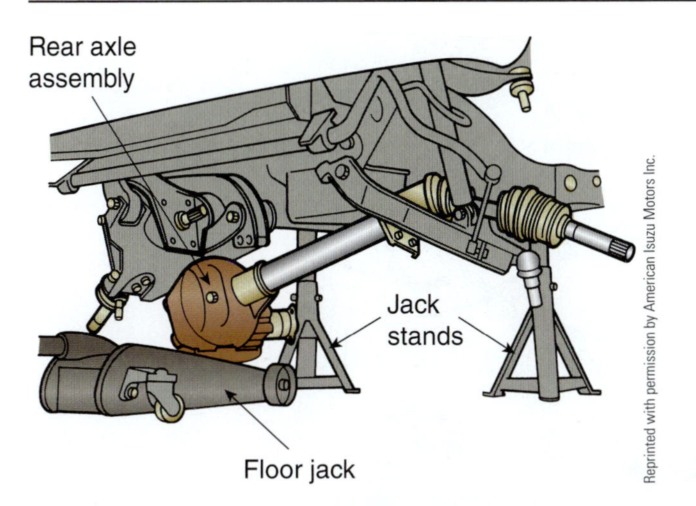

Figure 21-35 A floor jack is commonly used to lower and remove the rear axle housing from a vehicle. It can be very heavy.

Tie-rod ends attach the linkage to the steering knuckles. They are the final wearable pivots of the system. Looseness in the ball sockets causes steering play. If you can wiggle the tires sideways during the inspection (car off the ground, hand pressure only), the tie-rod ends are worn or damaged and should be replaced.

The rack-and-pinion steering system is the most common type of system found on unibody vehicles (Figure 21–37).

Rack-and-pinion steering uses a small pinion gear attached to the steering column and the rack gear in the steering gear housing. This rack gear is moved right to left within the housing by the rotation of the pinion gear. The ends of the steering rack are attached to the front wheel spindles by tie-rods. In unibody construction, the rack-and-pinion steering gear assembly on some cars is mounted to the cowl panel. In other cases, the rack-and-pinion steering gear is mounted to the front suspension cross member or the engine cradle assembly. The rack-and-pinion steering gear must be mounted securely because any movement will cause the car to wander as it travels down the road.

The automobile industry is now offering four-wheel steering systems, which allow the rear wheels to also help turn the car. This is done by either electrical or mechanical means.

Steering Inspection

To check a rack-and-pinion system, begin by raising the car and taking the weight off the front suspension. Visually inspect the steering system for any physical damage. Check the boots for leaks, inspect the tie-rods, and examine the mounting points for any distortion. Inspect the tie-rod ends. Grab the tie-rod near the tire and try pushing it up and down. Any vertical looseness indicates damage or wear.

Check the inner tie-rod socket by squeezing the bellows until you can feel the socket. With your other hand, push and pull on the tire. Looseness in the socket indicates damage or wear. Take a front tire in each hand to see whether they can be moved back and forth in opposite directions. If excessive movement is noted, wear or damage is likely. Observe the rack and pinion at the same time. Any movement might indicate a problem.

If you suspect damage to the steering system, check steering wheel rotational play and measure steering effort or force.

A *steering play check* involves measuring how far the steering wheel can be rotated without causing front wheel or tire movement.

Start the engine and rotate the steering wheel back and forth without causing the front wheel to turn. Compare how far the steering wheel can be turned without causing steering action at the tires. Typically, steering wheel play should not exceed about ½ inch (12 mm). However, always refer to the manufacturer's specifications.

A *steering effort check* involves using a spring scale to measure the force needed to turn the steering wheel. This is shown in Figure 21–38.

If the amount of effort to turn the steering wheel is higher than the manufacturer's specifications, something

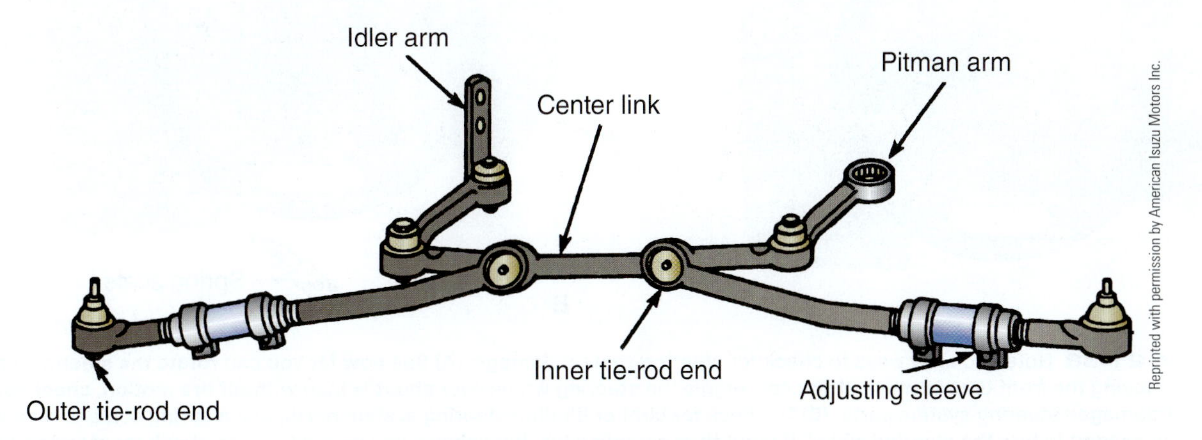

Figure 21-36 Memorize the parts of a parallelogram steering system. It is still found on many large pickup trucks.

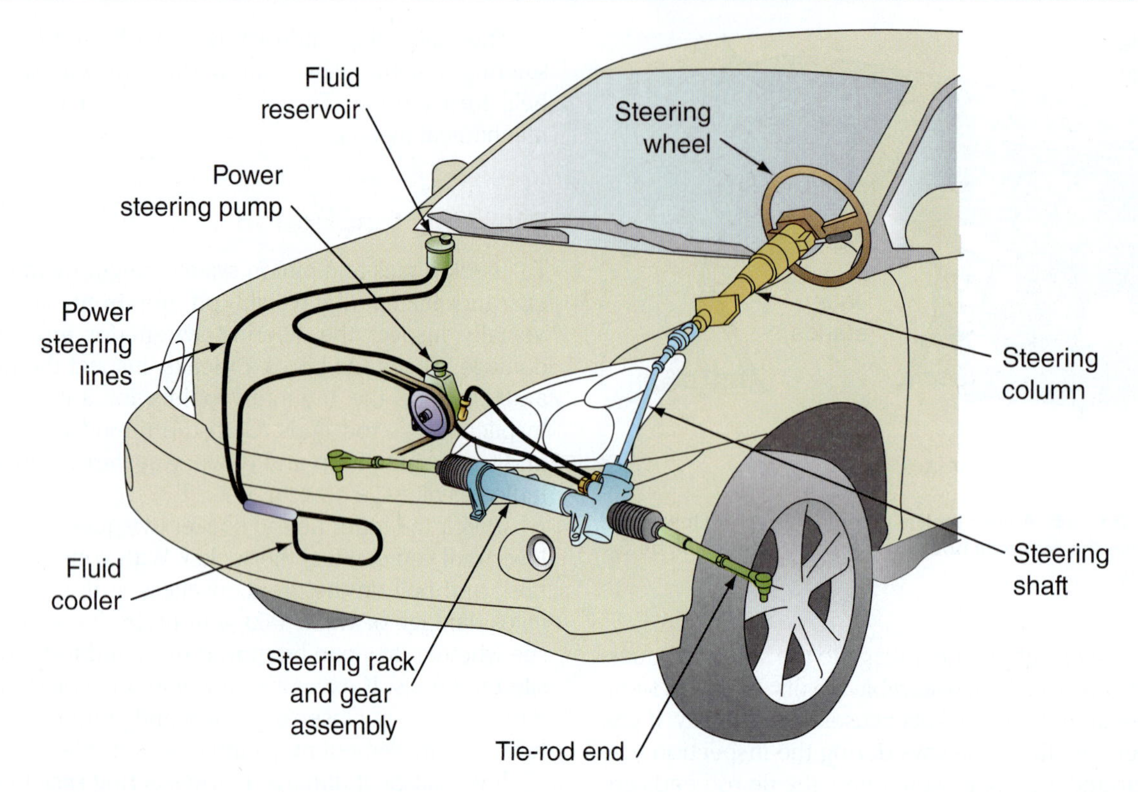

Figure 21-37 Note the parts of a rack-and-pinion steering system. The steering gear or rack, steering column, hoses, lines, and pump can be damaged in a high-speed collision or when a tire hits a curb.

was probably damaged in the collision. Quite often, the steering rack assembly is bent, which increases steering effort. The steering rack would normally have to be replaced with a new or rebuilt unit.

Misalignment of the rack and pinion will cause changes in the steering geometry during jounce or rebound, which are discussed later in this chapter. This condition cannot be corrected by changing the length of the tie-rods.

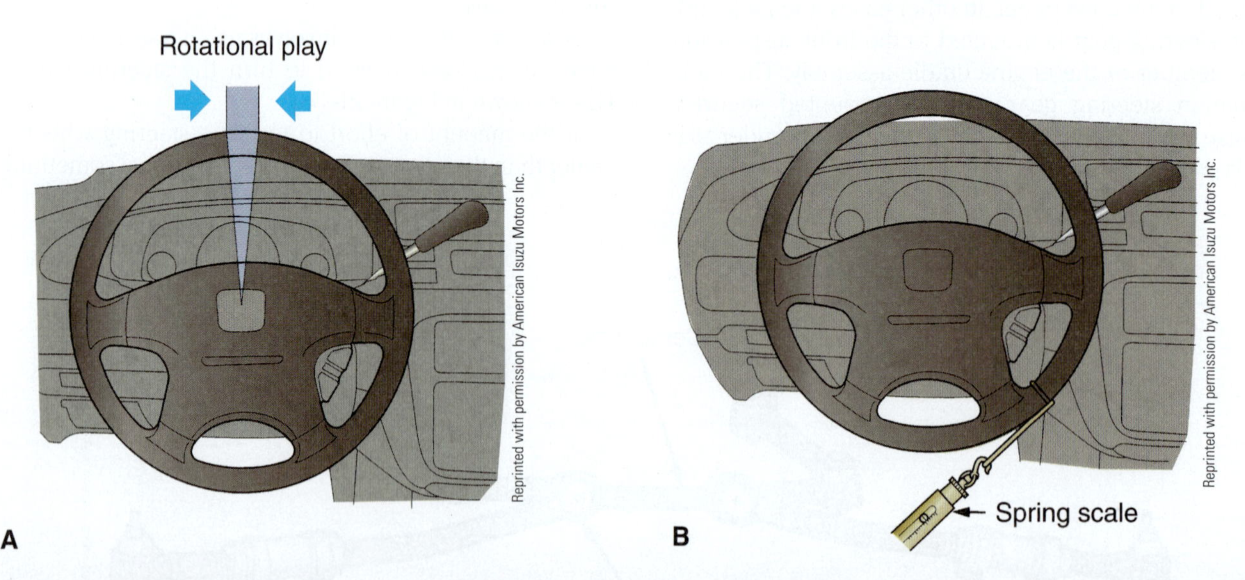

Figure 21-38 Note two easy ways to check for steering system damage: (A) See how far you can rotate the steering wheel before causing the front tires to move. If you can wiggle the steering wheel over about ½ inch without tire motion, check for worn or damaged steering system parts. (B) To check for bent or binding steering system parts, use a spring scale to measure the force needed to turn the steering wheel. If more than specified for that vehicle, you would know to check for steering system problems caused by collision damage.

Power Steering

Power steering uses hydraulic or electric energy to reduce the amount of effort required to turn the steering wheel. It also reduces driver fatigue on long drives and makes it easier to steer the vehicle at slow road speeds, particularly during parking.

Power steering can be broken down into two design arrangements: hydraulic and electrically controlled.

In the conventional arrangement, hydraulic power (fluid under pressure) is used to assist the driver. Power steering hoses carry the oil to and from the pump. A hydraulic piston on the steering linkage or in the gearbox helps turn the wheels. Hydraulic valves control power assist.

With the electric-type assist, a motor and electronic controls provide power assistance in steering.

Power Steering Service

Here are some power steering service tips that you should keep in mind:

- *Protect the system.* Protect the system from invasion by dirt and moisture. If the system must be open, be sure to plug or tie off all openings with a plastic sheet or rubber plugs.
- *Use recommended fluid.* Always replace the fluid lost with the manufacturer's recommended type to protect the warranty. Most vehicles require either Dexron or Type F fluid. Some fluids claim to meet the specifications for both of these types.
- *Bleed the system.* Many systems are self-bleeding. Some have specific bleeding procedures, outlined in the service manual, to eliminate air. Air in the system can cause noise, vibration, and erratic performance.
- *Check the hose routing.* Check the hose routing when reassembling power steering systems. Always route and hang hoses the same way as in the factory installation. Avoid contact with other parts. In particular, look for rubbing against moving parts.
- *Check for leaks.* After making repairs, always check for fluid leaks before releasing the vehicle to the customer.
- To perform a diagnostic check of possible steering problems, see Table 21–2.

Steering Rack Replacement

To remove a rack-and-pinion steering gear, separate the outer tie-rod ends from the steering knuckles. Then, unbolt the steering gear mounting brackets from the frame, unibody, or cross member. Also, disconnect the steering column coupler or U-joint. Rotate the steering gear and slide it out of the chassis. Refer to Figure 21–39.

Refer to service literature for additional information. You may have to remove the front wheels or slide the gear out one particular side of the vehicle.

Wheel and Tire Service

A cut in a tire can result in a high-speed blowout (rapid loss in air pressure). A blowout can cause a driver to lose control of the vehicle, possibly causing a serious accident. Inspect tires carefully for damage (Figure 21–40).

Wheels can be bent and cracked in a collision. Closely inspect the wheels for problems (nicks, dents, bent lips, cracks, and other damage). A bent, damaged rim causes **wheel runout**. *Radial runout* causes the diameter of the wheel to change as it is rotated. **Axial runout** causes the wheel to wobble sideways as it rotates. A dial indicator can be used to quickly check for wheel runout (Figure 21–41).

If you find problems with the wheels or tires, have them replaced. Most body shops send them to a tire shop for replacement. These specialty shops have tire changers, wheel balancers, and other equipment (Figure 21–42).

When installing wheels, especially lightweight aluminum ones, use a torque wrench to tighten lug nuts in a star or crisscross pattern. This will prevent you from warping the hub and wheel and causing runout vibration (Figure 21–43).

Collapsible Steering Columns

To reduce the chance of injury, automotive engineers have designed *collapsible steering columns* that crush when hit by the driver's body during a collision.

The steering wheel is often held on by a large nut and a press-fit. A wheel puller is used to remove a steering wheel. This is a common task in a body shop because steering wheels are frequently damaged. Use hardened bolts to hold the puller into the steering wheel. Then, tighten down the large bolt in the center of the puller to force the steering wheel off its shaft.

Steering Column Service

Steering column service is needed after a collision (crushing of collapsible steering column) or when internal parts of the column fail. Most steering column repairs can be done with the column mounted in the vehicle. However, some repairs require steering column removal.

A **wheel puller** is used to remove a steering wheel from its shaft. After removing the horn button or air bag and steering shaft nut, scribe alignment marks on the steering wheel and steering shaft. This will help you position the steering wheel correctly during assembly.

Mount the wheel puller as shown in Figure 21–44. Screw the bolts into the threaded holes in the wheel. Make sure the bolts have the correct thread type. Using a wrench or ratchet, tighten the puller down against the steering shaft.

The steering column is bolted to the bottom of the dash panel and to the firewall. By removing these bolts and disconnecting the column shaft from the steering gear, you can remove the steering column. Always follow service manual directions when servicing a steering column, and assemble everything in reverse order.

TABLE 21–2 STEERING PROBLEM DIAGNOSIS

	Problem					
Check	**Noise**	**Instability**	**Pulls to One Side**	**Excessive Steering Play**	**Hard Steering**	**Shimmy**
Tires/wheels	Road/tire noise	Low/uneven tire pressure; radial tire lead	Low/uneven tire pressure; radial tire lead	Low/uneven air pressure	Low/uneven tire pressure	Unbalanced wheel; uneven tire wear; overworn tires
Tie-rods	Squeal in turns: worn ends	—	Incorrect toe: tie-rod length	Worn ends	Worn ends	Worn ends
Mounts/ bushings	Parallelogram steering: steering gear mounting bolts, linkage connections; rack-and-pinion steering: rack mounts	Idler arm bushing	—	Parallelogram steering: steering gear mounting bolts, linkage connections; rack-and-pinion steering: rack mounts	Parallelogram steering: steering gear mounting bolts, linkage connections; rack-and-pinion steering: rack mounts	Parallelogram steering: steering gear mounting bolts, linkage connections; rack-and-pinion steering: rack mounts
Steering linkage components	Bent/damaged steering rack	Incorrect center link/rack height	Incorrect center link/rack height	Worn idler arm, center link, or pitman arm studs; worn/ damaged rack	Idler arm binding	Worn idler arm, center link, or pitman arm studs
Steering gear	Improper yoke adjustment on rack-and-pinion steering	—	—	Improper yoke adjustment on rack-and-pinion steering: worn steering gear/ incorrect gear adjustment on parallelogram steering; loose or worn steering shaft coupling	Parallelogram steering: low steering gear lubricant, incorrect adjustment; rack-and-pinion: bent rack, improper yoke adjustment	—
Power steering	—	—	—	—	Fluid leaks, loose/worn/ glazed steering belt, weak pump, low fluid level	—
Alignment	—	—	Unequal caster/ camber	—	Excessive positive caster, excessive scrub radius (incorrect camber and/or SAI)	Incorrect caster

Wear eye protection when tightening a wheel puller. If one of the bolts were to break, bits of metal could fly into your face.

21.3 WHEEL ALIGNMENT

In collision repair, **wheel alignment** involves adjusting the vehicle's tires so that they roll properly over the road surface. Wheel alignment is essential to safety, handling, fuel economy, and tire life.

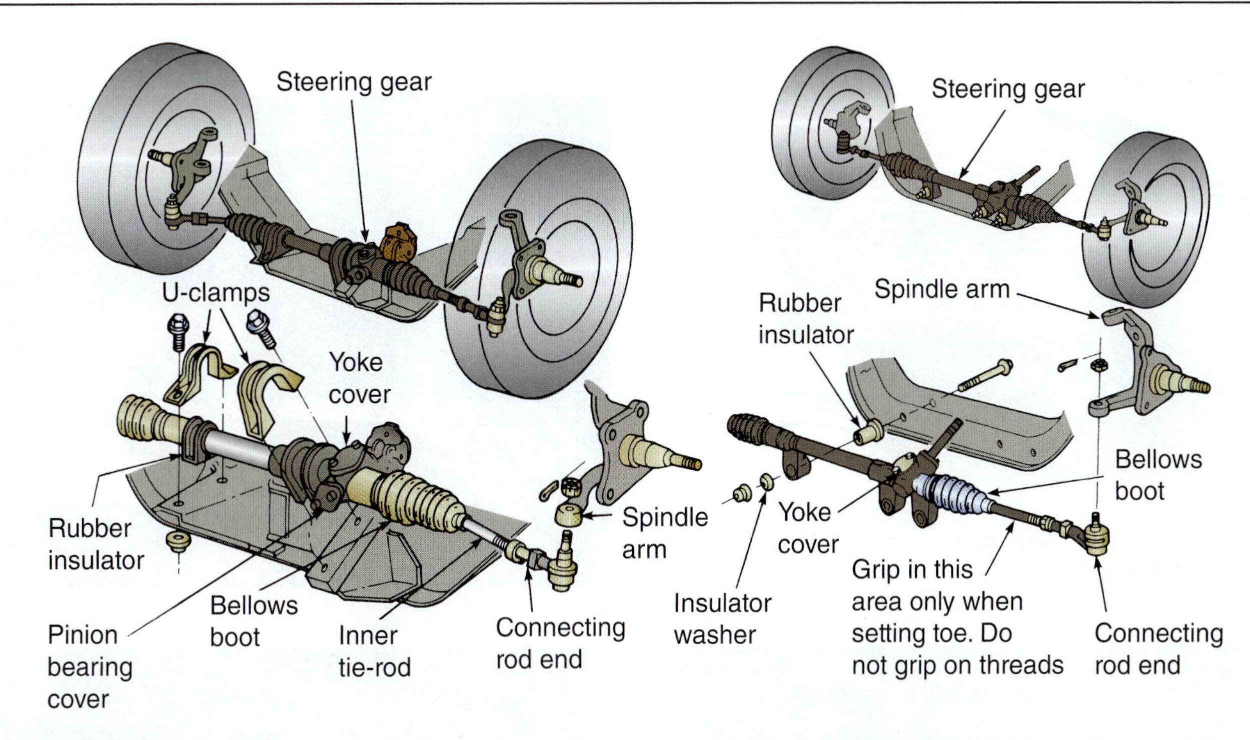

Figure 21-39 Here are the typical attaching methods for a manual rack-and-pinion assembly. Refer to vehicle-specific service information when in doubt.

Following a collision, a vehicle requires an alignment if any of the following conditions exist:

- Damage to any steering or suspension parts
- Damage to any steering or suspension mounting locations
- Engine cradle damage or a position change
- Removal of suspension or steering parts for access to body parts

REMEMBER *Wheel alignment is done to fine-tune body/frame adjustments. The job of the collision repair technician is to make sure that everything can be fine-tuned and the wheels can be aligned properly.*

The proper alignment of a suspension/steering system centers around the accuracy of seven control angles:

1. Camber
2. Caster
3. Steering axis inclination (SAI)
4. Scrub radius
5. Toe (in and out)
6. Thrust line
7. Turning radius

Camber

Camber is the angle represented by the vertical tilt of the wheels inward or outward when viewed from the front of the vehicle. It ensures that all of the tire tread contacts the road surface. Camber is measured in degrees. It is usually the second angle adjusted during a wheel alignment (Figure 21–45).

Camber is usually set equally for each wheel. Equal camber means each wheel is tilted outward or inward the same amount.

Positive camber means the top of the wheel is tilted out when viewed from the front. The outer edge of the tire tread contacts the road.

Negative camber means the top of the wheel is tilted inward when viewed from the front. The inner tire tread contacts the road surface more. Note how camber changes when turning.

Camber is controlled by the control arms and their pivots. It is affected by worn or loose ball joints, control arm bushings, and wheel bearings. Anything that changes chassis height will also affect camber (Figure 21–46).

Camber is adjustable on most vehicles. Some manufacturers prefer to include a camber adjustment at the spindle assembly. Camber adjustments are also provided on some strut suspension systems at the top mounting position of the strut. Remember that camber adjustment also changes SAI or the included angle.

Very little adjustment will be required if the strut tower and lower control arm positions are in their proper place. If you find serious camber error but suspension mounts have not been damaged, it is an indication of bent suspension parts. In this case, diagnostic angle and dimensional checks should be made to the suspension parts. Damaged parts must be replaced.

Figure 21-40 **Inspect tires for damage; cuts, splits, tears, and broken plies or cords show up as small bulges in the rubber. (A) If a tread wear gauge shows a tire is worn, inform the customer. (B) This tire has bulging sidewalls, which indicate low air pressure. (C) Use an accurate tire gauge to check inflation pressure. (D) Inflate the tire a few pounds below the maximum inflation pressure printed on the tire sidewall.**

Caster

Caster is the angle of the steering axis of a wheel from true vertical, as viewed from the side of the vehicle. It is a directional stability adjustment. Caster is measured in degrees (Figure 21–47).

Caster has little effect on tire wear. Caster affects where the tires touch the road compared to an imaginary centerline drawn through the spindle support. Caster is the first angle adjusted during an alignment.

Positive caster tilts the tops of the steering knuckles toward the rear of the vehicle. It helps keep the vehicle's wheels traveling in a straight line. The wheels resist turning and tend to return to the straight-ahead position.

Negative caster tilts the tops of the steering knuckles toward the front of the vehicle. Negative caster makes the wheels easier to turn. However, it produces less directional stability. The wheels tend to follow imperfections in the road surface.

Caster is designed to provide steering stability. The caster angle for each wheel should be almost equal. Unequal caster angles will cause the vehicle to steer toward the side with less caster. Too much negative caster can cause the vehicle to have "sensitive" steering at high speeds. The vehicle might wander as a result of too much negative caster.

Caster is measured in degrees from true vertical. Specifications for caster are given in positive or negative degrees. Typically, more positive caster is used with power steering. More negative caster is used with manual steering to reduce steering effort. Also, a vehicle

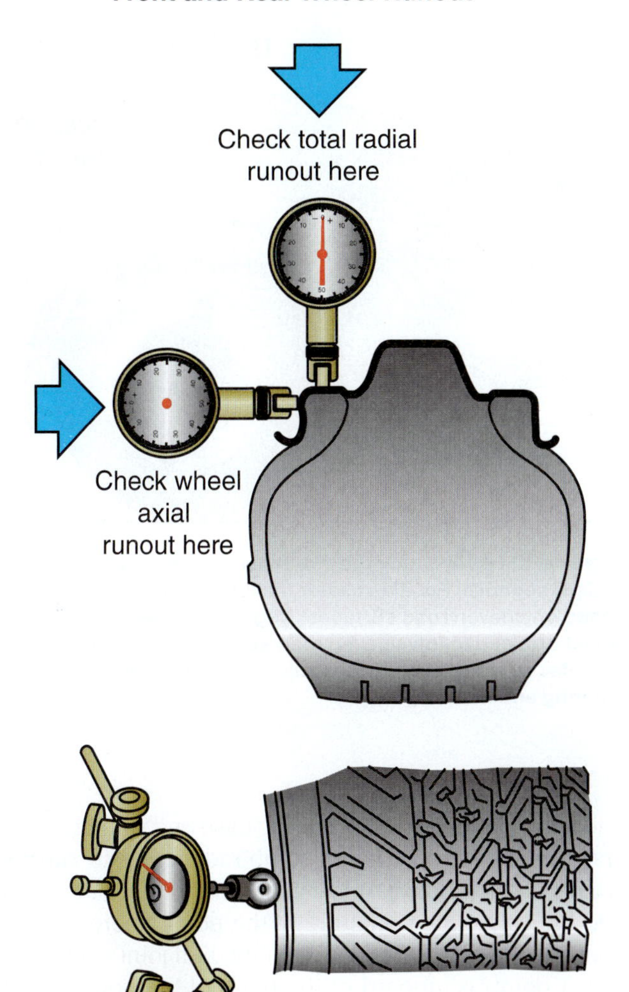

Figure 21-41 Always check for wheel damage when the vehicle has suffered major structural damage. Check for both axial and radial runout. Specs shown here are typical, but refer to the service manual for a specific vehicle.

pulls to the side with the least amount of caster (Figure 21–48).

With some vehicle designs, caster is adjusted by loosening and moving the subframe or cradle bolted to the unibody. Pushing the entire cradle forward or rearward can be done to change caster settings.

Steering Axis Inclination

Steering axis inclination (SAI) is the inward tilt of the steering axis at the top. It also contributes to directional stability. Because the steering axis is inclined, the spindle is forced to move in a downward arc as the wheel is turned. This action causes the vehicle to rise as the wheel is turned in either direction, so the weight of the car forces the wheels back to the straight-ahead position (Figure 21–49).

Figure 21-42 A high-speed balancer is being used to locate wheel weights after tire and wheel replacement. Most body shops send damaged wheels out for replacement.

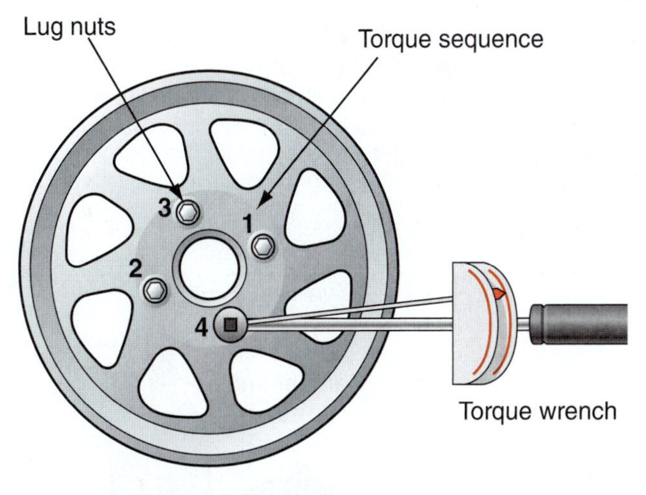

Figure 21-43 Always use a torque wrench to properly tighten lug nuts to specs. This is critical with today's lightweight aluminum "mag wheels." It ensures the vehicle is safe to drive and that soft aluminum wheels are not warped by excess torque. Run the lug nuts down lightly with air impact. Then use a torque wrench to final tighten them.

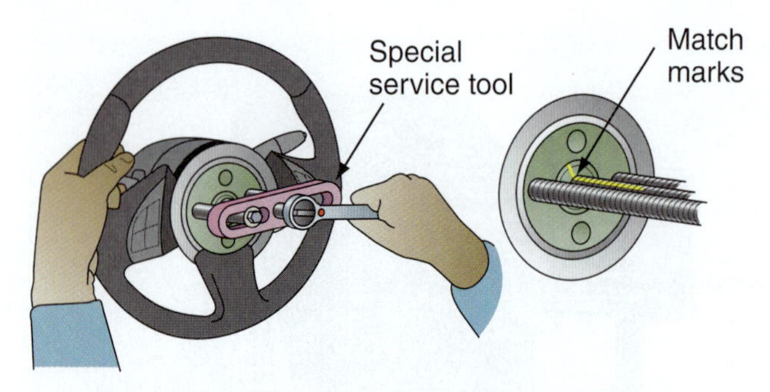

Figure 21-44 A wheel puller is needed to remove a press fit steering wheel from its shaft. If seat belts were not being worn during a head-on collision, the steering wheel may have been bent by the driver's body being thrown into it.

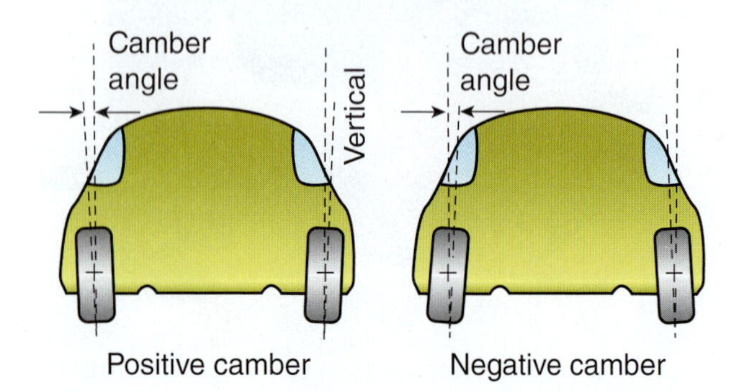

Figure 21-45 Compare the two types of camber: (A) positive and (B) negative. Improper camber adjustment will wear the inner or outer edges of a tire tread.

SAI is not generally considered a tire wear factor unless there is an extreme change. The amount of inclination is preset and should not change unless there is damage to the spindle support arm.

Camber and SAI are sometimes measured together as the "included angle." The amount of tilt is measured in degrees from vertical.

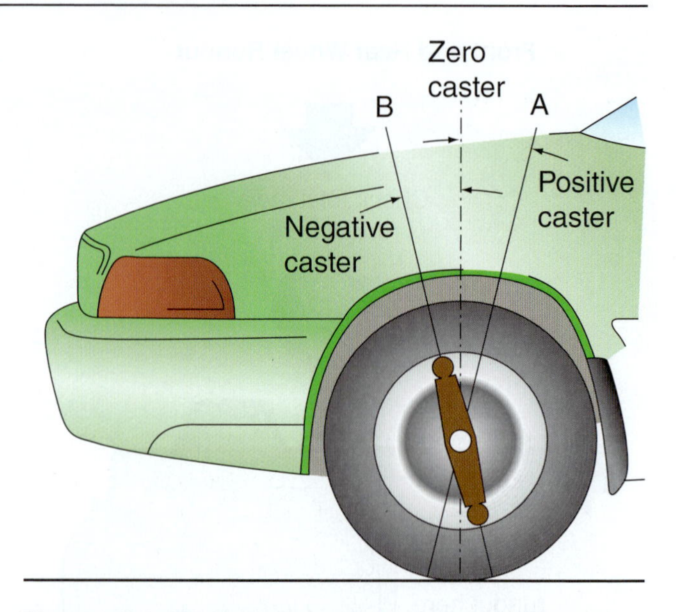

Figure 21-47 Study the two types of caster: (A) positive and (B) negative. Positive caster makes tires travel straight ahead on uneven road surfaces. Negative caster makes the wheel turn and follow uneven road surfaces. The vehicle's manufacturer supplies caster adjustment for proper vehicle steering and handling.

Scrub Radius

The importance of SAI to steering ease and stability centers around the reduction of scrub radius. **Scrub radius** is the distance between the centerline of the ball joints and the centerline of the tire at the point where the tire contacts the road surface. When the ball joint centerline (pivot point) is inboard of the point of tire contact, the tire does not pivot where it touches the road. Instead, it has to move forward and backward to compensate as the driver turns the steering wheel. Steering effort is greatly increased as the tires scrub against the road during turns.

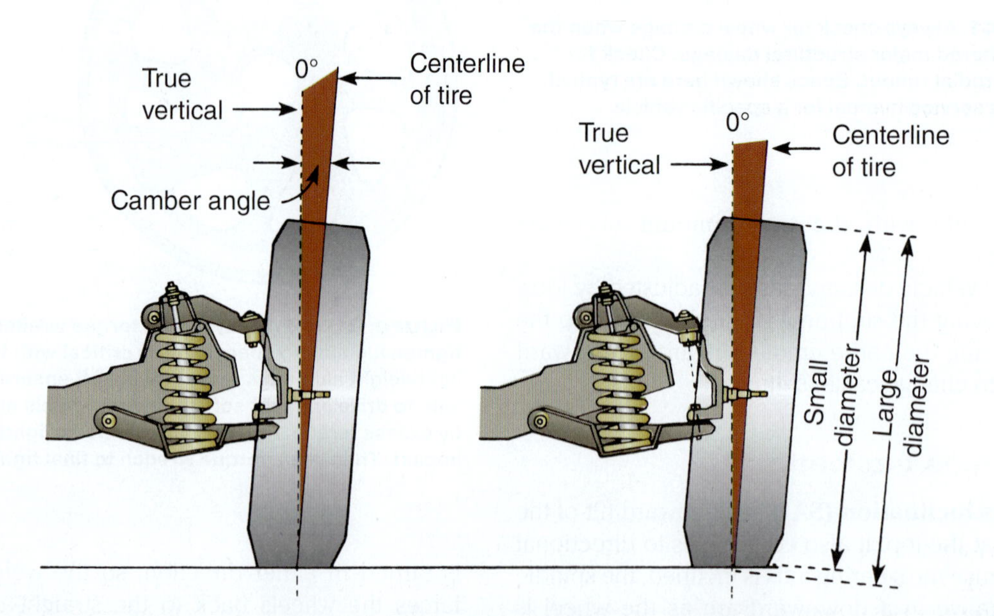

Figure 21-46 When camber is out of specifications, it can cause the inner or outer edge of the tire to contact the road.

Slide the plate toward the front or rear of the car until the desired caster reading is obtained

Engine

Add shims here to increase caster

Frame

Subtract shims here to increase caster

Figure 21-48 On some struts, the caster and camber can be adjusted by sliding a plate to relocate the top of the strut rod and steering knuckle. With older suspension systems, the caster and camber are changed by adding or subtracting shims or by turning the adjustment nuts or screws.

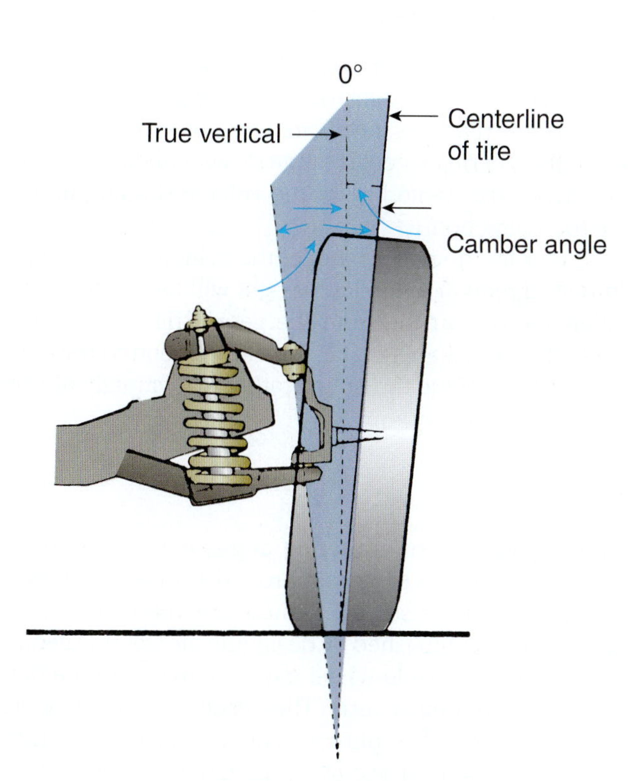

Figure 21-49 This diagram illustrates SAI.

If the control arm assembly were designed with no SAI, scrub radius would be quite large.

Both positive camber and steering axis inclination combine to reduce scrub radius to a minimum.

Toe

Toe is the difference in the distance between the front and rear of the left- and right-hand wheels. Toe can be measured in inches (or millimeters) or degrees, depending on the equipment used. Toe should be the last wheel alignment adjustment made (Figure 21–50).

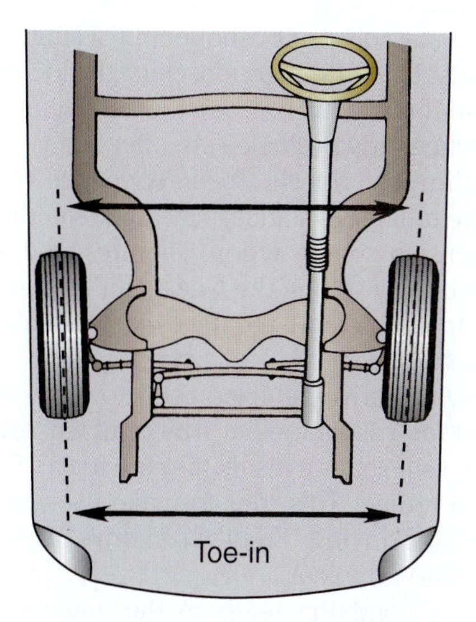

A

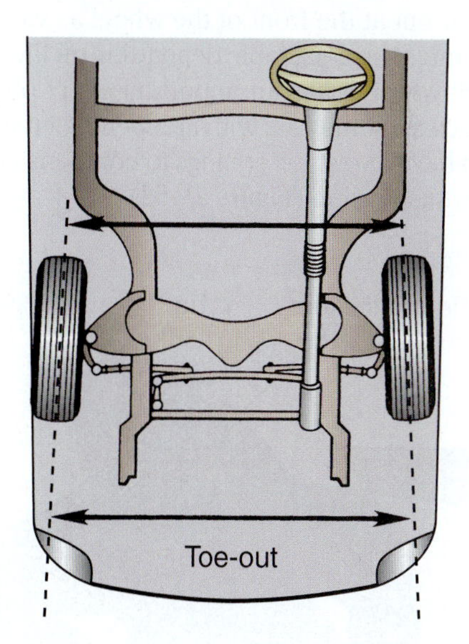

B

Figure 21-50 Improper toe adjustment will cause rapid tire wear. Compare toe conditions: (A) Toe-in is the amount that wheels are closer together at the extreme front of the tires than they are at the extreme rear. It is often used on RWD vehicles because tires are pushed rearward when driving. (B) Toe-out is just the opposite of toe-in, with a greater measurement in front than in the rear. It is used on FWD vehicles because the drive wheels push forward when driving.

Toe adjustment is critical to tire wear. If properly adjusted, toe makes the wheels roll in the same direction. If toe is not correct, the misaligned wheels will scuff or drag the tires sideways, causing rapid tire wear.

REMEMBER *Excessive toe—in or out—will cause a sawtooth edge on the tire tread from dragging the tire sideways.*

Toe-in results when the front of the wheels is set closer than the rear. The wheels point in at the front. *Toe-out* has the front of the wheels farther apart than at the rear. It is the opposite of toe-in—the wheels point out at the front.

Toe is a critical tire-wearing angle. Wheels that do not track straight ahead have to drag as they travel forward. Toe is normally adjusted by shortening or lengthening the tie-rod ends. This is done by loosening the locknut on the tie-rod and rotating the rod, as shown in Figure 21–51.

Rear-wheel-drive vehicles are often adjusted to have toe-in at the front wheels. Toe-in is needed to compensate for tire rolling resistance, play in the steering system, and suspension system action. The tires tend to toe-out while driving. By setting the wheels for a small toe-in of about 1⁄16 inch (1.5 mm), the tires will roll straight ahead over the road surface.

Front-wheel-drive vehicles need to have their front wheels set for a slight toe-out. The front wheels pull and propel the vehicle. As a result, they are forced forward by drive train torque. This tries to make the wheels point inward while driving. Front-wheel-drive toe-out of 1⁄16 inch (1.5 mm) is typical.

Rear toe condition refers to the angle of the rear wheel in or out at the front of the wheel as viewed from the top. It might be adjustable depending on the design of the car. However, it has an important effect on the handling of the car. Some cars with independent rear suspensions also have at-rest toe settings to compensate for play in the rear suspension (Figure 21–52).

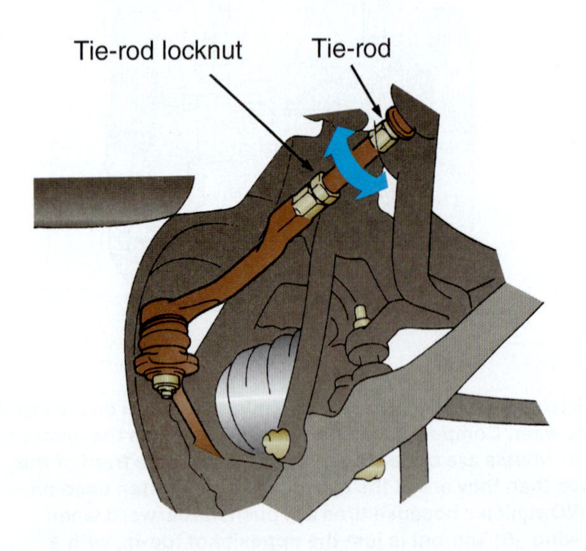

Figure 21–51 Toe-in or toe-out is adjusted by turning the threaded tie-rod ends. Make sure to retighten the locknut after adjustment.

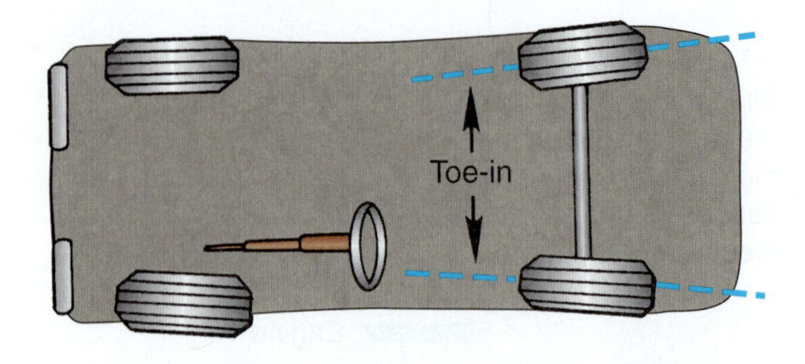

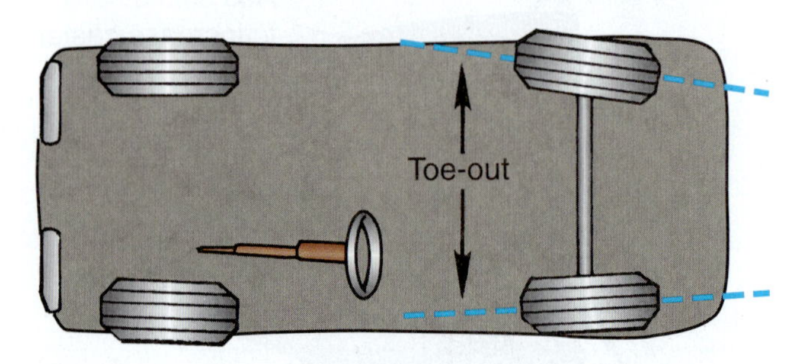

Figure 21–52 These are typical rear toe conditions that should be checked after repairing a collision-damaged vehicle.

Rear camber refers to the position of a rear wheel in or out at the top as viewed from the rear of the rear wheel. It might be adjustable, depending on the design of the car. It too has an important effect on the handling of the car.

Thrust Line Alignment

A main consideration in any alignment is to make sure the vehicle runs straight down the road. With *proper tracking*, the rear tires travel directly behind the front tires when the steering wheel is in the straight-ahead position. The geometric centerline of the vehicle should parallel the road direction.

If rear toe does not parallel the vehicle centerline, a "thrust" direction to the left or right will be created. This difference of rear toe from the geometric centerline is called the thrust angle. The vehicle will tend to travel in the direction of the thrust line rather than straight ahead (Figure 21–53).

Turning Radius

Turning radius, or cornering angle, is the amount of toe-out on turns. As a car goes around a corner, the inside tire must travel in a smaller radius circle than the outside tire. This is accomplished by designing the steering geometry to turn the inside wheel more sharply than the outside wheel during a turn. The result can be seen as toe-out on turns. The purpose is to eliminate tire scrubbing on the road surface by keeping the tires pointed in the direction they have to move (Figure 21–54).

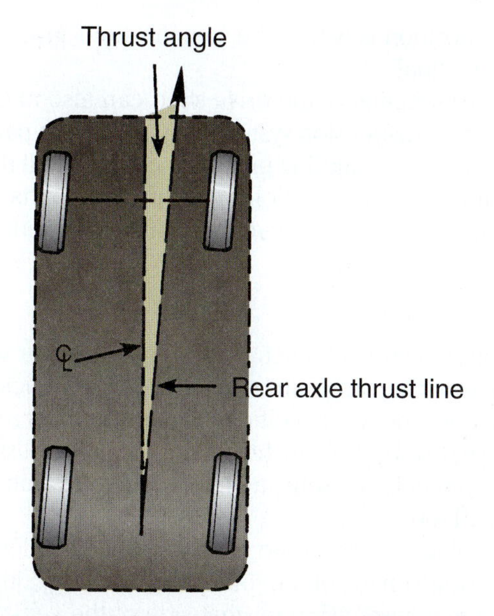

Figure 21-53 If the rear axle or wheels are not in alignment, it will affect thrust line alignment and tracking of the vehicle down the road.

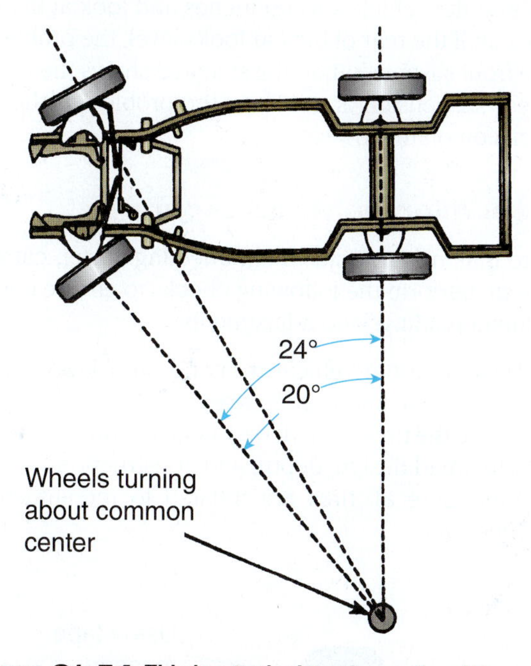

Figure 21-54 This is a typical turning radius. It is measured by reading the scales on a turning radius gauge or wheel alignment machine.

One of the very useful diagnostic checks that can be made with a minimum of equipment is a jounce–rebound toe-in change check, which can help determine the condition of the suspension system.

Jounce is the motion caused by a wheel going over a bump and compressing the spring. During jounce, the wheel moves up toward the chassis. Jounce can be simulated for in-shop testing by pushing down on the bumper. The car must be jounced equally on both sides.

Rebound is the motion caused by a wheel going into a dip or returning from a jounce and extending the spring.

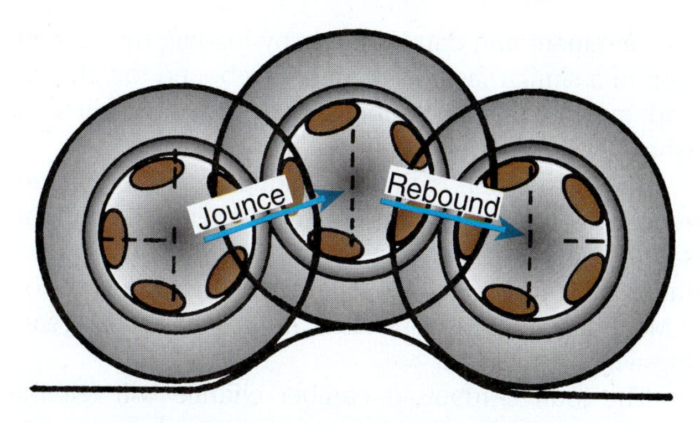

Figure 21-55 Jounce and rebound refer to the reactions of the suspension to an irregular road surface.

During rebound, the wheel moves down away from the chassis. Rebound can be simulated for in-shop testing by lifting up on the fender. The car must be lifted equally on both sides.

This jounce–rebound check will determine whether there is some misalignment to the rack-and-pinion gear. For a quick check, unlock the steering wheel and see whether it moves during the jounce and/or rebound (Figure 21–55).

The next diagnostic check is for cornering angle. The cornering angle check evaluates the proper relationship of the two front wheels as they are turned through a steering arc. To measure cornering angle, one wheel is turned a given amount on a turn plate or protractor. The amount of rotation of the opposite wheel is measured in a similar manner. The results are compared right to left to determine whether the two front wheels are rotating through the same arc.

During a cornering angle check, the left front wheel should be turned out 20 degrees. Then the right wheel rotation is measured. The right wheel should turn in the same amount or about 2 degrees less. The difference accounts for the turning radius difference between the inside and outside wheels during cornering.

The process is repeated with the right wheel; the right wheel is turned out 20 degrees. The movement of the left wheel is measured on the protractor or turn plate. The left wheel should turn in the same amount or about 2 degrees less.

By design, a vehicle might use a different turning radius from one side to the other. If in doubt, refer to the manufacturer's specifications. If these measurements do not repeat within 2 degrees, damage to the steering arms or gear is indicated. Cornering angle measurements are especially useful in determining whether improper toe conditions are caused by poor wheel alignment or damaged suspension components.

Camber Checks

Some camber checks can be made to diagnose the condition of a strut and can be measured easily with a camber gauge. One is called a jounce–rebound camber measurement and can be made by loading the suspension in a similar fashion to jounce–rebound toe change and measuring the camber angle from an individual wheel.

The suspension is then unloaded as in the jounce–rebound toe check and a second camber reading of the same wheel is made. The two readings are compared; these readings should not differ more than 2 degrees on a MacPherson strut suspension. In most cases, the readings will be the same.

The jounce–rebound camber change will tell the technician whether the strut is bent either inboard or outboard. Check each wheel individually before deciding whether one strut is bad based on the readings. If the readings differ between wheels more than 2 degrees, a bent strut is indicated.

A swing camber measurement is made by turning the front wheel "in" a given amount and performing a jounce–rebound camber check.

The front wheel is then turned out the same amount and the camber angle is measured again. If the camber angle change differs more than 3 degrees from the left wheel to the right wheel, it is likely that either the strut is bent forward or rearward of its normal position or the caster angle is incorrect. As a further test for a bent strut, perform a jounce–rebound check while the wheels are turned in and while they are turned out. Check each wheel and compare the readings. These diagnostic angles are especially helpful in determining the cause of vehicle handling and tracking problems.

Engine Cradle Position

Proper positioning of the engine cradle can affect the steering angles. Because the cradle provides the lower pivot point, movement of the cradle will cause a camber change. Both wheels will show an equal camber change: one side negative and one side positive. It will also cause an SAI, but not an included angle change. Make sure the cradle's position is within the specifications given in the service manual.

The positioning of the drive shaft can also affect the steering and suspension systems. If any of these parts are bent, shimmy or handling problems can occur. If there is any doubt about the positioning of the drive shafts, measure them following factory-recommended procedures.

Curb Height

For proper alignment, each of the front and rear wheels must carry the same amount of weight. Vehicles are designed to ride at a specific height, sometimes referred to as **curb height**. Curb height specs are published in service manuals and some alignment specification books (Figure 21–56).

If a vehicle leans to one side or seems to be lower on one side than on the other, something is wrong. Either the front or rear suspension on that side of the vehicle can cause the condition.

To isolate the height problem, place a jack in the center of the main cross member in the front of the vehicle. Raise the vehicle several inches and look at the rear of the car. If the rear of the car looks level, the problem is in the front suspension on the side that shows the lean. If the rear suspension is not level, the problem is the rear suspension on the low side.

Wheel Alignment Procedure

Before making any adjustment affecting caster, camber, or toe-in, perform the following checks to ensure correct alignment readings and adjustments:

1. Make sure the vehicle is sitting on a level surface (side to side and front to rear).
2. Rotate the tires if needed. (Check the tires for similar size, tread design, depth, and construction.)
3. Make sure all tires are inflated to recommended pressure.

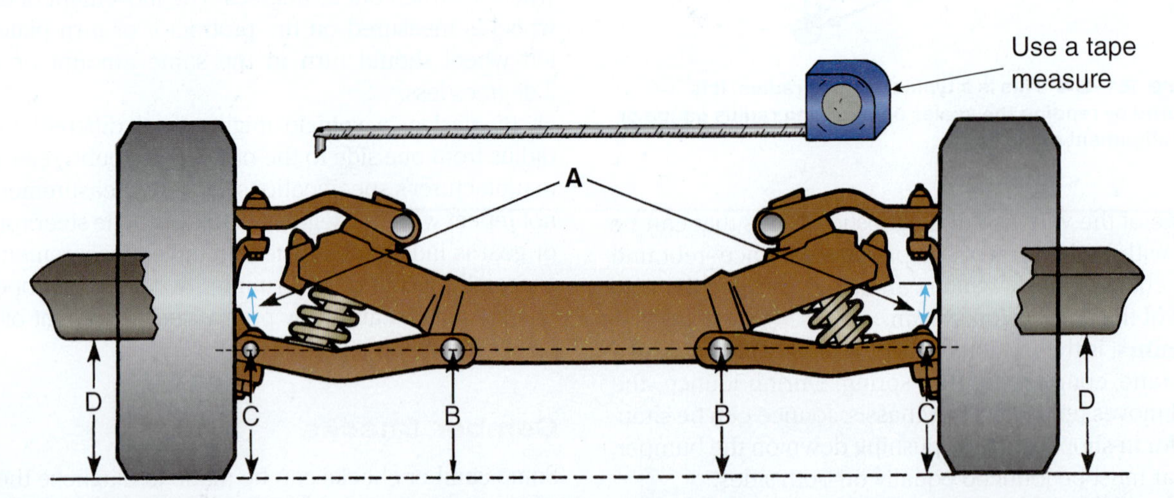

Figure 21–56 Note the typical locations for measuring curb, or ride, height. Service literature will give details for specific makes and models of vehicles.

4. Inspect for worn or bent parts and replace. Much of this should be checked during body/frame correction.
5. Check and adjust wheel bearings if necessary; spin tires and check for looseness or unusual noises.
6. Check for unbalanced loading (proper chassis height). This should be checked after body/frame correction.
7. Check for loose ball joints, tie-rod ends, steering relay rods, control arms, and stabilizer bar attachments.
8. Check for runout of wheels and tires.
9. Check for defective shock absorbers.
10. Consider excess loads, such as toolboxes.
11. Consider the condition and type of equipment being used to check alignment, and follow the manufacturer's instructions.

Caster and camber angles are measured with gauges available from specialty tool manufacturers. They must be used as directed to get proper measurements. Figure 21–57 shows a technician using one type of alignment gauge.

The adjustment sequence of caster first, then camber, and finally toe is recommended regardless of the vehicle make or its type of suspension. Methods of adjustment vary from vehicle to vehicle and, in some cases, from year to year of the same make car. Refer to the manufacturer's service manual for details.

A typical wheel alignment procedure follows:

1. Obtain manufacturer's specifications.
2. Check camber—tilt of wheel inward and outward.
3. Check caster—forward or rearward tilt of steering axis.

Courtesy of Hunter Engineering Company

Figure 21–57 A wheel alignment machine will help you quickly and accurately realign wheels after collision repairs. This technician is doing four-wheel alignment.

Courtesy of Hunter Engineering Company

Figure 21–58 This wheel alignment machine can be used with the vehicle still mounted on the frame rack.

4. Check steering axis inclination—inward tilt of steering axis at the top.
5. Check turning radius—wheel angles while turning.
6. Check toe—difference in distance between the front and rear of the tire.

Many of today's wheel alignment machines are computerized. They provide exact specs, specify where adjustments are needed, and may even show a picture of what is wrong (Figure 21–58).

Turning radius gauges measure how many degrees the front wheels are turned right or left. They are commonly used when measuring caster, camber, and toe-out on turns (Figure 21–59).

A **caster–camber gauge** is used with the turning radius gauge to measure caster and camber in degrees. The gauge either fits magnetically on the wheel hub or

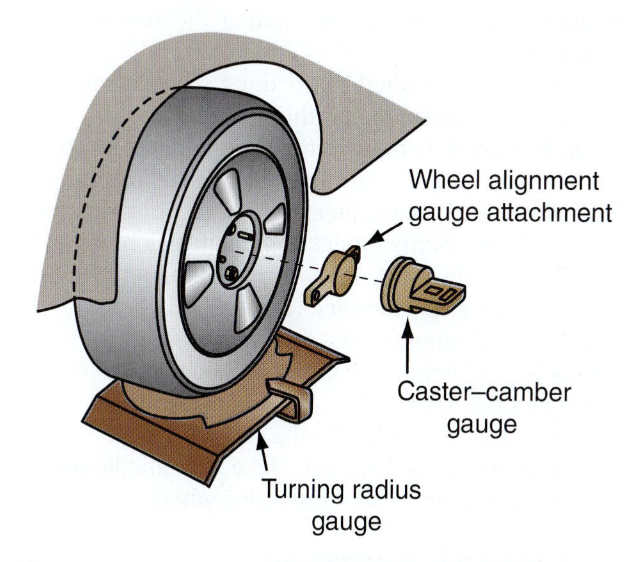

Figure 21–59 Turning radius gauges have scales so you can measure how far each wheel turns in degrees when doing wheel alignment. Camber–caster gauges have a bubble float for showing readings.

may fasten on the wheel rim. Normally, caster and camber are adjusted together, because one affects the other.

To measure caster with a caster–camber gauge, turn one of the front wheels inward until the radius gauge reads 20 degrees. Turn the adjustment knob on the caster–camber gauge until the bubble or indicator is centered on zero. Then, turn the wheel out 20 degrees.

The degree marking next to the bubble or indicator will equal the caster of that front wheel. Compare your reading to specifications and adjust as needed. Repeat this operation on the other side of the car.

To measure camber with a bubble-type caster–camber gauge, turn the front wheels straight ahead (radius gauges on zero). The car must be on a perfectly level surface or on an alignment rack.

Read the number of degrees next to the bubble on the camber scale of the gauge. It will show camber for that wheel. If not within specs, adjust camber.

If shims are used, add or remove the same number of shims from the front and rear of the control arm. This will keep the caster set correctly. Double-check caster, especially when an excessive amount of camber adjustment is needed.

To measure toe with a tram gauge, raise the wheels and rub a chalk line all the way around the center rib on each tire. Then, using a scribing tool, rotate each tire and scribe a fine line on the chalk line. This will give you a very thin reference line for measuring the distance between the tires. Lower the car back on the radius gauges.

First, position the tram gauge at the back of the tires. Move the pointers until they line up with the lines you scribed on the tires. Then, without bumping the gauge, position the gauge at the front of the tires. The difference in the distance between the lines on the front and rear of the tires shows toe.

For example, if the lines on the front of the tires are closer together than on the rear, the wheels are toed-in. If the lines are the same distance apart at the front and rear, toe is zero.

Using service manual instructions, adjust the tie-rods until the tram gauge reads within specs.

Modern equipment saves time and requires less training. Many alignment machines will give instructions, illustrations, and measurements on a computer monitor (Figure 21–60). Some wheel alignment machines even provide a graphic showing which alignment specifications are not within tolerance (Figure 21–61).

Table 21–2 summarizes a typical steering problem diagnosis. It is important to remember that the typical customer judges the quality of a wheel alignment by the position of the "fifth wheel"—the steering wheel in his or her hands. It must be straight. Make sure all alignments end with a properly centered steering wheel.

21.4 BRAKE SYSTEMS

The *brake system* uses hydraulic pressure to slow or stop wheel rotation with brake pedal application (Figure 21–62).

Figure 21-60 Always inspect all rubber boots on tie-rods and ball joints for damage or failure. Replace any with split or cracked boots.

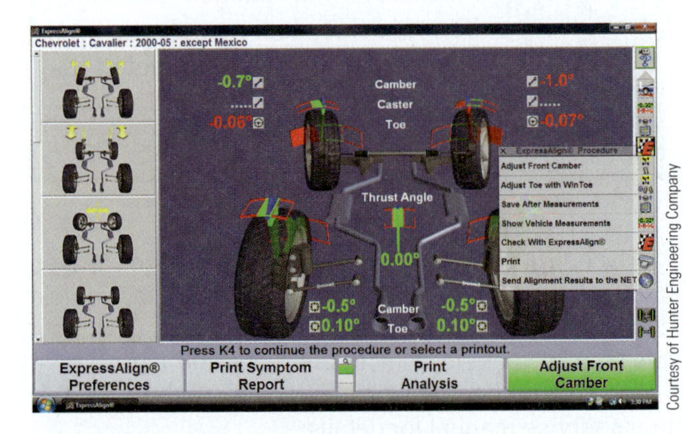

Courtesy of Hunter Engineering Company

Figure 21-61 The screen of this computerized wheel alignment machine gives specs and instructions for adjustment. Live readings allow you to watch your progress as you adjust caster, camber, and toe to within factory tolerances.

The brake pedal transfers the driver's foot pressure into the **master cylinder**. The master cylinder develops hydraulic pressure (oil pressure) for the system. Brake lines and hoses carry fluid out to the wheel cylinders. The wheel cylinders use hydraulic pressure to push the brake pads or shoes outward (Figure 21–63).

The *brake pads*, or brake shoes, have a friction lining for rubbing on the brake rotor or drum. The brake drums provide heavy metal friction surfaces bolted between the hub and wheel. A caliper holds the piston(s) and brake pads on disc brakes.

A power brake system is a standard hydraulic brake system with a vacuum, hydraulic, or electric assist. A booster unit is added to help apply the master cylinder and brakes.

Two basic types of hydraulic brakes are used in unibody vehicles. They are drum brakes and disc brakes.

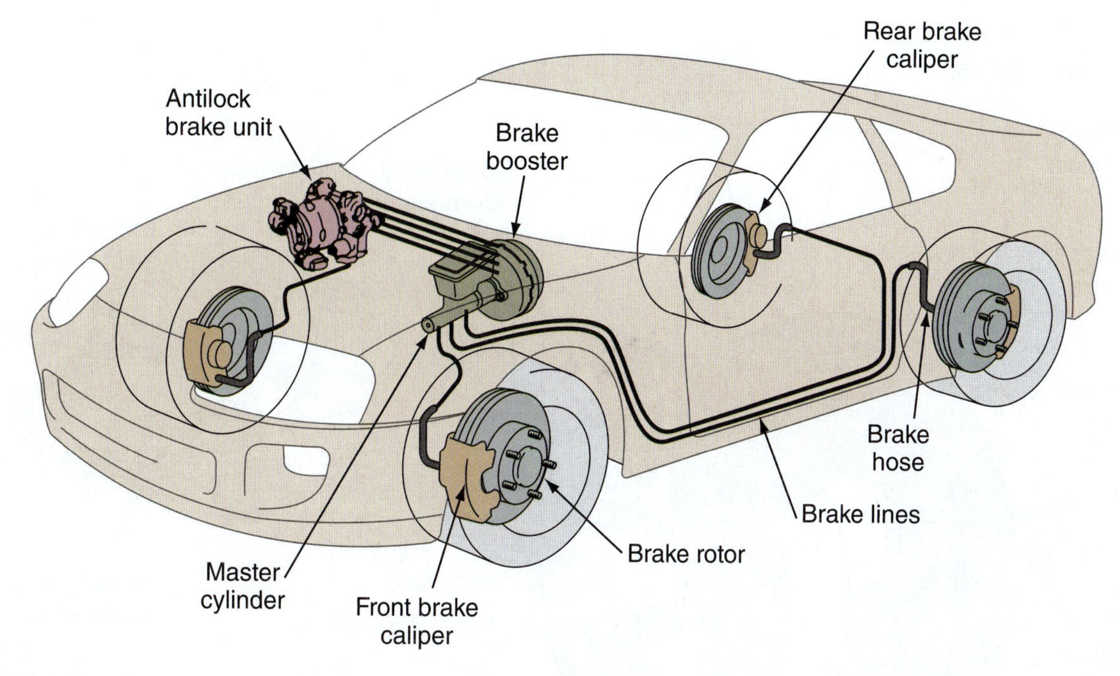

Figure 21-62 **Study the major parts of a modern brake system.**

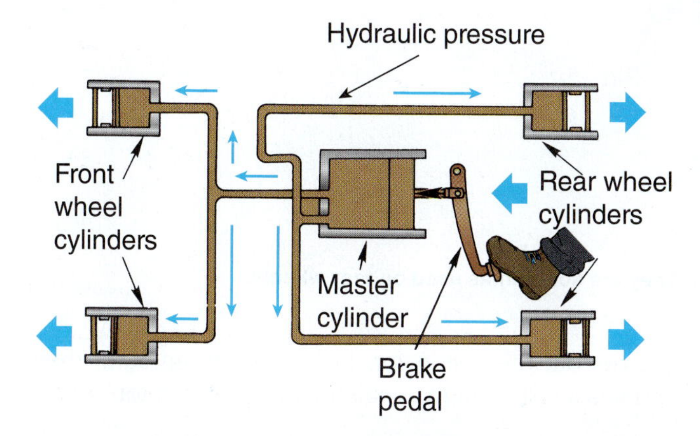

Figure 21-63 **This simple automotive hydraulic system diagram shows how a brake system operates. When you press the brake pedal, the piston in the master cylinder is moved to develop hydraulic pressure. Pressure is sent out through the brake lines to the wheel cylinders. The wheel cylinders then slide out to activate the wheel brake assemblies.**

Drum Brakes

A *drum brake* assembly consists of a cast-iron drum bolted to the vehicle axle. A fixed brake backing plate holds the shoes and other components—wheel cylinders, automatic adjusters, linkages, and so on. Additionally, there might be some extra hardware for parking brakes (Figure 21–64).

The *brake shoes* are surfaced with frictional linings, which contact the inside of the drum when the brakes are applied. The shoes are forced outward against the action of the return springs by pistons or *wheel cylinders*, which are actuated by hydraulic pressure. As the drum rubs against the shoes, the energy of the moving drum is transformed into heat, and this heat energy is passed into the atmosphere.

When the brake shoe is engaged, the frictional drag acting around its circumference tends to rotate it about its hinge point, the brake anchor. If the rotation of the drum corresponds to an outward rotation of the shoe, the drag will pull the shoe tighter against the inside of the drum and the shoe will be self-energizing.

Disc Brakes

Disc brakes resemble the brakes on a bicycle: The friction elements are in the form of pads, which are squeezed or clamped about the edge of a rotating wheel. With automotive disc brakes, this wheel is a separate unit, called a rotor, inboard of the vehicle wheel (Figure 21–65).

The **rotor** is made of cast iron and, because the pads clamp against both sides of it, both sides are machined smooth. Usually the two surfaces are separated by a finned center section for better cooling. The pads are attached to metal shoes, which are actuated by pistons, as in drum brakes. The pistons are contained within a caliper assembly, a housing that wraps around the edge of the rotor. Usually two large bolts secure the caliper to the steering knuckle to keep it from moving when the brakes are applied.

The **caliper** is a housing containing the pistons and related seals, springs, and boots, as well as the cylinders and fluid passages necessary to force the friction linings or pads against the rotor. The caliper resembles a hand in the way it wraps around the edge of the rotor. It is attached to the steering knuckle. Some models employ light spring pressure to keep the pads close against the rotor; in other caliper designs this is achieved by a unique seal that

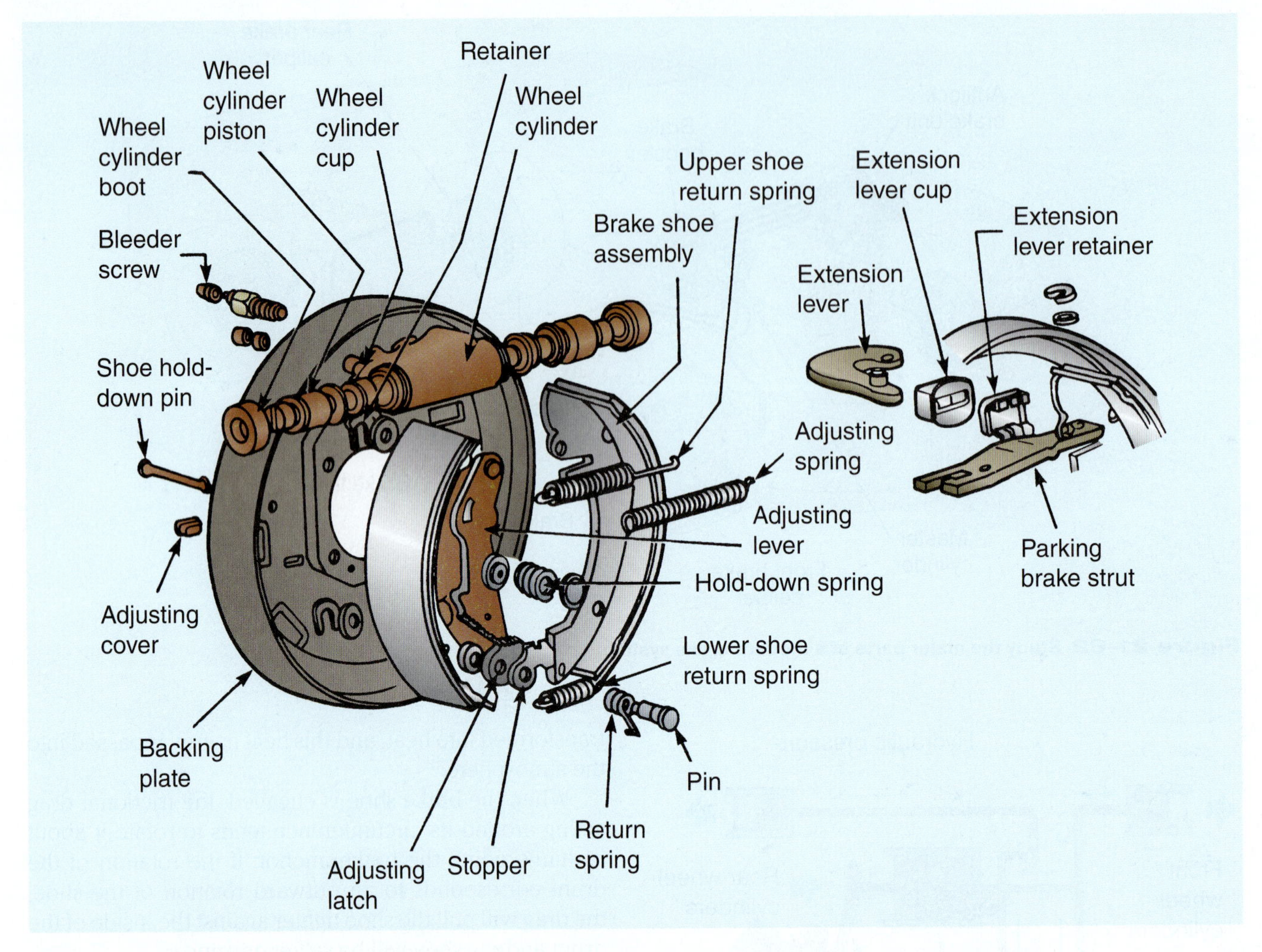

Figure 21-64 Study the parts of a rear drum brake assembly. They are sometimes used on rear wheels.

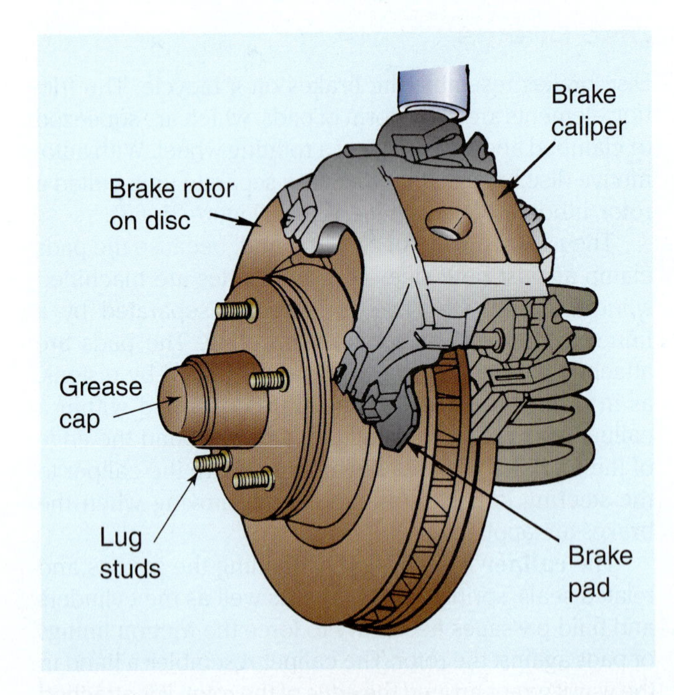

Figure 21-65 The typical disc brake has a caliper that clamps around the rotor. Front and rear disc brakes are now common because of their superior stopping power over drum brakes.

pushes out the piston for the necessary amount, then retracts it just enough to pull the pad off the rotor.

Brake Pad Service

To replace worn brake pads on a floating caliper, first loosen the lug bolts. Place the vehicle on jack stands and remove the wheels and tires. Before caliper removal, use a large C-clamp to push the piston back into its cylinder. Then the piston will be retracted and out of the way, allowing the new, thicker pads to fit into the caliper.

Unbolt and slide the caliper off the disc. To prevent brake hose damage, hang the caliper by a piece of mechanic's wire if the caliper is not to be removed (Figure 21–66).

Remove the old pads. Install anti-rattle clips on the new pads, and fit them back into the caliper. Slide the caliper assembly over the disc. Assemble the caliper mounting hardware in reverse order of disassembly. Make sure all bolts are torqued properly. Install wheels and tighten wheel lug nuts to specs. Repeat these operations on the other disc brake assemblies as needed.

Reassemble the disc brake in the opposite order of disassembly. After installing the brake disc, fit the caliper

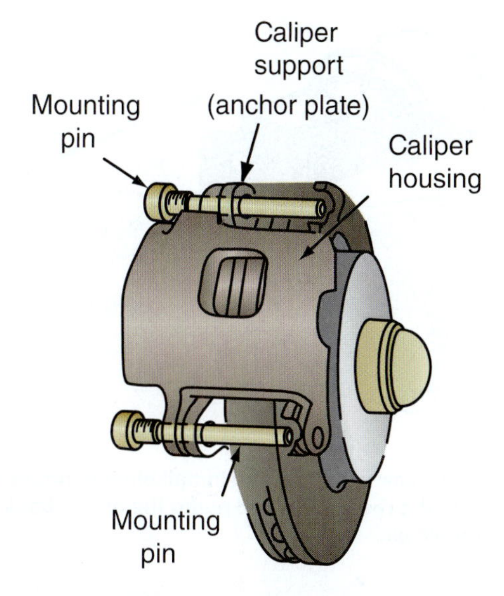

Figure 21–66 Two large bolts normally secure the brake caliper to the knuckle. Caliper removal is necessary for rotor and CV-axle removal. Hang the caliper on a piece of wire so that you do not damage the brake hose.

assembly into place. Make sure the new pads are properly installed. Torque all fasteners to specs.

Master Cylinder

The *master cylinder* is the heart of the hydraulic system. It is located in the engine compartment, usually on the driver's side, and is connected to the brake pedal by a special rod. The master cylinder initiates braking when the brake pedal is depressed by pushing out a piston inside the cylinder, exerting pressure that is transferred through the system. To protect against total failure of the system, all cars are now required to have two hydraulic systems (Figure 21–67).

The master cylinder must be checked before the car is put back in service. To check the fluid level, clean all dirt and grease off the unit, then simply pop the wire bracket (or whatever locking device is on the top) and remove the lid. The level should not be more than ¼ inch (6.3 mm) below the top of the reservoir. If the level is below this, check the brake line connections and refill the reservoir with fluid. If the system is leaking anywhere but at the brake line connections, the master cylinder should be replaced or rebuilt by a brake specialist.

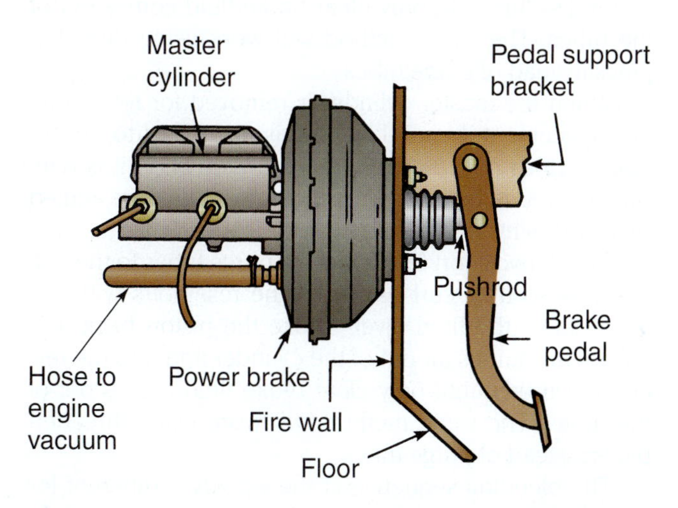

Figure 21–67 The master cylinder normally bolts to the firewall. Note the part names.

Never use the master cylinder and other brake mounts as a pulling attachment. If a pull must be made to correct cowl damage, use a plate and different bolts to anchor the pull.

Brake Fluid

When brake fluid absorbs moisture, its boiling point is drastically reduced. This effect is even more pronounced in high-temperature brake fluids that are used in heavy-duty and disc brake service.

To prevent contamination, strictly observe the following precautions when handling brake fluid:

- Keep the master cylinder tightly covered.
- Always recap it immediately after filling.
- Use the smallest possible can of fluid, and use it all if possible. For instance, if you have a choice between using two small cans or a portion of a large can, use the two small ones.
- Tightly cap the fluid container after use.
- If using a pressure brake bleeder, keep its fluid reservoir tightly closed, just like the master cylinder.
- If any fluid has become contaminated, throw it out.
- Do not reuse old brake fluid.
- Do not reuse an old brake fluid container, because it is not possible to know what else might have been in the can.
- Do not transfer brake fluid from its original container to anything other than a container specifically designed to hold brake fluid, such as a pressure bleeder.

Brake Lines

The *brake lines* carry fluid pressure between the master cylinder and wheel cylinders and related parts. They are generally the major brake component that a body technician must repair.

When making a collision inspection, check the brake lines for chafing, crimps, loose or missing tube clips, kinks, dents, and leakage. Leaks are evidenced by fluid seepage at the connections or stains around hose ends. Blockages are not so readily apparent but are just as detrimental to brake system functioning, often acting as a check valve to prevent proper release of the brakes. During a brake application, the pressure forces the fluid past the obstruction, but when the pressure is relaxed, the fluid will not readily flow back past the blockage, and the brakes drag. Brake lines are usually steel, except where they have to flex—between the chassis and the front wheels, and the chassis and the rear axle. At these locations, flexible hoses are used.

When replacing damaged brake lines, use the same type of material as the original factory installation. This includes stainless steel, armor plate tubing, or ribbed hose. Local availability of special types might be limited, but it is important to try to match the factory materials. Never use a weaker material to make a brake line or catastrophic brake failure may result.

Most cars use a double flare connection, so check for details carefully. Do not use compression fittings in brake line repairs. Replace all supporting clamps removed during the repair. Support springs prevent kinking and serve a very important role. Be sure to replace them just as they were, and install new ones if the original ones are damaged. Always replace brake lines in the original routing to avoid later damage to the lines.

Most brake hoses have a male fitting on one end and a female fitting on the other. Disconnect the female end first, remove the clip or jam nut holding it down, then unscrew the male end. Install the new hose by connecting the male end. If a copper gasket was used, make sure it is reinstalled. When the male end is tight, connect the female end. Tighten it in such a way as to keep the hose from touching any part of the chassis or suspension.

Check for interference during suspension deflection and rebound and turning of the front wheels.

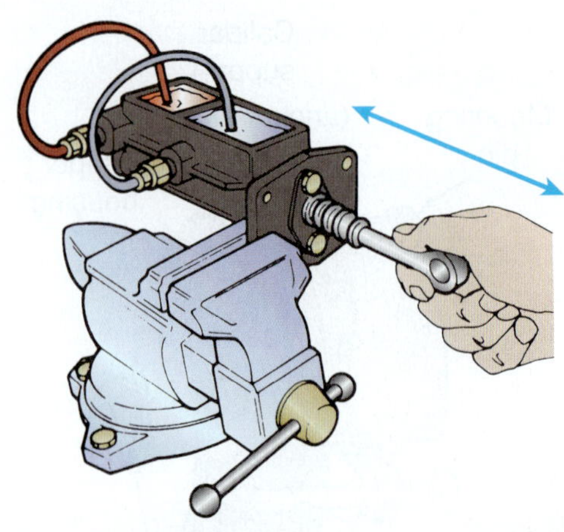

A **Fill the reservoir with fluid and install bleeder hoses from the outlets to the reservoir. Then pump the piston back and forth to remove air.**

B **Here the technician is using a vacuum and a large container to bleed the brake lines after repairs.**

Figure 21–68 **Before installing a new master cylinder, bleed it.**

Bleeding Brakes

To remove or replace a brake component, follow the instructions in the service manual. Remember that any time the brake system is open, it must be bled. Keep the system open for as short a time as possible to prevent moisture from entering and causing sludge and corrosion.

Bleeding removes air from the brake system. Air is lighter than liquid, and it seeks high points in the hydraulic system. Bleeder screws are provided at each of these collecting points: calipers, wheel cylinders, and on some master cylinders. Bleeding involves opening up these screws in a specific order to let the trapped air escape. Fluid is added to the master cylinder to replace whatever is lost in bleeding.

When the master cylinder is removed for rebuilding or replacement, bench bleeding is necessary to ensure that air does not remain in the cylinder when it is reinstalled. Mount the cylinder in a vise with the bore angled slightly downward (Figure 21–68),

Attach two short brake lines or purge tubes to the outlet ports so they curl back into the reservoirs with the ends below the fluid level. Stroke the piston back and forth. This pumps air out of the cylinder and into the reservoir. Do this until only clear brake fluid comes out of the tubes. The same method will work using threaded plugs instead of purge tubes.

When the master cylinder is removed for rebuilding or replacement, bench bleeding is necessary to ensure that air does not remain in the cylinder when it is reinstalled. Mount the cylinder in a vise with the bore angled slightly downward (Figure 21–68).

Attach two short brake lines or purge tubes to the outlet ports so they curl back into the reservoirs with the ends below the fluid level. Stroke the piston back and forth. This pumps air out of the cylinder and into the reservoir. Do this until only clear brake fluid comes out of the tubes. The same method will work using threaded plugs instead of purge tubes.

The bleeding sequence at the wheels is different for dual front/rear systems than for dual diagonal systems. In addition, each manufacturer might have a preferred sequence for any given model design. Check the service manual for each vehicle.

Some four-piston calipers have two bleeder screws. In this case, bleed the lower one first. On diagonal systems, bleed one system at a time. Do one front disc brake first, then the diagonally connected rear drum.

Always check the master cylinder first. If the brake fluid falls below the level of the intake ports, air will get into the system. Refill the reservoir and pump the brake pedal slowly a number of times. Often this will purge it of all unwanted air; if this does not work, bleed the system.

When bleeding modern antilock brake systems, refer to the service manual. It will give the detailed instructions needed to do good work.

Manual Bleeding of Brakes

Manual bleeding should be done only if a pressure bleeder is not available. Begin at the master cylinder. Clean the cover before removing it and the diaphragm gasket. Fill the reservoir to ¼ inch (6.3 mm) from the top. Apply pressure to the brake pedal slowly and with a smooth action. Open the bleeder screw on the first wheel in the sequence. Drain the aerated fluid through the bleeder hose into a jar partially filled with clean brake fluid (Figure 21–69).

Keep up pedal pressure while the bleeder screw is open. When the pedal bottoms out, close the screw and release the pedal. If all the air is not yet purged and air bubbles can be seen in the fluid, repeat the process. When you see only clear fluid with no bubbles, go on to the next wheel in the sequence.

While bleeding the brakes, watch the fluid level in the reservoir. After about every six pedal applications, more fluid will have to be added so it does not fall below the level of the intake port. If it does, more air will enter into the system.

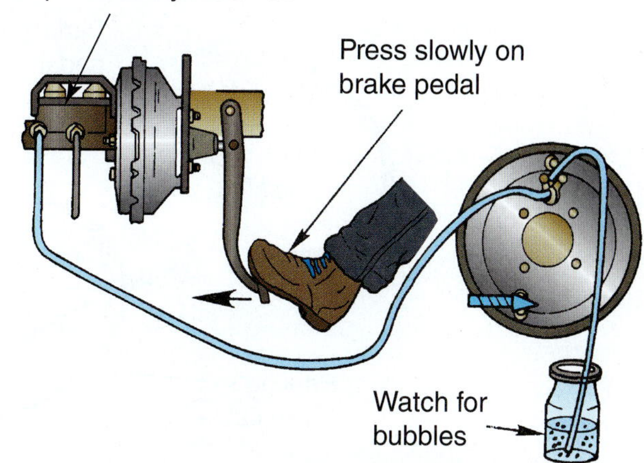

A To manually bleed brakes, open the wheel cylinder bleeder screw. Connect a hose from the screw to a container of brake fluid. By pumping the brake pedal, you will force air from the system. Keep the reservoir full.

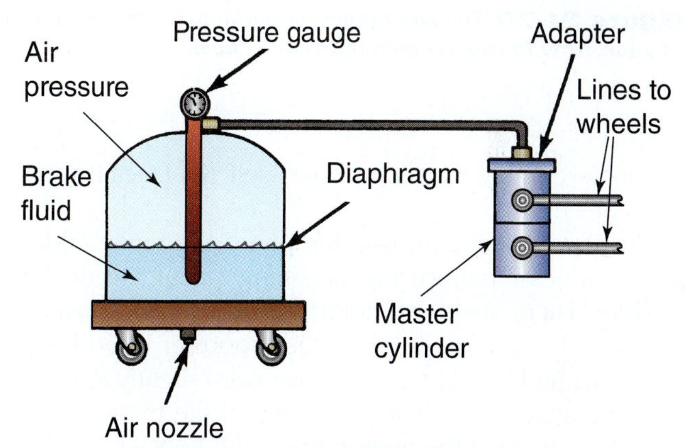

B Pressure bleeding is needed with some brake systems. Refer to vehicle-specific service instructions when in doubt.

Figure 21–69 Whenever a brake system hose or line has been disconnected for repairs, you must bleed air out of the hydraulic system.

Pressure Bleeding Brakes

This procedure is sometimes used to rid a hydraulic system of air. It is the most efficient bleeding procedure, requiring only one person to perform it.

The pressure unit used in this process is a tank divided into two sections by a flexible diaphragm. Bring the pressure unit up to 15–20 psi (103–138 kPa). Make sure the master cylinder cover is clean so that no loose particles of dirt fall into the reservoir. Remove the gasket and clean the gasket seat. Fill the reservoir and attach the adapter cap and hose. Check the coupling sleeve to make sure it is fully engaged before opening the fluid supply valve.

Follow the sequence for bleeding as recommended by the service manual. Allow the aerated fluid to flow out of the bleeder screws through a short bleeder hose into a jar. Once completed, close the supply valve of the pressure unit.

The recommended method of pressure bleeding a brake system is with a vacuum-type bleeder. This technique withdraws the fluid from the system rather than pumping it. Check the connection pattern of the system's wheels. Some cars have the two front and two back wheels connected; some are crossed diagonally. Others combine the front two with one rear and the rear two with one front wheel. Always check the service manual.

Power Brakes

As mentioned earlier, power brakes are nothing more than a standard hydraulic brake system with a vacuum assist or booster unit between the pedal and the master cylinder to help activate the brakes.

When a unibody car is involved in a collision, the power brake booster should be carefully inspected. Pay particular attention to vacuum hoses, check valves,

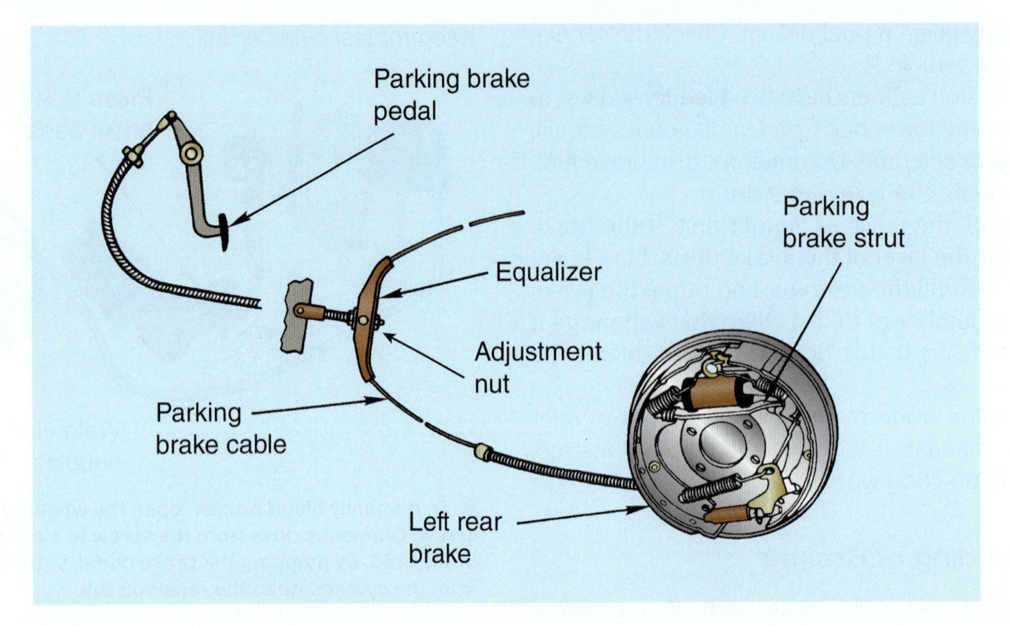

Figure 21-70 **The emergency, or parking, brake is simply a mechanical system of cables and levers that applies rear shoes of pads. Note the adjustment nut on the cable.**

fasteners, and the master cylinder itself. Replace all damaged pieces.

To test the vacuum booster, pump the brake pedal several times to remove any remaining vacuum inside the booster. Then, press and hold the brake pedal down as you start the engine. If the vacuum booster is working, the brake pedal will be pulled downward slightly as soon as the engine starts. If you cannot feel the brake pedal move down when the engine starts, the booster may be faulty.

Parking Brakes

The parking, or *emergency, brake* uses a steel cable to physically apply the rear brake shoes or pads. The rear wheel brakes act to hold the car stationary when not in use. Parking brakes are mechanical, not hydraulic. When you press down on the parking brake pedal or pull on the parking brake handle, you move a steel cable that mechanically applies the rear brakes, as shown in Figure 21–70.

To adjust the parking brakes, you must normally tighten a small nut on the parking brake cable assembly. This will shorten the cable assembly to apply the brakes more. Be careful not to tighten the cable adjustment too much or the parking brake may not release and could burn up the pads or shoes during driving.

21.5 COOLING SYSTEMS

A *cooling system* helps maintain the correct engine operating temperature. It is often damaged in a frontal collision and must be restored to its pre-accident condition. The basic parts of a cooling system are the radiator, water pump, and heater core (Figure 21–71).

The *radiator* transfers coolant heat to the outside air. The radiator pressure cap prevents the coolant from boiling. A radiator fan draws outside air through the radiator to remove heat.

The *water pump* circulates coolant through the inside of the engine, hoses, and radiator. The water jackets are passages in the engine for coolant. The thermostat regulates coolant flow and system operating temperature.

A heater system uses coolant heat and a **heater core** (a small radiator under the dash) to warm the passenger compartment. The automatic transmission cooler uses the radiator to reduce transmission fluid temperature.

Antifreeze is used to prevent freeze-up in cold weather and to lubricate moving parts. Antifreeze also prevents engine overheating. A coolant recovery system stores an extra supply of coolant for the system.

Coolant Service

One of the most frequently overlooked areas of the cooling system is the strength of the antifreeze. A common idea is that the stronger the concentration, the better. This is not so. Pure water transfers heat better than pure antifreeze, but it does not protect the system from freezing or corrosion.

In addition, water has a boiling point of only 212°F. Pure antifreeze has a higher boiling point (330°F) than pure water, but, due to its lack of heat transferability, it can cause an engine to overheat.

The ideal antifreeze-to-water ratio is 50:50. This ratio provides freezing protection to –34°F, while increasing

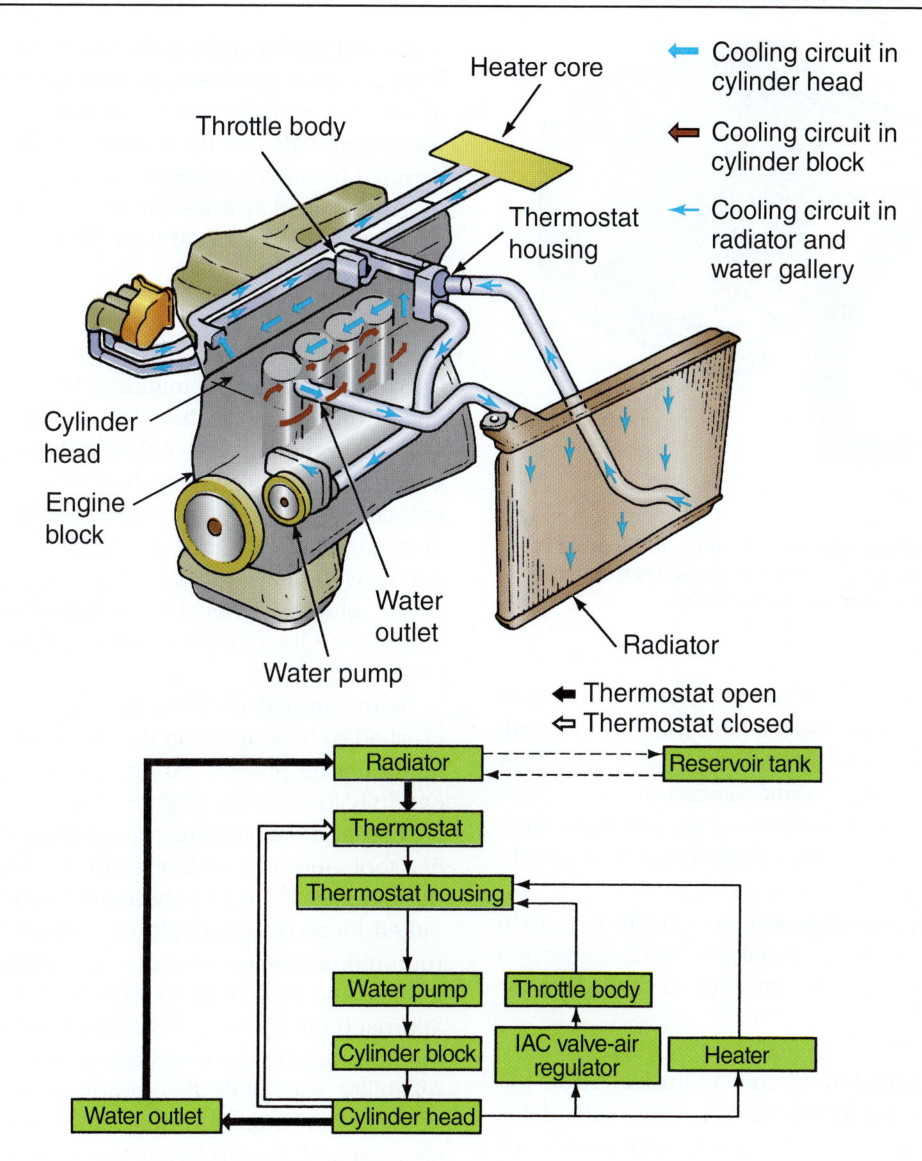

Figure 21-71 Study the cooling system components and flow. A belt-driven water pump forces coolant through the internal passages in the engine as well as through the hoses, radiator, and heater core.

the coolant's boiling point to 224°F. A combination of half water–half antifreeze also provides the best mix for preventing system corrosion.

An antifreeze tester, commonly called a hydrometer, is used to determine the freeze-up protection of the coolant mixture. Pull a sample of the vehicle's coolant solution into the tester. Then, read the lowest temperature the coolant will withstand without freezing. Add more coolant, if needed.

Do not let antifreeze, brake fluid, or other chemicals drip on painted surfaces. They can discolor or damage the vehicle's paint or finish!

Some systems, such as mid-engine cars and vans and those having dual heaters, can require up to 3 or 4 gallons of antifreeze. Check the manufacturer's recommended coolant capacity and change schedule for the specific vehicle.

Coolant Leaks

Coolant leakage often results from frontal collisions. A low coolant level reduces cooling capacity and can cause engine overheating.

An external coolant leak will cause coolant and steam to spray out of any leakage point after the engine warms to operating temperature. External leakage is often due to a split hose, damaged radiator, crushed water pump, or similar troubles.

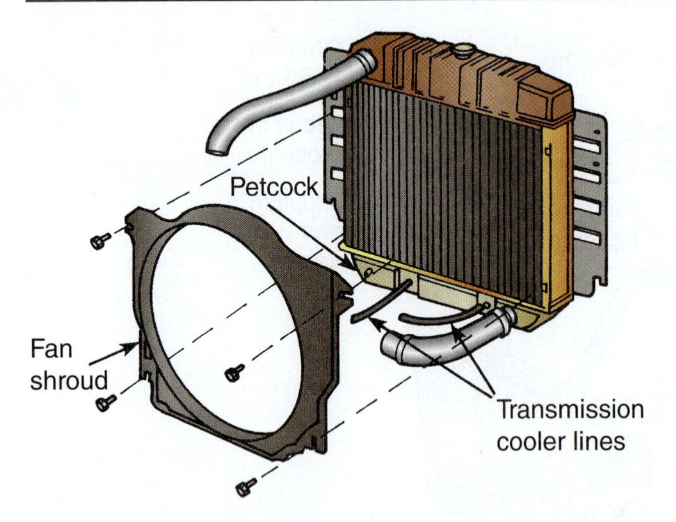

Figure 21-72 Transmission cooler lines often attach to the lower tank on the radiator. The fan shroud is frequently cracked and broken during a frontal collision.

Another type of coolant leak is internal and occurs when the automatic transmission fluid (ATF) cooler tank leaks from the inside into the radiator (Figure 21–72). A sure sign of this is a thick, pink solution in the radiator caused by the mixing of coolant and transmission fluid. Due to the impact forces in a collision, this area should be a high priority in any post-repair inspection.

Internal cooling system leaks can result from a blown head gasket, leaking intake manifold gasket, or warped cylinder heads. Internal leaks are seldom caused by collision damage.

Another type of internal leak occurs when the automatic transmission fluid (ATF) cooler tank leaks from the inside into the radiator (Figure 21–72). A sure sign of this is a thick, pink solution in the radiator caused by the mixing of coolant and transmission fluid. Due to the impact forces in a collision, this area should be a high priority in any post-repair inspection.

A **coolant recovery bottle** is normally plastic and can be easily damaged. Check for cracks or abrasions in the bottle, and make sure the hose leading to the radiator is connected and in good shape. These plastic tanks are normally not repaired. If cracked or distorted, replace the recovery bottle with a new or salvage unit.

Before making any replacements in the cooling system, drain the coolant from the system. Properly dispose of any old coolant (check local regulations). When refilling the system, make sure that the proper coolant is used in vehicles with aluminum engines or radiators. Some warranties will not be honored if the coolant recommended by the manufacturer is not used.

After installation of the engine coolant, bleed the cooling system of air. Always follow the manufacturer's recommendations.

SHOP TALK

If the vehicle is not equipped with a mechanism for bleeding air, you might be able to jack up the front of the vehicle, which raises the fill point or upper radiator hose higher than the rest of the system. This will allow the system to be completely filled and bled of all air.

Radiator Construction

Coolant flows from the engine to tubes located inside the fins of the radiator, where the airflow cools it. If these tubes become plugged, either by being bent or through maintenance neglect, the flow of coolant through the radiator is reduced and engine overheating can result. This condition is more noticeable at highway speeds and/or with heavier loads. If the vehicle is not air conditioned, plugged areas of the radiator can be identified by cold spots felt on the front of the radiator after the vehicle is warmed up.

Many frontal collision-damaged vehicles will have crushed or bent areas on the radiator even though leaks might not be present. So always check the radiator fins carefully for damage (Figure 21–73).

Crushed radiator fins can be straightened with a special tool, and tubes that are not too badly mangled can be soldered. But if large hunks of cooling fins have been pulled loose or if multiple tubes have been crushed or ruptured, a new radiator core is recommended. Time is money, and if the cost to repair core damage begins to approach the cost of a new core, most shops will opt for the new radiator core. Besides, a new core offers greater reliability, especially if the radiator is showing its age. Radiators are normally sent out to a specialty radiator shop for repairs or to have a new core installed.

There are two basic types of radiator designs. They are distinguished from one another by the direction of the coolant flow and the location of the two tanks.

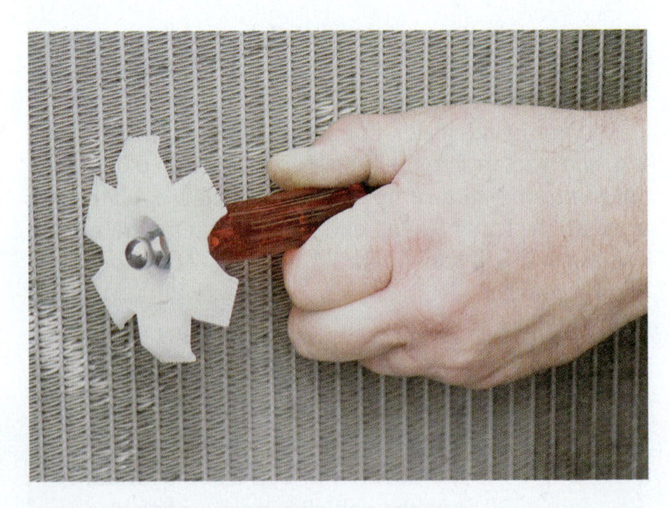

Figure 21-73 The radiator is often damaged in a frontal collision. If it is not leaking, you can sometimes straighten the soft metal fins on the radiator core with a stiff plastic comb.

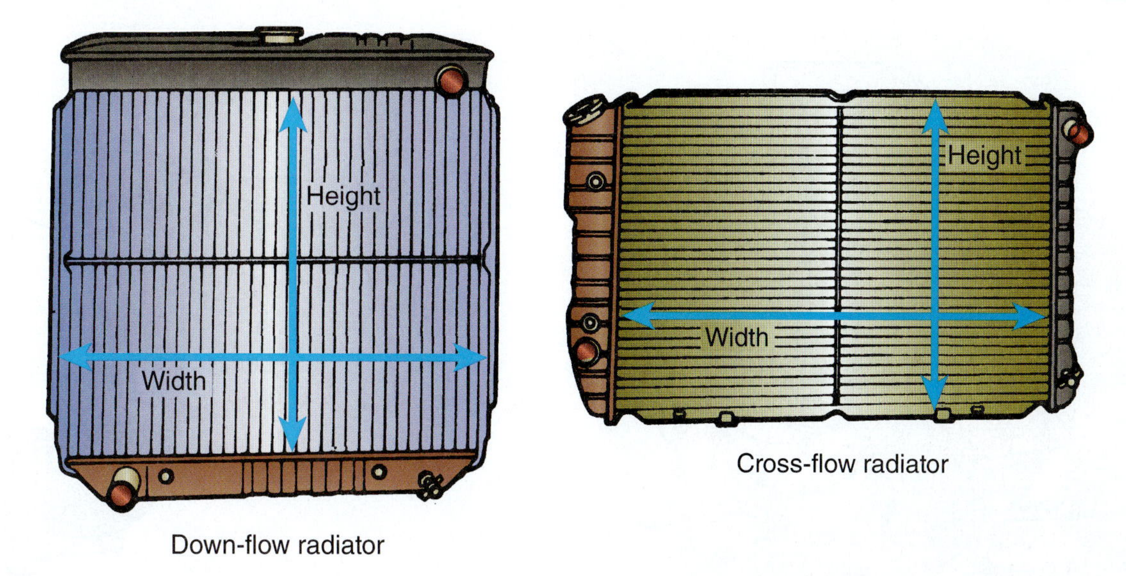

Figure 21–74 Measuring a radiator will allow you to order a new one. Note the tank locations on down-flow versus cross-flow designs.

In the *down-flow radiator*, the coolant flows from the top tank downward to the bottom tank. In the *cross-flow radiator*, the tanks are located at either side, and the coolant flows across the radiator core from tank to tank (Figure 21–74).

Replacing a Radiator

When replacing a radiator, make sure the new or salvage radiator is the same size as the old one. Radiator height, length, and thickness must be the same.

Be careful not to hit and damage the new radiator when lowering it down into place. The core tubes are very thin and can leak if hit even lightly (Figure 21–75).

Use care in reconnecting the transmission lines, because they are easy to cross thread. Align the transmission line straight into the radiator fitting and then hand-start the fitting threads. Once they are started by hand, properly tighten the transmission cooling lines.

Figure 21–75 Be very careful when working with the radiator, condenser, and oil coolers. Their fins and the core are very soft and can be easily damaged with rough handling.

Thermostat

The **thermostat** controls the engine's operating temperature by controlling coolant flow through the system. When the engine is cold, the thermostat stays closed to keep the coolant from circulating through the radiator. Once the proper engine operating temperature (usually 180°–195°F) has been reached, the thermostat opens up, allowing the coolant to flow to the radiator for cooling.

A stuck thermostat can cause the engine to overheat or run too cool. A quick way to tell whether this is happening is to feel the upper radiator hose. If it is cold but the engine is hot, the thermostat is probably stuck closed. On the other hand, if the thermostat is stuck open, the engine will take a long time, or possibly fail, to reach proper operating temperature.

Water Pump

The *water pump* circulates the coolant throughout the cooling system. It has internal blades called *impellers* that push the fluid through the system. The major parts of a water pump are shown in Figure 21–76.

A bad water pump will normally have a worn or damaged impeller shaft or shaft bushing. The water pump shaft will be loose in its housing and will make noise. Quite often with a damaged water pump, coolant will leak out a small hole in the bottom of the water pump housing.

Cooling System Hoses

While working on the cooling system during collision repair, check that the hoses are in good condition and are securely clamped. The lower radiator hose routes coolant

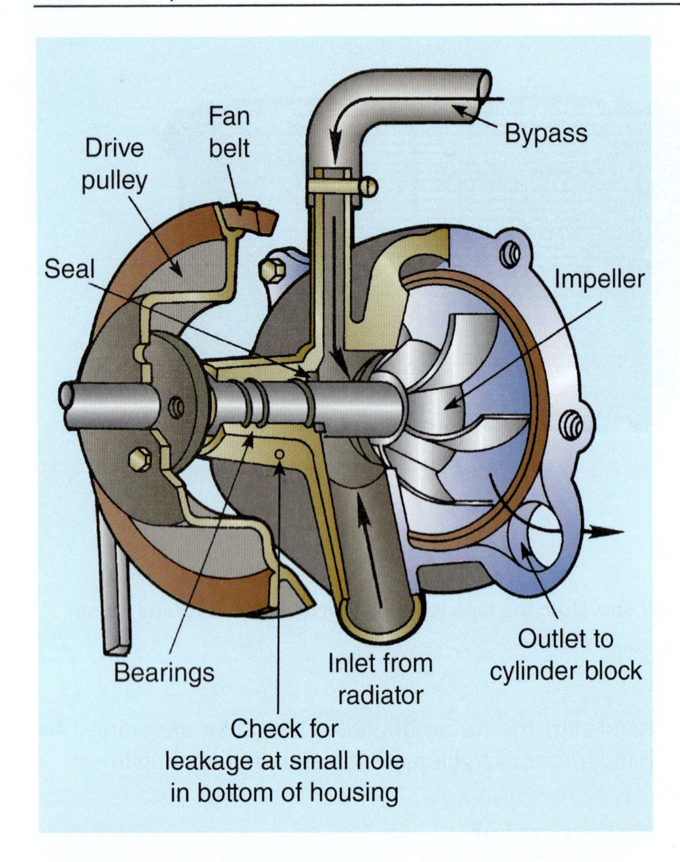

Figure 21–76 Note the basic parts of a water pump. The housing can be cracked or the bushing damaged by frontal impact. Coolant leakage out of a small hole in the bottom of the housing means that the bearings and seals are bad and water pump replacement is required.

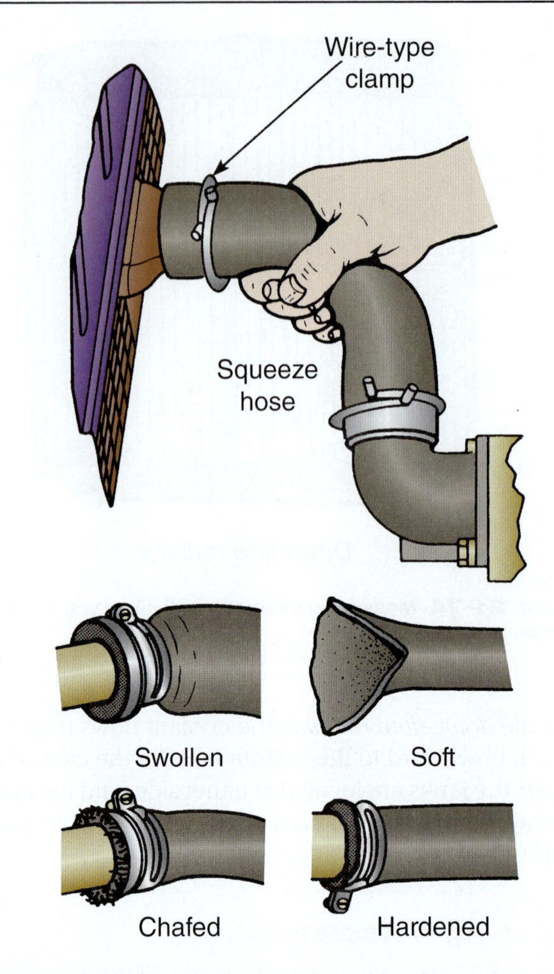

Figure 21–77 Always check hoses for damage or deterioration. Hoses should not feel too hard or too soft when squeezed.

from the radiator to the water pump. The turning water pump draws coolant, creating a low-pressure area in the lower hose. A coiled spring inside the lower hose keeps it from collapsing. The lower hose should not show signs of collapse during engine operation. Refer to Figure 21–77.

Cooling System Belts

Always make sure engine belts are not cut or damaged before installation. During a frontal collision, metal parts can smash into spinning belts. This can cut the edges of the belts, requiring their replacement.

If it does not use a spring tensioner, you must adjust belt tension during engine belt installation. Always refer to factory instructions for adjusting belt tension. Belt tension gauges are available that will measure the tightness of the belt.

Generally, the belts should be as loose as possible without slipping. This will place minimal load on the water pump or other bearings to avoid bearing damage.

Cooling System Fans

On vehicles in which the cooling fan is driven by belts from the engine, proper tension must be maintained or belt slippage will occur. When this happens, the cooling fan does not turn at full speed, resulting in reduced airflow through the radiator at idle. The engine may overheat at idle. At highway speeds, the airflow is sufficient to maintain cooling.

On vehicles with electric fans, check for loose electrical connections and bare, burnt, or cut wires. Check to make sure the fan blades turn with no interference and that the fan mounts are not rubbing against the radiator or body. On some cooling systems, there might be two or more fans. Look at Figure 21–78.

Some vehicles are equipped with a special clutch fan. Fan clutches either can be filled with a fluid or use a thermostat spring. The clutch allows the fan to slip at highway speeds when the fan is not needed for added airflow. This reduces drag on the engine, resulting in increased fuel economy.

Always make sure a clutch fan is working. Allow the engine to warm up, and check to be sure that the fan kicks in and starts blowing air through the radiator. If a fan clutch does not work, it could have been damaged during the collision. It might be leaking fluid or have a broken thermostatic spring.

If the vehicle has an electric fan, make sure it turns on after the engine warms up.

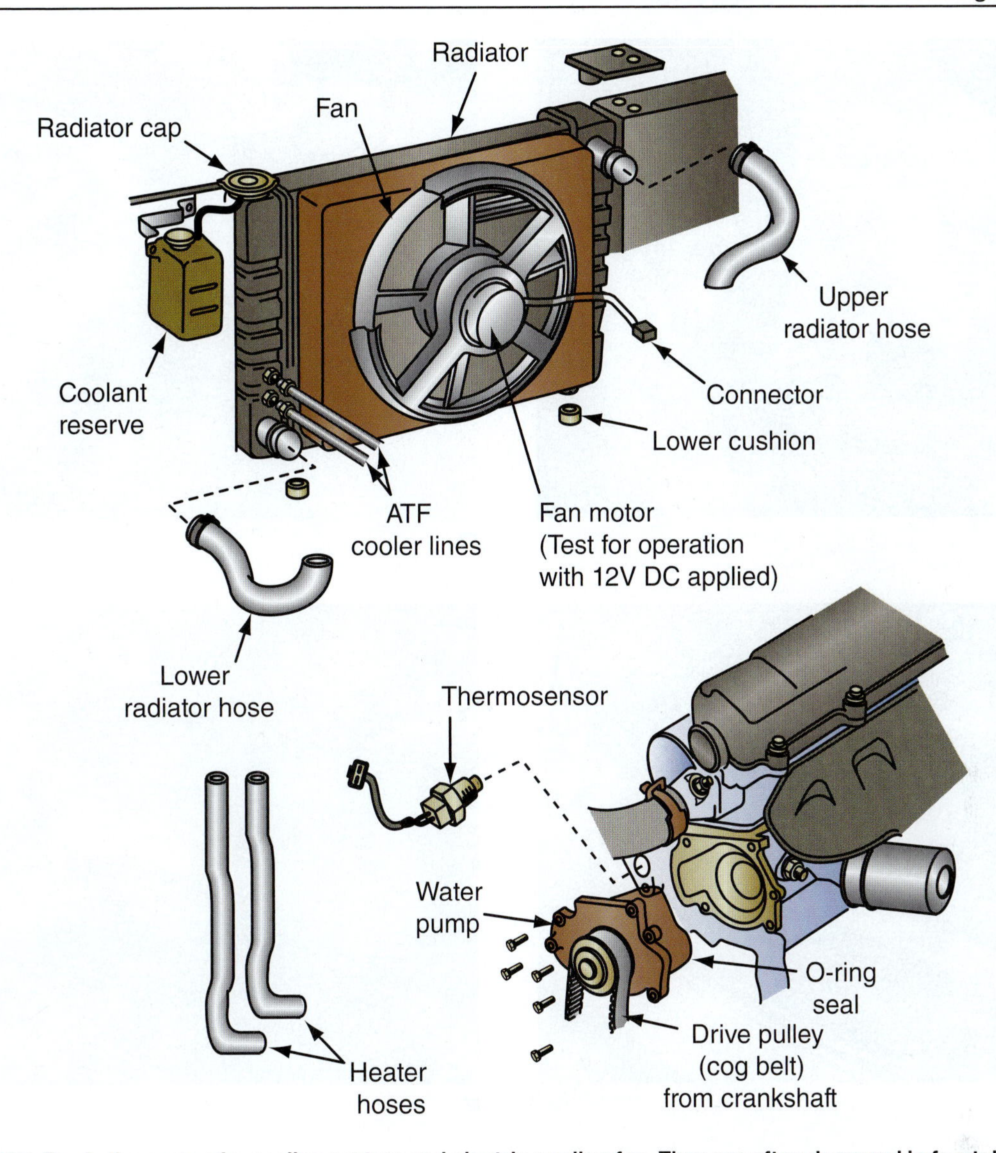

Figure 21–78 **Study the parts of a cooling system and electric cooling fan. They are often damaged in frontal collisions.**

DANGER Be careful when working around an electric cooling fan. It can suddenly turn on, even if the engine is turned off, and cause serious hand injury. Keep your hands away from the blade at all times. Disconnect the fan, if needed.

Radiator Cap Operation

Most cooling systems operate under a pressure of about 15 psi. This is because increased pressure on a liquid raises its boiling point. With a 50–50 mixture of antifreeze and water, a boiling point of 263°F is achieved. To maintain the correct pressure in the system, the radiator cap must be able to hold the required pressure.

A defective radiator cap lowers the boiling point of the coolant. The coolant could boil, even though the engine is not actually overheated. Dried calcium deposits in a radiator cap can make the radiator inoperative. Always inspect the radiator cap seal for deterioration or damage.

Pressure Testing Cooling System

A *cooling system leak test* is performed by installing a pressure tester on the radiator neck to check for coolant loss or leakage. Pump the tester handle until its gauge equals the cap pressure rating. With the system pressurized, look around the radiator, hoses, water pump, heater hoses, and engine to find any coolant leakage. A loss of pressure or coolant leakage means there is a problem. Refer to Figure 21–79.

The radiator cap pressure rating is stamped on the cap. If the cooling system is disassembled during repair or parts are replaced, a pressure test should be performed on the cap.

A *radiator cap pressure test* makes sure the cap will hold the rated pressure but release above normal pressure

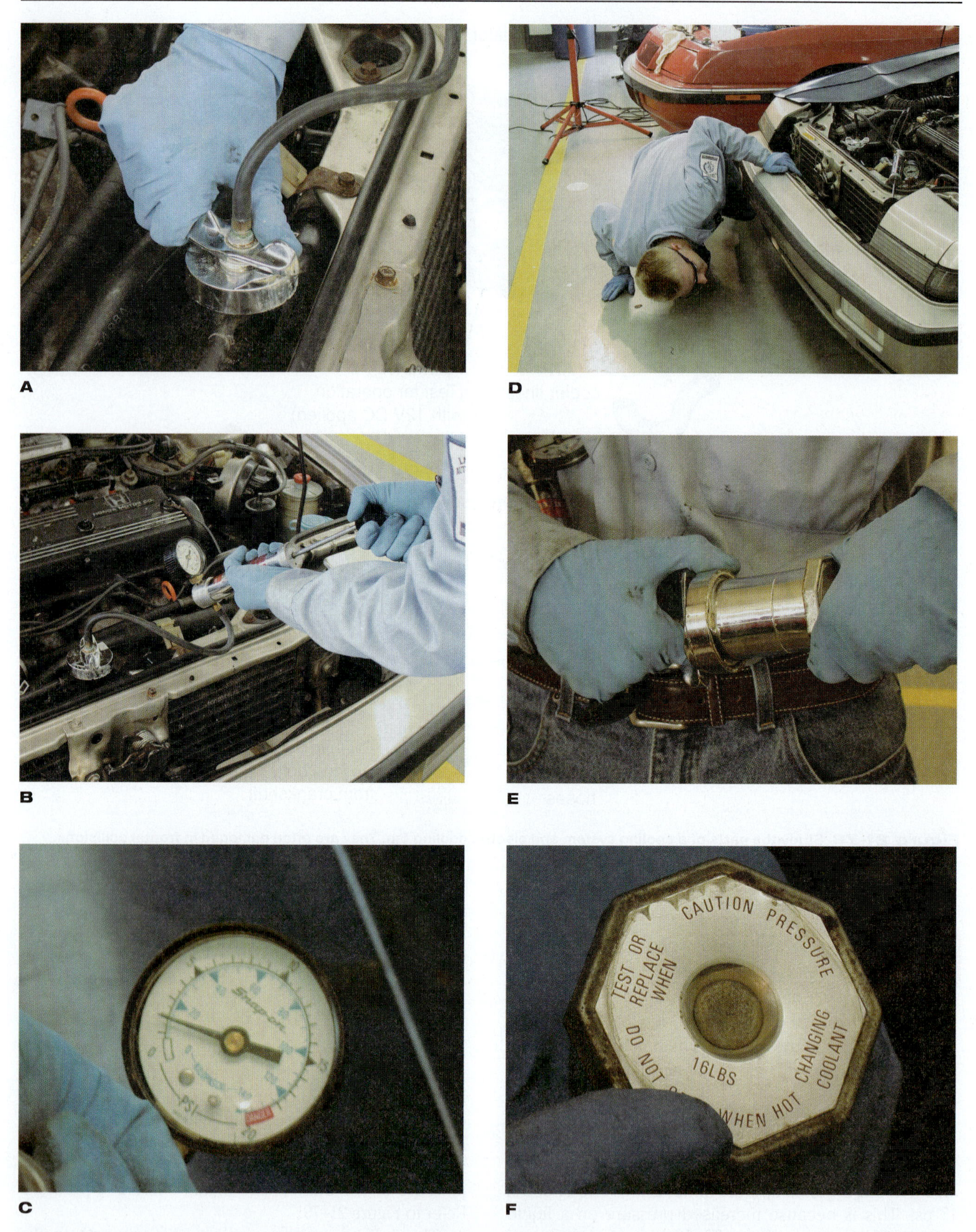

Figure 21-79 A pressure tester should be used to check the cooling system after repairs. (A) To check for coolant leaks, install a tester on the radiator filler neck. (B) Pump until the gauge reads at the pressure cap rating. (C) The gauge should hold pressure without the needle dropping. (D) With pressure in the system, look for leakage or loss of coolant under the engine and radiator. (E) Test the radiator cap if needed. Install the cap on the adapter and pump the handle of the tester. (F) Cap should build specified pressure and then release pressure at the cap rating. Cap pressure is stamped or printed on the top of the cap.

as needed. Use the tester adapter to mount the radiator cap on the pressure tester. Refer to Figure 21–79B.

Pump the tester handle while watching the pressure gauge. The tester gauge should stop increasing the pressure reading when the cap rating is reached. If the cap leaks or does not open at its rated pressure, replace it.

21.6 HEATER OPERATION

A *heater* uses warm engine coolant flowing through a heater core to warm the passenger compartment. The heater core is a miniature version of the radiator mounted under the dash or on the firewall (Figure 21–80).

As hot coolant flows through the heater core, a fan blows air over the tubes, warming it and delivering it to the passenger compartment. The blower fan, located in the heater housing, forces air through the heater core and into the passenger compartment.

The air heating distributor system is a duct system. Outside air enters the system through a grille, usually located directly in front of the windshield, and goes into a plenum chamber where rain, snow, and some dirt are separated from it. The air from the plenum is directed through the car's heater core, through the A/C evaporator, or into a duct that runs across the firewall of the car. Outlets in the duct direct the airflow into the passenger compartment.

21.7 AIR-CONDITIONING SYSTEMS

Proper handling of the air-conditioning (A/C) system during collision repair is one of the most important and least understood aspects of working with mechanical components. Many needless repairs are caused unknowingly by service technicians who do not understand the importance of following some strict rules for working with A/C systems. What compounds the problem is that malfunctions often occur several months after the collision repair work is completed, so the customer is unaware of who caused the problem.

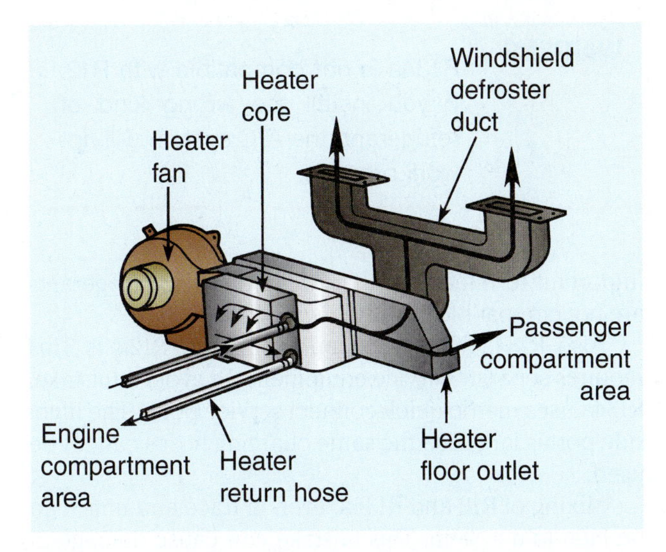

Figure 21-80 The heater core is a small radiator that transfers coolant heat to warm the passenger compartment. A fan blows air through the core, ducts, and vents.

A/C Operation

An *air-conditioning system* is designed to cool the passenger compartment. Although designs vary somewhat, all automotive A/C systems use similar operating principles. The major parts of a modern A/C system are shown in Figure 21–81.

An *air-conditioning compressor* is an engine-driven pump that forces the refrigerant through the system. Lines and hoses carry the refrigerant to the different parts of the system.

The *condenser* transfers refrigerant heat to the outside air. It is normally located in front of the radiator. It is constructed like a radiator, with core tubes and metal fins.

The *evaporator* draws warm air out of the passenger compartment to provide a cooling effect. It is normally located under the dash. It is also like a small radiator.

A *receiver/drier* or an accumulator uses a desiccant bag to remove moisture from the system. The only difference between them is their location. The accumulator is found between the evaporator and compressor. The receiver/drier is located between the condenser and expansion device. Both act as storage tanks.

An *expansion valve*, or orifice tube, causes the refrigerant pressure to drop to produce a cooling action in the evaporator.

Air-conditioning systems are divided into two sides: high and low. The dividing points are the compressor and the expansion device.

A/C High- and Low-Sides

The *A/C high-side* contains high-pressure/high-temperature refrigerant. Its hoses feel hot to the touch. High-side hoses are generally smaller in diameter than low-side hoses (Figure 21–82).

The *A/C low-side* contains low-pressure/low-temperature refrigerant. Its hoses feel cold to the touch. Low-side hoses are generally larger in diameter than the high-side hoses.

In the basic A/C system, the heat is absorbed and transferred in the following six steps:

1. Refrigerant leaves the compressor as a high-pressure, high-temperature vapor.
2. By removing heat via the condenser, the vapor becomes a high-pressure, lower-temperature liquid.
3. Moisture and contaminants are removed by the receiver/dryer, where the cleaned refrigerant is stored until it is needed.
4. The thermostatic expansion valve converts the high-pressure liquid into a low-pressure liquid by controlling its flow into the evaporator.

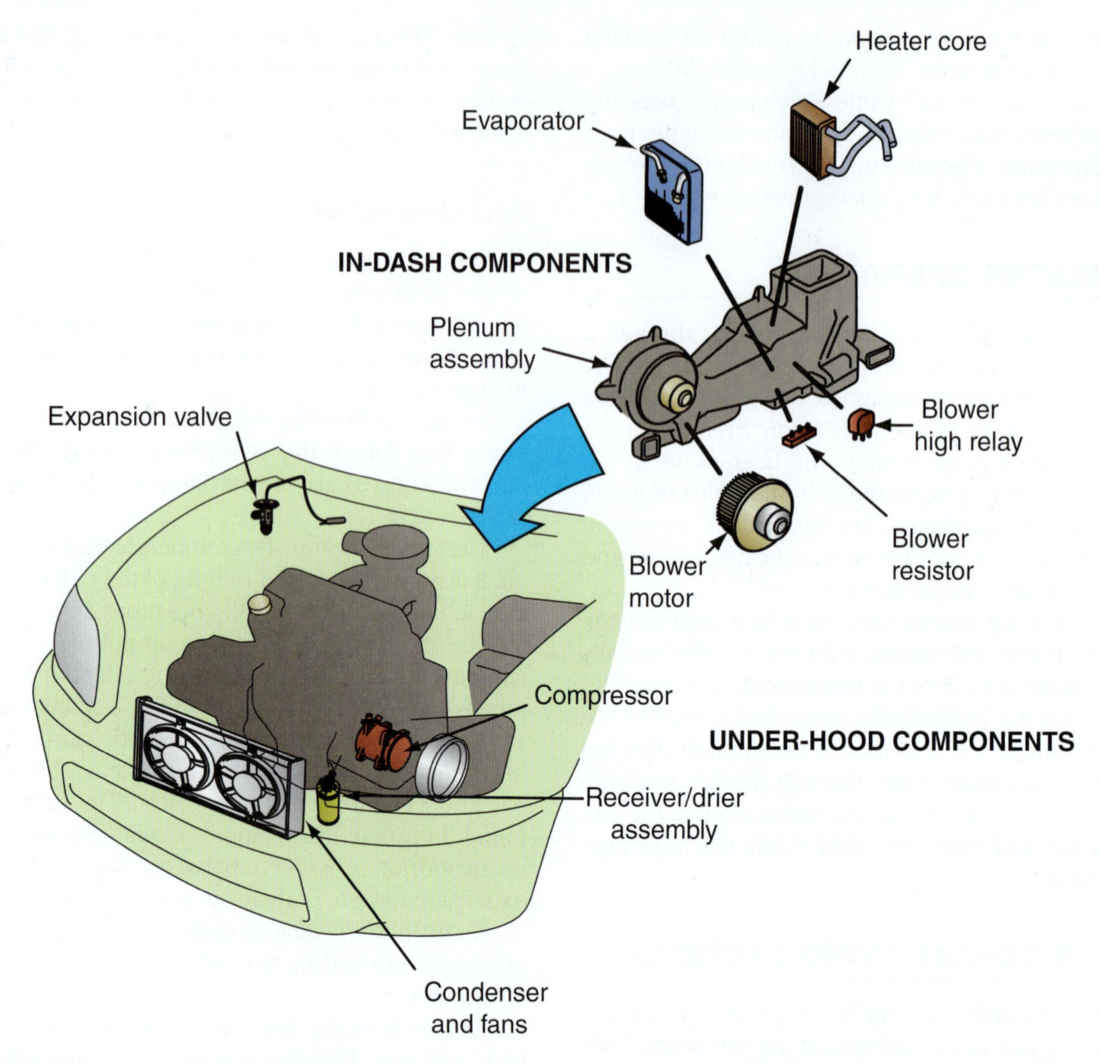

Figure 21-81 Study the major parts of an air-conditioning system. An engine-driven compressor forces refrigerant through the system. A condenser transfers heat to outside air. An evaporator collects heat to cool the air blowing into the passenger compartment.

5. Heat is absorbed from the air inside the passenger compartment by the low-pressure, low-temperature refrigerant, causing the liquid to vaporize.
6. The refrigerant returns to the compressor as a low-pressure, higher-temperature vapor.

Refrigerants

Due to its possible depleting effect on the Earth's ozone layer, R12 refrigerant is being phased out. Environmental regulations call for a gradual phase-out of most ozone-depleting substances. They allow R12 systems to be serviced using recovered and recycled refrigerant. After filtering, R12 can be used again in another vehicle. Such reuse is designed to extend the supply of refrigerant.

R134a is the present replacement refrigerant for R12. It is less harmful to the ozone layer. New vehicles are being designed to run on this new refrigerant. The compressor and other parts are designed to be used with R134a. It is important to remember that R134a and R12 refrigerants are not compatible. Refer to Figure 21–83.

R134a is *not* compatible with R12. If you install the wrong kind of refrigerant, the A/C system will not work properly.

Also, R134a oils are not compatible with R12 oils. This requires separate service equipment. To avoid a mistake, R134a uses metric quick-connect service ports. The high-side port is larger, so the same charging hoses cannot be used.

Mixing of R12 and R134a, even in trace amounts, can be fatal to a system. This mistake can cause damage to seals, bearings, compressor reed valves, and pistons. Mixing refrigerants can also cause desiccants used in R12 systems to break down and form harmful acids.

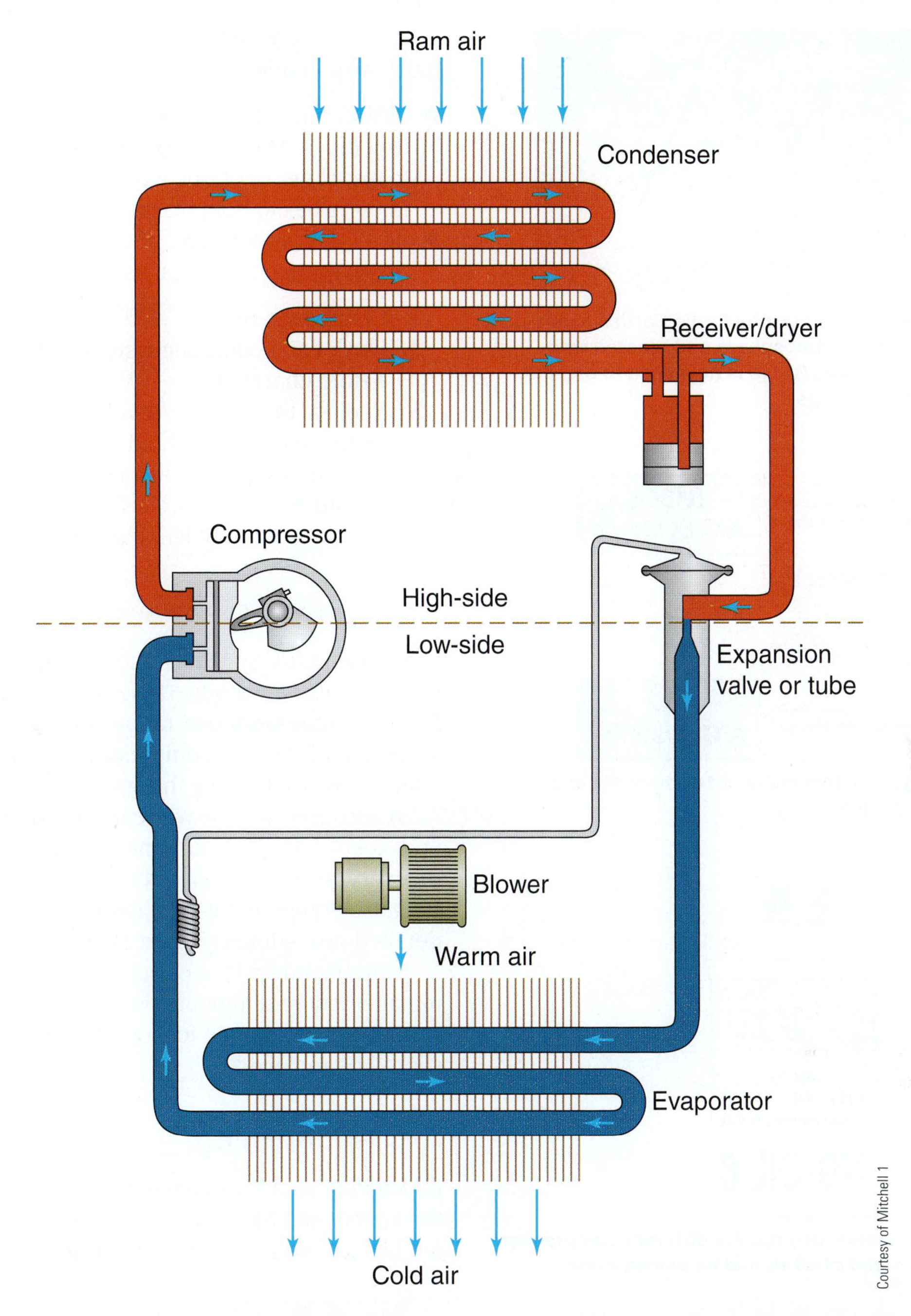

Figure 21-82 **Note the refrigerant flow cycle. High-side is from the outlet of the compressor to the refrigerant flow control device. Low-side is from the outlet of the flow control device (expansion valve) to the inlet of the compressor. Hoses and lines are large diameter on the low-side.**

Wear hand and face protection when working on an A/C system. When refrigerant escapes from the system or if you touch the supply tank, you can get severe frostbite burns.

An A/C system *sight glass* is used to check the amount of refrigerant in the system. It is often located on the receiver/drier or in a refrigerant line. Before starting collision repairs, you might want to start the system and inspect the sight glass. This will let you know whether the system has been damaged and has lost its refrigerant charge. Refer to Figure 21–84.

When looking through the A/C system sight glass, one of the following four conditions will be evident:

1. *Clear sight glass*—refrigerant completely full or completely empty
2. *Oil streaks on sight glass*—no refrigerant

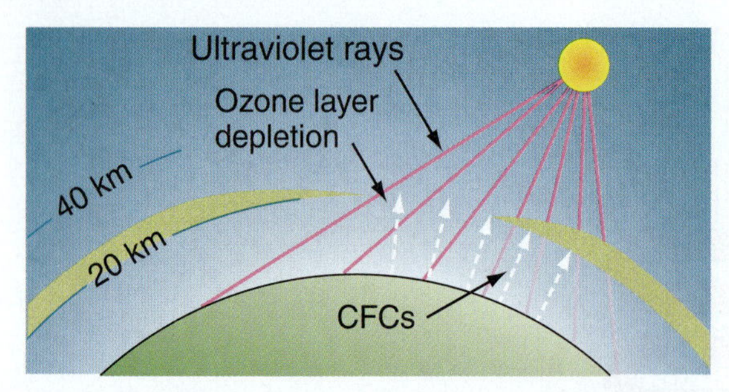

A Older R12 refrigerant was harmful to the Earth's ozone layer. It floated up into the atmosphere and reacted with ultraviolet rays from the sun. R134a is now used to prevent further environmental damage.

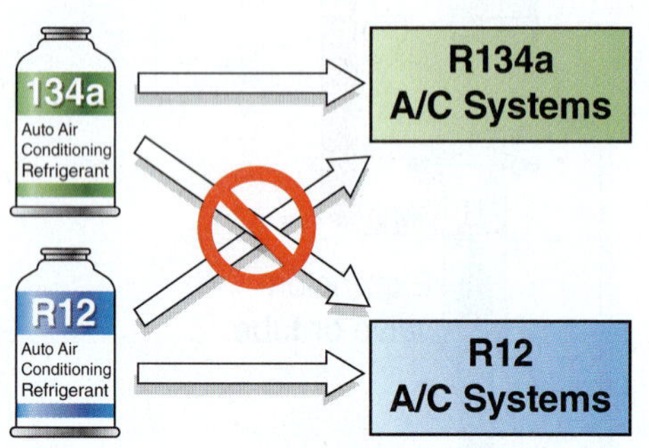

B Older R12 and modern R134a are not compatible and systems are designed differently.

C R12 and R134a systems also require different compressor oil. The specified amount of oil should be placed in the system as needed.

Figure 21-83 **Study the basic refrigerant information.**

3. *Foam or constant bubbles in sight glass*—low refrigerant charge
4. *Clouded sight glass*—desiccant being circulated through the system

A/C Service Tips

Although many A/C repairs and service jobs are done in specialty A/C shops, some larger body shops do basic service work. Many collision repair shops recover refrigerant, remove and install A/C parts, and recharge A/C systems.

Here are pointers that the collision repair technician must keep in mind:

- When removing or opening up the A/C unit, seal all openings. These can be synthetic rubber, tight-fitting caps, plugs, or plastic wraps. Use sturdy rubber bands or wire ties to hold plastic wraps in place securely.
- If an A/C system has been open to the atmosphere more than a few hours, use the following procedure:

1. Change the oil.
2. Flush each component separately with nitrogen gas before charging.
3. Replace the receiver/dryer or accumulator.
4. During evacuation, hold the system at high vacuum for a minimum of 30 minutes to pull out air and moisture.
5. Recharge without leaking refrigerant.

Refrigerant Discharging/Recovery

Discharging A/C systems remove refrigerant from the system and must always be done before parts are removed. Some compressors use a special back seating service valve that allows the compressor to be removed without completely discharging the system.

A *recovery system* will capture the used refrigerant and keep it from contaminating the atmosphere. Most recovery systems also filter the refrigerant for reuse. Because equipment varies, refer to the user's manual for detailed procedures (Figure 21–85).

Manufacturers' receiver/drier and accumulator replacement recommendations vary. Generally, if an A/C system has been open for several days, the receiver/drier should be replaced.

A/C Recharging

Evacuating an A/C system removes air and moisture from the system and allows you to check for leakage. Any time air has entered an A/C system, it must be evacuated.

The release of R12 into the atmosphere is prohibited by current environmental regulations. Never vent the refrigerant into open air. Use a recovery/recycling machine.

Evacuating is done by connecting a vacuum pump to the vehicle's A/C system. After pulling a vacuum, you close off the pump to see whether the system will hold the vacuum. If the vacuum drops, the system is leaking and should not be charged. Find and correct the leak first (Figure 21–86).

Before *charging* (filling) the system with refrigerant, determine the amount and type of refrigerant used. This

Item	Symptom	Amount of refrigerant	Remedy
1.	Bubbles present in sight glass.	Insufficient*	1. Check for gas leakage with gas leak tester and repair if necessary. 2. Add refrigerant until bubbles disappear.
2.	No bubbles present in sight glass.	None, sufficient, or too much	Refer to items 3 and 4.
3.	No temperature difference between compressor inlet and outlet.	Empty or nearly empty	1. Check for gas leakage with gas leak tester and repair if necessary. 2. Add refrigerant until bubbles disappear.
4.	Temperature between compressor inlet and outlet is noticeably different.	Correct or too much	Refer to items 5 and 6.
5.	Immediately after air conditioning is turned off, refrigerant in sight glass stays clear.	Too much	1. Recover refrigerant. 2. Evacuate air and charge proper amount of purified refrigerant.
6.	When air conditioning is turned off, refrigerant foams and then stays clear.	Correct	—

*Bubbles in the sight glass with ambient temperatures higher can be considered normal if cooling is sufficient.

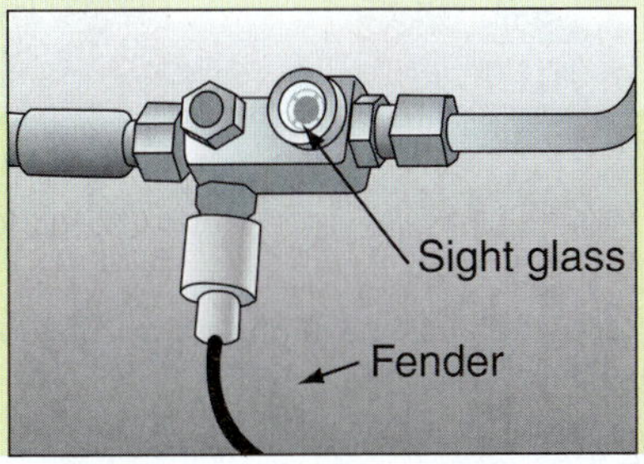

Sight glass

Fender

Figure 21-84 A sight glass will let you check the general charge in an A/C system. Run the engine at a fast idle and turn the A/C system on high cool while viewing the sight glass. Note symptoms and remedies in the chart.

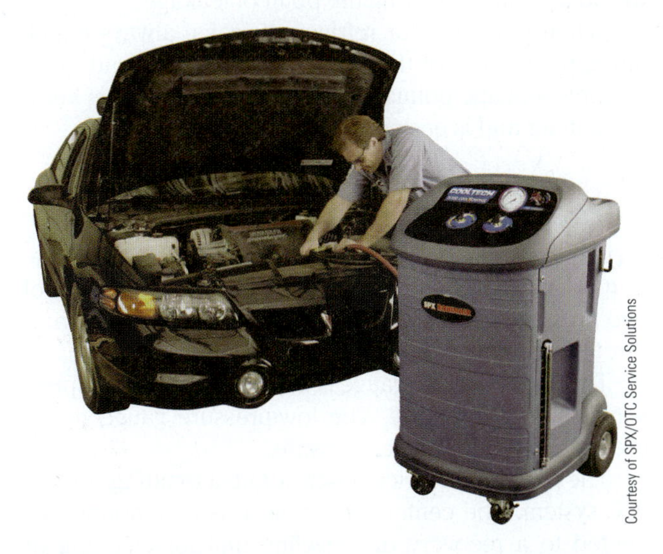

Courtesy of SPX/OTC Service Solutions

Figure 21-85 Here a body technician is removing old refrigerant from a system before disconnecting the lines and hoses. Never release refrigerant into the atmosphere.

information is found in the service manual or on the label on the radiator support or compressor. Do *not* mix different types of refrigerants. Charging can be done with a gauge set or with a charging station. Refer to Figure 21–87.

Purging uses refrigerant to push air and dirt out of the hoses. It prevents air and other contaminants from being pushed into the A/C system. Always purge the gauge hoses before charging.

Refrigerant oil lubricates moving parts in the A/C system. Use only refrigerant oil. Do not use any other type of oil. Make sure to use the type recommended for the system being serviced. For example, some oils are designed only to be used with specific types of refrigerant. Using a different type can result in damage to the compressor, seals, and other parts.

General rules are to add the amount of oil that was removed during discharge. There are adapters available to use refrigerant pressure to add oil during recharging.

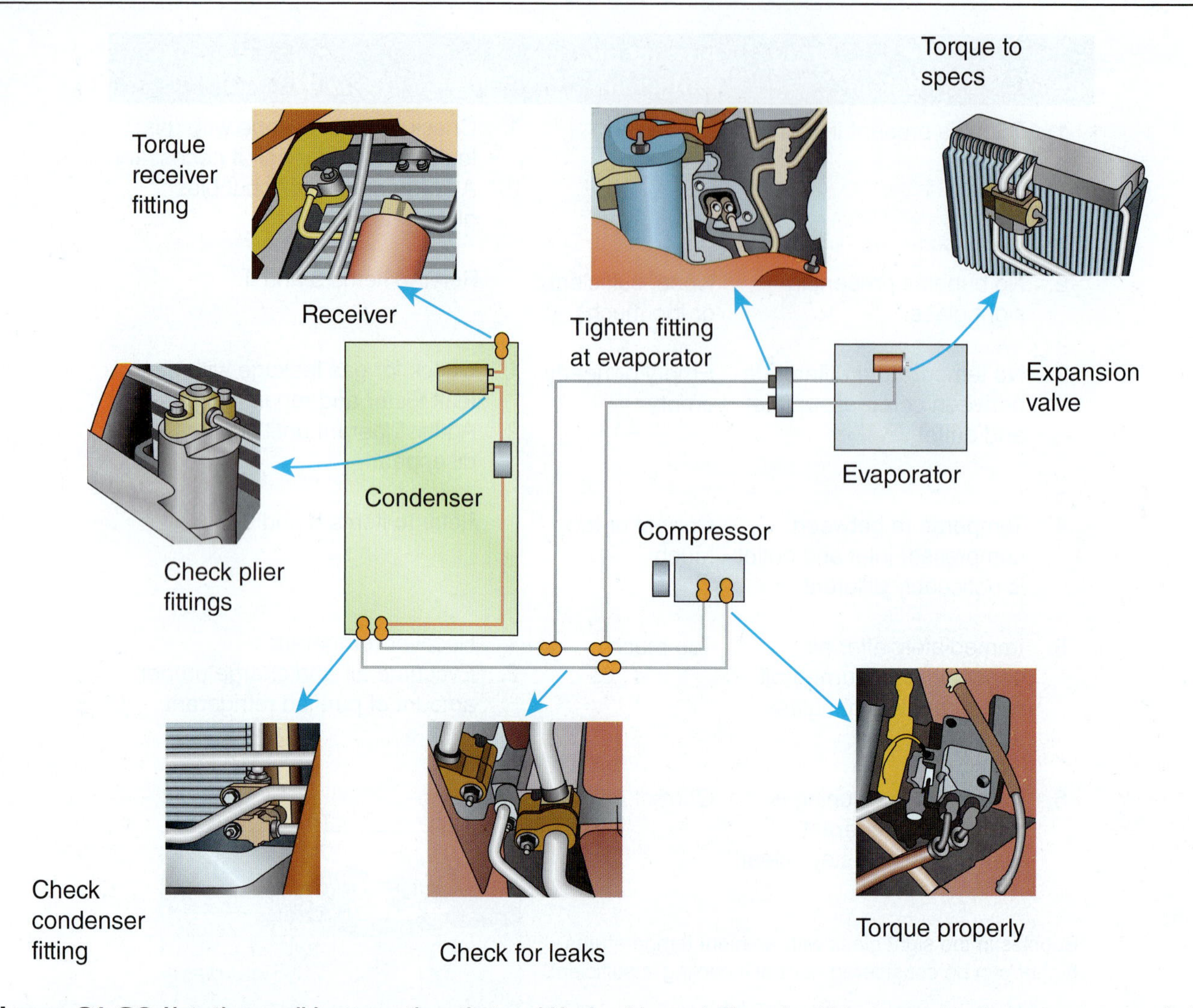

Figure 21-86 **Note the possible connections that could leak refrigerant. They should be torqued to factory-recommended specifications to avoid leakage.**

Too much oil causes reduced cooling, because the oil takes up space normally used by the refrigerant. It also damages the compressor and seals. Too little oil causes poor lubrication of the system, premature compressor wear, and poor system performance.

An open can of refrigerant oil can collect dirt and moisture. Adding contaminated refrigerant oil to the system causes corrosion, which results in the failure of the compressor and other parts.

There are three ways to find refrigerant leaks:

1. Electronic leak detector
2. Refrigerant cans with dye
3. Soap and water solution in a spray bottle

Electronic leak detectors are battery-operated instruments that use a sound to alert you to a gas leak. They are designed to detect different types of gases.

Refrigerant dye can be injected into the A/C system to help find leaks. The dye will stain any point of leakage and some dyes can be illuminated with a "black light," or ultraviolet light.

Soap and water is an old-fashioned method of finding refrigerant leaks. Any leaking gas will form bubbles in the soapy water, showing the point of leakage.

When checking for refrigerant leaks, always check along the bottom of the hoses, fittings, seals, and other possible leakage points, because the refrigerant is heavier than air and is easier to detect below these parts.

A/C Troubleshooting

A *pressure gauge set* is used to troubleshoot the operation of an A/C system. The gauge set typically consists of two pressure gauges, a manifold, two on–off valves, and three service hoses. One is shown in Figure 21–88.

The high-pressure gauge is used to measure compressor discharge pressure. The low-pressure gauge measures suction, or low-side, pressure.

The two outer service hoses connect to fittings on the A/C system. The center service hose is commonly connected to a recovery or recycling unit for cleaning or evacuating or to a refrigerant container for charging (filling) the system.

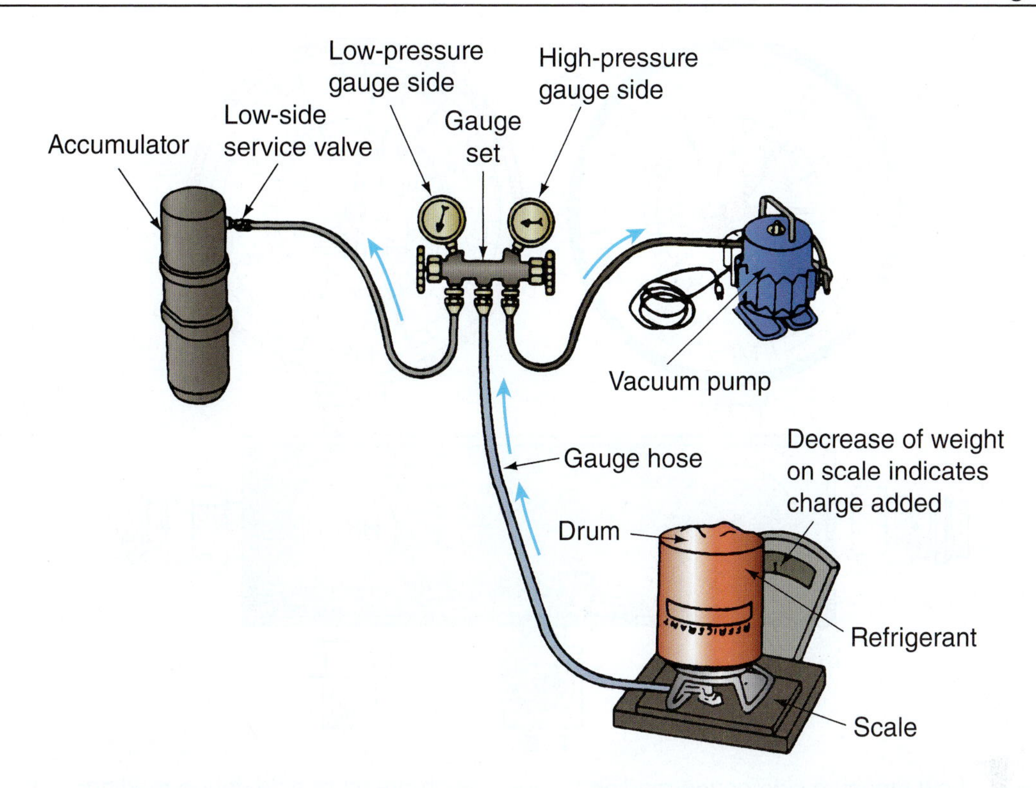

Figure 21-87 **This setup will allow evacuation and recharging. First, the gauge set valves are opened to allow the vacuum pump to pull all air out of the system. Then the gauge valve to the vacuum pump is closed to see whether the vacuum drops, which indicates leakage. Next, the valve to the vacuum pump is closed and the valve on the refrigerant tank is opened to allow a recommended amount of refrigerant to flow into the system. The scale will show how much refrigerant has been added.**

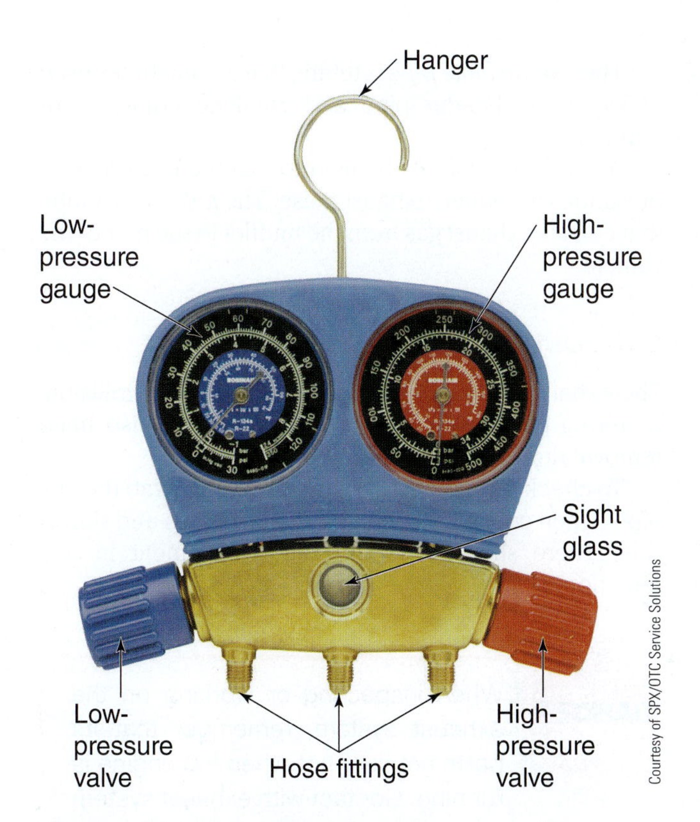

Figure 21-88 **Note the parts of the gauge set for servicing an R134a A/C system.**

A/C service valves provide a means of connecting the pressure gauge assembly for testing, discharging, evacuating, and charging (filling) the A/C system. Most systems have two service valves. The service valves may be located on the compressor fittings or in the refrigerant lines. The service valves for a system using R12 are different from those for a system using R134a. The R134a fittings are larger.

A static A/C pressure reading will indicate how much refrigerant is in the system. With the engine off, read the high-side pressure gauge. If the high-pressure gauge shows approximately 50 psi (345 kPa), then the system should have an adequate charge. If the pressure gauge reads below 50 psi (345 kPa), some of the refrigerant charge has leaked out and the system should *not* be operated. Correct any leak. Add refrigerant before making other tests. Refer to Figure 21–89.

An *A/C performance test* evaluates a system's condition by measuring system pressures with the engine running. Start and fast idle the engine at approximately 1,500 rpm. Set the system for maximum cooling for about 10 minutes to allow pressures to stabilize. Close the car's doors and windows. Leave the hood fully open.

Place a temperature gauge in one of the air outlets in the passenger compartment. Place another temperature gauge at the condenser to measure ambient (outside) air

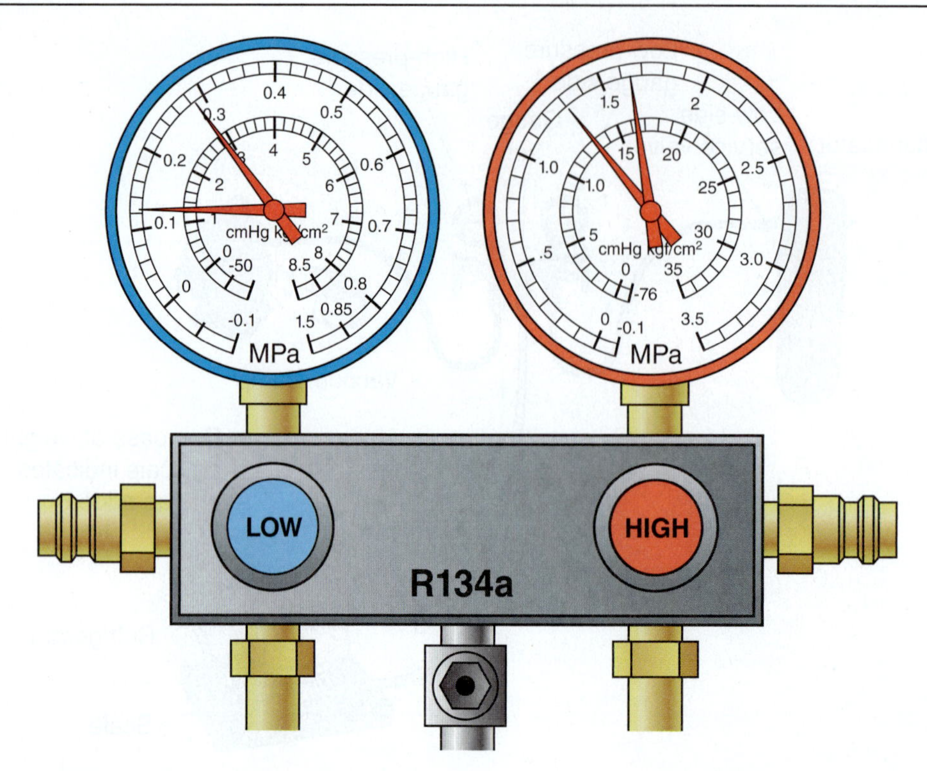

Low-pressure side gauge reading:
0.15–0.25 MPa (1.5–2.5 kgf/cm^2)

High-pressure side gauge reading:
1.37–1.57 MPa (14–16 kgf/cm^2)

Figure 21-89 These are typical gauge set readings for properly operating an R134a A/C system on a typical day, with the engine at fast idle and the A/C set to maximum cool.

temperature. Both temperatures are usually needed to analyze system performance.

Read the pressure gauges and compare them to factory specifications. Some typical A/C system pressure readings with possible causes and remedies are shown in Figure 21–90.

21.8 EXHAUST SYSTEMS

The **exhaust system** collects and discharges exhaust gases caused by the combustion of the air–fuel mixture within the engine. It also quiets the noise of the running engine. The major parts of an exhaust system are shown in Figure 21–91.

The *header pipe* is steel tubing that carries exhaust gases from the engine's exhaust manifold to the catalytic converter.

The *catalytic converter* is a thermal reactor for burning and chemically changing exhaust by-products into harmless gases (Figure 21–92).

Modern vehicles often have one or two catalytic converters. A small catalytic converter is used next to the engine exhaust manifold. It heats up rapidly and starts working before the main converter in the intermediate pipe.

The *intermediate pipe* is tubing that is sometimes used between the header pipe and catalytic converter or muffler.

A *muffler* is a metal chamber for dampening pressure pulsations to reduce exhaust noise. The *tailpipe* is a tube that carries exhaust gas from the muffler to the rear of the vehicle.

Exhaust System Service

The exhaust system can be damaged during a collision, requiring partial replacement. Its parts may also need removal during major collision repairs.

To check the exhaust system's condition, grab the tailpipe (when cool). Try to move it up and down and side to side. There should be only slight movement in any direction.

When inspecting or working on the exhaust system, remember that its parts get very hot when the engine is running. Contact with exhaust system parts can result in a severe burn.

Condition:
Periodically cools and then fails to cool

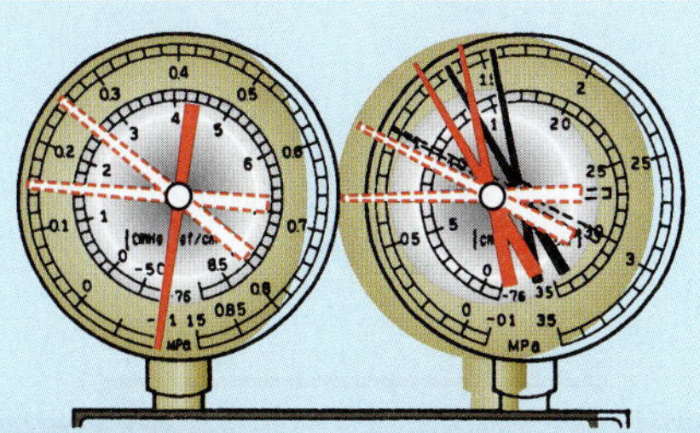

Symptom	Probable Cause	Diagnosis	Remedy
• During operation, pressure on low-pressure side sometimes becomes a vacuum and sometimes is normal	• Moisture entered in refrigeration system freezes at expansion valve orifice and temporarily stops cycle, but normal state is restored after a time when the ice melts	• Drier in oversaturated state ↓ • Moisture in refrigeration system freezes at expansion valve orifice and blocks circulation of refrigerant	1. Replace receiver-drier 2. Remove moisture in cycle through repeatedly evacuating air 3. Charge new refrigerant to proper amount

Condition:
Insufficient cooling

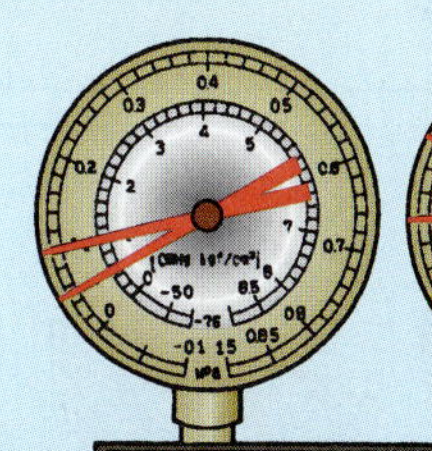

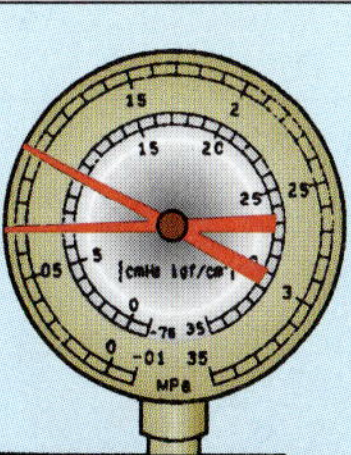

Symptom	Probable Cause	Diagnosis	Remedy
• Pressure low on both low- and high-pressure sides • Bubbles seen in sight glass continuously • Insufficient cooling performance	• Gas leakage at some place in refrigeration system	• Insufficient refrigerant in system ↓ • Refrigerant leaking	1. Check for gas leakage with leak detector and repair if necessary 2. Charge refrigerant to proper amount 3. If pressure indicated value is near 0 when connected to gauge, create the vacuum after inspecting and repairing the location of the leak

Condition:
Insufficient cooling

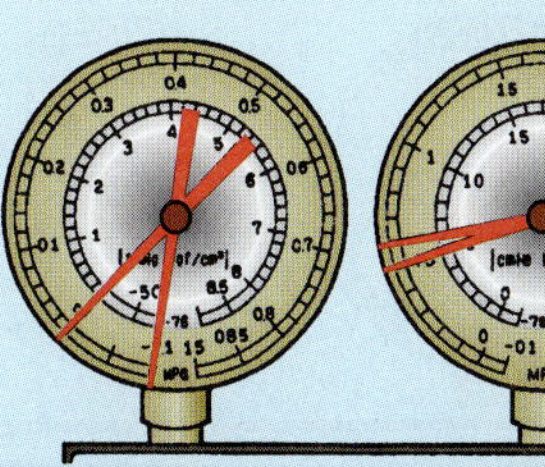

Symptom	Probable Cause	Diagnosis	Remedy
• Pressure low on both low- and high-pressure sides • Frost on tubes from receiver to unit	• Refrigerant flow obstructed by dirt in receiver	• Receiver clogged	• Replace receiver

Figure 21-90 **Study the typical gauge readings for common air-conditioning system problems or conditions.**

(continued)

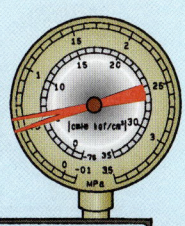

Condition:
Does not cool (cools from time to time in some cases)

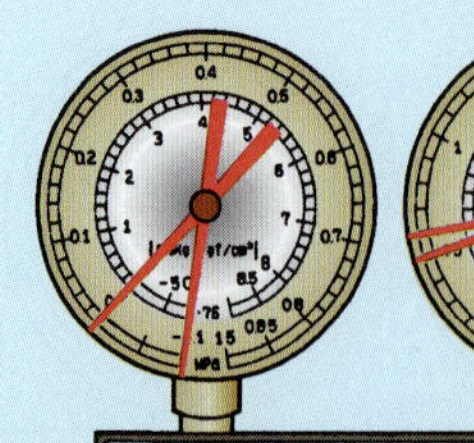
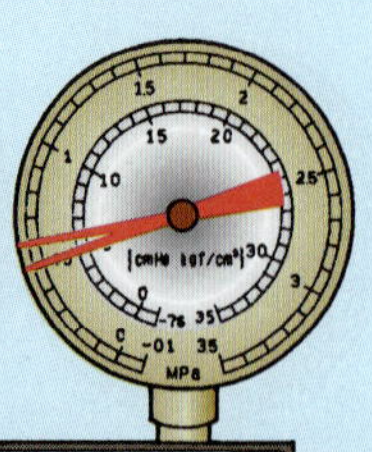

Symptom	Probable Cause	Diagnosis	Remedy
• Vacuum indicated on low-pressure side, very low pressure indicated on high-pressure side • Frost or dew seen on piping before and after receiver/drier or expansion valve	• Refrigerant flow obstructed by moisture or dirt in refrigeration system • Refrigerant flow obstructed by gas leakage from expansion valve heat sensing tube	• Refrigerant does not circulate	1. Check heat sensing tube, expansion valve, and EPR 2. Clean out dirt in expansion valve by blowing with air. If not able to remove dirt, replace expansion valve 3. Replace receiver 4. Evacuate air and charge new refrigerant to proper amount. For gas leakage from heat sensing tube, replace expansion valve

Condition:
Does not cool sufficiently

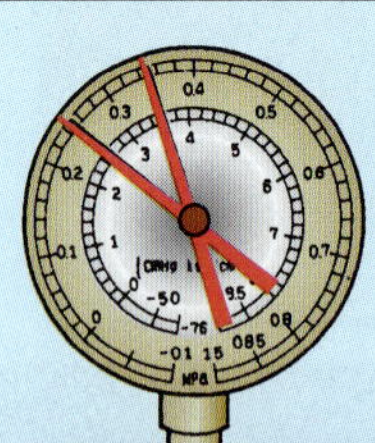
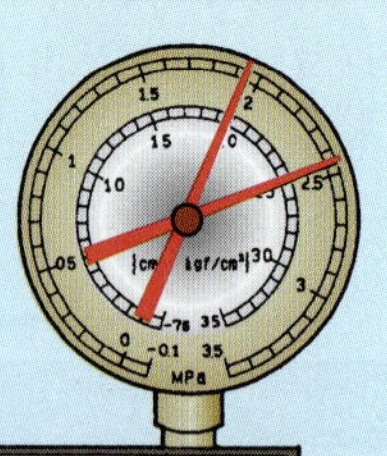

Symptom	Probable Cause	Diagnosis	Remedy
• Pressure too high on both low- and high-pressure sides • No air bubbles seen through the sight glass, even when the engine rpm is lowered	• Unable to develop sufficient performance due to excessive refrigerant in system • Insufficient cooling of condenser	• Excessive refrigerant in cycle → refrigerant overcharged • Condenser cooling insufficient → condenser fins clogged or fan motor faulty	1. Clean condenser 2. Check fan motor operation 3. If (1) and (2) are in normal state, check amount of refrigerant. Charge proper amount of refrigerant

Condition:
Does not cool down sufficiently

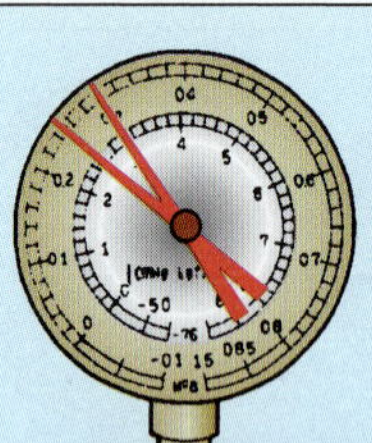
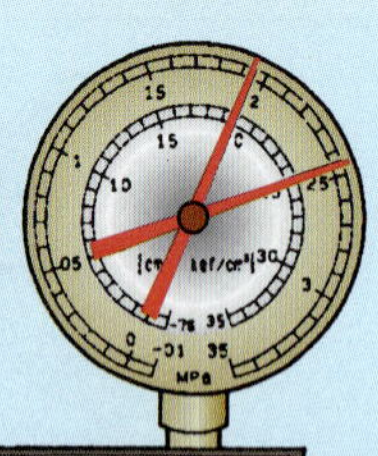

Note: These gauge indications are shown when the refrigeration system has been opened and the refrigerant charged without vacuum purging

Symptom	Probable Cause	Diagnosis	Remedy
• Pressure too high on both low- and high-pressure sides • The low-pressure piping is hot to the touch • Bubbles seen in sight glass	• Air entered in refrigeration system	• Air present in refrigeration system ↓ • Insufficient vacuum purging	1. Check compressor oil to see if dirty or insufficient 2. Evacuate air and charge new refrigerant

Figure 21-90 ***(Continued)***

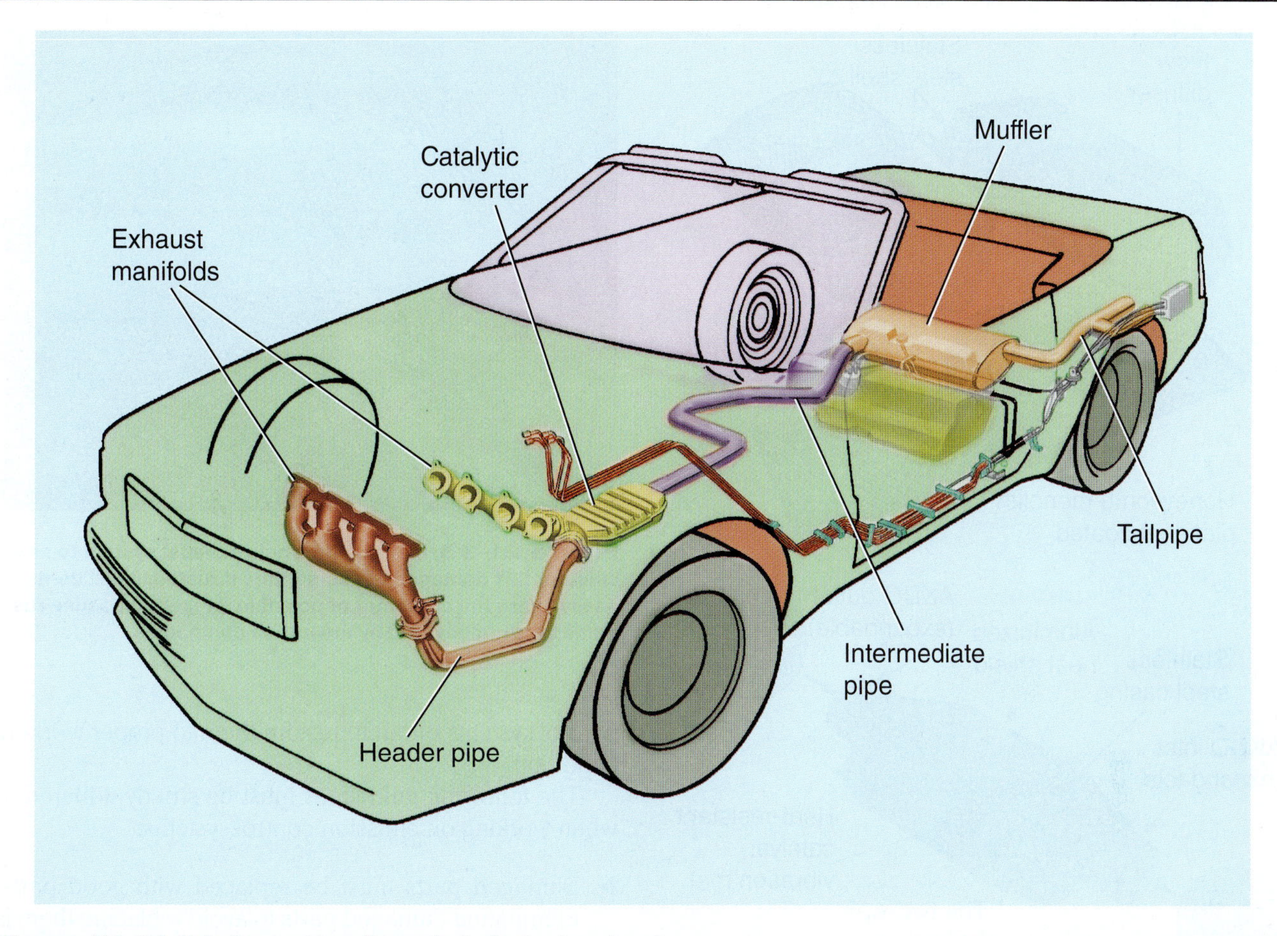

Figure 21-91 Note the part locations of a typical exhaust system.

To check further, start the engine (never in a closed shop). Stuff a rag in the tailpipe and feel around every joint for leaks. If one is found, try tightening the clamp. If this does not stop the leak, it must be repaired.

If necessary, raise the vehicle. Check the clamps and hangers that fasten the exhaust system to the underbody. Also, jab at all rusted areas in the system with an old screwdriver. If the blade sinks through the metal at any point, that part is badly rusted. You can also tap on parts with a hammer or mallet. A ringing sound indicates that the metal is good. A badly corroded part produces a dull thud from thinned metal.

If a loud ticking, clicking, or puffing sound can be heard, there is probably a large exhaust leak in the system. Make sure that fittings are not loose and leaking. If parts are damaged, loosen the clamps or fittings and separate each part for replacement (Figure 21–93).

Because of constant changes in recommended catalytic converter servicing and installation requirements, check with the vehicle manufacturer for the latest data regarding replacement.

21.9 EMISSION CONTROL SYSTEMS

Emission control systems are used to prevent potentially toxic chemicals from entering our atmosphere (Figure 21–94).

The *exhaust gas recirculation* (EGR) valve opens to allow the engine vacuum to siphon exhaust into the intake manifold and combustion chambers. The EGR valve consists of a poppet and a vacuum-actuated diaphragm. When vacuum is applied to the diaphragm, it lifts the poppet off its seat. Exhaust gas then flows back into the engine. The exhaust entering the combustion chambers lowers peak combustion temperatures. This reduces nitrogen oxide air pollution.

The *positive crankcase ventilation* (PCV) system channels engine crankcase blowby gases into the engine intake manifold. The gases are then drawn into the engine and burned. This prevents crankcase fumes from entering the atmosphere (Figure 21–95).

The *fuel evaporative system* pulls fumes from the gas tank and other fuel system parts into a charcoal canister. The charcoal canister absorbs and stores vaporized fuel (Figure 21–96). When the engine is started, these vapors are drawn into the engine and burned, which prevents this source of pollution from entering the Earth's atmosphere.

Emission Control System Service

Many times emission control systems are damaged in a collision and must be serviced as part of the repair. The Clean Air Act, which is a federal law, makes the body

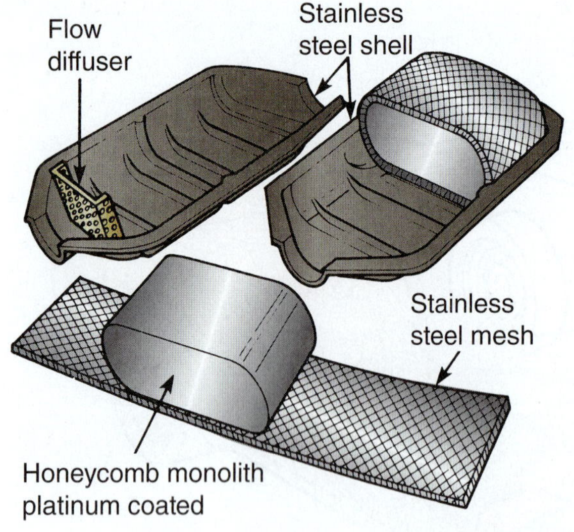

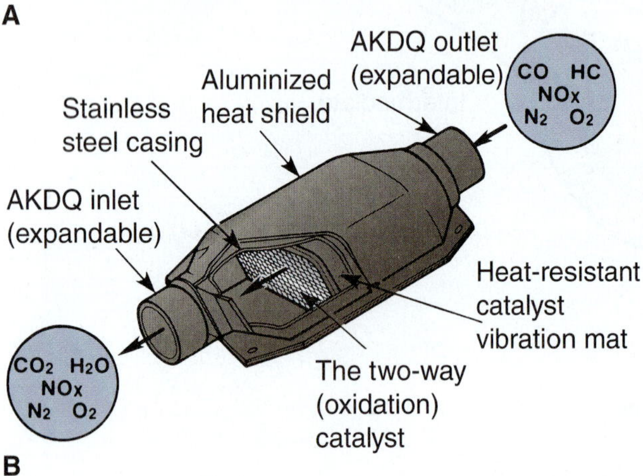

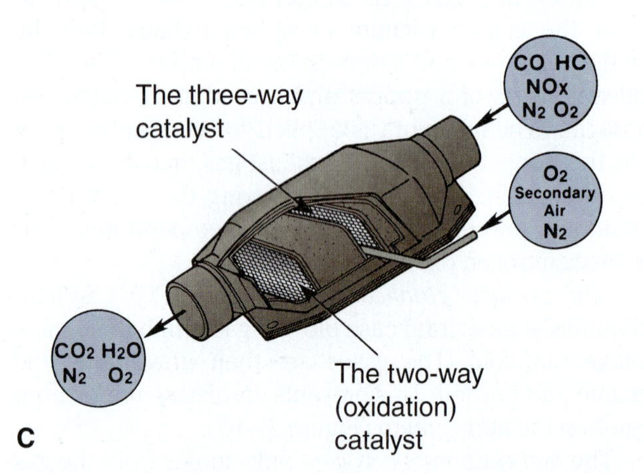

Figure 21-92 **Compare the types of catalytic converters: (A) single-bed, (B) dual-bed, and (C) three-way. Note how they convert harmful exhaust gases into harmless water, oxygen, and carbon dioxide.**

shop responsible for the emission control system. The law requires technicians to restore emission control systems to their original condition. The law also imposes penalties for shops and technicians who alter emission control systems or fail to restore them to proper working condition.

Figure 21-93 **Inspect the whole exhaust system to see whether it is damaged or rusted thin. If rusted, it will dent easily. Warn the customer of possible dangers, because rust damage is not covered by insurance companies.**

The following guidelines must be strictly adhered to when working on emission control systems:

- Damaged parts must be replaced with good parts. Eliminating damaged parts to avoid replacing them is against the law.
- Using parts that prevent proper operation of the emission control system is against the law.
- Proper repairs to the emission control system must be made to the manufacturer's specifications.
- All replacement parts for emission control systems must satisfy the original design requirements of the manufacturer.

To make it easier for the technician, manufacturers are required by law to install emission control identification labels and labels supplying vacuum routing and connection information and adjustment specifications when vehicles are built. These labels are considered part of the emission control systems under the law. The labels must be replaced when collision repair services require their removal or cause damage to the labels. Part numbers appear on the labels as required by law (Figure 21–97).

21.10 HOSE AND TUBING INSPECTION

There are a number of hoses of various types and sizes in a car. Before the vehicle is returned to the customer, these should be checked. If any hoses, tubing, or clamps appear damaged, they should be replaced. This is especially true of the fuel lines (Figure 21–98).

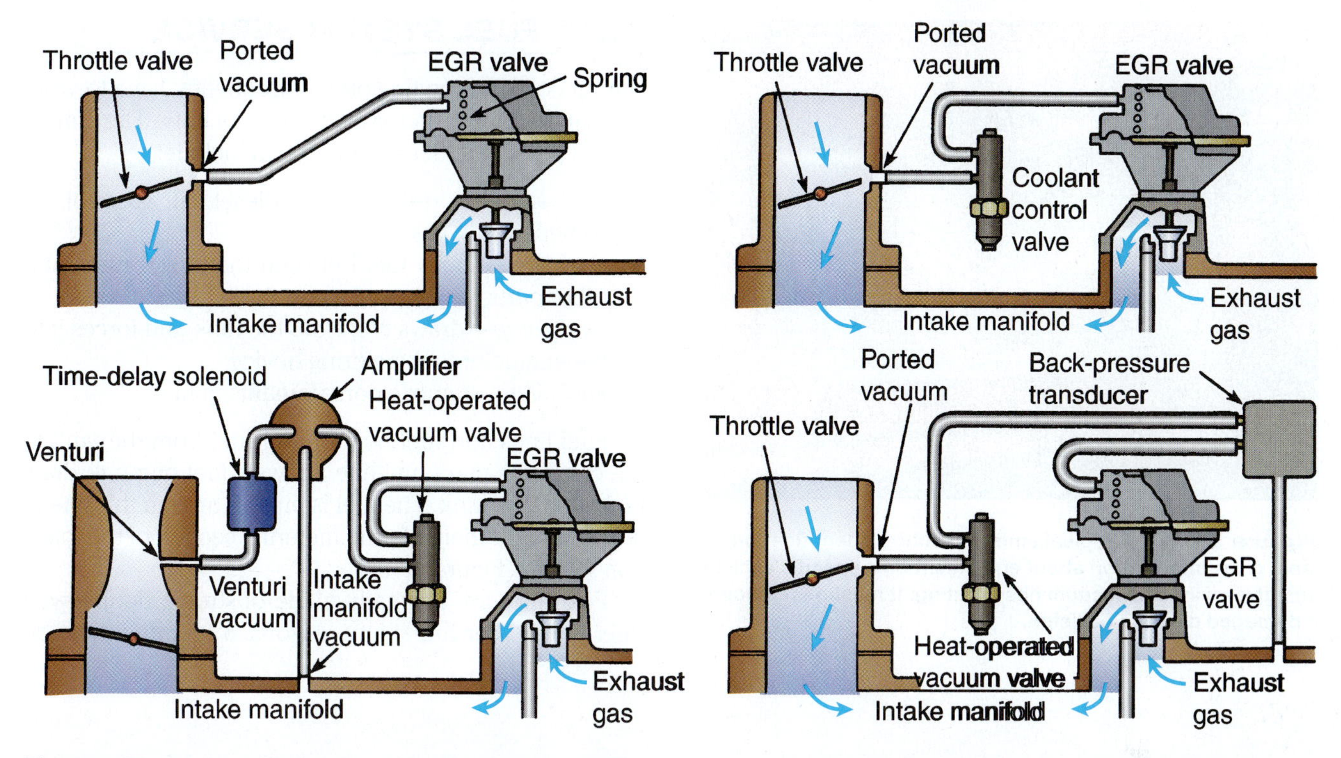

Figure 21-94 An exhaust gas recirculation system (EGR) routes burned exhaust gases back into the engine intake manifold to lower peak combustion temperatures to reduce NOx emissions. Note how the vacuum hose routing can vary with vehicle make and model.

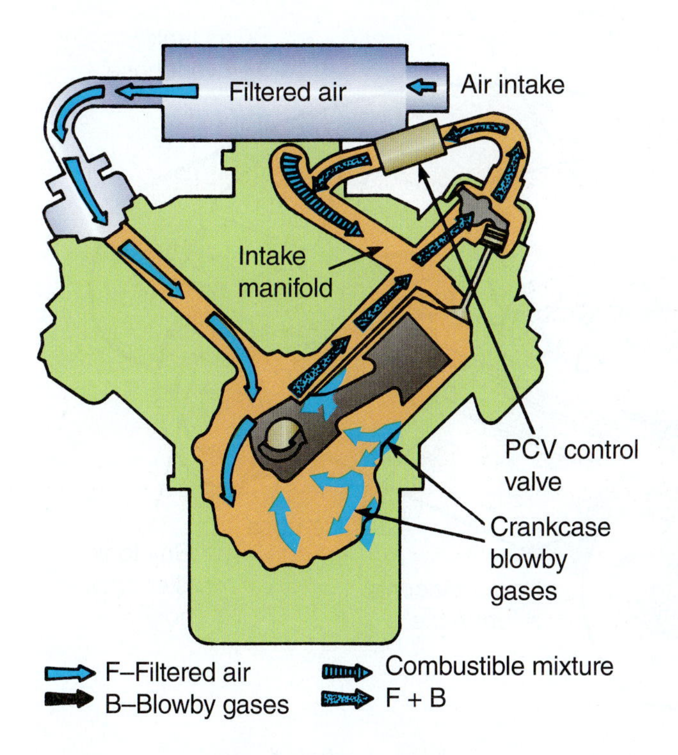

Figure 21-95 The PCV system prevents crankcase fumes from entering and polluting the Earth's atmosphere. Fumes are drawn into the engine intake manifold for burning.

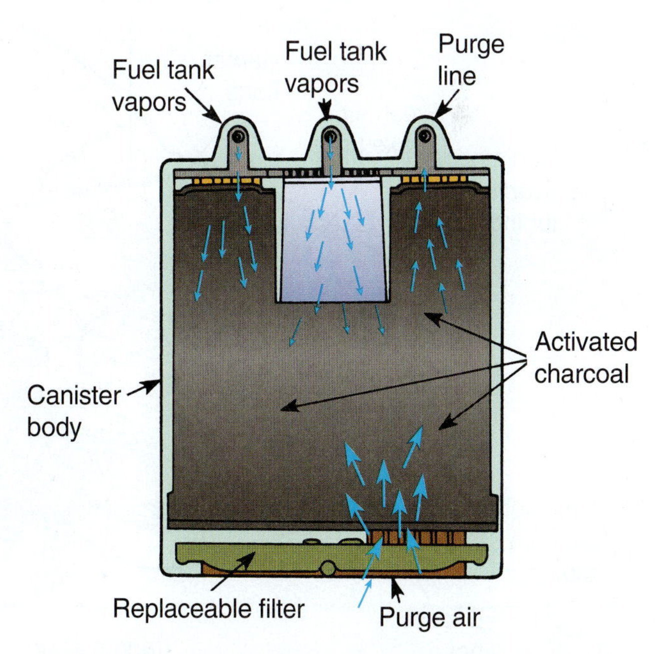

Figure 21-96 The charcoal canister captures and stores fuel system fumes so they can be burned in the engine upon startup. The plastic canister can be damaged and broken during a collision. Always inspect it for cracks and breakage while doing body work.

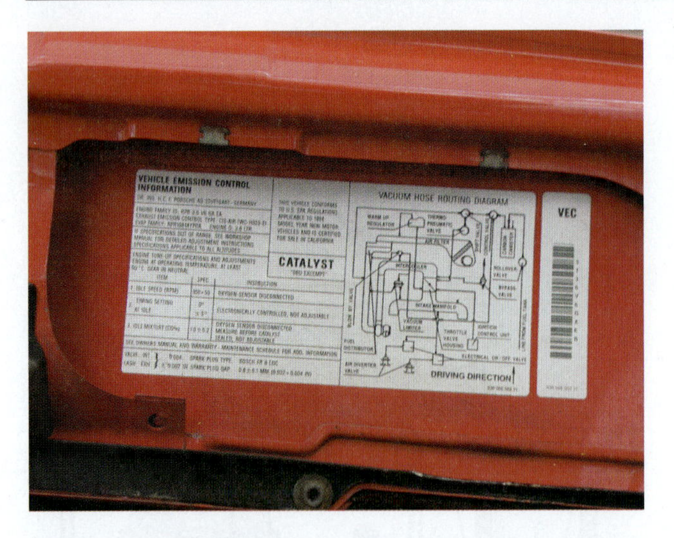

Figure 21-97 A typical emission control identification label gives information about equipment on a specific vehicle and gives wire and vacuum hose routing. It must be replaced if damaged during a collision.

21.11 FUEL SYSTEM SERVICE

During collision repair, you will frequently have to work with and around fuel system components. The fundamental parts of a fuel supply system include:

- *Fuel tank*—stores gasoline, diesel oil, gasohol, or sometimes LP gas
- *Fuel lines*—carry fuel between the tank, pump, and other parts
- *Fuel pump*—draws fuel from the tank and forces it to the engine or fuel metering device
- *Fuel filters*—remove contaminants in fuel

Fuel lines and fuel hoses carry fuel from the tank to the engine. A main fuel line allows a fuel pump to draw fuel out of the tank. The fuel is pulled through this line to the pump and then into the metering section of the injection system (Figure 21–99).

Fuel lines are normally made of strong, double-wall steel tubing. For fire safety reasons, a fuel line must be

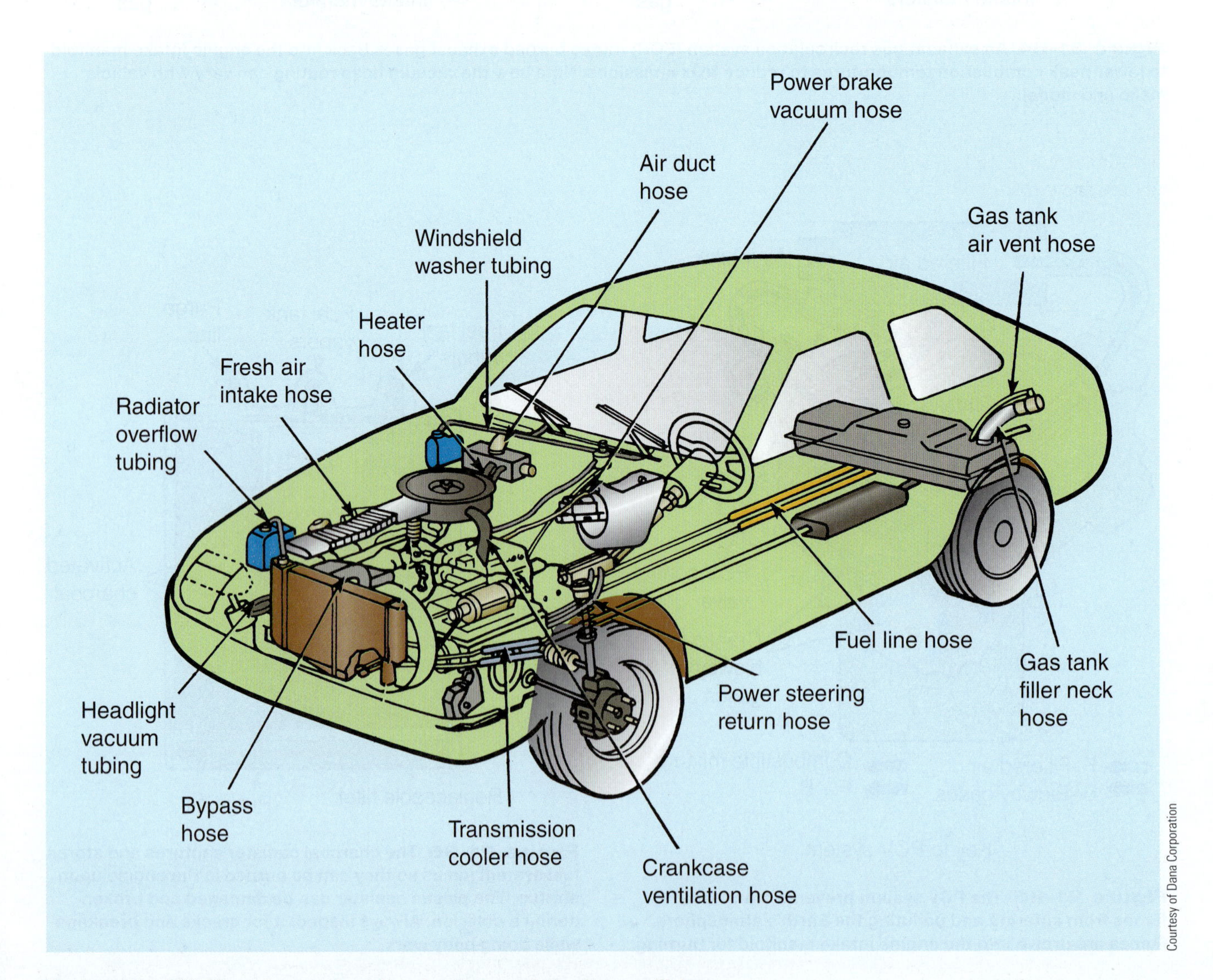

Figure 21-98 These hoses should be checked carefully after a collision.

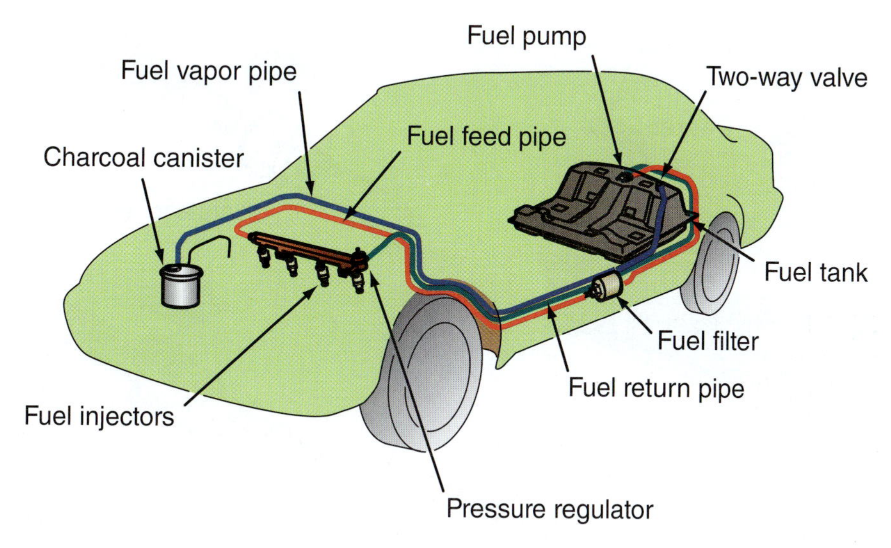

Figure 21-99 Note the basic parts of a fuel system. An electric pump in the tank forces fuel through lines and hoses to the engine. Injectors spray fuel into the engine intake manifold.

able to withstand the constant and severe vibration produced by the engine and road surfaces.

Fuel hoses made of synthetic rubber are needed where severe movement occurs between parts. For example, a fuel hose is used between the main fuel line and the engine. The engine is mounted on rubber motor mounts. The soft mounts allow the engine some movement in the car frame or body.

DANGER Do not try to repair a damaged fuel tank. If it is not badly rusted, send the leaking tank to a well-trained specialist. Even an empty tank can explode when fuel gum melts, vaporizes, and ignites from the heat of soldering or welding.

A fuel tank can be located under the trunk, in a body panel, or under the rear seat. It may be held in the vehicle by large metal straps or by bolts passing through the tank flange. On rear-engine vehicles, the fuel tank can be located in the front.

DANGER Before servicing a fuel tank, empty it. A full tank is very heavy and can rupture if dropped. This can result in injuries and fire.

To remove fuel from the tank, unscrew the drain plug and drain the fuel into an approved safety can. If a drain is not provided, use an approved pumping method to draw the fuel out of the tank.

After draining, you can remove the tank from the vehicle. Disconnect the filler neck, fuel lines, wires, and other components. Then remove the tank straps or bolts securing the tank to the body or frame. Slowly lower the tank without spilling any fuel.

When installing a fuel tank, make sure you replace the rubber insulators. Check that all fuel lines are properly secured. Replace the fuel in the tank and check for leaks. Vehicle-specific service information will detail exact tank installation procedures (Figure 21–100).

SHOP TALK Some late model fuel lines have a special snap-type fitting. Do not try to pry the fitting apart or it will be damaged. You need an inexpensive tool to release the fuel line fitting for service. This tool can be purchased at most auto parts stores.

Remember these rules when working with fuel lines and hoses:

- Place a shop rag around the fuel line fitting during removal. This will keep fuel from spraying on you or on the hot engine. Use a flare nut or tubing wrench on fuel system fittings.
- Only use approved double-wall steel tubing for fuel lines. Never use copper or plastic tubing.
- Make smooth bends when forming a new fuel line. Use a bending spring or bending tool.
- Form double-lap flares on the ends of the fuel line. A single-lap flare is not approved for fuel lines.

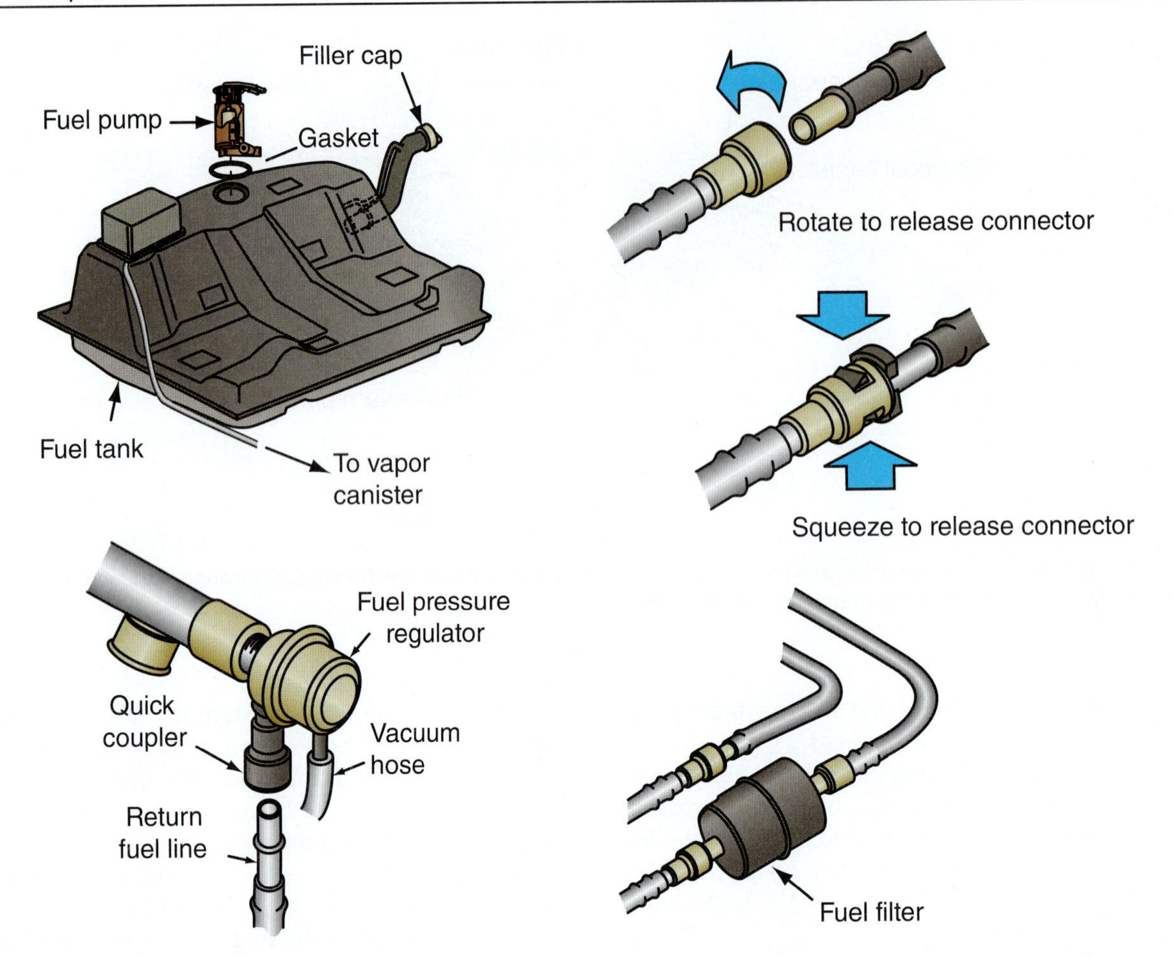

Figure 21-100 **Fuel line tubing should be visually checked for leakage, damage, and looseness of clamps. The fuel tank should always be drained before removal.**

- Reinstall fuel line hold-down clamps and brackets. If not properly supported, the fuel line can vibrate and fail.
- Route all fuel lines and hoses away from hot or moving parts. Double-check clearance after installation.
- Only use approved synthetic rubber hoses in a fuel system. If vacuum-type rubber hose is accidentally used, fuel can chemically attack and rapidly ruin the hose. A dangerous leak could result.
- Double-check all fittings for leaks. Start the engine and inspect the connections closely.

Most fuel injection systems have very high fuel pressure. Follow recommended procedures for bleeding or releasing pressure before disconnecting a fuel line or fitting. This will prevent fuel spray from possibly causing injury or a fire!

SUMMARY

1. The powertrain includes all of the parts that produce and transfer power to the drive wheels.
2. The engine provides energy to move the vehicle and power all accessories.
3. When starting an engine before or after repairs, check the oil level with the oil dipstick.
4. When a CV-joint boot is torn or missing, there is often damage or wear in the joint.

5. When more time is saved in the repair of adjacent panels than is necessary to remove and reinstall the drivetrain, the drivetrain should be removed.

6. An engine-transaxle or transmission assembly is very heavy. If dropped, it can easily chop off toes and fingers or crush bones. A coil spring has deadly force when compressed!

7. Visually inspect the steering system for any physical damage. Check the boots for leaks, inspect the tie-rods, and examine the mounting points for any distortion.

8. In collision repair, wheel alignment involves adjusting the vehicle's tires so that they roll properly over road surfaces.

9. Camber is the angle represented by the vertical tilt of the wheels inward or outward when viewed from the front of the vehicle.

10. Caster is the angle of the steering axis of a wheel from true vertical, as viewed from the side of the vehicle.

11. Steering axis inclination is the inward tilt of the steering axis at the top.

12. Toe is the difference in the distance between the front and rear of the left- and right-hand wheels.

13. With proper tracking, the rear tires travel directly behind the front tires when the steering wheel is in the straight-ahead position.

14. The brake system uses hydraulic pressure to slow or stop wheel rotation with brake pedal application.

15. R134a is the present replacement for R12 refrigerant. It is less harmful to the ozone layer. A recovery system captures used refrigerant and keeps it from contaminating the atmosphere.

16. Emission control systems are used to prevent potentially toxic chemicals from entering our atmosphere. The most common of these are the exhaust gas recirculation (EGR), catalytic converter, air injection, and positive crankcase ventilation (PCV) systems.

EXERCISES

On a separate sheet of paper, complete the following learning activities for this chapter. Write definitions for the key terms and answer the ASE-style review questions, essay questions, critical thinking problems, and math problems. You can also do the outside activities, possibly for extra credit.

Key Terms

antifreeze
axial runout
caliper
camber
caster
caster–camber gauge
clutch
coolant recovery bottle
curb height
CV axles
differential
drive shaft
emission control system
engine
exhaust system
heater core
jounce
master cylinder
motor mounts
parallelogram steering system
powertrain
rear axle assembly
rebound
rotor
scrub radius
shock absorbers
steering axis inclination (SAI)
steering gearbox
suspension system
thermostat
transaxle
transmission
turning radius
turning radius gauge
wheel alignment
wheel puller
wheel runout

ASE-Style Review Questions

1. To begin a caster angle check, Technician A turns the left front wheel out 20 degrees. Technician B turns the left front wheel out 70 degrees. Who is correct?
   A. Technician A
   B. Technician B
   C. Both A and B
   D. Neither A nor B

2. In unibody construction, what provides the critical mounting positions for the suspension and steering systems?
   A. Structural panels
   B. Drivetrain
   C. Upper and lower control arms
   D. Cradle assembly mounting biscuits

3. What is the definition of camber?
   A. The forward or backward tilt of the steering axis
   B. The distance between the centerline of the ball joints and the centerline of the tire at the point where the tire contacts the road surface
   C. The amount of toe-out present on turns
   D. Represented angle by the vertical tilt of the wheels when viewed from the front of the vehicle
4. In order to correct toe, Technician A adjusts the tie-rod; Technician B makes a jounce–rebound check. Who is correct?
   A. Technician A
   B. Technician B
   C. Both A and B
   D. Neither A nor B
5. Which parts mount on the outer end of the control arms to allow the steering knuckles to swivel and turn?
   A. Bushings
   B. Sleeves
   C. Ball joints
   D. Ball sockets
6. Which of the following statements concerning brake systems is incorrect?
   A. All cars are now required to have two hydraulic systems
   B. The major brake component that a technician must repair is the brake line
   C. A hose that is blistered does not necessarily have to be replaced
   D. The wheel cylinder converts hydraulic pressure to mechanical force
7. The _______ is the foundation of the engine; all the other engine parts are either housed in or attached to it.
   A. block
   B. head
   C. rocker arm
   D. piston
8. Which emission control subsystem is responsible for channeling crankcase blowby gases into the fuel intake area?
   A. Engine control
   B. Positive crankcase ventilation
   C. Evaporative
   D. Exhaust gas recirculation
9. Which assembly has a transmission and differential combined into a single housing or case?
   A. Differential
   B. Engine
   C. Transfer case
   D. Transaxle
10. After recovery from an A/C system, Technician A uses hydraulic caps to keep out moisture. Technician B uses plastic wrap held in place with wire ties. Who is correct?
    A. Technician A
    B. Technician B
    C. Both A and B
    D. Neither A nor B
11. Which type of camber has the top of the wheel tilted inward when viewed from the front so the inner tire tread contacts the road surface more?
    A. Reverse camber
    B. Forward camber
    C. Positive camber
    D. Negative camber
12. What should be done before making any adjustments affecting caster, camber or toe-in?
    A. Verify tire pressure is correct
    B. Check for loose ball joints
    C. Check for runout of wheels and tires
    D. All of the above

Essay Questions

1. Describe the three functions of a vehicle's suspension system.
2. List five power steering service tips that should be kept in mind.
3. What is the A/C system low-side?
4. Explain powertrain inspection.

Critical Thinking Problems

1. A steering wheel has been badly bent by the driver's body flying forward without a seat belt. What should be done?
2. You must replace the A/C system compressor. How should you proceed?
3. Explain the seven major parts of an engine.

Math Problems

1. A worn ball joint moves up and down 1/32 inch (0.8 mm). The ball joint moves 1/64 inch (0.4 mm). What, if anything, should be done? Give measurements.
2. The front wheels of a car are toed-in 1/16 inch (1.5 mm). Specs call for 1/16 inch (1.5 mm) toe-out. How much adjustment is needed?
3. A body shop may spend $30 in preparation supplies to clean, mask, prime, and paint each vehicle. If the shop repairs and paints 134 vehicles per year, how much will the shop have spent to cover those costs?

Activities

1. Inspect several badly damaged vehicles. Make a list of the mechanical parts that have been damaged in the collision.
2. Make estimates of the costs to repair the damaged mechanical parts you identified in the previous activity.

CHAPTER 22

Electrical/Electronic System Operation and Service

OBJECTIVES

After reading this chapter, you should be able to:

- Explain basic electrical values: volts, ohms, and amps.
- Use Ohm's Law to calculate circuit values.
- Describe the difference between series and parallel circuits.
- Describe the test procedures used to repair electrical and electronic systems.
- Repair vehicle wiring harnesses damaged during a collision.
- Use a scan tool to check for electrical troubles before and after collision repairs.
- Replace damaged computer system components.
- Connect and use a multimeter to perform basic electrical tests.
- Correctly answer ASE-style review questions relating to electrical/electronic systems.

INTRODUCTION

Electrical repairs include tasks like repairing severed wiring, replacing engine sensors, and scanning for computer or wiring problems. A collision impact and the resulting metal deformation can easily crush a vehicle's wires and electrical components. For this reason, today's auto body technician must have the basic skills needed to work with and replace electrical/electronic components.

Most modern vehicle mechanical systems are monitored or controlled by on-board computers. This makes electrical/electronic knowledge important to today's collision repair technician. Modern body shops commonly use a scan tool to check and repair electrical problems caused by collision damage. This chapter will give you the basic information needed to do body shop electrical repair work. See Figure 22–1.

22.1 ELECTRICAL TERMINOLOGY

To understand electricity, you must become familiar with four electrical terms: *current, voltage, resistance*, and *conductor.*

Current is the movement of electricity (electrons) through a wire or circuit. It is measured in amperes or amps, using an ammeter. The common electrical symbol for current is A or I.

Direct current (DC) only flows in one direction through a circuit. Most automotive circuits use DC from the battery or charging system alternator. *Alternating current* (AC) flows in one direction, reverses, and then flows in the other direction. The charging system alternator generates AC and it is then rectified (changed) into DC before flowing into the vehicle's electrical system.

Voltage is the pressure that pushes the electricity through the wire or circuit. The power source—the battery or alternator—generates the voltage that causes current flow in a vehicle. Voltage is measured in volts, using a voltmeter. The symbol for voltage is E or V.

Resistance is a restriction or obstacle to current flow. It tries to stop the current caused by the applied voltage. Circuit or part resistance is measured in ohms, using an ohmmeter. The symbol for resistance is R or V. An open switch set to off would have infinite (maximum) resistance to shop current flow. A closed switch set to ON would have little or no resistance, so current can flow through and operate the circuit.

A *conductor* carries current to the parts of a circuit. "Hot" wires connect the battery positive to the components of each circuit. Insulation stops current flow and

Figure 22–1 Today's auto body technician should have a solid understanding of electricity and electronics. Many minor electrical repairs are performed while doing collision repairs.

keeps the current in the metal wire conductor. The vehicle's body structure provides the ground conductor back to the battery negative cable.

Ohm's Law

Ohm's Law is a mathematical formula for calculating an unknown electrical value (amps, volts, or ohms) when two values are known. If you know two values, you can mathematically calculate the third unknown value (Figure 22–2).

For instance, if you know that a circuit has 12.6 volts and 2 amps, you can calculate circuit resistance. Plug the two known values into Ohm's Law and you can calculate the unknown value (12.6 divided by 2 equals 6.3 ohms). Understanding Ohm's Law will help you visualize what might be causing an electrical problem.

Magnetism

Magnetism involves the study of how electrical fields act on ferrous (iron-containing) objects. Many electronic parts utilize magnetism.

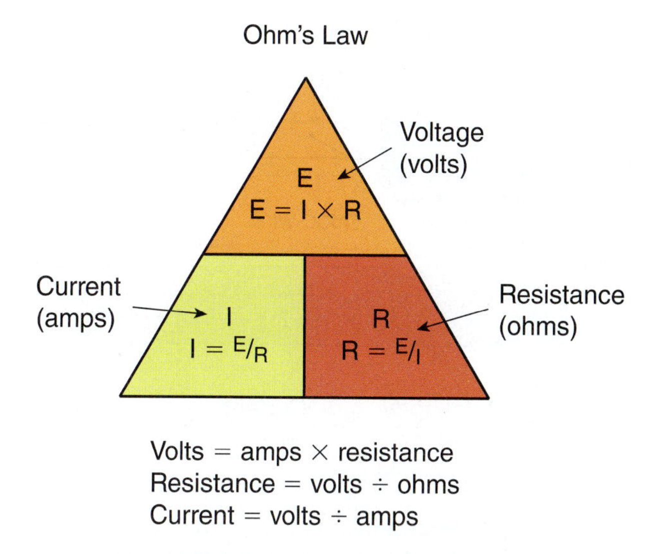

Figure 22–2 Ohm's Law is a simple formula for mathematically finding an unknown electrical value when two values are known.

A flux, or **magnetic field**, is present around permanent magnets and current-carrying wires. This invisible energy is commonly used to move metal parts. An *electromagnet* is a set of windings (wrapped wires) around an iron core. When current flows through the windings, a powerful magnetic field is produced. Electric motors, solenoids, relays, and other parts use this principle.

Electric Circuits

An **electric circuit** contains a power source, conductors, and a load. Some resistance is designed into a circuit in the form of a load. The load is the part of a circuit that converts electrical energy into another form of energy (light, movement, heat, and so on). Other parts are added to this simple circuit to protect it from damage and to do more tasks.

A *frame ground circuit* uses the metal body as a return wire to the negative terminal of the battery. Motor vehicles use frame ground circuits to reduce the number of wires needed for circuit operation. Refer to Figure 22–3.

A **series circuit** has only one conductor path, or leg, for current through the circuit. Current must flow through the wires and components one after the other. If any part of the series circuit is opened (disconnected), all of the circuit stops working (Figure 22–4).

A **parallel circuit** has two or more legs, or paths, for current. Current can flow independently through either leg. One path can be closed (electrically connected) and the other opened, and the closed path will still operate.

A *series-parallel circuit* has both series and parallel branches in it. It has characteristics of both circuit types.

To better understand an electric circuit, it is useful to compare it to a hydraulic circuit. The battery is

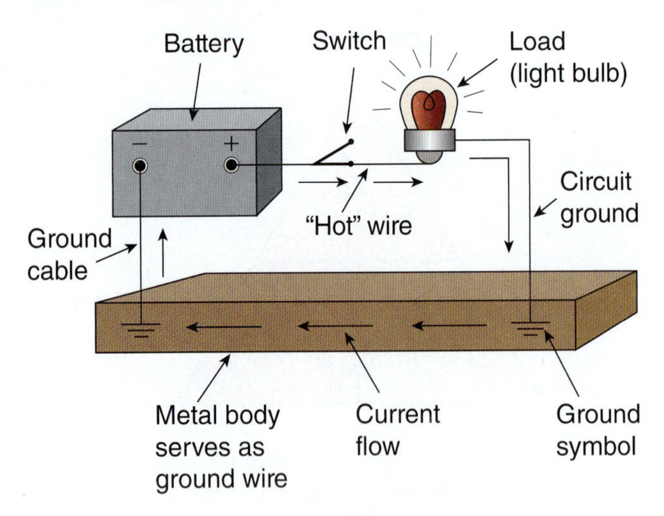

Figure 22-3 Most circuits in a car or truck are frame ground circuits. The frame or unibody is used to return electrical current to the battery. This reduces the amount of wiring needed to complete circuits. Note ground symbols.

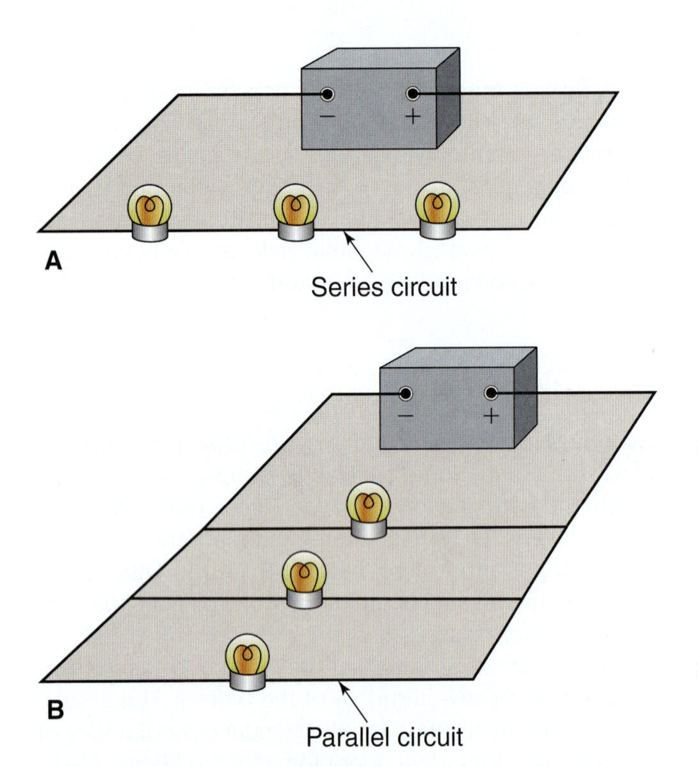

Figure 22-4 Compare basic series and parallel circuits. (A) With a series circuit, if one light bulb or load burns out, the whole circuit will stop working. (B) With a parallel circuit, other legs of the circuit will keep it working if one bulb or load burns out.

comparable to a hydraulic pump that develops pressure and flow. A hydraulic pressure relief valve is similar to a fuse that protects the circuit from damage. An electric switch is comparable to a flow control valve. An electrical load (for example, an electric motor) is comparable to the hydraulic motor that does the work. This is illustrated in Figure 22–5.

Wiring Diagrams, Schematics, and Trees

To determine and isolate electrical problems, it is often necessary to trace through the electric circuit using a wiring diagram. A *wiring diagram* is a graphic representation of all the parts and wires in a circuit (Figure 22–6).

Electrical abbreviations are used on wiring diagrams so that more information can be given. Service manuals provide charts to explain electrical abbreviations, numbers, and symbols on their wiring diagrams.

Table 22–1 lists some common electrical abbreviations.

Electrical symbols are graphic representations of electrical/electronic components. Symbol charts can be found at the beginning of the wiring diagram section of a service manual. Figure 22–7 shows some common electrical symbols.

The symbol used to identify a part is either a universal symbol or an auto manufacturer's symbol. Because manufacturers use different symbol designs for some parts, it is important to follow the symbol chart for the specific wiring diagram you are using.

Most wires on wiring diagrams are identified by their insulation color. **Wire color coding** allows you to find a specific wire in a harness or connector. The color-code abbreviation chart can also be found at the beginning of the wiring diagram section of the service manual. Remember, different auto makers use different color-code abbreviations on their wiring diagrams.

A wiring diagram is more like a book than a picture. You cannot understand a wiring diagram just by glancing at it. As with a book, you must read the diagram carefully all the way through for complete understanding.

Typical color codes are given in Table 22–2. The first letter in a combination of letters usually indicates the base color. The second letter usually refers to the stripe color (if any). Tracing a circuit through a vehicle is basically a matter of following the colored wires.

Many wiring diagrams found in service manuals also have numbered circuits and connectors. Circuit numbering is used to specify exactly which part of the circuit the service manual is describing.

For example, Figure 22–8 shows how harness circuits are numbered for reference when doing repairs. Study the electrical repair information.

A **wiring harness** has several wires enclosed in a protective covering. A vehicle has several wiring harnesses, usually named after their location in the vehicle. The vehicle's service manual will give illustrations with code numbers for locating parts and connections.

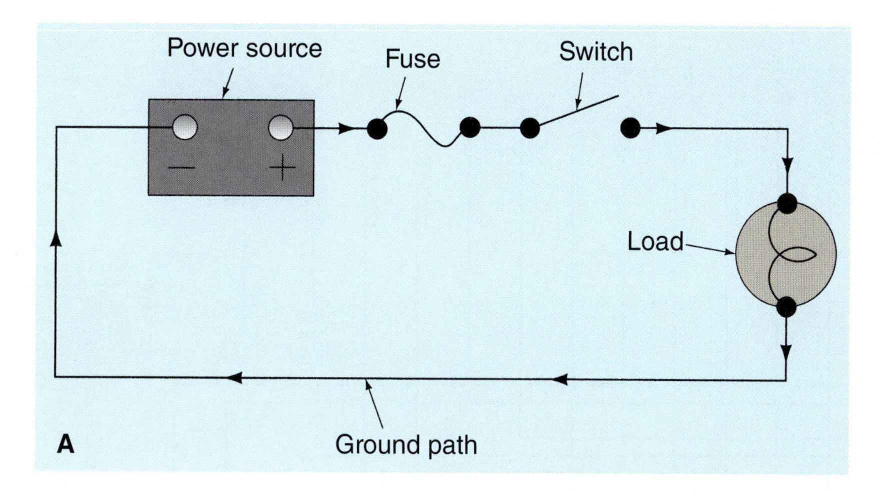

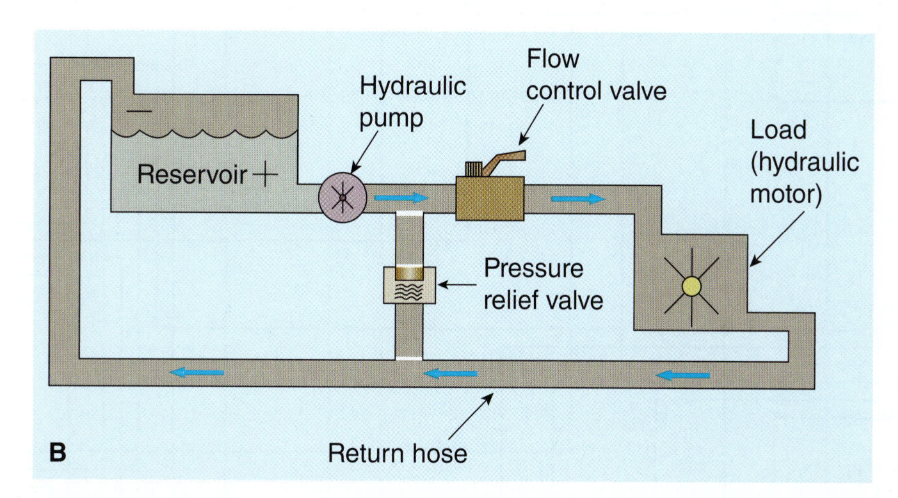

Figure 22-5 Note the similarities between an electric circuit and a hydraulic circuit. (A) The electrical circuit has a battery as a power source. A fuse protects the circuit from too much current flow. A switch turns the circuit on and off. The load uses electricity to perform work. (B) The pump is the source of power fluid pressure, like a battery that generates voltage. A pressure relief valve limits fluid pressure like a fuse. A flow control valve limits fluid flow like a switch. The hydraulic motor is the load.

TABLE 22-1 COMMON ELECTRICAL ABBREVIATIONS

A	ampere	POS	positive
AC	alternating current	PRES	pressure
ACC	accessory	SOL	solenoid
BAT	battery	SPDT	single-pole double-throw
C/B	circuit breaker	SPST	single-pole single-throw
DC	direct current	TEMP	temperature
DPDT	double-pole double-throw	TOG	toggle (switch)
MOM	momentary	V	volt
MOT	motor	W	watt
(n)	none	–	negative
NC	normally closed	Ω	ohm
NEG	negative	+	positive
NO	normally open	±	plus or minus
PB	push button	%	percent

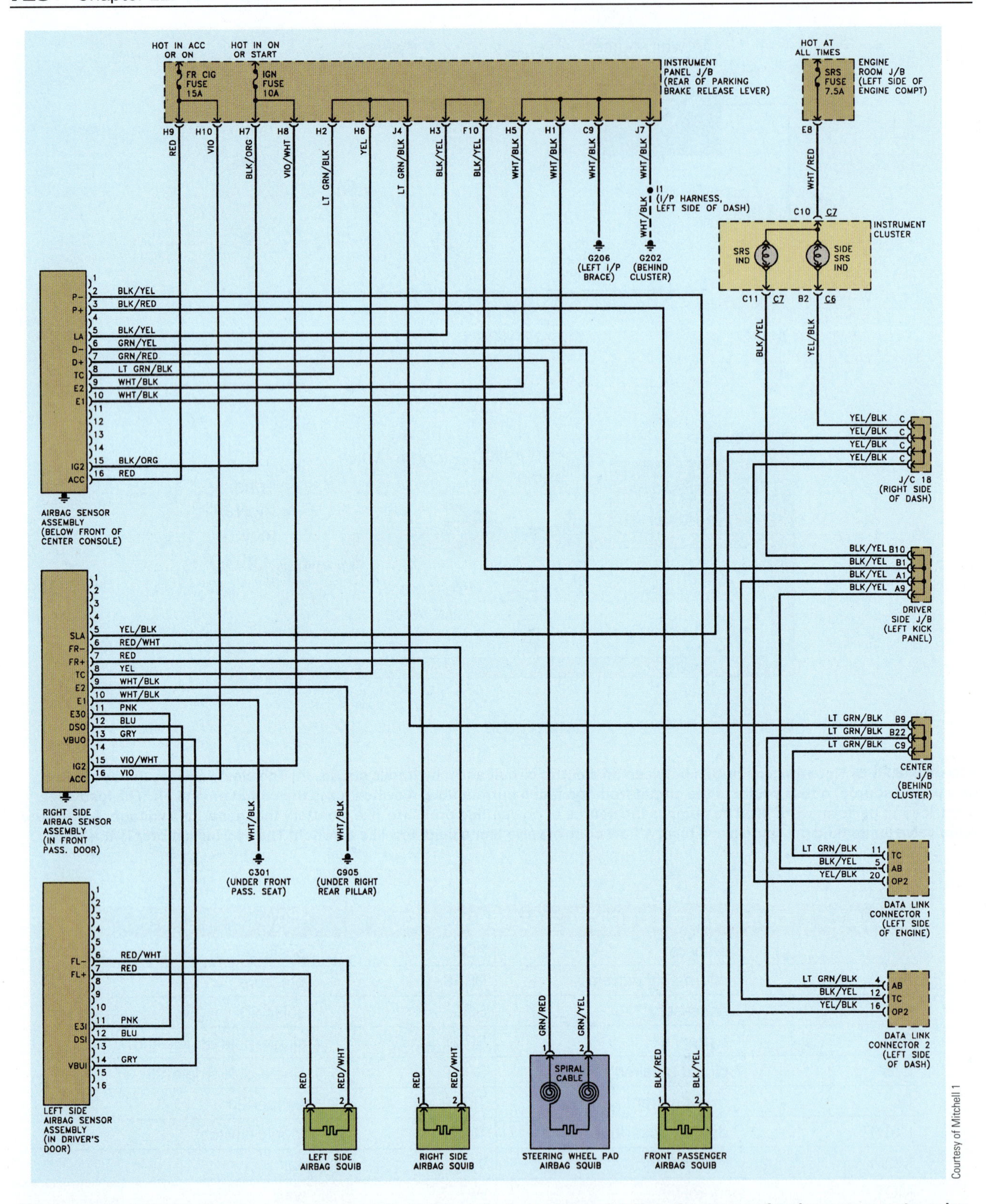

Figure 22-6 **Wiring diagrams must be carefully read and studied to be useful. Note the names of major components, color code abbreviations, and component locations. Can you find the instrument cluster, main fuse, air bag sensors, and data link connector?**

Harness connectors allow you to disconnect or unplug wiring connections for service. There are several ways to disconnect wiring harness connectors. Make sure you use the recommended technique to prevent connector damage.

Remember when working on a vehicle's wiring, use correct procedures to release and free electrical connectors in a wiring harness. If you break a plastic connector, they can be difficult to order and time consuming to replace. Always move the correct tab or push on the correct

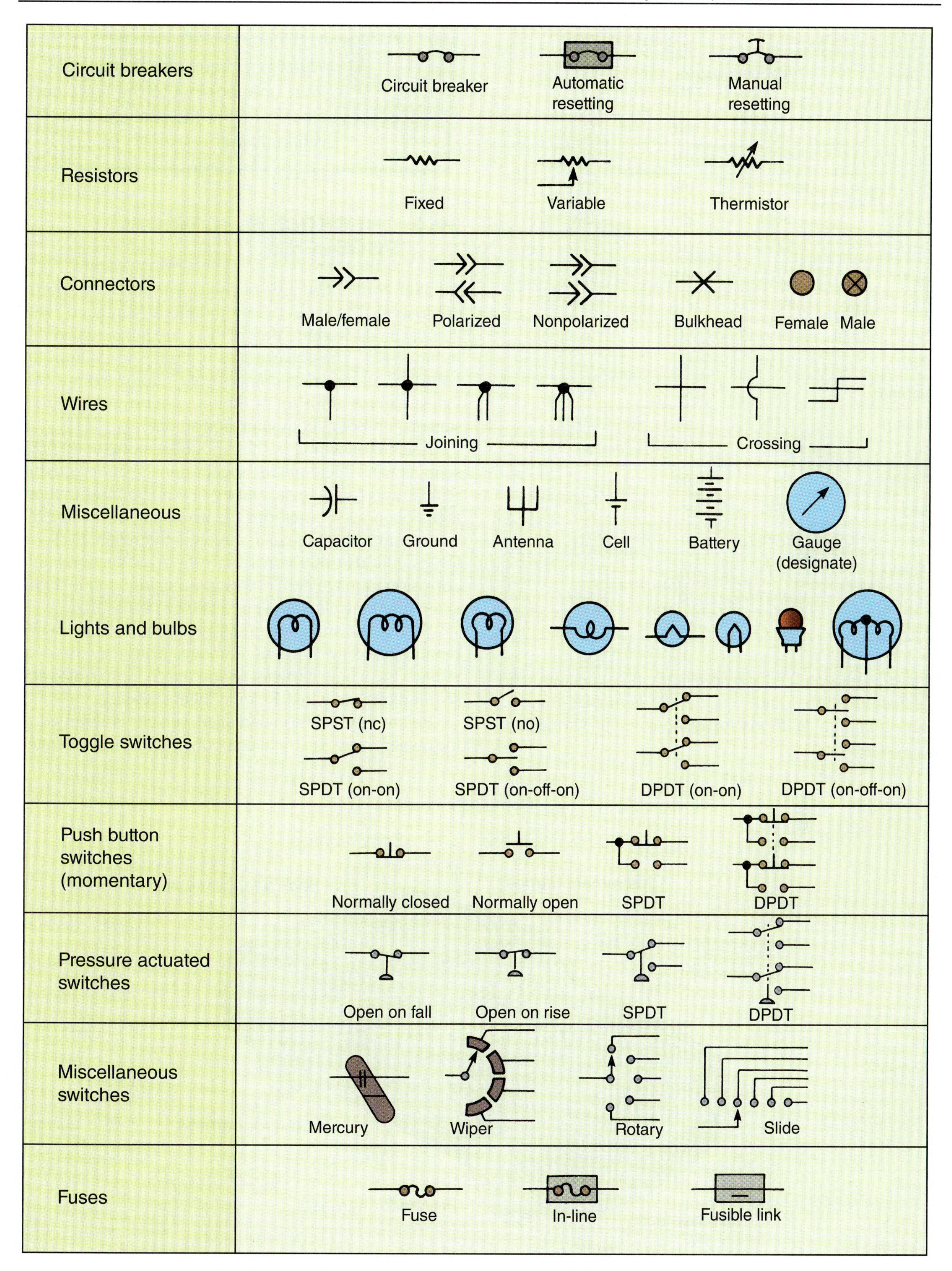

Figure 22-7 Memorize the common symbols used on automotive wiring diagrams.

TABLE 22–2 COMMON WIRE COLOR CODES

Color	Abbreviations		
Aluminum	AL		
Black	BLK	BK	B
Blue (Dark)	BLU DK	DB	DK BLU
Blue (Light)	BLU LT	LB	LT BLU
Brown	BRN	BR	BN
Glazed	GLZ	GL	
Gray	GRA	GR	G
Green (Dark)	GRN DK	DG	DK GRN
Green (Light)	GRN LT	LG	LT GRN
Maroon	MAR	M	
Natural	NAT	N	
Orange	ORN	O	ORG
Pink	PNK	PK	P
Purple	PPL	PR	
Red	RED	R	RD
Tan	TAN	T	TN
Violet	VLT	V	
White	WHT	W	WH
Yellow	YEL	Y	YL

button to release the lock on electrical connectors. The connector will then slide apart easily. Figure 22–9 shows some common methods for disconnecting wiring harness connections.

Wires in a circuit can change color from one terminal to the next, so closely follow the manufacturer's wiring diagram.

22.2 CHECKING ELECTRICAL PROBLEMS

An often overlooked area of collision repair is the electrical system. The modern automobile is "threaded" with literally miles of wires. Most of these are bundled together in harnesses. These harnesses route the wires from the battery to all electrical components—dome lights, headlights, electric door locks, remote control side mirrors, sensors, on-board computer, and so on.

A vehicle's wiring harnesses snake along body parts such as windshield pillars, rocker panels, doors, quarter panels, and roof panels, among others. Damage in these areas often cuts or abrades the insulation protecting the wires, and a short or open circuit is the result. Collision forces can also pull wires from their connections, and corrosion damage can loosen ground wire connections, again breaking electrical circuits (Figure 22–10).

Follow the manufacturer's recommendations when repairing wiring harness damage. You may have to replace the whole harness, or you may have to splice and solder damaged wires. Refer to Figure 22–11.

Before any collision-damaged vehicle is returned to the owner, every electrical component should be operated

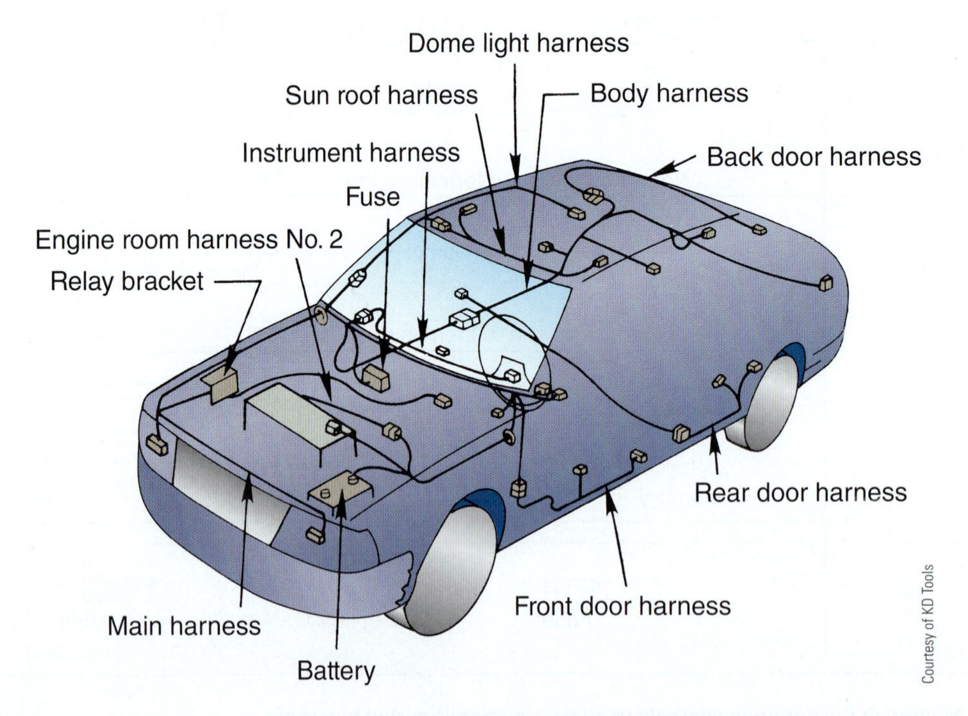

Figure 22–8 The electrical wiring harness must be inspected for cuts, splits, kinks, broken wires, and loose terminals. Collision damage can pull a harness out of its hold-down clamps and stretch and break wires inside the harness. Study the general names of the wiring harness sections.

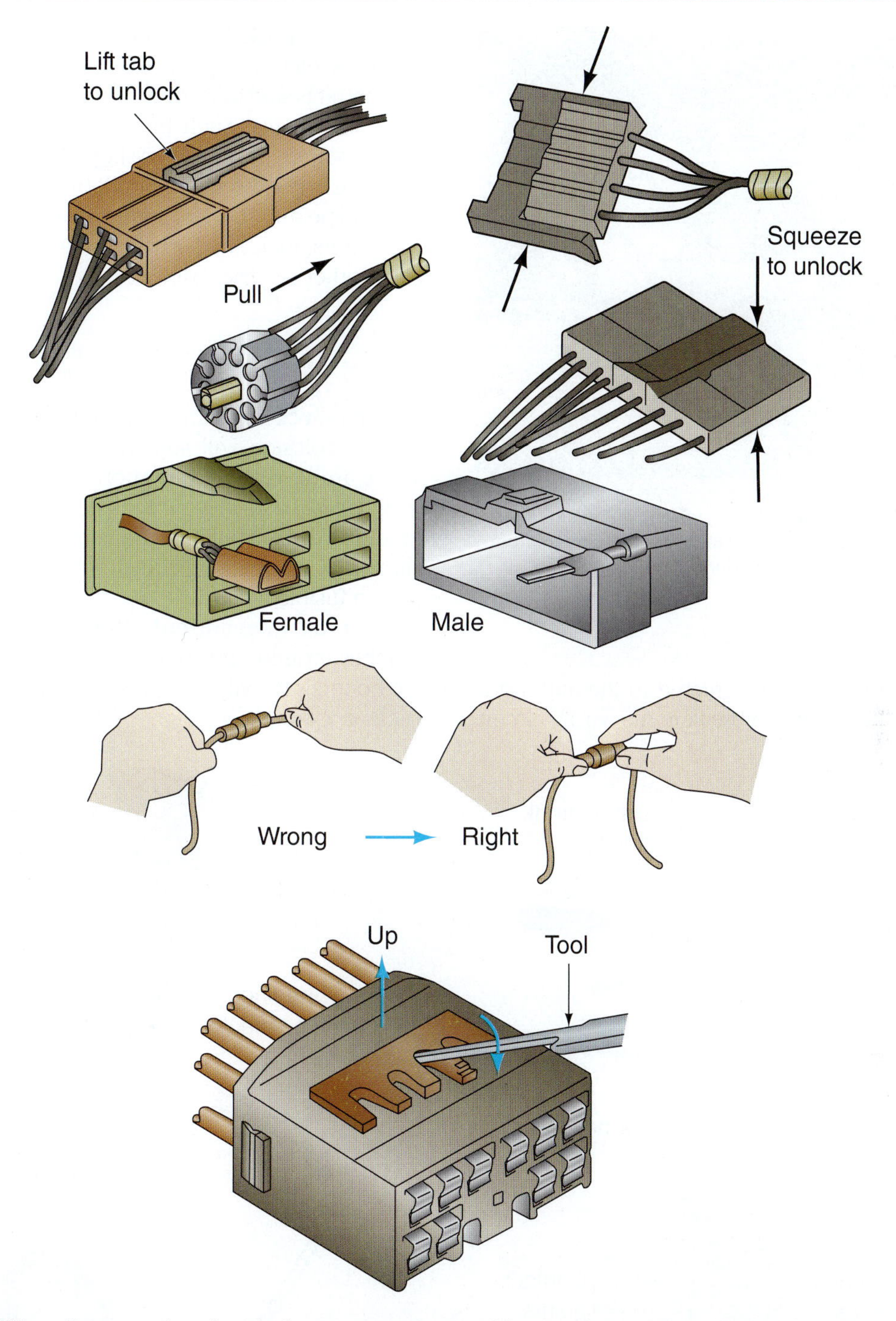

Figure 22-9 **When disconnecting electrical connectors, be careful not to damage them. You might have to lift a small tab, insert a pointed tool, or squeeze the connector together to release and disconnect it.**

to verify that it works and that it stops working when turned off. Any problems must be traced to their source and the faulty wires or loose connections repaired. Oftentimes, the problem is simply a blown fuse or a loose connection.

Consider, for example, what happens when you must replace an outer door skin. The door, of course, is removed from the vehicle and any electrical components are disconnected from the related wiring harness. If the connectors are not properly connected when the door is replaced, the windows, lock, or exterior mirror will not be operable. Always double-check connections. Make sure harnesses are secured in their clips and that ground wires are tightly secure.

Some body shops send vehicles with major or complex electrical problems to a specialty shop for repair, but there are many minor problems in the lighting and accessory circuits that could be quickly diagnosed and repaired in the body shop. Minor electrical troubleshooting requires only basic knowledge of electrical theory and a few simple diagnostic tools.

Figure 22-10 During a collision, wires can be smashed and shorted. You or a specialized electrical technician must find and correct wiring problems during collision repair.

Open Circuits

Most electrical problems encountered in the auto body repair shop are the result of a break in the wiring circuit. An **open circuit** is disconnected and does not have a complete electrical path. When the continuity is broken, the circuit is said to be open. Power does not flow to vehicle components. Such problems should be uncovered and repaired in the body shop. Too often, however, a problem is not discovered until the owner has picked up the vehicle and (days or weeks later) attempts to turn on the wipers, lock the doors, use the map light, or operate some other electrical device.

Short Circuits

A **short circuit** has an unwanted path for electrical current. A collision will sometimes create a short circuit. There are several kinds of short circuits.

A *dead short to ground* is usually a bare conductor touching directly against the vehicle frame. This type of short always opens a circuit breaker, blows a fuse, or pops a fusible link.

With an **intermittent short**, the shorted wire only touches ground and shorts momentarily when the vehicle bounces heavily or is jarred. A flickering dash lamp is such an example.

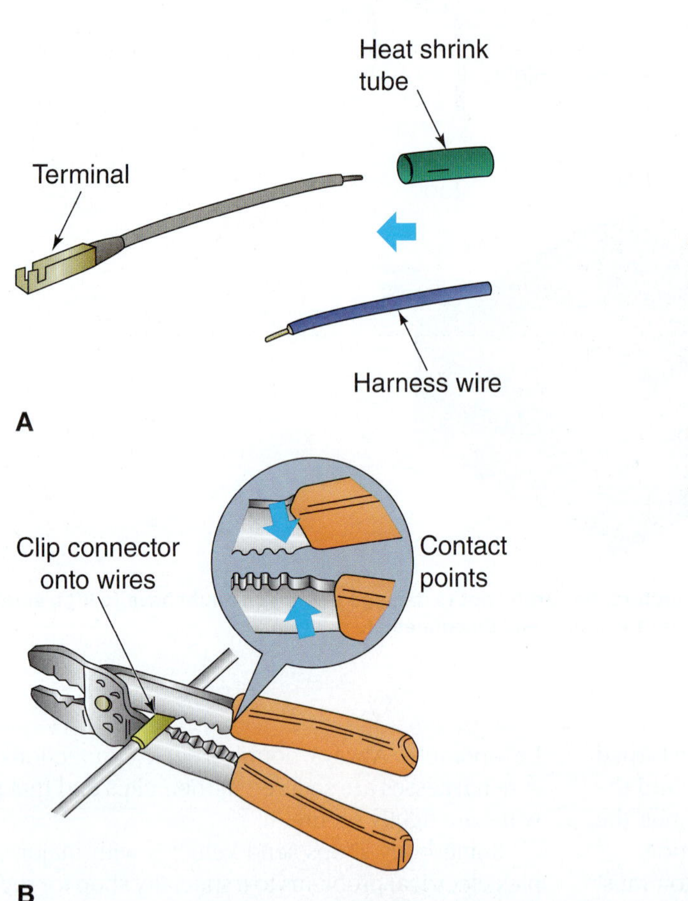

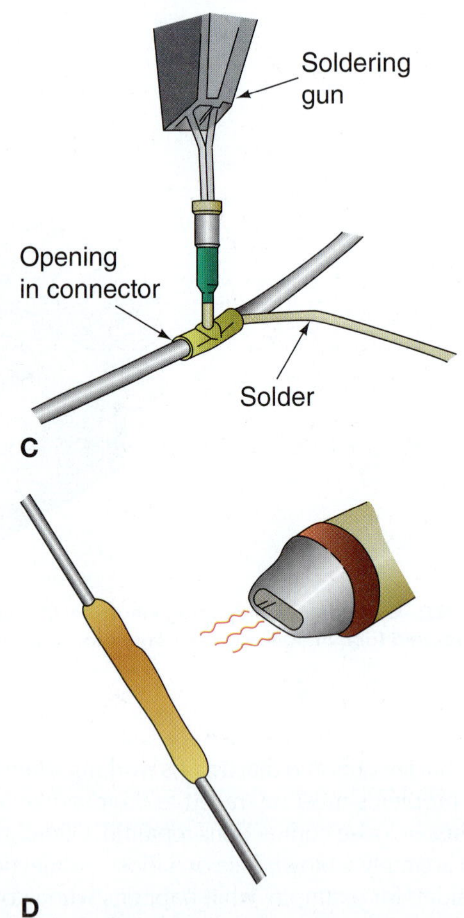

Figure 22-11 Note the basic steps for repairing an individual wire using a crimp and solder connector. (A) Cut away the smashed area of wire. Strip off a short piece of plastic insulation. Slide a heat shrink tube over the wire. (B) Insert the wires into the metal crimp connector. Use electrical pliers to crush the connector down over the stripped wire ends. (C) Use a soldering gun to solder the wires and connector together. (D) Slide the heat shrink tube down over the connector. Heat it with a heat gun. The plastic will melt and shrink down over the connector.

A *cross circuit short* occurs when two hot wires come in contact. This is usually caused by abrasion of the protective plastic coverings or by an overload that causes the coverings to melt. Such a short can cause more than one component to operate on a single switch. For example, when the lights are turned on, the windshield wipers operate. This is because the shorted wires are sharing current—supplying current to one (by actuating a switch) supplies current to both.

The last type of short circuit is called a *high-resistance short to ground.* The circuit is not broken, but contact is present between the hot wire and the ground. A high-resistance ground might not blow a fuse until the circuit is loaded to full capacity, or it might not blow a fuse at all but will slowly drain the battery.

22.3 BATTERY

The *battery* stores electrical energy as chemical energy. Therefore, it must be treated with respect as well as protected. Because of the potential electrical problems incurred in a collision, always disconnect the battery ground cable as soon as a vehicle is received at the shop.

There are several precautions that must be taken when recharging a battery. The following precautions protect the battery, vehicle, and technician:

- Always disconnect the battery ground wire before charging a battery on the vehicle. Charging the battery while connected can damage a vehicle's electrical components. The transistors and microcircuits on many cars are very sensitive to current levels. The high-amperage chargers can burn out such expensive parts as the regulator, alternator, power transistors, computerized control modules, and so on.
- Never disconnect the battery with the ignition switch in the "on" position. The resulting voltage "spike" will destroy many microcircuits in today's electronic systems.
- Check the battery carefully before charging. Look for conditions such as low water level, a cracked case, and so forth. Add distilled water or electrolyte, if needed. Use a *battery hydrometer* to check the state of charge (specific gravity) or read the built-in hydrometer (Figure 22–12).
- Table 22–3 shows the state of charge that corresponds to a specific gravity reading.
- Follow the manufacturer's instructions carefully. Do not overheat a battery by charging it for too long. A standard battery charging guide is usually given in the owner's manual. Test the specific gravity once an hour. Once no change is noticed in readings, disconnect the charger. Overcharging the battery will destroy the active material on the plates.
- Never charge the battery near any welding operations, open flames, or other heat sources. Do not smoke near the charging battery. The battery gives off very flammable hydrogen gas while charging. If the gas ignites, the battery will explode.
- Protect yourself from battery acid. Wear eye protection when handling a battery and immediately wash

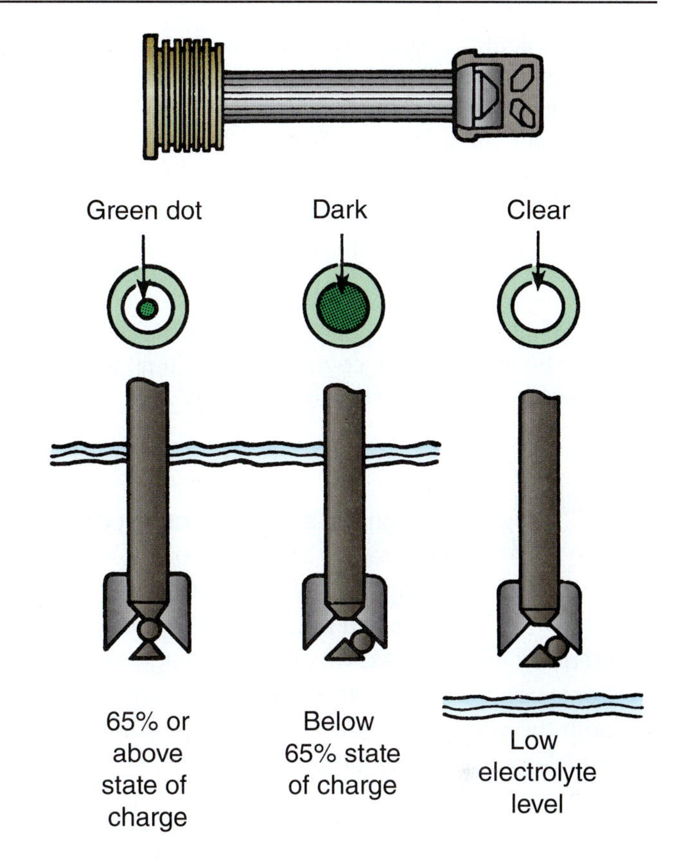

Figure 22–12 Most batteries have a built-in hydrometer that will indicate the state of the battery's charge and condition.

TABLE 22–3 SPECIFIC GRAVITY VERSUS STATE OF CHARGE AT 80°F

Specific Gravity	State of Charge	Open Circuit Cell Voltage
1.260	100%	2.10
1.230	75%	2.07
1.200	50%	2.04
1.170	25%	2.01
1.110	0	1.95

DANGER Batteries are filled with acid and emit explosive gas. Improper handling of batteries can cause severe acid burns, and ignited gas can cause the battery to explode. Always use care when working with car batteries.

off any acid that splashes on clothing or skin. If battery acid does get into your eye, hold the eye open and flush it with water at room temperature. See a physician immediately. Battery acid can cause blindness.

Using Jumper Cables

In an emergency, **jumper cables** may be needed to start an engine in order to bring the vehicle into the shop. If a car must be jump-started, connect the red jumper cable to the positive terminal of the dead battery.

Although it is a common practice in some repair shops, avoid jump-starting whenever possible. The discharged battery can explode or create voltage spikes, which can damage electronic components. Voltage spikes can damage both the vehicle you are trying to start and the vehicle providing the jump.

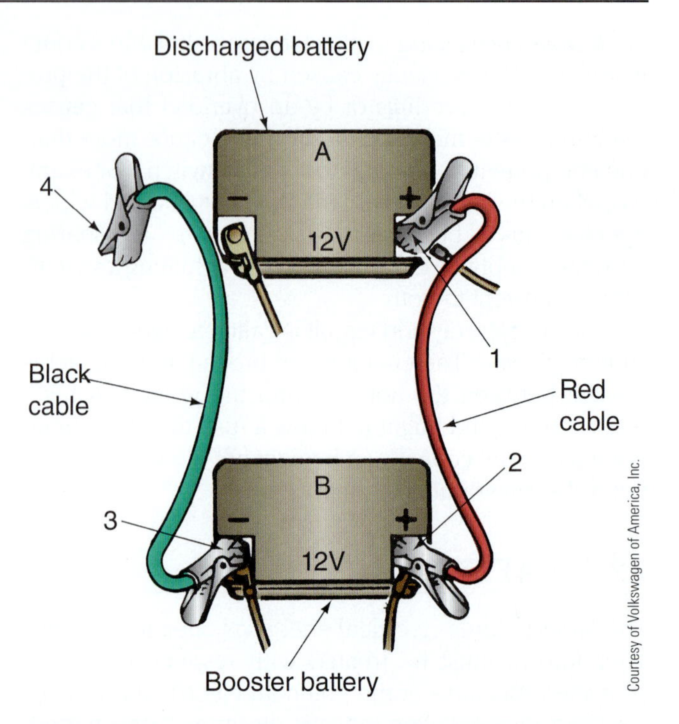

Figure 22-13 To prevent battery explosions, connect the jumper cables carefully to prevent sparks around the dead battery. Connect in the number sequence shown. The last cable connecting is to the good frame ground away from both batteries.

Connect the other end to the positive terminal of the good battery. Next, connect the black jumper cable to the negative terminal of the dead battery. Connect the other end of the black cable to the running car's chassis ground or to the good battery's negative terminal. If using another car for the jump, make sure the two vehicles do not touch each other (Figure 22–13).

In addition, consider the following to avoid damage from voltage spikes in the electronic circuits of a vehicle with a dead battery:

- Make sure every electrical device in that car, including the dome light, is turned off before connecting the batteries.
- Turn the key in the "dead" car to get it started *only* after the hookups are properly made.
- Once the "dead" car is running, remove all jumper connections before turning on any electrical devices.
- Remove clamps in reverse order, starting with the negative to the car body. This ensures that a spark will be well away from any gases produced by the charging battery.

Since most new batteries are sealed and you cannot check specific gravity (percentage of acid compared to water), use a voltmeter to check the state of charge (Figure 22–14). A fully charged 12-volt battery should show about 12.6 volts. Anything below this value shows battery drain, requiring a charge or possibly a new battery.

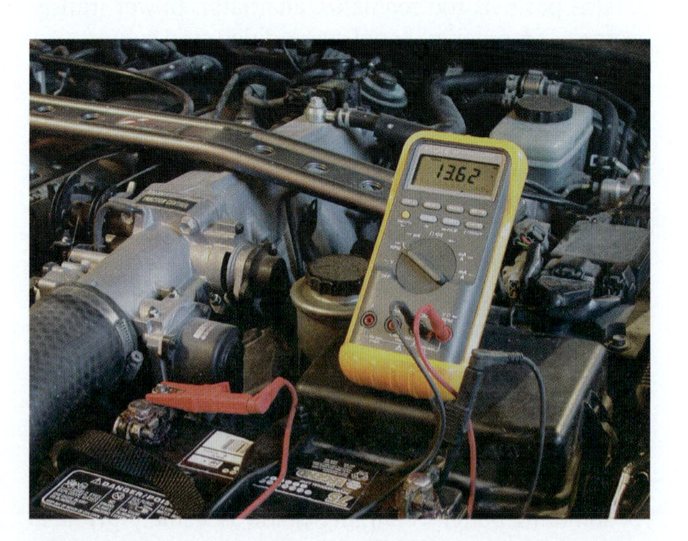

Figure 22-14 An accurate voltmeter will let you check the general state of the charge on modern sealed batteries. A reading of 12.6 volts normally indicates a fully charged battery. A lower reading means the battery should be recharged. If the reading is still low after charging, the battery may be defective. When you start the engine, the alternator should put out 13 to 14 volts to charge the battery.

22.4 ELECTRICAL DIAGNOSTIC EQUIPMENT

Locating an electrical fault is not possible without using diagnostic tools (meters, test lights, jumper wires, etc.). Keep in mind that today's delicate electronic systems can be damaged if the wrong methods and equipment are used.

During a collision, wiring damage is common. Wires can be severed, torn apart, or abraded. You may be required to find and fix these damaged wires.

Testlight and Jumper Wires

An externally powered **testlight** is often used to determine whether there is a complete, unbroken circuit. One lead of the test lamp is connected to a good ground. The other lead connects to a point in the circuit. If the lamp lights, current is flowing through the circuit (Figure 22–15).

Whenever a technician has an electrical system problem that does *not* directly concern the computer, a testlight is handy.

Connect the light to the voltage source and to ground. If the testlight does *not* glow, there is an open circuit somewhere. If the light is on but the part does not work, the part is probably bad (Figure 22–16).

Figure 22–17 shows how to test for a short with a testlight.

Figure 22–18 and Figure 22–19 demonstrate how to use a testlight to check wire continuity, that is, if a wire is broken or has a complete path.

Jumper wires are used to temporarily bypass circuits or components for testing. They consist of a length of wire with an alligator clip at each end. They can be used to test circuit breakers, relays, lights, and other components. Refer to Figure 22–20 and Figure 22–21.

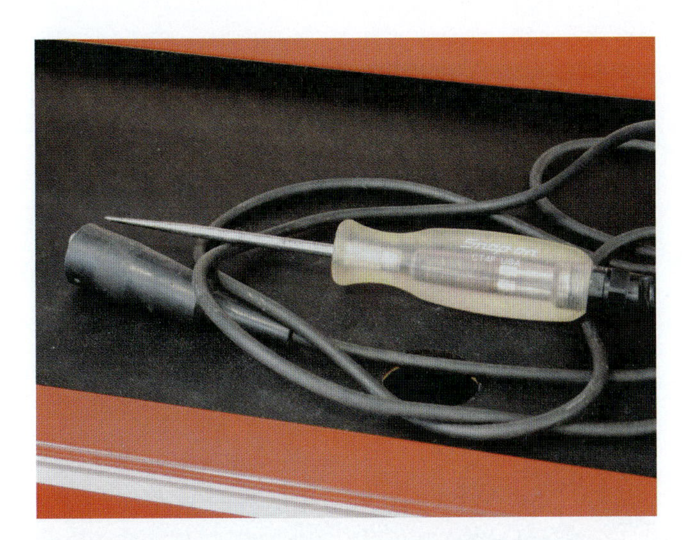

Figure 22–15 A testlight can check for power to electrical, but not electronic, devices. Ground the testlight and touch the pointed probe on the wire. The light will glow if power is being sent through the wiring.

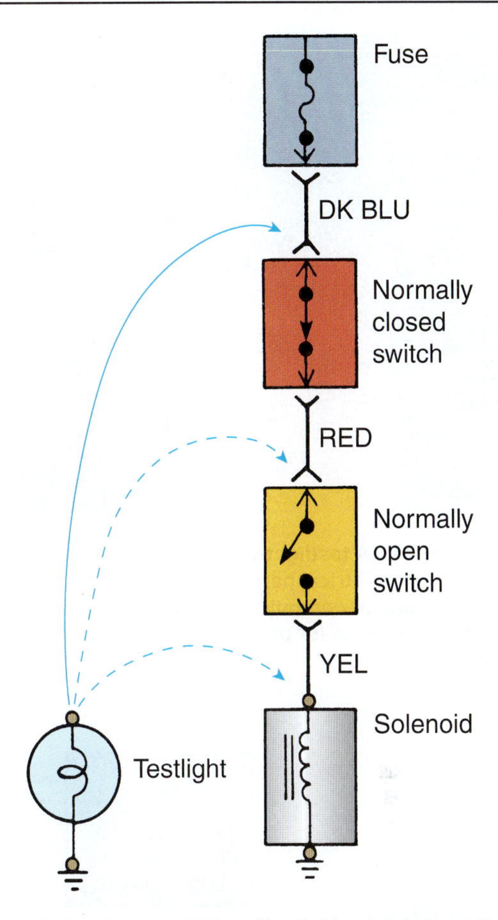

Figure 22–16 A testlight will quickly check for voltage or power at different locations in a circuit to isolate any trouble areas. When you touch a test point and the light does not glow, you have found a fault in the circuit.

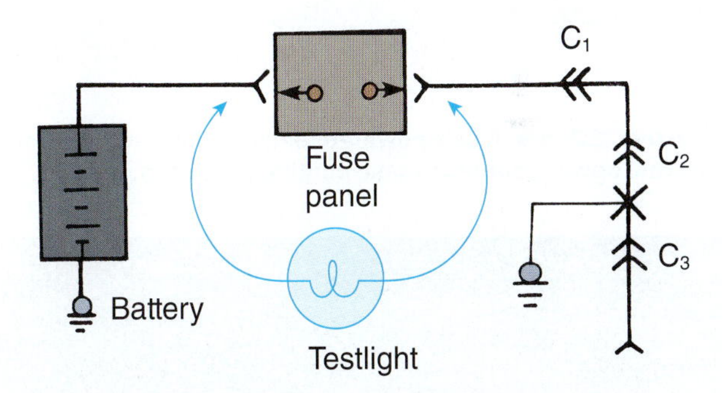

Figure 22–17 If a fuse keeps blowing from a short, remove the fuse and connect a testlight across the terminals. The light will glow as long as the short remains. If you disconnect or fix the shorted circuit, the testlight will go out.

Testlights should *not* be used randomly, only when specified in the service manual procedures or on isolated wires. Do not use a testlight on electronic circuits. They can damage electronic parts if their impedance (internal resistance) is too low.

Figure 22–18 A testlight will quickly check the continuity of an electrical harness. Never use a testlight on computer circuits or computer damage may occur. Here the technician is checking for power to the door window motor.

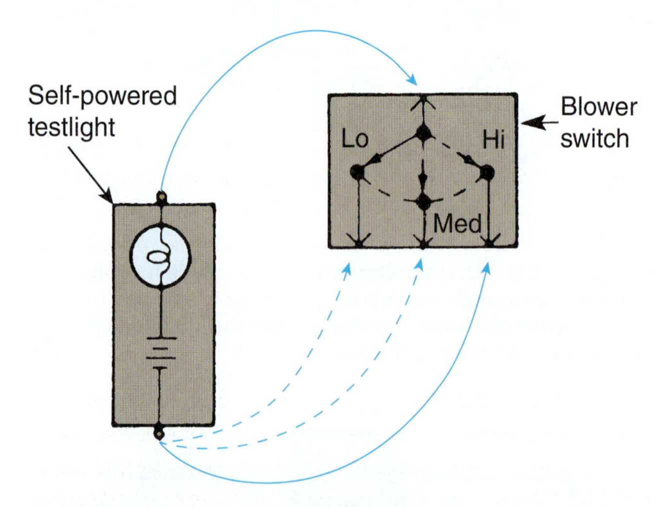

Figure 22–19 A self-powered testlight does not need an external power source to make continuity checks.

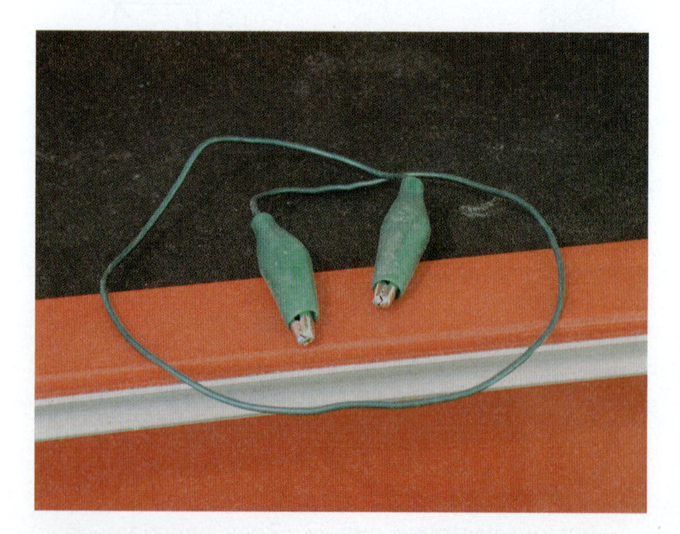

Figure 22–20 Jumper wires provide another way to make quick checks of circuits and parts. They can bypass switches or connect power directly to an electrical component to see whether it works.

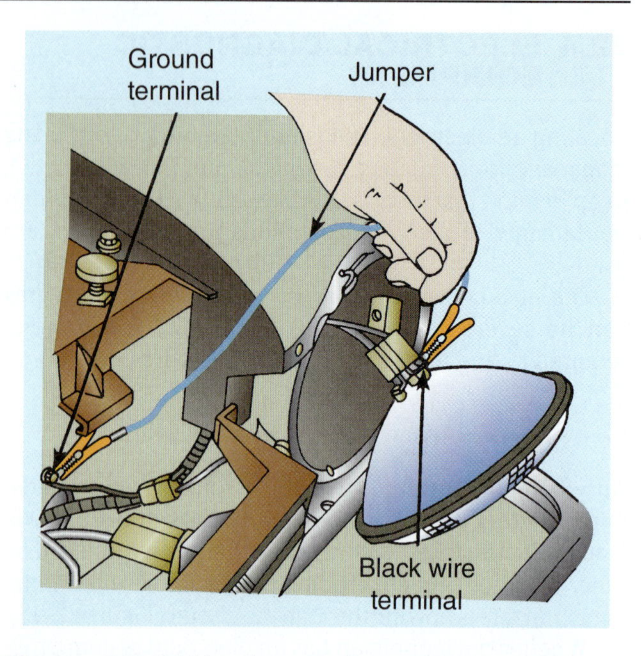

Figure 22–21 Here jumper wires are being used to check for a bad ground wire. If the bulb starts working with the jumper, you have found high circuit resistance or an open in the ground circuit.

Multimeters

A **multimeter** is a voltmeter, ohmmeter, and ammeter combined into one case. Also called a volt-ohm-ammeter (VOM), it can be used to measure actual electrical values for comparison to known good values (Figure 22–22).

A *digital multimeter* (DVOM) has a number readout for the test value. This type of multimeter is recommended by auto manufacturers because it will not damage delicate electronic components. A high-impedance (10 mega-ohm input) DVOM is recommended to avoid damaging sensitive components. Digital readouts give

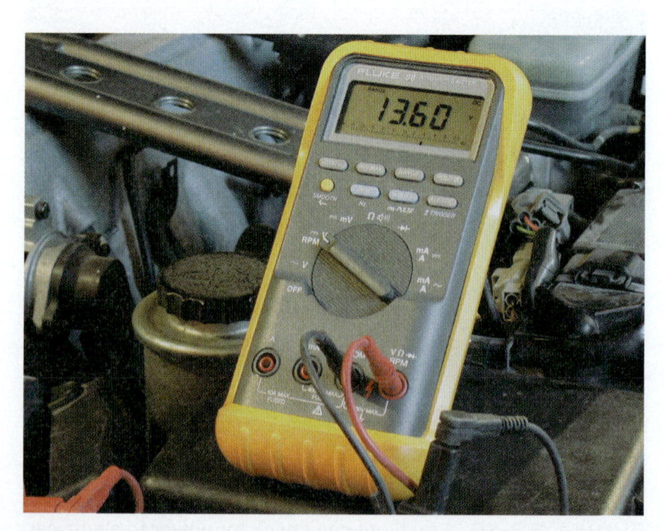

Figure 22–22 A multimeter is an important tool for today's repair technician. It reads voltage, current, and resistance.

Figure 22–23 A digital multimeter with high internal resistance is needed when testing a computer-controlled circuit to avoid damaging delicate electronic components.

the precise measurement needed for proper diagnosis. DVOMs should be used in conjunction with the vehicle's service manual (Figure 22–23).

An *analog multimeter* has a pointer needle that moves across the face of a scale when making electrical measurements. Analog VOMs can damage sensitive electronic components. They should only be used when testing electrical, not electronic, circuits. They help show a fluctuating or changing reading, such as from an intermittent problem.

Measuring Resistance

When measuring resistance, always disconnect the circuit from the power source. The multimeter must never be connected to a circuit that has voltage applied. Doing so can damage the meter.

Use the service manual to determine the normal resistance of the part being checked. For example, a computer system sensor may have a normal resistance reading of 45–55 ohms. You would select the 200 range on the multimeter because the resistance reading of 44–55 ohms will be 200 ohms or less.

Always refer to the equipment manual before using the multimeter for diagnostic purposes. You must make sure that you understand how to use the multimeter before performing diagnostics. All multimeters are fairly similar, but there are differences between makes.

To measure resistance, use the following general procedure:

1. Set the range selector switch to the highest range position. This will protect the meter from damage.
2. Connect the multimeter test leads to opposite ends of the circuit or to the wires being tested.
3. Reduce the range setting until the meter shows a reading near the middle of its scale. Some DVOMs have an auto ranging function that adjusts the settings automatically.

Measuring Voltage

The multimeter allows you to select either alternating current voltage (ACV) or direct current voltage (DCV). ACV is selected when measuring alternating current voltage. AC is the type of current that is found in your home wiring.

DCV is selected when measuring direct current voltage. DC is what is normally measured in an automobile. Signals from sensors can be AC or DC.

When using the multimeter to measure voltage, the selection of the range scale is very important. If the voltage reading will be 12–14 volts, select the 20V range on the multimeter. This range is selected because the voltage reading will be less than 20 volts.

If the lights do not work, check the voltage available to the fuse. If approximately 12 volts are not displayed on the meter, a short exists between the battery and the fuse box. If 12 volts are available, check the voltage between the switch and the first lamp (Figure 22–24).

If voltage is not available, bridge the switch with a lead wire. If the light comes on or the voltmeter reads 12 volts, the switch is defective. If no voltage shows on the gauge, bridge the fuse box with the jumper wire. The short might be either in the box or in the wiring between the fuse box and the switch.

Assume that after repairing the shortage in the same circuit, only the first two lights come on. Check the voltage available to the third and fourth lights. If the meter reads 0 volts, the wire between the two lights is probably broken or disconnected. If 12 volts register on the meter, the problem must be narrowed down to one of the following:

- Defective ground connection
- Defective light socket
- Burned-out bulbs

Try new bulbs in each of the sockets. If that does not solve the problem, test the ground wires of the third and fourth lights with a multimeter set to read voltage. Connect the leads to the ground side terminal of the socket

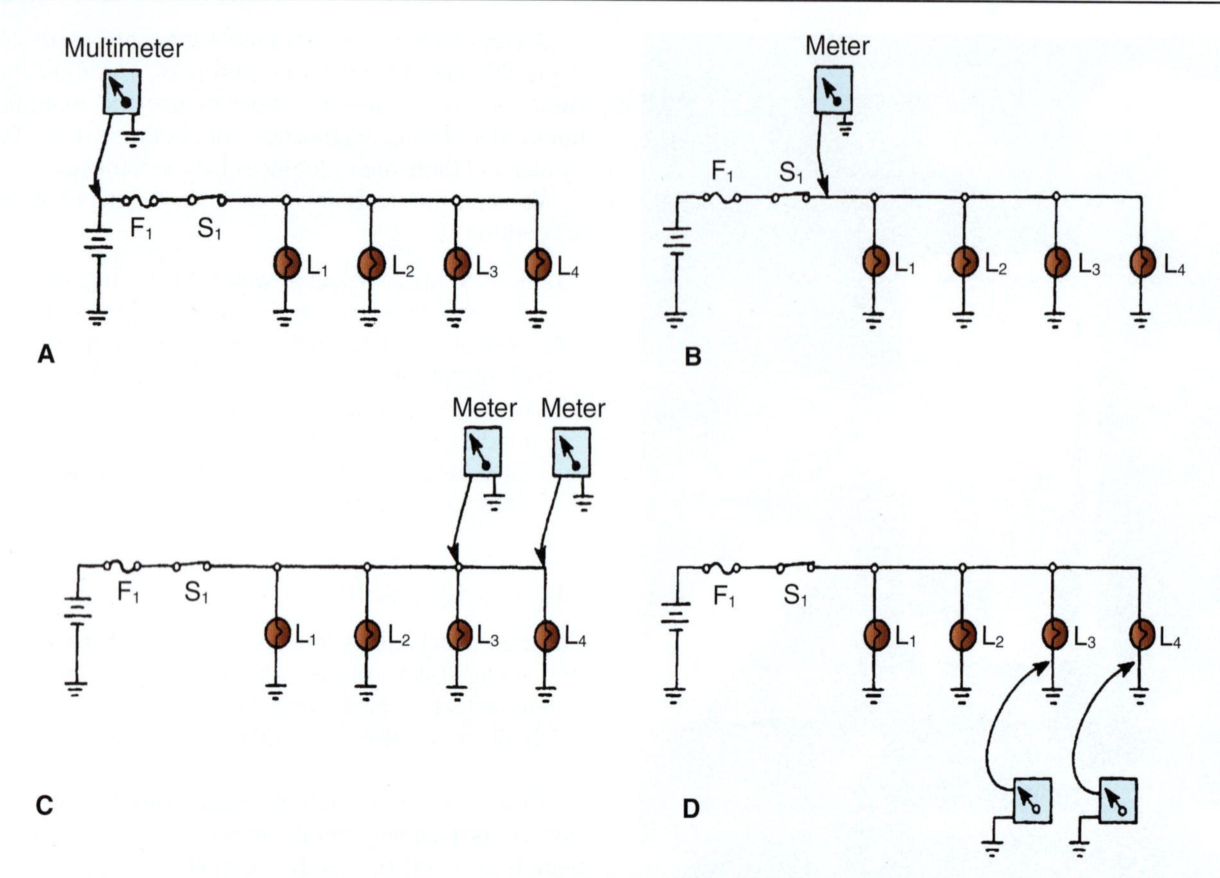

Figure 22-24 Note the various checks that can be made with a multimeter to determine taillight circuit problems. This type of analysis also applies to other types of circuits. (A) Check for power to the fuse panel. (B) Check for voltage after the main switch and fuse. (C) Check for voltage at taillight sockets. (D) An ohmmeter will check for grounds at taillights.

and to a suitable ground. If the meter reads 12 volts, this indicates a defective ground. If 0 volts are read, the sockets are probably bad. Verify that no continuity exists in the bulbs by setting the multimeter to read ohms. Connect the leads to either side of a socket. If the ohmmeter reads "infinity," no continuity exists and the sockets are defective.

If the problem with the taillights is narrowed down to a fuse that "blows" whenever the lights are turned on, the lights are drawing too much current. This indicates a high-resistance short somewhere in the circuit. To determine where the short is, connect the multimeter in series with the circuit. Then, disconnect the hot wire from the first light. If current registers on the ammeter, a shortage somewhere between the switch and the first taillight is grounding the circuit.

If no current flow registers, remove the light bulbs and replace them one at a time. If the circuit is designed to draw 5 amperes, each bulb should draw 1.25 amperes (5 ÷ 4 = 1.25). Replace the first bulb. The ammeter should read 1.25 amperes. When the second lamp is added, the reading should be 2.5 (1.25 + 1.25). Add the third lamp to the circuit, and the reading should be 3.75 amperes; add the fourth lamp, and the reading should be 5 amperes. If a higher than normal reading is achieved at any light, the short drawing the additional current lies between that light and the preceding one. This type of analysis will enable you to find any circuit problem.

Measuring Current

Current or amperage is sometimes measured to check the consumption of power by a load. For example, current draw is often measured when checking the condition of a starting motor.

Modern ammeters have an inductive pickup that slips over the wire or cable to measure current. With older ammeters, the circuit had to be disconnected and the meter connected in series to measure current.

A high-current draw indicates a low resistance, as from a dragging or partially shorted motor. A low-current draw means there is a high resistance in the circuit, as from a bad connection or dirty motor brushes.

Do not connect an ammeter in series with a source of high-current draw. High-current flow through the ammeter could damage it or blow its internal fuse. If high-current draw is possible, use an inductive or clip-on ammeter.

22.5 ELECTRIC COMPONENTS

Service literature will have illustrations showing the locations of electrical and electronic components. Two examples are shown in Figure 22–25 and Figure 22–26.

A **switch** is used to turn a circuit on or off manually (by hand). When the switch is closed (on), the circuit is complete (fully connected) and will operate. Various types of switches can be found in today's vehicles. A bad switch will often be open in both the closed and open positions.

A **solenoid** is an electromagnet with a movable core or plunger. When energized, the plunger is pulled into the magnetic field to produce motion. Solenoids are used in

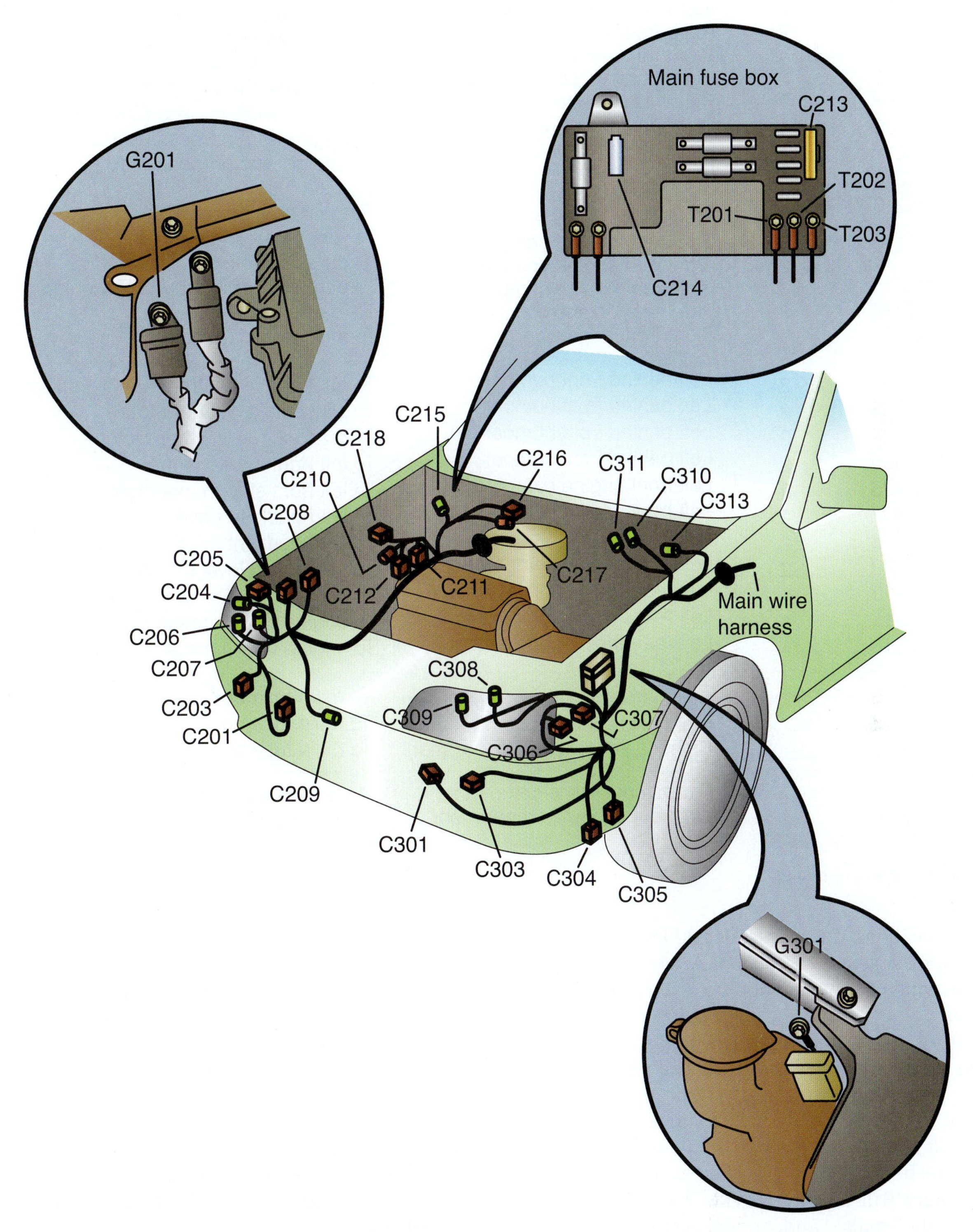

Figure 22-25 Service illustrations show locations of relays and other components in the engine bay for specific vehicles. Note how the accompanying chart in this illustration explains the number of pins, locations, and where wires go from each connector.

(continued)

Connector and terminal number	Number of pins	Location	Where the wire goes
C201	2	Right behind front bumper	To right horn
C203	2	Right behind front bumper	To right front turn signal light
C204	2	Right front engine compartment	To right front marker light
C205	2	Right front engine compartment	To right front side marker light
C206	2	Right front engine compartment	To right headlight (low)
C207	2	Right front engine compartment	To right headlight (high)
C208	4	Right front engine compartment	To radiator fan motor
C209	2	Right front engine compartment	To A/C wire harness (291)
C210	2	Right front engine compartment	To engine wire harness (C101 or C151)
C211	14	Right center engine compartment	To engine wire harness (C201 or C152)
C212	8	Right center engine compartment	To engine wire harness (C153)
C213	6	Right center engine compartment	To main fuse box (C901)
C214	4	Right center engine compartment	To main fuse box (C902)
C215	3	Right center engine compartment	To windshield wiper motor
C216	5	Right rear engine compartment	To purge cut-off SOL.V
C217	2	Right rear engine compartment	To purge cut-off SOL.V/EGR control
C218	4	Right rear engine compartment	To map sensor
	3	Right rear engine compartment	To auto shoulder seat belt fuse
	2	Right front engine compartment	
C301	2	Left behind front bumper	To left horn To left front turn signal light
C303	2	Left behind front bumper	To windshield washer motor
C304	2	Left behind front bumper	To rear window washer motor
C305	2	Left behind front bumper	To left front marker light
C306	2	Left front engine compartment	To left front side marker light
C307	2	Left front engine compartment	To left headlight (low)
C308	2	Left front engine compartment	To left headlight (high)
C309	2	Left front engine compartment	To brake fluid level switch (+)
C310	1	Left rear engine compartment	To brake fluid level switch (-)
C311	1	Left rear engine compartment	To engine wire harness (C122)
C313	14	Left rear engine compartment	To engine wire harness (C175)
T201		Right center engine compartment	To main fuse box
T202		Right center engine compartment	To main fuse box
T203		Right center engine compartment	To main fuse box
G201		Right front engine compartment	To body ground, via main harness
G301		Left front engine compartment	To body ground, via main harness

Figure 22–25 ***(continued)***

many applications: door locks, engine idle speed, emission control systems, and so on. A bad solenoid can develop winding opens, shorts, or a high-resistance problem.

A **relay** is a remote control switch. A small switch can be used to energize the relay. The relay coil then acts on a moveable arm to close the relay contacts. This allows a small switch to control high current going to a load. A relay is commonly used with electric motors because they draw heavy current. The service manual will usually provide relay locations for the specific make and model vehicle. Refer to Figure 22–27.

Bad relays will often have burned points that prevent current flow to the load. They can also develop coil opens and shorts that keep the points from closing.

An *inertia switch* is designed to shut the electric fuel pump off after a collision. It must be reset for the engine to restart after a collision. A button on the side of the inertia switch must be pressed to reclose the fuel pump circuit (Figure 22–28).

Motors use permanent magnets and electromagnets to convert electrical energy into a rotation motion for doing work. Some examples are the electric starting motor for the engine and stepper motors for computer control of parts.

Faulty motors can have worn bushings and brushes that decrease efficiency. They can also have winding shorts and opens that prevent motor operation.

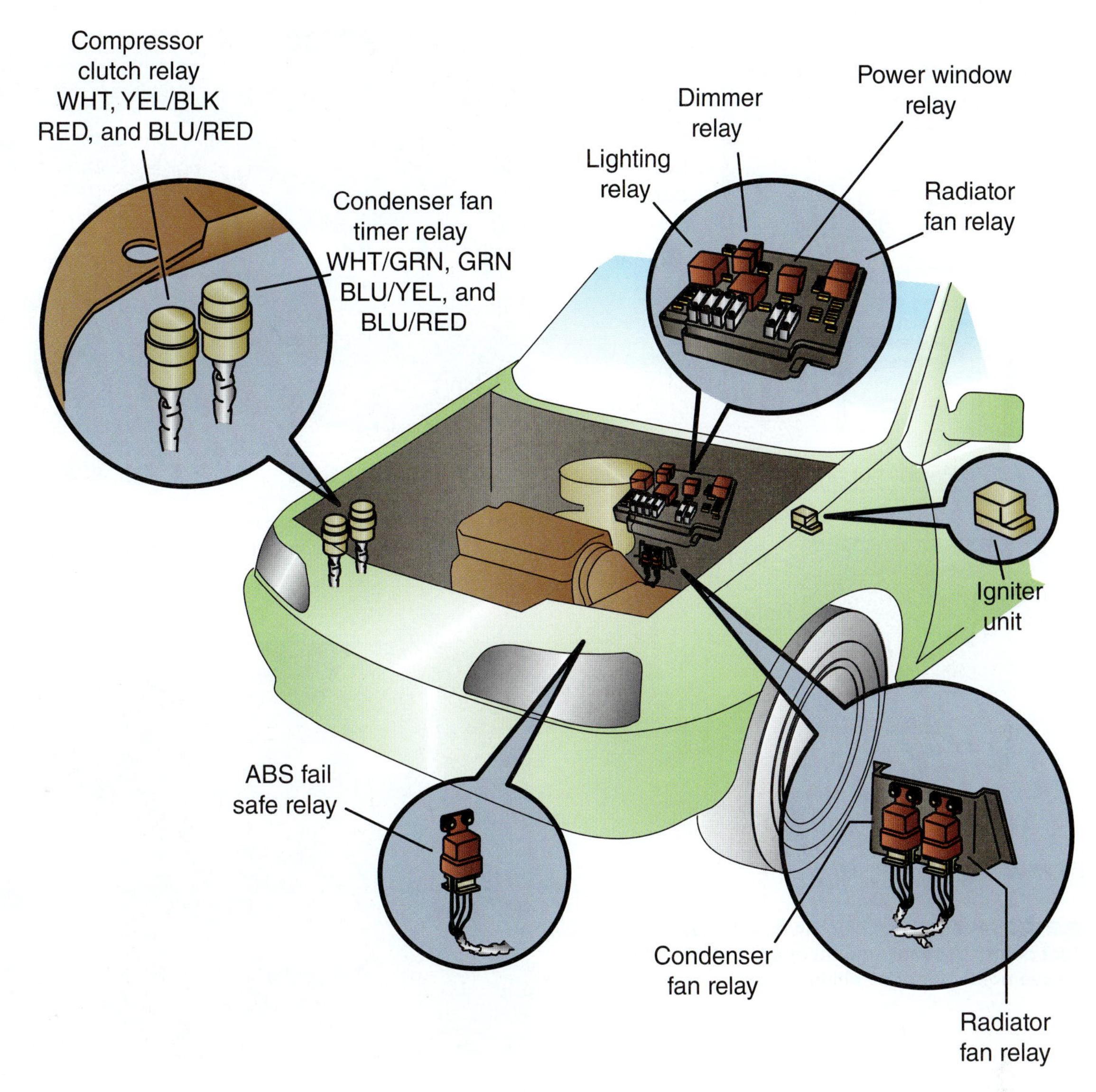

Figure 22–26 This illustration shows the electrical components found in the engine bay of one vehicle. During a hard frontal collision, these parts can be smashed and damaged, requiring replacement.

22.6 CIRCUIT PROTECTIVE DEVICES

Circuit protection devices prevent excess current from burning wires and components. With an overload or short, too much current tries to flow. Without a fuse or breaker, the wiring in the circuit would heat up. The insulation would melt and a fire could result.

A **fuse box**, or fuse panel, holds the various circuit fuses, breakers, and sometimes the flasher units for the turn and emergency lights. It is often located under the instrument panel, behind a panel in the foot well, or in the engine compartment.

Many circuit breakers and main fuses are housed in the main fuse box. The fuse panel can be located under the dash, in the engine compartment, under the back seat, and in other locations. Usually, an illustration on the lid of the fuse panel gives the names and ratings for each fuse and circuit breaker (Figure 22–29).

Fuses

Fuses burn in half with excess current to protect a circuit from further damage. They are normally wired between the power source and the rest of the circuit. There are three types of fuses in automotive use: cartridge, blade, and ceramic (Figure 22–30).

The cartridge fuse is found on most older domestic vehicles and a few imports. It is composed of a strip of metal enclosed in a glass or transparent plastic tube. To check the fuse, look for a break in the internal wire or metal strip. Discoloration of the glass cover or glue bubbling around the metal end caps is an indication of overheating.

Late model domestic vehicles and many imports use blade, or spade, fuses. To check the fuse, pull it from the fuse panel and look at the element through the transparent plastic housing. Look for internal breaks and discoloration.

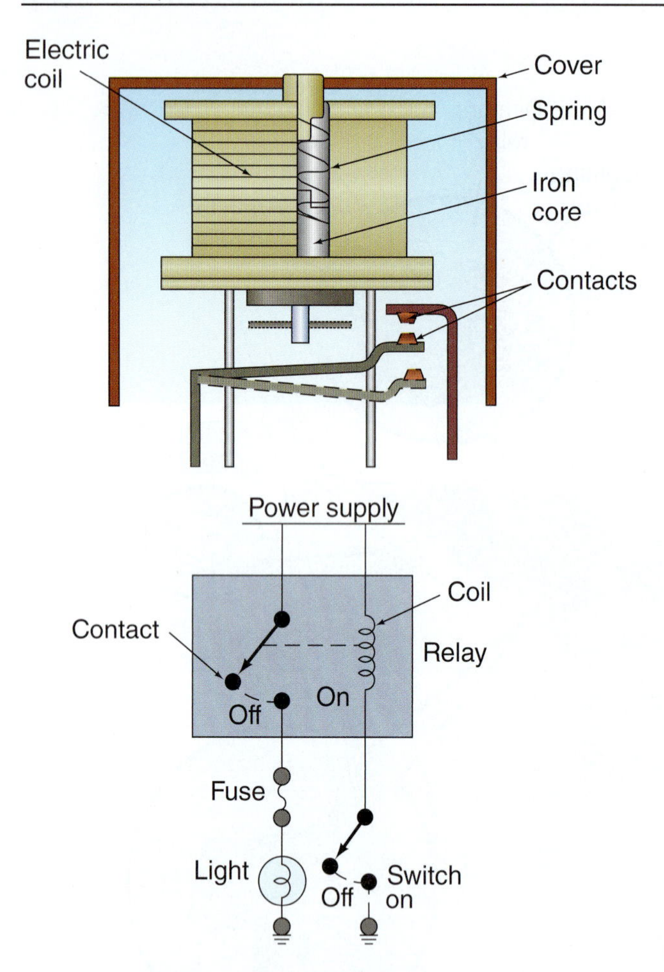

Figure 22–27 A relay is an electric coil that operates small contact points. A small amount of current will cause the coil to close points for controlling much larger current flow.

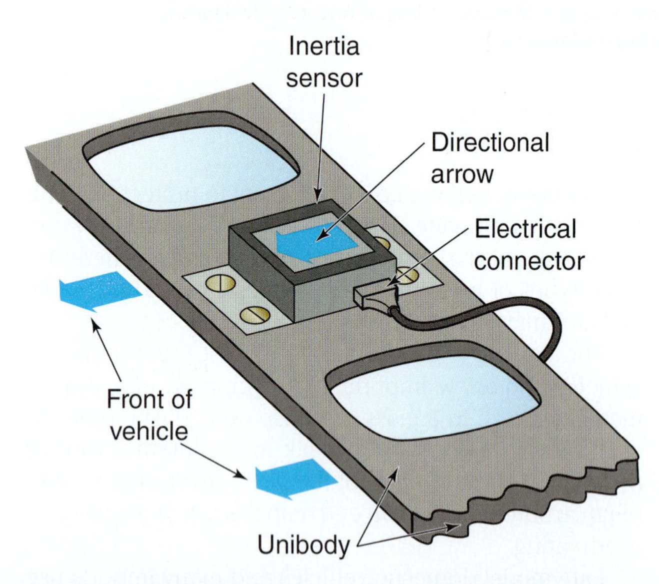

Figure 22–28 The inertia switch in the trunk must often be reset after a collision. It will open during a collision to shut off the electric fuel pump to prevent a possible gasoline fire. When you press the reset button, the fuel pump and engine will then operate.

Figure 22–29 The fuse box can be located under the dash or rear seat cushion or in the engine compartment.

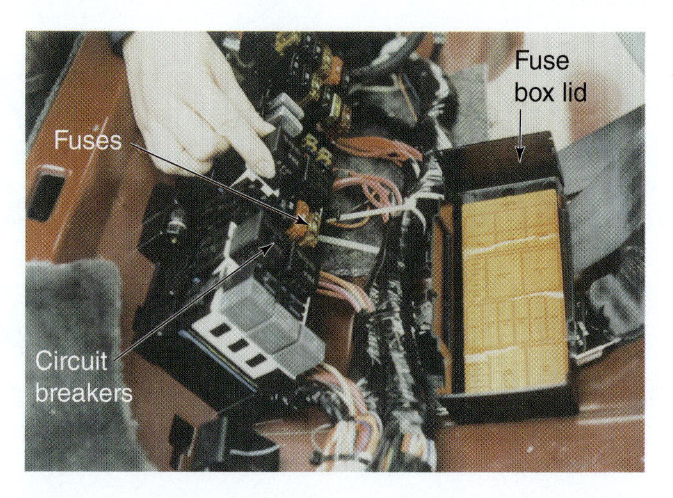

Figure 22–30 The fuse box lid normally provides a key that explains what each fuse protects.

The ceramic fuse is used on many European imports. The core is a ceramic insulator with a conductive metal strip along one side. To check this fuse, look for a break in the contact strip on the outside of the fuse.

All fuse types can be checked with a circuit tester or multimeter (Figure 22–31). A blown fuse will have infinite resistance.

Fuse ratings are the current at which the fuse will blow. Fuse ratings are printed on the fuse. Always replace a fuse with one of the same amp rating; if you do not, part damage can result from excess current flow.

Never permanently bypass a fuse or circuit breaker with a jumper wire. Do it only for test purposes. An electrical fire can result from excess current flow.

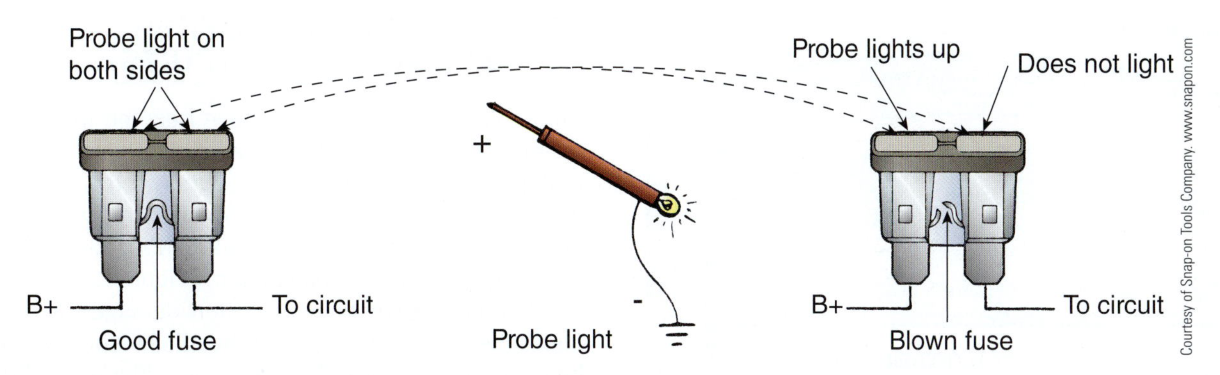

Figure 22-31 A testlight can be used to quickly find dead fuses. The testlight should glow when touched on both sides of a fuse. A dead fuse will only light the testlight on the hot side.

Fuse Links

Fuse links, or fusible links, are smaller diameter wires spliced into the larger circuit wiring for overcurrent protection. Fuse links are normally found in the engine compartment near the battery. They are often installed in the positive battery lead that powers the headlights and other circuits that are live with the key off (Figure 22–32).

Fuse link wire is covered with a special insulation that bubbles when it overheats. This indicates that the fuse link has melted. If the insulation appears good, pull lightly on the wire. If the fuse link stretches, the wire has burned in half. When it is hard to determine whether the fuse link has burned out, perform a continuity check.

When replacing fuse links, first cut the protected wire where it is connected to the fuse link. Then, solder a new fuse link of the same rating in place. The insulation on manufactured fuse links is flameproof. Never fabricate a fuse link from ordinary wire, because the insulation may not be flameproof. Refer to Figure 22–32.

A number of new electrical systems use maxi-fuses in place of traditional fuse links. Maxi-fuses look and operate like two-prong blade fuses, except they are much larger and can handle more current. Maxi-fuses are often located in their own underhood fuse block.

Maxi-fuses are easier to inspect and replace than fuse links. To check a maxi-fuse, look at the fuse element through the transparent plastic side housing. If there is a break in the element, the maxi-fuse has blown. To replace it, pull it from its fuse box or panel. Always replace a blown maxi-fuse with a new one of the same amp rating.

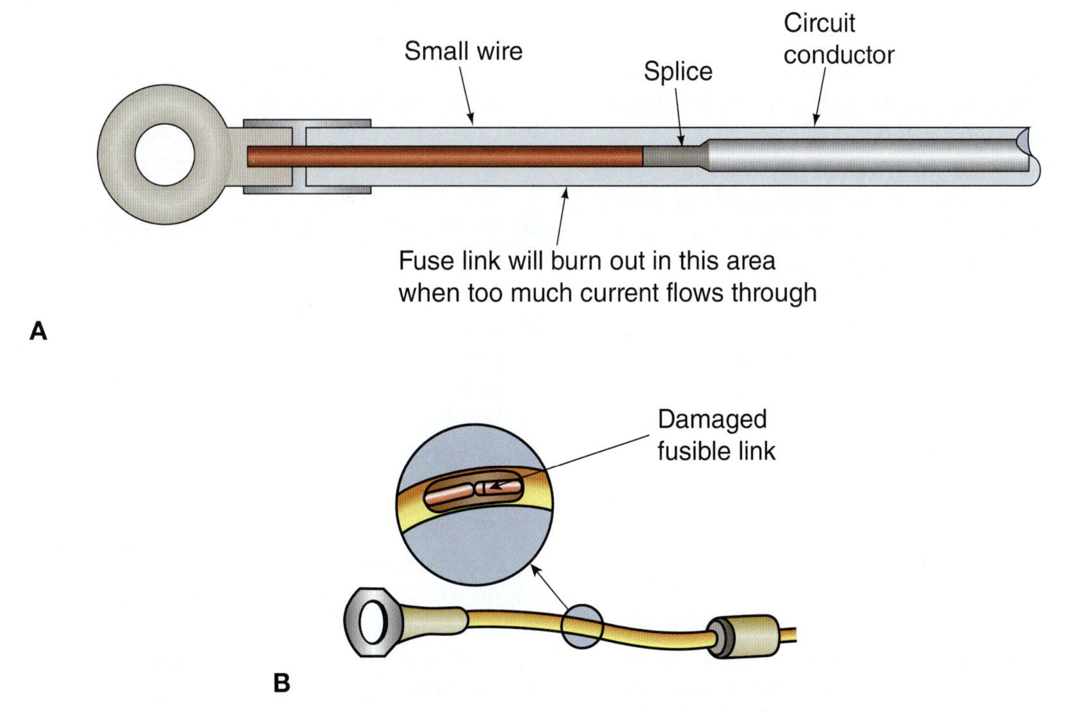

Figure 22-32 Note the construction of a fuse link. (A) A fuse link has a small wire in series with a larger wire. (B) The small wire will melt and break the circuit connection if the current goes too high.

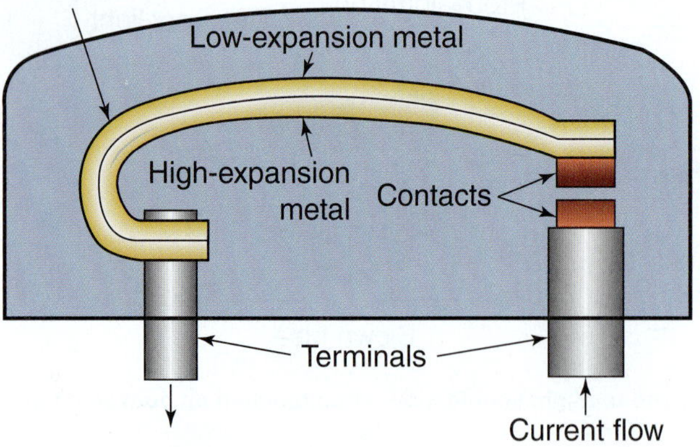

Figure 22-33 **A circuit breaker uses a bimetal arm that bends with heat-generated current flow. If the current goes too high, heating of the bimetal arm causes it to bend and open circuit contacts. When the bimetal arm cools, it bends back to reconnect the circuit.**

Circuit Breakers

Circuit breakers heat up and open with excess current to protect the circuit. They do not suffer internal damage as a fuse does. Many circuits are protected by circuit breakers. They are normally fuse panel-mounted. Like fuses, they are rated in amperes.

Each circuit breaker conducts current through an arm made of two types of metal bonded together, known as a bimetal arm. If the arm starts to carry too much current, it heats up. As one metal expands farther than the other, the arm bends, opening the contacts and breaking the current flow. A simplified circuit breaker is shown in Figure 22–33.

In the cycling breaker, the bimetal arm will begin to cool once the current to it is stopped. When it returns to its original shape, the contacts close and power is restored. If the current is still too high, this cycle of breaking the circuit will be repeated.

Cycling circuit breakers are generally used in circuits that are prone to occasional overloads, such as power windows. A jammed or sticking window can overwork the motor, so a circuit breaker prevents the motor from burning out.

Noncycling breakers open and must be manually reset to close the circuit. In automotive work, two types of noncycling circuit breakers are used. One is reset by removing the power from the circuit. The other type is reset by depressing a reset button.

When replacing fuses and circuit breakers, always install a replacement with the same amp rating. A higher rated unit could cause an electrical fire.

22.7 LIGHTING AND OTHER ELECTRICAL CIRCUITS

Automotive lighting systems have become increasingly sophisticated. Headlights and taillights have evolved into multiple-light systems. Indicator lights on the dashboard commonly warn of failure of the charging system, seat belts, brake systems, parking brakes, door latches, directional lights, and computer system (Figure 22–34).

Headlight assemblies are often damaged in frontal collisions. Consequently, new headlight housings must be installed to replace damaged ones (Figure 22–35).

Headlights have both high and low beams. Two circuits feed out to the bulbs. A switch in the turn signal cluster often operates the high and low beams. A few trucks still have a floor-mounted switch.

Make sure you check the operation of both high and low beams after repairs. Many late model cars use a cartridge-type headlight bulb (Figure 22–36).

Aiming Headlights

If the vehicle was involved in a front-end collision, the headlights should be adjusted after needed parts have been replaced. Most modern headlight bulbs are a small cartridge type that snaps into a plastic housing. A glass or plastic lens fits over the front of the housing.

Before making adjustments to a vehicle's headlights, make the following inspections to ensure that the vehicle is level. Any one of the adverse conditions listed here can result in an incorrect setting.

- If the vehicle is heavily coated with snow, ice, or mud, clean the underside with a high-pressure stream of water. The additional weight can alter the riding height.
- Ensure that the gas tank is half full. Half a tank of gas is the only load that should be present on the vehicle.
- Check the condition of the springs or shock absorbers. Worn or broken suspension components will affect the setting.
- Inflate all tires to the recommended air pressure levels. (Take into consideration cold or hot tire conditions.)
- If collision damage requires straightening of the frame, make sure that the wheel alignment and rear axle tracking path are correct before adjusting the headlights.
- After placing the vehicle in position for the headlight test, bounce the vehicle by pushing down on the bumper or front fenders to settle the suspension.

Normally, the body shop will have a wall of the shop set up to make correct headlight adjustments. Once the vehicle is properly interfaced with the alignment unit, the horizontal and/or vertical adjustment screws on the headlight are adjusted for the proper reading or indication on the unit. The headlights should be aimed or adjusted after a frontal collision (Figure 22–37).

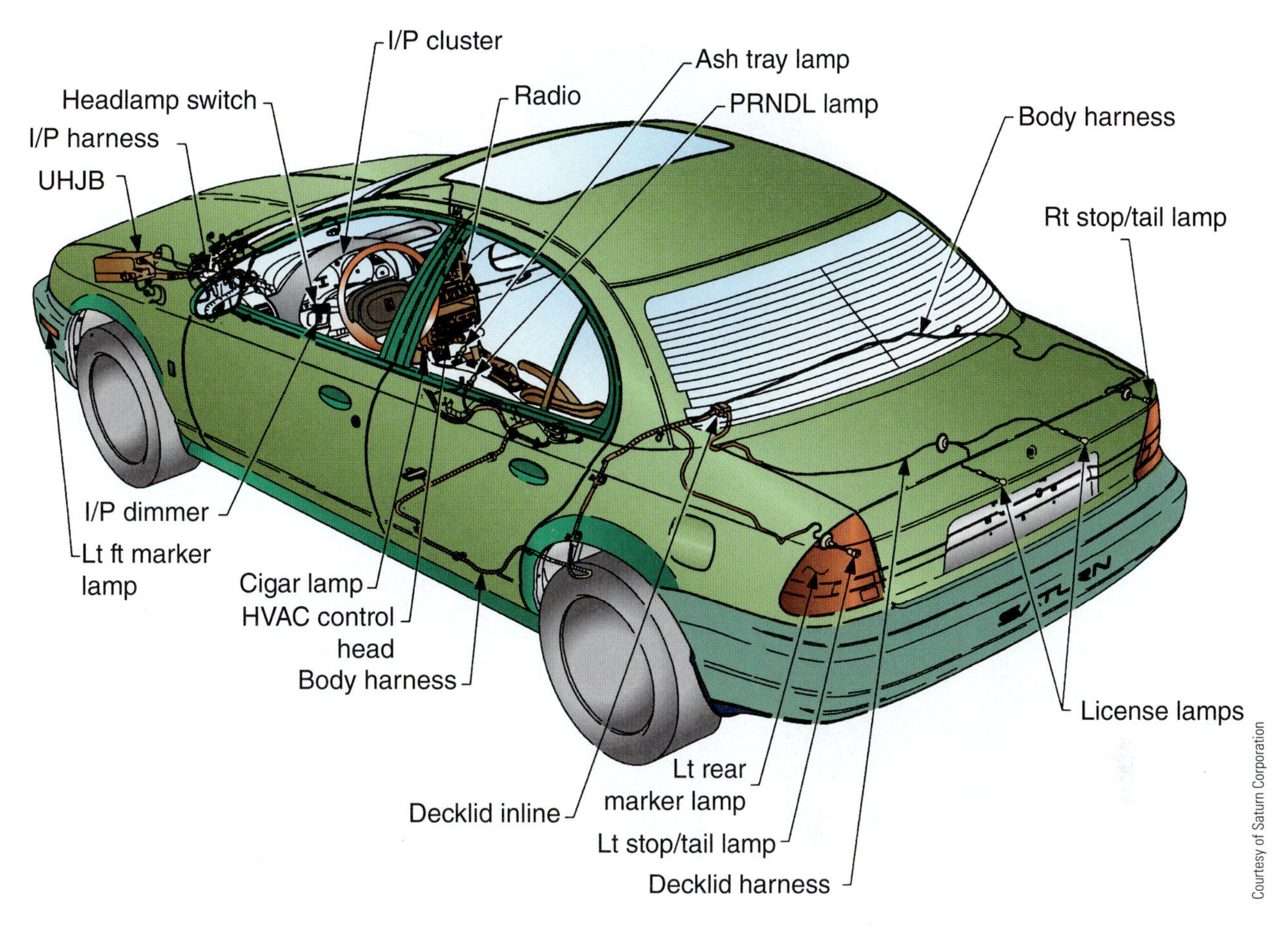

Figure 22-34 Study the typical automotive lighting systems.

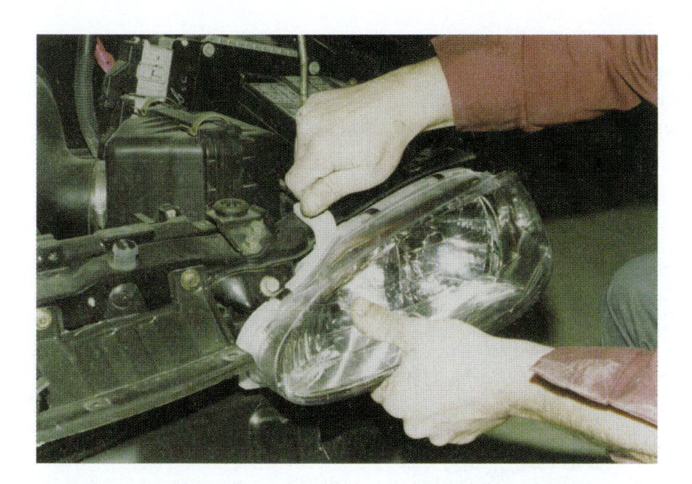

Figure 22-35 When modern headlights are damaged, new plastic lenses and housing assemblies must be installed. Reconnect wires before bolting the headlight into place.

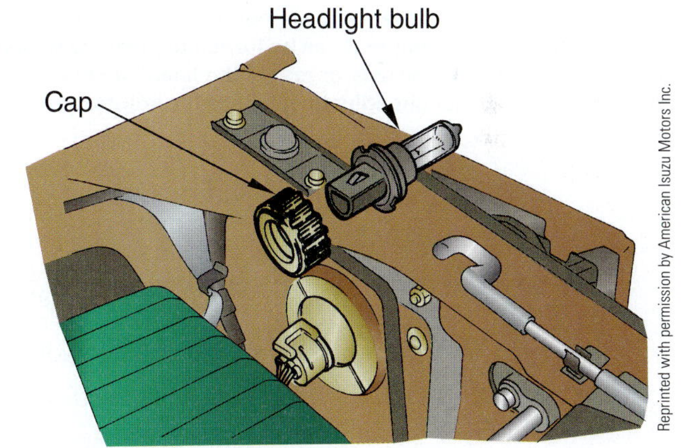

Figure 22-36 Most modern headlight bulbs are small cartridge-type bulbs that snap in from the rear. Do not touch the surface of the bulb with your hands during installation. Oil on your skin can cause the bulb to break at operating temperature.

Some late model vehicles have built-in headlight leveling bubbles for aiming the headlight beams. They are often mounted on top of each headlight assembly.

Headlight adjustment screws can be turned to tilt the headlight lens up or down or right or left for adjustment.

With modern headlights, use the headlight aiming setup shown in Figure 22–38. You can measure and mark the shop floor and a wall so that the headlight beams strike the wall at specified heights. The vehicle must be located the correct distance from the wall.

Tail, Backup, and Stoplights

Failure of a tail, backup, stop, or directional light on a vehicle can usually be attributed to a faulty bulb. Often

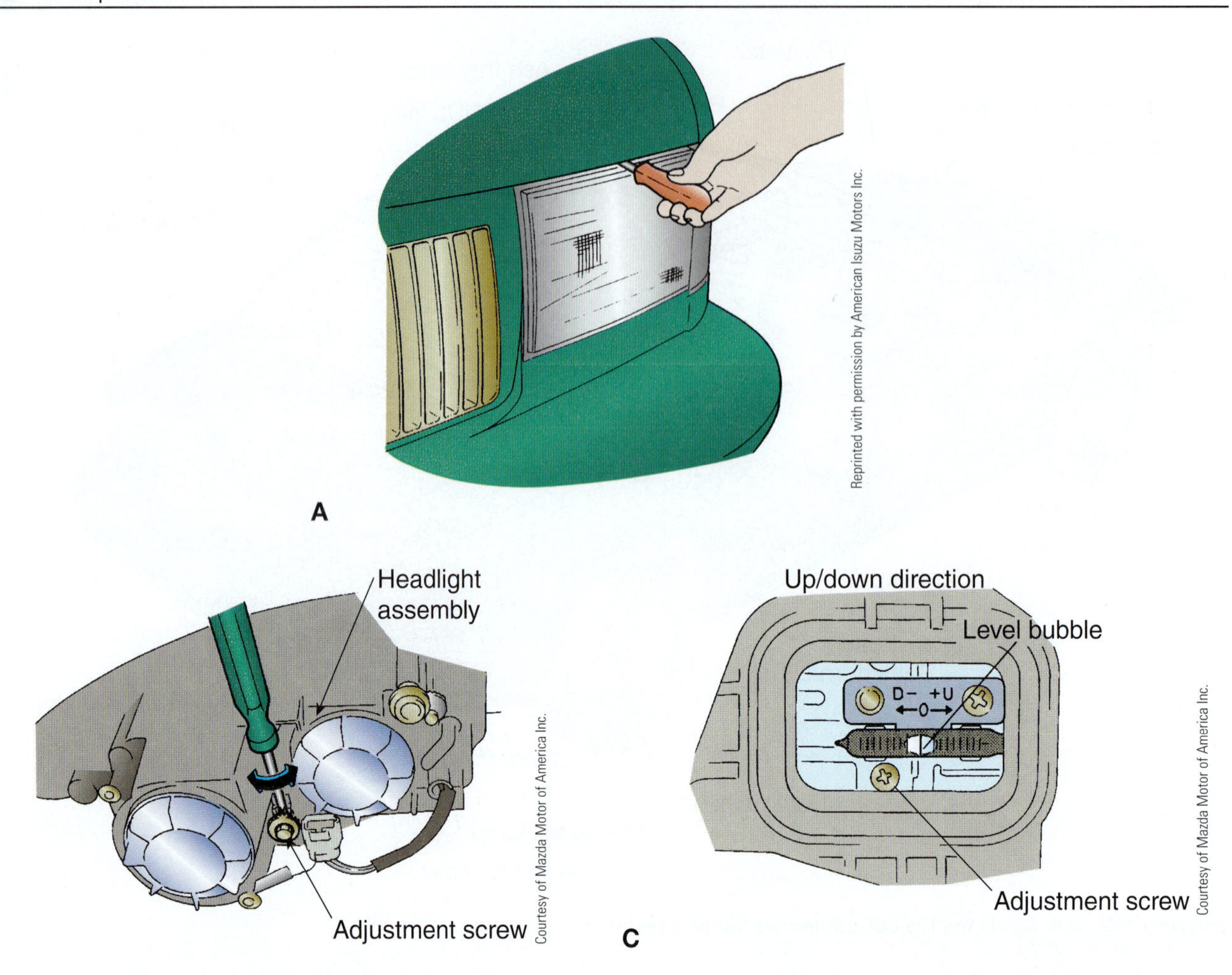

Figure 22-37 Note the basic method for adjusting headlight beams. (A) Many vehicles, especially older ones, have small screws under the trim ring that can be turned to pivot the headlight bulb in its housing. (B) Late model vehicles often have adjustment screws on the side or rear of the headlight housing. (C) Some new vehicles have a leveling bubble on the headlight housing to aid in headlight adjustment. Turn the adjustment screws until the bubbles are centered next to the lines on the scales.

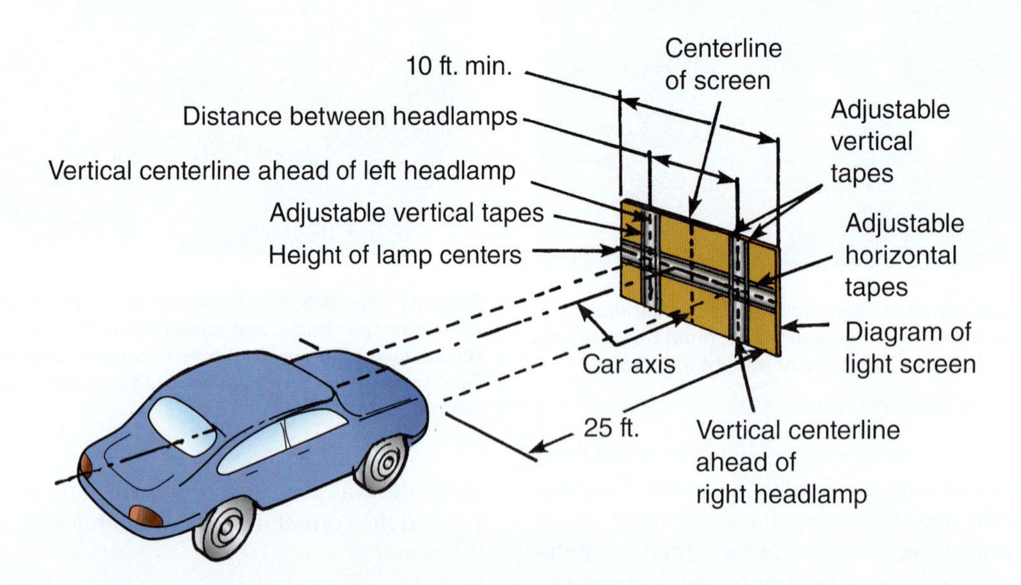

Figure 22-38 Note how the shop wall and floor have been marked for aiming headlights. Pull a vehicle up to the line drawn 25 feet from the wall. Measure the headlight heights, the width between bulbs, and the centerline. Draw lines on the wall or mark them with masking tape. When the headlights are turned on, they should hit the wall on your alignment marks.

moisture gets into the bulb socket and causes corrosion of the electrical contacts and the bulb. Corrosive conditions can be repaired by using sandpaper on the affected areas. For severe cases, replace the socket and/or bulb.

After any repair, always attempt to waterproof the assembly to prevent future problems. Sometimes, *dielectric grease* (special electrically conductive grease) is recommended in connectors and sockets.

Backup, tail, stop, and directional lights are also contained within a lens or bezel-type assembly, usually amber or red in color. Cracked or broken assemblies are easily replaced; they are secured by attaching hardware readily accessible to the body technician.

The turn, brake, and running lights use bulbs that fit into removable sockets. The vehicle service manual will give directions for accessing and changing bulbs. Quite often, you must reach in behind the bulb and turn to pop out the bulb and socket for service (Figure 22–39).

Before releasing any vehicle to the customer, make sure all light bulbs are working properly. Turn all lights on and walk around the car. If any bulbs are burned out or not functional, find out why. You may have to install a new bulb or fix wiring problems.

Figure 22–40 shows a wiring diagram for an interior light circuit. Note the symbols and connections in the circuit.

A few of the other circuits in an automobile include power seats, windows, door locks, mirrors, and cruise control. Other electrical devices include radios, cassette players, speaker systems, chimes, buzzers, graphic displays, analog instruments, and computer commands. Each can be fixed if the basic testing methods described earlier are followed.

When working on any electrical circuit, make sure all wires are routed in their original locations. Also, make sure all clips that hold wires to the body are reinstalled. This will prevent the wires from moving and possibly being damaged by contact with hot or moving parts. See Figure 22–41.

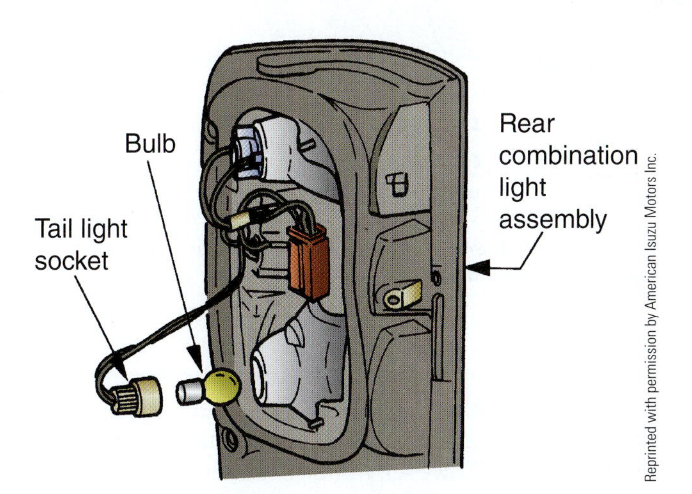

Figure 22–39 **Note the typical mounting of bulbs for tail and brake lights. They should be checked before releasing the vehicle. Collision impact can fracture bulb elements.**

Windshield Wipers and Washers

A typical windshield wiper operates on a small single- or multispeed electric motor. A switch on the steering wheel assembly or dashboard activates the motor. The spray washer generally has its own motor, plastic container, or reservoir and pump, which forces liquid through plastic tubing to the nozzles. The nozzles spray the liquid washer fluid on the windshield (Figure 22–42).

Horn

Most horn systems use the steering wheel switch and relay for sounding the horn. When the horn button, ring, or padded unit is depressed, electricity flows from the battery through a horn lead, into an electromagnetic coil in the horn relay to the ground. A small flow of electric current through the coil energizes the electromagnet, pulling a movable arm. Electrical contacts on the arm touch, closing the primary circuit and causing the horn to sound (Figure 22–43).

Starting and Charging Systems

The **starting system** has a large electric motor that turns the engine flywheel. This spins, or "cranks," the crankshaft until the engine starts and runs on its own power. Figure 22–44 shows a typical starting system wiring diagram.

The ignition switch in the steering column is used to connect battery voltage to a starter solenoid or relay. Other ignition switch terminals are connected to other electrical circuits. A starter solenoid, when energized, connects the battery and starting motor.

The *starting motor* is a large DC motor for rotating the engine flywheel. It normally bolts to the lower rear of an engine. On a few models, the starting motor is mounted inside the engine under the intake manifold. The flywheel ring gear meshes with the starter-mounted gear while cranking.

The **charging system** recharges the battery and supplies electrical energy when the engine is running. An alternator or belt-driven DC generator produces this electricity. Figure 22–45 shows a wiring diagram for a typical charging system.

A *voltage regulator*, usually mounted on the alternator, controls alternator output. Charging system voltage is typically 13–15 volts.

To quickly check the condition of a charging system, connect a voltmeter across the battery. With the engine off, you will read battery voltage. It should be above 12.6 volts. If not, the battery needs charging or is defective. When you start the engine with all electrical accessories on (lights, radio, and so on), the voltage must stay

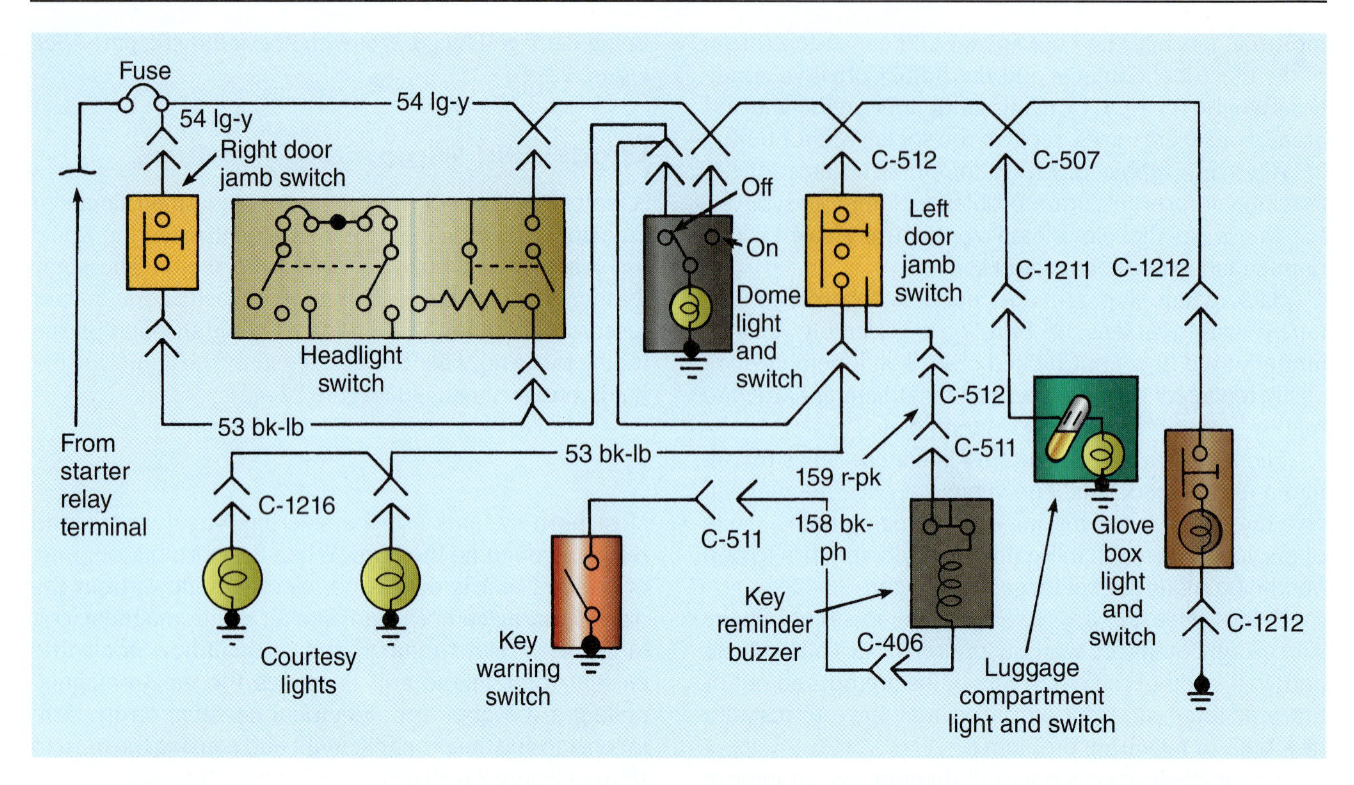

Figure 22–40 Study a typical interior light wiring diagram. Can you find the dome light switch, right door jamb switch, and mercury switch?

Figure 22–41 Be sure wiring harness groupings are secure in their clips or holders. When doing collision work, always check direct damage areas for small cuts in the harness that could mean major wiring repairs are needed.

above battery voltage. If not, there is something wrong with the charging system.

Electrical Troubleshooting Charts

A **troubleshooting chart** provides diagnostic procedures that are read from the top to the bottom in a specific sequence. One is shown in Figure 22–46.

The most efficient way to locate an electrical fault is to follow the manufacturer's diagnostic steps. These step-by-step procedures are used hand in hand with the wiring diagrams. If, however, troubleshooting charts are not available, begin at the battery and systematically trace the circuit from there. Do not jump back and forth in the circuit or between circuits. Be patient and narrow down possible trouble spots by a process of elimination.

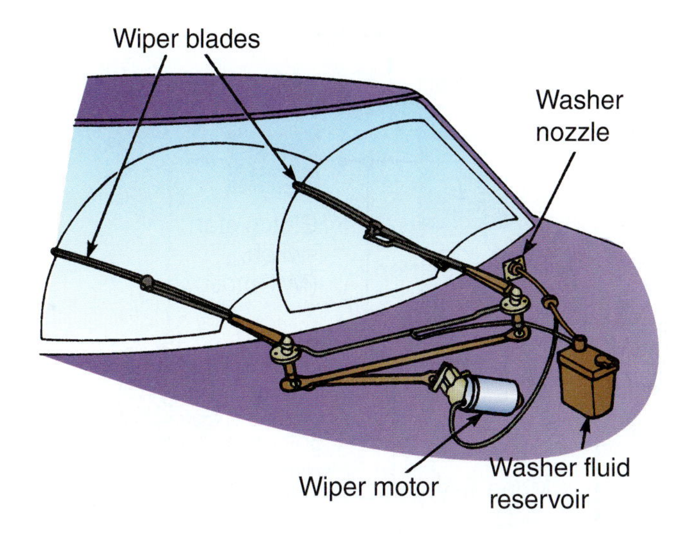

Figure 22-42 This is a typical configuration of windshield wipers and washers. The wiper motor normally attaches to the firewall, either in the engine compartment or under the dash. The washer bottle is often cracked during frontal collisions and should be inspected for leakage and normal operation.

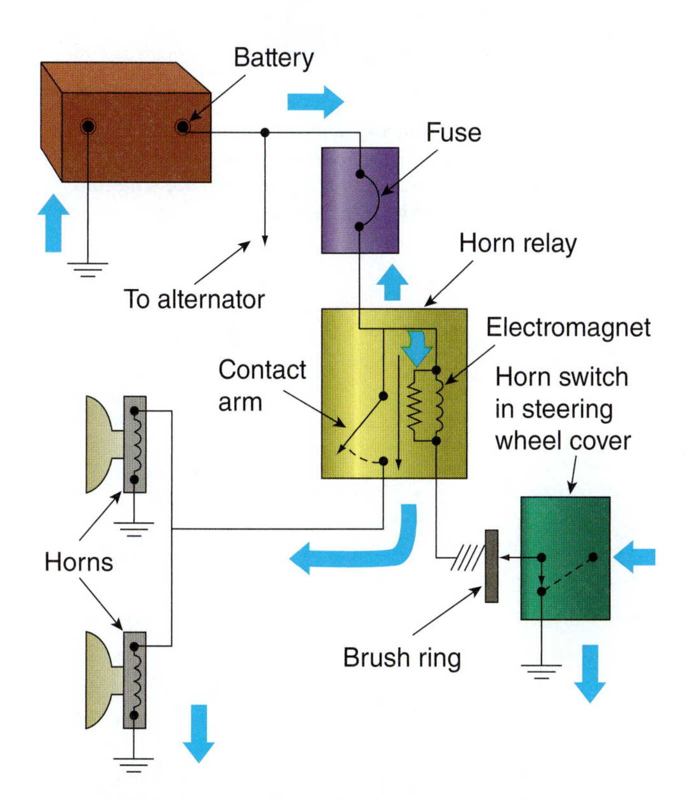

Figure 22-43 Study a basic horn circuit. When you press the horn button, you close the switch in the steering wheel cover. Current passes through the sliding brush ring in the steering column and to the relay. A small current flow energizes the coil in the relay to pull the contacts closed. The relay then allows a high current flow through the fuse, relay, and wiring to the horns to make a loud "beep" sound.

Faulty grounds are another common source of circuit problems (Figure 22–47). A bad ground can prevent current from returning to the battery, causing a dead circuit. The service manual will give ground locations for circuit testing with a multimeter.

Figure 22–48 shows a technician checking for a poor battery cable ground. While cranking the engine, a voltage drop reading was taken across the cable end and ground. A high voltage drop indicates a poor electrical connection.

The service manual may also provide **component location diagrams** for finding electrical parts (harnesses, sensors, switches, computers, and so forth). It is helpful when you are having trouble finding something.

An example of a component location diagram is shown in Figure 22–49.

Battery Drain

A *battery drain* is a problem that causes current to flow out of the battery when everything should be off. For example, a low-amperage short from a severed hot wire can gradually drain a vehicle's battery. Too often when this happens, the body shop's remedy is to simply recharge the battery. Charging the battery can get the vehicle back to the owner, but it is only a temporary solution. The owner will have a dead battery in a day or two.

To find a battery drain, connect an ammeter between one of the battery cables and the battery. Turn everything off. Disconnect the fuse for the clock; the meter should show little or no current draw. If you still have a drain, keep removing fuses until the meter shows zero. This will tell you which circuit has the short. Pinpoint test this circuit until the shorted wire is found and repaired. When it is corrected, the ammeter should show only a tiny current flow to the clock.

Engine No-Start

If an engine cranks but fails to start, check for "spark" and "fuel." Both are needed for an engine to operate.

To *check for spark*, pull off one of the spark plug wires. Install an old spark plug into the wire and lay the spark plug on an engine ground. When you crank the engine, a bright spark should jump across the spark plug gap. If not, something is wrong with the ignition system (blown fuse, damaged wires, crushed components, or so forth).

If you have spark, *check for fuel*. This can often be done by installing a pressure gauge on the engine's fuel rail. A special test fitting is usually provided for a pressure gauge. With the engine cranking or the key on, the gauge should read within specs. If it is not, something is keeping the electric fuel pump(s) from working normally. Check for a clogged fuel filter, blown pump fuse, or wiring problem.

With older carbureted engines, you can look inside the carburetor to check for fuel. With the engine and air cleaner off, move the throttle open and closed. This should make fuel squirt into the carburetor and engine. If it does not, something is preventing normal fuel flow from the tank to the engine.

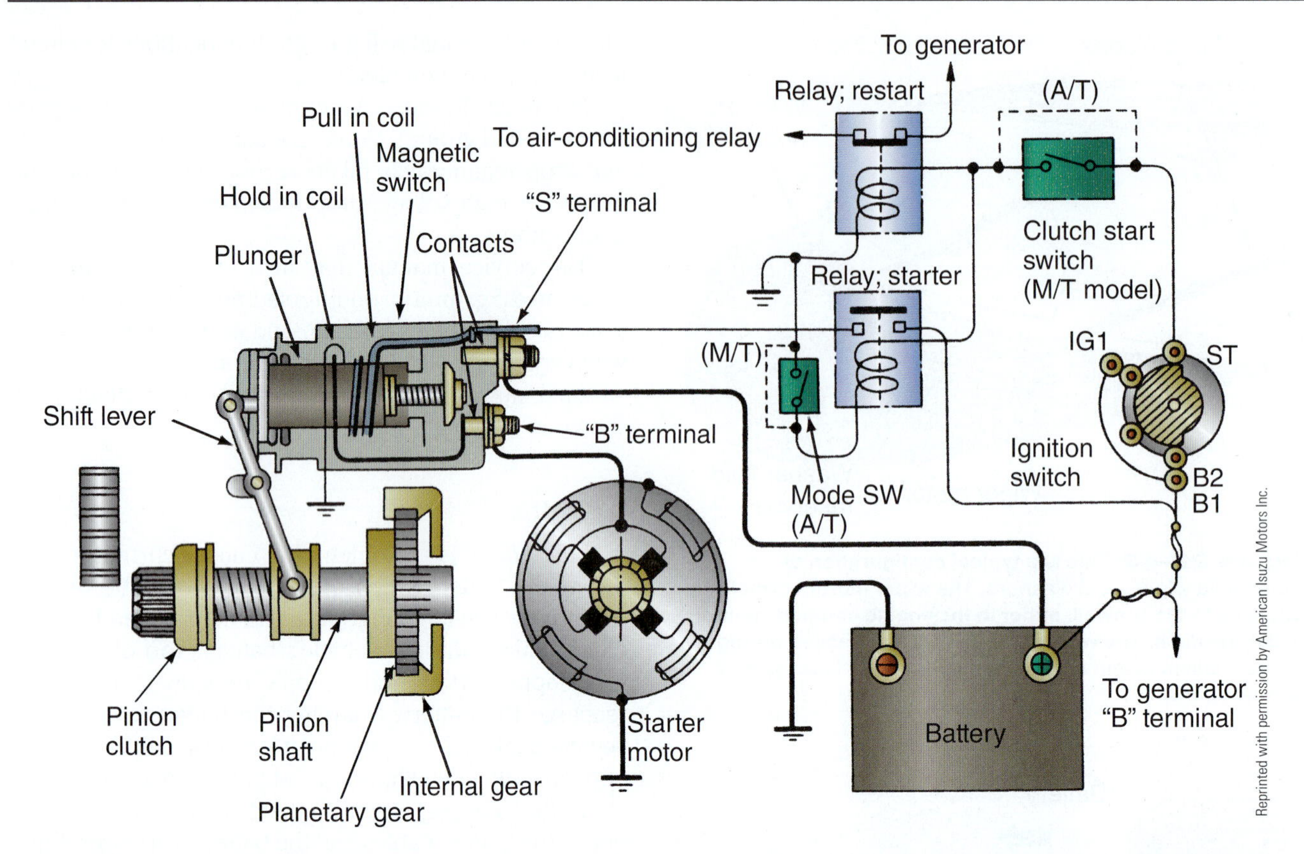

Figure 22-44 Study the wiring of a starting system. To turn the engine over, the ignition key switch sends a small current to the starter relay only when the clutch start and transmission safety switches are closed. Relay current then energizes solenoid windings that pull on a plunger to engage the pinion gear with the flywheel ring gear. At the same time, the solenoid plunger makes a high-current electrical connection from the battery to spin a high-torque starting motor.

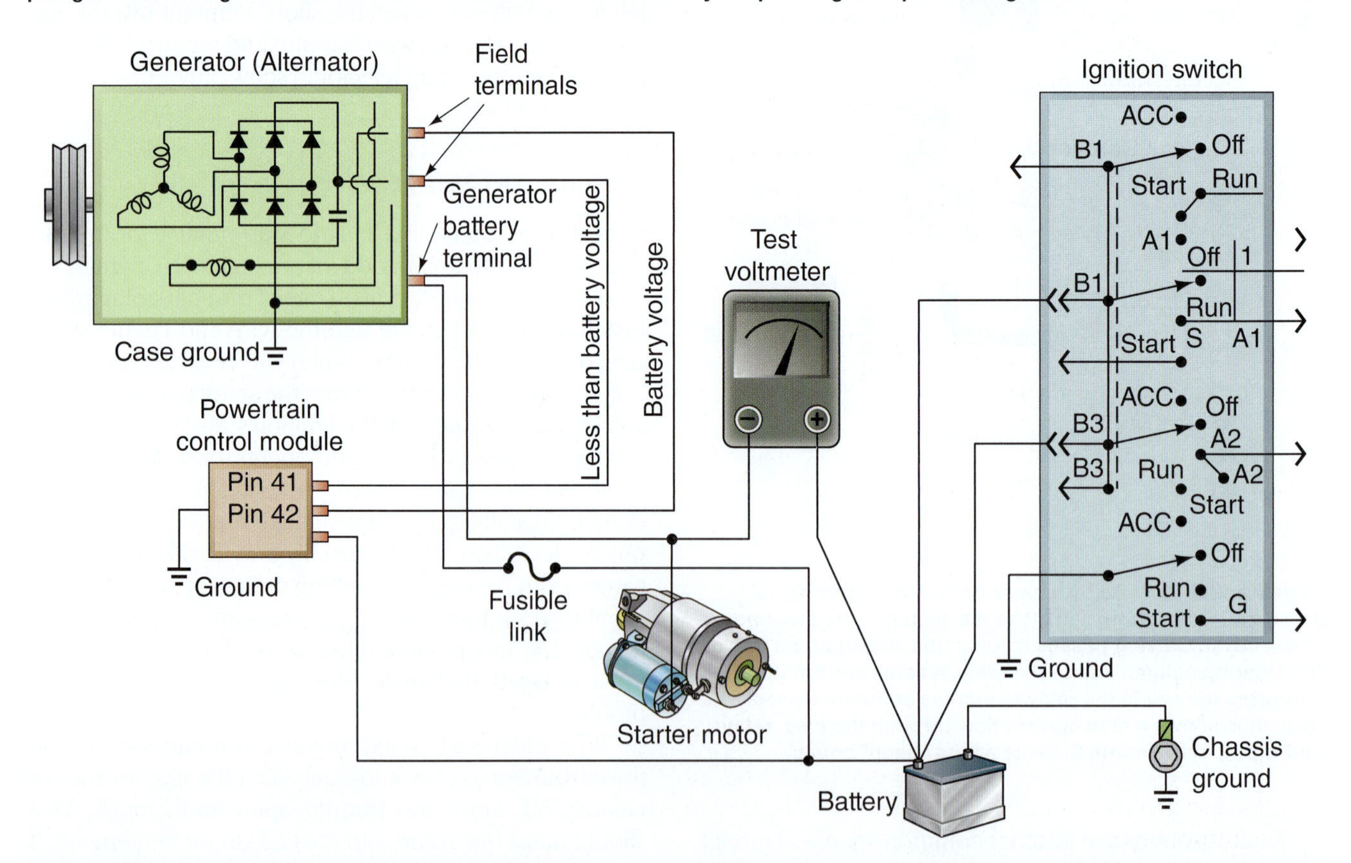

Figure 22-45 Note how an alternator or generator wires into the starting circuit. The ignition switch controls current flow to the powertrain power module and other components.

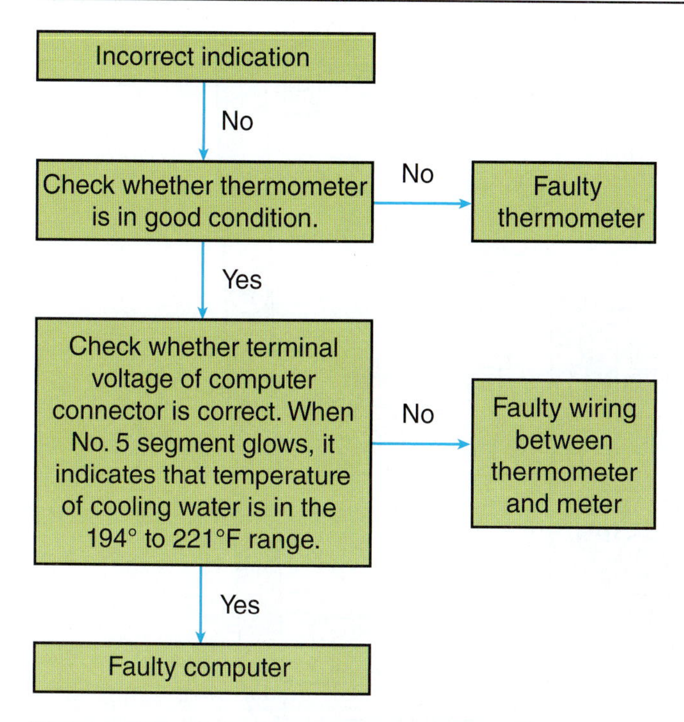

Figure 22-46 Manufacturers' troubleshooting charts vary. This one is a flow chart for an engine temperature gauge. Follow the arrows from top to bottom as you read the information in each box.

22.8 ELECTRONIC SYSTEM SERVICE

Many electronic diagnostic and repair procedures are not that difficult. Technicians who go back to the basics of electrical circuits and follow a few simple rules can

Figure 22-48 Auto body technicians must understand the principles of electrical testing and repair.

find and fix a lot of electronic problems. Knowledge of electronics is necessary for technicians to perform their jobs.

The use of the word *electronics* in this book refers to on-board computers and other "black box" items, while *electrical systems* means wiring and electrical components such as alternators, lights, heater motors, and so on.

Figure 22-47 Bad grounds can cause electrical malfunctions. Grounds can be damaged or loosened during a major collision. The typical locations for major grounds are shown here. However, always refer to service reference material for exact locations on specific vehicles.

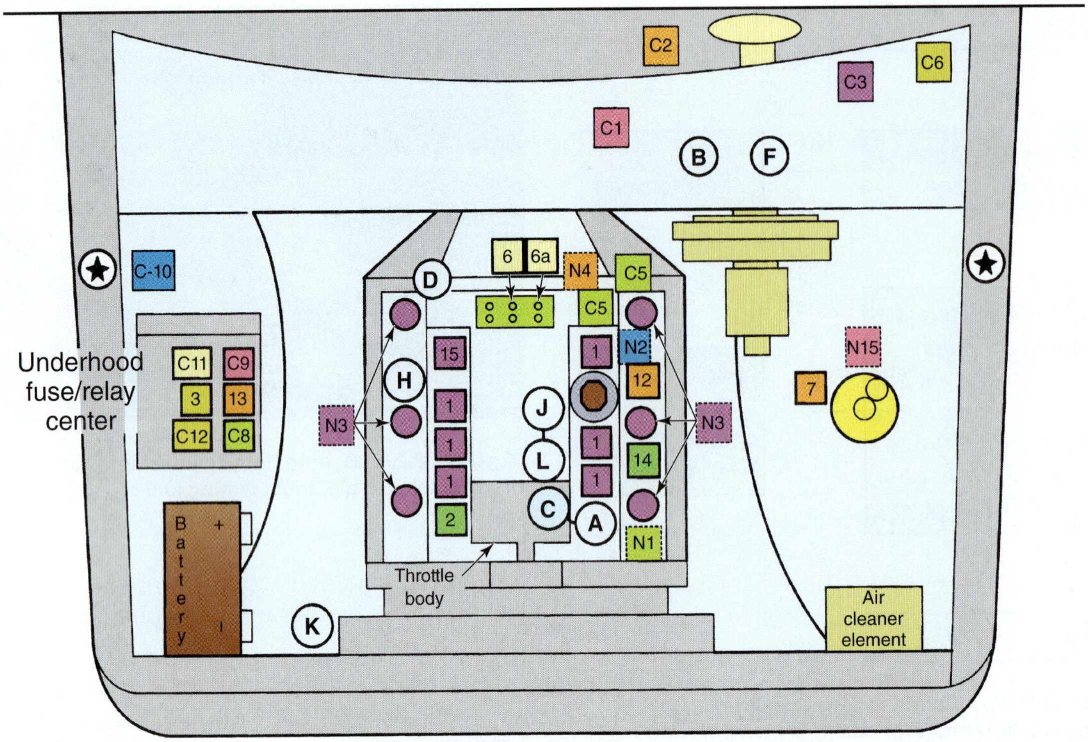

Computer harness

C1	Engine control module (ECM)*
C2	DLC diagnostic connector
C3	"Check engine" malfunction indicator lamp
C5	ECM harness ground
C6	Fuse panel
C8	ECM main relay
C9	Fuel pump fuse 15A
C10	Injector resistor
C11	Oxygen 10A heater fuse
C12	30A ECM main fusible link

ECM information sensors

A	Manifold absolute pressure
B	Heated oxygen sensor
C	Throttle position sensor
D	Engine coolant temperature
F	Vehicle speed sensor
H	Crank angle sensor
J	Knock sensor (under intake assembly)
K	Power steering pressure switch (in-line)
L	Intake air temperature

ECM controlled components

1	Fuel injector
2	Idle air control
3	Fuel pump relay
6	Ignition control module ignition control
6a	Ignition coils
7	Knock sensor module under charcoal canister
12	Exhaust gas recirculation (EGR) VSV
13	Air-conditioning relay
14	Evaporate emission canister purge VSV
15	Induction air control plate system VSV

Emission components (not ECM controlled)

N1	Crankcase vent valve (PCV)
N2	Exhaust gas recirculation valve back pressure transducer
N3	Spark plugs
N4	Fuel rail test fitting (for fuel pressure test)
N15	Fuel vapor canister

*The ECM is located behind the console in the lower dash.

EGR valve ★ Chassis grounds

Figure 22-49 A component location diagram such as this one is handy when you cannot find a part on a vehicle. Service literature will include location diagrams for different sections of each make and model vehicle. Note how the alphanumeric codes correspond to locations.

22.9 ELECTRONIC DISPLAYS

Electronic instrument displays are becoming more and more popular. The technology has become less expensive and more reliable, and many customers like the high-tech, state-of-the-art image of electronic displays. But in collision repair, electronic displays call for some special cautions. These complex and expensive parts must be handled carefully to avoid damage.

Most dashboard gauges are driven by some type of electrical signal. These gauges can be either analog or digital (Figure 22–50).

A digital display uses numbers instead of a needle or graphic symbol. In an analog display, an indicator moves in front of a fixed scale to give a variable readout. The indicator is often a needle, but it can also be a liquid crystal or graphic display. An example is a speedometer in which the speed is shown by a set of vertical bars that light up or dim as the speed changes.

The advantage of analog displays is that they show relative change better than digital displays. They are useful when the driver must see something quickly, and the exact amount of change is not important. For example, an analog tachometer shows the rise and fall of the engine speed better for shifting than a digital display does. Here the driver does not have to know at exactly how many rpm the engine is running. The most important thing is how fast the engine is reaching the red line on the gauge.

A digital display is better for showing exact data such as miles or operating hours. Many speedometer/odometer displays are both analog (speed) and digital (distance).

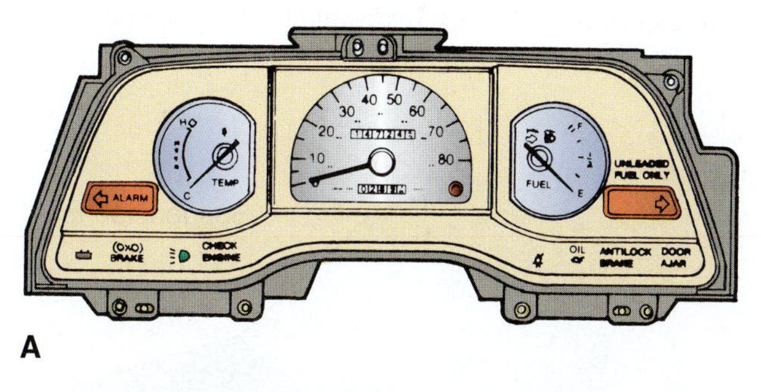

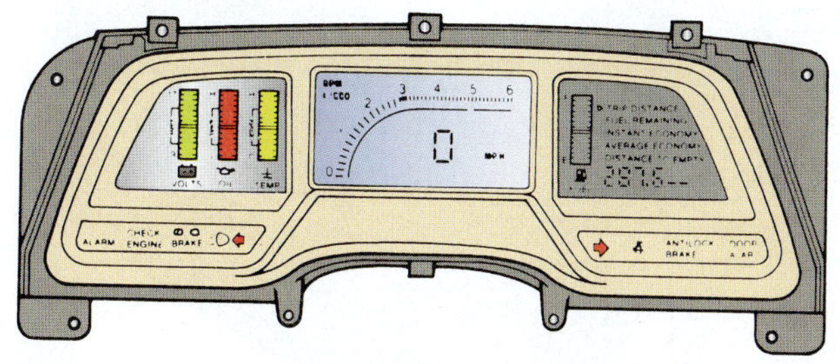

Figure 22-50 (A) An analog display has gauge needles that sweep around. (B) A digital display has number displays and sometimes analog indicators that move or slide across to show vehicle speed, for example.

The choice of display types is a matter of designer and buyer preferences.

An *analog signal* is continuously variable. An analog current is like the water flowing from a faucet that is gradually turned up and down. Sometimes it flows a lot, sometimes only a little, and sometimes not at all.

For example, a temperature sensor causes the current to change as the temperature changes. As the temperature rises, the resistance decreases. This causes a gradual increase in the circuit current. As the sensor cools, the current steadily decreases.

The changing current is used to drive a gauge. The higher the temperature and pressure, the more current flows in the gauge circuit. The current creates a magnetic field that moves the pointer. In a temperature gauge, the higher the current (temperature), the greater the magnetic field and the more the pointer moves. Such magnetic gauges are widely used.

A *digital signal* has only two states—on or off. If a switch is turned on and off many times, the number of pulses can be counted. For example, a sensor can be made to turn on and off each time a wheel moves a certain distance. The number of pulses that are counted in a given period of time allows the computer to display the speed. The pulses can also be used by the computer to change the odometer reading. This is the principle of computer system operation.

Display Types

Three types of electronic displays are used today:

1. *Light-emitting diode (LED).* These are used as single indicator lights, or they can be grouped to show a set of letters or numbers. LED displays are commonly red, yellow, or green. LED displays use more power than other displays. They can also be hard to see in bright light.
2. *Liquid crystal diode (LCD).* These displays have become very popular for many uses, including watches, calculators, and dash gauges. They are made of sandwiches of special glass and liquid, which is why the term *liquid* is used. A separate light source is required to make the display work. The display has wires on the glass. When there is no voltage, light cannot pass through the fluid. When voltage is applied, the light passes through the segment. LCDs do not like cold temperatures, and the action of the display slows down in cold weather. These displays are also very delicate and must be handled with care. Any rough handling or force on the display can damage it.
3. *Vacuum fluorescent diode (VFD).* These displays use glass tubes filled with argon or neon gas. The segments of the display are little fluorescent lights, like the ones in a fluorescent fixture. When current is passed through the tubes, they glow very brightly. These displays are both durable and bright.

All gauges require input from a sensor or sending unit. However, with modern computer-controlled displays, the sensor's output is used in two ways. The engine control computer needs the same information as the electronic display, so the information first passes through the computer. It then travels to the gauge. As an example, compare the temperature sensor on the vehicle of ten years ago with a present-day vehicle. On the ten-year-old car, the temperature gauge was connected directly to a sensor that checked the engine temperature. A rise in temperature resulted in increased current in the gauge circuit. This caused the pointer and magnetic gauge to move, showing the temperature to the driver on an analog scale.

On present-day vehicles, the system works basically the same, with one very important exception. The information from the sensor is first fed through the vehicle's engine control computer. The computer uses the information to manage a variety of systems, including air–fuel ratio, spark timing, and switching of emission control system components. In addition, the computer uses the information to operate the temperature gauge—digital, analog, or just a temperature warning lamp (a form of digital display).

Air Bags

The air bag system uses impact sensors, the vehicle's onboard computer, an inflation module, and a nylon balloon in the steering column to protect the driver during a head-on collision. These systems can be expensive to service because the bag and all sensors often require replacement after inflation.

For details of air bag system operation and service, refer to Chapter 23 on restraint system service.

22.10 COMPUTER SYSTEMS

Almost all vehicle systems are now controlled by computer. These include the fuel, ignition, charging, suspension, brake, climate control, air bag, and other systems. See Figure 22–51.

A basic computer system consists of:

- Sensors (input devices)
- Actuators (output devices)
- Computer (electronic control unit)

The **sensors** are devices that convert a condition (temperature, pressure, part movement, and so on) into an electrical signal. They send an electrical input signal back to the computer. Once the computer analyzes the sensor data, it produces a preprogrammed output that is sent to system actuators.

Actuators are devices (for example, solenoids or servo motors) that move when responding to electrical signals from the computer. In this way, a computer system can react to sensor inputs and then act on these conditions by operating the motors or solenoids.

The **computer** is a complex electronic circuit that produces a known electrical output after analyzing electrical inputs. Today's vehicles can have one or more computers that monitor and control the operation of electrical systems.

Antilock Brakes

The modern *antilock brake system (ABS)* can be thought of as an electronic/hydraulic "pumping" of the brakes for straight-line stopping under panic conditions. That is, this system is another control arrangement that is used in conjunction with a basic hydraulic braking operation.

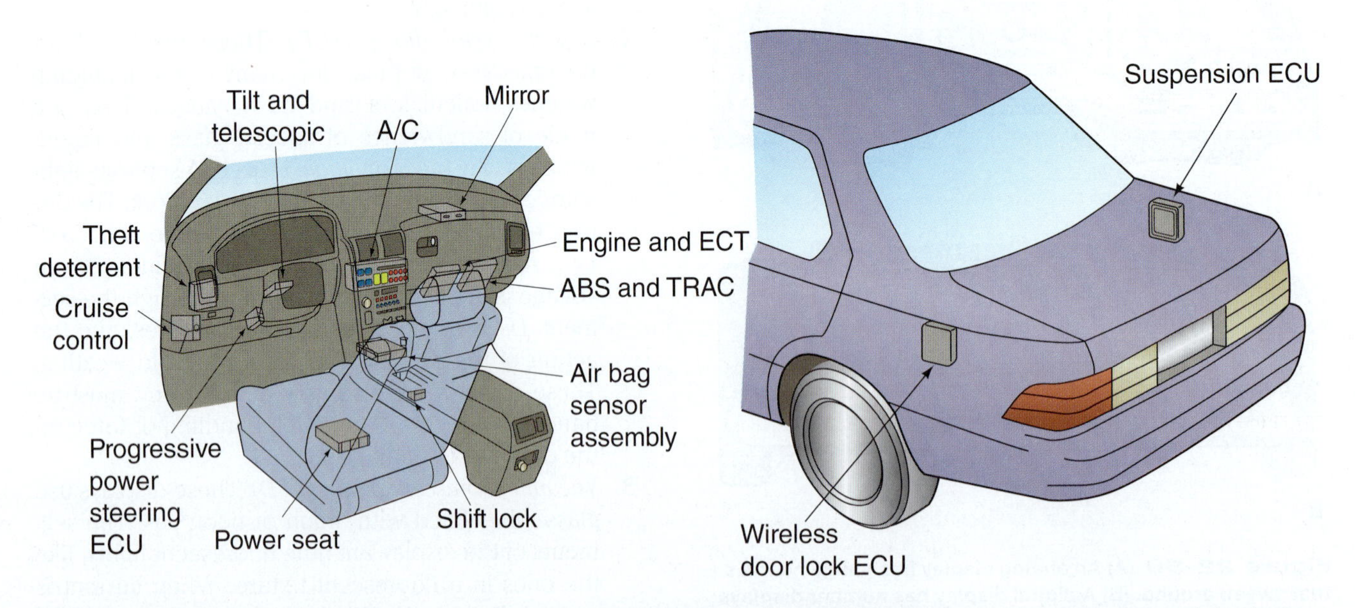

Figure 22–51 Computers, also called electronic control modules, can be found throughout the vehicle. When doing body repairs that require welding, computers should be protected from heat or removed.

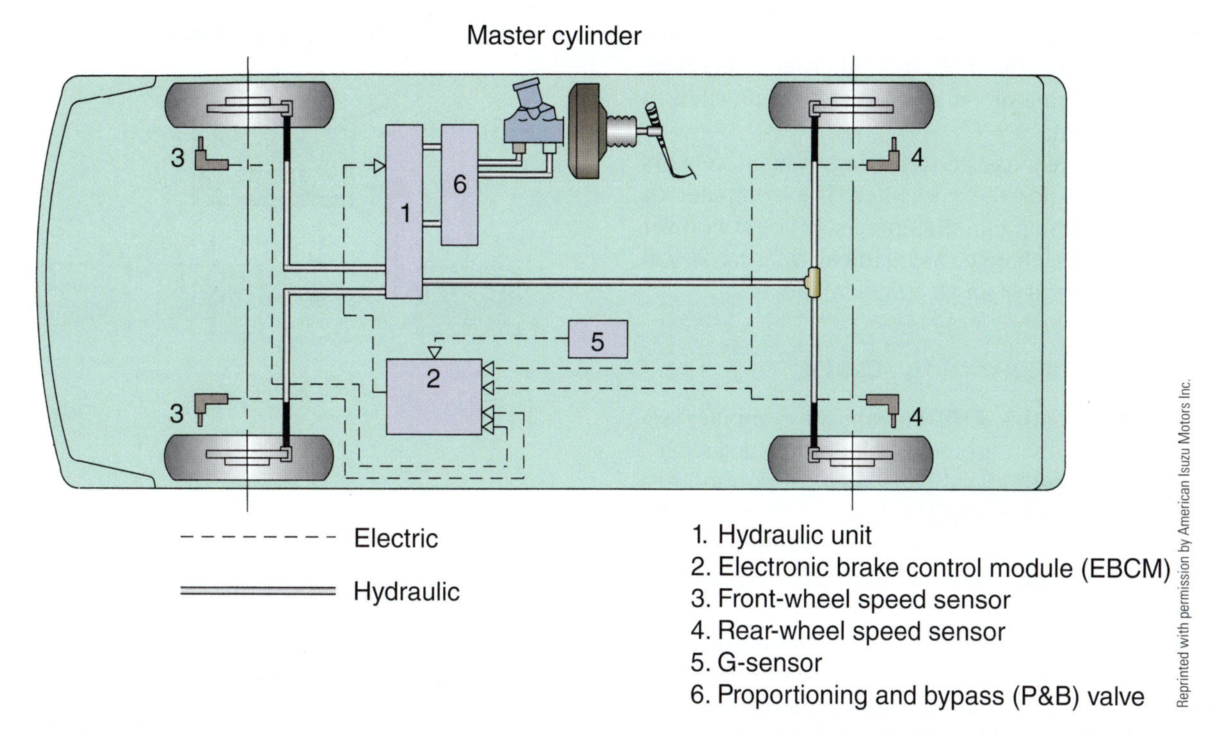

Figure 22-52 **Study the parts of this computer system for controlling the antilock brake system. The brake control module in the engine compartment is often damaged and requires replacement after a major frontal collision. Wheel speed sensors can also be damaged in a front side hit.**

During hard braking conditions with a conventional hydraulic system, it is possible for the wheels of a vehicle to lock, resulting in reduced steering as well as braking. On vehicles equipped with ABS, however, an electronic sensor constantly monitors wheel rotation (Figure 22–52).

If one or more of the wheels begins to lock, the system opens and closes solenoid valves, cycling up to ten times per second. This applies and releases the brakes rapidly and repeatedly, so that the front wheels alternately steer and brake. This makes it possible for vehicles equipped with ABS to avoid skidding under conditions that might cause other vehicles to handle differently.

The antilock or antiskid brake system has a controller that senses rotation at each of the wheels through wheel sensors. It can apply the ABS to each of the front wheels independently, to the rear wheels as a pair, or to any combination of these three, as the need arises. Because there are several antilock or antiskid systems, it is important to check service manuals for diagnostic and service procedures.

Servicing antilock brakes is similar to servicing conventional brakes. However, electronic parts are added to operate the system. Most ABS brakes have self-diagnosis. The computer will output a trouble code if an electrical/electronic malfunction develops. You can refer to charts in the service manual to see what each number code means. The code will pinpoint which part might be at fault.

For example, if a trouble code indicates a problem with one of the wheel speed sensors, make sure it is adjusted properly and undamaged. You may also need to test the sensor and its wiring.

Computerized Suspension

Computer suspension systems use sensors and an electronic control unit to adjust height, leveling of the car, and ride firmness in some models. Repairing these systems will become more common as more vehicles come equipped with them. Anyone involved in the collision repair industry needs a good basic knowledge of how these systems work. The best source of information about computer suspension systems is vehicle-specific service manuals or computer software.

An air spring replaces the coil spring used in a conventional independent suspension system. The computer controls the car's height either by telling a battery-driven air compressor to pump air into an air spring or telling a valve to let air out. Height sensors tell the computer whether the car is too high or too low. When the car reaches the right height, the sensors send a trim signal. The pump is shut off, the air spring valve is closed, and the air is trapped in the spring.

Computer-Assisted Steering

In an electronically controlled power steering arrangement, an electric motor replaces the hydraulic pump, hoses, and fluid associated with conventional power steering systems. The housing and rack are designed so that the rotary motion of the electric motor can be transferred to linear movement to assist steering wheel rotation. The motor armature is mechanically connected to the steering system.

With a computer-assisted steering system, sensors provide feedback for the computer. The computer or electronic control unit can then precisely control power assist as variables change. Mechanical steering is still provided in the event of an electrical failure.

On-Board Diagnostics (OBD)

On-board diagnostics (OBD) means the computer system can detect its own problems. Modern vehicles have built-in, self-diagnostic capabilities that check hundreds of operating parameters. OBD makes sure that all of the wires, fuses, sensors, actuators, and electronic control modules are working, such as the ABS and air bag systems.

A poorly tuned, damaged, or malfunctioning automobile can produce several times the normal amount of emissions or pollution. For this reason, the Environmental Protection Agency (EPA) has passed regulations that require auto manufacturers to install performance monitoring systems on their cars and light trucks.

Early OBD systems simply illuminated an indicator lamp in the dash if a circuit stopped working. On-board diagnostics II (OBDII) systems go a step further by monitoring how well or efficiently each part of the system is operating. For example, if a sensor becomes lazy or remains in the low end of its operating parameter, this potential problem is stored in the vehicle's computer memory for retrieval at a later date. If the low spec output continues, the system will turn on the dash *malfunction indicator lamp (MIL)*.

OBDII is designed to more efficiently monitor the condition of components that affect emissions. New vehicle diagnostics detect part wear (change in performance), not just complete part failure.

If a problem exists, the computer will turn on the MIL in the instrument panel, which prompts you to use OBD to find the source of the trouble. The computer performs a self-check each time the ignition key is turned to the on position (Figure 22–53).

If the OBD system finds a problem, it will also store a **fault code** (code representing a specific circuit or part with a problem) for later servicing. A fault code is recorded into the computer's memory whenever the system malfunctions.

Computer trouble codes are pulsing signals (number codes) produced by the computer when an operating parameter is exceeded. An operating parameter is an acceptable minimum and maximum electrical value. It might be an acceptable voltage range from the oxygen sensor, a resistance range for a temperature sensor, or an acceptable current draw from an injector coil. In any case, the computer "knows" the limits for most input and output levels. If an electrical valve is too weak or too strong for known parameters, the computer is programmed to store a trouble code and turn on a dashboard warning light.

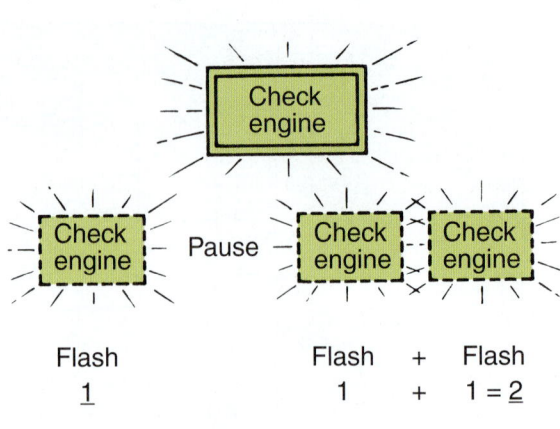

Before reading fault codes, do a visual check to make sure the problem is not the result of wear, loose connections, or faulty vacuum hoses. Inspect all wire and vacuum-hosed connections. Remember that the low-level electrical signals in today's electronic circuits cannot tolerate the increased resistance caused by corrosion in connector contacts.

An *assembly line diagnostic link (ALDL)* connector is provided for reading fault or trouble codes. This connector can be located in the passenger or engine compartment (Figure 22–54).

Diagnostic code display

Check engine

Check engine — Pause — Check engine — Check engine

Flash 1 — Flash + Flash: 1 + 1 = 2

1 and 2 = Code 12

Figure 22–53 Older vehicles use a flashing signal code on the dash to indicate a problem. For example, indicator lights may flash, pause, and then flash two more times to indicate a Code 2. Codes vary by vehicle make and model.

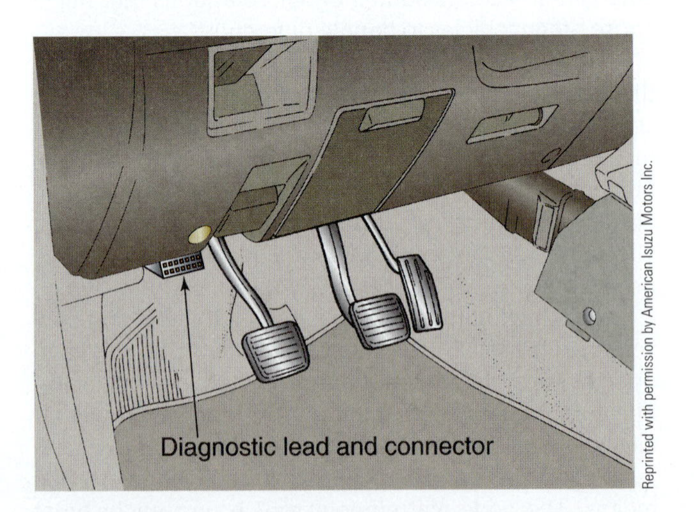

Figure 22–54 The diagnostic connector is normally located under the dash. A scan tool can be plugged into this connector to pull or retrieve trouble code data from the vehicle's computer.

Scan Tool Use

A **scan tool** is an electronic instrument for reading trouble codes and converting them into problem explanations.

To read computer fault codes, plug a scan tool into the ALDL connector. Turn the ignition key to the on position. The scan tool will then communicate with the vehicle's computer. The scan tool will retrieve the trouble code number and translate it into an explanation of the problem. This is the best way to find electrical problems on modern vehicles (Figure 22–55).

Because fault codes vary from model to model, the vehicle's service manual and scanner owner's manual must be consulted. Different cartridges usually must be installed in the scan tool to match the year, make, and model of the vehicle (Figure 22–56).

Plug the scan tool connector into the vehicle's diagnostic connector. After you follow instructions given by the scan tool, the tool will then be able to communicate with the vehicle's on-board computer to analyze possible problems. Refer to Figure 22–57.

Figure 22–58 shows how a scan tool connects to and can communicate with a vehicle's computer.

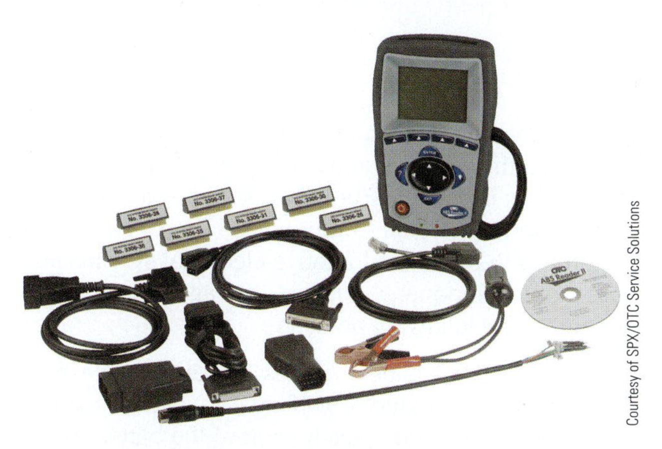

Figure 22–55 Scan tools should be used to check wiring after making any electrical repairs. A scan tool can more quickly access and analyze computer system trouble codes and the operating parameters of circuits.

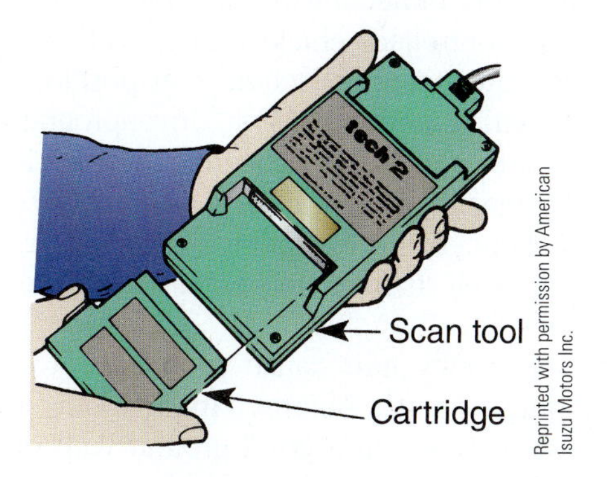

Figure 22–56 A scan tool will normally have different cartridges for different years and makes of vehicles. Some scan tools will actually tell you what to do while troubleshooting.

Figure 22–57 It only takes a couple of minutes to scan a vehicle after collision repair. You do not want the vehicle to leave the shop only to have it trip a trouble code and cause the customer to have to return for further repairs.

If you get a fault code, the scan tool may be able to describe the problem or it may only give a fault code number. If you only get a number, refer to the service manual fault code chart. It will explain the number code

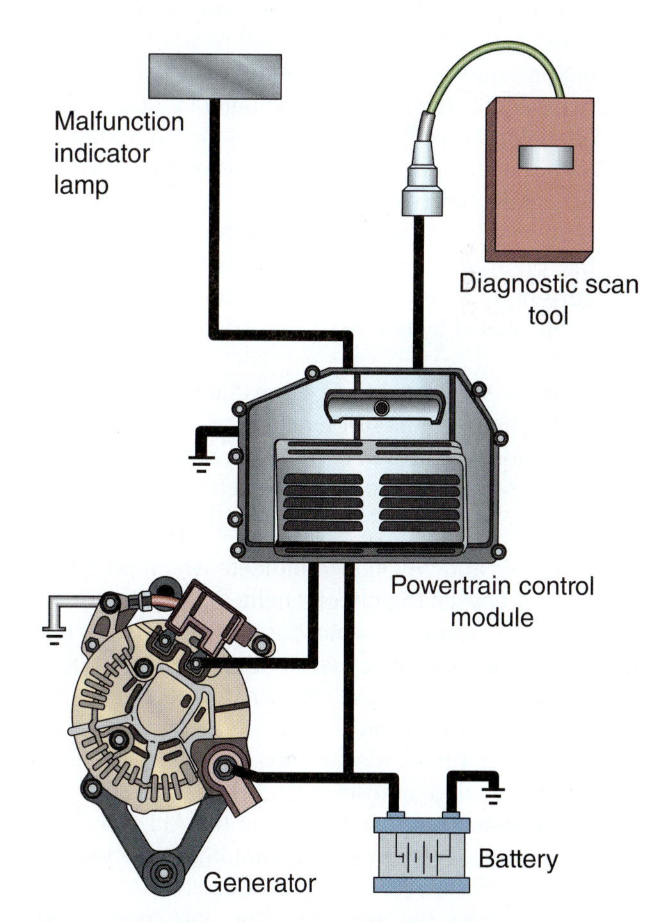

Figure 22–58 When connected to the diagnostic connector under the dash, a scan tool will quickly pull trouble codes and electrical operating parameters out of the control module.

and describe which parts might be causing the trouble (Figure 22–59).

Modern OBD systems detect wear, not just failure, of over 50 emission-related parts. They check the operating parameters of switches, sensors, actuators, and their related wiring.

With OBDII, engine misfire will prompt a warning light in the dash to blink on and off. This is to warn that the misfire could overheat and damage the catalytic converter. This alerts the driver to the fact that the vehicle is being damaged and should not be driven.

OBDII has also helped to standardize data link connections, codes, and scan tool capabilities. In the past, one manufacturer required over a dozen different connectors for just its vehicle computer systems. This made it almost impossible for a small repair shop to scan test all makes of vehicles. To solve this repair problem, OBDII standardization has made it more economical and practical to scan troubles on any manufacturer's vehicles.

OBDII trouble codes can be broken down into four general categories:

1. A *general circuit failure* has a fixed value or no output. This is the most severe and easy to locate failure. It is caused by disconnected wires, high-resistance connections, shorts, and similar problems.
2. A *low input failure* is one that produces a voltage, current, resistance, or signal below normal operating parameters. A weak or abnormally low signal is being sent to the on-board computer module. This is often caused by high circuit resistance, a poor electrical connection, a contaminated or failed sensor, or a similar problem.
3. A *high input failure* results when the signal reaching the on-board computer has more voltage, more current, or a higher frequency than normal. This is often caused by failure of a sensor or a mechanical fault.
4. An *improper range failure* is one that is slightly low or high. The circuit is still functioning, but not as well as it should. This can be due to a damaged sensor, partial sensor failure, a poor electrical connection, or similar causes.

Scan tools and OBDII also indicate which on-board control unit causes the check engine light to come on. This can aid further troubleshooting because it narrows down the possible causes of the problem to parts related to that control unit (the air bag controller, the ABS controller, and so on).

If you do not have a scan tool, you can sometimes use computer self-diagnosis to find circuit problems. By jumping across specific terminals on the ALDL connector or turning the ignition key on and off as directed, the computer will flash a Morse-type code in the instrument panel, indicating the problem. You can note these flashes and compare them to service manual details to find a problem.

Computer Problems

The computer itself is a highly reliable electronic device. Trouble is more often caused by damaged wiring, connectors, or sensors. The computer is one of the last parts to suspect with computer system malfunctions.

Computers can be found in just about any location on the vehicle, including:

- Under the dash
- Under the seats
- Under the hood
- Behind kick panels
- In the trunk

A PROM is a programmable read-only memory chip mounted in the computer. If you have to replace a computer, you may have to install the PROM from the old computer into the new one. This will ensure that the computer is programmed for the equipment on the vehicle being serviced.

Disconnecting the battery can erase stored electrical electronic system fault codes on some vehicles. Sometimes you can pull and reinstall the computer fuse to erase fault codes. Always refer to the service manual for specific information about computer system service.

Protecting Electronic Systems

The last thing a technician wants to do when a vehicle comes into the shop for collision repair is to create problems. This is especially true when it comes to electrical systems and electronic components. There are proper procedures to protect automotive electrical systems and electronic components during storage and repair. Remember the following pointers:

- Disconnect the battery cables (negative cable first) before doing any kind of welding. To avoid the possibility of explosion, completely remove the battery when welding under the hood or on the front end. Also, make sure the ground connection is clean and tight. Position the ground clamp as close as possible to the work area to avoid current seeking its own ground.
- Whenever disconnecting or removing the battery on a computer-controlled vehicle, remember that the memory for radio station selection, seat position, climate control setting, and any other "driver-programmable" options is erased. When delivering the vehicle, advise the customer to reprogram these settings. Or better yet, record them before beginning work and reprogram them yourself before delivering a vehicle to the customer.
- Static electricity can cause problems. Avoid it by grounding yourself before handling new displays. One way is to touch a good ground with one hand before handling the display with the other hand. Another way is to use a grounding strap that attaches to the wrist and then to the vehicle.

Diagnostic trouble code	DRB scan tool display	Description of diagnostic trouble code
11*	No crank reference signal at PCM	No crank reference signal detected during engine cranking
15**	No vehicle speed sensor signal	No vehicle distance (speed) sensor signal detected during road load conditions
34*	Speed control solenoid circuit	An open or shorted condition detected in the speed control vacuum or vent solenoid circuits
	or	
	Speed control switch always low	Speed control switch input below the minimum acceptable voltage
	or	
	Speed control switch always high	Speed control switch input above the maximum acceptable voltage
55*	N/A	Completion of fault code display on check engine lamp

* Check engine lamp will not illuminate at all times if this diagnostic trouble code was recorded. Cycle ignition key as described in manual and observe code flashed by check engine lamp.

** Check engine lamp will illuminate during engine operation if this diagnostic trouble code was recorded.

A

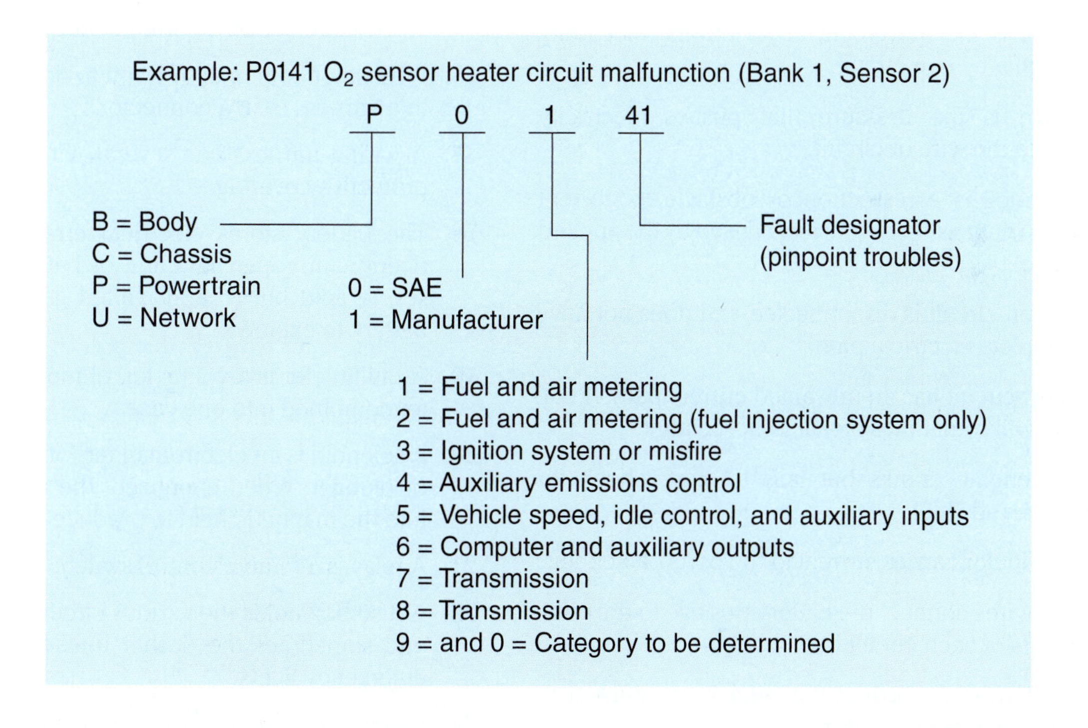

B

Figure 22-59 Trouble code numbers indicate specific problems. Compare early trouble codes with modern OBDII codes. (A) Trouble code numbers displayed on the scan tool are explained in the service manual chart, which provides added descriptions. (B) Study the meanings of an OBDII trouble code alphanumeric display. The first letter denotes whether it is a body, chassis, powertrain, or network problem; the second to last number specifies the general circuit types involved. The last two-digit number is the trouble code. Note how 41 means there is a problem with a specific sensor circuit.

- Avoid touching bare metal contacts. Oils from your skin can cause corrosion and poor contacts.
- Be careful about the placement of welding cables. Keep the electrical path as short as possible by placing the ground clamp near the point of welding. Also, do not let welding cables run close to electronic displays or computers.
- Take care when handling electronic displays and gauges. Never press on the gauge face, because this can damage it.
- If a fault code indicates a problem with the oxygen sensor, extra caution is required. The oxygen sensor wire carries a very low voltage and must be isolated from other wires. If it is not, nearby wires can add more induced voltage. This gives false data to the computer and can result in a driveability problem. Some manufacturers use a foam sleeve around the oxygen sensor wire to keep it separate and insulated from other wires.
- The sensor wires that connect to the computer should never be rerouted. The resulting problem might be impossible to find. When replacing sensor wiring, always check the service manual and follow specified routing instructions.
- Remove any computer that could be affected by welding, hammering, grinding, sanding, or metal straightening. Be sure to protect the removed computer and its connectors by wrapping them in plastic bags to shield them from moisture and dust.
- Be careful not to damage wiring when welding, hammering, or grinding.
- Be careful not to damage connectors and terminals when removing electronic components. Some may require special tools to remove them.
- Always route wiring in its original position. If you do not, electronic crossover from the current-carrying wires can affect the sensing and control circuits. Reuse or replace all electrical shielding for the same reason.

SUMMARY

1. Electrical repairs include tasks such as repairing severed wiring, replacing engine sensors, and scanning for computer or wiring problems.
2. Current is the movement of (electrons) electricity through a wire or circuit.
3. Voltage is the pressure that pushes electricity through the wire or circuit.
4. Resistance is a restriction or obstacle to current flow. It tries to stop the current caused by the applied voltage.
5. An open circuit is disconnected and does not have a complete electrical path.
6. A short circuit has an unwanted current path. A collision will sometimes create a short circuit.
7. If an engine cranks but fails to start, check for "spark" and "fuel."
8. A conductor carries current to the parts of a circuit.
9. "Hot" wires connect the battery positive to the components of each circuit.
10. Insulation stops current flow and keeps current in the metal wire conductor.
11. A frame ground circuit uses the metal body as a return wire to the negative terminal of the battery.
12. A series circuit has only one conductor path, or leg, for current through the circuit.
13. A parallel circuit has two or more legs, or paths, for current.
14. A wiring diagram is a graphic representation of all parts and wires in the circuit.
15. Electrical symbols are graphic representations of electrical/electronic components.
16. Wire color coding allows you to find a specific wire in a harness or in a connector.
17. A wiring harness has several wires enclosed in a protective covering.
18. The battery stores electrical energy as chemical energy. Improper handling of batteries can cause severe acid burns, and ignited gas can cause the battery to explode.
19. A multimeter is a voltmeter, ohmmeter, and ammeter combined into one case.
20. A solenoid is an electromagnet with a movable core or plunger. When energized, the plunger is pulled into the magnetic field to produce motion.
21. A relay is a remote control switch.
22. A fuse box holds the various circuit fuses, breakers, and sometimes the flasher units for the turn and emergency lights.
23. Fuses burn in half with excess current to protect a circuit from further damage.
24. A fuse link, or fusible link, is a smaller diameter wire spliced into larger circuit wiring for overcurrent protection.
25. Circuit breakers heat up and open with excess current to protect the circuit. They do not suffer internal damage like a fuse.

26. The starting system has a large electric motor that turns the engine flywheel.
27. The charging system recharges the battery and supplies electrical energy when the engine is running.
28. A battery drain is a problem that causes current to flow out of the battery when everything should be off.
29. Sensors are devices that convert a condition (temperature, pressure, part movement, and so on) into an electrical signal.
30. Actuators are devices (solenoids or servo motors, for example) that move when responding to electrical signals from the computer.
31. A computer is a complex electronic circuit that produces a known electrical output after analyzing electrical inputs.
32. Modern antilock brake systems (ABS) can be thought of as electronic/hydraulic "pumping" of the brakes for straight-line stopping under panic conditions.
33. On-board diagnostics (OBD) means the computer system can detect its own problems.
34. A scan tool can convert the computer number code into an explanation of the problem.

EXERCISES

On a separate sheet of paper, complete the following learning activities for this chapter. Write definitions for the key terms and answer the ASE-style review questions, essay questions, critical thinking problems, and math problems. You can also do the outside activities, possibly for extra credit.

Key Terms

actuators
charging system
circuit breakers
component location diagrams
computer
current
electric circuit
fault code
fuse box
fuse links
fuse ratings
intermittent short
jumper cables
jumper wires
magnetic field
magnetism
multimeter
Ohm's Law
on-board diagnostics (OBD)
open circuit
parallel circuit
relay
resistance
scan tool
sensors
series circuit
short circuit
solenoid
starting system
switch
testlight
troubleshooting chart
voltage
wire color coding
wiring harness

ASE-Style Review Questions

1. Technician A says that voltage is the pressure that pushes electricity through the wire or circuit. Technician B says that voltage is measured by using a voltmeter. Who is correct?
   A. Technician A
   B. Technician B
   C. Both A and B
   D. Neither A nor B
2. Technician A says that an inductive ammeter should be used to check starter current draw. Technician B says that a small ammeter in series will work better. Who is correct?
   A. Technician A
   B. Technician B
   C. Both A and B
   D. Neither A nor B
3. A fuse box is often located ________.
   A. Under the instrument panel
   B. Behind a panel in the foot well
   C. In the engine compartment
   D. All of the above
4. Technician A says wire color coding allows you to find a specific wire in a harness or connector. Technician B says different auto makers use different color-code abbreviations on their wiring diagrams. Who is correct?
   A. Technician A
   B. Technician B
   C. Both A and B
   D. Neither A nor B

5. Technician A says a cheap testlight can be used to troubleshoot computer circuits. Technician B says that this could damage the computer circuitry. Who is correct?
   A. Technician A
   B. Technician B
   C. Both A and B
   D. Neither A nor B
6. If you were going to measure approximately 12 volts with a multimeter, what meter range setting would you use?
   A. 5V
   B. 10V
   C. 20V
   D. 100V
7. A fully charged battery should produce how much voltage when read with a multimeter?
   A. 12V
   B. 12.5V
   C. 12.6V
   D. 13V
8. A fuse blows when replaced. Technician A says it is okay to use a fuse with a higher rating to keep it from blowing. Technician B says this could start an electrical fire. Who is correct?
   A. Technician A
   B. Technician B
   C. Both A and B
   D. Neither A nor B
9. A new battery goes dead overnight. Technician A says to check for a battery drain. Technician B says to replace the battery. Who is correct?
   A. Technician A
   B. Technician B
   C. Both A and B
   D. Neither A nor B
10. A check engine light in the dash glows after doing body repairs. Technician A says that this is not a concern. Technician B says that a scan tool should be connected to the vehicle before releasing it to the customer. Who is correct?
   A. Technician A
   B. Technician B
   C. Both A and B
   D. Neither A nor B

Essay Questions

1. Explain the difference between voltage, current, and resistance.
2. Summarize the three major parts of a computer system.
3. Before making adjustments to a vehicle's headlights, what should you do to make sure the vehicle is level?
4. Explain the difference between direct current and alternating current.
5. Define a series circuit and a parallel circuit.

Critical Thinking Problems

1. A battery will not crank and start an engine. Describe what you should do to the vehicle.
2. A headlight circuit is dead and will not work. How would you find the problem?

Math Problems

1. When adjusting headlights, you find the front curb height measurement to be 12.6 inches (320 mm). Specs call for a minimum ride height of 13.2 inches (335 mm). How much is the vehicle out of spec?
2. A circuit has 12.6 volts applied to it from the battery. An ohmmeter says the circuit has 10 ohms resistance. Using Ohm's Law, how much current would flow through this circuit?

Activities

1. Look up the headlight wiring diagram for a specific vehicle. Use the diagram to find specific wire color codes and connectors at various locations on the vehicle.
2. Use a voltmeter to read voltage across the battery terminals of a vehicle. Take a reading with the engine off, with the engine running, and with the engine running with all accessories on. Make a report explaining why the voltmeter readings changed.

Restraint System Operation and Service

CHAPTER 23

OBJECTIVES

After studying this chapter, you should be able to:

- Define the term *restraint system*.
- Compare active and passive restraint systems.
- Explain the operation of retracting seat belts.
- Inspect and replace seat belts and retractors.
- Summarize the operation of air bags.
- List and describe the types and locations of air bags.
- Troubleshoot air bag problems.
- Properly replace air bags and related components.
- Safely work with undeployed air bags.
- Summarize the operation and service of automatic roll bars.
- Answer ASE-style review questions relating to seat belts and air bag service.

INTRODUCTION

The National Highway Traffic Safety Administration (NHTSA) requires that all new vehicles sold in the United States be equipped with seat belts and air bags.

A **restraint system** is designed to help hold people in their seats and prevent them from being injured during a collision. Without restraint devices, the people in a vehicle would be thrown violently around in the passenger compartment during a high-speed crash. Their bodies would fly into the dash, steering wheel, and windshield, often resulting in very serious injury (Figure 23–1).

If it were not for today's advanced restraint systems, thousands more people would be killed in auto accidents. For this reason, many states have laws requiring people to buckle their seat belts.

An **active restraint system** is one that the occupants must make an effort to use. For example, in most vehicles the seat belts must be fastened by hand for crash protection. Conventional, manually operated seat belts are classified as an active restraint system.

A **passive restraint system** is one that operates automatically. No action is required to make it functional. Two types are automatic seat belts and air bags (Figure 23–2).

This chapter will summarize the operation and repair of seat belts, air bag systems, and automatic roll bars. Both systems can be deployed and/or damaged during a collision. Because seat belts and air bags are crucial to preventing collision injuries, you must know how to properly service restraint systems.

23.1 SEAT BELT SYSTEMS

Seat belts are strong nylon straps with special ends attached for securing people in their seats. In today's vehicles, seat belts work in conjunction with the air bags. These restraint devices keep the people in the vehicle from flying around or from being ejected out of the vehicle during the collision (Figure 23–3).

Lap belts are the belts that extend across a person's lap. *Shoulder belts* extend over a person's chest and shoulder.

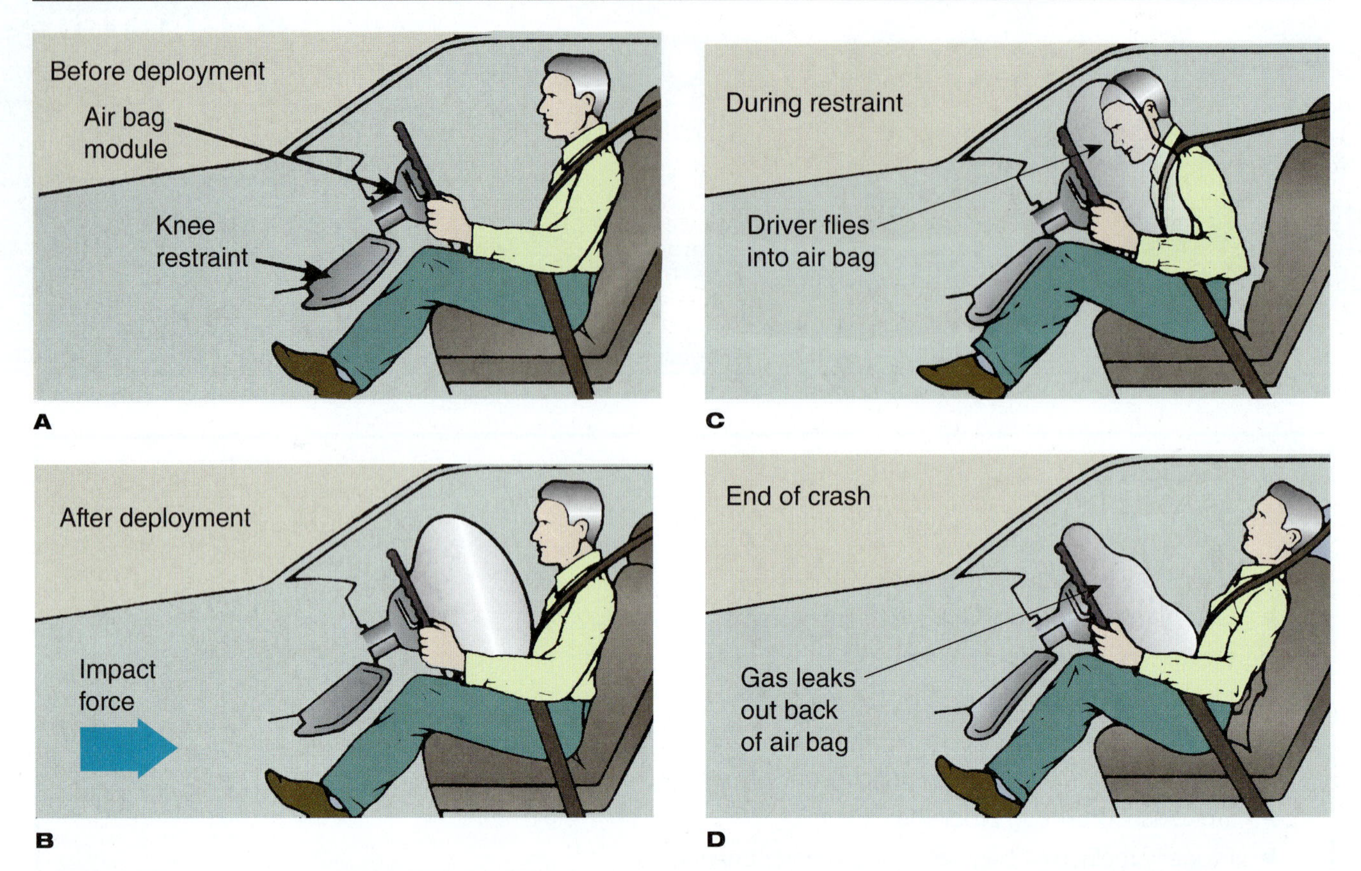

Figure 23-1 Note the sequence of what happens to a typical air bag system during a crash. (A) The air bag and knee diverter or restraint are located in front of the driver. (B) During a collision, the air bag deploys to form a relatively soft cushion in front of the driver. (C) The driver's body flies forward from rapid deceleration caused by the collision. The driver's body is slowed and stopped by the combined action of seat belts and the air bag. (D) The driver's body falls back. Holes in the air bag allow gas pressure to leak out so the bag deflates.

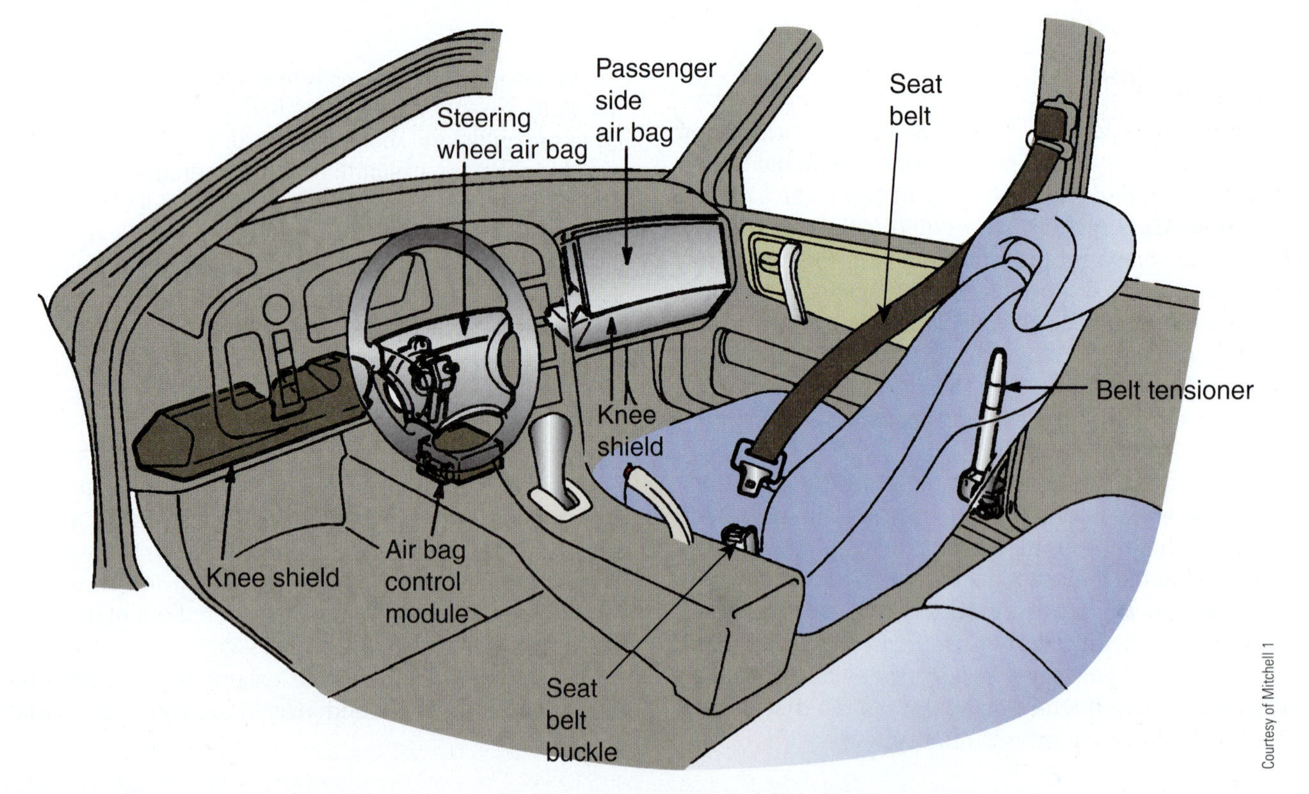

Figure 23-2 The modern vehicle restraint system uses seat belts, air bags, and a knee diverter shield under the dash. They work together to help prevent people from being seriously injured or killed in automobile accidents.

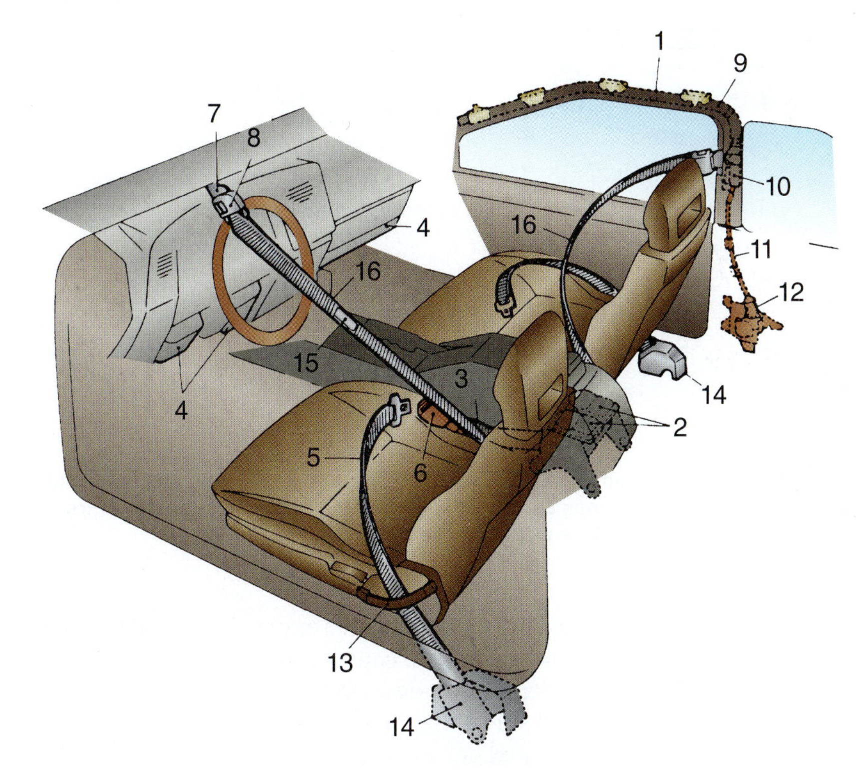

1 Rail and motor assembly
2 Emergency locking retractor assembly
3 Belt guide
4 Knee panel
5 Outer belt assembly (manual lap belt)
6 Inner belt assembly (manual lap belt)
7 Shoulder anchor
8 Emergency release buckle
9 Rail
10 Locking device
11 Tube
12 Motor
13 Belt holder
14 Emergency locking retractor assembly (manual lap belt)
15 Caution label
16 Shoulder belt

Figure 23-3 Study the parts of a passive seat belt system.

A **seat belt buckle** mechanism allows you to put the seat belt on and take it off. It has a button that can be pressed to release the buckle for seat belt removal.

Seat belt anchors allow one end of the belt to be bolted to the body structure. They are hardened metal ends attached to the lower end of the seat belt strap. Specially shaped, case-hardened bolts go through the anchor holes and thread into a nut welded into the body structure.

Two-point seat belts only use a lap belt and are only found in pre-1960s vehicles. *Three-point belts* have three anchor points and one buckle for the lap and shoulder belts. They are found in most motor vehicles. *Four-point belts* with four anchors, two buckles, and two shoulder belts are used in some high-performance sports cars.

A **belt retractor**, or tensioner, is used to remove slack from seat belts so they fit snugly. Various mechanisms are used in belt retractors. One is shown in Figure 23–4.

The *active belt system* consists of a single continuous length of webbing. The webbing is routed from the anchor (at the rocker panel), through a self-locking latch plate (at the buckle), around the guide assembly (at the top of the center pillar), and into a single retractor in the lower area of the center pillar.

The *passive seat belt system* for coupes and late model sedans differs from the active system in that two retractors are used. One is provided for the seat belt and a second for the shoulder belt.

Some late model vehicles use a force-limiting tensioner to prevent injury. It takes up seat belt slack during a collision. Then, a constant, controlled load is maintained on the belt while allowing the belt to extend back out slightly. This reduces the force applied to the chest when a person is thrown forward and snapped to a stop by the shoulder belt (Figure 23–5).

A **pyro-technique retractor** uses a gas-generating retractor to develop pressure for quickly taking up slack in a seat belt when a collision is detected. This type of seat belt retractor often works in conjunction with, and similar to, an air bag system. Pyro-technique retractors must usually be replaced after air bag deployment.

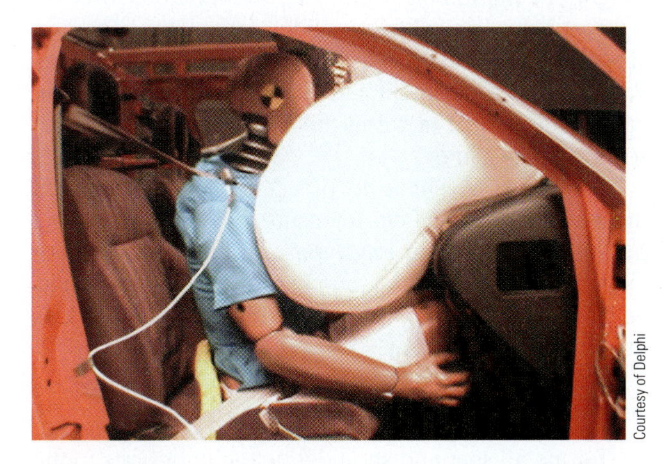

Figure 23-4 During a hard frontal impact, the driver's body is thrown forward with tremendous force. The air bag deploys and the seat belt retractor mechanism is energized. This slows forward motion of the driver's body and keeps it from impacting the steering wheel and dash.

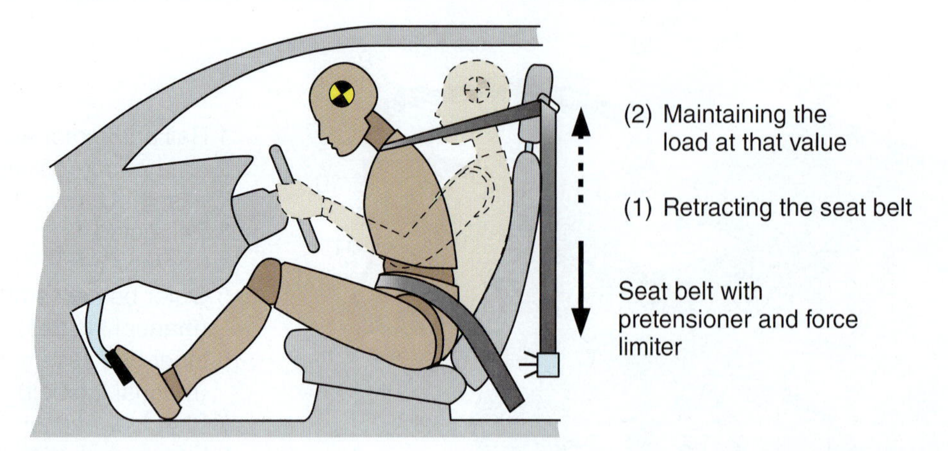

Figure 23-5 Some seat belt retractors are force limiting. When engaged, the retractor first takes up any seat belt slack. It then slowly releases while maintaining a load on the belt. This cushions the blow to the driver's chest when the driver first hits the shoulder belt during a collision. The tensioner then locks to hold the passenger in the seat.

Seat Belt Reminder Systems

A **seat belt reminder system** uses sensors and a warning system to remind the driver to fasten the seat belt. On active systems, the driver's side front seat belt uses a 4- to 8-second fasten seat belt reminder light and sound signal. This is designed to remind the driver if the lap and shoulder belts are not fastened when the ignition is turned on. If the driver's seat belt is not buckled, the reminder light and sound signal will automatically shut off after a few seconds.

On a passive system, the belt warning light will glow for a few seconds. An audible signal will sound if the driver's lap and shoulder belt are not buckled. The system will also signal if the ignition is on and the driver's door is open or if a system failure occurs.

23.2 SEAT BELT SERVICE

Proper seat belt service is critical after a major collision. During the rapid deceleration in a head-on collision, the weight of a human body can exert tremendous force on the seat belts. In a rollover accident, seat belts keep passengers from being ejected through the windows. Passenger compartment intrusion from a side hit can cause sharp sheet metal in the doors and pillars to cut seat belts like a knife.

To restore a vehicle to preaccident condition, you must inspect all seat belts to ensure that they will protect the occupants during any type of future collision. For example, if the seat belt fabric has even a small cut in its edge, the belt fabric could tear and snap if the vehicle is in another bad accident.

When servicing or replacing lap and shoulder belts, keep the following pointers in mind:

- Torque all seat and shoulder belt anchor bolts to the manufacturer's specifications with a torque wrench. Undertightening or overtightening could result in anchor failure during a future accident. An undertightened anchor bolt could loosen and unscrew itself. An overtightened anchor bolt could be stretched and weakened, causing failure and injury.
- Keep sharp edges and damaging objects away from belts. The slightest nick in the fabric could cause tearing and belt breakage during a collision.
- Avoid bending or damaging any portion of the belt buckle or latch plate. Always check for normal engagement and disengagement of all buckle mechanisms during inspection.
- Do not attempt repairs on lap or shoulder belt retractor mechanisms or lap belt retractor covers. Replace with new or salvage replacement parts.
- Do not intermix types of seat belts on front or rear seats.
- Only use original equipment fasteners on seat belt anchors. They usually have a special shape to work with the shape of the anchor plate. Also, the anchor fasteners have very high tensioning strength to avoid being snapped off during a collision.

Seat Belt Inspection

A visual and functional inspection of the belts is critical to ensure maximum protection for vehicle occupants. During seat belt inspection, you should:

- Check for twisted webbing due to improper alignment when connecting the buckle.
- Fully extend the webbing from the retractor. Inspect the webbing and replace with a new assembly if the following conditions are noted:
  - Twists
  - Cuts or damage
  - Broken or pulled threads
  - Cut loops
  - Color fading or stains
  - Binding in guide plates

Any of these defects can cause seat belt webbing to be weakened and possibly fail during a collision. Check for the problems shown in Figure 23–6.

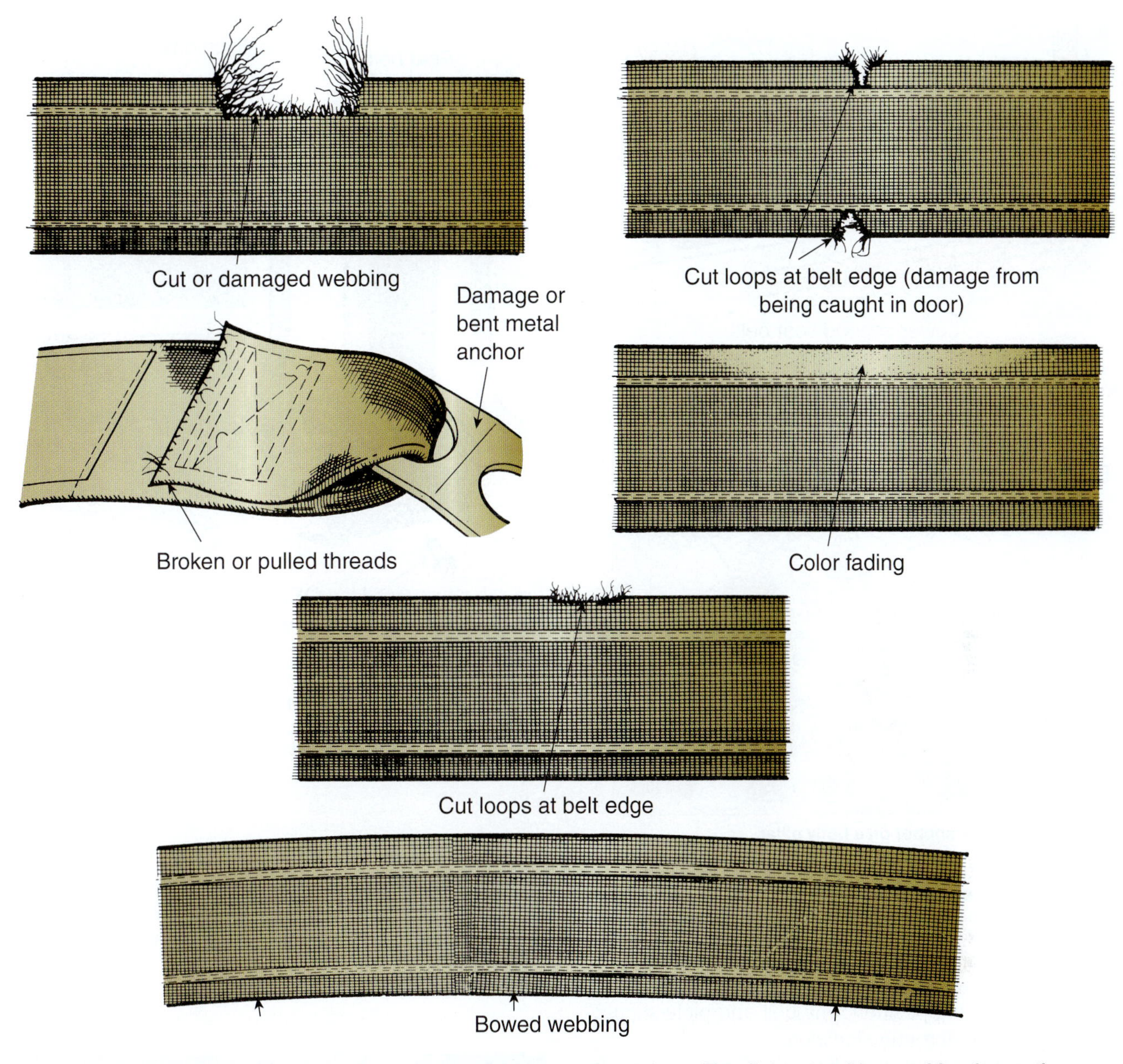

Figure 23-6 **Always inspect seat belts for these kinds of webbing defects. The slightest problem could endanger the people using the seat belt. Install new seat belts when any of these problems are found.**

Do not bleach or dye seat belt webbing. This could weaken the belt fabric. If needed, clean with a mild soap solution and water.

Buckle Service

To inspect seat belt buckles, perform the following procedures:

1. Insert the tongue of the seat belt into the buckle until a click is heard. Pull back on the webbing quickly to ensure that the buckle is latched properly.
2. Replace the seat belt assembly if the buckle will not latch.
3. Depress the button on the buckle to release the belt. The belt should release with normal finger pressure.
4. Replace the seat belt assembly if the buckle cover is cracked or the push button is loose. Also replace the unit if the pressure required to release the buckle is too great.

Anchor Service

To inspect seat belt anchors, remove any plastic trim pieces over the metal anchor plates. Check the seat belt anchors and bolts for signs of movement or deformation. Replace if necessary. Position the replacement anchor

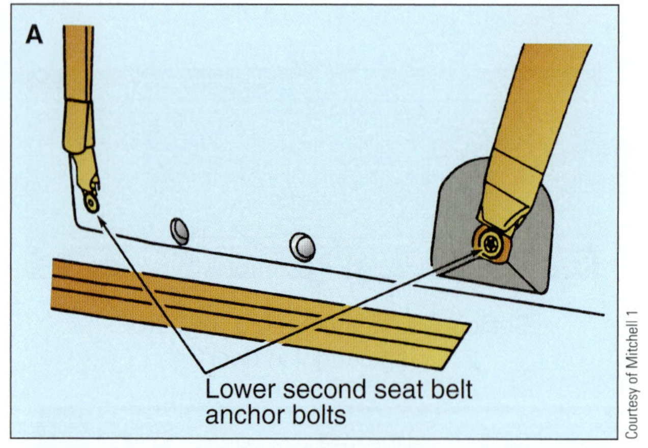

A These are lower seat belt anchor bolts.

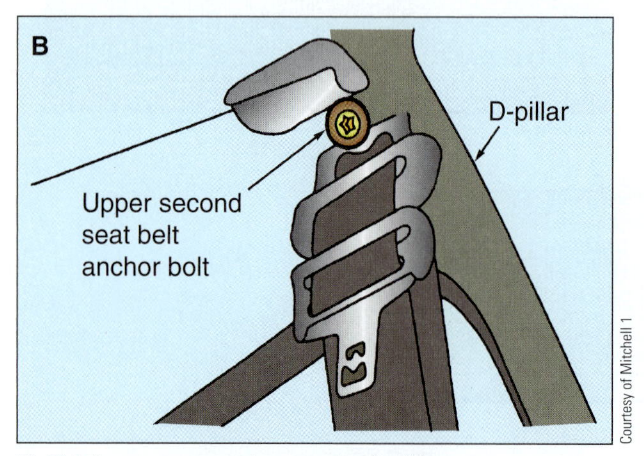

B This is an upper anchor on a body pillar.

Figure 23–7 Use only factory fasteners on seat belt anchors. After positioning the belt plate in the proper direction, use a torque wrench to tighten the anchor bolts to specifications.

plate in the original position. The belt and plate should face the seat or belt routing location.

Look up the torque specifications for the seat belt anchor bolts. Anchor bolt tightening specs can be found in the service manual or computer-based information for that vehicle. Refer to Figure 23–7.

Retractor Service

To inspect a belt retractor assembly, remember the following steps:

1. Grasp the seat belt webbing. While pulling from the retractor, give the belt a fast jerk. The belt should lock up.
2. Drive the vehicle in an open area away from other vehicles. Drive at about 5–15 mph (8–24 km/h). Quickly apply the foot brake. The belt should lock up.
3. If the retractor does not lock up under these conditions, remove and replace the seat belt assembly (Figure 23–8).

Figure 23–8 If a retractor does not work properly or is a one-time deployment mode, replace it. Unplug any wires without damaging the connectors. Then unbolt and replace the unit.

If the vehicle is equipped with pyro-technique seat belt retractors, they will have to be replaced after air bag deployment. The propellant charge will have been depleted by seat belt tensioning during the collision. See Figure 23–9.

Figure 23–9 Inspect all seat belts when repairing a vehicle after a major collision. Check their webbing and make sure buckles work normally and that retractors work.

Child Seats

A *child seat* is a small, supplemental seat designed to protect small children from collision injuries. There are three types of child car seats: rear facing, forward facing, or a combination.

The child safety seat is often secured with a lap and shoulder seat belt (using a locking clip provided by the car seat manufacturer) or just a lap belt. The vehicle manufacturer sometimes builds an integral child seat into the rear seat.

DANGER Small children should not sit in the front seat of air bag-equipped vehicles. This is because the large passenger side air bag can cause the injury or death of the child. Small children can sustain serious head and neck injuries from air bag deployment even during a minor collision.

No one can be sure a child car seat will prevent injury in an accident. However, the proper use of a child safety seat should reduce the risk. Aftermarket child seats may not be suitable for use in the front seat with an air bag system. Some seats locate the child too close to the air bag. Injury to the child could result from bag deployment. Warn your customers of this danger.

23.3 AIR BAG SYSTEM OPERATION

All new cars, vans, SUVs, and light trucks are now equipped with multiple air bags. At first, vehicles only had a driver's side air bag in the steering column. A few years ago manufacturers started using passenger side air bags as well in the right side of the dash. Now, new vehicles may even have air bags all the way around the perimeter of the passenger compartment.

An **air bag system** automatically deploys and inflates one or more nylon cushions during a collision. The system detects the rapid deceleration of an impact and blows up air bags to keep the driver and passengers from hitting the windshield, dash, and metal pillars.

While the location and design of air bag systems vary from manufacturer to manufacturer, all air bag systems have similar parts:

- An *air bag module* contains the inflator mechanism and tough nylon pouch. When triggered electrically, the inflator generates a gas that blows into and fills the air cushion during a collision. Refer to Figure 23–10.
- *Air bag system sensors* electrically signal the computer of a major collision. With front air bags only, a hit to the frontal area must be sufficient to close the switches in two impact sensors. This causes the control unit to fire the air bags (Figure 23–11).
- The *air bag control unit* is a dedicated computer that operates the system. If two sensors send a signal to the control unit, only then does it send an electrical current to the air bag module to ignite and inflate the air bag (Figure 23–12).
- The *air bag harness* includes the wiring and connectors that link the impact sensors, control unit, and air bag modules.
- An *air bag warning lamp* in the dash is turned on to warn of system problems.

Air Bag Sensors

Two or more sensors are used in air bag systems: impact sensors and arming sensors. Some late model vehicles also have impact sensors mounted in the system electronic control unit.

Impact sensors are the first sensors to detect a collision because they are often mounted at the front of the vehicle. Impact sensors are usually located in the engine compartment, while the safing sensor is usually located in the passenger compartment.

The **arming sensor** ensures that the particular collision is severe enough to require that the air bag be deployed. The arming sensors are usually located near the front bumper or radiator support of the vehicle. This frontal location helps them instantly trigger or close with a severe collision.

Some vehicles also use a **safing sensor** as a third fail-safe sensor to prevent accidental deployment of the air bags. All three sensors must typically be closed to signal the computer or electronic control unit that a collision is taking place.

Both impact and arming sensors are inertia sensors. *Inertia sensors* detect a rapid deceleration or g-forces to produce an electrical signal. Some air bag sensors have a small metal ball held in place by a permanent magnet. The sensor ball is thrown forward and away from the magnet by the inertia or rapid deceleration of the collision. It then touches two electrical terminals that close the sensor circuit to the system computer. One is illustrated in Figure 23–13.

Another air bag sensor design uses a weight attached to a coil spring. During impact, the weight is thrown forward. This overcomes spring tension to close the sensor contacts, which completes the sensor circuit so a small electric current can flow into the computer to signal a possible collision.

Both the impact sensor and a safing sensor must close at the same time for air bag inflation. They work together to provide a fail-safe system to prevent accidental air bag deployment. When both an impact sensor and a safing sensor close, the diagnostic control module sends a signal to the igniter, which starts a chemical reaction to inflate the bag.

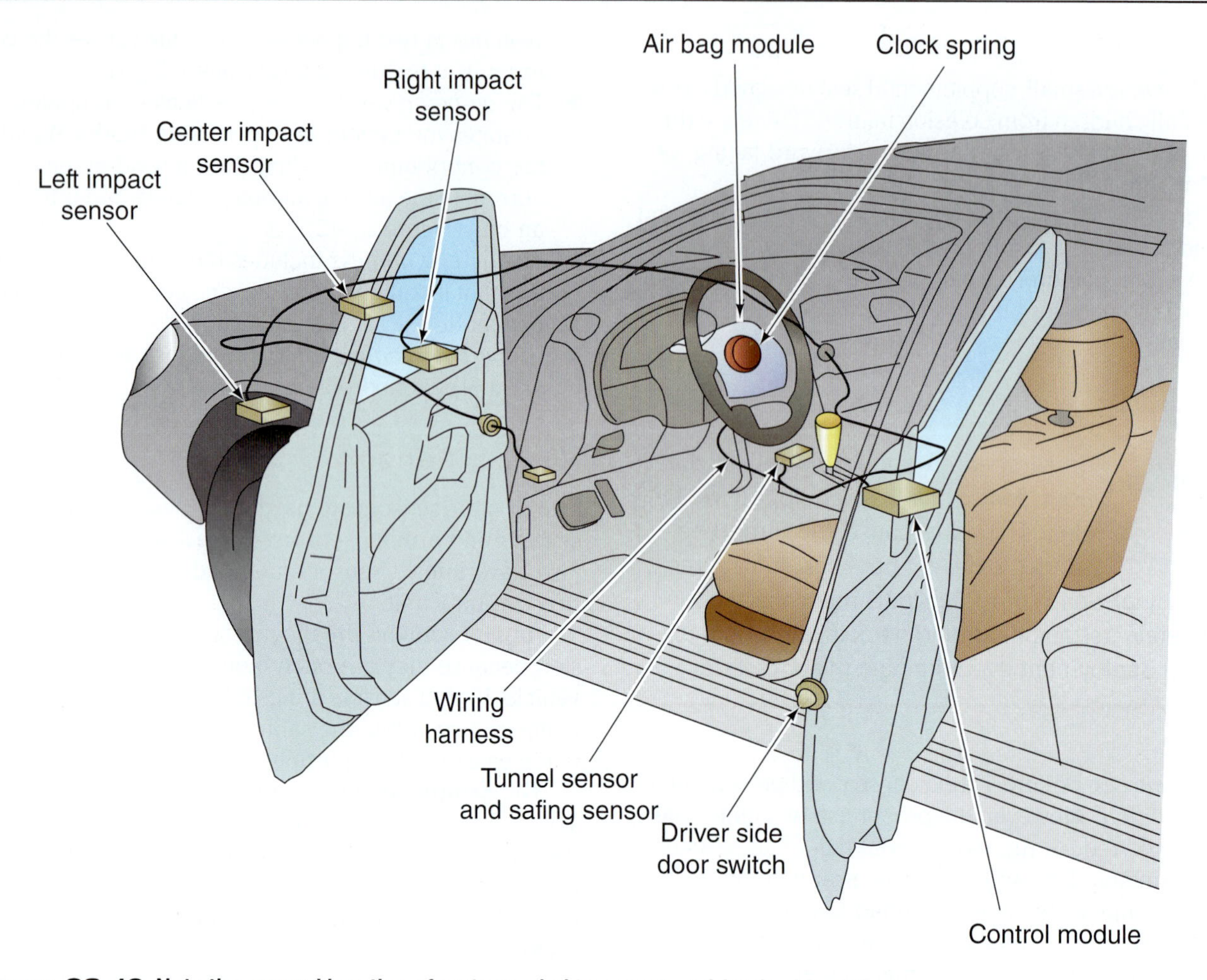

Figure 23-10 Note the general location of parts needed to operate a driver's side air bag.

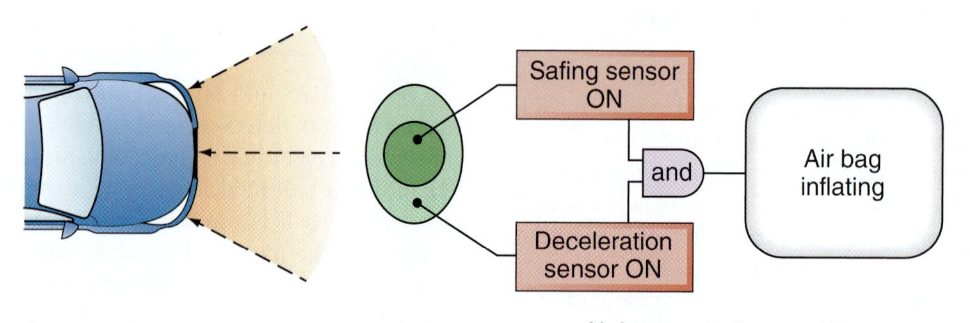

Figure 23-11 When a vehicle takes a hard hit toward the front, at least two air bag sensors must close because of rapid deceleration. The air bag control assembly can then produce an electrical signal to fire the air bag.

It is important to remember that the tandem action of at least one main sensor and a safing sensor will activate the system. Refer to Figure 23–14.

Deformation sensors are sometimes used to operate side air bags. To prevent accidental deployment, deformation sensors detect intrusion or structural impact damage into the doors. They do not detect inertia or g-force loads. Metal deformation from a side hit will change an air gap at the sensor, which triggers the sensor. The sensor sends a shock wave through plastic tubes going to the side air bag. When the air bag module is activated by the shock wave, it ignites a gas-generating charge that inflates the side air bag.

Air Bags

The **driver's side air bag** deploys from the steering wheel center pad. This type of air bag can also be found on older model cars. The air bag assembly replaces the horn button used on pre-air bag vehicles. The inflator

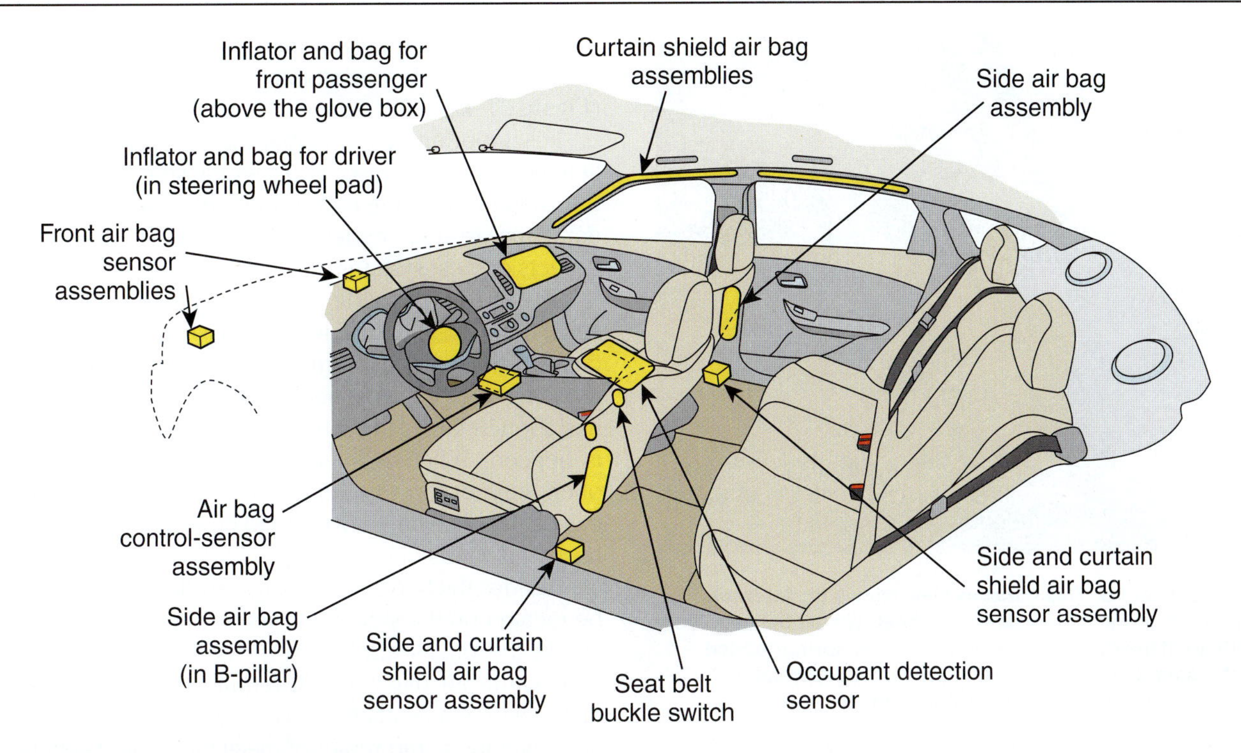

Figure 23–12 **Study the component locations in a modern restraint system. Note that this car has curtain air bags for head protection during a collision from the side or during a rollover crash.**

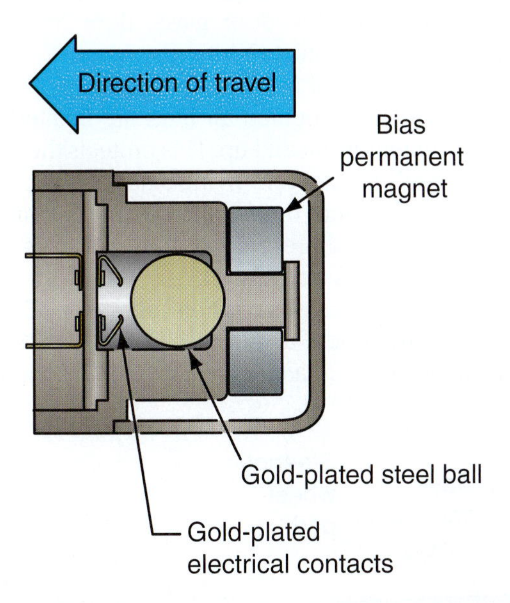

Figure 23–13 **This is one type of inertia sensor. A steel ball is normally held on one end of the cylinder by a small magnet. The impact of a collision causes the steel ball to fly forward away from the magnet and into gold-plated electrical contacts. This closes the sensor circuit so that the computer and electrical signal indicate a major collision is occurring.**

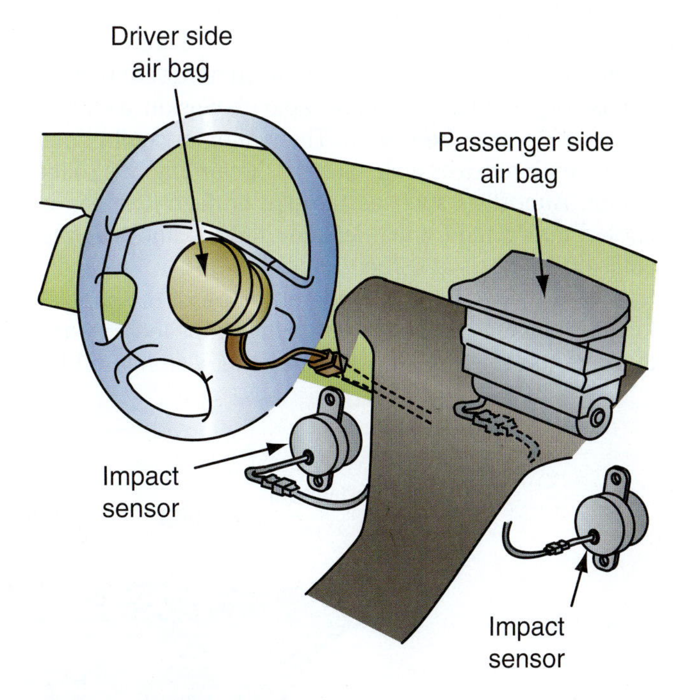

Figure 23–14 **Impact sensors for this vehicle's air bag system are located under the dash.**

charge is mounted behind the bag. The air bag is folded over the inflator and behind the steering wheel cover (Figure 23–15).

The **air bag igniter**, also called a *squib*, produces a small spark when an electrical signal is sent from the electronic control unit. When the electric current is applied, the igniter produces an electric arc across two small pins. This spark ignites a propellant charge in the squib that goes off like a firecracker (Figure 23–16).

The squib ignites a larger **propellant charge** that burns to produce the gas expansion that inflates the air bag. The inflator charge is usually pellets of sodium

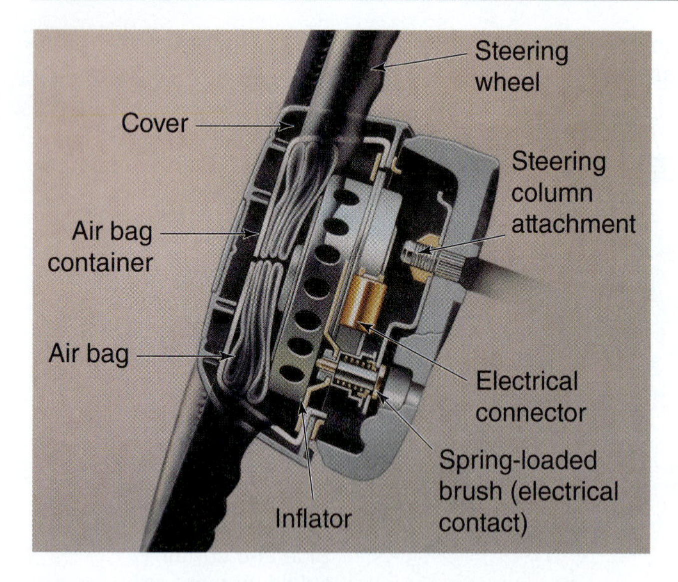

Figure 23-15 This cutaway view shows the internal parts of an air bag in the steering wheel. The air bag is folded tightly to fit inside the cover. Note how the spring-loaded brush contacts the metal ring to provide a dependable electrical connection into the rotating steering wheel assembly.

azide. When these flammable pellets are ignited, the sodium azide burns rapidly and is converted to nitrogen gas. Heat causes the chemicals in the unit to produce a large amount of nitrogen gas. Refer to Figure 23–17. Inflation of the air bag is caused by an almost explosive release of gas. To allow for this rapid expansion, a chemical reaction must be started. The igniter, or squib, does this when it receives a signal from the air bag controller circuit. Almost as soon as the bag is filled, the gas is cooled and vented, thus deflating the assembly as the collision energy is absorbed (Figure 23–18).

The driver is cradled in the envelope of the supplemental restraint bag instead of being propelled forward to strike the steering wheel or be otherwise injured by follow-up inertia energy from seat belts. In addition, there is some facial protection against flying objects (Figure 23–19).

It is important to remember that the tandem action of at least one main sensor and a safing sensor will activate the system. The microcontroller also provides failure data and trouble codes for use in servicing various aspects of most systems.

The **passenger side air bag** deploys from behind a small door in the right side of the dash. It is a much larger cushion that extends from the center of the seating area to the right door during deployment. Because this air bag has a much larger internal volume, a large pellet charge is needed. When inflated, the passenger side air bag often tears or forces open a hinged door formed in the dash cover. The bag blows out to cover the left side of the dash and windshield (Figure 23–20).

Besides the two front air bags, additional air bags can be located on the side of the seats, in the door trim panels, in the pillars, and even in the rear seat area. The increase in air bag use will require the collision repair technician or adjuster to become familiar with the operation, diagnosis, and repair of these vital safety systems.

Side air bags may be located in the door panels, roof, pillars, or seats to protect against side impact injury. Without side air bags, passengers' heads are thrown violently sideways into the door glass, door frame, and strong pillars when a vehicle is hit in the side, or "T-boned." A seat-mounted side air bag is shown in Figure 23–21.

When most seat-mounted air bags are inflated, each side air bag expands out and up. This protects the driver's and passenger's heads from flying sideways and hitting the door glass or steel door frame. Many human head and neck injuries are prevented by side air bags during high-speed collisions from the side.

Imagine what happens when one car "T-bones" another from the side. Car A smashes into Car B's doors at high speed. The impacting vehicle crushes the doors

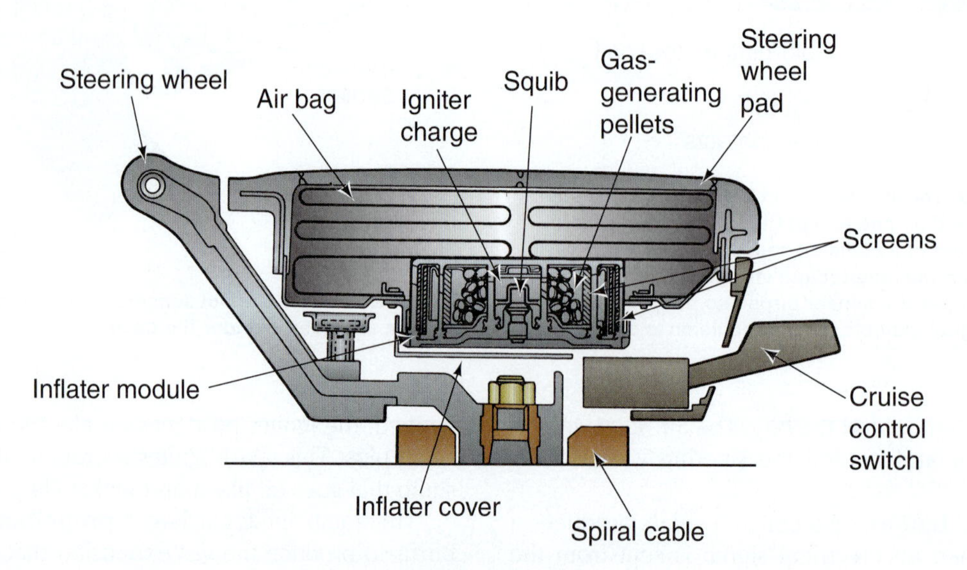

Figure 23-16 Study the details of the squib and gas-generating pellets.

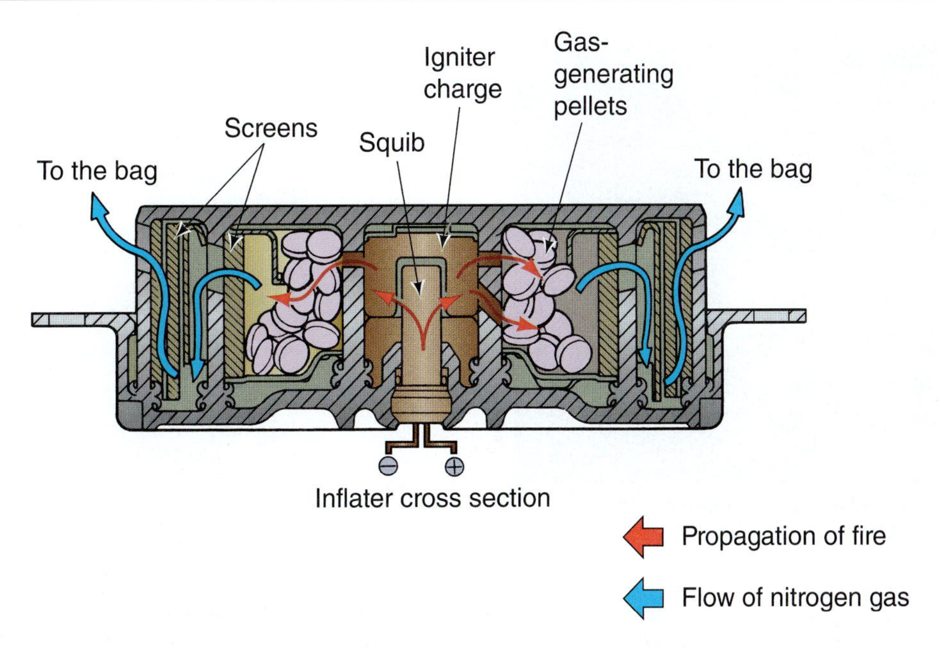

Figure 23-17 When electric current is sent to the squib from the electronic control unit, the squib explodes like a firecracker. This ignites the charge. The pellets rapidly burn and produce a hot gas that blows out of the module and into the air bag.

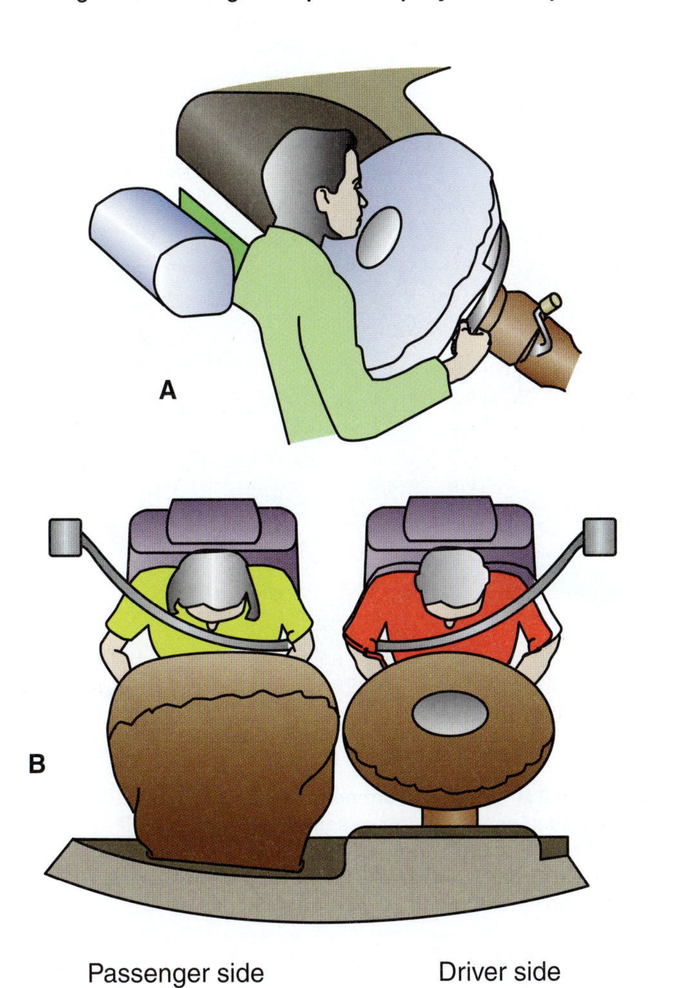

Figure 23-18 (A) Air bag deployment only takes about 1/20 of a second, or the "blink of an eye." The driver's body hits the soft air bag instead of flying farther forward, possibly into the steering wheel. (B) Note the differences between driver and passenger air bags.

and pillars inward and toward Car B's centerline. Human bodies in Car B will be thrown sideways violently against the direction of impact. Side air bags help protect human life during these kinds of auto accidents. They "pop open" to keep humans from being injured by hitting the imploding steel doors, pillars, and side body panels of the struck car.

Side air bags can also be mounted inside the door trim cover. When deployed, this type of side air bag deploys through the door panel. It protects the driver or passenger from being injured on the door frame when that person is thrown sideways.

A **curtain air bag** deploys from the front pillar and roof trim. It is a variation of the side air bag (Figure 23–22).

The curtain air bag is longer and thinner, as shown in Figure 23–22 and Figure 23–23.

Diagrams showing when and how front and side air bags are typically deployed are presented in Figure 23–24. Compare the conditions for front and side air bag deployment.

Another vehicle part relating to an air bag is the knee diverter. A **knee diverter** cushions the driver's knees from impact and helps prevent the driver from sliding under the air bag during a collision. It is located underneath the steering column and behind the steering column trim.

Multiple Threshold Deployment

Multiple threshold deployment means the air bag system can deploy the air bag at different speeds. Systems with multiple threshold deployment use special sensors that measure the rate of deceleration. The computer detects the power of the collision impact and ignites different amounts of propellant charge in the air bag

A The air bag is fired and deploys by opening a hinged cover formed in the steering wheel pad.

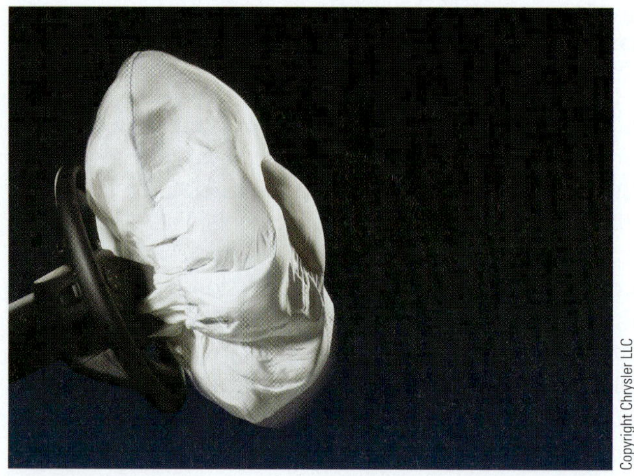

B Gases fill the air bag to form a large cushion in front of the driver.

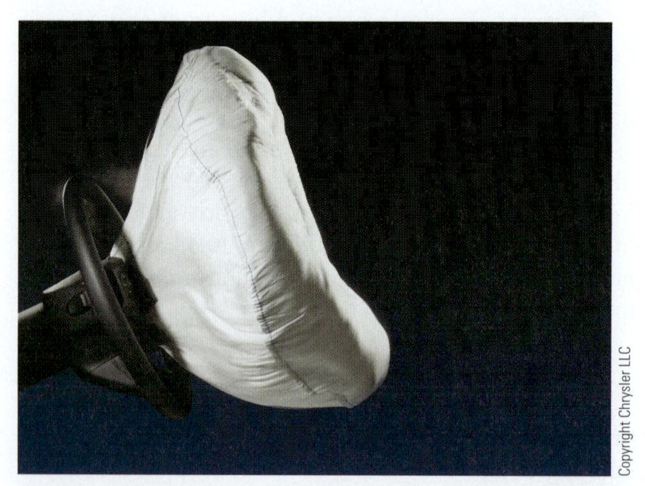

C Holes in back of the air bag allow deflation as soon as a body hits and compresses the air bag.

Figure 23-19 Study the phases of air bag deployment, which takes about $1/20$ of a second.

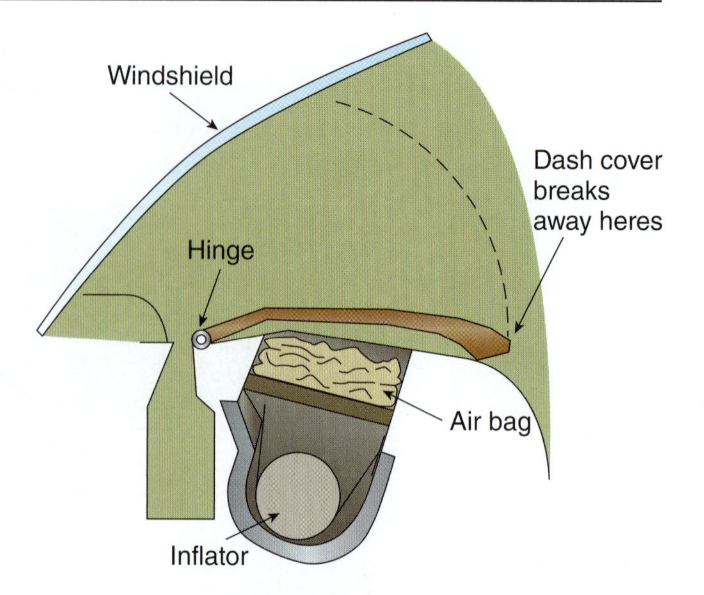

Figure 23-20 A passenger side air bag fits under the door in the right side of the dash. The air bag is folded into a small area. When the computer fires the propellant in the inflator cartridge, hot gas makes the deploy in a fraction of a second.

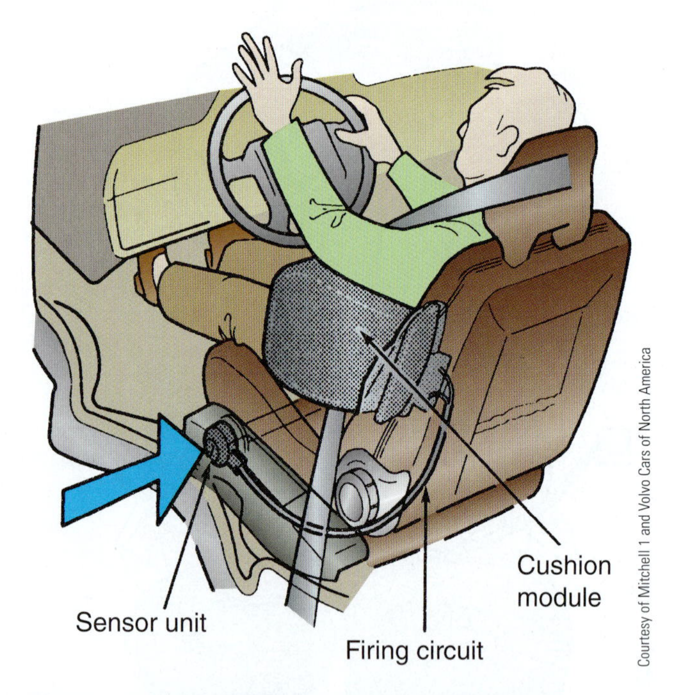

Figure 23-21 This front seat side air bag has a dedicated sensor-actuator assembly. A hard side hit triggers a firing mechanism in the sensor-actuator. The firing circuit then ignites the air bag module so gas expansion can inflate the side air bag.

module. The propellant charge is separated into compartments in the module.

For example, during a low-speed accident, the sensors would signal a smaller, less violent change in inertia. The computer would then only fire some of the propellant

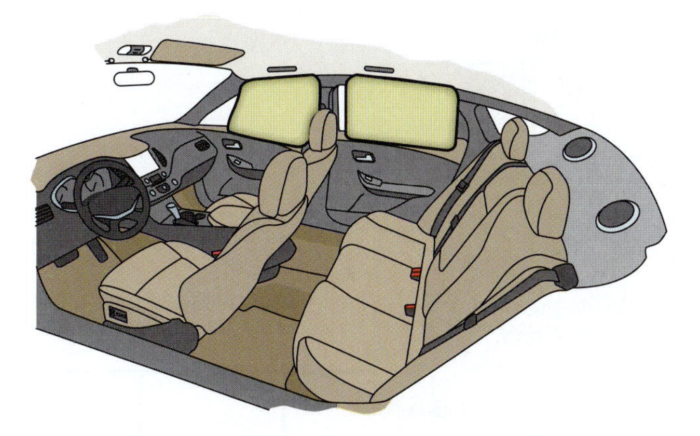

Figure 23-22 This vehicle is equipped with a drop-down curtain-type air bag for better head protection from side hits during a collision. Air curtains are more expensive and time-consuming to replace, however.

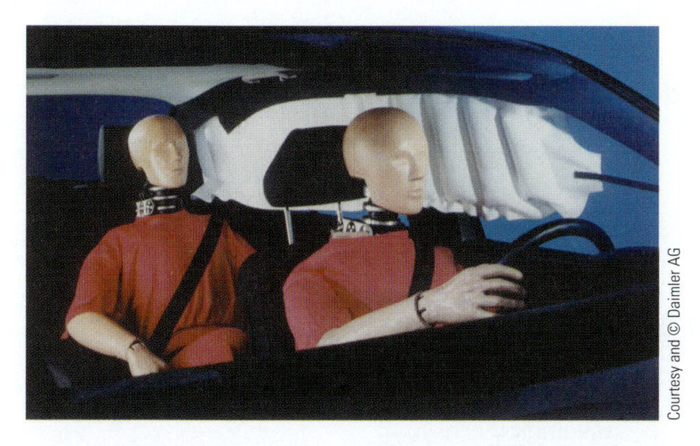

Courtesy and © Daimler AG

Figure 23-23 During a hard side hit, an air curtain will provide excellent protection. It keeps a passenger's head from hitting and being seriously injured on the body structure or door glass.

charge in the air bag module. The air bag would inflate more slowly to prevent air bag-inflicted injuries.

In a high-speed collision, the sensor would signal rapid deceleration. The computer would then ignite all of the propellant charge for more rapid air bag deployment. During a high-speed crash, you want the air bag to deploy much more quickly to keep the driver and passengers from flying into the deploying air bag. The air bag should be fully deployed when the driver's body hits it.

Air Bag Controller

The **air bag controller** analyzes inputs from the sensor to determine whether bag deployment is needed. If at least one impact sensor and the arm sensor are closed, the controller sends current to the air bag module. This "fires," or deploys, the air bag(s). The electronic control unit also provides failure data and trouble codes for use in troubleshooting and servicing circuits and components.

Passenger side air bags are very similar in design to those in the driver's unit. However, the capacity of gas required to inflate the bag is much greater, because the bag must span the extra distance between the occupant and the dashboard at the passenger seating location. The steering wheel and column make up this difference on the driver's side. Conversely, side air bags are smaller and require less propellant.

23.4 SERVICING AIR BAG SYSTEMS

Before servicing a vehicle equipped with an air bag, the system must be disarmed. An air bag system is disarmed by disconnecting all sources of electricity that could fire the air bags. Procedures for disarming air bag systems vary.

Many vehicles' service instructions require you to disconnect the negative battery cable and wrap tape around the cable end. The tape insulates the metal cable end so that it cannot accidentally touch the battery terminal.

Manufacturers may also specify removal of the system fuse or disconnection of the module. Always refer to the service manual for exact procedures for disarming the system. This will help you prevent electrical system damage and accidental deployment of the new air bag(s).

Air bag systems may be equipped with an **energy reserve module** that allows the air bag to deploy in the event of a power failure. It must be removed from the system or allowed to discharge for a period of time ranging from a few seconds to 30 minutes after disconnecting the battery. Refer to vehicle-specific service information if in doubt about the use of an energy reserve module.

Even with the battery disconnected, the energy reserve module can fire an air bag. If you are working near that bag, you can be seriously injured. Make sure that the air bag system is properly disarmed before working around undeployed air bags.

Air Bag Sensor Replacement

Air bag system parts replacement after a deployment varies depending on the vehicle. Check for specific

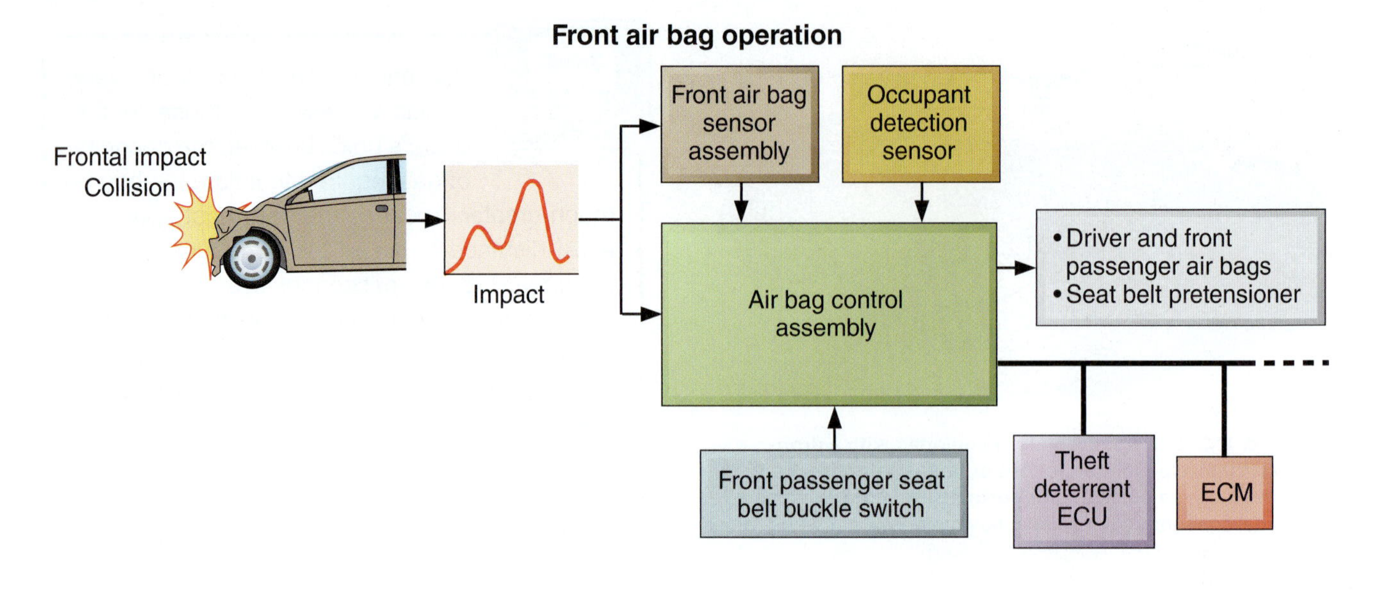

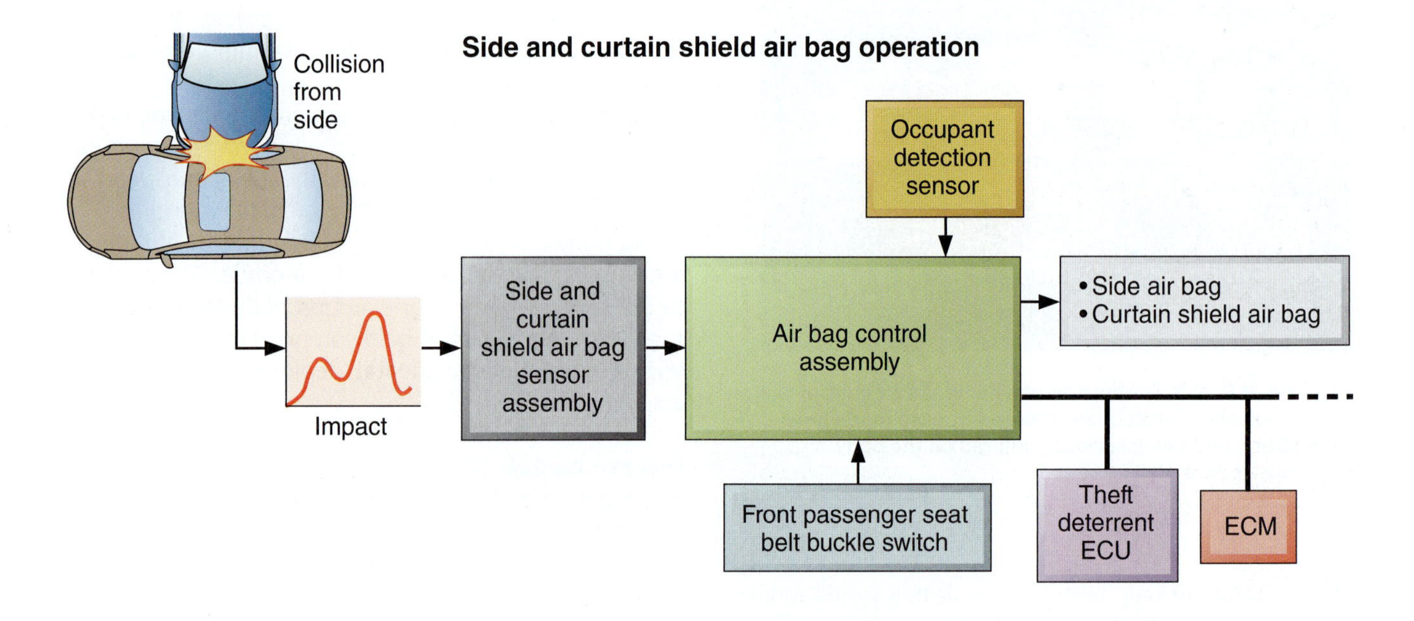

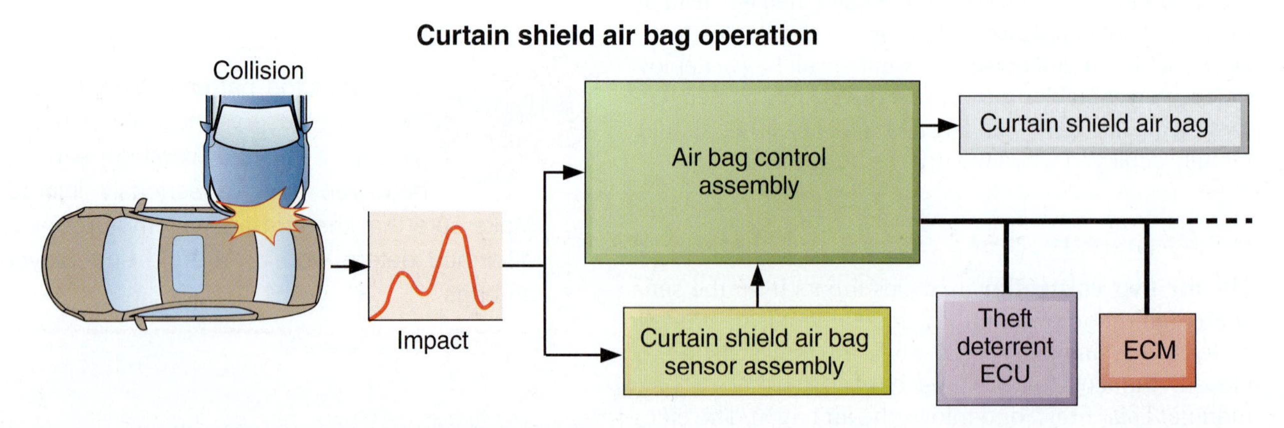

Figure 23-24 Study what must happen to deploy front, side, and curtain-type air bags. Only the air bags that are needed for each direction and location of the impact are deployed.

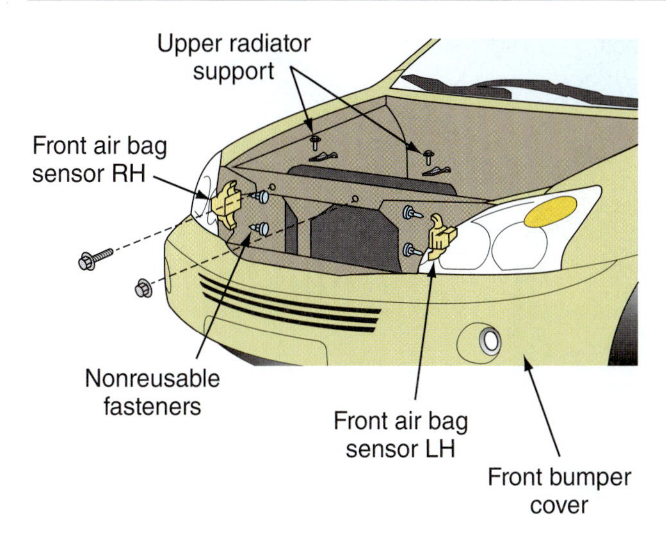

Figure 23-25 If damaged or recommended by the auto maker, air bag sensors should be replaced. All wiring in the air bag circuit must also be inspected and repaired before installing new air bags. Make sure the sensor arrows point toward the front of the vehicle during replacement.

manufacturer recommendations on parts replacement. Many auto makers recommend replacement of all sensors and sometimes the electronic control unit when servicing a deployed air bag. Sensors can be damaged internally from a severe collision. Refer to Figure 23–25.

When replacing air bag system sensors, double-check that the system is disarmed before removing any sensor. The service manual will give sensor locations. Make sure you have the correct replacement sensor. During air bag sensor installation, check that the **sensor arrow** (directional arrow stamped on the sensor) is facing forward. If a sensor is installed backward, the air bag will not deploy during a future accident.

Always obtain the correct replacement parts from the manufacturer. Also, refer to the service manual for exact procedures, because system designs vary.

Deployed Air Bag Removal

When servicing a vehicle after air bag deployment, use a shop vacuum to clean the passenger compartment. Residual powder, which is an eye and skin irritant, can be present. The powder is added during manufacture to reduce friction during air bag deployment. Vacuum the dash, vents, seats, carpet, and other surfaces contaminated with this powder (Figure 23–26).

To remove the deployed driver's air bag, remove the small screws from the rear of the steering wheel. You can then lift out the used module and disconnect

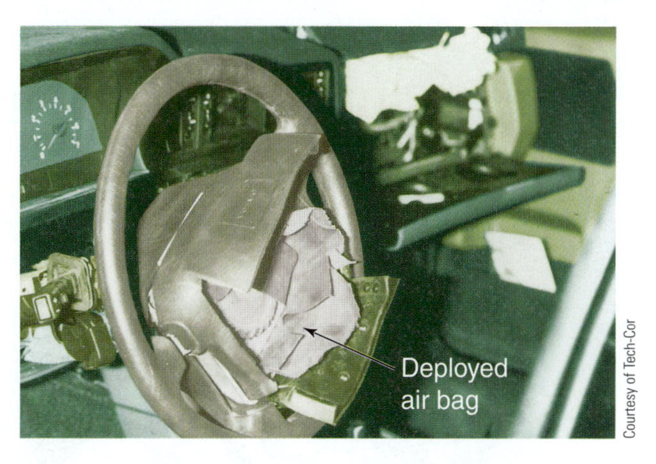

Figure 23-26 When an air bag is deployed, use a vacuum to clean up all powder. Wear a respirator, gloves, and eye protection, because powder can be an irritant.

DANGER Wear safety goggles, gloves, and a respirator while vacuuming and removing the deployed bag. This will protect you from the residual powder. Deployment of a passenger side air bag will frequently break the car's windshield, sending glass bits and slivers into the passenger area. Be careful not to cut yourself on these bits of sharp glass. See Figure 23–27.

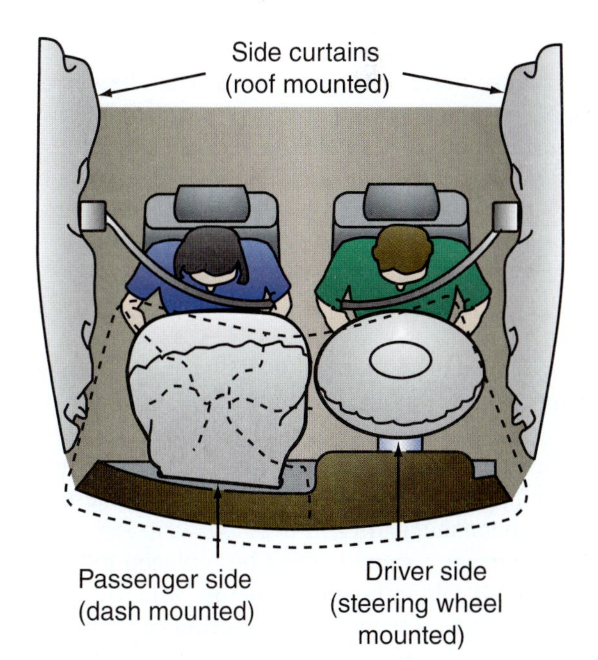

Figure 23-27 Deployment of a passenger side air bag sometimes breaks the windshield.

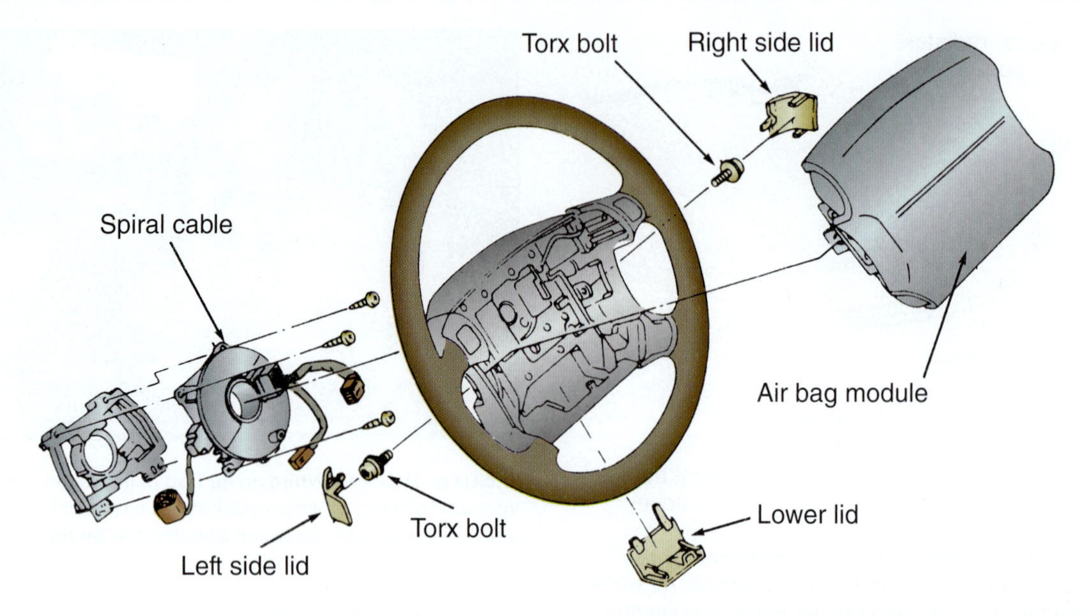

Figure 23-28 Hardened bolts hold the driver's side air bag assembly in the steering wheel. Note the small trim covers that must be removed to access these bolts. The spiral cable or clock spring must normally be replaced after air bag deployment.

its wires, following the manufacturer's procedures (Figure 23–28).

Inspect all parts for damage. Parts that have visible damage should be replaced, including the steering wheel, steering column, clock spring, and related parts. Damage to the electrical wiring may also require wiring harness replacement or careful wiring repair. Refer to Figure 23–29.

Air Bag Installation

When carrying a live (undeployed) air bag module, make sure the bag and trim cover are pointed away from your body. This helps to reduce the chances of serious injury if the bag accidentally inflates. When laying a module down on a work surface, make sure the bag and trim cover are face up to minimize the "launch effect" of the module if the bag suddenly inflates. Follow manufacturer's policies for replacing parts following a deployment.

Do *not* carry any system parts by the wire harness or pigtails. Follow the manufacturer's policies if any part is dropped or shows visible signs of damage. Do not attempt to repair any parts unless specified by the manufacturer. Do not apply electrical power to any part unless specified by the manufacturer.

The air bag assembly must be replaced following a deployment. Often the **clock spring**, which is the electrical connection between the steering column and the air bag module, will also have to be replaced. Refer to Figure 23–30.

During replacement of the clock spring, also called a spiral cable, you must often align marks for proper assembly. This is shown in Figure 23–31.

To install a driver's side air bag, double-check that all sensors are replaced or undamaged and that the battery is still disconnected. Plug in the air bag's electrical connector. Then fit the air bag down onto the steering wheel. Install and torque the fasteners that secure the air bag.

Installing a new passenger side air bag involves a similar process. Sometimes the glove box and a heating duct must be removed to gain access to the fasteners holding this air bag. Service literature will give detailed information if you are not sure how to access these bolts (Figure 23–32).

You must also disconnect the wiring harness connector from the passenger side air bag. Harness clips may also secure the wires going to the air bag (Figure 23–33).

Be careful when unplugging and reconnecting air bag electrical connectors. The plastic connectors are easily broken if mishandled. One type of double-lock connector used with air bag wiring harnesses is pictured in Figure 23–34.

With the door trim panel removed, the technician can then easily access the side air bag mounting bolts. These bolts should be final tightened with a small torque wrench.

Procedures for replacing seat- and roof-mounted air bags are similar. Again, refer to factory service literature for detailed instructions. A technician installing a new door-mounted air bag is shown in Figure 23–35.

REMEMBER *There is no need to fear working on air bag-equipped vehicles, but they must be treated with respect. When working, keep your arms and head out from directly in front of the air bag in case it accidentally fires. Your arms can be broken easily if they are too close to a deploying air bag. You could be killed if your face is too close to an air bag during accidental deployment.*

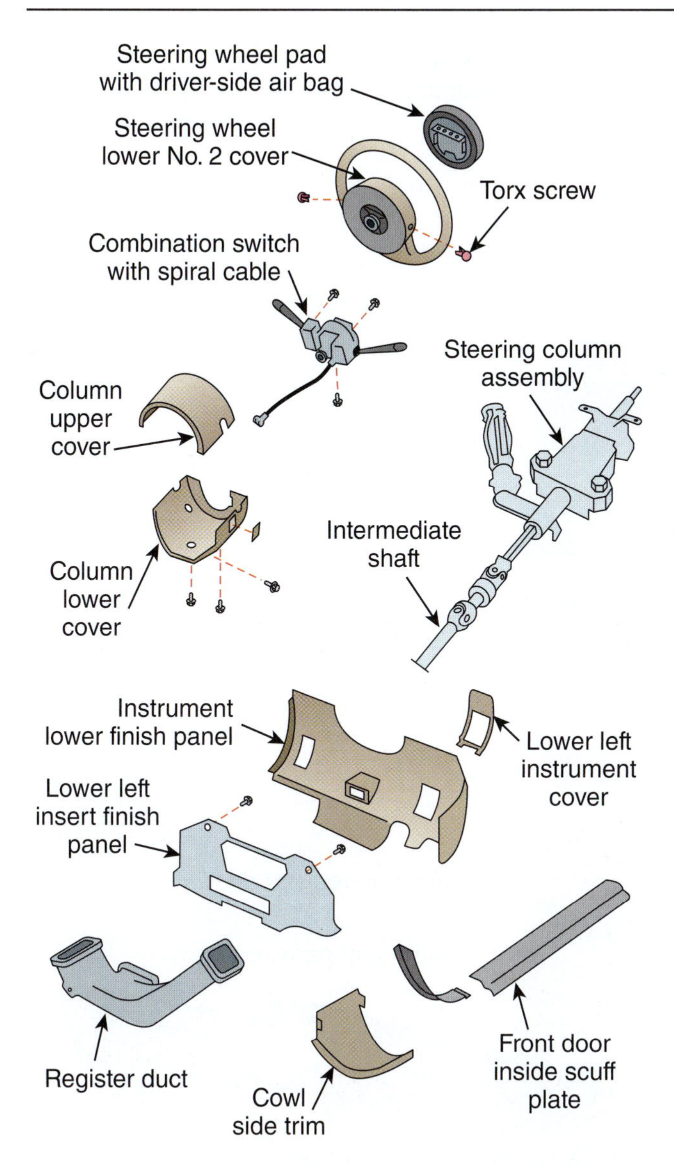

Figure 23-29 If you find damage to the steering column, it should be repaired during air bag replacement. Note how large nuts secure the steering column to the bottom of the dash. The combination switch should also be checked for normal operation at this time.

Servicing an Air Bag Control Unit

A few auto manufacturers recommend that you replace the electronic control unit any time the air bags have deployed. Other auto makers allow reuse of the control unit if it passes diagnostic tests. Normally, the air bag electronic control unit is mounted under the center of the dash, under a seat, or under the center console (Figure 23–36).

Make sure you have the correct replacement air bag control unit. Tighten the mounting fasteners properly. Also, make sure the electrical connectors are fully engaged and locked (Figure 23–37).

Wiring diagrams for two different air bag systems are shown in Figure 23–38 and Figure 23–39. Study how the wires are routed to each component.

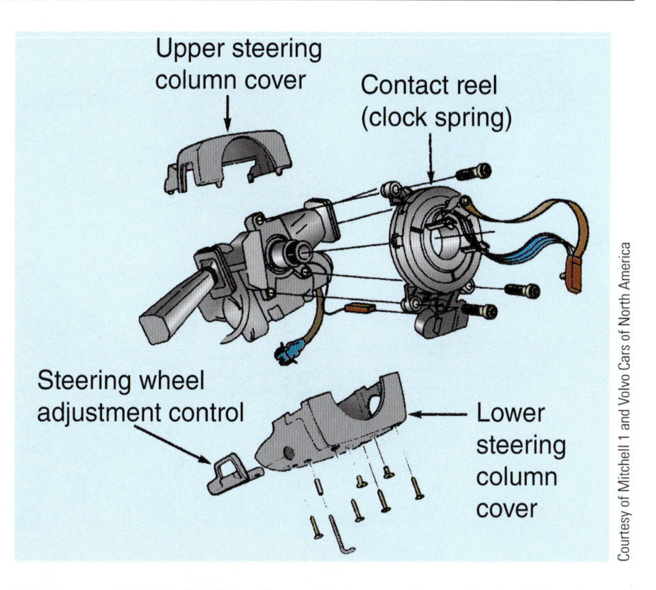

Figure 23-30 Note how the contact reel is held in the steering column.

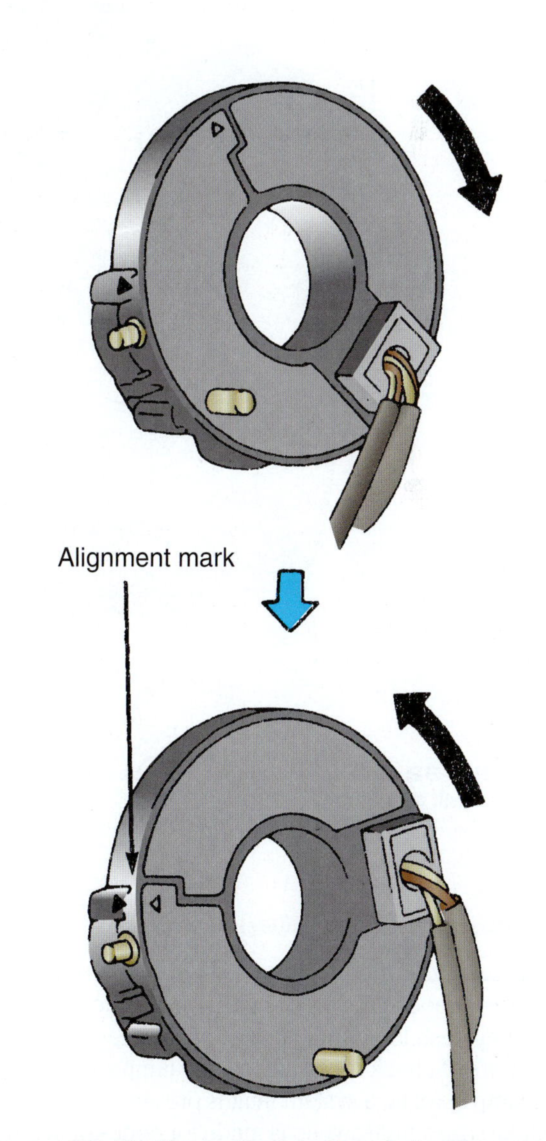

Figure 23-31 You must rotate the spiral cable assembly back and forth during installation. Arrows must align before installing fasteners.

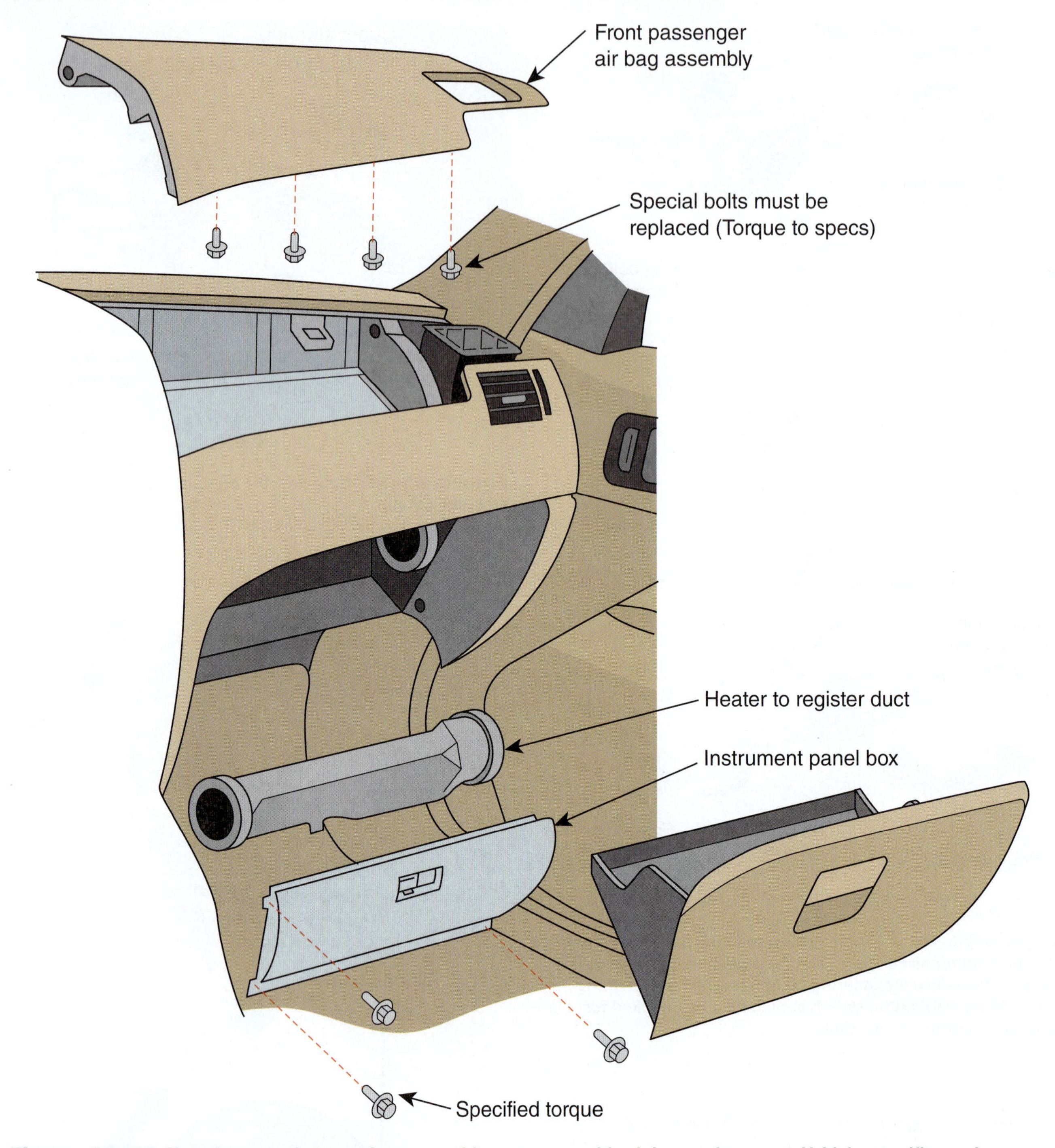

Figure 23-32 **Note the parts that must be removed for passenger side air bag replacement. Vehicle-specific service information will give details.**

Checking Air Bag Repairs

The air bag system performs a self-check every time the ignition is turned to the on position. During the self-check, the air bag dash lamp indicator will light steadily or blink. When the self-check is completed, the lamp should go off. If the lamp stays lit, a system fault is present.

Make sure a final sweep is made for codes or accident information, using the approved scan tool. Carefully recheck the wire and harness routing before releasing the vehicle to the customer (Figure 23–40).

DAB refers to the driver's side air bag and PAB refers to the passenger side air bag. Some OBDII trouble codes for an air bag system are given in (Figure 23–41).

A final inspection of the job should include checking to make sure the sensors are firmly fastened to their mounting fixtures with the arrows on them facing forward. Be certain all the fuses are correctly rated and replaced.

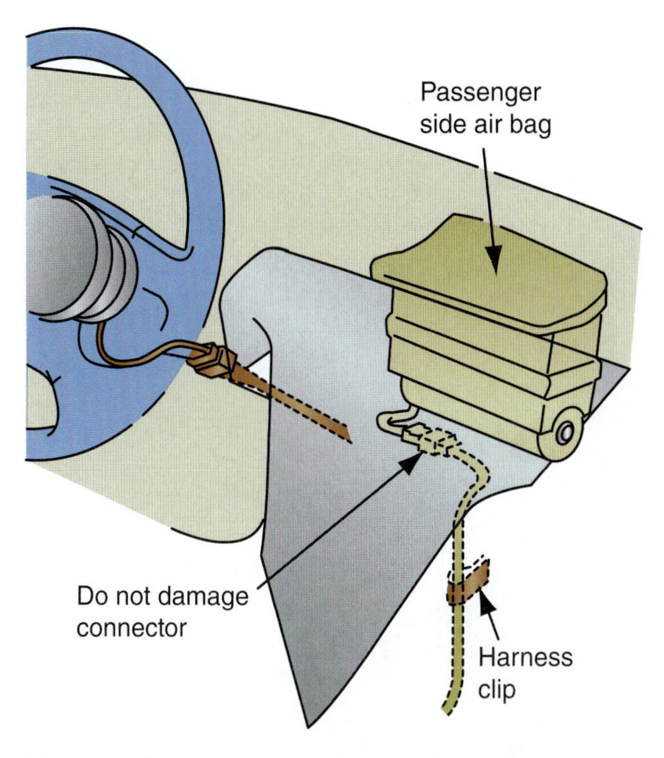

Figure 23-33 Make sure all air bag harness wires fit back into their clips. Route the wires in their original locations.

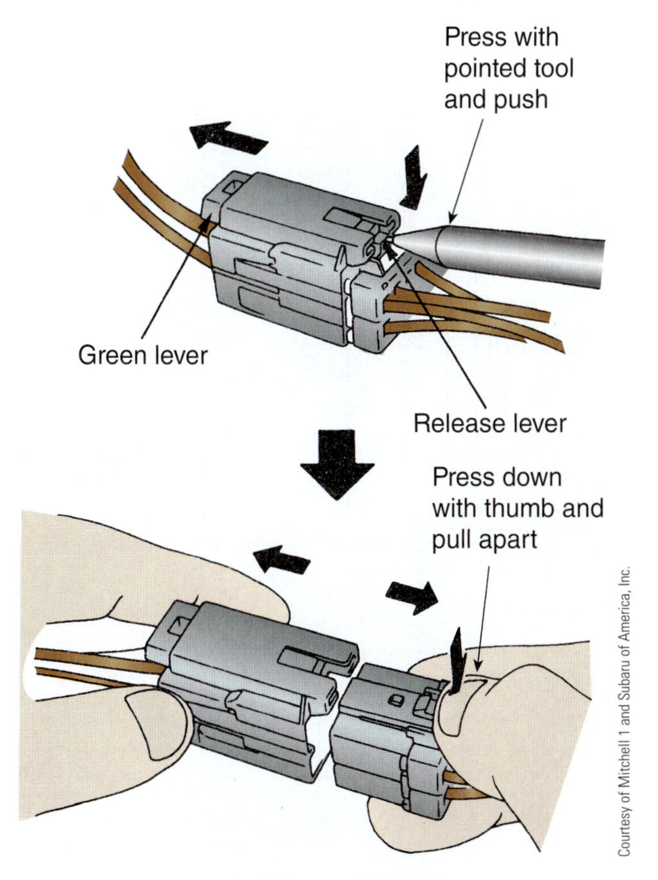

Figure 23-34 This is a double-locking electrical connector for an air bag circuit. You must depress both levers to disconnect the harness.

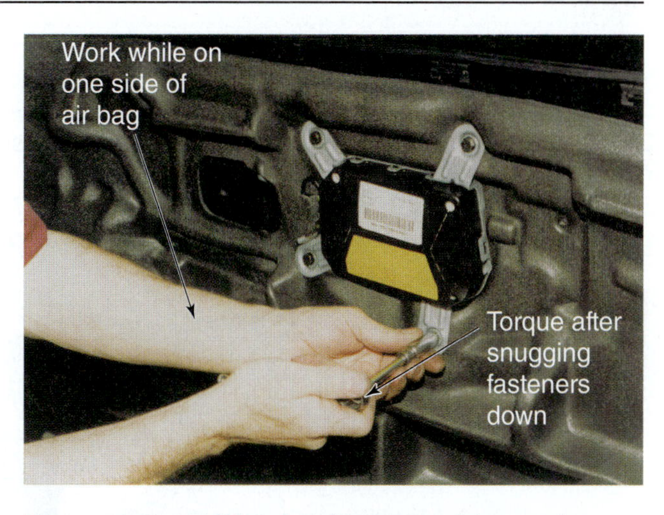

Figure 23-35 When installing a new air bag, such as this door-mounted side air bag, keep your face and body to one side. This will help protect you if accidental deployment occurs.

Air Bag Service Rules

Common sense can go a long way during work on air bag systems. By following some simple rules, air bags can be safely serviced.

- Always have the service manual on hand when working on an air bag-equipped vehicle.
- When servicing a vehicle that has an undeployed air bag, follow the manufacturer's instructions for disarming the system. You should also disarm the system when installing a new air bag.
- Wear rubber gloves, a respirator, and eye protection when servicing an air bag following a deployment. In case of skin or eye irritation, wash the affected area thoroughly with water and seek medical attention.
- Disarm the air bag system prior to performing any welding operations.
- Keep arms out of the steering wheel spokes when working on an air bag. The air bag can shatter bones if accidentally deployed. Also keep your head to one side of the bag during installation.
- Follow the manufacturer's guidelines on force drying paint on vehicles equipped with an air bag.
- Air bag disposal procedures vary depending on whether the air bag has been deployed.
- If the air bag module is defective or the vehicle is to be scrapped, the air bag should be manually deployed using the procedures described in the service manual. Do not dispose of an undeployed air bag.
- On air bag modules that cannot be manually deployed, the disposal procedure is to ship it back to the manufacturer, using the packaging that the replacement module came in. By using the replacement part's packaging, all of the needed warning labels are already on the package.

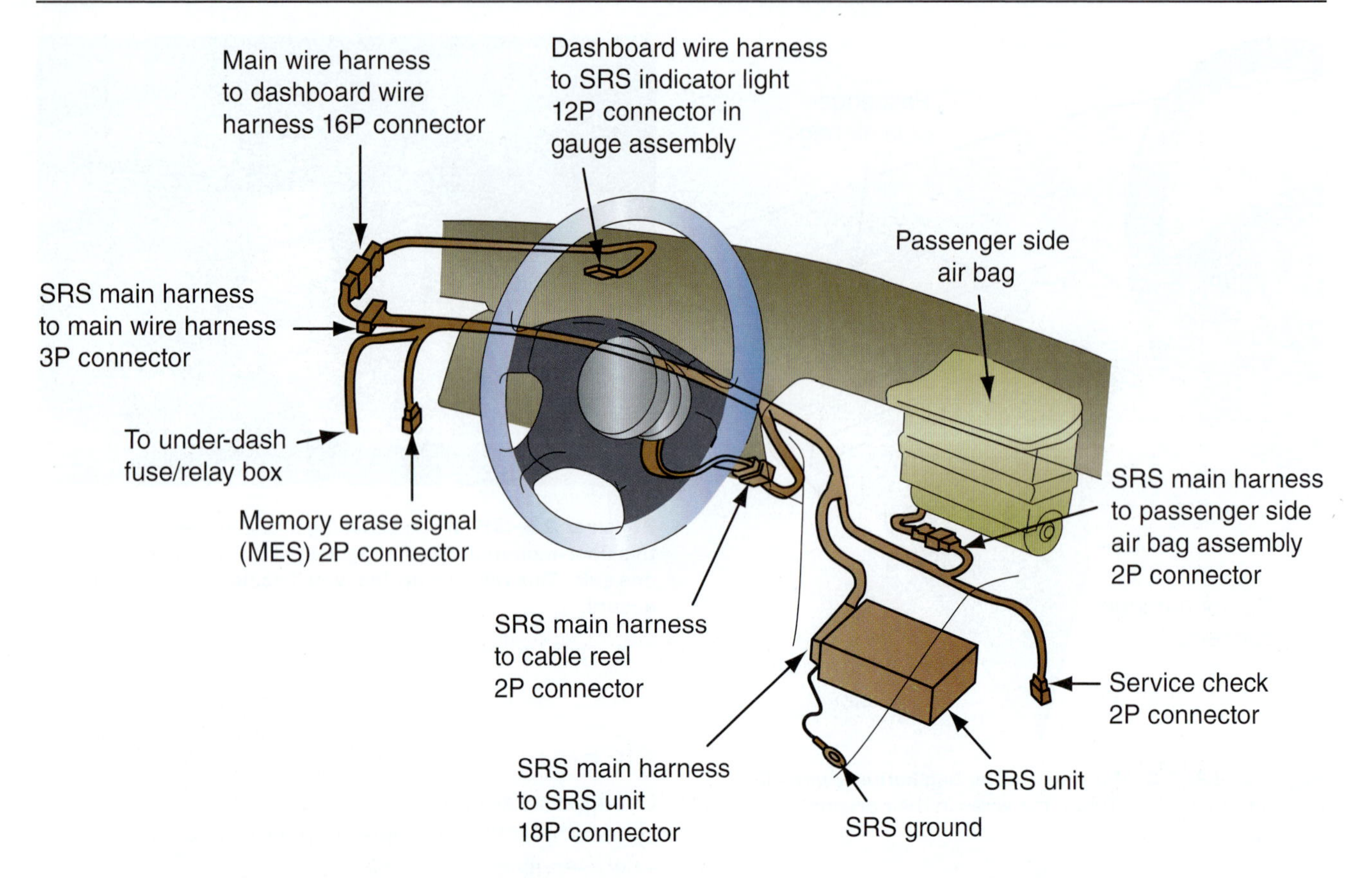

Figure 23-36 With major dash damage, make sure all harness connectors are undamaged and plugged in.

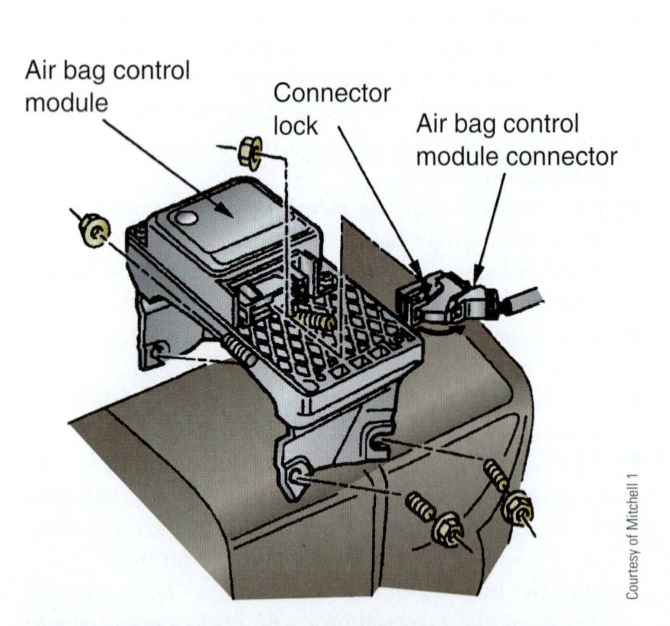

Figure 23-37 The air bag electronic control unit is usually located under the dash, often near the center of the vehicle. Make sure you have the correct replacement unit part numbers.

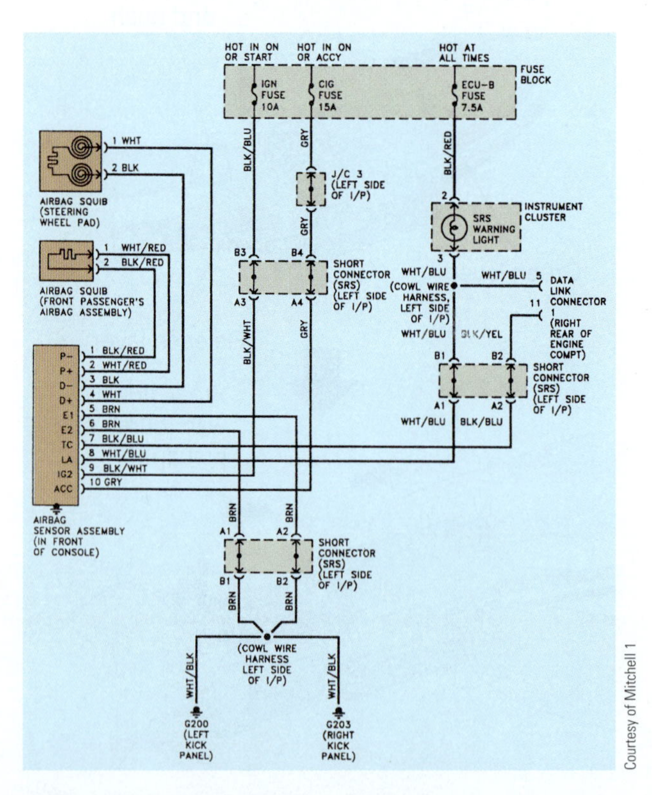

Figure 23-38 Study this wiring diagram for an air bag system.

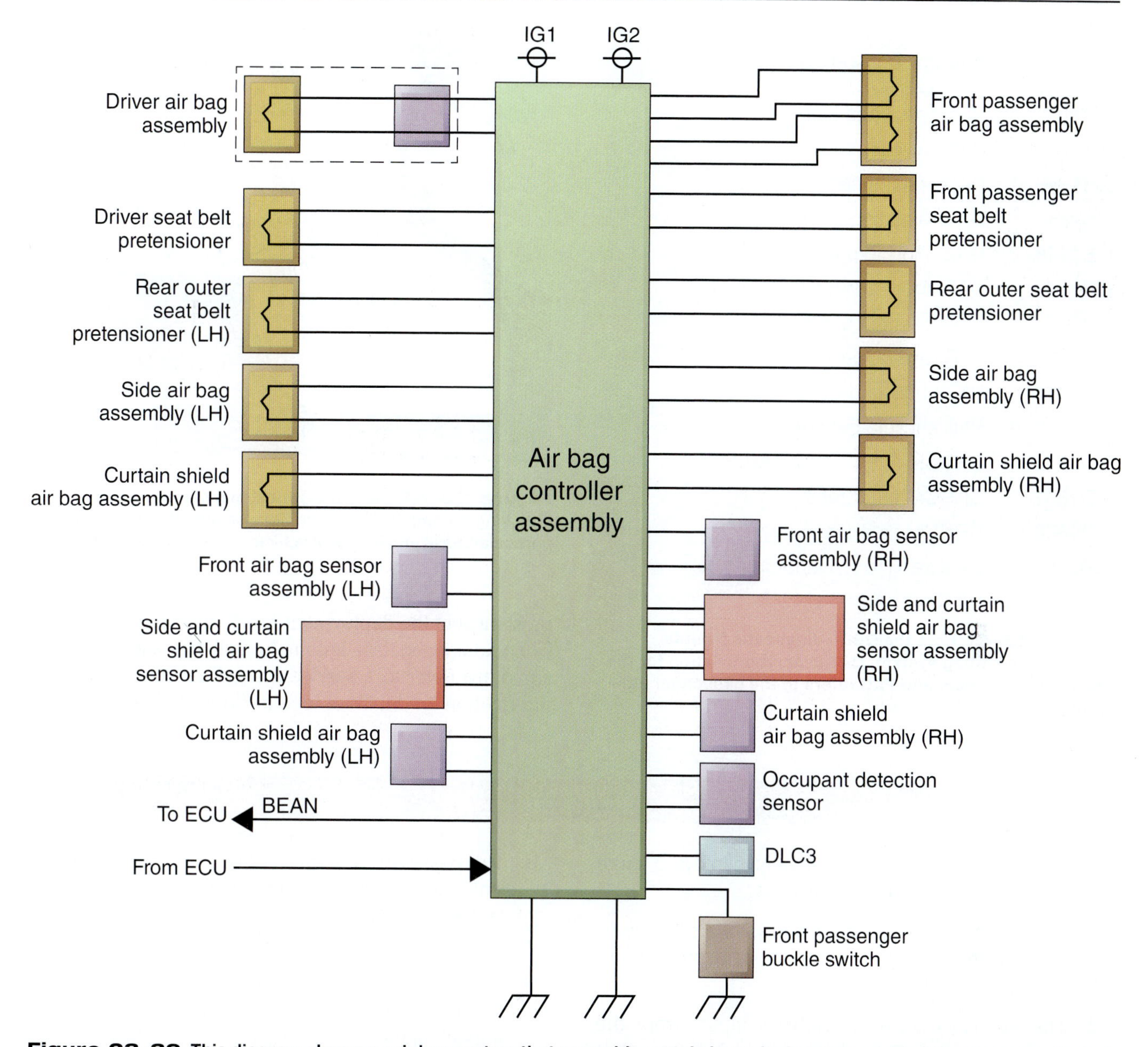

Figure 23-39 **This diagram shows an air bag system that uses side curtain bags. It also uses seat belt tensioners activated by the computer during air bag deployment. Note that the buckle switch is wired to the controller. The air bag controller assembly works in conjunction with the main engine control unit (ECU) or body computer.**

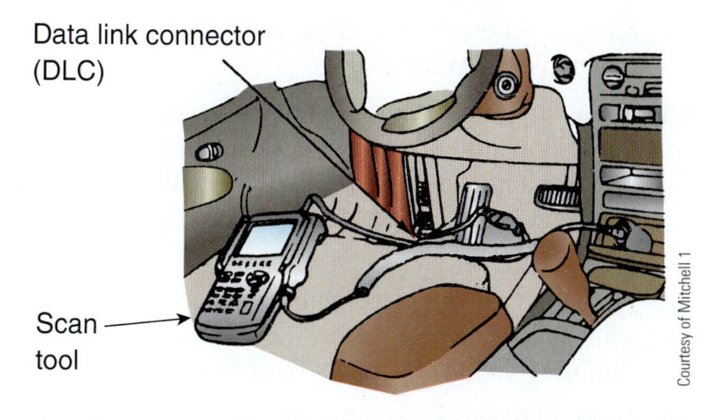

Figure 23-40 **A scan tool can be used to quickly check air bag circuits for problems.**

23.5 ROLL BAR SERVICE

A few convertible sports cars are equipped with either a fixed or automatic roll bar.

With an automatic roll bar, the vehicle's on-board computer system uses a yaw sensor to trigger activation of the roll bar. As soon as the vehicle angle signals that a rollover accident is imminent, the computer sends current to the actuators that allow the roll bar to slide up and lock in place.

You must make sure the roll bar has not been damaged in the collision. A fixed roll bar must not be bent, cracked, or damaged. Take measurements to make sure the roll bar is in alignment. With an automatic roll bar, you

DTC	Scan Tool Display
B1111	Battery voltage high
B1112	Battery voltage low
B1346	DAB resistance high
B1347	DAB resistance low
B1348	DAB short to ground
B1349	DAB short to battery
B1352	PAB resistance high
B1353	PAB resistance low
B1354	PAB short to ground
B1355	PAB short to battery
B1372	SRSCM firing circuit DAB-PAB
B1661	SRSCM parameter
B1650	SRSCM crash recorded
B2500	SRS service reminder indicator

Courtesy of Mitchell 1

Figure 23-41 Note the OBDII diagnostic trouble codes for an air bag system. DAB is an abbreviation for the driver's side air bag, and PAB refers to the passenger side air bag.

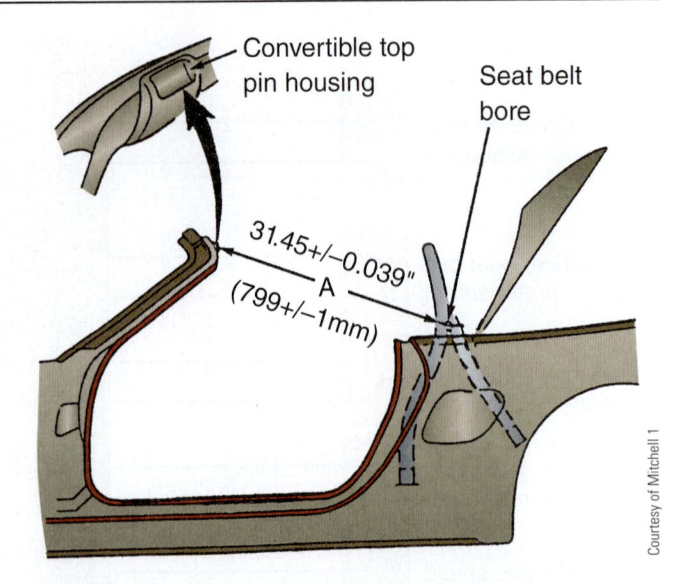

Figure 23-42 Note the critical measurements for checking damage to a roll bar. Service specifications for a particular make and model of vehicle should be used, however.

must inspect the deployment mechanism to make sure it is not damaged. An automatic roll bar should retract smoothly. Refer to a service manual or computer-based information for details of repairs. Look at Figure 23–42.

SUMMARY

1. An air bag system automatically deploys a large nylon bag during frontal collisions.
2. Both impact and arming sensors are inertia sensors, which detect a rapid deceleration to produce an electrical signal.
3. The air bag controller analyzes inputs from the sensor to determine whether bag deployment is needed.
4. The arming sensor ensures that a collision is severe enough to require that the air bag be deployed.
5. Some vehicles use a safing sensor as a third fail-safe sensor to prevent accidental deployment of the air bag(s).
6. Deformation sensors are sometimes used to operate side air bags. To prevent accidental deployment, deformation sensors detect intrusion or structural impact damage into the doors. They do not detect inertia or g-force loads.
7. The air bag igniter, also called a squib, produces a small spark when an electrical signal is sent from the electronic control unit. The squib ignites a larger propellant charge that burns to produce the gas expansion that inflates the air bag.
8. The passenger side air bag deploys from behind a small door in the right side of the dash. It is a much larger cushion that extends from the center of the seating area to the right door during deployment.
9. Side impact air bags may be located in the door panels, roof, pillars, or seats to protect against side impact injury.
10. A knee diverter cushions the driver's knees from impact and helps prevent the driver from sliding under the air bag during a collision.
11. Small children should not sit in the front seat of air bag-equipped vehicles.
12. Before servicing a vehicle equipped with an air bag, the system must be disarmed; all sources of electricity for the igniter must be disconnected. Even with the battery disconnected, the reserve module can fire the air bag.
13. Most auto makers recommend replacement of all sensors and sometimes the electronic control unit during servicing of a deployed air bag.
14. During air bag sensor installation, check that the sensor arrow, the directional arrow stamped on the sensor, is facing forward.

15. Wear rubber gloves, a respirator, and eye protection when servicing an air bag.
16. During the rapid deceleration of a head-on collision, the weight of a human body can exert tremendous force on the seat belts.
17. Proper seat belt service is critical after a major collision.

EXERCISES

On a separate sheet of paper, complete the following learning activities for this chapter. Write definitions for the key terms and answer the ASE-style review questions, essay questions, critical thinking problems, and math problems. You can also do the outside activities, possibly for extra credit.

Key Terms

active restraint system
air bag controller
air bag igniter
air bag system
arming sensor
belt retractor
clock spring
curtain air bag
deformation sensors
driver's side air bag
energy reserve module
impact sensors
knee diverter
multiple threshold deployment
passenger side air bag
passive restraint system
propellant charge
pyro-technique retractor
restraint system
safing sensor
seat belts
seat belt anchors
seat belt buckle
seat belt reminder system
sensor arrow
side air bags

ASE-Style Review Questions

1. Technician A attempts to repair lap and shoulder belt retractor mechanisms before replacing them. Technician B automatically replaces defective lap and shoulder belt retractor mechanisms. Who is correct?
   A. Technician A
   B. Technician B
   C. Both A and B
   D. Neither A nor B
2. When servicing an air bag system, Technician A probes the electrical connectors on the air bag module with a test light; Technician B does not. Who is correct?
   A. Technician A
   B. Technician B
   C. Both A and B
   D. Neither A nor B
3. Technician A disconnects the battery negative terminal and wraps it with tape before installing new air bags. Technician B says this is not necessary for safety. Who is correct?
   A. Technician A
   B. Technician B
   C. Both A and B
   D. Neither A nor B
4. Technician A uses an air ratchet to tighten air bag mounting bolts. Technician B uses a small torque wrench to tighten them. Who is correct?
   A. Technician A
   B. Technician B
   C. Both A and B
   D. Neither A nor B
5. Technician A says that you should vacuum up the powder dust remaining after air bag deployment. Technician B says this powder can be a lung and skin irritant. Who is correct?
   A. Technician A
   B. Technician B
   C. Both A and B
   D. Neither A nor B
6. Technician A uses strong bleach to clean bloodstains off seat belt webbing. Technician B says that this could weaken the belt fabric. Who is correct?
   A. Technician A
   B. Technician B
   C. Both A and B
   D. Neither A nor B

7. Technician A says you should carefully inspect seat belts for signs of damage. Technician B says even a small cut in the belt fabric would require seat belt replacement. Who is correct?
   A. Technician A
   B. Technician B
   C. Both A and B
   D. Neither A nor B

8. Technician A says that some auto manufacturers recommend that all air bag sensors be replaced after a deployment. Technician B says that some manufacturers allow you to reuse sensors. Who is correct?
   A. Technician A
   B. Technician B
   C. Both A and B
   D. Neither A nor B

Essay Questions

1. Explain the difference between active and passive restraint systems.
2. Summarize the five major parts of an air bag system.
3. What happens when an air bag inflates?
4. After servicing an air bag system, what should you do to check for problems?
5. How does a seat belt reminder system operate?
6. What five things must you check when inspecting seat belt webbing?
7. List nine rules for working with air bag systems.

Critical Thinking Problems

1. If you have your arm through the spokes of a steering wheel and the air bag inflates, what can happen?
2. What would happen if you install an air bag sensor with its arrow facing to the rear?
3. How does gas cause an air bag to inflate?
4. Explain the operation of air bag system sensors.
5. How do you handle a live air bag module?
6. A visual and functional inspection of the belts is critical to ensure maximum protection for vehicle occupants. During seat belt inspection, what should you do?
7. Please list safety rules for working on air bag systems.

Math Problems

1. If an air bag shoots out at 100 miles per hour, how long will it take the air bag to deploy and shoot out 1 foot?
2. A new seat belt can hold 1,000 pounds without breaking. If a seat belt has been cut through about 10 percent of its webbing, how much might it be able to hold without breaking?

Activities

1. Inspect a wrecked vehicle's restraint systems. List the types of restraints installed in the vehicle. Does it have shoulder harnesses and air bag(s)? List your findings.
2. Refer to the repair manual for a specific make and model vehicle. In your own words, write a report summarizing the safety procedures for replacing deployed air bags.

SECTION

6

Refinishing

CHAPTER 24

Refinishing Equipment Technology

OBJECTIVES

After reading this chapter, you should be able to:

- Identify the various types of equipment used in auto refinishing.
- Explain how a spray gun works.
- Adjust spraying equipment to test for and develop a good spray pattern.
- Implement the stroke technique procedure for single- and double-coat application of refinishing materials, and recognize common errors made by apprentice refinishers.
- Identify the various types of spray coats.
- Determine when and how to make spot repairs.
- Clean and properly care for a spray gun.
- Compare modern HVLP spray guns with other types of guns.
- Explain the operation of spray booths and respirators.
- Correctly answer ASE-style review questions relating to refinishing equipment.

INTRODUCTION

This chapter summarizes the steps needed to prepare refinishing equipment and the paint area so you can paint vehicles. There are a number of shop and equipment variables that affect the refinishing operation. These variables include the painting environment, as well as the painting equipment and their adjustments. All are important. You must pay close attention to these variables because they can affect the quality of your paint work.

To do a good paint job, your shop and equipment must be in perfect condition. A dirty spray booth, a poorly maintained paint spray gun, a contaminated air supply, and other such avoidable situations can ruin your work. Sloppy shop conditions will usually result in a horrible paint job.

Automotive paint must be kept free of dust and dirt during spraying and while drying. In fact, today's clearcoat finishes magnify any dust, dirt, or other flaws in the finish.

A professional painter spends more time on the shop and equipment maintenance than on any other single task. Spraying the vehicle only takes a very short amount of time by comparison.

The proper painting environment must address six variables:

1. Cleanliness
2. Temperature/humidity
3. Light
4. Compressed air
5. Controlled ventilation
6. Fire safety

NOTE *Chapter 5 introduced basic refinish equipment. Selection and use of air compressors, air control equipment, and air hose connectors was thoroughly described in Chapter 6. Chapter 7 explained the selection and qualities of various refinish materials. You should study these chapters thoroughly so you can fully understand the information in this chapter.*

There are many ways to keep dirt from becoming a problem in a finish during body repairs:

- Using a dustless, or vacuum, sanding system
- Cleaning the vehicle before bringing it into the shop
- Having the paint prep area separate from the body repair area

- Using downdraft prep stations
- Maintaining the compressed air system's filters and water traps

24.1 SPRAY GUNS

A *spray gun* breaks the liquid sealer, primer, paint, and so on into a fine mist and forces it onto the surfaces of the vehicle. It is the key component in a refinishing system. A quality spray gun is a precision-engineered and manufactured tool (Figure 24–1).

There are many spray gun types and sizes. Each is specifically designed to perform certain tasks. Even though all spray guns have common parts and components, each gun type or size is suited only for a defined range of jobs.

Many professional painters have two or three spray guns. Each is set up and used with a specific type of refinish material or for a specific task. One might have a tip or cap size used for only primers and sealers. Another might be set up and used for misting on colorcoats. A third spray gun might be adjusted for the wet application of clearcoats. Having more than one spray gun allows you to preset each gun for the type of material needed for typical repairs.

Ideally, you would have separate spray guns for each type of material to be sprayed. Each would be cleaned, adjusted, and ready to use for:

- Applying primer surfacers
- Applying primer sealers
- Applying basecoats
- Applying clearcoats

By having a gun adjusted for each material, you will save time. Also, your paint work will be of better quality because you will know how each gun setting applies its own material.

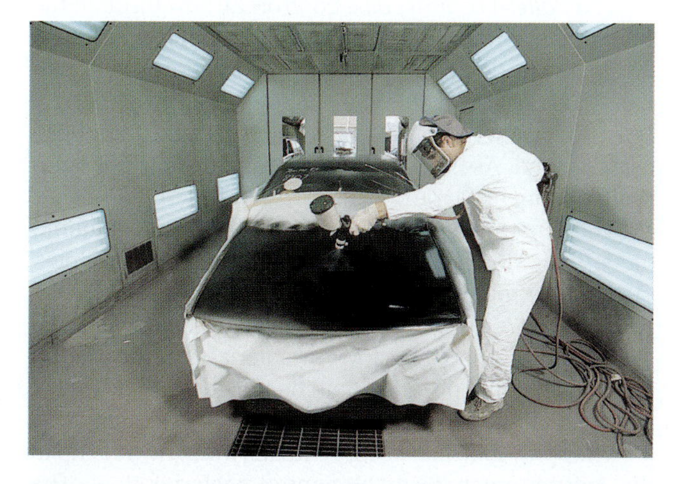

Figure 24–1 Modern spraying or refinishing equipment is designed to protect the painter and prevent waste of materials while producing a glamorous paint job.

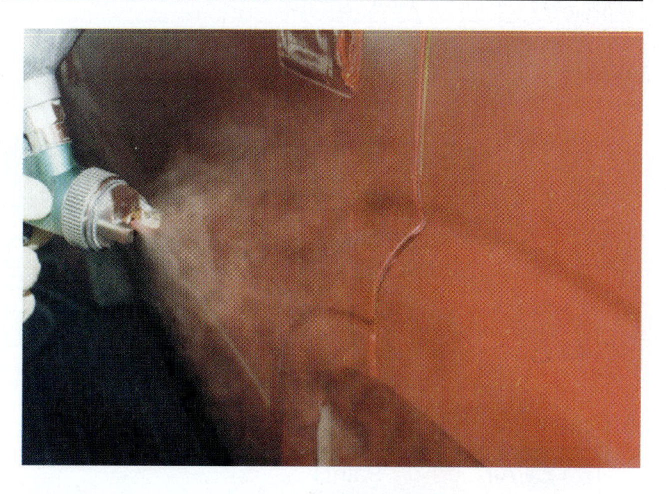

Figure 24–2 A spray gun is designed to atomize liquid material into a fine mist of tiny droplets that smoothly deposit on the vehicle body surface.

As in most other areas of refinishing work, having the right tool for the job goes a long way toward getting a professional job done right in the minimum time.

Spray Gun Atomization

A thorough understanding of atomization is the key to using a spray gun correctly. **Atomization** breaks the liquid material into a fine mist spray of tiny, uniform droplets. When properly applied to the vehicle's surface, these droplets flow together to create an even film thickness with a mirror-like gloss (Figure 24–2).

Proper atomization is essential when working with today's basecoat/clearcoat finishes because the basecoat is so thin. Basecoat/clearcoat finishes will also not achieve proper hiding and coverage if they are not correctly atomized. Also, clearcoats will show any surface roughness more easily if not sufficiently atomized.

Atomization takes place in three basic stages (Figure 24–3):

1. In the first stage, the paint siphoned from the fluid tip is immediately surrounded by air streaming from the annular ring. This turbulence begins breaking up the paint.
2. The second stage of atomization occurs when the paint stream is hit with jets of air from the containment holes. These air jets keep the paint stream from getting out of control and aid in paint breakup.
3. In the third phase of atomization, the paint is struck by jets of air from the air cap horns. These air streams hit the paint from opposite sides, causing the paint to form into a fan-shaped spray.

Spray Gun Parts

The external parts of a typical air spray gun are illustrated in Figure 24–4. This is a gravity-fed spray gun, which is the most common type found in today's body shops.

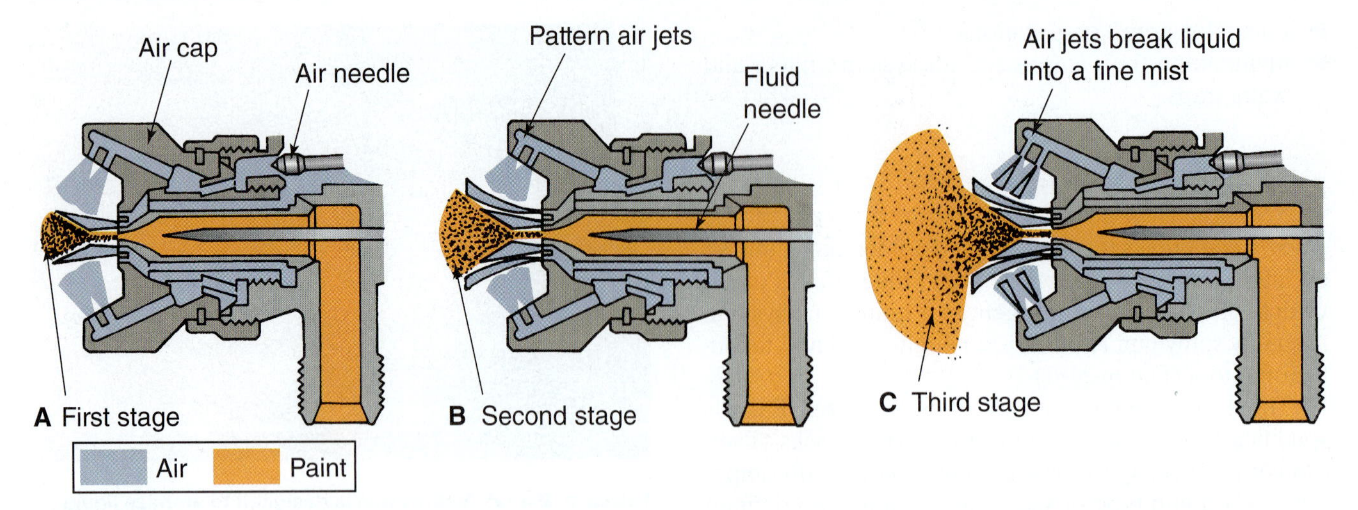

Figure 24–3 Study the three stages of spray gun liquid atomization. (A) With the trigger pulled all the way back, air and paint are sprayed out of the nozzle. (B) Air jets mix with liquid paint to continue atomization. (C) A fan of atomized paint flows out from the nozzle in tiny droplets.

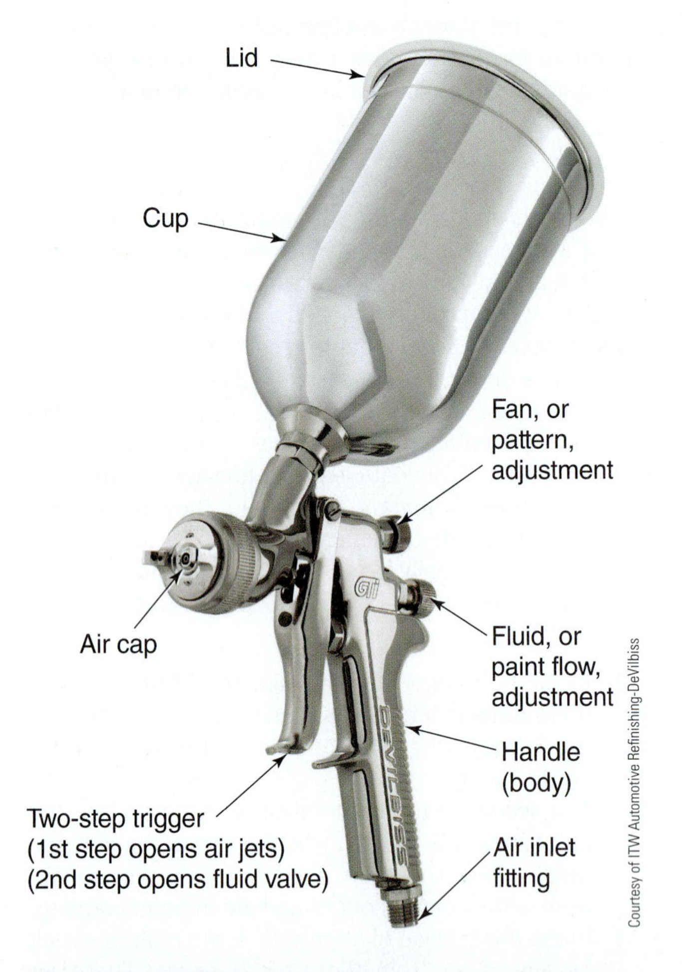

Figure 24–4 This is the most common type of spray gun used by professional painters. It handles well, is balanced, and has a high-efficiency HVLP design to waste less refinish material.

The major parts of a typical professional spray gun are shown in Figure 24–5.

The *air cap* directs compressed air into the material stream to atomize it and form the spray pattern. The air cap threads onto the front of the spray gun body and holds the fluid tip in place.

There are three types of orifices (holes) in an air cap: the center orifice, the side orifices or ports, and the auxiliary orifices. Each of these holes has a different function.

1. The *center orifice* located at the nozzle tip creates a vacuum for the discharge of the paint.
2. The *side orifices* in the air cap horn determine the spray pattern by means of air pressure.
3. The *auxiliary orifices* promote atomization of the paint.

The relationship between the auxiliary orifices and the gun's performance is pictured in Figure 24–6.

Large orifices increase the gun's ability to atomize more material for painting large objects with great speed. Fewer or smaller orifices usually require less air, produce smaller spray patterns, and deliver less material to conveniently paint smaller objects or apply coatings at lower speeds.

The *pattern control valve* controls airflow through the side orifice to control the shape of the paint mist. Air also flows through the two side orifices in the horns of the air cap. This air flow forms the shape of the spray pattern. When the pattern control valve is closed, the spray pattern is round. As the valve is opened, the spray becomes more oblong in shape.

The fluid needle valve and fluid tip work together to meter the amount of material leaving the gun and entering the air stream.

The *fluid tip* forms an internal seat for the fluid needle to shut off or allow the flow of material. The needle tip contacts the fluid tip, and it can be pulled open by trigger action to allow the material to spray out of the gun.

The spray gun *fluid needle* extends from the fluid tip to the trigger mechanism to control material flow. The amount of refinishing material that leaves the gun depends on the needle valve adjustment.

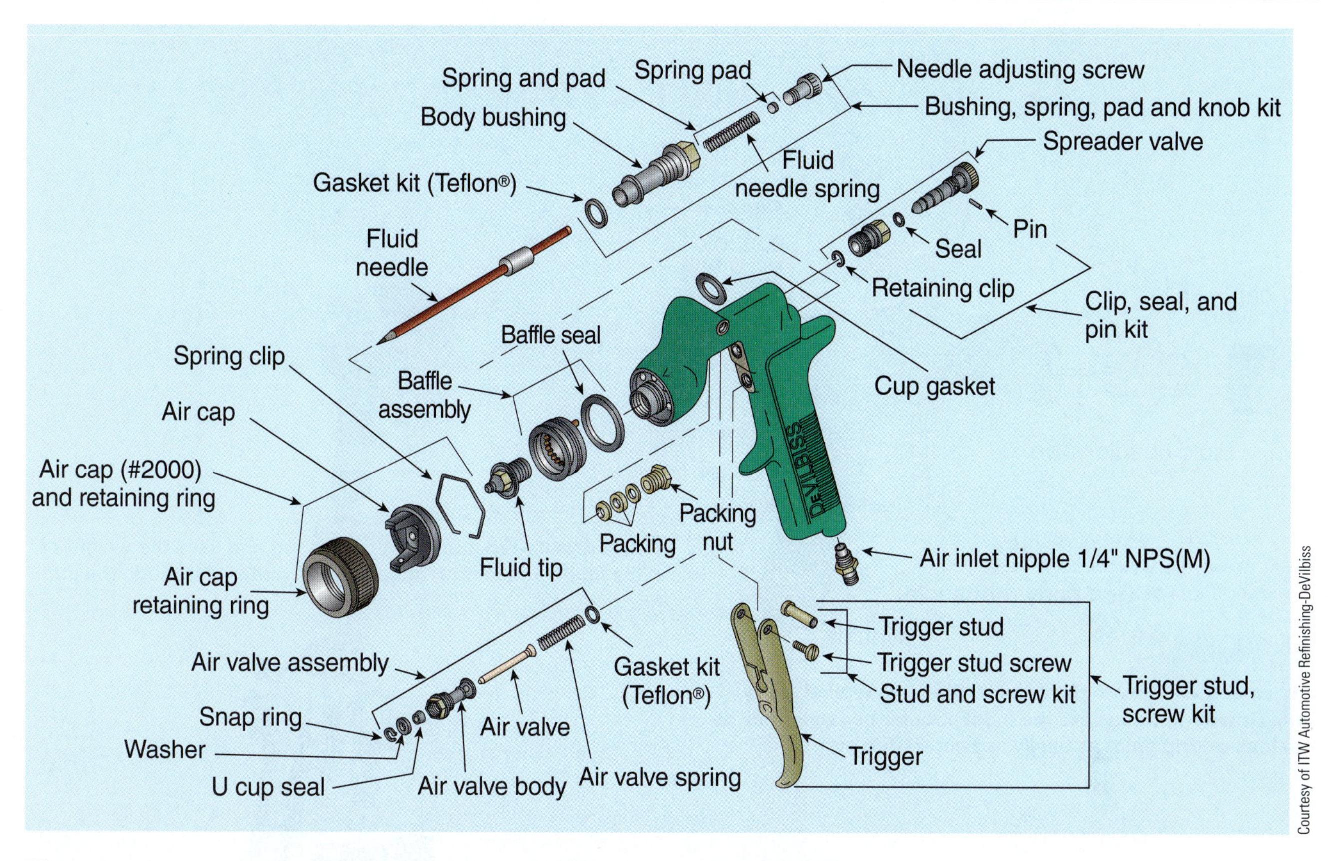

Figure 24-5 Study this exploded view of a typical HVLP gravity-fed spray gun. Memorize the names of parts.

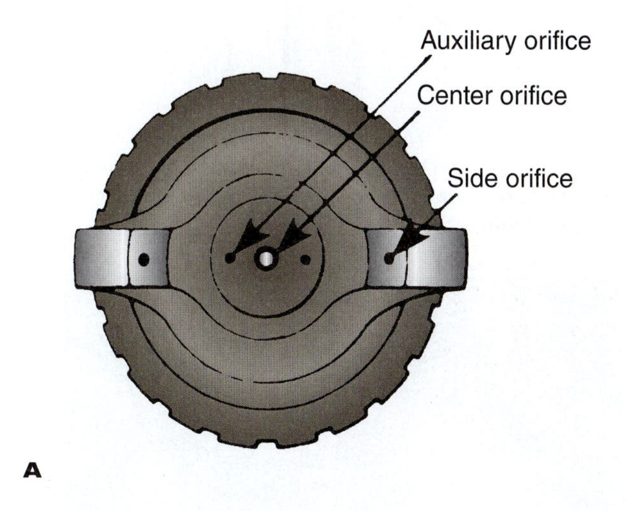

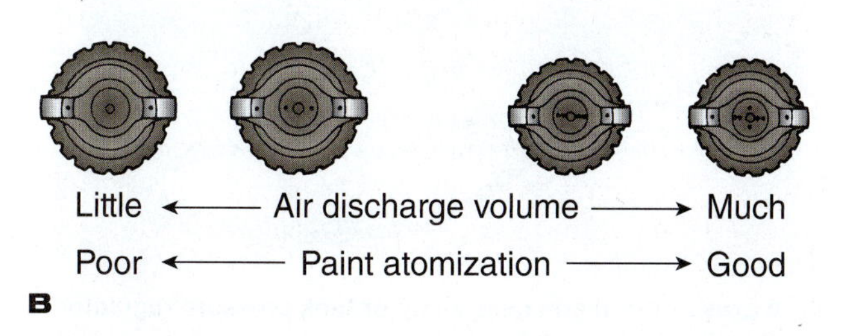

Figure 24-6 (A) An air cap has three types of orifices, or holes. (B) Quality spray guns generally have more holes for better atomization.

Fluid tips and air nozzles are available in a variety of sizes to accommodate materials of various types and viscosity. Each passes the required volume of material to the cap for different speeds of application. See Figure 24–7.

The **fluid control knob** (valve) changes the distance the fluid needle valve moves away from its seat in the nozzle when the trigger is pulled.

The **air valve**, like the fluid valve, is opened by moving the trigger. When the trigger is pulled part way, the air valve opens. When it is pulled a little farther, the fluid valve opens. Conventional guns have an air needle valve. Newer high-efficiency guns may use an adjustable air cap to control airflow.

HVLP Spray Guns

The **high-volume, low-pressure (HVLP)** spray gun (also known as the *high-solids* system) uses a high volume of air delivered at low nozzle pressure to atomize paint into a pattern of low-speed particles. This type of system is required in many areas to pass strict air emission or pollution standards. The most important way it differs from conventional spray systems is its high transfer efficiency. See Figure 24–8.

High transfer efficiency means that more of the material leaving the gun stays on the surface being refinished.

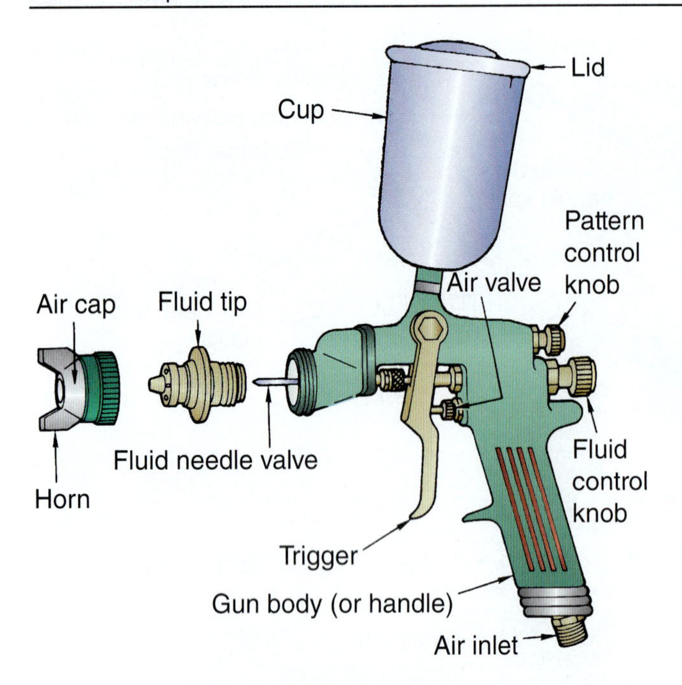

Figure 24-7 Note the construction of a typical spray gun. Gravity-fed guns are the most popular because they do not leak or drip paint as easily as suction-fed guns.

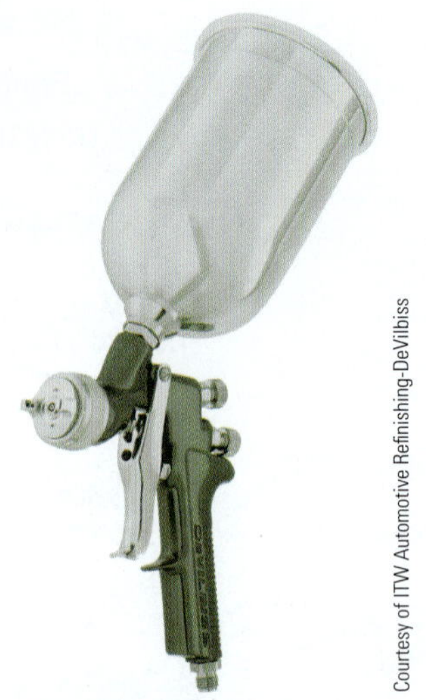

A A gravity-fed gun has a cup on top and uses the weight of the liquid and venturi action to pull material through the gun.

B A suction-fed gun has a cup on the bottom. Suction at pickup and pressure on top of the liquid feed material through the gun.

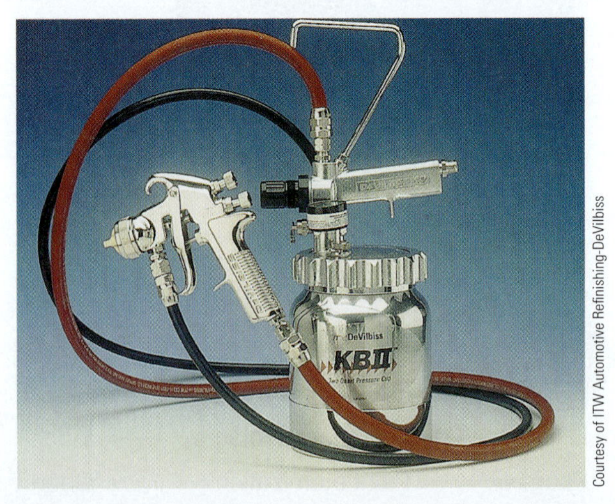

C A pressure-fed gun uses a cup or tank pressure regulator to force liquid through the gun.

Figure 24-8 High-efficiency HVLP spray guns are now available in several configurations.

Less material is wasted, and more is prevented from entering the atmosphere as air pollution. This is the primary purpose of HVLP guns (Figure 24–9).

The high pressure of conventional spray guns tends to "blast" the paint into small particles. In the process, it creates a fair amount of overspray. The transfer efficiency of high-pressure systems suffers as a result of overspray, particle "bounce," and blowback.

In contrast, HVLP relies on air delivered to the tip at 10 psi (69 kPa) or less to break the paint into small particles. As the material flows into the air stream, far less is lost in overspray, bounce, and blowback, hence the dramatic improvement in transfer efficiency. HVLP will work with any material that can be atomized by a spray gun, including two-component paints, urethanes, acrylics, epoxies, enamels, lacquers, stains, primers, and so on.

HVLP spray guns have a thicker body to allow for larger internal air passages. Larger air passages are needed because HVLP spray guns are designed to operate on lower line or hose pressures. Refer to Figure 24–10.

Some HVLP guns, especially retrofit guns, require normal inlet pressure (45–60 psi, or 82–207 kPa) with 10 psi outs. Most complete redesign HVLP guns use conventional inlet pressure to help atomization. These guns lower nozzle pressure to 10 psi (69 kPa) internally while allowing a high volume of air and paint to pass through the gun. This increases gun efficiency so that more paint is applied to the body surface and less is wasted as overspray. Unlike early gun designs, modern HVLP guns are easy to use and will produce an excellent paint finish.

High transfer efficiency is attractive for several reasons. It reduces air pollution and paint waste. In many

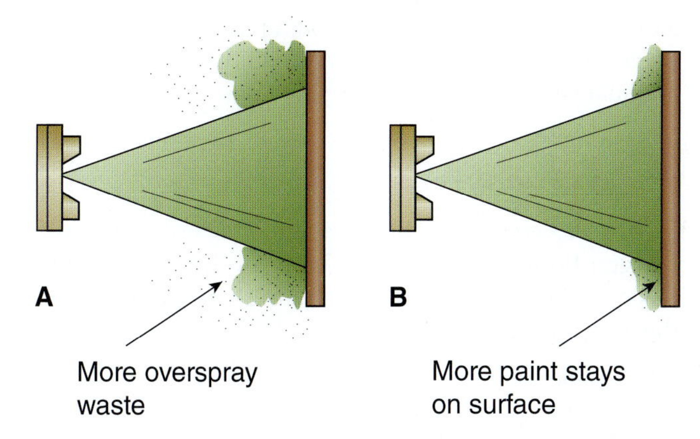

Figure 24-9 **(A) Older low-efficiency guns caused much more overspray waste. More paint was blown off the body surface by high pressure. (B) Modern high-volume, low-pressure, or HVLP, spray guns use less material to cover the same area. They waste less paint material due to reduced overspray.**

states, California, for example, new laws require the use of spray equipment that has at least 65 percent transfer efficiency. Low-pressure spray (up to 10 psi, or 69 kPa, at the nozzle) and electrostatic spray methods have been approved by this legislation. Similar legislation has been passed in many other locations. There are other good reasons for HVLP, however. Higher transfer efficiency improves the quality of both the workplace and the finished product. Overspray not only makes painting work less desirable, it also reduces visibility, which contributes to mistakes and low productivity. Overspray is one of the main reasons for paint booth downtime necessitated by maintenance. All paint spraying equipment can be affected by overspray, but the booth and its filters are affected the most.

To illustrate how much of a difference transfer efficiency makes in booth maintenance, consider that HVLP can be two to three times as efficient as conventional spray equipment. Depending on how it is used, conventional spray equipment is as little as 20–30 percent efficient. That means for every 3 gallons (11.4 liters) of paint sprayed, more than 2 gallons (7.6 liters) are wasted.

Also, if a conventional gun and an HVLP gun are used side by side to paint identical surface areas, the older gun will run out of paint while the HVLP gun will have enough paint to finish the job. This saves on the cost of paint and the time needed to refill the gun.

Some HVLP gun designs place the air valve in the handle, replacing the conventional air needle. The gun handle and body are larger to allow a high volume of air to flow to the tip. Other designs only use one needle valve to control paint flow. The air valve is in the handle right at the inlet fitting.

Except for a few subtle differences, the HVLP and conventional air spray gun operate in basically the same manner. For instance, the HVLP gun should be held closer to the surface of the workpiece because of the

Courtesy of ITW Automotive Refinishing-DeVilbiss

Courtesy of ITW Automotive Refinishing-DeVilbiss

Figure 24-10 **Compare HVLP high-efficiency suction-fed and gravity-fed spray guns.**

lower speed of the particles. A rule of thumb would be to hold the gun 6–8 inches away when spraying with HVLP, compared to 8–10 inches for a conventional gun. Greater distances result in excessive dry spray and lack of film buildup.

HVLP high-efficiency spray guns are now the industry standard. Older low-efficiency guns are even outlawed in some states, such as California. The body shop can be fined if even one low-efficiency gun is found on the premises.

Use the recommended hose size for HVLP spray guns. The manufacturer will give a minimum hose inside diameter to allow for sufficient airflow. High airflow is needed through the HVLP gun body to produce ideal atomization, spray fan, and material deposit on the surface.

Always use the manufacturer-recommended size air hoses, air nipples, and air couplers with HVLP spray guns. Most HVLP guns require air fittings with an air hole of at least 5/16-inch or even 3/8-inch inside diameter (ID). Because HLVP spray guns operate on such low air pressures, any line restriction (smaller than recommended air hole in an air fitting or hose) can adversely affect their operation. The primer or paint may not atomize and spray out properly. Also, make sure that your gun-mounted pressure regulator has these larger diameter fittings.

The most common mistake when first using an HVLP spray gun is insufficient airflow. Initially, novice technicians may have trouble painting with HVLP guns because they may not follow directions for providing adequate airflow to the gun.

Spray Gun Feeds

Spray gun feed refers to how the liquid material is fed into the gun body. As detailed in Table 24–1, there are four basic methods of feeding liquid refinish material through the air spray gun (Figure 24–11):

1. Gravity feed
2. Suction (siphon) feed
3. Pressure feed
4. Pressure-assist feed (gravity or suction cup spray guns)

Gravity-Feed Spray Guns

You can identify a **gravity-fed spray gun** easily because the cup is on top of the gun, not under it. This high-efficiency system is ideal for all spraying operations.

With a gravity-fed spray gun, the refinish material (primer, sealer, or paint) is initially fed into the gun by gravity and then suction-forced to the nozzle tip.

Gravity-fed spray guns are the most common type found in modern collision repair facilities. Because their design provides a consistent, spatter-free method of feeding liquid into the gun, they are popular in auto refinishing shops for all types of work (spot, panel, and overall). Gravity-fed guns can be used for basecoat/clearcoat work and to spray undercoat refinishing materials, such as primers and sealers, as well as some lighter spray-on fillers.

TABLE 24–1 TYPES OF AIR SPRAY GUNS

Type	Paint Feed Method	Advantages	Disadvantages
Gravity feed	As the paint cup is installed above the spray nozzle, paint is supplied by gravity and a suction force at the nozzle tip.	Because there is no change in paint viscosity, there is no variation in the injection volume. The position of the cut can be changed according to the configuration of the painted item.	Because the cup is installed above the injection nozzle, it adversely affects gun stability. Cup capacity is small, so it is not useful for painting larger surfaces.
Suction (Siphon) feed	Paint container is installed below the spray nozzle and paint is supplied by suction force alone.	Stable gun operation. Easy to refill container or make color changes.	Difficult to spray on horizontal surfaces and some variations occur in discharge volume due to variations in viscosity. Has a larger paint container than gravity-fed type, but this causes quicker painter fatigue.
Pressure feed	Paint is pressurized by a compressed air tank or pump.	Large surfaces can be painted without stopping to refill container. A paint with a high viscosity can also be used.	Not suitable for painting small areas. Color changes and gun cleaning take time.
Pressure-assist feed	Low air pressure in cup helps gravity force material through gun.	Low paint level in cup does not affect paint flow as much as with gravity-fed-type spray guns.	Spray gun lid is more prone to leakage if not sealed properly.

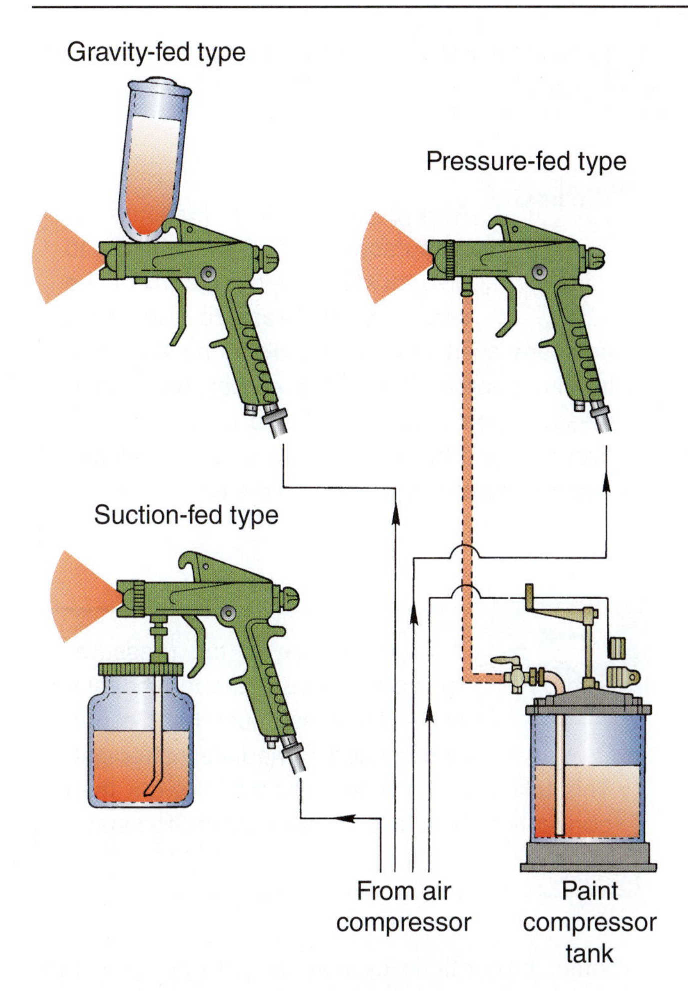

Figure 24–11 Study the paint feed methods of modern air spray guns. Note that HVLP pressure-fed guns require air pressure lines to the gun and to the tank.

The operation and adjustment of the gravity spray gun is about the same as a suction-fed gun, but a gravity-fed gun is easier to handle because of its better balance. The cup is also up and out of the way when spraying, and it is less likely to touch and damage a painted surface.

A gravity-fed gun also is less likely to leak and drip paint when spraying, which can ruin the paint job. Dripping from a spray gun can be a particular problem when spraying large horizontal surfaces, such as a roof or hood.

Some high-efficiency spray guns use a combination gravity-pressure feed cup design. It offers the benefits of both types of guns. Cup leakage can sometimes be a problem, however.

The main requirement of gravity feed is that the container be vented so that atmospheric air can replace the material as it is being sprayed. The vented cup lid can be made of flexible synthetic rubber or metal with a plastic or synthetic rubber O-ring seal. Gravity-fed cups come in a variety of convenient sizes, with the ½- and 1-quart sizes the most common in body shops.

Viscosity and flow characteristics of the material directly affect rate of flow to the gun, as do hose size, hose length, and nozzle size.

Some modern HVLP guns look like gravity-fed guns, but they are the pressure-gravity-fed type. The cup is sealed and gun pressure is used to help move the material out of the cup and into the gun.

Suction-Feed Spray Guns

Suction spray guns use airflow through the gun head to form a siphoning action (negative pressure) in the cup that pulls the refinish liquid through the gun body and into the air stream. When negative pressure or suction is formed at the bottom of the cup outlet, atmospheric pressure enters the vent hole to make the liquid flow out of the cup and through the spray gun. With a suction gun design, the paint material is often held in a 1-quart (0.94-liter) cup attached to the bottom of the gun.

Both low-efficiency and HVLP high-efficiency suction-fed spray guns are available. The body of an HVLP gun will be thicker for added airflow at lower line pressures.

The suction-fed spray gun was once the most common type of gun, but most paint technicians now prefer the gravity-fed type. It is easier to hit the vehicle with the suction feed gun's bottom-mounted cup when spraying.

An **air vent hole** and hose on the siphon spray gun allow atmospheric pressure to enter the cup. This vent can become clogged with dry primer or paint. If the vent is plugged, paint will not flow out of the gun.

When the spray gun trigger is partially depressed, the air valve opens and air rushes through the gun. As the air passes through the openings in the air cap, a partial vacuum is created at the fluid tip. Further squeezing of the trigger withdraws the fluid needle from the fluid tip. The vacuum sucks paint from the cup, up the fluid inlet, and out through the open fluid tip. Air enters through the air hole and replaces the siphoned paint. The inlet air vent holes in the cup lid must be open (Figure 24–12).

Pressure-Feed Spray Guns

Pressure-fed spray guns use air pressure inside the paint cup (pot) or tank to force the material out of the gun. No vent hole is needed in the cup lid. HVLP pressure-fed guns are available.

Pressure pot or pressure cup guns provide possible advantages over siphon cup guns. They allow more paint to speed through a smaller nozzle, and smaller paint streams atomize better.

Having a remote cup makes the pressure-fed gun lighter and easier to handle. It also permits spraying while holding the gun horizontally for painting under flared

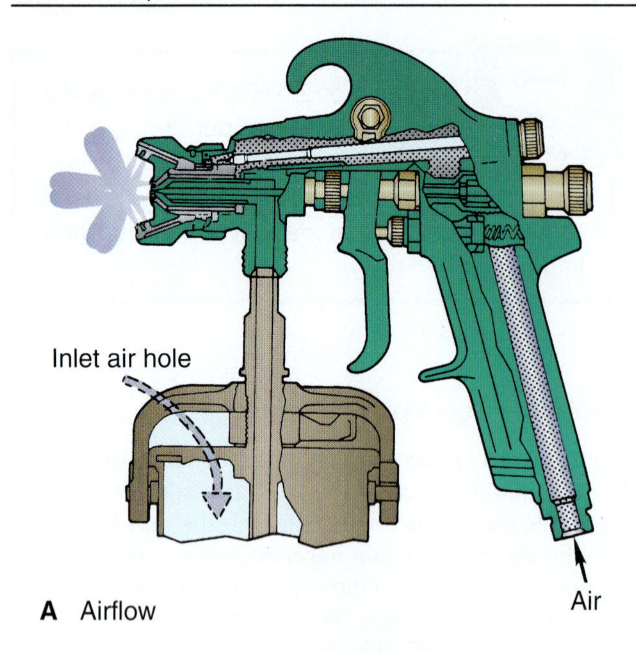

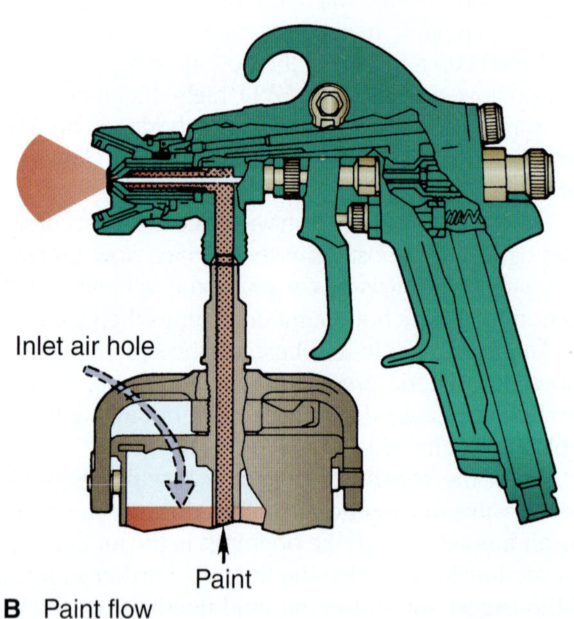

Figure 24-12 **Study paint flow and airflow paths through a spray gun.**

parts without danger of "spitting." With the tank or cup away from the vehicle, paint dripping from the cup and onto the vehicle is not a concern.

Pressure-tank spray guns use a much larger storage container for paint materials. They hold enough paint for a complete paint job, which saves time. You can also be sure that the paint will match throughout the whole job. This is particularly helpful when spraying hard-to-match metallic or pearl paints. Pressure-tank spray guns are often used to paint large trucks and motor homes.

Remember that pressure cups have seals that must be kept clean and regularly inspected for damage. A loss of cup pressure affects the delivery of fluid to the spray gun.

WARNING

Always adjust line pressure to manufacturer's specifications to prevent damage or rupture of the cup or tank. Pressure cups hold pressure even after being disconnected from the air source. It can be embarrassing and messy if you open the lid and paint blows all over you and the shop. Make sure you release cup pressure before opening the lid.

A disadvantage of the pressure-fed spray gun is cleanup. The tank and fluid lines must be cleaned and flushed immediately after use. This takes a little more time and solvent than conventional suction-fed and gravity-fed spray guns.

Some cups or tanks have an *agitation paddle* system to keep the pigments and solids thoroughly mixed at all times, ensuring color uniformity. Some siphon gun cups also have an agitator system. These cups provide constant mixing of all automotive finishes and primers; they even keep metal flakes and metallics in total suspension and complete dispersion (Figure 24–13).

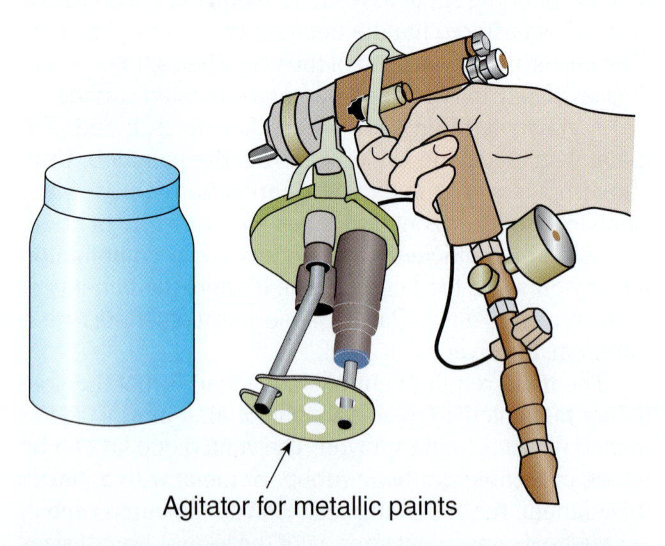

Figure 24-13 **An agitator-type paint cup has a paddle that moves up and down to mix the paint. This is particularly important with metallic paints because they tend to settle in the cup.**

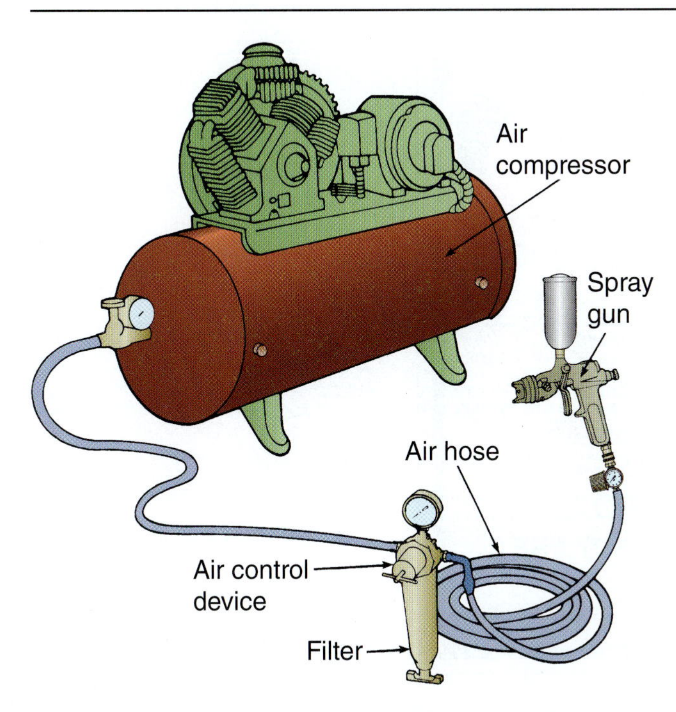

Figure 24–14 Study the basic equipment needed to operate a spray gun. Two pressure gauges provide more accurate adjustment.

Spray Gun Air Supply

Connect a spray gun to its air hose using the following steps (Figure 24–14):

1. Adjust the main pressure regulator at the compressor to provide the manufacturer-recommended line pressure. Pressure will drop as it flows through the hose and to the gun. Make sure excess pressure is not fed to suction and pressure-fed spray guns, because they could be damaged.
2. Pull back on the quick disconnect fitting and slide it over the male fitting on the spray gun. Make sure no air is leaking out of the fittings. Tighten fitting nuts as needed.
3. If used, adjust the spray gun-mounted regulator to the air pressure recommended by the gun manufacturer. This will vary slightly with the nozzle size and viscosity of the material to be sprayed.

24.2 EQUIPMENT AND MATERIAL PREPARATION

Spraying a vehicle is a skilled job. It calls for considerably more experience and knowledge than just holding down the trigger and moving the gun. There are several variables contributing to the quality of a spray finish, including spraying material viscosity, spray booth temperature, film thickness, and spray method.

The proper painting environment must address six variables:

1. Cleanliness—to keep dirt out of paint
2. Temperature/humidity—to provide proper paint curing or drying conditions
3. Light—to properly illuminate the vehicle and paint as it is applied
4. Compressed air—to send clean air at the right pressure to the gun
5. Controlled ventilation—to ensure the health of workers
6. Fire safety—to protect the shop and employees

There are many ways to keep dirt and other contaminants from becoming a problem in a paint or finish. Before beginning a refinish operation, you must prepare (prep) the equipment, refinish materials, and paint area.

DANGER Make sure you are wearing an approved respirator for the type of refinish material being applied. The instructions and warnings on the container will give safety rules for proper use. Many of today's catalyzed paint products require a fresh air-supplied respirator for adequate protection of your throat and lungs. Refer to Figure 24–15.

Spraying Material Viscosity

Viscosity is the measurement of thickness, fluidity, or flow resistance of a liquid. High viscosity means a liquid is very thick and resists flow, like honey. Low viscosity means the material is very thin and runs easily, like water (Figure 24–16).

The viscosity or flow characteristics of liquids relate directly to the degree of internal friction formed in the material. Therefore, anything that will influence the internal friction (such as solvents, thinners, or temperature change) will influence flow. Similarly, it is the flow characteristics that determine how well a material will atomize, how well it will "flow out" on the body surface, and the type of equipment needed for application.

Using an incorrect viscosity paint will result in various paint finish defects. Paint must be thoroughly mixed and properly reduced with the correct amount of solvent added or a good-quality paint finish is impossible to attain.

Improper viscosity can cause excessive overspray, increased booth maintenance, runs or sags, pebble-dry finishes, and color mismatches. Always mix paint according to the manufacturer's recommendations and check the viscosity with a Zahn cup and stopwatch as described later in this chapter.

A

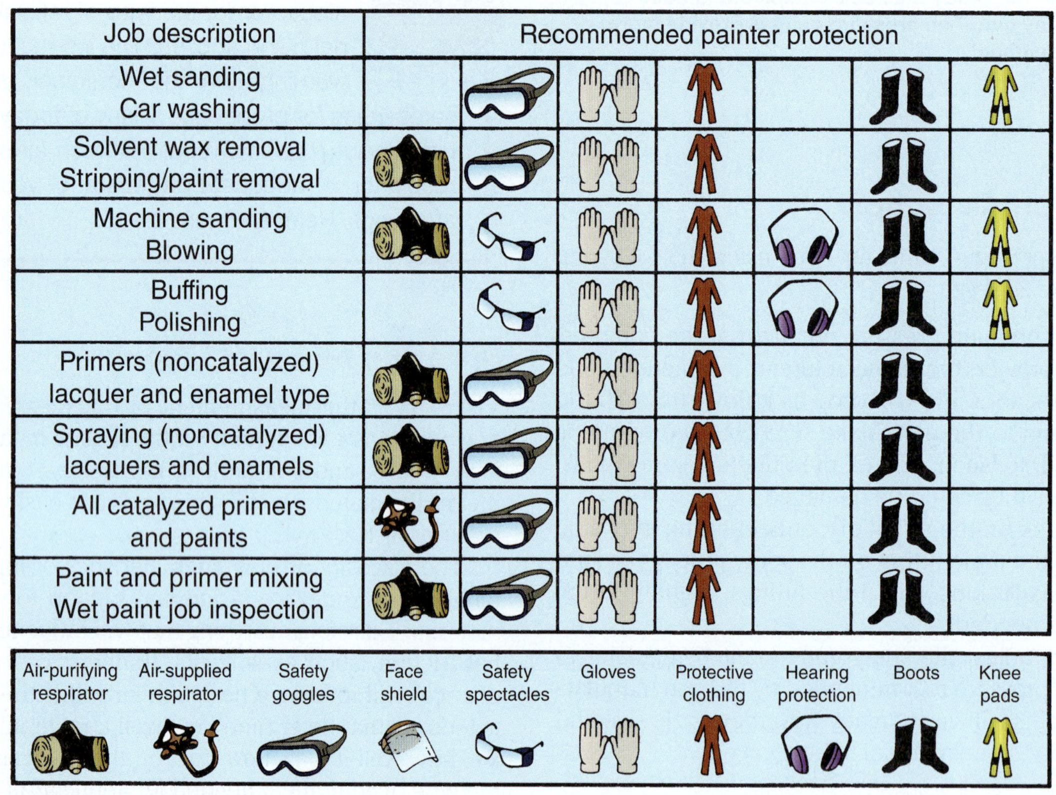

Job description
Recommended painter protection
Wet sanding
Car washing
Solvent wax removal
Stripping/paint removal
Machine sanding
Blowing
Buffing
Polishing
Primers (noncatalyzed)
lacquer and enamel type
Spraying (noncatalyzed)
lacquers and enamels
All catalyzed primers
and paints
Paint and primer mixing
Wet paint job inspection
Air-purifying respirator
Air-supplied respirator
Safety goggles
Face shield
Safety spectacles
Gloves
Protective clothing
Hearing protection
Boots
Knee pads

B

Figure 24–15 Many of today's repainting products contain harmful chemicals that require the use of an air-supplied respirator. (A) A shop warning sign explains the dangers associated with paint materials containing isocyanates. (B) A chart shows what safety gear should be worn when spraying different materials. Note when use of an air-supplied respirator is recommended.

If a paint's viscosity is too high when sprayed, it will be dry and will not lay down smoothly on the body surface. If a paint is too thin, it will not provide good coverage and will run and sag easily.

There are several ways to mix paint or other materials to the desired viscosity before spraying.

Paint Mixing Sticks

Graduated *paint mixing sticks* have conversion scales that allow you to easily convert ingredient percentages into part proportions. They are used by painters to help mix paints, solvents, catalysts, and other additives right

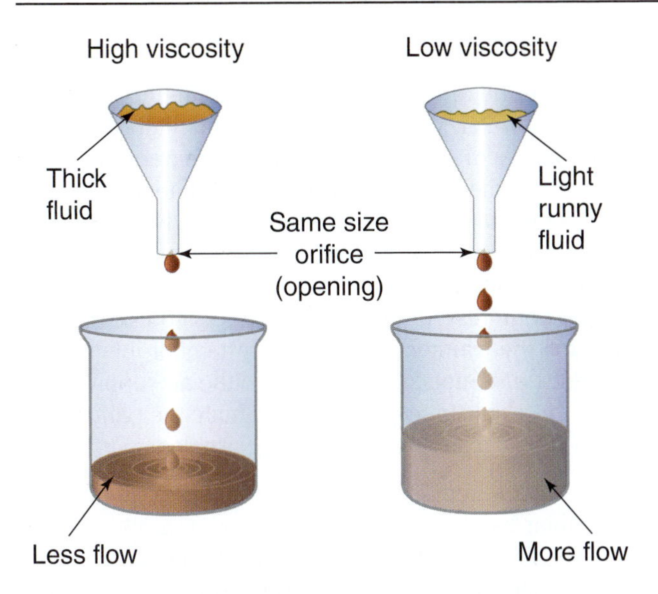

Figure 24–16 Viscosity refers to thickness of liquid. If poured through the same size orifice or hole, a high-viscosity liquid will pour out much more slowly. Primer and paint materials must be the right viscosity to spray out properly onto surfaces.

before spraying. Almost all shops have paint mixing sticks (Figure 24–17).

Detailed instructions for using paint mixing sticks are given in Chapter 26.

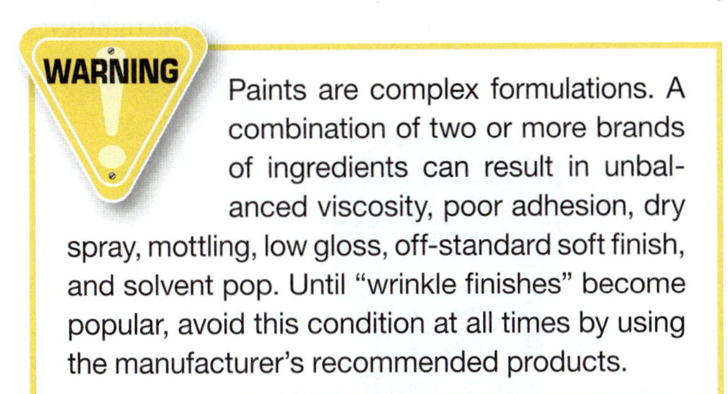

Paints are complex formulations. A combination of two or more brands of ingredients can result in unbalanced viscosity, poor adhesion, dry spray, mottling, low gloss, off-standard soft finish, and solvent pop. Until "wrinkle finishes" become popular, avoid this condition at all times by using the manufacturer's recommended products.

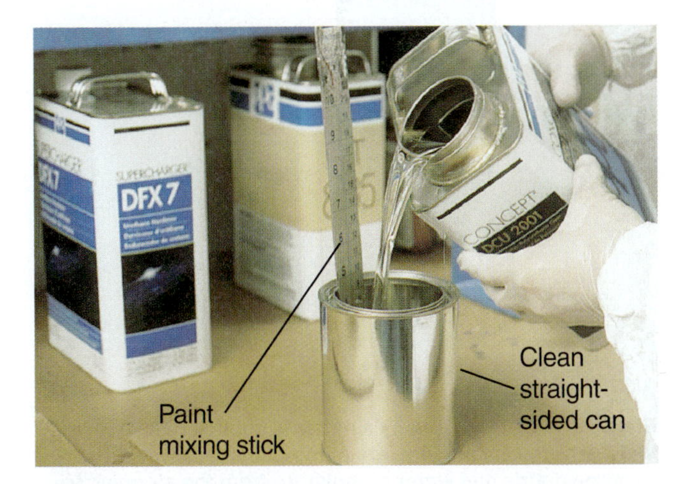

Figure 24–17 Paint mixing sticks are sometimes used to add the correct amount of ingredients: paint, reducer, and hardener.

Viscometers

The most accurate way to measure material viscosity is with a viscometer, or **viscosity cup**. The two types of viscometers used for automobile painting are the Ford cup and the Zahn cup. Although the Ford cup is very accurate, it is too expensive for the average collision repair shop. The Zahn cup is less expensive and almost as accurate.

When preparing a refinish material for spraying, thin it to the proper viscosity according to the directions on the can, using the reducer or thinner best suited for the shop temperature and conditions.

Various automotive finish materials are manufactured to spray at ideal viscosities. Refer to the label on the material container to find the recommended viscosity.

Using a Viscometer

Viscosity of sprayable materials is measured in seconds. This *viscosity in seconds* specification is determined by timing how long it takes the paint material to leak out of a viscosity measurement tool at a specific temperature.

When measured with a Zahn cup, these are the general viscosities in seconds for various materials:

- Very thin materials—wash primers, dyes, and stains—14–16 seconds
- Thin materials—sealers, primers, zinc chromates, and acrylics—16–20 seconds
- Medium materials—synthetic enamels, primer-surfacers, epoxies, urethanes, basecoat/clearcoat, and so on—19–30 seconds

Finish Material Temperature

The temperature at which material is sprayed and dried has a great influence on the smoothness of the finish. This involves not only the air temperatures of the shop, but the temperature of the work as well. A vehicle or panel should be brought into the shop long enough ahead of spraying time to arrive at approximately the same temperature as the shop. Spraying warm paint on a cold surface or spraying cool material on a hot surface completely upsets flow characteristics. The rate of evaporation on a hot summer day is approximately 50 percent faster than it is on an average day with a shop temperature of 72°F (22°C). Appropriate thinners or reducers should be used for warm and cold weather applications (Figure 24–18).

Finish Film Thickness

When spraying on a refinish material, you must try to produce the recommended film thickness. The thickness of the paint film applied when spraying has several effects on the finish.

Never apply today's enamels in a coat that is too thick or too wet. With enamel paints, "thin is in and thick is out."

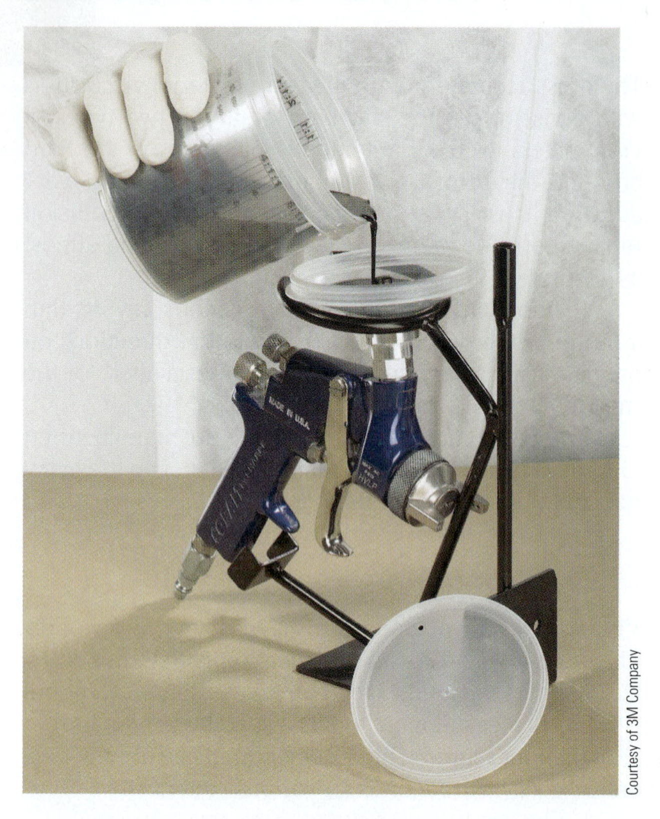

Figure 24-18 **Paint must be mixed correctly before being sprayed. Here the technician has mixed and filtered a small amount of paint. A small plastic cup on the gun prevents waste of paint materials when only a tiny amount of colorcoat is needed.**

You must use normal medium-wet coats with proper flash times between coats. This will prevent problems.

As enamel paint dries, reducer or solvent gases must pass up through the paint surface. If too much paint is sprayed on at one time, the paint will not dry or cure normally. If the surface of the paint flashes or "skins over," the escaping gas bubbles can lift or blister the fresh paint film.

You should develop your spraying technique to apply a coat onto a surface that will remain wet long enough for proper flow-out and no longer. Heavier coats are not necessary. They can produce sags, curtains, or wrinkles, as well as strongly influence metallic color when matching.

The amount of material sprayed on a surface with one stroke of the gun will depend on the width of the fan, distance from the gun, air pressure at the gun, amount of reduction, speed of the stroke, and selection of a reducer.

24.3 SPRAY GUN SETUP

Before attempting to paint a vehicle, it is critical that the spray gun and its air supply system be set up and adjusted properly. Clean, dry air at the correct pressure must be connected to the spray gun. The spray gun must also be adjusted to produce a correct spray pattern for uniform application of the new paint film.

Spray Gun Air Supply

Make sure the air compressor and the in-line filters are drained of water. Many compressors and filters now have an automatic drain that does not require periodic maintenance. Some air compressors and filters, however, require manual bleeding. You must open the drain valve on the bottom of the compressor tank or filter to bleed out condensation or moisture (Figure 24–19).

The compressed air system can be a source of problems for the painter. Air supply system problems can introduce dirt, moisture, and oil into the air supply. From there, these substances can contaminate the paint.

To avoid air supply and painting problems:

- Check and replace oil and water filters and traps on a regular basis.
- If the system is not automatic, drain moisture from it daily. Draining the system in the morning allows more moisture to be removed because it is cool and has condensed.
- Replace air hoses as necessary. Deteriorated air hoses can also introduce dirt into the system.

Spray Gun Adjustments

A good paint spray pattern depends on the proper mixture of air and paint droplets. The process of tuning a spray pattern is much like fine-tuning a race car engine. It depends on adjusting the air pressure to just the right level.

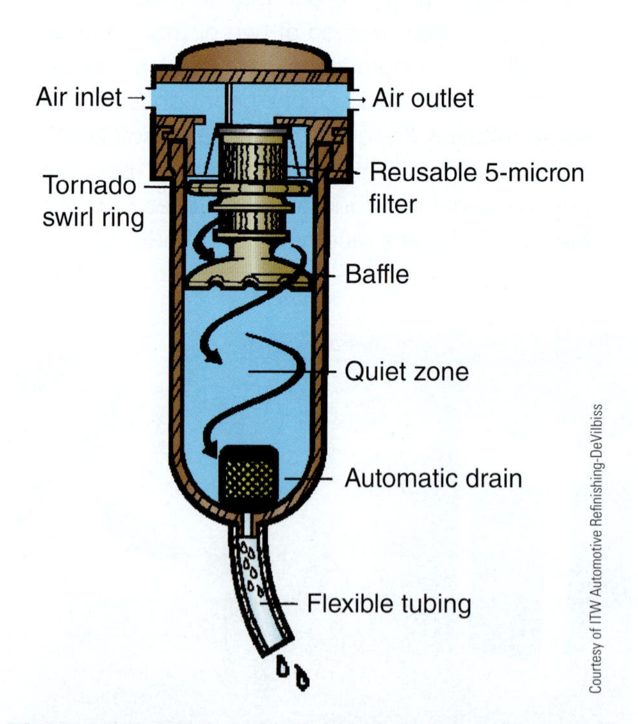

Figure 24-19 **A quality air filter is important if you want to do a professional paint job. A good filter will remove airborne debris and water or moisture from the air going to the spray gun. This unit has an automatic drain valve. Air filters and dryers should be serviced if there is any sign of moisture coming out of air hoses.**

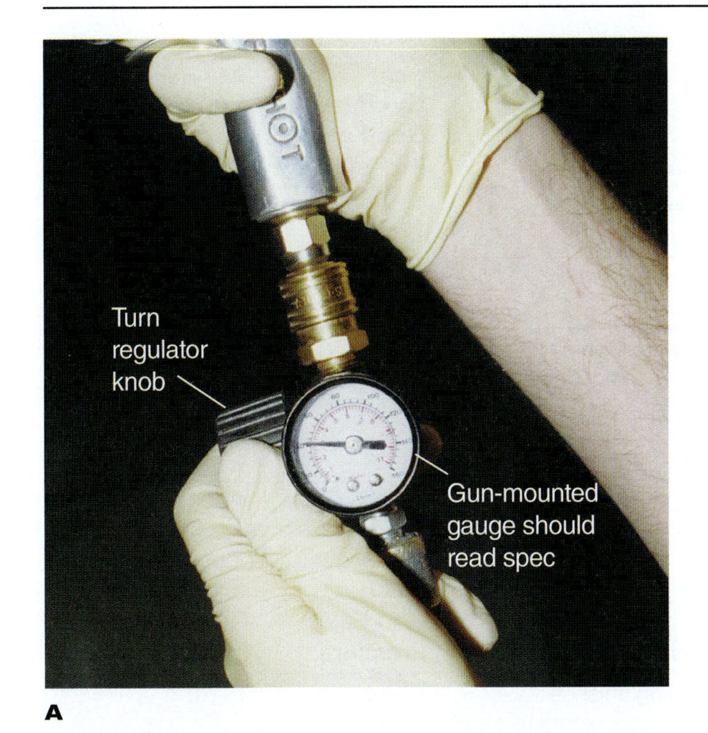

A

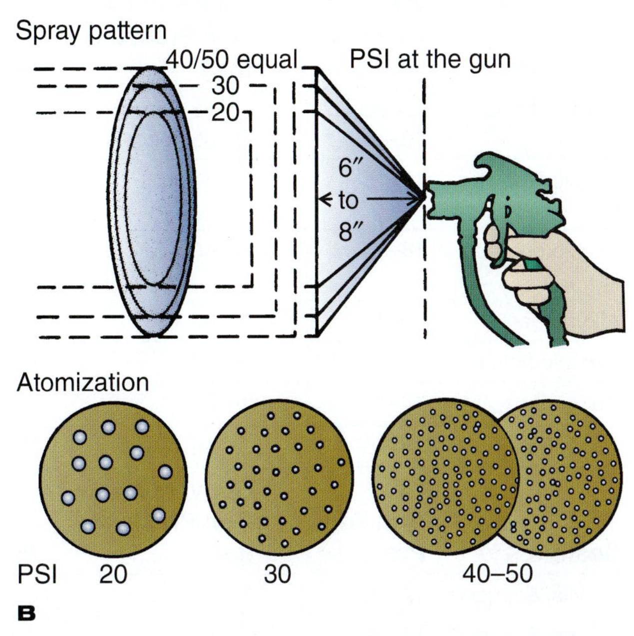

B

Figure 24–20 **With today's low-pressure, high-transfer efficiency spray guns, accurate inlet air pressure adjustment is critical to a consistent, quality paint job. (A) A small air pressure regulator and gauge assembly fits between the spray gun and air hose. This will let you maintain ideal spray gun feed pressure with changes in pressure in the main shop air line. The gun should be set to the manufacturer's recommendations or what works best for your painting technique. (B) Note the effects of gun pressure on spray pattern and atomization. Note how high pressure improves dispersion of the paint mist and enlarges the spray pattern size. To protect the Earth's atmosphere from overspray pollution, use the lowest pressure that produces good atomization.**

You adjust gun pressure while pulling the trigger to allow air flow through the gun. See Figure 24–20.

HVLP spray gun manufacturers often provide a gun-mounted pressure regulator-gauge assembly with their guns. It fits between the gun inlet fitting and the air hose fitting. It allows you to precisely adjust air feed pressure without worrying about any loss of pressure through the hose.

The optimum spraying pressure is the lowest needed to obtain proper atomization, flow rate, and pattern width. Proper spray gun air pressure varies with the kind of material sprayed and type of gun. Many low volatile organic compound (VOC) recommendations require 10 psi or less at the air cap. Always follow the spray gun manufacturer's air pressure recommendations for the type of material to be sprayed.

High spray gun pressure results in excessive paint loss through overspray and poor flow due to high solvent evaporation before the paint reaches the surface. A *low spray gun pressure* produces poor drying characteristics due to high solvent retention and makes the paint film prone to bubbling and sagging.

The recommended pounds of air pressure vary with the kind of material to be sprayed and the type of gun. The typical pressure ranges for conventional and HVLP guns are given in Table 24–2. Remember that gun inlet pressures vary among manufacturers.

Also note that each nozzle requires a different inlet pressure. Generally, the thicker the material being sprayed, the higher the needed inlet pressure and the larger the gun tip needed. If you change material thickness (for example, use VOC, high-solids paint), you may have to use a larger nozzle and higher pressure to get a smooth paint film without orange peel (Table 24–3).

Spray Gun Distance

Spray gun distance is measured from the spray gun nozzle to the surface being painted. Most spray gun manufacturers recommend that you keep the spray gun about 6–10 inches (203–254 mm) from the surface being sprayed.

HVLP guns should be kept 6–8 inches from the body surface. Older higher pressure spray guns should be kept a little farther from the body surface, 8–10 inches. This is true when doing test spray patterns and when painting a vehicle. Look at Figure 24–21.

If you hold the spray gun too close to the surface, paint will pile up unevenly on the surface and the paint film can run or sag. If you hold the spray gun too far away, the material will partially dry before it hits the surface. The paint will look dull and will not flow down smoothly. This is illustrated in Figure 24–22.

Difficult areas such as corners and edges should be sprayed first. Aim directly at the area so that half of the spray covers each side of the edge or corner. Hold the gun 1–2 inches (25–50 mm) closer than usual, or screw the pattern control knob in a few turns. Either technique will

TABLE 24–2 TYPICAL AIR PRESSURE RANGES

Topcoats	HVLP Gun Feed Pressure	Conventional Gun Pressure (psi)	Undercoats	HVLP Gun Feed Pressure	Conventional Gun Pressure (psi)
Polyurethane enamel	18–20 20–30	50–55 (solids) 60–65 (metallic)	Lacquer primer-surfacers	15–18 16–20	25–30 (spot) 35–45 (panel)
Acrylic lacquer	12–18	20–45	Multipurpose primer-surfacers	16–20	30–40 as primer-surfacer
Acrylic enamel	18–20	50–60	Multipurpose primer-surfacers as nonsanding	17–20	35–40
Alkyd enamel	18–30	50–60	Nonsanding primer-sanders	18–20	45
Flexible finishes	14–28	35–40	Enamel primer-surfacers	18–20	45
Basecoat	14–16	30–35	Epoxy primer	18–20	45
Clearcoat	18–20	35–40	Zinc chromate primer	18–20	45
Sealers	**HVLP Gun Feed Pressure**	**Conventional Gun Pressure (psi)**	**Miscellaneous**	**HVLP Gun Feed Pressure**	**Conventional Gun Pressure (psi)**
Acrylic lacquer	12–16	25–30	Uniforming finishes	12–14	15–20
Universal sealer	14–18	35–45			
Bleederseal	14–16	35–40			

NOTE: Spot repairs should be made at the low end of the air pressure range.

reduce the pattern size. If you only hold the gun closer, the stroke will have to be faster to compensate for a normal amount of material being applied to the smaller areas. After all of the edges and corners have been sprayed, the flat or nearly flat surfaces should then be sprayed.

Spray Gun Adjustment Knobs

A spray gun normally has two adjustment knobs: fluid adjustment and pattern adjustment. These knobs allow you to set the spray gun for the viscosity of the material to be sprayed.

TABLE 24–3 ESTIMATED AIR PRESSURES AT THE GUN

	Pressure Reading (lbs.) at Gauge	Pressure at the Gun for Various Hose Lengths					
		5 feet	10 feet	15 feet	20 feet	25 feet	50 feet
¼-Inch hose	30	26	24	23	22	21	9
	40	34	32	31	29	27	17
	50	43	40	38	36	34	22
	60	51	48	46	43	41	29
	70	59	56	53	51	48	36
	80	68	64	61	58	55	43
	90	76	71	68	65	61	51
5⁄16-Inch hose	30	29	28½	28	27½	27	23
	40	38	37	37	37	36	32
	50	48	47	46	46	45	40
	60	57	56	55	55	54	49
	70	66	65	64	63	63	57
	80	75	74	73	72	71	66
	90	84	83	82	81	80	74

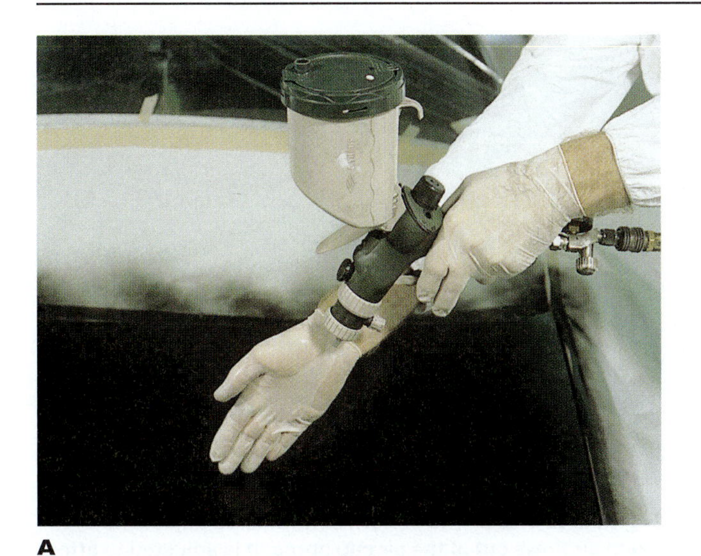

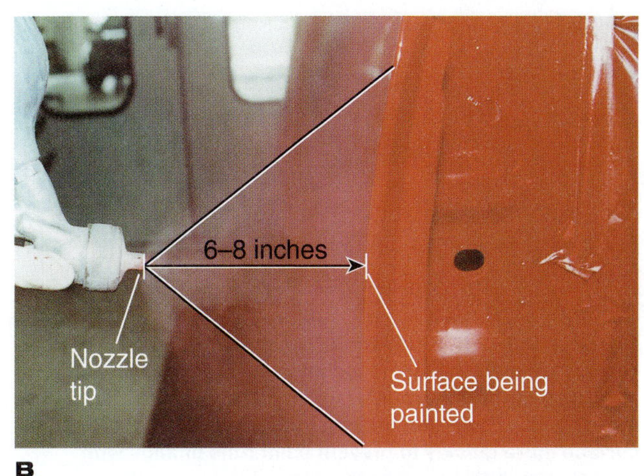

A B

Figure 24–21 When painting, the spray gun must be held at a constant distance from the body surface. (A) The length of your hand is a good visual reference for maintaining the correct nozzle distance from work. (B) Your hand motion should keep the spray gun the same distance from the body surface. Here a paint technician is spraying the edge of a panel. The spray gun is held level and moved straight down along the edge.

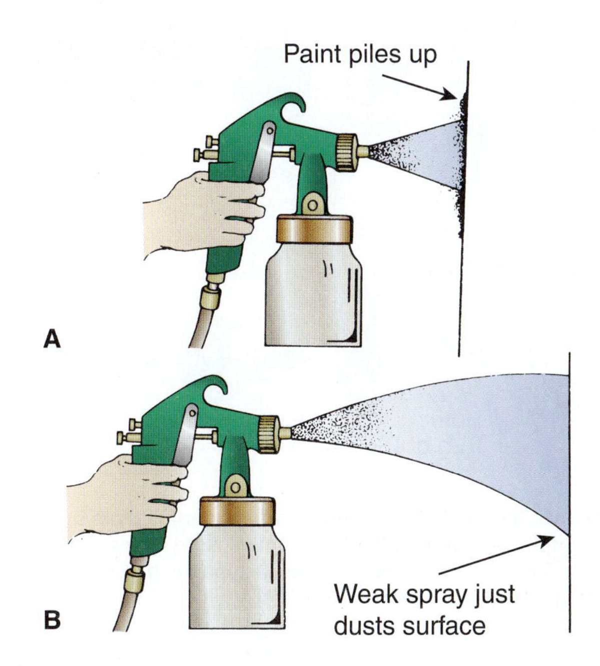

Figure 24–22 The correct gun-to-work distance is important to prevent runs or dry spray problems. (A) If the gun is too close, the finish material piles up and causes runs and sags. (B) When the gun is too far away, material tends to dry into dust before reaching the surface. Adjust distance accordingly.

The *fluid control knob* allows you to adjust the amount of paint or other material leaving the spray gun. Back the fluid knob out to increase the paint flow. Screw the knob inward to decrease the flow (Figure 24–23).

When adjusting your spray gun, regulate the volume of paint according to the desired pattern size and speed at which you will move the gun.

Figure 24–24 shows what happens when you turn the spray gun fluid control knob in and out.

Experienced painters adjust the fluid control knob out to almost the full open position. By moving or passing the spray gun over the vehicle body more quickly, they can apply the right amount of paint on the vehicle in less time without paint runs or excessively wet coats. This allows the experienced paint technician to refinish a vehicle in the minimum time.

If you are less experienced, you must adjust the fluid control knob so that paint flow matches the speed at which you move the spray gun. If you move the gun more slowly when painting, you must screw the fluid control valve in to reduce paint flow and to avoid spraying the paint on too wet.

Adjust the fluid control valve so that your test panel or spray-out panel matches the look of the paint film on the vehicle. If the paint on the vehicle is very smooth,

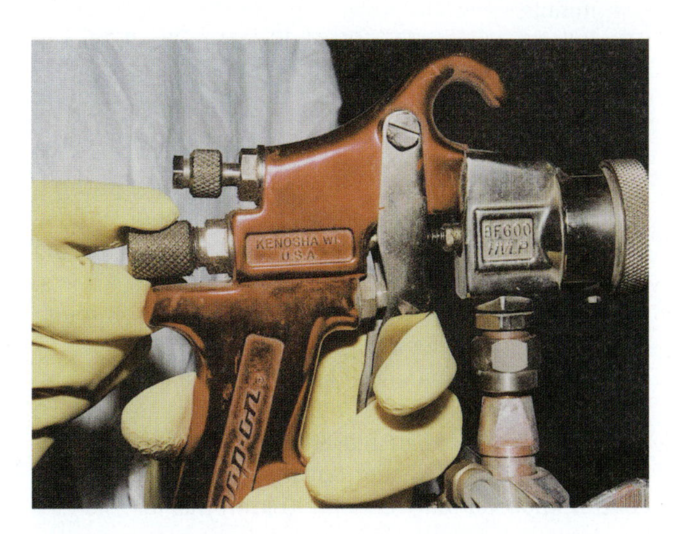

Figure 24–23 Here a technician is adjusting the lower fluid flow control knob, which controls how fast liquid is sprayed out of the gun.

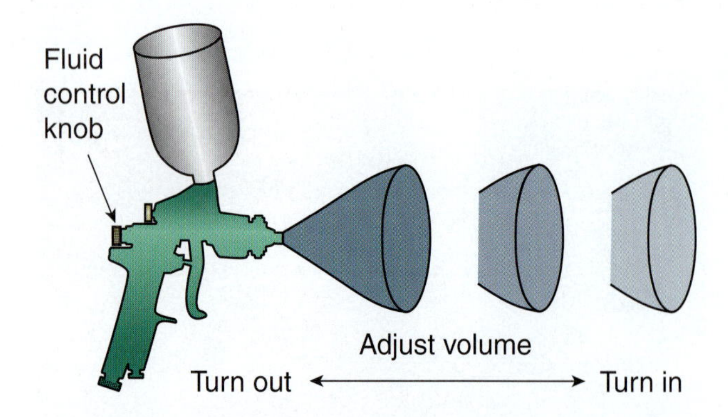

Figure 24-24 **Study what happens by turning the fluid control knob. Professional technicians open the fluid valve almost all of the way counterclockwise so they can paint more quickly. However, the gun must be moved over the body surface more quickly to prevent paint runs or sags with maximum paint flow out of the gun.**

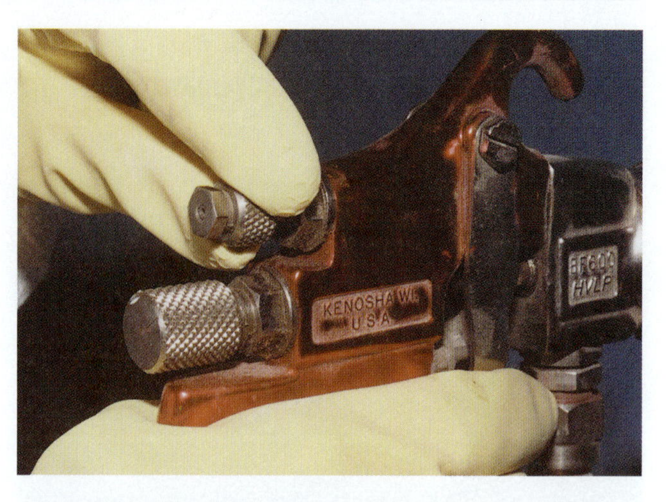

Figure 24-25 **The pattern control knob adjusts how much air flows out of the air cap horns. It is adjusted to affect the shape of the paint mist when it leaves the gun.**

use full wet coats of clearcoat that will flow out more smoothly.

To adjust the flow rate from a spray gun, use the following procedure:

1. Fill the gun with properly mixed and reduced material to be sprayed. Make sure the correct pressure is being fed into the gun from the air line. Check pressure at the gun-mounted pressure gauge.
2. Remove the air cap from the spray gun. Obtain a clean graduated container for measuring liquid volume.
3. While wearing protective gear, aim the spray gun into the container. Pull the trigger for 10 seconds. Measure the amount of material that flows into the container in that time. Multiply the volume of liquid by 6. This is the fluid flow rate in ounces per minute.
4. For standard refinishing, flow out of the spray gun should be about 14–16 ounces (413–472 mm) per minute.
5. If the flow rate is less than this, open the fluid control valve more and repeat. If flow rate is faster than this, turn the fluid valve in a little and try again. When the flow rate is correct, reinstall the air cap.

You set the size of the spray pattern using the **pattern control knob**. This knob adjusts the amount of air flowing through the air cap horns or side air nozzles. Turning the knob in restricts airflow through the side air jets. Unscrewing the pattern knob increases airflow through the air horn jets. Refer to Figure 24–25.

With the pattern control knob turned all the way in, the paint spray shape will be almost round. As you turn the pattern knob out, increased air flow will reshape the pattern into a tall, slender fan shape. Turn the pattern control knob out only enough to form this tall, slender fan shape.

Figure 24–26 represents the adjustment of spray pattern from all the way in to all the way out. After mixing the primer or paint to its recommended viscosity (thickness), check spray gun adjustment by testing the gun on a sheet of masking paper (Figure 24–27).

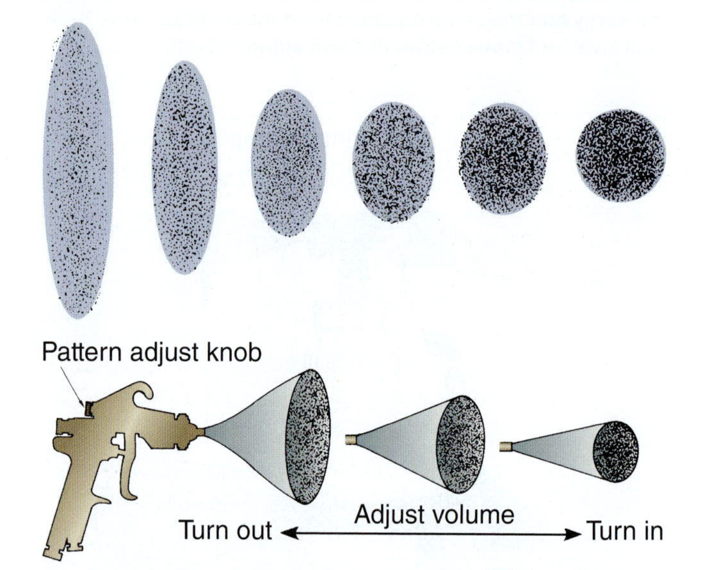

Figure 24-26 **Pattern width adjustment is done by turning the adjustment knob as shown. Note how screwing the air knob out, or counterclockwise, makes a round pattern form into the desired fan shape.**

When making your spray gun pattern adjustments, tape a large sheet of masking paper to the wall of the spray booth. This will let you see how the spray pattern is forming on a test surface (Figure 24–27).

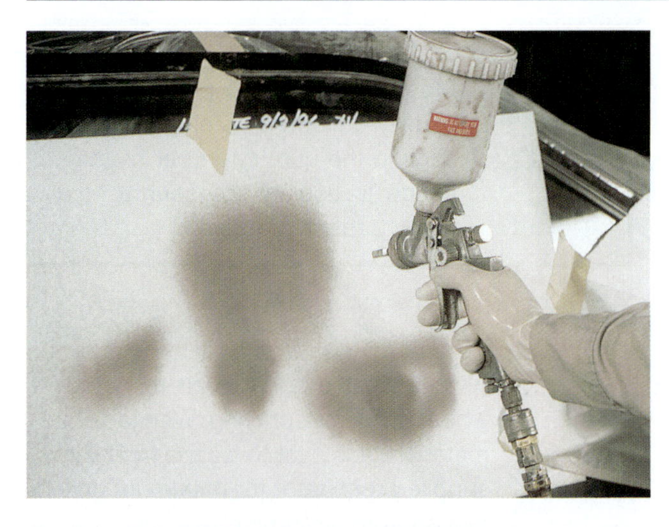

Figure 24–27 The spray gun should always be checked for normal operation on a piece of masking paper before starting to refinish the vehicle.

Testing Spray Pattern

A **spray pattern test** checks the operation of the spray gun on a piece of paper. Before attempting to paint the vehicle, it is very important to test the spray pattern. You must check the spray pattern for proper atomization, adequate paint flow, proper pattern shape, and even distribution (Figure 24–28).

SHOP TALK

If you recycle and use old newspaper as a spray pattern test sheet, do not attach it to the vehicle. The newspaper ink can bleed through the newspaper, contaminating the body surface. Attach the test sheet to a fabricated stand or holder or to a wall in the spray booth.

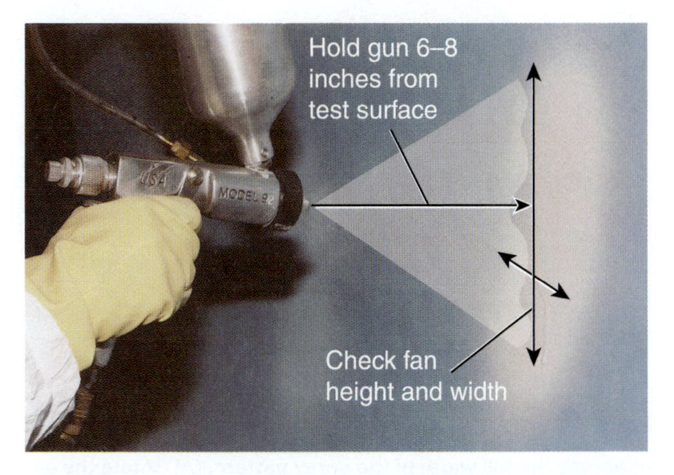
Hold gun 6–8
inches from
test surface
Check fan
height and width

Figure 24–28 To check spray gun adjustment, hold the gun at the correct distance from the vehicle surface. Pull the trigger and check the height, width, and wetness of the spray pattern.

Typically, you should hold a conventional gun 8–10 inches (203–254 mm) away from the surface. If using an HVLP gun, hold it 6–8 inches (152–203 mm) away from the paper. Pull the trigger all the way back and release it immediately. This burst of paint should leave a long, slender pattern on the test paper. Make a couple of spray passes over the test sheet to make sure the gun is not spitting, leaking, or having other problems.

To narrow the pattern, adjust the air valve inward (clockwise). To widen the spray pattern, turn the air valve outward (counterclockwise). Remember the following tips when evaluating a spray pattern test:

- A pattern that is heavy in the middle indicates too little airflow.
- A pattern that is divided in the middle indicates too much airflow.
- Too much paint at top or bottom may be caused by a restriction at the fluid needle or air cap horn.
- A pattern that leans to one side could be the result of a restriction at the fluid needle or air cap horn.
- If the pattern is heavy on one side or on the top or bottom, try turning the air cap 180 degrees. If the pattern remains the same, clean or replace the fluid needle and fluid nozzle. If the pattern rotates 180 degrees, then the problem is in the air cap horns.

Spraying primer-surfacer usually requires a smaller spray pattern. Turn in the pattern control knob until the spray pattern is 6–8 inches (152–203 mm) wide. For spot repair, the pattern should be about 5–6 inches (127–152 mm) from top to bottom.

If the paint droplets are coarse and large, close the fluid control knob about one-half turn or increase the air pressure 5 psi (34 kPa). If the spray is too fine or too dry, either open the fluid control knob about one-half turn or decrease the air pressure 5 psi (34 kPa). Make sure the test spray pattern has even distribution and that the paint is atomizing into tiny droplets. If the paint does not atomize properly, increase air pressure slightly or check the paint viscosity. Refer to Figure 24–29.

Next, test the spray pattern for uniformity of paint distribution. Loosen the air cap retaining ring and rotate the air cap so that the horns are straight up and down. In this position, you will get a horizontal spray pattern instead of a vertical one.

Spray again on your test paper. However, hold down the trigger until the paint begins to run; this is known as *flooding the pattern*. Inspect the lengths of the runs. If all adjustments are correct, the runs will be almost equal in length (Figure 24–30).

The uneven runs in the split pattern shown are a result of setting the spray pattern too wide or the air pressure too high. Turn the pattern control knob in one-half turn or raise the air pressure 5 psi (36 kPa). Alternate between these two adjustments until the runs are even in length.

If paint runs are longer in the middle than on the edges, too much paint is being discharged. Turn the fluid control knob in until the runs are even in length.

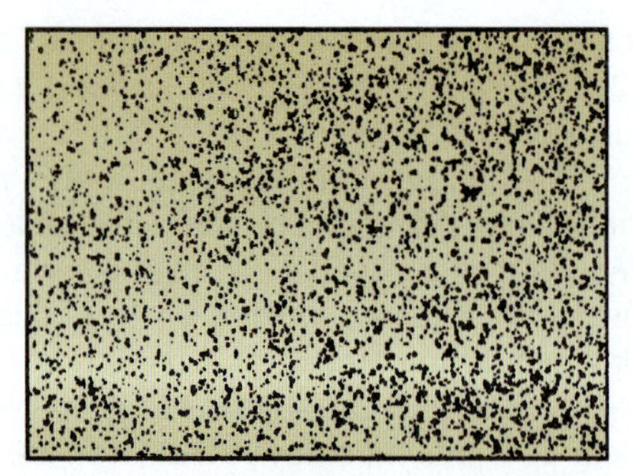

A Dry spray and uneven distribution result from poor atomization. With improper atomization, paint will go on coarse and dry. Small bumps will be evident in one or more areas of the spray pattern, with tiny spaces between paint droplets.

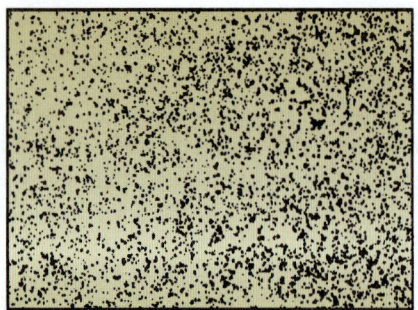

B With even distribution and good atomization, tiny paint droplets will be deposited across the spray pattern with full coverage of the surface.

Figure 24-29 Closely inspect a test spray pattern for good atomization and distribution.

Most spray pattern problems are usually caused by a clogged passage in the gun. Improper maintenance is usually the source of a spray pattern problem.

Always wear a suitable air respirator when doing any spraying. With today's materials, it is best to wear a fresh air-supplied respirator for maximum protection from airborne chemicals.

24.4 USING A SPRAY GUN

There are many factors that affect the use of a spray gun. Even when the spray gun is properly set up and adjusted, it takes great skill to paint a vehicle. You must learn how to move, trigger, aim, and handle a gun properly when spraying.

Spray Gun Stroke

Spray gun stroke refers to the hand motion used to move the gun while spraying. The proper stroke is

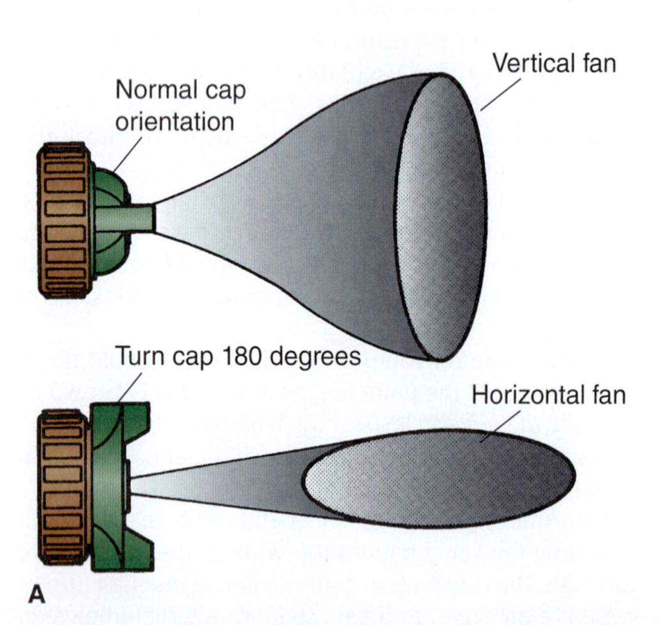

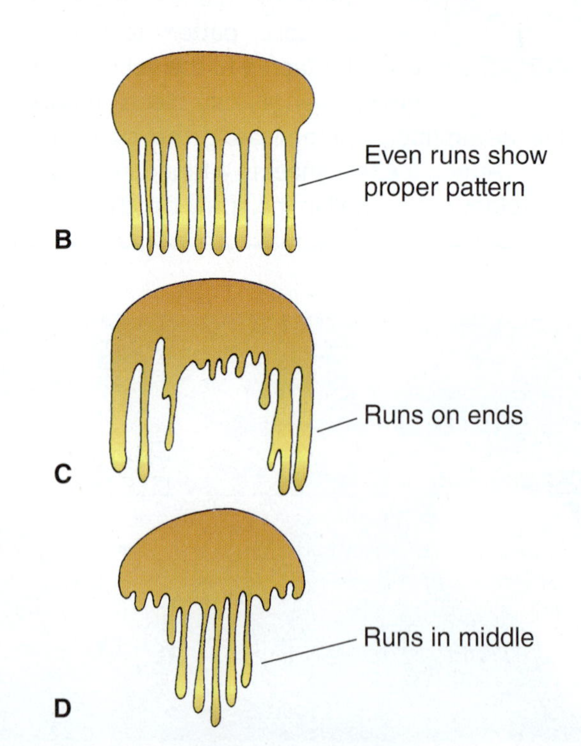

Figure 24-30 A flood test checks for equal distribution of paint across the full width of the spray pattern. (A) Rotate the air cap sideways so that the pattern is sideways or horizontal. Pull the gun trigger and keep spraying a test pattern until the paint starts to run or sag. (B) When flooding the test sheet, a balanced spray pattern should result. Runs should be equal across the pattern. (C) A split pattern shows too much spray on each end of the pattern. In this case, airflow from the sides of the cap is insufficient. (D) A heavy center pattern means too much air is being metered out of the side orifices in the cap.

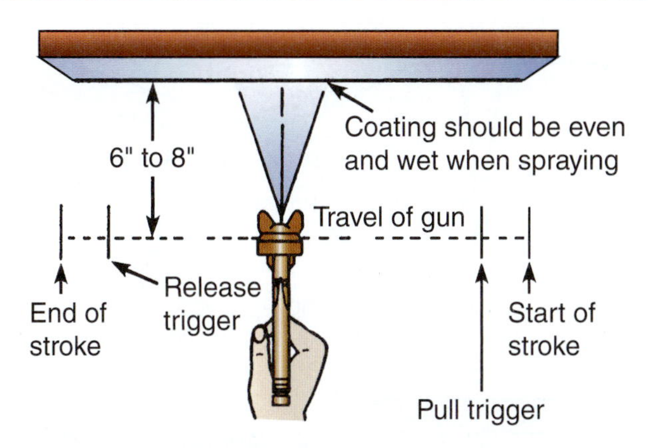

Figure 24–31 Handling a spray gun takes concentration. You must constantly watch the body and keep aiming the gun straight at the surface contour. Try to hold the gun parallel with the surface. When applying a full wet coat to a flat surface, keep the gun nozzle aimed straight at the surface; do not fan or swing the gun sideways.

important in obtaining a good paint job. To obtain a good stroke technique, hold the spray gun at the proper distance from the surface—8–10 inches (203–254 mm). If the humidity or temperature is high, a slightly shorter distance may be necessary.

If spray gun distance is too close, the high velocity of the spraying air tends to ripple the wet paint film. If the spray gun distance is too great, more reducer will evaporate, resulting in orange peel, dry film, or poor color match.

A lower evaporating thinner will permit more variation in the distance of the spray gun from the job but will produce runs if the gun gets too close. Excessive spraying distance also causes a loss in materials due to overspray.

Generally, you should hold the gun perpendicular to the surface being sprayed (Figure 24–31).

Spray Gun Speed

Spray gun speed is how fast you move the gun sideways over the surface. Proper spray gun speed will deposit a smooth film of paint or other material on the body surface.

Move the gun in a steady deliberate pass, about 1 foot (0.3 m) per second. Moving the gun too quickly will produce a thin film, whereas moving it too slowly will result in the paint running. The speed must be consistent or it will result in an uneven paint film. Never stop in one place or the sprayed coat will drip and run!

Figure 24–32 illustrates what happens with proper and improper gun speeds.

Spray Gun Triggering

Spray guns have a two-step trigger mechanism. **Spray gun triggering** involves keeping constant airflow out of the nozzle and only triggering paint flow when depositing paint materials on the body.

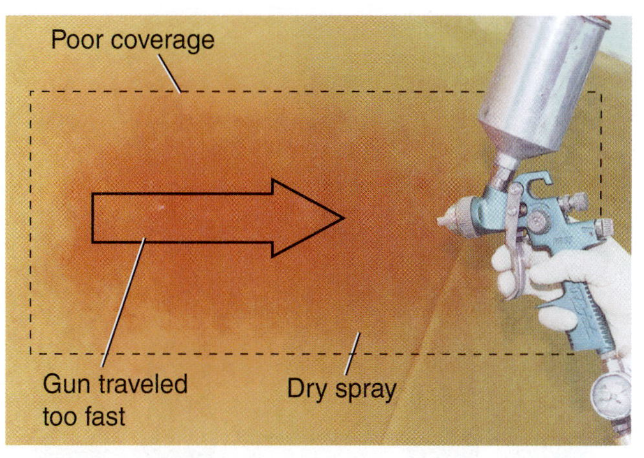

A The spray gun stroke speed or side movement is too fast, so the paint is going on too thinly with poor coverage.

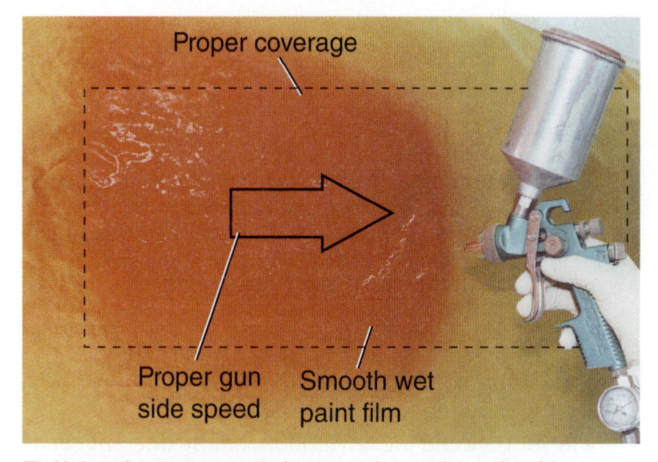

B Using the proper speed across the surface should produce good, even coverage.

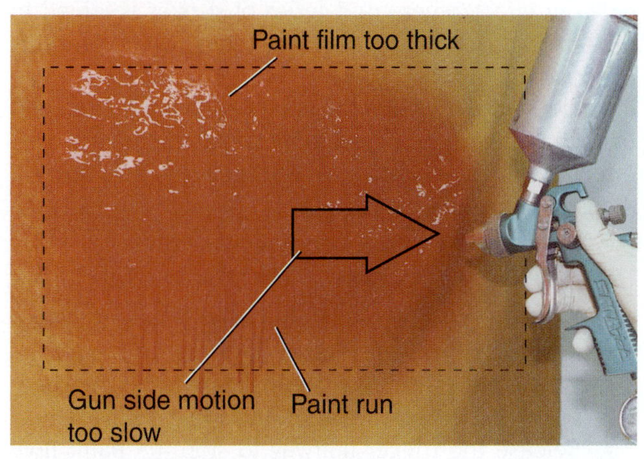

C If a spray gun is moved too slowly, even for a second, the paint will be too thick and can run or sag.

Figure 24–32 The stroke or side motion of a spray gun must be at the correct speed.

When you pull the trigger halfway back, only air flows out of the gun. This keeps the gun passages clean and helps prevent spitting. When you pull the spray gun trigger all the way back, paint material flows out of the nozzle. You should only pull the trigger all the way back

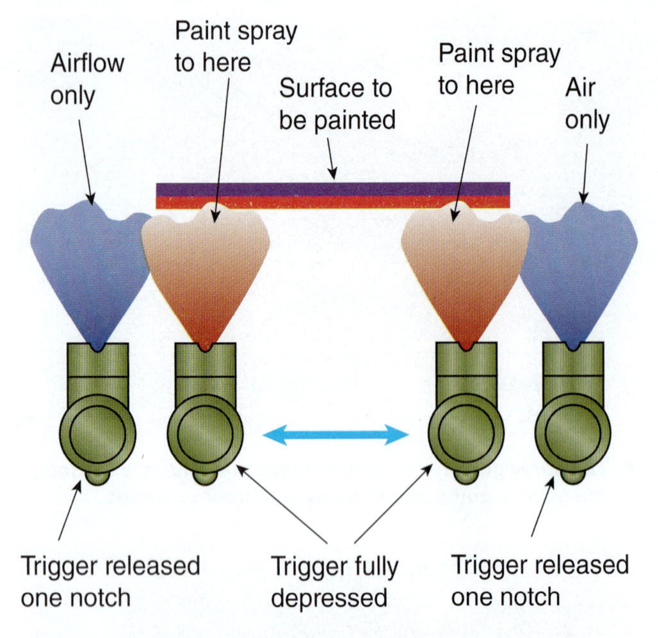

Figure 24–33 Gun triggering involves releasing halfway up on the spray gun trigger at the end of each pass to blow only air out of the nozzle. After you change the pass direction and the gun is moving back over the surface, press all the way down on the trigger to evenly apply the paint. Triggering prevents excess paint application and runs at the ends of a pass.

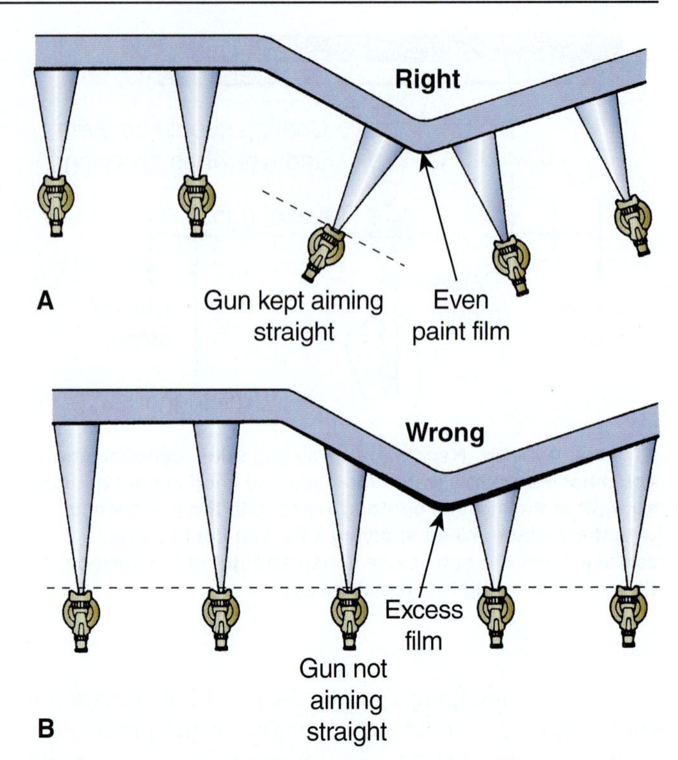

Figure 24–34 Note how spray gun direction must be changed during each pass if the vehicle body shape is not flat. (A) Proper spray gun movement keeps the gun tip aimed straight at an angled surface. As the curve in the panel is reached, the gun must be rotated so that the spray hits the surface head on. (B) If the spray gun is not kept at a right angle on angles or curves in the body, an uneven paint film will result.

when you are applying material to a body surface with the spray gun in motion (Figure 24–33).

The gun should be in motion before the trigger is pulled, and the trigger should be released before the gun motion stops. This technique produces a fade-in and fade-out effect, where one series of strokes is joined to the next by overlapping the stroke ends, which prevents overloading.

Release the trigger at the end of each pass. Then pull back the trigger when beginning the pass in the opposite direction. In other words, "trigger" the gun and turn off the gun at the end of each sweep. This avoids runs, minimizes overspray, and saves paint. Proper triggering involves four steps:

1. Begin the stroke over the masking paper, triggering the gun halfway to release only air.
2. When the starting edge of the panel is reached, squeeze the trigger all the way to release the paint.
3. Release the trigger halfway to stop the paint flow when directly over the finishing edge.
4. Continue the stroke several more inches before reversing the direction and repeating the sequence.

Spray Gun Direction

Spray gun direction involves keeping the gun aimed so that an even film of paint deposits on the body surface. If the spray gun is not kept aimed straight at the surface, an uneven paint film will result. Refer to Figure 24–34.

On flat surfaces, such as the hood or roof, the gun should be pointed straight down and moved straight across the flat surface. The spray gun should be kept aimed straight at flat surfaces all the way across the panel.

Do not fan the gun if a uniform film is desired when full paint coverage is needed. The only time fanning is permissible is on small spot repair areas where the paint film at the edges of the spot should be thinner than the center portion.

Spray Overlap

Spray overlap ensures full, even coverage without applying too much paint at once. Each pass of the spray gun should paint over or cover about half of the previous paint stroke for proper spray overlap.

Generally, you should start at the top of an upright surface such as a door panel. The spray gun nozzle should be level with the top of the surface. This means that the upper half of the spray pattern will hit the masking material. Refer to Figure 24–35.

The second pass is made in the opposite direction, with the nozzle level at the lower edge of the previous pass. Thus one-half (50 percent) of the pattern overlaps the previous pass, and the other half is sprayed on the unpainted area.

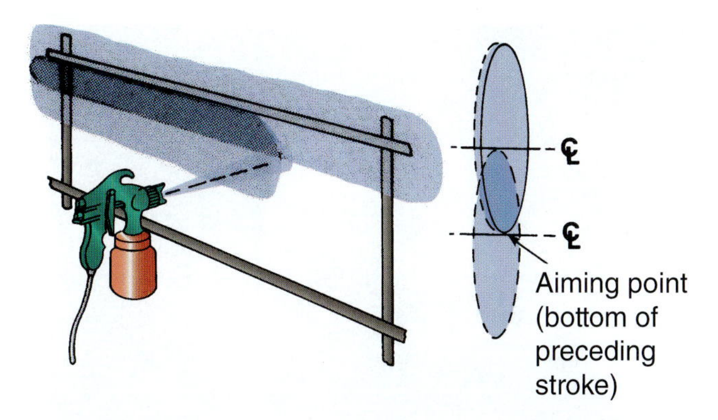

Figure 24–35 The first pass should be centered at the top of a panel—half on the masking paper and half on the body surface. The next paint pass or stroke should overlap and cover half of the previous pass. The pass should extend a bit off the sides of the panel and onto the masking paper a little. Trigger the paint off but leave air on at the end of each pass.

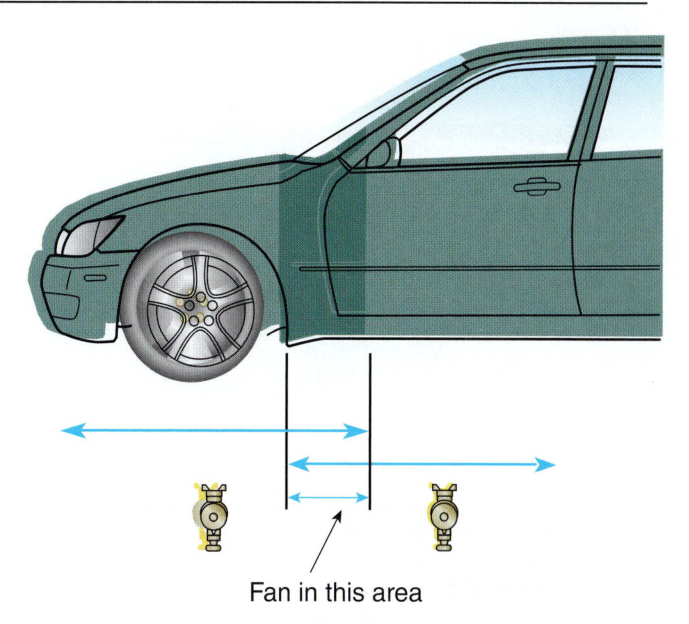

Figure 24–37 Remember that the gun overlap area is where runs are likely to occur in overall painting jobs or multipanel spot repairs. Triggering and fanning the gun slightly at the end of each pass at panel gaps will help avoid paint run problems.

Continue back and forth passes, triggering the gun at the end of each pass and lowering each successive pass one-half the top-to-bottom width of the spray gun pattern.

The last pass should be made with the lower half of the spray pattern below the surface being painted. If it is a door, the pattern would shoot off into space below it.

Always blend into the wet edge of the previous section sprayed (Figure 24–36).

Proper triggering technique at the area where the sections are joined will avoid the danger of a double coat at this point and the possibility of getting a sag (Figure 24–37).

The procedure just outlined is called a *single coat*. For a *double coat*, immediately repeat the single coat procedure. Two or three single coats are normally required for enamel topcoats. Allow the first coat to set up—become tacky, or partially dry or cured—before applying additional coats.

Gun Handling Problems

The inexperienced painter is prone to several spraying errors.

Heeling is when the painter allows the gun to tilt. Because the gun is no longer perpendicular to the surface, the spray produces an uneven layer of paint, excessive overspray, dry spray, and orange peel. Look at Figure 24–38.

Arcing is a gun handling problem that occurs when the spray gun is not moved parallel with the surface. At the outer edges of the arced stroke, the gun is farther away from the surface than at the middle of the stroke. The result is uneven film buildup, dry spray, excessive overspray, and orange peel (Figure 24–39).

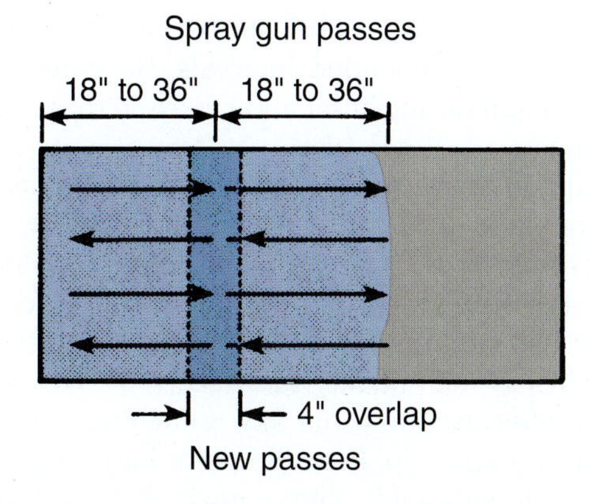

Figure 24–36 Always blend into the wet edge. Blend where body areas meet by fanning the gun over the edges of panels. This will thin the paint film between the two areas. Because they will be sprayed twice, a thinner blend edge is needed to avoid excess paint film thickness.

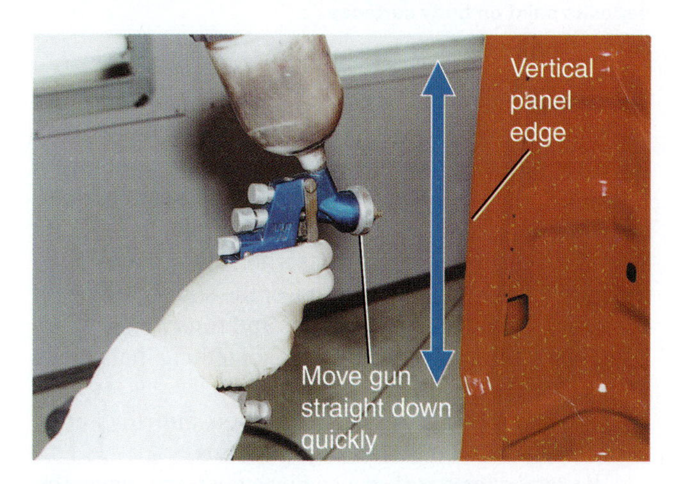

Figure 24–38 Always make sure you paint the edges of panels. Hold the gun parallel and aim straight at the edge to prevent heeling of the gun. A quick stroke of the gun will provide adequate paint film thickness on sharp edges.

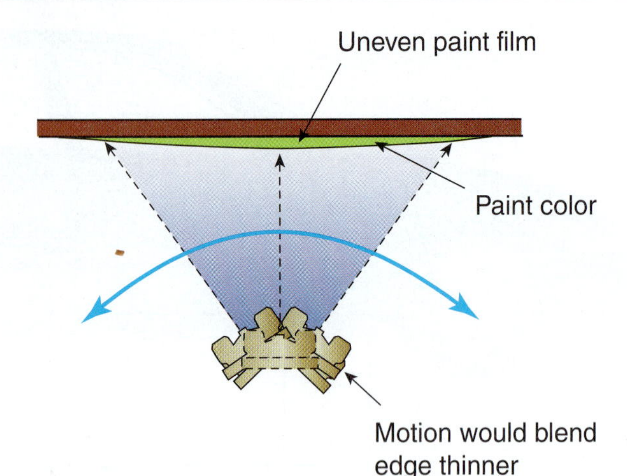

A If you arc or fan the gun on a flat surface, the paint film will be thicker in the middle and thinner out to the sides. This is not desirable unless you want to blend paint into an area that is not being painted.

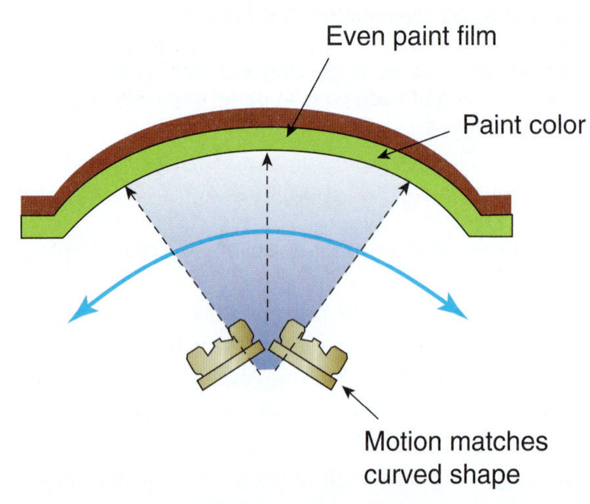

B If you arc or fan the gun to match the shape of a body contour, you can maintain equal paint film thickness. Again, you normally want to keep the spray pattern hitting straight onto the body surface.

Figure 24-39 Note how arcing or fanning a spray gun deposits paint on body surfaces.

Fanning involves moving your wrist sideways to swing the spray gun into an angle to the surface being sprayed. Fanning should only be done when you intend to blend the paint thinner.

Improper stroke speed results when the stroke is made too quickly; the paint will not cover the surface evenly. If the stroke is made too slowly, sags and runs will develop. Using the proper stroking speed is something that comes with experience.

Improper overlap results in uneven film thickness, contrasting color hues, and sags and runs.

Wasteful overspray is caused by a failure to trigger the gun before and after each stroke. Wasteful overspray results in excessive buildup of paint at the beginning and end of each stroke.

Figure 24-40 This technician is using a very long pass of the spray gun across a trunk lid. The spray gun is held straight down on each pass. When going down the rear of the trunk lid, the gun would have to be aimed straight at the side of the panel surface. Note the use of a plastic cup with a disposable liner to reduce paint waste.

Improper coverage from triggering at the wrong time is another common error. Failure to trigger exactly over the edge of the panel results in uneven coverage and film thickness. Look at Figure 24–40.

Table 24–4 summarizes the variables that control quality when spray painting.

24.5 SPRAY GUN MAINTENANCE

Neglect and poor maintenance are responsible for the majority of spray gun difficulties. Proper care of a gun requires little time and effort. Thorough cleaning of a gun and its accessory equipment *immediately after use* is critical. You must also lubricate bearing surfaces and packings at recommended intervals. Be careful when handling a gun during lubrication to avoid dropping and damaging it.

A buildup of overspray can collect on the air cap and turn into a kind of fuzz. Clean the gun frequently to prevent this fuzz from blowing off and ruining a paint job. Paint will set up in and on the gun. If the dried paint flakes, it will land on the job and cause a defect. Clean the gun inside and out after each job.

Right after use, pour out any remaining paint material from the gun cup. Pour the paint into an approved storage tank with a lid. With a gravity-fed gun, you can depress the trigger so that any paint remaining in the gun body squirts out. Fill the cup with clean solvent and spray it through the gun and into the waste container (Figure 24–41).

TABLE 24–4 SUMMARY OF VARIABLES CONTROLLING QUALITY IN SPRAY FINISHING

Atomization	1. Fluid viscosity 2. Air pressure 3. Fan pattern width 4. Fluid velocity or fluid pressure 5. Fluid flow rate 6. Distance of spray gun from work
Evaporation stages	1. Between spray gun and part 2. From sprayed part
Evaporation variables between spray gun and sprayed part	1. Reducer temperature spec 2. Atomization pressure 3. Amount of reducer 4. Temperature in spray area 5. Degree of atomization
Evaporation variables affecting	1. Physical properties of solvents (that is, fast or slow evaporation) 2. Temperature a. Fluid b. Work c. Air 3. Exposed area of the surface sprayed
Evaporation variables from the sprayed part	1. Surface temperature 2. Room air temperature 3. Air pressure velocity 4. Flash time between coats 5. Flash time after final coats 6. Physical properties of the solvents (that is, fast or slow evaporation)
Operator variables	1. Distance of spray gun from the work surface 2. Stroking speed over the work surface 3. Pattern overlap 4. Spray gun attitude a. Heeling b. Arcing (fanning) 5. Triggering

Avoid spraying solvent into the air when cleaning a spray gun. This pollutes our atmosphere. If a gun washer is not available, direct the flushing spray into a closed container of solvent. Then, you will catch most of the solvent spray in the container and produce less air pollution. Areas in the United States with air pollution problems, such as southern California, require the use of an enclosed spray gun cleaning tank.

Figure 24–41 **Always empty unused paint material into an approved storage container with a lid. Fill the cup with solvent and spray it through the gun into the container. Always keep the lid closed to prevent solvent evaporation and air pollution. Used paint can be processed in a recycling machine to remove pigments and separate resins and solvents.**

If you ever forget to clean a spray gun filled with a curing-type material (for example, high-build primer or epoxy primer), you will probably have to completely disassemble the gun and tediously clean each part separately, consuming an enormous amount of time and energy.

Spray Gun Cleaning Tank

A **spray gun cleaning tank**, also called a gun washer/recycler, is a pressurized container for flushing the gun and other tools with cleaning solution. It is used by most modern body shops because it saves time and keeps spray guns in good, clean condition.

A spray gun washer directs a cleaning solution into the cup and over and through the spray gun body to remove paint or other materials (Figure 24–42).

Paint-covered equipment (guns, cups, stirrers, and strainers) is placed in the larger tub of the gun washer/recycler. The lid is closed, then the pump recirculates the solvent into the upper portion of the tub. In less than 60 seconds, the equipment is clean and ready for use.

The automatic gun washer/recycler saves the technician time: Compared with traditional manual cleaning methods, it saves 10 minutes on each color change. The cleaning system also offers increased safety, because solvent fumes are contained in the tank.

The spray gun cleaning tank is designed so that sludge from the cleaning action settles to the bottom for easy drainage and disposal with other shop wastes. Check the owner's manual for complete operational details and the proper solvents to use.

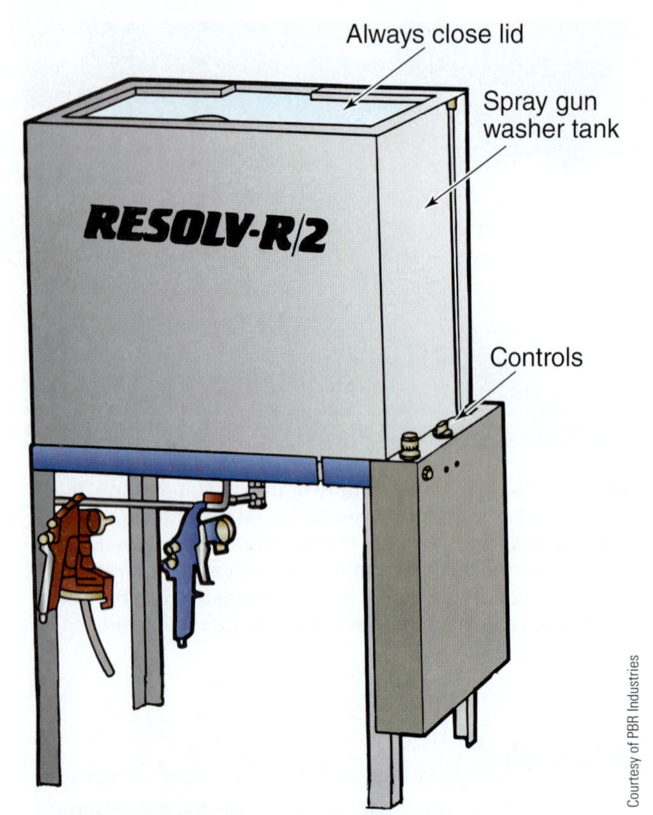

A A spray gun cleaning tank or washer saves time in high-production shops.

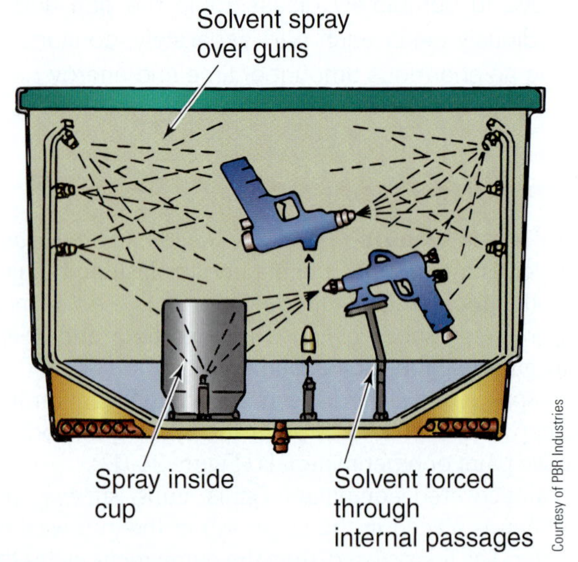

B Note how cleaning solvent is forced through and over a gun and into the cup for rapid removal of finish materials.

Figure 24-42 A spray gun washer or cleaning tank can be found in most professional body shops.

To use a gun cleaning tank, remove any parts that might be damaged by the cleaning solvent. These parts would include the pressure gauge-regulator on HVLP guns and any plastic vent hoses. Place the cup into the tank. Then place the gun into the cleaning mechanism so that the trigger is engaged and the inlet tub is in place over the cleaning nozzle.

Close the lid and turn the cleaning machine on. This will force solution through all passages in the gun. When it is cleaned, remove the gun and wipe it off with a clean rag.

After removing the gun from the spray gun washer, you may need to use soft bristle brushes to final clean around the air cap, trigger, and other hidden pockets on the outside of the gun. Use a clean, lint-free rag to wipe the cup and body.

With a solvent-soaked rag, wipe the inside and outside of the cup, the gun body, the air cap, and all external parts. Remove all traces of paint film.

Manual Spray Gun Cleaning

Even if you use a spray gun cleaning tank, you should periodically disassemble the gun for thorough service. To manually clean a suction-fed gun, first loosen the cup from the gun. With a gravity-fed gun, remove the lid. Pour out any remaining, unused material into an approved container for proper disposal. Pour some spray gun cleaning solvent into the gun. Slosh it around to partially remove the paint film in the cup. Spray the solvent through the gun to remove most of the paint.

Following manufacturer's instructions, remove any parts that require further cleaning (air cap, nozzle, needles, vent tube, and so on). While wearing plastic or rubber gloves, wipe them clean with a solvent-soaked rag. When blowing off a spray gun, use very low pressure (5 psi or 34 kPa).

If needed, clean small, hard-to-reach areas on the gun with a thin, soft bristle brush. Wipe off residue with a clean rag soaked with solvent. Then, pour 1 inch of clean solvent in the cup again. Spray the solvent through the gun to clean out the fluid passages. See Figure 24–43.

When cleaning a spray gun, it is best to use a recommended spray gun cleaning solvent. It will not damage gun parts while removing deposits. You should also use a special spray gun oil to lubricate parts as needed. Spray gun oil is formulated to not contaminate the paint and will not cause fish-eyes like conventional oils will.

To avoid gun damage, *never* use wires or nails to clean the precision-drilled openings. Clean the fluid tip with a gun brush and solvent. With a clean rag soaked in thinner, wipe the outside of the gun to remove all traces of paint.

Packings, springs, needles, and nozzles must be replaced periodically due to normal wear and tear.

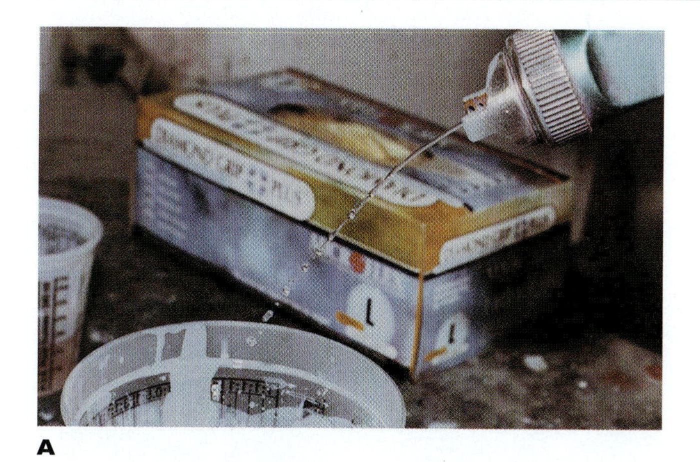

A

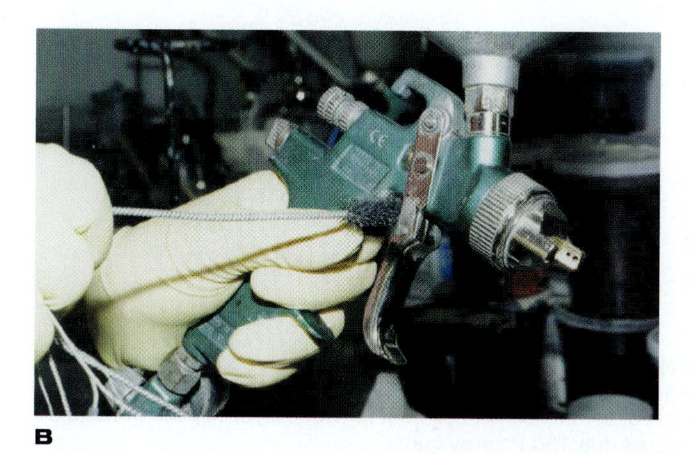

B

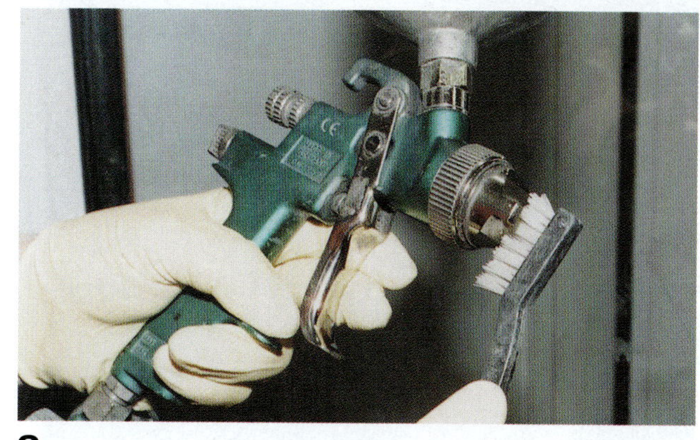

C

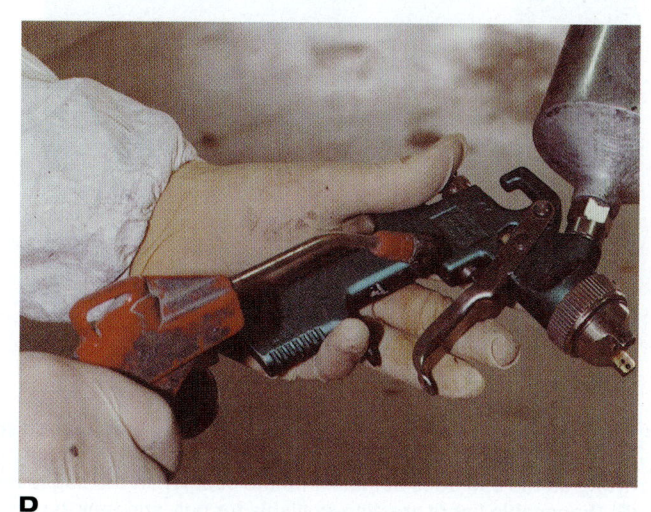

D

Figure 24-43 **It is easy for paint to dry on the outside of a spray gun. Keep your spray gun perfectly clean to keep dry paint flakes from falling into your paint job. (A) With a gravity gun, press the trigger to run clean solvent through the gun. You do not have to use air pressure with a gravity-fed gun. (B) Use solvent or paint remover to clean paint off the outside of a gun. Get into tight areas such as behind the trigger with a soft bristle brush. (C) Check the air nozzle frequently for paint buildup. If one bit of dry paint falls off your spray gun into a paint job, you will need to spend a lot of time repainting. (D) Use compressed air and a blow nozzle to final clean debris off the outside of a gun before use.**

This should be done only in accordance with the manufacturer's instructions.

Note that some manufacturers provide plastic bags that can be installed in the spray gun cup or tank. The plastic bag helps ease gun cleanup. After you remove the paint-coated bag, the cup or tank remains relatively clean. This is shown in Figure 24–44.

Spray Gun Lubrication

Most spray gun manufacturers recommend lubricating the parts shown in Figure 24–45 with spray gun oil. **Spray gun oil** is compatible with paint and will not contaminate the gun.

Avoid using conventional oil to lubricate a spray gun. Excess oil can overflow into the paint and air passages, mixing with the paint and resulting in a defective paint film. Conventional oil and paint do not mix; cross-contamination will result in fish-eyes in the fresh paint.

24.6 SPRAY GUN TROUBLESHOOTING

If an air spray gun is not adjusted, manipulated, and cleaned properly, it will apply a defective coating to a surface. Fortunately, defects from incorrect handling and improper cleaning can be tracked down and corrected without much difficulty. Refer to Figure 24–46.

NOTE *The most common spray gun application problems, with their possible causes and suggested remedies, are given in Chapter 28.*

If not properly maintained, the air spray gun itself can also create some problems.

Table 24–5 contains the causes of and possible solutions to some of the more common spray gun difficulties.

Failure of the compressed air supply system to perform properly can cause the paint problems shown in Table 24–6.

A

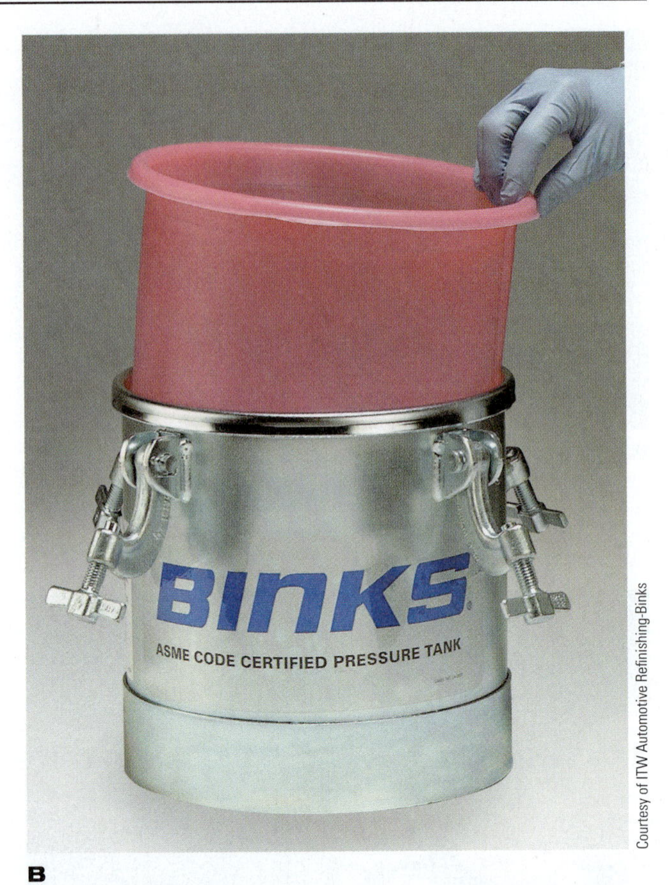

B

Figure 24-44 Disposable plastic containers can reduce spray gun cleanup time. (A) Here a technician is installing a small disposable plastic bag into a gravity-fed cup. A special tool is being used to snap the hole in the bottom of the bag over the fitting so the bag does not leak into the cup. When the bag is removed, the cup will not be coated with a layer of paint material. (B) Disposable liners are also available for pot- and tank-type pressure-fed, HVLP spray guns.

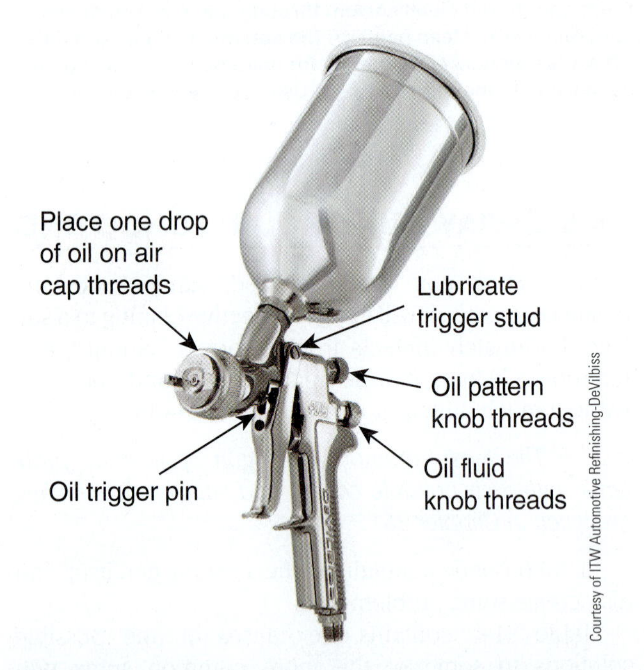

Figure 24-45 Note the locations on a spray gun that should be oiled periodically. Use manufacturer-recommended spray gun lubrication that will not contaminate paint materials. Only use a drop of oil on each point, to avoid drips. Refer to the owner's manual for details.

24.7 OTHER SPRAY SYSTEMS

There are three other types of spray systems that can be found in some shops: airless spray gun, electrostatic, and airbrush. Their operation is basically the same as the air spray equipment just described. Even though they are not as common as HVLP spray guns, you should be familiar with them.

Airless Spray Gun System

Airless spraying equipment uses hydraulic pressure rather than air pressure to atomize paint material. With the airless spray method, pressure is applied directly to the paint, which is ejected at a high speed through small holes in the nozzle and formed into a mist (Figure 24–47).

Electrostatic Spraying System

Electrostatic spraying utilizes the principle that positive (+) and negative (–) electrical charges mutually attract each other but oppose a like charge. Therefore, when paint particles are given a negative charge by a high-voltage generator, the particles oppose each other, causing them to become atomized. This causes more of the

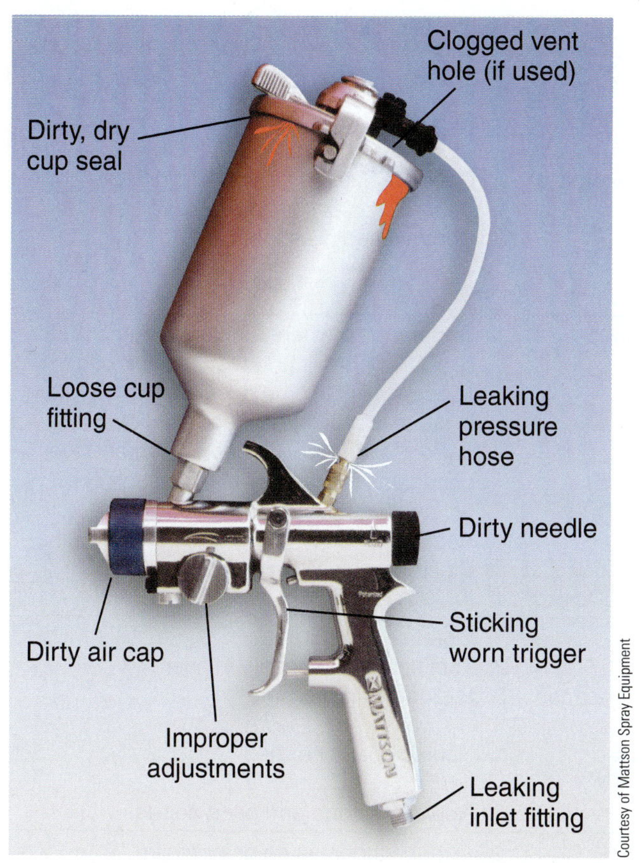

A A leaking pressure hose on a can lid is a common problem that can ruin a paint job.

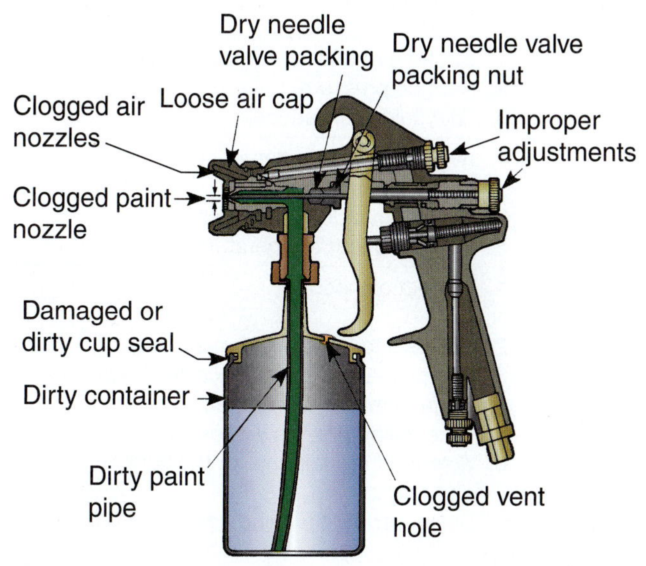

B A clogged vent hole is the most common problem with suction-fed spray guns.

Figure 24-46 Study possible problem points on a spray gun.

paint overspray to be attracted and pulled against the vehicle body. Less paint is wasted as overspray. It should be mentioned that this type of spray equipment is generally not used in the auto body shop (Figure 24–48).

As for portable electrostatic painting equipment, there are both the air spray type (Figure 24–49) and the airless spray type.

Table 24-7 details some of the features of the six types of spray painting.

It is best to connect the body of the vehicle to an earth ground before final tack cloth cleaning. This will help cut down on static electricity that can draw dust onto the vehicle (Figure 24–50).

Touch-Up Guns

Touch-up guns are like normal spray guns, but they are much smaller and hold less paint material. They are ideal for spraying small or confined areas. They are easier to use and clean when you only have to spray a very small area, as when spraying a small repair area with colorcoat. See Figure 24–51.

Airbrushes

An **airbrush** is a small spray gun designed to do custom painting. It can be used to apply a fine line of paint or to blend paint over a custom pattern when doing complex shapes in a paint job. Refer to Figure 24–52.

It is important to select the correct airbrush for the type of work to be performed. Consider the size and type of the work to be done, the fineness of the line desired, and the fluids to be sprayed. Airbrushes used for custom auto finishing are generally one of two types: double-action and single-action (Figure 24–53).

Double-action brushes are commonly found in most custom paint shops. They are available with a choice of tips to further increase their versatility. The double-action airbrush is usually recommended for projects that require very fine detailing. They produce a variable spray that works by depressing the finger-controlled front lever for air and pulling back on the same lever for the proper amount of color to be sprayed.

With single-action airbrushes, air is released by depressing the finger lever, while the amount of color desired is controlled by rotating the rear needle adjusting screw. While working, it is not possible to change the amount of color being sprayed, because the operator must stop spraying to rotate the needle adjusting screw in the rear.

Airbrushes operate on a range of 5–50 psi (35–345 kPa), with a normal operating pressure of approximately 30 psi (207 kPa). Compact compressors are very popular with custom auto painters for use with airbrushes.

After use, airbrush spray guns must be completely disassembled for cleaning.

TABLE 24–5 TROUBLESHOOTING AN AIR SPRAY GUN

Trouble	Possible Cause	Suggested Correction
Spray pattern top heavy or bottom heavy	1. Horn holes partially plugged (external mix). 2. Fluid tip clogged, damaged, or not installed properly. 3. Dirt on air cap seat or fluid tip seat.	1. Remove air cap and clean. 2. Clean, replace, or reinstall fluid tip. 3. Remove and clean seat.
Spray pattern heavy to right or to left	1. Air cap dirty or orifice partially clogged. 2. Air cap damaged. 3. Paint nozzle clogged or damaged. 4. Too low a setting of the pattern control knob.	1. To determine where buildup occurs, rotate cap 180 degrees and test spray. If pattern shape stays in same position, the condition is caused by fluid buildup on fluid tip. If pattern changes with cap movement, the condition is in the air cap. Clean air cap, orifice, and fluid tip accordingly. 2. Replace air cap. 3. Clean or replace paint nozzle. 4. Adjust setting.
Spray pattern heavy at center	1. Atomizing pressure too low. 2. Fluid of too great viscosity. 3. Fluid pressure too high for air cap's normal capacity (pressure feed). 4. Caliber of paint nozzle enlarged due to wear. 5. Center hole enlarged.	1. Increase pressure. 2. Thin fluid with suitable thinner. 3. Reduce fluid pressure. 4. Replace paint nozzle. 5. Replace air cap and paint nozzle.
Spray pattern split	1. Not enough fluid. 2. Air cap or fluid tip dirty. 3. Air pressure too high. 4. Fluid viscosity too thin.	1. Reduce air pressure or increase fluid flow. 2. Remove and clean. 3. Lower air pressure. 4. Thicken fluid viscosity.
Pinholes	1. Gun too close to surface. 2. Fluid pressure too high. 3. Fluid too heavy.	1. Stroke 6 to 8 inches from surface. 2. Reduce pressure. 3. Thin fluid with reducer.
Blushing or a whitish coat	1. Absorption of moisture. 2. Too quick drying.	1. Avoid spraying in damp, humid, or too cool weather. 2. Correct by adding retarder. 3. Reduce booth temperature.
Orange peel (surface looks like orange peel)	1. Too high or too low an atomization pressure. 2. Gun too far or too close to work. 3. Fluid not thinned. 4. Improperly prepared surface. 5. Gun stroke too rapid. 6. Using wrong air cap. 7. Overspray striking a previously sprayed surface. 8. Fluid not thoroughly dissolved. 9. Drafts (synthetics and lacquers). 10. Humidity too low (synthetics).	1. Correct as needed. 2. Stroke 6 to 8 inches from surface. 3. Use proper reducing process. 4. Surface must be prepared. 5. Take deliberate, slow strokes. 6. Select correct air cap for the fluid and feed. 7. Select proper spraying procedure. 8. Mix fluid thoroughly. 9. Eliminate excessive drafts. 10. Raise humidity of room.
Excessive spray fog or overspray	1. Atomizing air pressure too high or fluid pressure too low. 2. Spraying past surface of the product. 3. Wrong air cap or fluid tip. 4. Gun stroked too far from surface. 5. Fluid thinned out too much.	1. Correct as needed. 2. Release trigger when gun passes target. 3. Ascertain and use correct combination. 4. Stroke 6 to 8 inches from surface. 5. Add correct amount of thinner.

TABLE 24–5 *(continued)*

No control over size of pattern	1. Air cap seal is damaged. 2. Foreign particles are lodged under the seal.	1. Check for damage; replace if necessary. 2. Make sure surface that this sets on is clean.
Sags or runs  	1. Dirty air cap and fluid tip. 2. Gun manipulated too close to surface. 3. Not releasing trigger at end of stroke (when stroke does not go beyond object). 4. Gun manipulated at wrong angle to surface. 5. Fluid piled on too heay. 6. Fluid thinned out too much. 7. Fluid pressure too high. 8. Operation too slow. 9. Improper atomization.	1. Clean cap and fluid tip. 2. Hold the gun 6 to 8 inches from surface. 3. Release trigger after every stroke. 4. Work gun at right angle to surface. 5. Learn to calculate depth of wet film of fluid. 6. Add correct amount of fluid by measure. 7. Reduce fluid pressure with fluid control knob. 8. Speed up movement of gun across surface. 9. Check air and fluid flow; clean cap and fluid tip.
Streaks	1. Dirty or damaged air cap and/or fluid tip. 2. Not overlapping strokes correctly or sufficiently. 3. Gun moved too quickly across surface. 4. Gun held at wrong angle to surface. 5. Gun held too far from surface. 6. Air pressure too high. 7. Split spray. 8. Pattern and fluid control not adjusted properly.	1. Same as for sags. 2. Follow previous stroke accurately. 3. Take deliberate, slow strokes. 4. Same as for sags. 5. Stroke 6 to 8 inches from surface. 6. Use least air pressure necessary. 7. Reduce air adjustment or change air cap and/or fluid tip. 8. Readjust.
Gun sputters constantly 	1. Connections, fittings, and seals loose or missing. 2. Leaky connection on fluid tube or fluid needle packing (suction gun). 3. Lack of sufficient fluid in container. 4. Tipping container at an acute angle. 5. Obstructed fluid passageway. 6. Fluid too heavy (suction feed). 7. Clogged air vent in canister top (suction feed). 8. Dirty or damaged coupling nut on canister top (suction feed). 9. Fluid pipe not tightened to pressure tank lid or pressure cup cover. 10. Strainer is clogged up. 11. Packing nut is loose. 12. Fluid tip is loose. 13. O-ring on tip is worn or dirty. 14. Fluid hose from paint tank loose. 15. Jam nut gasket installed improperly or jam nut loose.	1. Tighten and/or replace as per owner's manual. 2. Tighten connections; lubricate packing. 3. Refill container with fluid. 4. If container must be tipped, change position of fluid tube and keep container full of fluid. 5. Remove fluid tip, needle, and fluid tube and clean. 6. Thin fluid. 7. Clean. 8. Clean or replace. 9. Tighten; check for defective threads. 10. Clean strainer. 11. Make sure packing nut is tight. 12. Tighten fluid tip. Torque to manufacturer's specifications. 13. Replace O-ring if necessary. 14. Tighten. 15. Inspect and correctly install or tighten nut.
Uneven spray pattern	1. Damaged or clogged air cap. 2. Damaged or clogged fluid tip.	1. Inspect air cap and clean or replace. 2. Inspect fluid tip and clean or replace.

(continued)

TABLE 24–5 TROUBLESHOOTING AN AIR SPRAY GUN *(continued)*

Fluid leaks from spray gun **Nozzle drip**	1. Fluid needle packing not too tight. 2. Fluid needle packing dry. 3. Foreign particle blocking fluid tip. 4. Damaged fluid tip or fluid needle. 5. Wrong fluid needle size. 6. Broken fluid needle spring.	1. Loosen nut; lubricate packing. 2. Lubricate needle and packing frequently. 3. Remove tip and clean. 4. Replace both tip and needle. 5. Replace fluid needle with correct size for fluid tip being used. 6. Remove and replace.
Fluid leaks from packing nut **Packing nut leak**	1. Loose packing nut. 2. Packing is worn out. 3. Dry packing.	1. Tighten packing nut. 2. Replace packing. 3. Remove and soften packing with a few drops of light spray gun oil.
Fluid leaks through fluid tip when trigger is released	1. Foreign particles lodged in the fluid tip. 2. Fluid needle has paint stuck on it. 3. Fluid needle is damaged. 4. Fluid tip has been damaged. 5. Spring left off fluid needle.	1. Clean out tip and strain paint. 2. Remove all dried paint. 3. Check for damage; replace if necessary. 4. Check for nicks; replace if necessary. 5. Make sure spring is replaced on needle.
Excessive fluid	1. Not triggering the gun at each stroke. 2. Gun at wrong angle to surface. 3. Gun held too far from surface. 4. Wrong air cap or fluid tip. 5. Depositing fluid film of irregular thickness. 6. Air pressure too high. 7. Fluid pressure too high. 8. Fluid control knob not adjusted properly.	1. It should be a habit to release trigger after every stroke. 2. Hold gun at right angles to surface. 3. Stroke 6 to 8 inches from surface. 4. Use correct combination. 5. Learn to calculate depth of wet film of finish. 6. Use least amount of air necessary. 7. Reduce pressure. 8. Readjust.
Fluid will not come from spray gun	1. Out of fluid. 2. Grit, dirt, paint skin, and so on, blocking air gap, fluid tip, fluid needle, or strainer. 3. No air supply. 4. Internal mix cap using suction feed.	1. Add more spray fluid. 2. Clean spray gun thoroughly and strain spray fluid; always strain fluid before using it. 3. Check regulator. 4. Change cap or feed.
Fluid will not come from fluid tank or canister	1. Lack of proper air pressure in fluid tank or canister. 2. Air intake opening inside fluid tank or canister clogged by driedup finish fluid. 3. Leaking gasket on fluid tank cover or canister top. 4. Gun not converted correctly between canister and fluid tank. 5. Blocked fluid hose. 6. Connections with regulator not correct.	1. Check for air leaks or leak of air entry; adjust air pressure for sufficient flow. 2. This is a common problem; clean opening periodically. 3. Replace with new gasket. 4. Correct per owner's manual. 5. Clear. 6. Correct per owner's manual.
Sprayed coat short of liquid material	1. Air pressure too high. 2. Fluid not reduced or thinned correctly (suction feed only). 3. Gun too far from work or out of adjustment.	1. Decrease air pressure. 2. Reduce or thin according to directions; use proper thinner or reducer. 3. Adjust distance to work; clean and adjust gun fluid and spray pattern controls.
Spotty, uneven pattern, slow to build	1. Inadequate fluid flow. 2. Low atomization air pressure (suction feed only). 3. Gun motion too fast.	1. Back fluid control knob to first thread. 2. Increase air pressure, rebalance gun. 3. Move at moderate pace.
Unable to get round spray	4. Pattern control knob not seating properly.	5. Clean or replace.
Dripping from fluid tip	1. Dry packing. 2. Sluggish needle. 3. Tight packing nut. 4. Spray head misaligned, causing needle to bind.	1. Lubricate packing. 2. Lubricate. 3. Adjust. 4. Tap all around spray head with wood and rawhide mallet and retighten locking bolt.

TABLE 24–5 *(continued)*

Excessive overspray	1. Too much atomization air pressure. 2. Gun too far from surface. 3. Improper stroking (for example, arcing, moving too fast).	1. Reduce. 2. Check distance. 3. Move at moderate pace, parallel to work surface.
Excessive fog	1. Too much or quick drying thinner. 2. Too much atomization air pressure.	1. Remix. 2. Reduce.
Will not spray on pressure feed	1. Control knob on canister cover not open. 2. Canister is not sealing. 3. Spray fluid has not been strained. 4. Spray fluid in canister top threads. 5. Gasket in canister top worn or left out. 6. No air supply. 7. Fluid too thick. 8. Clogged strainer.	1. Set this knob for pressure spraying. 2. Make sure canister is on tightly. 3. Always strain before using. 4. Clean threads and wipe with grease. 5. Inspect and replace if necessary. 6. Check regulator. 7. Thin fluid with proper thinner. 8. Clean or replace strainer.
Will not spray on suction feed	1. Spray fluid is too thick. 2. Internal mix nozzle used. 3. Spray fluid has not been strained. 4. Hole in canister cover clogged. 5. Gasket in canister top worn or left out. 6. Plugged or clogged strainer. 7. Fluid control knob adjusted incorrectly. 8. No air supply.	1. Thin fluid with thinner. 2. Install external mix nozzle. 3. Always strain before use. 4. Make sure this hole is open. 5. Inspect and replace if necessary. 6. Clean or replace strainer. 7. Correct adjustment. 8. Check regulator.
Air continues to flow through gun when trigger has been released (on nonbleeder guns only)	1. Air valve leaks. 2. Needle is binding. 3. Piston is sticking. 4. Packing nut too tight. 5. Control valve spring left out.	1. Remove valve, inspect for damage, clean valve, and replace if necessary. 2. Clean or straighten needle. 3. Clean piston, check O-ring, and replace if necessary. 4. Adjust packing nuts. 5. Make sure to replace this spring.
Air leak at canister gasket	1. Canister not sealing on canister cover.	1. Check gasket, clean threads, and tighten canister.
Leak at setscrew in canister top	1. Screw not tight. 2. Damaged threads on setscrew.	1. Clean threads and tighten screw. 2. Inspect and replace if necessary.
Leak between top of canister cover and gun body	1. Retainer nut is not light enough. 2. Gasket or gasket seat damaged.	1. Check nut to make sure it is tight. 2. Inspect, clean, and replace if necessary.
	Pressure Fluid Cup or Tank Problems (HVLP Included)	
Leaks air at the top of the tank lid	1. Gasket not seating properly or damaged. 2. Wing screws not tight enough. 3. Fittings leak. 4. Air pressure too high.	1. Drain off all of the air from fluid tank, thus allowing the gasket to seat. Retighten wing nut and fill with air again. Lid will seat tightly. 2. Make sure all wing screws are tight. By following remedy #1 (above), wing screws can be pulled down even tighter. 3. Check all fittings and apply pipe dope if necessary. 4. Maximum 60 psi. Normal w.p. 25–30 psi.
No fluid comes through the spray gun	1. Not enough pressure in tank. 2. Out of fluid. 3. Fluid passages clogged.	1. Increase regulator setting until fluid flows; do not exceed 60 psi. 2. Check fluid supply. 3. Check tube, fittings, hose, and spray gun. Clean out fittings, hose, tube, and spray gun, making sure all residual fluid is removed.

TABLE 24–6 TROUBLESHOOTING A COMPRESSED AIR SUPPLY

Fault	Result	Blistering	Nondrying	Poor Adhesion	Contamination	Poor Atomization	Poor Flow	Overloading	Sags	Popping	Slow Application	Off-shade Metallic	Uneven Application	Dry Spray	Dirt	Remedy
Oil or water not adequately condensed out.	Oil/water at spray gun	A	C	A	C											Ensure regular drainage of air receiver, separator, and transformer. Site transformers of adequate capacity in cool places. Lubricate compressors with recommended grade of mineral oil of good emulsifying properties.
Long air line; inadequate internal bore of air line; connectors, fittings, compressor, air transformers, and regulators of inadequate capacity.	Pressure drop					B	C	C	A	A	C	A				Ensure adequate air supply with 30 feet of 5/16-inch (8 mm) internal bore air line with appropriate fittings. NOTE: Reduction of viscosity to give improvement may produce other defects.
Inadequate compressor capacity. No pressure regulator. Regulator diaphragm broken.	Pressure fluctuation							A	A	A		A	A	A		Increase capacity. Use pressure regulator. Replace regulator diaphragm.
Compressed air intake filter breached. Transformer filter not properly maintained. Compressor sited in dusty area.	Dirt in compressed air														A	Repair air intake filter. Replace transformer filter. Clean dust and dirt from compressor site.

A Most likely failure to be associated with the fault

B Likely failure

C Failure less likely to be associated with the fault

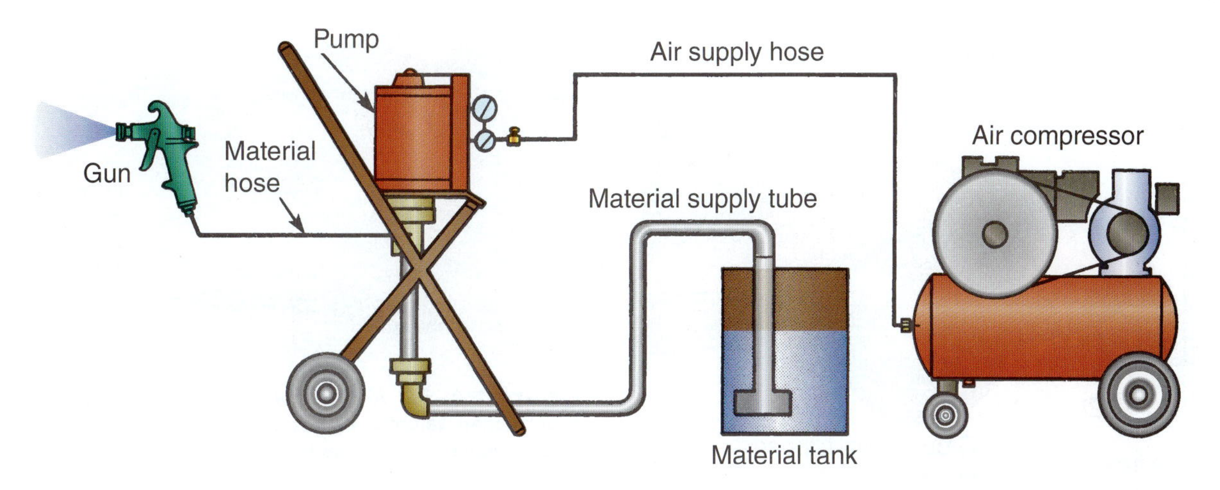

Figure 24-47 An airless spray equipment setup is fairly simple.

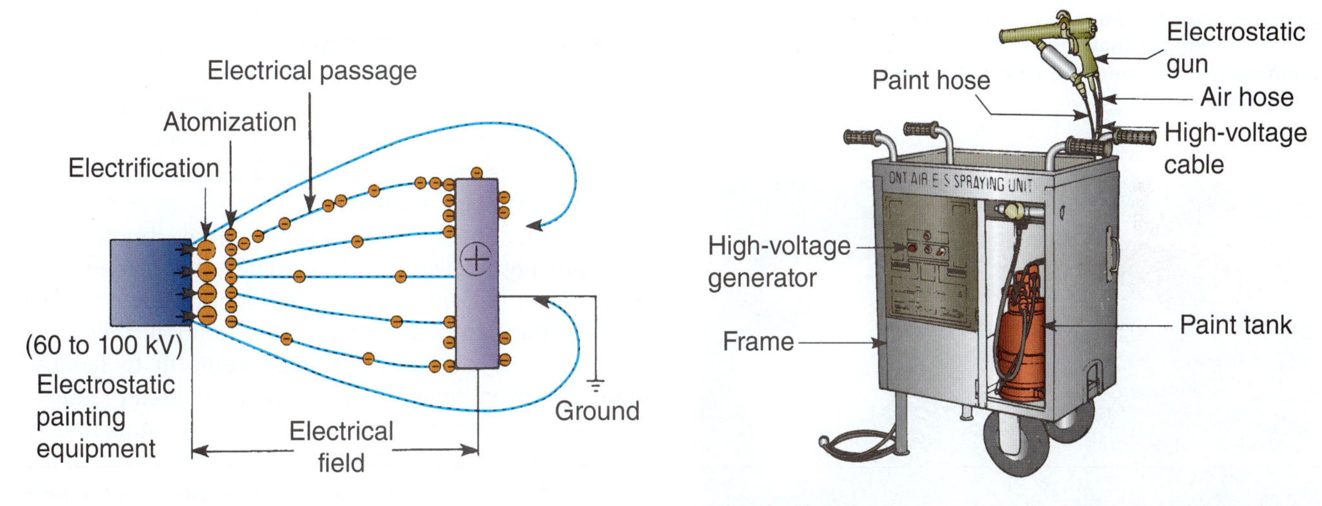

Figure 24-48 Note the principles of electrostatic painting.

Figure 24-49 Note the parts of air-type electrostatic painting equipment.

TABLE 24–7 COMPARISON OF VARIOUS SPRAYING SYSTEMS

	Conventional Air Spraying	Conventional Airless Spraying	Air-Assisted Airless Spraying	Air-Assisted Electrostatic Spraying	Airless-Type Electrostatic Spraying	HVLP Spraying
Adhesion of spray efficiency	20 to 40%	50 to 60%	40 to 60%	60 to 70%	70 to 80%	65 to 90%
Quality of finished surface	Excellent	Poor	Fair to good	Good	Fair	Good to excellent
Work environment (paint mist dispersion)	Poor	Good	Good	Excellent	Excellent	Excellent
Paint speed	Slow	Fast	Fast	Very fast	Very fast	Slow
Paint of depressed areas	Excellent	Poor	Fair	Good	Fair	Excellent
Gun handling (partial repainting and touch-up)	Excellent	Fair	Fair	Fair	Good	Excellent

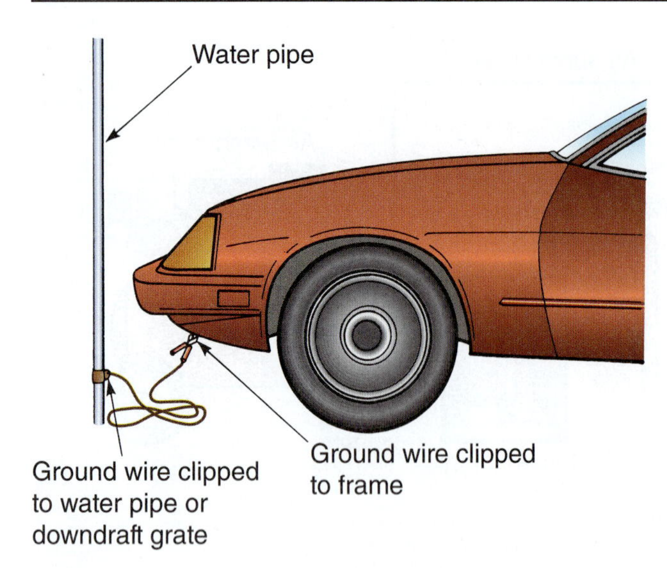

Figure 24–50 For safety reasons and to avoid static electrical charges that can attract dirt, you should use a jumper wire to ground a vehicle in the spray booth before spraying.

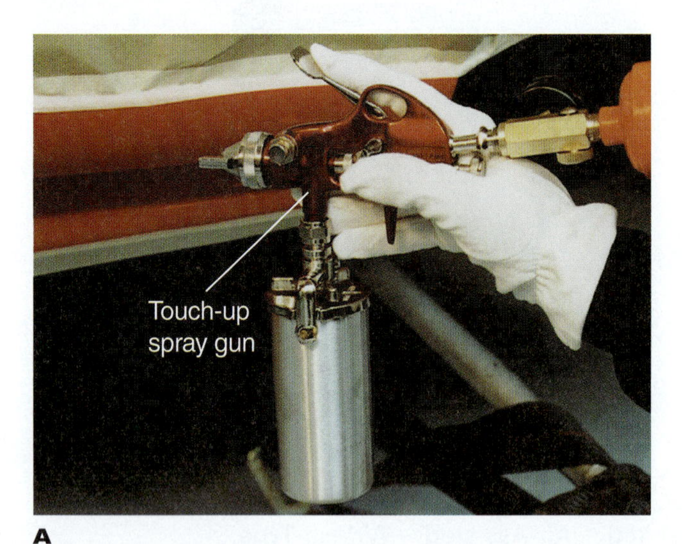

A

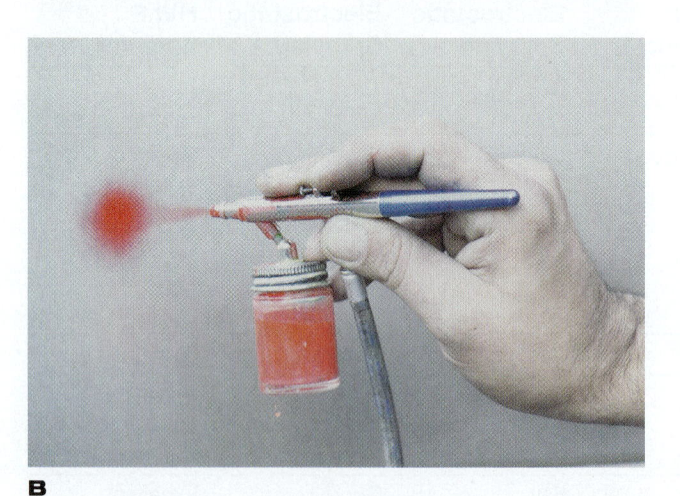

B

Figure 24–51 (A) A touch-up gun is sometimes used to spray very small areas with colorcoat. (B) An airbrush is used to do custom paint work.

Figure 24–52 Airbrushes are commonly used by experienced, talented painters to do custom finishes. This vehicle's painter was also an artist!

24.8 SPRAY BOOTHS

A **spray booth** is designed to provide a clean, safe, well-lit, and well-ventilated enclosure for painting vehicles. A spray booth isolates the painting operation from dirt- and dust-producing activities and confines and exhausts the volatile fumes created by spraying automotive finishes. Please refer to Figure 24–54.

Modern one- and two-room spray booths are scientifically designed to create the proper air movement, provide necessary lighting, and safely enclose the painting operation. In addition, their construction and performance must conform to federal, state, and even local safety codes, not to mention those of insurance underwriters. In most areas, automatically operated fire extinguishers are required because of the highly explosive nature of the refinishing materials (Figure 24–55).

Cleaning preparation should be done outside the booth area. Steam clean the underbody of a vehicle thoroughly and air dust the entire vehicle before moving it into the spray booth. After the vehicle is in the booth, close the booth doors tightly and wipe the entire vehicle again with a tack cloth before proceeding with the painting operation.

All spray booth doors must be kept tightly closed during painting. If it becomes necessary to open a door, be sure the air supply is turned off. In fact, many spray booths are equipped with door switches that shut off the air supply when the doors are opened. The air compressor should be outside the booth with the air delivery pipes slanting back toward the compressor.

An *air makeup* or *air replacement system* is important because of the large volume of air exhausted from a spray booth. During the winter, the spray area can become cold and uncomfortable. Finish problems can arise from spraying cold materials on cold vehicles in cold air. An air makeup system will provide even temperatures and clean filtered air as well as ensure proper booth performance.

Sometimes paint shops employ an independent air replacement system specifically designed for the spray

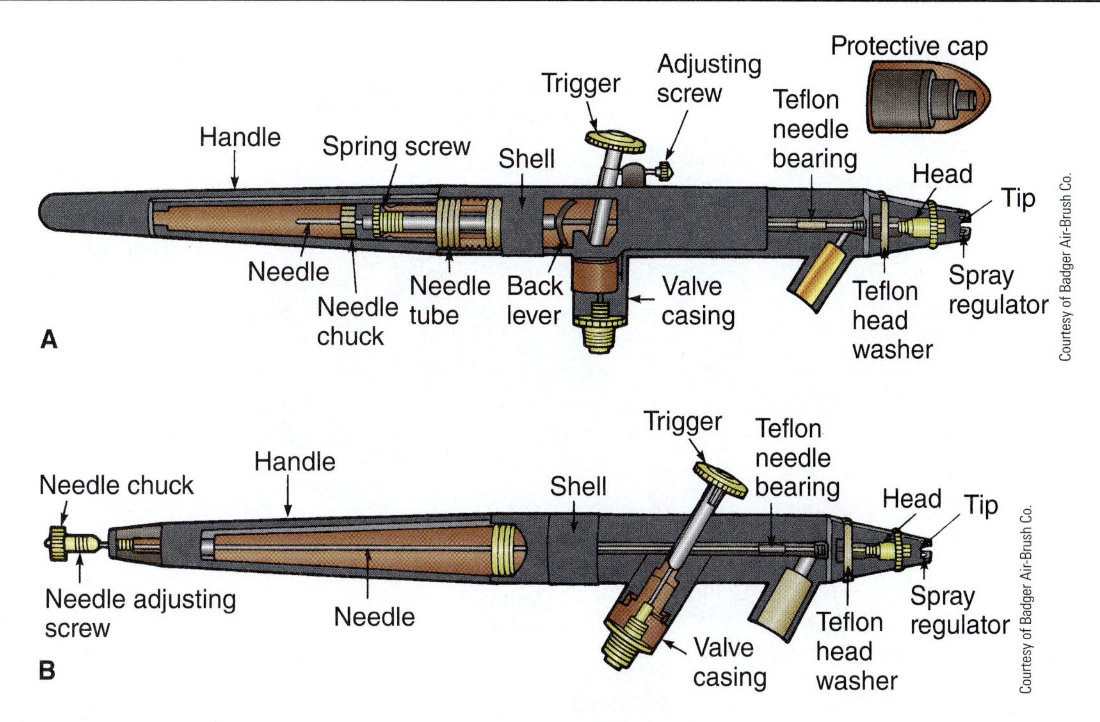

Figure 24-53 Compare a (A) double-action and a (B) single-action airbrush.

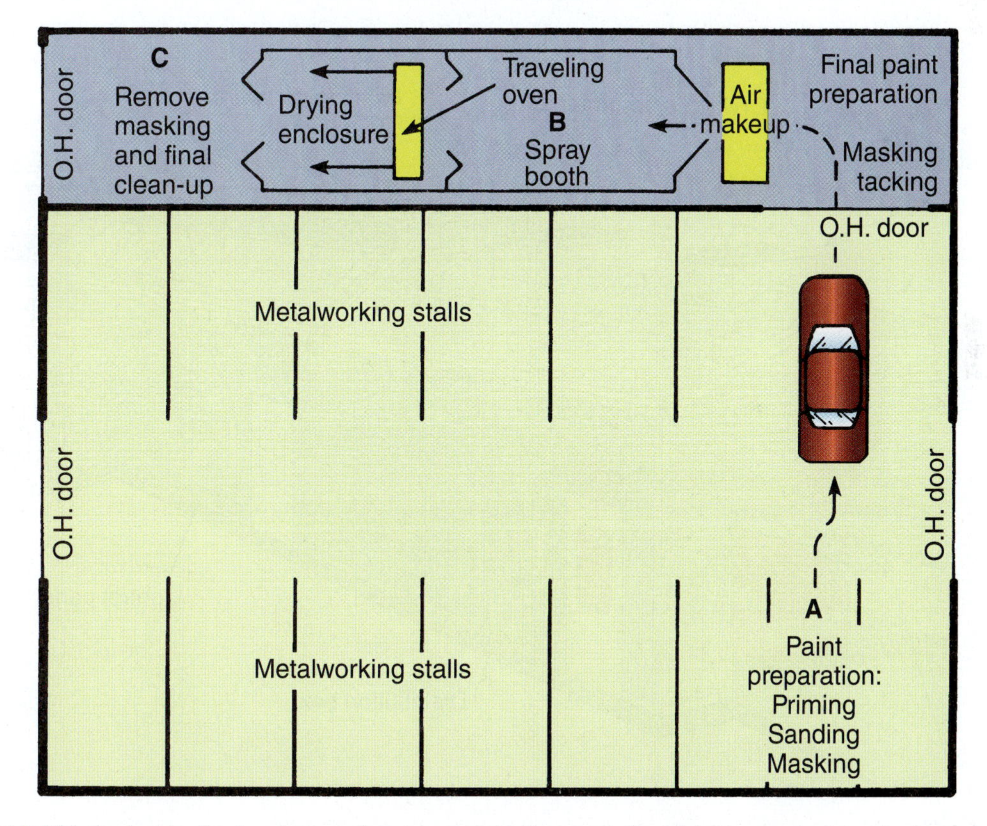

Figure 24-54 This typical body shop layout shows a straight-line work flow finishing operation. The important steps in such an arrangement are (A) paint preparation, (B) spray booth, and (C) final clean-up.

booth. This provides clean, dry, filtered air from the outside to the booth, heating the air in colder weather. Replacement air can be delivered to the general shop area or directly into the booth for a completely closed system.

There are four air makeup systems in use today:

1. Regular flow booth
2. Reverse flow booth
3. Crossdraft booth
4. Downdraft booth

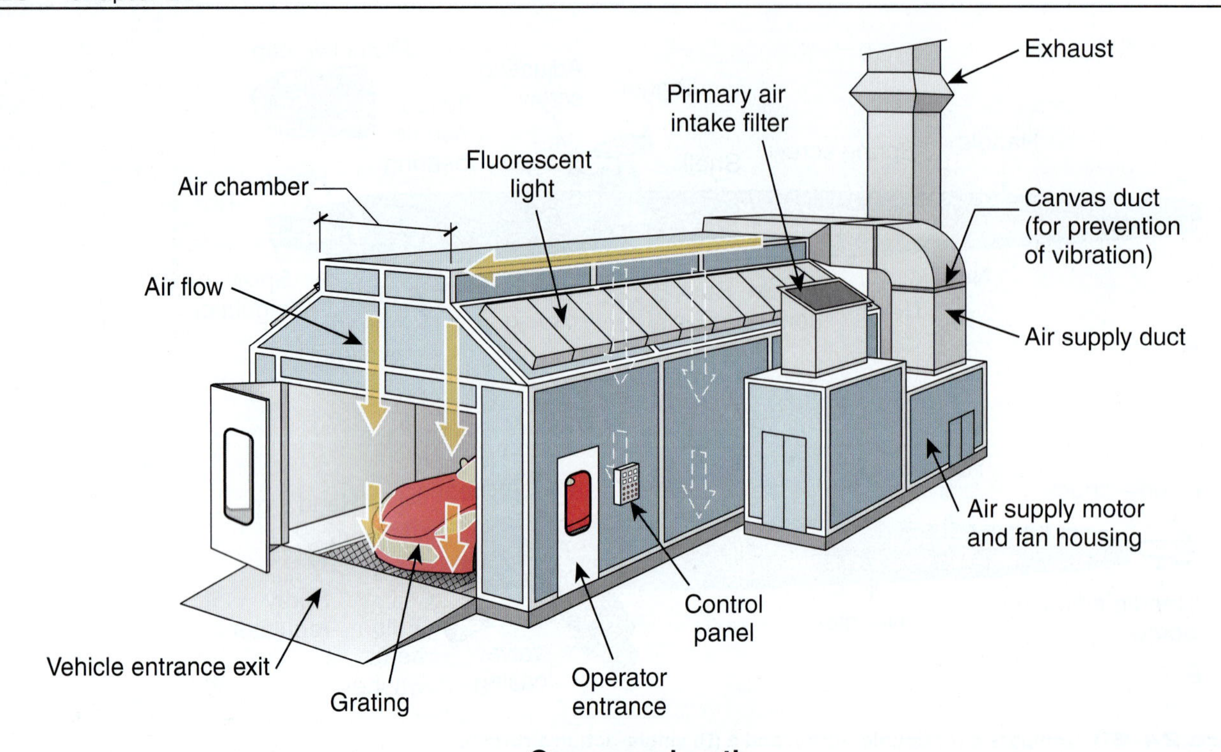

One-room booth

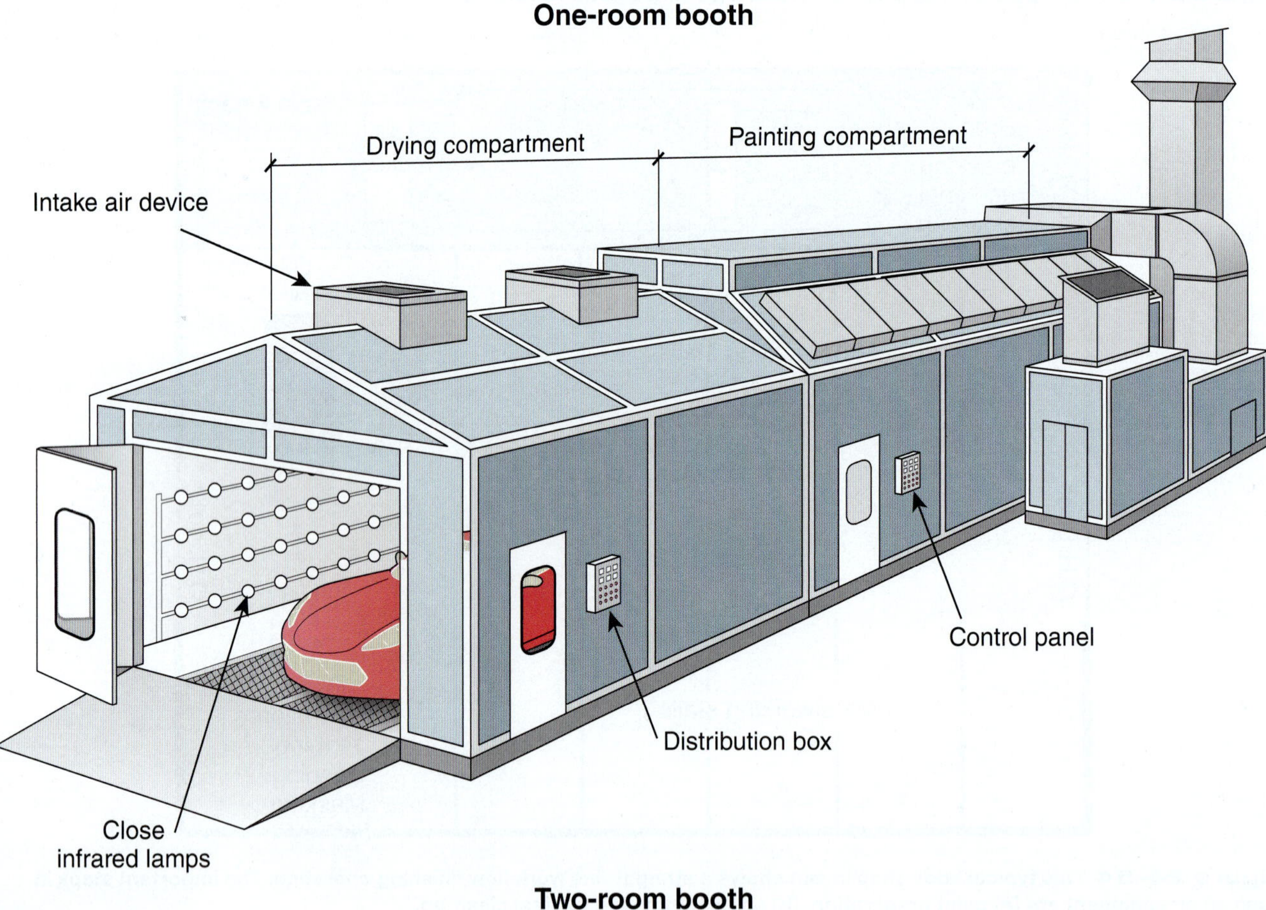

Two-room booth

Figure 24-55 Compare one- and two-room spray booths.

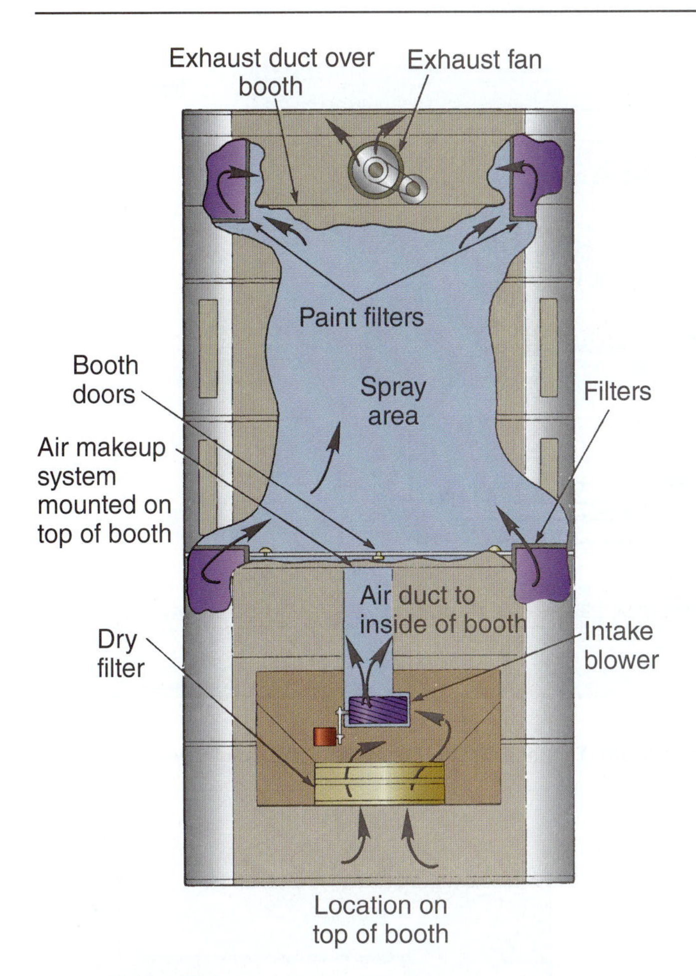

Figure 24–56 This is a typical makeup air system for a spray booth.

Both the regular and reverse flow types of booths were once considered standards in spray booth construction. However, since the late 1970s they have been replaced to a great degree by the downdraft airflow booth.

As shown in Figure 24–56, in the regular system the airflow is from back to front. In the reverse flow process, the airflow is from front to back. The reverse flow type of booth generally has a solid back, whereas a regular flow booth usually is of the drive-through style. It is interesting to note that a good number of vehicles that were sprayed in a reverse flow booth were backed in (Figure 24–57).

The *downdraft spray booth* forces air from the ceiling down through exhaust vents in the floor. It is the most popular air movement system used today. The downward flow of air from the ceiling to the floor pit creates an envelope of air passing by the surface of the vehicle. Look at Figure 24–58.

Downdraft booths are available in raised platform models and floor models and are usually of the drive-through type. Refer to Figure 24–59.

A *side draft* or *crossflow spray booth* moves air sideways over the vehicle. An air inlet in one wall pushes fresh air into the booth. A vent on the opposite wall removes booth air.

Air Filtration Systems

The most important safety feature of spray booths is the filtration system. Currently there are two common types in use: the wet filtration system and the dry filtration system (Figure 24–60).

A wet, or wash, filtration system has a higher initial cost than a dry filter system. There can also be an additional cost associated with waste disposal. However, wet filtration does an excellent job of removing paint particles from exhaust air regardless of the paint viscosity or drying speed. The air that emerges is as clean or cleaner than that achieved by a quality dry filtration system (Figure 24–61).

Dry filtration systems come in various configurations and use different filter media (paper, cotton, fiberglass, and polyester). These systems mechanically filter out particles of paint and dirt by trapping the particles as air flows through the filter. Some are also coated with a tacky substance so particles will readily adhere to the surface of the fibers.

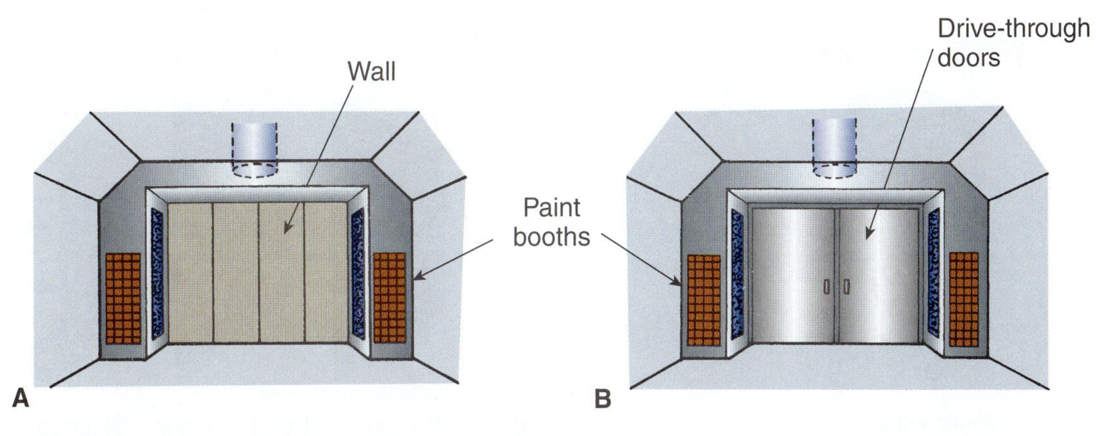

Figure 24–57 Note the two designs of spray booths: (A) solid back and (B) drive-through. The solid design is generally found in small shops.

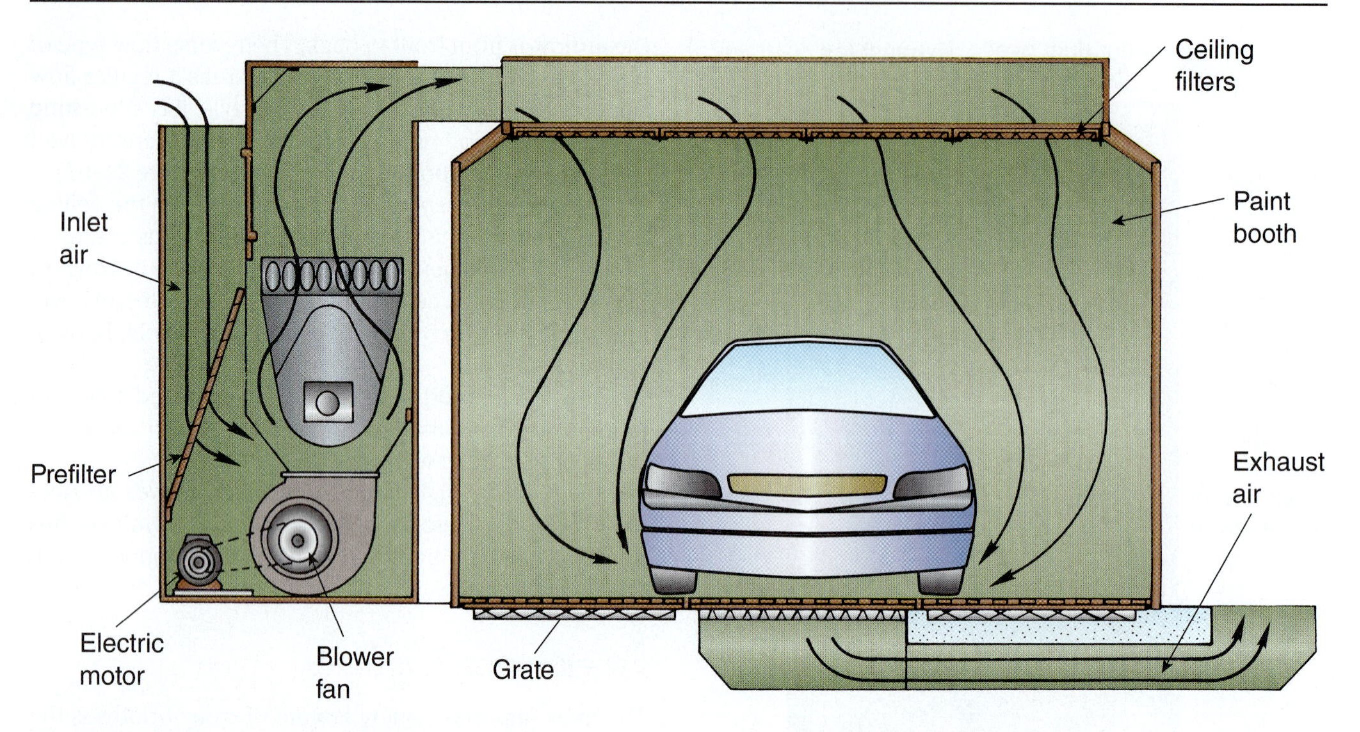

Figure 24-58 Follow the airflow pattern of a downdraft spray booth. The location of the intake and exhaust system will depend on the system manufacturer.

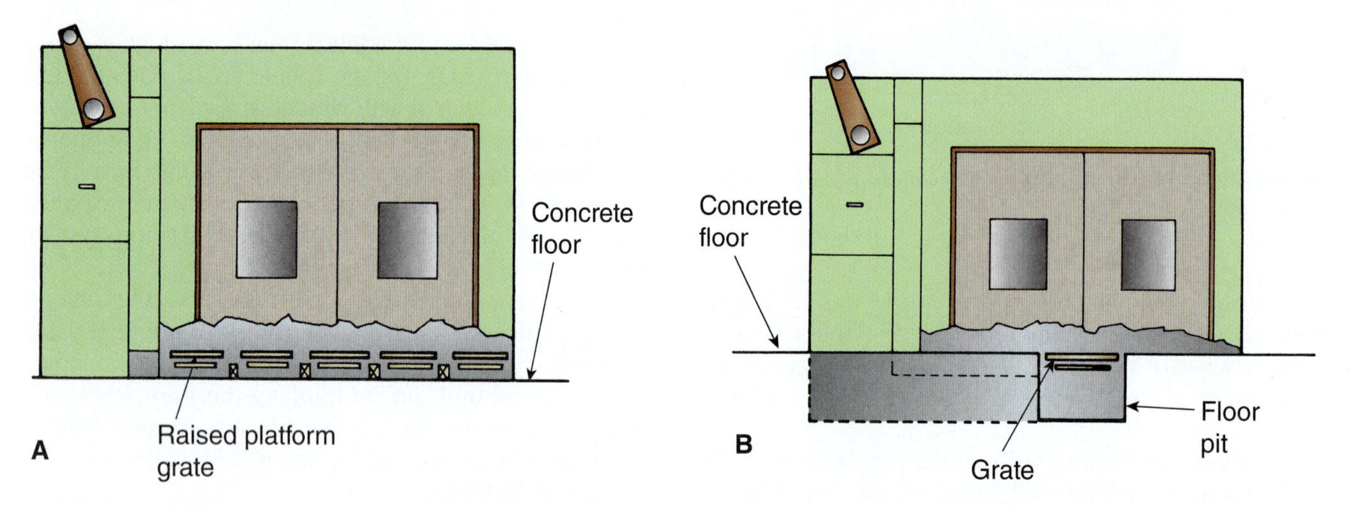

Figure 24-59 Note the two models of downdraft booths: (A) the raised platform model and (B) the floor model with underfloor pit.

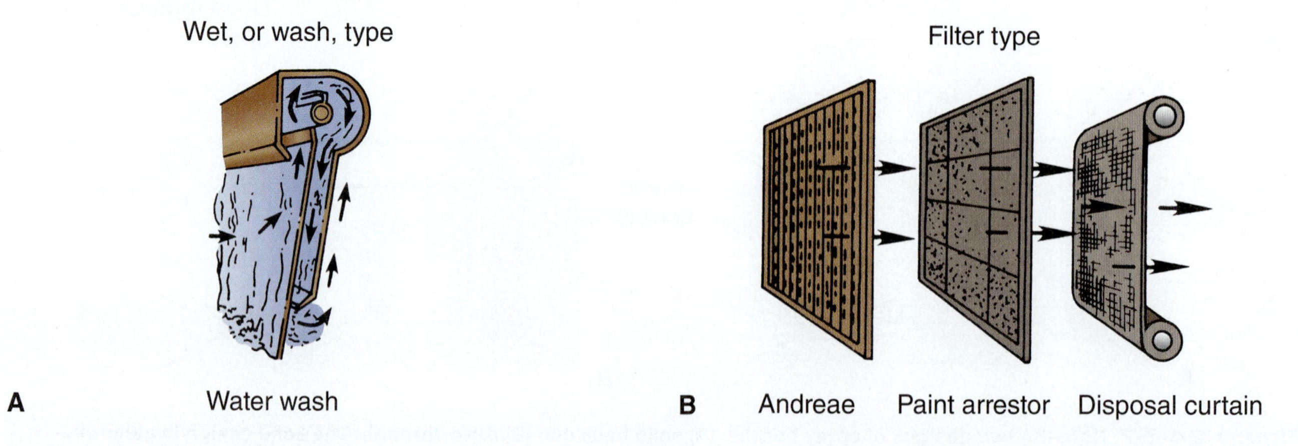

Figure 24-60 Here are different methods of spray booth filtration: (A) wet filtration and (B) dry filtration.

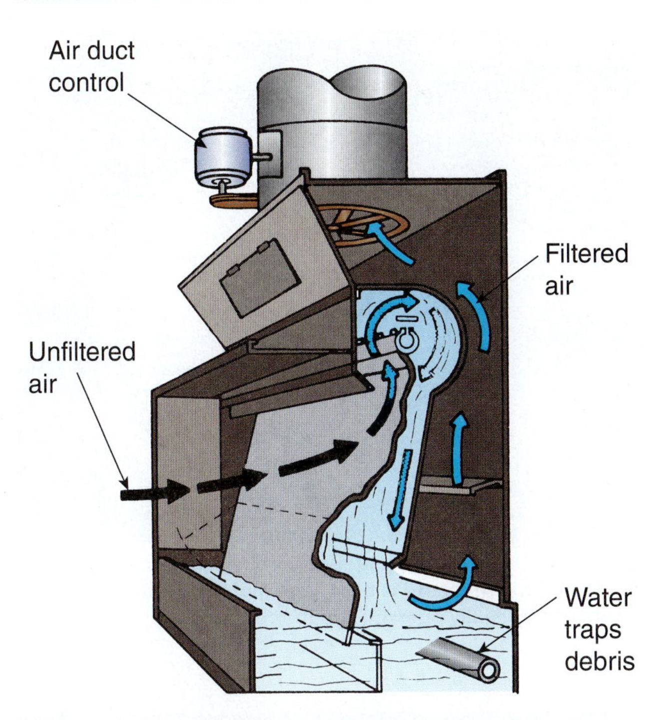

Figure 24-61 A wet filtration system must remove contaminants from air circulating through the spray booth. Note how the airflow moves over water to help trap bits of dirt and dust.

The best way to judge a filter's condition is to measure its air resistance with a water column pressure differential gauge (manometer gauge). Some booths have built-in gauges, whereas others do not. Comparing the air pressure upstream of the filter to that which is downstream is a good indication of whether it is time to replace filters. The amount of restriction that is considered acceptable will vary according to filter construction and media, spray booth construction, and air volume.

24.9 SPRAY BOOTH MAINTENANCE

Regardless of the type of filtration system that a paint shop employs, spray booth maintenance is a prime consideration, not only from the standpoint of cost and convenience but also because it is essential to achieving quality paint jobs. The best air filtration system in the world will not be able to do its job if it is poorly maintained.

The first task in learning how to avoid dirt is to understand where it comes from. Anything that is brought into the booth can bring in dirt with it. Potential sources of dirt include the air, the vehicle, the painter, the equipment and supplies, and even the paint.

Incoming air is a prime source of dirt. Dirt is generated by dirty filters, imbalanced air pressures, and open doors. Check the intake filters daily and change them as soon as the manometer indicates. When dust and dirt start to clog filters and restrict airflow, the velocity of air passing through the filters begins to climb. Increased velocity increases the likelihood of pulling dirt through the filters. Balance the input air pressure against the exhaust air to provide slightly positive pressure in the booth. This balance can change as filters load up, and it also differs from car to car. Therefore, check and adjust it with each new job.

Enter the booth only when the fans are running. The positive pressure helps keep the dirt out. Once a vehicle is inside, keep traffic flow in and out of the booth to an absolute minimum. Also, make sure that the body shop doors are closed at all times during a painting operation. Opening and closing these doors can cause the booth balance to fluctuate, creating turbulence and dirt inside the booth (Figure 24–62).

A *booth manometer* (Figure 24–63) or a *booth pressure gauge* (Figure 24–64) indicates how much airflow is going through a paint booth. High booth pressure indicates an airflow restriction from clogged exhaust filters. Low booth pressure indicates a booth air leak, as when the booth doors are not fully closed. For example, when you open a booth door, the booth pressure gauge will drop to zero.

The booth itself can be a main contributor to dirt problems due to air leaks, poor housekeeping habits, exhaust air, and floor coverings. There are recommended seals for door frames, light openings, and panel seams that must be properly installed and periodically replaced. Heavy usage and temperature extremes quickly destroy these seals. Use caulking as an inexpensive gap sealant to keep dirt out of the air stream.

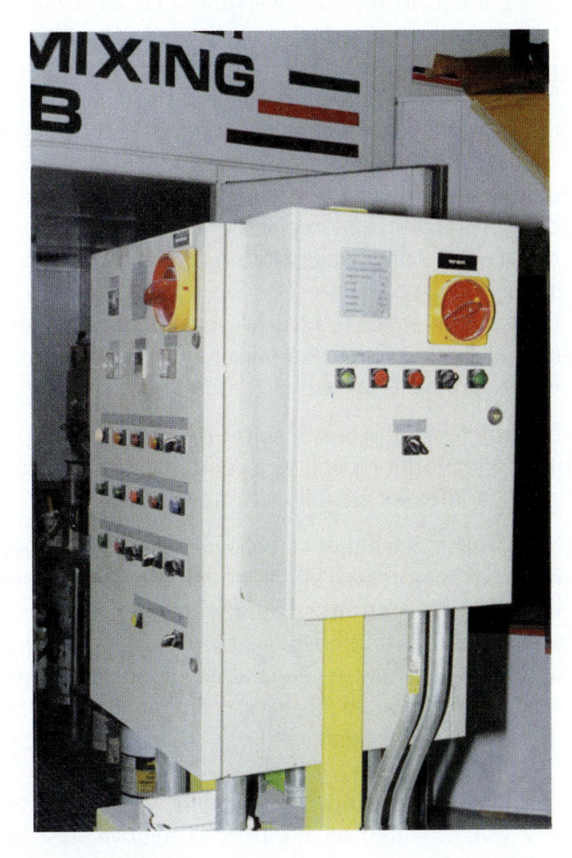

Figure 24-62 A paint booth control station will have knobs and indicators for airflow, lights, and temperature.

Figure 24-63 A paint spray booth should be like a clean room in a hospital. It provides clean, filtered air so that no airborne dirt or dust can get into the fresh paint.

A The booth pressure gauge serves the same function as a manometer. If pressure is high, you might have clogged filters or a partially open booth door.

When operating the spray booth, keep the following points in mind:

- Follow the manufacturer's recommendations for the minimum velocity needed to properly exhaust spray vapors. If that recommendation is exceeded, turbulence cancels out the screening performed by the filters. If the velocity is too low, the air will not move fast enough to remove overspray and airborne dirt before they cause defects.
- Paint arresters are a high-consumption item requiring frequent changing. Check filter resistance daily on the manometer. When paint accumulation builds up, velocity goes down, and air movement is too slow.
- In a dry filtration system, the filters must be periodically inspected and replaced. And when they are replaced, the multistage filters designed for the booth should be used (Figure 24–65).
- Be sure the water level in a wet filtration system is kept at its proper working level and that the correct water additive is used.

In order to get the best results from any type of spray booth, it is important to follow a good housekeeping program that addresses the following items:

- Maintain air line filters. Unbelievable as it might seem, dirt from compressed air lines often causes blemishes in paint jobs. Air transformers, with properly cleaned and regularly drained filters, keep booth air clean and dry. Oil and water separators are absolutely necessary to eliminate dirt and contamination. Drain them of water and replace them periodically (Figure 24–66).
- Periodically wash down the booth walls, floor, and any wall-mounted air controls to remove dust and paint particles. Many shops require that floors and walls be wiped down after every job (Figure 24–67).

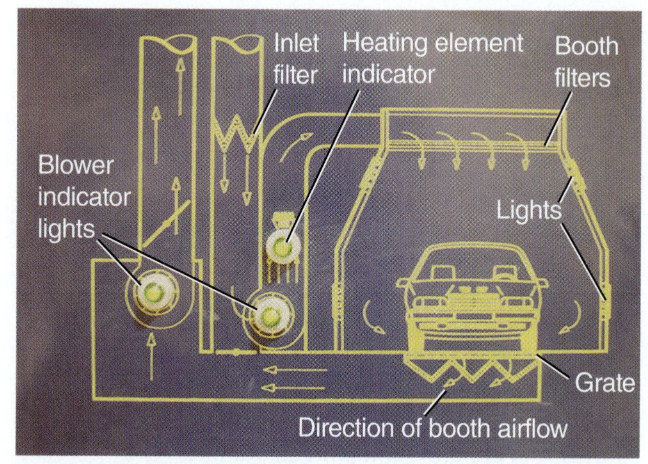

B Indicator lights on a diagram of a booth control station show whether the blowers and heating element are working.

C To keep dirt and dust out of a paint booth, always keep the doors completely shut.

Figure 24-64 This booth control station has gauges, meters, and adjustment knobs for airflow and baking.

Figure 24–65 Always change booth air filters when restricted. Good airflow is needed through the booth to keep dirt and dust out of paint jobs.

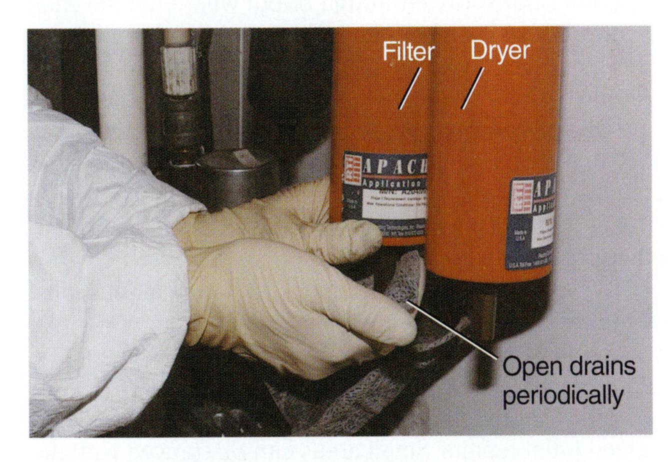

Figure 24–66 If they do not have automatic drain valves, open the filter and dryer drains to remove condensed moisture. These are for the fresh-air supply to the respirator.

Figure 24–67 Paint spray booths must be perfectly clean to produce consistently flaw-free paint work.

- Pick up all scraps, rags, and so forth. Blow dust out from around the walls and seals on the bottom of the booth door frame (Figure 24–68).

Figure 24–68 Here a technician is using an air blow nozzle to remove dust and dirt from under the door frame in a spray booth. Booths should be swept and blown out periodically.

- A spray booth is no place to store parts, paint, trash cans, or workbenches because dirt will accumulate on these things and will eventually land on the vehicle. Keep these items in a sealed, ventilated storage area.
- Be sure that all bodywork and most paint preparation procedures are done outside of the spray booth. Make certain no sanding or grinding operations are performed in or near the spray booth. The dust created will spread all over and ruin not only a present job but many future jobs.
- Water is most often used to contain dirt. It is relatively inexpensive and very effective at trapping dirt, but it can splash on a vehicle midway through the job or, in a heated booth, dry out before the paint job is finished. If water is sprayed on the floor to keep any stray dust down, eliminate all puddles to prevent splashes. Water can also rust the walls of a spray booth, resulting in premature deterioration.
- Spray guns, masking paper, paint cans, tape, wheel covers, air transformers, hoses, respirators, coveralls, tack rags, and various other supplies can all collect dirt if stored in a dirty environment. All of these items should

be kept in a filtered, ventilated storage/mix room. If subject to sanding dust, they will quickly ruin a paint finish. Avoid using dirt-collecting cloth wheel covers. It is better to back mask the wheel wells with masking paper or cover them with plastic film wheel covers.

- Oil the fan pulley and motor bearings of the spray booth regularly, if required. Always switch off the main fan power supply before oiling the fan. If the spray booth is not properly maintained, it can cause finish problems.

The vehicle itself is often the greatest source of dirt in the spray booth. Dirt hides in cracks and crevices, behind bumpers, and in the engine compartment. Even a thoroughly cleaned vehicle collects dirt when left in the general sanding area before being brought into the booth. When the spray gun hits this dirt, it kicks it out of its hiding places and deposits it into the finish. That is why a good prep job is so important.

Cotton clothing is perhaps the greatest source of contamination and should never be worn in a spray booth. Lint-free paint suits, rubber form-fitting gloves, a dirt-free head cover, and the appropriate respirator should be worn inside the booth. Remain in the booth between the application of coats rather than risk dragging dirt back inside.

Technicians not wearing the proper attire should not enter a paint booth. They should view the work through a booth window rather than risk contaminating the paint.

Table 24–8 summarizes how to troubleshoot paint booth problems.

24.10 DRYING ROOM

A dust-free drying room will speed up drying, turn out a cleaner job, and increase the volume of refinishing work that can be handled. The **drying rooms** of more sophisticated paint shops have permanent infrared or sodium quartz units for the forced drying of paint, particularly enamels. These oven-like units can speed up the drying time of enamels by as much as 75 percent (Figure 24–69).

The use of forced drying on putty, primer, and sealer coats will reduce waiting time between operations and can also be used for fast drying spot and panel finish coats.

Infrared or sodium quartz drying equipment is available as portable panels for partial or sectional drying. Drying equipment also takes the form of large traveling ovens capable of moving automatically on a track over the vehicle to dry an overall job (Figure 24–70).

There are two types of infrared drying equipment:

1. *Near drying equipment.* Because drying equipment uses lamps as the heat source, this type of equipment is easy to handle. The radiation angle can be easily varied. The construction, relocation, and assembly are simple, so it is the most common type used for automobiles. There are several shapes and sizes of this equipment, depending on what it is used for.
2. *Far drying equipment.* Far drying or sodium quartz equipment affects paint drying by means of heat radiated from a tubular or plate-type heater. The heat source is either gas or electricity. Far drying equipment also comes in various types and sizes, depending on its intended use.

Drying can best be accomplished in a separate drying chamber attached to the back of a downdraft system or conventional drive-through booth where the traveling oven is housed and operated. This configuration achieves the highest production because both the painting and drying operations can be performed simultaneously.

Drying can also be performed directly in the spray booth after painting. A storage vestibule is used to store the traveling oven until it is needed. After the vehicle is painted, the oven is rolled out of the vestibule and into the spray booth for the drying operation.

When using a drying room, certain precautions must be taken not to destroy the finish.

Table 24–9 gives the common difficulties that can be caused in the drying room.

Many shops are now using a UV (ultraviolet) primer to speed paint repairs. Small areas can be sprayed with the UV primer and cured in a couple of minutes by exposure to a UV heat lamp. The primed area can then be painted after very little drying time.

24.11 AIR-SUPPLIED RESPIRATORS

An *air-supplied respirator* circulates fresh air over the painter's face. It is the most common type of respirator used by painters and refinish technicians. When fresh air is forced over the painter's nose and mouth, harmful paint contaminants cannot pass around or through the respirator.

There are two common types of air-supplied respirators: the hood type and the face shield type. Either type must be worn when spraying catalyzed paint materials.

A *full hood fresh air respirator* covers a painter's head and face. A fresh air hose connects to the hood. One is shown in Figure 24–71.

A *face shield fresh air respirator* uses a clear plastic shield with a small air nozzle to direct breathing air over the painter's face. As with the hood type, an extra fresh air supply hose is attached to a small hose going to the air nozzle. See Figure 24–72.

Note how the fresh-air respirator connects to shop air line pressure. A carbon filter is needed before the

TABLE 24–8 TROUBLESHOOTING SPRAY BOOTH PROBLEMS

Fault		Result	Dirty Job	Thin Coats	Poor Opacity	Sags	Overloading	Popping	Softness	Overspray	Uneven Application	Recoat Failure	Fire Hazard	Water Splashes
Dirty filters		Vacuum in booth (hot air drawn from oven)	C				B	A	A	C	C,D	A		
		OR												
		Not pressurized (low air movement and dirty air drawn in from preparation Area)	A	A*	C	B					C,D			
Breached or damaged filter		Turbulence								B	B,D			
		Overpressurized				A†	A	A	A**	B	C,D	A		
Water level	**Low**	Increased extraction					A	A	A		C,D	A		A‡
	High	Restricted extraction		A*	C	B				C	C,D			A‡
	Empty	Increased extraction with buildup of dry paint in reservoir	A										A	
Use of incorrect water additive or incorrect use of water additive.		Blocked water jets and filters. Formation of dry powder on antisplash panels.	A										C	A
		E Bacteria/germ cultivation (unpleasant smell)												
		Corrosion of paint.											A	
		Paint deposits difficult to remove.												
Rags, masking paper, old cans, and so on in booth.		Dirt accumulation	A										A	
Spraying on walls of booth.		Poor light reflection									C,D			
Loose deposits of dirt, dry spray, rust, and so forth on booth walls.		Dirt in atmosphere	A											

A Most likely failure to be associated with the fault
B Likely failure
C Failure less likely to be associated with the fault
D Will affect color of metallics
E Health hazard

* Poor build
† In oven
** Cold air forced into oven
‡ Alkali contamination

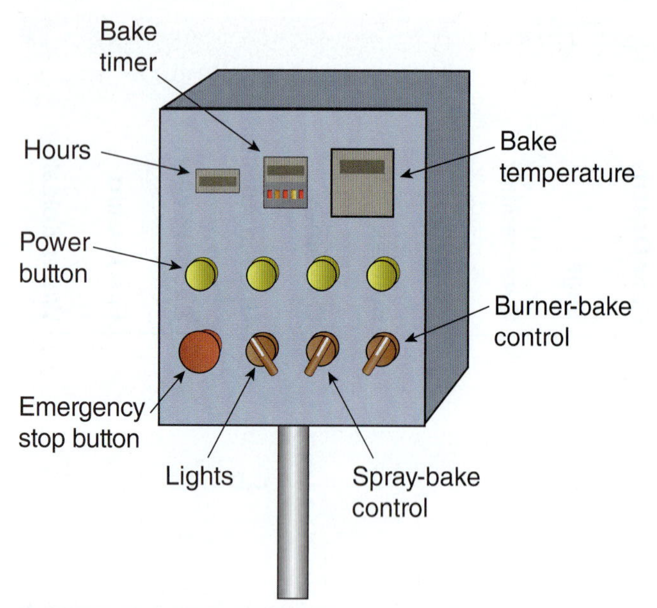

A Note the readings and controls on this booth.

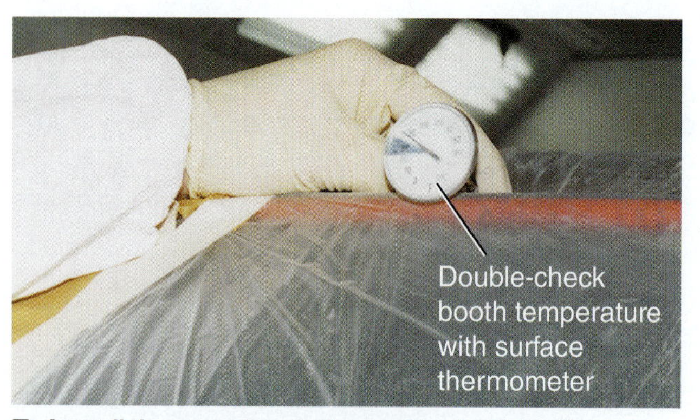

B A small thermometer can be used to check booth temperature, if needed. The temperature rating of the reducer must match the actual temperature of the vehicle being painted.

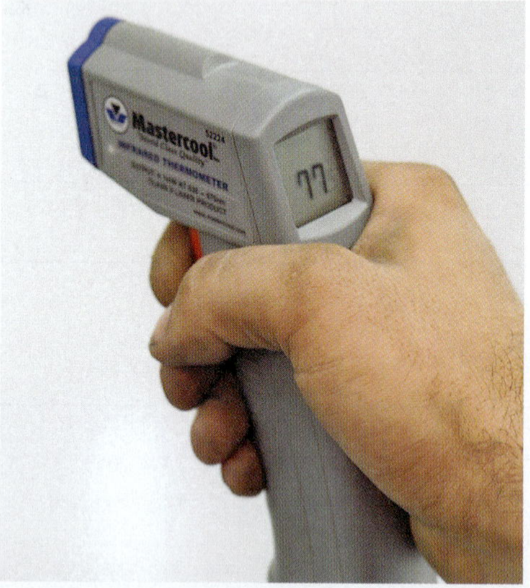

C An infrared thermometer is handy because you can aim it at surfaces to read the temperature.

Figure 24-69 Paint drying booths have controls for heat and airflow.

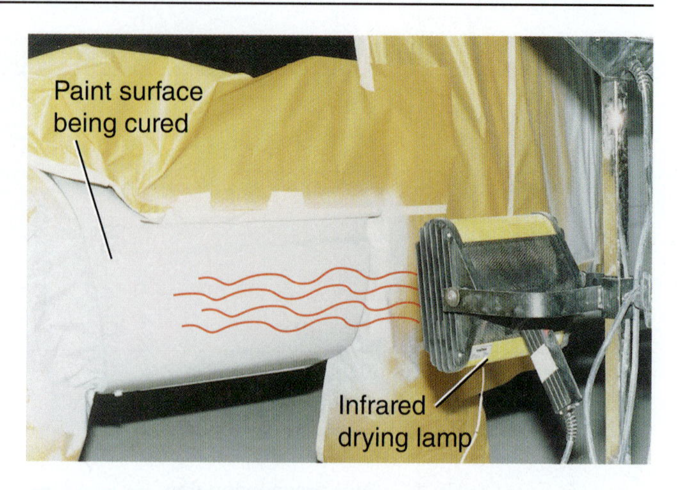

Figure 24-70 Portable lamps are often used when a booth oven is not available or when curing UV primer or other paint material outside of a booth.

respirator to make the air breathable. Clean air can then flow through the respirator feed tube.

A simplified booth layout for a fresh air-supplied respirator is shown in Figure 24–73.

Note that separate air lines are provided for the spray gun and respirator. The air intake for the respirator is normally located on the outside of the shop. This keeps fumes and dust from being pulled into the system. Also note that a small dedicated air compressor can be used to pump air to the respirator.

It is very important that you periodically replace the fresh-air respirator filter. It can normally be replaced by unscrewing a cover over the element. Refer to Figure 24–74.

If shop air pressure is going to be used for an air-supplied respirator, additional in-line filters and a desiccant drying unit are needed. They will clean the shop air enough for breathing. See Figure 24–75.

There are two good reasons for wearing a respirator. First, some sort of respiratory protection is dictated by OSHA/NIOSH regulations. Second, even if the first reason were not true, common sense would indicate that inhaling overspray is not healthy. Overspray can contain particles of toxic paint pigments, harmful dust, and vapor fumes that can be harmful to your health. Depending on design, a respirator can remove some or all of the dangerous elements from the air around a spray finishing operator.

24.12 OTHER PAINT SHOP EQUIPMENT AND TOOLS

There are several pieces of paint shop equipment that can help refinishing technicians provide better paint jobs.

A *wet sanding stand* is used for wet sanding individual components or small parts. These cabinets are made by individual paint shops, with the size and installation location dictated by shop requirements and conditions (Figure 24–76).

TABLE 24–9 TROUBLESHOOTING DRYING ROOM PROBLEMS

Fault	Result	Popping	Softness	Dirty Job	Overspray	Impaired Durability	Polishing Impaired	Fire and Explosion Hazard	Loss of Gloss	Recoat Failure	Discolorate
Dirty filters	Diminished air velocity	A[1]	A[3]			C	B			A	
	Diminished oven pressure		A[4]	C	B[5]	C	B				
	Spray booth/oven pressure imbalance			B							
Filters damaged or breached	High velocity jet streams and turbulence	A[2]	B[2]	A		C	C				
Thermostat probe not correctly sited in moving airstream and/or insufficiently sensitive.	Excessive high/low temperature modulation	A	A			C	C				
10 percent bleed duct closed 10 percent make up filter clogged	Foul oven Excessive fumes					B		A	A[6]		A[7]
Failure to remove deposits of rust, dust, and flaking paint from oven surfaces	Excessive dirt circulation			A							
Failure to clean unpainted areas on vehicles. Failure to clean masking or remask. Operators entering oven with dirty overalls.	Unnecessary dirt introduced into oven			A							

A Most likely failure to be associated with the fault
B Likely failure
C Failure less likely to be associated with the fault
D Will affect color of metallics
E Health hazard

Note: Repair ill-fitting or damaged oven doors immediately.

[1] Upper parts
[2] Local
[3] Lower parts
[4] Cold air drawn from booth
[5] Drawn from spray booth
[6] Microshrivel
[7] Chemical reaction

Paint hangers are used to suspend or secure individual components or small parts for spray painting. As with the wet sanding stands, these are made by the individual shop in accordance with the shape of the item to be painted, the quantity required, and so on. Paint hangers keep panels from dropping during painting. They must be made of a material that will withstand heat during paint drying. An example is shown in Figure 24–77.

Panel drying ovens are small ovens used to dry test pieces. There are various types—from a very simple kind using infrared lamps to more complicated units with an electric heater, vent fan, and a timer for controlling the temperature and drying time.

Paint shakers will vibrate a can of paint to mix it before opening. For a good refinishing job, it is very important that the paint be thoroughly mixed, or agitated. In fact, it is essential with metallic paint topcoats. These paints contain metallic particles that are heavier than the paint itself and quickly settle to the bottom of the container. The quickest method to achieve a proper mixing job is with a paint shaker.

A *blade agitator* is a stirring paddle used with paint mixing systems. The blades of the agitator are dipped into the paint and the paint can is sealed by the agitator cover. The cover locks over the can opening by spring action. These agitators usually come in 1- and 4-quart

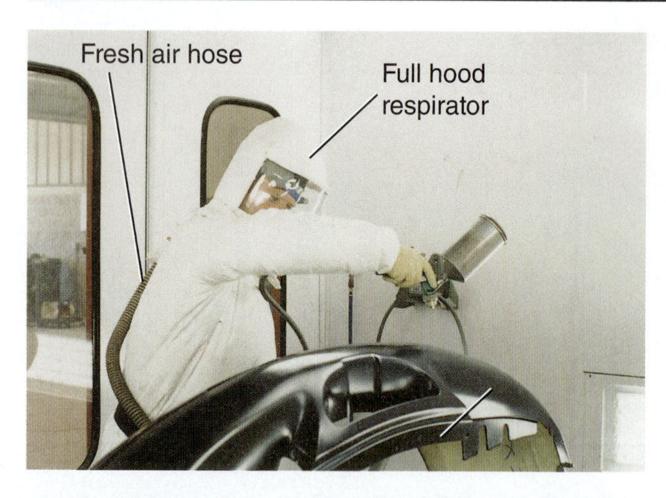

Figure 24–71 **This technician is wearing a full hood air-supplied respirator. It is the safest type to use when spraying toxic materials, such as today's catalyzed paint materials. Note the fresh-air hose attached to the technician's belt.**

sizes and are the types of mixers used by color mixing centers (Figure 24–78).

Churning knives are also used to stir paint. The handle tip is designed as a paint can lid opener. Some churning knives have a scale for measuring paint or hardening agents.

Paint scales are used to weigh paint ingredients when mixing the paint to match the original color or tone (Figure 24–79).

Paint cabinets are used for storage and stock control of paint, thinner, and putties. These cabinets should be selected for the amount of paint and thinner normally stored, conditions of the shop layout, and local fire codes.

It is important that the areas of a vehicle to be painted are clean. Most paint shops provide clean cloths or special disposable paper wipes. Whichever is used, these simple tips will help:

- Use clean, dry cloths folded into a pad.
- When using a cleaning solvent, be sure to pour enough onto the pad being used to thoroughly wet the surface to be cleaned.
- Do not wait for the solvent to dry. Wipe it dry with a second clean cloth.
- Refold the cloth often to provide a clean section.
- Change cloths often.
- Once an area is clean, do not touch it with the hands, as it might affect adhesion.

Tack cloths are porous cheesecloth fabric coated with a "sticky varnish" for wiping dust off the body surface right before spraying. While blowing the surface with compressed air, lightly wipe the body surface with the tack cloth to remove all dust particles and dirt from the surface to be painted. A tack cloth will pick up fine particles that are invisible to the naked eye. Tack cloths should be stored in an airtight container to conserve their tackiness.

Made of wood, metal, or plastic, *paint paddles* are used to stir the paint material.

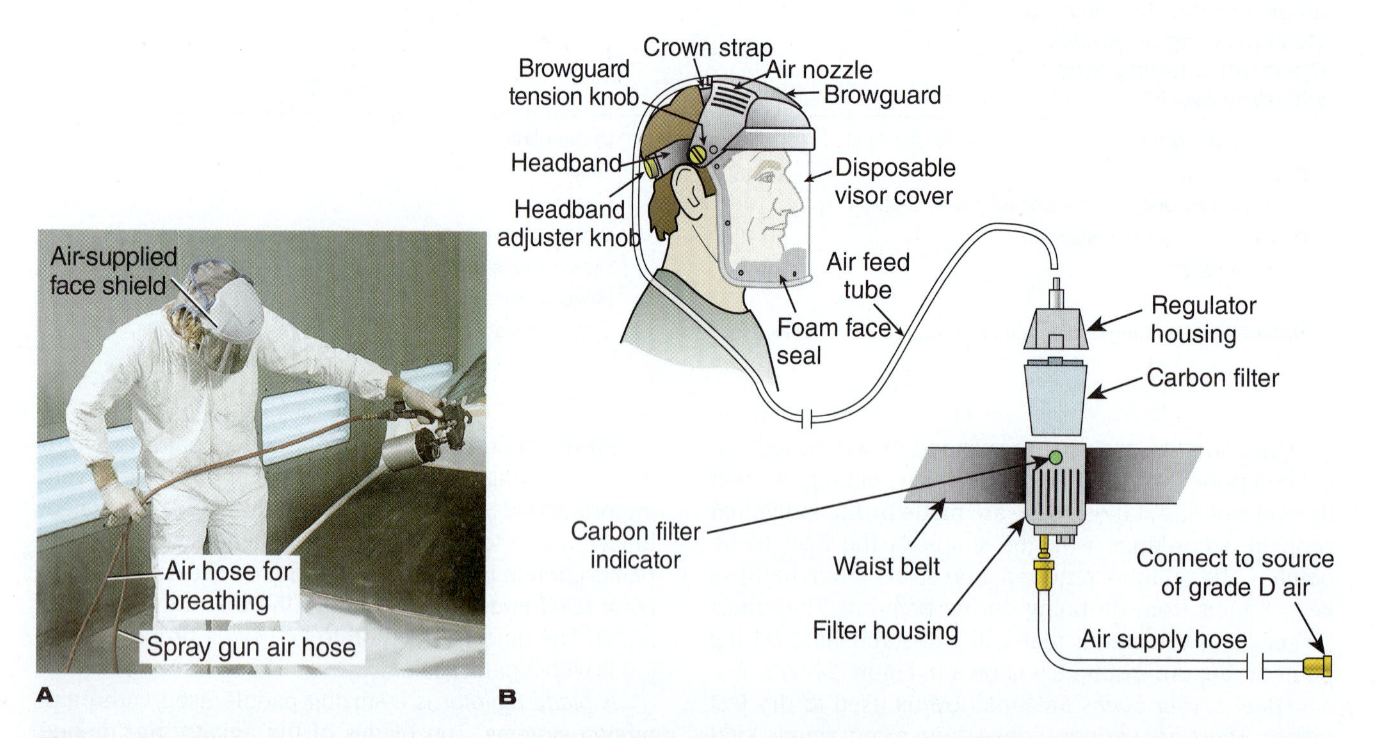

Figure 24–72 **An air-supplied respirator is cool on your face in hot weather and is the only safe way to spray today's paint products. (A) This technician is using an air-supplied face shield, or visor. (B) Study the parts of an air-supplied paint visor. It is lightweight and has disposable visor covers. You can replace a scratched or overspray-coated clear plastic visor cover with a new one so you can see well when painting.**

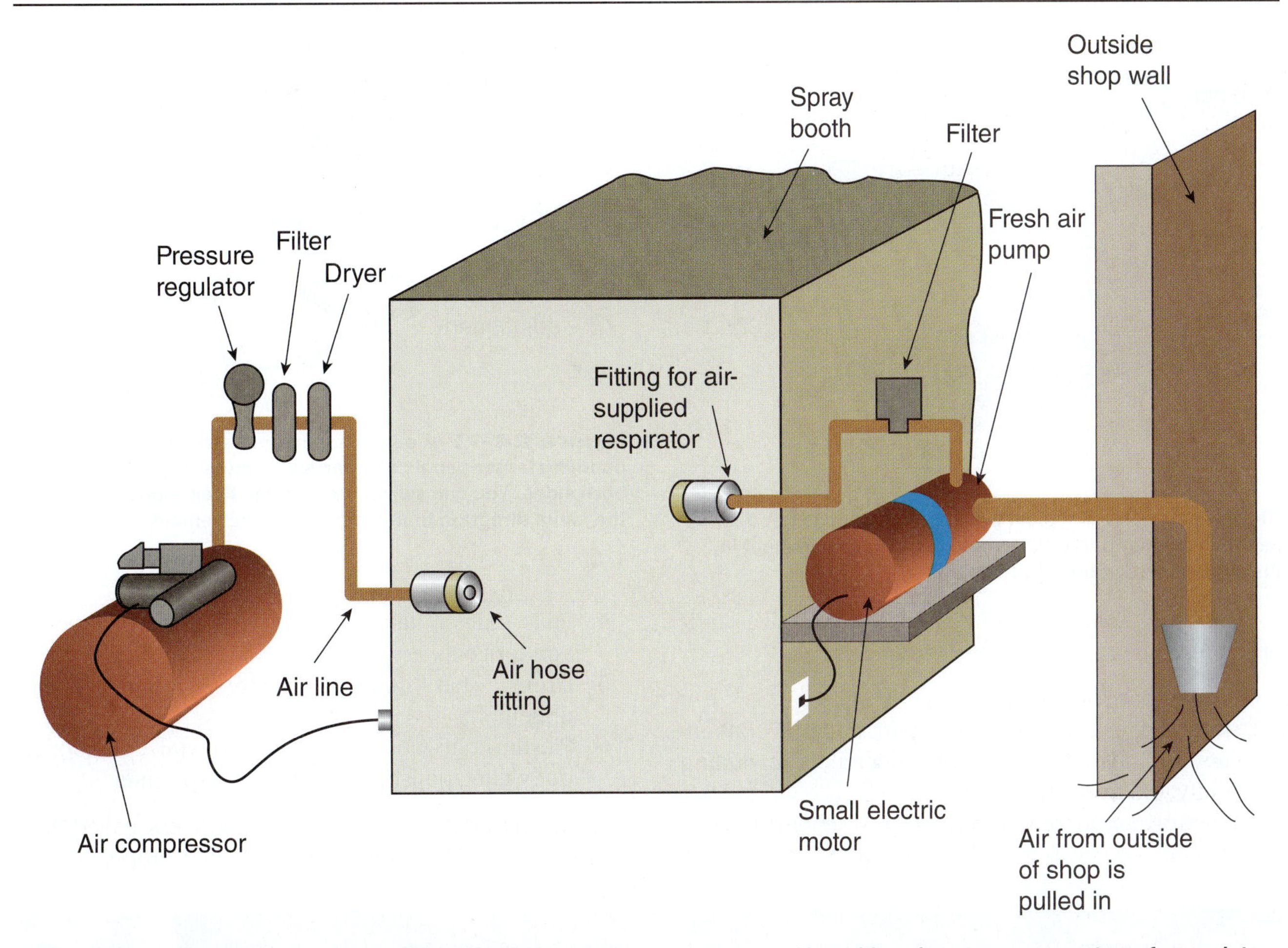

Figure 24-73 Note the components of simplified main air compressor and breathing air compressor systems for a paint booth. Breathing air should be drawn from outside of the shop so dust and fumes are not pulled into the air-supplied respirator.

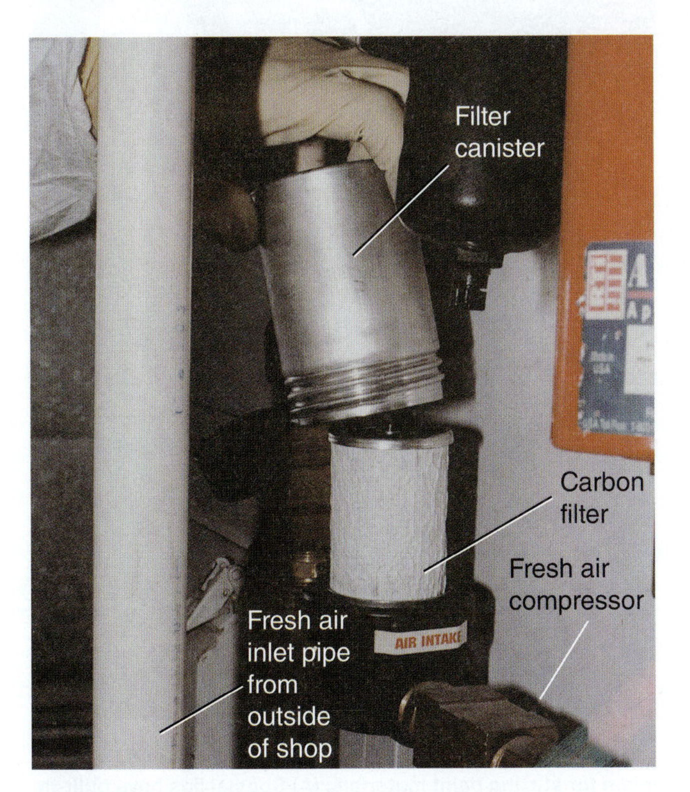

Figure 24-74 This technician is installing a new carbon filter so the breathing air is not contaminated.

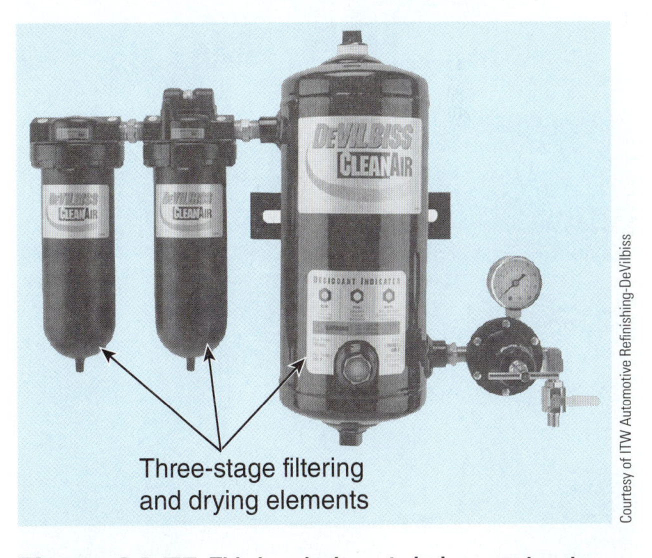

Figure 24-75 This is a desiccant air dryer system to purify and produce breathable air for an air-supplied respirator.

Consisting of a cardboard funnel with cotton mesh, a **strainer** is used when pouring thinned topcoat and other materials into a spray cup. This is done to make sure it is free of any dirt or foreign material.

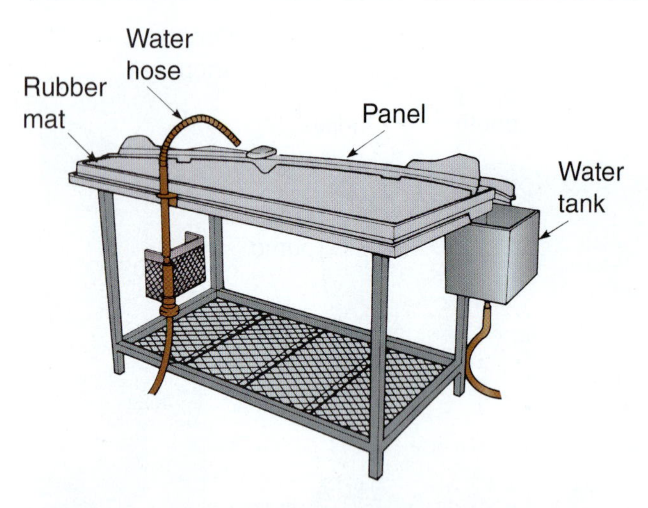

Figure 24-76 **A typical wet sanding stand is good for handling smaller parts. Water and sanding grit are kept in the system, not on the shop floor.**

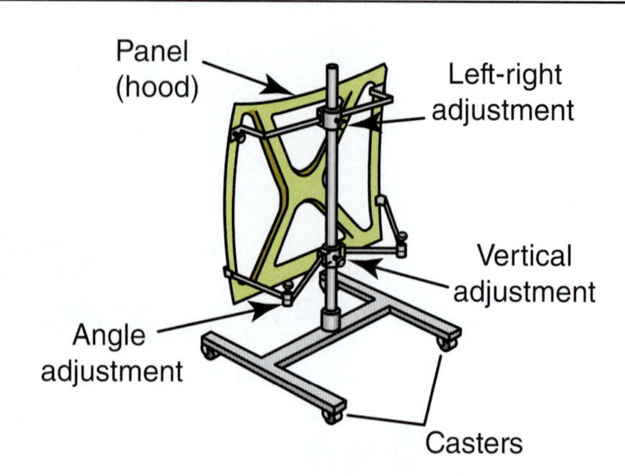

Figure 24-77 **A paint shop hanger stand is good for handling larger panels and parts that must be painted on both sides. You can swivel the panel to keep it oriented in the same direction that it mounts on the vehicle.**

Paint shop containers come in six common sizes and/or shapes and contain various materials:

1. Tubes that contain putty
2. Round cans such as gallon, quart, and pint containers for topcoats and some undercoats, including putty and body filler
3. Square cans for thinners, reducers, primer-surfacers, sealers, and clear topcoats.
4. Pails that contain thinners, reducers, undercoats, and topcoats
5. Drums that contain thinners, reducers, and undercoats
6. Plastic containers that contain metal conditioner, body filler, polish, and buffing compounds

Lids on round containers of a gallon or less, called "friction lids," should be carefully opened with a proper opener.

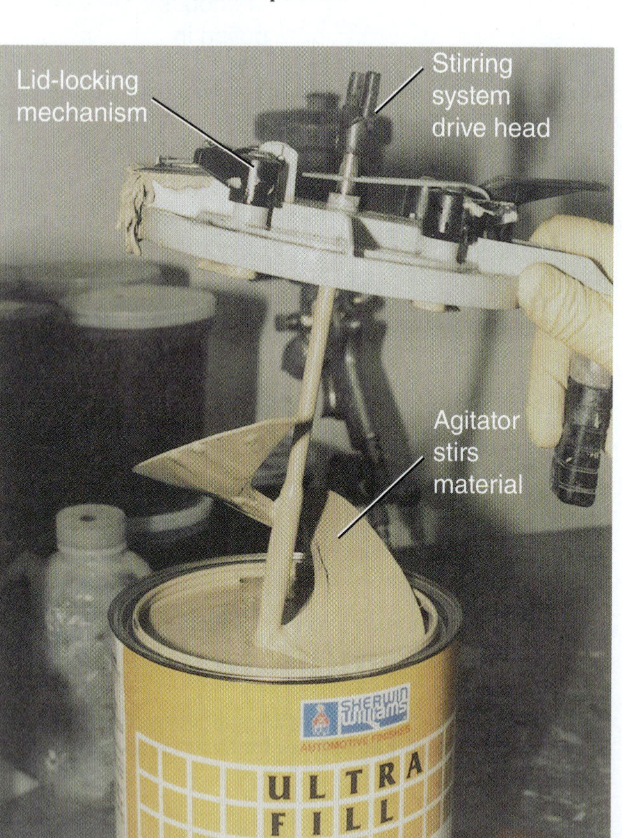

A

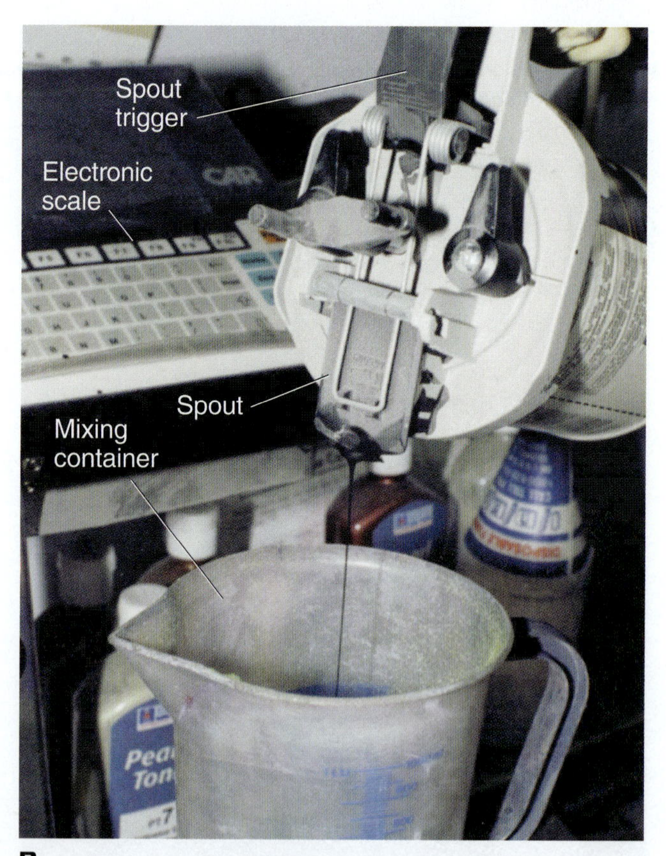

B

Figure 24-78 **Paint-mixing system shelves have a drive mechanism for stirring paint materials. (A) Special lids have built-in agitators and spring-loaded spouts. (B) Mixing system lids also have a spring-loaded spout that can be opened when pouring paint ingredients.**

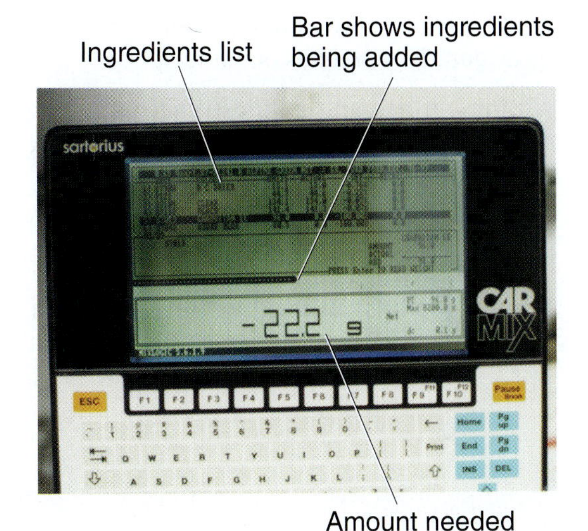

A

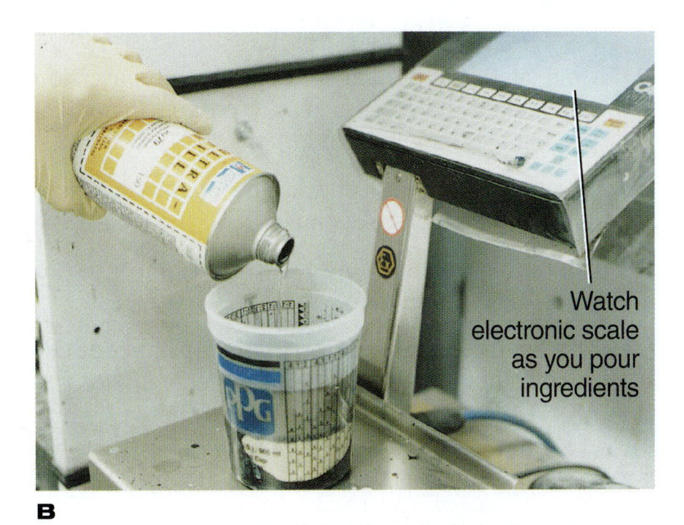

B

Figure 24-79 Electronic scales are now used in more medium- to high-volume body shops. (A) Electronic mixing scales will pull up a paint formula and tell you how much of each ingredient is needed to properly mix the material. (B) As you pour each ingredient into the container, watch the readout on the scale so you do not add too much material.

After pouring off whatever amount of material is needed from a round can, the lip should be wiped and the lid replaced tightly to form a good seal.

Plastic measuring cups are used for matching or thinning paint. Generally their sizes range from 1 quart to 5 quarts, and they are made of easy-to-use, easy-to-clean plastic.

Masking paper is used to cover surrounding areas that will not be painted so that paint mist does not settle on them. Masking paper must be capable of preventing solvent in the paint mist from bleeding through to the surface of the object. Some paint shops use newspapers for this purpose, but this is not recommended because thin fibers from the paper come off and adhere to the painted surface, resulting in dirt. Solvent will also seep through newspaper or even transfer newsprint onto the covered area.

Masking tape is used to stick the masking paper to the areas to be covered, or it can be used by itself. Masking tape is made of different types of material, such as paper, cloth, and vinyl, so that adhesion performance is ensured regardless of the season or weather. The adhesive performance of masking tape does not change when heat is applied, and traces of adhesive will not be left when the tape is removed. It is also easy to cut or tear off masking tape.

There are several types of paper masking tape that are classified by their intended use. General masking tape is used for air drying, and heat-resistant tape is used for baking enamel. The proper tape for the job should always be used. Full information on masking paper and tape and their proper use is provided in Chapter 25.

SUMMARY

1. To do a good paint job, your shop and equipment must be in perfect condition!
2. The spray gun atomizes liquid primer or paint into a fine mist and forces it onto the surfaces of the vehicle.
3. Gravity-fed, high-efficiency spray guns are now the most common type used in today's body shops.
4. Suction (siphon) spray guns use airflow through the gun head to form a suction that pulls paint into the air stream.
5. In the gravity feed system, the paint is supplied by gravity and the material is pressure- or suction-forced at the nozzle tip.
6. Some high-efficiency spray guns use a combination gravity–pressure feed cup design. It offers the benefits of both types of guns.
7. The high-volume, low-pressure (HVLP) spray system, also known as the "high-solids" system, uses a high volume of air delivered at low nozzle pressure to atomize paint into a pattern of low-speed particles.
8. High transfer efficiency means that more of the paint leaving the gun stays on the surface being painted.
9. Touch-up spray guns are very small and are ideal for painting small repair areas. Often called "airbrush" or "door jamb" guns, they have a tiny cup for holding a small amount of material.

10. Many low volatile organic compound (VOC) regulations require 10 psi (69 kPa) or less at the air cap.
11. Graduated paint mixing sticks have conversion scales that allow you to easily convert ingredient percentages into part proportions.
12. A spray pattern test checks the operation of the spray gun on a piece of paper.
13. Gun stroke refers to the hand movement used to move the gun while spraying.
14. Thorough cleaning of a spray gun and its accessory equipment immediately after use is critical.
15. Avoid spraying solvent into the air when cleaning a spray gun. This pollutes our atmosphere.
16. A spray gun cleaning tank, also called a gun washer–recycler, is a pressurized container for flushing spray guns and other tools with cleaning solution.
17. Most spray gun manufacturers recommend lubricating the moving parts with spray gun oil.
18. A spray booth is designed to provide a clean, safe, well-lit, and well-ventilated enclosure for painting.

EXERCISES

On a separate sheet of paper, complete the following learning activities for this chapter. Write definitions for the key terms and answer the ASE-style review questions, essay questions, critical thinking problems, and math problems. You can also do the outside activities, possibly for extra credit.

Key Terms

airbrush
air valve
air vent hole
arcing
atomization
drying rooms
fluid control knob
gravity-fed spray gun
heeling
high-volume, low-pressure (HVLP)
paint shakers
pattern control knob
pressure-fed spray guns
spray booth
spray gun cleaning tank
spray gun direction
spray gun distance
spray gun oil
spray gun speed
spray gun stroke
spray gun triggering
spray overlap
spray pattern test
strainer
suction spray guns
viscosity
viscosity cup

ASE-Style Review Questions

1. Which of the following promote atomization of the paint?
   A. Center orifices
   B. Side orifices
   C. Ports
   D. All of the above
2. Technician A says that gravity-fed spray guns are the most common type found in body shops today. Technician B says that electrostatic spray guns are more common. Who is correct?
   A. Technician A
   B. Technician B
   C. Both A and B
   D. Neither A nor B
3. Technician A says that HVLP spray guns should be held a little closer than older low-efficiency spray guns. Technician B disagrees. Who is correct?
   A. Technician A
   B. Technician B
   C. Both A and B
   D. Neither A nor B
4. If you hold a spray gun too close to the vehicle surface, which problem will result?
   A. Poor coverage
   B. Dry spray
   C. Runs and sags
   D. Fish-eyes
5. A spraying pressure that is too high results in which of the following?
   A. Poor flow
   B. Poor drying characteristics
   C. Bubbling
   D. Sagging
6. What happens when testing the spray pattern for uniformity of paint distribution?
   A. A vertical spray pattern is used
   B. The trigger is pulled all the way back and released immediately
   C. The pattern is flooded
   D. None of the above
   E. Both A and B

7. This spray system (also known as the "high-solids" system) uses a high volume of air, delivered at low pressure, to atomize paint into a pattern of low-speed particles.
   A. LVLP
   B. HVLP
   C. HVHP
   D. LVHP
8. Which type of spray booth is the most popular air movement system used today?
   A. Regular flow booth
   B. Reverse flow booth
   C. Downdraft booth
   D. Crossdraft booth
9. Technician A says that a cartridge respirator will protect you from catalyzed paint. Technician B says that an air-supplied respirator is needed with catalyzed materials. Who is correct?
   A. Technician A
   B. Technician B
   C. Both A and B
   D. Neither A nor B
10. Technician A monitors the booth intake filters and manometer readings hourly. Technician B monitors the booth intake filters and manometer readings daily. Who is correct?
    A. Technician A
    B. Technician B
    C. Both A and B
    D. Neither A nor B
11. What is the purpose of a tack cloth?
    A. Wiping off solvent cleaner
    B. Removing dirt from fresh clearcoat
    C. Wiping dust off the surface before refinishing
    D. None of the above
12. Technician A uses the same spray gun for the topcoat that was used for the undercoat. Technician B uses a different spray gun. Who is correct?
    A. Technician A
    B. Technician B
    C. Both A and B
    D. Neither A nor B

Essay Questions

1. What happens if the air vent hole becomes plugged in a spray gun cup?
2. Summarize some advantages of an HVLP spray gun.
3. How do you do a spray pattern test?
4. Describe five problems often found during a spray pattern test.
5. What are some common spray gun handling problems experienced by an inexperienced painter?

Critical Thinking Problems

1. What will happen if you hold a spray gun too close to a vehicle when spraying?
2. When spraying, if the paint viscosity is too thick, what will be the result?
3. Name six variables that must be addressed before painting.

Math Problems

1. An HVLP spray gun requires air delivered at 10 psi (69 kPa). Your gauge only shows 6.3 psi. By what percentage is the pressure too low?
2. If you are holding a spray gun 4.2 inches (106 mm) away when spraying enamel and the specs call for a distance of 3.4 inches (87.2 mm), how much closer do you have to move your gun in?
3. If you spray fluid out of your spray gun into a graduated container for 30 seconds and 7 ounces flow out, what is your spray gun adjustment in ounces per minute?

Activities

1. Make a spray pattern test with a spray gun. Try adjusting the gun knobs for making good and poor spray patterns. Use waste paint from a storage container so you do not waste shop materials.
2. Make a wall chart showing how to use one type of paint mixing stick. Place callouts and steps on the chart for display in the classroom.
3. Inspect the condition of your spray booth. Is it ready for use? Make a report on its condition.

CHAPTER 25

Vehicle Surface Preparation and Masking

OBJECTIVES

After reading this chapter, you should be able to:

- Determine whether an existing finish is defect free and adheres soundly to a vehicle.
- Select the correct abrasive and sanding techniques for specific final sanding operations.
- Prepare existing paint films and bare metal substrates for refinishing.
- Describe the three methods of removing a deteriorated paint film.
- Determine when to apply a primer, a primer-sealer, a primer-surfacer, or glazing putty.
- Prepare plastic parts for refinishing.
- Mask a car, panel, or spot repair for refinishing.
- Comply with EPA 6H rules.
- Correctly answer ASE-style review questions relating to vehicle surface preparation and masking.

INTRODUCTION

After leaving the metalworking area, a vehicle is sent to the surface preparation area. Here, the painter and/or a helper must ready the vehicle for painting or refinishing. All surfaces to be painted must be closely inspected for surface problems, cleaned, scuffed or sanded, and recleaned. Body surfaces and parts that will not be painted must be masked.

Any scratches and existing paint problems must be fixed prior to panel refinishing. Final sanding, surface scuffing, part removal, and masking takes place in the surface preparation area. Most surface prep and masking operations are done before pulling a vehicle into the paint booth (Figure 25–1).

The term **surface preparation** refers to getting a body surface clean, smooth, primed, puttied, sanded, scuffed, and wiped clean to ready it for painting. To get a smooth, level surface often requires minor putty filling and sanding operations besides those done during body filler application.

The life and appearance of a finish depend on the condition of the surface on which the paint is applied. In other words, proper surface preparation is the foundation of a good paint job. Without it, there will be a weak base for the topcoat and the paint will fail or not look good.

Figure 25–1 After sheet metal work is done, a vehicle is moved to the surface prep/masking area.

25.1 EVALUATE SURFACE CONDITION

Vehicle surface evaluation involves determining what must be done before painting a body surface. Is the old paint in good condition, requiring only scuffing before painting? Is the old paint badly deteriorated, requiring complete removal down to bare metal? You must ask yourself these kinds of questions during vehicle surface evaluation.

Before painting, you must first identify the type of paint and overall condition of the existing paint. Failure to identify defects at this stage can be very expensive to correct. Missing and painting over even a tiny surface flaw could involve resanding and repainting the whole panel.

To evaluate the surface condition, first clean the areas to be inspected. Dust will quickly collect on surfaces while a vehicle is in the body shop.

Look carefully for any signs of paint damage on all panels to be painted. Also check old paint for *film breakdown* problems, such as checking, cracking, and blistering. Horizontal surfaces usually show the greatest film deterioration. Careful inspection of the hood and trunk areas will give a good indication of the overall condition of the paint system (Figure 25–2).

In particular, note the **gloss level**, or how much shine remains in the paint surface. Low paint gloss will often indicate surface irregularities caused by defects such as checking or micro cracking, which will need a more thorough investigation with a magnifying glass. Any signs of disfigurement or discoloration of the paint film due to industrial fallout or acid rain must be completely removed.

Determine whether the old finish has *good adhesion*—that is, whether the old paint is still bonded tightly to the vehicle body. To test adhesion, sand through the finish and featheredge a small spot. If the thin edge does not break or crumble, it is reasonable to assume that the old paint will stay on when the refinish color is applied over it.

Developing *surface* rust can be detected by roughness, bubbling, or pitting of a paint surface. On those areas where poor adhesion or rust is found, the paint must be removed down to bare metal. Close-ups of minor surface rust damage are shown in Figure 25–3.

Even if the original paint finish is in good condition, it should be thoroughly sanded or scuffed after washing to remove dead film and to smooth out imperfections. Sanding also reduces the existing paint thickness to avoid too much paint thickness after refinishing. If the surface is in poor condition, you should remove all the paint—right

A This door has been keyed down through the colorcoat. The area above the scratch will be wet sanded and featheredged. The existing paint is still in excellent condition and will provide a good base for new paint.

B This missing flake of paint indicates there could be a serious paint adhesion problem or that the paint or undercoat is failing or deteriorating. All of the old paint may have to be removed before repainting.

Figure 25–2 To begin surface prep, closely inspect all surfaces to be painted. You do not want to miss any surface flaw that would blemish the vehicle's new finish.

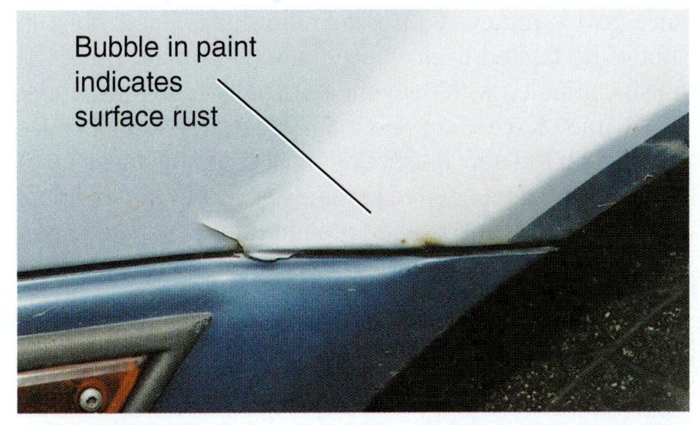

A A bubble in the finish often hides surface rust between the paint and metal.

B With the bubble peeled off, you can see surface rust. Rust should be coarse sanded or blasted off carefully without further thinning the metal.

Figure 25–3 Surface rust can be a problem when repairing panels on older vehicles.

down to the bare metal. In this way, you achieve a good foundation.

Professional painters know that today's thin film colorcoat/clear-coat paints do little to hide or fill rough areas or minor surface imperfections. They know that their paint jobs can be no better or smoother than the surface below them.

Checking Paint Thickness

Paint thickness is measured in mils, or thousandths of an inch (hundredths of a millimeter). Original OEM paints are typically about 3–6 mils thick. With basecoats/clearcoats, the basecoat is approximately 1 mil thick and the clearcoat is about 2 mils thick. This is approximately the thickness of a piece of typing paper.

If a panel has been repainted, paint thickness will increase. If too much paint is already on the vehicle, it may have to be removed prior to refinishing. Paint buildup should be limited to no more than 12–14 mils. The OEM finish combined with one refinish usually equals just under 12–14 mils. Exceeding this paint thickness could cause cracking in the new finish. Chemical stripping, blasting, or sanding would be needed to remove the old paint buildup.

A mil gauge, also called a paint thickness gauge, can be used to measure the thickness of the paint on a vehicle. If the paint is too thick (over approximately 14 mils), the paint should be chemically removed or abraded off with coarse sandpaper (Figure 25–4).

The *pencil mil gauge* has a spring-loaded magnet with a ruler-type scale for checking paint thickness. The tool's magnet sticks to a vehicle when placed against a painted steel body panel. It is pulled away from the body and a reading is taken in mils (Figure 25–5).

To use this type of mil gauge, place the tool against the steel body surface. Slowly pull the tool away from the vehicle body while watching the tool scale. The highest mil reading exposed on the tool before the magnet snaps away from the body equals paint thickness in mils.

An *electronic mil gauge* is similar but has a digital read-out showing paint thickness. When placed and held on the surface, the tool will automatically register paint thickness.

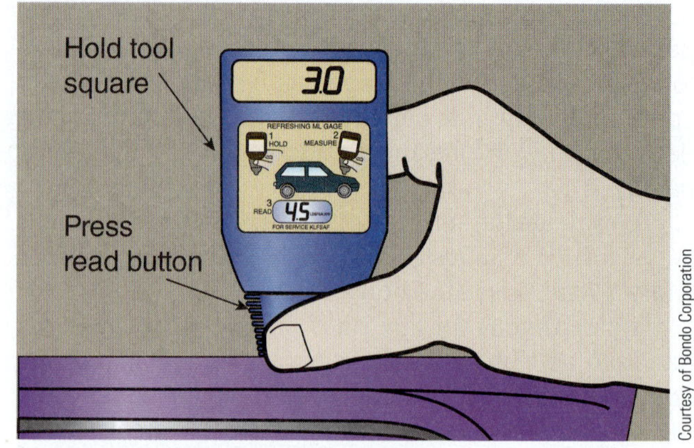

A To use a typical electronic mil gauge, hold the tool square on the surface and press the button.

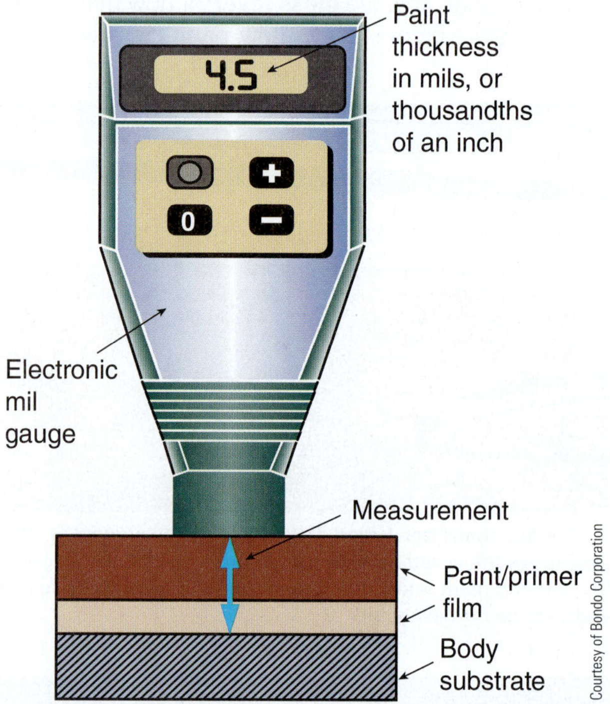

B An electronic mil gauge measures the distance between the substrate and the top of the paint film. This gauge shows 4.5 mils, which is about the thickness of a piece of paper.

Figure 25–4 An electronic mil gauge can be used on steel body panels to measure paint thickness. If the paint is too thick, you should not paint the panel without removing the old paint.

25.2 PAINT REMOVAL

Most forms of paint failure are progressive. They are conditions that cannot be stopped by any form of repair. In fact, repairing will usually accelerate the deterioration of the original finish. If the old finish is badly weathered or scarred, it is not suitable for recoating and should be completely removed. This can be a labor-intensive and time-consuming process.

There are three common ways to strip paint from metal surfaces:

1. Chemical stripping
2. Media blasting
3. Sanding or grinding

Chemical Paint Removal

A **chemical paint remover** can be used for stripping large areas of paint if environmental regulations allow. It is an alternative to blasting in those areas that a power

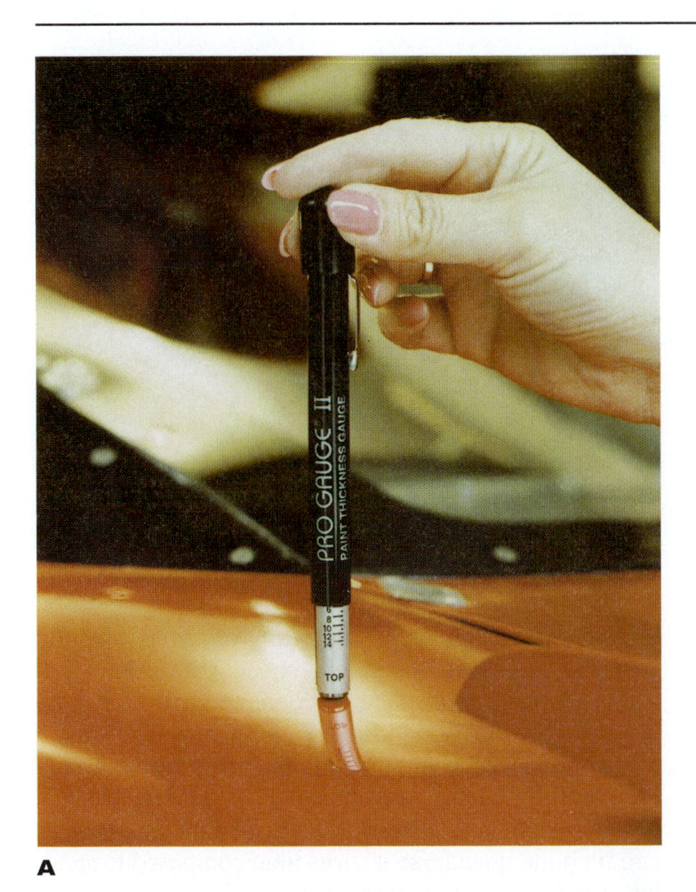

A

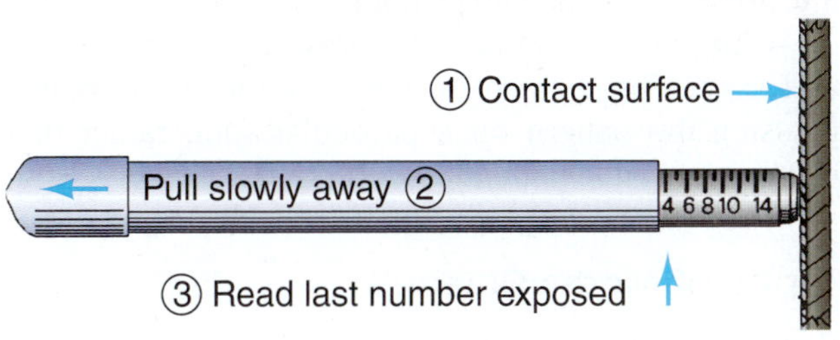

B

Figure 25–5 **(A) A Tinsley gauge, also called a pencil mil gauge, is an inexpensive but effective way to check paint thickness in three steps. (B) By pulling the gauge, a magnet will stick to a steel body but will not work on aluminum or plastic panels. Thicker paint will allow you to pull the tool off the vehicle body with less effort.**

sander cannot reach. One advantage of chemical stripping is that there is no danger of the metal warping, which can happen with media blasting.

Before applying paint remover, mask off the area to ensure that the remover does not get on any area that is not to be stripped. Use two or three thicknesses of masking tape to give adequate protection. Cover any crevices to prevent the paint remover from seeping to the undersurface of a panel.

Slightly scoring or scratching the surface of the paint to be stripped will help the paint remover to penetrate more quickly.

Paint remover should be applied following the manufacturer's instructions. Pay attention to warnings regarding ventilation, smoking, and the use of protective clothing such as PVC or rubber gloves, long-sleeved shirts, and safety glasses or goggles. Chemical remover will cause irritation and burning of the skin or eyes.

> **DANGER** Chemical paint remover normally contains a powerful acid that can cause severe chemical burns. If paint remover gets on your skin or in your eyes, immediately flush off the acid with water using an eye flushing station or a water hose. Wear a full face shield and rubber gloves when working with paint removers.

To apply, brush a heavy coat of paint remover in one direction only onto the area being treated. Use a soft bristle brush, but do not brush the material out. Allow the paint remover to stand until the finish is softened (Figure 25–6).

Although paint remover is effective on most vehicle topcoats, some modern car undercoats can prove stubborn. If a finish resists the remover, more than one application may be needed.

Caution should be taken when removing the loosened paint coatings. Some paint removers are designed to be neutralized by water. Remove the dissolved paint with a squeegee or scraper. Be sure to rinse off any residue that remains on the body, using cleaning solvent and steel wool. Follow by wiping with a clean rag. This rinsing operation is essential. Many paint removers contain wax, which, if left on the surface, will prevent the refinish paint from properly adhering, drying, and hardening.

Rusting occurs very rapidly on metal that has been chemically stripped. In fact, any bare metal substrate should be treated immediately. Before selecting the type of metal treatment or conditioning system, first consider the types of rust. The least amount of rust is microscopic rust, which is not really visible to the eye but can be a hazard to the performance of a refinish job. The second type of rust is called flash rust, which usually develops when there is moisture or humidity present. The other types of rust are very visible and might even be large and scabby.

The decision about which metal conditioning system to use depends on the type of rust and the type of substrate.

Blasting Off Paint

Blasters are air-powered tools for forcing sand, plastic beads, or another abrasive onto surfaces for paint removal. Blasting is preferred when trying to remove surface rust from body panels. Blasting will clean out rust pits without further thinning the panel (Figure 25–7).

With today's thin-gauge, high-strength steel, blasting is often recommended over grinding to clean out rust pockets. Grinding thins the metal and makes it weaker. Grinding also may not get down into the bottom of rust pits completely.

Spread chemical stripper on thickly

A **Brush the chemical stripper onto the paint without splashing it. Spread the paint stripper to form a thick coating on the old paint.**

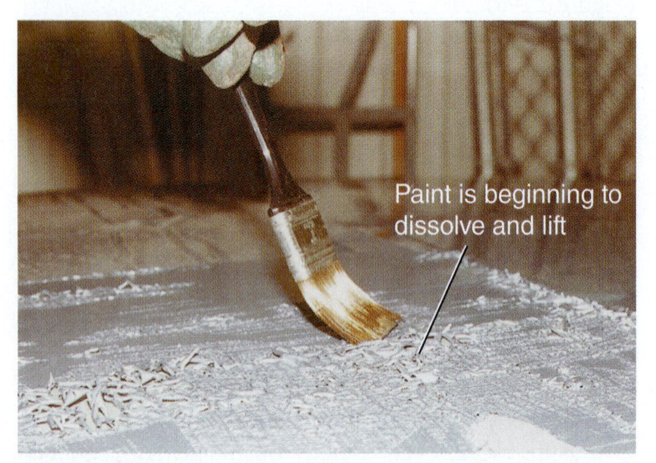
Paint is beginning to dissolve and lift

B **Here the paint is starting to soften. If any areas fail to soften and lift, apply a second coat of paint stripper.**

Scrape off old paint without scratching panel

C **Scrape off softened paint with a flexible putty knife. Old paint should be treated as a hazardous material and must be disposed of properly.**

Figure 25-6 **If a panel has been painted several times, you might want to chemically strip off the paint. Be sure to wear face, hand, and body protection and a respirator because this chemical contains strong acids.**

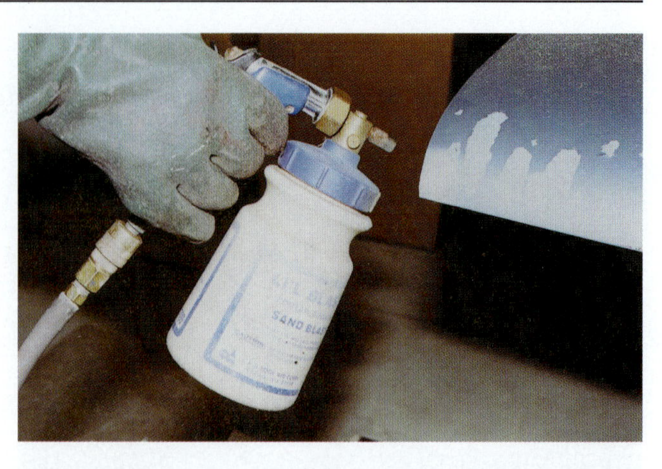

Figure 25-7 **Media blasting is a fast and effective way to remove paint from smaller areas.**

Blasting can be done on nearly all types of body construction—even aluminum sheet, with caution. It leaves a clean, dry surface, an ideal condition for refinishing. It has the added advantage of revealing rusted areas and places where hidden rust would result in scaling after the job has been refinished. In addition, it makes hard-to-reach areas accessible to the technician. Blasting can be done quickly, so it saves time compared to sanding, grinding, or chemical stripping.

A blaster concentrates the pressure and flow of air and sand. The technician can vary the blast volume, focusing the pattern on a particular spot, rather than blasting in a wide pattern.

Types of Media Blasters

Pressure blasters are pressurized containers filled with abrasive material (such as silica sand or plastic beads). The material travels through one hose while high-velocity air travels through another hose. Both meet at a third hose and travel out toward the surface together at tremendous speed and force.

In a *siphon blaster*, compressed air draws the abrasive from the reservoir by producing a suction, or vacuum, at the blaster head. The abrasive accelerates and is shot out of the nozzle at the intended surface. Small bottle blasters are available for spot jobs.

A portable blaster can be carried or is mounted on a rolling cart. *Blasting rooms* are metal enclosures that are similar to paint booths. They have sealed doors, wall and ceiling lighting, forced ventilation, media recovery capability, and a dedicated air supply for the blasting equipment.

Types of Blasting Media

There are three types of media used to blast paint off: plastic media, sand, and soda media.

Plastic media blasting uses small plastic pebbles or beads to remove the old finish. It is the method of paint removal preferred by most professionals and for stripping an entire vehicle. It requires large, heavy-duty blasting equipment. Plastic media blasting is fast and will not warp metal as easily as sand blasting.

Sand blasting uses hard silica sand to abrade off the old paint. It should not be used on thin-gauge steel panels or on soft aluminum panels. Heavy, hard bits of sand can generate heat and warp thin-gauge sheet metal.

Soda blasting uses extremely high air pressure and a fine powder or baking soda as the blasting medium. The blasting gun can be equipped with a water hose to inject water into the air–soda stream to aid in paint removal and to keep the dust down in the media blasting booth or area. Soda blasting can remove specific layers of paint. It will leave undercoats of primer intact to save surface prep time.

Some metal conditioners should not be applied over blasted, pitted steel. The instructions for some metal conditioners require you to wipe the conditioner off the metal after application. If the metal is deeply pitted, you will not be able to remove the caustic conditioner from the pits. This can cause paint problems, such as blistering and popping after repainting.

Using a Blaster

The basic procedure for operating a blaster is as follows:

1. Mask off the areas that will not be affected by the spot repair. For instance, when spot repairing a quarter panel, mask the wheel covers, glass, trim, and the top of the vehicle. Thick duct tape and a plastic tarp can be used for added protection where needed.
2. Put on the necessary safety gear. Wear gloves, eye protection, a helmet, and a respirator. A respirator must be worn because dust can build up in the lungs over an extended period of time.
3. Before blasting, check the manufacturer's instructions for proper blasting pressures, sand load procedures, and setup arrangements. When ready to blast, apply the abrasive material directly onto the area to be blasted. Eventually, the area will turn a gray or white color. Blasting has textured the surface by opening the pores of the metal in these areas. This etched texture makes an excellent surface for primer adhesion. When the area shows no signs of brown rust, remove the pressure.

   Pressure should be applied by holding the nozzle the recommended distance away from the area being repaired. It should hit the surface at a 20- to 30-degree angle. That way, the media will fly sideways and not back at you.
4. Watch the surface carefully. Blasting may reveal a rust hole. If you find major corrosion, blast as much of the hole out as possible. Blasting is designed to reveal weak spots like these. Before priming the rusted out area, cut out all corrosion-thinned metal and weld a metal patch on it.
5. After the paint has been removed, use an air blowgun to remove the sand from all areas of the vehicle. If it is not removed, it can get into the paint. The abrasive could also get stuck in windshield wiper blades or window slots and scratch the windows.
6. It is advisable to prime the metal as soon as possible after any stripping process. Blasting requires that the job be primed almost immediately because the metal is in a raw state following this treatment. The bare metal will start rusting if allowed to stand overnight.

Sanding Off Paint

Orbital sanding with a flexible backing pad is suitable for removing old finish from small flat areas and gently curved areas. Start finish removal with a coarse-grit sandpaper (about #24–#36) on an open-coat flexible disc. Holding the face of the disc at a slight angle to the surface, work forward and backward evenly over the area to remove the bulk of the old finish down to the metal. Using a #50 or #80 closed-coat disc, go over the entire area, and slightly out on the surrounding surface, to clean up the work and eliminate the troughs and deep sand scratches caused by the coarse grit. Refer to Figure 25–8.

After all of the paint is removed with the coarse grit disc, resand the area with the orbital or DA sander and #100-grit sandpaper to remove the metal scratches. Then finish sand the panel using #180-grit sandpaper. In this way most of the scratches created by the stripping operation will be eliminated. Featheredge any bare metal areas into areas that still have paint on them (Figure 25–9).

Table 25–1 reviews which grit, or coarseness, of sandpaper should be used for each surface prep operation. Logically, you use coarse grits first for fast material removal and surface leveling. Progressively finer grits are then used until the surface is smooth enough for priming.

Never use a grinder to remove paint from a very large surface area on a panel or from the whole vehicle. Heat from grinding will usually warp the panel and thin the metal too much.

For more information on filling and priming bare metal surfaces, refer back to Chapter 11 and Chapter 12.

25.3 PREPARING BARE METAL

Proper preparation of bare metal is a critical step in restoring a vehicle to its pre-accident condition. Yet it is often ignored or carried out in a haphazard manner. This can only result in poor adhesion, corrosion, and customer complaints.

When a vehicle is made at the factory, auto manufacturers devote as much if not more attention to metal treatment than they do to priming or topcoating. Auto manufacturers basically use a multistage zinc phosphate metal treatment process to ensure adhesion of primers to the metal

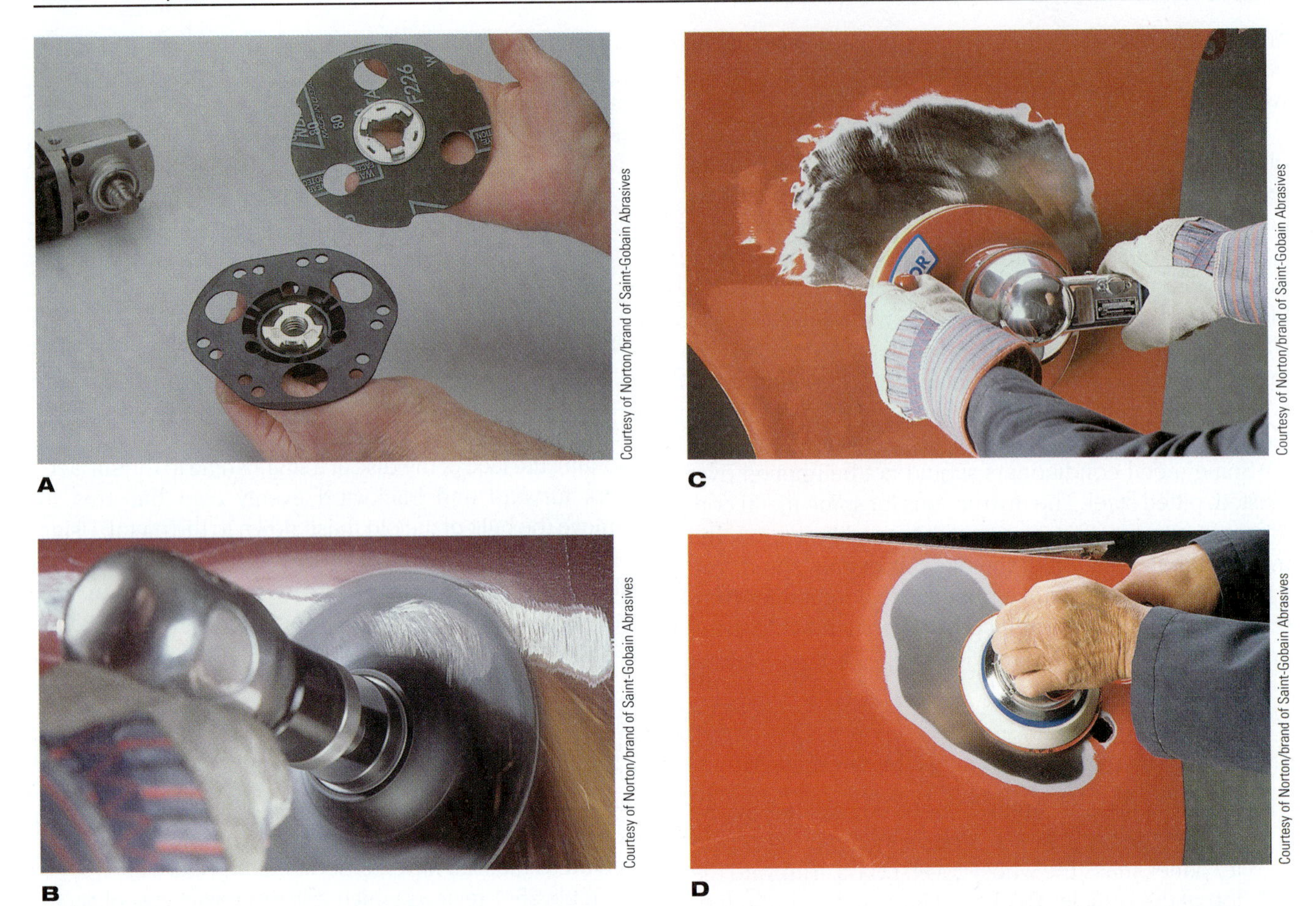

Figure 25-8 **Paint can also be stripped with a flexible grinding disc (for heavy paint buildup) or with coarse sandpaper and a DA sander (for light paint buildup). (A) With today's thin-gauge steel body panels, use a flexible grinding disc or coarse scuff wheel to prevent metal damage. (B) Grind only to bare metal and keep the disc moving to avoid overheating the metal. (C) If it is not too thick or hard, coarse sandpaper in a DA will remove paint safely with minimum abrasion to a body panel. (D) Be careful not to oversand or overgrind when removing old paint, either of which can thin, weaken, and possibly warp the panel.**

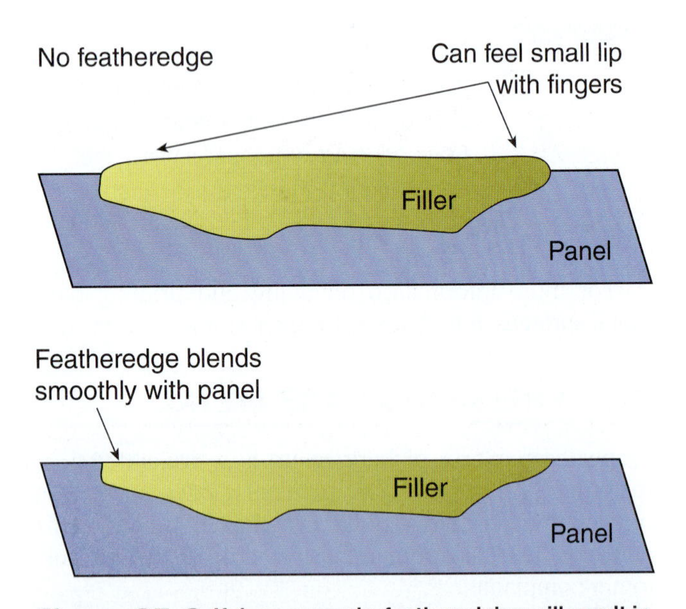

Figure 25-9 **If done properly, featheredging will result in a level surface that blends outward without bumps or dips. During surface prep, check for proper featheredging of any metalworking repairs. Feel or rub over the repair surface with a clean glove or rag. Any surface irregularity must be found now, before painting.**

body surface. Though the techniques and equipment used at the OEM level are not adaptable to body shops, some metal treatment products on the market today enable you to simulate original equipment metal treatment.

Why is bare metal treatment so important? Water vapor penetrates all paint films. The fresher the paint and the more humid the weather, the further the vapor penetrates, sometimes reaching the bare metal. Once water droplets form under paint film, pressure starts to build, causing bubbling, blistering, and loss of adhesion. Rust can also begin to form, further pushing the paint film away from the metal.

The only way to prevent this potential problem is to create such a strong bond between the undercoat and the metal that water vapor cannot penetrate down to the substrate. If the water vapor is not allowed to condense under the paint film, it will return to the surface of the finish and eventually evaporate.

Using Metal Conditioners

A *metal conditioner* is an acid-type chemical wash that eats away any material that might prevent a good

TABLE 25–1 TYPE OF GRIT AND NUMBERING SYSTEM

Grit	Aluminum Oxide	Silicone Carbide	Zirconia Alumina	Primary Use for Auto Body Repair
Ultra fine	—	2,000 1,500 1,250 800	—	Used for basecoat/ clearcoat paint system or for colorcoat sanding.
Very fine	—	600	600	Used for colorcoat sanding. Also for sanding the paint before polishing.
	400 320 280 240	400 320 280 240	400 — 280 240	Used for sanding primer-surfacer and old paint prior to painting.
	220	220	—	Used for sanding of topcoat.
Fine	180 150	180 150	180 150	Used for final sanding of bare metal and smoothing old paint.
Medium	120 100 80	120 100 80	— 100 80	Used for smoothing old paint body filler.
Coarse	60 50 40 36	60 50 40 36	60 — 40 —	Used for rough sanding of body filler.
Very coarse	24 16	24 16	24 —	Used on sander or grinder to remove paint.

primer-to-metal bond. It was commonly recommended by paint manufacturers in the past. However, with new self-etch primers, wipe-on metal conditioners are seldom used today. Nevertheless, metal conditioners or an acid wash might still be used when surface rust could pose a problem, as when restoring an antique car.

To use a metal conditioner, mix the appropriate amount of conditioner with water in a plastic bucket according to the instruction label. Apply to the metal body surface with a cloth, sponge, or spray bottle. Before the conditioner dries, wash the conditioner off the body with clean water. Again, follow manufacturer instructions because procedures vary.

A metal conditioner with phosphoric acid reducer not only cleans, it also etches the metal and promotes the adhesion of the paint film. It helps prevent the occurrence of rust and also eases sanding marks.

Preparing Hard Chrome Surfaces

Sometimes chrome parts, such as bumpers and trim, are painted when customizing a vehicle. Chromium is very hard. Primer and paint will not adhere to chrome very well. When painting is desired, prepare the surface as follows:

1. Clean the metal thoroughly with a wax and grease remover.
2. Thoroughly sand the metal, using #320 wet or dry sandpaper.
3. Reclean with wax and grease remover.
4. Apply any of the metal treatments described earlier in this chapter.
5. Spray two coats of primer-surfacer. Allow adequate drying time before dry sanding.
6. Blow out cracks, then use a tack cloth on the entire surface. The final coat can now be applied.

Regardless of the cleaning procedure, once the metal is clean and prepared, it must not be contaminated by fingerprints, so wear clean plastic gloves when handling the vehicle.

Preparing Metal Replacement Parts

Some aftermarket panel or part suppliers protect new panels with a nonpaintable **shipping coating**. The function of this coating is to protect the metal against corrosion and rust during storage and shipping. Some new parts must have this shipping coating, which looks like primer, removed before painting. Always refer to the manufacturer's instructions for preparing the new part for painting.

Most auto manufacturers supply new panels and parts with an electro-coat primer. This **E-coat** is a paintable primer and is an essential part of the factory corrosion protection warranty on new factory panels and parts. Factory E-coats on new parts should not be removed unless directed by the manufacturer. Modern E-coats should be cleaned, scuffed, and repaired to prepare them for the painting process.

If a non-OEM replacement panel or older OEM panel has a shipping coat rather than a paintable E-coat, you will have to sand or chemically remove the coating completely. Then you can spray the part with self-etch primer and primer-surfacer to provide a smooth, solid base for the new paint. If not cleaned and sanded properly, colorcoat will not stick to the new part.

Remember that replacement parts with various types of coatings are provided by manufacturers. Some have a protective coating for shipping and others are primed for painting. Refer to published information to determine what must be done to prepare the new part for painting.

If the part is already factory primed and ready for paint, simply clean it with wax and grease remover. Then examine the part for imperfections such as drips or scratches. Sand any imperfections until smooth. You may not have to completely remove the coating. Scuff sand the entire panel, then apply primer or primer-surfacer before painting.

If in doubt about the quality of a new part's protective coating, check with the manufacturer of the part for the recommended finishing procedures. This will ensure a long-lasting, quality paint job.

Using Self-Etch Primer

In the past, metal conditioning and priming were considered separate surface preparation steps. Today, however, self-etch primer makes it possible to combine these steps. *Self-etch primers* chemically eat into the bare metal to improve paint adhesion and corrosion resistance while also priming the surface. Self-etch primers are now the most common material used to prepare bare metal for undercoats.

A self-etch primer should be applied to all bare metal surfaces before applying primer-surfacer, sealer, or any other type of primer. Self-etch primer comes in aerosol cans for small areas of exposed metal. To coat larger bare metal areas, self-etch primer can also be mixed and applied with a spray gun.

To use a self-etch primer, follow these basic steps:

1. Sand metal thoroughly. Remove all visible scale or rust.
2. Clean the surface with wax and grease remover and wipe dry.
3. Spray on a full wet coat of self-etch primer to any exposed bare metal. See Figure 25–10.
4. Apply an undercoat (primer or primer-surfacer) or sealer. If a self-etching primer-filler is used, this step might not be necessary (Figure 25–11).
5. Once the undercoat refinish system is dry and sanded, wipe with a tack cloth. The surface is ready for the colorcoat.

If filling is required, apply an epoxy primer and allow it to cure a minimum of 1 hour and then apply a primer-surfacer. Sand the primer-surfacer after it has cured properly. Once the undercoat system is completed, the surface is ready for the topcoat.

Apply Seam Sealer

In the prep area, you may also have to apply seam sealer to newly welded panels. *Seam sealer* is needed anywhere water leakage might be a problem between panel joints—often on welded panels forming the passenger compartment and trunk areas.

After application and drying of self-etch primer, use a caulk gun to apply the seam sealer to the panel joints. Many modern sealers are two-part materials. A caulk gun with an intermix tip is used to mix and apply the seam sealer ingredients. This is shown in Figure 25–12.

Refer to the vehicle service manual or computer-based data to find out where and how seam sealer should be applied. Service illustrations will give panel locations and bead width specifications for applying seam sealer properly.

25.4 PRIMECOAT SELECTION

The decision to apply a primer, a primer-sealer, an adhesion promoter, or a primer-surfacer depends on three factors. These are:

1. Condition of the substrate—smooth or rough, bare or painted
2. Type of finish on the substrate—if painted
3. Type of finish to be used for the topcoat

Generally, use primecoats as follows:

- Use a *self-etch primer* or *epoxy primer* on bare metal.
- Use a *sealer* over repair areas to prevent repair material ingredients from bleeding or showing through the new paint.
- Use a *primer* to improve adhesion and to help cover repaired surfaces with dissimilar materials on them.
- Use a *primer-sealer* for covering dissimilar materials and for preventing bleedthrough in the new paint.
- Use a *primer-surfacer* or *primer-filler* to help smooth and level surfaces with a large area of minor surface imperfections or to help featheredge repair areas.
- Use an *adhesion promoter* on hard, baked-on OEM finishes to help prevent peeling or a poor bond between the new paint and the existing finish.

Table 25–2 summarizes the use of common primecoat materials for refinishing.

Applying Primecoats

Reduce the primecoat chosen according to the manufacturer's instructions. Be careful to select the proper solvent for the weather conditions, and thoroughly mix the material.

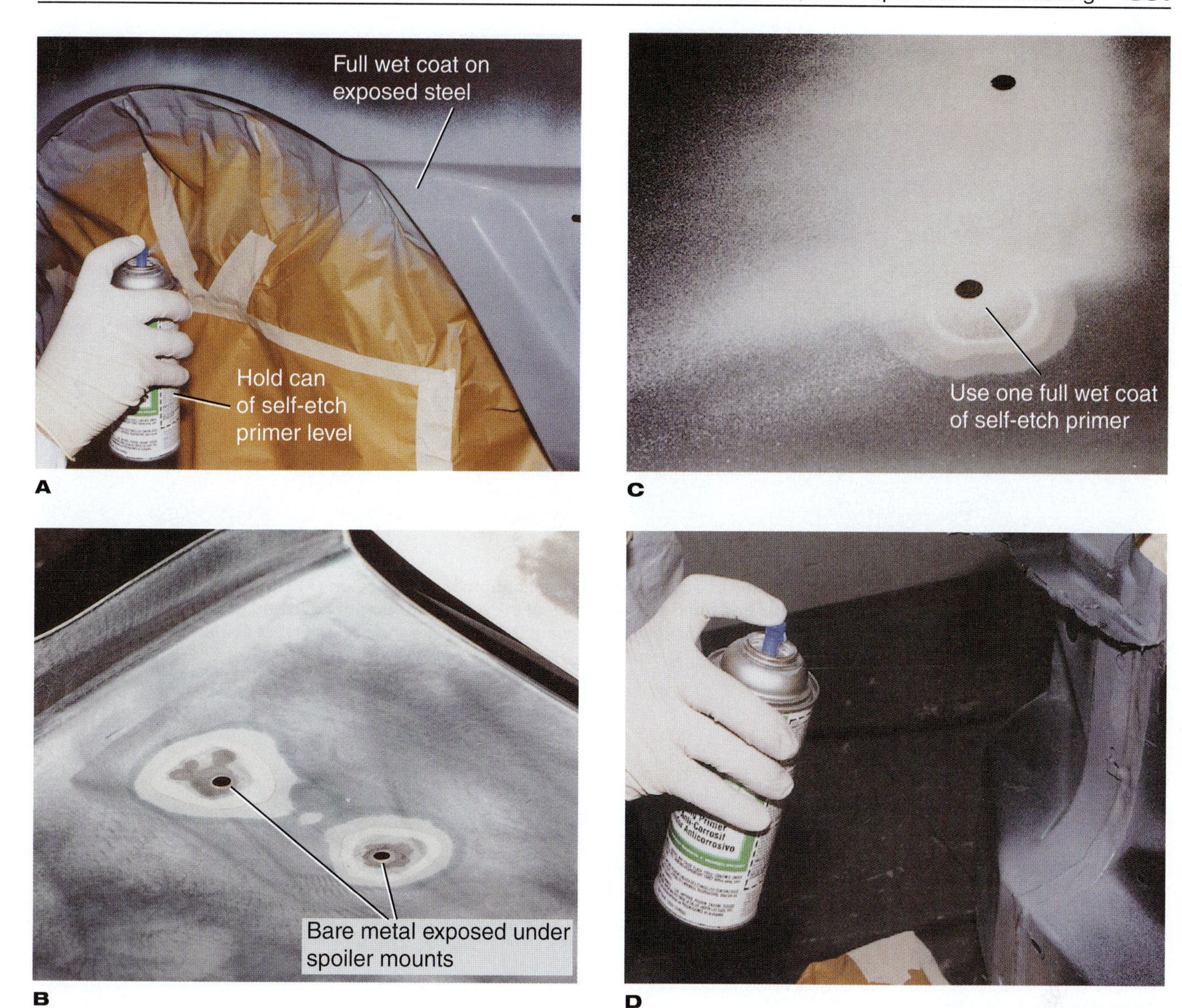
Full wet coat on exposed steel
Hold can of self-etch primer level
A
Use one full wet coat of self-etch primer
C
Bare metal exposed under spoiler mounts
B
D

Figure 25-10 Most modern shops now spray self-etch primer over all bare metal surfaces. It will eat into metal to form a strong bond. (A) Only spray self-etch primer where needed or on bare steel. Apply a good wet coat. (B) After removing a rear spoiler, surface rust was found under the mounts. The surface was blasted and then sanded to remove all traces of rust. (C) Close-up shows self-etch or bare metal primer depositing over a repair area. (D) If any new panels have been welded on a vehicle, make sure their flanges are fully coated with self-etch primer.

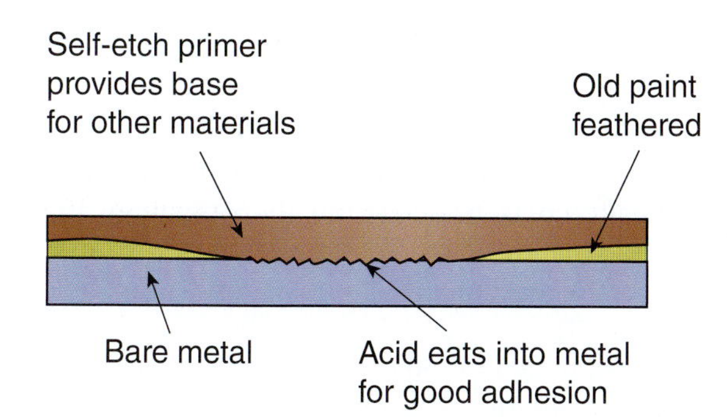
Self-etch primer provides base for other materials
Old paint feathered
Bare metal
Acid eats into metal for good adhesion

Figure 25-11 Self-etch and epoxy primers are commonly used to provide proper adhesion to bare metal. They also provide for some filling and leveling.

Generally, only one or two coats of primer or primer-sealer are required. Primer-surfacer and primer-filler also require one or two coats for proper buildup. Apply the first coat of primecoat. Allow this coat to flash dry, following the recommended flash time on the label.

Flash time is the time needed for a fresh coat of sprayed material to partially dry or cure. Flash time is needed to prevent the material from sagging, running, cracking, or experiencing other problems when another coat is applied.

After the recommended flash time has passed, apply the next coat or coats as medium wet coats for additional film buildup, with sufficient flash time between each application. When making a spot repair, extend the primecoat (primer material) several inches or millimeters around the first coat.

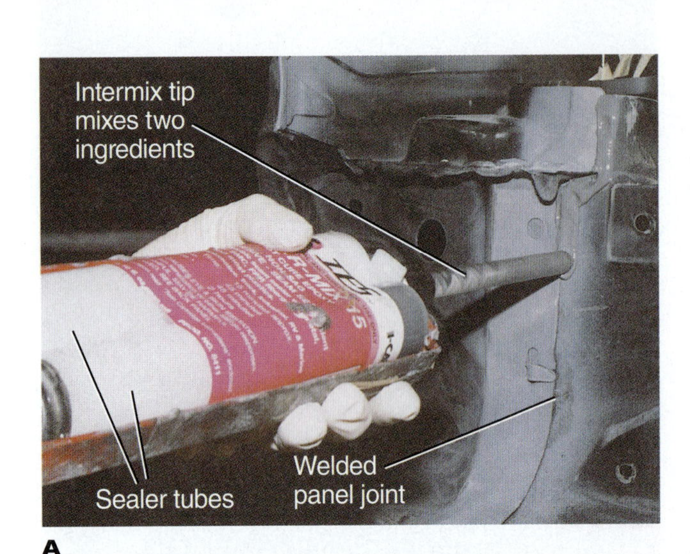

A

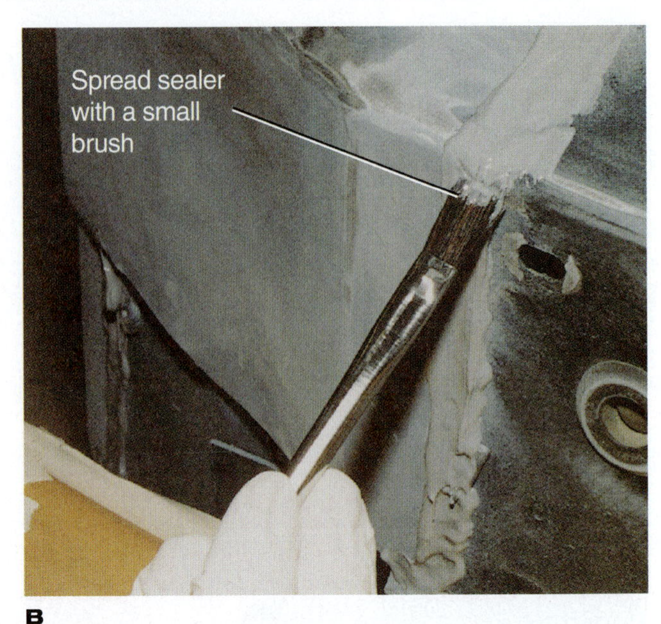

B

Figure 25-12 In the prep area, you must also check for correct application of seam sealer. (A) The sheet metal technician forgot to apply seam sealer here. If the vehicle had been painted and leaked, the painter might have had to repaint the vehicle for free. (B) Use a small brush to smoothly spread seam sealer. You want it to look like the original OEM sealer. Brushing will also improve bonding of sealer.

TABLE 25–2 FUNCTIONS OF PRIMECOATS

Primecoat Function	Self-Etch Primer	Primer	Primer-Surfacer	Primer-Sealer	Sealer
Apply to bare metal	Yes	No	No	No	No
Resists rust and corrosion	Yes	Yes	Yes	Yes	No
Makes topcoat adhere better	Yes	Yes	Yes	Yes	Yes
Fills scratches and nicks	No	No	Yes	No	No
Provides uniform holdout of the topcoat	No	No	No	Yes	Yes
Prevents show-through of sand scratches	No	No	No	Yes	Yes

SHOP TALK

A common mistake is to apply primer materials that are too dry or too thick. Both conditions can cause problems when painting. Spray the primer on wet in thin coats. Allow each coat to flash before applying another.

Allow the primecoat to dry thoroughly. Do not apply extra heavy coats to speed up the operation. Primer applied too thickly will require more time to dry and can lead to cracking, crazing, pinholes, and poor holdout (Figure 25–13).

It is also difficult to tell when a thick coat of primer-surfacer is really dry. The surface will appear dry while there is still a lot of solvent trapped below the surface. The lower layer of primer-surfacer is still trying to dry and shrink. Follow the manufacturer's guidelines to avoid problems!

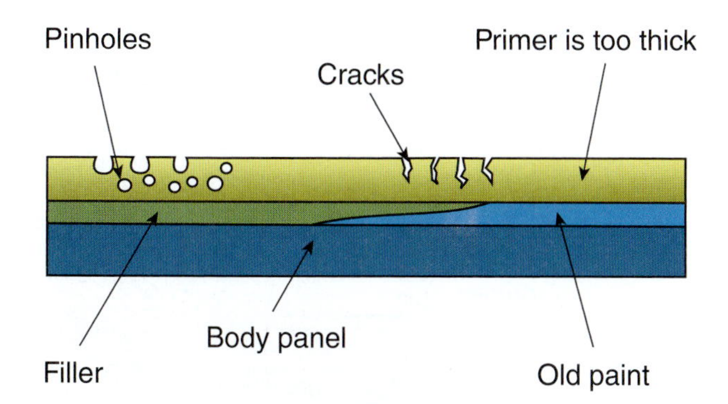

Figure 25–13 If primer is put on too thickly in one coat, pinholes and cracks can result. Allow primer to flash for a few minutes before applying the next coat.

If primer-surfacer is sanded before all of its solvent has evaporated, the material in the scratches will continue to shrink down in the scratches. They will show up in the final finishing color topcoats as sanding scratches.

At the other extreme, thin dry coats of primer-surfacer can cause loss of adhesion, not only to the substrate but also to the topcoat color. Always spray wet coats of primer-surfacer.

After the material is fully dry, block sand the area until it is smooth.

As discussed in Chapter 12, some technicians like to apply a guide coat. Using a guide coat, you can easily find high and low spots by sanding the area. If the second guide coat does not sand off, you have found a low spot. If it sands off too quickly, you have found a high spot.

Ideally, the guide coat should sand off at the same time. This shows that the surface is flat and ready for sealer, a colorcoat, and other operations.

Applying Spot Putty

Once the primer is dry, any remaining small pinholes and scratches must be filled with *spot putty*, or glazing putty. Closely inspect all panels to be painted for remaining surface problems: paint chips, pits, sanding marks, or other imperfections. It can be very difficult to see some of the flaws, so make sure the surface is well lit. Use a hand-held shop light for your inspection if needed.

Place a small amount of properly mixed putty onto a clean rubber squeegee or tiny putty spreader. Wipe a thin coat over the primer imperfections. Use single strokes and a fast scraping motion (Figure 25–14).

Putty dries very quickly. Use a minimum number of strokes when applying putty. Repeated passes of the spreader might pull the putty away from the primer. Go to the next surface imperfection and repeat putty application.

After curing, block sand the putty flush with the surrounding surface. Although wet sanding works well, it is not recommended by some putty manufacturers because moisture can soak into the putty. Refer to label instructions for details. Dry sand the putty if needed (Figure 25–15).

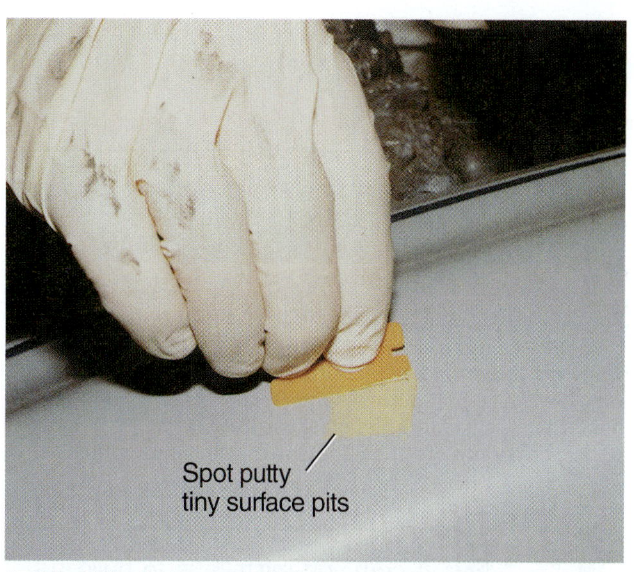

A It is difficult to see small dimples or depressions in dull primer, so look closely. Use a small flexible plastic mini-spreader or rubber squeegee to apply spot putty.

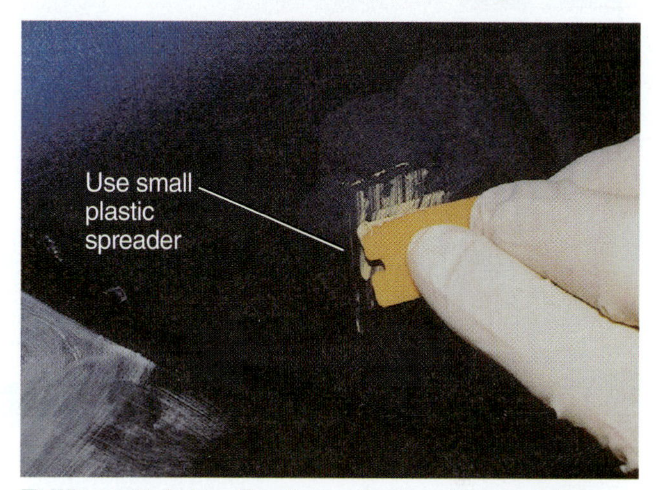

B Wipe putty in two directions only. Do not keep rubbing spot putty because it will tend to dry and may not bond properly, resulting in imperfections.

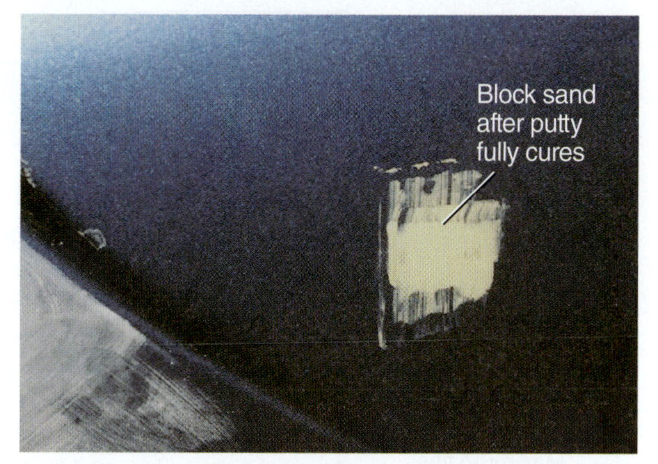

C Some spot putties can be used on painted surfaces. Use your finger to scuff the paint chip before putty application. After thorough curing, you can featheredge putty into existing paint with less sanding.

Figure 25–14 Spot putty should be applied only to small surface imperfections. Never use it like plastic body filler.

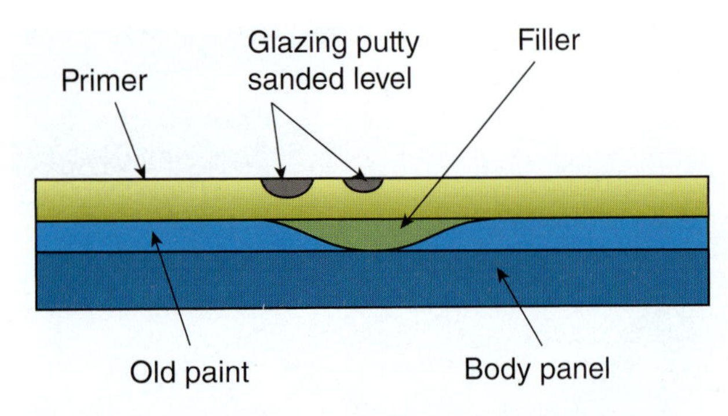

Figure 25-15 Glazing putty is like a very thick primer, only it will fill minor surface imperfections such as pinholes. If a large area has tiny pinholes, you might want to spray the panel with primer-surfacer instead of using spot putty.

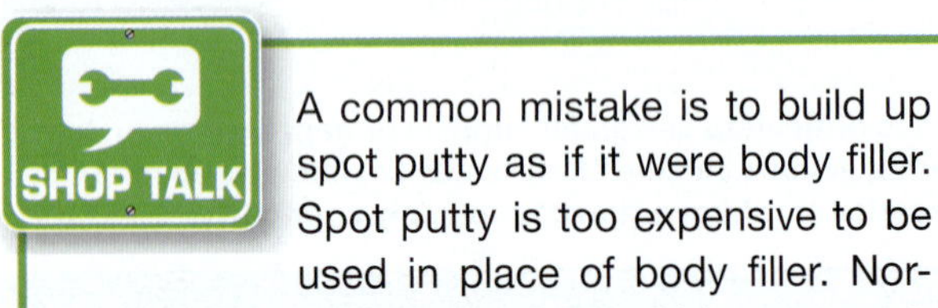

A common mistake is to build up spot putty as if it were body filler. Spot putty is too expensive to be used in place of body filler. Normally, only use two-part putty. Older one-part, lacquer putty is seldom used because it dries so slowly that shrinkage problems can result (Figure 25–16).

Spot putty is normally applied over the primer or primer-surfacer (Figure 25–17).

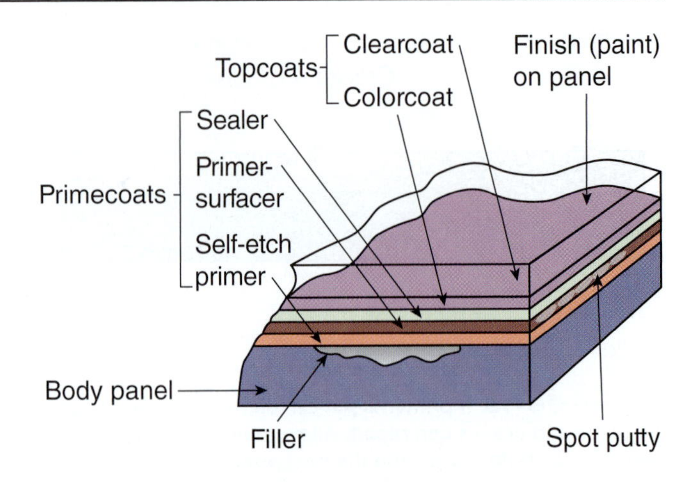

Figure 25-17 Note how different layers of repair materials are often applied during panel repair.

When featheredging and sanding spot putty, use a very fine grit dry sandpaper or wet sand with very fine wet sandpaper. Many paint technicians use #220- to #600-grit sandpaper to featheredge spot putty.

If the successive layers of paint are not properly tapered, a depression called a **bull's-eye** will show up under the freshly painted finish. This condition can usually be corrected by extending each paint and primer ring farther from the bare metal. Do this until the depression can no longer be felt when a hand is run over the featheredged area. Occasionally, when featheredging areas with several layers of paint, primer and putty might

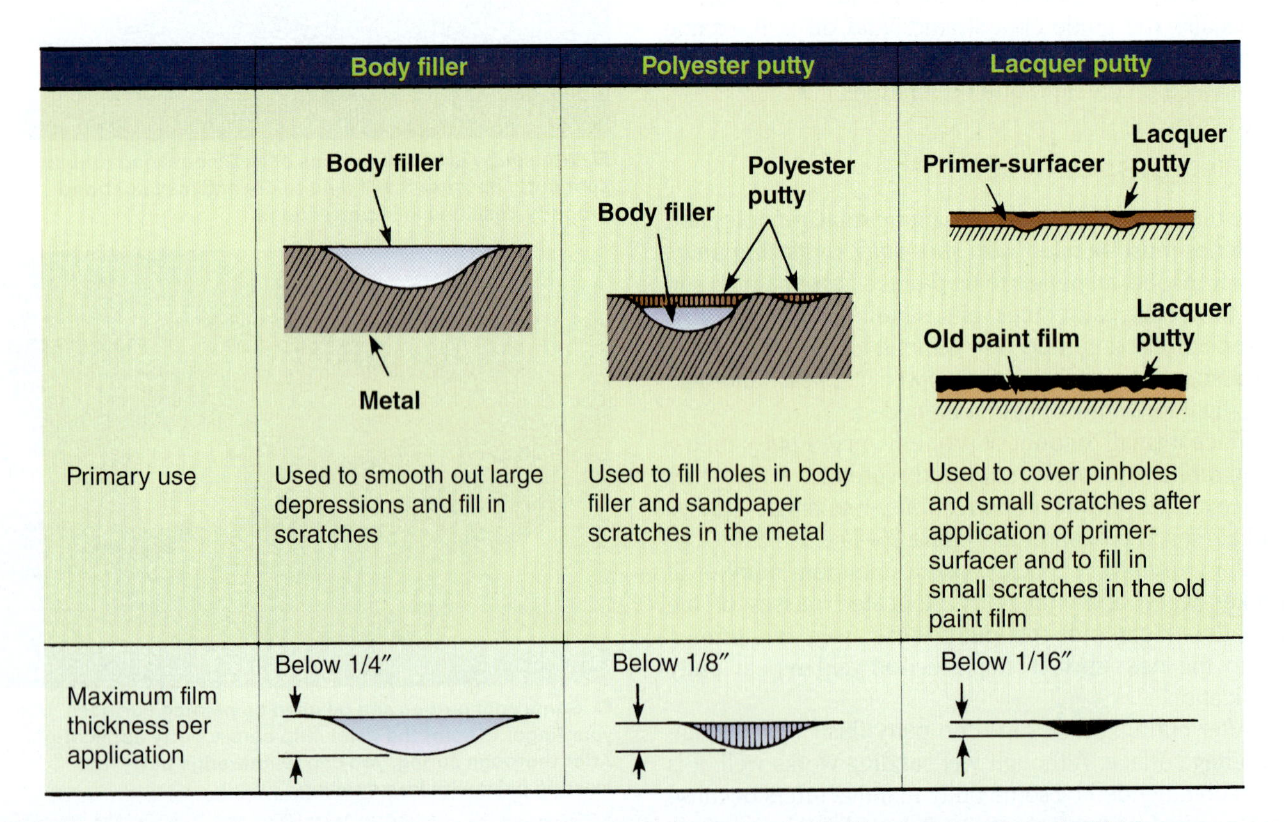

	Body filler	Polyester putty	Lacquer putty
Primary use	Used to smooth out large depressions and fill in scratches	Used to fill holes in body filler and sandpaper scratches in the metal	Used to cover pinholes and small scratches after application of primer-surfacer and to fill in small scratches in the old paint film
Maximum film thickness per application	Below 1/4″	Below 1/8″	Below 1/16″

Figure 25-16 Compare the uses of filler and putty. Materials must not be applied beyond recommended thickness or problems will result.

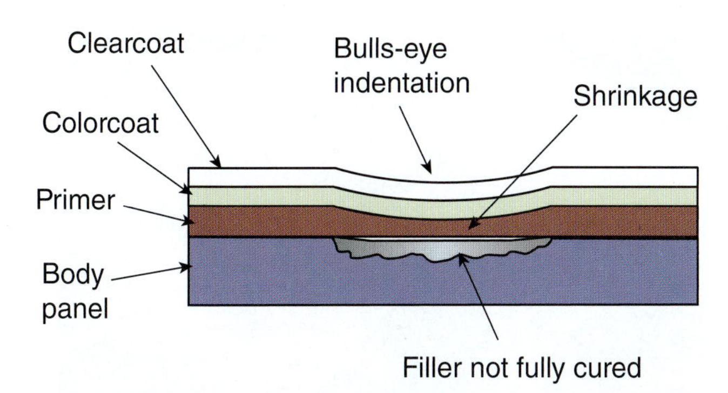

Figure 25–18 A bull's-eye is a depression formed over a repair area. It is caused by shrinkage of repair materials that were sanded too soon or from improper featheredging when sanding. You have to spray on primer-surfacer and block sand larger areas to correct a bull's-eye problem.

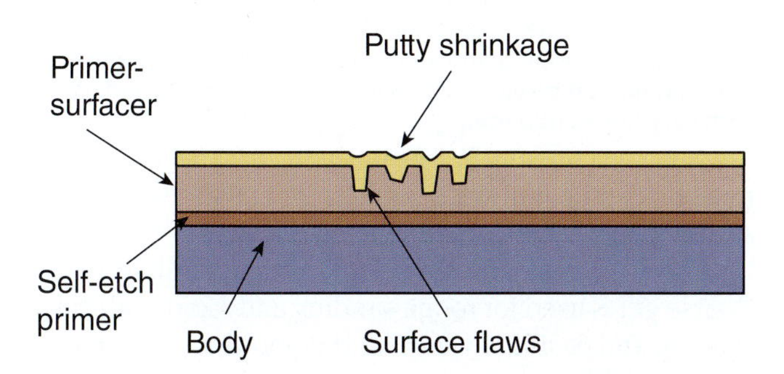

Figure 25–19 Even when using two-part putty, allow glazing putty enough time for proper curing or drying before sanding. If you sand the putty too soon, it can shrink more over thicker areas and cause indentations in the paint surface.

be necessary to fill the bull's-eye to the level of the existing film buildup (Figure 25–18).

If a large surface area has small pits, spray primer-surfacer over that area. Primer-surfacer will coat and fill in tiny imperfections. You can block sand the area to make it level and smooth.

Allow the putty to air dry or cure until it is hard. If it is sanded too soon, the putty will continue to shrink, leaving part of the scratch unfilled (Figure 25–19).

Once it hardens, the putty should be dry sanded with #220-grit paper. After sanding the puttied area, clean the surface and then reprime if needed. If the putty has been wet sanded, make sure to dry the surface thoroughly before applying sealer.

Figure 25–20 shows that the shrinkage and swelling of primecoats is an important point to consider in the elimination of sand scratches. If the primecoat is not allowed

Using too much glazing putty indicates a substandard repair. Spot putty should only be used on very small pinholes and other tiny surface problems.

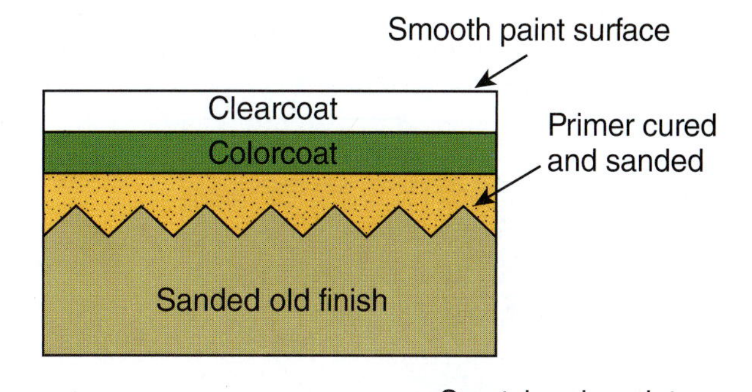

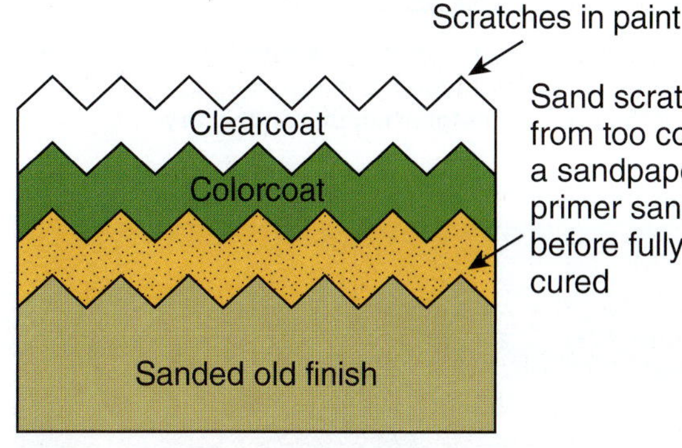

Figure 25–20 The best way to prevent sand scratches from showing is to use the correct grit sandpaper and proper sanding technique. When hand sanding, always sand in a straight line, never in a circular motion.

Figure 25–21 This quarter panel has been straightened, body filled, primed, and spot puttied. It is now ready for final fine sanding to prep it for topcoats of paint.

to dry down to its final position before sanding or applying finish coats, scratches are likely to result.

Figure 25–21 shows a quarter that has been straightened and properly prepared for final sanding according to recommended procedures.

25.5 FINAL SANDING

Final sanding involves using fine and very fine grits of material to prepare body surfaces for painting. Final sanding is one of the most important steps in surface preparation. In fact, this operation is a standard part of most surface preparation procedures. Refer to Figure 25–22.

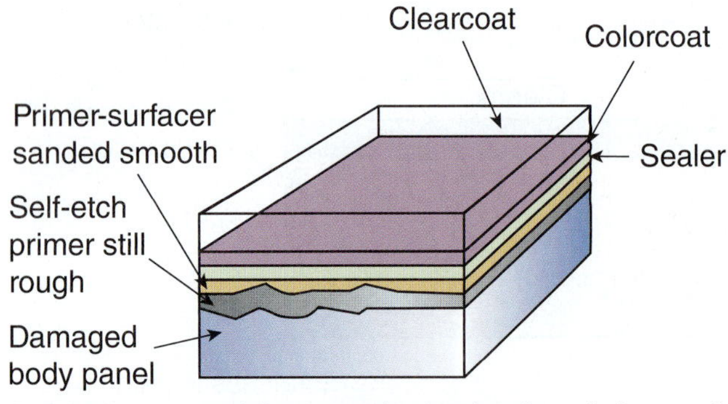

A **Primecoat materials are used to level and repair damaged metal. The surface of primecoats follows the approximate contour of the original metal when thoroughly dry.**

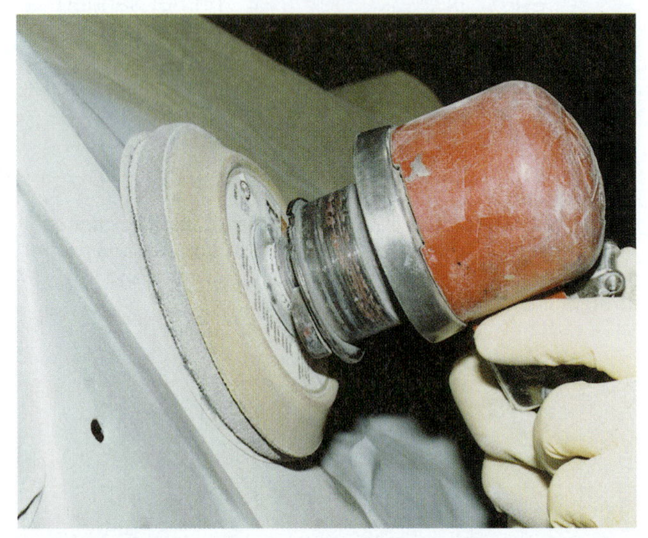

B **Final sanding levels off any remaining high or low spots, producing a flat, smooth surface. Uneven or rough surfaces are hard to see in dull primer undercoats. They will be more evident when sprayed with a shiny topcoat.**

Figure 25-22 **Study this enlarged cross section of rough metal.**

Courtesy of Dynabrade Inc.

Figure 25-23 **When final sanding, only use a power sander when needed. Hand-sand when possible to better control featheredging and smoothing of the last primecoat. This sander can be used with water for power wet sanding with very fine sandpaper.**

The entire surface to be refinished must be scuff sanded to improve adhesion of the new paint. A clean, scuffed surface is very important for proper bonding of the new topcoat. Look at Figure 25–23.

Because coated abrasives (sandpaper) perform the actual cutting and leveling in the sanding operation, selecting the correct abrasive is critical to the quality of the finished work.

Use the Right Grit

REMEMBER *When sanding during surface preparation, always use the correct grit or sandpaper coarseness. The lower the number on the back of the sandpaper, the coarser the grit.*

Very coarse grit includes #16 to #24 grinding discs and sanding discs. Very coarse grit is used for rapid removal of paint down to bare metal. Very coarse grit abrasives must be used carefully to avoid thinning and weakening a body panel. They are dry abrasives.

Coarse grit includes #36–#60 rough sanding discs. Coarse grit is used for rough sanding and rapid body filler shaping and to initially level and contour a large area of body filler. They are normally dry abrasives.

Medium grit of #80–#120 grit is often used to continue sanding body filler and for quickly sanding off old paint. Medium grits will remove material quickly but will not leave sand scratches as deep as coarse grits. They are usually dry sandpapers.

Fine grit of #150–#180 is normally used to continue smoothing filler or painted surfaces. Fine grits are also used to final sand body filler and to featheredge paint and filler. They are usually dry sandpapers.

Very fine grit ranges from #220 to about #600 grit and is used for numerous final smoothing operations. Larger grits of #220–#360 are for the first phase of smoothing and final leveling of body filler and old paint. Smaller grits of #400–#600 are for removal of sand scratches to prepare softer primer and primer-surfacer before painting. They can be either dry or wet sandpapers.

Ultrafine grit sandpaper ranges from #800 to #2000 grit. Ultrafine grits are often used to remove minor paint flaws. They are for final wet sanding before machine buffing. Ultrafine grits are usually wet sandpaper to keep the paper from becoming clogged with paint.

Ultrafine #800 grit might be used to knock the high spots off a paint run. Then, even finer grits would be used to final smooth and level the paint flaw.

Also called *compounding sandpapers*, the #1200, #1500, and #2000 grits are used to solve problems on basecoat or clearcoat paint surfaces before compounding or machine buffing. Sandpaper with #4000 grit is now available for color shading.

Generally, start your work with the coarsest grit practical. This will remove and smooth the work quickly. Then,

Figure 25–24 When scuff sanding, use very fine sandpaper to knock the gloss off the old paint.

gradually go to finer paper to achieve the desired surface smoothness. Refer to Figure 25–24.

Surface Sanding Methods

Refinishing sanding can be done by hand or by using power equipment. Most heavy sanding, such as removing the old finish, is done with power sanders, but some conditions, particularly the delicate operations, dictate hand sanding.

To the untrained eye, a dull primer surface may look smooth and ready for paint. You must remember that the shiny paint will act like a magnifying glass to exaggerate any scratches or irregularities. If you ever paint over a flawed surface, you will never make this mistake again. It is time-consuming, embarrassing, and frustrating to resand and repaint after making surface preparation mistakes.

Power Sanding

An orbital sander, or DA sander, moves in two directions at the same time. This produces a much smoother surface finish. A DA sander is used to featheredge a repair area and to final sand larger panel areas. It is the workhorse of body technicians.

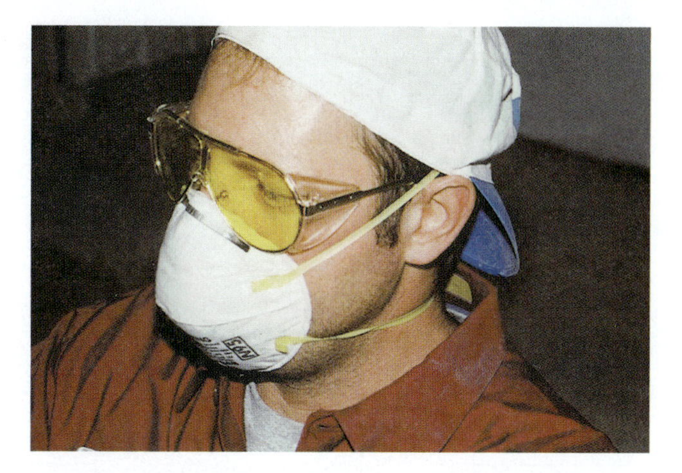

Figure 25–25 When sanding is being done in your surface prep area, wear a dust mask. You would be amazed at how much paint dust a cloth respirator will keep out of your lungs in just one day.

DANGER

Remember to wear a dust respirator when sanding. Sanding dust can contain toxic paint chemicals. If you do not wear a respirator, you may not feel sick right away, but long-term exposure can result in serious disease and illness (Figure 25–25).

To operate an air sander, set the air pressure at the equipment manufacturer's specifications (typically about 70 psi, or 476 kPa). If you are right-handed, hold the handle of the sander in your right hand while using your left hand to apply light pressure and guide the tool.

Table 25–3 shows which type of sander should be used for various areas and operations.

To protect chrome and emblems from damage, do not sand too close to trim and moldings. Mask nearby trim, decals, glass, handles, and emblems to prevent metal sparks from pitting these surfaces. In fact, it is a good idea to double-tape all moldings and trim on the panel before sanding.

When using any mechanical sander, particularly a disc grinder, keep it moving so that no deep scratches, gouges, or burn-throughs develop. Except when sanding bare metal, do not power sand styling lines as this will quickly distort the styling edge.

When power sanding, replace the sandpaper as soon as paint begins to cake or "ball up." This paint buildup can scratch the surface and reduce the sanding action of the disc. Slowing down the speed of the sander helps prevent paint buildup on the sanding disc and prolongs sandpaper life. Generally, six to eight sanding discs or pads are required to featheredge the chips and scratches on the average automobile.

A **foam backing pad** should be used when power sanding crowned surfaces or when final sanding. With a Velcro sandpaper attachment, the foam backing pad fits

TABLE 25–3 USE OF SANDERS

Sander Type	Normal Area of Operation	Normal Use: Paint Stripping	Featheredging	Rough Sanding of Solder	Rough Sanding of Metal Putty	Rough Sanding of Poly Putty	Sanding of Metal Putty	Sanding of Poly Putty
Disc sander	Suitable for narrow areas	A	C	B	C	C	C	C
Dual action sander		B	A	C	A	A	A	A
Orbital sander		B	B	C	A	A	A	A
Straight line sander	Suitable for wide open spaces	B	C	C	A	B	A	B
Long orbital sander		B	C	C	A	B	A	B

Note: It is important that the correct type of sander and abrasive paper be used for each type of job. Also, always wear a mask or use some sort of dust arrester when using a sander.

A Preferred

B Acceptable

C Least preferred

between the sander pad and the sandpaper. It helps soften the sandpaper backing to allow the sandpaper to better conform to a curved surface. It also help prevents the sandpaper from digging into the surface if you accidentally tilt the sander too much. Refer to Figure 25–26.

When final sanding with a DA, the very fine grit sandpaper can become clogged. You can clean out the paint by holding the clogged DA paper against a scuff pad. Turn the DA on and hold the sandpaper against the scuff pad. The abrasive action will clean the sandpaper and get it ready for use again.

Use a thorough sanding procedure before painting for these reasons:

1. To remove paint gloss or shine to texture old paint
2. When the old finish is rough or in poor shape
3. To level and smooth primed areas
4. To reduce paint mil thickness before refinishing

Light sanding or scuffing should be done on all panel surfaces to be repainted. You must sand or scuff to dull the paint gloss or shine. This will texture the paint surface so the new paint will adhere securely to the old paint. Very fine sand or scuff all areas to be painted with the correct grit sandpaper or scuff pad.

The grit sandpaper used for final sanding before painting will depend on several variables. You may have to sand the vehicle's existing finish with grit ratings from #320 all the way to #1000.

Use a larger #320–#600 grit to remove orange peel or any roughness in the old paint. Use #600 grit to final sand old paint in repair areas.

In a blend area where you are only scuffing the clearcoat to accept more clearcoat, wet sand the area with #1000 grit wet sandpaper. This will prevent sand scratches from showing up in the blended clearcoat where the new and old clearcoats overlap.

Hand Sanding

Hand sanding is a simple back and forth scrubbing action with the sandpaper flat against the surface. Hand sanding can be done with your fingers on sharply curved surfaces. However, on flat or large curves use a sanding block to support the sandpaper when final sanding (Figure 25–27).

Do not sand in a circular motion. This creates sand scratches that might be visible under the paint finish. To achieve the best results, always sand in the same direction as the body lines on the vehicle (Figure 25–28).

When sanding sharply curved surfaces, wrap your sandpaper around a soft sanding sponge or around a round rubber sanding block. The shape of the sanding aid should match the shape of the body contour.

Although you should avoid sanding with your fingers or hands whenever possible, it is often necessary to reach into sharply curved body lines. Sandpaper is also sometimes folded over and held in your fingers to reach into tight openings between parts. Again, the shape of sandpaper backing (in this case, your fingers or hands) must match the shape of the curved surface being sanded.

Avoid using only your hands to sand flat or gently curved body surfaces. Use a sanding block. Your hand is not a flat surface, so your fingers do the sanding.

Lightly sand
sharp contours,
such as
on pillars

A **Move the sander quickly over a crowned or sharp edge, such as on this A-pillar. Avoid cutting through paint or primer.**

B **Here a technician has installed a foam pad between the sandpaper and the backing pad on the tool. It will help avoid burn-through when sanding.**

Figure 25-26 **When sanding crowned panels or sharp edges, it is easy to oversand and accidentally cut through to bare metal.**

This results in uneven pressure being applied in the spaces between your fingers.

Be sure to carefully sand around unremoved parts, such as trim, moldings, door handles, radio antennas, and behind bumpers. These parts are easily damaged during sanding. Protect them with masking tape when needed.

Dry Sanding

This is basically the back-and-forth procedure just described. One of the problems with dry sanding is that the paper tends to clog with paint or metal dust. Tapping the paper from time to time or spinning it against a scuff pad will remove some of the dust.

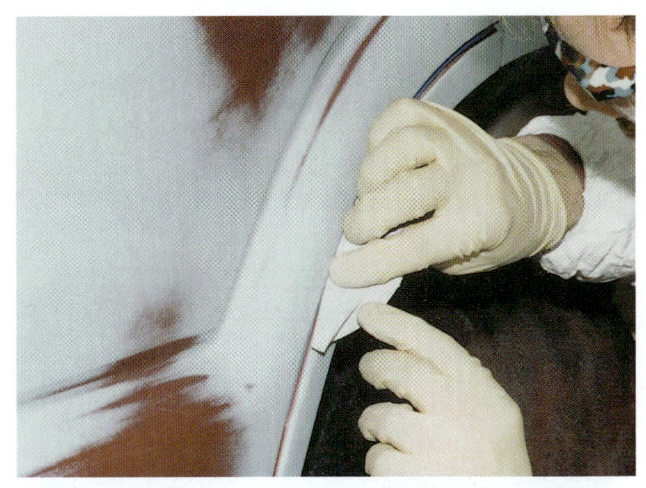

A **This paint technician is hand sanding the edge around a rear quarter panel. Small indentations were found on the edge from paint chips.**

Courtesy of Dynabrade Inc.

B **This sander is designed to sand inside louvers and similarly tight places. It can save enough time on one job to pay for itself.**

C **Here a technician is using a handy sanding pen to scuff inside recessed letters on a bumper. The tool has a tiny sanding tip made of abrasive fiberglass material. The small sanding tip will fit down into grooves to scuff, clean, and prepare them to hold paint.**

Figure 25-27 **Make sure you sand all edges and small indentations on panels to be painted.**

Figure 25-28 Generally, sand with the body lines of a panel. This will speed your work and help to prevent paint runs on the side of the vehicle.

Figure 25-29 Many paint technicians like to wet sand during final sanding. Wet sanding lets you feel surface smoothness, and it does not produce airborne dust.

Wet Sanding

Wet sanding solves the problem of paper clogging when fine sanding. It is basically the same action as dry sanding except that water, a sponge, and a squeegee are used in addition to the sanding block. Sandpapers are available in dry, wet, or wet-or-dry abrasive types (Figure 25–29).

When wet sanding, dip the paper in the water or wet the surface with the sponge. Use plenty of water to flush away old paint and sanding grit (Figure 25–30).

Most sandpaper manufacturers recommend that you soak wet sandpaper overnight before color sanding. This will soften the paper backing and help prevent tiny scratches in the paint.

Use long, smooth strokes and light pressure when wet sanding large areas. A sanding block is needed to keep the surface level and to keep your fingertips from digging troughs in the surface (Figure 25–31).

If wet sanding a small area, use a circular motion only on the specific problem area. Small dirt nibs or pieces of dust in the primer can often be removed with wet sanding in this manner (Figure 25–32).

Never allow the surface to dry during the wet sanding operation. Also, do not allow paint residue to build up on the abrasive paper.

It is possible to tell how well the paper is cutting by the amount of drag felt as it moves across the surface being sanded. When the paper begins to slide over the surface too quickly and easily, it is no longer cutting. The grit has become filled with paint particles, or sludge. Rinse the paper in water to remove the paint, and sponge the surface to remove the remaining particles. Then the sandpaper will again cut the surface.

Avoid wet sanding body filler! If not waterproof (fiberglass-impregnated filler), body filler can absorb and hold the moisture. The water can "come back to haunt you" when it shows up in the fresh paint! Power sand the repair area level with dry paper and prime the area before wet sanding.

Check your work periodically by sponging the surface off and wiping it dry with a squeegee. This will remove all excess water so that it is easier to evaluate the surface condition. It is usually wise to complete one panel or body section at a time. Then remove the sanding residues with a sponge and dry off with a squeegee before sanding the next panel.

When the surface is wet, view it from an angle or from the side. This will help magnify the surface so you can find any remaining flaws that must be wet sanded. This is shown in Figure 25–33.

Once the wet sanding operation is completed, be sure that all surfaces are dry. Blow out the seams and molding with compressed air at a lower pressure and tack cloth the entire surface. Do not touch the body surface with your bare hands (Figure 25–34).

A comparison of the advantages and disadvantages of wet and dry sanding is given in Table 25–4.

A Wet sanding does not generate dust and you can feel the smoothness of a body panel as you sand.

B Use plenty of water when wet sanding. Water speeds the sanding action by washing away debris that can clog ultrafine grit.

Figure 25-30 Wet sanding is commonly done after priming or using primer-surfacer.

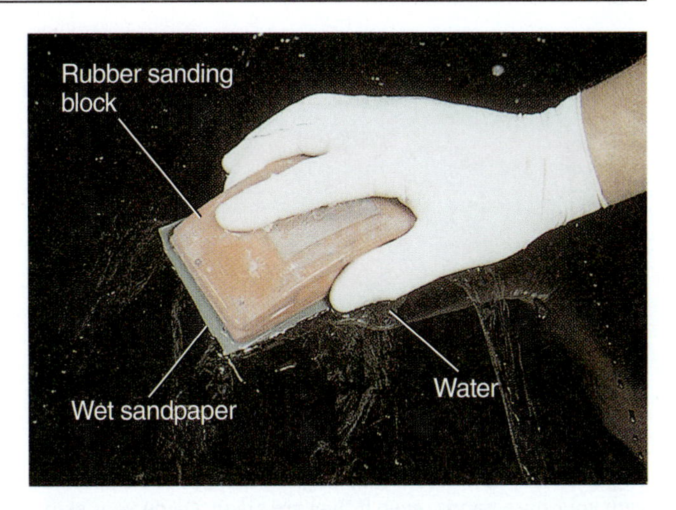

Figure 25-31 When final wet sanding flat or gently curved surfaces, use a sanding block. The block will help plane down and level minor surface imperfections much better than just using your hand.

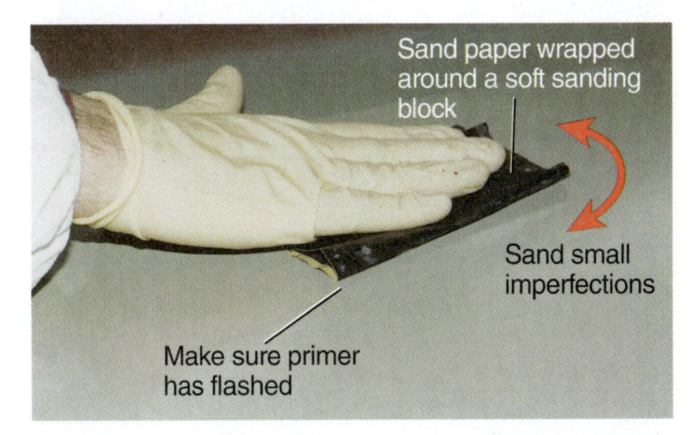

Figure 25-32 This technician has found a minor surface blemish in a freshly applied primer. After primer has flashed enough, the small blemish is wet sanded out before applying the next coat of refinish material.

Figure 25-33 After wet sanding, pour water over the repair area and view the area from an angle. The water acts like a shiny clearcoat to help you find any remaining flaws in the body surface.

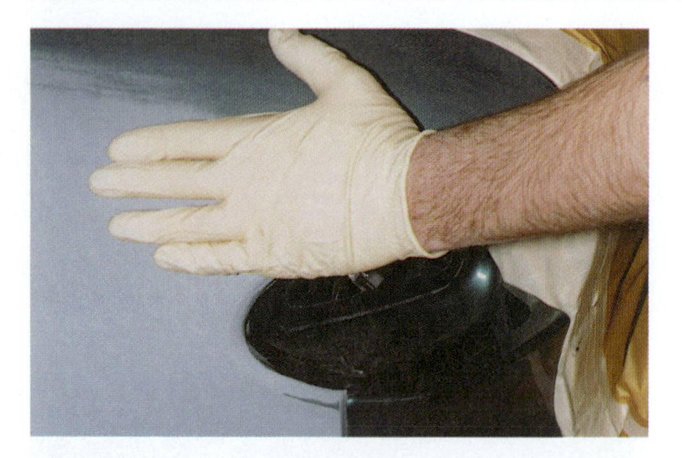

Figure 25–34 Avoid touching any surface to be painted with your bare hands, even if they are clean. Oil on your skin can contaminate the surface and cause problems when you spray topcoats of paint.

TABLE 25–4 COMPARISON OF WET AND DRY SANDING

Item	Wet Sanding	Dry Sanding
Work speed	Slower	Faster
Amount of sandpaper required	Less	More
Condition of finish	Very good	Final finish difficult
Workability	Normal	Good
Dust	Little	Much
Facilities required	Water drain necessary	Dust collector and exhaust necessary
Drying time	Necessary	Not necessary

Surface Scuffing

Surface scuffing involves using a very fine or ultrafine abrasive (paste material, sandpaper, or a scuff pad) to cut microscopic scratches in the body surface to be painted. This finely scratches the surface to aid proper paint adhesion. For example, if you paint over a fully cured, glossy paint without fine sanding or scuffing it, the new paint will probably peel and flake off.

Sanding or an abrasive paste is used to scuff relatively large, flat surfaces. Very fine sandpaper on a DA will quickly scuff large, flat surfaces. Scuffing paste can also be used. The abrasive paste can be applied by hand on small areas or applied with a power buffer on larger areas (Figure 25–35).

Scuff pads, made of a synthetic plastic sponge-like material, are commonly used for texturing or scuffing surfaces that will be painted. Scuff pads come in different stiffnesses, which relate to coarseness. A stiffer scuff pad will produce slightly deeper scuff marks than a softer scuff pad.

Courtesy of Norton/brand of Saint-Gobain Abrasives

A Wall cabinets should be kept stocked with various types of final sanding products.

Courtesy of Norton/brand of Saint-Gobain Abrasives

B A rolling cart is handy because you can take a supply of abrasives and other final prep products to the vehicle you are working on.

Figure 25–35 Many types of abrasives are used during final surface preparation: fine sandpapers, scuff wheels, scuff pads. You should know when and how to use each type.

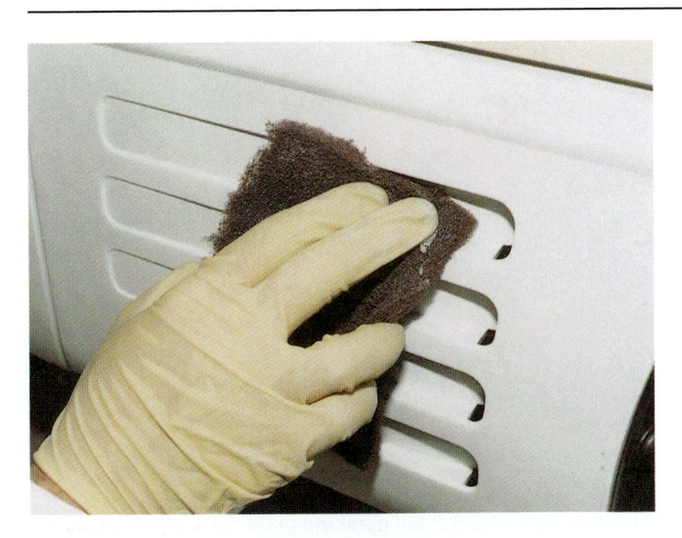

Figure 25-36 **Rub a scuff pad over all surfaces that cannot be easily sanded. Scuff pads work well on hard-to-reach surfaces around edges of panels. The louvers in this front fender must not be overlooked. Make sure all surfaces are scuffed and dulled to ready them for paint.**

All surfaces to be painted must be sanded or scuffed until they are dull

Figure 25-37 **Note how sanding and scuffing have removed the shine from all surfaces to be painted.**

Scuff pads are commonly used to knock the gloss off hard-to-reach areas on a vehicle body. Use a scuff pad on panel edges, on deep contours, around openings in panels, and on other surfaces that cannot be easily reached. Some examples of areas where a scuff pad might come in handy are shown in Figure 25–36.

All surfaces that will be painted must be dull and nonreflective. Any shiny paint areas that have not been scuffed will not take and hold the new paint securely. Double-check all body surfaces to make sure they are scuffed and ready for good paint adhesion. Look at Figure 25–37.

25.6 MASKING

Masking keeps paint from contacting areas that are not meant to be refinished or painted. It is a very important step in the vehicle surface preparation process (Figure 25–38).

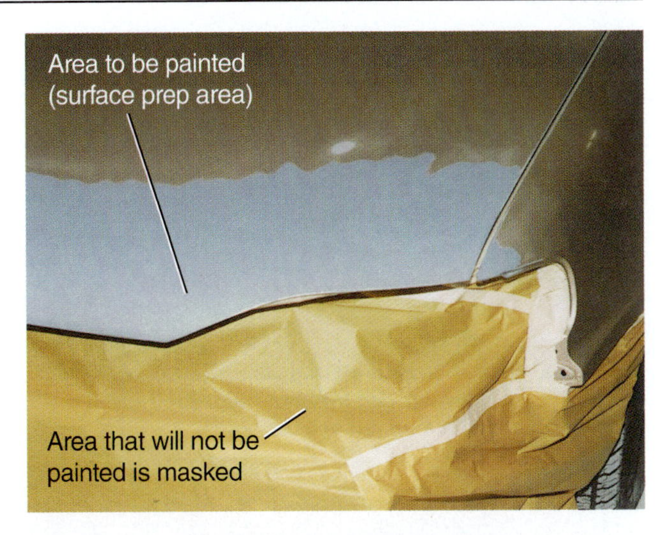

Figure 25-38 **Masking materials are used to cover panels and parts that will not be painted.**

Masking has become even more important with the advent of two-part paints. Once these paints dry, the overspray *cannot* be removed with a thinner or other solvent. Any overspray must be removed with a rubbing compound or by other time-consuming means. Proper masking takes great skill, as in Figure 25–39.

There are four basic ways to mask the parts of a vehicle:

1. With masking paper and masking tape
2. With plastic sheeting and masking tape
3. With specially shaped cloth or plastic covers (for wheels, the antenna, and rearview mirrors)
4. With liquid masking material

Traditional masking products may not work with water-based paints because of their high water content. If using water-based products, check with your supplier to determine which masking papers and tapes are made for use with these paints.

Clean Before Masking

Before any types of masking materials are applied, the vehicle must be completely cleaned and all dust blown away. Masking tape and other masking materials will not stick to surfaces that are not clean and dry. It is important that the tape is pressed down firmly and adheres to the surface. Otherwise, paint will creep under it.

It is wise to completely clean and detail the vehicle before masking and again after the refinishing job is completed. This is because masking over a dirty vehicle can cause a "dirty paint job."

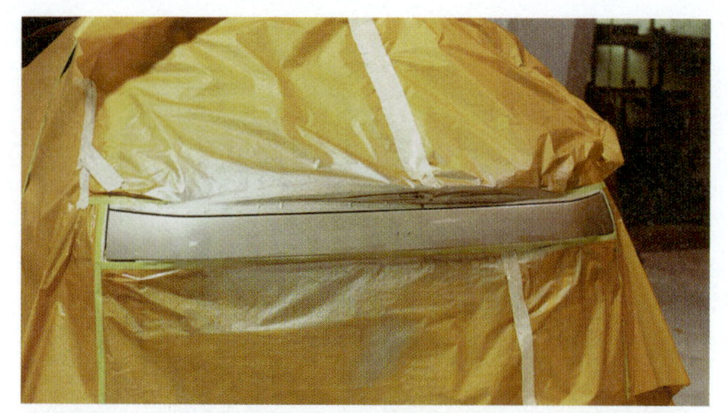

A Look at all of the masking material needed to repaint the small panel on the rear of this vehicle.

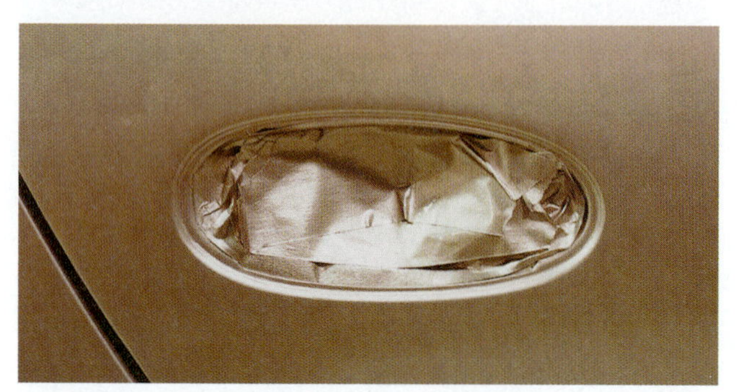

B After the door handle was removed, the hole in the panel was masked to keep paint spray out of the door.

Figure 25–39 Proper masking takes great skill. It can be a simple task for some jobs or very time-consuming for others.

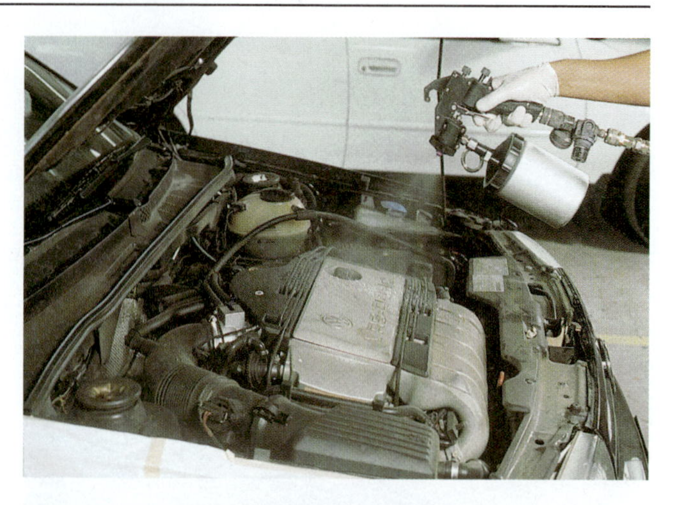

Figure 25–40 Liquid masking material is often used on large surfaces that will not be painted. After painting, liquid mask material washes off with soap and water or with a pressure washer. Here a spray-on mask is being applied to the engine compartment to protect the engine and other parts from overspray.

Local regulations may require that liquid masking residue be captured in a floor drain trap and not put into the sewer system.

Liquid Masking Material

Liquid masking material seals off large, complex surfaces of the vehicle to protect them from paint overspray. It is the newest masking system available to painters. Liquid masking is used on areas where masking is necessary, but difficult to apply, including wheel wells, headlights, the grille, the underbody chassis, and even the engine compartment.

Masking liquid, also called masking coating, is usually a water-based sprayable material for keeping overspray off body parts. Some are solvent-based. Masking liquid comes in large, ready-to-spray containers or drums. These materials are sprayed on and form a paint-proof coating over the vehicle (Figure 25–40).

Some masking coatings are tacky and are used only during priming and painting. They form a film that can be applied when the vehicle enters the shop. Others dry to a hard, dull finish.

Liquid masking material is usually left on the vehicle until after buffing or compounding. When buffing compound sprays off the buffing pad, the masking coating will keep the compound from sticking to body surfaces. Because the masking coating washes off with soap and water, this will ease final cleanup. The masking coating is often removed right before the vehicle is ready to be returned to the owner or customer.

To mask a vehicle using the liquid masking system, proceed as follows:

1. Partially mask the area to be painted by going around it with masking paper. Fold the paper over onto the area to be painted. Secure the paper with masking tape.
2. Apply the liquid masking material. Use a heavy, single, overlapping coat. Apply the material to all surfaces that are not meant to be painted. This would include bumpers, grilles, doors, windshields, body panels, wheels, wheel wells, door jambs, and even the engine compartment. An airless spray system is generally recommended for applying the masking material.
3. Fold the masking paper back over the liquid masking material. Wipe away any material from the area to be painted with a damp sponge. Allow the surface to dry.
4. Prepare the surface. Then apply primer and paint according to the manufacturer's instructions.
5. Allow the paint to dry, then unmask the vehicle. Liquid masking may be used in both air dry or bake conditions.
6. After the paint is cured, wash off the dried liquid masking material with a garden hose or pressure wash.

Plastic Sheet Masking

Plastic sheet masking is very thin clear plastic that can be used to cover and protect large body surfaces. If one end or section of a vehicle is going to be painted, plastic sheet masking can be used to cover the other end of the vehicle to protect it from overspray. Refer to Figure 25–41.

When using plastic sheeting, drape it over the vehicle. Do not let the plastic touch the ground and get dirty. Open the sheeting over the top of the vehicle and allow it to drape down over the sides.

Lap the plastic sheeting under any masking paper. The masking paper should be applied over the masking plastic sheet. See Figure 25–42.

Non-corona-treated plastic sheet should be used next to a panel that will be painted, however. The soft, thin plastic does not absorb paint, which can allow dripping onto the fresh paint. Masking paper absorbs wet paint to help avoid dripping.

Mono-linear plastic sheeting is also so thin that it can blow onto the freshly painted surface more easily than masking tape and paper. However, newer paintable, tri-layer, corona-treated sheeting can be used right up to the paint line, won't flake paint, can be baked, and eliminates paper altogether (Figure 25–43).

Poke a hole in the plastic as needed for an antenna or other obstruction so that the sheet lays flat. Tape down the corners of the plastic sheet so it cannot blow around with the wind or when spraying. Also mask any part sticking through the plastic sheeting.

Figure 25–44 shows a technician masking an antenna that is sticking through the plastic.

Figure 25-41 **Because only the trunk area of this car is going to be painted, the technician is covering the front section with masking plastic. Try to keep the plastic from touching the floor.**

Figure 25-42 **The masking paper should lap over the plastic. Tape the plastic to the body and then the paper to the plastic.**

Masking Paper and Tape

Automotive masking paper is heat-resistant so that it can be used safely in baking ovens. It also has good wet strength, freedom from loose fibers, and resistance to solvent penetration. Automotive masking paper comes in various widths—from 3 to 36 inches (76–914 mm).

Masking paper comes in different grades and colors: green, gold, gray, yellow, blue, and white. Some types of paper are designed for masking when priming. Others are designed for masking when painting. Make sure you use the correct type of masking paper as recommended by the manufacturer.

Primer masking paper (often green or gray) is suitable for masking off primer spray, but not paint spray. It is less expensive, more porous masking paper. If you use it to mask off when painting, the paint solvent can soak through the paper and get onto the unpainted surfaces, possibly damaging the finish.

Paint masking paper (often gold, yellow, blue, or white) is a nonporous paper designed for masking off paint spray. It is a more expensive paper that will keep paint from soaking through the paper and onto body surfaces.

Never use newspaper for masking a vehicle since it does not meet any of the requirements for properly blocking out primers and paints. Newspaper has the added disadvantage of containing printing inks that are soluble in some paint solvents. These inks can be transferred to the underlying finish, causing paint staining.

A

B

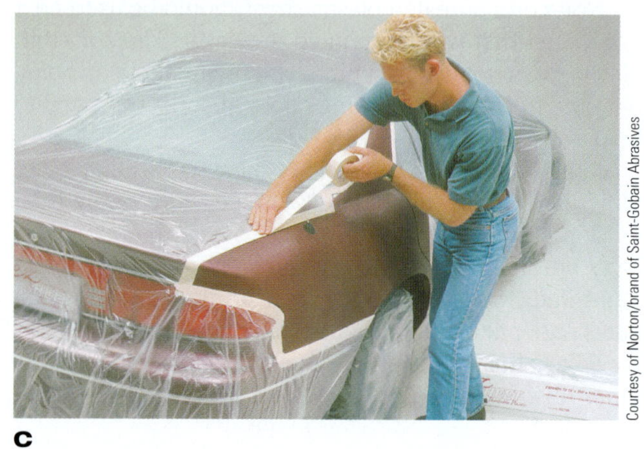

C

Figure 25–43 Here is a trick for quickly masking a small area or panel during a spot repair: (A) Apply fine line and then regular masking tape around the area to be painted. (B) Spread paintable, corona-treated, tri-layer plastic sheet over the vehicle. Stretch the plastic and tape it in place around the perimeter of the car. (C) Use a single-edge razor blade to cut the plastic around the area to be painted. Then use several passes of masking tape to secure the cut edge of the plastic to the vehicle. Make sure the plastic cannot flop and hit wet paint.

Masking tape is a very sticky paper tape designed to cover small parts and also to hold masking paper in place. Automotive masking tape comes in various widths—from 1⁄16 to 2 inches (2–51 mm). Larger width tapes are used only occasionally because they are expensive and difficult to handle (Figure 25–45).

It is interesting to note that an average-sized vehicle takes 2–2½ rolls of tape to be completely masked.

Fine line masking tape is made of thin, flexible plastic with an adhesive on one side. It is designed to be used when extreme accuracy is needed along an edge. Fine line masking tape should be used when masking around parts that cannot be removed.

Also called flush masking tape, fine line masking tape can be used to produce a better paint edge—that is, the edge where old paint and new paint meet. When the fine line tape is removed, the edge of the new paint will be straighter and smoother than if conventional masking tape were used.

Fine line tape can be used to protect existing stripes from overspray. Also use fine line tape for precise color separation in two-tone painting and for creating vivid, clean stripes. Its added flexibility makes painting curved lines easier, with less reworking.

Fine line tape is often used at the paint parting line (line separating surfaces to be painted and not to be painted). Then, conventional tape is placed over half of the fine line tape to secure the masking paper. In the case of a two-tone finish where the color break is not hidden by a stripe or molding, use fine line masking tape and

Automotive masking tape should not be confused with masking tape purchased at hardware or paint stores for home use. Regular masking tape will *not* hold up to the demanding requirements of automotive refinishing.

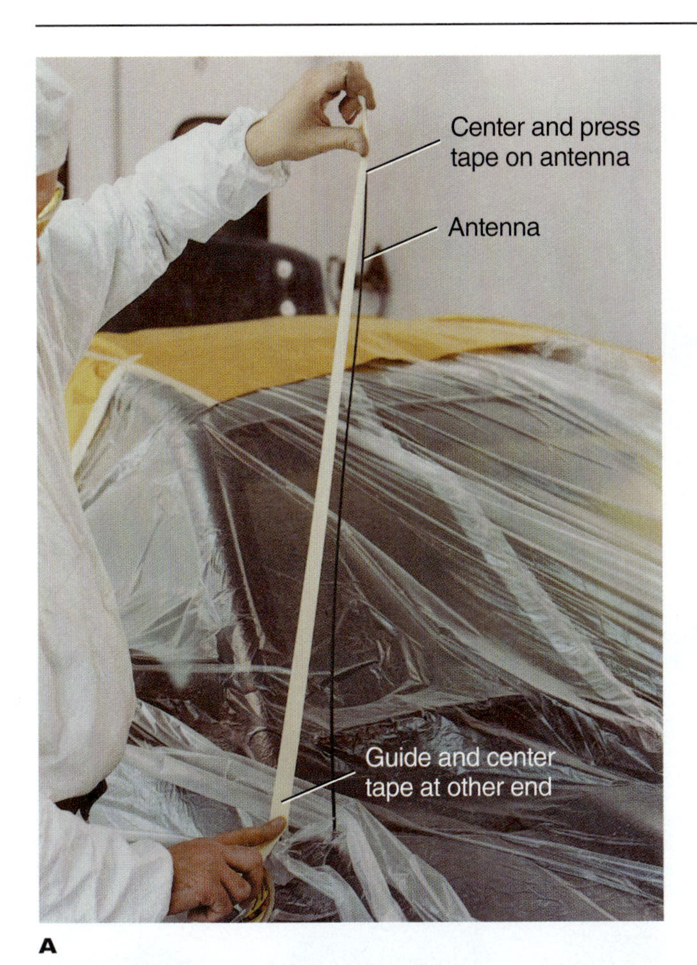
Center and press
tape on antenna
Antenna
Guide and center
tape at other end

A

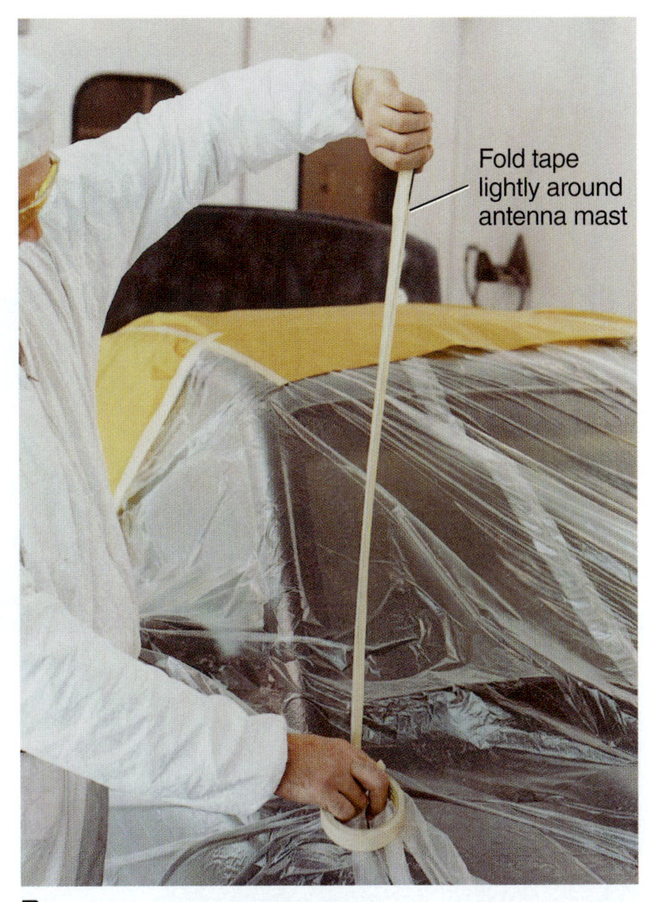
Fold tape
lightly around
antenna mast

B

Figure 25-44 Force the plastic sheeting down over the antenna and then mask the antenna. (A) Attach tape to the top and then the bottom of the antenna by pulling it straight with your thumb and finger. The center of the tape should attach to the antenna. (B) Gently fold the sides of the tape around the antenna. Do not press the tape down too hard or it will be difficult to remove.

press its edge down firmly. Do not use conventional masking tape on paint parting lines.

A *masking paper dispenser* automatically applies tape to one edge of the masking paper as the paper is pulled out. This saves time when masking a vehicle. The average vehicle takes 2–2½ rolls of tape to be completely masked. The use of masking paper and tape-dispensing equipment makes it easy to tear off the exact amount needed (Figure 25–46).

To use a masking paper dispenser, make sure the tape is lined up with the edge of the paper. One-half of the masking tape should extend out beyond the masking paper. Adjust the arm holding the masking tape as needed to align the tape for proper feed onto the paper. Pull straight out on the masking paper to the desired length. Then, jerk the paper down and sideways over the serrated jaw to tear the paper and tape straight off the roll.

Masking Versus Removing a Part

There are no clear ground rules on when to mask and when to remove parts. This includes decisions about parts such as trim, moldings, door handles, antennas, and so on. The decision to remove or mask depends on the design of the vehicle, insurance company stipulations, and the expectations of the customer. See Figure 25–47.

If the insurance company or customer will pay for it, it is always better to remove rather than to mask parts. Removal allows for new paint to be deposited underneath where the part will mount. You can clean and scuff the surface under the part before painting.

If you mask around the part, the paint will stop right next to the fine line masking tape. Because the surface under the part cannot be properly cleaned and scuffed, the paint may someday start to peel next to the masked part.

Even when you remove a part, you must back mask the hole to keep overspray out of the openings. Masked and removed door handles are pictured in Figure 25–48.

If it can be done easily, it is better to remove a part than to try to mask around the part. Also, if you cannot sand and clean right up to the part, the part should be removed. If it will be difficult to mask the part, you might save time removing it from the vehicle. Each part requires an individual decision. Some trim parts, such as windshield or glass moldings, are easily damaged and often ruined during removal and this should be accounted for in the estimate (Figure 25–49).

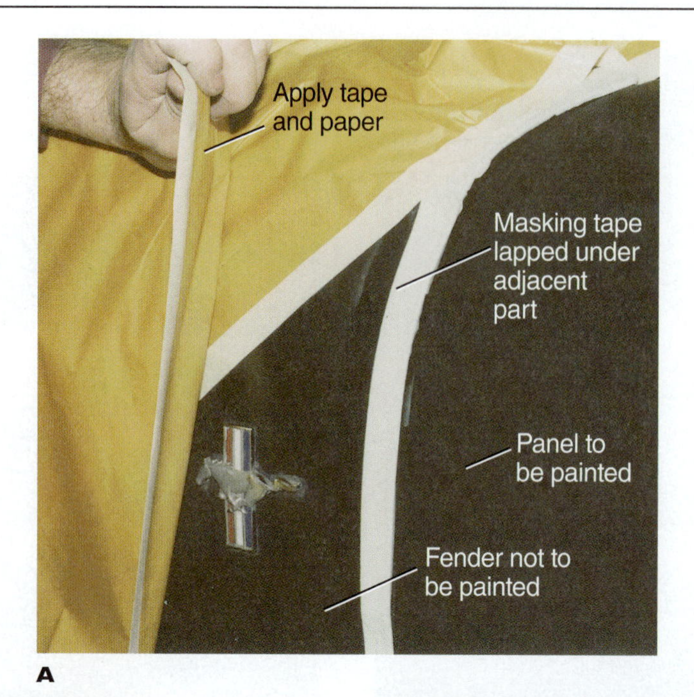

A

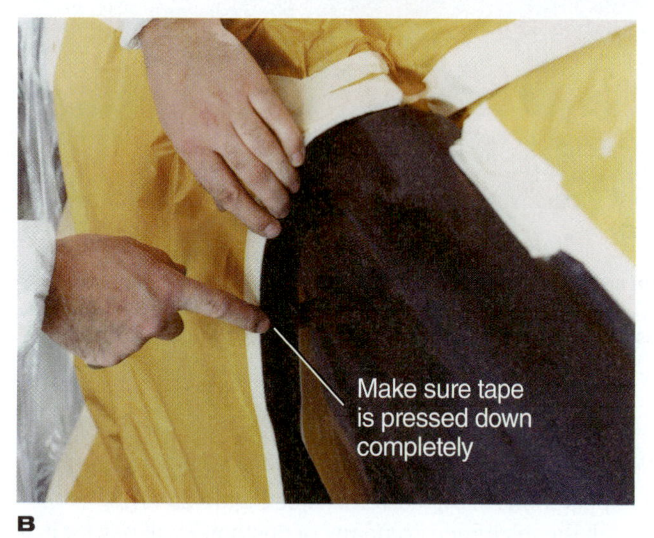

B

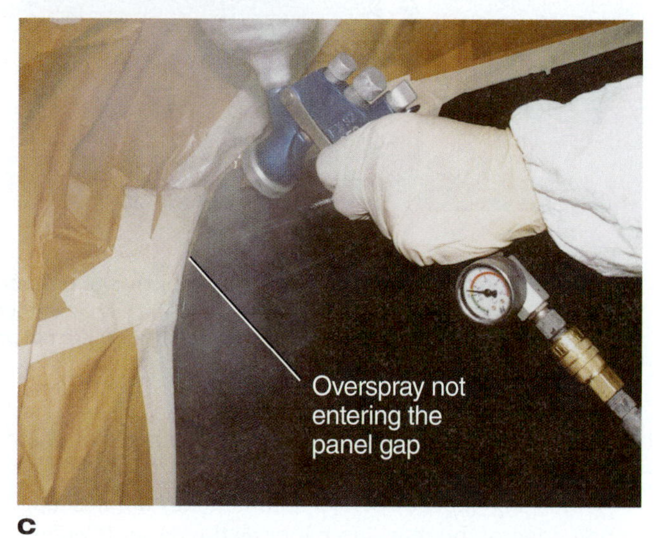

C

Figure 25-45 Here a technician is masking a front fender so a door can be color blended and then sprayed with clearcoat. (A) This technician applies masking tape to the edge of the fender, then applies masking paper over the tape on the fender. (B) The technician makes sure the edge of all masking tape is stuck down securely by running a finger over the masking tape to make sure it is securely adhered. (C) The technician is spraying clearcoat over a door. Each spray gun pass should hit the masking material to ensure an even coat on the door.

Removal of trim is more often necessary when using a basecoat/clearcoat system than with a single-stage finish. The film buildup can be greater with basecoat/clearcoat. This additional thickness makes the paint edge more likely to crack or chip.

Decal and Tape Stripe Removal

To remove small decals or tape pinstripes from a painted surface, first try using a *rubber scrub wheel* in an air drill. The spinning rubber wheel will usually heat, abrade, and lift the tape or decal without damaging the painted surface.

On larger or stubborn decals or stripes, slip a razor blade under the edge. This will start removal of a small area. Then the decal or stripe tape can be pulled up and peeled off entirely.

If the decal or stripe will not peel or abrade off, use a heat gun to warm and soften its adhesive. Heat the decal or stripe and surrounding surface to soften the adhesive, then peel it off. There are also chemical decal removers on the market. However, they must be used with care because they can damage the surface on which the decal or stripe is applied.

Decal installation includes preparation, application, and refinishing of decals. Before installing new decals, clean the body surface with wax and grease remover. Use a tape measure and water-soluble marker to mark the body surface for proper decal location and alignment.

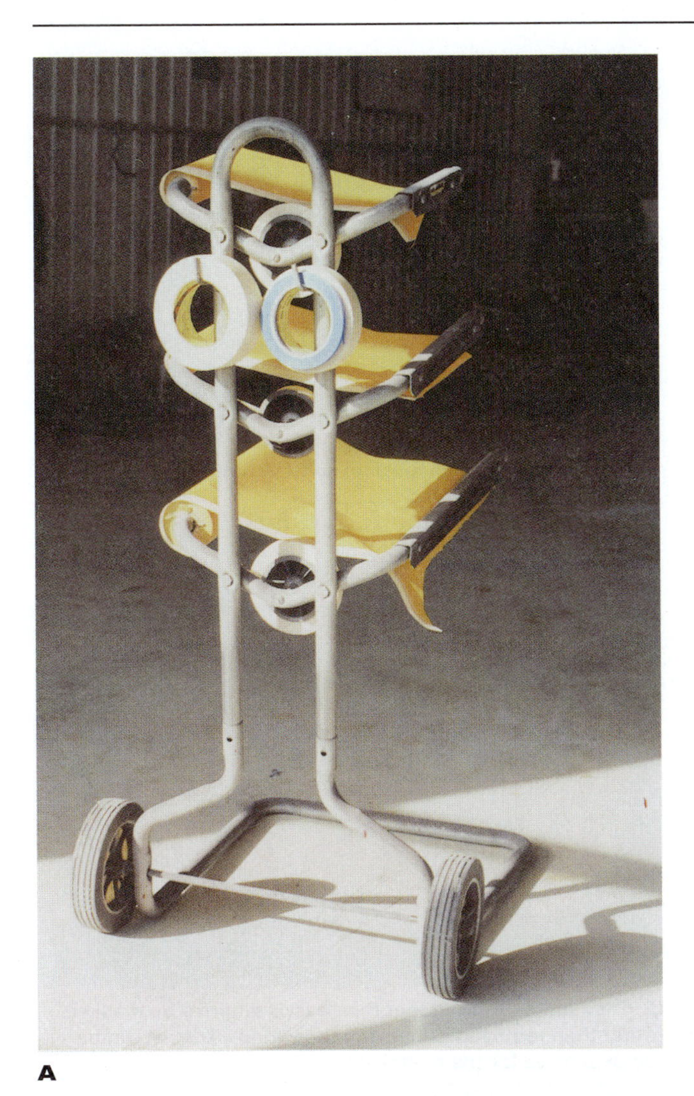

A

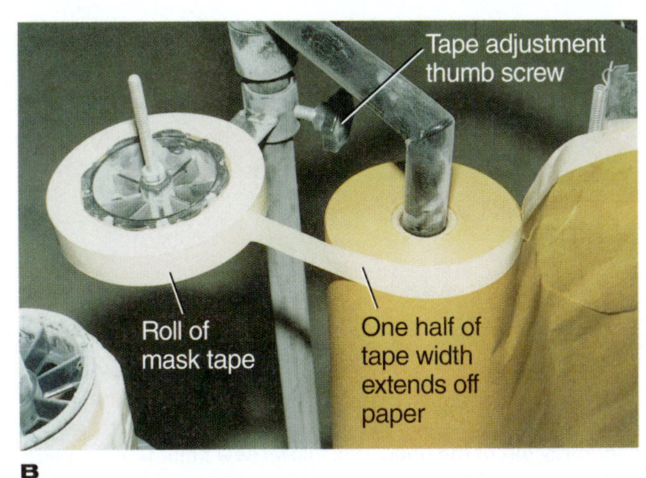

B

C

Figure 25-46 A masking machine or dispenser saves time when prepping a vehicle for painting. (A) A masking dispenser has several widths of masking paper. Use the appropriate width to avoid waste. (B) Make sure the masking tape is aligned with the paper edge so that the tape properly adheres to the paper. Half of the tape width should hang off the edge of the paper. (C) Pull masking paper straight off the roll. Quickly jerk paper across the serrated blade to cut off the paper evenly.

Figure 25-47 Skill and knowledge are needed to properly mask a vehicle. Mistakes in masking can cause costly paint problems.

You might want to apply masking tape over your marks as a guide for aligning very large decals.

Mix a couple of teaspoons of dishwashing liquid into a gallon of water. Use a clean sponge to wipe the soapy water solution onto the body surface. This will keep the decal from sticking right away so the decal can be moved and spread out flat on the body. This is very important for large decals. Small decals can sometimes be installed without soapy water.

Tape the top of the decal into position over your alignment marks. Pull off the decal backing to expose the adhesive. Pull the decal out flat and down onto the body surface. Use a rubber squeegee and rag to smooth out the decal. Rub the squeegee toward the edges of the decal to remove any trapped air bubbles and wrinkles. The soapy water will let you shift and move the decal as needed. Keep wiping from the middle outward until the soapy water dries and the decal bonds to the surface. Peel off the carrier paper on the front of the decal carefully without tearing and damaging the vinyl decal.

Continue wiping the decal as needed. If you have any stubborn, trapped air bubbles, pierce the decal with a needle. Rub the air bubble with a rag to force the trapped air out from under the decal. Double-check that the

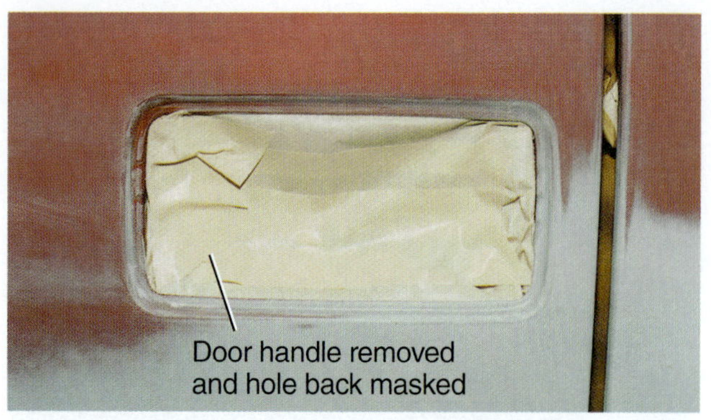

A Here the insurance company approved removal of the door handle. The hole in the door was back masked to keep overspray out of the door. This allows new paint to be applied under the door handle.

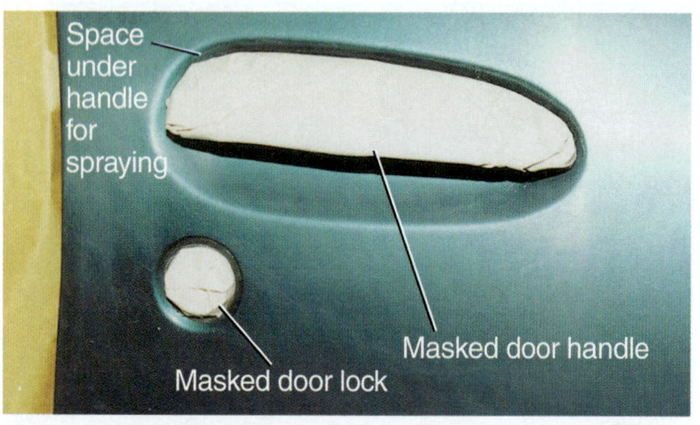

B This door handle had to be masked because the customer did not want to pay to have it removed before painting the door.

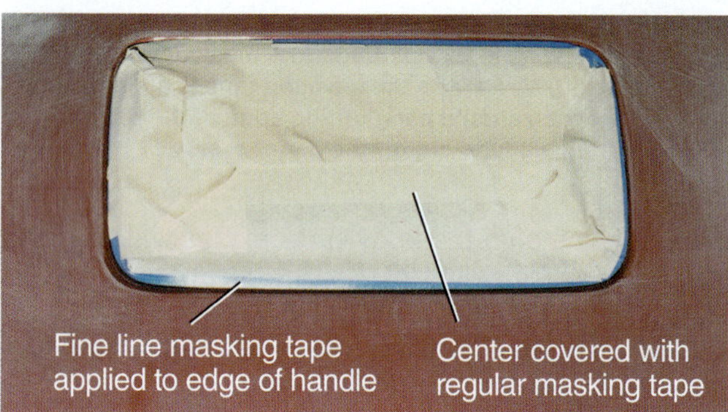

C When paint will be applied up to a masked part, first use fine line masking tape around the part. Then apply regular masking tape over the fine line tape. Extreme precision is needed when applying fine line tape.

Figure 25–48 It is always better to remove a part for painting than to mask it. However, labor costs sometimes prevent part removal, so the parts must be carefully masked.

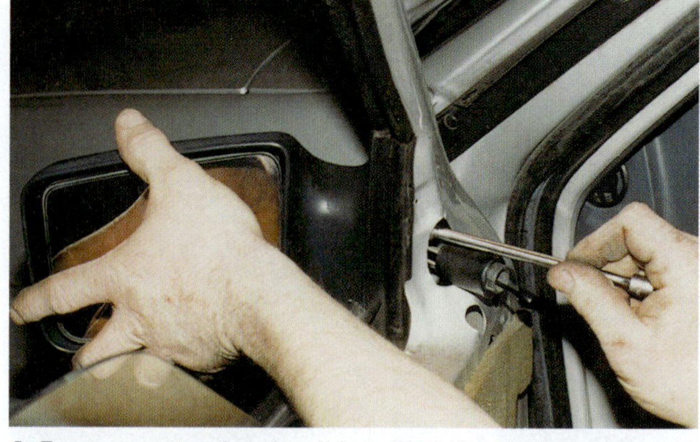

A Remove parts when feasible so that the whole panel surface can be repainted.

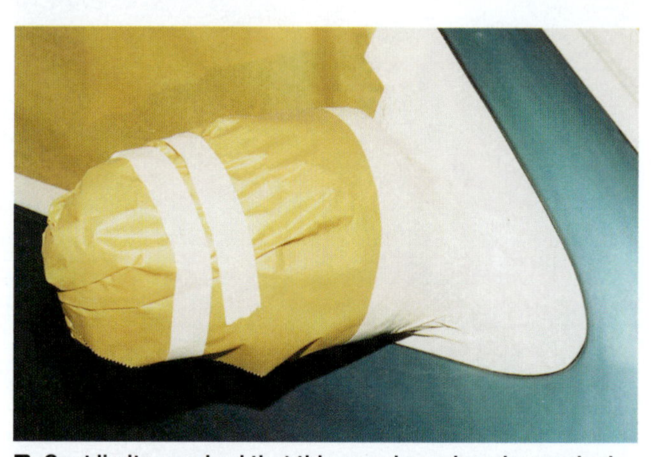

B Cost limits required that this rearview mirror be masked rather than removed. Removing the window trim would be too expensive for the customer.

Figure 25–49 The estimate and work order will determine whether parts should be removed or masked.

edges of the decal are adhered properly, and keep rubbing them outward until dry.

If you have any old, existing decals that have a dulled finish, you may need to refinish them. Rub them with glazing compound or polish so that their gloss matches the new decal.

Applying Masking Tape and Paper

If the painting environment is cold and damp, masking tape may not stick to glass or chrome parts. Condensation on them can prevent the tape from sticking properly. If the vehicle is not in an enclosed paint booth, dust will collect quickly on body surfaces. Blow and wipe the parts off before masking them to ensure good tape adhesion.

When applying masking tape, hold and peel the tape with one hand. Use your other hand to guide and secure the tape to the vehicle. This provides tight edges. This also allows you to change directions and go around corners. This is shown in Figure 25–50.

Although masking tape is elastic, do not stretch the tape too much when making a straight line. Overstretching can make the tape edges pull back up and leak paint spray. Only stretch masking tape when going around curved surfaces. Even then you should stretch the masking tape as little as possible while still following the curve of the part.

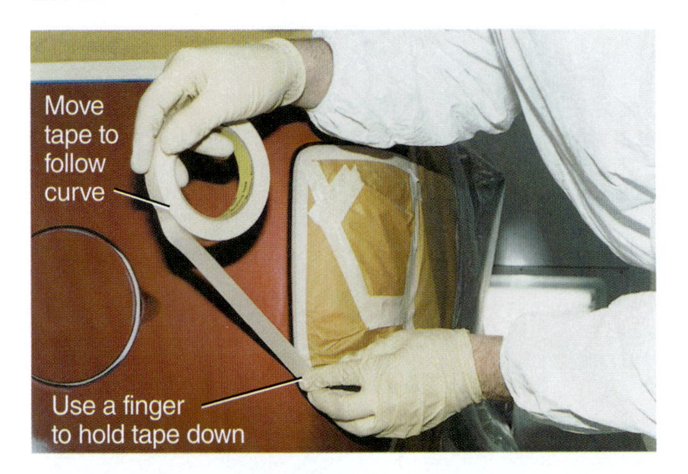

Figure 25-50 When masking around curves, guide the roll around the part with one hand while pressing the tape down with your other hand.

To cut the tape easily, quickly tear upward against your thumbnail. This permits a clean cut of the tape without stretching.

In Figure 25–51, note how fine line masking tape has been carefully applied to a body trim piece. The tape must be carefully aligned so that a tiny gap still exists along the bottom of the trim piece. This small gap along the bottom of the masked part will allow some paint to spray under the part. Regular masking tape is then applied over about half of the fine line tape. Masking paper can then be applied to cover the rest of the area to be protected from overspray.

Fine line tape can also be applied after the rest of the part has been masked. The painter can then remove the fine line tape while the paint is still wet for a superior edge.

Tape down all loose paper and plastic to keep it from blowing around when spraying on the new finish. Use masking tape to hold down large areas of masking paper. Tape along the bottom of the vehicle to hold the paper on the bottom of the body (Figure 25–52).

A *masking leak problem* is an opening in the masking material that allows paint overspray to hit and coat parts that were not supposed to be painted. Make sure all tape edges are tightly pressed down.

If you leave any of the tape loose, overspray can blow under the masking paper and get onto glass, chrome, and other surfaces. You will then have to waste time using solvent or polishing compound to remove the overspray problem after the new paint cures.

An *overmask problem* results when you accidentally place the edge of masking tape over a surface that is

Be careful that the masking tape does not overlap any area to be painted. After painting, you can remove paint from parts, but you cannot add missing paint to the body.

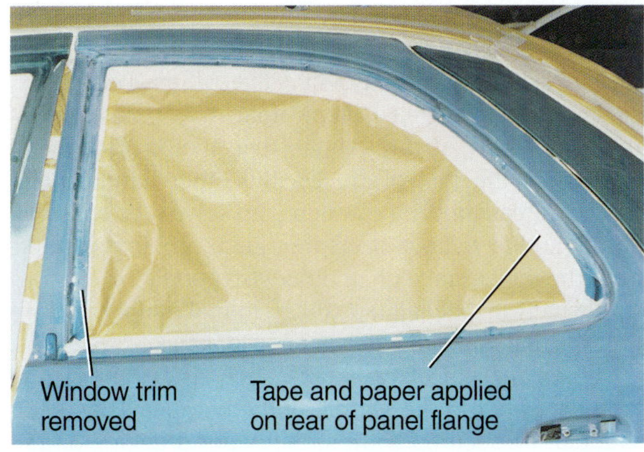

A Here door trim was removed so that the opening could be back masked. New paint can be applied under where the trim fits to the body panel.

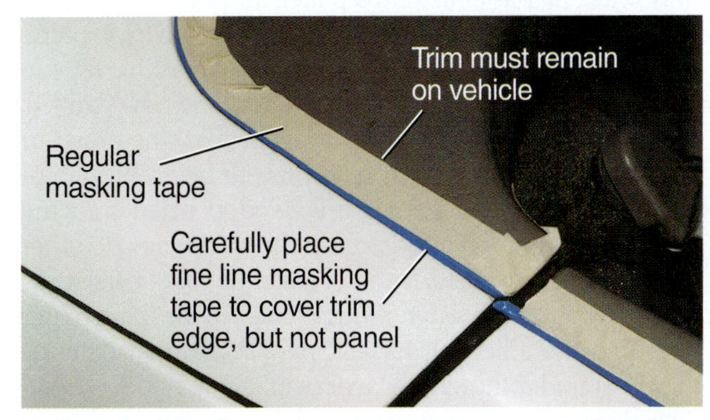

B Here the labor for part removal was not approved, so fine line masking tape was carefully applied right up to the lower edge of the trim.

Figure 25-51 Note masking differences that depend on insurance company and customer wishes.

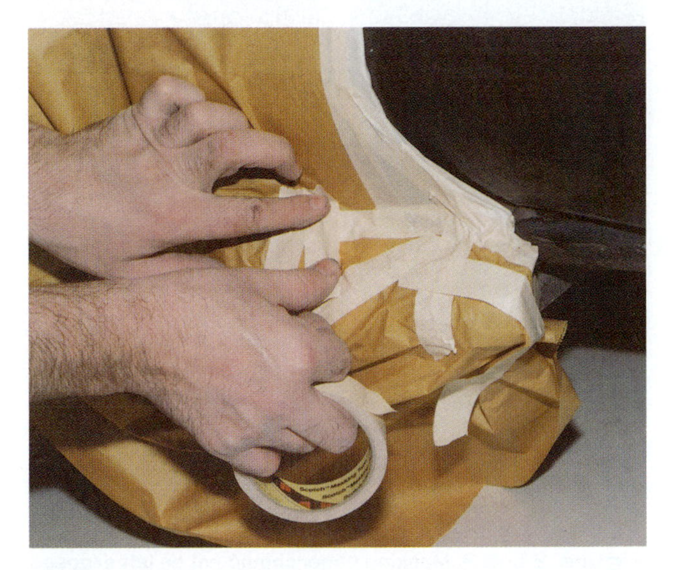

Figure 25-52 Make sure loose masking paper cannot flop and hit wet paint. Airflow from a spray gun can make loose paper blow around. Tape around or over all loose ends.

supposed to be painted. Because the tape is positioned down over the panel, it will get painted. When you remove the tape, a flaw will be evident in the overmasked area.

Overmasked areas mean that the painter must retouch a part of the vehicle that should have been painted in the first place. On the other hand, undermasked areas must be cleaned with solvent or polishing compound to remove overspray. Each problem detracts from the appearance of an otherwise good job.

Visually inspect all tape edges to make sure there are no gaps that could allow overspray leakage.

When masking large areas, such as bumpers, first tape the paper to the middle of the bumper. Then each side can be secured without the paper dragging on the floor and getting in the way.

When masking next to panels that will be painted, make sure masking tape covers the paper. If you leave only masking paper next to a panel, air pressure from the spray gun can make the paper flap and hit the fresh paint. Cover the paper with a couple of strips of masking tape to hold it down securely and keep it out of the new paint.

Double masking uses two layers of masking paper or a layer of tape over the paper to prevent bleedthrough or finish-dulling from solvents. It is needed when spraying horizontal surfaces (hood, trunk, and so on) next to other masked horizontal surfaces (Figure 25–53). Overspray will tend to soak through a single layer of masking paper and onto the adjacent masked panel. This can damage the paint under the single layer of masking paper.

Masking Aids

Wheel masks are preshaped plastic or cloth covers that fit over the vehicle's wheels and tires. Plastic wheel covers are disposable and should only be used once. Cloth covers are reusable but should be cleaned off periodically. Refer to Figure 25–54.

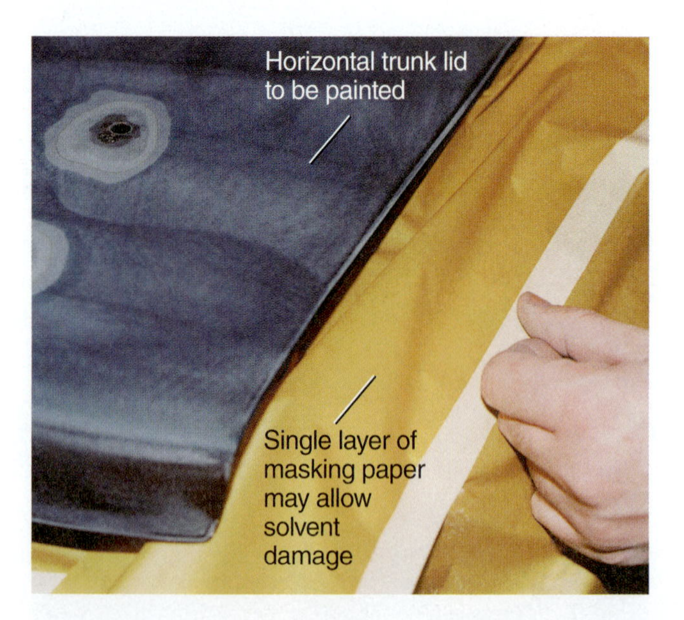

Figure 25–53 Masking paper should not be left exposed when painting horizontal panels such as a trunk lid or hood. This quarter panel is being masked so the trunk lid can be repainted.

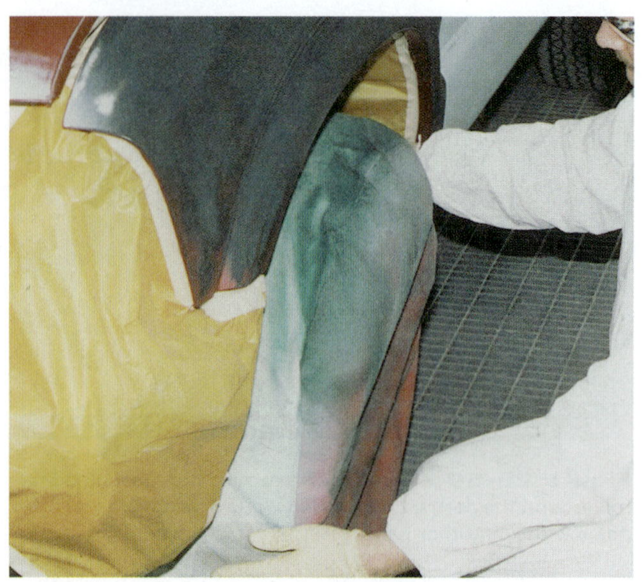

A Cloth wheel masks must be kept clean to avoid dust contamination of new paint.

B Disposable plastic wheel masks provide for a cleaner paint booth. They are discarded after use.

Figure 25–54 Wheel masks can save time.

It is very easy for cloth wheel covers to collect debris that can be blown into wet paint. For this reason, plastic, disposable wheel masks are preferred over reusable cloth ones.

Preshaped plastic antenna, headlamp, and mirror covers are also available. They can be quickly slid down or fit over these parts to reduce masking time.

Masking Panel Gaps

Back masking involves applying tape to the rear or inside edges of panels so that only the front of the panel is painted. When painting several panels, you often do not want paint to spray into gaps between panels. Spray would deposit on the panel surfaces and parts behind the gap or space between the panels.

To back mask a panel gap, apply the tape along the back of the panel so that the tape sticks out enough to cover the gap or space between the adjacent panels (door, fender, hood, trunk lid). Then, close the door or hood and push the tape back into the part gap. You can then paint the panels, and an almost invisible parting line will be hidden on the inner edge of the panels. Paint will spray on the edges of the panels, but no overspray will enter the panel gaps.

Masking Openings

When masking openings in panels, apply the masking tape to the back edge of the opening. Apply short pieces of masking tape to the back of the opening. Then, apply tape to the center of the hole to seal off overspray. This is shown in Figure 25–55.

In many instances, it is better to use tape and paper to mask wheels and wheel openings. If you are painting a bumper, you can simply place masking paper over the quarter panel and wheel. The location of masking depends on which parts are to be repainted. See Figure 25–56.

Figure 25–57 shows how to correctly mask a wheel opening when painting a fender or quarter panel.

First, apply wide masking tape to the inside edge of the panel flange. Press the tape back against the flange edge so that it hangs down into the wheel opening. You can then use this sticky surface to attach the masking paper.

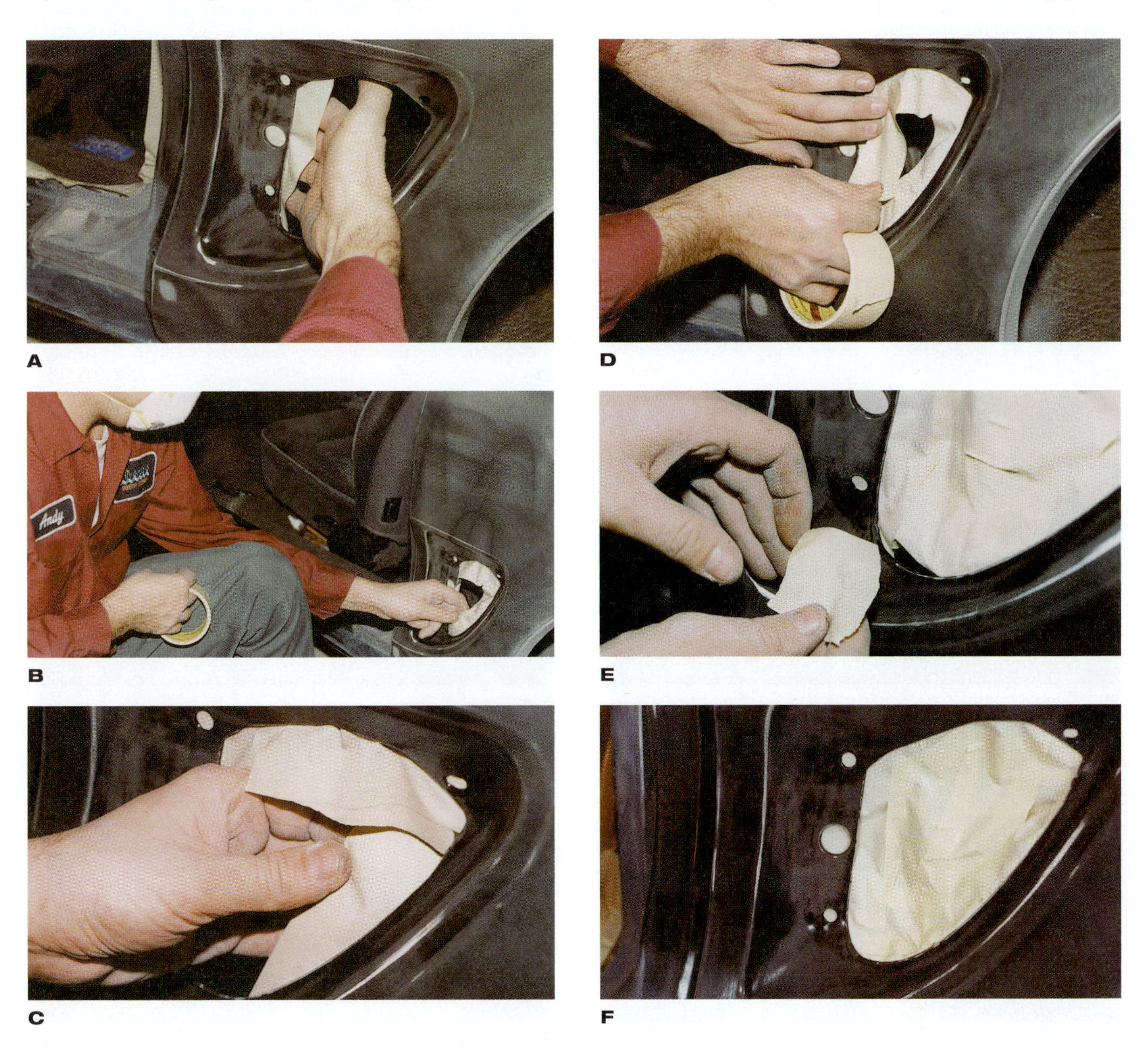

Figure 25-55 Note the basic steps for back masking an opening in a panel. (A) Tear off the correct length of masking tape. Apply it to the inner edge of the hole. (B) Continue applying short pieces of masking tape around the inside edge of the opening. (C) Use wider masking tape so it sticks well to the rear of the panel opening. (D) Once tape is applied all the way around the inside of the opening, apply more tape to close the center of the hole. (E) Check back masking for leaks. Small pieces of tape may be needed to close the "leakers." (F) Note how mounting holes are also back masked. Back masking allows paint to be deposited over the whole surface of a panel for superior repair.

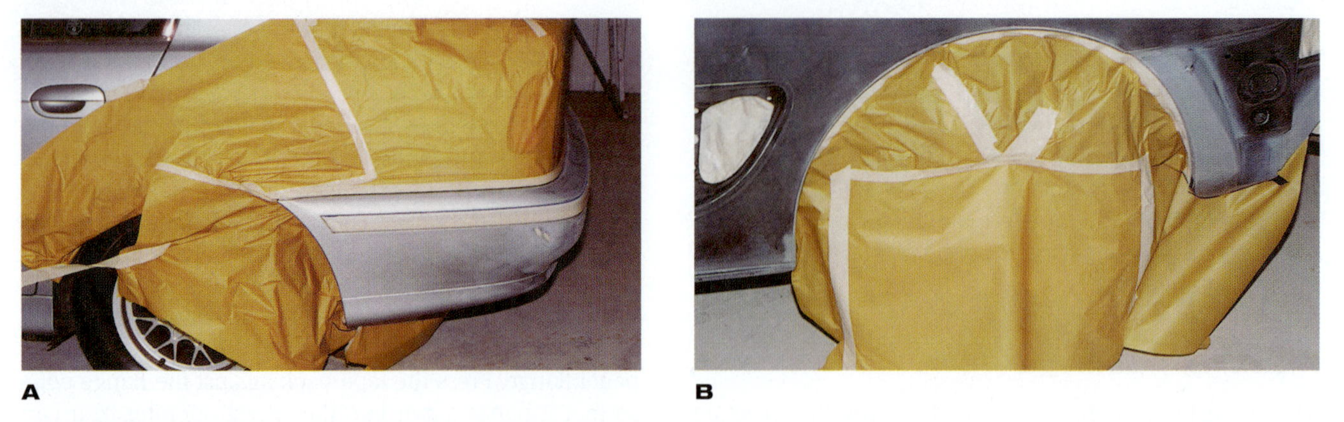

Figure 25-56 Note the two methods for masking a wheel opening. (A) Because only the rear bumper is to be painted, the whole wheel opening did not have to be fully masked. (B) Full wheel opening masking is needed when painting the panel area over or around the opening.

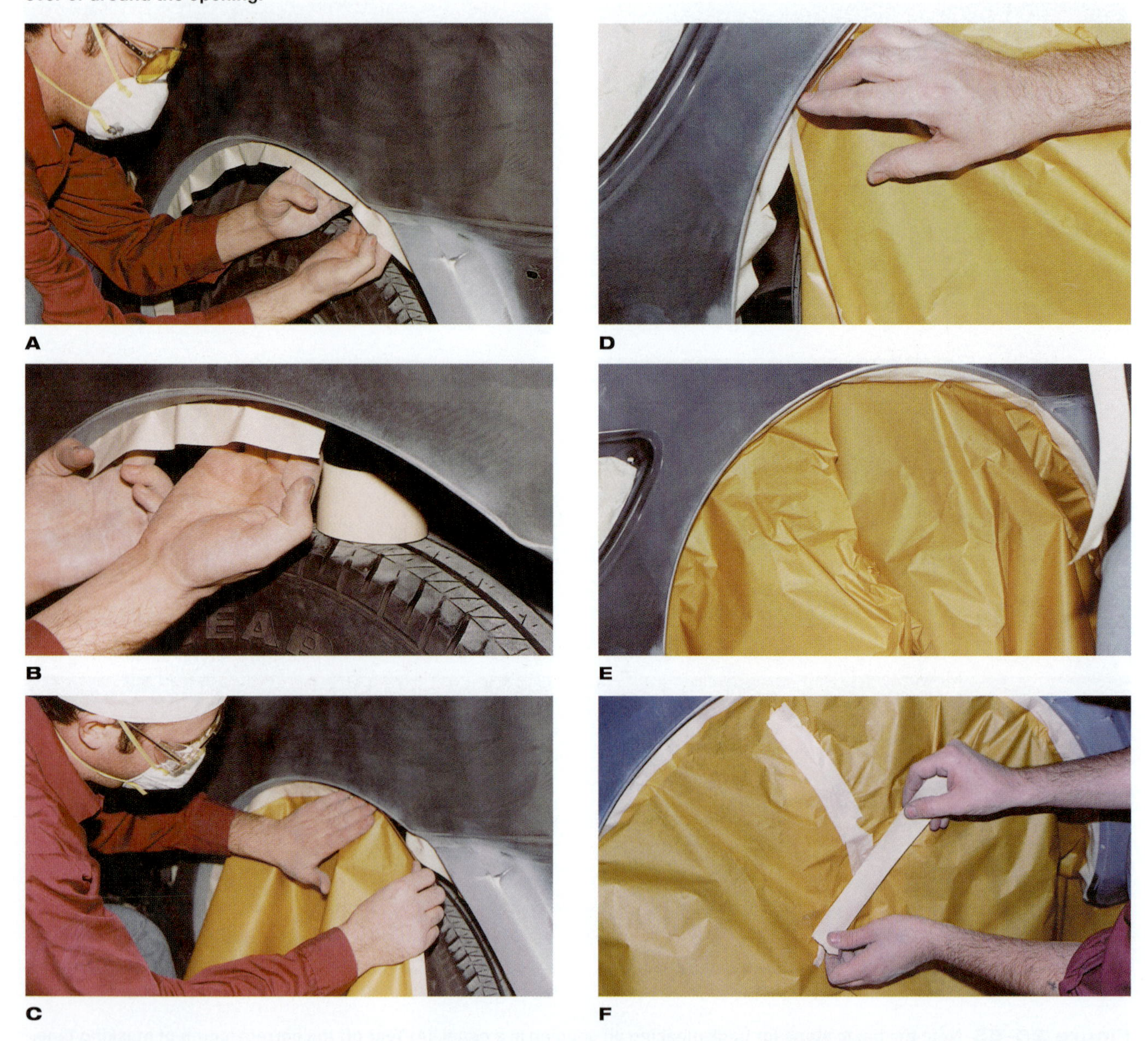

Figure 25-57 Note the major steps for fully masking a wheel opening. (A) Start by applying masking tape to the flange around the wheel well. (B) Press 2-inch-wide masking tape to the inner edge of the flange. (C) Apply masking paper to the back masked tape hanging down from the wheel opening. (D) Press the tape on the masking paper to the tape on the flange. (E) Fold the paper in the center so that it can be mounted to the curved flange. (F) Tape over the bottom of the paper so that loose ends do not blow up and get into the wet paint.

Align and press the masking paper tape against the flange-mounted tape. Work the paper around the wheel opening until it is fully masked. Tape the bottom of the paper down to keep it from blowing around when spraying.

Proper wheel opening masking will allow you to paint the fender or quarter panel flange. It will also keep overspray off the inner fender panel and tire–wheel assembly.

When masking a hood or trunk lid, you often want to be able to open it for painting. You must mask the hinges so that they can move without tearing the masking tape and paper (Figure 25–58).

Figure 25–59 shows the major steps for masking a trunk opening. Again, apply masking tape to the back edge of the trunk opening so that paint can be applied to all flanges around the opening. Press the tape tightly on the back side of the trunk flange. Leave about half of the tape sticking out. This will let you apply masking paper to the tape on the flange. Use extra masking tape as needed to secure the paper to the tape on the trunk opening flange.

Figure 25–60 shows how to mask a trunk shock or lift cylinder to allow the trunk to be opened and closed while painting. You may want to use a screwdriver to engage the trunk latch so that it will not fully close and lock. The engaged latch will hold the lid open slightly so that painted surfaces will not touch when the lid is closed. Refer to Figure 25–61.

Never allow paint overspray to enter electrical connections or to cover wiring harnesses. If paint gets inside an electrical connector, the circuit may not work properly. The paint can insulate the metal connectors and prevent normal current flow (Figure 25–62).

When masking a door opening, use the same general technique described for other openings. Apply the tape to the back of the opening flange. Then attach the masking paper to the exposed tape on the flange. This is shown in Figure 25–63.

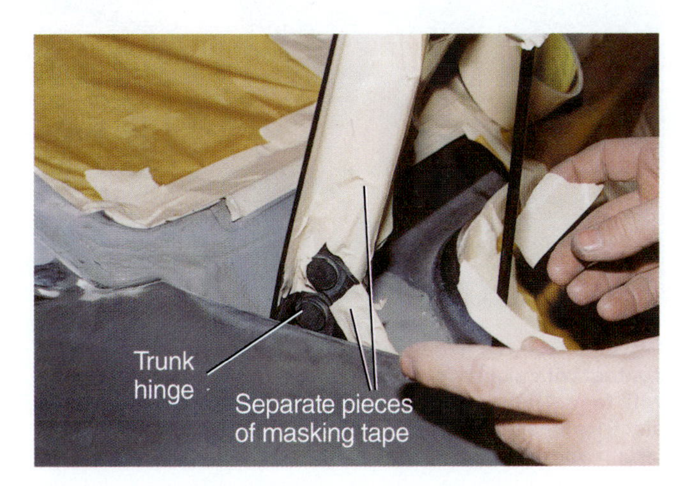

Figure 25-58 **Sometimes hinged parts must be opened and closed while painting. You must mask hinges so that the tape and paper will not tear when the panel is opened. Note how small pieces of masking tape are applied to the hinge arms. More tape will be applied over the hinge pin so each will not tear when the trunk lid is opened.**

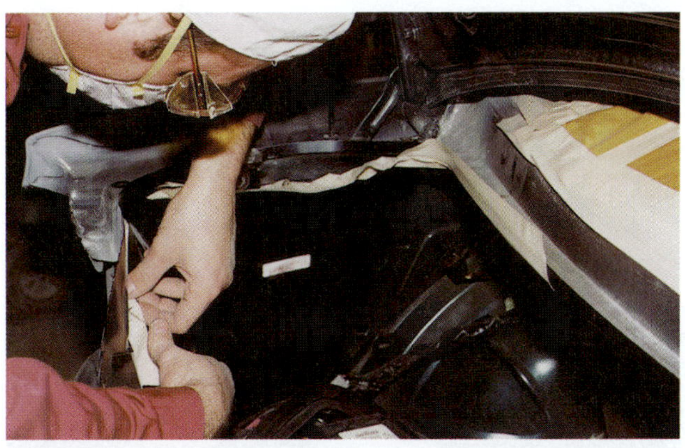

A Apply masking tape under the flange on the trunk opening so that half of the tape sticks out from under the flange.

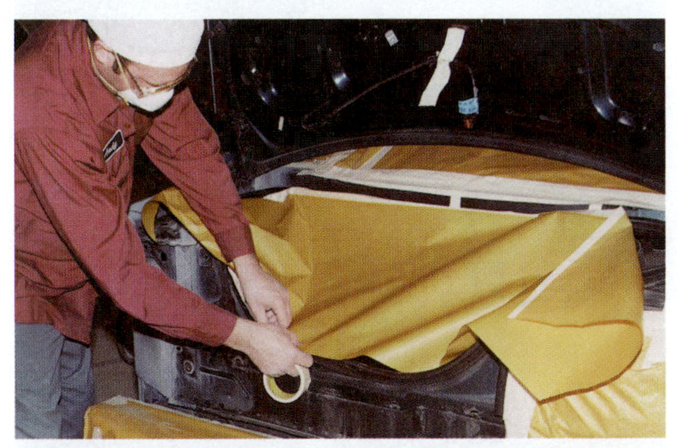

B Apply masking paper to the tape already applied to the opening. Tape the edge of the paper all the way around the opening to seal off the interior of the trunk.

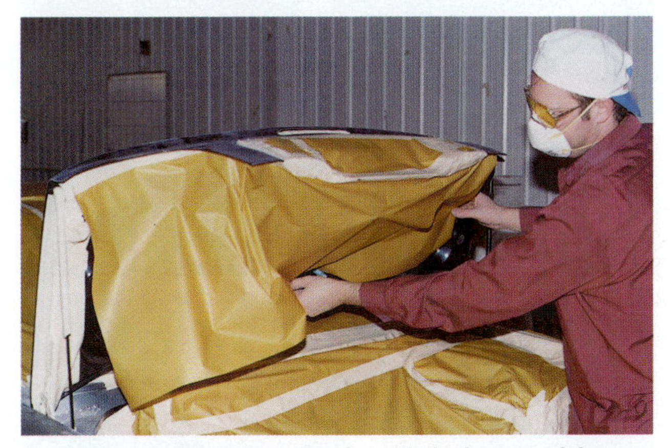

C If it is not going to be painted, mask the bottom of the trunk lid using tape and paper.

Figure 25-59 **Note the basic steps for back masking a trunk opening.**

When masking door jambs, be sure to cover both the door lock assembly and the striker bolt if it has not been removed. They can become filled and clogged with paint if not masked.

Figure 25–64 shows how to mask a door opening. Note that the technician is only going to paint the rocker panel. The door opening, roof, and door are being masked to prevent damage to the wet paint.

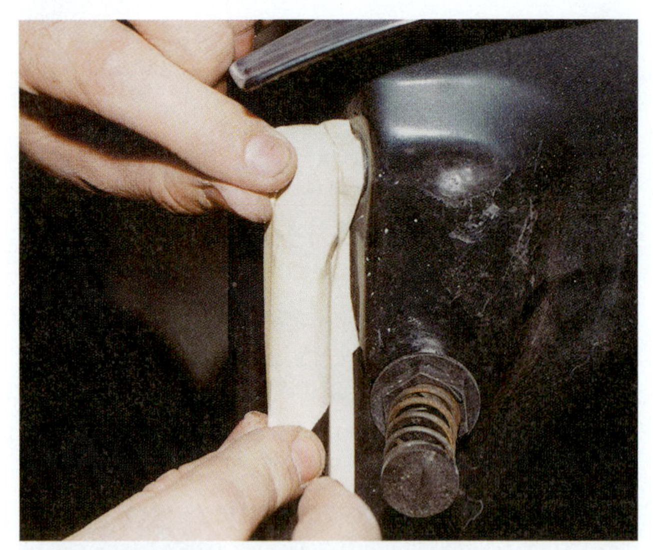

A To allow the lid to be opened while painting, the technician uses separate pieces of masking tape on moving parts, wrapping each piece carefully around parts.

B The technician uses tape wide enough to go all the way around the trunk lid spring-shock.

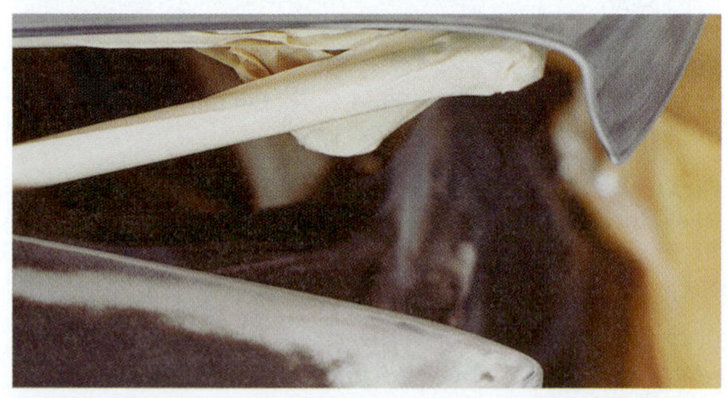

C Note how separate pieces of masking tape have been used on the shock and on the mounting bracket to allow the trunk lid to be opened while painting.

Figure 25-60 Note how this technician masks the spring-shock on a trunk lid.

Engage latch with screwdriver to keep trunk open while painting

Figure 25-61 This trunk lid must be opened so the inner body flange can be painted. A trick of the trade is to engage the trunk lid latch by pushing on it with a screwdriver. This will keep the trunk from closing fully while you paint. It might also turn off the trunk light so you do not drain the battery while painting.

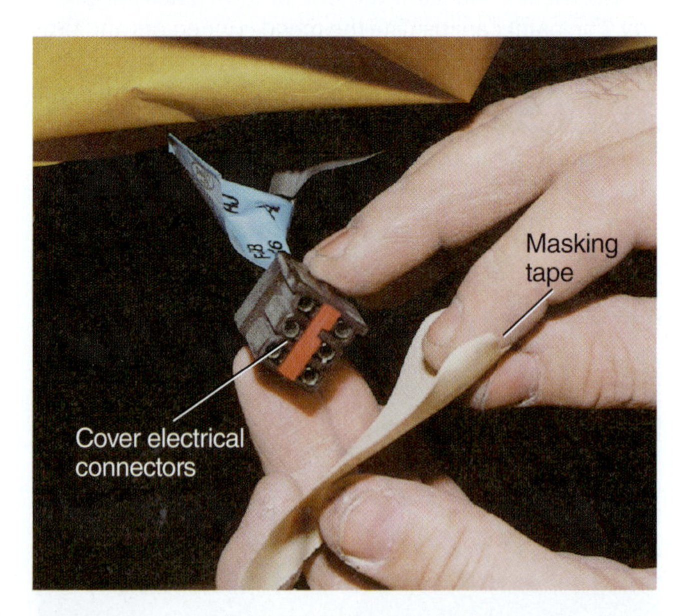
Masking tape
Cover electrical connectors

Figure 25-62 When overspray might come into contact with electrical connectors, cover the terminals with tape. The slightest amount of overspray can increase electrical resistance in the connector, possibly affecting circuit operation.

Reverse Masking

Reverse masking, or *blend masking*, is done by rolling the tape over and into a curved shape to prevent a visible paint parting line along the masking tape. It requires you

Figure 25-63 Masking a door opening is similar to masking other openings. This one is being back masked along the bottom of the opening and top masked along the top of the opening. This will allow for coverage of the rocker panel, but not of the upper roof rail.

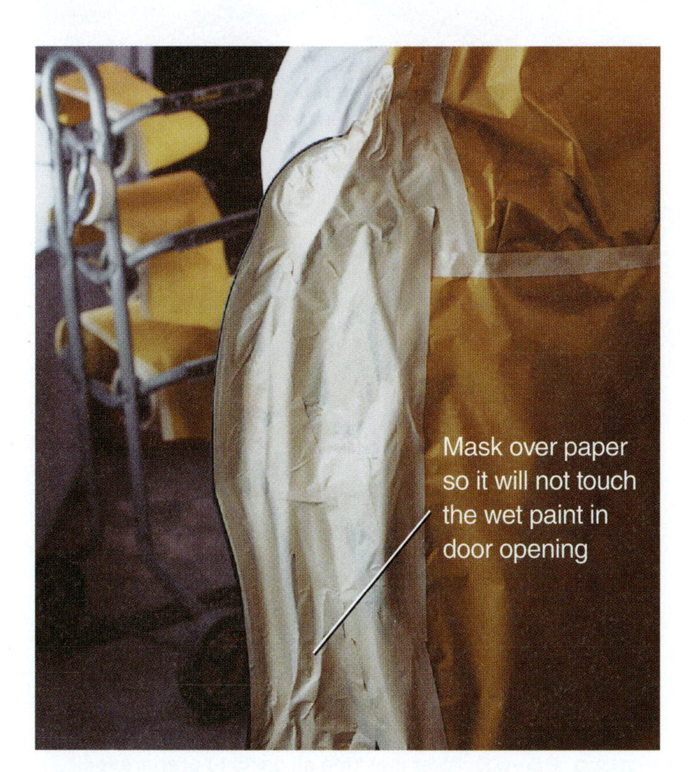
Mask over paper
so it will not touch
the wet paint in
door opening

Figure 25-64 Note how extra masking tape has been applied to a door frame. This will keep the paper from touching and ruining paint applied to the door opening around the striker plate.

to fold the masking paper back and over the masking tape. The rolled edge of the tape will block most of the overspray, but a blended paint edge will be formed under the curved up tape. Please refer to Figure 25–65.

This technique is often used during spot repairs to help blend the painted area and make it less noticeable. This also helps prevent bleedthrough. The paper is taped on the inside and allowed to billow slightly, which keeps it lifted up a bit off the surface.

Masking rope, also called *aperture tape*, is self-stick, foam rubber cord designed for quickly masking behind panels. Masking rope can be applied behind door, hood, gas cap lid, and other panels to block overspray. It provides a quick and easy way of blending paint behind panels.

Figure 25–66 shows how masking rope can be used behind parts like a fuel fill door. The curved edge of the rope will blend paint spraying between the gap at the door and quarter panel. No sharp paint parting line will be produced inside the fuel filler opening.

Masking rope is being applied to the bottom of a hood in Figure 25–67. This will keep overspray from going down into the gap between the hood and fender. It will produce a soft edge of spray next to the self-stick foam rope.

Figure 25–68 shows how to use masking tape and paper to form a blend area. You can form a pocket over an area where you want the new paint to blend out into the old paint. This is commonly done on inner body surfaces.

When everything is masked, walk around the vehicle while inspecting all of the masking material. Make sure there is no loose paper or plastic that could blow into the wet paint. Check for masking leaks or overmask mistakes.

SHOP TALK

Sometimes people rub their bare hands over an area to determine the effect of sanding without realizing they are contaminating the surface. There is oil on your skin as well as on shop tools. Even after washing your hands, a fine oily film remains on your skin. If you touch a surface that is to be painted, even a tiny deposit of oil can make the paint fish-eye or not lay down smoothly.

Inspect where masking tape has been curved around parts. When masking tape is stretched and curved, it can pull up and allow a leak under the curve. Press the tape back down or apply new tape to hold the curved masked surface tightly on the part. Refer to Figure 25–69.

25.7 SURFACE CLEANING

It is unwise to apply any kind of finish to a surface that has not been properly prepared. Quality suffers, customer dissatisfaction is inevitable, and costs increase because the job usually has to be done over.

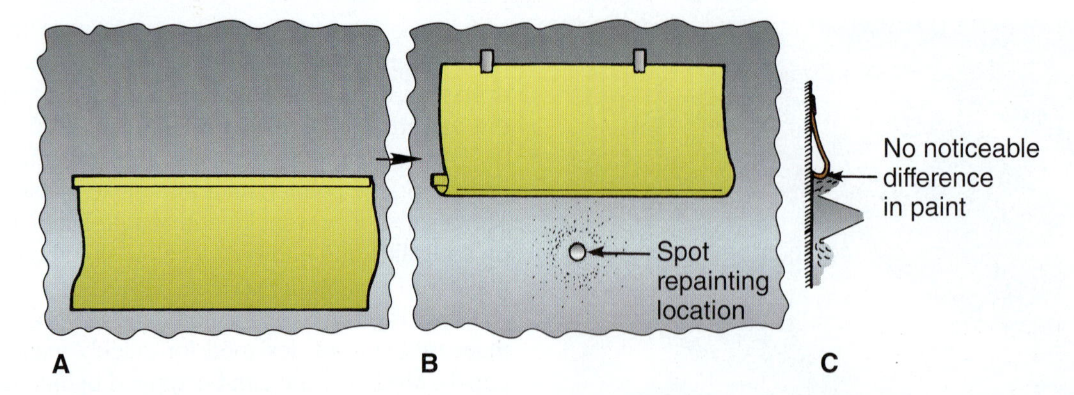

Figure 25-65 Reverse masking is used to prevent a sharp paint edge. It will help blend paint out for a less noticeable difference between new and old paint. (A) Place masking tape and paper hanging down over the area to be blended. (B) Lift the paper up over the tape and secure. (C) The curved edge of the tape will catch and trap paint spray so a blended paint film will deposit next to the tape.

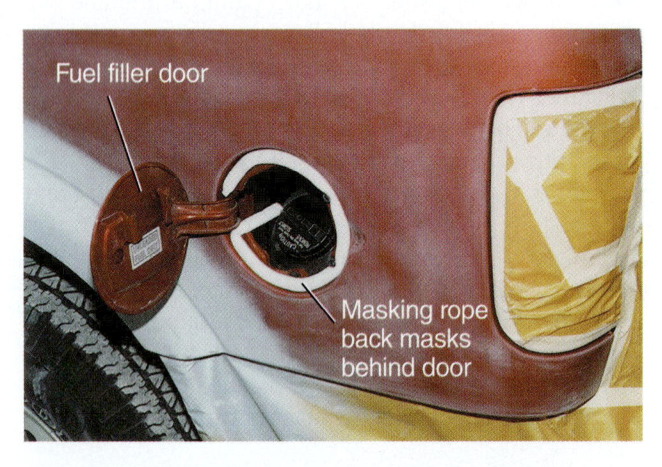

Figure 25-66 Masking rope or aperture tape will produce a soft paint edge like reverse masking. It has been used to mask behind this gas cap door. It will mask behind the door and produce a soft paint edge around the perimeter or opening in the quarter panel.

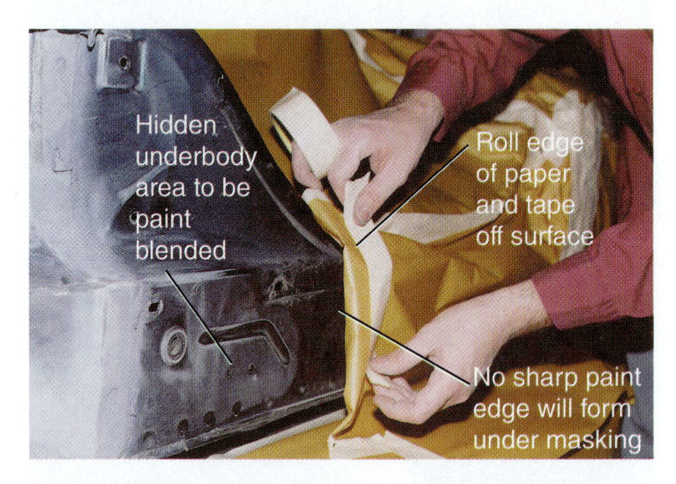

Figure 25-68 Even though this area will not be visible after vehicle reassembly, the paint technician is forming a blended area with masking paper and tape. When this area is painted, the new paint will blend into and better match the existing paint.

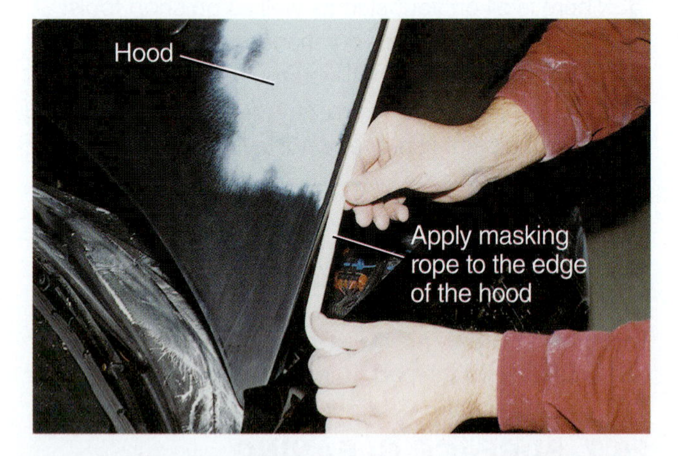

Figure 25-67 Masking rope or aperture tape is handy for quickly back masking gaps between panels. Here it is being aligned over the lip of a hood. When the hood is closed, the soft foam will block the gap between the fender and hood while producing a soft, blended edge along the edges of the panels.

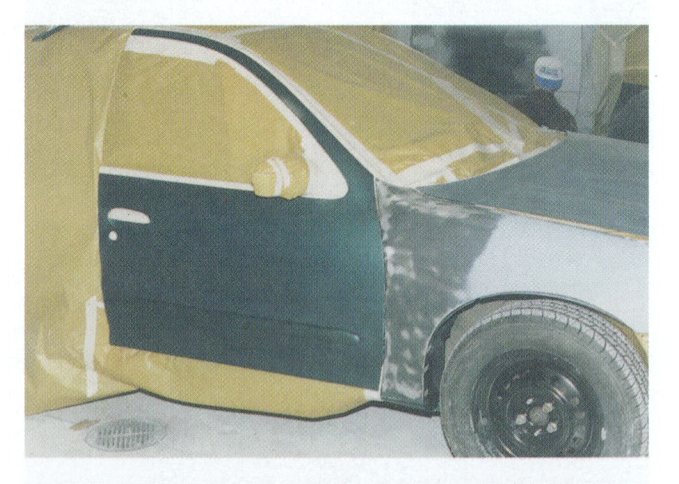

Figure 25-69 After masking all parts to prepare for painting, walk around the vehicle to check for masking problems. Retape as needed because a "leaker" or an "overmask" can necessitate painting the vehicle again at your expense.

Figure 25–70 **Before pulling a vehicle into the paint booth, use an air blowgun to clean the whole vehicle, including the masking material. You do not want the vehicle to carry any debris into the paint booth.**

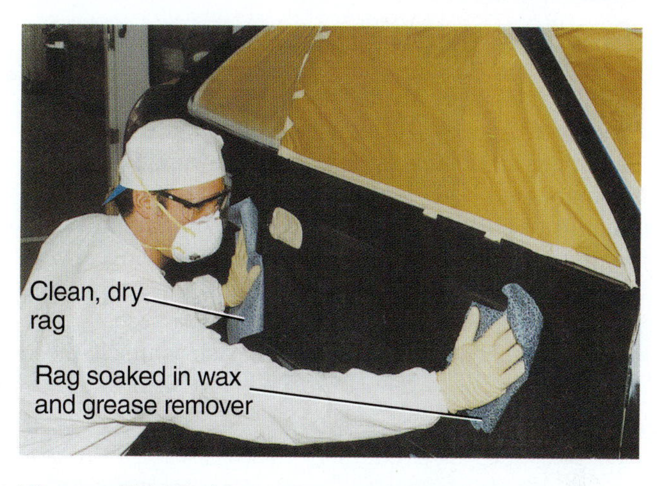

Figure 25–71 **After surface preparation, the vehicle should again be wiped down with wax and grease remover. Use one clean rag to apply the cleaning agent. Use another dry rag to wipe the cleaner off right away before it dries.**

Before pulling a vehicle into the paint booth, blow it off again (Figure 25–70).

Even while masking, a lot of dust can collect on the body and masked surfaces. Starting on the roof, blow dust off the whole vehicle. Direct compressed air over the masking paper, inside openings, along the windshield, and anywhere dust and debris can hide. This will reduce the amount of foreign matter that enters the paint booth. You should keep the paint booth as clean as a hospital operating room.

If you fail to remove contaminants, especially tar and grease, from the surface before final sanding, this debris can be buried in the filler or primer. If you fail to clean the surface again right before painting, oil from air tools or your hands can contaminate the vehicle surface. Problems can then result when spraying on the topcoats.

Using Wax and Grease Remover

It is simple to repaint over an existing paint film in good condition, providing it is stable and does not react to the solvent of the refinish paint. Before and after the job is sanded, use a wax and grease remover or recommended solvent to thoroughly clean the surface. Be sure to clean areas where a heavy wax buildup can be a problem, such as around trim, moldings, door handles, radio antennae, and behind the bumpers. Paint will not adhere properly to a waxy surface.

Some technicians like to use one hand to wipe on a cleaning agent and the other to dry the same area. By using both hands and two rags, you can more easily wipe off the solvent while it is still wet to remove surface debris. Refer to Figure 25–71.

Many shops like to use wax and grease remover before sanding and again right before painting. This procedure double-checks that all contaminants have been removed from a vehicle's surface before it is painted.

When wiping down a vehicle body, use a lint-free rag or towel. Special lint-free cleaning towels are available and recommended. They come in box dispensers or rolls. Avoid using rags with frayed edges that will deposit lint on the vehicle surface. When painting, lint can be a "nightmare."

Work small areas (a few feet or meters), liberally wetting the surface. Never attempt to clean too large an area. The solvent will dry before the surface can be wiped. Maximum effectiveness is achieved by wiping up wax and grease remover while it is still wet. Always use new wiping cloths because laundering may not remove all oil or silicone residue.

To remove any last trace of moisture and dirt from seals and moldings, carefully blow them out with compressed air at low pressure. Wax and silicone can penetrate beneath the surface. This contamination is not easily detectable. It is wise to assume that it is present, so always include some wax and grease cleaner or detergent in the sanding water.

It is a common mistake to use paint reducer or solvent to clean vehicle surfaces. Reducers and thinners can absorb into the paint film. Blistering or lifting can result from improper cleanup with these solvents. Use only a recommended wax and grease remover when cleaning before painting.

25.8 EPA 6H RULES

The 6H rules were written by the EPA (Environmental Protection Agency) to reduce the public's exposure to hazardous air pollutants (HAPs). While working as a professional auto body technician, always try to lessen the amount of HAPs released into the environment. When paint stripping, spray priming, and paint spraying, always do everything you can to prevent the release of volatile organic compounds into the earth's atmosphere.

To comply with EPA 6H rules, ask yourself these questions:

1. Are spray booths and prep stations clean and in good condition?
2. Are the paint booth filters and exhaust systems 95 percent efficient?
3. Does the water wash spray booth operate up to specs?
4. Are all of your spray guns HVLP (high-volume, low-pressure), electrostatic, airless, or air-assisted airless?
5. Does your spray gun inlet pressure not exceed 10 psi at the air cap?
6. Do you have a properly functioning spray gun cleaner machine and a paint/solvent recycling machine?

SUMMARY

1. The term *surface preparation* refers to getting a vehicle's surface clean, smooth, and ready for the application of the final colorcoats.
2. A dull primer surface may look smooth and ready for paint to the untrained eye. You must remember that the paint will actually work like a magnifying glass to exaggerate any scratches or irregularities.
3. Before painting, you should identify the type of paint and overall condition of the existing paint system.
4. Paint thickness is measured in mils or thousandths of an inch. OEM paints are typically about 3–6 mils thick.
5. Blasters are air-powered tools for forcing sand, plastic beads, or another material onto surfaces for paint removal. A chemical paint remover can also be used for stripping large areas of paint if environmental regulations allow.
6. Sanding uses an abrasive-coated paper or plastic backing to level and smooth a body surface.
7. Grit sizes vary from coarse to ultrafine grades and are rated by number.
8. Power grinding is done to quickly remove large amounts of old paint and other materials.
9. When sanding, make sure you are using the correct grit number for the job.
10. An orbital sander or DA sander moves in two directions at the same time, an action that produces a much smoother surface finish.
11. Scuff sanding removes any trace of contaminants on the existing finish and textures the surface so paint will adhere properly.
12. If the primer-surfacer is sanded before all of the solvent has evaporated, the material in the scratches will continue to shrink down in the scratches.
13. After the primecoat is dry, block sand the area until it is smooth.
14. A guide coat helps to point out unlevel surface areas.
15. Once the primer is dry, small pinholes and scratches can be filled with spot putty or glazing putty.
16. Masking keeps paint from contacting areas other than those to be refinished or painted.
17. Automotive masking paper is heat-resistant so that it can be used safely in baking ovens.
18. Liquid masking material seals off the entire vehicle to protect undamaged panels and parts from paint overspray.
19. Masking tape is very sticky paper tape designed to cover small parts and also to hold the masking paper in place.
20. Fine line masking tape is a very thin, smooth-surface plastic masking tape.
21. Masking covers are specially shaped cloth or plastic covers for masking specific parts.

EXERCISES

On a separate sheet of paper, complete the following learning activities for this chapter. Write definitions for the key terms and answer the ASE-style review questions, essay questions, critical thinking problems, and math problems. You can also do the outside activities, possibly for extra credit.

Key Terms

blasters
bull's-eye
chemical paint remover
E-coat
final sanding
flash time
foam backing pad
gloss level
masking
masking rope
reverse masking
shipping coating
surface preparation
surface scuffing

ASE-Style Review Questions

1. Technician A uses a DA sander to remove heavy surface rust, whereas Technician B says it is better to use a media blaster. Who is correct?
   A. Technician A
   B. Technician B
   C. Both A and B
   D. Neither A nor B
2. If the original paint surface is in good condition, Technician A simply washes the car to prepare the surface for painting. In the same instance, Technician B also scuff sands the surfaces to be painted. Who is correct?
   A. Technician A
   B. Technician B
   C. Both A and B
   D. Neither A nor B
3. A body surface has been sanded down to bare metal. Technician A first applies a metal conditioner and then body filler. Technician B applies a self-etch primer to the bare metal. Who is correct?
   A. Technician A
   B. Technician B
   C. Both A and B
   D. Neither A nor B
4. Which of the following methods can be used to strip paint from the metal surfaces of a vehicle?
   A. Sanding
   B. Blasting
   C. Chemical stripping
   D. All of the above
5. A fender has tape stripes that must be removed before painting. Technician A says to use a rubber scrub wheel in an air drill. Technician B says that a heat gun might aid stripe removal. Who is correct?
   A. Technician A
   B. Technician B
   C. Both A and B
   D. Neither A nor B
6. Technician A uses #600-grit sandpaper to remove paint to bare metal. Technician A says to use #36-grit sandpaper for rapid paint removal. Who is correct?
   A. Technician A
   B. Technician B
   C. Both A and B
   D. Neither A nor B
7. Technician A says that modern paint removers are not harmful to your skin and eyes. Technician B says to wear a face shield and rubber gloves when using chemical paint removers. Who is correct?
   A. Technician A
   B. Technician B
   C. Both A and B
   D. Neither A nor B
8. Which product should be used to mask next to a part that is being painted around?
   A. Narrow masking tape
   B. Wide masking tape
   C. Fine line masking tape
   D. Duct tape
9. Technician A says when wiping down a vehicle body, use lint free towels. Technician B says to use one lint free towel to clean each vehicle panel before refinishing. Who is correct?
   A. Technician A
   B. Technician B
   C. Both A and B
   D. Neither A nor B
10. A new metal bumper is coated with a factory E-coat. Technician A says to remove all of the coating before priming and painting. Technician B says that an E-coat normally can be prepped and painted. Who is correct?
    A. Technician A
    B. Technician B
    C. Both A and B
    D. Neither A nor B

11. Technician A says spot putties can be painted over without sanding them. Technician B says that a two-part spot putty dries very quickly. Who is correct?
    A. Technician A
    B. Technician B
    C. Both A and B
    D. Neither A nor B
12. Technician A uses a sanding block and wet sandpaper to scuff and prepare existing painted surfaces for repainting. Technician B uses his fingers and a scuff pad to get inside restricted or inner surfaces of panels. Who is correct?
    A. Technician A
    B. Technician B
    C. Both A and B
    D. Neither A nor B
13. Technician A says that newspaper can be used to mask right next to panels being painted. Technician B says that a plastic sheet should be used to mask right next to panels being painted. Who is correct?
    A. Technician A
    B. Technician B
    C. Both A and B
    D. Neither A nor B

Essay Questions

1. How do you evaluate the surface condition of a vehicle?
2. List some points that must be kept in mind when working with abrasives.
3. Describe the purpose of wet sanding.
4. Explain the use of a guide coat.
5. Explain fine line masking tape.
6. How do you mask a vehicle using the liquid masking system?
7. Vehicle surface evaluation involves determining what? Please answer as thoroughly as possible.

Critical Thinking Problems

1. If you are going to spray a two-part paint on a vehicle and there are leaks in the masking paper, what are the implications for the vehicle, for you, and for the shop?
2. What happens if a technician uses too coarse a sandpaper before painting?
3. If measurements show a paint to be 24 mils thick, what does that tell you?

Math Problems

1. If a paint measures 6 mils thick and paint buildup should be limited to no more than 12 mils, how much more paint can be applied to the body?
2. If a pint of material will cover 3 square feet (1.8 square meters), how much material would be needed for 23 square feet (6.9 square meters)?

Activities

1. Inspect the paint or finish on an older vehicle. Report on the condition of the old paint and what would have to be done to prepare it for painting.
2. Measure paint film thickness in various locations on different vehicles. From your measurements, can you find any panels that have been repainted or painted more than once? Would any panels require removal of the old paint before repainting?

CHAPTER 26

Refinishing Procedures

OBJECTIVES

After reading this chapter, you should be able to:

- Explain when primers (self-etch primer, primer-surfacer, sealer, and adhesion promoter) should be used before painting.
- Compare OEM or original factory paint jobs with those done in a body shop.
- Name the types of colorcoats and clearcoats, including water-based colorcoats or basecoats.
- Explain the advantages and disadvantages of basecoat/clearcoat finishes when compared to single-stage finishes and water-based colorcoats.
- Describe the role of solvents, distilled water, and the variables that affect their spraying capabilities.
- Select and mix paint solvents and distilled water.
- Determine the type of paint on a car and whether the car has been repainted or is a water-based paint
- Use a spray gun properly.
- Describe the different kinds of spray gun coats, including blend coats, often required with colorcoat/clearcoat paints.
- Summarize common spray gun handling problems.
- Properly complete spot repairs, panel repairs, and an overall paint job.
- Professionally apply single-stage finishes, water-based paints, as well as colorcoat/clearcoat systems.
- Describe the paint finishing systems applicable to plastic parts.
- Summarize the methods for applying a smooth, glossy finish.
- Correctly answer ASE-style review questions relating to refinishing procedures.
- Summarize the differences required to properly prepare and spray water-based paints.

INTRODUCTION

From the customer's standpoint, the look of the new paint is the most important aspect in determining the quality of the collision repair. Many customers do not understand all of the work that goes under, and into, a paint job. Paint chemistry and spray equipment are much more complex than in the past. Proper material selection, skill, and knowledge are needed to produce a durable, long-lasting finish.

The expert refinisher takes special pride in producing a beautiful topcoat of paint that matches both the color and the texture of the original, or OEM, finish. It is your job to satisfy the customer with a professional paint

HISTORY NOTE

During the "olden days" of motor car manufacturing, auto makers applied their new car paint finishes with a paint brush. Linseed oil-based paint was brushed on by hand in several coats. By today's standards these new car finishes did not look very good, and the linseed oil paint deteriorated in a couple of years. This is a far cry from the high-tech paint systems used in today's collision repair shops.

Figure 26-1 As a paint technician, you must try to match the original paint of the vehicle as closely as possible. The technician is blowing off all surfaces while wiping with a tack rag.

application that not only looks good but also will last in all kinds of weather (Figure 26–1).

With today's high-solids, low volatile organic compound (VOC) paints and high-efficiency spray guns, refinishing procedures have changed. The industry is now using high-volume, low-pressure (HVLP) spray equipment to reduce paint waste and emissions, and paint products that have more solid content to reduce solvent waste are now industry standards. These changes have made collision repair even more challenging.

This chapter summarizes the basics of spraying today's refinishing materials and using high-efficiency spray equipment.

26.1 PURPOSE OF REFINISHING

Many customers see and appreciate only the top layers of paint. They judge the quality of the refinisher's work on the finish appearance alone. For some customers, there is little appreciation of all the work done underneath the topcoats. Customers often do not know the skill it takes to do metalwork, filler sanding and shaping, surface prep, and the application of primers, sealers, and other primecoats.

As discussed in previous chapters, the cleaning, filling, and sanding of the substrate are painstaking processes. A perfectly smooth surface must be readied before the topcoat can be applied. Otherwise, any surface imperfection—even the smallest—will show in the topcoat and could require repainting to fix the surface flaw and resulting paint problem.

Functions of Paint

Automobile finishes or paints perform four extremely important functions:

1. Paint provides a "skin" to protect the body substrate (steel, aluminum, and plastics) from the elements. Most motor vehicles are constructed primarily of steel sheet metal. If this steel were left uncovered, the reaction of oxygen and moisture in the air would cause it to rust. Painting serves to prevent rust, therefore protecting the body.
2. Paint improves the appearance of the body. The shape of the body is made up of several types of surfaces and lines, such as elevated surfaces, flat planes, curved surfaces, straight and curved lines, and so forth. Therefore, another objective of painting is to improve the body appearance by giving it a three-dimensional color effect.
3. Paint increases the value of the vehicle. When comparing two vehicles of identical shape and performance capabilities, the one with the more beautiful paint finish will have a higher market value. Hence another object of painting is to increase resale value.
4. Paint color makes vehicles of the same make and model different or distinguishable. Painting automobiles also makes them easily distinguishable by application of certain colors or markings. Examples are police car paint schemes, yellow taxicabs, and red fire trucks. If all passenger cars were the same color, as when all Model T Fords were only painted black, cars would be dull and drab.

To achieve a high-tech finish, the typical automotive finishing system consists of several coats of two or more different materials:

1. **Primecoats** are surface preparation coats (self-etch primers, primer-surfacer, sealer, and adhesion promoter) sprayed on the repair area first. These initial primecoats ready the body surface for the topcoats of paint.
2. **Topcoats**, or glamour coats, are coats of color and clear paint sprayed over the primecoats (usually urethane colorcoat/clearcoat paint) to cover the vehicle with a shiny, bright color. Refer to Figure 26–2.

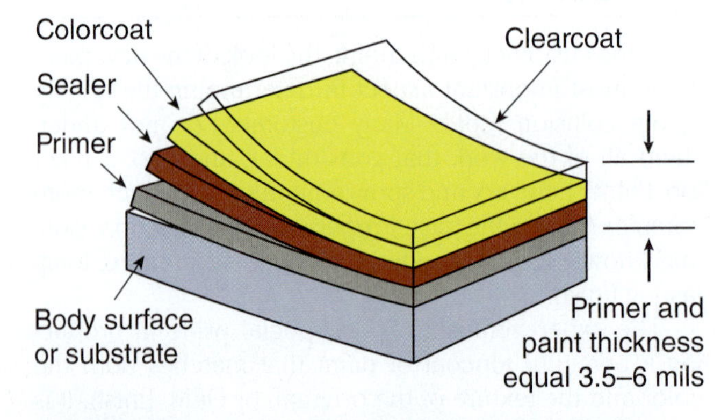

Figure 26-2 Note the basic layers of primer and paint used to produce a vehicle finish. Transparent clearcoat is applied over colorcoat on most late model finishes.

26.2 TOPCOATS

There are several types of OEM paints (solid colors, metallics, micas, and pearls) used on new cars. Various types of paint products are used when refinishing in a body shop. You must be familiar with modern paint technology to become a good paint technician.

OEM Finishes

An **OEM finish** is the original factory paint job. It is a hard, baked-on finish.

Today, most manufacturers use similar processes when priming and painting their new vehicles on the assembly line. During vehicle manufacturing at the factory, several automated processes are used to apply the finish. The vehicle body is painted before most parts are assembled onto the body. Because nonmetal parts, interior seats and trim, the drivetrain, wiring, and rubber seals have not yet been installed, OEM methods and materials are difficult and sometimes impossible to duplicate in a collision repair shop.

Typically, the unibody panels are robotically spot welded together. The steel unibody is then carried by conveyor to a "bathing room." The whole unibody is lowered and dipped into several solutions to prepare all metal surfaces, especially the welds, to accept and hold the paint materials.

First the new car body is dipped in an acid bath to dissolve all contaminants from welding and assembly. The acid bath cleans and etches the steel so that primer will adhere securely to the body. Then the body is dipped in a solution to neutralize this acid. Many new vehicle bodies get a metal treatment (phosphate coating) to provide a barrier coat that protects steel body panels from corrosion or rusting.

Using automatic equipment, the unibody and other body parts are *E-coated* by dipping them into an electrically charged bath of corrosion protection material. Called electro-deposition, the materials are attracted to the body parts by the static electrical charge. After leaving the corrosion protection bath, the vehicle body is dried. Then robotic spray equipment is used to apply sealer, colorcoat, and clearcoat to the vehicle body as it moves down the assembly line.

Today's paints are usually "thermoset" furnace-hardened, acrylic urethane, high-solids basecoat/clearcoat enamels. This allows the new finish to be baked at extremely high temperatures to cure and harden the paint materials to a durable, glossy finish.

OEM primecoat and topcoat ingredients are a slightly different formulation than the refinish materials used in a body shop. OEM paint finishes are normally harder and more durable than those done in a body shop because of these differences in materials and methods.

Higher OEM baking temperatures (over 350°F) can be used at the factory because nonmetal parts have not yet been installed on the car. The unibody is baked in huge ovens to shorten the drying times and to better cure the acrylic urethane enamel paint. Higher temperatures like those used during vehicle manufacturing could melt rubber or plastic parts, damage vehicle computers, or cause paint solvent popping, blistering, and other curing problems.

When a typical body shop bakes a new paint job they use much less heat—about 150°F.

Vehicle manufacturers use slightly different types of finish materials, coating processes, and application processes. Each type of finish requires different planning and repair steps.

The most common types of OEM coating processes include:

- Single-stage paint—color without clear
- Two-stage paint—colorcoat/clearcoat
- Three-stage paint—tri-coat that involves applying basecolor, mid-coat with flakes, and then a clearcoat
- Multistage—several coats of different paint materials must be applied

These types of paint will be explained fully in Chapter 27.

OEM paint thickness averages about 4–8 mils (0.004–0.008 inch) thick. The topcoat thickness on a new car when it comes from the factory can be as thin as 2.5 mils but can also be as thick as 8 mils. Horizontal panels such as the hood and trunk lid will usually have more paint on them than vertical panels such as doors, fenders, and quarter panels. Remember, 1 mil equals approximately one-thousandth of an inch, which is extremely thin.

Refinishing materials and methods are constantly changing. The refinish technician or painter must respond to these changes to provide the perfect matches demanded by the customer on a refinishing job. Keeping up-to-date on the changes is crucial.

Refinishing Topcoats

The repainted topcoat is the finish that is seen on the car. From an appearance standpoint, it must be smooth, glossy, and eye-catching. Functionally, it must be tough and durable.

The expert refinisher takes special pride in producing a beautiful finish that matches both the color (or color effect) and the texture of the original finish. Therefore, it is of great importance to fully understand the topcoat materials and how they are applied.

There are several types of paint for topcoat refinishing:

1. Lacquer paints—seldom used
2. Single-stage enamel paints—sometimes used
3. Water-based paints—sometimes used to correct paint compatibility problems or as OEM colorcoat under the urethane clearcoats
4. Basecoat/clearcoat enamel paints—most commonly used for refinishing
5. Multicoat paints—often used to match OEM paint
6. Solid color paints—do not have metal flakes mixed in

TABLE 26–1 COMPARING DURABILITY OF TOPCOATS

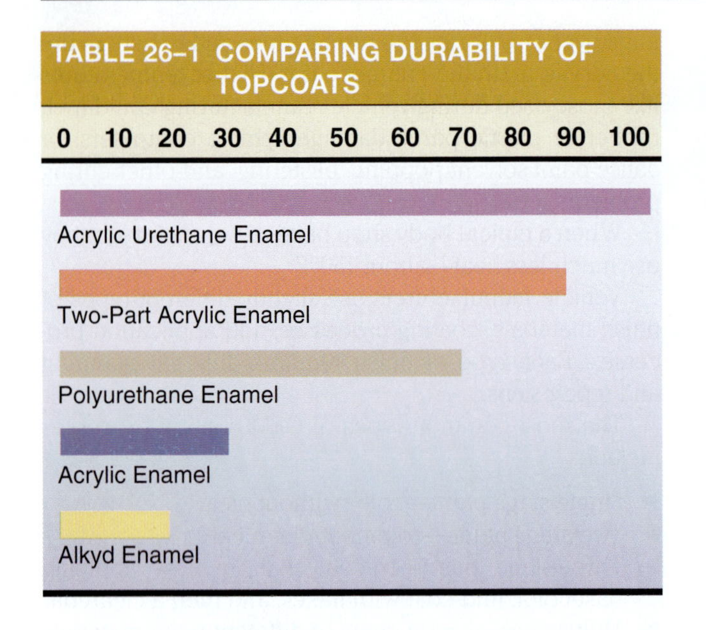

7. Metallic paints—small reflective flakes mixed with paint color
8. Pearl paints—tiny reflective mica chips or flakes are mixed in mid-coat clear

As you will learn, there are variations within these categories. It is important that you know what types of finishes manufacturers use because there are slightly different methods required when mixing and spraying them.

Table 26–1 shows the relative durability of the various types of topcoat systems, comparing them to OEM high-baked thermoset acrylic enamels. Table 26–2 provides a general summary of the properties of paint used for refinishing.

Waterborne Paint

Waterborne paint has been used successfully in Europe since the mid-1990s. In 2009 the state of California mandated that collision repair facilities replace solvent basecoat with waterborne basecoat in order to reduce the emission of VOCs. Waterborne paint (which refers to waterborne basecoat) has been growing in popularity in the North American collision repair industry ever since its adoption by California.

If you compare one quart of ready-to-spray solvent basecoat with one quart of ready-to-spray waterborne basecoat, this is what you will find:

	Waterborne Basecoat	Solvent Basecoat
Pigments	15%	20%
Metallic flake	5%	5%
Solvent reducer	10%	75%
Cosolvent (alcohol)	5%	0%
Water	65%	0%

The formulation of waterborne basecoat replaces much of the traditional solvent (paint thinner) with water. Alcohol used as a cosolvent allows the water-based and oil-based chemistry to mix together. Water evaporates instead of solvent as waterborne paint dries; this is better for the environment and can increase shop productivity in regions with dry climates.

Tips for Success Using Waterborne Basecoat

Many repair facilities have switched to waterborne paint because it is the type of paint used in the vehicle manufacturing process. Some vivid colors that are painted originally with waterborne basecoat can only be matched by using waterborne basecoat. Certain rules must be followed in order to use waterborne paint with success. Those rules are:

- Keep it clean.
- Prep your panels to a higher grit.
- Dry the basecoat thoroughly before clearcoat application.

Keep It Clean

Vehicle panels that will be refinished with waterborne paints must be cleaned with a solvent cleaner before repairs begin. After repairs are completed, the panels must be cleaned with a solvent cleaner for the second time, and after this step the panels must be cleaned with a waterborne cleaner and dried with disposable lint-free towels.

Waterborne paints must be applied with a compressed air supply that is absolutely free of oil residue and moisture. Most repair facilities use three-stage air filtration systems to clean the compressed air that will be used for refinishing. The first two stages remove oil and water; the third-stage desiccant canister serves as a backup to capture any contaminants that may have passed the first two stages. Waterborne paints also require special tack-rags to remove dust particles before the refinishing process and between coats during the process.

Prep the Panels to a Higher Grit

Waterborne paint will not hide as many surface imperfections as solvent paint will, so prep work has to be taken to a higher level. Vehicle panels that would normally be sanded with 400 grit for solvent refinish need to be sanded with 600 grit for refinish with waterborne paint. Blend panels that would be sanded with 600 grit for solvent need to be prepped to 800 grit for blending with waterborne paint. The proper preparation procedures must be followed if a shop is to be successful using waterborne paint.

TABLE 26–2 SUMMARY OF TOPCOAT FEATURES

Nomenclature		One-Part Type			Two-Part Type	
		Alkyd Enamel	Acrylic Lacquer	Acrylic Enamel	Polyurethane	Acrylic Urethane Enamel
Spray characteristics		Excellent	Excellent	Good	Good	Good
Possible thickness per application		Fair	Fair	Good	Excellent	Excellent
Gloss	without polishing	Fair	Good	Good	Excellent	Excellent
	after polishing	Good	Good	Good	—	Good
Hardness		Good	Good	Good	Excellent	Excellent
Weather resistance (frosting, yellowing)		Fair	Fair	Good	Excellent	Excellent
Gasoline resistance		Fair	Fair	Fair	Excellent	Good
Adhesion		Good	Good	Fair	Excellent	Excellent
Pollutant resistant		Fair	Fair	Fair	Excellent	Excellent
Drying time	to touch	68°F 5–10 minutes	68°F 10 minutes	68°F 10 minutes	68°F 2–30 minutes	68°F 10–20 minutes
	for surface repair	68°F 6 hours 140° 40 minutes	68°F 8 hours 158°F 30 minutes	68°F 8 hours 158°F 30 minutes	—	68°F 4 hours 158°F 15 minutes
	to let stand outside	68°F 24 hours 140°F 40 minutes	68°F 24 hours 158°F 40 minutes	68°F 24 hours 158°F 40 minutes	68°F 48 hours 158°F 1 hour	68°F 16 hours 158°F 30 minutes

Dry the Basecoat Thoroughly Before Clearcoat Application

Both solvent basecoat and waterborne basecoat use the same type of solvent-based clearcoat. It is important to make sure all the water has evaporated out of the paint film before applying clearcoat over waterborne basecoat. This can be done by using an infrared no-touch thermometer and checking the temperature of the area that has been refinished; compare the temperature with that of another part of the vehicle that has not been refinished. If the refinished area is cooler, that means that there is still water present in the paint film. The film with water present is cooler because as the water evaporates, it takes heat with it out of the panel. If the refinished panel and the panel that has not been refinished are the same temperature, there is no longer water present, and the refinished panel is ready for the application of clear topcoats.

Urethane Enamel Topcoats

Acrylic enamel and acrylic urethanes are two variations of paint used in the collision repair industry. However, acrylic urethane is recommended because of its improved durability.

Acrylic urethanes are slightly harder and more durable than plain acrylic enamels. Each is available in a variety of colors as solid color, metallic, and pearl paints. These are the most common type of paint used in modern body shops.

Solid color paints do not have reflective flakes or mica particles in them. Generally, solid colors are easier to spray and match than metallic and pearl paints. Compare the solid color and metallic paints shown in Figure 26–3.

Metallic paints have medium to large metal flakes in them. Metallic finishes are desirable because they sparkle in the sunlight. As you will learn, there are things you should do differently when spraying metallic paints compared to solid color paints.

The size, shape, color, and material in the flakes can vary. Often called *metal flakes*, the flakes can be made of tiny but visible bits of aluminum or polyester. When light strikes the flakes, it is reflected at different angles, making the flakes appear like tiny, glittering stars inside the paint (Figure 26–4).

Pearl paints have very small reflective pigment particles, usually shiny mica bits, in them. Pearl paints produce a luster or shine that tends to change color with different viewing angles. Pearl paints are the most difficult to match when repainting.

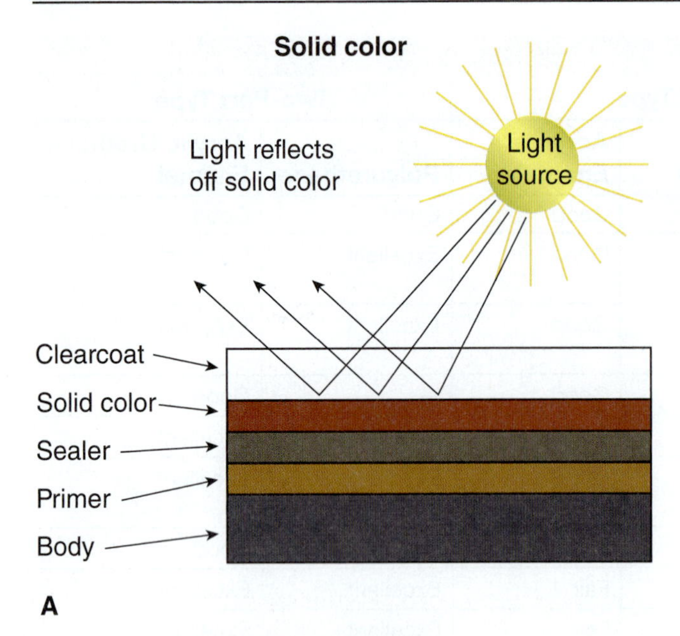

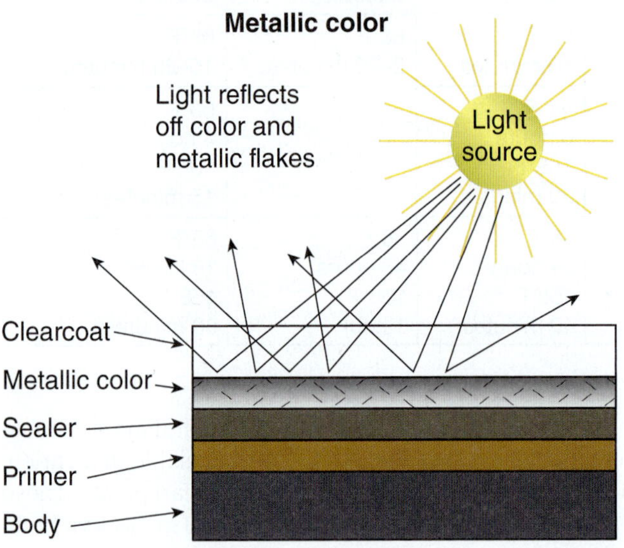

Figure 26-3 Compare how light reflects off solid color paints and metallic paints. (A) With a solid color finish, color is seen as light reflecting off the basecoat only. The basecoat color shows through the transparent clearcoat. (B) With metallic color paints, color is seen as light reflecting off the basecoat and also off metal flakes. The base color and metal flakes both reflect light to produce the paint color.

If this is new to you, start looking at paint jobs more closely. See whether you can tell the difference between a solid color, a pearl, and a metallic.

NOTE *For more information on metallic and pearl paints, refer to Chapter 27, which explains color matching in detail.*

Lacquer Topcoats

Lacquer is an older paint that dries quickly because of solvent evaporation. Generally, lacquers have been phased out in favor of the more durable acrylic urethane

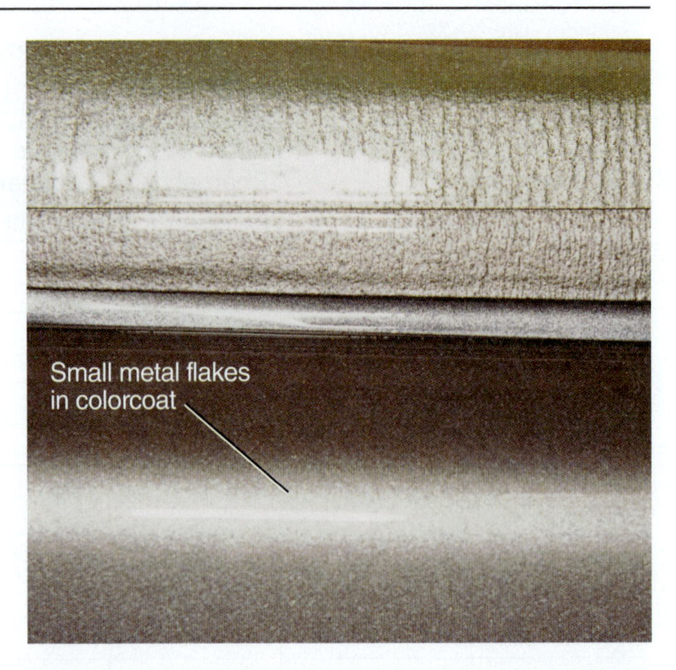

Figure 26-4 This close-up shows the tiny metal flakes in a metallic colorcoat. As you will learn, metallic paints require slightly different spraying methods.

enamel paints. When used, lacquer topcoats usually must be buffed or compounded to bring out their gloss.

Avoid spraying lacquer paint over an enamel paint, because the enamel can lift and cause problems. You can spray enamel over an old lacquer finish, however.

26.3 PRIMECOATS

Repair primecoats provide a sound foundation for the topcoats. If you sprayed paint directly onto the bare body substrate (metal, fiberglass, or plastic) without primecoats, the paint would probably peel off or look rough. This is why primecoats are sandwiched between the substrate (body panel) and the topcoat (paint). The primecoats also protect against rusting and help fill tiny scratches and other flaws in the panel (Figure 26–5).

Primecoating Sequence

A typical sequence for application of primecoats before topcoating is:

1. Spray *self-etch primer or epoxy primer* over all bare metal body surfaces to improve adhesion and add to corrosion protection. Only apply etching or epoxy primer to bare metal, not body filler or plastic parts.

Primecoats

A This technician is spraying flexible plastic primer-sealer onto a front bumper.

Topcoats

B Then a colorcoat with flex additive is sprayed over the plastic bumper.

Figure 26-5 Repair primecoats provide a sound foundation for the topcoats.

Apply the first coat over the whole repair area. Then apply additional coats to smaller areas to build up the primer and fill the repair area.

2. Apply *primer-surfacer* over the whole repair area (body filler, self-etch primer, spot putty) if needed to help build and smooth the heavily sanded repair area up level with the original paint surface.
3. Apply *sealer* over all primers and putties to keep them from showing or bleeding into the topcoats of paint. If possible, the sealer should be tinted to about the same color as the vehicle's paint color.
4. Apply clear *adhesion promoter* over original, unrepaired, but scuffed, OEM paint areas to keep the new topcoats from peeling or flaking. Adhesion promoter does not have to be applied over sealed areas.

The application of these primecoats will ready the area for topcoats. You can then spray on the colorcoats and clearcoats.

Water-Based Paints

Water-based paint, as implied, uses water (H_2O) to carry the pigment. This type of paint is like latex house paint and dries through evaporation of the water. Drying times for water-based colorcoats are much longer than those for mineral oil-based paints.

Some manufacturers have used water-based paints on new vehicles to help satisfy stricter emission, or air pollution, regulations on factories in some geographic areas. The aftermarket is bringing back the use of water-based materials. These new water-based materials have been refined to help reduce emissions of volatile organic compounds (VOCs) into the atmosphere.

Water-based primers have been used for years as a fix for paint lifting problems. Water-based primers serve as an excellent barrier coat when there are paint incompatibility problems. Water-based primers will not react with existing lacquer or enamel paints. If you ever run into difficulty with old paint or primer that crinkles up or lifts right after spraying, you can apply water-based primer to solve the chemical incompatibility problem. Water-based primers also dry to a hard, rock-like surface.

Water-based primers and paints generally come *premixed* (ready to spray), and normally they are not reduced. In an emergency, distilled water can be added to make a thinner, more liquid solution.

When using a water-based primer, do not wet sand it. Logically, the water-based material is water soluble. Dry sand water-based paints only! Water-based primers and sealers are sometimes used as a barrier coat. They help seal a paint problem, such as lifting when trying to repaint over a previously repaired, featheredged area on the body.

26.4 PREPARING REFINISH MATERIALS

When preparing to paint a vehicle, you must first decide what type of repair is called for: spot repair, panel repair, or overall repainting of the whole vehicle. You must order or mix all refinish materials needed to complete the repair. You must also check what type of paint is already on the vehicle and check whether the vehicle has been repainted before.

If spot or panel repair is planned, it is important to purchase or mix the topcoat color to accurately match the original paint color. When planning an overall refinish, the customer may want to match an old finish or choose a completely new color.

Has the Vehicle Been Repainted?

There are three ways to determine whether an automobile has already been repainted. They are:

1. *Sanding method.* Sand an edge on a panel that will be repainted until the bare metal appears. There will be more layers of paint exposed in the sanded

area if the vehicle has been repainted previously. You will see an extra layer of primer and paint over the top of the original paint film.

2. *Paint film thickness measurement method.* A thicker-than-normal paint coating generally indicates that a vehicle has been repainted. The standard paint film thicknesses of new vehicles are approximately:
   - Domestic vehicles 4–6 mils
   - European vehicles 5–8 mils
   - Asian vehicles 4–6 mils

   As discussed in previous chapters, you can use a mil gauge to measure paint thickness. Check all panels that will be repainted. If your measurements show a paint thickness about twice as thick as normal (over about 12–15 mils), the vehicle has probably been repainted.

   Reduction of mil thickness by paint removal would be needed before painting again. This was detailed in Chapter 25 on vehicle surface preparation.
3. *Inspection method.* Inspect the vehicle closely for signs of repainting. Look for masking tape-created paint lines, overspray, and other signs of repair. If a true professional repaired the vehicle, it can be difficult to tell whether the car has been refinished, because all signs of repainting will be hidden.

Determining the Existing Paint Finish

When planning how to refinish a vehicle, you must find out what type of paint is already on the vehicle. The vehicle might have its original paint or it could have been repainted with a different type of paint product.

Methods for finding out the type of paint on a vehicle include solvent application method, hardness method, and clearcoat method.

With a *solvent application method*, rub the paint with a white cloth soaked in lacquer thinner to see how easily the paint dissolves. If the paint film dissolves and leaves a mark or color stain on the rag, it is some type of air-dried paint. If it does not dissolve, it is either an oven-dried or a two-part paint. An acrylic urethane paint film will not dissolve as easily as an air-dried paint, but sometimes the thinner will penetrate sufficiently to blur the paint gloss.

With the *hardness method*, you must check the general hardness of the paint. Paints do not dry or cure to the same hardness. Generally, two-part and oven-dried paints dry to a harder film than noncatalyzed air-dried paint.

With the *clearcoat method* of paint identification, you can tell whether the panel is painted with a clearcoat. Dry sand a small area along the bottom of a panel. If the dust is white, the car has a basecoat/clearcoat finish. If the dust is the color of the car, it is a solid or single-stage color. This method works with all but white single-stage colors. Both single-stage white and clearcoat will produce a white dust when sanded.

Table 26–3 lists the types of previously applied paints and those topcoats that can be applied over them.

Do *not* spray lacquer paint over enamel paint. However, you can spray enamel over lacquer without problems. If you apply lacquer over enamel, incompatibility problems, such as paint lifting, can develop.

Some limited repairs can be done to vehicle finishes that do not require repainting. Minor scratches can be sanded with a fine grit of sandpaper and buffed out to a shiny, like-new appearance. For this repair to succeed, the scratch must not cut into the lower layers of paint. If a scratch can be felt with a fingernail, it is too deep to be repaired by sanding and buffing. One way to determine if a scratch can be removed successfully is to spray the scratched area with a solvent-based cleaner; if the scratch disappears when the panel is wet with the solvent, this indicates the scratch can be removed by sanding and buffing.

Color Mixing

To order or mix a matching topcoat color, first locate the vehicle identification plate (VIP). Write down the car manufacturer's paint code shown on the plate.

TABLE 26–3 APPLICATION CHART—PREVIOUSLY APPLIED PAINT AND REPAINTING CHART

	Previously Applied Paint				
Topcoat	Alkyd Enamel	Acrylic Lacquer	Acrylic Enamel	Polyurethane Enamel	Acrylic Urethane Enamel
Alkyd enamel	A	B	A	A	A
Acrylic lacquer	A	A	B	B	B
Acrylic enamel	A	A	A	A	A
Polyurethane enamel	A	A	A	A	A
Acrylic urethane enamel	A	A	A	A	A

A Okay to repaint with

B Okay if primer-surfacer or sealer specified by paint manufacturer is used

A chart that will help you locate paint code numbers on most vehicles is given in Figure 26–6.

An example of a service parts label for one make of vehicle is shown in Figure 26–7. Note the breakdown of information about the vehicle's paint.

Most auto refinishing shops have a color book. The paint color directory contains color chips and color information for almost all makes and models of vehicles worldwide. Color locations for each vehicle can also be found in the front of crash estimating guides.

If the vehicle's original paint is in good condition (not faded or chalked), mixing or ordering paint that matches the paint code color will normally provide good results. However, keep in mind that tinting and blending might be needed for a perfect color match. For more information on this subject, refer to Chapter 27 on color matching.

As a double-check, it is wise to compare the color chip with the actual vehicle color. There is always the chance that the car has been repainted with a different color. Place the chip on the paint and compare them. Remember that paint chips are not perfectly accurate, because they are placed on paper, not over a primecoat.

If the color match is correct, order the topcoat from a local supplier or mix the paint yourself using the stock color code number. Refinish suppliers supply topcoat colors in three ways:

1. If it is a recent model or a popular color, chances are they will have it ready-mixed in pint, quart, and gallon cans. These ready-mixed colors are called *factory-packaged.*
2. If it is an older or less popular color, they might have to mix it in pint, quart, or gallon quantities.

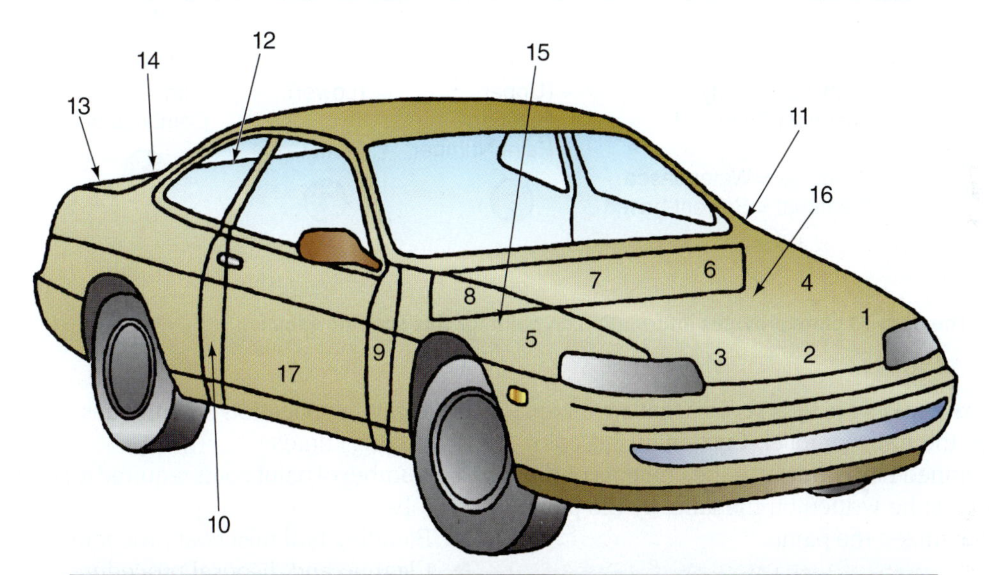

Model	Position
Acura	9
Alfa Romeo	4,13
AMC	9,10
Audi	12,13
Austin Rover	17
BMW	4,5
Chrysler	3,5,16
Chrysler Corp	3,5
Caravan / Voyager / Ram Van	6
Chrysler Imports	1,2,4
Colt Vista	16
Conquest	7
Diahatsu	1,6,7
Datsun	2
Dodge D50	3
Ford	10
Ford Motor Co.	10
General Motors	
A, J and L Bodies	14
E and K Bodies	12
B,C,H and N Bodies	13
GM Imports	2,12,13,14
Honda	8,10
Hyundai	6,7
Isuzu	2,10
Lexus	7,8
Mazda	1,2,3,4,6,8
Mercedes	2,7,9
Mitsubishi	7
Montero / Pickup	3
Cordia / Tredia	4
Others	1,2,3
Nissan	1,3,4,6,8,15,*
Peugeot	2,3,4,5,8
Porsche	9
Renault	1,3,4,5,8
Rover	1,3,4,5
Saab	5,6,8
Subaru	2
Suzuki	7,11
Toyota Passenger	7,8,14
Truck	4
Volkswagen	2,11
Volvo	6,7,8
Yugo	12

* Under Right Front Passenger Seat

Figure 26–6 This chart can be used to locate the service label that provides paint codes for most makes of vehicles.

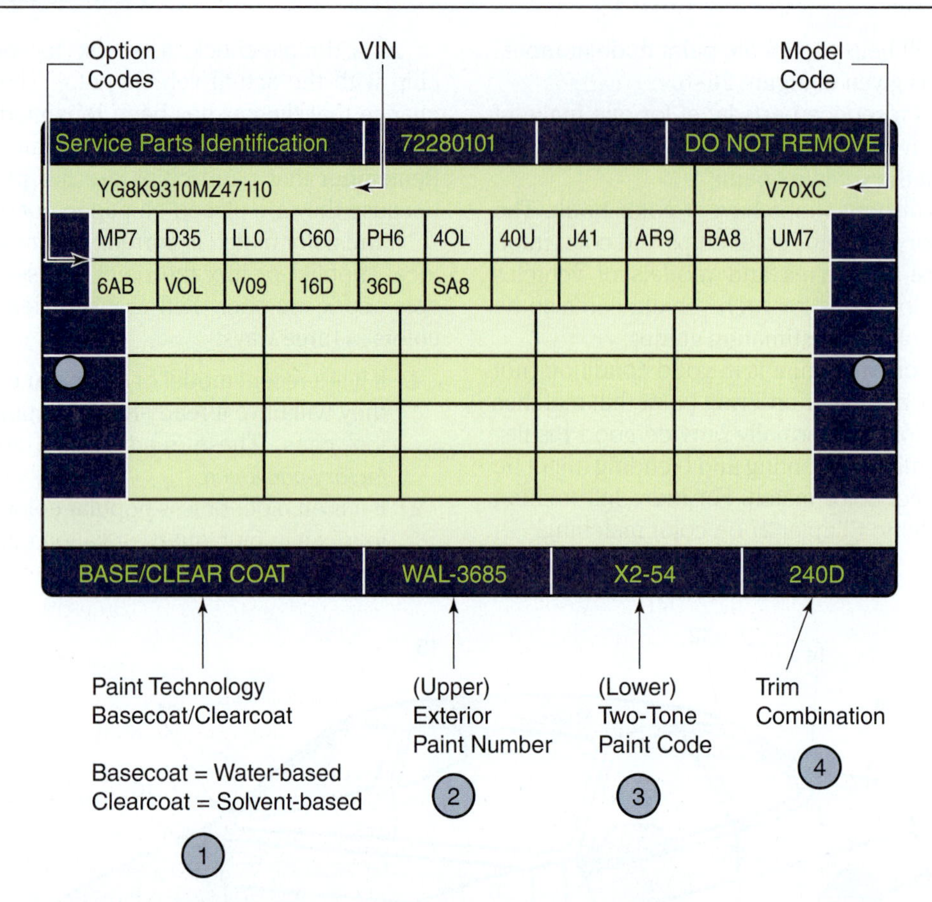

Figure 26-7 The body ID plate provides information on paint for this specific vehicle.

3. *Custom-mixed colors* are those colors that are mixed to order by the paint supplier. Custom-mixed color can be identified easily because the contents of the container must be written on the label by the paint supplier that mixed the paint.

Larger body shops often have their own paint mixing room. The paint mixing room will contain different color pigments and other ingredients so the body shop can mix its own paint color. This saves time and money over having to order mixed paint from an outside supplier.

Read Label Directions

Before applying any refinish product, carefully read the manufacturer's directions on the paint container label. While different types of paints might have the same general characteristics, each manufacturer has specific formulations for its products. For this reason, the best source of data on how to apply a specific brand of paint is the container label.

When reducing the paint, refer to the product bulletin for the proper procedures. As explained in previous chapters, some of the more important label and literature data that should be checked include (Figure 26–8):

- Proper reduction viscosity of paint material
- Use of paint additives (hardeners or catalyst, flex agents, and so forth), when necessary
- Spray application techniques and typical flash (drying) times
- Number of paint coats required for different refinishing jobs
- Blending and mist coat procedures, if necessary
- Cleanup and disposal procedures
- Safety warnings

Selecting Paint Solvents

Solvents (reducers or thinners) are added to the primer, sealer, paint, or other liquid material to lower the

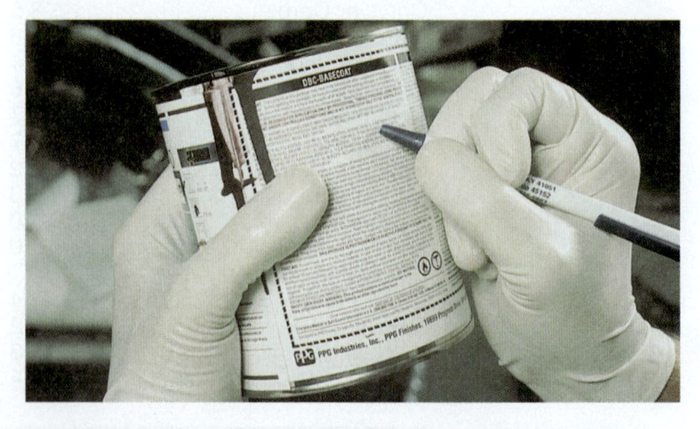

Figure 26-8 Product label directions are "the bible" for how to do things right when it comes to painting properly. Always read and follow the manufacturer's directions.

viscosity (thickness or fluidity) so it will flow through and out of the spray gun properly. Solvents or reducers also affect how fast paint cures or dries in existing temperature and weather conditions.

There are two basic types of paint solvent:

1. *Reducer*—used to thin today's urethane enamel-based materials
2. *Thinner*—used in the past to thin old lacquer-based materials

Note that some manufacturers call reducer "thinner" even though it is used with urethane paint.

Topcoat paint colors are usually shipped at as high a viscosity as practical to slow down the rate of settling. In order to apply these thick, honey-like paint materials, they must be reduced to a viscosity "runny" enough to be properly atomized by the spray gun.

REMEMBER *Some paint products are "ready to spray." They do not require reduction and can be sprayed right out of the can.*

With some new paint products, a basecoat stabilizer-reducer replaces the standard reducer. A **stabilizer-reducer** contains a special basecoat resin designed to allow a faster recoat time and better metal flake control. This is especially important if any blending is required. The stabilizer is formulated to prevent a washed out, halo-like effect at the edge of the repair. If you use plain reducing solvent without the stabilizer, you can have trouble matching your color to the original color.

Spraying Temperature and Humidity

There are two vital variables that affect painting a car: (1) *temperature* and (2) *humidity*. Of the two, temperature is the more critical. If your shop does not have year-round shop and booth temperature control, you must chemically compensate for the effects of temperature and humidity by using a different reducing solvent when mixing the paint.

Temperature and humidity affect sprayed material in five important ways:

1. Hot, dry weather produces the fastest drying time. If the paint dries too fast, numerous paint problems can result.
2. Hot, humid, or moist weather produces a fast drying time, but a little slower than hot, dry weather. Hot, humid weather conditions do not cause problems if a quality paint and the right temperature reducer are used.
3. Normal weather—70°F (21°C) with 45–55 percent relative humidity—produces a normal drying time. This is the ideal temperature for painting.
4. Cold, dry weather produces a slow drying time. If it is excessively cold, you will have to wait just a little longer between flash times before spraying the next coat.
5. Cold, wet, or humid weather produces the slowest drying time. High humidity can cause occasional paint problems.

Solvent Temperature Ratings

If you live and work in a geographic location with severe changes in winter and summer weather, up to four different temperature solvents should be used to avoid paint problems:

1. Slow-drying solvent is formulated for very hot, dry summer weather above 75°F. A slow-drying paint reducer is required to keep paint from curing too quickly in hot, dry summer weather.
2. Medium-drying solvent is formulated for average temperatures (60° to 75°F and average humidity). If the outside and booth temperatures are within 60°–75°F, use this medium-drying reducer to thin your paint.
3. Fast-drying solvent is for cooler, moister weather conditions below 60°F. The quicker flash times of a fast-drying reducer keep you from having to wait for long periods between coats.
4. Retarder is used to slow drying even more than a slow-drying reducer in extremely hot, dry weather conditions above 85°F. This is a very slow-drying solvent sometimes needed on "super warm" or hot summer days.

A good general rule to follow when selecting the proper solvent is the faster the shop drying conditions, the slower drying the solvent you should use. In hot, dry weather, use a slow-drying solvent. In cold, wet weather, use a fast-drying solvent. If you use the wrong temperature solvent, several paint curing problems can result. Paint problems are detailed in Chapter 28.

Blending Solvent

A **blending solvent** is formulated to help fade the new paint into the original paint when spraying. It is often used when spraying clearcoats over a blended repair area. The blending solvent has stronger ingredients that etch and eat into the old paint more than conventional solvents can.

The blending solvent will help dissolve the original paint film so it matches the new clearcoat more closely. The blending solvent will help the two finishes flow together more smoothly.

For example, when blending a quarter panel into a roof panel, you should use blending solvent in your clearcoat on the sail panel or where the new and existing clearcoat will meet. Blending solvent will help you feather and fade the clears together, preventing a difference in paint smoothness in the blend area.

Follow Mixing Instructions

Material mixing instructions, printed on the container label or product bulletins, state how much of each ingredient (solvent, hardener, flex agent, and so on) must be added to the paint product. A percentage of each ingredient might be called for. A simple example is given in Figure 26–9.

A *percentage reduction* means that each ingredient must be added in certain proportions or parts. For instance, if paint requires a 50 percent reduction, this means that one part reducer (solvent) must be mixed with two parts of paint.

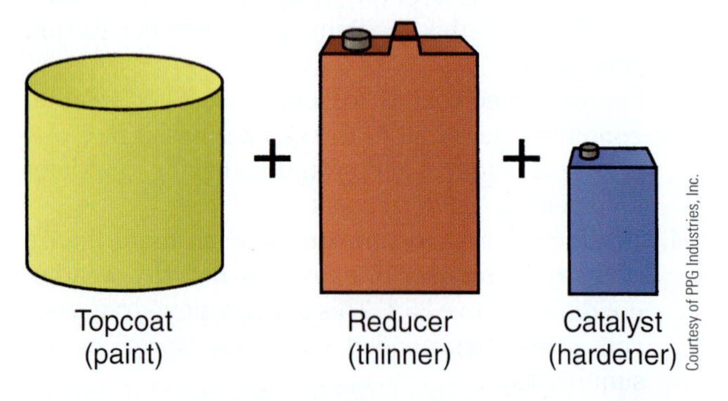

Figure 26-9 **Many priming and painting products require you to add reducing solvent and a hardener or catalyst. It is important that the right amount of each ingredient be added according to the manufacturer's directions.**

Mixing by parts means that for a specific volume of paint or other material, a specific amount of another material must be added. For example, if label directions call for a 25 percent reduction, you would add 1 quart of reducer to 1 gallon of paint. Because there are 4 quarts in a gallon, 1 quart is 25 percent of a gallon. You would mix one part (25 percent) reducer for each four parts (100 percent) of paint.

Proportional mixing numbers compare how much of each ingredient must be mixed regardless of quantity. The first number is the parts of paint needed. The second number is usually the hardener. A third number might be used to denote the amount of reducer required. The sequence of ingredients can vary. Some instructions list four ingredients to mix, so read the paint manufacturer's instructions carefully.

For example, if the mixing instruction is 2:1:1, the first digit (2 in 2:1:1) would normally mean two parts of paint. The next digit (the first 1 in 2:1:1) would mean to add one part hardener. The last digit (the second 1 in 2:1:1) would tell you to add one part reducer when mixing the paint material.

A **mixing chart** converts a percentage into how many parts of each material must be mixed. This kind of chart allow you to study the percentages and parts of each material that must be mixed. One mixing chart is shown in Figure 26–10.

As explained in several other chapters, many shops now use an electronic or computerized scale when mixing paint materials (Figure 26–11). The scale will prompt

Reduction/Thinning Percentage		Reduction/Thinning Proportions	Paint (Color)	Solvent
20%	=	5 parts paint / 1 part solvent		10%
25%	=	4 parts paint / 1 part solvent		25%
33%	=	3 parts paint / 1 part solvent		33%
50%	=	2 parts paint / 1 part solvent		50%
75%	=	4 parts paint / 3 parts solvent		75%
100%	=	1 part paint / 1 part solvent		100%
125%	=	4 parts paint / 5 parts solvent		125%
150%	=	2 parts paint / 3 parts solvent		150%
200%	=	1 part paint / 2 parts solvent		200%
250%	=	2 parts paint / 5 parts solvent		250%

Figure 26-10 **This chart shows conversions for different reduction/thinning percentages. Study them.**

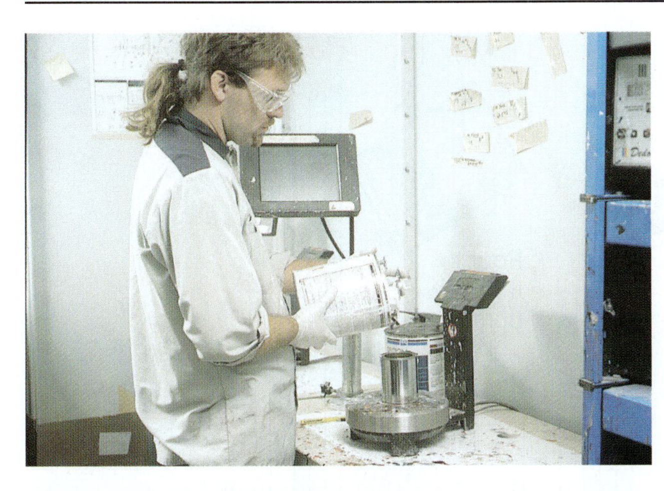

Figure 26–11 Electronic mixing scales are often used when adding ingredients to mix or formulate repair products. This is explained further in the next chapter.

you to add each ingredient and also show you how much of each ingredient to add.

Using Paint Mixing Sticks

Paint mixing sticks have graduated scales that allow you to easily convert ingredient percentages into part proportions. They are used by painters to help mix paints, solvents, catalysts (hardeners), and other additives before spraying. They are often provided by the paint manufacturer. Several different ones are needed to provide graduated scales for each type of paint (Figure 26–12).

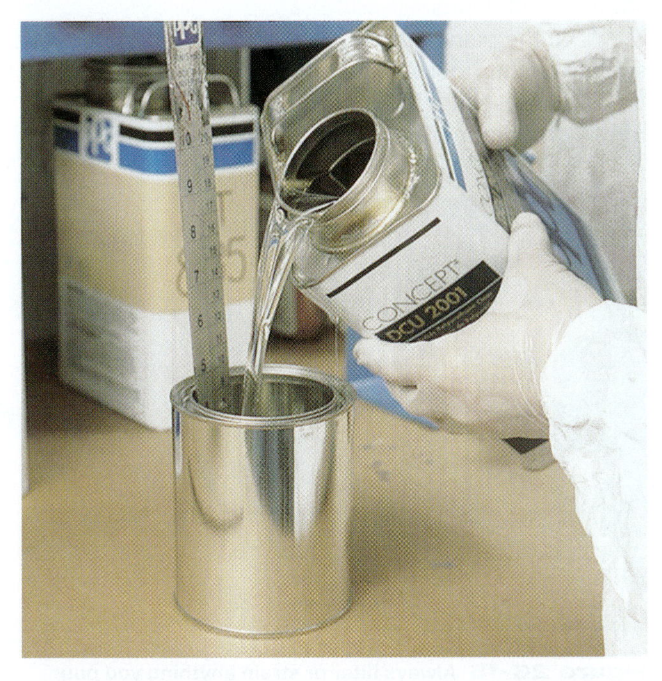

Figure 26–12 A paint mixing stick can be used to add repair material ingredients in the correct amounts. Choose the correct stick for the paint material you are using. Only use perfectly clean containers with horizontal sides when mixing paint products.

Paint mixing sticks should not be confused with paint stirring sticks (wooden or metal sticks). Stirring sticks do not have a ruler-type scale. However, a paint mixing stick can be used to stir a liquid before pouring it into the spray gun cup.

All mixing sticks have a ratio or percentage printed at the top. Marks and numbers are placed along the mixing stick for specific ingredients. You have to pour out the correct quantity of each material until it is even with the appropriate mark on the mixing stick. For example, say that the paint mixing stick is for a mixing ratio of 4:1:1. This means that you need four parts color, one part hardener, and one part solvent.

To use a mixing stick, first select the right mixing stick to match the label mixing ratio and type of paint product. Obtain a clean tin can or plastic pail with straight sides. Tapered sides on the container will upset your measurements.

The mixing container should be big enough to hold *all* paint, hardener, and solvent needed for the job. A gallon or several liter container saves mixing time for an overall or complete paint job.

Place the correct paint mixing stick for the type of paint into the straight-sided container. Pour paint or primer into the bucket. Stop pouring when the material is even with the number corresponding to the type of product. The material must usually be even with a number on the left column, which might be 1 through 7, depending on the quantity of paint needed (Figure 26–13).

For a larger repair, you might fill to lines 6 or 7; for a spot repair, you might only need to pour material even with lines 2 or 3. Make sure the paint is perfectly even with one of the numbers on the mixing stick. If color is listed on the left, the paint should be even with a number in that column that would provide enough paint for the size of the repair.

Next, pour in hardener until even with the same number in the next column on the stick. If the paint already in the container is aligned with a 3, pour in hardener until it aligns with 3, but in the appropriate column printed on the mixing stick.

Finally, pour in the last ingredient (usually reducer) until it aligns with the same number in the last column on the mixing stick. If the paint and hardener are aligned with 3, pour reducer in until it aligns with the 3 in its mixing stick column.

After adding the correct amounts of each material, thoroughly mix them. You can then strain and fill your spray gun with properly mixed paint. Some graduated mixing cups are used in the same way as mixing sticks. They have graduated scales for adding ingredients, as shown in Figure 26–14.

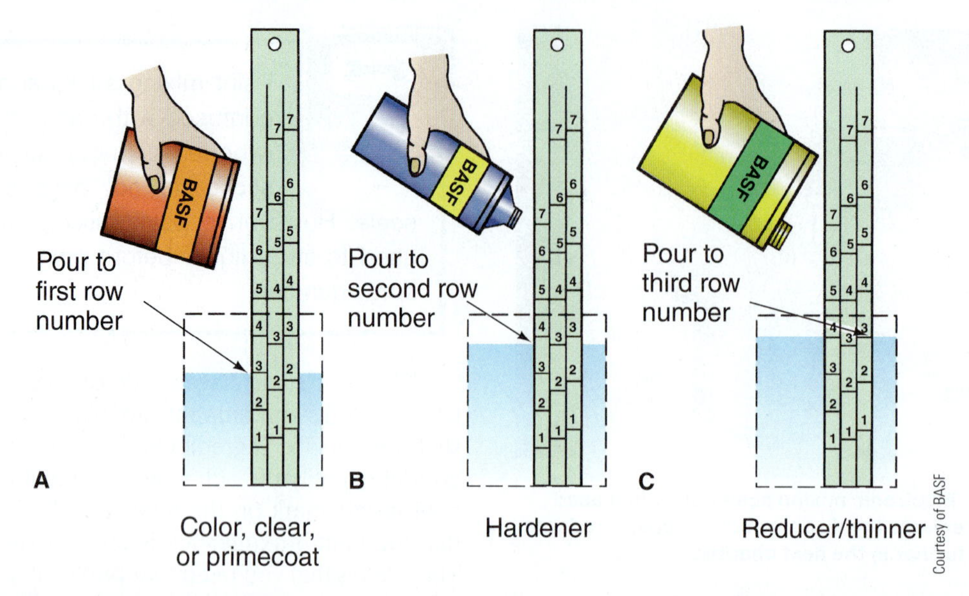

Figure 26–13 Study the basic steps for using a mixing stick. (A) Depending on information on the top of the mixing stick, you must usually pour in paint or primer first. Pour in the amount of material needed for your job. Stop when you reach any of the numbers on the mixing stick. (B) If used, pour hardener or catalyst in next. Pour material into the can until it is even with the same number on the mixing stick, but in the next column. If the paint was even with number 5, pour in hardener to number 5 on the next, or center, column on the stick. (C) Pour in solvent until the liquid is even with 5 on the stick in the right column. Pour all materials slowly so you do not add too much. Stir materials with a metal stick.

Figure 26–14 This graduated mixing cup will allow you to add exactly the right amount of each ingredient when preparing repair products.

Paint Straining

Every time you fill your spray gun, you should use a paint strainer over the cup. The *paint strainer* is a paper funnel with a fine mesh opening in the bottom that traps debris and keeps it out of the spray gun. It is a major mistake to pour any paint material into your spray gun without straining out debris.

Place the paint strainer in the spray gun cup or use a strainer holding tool. A paint strainer or funnel holding tool is desirable because it helps keep paint from dripping onto the sides of the spray gun cup, as shown in Figure 26–15.

Figure 26–15 Always filter or strain anything you pour into your spray gun cup. One bit of hardened paint, lint, woodchip, or dust could ruin your paint job. This technician is filtering clearcoat into the spray gun cup. Lift filter off of cup to see level of paint in cup so you do not overfill.

Use a paint strainer whenever you pour material into your spray gun cup! If you do not want dirt in your paint job, filter or strain everything that you pour into the spray gun. There can be hardened debris in the primer or paint can. Dust and dirt can also fall off the can or lid when pouring.

Paint Mixing/Spraying Considerations

Before applying any topcoat material, ask yourself four questions:

1. Is the paint properly stirred or mixed?
2. Is the paint properly reduced with the correct temperature solvent to the desired viscosity for booth temperature and humidity conditions?
3. Is the vehicle's surface temperature correct?
4. Are all body surfaces clean, smooth, and ready for painting?

Failure to properly stir all of the settled pigment into the liquid is a principal cause of paint problems. Stirring or mixing can be done by hand or by machine.

Pigment gives paint its color, opacity, and specific performance properties. The weight of pigments varies greatly. Some of the commonly used pigments are seven to eight times as heavy as the liquid part of paint. Because of their weight, the heavy pigments slowly settle.

Some of the pigments are light and fluffy and have very little tendency to settle. However, commonly used pigments that settle quite rapidly are whites, chrome yellows, chrome oranges, chrome greens, and red and yellow iron oxides.

If a color that contains one or more of these heavy pigments is reduced to spraying consistency and allowed to stand 10–15 minutes without being stirred, it will have settled enough in that time to be off-color when sprayed.

Do not use sharp sticks or screwdrivers for stirring. They will not adequately stir the heavier material that has settled to the bottom of the paint can. Use a flat-bottomed, clean stirring paddle or steel spatula that is at least 1 inch (25 mm) wide.

There are four important points to remember when mixing basecoat color:

1. Always read the directions first.
2. Use only the manufacturer's recommended hardeners and reducers (or basecoat stabilizer instead of plain reducer, if recommended).
3. Use the proper reducer or stabilizer temperature for the shop conditions and size of the job.
4. Use only the proper mixing ratios.

Most major paint manufacturers offer a choice of reducers to offset atmospheric conditions that can cause color shifting. Problems such as soak-in (too slow a reducer) or dry overspray (too fast a reducer) can result from selecting the wrong reducer temperature rating.

The temperatures of concern when refinishing are the spray booth temperature, the temperature of the surface of the car, and the temperature of the paint.

Be careful when bringing in a vehicle that has been outside in the cold. If the body surface is too cold, the solvents will not evaporate as quickly as they should. Color matching and paint curing problems can result. Check booth temperature and body surface temperature in cold weather. Allow the vehicle enough time to warm up if it has been sitting outside in the cold weather.

26.5 PREPAINTING PREPARATIONS

Before spraying the vehicle, there are several things you should do to prepare your equipment and the vehicle for repainting. If you fail to do even one of these steps, your paint job could be inferior and require finesse wet sanding or even repainting.

Booth Prep

After properly masking a vehicle and preparing its surfaces for repainting, pull the vehicle into a clean paint booth. Drive slowly, especially if masking materials are blocking your view out the windows. Center the vehicle in the spray booth so that you have plenty of room to work all the way around (Figure 26–16).

Figure 26-16 A paint booth should be perfectly clean before pulling any vehicle into the booth. Drive carefully, especially if some of the windows have been masked over.

Turn on the booth blower fan and close the booth doors so that dust, dirt, and fumes are pulled out of your work area. Also check the booth temperature and adjust it to match the temperature of your reducer.

Make sure the shop's main air supply system is drained of moisture and working properly. If needed, also service and turn on the separate compressor for your fresh air-supplied respirator.

Even if you turn on the circulation fans in a paint booth, you must close the booth doors. This is the only way that clean filtered air can be pulled through the booth. Even if you only leave the doors cracked open, you will draw outside dust and other airborne debris into the paint booth. This foreign matter could ruin your paint job. Keep the booth doors closed tightly all the time! Treat your paint booth like a hospital operating room.

Final Check of Masking Materials

Once in the paint booth, carefully inspect all masking tape and paper one last time. Make sure none of the tape has pulled up. Replace or remask any masking paper that has been torn while moving the vehicle.

In particular, closely inspect all fine line tape edges next to all panels that will be painted. Check for a crooked tape edge or a missed surface that should be masked. If even one piece of tape has been moved or shifted, it could damage your paint job.

Also inspect all masking paper and plastic closely for openings that could allow overspray leaks. After driving the vehicle into the booth, you might have to mask the driver's door opening (Figure 26–17).

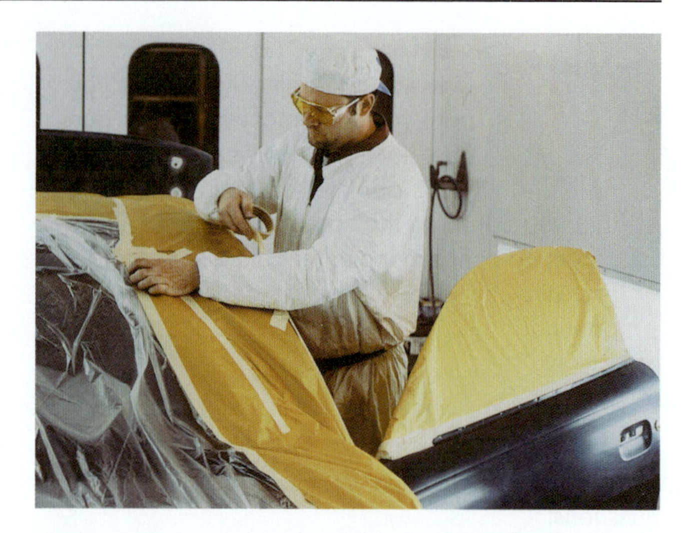

Figure 26–17 **If needed, mask the door opening after driving the car into the booth. Also, check all masking for leaks that could affect the paint job.**

Prepainting Surface Cleaning

Once in the spray booth, clean all surfaces to be painted with compressed air and a prepaint cleaning agent again. Someone could have touched the vehicle with bare hands, or an air tool could have thrown oil on body surfaces. You must carefully remove anything that could contaminate your new paint (Figure 26–18).

Figure 26–18 **After the vehicle is moved into the paint booth, any surface to be painted will need a final cleaning. One fingerprint on the body can cause a paint problem after spraying.**

While wearing plastic gloves, give the vehicle body a final cleaning with a low VOC wax and grease remover or an approved cleaning solution. Use one rag soaked in cleaning liquid to wipe down all body surfaces to be painted. Use a second clean, dry, lint-free rag to wipe off the cleaning solution while it is still wet. Do not allow the cleaning solution to dry or debris can remain on the body surface (Figure 26–19).

Most shops use special lint-free disposable cloths to wipe down vehicles right before painting. Disposable towels or rags ensure that no paint contaminants are on the cleaning rags.

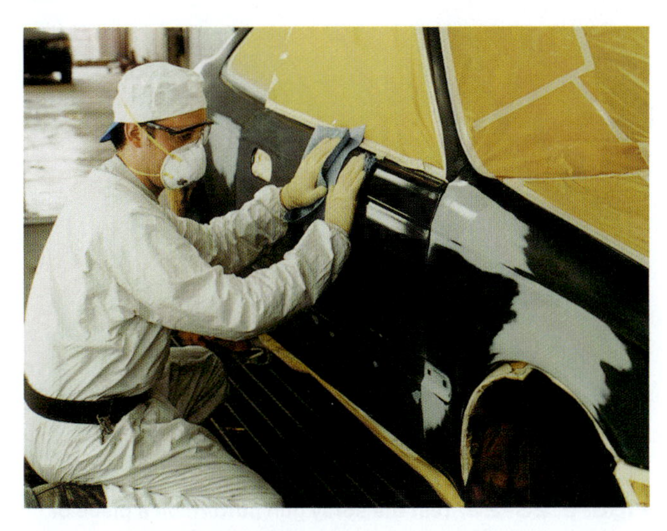

Figure 26-19 Use an approved prepainting cleaning solution to wipe down all body surfaces that will be painted. Use one clean rag to apply the cleaner; use another clean rag to wipe off the cleaning solution before it dries.

Never use paint solvent (reducer or thinner) to clean body surfaces. Reducers and thinners can cause serious paint problems if used to final clean body surfaces before painting. Paint solvents can leave a residue on the surface or damage primecoats. Use only a specially formulated wax and grease remover designed for washing surfaces before painting.

Any static electrical charge on a vehicle body will act like a dust magnet. For this reason, some paint technicians like to ground a vehicle before the prepainting wipe down. A metal wire is connected from the vehicle chassis to an earth ground (metal pipe or metal rod in booth floor). This helps to discharge any static electricity on the body so that it will not attract airborne dust.

Put on your painting coveralls, a pair of clean gloves, and an air-supplied respirator hood or face shield. Connect your spray gun to the booth air hose and your respirator to its fresh air supply hose. Blow off any remaining dust on the vehicle with an air nozzle. An air nozzle will blow more air than your spray gun. First, blow down into hidden pockets that could be holding dirt and dust. Next, blow off the masking paper and then the body panels that will be painted.

Using a Tack Cloth

As you blow debris off body surfaces, wipe the vehicle down with a tack cloth. A tack cloth is a disposable rag with a sticky coating on it for removing dust and debris from vehicle surfaces before painting.

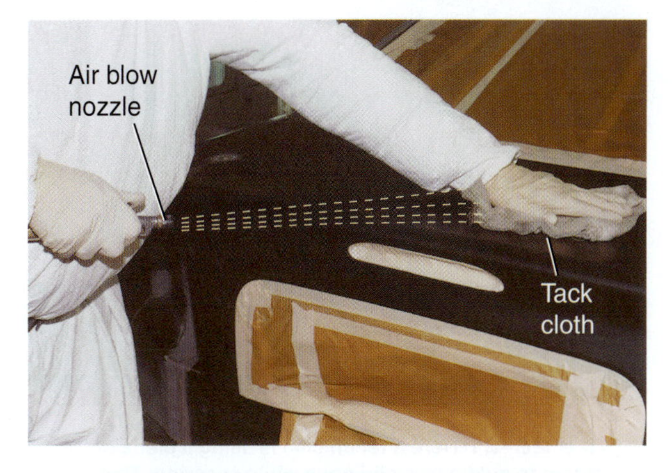

Figure 26-20 Wipe vehicle surfaces down with a tack cloth while blowing with an air nozzle. The tack cloth will help lift and hold any dirt and dust that could flaw the new paint.

Wipe all surfaces gently with the tack cloth as you direct the nozzle's airstream over the rag. The tack cloth will lift and hold any dust particles that could show up in the paint. This procedure is shown in Figure 26–20.

As the tack cloth gets dirty, fold it over to expose a clean surface. Methodically wipe every square inch of the body surface to be painted. If you miss an area, dirt or pieces of lint could mar your fresh paint finish.

To further avoid dust and dirt problems in the paint, wipe off your spray gun air hose with the used tack cloth. In particular, wipe off the end of the hose that will be picked up and moved around in the air near the vehicle. It is very easy for the hose to collect and then deposit debris into the fresh paint.

After wiping with a tack cloth, be careful not to touch the surface being refinished or to stir up dust in the booth. Do not touch the paint or lay anything, such as masking tape, on body surfaces. Keep the paint booth doors closed. When necessary, open and close them gently.

Final Prepainting Check

To help prevent problems during and after painting, there are several prepaint checks you should make before spraying the new finish.

For example, make sure everything is off the paint booth floor. Remove the masking machine or anything that might get in your way while painting.

Figure 26–21 shows a technician placing a piece of Styrofoam™ under the trunk lid, which helps to hold the trunk lid partially open while painting. This is a handy tip if you have to paint under a hinged panel.

Check your spray gun air supply pressure, preferably with a gun-mounted pressure regulator. Operating pressures will vary with gun design and the type of paint material being sprayed. Refer to the spray gun and paint product recommendations for inlet pressures if in doubt. See Figure 26–22.

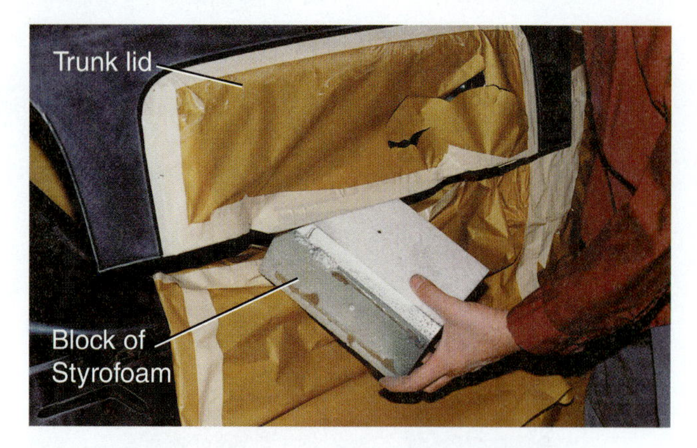

Figure 26-21 Here a technician is using a piece of Styrofoam to hold a trunk lid open while spraying. This technique keeps the lid from closing fully so wet paint surfaces do not touch.

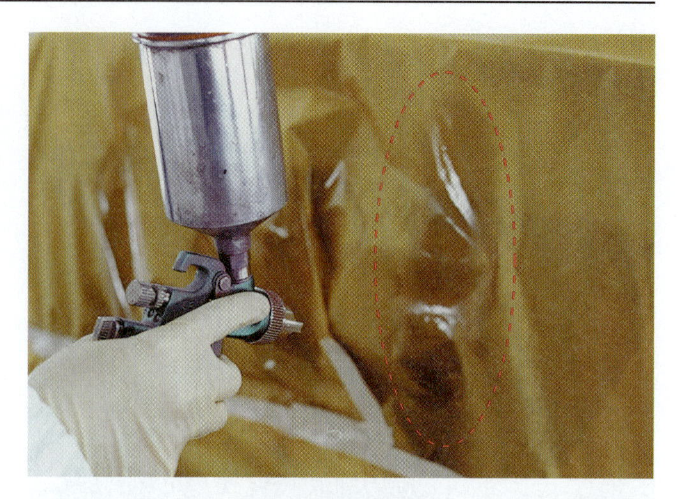

Figure 26-23 Test the spray gun pattern on a piece of masking paper. If the spray gun is not operating normally, you should find out before trying to paint the vehicle.

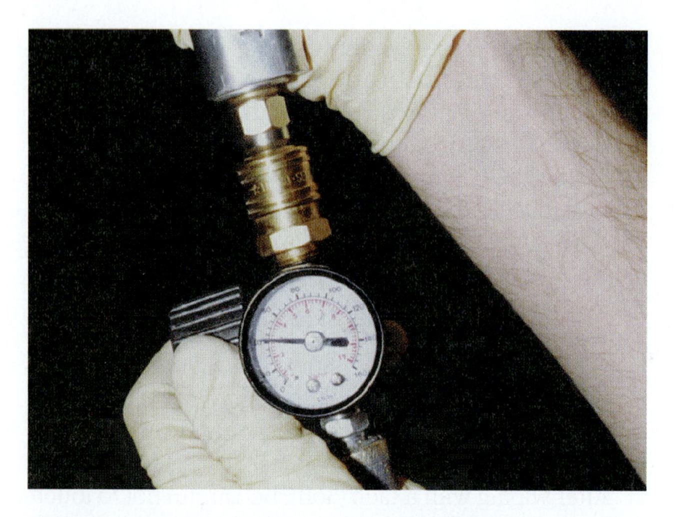

Figure 26-22 Check spray gun operation right before painting. While pressing the trigger halfway to get airflow through the gauge, check inlet pressure, preferably with an in-line pressure regulator-gauge assembly.

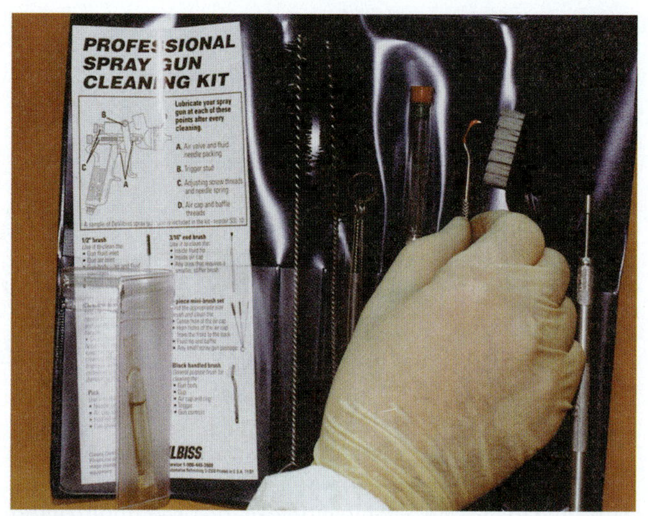

A Spray gun cleaning kits have all the tools needed to clean and lubricate a spray gun.

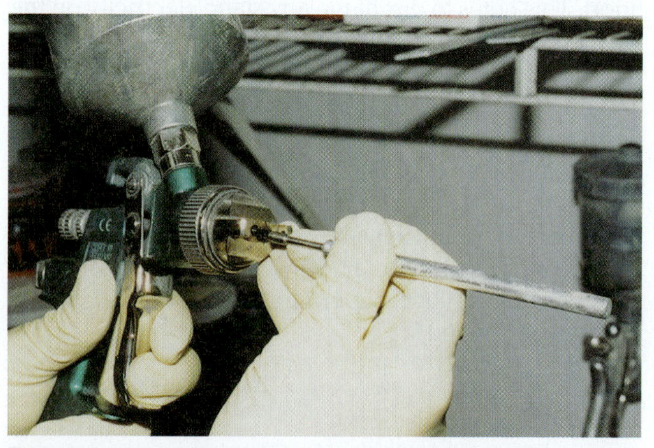

B A common problem area is around the air cap. Dry spray can collect on the cap and affect atomization. Clean the air cap frequently.

Figure 26-24 If a spray gun spits or does not properly atomize the repair material, it needs to be serviced.

Test and adjust your spray gun pattern, as described in Chapter 24. Adjust fluid flow and the spray pattern. Make sure the spray gun is operating properly before starting to spray the vehicle. Look at Figure 26–23.

If needed, use a spray gun cleaning kit to service your spray gun. These kits have special rifle-type brushes, special oil, and other tools for servicing a spray gun. Your spray gun must be perfectly clean, lubricated with noncontaminant oil, and properly adjusted before starting to refinish a vehicle (Figure 26–24).

Trigger the spray gun a few times to check its operation. Make sure a smooth, well-atomized mist flows out of the nozzle when fully triggered. Spray the gun on masking paper before attempting to paint vehicle body panels. If there is a spray gun or paint problem, you should find and correct it before it ruins your paint job (Figure 26–25).

A **spitting spray gun** fails to atomize all of the liquid properly. Large droplets of liquid will spatter on the test

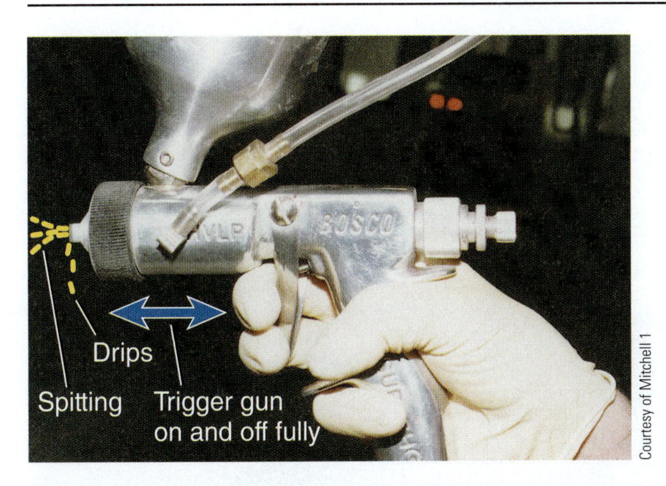

Figure 26-25 **Trigger the spray gun on and off when testing the spray pattern. Make sure the gun does not spit or leak.**

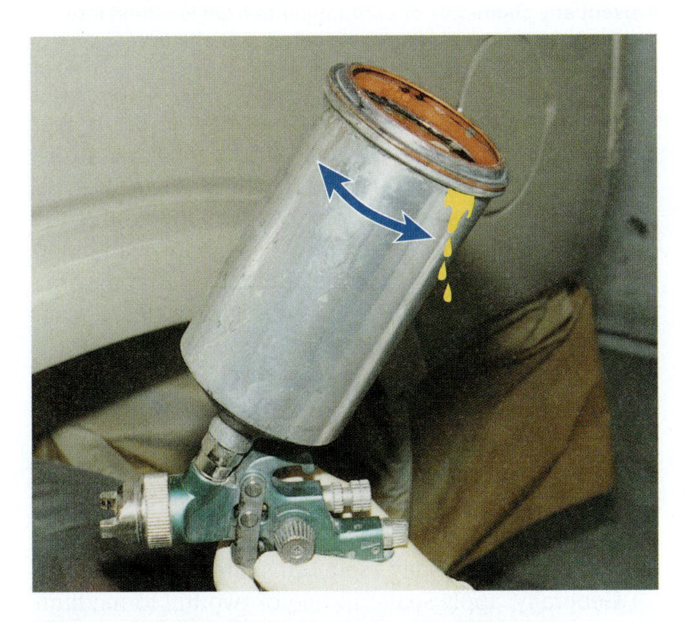

Figure 26-26 **Shake the spray gun a little while watching for leakage around the cup. If the lid on the cup leaks, paint could drip onto the vehicle body and cause paint problems that are time consuming to fix.**

spray pattern or on the vehicle. This is normally due to air getting into the paint stream as it moves through the spray gun body. When the spray gun cup is almost empty, air can be pulled into the gun and cause unatomized paint to spit out of the nozzle. Loose spray gun parts (for example, fluid nozzle or worn, dry packing around the fluid needle) can also cause spray gun spitting.

A **leaking spray gun** allows paint to seep out of the gun lid or another part. A leaking spray gun lid or cap is a common problem (Figure 26–26).

Dry paint can quickly collect around the plastic seal on the spray gun lid. Any dry paint on the sealing surfaces around the lid can allow liquid to drip into your paint job. Always keep the spray gun lid seal and other parts perfectly clean.

Your spray gun must be working perfectly if you want a perfect paint job. Maintain your spray gun and test it before each refinish job. You do not want your spray gun to "drip and spit" in your fresh finish!

26.6 APPLYING PRIMECOATS

As discussed in previous chapters, a primer is generally the first primecoat in any finishing system. Primers are designed to prepare the bare substrate (steel, aluminum, SMC, fiberglass, or plastic) to accept and hold the color topcoat. Primers should be selected to match the substrate. Epoxy and etching primers provide maximum adhesion on metal panels and produce a corrosion-resistant foundation to prevent rusting.

There are several types of primers or primecoat materials available. Make sure you use the right kind for the body substrate or panel being painted. See Chapters 7, 12, 24, and 25.

Always read the label directions on paint materials before using them. The directions are specific to the product and include detailed application, mixing, and safety information.

The following section provides a quick review of the different primecoats and how they should be applied. See Figure 26–27.

Applying Primers to Bare Metal

Apply *etching primer* or *two-part epoxy primer* to all bare metal surfaces. On smaller areas sanded to bare steel, you can use a spray can of etching primer. On larger areas of bare metal, mix and apply the primer with your spray gun. Generally, use one full wet coat over all exposed steel areas on the panel. Blend or fan the primer thinner around the perimeter of the repair area. Apply this material following label directions, usually as full wet or medium wet coats. One or two coats are generally recommended, with proper flash time between coats.

Etching primer-surfacers are also available that can be applied to bare metal. They should be used if the metal has surface rust or surface pits. The thicker etching primer-surfacer will bond to the metal while also filling small pits in the surface. Plain primer is often not thick enough to fill rust pits in metal.

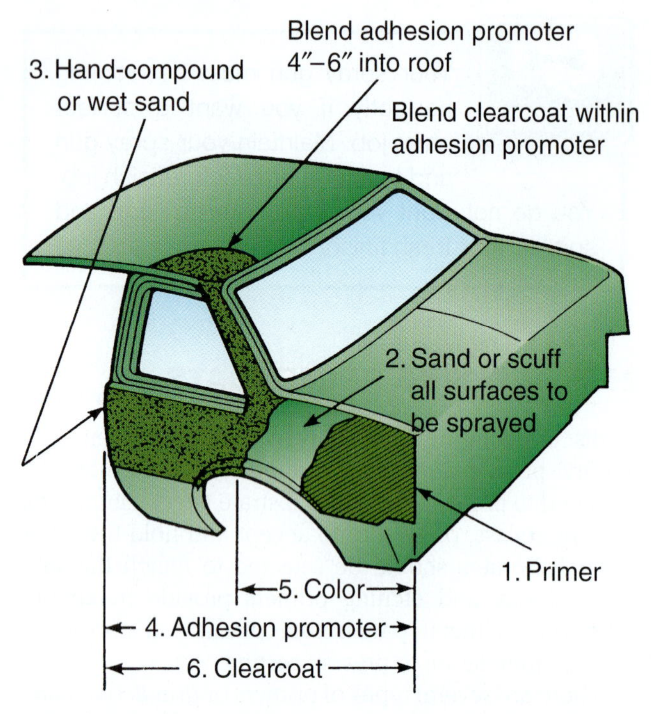

Figure 26–27 **Study the major steps of applying primecoats during panel repair. Clear adhesion promoter is often recommended over hard-baked original paint, because the paint is so hard that clearcoat alone may not properly adhere or stick.**

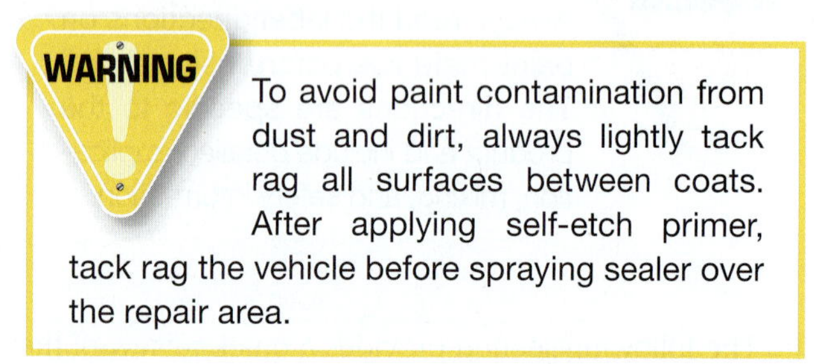

To avoid paint contamination from dust and dirt, always lightly tack rag all surfaces between coats. After applying self-etch primer, tack rag the vehicle before spraying sealer over the repair area.

After spraying etching primer-surfacer over the surface, allow it to fully cure. You should then block sand to level the primer-surfaced area and prepare it for sealer and paint.

Most professionals tack cloth the vehicle twice or more. First, tack cloth the body surfaces while blowing with an air nozzle. Then, right before you paint, tack cloth again while blowing air with your spray gun. When painting a vehicle, remember that cleanliness is critical.

Applying Sealers

Sealers are sprayed over primers, spot putties, and old finish to provide a barrier layer for the topcoats. Paint sealer prevents primecoat solvents from penetrating the primer and affecting the color. They provide uniform holdout so discolorations do not show up in the new paint (Figure 26–28).

Figure 26–28 **Sealer should be applied over all repair areas to provide a barrier layer. The sealer barrier helps prevent any chemicals or contaminants from leaching into the topcoats and affecting or discoloring the finish.**

Sealer provides a solid color over which to apply the colorcoat. If there have been previous repairs with different primers, the sealer will make the whole repair area the same color. This will reduce the amount of topcoat needed, especially with transparent colorcoats.

If the old finish is good and hard, a sealer is not mandatory. Only spray the sealer over the repaired areas on panels.

Mix and apply sealers following label directions. Sealers can be clear or colored, or you can tint them to match the color of the vehicle. A tinted sealer should be used with some of today's semitransparent colorcoats.

Generally, apply sealer in one or two full to medium wet coats. Apply full coats of sealer to build the sealer over primer, spot putty, and body fillers. Blend the sealer out by fanning the spray gun sideways around the perimeter of repair areas. This will feather the paint film thinner so it blends into the existing paint.

Applying Adhesion Promoter

Adhesion promoter, also called a mid-coat primer, is designed specifically to aid adhesion on hard-baked basecoat/clearcoat OEM finishes. This product is often a premixed, clear primer with good durability and excellent adhesion to very hard clearcoats. It is recommended on scuff-sanded OEM finishes to avoid lifting and peeling of the new paint.

Adhesion promoter should be applied over factory-baked finishes before repainting them. It does not have to be applied over the primed and sealed repair areas. Clear adhesion promoter can be sprayed with clearcoat paint without affecting the color of the original finish. Usually one medium to full wet coat is recommended.

26.7 REFINISHING PLASTIC PARTS

After plastic parts have been repaired as described in Chapter 13 and “surface prepped” as detailed in Chapter 25, primecoats and topcoats can be applied.

Plastic parts often require a special plastic primer. If you do not apply a plastic primer or conventional primer, you may have paint peeling or lifting problems. Semirigid plastic parts, such as bumpers, also require that you add a flex additive to the paint to prevent paint cracking.

Follow the manufacturer’s recommendations to determine whether a particular paint system can be used on a specific type of plastic, or whether a plastic primer or flex agent is required. Automotive plastics can generally be topcoated using conventional paint systems.

Table 26–4 lists the more popular automotive plastics and suggested finishing systems.

Using Paint Flex Agent

Semirigid (flexible) plastics often require the addition of a “flex agent” additive into the paint. The additive is needed because soft, flexible plastic parts vibrate and bend easily.

A **paint flex additive** keeps the paint film softer and more pliable so the cured paint film will not crack when bent or flexed. A flex additive should be added to paint that will be applied to flexible bumpers and similar parts. Refer to paint manufacturer recommendations for when and how to use flex additive.

As always, it is best not to mix products from different manufacturers. The flex additive, primecoat materials, paint, hardeners, and reducer should be from the same manufacturer. Mixing products from different manufacturers on the same job can result in poor performance.

Basecoat/clearcoat material is usually urethane-based. Some manufacturers recommend flex additives in their basecoats of color but not in their clearcoats, because the additive can reduce paint gloss and shine. Other paint manufacturers recommend just the opposite. Therefore, always refer to label directions for flex additive use.

To apply a flexible (elastomeric) finish, proceed as follows:

1. Thoroughly scuff sand the entire part with the recommended grit abrasive paper (typically #400 to #600 grit). Clean the surface with a prepainting cleaning solvent.

TABLE 26–4 FINISHING SYSTEM FOR POPULAR PLASTICS

KEY I Interior E Exterior P Primer NP No primer SP Special primer/adhesion promoter NA None approved * Flexible primer and/or additive recommended		**Standard Lacquer System**	**Flexible Lacquer/ Enamel System**	**Polypropylene System**	**Vinyl System**	**UrethaneSystem**
ABS	Acrylonitrile-Butadiene-Styrene	I/NP E/NP				
ABS/PVC	ABS/Vinyl (soft)		I/NP E/NP		I/NP	
EP I, EP II, or TPO	Ethylene Propylene			E/SP*		
PA	Nylon	E/P				
PC	Lexan	I/NP				
PE	Polyethylene	NA	NA	NA	NA	NA
PP	Polypropylene			I/SP		
PPO	Noryl	I/NP				
PS	Polystyrene	NA	NA	NA	NA	NA
PUR, RIM, or RRIM	Thermoset Polyurethane		E*			E
PVC	Polyvinyl Chloride (vinyl)		E/NP I/NP		E/NP	E I/NP
SAN	Styrene Acrylonitrile	I/NP				
SMC	Sheet Molded Compound (polyester)	E/P				
UP	Polyester (fiberglass)	E/P				
TPUR	Thermoplastic Polyurethane		E*			E
TPR	Thermoplastic Rubber		E*			E

2. Following the manufacturer's instructions, mix the base color and flex additive thoroughly before adding the type of solvent best suited for the shop temperature. Remember to mix only the amount needed for the job, because the catalyzed material cannot be stored.
3. Using the recommended air pressure at the gun, apply a sufficient number of coats, as directed by the manufacturer. Typically, apply double coats to achieve complete hiding and the proper color match on flexible parts.
4. Wet double coats are applied as follows: Spray the first pass left to right. Spray the second pass right to left, directly over the first pass. Drop the nozzle so that 50 percent of the pattern overlaps the bottom half of the initial double coat. Continue the pattern until complete. Be sure to allow the recommended flash time between coats.
5. Allow the basecoat adequate drying time before applying the clearcoat (typically 30–60 minutes). Avoid sanding the basecoat before applying the clearcoat.
6. If colorcoat or basecoat sanding is needed to remove paint film imperfections, such as dirt or lint in the paint, wet sand with very fine grit sandpaper (about #600 or finer grit). Reclean the area(s). Apply one additional mist coat of colorcoat over the sanded area and let it flash.
7. Sometimes flex additive should be mixed with the clearcoat. However, with some paints you should not mix flex agent into the top clearcoats. This can make them a little dull and keep them from matching the OEM finish.

Compounding is not necessary when a flex additive is used in the topcoat; the mixture should cure with acceptable gloss. Compounding can dull the gloss of paint with a flex additive, creating a flat appearance. If you compound and dull a flexible paint, the finish cannot be brought back to the same gloss level without applying more paint. Make sure you produce a glossy finish when spraying flexible paints.

Flexible replacement panels are factory primed with a flexible enamel-based primer. The only preparation required prior to topcoating is cleaning with wax and grease remover, fine sanding, and a second cleaning after the sanding.

WARNING

Keep flexible paint material overspray off regular, rigid body surfaces. If applied to them, you could have problems color matching the gloss differences between the flexible paint and the rest of the paint.

26.8 FLASH TIMES

Flash time is the amount of time needed for the fresh coat of primer, sealer, or paint to partially dry before the next coat can be sprayed. Proper flash time is needed so the next coat will apply to the previous coat without problems.

Flash times vary with the type of repainting material, booth temperature, and humidity. Again, refer to label directions for specific flash times for each type of paint product.

Label directions often recommend flash times from 10 to 20 minutes for modern urethane and catalyzed paint products. A cold spray booth, paint material, or vehicle surface will increase the necessary flash times. Warm temperatures reduce flash times between coats.

Drying/Flashing Temperature

Flash times can be made shorter with today's climate-controlled paint booths. Painters often increase booth temperature between coats.

A *finish drying temperature* of 100°F is used by many professional painters to reduce flash times between coats. This warmer booth temperature can cut flash times to about 5–10 minutes for most paint products. This can save considerable time over having the paint dry at room or shop temperature.

Incorrect Flash Times

A *short flash time*, or not waiting long enough between spray coats, can cause paint to run or sag. This often happens when a painter gets in a hurry and does not wait long enough between coats. The paint will often run when the last heavy coats of clearcoat are applied. "Solvent popping" is an even greater concern if you cheat on the flash time.

An *excessive flash time*, or waiting too long between spray coats, can lead to paint adhesion and flow-out problems. You do not want the previous coat to dry too much before spraying the next coat. If the paint cures completely, you have to scuff sand it for proper adhesion of the next coat.

WARNING

Never adjust a spray booth temperature higher than about 100°F between coats. A high temperature can cause blistering and other paint problems. Never use higher baking temperatures (150°F) to reduce flash times or the solvent could pop and blister the paint film.

Spray Schedule

A good idea is to write down a **spray schedule** that lists the times when each material (primer, sealer, adhesion promoter, colorcoats, clearcoats) has been sprayed. Write down spray times on a small notepad kept in the paint mixing room. This will help you keep track of the exact time between spray coats and ensure proper, consistent flash times (Figure 26–29).

Finger Flash Testing

A *finger flash test* is done by touching the partially cured paint with a gloved finger to see whether it is dry enough for another coat. Touch on a piece of masking tape, not on the body panel. If the paint has flashed properly, small strings of semicured paint will stick to your gloved finger.

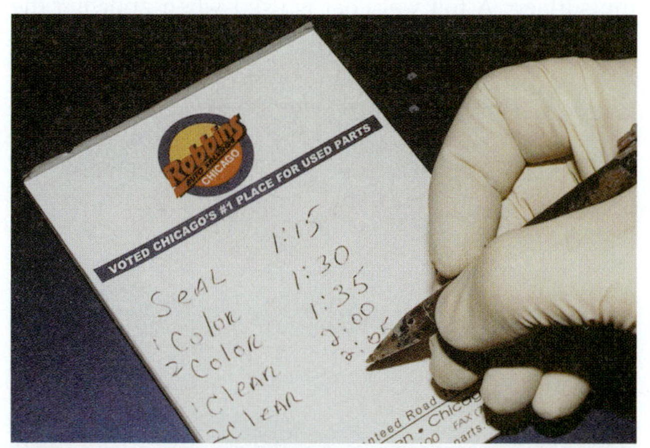

A Write down the times each coat is sprayed on a spray schedule sheet. This will help you remember when the next coat should be sprayed.

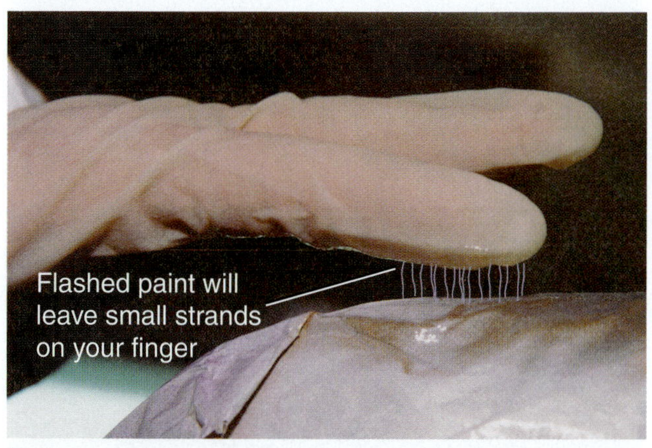

B You can also touch fresh paint on masking paper to check for proper flashing. Paint should stick to your finger and form "spider webs" if flashed and ready for another coat. Wet paint will deposit on your finger as a liquid and has not flashed and is not ready for another coat.

Figure 26–29 Always make sure the previous coat has dried before applying the next coat. Label directions will give typical flash times for the specific product.

If the paint has not had enough time to flash and is still too wet for another coat, no paint strings, only liquid paint, will stick on your plastic glove.

If you accidentally repaint a car too quickly, without proper flash times, huge paint sags and runs can flow down and ruin the finish, especially on vertical panels. Because the first coats were not flashed and cured enough, the heavy coats of wet paint can be pulled down by gravity. This is an embarrassing mistake that requires resanding and possibly repainting the vehicle.

26.9 BASIC SPRAY COATS

You can use a spray gun to apply slightly different kinds of coats of primer or paint. Sometimes a paint product should be thicker or thinner, depending on its purpose. There are varying degrees of thickness and wetness for sprayed coats. For example, a colorcoat should be applied more thinly than a primer-surfacer.

The different kinds of spray gun coats are mist coat, light coat, medium coat, full wet coat, and blend coat.

The easiest way to control the thickness of a spray coat is by varying the speed with which the gun is moved. That is, the slower the speed of the spray gun movement, the heavier and wetter the coat. If you move the spray gun more quickly over the panel, a thinner, dryer coat will result. Fanning the gun sideways will also thin the coat for blending.

Spraying Mist Coats

A **mist coat**, also called a *dust coat, drop coat, or tack coat*, is a very light, thin coat. Spray gun pressure is reduced and the gun is held a little farther from the surface. The spray gun is moved more quickly from side to side. The mist coat cures or dries in a short period of time to bond and form a lightly textured paint film.

Mist coats are sometimes used to apply metallic colorcoats. The last coat of metallic color is misted onto the panel to keep the metal flakes suspended in the paint for a better color match. If you apply metallic paints in coats that are too wet, the metal flakes can settle to the bottom of the paint film and not match the factory color.

Mist coats can be used, and are often recommended, to help avoid problems like mottling and blotching with difficult-to-spray metallic colors such as gold and silver. For example, some paint manufacturers recommend a mist coat as the last colorcoat with some metallics. Some painters also mist coat the first layer of clearcoat on troublesome metallics to prevent movement of the metal flakes.

Spraying Light Coats

A **light coat** is usually produced by moving the spray gun a little more quickly than normal across the surface of the vehicle. A thinner than normal coating of paint film will be deposited on the surface. A light coat is sometimes used when applying the first colorcoats to get good coverage.

The basecoat or colorcoat can go on light, without a gloss. This helps you blend the new finish into the existing paint with a less dramatic change of color. A light spray coat of color also distributes any metallic flakes more evenly. The wetter clearcoats will give the colorcoat its gloss. See Figure 26–30.

Generally speaking, you should aim for the thinnest coat possible that produces complete coverage and good flow out. Thinner spray coats will be more durable and less prone to problems than thicker coats. Thicker coats of primer and paint crack and chip more easily.

If you need to build up primer or primer-surfacer thickness to fill tiny imperfections, apply more coats. Do not try to spray the material on thicker in one pass or coat. Several thin coats are always better than one thick coat.

Spraying Medium Wet Coats

A **medium wet coat** is produced by moving your spray gun at a normal speed over the surface being refinished. It produces normal gloss, provides adequate coverage, and helps prevent runs and sags. A medium wet coat is the most common coat recommended by paint manufacturers and used by professional painters. Refer to Figure 26–31.

Spraying Full Wet Coats

A **full wet coat** is done by moving the spray gun slightly more slowly than normal. It will deposit more paint on the surface. A full wet coat is used when applying the

A Basecoat or colorcoat is sprayed on in light to medium coats.

B The colorcoat can go on without a gloss. It will often look dull until the clearcoats are applied.

Figure 26–30 Colorcoats of basecoat/clearcoat finish are being sprayed by a professional painter.

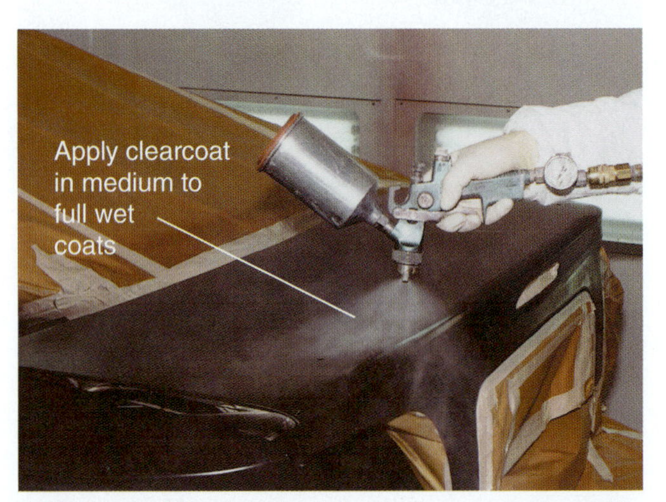

A Medium to full wet coats of clearcoat are normally used to produce a smooth, shiny finish.

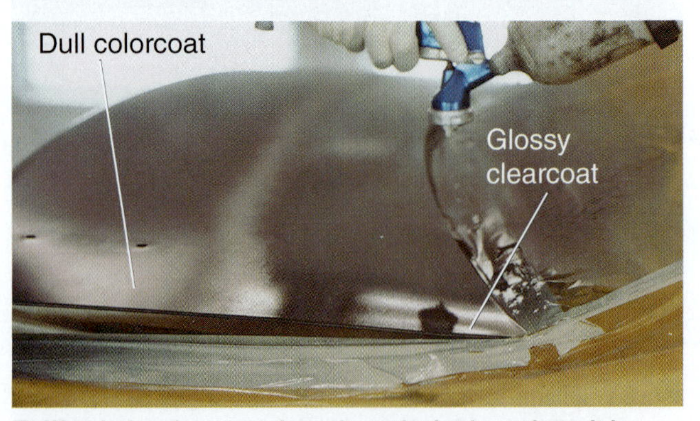

B Watch the clearcoat deposit on the body surface. It is easy to apply too much transparent clearcoat, which produces paint runs or sags. Apply uniform medium wet to full wet coats, and do not flood the body with too much clearcoat.

Figure 26–31 After the colorcoats have flashed enough, apply the clearcoats to bring out the gloss or shine in the finish.

final layer of clearcoat or single-stage colors. It is important that the last coat of clearcoat goes on wet to produce a high gloss or shine. A full wet coat is also recommended when building a surface with primer-surfacer or primer-filler.

Spraying a full wet coat requires skill and practice to prevent runs. The wet coat makes the paint lie down smooth and shiny. It prevents a dull, textured "orange peel" paint film. However, a full wet coat can be almost ready to run and sag. Look at Figure 26–32.

With today's basecoat/clearcoat paint systems, professional painters often use light to medium wet coats of basecoat for good coverage, followed by a mist, almost dry, coat of color. Finally, full wet coats of clear are applied over the colorcoat to produce a high-gloss finish.

Spraying Blend Coats

Blend coats, also called *shading coats*, are applications of paint done progressively so the paint film becomes thinner. Blend coats are normally used on the boundary of spot repair areas so that a color difference is not noticeable.

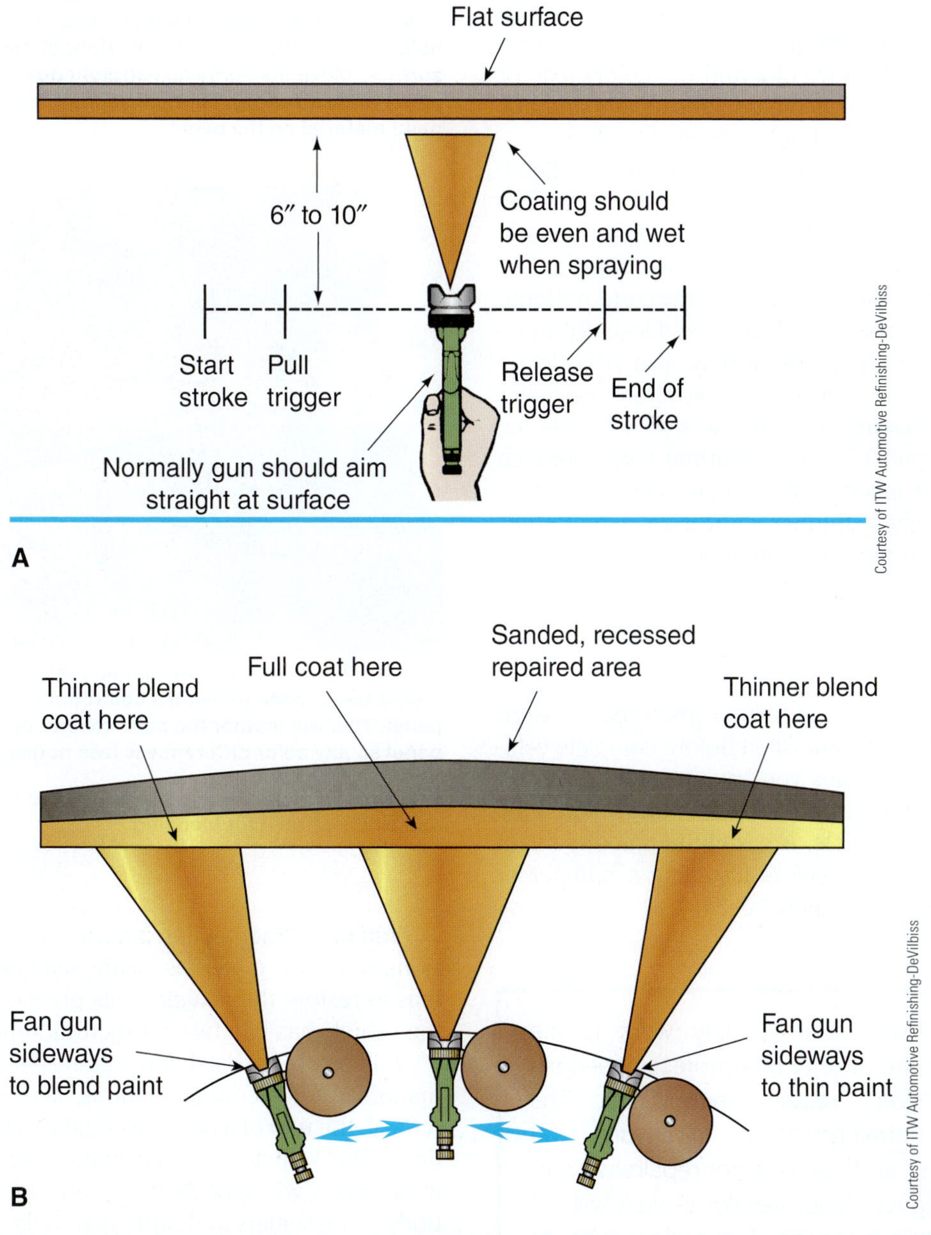

Figure 26-32 Note the different ways to spray coats of material onto a vehicle body. (A) When spraying a whole panel or large area, the paint coat should be the same thickness. Aim the gun straight at the vehicle and keep it the same distance from the body as you move the spray gun sideways. Release the trigger right before the end of each stroke when changing gun directions. (B) Only when blending paint should you swing the spray gun so it is aiming away from the body surface. By fanning the gun sideways, the paint film will be thinner near the edges of the repair area. This is often done when blending colorcoats into original paint during spot repairs.

With most brands of paint products, a special blending solvent should be mixed with the paint when trying to fan and fade a repair area. This type of solvent is commonly used when blending clearcoats so the old and new paints flow out and match each other better.

Blend coats are applied in two or more coats to fade or shade the new paint color into the existing color during spot repairs. The second and third coats are thinner and sprayed over a wider area than the first. This helps make the new paint gradually blend or fade into the original paint color and texture.

Blend coat application involves fanning the gun sideways to deposit less paint around the perimeter of a repair area. This is often done when applying the basecoat of color over the repair area. The blend coat will make the color become progressively thinner away from the repair area so the new color and old colors fade or shade into each other. This makes spot repairs and any color variation less visible. Look at Figure 26–33.

Blending is the key to a successful spot repair. Blending involves tapering the new paint gradually into the old paint. It helps hide slight differences in color or texture, making any difference in the new paint less noticeable.

When spraying a blend coat, the spray pattern should be broadened and the fluid delivery reduced. To minimize overspray, the air pressure may also have to be reduced. Apply the finish in short strokes from the center outward. Again, extend each coat so that it blends out farther than the previous one. Blending coats produce full coverage and result in more paint over the repair area. However, lighter, thinner, more transparent coats are applied around the repair area so the original finish can show through the new colorcoat.

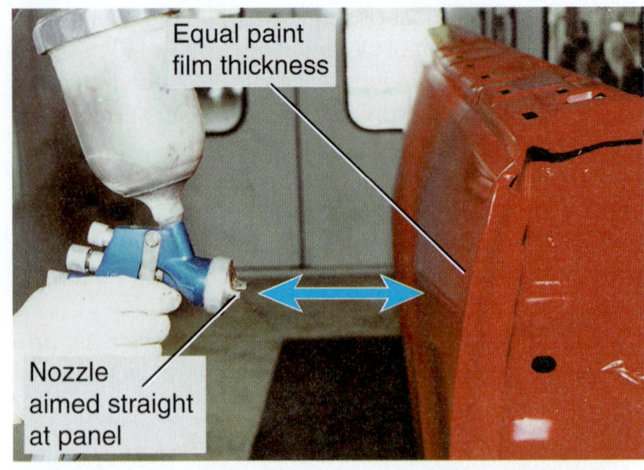

A When spraying the full surface of a relatively flat panel, hold the spray gun level and the right distance from the surface. Move the spray gun straight down the side of the panel while spraying. This will deposit the same amount of paint material on the panel.

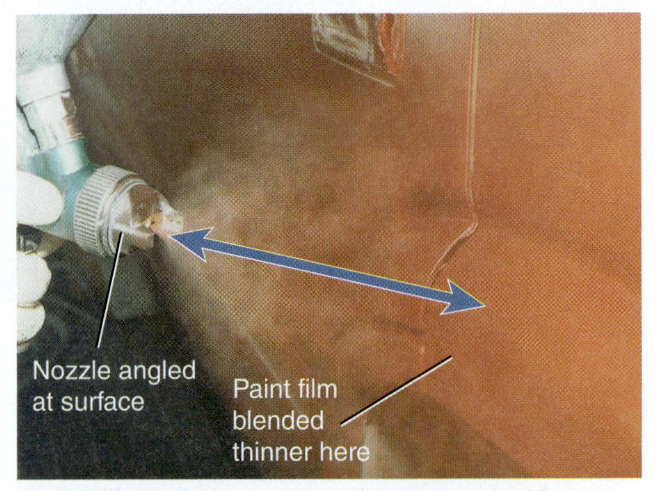

B Here the technician has fanned the spray gun sideways. This is being done to blend a colorcoat into an adjacent panel. This will feather the paint thinner on the adjacent panel so any color difference is less noticeable.

Figure 26–33 **Note when to aim the spray gun straight and when to fan it for blending.**

Underbody Refinishing

When doing paint repairs, hidden or restricted areas on the body must often be refinished before complete vehicle reassembly. Sometimes you have to paint the back sides of panels (fenders, hoods, doors, lids) before they are installed on the vehicle. Areas of the unibody might also have to be painted before installing a bumper or SMC panel, for example, Figure 26–34.

SHOP TALK

Remember! Quality work is the first priority of a professional collision repair technician. The untrained novice or anyone who "cuts corners" and does inferior repairs will suffer in the long run. Unprofessional work will not generate return business. Technicians who do substandard repairs will not make as much money as professional body technicians who take pride in their work.

You or another technician will have to apply all corrosion-protection materials, seam sealers, and paint coatings to restore the vehicle to its preaccident condition if the vehicle has had major structural repairs.

All corrosion-protection coatings must be applied per manufacturer recommendations. If you forget or fail to refinish the rear of a panel, it could rust out in a few years' time. The vehicle's service manual or computer-based information will give the types and locations for underbody seam sealers and other materials.

Applying Band Coats

A **banding coat** is done to deposit enough paint on panel edges, corners, or near the ends of body panels.

A **Underbody surfaces have been painted prior to bumper installation and painting.**

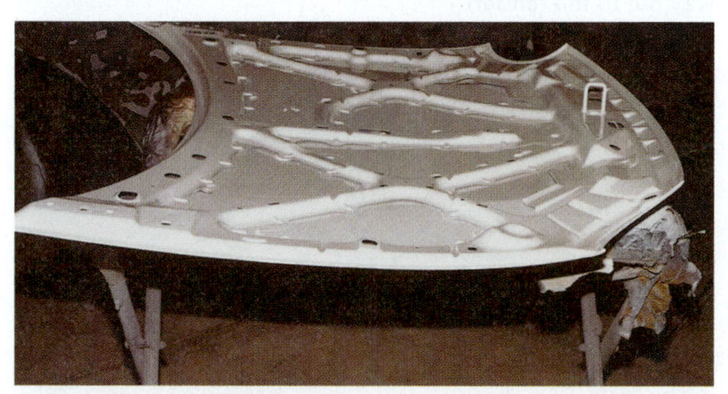

B **The bottom of this hood has been painted first. After curing, the hood will be installed on the vehicle and painted with the fenders and other panels.**

Figure 26–34 **You will have to paint hidden or hard-to-reach surfaces before painting the front of panels.**

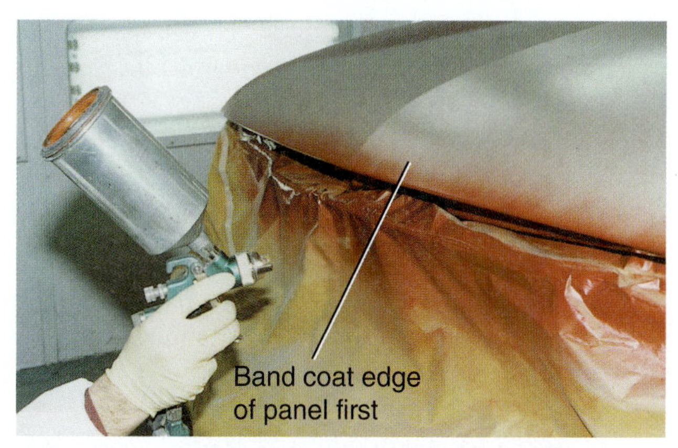

A **Banding first ensures proper coverage along the edge and end of the panel.**

B **After banding, the face of the hood or panel can be sprayed with overlapping passes.**

Figure 26–35 **Banding, or edge coating, is often done before spraying the face or front of the panel. This procedure ensures a problem-free paint application to edges.**

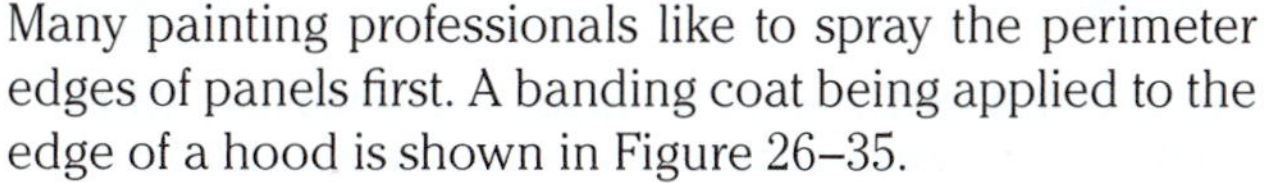

Many painting professionals like to spray the perimeter edges of panels first. A banding coat being applied to the edge of a hood is shown in Figure 26–35.

To apply a band coat, hold the gun an inch or two closer than usual while moving it along the panel edge. You might also want to adjust the spray gun's air control knob for a slightly smaller spray pattern. This helps to concentrate the spray on the panel edge and avoid overspray waste.

To make a narrow banding coat on the edge of a panel, aim the spray gun nozzle directly at the corner or edge. You want the center of the spray pattern to hit the panel edge. Move the spray gun along the edge at normal speed and only apply one coat. Edges of panels cover quickly with a single band coat. See Figure 26–36.

After all of the edges and corners have been sprayed, spray the face or front of the panel. The spray should extend off a panel and onto the masking paper before releasing the gun trigger one detent (notch) (Figure 26–37).

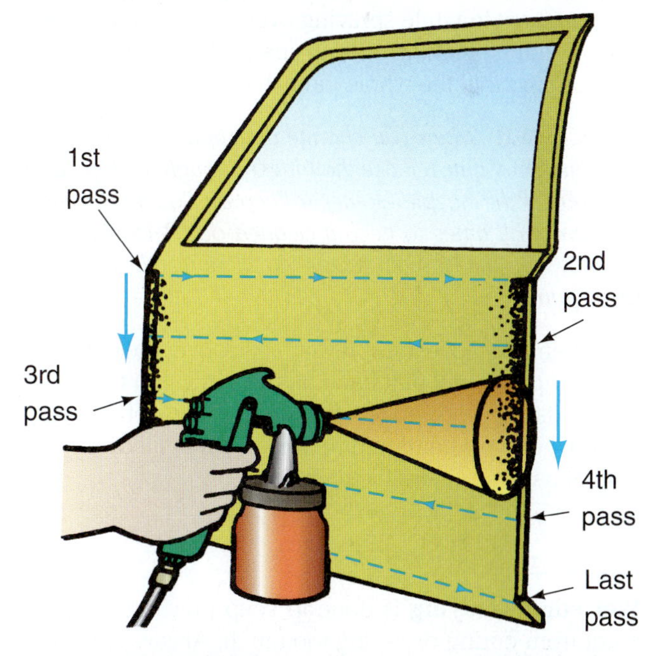

Figure 26–36 **Make your first pass of a banding coat along one side of the panel. Make the second pass along the other side of the panel. Then repeat this process so that panel ends get two coats with a little flash time between each coat. Finally, make overlapping spray passes across the front of the panel.**

Overlap Coats

Overlap coats are done by aiming the spray gun nozzle so that the next pass of paint spray overlaps half of the previous one. Overlapping coats ensures even paint coverage

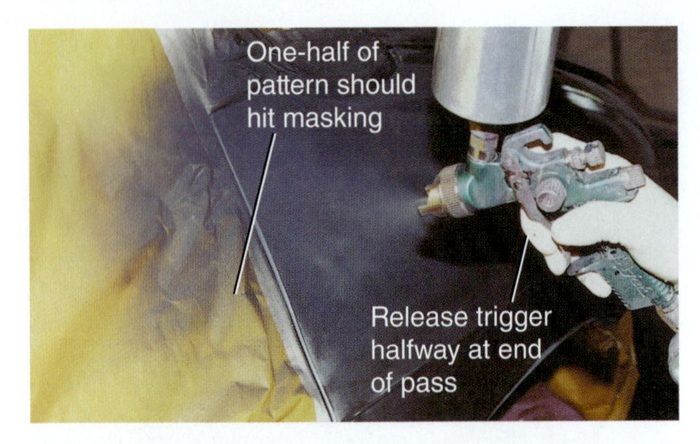

Figure 26–37 **When triggering your spray gun, let up on the trigger one notch, or detent, as soon as the paint spray goes off the edge of the panel. Some of the spray pattern should hit the masking paper. After changing directions and just before moving back over the panel again, trigger the gun fully to again start the paint spray onto the vehicle.**

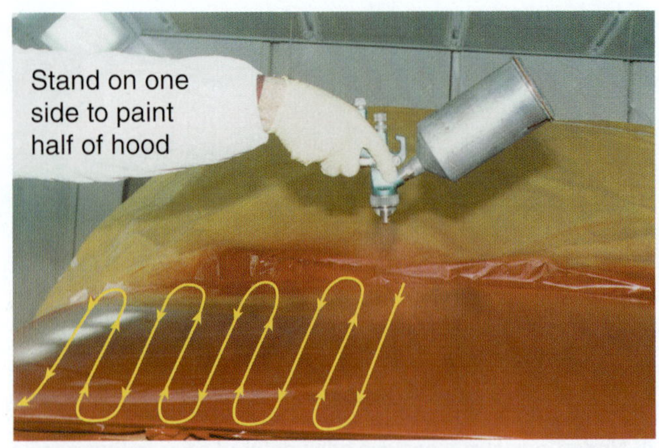

A Stand to one side of the vehicle to apply colorcoat. Start in the middle and use overlapping coats while working your way out to the fender.

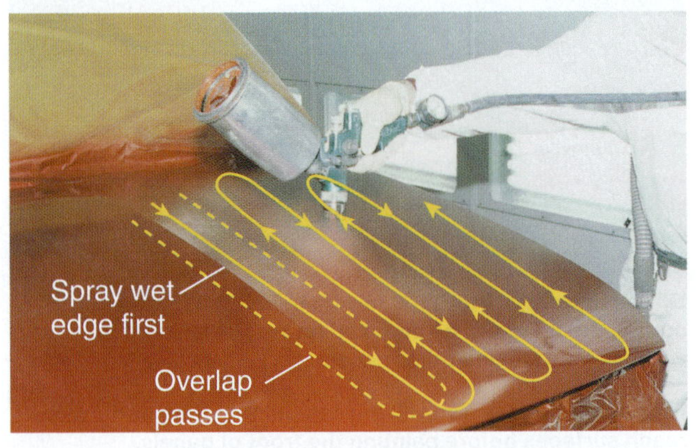

B Go to the other side of the hood and spray the wet edge first. Again, use overlapping coats, working your way outward.

Figure 26–38 **Always spray your wet edge first. When painting large areas, such as a hood or roof panel, keep spraying the wet edge so it does not dry or cure too much before being finished.**

and equal paint film thickness. Overlap coats should be sprayed methodically over the top of each other, using an even, back-and-forth motion.

On a relatively flat surface, hold the gun the same distance from the panel and keep the spray nozzle aiming straight at the surface. Normally, you should start at the top of a panel, a door for example, and move the spray gun along the top while spraying. On the first pass, half of the paint mist should hit any masking material and the rest should coat the top of the panel.

On the next pass, move the gun down about half the width of the first coat. Move the spray gun straight across the panel again while spraying over half of the previous coat. Continue making overlap coats all the way down to the bottom, until the whole panel has been coated.

REMEMBER *When you change directions for each pass with the spray gun, release the trigger enough to stop paint flow out of the nozzle. However, keep air flowing through the gun at all times to keep it cleaned out. Releasing only halfway on the trigger keeps air flowing through the spray gun.*

When you move the spray gun back over the panel for the next pass, fully pull back on the spray gun trigger so paint hits the very edge of the panel. By not spraying paint when you change directions, you avoid a double coat of paint and possible paint runs.

Wet Edge Spraying

Wet edge spraying is done to keep previous coats of paint from curing or drying too much. Always spray the wet edge first when painting large panels (hood, lid, roof) from two sides of the vehicle. This is illustrated in Figure 26–38.

For example, when painting a hood, spray the front edge first, especially if the hood curves down at the front. Then stand on one side and lean out over the fender while spraying the top of the hood. Spray next to the fender first. Work your way to the middle of this large panel surface while leaning out carefully. Make sure you keep the gun aiming straight into the panel surface. Make back-and-forth passes, working your way inward to the middle of the hood.

Walk to the other side of the vehicle and spray the wet edge in the center of the hood first. This will allow the two coats to melt into each other right away. By again starting in the center of the hood, your first coat will hit and overlap the wet edge of paint already sprayed. This will prevent the paint already on the panel from drying too much before getting fully painted.

If you painted somewhere else on the vehicle and forgot to go right to the wet edge, this paint can cure too much. When sprayed later, the wet edge will be too tacky

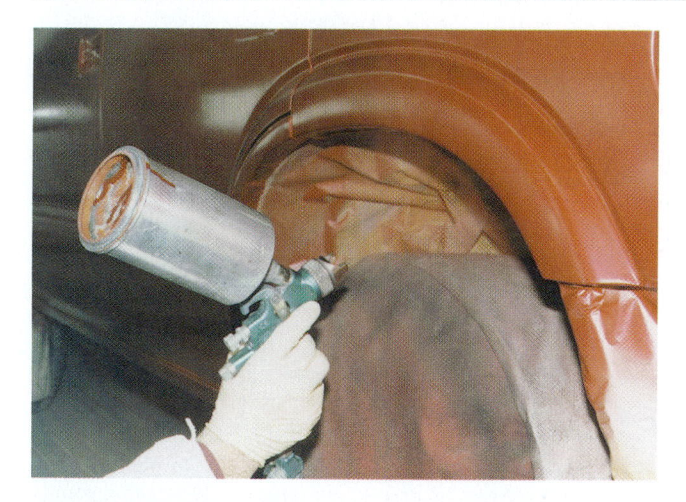

Figure 26–39 To spray a round surface, such as this fender well opening, you must rotate the spray gun around to keep the nozzle aiming straight at the surface being painted.

and might not flow out like the rest of the paint. Orange peel or dry spray can ruin the appearance of the finish in the center of the hood or other panel being painted.

Spraying Curved Surfaces

When spraying curved or contoured surfaces, remember to move your spray gun so it follows the shape of the panel. This is one of the few times that you should use wrist action when handling your spray gun.

For example, when spraying the wheel openings on fenders and quarter panels, aim your gun nozzle directly at the flange. Then swing your gun around inside of the wheel well so that an equal coat of material deposits on the round flange. Use wrist action to pivot the gun around the center of the opening to keep the nozzle the same distance from the wheel well flange while spraying (Figure 26–39).

26.10 METHODS OF REFINISHING

How much of the vehicle should be repainted is determined by several factors:

- Location of repair areas
- Amount and number of panels repaired or replaced
- Extent of paint film deterioration on older vehicles
- Whether the vehicle has been previously repainted
- The type of paint used
- How well the colorcoat matches the existing paint color
- Funds available to pay for repairs

In turn, these factors determine how much of each panel should be repainted. There are three general types of refinishing repairs:

1. **Spot repairs**—only some of a panel is colorcoated
2. **Panel repairs**—whole panel is colorcoated
3. *Overall refinish*—entire vehicle is repainted

26.11 BASECOAT/CLEARCOAT REPAIRS

Basecoat/clearcoat topcoats are now the most common type of finish used in today's body shops. Because almost all late model vehicles now have colorcoat/clearcoat finishes, it is very important to become familiar with them.

When estimating a basecoat/clearcoat repair, carefully examine the finish on the area adjacent to the damage. If it is chalked, dulled, or otherwise impaired, matching the old finish might prove impossible.

Try compounding a small area of the old paint to see whether you can bring out the color and shine. If not, such jobs should be performed as overall repaints of the whole vehicle. This approach will eliminate many problems in repairing colorcoat/clearcoat finishes that are severely weathered.

Basecoat/Clearcoat Spot Repairs

A typical basecoat/clearcoat spot repair involves spraying the colorcoat only over the repair area and not the whole panel. The basecoat color is only sprayed where needed over the primer and sealer. Basecoat or colorcoat does not have to be sprayed over existing paint as long as it is in good condition.

The colorcoat must be blended out around the edges of the repair area to help match the existing color finish. The whole panel is then clearcoated. This is the most common type of spot repair in today's body shops.

Spot repairs with a colorcoat/clearcoat paint system generally involves the following steps:

1. Minor panel repair using body filler.
2. Proper sanding and scuffing of all surfaces to be painted.
3. Application of a primecoat system (primers and sealer) over repair area (Figure 26–40).
4. Spraying adhesion promoter over OEM paint, but not over sealer.
5. Spraying two to three coats of color over repair areas, blending the color into the surrounding original finish.
6. Spraying two to three coats of clear urethane over the whole panel.

Even though the whole panel must be sprayed with clear, this is still considered a spot repair. Spot repairs are recommended where a complete panel repair is either uneconomical or impractical. Spot repairs are often used when only a small area of a panel is damaged. With colorcoat/clearcoat repairs, they allow as much of the original color as possible to remain visible after the repair. Spot repairs are also used when there is a break line (molding or sharp body contour) that hides any paint difference (Figure 26–41).

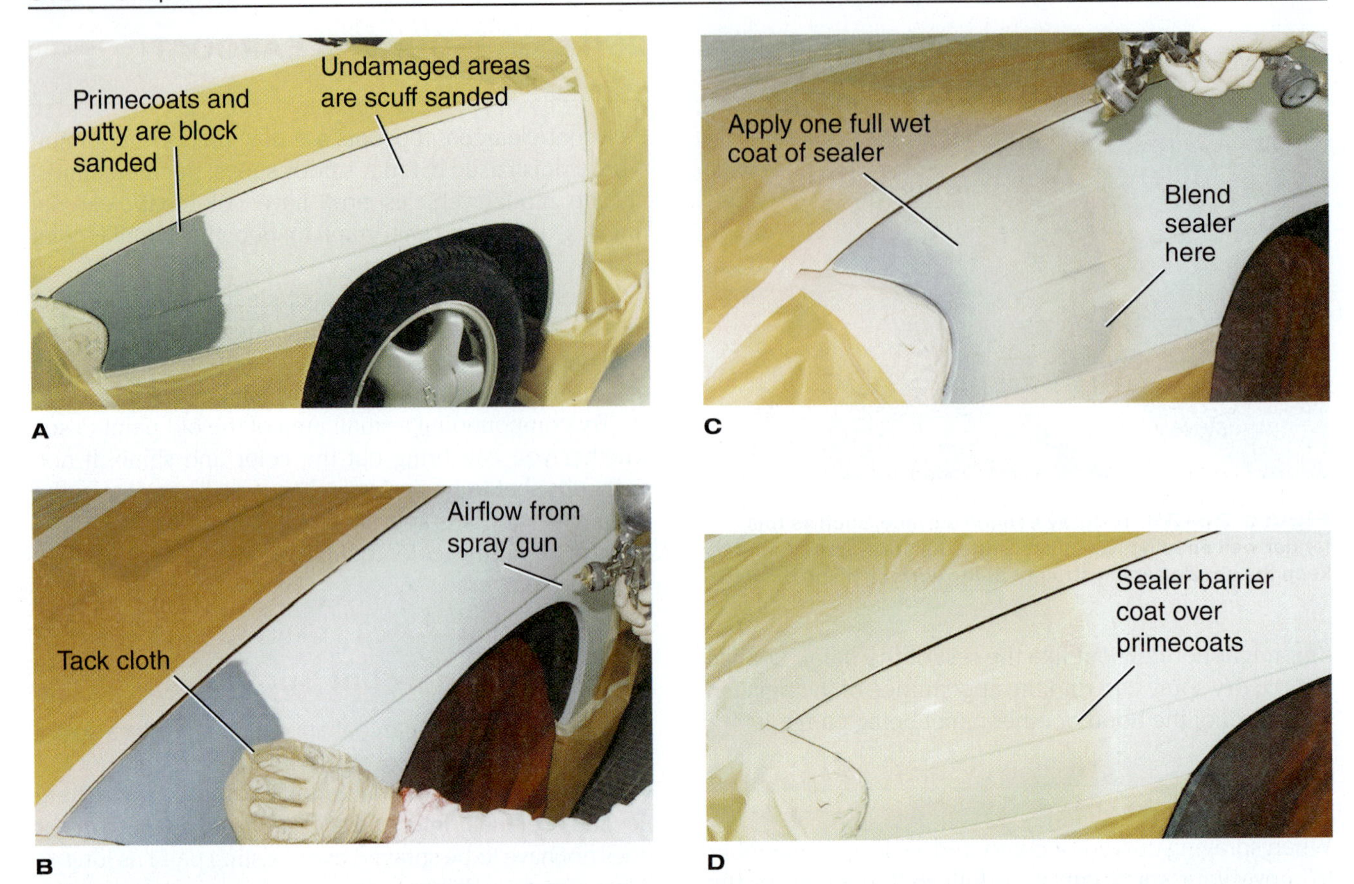

Figure 26-40 Note the steps for applying sealer during the spot repair of this fender. (A) The area around the fender has been masked off. (B) The fender is wiped down with a tack cloth to remove dust and dirt. (C) Sealer is applied only over the primer area to ready it for topcoats of paint. (D) Sealer has been blended slightly so that its thickness is thinner around the edge of the repair area.

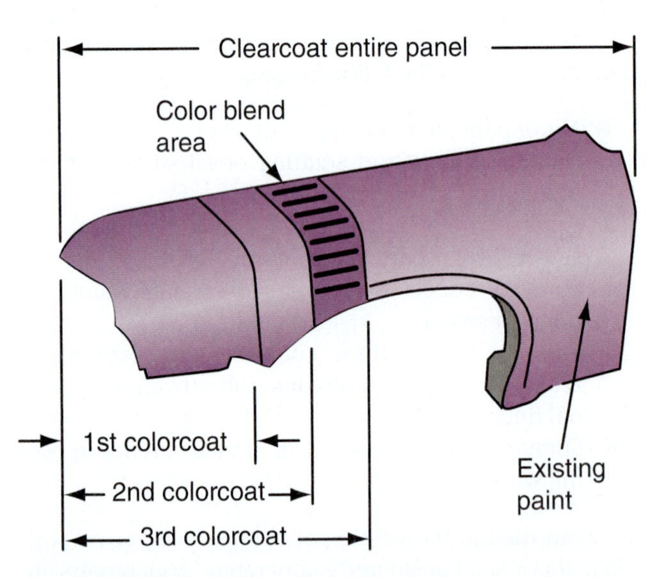

Figure 26-41 When doing a spot repair, each coat of paint material should be blended out over a larger area. The last coat will be the thinnest so that the new paint and old paint blend together smoothly.

Spraying Colorcoats

When spraying basecoats or colorcoats, two to three coats of color should be applied for good coverage of primecoats. Generally, colorcoats should be sprayed more lightly than clearcoats, keeping the following points in mind:

- The first coat of colorcoat should be a medium wet coat to provide good coverage over the damaged area.
- The second coat of colorcoat should be a little lighter coat or a medium coat that blends a few inches beyond the first full colorcoat.
- The third coat of colorcoat should be an even thinner light coat that blends a few more inches beyond the second coat of color. This is shown in Figure 26–42.

A line drawing illustrating how to apply basecoat/clearcoat paint to a front fender is shown in Figure 26–43.

The colorcoat does not need to be glossy, and only enough material should be used as needed to achieve hiding and full color coverage. When spraying metallic colorcoats, a very light final mist coat is often used. This helps to disperse the reflective flakes in the paint better and match the existing finish (Figure 26–44).

Blending Colorcoats

To match the existing color on the vehicle, you must exercise care to blend the color out more thinly near the

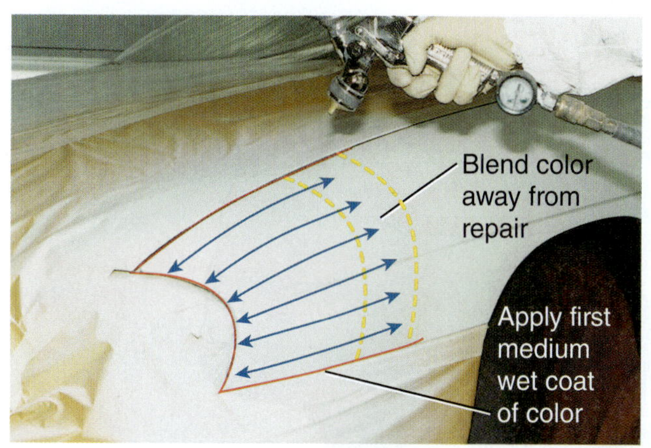

A The first coat of color should only be applied over sealer. Spray a medium wet coat for full coverage of the sealer. Blend or shade the colorcoat by wrist fanning the spray gun sideways and away from the repair area over the sealer edge.

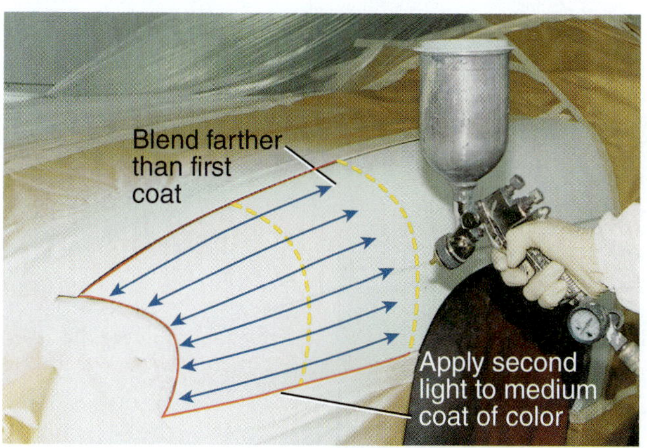

B A second coat of color is applied over a larger area than the first. Wrist fan the gun to blend the outer edge of the colorcoat more thinly at the perimeter of the repair. Use a medium coat blended toward the middle of the fender panel.

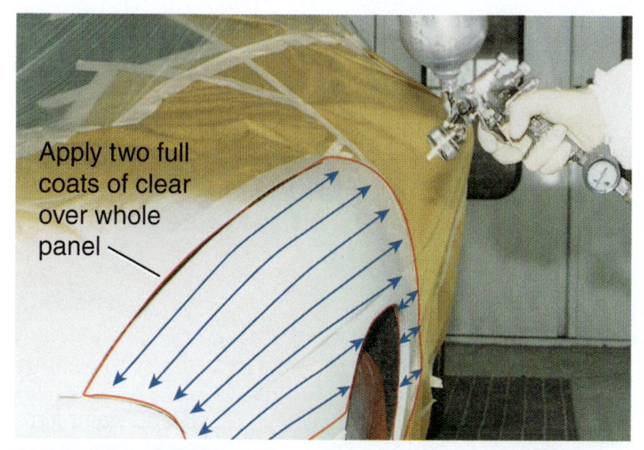

C Apply two full wet coats of clear over the whole panel. Spot repair will leave most of the original color showing on the panel so the fender color better matches other panels (hood and door).

Figure 26–42 Study the major steps for making a spot repair in a solid, nonmetallic fender panel. Only the damaged area is sprayed with colorcoats or basecoats, but the whole panel is clearcoated.

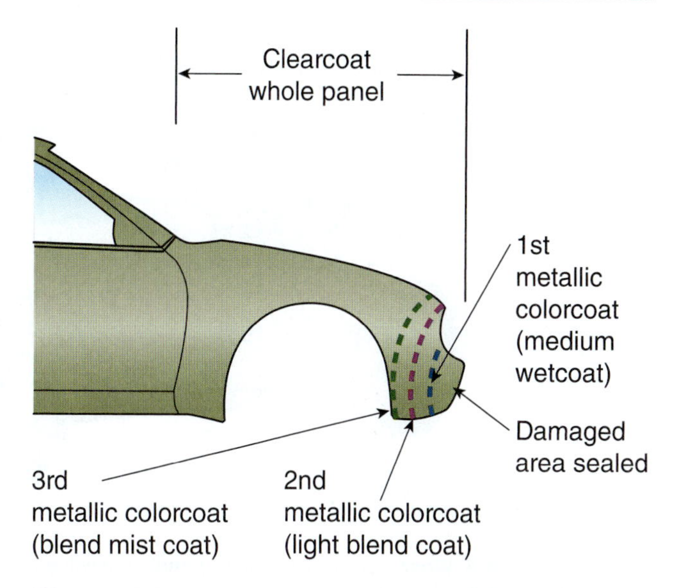

Figure 26–43 Repainting with basecoat/clearcoat paint is best done by colorcoating the smallest area possible. Then, clearcoat the whole panel.

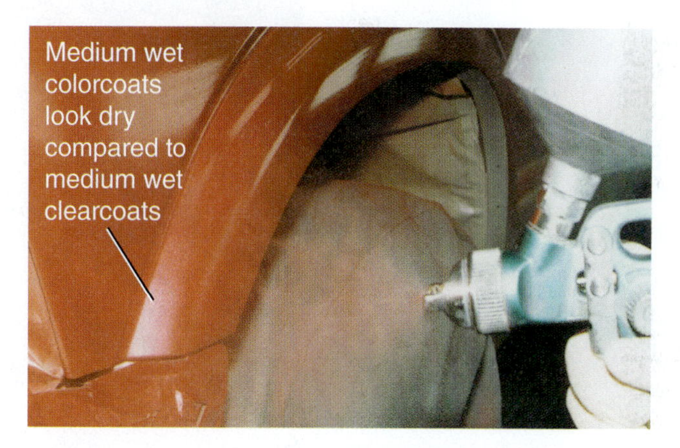

Figure 26–44 Colorcoats or basecoats do not have to be sprayed with a high gloss. In fact, with metallic and pearl paints it is better to mist the color onto the panel. This will distribute the metal or mica flakes to more closely match the existing paint.

perimeter of the repair area. Again, keep as much of the original color showing as possible. Only apply colorcoats where needed. Paint blending on the surface of a car's quarter panel is illustrated in Figure 26–45.

To blend colorcoats, fan the gun sideways near the perimeter of the repair area. Intentionally make the paint film thinner so it fades into the original finish. To do this, each colorcoat should be sprayed a little more lightly and farther out than the previous coat.

Each coat of color is blended more thinly and farther away from the repair area. Because the rear corner of the quarter panel was the only area damaged in this example, colorcoat is applied only over this area.

Figure 26–46 summarizes how to apply colorcoats when repairing a quarter panel.

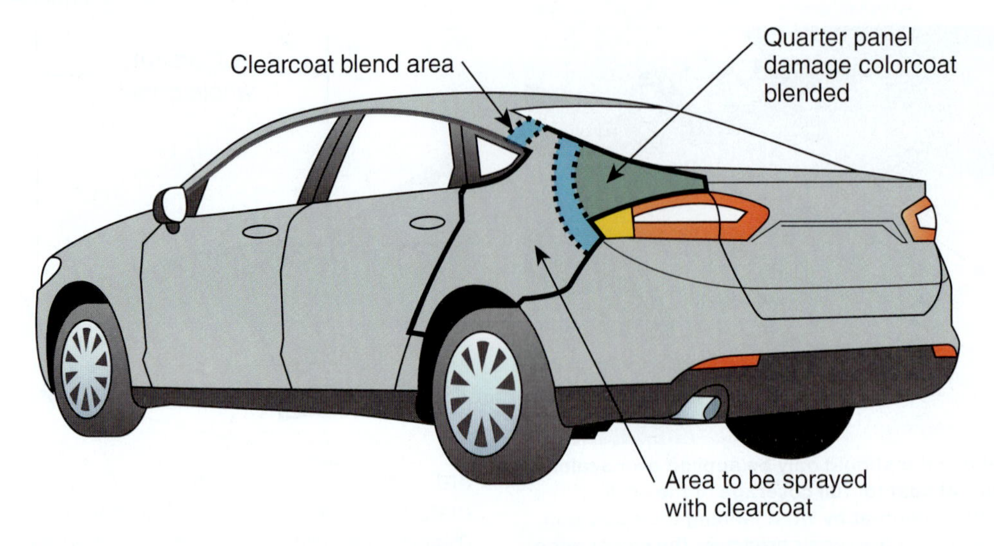

Figure 26-45 When painting a quarter panel, you will normally need to blend the paint at the sail panel between the roof and quarter panel.

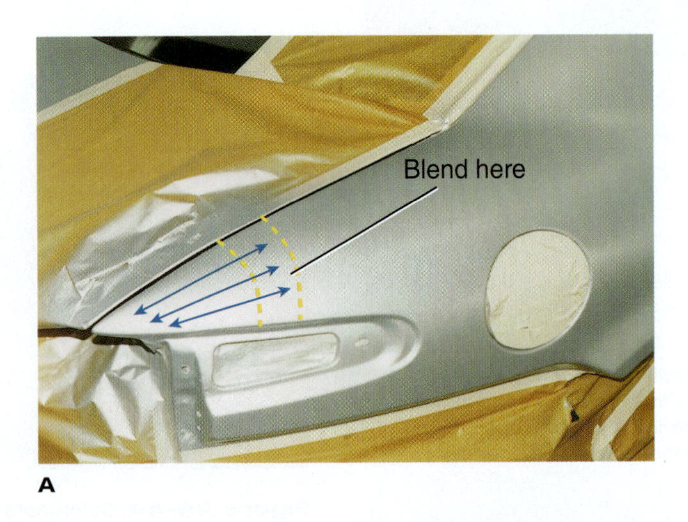

A

B

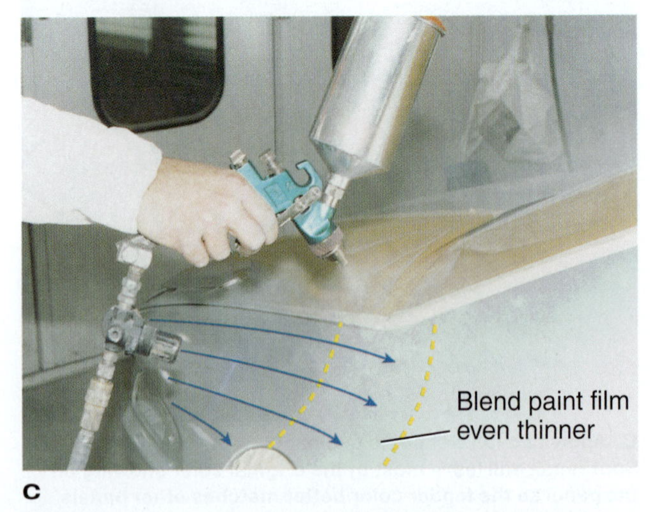

C

Figure 26-46 Note the major steps for colorcoating a quarter panel with minor damage. (A) First spray a medium wet coat only over the repair area for full coverage. (B) After the proper flash time, apply a light second coat of color a little farther out than the first coat. (C) The third coat of color should blend out several inches beyond the second coat. Use a very light mist coat of color so the old and new finishes blend into each other. Note how this colorcoat is dull and lacks shine. This is fine because clearcoat will bring out the color and gloss.

Clearcoats are then applied to the whole quarter panel. The clear must be blended into the existing finish near the roof panel. This sail panel area between the quarter panel and roof panel is smaller and will not show paint differences as much. See Figure 26–47.

Proper blending is critical if you do not want the new paint to show in the existing paint. When blending a colorcoat, you should blend the color down and away from the repair area instead of straight across. This will make any difference in color more difficult to notice because the color variance is not straight up or down or across the panel (Figure 26–48).

Figure 26–49 shows how a small spot repair may require that you blend the colorcoats into an adjacent panel. This is commonly required with metallic and pearl paints. By blending the color in a larger area, there will be less noticeable differences in color.

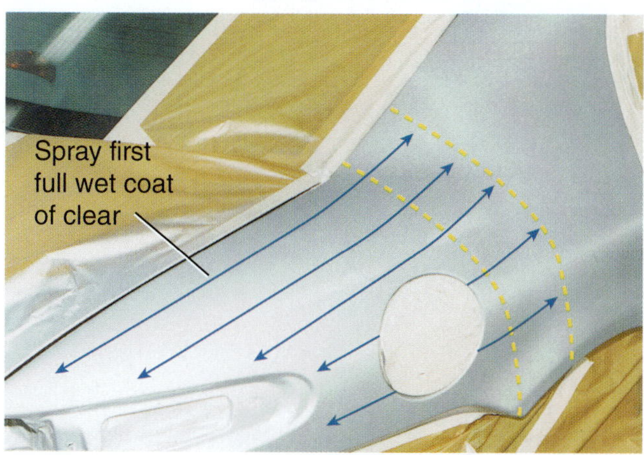

A First, spray a full wet coat of clear over new basecoat only.

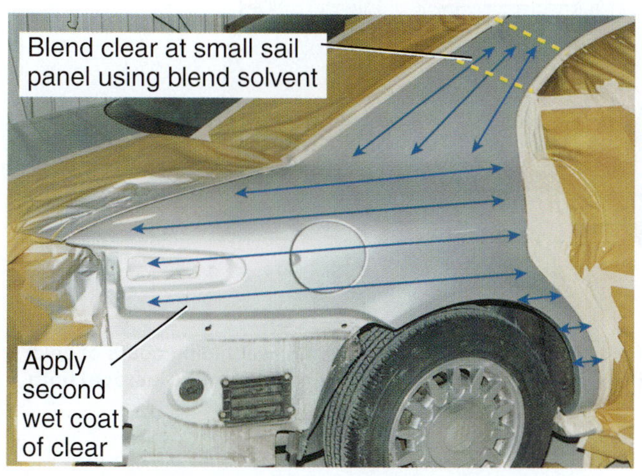

B Clear the remainder of the quarter panel. Blend the clear at the narrow area of the sail panel. Use a blend solvent where the old and new finishes meet. A blend solvent will help dissolve the original finish into the new finish for a better match.

Figure 26–47 After colorcoats have flashed, clearcoat the quarter panel.

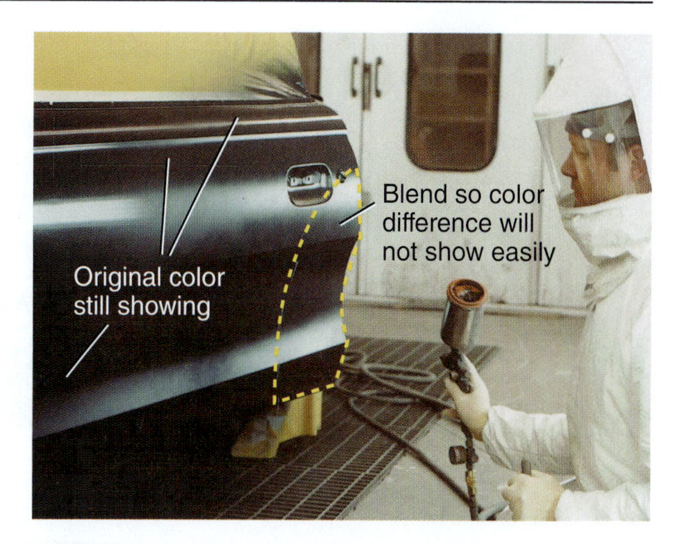

Figure 26–48 When blending a colorcoat, avoid blending straight across a panel. Here the technician is blending the color on this door down and away from the repair area. This will trick the eye, making any color difference less noticeable.

Spraying Metallic Colors

Matching metallic colors is complicated because you have to match both the color and the density of the metal flakes in the paint. Spot repairs with metallic color require skill to match the color and achieve proper distribution of the metallic flakes. The blend area will be less noticeable if it is angled away from any body line.

When spraying metallic colorcoats, spray them on with light coats to keep the flakes suspended near the surface of the paint film. Avoid spraying full wet coats of metallic paint because the metal flakes can settle or mottle (float together) in the paint film. Between coats, pour the paint out of the gun cup and into a clean container. This will prevent the flakes from settling in the gun and affecting color. You can then remix the paint before the next coat.

Spraying Clearcoats

Clearcoats are normally sprayed on in medium to full wet coats. Two or three coats of clear are normally applied over a whole panel that is being repaired. It is better to mask and clear an entire panel when possible.

Always remember to tack cloth the body surfaces after colorcoating and before clearcoating. After the previous colorcoat has flashed, use your spray gun to blow off the body as you wipe it gently with your tack rag (Figure 26–50).

Do not rub too hard with the tack cloth on colorcoats, because you could mar the fresh paint. Concentrate your tack cloth on horizontal surfaces (roof, hood, deck lid, and the tops of fenders, bumpers, and quarter panels). More dust falls on horizontal body panel surfaces than vertical panels.

Vertical surfaces (sides of vertical panels) do not collect dust as much. Airborne dust and lint, unless blown sideways, will settle straight down onto the top of wet paint.

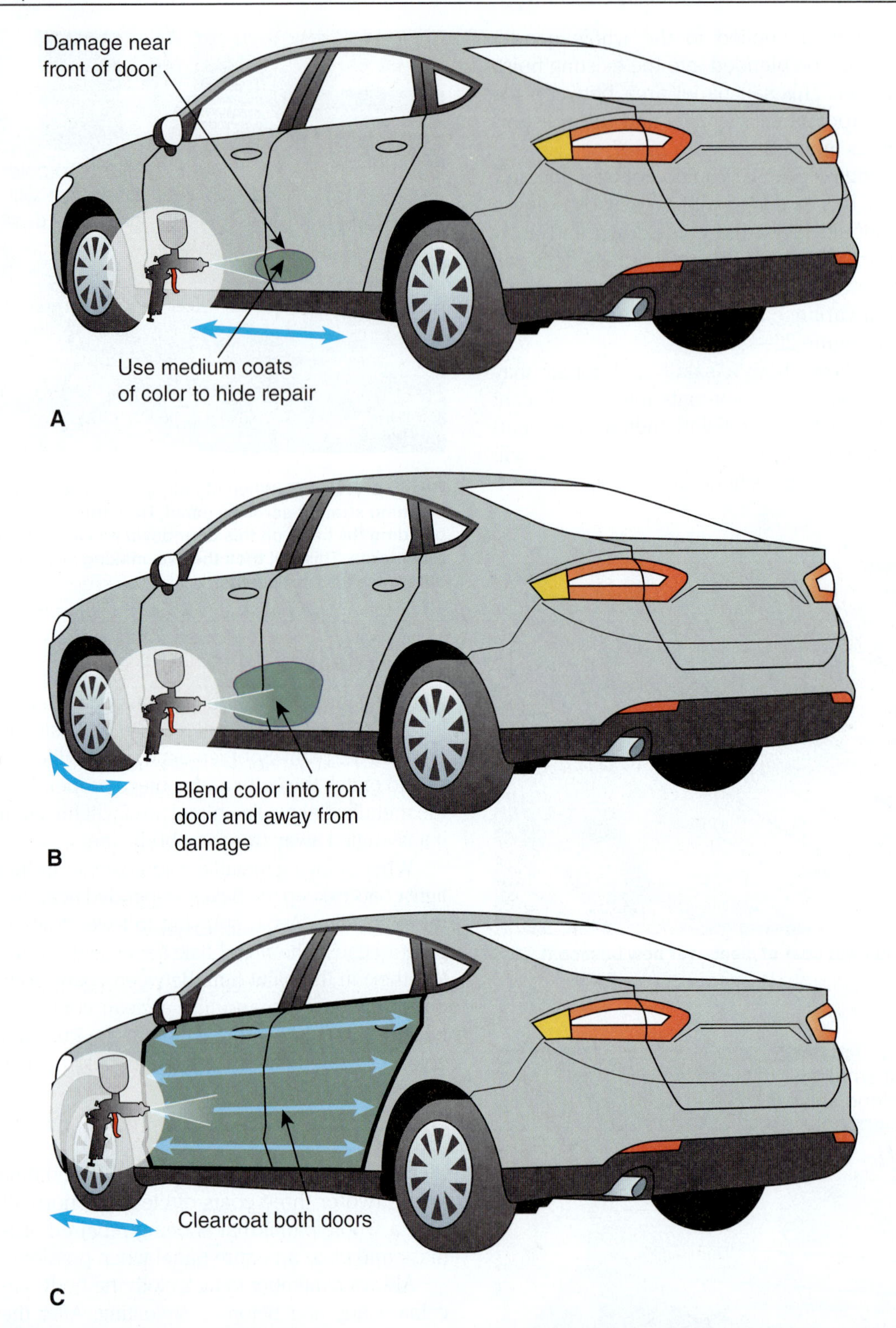

Damage near front of door
Use medium coats of color to hide repair
A
Blend color into front door and away from damage
B
Clearcoat both doors
C

Figure 26-49 Note the steps for blending color on the side of this vehicle. (A) Apply full, medium wet hiding coats of color over the repair area. (B) Blend color out into the existing paint so it fades at an angle into the existing finish. (C) Clearcoat both doors as in a full panel repair.

NOTE *Chapter 28 explains paint problems, such as contaminants, and how to prevent them.*

Two or three medium wet coats of clear should be applied, with at least 10–15 minutes flash time between coats. Do not let a colorcoat dry completely before clearcoating. Follow the manufacturer's recommended flash time. Many companies suggest 15–30 minutes of flash time before clearcoating.

Several things can be done to help you properly apply clearcoats:

- Do not load clearcoats on heavily. Because these finishes are clear, refinishers have a tendency to use too

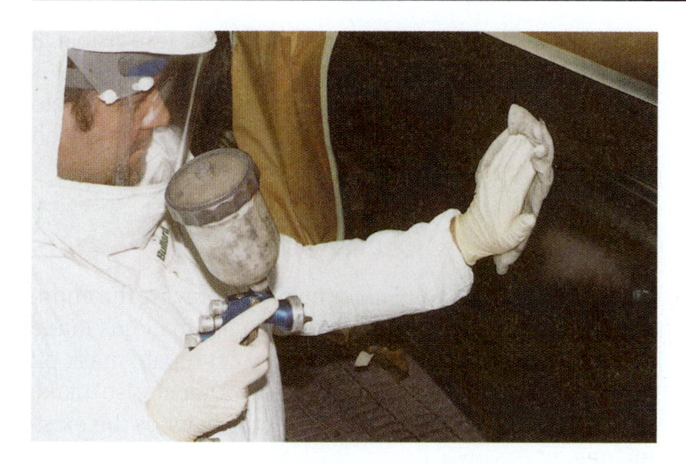

Figure 26-50 **Always remember to lightly tack cloth panels before applying colorcoats and clearcoats. Rub lightly because paint is only flashed and is not fully dry.**

much in an attempt to increase the desired glamour effect. As a result, they "bury" the colorcoat. Remember, clearcoats are not perfectly clear and they tend to alter the color.

- Do not use overreduced clearcoats. Contrary to some opinions, clearcoats do not shine more when they are underreduced. Reduce the clearcoat with the recommended amount of solvent according to the label instructions.
- Do not use cheap generic reducers when spraying clearcoats. Use a quality thinner/reducer recommended for the shop temperature conditions. Also, thinner or reducer that is too fast weakens the performance of clearcoats by trapping solvents and hurting the flow and leveling characteristics. Use the correct speed solvent for shop conditions. Let each coat flash thoroughly before applying the next one.
- When blending clear over existing paint that is not to be clearcoated, use a special blending solvent to reduce the clearcoat. The blending solvent will help dissolve the new clear into the original finish. This strong solvent will help you blend spray the clear so that there is no noticeable difference in the new and old finishes.

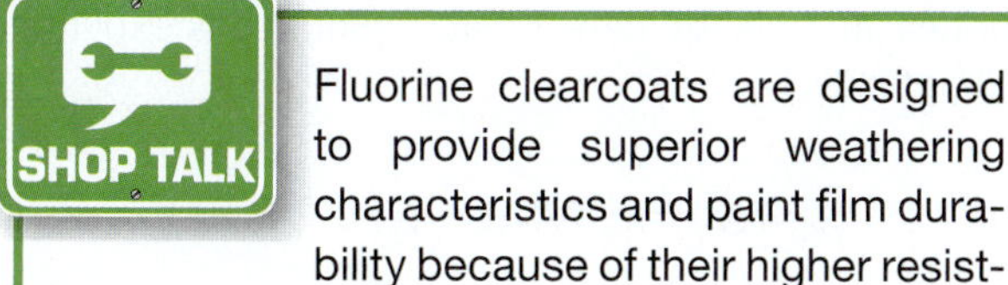

Fluorine clearcoats are designed to provide superior weathering characteristics and paint film durability because of their higher resistance to ultraviolet rays. If the vehicle has an OEM fluorine clear topcoat, the refinish system must also use a fluorine clearcoat. Read the label directions. Some might require slightly different application and blending procedures.

Spraying Tri-Coat Finishes

Tri means "three." In auto repair terminology, it generally refers to a three-stage finish. Repainting a tri-coat finish requires multiple coats of three different paint materials to form the glamour topcoat finish. First, you must apply a basecoat color without mica flakes. Second, semitransparent mica, or pearl, mid-coats are sprayed over the basecolor. Finally, coats of clear urethane enamel are sprayed over the mica coats.

Three-stage paints produce a three-dimensional look in the finish. Instead of being partially submerged and buried in the paint color, the shiny mica flakes float above the layer of color. When you look at pearl paint, the color and mica flakes seem to glisten and change colors as you walk by.

Although tri-coats require somewhat different refinish procedures and techniques, they are essentially the same as the repairs done on colorcoat/clearcoat finishes. Newly applied mica colors are difficult to match with the OEM finish, however. Matching three-stage and other hard-to-color paints is explained in Chapter 27.

Following are some key points to keep in mind when performing a tri-coat refinish:

- Follow the recommendations furnished by the paint manufacturer for this type of repair.
- Pay close attention when the instructions call for the use of adhesion promoters, antistatic materials, and so on.
- Make a test or let-down panel, as described in Chapter 27. A mismatch in the colorcoat, mica coats, or clearcoat can affect the overall finish match.
- Keep the repair area as small as possible.
- Avoid a halo effect by applying the first coat of mica to the colorcoat only.
- The more intermediate coats that are applied, the darker the finish will appear.
- Allow a larger area in which to blend the mica intermediate coats. They require more room to blend than a standard colorcoat.
- Do not try to substitute another type of paint for the recommended colorcoat.
- Always check the basecoat color against the OEM colorcoat. To do this, find an uncleared mica-free area of the vehicle. Some car companies leave an exposed portion of basecoat beneath the right and left sill plates that is perfect for this check.

For more information on applying tri-coat paints, refer to Chapter 27.

26.12 APPLYING SINGLE-STAGE PAINTS

A single-stage paint has a gloss or shine without a clearcoat. Single-stage paints go on shiny, unlike colorcoats, which go on dull. Single-stage paints are sometimes used to speed repairs and to keep costs down. You do not have to spray clearcoat over a single-stage color.

Single-stage urethanes are now being used in some shops to repaint solid colors. Single-stage paints often can be mixed to match basecoat/clearcoat solid colors, but not metallics. Generally, they are used with solid colors that are easy to match or on parts that are not easily seen.

Single-stage urethanes provide good durability and eliminate white scratching of clearcoats. Single-stage paints save time compared to colorcoat/clearcoat finishes, and still have low VOCs. There is also less waste of unused material, and less cleanup time is required.

Single-stage color repairs are not as common as colorcoat/clearcoat repairs. However, some shops still use solid color spot repairs. Blending with a single-stage paint is sometimes used for repairing light damage or when the estimate will not allow the cost of painting the whole panel.

When blending a solid color at a body line, you do not have to paint the upper portion of the fender where paint banding shows more. This can help avoid problems with paint color and texture differences.

When spot repairs are made with a single-stage paint, the blend area should be scuffed with rubbing compound or sanded with #1000 grit sandpaper before refinishing. Apply full wet coats of single-stage paint over the sealer, and fan the gun or blend the paint out over undamaged areas.

You should only blend single-stage paint on small areas or areas that cannot be easily seen (along the bottom of the body or next to body contours). A paint band where the new blended paint and old paints meet may be slightly visible, even after buffing.

26.13 PANEL REPAIRS

Basecoat/clearcoat panel repair involves painting the whole body part with color and clear. The whole panel is refinished without blending the colorcoats. For example, if a new fender has been installed, the whole fender would have to be sprayed with primer primecoats and sealer, then color basecoat, and finally clearcoat.

Most solid colors can be matched with panel repairs. With hard-to-match colors (pearls and metallics, especially golds and silvers), sometimes adjacent panels must have the colorcoat blended where the repaired and unrepaired panels meet.

Metallic Color Panel Repairs

New and old paint differences tend to be very noticeable with bright metallic colors. It can be difficult to match the new metallic finish exactly with the previous one when repainting the whole panel. When unmasked, the new metallic paint may be a slightly different color than the original finish.

You must extend the blending over a wide area, possibly into undamaged panels, to help hide the paint differences. This will make the panel repair less visible. If a panel has damage at both ends or if the whole panel is to be refinished, the blending might have to extend onto both adjacent panels even though they are undamaged and do not require repainting.

Solid Color Panel Repairs

Panel repair with a solid color involves spraying the entire panel (door, hood, and so on). The paint match is made at the panel joints. For a complete panel repair, mask off the area that will not be painted. You must also make sure that the new color almost exactly matches the existing one.

Panel repainting is done to repair complete panels separated by a definite boundary, such as a door or a fender gap. The whole panel is given normal medium wet coats of paint. However, you may have to blend or shade the paint film gradually more thinly over areas such as between the quarter panel and roof panel where the whole area cannot be painted.

26.14 OVERALL REFINISHING

Overall refinishing is just what its name implies: The whole vehicle is sprayed with a fresh coat of paint. Overall refinishing is a costly, major repair that takes considerable time and skill. Some reasons for refinishing the entire vehicle include:

- If over half of all body surfaces require repainting, the estimate and your work order may stipulate overall refinishing.
- If the vehicle has a deteriorated finish (dull, cracked, peeling, flaking, or worn paint), overall refinishing may be stipulated.
- If a car owner wishes to change the vehicle's color, he or she will request a new paint job.
- When several new panels have been welded in place after a major collision, the insurance company or owner may okay an overall paint job so all panels match perfectly.

All body shops do overall refinishing. However, there are also specialty paint shops that do nothing but overall painting at reduced prices. Custom auto body shops specialize in glamour finishes for custom vehicles, antiques, and classic cars. Custom painting is explained in Chapter 27.

Overall Spraying Methods

Generally, most technicians start a complete paint job at the roof. Spray at the highest panels on the vehicle and work your way down, always going to the wet edge. Because overspray and dry spray are pulled down by gravity, this technique prevents overspray dust from settling onto areas that have already started to dry.

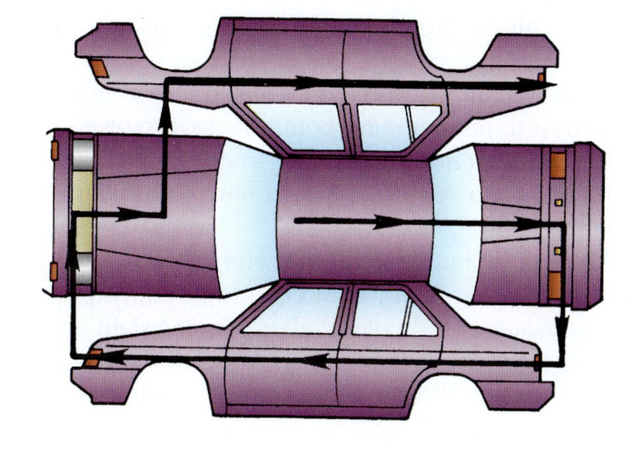

Painting order for one person

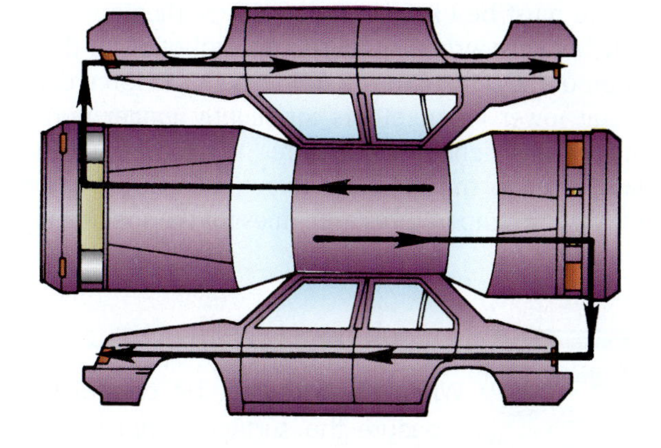

Painting order for two people

Figure 26–51 **Study the sequence for overall painting procedures for one and two people by following the arrows from panel to panel.**

There is no single, most perfect procedure for overall repainting of a vehicle. However, most refinishers will agree that the diagrams in Figure 26–51 illustrate the best panel sequence.

For overall spraying, always maintain a wet edge. Do not paint half of a large panel (roof, hood, trunk lid) and then forget to paint the other half right away. If the paint edge dries, a rough textured surface will be formed where the fresh paint covers the partially cured paint. The partially dry paint will not melt and flow together with the fresh paint. A dull line in the paint finish will form down the middle of the panel.

Avoid sags in the overlap line by changing the point of overlapping. Methodically overlap each pass of the spray gun. Trigger the gun to stop the paint flow as you change directions while spraying.

Some painters like to *walk spray* the sides of a vehicle. While holding the spray gun steady and the right distance from the body, they walk down the side of the vehicle while spraying. They paint the fenders, doors, and quarter panels all at once, in long passes. This helps avoid uneven paint film thickness, because the whole side of the vehicle is painted as if it were one large panel. You do not have to trigger the gun on and off as you change directions over each panel. This is a fast, efficient way of painting the whole side of a vehicle.

To quickly spray while walking, the spray gun must be adjusted for almost maximum fluid flow. Walk and move the gun at normal speed to apply a medium wet coat, unless otherwise specified by product directions. As you walk back and forth while spraying, overlap each coat about halfway as you would with spot or panel repairs.

Sometimes with an overall paint job, panel gaps are back masked and underbody surfaces are not painted. However, when restoring a classic vehicle or when changing to a different color, all trim, glass, and many fastened parts must be removed. Part removal is needed so that all surfaces (both the front and rear of panels and panel flanges) can be repainted. This is an expensive repair, usually costing the customer several thousand dollars.

Force Drying Enamel Topcoats

Force drying fresh paint by means of spray booth heat convection ovens or infrared lights will greatly reduce waiting time between coats. To help force-flash paints between coats, warm the spray booth only to about 100°F or to paint manufacturer-specified temperatures. This is enough heat to speed paint flashing between coats, but it will not affect the paint. Warming the booth between coats can greatly reduce the wait time between coats and speed shop production.

After painting, you must allow the paint to flash or partially cure before baking. Baking is done at higher booth temperatures. If you accidentally bake fresh paint too quickly or before enough solvent has evaporated, paint problems can develop.

Typically, **paint baking** is done for 15–20 minutes at 160°F (72°C) until partially cured. Again, allow sufficient flash time before baking for the solvents to escape so the paint will not blister. Generally speaking, pastel colors are more heat sensitive, and extreme caution must be used in force drying or baking them to avoid discoloration.

Never bake new paint until it has had sufficient time to flash (called solvent purge time). If you turn the heat up to 160°F for rapid curing before the paint solvents have evaporated enough, several paint problems can develop.

Care must be exercised to avoid overheating a new finish, because wrinkling, blistering, pinholing, or discoloration of the paint can result. It is better to force dry or bake at lower temperatures for slightly longer periods than to run higher temperatures for shorter periods. Again, refer to the paint manufacturer's directions for force drying temperatures and times for the specific type of paint.

When force drying, be careful to measure the surface temperature of the vehicle, not the air temperature. Generally, do not bake at too high a temperature (above 160°F or 72°C), because nonmetal part damage and paint problems can occur.

Water-Based Paint Characteristics

Waterborne or water-based paints use distilled water as the solvent or carrier. Water-based automotive paint has been around for years as a barrier coat. Water-based paints are still not used very much in the auto repair field. However, some states with a more serious pollution problem have made water-based colorcoats mandatory.

Water-based or latex clearcoats cannot be used in warranty repair of factory vehicles. They are not durable enough to protect the metal or plastic body panels from moisture, acid rain, and the elements.

Most shops are still using solvent-based paints because of their increased service life and durability. A paint chip in the clearcoat could expose the water-based paint to moisture. The small chip can allow the elements to attack the water-based color film. The water-based paint could continue to dissolve under the clear. After enough time, the small chip in the waterproof clear can result in catastrophic paint failure of a larger paint area.

Since you may have to apply either type of paint someday, you must understand the differences and similarities between water-based and solvent-based paints. Water-based paints might be the future of automotive repair and painting. Several states have already passed laws requiring the use of less toxic water-based colorcoats.

Paint manufacturers are selling water-based products into both the OEM and refinishing markets. Water-based colorcoat is available in all solid and metallic colors. Since solvent-based and water-based paints use the same pigments and similar binders, it is not difficult to match paint colors.

Most shops are still using solvent-based paints because of their increased service life and durability. A paint chip in the clearcoat could expose the water-based paint to moisture. The small chip can allow the elements to attack the water-based color film. The water-based paint could continue to dissolve under the clear. After enough time, the small chip in the waterproof clear can result in catastrophic paint failure of a larger paint area.

Many paint manufacturers premix their water-based paints so they are ready to filter and spray. If you have to mix the water-based colorcoat, mix in the right amount of each toner or color pigment by weight, add the correct amount of distilled water recommended, and test spray.

Water-based colorcoats do not have a "pot life" and can be stored in closed containers for later application. You might also give the extra water-based paint to the customer since it matches the vehicle.

Water-based paints do not emit the strong mineral oil odor, as do solvent-based paints. This strong varnish-type odor is caused by organic chemicals evaporating into the atmosphere as VOCs. Solvent odor and VOC emissions can be harmful to your respiratory system and the environment. This is the reason some states are requiring that body shops convert to water-based refinishing products. Water-based primers and colorcoat paints are desirable because they emit less VOC pollution when drying from the evaporation of water out of the binder and pigments.

The water-based colorcoats must be protected from the elements by layers of oil-based, solvent-based clearcoats. After applying the water-based colorcoats to cover the repair, conventional oil-based clearcoats must be applied over the water-based paint to protect it from rainwater and the environment.

Water-based paints use the same pigments as urethane paints. Therefore both types of colorcoats will produce the same light-fastness and color matching.

The main disadvantage of water-based paints is their very slow drying times. They can take three times as long to dry. Solvent, 2K (paint and hardener) urethanes cure through a chemical action rather than by slower water (H_2O) evaporation.

Another disadvantage of water-based paints is that they are not as hard when dry as urethane paints. They will scratch more easily than a harder film of all oil-based paint.

Water-based paint manufacturers are developing mild oil-based co-solvents to reduce water-based paint drying times. However, these co-solvents contribute to VOC emissions.

Even though water-based paints are not as toxic as urethane paints, you still must wear an approved respirator when spraying. Even breathing airborne pigments from water-based paints can be very detrimental to your health. Fumes from water-based paints are known to cause allergies and/or asthma, especially if you are a smoker.

Do not let the idea of water-based paint cause you to ignore the dangers of prolonged exposure to chemically hardening substances. It is very important to always use proper respiratory protection when spray applying any type of primer or paint.

Both solvent- and water-based paints contain harmful isocyanates. They are very harmful to your respiratory system and must never be inhaled. Wear an approved, properly fitting respirator mask, preferably an air-supplied respirator.

Water-based clear is an "intercoat," not a final clearcoat. It can be used to decrease VOCs to build up clear over the colorcoats before final coats of non–water soluble oil-based clearcoats.

Water-based paints never use an activator or hardener. They dry softer and more porous than oil-based paints while reducing air pollution.

Water-based paint cannot be used as the top clearcoat! The latex or water-based paint can slowly dissolve from exposure to rainwater and the elements. Water-based paint (colorcoats) must be fully coated with conventional urethane-, solvent-, or oil-based clearcoats or even old clear lacquer to protect it.

Water-based and solvent-based paints are typically compatible. Most urethane primers, paints, and clearcoats work perfectly well with water-based paints.

At present, water-based paint systems require a urethane primer and urethane clearcoat to produce a durable finish. Water-based paints are nonreactive. This means that water-based paints can accept and bond with any type of existing paint. In fact, water-based paint was originally used as a "barrier coat" (sealer) between old lacquer and newer urethane paints, to prevent cracking, lifting, and peeling of the new finish.

Always use a complete water-based paint system to assure proper bonding of all spray coats. For example, enamels and lacquers that use a very harsh solvent may degrade and chemically damage the water-based paint.

Although water-based paint cannot be used as a topcoat or clearcoat, by using water-based primers and colorcoats, you will reduce your exposure to toxic emissions while still producing a beautiful, long-lasting finish.

Water is not as good as solvents as a carrying agent for pigments. Also, shine and "depth" may not be as attractive as with solvent-based paints. An oil-based colorcoat will have more shine than a water-based basecoat. Paint manufacturers are constantly working on making their water-based colorcoats better match oil-based paints. If your shop has gone to all water-based colors, keep this in mind when trying to repair and match the color of a water-based colorcoat with an existing solvent-based colorcoat.

Water-based paint viscosity or thickness can vary by manufacturer. Most mix their water-based colorcoats so that they are ready to spray; no water reduction is required. If water-based paint has been on the shelf for a while, distilled water reduction may be required to achieve the correct spray viscosity.

Spraying Water-Based Paints

Any modern, HVLP gravity feed spray gun will work with water-based paints. Make sure your needle and tip are made of stainless steel and not steel. Steel needles will quickly corrode and rust when repeatedly spraying water or latex paints.

Some painters like to use one size smaller spray gun tip set to improve the atomization of water-based paints. Better atomization will help pull some of the water out of the spray mist before it hits the body surface to speed drying. It will also help smooth and level the water-based paint during spray application.

Adjust your spray gun before spraying water-based paint on the vehicle. Test your spray pattern on masking paper. Adjust the rate of flow and fan adjustments until an equal-sided elliptical shape is shooting out of your spray gun and onto the paper. Again, your test pattern will become wet very quickly as the water, binder, and pigments deposit on the paper sheet.

When spraying water-based paint in very hot, dry conditions, you may need to pour a slow-drying additive to the paint. Refer to the paint manufacturer information for details on mixing, additives, spraying, and drying temperatures.

When clearcoating a water-based paint, be very careful to make full wet coats over all surfaces, including along the bottoms of doors, fenders, and aprons. If you fail to spray the clear fully in any area of repair, the clear protective coating can be easily fractured to expose it to the elements. The area without enough clearcoat could be very soft and fail to provide proper protection from even minor impacts that could penetrate the topcoat, exposing the latex colorcoat.

A car painted or refinished with water-based paint should be waxed more often than conventional oil-based finish. The wax will help protect the thin urethane film that protects the water-based color from the environment.

Water-based paints go on very wet when sprayed compared to solvent-based colorcoats. This is fine, and the high pigment ratio will help the water-based paint fully cover very quickly. When first spraying water-based paints, it is very easy to overapply the liquid paint material to the point of paint running and sagging, ruining the paint job. Apply water-based paint a little more slowly, and stop applying when the new paint film looks wet to avoid ugly runs!

Water-based paints should be baked under heat lamps at approximately 140–160°F (60–70°C).

Water-based paints should not be run through an automated car wash because the abrasive brushes could break through the thin urethane clearcoat and into the softer, more porous water colorcoats. This could result in rapid paint film deterioration, discoloring, fade, paint breakdown, and other problems that lead to rusting.

At the proper drying temperature, oil or solvent-based colorcoats might take 15 minutes to cure and solidify. Water-based colorcoats can take up to 45 minutes to evaporate the water and dry. This can adversely affect body shop production since the painter has to wait so long before moving another car into the paint booth. Some shops are buying or building ovens so that vehicles can be moved into the drying room and free up the paint booth.

Once water-based paint has been sprayed, large fans can be used to increase airflow over the wet paint to speed drying times. This can reduce drying times down to those of conventional urethane-based paint systems.

Blend water-based colorcoats into any existing color to produce a gradual fade between the old and new paints. This will help hide any slight differences in the two colors.

26.15 REMOVAL OF MASKING MATERIALS

Removal of masking materials is a fairly simple task. If you do not get careless and hit or bump the fresh paint, demasking can be a very rewarding task. As you remove masking materials, the beauty of the new paint job becomes evident and you can begin to admire your work.

Demasking Wait Time

Never remove paint masking materials until the finish has had enough time to fully flash. The very top of the paint must be cured enough that it is not easily damaged by a light touch. Again, wait time before demasking will vary with shop temperature, type of paint product, and other factors. Refer to Figure 26–52.

Figure 26–52 Masking materials are removed after paint has fully flashed. Be careful not to bump into and damage fresh paint, however.

To make sure the paint is flashed enough for masking material removal, touch a piece of masking tape that has been covered with paint. A light touch should not leave a fingerprint in the paint film. If the paint film is not marred when touched lightly, you can carefully remove the masking materials.

If the finish has been force dried, remove the masking tape while the finish is still warm. If the finish is allowed to cool, the tape will stick and be difficult to remove. It can also leave adhesive behind or lift paint edges along fine line tape.

When removing fine line tape, pull the tape back and straight out away from the body. This will make the fine line tape shear or cut the paint film cleanly without lifting or pulling up any new paint. Normally, you should pull fine line tape off first. Avoid pulling fine line tape sideways, which could damage the new paint.

For a superior edge, fine line tape can be removed when the paint has partially flashed. This must be performed carefully by the painter right after painting. Tabs can be left sticking out on the tape at the ends so the fine line tape can be pulled off easily. Extreme care must be used so that the fresh, soft paint is not damaged.

Pull the tape slowly so that it comes off evenly. Take care not to touch any painted areas, because the paint will still be soft under the flashed paint film. Accidental impact marks in the fresh paint will take time to fix. Removal of masking materials is shown in Figure 26–53.

If you used liquid masking material, wash it off with soap and water. Do *not* wash freshly painted surfaces until they are fully cured, however.

Never mask a vehicle and let it sit for a prolonged period. Also, do not let masking paper and tape get wet. Either situation will make demasking difficult and can cause painting problems.

If masking tape is left on a body surface for too long, the tape adhesive can dry out. Dried adhesive can make it impossible to pull the tape back off. In that case, you have to gently scrape off the old masking tape. Then the old tape adhesive has to be cleaned off with an adhesive solvent.

If the masking material is on too long before painting, the tape edge can also roll up and allow paint to spray onto parts that are not intended to be painted.

A

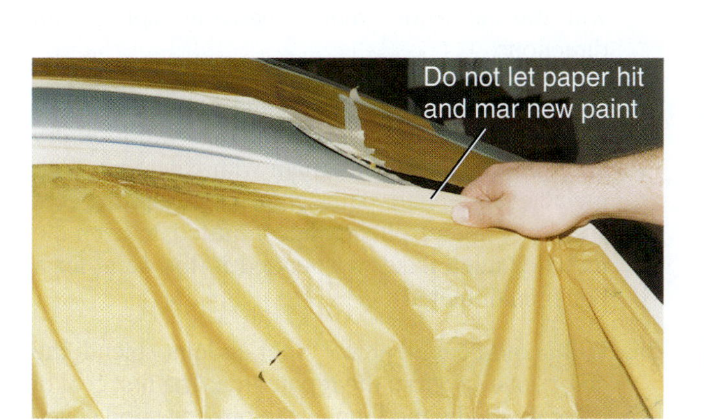

B

C

Figure 26-53 Note the steps for removing masking material. (A) Remove fine line masking tape first. Pull the tape straight out and away from the body. This will help shear off flashed paint so the paint does not peel up with the tape. (B) When removing masking paper, make sure it does not fall down and hit the new paint. (C) Remove masking rope to reveal the blended paint edge formed in the gap between the edge of the rope and the edge of the panel.

SUMMARY

1. Vehicle surfaces must be straight and ready for painting. A common mistake is to overlook a surface problem before painting. Because the surface is usually sanded dull, imperfections are easy to overlook. Double-check metal straightening work, body filler, and primer before proceeding. Overlooking even a small paint chip can ruin a paint job.
2. Surfaces must be perfectly clean and sanded or scuffed dull. Paint will not adhere to a dirty or glossy, smooth surface.
3. It is very easy for dirt and dust to collect on a vehicle prior to painting. You must methodically blow off all surfaces that can hold debris. When spraying, do not blow dirt onto wet paint.
4. Use an approved wax and grease remover or similar product to wipe down the vehicle. Use one clean, lint-free rag to apply the cleaning agent and a clean, dry one to wipe the solvent off before it dries. Do not use paint reducers or thinners or synthetic rags because they can leave a residue.
5. Tack cloth all body surfaces while blowing air with your spray gun to lift off any remaining dust. You might also want to tack cloth lightly between coats.
6. Paint in an ideal environment. Make sure the spray booth and vehicle are the right temperature. Are the booth filters and blower working properly? Do not try to paint a vehicle in an open shop.

7. Be sure to use the correct paint product and mix it correctly. Use an etching primer or epoxy primer on all bare metal. Apply tinted sealer over all repair areas. Apply clear adhesion promoter over scuff-sanded OEM finishes. Is the type of paint and its color correct?

8. Mix all required additives into the paint. If you forget to add hardener, the paint will not normally reach a full chemical cure. You could have trouble wet sanding and buffing out any paint imperfections. If painting a flexible bumper, add flex agent so the paint will not crack.

9. Test and adjust the spray gun spray pattern on a sheet of paper or an old part. If the gun is spitting or not working properly, do *not* use it to paint the vehicle.

10. A leaking spray gun cup can drip and ruin the paint. You must tilt the gun down when spraying the roof, hood, and trunk lid. If the cup is leaking, paint might drip out and onto these surfaces. Shake the spray gun before spraying to make sure the cup is not leaking around the lid.

11. Closely watch the paint as it deposits on the surface being painted. Check for application problems such as dry coat, excessive wet coat, spitting, and so on. Constant inspection of the wet paint as it hits the surface is critical to doing good paint work.

12. Determine whether the inside or rear of panels has been prepainted. The bottom of hoods, deck lids, and the underbody are often painted before the panel is installed and painted on the vehicle.

13. Band coat the edges of panels with paint first. Edging a panel involves painting around the perimeter of the panel before painting the front of the panel.

14. Apply paint properly. Hold the spray gun at the correct distance from the surface, typically 8–10 inches (203–254 mm). Aim the gun directly at the surface while moving it at the correct speed. Do *not* fan the gun unless you want to blend the paint film thinner.

15. Apply the correct film thickness. A common mistake is to apply too much paint. You should apply only enough paint to provide good color coverage and gloss. Film thicknesses are much less than in the past. Use the number of coats recommended by the paint manufacturer.

16. Spray to the wet edge. After painting half a panel, start spraying where you last stopped. Stand on one side to paint half the panel. Walk around to the other side and finish spraying the panel, starting in the middle. This will cause both coats to merge and flow out smoothly.

17. Keep air hoses a safe distance from the wet paint. Do not brush up against the wet paint and damage it.

18. Ensure that you get proper paint coverage along the edges of panels. It is easy to apply too much or not enough paint on panel ends. If you fail to trigger the liquid spray off at the end of each pass, double coats will be applied when changing spray gun directions.

19. Colorcoats should be applied only to get good color coverage over the sealer. Colorcoats will look dry and dull until you spray them with clearcoat. With metallic and pearl paints, the color or mid-coat should be sprayed light, almost misted, to keep the metal or mica flakes suspended in the paint for a better color match.

20. With colorcoat/clearcoat spot repairs, blend the color out away from the repair area. The first coat of color should be medium wet to provide good coverage over the damaged area. The second coat should be a little lighter or a medium coat that blends a few inches beyond the first colorcoat. The third coat of color should be an even thinner light or mist coat that blends a few more inches beyond the second coat of color. This technique will make any difference between the new color and original paint color less noticeable.

21. Allow enough flash time between coats. Flash time is the time needed for a fresh coat of paint to partially dry. Flash time is needed to prevent the paint from sagging or running. Always refer to the material manufacturer's guidelines. Increasing booth temperature to about 100°F will dry the paint and reduce flash times.

22. Use proper methods and materials when refinishing plastics. They must be treated differently than metal.

EXERCISES

On a separate sheet of paper, complete the following learning activities for this chapter. Write definitions for the key terms and answer the ASE-style review questions, essay questions, critical thinking problems, and math problems. You can also do the outside activities, possibly for extra credit.

Key Terms

adhesion promoter
banding coat
blend coats
blending solvent
force drying
full wet coat
leaking spray gun
light coat
medium wet coat
mist coat
mixing chart
OEM finish
paint baking
paint flex additive
panel repairs
primecoats
repair primecoats
solid color paints
spitting spray gun
spot repairs
spray schedule
stabilizer-reducer
topcoats
wet edge spraying

ASE-Style Review Questions

1. Technician A says you can pour solvent out of the can into a spray gun without using a strainer. Technician B says you should strain everything you pour into your spray gun. Who is correct?
   A. Technician A
   B. Technician B
   C. Both A and B
   D. Neither A nor B
2. Which of the following statements is incorrect?
   A. The objective of blending is to create the illusion that only one color is being seen
   B. It is not necessary to color match when blending
   C. When preparing the rest of the car, always prepare the adjacent panels for blending
   D. None of the above
3. Which product should be sprayed over an original OEM finish before clearcoating?
   A. Primer
   B. Sealer
   C. Adhesion promoter
   D. Epoxy
4. When spot repainting with solid colors, Technician A applies the topcoat with a circular motion from the center outward. Technician B applies the topcoat in short strokes from the center outward. Who is correct?
   A. Technician A
   B. Technician B
   C. Both A and B
   D. Neither A nor B
5. Technician A says to fan the spray gun to blend the colorcoats more thinly away from the repair area during spot repairs. Technician B says you should never fan your spray gun sideways. Who is correct?
   A. Technician A
   B. Technician B
   C. Both A and B
   D. Neither A nor B
6. Which of the following statements concerning color-coat/ clearcoat applications is incorrect?
   A. Never sand a colorcoat for any reason
   B. Never thin or reduce clearcoats before applying them
   C. Only one mist coat of colorcoat is needed
   D. All of the above
7. Technician A says to write down a spray schedule as you paint a vehicle to keep track of flash times. Technician B says you can touch the wet paint on a piece of masking tape to check flashing. Who is correct?
   A. Technician A
   B. Technician B
   C. Both A and B
   D. Neither A nor B
8. Which of the following would be a blend coat of paint?
   A. A full wet coat applied evenly
   B. A light coat applied evenly
   C. A coat faded more thinly near the perimeter of the repair
   D. A coat built up more thickly around the perimeter of the repair
9. Which of the following would be considered a panel repair?
   A. Paint is blended
   B. Whole panel is painted
   C. Half of a panel is painted
   D. Several panels are painted
10. Technician A says that OEM paints once used a water-based color. Technician B says that water-based primer can help solve paint compatibility problems. Who is correct?
    A. Technician A
    B. Technician B
    C. Both A and B
    D. Neither A nor B

11. A flexible bumper is going to be painted. Technician A says a flex additive is not needed. Technician B says it is best to use a flex additive on flexible panels to prevent the paint from cracking. Who is correct?
    A. Technician A
    B. Technician B
    C. Both A and B
    D. Neither A nor B

12. Technician A says to set the spray booth temperature at about 100°F to help flash paint between coats. Technician B says you can use a baking temperature of about 160°F after the paint has flashed and partially cured. Who is correct?
    A. Technician A
    B. Technician B
    C. Both A and B
    D. Neither A nor B

Essay Questions

1. Summarize the four important functions of a topcoat.
2. Describe some reasons for overall refinishing.
3. Summarize some characteristics of sealers.
4. What are some methods for determining the type of paint on a vehicle?
5. Summarize some of the more important paint label and literature data that should be checked.
6. In your own words, describe how to use paint mixing sticks.

Critical Thinking Problems

1. If you mix the paint too thin or too thick, what can happen?
2. If the first coat you spray on the vehicle is a very wet coat, what can happen after the next two coats?

Math Problems

1. If a spec temperature is given as 75° Fahrenheit, what is its equivalent in degrees Celsius? (Use the formula degrees F minus 32 multiplied by 0.556 equals degrees C.)
2. If directions say to reduce a paint 33 percent and you need 1 gallon of paint, how many parts of both reducer and paint do you need?

Activities

1. Practice spraying on an old body part. Prepare its surface properly using information in previous chapters and this chapter. You might be able to obtain extra paint for free from a body shop. Write a report on what you learned. Discuss any problems that you encountered. List the materials used for the job.
2. Visit a local body shop. Get permission to watch a professional painter at work. Write a one-page report about what you learned.
3. Practice spraying a metallic paint onto an old part or a sheet of masking paper. Intentionally apply so much paint that it runs or sags in one location. In another location, spray a dry mist of metallic by holding the spray gun too far away from the surface. Then adjust the gun properly and apply a medium wet coat. After drying, inspect the difference in the metallic finishes. Report your findings to the class.

CHAPTER 27

Color Matching and Custom Painting

OBJECTIVES

After studying this chapter, you should be able to:

- Describe color theory and how it relates to matching paint colors.
- Define terms relating to color.
- Describe the use of a computerized color matching system.
- Make spray-out and let-down test panels.
- Explain how to tint solid and metallic colors.
- Summarize the repair procedures for multistage finishes.
- Describe basic methods of doing custom paint work.
- Correctly answer ASE-style review questions relating to color matching.

INTRODUCTION

Color matching involves the steps needed to make a new finish match the existing finish on a vehicle. Even if you use the correct body color code numbers and correct paint mixing formula, the new paint may still not be exactly the same color. With today's metallics, three-stage pearl paints, and high-tech factory robotic painting, it can be very difficult to match colors when making both spot colorcoat paint repairs and full panel repainting.

Multistage paints, such as three-stage pearl paints, for example, are very difficult to match. Even if you have a new vehicle with a fresh OEM finish, it can be a challenge to make your repair look like the paint already on the vehicle. More than ever, there are many variables that affect the color and appearance of your paint work (Figure 27–1).

Oddly enough, as you learned in the previous chapter, a color mismatch in a panel repair will usually show up more than a spot repair. This is because a panel—such as a car door—has a distinct edge. And the repair, obviously, cuts off at that edge. Any color difference between the new paint and the existing paint will be visible because the panels are right next to each other, separated only by the door gap. For this reason, the color mix must be as good, if not better, when spraying a whole panel with colorcoat.

A spot repair, on the other hand, is performed by blending the repair into the surrounding area. You can hide a slight color difference better with a spot repair. The first coat is applied to the immediate area being repaired. The colorcoats that follow gradually extend beyond the repair area. A final light blend coat extends beyond the previous color coats. Thus, if there is a slight mismatch, the blend coat and the last colorcoat allow enough of the old finish to show through to make the color difference a gradual one.

This chapter will help you develop the skills needed to match the color of any type of paint. It will summarize color theory, color evaluation, color matching, computer analysis of paint, tinting, and other factors.

Figure 27–1 The beautiful finish found on today's vehicles is much more difficult to match than in the past. Multistage metallic and pearl paints can be a challenge when repainting.

27.1 COLOR THEORY

Color results from how objects reflect light at different frequencies, or wavelengths, into our eyes. The color seen depends on light waves reflecting off the surface. When these light waves bounce off a surface and strike the retina, our brains process the different frequencies into color.

What the eye sees as color is the absorption of all of the light except the color that it appears to be. Red has the lowest frequency, or wavelength, of light. A red object only reflects the lowest frequency waves in the light. A red ball appears red because the ball absorbs all of the colors in the light shining on it except for the reds. In contrast, a black object absorbs almost all light, while a white object absorbs none of the light wavelengths.

White light is actually a mixture of various colors of light. By passing light through a *prism* (piece of glass with a pyramid-like shape), light can be broken down into its separate colors, called the **color spectrum**. This is the same principle by which raindrops sometimes break up sunlight to form a colorful rainbow arc in the sky (Figure 27–2).

The colors in the spectrum are easily remembered with the phrase "Roy G Biv," which stands for the colors:

Red
Orange
Yellow
Green
Blue
Indigo
Violet

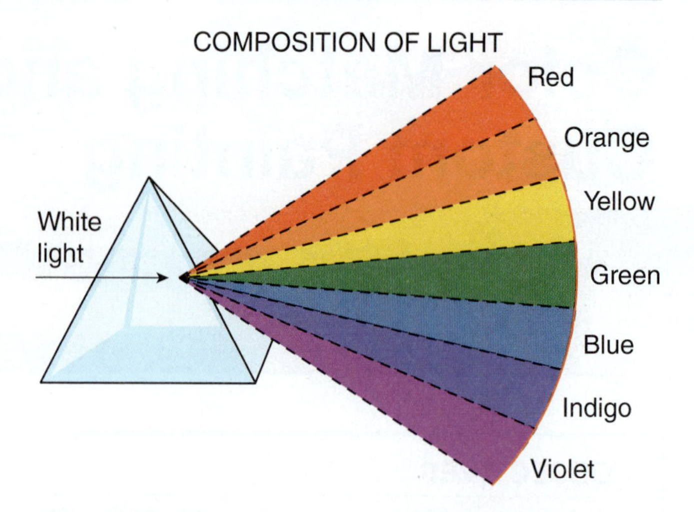

Figure 27–2 White light contains all of the colors of the spectrum. When white light shines through a glass prism, these colors are separated and can be seen individually. The same principle applies when you see a rainbow.

Lighting

Sunlight contains the entire visible spectrum (all wavelengths) of light. It is the standard by which other light sources are measured. Because a vehicle will be seen in sunlight, you should always use sunlight, or daylight-corrected indoor lighting, when making color evaluations.

There are different kinds of light bulbs used in a body shop. Each affects how you see color. Each light source has a different mixture of colored light, as shown in Figure 27–3.

Compared to daylight, *incandescent light* has more yellows, oranges, and reds. *Fluorescent light* has more violets and reds. They should not be used when analyzing a color. The same color of paint will look very different under different kinds of light. That is why it is so important to check the color match in daylight or under a balanced or color-corrected artificial light.

Color-corrected light bulbs, used by photographers, most closely match the full spectrum of sunlight. Color-corrected light bulbs are helpful when analyzing difficult-to-match paint colors inside the shop. They can be purchased at photo or camera supply stores and installed in a drop light.

What the eye sees as color is really light reflected from an object. The eye might see different shades of a color depending on the type of light source used. This is shown in Figure 27–4.

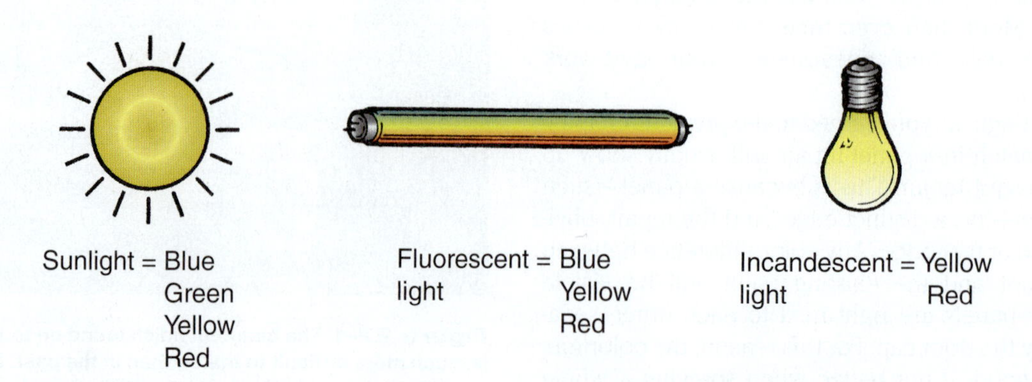

Figure 27–3 When evaluating paint color, use sunlight or color-corrected lighting. Colors look different under fluorescent and incandescent lighting.

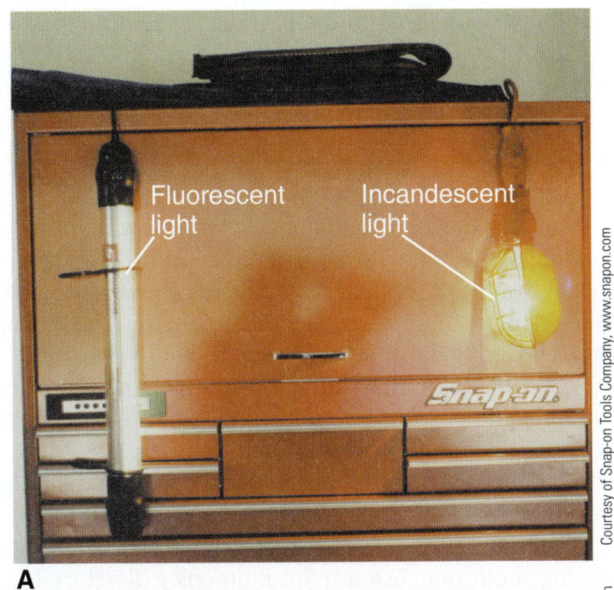

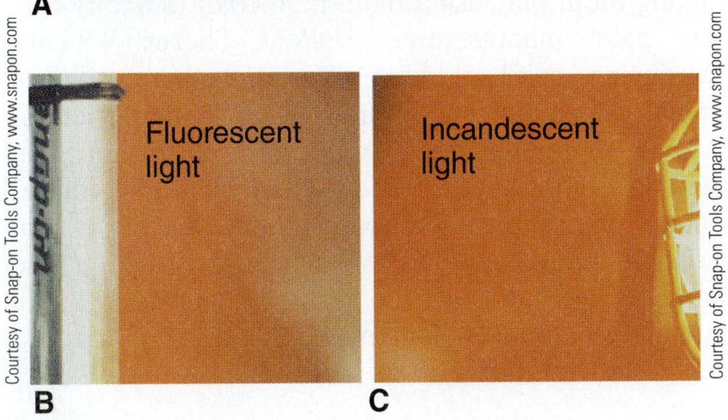

Figure 27-4 **(A) Fluorescent and incandescent drop lights have been hung on the front of this red toolbox. In (B) and (C), note how the close-up views of the red look different. This is why sunlight or color-corrected light bulbs must be used when matching paint colors.**

The index for measuring how closely indoor lighting matches actual daylight is the *Color Rendering Index (CRI)*. A CRI of 100 duplicates daylight. A range of 85–100 is acceptable for spray booth lighting.

Lamps are rated for *light temperature*, as measured in kelvin units. Daylight is 6,200 kelvin. For painting, a lamp rating of 6,000–7,000 kelvin is recommended to enable you to see the real color of the vehicle.

Lamps may also have a **lumen rating** for brightness. Lamps are normally between 1,000 and 2,000 lumens, with a higher lumen rating producing a brighter light.

Always evaluate and match paint colors in daylight or while using daylight-corrected lighting. The characteristics of the light in the shop affect how you perceive the color of paint on a vehicle. The color of the paint on the panel does not change. What changes is the amount of colored light from each light source reflected from the panel. For this reason, choose lamps or bulbs that most closely simulate actual sunlight. The spray booth manufacturer or representative may have recommendations (Figure 27–5).

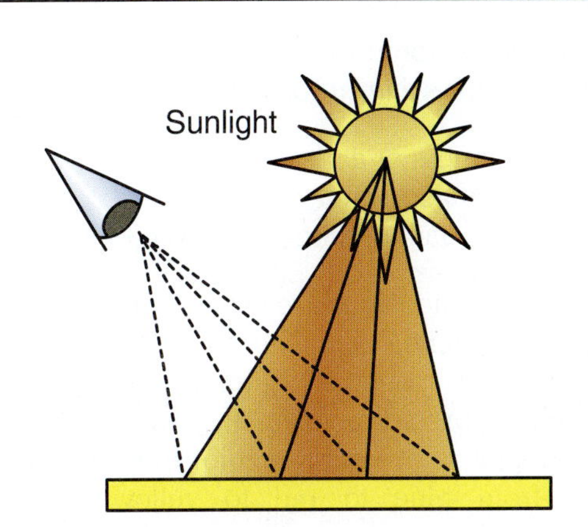

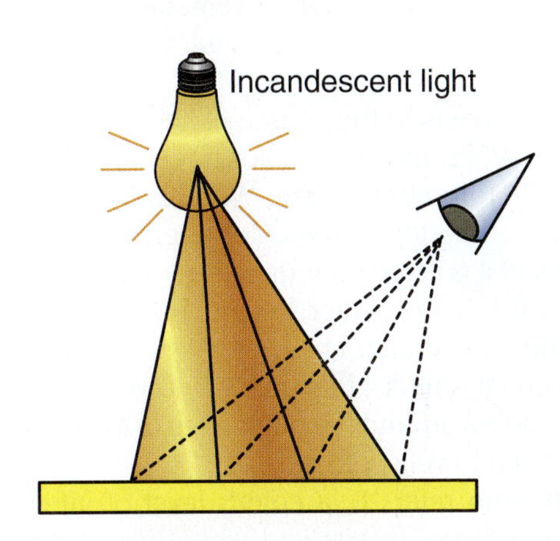

Figure 27-5 **Sunlight is the best lighting to use when evaluating color.**

Color Blindness

Color blindness means a person cannot see all colors of the spectrum normally. Color blindness makes it difficult to see paint colors accurately.

If you have problems matching certain colors, it might be wise to have your eyes checked for color vision. Blues and greens are most often the problem, or it may be difficult to discern between similar colors.

Nearly 10 percent of all men have trouble seeing one or more colors. However, almost no women have this difficulty. This is one of the reasons women do much of the touch-up paint work in auto manufacturing plants.

If you have a color vision problem, ask someone in the shop to help you match colors. A coworker can help determine what color pigments must be added to the paint to change the paint color to match the finish on the vehicle. Color-blind people can be good painters, but they often need help checking the color match before painting.

Dimensions of Color

Many people describe color in terms of what they see. For example, you might have heard these descriptions: sky blue, ruby red, emerald green, or midnight blue. Such terms can cause confusion when matching colors.

To minimize the confusion when painting, color should be based on the three *dimensions of color*:

1. Value—lightness or darkness
2. Hue—color, cast, or tint
3. Chroma—saturation, richness, intensity, or muddiness

These three dimensions are used to organize colors into a logical sequence on a color tree. The **color tree**, also called "color space" or "color wheel," is used to locate colors three-dimensionally when matching colors. Colors move around the color tree in a specific sequence—from blue to red to yellow to green. This sequence is easier to remember if you think of "BRYG." These are the first letters of blue, red, yellow, and green.

Value refers to the degree of lightness or darkness of a color. When using the color tree, the value scale runs vertically through the tree. White is the brightest value and it is at the top of the color tree. Black is the darkest value and it is at the bottom of the color tree. Neutral gray is the value at the center. Refer to Figure 27–6.

Hue, also called color, cast, or tint, describes what we normally think of as color. Hue is the color that we see. It moves around the outer edge of the color tree from blue to red to yellow to green.

Chroma refers to a color's level of intensity or the amount of gray (black and white) in a color. It is also called saturation, richness, intensity, or muddiness. Chroma moves along the spokes that radiate outward from the central gray axis of the color tree. Weak, washed-out colors with the least chroma are closest to the core of the color tree. Highly chromatic colors that are rich, vibrant, and intense are at the outer edge. When using the color tree, chroma increases as it moves outward from the neutral gray center and decreases as it moves closer.

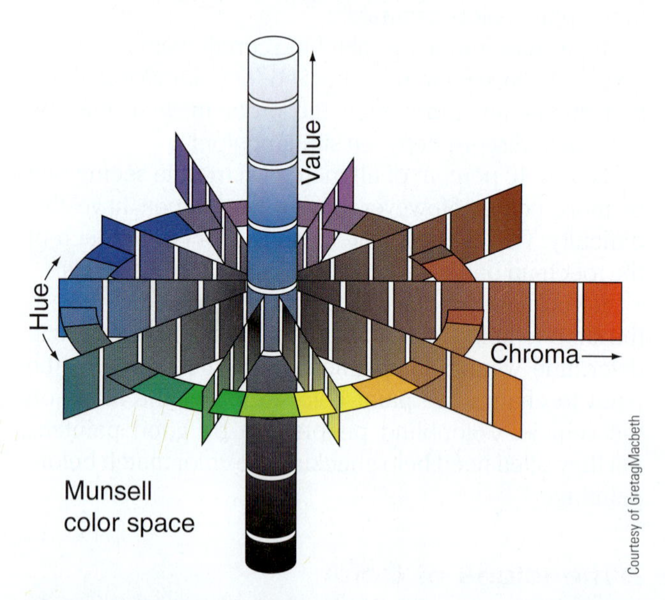

Figure 27-6 A Munsell graphic representation shows the three dimensions of color. Hue or color moves around the outside of the color wheel. Chroma moves in and out on the wheel spokes to change the intensity of a color. Value moves up and down on the wheel axle from black to white.

27.2 USING A PAINT COLOR DIRECTORY

The first step in color matching is to write down a vehicle's original paint color code. The paint color code number is located on a small plate or sticker on the vehicle body (see Chapter 26). This will give you a starting point for mixing or ordering the paint needed to repair the vehicle.

A **paint color directory** is a book containing information to help you match the paint color on the vehicle. This book normally contains color chips, paint formulas, and paint tinting instructions for most makes and models of vehicles. See Figure 27–7.

Collision repair and refinishing shops with an in-shop mixing room only use a refinishing color directory from one paint manufacturer (DuPont, Sherwin-Williams, Martin Senour, ICI, and so on). The paint mixing room will only have paint ingredients from that paint manufacturer.

Smaller body shops may not have an in-shop paint mixing room. They order and purchase paint from an outside paint supplier or mixing house. This allows them to use different color directories from several paint companies. They can order the brand of paint used by the auto maker or OEM to help match the color.

Paint Matching Chips

A *paint matching chip* is a patch of color printed on coated paper to closely match the color of a paint. A color directory will have hundreds of paint chips for almost every color of paint from every auto maker.

To use a color directory, first look up the make and model vehicle being repainted. Then, find the sample color chip that corresponds to the vehicle's paint code number. You can then compare the color on the paint color chip with the existing color on the vehicle.

Figure 27-7 Color manuals provide paint color chips, paint formulas, and instructions for mixing paint products.

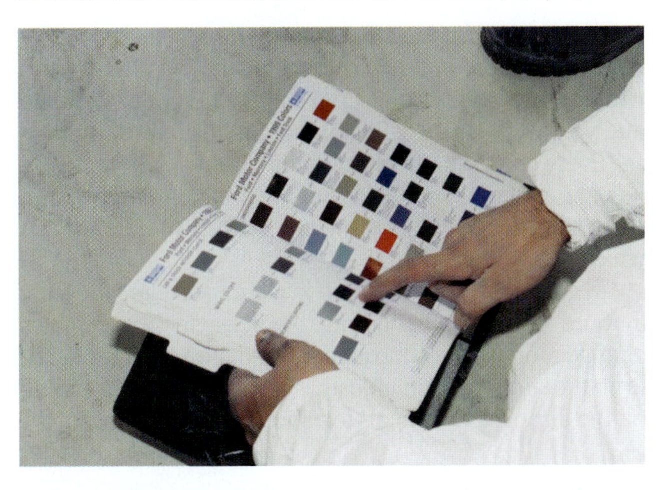

Figure 27–8 After obtaining the paint color code on a vehicle, find the paint color chip for that paint color. The paint chip for that body color code is compared to the paint color to check for a basic match.

Make sure the color on the sample chip matches the color on the vehicle. There is always the chance that the vehicle has been repainted a different color by another body shop (Figure 27–8).

Standard Color Chips

The **standard color chip** represents the intended color used when the vehicle was painted at the factory. It is the paint formula or color mix specified by the vehicle manufacturer. If everything goes as planned on the new car assembly line, the standard color chip will match the paint color on the vehicle body.

The standard color chip will often be the closest to matching the paint color on the vehicle. It can normally be used to mix or order the paint for your repair work.

Alternate Color Chips

Alternate color chips, also called *variance color chips*, are color chips that have been tinted a slightly different color than the standard color chip. These pretinted chips are used to help match minor paint color variations. Three or more alternate color chips are often given next to the standard color chip. They can be used to adjust the paint formulation more precisely than by using just one chip.

Paint manufacturers formulate these alternate colors when their research uncovers common problems in trying to match original paint colors during body shop repairs. The alternate chips give you examples of tinted colors that can be mixed to their specific formula. You do not have to experiment as much with tinting to find a closer color match than the standard color chip. Refer to Figure 27–9.

To use alternate or variance chips, lay the chips on the vehicle under proper color-corrected lighting. If the standard color chip does not match the vehicle's color, find the variance chip that best matches the color on the vehicle (Figure 27–10).

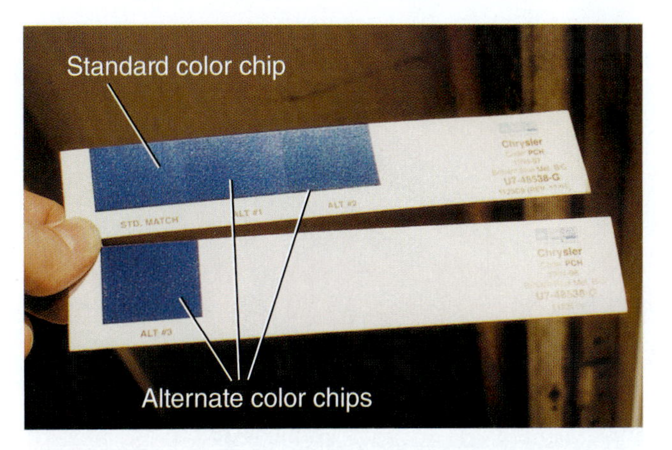

Figure 27–9 Most paint manufacturers provide both a standard paint chip color and alternate or variant paint chip colors.

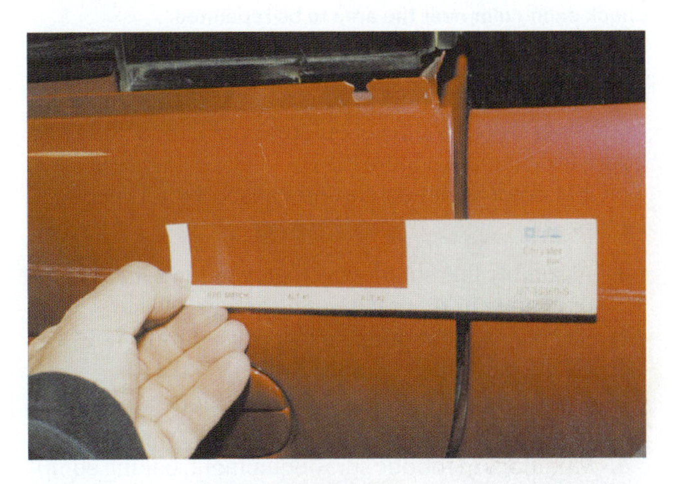

Figure 27–10 After cleaning the surface, paint color chips are held next to the vehicle body to check for a color match. The chip that best matches the actual color on the vehicle is used to find the paint formula for mixing the right color paint.

With weathered or dull paint, compound and polish a small area to bring it up to its original color and shine. You can also use a less deteriorated area of the finish, inside a door opening, for example. This will allow you to better compare the color chips with the actual color on the vehicle.

If the original finish is in good condition, many technicians still like to wet the vehicle finish with a cleaning agent when using color chips. A wet vehicle surface helps magnify and show off the real color.

Color Variations

When using a paint color directory, you must be on the lookout for color variations, regardless of the cause.

OEM Color Variance

An **OEM color variance** is a color difference on original manufacturer paint jobs from vehicle to vehicle. It should be noted that color variances from panel to panel exist even with original factory finishes. You must check for

Figure 27–11 **Paint color can vary on different areas of a vehicle, even with an OEM finish. For this reason, always check paint color near the area to be repainted.**

these factory differences when trying to match the color. A bumper may require a different color mix than the hood, for example, Figure 27–11.

OEM color variance can be caused by a number of factors:

1. Assembly line shutdowns that result in settled paint or change in robotic painting action.
2. Switching paint suppliers or vats during new car painting on the assembly line.
3. If two vehicles with the same color code are sprayed at different factories, the color can vary.
4. Variations in robotic spray equipment, baking equipment, and so on at factory.
5. Part may not be oriented the same as during spraying operations at the factory.
6. Material used to make the part (steel, aluminum, or plastics) may cause paint, especially metal flakes, to flow out differently. If two abutting parts are made of different materials (one steel and one plastic), it can be difficult to avoid a slight variance of color.

Whatever the reason for color variance, you must match the only "color measuring stick" that really matters—the vehicle itself. For better or worse, the paint color on the vehicle is the standard!

Finish Weathering

It must be remembered that the color code on the vehicle might not be exactly the same color, because all automotive finishes gradually change color when exposed to light and weather. Some colors fade; others become darker.

Yellow, for instance, fades fairly rapidly. If the yellow fades from a cream, the color will usually go lighter and whiter. If the yellow fades from a green composed of blue and yellow, the color will go bluer and usually darker.

Every color weathers a little differently, depending on its pigment composition. Other factors that affect weathering are owner upkeep and the part of the country in which the vehicle is driven.

In general, a vehicle stored in a garage suffers the least amount of paint weathering. Vehicles in hot climates change more rapidly due to increased ultraviolet radiation from the sun. Those in industrial areas change depending on the airborne chemicals to which they are exposed.

Paint Formula

The **paint formula** gives the percentage of each ingredient that is needed to match the original color on a vehicle. If the standard color chip matches the paint on the vehicle, use the paint formula for the standard paint color. If one of the alternate chips better matches the vehicle's paint, use the paint formula that corresponds to that paint color chip.

If you have to order the paint, paint suppliers provide topcoat colors in two ways: (1) factory colors and (2) custom colors. If the vehicle is a recent model or is finished with a popular color, chances are they will have the paint ready-mixed. These ready-mixed paints are called *factory-packaged colors.*

If it is an older color, the paint supplier may have to mix it. *Custom-mixed colors* are those colors that are mixed to order by the paint supplier. Custom-mixed colors can always be identified easily. The contents of the can must be written or typed on the label by the person who mixed the paint.

An *intermix system* is a full set of paint pigments and solvents that can be mixed at the body shop. Large body shops often purchase all of these ingredients and an automatic stirring shelf for the paint mixing room. With an intermix system, it is possible to mix thousands of paint colors in-house. This can save money over having to order paint from an outside paint supplier.

An in-shop intermix system also allows you to remix the paint to better and more easily match the color of the car. If the mixed paint does not match the color on the vehicle properly, you do not have to go back to the paint supplier for a remix.

27.3 MATCHING BASIC PAINT COLORS

Incorrect color matching is one of the single most common problems in the automotive refinishing industry. Most of the color matching problems are experienced when attempting to match metallic and pearl three-stage paints. Although some problems are encountered with solid colors, they cause the fewest color matching problems for the average technician.

Color Test Panels

Most paint colors will look different as a liquid compared to after they have been sprayed on a panel. The liquid paint on a mixing stick gives a general indication of how well your color matches the existing color on the vehicle. This will only work, though, if you have an easy color to match, such as a solid color, or if you have had previous experience with the specific color (Figure 27–12).

Checking your color match is much more accurate if you spray a color test panel to see how well your paint matches the original finish. A **color test panel** is a black-and-white coated sheet of cardboard with a wooden handle. Color test panels can be purchased from paint suppliers or from your paint salesperson (Figure 27–13).

A test panel can be sprayed and compared to the color on the vehicle. Using a color test panel is recommended with metallic paints, pearl paints, and some hard-to-match solid colors.

The extra time spent spraying one or more test panels will be rewarded with a satisfied customer. You do not want the customer to complain that your paint repair looks like a different color than the original finish.

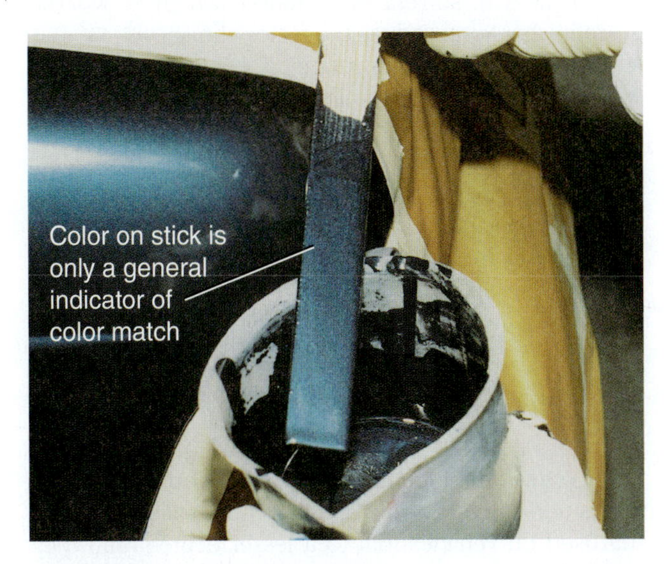

Figure 27–12 The color on your paint stick can be used to check the general match of the paint color. However, the paint color will change slightly after it has been sprayed.

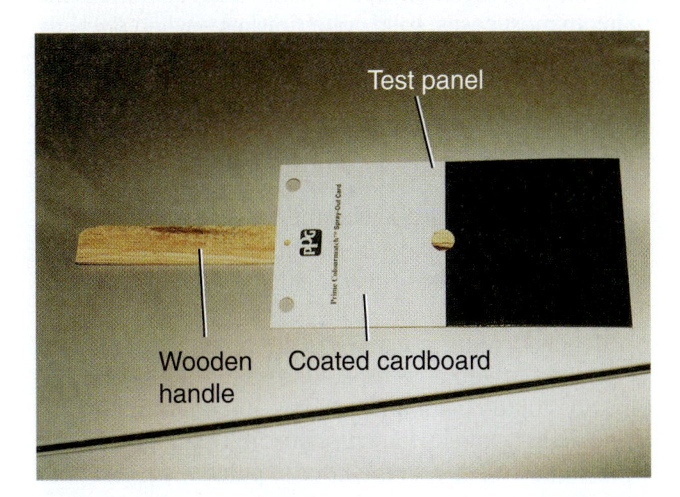

Figure 27–13 A color test panel is a black-and-white coated sheet of cardboard with a wooden handle. Color test panels can be purchased from paint suppliers or from your paint salesperson.

Spray-Out Test Panel

The **spray-out test panel** is used to check the paint match of solid and metallic colors, but not three-stage pearl paints. It typically involves spraying the test panel with sealer, colorcoat, and clearcoat. With a solid color, spray the test panel with sealer and then paint it.

Spray-out test panels are prepared by applying the paint as near to actual spraying conditions as possible. When done properly, a spray-out test panel shows the paint almost exactly as it will look when sprayed on a vehicle.

It is best to tape the color test panel to a piece of masking paper. However, some technicians like to hold the test panel by its handle while spraying (Figure 27–14).

You spray a color test panel just as you plan to paint the vehicle. The spray gun adjustments must be correct. Spray gun distance and speed must be normal. Apply the same number and types of coats that will be used when painting the vehicle. Allow proper flash times between coats.

To make a spray-out test panel for basecoat and clearcoat paints, proceed as follows:

1. Spray a full wet coat of sealer to the test panel. Both the black and white sides of the color test panel must be fully hidden. Allow the sealer to flash per label instructions.
2. Spray your coats of color onto the test panel, with proper flash times between coats. With solid colors, normally use two medium coats. With metallic paints, use three coats—two medium coats and a third light coat.
3. Apply your full wet coats of clear next. Allow proper flash times.

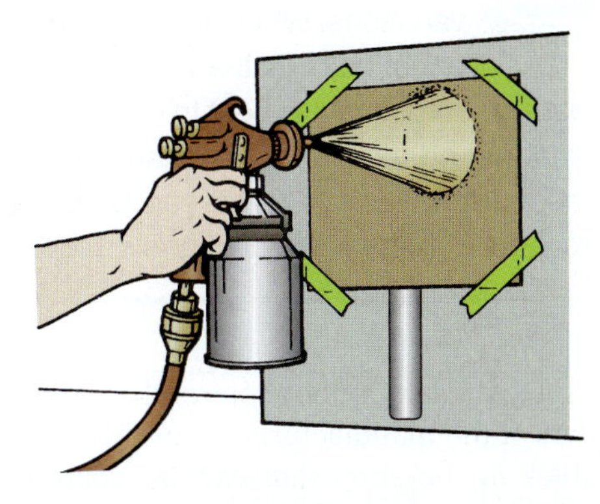

Figure 27–14 When spraying a spray-out test panel, tape it to a piece of masking paper. This will let you use normal spray gun action to apply proper coats to the test panel. The test panel should be sealed, colorcoated, and clearcoated using the exact methods that will be used when painting the vehicle.

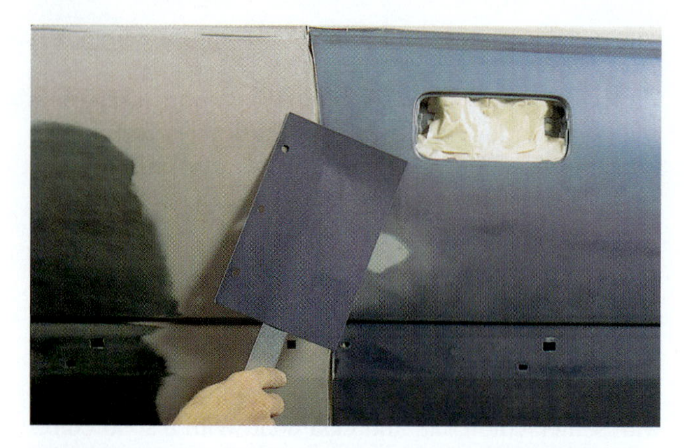

Figure 27-15 Once the spray-out test panel has dried, hold it next to the original finish. The test panel will show you exactly how well the new paint color will match the existing paint color.

The spray-out test panel should be fully dry or cured prior to evaluating the color match. Hold the color test panel against the cleaned vehicle body to see whether the existing color and your color match. If the test panel does not look like the same color, you will have to remix or tint the color for an exact match (Figure 27–15).

A number of adjustments can be made to make slight changes in the paint's color. Several test panels might be needed in order to achieve a good match. The ability to match is mostly a matter of experience and can be honed with practice.

Color Spraying Variables

A *variable* is a painting condition (temperature, humidity, or ventilation) or a painting process (reduction, evaporation, speed of solvents, air pressure, or type of equipment).

Even late model vehicles with an OEM factory finish can have color variations. When holding a color test panel next to the existing finish to compare color match, it is best to check near the area that will be repainted. For example, a hood or trunk lid that lies horizontal can be a slightly different color than a vertical door side or fender.

A paint technician cannot control the variables that affect colors at the manufacturing plant. However, there are variables in the shop that can be controlled. For example, a painter can control paint color through:

- Agitation of the paint
- Application techniques
- Amount of material applied
- Spray gun adjustment
- Atomizing air pressure
- Type and amount of solvent used for reduction
- Identification of the correct paint code and mixing formula

All of these factors can be changed to help you make the new paint match the existing finish.

Varying the spraying technique can affect color. In other words, the application technique can cause the color to vary. Painters who spray wet end up with a darker color than those who spray drier, especially with metallic colorcoats.

Variables are divided into two categories: positive and negative. *Positive variables* are those things that a painter does to duplicate the original finish, which in turn result in a good color match. They are:

- Slowness of solvent evaporation, which allows the refinisher to reproduce the factory finish
- Wetness of color application
- Proper spraying technique and the correct air pressure

Negative variables are those that cause the shade of a color to be off-standard. The most common are:

- Improper reduction
- Improper agitation
- Improper application, primarily too high or too low air pressure

Matching Solid Colors

For many years all vehicles were solid colors, such as black, white, tan, blue, green, maroon, and so on. Solid colors reflect light in only one direction. Solid colors are still used on vehicles, but to a lesser degree than they were a few years ago.

Matching solid colors is easier than matching metallic or mica paints. You only need to match the color pigment and not the metal or mica flakes suspended in the paint. In most cases, solid color finishes—when properly prepared, reduced, and sprayed—will provide a good color match.

Matching Metallic Finishes

Matching a metallic (polychrome) finish takes more skill than matching a solid color. This is because metallic colors are made with both color pigments and aluminum flakes.

Metallic paints contain small flakes of aluminum suspended in liquid. The position of the flakes and the thickness of the paint affect the overall color. The flakes reflect light, whereas the color absorbs a higher amount of the light. The thicker the layer of paint, the greater the light absorption.

As the position or orientation of tiny metal flakes changes in a color film, the color shade changes accordingly. Each metallic particle is like a tiny mirror that changes the appearance of the color when viewed from different angles. When viewing a metallic color at right angles or perpendicular to the surface, you are seeing the **color face**. When viewing the metallic finish at a 45-degree angle or less, you see the *color pitch* or **side tone** of a color. Metallic color also appears to vary when viewed under different kinds of light, such as daylight, shade, sunlight, or artificial light.

Spraying Metallic Colors

How wet you spray metallic coats affects the paint color. Normally, three coats of a metallic color are sprayed over the sealer. The first two coats are sprayed medium wet for good coverage. The last coat is a light mist coat to help distribute and orient the metal flakes on the top of the paint film.

A *medium metallic* spray will disperse the metal flakes evenly throughout the paint film. Medium coats of metallic color are normally recommended by paint manufacturers. See Figure 27–16.

If a paint technician is to get good color matches, it is important to understand how certain paint variables affect the shades of metallic colors. In summary, the shades of metallic colors are controlled by:

- Choice of solvents
- Color reduction
- Air pressure
- Wetness of application
- Spraying techniques

Wet Coats of Metallic Color

A *wet metallic paint spray* makes the color appear darker and less silver. Because the flakes have sufficient time to settle down in the wet paint, they lie parallel to and deeper within the paint film. Light reflection is uniform and, because the light has to go further into the paint film, light absorption is greater. The result is a painted surface that appears deeper and darker in color.

Normally, metallic colors, both single-stage and basecoat/clearcoat, should never be sprayed as full wet coats. Use only medium to light spray coats to apply metallic color to achieve the best color match. If your first colorcoat of metallic looks too light, spray the next coat in a medium wet coat to darken the color.

Dry Coats of Metallic Color

A *dry metallic spray coat* makes the paint appear lighter and more silver. The aluminum flakes are trapped at various angles near the top surface of the paint film. The light has less paint film to travel through before hitting the metal flakes, so little of it is absorbed. The result is nonuniform light reflection and minimum light absorption.

A dry metallic spray coat allows more of the metal flakes to remain on the top of the paint film. The paint spray will dry quickly, suspending and angling the flakes

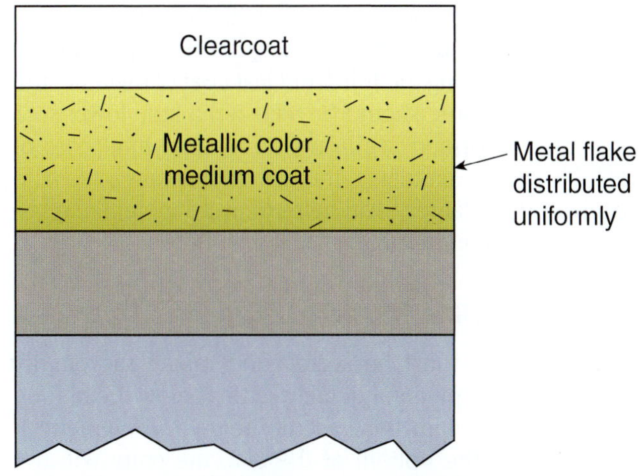

A A medium coat of metallic color will distribute metal flakes evenly throughout paint film.

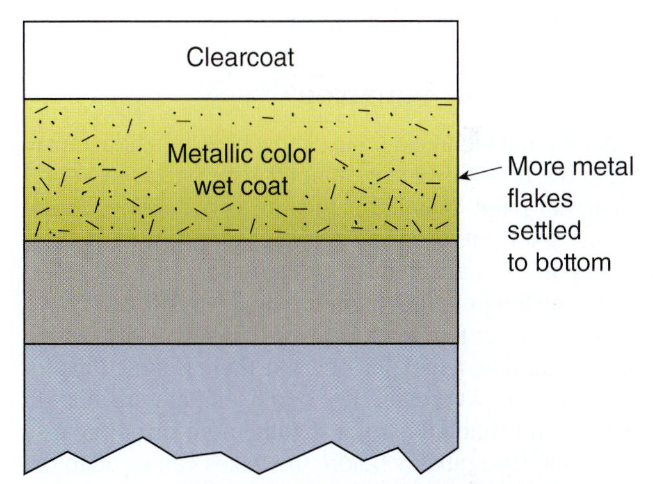

B A wet coat of metallic color allows heavy metal flakes to sink to the bottom of the paint, darkening the color.

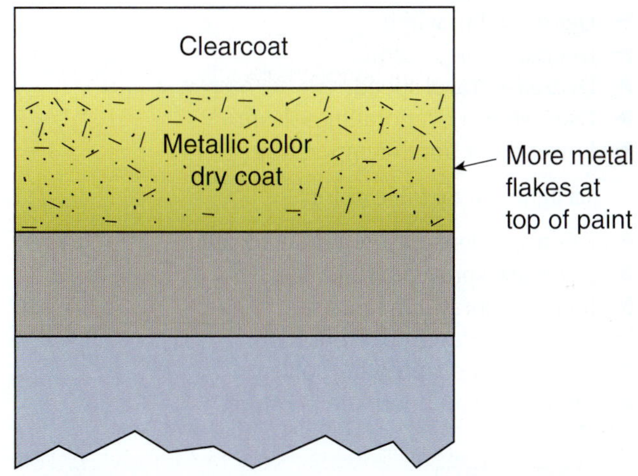

C A light coat of metallic color keeps more of the metal flakes near the top of the paint film, lightening the color and making it appear more silver.

Figure 27-16 Study how metallic paint is affected by the type of coat used when spraying.

at different angles near the paint film's surface, which gives the colorcoat a brighter, more silvery appearance.

Before spraying the third and last coat of metallic, the spray gun pressure should be adjusted a little lower so a finer mist is formed at the nozzle. A light final coat of metallic color simulates the robotic spray a vehicle receives at the factory and helps to avoid metal flake settling and a color mismatch.

Keep Metallics Mixed

Metallic colors must be stirred and mixed thoroughly before use. The pigment quickly settles below the binder. Also, the aluminum flakes settle below the pigment. If flakes stay at the bottom of the can, the paint will not match the color on the vehicle being refinished.

An *agitator paint cup* will help keep the metallic flakes mixed and evenly distributed in the cup. A small air motor in the gun moves a mixing paddle up and down. This helps keep the paint stirred.

Metallic Color Variables

A good paint technician must know how to handle metallic colors, both single-stage paint and two-stage basecoat/clearcoat paint. Metallic paints are very sensitive to the reducing solvents used and spray gun air pressure. Metallic colors are also affected by a number of other variables, including temperature, humidity, and ventilation.

Cold booth temperature slows paint flashing, which tends to darken a metallic because there is more time for the metal flakes to settle. High booth temperature tends to lighten and shade the color slightly more silver because the paint dries quickly before the flakes can settle to the bottom of the paint film.

To darken a metallic color:

- Use a larger spray gun fluid nozzle.
- Increase fluid flow.
- Decrease fan width.
- Decrease air pressure.
- Decrease travel speed.
- Use a slower evaporating solvent.
- Lower paint booth temperature.

To lighten a metallic color:

- Use a smaller spray gun fluid nozzle.
- Decrease spray gun fluid flow.
- Increase spray gun fan size.
- Increase spray gun air pressure.
- Use a faster evaporating solvent.
- Increase paint booth temperature.

Metamerism

Metamerism occurs when different light sources affect paint pigments differently. A paint may have some blue in it that is not noticeable in daylight but that becomes very evident under streetlights. The problem then is that the refinish and OEM paint formulas are not made of the same pigments. This causes the pigments to look different under different light sources.

Paint manufacturers formulate refinish paints to minimize the effect of metamerism. Metamerism is most often a problem when a painter varies from the paint formula during tinting operations. That is why it is important to use only the tints called for in the formula.

Color Flip-Flop

Color flip-flop makes a multistage paint color appear a different tone when viewed at different angles. This color-changing "chameleon" effect results from the positioning of the mid-coat particles (aluminum metal flakes, mica, and other reflective materials). Different colors of light reflect off the mid-coat flakes or particles to make the paint seem to change color with viewing angle (Figure 27–17).

Flip-flop results from the percentage of mid-coat particles oriented in a specific direction and their depth in the paint film. The direction and intensity of the light being reflected back through the paint film cause the flip-flop phenomenon.

The first approach to correcting the problem is to adjust your spraying technique to compensate for this effect. Spraying the panel a little wetter will slightly darken the appearance when looking directly into the panel. When viewed from an angle, the resulting appearance is lighter. This occurs because the aluminum particles are positioned flatter and deeper in the paint film.

Spraying the panel slightly dryer reverses the effect, giving a light appearance when looking directly at the panel. This is because the aluminum particles are closer to the surface. The result is a darker appearance viewed at an angle, as light becomes trapped. Both of these techniques are a compromise and should be used to correct minor conditions of flip-flop.

If spray techniques cannot correct this condition, the addition of a small amount of white toner will eliminate

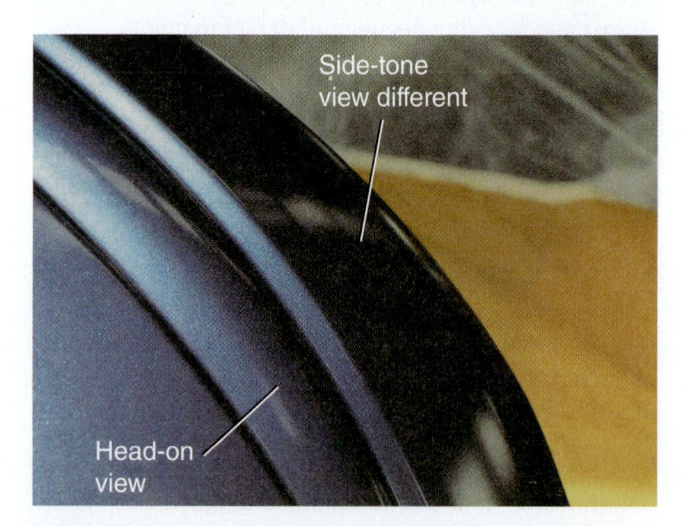

Figure 27–17 Color flip-flop results when a metallic or pearl paint looks like a different color when viewed from different angles.

the sharp contrast from light to dark when the surface is viewed at various angles. The white acts to dull the transparency, giving a more uniform, subdued reflection through the paint film. Care should be taken when adding white because the change occurs quickly. Once too much white is added, recovering the color match becomes virtually impossible.

When confronted with an extremely difficult flip-flop condition, the best method to solve the problem is to add white and blend the color into the adjacent panels. When blending, extend the color in stages. In a basecoat/clearcoat system, for example, spray the blend into the adjacent panel as needed. Then apply the clear to both panels.

27.4 MATCHING BASECOAT/ CLEARCOAT FINISHES

The technology for basecoat/clearcoat finishes was developed in Europe. The durability and popularity of these finishes prompted Japanese and American automobile manufacturers to begin using them as well. In fact, most automotive experts agree that basecoat/clearcoat paints will continue to be used on the vast majority of new and refinished vehicles for many years. The clear sprayed over the colorcoat provides a brighter shine than single-stage paints.

Spot repair is made easier because basecoat/clearcoat finishes make it simple to blend in an edge. Also, the need for buffing is just about eliminated. Most important to body shop paint technicians, however, is the outstanding color match capability that state-of-the-art basecoat/clearcoat systems provide.

When working with basecoat/clearcoat finishes, remember:

- Clearcoats are not all perfectly clear and they may change the appearance of a color.
- Blend the basecoat and clearcoat the entire panel.
- Avoid blending the clearcoat. If needed, blend the clear into the smallest area possible to hide the repair.

Fluorine Clearcoat Spot Repairs

The basic steps for spot repair with a fluorine clearcoat system are as follows:

1. Compound or sand all surfaces to be sprayed with #1200–#1500 grit sandpaper for proper adhesion of the new paint (Figure 27–18).
2. Apply the first coat of basecoat as a medium wet coat over only the area needing color.
3. Apply the second coat of basecoat as a medium coat and blend it out a little farther than the first coat. Two coats of color will normally be enough paint with solid, nonmetallic colors.
4. If needed, or with a metallic basecoat, apply a light coat to obtain a good color match of the metal flakes.
5. Use a blending solvent in blend areas if necessary. Dry at 140°F (60°C) for 20 minutes.

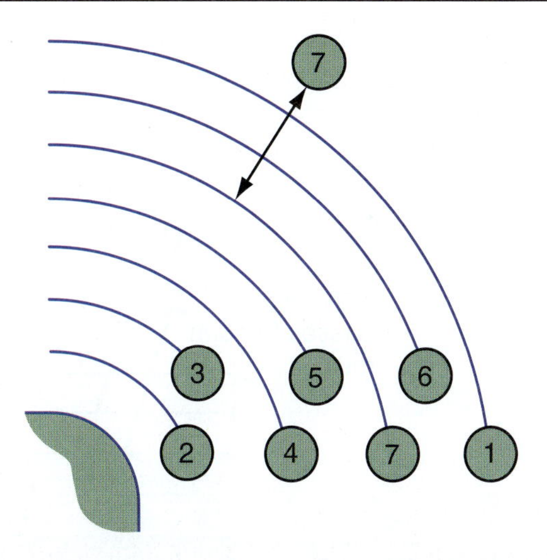

Figure 27–18 Compare the numbers on this illustration to the following to see how to do a basecoat/clearcoat spot repair: 1. Compound or sand with #1200–1500 grit sandpaper. 2. Apply first coat of basecoat. 3. Apply second coat of basecoat. 4. Apply third coat of basecoat or until hiding is obtained. 5. Apply color blender, if necessary; dry at 140°F (60°C) for 20 minutes. 6. Apply three to four coats of fluorine clearcoat; dry properly between coats. 7. Polish with fine compound.

6. Apply two to three coats of fluorine clearcoat. Clearcoat the entire panel when possible. As shown in Figure 27–18, the area between 5 and 6 is faded out or blended if required. Flash dry at 70–80°F (16–21°C) for 10 minutes between coats.
7. After applying the final coat, bake the finish at 160°F (75°C) for 45 minutes.

27.5 MATCHING THREE-STAGE PAINTS

In an effort to attract buyers who like a flashy paint job, auto manufacturers now offer highly iridescent colors or pearl paints applied in three stages. The first stage is a basecoat of color, the second stage of paint is a mica, or "pearl," mid-coat. The final stage is the clear topcoat (Figure 27–19).

Figure 27–19 Three-stage pearl paints are becoming more popular because of their brilliance.

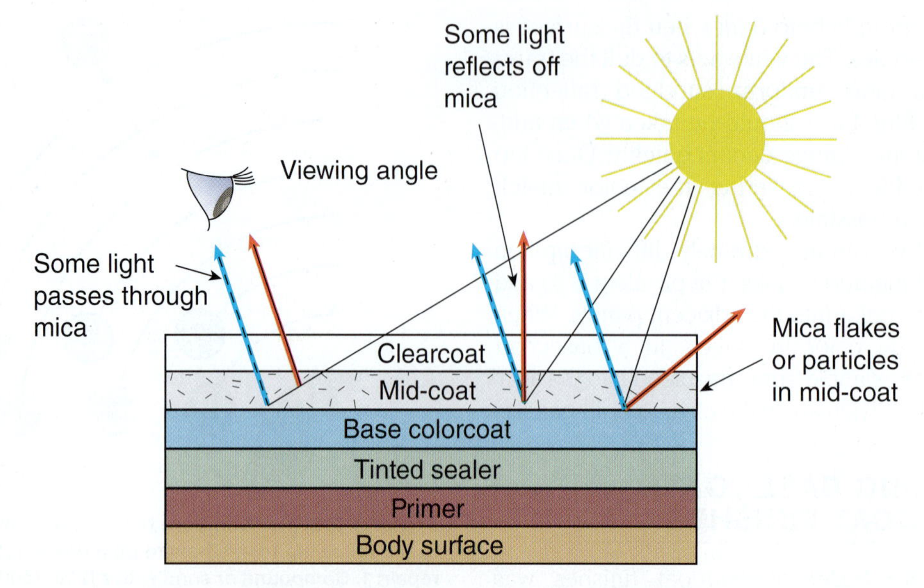

Figure 27-20 **The mica particles in a mid-coat act like tiny diamonds or rubies that reflect, refract, and pass light. Basecoat color and mica particles in the middle layer of a clearcoat produce a pearl effect.**

These paints often use a synthetic pearl luster pigment of mica particles covered with a thin layer of titanium dioxide. The coated mica particles are reflective, but they are also transparent, like diamonds or rubies. The titanium dioxide layer provides the rainbow or pearl effect, as light reflects but also passes through the mica particles. This is shown in Figure 27–20.

Transparent pearl luster pigments allow a much higher reflective brilliance than aluminum flakes. Aluminum flakes act as miniature mirrors that reflect light. Mica particles act like tiny "see-through" diamonds in the paint.

The repaint formulas available for the pearl colors on new vehicles usually have colored pigments and mica pigments combined. More work is involved because more spray coats are needed than with single-stage or basecoat/clearcoat paints. Several additional mid-coats are required with the proper flash times between them. Allow more spraying time when scheduling jobs with mica finishes.

There are a number of factors to keep in mind when working with pearl luster paints:

1. Mica flakes are heavy. Keep the paint agitated to ensure even distribution.
2. Spray test panels before painting the vehicle.
3. Continually blend on spot repairs.
4. Do not rush. Allow enough flash time between coats.
5. Spray in a well-lit booth.
6. Ultraviolet light can help in checking the pearlescent effect.
7. Direct sunlight is the best source of light for evaluating touch-ups.
8. Check the paint match from three angles in direct sunlight: straight on and standing on both sides of the repair area.

In a three-stage finish, the mid-coat is the layer that contributes the most to the final appearance of the color. Particles making up the mid-coat can be designed to reflect, absorb, and refract differing amounts of light, as in pearl luster finishes. Changing the amount or color of the mica flake coating drastically alters the color of the finish when viewed from straight on or from an angle.

While most of the tri-coat finishes currently used by auto makers are pastels, darker shades are also feasible. Tri-coat finishes have given automotive stylists an exciting new palette of available colors. Anywhere from one to five coats of pearl luster mid-coat need to be applied to achieve the desired effect.

When working with tri-coat and multistage finishes, a technician must match the basecoat color prior to applying the pearlcoats and clearcoats. Some vehicle manufacturers leave an area of basecoat that is not coated with mica or clear for comparing the basecoat spray-out to the vehicle.

Let-Down Test Panels

A **let-down test panel** is used to evaluate color match on three-stage, tri-coat, and multistage paint systems. Directions for making a let-down test panel are available from the manufacturer of the paint.

Here is the procedure for a typical let-down test panel for a three-stage finish:

1. Prepare the let-down test panel with the same sealer being used on the vehicle. Many painters like to tint the sealer to match the color of the basecoat to aid in matching these difficult colors (Figure 27–21).
2. After the sealer has flashed, apply the basecoat color using the same air pressure and spray pattern that will be used on the vehicle. Duplicating the actual spray techniques when preparing the let-down test panel is critical. Make sure you do *not* vary your procedures.

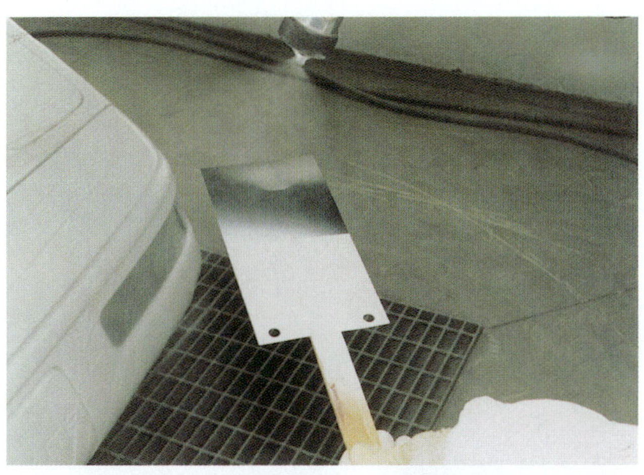

A First spray the let-down test panel with a sealer tinted to match the base color. Apply enough tinted sealer to completely cover the black and white halves of the test panel.

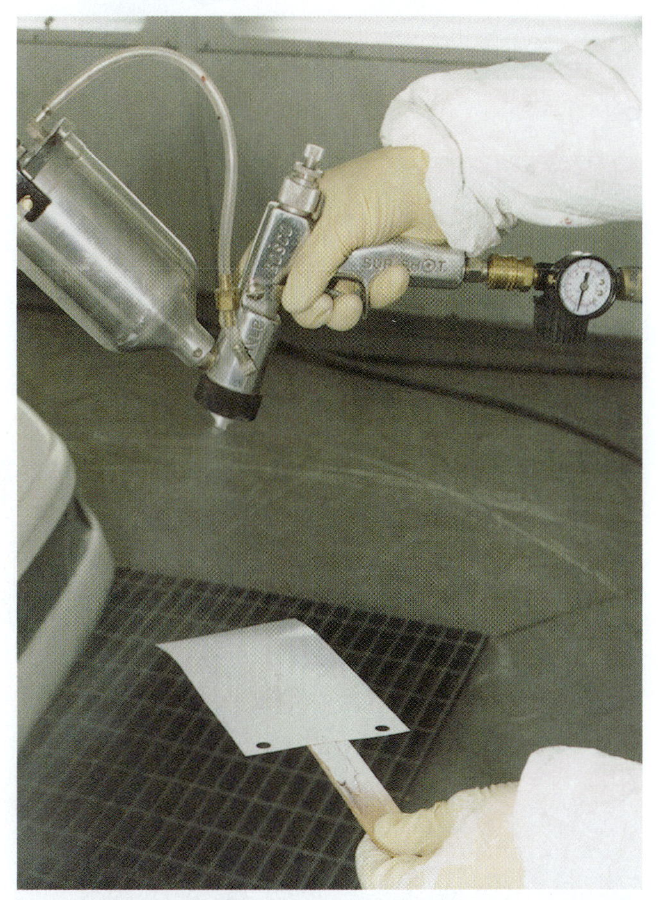

B After allowing the tinted sealer to flash properly, spray the whole test panel with the correct base color for full hiding of the sealer.

Figure 27-21 A let-down test panel is needed to properly match three-stage paints.

3. After the test panel has dried, mask it into four equal sections with small pieces of masking paper and tape. Leave the top fourth of the test panel exposed (Figure 27–22).
4. Apply one coat of semitransparent mid-coat color (usually mica) over the top quarter of the card.
5. After the mid-coat has flashed, remove the top layer of masking paper, exposing more of the test panel.
6. Apply another coat of mid-coat color over the exposed top half of the test panel.
7. After this second coat has flashed, remove the masking paper to expose three-quarters of the panel.
8. Apply another coat of mid-coat color over the exposed three-quarters of the panel.
9. After flashing, remove the masking paper entirely.
10. Apply a fourth coat of mid-coat color. As always, spray the coating in the same way as you plan to on the vehicle.
11. After the entire let-down test panel has dried, mask off half of the panel lengthwise.
12. Apply the manufacturer's recommended number of clearcoats to the exposed half.

Compare the different shades on the let-down test panel with the paint on the vehicle. Each layer of the intermediate or mid-coat will darken the appearance of the finish. The let-down test panel serves as a handmade set of alternate paint color chips. You must find which area on the let-down test panel best matches the color of the vehicle.

When repainting the vehicle, use the same number of coats used on the matching section of the let-down test panel. This will help you achieve the correct paint match. Refer to Figure 27–23.

A *basecoat-only patch* is a small area on the vehicle's surface without clearcoat that enables a technician to check for color match with tri-coat, or three-stage, colors. The manufacturer masks the patch before clearcoating to help you match the color more easily. The basecoat-only patch is sometimes located on the driver's side rocker panel. Other common locations for the exposed basecoat are under the deck lid or hood. Refer to the vehicle's service manual to find the exact location of the basecoat-only patch.

You can hold your spray-out test panel next to this small area without clearcoat on the vehicle body when checking your color mix.

Once made, the test panel can be kept and used on vehicles with the same color code. On the back of the panel note the color code, gun settings, and paint technician's name. One must be made for each different multi-stage color and each painter.

Matching Three-Stage Spot Repairs

A **halo effect** is an unwanted shiny ring, or halo, that appears around a pearl or mica paint repair. It is caused by the paint being wetter in the middle and drier near the outer edges of the repair.

Avoid a halo effect by applying the first coat of mica to the basecoat only. The more intermediate mica coats that are applied, the darker the finish will appear. Allow a larger area in which to blend the intermediate mica coats. They require more room to blend than a standard basecoat. Keep the tri-coat repair area as small as possible.

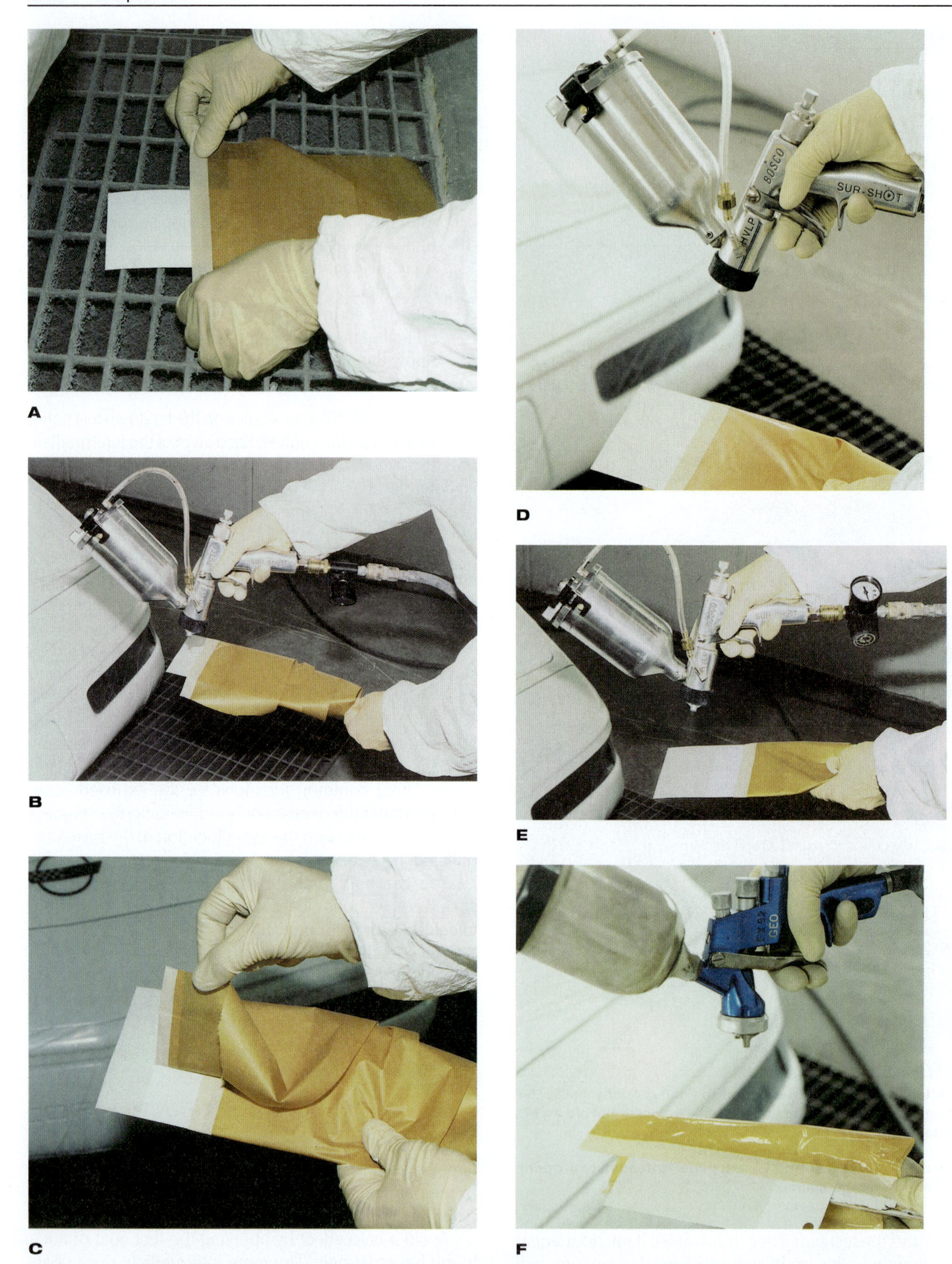

Figure 27-22 Study the basic steps for masking a let-down test panel and spraying it with the mica mid-coats and clearcoats. (A) Evenly overlap four pieces of masking paper and tape to cover all but the top of the color test panel. (B) Spray the top exposed portion of the test panel with one normal coat of pearl. (C) Remove the top masking paper and tape to expose another section of the test panel. (D) Spray the exposed area of the test panel with another coat of pearl. (E) Repeat this on the rest of the test panel. (F) Mask vertically down the middle of the test panel. Then, apply full wet coats of clear to the exposed portion of the let-down panel.

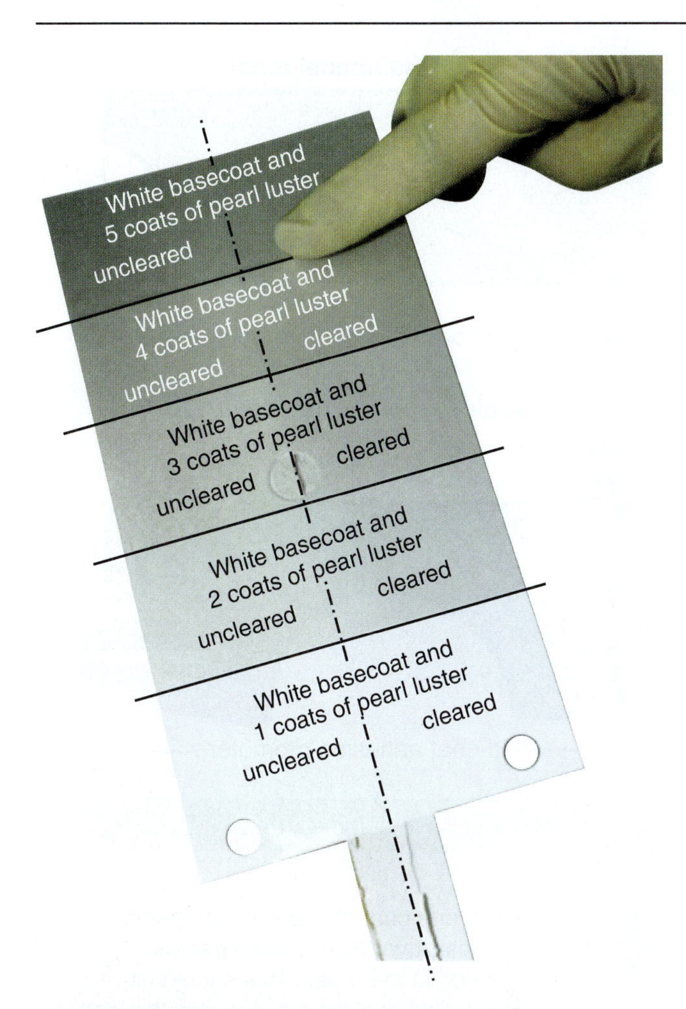

Figure 27–23 Having a let-down test panel is like having several pearl paint color chips. The area or "chip" on the panel that best matches the color on the vehicle will tell you how many coats of pearl or mica should be applied. For example, if the area on the let-down test panel with three mid-coats of pearl or mica best matches the existing finish, then use three mid-coats of pearl or mica when painting the car.

This is one manufacturer's method for a spot or partial repair on a three-stage paint system:

1. Apply adhesion promoter to all unsealed panels. Adhesion promoter should extend beyond the repair area.
2. Apply primer and tinted sealer to the area over the body filler.
3. Apply two or more coats of basecoat to the area to provide full hiding. Extend each coat slightly beyond the previous one, allowing time to dry between coats (Figure 27–24).
4. Check the let-down test panel for the total number of intermediate mica coats needed to match the OEM finish. Apply the intermediate coats to the repair area, extending each coat beyond the last. Allow adequate flash time between coats.
5. Apply two coats of clear over the entire panel. The clearcoat may have to be blended into the sail panel.

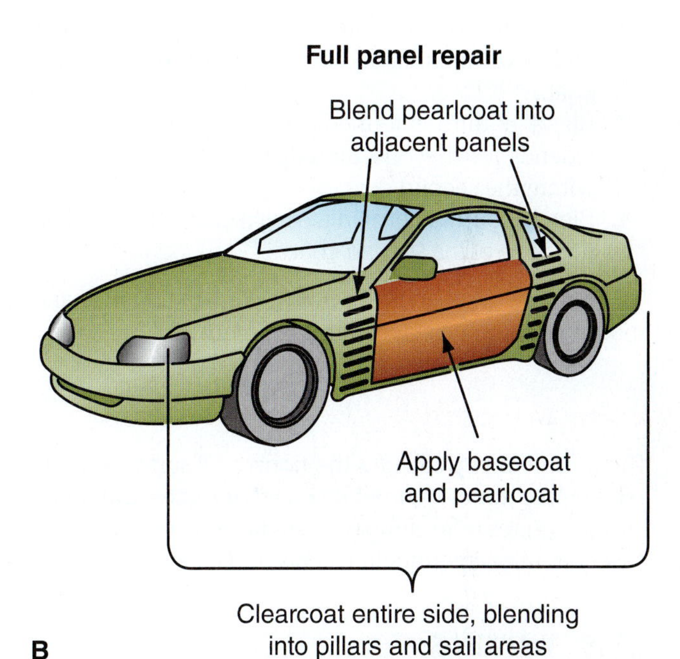

Figure 27–24 Note how to do spot and panel repairs with a three-stage finish. Three-stage pearl paints are very difficult to match. (A) With spot repair of three-stage pearl, you must blend a pearlcoat over an area larger than the repair, then apply clearcoat over the whole panel. (B) For panel repair with a multistage pearl finish, you will usually have to blend pearl into adjacent panels, then apply clearcoat over the whole side of the vehicle to get a good color match.

Matching Three-Stage Panel Repairs

These are typical steps for a panel repair with a multistage mica or pearl finish:

1. Apply primer, sealer, and adhesion promoter as needed to the area.
2. Apply two or more coats of basecoat to the area to provide full hiding. Extend each coat slightly

beyond the previous one, allowing time to dry between coats.

3. Check the let-down test panel for the total number of intermediate mica coats needed to match the OEM finish. Extend each coat beyond the previous one, with only the last coat extending into the adjacent panel. Allow adequate flash time between coats.
4. Apply two coats of clear to both doors.

Blending Mica Coats

This is a typical mica mid-coat blending procedure for a three-stage paint:

1. After allowing the base color to flash properly, blend the first mica or pearl mid-coat over the primer repair area.
2. Blend the second intermediate mica coat a few inches or millimeters beyond the edge of the first coat.
3. Blend a third intermediate mica coat so that it extends just beyond the edge of the first coat, but within the second coat.
4. Blend a fourth intermediate mica coat to just beyond the edge of the second coat.

Compare what must be done with multipanel and spot repair of three-stage paints in Figure 27–25.

Zone Concept

The **zone concept** divides the horizontal surfaces of the vehicle into zones defined by character lines and moldings. It requires refinishing of an entire zone or zones with basecoat, mica intermediate coats, and clearcoats.

27.6 TINTING

Tinting involves altering a paint color by adding small amounts of color pigments to better match an existing paint color. Tinting may be one of the least understood procedures in finish color matching.

There are six basic reasons for tinting a paint color:

1. To alter color when alternate color chips in the paint color directory do not match the vehicle paint color
2. To adjust color on an aged or weathered finish
3. To make a color for which there is no available formula or paint codes
4. To adjust the shade to match a nonstandard color from the manufacturer
5. When your spray-out test panel does not match the paint color on the vehicle
6. If your first coat of color does not match the existing paint as well as it should

To do color matching, the refinisher must be able to see the qualities of color. It is important not only to see the main color of the paint, but also the overtones within that color. A good painter must see the shades of darkness or lightness and the richness or fullness of the color.

Typical multipanel repair

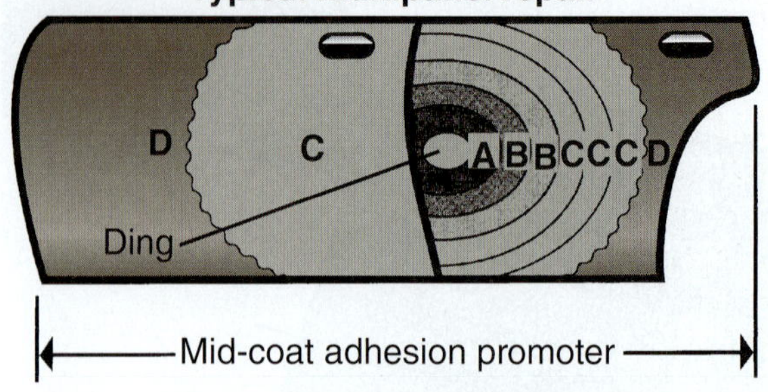

Typical spot/partial repair

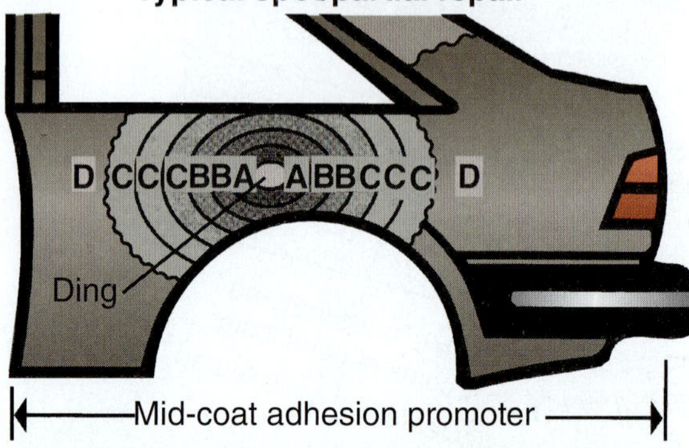

A. Primer B. Base color C. Pearl D. Clear

1. Apply one coat of mid-coat adhesion promoter over all unsealed panels involved in the repair. This should be applied beyond the areas where the last coat of clear will go in order to help the material adhere to the slick OEM surface.

2. Apply the basecoat to full hiding, extending each coat slightly beyond the previous one. Allow adequate flash time between coats.

3. Consult the let-down panel to decide how many coats of pearl are needed for a match.

4. Apply the number of pearl coats indicated by the let-down panel. Be sure to fully cover the base color before tapering the blend edge of each coat of pearl. Blend each coat beyond the next and allow sufficient flash time between coats.

5. Clear all of the panels with one of the recommended clearcoats.

Courtesy of DuPont Automotive Finishes

Figure 27-25 Compare the procedures for multipanel and spot repairs with three-stage paint.

Use three angles to determine whether a color adjustment is necessary:

1. Head-on—viewing the repaired area from an angle that is perpendicular to the vehicle
2. Near specular—viewing the repaired area from an angle just past the reflection of the light source
3. Side tone—viewing the repaired area at an angle of less than 45 degrees

The color of a repaired area should be the same as the rest of the vehicle. If not, correct it by adding tinting colorant or toner until it is the same when viewed from all three angles.

Why a Color Mismatch?

Tinting should only be used as a last resort. When the paint does not match the vehicle, you must determine whether the:

- Paint code was properly identified
- Proper paint code was used
- Spray-out test panel was made properly

You also must determine whether the spray-out test panel was checked:

- Against a clean vehicle
- In the proper light
- On both face and side tone views

Finally, check the paint company variance chips for a formula that may obtain a better match. If there is not a variance formula, determine whether the color is close enough to blend. If not, the paint must be tinted to create a blendable match.

A paint mismatch does not automatically mean tinting is needed. Tinting should only be done to move the paint close enough for blending. Do not try to tint to a perfect match. Blending the paint out over a larger area takes care of any final color variations.

It is important to remember that the color might not be exactly right, because all automotive finishes gradually change color over time. Some colors fade lighter; others go darker.

Shading Adjustments

Here is a list of shading adjustments to help avoid the need for tinting the paint:

To make color appear darker, proceed as follows:

1. Open the fluid valve more.
2. Reduce the size of the fan pattern.
3. Decrease the gun distance.
4. Slow down the gun stroke.
5. Allow less flash time.

To make a color appear lighter, follow these steps:

1. Close the fluid valve slightly.
2. Increase the size of the fan pattern.
3. Increase the gun distance.
4. Speed up the gun stroke.
5. Increase flash time.

Tinting Questions

If your paint color does not match the original finish, check the following possible reasons for the mismatch before deciding to tint the paint:

- The original finish may have faded. Check the paint on unexposed areas such as door jambs or under the trunk lid or hood to determine whether the finish has faded. If this is the case, restore the paint's luster by compounding the old finish well beyond the repair area.
- Has the clearcoat oxidized and turned white? If using a clearcoat, remember that compounding the clearcoat will make the paint appear darker. Cleaning and compounding will remove any whitish, oxidized layer on the finish to deepen the color.
- Was the wrong color used? Check the auto manufacturer's code and the paint company's stock number for the color being used to make sure that it is the right one.
- Was the paint poorly mixed before spraying? The pigment and/or flakes may not have been mixed thoroughly. Leaving pigment, flake, or pearl in the bottom of the can can cause a mismatch, so agitate thoroughly.
- Has the amount of solvent or reducer been measured carefully? Overreducing will lighten or desaturate a color. Remember that it is easy to add more solvent, but it cannot be taken out.
- When spraying a color test panel, did you do everything properly? Did you mix the paint formula precisely? Did you use normal spray gun adjustments and spraying methods? Did you allow proper flash times between coats?
- If using a pearl paint, did you make an accurate letdown test panel and use it properly? Three-stage pearl paints are the most difficult to match. Did you use the correct number of pearl coats and clearcoats? Each will slightly change the color of a three-stage pearl paint.
- Did you clearcoat your test panel before comparing it to the vehicle's color? If testing for a basecoat/clearcoat finish, color matching cannot be judged until the clear is applied to the basecoat.
- Did you vary your spraying technique to help match the color?

Plotting Color

Plotting color is the process of identifying paint color in a graphic way based on value, hue, and chroma. Although seldom used, this knowledge will help you become familiar with the refinishing evaluation process.

Plotting is *not* an exact science. It only has meaning when it is compared to the color on the vehicle. Plotting

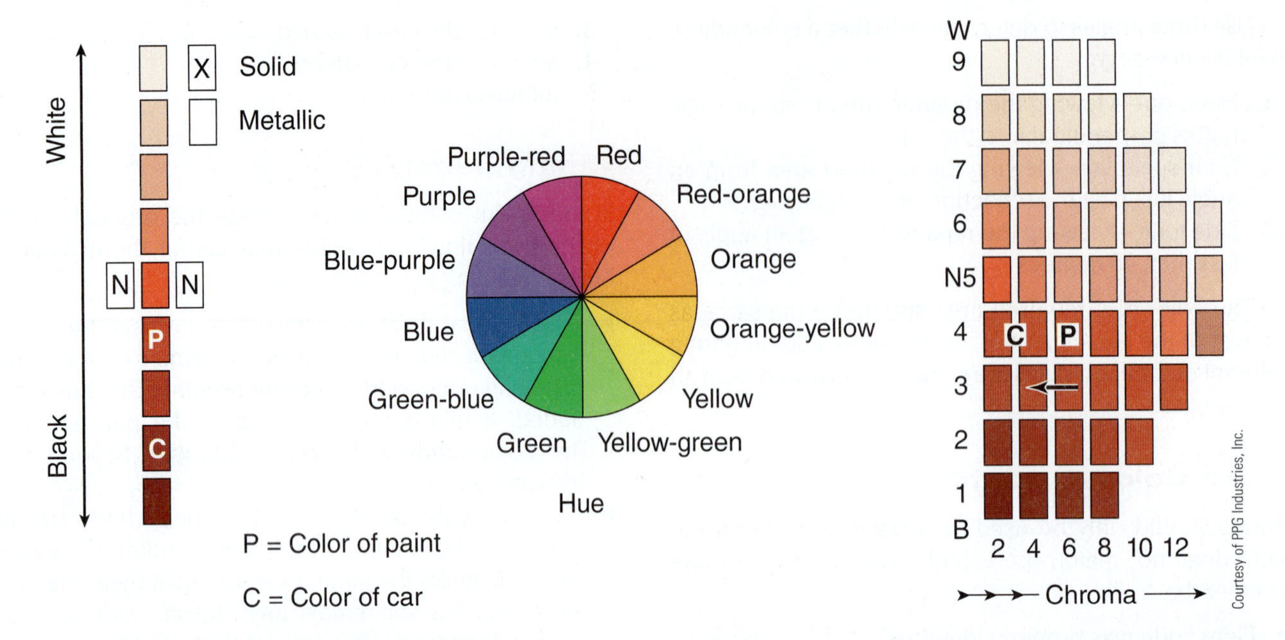

Figure 27-26 When plotting color, the letter C stands for the car or vehicle and P stands for paint. Draw an arrow from P to C to show the proper direction of tint. First, you would tint the paint to adjust its value. You would adjust the hue second by adding color to move the hue around the wheel. Chroma is rarely adjusted.

will help you recognize changes in value, hue, and chroma and to adjust color as needed.

Evaluate colors in this order:

1. Value (lightness or darkness)
2. Hue (color, cast, or tint)
3. Chroma (saturation, richness, intensity, muddiness, or grayness)

Develop a tinting plan that makes sense for all three dimensions of the color. In most cases, only one dimension of a color will require adjustment. Please refer to the plotting chart in Figure 27–26.

The procedure for plotting and tinting a color is as follows:

1. Adjust value first. Add white to increase value. Add black or color pigments to decrease value. Remember that changing value affects chroma and changing hue affects value.
2. Adjust hue second. Adding color pigments moves the directed color around the color wheel. Adding color pigments affects chroma.
3. Chroma is changed last but is seldom adjusted. Increasing chroma decreases value. Use gray to decrease chroma with minimal effect on value.

Diagrams that illustrate the sequence of steps for tinting both solid and metallic colors are given in Figure 27–27.

When working with colors that are between the four main colors on the color wheel, a painter should remember that:

- Orange, bronze, and gold colors can move toward red or yellow.
- Maroon and purple can move toward blue or red.

Figure 27-27 The paint mixing room will have color for tinting so your paint will match the existing paint on the vehicle.

- Lime can move toward green or yellow.
- Aqua and turquoise can move toward green or blue.

When adjusting hue, a paint can be moved to either side of the dominant color, or toward one of two dominant colors if the paint is between major colors. Always use the terms *bluer, redder, yellower,* or *greener* to describe how hue is to be moved.

Through various application techniques, color cast, depth, and grayness can be adjusted so that the paint technician can achieve a good match (Table 27–1).

Before adjusting begins, the color must be checked to see whether the finish sprayed on the test panel is brighter or grayer than the original. The sprayed portion must always be allowed to dry before any adjustments are made.

TABLE 27–1 HOW COLORS ARE DESCRIBED

Color depth (lighter or darker)	1. Direct look (panel to panel) 2. Side angle look (panel to panel)
Color cast	1. Redder 2. Bluer 3. Greener 4. Yellower
Color cleanliness	1. Grayer (dirtier or more muddy) 2. Brighter (cleaner appearance)

Adjustments for lightness or darkness rely primarily on shop conditions, spraying techniques, and solvent usage (Table 27–2). Other variables include the amount of paint applied, the air pressure at the spray gun, and the amount of color added to the mix.

Once the lightness or darkness has been adjusted, tinting might be required to get the right cast or hue. Each color can only vary in cast in two directions.

- Colors that are either greener or redder in cast include:
  Blues Purples
  Yellows Beiges
  Golds Browns
- Colors yellower or bluer in cast are:
  Greens Blacks
  Maroons Grays or silvers
  Whites
- Colors yellower or redder in cast are:
  Bronzes Reds
  Oranges
- Colors bluer or greener in cast include:
  Aquas Turquoises

Tinting manuals, available from paint manufacturers, give instructions for using their tinting products. They can help you decide which tint color to use.

Once the color necessary to correctly adjust the hue or cast is determined (refer to Table 27–3), you must add only a drop or two at a time. After you have added a drop or two of the tinting color, test the color again on a test panel. It is very easy to add too much toner color, requiring you to start over with a new mix of paint (Figure 27–28).

The tinting manual describes when and how to use each tinting toner. The various colors found in one paint manufacturer's tinting manual are shown in Figure 27–29.

After the right amount of tinting material has been added, the paint must be thoroughly mixed. The spray gun must be triggered to clear internal passages of remaining untinted paint. Then another test panel can be sprayed, allowed to dry, and checked against the original color.

Hints on Color Tinting

Here are some additional tips that will prove helpful when tinting a color:

- Check the color in daylight as well as artificial light. It might not look the same in both lights. When a refinish color matches in one light but not another, it often indicates that the same pigments were not used in the refinish material as in the original finish.
- Be sure the panel to be matched is thoroughly cleaned and compounded so the true color is clearly visible.
- Do all tinting systematically by keeping a list of tinting colors used and a record of the amount used.
- Do not use two tints at one time. Tint the color using the right sequence: Change cast first, then depth, and finally grayness as needed.
- Determine what the color problem is and select the proper tinting colors. Adjust the color to make the hue redder, greener, bluer, or yellower.
- Never use premixed topcoat colors for tinting. Use only pure, base colorants.

TABLE 27–2 ADJUSTING COLOR DEPTH

Variable	To Make Colors	
	Lighter	**Darker**
Shop Conditions		
1. Temperature	1. Increase	1. Decrease
2. Humidity	2. Decrease	2. Increase
3. Ventilation	3. Increase	3. Decrease
Spraying Techniques		
1. Gun distance	1. Increase distance	1. Decrease distance
2. Gun speed	2. Increase speed	2. Decrease speed
3. Flash time between coats	3. Allow more flash time	3. Allow less flash time
4. Mist coat	4. Will not lighten color	4. Wetter mist coat
Solvent Usage		
1. Type solvent	1. Use faster evaporator solvent	1. Use slower evaporator solvent
2. Reduction of color	2. Increase amount of solvent	2. Decrease amount of solvent
3. Use of retarder	3. Do not use retarder	3. Add retarder to solvent

TABLE 27–3 METHOD OF CHANGING CASTS

Color	Add		Cast
Blue	Green	to kill	Red
Blue	Red	to kill	Green
Green	Yellow	to kill	Blue
Green	Blue	to kill	Yellow
Red	Yellow	to kill	Blue
Red	Blue	to kill	Yellow
Gold	Yellow	to kill	Red
Gold	Red	to kill	Yellow
Maroon	Yellow	to kill	Blue
Maroon	Blue	to kill	Yellow
Bronze	Yellow	to kill	Red
Bronze	Red	to kill	Yellow
Orange	Yellow	to kill	Red
Orange	Red	to kill	Yellow
Yellow	Green	to kill	Red
Yellow	Red	to kill	Green
White	White	to kill	Blue
White	White	to kill	Yellow
Beige	Green	to kill	Red
Beige	Red	to kill	Green
Purple	Blue	to kill	Red
Purple	Red	to kill	Blue
Aqua	Blue	to kill	Green
Aqua	Green	to kill	Blue

- To understand what the overcasts are to a tinting color, put a few drops on a quart lid with a few drops of white, then mix these two and make a finger smear on the lid. This will enable you to determine what the overcast of that tinting color is.
- The original formula for the paint color is a help in tinting. The paint formula shows the original base colors so you can determine which color pigments have faded out. The faded pigments have to be toned down when mixing the refinish material.
- Be sure to mix all tinting colors thoroughly before using.
- Add tinting colors in small amounts, because it is very easy to overtint. Keep in mind that more color can be added, but it cannot be taken out.
- Do not tint an entire can of paint at one time. Make progressive tryouts with small samples until a color match is achieved.
- Be conservative when tinting near the limits of the color range. Correct the most noticeable color differences first.
- Use caution when adding white to metallics or pearls, and always use low-strength whites.

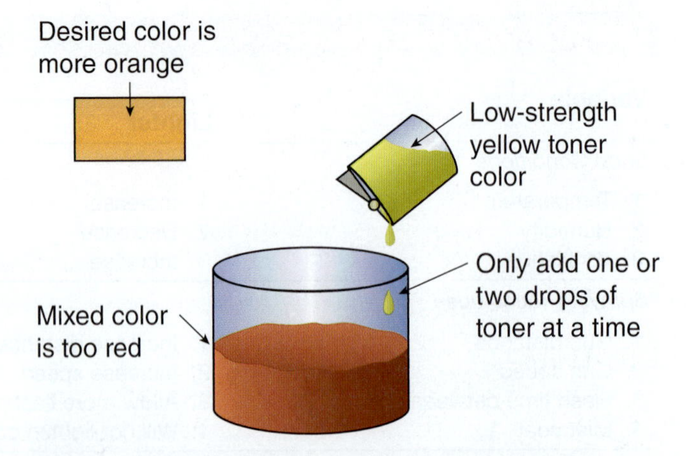

Figure 27–28 **Tinting is done by adding a very small amount, often just a few drops, of a low-strength toner color to help better match the paint. For example, if your color is red and you need it to be orange, you would add yellow.**

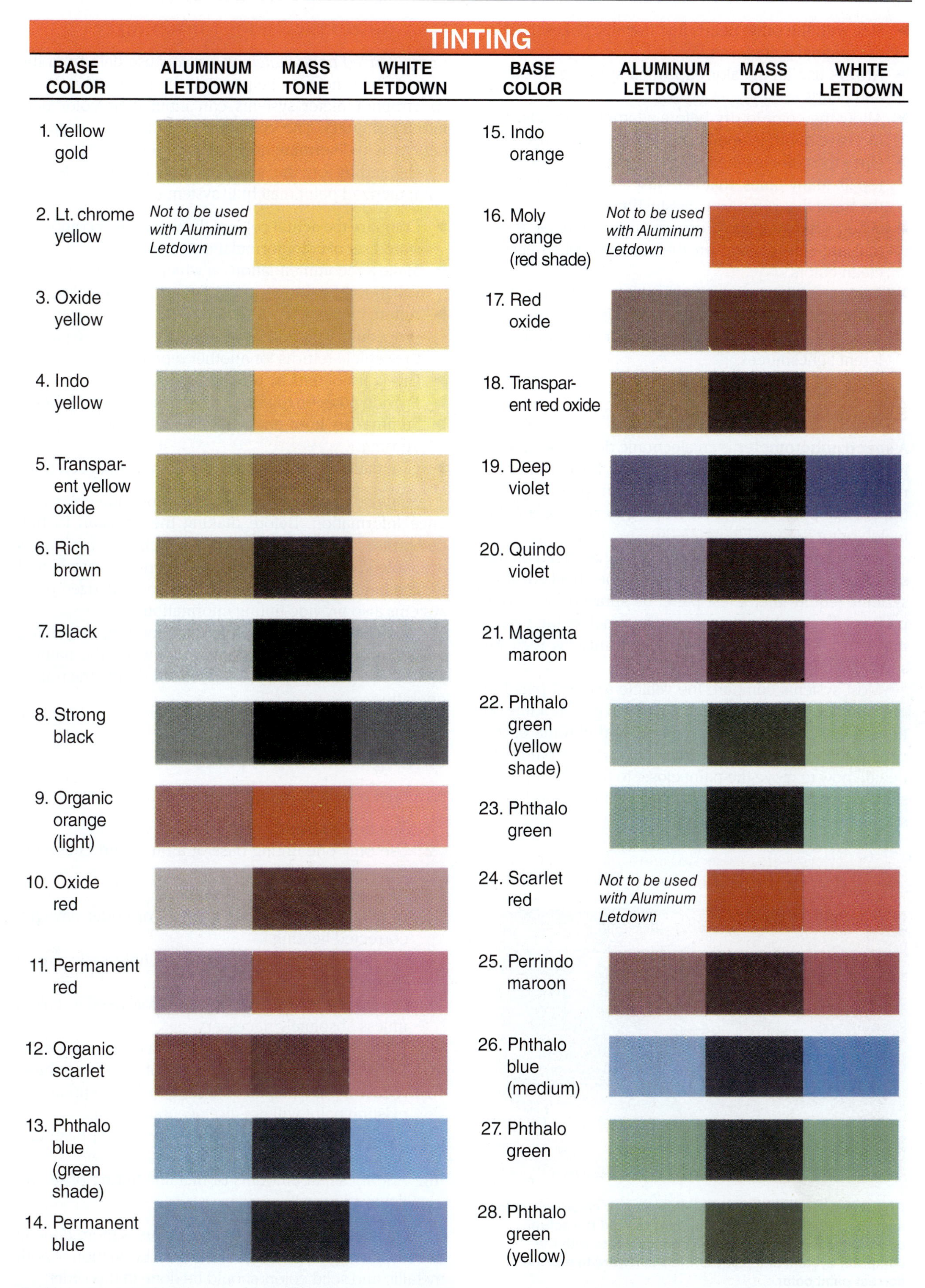

Figure 27-29 A color manual will provide tinting instructions and note different toners for altering paint color.

- Stay with the same pearls and metallic flakes used in the original color formula.
- Always use an agitator cup when spraying metallics and pearls.
- Allow the color to dry before attempting to adjust it. Increase booth temperature or use a heat lamp.
- Once the color is tinted "close enough," complete the repair. Many times, that last "just a little bit closer" is the thing that ruins a successful tinting.
- When changing cast, use partner colors or those that are side by side on the color wheel to ensure clean colors.
- Avoid using opposing colors, which results in a dirty or grayed color. Opposing colors also contribute to metamerisms, so the color will not match under different light sources.

Spectrophotometer

A **spectrophotometer** is an electronic device for analyzing the color of the paint on a vehicle. It electronically reads the color frequencies in the finish to quickly find the correct paint formula for helping to achieve a color match or for tinting (Figure 27–30).

The spectrophotometer wand or box is placed on the surface (either vehicle or test panel) to be checked. Most systems require that a test panel be sprayed for a comparison. The multiangle spectrophotometers take readings at 25, 45, and 75 degrees. Each angle is read for several variables.

Most systems compare the vehicle and test panel to one another. The refinish technician will get a reading on the relative lightness/darkness, hue, and chroma of the vehicle to the panel checked. It is still up to the painter to decide how to move the paint closer to the vehicle. Decisions on which tint and how much will be added must still be made using human judgment.

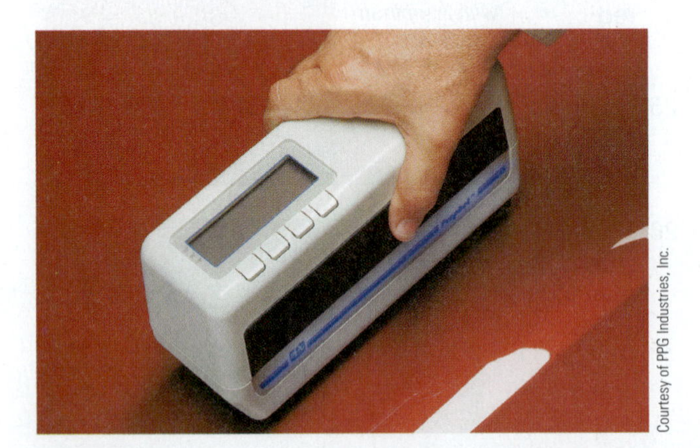

Courtesy of PPG Industries, Inc.

Figure 27-30 **A spectrophotometer, or electronic color analyzer, uses electronic technology to read the actual color of the vehicle. It can then communicate with the computer system and paint formula software to mix the correct paint color.**

Computerized Paint Matching

Computerized paint matching systems use data from the spectrophotometer to help match the paint color. Many spectrophotometer systems can input their color data into a computer. The computer can then use its stored data to help determine how to mix or tint the paint.

Depending on the sophistication of the system, a computerized paint matching system may be able to:

- Compare the actual color of the vehicle to a computer-stored set of color formulations.
- Make a recommendation on which tint in the formula will move the sample panel closer to the vehicle color.
- Automatically keep a record of the mixing or tinting procedure. This will let you quickly match the paint if the vehicle returns for another repair.
- Give a list of tints by number and name.
- Provide notes on tint strength or hiding characteristics.
- Summarize how each tint affects value, hue, and chroma.
- Give cautionary notes, if needed, for using each tint.

Some computerized paint systems provide color variance information. Before making the decision to tint, determine whether a color variance chip or formula is available. These may provide a blendable match and reduce or eliminate the need to tint. Computerized paint systems also provide tinting information.

This computer matching information can be printed out with a copy of the paint formula when the paint is mixed. The printout can be referenced during the tinting operation.

Tinting Dos and Don'ts

As a brief review of tinting, remember these ten points:

1. Tint only to blend.
2. Use only one tinting base at a time, and check the color after every "hit."
3. Always tint within the formula.
4. Evaluate color match in daylight or under daylight-corrected lighting.
5. Do *not* use a black tinting base unless absolutely necessary.
6. Do *not* use a white tinting base for metallic colors unless absolutely necessary.
7. Evaluate the color match against a clean vehicle.
8. Do *not* use the "dab method" of color matching. A dab of paint placed on a surface and compared to existing paint will not give accurate results.
9. Adding small amounts of metallic tinting base can darken side tone.
10. Adding small amounts of metallic tinting base can lighten a metallic color.

Evaluating and tinting metallic colors is more difficult than evaluating and tinting solid colors. Evaluation of both metallic and solid colors should be done in this order:

1. Verify the formula.
2. Plot the color on the plotting chart.
3. Compare the test panel to the vehicle.
4. Check and adjust value.
5. Check and adjust hue.
6. Check and adjust chroma.

Using kill charts and adding tints not in the paint formula will move the color, but not around the outside of the color wheel. Adding tints suggested by a kill chart will move the paint more directly toward the gray center of the color wheel, affecting both hue and chroma.

There may be situations where this is desirable, but as a rule of thumb do not use kill charts to change hue. The proper tint for changing hue is already in the formula. Adding tints that are not in the formula may also cause problems with metamerism.

Adding paint instead of tints will add the dominant tint and all the other tints in that paint formula. Again, use only tint bases and do *not* add paint.

Some paint companies produce metallic tinting bases designed to correct a specific problem, usually having to do with changes in side tone. Use these products according to the manufacturer's recommendations. Using the paint formula and tinting guide, select the tinting base that will move the paint in the right direction.

When tinting, a painter should remember to:

- Use only half the can of paint.
- Use only tint bases; never add paint.
- Use only tints that are in the paint formula.
- Add tint bases in small amounts, and check the paint following each addition.
- Keep records of each tint base and the amount added, which is useful if more paint needs to be mixed.

A chart summarizing the general tinting rules is given in Table 27–4. Read through it carefully.

27.7 CUSTOM PAINTING

Custom painting involves using multiple colors, metal flake paints, multilayer masking, and special spraying techniques to produce a personalized paint job. Multicolored stripes, flames, murals, landscapes, names, and other artwork can be added to the finish.

Some custom paint jobs require the use of a small airbrush. When spraying with an airbrush, you should overreduce or overthin the paint material about 20 percent. This allows the paint to flow smoothly through the small orifices or passages in the airbrush.

A double-action airbrush has a small spring-loaded lever that acts as a throttle. The more you pull back on the fluid lever, the more paint sprays out of the airbrush nozzle.

Complex images require you to have special artistic talent. If you cannot paint an attractive image on paper, you will not be able to do it on a vehicle (Figure 27–31). However, simple custom paint schemes can be done by even nonartistic painters.

Custom painting requires considerable talent, skill, and knowledge. You need to plan the custom job carefully. This will let you determine how to mask and spray or apply each color. Custom painters are good at using airbrushes, striping tools, and masking materials. See Figure 27–32.

Before custom painting, make sure the base finish is in good condition. You do not want to waste your time trying to paint over a weathered or problem finish. Wet sand and clean the area to be custom painted. Use surface preparation methods detailed in other chapters.

Custom masks can be made by drawing designs on thin poster board and then cutting them out. A design is taped onto the vehicle and spray painted. Using an airbrush and translucent paint, various attractive effects can be produced.

Card masking involves using a simple masking pattern to produce a custom paint effect. Usually, an airbrush is used to mist the paint over the edge of the masking card. The card can be moved to repeat the pattern and produce a wide range of paint effects (Figure 27–33).

Lace painting involves spraying through lace fabric to produce a custom pattern in the paint. Various lace designs can be purchased at fabric stores. The cloth pattern will allow the paint to pass through the holes in the lace, but mask it in other areas (Figure 27–34).

A marble effect can be made by forcing crumpled plastic against a freshly painted stripe or area. You might want to spray the area with two colors to reveal a desired color under the wet coat. When the plastic is lifted off, it will remove the top layer of paint in random areas, creating a marble-like effect (Figure 27–35).

Spider webbing is done by forcing paint through an airbrush in a very thin, fibrous spray. Air pressure from the gun can also be used to spread and smear the wet paint to alter the special paint effect (Figure 27–36).

Painted flames are a custom painting technique often used on hot rods and race cars. First, fine line masking tape is used to form the outline of the flames. Then the area around this shape is covered with masking paper or plastic.

First, the base color for the flames is applied. A second translucent color is blended inside the flame area. A third color may be used to darken the outer edges and center area of the flames. The flames are finally wet sanded and polished when dry (Figure 27–37).

Painted lettering involves masking off letters on the finish and spraying or brushing them with a different color. This can be time-consuming but is sometimes requested by customers.

When doing custom paint work, do not "bite off more than you can chew." Start out simple with jobs of minor complexity. As you learn to do custom work successfully, you can progress to more complex paint work. Experience is the best teacher with custom paint work. To be a

TABLE 27–4 GENERAL RULES FOR TINTING

Base Color	General Usage
1. Yellow Gold	Use for reddish-yellow tint in solids and metallics. Lightens flop in metallics.
2. Lt. Chrome Yellow	Use in substantial amounts to give bright greenish-yellow hue. Not used in metallics.
3. Oxide Yellow	Use for reddish-yellow tint in all colors. Lightens flop in metallics.
4. Indo Yellow	Use for greenish-yellow tint in all colors. Lightens flop in metallics.
5. Transparent Yellow Oxide	Use for yellow tint in solids and metallics. Lightens flop in metallics.
6. Rich Brown	Use to give clean golden tint to metallics and clean beige to solid colors.
7. Black	Use where small amounts of black are needed. Has brown or yellow undertone.
8. Strong Black	Use where a large amount of black is needed. Has brown or yellow undertone.
9. Organic Orange (Light)	Use as reddish-orange tint in pastel solid and metallic colors.
10. Oxide Red	Use for clean red tint in beige colors. Not commonly used in metallics.
11. Permanent Red	Use for blue-red tint in pastel solids and metallics. Lightens flop in metallics.
12. Organic Scarlet	Use as red in pastel solids and metallics.
13. Phthalo Blue (Green Shade)	Use as blue tint in pastel and metallics. Has a very green-blue tint.
14. Permanent Blue	Use as blue tint in pastel solids and metallics. Has a very red-blue tint.
15. Indo Orange	Use as orange tint in pastel solids and metallics.
16. Moly Orange (Red Shade)	Use as reddish-orange tint in solids only.
17. Red Oxide	Use for clean red tone in beige solids, not commonly used in metallics.
18. Transparent Red Oxide	Use to give reddish-gold tint in metallics. Also beige tint in solids.
19. Deep Violet	Use as purple tint in pastel solids and metallics. Also use in blues and grays for violet tones.
20. Quindo Violet	Use as blue-red tint in pastel solids and metallics.
21. Magenta Maroon	Use as blue-red tint in pastel solids and metallics.
22. Phthalo Green (Yellow Shade)	Use as green tint in pastel solids and metallics. Has yellow-green tint.
23. Phthalo Green	Use as green tint in all colors.
24. Scarlet Red	Use in bright red solid colors.
25. Perrindo Maroon	Use as rich brown-maroon tint in solids and metallics.
26. Phthalo Blue (Medium)	Use as blue tint in pastels and metallics. Has a clean red-blue tint.
27. Phthalo	Use as green tint in all colors.
28. Phthalo Green (Yellow)	Use as yellow-green tint in pastel solids and metallics.

1. Base colors have two tones: mass tone (as appears in can) and tint tone (small amount mixed with white or aluminum).
2. Colors darken as they dry. Always match on the light side.
3. The same color arrived at with two different formulations (using different pigments) might vary in color under different lights and might weather differently.
4. Metallic colors have varying degrees of flop. Colors with a deep rich flop contain coarse aluminum and tinting colors with greater transparency of depth.
    a. To maintain a rich flop, use coarse aluminum and, if required, tinting colors with greater transparency.
    b. When the flop requires a grayer appearance, use finer aluminum.

good custom painter, you have to be a true artist with the "car as your canvas."

A good idea is to practice techniques on old parts using leftover paint. Then, you can learn from your mistakes and successes without taking a chance with a customer's vehicle (Figure 27–38).

Spatter Finishes

The interior of luggage compartments may be painted with a special spatter finish. The material is often water-reducible and can be applied in one heavy or two medium coats. Do not shake spatter finish on a paint

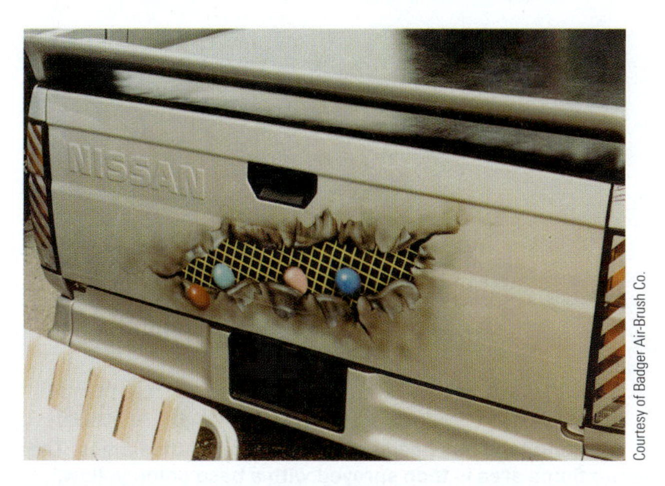

Figure 27-31 Experience and artistic abilities are needed to do custom paint work. It is best to learn on small projects first, such as motorcycle gas tanks. Then, with experience, you can do custom paint work on cars and trucks.

Figure 27-33 An airbrush is used to mist paint over the edge of a masking card. The card is then moved to repeat the pattern for a custom paint effect.

A Carefully cut poster board or masking tape can be used to form a mask over the area to be custom painted. An airbrush and translucent paints are often used to produce a colorful effect inside the masked area.

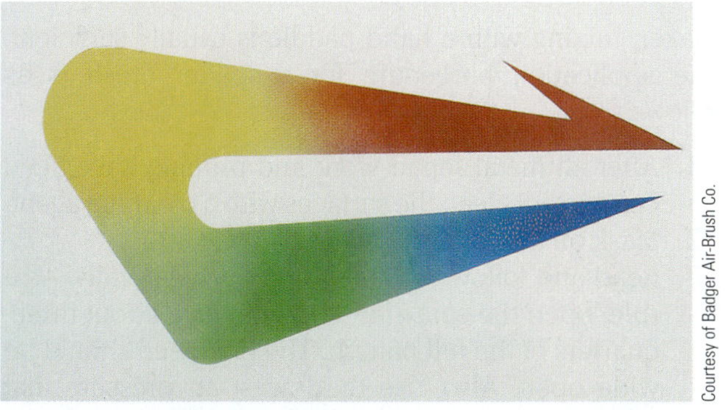

B After removing the custom mask, a colorful, one-of-a-kind image remains on the unmasked area.

Figure 27-32 Note how a basic custom mask can be used to produce a simple but good-looking custom design.

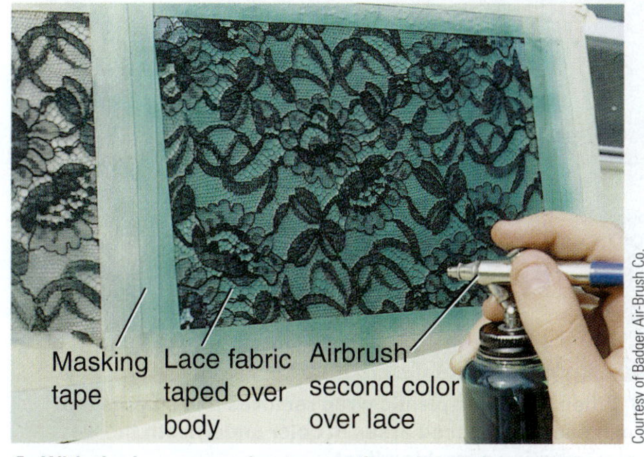

A With the lace spread out smooth and taped over the painted body, spray the second color over the lace.

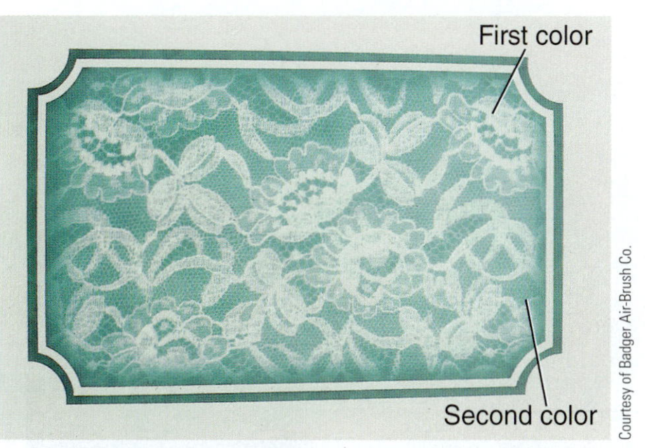

B When the lace is removed, the second color will be overlayed with the lace pattern.

Figure 27-34 Lace painting involves spraying through lace fabric to produce a custom pattern in paint. Various lace designs can be purchased at fabric stores.

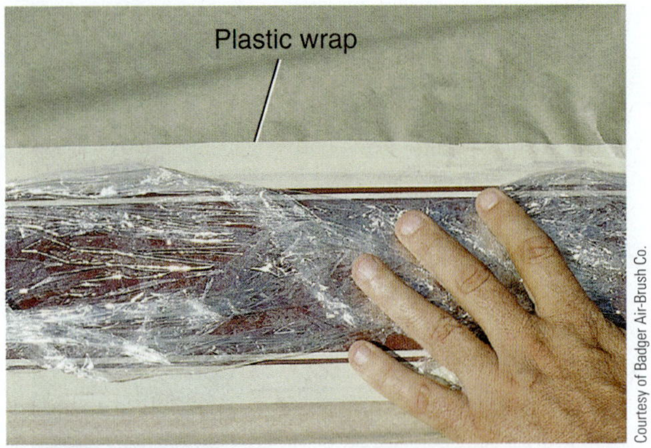

A **After spraying a different color onto the body, crumpled plastic wrap is forced against the wet coat of paint and then removed.**

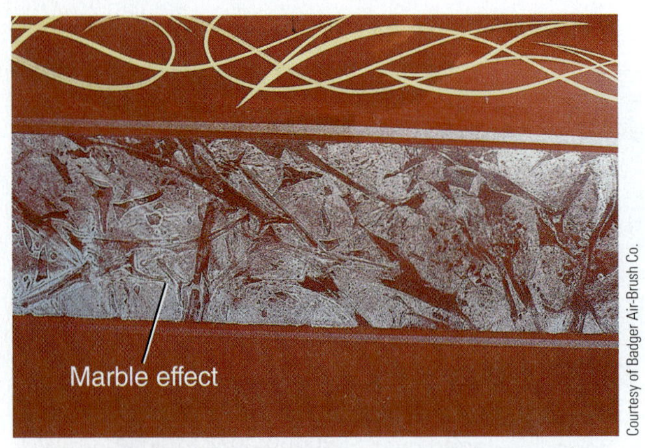

B **When lifted, the plastic wrap removes and smears the fresh paint to exposed areas of color underneath the wet paint, creating a marble effect.**

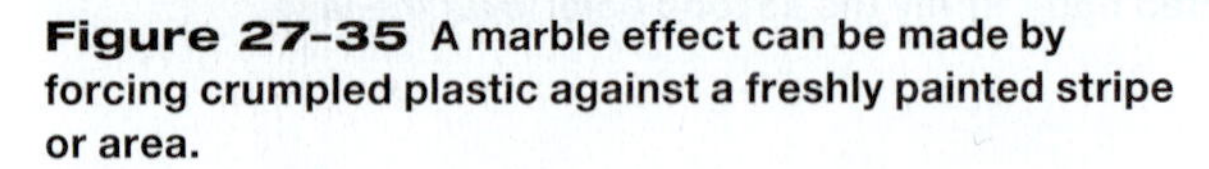

Figure 27-35 **A marble effect can be made by forcing crumpled plastic against a freshly painted stripe or area.**

A **Masking tape has been cut to the shape of the flames. The entire flame area is then sprayed with a base color, yellow, for example. An airbrush is used to feather spray translucent red paint around the perimeter of the flame.**

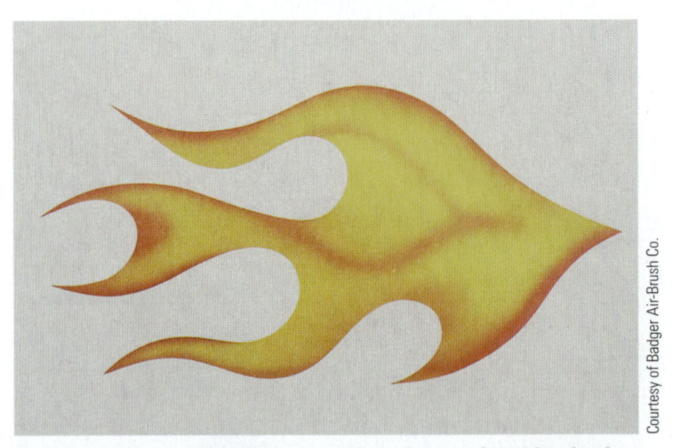

B **When the masking is removed, a sharp edge of color is formed on the outside edge of the flame design. However, a soft edge of color blends into the center of the flame to represent changes in flame temperature.**

Figure 27-37 **Painted flames are a custom painting technique often used on "hot rods."**

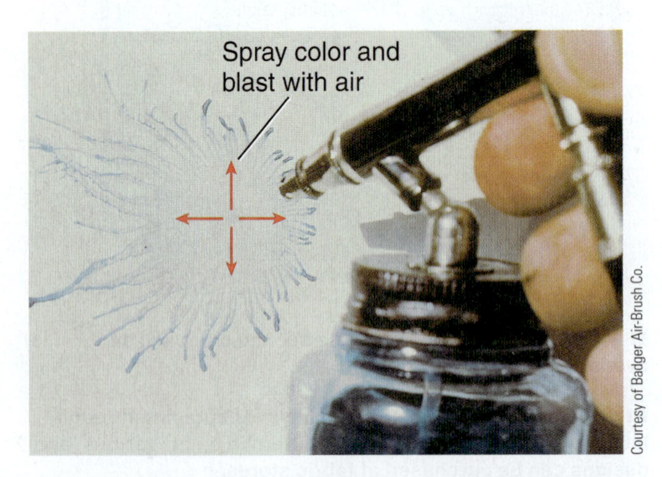

Figure 27-36 **Spider webbing is done by forcing paint through the airbrush in a very thin, fibrous spray.**

shaker; mixing with a hand paddle is usually sufficient. The application procedure for a spatter finish is as follows:

1. After all metal repair work and priming have been completed, clean the surfaces with a cleaning agent.
2. Mask off the area, as required.
3. Read and follow the label directions carefully. As a rule, open the spray fan nozzle to give about three-quarters of the full pattern. The fluid feed should be wide open. Also, use the lowest air pressure that causes the desired spray pattern:
   - For smaller spatters, increase the air pressure.
   - For larger spatters, reduce the air pressure.
4. Apply the coating. If two coats are needed, allow for the recommended flash time.

Figure 27-38 **Note the more detailed steps for producing a custom paint job on a motorcycle fairing. (A) The fairing has already been custom masked and painted with the main colors of black, white, and yellow. (B) Masking tape is removed to expose the checkered flag pattern. A full-size spray gun was used to apply these colors. (C) A small airbrush is used to feather color along the design's edges. (D) The masking tape is removed to form a ripped hole look in the flag. Note how an airbrush has been used around the perimeter of the "hole." (E) Finesse sanding is often needed to smooth edges where masking tape has been removed. Fine sand carefully so you do not cut through the base colors. (F) Clearcoats are finally applied to bring out the gloss of colorcoats used on the custom paint job.**

SUMMARY

1. Color matching involves the steps needed to make a new finish match the existing finish on a vehicle.
2. Color is caused by how objects reflect light at different frequencies or wavelengths into our eyes.
3. Because the vehicle will be seen in sunlight, you should always use sunlight or daylight-corrected lighting when making color evaluations.
4. To minimize the confusion when painting, color matching should be based on the three dimensions of color: value, hue, and chroma.

5. A color directory is a book that provides color chips and other paint-related information for most makes and models of vehicles.
6. A spectrophotometer is an electronic device for analyzing the color of paint on a vehicle.
7. Computerized paint matching systems use data from a spectrophotometer to analyze how to match the paint color.
8. Tinting involves altering the paint color slightly to better match the new finish with the old finish.
9. A dry metallic paint spray makes the paint appear lighter and more silver.
10. A wet metallic paint spray makes the color appear darker and less silver.
11. Flip-flop is a condition that occurs in metallics involving the positioning of the aluminum particles and the manner in which light is reflected to the observer.
12. Clearcoats are not all perfectly clear, and they can change the appearance of a color.
13. A halo effect is an unwanted shiny ring that appears around a pearl paint repair.
14. A variable is a condition (temperature, humidity, or ventilation) or a process (reduction, evaporation, speed of solvents, air pressure, or type of equipment) that affects a paint job.
15. The spray-out test panel checks the paint color and also shows the effects of the paint technician's technique. A let-down test panel is often needed to match three-stage pearl paints.
16. Custom painting involves using multiple colors, metallic paints, multilayer masking, and special spraying techniques to produce a personalized paint job.

EXERCISES

On a separate sheet of paper, complete the following learning activities for this chapter. Write definitions for the key terms and answer the ASE-style review questions, essay questions, critical thinking problems, and math problems. You can also do the outside activities, possibly for extra credit.

Key Terms

alternate color chips
chroma
color-corrected light bulbs
color face
color flip-flop
color matching
color spectrum
color test panel
color tree
halo effect
hue
let-down test panel
lumen rating
metamerism
OEM color variance
paint color directory
paint formula
plotting color
side tone
spectrophotometer
spray-out test panel
standard color chip
sunlight
tinting
value
zone concept

ASE-Style Review Questions

1. A spray-out test panel does not match the metallic color on the vehicle. The color is a little too dark. Technician A says to tint the paint first. Technician B says to try to lighten the color by misting the last coat of metallic. Who is correct?
   A. Technician A
   B. Technician B
   C. Both A and B
   D. Neither A nor B
2. When evaluating the color of a finish, under which type of light should it be viewed?
   A. Sunlight
   B. Incandescent light
   C. Fluorescent light
   D. Drop light
3. Technician A says to use a spectrophotometer when trying to match the color on a vehicle. Technician B says to use a computerized paint matching system. Who is correct?
   A. Technician A
   B. Technician B
   C. Both A and B
   D. Neither A nor B
4. What term refers to the degree of lightness or darkness of a color?
   A. Value
   B. Hue
   C. Tint
   D. Shade

5. Technician A says to use a spray-out test panel to help check the match of a pearl paint. Technician B says to use a let-down test panel. Who is correct?
   A. Technician A
   B. Technician B
   C. Both A and B
   D. Neither A nor B
6. Which term refers to a color's level of intensity, or the amount of gray (black and white) in it?
   A. Hue
   B. Chroma
   C. Tint
   D. Value
7. A dry metallic paint spray makes the paint appear
   A. darker.
   B. lighter.
   C. more red.
   D. more blue.
8. When making a spray-out test panel, Technician A says not to clearcoat the panel. Technician B says you should clearcoat the panel. Who is correct?
   A. Technician A
   B. Technician B
   C. Both A and B
   D. Neither A nor B
9. What are the basic reasons for tinting?
   A. To adjust color variations in shades to match the color from the manufacturer
   B. To adjust color on an aged or weathered finish
   C. To make a color for which there is no formula or for which there are no paint codes available
   D. All of the above
   E. None of the above
10. What is the name for a full set of paint pigments and solvents that can be mixed at the body shop?
    A. Factory-packaged colors
    B. Custom-mixed colors
    C. Intermix system
    D. Manufacturer system

Essay Questions

1. What is a paint color directory?
2. What is a spectrophotometer?
3. List three tasks that a computerized paint matching system is generally able to do.
4. Explain alternate color chips in detail.
5. How do we see the color of a finish or any object?
6. List six ways to darken a metallic color.
7. Define the term *flip-flop*.
8. Summarize the basic steps for spot repair with a fluorine clearcoat system.

Critical Thinking Problems

1. If the color of a refinish job varies from the original, what are some possible reasons for the mismatch?
2. Why should you view a color at different angles when evaluating a color match?
3. What colors do the phrase "Roy G Biv" stand for?
4. Name the seven basic steps for spot repair with a fluorine clearcoat system.

Math Problems

1. A paint technician uses a mil gauge to measure the paint thickness on a vehicle. The gauge shows a thickness that varies from 8–10 mils. If the new paint must be applied with a thickness of 2–4 mils, how thick can the paint become?
2. If a three-stage paint requires six coats total and flash time is 6 minutes, how much total time must you wait while painting?

Activities

1. Inspect the paint on every panel of several vehicles. See whether you can find any paint mismatch.
2. Mask a small area on an old panel. Try to spray and match the paint already on the panel.
3. Read the instructions in a paint color directory and a tinting manual. Write a report on what you learned in each.

CHAPTER

28 Paint Problems and Final Detailing

OBJECTIVES

After studying this chapter, you should be able to:

- Visually identify and define automotive paint problems.
- Explain the causes and symptoms of finish flaws.
- Use a logical sequence of operations to repair a finish.
- Detail or finesse sand paint flaws using the right tool and abrasive.
- Explain how to avoid damaging a paint job by cutting through the clearcoat when sanding and compounding.
- Summarize things you should do to protect unpainted parts from being damaged during paint repair.
- Hand rub and machine compound a finish using different types of products.
- Describe how to operate a buffing machine without burning through the clearcoat.
- Touch up small chips in the finish on unrepaired panels.
- Final clean and detail a vehicle for improved customer satisfaction.
- Answer ASE-style review questions relating to paint problems and final detailing.

INTRODUCTION

Paint problems include a wide range of defects that can be found before or after painting. To maintain repair quality and satisfy customers, you must be able to analyze and correct finish problems efficiently.

If all technicians in the shop do their jobs, there will seldom be a reason to fix costly, time-consuming paint problems. Ideally, every vehicle can be released to the customer after paint baking and a minor clean-up.

Regretfully, even the best, most professional collision repair businesses will encounter minor paint flaws that must be fixed. On rare occasions, shop and paint company personnel will have to solve major paint problems on existing and freshly painted surfaces.

The technical information provided in previous chapters will help you avoid mistakes that result in paint problems and repainting. This chapter will continue your study of collision repair by teaching you about paint problem conditions, causes, prevention, and correction. The last section of the chapter summarizes how to clean a vehicle before it is released to the customer.

28.1 REPAIRING PAINT PROBLEMS

Most refinishing problems can usually be repaired, but this reworking requires time and money. Smart technicians take the time to prevent paint problems before they occur. Unfortunately, there are a variety of causes for defects in a vehicle's finish. They usually originate as a result of problems in the preparation of the body surface, painting procedure, environment, paint ingredients, and other sources.

Problems in Wet Paint

If you see paint defects while spraying, you must decide whether to stop work immediately or whether the problem can be fixed so you can continue painting. This depends on the type and extent of the paint problem.

For example, if the problem is poor, wavy bodywork, you should stop right away. A rough body surface will look even rougher after being sprayed with a shiny coat of paint. With an improper body repair, perhaps someone forgot to properly block sand and featheredge a small area of body filler. The metalworking, body filler, or other work problem would have to be corrected before continuing to paint the vehicle.

When painting in even the cleanest paint booth, tiny particles of dust, dirt, hair, and so on can sometimes fall or blow into the paint. There are several things you can do to keep foreign matter out of your freshly applied paint. Keep the spray booth doors closed for several minutes before starting to tack rag and blow off vehicle surfaces. This gives any airborne dust and dirt enough time to settle out of the air.

Do not let anyone open the spray booth doors while you are spraying. If someone opens the booth doors, wind can blow dust, hair, lint, and dirt into the spray booth and onto wet paint. Place a "Keep Out" or "Do Not Open!" sign on the spray booth doors when you are painting.

Removing Foreign Matter in Wet Paint

Paint foreign matter includes anything you see in the paint that will adversely affect the finish (dust, lint, hair, and so forth). Sometimes you can remove a tiny bit of foreign matter while the paint is still wet, or you may have to wait until the paint dries to sand out the flaw before continuing.

If you notice something in the wet paint, try to remove it right away—the sooner the better because today's catalyzed paints flash so quickly. A tiny piece of lint or dust can often be lifted out of the wet paint so you can continue painting. Depending on the type of matter in the paint, there are several ways to remove debris from wet paint.

Sharp tweezers can sometimes be used to grab and remove lint and hair from wet paint. Be careful to only touch the debris without disturbing the paint surface.

A piece of fine wire or a toothpick can also be used to lift and remove small flakes of dust from wet paint. Very fine wire can reach into the wet paint and reach under the piece of dust and lift it out of the finish. If done quickly, this will allow the wet paint to flow back out and partially fill the paint surface imperfection.

After removing debris in the paint, blend spray another coat of paint over that area right away. This will help fill in any paint surface flaw that remains where the hair, dust, or lint was removed.

Wet Sanding Between Coats

If you notice small particles or imperfections in the colorcoat, repair them before spraying the clearcoats. Try lifting the particles out of the colorcoat while it is still wet. If the paint still looks too flawed to be hidden by another coat, you will have to let the basecoat flash enough to wet sand the surface flaw.

Carefully wet sand right over the top of the flawed paint with ultrafine #1000–#1200 or finer grit sandpaper. Wrap the wet sandpaper around a soft sponge-type sanding block. Concentrate your wet sanding action right over the paint flaw. Sand lightly because the paint is flashed, but not fully dry.

After you have sanded the flaw level, wipe the area dry. Clean and tack rag the area before giving it another coat of color. Lightly mist and blend the basecoat over the surface flaw to cover any visible problem in the color.

Do not allow the colorcoat to fully dry or cure before applying the clearcoat. Most clearcoats are designed to be applied before the basecoat fully cures. If the colorcoat dries fully, you should scuff sand the surface before clearcoating to provide good adhesion. This is why you must work quickly when trying to remove foreign matter from newly applied paint.

If the piece of dust or dirt in the clearcoat is too small to be lifted out, you can usually repair the problem after the finish dries. As you will learn later, the clearcoat can normally be finesse sanded and compounded to repair minor paint problems.

Paint Color Mismatch

A **paint color mismatch** causes a repair area to look a different color from the original color on the vehicle. The value (lightness or darkness), hue (color, cast, or tint), or chroma (cleanliness, grayness, or muddiness) may not be exactly the same in the two paints (Figure 28–1).

Causes of Paint Color Mismatch

As detailed in the previous chapter, there are several reasons for a paint color mismatch. The most common causes of a color difference include improper paint mixing and not spraying the paint on properly.

Preventing Color Mismatch

To prevent a color mismatch, always use spray-out test panels and let-down test panels. Use a spray-out panel with two-stage, basecoat/clearcoat paints; use a let-down test panel with three-stage pearl paints.

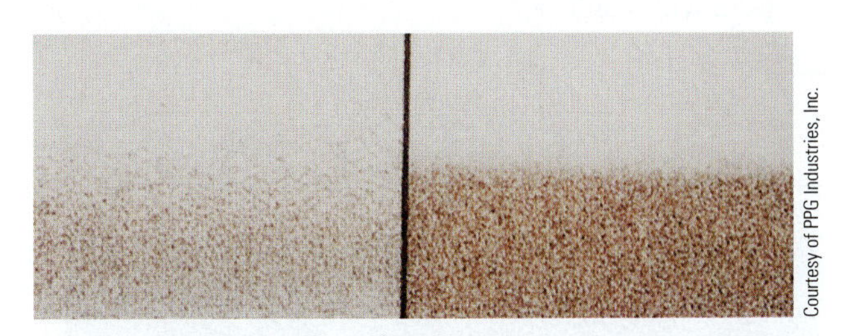

Courtesy of PPG Industries, Inc.

Figure 28–1 Note the difference in paint color after this improper panel repair. The metallic paint on the right looks much darker. This could be due to spraying the metallic too wet or to improper tinting.

Correcting Color Mismatch

To correct a paint mismatch, you must repaint the area. You might have to tint the paint a slightly different color or use different spraying techniques.

Orange Peel

Orange peel is an uneven surface formation, much like that on the skin of an orange. When viewed under a magnifying glass, the paint surface looks rough, bumpy, or textured. Orange peel is caused by poor fusion of atomized paint droplets. Paint droplets dry out before they can flow out and level smoothly together. See Figure 28–2.

Note that some degree of orange peel can be found in most finishes, both OEM and repainted. It is when the orange peel becomes obvious or offensive that it becomes a paint problem.

Causes of Orange Peel

- Improper gun adjustment and spraying techniques often cause orange peel. Too little air pressure, wide fan patterns, and spraying at excessive gun distances cause droplets to become too dry during their travel time to the work surface. Improper adjustment does not let the paint flow out smoothly.
- High paint booth temperature can cause orange peel. When air temperature is too high, droplets lose more solvent and dry before they can flow out and level properly on the body surface.
- Improper flash or recoat time between coats can cause orange peel. If the first coat is allowed to flash too much, solvent in the paint droplets of subsequent coats will be absorbed into the first coat before proper flow is achieved.
- Using the wrong reducer can cause orange peel. Underdiluted paint or paint thinned with fast-evaporating thinners or reducers causes the atomized droplets to become too dry before reaching the surface.
- If you improperly mix in too little thinner or reducer, the paint can be too thick and will not flow out smoothly, causing orange peel.
- Materials not uniformly mixed can also cause orange peel. Many finishes are formulated with components that aid fusion. If these are not properly mixed, orange peel will result.

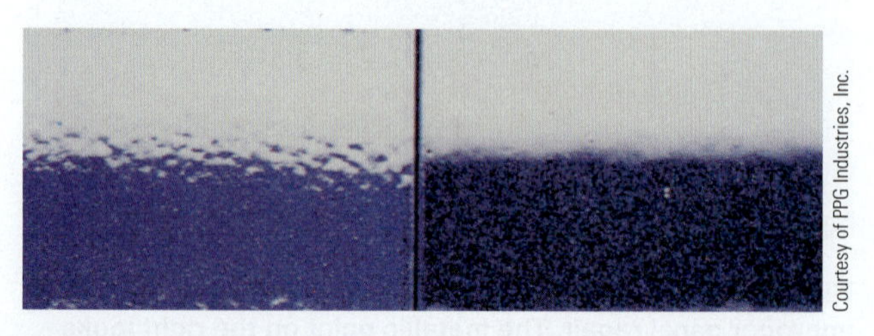

Figure 28-2 Orange peel is excessive on the left and normal on the right.

Preventing Orange Peel

- Use proper gun adjustments, techniques, and air pressure.
- Schedule painting to avoid temperature and humidity extremes. Select the reducer that is suitable for existing conditions. The use of a slower evaporating thinner or reducer will overcome an orange peel problem.
- Allow sufficient flash and dry time. Do not dry by fanning.
- Allow proper drying time for undercoats and topcoats (not too long and not too short).
- Reduce to recommended viscosity with proper thinner/ reducer.
- Stir all pigmented undercoats and topcoats thoroughly.

Correcting Orange Peel

Two full wet coats of clear, properly applied, with the correct flash times between each coat, will normally correct an orange peel problem.

Minor orange peel can be corrected by machine buffing or compounding the finish after it has dried. In extreme cases, wet sand the orange peel area before compounding.

Runs and Sags

Paint runs occur when gravity produces a mass slippage of an overwet and thick paint film. The weight of the uncured paint causes it to slide or flow down the surface. A large area of paint may flow down and form large globules of paint. Large bumps form in the paint surface where the run stops flowing.

A *paint sag is* a partial slipping down of the paint created by a film that is too heavy to support itself. It appears like a curtain. Runs and sags are more of a problem on vertical panels (fenders, doors, and quarter panels) because of gravity. Runs and sags are not as much of a problem on horizontal panels (roof, hood, and trunk lid). See Figure 28–3.

Causes of Runs and Sags

- Applying too much paint in one coat
- Triggering paint spray incorrectly when changing spray gun directions
- Not allowing enough flash time between coats
- Wrong temperature rating of solvent (reducer or thinner)
- Low air pressure, causing lack of atomization
- Holding gun too close or making too slow a gun pass
- Shop or surface too cold

Preventing Runs and Sags

- Use proper spray gun motion, distance, and speed of pass with equal overlaps of coats.
- Select the proper thinner/reducer.

A A paint run in clearcoat can sometimes be wet sanded out to fix the problem.

B A metallic sag in colorcoat requires repainting with a lighter mist coat.

Figure 28–3 Paint runs and sags are caused by spraying on too much paint at once or by not allowing enough flash time between coats.

A Sand scratch swelling is enlarged sand scratches caused by the swelling action of topcoat solvents.

B A bull's-eye is normally caused by sanding filler or putty before it has fully cured. It will shrink and form an indentation in the paint film.

Figure 28–4 If materials are not allowed to cure properly before sanding and painting, problems like these can develop.

- Do not pile on paint coats too thickly. Allow sufficient flash or dry time in between coats.
- Use proper gun adjustment, techniques, and air pressure.
- Allow the vehicle surface to warm up to at least room temperature before attempting to refinish. Try to maintain an appropriate shop temperature for paint areas.

Correcting Runs and Sags

On a small area of wet paint, you can use solvent to wash off the run or sag before repainting. If the run or sag is on a larger panel, allow the paint to dry enough for wet sanding. Use a relatively coarse #600-grit wet sandpaper and a stiff rubber sanding block to level off the run or sag. Then block sand the area again with finer #1000-grit wet sandpaper to avoid sand scratches.

If the run or sag is in the clearcoat, try to wet sand the area without cutting through the clear. If you cut into the colorcoat, the panel will have to be sprayed again with clear.

Sand Scratch Swelling

Sand scratch swelling is enlarged sand scratches caused by the swelling action of topcoat solvents. This problem is shown in Figure 24–4A.

Causes of Sand Scratch Swelling

- Improper surface cleaning or preparation. Use of too coarse a sandpaper or omitting a sealer in panel repairs greatly exaggerates swelling caused by thinner penetration.
- Improper solvent (reducer or thinner), especially a slow-drying solvent when sealer has been omitted.
- Underreduced or too fast a solvent (reducer or thinner) used in primer-surfacer causes "bridging" of scratches.

Preventing Sand Scratch Swelling

- Use appropriate grits of sandpaper.
- Apply a sealer over the primer to eliminate sand scratch swelling. Select thinner or reducer suitable for existing shop conditions.
- Use proper thinner and reducer for primer-surfacer.

Correcting Sand Scratch Swelling

Sand the affected area down with ultrafine sandpaper and apply appropriate sealer before applying paint.

Bull's-Eye Featheredge

A *bull's-eye featheredge* is an indented area that results from shrinkage of spot putty or filler, producing an area that is lower over the top of the putty or filler. This problem is shown in Figure 28–4B.

Causes of Bull's-Eye Featheredge

- Not allowing the spot putty to cure enough before block sanding. The use of older, slow-drying, one-part, lacquer-based spot putty is another cause.
- Not allowing body filler to cure fully.
- Improper mixing of two-part filler or putty.

Preventing Bull's-Eye Featheredge

Bull's-eye featheredge can be prevented by only using two-part spot putty and allowing proper time for putty or filler to cure. Also be sure to mix putty and filler thoroughly.

Correcting Bull's-Eye Featheredge

Correcting bull's-eye featheredging requires that you sand and refinish the affected areas.

Featheredge Splitting

Featheredge splitting appears as stretch marks, or cracking, along a featheredge.

Causes of Featheredge Splitting

- Using too much primer or primer-surfacer in heavy and wet coats over a repair area can cause featheredge splitting. Solvent is trapped in lower layers that have not had sufficient time to set up.
- Material has not uniformly mixed. Because of the high pigment content of primer-surfacers, it is possible for settling to occur after it has been thinned. Delayed use of the material without frequent stirring results in applying a film with loosely held pigment containing voids and crevices throughout. This causes the film to act like a sponge.
- The wrong solvent is used. For example, if you use lacquer thinner as an enamel reducer, featheredge splitting and other problems can result.
- Improper surface cleaning or preparation. When not properly cleaned, primer-surfacer coats can crawl or draw away from the edge because of poor wetting and adhesion.
- Excessive putty use and film buildup.

Preventing Featheredge Splitting

- Apply properly reduced primer-surfacer in medium to full wet coats with enough flash time between coats. This will allow solvents and air to escape.
- Stir all pigmented undercoats and topcoats thoroughly. Select a paint solvent temperature that is suitable for existing shop conditions.
- Select only reducers that are recommended for existing shop conditions.
- Thoroughly clean areas that will be painted before sanding.
- Spot putty should be limited to filling minor imperfections. Putty applied too heavily or too thickly will eventually shrink, causing featheredge splitting.

Correcting Featheredge Splitting

To correct featheredge splitting, you must remove the finish from the affected areas and refinish.

Water Spotting

Water spotting is the general dulling of gloss in spots or masses of spots.

Causes of Water Spotting

- Water evaporating on the finish before it is thoroughly dry
- Washing the finish in bright sunlight

Preventing Water Spotting

- Do not apply water to a fresh paint job, and try to keep a newly finished vehicle out of the rain and snow. Allow sufficient drying time before delivering the vehicle to your customer.
- Wash the car in the shade and wipe it completely dry.

Correcting Water Spotting

Compound or polish the vehicle with rubbing or polishing compound. In severe cases, sand the affected areas and refinish.

Chemical Spotting

Chemical spotting, such as acid and alkali spotting, causes an obvious discoloration of the paint surface. Various paint pigments react differently when they come into contact with acids or alkalies (Figures 28–5A, B, and C).

Causes of Chemical Spotting

The cause of acid and alkali spotting is a chemical change of pigments. This chemical change results from atmospheric contamination or the presence of moisture. This problem is often found on older finishes that have been exposed to industrial pollution.

Preventing Chemical Spotting

- Keep the finish away from a contaminated atmosphere if possible.
- Immediately following contamination, the body surface should be vigorously flushed with cool water and detergent.

Correcting Chemical Spotting

- Wash with detergent and water and follow with a vinegar bath.
- Sand and refinish. You might try wet sanding and compounding if there is only minor spotting.
- If contamination has reached the metal or substrate, the spot must be sanded down to the metal before refinishing.

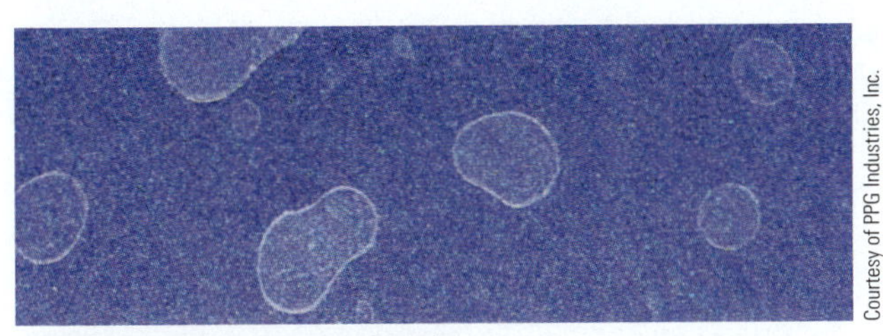

Courtesy of PPG Industries, Inc.

A **Water spotting on fresh paint.**

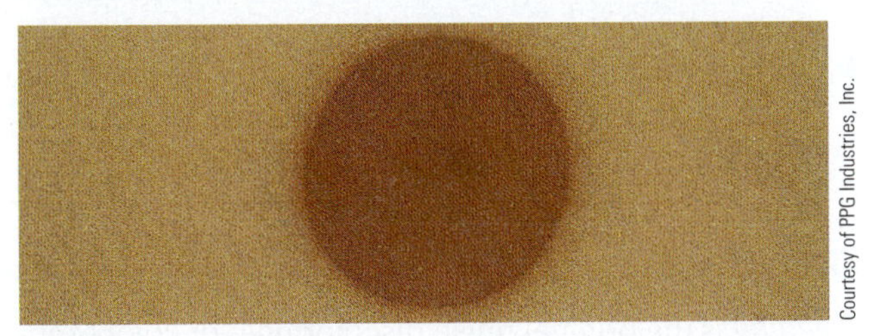

Courtesy of PPG Industries, Inc.

B **Chemical spotting is normally due to contamination of paint.**

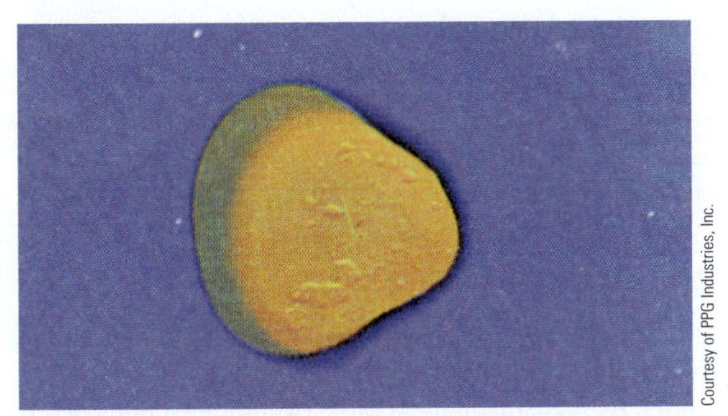

Courtesy of PPG Industries, Inc.

C **Acid or alkali spotting is due to a chemical change in the paint pigments.**

Figure 28-5 **Note the different types of paint spotting.**

Curing or Drying Failure

A *curing or drying failure* is the abnormally slow hardening of a refinish product. The material remains wet or soft for a prolonged period of time. This might involve a body filler, spot putty, primer, sealer, paint, or corrosion-protection material.

Causes of Curing or Drying Failure

- Improper stirring or mixing of product ingredients
- Sloppy surface cleaning and preparation, allowing chemical contamination of refinish materials, splashing chemical paint remover, for example
- Wet sanding with contaminated water
- Faulty refinish product; product ingredients were mixed incorrectly during manufacturing
- Shelf life of product exceeded

Preventing Curing or Drying Failure

- Thoroughly clean all areas to be repaired with a wax and grease remover to avoid chemical contamination.
- Finger knead (mix) cream hardener for spot putty and body filler.
- Do not forget to add a hardener or catalyst.

Correcting Curing or Drying Failure

- Wash or sand all affected areas thoroughly as needed and then refinish.
- Properly dispose of aged refinish products.

Paint Fish-Eyes

Paint fish-eyes are small, BB-sized dimples or craters that form in the liquid paint film right after spraying. If watched closely when forming, the paint will actually flow up and out of the small dimple or crater. Contaminants mixed with the paint are pushing the paint out of the area. If these flaws in the paint are deep enough, you can sometimes see the primer or sealer under the colorcoat (Figure 28–6).

Causes of Paint Fish-Eyes

- Fish-eyes are commonly caused by improper body surface cleaning or preparation. Many waxes and polishes contain silicone, the most common cause of fish-eyes. Silicones adhere firmly to the paint film and require extra effort to remove. Even small quantities in sanding dust or rags, from cars being polished nearby, or from touching the vehicle body with dirty hands can cause this paint film failure.
- The old finish or a previous repair can contain excessive amounts of silicone from additives used during their application. Even washing with a prepainting cleaning solution will not remove embedded silicone.
- Contaminants in shop air lines and hoses can blow out of air sanders, a spray gun, or other power tools onto the car body.
- Using the wrong type of air tool lubricating oil. It is best to use spray gun oil in all shop air tools, even grinders and sanders. Spray gun lubricant is formulated not to contaminate a paint job and cause fish-eye and other chemical interaction problems.

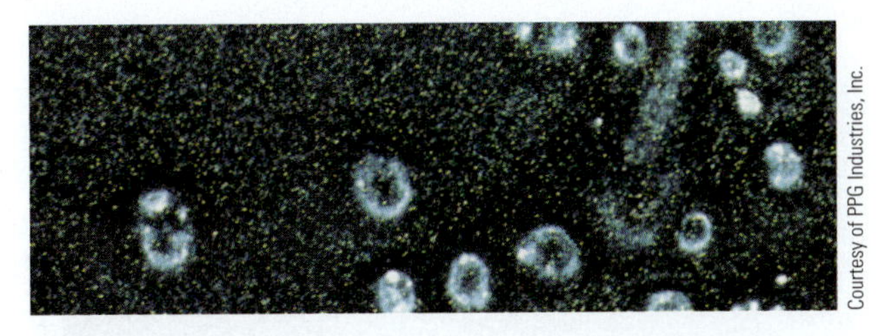

Courtesy of PPG Industries, Inc.

Figure 28-6 **Fish-eyes can form when painting a silicone- or oil-contaminated surface. Paint additives can help correct fish-eye problems if sprayed again before full flashing.**

Preventing Paint Fish-Eyes

- Do not touch the body surface with anything that could have silicone or another contaminant on it. Wear clean gloves when touching body surfaces.
- Use only clean disposable cloths when wiping body surfaces that will be painted. Clean off all traces of silicone and other contaminants by thoroughly washing all surfaces with wax and grease remover. Wipe the cleaning solution off the body before it dries.
- Routinely drain and clean air filters and driers. The shop air compressor tank should be drained daily. Also replace air line filters and driers periodically.

Correcting Paint Fish-Eyes

- Mix in a small amount of fish-eye eliminator additive with the paint. Spray another coat over the affected area as soon as possible to see whether the paint film will flow out smoothly over the fish-eye dimples.
- If the area cures too much or if the problem is too severe, allow the paint to cure or dry. Then, wet sand or power dry sand the area to level the dimples in the paint surface. Repaint the spot or panel as needed.

Blushing

Blushing is a problem that makes the finish turn white or milky looking (Figure 28–7).

Causes of Blushing

- In hot, humid weather, moisture droplets can become trapped in the wet paint film. Air currents from the spray gun and the evaporation of the paint solvent tend to lower the temperature of the surface being sprayed below that of the surrounding atmosphere. This causes moisture in the air to condense on the wet paint film.
- Excessive air pressure
- Using too fast a thinner or reducer

Preventing Blushing

- In hot, humid weather, try to schedule painting early in the morning when temperature and humidity conditions are more suitable.
- Use proper gun adjustments and techniques.
- Select the thinning solvent or reducer that is suitable for existing shop conditions. Add retarder to the thinned or reduced color and apply additional coats.

Courtesy of PPG Industries, Inc.

Figure 28-7 **Blushing is a whitish blotch in paint, normally due to moisture buildup on the painted surface.**

Correcting Blushing

- If the blushing problem is in the colorcoat, recoat the area, using the proper booth temperature, the correct reducer, and the recommended spray methods.
- If the problem is in the clearcoat or if the paint has cured, scuff sand and repaint the area or panel as needed.

Bleeding

Bleeding is the original finish discoloring—or color seeping through—the new topcoat color.

Causes of Bleeding

Bleeding is usually caused by not using a sealer before painting. Contamination can also cause bleeding—usually in the form of soluble dyes or pigments on the older finish before it was repainted.

Preventing Bleeding

Thoroughly clean areas to be painted before sanding, especially when applying lighter colors over darker colors. Avoid using lighter colors over older shades of red without sealing first.

Correcting Bleeding

Apply two medium coats of paint sealer. Seal and flash dry according to label instructions. Then reapply the colorcoat.

Primecoat Show-Through

Primecoat show-through is a variation in the surface color.

Causes of Primecoat Show-Through

- Insufficient colorcoats used
- Sealer not used or not tinted to match basecoat

Preventing Primecoat Show-Through

- Apply a good coverage of color.
- Tint the sealer to match the vehicle color.

Correcting Primecoat Show-Through

To correct primecoat show-through, you must sand and refinish.

Blistering

Blistering shows up as small, swelled areas on the finish that look like a water blister on human skin. There will be a lack of gloss if blisters are small. You will find broken-edged craters if the blisters have burst (Figure 28–8).

Causes of Blistering

- Improper surface cleaning or preparation. Tiny specks of dirt left on a surface can act like a sponge and hold moisture. When the finish is exposed to the sun or abrupt changes in atmospheric pressure, moisture expands and builds up pressure. If the pressure is great enough, blisters form.
- Use of the wrong thinner or reducer, such as using a fast-drying thinner or reducer, especially when the material is sprayed too dry or at an excessive pressure. Air or moisture can be trapped in the film.
- Excessive film thickness. Insufficient drying time between coats or too heavy an application of primecoats can trap solvents that escape later and blister the colorcoat.
- Contamination of compressed air lines, such as with oil, water, or dirt.

Preventing Blistering

- Thoroughly clean areas to be painted before sanding. Be sure the surface is completely dry before applying either primecoats or topcoats. Do not touch a cleaned area because the oil on your hands can contaminate the surface.
- Select the thinner or reducer most suitable for existing shop conditions.
- Allow proper drying time for primecoats and topcoats. Be sure to let each coat flash before applying the next coat.
- Drain and clean the air pressure regulator on a daily basis to remove trapped moisture and dirt. The air compressor tank should also be drained daily.

Correcting Blistering

If damage is extensive and severe, the paint must be removed down to primecoat or metal, depending on the depth of the blisters. Then the surface can be refinished.

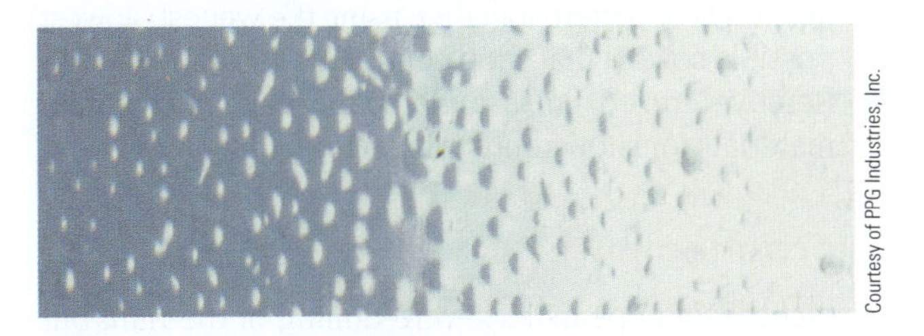

Figure 28–8 Paint blistering is normally caused by excessively fast solvent evaporation.

In less severe cases, blisters can be sanded out, resurfaced, and topcoated again.

Solvent Popping

Solvent popping is "blisters" or "pimples" on the paint surface caused by the paint topcoats trapping evaporating solvent gases during curing and drying. The gas bubbles try to escape by pushing up small blisters in the wet paint or topcoat. Solvent popping is further aggravated by forced drying and baking, because the trapped solvents evaporate even more quickly.

Causes of Solvent Popping

- Surface is improperly cleaned or prepared.
- The wrong solvent or reducer is used. Use of fast-drying solvent or reducer, especially when the material is sprayed too dry or at excessive pressure, can cause solvent popping by trapping air in the film.
- Excessive film thickness formed by too many layers of material. Insufficient drying time between coats and too heavy an application of undercoats can trap solvents, causing popping of the colorcoat as they later escape.

Preventing Solvent Popping

- Thoroughly clean areas to be painted.
- Select the thinner or reducer suitable for existing shop conditions.
- Do not pile on primecoats or topcoats. Allow for sufficient flash and dry times. Allow proper primer and sealer drying time by spraying topcoats. Allow each coat of primer-surfacer to flash naturally. Do not fan.

Correcting Solvent Popping

If damage is extensive and severe, the paint must be removed down to an unaffected layer or to bare metal, depending on the depth of the blisters. Then the affected area must be refinished using proper flash times and drying or baking temperatures.

Paint Cracking

Paint cracking is a series of deep cracks resembling mud cracks in a dry pond or lake bed. Often taking the form of three-legged stars in no definite pattern, they usually go through the topcoat and sometimes the primecoat as well. Different kinds of paint cracking are shown in Figure 28–9.

Causes of Paint Cracking

- Excessive film thickness. Excessively thick topcoats magnify normal stresses and strains that can result in cracking even under normal conditions.

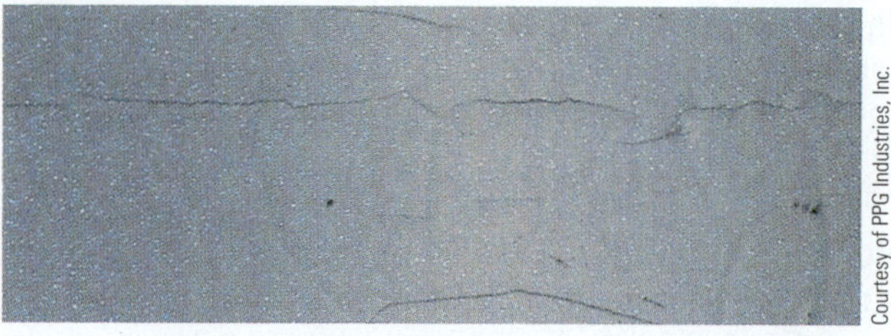

A

B

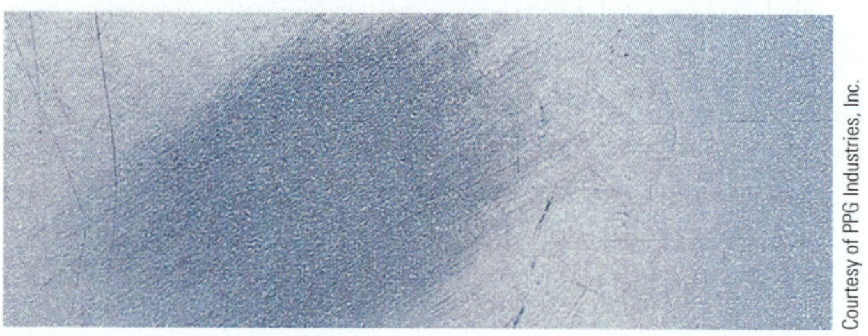

C

Figure 28-9 Note the different kinds of paint cracking: (A) line cracking, (B) crow's feet cracking, and (C) cold cracking.

- Materials not uniformly mixed
- Insufficient flash time
- Incorrect use of additive

Preventing Paint Cracking

- Do not pile on topcoats. Allow sufficient flash and drying times between coats. Do not dry by gun fanning.
- Stir all pigmented undercoats and topcoats thoroughly. Strain and add fish-eye eliminator to topcoats, when necessary.
- Read and carefully follow label instructions. Additives not specifically designed for a colorcoat can weaken the final paint film and make it more sensitive to cracking.

Correcting Paint Cracking

The affected areas must be sanded to a smooth finish or, in extreme cases, removed down to the bare metal and refinished.

Line Checking

Line checking is similar to cracking, except that the lines or cracks are more parallel and range from very short to very long.

Causes of Line Checking

- Excessive film thickness
- Improper surface preparation, often due to the application of a new finish over an old film that had cracked and was not completely removed

Preventing Line Checking

- Do not pile on topcoats. Allow sufficient flash and drying times. Do not dry by gun fanning.
- Thoroughly clean all areas that will be painted before sanding. Be sure the surface is completely dry before applying any undercoats or topcoats.

Correcting Line Checking

Remove clearcoat and colorcoat down to the sealer or primer and apply new topcoats.

Crazing

Crazing results in fine splits or small cracks—often called "crow's-feet"—that completely checker an area in an irregular manner. This problem was common with older lacquer finishes.

Causes of Crazing

- Shop temperature is too cold.
- Surface tension of the original material is under stress, and it literally shatters under the softening action of the solvents being applied.
- OEM lacquer crazes due to age and temperature extremes.

Preventing Crazing

- Select the thinner or reducer that is suitable for existing shop conditions.
- Schedule painting to avoid temperature and humidity extremes in the shop or between the temperature of the shop and the job.
- Bring the vehicle to room temperature before refinishing.

Correcting Crazing

- Continue to apply wet coats of topcoat to melt the crazing and flow pattern together, using the wettest/slowest possible solvent that shop conditions will allow.
- Remove crazed finish and repaint with appropriate materials for shop conditions.

Microchecking

Microchecking appears as severe dulling of the film, but when examined with a magnifying glass, it contains many small, microscopic cracks.

Causes of Microchecking

Microchecking is the beginning of film breakdown and might be an indication that film failures, such as cracking or crazing, will develop.

Preventing Microchecking

- Do not pile on topcoats. Allow sufficient flash and drying times. Do not dry by gun fanning.
- Thoroughly clean all areas that will be painted before sanding. Be sure the surface is completely dry before applying any undercoats or topcoats.

Correcting Microchecking

Sand off the colorcoat to remove the cracks, then recoat as required. Make sure you sand down deep enough to remove all microscopic cracks in each refinish material.

Lifting

Lifting is a condition that causes surface distortion or shriveling, while the topcoat is being applied or while drying (Figure 28–10).

Causes of Lifting

- Use of incompatible materials. Solvents in the new topcoat attack the old surface, which results in a distorted or wrinkled effect.
- Insufficient flash time. Lifting will occur when the paint film is an alkyd enamel and is only partially cured. The solvents from the coat being applied cause localized swelling or partial dissolving that later distorts the final surface.

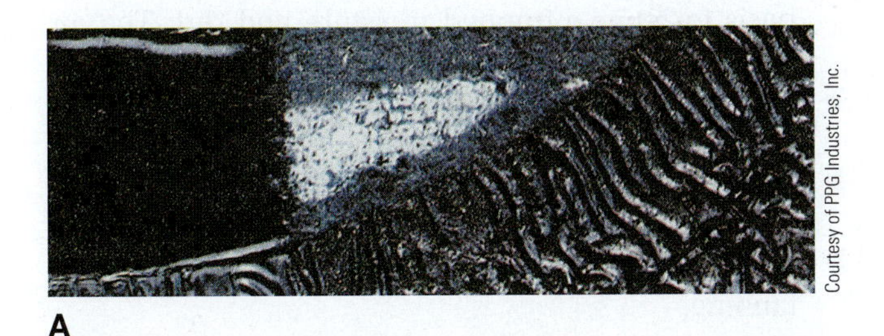

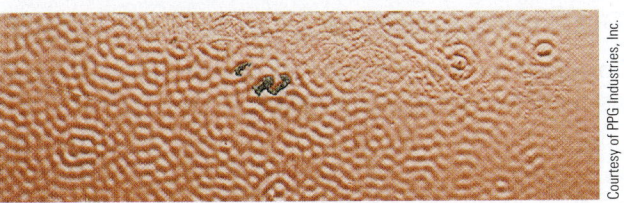

Figure 28–10 (A) Lifting and (B) wrinkling are usually caused by incompatible chemicals or by using the wrong paint ingredients when mixing.

Courtesy of PPG Industries, Inc.

- Improper drying. When synthetic enamel-type undercoats are not thoroughly dry, topcoating with lacquer can result in lifting.
- The effect of an old finish or a previous repair. Lacquer applied over a fresh air-dried enamel finish will cause lifting.
- Improper surface cleaning or preparation. Use of an enamel-type primer or sealer over an original lacquer finish that is to be topcoated with a lacquer will result in lifting due to a sandwich effect.
- Wrong thinner or reducer. The use of lacquer thinners in enamel increases the amount of substrate swelling and distortion, which can lead to lifting, particularly when two-toning or recoating.

Preventing Lifting

- Avoid incompatible materials, such as a thinner with enamel products or incompatible sealers and primers.
- Do not pile on topcoats. Allow sufficient flash and drying times. The final topcoat should be applied when the previous coat is still soluble or after it has completely dried and is impervious to topcoat solvents.
- Select the correct thinner or reducer for the finish applied and suitable for existing shop conditions.

Correcting Lifting

To correct a lifting problem, remove the finish from all affected areas and refinish.

Paint Wrinkling

Paint wrinkling is a severe puckering of the paint film that appears like the skin of a prune. It is more common with enamel paints. There is a loss of gloss as paint dries. Minute wrinkling may not be visible to the naked eye.

Causes of Wrinkling

- Improper drying. When a freshly applied topcoat is baked or force dried too soon, softening of the undercoats can occur. This increases topcoat solvent penetration and swelling. In addition, baking or force drying causes surface layers to dry too soon. The combination of these forces causes wrinkling.
- Using too many heavy or wet coats. When enamel coats are too thick, the lower wet coats are not able to release their solvents and set up at the same rate as the surface layer, which results in wrinkling.
- Improper reducer or incompatible materials. A fast-drying reducer or the use of a lacquer thinner in enamel can cause wrinkling.
- Improper or rapid change in shop temperature. Drafts of warm air cause enamel top to "skin," or set up and shrink, before sublayers have released their solvents. This results in abnormal surface drying and wrinkling in uneven patterns.

Preventing Wrinkling

- Allow proper drying time for undercoats and topcoats. When force drying alkyd enamel, baking additive is required to retard surface setup until the lower layers harden. Lesser amounts can be used in hot weather. Read and carefully follow product instructions.
- Do not use too many topcoats. Allow sufficient flash and drying times.
- Select the proper reducer and avoid using incompatible materials such as a reducer with lacquer products or thinner with enamel products.
- Schedule painting to avoid temperature extremes or rapid temperature changes.

Correcting Wrinkling

To correct this paint problem, you must remove the wrinkled enamel and refinish the area.

Mottling

Paint mottling occurs only in metallics when the metal flakes float together to form a more silver or streaked appearance in the paint color.

Causes of Mottling

- Wrong solvent (reducer or thinner)
- Materials not uniformly mixed
- Spraying too wet
- Holding spray gun too close to work
- Uneven spray pattern
- Low shop temperature

Preventing Mottling

- Select the paint solvent that is suitable for existing shop conditions and mix properly. In cold, damp weather, use a faster drying solvent.
- Stir all pigmented topcoats—especially metallics—thoroughly.
- Use proper gun adjustments, techniques, and air pressure.
- Keep your spray gun clean (especially the needle fluid tip and air cap) and in good working condition.

Correcting Mottling

To correct mottling, first spray two medium coats of metallic color. Apply a lighter third mist coat of color to help distribute the metal flakes evenly throughout the paint.

Pinholing

Paint pinholing is tiny holes in the finish, which are usually the result of trapped solvents, air, or moisture. Refer to Figure 28–11.

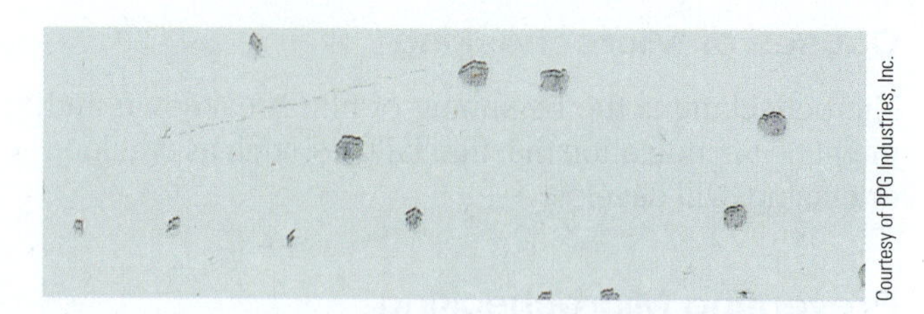
Courtesy of PPG Industries, Inc.

Figure 28–11 Pinholes look a lot like chips in the paint.

Causes of Pinholing

- Improper surface cleaning or preparation. Moisture left on primer or sealer will pass through the wet topcoat to cause pinholing.
- Contamination of air lines. Moisture or oil in air lines will enter the paint while it is being applied and cause pinholes when released during the drying stage.
- Wrong gun adjustment or technique. If adjustments or techniques result in an application that is too wet, or if the gun is held too close to the surface, pinholes will occur when air or excessive solvent is released during drying.
- Wrong solvent. The use of a solvent that is too fast for shop temperature tends to make the refinisher spray too close to the surface in order to get adequate flow. When the solvent is too slow, it is trapped by subsequent topcoats.
- Improper drying. Fanning a newly applied finish can drive air into the surface or cause a dry skin, both of which result in pinholing when solvents retained in lower layers come to the surface.

Preventing Pinholing

- Thoroughly clean all areas to be painted. Be sure the surface is completely dry before applying undercoats or topcoats.
- Drain and clean the air pressure regulator on a daily basis to remove trapped moisture and dirt. The air compressor tank should also be drained daily.
- Use proper gun adjustments, techniques, and air pressure.
- Select the solvent (reducer or thinner) that is suitable for existing shop conditions.
- Allow sufficient flash and drying times. Do not dry by fanning.

Correcting Pinholing

To correct pinholes, sand the affected area down as deep as needed and refinish the area.

Peeling

Paint peeling is caused by a loss of adhesion between refinish products (primer, sealer, or topcoats). The different coats of paint materials separate and one peels off another (Figure 28–12).

Courtesy of PPG Industries, Inc.

Figure 28-12 Peeling is a catastrophic paint failure of older finishes.

Causes of Peeling

- Improper cleaning or preparation. Failure to remove sanding dust and other surface contaminants will keep the finish coat from coming into proper contact with the substrate.
- Metal is not treated properly.
- Materials are not uniformly mixed.
- The proper sealer is not used.

Preventing Peeling

- Thoroughly clean areas to be painted. It is good shop practice to always wash the sanding dust off the area to be refinished with a cleaning solvent.
- Use the correct metal conditioner and conversion coating.
- Stir all pigmented undercoats and topcoats thoroughly.
- In general, sealers are recommended to improve adhesion of topcoats.

Correcting Peeling

Remove the finish from an area slightly larger than the affected area and refinish.

Chalking

Paint chalking is a problem that produces a lack of gloss on the paint surface. Extreme cases show up as a powdery surface. Chalking is also used to refer to an old finish that has deteriorated over time.

Causes of Chalking

- Wrong thinner or reducer, which can harm topcoat durability
- Materials not uniformly mixed
- Excessive mist coats when finishing a metallic color application
- Paint surface exposed to bright sunlight and the elements for too long

Preventing Chalking

- Select the thinner or reducer that is best suited for existing shop conditions.
- Stir all pigmented undercoats and topcoats thoroughly.
- Meet or slightly exceed the minimum film thickness.
- Apply metallic color as evenly as possible so that misting is not required. When mist coats are necessary to even out flake, avoid using straight reducer.
- Store the vehicle in an enclosed garage to protect paint from sunlight damage.

Correcting Chalking

To correct chalking, remove the surface in the affected area by sanding, and then clean and refinish.

Paint Color Fade

Paint color fade means color pigments have changed after exposure to prolonged sunlight. Even with compounding, the old finish that has been exposed to sunlight will look like a weaker hue or color than unexposed surfaces.

Causes of Paint Color Fade

- Finish is old and has been in bright sunlight too much.
- Inferior paint product was used when the vehicle was painted, or there is a paint pigment problem.

Preventing Paint Color Fade

- Keep paint protected from bright sunlight when the vehicle is not in use.
- Only use quality paint products from reputable manufacturers.

Correting Paint Color Fade

- Refinish the faded area.
- If the vehicle is a solid color, machine buff the paint to remove the faded surface layer.

Dulled Finish

A *dulled finish* means the paint does not have, or no longer has, its normal gloss or shine. Dulled finishes are the result of improper repair procedures or paint deterioration.

Causes of Dulled Finish

- Compounding or buffing before paint has cured fully
- Using too coarse a compound (for example, rubbing compound)
- Poorly cleaned surface
- Topcoats put on wet subcoats
- Washing with caustic cleaners

Preventing Dulled Finish

- Allow paint to cure properly before buffing.
- Use recommended materials.

- After using rubbing compound, follow up by using a finer glazing compound.
- Never use hand-rubbing compound with a buffing machine.

Correcting Dulled Finish

Allow the finish to dry hard, and use the correct methods to hand or machine compound the paint to a high gloss.

Debris in the Finish

Debris in the finish simply means foreign particles have gotten into the paint film. Debris includes dust particles, human hair, sand, dirt, lint from rags, and other small pieces of foreign matter. This is one of the most common and easily avoidable paint problems (Figure 28–13).

Causes of Debris in the Finish

- Improper surface cleaning, blowing off, and tack rag wiping of the surface to be painted
- A dirty or failed air line filter
- A dirty spray booth
- Defective or dirty air booth inlet filters
- A dirty spray gun; not straining paint material into the spray gun cup
- Wearing improper clothing or dirty or dusty shop coveralls
- Opening and closing spray booth doors right before or while painting vehicle

Courtesy of PPG Industries, Inc.

A Dust particles in the basecoat require repainting. Dust in the clearcoat can often be wet sanded out of the finish.

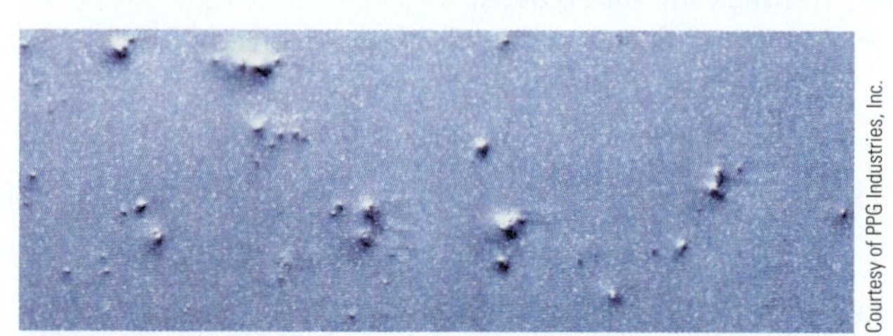

Courtesy of PPG Industries, Inc.

B Dirt in the finish results from larger pieces of lint and larger flakes of dust. This is a highly magnified view.

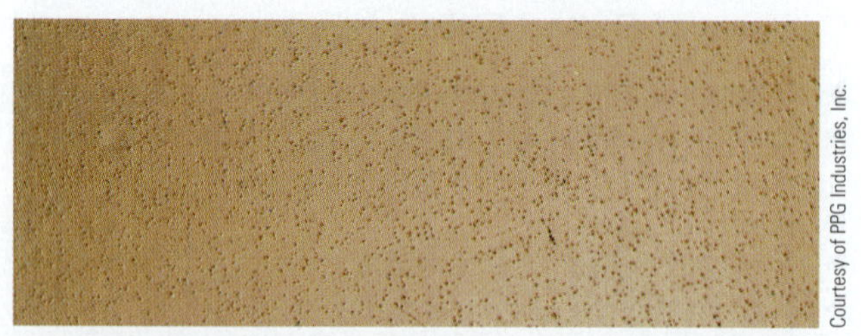

Courtesy of PPG Industries, Inc.

C Sand in the finish can occur if areas were sandblasted during rust repair and someone did not blow out all of the repair areas to remove all sand before painting.

Figure 28–13 Compare different types of foreign matter that can get into the finish when painting.

Preventing Debris in the Finish

- Blow out all cracks and body joints with an air nozzle while wiping with a tack cloth.
- Service air line filters and driers regularly.
- Sweep and blow out the spray booth on a regular basis. Maintain spray booth door seals and blow out cracks around booth doors and the bottoms of walls.
- Periodically replace booth and air line filters.
- Clean dry paint off the spray gun with paint remover.
- Wash coveralls and the paint hood when dirty. Wear a paint suit and head or hair covering.
- Use a paint strainer when pouring material into the spray gun. Keep all containers closed when not in use to prevent contamination.
- Close paint booth doors. Wait several minutes, then blow off the body with the spray gun while wiping with a tack rag.

Correcting Debris in the Finish

- Use tweezers, a piece of fine wire, or a toothpick to remove debris while the paint is still wet.
- If the paint has cured, wet sand the debris area level with the paint surface. Then hand or machine compound the area to return gloss.
- If dirt is deep in the finish, the area may have to be repainted.

Rust Under the Finish

Rust under the finish will show up as raised surface spots or peeling or blistering in the paint. In this instance, rusting steel has oxidized, started to flake, and pushed the paint upward and away from the body.

Causes of Rust Under the Finish

- Poor corrosion protection or forgetting to apply weld-through and self-etch primer is the main reason for rusting. Was weld-through primer sprayed on panel flanges to be welded? Were corrosion-protection materials injected into the closed box sections on the steel unibody panels?
- Broken paint film allows moisture to creep under the surrounding finish.
- Water in air lines can blow out of air tools to contaminate surfaces and cause rusting.
- Fingerprints (moisture on the skin) also cause rusting over time.

- Not applying seam sealer to newly welded panel flanges leads to rusting.
- The vehicle is very old and has been subjected to road salt in cold climates.

Preventing Rust Under the Finish

- Coat bare metal with self-etch primer. Do not leave bare metal exposed in the shop; spray it with self-etch primer right away.
- When replacing ornaments or molding, be careful not to break the paint film and allow dissimilar metals to come in contact with each other.
- Periodically drain the air compressor tank, air line filters, and driers.
- Wear plastic or rubber gloves when touching body surfaces.
- Apply all corrosion-protection materials according to the vehicle manufacturer's recommendations.
- Keep the vehicle cleaned and waxed. Occasionally pressure wash the underbody to remove road salts.

Correcting Rust Under the Finish

- Seal off possible entrance points for moisture from the inner parts of panels.
- Sand down to bare metal, prepare the metal, and treat it with phosphate before refinishing.
- Cut off the rusted panel and install a new or salvage one.

28.2 MASKING PROBLEMS

After the topcoat has dried, the masking paper and tape must be removed. If the finish has been force dried, the masking should be removed while the paint finish is still warm. If the finish is allowed to cool, the tape is more difficult to remove and may leave adhesive residue on the vehicle. You will then need to take extra time to clean off the adhesive with an adhesive solvent.

The tape should be removed slowly so that it comes off evenly. Pull the tape away from the paint edge—never across it. If the vehicle was not force dried, take care not to touch any painted areas, because the paint might not be completely dry. Fingerprints or tape marks can result if the surface is touched. Also, be aware of loosely fitting clothing or belt buckles that can accidentally rub against the paint.

28.3 FINAL DETAILING

Final detailing involves a series of steps to properly clean and shine all visible exterior and interior surfaces of the vehicle, taking special care not to harm newly painted surfaces. Basically, surfaces without new paint are hand washed and the interior is vacuumed.

Some shops or work orders stipulate a complete vehicle detailing, including paint touch-up work to unrepaired body panels. If minor paint problems were found in the new paint, such as debris in the finish, these problems must also be repaired before the vehicle is released to the customer.

Corrective steps for paint detailing include the following:

1. Detail wet sand the flaws in new paint.
2. Machine compound with an abrasive liquid and high-speed buffer equipped with a wool pad.
3. Machine glaze with a finer abrasive liquid and a buffer equipped with a foam pad.
4. Hand-rub and glaze small areas that cannot be machine buffed.
5. Clean all interior and exterior surfaces.

Each of these steps has its own requirements. As a general rule, increasingly finer grades of products—wet sandpaper and compounds—are used for all of these steps. Also, a single product line should be used throughout the repair, and the manufacturer's recommendations should be followed.

Inspecting Painted Surfaces

After all masking materials have been removed, walk around the vehicle to closely inspect all repair work and repainting. Try to find anything that may offend the customer (dirt in paint, paint run, overspray, and so on). Use a drop light and check reflections off the painted surface to find any finish surface flaws, as shown in (Figure 28–14).

If everything went as planned during all shop repair operations, you will not find any paint flaws. The vehicle can be released to the customer after an interior and exterior cleaning.

A **paint protrusion** is a particle of debris (such as dust, dirt, or lint) sticking out of the paint film after refinishing. These problems result from a lack of cleanliness during body repair and repainting processes. Something or someone was not clean enough when doing the repair work.

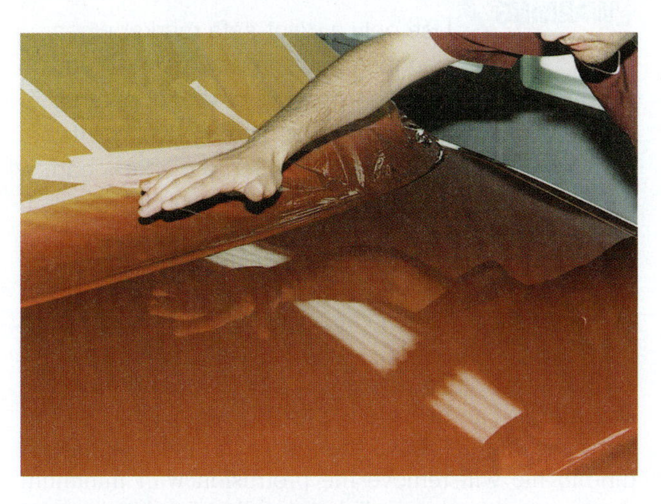

Figure 28-14 Look closely to check that there are no problems in your repair work.

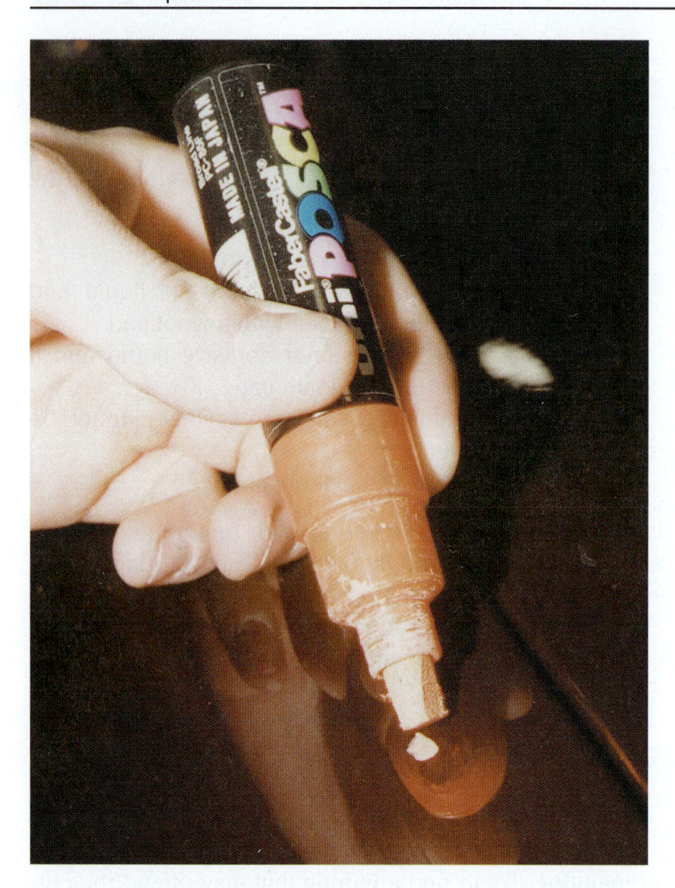

Figure 28–15 **Use only a paint-safe marker to identify surface imperfections.**

Use a **paint-safe marker** to identify any flaws in the new paint film. Touch the marker on the piece of dust or dirt so you will know where to detail sand without having to search closely again. Take your time when marking flaws. Methodically look back and forth across each panel to make sure you find every tiny surface imperfection. See Figure 28–15.

> **WARNING**
>
> Use only a paint-safe marker when denoting flaws in new paint. If you use a permanent marker or one not formulated for automotive paints, you can damage the finish. Some markers can bleed down and leave a mark deep in the paint that is difficult to detail sand out of the finish.

Dirt-Nib Filing

Dirt-nib files can be used to partially remove tiny protrusions sticking up above the rest of the paint surface. A dirt-nib file will remove the protrusion with minimum damage to the surrounding paint film. Dirt-nib files are available commercially or you can make your own.

To use a dirt-nib file, place the file lightly on the paint film. Do not push down or you could scratch the fresh paint. Use short, straight strokes in one direction only. Make two or three light passes to remove most but not all of the defect. After gentle filing, the area must be sanded with an ultrafine grit to further level out the paint flaw.

Detail Sanding

Detail sanding involves using a small dirt-nib sanding block and ultrafine sandpaper to level and smooth small specks of debris (dust, dirt, hair, and so on) in the paint. A small detail sanding block held in your fingers is used on dust and dirt in the finish. A larger handheld sanding block is used on larger paint flaws, such as paint runs (Figure 28–16).

In the past, detail technicians used a dirt-nib file and a whetstone to rub and abrade off minor paint flaws. Now, technicians use a detail sanding block and very fine grits of wet sandpaper to repair paint flaws.

A *detail sanding block* uses Velcro to attach and hold small round, wet sanding mini-discs. This allows you to change to new sandpaper when they become clogged with paint. You can also use different grits of mini-discs for each repair situation. Detail sanding mini-sandpaper discs commonly used during minor paint repair range from #600 grit up to #1500 grit.

Some technicians like to dry sand small dirt particles in the paint first. Others prefer to wet sand them through the whole paint repair process (Figure 28–17).

Dry sanding allows you to see which areas have been sanded and which areas have not been sanded enough. Dry sanding turns the clearcoat white so you can see where clearcoat has been removed. When you wet sand a clearcoat, the sanding marks and dust will stay clear and transparent. It might be best to dry sand dirt in the paint first. Then wet sand to help smooth and level the surface. Refer to Figure 28–18.

Attach the appropriate grit sandpaper using the Velcro backing. The larger the surface flaw, the more

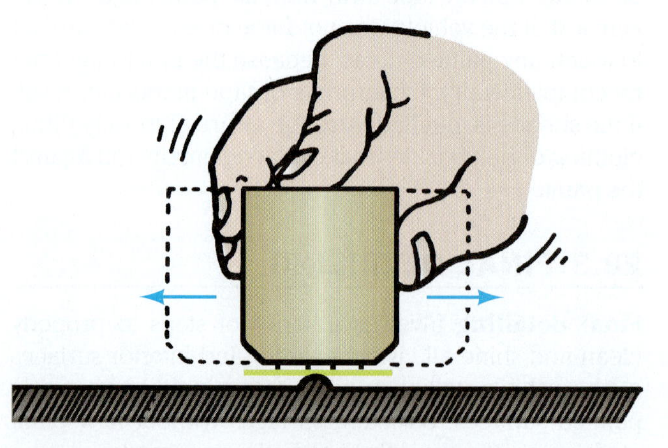

Figure 28–16 **A detail sanding block will concentrate sanding action over the top of paint protrusions. Press lightly to remove only the high spot on the nib.**

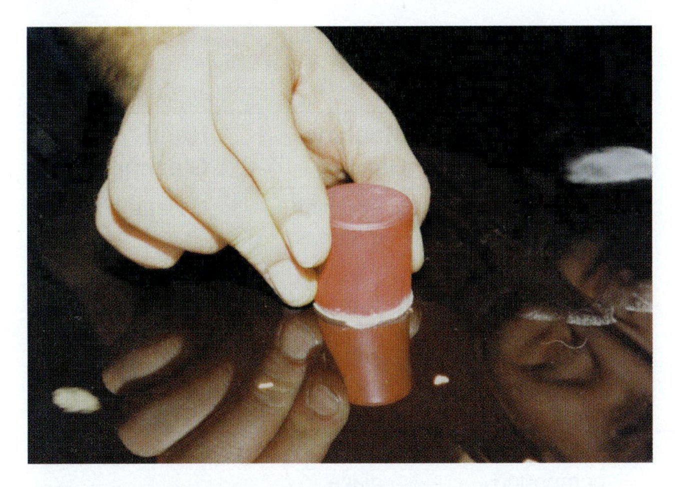

Figure 28–17 Here a detail technician is dry sanding dirt in the paint finish with a detail sanding block and ultrafine sandpaper. Dry sanding will more easily let you see how much you have sanded off.

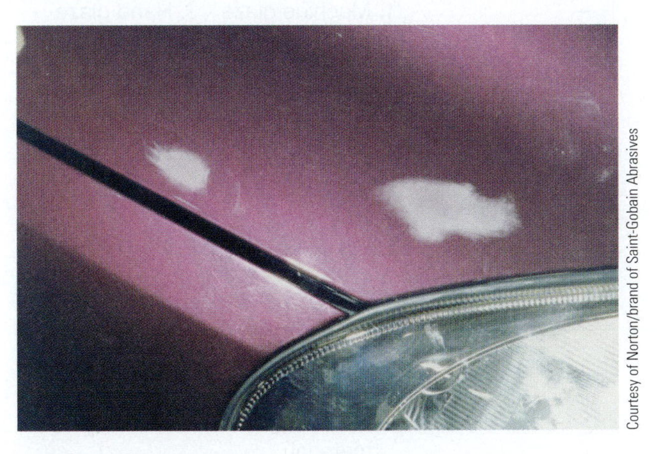

Courtesy of Norton/brand of Saint-Gobain Abrasives

Figure 28–18 Only sand enough to knock off the dirt-nib. This will dull the finish.

coarse a grit sandpaper you should use. For example, on a small piece of dust, use #1000 grit first and then finer #1500 grit as you final level the flaw. This finer grit will make it easier to buff out the sand scratches. Place the detail sanding block over the protrusion. Move it back and forth gently in different directions. Do not press down very hard. Only sand where needed—on the high spot or protrusion in the paint.

When the protrusion has all but disappeared, inspect the paint flaw closely. Sand as little as possible. Remember that the clearcoat is only a few thousandths of an inch (mils) thick. You must not cut through the clearcoat or the panel will have to be repainted. You must leave enough clearcoat on the area for buffing, compounding, or polishing to return the paint gloss or shine.

After the paint flaw or protrusion has been sanded level, wet sand with #1500-grit or #2000-grit wet sandpaper to remove sand scratches in the paint. Some paint manufacturers recommend even finer sandpaper (#3000 to #4000 grit) before buffing. This will ready the area to be machine buffed back to a high gloss (Figure 28–19).

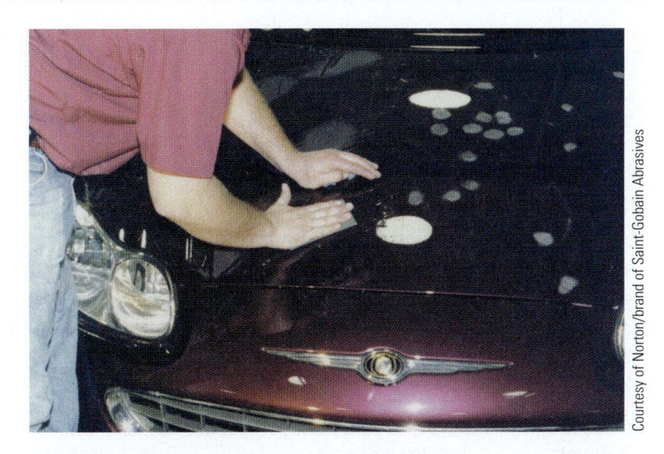

Courtesy of Norton/brand of Saint-Gobain Abrasives

Figure 28–19 After dry sanding dirt-nibs, wet sand them with #1500 grit sandpaper and water. Wet sand the scratches left from dry sanding.

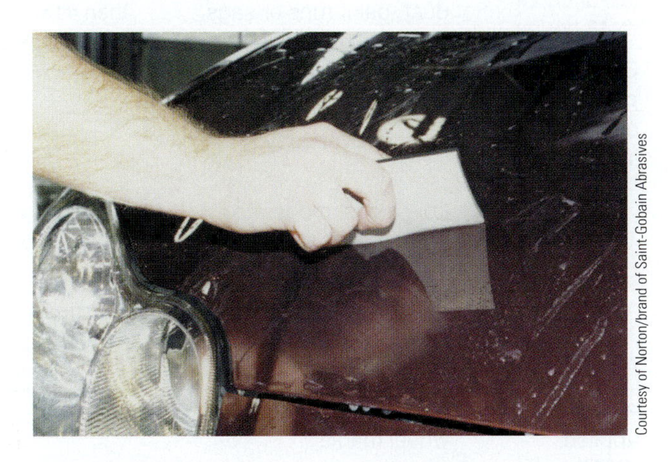

Courtesy of Norton/brand of Saint-Gobain Abrasives

Figure 28–20 When wet sanding, squeegee off the water so you can see whether the flaw has been leveled.

The water on the paint will let you find any remaining surface flaws. Use a soft rubber sanding pad or block to squeegee the water off the flawed area. See Figure 28–20.

Table 28–1 summarizes detail or finesse sanding procedures.

If you fail to catalyze enamel paints with a hardener, you may not be able to wet sand and repair minor paint problems right away. If slight defects in the finished surface appear, they should not be compounded until the paint has had a chance to cure. This can involve a period of several days if the material is not catalyzed.

Repairing Paint Runs

To repair a paint run without repainting, you must carefully wet sand the run while trying not to sand through

TABLE 28–1 DETAIL SANDING PROCEDURES

Paint Type	Paint Condition	Procedure: Wet Sanding	Procedure: Compounding	Procedure: Machine Glazing	Procedure: Hand Glazing
Refinish paints: cured enamels and urethanes* (air-dried more than 48 hours or baked)	1. Minor dust or mismatched orange peel (light sanding)	1. Ultrafine #1500	1. —	1. Finishing material	1. Hand glaze
	2. Heavy orange peel, dust, paint runs or sags	2. #1000 then #1500	2. Microfinishing compound	2. Finishing material	2. Hand glaze
Refinish paints: fresh enamels/ urethanes* (air-dried 24 to 48 hours)	1. Minor dust or mismatched orange peel (light sanding)	1. Ultrafine #1500	1. Microfinishing compound	1. Microfinishing glaze	1. Hand glaze
	2. Heavy orange peel, dust, paint runs or sags	2. #600, #1000, then #1500	2. Microfinishing compound	2. Microfinishing glaze	2. Hand glaze
Refinish paints: acrylic lacquer	1. Low gloss or overspray	1. —	1. —	1. Machine glaze	1. Hand glaze
	2. Low gloss, minor orange peel, or overspray	2. —	2. Paste or rubbing compound (heavy cut)	2. Machine glaze	2. Hand glaze
	3. Low gloss, moderate orange peel, or dust nibs	3. Ultrafine #1200	3. Microfinishing compound (medium cut)	3. Machine glaze	3. Hand glaze
	4. Low gloss, heavy orange peel, paint runs or sags	4. Ultrafine #1000	4. Paste or rubbing compound (heavy cut)	4. Machine glaze	4. Hand glaze
All factory applied (OEM)	1. New car prep or fine wheel marks	1. —	1. —	1. —	1. Hand glaze liquid polish
	2. Coarse swirl marks, chemical spotting, or light oxidation	2. —	2. —	2. Finishing material	2. Hand glaze
	3. Overspray or medium oxidation	3. —	3. Microfinishing compound (medium cut)	3. Finishing material	3. Hand glaze
	4. Heavy oxidation or minor acid rain pitting	4. —	4. Rubbing compound (heavy cut)	4. Finishing material	4. Hand glaze
	5. Dust, minor scratches, or major acid rain pitting	5. Ultrafine #1500	5. —	5. Finishing material	5. Hand glaze
	6. Orange peel, paint runs or sags	6. Ultrafine #1200 or #1500	6. Microfinishing compound (medium cut)	6. Finishing material	6. Hand glaze

*Enamels and urethanes—as referred to in this table—are catalyzed paint systems (including acrylic enamel, urethane, acrylic urethanes, acrylic urethane enamels, polyurethane enamels, and polyurethane acrylic enamels) and nonisocyanate-activated paint systems used in color or clear coats.

the clearcoat and into the colorcoat. Use a full-size stiff rubber sanding block and coarser #600-grit wet sandpaper first. Then final sand the paint run with finer #1500 or finer wet sandpaper until it is smooth and level. If you get lucky, you will not sand through to the colorcoat. If you do, the area will have to be repainted.

With a paint run, you want to plane off the high points without cutting through the clearcoat. If you initially try to block sand a run with fine sandpaper, you will cut too deep in the low spots and "wallow out" the low spots, usually cutting through the clearcoat.

The low spots in a paint run will be softer than the high spots. This is because softer resins will tend to flow down and collect in the low spots of a paint run.

Fill a bucket with clean water. Some technicians like to add a mild soap to the water. The soap-and-water solution will help keep the wet sandpaper from sticking and digging into the fresh paint.

First, block sand the paint run using #600 sandpaper on a stiff rubber sanding block. Only use this grit sandpaper to cut the tops off the run. Use plenty of water and watch how the surface dulls to see where you are sanding and removing clearcoat. Wipe the area with a rubber squeegee so you can see the run. Stop as soon as the high spots of the run are sanded level. A few more passes with #600-grit sandpaper could cut into the colorcoat and require repainting.

A good detail technician can "feel" the sandpaper and how deep the wet sanding process has cut into the clearcoat. When the wet sandpaper begins to feel "sticky" as it is moved back and forth, you have cut down into the softer, less cured clearcoat. This is when you should wet sand very lightly and be ready to stop sanding. If you do not cut through the clearcoat when wet sanding and buffing a run, you have saved yourself hours of work.

Next, lightly block sand the paint run area with #1000-grit wet sandpaper. Use the finer grit wet sandpaper to feather and level the run with the surrounding paint surface.

As soon as the run is sanded level, use the sanding block with finer #1500-grit wet sandpaper to smooth the surface and prepare it for machine buffing. Sand very gently, because the clearcoat will be very thin at this point.

Repairing Chipped Paint

Chipped paint results from mechanical impact damage to the paint film. It is a condition where small flakes or areas of paint have been crushed and damaged. The areas around the missing paint chip have lost adhesion with the substrate.

Chipped paint is normally caused by the impact of stones or hard objects. Chipped paint also happens when someone opens a car door and it hits another car or object. Chips in the paint are most common on the front bumper, front edge of the hood, doors, and around the rear of wheel openings on fenders and quarter panels.

If the whole vehicle is not refinished, you may need to touch up chipped paint on panels that have not been painted. Use the paint mixed for the repair. It will usually have hardener in it to speed curing.

Degrease the area with wax and grease remover. If you use a small paintbrush, slowly move the touch-up paint straight into each chip. On smaller chips, a toothpick will reach into the chip more efficiently. If you are using a solid color, use a thicker viscosity touch-up paint to fill the chip in one application. If you have metallic paint, use thinner touch-up paint and several coats to help match the color.

Allow the paint to cure sufficiently before wet sanding and polishing the chip repairs to level the repair (Figure 28–21).

Panel Detail Sanding

Panel detail sanding can be done to smooth the paint surface on larger areas, as when removing orange peel. It is detail sanding, but over a large surface area, using a larger sanding block and sandpaper.

Panel detail sanding should normally be done with a backing pad or rubber sanding block to avoid crowning of the paint surface. A pad or block will help keep large, relatively flat surfaces level and uncrowned. On restricted and curved surfaces, you can use only your hand to color sand.

Sanding blocks and sandpapers are available in a variety of grit sizes. For major surface repairs, use coarser wet sandpapers, #400–#600. For detail sanding, use #1000, #1500, and finer grits of wet sandpaper.

Wet sand in a back-and-forth or small circular motion, depending on the shape of the surface problem and the contour of the body panel. Use plenty of water to flush away paint debris. Dip the block in a bucket of water. You can also use a sponge, garden hose, or spray bottle to flow water over the area. Some air sanders are equipped with a wet sanding attachment that uses a small plastic hose to feed flushing water up to the sanding pad.

Check the defect often when using a sanding block. You do not want to cut too deeply into the finish. If you cut through the clearcoat or color, repainting will be necessary. Wash surfaces thoroughly with clean water and a sponge after panel detail sanding.

28.4 PAINT COMPOUNDING

Paint compounding involves using different abrasive pastes and liquids to hand and machine polish a surface to a high gloss or shine. You must be familiar with each type of compounding material to select and use them correctly.

Rubbing Compound

Rubbing compounds, also called hand compounds, generally contain the coarsest grit abrasive. They are used to abrade and smooth a surface film by hand to level minor surface imperfections. Rubbing compounds remove the surface gloss and must be followed up with a hand-glazing compound to restore paint shine. They are commonly used on smaller parts or areas that cannot be compounded with a buffing machine.

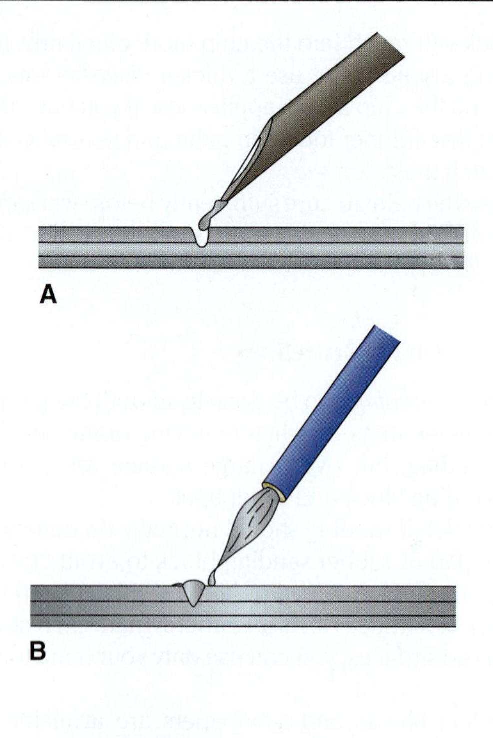

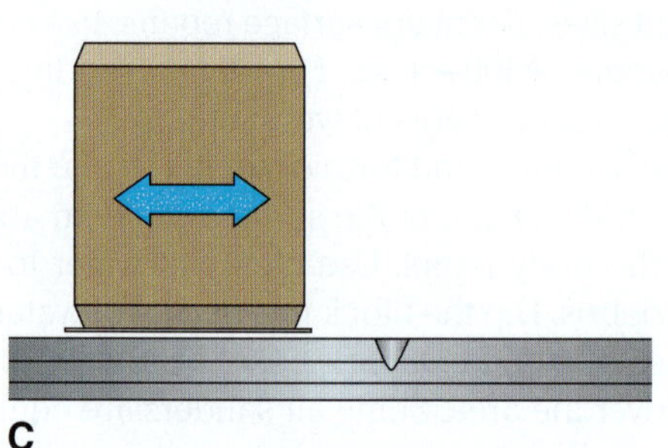

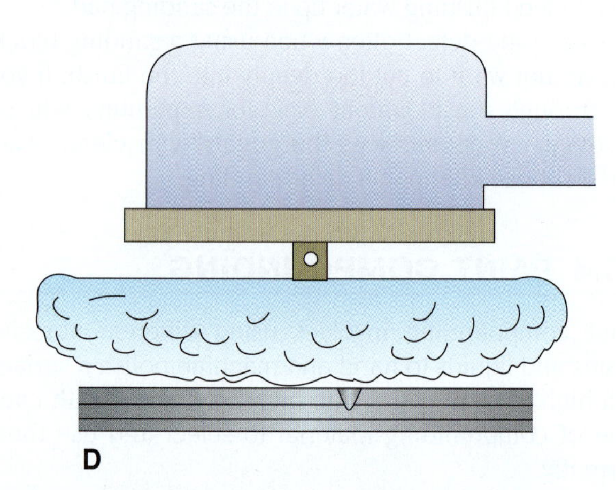

Figure 28-21 Study the basic steps for touching up a paint chip. (A) Use a toothpick or small brush to apply catalyzed paint into the chip. (B) Let the first coat flash and apply another drop of paint over the chip. (C) After allowing the paint to cure, detail wet sand the paint flush. (D) Hand- or power-buff the repair area to return the gloss.

Rubbing compounds are available in various cutting strengths. Hand compounds are oil-based to provide lubrication. Small or blended areas are best treated by hand-compounding. On large surfaces, machine compounding is recommended.

Rubbing compounds are used to:

- Eliminate fine sand scratches around a repair area
- Correct a gritty surface
- Smooth and bring out some of the gloss of lacquer topcoats
- Repair paint on areas that cannot be buffed with a machine

Hand-Compounding

Fold a soft, lint-free cloth into a thick pad or roll it into a ball and apply a small amount of hand compound to it. Use straight, back-and-forth strokes and medium-to-hard pressure until the desired smoothness is achieved.

Hand-compounding takes a lot of elbow grease and is time-consuming. To keep the compounding of topcoats to a minimum, it is important to apply the clearcoats as full wet coats, without sags or runs.

When using hand polishes or glazes, apply the glaze to the surface using a clean dry cloth. Rub the glaze thoroughly into the surface. Then wipe it dry.

Table 28–2 shows some applications for different rubbing and polishing compounds.

Machine Compounding

Machine compounds are water-based to disperse the abrasive while using a power buffer. Some product manufacturers rate their compounds or liquids and pastes by a grit rating system: #1000, #1500, #2000, and finer. Just like

TABLE 28–2 POLISHING AND RUBBING COMPOUNDS

Grade	Liquid	Paste	Use and Application
Very fine	Machine or hand	–	Used to remove swirl marks on topcoat. Spread material evenly with buffing wheel pad before starting compounding.
Fine	Machine or hand	Hand (add water for machine use)	Used to level orange peel. Can also be used to clean, polish, and restore older finishes leaving no wheel marks or swirls.
Medium	Machine or hand	Paste (add water for machine use)	Used for quick-leveling orange peel. Can be used to repair other minor paint defects.
Coarse	Machine	Machine	Used for compounding before final topcoating.

Figure 28-22 Various materials are available to help you repair paint problems: dirt-nib file, soft sanding blocks, hand glaze, machine glaze, swirl mark remover, and so on.

sanding, you would start out with a coarser rated compound and follow up with finer compounds to bring the paint to a high gloss and shine (Figure 28–22).

Refer to the label directions for the machine compound or glaze to learn about its cutting and polishing characteristics. The label directions will give instructions for buffer speed, surface temperature, and so on for properly using their buffing product (Figure 28–23).

A **buffing pad** is rotated by an electric or air buffer to force the compound over the paint surface. If done properly, this will quickly bring the wet sanded paint surface back to a glossy shine. There are different kinds of machine buffing compounds and pads.

A **buffing machine** uses a spinning or rotating action to level and quickly smooth a paint surface. Machine buffing can be done with either a soft wool pad or a foam rubber pad to apply abrasive compound to the paint. Most paint repair technicians use the wool pad first and the softer foam pad second. A **polishing machine** uses an orbital action to bring out the full paint gloss or shine.

Figure 28-23 Compounding removes a thin layer of paint and is needed to return the paint gloss after wet sanding.

Instead of spinning the pad in a circle, the pad is spun and moved sideways by the dual action of the machine. An orbital action polishing machine is needed to bring out a "show-type finish" in a paint. It will remove swirl marks and the finish will look like it has been hand-polished. Compare buffing and polishing machines results in Figure 28–24.

Paint swirl marks are patterns of very fine scratches produced when power buffing or compounding. They are caused by a dirty, worn buffing pad, too much pressure on the buffer, or using too coarse a compound. Always keep your buffing pad clean and replace it when worn.

To avoid swirl marks, most detail technicians first buff the surface with a wool pad and a coarser machine compound. Then they follow with a foam pad used with finer glazing compound. The wool pad buffs more quickly and the foam pad smooths the surface to take out any

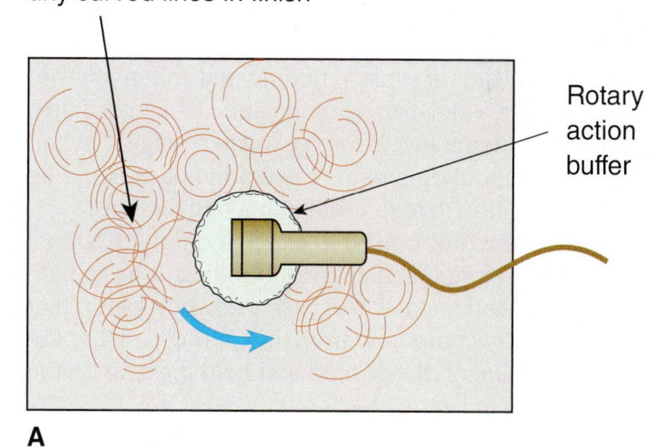

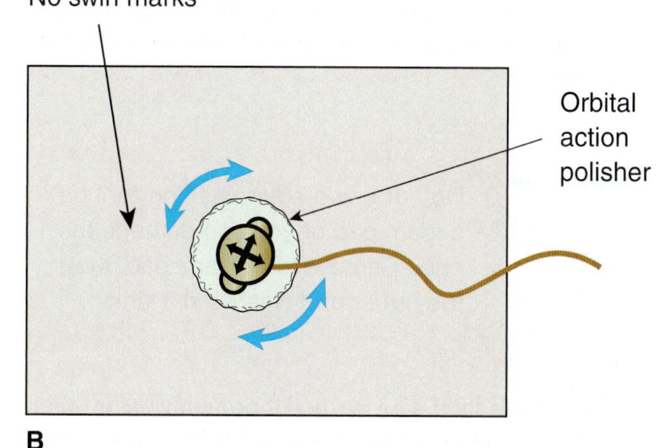

Figure 28-24 Compare the action of buffing and polishing machines. (A) A buffing machine spins a pad in a rotary motion for fast action. Swirl marks can remain after using a single-action buffer. (B) A polishing machine uses dual actions to avoid swirl marks. Though slower, it is often used as a final way to bring out paint shine.

remaining swirl marks. Some shops like to follow the wool and foam pads with an orbital action polisher to remove any remaining trace of swirl marks.

Using Buffers and Polishers

When using buffing and polishing pads, there are several things you must remember to avoid paint damage. In untrained hands, an air or electric buffer can quickly cut through and damage a paint job—a costly mistake.

Wear eye protection when machine compounding or buffing. It is very easy for the abrasive liquid to fly into your face and eyes. It is also possible for chunks of buffing pad to fly into your face. Buff so that any debris flies away from your body and face, not toward it.

Inspect, clean, or replace buffing pads often to avoid problems. If a wool pad is worn and no longer fluffy, replace it. If a foam pad is torn or worn, replace it. When machine compounding paint, the buffing pad should be in good condition to help avoid damaging the finish.

A **pad spur tool** is used to clean and fluff up a wool buffing pad before machine compounding. It has a handle and a spoked metal wheel (like a cowboy's spur) that will remove dry compound from the buffing pad. Lay the buffing machine on the ground and hold the spur on the pad as shown in Figure 28–25.

While wearing eye protection, turn the buffer on and move the spur over the pad while pressing lightly. Hold the spur on the side of the pad that is spinning away from you. This will keep the debris from flying in your direction. Spin spur on the pad until all dried compound is forced out of the pad and the wool fibers are fluffed up and soft.

Do not use a spur cleaning tool on a foam pad or you can damage the pad. Remove the foam pad from the buffer and wash it in a sink.

Foam pads should be washed thoroughly before each use. You should clean the pores of the foam pad out so they can hold compound and polish properly.

Avoiding Paint Burn-Through

Paint burn-through is damage caused by the machine buffing pad removing too much clearcoat. This is a costly, time-consuming mistake that necessitates repainting of the panel. There are several things you can do to avoid burn-through and paint damage. Refer to Figure 28–26 and Figure 28–27.

Edge masking involves taping over panel edges and body lines prior to machine buffing to protect the paint from burn-through. Masking tape is applied to these surfaces to protect them. Buff right up to the tape. After buffing surfaces and removing tape, hand-compound

A This technician is using a spur tool to clean and fluff a wool buffing pad.

B Foam pads should be washed and wetted before use.

Figure 28–25 Your buffing pad should be clean before using it on any vehicle.

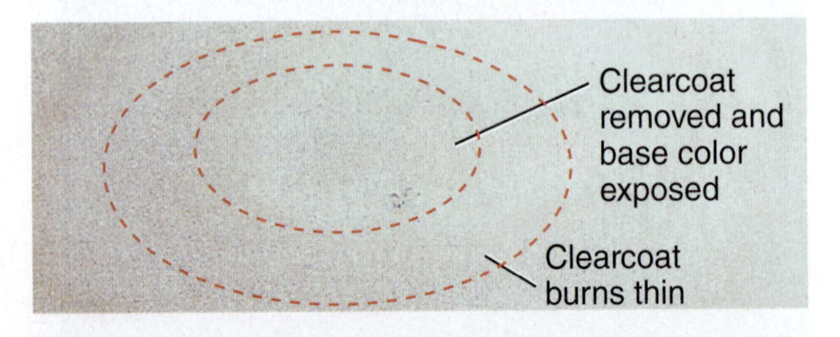

Figure 28–26 Clearcoat burn-through is hard to see. The exposed colorcoat area will be slightly duller than the clearcoat. The clearcoat was cut through into the basecoat in the center area of this photo.

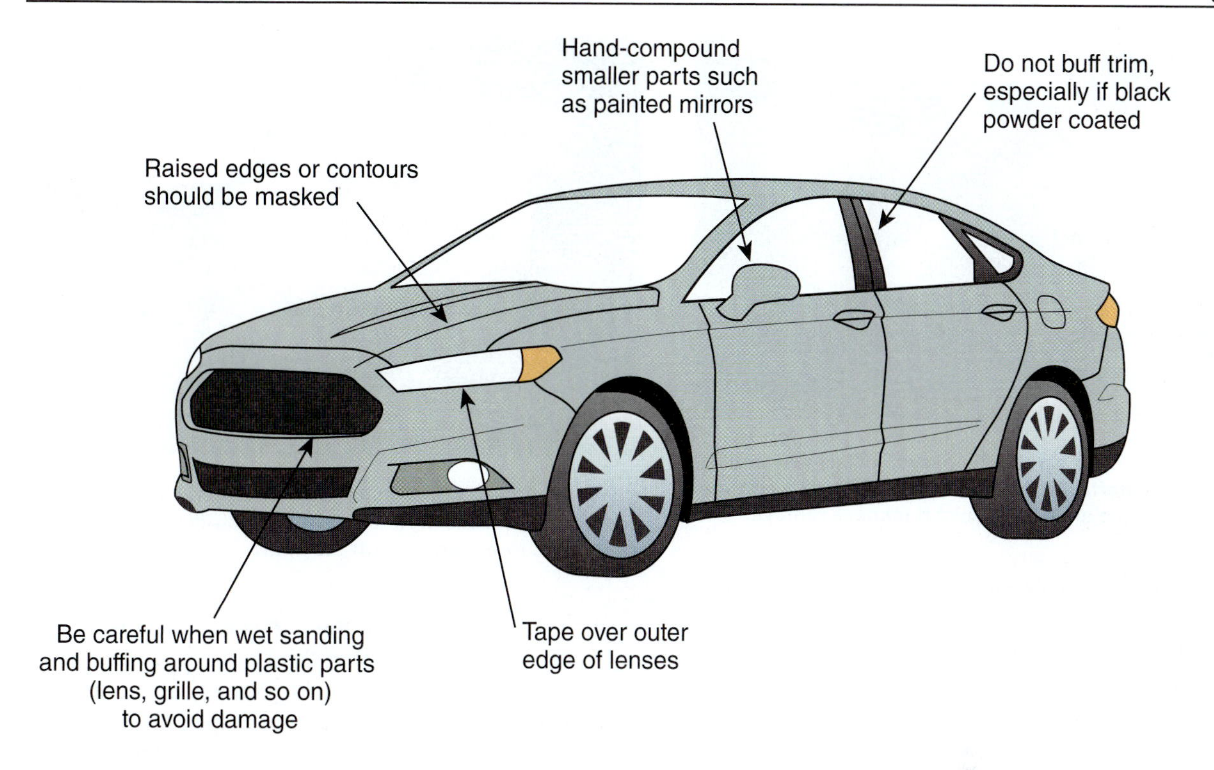

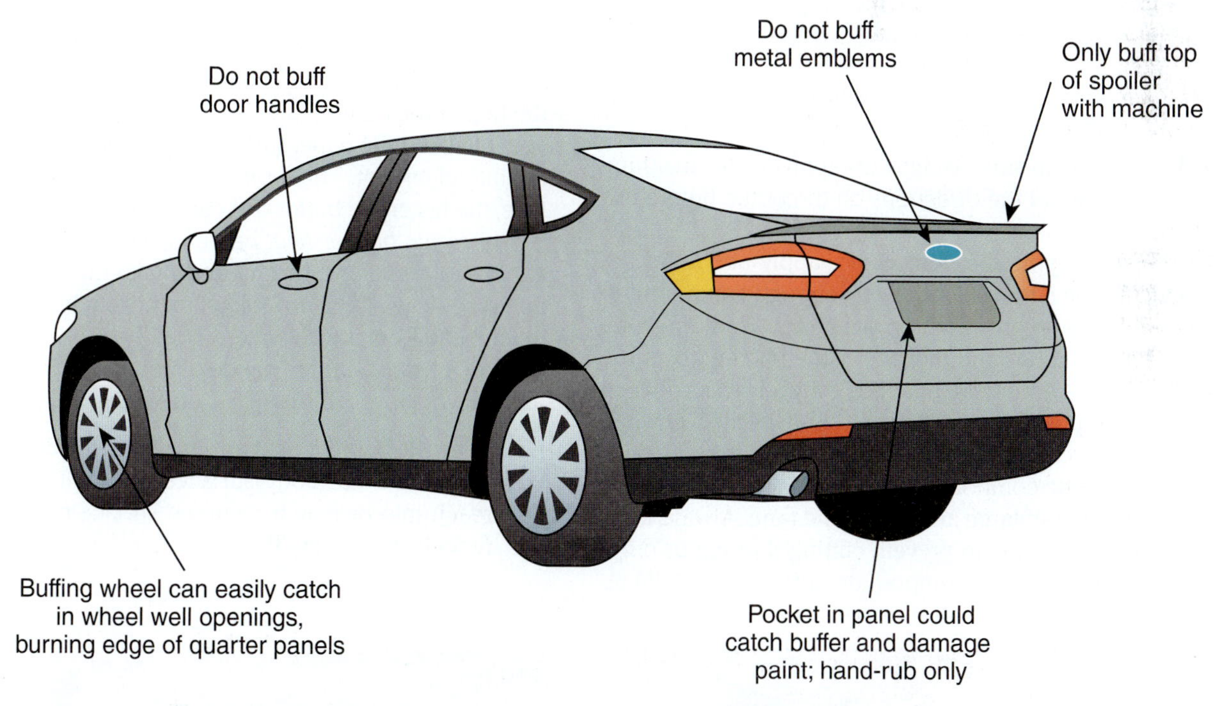

Figure 28-27 Study things you can do to keep from burning through the paint when machine buffing.

these edges. Place masking tape over all sharp body edges. A buffing pad can burn the paint on sharp edges almost instantly. Place masking tape on panel edges so you can buff over the tape without danger of burning through the paint.

You should also mask door handles, emblems, trim, plastic headlight and taillight lenses, and similar parts. The spinning buffing pad can easily damage these parts, requiring their replacement. Today's flat black trim pieces are very prone to damage from a buffer. The black coating can be instantly marred if touched with a high-speed buffing pad. Tape over any part that could be damaged when buffing the paint.

Use separate pads for different grades and types of products. One wool pad should be used for initial buffing with a coarser machine compound. Another foam pad should only be used for applying a finer machine glaze.

To avoid paint burn-through, always move the buffer in even passes over the body surface, as when you paint a car. Make one pass across the panel and then move

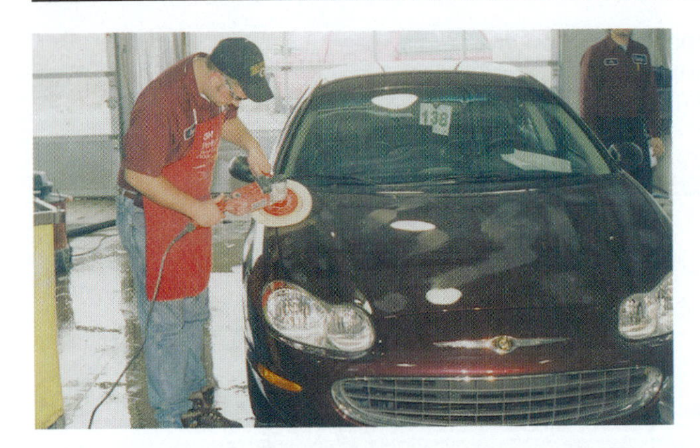

Figure 28-28 Buff a panel in passes, just as you would paint it. Move the buffer back and forth in long passes so you can keep track of the amount of paint thickness removal.

Figure 28-30 Before hitting the trigger, spread compound over the area with a pad. This will keep the compound from spraying off the pad.

down a little. Buff in passes across the panel so you can keep track of how much paint has been buffed. Avoid buffing too much in one location. Do not press down on the buffing machine. Let the weight of the machine do the work. Stay off crowned body contours and sharp edges with the buffing pad. If needed, these surfaces can be hand compounded quickly after you are done machine compounding (Figure 28–28).

Machine Buffing Procedures

Make sure you are using the appropriate machine compound. Read the directions on the bottle before use (Figure 28–29).

When applying the compound, apply an "X" of the product to the surface. Only apply enough compound to buff an "arm's length" area on the panel. Work the compounding liquid around the face of the pad and over the surface before hitting the machine's trigger. This will help prevent compound from flying and spraying all over when you first turn on the buffing machine. See Figure 28–30.

Because the compound has a tendency to dry out, do not try to buff too large an area at one time. Always keep the machine moving to prevent cutting through or burning the topcoat. As the compounds start to dry out, lift up a little on the machine so the pad speed increases. This will make the surface start to shine.

Buffer speed and pressure have an effect on the paint cutting and polishing action. For example, the higher the rpm, the higher the cutting rate; the lower the rpm, the lower the cutting rate.

The faster the buffer is moved across the panel, the slower the cutting rate. The slower the buffer is moved, the higher the cutting rate.

The flatter the panel surface, the slower the buffer will cut into clearcoat. The more round or sharp a panel surface, the faster the buffer will cut into the paint.

Excessive buffing heat can cause swirl marks, warping, discoloring, and hazing and make the material dry out too quickly. If the area is hot to the touch, there is too much heat. Cool it with water.

When using a buffer, the detail technician should use the following procedures:

1. Keep the pad flat or at about a 5-degree angle to the surface on flat body surfaces. Only tilt the pad to reach into or match a curved surface on the body. Refer to Figure 28–31.

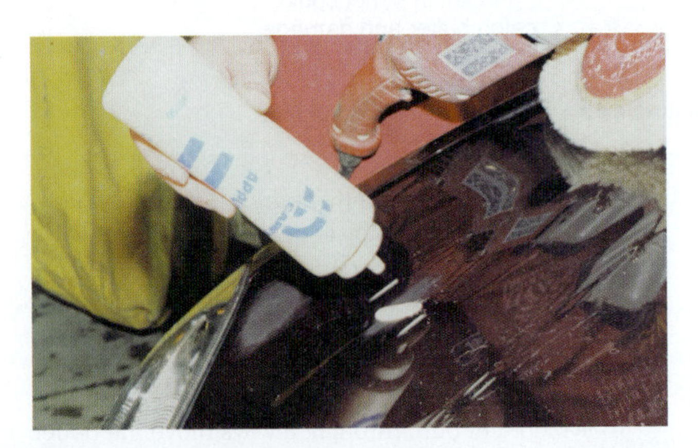

Figure 28-29 Only apply enough compound to buff a small area at a time.

Figure 28-31 Normally, you should hold the buffing pad square on the body surface to avoid cutting too deeply into the clearcoat.

2. Let the weight of the machine do the work. If you press down on a buffing machine, it can quickly cut through the clearcoat.
3. Use care around panel edges and body lines to avoid burn-through. Do not let the edge of the buffing pad get down into panel gaps, or you could even burn through your protective masking tape and the paint (Figure 28–32).
4. Check the repair area often and apply more product as needed. Buff as little as possible to smooth and shine the paint surface.
5. Compound until the product begins to dry. Do not keep buffing if the compound has dried, because you will burn through the paint surface.
6. Never lay the face of a buffing pad on a workbench or any surface that could contaminate the pad with dirt and debris. One speck of sand in the pad can badly scratch the paint.
7. Never use a power buffer with a hand-rubbing compound. This will cause deep scratches, swirl marks, and burn-through. Only use machine compounds when power buffing.
8. Place masking tape over gaps in panels. This will keep compound out from behind panels so time is not wasted cleaning these areas after buffing.
9. Hand-rub small parts and internal pockets in panels that could be easily damaged by the spinning buffing pad. Hand-compound these areas to avoid burn-through.
10. After initial compounding with a wool pad, buff again lightly with a foam pad and finer glazing compound. This will help remove swirl marks and bring out the paint gloss.
11. After the machine compounding, remove the tape and hand-compound all edges and contours just enough to produce a smooth finish. Keep in mind that body lines usually retain less paint than flat surfaces and thus should get only minimal compounding.

A common mistake for the beginner is to burn through new paint while machine compounding. In an effort to make the paint job really shine, the technician may cut right through the paint to the basecoat. The result is usually a time-consuming repaint of the panel.

Hand and Machine Glazing and Polishing

Glazing or polishing involves using very fine grit compound to bring the paint surface up to full gloss. It is usually done after compounding. You can hand-polish small or hard-to-reach areas. Machine polish larger areas to save time.

Slight defects in the topcoat can be repaired by polishing. The choice of compounds depends on the extent of the damage. Final polishing should always be done with an ultrafine polishing compound.

When using rubbing compounds and machine glazes, be sure to follow these procedures:

1. Use a single manufacturer's product line.
2. Follow the manufacturer's recommendations for use.
3. Use the materials sparingly.
4. Use the buffing wheel to evenly distribute the material over the area that is being repaired (Figure 28–33).
5. Keep the pad flat and directly over the surface being repaired.
6. Use a slow, methodical motion so you can keep track of how much area has been buffed (Figure 28–34).
7. Use the finest grit product possible last. Using a finer grit product may take a little longer initially but will generally require less time to complete the repair.
8. Reduce swirl marks by avoiding coarse products and worn buffing pads.

Figure 28–32 Be careful when buffing next to the edges of panels. The buffing pad can catch on sharp edges and damage paint or trim pieces.

Figure 28–33 After buffing with a wool pad, many detail technicians like to machine glaze the paint with a softer foam pad. This will bring out more of the paint shine or gloss.

Figure 28–34 Note how the plastic headlight lens, plastic parts on the bumper, and sharp body contours have been protected with masking tape.

Instead of a circular action buffer, you should use an orbital action machine for final polishing. An orbital action polisher will move the polishing compound in a random manner to remove swirl marks from buffing.

28.5 FINAL CLEANING

Final cleaning or get ready is the last, thorough cleanup before returning a vehicle to a customer. You must do all the "little things" that make a big difference to customer satisfaction. The interior and exterior of the vehicle should be cleaner than when the customer brought it in.

Vacuum the interior of the vehicle carefully. Clean the seats, door panels, seat belts, and carpets. If dusty, clean and treat vinyl surfaces with a conditioner. Be sure to remove all excess cleaner/conditioner from the seat crevices and folds. Stubborn stains, such as blood, should be cleaned with a recommended cleaning solution (Figure 28–35).

Carefully remove any overspray that may have been left on windows or chrome. If it can be done without

Courtesy of Dynabrade Inc.

Figure 28–35 Here a detail technician is using an air-powered scrubber to clean bloodstains out of carpeting.

Avoid using strong cleaning agents on the plastic parts in the dash panel. Some cleaners will dissolve and damage plastic, a costly mistake. You should also avoid having any product with silicone in it in the body shop.

dripping on the new finish, use paint solvent (thinner or reducer). Clean and polish chrome, moldings, and bumpers. Thoroughly clean all the glass, including windows, mirrors, and lights.

Use a brush with soap and water to clean the tires and wheels. Do not let dirty wheels and tires spoil the appearance of an otherwise quality job. Coat them with a conditioner.

Steel wool should not be used to polish chrome, because pieces of the wool can easily become embedded in the new finish. Instead, use a commercial chrome polish.

Spray rubberized undercoating to blacken wheel openings and any other exposed undercarriage parts, because color overspray often gets on these areas. The customer generally will not necessarily notice the undercoating, but certainly will notice if it is not done.

Replace wipers, moldings, and emblems that were removed before finishing. Take the time to clean off these items and be certain that everything is replaced. Make sure all weather stripping is installed properly.

If the vehicle has a vinyl top, do not forget to wipe it with a damp cloth or a commercial vinyl cleaner.

As a finishing touch, clean the engine compartment using a pressure washer. A clean engine compartment usually makes a big impression on the customer. Be careful not to damage any new paint on the fenders when pressure washing. Keep strong engine degreasing agents off the paint.

Finally, inspect the vehicle with a careful eye for details. If a window is smeared, clean it again. If a piece of masking tape remains, remove it. If an emblem is missing, replace it before the customer asks where it is.

If the vehicle gets dirty while waiting to be picked up, wipe it down with a clean cloth. The number one objective should always be a satisfied customer.

28.6 CARING FOR A NEW FINISH

A newly refinished vehicle must receive special care, as the paint can still be curing for several days or months. Each paint manufacturer will have specific recommendations for caring for a new finish. Explain all of these precautions to the vehicle owner.

To care for a new finish, you and the customer should:

- Avoid commercial car washes and harsh cleaners for one to three months.
- Hand-wash using only water and a soft sponge for the first month. Dry with cotton towels only. Do not use a chamois.
- Avoiding waxing and polishing for up to three months. After that time, use a wax designed for basecoat/clearcoat finishes, as they are the least aggressive.
- Avoiding scraping ice and snow near newly refinished surfaces.
- Flush gas, oil, or fluid spills with water as soon as possible for the first month. Do not wipe off.

SUMMARY

1. Paint problems include a wide range of troubles that can be found before or after painting. You must be able to efficiently analyze and correct paint problems.
2. If you see paint defects while spraying, you must decide whether to stop work immediately or wait until the painting is finished to correct the problem. Your decision will depend on the type and extent of the problem.
3. The objective in final detailing is to locate and correct any defect that may cause customer complaints.
4. Paint surface chips result from mechanical impact damage to the paint film: door dings, damage from road debris, and so on.
5. A paint surface protrusion is a particle of paint or other debris sticking out of the paint film after refinishing.
6. A detail sanding block is commonly used to remove any defect on or above the surface of the paint.
7. Wet block sanding can be done to smooth the paint surface on larger areas, as when removing orange peel.
8. Hand-rubbing compounds contain coarser grit abrasives than machine compounds and glazes.
9. Machine compounds are water-based to disperse the abrasive while using a power buffer.
10. Edge masking involves taping over panel edges and body lines prior to machine buffing or polishing to protect the paint from burn-through.

EXERCISES

On a separate sheet of paper, complete the following learning activities for this chapter. Write definitions for the key terms and answer the ASE-style review questions, essay questions, critical thinking problems, and math problems. You can also do the outside activities, possibly for extra credit.

Key Terms

blistering
blushing
buffing machine
buffing pad
chipped paint
detail sanding
final detailing
lifting
line checking
machine compounds
orange peel
pad spur tool
paint chalking
paint color mismatch
paint fish-eyes
paint mottling
paint protrusion
paint-safe marker
paint wrinkling
polishing machine
sand scratch swelling
solvent popping

ASE-Style Review Questions

1. After a panel repair and repainting, the masking paper is removed and it is discovered that the new metallic paint is darker than the original color. Technician A says that the last color coat might have been sprayed too wet. Technician B says that the painter might not have made a spray-out test panel to check the new paint color. Who is correct?
   A. Technician A
   B. Technician B
   C. Both A and B
   D. Neither A nor B

2. A new spot repair has a small paint run in it. Technician A says that the spray gun may have been moved too slowly or not moved in controlled passes over this area. Technician B says that a slower drying solvent should have been used. Who is correct?
   A. Technician A
   B. Technician B
   C. Both A and B
   D. Neither A nor B
3. A small paint run is being repaired. Technician A says to block sand the run with #600-grit wet sandpaper to quickly plane the area level. Technician B says to start by sanding with much finer sandpaper, such as #1500 grit. Who is correct?
   A. Technician A
   B. Technician B
   C. Both A and B
   D. Neither A nor B
4. A paint job shows a few small fish-eyes. Technician A says to let the paint cure, sand, and refinish the area. Technician B says mixing a little fish-eye eliminator additive in the paint may correct the problem. Who is correct?
   A. Technician A
   B. Technician B
   C. Both A and B
   D. Neither A nor B
5. Which condition can sometimes be found on OEM finishes and repair finishes, to a certain degree?
   A. Bull's-eye
   B. Lifting
   C. Orange peel
   D. Chalking
6. Technician A says to use a light mist coat when applying the last color coat. Technician B says to always use a full wet coat with metallics. Who is correct?
   A. Technician A
   B. Technician B
   C. Both A and B
   D. Neither A nor B
7. Which should not be done when machine buffing a finish?
   A. Mask over sharp edges
   B. Mask over plastic parts, trim, and so forth
   C. Keep buffing pad level on surface
   D. Push down on buffing machine
8. Technician A says to try panel detail sanding and polishing if acid spotting is not too severe. Technician B says that sanding and refinishing might be needed. Who is correct?
   A. Technician A
   B. Technician B
   C. Both A and B
   D. Neither A nor B
9. Which of these types of buffing pads is used last when machine compounding a finish?
   A. Wool
   B. Foam
   C. Rubber
   D. Steel
10. A new paint job shows signs of blushing. Technician A says to lower air pressure at the gun. Technician B says to use a slower reducer. Who is correct?
   A. Technician A
   B. Technician B
   C. Both A and B
   D. Neither A nor B

Essay Questions

1. Explain how to prevent paint cracking.
2. Describe some causes of crazing.
3. How can you prevent orange peel?
4. Summarize final detailing.
5. Explain the causes of orange peel.

Critical Thinking Problems

1. Explain when a hand-rubbing compound should be used.
2. How can you prevent burn-through on edges when machine polishing?
3. How do you care for a new finish?
4. Most imperfections in paint originate from what?
5. Explain how to repair a paint run without repainting.

Math Problem

1. Instructions for a decal overlay state that it should be trimmed to extend off each end of a panel by ¾ inch. The panel measures 21 ¼-inches long. How long should the overlay be cut to fit?

Activities

1. Inspect several vehicles to analyze the condition of the paint. Try to find as many paint problems as you can. Write a report on the causes and corrections of the paint problems you find.
2. Visit a body or detail shop. After receiving permission from the shop owner, observe a professional detail technician at work. Report on things you learned by watching this person work.

SECTION 7

Collision Repair Professionalism

Chapter 30 Collision Repair for Hybrid and Electric Vehicles

CHAPTER 29 Job Success, ASE Certification, and Future Technology

OBJECTIVES

After studying this material, you should be able to:

- Get and keep a good job in collision repair.
- Describe desirable and undesirable worker traits.
- Define the term *entrepreneur*.
- Advance into other automotive-related professions.
- Explain the benefits of ASE certification.
- Summarize the purpose of I-CAR.
- Know the sources of professional training and certification available to collision repair shop personnel.
- Correctly answer ASE-style review questions relating to job success and ASE certification.
- Discuss the future of auto body repair.

INTRODUCTION

You have selected an excellent profession—collision repair technology. With the knowledge base gained from reading this book, from your teacher, and from hands-on learning experiences, you should be able to become a good auto body technician. You will then earn a good living for as long as there are motor vehicles on our roads and highways (Figure 29–1).

According to economic predictions, there will be a strong demand for collision repair technicians in the future.

This chapter will utilize everything you have learned about collision repair for maximum productivity and job success. It will give you tips on how to be happy and hard-working on the job. You will learn how to communicate and work well with others in the shop. You will also see how to keep up-to-date with new repair technology. The last section of the chapter summarizes the ASE testing program to gain certification in collision repair.

29.1 CAREER GOALS

What is your ultimate goal in becoming an auto body technician? How far do you want to go in your professional life? These are important questions to consider.

Many technicians find auto body repair extremely rewarding. They take a "crumpled mass of twisted metal" and restore it to a beautiful, functional vehicle. They find working with their hands and their minds to be rewarding.

Courtesy of Liss CarStar Collision Center

Figure 29–1 **The collision repair industry is a fast-paced, highly technical field. Only highly trained professionals can hope to prosper in this industry.**

The Collision Repair Technician

As you have learned, a collision repair technician is a highly skilled individual capable of doing a wide variety of tasks. This person must have a working knowledge of mechanics, electronics, measurement, welding, cutting, straightening, painting, environmental dangers, and so on (Figure 29–2).

Collision repair technicians earn good money. There is also a constant demand for well-trained body technicians. You should be proud of becoming a collision repair technician. See Figure 29–3.

Specialized technicians concentrate their study in one area. Some shops have specialized paint technicians, sheet metal technicians, masking technicians, frame rack technicians, and detailing technicians. This allows technicians to be more efficient at their own skill area. Thus, more vehicles are repaired and more profit is gained by the shop and its workers.

Figure 29-2 There are almost always job openings in body shops. Check your local newspaper for job openings.

Figure 29-3 Collision repair technicians who work hard and have a good knowledge of their trade earn very good money!

Professionalism

Professionalism is a broad trait in a worker that includes everything from being able to follow orders to taking pride in workmanship. A **professional** does everything "by the book." Professionals never cut corners (for example, leave out the hard-to-reach bolt) when trying to get the repair done. The professional always take the time to do everything the right way. Always think, dress, and act like a professional.

Professionals are:

- Customer-oriented
- Up-to-date on vehicle and product developments
- Aware that vehicle safety and integrity depend on repair quality
- Attentive to detail
- Certain that their work is up to specs
- Active in trade associations
- ASE-certified
- Always eager to learn something new to improve their skills
- Clean and organized, especially when it comes to their tools
- Helpful to other technicians when they need it

Average Versus Professional

Do you want to be an "average technician," only doing enough to get by, or do you want to be a "professional technician" who takes pride in doing the very best work? The professional technician will be rewarded by happy customers who return again and again to get their cars and trucks fixed properly and with pride.

An unprofessional technician will be happy if the customer does not notice a bad paint job. If you try to cut corners and get too much done too quickly, quality will suffer. You must use proper work habits, common sense, and hard work to be productive. Unprofessional technicians cut corners trying to make more money; however, in the long run, they will suffer with smaller paychecks. Their poor work habits will often result in comebacks.

A **comeback** is a repaired vehicle that is returned to the shop by an unsatisfied customer. The technician who did the original repair as well as other workers in the shop may have to do the repair over again. Many insurance companies and customers will not pay their bill until after they are satisfied with repair quality.

29.2 WORKER TRAITS

Work traits are the little things, besides skills, that make a good or bad employee. Work traits often determine whether you keep a good job or get fired. The most important work traits are detailed in the following section.

Reliability

Reliability means being at work on time, being at work every day, and doing a job right. This is the most important job trait. Without it, you will have a difficult time keeping a job.

If you always miss work, you negatively affect everyone in the shop. If a vehicle has been promised on Friday and you are not there, either someone else has to do your work, or the vehicle may not get repaired. Then the customer will be unhappy and will not return to your shop again. If this continues, word of mouth will ruin the reputation of the shop, and profits will decline.

Social Skills

Social skills are important so that other workers and customers like you. Many times you will need the help of another worker to complete a difficult, two-person task. If you are unliked by coworkers, you will find many tasks almost impossible. Look at Figure 29–4.

A good body shop will have a team of workers who help each other succeed and prosper. They will exchange information, help each other with small tasks, and enjoy working together. You spend much of your life on the job, so why not enjoy it?

It is important for body shop personnel to have good communication and cooperation skills. A wise old saying goes, "A chain is only as strong as its weakest link." This applies to the smooth operation of a body shop. If anyone, from the estimator to the employee responsible for the final get-ready, does not do the job right, everyone suffers. Customers will not return to the shop, and everyone's paycheck will go down.

Productivity Versus Quality

Productivity is a measure of how much work gets done. A highly productive technician will turn out a large amount of work. This will result in higher pay for the technician and more profits for the shop.

Courtesy of ASE

Figure 29–4 To do well in any profession, you must get along with your coworkers and customers. Professionalism pays off!

Quality means that you do every step of a repair correctly so that the repair is strong, reliable, and attractive. You want to balance productivity and quality.

Systematic Approach

It is important that you develop a systematic approach for doing your job. A **systematic approach** involves organizing and using a logical sequence of steps to accomplish a task or job. A systematic approach will result in more efficient ways of working. It will also help prevent the kinds of mistakes that can affect repair quality.

Ask yourself these kinds of questions:

- Did I take all needed tools to the vehicle?
- Is there a better tool for a specific task?
- Have I been reading the manufacturer's instructions (paints, body filler, equipment, and so on)?
- Is my body in the right position to protect myself and perform the repair comfortably?

29.3 ENTREPRENEURSHIP

An **entrepreneur** is a person who starts his or her own business. This business might be a body shop, materials supply house, parts warehouse, or similar endeavor. You might want to consider being an entrepreneur someday.

If you want to start your own successful business, you need to understand personnel management, bookkeeping, payroll, and state and local laws controlling the industry.

After mastering the knowledge needed to do collision repair work, you will be more marketable. The skills a body technician develops can help when applying for other jobs or starting a business.

Cooperative Training Programs

Some schools offer **cooperative education programs** that allow you to have a job and get school credit at the same time. If available, you could be hired to work and train in a real body shop. Some "co-op" programs pay a small wage while you learn.

During cooperative training, you might be a helper technician. You would work under and be further trained by an experienced professional. This is an excellent way of gaining practical, hands-on experience, and remember, experience is the best teacher.

Automotive-Related Professions

Your career potential in automotive-related fields is vast. Besides the thousands of people who repair motor vehicles, there are countless other related fields that support our nation's transportation industry. Automobile-related professions are jobs that require an understanding of auto body repair. There are many related professions that you can advance into after gaining a working knowledge of body repair. By understanding body repair and furthering your education, you might be able to get a job at a variety of businesses:

- Insurance company—adjuster, agent
- Materials manufacturer—paint, filler, or other repair products
- Parts suppliers—new, used, and remanufactured parts
- Auto manufacturer—new or used car sales, factory representative, designer, or engineer
- Instructor—teaching collision repair classes

Keep Learning

Try to learn something new that is job-related every day. Always learn from your mistakes. Read technical magazines and other publications. While accomplishing tasks, try to think of more efficient ways of working. Consider new tools and techniques. Participate in trade associations, such as VICA, NACE, ASE, and so on. This kind of attitude will help you become more productive.

To stay productive, you must keep track of new vehicle construction technology. You must learn about new repair products and methods to try to improve repair speed and repair quality.

In Canada, compulsory training is required before you can get a job as an auto body technician. In most areas of Canada, untrained people cannot start work until after they have taken state-mandated courses in collision repair. Upon receiving journeyman status and a diploma, certified graduates are then allowed to work in the Canadian collision repair industry.

Career Advancement

Even if you have a good job as a body technician, always consider career advancement. Maybe you can find a better paying job or one with better working conditions.

Job openings are common, as people move up into other positions and retire. Always read newspapers to stay aware of job prospects. Some may offer a higher salary or better benefits (for example, insurance, vacation time, holiday pay) than your present job.

Even if you work as a technician, you still might want to take college courses. Take subjects that might help you in your career.

29.4 CERTIFICATION PROGRAMS

Various companies and professional associations offer certification programs. They offer training and/ or tests in collision repair subjects. The most important and well-known certification program in the United States is the Automotive Service Excellence program, or ASE.

ASE Certification

ASE certification is a testing program to help prove that you are a knowledgeable auto body technician. Just as doctors, accountants, electricians, and other professionals are licensed or certified to practice their profession, collision repair and refinishing technicians can also be certified. Presently in the United States, unlike in Canada, certification is voluntary and is not required by law.

Certification protects the general public and the professional. It assures the general public and the prospective employer that certain minimum standards of performance have been met. See Figure 29–5 and 29–6.

Many employers now expect their collision repair and refinishing technicians to be certified. The certified technician is recognized as a professional by the public,

SCRS
Society of Collision Repair Specialists
ASE
CERTIFIED
We employ technicians certified by the National Institute for
AUTOMOTIVE SERVICE EXCELLENCE
Let us show you their credentials
ASA

Figure 29–5 Various trade associations offer training and certification programs, as do large collision repair product manufacturers. Any of these programs can help you improve your skills and knowledge.

Figure 29–6 If you pass one of the ASE tests in collision repair, you will be given a patch that can be sewn on your work shirt. This will show everyone that you are a certified and knowledgeable collision repair professional.

employers, and peers. For this reason, the certified technician usually receives higher pay than one who is not certified. Refer to Figure 29–7.

ASE Tests

There are five ASE tests, four for technicians and one for estimators. The titles of the ASE tests in auto body repair are:

1. Painting and Refinishing
2. Nonstructural Analysis and Damage Repair
3. Mechanical and Electrical Components
4. Structural Analysis and Damage Repair
5. Damage Analysis and Estimating

Table 29–1 gives a breakdown of the ASE tests. Table 29–2 summarizes the test for becoming a certified estimator.

Courtesy of ASE

Figure 29-7 Here technicians are taking one of the ASE certification tests. They want to prove to themselves, their employers, and the public that they are competent to repair collision-damaged vehicles.

TABLE 29–1 ASE TECHNICIAN CERTIFICATION TESTS

Test Specifications — Painting and Refinishing

Content Area	Questions in Test	Percentage of Test
A. Surface preparation	13	24%
B. Spray gun operation and related equipment	8	14%
C. Paint mixing, matching, and applying	14	26%
D. Solving paint application problems	9	16%
E. Finish defects, causes, and cures	6	11%
F. Safety precautions and miscellaneous	5	9%
Total	**55**	**100%**

Test Specifications — Nonstructural Analysis and Damage Repair

Content Area	Questions in Test	Percentage of Test
A. Preparation	5	9%
B. Outer body panel repairs, replacement, and adjustments	17	31%
C. Metal finishing and body filling	7	13%
D. Glass and hardware	5	9%
E. Welding and cutting	12	22%
F. Plastic repair	9	16%
Total	**55**	**100%**

Test Specifications — Structural Analysis and Damage Repair

Content Area	Questions in Test	Percentage of Test
A. Frame inspection and repair	12	24%
B. Unibody inspection, measurement, and repair	19	38%
C. Stationary glass	5	10%
D. Metal welding and cutting	14	28%
Total	**50**	**100%**

Test Specifications — Mechanical and Electrical Components

Content Area	Questions in Test	Percentage of Test
A. Suspension and steering	12	26%
B. Electrical	8	18%
C. Brakes	5	11%
D. Heating and air conditioning	4	9%
E. Engine cooling systems	4	9%
F. Drive train	4	9%
G. Fuel, intake, and exhaust systems	3	7%
H. Restraint systems	5	11%
1. Active restraint systems (1)		
2. Passive restraint systems (1)		
3. Supplemental restraint systems (SRS) (3)		
Total	**45**	**100%**

TABLE 29–2 ASE ESTIMATOR CERTIFICATION TEST

Test Specifications Damage Analysis and Estimating		
Content Area	**Questions in Test**	**Percentage of Test**
A. Damage analysis	10	20%
B. Estimating	13	26%
C. Legal and environmental practices	4	8%
D. Vehicle construction	6	12%
E. Vehicle systems knowledge	10	20%
1. Fuel, intake, ignition, and exhaust systems (1)		
2. Suspension, steering, and powertrain (2)		
3. Brakes (1)		
4. Heating, cooling, and air conditioning (2)		
5. Electrical/electronic systems (1)		
6. Safety systems (2)		
7. Fasteners and materials (1)		
F. Parts identification and source	4	8%
1. New original equipment manufacturer (OEM) (1)		
2. New aftermarket (1)		
3. Salvage/used (1)		
4. Remanufactured/rebuilt/reconditioned (1)		
G. Customer relations and sales skills	3	6%
Total	**50**	**100%**

If you pass any one of these tests and have two years of work experience, you will become ASE-certified.

ASE tests include multiple-choice questions pertaining to the service and repair of a vehicle. They do not include totally theoretical questions. To prepare for ASE auto body tests, study the material in this book carefully. To help you prepare for the Body Repair and/or Painting and Refinishing tests, some test questions at the end of service chapters in this text are similar to those used by ASE. The Body Repair and Painting and Refinishing tests contain 40 questions in various areas.

Craftspeople with proper work experience who pass the written tests are awarded a certificate and a shoulder emblem for their work clothes.

For further information on the ASE certification program or to get their test preparation booklet, write:

ASE
1305 Dulles Technology Drive
Herndon, VA 22071-3145

You can also visit ASE on the World Wide Web at <**www.asecert.org**>.

Other Trade Associations

In addition to ASE certification, a professional body/paint technician or the shop should consider membership in the various trade associations. The growth and influence of these trade associations within the industry has led to increased communication, training, and the sharing of knowledge and ideas from all sides.

The Society of Collision Repair Specialists (SCRS), the Automotive Service Association (ASA), the Mitchell Certification of Achievement Program, the International Autobody Congress and Exposition (NACE), and the **Inter-Industry Conference on Auto Collision Repair (I-CAR)** are examples of professionals promoting consumer and technician education.

Future Technology

What automotive design changes will we see in the next century? Will we see the all-plastic car? Will hybrids replace conventional gas-engine-only propulsion systems? Will another power source replace the piston engine?

Trace where the human race has gone in the last 100 years! We have progressed from the horseless carriage, to the motor car, to the moon in one human lifetime. Since human progress is increasing exponentially (by multiples), we should see even more rapid progress in the future of collision repair.

We have the technology to allow motor cars to clean up or even modify our atmosphere as desired. Just as a change from R-12 refrigerant to R-134a has had a desired effect on the earth's atmosphere, we can do many other things to help control the atmosphere and natural resources that support life on earth.

More Recycling

Recycling will become even more important in the auto body repair industry. Professional, reputable shops now recycle motor oil, antifreeze, refrigerant, solvents, primer, and paint.

Someday, auto makers might use recyling to the point that the whole car could be made from recycled aluminum or plastic materials. A new car body could someday be made out of "old soda pop cans" and "old plastic milk jugs." In this example, recycling would result in a lighter, easier-to-propel vehicle that will resist corrosion much better than conventional steel.

Auto parts are now being made using recycled aluminum and plastic. To help protect our environment, always purchase and use new parts manufactured from recycled materials.

Lighter Vehicle Construction

Vehicle body construction will involve the use of more lightweight materials, such as aluminum, plastics, carbon fiber, Kevlar, magnesium, and titanium. To increase gas mileage, vehicle weight must be reduced as much as possible. It takes about half the energy to propel a car that weighs 2,000 pounds compared to a full-size car that weights closer to 4,000 pounds.

Several supercar space frames are now made of carbon fiber. Their space frames or unibody structures are made of epoxy resin impregnated with directional glass strands. A carbon fiber frame is very light yet also very strong and rigid.

A carbon fiber space frame would be very difficult, if not impossible, to repair after a major collision with structural damage. All of the useable parts and panels on the damaged vehicle must be transferred over to a new "carbon fiber tub" or space frame structure. A very, very expensive collision repair, if not the most expensive!

Easier Diagnosis

Problem diagnosis should get even easier as on-board diagnostic systems are improved to provide even more feedback on potential electrical as well as mechanical problems in the test vehicle. Finding problems before releasing the car to the customer should become even easier in the future.

Solar Panel Paint

What about a paint that can serve as a solar collector or solar panel? What if dissimilar silica (sand) plates could be sprayed onto the vehicle body to serve as a solar panel with positive and negative terminals? The whole surface of the car body could help recharge the battery stack for your drive home from work.

Polymer (Plastic) Paints

Polymer paints are experimental self-healing paints made of polymers (plastics) that can actually flow out to fill minor scratches in the paint film. Paints containing these special polymers can "heal themselves, and in a few seconds. Maybe the industry will develop self-healing paints in the future to help a vehicle's paint to look new and shiny for a longer period of time.

SUMMARY

1. According to economic predictions, there will be a strong demand for collision repair technicians in the future.
2. Collision repair technicians earn good money. There is also a constant demand for well-trained professionals. You should be proud of becoming a collision repair technician.
3. A professional is a worker with a broad range of skills that include everything from being able to follow orders to taking pride in workmanship.
4. A "comeback" is a repaired vehicle that is returned to the shop by an unsatisfied customer.
5. Reliability means being at work on time, being at work every day, and doing the job right. This is the most important job trait.
6. Social skills are important so that other workers and customers like you. Many times you will need the help of another worker to complete difficult tasks.
7. An entrepreneur is a person who starts his or her own business successfully.
8. Try to learn something new that is job-related every day. Read technical magazines and other publications.
9. Always learn from your mistakes. Ben Franklin once said, "Everything that hurts instructs!"
10. Various companies and professional associations offer certification programs and training in collision repair subjects.

EXERCISES

On a separate sheet of paper, complete the following learning activities for this chapter. Write definitions for the key terms and answer the ASE-style review questions, essay questions, critical thinking problems, and math problems. You can also do the outside activities, possibly for extra credit.

Key Terms

ASE certification
comeback
cooperative education programs
entrepreneur
Inter-Industry Conference on Auto Collision Repair (I-CAR)
productivity
professional
reliability
social skills
systematic approach
work traits

ASE-Style Review Questions

1. Technician A gets along with his or her coworkers. Technician B keeps to himself and will not help others with complex two-person tasks. Which technician will be more successful?
   A. Technician A
   B. Technician B
   C. Both A and B
   D. Neither A nor B
2. Which term refers to a worker who can follow orders and shows pride in workmanship?
   A. Certified
   B. Professional
   C. Average
   D. Comeback
3. Workers who fail to get to work on time do not exhibit which desired worker trait?
   A. Quality
   B. Productivity
   C. Reliability
   D. Systematic approach
4. If a technician uses a systematic approach to doing repairs, which question would he or she not ask?
   A. Did I take all needed tools to the vehicle?
   B. Is there a better tool for a specific task?
   C. Should I call my wife or girlfriend/husband or boyfriend?
   D. Am I protecting myself with safety glasses?
5. How many total tests does ASE offer to collision repair personnel?
   A. 3
   B. 4
   C. 5
   D. 6

Essay Questions

1. List ten traits of a professional auto body technician.
2. Write a paragraph on how being late to work can hurt you as well as your coworkers.
3. In your own words, what is a cooperative education program?

Critical Thinking Problems

1. Why is cooperation between employees in a body shop important?
2. List some other professions where knowledge of collision repair would be helpful.
3. What is the official World Wide Web name for the ASE website?
4. Why is reliability such an important factor in a successful workplace?

Math Problems

1. By learning how to do her job more efficiently from continued study, a technician increased her income by 20 percent. If this technician was earning $35,000 dollars per year before, how much is she earning per year now?
2. A technician named Tom constantly mixes 20 percent too much paint for his paint work compared to other painters. If the average technician wastes $500 of materials per year, how much would Tom waste per year in dollars?

Activities

1. Visit several body shops and observe technicians doing repairs. Try to find a technician you would call professional and one you might call unprofessional. Compare their work habits: stall cleanliness, condition of tools and toolbox, work clothes, and quality of repairs. Write a report on your findings.
2. Go to the library. Look up job descriptions for auto body-related professions: insurance adjuster, automotive engineer, salvage yard worker, auto salesperson, and so on. Draw a chart that shows these professions and how they relate to collision repair.

CHAPTER 30 Collision Repair for Hybrid and Electric Vehicles

OBJECTIVES

After studying this chapter, you should be able to:

- Compare hybrid vehicles and their unique drive systems.
- Identify electric and alternative fuel vehicles and compare their designs.
- Identify the major parts of hybrid and electric vehicles.
- Understand the unique challenges and safety precautions while performing collision repairs to hybrid and electric vehicles.
- Identify the safety measures required to repair vehicles equipped with high-voltage systems.
- Inspect drive systems and components unique to hybrid and electric vehicles.
- Safely remove and replace high-voltage batteries and other components necessary to perform collision repairs to hybrid and electric vehicles.
- Perform final checks required for hybrid and electric vehicles before delivery to the customer.

INTRODUCTION

The task of repairing any vehicle with body and structural damage is much different than servicing or replacing parts on an undamaged vehicle. The purpose of this chapter is to assist you, the future collision repair technician, as you face the reality of repairing damaged hybrid, electric, and alternative fuel vehicles (Figure 30–1). Safety is an overwhelming theme throughout this chapter. As you read, you, notice warning and safety boxes regarding a number of hazards specific to repairing hybrid and electric vehicles. The warning boxes are not to invoke fear but rather to remind you of the danger to you and your coworkers. Hybrid vehicles damaged in a collision could become energized (electrically hot) by simply pushing them through the shop. Safety must be the first priority when working on vehicles equipped with high voltage drive systems.

Figure 30–1 A damaged second-generation Toyota Prius awaiting repair.

There are new models of hybrid and electric vehicles emerging constantly from the auto manufacturers. It has been our goal to focus on the models and technologies you will most likely encounter while working in the collision repair industry. It is important for you to understand the basics of hybrid vehicle operation so that you may be able to recognize and troubleshoot problems when they occur. As you study this chapter, you will learn about the basic operation and recommended repair procedures for the most common hybrid models on the road today.

A review of all-electric and alternative fuel vehicles will also be provided. Accepted safety and repair procedures for all hybrid and electric vehicles will be summarized.

30.1 WHAT IS A HYBRID VEHICLE?

A hybrid is a vehicle that has more than one power source. Common power sources for a hybrid vehicle are an internal combustion engine and an electric motor-generator. There are different types of hybrid designs, some having more parts and complexity than others. Hybrids have been put into at least three classifications:

1. Assist hybrids
2. Power-split-type full hybrids
3. Series, parallel, and other full hybrid designs

How can one make sense of all this hybrid terminology? Let us start by looking at some examples of vehicles in each category.

Assist Hybrids

Honda Motor Company was the first manufacturer to get a hybrid vehicle to the North American marketplace in December 1999. The vehicle was called the Insight (Figure 30–2), and it had the first generation of Honda's **Integrated Motor Assist (IMA)**. The Insight is classified as an assist hybrid. The heart of the Insight's IMA system was a 144-volt electric motor-generator connecting the engine to the transmission. The Insight had other parts unique to hybrid vehicles, such as a **battery pack**, **powertrain control module (PCM)**, and **high-voltage (HV)** cables, but the use of only one electric motor-generator made Honda's IMA distinctive. The original IMA system was not designed to propel the vehicle with electric power alone but rather to yoke the gasoline engine and electric motor together for acceleration. The electric motor could also act as a "generator brake" producing drag to slow the vehicle. At the same time during braking, the generator also recharged the battery pack. Known as **regenerative braking**, this recaptures and saves a huge amount of electrical energy during deceleration.

Power-Split-Type Full Hybrids

A full hybrid is a vehicle that has the ability to be propelled by an electric motor alone. It is also important to note that these full hybrid vehicles are normally powered by a gasoline engine, any number of electric motors, or by any combination of these power sources (Figure 30–3).

Toyota was the first to manufacture a production hybrid, the Prius in 1997. Along with all the expected hybrid components, Toyota's first generation of the Prius (2000–2003) was fitted with two electric motors housed inside the vehicle's drivetrain. The larger of the two motors is called the traction motor, while the smaller is called the starter motor. Both motors can also generate electric power.

Here is how it works: Let us assume the Prius is stopped at a red light that just turned to green. As the brake is released and the driver pushes the gas pedal, the large electric traction motor propels the vehicle into motion. If the driver calls for rapid acceleration by pushing the accelerator pedal farther down, the starter motor will start the **internal combustion engine (ICE)**. At this point, the engine will work together with the traction motor to provide acceleration as long as it is called for. There is a power-splitting device in the Prius drivetrain that allows propulsion from the gasoline engine, the traction motor, or both at the same time. If the high-voltage battery pack needs to be recharged while driving, the starter motor can start the engine, which will spin the starter motor (now acting as a generator), which replenishes electrical power to the high-voltage battery pack.

Figure 30–2 This second-generation Honda Insight has a more powerful electric motor when compared to the original Insight.

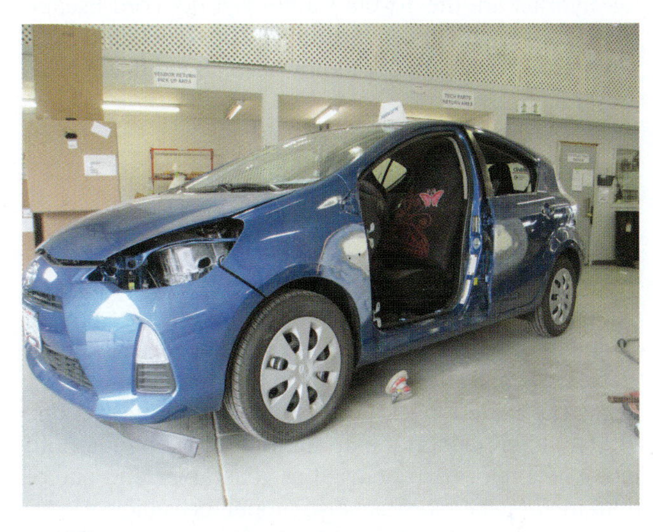

Figure 30–3 A third-generation Toyota Prius ready to go to the paint department.

Figure 30-4 On later model Toyotas the hybrid propulsion system is called the Hybrid Synergy Drive.

As the Prius slows down, the fossil fuel engine is turned off and the traction motor becomes a generator. The hybrid uses the mass of the vehicle (almost two tons) to rotate the motor-generator (regenerative braking) to produce power stored in the high-voltage battery pack.

As long as there is enough power in the high-voltage battery pack, the engine will remain off to save fuel. A fully charged battery pack can operate all vehicle electronics and the air conditioning system. All of the functions of Toyota's hybrid system are controlled by the **hybrid vehicle control unit (HVCU)** (Figure 30–4).

Today, most manufacturers are now manufacturing plug-in hybrids. These hybrid vehicles have a lithium-ion battery pack instead of the nickel-metal hydride (NiMH) battery pack found in earlier models. The lithium-ion battery pack allows a plug-in electric drivetrain to charge faster and to have a greater driving range. Most can be driven in the all-electric mode for up to 40 miles without starting the engine.

Other vehicles that are considered power-split-type full hybrids are the Toyota Camry Hybrid, Ford Escape Hybrid, and Ford Fusion Hybrid. The operation and components of Toyota and Ford power-split-type full hybrids are similar.

Series and Other Full Hybrid Designs

The Chevrolet Volt can be considered a series hybrid. The Volt has more than one power source (gas/electric), even though it is propelled most of the time by electric motors. Under the hood of the Volt, you will find a conventional 1.4 liter gasoline engine rated at 84 horsepower. The Volt also has a 16.5-kilowatt-hour (kWh), lithium-ion battery pack. The battery pack powers two electric motors one 149 HP main traction motor and one 74 HP motor. Both motor-generators can be used as generators, and one or both motor-generators can also propel the vehicle.

The Volt can be charged by plugging it into a 120- or 240-volt residential power supply. When it is fully charged, the Volt is designed to drive with only electrical power for the first 38 miles before the gasoline engine starts. The gas engine stays off until the battery pack is almost depleted. The Volt's larger traction motor-generator acts as a huge generator or alternator, providing power to the battery pack during deceleration (regenerative braking) as in other hybrid designs. The Volt has a range of 380 miles, thanks to its ability to generate power by running the gasoline engine. Two more examples of series-type hybrids are the Fisker Karma and Mercedes Benz Sprinter, both of which have the ability to be charged by residential power sources.

Honda improved the design of its IMA to the point that the 2012 Civic is no longer considered an assist hybrid, but rather a full hybrid because its electric motor can propel the vehicle with no assistance from the ICE. Honda's new Insight and CR-Z are also considered full hybrid vehicles for the same reason.

Honda also offers a plug-in Civic and Accord that are full hybrids with two electric motors. Similar to a power-split-type hybrid design, one motor is used for propulsion and the other is used for generating power and replenishing the battery pack. The plug-in Hondas can run on battery power for a range of about 15 miles; after that the engine and traction motor work together to propel the vehicle. At high speeds, the vehicles are powered only by the engine.

The Hyundai Sonata Hybrid uses a **transmission-mounted electrical device (TMED)** that is mounted between the engine and transmission. The TMED contains a 40 HP electric motor that can be used for propulsion or regenerative braking. The Sonata Hybrid has a conventional 12-volt battery in addition to its 270V battery pack.

The vehicles described so far are the most common hybrid types on the road today. You will now be able to compare different hybrid designs and understand their basic operation.

30.2 HISTORY OF ELECTRIC-POWERED VEHICLES

It may be surprising that hybrid and electric vehicles are not new ideas; in fact they have a very interesting history. At the beginning of the 20th century, electric vehicles (some of them hybrids) actually outsold internal combustion or gasoline-only powered vehicles. One early hybrid was the 1903 Kreiger; it was front wheel drive, had a gasoline-electric powertrain, and even had power steering. In the early 1900s electric cars were appealing to many people for their ease of operation and lack of noise and fumes. Hybrid and electric cars had limitations, but in the early days of the automobile people did not need to travel long distances or at high speed.

By 1912 there were over 34,000 electric cars registered in the United States with over 50 manufacturers producing them. By the mid-1920s the electric car had nearly disappeared. What killed the early electric car? History reveals it was simply public interest. Factors in the demise of the electric car may have included improved roads and the availability of gasoline along those roads and even the refinement of the gasoline-powered car. Henry Ford's mass-produced Model T was less than half the cost of similar electric or even alternative fuel steam-powered cars, and many felt the Model T was safer and more reliable as well.

Gasoline-powered cars continued to be refined and improved through the 20th century. It was not until the late 1980s that the same thing that led to the demise of early electric and alternative fuel vehicles would bring them back to life: public interest. Oil crisis concerns and smog problems in large cities during the 1970s and 1980s caused people to look again for alternatives to crude oil derived gasoline-powered vehicles.

Modern Electric Vehicles

In recent years, new lithium battery cell technology has removed some limitations of all-electric vehicles. The Nissan Leaf was introduced to the North American market at the end of 2010, and it can travel up to 100 miles on a fully charged battery. The Leaf is a **battery electric vehicle (BEV)** equipped with a 110 HP electric motor and 24 kWh lithium-ion battery pack. The Leaf is designed to be charged by residential power, and it has an option for high-voltage charging.

The Chevrolet Spark EV **(electric vehicle)** is another plug-in BEV available in limited quantities to participating dealers in California and Oregon at the time of this writing. The Spark EV uses much of the same electric technology used in the Chevrolet Volt, such as regenerative braking and a lithium-ion battery pack.

Alternative Fuel Vehicles

The Honda Civic GX is a vehicle powered by **compressed natural gas (CNG)** and is considered an alternative fuel vehicle. Looking under the hood of a Civic GX, the engine appears to be conventional; however the fuel lines are pressurized with up to 10,000 psi of CNG. The CNG fuel tank is located behind the rear seat, which makes the trunk compartment smaller than that of a gasoline or hybrid model Civic. The CNG-powered Civic GX is the cleanest internal combustion vehicle ever tested by the U.S. Environmental Protection Agency. Civic GX models can be identified by a CNG emblem on the rear deck lid. Other manufacturers are also producing CNG-powered vehicles, including trucks from General Motors, Ford, and Dodge. Now that we have reviewed the most common electric and alternative fuel vehicles, we can start thinking about repairs.

30.3 BEGINNING REPAIRS

Starting the repair process for a damaged hybrid, electric, or alternative fuel vehicle should begin with a careful walk around (Figure 30–5). As you walk around the vehicle, ask yourself the following questions:

Figure 30-5 It is important for your safety to carefully inspect damaged vehicles equipped with high-voltage systems.

- Was the vehicle driven to the repair facility or hauled on a flatbed truck?
- Does this vehicle have a high-voltage electrical system?
- Is the vehicle damage severe enough to affect any high-voltage cables, hybrid and electric vehicle control systems, battery packs, or electric drivetrain components?

Hybrid and electric vehicles can become energized (electrically hot) as damaged steel vehicle structure could be in contact with high-voltage components. High-voltage cables (recognized by an orange protective covering) can easily become pinched or severed in a collision.

As you are starting repairs or even pushing the damaged vehicle into the repair facility in order to start repairs, only touch the vehicle if you are wearing insulated lineman's gloves. Damaged vehicles with high-voltage systems should be placed on wheel dollies in order to move them; this prevents them from producing electrical power by turning the drive wheels. Safety must be the first priority when working on vehicles equipped with high voltage drivetrains.

Insulated lineman's rubber gloves are available for protection against different levels of voltage. A class zero lineman's glove is a popular choice for working on hybrid and electric vehicles as it will protect against 1,000 volts AC. Class 1 through 3 lineman's gloves protect against progressively higher amounts of voltage, class 3 being the highest rated (Figure 30–6).

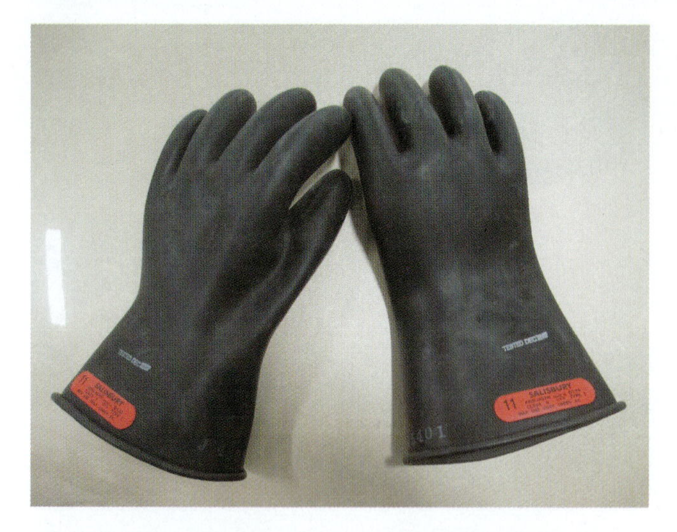

Figure 30-6 Check your lineman's gloves before each use to make sure there are no cuts, rips, or tears in them. Damaged lineman's gloves offer no protection from high voltage.

Wearing the proper lineman's gloves, you will be ready to test and replace the damaged high voltage components. Use extreme caution not to snag or tear the lineman's gloves on sharp metal in damaged areas of the vehicle. Lineman's gloves with holes or tears offer no protection against electric shock. There are leather covers available to help protect the soft lineman's gloves. Newer hybrids like the Chevrolet Volt and Toyota Prius have a high-voltage disconnect that occur automatically in the event of a collision; however, always treat every damaged vehicle equipped with high voltage as if it were hot.

A **digital volt-ohm-ammeter (DVOM)** designed for measuring high voltage can be used to make sure there is no stray electrical energy present in the vehicle structure.

Once you have verified that a damaged vehicle with a high-voltage system is not energized (electrically hot) and does not have hazards from damaged HV cables or components, before starting repairs, shut down the high-voltage system. For collision repair scenarios where straightening, welded-on parts replacement, or service to restraint systems is necessary, the vehicle's high-voltage system must be shut down. This is a safety precaution that must not be skipped.

Depending on the extent of the damage to the vehicle, you may want to retrieve **diagnostic trouble codes (DTCs)** before shutting down the high-voltage system. Retrieving DTCs before completing repairs is only an option if the vehicle is in a safe condition; if there are any hazards from HV electrical cables or components, the vehicle must be shut down immediately. In the latter scenario, DTCs can be retrieved after repairs are complete.

Disabling High-Voltage Electrical Systems

You must shut down hybrid and electric vehicle high-voltage systems for almost any collision repair procedure. If you need to open the hood to complete the needed repairs, then you need to shut down the vehicle's high-voltage electrical system. Although this may seem an unnecessary inconvenience, it is much better than electrocution. Remember to wear your protective gloves when shutting down any high-voltage electrical system.

Disabling a Toyota Prius

The first generation of the Toyota Prius can be shut down by the following steps:

1. Place the shifter in park and set the parking brake.
2. Turn the ignition to off and remove the key.
3. Disconnect the 12-volt battery found on the left side of the trunk compartment.
4. Wearing the required **personal protection equipment (PPE)**, remove the service plug from the high-voltage battery pack located in the trunk compartment and keep it in a safe place (so it is not restored without your knowledge).
5. Wait a minimum of 5 minutes to allow residual electrical power to discharge before beginning repairs.

There is also an alternative method of disabling the high-voltage system on a first-generation Prius, which consists of removing the ignition circuit relay and high-voltage fuse from separate boxes under the hood.

The second (2004–2009) and third (2010–2015) generations of the Toyota Prius have a similar procedure. Later model vehicles may be equipped with a key fob that within 20 feet of the vehicle enables drivers to push a start button to "power up" a hybrid or electric vehicle. Make sure the key fob is out of range or remove the fob battery during shutdown on models so equipped.

The second and third generations of Prius have a slide, flip, and pull style service plug. There is an orange handle on these models that is removable for safety (Figure 30-7). Most manufacturers specify an extended wait time of 10 minutes on newer models to allow the discharge of residual electrical power from the high power control unit, which is part of the HV system.

Specific service information for all Toyota and Lexus hybrid vehicles can be accessed through the manufacturer's website.

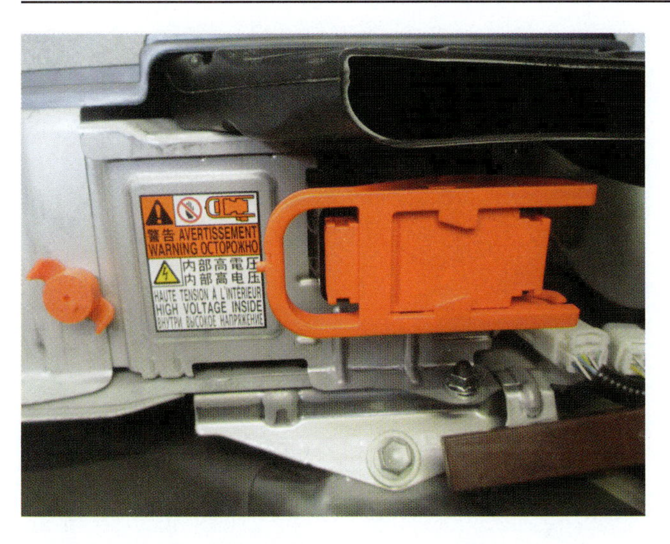

Figure 30-7 **The Prius manual service disconnect is located behind the rear seats of the vehicle.**

Disabling Honda's IMA Vehicles

The following procedure can be used for shutting down first-generation Honda Civic and Accord hybrids:

1. Place the shifter in park and set the parking brake.
2. Turn the ignition off and remove the key (if the vehicle is equipped with a key fob make sure it is at least 20 feet from the vehicle or has its battery removed).
3. Access the Intelligent Power Unit (IPU), which contains the HV battery pack and control modules, by removing the rear seat and seat back.
4. Wearing the appropriate PPE, remove the battery module switch cover from the lid of the IPU and remove the locking cover from the switch.
5. Turn the battery module switch off and reinstall the locking cover.
6. Disconnect the 12V battery by removing the negative terminal.
7. Wait a minimum of 5 minutes to allow residual electrical power to discharge.
8. After the wait time, check for any electrical voltage with a DVOM before beginning repairs.

Specific service information for all Honda and Acura hybrid vehicles can be accessed through the manufacturer's website.

Disabling a Chevrolet Volt

The Chevrolet Volt is equipped with a collision detection system. This system uses input from the **supplemental inflatable restraint (SIR)** sensors to identify the severity of a collision and from what direction a collision has occurred. The SIR sensors typically detect collision conditions for occupant impact-protection reasons. Even if the high voltage has been disabled by the hybrid powertrain control module in the event of a collision, the **manual service disconnect (MSD)** must be used before any repair work is done to the vehicle.

To isolate the high-voltage battery on a Chevrolet Volt, the MSD can be found under the center console. General Motors requires a complete inspection of the high-voltage system of any vehicle that has been involved in a major collision. This inspection procedure for the Volt's high-voltage system is several pages in length.

The MSD instructions can also be downloaded at: www.genuinegmparts.com, which is a free website provided by GM. Once the website has been accessed, the collision repair tab will give you access to repair information for the Volt and many other late-model GM vehicles such as the Spark EV. In order to maintain a high level of customer satisfaction, collision repair technicians are encouraged to use manufacturer resources and repair recommendations.

Some of the most important tools that you can have for working on hybrid, electric, and alternative fuel vehicles are a computer with Internet access and a printer. These tools should be right on top of your toolbox or at least at a technician station in the repair facility. It is impossible for any technician to memorize all the repair procedures for every vehicle.

Manufacturers have been known to change or update repair recommendations, and the only way to know this is by using vehicle maker websites to get the most current and up-to-date information for the vehicle you are working on.

30.4 REPAIR STRATEGIES FOR HYBRID AND ELECTRIC VEHICLES

While many of the same collision repair methods apply to hybrid and electrical vehicles as conventional ICE only vehicles, there are some common precautions for any vehicle equipped with a high-voltage system. After the high-voltage electrical system has been disabled and the proper wait time has been observed, follow the recommended vehicle maker procedure to make sure there is no electrical shock hazard. At this point, we can safely proceed with the repair of the vehicle, starting with the vehicle teardown.

Inspect the Major Parts and Assemblies

Take the opportunity to inspect all the hybrid and electric assemblies and components as damaged parts are removed from the vehicle (Figure 30–8). If damage exists on any part in a hybrid or electrical system, it must be

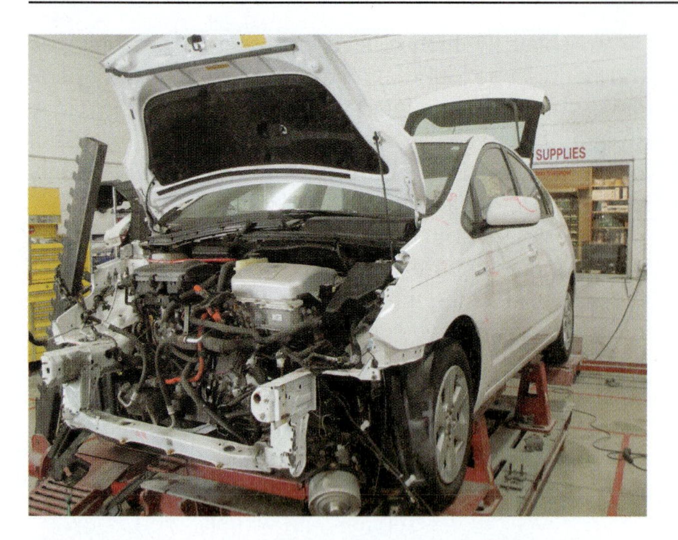

Figure 30-8 **Collision-damaged parts may need to be removed in order to properly inspect components.**

Figure 30-9 **The inverter/converter under the hood of a third-generation Prius.**

removed and replaced. The following is a list of assemblies from a third-generation (2010–2015) Prius to inspect for collision damage during vehicle teardown. There are similar assemblies in all hybrid vehicles, although they may be in different locations. Some of the technical information is included in this list to give you a better understanding of how hybrids work and what to look for during inspection.

Inspecting the Toyota Prius

1. The *hybrid vehicle battery pack* is located in the cargo area, mounted to the cross member behind the rear seat. It is a 201.6 volt Nickel Metal Hydride (NiMH) battery pack made up of 28 low voltage (7.2 Volt) modules connected in series. Because of its location, it may be unlikely that the battery pack would sustain damage from objects outside the vehicle, however the forces present during a collision could cause unsecured items in the cargo area to damage the battery pack. This type of damage is called secondary damage or inertia damage. If any damage such as bends or dents are found in the battery pack it must be replaced. Also look for broken or cracked paint in the vehicle structure around the battery pack mounting locations. The cracked paint would indicate movement in the vehicle structure during the collision which could have resulted in damage to the battery pack.
2. The *12-volt auxiliary battery* is located in the passenger side of the cargo area. It is a lead-acid battery that supplies power to the low-voltage devices. Carefully check the 12V battery, especially if collision damage exists in the right rear of the vehicle. Look for any signs of damage, bulging, or leakage; if any of these are found the battery must be replaced.
3. There are several *high-voltage power cables* present in the engine compartment and under the vehicle body. These orange-colored power cables carry high-voltage direct current (DC) between the HV battery pack, the inverter/converter, and the A/C compressor. These cables also carry three-phase alternating current (AC) between the inverter/converter, the electric motor (traction motor), and the generator (starter motor). Inspect the HV power cables for any damage to the orange insulators; if any damage is found, the HV power cable must be replaced. Splicing or any type of electrical repairs are not permitted on HV power cables.
4. Toyota states that only certified hybrid service technicians are qualified to replace high-voltage system wiring.
5. The *inverter/converter* (Figure 30–9) is in the engine compartment directly in front of the driver position. The inverter/converter changes the high-voltage DC electricity from the HV battery pack to three-phase AC electricity that drives the electric motor. In this process the 201.6V battery pack DC voltage is increased to 650 volts AC. The inverter/converter also converts AC electricity from the electric generators into DC that recharges the HV battery pack. The inverter/converter can be easily damaged in a frontal collision because of its location. The inverter/converter is water cooled, so carefully inspect the coolant lines for any damage. Replace any parts that are damaged. Some individual parts may be serviceable (meaning that they can be purchased individually) such as the cover for the inverter/converter, while other parts are only serviced as an assembly.

The *internal combustion engine (ICE)* powers the vehicle and powers the generator in order to recharge the HV battery pack. The engine is controlled by the hybrid vehicle control unit (HVCU), which is one of the vehicle's computers. Carefully inspect the engine for damage, paying close attention to fuel lines,

Figure 30–10 This coolant tank on the second-generation Prius can be easily damaged in a collision.

electrical wiring, engine mounts, pulleys, and coolant hoses. Inspect the exhaust system carefully and look for signs of leaking engine coolant. Starting with the 2010 Prius, coolant was routed through the exhaust system of the ICE for faster preheating in order to reduce engine warm up times and emissions. In the second-generation Prius (2004–2009), there is a coolant tank (Figure 30–10) behind the bumper cover on the left side of the vehicle that is frequently damaged in a collision and requires replacement.

6. The *engine radiator, A/C condenser, and inverter/converter radiator* are frequently damaged in collisions because of their location behind the front impact bar (bumper). These three liquid-to-air cooling units are mounted together and are made of soft aluminum that are easily damaged. Damage to these radiator/coolers could cause overheating, so they must be replaced if any distortion or bent cooling fins are discovered.
7. The *A/C compressor* (with inverter) is mounted to the engine in the right front of the vehicle. The A/C compressor is powered by high voltage so that it may run while the engine is off. Because of its location, the A/C compressor can be easily damaged in a collision. Carefully inspect its housing for cracks or distortion, and check the high-voltage cable connections to make sure they are tight. Always use tools made of nonconductive materials such as carbon fiber when checking HV electrical connections, even when the HV system is disabled.
8. The *electric motor* (traction motor) is a three-phase high-voltage AC permanent magnet electric motor contained in the front transaxle. It is used to power the front wheels, and during regenerative braking it is used as a power source to replenish the HV battery pack. Check the vehicle transaxle housing for cracks, and look carefully at any place on the assembly that may have been contacted by the vehicle body structure as a result of a collision. If the housing is damaged, it may affect the operation of the electric traction motor and should be replaced.
9. The *electric generator* is a three-phase high-voltage AC generator that is contained in the transaxle. The electric generator starts the engine and also recharges the HV battery pack as it is turned by the engine crankshaft. Inspect the transaxle for damage in the same way as for the electric traction motor.
10. The engine *fuel tank and fuel lines* should be inspected carefully after a collision. Look for deep scratches or gouges in the fuel tank assembly, and inspect the fuel lines for signs of leakage. Inspect the fuel lines for bends and kinks, especially if the vehicle left the road in the accident (signs of this are debris between the tires and wheels and under the vehicle undercarriage). If the fuel lines need to be replaced, the high pressure in the fuel system needs to be relieved before they are disconnected.

Inspecting Honda's Hybrids

While Hondas have many of the same hybrid parts as the Toyota Prius, there can be some differences in the names and locations of those parts. Honda models equipped with Intelligent Power Units (IPU) house many of the necessary hybrid parts in one central location. On models so equipped, the IPU is located behind the rear seat back or in the rear cargo area. Depending on the vehicle model and year, the IPU can contain the HV battery, AC inverter, DC-DC converter, A/C compressor driver, and IPU cooling fan.

Because of its location (similar in location to the HV battery pack of a Prius) it may be unlikely that the IPU would sustain damage from objects outside the vehicle. However, unsecured items in the cargo area could damage the IPU in a collision. If any damage such as bends or dents are found in the IPU, it must be replaced. The IPU is

WARNING

The high-voltage electrical system is disabled on a Honda Intelligent Power Unit (IPU) by deactivating/turning off the manual service disconnect. Honda recommends that you disable the 12V system (if the vehicle is so equipped) after the HV system is disabled. Even after the HV and 12V systems have been disabled, the IPU still contains high voltage. The protective cover of the IPU must be left in place. The components and orange cables within the IPU are charged with high voltage, and if the cover is removed they can cause electrocution even if the system has been disabled by turning off the manual service disconnect.

air cooled by an electric fan or outside air sources, so it is important to look for broken or cracked cooling ducts. If the air cooling system for an IPU is damaged in a collision, the IPU may overheat and malfunction.

Welding on Hybrid and Electric Vehicles

Repair recommendations from vehicle manufacturers allow welding in order to repair structural damage to hybrid and electric vehicles (Figure 30–11). The 12V and high-voltage electrical systems of these vehicles must always be disabled before welding. HV parts and assemblies may need to be removed before welding in certain areas of the vehicle, especially if welding will be done in close proximity to HV parts. Look up the specific manufacturer repair procedures for the parts that you will be welding on vehicles equipped with high-voltage systems.

It is important to have the ground for welders and some other tools attached to the vehicle in close proximity to the area that is to be welded (Figure 30–12). The welding ground needs to have good contact with bare metal in order for the electrical current to flow freely and to make a good weld. If a welding ground is placed on a painted surface, the current will not be able to flow freely, and making a good weld will not be possible. Never place a welding ground on parts of the engine or drivetrain if you are welding the vehicle body. Electricity follows the path of least resistance, so the farther away the ground is from the weld, the more likely it is that the welding current will find a path through a computer or other sensitive component. Many vehicle computers and electrical systems have been ruined from either not disabling the electrical system before welding or attaching the work clamp too far away from the welding operation.

Keep an ABC fire extinguisher nearby while welding on hybrid and electric vehicles. When welding underneath the vehicle, it is important to have another technician act as a spotter, looking on the top side of the panel for unexpected fires or heat damage. While welding inside a vehicle on a floor or trunk compartment area, the spotter should watch the underside closely for fires and heat damage as well.

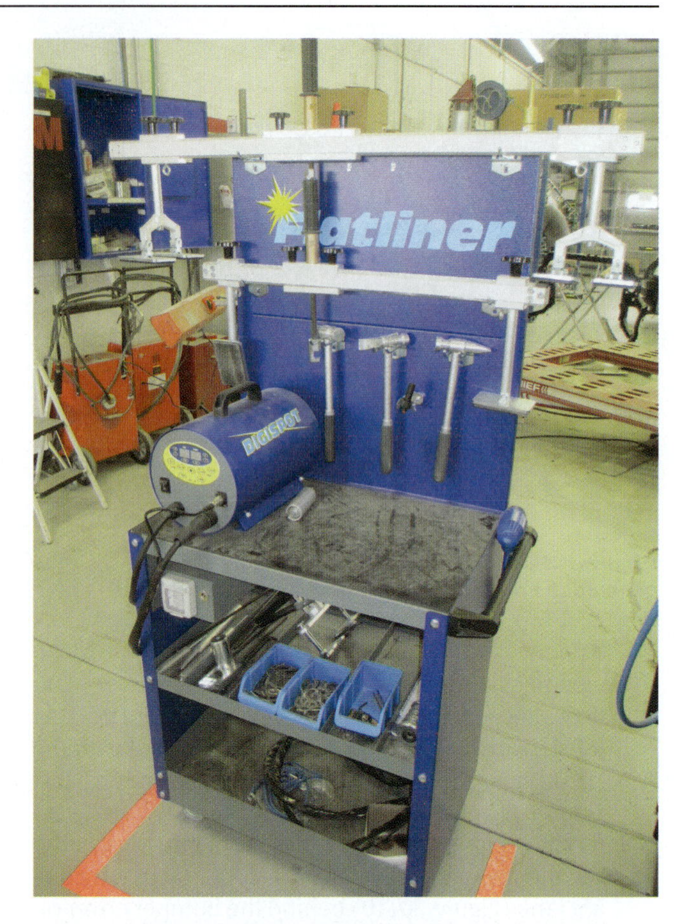

Figure 30–12 Similar to welders, some dent removal tools may require a good ground location.

Figure 30–11 MIG welders are commonly used for structural repairs.

Use the Correct Air Conditioning Compressor Oil

Hybrid and electric vehicles that have electric-powered air conditioning compressors require a specific type of oil. After damaged A/C parts are replaced and the system is ready to be recharged with refrigerant, the manufacturer-specified oil must be used. Using the wrong type of oil can cause moisture to form in the electric A/C compressor, changing resistance. If the wrong oil type is used in an electrical air conditioning system it can cause a DTC to be displayed. After contamination with the wrong oil type, all affected A/C components must be replaced.

30.5 SAFE PRACTICES AND GENERAL REPAIRS FOR HYBRID AND ELECTRIC VEHICLES

Before doing structural repairs to a hybrid or electric vehicle, you must sometimes remove hybrid/electric drivetrain assemblies in order to gain access to the damaged

Removing paint coatings is a good way to prepare an area for a welding ground; unfortunately in some cases this causes extra work for collision repair technicians who after the welding is complete must sand, prime, and refinish the ground area. Working in tight spots where removing paint coatings might be considered too intrusive, there is another option to get a good welding ground. Find a medium-sized body bolt near the area where the welding will take place. Remove the body bolt, and if there are no protective coatings on the threads (which could act as an insulator), you can use the threaded hole. Use a longer replacement bolt (that is free of any coatings) with the same thread size that has a nut threaded on it about 10–15 mm from the end that you will thread into the vehicle body. Turn the replacement bolt into the vehicle body; when it has threaded in 10–15 mm, tighten down the nut against the vehicle body. The replacement bolt should be sticking out 25–45 mm and can be used to attach the welding ground. Because the bolt has a nut holding it tightly against the threads in the vehicle body, it will allow welding current to flow freely, helping to achieve a good weld.

After the welding is complete, the replacement bolt and nut used as a ground can be removed and the factory bolt reinstalled in the threaded hole.

WARNING

The first generation of the Prius (2000–2003) would not shut down the internal combustion engine if the climate control was set to MAX A/C; this is because the engine drives the A/C compressor in that model. Starting with the 2004 model year, the Prius was modified to have an electrically powered A/C compressor. This allows the hybrid vehicle control unit to shut off the internal combustion engine without affecting A/C operation. Use extreme caution when servicing the A/C on 2004 and newer Priuses to avoid the possibility of electrical shock.

body structure. Trained collision repair technicians use manufacturer-recommended procedures for working with the high-voltage circuits found in modern electric drivetrains.

Follow manufacturer instructions when disabling HV systems; these systems must be disabled before performing repair work. Not properly shutting down the vehicle's electric drivetrain could be a fatal mistake. If the orange-colored HV cables were damaged in the collision, a short could unexpectedly energize the vehicle body and become a source of electrocution. Avoid pushing a hybrid around the shop on its own wheels; always use wheel dollies. Be careful when disconnecting the high- and low-voltage power cables and harness connectors from the electric drive assemblies. Properly disengage all plastic electrical connectors without damaging them.

Never use a conventional 12V DVOM to try to measure hybrid drive circuit voltages. If connected to high voltage, the meter will blow its internal fuse or can be ruined. Only use voltmeters rated for high voltage when checking for high voltage.

Avoid jump-starting a hybrid vehicle, unless it has a 12V battery and 12V electrical system. If attempting to jump-start, never connect directly to the HV battery. Most late model full electric drivetrains (without 12V battery) cannot be jump-started.

Avoid removing the protective housing from any damaged electric drive assembly, including the HV-PCM and HV battery. Removing and unsealing the cover or lid could void any warrantee and increase the chance of electric shock injury.

If an electric **motor/generator** requires replacement, the vehicle is normally sent to a certified transmission shop or authorized dealership for repair. Any damaged or faulty hybrid electric drive assembly must be replaced with a new or rebuilt unit. Damaged HV batteries can be repaired by aftermarket companies. If not repairable, components can be recycled and used to rebuild other HV batteries, so they should always be disposed of properly.

Many of the electrical/electronic service methods you learned in previous chapters apply to hybrid vehicles. However, now you must remember how to work safely with the high voltage present in modern hybrid and electric vehicles. Although they can be cumbersome and uncomfortable, always take the precaution to wear your PPE, especially lineman's rubber gloves, rubber apron, and eye protection. Remember to use the recommended insulated tools when servicing high-voltage electric driveline circuits.

When replacing an HV-PCM (powertrain control module), unplug all electrical connectors without damaging them. They are water-tight connectors that must be clipped together securely during reassembly. Make sure you identify the positive and negative markings on the HV battery cables, and where they connect to the HV-PCM, mark them before disassembly. Also label all of the power cables running between electric motor/generators

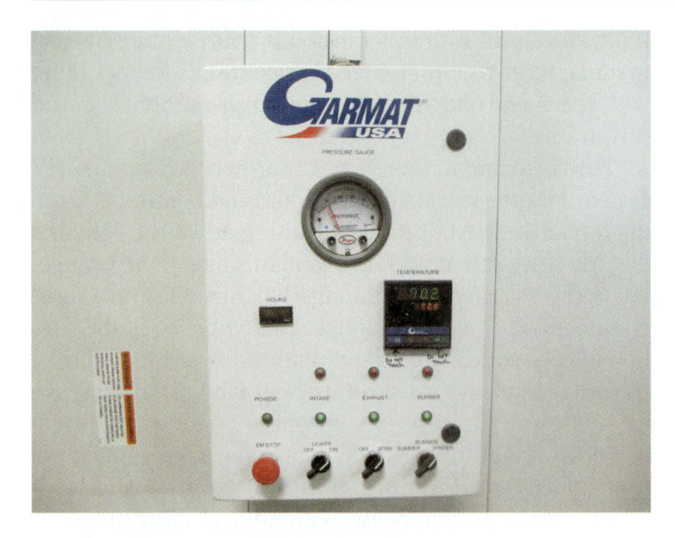

Figure 30-13 **Paint booth controls must be set not to overheat Hybrid components.**

and the HV-PCM. They must be installed in their original locations during reassembly.

30.6 PAINT BOOTH CONCERNS

Excessive heat can damage HV batteries, so some vehicle manufacturers have given recommendations concerning paint booth bake temperatures (Figure 30–13). Honda states that vehicles equipped with HV systems should not be heated to more than 150°F. Ford's recommendation is a maximum temperature of 140°F for no more than 45 minutes. Toyota states that the spray booth bake cycle is not a concern because the internal temperature of the vehicle is not nearly as high as the external temperature during bake cycles.

Honda states that the CNG-powered Civic GX needs to have an empty fuel tank with the manual relief valve in the open position before baking the vehicle. The manual relief valve on a Civic GX is located behind the left rear tire.

Any concerns of the HV battery being damaged by excessive heat can be avoided by allowing the vehicle to air dry.

30.7 GENERAL RECOMMENDATIONS FOR HV BATTERY REMOVAL AND REPLACEMENT

If an HV battery pack is damaged in a collision, it must be replaced. Replacement options may include a new battery from the manufacturer, a remanufactured battery from an aftermarket company, or a used battery from a recycling center. The following steps are general guidelines; refer to manufacturer-specific procedures for the vehicle you are working on.

1. Shut down the HV electrical system following vehicle manufacturer instructions before performing electric drive service operations. Normally, you must pull a safety disconnect plug or move a battery disconnect switch to disconnect the HV battery from the vehicle's electrical system.
2. To remove an HV battery, refer to factory service information showing battery location, electrical connections, and mounting. Hybrid HV batteries can be located in the trunk area, under a rear seat cushion, under the hood, or under the vehicle body.
3. If the HV battery is mounted under the rear seat, remove the interior trim, seats, and carpeting over the top battery module. If in the trunk, remove any carpet, spare tire, or cover over the HV battery. If under the vehicle, a special hydraulic jack is needed to lower the heavy battery pack from the floorpan.
4. Avoid removing protective covers of HV batteries. If removing the HV battery cover is necessary, use extreme caution, and never lay tools or test equipment in the area where HV is a potential hazard. If a protective cover is removed, HV can be present even if the service disconnect has been switched to the off position.
5. Some modern electric and gas-electric powertrains locate the HV battery stack under the center console of the vehicle. The HV battery fits up into the center hump under the floor and bolts to the metal frame. The heavy battery placement lowers the vehicle's center of gravity. It also isolates HV battery heat and any expansion gas outside the passenger compartment for driver and occupant safety.
6. When removing or installing an HV battery, use an engine crane or extension jack to avoid dropping the heavy HV battery. On certain hybrid models, two people can lift smaller HV batteries out of the vehicle for replacement. Larger batteries require a hydraulic lift to safely raise and move the bulky, heavy series of battery packs. If the HV battery is mounted under the floor pan, you must raise the vehicle on a two- or four-post lift. A single-post lift will not let you access the HV battery. Unfasten and disconnect any wires, hoses, or ducts that prevent HV battery removal.
7. Place a transmission or HV battery jack under the HV battery. Position the lifting arms on the jack to balance and support the weight of the unit. Remove the bolts or nuts that secure the heavy HV battery to the unibody structure.
8. While wearing insulated rubber gloves, remove the large fasteners or connectors that secure the high-voltage power cables to the HV battery. To prevent a high-voltage short, wrap electrical tape around the metal terminals on the ends of the orange power cables if needed. You do not want them to accidentally contact anything and conduct high-voltage electricity into anyone or anything! Remember to

treat the HV electrical system as if it were hot—even while in a disabled condition!

9. Unbolt the HV battery hold-down flange from the vehicle frame. If it weighs more than two people can lift (some late model hybrid batteries can weigh over 600 pounds) use an engine crane and nylon straps to raise the large assembly up and out of the vehicle. The HV battery should be handled carefully since it contains a tremendous amount of stored chemical-electrical energy and is fragile when not bolted in place in the vehicle.
10. Even a small hybrid battery assembly can weigh 100–300 pounds. Make sure you and your helper are wearing eye protection and thick rubber insulated gloves during HV battery removal and installation.

DANGER If you accidentally drop an HV battery, especially a modern lithium-acid battery, it can rupture several of the lithium cells and short them out. The damaged HV battery cells can start to smoke and hiss as if catching fire. Open the shop overhead doors and stay clear of the damaged HV battery for a few minutes until it stops smoking (gassing). Use fans to ventilate the area immediately. Never spray water on an HV battery because this will intensify and feed the fire!

11. If an HV battery is to be replaced, secure it on a wooden pallet for shipment to an authorized recycler. Again, wrap electrical tape over the HV battery cables or connectors. Cover the HV battery with a thick plastic box and place it in a shipping container. Place warning labels on the outside of the battery module or shipping container to warn others of the high-voltage and chemical dangers inside.
12. To install a new HV battery or HV components follow the removal procedure in reverse order. Vehicle manufacturer service information is necessary in order to perform these repairs with efficiency and confidence. If the proper recommendations are not followed, the procedure may have to be redone, wasting a great deal of time that could have been used performing other repairs.

30.8 FINAL CHECKS

After repairs are complete, the HV system of the vehicle must be reconnected or "powered up"; with some vehicles this procedure is the same as the disassembly procedure only in reverse order. Follow the vehicle maker's specific procedures for enabling HV systems, and do not forget to wear the proper PPE, especially lineman's gloves, for this operation.

Once the vehicle is powered up, check for DTCs using a scan tool. Clear DTCs that have been corrected by the replacement of parts and the repair of the vehicle. If there are unexpected DTCs, make the proper repairs in order to correct the malfunction and clear the code. Test drive the vehicle in order to verify that all its systems are working properly. It is recommended to drive several miles in varying conditions in order to check all the modes in a hybrid vehicle. Taking the steps to verify a professional repair can help you return the vehicle to the customer with confidence.

SUMMARY

1. A hybrid vehicle has more than one power source, such as an internal combustion engine and an electric motor, to provide energy for propulsion.
2. A battery electric vehicle (BEV) has one power source, which is electricity. BEVs have no method of recharging the HV battery while driving and are generally recharged by residential power sources. BEVs are also called EVs (electric vehicles)
3. Alternative fuel vehicles run on fuels other than traditional gasoline or diesel. An example of an alternative fuel is CNG.
4. A high-voltage battery consists of a large number of voltaic cells wired or stacked in series to produce a high-voltage, high-power storage cell.
5. A high-voltage motor/generator (M/G) is part of a hybrid vehicle system. The M/G can propel the vehicle alone or with the ICE (internal combustion engine), start the ICE, generate electrical power, and slow down the vehicle.
6. The hybrid vehicle control unit (HVCU) is what Toyota calls a vehicle computer that monitors driving conditions to help control the operation of the Hybrid System or Hybrid Synergy Drive.
7. The HV power cables are large, orange-colored, insulated conductors that electrically connect the battery pack, HVCU, and motor-generator assemblies together.
8. In all-electric drive mode, a hybrid vehicle operates just like an all-electric car. The HV battery provides all of the energy needed to propel the vehicle. The engine is shut off but is ready to start up when the HV battery becomes discharged.
9. When the driver applies the brakes, the traction motor/MG helps slow and stop the vehicle while also

recharging the battery pack; this is called regenerative braking.

10. Plug-in hybrids and BEVs can be connected to a residential power supply in order to fully recharge the HV battery pack.
11. Some hybrids have a high-voltage electric air conditioning compressor.
12. When repairing high-voltage electric air conditioning systems, it is important to replace the A/C oil with the proper type to avoid causing a DTC.
13. Hybrid water cooling circulates engine coolant through one or more of the main assemblies of the electric drivetrain.
14. Hybrid air cooling circulates outside air over the HV battery to dissipate heat. Electric fans can also be used to force air through the battery pack housing.
15. A hybrid power display in the instrument cluster informs the driver of the hybrid drive operating conditions. Many hybrid instrument clusters show how electrical energy is being used and how much energy remains in the HV battery. This system can be helpful while performing final vehicle checks before delivery to the customer.
16. Avoid jump-starting a hybrid vehicle, unless it has a 12V battery and 12V electrical system. If attempting to jump-start, never connect directly to the HV battery. Most late model full electric drivetrains (without 12V battery) cannot be jump-started.
17. Use your scan tool or laptop connected to the hybrid OBD port in order to find DTCs. Stored trouble codes will quickly help identify most hybrid/electric drivetrain problems.
18. Always be aware of the deadly energy levels present in high-voltage hybrid circuits! Modern hybrid electric drive assemblies and their orange power cables can carry 650 volts.
19. Always wear insulated gloves rated for high voltage when working on HV components and cables. Wear all the proper PPE when performing repairs.
20. Wrap electrical tape around any exposed power cable end after removal. Also wrap electrical tape around any damaged HV cables (after the HV system has been disabled) just as a precaution until the cable can be replaced.
21. Before disconnecting any hybrid power cable, set up a buffer zone around the vehicle by moving all carts, tools, and metal equipment away from the car body.
22. HV battery power must be disabled before doing any body shop repairs.
23. When a vehicle has been in an accident suffering any sheet metal damage, always inspect all high-voltage electric drive assemblies for impact damage.
24. Do not weld too close to the hybrid system components, hybrid power control modules, and orange power cables. Too much welding heat can damage delicate electronic circuits and melt wire insulation. Follow vehicle makers' recommendations when welding on hybrid vehicles.
25. The most common reason for HV battery replacement in a collision is secondary damage or inertia damage. In rare cases a severe collision or auto accident can rupture the battery pack elements. Refer to the hybrid and emergency response guides on the manufacturer websites for more information about badly damaged battery packs.
26. Although it is unlikely that a high-voltage battery pack would catch fire in a repair facility, there is a possibility of such a scenario. If a high-voltage battery pack begins to smoke or catch on fire, leave the area immediately and call the fire department. Repair facilities generally do not have the required safety equipment to deal with such scenarios. Never pour water on a smoking battery pack or you will intensify the fire.

EXERCISES

On a separate sheet of paper, complete the following learning activities for this chapter. Write definitions for the key terms and answer the ASE-style review questions, essay questions, critical thinking problems, and math problems. You can also do the outside activities, possibly for extra credit.

1. Look up and print vehicle maker recommendations from vehicle maker websites for disabling the high-voltage systems on current year Toyota Prius, plug-in Honda Accord, Ford, and Chevrolet Spark EV.
2. Find and print front and rear frame rail replacement recommendations for one of the vehicles listed in Activity 1 from the vehicle maker's website.
3. Look up factory position statements on replacement hybrid parts for one of the vehicles listed in Activity 1. What does the position statement say?

Key Terms

battery electric vehicle (BEV)
compressed natural gas (CNG)
diagnostic trouble code (DTC)
digital volt-ohm-ammeter (DVOM)
electric vehicle (EV)
high voltage (HV)
Hybrid Synergy Drive (HSD)
hybrid vehicle control unit (HVCU)
Integrated Motor Assist (IMA)
internal combustion engine (ICE)
manual service disconnect (MSD)
motor/generator
personal protection equipment (PPE)
powertrain control module (PCM)
regenerative braking
transmission mounted electrical device (TMED)

ASE-Style Review Questions

1. Technician A says that hybrids can only have one motor/generator. Technician B says that hybrids can have more than one motor/generator. Who is correct?
   A. Technician A
   B. Technician B
   C. Both A and B
   D. Neither A nor B
2. The term *hybrid* means:
   A. It is a crossover vehicle.
   B. A vehicle with more than one power source.
   C. Only one power source.
   D. Only electricity is used for power.
3. What color are the high-voltage power cables used in a car's electric drivetrain?
   A. Blue
   B. Red
   C. Orange
   D. Green
4. This hybrid component is more likely than other components to be damaged by secondary damage or inertia damage in a collision.
   A. HV battery
   B. HVCU
   C. HV power cables
   D. HV motor/generator
5. Technician A says that hybrid vehicles need to have their HV electrical systems disabled before performing most collision repairs. Technician B says that most hybrids have a manual service disconnect located in an area that is easy to access. Who is correct?
   A. Technician A
   B. Technician B
   C. Both A and B
   D. Neither A nor B
6. Today's hybrid vehicles can do all of the following EXCEPT:
   A. Crank the engine over for starting almost silently.
   B. Generate power by running the engine.
   C. Allow the car owner to plug the vehicle into a special charging station.
   D. Never require gasoline.
7. How many volts drive the traction motor on a third-generation Toyota Prius?
   A. 201.6V
   B. 144V
   C. 650V
   D. 288V
8. Technician A says you cannot jump-start most modern hybrids. Technician B says all hybrids still have a 12V battery and 12V starting motor, so they can be jump-started with another, fully charged 12V battery. Who is correct?
   A. Technician A
   B. Technician B
   C. Both A and B
   D. Neither A nor B
9. This equipment would be required for large, plug-in battery replacement from under the vehicle.
   A. Engine crane
   B. Hydraulic scissor jack
   C. Both of the above
   D. None of the above
10. Technician A says to wrap electrical tape around any exposed electrical terminals on the ends of the hybrid power cables and other exposed high-voltage conductors. Technician B says that any collision-damaged HV battery or other assembly must be properly packed and crated for recycling or rebuilding. Who is correct?
    A. Technician A
    B. Technician B
    C. Both A and B
    D. Neither A nor B

Essay Questions

1. According to Toyota, who is authorized to service hybrid power cables?
2. Summarize how to inspect a hybrid vehicle before repairs are started. What safety precautions need to be considered?

3. Basically, how do you remove a large, heavy HV battery assembly?
4. What are the concerns of welding on a hybrid/electric vehicle?

➤ Critical Thinking Problems

1. Can you describe the main differences between popular hybrid drive systems?
2. What do CNG and BEV stand for?
3. What are the major assemblies in a hybrid vehicle?
4. What is the primary benefit of a hybrid gas-electric vehicle over an ICE-only powered vehicle?

➤ Math Problems

1. If a hybrid vehicle contains 100 battery packs of 3.6 volt DC cells in series, how many volts does the whole HV battery generate?
2. If each hybrid battery cell pack generates 7.2 volts and the battery has 40 cell packs, the HV battery would generate how much open circuit voltage when fully charged?

➤ Activities

1. Select any make and model hybrid vehicle. Look up its operating specs and service procedures on your shop computer. Make a top view sketch of component placement and hybrid power cable routing.

Auto Body Shop Terms—Términos del Taller de Carrocería

APPENDIX

A

abrasive Material such as sand, crushed steel grit, aluminum oxide, silicon carbide, or crushed slag, used for cleaning or surface roughening.
abrasivo Una material tal como la arena, la granalla de acero para bruñir, el carburo de silicio, o la escoria machacada que sirve para limpiar o deslustrar una superficie.

accelerator Fast-evaporating thinner or reducer used to speed up drying time; used in very cold weather.
acelerador Diluyente o reductor de evaporación rápida utilizado para acelerar el tiempo de secado; se utiliza en clima muy frío.

access time Time required to remove extensively damaged collision parts by cutting, pushing, pulling, and so on.
tiempo de acceso El tiempo requerido para remover las partes dañadas extensivamente en una colisión por medio de cortar, empujar, jalar y etcétera.

accident report Summary of what happened during the collision; lists information about the drivers, their insurance companies, and their vehicles.
informe de accidente Resumen de lo que sucedió durante el choque y que contiene información sobre los conductores, sus compañías de seguro y sus vehículos.

acid rain An air pollutant that can cause damage to an automotive finish, created when sulfur dioxide or nitrogen oxide is released into the atmosphere.
lluvia ácida Contaminante del aire que puede causar daños al acabado del vehículo; se crea al liberar a la atmósfera dióxido de sulfuro u óxido de nitrógeno.

active restraint system One that the occupants must make an effort to use.
sistema de sujeción activo Sistema que los ocupantes deben utilizar.

actuators Devices that move when responding to electrical signals from the computer.
impulsores Dispositivos que se mueven en respuesta a señales eléctricas de la computadora.

adapter A connection that is male on one end and female on the other end; used to convert the connections on a hose and other equipment from one thread size to another.
adaptador Conexión con un extremo macho y otro extremo hembra, utilizada para convertir los tamaños de rosca de las conexiones en una manguera y otros equipos.

adhesion Ability of one substance to stick to another.
adhesión La habilidad de una substancia de pegarse a otra.

adhesion promoter A clear, ready-to-spray material that provides a chemical etch to OEM finishes.
promovedor de adhesión Una material de laca blanca soluble en agua, listo para atomizar que provee un grabado químico en los acabados OEM.

adhesives Special glues used to bond parts together.
adhesivos Pegamentos especiales utilizados para unir piezas.

adjustable wrench A wrench with one fixed jaw and one moveable jaw for fitting different-sized fasteners.
llave ajustable Llave con un lado de la abertura fija y uno movible para su uso con sujetadores de distintos tamaños.

aerodynamic Shape with low wind resistance.
aerodinámico Una forma de poca resistencia al viento.

aftercooler Used to reduce the temperature of compressed air. Heat and other impurities like oil residue or water can also be removed with an aftercooler.
posrefrigerador Utilizado para reducir la temperatura del aire comprimido. También se puede remover impurezas como residuos o agua con un posrefrigerador.

aftermarket replacement panels Manufactured by smaller companies, not the OEM.
Paneles de reposición de posventa Fabricados por compañías más pequeñas y no por la OEM.

air bag controller Analyzes inputs from the sensor to determine whether air bag deployment is needed.
controlador de bolsa de aire Analiza los datos del sensor para determinar si es necesario el despliegue de la bolsa de aire.

air bag igniter Produces a small spark when an electrical signal is sent from the electronic control unit; also called a squib.
ignición de la bolsa de aire Produce una chispa pequeña cuando la unidad de control electrónica envía una señal eléctrica.

air bag system System that uses impact sensors, vehicle's on-board computer, an inflation module, and a nylon bag in the steering column and dash to protect the driver and passenger during a head-on collision.
sistema de bolsa de aire Un sistema que usa los sensores, la computador a bordo del vehículo, un módulo inflador, y las bolsas de nylon en la columna de dirección y en el tablero para proteger al conductor y al pasajero en caso de una colisión frontal.

airbrush A small spray gun with a tiny cup for holding a small amount of material that operates like a conventional siphon gun; also called a door jamb gun.
pistola pulverizadora Tiene una taza pequeña para la colocación de una pequeña cantidad de material y funciona como una pistola-sifón convencional; también llamada pulverizador neumático.

air compressor Used to raise the pressure of air from normal to some higher pressure, up to 200 psi.
compresor de aire Utilizado para aumentar la presión del aire de normal a una presión más alta de hasta 200 psi.

air drill A drill that applies shop air pressure to spin a drill bit out. It is smaller, lighter, and more compact than an electric drill.
taladro neumático Un taladro que aplica presión de aire para retirar una broca de taladro girándola. Es más pequeño, más leve y más compacto que un taladro eléctrico.

air file A long, thin air sander for working large flat surfaces on panels; also called a long air sander.
Lija neumática Una lija neumática larga y fina que se utiliza para trabajar sobre superficies planas grandes en paneles.

air leaks Cause a whistling or hissing noise in the passenger compartment while the vehicle is being driven.
pérdidas de aire Provocan un silbido o siseo en el compartimiento de pasajeros cuando el vehículo está en movimiento.

air line lubricator Installed on the leg or branch line furnishing air to air-powered tools to help maintain them and keep them running smoothly.
lubricador de línea de aire Instalado en la ramificación o línea principal, provee aire a herramientas neumáticas para su buen mantenimiento y funcionamiento.

air polisher A tool that uses a soft buffing pad for smoothing and shining painted surfaces.
pulidor neumático Herramienta para suavizar y dar brillo a superficies pintadas, para lo que usa una almohadilla pulidora suave.

air pressure regulator A device used to reduce main line air pressure as it comes from the compressor. Maintains required air pressure. Used with air condenser or filter.
regulador de presión del aire Dispositivo utilizado para reducir la presión del aire del conducto principal mientras sale del compresor. Mantiene la presión de aire requerida. Se utiliza con un condensador de aire o filtro.

air ratchet An angled wrench used to work in hard-to-reach places and speed part replacement.
llave de trinquete neumática una llave en ángulo utilizada para trabajos en lugares difíciles de alcanzar.

air sander A tool that uses abrasive action to smooth body surfaces. Disc sanders are used for rougher automotive work, whereas orbital, or dual-action, sanders are used for buffing.
lijadora neumática Herramienta que utiliza acción abrasiva para suavizar superficies de la carrocería. Para trabajos automotrices más pesados se utilizan lijadoras a disco, mientras que para pulir, se utilizan lijadoras orbitales o de doble acción.

air-supplied respirator A hood with a clear visor and an external air supply hose that provides clean air for breathing.
respirador con provisión de aire una capucha con visor transparente y manguera de provisión de aire externo que provee aire puro para la respiración.

air tank shutoff valve A hand valve that isolates the tank pressure from shop line pressure. If not closed properly, the compressor may run all night.
válvula de cierre de tanque de aire Válvula manual que aísla la presión del tanque de la presión de la línea del taller. Si no se la cierra correctamente, el compresor puede quedar prendido toda la noche.

air transformer A device that filters and regulates the compressed air by removing dirt, oil, and water. A gauge shows the regulated air pressure. May also provide air outlets for spray guns, blowguns, and air-operated tools.
transformador de aire Dispositivo que filtra y regula el aire comprimido al retirándole la tierra, aceite y agua. Un medidor exhibe la presión del aire regulado. También puede proveer salidas de aire para pistolas pulverizadoras, pistolas de aire y herramientas neumáticas.

air valve Opened by pulling the trigger part way.

válvula de aire Se abre al tirar del gatillo parcialmente.

air vent hole Allows atmospheric pressure to enter the spray gun cup.
orificio de ventilación Permite que la presión atmosférica ingrese en la taza de la pistola pulverizadora.

aligning punch A tool with a long tapered punch used to align body parts for welding and for starting bolts.
punzón de alineamiento Herramienta con un punzón largo cónico utilizada para alinear piezas de la carrocería para la soldadura y para la colocación inicial de tornillos.

Allen wrench A hexagon, or six-sided, wrench used to remove setscrews.
llave allen Llave hexagonal o con seis lados utilizada para retirar tornillos de fijación.

alternate color chips Color chips that have been tinted a slightly different color than the standard color chip; also known as variance color chips.
Chips de colores alternativos También conocidos como chips de colores variados. Son chips de colores levemente distintos al color estándar del chip.

aluminum electrode wire Classified by series according to the metals the aluminum is alloyed with, and whether the aluminum is heat treated. Does not indicate the strength of the electrode.
cable de electrodo de aluminio Clasificado por serie de acuerdo con los metales con los que se ha aleado el aluminio y si el aluminio ha recibido tratamiento térmico. No es indicación de la fuerza del electrodo.

antichip coating A rubberized material put on a vehicle's lower panels and on the front edge of hoods and fenders; helps the finish to resist chips from rocks and other flying debris thrown up from the tires.
revestimiento contra el astillado Material encauchado colocado en los paneles inferiores del vehículo y en el borde delantero del capó y guardabarros; hace que el acabado sea resistente a astillas provenientes de piedras y otros detritos lanzados al aire por los neumáticos.

anticorrosion material Additional coating applied over and under the paint film.
material anticorrosivo Revestimiento adicional aplicado sobre y debajo la película de pintura.

antifreeze Used to prevent freeze-up in cold weather and to lubricate moving parts.
anticongelante Se utiliza para evitar el congelamiento en clima frío y lubricar piezas móviles.

A-pillars Steel box members that extend down from the roof panel to the main body section; also called "doghouse."
pilares A Miembros de la caja de acero que bajan del panel del techo a la sección principal de la carrocería.

arcing A gun handling problem that occurs when the spray gun is not moved parallel with the surface. Also an electric arc or spark that occurs when welding or when an electrical hot wire is shorted to grounds.
arqueo problema en la manipulación de la pistola que ocurre cuando el movimiento de la pistola pulverizadora no es paralelo a la superficie.

arming sensor Ensures that the particular collision is severe enough to require that an air bag be deployed.
sensor de armado Asegura que un choque específico sea lo suficientemente grave para requerir el despliegue de la bolsa de aire.

asbestos dust Cancer-causing agent used in the manufacture of older brake and clutch assemblies.
polvo de amianto Un agente carcígeno usado en la fabricación de las asambleas antiguas de frenos y embragues.

ASE certification Testing program to help prove that you are a knowledgeable collision or mechanical repair technician.
certificación ASE Una programa de examenes para certificar que es Ud. un técnico de reparación de colisión experimentado.

asphyxiation Anything that prevents normal breathing. There are many mists, gases, and fumes in a body shop that can damage your lungs and affect your ability to breathe.
asfixiación Se refiere a cualquier cosa que impide la respiración normal. Hay muchas brumas, gases y humos en un taller de carrocería que pueden dañar sus pulmones y afectar su habilidad de respirar.

assembly A number of parts that are either bolted or welded together to form a single unit.
asamblea Una cantidad de partes que se empernan o se sueldan juntos para formar una sola unedad.

atomization Breaks the liquid material into a fine mist spray of tiny, uniform droplets.
atomización Convierte el material líquido en un fina pulverización de gotas diminutas y uniformes.

automatic unloader A device designed to maintain a supply of air within given pressure limits on compressors when it is not practical to start and stop the motor.
descargador automático Dispositivo diseñado para mantener la provisión de aire dentro de ciertos límites de presión en los compresores cuando resulta inconveniente prender y apagar el motor.

axial runout Causes the wheel to wobble sideways as it rotates.
desviación axial Hace que la rueda bambolee hacia los lados al girar.

backing strip Backing patch bonded to the back of a repair area to restore the reinforced plastic's strength.

tira de refuerzo Parche de refuerzo que se une a la parte posterior de área de reparación para reforzar la fuerza del plástico.

ball peen hammer A multipurpose tool used to straighten, smooth, or shape sheet metal.
martillo de bola Herramienta multiuso utilizada para enderezar, alisar o dar forma a la lámina metálica.

banding coats Spraying the perimeter to deposit enough paint on panel edges, corners, or near the ends of body panels.
ligado de manos Se realizar para depositar pintura suficiente en bordes de paneles, esquinas o cerca de las extremidades de los paneles de la carrocería.

basecoat/clearcoat A paint system in which the color effect is given by a highly pigmented basecoat. Gloss and durability are given by a subsequent clearcoat. The basecoat can be either a solid color or metallic.
capa de fondo/capa transparente Un sistema de pintar en el que el efecto del color se logra por medio de una capa de fondo con mucho pigmento. Una capa transparente subsiguiente produce la brillantez y la durabilidad. La capa de fondo puede ser de un color sólido o metálico.

belt retractor Used to remove slack from seat belts so they fit snugly.
retractor de cinturón Utilizado para ajustar los cinturones de seguridad.

battery electric vehicle (BEV) A vehicle whose sole power source is electricity.
vehículo eléctrico a batería (BEV) Vehículo cuya única fuente de alimentación es la electricidad.

blasters Air-powered tools for forcing sand, plastic beads, or another abrasive onto surfaces for paint removal.
sopladores herramientas neumáticas utilizadas para remover pintura en superficies a través de la liberación de arena, cuentas plásticas u otros abrasivos.

bleeding Original color showing through after a new topcoat has been applied; also called bleed-through.
color sangrado El color original se puede ver al través de la capa superior que se ha depositado.

blend coats Progressive application of paint so the paint film becomes thinner; also called shading coats.
mezcla de manos Aplicación progresiva de pintura para reducir el espesor de la película de pintura.

blending solvent Formulated to help fade the new paint into the original paint when spraying.
solvente de mezclado Formulado como ayuda para fundir la pintura nueva en la pintura original al pulverizar.

blistering Formation of hollow bubbles or water droplets in a paint film. Blistering is usually caused by expansion of air or moisture trapped beneath the paint film. Blisters can form around salt crystals trapped under a film because salt attracts moisture and is dissolved.
formar ampollas La formación de las burbujas o las gotitas de agua dentro de una película de pintura. Las ampollas suelen ser causadas por una expansión del aire o la humedad que se ha atrapado debajo de la película de pintura. Las ampollas pueden formarse alrededor de los cristales de sal atrapados debajo de la película porque el sal atrae la humedad y se disuelve.

blushing Hazing or whitening of a film caused by absorption and retention of moisture in the drying paint film.
aspecto lechoso Una película que tiene un haz o se blanquea por causa de la absorción y retención de la humedad en una película de pintura que se esta secando.

body center marks Often stamped into the sheet metal in both the upper and lower body areas of the vehicle.
marcas centrales de la carrocería Con frecuencia, gravadas en la lámina metálica en las áreas superior e inferior de la carrocería del vehículo.

body dimensions The known correct body measurements of an undamaged vehicle.
dimensiones de la carrocería Son las medidas correctas conocidas de la carrocería de un vehículo sin daños.

body dimensions chart Used to make accurate damage assessment of the vehicle body.
cuadro de dimensiones de la carrocería Utilizado para realizar una evaluación precisa de daños en la carrocería del vehículo.

body filler A heavy-bodied polyester plastic material that cures very hard and is used to fill small dents in metal.
masilla Una material espesa de plástico que se endurece mucho al curarse y sirve para rellenar las abolladuras pequeñas en el metal.

body hammer A tool designed to strike and rebound off the surface of sheet metal to straighten minor bumps and dents.
martillo de carrocería Herramienta proyectada para golpear y rebotar sobre la superficie de la lámina metálica; se usa para enderezar golpes y abolladuras de poca importancia.

body ID number Gives information about the finish and the trim or moldings used on a specific vehicle; also called the service part label.
número de identificación de la carrocería Brinda información sobre el acabado y los bordes o molduras utilizados en un vehículo específico; también denominado etiqueta de pieza de servicio.

body jack A portable hydraulic pump and ram for body and minor frame-straightening operations; also called a hydraulic power set or portable power unit.

elevador de carrocería Bomba hidráulica portátil y pisón para la carrocería y para operaciones menores de enderezado de bastidor; también denominado kit de reparación de carrocería o unidad eléctrica portátil.

body line sanding guide A tool used for sanding perfectly straight contour lines on panels.

Guía de lijado de línea de la carrocería Herramienta utilizada para el lijado en líneas perfectamente rectas en los paneles.

body-over-frame construction A vehicle that has seperate body and chassis parts bolted to the frame.

construcción carrocería-sobre-bastidor Vehículo con piezas de carrocería y chasis separadas, atornilladas al bastidor.

body shim Thin U-shaped piece of metal for making part adjustments.

cuña de carrocería Pedazo de metal en forma de U utilizado para ajustes de piezas.

body specifications Normal measurements on an undamaged vehicle.

especificaciones de la carrocería Medidas normales en un vehículo sin daños.

body spoon A tool used like a hammer or dolly, particularly useful on creases and ridges. The flat surfaces of a spoon distribute the striking force over a wide area.

cucharas de carrocería Herramienta que se usa como un martillo o remachador, especialmente útil para estrías y rebordes. Las superficies planas de la cuchara distribuyen la fuerza del golpe en un área amplia.

bolt grade markings Lines or numbers on the top of the head of the bolt used to identify bolt hardness and strength.

marcas de clase de tornillos Líneas o números en la cabeza del tornillo para identificar la dureza y fuerza del tornillo.

box-end wrench A wrench with a closed end, allowing for better holding power without slippage.

llave de cubo Llave con una extremidad cerrada; permite una mayor fuerza de sujeción sin resbalar.

B-pillars Roof supports between the front and rear doors; also called center pillars.

Pilares B Soportes del techo entre las puertas delanteras y traseras; también llamados pilares centrales.

brazing Welding in which a nonferrous metal with a lower melting point than the base metal is melted and spread between two base metals to form a strong bond.

soldadura fuerte Soldadura en la que se derrite un metal no ferroso con un punto de fusión más bajo que el del metal-base, el que luego se desparrama entre dos metales-base para formar una fuerte unión.

breaker bar An extra-long handle used as a socket wrench accessory. When held at a 90-degree angle, it provides the torque needed to loosen stubborn fasteners.

prolongador Manija extra larga que se usa como accesorio de llave tubular. En un ángulo de 90 grados, proporciona el esfuerzo de rotación necesario para aflojar sujetadores demasiado ajustados.

brushable seam sealers Used on interior body seams where appearance is not important.

selladores a escobilla para uniones Se usan en las uniones del interior de la carrocería, donde la apariencia no tiene importancia.

buckles Deformations in metal caused when metal bends beyond its elastic limit and will not return to its original shape.

hebillas Deformaciones en el metal causadas cuando el metal se dobla más allá de su límite de flexibilidad y no regresa a su forma original.

buffing machine Uses a spinning or rotating action to level and smooth the paint surface quickly.

pulidora Utiliza una acción giratoria o rotativa para nivelar y emparejar rápidamente la superficie de la pintura.

buffing pad Rotated by an electric or air buffer to force compound over the paint surface.

almohadilla de pulido La hace girar una pulidora eléctrica o neumática para que coloque el compuesto sobre la superficie de la pintura.

bull's-eye A depression underneath fresh paint due to successive layers of paint not being properly tapered.

ojo de buey Depresión debajo de la pintura fresca debido a la colocación de capas sucesivas de pintura sin reducción apropiada.

bumper energy absorbers Used to cushion some of the impact of a collision and reduce damage.

Amortiguadores de energía del paragolpes Se utiliza para amortiguar parte del impacto de un choque y reducir los daños.

bumping hammer A round- or square-faced hammer used on large dents. The faces are large so that the force of the blows is spread over a large area.

martillo desabollador Martillo con punta redonda o cuadrada utilizado en abolladuras grandes. Sus caras son grandes para que la fuerza de los golpes se extienda en un área extensa.

burn mark An indication of good weld penetration found on the back of a weld. Also mark on paint from improper machine buffing.

marca de quemadura Indicación de buena penetración de soldado encontrada en la parte de atrás de una soldadura.

burn-through A hole through the back side of the weld, which indicates too much penetration into the lower base metal. Also unwanted mall in paint from improper use of high-speed buffer.
quemadura excesiva Un orificio que atraviesa la parte posterior de la soldadura, indicando penetración excesiva en el metal inferior.

caged plate Heavy, thick steel plate with threaded holes to accept large bolts.
lámina para tornillos Lámina pesada y gruesa con orificios con rosca para la colocación de tornillos grandes.

caliper Housing containing the pistons and related seals, springs, boots, cylinders, and fluid passages necessary to force the friction linings or pads against the rotor.
calibrador Caja que contiene los pistones y sellos correspondientes, y fundas, cilindros y conductos de fluido necesarios para provocar la fricción de revestimientos o almohadillas contra el rotor.

camber Inward or outward tilt of a wheel at the top from true vertical. It is the tire-wearing angle in degrees.
camber (comba) La inclinación fuera de vertical perfecto hacia adentro o hacia afuera en la parte superior de una rueda. Es el ángulo en grados que desgasta las llantas.

carbon monoxide (CO) An invisible, odorless, and potentially fatal gas produced by car and truck engines.
monóxido de carbono Gas invisible, inodoro y potencialmente fatal producido por los motores de auto-móviles y camiones.

carburizing flame A flame obtained by mixing slightly more acetylene than oxygen, used for welding nickel, aluminum, and other alloys; also called a surplus or reduction flame.
Llama carburizadora Llama obtenida a través de la mezcla de un poco más de acetileno que oxígeno, utilizada para soldar níquel, aluminio y otros aleados; también llamada llama de reducción.

carcinogens Cancer-causing agents.
carcinógenos Agentes que provocan el cáncer.

carpeting Woven fabric cover, often with a sound-deadening backer, that fits over the floor panels.
alfombrado: Forro de tela que se coloca sobre los paneles del piso; con frecuencia, cuenta con un revestimiento posterior reductor de sonido.

cartridge filter respirator Protects against vapors and spray mists given off by one-part enamel or lacquers. Used in well-ventilated areas.
filtro ventilador de cartucho Protege de vapores y pulverizaciones despedidos por esmaltes o lacas de una parte. Debe utilizarse en áreas con buena ventilación.

caster Backward or forward tilt of kingpin or spindle support arm at top from true vertical. It is the directional control angle measured in degrees.
angulo de caster La inclinación fuera de vertical perfecto hacia atrás o hacia afrente en la parte superior de un pivote o el brazo de soporte del husillo. Es el ángulo de control direccional que se mide en grados.

castor–camber gauge Used with the turning radius gauge to measure caster and camber in degrees.
medidor de ángulo de avance y de inclinación Se usa con el medidor de radio giratorio para medir el ángulo de avance y el de inclinación en grados.

catalyst A substance that causes or speeds up a chemical reaction when mixed with another substance but does not change itself.
catalizador Una substancia que causa o acelera una reacción química al ser mezclada con otra substancia pero que no cambia en si misma.

center plane Divides the vehicle into two equal halves, the passenger side and the driver's side; also called the centerline.
plano central Divide el vehículo en dos mitades iguales, el lado del pasajero y el lado del conductor; también llamado línea central.

center punch A tool used to mark parts before they are removed or for marking a spot for drilling. The punch mark keeps the drill bit from wandering.
punzón de centrar Herramienta utilizada para marcar piezas antes de retirarlas o para perforación. La marca de punzón evita que la broca de la perforadora se mueva.

certified crash tests Use real vehicles and sensor-equipped dummies to estimate how much impact people and vehicles would suffer in collisions.
pruebas de colisión certificadas Utilizan vehículos verdaderos y muñecos equipados con sensores para estimar el nivel del impacto que las personas y el vehículo sufrirían en una colisión.

charging system Recharges the battery and supplies electrical energy when the engine is running.
sistema de carga Recarga la batería y suministra energía eléctrica cuando el motor se encuentra en funcionamiento.

chemical paint remover Used for stripping large areas of paint if environmental regulations allow.
removedor de pintura químico Se utiliza para quitar grandes áreas de pintura cuando los reglamentos ecológicos permiten su uso.

chipped paint Results from mechanical impact damage to the paint film.
pintura lascada Provocada por daños causados por el impacto mecánico en la película de pintura.

chisel A tool that is a steel bar with a hardened cutting edge for shearing steel.
cincel Herramienta que es una barra de acero con un borde afilado endurecido para cortar acero.

chroma Strength or intensity of a color; the amount a color differs from the white, gray, or black of the neutral axis of the color tree. Often referred to as saturation or desaturation.
saturación cromático La fuerza o intensidad de un color; la cantidad que difere un color del blanco, gris o negro en el eje neutro de una escala de colorimetría. Esta intensidad suele referirse como la saturación/desaturación.

circuit breakers Heat up and open with excess current to protect a circuit.
disyuntores Se calientan y se abren al recibir corriente excesiva para proteger un circuito.

clip pullers A tool used to remove the clip from an interior door handle.
extractor de perno Herramienta utilizada para retirar el perno de una manija interior de la puerta.

clock spring The electrical connection between the steering column and the air bag module; also called a spiral cable.
resorte de reloj La conexión eléctrica entre la columna de dirección y el módulo de la bolsa de aire.

clutch Device used to couple and uncouple engine power to a manual transmission or transaxle.
embrague Dispositivo utilizado para acoplar y desacoplar la fuerza del motor a una transmisión manual o eje de la transmisión.

collision Accidental damage caused by an impact on a vehicle body and chassis; commonly called a "crash" or "wreck."
colisión Daño accidental ocasionado por un impacto en la carrocería y chasis de un vehículo; comúnmente denominado "choque".

collision estimating guides Provide information for calculating labor costs and part costs, such as part numbers, part prices, and labor times per repair task.
guías de estimación de colisiones Suministran información para el cálculo del costo de reparaciones de colisiones, tales como números de piezas, precios de piezas y mano de obra por tarea de reparación.

collision repair Fixing a vehicle that has been in an accident; also called auto body repair.
reparación de colisión Reparación de un vehículo que ha tenido un accidente; también reparación de carrocería.

collision repair facility A place with well-trained workers, specialized tools, and equipment for restoring damaged vehicles; also called a body shop.
establecimiento de reparación de colisiones Un lugar con trabajadores capacitados, herramientas especializadas y equipos para la restauración de vehículos dañados; también llamado taller de auto-móviles.

collision repair guides Provide instructions, safety warnings, and technical illustrations for various makes and models of vehicles; summarize procedures for removing parts, cutting and welding structural body panels, and similar types of repair.
guías de reparación de colisiones Provee instrucciones, advertencias de seguridad e ilustraciones técnicas para diversas marcas y modelos de vehículos; resume los procedimientos para la extracción de piezas, corte y soldadura de paneles estructurales de la carrocería y tipos de reparaciones similares.

colorcoat Paint applied over the primecoat; also called the topcoat.
mano de color Pintura que se aplica sobre la capa de imprimación; también denominada mano superior.

color-corrected light bulbs Bulbs used by photographers because they most closely match the full spectrum of light.
bombillas de luz con corrección cromática Bombillas utilizadas por los fotógrafos por asemejarse más al espectro completo de la luz.

color face The side seen when a metallic color is viewed at right angles or perpendicular to the surface.
faz de color Cuando se mira un color metalizado en ángulos rectos o perpendicular a la superficie.

color flip-flop A condition that occurs in metallic paints involving the positioning of the aluminum particles and the manner in which light is reflected to the observer.
flip-flop de color Condición que ocurre en las pinturas metalizadas asociada al posicionamiento de las partículas de aluminio y la forma en que refleja la luz para el observador.

color matching The steps needed to make the new finish on a vehicle match the existing finish.
combinación de colores Los pasos necesarios para que el acabado nuevo del vehículo combine con el acabado existente.

color-matching guides Provide information needed for repainting panels so that the repair has the same appearance as the old finish.
guías de combinación de colores Proporciona la información necesaria para volver a pintar paneles de modo que el arreglo tenga el mismo aspecto del acabado anterior.

color spectrum Light that has been broken down into its separate colors by passing through a prism.

espectro de colores Luz separada en sus distintos colores al atravesar un prisma.

color test panel A black-and-white coated sheet of cardboard with a wooden handle.

panel de prueba de color Plancha de cartón revestida en blanco y negro con una manija de madera.

color tree Used to locate colors three-dimensionally when matching colors.

árbol de colores Se utiliza para ubicar colores en tres dimensiones al combinarlos/compararlos.

combination pliers The most common type of pliers with both flat and curved surfaces and two jaw-opening sides. One jaw can be moved up and down on a pin attached to the other jaw to change the opening.

pinzas de combinación El tipo más común de pinza con superficies tanto planas como curvas y dos lados con mordazas que se abren. Una de las mordazas puede moverse hacia arriba y hacia abajo sobre un pino agarrado a la otra mordaza para modificar la abertura.

combination wrench A tool with an open-end jaw on one end and a box-end on the other. Both ends are the same size.

llave combinada Herramienta con una mordaza de punta abierta en una extremidad y una punta cuadrada en la otra. Ambas extremidades son del mismo tamaño.

comeback A repaired vehicle that is returned to the shop by an unsatisfied customer.

devolución Vehículo reparado que ha sido devuelto al taller por un cliente insatisfecho.

component location diagrams Found in a service manual and used for locating electrical parts.

diagramas de ubicación de componentes Se encuentran en los manuales de servicio; utilizados para ubicar piezas eléctricas.

composite force Force of all pulls combined.

fuerza compuesta La fuerza de todos los esfuerzos combinados.

composite unibody A frame made almost entirely of plastics or other materials like carbon fiber; helps improve vehicle performance and fuel economy.

'unibody' compuesto Bastidor hecho prácticamente en su total de plásticos y otros materiales como fibra de carbón; ayuda a mejorar el desempeño del vehículo y la economía de combustible.

compounding Action of using an abrasive material either by hand or machine to smooth and bring out gloss of the applied topcoat.

pulimentación El uso de una material abrasiva aplicada sea por mano o por máquina para que la capa superficial queda lisa y brillante.

compressed air supply system A system designed to provide an adequate supply of clean, dry air at a predetermined pressure for the operation of all air-powered tools.

sistema de suministro de aire comprimido Sistema diseñado para proveer un suministro adecuado de aire seco limpio, a una presión predeterminada, para el funcionamiento de toda herramienta neumática.

compressed natural gas (CNG) Natural gas stored at up to 10,000 psi on a vehicle to provide a source of clean-burning fuel for an internal combustion engine.

gas natural comprimido (GNC) Gas natural almacenado a presiones de hasta 3600 psi en un tanque de combustible de GNC y presurizado hasta 10 000 psi en las líneas de combustible de GNC. El GNC es un combustible de combustión limpia para motores de combustión interna.

compressor air tank Heavy-gauge steel tank for holding an extra supply of compressed air.

tanque de aire comprimido Tanque de acero pesado donde se almacena una provisión adicional de aire comprimido.

compressor drain valve Because compressing air makes moisture condense, this valve allows you to drain water from a compressor air tank. The valve is found at the bottom of the tank.

válvula de drenaje del compresor Debido a que el aire comprimido provoca la condensación de la humedad, esta válvula permite el drenaje del agua de un tanque de aire comprimido. La válvula está ubicada en la base del tanque.

compressor oil plugs Used for filling and changing air pump oil.

tapones del aceite del compresor Se utilizan para cargar y cambiar el aceite de la bomba neumática.

computer Electronic device for storing and manipulating information.

computadora Un dispositivo electrónico que sirve para almacenar y manipular la información.

computer-based estimating Provides more accurate and consistent damage reports.

estimativas computarizadas Proporcionan informes de daños más precisos y compatibles.

computer database Contains all service information (estimating guides, labor rates, and part and material vendor information) in electronic form.

base de datos informatizada Contiene toda la información de servicio (guías de presupuesto, tarifas de mano de obra, e información sobre proveedores de piezas y materiales) en forma electrónica.

computerized laser measuring system Uses light beams, reflective targets, a laser-receiver unit, and a PC to speed and simplify damage analysis.

computerized measuring system Uses a PC and specialized electronic hardware to evaluate structural damage.

sistema de medición computarizada Utiliza un PC como ayuda para calcular medidas del vehículo.

conductor Carries current to the parts of a circuit.

conductor Lleva corriente a las piezas de un circuito.

console Located between the bucket seat for the gear shift mechanism, electric controls, a console lid and compartment, and other items.

consola Ubicada entre el asiento individual para el mecanismo de cambios, controles eléctricos, tapa y compartimiento de la consola y otros artículos.

contaminants Foreign substances on a surface to be painted (or in paint) that adversely affect the finish.

contaminantes Las materias extrañas en la superficia de acabado (o en la pintura) que pueden afectar negativamente al acabado.

continuous weld Welding with an uninterrupted seam in a slow, steady, ongoing movement.

soldadura continua Soldadura con juntura sin interrupción con un movimiento lento, constante y continuo.

contour sanding stick A sanding stick with a specific shape to match the body line being sanded.

Vara de lija de contorno Vara de lija con forma específica que combina con la línea de la carrocería que se está lijando.

control points Points on a vehicle, including holes, flats, or other identifying areas, used to position panels and rails during manufacturing of the vehicle.

puntos para comprobación Los puntos de un vehículo, incluyendo los agujeros, las areas planas, u otras áreas de identificación que sirven para posicionar los paneles y los largueros durante la fabricación del vehículo.

conversion charts Charts that help convert from one measuring system to another or from one value to another.

cuadros de conversión Cuadros que ayudan a convertir de un sistema de medición a otro o de un valor a otro.

coolant recovery bottle Aid for the replacement of coolant in the cooling system.

botella de recuperación de refrigerante Se utiliza para reemplazar el refrigerante en el sistema de refrigeración.

cooperative education programs A program offered by a school that allows a student to have a job and obtain school credit at the same time.

programas educativos cooperativos Un programa ofrecido por una escuela y que permite que el alumna tenga un empleo y obtenga crédito escolar al mismo tiempo.

Corporate Average Fuel Economy (CAFE) Set by the Environmental Protection Agency to help reduce U.S. fuel consumption and air pollution; requires automotive manufacturers to start building cars with improved fuel economy.

Corporate Average Fuel Economy (CAFE—Promedio de Economía de Combustible Corporativa) Establecida por la Agencia de Protección Ambiental de los Estados Unidos para ayudar a reducir el consumo de combustible y la polución del aire en los EE.UU.; exige que los fabricantes de vehículos automotores comiencen a construir vehículos con mayor economía de combustible.

corrosion Chemical reaction of air, moisture, or corrosive materials on a metal surface; usually referred to as rusting, or oxidation.

corrosión La reacción química del aire, la humedad, o las materias corroídas en una superficie metálica; normalmente se refiere como comido por la herrumbre o la oxidación.

corrosion prevention Implies a lifetime vehicle maintenance responsibility to the consumer.

prevención de la corrosión Implica la responsabilidad vitalicia del consumidor por el mantenimiento del vehículo.

corrosion protection Involves using various methods to protect steel body parts from rusting.

protección de la corrosión Involucra el uso de varios métodos para proteger las partes de acero de la carrocería de la oxidación.

coupling A connection that is male on both ends. Used to pair two pieces of hose together or to convert a female connection of one thread size to a male connection of another thread size.

pieza de conexión Conexión en la que ambas extremidades son macho. Se utiliza para unir dos pedazos de manguera o para convertir una conexión hembra con un tamaño de rosca a una conexión macho con otro tamaño de rosca.

cowl The panel at the rear of the front section, right in front of the windshield.

cubretablero El panel ubicado en la parte trasera de la sección delantera, justo delante del parabrisas.

C-pillars Pillars that extend up from the quarter panels to hold the roof and rear window glass; also called rear pillars.

pilares C Pilares que se extienden hacia arriba, a partir de los paneles del cuarto trasero, para sujetar el techo y el vidrio de la ventana trasera; también denominados pilares traseros.

cream hardener Body filler catalyst that must be proportionally mixed with filler or numerous problems may occur; used to cure body filler by making the filler heat up and harden.
crema endurecedora Catalizador de relleno que debe ser compatible con un endurecedor para ser mezclado; de lo contrario, pueden ocurrir numerosos problemas. Se usa para curar el relleno de carrocería al calentar y endurecer el relleno.

crush zones Sections built into the frame or body designed to collapse and absorb some of the energy of a collision.
zonas de impacto Las secciones incorporadas en el diseño del bastidor o la carocerría que se hunden así absorbiendo algo de la energía producida por una colisión.

cubic feet per minute (cfm) The volume of air being delivered by the compressor to the air tool. Used to measure the compressor's capabilities. Compressors with higher cfm ratings provide more air through the tools, making them more practical for larger jobs.
pies cúbicos por minuto (pcm) Volumen de aire enviado a la herramienta neumática por el compresor. Se utiliza para medir las capacidades del compresor. Los compresores con índices más altos de pcm envían más aire a través de las herramientas, lo que las hace más convenientes para trabajos grandes.

curb height The height at which a car is designed to ride.
altura de contén Cuando se diseña el automóvil a una altura específica.

curing Chemical reaction in paint or other material that causes hardening.
endurecimiento Una reacción química de secado de las pinturas que se secan por un cambio químico.

current Movement of electricity through a wire or circuit.
corriente Movimiento de la electricidad a través de un cable o circuito.

curtain air bag Deploys from the front pillar and roof trim.
bolsas de aire de cortina Al abrirse, salen del pilar delantero y el tapizado del techo.

CV axles Constant velocity axles transfer torque from the transaxle to the wheel hubs.
ejes de velocidad constante Transfieren el esfuerzo de rotación del eje de la transmisión a los cubos de las ruedas.

DA Disc orbital sander used to remove scratches.
DA Lija orbital de disco o neumática que puede usarse para remover rayas.

damage analysis form Helps you organize specs, actual measurements, and differences between good and actual measurements.
formulario de análisis de daños Ayuda a organizar especificaciones, medidas reales y diferencias en medidas buenas y reales.

damage estimate A calculation of the cost of repairs.
estimativa de daños Cálculo del costo de reparaciones.

dash assembly Assembly including the metal frame, soft dash pad, instrument cluster, radio, heater and air-conditioning controls, vents, and other similar parts; also termed instrument panel.
conjunto del tablero Conjunto que incluye la almohadilla blanda del tablero, el conjunto de instrumentos, la radio, los controles de calefacción y aire acon-dicionado, respiraderos y otras piezas similares; también llamado panel de instrumentos.

dash panel Fits between the A-pillars, or windshield pillars, to hold the instrument cluster, air-conditioning system vents, passenger side air bag, glove box door, stereo, and other items.
tablero Ubicado entre los pilares A, o pilares del parabrisas, para sujetar el conjunto de instrumentos, los respiraderos del sistema de aire acondicionado, la bolsa de aire del lado del pasajero, la puerta de la guantera, el equipo de música y otros artículos.

datum plane An imaginary line that appears on frame blueprints or charts to help determine correct height dimensions.
plano de datos Línea imaginaria que aparece en planos o cuadros del chasis como ayuda para determinar la altura correcta.

DC reverse polarity When the electrode is positive and the workpiece is negative; the greatest welding penetration is produced at this connection.
polaridad invertida DC Cuando un electrodo es positivo y la pieza de trabajo es negativa; se produce la mayor penetración de soldadura en esta conexión.

dead blow hammer A hammer with a metal face filled with lead balls to prevent it from rebounding after it has been struck.
martillo de golpe seco Martillo con faz metálica llena de bolitas de plomo para evitar que rebote después del golpe; es bueno para evitar dañar las piezas.

dealership body shop A shop owned and managed under the guidance of a new car dealership.
taller de la concesionaria Taller de propiedad de la concesionaria de automóviles 0 km, la que supervisa su administración.

deductible The amount the vehicle owner has agreed to copay on the insurance policy.
deducible Suma que el propietario del vehículo ha acordado pagar y a partir de la cual comienza la cobertura de su póliza de seguro.

deductible clause Makes the owner responsible for a given amount of a repair estimate (usually

$100–$500), with the remaining cost paid by the insurance company.

cláusula de deducible Responsabiliza al propietario por un valor determinado de una estimativa de reparación (generalmente, entre $100 y $500); el costo restante lo paga la compañía de seguros.

deep socket A longer socket used for reaching over stud bolts.

cubo hondo Cubo más largo utilizado para alcanzar por arriba de tornillos prisioneros.

deformation New, undesired bent shape a metal takes after an impact or collision.

deformación Forma doblada, nueva y no deseada, que toma el metal después de un impacto o colisión.

deformation sensors Used to operate side air bags.

sensores de deformación Se usan para el funcionamiento de las bolsas de aire laterales.

destructive test A way to examine whether a weld is satisfactory by working on a test piece of the same metal and thickness as the panel, and then separating it to see how cleanly the weld comes apart.

verificación por destrucción Forma de fijarse si una soldadura es satisfactoria, en la que se trabaja en una muestra de un pedazo del mismo metal de igual espesura que el panel, y luego se lo separa para ver si la soldadura se desprende de manera prolija.

detail sanding Involves using a small dirt nib sanding block and extra-fine sandpaper to level and smooth specks of debris.

lijado en detalle Utiliza un bloque de lijado con punta pequeña y papel de lija extrafino para nivelar y suavizar manchas de detritos.

diagnosis charts Found in service manuals, these charts give the logical steps for finding the sources of problems; also called troubleshooting charts.

cuadros de diagnóstico Encontrados en los manuales de servicio, estos cuadros proporcionan los pasos lógicos para encontrar la fuente de problemas; también denominados cuadros de resolución de problemas.

diagnostic trouble code (DTC) Malfunction data stored in a vehicle's computer that can be retrieved by a scan tool.

código diagnóstico de falla (DTC, por sus siglas en inglés) Datos de fallas almacenados en la computadora de un vehículo que pueden recuperarse mediante una herramienta de exploración.

diamond damage A damage condition where one side of the vehicle has been moved to the rear or front, causing the frame/body to be out of square.

daño de diamante Una tipo de daño en el que una lateral del vehículo se ha movido hacia atrás o adelante, haciendo que el chasis/carrocería quede fuera de escuadra.

die A tool used to straighten damaged threads on bolts or studs.

troquel Herramienta utilizada para enderezar las roscas dañadas de pernos o tornillos prisioneros.

differential Assembly within the drive axle assembly that uses gears to allow a different amount of torque to be applied to each drive wheel while the vehicle is making a turn.

diferencial Conjunto dentro del conjunto del eje de la transmisión que, al girar con el vehículo, utiliza engranajes para permitir la aplicación de distintos niveles de torsión a cada rueda impulsora.

digital volt-ohm-ammeter (DVOM) A digital device used to troubleshoot electrical systems.

voltímetro, ohmímetro y amperímetro digital (por su sigla en inglés, DVOM) Dispositivo digital utilizado para resolver problemas en sistemas eléctricos

displacement The theoretical amount of air in cubic feet that a compressor can pump in one minute.

desplazamiento La cantidad teórica de aire en pies cúbicos que un compresor logra bombear en un minuto.

dolly Heavy steel block with various shapes on each side; used like a small anvil held on the back side of a panel being struck by a hammer to work out body damage.

remachador Bloque de acero pesado con varias formas en cada lado; se lo usa como un yunque pequeño que se sujeta en la parte trasera del panel que se está martillado para arreglar daños en la carrocería.

door dust cover Fits between the inner trim panel and door frame to keep out wind noise.

funda contra el polvo Ubicado entre el panel de revestimiento interior y el bastidor de la puerta para aislar el viento y los ruidos.

door frame Main steel frame of the door.

bastidor de la puerta Bastidor principal de acero de la puerta.

door hinge check Test to move the door assembly up and down on its hinges.

prueba de bisagras de la puerta Prueba realizada para mover el conjunto de la puerta hacia arriba y hacia abajo sobre sus bisagras.

door sagging Results when the rear of the door is lower than the front; a common problem often due to badly worn hinge pins.

puerta vencida Cuando la parte trasera de la puerta está más baja que la delantera.

door skin Outer panel over the door frame that is clinched to the pinch weld flange.

piel de la puerta Panel exterior que cubre el bastidor de la puerta que está remachado al reborde del doblez.

door trim panel Attractive padded trim piece that covers the inner door frame.
panel de revestimiento de la puerta Pieza de revestimiento acolchada y decorativa que recubre el bastidor interior de la puerta.

door weatherstripping Fits around the door or door opening to seal the door-to-body joint.
burlete de la puerta Se coloca alrededor de la puerta o de la abertura de la puerta para que la puerta quede sellada a la junta de la carrocería.

drive shaft A long tube that transfers power from the transmission to the rear axle assembly.
eje de transmisión Tubo largo que transfiera fuerza de la transmisión al conjunto del eje trasero.

drive sizes Fraction sizes of drive post for sockets: ¼, ⅜, ½, and ¾ inch. When purchasing socket wrench sets, smaller drive sizes are used for turning small fasteners on emblems and trim where little torque is required. Larger drive sizes are used where greater torque is needed.
tamaño de extremidad de llave de cubo Al comprar conjuntos de llaves de cubo, los tamaños de extremidad más pequeños se usan para girar sujetadores pequeños en emblemas y adornos para los que se necesite poca torsión. Cuando se necesita una torsión mayor, se usan tamaños más grandes.

driver's side air bag Deploys from the steering wheel center pad.
bolsa de aire del lado del conductor Se abre a partir de la almohadilla central del volante.

dry sandpaper Sandpaper designed to be used without water, which will ruin the paper.
papel de lija seco Papel de lija diseñado para su uso sin agua, la que arruinaría el papel.

drying Process of changing a coat of paint from a liquid to a solid state due to evaporation of the solvent, a chemical reaction of the binding medium, or a combination of these causes.
secado El proceso de cambiar una capa de pintura de un líquido a un estado sólido debido a la evaporación del solvente, una reacción química del medio aglomerante, o una combinación de éstas causas.

drying rooms Permanent infrared or sodium quartz units for the forced drying of paint in more sophisticated paint shops.
salas de secado Los talleres de pintura más sofisticados cuentan con unidades permanentes de infrarrojos o cuarzo de sodio para el secado de la pintura.

dust respirator A paper filter that fits over the nose and mouth to block small airborne particles.
máscara contra el polvo Filtro de papel con el que se cubre la nariz y la boca para bloquear la entrada de pequeñas partículas en el aire.

dye penetrant A sprayed-on material used to check for cracks in aluminum panels and components.
tinta penetrante Material que se pulveriza utilizado para detectar rajaduras en paneles y componentes de aluminio.

eccentric An egg-shaped part, mounted on the motor shaft, that acts on the diaphragm plate to pull the diaphragm up and down.
excéntrico Pieza con forma ovoide.

E-coat Paintable primer; an essential part of the factory corrosion protection warranty on new factory panels and parts.
E-coat Pintura preliminar pintable; parte esencial de la garantía de fábrica contra la corrosión en paneles y piezas nuevos.

electrical fires Fires caused when excess current causes wiring to overheat, melt, and burn.
incendios eléctricos Incendios ocasionados cuando el exceso de corriente hace que los cables se recalienten, se derritan y se prendan fuego.

electric circuit Contains a power source, conductors, and a load.
circuito eléctrico Contiene una fuente de energía, conductores y una carga.

electric vehicle same as battery electric vehicle A vehicle whose primary power source is an electric motor.
vehículo eléctrico, igual a vehículo eléctrico a batería Vehículo cuya fuente principal de alimentación es un motor eléctrico.

electrocution Electricity passes through a human body. Severe injury or death can result.
electrocución La electricidad pasa por el cuerpo humano. Puede causar los daños severos o la muerte.

electronic leak detector Uses a high-frequency sending unit and a receiver unit to find openings or leaks between parts.
detector de pérdidas electrónico Utiliza una unidad emisora y una unidad receptora de alta frecuencia para ubicar aberturas o pérdidas entre piezas.

electronic shielding Protection used when welding near on-board computers and sensor wiring.
protector electrónico Protección que se utiliza al soldar cerca de computadoras de a bordo y cableado de sensores.

electronic stethoscope Instrument used to find rattles and other mechanical noises.
estetoscopio electrónico Instrumento que se utiliza para ubicar rechinamientos y otros ruidos mecánicos.

elements Moisture, road salt, mud, or other outdoor substances that can damage vehicle parts.

elementos Humedad, sal en los caminos, barro u otras sustancias exteriores que pueden ocasionar daños en las piezas del vehículo.

emission control systems Used to prevent potentially toxic chemicals from entering our atmosphere.
sistemas de control de emisión Se utilizan para evitar el ingreso a la atmósfera de sustancias químicas potencialmente tóxicas.

enamel A type of paint that dries in two stages: first, by evaporation of the solvent and then by oxidation of the binder.
esmalte Un tipo de pintura que se seca en dos etapas: primero por evaporación del solvente y luego por la oxidación del aglomerante.

energy reserve module Allows the air bags to deploy in the event of a power failure.
módulo de reserva de energía Permite que las bolsas de aire se abran en caso de falla eléctrica.

engine Provides energy to move the vehicle and power all accessories.
motor Suministra la energía para el movimiento del vehículo y el funcionamiento de todos los accesorios.

entrepreneur A person who starts his or her own business.
empresario Persona que abre su propia empresa.

Environmental Protection Agency (EPA) Established in 1970 as an independent agency in the executive branch of the U.S. government, this agency coordinates government and industry action to protect Earth's environment from pollution damage.
Agencia de Protección Ambiental (EPA Environmental Protection Agency) Establecida en 1970 como organismo independiente del poder ejecutivo del gobierno de los EE.UU., esta entidad coordina las acciones del gobierno y de la industria para la protección del medio ambiente de la tierra contra daños ocasionados por la polución.

environmental safety Procedures that protect people and the Earth's resources from toxic chemicals.
seguridad ambiental Procedimientos que protegen a los recursos de la tierra contra las sustancias químicas tóxicas.

epoxy A class of resins characterized by good chemical resistance.
resina epósica Una clase de resinas caracterizadas por su buena resistencia química.

epoxy primer A two-part primer that cures fast and hard and increases body filler adhesion and corrosion resistance.
pintura preliminar de epoxi Pintura preliminar de dos parte que se cura rápidamente quedando dura y aumenta la adhesión del relleno de la carrocería.

estimating Analyzing damage and calculating how much it will cost to repair a vehicle.
dar presupuesto estimativo Analizar los daños y calcular los costos de efectuar las reparaciones del vehículo.

estimating program Software that automatically helps to find parts needed and determine labor rates and the total cost of repairs.
programa de estimativas Software que automática ayuda a ubicar las piezas necesarias y determinar tarifas de mano de obra y el costo total de reparaciones.

estimator Person who makes an appraisal of vehicle damage and determines what needs to be repaired.
estimador Persona que realiza una evaluación de los daños en un vehículo y determina qué necesita ser reparado.

exhaust system Collects and discharges exhaust gases caused by the combustion of the air–fuel mixture within the engine.
sistema de escapa Junta y libera gases de escape provocados por la combustión de la mezcla de aire y combustible dentro del motor.

extension Used on a socket wrench to reach otherwise inaccessible places.
extensión Se utiliza en la llave de cubo para alcanzar lugares que de otra forma serían inaccesibles.

eye flushing station A station equipped with fresh water and nozzles for washing out eyes.
estación de lavado de argollas Estación equipada con agua limpia y boquillas para el lavado de argollas.

fabricated panels Hand-made repair parts to fix small problems.
paneles fabricados Piezas de reparación hechas a mano utilizadas para el arreglo de problemas pequeños.

fastened parts Parts held on a vehicle by bolts, nuts, screws, clips, and adhesives.
piezas sujetas Piezas sujetas al vehículo por tornillos, tuercas, pernos, sujetadores y adhesivos.

fasteners The bolts, nuts, screws, clips, and adhesives that hold a vehicle together.
fijadores Los tornillos, tuercas, pernos, sujetadores y adhesivos que mantienen el vehículo armado.

fault code A code representing a specific circuit or part with a problem.
código de falla Un código que representa un circuito o pieza específico(a) con un problema.

featheredging Tapering edges of the damaged area with sandpaper or special solvents.
biselado Reducción de los bordes del área dañada con papel de lija o solventes especiales.

fiberglass cloth Made by weaving fiberglass strands into a stitched pattern.
paño de fibra de vidrio Paño hecho a través del entretejido de hilos de fibra de vidrio formando un calado.

fiberglass mat A series of irregularly distributed fiberglass strands that form a patch for the resin liquid.
malla de fibra de vidrio Una serie de hilos de fibra de vidrio distribuidos de manera irregular y que forman un parche para el líquido de resina.

fiberglass resin A thick resin liquid used for plastic body repair. Must be mixed with its own type of hardener to cure.
resina de fibra de vidrio Líquido de resina espeso que se usa para arreglos de carrocerías plásticas. Se lo debe mezclar con su propio tipo de endurecedor para la cura.

fiberglass-reinforced body filler A strong plastic filler with fiberglass in it that is used for rust repair.
relleno de carrocería reforzado con fibra de vidrio Relleno plástico fuerte con fibra de vidrio que se usa para reparar la oxidación.

filler Any material used to fill (level) a damaged area.
relleno Cualquier material que sirve para rellenar (aplanar) una área dañada.

filler grating Method that uses a coarse body file to remove high spots or edges that stick up on freshly applied filler.
rallado de relleno Método que utiliza una lima gruesa de carrocería para remover partes altas o bordes que sobresalen en el relleno recién aplicado.

final detailing Involves a series of steps to properly clean and shine all visible exterior and interior surfaces of the vehicle.
detalles finales Una serie de pasos para limpiar y dar brillo correctamente a toda la superficie exterior e interior del vehículo.

final sanding Using fine and very fine grits of material to prepare body surfaces for painting.
lijado final El uso de arenilla de material para preparar las superficies de la carrocería para la pintura.

fine line masking tape Thin, smooth surface masking tape. When removed, will produce a straighter and smoother place where old and new paint meet.
galón de pasamano de línea fina Galón de pasamano de superficie fina y lisa. Al retirarse, hace que el punto en que se encuentran la pintura antigua y la nueva quede más derecho y más liso.

finishing hammer A hammer with a smaller face used to achieve final contour on sheet metal. The surface of the face is crowned to concentrate the force on top of the ridge or high spot.
martillo de acabado Martillo con faz más pequeña que se usa para lograr el contorno final en láminas de metal. La superficie de la faz es bombeada para concentrar la fuerza sobre la protuberancia o punto alto.

first-aid kit A kit containing medical items, such as sterile gauze, bandages, scissors, or band-aids, to treat minor injuries.
kit de primeros auxilios Kit que contiene artículos médicos tales como gasa esterilizada, vendas, tijeras o curitas para el tratamientos de lesiones menores.

fish-eye eliminator An agent added to paint that helps smooth the paint when there are small craters or holes. Because it is oil-based, the paint flows over the tops of the craters or holes.
eliminador de orificios Agente que se agrega a la pintura y que ayuda a alisar la pintura cuando hay pequeñas crateras u orificios. Debido a que es a base de aceite, la pintura fluye sobre las crateras u orificios.

flash First stage of drying, where some of the solvents evaporate, which dulls the surface from an exceedingly high gloss to a normal gloss.
vaporización instantánea La primera etapa del secado en la cual algo de los solventes se evapora, lo que deslustra la superficie de una brillantez fuertísima a una brillantez normal.

flash time Time between coats or paint application and/or baking.
tiempo de vaporización instantánea El tiempo entre las capas o aplicaciones de pintura y/o el horneo.

flat rate A preset amount of time required and money charged to perform a specific repair operation.
tarifa fija Una cantidad establecida del tiempo requerido y del dinero cobrado para efectuar una reparación específica.

flat welding Welding with pieces of metal parallel with the bench or shop floor; fast and easy welding with the best penetration.
soldadura plana Soldadura con pedazos de metal paralelos con el banco o piso del taller; soldadura rápida y fácil con la mejor penetración.

flattener An agent added to paint to lower gloss or shine.
aplanador Agente que se agrega a la pintura para reducir el brillo.

flex agent An agent added to paint that allows primers and paints to flex or bend without cracking. Commonly used for plastic bumper covers; also called an elastomer.
agente de flexibilidad Agente que se agrega a la pintura y que permite que las manos preliminares y pinturas se doblen sin quebrarse. Normalmente, se usa para el revestimiento de paragolpes de plástico.

floor jack A tool that uses a saddle to raise the front, sides, or rear of a vehicle.
gato de piso Herramienta que usa un soporte para elevar la delantera, laterales o trasera de un vehículo.

floor tunnel A space under the floor pan for the rear-wheel shaft.
túnel del piso Espacio debajo del piso de la carrocería donde está ubicado el eje de las ruedas traseras.

flow diagrams Shows how various fluids or air moves through a hydraulic (oil-filled) or pneumatic (air-filled) circuit. Various vehicle systems use hydraulic circuits for operation. Shop air tools use a pneumatic system or a high-pressure air supply system for power.
flujogramas Indican cómo se mueven diversos líquidos o aire a través de un circuito hidráulico (con aceite) o neumático (con aire). Diversos sistemas de vehículos usan circuitos hidráulicos para su funcionamiento. Las herramientas neumáticas de los talleres usan un sistema neumático o un sistema de alta presión de suministro de aire para energía.

fluid control knob Allows you to adjust the amount of paint or other material leaving the spray gun by changing the distance the fluid needle valve moves away from its seat in the nozzle when the trigger is pulled.
perilla de control de flujo Permite el ajuste de la cantidad de pintura u otro material que fluye del pulverizador cambiando la distancia que se aleja la válvula de aguja de su lugar en la boquilla al apretar el gatillo.

foam backing pad Used when power sanding crowned surfaces or for extra-fine final sanding.
almohadilla de espuma de refuerzo Se utiliza al lijar a máquina superficies bombeadas o para el lijado final extra fino.

foam fillers Used to add rigidity and strength to structural parts and to reduce noise and vibrations.
rellenos de espuma Se usan para agregar rigidez y fuerza a piezas estructurales.

force drying A means of reducing the time you must wait between coats by spray booth heat convection or infrared lights.
secado forzado Medio de reducir el tiempo de espera entre capas a través de convección o luces infrarrojas en la cabina de pulverización.

frame machine Large framework with hydraulic equipment for pulling out major structural damage; also called frame rack or frame bench.
máquina de bastidor Bastidor estructural grande con equipos hidráulicos para arrancar daños estructurales grandes; también denominada banco de bastidor.

frame rails The box members extending out near the bottom of the front section.
largueros de bastidor Miembros de la caja que se extienden hacia afuera cerca de la base de la sección delantera.

front clip Refers to the front body section; also called the "doghouse."
abrazadera delantera Se refiere a la sección delantera de la carrocería.

front clip assembly Includes all body parts from the front bumper to the rear of the fenders.
conjunto de la abrazadera delantera Incluye todas las piezas de la carrocería, desde el paragolpes delantero hasta la parte trasera de los guardabarros.

full body sectioning Replacing the entire rear section of a collision-damaged vehicle with the rear section of a salvage vehicle.
corte completo de carrocería El reemplazo de la sección trasera completa de un vehículo chocado por la sección trasera de un vehículo de chatarra.

full face shield Used during dangerous operations to protect the face from deep lacerations and scars.
protección de rostro entero Se usa durante operaciones peligrosas para proteger al rostro de laceraciones profundas y cicatrices.

full wet coat Produced by moving the spray gun slightly more slowly than normal.
capa húmeda completa Producida moviendo el pulverizador a una velocidad levemente más lenta que la normal.

fuse box Holds the various circuit fuses, breakers, and the flasher units for the turn and emergency lights.
caja de fusibles Contiene los diversos fusibles de circuitos, disyuntores y las unidades de luces intermitentes de las luces de giro y de emergencia.

fuse links Smaller diameter wire spliced into the larger circuit wiring for overcurrent protection.
enlaces de fusibles Cable de diámetro pequeño empalmados en un cableado de circuito más grande para protección contra sobrecarga de corriente.

fuse ratings The current at which a fuse will blow.
clasificaciones de fusibles La corriente en la que se quema un fusible.

fusion welding Pieces of metal are heated to the melting point, joined together with a filler rod, and allowed to cool.
soldadura por fusión Metal calentado a su punto de fusión, unido por una varilla de llenado, y que luego se deja enfriar.

galvanic corrosion Occurs when two dissimilar metals are placed in contact with each other.
corrosión galvánica Ocurre cuando dos metales disimilares se colocan en contacto entre sí.

galvanizing Process of coating steel with zinc.

galvanizado El metal cubierto del zinc.

gauges Indicate regulated air pressure and sometimes main line pressure.
indicadores Indican la presión del aire regulada y, a veces, la presión de la línea principal.

glass trim panels Fit over the edges of the windshield and back glass.
paneles de tapizado de las ventanas Se colocan sobre los bordes de parabrisas y la ventana trasera.

glazing putty A material for filling small surface imperfections that is applied over an undercoat of primer-sealer.
masilla de vidriero Material para rellenar imperfecciones pequeñas en la superficie, que se aplica sobre una base de pintura preliminar selladora.

gloss level The amount of shine remaining on the paint surface.
nivel de brillo La cantidad de brillo que permanece sobre la superficie de la pintura.

goggles A device with colored lenses or clear safety glass that protects the eyes from harmful radiation during welding and cutting operations.
gafas protectivas Un dispositivo con lentes obscurecidos o de vidrio inastillable transparentes para proteger los ojos de la radiación dañosa al efectuar la soldadura o los cortes.

gouge A sharp dent or crease in a panel produced by a focused impact that causes the metal to be stretched.
gubia Abolladura o arruga grave en un panel, producida por un impacto directo que estira el metal.

gravity-fed spray gun Provides a consistent, spatter-free method of feeding liquid into the gun; used for spot, panel, and overall work.
pulverizador por gravedad Proporciona un método homogéneo y libre de salpicaduras para la colocación de líquido en la pistola pulverizadora; se usa para trabajos en puntos determinados, paneles y globales.

grinder Designed for fast removal of material, this tool is often used to smooth metal joints after welding and to remove paint and primer.
esmerilador Diseñada para la remoción rápida de material, esta herramienta se usa con frecuencia para alisar juntas metálicas después de la soldadura y para remover pintura y pintura preliminar.

grinding discs Round, very coarse abrasives used for preliminary removal of paint, plastic, and metal.
discos esmeriladores Abrasivos redondos y muy gruesos que se usan para la remoción preliminar de pintura, plástico y metal.

grit A measure of size of particles on sandpaper or discs.
grano Una medida del tamaño de los partículos en el papel o los discos de lijar.

grit numbering system A measurement of how coarse or fine an abrasive is.
sistema de numeración de arenilla Medida de la espesura de un abrasivo.

guide coat Thin layer of different color primer or a special powder applied to a repair area; used to check for high and low spots.
mano guía Capa fina de pintura preliminar de distinto color o polvo especial que se aplica sobre el área de reparación; se usa para encontrar puntos altos o bajos.

halo effect An unwanted shiny ring, or "halo," that appears around a pearl or mica paint repair.
efecto halo Un aro brilloso o halo no deseado que aparece alrededor de reparaciones de pintura perlada o de mica.

hammer-off-dolly Method used to simultaneously raise low spots and lower high spots.
hammer-off-dolly Método usado para elevar puntos bajos y reducir puntos altos simultáneamente.

hammer-on-dolly Method used to exert a powerful, concentrated force to a small area on a damaged panel.
hammer-on-dolly Método utilizado para ejercer fuerza poderosa y concentrada en un área pequeña de un panel dañado.

hardener A curing agent used in plastics.
endurecedor Un agente para curar que se usa en los plásticos.

hardener kneading Method of thoroughly mixing the contents inside a tube of hardener by squeezing them back and forth.
sobado de endurecedor Método a través del cual se estruja a fondo el contenido de un tubo de endurecedor de una punta del tubo a la otra.

hazardous waste A solid or liquid determined by the EPA to harm people and the environment.
detritos peligrosos Sólido o líquido que la EPA haya determinado que es dañino para las personas y el medio ambiente.

headliner assembly Cloth or vinyl cover for the inside of the roof panel.
conjunto de forro del techo Revestimiento de tejido o vinílico para el interior del panel del techo.

head-on collision A very bad accident where two vehicles ram their front bumpers together while driving.
choque de frente Un accidente muy grave en el que dos vehículos chocan sus respectivos paragolpes delanteros al desplazarse.

headrest Padded frame that fits into the top of the seat back.
apoyacabeza Armazón acolchado que se coloca en la parte superior del respaldo del asiento.

heat crayon Thermal paint used to determine the temperature of aluminum or other metal being heated.
crayón de calor Pintura térmica utilizada para determinar la temperatura del aluminio u otro metal que se esté calentando.

heat effect zone Area around the weld that becomes hot; should be kept to a minimum to prevent panels from warping or parts from being damaged.
zona de efecto de calor Área alrededor de la soldadura que se calienta; se debe mantener a un mínimo para evitar que los paneles se deformen o las piezas se dañen.

heat gun A tool used when controlled heat is needed, such as when repairing vinyl roofs, repairing other plastics, shrinking panels, or speeding up drying times.
pistola de calor Herramienta que se utiliza cuando se necesita calor controlado, tal como al reparar techos vinílicos, reparar otros plásticos, encoger paneles o acelerar el tiempo de secado.

heat sink compound A clay-like material that reduces the heat effect zone and prevents warping.
compuesto absorbente de calor Material similar a la arcilla que reduce la zona de efecto de color y evita la deformación.

heater core Small radiator under the dash used to warm the passenger compartment.
núcleo de calorífero Radiador pequeño debajo del panel de instrumentos para calentar el compartimiento de pasajeros.

heavy-bodied sealers Used to fill seams from ⅛ to ¼ (3.2–6.4 mm) wide.
selladores de cuerpo pesado: Se utilizan para rellenar uniones de ⅛ a _ (3,2 a 6,4 mm) de ancho.

heeling A gun handling problem that occurs when the painter allows the gun to tilt.
ladeado Problema en la manipulación de la pistola pulverizadora en que el pintor deja que la pistola se incline.

height damage Results from impact damage that forces parts or panels up or down.
daños de altura Consecuencia de daños de impacto que fuerzan las piezas o paneles hacia arriba o hacia abajo.

height dimensions Taken by the vehicle manufacturer in reference to the datum plane and used to measure the vehicle during repair.
dimensiones de altura Tomadas por el fabricante del vehículo, se usan para medir el vehículo durante su reparación.

high volume, low pressure (HVLP) Spray gun that uses a high volume of air delivered at low nozzle pressure to atomize paint into a pattern of low-speed particles.
volumen alto, presión baja (HVLP, por sus siglas en inglés) Pistola pulverizadora que usa un volumen alto de aire que se suelta a presión baja de la boquilla para pulverizar pintura en un padrón de partículas de baja velocidad.

hood hinges Allow the hood to open and close while staying in alignment.
bisagras del capó Permiten que el capó se abra y se cierre manteniéndose alineado.

hood latch Mechanism that keeps the hood closed and releases the hood when activated.
traba del capó Mecanismo que mantiene al capó cerrado y se abre el capó cuando se lo acciona.

hood prop tool Rubber-tipped extension rod for holding a hood open as you remove the hood shocks and other parts.
herramienta de sostén del capó Varilla de extensión con punta de goma cuya finalidad es mantener el capó abierto mientras retira amortiguadores del capó y otras piezas.

hood stop Controls the height of the front of the hood.
tope del capó Controla la altura de la parte delantera del capó.

hood striker Bolts to the hood and engages the hood latch when closed.
dispositivo de cierre del capó Se atornilla al capó y acciona la traba del capó al cerrarlo.

hood struts Spring-loaded rods used to hold the hood open while working in the engine compartment; also called shocks.
tirantes del capó Varillas con resortes que usen para mantener el capó abierto al trabajar en el compartimiento del motor.

horizontal welding Welding with the pieces of metal turned sideways.
soldadura horizontal Soldadura con los pedazos de metal colocados de lado.

hue Characteristic by which one color differs from another, such as red, blue, green, and so forth.
matiz La característica por la cual un color es distincto de otro, tal como el rojo, el azul, el verde, etc.

Hybrid Synergy Drive (HSD) Toyota's current version of a hybrid vehicle drive system.
Transmisión de sinergia híbrida (por su sigla en inglés, HSD) La versión actual de Toyota de un sistema de transmisión en un vehículo híbrido.

hybrid vehicle control unit (HVCU) Computer that controls Toyota's hybrid vehicle drive system.

unidad de control del vehículo híbrido (por su sigla en inglés, HVCU) Computadora que controla el sistema de transmisión de un vehículo híbrido Toyota.

hydraulic lift Used to raise an entire vehicle in the air so that the vehicle is easier to work on.
elevador hidráulico Se usa para elevar el vehículo entero en el aire para facilitar el trabajo.

hydroformed frame Frame created out of thinner gauge steel by using water under high pressure to force frame rails into the desired shape and contour.
bastidor hidroformado Bastidor creado con acero fino utilizando agua bajo alta presión para que los largueros de bastidor resulten con la forma y contorno deseados.

impact A hit from another vehicle or object.
impacto Golpe proveniente de otro vehículo u objeto.

impact sensors The first sensors to detect a collision because they are often mounted at the front of a vehicle.
sensores de impacto Los primeros sensores en detectar un choque, ya que, con frecuencia, están montados en la delantera del vehículo.

impact wrench An air-powered tool used for rapidly turning bolts and nuts.
llave de choque Herramienta neumática utiliza para girar tornillos y tuercas rápidamente.

impervious gloves Latex or synthetic rubber gloves worn to prevent skin exposure when working with harmful chemicals.
guantes impermeables Guantes de látex o goma sintética para evitar la exposición de la piel a sustancias químicas dañinas al trabajar.

included operations Jobs that can be performed individually but are also part of another more complex operation.
operaciones incluidas Tareas que pueden realizarse individualmente, pero que también son parte de otra operación.

independent body shop A shop owned and operated by a private individual not associated with other shops or companies.
taller independiente Taller de propiedad de una persona que se encarga de sus operaciones; no está asociado a otros talleres o compañías.

industrial fallout Caused when small, airborne particles of iron fall on and stick to the vehicle's surface.
caída de polvillo industrial Ocurre cuando pequeñas partículas de hierro transportados por el aire se caen sobre la superficie del vehículo, adhiriéndose a la misma.

in-line oiler An attachment for air tools that will automatically meter oil into the air lines.
lubricador en línea Accesorio para herramientas neumáticas que mide automáticamente el aceite en las líneas neumáticas.

insert A piece of metal made of the same metal as the base metal that can be placed behind the weld; also called a backing strip.
inserción Pedazo de metal hecho del mismo metal que el metal base y que se puede colocar detrás de la soldadura; también se lo llama tira de respaldo o pletina de respaldo.

instrument cluster Part of the dash assembly that normally contains the warning and indicator lights, gauges, and speedometer head.
grupo de instrumentos Parte del conjunto del tablero que normalmente contiene las luces de advertencia e indicadoras, medidores y velocímetro.

insurance adjuster Reviews estimates and determines which one best reflects how the vehicle should be repaired.
ajustador de aseguranza Repasa los presupuestos estimativos y determina cual refleja mejor la manera en que se debe efectuar las reparaciones del vehículo.

insurance rating system Uses a numerical scale to rate a vehicle's accident survivability during a collision.
sistema de clasificación de seguros Usa una escala numérica para clasificar la posibilidad de supervivencia de accidente de un vehículo durante un choque.

intake air filter Filter found in a compressor. All air going into the compressor must pass through this filter so that no grit or dust passes into the cylinders. The filter prevents excessive wear on cylinder walls, piston rings, and valves.
filtro de aire de admisión Filtro ubicado en el compresor. Todo el aire que entra al compresor debe pasar por este filtro para evitar la entrada de arenilla o polvo en los cilindros. El filtro evita el desgaste excesivo de las paredes del cilindro, anillos de pistón y válvulas.

Integrated Motor Assist (IMA) Honda's version of an assist hybrid drive system.
Asistencia integrada del motor (por su sigla en inglés, IMA) La versión de Honda del sistema de transmisión híbrida.

Inter-Industry Conference on Auto Collision Repair (I-CAR) An advanced training organization dedicated to promoting high-quality practices in the collision repair industry.
La Conferencia Industrial Para las Reparaciones de Colisiones Vehiculares (I-CAR) y la organización de entrena-miento avanzado dedicada a la promoción de las practicas de alta calidad en la industra de reparaciones de colisión.

interior trim All upholstery and moldings on the inside of the vehicle.
moldura interior Todas las guarniciones y molduras que se hayan en el interior de un vehículo.

interior trim panels Used in the passenger compartment for appearance and safety.
paneles de revestimiento interior Se usan en el compartimiento de pasajeros con fines estéticos y de seguridad.

intermittent short A circuit where a shorted wire only touches ground and shorts momentarily when the vehicle bounces heavily or is jarred.
corto intermitente Un cable en corto sólo toca tierra y realiza un corto de poca duración cuando el vehículo se sacude con fuerza.

internal combustion engine (ICE) The gas-fueled engine that complements the electric motor in a hybrid drive system.
motor de combustión interna (por su sigla en inglés, ICE) Motor alimentado por combustible que complementa el motor eléctrico en un sistema de transmisión híbrida.

jack stands Used to support the vehicle when working under it, after it has been raised by a floor jack.
torres Sirven para apoyar el vehículo al trabajar abajo de él.

job overlap When replacement of one part duplicates some labor operations required to replace an adjacent or attached part.
superposición de tareas Cuando la sustitución de una pieza duplica algunas operaciones de mano de obra necesarias para reemplazar una pieza adyacente o anexa.

joint fit-up Holding work pieces in alignment before welding.
alineación de juntas Sujetar piezas de trabajo de manera alineada antes de soldar.

jounce Compression of the springs, or an upward movement of the wheel and a downward movement of the frame.
sacudo Una compresión de los resortes, o un movimiento hacia arriba de la rueda y un movimiento hacia abajo del bastidor.

jumper cables Used to start the engine in an emergency.
cables puente Se usan para dar arranque al motor en una emergencia.

jumper wires Used to temporarily bypass circuits or components for testing.
alambres de cierre de circuito Se usan para puentear circuitos o componentes para pruebas.

knee diverter Cushions the driver's knees from impact and helps prevent the driver from sliding under the air bag during a collision.
desviador de rodillas Protege las rodillas de conductor de impactos y ayuda a evitar que el conductor se deslice por debajo de la bolsa de aire durante un choque.

laminated plate glass Two thin sheets of glass with a thin layer of clear plastic between them.
vidrio laminado Dos láminas finas de vidrio con una capa fina de plástico transparente entre las mismas.

lap weld A weld made along the top edge of an overlapping piece.
soldado de recubrimiento Una soldadura hecha en el borde de una pieza solapada.

laser measuring system Type of universal measuring system that uses laser optics in partial or total vehicle demensioning.
sistema de medición a láser Tipo de sistema de medición universal que usa la óptica de láser para calcular las medidas parciales o totales de un vehículo.

lateral dimensions Measured from the center of a vehicle. For example, the measurement from the centerline to a specific point on the right side will be exactly the same as the measurement from the centerline to the same point on the left side.
dimensiones laterales Se miden a partir del centro del vehículo. Por ejemplo, la medida desde la línea central a un punto específico del lado derecho será exactamente igual a la medida desde la línea central al mismo punto en el lado izquierdo.

leaking spray gun A spray gun that allows paint to seep out of the lid or other part of the gun.
pistola pulverizadora con pérdidas Permite que pase pintura por la tapa u otra parte de la pistola pulverizadora.

left-to-right symmetry check Compares measurements on the undamaged side of the vehicle to measurements on the damaged side.
verificación de simetría de izquierda a derecha Compara las medidas de la lateral del vehículo sin daños con la lateral dañada.

length damage Results from a front or rear hit that pushes body panels toward the center of the vehicle.
daños longitudinales Resultado de un choque delantero o trasero que empuja los paneles de la carrocería hacia el centro del vehículo.

let-down test panel Used to evaluate color match on three-stage, tri-coat, and multistage paint systems.
panel de prueba de disminución Se usa para evaluar si los colores son parejos en sistemas de pintura de tres etapas, tres manos y muchas etapas.

lifting Attack on a primecoat by the solvent in a topcoat that results in distortion or wrinkling of the primecoat.

despegue El ataque en una capa de imprimación por el solvente de una capa superior que resulta en la distorsión o las arrugas en la capa de imprimación.

light body filler Used as a thin topcoat in final leveling, or can be spread thinly over large surfaces for easy sanding and fast repairs.
relleno leve para carrocerías Una mano fina superior que se da en el emparejamiento final o que puede colocarse sobre superficies grandes para facilitar el lijado y las reparaciones rápidas.

light coat Produced by moving the spray gun a little more quickly than normal across the surface of the vehicle.
mano leve Se produce moviendo la pistola pulverizadora sobre la superficie del vehículo con velocidad un poco mayor que la normal.

line checking Similar to cracking, but the lines or cracks are more parallel and range from very short to very long.
verificación de líneas Similar a las grietas, pero las líneas o grietas son más paralelas y pueden ser muy cortas.

long-hair fiberglass filler Contains long strands of fiberglass for added strength; used when filler strength is vital to repair.
relleno de fibra de vidrio de fibras largas Contiene hilos largos de fibra de vidrio como refuerzo; se usa cuando la fuerza del relleno es crucial para la reparación.

longitudinal engine Engine that is mounted with the crankshaft centerline extended front to rear.
motor longitudinal Motor montado con línea central de cigüeñal que se extiende de la delantera a la trasera.

lumen rating A measure of a lamp's brightness. Lamps are normally between 1,000 and 2,000 lumens, with a higher lumen producing a brighter light.
clasificación de lúmenes Medida de la luminosidad de una lámpara. En general, las lámparas son de 1.000–2.000 lúmenes; cuanto más alta la clasificación de lúmenes, más luminosa será la luz que emite.

machine compounds Water-based compound to disperse the abrasive while using a power buffer.
compuestos mecánicos Con base de agua para la dispersión del abrasivo al usar una pulidora eléctrica.

magnetic field Present around permanent magnets and current-carrying wires; also called flux.
campo magnético Presente alrededor de imanes permanentes y cables con corriente.

magnetism Involves the study of how electric fields act upon iron-containing objects.
magnetismo Se trata del estudio de la acción de los campos eléctricos sobre objetos que contienen hierro.

major body damage When a vehicle requires large body parts to be replaced or repaired before repainting.
daños mayores en la carrocería Cuando un vehículo requiere la sustitución o reparación de piezas grandes de la carrocería antes de ser pintado.

manual service disconnect (MSD) The disconnect feature on a high-voltage hybrid power system that is used to manually cut the power for service.
desconexión del servicio manual (MSD) Función de desconexión en un sistema de energía híbrida de alto voltaje que se usa para cortar manualmente la alimentación durante el mantenimiento.

manufacturer's instructions Detailed procedures for a specific product.
instrucciones del fabricante Procedimientos detallados para un producto específico.

mash damage Vehicle body damage condition in which the length of any section or frame member is shorter than factory specifications.
daños por aplastamiento Condición de daños en la carrocería del vehículo en la que el largo de cualquier sección o miembro de bastidor es inferior al de las especificaciones de fábrica.

masking Paper or plastic that protects surfaces and parts from paint overspray.
protección El papel o plástico que protege las superficies y las partes del desparrame de pintura.

masking liquid A water- or solvent-based spray used to keep overspray off body parts; also called masking coating.
líquido para cubrir Líquido pulverizado de base agua o de base solvente que se usa para mantener alejado de las partes de la carrocería al exceso de la pulverización; también llamado capa para cubrir.

masking materials Items used to cover and protect body parts from paint overspray.
materiales para cubrir Elementos utilizados para cubrir y proteger las partes de la carrocería contra el exceso de la pulverización.

masking paper Special paper designed to cover panels and parts that will not be painted.
papel para cubrir Papel especial diseñado para cubrir paneles y piezas que no van a pintarse.

masking plastic Used like paper to cover larger areas and to protect panels and parts from overspray.
plástico para cubrir Se utiliza como papel para cubrir grandes áreas y proteger paneles y piezas contra el exceso de la pulverización.

masking rope Self-stick, foam rubber cord designed for quickly masking behind panels.
cuerda para cubrir Cordón de hule espuma autoadhesivo diseñado para cubrir rápidamente detrás de los paneles.

masking tape Used to hold masking paper to the areas to be covered, or it can be used by itself.
cinta para cubrir Se usa para sujetar el papel para cubrir en las áreas que serán cubiertas, también se le puede usar sola.

master cylinder The heart of the hydraulic system, it initiates braking when the brake pedal is depressed.
cilindro maestro Corazón del sistema hidráulico.

material safety data sheets (MSDS) Available from all product manufacturers, they detail chemical composition and precautionary information for all products that can present a health or safety hazard.
hojas de dato de seguridad de los materiales (MSDS) Disponibles de todos los fabricantes de productos, detallan la composición química y la información de precaución para cada producto que puede causar daños al salud o peligro a la seguridad.

measurement gauges Special tools used to check specific frame and body points.
Calibradores de medición Herramientas especiales utilizadas para verificar puntos del armazón y de la carrocería.

measurement tolerance A difference of 2–3 millimeters between the spec dimensions and your actual dimensions.
tolerancia de la medición Diferencia de 2 a 3 milímetros entre las dimensiones especificadas y las dimensiones reales.

mechanical measuring systems One of three universal measuring systems.
sistemas de medición mecánica Uno de los tres sistemas de medición universal.

medium wet coat Produced by moving the spray gun at a normal speed over the surface being refinished.
mano húmeda media Se produce al desplazar la pistola pulverizadora a velocidad normal sobre la superficie que está recibiendo el nuevo acabado.

melt-flow plastic welding The most commonly used airless welding method.
soldadura plástica de flujo por fusión Es el método de soldadura sin aire más comúnmente utilizado.

metal conditioner Chemical cleaner that removes rust and corrosion from bare metal and helps prevent further rusting.
acondicionador de metal Una limpiador química que remueva la oxidacion y la corrosión del metal descubierto y previene que se oxide más.

metal inert gas (MIG) welding Gas-shielded metal arc welding.
soldadura de metal en gas inerte (MIG) soldadura de arco para metales en gas protector.

metal treating Four-step process for ensuring long-term corrosion resistance of body panels and other metal parts.
tratamiento de los metales Proceso de tres etapas para asegurar la resistencia contra la corrosión a largo plazo de los paneles de la carrocería y de otras piezas metálicas.

metallic paints Paint colors that contain metallic flakes in addition to pigment.
pinturas metálicas Colores para pintar que contienen hojuelas metálicas además del pigmento.

metalworking areas Shop locations where damaged parts are removed, repaired, and installed.
áreas de metalistería Lugares del taller donde las piezas dañadas se quitan, reparan e instalan.

metamerism A term used to describe two or more colors that match when viewed under one light source, but do not match when viewed under a second light source.
metámero Un término para describir dos o más colores que parecen iguales al verlos bajo una fuente luminosa, pero que no parecen iguales cuando se ven bajo otra fuente luminosa.

mil gauge Used to measure the thickness of the paint on the vehicle.
medidor de milésimas Se usa para medir el espesor de la pintura del vehículo.

mill and drill pads Used to help the factory hold panels in place while the adhesive cures.
almohadillas de fresa y taladro Se usan para ayudar en la fábrica a mantener los paneles en su lugar mientras cura el adhesivo.

minor body damage When a vehicle only requires a few parts to be replaced or repaired before being repainted.
daños menores en la carrocería Cuando un vehículo sólo necesita la sustitución o reparación de unas cuantas piezas antes de que se vuelva a pintar.

mist coat A light spray coat of high-volume solvent for blending and/or gloss enhancement.
capa de neblina Una capa de rocío ligera de un volumen muy alto de solvente para casar colores o mejorar la brillantez.

mixing board Flat surface used for mixing body filler and hardener.
tablero de mezclado Superficie plana utilizada para mezclar el endurecedor y el relleno de carrocería.

mixing by parts Refers to the fact that for a specific volume of paint or other material, a specific amount of another material must be added.
mezclar por partes Se refiere al hecho de que para un volumen específico de pintura o de un material

diferente, debe agregarse una cantidad específica de otro material.

mixing chart Converts a percentage into how many parts of each paint material must be mixed.
cuadro de mezclado Convierte el porcentaje en el número de partes que deben mezclarse de cada material.

mixing scales Used by paint suppliers and technicians to weigh the various ingredients when combining paint materials.
escalas de mezclado Son las que utilizan los técnicos y los proveedores de pinturas para pesar los diversos ingredientes cuando combinan los materiales de las pinturas.

molded core Curved body repair part made by applying plastic repair material over a part and then removing the cured material.
núcleo moldeado Pieza curva para la reparación de la carrocería que se fabrica mediante la aplicación del material plástico de reparación sobre una parte y la posterior remoción del material curado.

molding tool Removes adhesive-held moldings using a thin blade that will not cause part damage.
herramienta para molduras Retira las molduras fijadas con adhesivo usando una hoja delgada que no daña a la pieza.

motor mounts Prevent minor engine vibrations and noise from being transferred into the body.
soportes del motor Evitan que el ruido y las vibraciones menores del motor se transfieran hacia la carrocería.

motor/generator The high-voltage electric motor in a hybrid drive system that can propel the vehicle, generate electrical power, or slow the vehicle.
motor/generador Motor eléctrico de alto voltaje en un sistema de transmisión híbrida que puede impulsar el vehículo, generar corriente eléctrica o desacelerar el vehículo.

multimeter A voltmeter, ohmmeter, and ammeter combined into one case.
multímetro Voltímetro, ohmetro y amperímetro combinados en una sola caja.

multiple-pull method Uses several pulling directions on many panels with major damage.
método de tracción múltiple Utiliza varias direcciones de tracción en los paneles que tienen daños mayores.

multiple threshold deployment Means the air bag system can deploy the air bags at different speeds.
despliegue de múltiples umbrales Significa que el sistema de bolsas de aire puede desplegar las bolsas de aire a diferentes velocidades.

National Institute of Occupational Safety and Health (NIOSH) Agency that conducts scientific and technical research to establish basic safety standards.
Instituto Nacional de Seguridad e Higiene Ocupacional (NIOSH) Agencia que realiza investigación científica y técnica para establecer normas de seguridad básicas.

needle nose pliers Pliers with long, tapered jaws used for grasping small parts or reaching into tight places; excellent for electrical work.
pinzas de punta Pinzas con mordazas largas y de forma cónica que se usan para agarrar piezas pequeñas o para llegar a lugares cerrados; son excelentes para trabajos eléctricos.

neutral flame A standard flame achieved by mixing acetylene and oxygen in a 1:1 ratio by volume.
llama neutra Llama estándar que se obtiene al mezclar acetileno y oxígeno en una proporción de 1 a 1 en volumen.

New Car Assessment Program (NCAP) Uses star rating scale to indicate how well vehicles do on various types of crash tests.
Programa de Evaluación de Coches Nuevos (NCAP) Utiliza una escala de calificación con estrellas para indicar que tan bien se comportan los vehículos en varios tipos de pruebas de choques.

nondestructive test Procedure to confirm a spot weld after it has been made by using a chisel and a hammer to confirm its quality.
prueba no destructiva Procedimiento para comprobar una soldadura de punto, después de que esta se ha realizado, mediante el uso de un cincel y un martillo para verificar la holgura.

nonparallel body gaps Indicate panel misalignment from structural damage, shifted panel fasteners, or worn mechanical parts.
espacios no paralelos en la carrocería Indican la desalineación del panel resultado del daño estructural, sujetadores de panel desplazados o piezas mecánicas gastadas.

nonsectionable areas Locations where you must not cut through parts when making structural repairs.
áreas no divisibles Lugares que no se deben cortar por las piezas cuando se efectúan reparaciones estructurales.

Occupational Safety and Health Administration (OSHA) Established in 1970 to create and enforce standards that ensure every American worker has safe and healthy working conditions.
Asociación para la Seguridad y la Salud Ocupacional (OSHA) Establecida en 1970 para la creación y la puesta en vigor de normas que aseguren

que todo trabajador estadounidense tenga condiciones laborales sanas y seguras.

OEM color variance A color difference on an original paint or finish.
variación de color del OEM Diferencia de color en una pintura o acabado original.

OEM finish The original factory paint job.
acabado del OEM *El trabajo de pintura original de fábrica.*

OEM replacement panels Purchased from the original equipment manufacturer.
paneles de reposición del OEM Comprados al fabricante de equipo original.

Ohm's Law A math formula for calculating an unknown electrical value when two values are known.
Ley de Ohm Fórmula matemática para calcular un valor eléctrico desconocido cuando se conocen los otros dos valores.

on-board diagnostics (OBD) Means the computer system can detect its own problems.
diagnóstico de a bordo (OBD) Significa que el sistema de la computadora puede detectar sus propios problemas.

one-part spot putty Material designed for filling minor surface imperfections to produce a smoother surface.
masilla de una parte para zonas pequeñas Material diseñado para rellenar imperfecciones superficiales menores y producir una superficie más lisa.

open circuit A circuit that is disconnected and does not have a complete electrical path.
circuito abierto Circuito que está desconectado y no tiene una vía eléctrica completa.

open-end wrench A tool used when there is not enough space for a box wrench. The jaws slide around bolts or nuts, but there is a greater tendency for slippage or injury. Fits both square head and hex head nuts.
llave de tuercas abierta o llave española Herramienta que se usa cuando no hay suficiente espacio para una llave de caja. Las mordazas se deslizan ajustándose alrededor de tornillos o tuercas, aunque la tendencia a las lesiones o a patinarse es mayor. Se ajustan tanto a tuercas de cabeza cuadrada como hexagonal.

orange peel An irregularity in the surface of a paint film resulting from the inability of a wet film to "level out" after being applied. Orange peel characteristically appears to the eye as an uneven or dimpled surface, but it usually feels smooth to the touch.
piel de naranja Una irregularidad en la superficie de una película de pintura causada por la inabilidad de una película húmeda en "aplanarse" después de su aplicación. Típicamente, el piel de naranja tiene un aspecto de una superficie desigual o con hoyitos, pero suele tener una sensación lisa al tacto.

overhaul (O/H) To remove an assembly from the vehicle; disassemble, clean, inspect, and replace parts as needed; then reassemble, install, and adjust (except wheel and suspension alignment).
Reacondicionamiento (O/H) Quitar del vehículo un conjunto, desarmarlo, limpiarlo, inspeccionarlo, sustituir piezas según sea necesario y después volver a armar, instalar y ajustar (con excepción de las ruedas y la alineación de la suspensión).

overhead welding Welding with the pieces of metal upside down or over your head, the most difficult position.
soldadura sobre la cabeza Soldadura con las piezas de metal colocadas al revés.

overpulling Stretching metal too much, resulting in a need to replace the part.
sobrejalar Estirarse demasiado al acero, lo que resulta en una necesidad de reemplazar un elemento.

oxidizing flame A flame obtained by mixing slightly more oxygen than acetylene, used for welding brass and bronze.
llama oxidante Llama que se obtiene mezclando un poco más de oxígeno que de acetileno, se usa para soldar latón y bronce.

pad spur tool Tool with a star wheel to clean and fluff up a wool buffing pad before machine compounding.
herramienta de espuela para almohadilla Se usa para limpiar y esponjar una almohadilla pulidora de lana antes de la aplicación de los compuestos mecánicos.

paint The visible topcoat of color.
pintura La capa o mano de color superior visible.

paint baking Partially curing the paint at 160°F (72°C) for 15–20 minutes.
horneado de la pintura Curado de la pintura a 160°F (72°C) durante 15 a 20 minutos.

paint binder Ingredient in a paint that holds the pigment particles together.
aglutinantes para pintura Ingredientes de una pintura que mantienen unidas las partículas de pigmento.

paint chalking A problem that causes a lack of gloss on the paint surface.
polveado de la pintura Problema que ocasiona la pérdida de brillo de la superficie pintada.

paint code An alphanumeric code that states the type and color of paint used during vehicle manufacturing.
código de la pintura Código alfanumérico que establece que tipo y color de pintura se utilizó durante la fabricación de un vehículo.

paint color directory A book containing information to help you match the paint color on the vehicle.
directorio de color de pinturas Libro que contiene información que le ayudará a igualar el color de la pintura de un vehículo.

paint color mismatch Occurs when the repair area looks a different color from the original color on the vehicle.
desigualdad en el color de la pintura Se presenta cuando el área de reparación ve un color diferente al color original del vehículo.

paint fish-eyes Small, BB-sized dimples or craters that form in the liquid paint film right after spraying.
ojos de pescado en la pintura Pequeñas depresiones o cráteres "de tamaño BB" que se forman en la película de pintura líquida exactamente después de la pulverización.

paint flex additive Keeps the paint film softer and more pliable so the cured paint film will not crack when bent or flexed.
aditivo para la flexibilidad de la pintura Mantiene a la película de pintura más suave y flexible, de modo que la película de pintura curada no se agriete cuando se doble o flexione.

paint formula Gives the percentage of each ingredient that is needed to match an OEM, or original color, on the vehicle.
fórmula de pintura Proporciona el porcentaje de cada ingrediente que se necesita para igualar el color original o del OEM de un vehículo.

paint masking paper Nonporous paper used for masking off paint spray.
papel de enmascarar Papel no poroso utilizado para evitar que la pintura en spray manche otras partes.

paint mottling Occurs only in metallics when the metal flakes float together to form a more silver or streaked appearance in the paint color.
jaspeado de la pintura Se presenta solamente en las pinturas metálicas cuando las hojuelas metálicas flotan para formar en el color de la pintura una apariencia más plateada.

paint protrusion A particle of debris sticking out of the paint film after refinishing.
proyección de la pintura Partícula residual que se desprende de la película de la pintura después de la aplicación del nuevo acabado.

paint runs When excess paint flows down or runs.
corrido de la pintura Cuando hay un exceso, la pintura se corre o chorrea.

paint-safe marker Used to identify any flaws in the new paint film.
marcador de seguridad para pintura Se usa para identificar cualquier defecto en la nueva película de pintura.

paint sealer An innercoat between the topcoat and the primer or old finish that prevents bleeding.
sellador de pintura Capa o mano interna entre la mano superior y la pintura preliminar o el acabado anterior que evita la exudación.

paint shaker A device that vibrates a can of paint to mix it before opening.
agitadores de pintura Ponen a vibrar una lata de pintura para mezclarla antes de abrirla.

paint solvent The liquid solution that reduces and thins the pigment and binder so the mixture can be sprayed.
solvente para pintura Solución líquida que rebaja y diluye el aglutinante y transfiere el pigmento y el aglutinante por la pistola pulverizadora hacia la superficie que va a pintarse.

paint stripper A powerful chemical that dissolves paint for fast removal.
disolvente de pintura Un compuesto químico poderoso que disuelve la pintura para su rápida remoción.

paint thickness Paint thickness is measured in mils or thousandths of an inch (hundredths of a millimeter).
espesor de la pintura El espesor de la pintura se mide en milésimas de pulgada (cien milésimas de milímetro).

paint wrinkling Severe puckering of the paint film that appears like the skin of a prune. Most common with enamel paints.
arrugamiento de la pintura Fruncido grave de la película de pintura que se parece a la piel de una ciruela pasa. Muy común en pinturas de esmalte.

panel A large plastic or metal body part, such as a fender, hood, or roof.
panel Una pieza de carrocería grande de plástico o de metal, como una defensa, el capó o el techo.

panel adjustment Moving or shifting a panel to align it with other panels.
ajuste del panel Movimiento o desplazamiento de un panel para alinearlo con los demás paneles.

panel clamping pliers A tool with wide U-shaped jaws for reaching around and holding panels at their flanges.
pinzas para sujetar paneles Herramienta que tiene mordazas anchas con forma de U para alcanzar a rodear y sostener paneles por sus rebordes.

panel repair A type of refinish repair job in which a complete section (door, hood, deck lid, and so on) is repainted.

reparación del panel Un tipo de reparación del acabado en que una sección completa (puerta, capót delantera y trasera y etc.) se vuelve a pintar.

panel replacement illustration Shows the type, number, and location of cuts and welds needed to properly install a structural panel.
ilustración de reposición de paneles Muestra el tipo, número y ubicación de los cortes y las soldaduras necesarias para instalar adecuadamente un panel estructural.

panel welding illustrations Diagrams that provide the factory-recommended weld count, weld type, and weld locations for replacing structural panels.
ilustraciones de soldadura del panel Diagramas que indican los lugares de la soldadura, el tipo de soldadura y la cuenta de las soldaduras recomendadas para la reposición de paneles estructurales

parallax error The error that results when you read a rule or scale from an angle instead of looking at it straight down.
error de paralaje Error que resulta cuando se lee una regla o escala con un cierto ángulo, en lugar de mirarla exactamente en ángulo recto.

parallel circuit Has two or more legs, or paths, for current.
circuito paralelo Tiene dos o más ramas o vías para la conducción de la corriente.

parallelogram steering system Still used on some full-frame cars and large pickup trucks.
sistema de dirección en paralelogramo Todavía utilizado en algunos coches de armazón completo y en grandes camiones de caja

partial replacement panels Designed to replace only a section or area of a large panel.
paneles de reposición parcial Diseñados para reponer o sustituir sólo una sección o área de un panel grande.

parts interchange guides Provide information about which model vehicles from the same manufacturer can use the same parts.
guías para intercambio de piezas Proporcionan información sobre cuales modelos de vehículos del mismo fabricante pueden usar las mismas piezas.

passenger side air bag Deploys from behind a small door in the right side of the dash.
bolsa de aire del lado del pasajero Se despliega desde detrás de una pequeña puerta en el lado derecho del tablero.

passive restraint system One that operates automatically.
sistema de sujeción pasivo Aquel que funciona automáticamente.

pathogens Microorganisms that can cause illness and disease.
patógenos Microorganismos que pueden provocar trastornos y enfermedades.

pattern control knob Adjusts the amount of air flowing through the air cap horns or side air nozzles of a spray gun.
perilla de control del patrón Ajusta la cantidad de aire que fluye a través de las bocinas de tapa de aire o de las boquillas de aire lateral.

pearl paints Paint with luster or shine (through the addition of mica or other reflective pigment particles) that changes color at different angles.
pinturas perladas Pintura que tiene un lustre o brillo (mediante el uso de mica u otras partículas de pigmento reflexivas) que cambia de color a diferentes ángulos.

personal protection equipment (PPE) Safety equipment worn and used by service technicians to avoid injury while servicing vehicles.
equipo de protección personal (EPP) Equipo de seguridad que usan los técnicos de servicio para prevenir lesiones al realizar tareas de mantenimiento en los vehículos.

Phillips screwdriver A tool with four prongs that fit the four slots in the Phillips screw head, making it less likely to slip off the fastener; often used in the automotive field.
destornillador Phillips o de cruz Herramienta con cuatro patillas que se ajusta en las cuatro ranuras de la cabeza de un tornillo Phillips o de cruz, lo que hace menos probable que se patine del tornillo; se usa con frecuencia en el campo automotriz.

picking hammer A tool with a pointed tip on one end and a flat head on the other used to remove small dents.
martillo para picar Herramienta que en un extremo tiene una punta afilada y en el otro tiene una cabeza plana, que se usa para la eliminación de abolladuras pequeñas.

picks Tools with a U-shaped end used to reach into confined spaces and pry up low spots on body sections without having to drill or weld.
picas Herramientas que tienen un extremo con forma de U, se usan para llegar a espacios cerrados y para levantar haciendo palanca áreas bajas en secciones de carrocería si tener que perforar o soldar.

pigments Material in the form of fine powders used to impart color, opacity, and other effects to paint.
pigmento Una materia en forma de los polvos finos que sirve para conferir el color, la opacidad u otros efectos a la pintura.

pillar trim panels Panels that fit over the upper sections of the A-, B-, and C-pillars.

paneles de revestimiento de pilares Paneles que se ajustan sobre las secciones superiores de los pilares A, B y C.

pinch weld clamps Large bottom clamps to anchor a unibody vehicle on a frame rack.

prensas para soldadura de apriete Se usan para anclar vehículos de monocarrocería.

pipe wrench A tool most commonly used for grabbing and turning round objects, such as pipes or studs.

llave para tubos Herramienta que se usa muy comúnmente para sujetar y dar vuelta a objetos redondos como tubos o espárragos.

pivot measure system Uses rotating rods and pointers to measure vehicle damage.

sistema de medición de pivote Utiliza varillas y punteros giratorios para medir el daño en el vehículo.

plasma arc cutting Cutting process in which metal is severed by melting a localized area with a constructed arc and removing molten material with a high-velocity jet of hot, ionized gas issuing from an orifice.

corte con arco de plasma Un proceso de cortar en que el metal se quita al derritir una área localizada con un arco construido y se remueva al material derritido con un chorro caliente de alta velocidad del gas ionizado que sale de un orificio.

plastic flexibility test Test used to help identify what type of plastic is used in a part.

prueba de flexibilidad del plástico Prueba utilizada para ayudar a identificar qué tipo de plástico se utiliza en una pieza.

plastic memory The tendency of material to keep or return to its original molded shape.

memoria plástica La tendencia del material a mantener o regresar a su forma moldeada original.

plastics Reference to a wide range of manufactured materials synthetically compounded from crude oil, coal, natural gas, and other natural substances.

plásticos Se refiere a una extenso número de materiales fabricados de forma sintética compuestos de petróleo crudo, carbón, gas natural y otras substancias naturales.

plastic speed welding Uses a specially designed tip to produce a more uniform weld at a higher rate of speed.

soldadura plástica rápida Utiliza una punta especialmente diseñada para producir una soldadura más uniforme a una mayor velocidad.

plastic stitch-tamp welding Used on hard plastics to ensure a good base and rod mix.

soldadura plástica apisonada por puntos Se utiliza en plásticos duros para asegurar una buena mezcla entre la base y la varilla.

plastic welding Uses heat to repair or join plastic parts.

soldadura plástica Utiliza calor para reparar o unir piezas de plástico.

pliers An all-purpose grabbing and holding tool for working with wires, clips, and pins.

pinzas Una herramienta para sujetar y sostener, para cualquier propósito y al trabajar con alambres, sujetadores y pasadores.

plotting color The process of identifying paint color in a graphic way based on value, hue, and chroma.

colores de gráfica Proceso de identificación del color de la pintura en una forma gráfica, basado en el valor, matiz e intensidad.

plug weld Adding metal into a hole and fusing all metal.

soldadura de tapón Colocando el metal en un agujero y uniendo todo el metal.

plug weld holes Needed to achieve an adequate nugget and acceptable weld strength.

orificios de soldadura de tapón Necesarios para lograr una soldadura de costura adecuada y una aceptable resistencia de soldadura.

polishing compound A fine grit compound applied with machines that makes paint shiny and smooth.

compuesto pulidor Compuesto de arenilla fina aplicado con máquinas y que hace que la pintura sea brillante y lisa.

polishing machine Power tool that uses an orbital action to bring out the full paint gloss or shine.

máquina pulidora Herramienta eléctrica que utiliza un acción orbital para hacer resaltar plenamente el brillo o lustre de la pintura

polyester glazing putty Fine-grained filler material designed not to shrink, sink, bleed through, or cause a flaw in paint or topcoat on small surface repairs; also called two-part spot putty.

masilla de vidriado de poliéster Material de granulado fino diseñado para no contraerse, chuparse, sudar u ocasionar una mancha o desperfecto en la pintura o mano superior.

post-painting operations Tasks that must be done after painting and before returning the vehicle to the customer, such as removing masking tape, reinstalling parts, and cleaning the vehicle.

operaciones posteriores a la pintura Estas tareas deben realizarse después de pintar y antes de regresar el vehículo al cliente y son: remover la cinta para cubrir, reinstalar las piezas y lavar el vehículo.

pounds per square inch (psi) A measurement of the air pressure or force delivered by the compressor.
libras por pulgada cuadrada (psi) Medida de la presión del aire o fuerza liberada por la compresora.

power tools Tools that use air pressure, oil pressure, or electrical energy to help in making repairs.
herramientas eléctricas Herramientas que utilizan presión de aire, presión de aceite, o energía eléctrica para ayudar en las reparaciones.

power train Includes all of the parts that produce and transfer power to the drive wheels to propel the vehicle.
tren motriz Incluye todas las partes que producen y transfieren potencia a las ruedas de tracción para impulsar el vehículo.

powertrain control module (PCM) The computer that controls a Honda hybrid drive system.
módulo de control del tren de fuerza (por su sigla en inglés, PCM) Computadora que controla un sistema de transmisión híbrida.

prep solvent A fast-drying solvent used to remove wax, oil, grease, and other debris that could ruin a paint job.
solvente primario Solvente de secado rápido utilizado para remover cera, aceite, grasa y otros detritos que pudiesen arruinar un trabajo de pintura.

pressure switch An electric switch controlled by air pressure used for starting and stopping electric motors at preset minimum and maximum pressures.
interruptor de presión Interruptor eléctrico controlado por la presión de aire, utilizado para arrancar y detener motores eléctricos a presiones mínimas y máximas preestablecidas.

pressure welding Heating metal with electrodes and joining them by applying pressure.
soldadura de presión Calentamiento de metal con electrodos y unión de los mismos mediante la aplicación de presión.

pressure-fed spray guns Using air pressure inside the paint cup or tank to force the material out of the gun.
pistolas pulverizadoras alimentadas mediante presión Utilizan presión de aire dentro de la lata o tanque de pintura para forzar al material a salir de la pistola.

primary damage Damage that occurs at the point of impact on the vehicle.
daños primarios Los daños que ocurren en el punto de impacto del vehículo.

primecoat First coat in a paint system. Its main purpose is to improve adhesion and provide corrosion protection.
capa de imprimación La primera capa en un sistema de pintura. Su propósito principal es de mejorar la adhesión y proveer la protección contra la corrosión.

primer A type of paint applied to a surface to increase its compatibility with the topcoat or to improve the adhesion or corrosion resistance of the substrate.
imprimación Un tipo de pintura aplicada a una superficie para mejorar su compatabilidad con una capa superior o para mejorar la adhesión o la resistencia a la corrosión de un substrato.

primer masking paper Less expensive, porous masking paper used for masking off primer spray but not paint.
papel de enmascarar para pintura preliminar Papel económico y poroso de enmascarar utilizado para evitar que la pintura preliminar manche partes no deseadas; no sirve para pintura.

primer-surfacer A high-solids primer that fills small imperfections in the substrate and usually must be sanded.
imprimación-preparación Una imprimación de muchos sólidos que rellena las imperfecciones pequeñas en un subestrato y que se tiene que lijar.

procedure pages Found in collision estimating guides, "P pages" provide information such as the arrangement of material in the guide, an explanation of the symbols, definitions of terms, how to read and use the parts illustrations, procedure explanations, information about discontinued and interchangeable parts, additions to labor times, labor times for overlap items, and how to identify structural and mechanical operations.
páginas de procedimiento Encontradas en guías de estimación de colisión, las "Páginas P" proporcionan información como: la secuencia que tiene el material en la guía, una explicación de los símbolos, definición de los términos, cómo leer y utilizar las ilustraciones de las piezas, explicación de los procedimientos, información sobre piezas discontinuadas o intercambiables, adiciones a los tiempos de trabajo, tiempos de trabajo para artículos de superposición, y cómo identificar las operaciones mecánicas y estructurales.

productivity Measure of how much work gets done.
productividad Medición sobre cuánto trabajo puede realizarse.

professional A worker with a broad range of skills that include everything from being able to follow orders to pride in workmanship.
profesional Característica general de un trabajador, la cual incluye todo, desde poder seguir órdenes para hasta enorgullecerse por el trabajo.

propellant charge Burns to produce gas expansion that inflates the air bag.

carga propelente Se quema para producir gas expansivo que infla la bolsa de aire

proportional mixing numbers Denotes the amount of each material needed by volume. The first number usually refers to the parts of paint needed. The second number is usually the solvent. A third number might be used to denote the amount of hardener or other additive.

números de mezclado proporcionales Denota la cantidad que se necesita de cada material por volumen. El primer número se refiere por lo regular a las partes necesarias de pintura. El segundo número es por lo regular el solvente. Deberá utilizarse un tercer número para denotar la cantidad de endurecedor u otros aditivos.

pulling Using hydraulic equipment to apply force to a part to make a change.

jalando Aplicando la fuerza en una parte para efectuar un cambio.

pyro-technique retractor Uses a gas-generating retractor to develop pressure for quickly taking up slack in the seat belts when a collision is detected.

retractor pirotécnico Utiliza un retractor generador de gas para desarrollar presión para efectuar un rápido ajuste de los cinturones de seguridad cuando se detecta una colisión.

quarter panels Side panels, which extend from the rear doors to the end of a car.

aleta trasera El panel del lado, que extienda de la puerta trasera al posterior del vehículo.

ratchet handle The most commonly used handle; allows for removing or tightening without removing the socket from the fastener.

manija de trinquete La manija que se utiliza más comúnmente; permite aflojar o apretar sin retirar el cubo del sujetador.

rear axle assembly Housing that contains the ring gear, pinion gear, differential assembly, and axles.

ensamble de eje trasero Caja que contiene el engranaje de anillo, el engranaje de piñón, un ensamble diferencial y los ejes.

rebound The motion caused by a wheel going into a dip or returning from a jounce and extending the spring.

rebote Efecto causado por una rueda al caer en un bache, o al regresar de una amplia sacudida y extender el resorte.

recycled assemblies Undamaged parts from another damaged vehicle that are used for repairs.

conjuntos reciclados Piezas no dañadas de otro vehículo dañado, las cuales se utilizan para reparaciones.

reducer Solvent combination used to thin enamel.

reductor Se refiere a una combinación de solventes que sirven para adelgazar el esmalte.

reference points Points on a vehicle, including holes, flats, or other identifying areas, used to position parts during repairs.

puntas de comprobación Las puntas en un vehículo, incluyendo los agujeros, los planos u otras áreas de identificación, que sirven para colocar las partes durante la reparación.

refinishing guides Explain paint codes, types of paints, and how to apply and buff paints, and provide other paint-related information.

guías de acabado final Explican códigos de pinturas, tipos de pinturas, cómo aplicar y pulir pinturas, y proporcionan otra información relacionada con pintura.

refinishing materials Products used to repaint vehicles.

materiales de acabado final Productos utilizados para repintar vehículos.

reinforced plastic Provides a durable plastic skin over a steel unibody.

plástico reforzado Proporciona una película de plástico duradera sobre un cuerpo unitario de acero.

relay Remote control switch.

relé Interruptor de control remoto.

reliability Means being to work on time, coming to work every day, and doing the job right.

fiabilidad Se refiere a llegar temprano a trabajar, ir a trabajar todos los días y hacer el trabajo de forma correcta.

remove and install (R&I) To remove an item as an assembly, set it aside, and later reinstall and align it for a proper fit.

remover e instalar (R&I) Retirar un accesorio, por ejemplo un conjunto, reservarlo y más tarde reinstalarlo y alinearlo para un ajuste adecuado.

remove and replace (R&R) To remove the old parts, transfer necessary items to new parts, replace, and align.

remover y reparar (R&R) Retirar las piezas viejas, transferir los accesorios necesarios a las piezas nuevas, reemplazar y alinear.

repair order A printed sheet with exact repair instructions.

orden de reparación Un hoja impresa con las instrucciones exactas de reparación.

repair primecoats Provide a sound foundation for the topcoats.

mano primaria de reparación Proporciona una base firme para las manos superiores.

replacement panels Whole, unsectioned panels as designed by the auto manufacturer.
paneles de reemplazo Los paneles no seccionados completos, conforme al diseño del fabricante del vehículo.

resistance A restriction or obstacle to current flow.
resistencia Una restricción u obstáculo al flujo de corriente.

respirator fit test A way to test a respirator for air leaks.
prueba de ajuste del respirador Una forma de probar un respirador para fugas de aire.

restraint bar Used to hold or maintain a dimension in an opening when pulling.
barra de sujeción Se utiliza para conservar o mantener una dimensión en una abertura, al jalar.

restraint system Designed to help hold people in their seats and prevent them from being injured during a collision.
sistema de sujeción Diseñado para ayudar a sujetar a las personas en sus asientos y prevenir lesiones durante una coalición.

retarder Slow-evaporating thinner or reducer used to slow drying.
retardador Adelgazador o reductor de lenta evaporación, utilizado para secado lento.

reverse masking Blend masking done by rolling the tape over and into a curve shape to prevent a visible paint parting line along the masking tape.
cubierta invertida Mezcla de cubierta hecha enrollando la cinta sobre y dentro una forma curva para prevenir una línea de pintura visible a lo largo de la cinta para cubrir.

Right-to-Know Laws They give essential information and stipulations for safely working with hazardous materials.
leyes de Derecho a la Información Proveen la infor-mación esencial y las estipulaciones para los proce-dimientos de seguridad para trabajar con las materias tóxicas.

robot arm measuring system Uses a movable arm to send electronic measurement data back to a computer.
rocker panels Narrow, outer panels attached below the car door. Also called door sills, these strong beams fit at the bottom of door openings.
estribo Un panel estrecho exterior conectado debajo de la puerta del coche. Los umbrales de las puertas o las vigas fuertes que quedan en la parte inferior de los marcos de las puertas.

roll bar Steel framework designed to protect people during a rollover accident in a vehicle with a convertible roof.
Barra de refuerzo Armazón de acero diseñado para proteger a los pasajeros durante un accidente de volcadura con techo convertible.

rotor Brake part made of cast iron that is clamped against both sides by pads.
rotor Pieza del freno hecha de hierro fundido y que se encuentra afianzada contra ambos lados por almohadillas.

rubbing compound Coarsest type of hand compound; may leave scratch marks.
compuesto friccionador Tipo más áspero de compuesto que se aplica con la mano, puede dejar marcas de rayas.

rust Corrosion product which forms on iron or steel when exposed to moisture.
oxidación Un producto de la corrosión que forma en el hierro o en el acero cuando se exponen a la humedad.

rust converters Change ferrous (red) iron oxide to ferric (black/blue) iron oxide.
convertidores de óxido Cambian el óxido ferroso (rojo) a óxido férrico (negro/azul).

rustout Corrosion that has penetrated completely through a steel panel.
oxidación Corrosión que ha penetrado por completo a través del panel de acero.

safing sensor Used as a third fail-safe sensor to prevent accidental deployment of air bags.
sensor de seguridad Se utiliza como un tercer sensor de seguridad contra fallas para prevenir el despliegue accidental de las bolsas de aire.

sag damage A condition in which one area of a vehicle is lower than normal.
daño por hundimiento Condición en la cual un área del vehículo está más baja de lo normal.

salvage parts Used parts in good condition that were removed from totally wrecked vehicles by salvage yards.
partes recuperadas Partes usadas y en buenas condiciones que fueron retiradas de vehículos totalmente destruidos, en los deshuesaderos.

salvage parts software Provides easy access to a user-customized database of recycled parts and suppliers in North America and Canada so that technicians can quickly search and find quality recycled parts.
software de partes recuperadas Proporciona fácil acceso a la base de datos personalizada por el usuario sobre las partes recicladas y proveedores en Estados Unidos y Canadá, con la cual los técnicos pueden buscar y encontrar rápidamente partes recicladas.

salvage replacement panels Undamaged panels cut off a wrecked vehicle at a salvage yard.
paneles de reemplazo recuperados Paneles no dañados extraídos en un deshuesadero a partir de un vehículo chocado.

salvage yard A business that resells parts from a damaged car.
deshuesadero Negocio que revende las partes de un vehículo dañado.

sand scratch swelling Enlarged sand scratches caused by the swelling action of topcoat solvents.
protuberancias de rayas de arena Rayas de arena alargadas ocasionadas por la acción protuberante de los solventes de mano superior.

sander pad A soft mounting surface for sandpaper. Disc adhesive, self-stick paper, or hook and latch paper secures the sandpaper to the pad.
almohadilla lijadora Superficie de montadura suave para papel de lija Disco adhesivo, papel autoadherible o papel de bucles y ganchillos, que asegura el papel de lija a la almohadilla.

sandpaper Heavy paper coated with abrasive grit.
papel de lija Papel resistente revestido con arenilla abrasiva.

scan tool An electronic instrument for reading trouble codes and converting them into problem explanations.
herramienta de escaneo Instrumento electrónico para la lectura de códigos de problema y su conversión a explicaciones del problema.

scrub radius The distance between the centerline of the ball joints and the centerline of the tire at the point where the tire contacts the road surface.
separación del radio Distancia entre la línea central de la unión de las juntas de la bola y la línea central de la llanta en el punto en donde la llanta hace contacto con la superficie del camino.

scuff pad Tough synthetic pads used to clean and lightly scratch the surface of paints so that new paint will stick; useful for scuffing irregular surfaces.
almohadilla de tallado Almohadillas sintéticas resistentes que se utilizan para limpiar y raspar ligeramente la superficie de las pinturas para que la nueva pintura se adhiera, ideal para frotar sobre superficies irregulares.

sealer An intercoat between the topcoat and the primer or old finish, giving better adhesion.
sellante Una capa intermedia entre la capa superior y la imprimación o el acabado antiguo, dándole mejor adhesión.

seam sealer Used to make a leakproof joint between body panels. Needed when two panels overlap each other.
sellador de unión Se utiliza para realizar una unión hermética entre los paneles de la carrocería Se requiere cuando dos paneles se superponen uno con otro.

seat anchor bolts Case-hardened cap screws that secure the seat track to the floor structure.
tornillos de anclado del asiento Tornillos de casco con superficie endurecida que aseguran el carril del asiento a la estructura el piso.

seat assembly Includes the seat track, seat cushions, headrest, trim pieces, and often the power seat accessories.
conjunto del asiento Incluye al carril del asiento, cojín del asiento, el apoyacabeza, piezas de moldura y, por lo regular, los accesorios eléctricos del asiento.

seat back Rear assembly that includes cover, padding, and metal frame.
respaldo del asiento Conjunto trasero que incluye la cubierta, el relleno y el armazón de metal.

seat belt anchors Allow one end of the belts to be bolted to the body structure.
anclas de cinturones de seguridad Permiten que un extremo de los cinturones de seguridad esté atornillado a la estructura de la carrocería.

seat belt buckle Mechanism that allows you to put the seat belt on and take it off.
hebilla del cinturón de seguridad Mecanismo que le permite ponerse y quitarse el cinturón de seguridad.

seat belt reminder system Uses sensors and a warning system to remind a driver to fasten the seat belt.
sistema recordatorio del cinturón de seguridad Utiliza sensores y un sistema de advertencia para recordarle a los conductores que se pongan el cinturón de seguridad.

seat belts Strong nylon straps with special ends attached for securing people in their seats.
cinturones de seguridad Correas de nylon resistentes que tienen extremos especiales sujetados para asegurar a las personas en sus asientos.

seat cushion Bottom section of the seat that includes the cover, padding, and frame.
cojín del asiento Sección inferior del asiento que incluye la cubierta, el relleno y el armazón.

seat track Mechanical slide mechanism that allows the seat to be adjusted forward or rearward.
carril del asiento Mecanismo de corredera mecánica que permite ajustar el asiento deslizándolo hacia adelante o hacia atrás.

secondary damage Indirect damage that occurs due to misplaced energy that causes stresses in suspension and/or body dimensions at areas other than the primary impact zone.

daños secundarios Los daños indirectos que ocurren debido a la energía mal dirigido causando las tensiones en la suspensión y/en las dimensiones de la carrocería en las áreas que no son de la zona del primer impacto.

sectioning A means of replacing partial areas of a vehicle.

reemplazar por elementos Un metodo de remplazar las áreas parciales de un vehículo.

self-etching primer An undercoat that contains acid; used to treat bare metal so that the primer will adhere properly.

pintura preliminar automordente Un recubrimiento base que contiene ácido y que se utiliza para tratar metal sin recubrir, de modo que permite que la pintura preliminar se adhiera en forma adecuada.

sensor arrow The directional arrow stamped on the sensor.

sensors Devices that convert a condition (temperature, pressure, part movement, and so on) into an electrical signal.

sensores Dispositivos que convierten alguna condición en un sensor eléctrico.

series circuit Has only one conductor path, or leg, for current through the circuit.

circuito en serie Tiene una sola vía o rama para conducir la corriente a través del circuito.

service manuals Books published annually by vehicle manufacturers that list specifications and service procedures for each make and model of vehicle; also called shop manuals.

manual de servicio Un libro publicado cada año por cada fabricante de vehículos que registra las especificaciones y los procedimientos de reparación para cada marca y modelo de vehículo; se puede denominar también un manual de taller.

sheet-molded compound (SMC) Fiber-reinforced composite plastic panel.

compuestos moldeados en láminas Paneles de plástico compuestos reforzados con fibra.

shipping coating Protects metal panels or parts against corrosion during storage and shipping.

revestimiento de embarque Protege al metal contra la corrosión durante el almacenamiento y el embarque.

shock absorbers Dampening devices that absorb spring oscillations to smooth the vehicles ride quality.

amortiguadores Dispositivos amortiguadores que absorben las oscilaciones de los resortes para homogeneizar la calidad con la que viajan los vehículos.

shock towers Reinforced body areas for holding the upper parts of the suspension system; coil springs and struts or shock absorbers fit into the shock towers. Also called strut towers.

torres apoyaderos (torres de los postes) Las áreas de la carrocería reinforzadas que sirven para apoyar las partes superiores del sistema de suspensión: los muelles de embrague y los postes o los amortiguadores quedan dentro de los torres apoyaderos.

shop management software Designed to help enhance a collision repair facility's productivity, profitability, and customer service by coordinating all aspects of collision repair by converting electronic estimates into repair orders.

software para la administración del taller Diseñado para ayudar a mejorar la productividad, rentabilidad y el servicio a los clientes de los establecimientos del reparación de colisiones al coordinar todos los aspectos de la reparación de colisiones mediante la conversión de las estimaciones electrónicas en órdenes de reparación.

short circuit A circuit that has an unwanted path for electrical current.

corto circuito Es una vía indeseable para la corriente.

short-hair fiberglass filler Filler that contains tiny particles of fiberglass and is much stronger than conventional filler.

relleno de fibra de vidrio de fibra corta Contiene partículas diminutas de fibra de vidrio y es mucho más fuerte que el relleno convencional.

shrinking metal Needed to remove strain or tension on a damaged, stretched sheet metal area.

contracción metálica Necesaria para eliminar el esfuerzo o la tensión de una área de lámina metálica estirada dañada.

side air bags May be located in the door panels, roofs, pillars, or seats to protect against impact injury.

bolsas de aire laterales Pueden estar ubicadas en los paneles de las puertas, techos, pilares o asientos para proteger contra las lesiones por impacto.

side tone A metallic finish viewed at a 45-degree angle or less; also called color pitch.

tono lateral Es cuando se observa un acabado metálico con un ángulo de 45 grados o menor.

side-sway damage Collision impacts that occur from the side.

daño lateral Impactos por colisión que están en el costado.

sill plates Cover the rocker panels and hold edges of the carpeting; also called scuff plates.

placas de larguero Cubren los paneles de balancín y sostienen los bordes de la alfombra; también llamadas placas de nuca.

single-pull method Uses only one pulling chain to repair minor damage on one part.

método de un sólo tirón Utiliza sólo una cadena de tracción para reparar daños pequeños en una pieza.

single-stage compressor A compressor in which air is drawn from the atmosphere and compressed in a single step.
compresor de una etapa Nombre que se da a un compresor cuando toma aire de la atmósfera y lo comprime en un sólo paso.

sledgehammer A very large, heavy driving tool.
marro Herramienta para clavar grande y pesada.

social skills Important skills that are developed to create a harmonious working environment between coworkers.
habilidades sociales Importantes habilidades que se desarrollan para generar un entorno de trabajo armonioso entre usted y su compañeros de trabajo.

socket wrench A wrench that is much faster and easier to use than an open-end or box-end wrench. Made up of different type handles and barrel-shaped sockets that fit over and around a given nut size or wrench.
llave de cubo Llave más rápida y de uso más fácil que el de una llave de tuercas abierta, también llamada española, o de tuercas cerrada, también llamada de estrías. Formada por diferentes tipos de manerales y cubos de forma cilíndrica, también llamados dados, que se ajustan en un tamaño determinado de tuerca o llave.

solenoid An electromagnet with a movable core or plunger.
solenoide Un electroimán que tiene un núcleo o émbolo móvil.

solid color paints Paints that do not contain reflective flakes or mica particles.
pinturas de color sólido Pinturas que no contienen hojuelas reflexivas ni partículas de mica.

solid seam sealers Contain 100 percent solids and are used to fill larger voids at panel joints or holes.
selladores de unión sólida Contienen 100 porciento de sólidos y se utilizan para rellenar grandes huecos en uniones de paneles u orificios.

solvent popping Blisters, caused by trapped solvents, that form on a paint film.
granos de solvente Las ampollas que forman en una película de pintura, causadas por los solventes atrapados.

sonic measuring systems Use sound waves to measure vehicle dimensions for evaluating damage.
sistema de medición sónica Utiliza ondas de sonido para medir daños en vehículos.

sound-deadening materials Insulation materials that help quiet the passenger compartment.
materiales de reducción de sonido Materiales aislantes que ayudan a insonorizar el compartimento de pasajeros.

space frame A metal body structure covered with an outer skin of plastic or fiberglass panels; used on some vans and economy vehicles.
armazón espacial Estructura metálica de carrocería revestida con una película externa de plástico o con tableros de fibra de vidrio; utilizado en vehículos económicos.

specialty shop Shop that performs only specific kinds of repairs.
taller especializado Taller que únicamente realiza tipos de reparaciones específicas.

specifications Data as supplied by the manufacturer covering all measurements and areas of the vehicle.
especificaciones La información proveido por el fabricante que toma en cuenta todas las medidas y las áreas de un vehículo.

spectrophotometer An instrument to measure color. Compares reflectance of a test sample to reflectance of a standard at all points of a visible spectrum.
espectrofómetro Un instrumento para medir el color. Compara lo reflectivo de un modelo en todos puntos de un espectro visible.

spitting spray gun Fails to atomize all of the liquid properly.
pistola pulverizadora que escupe Pistola que no logra pulverizar todo el líquido en forma adecuada.

spontaneous combustion Fire that starts by itself.
combustión espontánea Un incendio que se encienda si mismo.

spot putty A material made for filling small holes or sand scratches; also called glazing putty.
masilla de retoque Una materia hecha para rellenar los agujeros chicos o las rayas del limado.

spot repair Type of refinish repair job in which a section of a car smaller than a panel is refinished (often called ding and dent work).
reparación localizada Un tipo de reparación del acabado en que una seccion de un coche que es más pequeña del panel se reacaba (muchas veces se refiere como trabajo de repiques y abolladuras).

spot weld Weld in which an arc is directed to penetrate both pieces of metal, while triggering a timed impulse of wire feed.
soldadura por puntos Una soldadura en que un arco dirigido penetra ambos pedazos de metal, mientras que distribuya una alimentación de alambre sincronizado.

spot weld drill Includes a lever and clamp for pushing the drill into the panel flange to rapidly remove panel welds.

taladro de soldadura de puntos Incluye una palanca y una prensa para empujar el taladro en el reborde del panel para retirar rápidamente las soldaduras de los paneles.

spray booth A clean, safe, well-lit, and properly ventilated enclosure for painting a vehicle.
cabina de pulverización Diseñada para proporcionar un recinto limpio, seguro, bien iluminado y con la ventilación adecuada para pintar un vehículo.

spray gun A painting tool powered by air pressure that atomizes liquids. Spray paint is atomized in a spray gun, and a stream of atomized paint is directed at the part to be painted.
pistola de atomización Una herramienta de pintar neumática que atomiza los líquidos. La pintura rociada se atomiza en la pistola, y un chorro de pintura atomizada se dirige hacia la parte para pintar.

spray gun cleaning tank A pressurized container for flushing the gun and other tools with cleaning solution; also called a gun washer–recycler.
tanque para limpiar la pistola pulverizadora Un recipiente presurizado para lavar con un flujo abundante de solución limpiadora la pistola y otras herramientas.

spray gun direction Involves keeping the gun aimed so that an even film of paint deposits on the body surface.
dirección de la pistola pulverizadora Implica mantener la pistola apuntada en forma tal que sobre la superficie de la carrocería se deposite una película uniforme de pintura.

spray gun distance The distance measured from the spray gun nozzle to the surface being painted.
distancia desde la pistola pulverizadora Es la distancia medida desde la boquilla de la pistola hasta la superficie que se va a pintar.

spray gun oil A spray gun lubricant that is compatible with paint and will not contaminate the gun.
aceite para pistola pulverizadora Aceite compatible con la pintura y que no contaminará la pistola.

spray gun speed How fast you move the gun sideways over the surface.
velocidad de la pistola pulverizadora Rapidez con la que el operador mueve la pistola lateralmente sobre la superficie.

spray gun stroke The hand motion used to move the gun while spraying.
carrera de la pistola pulverizadora Se refiere al movimiento de la mano utilizado para mover la pistola mientras se pulveriza.

spray gun triggering Involves keeping constant air flow out the nozzle and only triggering paint flow when depositing paint materials on the body.
disparo de la pistola de pulverización Implica mantener un flujo constante de aire a través de la boquilla y disparar el flujo de pintura sólo al momento de depositar la pintura sobre la carrocería.

spray overlap Refers to how each pass of the spray gun should paint over or cover about half of the previous paint stroke.
superposición de la pulverización Se refiere a que cada pasada de la pistola de pulverización debe pintar o cubrir aproximadamente la mitad de la pasada de pintura anterior.

spray-out test panel Used to check the paint match of solid and metallic colors, but not three-stage or pearl paints.
panel de pruebas de pulverización Se utiliza para verificar que la pintura iguala los colores sólidos o metálicos, pero no con pinturas de tres etapas ni perladas.

spray pattern test Checks the operation of the spray gun on a piece of paper.
prueba del patrón de pulverización Verifica la operación de la pistola de pulverización sobre un pedazo de papel.

spray schedule A list of the times that each material has been sprayed.
programa de pulverización Listado de las horas en las que se ha pulverizado cada uno de los materiales.

spreaders A rigid plastic tool used to apply body filler during auto body resurfacing.
espátula aplicadora Herramienta de plástico rígido utilizada para aplicar el relleno para carrocerías durante reparaciones superficiales de carrocerías automotrices.

spring-back The tendency for metal to return to its original shape after deformation.
recuperación Tendencia de un metal a regresar a su forma original después de la deformación.

squeegee A flexible rubber pad or block used to apply glazing putty and lights coats of body filler during auto body resurfacing. Also used to skim water and sanding grit from the repair area when wet sanding.
barredera de hule Bloque de hule flexible utilizado para aplicar la masilla de vidriero y revestimientos ligeros de relleno para carrocería durante la reparación superficial de carrocerías automotrices. También se utiliza para retirar del área de la reparación el agua y los granos de la lija cuando se lija en húmedo.

stabilizer-reducer Contains a special basecoat resin designed to allow a faster recoat time and better metal flake control.
estabilizador-reductor Contiene una resina especial para capa base diseñada para obtener un menor

tiempo recubrimiento y un mejor control de la hojuela metálica.

standard color chip The intended color used when the vehicle was painted at the factory, printed on a patch of coated paper.
chips de color estándar El color utilizado cuando el vehículo se pintó en la fábrica impreso en un parche de papel revestido.

standard screwdriver A tool with a single blade that fits a screw with a slotted head. The blade tip width and thickness must fit the screw head perfectly.
destornillador estándar Herramienta de una sola hoja que se inserta en la ranura de la cabeza de un tornillo. La anchura y el espesor de la punta de la hoja debe ajustar perfectamente en la cabeza del tornillo.

starting system A system for starting an engine that features a large electric motor that turns the engine flywheel.
sistema de arranque Tiene un motor eléctrico grande que hace girar el volante del motor.

steering axis inclination The inward tilt of the steering axis at the top.
inclinación del eje de la dirección Inclinación hacia dentro del eje de dirección en la parte superior.

steering column assembly Uses a long steel shaft to transfer steering wheel rotation to the steering gear assembly.
conjunto de la columna de la dirección Utiliza un eje largo de acero para transmitir la rotación de la rueda de dirección al conjunto de engranes de la dirección.

steering gearbox Either a worm or rack-and-pinion type, it changes the steering wheel rotation into side movement for turning the wheels of the vehicle.
caja de engranes de la dirección Ya sea de tipo cremallera o piñón gastado, convierte la rotación en movimiento lateral para hacer girar las ruedas.

stitch weld Intermittent welds used to join two or more parts.
soldadura de puntadas Soldaduras intermitentes usadas para unir dos o más piezas.

straightening equipment Hydraulic equipment used to apply tremendous force to move the frame or body structure back into alignment; frame rack, porta power, etc.
equipo de enderezado Se utiliza para aplicar una gran fuerza para volver a alinear el armazón o estructura de la carrocería, llamado también bastidor de armazón.

strainer Fine strainer used when pouring thinned primer and paint materials into a spray cup.
cedazo Se utiliza cuando se vacía la mano superior diluida y otros materiales dentro de la copa del pulverizador.

stress cracks Cracks that indicate minor panel damage and movement.
grietas por esfuerzo Grietas que indican movimiento y daño menor en el panel.

stress relieving Uses hammer blows, and sometimes carefully controlled heat, to help return damaged metal to its original shape and state.
liberación de esfuerzos El uso de golpes de martillo para ayudar a que el metal dañado regrese a su forma y estado originales.

stretched metal Metal that has been forcibly made thinner in thickness and larger in surface area by impact.
metal estirado Metal que mediante impactos se ha hecho de un espesor más delgado y de mayor área superficial

structural panel A panel used in a unibody that becomes part of the whole unit and is vital to the strength of the body.
paneles estructurales Los paneles de un monocasco que son partes íntegros de la unedad y que son esenciales para la fuerza de la carrocería.

structural panel replacement Cutting, measuring, and welding a new panel in place of an old or damaged one.
reposición de panel estructural Corte, medición y soldadura de un panel nuevo en lugar de un panel viejo o dañado.

strut plate An accessory that allows you to attach a pulling chain to the top of a shock tower.
placa de tirante Accesorio que le permite conectar que una cadena de tracción a la parte superior de una torre de impacto.

strut tower gauge Shows misalignment of the strut tower/upper body parts in relation to the centerline and datum line plane.
medidor de torre de tirante Muestra la desalineación de la torre de tirante/piezas superiores de la carrocería con respecto a la línea central y al plano de la línea de datos.

suction spray gun A tool that uses airflow through the gun head to form a siphoning action that pulls the refinish liquid into the airstream.
pistola pulverizadora de succión Herramienta que usa un flujo de aire a través del cabezal de la pistola para formar un efecto de sifón que succiona el líquido de acabado final hacia la corriente de aire.

sunlight Contains all frequencies, or wavelengths, of light.
luz de sol Contiene todas las frecuencias o longitudes de onda de la luz.

surface preparation Also called "surface prep," it involves inspection and treatment of the old surface to prepare it for refinishing or painting.

preparación de la superficie Se puede denominar también "preparación." Involucra la inspección y el tratamiento de la superficie antigua para preparla para el reacabado o la pintura.

surface rust Early stage of corrosion that has not penetrated the steel panel.
oxidación superficial Primera etapa de la corrosión que no ha penetrado al panel de acero.

surface scuffing Using an ultrafine abrasive to cut microscopic scratches in the body surface to be painted.
tallado superficial Uso de un abrasivo extra fino para eliminar las rayas microscópicas de la superficie de la carrocería que se va a pintar.

surface thermometer Used to measure a vehicle's body temperature when painting.
termómetro de superficie Se usa para medir la temperatura de la carrocería de un vehículo cuando se pinta.

surform file A tool used to shape body filler while it is semihard.
lima de sobreforma Herramienta que se usa para dar forma al relleno de carrocería cuando está semiduro.

suspension system Springs and other parts that support the upper part of a vehicle on its axles and wheels.
sistema de suspensión Los muelles u otras partes que soportan la parte superior del vehículo en sus ejes y ruedas.

swirl marks Round or curved lines in the paint caused by buffing; they can be prevented by keeping the polishing pad clean and free of compound.
marcas onduladas líneas redondeadas o curvas en la pintura provocadas por el pulido, pueden evitarse si la almohadilla de pulido se mantiene limpia y sin compuestos.

switch Used to turn a circuit on or off manually.
interruptor Se utiliza para encender o apagar manualmente un circuito.

symmetrical The dimensions on the right side of the vehicle are equal to the dimensions on the left side of the vehicle.
simétrico Las dimensiones del lado derecho del vehículo son iguales a las dimensiones del lado izquierdo del vehículo.

systematic approach Organizing and using a logical sequence of steps to accomplish a task or job.
aproximación sistemática Organización y uso de una secuencia lógica de pasos para cumplir una tarea u objetivo.

tack cloth A cheesecloth that has been treated with nondrying varnish to make it tacky. Used to pick up dust and lint from the surface to be painted.
trapo de limpieza Una estopilla impregnada con una laca que no se seca para hacerla muy pegajosa. Sirve para recojer el polvo y la borra de la superficie que se tiene que pintar.

tack weld A temporary spot weld used before making a permanent weld instead of a clamp or metal screw.
soldadura puntual Soldadura temporal de una zona usada en lugar de una prensa o tornillo metálico mientras se pone la soldadura permanente.

tap A tool used to clean and rethread holes.
machuelo Herramienta para limpiar y renovar la rosca de orificios.

tempered glass Heat-treated glass that has more resistance to impact than regular glass of the same thickness.
vidrio templado vidrio con tratamiento térmico que tiene mayor resistencia al impacto que el vidrio normal del mismo espesor.

tensile strength A measurement of the maximum force per unit of area that causes a complete fracture or break in the material.
resistencia a la tensión La resistencia de una película a distorcionarse.

test weld Done on scrap pieces of metal the same thickness and type metal as the parts to be repaired.
soldadura de prueba Se efectúa en piezas metálicas de desecho que son del mismo espesor y tipo de metal que las piezas que serán reparadas.

testlight Used to determine whether there is a complete, unbroken circuit.
luz de prueba Se usa para determinar si hay un circuito completo y sin interrupciones.

thermoplastic A polymer or other solid material that becomes soft or fluid when it is heated and hardens again when it is cooled.
termoplásticos Un polímero u otra material sólida que se ablanda o se pone flúido al ser calentado y se endurece de nuevo al enfriarse.

thermosetting plastics Plastics that undergo a chemical change by the action of heating, a catalyst, or ultraviolet light, and are hardened into a permanent shape that cannot be altered by reapplying heat or catalysts.
plásticos termofijos plásticos que sufren un cambio químico por medio de la acción de calor, de un catalizador o de luz ultravioleta y se endurecen adoptando una forma permanente que no puede alterarse cuando se vuelve a aplicar calor o catalizadores.

thermostat Controls the engine's operating temperature by controlling coolant flow through the system.
termostato Controla la temperatura de funcionamiento del motor al controlar el flujo de refrigerante a través del sistema.

thin-bodied sealers Designed to fill seams under ⅛ (3.2 mm) inch wide.

selladores para carrocería delgada Diseñados para rellenar uniones menores de ⅛ de pulgada (3.2 mm).

three zero plane sections Break the vehicle into three areas: front, center, and rear.

secciones de plano de tres ceros Dividen al vehículo en tres áreas: delantera, central y trasera.

tinting Altering a paint color by adding small amounts of color pigments to better match an existing paint color.

coloreado Alteración del color de la pintura mediante la adición de pequeñas cantidades de pigmentos de color para mejorar la coincidencia o igualado del color de una pintura existente.

topcoat Final layer of paint applied to a substrate. Several coats of topcoat may be applied in some cases.

capa superior La última capa que se aplica en un substrato. Varias capas superiores pueden aplicarse en algunos casos.

torque pattern A crisscross pattern that clamps down all parts evenly to prevent warping; also called a tightening sequence. Recommended for wheels.

patrón de torsión Patrón cruzado que aprieta todas las piezas uniformemente para evitar la deformación; también llamado secuencia de apriete. Recomendado para las ruedas.

torque specifications Tightening values for a specific bolt or nut.

especificaciones de torsión Valores de apriete para una tuerca o un tornillo específico.

torque steer Problem where engine torque is transmitted into the steering wheel under hard acceleration.

conducción con torsión Problema cuando el esfuerzo de torsión del motor se transmite al volante de dirección con una fuerte aceleración.

Torx fastener A six-point fastener that is easier to grip and drive without slippage; also called a star fastener. Used on most late model cars for luggage racks, headlights, taillight assemblies, mirror mountings, door strikers, seat belts, and exterior trim.

sujetador Torx un sujetador o tornillo de 6 puntas que puede agarrarse y apretarse más fácilmente sin patinarse; también llamado "tornillo de estrella". Se utilizan en los coches de modelo más reciente en las canastillas para equipaje, faros, conjuntos de luces traseras, monturas de espejos, enganches de puertas, cinturones de seguridad y adornos exteriores.

total loss When repairing the vehicle would be more costly than replacing it.

pérdida total Se presenta cuando la reparación del vehículo costaría más que sustituirlo.

tower collars Fixtures on pulling towers that allow you to adjust the height of the traction directions.

collarines de torre Permiten ajustar la altura de las direcciones de tracción.

toxic Basic biological property of a material reflecting its inherent capacity to produce injury; adverse effects resulting from overexposure to a material, generally via mouth, skin, eyes, or respiratory tract.

toxico Una propriedad básica biológica de una materia que refleja su capacidad inherente de causar daños; los efectos adversos que resultan de exponerse demasiado a una materia, generalmente por la boca, el piel, los ojos o el sistema respiratorio.

Toyota Hybrid System (THS) Toyota's hybrid drive system, now known as Hybrid Synergy Drive.

Sistema híbrido Toyota (por su sigla en inglés, THS) El sistema de transmisión híbrida de Toyota, conocido ahora como transmisión de sinergia híbrida.

traction Applying force in the direction opposite to the impact force when realigning a vehicle.

tracción Aplicación de fuerza en la dirección opuesta a la fuerza del impacto cuando se vuelve a alinear un vehículo.

tram gauge A special body dimension-measuring tool whose pointers can be aligned with body dimension reference points to determine the direction and amount of body misalignment damage.

medidor de ajuste Herramienta especial para medir las dimensiones de la carrocería, utilizada para determinar la dirección y cantidad de daño por desalineación de la carrocería.

transaxle A transmission and differential combined into a single housing or case.

transeje Transmisión y diferencial combinados en una sola caja o alojamiento.

transmission An assembly with a series of gears for increasing torque to the drive wheels so the car can accelerate properly.

transmisión Conjunto que tiene una serie de engranes para aumentar la torsión hacia las ruedas de tracción, de modo que el coche puede acelerar en forma adecuada.

transverse engine Engine that is mounted sideways in the engine compartment. The crankshaft centerline extends toward the right and left of the body.

motor transversal Motor que está montado en forma lateral dentro del compartimento del motor. La línea central del cigüeñal se extiende hacia el lado derecho e izquierdo de la carrocería.

troubleshooting chart Gives diagnosis procedures that are read from the top to the bottom in a specific sequence.

cuadro de resolución de problemas Proporcionan procedimientos de diagnóstico que se leen de arriba hacia abajo en una secuencia específica.

tungsten inert gas (TIG) Gas-shielded tungsten electrode arc welding.
tungsten con gas inerte (TIG) La soldadura de arco con un electrodo de tungsten protegido por gas.

turning radius A tire-wearing angle measured in degrees; the amount that one front wheel turns more sharply than the other.
radio de viraje Un ángulo del desgaste de las llantas que se mide en grados; la cantidad que vira más una rueda delantera que la otra.

turning radius gauges Measure how many degrees the front wheels are turned right or left.
medidores del radio de giro Miden cuantos grados giran las ruedas delanteras hacia la derecha o hacia la izquierda.

twist damage A condition in which one corner of a vehicle is higher than normal; the opposite corner might be lower than normal.
daño por torsión Estado en el que una esquina del vehículo está más elevada de lo normal; la esquina opuesta quizá esté más abajo de lo normal.

two-part adhesive system A base resin and a hardener are mixed to form an adhesive that cures into a plastic material similar to the base material of a part.
sistemas adhesivos de dos partes Mezcla de una resina base y un endurecedor para formar un adhesivo que cura formando un material plástico similar al material base de una pieza.

two-part putty Used for faster curing, this material comes with its own hardener.
masilla de dos partes Se usa para un curado más rápido, este material viene con su propio endurecedor.

two-stage compressor A compressor in which air is first compressed to an intermediate pressure and then compressed to a higher pressure.
compresor de dos etapas Nombre que se da a un compresor cuando el aire se comprime primero hasta una presión intermedia y después se comprime hasta una presión mayor.

two-stage paint Consists of two distinct layers of paint: basecoat and clearcoat.
de dos etapas pintura Consiste de dos capas distinctas de pintura: la capa de fondo y la capa transparente.

ultrasonic plastic welding Method of repairing auto plastics in which welding time is controlled by the power supply. This method is best suited to rigid plastics.
soldadura ultrasónica de plástico Un método de reparación de los plásticos de auto en que el tiempo de la soldadura se controla por la toma del corriente. Este método se adapta mejor en los plásticos rígidos.

undercoating Asphalt-based product used for rustproofing.
recubrimiento o capa base producto de base asfalto que se utiliza para proteger contra la oxidación.

undercutting Welding the surface a level lower than the base metal, thereby creating a groove.
socavado Soldadura de una superficie hasta un nivel inferior al del metal base para crear una acanaladura.

unibody Vehicle in which parts of the body structure serve as support for overall vehicle parts.
monocasco Un vehículo en el cual las partes de la estructura de la carrocería sirve de soporte para todas las partes del vehículo.

universal joint On a hand tool, a feature that allows a handle or extension to be placed at an angle other than 90 degrees to the fastener.
junta universal Herramienta que permite que un maneral o extensión se coloque en un ángulo con el suje-tador.

value Lightness or darkness of a color.
valor Lo claro o obscuro de un color.

vehicle anchoring Clamping down a vehicle so that it will not move during the straightening process.
anclado del vehículo Sujeción de un vehículo, de modo que este no se moverá durante el proceso de enderezado.

vehicle chassis The frame, engine, suspension system, and other mechanical parts with the body removed.
chasis del vehículo El armazón, motor, sistema de suspensión y otras piezas mecánicas a los que se ha quitado la carrocería.

vehicle dimension illustrations Diagrams that provide measurements across specific points on a vehicle to help determine the extent of damage.
ilustraciones de las dimensiones del vehículo Diagramas que proporcionan mediciones a través de puntos específicos de un vehículo para ayudar a determinar la extensión del daño.

vehicle dimension manuals Provide the body and frame measurements of undamaged vehicles.
manuales de dimensiones del vehículo Proporcionan las mediciones de la carrocería y del armazón de vehículos no dañados.

vehicle identification number (VIN) Number assigned to each vehicle by its manufacturer, primarily for registration and identification purposes.
número de identificacíon del vehículo El número asignado a cada vehículo por su fabricante, primariamente con el propósito de la registración y la identificación.

vehicle lift points Exact positioning locations that locate a vehicle's center of gravity for safely raising the vehicle off the ground.

puntos para elevar un vehículo Localización exacta de los lugares que ubican el centro de gravedad del vehículo para levantarlo del suelo en forma segura.

vehicle measurement Using specialized tools and equipment that locate reference points on a vehicle to measure the extent and direction of major damage.

medición del vehículo Uso de herramientas y equipos especializados que localizan los puntos de referencia de un vehículo para medir la extensión y dirección del daño importante.

vertical welding Welding with the pieces of metal turned upright or in up and down position.

soldadura vertical Soldadura con las piezas de metal colocadas verticalmente.

vinyl A soft, flexible, thin plastic material often applied over a foam filler, commonly used for dash pads, armrests, indoor trim, and roof coverings.

vinilo Un clase de los monómeros que se pueden combinar para formar los polímeros vinílicos. Los polímeros vinílicos son polímeros de agregación, se usan extensamente para crear los acabados de resistencia química, los artículos de plástico durables, los disco fonográficos, y las losas para el piso.

viscosity Consistency or body of a paint; thickness or thinness of a liquid.

viscosidad La consistencia o cuerpo de una pintura; lo espeso o líquido de un líquido.

viscosity cup Most accurate way to measure material viscosity.

copa de viscosidad La forma más precisa de medir la viscosidad de un material.

Vise-Grips Pliers that lock closed with a very tight grip; used for holding parts together or for getting a firm grip on badly rounded fasteners.

pinzas de presión Pinzas que tienen un agarre muy fuerte y se traban al cerrarse; se usan para mantener unidas varias piezas o para agarrar con firmeza sujetadores mal redondeados.

voltage The pressure that pushes electricity through a wire or circuit.

voltaje Presión que impulsa a la electricidad a través de un alambre o circuito.

wash-up Cleaning the entire vehicle with soap and water.

lavado Limpieza de todo el vehículo con agua y jabón.

water-based paints Water is used to carry the pigment in this type of paint, which dries through the evaporation of the water.

pinturas de base acuosa En este tipo de pintura se usa agua para transportar el pigmento; la pintura se seca cuando se evapora el agua.

water leaks Occur when moisture or rain enters the passenger compartment and collects on the carpeting.

fugas de agua Se presentan cuando al compartimento de pasajeros entra humedad o lluvia y es recolectada por la alfombra.

weld face The exposed part of a weld on the welded side.

cara de la soldadura La parte expuesta de una soldadura por el lado soldado.

welding Joining two metal or plastic pieces together by bringing them to their melting points, often involving the use of a welding rod to add metal or plastic to a joint.

soldadura Unión de dos piezas metálicas o de plástico llevándolas hasta sus puntos de fusión, que con frecuencia involucra el uso de una varilla o electrodo de soldadura para agregar metal o plástico a la unión.

welding blankets Thick covers made of fire-resistant cloth for protecting vehicle surfaces from heat, sparks, and weld splatter.

mantas para soldar Cubiertas gruesas hechas de tela pirorresistente para proteger las superficies del vehículo contra el calor, las chispas y las salpicaduras de soldadura.

welding current As electrical current increases, the penetration depth, melting speed, arc stability, amount of weld spatter, excess metal height, and bead width also increase.

corriente para soldar Conforme aumenta la corriente eléctrica, se afectan la profundidad de penetración, la velocidad de fusión, la estabilidad del arco, la cantidad de salpicado de la soldadura, el exceso de la altura del metal y la anchura del cordón.

welding filter lens A shaded glass welding helmet insert used to protect eyesight; also called a filter plate.

lentes filtrantes para soldar Inserto para casco de soldadura con vidrios de color que se usan para proteger la vista; también llamados placa filtrante.

welding helmet A helmet that covers the entire head and has a dark, shaded lens for the eyes; used when electric welding.

casco para soldar Casco que cubre toda la cabeza y tiene lentes de color oscuro para los ojos, se usa para la soldadura eléctrica.

welding respirator Has special cartridge inserts designed to trap welding fumes.

mascarilla para soldar Tiene insertos de cartucho especiales, diseñados para atrapar los humos de la soldadura.

weld inserts Small pieces cut to fit behind or inside a box section of a vehicle body.
insertos para soldar Pequeñas piezas cortadas para que ajusten debajo o dentro de una sección de caja.

weld legs Width and height of a weld bead.
catetos de soldadura Anchura y altura de un cordón de soldadura.

weld penetration Indicated by the height of the exposed surface of a weld on the back side.
penetración de la soldadura Se indica por medio de la altura de la superficie expuesta de la soldadura por la parte posterior.

weld root The part of a joint where the electrode is directed.
raíz de la soldadura Parte de una junta hacia donde se dirige el electrodo.

weld throat The depth of the triangular cross section of a weld.
cuello de la soldadura Profundidad de una sección triangular de la soldadura.

weld-through primer Provides anticorrosion protection to weld zones.
pintura preliminar para soldadura Brinda protección anticorrosiva a las zonas de soldadura.

wet edge spraying A spraying technique used to keep previous coats of paint from curing or drying too much.
pulverización de borde húmedo Se realiza para evitar que las manos de pintura se curen o sequen excesivamente.

wet sandpaper Works with water to flush away debris that might clog fine grits; used for final smoothing.
papel lija de agua Se usa con agua para arrastrar los residuos que pudieran tapar los granos finos; se usa para el alisado final.

wheel alignment Positioning suspension and steering components to assure a vehicle's proper handling and optimum tire wear and fuel economy.
alineamiento de las ruedas Posicionar la suspensión y los componentes de dirección para asegurar el manejo satisfactorio del vehículo y el rendimiento óptimo de las llantas.

wheel masks Plastic or cloth covers that fit over wheels and tires.
mascarillas para ruedas Cubiertas de plástico o tela que se colocan sobre ruedas y neumáticos.

wheel puller Used to remove a steering wheel from its shaft.
extractor de ruedas Se usa para quitar el volante de dirección de su eje.

wheel runout An irregularity in a wheel caused by a bent, damaged rim.
descentrado de rueda Provocado por un aro o rin doblado y dañado.

width damage Results when parts are pushed to the center by a side or angled impact.
daño en la anchura Resulta cuando las partes son empujadas hacia el centro por un impacto lateral o en ángulo.

wind noises High-frequency swishing sounds heard when the vehicle is being driven.
ruidos por viento Sonidos silbantes de alta frecuencia que se escuchan cuando se conduce.

window regulator Gear and arm mechanism for moving the window glass.
regulador de la ventana mecanismo de engrane y brazo para desplazar el vidrio.

wire color coding Allows you to find a specific wire in a harness or in a connector.
codificación de cables por colores Permite encontrar un cable específico en un arnés o en un conector.

wiring diagrams Drawings that show the location of electrical components. They show how wires are routed to electronic parts (motor, solenoid, sensors), fuses or circuit breakers, electronic control units, and other components.
diagramas de cableado o de conexiones Dibujos que muestran la ubicación de los componentes eléctricos. Muestran por donde pasan los cables hacia las partes eléctricas o electrónicas (motor, solenoide, sensores), hacia los fusibles o cortacircuitos, a las unidades electrónicas de control y a otros componentes.

wiring harness Has several wires enclosed in a protective covering.
arnés de cables Tienen varios cables contenidos dentro de una cubierta protectora.

work hardening Upper limit of plastic deformation causing a metal to become very hard in the bent area.
endurecimiento por medios mecánicos Límite superior de la deformación plástica que provoca que un metal se endurezca en el área doblada.

work order A printed form that outlines the procedures which should be taken to repair the vehicle.
orden de trabajo Formulario impreso que describe los procedimientos que deben seguirse para reparar un vehículo.

work traits The things besides skills that make a person a good or bad employee.

características para el trabajo Características adicionales a las habilidades que hacen de usted un buen o un mal empleado.

wrecker A truck equipped with special lifting equipment for raising and transporting a damaged vehicle.

vehículo de auxilio Camión equipado con equipo especial de elevación para levantar y transportar un vehículo dañado.

wrench size Size determined by the width of the jaw opening of a wrench.

tamaño de la llave Tamaño determinado por la anchura de la abertura de las mordazas de una llave.

yield strength A measurement of the minimum force per unit of area that causes a material to begin to permanently change its shape.

límite elástico Medición de la fuerza mínima por unidad de área que provoca que un material empiece a cambiar permanentemente de forma.

zone concept Divides the horizontal surfaces of a vehicle into zones defined by character lines and moldings.

concepto de zona Divide las superficies horizontales de un vehículo en zonas definidas por líneas características y molduras.

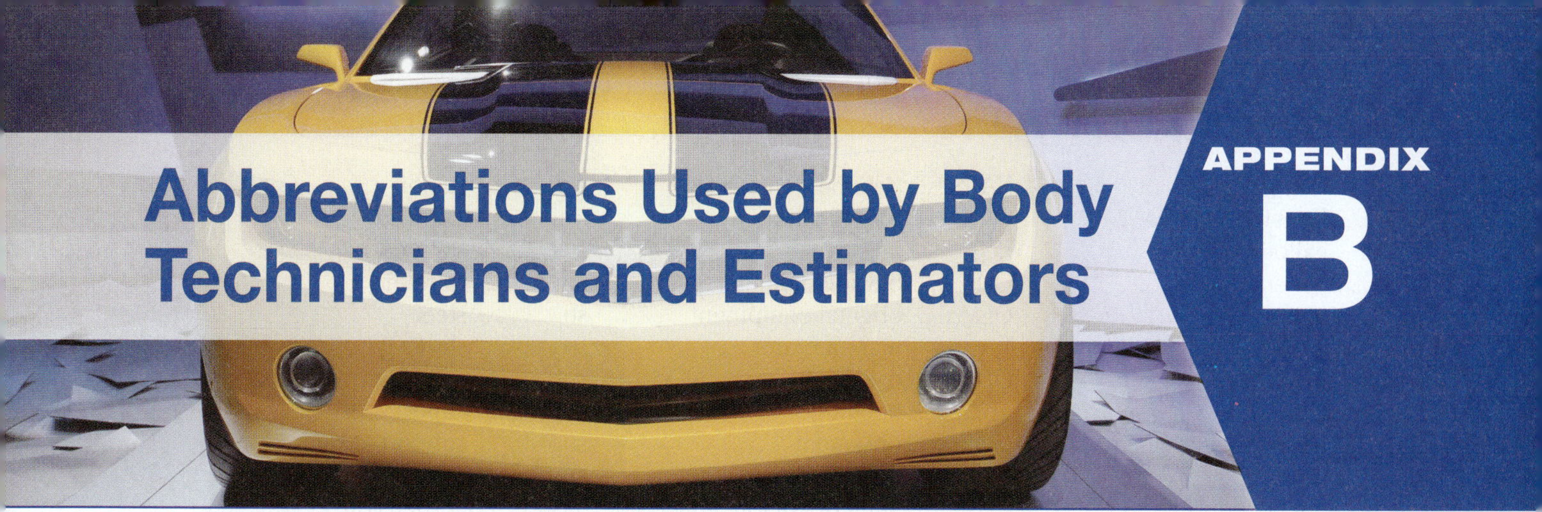

Abbreviations Used by Body Technicians and Estimators

APPENDIX B

It is important that the estimator and body shop technician be able to communicate verbally as well as in writing. Both in estimates and work procedure reports, most estimators use abbreviations. Generally, these abbreviations are the same as those used in the estimating crash guides. Some abbreviations are even used verbally. For example, three of the most commonly used abbreviations in a body shop are:

- **R&I: Remove and Reinstall**. The item is removed as an assembly, set aside, and later reinstalled and aligned for a proper fit. This is generally done to gain access to another part. For example, R&I bumper would mean that the bumper assembly would have to be removed to install a new fender or quarter panel.
- **R&R: Remove and Repair**. Remove the old parts, transfer necessary items to new parts, replace, and align.
- **O/H: Overhaul**. Remove an assembly from the vehicle, disassemble, clean, inspect, replace parts as needed, then reassemble, install, and adjust (except wheel and suspension alignment).

In addition to these abbreviations, the following terms are those accepted by most estimating guides, shop manuals, and estimators. They are the ones used in most written forms.

A .... Manufacturer has no list price for the part
ABS .... antilock brake system
AC .... alternating current
A/C .... air conditioner
ACRS .... air cushion restraint system
adj .... adjuster or adjustable
AIR .... air injector reactor
alt .... alternator
alum .... aluminum
amp .... ampere
assy .... assembly
AT .... automatic trans
ATF .... automatic transmission fluid
auto .... automatic
aux .... auxiliary
bbl .... barrel
bk .... back
blwr .... blower
bmpr .... bumper
brg .... bearing
brkt .... bracket
Bro .... Brougham
btry .... battery
btwn .... between
B-U .... back-up
bush .... bushing
Calif .... California
chnl .... channel
c/mbr .... cross member
cntr .... center
col .... column
comp .... compressor
compt .... compartment
cond .... conditioning or conditioner
cont .... control
conv .... converter or convertible
cor .... corner
cov .... cover
cpe .... coupe
C/R .... customer request
crossmbr .... cross member
c/shaft .... crankshaft
ctl .... control
ctry .... country
cust .... custom
cyl .... cylinder
D .... discontinued part
d .... drilling operational time
DAB .... driver's side air bag
dbl .... double
DC .... direct current
def .... deflector
dehyd .... dehydrator
desc .... description
dia .... diameter
diag .... diagonal
dist .... distributor
div .... division
dlxe .... deluxe
dr .... door
ea .... each
elec .... electric
emiss .... emission
eng .... engine
EP .... exhaust purging
equip .... equipment
evap .... evaporator
exc .... except
exh .... exhaust
extn .... extension
flr .... floor
fndr .... fender
fr & rr .... front and rear
frm .... from

ft .... foot
FWD .... front-wheel drive
gal .... gallon
gen .... generator
grds .... guards
grv .... groove
H'back .... hatchback
HD........... heavy duty
HDC .... heavy-duty cooling
hdr........... header
HEI .... high-energy ignition
Hi Per ...... high performance
H/L .... headlights
horiz........ horizontal
HP .... high-performance
hsg........... housing
HSLA....... high-strength alloy steel
HSS.......... high-strength steel
HT .... hard top
h'top........ hard top
hyd .... hydraulic
hydra....... hydramatic
ign .... ignition
in .... inch
incl .... includes
inr............ inner
inst .... instrument
inter......... intermediate
l .... left
lic .... license
lp .... lamp
lwr .... lower
max .... maximum
mdl.......... model
mldg........ molding
mt .... manual transmission
mtd.......... mounted
mtg .... mounting
muff......... muffler
NAGS....... National Auto Glass Specification
neg .... negative
OD........... outside diameter
OEM .... original equipment manufacturer
O/H .... overhaul
opng........ opening
orna .... ornament
otr............ outer
p .... paint operational time
PAB.......... passenger's side air bag
pass .... passenger
pkg .... package
plr............ pillar
pnl .... panel
pos .... positive
PS .... power steering
pwr.......... power
qtr............ quarter
r .... right
rad........... radiator
R&R......... remove and reinstall
R-L........... right or left
rec .... receiver
refl .... reflector
reg .... regulator
reinf......... reinforcement
RO .... repair order
reson .... resonator
rr.............. rear
RWD........ rear-wheel drive
sed........... sedan
ser............ serial or series
shld .... shield
sidembr .. side member
sig .... single
S/M .... side marker light
spd .... speed
spec......... special
sta............ station
stab.......... stabilizer
stat........... stationary
std .... standard
stl............. steel
strg .... steering
sub .... suburban
sup .... super
supt .... support
surr.......... surround
susp......... suspension
SW........... station wagon
tach .... tachometer
t & t.......... tilt and telescope or tilt and travel
TE .... thermactor emission
tel .... telescope
t/l .... taillight
trans .... transmission
t/s .... turn signal
upr........... upper
vent .... ventilator
vert .... vertical
vib .... vibration
VIN .... vehicle identification number
VIR .... valve-in-receiver
w/ .... with
WB .... wheelbase
WD .... wheel drive
wgn .... wagon
whl .... wheel
whlse........ wheelhouse
w'house ... wheelhouse
wndo........ window
wo/ .... without
w/o.......... wheel opening
wshd .... windshield
w'strip ...... weatherstrip
xmember. cross member

Decimal and Metric Equivalents

APPENDIX C

Decimal and Metric Equivalents					
Fractions	Decimal (in.)	Metric (mm)	Fractions	Decimal (in.)	Metric (mm)
1/64	.015625	.397	33/64	.515625	13.097
1/32	.03125	.794	17/32	.53125	13.494
3/64	.046875	1.191	35/64	.546875	13.891
1/16	.0625	1.588	9/16	.5625	14.288
5/64	.078125	1.984	37/64	.578125	14.684
3/32	.09375	2.381	19/32	.59375	15.081
7/64	.109375	2.778	39/64	.609375	15.478
1/8	.125	3.175	5/8	.625	15.875
9/64	.140625	3.572	41/64	.640625	16.272
5/32	.15625	3.969	21/32	.65625	16.669
11/64	.171875	4.366	43/64	.671875	17.066
3/16	.1875	4.763	11/16	.6875	17.463
13/64	.203125	5.159	45/64	.70312	17.859
7/32	.21875	5.556	23/32	.71875	18.256
15/64	.234275	5.953	47/64	.734375	18.653
1/4	.250	6.35	3/4	.750	19.05
17/64	.265625	6.747	49/64	.765625	19.447
9/32	.28125	7.144	25/32	.78125	19.844
19/64	.296875	7.54	51/64	.796875	20.241
5/16	.3125	7.938	13/16	.8125	20.638
21/64	.328125	8.334	53/64	.828125	21.034
11/32	.34375	8.731	27/32	.84375	21.431
23/64	.359375	9.128	55/64	.859375	21.828
3/8	.375	9.525	7/8	.875	22.225
25/64	.390625	9.922	57/64	.890625	22.622
13/32	.40625	10.319	29/32	.90625	23.019
27/64	.421875	10.716	59/64	.921875	23.416
7/16	.4375	11.113	15/16	.9375	23.813
29/64	.453125	11.509	61/64	.953125	24.209
15/32	.46875	11.906	31/32	.96875	24.606
31/64	.484375	12.303	63/64	.984375	25.003
1/2	.500	12.7	1	1.00	25.4

Viscosity Conversion Chart

VISCOSITY CONVERSION CHART

For Materials at 77°F, without Special Thixotropic Characteristics

	LIGHT CONSISTENCY									MEDIUM CONSISTENCY										HEAVY CONSISTENCY						
	Water or light oil-type materials with translucent or very fine grind color									Light creamy or thin syrup-type materials with medium to fine color or filler grind										Fluffy cream or slow pouring syrup-type materials with medium to coarse grind color or filler and highly filled materials						
Poise	.1	.2	.3	.4	.5	.6	.7	.8	.9	1	1.2	1.5	1.7	2.0	2.5	3.0	3.5	4.0	4.5	5	10	25	50	75	100	150
Centipoise (CPS)	10	20	30	40	50	60	70	80	90	100	120	150	170	200	250	300	350	400	450	500	1000	2500	5000	7500	10,000	15,000
Brookfield (CPS)	10	20	30	40	50	60	70	80	90	100	120	150	170	200	250	300	350	400	450	500	1000	2500	5000	7500	10,000	15,000
Fisher #1 (sec.)	20	30	39	50																						
Fisher #2 (sec.)		15	18	21	24	29	33	39	44	50	62															
Ford #3 (sec.)		12	19	25	29	33	36	41	45	50	58	70														
Ford #4 (sec.)	5	10	14	18	22	25	28	31	32	34	41	47	52	58	67	74										
Gardner-Holdt	A-4	A-3	A-I	A		B		C		D	E	F	G	H	J	L	N	P	Q	S	W	Z-I	Z-3	Z-4	Z-5	Z-6
Bubble Units (sec.)																										
Krebs Unit (sec.)					30	33	35	37	38	40	43	47	49	52	57	60	62	64	66	68	85	114	140			
Parlin #7 (sec.)	27	32	43	50	57	64																				
Parlin#10(sec.)	11	13	15	16	17	18	20	22	23	25	30	35	40	45												
Parlin #15 (sec.)														10		15		20		25	47	135	232	348	465	697
Parlin #20 (sec.)																				8	17	55	83	125	167	250
Parlin #30 (sec.)																							19	29	38	58
Saybolt (Universal) (SSU) (sec.)	60	100	160	210	260	320	370	430	480	530	580	740	345	1000	1240	1330	1475	1950	2215	2480	4600	11,600	23,500	35,000	46,500	69,500
Stormer (150 Gr.) (sec.)		10	12	14	16	18	20	22	25	27	32	38	44	49						114	223	450	1090	1635	2180	
Zahn #1 (sec.)	30	37	44	52	60	68																				
Zahn #3 (sec.)									10	12	14	17	19	23	29	34	40	46	51	57						
Zahn #4 (sec.)										10	11	13	15	17	21	24	27	30	34	37						
Zahn #5 (sec.)														10	13	15	18	20	22	25	49					
Sears				19	20	21	23	24	26	27	31	36	39	44												
Craftsman Cup (sec.)																										
Dupont M-50 (sec.)	16	18	20	22	24	27	30	34	37	41	49	62	70	82												

CAUTION

Your viscosity cup is a precision instrument requiring careful handling, cleaning, and storage. Improper care will adversely affect its accuracy.

Index

C

D

E

G

H

I

J

K

L

N

O

P

Q

R

S

T

U

V